ERRATA

VOLUME 22

Page	Line	For:	Read:
105	ref. 20	delete reference (limited circulation, extremely difficult to find).	
213	6	8 kg	2 kg
	7	8 lb	4 lb
394	last 2 lines	transpose to column headed Bath ratio	
748	structure for terpinene-4-ol	CH_2 etc	CH_3 etc

KIRK-OTHMER

ENCYCLOPEDIA OF CHEMICAL TECHNOLOGY

Third Edition

VOLUME 22

Sulfonation and Sulfation
to
Thorium and Thorium Compounds

KIRK-OTHMER

ENCYCLOPEDIA OF CHEMICAL TECHNOLOGY

THIRD EDITION

VOLUME 22

SULFONATION AND SULFATION
TO
THORIUM AND THORIUM COMPOUNDS

A WILEY-INTERSCIENCE PUBLICATION

John Wiley & Sons

NEW YORK • CHICHESTER • BRISBANE • TORONTO • SINGAPORE

Library of Congress Cataloging in Publication Data:

Main entry under title:
Encyclopedia of chemical technology.

At head of title: Kirk-Othmer.
"A Wiley-Interscience publication."
Includes bibliographies.
1. Chemistry, Technical—Dictionaries. I. Kirk, Raymond Eller, 1890–1957. II. Othmer, Donald Frederick, 1904—
III. Grayson, Martin. IV. Eckroth, David. V. Title: Kirk-Othmer encyclopedia of chemical technology.

TP9.E685 1978 660′.03 77-15820
ISBN 0-471-02075-3

Printed in the United States of America

CONTENTS

Sulfonation and sulfation, 1
Sulfonic acids, 45
Sulfoxides, 64
Sulfur, 78
Sulfur compounds, 107
Sulfur dyes, 168
Sulfuric acid and sulfur trioxide, 190
Sulfuric and sulfurous esters, 233
Sulfurization and sulfurchlorination, 255
Sulfur recovery, 267
Superconducting materials, 298
Surfactants and detersive systems, 332
Sutures, 433
Sweeteners, 448
Synthetic and imitation dairy products, 465
Syrups, 499
Talc, 523
Tall oil, 531
Tantalum and tantalum compounds, 541
Tar and pitch, 564
Tar sands, 601
Tea, 628
Technical service, 645
Tellurium and tellurium compounds, 658
Temperature measurement, 679
Terpenoids, 709
Textiles, 762
Thallium and thallium compounds, 835
Thermal pollution by power plants, 846
Thermodynamics, 868
Thermoelectric energy conversion, 900
Thiazole dyes, 918
Thioglycolic acid, 933
Thiols, 946
Thiophene, 965
Thiosulfates, 974
Thorium and thorium compounds, 989

EDITORIAL STAFF FOR VOLUME 22

CONTRIBUTORS TO VOLUME 22

Michael M. Abbott, *Renssalaer Polytechnic Institute, Troy, New York,* Thermodynamics
Herbert G. Arlt, Jr., *Arizona Chemical Company, Fair Lawn, New Jersey,* Tall oil
D. W. Bixby, *The Sulphur Institute, Washington, D.C.,* Sulfur
Bernard Bucholz, *Pennwalt Corporation, King of Prussia, Pennsylvania,* Thiophene
Arno Cahn, *Arno Cahn Consulting Services, Inc., Pearl River, New York,* Surfactants and detersive systems
G. T. Clark, *Eastman Kodak Co., Kingsport, Tennessee,* Disperse dyes under Thiazole dyes
Charles C. Coutant, *Oak Ridge National Laboratory, Oak Ridge, Tennessee,* Thermal pollution by power plants
John M. Derfer, *SCM Corporation, Jacksonville, Florida,* Terpenoids
Marian M. Derfer, *SCM Corporation, Jacksonville, Florida,* Terpenoids
J. R. Donovan, *Monsanto Enviro-Chem Systems, Inc., St. Louis, Missouri,* Sulfuric acid and sulfur trioxide
Robert E. Droegkamp, *Fansteel, Inc., North Chicago, Illinois,* Tantalum and tantalum compounds

E. M. Elkin, *Noranda Mines, Ltd., Montreal, Quebec, Canada,* Tellurium and tellurium compounds

Russell E. Farris *Sandoz, Inc., Martin, South Carolina,* Direct dyes under Thiazole dyes

H. L. Fike, *The Sulphur Institute, Washington, D.C.,* Sulfur compounds

J. G. Fisher, *Eastman Kodak Co., Kingsport, Tennessee,* Disperse dyes under Thiazole dyes

Alfred C. Glatz, *General Foods Corporation, Tarrytown, New York,* Thermoelectric energy conversion

Harold Graham, *Thomas J. Lipton, Inc., Englewood Cliffs, New Jersey,* Tea

R. A. Guest, *James Robinson & Company, Ltd., Hudderfield, Yorkshire, United Kingdom,* Sulfur dyes

W. James Harper, *New Zealand Dairy Research Institute, North Palmerston, New Zealand,* Synthetic and imitation dairy products

R. E. Hebeda, *CPC International, Summit-Argo, Illinois,* Syrups

Benjamin C. Hui, *Alfa Products, Danvers, Massachusetts,* Thallium and thallium compounds

Karl Kamman, *Keil Chemical Division, Ferro Corporation, Hammond, Indiana,* Sulfurization and sulfurchlorination

Leonard I. Katzin, *Consultant, Port Hueneme, California,* Thorium and thorium compounds

Otto S. Kauder, *Argus Chemical Corporation, Brooklyn, New York,* Thioglycolic acid

E. A. Knaggs, *Stepan Chemical Company, Northfield, Illinois,* Sulfonation and sulfation

John B. Lambert, *Fansteel, Inc., North Chicago, Illinois,* Tantalum and tantalum compounds.

B. C. Lawes, *E. I. du Pont de Nemours & Co., Inc., Wilmington, Delaware,* Sulfuric and sulfurous esters

Hillel Lieberman, *Betz Laboratories, Trevose, Pennsylvania,* Technical service

R. P. Louthan, *Phillips Petroleum Co., Bartlesville, Oklahoma,* Thiols

Jesse L. Lynn, Jr., *Lever Brothers Company, Edgewater, New Jersey,* Surfactants and detersive systems

W. S. MacGregor, *Crown Zellerbach Corp., Camus, Washington,* Sulfoxides

Robert Mazur, *GD Searle and Co., Chicago, Illinois,* Sweeteners

W. B. McCormack, *E. I. du Pont de Nemours & Co., Inc., Deepwater, New Jersey,* Sulfuric and sulfurous esters

Donald McNeil, *Technical consultant, Wilsden, Bradford, West Yorkshire, United Kingdom,* Tar and pitch

James B. McPherson, *American Cyanamid Company, Danbury, Connecticut,* Sutures

Frank P. Missell, *Universidade de São Paulo, São Paulo, Brazil,* Superconducting materials

H. T. Mulryan, *Cyprus Industrial Minerals, Englewood, Colorado,* Talc

John Norell, *Phillips Petroleum Co., Bartlesville, Oklahoma,* Thiols

M. L. Nussbaum, *Stepan Chemical Company, Northfield, Illinois,* Sulfonation and sulfation

J. V. Orle, *Crown Zellerbach Corporation, Camus, Washington,* Sulfoxides

Ludwig Rebenfeld, *Textile Research Institute, Princeton, New Jersey,* Survey under Textiles

J. M. Salomone, *Monsanto Enviro-Chem Systems, Inc., St. Louis, Missouri,* Sulfuric acid and sulfur trioxide

Mortimer Schussler, *Fansteel, Inc., North Chicago, Illinois,* Tantalum and tantalum compounds

Brian B. Schwartz, *Brooklyn College, CUNY, Brooklyn, New York,* Superconducting materials

J. E. Shelton, *U.S. Bureau of Mines, Washington, D.C.,* Sulfur compounds

A. Shultz, *Stepan Chemical Company, Northfield, Illinois,* Sulfonation and sulfation; Sulfonic acids

Robert W. Singelton, *University of Connecticut, Storrs, Connecticut,* Testing under Textiles

Henry E. Sostman, *Yellow Springs Instrument Co., Inc., Yellow Springs, Ohio,* Temperature measurement

James W. Swaine, Jr., *Allied Corp., Solvay, New York,* Thiosulfates

Donald F. Taylor, *Fansteel, Inc., North Chicago, Illinois,* Tantalum and tantalum compounds

Donald Towson, *Petro-Canada, Alberta, Canada,* Tar sands

Sidney L. Vail, *Southern Regional Research Center, United States Department of Agriculture, New Orleans, Louisiana,* Finishing under Textiles

Hendrick C. Van Ness, *Renssalaer Polytechnic Institute, Troy, New York,* Thermodynamics

Guy E. Verdino, *Keil Chemical Division, Ferro Corporation, Hammond, Indiana,* Sulfurization and sulfurchlorination

Edward D. Weil, *Stauffer Chemical Co., Dobbs Ferry, New York,* Sulfur compounds

James R. West, *Texasgulf Inc., South Hempstead, New York,* Sulfur recovery

T. K. Wiewiorowski, *Freeport Minerals Co., Belle Chasse, Louisiana,* Sulfur

W. E. Wood, *James Robinson & Company, Ltd., Huddersfield, Yorkshire, United Kingdom,* Sulfur dyes

Robert J. Ziegler, *Betz Laboratories, Trevose, Pennsylvania,* Technical service

NOTE ON CHEMICAL ABSTRACTS SERVICE REGISTRY NUMBERS AND NOMENCLATURE

Chemical Abstracts Service (CAS) Registry Numbers are unique numerical identifiers assigned to substances recorded in the CAS Registry System. They appear in brackets in the *Chemical Abstracts* (CA) substance and formula indexes following the names of compounds. A single compound may have many synonyms in the chemical literature. A simple compound like phenethylamine can be named β-phenylethylamine or, as in *Chemical Abstracts*, benzeneethanamine. The usefulness of the *Encyclopedia* depends on accessibility through the most common correct name of a substance. Because of this diversity in nomenclature careful attention has been given the problem in order to assist the reader as much as possible, especially in locating the systematic CA index name by means of the Registry Number. For this purpose, the reader may refer to the CAS Registry Handbook-Number Section which lists in numerical order the Registry Number with the *Chemical Abstracts* index name and the molecular formula; eg, **458-88-8,** Piperidine, 2-propyl-, (*S*)-, $C_8H_{17}N$; in the *Encyclopedia* this compound would be found under its common name, coniine [*458-88-8*]. The Registry Number is a valuable link for the reader in retrieving additional published information on substances and also as a point of access for such on-line data bases as Chemline, Medline, and Toxline.

In all cases, the CAS Registry Numbers have been given for title compounds in articles and for all compounds in the index. All specific substances indexed in *Chemical Abstracts* since 1965 are included in the CAS Registry System as are a large number of substances derived from a variety of reference works. The CAS Registry System identifies a substance on the basis of an unambiguous computer-language description of its molecular structure including stereochemical detail. The Registry Number is a machine-checkable number (like a Social Security number) assigned in sequential order to each substance as it enters the registry system. The value of the number lies in the fact that it is a concise and unique means of substance identification, which is independent of, and therefore bridges, many systems of chemical nomenclature. For polymers, one Registry Number is used for the entire family; eg, polyoxyethylene (20) sorbitan monolaurate has the same number as all of its polyoxyethylene homologues.

Registry numbers for each substance will be provided in the third edition cumulative index and appear as well in the annual indexes (eg, Alkaloids shows the Registry Number of all alkaloids (title compounds) in a table in the article as well, but the intermediates have their Registry Numbers shown only in the index). Articles such as Analytical methods, Batteries and electric cells, Chemurgy, Distillation, Economic evaluation, and Fluid mechanics have no Registry Numbers in the text.

Cross-references are inserted in the index for many common names and for some systematic names. Trademark names appear in the index. Names that are incorrect, misleading or ambiguous are avoided. Formulas are given very frequently in the text to help in identifying compounds. The spelling and form used, even for industrial names, follow American chemical usage, but not always the usage of *Chemical Abstracts* (eg, *coniine* is used instead of (*S*)-*2-propylpiperidine*, *aniline* instead of *benzenamine*, and *acrylic acid* instead of *2-propenoic acid*).

There are variations in representation of rings in different disciplines. The dye industry does not designate aromaticity or double bonds in rings. All double bonds and aromaticity are shown in the *Encyclopedia* as a matter of course. For example, tetralin has an aromatic ring and a saturated ring and its structure appears in the

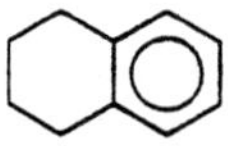

Encyclopedia with its common name, Registry Number enclosed in brackets, and parenthetical CA index name, ie, tetralin, [*119-64-2*] (1,2,3,4-tetrahydronaphthalene). With names and structural formulas, and especially with CAS Registry Numbers the aim is to help the reader have a concise means of substance identification.

CONVERSION FACTORS, ABBREVIATIONS, AND UNIT SYMBOLS

SI Units (Adopted 1960)

A new system of measurement, the International System of Units (abbreviated SI), is being implemented throughout the world. This system is a modernized version of the MKSA (meter, kilogram, second, ampere) system, and its details are published and controlled by an international treaty organization (The International Bureau of Weights and Measures) (1).

SI units are divided into three classes:

BASE UNITS

length	meter[†] (m)
mass[‡]	kilogram (kg)
time	second (s)
electric current	ampere (A)
thermodynamic temperature[§]	kelvin (K)
amount of substance	mole (mol)
luminous intensity	candela (cd)

SUPPLEMENTARY UNITS

plane angle	radian (rad)
solid angle	steradian (sr)

[†] The spellings "metre" and "litre" are preferred by ASTM; however "-er" are used in the Encyclopedia.

[‡] "Weight" is the commonly used term for "mass."

[§] Wide use is made of "Celsius temperature" (t) defined by

$$t = T - T_0$$

where T is the thermodynamic temperature, expressed in kelvins, and T_0 = 273.15 K by definition. A temperature interval may be expressed in degrees Celsius as well as in kelvins.

SUPPLEMENTARY UNITS

plane angle	radian (rad)
solid angle	steradian (sr)

DERIVED UNITS AND OTHER ACCEPTABLE UNITS

These units are formed by combining base units, supplementary units, and other derived units (2–4). Those derived units having special names and symbols are marked with an asterisk in the list below:

Quantity	*Unit*	*Symbol*	*Acceptable equivalent*
*absorbed dose	gray	Gy	J/kg
acceleration	meter per second squared	m/s^2	
*activity (of ionizing radiation source)	becquerel	Bq	1/s
area	square kilometer	km^2	
	square hectometer	hm^2	ha (hectare)
	square meter	m^2	
*capacitance	farad	F	C/V
concentration (of amount of substance)	mole per cubic meter	mol/m^3	
*conductance	siemens	S	A/V
current density	ampere per square meter	A/m^2	
density, mass density	kilogram per cubic meter	kg/m^3	g/L; mg/cm^3
dipole moment (quantity)	coulomb meter	C·m	
*electric charge, quantity of electricity	coulomb	C	A·s
electric charge density	coulomb per cubic meter	C/m^3	
electric field strength	volt per meter	V/m	
electric flux density	coulomb per square meter	C/m^2	
*electric potential, potential difference, electromotive force	volt	V	W/A
*electric resistance	ohm	Ω	V/A
*energy, work, quantity of heat	megajoule	MJ	
	kilojoule	kJ	
	joule	J	N·m
	electron volt†	eV†	
	kilowatt-hour†	kW·h†	

† This non-SI unit is recognized by the CIPM as having to be retained because of practical importance or use in specialized fields (1).

Quantity	*Unit*	*Symbol*	*Acceptable equivalent*
energy density	joule per cubic meter	J/m^3	
*force	kilonewton	kN	
	newton	N	$kg{\cdot}m/s^2$
*frequency	megahertz	MHz	
	hertz	Hz	1/s
heat capacity, entropy	joule per kelvin	J/K	
heat capacity (specific), specific entropy	joule per kilogram kelvin	$J/(kg{\cdot}K)$	
heat transfer coefficient	watt per square meter kelvin	$W/(m^2{\cdot}K)$	
*illuminance	lux	lx	lm/m^2
*inductance	henry	H	Wb/A
linear density	kilogram per meter	kg/m	
luminance	candela per square meter	cd/m^2	
*luminous flux	lumen	lm	cd·sr
magnetic field strength	ampere per meter	A/m	
*magnetic flux	weber	Wb	V·s
*magnetic flux density	tesla	T	Wb/m^2
molar energy	joule per mole	J/mol	
molar entropy, molar heat capacity	joule per mole kelvin	$J/(mol{\cdot}K)$	
moment of force, torque	newton meter	N·m	
momentum	kilogram meter per second	kg·m/s	
permeability	henry per meter	H/m	
permittivity	farad per meter	F/m	
*power, heat flow rate, radiant flux	kilowatt	kW	
	watt	W	J/s
power density, heat flux density, irradiance	watt per square meter	W/m^2	
*pressure, stress	megapascal	MPa	
	kilopascal	kPa	
	pascal	Pa	N/m^2
sound level	decibel	dB	
specific energy	joule per kilogram	J/kg	
specific volume	cubic meter per kilogram	m^3/kg	
surface tension	newton per meter	N/m	
thermal conductivity	watt per meter kelvin	$W/(m{\cdot}K)$	
velocity	meter per second	m/s	
	kilometer per hour	km/h	
viscosity, dynamic	pascal second	Pa·s	
	millipascal second	mPa·s	
viscosity, kinematic	square meter per second	m^2/s	

Quantity	*Unit*	*Symbol*	*Acceptable equivalent*
	square millimeter per second	mm^2/s	
volume	cubic meter	m^3	
	cubic decimeter	dm^3	L(liter) (5)
	cubic centimeter	cm^3	mL
wave number	1 per meter	m^{-1}	
	1 per centimeter	cm^{-1}	

In addition, there are 16 prefixes used to indicate order of magnitude, as follows:

Multiplication factor	*Prefix*	*Symbol*	*Note*
10^{18}	exa	E	
10^{15}	peta	P	
10^{12}	tera	T	
10^{9}	giga	G	
10^{6}	mega	M	
10^{3}	kilo	k	
10^{2}	hecto	h[a]	[a] Although hecto, deka, deci, and centi are SI prefixes, their use should be avoided except for SI unit-multiples for area and volume and nontechnical use of centimeter, as for body and clothing measurement.
10	deka	da[a]	
10^{-1}	deci	d[a]	
10^{-2}	centi	c[a]	
10^{-3}	milli	m	
10^{-6}	micro	μ	
10^{-9}	nano	n	
10^{-12}	pico	p	
10^{-15}	femto	f	
10^{-18}	atto	a	

For a complete description of SI and its use the reader is referred to ASTM E 380 (4) and the article Units and Conversion Factors which will appear in a later volume of the *Encyclopedia.*

A representative list of conversion factors from non-SI to SI units is presented herewith. Factors are given to four significant figures. Exact relationships are followed by a dagger. A more complete list is given in ASTM E 380-79(4) and ANSI Z210.1-1976 (6).

Conversion Factors to SI Units

To convert from	*To*	*Multiply by*
acre	square meter (m^2)	4.047×10^3
angstrom	meter (m)	1.0×10^{-10}†
are	square meter (m^2)	1.0×10^{2}†
astronomical unit	meter (m)	1.496×10^{11}
atmosphere	pascal (Pa)	1.013×10^{5}
bar	pascal (Pa)	1.0×10^{5}†
barn	square meter (m^2)	1.0×10^{-28}†

† Exact.

To convert from	*To*	*Multiply by*
barrel (42 U.S. liquid gallons)	cubic meter (m^3)	0.1590
Bohr magneton (μ_β)	J/T	9.274×10^{-24}
Btu (International Table)	joule (J)	1.055×10^{3}
Btu (mean)	joule (J)	1.056×10^{3}
Btu (thermochemical)	joule (J)	1.054×10^{3}
bushel	cubic meter (m^3)	3.524×10^{-2}
calorie (International Table)	joule (J)	4.187
calorie (mean)	joule (J)	4.190
calorie (thermochemical)	joule (J)	4.184†
centipoise	pascal second (Pa·s)	1.0×10^{-3}†
centistoke	square millimeter per second (mm^2/s)	1.0†
cfm (cubic foot per minute)	cubic meter per second (m^3/s)	4.72×10^{-4}
cubic inch	cubic meter (m^3)	1.639×10^{-5}
cubic foot	cubic meter (m^3)	2.832×10^{-2}
cubic yard	cubic meter (m^3)	0.7646
curie	becquerel (Bq)	3.70×10^{10}†
debye	coulomb·meter (C·m)	3.336×10^{-30}
degree (angle)	radian (rad)	1.745×10^{-2}
denier (international)	kilogram per meter (kg/m)	1.111×10^{-7}
	tex‡	0.1111
dram (apothecaries')	kilogram (kg)	3.888×10^{-3}
dram (avoirdupois)	kilogram (kg)	1.772×10^{-3}
dram (U.S. fluid)	cubic meter (m^3)	3.697×10^{-6}
dyne	newton (N)	1.0×10^{-5}†
dyne/cm	newton per meter (N/m)	1.0×10^{-3}†
electron volt	joule (J)	1.602×10^{-19}
erg	joule (J)	1.0×10^{-7}†
fathom	meter (m)	1.829
fluid ounce (U.S.)	cubic meter (m^3)	2.957×10^{-5}
foot	meter (m)	0.3048†
footcandle	lux (lx)	10.76
furlong	meter (m)	2.012×10^{-2}
gal	meter per second squared (m/s^2)	1.0×10^{-2}†
gallon (U.S. dry)	cubic meter (m^3)	4.405×10^{-3}
gallon (U.S. liquid)	cubic meter (m^3)	3.785×10^{-3}
gallon per minute (gpm)	cubic meter per second (m^3/s)	6.308×10^{-5}
	cubic meter per hour (m^3/h)	0.2271
gauss	tesla (T)	1.0×10^{-4}
gilbert	ampere (A)	0.7958
gill (U.S.)	cubic meter (m^3)	1.183×10^{-4}
grad	radian	1.571×10^{-2}
grain	kilogram (kg)	6.480×10^{-5}
gram force per denier	newton per tex (N/tex)	8.826×10^{-2}

† Exact.

‡ See footnote on p. xiv.

To convert from	*To*	*Multiply by*
hectare	square meter (m^2)	$1.0 \times 10^{4\dagger}$
horsepower (550 ft·lbf/s)	watt (W)	7.457×10^{2}
horsepower (boiler)	watt (W)	9.810×10^{3}
horsepower (electric)	watt (W)	$7.46 \times 10^{2\dagger}$
hundredweight (long)	kilogram (kg)	50.80
hundredweight (short)	kilogram (kg)	45.36
inch	meter (m)	$2.54 \times 10^{-2\dagger}$
inch of mercury (32°F)	pascal (Pa)	3.386×10^{3}
inch of water (39.2°F)	pascal (Pa)	2.491×10^{2}
kilogram force	newton (N)	9.807
kilowatt hour	megajoule (MJ)	$3.6^{\dagger}$
kip	newton (N)	4.48×10^{3}
knot (international)	meter per second (m/s)	0.5144
lambert	candela per square meter (cd/m^2)	3.183×10^{3}
league (British nautical)	meter (m)	5.559×10^{3}
league (statute)	meter (m)	4.828×10^{3}
light year	meter (m)	9.461×10^{15}
liter (for fluids only)	cubic meter (m^3)	$1.0 \times 10^{-3\dagger}$
maxwell	weber (Wb)	$1.0 \times 10^{-8\dagger}$
micron	meter (m)	$1.0 \times 10^{-6\dagger}$
mil	meter (m)	$2.54 \times 10^{-5\dagger}$
mile (statute)	meter (m)	1.609×10^{3}
mile (U.S. nautical)	meter (m)	$1.852 \times 10^{3\dagger}$
mile per hour	meter per second (m/s)	0.4470
millibar	pascal (Pa)	1.0×10^{2}
millimeter of mercury (0°C)	pascal (Pa)	$1.333 \times 10^{2\dagger}$
minute (angular)	radian	2.909×10^{-4}
myriagram	kilogram (kg)	10
myriameter	kilometer (km)	10
oersted	ampere per meter (A/m)	79.58
ounce (avoirdupois)	kilogram (kg)	2.835×10^{-2}
ounce (troy)	kilogram (kg)	3.110×10^{-2}
ounce (U.S. fluid)	cubic meter (m^3)	2.957×10^{-5}
ounce-force	newton (N)	0.2780
peck (U.S.)	cubic meter (m^3)	8.810×10^{-3}
pennyweight	kilogram (kg)	1.555×10^{-3}
pint (U.S. dry)	cubic meter (m^3)	5.506×10^{-4}
pint (U.S. liquid)	cubic meter (m^3)	4.732×10^{-4}
poise (absolute viscosity)	pascal second (Pa·s)	$0.10^{\dagger}$
pound (avoirdupois)	kilogram (kg)	0.4536
pound (troy)	kilogram (kg)	0.3732
poundal	newton (N)	0.1383
pound-force	newton (N)	4.448
pound per square inch (psi)	pascal (Pa)	6.895×10^{3}
quart (U.S. dry)	cubic meter (m^3)	1.101×10^{-3}

† Exact.

To convert from	*To*	*Multiply by*
quart (U.S. liquid)	cubic meter (m^3)	9.464×10^{-4}
quintal	kilogram (kg)	$1.0 \times 10^{2\dagger}$
rad	gray (Gy)	$1.0 \times 10^{-2\dagger}$
rod	meter (m)	5.029
roentgen	coulomb per kilogram (C/kg)	2.58×10^{-4}
second (angle)	radian (rad)	4.848×10^{-6}
section	square meter (m^2)	2.590×10^{6}
slug	kilogram (kg)	14.59
spherical candle power	lumen (lm)	12.57
square inch	square meter (m^2)	6.452×10^{-4}
square foot	square meter (m^2)	9.290×10^{-2}
square mile	square meter (m^2)	2.590×10^{6}
square yard	square meter (m^2)	0.8361
stere	cubic meter (m^3)	$1.0^{\dagger}$
stokes (kinematic viscosity)	square meter per second (m^2/s)	$1.0 \times 10^{-4\dagger}$
tex	kilogram per meter (kg/m)	$1.0 \times 10^{-6\dagger}$
ton (long, 2240 pounds)	kilogram (kg)	1.016×10^{3}
ton (metric)	kilogram (kg)	$1.0 \times 10^{3\dagger}$
ton (short, 2000 pounds)	kilogram (kg)	9.072×10^{2}
torr	pascal (Pa)	1.333×10^{2}
unit pole	weber (Wb)	1.257×10^{-7}
yard	meter (m)	$0.9144^{\dagger}$

Abbreviations and Unit Symbols

Following is a list of commonly used abbreviations and unit symbols appropriate for use in the *Encyclopedia*. In general they agree with those listed in *American National Standard Abbreviations for Use on Drawings and in Text* (*ANSI Y1.1*) (6) and *American National Standard Letter Symbols for Units in Science and Technology* (*ANSI Y10*) (6). Also included is a list of acronyms for a number of private and government organizations as well as common industrial solvents, polymers, and other chemicals.

Rules for Writing Unit Symbols (4):

1. Unit symbols should be printed in upright letters (roman) regardless of the type style used in the surrounding text.
2. Unit symbols are unaltered in the plural.
3. Unit symbols are not followed by a period except when used as the end of a sentence.
4. Letter unit symbols are generally written in lower-case (eg, cd for candela) unless the unit name has been derived from a proper name, in which case the first letter of the symbol is capitalized (W,Pa). Prefix and unit symbols retain their prescribed form regardless of the surrounding typography.
5. In the complete expression for a quantity, a space should be left between the numerical value and the unit symbol. For example, write 2.37 lm, *not* 2.37lm, and 35 mm, *not* 35mm. When the quantity is used in an adjectival sense, a hyphen is often used, for example, 35-mm film. *Exception:* No space is left between the numerical value and the symbols for degree, minute, and second of plane angle, and degree Celsius.

6. No space is used between the prefix and unit symbols (eg, kg).

7. Symbols, not abbreviations, should be used for units. For example, use "A," not "amp," for ampere.

8. When multiplying unit symbols, use a raised dot:

$$\mathrm{N{\cdot}m} \text{ for newton meter}$$

In the case of W·h, the dot may be omitted, thus:

$$\mathrm{Wh}$$

An exception to this practice is made for computer printouts, automatic typewriter work, etc, where the raised dot is not possible, and a dot on the line may be used.

9. When dividing unit symbols use one of the following forms:

$$\mathrm{m/s} \textit{ or } \mathrm{m{\cdot}s^{-1}} \textit{ or } \frac{\mathrm{m}}{\mathrm{s}}$$

In no case should more than one slash be used in the same expression unless parentheses are inserted to avoid ambiguity. For example, write:

$$\mathrm{J/(mol{\cdot}K)} \textit{ or } \mathrm{J{\cdot}mol^{-1}\cdot K^{-1}} \textit{ or } \mathrm{(J/mol)/K}$$

but *not*

$$\mathrm{J/mol/K}$$

10. Do not mix symbols and unit names in the same expression. Write:

$$\text{joules per kilogram } \textit{or } \mathrm{J/kg} \textit{ or } \mathrm{J{\cdot}kg^{-1}}$$

but *not*

$$\text{joules/kilogram } \textit{nor } \text{joules/kg } \textit{nor } \text{joules}{\cdot}\mathrm{kg^{-1}}$$

ABBREVIATIONS AND UNITS

A	ampere
A	anion (eg, H*A*); mass number
a	atto (prefix for 10^{-18})
AATCC	American Association of Textile Chemists and Colorists
ABS	acrylonitrile–butadiene–styrene
abs	absolute
ac	alternating current, *n.*
a-c	alternating current, *adj.*
ac-	alicyclic
acac	acetylacetonate
ACGIH	American Conference of Governmental Industrial Hygienists
ACS	American Chemical Society
AGA	American Gas Association
Ah	ampere hour
AIChE	American Institute of Chemical Engineers
AIME	American Institute of Mining, Metallurgical, and Petroleum Engineers
AIP	American Institute of Physics
AISI	American Iron and Steel Institute
alc	alcohol(ic)
Alk	alkyl
alk	alkaline (not alkali)
amt	amount
amu	atomic mass unit
ANSI	American National Standards Institute
AO	atomic orbital
AOAC	Association of Official Analytical Chemists
AOCS	American Oil Chemists' Society
APHA	American Public Health Association

API	American Petroleum Institute
aq	aqueous
Ar	aryl
ar-	aromatic
as-	asymmetric(al)
ASH-RAE	American Society of Heating, Refrigerating, and Air Conditioning Engineers
ASM	American Society for Metals
ASME	American Society of Mechanical Engineers
ASTM	American Society for Testing and Materials
at no.	atomic number
at wt	atomic weight
av(g)	average
AWS	American Welding Society
b	bonding orbital
bbl	barrel
bcc	body-centered cubic
BCT	body-centered tetragonal
Bé	Baumé
BET	Brunauer-Emmett-Teller (adsorption equation)
bid	twice daily
Boc	*t*-butyloxycarbonyl
BOD	biochemical (biological) oxygen demand
bp	boiling point
Bq	becquerel
C	coulomb
°C	degree Celsius
C-	denoting attachment to carbon
c	centi (prefix for 10^{-2})
c	critical
ca	circa (approximately)
cd	candela; current density; circular dichroism
CFR	Code of Federal Regulations
cgs	centimeter–gram–second
CI	Color Index
cis-	isomer in which substituted groups are on same side of double bond between C atoms
cl	carload
cm	centimeter
cmil	circular mil
cmpd	compound
CNS	central nervous system
CoA	coenzyme A
COD	chemical oxygen demand
coml	commercial(ly)
cp	chemically pure
cph	close-packed hexagonal
CPSC	Consumer Product Safety Commission
cryst	crystalline
cub	cubic
D	Debye
D-	denoting configurational relationship
d	differential operator
d-	*dextro*-, dextrorotatory
da	deka (prefix for 10^1)
dB	decibel
dc	direct current, *n.*
d-c	direct current, *adj.*
dec	decompose
detd	determined
detn	determination
Di	didymium, a mixture of all lanthanons
dia	diameter
dil	dilute
DIN	Deutsche Industrie Normen
dl-; DL-	racemic
DMA	dimethylacetamide
DMF	dimethylformamide
DMG	dimethyl glyoxime
DMSO	dimethyl sulfoxide
DOD	Department of Defense
DOE	Department of Energy
DOT	Department of Transportation
DP	degree of polymerization
dp	dew point
DPH	diamond pyramid hardness
dstl(d)	distill(ed)
dta	differential thermal analysis
(*E*)-	entgegen; opposed
ϵ	dielectric constant (unitless number)
e	electron

ECU	electrochemical unit
ed.	edited, edition, editor
ED	effective dose
EDTA	ethylenediaminetetraacetic acid
emf	electromotive force
emu	electromagnetic unit
en	ethylene diamine
eng	engineering
EPA	Environmental Protection Agency
epr	electron paramagnetic resonance
eq.	equation
esp	especially
esr	electron-spin resonance
est(d)	estimate(d)
estn	estimation
esu	electrostatic unit
exp	experiment, experimental
ext(d)	extract(ed)
F	farad (capacitance)
F	faraday (96,487 C)
f	femto (prefix for 10^{-15})
FAO	Food and Agriculture Organization (United Nations)
fcc	face-centered cubic
FDA	Food and Drug Administration
FEA	Federal Energy Administration
fob	free on board
fp	freezing point
FPC	Federal Power Commission
FRB	Federal Reserve Board
frz	freezing
G	giga (prefix for 10^9)
G	gravitational constant = 6.67×10^{11} N·m^2/kg^2
g	gram
(g)	gas, only as in $H_2O(g)$
g	gravitational acceleration
gem-	geminal
glc	gas-liquid chromatography
g-mol wt; gmw	gram-molecular weight
GNP	gross national product
gpc	gel-permeation chromatography
GRAS	Generally Recognized as Safe
grd	ground
Gy	gray
H	henry
h	hour; hecto (prefix for 10^2)
ha	hectare
HB	Brinell hardness number
Hb	hemoglobin
hcp	hexagonal close-packed
hex	hexagonal
HK	Knoop hardness number
HRC	Rockwell hardness (C scale)
HV	Vickers hardness number
hyd	hydrated, hydrous
hyg	hygroscopic
Hz	hertz
^{i}(eg, Pr^i)	iso (eg, isopropyl)
i-	inactive (eg, *i*-methionine)
IACS	International Annealed Copper Standard
ibp	initial boiling point
IC	inhibitory concentration
ICC	Interstate Commerce Commission
ICT	International Critical Table
ID	inside diameter; infective dose
ip	intraperitoneal
IPS	iron pipe size
IPTS	International Practical Temperature Scale (NBS)
ir	infrared
IRLG	Interagency Regulatory Liaison Group
ISO	International Organization for Standardization
IU	International Unit
IUPAC	International Union of Pure and Applied Chemistry
IV	iodine value
iv	intravenous
J	joule
K	kelvin
k	kilo (prefix for 10^3)
kg	kilogram
L	denoting configurational relationship
L	liter (for fluids only)(5)
l-	*levo*-, levorotatory
(l)	liquid, only as in $NH_3(l)$

LC_{50}	conc lethal to 50% of the animals tested
LCAO	linear combination of atomic orbitals
LCD	liquid crystal display
lcl	less than carload lots
LD_{50}	dose lethal to 50% of the animals tested
LED	light-emitting diode
liq	liquid
lm	lumen
ln	logarithm (natural)
LNG	liquefied natural gas
log	logarithm (common)
LPG	liquefied petroleum gas
ltl	less than truckload lots
lx	lux
M	mega (prefix for 10^{6}); metal (as in *MA*)
M	molar; actual mass
$\overline{M}_w$	weight-average mol wt
$\overline{M}_n$	number-average mol wt
m	meter; milli (prefix for 10^{-3})
m	molal
m-	meta
max	maximum
MCA	Chemical Manufacturers' Association (was Manufacturing Chemists Association)
MEK	methyl ethyl ketone
meq	milliequivalent
mfd	manufactured
mfg	manufacturing
mfr	manufacturer
MIBC	methyl isobutyl carbinol
MIBK	methyl isobutyl ketone
MIC	minimum inhibiting concentration
min	minute; minimum
mL	milliliter
MLD	minimum lethal dose
MO	molecular orbital
mo	month
mol	mole
mol wt	molecular weight
mp	melting point
MR	molar refraction
ms	mass spectrum
mxt	mixture
μ	micro (prefix for 10^{-6})
N	newton (force)
N	normal (concentration); neutron number
N-	denoting attachment to nitrogen
n (as n_D^{20})	index of refraction (for 20°C and sodium light)
^{n} (as Bu^n), *n*-	normal (straight-chain structure)
n	neutron
n	nano (prefix for 10^{9})
na	not available
NAS	National Academy of Sciences
NASA	National Aeronautics and Space Administration
nat	natural
NBS	National Bureau of Standards
neg	negative
NF	*National Formulary*
NIH	National Institutes of Health
NIOSH	National Institute of Occupational Safety and Health
nmr	nuclear magnetic resonance
NND	New and Nonofficial Drugs (AMA)
no.	number
NOI-(BN)	not otherwise indexed (by name)
NOS	not otherwise specified
nqr	nuclear quadruple resonance
NRC	Nuclear Regulatory Commission; National Research Council
NRI	New Ring Index
NSF	National Science Foundation
NTA	nitrilotriacetic acid
NTP	normal temperature and pressure (25°C and 101.3 kPa or 1 atm)
NTSB	National Transportation Safety Board
O-	denoting attachment to oxygen

o-	ortho
OD	outside diameter
OPEC	Organization of Petroleum Exporting Countries
o-phen	*o*-phenanthridine
OSHA	Occupational Safety and Health Administration
owf	on weight of fiber
Ω	ohm
P	peta (prefix for 10^{15})
p	pico (prefix for 10^{-12})
p-	para
p	proton
p.	page
Pa	pascal (pressure)
pd	potential difference
pH	negative logarithm of the effective hydrogen ion concentration
phr	parts per hundred of resin (rubber)
p-i-n	positive-intrinsic-negative
pmr	proton magnetic resonance
p-n	positive-negative
po	per os (oral)
POP	polyoxypropylene
pos	positive
pp.	pages
ppb	parts per billion (10^9)
ppm	parts per million (10^6)
ppmv	parts per million by volume
ppmwt	parts per million by weight
PPO	poly(phenyl oxide)
ppt(d)	precipitate(d)
pptn	precipitation
Pr (no.)	foreign prototype (number)
pt	point; part
PVC	poly(vinyl chloride)
pwd	powder
py	pyridine
qv	quod vide (which see)
R	univalent hydrocarbon radical
(*R*)-	rectus (clockwise configuration)
r	precision of data
rad	radian; radius
rds	rate determining step
ref.	reference
rf	radio frequency, *n.*
r-f	radio frequency, *adj.*
rh	relative humidity
RI	Ring Index
rms	root-mean square
rpm	rotations per minute
rps	revolutions per second
RT	room temperature
^{s} (eg, Bu^s); *sec*-	secondary (eg, secondary butyl)
S	siemens
(*S*)-	sinister (counterclockwise configuration)
S-	denoting attachment to sulfur
s-	symmetric(al)
s	second
(s)	solid, only as in $H_2O(s)$
SAE	Society of Automotive Engineers
SAN	styrene–acrylonitrile
sat(d)	saturate(d)
satn	saturation
SBS	styrene–butadiene–styrene
sc	subcutaneous
SCF	self-consistent field; standard cubic feet
Sch	Schultz number
SFs	Saybolt Furol seconds
SI	Le Système International d'Unités (International System of Units)
sl sol	slightly soluble
sol	soluble
soln	solution
soly	solubility
sp	specific; species
sp gr	specific gravity
sr	steradian
std	standard
STP	standard temperature and pressure (0°C and 101.3 kPa)
sub	sublime(s)
SUs	Saybolt Universal seconds
syn	synthetic

t (eg, Bu[t]), *t*-, *tert*-	tertiary (eg, tertiary butyl)	Twad	Twaddell
		UL	Underwriters' Laboratory
		USDA	United States Department of Agriculture
T	tera (prefix for 10^{12}); tesla (magnetic flux density)	USP	*United States Pharmacopeia*
		uv	ultraviolet
t	metric ton (tonne); temperature	V	volt (emf)
		var	variable
TAPPI	Technical Association of the Pulp and Paper Industry	*vic*-	vicinal
		vol	volume (not volatile)
tex	tex (linear density)	vs	versus
T_g	glass-transition temperature	v sol	very soluble
tga	thermogravimetric analysis	W	watt
THF	tetrahydrofuran	Wb	Weber
tlc	thin layer chromatography	Wh	watt hour
TLV	threshold limit value	WHO	World Health Organization (United Nations)
trans-	isomer in which substituted groups are on opposite sides of double bond between C atoms	wk	week
		yr	year
TSCA	Toxic Substance Control Act	(*Z*)-	zusammen; together; atomic number
TWA	time-weighted average		

Non-SI (Unacceptable and Obsolete) Units		*Use*
Å	angstrom	nm
at	atmosphere, technical	Pa
atm	atmosphere, standard	Pa
b	barn	cm^2
bar†	bar	Pa
bbl	barrel	m^3
bhp	brake horsepower	W
Btu	British thermal unit	J
bu	bushel	m^3; L
cal	calorie	J
cfm	cubic foot per minute	m^3/s
Ci	curie	Bq
cSt	centistokes	mm^2/s
c/s	cycle per second	Hz
cu	cubic	exponential form
D	debye	C·m
den	denier	tex
dr	dram	kg
dyn	dyne	N
dyn/cm	dyne per centimeter	mN/m
erg	erg	J
eu	entropy unit	J/K
°F	degree Fahrenheit	°C; K
fc	footcandle	lx
fl	footlambert	lx
fl oz	fluid ounce	m^3; L
ft	foot	m
ft·lbf	foot pound-force	J

† Do not use bar (10^5Pa) or millibar (10^2Pa) because they are not SI units, and are accepted internationally only for a limited time in special fields because of existing usage.

Non-SI (Unacceptable and Obsolete) Units		*Use*
gf den	gram-force per denier	N/tex
G	gauss	T
Gal	gal	m/s^2
gal	gallon	m^3; L
Gb	gilbert	A
gpm	gallon per minute	(m^3/s); (m^3/h)
gr	grain	kg
hp	horsepower	W
ihp	indicated horsepower	W
in.	inch	m
in. Hg	inch of mercury	Pa
in. H_2O	inch of water	Pa
in.-lbf	inch pound-force	J
kcal	kilogram-calorie	J
kgf	kilogram-force	N
kilo	for kilogram	kg
L	lambert	lx
lb	pound	kg
lbf	pound-force	N
mho	mho	S
mi	mile	m
MM	million	M
mm Hg	millimeter of mercury	Pa
$m\mu$	millimicron	nm
mph	miles per hour	km/h
μ	micron	μm
Oe	oersted	A/m
oz	ounce	kg
ozf	ounce-force	N
η	poise	Pa·s
P	poise	Pa·s
ph	phot	lx
psi	pounds-force per square inch	Pa
psia	pounds-force per square inch absolute	Pa
psig	pounds-force per square inch gauge	Pa
qt	quart	m^3; L
°R	degree Rankine	K
rd	rad	Gy
sb	stilb	lx
SCF	standard cubic foot	m^3
sq	square	exponential form
thm	therm	J
yd	yard	m

BIBLIOGRAPHY

1. The International Bureau of Weights and Measures, BIPM (Parc de Saint-Cloud, France) is described on page 22 of Ref. 4. This bureau operates under the exclusive supervision of the International Committee of Weights and Measures (CIPM).
2. *Metric Editorial Guide* (*ANMC-78-1*) 3rd ed., American National Metric Council, 5410 Grosvenor Lane, Bethesda, Md. 20814, 1981.
3. *SI Units and Recommendations for the Use of Their Multiples and of Certain Other Units* (*ISO 1000-1981*), American National Standards Institute, 1430 Broadway, New York, N. Y. 10018, 1981.
4. Based on *ASTM E 380-82* (*Standard for Metric Practice*), American Society for Testing and Materials, 1916 Race Street, Philadelphia, Pa. 19103, 1982.
5. *Fed. Regist.*, Dec. 10, 1976 (41 FR 36414).
6. For ANSI address, see Ref. 3.

R. P. LUKENS
American Society for Testing and Materials

S continued

SULFONATION AND SULFATION

Sulfonation and sulfation are chemical methods for introducing the SO_3 group into organic entities (1); they are closely related and usually treated jointly. Although the terms sulfonation and sulfation are often used interchangeably, the two methods differ chemically.

In sulfonation, an SO_3 group is introduced into an organic molecule, to give a product with —SO_3— linkages (sulfonates) in the form of a sulfonic acid (—SO_3H), a salt (—SO_3Na), or a sulfonyl halide (—SO_3X). Sulfonic acids are generally produced directly by reaction of an aromatic hydrocarbon with sulfuric acid, sulfur trioxide, or chlorosulfuric acid. Sulfonation of unsaturated hydrocarbons with metal sulfites or bisulfites produces the metal sulfonate salts. These reactions are referred to as sulfitation or bisulfitation. In instances where the sulfur atom, at a lower valence, is attached to a carbon atom, the sulfonation process entails oxidation. Thus, the reaction of a paraffinic hydrocarbon with sulfur dioxide and oxygen is referred to as sulfoxidation, and the reaction of sulfur dioxide and chlorine as chlorosulfonation (2). Sulfoalkylation and sulfoarylation reactions involve the addition of sulfoalkyl or a sulfoaryl group to an organic molecule.

Sulfation is defined as any process of introducing an SO_3 group into an organic compound in which the reaction product (sulfate) exhibits the characteristic —OSO_3— molecular configuration. Unlike the sulfonates, which show remarkable stability even after prolonged heating, sulfated products are unstable toward acid hydrolysis. This difference forms the basis of an analytical method to identify mixtures of these two species.

In sulfamation, also termed *N*-sulfonation, compounds of the general structure R_2NSO_3H are formed as well as the corresponding salts, acid halides, and esters. The reagents are sulfamic acid (amidosulfuric acid), SO_3–pyridine complex, SO_3–tertiary amine complexes, SO_3–aliphatic amine adducts, and chlorine isocyanate–SO_3 complex (3).

Sulfonation and sulfation processes are important tools for the organic chemist in the design of specific molecules. For example, the introduction of a sulfonic acid group into a specific position in an aromatic nucleus and the subsequent substitution of a hydroxyl group by alkali fusion represents a basic industrial process. On the other hand, the product-development and the application chemists are most interested in the physical and functional properties contributed by the sulfonate moiety, such as solubility, emulsification, wetting, and foaming. Just as a sulfonated emulsifier concentrates at the interface between oil and water phases, contributing to a stable emulsion, so too, in a broader sense, sulfonation and sulfation processes provide the means to interface with, modify, or bridge aqueous and organic systems. This principle is utilized in the development of water-dispersible or water-soluble dyes, flavorants, and medicinals by introducing the water-solubilizing sulfonate group.

Sulfonation, sulfation, and sulfamation processes have led to the commercial development of new sulfonated and sulfated products. Sulfonated dyes were first developed in 1862, followed by the bisulfitation of lignin (qv) in 1866, and the oleum sulfonation of petroleum (qv) and natural occurring fats and oils in 1875 (see Fats and fatty oils). Since then, many new and improved processes have been developed contributing to today's highly efficient and complex technology. Although sulfuric acid-based sulfonation processes continue to dominate, SO_3-based processes have increased significantly over the last 15 years, including the development and utilization of continuous SO_3 film sulfonation and sulfation processes (see also Surfactants and detersive systems).

Detergent sulfonates or sulfates for use in household detergent and personal care products must have light color, little unreacted material or free oil, low inorganic salt content, and negligible odor.

Economic Aspects

It is estimated that in the United States during 1980, ca 1.6×10^6 metric tons of sulfonation, sulfation, and sulfamation products were manufactured with a market value in excess of $\$2.5 \times 10^9$, representing ca 4.5% of the total value of all synthetic organic chemical production (see Table 1). A detailed compilation of U.S. production of sulfonated and sulfated surface-active agents for 1980 is given in Table 2, whereas annual production of total surface-active agents from 1969 to 1980, as well as principal product classes are shown in Figure 1. Total surface-active agents show a modest growth during this period, with lignosulfonates and sulfated products growing faster than average (see Table 3). A comparative cost analysis for sulfonating linear detergent alkylate with sulfuric acid, oleum, or liquid SO_3 is given in Table 4. Sulfonation with SO_3 offers a 65% saving over sulfuric acid and a 39% saving over oleum. A similar comparison for the sulfation of commercial lauryl alcohol shows a 56.3% cost saving for liquid SO_3 against chlorosulfuric acid, exclusive of costs or credit created by HCl disposal or neutralization.

Table 1. Estimated U.S. Volume and Value of Sulfonation, Sulfation, and Sulfamation Products, 1980[a]

Product class	Metric tons	Volume, %	Value $	Value %
surfactants	1,338,488	82.6	1,273,278,000	50.8
dyes	56,105	3.5	398,150,000	15.9
pigments	15,966	1.0	688,950,000	27.5
medicinals	4,534	0.3	56,216,000	2.2
pesticides	68	0.004	3,299,000	0.1
cyclic intermediates	204,750	12.6	88,467,000	3.5
Total	*1,619,911*		*2,508,360,000*	

[a] Refs. 4–5, and other industry sources.

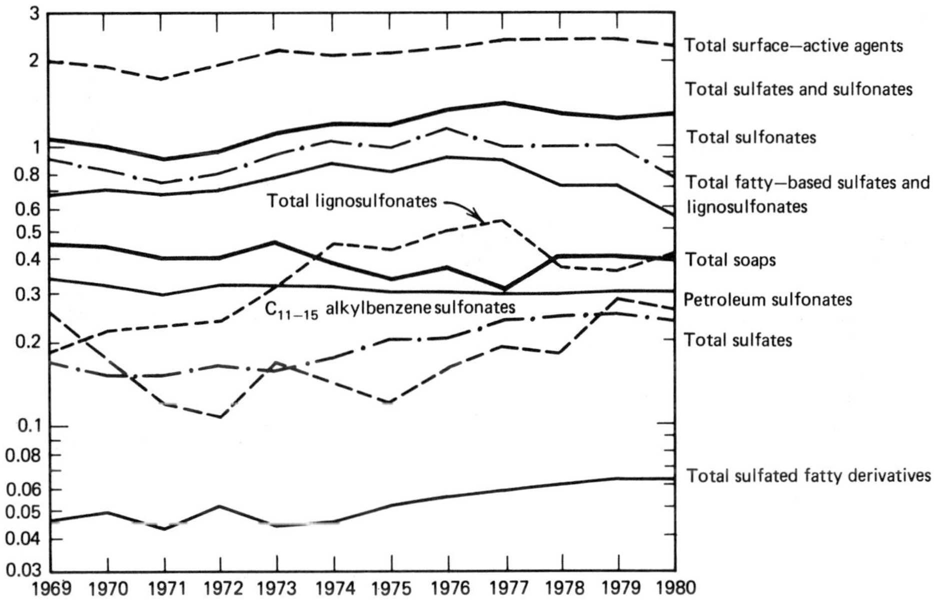

Figure 1. Annual U.S. production of surface-active agents, 10^6 metric tons.

Reagents

Sulfonation and sulfation reagents are listed in Table 5. In the 1860s, sulfuric acid was the only known sulfonating reagent; today, it and oleum are the most widely used reagents because of their convenience and low cost. Since the sulfonation of aromatic compounds with sulfuric acid produces 1 mol of water per mol of compound, 3–4 mol acid are required to obtain a suitably high conversion. This excess increases costs, but reduces viscosity and provides a good reaction medium, diluent, and heat sink. As a general rule, removal of the spent acid is required; neutralization, for example, with caustic soda results in a high sodium sulfate content which impairs the quality of the product.

Chlorosulfuric acid, usually employed in stoichiometric amounts, is another important low cost sulfonation reagent. It does, however, suffer from two drawbacks; namely, its corrosive nature, and the fact that it reacts with evolution of hydrogen

Table 2. U.S. Production of Sulfonated and Sulfated Anionic Surface-Active Agents, 1980[a,b]

Product	Metric tons	Percent of grand total	Unit value, $/t	Value, $
alkylbenzenesulfonates				
dodecylbenzenesulfonic acid	102,734		992.08	101,920,347
calcium salt	6,125		2,006.20	12,287,975
isopropylammonium salt	1,565		1,653.45	2,587,649
sodium salt	122,455		859.80	105,286,809
triethanolammonium salt	2,647		1,234.59	3,267,960
tridecylbenzenesulfonate, sodium salt	51,025		1,300.73	66,369,748
all other	7,648		1,653.47	12,645,739
Total	*294,199*	*22.00*		*304,366,227*
xylene and other aromatic sulfonates				
xylenesulfonate, ammonium salt	7,462		704.52	5,257,128
xylenesulfonate, sodium salt	15,211		947.99	14,419,876
all other	11,322		859.80	9,734,656
Total	*33,995*	*2.54*		*29,411,660*
ligninsulfonates				
calcium salt	281,631		88.18	24,834,222
iron salt	863		396.83	342,464
sodium salt	46,259		374.79	17,337,411
all other	70,394		308.65	21,727,108
Total	*399,147*	*29.85*	*154.32*	*64,241,205*
naphthalenesulfonates	9,689	*0.72*	*1,455*	*14,097,979*
naphthalenesulfonate–formaldehyde condensates[c]	*32,191*	*2.41*	*1,035*	*33,309,959*
sulfonic acids with amide linkages				
sulfosuccinamates	1,482		1,940.07	2,875,184
taurates	1,158		3,196.70	3,701,779
Total	*2,640*	*0.20*		*6,576,963*
sulfonic acids with ester or ether linkages				
sulfosuccinic acid esters, total	9,837		2,289.68	22,523,582
other sulfonic acids and salts, total	20,370		3,483.30	70,954,821
Total	*30,207*	*2.26*		*93,478,403*
other sulfonic acids and salts, total	*9,839*	*0.74*	*1,278.68*	*12,580,933*
petroleum sulfonates				
calcium salts	141,316		1,543.23	218,083,091
sodium salts	95,954		992.08	95,194,044
other petroleum salts	20,514		1,807.79	37,085,004
Total	*257,784*	*19.28*		*350,362,139*
sulfated acids, esters, and amides				
butyl oleate, sodium salt	478		992.08	474,214
isopropyloleate, sodium salt	17			
tall oil, sodium salt	1,453		617.29	896,922
all other	8,649		1,344.82	11,631,348
Total	*10,597*	*0.79*		*13,002,484*
sulfated alcohols				
dodecyl sulfonates				
ammonium salt	3,757		2,204.62	8,282,757
magnesium salt	96		3,284.88	315,348
sodium salt	8,186		2,292.80	18,768,861
triethylammonium salt	4,263		2,248.71	9,586,251
2-ethylhexyl sulfate, sodium salt	892		1,873.93	1,671,546

Table 2 (*continued*)

Product	Metric tons	Percent of grand total	Unit value, \$/t	Value, \$
mixed linear alcohol sulfates, [illegible]	24,412		2,182.57	53,280,899
sulfated [illegible] salts				
[illegible]um salt	123		3,130.56	385,059
alkylphenols, ethoxylated, [illegible]	67,213		2,270.76	152,624,592
[illegible]	*108,942*	*8.15*		*244,915,313*
dodecyl alcohol, ethoxylated, sulfated, ammonium salt	[illegible]		2,377.75	6,006,197
[illegible]			1,168.45	2,230,571
dodecyl alcohol, ethoxylated, sulfated, sodium salt	6,958		[illegible]11	13,652,361
mixed linear alcohols, ethoxylated, sulfated ammonium salt	38,897		837.76	32,586,351
all other	60,684		1,918.02	116,393,126
Total	*110,974*	*8.30*		*170,868,606*
natural fats and oils, sulfated				
castor oil, sodium salt	2,396		1,234.59	2,958,078
cod oil, sodium salt	746		573.20	427,607
herring oil, sodium salt	844		639.33	539,595
neat's-foot oil, sodium salt	613		683.43	418,943
tallow, sodium salt	1,260		595.25	750,015
Total	*5,859*	*0.44*		*5,094,238*
all other anionic surface-active agents sulfated and sulfonated, total	*31,061*	*2.32*	*771.62*	*23,967,289*
Grand total	*1,337,098*		*951.28 (avg)*	*1,366,273,398*

[a] Refs. 4–5.
[b] 100% pure.
[c] Tanning agents.

Table 3. U.S. Production Growth Rates for Surface-Active Agents, Soaps, and Sulfonated and Sulfated Products, 1969–1979

Products	%
surface-active agents, total	+14.6
soaps	−8.5
sulfonates, total	+10.7
alkylbenzenesulfonates	−12.1
lignosulfonates	+91.8
fatty acid-based sulfonates	+34.0
petroleum sulfonates	+5.3
sulfates, total	+47.6

chloride, which presents a disposal problem. Alternatively, the hydrogen chloride may be absorbed in a water scrubber and sold as a commercial grade or recycled.

Stabilized liquid sulfur trioxide, introduced commercially as a sulfonation/sulfation reagent in 1947 (7), is second only to oleum; in the United States, in fact, it is preferred to oleum for detergent manufacture. Because of its extreme reactivity, special

Table 4. Cost of Sulfonation/Sulfation Agents

Agent	kg required	\$/kg agent[a]	\$/t substrate	Total cost, \$
sulfonating agent				
for 1 t linear alkylate (LA)				
H_2SO_4	7292	0.0818	81.86	122.79
oleum	5590	0.0890	89.02	102.38
SO_3, stabilized	1730	0.2095	209.44	[illegible]5.77
sulfating agent				
for 1 t fatty alcohol				
chlorosulfuric acid	281.6	[illegible]	[illegible]	[illegible]
SO_3, stabilized	1944.4	[illegible]	[illegible]	[illegible]

[a] Based on quoted prices as [illegible]

handling and processing methods are required in order to avoid overreaction, oxidation, and charring. The extreme reactivity of liquid SO_3 can be moderated with solvents such as liquid SO_2, low boiling halogenated hydrocarbons, and paraffins. The reactivity of SO_3 is best controlled by dilution of gaseous SO_3 with nitrogen, SO_2, or dry air. Commercial sulfonation/sulfation processes generally employ a 2–8% SO_3–dry air mixture, or sometimes converter gas at 8% SO_3 concentration. Liquid SO_3 is continuously vaporized into a dry air stream to produce the desired mixture, or provided directly from a sulfur burner equipped with a catalytic converter (see below). Sulfur trioxide may also be obtained from oleum.

Sulfur trioxide occurs in three polymeric forms, known as alpha-, beta-, and gamma-SO_3 (see Table 6). Only the gamma polymer is liquid at room temperature; it can be stabilized to prevent conversion to the beta form, which is caused by small amounts of moisture. Unstabilized sulfur trioxide must be protected from freezing and moisture; otherwise, polymerization and solidification may create serious process problems.

The extreme reactivity of SO_3 may also be moderated by complexing with Lewis bases, eg, dioxane, DMF, pyridine, THF, dimethyl sulfoxide, trialkyl phosphates, and organic bases and amines such as trimethylamine; triethylamine; *N,N*-dimethylaniline; *N,N*-diethylaniline; morpholine; and quinoline. The reactivity of such complexes is inversely proportional to their stability, and consequently they can be selected for a wide variety of processing conditions. For example, the SO_3–trimethylamine complex is very stable, melting at 239°C. It is soluble in many solvents, including water, in which it is sufficiently stable to be used for sulfation. At present, SO_3-adducts are not commercially available. In general, the complexing agent must be removed after the reaction is completed.

A mixture of gaseous sulfur dioxide and chlorine is a useful reagent for reactions known as sulfochlorination. The actual reagent is believed to be sulfuryl chloride which also can be prepared by heating chlorosulfuric acid in the presence of catalysts, or from sulfur dioxide and chlorine in the presence of activated carbon or camphor. Sulfuryl chloride is soluble in glacial acetic acid and is readily decomposed by hot water or alkalies. Its bp is ca 69.2°C and it is extremely corrosive and toxic.

In sulfoxidation, sulfur dioxide and oxygen, catalyzed by uv radiation, convert paraffins to sulfonic acids:

$$2\,RH + 2\,SO_2 + O_2 \xrightarrow{h\nu} 2\,RSO_2OH \qquad (1)$$

Economic considerations heavily influence the choice of reagent. Although both sulfuric acid and oleum are cheaper than SO_3, a large excess of both reagents is required, which creates substantial quantities of spent acid that have to be discarded or recycled, and affects the economics of the process. In contrast, SO_3 and chlorosulfuric acid usually react stoichiometrically to give products containing very little sulfuric acid; consequently, the neutralized product contains only small amounts of sodium sulfate.

Sodium sulfite and bisulfite are also useful sulfonation agents; they offer economic advantages and versatility (see below).

Sulfonation

The sulfonation of benzene and short-chain alkylbenzenes, such as toluene, the xylenes, cumene, and ethylbenzene, is an important commercial process that is carried out in a number of ways, using sulfuric acid, oleum, or sulfur trioxide. The sulfur trioxide route presents potential fire and explosion hazards and produces by-product sulfones that impair product solubility. Precise process control is essential, and the sulfones are removed by extraction.

The reaction of toluene with 97% sulfuric acid is highly exothermic:

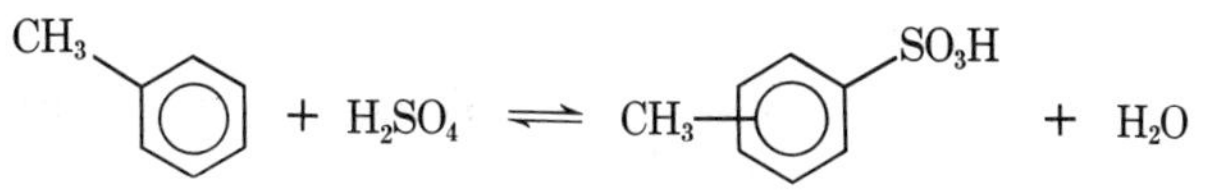

(2)

If excess acid is employed, the sulfonic acid can be extracted. However in other industrial processes, particularly in the sulfonation of xylenes, a large excess of the organic compound is used. The reaction is generally carried out under reflux and the water formed is distilled until the sulfuric acid has been completely consumed. The sulfonic acid is isolated by neutralization with caustic soda, and the excess xylene is separated and recycled. Toluene sulfonation yields 62% *para*-, 32% *ortho*-, and ca 6% *meta*-toluenesulfonic acid (8). The yield of meta isomer can be increased by raising the reaction temperature and increasing the strength of the sulfuric acid. Thus, 97.4% sulfuric acid at 100°C for 2.7 h gave a maximum meta isomer yield of 33%, with 47.3% para and 19.7% ortho isomers (9).

Sulfonated toluene, xylene, and cumene, primarily in the form of ammonium and sodium salts, are important hydrotropes or coupling agents in the manufacture of liquid cleaners and surfactant compositions, and serve as key crisping agents in drum and spray-drying operations.

Phenol from Benzene. For many years, phenol (qv) was manufactured principally via the sulfonation of benzene. The process was cumbersome and involved a large number of separate steps, and the production of 1 kg phenol, required 4.3 kg benzene:

$$C_6H_6 + H_2SO_4 \rightarrow C_6H_5SO_3H + H_2O \quad (3)$$

$$2\ C_6H_5SO_3H + Na_2SO_3 \rightarrow 2\ C_6H_5SO_3Na + H_2SO_3 \quad (4)$$

$$C_6H_5SO_3Na + 2\ NaOH \rightarrow C_6H_5ONa + Na_2SO_3 + H_2O \quad (5)$$

$$2\ C_6H_5ONa + H_2SO_3 \rightarrow 2\ C_6H_5OH + Na_2SO_3 \quad (6)$$

Table 5. Reagents for Sulfonation and Sulfation[a]

Reagent	Chemical formula	Physical form	Principal uses	Relative usage	Reactivity	General comments
sulfur trioxide	SO_3	liquid		very limited	extremely reactive	generally produces oxidation, charring, used with solvents to modify reactivity
		gas	widely used for organic compounds	used significantly	highly reactive generally mol per mol; instantaneous	used in dilution to 2–8% SO_3 in dry air or diluent gas; can be supplied by vaporizing liquid SO_3, stripping from oleum, sulfur burner or converter gas stream
from sulfur + air		gas	same as above; optional source of SO_3	moderate	same as above	
oleum, 20%, 30%, and 65%	$H_2SO_4.SO_3$	liquid	for alkylaryls for detergents, dyes	most widely used sulfonating, sulfating agent	high reactivity	equilibrium reaction, splits water; requires 3 mol available SO_3 or more per mol feedstock
chlorosulfuric acid	$ClSO_3H$	liquid	for alcohols, dyes	moderate	high reactivity; mol per mol reaction	generates HCl which must be recovered
sulfuryl chloride	SO_2Cl_2	liquid	sulfonation of acetylene groups, lab mainly	low; mainly lab research tool	moderate	produces sulfonyl chlorides
sulfamic acid	NH_2SO_3H	solid	sulfation of alkyl phenols, some specialties	limited	relatively low	requires substantial heat to effect reaction producing ammonium salt directly
sulfuric acid (96–100%)	H_2SO_4	liquid	used to sulfonate aromatics	significant	low	generally reacts under azeotropic heating to

Table 5 (*continued*)

Reagent	Chemical formula	Physical form	Principal uses	Relative usage	Reactivity	General comments
						remove H_2O
sulfur dioxide plus chlorine gases	$SO_2 + Cl_2$	gas mixture	chlorosulfonation of paraffinic hydrocarbons	very limited	low	requires a catalyst; produces sulfonyl chloride + HCl
sulfur dioxide and oxygen gases	$SO_2 + O_2$	gas mixture	sulfoxidation of paraffinic hydrocarbons	very limited	low	requires a catalyst; produces sulfonic acid
sodium sulfite	Na_2SO_3	solid	sulfonation of alkyl chlorides	substantial	low	requires heating in aqueous medium
sodium bisulfite	$NaHSO_3$	solid	sulfation of conjugated olefin species (succinates, etc), sulfonation of lignins	substantial	low	requires heating in aqueous medium
hydroperoxide–bisulfite (O_2 + $NaHSO_3$)	$NaHSO_3$ + O_2	gas and solid	sulfonation/sulfation of paraffinic petroleum fractions	very limited	low	aromatic compounds undergo cleavage to yield phenols which inhibit hydroperoxidation of most hydrocarbon species

[a] Reagents listed in the order of generally descending reactivity.
[b] U.S. raw material prices obtained from ref. 6; all prices fob producing point.
[c] Amidosulfuric acid.

This process was replaced by the alkaline hydrolysis of chlorobenzene under high pressure with a copper catalyst. Today, the principal U.S. process is based on the oxidation of cumene (qv), which gives acetone (qv) as the main by-product.

Resorcinol from Benzene. Resorcinol is prepared by the fusion of *m*-benzenedisulfonic acid with sodium hydroxide. Benzenesulfonic acid is first prepared using 100% sulfuric acid. The second sulfonic acid group is introduced by adding 65% oleum, whereas addition of sodium sulfate inhibits sulfone formation:

$$C_6H_6 + H_2SO_4 \rightarrow C_6H_5SO_3H + H_2O \quad (7)$$

$$C_6H_5SO_3H + H_2SO_4.SO_3 \xrightarrow{Na_2SO_4} m\text{-}C_6H_4(SO_3H)_2 + H_2SO_4 \quad (8)$$

$$m\text{-}C_6H_4(SO_3H)_2 + 4\ NaOH \xrightarrow{\text{Fusion}} m\text{-}C_6H_4(ONa)_2 + 2\ Na_2SO_3 + 2\ H_2O \quad (9)$$

Table 6. Polymers of Sulfur Trioxide

	Gamma form	Beta form	Alpha form
probable structure[a]	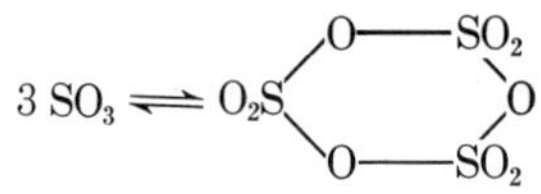	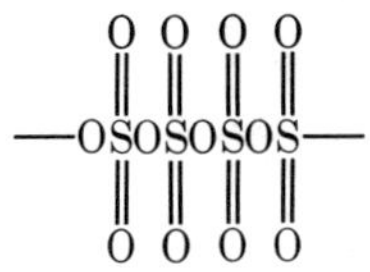	similar to beta form but with chains joined in a layered structure
physical form	liquid or vitreous	silky fibers	fibrous needles
equilibrium melting point, °C	16.8[b]	32.5	62.3
vapor pressure, kPa[c]			
at −3.9°C	28.3	24.1	3.4
23.9°C	190.3	166.2	62.0
51.7°C	908.0	908.0	699.1
79.4°C	3280.6	3280.6	3280.6

[a] As suggested by electron diffraction patterns, ir, and raman spectra.
[b] Boiling point, 44.7°C.
[c] To convert kPa to psi, multiply by 0.145.

Naphthalene and Derivatives. The sulfonation of naphthalene, naphthylamines, and naphthols is an important step in the manufacture of dyes and dye intermediates (qv) (10–13). Nearly all of these sulfonations are carried out with concentrated sulfuric acid or oleum. In general, mixtures are formed; however, satisfactory yields of preferred isomers can be achieved by careful control of conditions. As a rule, low temperatures (<100°C) and short reaction times favor sulfonation in the 1 position; substitution in the 2 position can be achieved by longer reaction time and higher temperature.

The monosulfonation of naphthalene is much faster than that of benzene; sulfone does not form. The effect of temperature on isomer distribution is shown below:

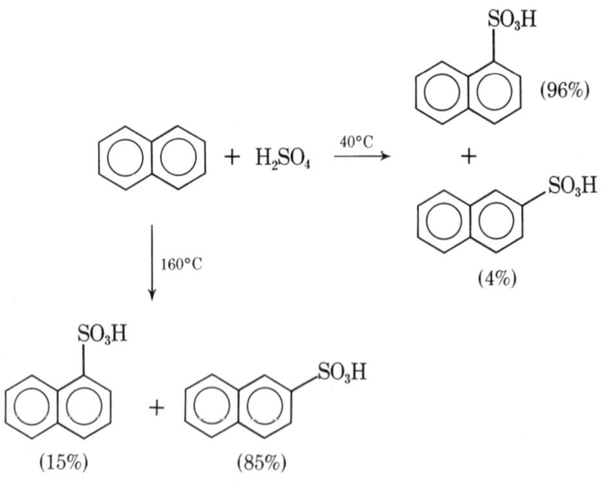

(10)

(11)

Heating naphthalene in a sealed tube for about 1450 h at ca 100°C with concentrated sulfuric acid at a mole ratio of 1.08, gives near a stoichiometric yield of 2-naphthalenesulfonic acid. Sulfur trioxide and its adducts are little used because they give undesirable by-products such as sulfones. Condensation of mono- and polysulfonated naphthalenes with formaldehyde gives products of commercial utility as dispersing agents.

The alkylnaphthalenesulfonates are used as dye-leveling agents (14), rubber stabilizers and antioxidants (15), crisping agents, anticaking agents (16), detergents and fertilizers (17), and heavy-duty synthetic detergents (18). In addition, they are effective in retarding the setting of cement, an important step in the cementing of oil wells (19). Other applications include stabilization of hydraulic fluids (qv) (20) and as additives for lubricating oils. Polymeric alkylnaphthalene sulfonates are used as tanning agents, dispersants, emulsifiers for latex polymerization (21), and as agents for improvement of wash fastness of dyes (22). High molecular weight polyalkylnaphthalenesulfonates improve the thermal stability of polymeric compositions containing inorganic fillers and conventional antioxidants or stabilizers (23).

Dyes and Dye Intermediates. In 1965, ca 1800 dyes were manufactured in the United States and 600 were imported. Since then, imports have increased as stringent environmental regulations and archaic facilities have weakened the competitive position of the U.S. dye industry.

Dyes bearing $—SO_3H$, $—SO_3^-$, $—SO_2NH_2$, and $—SO_3Na$ are grouped into 11 classifications that comprise ca 42.3% of all dye classes. They include nitroso, nitro, diazo, triazo, polyazo, stilbene, triarylmethane, quinoline, azine, oxazine, and phthalocyanine dyes (see also Azine dyes; Azo dyes; Dyes and dye intermediates; Pigments; Stilbene dyes; Sulfur dyes; Thiazole dyes; Triphenylmethane and related dyes). The 1980 U.S. production of these dyes and derived pigments was estimated at 142,755 t with a sales value of $\$1.266 \times 10^9$.

The cyclic intermediates for dyes are predominantly mono- and disulfonic acid derivatives, followed by sulfonamides and sulfonyl chlorides, eg, metanilic acid (*m*-aminobenzenesulfonic acid) and Schaeffer's acid (2-naphthol-6-sulfonic acid).

Long-Chain Alkylaromatic Compounds. The most widely used raw materials for the preparation of detergent sulfonates are the linear alkylates (LA) or linear alkylbenzenes (LAB) with a molecular weight range of 236–262 and their branched-chain counterparts, eg, dodecylbenzene (DDB). Sulfonation yields the sulfonic acid, LABS or DDBS (24). Sulfonates based on tetrapropylene and partially branched-chain paraffins (kerosene) do not meet the current standard for biodegradability exhibited by linear alkylatesulfonates.

The simplest alkylate sulfonation process uses 100% sulfuric acid (fp = 10.2°C). The by-product is water, which gradually dilutes the sulfuric acid. A large excess of sulfuric acid is therefore required. Oleum is employed in a smaller excess, giving a reaction product containing less spent acid. The yield of organic sulfonate can be increased to 85–90% by watering out the spent acid. Upon completion of sulfonation, the reaction mass is diluted with water to give a phase separation consisting of spent acid in the lower layer and organic sulfonic acid in the upper layer. Spent acid strengths of 72–78% give the quickest and most complete separation. Residues of spent acid in the alkylate sulfonic acid create the following problems: increased corrosiveness and shipping weight; increased caustic soda requirements for the neutralization step; and salts raising the cloud point of a liquid detergent.

In recent decades, batch sulfonation has given way to continuous oleum sulfonation units (25–26). By 1963, a total of 70 such units were in operation throughout the world (27). The Chemithon (25) and Ballestra (26) continuous oleum sulfonation units have been successfully utilized in cosulfonation of toluene or xylene with detergent alkylate.

The cosulfonation/sulfation of detergent alkylate with fatty alcohols or alkoxylated alcohols is an alternative for the postblending of neutralized salts (28). In the continuous Chemithon process, the alkylate is first converted to the sulfonic acid with oleum; in the second stage, the sulfonic acid is mixed with a fatty alcohol and additional oleum for sulfation. The mixture is conducted through a heat exchanger and then neutralized to prevent acid hydrolysis. In the Chemithon film process, two parallel reactors are used, one for sulfonation of the alkylate and the other for sulfation of the fatty alcohol. The reaction of the converging streams is completed in a digester that discharges to a continuous neutralization unit.

Detergent–alkylate still bottoms are usually sulfonated with 100% sulfuric acid and 20% oleum. Some of the products are used as controlled foaming surfactants in laundry detergents; others have limited application as calcium and barium salts in oil additives. The use of SO_3 for such sulfonations generally requires a solvent-like petroleum white oil and subsequent extraction with methanol.

By 1964, replacement of oleum by SO_3 was well underway; today, Stepan Chemical Company, Ultra Chemical Company (Witco), Tennessee Corporation, and Pilot Chemical Company operate SO_3–sulfonation processes with a combined capacity of ca 91,000 t/yr. By 1966, the sulfonation of alkylates had become a predominantly continuous process, with sulfur trioxide as the preferred agent. Thus, the alkylarenesulfonic acids currently produced are of high purity (97–99%) and exhibit good color; they are low in free oil and free sulfuric acid.

Since water is not a by-product, only a slight excess of SO_3 is required for near-stoichiometric conversion. A troublesome problem posed first by this process was the formation of the sulfonic anhydride $(RSO_2)_2O$, which caused pH drift in neutralized slurries. Studies showed that excess SO_3, and short reaction and digestion times favored anhydride formation. Hydrolysis after digestion eliminates this problem:

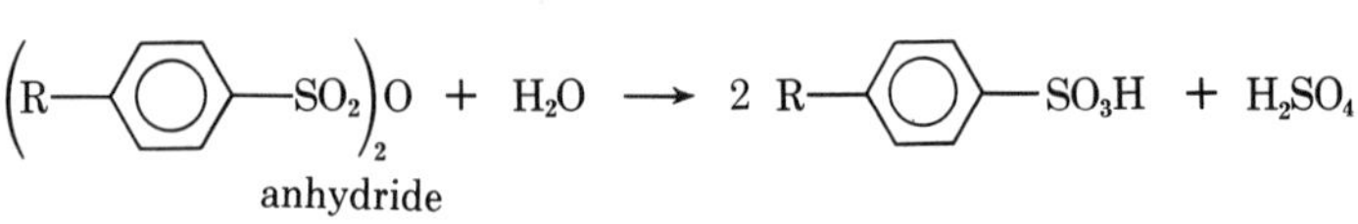

(12)

R = dodecyl. In addition, the water treatment prevents darkening of the sulfonic acid.

Sulfonation of purified linear alkylate bottoms containing substantial quantities of dialkylbenzenes gives lubricating-oil additives (29). The sodium, potassium, and ammonium salts of sulfonated styrene oligomer (styrene still bottoms) are useful as dispersing agents (30), viscosity reducers, and wetting agents. Mixtures (1:1) of nonyl- and octadecyl-*o*-xylenesulfonic acid salts of ethanolamine are used in oil-recovery processes (31), whereas low molecular weight linear alkylbenzenesulfonates are used as frothing agents in gypsum board-manufacture (32).

Sulfitation of Unsaturated Compounds. Sulfites react with unsaturated compounds such as olefins, olefin oxides, aldehydes, and ketones to give aliphatic sulfonates or hydroxysulfonates (30,33). These compounds can be used as intermediates in the synthesis of a variety of organic compounds.

Sulfosuccinates and Sulfosuccinamates. The principal starting materials for these compounds are bisulfite and maleic anhydride, maleic acid, and fumaric acid. Maleic anhydride reacts with a hydroxyl compound to give a mono- (or half-ester) or a diester:

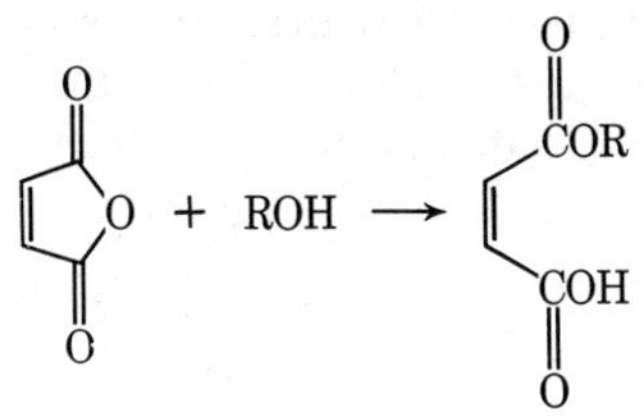

(13)

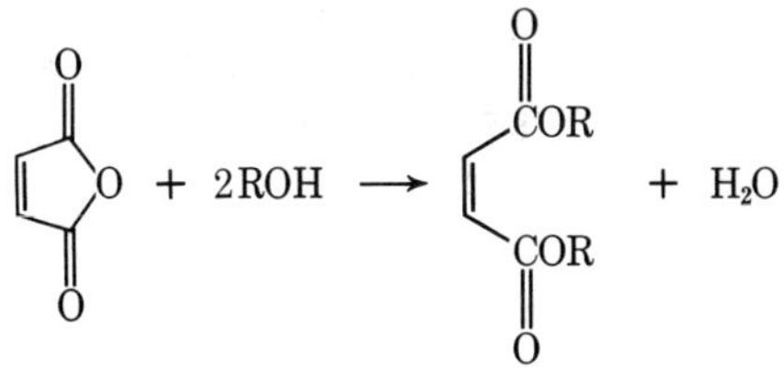

(14)

The water produced in the diester reaction must be removed to force the reaction to completion.

Treatment of these esters with an alkaline solution of sodium bisulfite gives the corresponding sulfosuccinates:

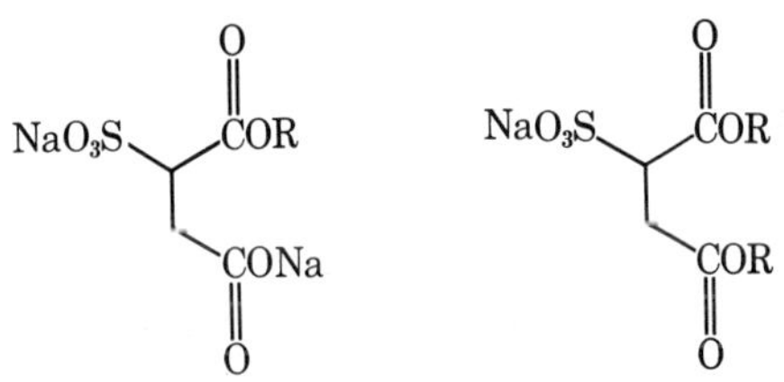

(15)

Sulfitation with bisulfite gives sulfosuccinamates. In both instances, the products are sulfonates and not sulfates, although sulfite was the reactant.

Alternatively, maleic anhydride and an amine gives a maleamic acid or maleic acid amide:

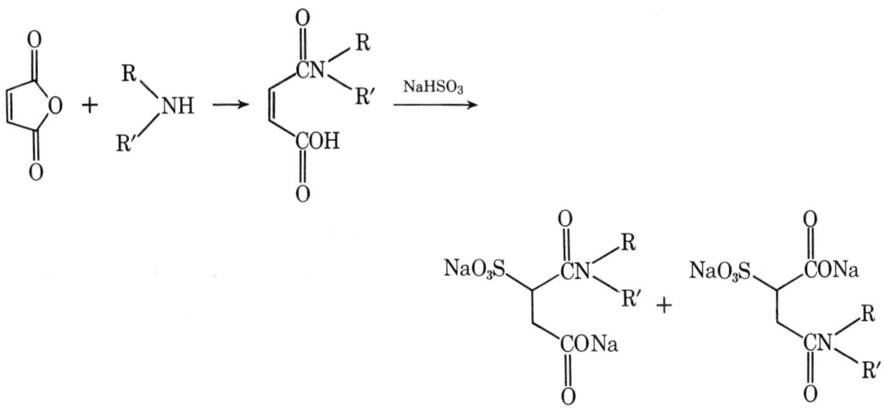

(16)

Variations in the choice of reactants can give a broad spectrum of sulfosuccinates and sulfosuccinamates of widely varying chemical and physical properties. These compounds find application in household, toiletry, and cosmetic products. Generally mild to the skin, they are used in shampoos and cleansing creams.

Bisulfitation of Alpha Olefins. Primary paraffin sulfonates (PPS) are prepared by radiation- or peroxide-catalyzed bisulfitation of linear α-olefins (34–35). The PPS products exhibit lower solubility and poorer performance than secondary alkanesulfonate (SAS) products, and are not produced commercially.

Other Sulfitation Reactions. The addition of sulfonate group, —SO_3Na, to alkenes and alkynes can be done by a free-radical chain process (36–38):

$$SO_3^{\cdot} + CH_2{=}CHR \rightarrow {-}SO_3CH_2\dot{C}HR \tag{17}$$

$$-SO_3CH_2\dot{C}HR + HSO_3^- \rightarrow {-}SO_3CH_2CH_2R + SO_3^{\cdot} \tag{18}$$

The reaction is used most often with terminal olefins. At atmospheric pressure, ethylene yields ethanesulfonic acid; at high pressure, ethylene and propylene yield telomeric sulfonates. Perfluorinated alpha olefins yield perfluoroalkanesulfonic acids.

Styrene yields three products (10):

$$C_6H_5CH{=}CH_2 \xrightarrow{NaHSO_3} C_6H_5CH_2CH_2SO_3Na + C_6H_5CH(OH)CH_2SO_3Na + C_6H_5CH{=}CHSO_3Na \tag{19}$$

Long-chain (C_{10}–C_{20}) alpha olefins form detergent sulfonates under the following conditions: air or peroxide as initiator; an alcoholic solvent (methanol, ethanol, or isopropyl alcohol); pH between 7 and 9; efficient mixing; and slow addition of bisulfite.

The sulfonation of allyl alcohol produces propanesultone, a powerful sulfoalkylating agent:

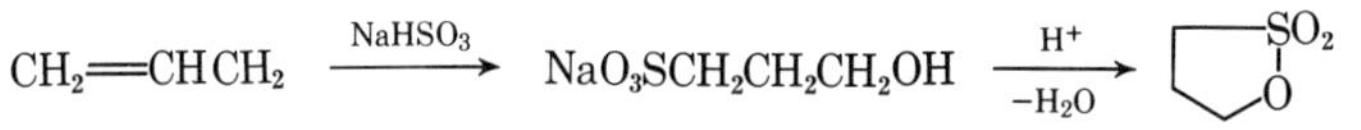

(20)

2-Methylallyl (methallyl) and crotyl alcohols, and 3-hydroxy-1-nonadecene have been similarly sulfonated; the first two products can be converted to sultones.

In contrast to the free-radical chain process, bisulfite reacts with olefins conjugated to carbonyl, cyano, nitro, or similar groups by ionic addition (39). Commercial long-chain, surface-active sulfonates are prepared by this reaction from amides and esters of acyclic, aconitic, citraconic, and maleic acids.

Bisulfite addition to alkynes is thought to be free radical in character (40), and gives a disulfonate:

$$RC{\equiv}CH + NaHSO_3 \rightarrow RCH{=}CHSO_3Na \xrightarrow{NaHSO_3} RCH(SO_3Na)\dot{C}H_2SO_3Na \tag{21}$$

Hydroxysulfonates are produced by adding bisulfite to aldehydes and ketones:

$$H_2CO + NaHSO_3 \rightarrow HOCH_2SO_3Na \tag{22}$$

Considerable research is being conducted on the reaction of sugars (glucose, galactose, mannose, rhamnose, and arabinose) with bisulfite, see refs. 41–45.

The reaction of sodium bisulfite with epoxides, cyclic imines, and episulfides results in ring cleavage and formation of the corresponding sulfonate. Thus, ethylene oxide reacts readily with sulfur dioxide to form ethylene sulfite, which is converted to sodium 2-hydroxyethanesulfonate by treatment with sodium bicarbonate (46):

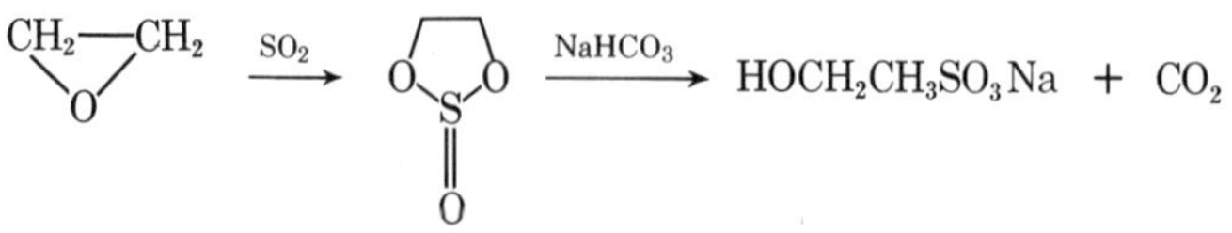

(23)

The product is a commercial sulfoethylating agent.

Ethyleneimine and sulfur dioxide give taurine a key intermediate in the synthesis of surfactants:

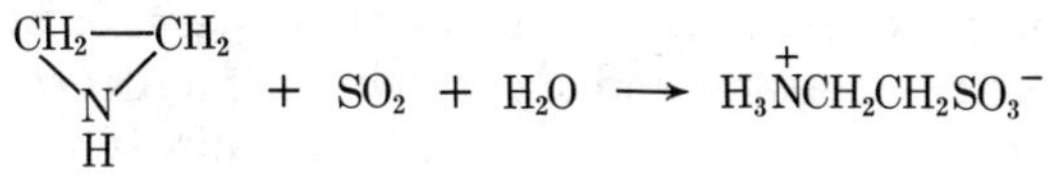

(24)

The reaction of hydroxy and amino compounds with bisulfite has led to the proposal of a new mechanism for the Bucherer reaction (47). This reaction has been conducted in the presence of an oxidizing agent or an oxidizing moiety in the compound being sulfonated. In some instances, the hydroxyl group is introduced during sulfonation and, on occasion, disulfonates are formed. For example, conversion of 1-nitroso-2-naphthol to 1-amino-2-hydroxynaphthalene-4-sulfonic acid which, together with its 6-methoxy derivative, has proved to be of commercial interest. This method has also been used for the synthesis of sulfamates.

Sodium sulfite reacts with nitro compounds:

$$RNO_2 + Na_2SO_3 \rightarrow RSO_3Na + NaNO_2 \qquad (25)$$

The nitro compound is heated or refluxed with an aqueous or alcoholic solution of sodium sulfite. This reaction is used to separate 1,2-dinitrobenzene from the 1,3-isomer which does not react; the 1,4-isomer, however, is sulfonated.

This reaction is also used for the purification of TNT; the 3-nitro groups in 2,3,4- and 2,4,5-trinitrotoluene are readily sulfonated, whereas the 2,4,6 isomer does not react (8).

The displacement of halogen with sulfite has been applied to a variety of brominated and chlorinated aromatic compounds; eg, the sulfonation of 2-bromo- and 2-chloro-1,4-diaminoanthraquinones for the preparation of dye intermediates (48).

The addition of bisulfite to a carbon–nitrogen double bond is illustrated by the reaction of 2-aminopyrimidine (49):

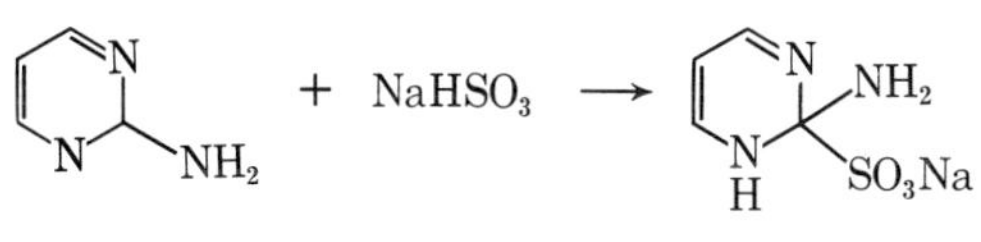

(26)

This synthesis has no significant industrial applications. Results have been similar for the products resulting either from the addition of bisulfite to heteroconjugate (50–51) systems or from the displacement of halogen in heterocyclics (52) with sulfite.

The use of an oxidizing agent to promote sulfonation with sulfite has been applied principally in extensive studies made of strychnine and brucine (53).

Alpha Olefins and SO_3. Alpha-olefin sulfonates, AO sulfonates or AOSs, are obtained by the treatment of α-olefins with SO_3 (54–58). It is estimated that ca 13,600 metric tons per year will be used in the United States from 1979–1984 (59) for personal care and household articles. Alpha-olefin sulfonates are less irritating to the skin than LAS; oral toxicity is also less (54).

Alpha-olefin sulfonates are the least toxic among the main anionic surfactants; LD_{50} is twice that of LAS (60). Eye irritancy is less than that of LAS. Alpha-olefinsulfonates have good solubility, are rapidly and completely biodegraded, have good detergency, and are stable. Industrial applications include fertilizer formulation (61), deinking of newsprint, emulsion polymerization (62), and admixture with concrete.

Processes. Some processes are based on solvent systems or SO_3 complexing agents (63–65). The latter are cumbersome and uneconomical. The development of continuous SO_3 film sulfonation technology made the commercialization of AOS possible (54–57,65–70). In this process, liquid AO and a gaseous mixture of SO_3 and dry air are introduced at the top of a vertical water-cooled tubular reactor unit. The reaction product is separated from the gaseous effluent, neutralized, and further processed to yield the AOS product. A slight excess of SO_3 is required to obtain high olefin conversion (54,71). A complex mixture of olefin mono- and disulfonic acids, sultones, and unreacted olefin is obtained. Some intermediates isomerize on acid aging (see Fig. 2) (55).

Acid Aging. The product obtained from the thin-film reactor contains a 1,2-sultone which yields 2-hydroxysulfonate, a less desirable component because of its limited solubility. A period of ca 2–15 min before neutralization allows the 2-sultone to isomerize to the 3-sultone which yields the more soluble 3-hydroxysulfonate (71). A longer period produces 4-sultone, which is much more difficult to hydrolyze. Freshly prepared AOS may contain oligomers (sulfonate esters) (58).

Neutralization and Saponification. The aged sulfonic acid product is neutralized with sodium hydroxide. A stoichiometric amount neutralizes the free acid; a slight excess is required to hydrolyze the sultone. In hyperalkalization, the base required for both neutralization and saponification is added (71). Allowing the neutralized product to age for some days, before saponification, produces noticeable differences in handling properties (58).

Hydrolysis of sultones yields alkenylsulfonates and hydroxyalkanesulfonates. The hydrolysis rate decreases in the order of 1,2 sultone > 1,3 sultone > 1,4 sultone. 1,3 Sultone is hydrolyzed after refluxing for 1 h in dilute NaOH; 1,4 sultone requires 34 h for 99% or better conversion. Complete hydrolysis is obtained in an autoclave after 30 min (58). Hydrolysis conditions affect the distribution of alkenylsulfonate and hydroxysulfonate produced (54,58,72–73). Mild conditions favor formation of hydroxyalkanesulfonate; elevated temperatures favor alkenylsulfonates (58). Overall composition of the final product following hydrolysis at 90–100°C is about 53% alkene sulfonate and 47% hydroxyalkanesulfonate (54).

Alpha-olefin sulfonic acid may contain an oligomer which hydrolyzes to alkenylsulfonates and hydroxyalkanesulfonates (58). Dermatological and feeding studies have found commercial AOS to be free of adverse effects (74–75).

Bleaching. The color produced in the continuous thin-film reactor is generally inferior to that of LAS or alcohol sulfates. Color depends on feedstock quality, process equipment, and operating conditions. Unbleached Klett color is usually in the 100–500 range (5% aqueous solution). This color may be improved, if desired, by preliminary treatment of the α-olefin with air blowing (76), absorbents (77), gaseous SO_3 (78), or H_2SO_4 (79). Alternatively, the AOS can be bleached with 0.5–3% sodium hypochlorite (80) or H_2O_2 (81), improving from 700 to 115–400 Klett (5% active basis) (54) (see Table 7). The TO-Reactor is reported to give light-colored AOS that may not require bleaching (72,82–83).

Saturated Fatty Esters. Alpha-sulfo fatty esters exhibit good detergency, biodegradability, and soap-dispersion properties. However, generally poor yields, quality, and purity, caused by production difficulties, have limited their acceptance in the United States. Much of the research on sulfoesters was done at the USDA (84–88).

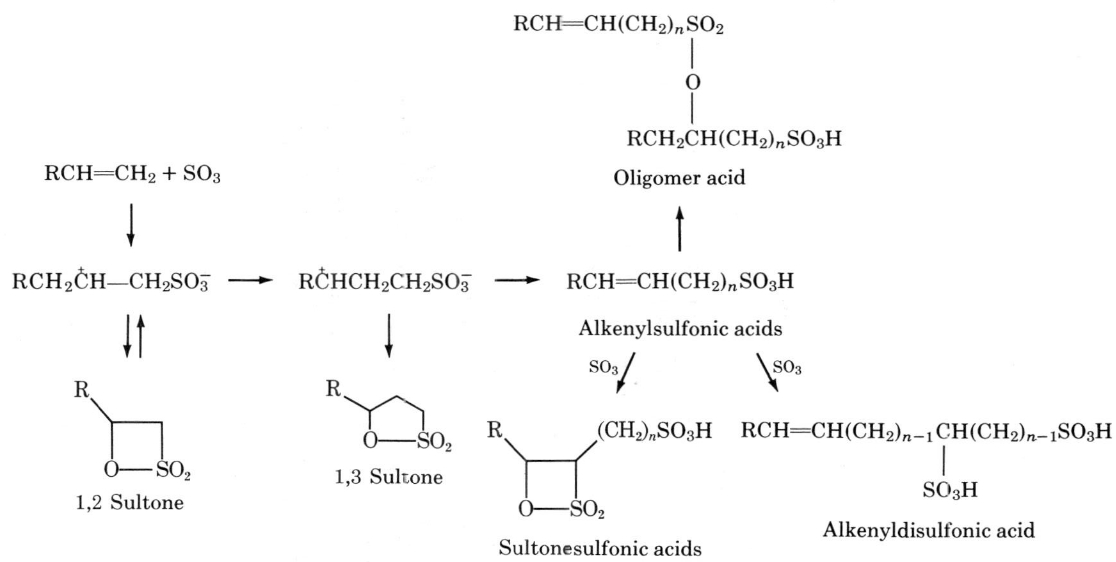

Figure 2. Sulfonation of α-olefins.

Table 7. Bleaching of α-Olefin Sulfonates with NaOCl

% NaOCl[a]	pH	Time, h[b]	Temperature, °C	Klett color[c]
0	12.2			450
1.0	12.3	5	50	120
	9.3	1		142
	8.6	<1/12		202
	6.7	<1/30		235
3.0	12.1	11	50	76
	9.5	1–2		122
	8.5	<1/12		158
	6.9	<1/30		190

[a] Based on active AOS (= α-olefin sulfonate).
[b] Until NaOCl content is <0.1%.
[c] 5% active components.

The falling-film-sulfonation technology has made high quality products available. Thus, SO_3 sulfonation of a methyl palmitate and methyl stearate mixture gave a sulfoester and sulfocarboxylic acid content of ca 90%. Reesterification of the sulfocarboxylic acid raises the sulfoester yield. High yields are claimed by postreactor digestion (89). Chemical and physical properties relating to stability, foaming, wetting and lime soap dispersion have been investigated (90). Sulfonation increases the stability of the ester linkage and in the pH range of 5–10, α-sulfomethyl tallowate is stable even when boiled. The isopropyl and *sec*-butyl esters are even more stable. Lime-soap dispersion was found to equal or surpass products based on taurates. Improved detergent properties are claimed for sulfoesters prepared in the presence of Na_2SO_4 (91).

Paraffins. Aliphatic sulfonates are prepared from paraffins by sulfochlorination or sulfoxidation, and from olefins by bisulfite addition. Although known for several decades, these processes attained commercial success only in the FRG where molecular-sieve technology has provided an economical source of linear C_{13-17} paraffins (see Adsorptive separation).

Sulfochlorination. Sulfonation of aliphatic hydrocarbons with a mixture of sulfur dioxide and chlorine in the presence of light was developed in 1936 (2):

$$Cl_2 \xrightarrow[\text{catalyst}]{h\nu} 2\,Cl\cdot \qquad (27)$$

$$Cl\cdot + RH \rightarrow R\cdot + HCl \qquad (28)$$

$$R\cdot + SO_2 \rightarrow RSO_2\cdot \qquad (29)$$

$$RSO_2\cdot + Cl_2 \rightarrow RSO_2Cl + Cl\cdot \qquad (30)$$

$$RSO_2Cl + 2\,NaOH \xrightarrow{\text{heat}} RSO_3Na + NaCl + H_2O \qquad (31)$$

This process attained only limited use in the United States, but was used by I. G. Farbenindustrie in Germany during World War II to produce Mersolates (92). Although the sulfochlorination of paraffins is random, polysulfonate formation can be controlled by limiting the sulfochlorination and recycling unreacted hydrocarbon (93–96).

Sulfoxidation. The direct sulfoxidation of paraffins catalyzed by uv radiation was discovered in Germany in 1940 (97). Several free-radical initiators were developed based on uv and gamma radiation and various organic accelerators (98–107):

$$RH \xrightarrow[\text{catalyst}]{\text{radiation}} R\cdot + H \tag{32}$$

$$R\cdot + SO_2 \rightarrow RSO_2\cdot \tag{33}$$

$$RSO_2\cdot + O_2 \rightarrow RSO_2O_2\cdot \tag{34}$$

$$RSO_2O_2\cdot + RH \rightarrow RSO_2O_2H + R\cdot \tag{35}$$

$$RSO_2O_2H + H_2O + SO_2 \rightarrow RSO_3H + H_2SO_4 \tag{36}$$

$$RSO_2O_2H \rightarrow RSO_2O\cdot + OH\cdot \tag{37}$$

$$OH\cdot + RH \rightarrow H_2O + R\cdot \tag{38}$$

$$RSO_2O\cdot + RH \rightarrow RSO_3H + R\cdot \tag{39}$$

Formation of a persulfonic acid, RSO_2O_2H, appears to be the rate controlling factor. Farbwerke Hoechst A.G., is probably the only commercial producer of paraffin sulfonates, and reportedly employs an ultraviolet catalyzed sulfoxidation process (108). This process, like sulfochlorination, produces random substitution with significant disulfonation. Paraffin sulfonates, sometimes referred to as secondary alkanesulfonates (SAS), have application in liquid and heavy-duty solid detergents in Western Europe (108).

Petroleum. Petroleum sulfonates are widely utilized in a variety of industrial products and applications. Petroleum sulfonates were the first petrochemicals produced in the United States in 1875, when petroleum fractions were refined by treatment with sulfuric acid, yielding sulfonated by-products. Uses were gradually developed and today petroleum sulfonates are important co-products, and in some instances represent the primary product (see Petroleum).

Mahogany Sulfonates. Natural petroleum sulfonates are by-products resulting from the manufacture of a variety of refined oils, most notably white (mineral) oils and lube oils. The feedstocks are selected on the basis of the desired characteristic of the refined oils or primary products and generally contain 15–30% aromatic material.

Oleum sulfonation. Petroleum sulfonates have traditionally been produced by both batch and continuous treatment of petroleum oils with oleum (109–110) (see Fig. 3). The oil is subjected to several successive oleum contacts at 60–65°C with sludge separation by gravity settling or centrifugation after each treatment. The reaction mixture forms two layers designated by color: the upper sulfonic acid–oil layer is a rich mahogany, whereas the lower acid sludge layer is green; hence, the terms mahogany acids and green acids. The green acids consist predominantly of water-soluble acids and sulfuric acid, whereas the mahogany acids are largely oil-soluble monosulfonic acids. Representative oil-oleum sulfonation process charge ratios (111–112), reaction products and derived mahogany sulfonate composition are assembled in Table 8. Sludge acids generally consist of a viscous sticky semisolid. They are difficult to handle and create a significant economic and disposal burden (55). After sludge separation, the crude mahogany acid is neutralized with caustic soda and extracted with isopropyl alcohol and purified by desalting from the solvent system.

As the demand for petroleum sulfonates grew over the past 40 years, numerous

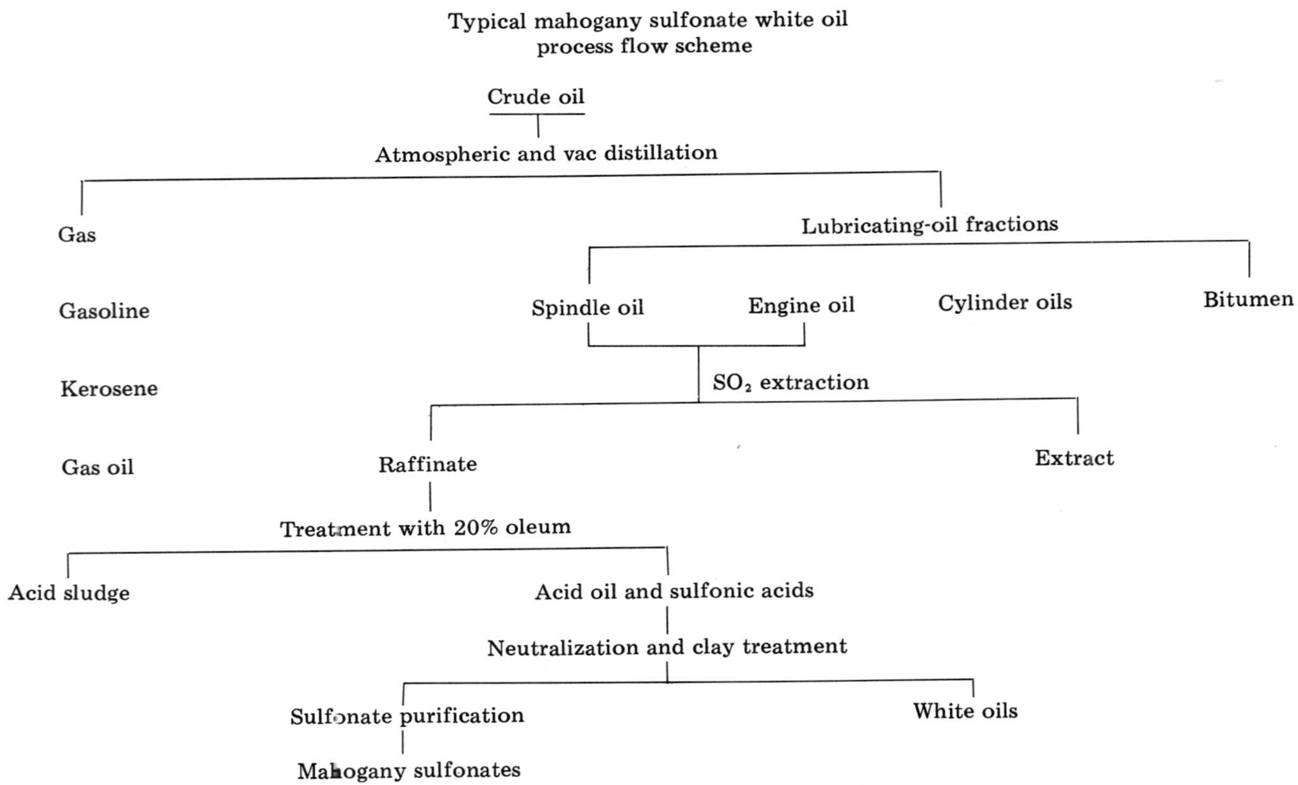

Figure 3. Typical mahogany sulfonate white oil process flow scheme.

Table 8. Representative White Oil and Mahogany Sulfonate Process Characteristics[a,b]

Property	Parts by weight
charge[c]	
lube oil raffinate	100
20% oleum	50
acid and oil products	
white oil, primary product	80
mahogany sulfonic acids	6
sludge acids	62
green acids	59
H_2SO_4	29
oil	12
refined sodium mahogany sulfonate	
petroleum sulfonate	62
mineral oil	33
water	4
inorganic salt	1

[a] Refs. 111–113.
[b] Sulfonate equivalent weight range = 400–550.
[c] Mole ratio of SO_3: aromatic compounds = 10:1.

process improvements were introduced. Since the sulfonation of petroleum oils with oleum is basically a two-phase liquid system, intimate contact is necessary and is generally accomplished by vigorous agitation (109–110,114–116). A continuous process employs a continuous high speed scraping blade in an annular unit equipped with a multiport gas-injection system and cooling jacket (117). Other processes claim centrifugal contacting systems (28,118). Various solvent systems have been disclosed utilizing SO_2 (118–119), chlorinated solvents (120), butane (120), and hexane (121) to reduce viscosity and thus allow operation at a lower temperature. A continuous sulfonation process was described in which the liquid sulfonating agent (conc H_2SO_4) is sprayed or atomized onto oil flowing down a heat-exchange surface (122) (see Heat-exchange technology). Oleum sulfonation of petroleum oils continues to be the predominant process employed, particularly for petroleum sulfonates that must meet strict specifications such as lube-oil additives. In general, oleum sulfonation gives light-colored products as the reaction is less severe and some of the color bodies are removed in the sludge.

SO_3 Sulfonation Systems. Sulfonation of aromatic compounds (including petroleum oils) is an instantaneous reaction. Liquid SO_3 without solvents results in overreaction, oxidation, and charring. With proper control, however, the high reactivity of SO_3 lends itself to continuous operation in combination with various solvents, eg, a liquid SO_2–SO_3 system (119), chlorinated solvents (123–127), and low boiling hydrocarbons (128). Viscosity is reduced and the reaction controlled at lower temperatures. Gas-phase procedures have also been described (129–157). Gaseous SO_3 has reportedly been used in the USSR for the production of white oils (158–159).

Gas-phase systems include the use of a continuous high speed rotating scraping blade (117,136), a device wherein oil is dispensed onto a cooling surface and subjected to centrifugal action while in contact with SO_3 gas or mist (160), and various other high shear or high velocity gas–liquid contacting systems (67,161–162). Cyclic batch SO_3-contactor systems employ a venturi mixing device (163), and a relatively high velocity gas-nozzle system for introducing the SO_3-gas into the circulating oil stream (138,164–167). Continuous SO_3 gas-sulfonation systems have been patented, based on the reaction of the oil in a mist or dispersed particulate state, followed by impingement onto a confining wall or heat-exchange surface (134,168–169). A more recent patent (157) claims a continuous air–SO_3 sulfonation process for gas oils in which water-soluble sulfonic acids are separated by decantation (settling), while the oil–sulfonic acid layer is subjected to saponification (neutralization). The recovered oil, rich in paraffins, is sent back to a reformer unit to generate more aromatic compounds. An extensive study comparing oleum and gaseous SO_3 batch sulfonation of specific petroleum fractions gave higher conversions of aromatic material in the SO_3 process and less sludge formation (170). For either sulfonating agent, multiple small treatments gave a better yield than a single treatment.

A number of continuous SO_3 sulfonation-reactor devices and processes utilize high velocity dilute SO_3 gas streams (66–70,83,162,166–167,171–175), especially for household-detergent sulfonates. Some patents claim sulfonation of high molecular weight aromatic oils. A wiped film reaction system for the continuous SO_3 gaseous sulfonation of high boiling petroleum oils for tertiary oil recovery was reported recently (150). Although high conversions of aromatic oils to sulfonates were achieved, polysulfonation occurred also. Decomposition was minimized by rapid neutralization. Film reactors in continuous gas-phase sulfonation of various oil fractions at 35–50°C, some using dichloroethane solvent, have been reported in the USSR (127,176–177).

Patents recently have been issued covering a number of SO_3 sulfonation processes for various petroleum materials, including whole and topped crude oils, to produce petroleum sulfonates primarily suitable for tertiary oil production (146,149,153–157). An example is a process where an SO_3–ethylene dichloride mixture is injected into a stream of gas oil under recycle (146,149). In another system whole or topped crude oil is continuously sulfonated in a recycle closed-loop tubular reactor (146,153–154). The resulting crude acid is generally very low in sulfonic acid. It is either phase-separated in water or extracted with alchohol, neutralized, formulated into a micellar sulfonate slug, and injected into an oil reservoir for tertiary oil recovery. A plant of 36,300 metric tons annual capacity started operation in 1978 (178). Substantially higher sulfonate yields and lower polysulfonate yields are claimed by an SO_3-continuous-sulfonation system incorporating small amounts of additives into the petroleum feedstock before sulfonation (155–156). A 45,400 t/yr plant for the direct production of 60% active petroleum sulfonates was put on stream in 1976 primarily for enhanced oil recovery (179). Increased use of petroleum sulfonate for tertiary oil recovery is forecast if improvements in technology and economics of current field applications materialize (180–181).

Petroleum Derivatives. As uses were developed for the natural petroleum mahogany sulfonates, synthetic alkylates were incorporated either as mixed feedstocks for cosulfonation (182), or by separate sulfonation (183–184). Detergent alkylate bottoms have substantial utilization in the production of synthetic petroleum sulfonates. Laboratory procedures for oleum and gas-phase SO_3 sulfonation of high boiling alkylated benzene have been described (183). The yields of sulfonate are related to the catalyst and the distribution of isomers. *p*-Dialkylbenzene isomers resist sulfonation; excess sulfonating agent results in dealkylation, and reduction in sulfonate equivalent weight. High molecular weight alkylates provide a good feedstock for synthetic petroleum sulfonates. These include long-chain mono and dialkylates based on polypropylene and polybutylene polymers as well as those derived from linear paraffins or olefins. Most of these are utilized in the production of overbased lube-oil sulfonates.

Synthetic petroleum feedstocks sulfonate easier than natural petroleum feedstocks. They give better yields and do not require oil extraction or product concentration. They may constitute ca 35% of the total petroleum sulfonate market but are rarely so identified.

Overbased Alkaline-Earth Petroleum Sulfonates. Overbased alkaline-earth petroleum sulfonates are used extensively in lubricant fluids as motor-oil additives, greases, and other industrial applications because of their excellent dispersing properties and their ability to neutralize engine acids. In general, they must meet stringent specifications and special processing methods are needed to remove all water-soluble sludge components. Sludge removal is aided by the use of solvents in the sulfonation process (118–120,123–127) or by solvent addition to the sulfonic acid after sulfonation, eg, xylene (137), hexane (185), heptane, naphtha (182–188), or kerosene (189). Extensive water washing (116) or extraction of the sulfonic acid from the sludge phase is also employed.

Alkaline-earth petroleum sulfonates are produced by a number of methods, eg, from the sodium sulfonate and alkaline-earth halides (55) or neutralization of the sulfonic acid with alkaline earth oxides, hydroxides, or alkoxides. Neutral and slightly overbased sulfonates are generally made by neutralization with the metal oxides or hydroxides in water, aqueous alcohol, or anhydrous organic solvents (184,190–200). Neutralization with ammonia, conversion to the sulfonamide by heating to 220°C followed by addition and heating of the metal oxide or hydroxide–sulfonate mixture is another route (201). Neutralization of the sulfonic acid and evaporation of an aqueous alkali medium in a sonic field has been reported (202).

The alkalinity of overbased petroleum sulfonates is generally determined by titration and is expressed as titratable base number (TBN), equivalent to the milligrams KOH per gram of sample. Overbased alkalinities range from 0–400 TBN, representing 0–50 wt % as CaO, $Ca(OH)_2$, or $CaCO_3$, typically suspended or complexed with 10–40% petroleum sulfonate. Super-overbased alkaline-earth petroleum sulfonates are produced by direct or indirect carbonation of excess or complexed alkaline-earth oxide or hydroxide in the petroleum sulfonate (187,198,203–242). Direct methods utilize gaseous CO_2, whereas indirect methods are based on thermal decomposition of ammonium carbamate (216,219,231) or urea (220). Overbased petroleum sulfonates are carefully filtered to remove particulates to give optically clear products. The most widely used overbased petroleum sulfonates are based on calcium (70%), followed by magnesium (20%), and barium (10%).

Lignin. Lignin (qv) occurs with cellulose as a component of woody plant tissue, and is one of the most abundant organic materials known. It consists of a group of high molecular weight polymeric amorphous materials based on *p*-hydroxyphenyl units (14%), syringyl units (6%), and guaiacyl units (80%). Sulfonation takes place on the α carbon of the side chain of the basic aromatic unit (see Lignin).

Lignin is obtained from spent wood-pulp liquors (see Pulp). It is recovered as free acid, sodium lignates, and lignosulfonate salts. Lignosulfonate technology dates back to 1866, when it was discovered that lignin material could be solubilized by digestion with sulfurous acid (243). Lignin by-products were usually discharged into waterways, but because of recent pollution regulations, they are now recovered. Lignosulfonates are produced by the acid sulfite process and the Kraft process.

Acid Sulfite Process. In this process, wood chips and sawdust are heated under pressure with a mixture of sulfurous acid and metal bisulfite, eg, calcium, magnesium, sodium, or ammonium bisulfite. Hydrolysis and sulfonation convert lignin into lignosulfonates of molecular weights between 200 and 100,000. Structures are linear at molecular weights below about 4000, and coiled and solvated at higher molecular weights. The degree of sulfonation appears to increase with decreasing molecular weight. In this reaction, hydrogen, oxygen, and sulfur are added in the proportion of 4:4:1, corresponding to $(SO_2 + 2\ H_2O)_n$ (244). In general, the wood material is digested at 125–145°C for 8–24 h during which the lignosulfonates are solubilized. The liquor is then filtered. Approximately half of the wood is converted into soluble lignosulfonates, which can be used as such or upgraded. Lignosulfonates are sold as 25–60% concentrates, or dried by spray or drum drying. The isolation of basic calcium lignosulfonate from sulfite liquor provides a means of producing essentially carbohydrate-free, high purity lignosulfonates (245). Various other processes based upon fermentation give up to 95% carbohydrate-free products. Other methods utilize ion-exchange, ion-exclusion, ultracentrifuging, and reverse osmosis.

Kraft Sulfate Process. Wood is heated with a 10–20% mixture of sodium hydroxide and sodium sulfite at 165–175°C for 4–6 h. Both batchwise and continuous processes are employed. The process is called the Kraft process because it imparts greater strength to the product. The sodium sulfite used to recharge spent liquors is subsequently reduced to sodium sulfide. The latter accelerates the delignification process which improves the pulp properties. Sodium lignosulfonate is recovered.

Lignosulfonate Uses and Derivatives. Lignosulfonates are used in animal feeds as a binder and molasses diluent; in adhesives; in oil-well drilling muds; and as surface-active agents, dispersants, and ore-flotation agents (see Flotation).

More recent uses as absorption agents and cosurfactants (246–247) for chemical oil recovery use are based on chemical modification by ozone (248), chloroacetic acid (249), and carboxylation (250).

Vanillin (qv), 3-methoxy-4-hydroxybenzaldehyde, is manufactured almost exclusively from lignosulfonate raw materials (251–252). Other derivatives include vanillic acid, ethyl vanillate, vanillil, and guaiacol (244).

Commercial applications account for only a small part of lignosulfonate production; the bulk is burned for fuel values or disposal. The U.S. International Trade Commission reports may overstate production of surface-active lignosulfonates and understate total production (253).

With escalating prices of crude oil and synthetic chemicals, lignin offers an alternative source for chemical intermediates.

Sulfoalkylation or Polymerization. In sulfoalkylation, low molecular weight sulfonates react with organic compounds to produce more complex sulfonates. Low molecular weight sulfonates are known as sulfoalkylating or organic sulfonating agents.

Sulfoalkylation affords a commercial route for the manufacture of sulfonated amines, amides, alcohols, and ketones. The following reaction is typical:

$$H_2N\overset{\overset{R}{|}}{C}HSO_3Na + R'COCl \longrightarrow R'CONH\overset{\overset{R}{|}}{C}HSO_3Na + HCl \qquad (40)$$

where R = H, methyl, or phenyl, and R′COCl = amino acid chlorides.

The products have wide application in the manufacture of polymers, dyes, drugs, surfactants, and photographic and agricultural chemicals (33).

Alkyl taurides, prepared by sulfoalkylation, possess most of the properties of soap, including foaming, emulsifying, and mildness. They function equally well in soft and hard water. Taurine (2-aminoethanesulfonic acid) or methyltaurine and a fatty acid chloride give the alkyl tauride:

$$\text{RC(=O)Cl} + \text{R}'\text{NHCH}_2\text{CH}_2\text{SO}_3\text{H} + 2\,\text{NaOH} \longrightarrow \text{RC(=O)N(R}')\text{CH}_2\text{CH}_2\text{SO}_3\text{Na} + \text{NaCl} + \text{H}_2\text{O} \qquad (41)$$

where R = hydrogen, methyl, or ethyl; R′ = alkyl. Alkyl taurides are also used in textile finishing and in the manufacture of agricultural wettable powders.

Alkyl isethionates resemble the taurides but are made from a fatty acid chloride and sodium isethionate (sodium 2-hydroxyethanesulfonate). The latter is made from ethylene oxide and sodium bisulfite:

$$\text{RC(=O)Cl} + \underset{\text{sodium isethionate}}{\text{HOCH}_2\text{CH}_2\text{SO}_3\text{Na}} \longrightarrow \underset{\text{alkyl isethionate}}{\text{RC(=O)OCH}_2\text{CH}_2\text{SO}_3\text{Na}} + \text{HCl} \qquad (42)$$

Alkyl isethionates are used mainly in cosmetics and toiletries.

Styrene and Styrene Polymers. 4-Styrenesulfonic acid (4-vinylbenzenesulfonic acid) is an important sulfonating agent used for the preparation of water-soluble and insoluble polymers. It is readily prepared by sulfonating 2 bromoethylbenzene, fol lowed by treatment with alkali to effect dehydrobromination and formation of the sulfonate salt.

Sulfonation imparts water solubility to polymers such as starch or cellulose. Likewise, polymerization of unsaturated sulfonates gives water-soluble polymeric sulfonates. On the other hand, copolymerization of the unsaturated sulfonates with water-insoluble monomers produces ion-exchange resins or fibers of enhanced receptivity.

Copolymers of acrylonitrile and other vinyl monomers with alkenylaromatic sulfonic acids improve tensile strength and dying quality of yarns. Terpolymers containing acrylonitrite and potassium styrenesulfonate are made commercially to enhance affinity resulting fibers for basic dyes.

Sulfation

Sulfation involves the reaction wherein a —COS— linkage is formed by the action of a sulfating agent on an alkene, alcohol, or phenol. Concentrated sulfuric acid forms a $\geqslant COSO_2OH$ moiety with an alkene, whereas, SO_3 yields a $\geqslant COSO_3H$ group with an alcohol or a phenol. Unlike the sulfonates, which exhibit excellent hydrolytic stability, the alcohol sulfates are readily susceptible to hydrolysis in acidic media. Sulfation of fatty alcohols and polyalkoxylates has produced a substantial body of commercial detergents and emulsifiers, and U.S. production of alkoxyalkyl sulfates increased from 75,900 metric tons in 1971 to 111,000 metric tons in 1980 (4,254). In 1977, an estimated 77,300 metric tons of alkyl sulfates was also produced.

Linear ethoxylates are the preferred raw materials for production of ether sulfates used in detergent formulations because of their uniformity, high purity and biogradability. The alkyl chain is usually in the C_{12}–C_{15} range with molar ethylene oxide: alcohol ratio of 3:1. In 1977, ca 18,200 metric tons of ether sulfate was used in emulsion polymerization and textile processing. Propoxylates, ethoxylates and mixed alkoxylates of aliphatic alcohols or alkyl phenols are sulfated for use in specialty applications.

Alkenes may be sulfated with sulfuric acid to yield primary or secondary alcohols, eg, ethanol is prepared commercially from ethylene.

$$CH_2{=}CH_2 + H_2SO_4 \rightarrow CH_3CH_2OSO_2OH \quad (43)$$

$$CH_3CH_2OSO_2OH + H_2O \rightarrow CH_3CH_2OH + HOSO_2OH \quad (44)$$

The detergent Teepol is produced in Europe by sulfation of C_{12}–C_{18} alkenes. The alkylsulfuric acid produced in this instance is neutralized with sodium hydroxide.

Sulfated natural fats and oils, sometimes erroneously referred to as sulfonated oils, made from glycerides such as castor, neat's-foot, olive, or rapeseed oils. Sulfated castor oil, used since ca 1875 in textile processing, was the first such commercial product. In turkey-red dyeing, it was known as turkey-red oil (255).

Alcohols and Alkoxylates. Alkyl and alkoxyalkyl sulfates can be produced from their corresponding alcohols or nonionics by reaction with a wide variety of reagents including chlorosulfuric acid, sulfur trioxide, sulfuric acid, and sulfamic acid (256). The products are similar in wetting time, detergency, and foam generation. Chlorosulfuric acid gives slightly better colors but requires disposal of HCl.

Sulfates of ethoxylated alkylphenols prepared with sulfamic acid or SO_3 show comparable performance. Ring sulfonation is appreciable (ca 12%) with SO_3 but not with sulfamic acid (257). Sulfonation in the presence of urea avoids ring sulfonation and discoloration (258). Urea–SO_3–H_2SO_4 is also employed as sulfating agent for alkoxylated derivatives of alkylphenol. The *in-situ* formation of sulfamic acid eliminates foaming; ring sulfonation is <1% (259).

Until recently, chlorosulfuric acid has been widely used for the sulfation of fatty alcohols (260). Rapid mixing of acid and alcohol, combined with vigorous air sparging to remove HCl improved the process. In continuous sulfation with chlorosulfuric acid, the reactants flow through concentric reactor tubes (261). The reaction mixture is quenched in recirculated product and effluent HCl gas is separated (261). Converter gas (a mixture of nitrogen, oxygen, and SO_3) has also been used and is an acceptable source of SO_3 (262). A quick reaction time at high temperatures gives the best results. For example, in film sulfonation, the reaction time is generally 0.1–0.5 s and the high temperature generated in the film is rapidly removed by the cooled reaction surface (66–70). In the United States, alkyl and alkoxyalkyl sulfates are mainly produced by this method. For sulfation of ethoxylated nonylphenol, this technique is modified (263). A heated cooling-medium reduces the film viscosity, accelerates heat removal, and gives a product with ca 6% ring sulfonation.

Fats and Oils. Fats and oils are traditionally sulfated with 98% H_2SO_4 (see Fats and fatty oils). A saturated or unsaturated hydroxy acid gives a sulfoxylated compound, whereas an unsaturated fatty acid yields a saturated fatty acid sulfate:

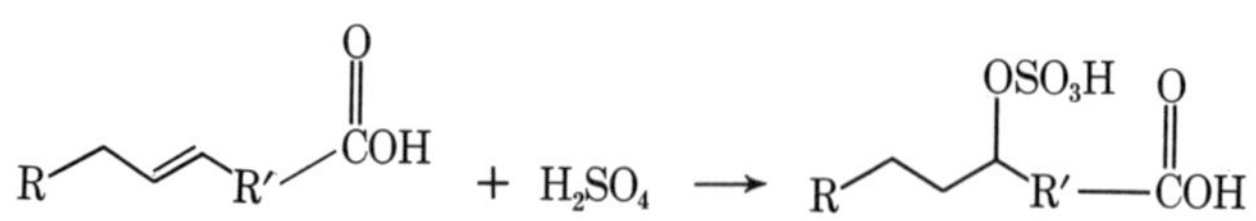

(46)

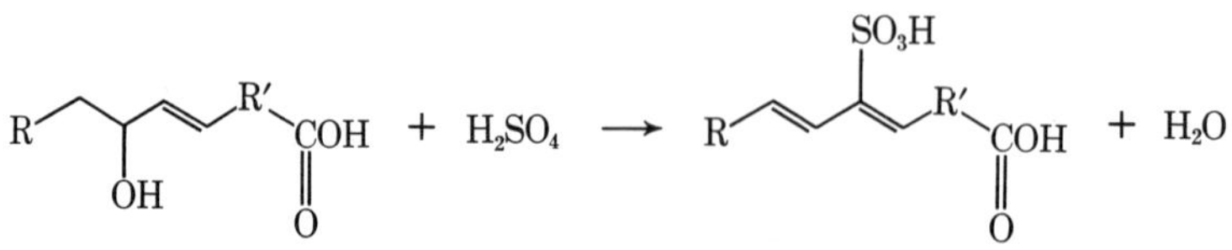

(45)

Reaction across the double bond is almost instantaneous (264), whereas sulfation of the hydroxyl group is much slower. Yet, 12-hydroxyoleic acid sulfates almost exclusively at the hydroxyl group (265–266).

These products have gradually been supplanted by sulfated fatty alcohols, sulfated monoglycerides, and sulfonated alkylbenzenes. From 1965 to 1980, U.S. sales of sulfated fats and oils decreased from ca 24,000 to ca 6,000 metric tons (4,254).

Other sulfating agents include chlorosulfuric acid (267) and sulfamic acid (268). Unsaturated hydroxy compounds yield unsaturated alkyl sulfates; turkey-red oil was produced batchwise with gaseous SO_3 (269). Processing time was much shorter than with H_2SO_4, and the product contained less inorganic salts and free fatty acids. Unsaturated hydroxy compounds and gaseous SO_3 in the falling-film reactor gave a 92% yield of sulfates with 95% double-bond retention (270).

Sulfated oils are used primarily in the emulsification of water-immiscible liquids such as mineral oil or pine oil. The products are used as cutting oils for metal working, oil sprays for insecticides, wetting agents, and disinfectants.

Carbohydrates. Sulfated cellulose is of interest as a possible substitute for naturally occurring gums and as an ion-exchange medium in biological studies. Treatment of cellulose with SO_3–pyridine yields a product with good transparency that is used as a thickening agent (271). Cation exchangers for lipoprotein absorption have been manufactured by sulfating hydroxyalkylated cellulose with SO_3–pyridine (272). Starch is sulfated with SO_3–triethylamine at 0°C (273) or at room temperature in aqueous media with SO_3–trimethylamine (274).

Alkenes. The manufacture of low molecular weight alcohols by sulfation of alkenes with 98% H_2SO_4 lends itself to continuous operation. Good agitation is required and the reaction is performed under mild pressure at ca 70–80°C; dialkyl sulfates also form. Longer-chain (C_{12}–C_{18}) alkenes yield detergent products. Addition of 96% H_2SO_4 to 1-dodecene at 0°C gives mainly dialkyl sulfate, but further addition gives an 80% yield of 2-monoalkyl sulfate (275). At −15°C with 98% H_2SO_4 in pentane, a 98% yield of monoalkyl sulfate is obtained. If the olefin is added to the acid, random isomers are produced.

Another process claims 80–95% conversion to monoalkyl sulfate; 75–95% of the product is the desired 2-alkyl sulfate (276). The temperature is controlled with a low boiling solvent, eg, isobutane or propane from which the alkyl sulfate crystallizes as it forms. An improved product is obtained using the corresponding secondary alcohol as solvent (277). Dialkyl sulfate formation is substantially reduced in the presence of excess alcohol.

Sulfamation

Sulfamation is the reaction of an amine in chloroform (278) with SO_3, an adduct, chlorosulfuric acid, or sulfamic acid

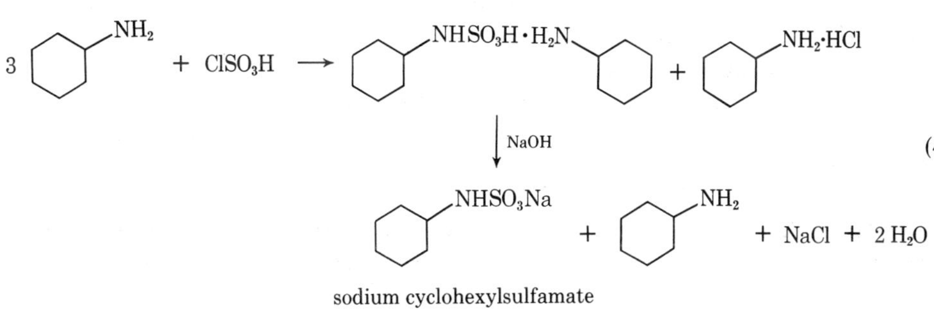

(47)

Sodium cyclohexylsulfamate is a sweetener. Because biological tests indicated a potential health hazard, the product has been removed from the U.S. market except under prescription (see Sweeteners).

Heating an amine with sulfamic acid is an alternative method.

Other processes make use of SO_3 or its adducts. Thus, refluxing cyclohexylamine with SO_3–triethylamine in methylene chloride, followed by treatment with aqueous NaOH, produced a 90% yield of sodium cyclohexylsulfamate (279). Direct addition of SO_3 to an amine is a highly exothermic reaction that can be used to prepare a sulfamate (280).

Sulfamic acid is prepared from urea and oleum (281). Sulfamation is little used commercially but sulfamic acid has various applications, eg, as dyeing aid (282–283), for bleaching wood pulp (284), and in metal cleaning (285); cigarette paper is treated with ammonium sulfamate to reduce the hazard of tumor formation from tobacco smoke (286).

Industrial Processes

Oleum Sulfonation. Typical oleum batch sulfonation of dodecylbenzene is disclosed (287–289). The temperature is generally kept between 32.2 and 46.1°C; 20% oleum is employed with a 3.6 mol ratio of available acid to 1.0 mol alkylate. The oleum is added over a 1.5-h period to the alkylate in a glass-lined kettle. Heat is removed through an external heat exchanger. Digestion is continued for two more hours at ca 46.1°C. The resultant acid is either neutralized by slow addition to a solution of caustic soda, or is first partially diluted with water followed by a gravity separation.

This process has been adapted to a number of continuous systems, including a scraping-blade heat-exchanger reactor (290). Procter and Gamble (291) utilized two loops or dominant baths; oleum and detergent alkylate are introduced into a stream of freshly produced sulfonic acid, then through a heat exchanger, and finally into a digestion zone for ca 5 min. Similar processes were patented by Chemithon (292) and Ballestra (293).

Sulfur Trioxide Sulfonation. ***Liquid SO_2 as Solvent.*** Liquid SO_2 was used in one of the first commercial processes at ca −10 to −11°C, where it served as a self-refrigerant and solvent. Product quality is high because of the low temperature, reduced viscosity, and a rapid assimilation of the reaction exotherm. The equipment consists of a jacketed glass-lined, corrosion-resistant kettle or reactor equipped with baffles, a high speed turbine impeller, optional internal cooling coils, condenser, and SO_3 sparger; an SO_2 recovery and recycle system is included. The use of liquid SO_2 as a solvent creates problems with regard to recovery, frothing, and purging of the reaction product. This process is normally operated batchwise, although it could be converted to continuous operation.

Pentane, hexane, heptane, methylene chloride, ethylene dichloride, and various fluorohydrocarbons, and fluorochlorohydrocarbon solvent systems can also be employed. Their main disadvantages are low productivity and the need for solvent recovery.

Gaseous SO_3. Dry air or inert gas is used as the carrier for the SO_3 (2–8%) which is introduced by a sparger in a conventional batch reaction. The reactor is jacketed

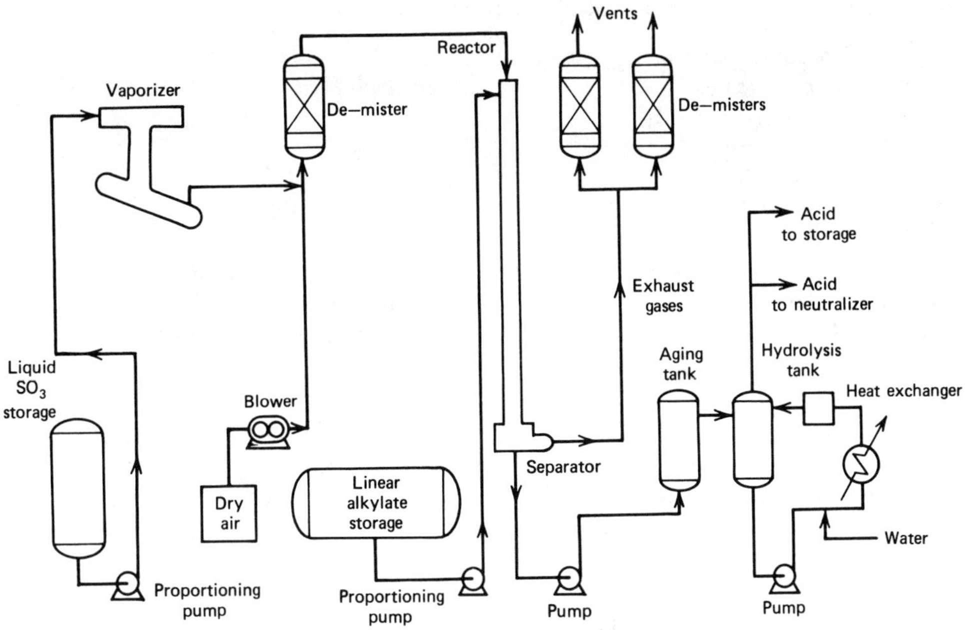

Figure 4. Process flow scheme for continuous SO_3 film sulfonation process.

and provided with internal cooling coils and baffles. Vigorous mechanical agitation minimizes local overreaction and overheating. Energy costs are high. In the batch process, the sulfonation cycle time is extended and thus the rate of gas sparging is reduced and the heat load is distributed over a longer time period. Generally, the batch reaction is held at the maximum temperature allowable to maintain suitable acid color stability (ca 60°C). This temperature limitation dictates a product viscosity profile which in turn determines the size of the heat exchanger.

The SO_3 for dry air–SO_3 mixtures is produced by: burning of elemental sulfur, followed by catalytic oxidation of SO_2 over a vanadium pentoxide catalyst; stripping of 65% oleum for direct use as gaseous SO_3; or vaporization of liquid SO_3. The dew point of the diluent air should be −28.9°C or preferably −40°C to prevent SO_3 loss by conversion to sulfuric acid.

Continuous SO_3 Processes. Starting in 1960, batch SO_3 sulfonation units were gradually replaced by continuous-film units which sharply reduced the disadvantages of batch systems, such as local overheating, prolonged exposure to SO_3, variable product quality and composition, limited productive capacity, fouling of complex spargers, and high energy costs.

Continuous sulfonation/sulfation units offer short residence time, precision distribution of reactants, rapid heat removal, and reduced energy costs.

In both batch and continuous SO_3 operations, the SO_3 gas must be diluted to a maximum concentration of 10% in order to moderate the reaction and obtain a satisfactory product color. For example, 1 mol of dodecylbenzene reacting with 1.03 mol SO_3 at a 10% concentration in dry air, is equal to ca 800 volumes of reaction gas to one volume of liquid organic reactant. The ratios of various reaction gases to liquid dodecylbenzene required for essentially complete reaction are given below.

SO_3 concentration in diluent gas, %	*Volume of reaction gas to liquid dodecylbenzene*
10	790
8	985
5	1576
2	3939

These SO_3 concentrations include the range generally employed in film–reactor systems. Gaseous residence time is generally <0.2 s. The sulfonation reaction is considered to be instantaneous and gas-diffusion limiting. Continuous-film reactors utilize high gas velocities to produce a turbulent gas–liquid contact with an extremely rapid and yet controlled reaction.

A typical continuous SO_3 falling-film sulfonation is shown schematically in Figure 4. The alkylate is pumped from storage to the distribution system at the top of the reactor. At this point, the film is established and progresses down the reactor walls. The ratio of SO_3 to organic is critical to the reaction; therefore, the liquid SO_3 is de-

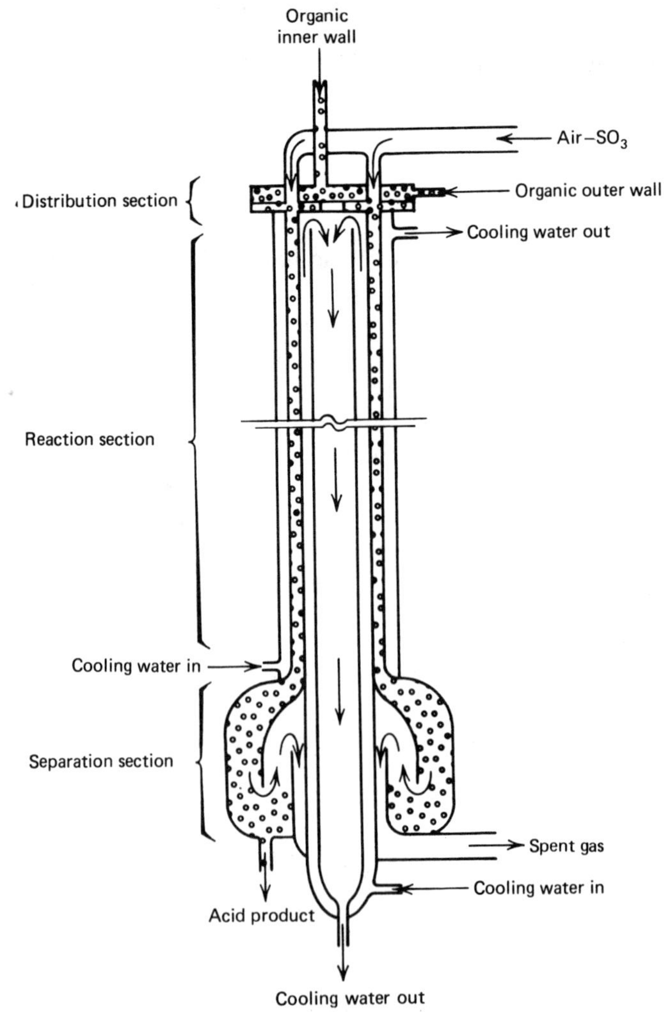

Figure 5. Allied continuous tubular SO_3 sulfonator.

livered to the vaporizer by a proportioning pump. The vaporized SO_3 is diluted with dry air from the blower to ca 4% by volume. The resulting SO_3–air mixture then enters the reactor, wherein the instant the gas stream contacts the organic film, the reaction begins and continues down the reactor walls to the gas-liquid separator located immediately below the reactor. After the acid product is separated, the spent gas is filtered and expelled or passed through a caustic scrubber to remove traces of sulfur dioxide by-product.

Before entering the aging tank, the sulfonic acid may optionally be pumped through a sensing device which controls the amount of sulfur trioxide entering the reactor. The sulfonic acid is aged for 5–15 min to reduce free sulfuric acid and unreacted organic material. It is then transferred to the hydrolysis loop, where ca 0.5% of water is administered to destroy any residual anhydrides (294).

Cost and profitability estimates for such a commercial reactor system are given in ref. 295.

The main commercial continuous SO_3 sulfonation units used by the surfactant industry are the Allied, Ballestra, Berol, Chemithon, Lion Fat and Oil, Mazzoni, and Stepan systems. The Allied Chemical, Chemithon Corporation, and Stepan Chemical units, often referred to as falling-film reactors, are perhaps the most widely used in the United States, whereas the others prevail in Canada, Europe, and Japan.

Allied continuous SO_3 falling-film sulfonation reactor. The Allied reactor (70,296) shown in Figure 5, consists essentially of a set of vertically mounted water-jacketed stainless steel concentric cylinders.

In the distribution section, an even film of organic feed is deposited continuously onto the inner and outer walls of the reaction section by pumping through small peripheral slots in the distributor. The air–SO_3 mixture is introduced above the dis-

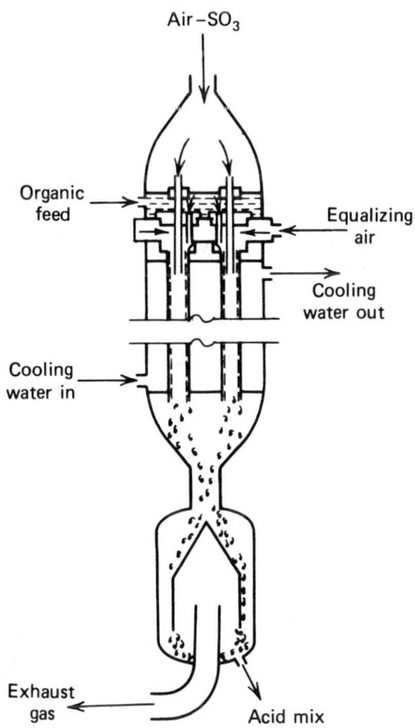

Figure 6. Mazzoni continuous SO_3 film sulfonator.

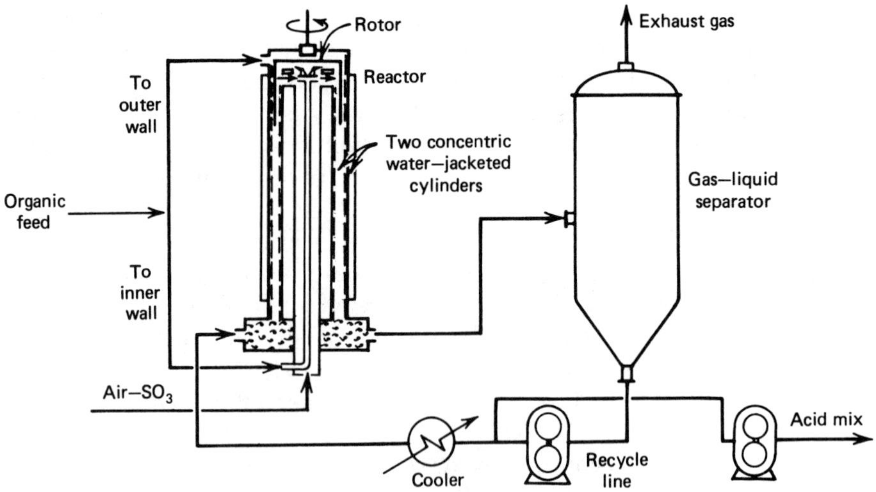

Figure 7. Chemithon continuous SO_3 film sulfonator.

tributor and passes through the annular space between the concentric cylinders in such a way that contact is made with the organic reactant as soon as the film is developed.

In the upper part of the reaction section, the air–SO_3 gas stream contacts the organic reactant. Moving down the reactor walls, the reaction approaches completion and the SO_3 concentration zero. The constantly moving film is in intimate contact with the water-jacketed reactor walls where heat is constantly removed and temperature control is provided.

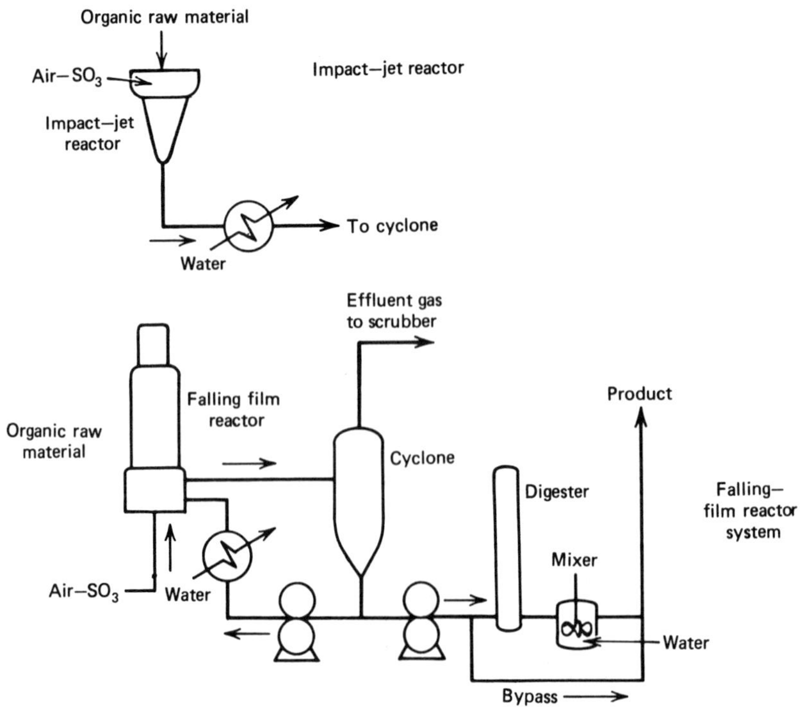

Figure 8. Process flow scheme for Chemithon continuous SO_3 sulfonators.

The internal section of the reactor is free of any moving parts that could disrupt the film and cause poor heat transfer. In the separation section, the spent gas is separated by sulfonic acid/sulfation products.

Stepan continuous SO_3 falling-film sulfonation reactor. Stepan's falling-film SO_3 sulfonation unit is based on a multi-tubular reactor design (67,289). The organic feed is distributed to the inner walls of a set of reaction tubes surrounded by a cooling medium. The SO_3-diluent gas mixture discharges from an inlet nozzle at high velocity and impinges on the organic film inducing turbulence and reaction in the falling film. Controlled reduced film viscosity and attendant heat exchange rapidly dissipate the reaction exotherm. The process is highly effective for a wide variety of organic feedstocks. Stepan operates a number of these units with annual capacities of ca 4,500–50,000 metric tons (100% active).

Mazzoni continuous SO_3 sulfonation reactor. The Mazzoni process, as depicted in Figure 6, is based upon the patent disclosure described in ref. 297. The reactor consists of a multi-tube reaction zone, a single feed chamber for air–SO_3, and a feed chamber for the organic feed. Both feeds are administered through metering nozzles. The organic feed is deposited as a film on the inner surface and the air–SO_3 is introduced in the central zone of each tube. A third zone or another single chamber supplies additional dry or equalizing air to the reaction tubes in order to equalize the pressure drops in the reaction tubes and downstream in the distributing nozzles. Uniform distribution of the reactants to the reactor tubes is thus ensured without limiting the pressure drop. The benefits of the process are minimal local overheating and charring and a tempering of the reaction severity.

Chemithon continuous SO_3 sulfonation reactor. The Chemithon film reactor utilizes a mechanical rotor at the top of a concentric tubular reactor which serves in part to distribute the organic feed to the reactor walls (see Fig. 7). This unit reportedly operates at high temperatures and gas velocities, and with short residence times

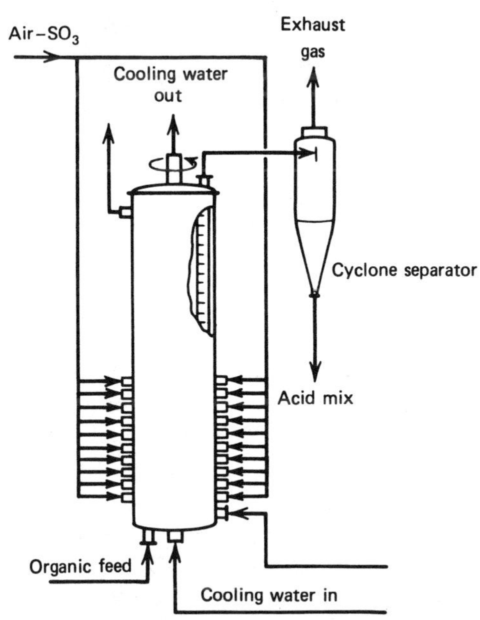

Figure 9. Berol continuous SO_3 sulfonator (174,301–302).

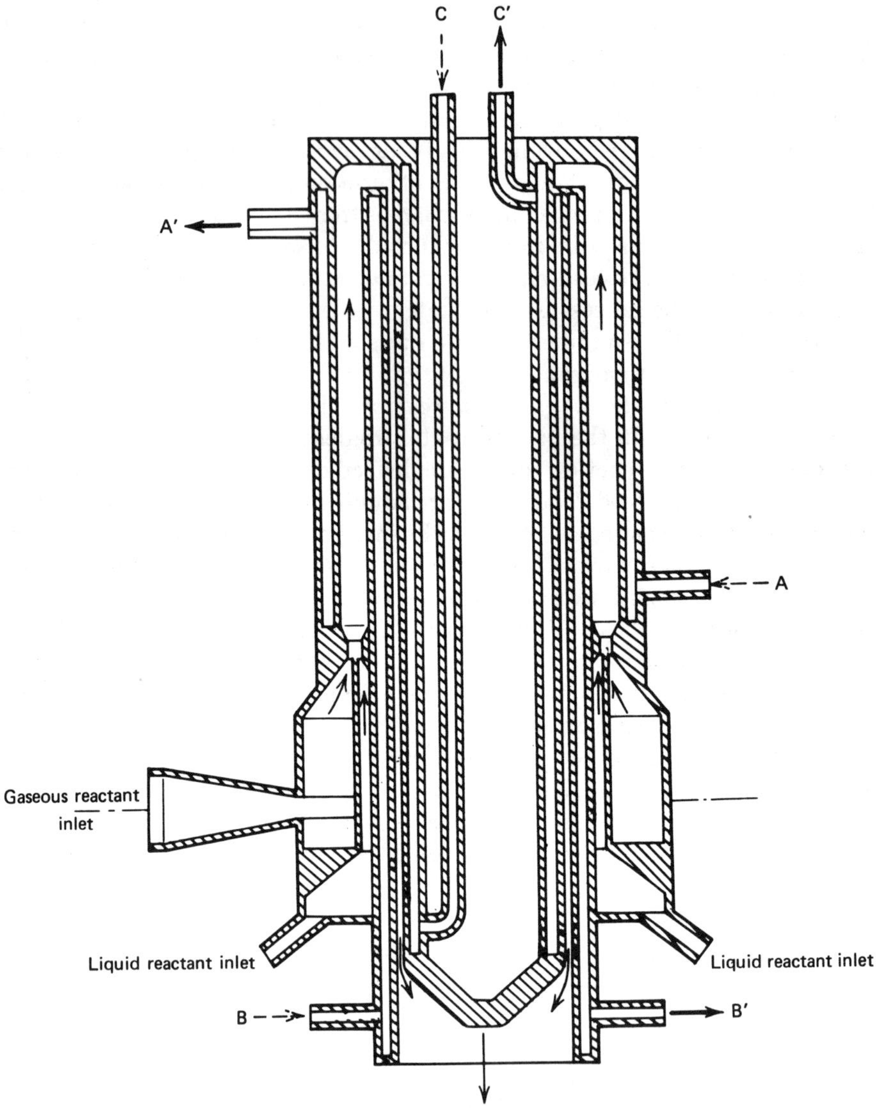

Figure 10. Lever continuous SO_3 film sulfonator (303). Cooling water in (A, B, C), ⇢; cooling water out (A′, B′, C′), ⟵; and reactant flow, →.

(69,298–300). The heat of reaction is first removed from the jacketed reactor and then by contacting the exiting reaction mixture with recycled, precooled reaction product. The reaction mixture leaves the reactor and enters the cyclone for gas–liquid separation. The liquid (acid) disengaged in the cyclone is pumped through a heat exchanger for cooling and then back to the heat-quenching zone of the reactor. A portion of the acid proceeds forward for further processing, eg, digestion, hydrolysis, etc. The SO_3 concentration is about 6%.

A number of these units are operating commercially throughout the world. Overall process flow diagrams for the falling-film reaction system and the newer impact-jet continuous SO_3 sulfonation system are both shown in Figure 8. Based upon patent disclosure (169), the jet-impact reactor utilizes a venturi mixing nozzle to mix and initiate a rapid mist reaction between a dilute SO_3 gas stream and fresh organic feed. The partially reacted mixture contacts (or impacts) a falling film of recycled liquid reaction product that quench cools the mixture and allows the sulfonation reaction to proceed essentially to completion.

The Lion isothermal sulfonation process appears to utilize processing equipment similar to that of Chemithon falling-film unit, with an inert gas stream introduced next to the flowing organic film in essentially parallel flow (83). This inert gas stream is claimed to improve process control. This system reportedly is beneficial in sulfonating and sulfating heat-sensitive feedstocks.

Procter and Gamble continuous SO_3 sulfonation reactor. A Procter and Gamble patent utilizes undiluted SO_3 vapor to contact the flowing film of organic reactant (169). The color quality of the products is not known.

The Berol continuous SO_3 sulfonation reactor utilizes a reaction process to introduce the organic reactant and the inert gas–SO_3 at the bottom of the reactor. Both travel upward through an annular reaction zone and discharge into a cyclone separator for collection of the organic acid and exhausting the effluent gas (170,301–302).

The organic feed is introduced through a bottom inlet and the air–SO_3 (6% SO_3) is distributed through two lateral rows of nozzles located in the lower section of the reactor (see Fig. 9). Owing to the movement of a water-cooled cylinder rotating within the shell of a stationary jacketed cylindrical vessel, the gas contacts the organic reactant and the mixture is forced upward in the annulus between the stator and the rotor. The upward driving force is imparted by the gas stream and the spiralling action of the rotor. The reaction mixture is withdrawn at the top of the reactor for gas–liquid separation.

Lever Brothers continuous SO_3 sulfonation reactor. The patent (303) discloses a process in which the flow of reactants is countergravitational in the first stage and gravitational, or falling-film type, in the second stage as shown in Figure 10. The degree of mixing increases with increased residence time since, at a fixed gas volume intro-

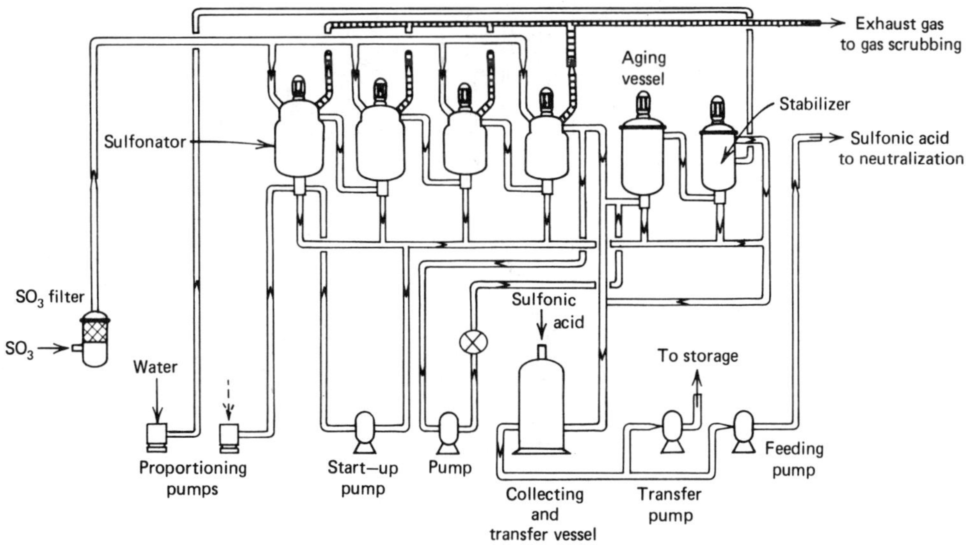

Figure 11. Sulfonation process based on Ballestra continuous SO_3 cascade reactor system. Free-oil control unit, ⊗; and organic raw material, →.

duced to the reactor, the superficial gas velocity in the first longitudinal reaction chamber is substantially less than that in the second longitudinal reactor, owing to the differences in the diameters of the two reaction zones. The greatest mixing is provided where the viscosity increase in the reaction product results in stoichiometric reaction. It is not known whether this process has been used commercially.

The Stratford Engineering continuous SO_3 sulfonation reactor (168) introduces the organic reactant as finely atomized microdroplets into the inert gas–SO_3 zone. The closed reactor vessel contains an upper reaction zone, reactive gas inlet, optional gas inlet, and a liquid atomizing rotor to disperse liquid in a lightly atomized condition. A vapor withdrawal conduit is positioned below. A second liquid atomizing rotor may be installed beneath the first rotor to provide a second gas-liquid or liquid-liquid contacting operation. A bottom product conduit permits withdrawal of the products.

The Ballestra continuous SO_3 sulfonation reactor relies on a multi-stage reaction carried out in two or more reaction vessels arranged in cascade sequence (304–306). It was the first commercial continuous air–SO_3 process that featured its own sulfur

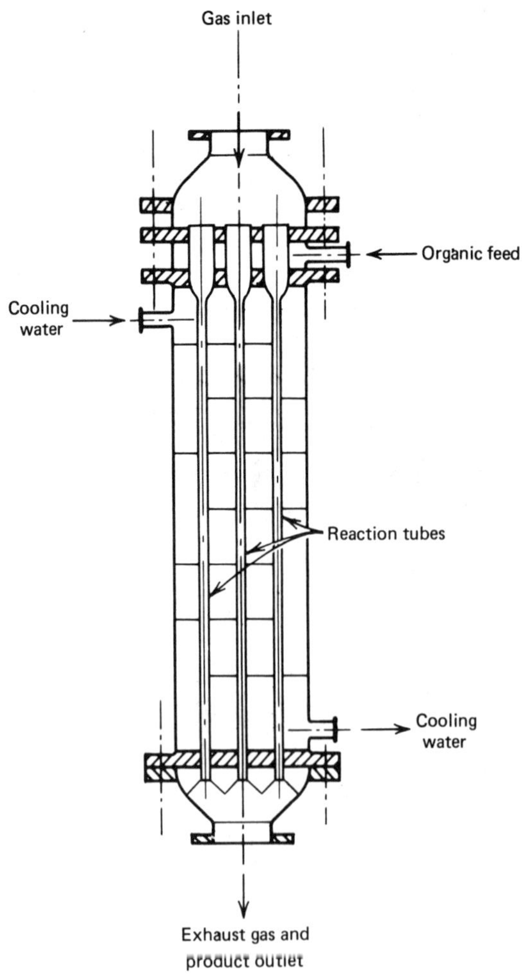

Figure 12. Ballestra continuous SO_3 film sulfonator.

burner–converter system for generating the 7–8% SO_3 reaction gas stream (see Fig. 11).

Each reactor is equipped with a high speed turbine impeller to disperse the gas and circulate the organic reactant. Cooling coils and baffles are strategically positioned in each reactor for efficient heat removal. The organic feed enters the first reactor and the liquid mixture overflows by gravity from one reactor to the next. The air–SO_3 mixture (7–8%) enters the reactors in parallel and in controlled amounts. The exhaust gas leaves each reactor from the top. A small portion of fresh organic reactant is administered to the last reactor via a feedback sensing system that controls the free oil content of the sulfonic acid.

The unique cascade arrangement of reactors permits distribution of the gas and heat loads over more reaction stages. Thus, the bulk of the reaction occurs in the relatively low viscosity medium of the first stage, which permits efficient removal of heat generated, and a minor amount in the higher viscosity medium of the last stage where the reaction exotherm is practically negligible. The cascade arrangement facilitates heat transfer and attendant control of temperature in each stage; it also reduces residence time in the final reaction stage. The capacity of a plant is governed by the number and size of the reactors.

Ballestra also offers a continuous SO_3 film sulfonation system (307). The liquid film is formed in small-diameter long tubes where the contact between the reactants occurs (see Figure 12). The unique features are allegedly embodied in process and equipment for maintaining the exact ratio of the two reactants in each tube. The tubes are followed by the usual demisters, cyclones, digesters, hydrolysis vessels, standardization tanks, and sulfonic acid storage tanks. Both batch and intricate continuous units are deployed.

Jergens solventless sulfonation process. In this process, substrate and gaseous SO_3 are combined in a venturi unit (163). The reaction mixture is cooled by a heat exchanger installed in the recirculating line. Operation is normally batchwise, but may be continuous.

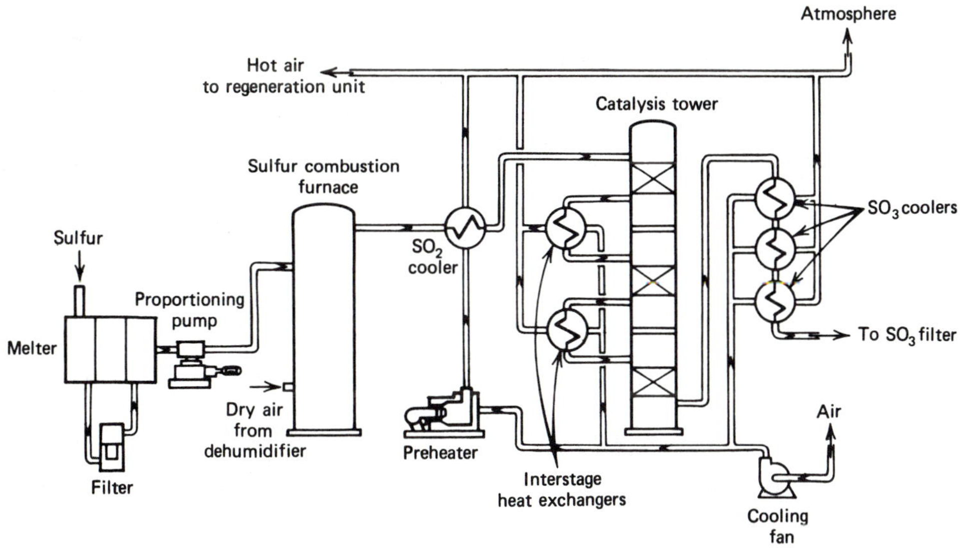

Figure 13. Typical sulfur burner–converter process for SO_3 generation (typically, 92% air and 8% SO_3).

Sulfur burner–converter SO_3 generator. A typical sulfur burner unit and converter system for the continuous generation of SO_3 reaction gas is shown in Figure 13. A proportioning pump supplies sulfur to the burner where it is burned to generate sulfur dioxide in an air stream. This, in turn, is fed into a converter unit usually containing a vanadium oxide catalyst and operating at ca 650°C. The resultant 7–8% air–SO_3 stream is cooled and filtered. This gas mixture can be utilized directly for various sulfonation and sulfation processes. Although this source offers some economic advantages, the disadvantages of limited flexibility associated with slow start up and shut down, and the supply of specific 7–8% SO_3 concentration, must be carefully considered.

The growth of SO_3 film-reactor technology will continue, and it is expected that such units will be installed in all new plants for the foreseeable future.

Current trends towards an increased shift from oleum to SO_3 are expected to continue and new surfactant plants will be based on SO_3. Hydrotrope (benzene, toluene, xylene, and cumene) sulfonate production will most likely continue to be based on sulfuric acid. Sulfur-burner-type processes will increase.

Increasing attention will be given to product biodegradability, especially for household and consumer surfactants. Increased utilization of AOS will displace some uses currently fulfilled by LAS and alkoxy sulfates. An important area of utilization of sulfonates and sulfates is tertiary oil recovery (see Petroleum, enhanced oil recovery).

BIBLIOGRAPHY

"Sulfonation and Sulfation" in *ECT* 1st ed., Vol. 13, pp. 317–337, by E. E. Gilbert, Allied Chemical Corporation; "Sulfonation and Sulfation" in *ECT* 2nd ed., Vol. 19, pp. 279–310, by E. E. Gilbert, Allied Chemical Corporation.

1. E. E. Gilbert, *Sulfonation and Related Reactions*, Interscience Publishers, a division of John Wiley & Sons, Inc., New York, 1965.
2. U.S. Pat. 2,046,090 (June 30, 1936), C. F. Reed (to Charles L. Horn).
3. R. Apel and W. Senkpiel, *Chem. Ber.* **91,** 1195 (1958).
4. United States International Trade Commission, *Synthetic Organic Chemicals, United States Production and Sales, 1969–1979*, Government Printing Office, Washington, D.C.
5. O. Kamatori and G. L. Hollis, *Chemical Economics Handbook*, Stanford Research Institute International, Menlo Park, Calif., Dec. 1980, pp. 583.5000C–583.5200C.
6. *Chem. Mark. Rep.* **221**(18), (May 3, 1982).
7. U.S. Pat. 3,160,474 (Dec. 8, 1964), W. G. Schnoor and A. W. Yodis (to Allied Chemical Corp.).
8. A. R. Broyles and R. E. Eckert, *Ind. Eng. Chem. Prod. Res. Dev.* **12,** 295 (1973).
9. V. S. Patwardhan and R. E. Eckert, *Ind. Eng. Chem. Prod. Res. Dev.* **20,** 80 (1981).
10. K. Venkataraman, *The Chemistry of Synthetic Dyes*, 8 Vols., Academic Press, New York, 1978.
11. J. C. Ambeland, J. A. Shotton, G. W. Gottschalk, H. P. Stevens, and G. E. P. Smith, Jr., *Ind. Eng. Chem.* **45,** 204 (1953).
12. U. S. Frank, *OTS Report, P.B. 7034412*, U.S. Department of Commerce, Washington, D.C., Aug. 1949.
13. K. Fukazumi and Y. Hirai, *J. Chem. Soc. Jpn. Ind. Chem. Sect.* **55,** 403 (1952).
14. U.S. Pat. 3,363,969 (Jan. 16, 1968), J. A. Brooks (to E. I. du Pont de Nemours & Co., Inc.).
15. Brit. Pat. 1,070,037 (May 24, 1967), E. Rogers, A. A. Jones, and K. R. Tay (to Monsanto Chemicals, Ltd.).
16. Ger. Offen. 2,402,927 (1973), J. C. Cheminaud, J. Schapira, and A. Azkienazy (to Compagnie Française de Products Industrials).
17. U.S. Pat. 3,640,697 (Feb. 8, 1972), E. E. Toops, Jr. (to Commercial Solvents Corp.).

18. K. I. Sadykhov, Z. D. Naibov, S. M. Aliev, and N. F. Tuzhikova, *Khim. Tekhnol. Topl. Mazel* **18**(3), 29 (1973).
19. U.S. Pat. 4,223,733 (Sept. 23, 1980), E. E. Bodor and J. T. Payton (to Texaco, Inc.).
20. U.S. Pat. 4,179,389 (Dec. 18, 1979), J. T. Mann (to Gulf Research and Development Co.).
21. Ger. Offen. 2,008,674 (1971), L. L. Harrell, Jr. (to E. I. du Pont de Nemours & Co., Inc.).
22. U.S. Pat. Appl. 820,120 (1977), S. A. Lipowski (to Diamond Shamrock Corp.).
23. U.S. Pat. Appl. 20,106 (1979), R. A. Suhoza (to R. T. Vanderbilt Co., Inc.).
24. U.S. Pat. 2,652,427 (Sept. 15, 1953), A. Shultz (to Allied Chemical Corp.).
25. W. Fricke, *Tenside* **4,** 317 (1967).
26. A. Davidson and B. M. Milwidsky, *Synthetic Detergents*, 5th ed., Leonard Hill, London, 1972, p. 137.
27. *Chem. Week*, 51 (June 29, 1963).
28. U.S. Pat. 3,058,920 (Oct. 16, 1962), R. Brooks and B. Brooks (to The Chemithon Corp.).
29. U.S. Pat. 3,677,934 (July 18, 1972), K. Majima and G. Morita (to Kao Soap Co., Ltd.).
30. U.S. Pat.4,096,089 (June 20, 1978), W. J. Shibe, Jr., and W. Wood (to Sybron Corp.).
31. U.S. Pat. 3,861,466 (Jan. 21, 1975), W. W. Gale (to Exxon Production Research Co.).
32. U.S. Pat. 3,577,248 (May 4, 1971), R. B. Dean and R. C. Taylor (to Atlantic Richfield Co.).
33. E. E. Gilbert and E. P. Jones, *Ind. Eng. Chem.* **45,** 2044 (Sept. 1953).
34. U.S. Pat. 3,048,531 (Aug. 7, 1962), E. L. Stogryn, F. Argabright, and A. Argabright (to Esso Research & Engineering Co.).
35. U.S. Pat. 3,084,186 (April 2, 1963), E. Clippinger (to California Research Corp.).
36. Research and development records, Stepan Chemical Company, Northfield, Ill.
37. Brit. Pat. 522,840 (June 28, 1940), (to E. I. du Pont de Nemours & Co., Inc.).
38. M. S. Karach, E. M. May, and F. R. Mayo, *J. Org. Chem.* **3,** 175 (1938).
39. M. Quaedvlieg in J. Houben and T. Weyl, eds., *Methoden der Organischen Chemie*, Vol. 9, 4th ed., Thieme Verlag, Stuttgart, 1955.
40. W. J. Wenisch, *Diss. Abstr.* **18,** 808 (1958).
41. J. B. S. Braverman, *J. Sci. Food Agric.* **4,** 540 (1953).
42. J. B. S. Braverman and J. Kopelman, *J. Food Sci.* **26,** 249 (1961).
43. H. Heiwinkel, *Sven. Papperstidn.* **47,** 265 (1944).
44. D. L. Ingles, *Aust. J. Chem.* **12,** 97 (1959).
45. *Ibid.*, **14,** 302 (1961).
46. U.S. Pat. 2,899,461 (Aug. 11, 1959), R. M. Smith.
47. A. Rieche and H. Seeboth, *Ann.* **638,** 66 (1960).
48. U.S. Pat. 2,517,613 (Aug. 8, 1950), S. V. Allmen and H. Eggenberger (to Sandoz, Ltd.).
49. I. Inoue and M. Kojima, *J. Pharm. Soc. Jpn.* **71,** 939 (1951).
50. Ger. Pat. 721,991 (May 14, 1942), K. Diehl.
51. E. Wawzonek, *Heterocyclic Compounds*, Vol. 2, John Wiley & Sons, Inc., New York, 1951.
52. W. Reppe and co-workers, *Ann.* **596,** 1 (1955).
53. P. Seidel, *Ber.* **83,** 20 (1950).
54. D. Marquis and co-workers, *J. Am. Oil Chem. Soc.* **43,** 607 (1966).
55. W. Linfield, *Anionic Surfactants*, Marcel Dekker, Inc., New York, 1976.
56. B. Webb, *Soap Chem. Spec.* **42**(11), 61 (1966).
57. R. Odioso, *Soap Chem. Spec.* **43,** 47 (1967).
58. A. Fort and co-workers, *Sulfonation of Detergent Range Olefin*, American Oil Chemists Society Meeting, Mexico City, April 29, 1974.
59. E. S. Strauss and O. Kamatori, *Chemical Economics Handbook*, SRI International, Menlo Park, Calif., Sept. 1980, p. 681.5032S.
60. K. Oba, *Yukagaku* **17,** 208 (1968).
61. U.S. Pat. 4,190,427 (Feb. 26, 1977), R. Ravallo (to General Foods Corp.).
62. Y. Chang, *Ying Yung Chieh Mien Hua Hsueh* **10,** 45 (1980).
63. F. G. Bordwell and co-workers, *J. Am. Oil Chem. Soc.* **67,** 827 (1945).
64. C. M. Suter and co-workers, *J. Am. Oil Chem. Soc.* **60,** 538 (1938); E. Gilbert, *Chem. Rev.* **62,** 549 (1962).
65. A. Turbak, *Ind. Eng. Chem.* **2,** 229 (1963).
66. U.S. Pat. 2,923,728 (Feb. 2, 1960), K. Falk and W. Taplin (to E. I. du Pont de Nemours & Co., Inc.).
67. U.S. Pat. 3,169,142 (Feb. 9, 1965), E. A. Knaggs and M. L. Nussbaum (to Stepan Chemical Co.).

68. U.S. Pat. 3,200,140 (Aug. 10, 1965), A. Sowerbey (to Marchon Products, Ltd.).
69. U.S. Pat. 3,259,645 (July 5, 1966), R. J. Brooks (to Chemithon Corp.).
70. U.S. Pat. 3,328,460 (June 27, 1967), J. Vander Mey (to Allied Chemical Corp.).
71. U.S. Pat. 3,488,384 (Jan. 6, 1970), A. Kessler and P. Pflaumer (to Procter & Gamble Co.).
72. I. Yamane, *J. Am. Oil Chem. Soc.* **55,** (Jan. 1978).
73. C. Kaiser, *Chem. Ber.* **97,** 2903, 2917, 2926 (1964).
74. B. Hunter and co-workers, *Toxicology* **5,** 359 (1976).
75. *Alpha Olefin Sulfonates, Soap and Detergent Association Report*, Vol. 2, Arthur D. Little, Cambridge, Mass., May 31, 1977, Chapt. 6.
76. U.S. Pat. 3,699,157 (Oct. 17, 1972), W. Sweeney (to Chevron Research Co.).
77. U.S. Pat. Offen. 876,002 (July 28, 1970), J. Smith, DEF Publ. (to Procter & Gamble Co.).
78. U.S. Pat. 3,660,472 (May 2, 1972), K. Majima and co-workers (to Kao Soap Co.).
79. U.S. Pat. 3,492,343 (Jan. 27, 1970), P. Garner and co-workers (to Lever Bros. Co.).
80. U.S. Pat. 4,052,431 (Oct. 4, 1977), T. Baker and co-workers (to Stepan Chemical Co.).
81. U.S. Pat. 3,531,518 (Sept. 29, 1970), T. Ohren and co-workers (to Procter & Gamble Co.).
82. U.S. Pat. 3,839,391 (Oct. 1, 1974), R. Suzuki, S. Tanimori, S. Toyoda, and T. Ogoshi (to Lion Fat and Oil Co., Ltd.).
83. U.S. Pat. 3,925,441 (Dec. 9, 1975), S. Toyoda and T. Ogoshi (to Lion Fat and Oil Co., Ltd.).
84. A. Stirton, *J. Am. Oil Chem. Soc.* **39,** 490 (1962).
85. U.S. Pat. 2,867,585 (Jan. 6, 1959), J. Weil, A. Stirton, and E. Maurer (to the Secretary of Agriculture).
86. J. Weil, A. Stirton and R. Bistline, *J. Am. Oil Chem. Soc.* **37,** 295 (1960).
87. A. Stirton, R. Bistline, J. Weil, W. Ault, and E. Maurer, *J. Am. Oil Chem. Soc.* **39,** 128 (1962).
88. J. Weil and A. Stirton, *J. Am. Oil Chem. Soc.* **41,** 355 (1964).
89. M. Nagayama, O. Okumura, T. Sakatani, and S. Hashimoto, *Yukagaku* 1975, **24**(6), 295 (to Lion Fat and Oil Co., Ltd.).
90. E. A. Knaggs, J. Yeager, L. Varenyi, and E. Fischer, *J. Am. Oil Chem. Soc.* **42,** 805 (1965).
91. Jpn. Kokai 76 43,716 (April 14, 1976), T. Ohkoshi and S. Sekiguchi (to Lion Fat and Oil Co., Ltd.).
92. F. Asinger, *Chemie und Technologie der Paraffin-Kohlenwasserstoffe*, Akademie Verlag, Berlin, 1956.
93. U.S. Pat. 2,193,824 (March 19, 1940), W. H. Lockwood and J. I. Richmond (to E. I. du Pont de Nemours & Co., Inc.).
94. U.S. Pat. 2,848,628 (Aug. 19, 1958), H. G. Boynton, E. W. Lewis, and A. T. Watson (to Esso Research & Engineering Co.).
95. H. G. Boynton, E. W. Lewis, and A. T. Watson, *Ind. Eng. Chem.* **51,** 267 (March 1959).
96. U.S. Pat. 2,528,320 (Oct. 31, 1950), J. B. Roberts, H. Buchanan, and C. H. Brautcheck (to E. I. du Pont de Nemours & Co., Inc.).
97. Ger. Offen. 735,096 (April 1, 1943), (to I. G. Farbenindustrier A.-G.).
98. Ger. Offen. 887,503 (Sept. 8, 1949), (to A. G. Farbwerke Hoechst).
99. L. Orthner, *Angew Chem.* **62,** 302 (Jan. 1950).
100. H. Pines and R. C. Wackher, *J. Am. Chem. Soc.* **68,** 599 (1946).
101. U.S. Pat. 2,503,280 (April 11, 1950), W. H. Lockwood (to E. I. du Pont de Nemours & Co., Inc.).
102. U.S. Pat. 2,507,088 (May 9, 1950), H. W. Bradley (to E. I. du Pont de Nemours & Co., Inc.).
103. Brit. Pat. 942,503 (Nov. 20, 1963), (to Esso Research & Engineering Co.).
104. U.S. Pat. 3,325,387 (June 13, 1967), J. F. Black (to Esso Research & Engineering Co.).
105. U.S. Pat. 3,260,741 (July 12, 1966), J. Mackinnon and A. A. Austin (to Esso Research & Engineering Co.).
106. Ger. Pat. 1,252,666 (Oct. 26, 1967), J. Mackinnon and A. A. Austin (to Esso Research & Engineering Co.).
107. U.S. Pat. 3,372,188 (March 5, 1968), T. G. Alston and W. D. Schaeffer (to Union Oil Co.).
108. *European Chem. News* **3,** 31 (June 14, 1963).
109. U.S. Pat. 1,937,521 (Dec. 5, 1933), H. Limberg (to Funtkote Corp.).
110. U.S. Pat. 2,035,106 (March 24, 1936), H. G. Vesterdal and E. W. Carlson (to Standard Oil Development Co.).
111. T. F. Brown, *Inst. Pet. Rev.* **9,** 314 (1955).
112. H. Kaye, E. Forsyth, and A. I. Mills, *The Use of Gaseous SO_3 in the Sulfonation of Petroleum Oils*, paper 25, proceedings of the 5th World Petroleum Congress, Section III, New York, 1959.

113. E. Meyer, *White Mineral Oil, Petroleum, and Related Products*, Chem. Publishing Co., New York, 1968, pp. 166–192.
114. U.S. Pat. 2,050,345 (Aug. 11, 1936), L. Liberthson (to L. Sonneborn Sons, Inc.).
115. U.S. Pat. 2,166,117 (July 18, 1939), M. Blumer (to L. Sonneborn Sons, Inc.).
116. U.S. Pat. 2,909,563 (Oct. 20, 1959), W. Whitney (to Phillips Petroleum Co.).
117. *Canadian Chemical Processing*, 72 (Aug. 1957).
118. U.S. Pat. 3,097,234 (July 9, 1963), W. Graham and H. Stratford (to Stratford Engineering Corp.).
119. U.S. Pat. 2,244,512 (June 3, 1941), R. Brandt (to Colgate-Palmolive-Peet Co.); Reissue U.S. Pat. 221,548 (Sept. 26, 1944).
120. U.S. Pat. 2,524,086 (Oct. 3, 1950), L. Schmerling (to Universal Oil Products Co.).
121. Neth. Pat. 7,400,177 (July 9, 1975), (to Witco Chemical Corp.).
122. Brit. Pat. 553,598 (May 27, 1943), K. Partisch (to Frazer & Frazer, Ltd.).
123. U.S. Pat. 2,783,273 (Feb. 26, 1957), G. M. Verley (to Sinclair Refining Co.).
124. U.S. Pat. 2,815,370 (Dec. 3, 1957), L. E. Hutchings and G. W. Crosby (to The Pure Oil Co.).
125. U.S. Pat. 2,769,836 (Nov. 6, 1956), E. E. Gilbert and B. Veldhuis (to Allied Chemical Corp.).
126. U.S. Pat. 2,828,331 (March 25, 1958), M. M. Marisic and L. E. Hutchings (to The Pure Oil Co.).
127. B. U. Gruzdev, L. A. Potolovskii, and N. A. Potapkina, *Tr. Vseš. Nauchno. Issed. Inst. Pererab. Nefti* **12,** 307 (1970).
128. U.S. Pat. 2,616,936 (Nov. 4, 1952), H. E. Mammon and G. L. Hervert (to Universal Oil Products Co.).
129. U.S. Pat. 2,203,440 (June 4, 1940), and U.S. Pat. 2,203,441 (June 4, 1940), T. C. Oliver (to Charlotte Chemical Laboratories, Inc.).
130. U.S. Pat. 2,448,184 (Aug. 31, 1948), N. E. Lemmon (to Standard Oil Co. of Indiana).
131. U.S. Pat. 2,523,582 (Sept. 26, 1950), R. W. Mattson (to Union Oil Co.).
132. U.S. Pat. 2,581,064 (Jan. 1, 1952), F. M. Archibald (to Standard Oil Development Co. of Delaware).
133. U.S. Pat. 2,680,716 (June 8, 1954), M. R. Lipkin and H. E. Reif (to Sun Oil Co.).
134. U.S. Pat. 2,697,031 (Dec. 14, 1954), G. L. Hervert (to Universal Oil Products Co.).
135. U.S. Pat. 2,691,040 (Oct. 5, 1954), H. S. Block and H. E. Mammen (to Universal Oil Products Co.).
136. U.S. Pat. 2,768,199 (Oct. 23, 1956), H. E. Luntz and D. O. Popvac.
137. U.S. Pat. 2,802,026 (Aug. 6, 1957), H. Hennig, W. J. Alford, and L. H. Hutchings (to The Pure Oil Co.).
138. U.S. Pat. 2,865,958 (Dec. 23, 1958), D. G. Davies and P. Peperzak (to Shell Development Co.).
139. U.S. Pat. 2,769,836 (Nov. 6, 1956), E. E. Gilbert and B. Veldhuis (to Allied Chemical Corp.).
140. L. A. Potolovskii, U. M. Rozhdestvenskii, T. K. Avaliani, V. K. Golikov, I. Z. Kotrenko, and N. V. Šemenova, *Tr. Vses. Nauchno. Issled. Inst. Pererab. Nefti* **13,** 131 (1970).
141. Jpn. Pat. 3,106 (Jan. 27, 1972), S. Okumura and N. Toyota (to Matsummura Petroleum Research Laboratory).
142. M. C. Dwivedi, *Indian Inst. Technology Petroleum Hydrocarbons* **7**(2), 35 (1972).
143. USSR Pat. 402,324 (Oct. 19, 1973), A. M. Manoilo, N. P. Kortunov, N. G. Chernov, and coworkers.
144. W. W. Gale and E. I. Sandvik, *Soc. Pet. Eng. J.* **13**(4), 191 (1973).
145. Jpn. Kokai 78,127 (Oct. 20, 1973), J. Sato (to Nisso Metal Co., Ltd.).
146. Ger. Pat. 2,256,758 (Oct. 11, 1973); Fr. Pat. 2,300,078 (Feb. 5, 1975); Can. Pat. 1,011,759 (June 7, 1977); Brit. Pat. 1,478,483 (June 29, 1977), M. A. Plummer, C. C. Zimmerman, Jr., and D. E. Schroeder (to Marathon Oil Co.).
147. U.S. Pat. 3,753,465 (Aug. 21, 1973), M. D. Denekas, M. A. Ashimov, M. A. Mursalova, Z. A. Dadasheva, and S. M. Aliev (to Esso Production Research Co.); *Azerb. Neft. Khoz.* **54**(3), 34 (1974).
148. Jpn. Pat. 6,306 (Feb. 13, 1974), N. Toyoda, T. Yagi, H. Inove, and K. Hirooka (to Matsumura Petroleum Research Lab.).
149. U.S. Pat. 3,956,372 (May 11, 1976), J. R. Coleman, Jr., M. A. Plummer, and C. C. Zimmerman, Jr. (to Marathon Oil Co.).
150. T. L. Ashcraft and R. K. Saunders, *Sulfonation of Petroleum Feedstocks in a Wiped Film Reactor*, paper No. 6119 presented Society of Petroleum Engineers of AIME meeting, Houston, Texas, Oct. 3–6, 1976.
151. A. N. Niyazou, E. A. Amanov, and T. L. Bososlavskaya, *Izv. Akad. Nauk Turkm. SSR* **2,** 115 (1976).
152. U.S. Pat. 4,087,456 (May 2, 1978), P. S. Ossip and L. M. Echelberger (to Marathon Oil Co.).

153. U.S. Pat. 4,144,266 (March 13, 1979), M. A. Plummer and D. E. Schroeder, Jr. (to Marathon Oil Co.).
154. U.S. Pat. 4,147,638 (April 3, 1979), M. A. Plummer (to Marathon Oil Co.).
155. U.S. Pat. 4,148,821 (April 10, 1979), M. L. Nussbaum and E. A. Knaggs (to Stepan Chemical Co.).
156. U.S. Pat. 4,177,207 (Dec. 4, 1979), M. L. Nussbaum and E. A. Knaggs (to Stepan Chemical Co.).
157. U.S. Pat. 4,240,987 (Dec. 23, 1980), R. C. Berg (to U.O.P. Inc.).
158. V. Razderishin and E. Dennemark, *Nefty. Khoz.* **20,** 95 (1931).
159. K. P. Likhushin, et al., *Azerbaidzhan. Neftyanone Khoz.* **11,** 67 (1932); **11,** 116 (1933); **1,** 63 (1934); **9,** 87 (1934).
160. Australia Pat. 243,951 (Oct. 22, 1959), (Berol Akiebolaget.).
161. U.S. Pat. 3,056,831 (Oct. 7, 1967), H. W. Stratford (to Stratford Engineering Corp.).
162. U.S. Pat. 3,188,183 (June 8, 1965), R. S. Logan (to Phillips Petroleum Co.).
163. U.S. Pat. 3,232,976 (Feb. 2, 1966), J. W. Lohr (to The Andrew Jergens Co.).
164. U.S. Pat. 3,270,038 (Aug. 30, 1966), P. Marshall and J. Blakeway (to Colgate-Palmolive Peet Co.).
165. U.S. Pat. 3,346,505 (Oct. 10, 1967), J. M. Blakeway (to Colgate-Palmovive Peet Co.).
166. U.S. Pat. 4,113,438 (Sept. 12, 1978), B. J. Brooks and R. J. Brooks (to The Chemithon Corp.).
167. Brit. Pat. 974,298 (Nov. 4, 1964), E. A. Knaggs and M. L. Nussbaum (to Stepan Chemical Co.).
168. U.S. Pat. 3,107,087 (Sept. 21, 1971), W. A. Graham (to Stratford Engineering Corp.).
169. U.S. Pat. 4,113,438 (Sept. 12, 1978), B. J. Brooks and R. J. Brooks (to The Chemithon Corp.).
170. E. E. Gilbert and B. Veldhuis, *Ind. Eng. Chem.* **49,** 31 (1957).
171. U.S. Pat. 3,328,460 (June 27, 1967); U.S. Pat. 3,501,276 (March 17, 1970); Fr. Pat. 1,375,974 (Sept. 20, 1963); Ger. Pat. 1,195,299 (June 24, 1965); Brit. Pat. 1,029,029 (May 11, 1966), J. Vander Mey (to Allied Chemical Corp.).
172. U.S. Pat. 3,427,342 (Feb. 11, 1969), R. J. Brooks and B. J. Brooks (to The Chemithon Corp.).
173. U.S. Pat. 3,535,339 (Oct. 20, 1970), H. H. Beyer and C. W. Molt (to The Procter and Gamble Co.).
174. U.S. Pat. 3,438,743 (April 15, 1969), H. Grunewald and O. Steijner (to Mooch Domsjo Aktiebolag).
175. U.S. Pat. 3,438,744 (April 15, 1969), D. Steijner (to Mooch Domsjo Aktiebolag).
176. L. A. Potolovskii, U. M. Rozhdestvenskii, T. K. Avaliani, V. K. Golikov, T. Z. Katrenko, and N. V. Semenova, *Tr. Vses. Nauchno Issled. Inst. Pererab. Nefti* **12,** 307 (1970); **13,** 350 (1970).
177. T. K. Avaliani, L. A. Potolouskii and T. I. Katrenko, *Tr. Vses. Nauchno Issled. Inst. Pererab. Nefti,* **12,** 299 (1970).
178. B. L. Knight, *A Current Review of Surfactant Tertiary Recovery Tests,* Marathon Oil Co., paper presented at ACS Chemical Marketing and Economics Symposium, New York, April 5–8, 1976, reprinted.
179. E. A. Knaggs, *The Role of the Independent Surfactant Manufacturer in Tertiary Oil Recovery,* Stepan Chemical Co., paper presented at ACS Chemical Marketing and Economics Symposium, New York, April 5–8, 1976, reprinted. (Economics of Enhanced Oil Recovery).
180. Lewin and Associates, Inc., DOE/BETC/2628-2, report for U.S. Department of Energy, May 1981.
181. *Chemicals for Chemical Flooding in Enhanced Oil Recovery,* Gulf Universities Research Consortium, Report for U.S. Department of Energy, March 11, 1980.
182. U.S. Pat. 3,591,498 (July 6, 1971), L. S. Benbury and U. Bray (to Bray Oil Co.).
183. E. E. Gilbert and B. Veldhuis, *Ind. Eng. Chem.* **50,** 997 (1958).
184. U.S. Pat. 3,476,800 (Nov. 4, 1969), V. P. Cease (to Continental Oil Co.); B. V. Gruzdev, L. A. Potolovskii, and N. A. Potapkina, *Tr. Vses. Nauchno. Issled. Inst. Pererab. Nefti* **12,** 307 (1970).
185. U.S. Pat. 2,779,784 (Jan. 29, 1957), M. L. Scharrah (to Continental Oil Co.).
186. U.S. Pat. 3,033,898 (May 8, 1962), U. Bray (to Bray Oil Co.).
187. U.S. Pat. 3,105,049 (Sept. 24, 1963), V. Voorhees (to Bray Oil Co.).
188. U.S. Pat. 3,250,710 (May 10, 1966), M. W. Hunt (to Continental Oil Co.).
189. U.S. Pat. 2,845,455 (July 29, 1958), L. V. Brown (to Sinclair Refining Co.).
190. U.S. Pat. 2,616,905 (Nov. 4, 1952), P. A. Asseff (to Lubrizol Corp.).
191. U.S. Pat. 2,270,577 (Jan. 20, 1942), R. F. Bergstrom and co-workers (to Shell Development Co.).
192. U.S. Pat. 2,279,086 (April 7, 1942), R. F. Bergstrom (to Shell Development Co.).
193. U.S. Pat. 2,402,325 (June 18, 1946), W. K. Griesinger and co-workers (to Atlantic Refining Co.).
194. U.S. Pat. 2,418,894 (April 15, 1947), J. G. McNab and T. R. Dilworth (to Standard Oil Development Co.).
195. U.S. Pat. 2,485,861 (Oct. 25, 1949), S. E. Campbell and I. S. Dellinger.
196. U.S. Pat. 2,616,904 (Nov. 4, 1952), P. A. Asseff, T. W. Mastin, and A. Rhodes (to Lubrizol Corp.).
197. U.S. Pat. 2,800,503 (July 23, 1957), G. W. Crosby (to The Pure Oil Co.).

198. U.S. Pat. 2,865,956 (Dec. 23, 1958), G. Ellis and co-workers (to Shell Development Co.).
199. U.S. Pat. 2,920,105 (Jan. 5, 1960), H. D. Kluge and co-workers (to Texaco, Inc.).
200. U.S. Pat. 3,242,080 (March 22, 1966), M. A. Wiley and H. D. Kluge (to Texaco, Inc.).
201. U.S. Pat. 2,924,617 (Feb. 9, 1960), O. L. Wright (to Continental Oil Co.).
202. Y. N. Shekhter and co-workers, *Neftepererab. Neftekkim (Kiev)* **10,** 16 (1970).
203. U.S. Pat. 3,155,616 (Nov. 3, 1964), V. Voorhees (to Bray Oil Co.).
204. U.S. Pat. 3,250,710 (May 10, 1966), M. W. Hunt (to Continental Oil Co.).
205. U.S. Pat. 2,501,731 (March 28, 1950), R. W. Mertes (to Union Oil Co.).
206. U.S. Pat. 2,723,234 (Nov. 8, 1955), P. A. Asseff and co-workers (to Lubrizol Corp.).
207. Can. Pat. 574,161 (April 14, 1959), R. L. McMillen (to Lubrizol Corp.).
208. U.S. Pat. 2,839,470 (June 17, 1958), G. E. Warren (to Shell Development Co.).
209. U.S. Pat. 2,861,951 (Nov. 25, 1958), R. L. Carlyle (to Continental Oil Co.).
210. U.S. Pat. 2,956,018 (Oct. 11, 1960), R. L. Carlyle and co-workers (to Continental Oil Co.).
211. U.S. Pat. 3,027,325 (March 19, 1962), R. L. McMillen (to Lubrizol Corp.).
212. U.S. Pat. 3,150,089 (Sept. 22, 1964), M. W. Hunt (to Continental Oil Co.).
213. U.S. Pat. 3,170,880 (Feb. 23, 1965), U. H. Bray (to Bray Oil Co.).
214. U.S. Pat. 3,172,855 (March 9, 1965), J. D. Rodgers and C. W. Stuehe (to Lubrizol Corp.).
215. U.S. Pat. 3,256,186 (June 14, 1966), R. Greenwald (to Lubrizol Corp.).
216. U.S. Pat. 3,258,426 (June 28, 1966), U. H. Bray and V. Voorhees (to Bray Oil Co.).
217. U.S. Pat. 3,277,002 (Oct. 4, 1966), M. W. Hunt (to Continental Oil Co.).
218. U.S. Pat. 3,312,618 (April 4, 1962), W. M. LeSuer and G. R. Norman (to Lubrizol Corp.).
219. U.S. Pat. 3,320,162 (May 16, 1967), W. N. Axe and S. Schiff (to Phillips Petroleum Co.).
220. U.S. Pat. 3,321,399 (May 23, 1967), J. Versteeg and M. W. Hill (to Esso Research and Engineering Co.).
221. U.S. Pat. 3,350,308 (Oct. 31, 1967), R. L. McMillen (to Lubrizol Corp.).
222. U.S. Pat. 3,384,585 (May 21, 1968), J. T. Gragson and D. W. Bosse (to Phillips Petroleum Co.).
223. U.S. Pat. 3,384,587 (May 21, 1968), E. H. Holst, R. S. Edwards, and C. E. Sibert (to Texaco, Inc.).
224. U.S. Pat. 3,428,564 (Feb. 18, 1969), C. Bluestein and B. Bluestein (to Witco Chem. Co.).
225. U.S. Pat. 3,470,097 (Sept. 30, 1969), J. B. Lavigne and M. F. Huges (to Chevron Res. Co.).
226. U.S. Pat. 3,474,035 (Oct. 21, 1969), J. G. Dadura (to Texaco, Inc.).
227. U.S. Pat. 3,488,284 (Jan. 6, 1970), W. M. LeSuer and G. R. Norman (to Lubrizol Corp.).
228. U.S. Pat. 3,523,896 (Aug. 11, 1970), J. T. Gragson, D. W. Bosse, and K. L. Mills (to Phillips Petroleum Co.).
229. U.S. Pat. 3,523,897 (Aug. 11, 1970), A. N. DeVault (to Phillips Petroleum Co.).
230. U.S. Pat. 3,523,899 (Aug. 11, 1970), A. N. DeVault (to Phillips Petroleum Co.).
231. U.S. Pat. 3,524,814 (Aug. 18, 1970), A. R. Sabol, E. W. Blaha, and C. Brannen (to Standard Oil Co., Chicago).
232. U.S. Pat. 3,535,242 (Oct. 20, 1970), M. Hashimoto, Y. Kono, and co-workers (to Nippen Oil Co., Japan).
233. U.S. Pat. 3,537,976 (Nov. 3, 1970), E. H. Hast, R. S. Edwards and J. E. May (to Texaco, Inc.).
234. U.S. Pat. 3,539,511 (Nov. 10, 1970), A. R. Sabol and E. W. Blaha (to Standard Oil Co., Chicago).
235. U.S. Pat. 3,544,463 (Dec. 1, 1970), E. Koft, Jr. (to Mobil Oil Corp.).
236. U.S. Pat. 3,595,790 (July 27, 1971), G. R. Norman and W. M. LeSuer (to Lubrizol Corp.).
237. U.S. Pat. 3,609,076 (Oct. 15, 1971), A. R. Sabol, E. W. Blaha, and C. G. Brannen (to Standard Oil Co., Chicago).
238. U.S. Pat. 3,658,703 (April 25, 1972), J. T. Grayson and D. W. Bosse (to Phillips Petroleum Co.).
239. U.S. Pat. 3,829,381 (Aug. 13, 1974), W. M. LeSuer (to Lubrizol Corp.).
240. K. I. Sadrkhov and co-workers, *Azerb. Neft. Khoz.* **5,** 34 (1978).
241. U.S. Pat. 4,129,589 (Dec. 12, 1978), T. L. Eliades, R. J. Muir, and J. D. Horner (to Surpass Chemical, Ltd.).
242. U.S. Pat. 4,165,291 (Aug. 12, 1979), J. T. Grayson (to Phillips Petroleum Co.).
243. Brit. Pat. 2,924 (1866); U.S. Pat. 70,485 (1867), B. C. Tilghman.
244. I. A. Pearl, *The Chemistry of Lignin*, Marcel Dekker, Inc., New York, 1967.
245. U.S. Pat. 1,856,558 (May 3, 1932), G. C. Howard.
246. G. A. Rasmussen and W. J. Detrat, *Lignin Utilization in Enhanced Oil Recovery*, paper presented March 27, 1980, American Chemical Society, Houston, Texas, Chemical Marketing and Economics Division reprints.
247. U.S. Pat. 4,006,779 (Feb. 8, 1977), G. Kalfoglov (to Texaco, Inc.).

248. U.S. Pat. 4,133,385 (Jan. 9, 1979), G. Kalfoglov (to Texaco, Inc.).
249. U.S. Pat. 4,249,606 (Feb. 10, 1981), G. Kalfoglov (to Texaco, Inc.).
250. U.S. Pat. 4,267,886 (May 19, 1981), G. Kalfoglov (to Texaco, Inc.).
251. U.S. Pat. 3,054,659 (Sept. 18, 1962), D. Craig and C. D. Logan (to Ontario Paper Co.).
252. U.S. Pat. 3,054,825 (Sept. 18, 1962), D. Craig and C. D. Logan (to Ontario Paper Co.).
253. *Chemical Economics Handbook*, Stanford Research Institute, International, Menlo Park, Calif., Nov. 1978, p. 583.4800A.
254. *Chemical Economics Handbook*, Stanford Research Institute International, Menlo Park, Calif., Aug. 1981, p. 583.8002.
255. J. P. Sisley, *Am. Dyest. Rep.* **43,** 741 (1954).
256. E. E. Gilbert and B. Weldhuis, *J. Am. Oil Chem. Soc.* **36,** 208 (1959).
257. *Ibid.*, **37,** 298 (1960).
258. U.S. Pat. 3,332,979 (July 25, 1967), C. E. Redemann (to Purex Corp.).
259. U.S. Pat. 3,392,185 (July 9, 1968), J. M. Walts and L. M. Schenck (to General Aniline Film Corp.).
260. P. Sosis and J. Dringoli, *J. Am. Oil Chem. Soc.* **47,** 229 (1970).
261. U.S. Pat. 3,069,242 (Dec. 18, 1962), R. Brooks and B. Brooks (to The Chemithon Corp.).
262. S. Holtzman and U. Zoller, *Soap and Chemical Specialties* **45,** 43 (1969).
263. U.S. Pat. 3,549,613 (Dec. 22, 1970), E. A. Knaggs and M. L. Nussbaum (to Stepan Chemical Co.).
264. A. Sunderlaid, *Soap* **2,** 63 (1935).
265. J. Levy, *J. Am. Chem. Soc.* **38,** 36 (1961).
266. Fr. Demande 2,342,782 (March 3, 1977), (to Dev. Finance Corp. of New Zealand).
267. Ger. Pat. 1,274,118 (Aug. 1, 1968), E. Koenig, P. Hahn, K. Stickdorn, and H. Braun (to Deutsches Hydriereveik Rodleler).
268. H. Paternak and S. P. Tluszye, *Kosmet. Parfum. Drogen Rundsch.* **11**(3), 158 (1967).
269. B. Milwidsky and S. Holtzman, *Soap Cosmet. Chem. Spec.* **48,** 44 (June 1972).
270. Jpn. Kokai 100,029 (Sept. 3, 1976), K. Yamamoto (to Lion Fat and Oil Co., Ltd.).
271. Jpn. Kokai 13,585 (1977), D. Saika, T. Yanagawa, and K. Nabazana (to Lion Fat and Oil Co., Ltd.).
272. L. Daradica, *Rev. Chim. Miner.* **31**(1), 14 (1980).
273. H. Roberts, R. Whistles, and E. Paschall, eds., *Starch: Chemistry and Technology*, Vol. 1, Academic Press, Inc., New York, 1965, p. 451.
274. U.S. Pat. 2,786,833 (March 26, 1957), O. Wurzburg, M. Rutenberg, and L. Ross (to National Products, Inc.).
275. E. Clippinger, *Ind. Eng. Chem. Prod. Res. Dev.* **3**(1), 3 (1964).
276. U.S. Pat. 3,234,258 (Feb. 8, 1966), R. Morris (to Procter and Gamble Co.).
277. U.S. Pat. 4,226,797 (Oct. 7, 1980), P. Bakker, C. Kind, and V. Anrich (to Shell Oil Co.).
278. L. F. Audrieth and M. Sveda, *J. Org. Chem.* **9,** 89 (1944).
279. Brit. Pat. 1,568,812 (1960), G. Smith and W. Laird (to Dawson International, Ltd.).
280. U.S. Pat. 909,011 (April 3, 1973), J. Schwier, Def. Publ.
281. Can. Pat. 802,106 (Dec. 24, 1968), A. Sowerby (to Electric Reduction of Canada Ltd.).
282. U.S. Pat. 3,308,012 (March 7, 1967), U. Ramon (to E. I. du Pont de Nemours & Co., Inc.).
283. Ger. Pat. 1,805,262 (Aug. 7, 1969), I. Michelson (to American Safety Equipment).
284. Brit. Pat. 1,395,497 (May 29, 1975), A. Calder and J. Whetstone (to Imperial Chem. Ind., Ltd.).
285. U.S. Pat. 3,226,430 (Dec. 28, 1965), M. Mhatre (to Abbott Laboratories).
286. R. Patterson, *Sugar J.* **38**(1), 28 (1975).
287. J. Kircher, *Ind. Eng. Chem.* **46,** 1925 (1954).
288. W. L. Faith, D. B. Keys, and R. L. Clare, *Industrial Chemicals*, 4th ed., John Wiley & Sons, Inc., New York, 1975, p. 60.
289. E. A. Knaggs and M. L. Nussbaum, *Soap Chem. Spec.* **38**(5), 237; (6), 165, 179; (7), 145, 149 (1962).
290. H. Huber, *J. Am. Chem. Soc.* **33,** 57 (1956).
291. B. Stain and W. Fider, *J. Ind. Eng. Chem.* **51,** 13 (1959).
292. U.S. Pat. 3,024,258 (March 6, 1962), R. J. Brooks and B. Brooks (to The Chemithon Corp.).
293. *Technical Bulletin of Societa Per Azonic Ing.*, Mario Ballestra & Co., Milano, Italy, 1979.
294. U.S. Pat. 2,928,867 (March 15, 1960), J. C. Kirk and E. L. Miller (to Continental Oil Co.).
295. P. Hoffenberg, *Seifen Ole Fette Wachse* **99,** 360 (1973).
296. F. J. Kremers and A. Shultz, *Soap. Chem. Spec.* **46,** 44 (1970).
297. U.S. Pat. 3,931,273 (Jan. 6, 1976), A. Lanteri (to G. Mazzoni S.P.A.).

298. U.S. Pat. 3,620,684 (Nov. 16, 1971), R. J. Brooks and B. Brooks (to The Chemithon Corp.).
299. U.S. Pat. 3,427,342 (Feb. 11, 1964), R. J. Brooks and B. Brooks (to The Chemithon Corp.).
300. *Sulfur Trioxide Detergent Process Equipment*, Technical Bulletin, Chemithon Corp., Seattle, Washington, 1980.
301. Ger. Offen. 2,523,875 (Dec. 11, 1975), A. Brostrom (to Berol Kemi A.B.).
302. *The Berol Kemi Sulfonation Unit*, Technical Bulletin, Berol Kemi A.B., Steungsund, Sweden.
303. U.S. Pat. 3,891,689 (June 24, 1975), A. Pryce (to Lever Brothers Co.).
304. B. R. Schaafsma, *Deterg. Age* **1**, 22 (1965).
305. U.S. Pat. 3,198,849 (Aug. 3, 1965), M. Ballestra (to Ing. M. Ballestra and C.S.P.A.).
306. S. J. Silvis and M. Ballestra, *J. Am. Chem. Soc.* **40**, 618 (1963).
307. A. Lanteri and G. Mazzoni, *Sulfonation and Sulfation Technology*, World Conference on Soaps and Detergents, Montreux, Switzerland, Oct. 1977.

E. A. KNAGGS
M. L. NUSSBAUM
A. SHULTZ
Stepan Chemical Company

SULFONIC ACIDS

The sulfonic acids are any of numerous acids that are characterized by the sulfonic group $—SO_3H$ (or $—SO_2OH$) and may be regarded as derived from sulfuric acid by replacement of a hydroxyl group by either an inorganic anion or univalent organic radical (see Sulfuric acid and sulfur trioxide). Thus, inorganic acids bearing the typical formula XSO_2OH, where X is Cl, F, or NH_2, are known as sulfuric acids, ie, chlorosulfuric, fluorosulfuric, and amidosulfuric acids, formerly called chlorosulfonic, fluorosulfonic, and amidosulfonic acids, respectively (see Chlorosulfuric acid). Organic sulfonic acids having the general formula RSO_2OH are classified as paraffin-based or alkane- and cycloalkanesulfonic acids, unsaturated cyclic or arenesulfonic acids, heterocyclic sulfonic acids, and unsaturated straight- or branched-chain sulfonic acids.

Physical Properties

The nature of the R group has a significant influence upon the physical and chemical properties of the specific acids. In general, they are strong acids comparable to sulfuric acid and are, for the most part, hygroscopic, nonvolatile, soluble in water and polar solvents, and chemically stable. Only in select instances can the sulfonic acids be distilled either at atmospheric pressure or even *in vacuo*, because they undergo thermal decomposition at elevated temperatures. Exemplary in stability are some hydrocarbon-derived alkanesulfonic acids (see Table 1), trifluoromethanesulfonic acid [*1493-13-6*] (F_3CSO_2OH), other perfluoroalkanesulfonic acids, benzenesulfonic acid [*98-11-3*], and *p*-toluenesulfonic acid [*104-15-4*] (1). In general, however, the arenesulfonic acids usually decompose upon distillation, even *in vacuo*.

Table 1. Physical Properties of Some Alkanesulfonic Acids

Product	CAS Registry No.	Boiling point (at 133 Pa = 1 mm Hg), °C	Melting point, °C	Density, d_4^{25}, g/cm^3
methane-	[*75-75-2*]	122	20	1.4844
ethane-	[*594-45-6*]	123	−17	1.3341
propane-	[*28553-80-2*]	136	7.5	1.2516
butane-	[*30734-86-2*]	149	−15.2	1.1906
pentane-	[*35452-30-3*]	163	−15.9	1.1226
hexane-	[*13595-73-8*]	174	16.1	1.1047

Contrary to expectations, trifluoromethanesulfonic acid differs quite dramatically from the alkanesulfonic acids cited in Table 1 by virtue of having a lower boiling point, 162°C at 101.3 kPa (= 1 atm), greater acid strength than sulfuric acid, and higher thermal stability as it undergoes no change up to 350°C (2). Trifluoromethanesulfonic acid forms a distillable hydrate. The other perfluoroalkanesulfonic acids exhibit similar properties; n-$C_8F_{17}SO_2OH$ [*1763-23-1*] boils at 260°C at 101.3 kPa. Benzene- and p-toluenesulfonic acids can be distilled at ca 13.3 Pa (0.1 mm Hg); their respective boiling points are 172°C and 187°C.

Chemical Properties

The sulfonic acids are strong acids like sulfuric acid. They ionize in water and in many polar organic solvents. Unlike the organic sulfates, they are moderately stable to hydrolysis and not prone to decomposition on heating. A brief review of some classical chemical reactions of sulfonic acids will serve to exemplify their chemical properties.

Transformation of the Sulfonic Acid Group. Replacement of the hydroxyl segment of the sulfo group with chlorine takes place by treatment of the acid or salt thereof with phosphorus pentachloride or chlorosulfuric acid to give sulfonyl chlorides. Notwithstanding, alkanesulfonyl chlorides are more often prepared by direct treatment of the parent aromatic compound with chlorosulfuric acid (see Sulfonation and sulfation). The sulfonyl chlorides are the intermediates generally employed to convert the sulfonic acids to the corresponding sulfonamides, sulfonate esters, sulfinic acids, and thiols, since (except for the thiols and certain aliphatic esters) these compounds cannot be prepared directly from the sulfonic acids.

Sulfonyl bromides are produced in a similar manner by treatment of the sulfonic acid or a salt with phosphorus tribromide and bromine. However, these products (compared to the chlorides) are only of limited commercial utility.

Sulfonyl fluorides can be made by reaction of fluorosulfuric acid with the parent aromatic compound, the sulfonic acid, or a salt thereof. The resulting sulfonyl fluoride exhibits remarkable stability of its fluorosulfonyl group under either acid or neutral conditions and allows its retention during various chemical reactions of the organic portion of the molecule.

Other transformations of the sulfonic acid group comprise conversion to

Sulfonic anhydrides by treatment of a sulfonic acid with dehydrating agents such as thionyl chloride or phosphorus pentoxide:

$$2\,RSO_2OH + \underset{\text{thionyl chloride}}{SOCl_2} \rightarrow \underset{\text{sulfonic anhydride}}{(RSO_2)_2O} + 2\,HCl + SO_2$$

Arenesulfonic anhydrides by direct treatment of the aromatic compound with a sulfur trioxide/nitromethane complex:

$$2\,ArH + 3\,SO_3 \rightarrow (ArSO_2)_2O + H_2SO_4$$

Sulfones by reaction of alkane- and arenesulfonic acids with aromatic compounds under a variety of dehydrating conditions:

$$RSO_2OH + R'C_6H_5 \rightarrow \underset{\text{sulfone}}{RSO_2C_6H_4R'} + H_2O$$

The reaction is thought to proceed via the formation of the sulfonic anhydride as intermediate (3) (see Sulfolanes and sulfones).

Secondary esters by reaction of alkanesulfonic acids at room temperature with olefins:

$$\underset{\text{methanesulfonic acid}}{CH_3SO_2OH} + RCH{=}CHR' \rightarrow CH_3SO_2OCH(R)CH_2R'$$

Thus, ethylene gives a 25% conversion to the ester at 100°C and 5.6 MPa (800 psig). In contrast, trifluoromethanesulfonic acid reacts rapidly with ethylene at room temperature, forming the ethyl ester and a low polymer of ethylene (4).

Sulfonic acids and their corresponding sulfonyl chlorides can be reduced with phosphorus and iodine or with powdered zinc or iron in the presence of aqueous mineral acids:

$$RSO_2OH + 2\,P \xrightarrow{I_2} RSH + P_2O_3$$

$$RSO_2Cl + 3\,H_2 \rightarrow RSH + 2\,H_2O + HCl$$

Although the former reaction proceeds with considerable difficulty, it has been the basis for the commercial manufacture of 4-chlorothiophenol for a number of years (5). The more common laboratory and preferred industrial route for thiol manufacture is via the sulfonyl chloride derivative (see Thiols).

Cleavage of the Carbon–Sulfur Bond. In contrast to the foregoing transformation reactions, an arenesulfonic acid group can be replaced by a variety of other substituents, eg, Cl, Br, I, CN, COOH, H, NH_2, NH_2NH, OH, OR, NO_2, aryl-N=N, RS, and SH. The reaction mechanism involves cleavage of the carbon–sulfur bond; the ease and degree of cleavage depend upon the geometrics of other structural substituents. Only a limited number of such reactions proceed with sufficient ease to have commercial utility.

Hydrolytic desulfonation is an important cleavage reaction that has had industrial applications for many years. The process involves steaming an arenesulfonic acid to recover the parent aromatic compound by reversal of the sulfonation reaction (6):

$$ArSO_2OH + H_2O \rightarrow ArH + H_2SO_4$$

High temperatures and pressures are generally required to complete desulfonation. Thus, in the sulfonation of naphthalene for ultimate manufacture of 2-naphthol, the 1-sulfonate group is removed by steaming at 165°C.

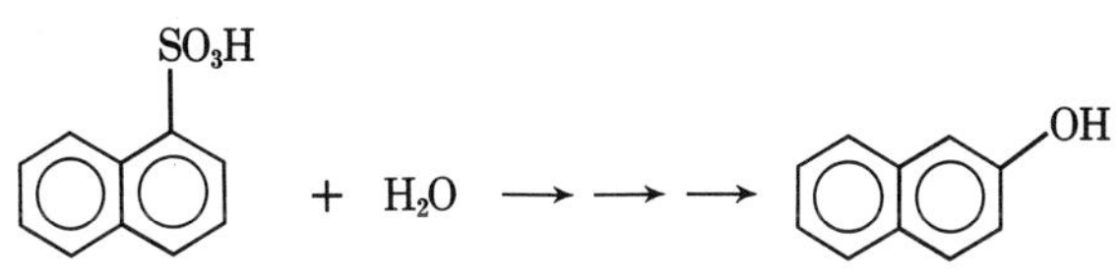

[85-47-2]

Fortunately, each sulfonic acid has a characteristic desulfonation temperature. The desulfonation process is used for synthesis and for the separation of isomers of xylene as well as other di- and polyalkylated benzenes and naphthalenes, chlorinated benzenes, and phenols.

Another cleavage reaction involves alkaline fusion of arenesulfonic acid salts to yield phenols:

$$ArSO_2ONa + 2\ NaOH \xrightarrow{\text{fusion}} ArONa + Na_2SO_3 + H_2O$$

The foregoing reaction is used for the commercial manufacture of a variety of phenolic compounds including phenol, *p*-cresol, 2-naphthol, resorcinol, hydroxyanthraquinones, 8-hydroxyquinoline, 4-hydroxybiphenyl, and many others (see Phenol; Alkylphenols).

The replacement of some arenesulfonic acid groups by halogens occurs by reaction with an oxidant such as sodium chlorate, which is used commercially to produce chlorinated anthraquinones and for the synthesis of chlorinated naphthalenes and heterocyclic derivatives:

$$3\ ArSO_2ONa + 6\ HCl + NaClO_3 \rightarrow 3\ ArCl + 4\ NaCl + 3\ H_2SO_4$$

Arenesulfonate groups can be exchanged for amino groups in some instances by heating with ammonia as well as mono- and dialkylamines and, in others, by fusion with sodamide.

Fusion with metallic cyanides yields nitriles (qv) (7):

$$RSO_2ONa + NaCN \rightarrow RCN + Na_2SO_3$$

The reaction is used for preparation of mono- and dicyano derivatives of benzene, naphthalene, biphenyl, and higher condensed ring systems.

Sulfonated phenols are converted to nitrophenols by treatment with nitric acid; ie, the sulfonic acid groups are replaced during nitration (qv).

By comparison, the instances of carbon–sulfur cleavage in the aliphatic series are few. Alkanesulfonyl chlorides are readily converted to the corresponding chlorides by heating to 150–250°C (8):

$$RSO_2Cl \xrightarrow{\Delta} RCl + SO_2$$

Arene- and Alkylarenesulfonic Acids

One of the simplest products, *p*-toluenesulfonic acid, which is made by sulfonation of toluene with 100 wt % sulfuric acid, is used commercially to effect aqueous solubility of disperse dyes (9) (see Dye carriers). Thus, azobenzene, *p*-aminoazobenzene, *p*-hydroxyazobenzene, *p*-dimethylaminoazobenzene, and 1-phenylazo-4-aminonaphthalene, are solubilized by the addition of one or more than two moles of *p*-toluenesulfonic acid to one mole of disperse dye (see Azo dyes). The solubilization mechanism appears to involve the formation of the sulfonic acid adduct with one mole of disperse dye by hydrophobic and nonpolar van der Waals interactions, which supply the necessary solubilizing groups to increase the water solubility of the disperse dye. A contrasting effect of the sulfate moiety, OSO_3H, on the substantivity of various acid dyes to nylon-6 is described in a study of the physicochemical properties of dye solutions and absorption of dyes (10–11). The investigation of the influence of sodium lauryl sulfate on mixed dyeing with certain acid dyes revealed that the absorption of each dye on nylon-6 decreases with increased concentration of the surfactant sulfate, probably because of the competing forces of the surfactant and dye anions for dye sites on the fabric.

Azosulfonic acids are potentially suitable as dyes for lipstick (12). They are more compatible with the skin than eosin-base dyes. They are soluble in oil and the common lipstick ingredients, and their salts are soluble in water so that lipsticks based on azosulfonic acids can be washed off with soap. The sodium salts of azosulfonic acids are used in food coloring and in cosmetics (qv) (see Food additives). The free sulfonic acids are prepared by treatment of the sodium salt with concentrated hydrochloric acid. 1-(4-Sulfo-1-naphthylazo)-2-naphthol-6,8-disulfonic acid [*7244-14-6*] typifies an azosulfonic acid used in lipstick. The azosulfonic acid is generally dissolved in a vehicle, eg, 1,2-propylene glycol, prior to kneading with the remaining formulation ingredients.

Both *p*-toluenesulfonic acid and 2-naphthalenesulfonic acid [*120-18-13*] are used as initiators for the catalytic polymerization of caprolactam (13). The initiation reaction is the acylation of the lactam through the amide salt (13). α-Methylstyrene can be dimerized at 40–120°C with *p*-toluenesulfonic acid as catalyst. The use of *p*-hydroxybenzenesulfonic acid [*98-83-9*] and methanesulfonic acid is also described (14).

Alkylarenesulfonic acids are used for the preparation of agricultural and industrial fungicides (15) (see Fungicides). Copper derivatives of alkylarylsulfonic acids are of particular commercial utility in fungicidal sprays (13,16). Aqueous *p*-toluenesulfonic acid is reported to be effective for the extraction of vanadium from petroleum and petroleum products (17).

Arenesulfonic acids, eg, 2-naphthalenesulfonic acid, 1,3-benzenedisulfonic acid [*98-48-6*], and 2,6-naphthalenedisulfonic acid [*581-75-9*], and disulfodiphenyl ether [*3232-24-4*] are suitable for stabilization of maleic anhydride–olefin copolymers (18). The nonvolatile sulfonic acids are used alone or as compounds that liberate the free sulfonic acid during the copolymerization step. 1,3-Benzenedisulfonic acid, 2,6-naphthalenedisulfonic acid, and 1,3,6,8-naphthalenetetrasulfonic acid [*33192-25-5*] are the best stabilizers of a wide variety of compounds studied (18).

Alkylated 2-naphthalenesulfonic acid is very effective for the controlled swelling of calf skins in pickling solutions irrespective of pH (19). The repression of swelling by certain aromatic sulfonic acids results not only from their dipole properties, but also from their tendency to enter compounds with hydrogen bonds, and from the possibility of salt formation between peptide chains. Reproducibility in the degree of swelling is of considerable practical importance in the processing of low quality skin materials (see Leather).

Another interesting application for alkylarenesulfonic acids is in the compounding of wax-flux compositions for low temperature soldering processes (20). The wax-soluble alkylarenesulfonic acids have the general formula R_nArSO_3H, where Ar is an arylene group, eg, phenylene, naphthalene, anthracenylene, etc; R is a wax-solubilizing alkyl group containing 1–22 carbon atoms; n is an integer from 1 to 3; and the total number of carbon atoms contained in the combined R groups is ca 8–30. The sulfonic acids can be produced any number of ways but preferably by continuous sulfonation of the alkylarene with gaseous SO_3 to give products of low free sulfuric acid content; free sulfuric acid causes staining and darkening of phenolic circuit boards. The wax-flux compositions make up a large amount of wax and an effective amount of the sulfonic acid to cause fluxing at or below the soldering temperatures of the piece to be soldered. A significant application of such wax-flux products is in printed circuit boards, wherein the molten wax-flux composition is applied to the metal pattern side of the board and allowed to solidify. The component leads are trimmed and then soldered to the printed circuit board (see Solders and brazing alloys).

Although not a recent development, application research continues on the use of linear alkylarenesulfonic acids (mol wt 318–345) for acid pickling of titanium components used in the construction of aircraft (see Metal surface treatments, cleaning, pickling, and related processes). The versatility of sulfonic acids in industrial applications is further exemplified by the use of sulfonated alkylphenol–formaldehyde resins or *p*-toluenesulfonic acid–cresol mixtures as accelerators for the curing of pressure-sensitive adhesive tape (21). Reduction in curing cycles can be as much as 20- to 30-fold.

Organic Sulfonic Acid Oligomers

Organic disulfonic acids are produced by heating a sulfonate monomer at above 110°C in the absence of water. Olefin sulfonation product mixtures, hydroxyalkanesulfonic acids, alkanesultones, alkenesulfonic acids, and mixtures thereof are oligomerized under these conditions. Thus, 1-hexadecenesulfonic acid [*30644-46-3*], prepared by SO_3 sulfonation of 1-hexadecene in a falling-film reactor, is oligomerized by heating at 150–153°C for 2.25 h. Esterification of the oligomerization product can be effected to yield the methyl ester, which is an important plasticizer for poly(vinyl chloride) resins and permits pressing of the resins into sheets (see Plasticizers).

The salts of oligomers of C_{14}–C_{20} sultones, after undergoing saponification processing, are useful surfactants in heavy-duty-detergent compositions. The disulfonic acids as unneutralized oligomers are commercially important as the foaming agents for the production of preformed acidic-well circulation fluids for oil wells (see Petroleum, drilling fluids). In the preparation of foamed-well circulation fluids, liquid concentrates of the surface-active agents are ordinarily used, provided that phase separation, especially solid formation, does not occur upon standing or at the low temperatures often encountered at the wellhead in the field. Thus, moderate activity continues in this important field of application. Liquid surfactant concentrates are being developed based on mixtures of the foregoing oligomers, ie, C_{14}–C_{19} olefins, with the precursor monomer or of the oligomers *per se* to give excellent low temperature characteristics.

Fluorinated and Chlorofluorinated Sulfonic Acids

The syntheses of fluorinated and chlorosulfonated sulfonic acids are described in refs. 2 and 22, respectively. The literature discloses the reaction of dialkyl sulfides and disulfides, sulfoxides and sulfones, alkanesulfonyl halides, alkanesulfonic acids, and alkanethiols with oxygen, hydrogen chloride, hydrogen fluoride, and oxygen–chlorine–hydrogen fluoride mixtures over metal halide catalysts, eg, $CuCl_2$–KCl and $MnBr_2$–$LaCl_3$, to give the respective title acids. The reaction of Cl_2CHSH with oxygen–hydrogen fluoride over $CuCl_2$–KCl gives CF_3SO_3H, $ClCF_2SO_3H$ [*73043-98-8*],

Cl_2CFSO_3H [*77801-23-1*], CHF_2SO_3H [*40856-11-9*], and $ClCHFSO_3H$ [*40856-08-4*].

Trifluoromethanesulfonic acid is useful in a number of commercial applications. It is used as a catalyst with sulfuric acid for the liquid-phase alkylation of low molecular weight isoparaffin hydrocarbons with olefin hydrocarbons. Thus, the liquid-phase alkylation of isoparaffin hydrocarbons, eg, isobutane, etc, with olefin hydrocarbons selected from the group consisting of propylene, butylene, and mixtures thereof can be carried out by use of a catalyst mixture consisting of a minor portion of trifluoromethanesulfonic acid and a major portion of sulfuric acid and containing ca 0–3 wt % water and ca 0–10 wt % acid oils. The resulting alkylate products, highly branched C_7–C_8 paraffin hydrocarbons with high octane values, are useful as motor fuels or in gasoline blending stocks (23) (see Gasoline and other motor fuels).

In another petroleum feedstock application, trifluoromethanesulfonic acid is used as a superacid catalyst (Magic acid) in combination with antimony pentafluoride (SbF_5) for either liquid- or vapor-phase hydroisomerization of C_4–C_{12}–paraffin hydrocarbons to give products useful in motor fuels (see Friedel-Crafts reactions) (24). The superacid catalyst, ie, SbF_5 and CF_3SO_3H, is applied in its supported form. The supports may consist of alumina, potassium fluoride on alumina, sodium fluoride, aluminum phosphate, aluminum fluoride, charcoal, etc. Either component of the catalyst can be added first or they can be added simultaneously to the support media in a stream of helium to facilitate distribution of the component(s) on the surface of the support. Upon deposition of the catalyst, the helium stream is continued only long enough to purge the system of any excess acid.

The versatility of perfluorosulfonic acids is also exemplified in the preparation of monopropellants, which are potentially useful as fuels in missiles and rockets (25) (see Explosives and propellants). Thus, 61.58 parts by weight of β,β,β-trifluoroethanesulfonic acid [*1827-97-0*] in admixture with 38.42 parts by weight of perchloric acid dihydrate ($HClO_4.2H_2O$) with cooling below 35–40°C, a chamber pressure of 5.5 MPa (800 psi) yields a monopropellant with a specific impulse of 125 s and a weight loss of 3.4% after 8 h at 249°C. Other mixtures that have been prepared are listed in Table 2 (25).

The use of fluorosulfuric acid and antimony pentafluoride for the polymerization or polycondensation of alkanes is applicable to butane, isobutane, ethane, and methane as a potential route for the synthesis of rubber or of extenders for elastomers and rubber (26). However, the products are highly branched, oily oligomers with molecular weights of 100–700. Research in this area continues toward the development of much higher molecular weight polymers. However, there is little published evidence to indicate that such oligomers and polymers have achieved any significant commercial utility. Also of interest is perfluorobutanesulfonic acid [*375-73-5*], which is used as a catalyst for the manufacture of chloropolydimethylsiloxanes (27).

Table 2. Perfluoro Monopropellants

Ingredients	Wt %	Specific impulse, s	Weight loss, % (8 h at 249°C)	Impact test
perchloric acid, dihydrate methanesulfonic acid	58.62 41.38	150	15	61.0 cm
perchloric acid, dihydrate ethanesulfonic acid	64.97 35.03	169	14.2	no explosion upon 15.2-m drop

Fluorocarbon sulfonic acids are used extensively in organic syntheses as, for example, in the preparation of anhydrides and sulfonyl halides (28). One route involving the reaction of phosphorus pentachloride with cyclic fluorocarbon sulfonic acids leads first, with loss of HCl, to a high melting solid, which decomposes upon heating to yield the sulfonic anhydride such as that shown below:

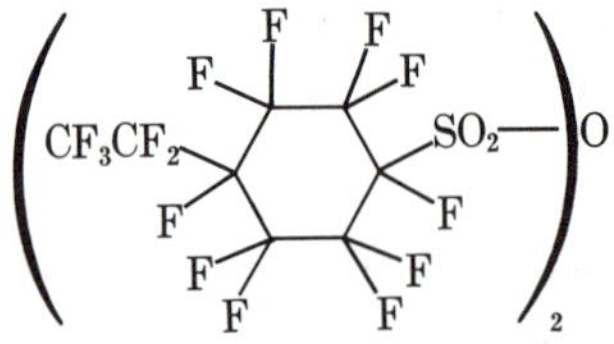

The noncyclic acids produce a mixture of anhydride and sulfonyl chloride. The higher boiling sulfonic anhydrides are cleaved above 200°C by metals, eg, NaF, $ZnCl_2$, and $AlCl_3$. Sulfonyl fluorides and chlorides can thus be prepared in good yield; however, only half of the anhydride is available for conversion.

Fluorinated sulfonic acids and derivatives thereof can be produced by a base-catalyzed addition reaction of the thiol R_fRSH (R_f = perfluoroalkyl) to an alkenylamidoalkanesulfonic acid or salt (29). The fluorinated alkenylamidoalkanesulfonic acids and salts are represented by the formula:

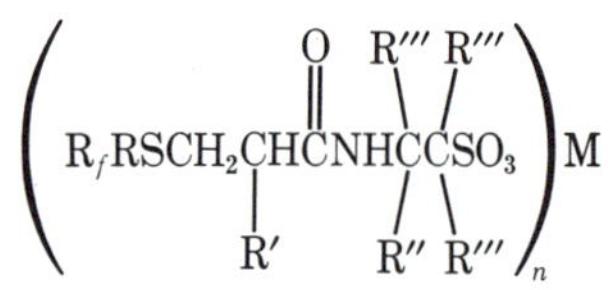

R_f = straight or branched-chain perfluoroalkyl, C_1–C_{18}; R = alkylene (usually ethylene); R′ = hydrogen or low alkyl; R″ = hydrogen, alkyl, aryl, or pyridyl; R‴ = hydrogen or hydrocarbon radicals. Such products are useful as surfactants and as leveling agents in floor polish (see Polishes; Surfactants and detersive systems). Thiols of the formula R_fRSH have been described in a number of U.S. patents (8,30–35).

The alkenylamidoalkanesulfonic acids and their salts are well known and have been described in numerous patents as well as in other literature references (36–49). Illustrative examples are listed in Table 3. The acrylamidoalkanesulfonic acids can be prepared by a two-step reaction sequence involving first the reaction of an olefin containing at least one allylic hydrogen atom with an acyl hydrogen sulfate to form a sulfonated intermediate and subsequently reacting this intermediate with water and acrylonitrile in the presence of sulfuric acid (38). Thus, 2-acrylamido-2-methylpropanesulfonic acid, which melts at 184–186°C, is prepared from isobutylene and acetyl hydrogen sulfate with acetic acid–methylene chloride as solvent in the first step and then treatment with water and reaction with acrylonitrile in the presence of 96 wt % sulfuric acid.

Of the alkenylamidoalkanesulfonic acids listed in Table 3, 2-acrylamido-2-methylpropanesulfonic acid has considerable commercial utility and is marketed by Lubrizol Corporation under the trademark of AMPS monomer (50). It is a white, crystalline solid with a molecular weight of 207, a melting point of 185°C, and an acid number of 270.7. It is moderately soluble in water and dimethylformamide, slightly soluble in methanol, ethanol, and isopropanol, and sparingly soluble in dioxane, tetrahydrofuran, acetonitrile, ethyl acetate, chloroform, benzene, and heptane. AMPS monomer is hygroscopic and polymerizes when wet but does not do so violently. The dry monomer is stable to autopolymerization at room temperature without inhibitors. Aqueous solutions of the monomer hydrolyze very slowly to give acrylic acid and β,β-dimethyltaurine. In contrast, aqueous solutions of the sodium salt at pH 9 resist hydrolysis for months. The homopolymerization and copolymerization of AMPS monomer have been studied extensively (50). Some of the applications of the AMPS monomer are described below.

Surfactants for aqueous solutions. Conjugate addition of a long-chain mercaptan or a primary amine to one or two moles of the sodium salt of AMPS monomer, respectively, produces surfactants that are effective in hard water. If the mercaptan is fluorinated, the conjugate adduct with AMPS monomer salts can generate aqueous fire-fighting foams.

Brass electroplating-bath additive. The addition of AMPS monomer to an aqueous alkaline cyanide brass-electroplating bath produces a semibright-to-bright brass plating (see Electroplating).

Dye receptivity in fibers and films. Blends of AMPS copolymers prepared with AMPS monomer provide good dye receptivity.

Textile sizes. Copolymers of AMPS monomer with vinyl acetate and acrylic acid are effective textile sizes for cotton and spun blends of cotton with polyester.

Permanent press and soil-release agents. AMPS monomer has been used to prepare copolymer compositions that impart permanent press and soil-release properties to spun blends of cotton with polyester and wrinkle and flame resistance to cotton fabrics (see Textiles, finishes).

Printing inks. Micronized aqueous suspensions of hydrophobic polymers containing AMPS monomer are impregnated with hydrophobic dye solutions. The dye solvents swell the polymers to produce nonsmearing jet printing inks (see Inks).

Clear antifog coatings. Homopolymers of AMPS monomer, when coated on dental mirrors or clear optical surfaces, render these surfaces fog-resistant.

Acrylimidoaminoethanesulfonic Acids

Acrylimidoaminoethanesulfonic acids and their derivatives are prepared by the reaction of acrylonitrile sequentially with SO_3, 2,2-dimethylpropane, and NH_3 at low temperatures (−30 to −75°C) (51). The products are useful as flocculating agents (qv) and in the preparation of polymers.

Amidoalkanesulfonic Acids

Amidoalkanesulfonic acids are prepared by the reaction of nitriles with olefins and concentrated sulfuric acid or oleum (39). They are useful as dispersing agents for calcium soaps.

Table 3. Alkenylamidoalkanesulfonic Acids

Name	CAS Registry No.	Empirical formula
2-acrylamidopropanesulfonic acid	[*33028-26-1*]	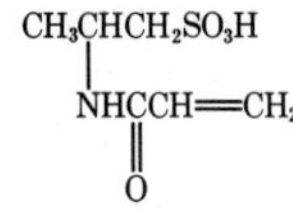
2-acrylamido-2-methylpropanesulfonic acid	[*15214-89-8*]	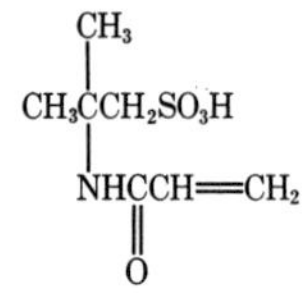
2-methacrylamido-2-methylpropanesulfonic acid	[*21838-63-1*]	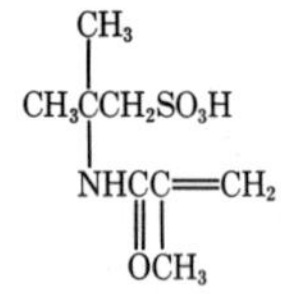
2-acrylamidobutanesulfonic acid	[*33028-25-0*]	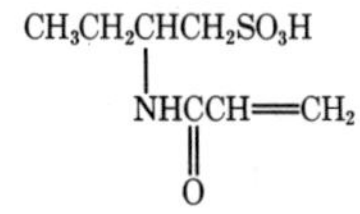
2-acryloamidobutane-2-sulfonic acid	[*79647-71-5*]	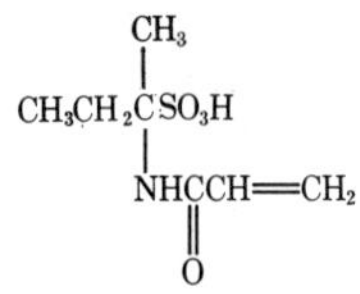
3-acrylamido-2,3-dimethylbutane-2-sulfonic acid	[*79840-59-8*]	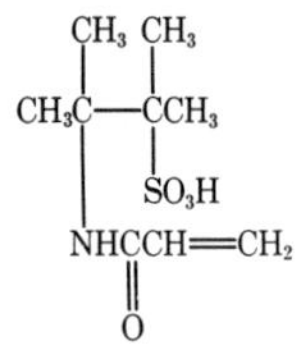
3-acrylamido-2,4,4-trimethylpentanesulfonic acid	[*79647-72-6*]	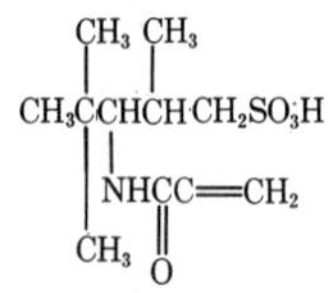
2-acrylamido-2-phenylethanesulfonic acid	[*32885-13-5*]	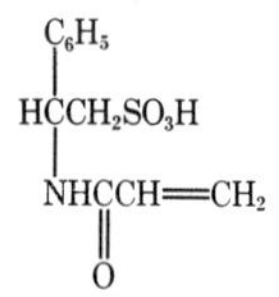
2-acrylamido-2-phenylpropanesulfonic acid	[*63314-84-1*]	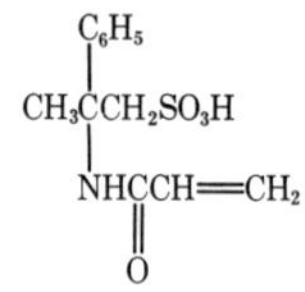
2-acrylamido-2-(*p*-tolyl)ethanesulfonic acid	[*79647-73-7*]	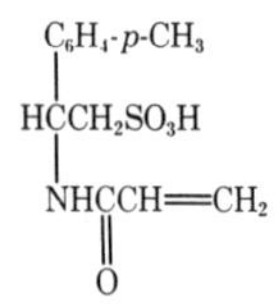

Table 3 (*continued*)

2-acrylamido-2-pyridylethanesulfonic acid	[*79647-74-8*]	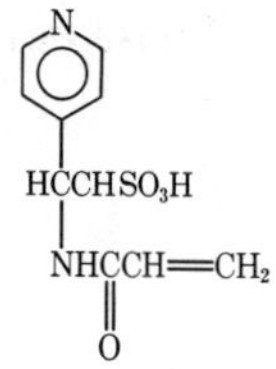

Perfluoroisoalkoxyalkanesulfonic Acids

Perfluoroisoalkoxyalkanesulfonic acids have a terminal polyfluorinated branched chain or cyclic perhaloisoalkoxy group linked through an ether or oxygen to a perfluoromethylene substituent (52). The perfluoroisoalkoxyalkanesulfonic acids can be prepared by conventional aqueous hydrolysis of the corresponding sulfonyl chlorides or fluorides (53). The hydrolysis is conveniently done by heating the sulfonyl halide with water at ca 100°C for 6–10 h and subsequently extracting the reaction mass with diethyl ether. The ether extract is then distilled to remove the ether and recover the sulfonic acid.

Alternatively, sulfonic acids with the —SO_3H group attached to a nonfluorinated methylene group can also be prepared by aqueous nitric acid oxidation of the corresponding fluorinated isoalkoxyalkyl thiocyanates (54). The nitric acid oxidation is generally carried out at ca 70–150°C with at least a stoichiometric proportion of nitric acid and preferably 200–230% of the stoichiometric requirement. The sulfonic acid is recovered from the reaction mass by extraction with diethyl ether followed by evaporation of the ethereal extract to dryness.

The perfluoroisoalkoxyalkanesulfonic acids are used as strong acid catalysts for organic reactions, as ingredients of dry-powder fire-extinguishing compositions (see Plant safety), and as intermediates in the manufacture of other polyfluoroisoalkoxyalkanesulfonyl-substituted compounds. The alkali metal salts of these sulfonic acids are readily soluble in water and are useful as aqueous dispersing agents, emulsifying agents, and detergents (see Surfactants). They are also used in organic lubricants and plasticizers (qv). The sulfonate salts of polyvalent metals, eg, divalent calcium, are highly effective in imparting oil and water repellancy to various substrates, eg, paper and textiles such as cellulose (see Waterproofing).

3,3′-Disubstituted Benzidine Derivatives

The 3,3′-disubstituted benzidine derivatives have in each of the 3,3′-positions the substituent

$$—O_n—A\overset{\overset{\displaystyle O}{\|}}{\underset{\underset{\displaystyle O}{\|}}{S}}—OW$$

where W is hydrogen or an alkali metal and where n is zero when A is the divalent branched-chain alkylene of 2–7 carbon atoms or one when A is a straight-chain divalent alkylene with three or four carbons (55–56).

The related sulfonic acids and their salts are useful as chromagen-reactive indicators. These new products have the following general formula:

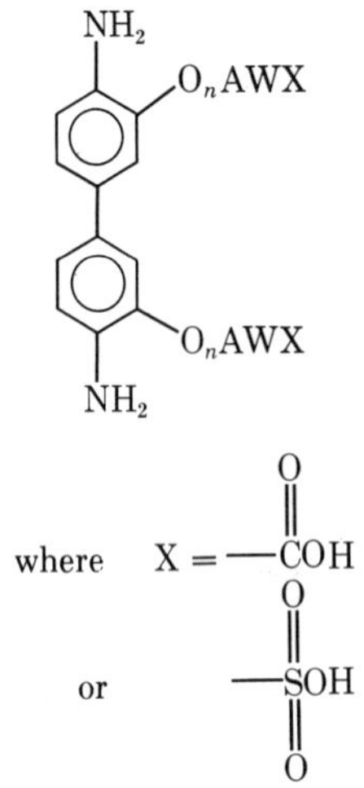

Two chromagen-reactive indicators are γ,γ'-(4,4′-diamino-biphenyl-3,3′-dioxy)dipropanesulfonic acid [*56716-06-4*] and δ,δ'-(4,4′-diamino-biphenyl-3,3′-dioxy)dibutanesulfonic acid [*34915-22-5*].

Alkanesulfonic Acids

Methanedisulfonic acid [*503-40-2*] and methanetrisulfonic acid [*54333-33-7*] are used as catalysts for the alkylation of phenols (57). Thus, 4-methylphenol and isobutylene react in the presence of catalyst to give 2,6-di-*tert*-butyl-4-methylphenol, (BHT) which is a commercially useful antioxidant and stabilizer for fats and oils. Methanesulfonic acid is also used for the manufacture of the unsaturated dimer of α-methylstyrene (37). *p*-Hydroxybenzenesulfonic acid is equally effective in this application. Methane- and ethanesulfonic acids are also used as the fluorosulfonic acids in the compounding of monopropellants (25).

Production and Economic Aspects

There are over 40 manufacturers of sulfonic acids and their various salts for in-plant use and the merchant market. Of these, Procter and Gamble, Colgate, and Lever Bros., and a few minor producers dominate the former field, which generally encompasses the manufacture of liquid and spray-dried household and institutional detergents. Manufacturers of sulfonic acids and their salts for the merchant market represent ca 22% of the total producers. The principal and minor producers of sulfonic acids are listed in Table 4 in order of decreasing productive capacity. Stepan Chemical Company is the largest single merchant manufacturer in the United States, with a productive capacity about equal to the total combined capacity of Onyx, Conoco, and Richardson. Their combined annual productive capacity is $>3.45 \times 10^5$ metric tons; this production is classified according to the sulfonation technology employed, as shown in Table 5.

Table 4. Manufacturers of Sulfonic Acids for the Merchant Market

Principal	Minor
Stepan Chemical Co.[a]	J. L. Prescott
Onyx Chemical Co./Gulf Oil	Westvaco
Conoco Chemical Co.	Continental Chemical Co.
Richardson Co.[a] (Witco)	Texize Division (Morton Norwich)
Pilot Chemical Co.[a]	E. I. du Pont de Nemours & Co., Inc.
Witco Chemical Corp.[a]	Lubrizol Corp.
Alcolac, Inc.[a]	

[a] Two or more manufacturing sites.

Table 5. Relative Production Capacities of Sulfonic Acids[a]

Productive capacity, t	Sulfonation process
283,675	continuous falling film SO_3-air
29,485	SO_3-solvent
22,680	chlorosulfuric acid
9,525	all other[b]

[a] Ref. 58.
[b] Low throughput continuous; batch SO_3, 100 wt % H_2SO_4, 20 wt % oleum, 30 wt % oleum, and sulfamic acid.

Production of sulfonic acids is treated in the literature in terms of their being intermediates for the manufacture of other products, eg, dyes, catalysts, electronic chemicals, polymers, oligomers, or chemicals for organic synthesis. Thus, the estimate of the total production of various important sulfonic acids is based on various industry sources and from a publication of the United States International Trade Commission (60). The statistics are detailed in Table 6. The difference in the quantities reported for sales and production reflects inventory changes and in-plant consumption of surface-active agents by synthetic rubber producers and by manufacturers of cosmetics, packaged detergents, and other formulated consumer products.

Health and Safety Factors, Toxicology

The sulfo or sulfonic acid group —SO_3H in some instances is moderately detoxifying; for example, toxic compounds, eg, aniline, 1-naphthylamine, or various nitro aromatics are relatively nontoxic when sulfonic acid substituents are introduced. The detoxifying effect upon the parent compound is exemplified by *m*-nitrobenzenesulfonic acid, *m*-aminobenzenesulfonic acid, and 1-naphthylamine-4-sulfonic acid; see Table 7 (60–61). The reason for this dramatic reduction in toxicity is thought to be the transition of the compound from one that is hydrophobic to one is hydrophilic upon introduction of the sulfo group. Thus, in the absence of the sulfo group, the hydrophobic compound easily penetrates the skin and is not eliminated through normal body functions. However, its conversion to the hydrophilic structure permits excretion in the urine as such or complexed with glycine or glucuronic acid (62). In the case of

Table 6. U.S. Production and Sales of Sulfonic Acids for 1980[a]

Products	Production, t	Percent of grand total	Sales: Quantity, t	Sales: \$/t	Sales: \$/kg	Value, $\$10^3$
p-dodecylbenzenesulfonic acid	102,734	38.58	48,623	992.08	0.99	101,920
benzene-, *p*-toluene-, xylene-, cumene-, *p*-hydroxybenzenesulfonic acids, etc	45,317	17.02	33,979	859.80	0.86	38,964
alkylnaphthalenesulfonic acids	9,675	3.63	7,740	1,300.71	1.30	12,584
naphthalene-based sulfonic acids	64,550	24.24	12,910	5,423.32	5.42	350,075
sulfonic acids having amide linkages		0.99	1,973	2,491.20	2.49	6,577
sulfosuccinamic acid derivatives	1,482	(0.56)	987	1,940.05	1.94	2,875
taurine fatty derivatives	1,158	(0.43)	986	3,196.68	3.20	3,702
Total	*2,640*					
sulfonic acids having an ester or ether linkage		11.85	14,457	3,068.80	3.06	96,873
sulfosuccinic acid esters	11,197	(4.20)	8,968	2,314.83	2.31	25,919
others	20,370	(7.65)	5,489	3,483.27	3.48	70,954
Total	*31,567*					
all other sulfonic acids and salts	9,839	3.69	9,752	1,278.67	1.28	12,581
Grand total	*266,322*	*100.00*	*145,864*			*723,024*

[a] Ref. 59.

benzenesulfonic acid, toxicity is nominal-to-high by the oral and inhalation route; the disaster hazard results when the product is heated to decomposition, yielding highly toxic fumes of SO_x.

The toxicity of *p*-toluenesulfonic acid is moderately high if the compound is ingested. Animal experiments indicate systemic toxicity. *o*-Toluenesulfonic acid [*88-20-0*] is highly irritating to the skin, eyes, and mucous membranes. It is also highly irritating when ingested or inhaled.

Indications are that benzene, toluene, xylene, and cumene are more toxic by inhalation than the corresponding sulfo derivatives primarily because of the higher volatility of the parent compound than its sulfonic acid derivative. This accounts for the narcotic effects sustained after exposure to high concentrations of vapor from the unsulfonated aromatic.

Addition of the hydroxyl group to the naphthalene nucleus tends to diminish the toxicity of the parent compound, naphthalene. On the other hand, introduction of the sulfonic acid group, as in 2-naphthol-6-sulfonic acid, dramatically increases the toxicity of the product. Addition of the amino group to naphthalene yields a reputedly carcinogenic compound that is significantly detoxified by introduction of the sulfonic acid group.

Fluorosulfuric acid is highly irritating to the skin, eyes, and mucous membranes;

it is also highly irritating by ingestion and inhalation (see Table 7). When heated to its decomposition temperature, it emits highly toxic fumes of sulfur fluorides and oxides. It generally reacts with water or steam to produce toxic and corrosive fumes.

Sulfonic acids used for the preparation of detergent bases are usually highly corrosive to the skin. Sulfonate salts widely used in detergents and as wetting agents, ie, those from C_{10}–C_{15} linear alkanes, C_{14}–C_{16} alpha-olefins, 1-isopropylnaphthalene, butylbiphenyl, butylphenylphenol, oleyl *N*-methyltauride, dodecyl diphenyl ether, dioctyl succinate, laurylsulfoacetic acid, and many others, prove nontoxic in both acute and chronic tests (67). The toxicity of individual sulfonic acids must be assessed separately, since they vary tremendously and the primary irritant effects do not correlate well with their ionization constants.

Hydrolysis of Oils and Fats

An important commercial application for sulfonic acids involves pure hydrolysis of oils and fats, which are glycerol esters of saturated and unsaturated fatty acids. The pure hydrolysis can be done with a current of live steam in the reaction mixture containing a sulfoaromatic compound, eg, Twitchell's reagent [*3055-92-3* and *82415-39-2*] (68). The reagent is used at ca 1.0 wt % of the hydrolyzing mixture.

Twitchell's reagent can be prepared by adding concentrated sulfuric acid to benzene or naphthalene and oleic acid and by keeping the reaction temperature preferably under 30°C:

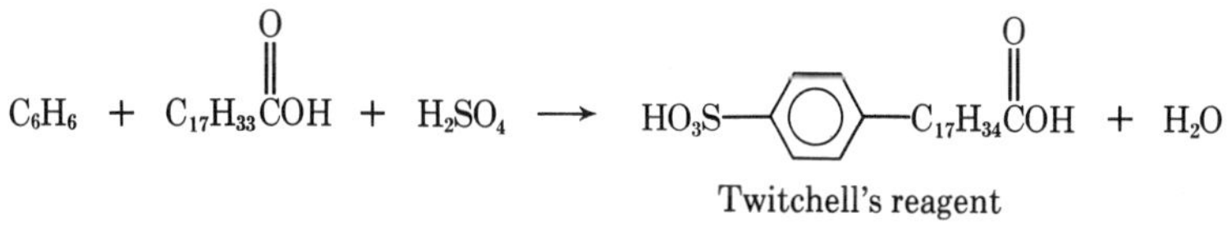

The resulting reagent is used for the pure hydrolysis, as exemplified below with glyceryl tristearate as one typical reactant:

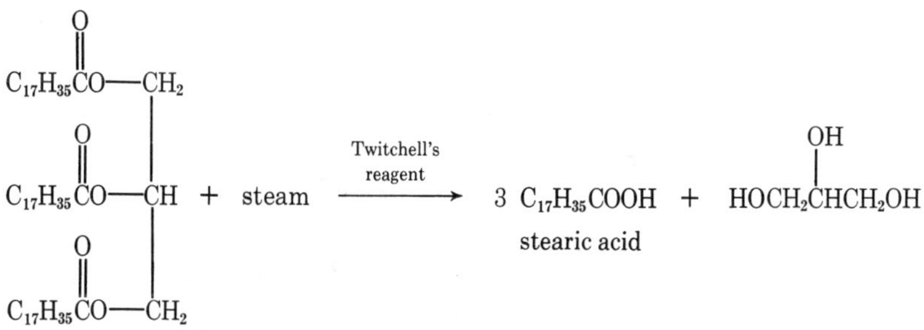

Table 7. Properties, Toxicity, and Uses of Sulfonic Acids[a]

Compound	CAS Registry No.	Bp, °C	Mp, °C	Toxicity	Uses
benzenesulfonic acid	*[98-11-3]*	110.6	50–51	highly irritating to skin, eyes, and mucous membranes; oral LD_{50} (rat) 890 mg/kg versus 3,800 mg/kg for benzene	manufacture of phenol by fusion with NaOH
m-nitrobenzenesulfonic acid	*[98-47-5]*		70	oral LD_{50} (rat) of sodium salt = 11,000 mg/kg versus 640 mg/kg for nitrobenzene	organic syntheses; sodium salt is a protective antireduction agent
m-aminobenzenesulfonic acid	*[121-47-1]*			oral LD_{50} (rat) = 12,000 mg/kg versus 440 mg/kg for aniline	manufacture of azo dyes; synthesis of certain sulfa drugs
p-toluenesulfonic acid	*[104-15-4]*	140 2.7 MPa	106–107	highly irritating to skin and mucous membranes; oral LD_{50} (rat) = 2,480 mg/kg versus 5,000 mg/kg for toluene	manufacture of dyes and oral antidiabetic drugs; conversion to sodium and ammonium salts for manufacture of hydrotropes
3,5-dimethylbenzenesulfonic acid (*m*-xylene-*m*-sulfonic acid)	*[18023-22-8]*			intraperitoneal LD_{50} (mouse) = 500 mg/kg	
dodecylbenzenesulfonic acid	*[27176-87-0]*			oral LD_{50} (rat) = 890 mg/kg versus 34,000 mg/kg for dodecylbenzene	electronic cleaning chemicals, pickling baths, and detergent manufacture

1-naphthalenesulfonic acid[a]	*[85-47-2]*		90[b]	oral LD_{50} (rat) = 420 mg/kg versus 1,780 mg/kg for naphthalene	manufacture of α-naphtholsulfonic acid and α-naphthylaminesulfonic acid; solvent for phenol in the manufacture of disinfectant soaps
2-naphthalenesulfonic acid[a]	*[120-18-3]*		91[c]	oral LD_{50} (rat) = 400 mg/kg versus 1,780 mg/kg for naphthalene	manufacture of beta-naphthol, beta-naphtholsulfonic acid, beta-naphthylaminesulfonic acid, and other intermediates
2-naphthol-6-sulfonic acid[a]	*[84-87-7]*		170 dec	oral LDL_o (mouse) = 250 mg/kg	preparation of azo dyes
1-naphthylamine-4-sulfonic acid[a]	*[84-86-6]*		170 dec	intraperitoneal LD_{50} (mouse) = 300 mg/kg	therapeutic category; Na–salt as hemostatic
fluorosulfuric acid	*[7789-21-1]*	165	−87	highly irritating to skin and mucous membranes; aquatic toxic rating, TL = 100–110 ppm after 96 h	fluorinating agent, catalyst in alkylation, acylation, polymerization, and condensation reactions; hydrofluorination of olefins; production of substituted pyridines
fluoromethanesulfonic acid	*[421-20-5]*	162		oral LD_{50} (mouse) = 112 mg/kg	alkylating catalyst; methylation of carbamates, octaethylporphyrin, and octaethylchlorine

[a] Refs. 60–61, 63–66.
[b] Dihydrate crystals.
[c] Monohydrate mp = 124–125°C.

BIBLIOGRAPHY

"Sulfonic Acids" in *ECT* 1st ed., Vol. 13, pp. 346–353, by A. A. Harban and C. E. Johnson, Standard Oil Company (Indiana); "Sulfonic Acids" in *ECT* 2nd ed., Vol. 19, pp. 311–319, by E. E. Gilbert, Allied Chemical Corporation.

1. Can. Pat. 1,093,581 (Jan. 13, 1981), R. F. Sweeney, J. O. Peterson, B. Sukornick, M. B. Berenbaum, H. R. Nychka, and R. E. Eibeck (to Allied Chemical Corp.).
2. R. E. Bank and R. N. Hazeldine, *The Chemistry of Organic Sulfur Compounds*, Vol. 2, Pergamon Press, Inc., New York, 1966, pp. 165–176.
3. W. A. Proell and C. E. Adams, *Ind. Eng. Chem.* **41,** 2217 (1949).
4. R. E. Banks and R. N. Hazeldine in N. Kharasch and C. Y. Meyers, eds., *The Chemistry of Organic Sulfur Compounds*, Vol. 2, Pergamon Press, Inc., New York, 1966, pp. 165–176.
5. U.S. Pat. 2,947,788 (Aug. 2, 1960), H. Pitt (to Stauffer Chemical Co.).
6. E. E. Gilbert, *Sulfonation and Related Reactions*, Interscience Publishers, a division of John Wiley & Sons, Inc., New York, 1965, pp. 425–442.
7. D. T. Mowry, *Chem. Rev.* **42**(2), 189 (1948).
8. F. Asinger, *Chemie und Technologie der Paraffin–Kohlenwasserstaffe*, Akademic Verlag, Berlin, 1956, pp. 428.
9. A. Katayama, T. Takagishi, K. Konishi, and N. Kuroki, *Kolloid Z.Z. Polym.* **206**(2), 162 (1965).
10. M. Mitsuishi, G. Aida, and E. Kosuge, *Sen'i Gakkaishi* **21,** 443 (1965).
11. *Ibid.*, 23 (1965).
12. Fr. Pat. 1,404,784 (July 2, 1965), (to G. Siegle & Co., G.m.b.H.).
13. G. Reinish, W. Jaeger, and I. Schopor, *Abh. Dtsch. Akad. Wiss. Berlin Kl. Chem. Geol. Biol.* (1) 283 (1963).
14. Jpn. Pat. 6335(66) (April 6, 1966), M. Suzumura and H. Yasui (to Yawata Chemical Industry Co., Ltd.).
15. Fr. Pat. 1,314,434 (Jan. 11, 1963), J. H. Duperray (to Chimiotechnic Union Chimique du Nord ed du Rhone).
16. U.S. Pat. 2,897,114 (July 28, 1959), T. W. Sauls (to Tennessee Corp.).
17. A. V. Kotova, S. V. Emel'yanova and V. G. Ben'kovskii, *Khim. Tekhnol. Topl. Masel* **10**(10), 29 (1965).
18. Neth. Appl. 6,405,547 (Nov. 23, 1964), (Imperial Chemical Industries, Ltd.).
19. D. Post, *Leder* **15,** 69 (1964).
20. U.S. Pat. 3,977,916 (Aug. 31, 1976), R. A. Stayner, R. C. Fox, and K. Jones (to Chevron Research Co.).
21. U.S. Pat. 3,231,419 (Jan. 25, 1966), R. Korpman (to Johnson & Johnson).
22. U.S. Appl. 642,833 (1975), R. F. Sweeney (to Allied Chemical Corp.).
23. U.S. Pat. 3,970,721 (July 20, 1976), J. W. Brockington and R. H. Bennett (to Texaco, Inc.).
24. U.S. Pat. 3,878,261 (April 15, 1975), L. E. Gardner (to Phillips Petroleum Co.).
25. U.S. Pat. 3,113,895 (Dec. 10, 1973), G. Barth-Wehrenalp and D. M. Gardner (to Pennsalt Chemicals Corp.).
26. D. T. Roberts, Jr., and L. E. Calihan, *Sci. Chem.* **7,** 1629 (1973).
27. Ger. Offen. 2,345,923 (March 20, 1975), K. F. Thom, A. DeMontigny, and R. Schliebs (to A. G. Bayer).
28. G. Van Dyke Tiers, *J. Org. Chem.* **28,** 1244 (1963).
29. U.S. Pat. 4,014,926 (March 29, 1977), R. E. A. Dear and E. K. Kleiner (to CIBA-GEIGY Corp.).
30. U.S. Pat. 2,894,991 (July 14, 1959), J.T. Barr and F. E. Lawlor (to Pennsalt Chemicals Corp.).
31. U.S. Pat. 2,961,470 (Nov. 22, 1960), W. A. Sheppard (to E. I. du Pont de Nemours & Co., Inc.).
32. U.S. Pat. 2,965,677 (Dec. 20, 1960), J. F. Harris, Jr. (to E. I. du Pont de Nemours & Co., Inc.).
33. U.S. Pat. 3,088,849 (May 7, 1963), W. S. Friedlander (to Minnesota Mining and Manufacturing Co.).
34. U.S. Pat. 3,544,663 (Jan. 30, 1967), M. Hauptschein, R. Bonner Hager, and T. Clark Allen (to Pennwalt Corp.).
35. U.S. Pat. 3,655,732 (April 11, 1972), C. Sciver Rondestvedt, Jr. (to E. I. du Pont de Nemours & Co., Inc.); U.S. Pat. 2,112,543 (March 29, 1938), R. F. Prescott (to The Dow Chemical Co.).
36. U.S. Pat. 2,983,712 (May 9, 1961), W. A. Wilkinson (to E. I. du Pont de Nemours & Co., Inc.).
37. U.S. Pat. 3,332,904 (July 25, 1967), E. M. LaCombe and W. P. Miller (to Union Carbide Corp.).

38. U.S. Pat. 3,506,707 (April 14, 1970), L. E. Miller and D. L. Murfin (to Lubrizol Corp.).
39. Brit. Pat. 1,090,779 (Nov. 15, 1967), D. Arlt.
40. Ger. Offen. 2,105,030 (Aug. 19, 1971), D. I. Hoke (to Lubrizol Corp.).
41. Can. Pat. 1,034,698 (July 11, 1978), M. V. Rostoker and L. Levin (to National Patent Development Corp.).
42. Ger. Offen. 2,426,452 (Jan. 2, 1975), W. R. Hollingshad (to Calgon Corp.).
43. Ger. Offen. 2,426,613 (June 4, 1973), W. R. Hollingshad and P. R. Ralston (to Calgon Corp.).
44. U.S. Pat. 3,926,718 (Dec. 16, 1975), L. J. Guilbault and J. R. Zakrewski (to Calgon Corp.).
45. U.S. Pat. 3,903,038 (Sept. 2, 1975), L. W. Neal (to E. I. du Pont de Nemours & Co., Inc.).
46. U.S. Pat. 4,026,856 (May 31, 1977), R. W. Jahnke (to Lubrizol Corp.).
47. R. L. Schild, M. S. El-Aasser, G. W. Pohlein, and J. W. Vanderoff in P. Bechner and M. N. Yudenfreund, eds., *Emulsions Latices, Dispersions*, Marcel Dekker, New York, 1978, pp. 99–128.
48. R. M. Reinhart and J. A. Harris, *Text Res. J.* **50**(3), 139 (1980).
49. T. J. Chen, *Research Disclosure* **159,** 62 (1977).
50. *AMPS Monomer*, Product Bulletin, Lubrizol Corp., Wickliffe, Ohio, Jan. 1982.
51. Ger. Offen. 2,347,627 (April 4, 1974), C. K. Steinhardt (to Lubrizol Corp.).
52. U.S. Pat. 3,821,290 (June 28, 1974), L. G. Anello and R. F. Sweeney (to Allied Chemical Corp.); R. B. Hodgdon, Jr., *J. Polym. Sci.* **A-1**(6), 171 (1968).
53. J. E. Millington, G. M. Brown, and F. L. M. Pattison, *J. Am. Chem. Soc.* **78,** 3846 (1956).
54. Fr. Pat. 1,561,360 (March 28, 1969), L. Foulletier and J. P. Lalu (to Ugine Kuhlmann).
55. U.S. Pat. 3,859,341 (Jan. 7, 1975), N. A. Jonsson, M. Ferenc, and L. E. Westlund (to A. B. Kabi, Sweden).
56. U.S. Pat. 4,008,267 (Feb. 15, 1977), N. A. Jonsson, M. Ferenc, and L. E. Westlund (to A. B. Kabi, Sweden).
57. U.S. Pat. 3,082,258 (March 19, 1963), W. V. McConnel and H. E. Davis (to Eastman Kodak Co.).
58. *Chemical Economics Handbook*, Stanford Research Institute, Menlo Park, Calif., Aug. 1980, p. 583.5001C–F.
59. *United States International Trade Commission Publication No. 1183 (1980)*, U.S. Government Printing Office, Washington, D.C., 1981.
60. *Registry of Toxic Effects of Chemical Substances*, Vol. 1, U.S. Government Printing Office, Washington, D.C., 1979, pp. 128, 134, 209, 217, 219, 221, 457.
61. *Registry of Toxic Effects of Chemical Substances*, Vol. 2, U.S. Government Printing Office, Washington, D.C., 1979, pp. 219, 623, 625, 708, 788.
62. D. W. Fassett, *Industrial Hygiene and Toxicology*, 2nd rev. ed., Vol. 2, 1963, pp. 1837–1845.
63. I. N. Sax and co-workers, *Dangerous Properties of Industrial Materials*, 5th ed., Van Nostrand Publishing Co., New York, 1979, pp. 403, 522, 1035, 1094.
64. Ref. 61, pp. 113, 122, 123, 127, 130.
65. Ref. 61, pp. 690, 693.
66. M. Windholz, ed., *The Merck Index*, 9th ed., Merck & Co., Inc., Rahway, N.J., 1976, p. 540.
67. R. D. Swisher, *The Soap and Detergent Association Scientific and Technical Report No. 4*, Soap and Detergent Association, New York, 1966.
68. P. H. Groggins, *Unit Processes in Organic Syntheses*, 3rd ed., McGraw-Hill Book Company, Inc., New York and London, 1947, pp. 689–691.

ANDREW SHULTZ
Stepan Chemical Co.

SULFOXIDES

Sulfoxides are compounds that contain a sulfinyl group covalently bonded at the sulfur atom to two carbon atoms. They have the general formula RS(O)R′, ArS(O)Ar′, and ArS(O)R. Sulfoxides represent an intermediate oxidation level between sulfides and sulfones. The naturally occurring sulfoxides often are accompanied by the corresponding sulfides or sulfones. The only commercially important sulfoxide is the simplest member, dimethyl sulfoxide [*67-68-5*] (DMSO, sulfinylbismethane).

Sulfoxides occur widely in small concentrations in plant and animal tissues, eg, allyl vinyl sulfoxide [*81898-53-5*] in garlic oil and 2,2′-sulfinylbisethanol [*3085-45-8*] as fatty esters in the adrenal cortex (1–2). Homologous methylsulfinylalkyl isothiocyanates, which are represented by the formula $CH_3SO(CH_2)_nNCS$, where $n = 3$ [*37791-20-1*], 4 [*4478-93-7*], 5 [*646-23-1*], 8 [*75272-81-0*], 9 [*39036-83-4*], or 10 [*39036-84-5*], have been isolated from a number of mustard oils in which they occur as glucosides (3). Two methylsulfinyl amino acids have also been reported: methionine sulfoxide [*454-41-1*] from cockroaches and the sulfoxide of *S*-methylcysteine, 3-(methylsulfinyl)alanine [*4740-94-7*]. The latter is the dominant sulfur-containing amino acid in turnips and may account in part for their characteristic odor (4).

Dimethyl sulfoxide occurs widely at levels of 3 ppm or less. It has been isolated from spearmint oil, corn, barley, malt, alfalfa, beets, cabbage, cucumbers, oats, onion, Swiss chard, tomatoes, raspberries, beer, coffee, milk, and tea (5). It is a common constituent of natural waters, and it occurs in seawater in the zone of light penetration where it may represent product of algal metabolism (6). Its occurrence in rainwater may result from oxidation of atmospheric dimethyl sulfide, which occurs as part of the natural transfer of sulfur of biological origin (7–8).

Properties

For the most part, sulfoxides are crystalline, colorless substances, although the lower aliphatic sulfoxides melt at relatively low temperatures (see Table 1). The lower aliphatic sulfoxides are water-soluble, but as a class the sulfoxides are not soluble in water. They are soluble in dilute acids and a few are soluble in alkaline solution. They dissolve to a variable extent in organic solvents, depending upon associated functional groups. Because of the very polar sulfoxide group, they generally are high boiling and, when distillable, require reduced pressure. Dimethyl sulfoxide (DMSO) is a colorless liquid and its properties are listed in Table 2. Dimethyl sulfoxide generally undergoes typical sulfoxide reactions and is used as an illustrative example in the following sections.

Thermal Stability. Dimethyl sulfoxide decomposes slowly at 189°C to a mixture of products that includes methanethiol, formaldehyde, water, bis(methylthio)methane, dimethyl disulfide, dimethyl sulfone, and dimethyl sulfide. The decomposition is accelerated by acids, glycols, or amides (30). This product mixture suggests a sequence in which DMSO initially undergoes a Pummerer reaction to give (methylthio)methanol, which is labile and reacts according to equations 1–3. Disproportionation (eq. 4) also occurs to a small extent:

$$CH_3SCH_2OH \rightleftharpoons CH_3SH + HCHO \qquad (1)$$

$$2\,CH_3SH + HCHO \rightleftharpoons CH_3SCH_2SCH_3 + H_2O \qquad (2)$$

Table 1. Melting and Boiling Points of Sulfoxides

Name	CAS Registry No.	Formula	Melting point, °C	Boiling point, °C	Ref.
sulfinylbismethane	[*67-68-5*]	$(CH_3)_2SO$	18.55	189.0	9
1,1′-sulfinylbisethane	[*70-29-1*]	$(C_2H_5)_2SO$	15	88–90 (at 2.0 kPa[a])	10
1,1′-sulfinylbispropane	[*4253-91-2*]	$(n\text{-}C_3H_7)_2SO$	18		10
1,1′-sulfinylbisbutane	[*2168-93-6*]	$(n\text{-}C_4H_9)_2SO$	32		11
1,1′-sulfinylbis(2-chloroethane)	[*5819-08-9*]	$(ClCH_2CH_2)_2SO$	110.2		12
1,1′-sulfinylbisbenzene	[*945-51-7*]	$(C_6H_5)_2SO$	70.5	340 (slowly decomposes)	13
methylsulfinylbenzene	[*1193-82-4*]	$C_6H_5S(O)CH_3$	30–30.5	139–140	14–15
phenylmethylsulfinylbenzene	[*833-82-9*]	$C_6H_5S(O)CH_2C_6H_5$	125.5		16
1,1′-sulfinylbis(methylenebenzene)	[*621-08-9*]	$(C_6H_5CH_2)_2SO$	135		17

[a] To convert kPa to mm Hg, multiply by 7.5.

$$2\,CH_3SH + (CH_3)_2SO \rightarrow CH_3SSCH_3 + CH_3SCH_3 + H_2O \quad (3)$$

$$2\,(CH_3)_2SO \rightarrow CH_3SO_2CH_3 + CH_3SCH_3 \quad (4)$$

Oxidation. Sulfoxides are readily converted to sulfones, and usually in high yield, by a number of strong oxidizing agents, eg, potassium permanganate, hypochlorites, hydrogen peroxide, ozone, selenium dioxide, or hot nitric acid (31). Side reactions producing sulfonic acids sometimes occur during oxidation with nitric acid. Treatment of DMSO under strongly alkaline conditions with either sodium hypochlorite or hypobromite results in oxidation accompanied by halogenation; the hexahalodimethyl sulfone forms in high yield (eq. 5) (32):

$$CH_3SOCH_3 + NaOCl \rightarrow CCl_3SO_2CCl_3 \quad (5)$$

The moderate resistance of DMSO to oxidation permits it to be used as a solvent for oxidations with lead tetraacetate or the 2-nitropropane anion (33–34). Dichromate oxidation and permanganate oxidation have been used for quantitative determination of DMSO (35–36).

Reduction. Dimethyl sulfoxide is reduced to dimethyl sulfide by a number of strong reducing agents, including aluminum hydrides, hydriodic acid, diborane, thiols, phosphine derivatives, and zinc in sulfuric acid (31). Quantitative procedures for determining DMSO have been based on its reduction by stannous chloride in hydrochloric acid, titanium trichloride in dilute hydrochloric acid, or sodium borohydride reduction followed by gas–liquid chromatography for low levels in aqueous systems (6,37–38). However, DMSO is sufficiently resistant to reduction to function as a solvent for polarography from a +0.3-V anode potential to a −2.8-V cathode potential, both being relative to a calomel electrode, with ammonium perchlorate electrolyte (39).

Dimethyl sulfoxide is reduced to dimethyl sulfide in a variety of organic reactions, which entail initial attachment of an electrophilic activating reagent to the oxygen atom of DMSO to give (1) as shown in equation 6. This is followed by displacement of the oxygen-containing groups from the sulfur by a nucleophile and leads to a sul-

Table 2. Properties of Dimethyl Sulfoxide

Property			Value		Ref.
boiling point, °C			189.0		9
conductivity (at 20°C), S/cm			3×10^{-8}		9
dielectric constant (at 25°C), 10 MHz			46.7		18
dipole moment, C·m[a]			1.4×10^{-29}		9
entropy of fusion, J/(mol·K)[b]			45.12		19
free energy of formation (*g*) (at 25°C), kJ/mol[b] C_{graph}, $S_{2(g)}$			115.7		20
freezing point, °C			18.55		9
heat capacity (liquid at 25°C), J/(mol·K)[b]			153.2		19
heat capacity (ideal gas) (T, K), C_p			$6.94 + 5.6 \times 10^{-2}\,T - 0.227 \times 10^{-4}\,T^2$		20
heat of formation (liquid at 18°C), kJ/mol[b] C_{graph}, S_{rhomb}			−199.6		21
heat of fusion, kJ/mol[b]			13.5		22
heat of vaporization (at 70°C), kJ/mol[b]			47.3		23
molal fp constant, °C/(mol·kg)			4.07		24
pK_a			35.1		25
pK_{BH^+} (aqueous sulfuric acid)			−2.7		26
refractive index, n_D^{25}			1.4768		27
flash point (open cup), °C			95		23
autoignition temperature in air, °C			300–302		23
flammability limits in air, vol %					
lower, 100°C			3–3.5		23
upper, 180°C			42–63		23
	25°C	*35°C*	*45°C*	*100°C*	
vapor pressure, kPa[c]	0.080	0.159	0.303	4.0	23, 28
density, g/cm³	1.0955	1.0855	1.0757		29
viscosity, mPa·s (= cP)	1.996	1.654	1.396	0.68	29

[a] To convert C·m to D, divide by 3.336×10^{-30}.
[b] To convert J to cal, divide by 4.184.
[c] To convert kPa to mm Hg, multiply by 7.5.

fonium intermediate (**2**) (eq. 7), which usually reacts further to give the products (40):

$$(CH_3)_2SO + \text{electrophile} \rightarrow (CH_3)_2\overset{+}{S}O\text{-electrophile} \quad (6)$$
(**1**)

$$\text{nucleophile} + (CH_3)_2\overset{+}{S}O\text{-electrophile} \rightarrow \text{nucleophile-}\overset{+}{S}(CH_3)_2 + (\text{O-electrophile})^- \quad (7)$$
(**2**)

Compounds, eg, phenacyl halides, benzyl halides, alkyl iodides, or alkyl esters of sulfonic acids, react with DMSO at 100–120°C to give aldehydes and ketones in 50–85% yields (eq. 8) (41):

$$C_6H_5CH_2Cl + (CH_3)_2SO \rightarrow [C_6H_5CH_2OS(CH_3)_2]^+Cl^- \rightarrow CH_3SCH_3 + HCl + C_6H_5CHO \quad (8)$$

By a suitable choice of activating reagents, primary and secondary alcohols can be selectively oxidized to carbonyl compounds in good yields at room temperatures. Typical activating reagents are acetic anhydride, sulfur trioxide–pyridine, dicyclohexyl carbodiimide, and phosphorus pentoxide (40).

In the alcohol oxidations, the sulfonium intermediate (**2**) loses a proton and dimethyl sulfide to give the carbonyl compound (42). The most common mechanism for the decomposition of (**2**) is attack by a mild base to remove a proton from one of the methyl groups. Subsequent cyclic collapse leads to the carbonyl compound and dimethyl sulfide (eq. 9):

$$\underset{(\mathbf{2})}{R_2C(OH)\overset{+}{S}(CH_3)CH_3} \xrightarrow{B^-} R_2C(OH)\overset{+}{S}(CH_3)CH_2^- \longrightarrow R_2C{=}O + (CH_3)_2S \qquad (9)$$

Because of the mild reaction conditions and the reaction path, these oxidations are very selective and often are used to oxidize hydroxyl groups in molecules containing sensitive groups that would also react with the common oxidizing agents.

Carbon–Sulfur Cleavage. The carbon–sulfur bond of DMSO is broken in a number of reactions. Attempts to form the DMSO anion by the reaction of DMSO with sodium result in cleavage accompanied by methane evolution (eqs. 10 and 11) (43):

$$CH_3\overset{O}{\overset{\|}{S}}CH_3 + 2\,Na \longrightarrow CH_3SONa + CH_3Na \qquad (10)$$

$$CH_3Na + CH_3\overset{O}{\overset{\|}{S}}CH_3 \longrightarrow CH_4 + CH_3\overset{O}{\overset{\|}{S}}CH_2Na \qquad (11)$$

Sulfoxides containing β-hydrogen atoms react with strongly basic systems, eg, potassium *t*-butoxide, in DMSO by sulfenic acid elimination to produce olefins (eq. 12) (44):

$$\underset{[2211\text{-}92\text{-}9]}{[(CH_3)_3C]_2SO} \xrightarrow[\text{DMSO}]{t\text{-}C_4H_9OK} (CH_3)_2C{=}CH_2 + (CH_3)_3CSO^- \qquad (12)$$

In other cases, sulfenic acid elimination can involve γ-hydrogen atoms with the formation of cyclopropane derivatives. γ-Elimination is favored when DMSO is the reaction solvent (eq. 13) (45):

$$\underset{[14198\text{-}15\text{-}3]}{C_6H_5CH_2CH(C_2H_5)CH_2SOCH_3} \xrightarrow[70°C]{NaCH_2SOCH_3,\ DMSO} C_6H_5CH\langle\begin{matrix}CH_2\\ |\\ CHC_2H_5\end{matrix} + CH_3SO^-\,Na^+ \qquad (13)$$

Pummerer Reactions. Acetic anhydride at 70°C converts ethyl phenylsulfinylacetate (**3**) to the α-acetoxy sulfide (**4**) in 70% yield (eq. 14) (46):

$$\underset{\substack{[54882\text{-}04\text{-}1]\\(\mathbf{3})}}{C_6H_5\overset{O}{\overset{\|}{S}}CH_2\overset{O}{\overset{\|}{C}}OC_2H_5}\ (CH_3\overset{O}{\overset{\|}{C}})_2O \longrightarrow \underset{(\mathbf{4})}{C_6H_5SCH(O\overset{O}{\overset{\|}{C}}CH_3)\overset{O}{\overset{\|}{C}}OC_2H_5} \qquad (14)$$

The reaction is quite general and usually results in 75–90% yields (47). All reactions in which a sulfoxide containing at least one α-hydrogen is reduced to the sulfide and also oxidized at the α-carbon are referred to as Pummerer reactions (48).

The initiating step in these reactions is the attachment of a group to the sulfoxide oxygen to produce an activated intermediate (**5**). Suitable groups are proton, acyl, alkyl, or almost any of the groups that also initiate the oxidations of alcohols with DMSO (40,48). In a reaction, eg, the one between DMSO and acetic anhydride, the second step is removal of a proton from an α-carbon to give an ylide (**6**). Release of an acetate ion generates the sulfur-stabilized carbonium ion (**7**), and the addition of acetate ion to the carbonium ion (**7**) results in the product (eq. 15):

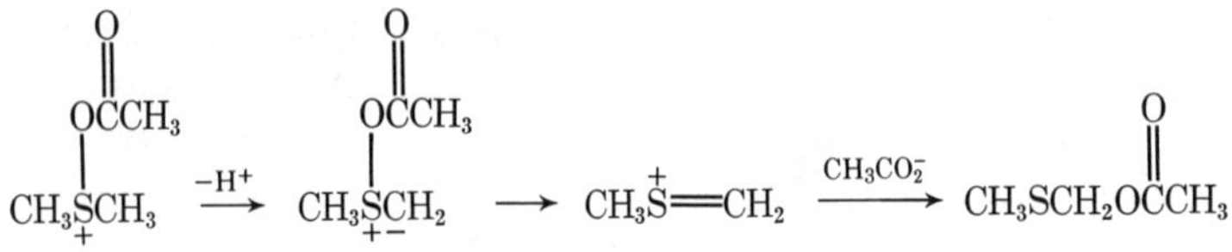

(**5**) (**6**) (**7**) (15)

Both inorganic and organic acid chlorides react very vigorously with DMSO to give α-chloromethyl methyl sulfide (**9**) when conditions are sufficiently controlled. In these reactions, a chloride ion displacement of the oxygen-containing group from the initial activated intermediate gives the labile chlorodimethylsulfonium ion (**8**). This sulfonium salt follows the reaction pattern shown above for the acetoxydimethylsulfonium ion (**5**) to give the product (**9**). The sequence involving thionyl chloride is shown in equation 16:

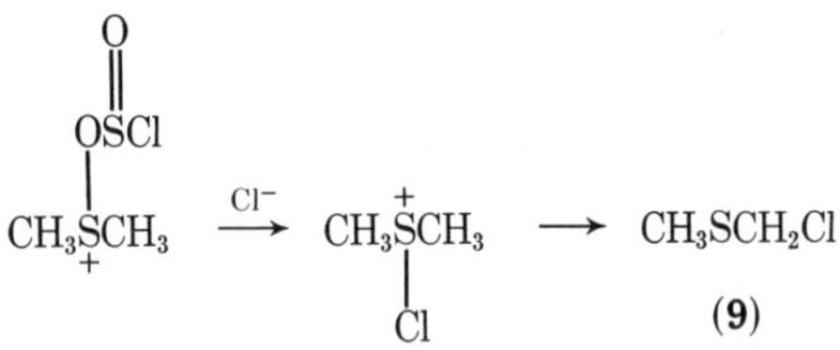

(**8**) (**9**) (16)

When DMSO is mixed with concentrated hydrochloric acid, protonated DMSO is in equilibrium with the chlorodimethylsulfonium ion. Pummerer reactions and subsequent reaction of the initial products give a complex mixtures of products including formaldehyde, bis(methylthio)methane, methanethiol, dimethyl disulfide, dimethyl sulfide, and others.

Methylsulfinyl Carbanion (Dimsyl Ion). The activating influence of the sulfinyl group on α-hydrogens is considerably less than that of a carbonyl group but is still sufficient to result in a pK_a of 35.1 for DMSO. Consequently, strong bases, eg, sodium hydride or sodium amide, react with DMSO producing solutions of methylsulfinyl carbanion [*15590-23-5*] (dimsyl ion) which are synthetically useful (49). The solutions also provide a strongly basic reagent for generating other carbanions. The dimsyl ion shows the expected nucleophilicity of carbanions and serves as a source of methylsulfinylmethyl groups (49). Thus, with alkyl halides or sulfonate esters, sulfoxides are obtained (eq. 17), carbonyl compounds yield β-hydroxysulfoxides (eq. 18), and esters give β-ketosulfoxides (eq. 19) (49):

$$n\text{-}C_4H_9Br + CH_3\overset{O}{\overset{\|}{S}}CH_2^- \longrightarrow \underset{[1561\text{-}74\text{-}6]}{C_5H_{11}\overset{O}{\overset{\|}{S}}CH_3} + Br^- \qquad (17)$$

$$(C_6H_5)_2CO + CH_3\overset{O}{\overset{\|}{S}}CH_2^- \xrightarrow{H^+} \underset{[2863\text{-}39\text{-}0]}{(C_6H_5)_2\overset{OH}{\overset{|}{C}}CH_2\overset{O}{\overset{\|}{S}}CH_3} \qquad (18)$$

$$C_6H_5\overset{O}{\overset{\|}{C}}OC_2H_5 + CH_3\overset{O}{\overset{\|}{S}}CH_2^- \longrightarrow \underset{[2813\text{-}22\text{-}1]}{C_6H_5\overset{O}{\overset{\|}{C}}CH_2\overset{O}{\overset{\|}{S}}CH_3} + C_2H_5O^- \qquad (19)$$

The dimsyl ion also adds to carbon–carbon double bonds, and if the mixture is heated for several hours, methanesulfenate is eliminated. The overall result is methylation, and with compounds such as quinoline or isoquinoline, yields are nearly quantitative (50). The reaction sequence is shown in equation 20:

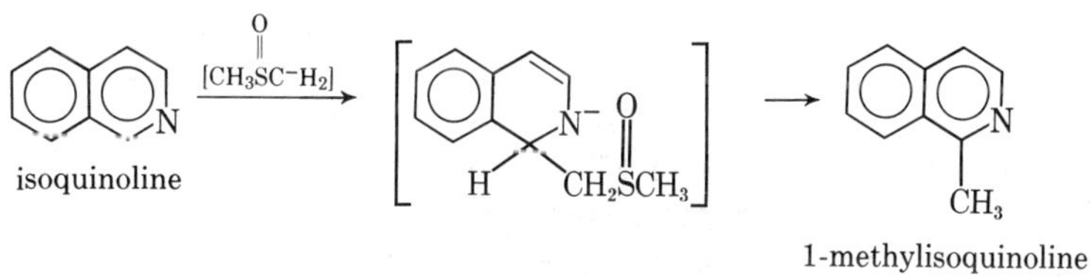

(20)

Care is required in running these reactions because the decomposition of the intermediate sulfoxide and of dimsyl sodium during the heating in the strongly alkaline system is exothermic and also produces a precipitate which can interfere with heat removal. Explosions have occurred (51).

Methoxydimethylsulfonium and Trimethylsulfoxonium Salts. Alkylating agents react with DMSO at the oxygen. For example, methyl iodide gives methoxydimethylsulfonium iodide (**10**) as the initial product. The alkoxysulfonium salts are quite reactive and, upon continued heating, either decompose to give carbonyl compounds or rearrange to the more stable trimethylsulfoxonium salts, eg, (**11**) (eq. 21) (52):

$$(CH_3)_2SO + CH_3I \longrightarrow \underset{(\mathbf{10})}{[(CH_3)_2SOCH_3]^+I^-} \longrightarrow \underset{(\mathbf{11})}{[(CH_3)_3SO]^+I^-} \qquad (21)$$

Trimethylsulfoxonium iodide (**11**) is of interest because treatment with sodium hydride or dimsyl sodium produces dimethylsulfoxonium methylide (**12**) (eq. 22), which is an excellent reagent for introducing a methylene group into a variety of structures (53):

$$\underset{(\mathbf{11})}{[(CH_3)_3SO]^+I^-} \xrightarrow{NaH} \underset{\substack{[5367\text{-}24\text{-}8] \\ (\mathbf{12})}}{(CH_3)_2\overset{O^+}{\overset{|}{S}}CH_2^-} \qquad (22)$$

Many aldehydes and ketones react with (**12**) to give better than 75% yields of epoxides (eq. 23):

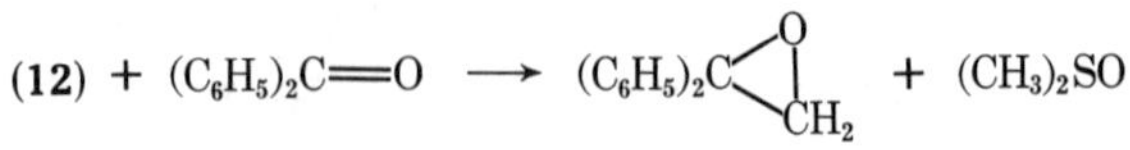

(23)

In similar reactions, (**12**) with carbon–carbon double bonds that are conjugated with carbonyl groups gives cyclopropane derivatives (eq. 24) (48):

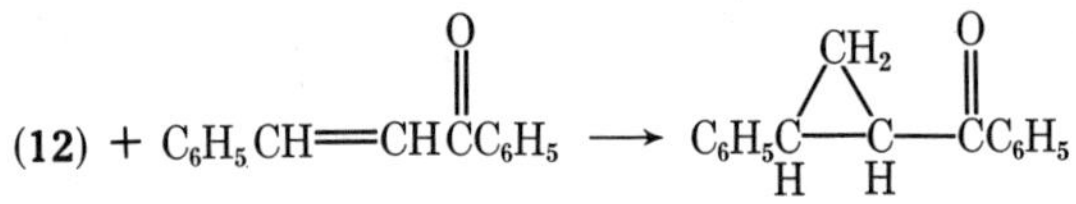

(24)

Complexes. The sulfoxides have a high dipole moment (ca 4), which is characteristic of the sulfinyl group and a basicity about the same as that of alcohols, and they are strong hydrogen-bond acceptors. They would be expected, therefore, to solvate ions with electrophilic character, and a large number of DMSO complexes of metal ions have been reported (54). The bonding to the metal is through the oxygen except for platinum(II), palladium(II), and rhodium(II) complexes where metal–sulfur bonds occur. The strength of the solvates is commonly about the same as that of the corresponding hydrate, and exchange of ligands is readily accomplished.

The strong tendency of the DMSO oxygen to act as a hydrogen-bond acceptor leads also to a number of organic complexes. Chloroform forms both 1:1 and 1:2 complexes (55). Pyrrole and phenol give 1:1 complexes (56–57). In solutions of monosaccharides in DMSO, the anomer with cis hydroxyls on the first and second carbons is stabilized (58). Complexes with organic molecules that do not involve hydrogen bonding also occur, eg, the 1:1 complex of DMSO and 4-chlorobenzonitrile and other nitriles (59–60). Strong 1:1 complexes form with nitrogen tetroxide and sulfur trioxide (61–62). Charge-transfer complexes occur with cyanogen iodide, tetracyanoethylene, and oxygen (63–65).

Synthesis and Manufacture. The sulfoxides are most frequently synthesized by oxidation of the sulfides (66–67). A broad group of oxidizing agents can be used and, since the oxidation to the sulfoxide is considerably more rapid than further oxidation to the sulfone, a proper choice of reagent quantity and conditions leads to high sulfoxide yields; eg, hydrogen peroxide in stoichiometric amounts can give 75–90% sulfoxide yields (68). Nitrogen tetroxide in a solvent, eg, carbon tetrachloride at temperatures below ca 0°C, selectively gives the sulfoxide in yields of up to 95% (69). Oxidations with sodium metaperiodate in aqueous or water–methanol solutions at ice-bath temperatures give high sulfoxide yields without sulfone formation (70). Oxidation with tertiary amine–bromine complexes in aqueous acetic acid gives yields above 70% and has been suggested as a convenient procedure for preparing ^{18}O-labeled sulfoxides when $H_2{}^{18}O$-acetic acid is used as the reaction solvent (71). There are a number of sulfoxide synthesis procedures in which the appropriate fragments are joined to give the product.

$$ROS(=O)OR + 2\,R'MgX \longrightarrow R'S(=O)R' + 2\,ROMgX \quad (25)$$

$$ArS(=O)OR + R'MgX \longrightarrow ArS(=O)R' + ROMgX \quad (26)$$

$$2\,CH_3OC_6H_5 + SO_2 \xrightarrow{AlCl_3} CH_3OC_6H_4S(=O)C_6H_4OCH_3 + H_2O \quad (27)$$

[*1774-36-3*]

$$2\,C_6H_6 + SOCl_2 \xrightarrow{AlCl_3} C_6H_5S(=O)C_6H_5 + 2\,HCl \quad (28)$$

[*945-51-7*]

The reaction of Grignard reagents with sulfite esters gives 40–70% yields (eq. 25), and with arylsulfinate esters the yields are ca 55% (eq. 26) (72). Optically active sulfoxides are synthesized in good yield from the reaction of optically active sulfinate esters with Grignard reagents (73). Sulfoxides are obtained in greater than 50% yield by alkylating sulfenates with alkyl bromides (74). Diaryl sulfoxides are also obtained by Friedel-Crafts syntheses, eg, the reaction of anisole with sulfur dioxide and aluminum chloride to obtain 40% bis(4-methoxyphenyl) sulfoxide (eq. 27) or the reaction of benzene with thionyl chloride and aluminum chloride to obtain 51% diphenyl sulfoxide (eq. 28) (75). A comprehensive review of procedures for synthesizing sulfoxides has been published recently (76).

Dimethyl Sulfoxide. Dimethyl sulfoxide is manufactured from dimethyl sulfide (DMS), which is obtained either by processing spent liquors from the Kraft pulping process or by the reaction of methanol or dimethyl ether with hydrogen sulfide. In the Kraft pulping process, the spent liquors are normally concentrated to ca 50% solids and burned to recover inorganic chemicals and heat values. The lignin in the liquor contains aromatic methoxyl groups, which are cleaved by sulfide ions to produce dimethyl sulfide when the concentrated liquor is processed in a reactor at 200–250°C (77). The synthesis of dimethyl sulfide from methanol and hydrogen sulfide is accomplished by a vapor-phase reaction over a catalyst at above 300°C.

Dimethyl sulfide has been oxidized to DMSO by several procedures. In pilot-plant quantities, the oxidation was accomplished with nitric acid, but this route has been supplanted by oxidation with nitrogen dioxide or oxygen containing minor amounts of nitrogen dioxide. The oxidation with nitrogen dioxide is diagrammed in Figure 1 (78). Dimethyl sulfide is oxidized with a DMSO solution of nitrogen dioxide in a reactor at 40–50°C. The reactor contents pass into a zone at 100°C where excess dimethyl sulfide is sparged from the crude DMSO with nitrogen; the crude DMSO is then neutralized and distilled. The flow of nitrogen dioxide into the reactor is kept insufficient to oxidize all of the dimethyl sulfide so that all the nitrogen dioxide is converted to nitric oxide, which is quite insoluble in DMSO and escapes in the exit-gas stream. The gas stream passes through a heat exchanger for condensation of some of the dimethyl sulfide which is recycled to the reactor. The gases then are conducted to a second reactor where an excess of nitrogen dioxide converts all of the remaining sulfide

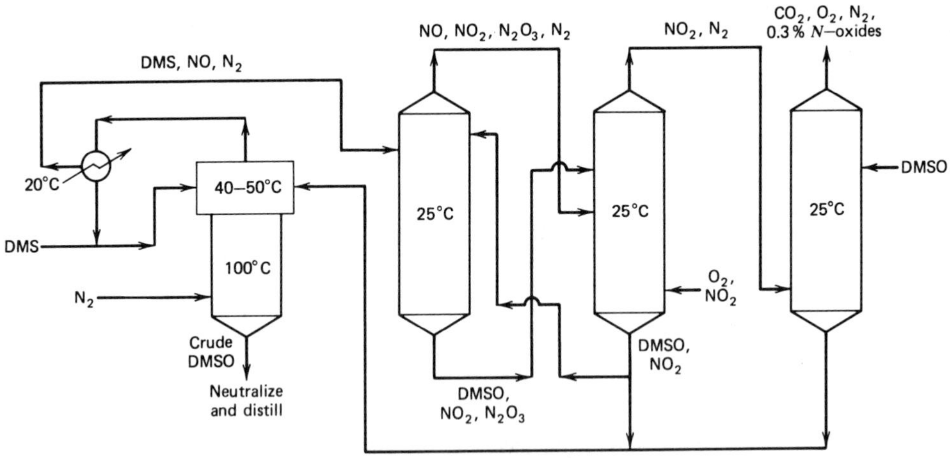

Figure 1. Dimethyl sulfoxide manufacture with nitrogen tetroxide.

to the sulfoxide. The gases from this reactor contain substantially no organic matter and are oxidized with oxygen in a third reactor to regenerate the nitrogen dioxide. The gases finally pass through a DMSO scrubber for removal of nitrogen dioxide before venting to the atmosphere.

Processes involving oxygen with nitrogen oxides as catalysts have been operated commercially with either vapor- or liquid-phase reactors. The vapor-phase reactors require particularly close control because of the wide explosive limit of dimethyl sulfide in oxygen (1–83.5 vol %); plants in operation use liquid-phase reactions. Figure 2 is a schematic diagram for the liquid-phase process. The product stream from the reactor is neutralized with aqueous caustic and is vacuum-evaporated, and the DMSO is dried in a distillation column to obtain the product.

Dimethyl sulfoxide is produced in commercial quantities in the United States by Crown Zellerbach Corporation. The 1982 sales price was $1.72/kg in bulk quantities. Typical properties of industrial DMSO are given in Table 3 (23).

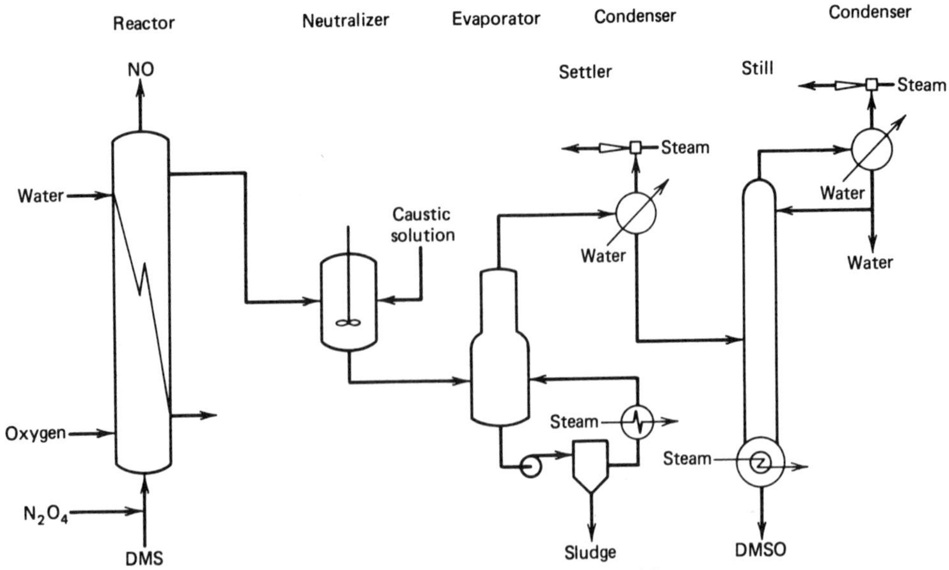

Figure 2. Dimethyl sulfoxide manufacture with oxygen.

Table 3. Typical Analysis of Industrial Dimethyl Sulfoxide

Properties	Analysis
assay, %	99.9
water content, wt %	0.1
color	water-white
other impurities	negligible
nonvolatiles	negligible

Health and Safety Factors

Dimethyl sulfoxide is a relatively stable solvent of low toxicity. The LD_{50} for single-dose oral administration to rats is ca 17,400–28,300 mg/kg. Dimethyl sulfoxide by itself presents less hazard than many chemicals and solvents commonly used in industry. However, DMSO can penetrate the skin and may carry with it certain chemicals with which it is combined under certain conditions. Normal protective measures should be followed in the laboratory. If large quantities are handled where splashing and accidental contact may occur, protective clothing is recommended including suitable gloves and eye protectants. Butyl rubber gloves are more suitable than other types of material in terms of resisting penetration by DMSO solutions (79).

Dimethyl sulfoxide has received considerable attention as a useful agent in medicine (80–81). Compositions containing it are marketed in the FRG and Austria as prescription items for analgesic uses. In the United States in 1981, the possible efficacy of DMSO as a drug was examined under jurisdiction of the FDA, but this agency has not released DMSO for drug use by humans other than for a painful bladder condition called interstitial cystitis. In veterinary medicine, DMSO is used for horses and dogs as a topical application to reduce swelling resulting from injury or trauma (see Veterinary drugs).

Uses of Dimethyl Sulfoxide

Polymerization and Spinning Solvent. Dimethyl sulfoxide is used as a solvent for the polymerization of acrylonitrile and other vinyl monomers, eg, methyl methacrylate and styrene (82–83). The low incidence of transfer from the growing chain to DMSO leads to high molecular weights. Copolymerization reactions of acrylonitrile with other vinyl monomers are also run in DMSO. Monomer mixtures of acrylonitrile, styrene, vinylidene chloride, methallylsulfonic acid, styrenesulfonic acid, etc are polymerized in DMSO–water (84). In some cases, the fibers are spun from the reaction solutions into DMSO–water baths.

Dimethyl sulfoxide can also be used as a reaction solvent for other polymerizations. Ethylene oxide is rapidly and completely polymerized in DMSO (85). Diisocyanates and polyols or polyamines dissolve and react in DMSO to form solutions of polyurethanes (86) (see Solvents, industrial).

Solvent for Displacement Reactions. As the most polar of the common aprotic solvents, DMSO is a favored solvent for displacement reactions because of its high dielectric constant and because anions are less solvated in it (87). Rates for these reactions are sometimes a thousand times faster in DMSO than in alcohols. Suitable nucleophiles include acetylide ion, alkoxide ion, hydroxide ion, azide ion, carbanions, carboxylate ions, cyanide ion, halide ions, mercaptide ions, phenoxide ions, nitrite ions, and thiocyanate ions (31). Rates of displacement by amides or amines are also greater in DMSO than in alcohol or aqueous solutions. Dimethyl sulfoxide is used as the reaction solvent in the manufacture of high performance, polyaryl ether polymers by reaction of bis(4,4′-chlorophenyl) sulfone with the disodium salts of dihydroxyphenols, eg, bisphenol A or 4,4′-sulfonylbisphenol (88). These and related reactions are made more economical by efficient recycling of DMSO (89). Nucleophilic displacement of activated aromatic nitro groups with aryloxy anion in DMSO is a versatile and useful reaction for the synthesis of aromatic ethers and polyethers (90).

Solvent for Base-Catalyzed Reactions. The ability of hydroxide or alkoxide ions to remove protons is enhanced by DMSO instead of water or alcohols (91). The equilibrium change is also accompanied by a rate increase of 10^5 or more (92). Thus, reactions in which proton removal is rate-determining are favorably accomplished in DMSO. These include olefin isomerizations, elimination reactions to produce olefins, racemizations, and H–D exchange reactions.

Extraction Solvent. Dimethyl sulfoxide is immiscible with alkanes but is a good solvent for most unsaturated and polar compounds. Thus, it can be used to separate olefins from paraffins (93). It is used in the Institute Français du Pétrole (IFP) process for extracting aromatic hydrocarbons from refinery streams (94). It is also used in the analytical procedure for determining polynuclear hydrocarbons in food additives of petroleum origin (95).

Solvent for Electrolytic Reactions. Dimethyl sulfoxide has been widely used as a solvent for polarographic studies and a more negative cathode potential can be used in it than in water. In DMSO, cations can be successfully reduced to metals that react with water. Thus, the following metals have been electrodeposited from their salts in DMSO: cerium, actinides, iron, nickel, cobalt, and manganese as amorphous deposits; zinc, cadmium, tin, and bismuth as crystalline deposits; chromium; silver; lead; copper; and titanium (96–103). Generally, no metal less noble than zinc can be deposited from DMSO.

Cellulose Solvent. Although DMSO by itself does not dissolve cellulose, the following binary and ternary systems are cellulose solvents: DMSO–methylamine, DMSO–sulfur trioxide, DMSO–carbon disulfide–amine, DMSO–ammonia–sodamide, DMSO–dinitrogen tetroxide, DMSO–paraformaldehyde, and DMSO–sulfur dioxide–ammonia (104–105). At least a ratio of three moles of active agent per mole of glucose unit is necessary for complete dissolution (104). Although only 80% of cellulose (qv) dissolves in DMSO–methylamine under cold anhydrous conditions, DMSO–dinitrogen tetroxide is a better solvent, particularly when a small quantity of water is added (106). Most of these systems can be used to produce cellulose fibers. The recently discovered DMSO–paraformaldehyde system does not degrade cellulose, and it can form solutions containing up to 10 wt % cellulose (107). It is believed that a methylolcellulose compound forms and is stable for extended periods of storage at ambient conditions (107). Regenerated cellulose articles, eg, films and fibers, can be prepared by contacting the DMSO–paraformaldehyde solutions with methanol and water (107–108).

Pesticide Solvent. The majority of organic fungicides, insecticides, and herbicides (qv) are soluble in DMSO, including such difficult-to-solvate materials as the substituted ureas and carbamates (see Fungicides; Insect control technology). Dimethyl sulfoxide forms cosolvent systems of enhanced solubility properties with many solvents (109).

Clean-up Solvent. Dimethyl sulfoxide is used to remove urethane polymers and other difficult-to-solvate materials from processing equipment.

BIBLIOGRAPHY

"Sulfoxides" in *ECT* 1st ed., Vol. 13, pp. 353–357, by E. G. Rietz, Chicago City Colleges, Wright Branch; "Sulfoxides" in *ECT* 2nd ed., Vol. 19, pp. 320–337, by W. S. MacGregor, Crown Zellerbach Corporation.

1. A. Zwergal, *Pharmazie* **7,** 245 (1952).
2. T. Reichstein and A. Goldschneidt, *Helv. Chim. Acta* **19,** 401 (1936).
3. A. Kjaer and B. Christensen, *Acta Chem. Scand.* **12,** 833 (1958).
4. C. J. Morris and J. F. Thompson, *J. Am. Chem. Soc.* **78,** 1605 (1955).
5. T. W. Pearson, H. J. Dawson, and H. B. Lackey, *J. Agric. Food Chem.* **29,** 1089 (1981).
6. M. O. Andreae, *Anal. Chem.* **52,** 150 (1980).
7. M. D. Bentley, I. R. Douglass, J. A. Lacadie, and D. R. Whittier, *J. Air. Pollut. Control Assoc.* **22,** 359 (1972).
8. J. E. Lovelock, R. J. Maggs, and R. A. Rasmussen, *Nature (London)* **237,** 452 (1972).
9. H. L. Schlafer and W. Schaffernicht, *Angew. Chem.* **72,** 618 (1960).
10. W. Strecker and R. Spitaler, *Chem. Ber.* **59,** 1754 (1926).
11. N. Grabowsky, *Ann. Chem.* **175,** 348, 351 (1875).
12. H. Mohler, *Helv. Chim. Acta* **20,** 1188 (1937).
13. F. Krafft and R. E. Lyons, *Chem. Ber.* **29,** 435 (1896).
14. D. Barnard, J. M. Fabian, and H. P. Koch, *J. Chem. Soc.* 2442 (1949).
15. H. Bohme, H. Fischer, and R. Frank, *Ann. Chem.* **563,** 54 (1949).
16. R. Pummerer, *Chem. Ber.* **43,** 1401 (1910).
17. H. Rheinboldt and E. Giescrecht, *J. Am. Chem. Soc.* **68,** 2671 (1946).
18. R. A. Hovermale and P. G. Sears, *J. Phys. Chem.* **60,** 1579 (1956).
19. H. L. Clever and E. F. Westrum, *J. Phys. Chem.* **74,** 1309 (1970).
20. H. Mackle and P. A. G. O'Hare, *Trans. Faraday Soc.* **58,** 1912 (1962).
21. T. B. Douglas, *J. Am. Chem. Soc.* **68,** 1072 (1946).
22. E. E. Weaver and W. Keim, *Proc. Indiana Acad. Sci. Monogr.* **70,** 123 (1961).
23. *Dimethyl Sulfoxide*, Technical Bulletin, Crown Zellerbach Corp., Chemical Products Division, Orchards, Washington, 1966.
24. R. Garnsey and J. E. Prue, *Trans. Faraday Soc.* **64,** 1206 (1968).
25. W. N. Olmstead, Z. Margolin, and F. G. Bordwell, *J. Org. Chem.* **45,** 3295 (1980).
26. P. Haake, D. A. Tysse, S. R. Alpha, J. Kleckner, and R. D. Cook, *Q. Rep. Sulfur Chem.* **3**(2), 105 (1968).
27. R. G. LeBel and D. A. I. Goring, *J. Chem. Eng. Data* **7,** 100 (1962).
28. T. B. Douglas, *J. Am. Chem. Soc.* **70,** 2001 (1948).
29. P. G. Sears, W. D. Siegfried, and D. E. Sands, *J. Chem. Eng. Data* **9,** 261 (1964).
30. V. J. Traynelis and W. L. Hergenrother, *J. Org. Chem.* **29,** 221 (1964).
31. J. Drabonicz, T. Numata, and S. Oae, *Org. Prep. Proced. Int.* **9**(2), 63 (1977).
32. U.S. Pat. 3,304,331 (Feb. 14, 1967), C. DiSanto (to Stauffer Chemical Co.).
33. V. Zitko and C. T. Bishop, *Can. J. Chem.* **44,** 1749 (1966).
34. B. H. Klanderman, *J. Org. Chem.* **31,** 2618 (1966).
35. K. Stelmach, *Chem. Anal. (Warsaw)* **11,** 627 (1966).
36. L. H. Krull and M. Friedman, *J. Chromatogr.* **26,** 336 (1967).
37. E. Glynn, *Analyst* **72,** 248 (1947).
38. R. R. Legault and K. Groves, *Anal. Chem.* **29,** 1495 (1957).
39. J. L. Jones and H. A. Fritsche, Jr., *J. Electroanal. Chem.* **12,** 334 (1966).
40. A. J. Mancuso and D. Swern, *Synthesis*, 165 (1981).
41. N. Kornblum, W. J. Jones, and G. J. Anderson, *J. Am. Chem. Soc.* **81,** 4113 (1959).
42. M. G. Burdon and J. G. Moffatt, *J. Am. Chem. Soc.* **89,** 4725 (1967).
43. D. E. O'Connor and W. I. Lyness, *J. Org. Chem.* **30,** 1620 (1965).
44. J. E. Hofmann, T. J. Wallace, P. A. Argabright, and A. Shriesheim, *Chem. Ind. (London)*, 1243 (1963).
45. R. Baker and M. J. Spillett, *Chem. Commun.*, 757 (1966).
46. R. Pummerer, *Chem. Ber.* **43,** 1401 (1910).
47. L. Horner and P. Kaiser, *Ann. Chem.* **631,** 198 (1960).

48. T. Durst in E. C. Taylor, H. Wynberg, eds., *Advances in Organic Chemistry: Methods and Results*, Vol. 6, Interscience Publishers, a division of John Wiley & Sons, Inc., New York, 1969, pp. 285–388.
49. E. J. Corey and M. Chaykovsky, *J. Am. Chem. Soc.* **87,** 1345 (1965).
50. G. A. Russell and S. A. Wiener, *J. Org. Chem.* **31,** 248 (1966).
51. F. A. French, *Chem. Eng. News* **44,** 48 (1966).
52. S. G. Smith and S. Winstein, *Tetrahedron* **3,** 317, 319 (1958).
53. E. J. Corey and M. Chaykovsky, *J. Am. Chem. Soc.* **87,** 1353 (1965).
54. W. L. Reynolds in S. J. Lippard, ed., *Progress in Inorganic Chemistry*, Vol. 12, Interscience Publishers, a division of John Wiley & Sons, Inc., New York, 1970, pp. 1–99.
55. A. L. McClellan, S. W. Nicksic, and J. C. Guffy, *J. Mol. Spectrosc.* **11,** 340 (1963).
56. D. M. Porter and W. S. Brey, Jr., *J. Phys. Chem.* **72,** 650 (1968).
57. R. S. Drago, B. Wayland, and R. L. Carlson, *J. Am. Chem. Soc.* **85,** 3125 (1963).
58. V. S. R. Rao and J. F. Foster, *J. Phys. Chem.* **69,** 656 (1965).
59. C. D. Ritchie and A. Pratt, *J. Phys. Chem.* **67,** 2498 (1963).
60. C. D. Ritchie and A. Pratt, *J. Am. Chem. Soc.* **86,** 1571 (1964).
61. C. C. Addison and J. C. Sheldon, *J. Chem. Soc.*, 2705 (1956).
62. R. L. Whistler, A. H. King, G. Ruffini, and F. A. Lucas, *Arch. Biochem. Biophys.* **121,** 358 (1967).
63. E. Augdahl and P. Klaeboe, *Acta Chem. Scand.* **18,** 18 (1964).
64. F. E. Stewart, M. Eisner, and W. R. Carper, *J. Chem. Phys.* **44,** 2866 (1966).
65. T. Sato, H. Inoue, and K. Hata, *Bull. Chem. Soc. Jpn.* **40,** 1502 (1967).
66. T. Durst in D. N. Jones, ed., *Comprehensive Organic Chemistry*, Vol. 3, Pergamon Press, Oxford, England, 1979, pp. 121–156.
67. E. E. Reid, *Organic Chemistry of Bivalent Sulfur*, Vol. 2, Chemical Publishing Co., New York, 1960, pp. 64–66.
68. D. Jerchel, L. Dippelhofer, and D. Renner, *Chem. Ber.* **87,** 947 (1954).
69. L. Horner and F. Hubenett, *Ann. Chem.* **579,** 193 (1953).
70. N. J. Leonard and C. R. Johnson, *J. Org. Chem.* **27,** 282 (1962).
71. S. Oae, Y. Ohmishi, S. Kozuka, and W. Tagaki, *Bull. Chem. Soc. Jpn.* **39,** 364 (1966).
72. H. Hepworth and H. W. Clapham, *J. Chem. Soc.* **119,** 1188 (1921).
73. K. K. Andersen, W. Gaffield, N. E. Papnikolaow, J. W. Foley, and R. I. Perkins, *J. Am. Chem. Soc.* **86,** 5637 (1964).
74. D. E. O'Connor and W. I. Lyness, *J. Org. Chem.* **30,** 1620 (1965).
75. R. L. Shriner, H. C. Struck, and W. J. Jorison, *J. Am. Chem. Soc.* **52,** 2060 (1930).
76. J. Drabowicz and M. Mikokajczyk, *Org. Prep. Proced. Int.* **14**(1–2), 45 (1982).
77. W. M. Hearon, W. S. MacGregor, and D. W. Goheen, *Tappi* **45**(1), 28A, 30A, 34A, 36A (1962).
78. H. Pruckner, *Erdoel Kohle Erdgas Petrochemie* **16,** 188 (1963).
79. Dimethyl Sulfoxide, *A Summary of Toxicological and Safety Information*, Crown Zellerbach Corp., Chemical Products Division, Orchards, Washington, 1972.
80. *Ann. N.Y. Acad. Sci.* **141,** 1 (Mar. 15, 1967).
81. S. W. Jacob, E. E. Rosenbaum, and D. C. Wood, eds., *Dimethyl Sulfoxide*, Vol. 1, Marcel Dekker, Inc., New York, 1971.
82. T. Ouchi, A. Tatsumi, and M. Imoto, *J. Polym. Sci. Polym. Chem. Ed.* **16,** 707 (1978).
83. C. H. Bamford and A. N. Ferrar, *Proc. R. Soc. London Ser. A* **321,** 425 (1971).
84. U.S. Pat. 3,781,248 (Dec. 25, 1973), H. Sakai, S. Hamada, Y. Yamanaka, I. Ito, Z. Izumi, H. Kitagawa, E. Mukoyama, Z. Suzuki, T. Kato, and S. Hosak (to Toray Industries, Inc.).
85. K. S. Kazanskii, A. S. Yanov, and S. A. Dubrovsky, *Makromol. Chem.* **179,** 969 (1978).
86. U.S. Pat. 3,658,731 (Apr. 25, 1972), T. Richardson and G. O. Hustad (to Wisconsin Alumni Research Foundation).
87. E. Buncel and H. Wilson in V. Gold and D. Bethell, eds., *Adv. Phys. Org. Chem.*, Vol. 14, Academic Press, New York, 1977, pp. 133–202.
88. U.S. Pat. 4,175,175 (Nov. 20, 1979), R. N. Johnson and H. G. Farnham (to Union Carbide Corp.).
89. *Dimethyl Sulfoxide Recovery and Environmental Engineering*, Crown Zellerbach Corp., Chemical Products Division, Orchards, Washington, 1974.
90. T. Takekoshi, J. G. Wirth, D. R. Heath, J. E. Kochanowski, J. S. Manello, and M. J. Webber, *J. Polym. Sci. Polym. Chem. Ed.* **18,** 3069 (1980).
91. R. Steward, *Q. Rep. Sulfur Chem.* **3**(2), 99 (1968).
92. D. J. Cram and L. Gosser, *J. Am. Chem. Soc.* **86,** 5457 (1964).

93. U.S. Pat. 4,267,034 (May 12, 1981), C. O. Carter (to Phillips Petroleum Co.).
94. P. J. Bailes, *Chem. Ind. (London)*, 69 (1977).
95. E. O. Haenni, F. L. Joe, Jr., J. W. Howard, and R. L. Leibel, *J. Assoc. Off. Agric. Chem.* **45**(1), 59 (1962).
96. J. A. Porter, *AEC Research and Development Report DP-389*, July 1959.
97. T. H. Handley and J. H. Cooper, *Anal. Chem.* **41,** 381 (1969).
98. S. Morisaki, N. Baba, and S. Tajima, *Denki Kagaku* **38,** 746 (1970).
99. U.S. Pat. 3,772,170 (Nov. 13, 1973), N. R. Bharucha.
100. S. Nakagawa, Z. Takehara, and Y. Yoshizawa, *Denki Kagaku* **41,** 880 (1973).
101. V. V. Kuznetov and co-workers, *Zashch. Met.* **1,** 631 (1975); *Chem. Abstr.* **84,** 10359X (1976).
102. V. V. Kuznetov and co-workers, *Izv. Sev. Kavk. Nauchno. Tsentra Vyssh. Shk. Ser. Estestv. Nauk.* **4,** 47 (1976); *Chem. Abstr.* **86,** 179414P (1978).
103. E. Santos and F. Dyment, *Plating (East Orange, N.J.)* **60,** 821 (1973).
104. B. Philipp, H. Schleicher, and W. Wagenknecht, *Chemtech*, 702 (Nov. 1977).
105. A. F. Turbak, R. B. Hammer, R. E. Davies, and H. L. Hergert, *Chemtech*, 51 (Jan. 1980).
106. U.S. Pat. 4,076,933 (Feb. 28, 1978), A. F. Turbak, R. B. Hammer, N. A. Portnoy, and A. C. West (to International Telephone and Telegraph Corp.).
107. U.S. Pat. 4,097,666 (June 27, 1978), D. C. Johnson and M. D. Nicholson (to The Institute of Paper Chemistry).
108. R. B. Hammer, M. E. O'Shaughnessey, E. R. Strauch, and A. F. Turbak, *J. Appl. Polym. Sci.* **23,** 485 (1979).
109. *DMSO—The Agricultural Chemical Solvent*, Crown Zellerbach Corp., Chemical Products Division, Orchards, Wash., Oct. 1973.

General References

References 9, 31, 48, 54, 66, 81, and 87 contain reviews of DMSO chemistry.
W. W. Epstein and F. W. Sweat, *Chem. Rev.* **67**(3), 247 (1967).
D. Martin and H. G. Hauthal, *Dimethyl Sulfoxide*, Halsted Press, a division of John Wiley & Sons, Inc., New York, 1975.
B. S. Thyagarajan and N. Kharasch, *Intrascience Sulfur Reports, Vol. 1, The Chemistry of DMSO*, Intrascience Research Foundation, Santa Monica, Calif., 1966.

W. S. MacGregor
J. V. Orle
Crown Zellerbach Corp.

SULFUR

Sulfur [*7704-34-9*], S, is a nonmetallic element and is the second element of group VIA below oxygen and above selenium. In massive elemental form, sulfur is often referred to as brimstone. Sulfur is one of the most important raw materials of the chemical industry. It is of prime importance to every sector of the fertilizer and industrial complexes of the world (see Fertilizers). Its consumption is generally regarded as one of the best measures of a nation's industrial development and economic activity (see also Sulfur compounds; Sulfur recovery; Sulfuric acid).

Sulfur has been known since antiquity. Early man used sulfur to color cave drawings, discovered that sulfur fumes could be used to kill insects and to fumigate, and knew about its color-removing or bleaching action. Mystical powers were attributed to the ethereal blue flame and pungent odor given off by burning the yellow rock. Medicinal use of sulfur was known to the Egyptians and Greeks. One contemporary use was developed in 500 BC, when the Chinese used sulfur as an ingredient of gunpowder. Although the modern history of sulfur may have begun with the proof in the late 18th century by Lavosier that sulfur is an element, the first commercial sulfur was produced in Sicily early in the 15th century. Sulfur production became Sicily's main industry when, in 1735, development of a process to make sulfuric acid from sulfur was commercialized. When in 1839 a French company gained an effective monopoly on the Sicilian deposits and tripled the price, other countries, particularly England and the United States, undertook the development of their own sources. They quickly learned that sulfuric acid could be made from sulfur dioxide obtained from the roasting of iron pyrites, which obviated the need for Sicilian sulfur.

The United States continued to depend on foreign sources of elemental sulfur, even after it was discovered in the United States by oil prospectors investigating a salt dome in Calcasieu Parish, Louisiana, in 1867. For the next 20 years, various unsuccessful attempts were made to sink mine shafts through perhaps 100 m of strata containing clay, gravel, and quicksand.

On realizing that conventional mining processes would be too uneconomical to compete with Sicilian sulfur, Dr. Herman Frasch conceived the idea of melting the sulfur underground by injecting superheated water into the formation and then lifting the melted sulfur to the surface with a sucker-rod pump. In 1894, the first flow of molten sulfur was pumped from the Calcasieu Parish deposit. In 1902, the Frasch process was successfully commercialized. This mining method later became important in the development and production of sulfur not only from the Texas–Louisiana salt dome area but also from areas in western Texas, Mexico, Poland, and Iraq.

Sulfur recovered as a by-product, ie, involuntary sulfur, accounts for a larger portion of world supply than the voluntary material. It is obtained from hydrogen sulfide, which evolves when natural gas, crude petroleum, tar sands, oil shales, and coal are desulfurized. Other sources of sulfur include metal sulfides, eg, pyrites; sulfate materials, eg, gypsum; and elemental sulfur in native and volcanic deposits mined in the traditional manner.

Sulfur constitutes ca 0.052 wt % of the earth's crust. The forms in which it is ordinarily found include elemental or native sulfur in unconsolidated volcanic rocks, in anhydrite over salt-dome structures, and in bedded anhydrite or gypsum evaporate basin formations; combined sulfur in metal sulfide ores and mineral sulfates; hydrogen sulfide in natural gas; organic compounds in petroleum and tar sands; and a combination of both pyritic and organic compounds in coal.

Properties

Allotropy. Sulfur occurs in a number of different allotropic modifications, that is, in various molecular aggregations which differ in solubility, specific gravity, crystalline form, etc. Like many other substances, sulfur also exhibits dynamic allotropy, ie, the various allotropes exist together in equilibrium in definite proportions, depending on the temperature and pressure. The formulas for the molecules of various allotropes is $S–S_n$, where n is a very large but unidentified number, ie, $n > 10^6$. The particular allotropes that may be present in a given sample of sulfur depend to a large extent upon its previous thermal history, the amount and type of foreign substance present, and the length of time that has passed for equilibrium to be attained.

In the solid and liquid states, the principal allotropes are designated traditionally as $S\lambda$, $S\mu$, and $S\pi$. Of these, only $S\lambda$ is stable in the solid state. Upon solidification of molten sulfur, $S\pi$ rapidly changes into $S\mu$, which is converted into $S\lambda$ although at a much slower rate. The molecular structure of $S\pi$ is that of an octatomic sulfur chain (1–2). The symbol $S\mu$ designates long, polymerized chains of elemental sulfur. $S\lambda$ is perhaps the most characteristic molecular form of sulfur, namely, that of a crown-shaped, octatomic sulfur ring designated in more recent literature as S_8^R (3). The allotropes differ in their solubility in carbon disulfide. $S\pi$ and $S\lambda$ are soluble in carbon disulfide, whereas $S\mu$ does not dissolve in this solvent.

Sulfur crystallizes in at least two distinct systems: the rhombic and monoclinic forms. Rhombic sulfur, $S\alpha$, is stable at atmospheric pressures up to 95.5°C, at which transition to monoclinic sulfur, $S\beta$, takes place. Monoclinic sulfur is then stable up to its natural melting point of 114.5°C. The basic molecular unit of both of these crystalline forms of sulfur is the octatomic sulfur ring S_8^R. Other forms of solid sulfur include hexatomic sulfur, as well as numerous modifications of catenapolysulfur (2,4).

The molecular constitution of liquid sulfur undergoes significant and reversible changes with temperature variations. These changes are evidenced by the characteristic temperature dependence of the physical properties of sulfur. In most studies of liquid sulfur, some striking changes in its physical properties are observed at ca 160°C. For example, the viscosity of purified sulfur at 120°C is about 11 mPa·s (= cP) drops to a minimum of 6.7 mPa·s at ca 157°C, and then begins to rise. At 159–160°C, the viscosity of liquid sulfur rises very sharply increasing to 30 mPa·s at 160°C and reaches a maximum of ca 93 Pa·s (930 P) at 187°C. Above this temperature, the viscosity gradually drops off again to ca 2 Pa·s (20 P) at 306°C. A qualitative exploration of these viscosity changes in terms of the allotropy of sulfur implies that, below 159°C, sulfur consists mainly of S_8 rings, and a normal decrease of viscosity with rising temperature is observed. The sudden increase in the viscosity of sulfur above 159°C is attributed to the formation of polymeric sulfur chain molecules. Then, as the temperature rises further, the concentration of polymeric sulfur continues to increase, but the opposing effect of decreasing chain length resulting from thermal sulfur–sulfur bond scission causes a gradual decrease in viscosity in the temperature range between 187°C and the boiling point of sulfur. The chemical equilibria between the various forms in molten sulfur have been extensively investigated (2–3,5). A recent critical review of the literature concerning the molecular composition of molten sulfur is given in ref. 6. Experiments that added much to the knowledge of the species present under different time–temperature parameters are described in refs. 7–8. Previous theories concerning the polymerization of S_8 were shown to be in disagreement with well-established experimental facts and are considered unsatisfactory.

The molecular composition of sulfur vapor is a complex function of temperature and pressure. Vapor pressure measurements have been interpreted in terms of an equilibrium between several molecular species (9–10). Mass spectrometric data for sulfur vapor indicate the presence of all possible S_n molecules from S_2 to S_8 and negligible concentrations of S_9 and S_{10} (11). In general, octatomic sulfur is the predominant molecular constituent of sulfur vapor at low temperatures, but the equilibrium shifts toward smaller molecular species with increasing temperature and decreasing pressure.

Constants and Chemical Properties. The constants of sulfur are presented in Table 1. Two freezing points are given for each of the two crystalline modifications. When the liquid phase consists solely of octatomic sulfur rings, the temperature ranges at which the various modifications form are called the ideal freezing points. The temperatures at which the crystalline forms are in equilibrium with liquid sulfur containing equilibrium amounts of $S\pi$ and $S\mu$ are called natural freezing points.

There are four stable isotopes of sulfur: ^{32}S, ^{33}S, ^{34}S, and ^{36}S with relative abundances of 95.1, 0.74, 4.2, and 0.016%, respectively. The relative abundance of the various isotopes varies slightly, depending on the source of the sulfur; the ratio of ^{32}S to ^{34}S is 21.61–22.60. Three radioactive isotopes of masses 31, 35, and 37 with half-lives of 2.6 s, 87 d, and 5 min, respectively, have been generated artificially. The normal orbital electron structure (17) is of the arrangement (2), (2,6), (2,4). Sulfur has valences of −2, +2, +3, +4, and +6.

Sulfur falls between oxygen and selenium in its group and resembles oxygen in its chemical reactions with most of the elements. An element closely related to sulfur is selenium, which has a similar group of valences and analogous allotropy. Sulfur is between phosphorus and chlorine horizontally in the periodic table, and its properties are generally those to be expected from its position in that table. An exception is that its melting point is higher than would be expected, probably because of its complex molecular structure (17).

Sulfur is insoluble in water but is soluble to varying degrees in many organic solvents, eg, carbon disulfide, benzene, warm aniline, warm carbon tetrachloride, and liquid ammonia (18). Carbon disulfide is the most commonly used solvent for sulfur.

Sulfur combines directly and usually energetically with almost all of the elements; exceptions include gold, platinum, iridium, and the inert gases (19). In the presence of oxygen or dry air, sulfur is very slowly oxidized to sulfur dioxide. When burned in air, it forms predominantly sulfur dioxide with small amounts of sulfur trioxide. In the presence of moist air, sulfurous acid and sulfuric acids are slowly generated.

Hydrochloric acid reacts with sulfur only in the presence of iron to form hydrogen sulfide. Sulfur dioxide forms when sulfur is heated with concentrated sulfuric acid at 200°C. Dilute nitric acid up to 40% concentration has little effect, but sulfur is oxidized by concentrated nitric acid in the presence of bromine with a strongly exothermic reaction (19).

Table 1. Physical Constants of Sulfur

Property	Value		Ref.
	Ideal	Natural	
freezing point of solid phase, °C			
rhombic, S	112.8	110.2	
monoclinic	119.3	114.5	
boiling point, °C		444.6	12
density of solid phase (at 20°C), g/cm^3			
rhombic		2.07	
monoclinic		1.96	
amorphous		1.92	
density of liquid, g/cm^3			
125°C		1.7988	
130°C		1.7947	
140°C		1.7865	
150°C		1.7784	
density of vapor at 444.6°C and 101.3 kPa (= 1 atm), g/L		3.64	
refractive index, n_D^{110}	1.929		
vapor pressure, p = Pa, T = K[a]			
rhombic (20–80°C)	$\log p = 16.557 - 5166/T$		13
monoclinic (96–116°C)	$\log p = 16.257 - 5082/T$		14
liquid (120–325°C)	$\log p = 19.6 - 0.0062238\ T - 5405.1\ T$		12
(325–550°C)	$\log p = 12.3256 - 3268.2/T$		
surface tension, mN/m (= dyn/cm)			
120°C		60.83	15
150°C		57.67	
critical temperature, °C		1040	
critical pressure, MPa[b]		11.75	
critical volume, mL/g		2.48	
specific heat, J/(kg·K)[c]			
rhombic (24.9 to 95.5°C)	$Cp = 468 + 0.814\ T$		
monoclinic (−4.5 to 118.9°C)	$Cp = 465 + 0.908\ T$		
liquid (Sλ) (118.9 to 444.6°C)	$Cp = 706 + 0.65\ T$		
gas (S) (25 to 1727°C)	$Cp = 709 - 0.034\ T - 3.5 \times 10^6\ T^{-2}$		
gas (S_2) (25 to 1727°C)	$Cp = 558 + 0.018\ T - 5.2 \times 10^6\ T^{-2}$		
heat of transformation (rhombic to monoclinic) at 95.5°C, J/g[c]		11.25	
heat of fusion, J/g[c]			
112.8°C (S) (rhombic) → Sλ (liquid)		49.8	
118.9°C (S) (monoclinic) → Sλ (liquid)		38.5	
thermal expansion of rhombic sulfur (linear)			
0–13°C	4.567×10^{-5}		
13–50°C	7.433×10^{-5}		
50–78°C	8.633×10^{-5}		
78–97°C	20.67×10^{-5}		
98–110°C	103.2×10^{-5}		
latent heat of vaporization, J/g[c]	L[d]	L[e]	12
200°C	308.6		
300°C	289.3		
400°C	286.4	278.0	
420°C	287.6	276.3	
440°C	290.1	274.6	
460°C	293.1	273.0	
electrical resistivity, ohm·cm			16
20°C	1.9×10^{17}		

Table 1 (*continued*)

Property	Value: Ideal	Value: Natural	Ref.
110°C	4.8×10^{12}		
400°C	8.3×10^{6}		
magnetic susceptibility, m^3/mol^f			16
rhombic at 18°C		1.539	
monoclinic at 112°C		1.539	
liquid at 220°C		1.539	
standard reduction potential, S/S^{2-}, V		−0.508	16

[a] To convert log P_{Pa} to log P_{psi}, subtract 3.8384 from the constant.
[b] To convert MPa to psi, multiply by 145.
[c] To convert J to cal, divide by 4.184.
[d] Including heat of dissociation to S_2 present in vapor.
[e] Minus heat of dissociation to S_2 in vapor.
[f] To convert m^3/mol to emu/mol (cgs unit), divide by $4\pi \times 10^6$.

Sulfur combines directly with hydrogen at 150–200°C to form hydrogen sulfide. Molten sulfur reacts with hydrogen to form hydrogen polysulfides. At red heat, sulfur and carbon unite to form carbon disulfide; this is a commercially important reaction in Europe, although natural gas is used in the United States. In aqueous solutions of alkali carbonates and alkali and alkaline-earth hydroxides, sulfur reacts forming sulfides, polysulfides, thiosulfates, and sulfites.

At room temperature, sulfur unites readily with copper, silver, and mercury and vigorously with sodium, potassium, calcium, strontium, and barium to form sulfides. Iron, chromium, tungsten, nickel, and cobalt react much less readily. In a finely divided state, zinc, tin, iron, and aluminum react with sulfur on heating (19).

Various sulfides of the halogens are formed by direct combination of sulfur with fluorine, bromine, and chlorine. No evident reaction occurs with iodine; instead, the elements remain as components of a mixture. Mixtures of sulfur and potassium chlorate, or sulfur and powdered zinc are highly explosive.

Sulfur is involved in numerous organic reactions (20). When dissolved in amines, chemical interaction between sulfur and the solvent results in the formation of colored species ranging from deep yellow to orange and green. Many organic reactions involving sulfur are commercially significant. Sulfur is important in the manufacture of lubricants, plastics, pharmaceuticals (qv), dyes, and rubber goods (see Lubrication and lubricants; Plastics processing; Dyes; Rubber chemicals).

Sulfur is not considered corrosive to the usual construction materials. Dry, molten sulfur is handled satisfactorily in mild steel or cast-iron equipment. However, acid-generating impurities which may be introduced in handling and storage, create corrosive conditions. The exposure of sulfur to moisture and air causes the formation of acids, which attack many metals. To combat such corrosion difficulties, protective coatings of organic compounds, cement, or sprayed resistant metals are often applied to exposed steel surfaces, including pipe and equipment used in handling liquid sulfur, and to structural members that come in contact with solid sulfur. Also practical in some applications is the use of resistant metal alloys, particularly those of aluminum and stainless steel. Neutralization of the generated acids by the addition of basic chemicals is sometimes employed.

Elemental Sulfur

Occurrence. ***Salt-Dome Sulfur Deposits.*** The sulfur deposits associated with salt domes in the Gulf Coast regions of the southern United States and Mexico have until recently been the primary sources of U.S. sulfur and remain an important segment of both U.S. and world sulfur supply. Although the reserves are finite, they are large and the voluntary nature of their large productive capacity assures the importance of these sources for some time to come. In 1980, the output from the salt domes in the United States and Mexico was ca 8.1×10^6 metric tons.

Salt domes of the U.S. Gulf Coast are vertical structures, usually circular in outline with steeply dipping flanks, and composed of coarsely crystalline halite (NaCl) interspersed with anhydrite ($CaSO_4$). The cap rock which surmounts the salt dome consists of anhydrite in contact with the salt and gypsum ($CaSO_4.2H_2O$) derived from the anhydrite. Limestone in the form of fine gray carbonate interspersed with vugs, seams, fissures, and cavities is frequently associated with the gypsum and anhydrite formations. It may be present as a stratum overlying these formations, as lenticular beds covering part of them, or as disseminated lenses and nodules included in the upper part of the cap rock.

The sulfur occurs as well-developed crystal aggregates in veins and vugs or as disseminated particles in the porous limestone and gypsum section of the cap rock. Several theories have been proposed for the occurrence of sulfur in salt domes. One theory suggests the formation of limestone and hydrogen sulfide from anhydrite in the presence of reducing agents. This reaction requires temperatures of ca 650°C and, although oil or other hydrocarbons may be present to act as reducing agents, the temperature actually attained is not sufficient to support this theory of the origin of sulfur. In 1946, the presence of anaerobic, sulfate-reducing bacteria was discovered in cap rock; their ability to promote reaction at normal temperatures is recognized as the more likely origin of sulfur. Anaerobic bacteria consume hydrocarbons as a source of energy but combine sulfur instead of oxygen with the hydrogen. The hydrocarbon-fueled bacteria reduce anhydrite to hydrogen sulfide, calcium carbonate, and water. The hydrogen sulfide remains dissolved in the formation waters until it precipitates as crystalline sulfur through various oxidation reactions, possibly initiated by oxygen and carbon dioxide dissolved in water percolating from the upper sediments. In 1966, 329 salt-dome structures were identified by the U.S. Bureau of Mines as existing in the U.S. Gulf Coast area and offshore tidelands. Of these, 27 have been commercial sulfur producers. There are nine salt-dome sulfur deposits being commercially mined by the Frasch process, of which six are in the United States, three are in Mexico, and 19 are no longer being mined although one is on stand-by. The latter and certain others could be reactivated if justified by economic circumstances. The 9 domes in operation as of April 1981, are listed below:

Dome	*Company*
Texas	
Long Point Dome	Jefferson Lake Sulphur Co. (division of Occidental Petroleum Co.)
Moss Bluff Dome	Texasgulf, Inc.
Newgulf (Boling Dome)	Texasgulf, Inc.

Dome	*Company*
Louisiana	
Caillou Island Dome	Freeport Minerals, Inc.
Garden Island Bay Dome	Freeport Minerals, Inc.
Grand Isle Dome	Freeport Minerals, Inc.
Mexico	
Salinas Dome (Coachapa)	Azufera Panamericana, S.A.
Jaltipan Dome	Azufera Panamericana, S.A.
Nopalapa Dome (Texistapec)	Compania Exploradadora del Istmo, S.A.

Evaporite Basin Sulfur Deposits. Elemental sulfur occurs in another type of subsurface deposit, which is similar to the salt-dome structures in that the sulfur is associated with anhydrite or gypsum. The deposits are sedimentary and occur in huge evaporite basins. It is believed that the sulfur in these deposits, like that in the Gulf Coast salt domes, was derived by hydrocarbon reduction of the sulfate material, assisted by anaerobic bacteria which permit the reaction to occur at ordinary temperatures. The sulfur deposits in Sicily, Poland, Iraq, the USSR, western United States, and in western Texas fall into this category.

Mining techniques similar to the Frasch salt-dome mining systems have been applied successfully. These developments and particularly those in west Texas, Poland, and Iraq have significantly contributed to world sulfur production and reserves. Hot-water mining of the Polish deposits was first begun in 1966 at Gryzbow. In 1979, production at Gryzbow and another deposit at Jeziorko was nearly 5×10^6 t. Production of sulfur from the Culberson mine (Duval Corporation, Division of Pennzoil Inc.) in west Texas began in 1968, and this mine is the world's largest sulfur producer, exceeding 2×10^6 t/yr. Other smaller mines in the area include Phillips Ranch (Duval Corp.), Comanche Creek (Texasgulf), and the Farmlands operation near Fort Stockton. In Iraq, the Mishraq deposit has the potential for contributing over 10^6 t/yr after overcoming logistical problems that limit it to about half of that rate. Salt domes and similar sulfur-bearing structures occur in regions other than the United States, Mexico, Poland, Iraq, and the USSR, but sulfur deposits that could be economically productive have not been discovered in any of them.

Volcanic and Native Sulfur Deposits. Elemental sulfur occurs in other types of surface or underground deposits throughout the world, but seldom in sufficient concentration to be commercially important. Sulfur of volcanic origin occurs in many parts of the world. These deposits originated from gases emitted from active craters, solfataras, or hot springs, which contain deposited sulfur in fractures of rock, by replacement in the rock itself or in the sediments of lake beds. Volcanic deposits usually occur in tufas, lava flows, and similar volcanic rocks but also in sedimentary and intrusive formations. Scattered deposits of this type have been discovered throughout the mountain ranges bordering the Pacific Ocean, particularly in Japan, the Philippine Islands, and Central and South America. Some volcanic deposits are worked profitably and are important in the countries in which they are found. The Japanese deposits are among this group because of their long productive history and the considerable tonnages that have been produced from them. Most volcanic deposits, however, are in isolated regions and at high elevations where production and transportation costs are prohibitive.

Extraction. ***Frasch Process.*** In the Frasch process, large quantities of hot water are introduced through wells drilled into buried deposits of native sulfur. The heat from the water melts the sulfur in the vicinity of the wells; the melted sulfur is then removed to the surface as molten elemental sulfur of high purity. Economical operation of a salt dome or a subsurface sulfur deposit by the Frasch process requires a porous sulfur-bearing limestone, a large and dependable supply of water, and a source of inexpensive fuel. A power plant is required, in which the necessary volume of hot water is produced, as is compressed air for pumping molten sulfur from the wells and electric power for drilling, lighting, operating maintenance equipment, loading sulfur for shipment, and similar operations.

A typical setting of equipment for a sulfur well and the principles of mining are illustrated schematically in Figure 1. First, a hole is drilled to the bottom layer of the salt-dome cap rock with equipment of the same type as that used in oil fields. Three concentric pipes within a protective casing are placed in the hole. A 20-cm pipe inside an outer casing is sunk through the cap rock to the bottom of the sulfur deposit. Its lower end is perforated with small holes. Then a 10-cm pipe is lowered to within a short distance of the bottom. Last and innermost is a 2.5 cm pipe, which is lowered more than halfway to the bottom of the well.

Water heated under pressure to 160°C is pumped down the space between the 20-cm and 10-cm pipes and, during the initial heating period, also down the 10-cm pipe. The superheated water flows out the holes at the bottom into the porous sulfur-bearing formation (Fig. 1**a**). When the temperature of the sulfur-bearing formation exceeds the melting point of sulfur, the liquid sulfur, being approximately twice as heavy as water, percolates downward through the porous limestone to form a pool at the bottom of the well. A heating period of 24 h or longer is required to accumulate a liquid sulfur pool of sufficient size, and then pumping of hot water down the 10-cm pipeline is stopped. Static pressure of the hot water pumped into the formation then forces the liquid more than 100 m up into the 10-cm pipe (Fig. 1**b**). Compressed air

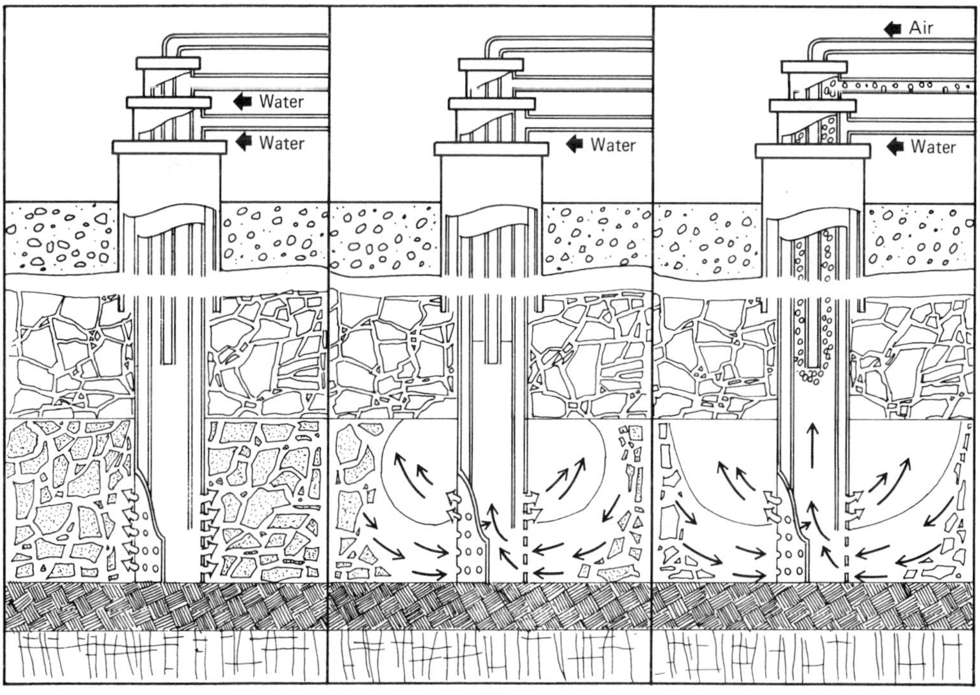

Figure 1. The Frasch process. Courtesy of Freeport Minerals Co.

forced down the 2.5-cm pipe aerates and lightens the liquid sulfur so that it rises to the surface (Fig. 1**c**). Injection of hot water is continued down the 20-cm pipe to maintain the sulfur melting process, and the compressed-air volume is adjusted to equalize the sulfur pumping rate with the sulfur melting rate. If the pumping rate exceeds the melting rate, the sulfur pool is depleted and the well produces water. At this point, the compressed-air flow is stopped, and hot water is again injected until the liquid sulfur pool is reestablished.

The sulfur-bearing cap rock, being an enclosed formation, is essentially the equivalent of a pressure vessel. Hot water, pumped into the formation to melt sulfur, must be withdrawn after cooling at approximately the same rate as it is put in, otherwise the pressure in the formation would increase to the point where further water injection would be impossible. Bleedwater wells, which are used to extract water from the formations, usually are located on the flanks of the dome away from the mining area where the water temperature is lowest. The water is treated to remove soluble sulfides and other impurities before being discharged to disposal ditches or canals.

On the surface, the liquid sulfur moves through steam-heated lines to a separator where the air is removed. Depending on the mine location, the liquid sulfur may be pumped to storage vats to be solidified, to tanks for storage as a liquid, to pipelines, or to thermally insulated barges for transport to a central shipping terminal.

Sulfur wells that are favorably located produce continuously over long periods. Some may last a year or more, whereas others may be abandoned within a few weeks because denseness of the rock formation may retard circulation of hot water and molten sulfur. The extraction of sulfur weakens the rock formation and subsidence may follow. This may break the pipes in the well and end its productivity. However, subsidence is desirable in mining even though wells may be lost as a result. The advantage of subsidence is that the volume of exhausted formation through which hot water can circulate is reduced. After caving, the crushed exhausted formation is relatively impervious and therefore confines the circulation of hot water to the more porous sulfur-bearing parts of the deposit.

Directional drilling techniques were an important advance in sulfur mining methods. Casings are placed in the hole in such a manner as to extend into the sulfur formation somewhat horizontally. Thus, substantial amounts of sulfur in the deposit overlie the hole. Also, the casings in the subterranean volume likely to be affected by subsidence are parallel to the expected earth movement and therefore are less affected by shear. The result is better utilization of heating water and longer well life; it has also permitted efficient reworking of areas exhausted to vertical mining techniques.

Another advance in sulfur mining technology has been the development of a method involving seawater in the Frasch process, making it feasible to mine deposits distant from freshwater supplies. In such a plant, seawater is first deoxygenated by bringing it in direct contact with combustion gases in a packed tower. The seawater is preheated by these gases and its temperature is raised to 106°C in indirect heat exchangers by means of steam furnished by high pressure boilers. Condensate from the heat exchangers is recycled to the boilers, which limits freshwater requirements to leaks and other small losses in the system. Production from several sulfur mines involves seawater both from stationary systems and portable, barge-mounted power plants.

Hydrodynamic Process. The hydrodynamic process is similar to the Frasch process in that superheated water is used to melt the underground sulfur. However, the techniques involved are sufficiently different to warrant a separate characterization. The process was developed in Poland to exploit sulfur deposits which, because of their thin bedding, wide dispersion, and frequent impermeability, did not appear to be amenable to production by the Frasch method which requires some degree of special isolation of the deposits. The techniques employed include the use of explosives to control permeability or to create sealed-off gases and the calculated manipulation of underground water pressure, temperature, and flow conditions by control at injection and breakwater points. The system requires a constant rate of fuel consumption throughout the life of the mine and improves the rate of sulfur recovery. Hydrodynamic mining is used in Poland and in Iraq where the mining area was developed with Polish assistance.

Extraction from Volcanic and Other Surface Deposits. Sulfur is recovered from volcanic and other surface deposits by a number of different processes, including distillation, flotation, autoclaving, filtration, solvent extraction, or a combination of several of these processes. The Japanese sulfur deposits are reached by tunnel, and mining is done by the room-and-pillar, chamber-and-pillar with filling, and cut-and-fill systems. Sulfur is extracted from the ore by a distillation process, which is performed in rows of cast-iron pots, each containing ca 180 kg of ore. Each row of pots is connected to a condensation chamber outside the furnace. A short length of pipe connects each pot with a condenser. Brick flues connect combustion gases under the pots. Sulfur vapor flows from the pots to the condensation chamber where the liquid sulfur is collected. Japanese ore contains 25–35 wt % sulfur. However, this method is mainly of historical interest, having been superseded by other sources of sulfur production.

The sulfur deposits in Sicily have been worked since ancient times. Originally, the sulfur was removed by piling the ore in central heaps, covering it with earth, and then igniting the pile. By this method, 30–50 wt % of the sulfur was burned to provide heat for melting the remainder of the sulfur in the ore. Less than 50 wt % of the sulfur originally contained in the ore was recovered. In about 1880, the first Gill gas furnace was installed. The original furnace had two chambers arranged so that the heat from burning ore in one chamber passed through ore in the other chamber to melt a considerable portion of the sulfur. When the sulfur in the first chamber burned, the chamber was refilled and when the partially extracted ore in the first chamber burned, it was refilled and the partially extracted ore in the other chamber was ignited. This method involved better utilization of the heat of combustion. Later, furnaces contained as many as six chambers and permitted up to 80% sulfur recovery (21).

Extensive experimentation has led to the granting of numerous patents for various thermal processes for extracting sulfur from ores, either as elemental sulfur or as SO_2, but very few of these processes have been operated commercially. The proposed processes involve shaft furnaces, multiple-hearth furnaces, rotary kilns, and fluidized-bed roasters. In all of these, ground ore is heated with oxygen-free, hot combustion gases to distill elemental sulfur, or the sulfur in the ore is burned with air to yield SO_2 for sulfuric acid production. In 1953, a commercial plant was brought on stream at the Yerrington, Nevada, copper mine for recovering the sulfur as SO_2 from the Leviathan deposit of low grade sulfur ore in Alpine County, California. The process consisted of four fluid-bed reactors, in which the ore was roasted in air to produce SO_2 for a contact sulfuric acid plant (22). For the production of elemental sulfur, the use of oxygen-free, hot combustion gases in a fluidized bed has been proposed to distill sulfur from the ore as a vapor, which is then condensed to liquid sulfur.

Various processes have been proposed and tested for the recovery of sulfur from native ores by solvent extraction, and many patents have been issued. Carbon disulfide, being the best solvent for sulfur, has often been suggested for extraction of sulfur from ore. Some plants in Italy, the FRG, South America, and the United States have used carbon disulfide for this purpose, but the cost of the solvent, the high losses, and its flammability detract from low operating costs. Many other solvents have been tried, including hot caustic solution, chlorinated hydrocarbons, ammonium sulfide, xylene, kerosene, and various high boiling oils. The sulfur is recovered either by volatilizing the solvent or by crystallizing the sulfur.

Various combinations of autoclaving, filtration, and centrifuging are used in some processes to recover sulfur from ore. One such process, involving continuous autoclaving, flotation, and filtration, was first used commercially at a plant in Columbia (23). The ore is finely ground (<625-μm or −28-mesh) and suspended in water to form a slurry of ca 30% solids, which is pumped continuously to a three-compartment, agitated autoclave. In the autoclave, the slurry is heated and the sulfur melts by steam injection into the bottom of each compartment. Agitation causes the sulfur to coalesce into globules which separate from the gangue. Hot slurry from the autoclave flows into a quench pot to be cooled by water injection, and the sudden cooling solidifies the separated sulfur particles. The cooled slurry is throttled to atmospheric pressure and flows into a 625-μm (28-mesh) vibrating screen. The oversize material passes directly to a sulfur melter, whereas the underflow from the screen passes to a flotation circuit for separation of the smaller sulfur particles; this gives a concentrate of 90–95 wt % sulfur which then passes to the sulfur melter. Melted sulfur is pumped through a filter for removal of gangue.

By similar techniques but on a much more massive scale, the Piaseczno and Machow open-pit mines in Poland produced over 2×10^6 t of sulfur during 1957–1971 and 1968–1979, respectively. A multibucket extractor was used to remove overburden after a two-year period of dewatering. The sulfur ore after precrushing was moved by conveyor belt to the plant where it was ground and the sulfur was floated off, dehydrated in centrifuges, melted, and filtered in Kelly-type filters. The filter cake was used in sulfuric acid production. Many natural and technical obstacles were successfully overcome, and much of the otherwise unexploitable (by Frasch techniques) sulfur deposit was processed with over 80% recovery. At this time, the Machow mine is not operating because of equipment problems and its fate is uncertain. It is believed, however, that grinding–flotation methods currently account for a substantial portion of sulfur production in the USSR.

One more variation to the many methods proposed for sulfur extraction is the fire-flood method. It is a modern version of the Sicilian method, by which a portion of the sulfur is burned to melt the remainder. It would be done *in situ* and is said to offer cost advantages, to work in almost any type of zone formation, and to offer better sweep efficiency. The recovery stream would be ca 20 wt % sulfur as SO_2 and 80 wt % elemental sulfur. The method was laboratory tested in the late 1960s and patents were issued, but it was not commercially exploited when sulfur prices dropped. It is, however, still considered to be a feasible process by Tor Developments, Inc. in Dallas, Texas, who were the originators of the process.

Sulfide Ores

Occurrence. The metal sulfides, which are scattered throughout most of the world, have been an important source of elemental sulfur. The potential for their recovery exists, although at present they are less attractive economically and technologically than other sources of brimstone. Nevertheless they are and will continue to be an important source of sulfur in other forms, ie, sulfur dioxide and sulfuric acid.

Some of the most important metal sulfides are pyrite, FeS_2; chalcopyrite, $CuFeS_2$; pyrrhotite, $Fe_{n-1}S_n$; sphalerite, ZnS; galena, PbS; arsenopyrite, $FeS_2.FeAs_2$; and pentlandite, $(Fe,Ni)_9S_8$. Sulfide deposits often occur in massive lenses, but may occur in tabular shape, in veins, or in a disseminated state. They may be of igneous, metamorphic, or sedimentary origin.

Pyrite is the most abundant of the metal sulfides. For many years, until the Frasch process was developed, pyrites were the main source of sulfur and for much of the first half of this century, over 50% of world sulfur production was derived from pyrite. In 1979, sulfur in various forms from pyrite represented ca 12.5% of the total sulfur from noncommunist countries. In the communist countries, the portion is considerably higher: 36.8%. Pyrite is usually obtained either as run-of-mine or beneficiated ore from straight pyrite deposits or it is recovered as a by-product flotation concentrate in the processing of ores for their base-metal values. Pyrite reserves are distributed throughout the world and deposits are being mined in ca 30 countries. Possibly the largest pyrite reserves in the world are located in southern Spain, Portugal, and the USSR. Large deposits are also in Canada, Cyprus, Finland, Italy, Japan, Norway, South Africa, Sweden, Turkey, the United States, and Yugoslavia. In 1981, the principal noncommunist producers were Spain, Italy, South Africa, and Canada. In 1981, there were no known plants producing elemental sulfur from pyrites in noncommunist countries. The pyrometallurgical processes are not economic for voluntary brimstone production under present circumstances.

Pyrometallurgical Processes. ***Orkla Process.*** A process for recovering sulfur from cuprous pyrite was developed by the Orkla Mining Company in Norway (24). The sulfur output of 80,000–100,000 t/yr furnished an important part of European requirements until 1962, when the smelter was shut down. The process was once used on a much smaller scale in Portugal and Spain and probably in the USSR (25). The Orkla process involves recovery of ca 80% of the sulfur contents of a pyritic copper ore by direct smelting in the presence of a carbonaceous reducing agent.

Noranda Process. When pyrites are heated to ca 540°C in the absence of oxygen, they evolve about half of their sulfur content in the elemental form. Noranda Mines Ltd. and Batelle Memorial Institute developed a process based upon this property to recover elemental sulfur from pyrite (26). The first commercial plant was built at Welland, Ontario, in 1954 but operated on an experimental basis for only a few years before being closed for economic reasons.

Outokumpu Process. Outokumpu Oy, Finland's largest mining and metallurgical company, discovered a complex ore body at Pyhasalmi, Finland, containing pyrite, sphalerite, chalcopyrite, barite, and small amounts of pyrrhotite, arsenopyrite, and molybdenite. The ore can be beneficiated by flotation to obtain pyrite concentrate as well as copper and zinc concentrates. A process was developed to treat the pyrite concentrate in a flash smelter for recovery of elemental sulfur and iron cinder. The commercial smelter located at Kokkolla began operating in 1962 (27). In 1977, production of elemental sulfur was stopped, although sulfur dioxide is still produced and sold for sulfuric acid production. Similarly, the Outokumpu process was used to recover elemental sulfur at a plant in Botswana, but the sulfur is now recovered as SO_2–H_2SO_4.

Hydrometallurgical Processes. Recovery of sulfur in the processing of nonferrous metal sulfides has been in the form of SO_2 and/or H_2SO_4, and this is still true of smelter (pyrometallurgical) operations. However, there have been recent accounts of processes, mainly hydrometallurgical, in which sulfur is recovered in the elemental form (see Extractive metallurgy).

One, called the CLEAR process, is operated by Duval Corporation near Tucson, Arizona (28). It involves leaching copper concentrated with a metal chloride solution, separation of the copper by electrolysis, and regeneration of the leach solution in a continuous process carried out in a closed system. Elemental sulfur is recovered. Not far from the Duval plant, Cyprus Mines Corporation (a division of Standard Oil of Indiana) is operating a process known as Cymet. Sulfide concentrates undergo a two-step chloride solution leaching and are crystallized to obtain cuprous chloride crystals. Elemental sulfur is removed during this stage of the process.

Another process, which also generates elemental sulfur as a by-product, has been patented by Envirotech Research Center in Salt Lake City (28). In the Electroslurry process, a ball mill finely grinds a chalcopyrite concentrate, which then reacts with an acidic copper sulfate solution for iron removal. The liquor is electrolyzed, with the iron oxidized to the ferric form which leaches copper from the copper sulfide for deposition on the cathode. Elemental sulfur is recovered at the same time. As of 1981, a demonstration plant had not been built.

A pressure leaching system to handle copper sulfide called the Sherritt-Cominco (S.C.) copper process was developed by these two Canadian firms; pilot-plant testing was completed in 1976 (28). It is likely to be commercialized when new copper deposits are developed. Plans have been made for a plant producing 200 t/d of copper.

Nine hundred and fifty metric tons per day of copper concentrate is fed to a roaster 7.5 m in diameter and 12 m high. In the upper hearths, some sulfur is driven from the pellets by the 690°C heat. Upon reaching the roaster bottom, the feed contacts hydrogen gas. More sulfur is removed from the feed as hydrogen sulfide. Sulfur vapor and hydrogen sulfide burn in the roaster, thereby producing sulfur dioxide and heating the roaster gases. The SO_2 content of the roaster gases can be used to make 360 t/d of sulfuric acid. Activated concentrate leaving the roaster is fed to a series of rubber-lined batch tanks containing sulfuric acid. As iron sulfide dissolves in the acid, hydrogen sulfide gas is liberated. Most of the gas is drawn off and sent to a small Claus plant and converted to 100 t/d of elemental sulfur.

By-Product Sulfur

The preceding discussions have focused upon those sources of sulfur that are voluntary, ie, sulfur is considered to be the principal product and often the only product. Sulfur has also been recovered as a by-product from various process operations for many years. Such sulfur is termed involuntary sulfur and accounts for the largest portion of world sulfur production (see Sulfur recovery).

Sulfates

Occurrence. The largest untapped source of sulfur occurs in the ocean as dissolved sulfates of calcium, magnesium, and potassium. The average sulfur concentration in seawater is 880 ppm; thus, 1 km^3 of seawater contains ca 0.86×10^6 t of elemental sulfur in the form of sulfates. Natural and by-product gypsum ($CaSO_4.2H_2O$) and anhydrite ($CaSO_4$) rank second only to the oceans as potential sources of sulfur. Mineral deposits of gypsum and anhydrite are widely distributed in extremely large quantities. Gypsum is a by-product waste material from several manufacturing processes; most notable is the waste gypsum produced in the manufacture of phosphoric acid from phosphate rock and sulfuric acid.

Extraction. Although many processes have been developed to recover elemental sulfur from gypsum or anhydrite, high capital and operating costs have precluded their widespread use and will continue to do so while less expensive sources remain available. Obtaining sulfur from gypsum processes has been attractive during periods when sulfur has been in short supply and energy costs remained relatively low. Because the processes are energy-intensive and energy costs are high, they are unlikely to be competitive. However, gypsum and anhydrite remain economical sources of sulfur in other forms, eg, sulfuric acid–cement and ammonium sulfate, in areas such as India where brimstone must otherwise be imported.

Thermal Reduction of Gypsum. The initial work involving the thermochemical technique was carried out in the FRG and more recently by the U.S. Bureau of Mines which did research work on two processes for the recovery of elemental sulfur from gypsum at their Salt Lake City Metallurgy Research Center in the late 1960s (29). Both processes involved reduction roasting of gypsum with coal or reducing gases at 900–950°C to produce calcium sulfide. Process 1 involved carbonation of a water slurry of calcium sulfide with CO_2-bearing flue gases from the reduction kiln to precipitate calcium carbonate and to evolve hydrogen sulfide. The latter could be converted to sulfur in a standard Claus unit. Process 2 made use of a countercurrent ion-exchange system and sodium chloride to produce by-product sodium carbonate and calcium chloride as well as elemental sulfur. Three metric tons each of the by-product were produced per metric ton of sulfur.

Neither of these processes has been commercialized, although some aspects of the methodology were incorporated into a plant operated for a short time by the Elcor Co. (30). This company, which operated briefly in west Texas in 1968 and used natural gypsum, is the only one known to have commercially attempted to recover elemental sulfur from this material by a two-step thermal process. The principal piece of equipment was a closed-shaft kiln which served as a reaction chamber. Gypsum was fed to the top of the kiln and the reducing gas, a mixture of carbon oxides and hydrogen produced by reforming methane, was fed to the bottom. In the reaction zone of the furnace, the reformed gas reacts with calcium sulfate forming calcium sulfide according to the following reactions:

$$CaSO_4 + 4\ H_2 \rightarrow CaS + 4\ H_2O$$

$$CaSO_4 + 4\ CO \rightarrow CaS + 4\ CO_2$$

These reactions are exothermic causing the gas temperature to rise to 1040°C from 1000°C. As the gas passes up the shaft, its temperature falls, reaction ceases at 815°C, and gases leave the reactor at ca 230°C. Solid calcium sulfide leaving the bottom of the reactor is quenched in water and then ground in a ball mill. After grinding, the calcium sulfide is hydrolyzed in a reactor upon contact with water and a countercurrent stream of carbon dioxide recycled from the shaft kiln:

$$CaS + CO_2 + H_2O \rightarrow H_2S + CaCO_3$$

The hydrogen sulfide formed is removed from the top of the decomposition reactor to a conventional Claus sulfur recovery unit and calcium carbonate is removed from the base.

The Elcor plant was shut down shortly after it began operation. Although most technical problems were said to have been solved, production costs were prohibitive.

Phosphogypsum. A significant technical achievement would be the economical recovery of sulfur from phosphogypsum, a by-product produced in tremendous quantities in the manufacture of phosphatic fertilizers. Like the Elcor plant, the process used by Fertilizer India, Ltd. (Planning and Development) involved a shaft kiln (30). Following bench-scale tests, tests on a larger scale were conducted at the Regional Research Laboratory at Jorhat (Assam Province). The feed to the top of the kiln consisted of a nodulized mixture of phosphogypsum, pulverized coke, and clay additives. Air was introduced to the bottom of the kiln such that the temperature in the hottest zone was maintained at 1100–1200°C. Under these conditions, calcium sulfate is reduced to sulfur dioxide which then reacts to yield elemental sulfur:

$$CaSO_4 + 4\ C \rightarrow CaS + 4\ CO$$

$$CaS + 3\ CaSO_4 \rightarrow 4\ CaO + 4\ SO_2$$

$$SO_2 + 2\ CO \rightarrow 2\ CO_2 + S$$

$$SO_2 + C \rightarrow CO_2 + S$$

Analysis of the gas stream leaving the kiln shows that 90% of the sulfur removed from the nodules is present as elemental sulfur and 10% as sulfur dioxide. The degree of desulfurization depends on the size of the kiln used. The larger of two kilns tried was the most efficient with a desulfurization efficiency of 75–80%. This kiln was ca 4.7 m in height and had a base diameter of 1 m widening to 1.6 m at the top.

Roasted nodules removed from the kiln can apparently be used after additional roasting at higher temperatures in the production of cement clinker, and this has a favorable effect on the overall economics of the process. Raw materials and utilities consumption data are listed in Table 2.

Bacteriological Sulfur. Anaerobic, sulfate-reducing bacteria burn hydrocarbons as a source of energy but combine sulfur instead of oxygen with the hydrogen to form hydrogen sulfide. Several experimenters have tried to develop this idea into a controlled process for producing sulfur from gypsum or anhydrite (31). This process requires a strain of sulfate-reducing bacteria, an organic substrate whose hydrocarbons provide food for the bacteria, and close control of environmental conditions in order to obtain maximum sulfur yields.

Finely ground gypsum is fed into a stirred reaction tank containing the organic substrate and the bacteria. The substrate can be a petroleum fraction, although sewage, spent sulfite liquor, molasses, or brewery waste can also be used. The advantage of a petroleum-based substrate is that its composition can be more closely controlled. Air must be excluded from the system because the bacteria are anaerobic. A hydrogen-purging system keeps air out and at the same time promotes fermentation.

Table 2. Raw Materials and Utilities Consumption per Metric Ton of Sulfur Recovered from Phosphogypsum

Property	Value
raw materials, t	
phosphogypsum	6.75
coke breeze	1.55
clay	0.345
natural gypsum	0.2
utilities	
power, MW·h	0.412
process water, m^3 [a]	1.6

[a] To convert m^3 to gal, multiply by 264.2.

Carbon dioxide generated by the fermentation process must be removed to help maintain the pH of the solution at pH 7.6–8.0. Carbon dioxide also inhibits the activity of the bacteria. The oxidation reduction potential is kept at 100–200 mV. The ideal temperature in the reactor varies with different strains in the bacteria but generally is 25–35°C.

As the reaction proceeds, a part of the mix is continuously withdrawn from the tank and is centrifuged, and the solids removed by centrifuging are resuspended in the reactor. Filtrate from the centrifuge goes to a stripping tower for removal of dissolved carbon dioxide and hydrogen sulfide, which is combined with the carbon dioxide and hydrogen sulfide gases passing from the top of the reactor. The combined gases are passed through a scrubbing tower for removal of the carbon dioxide and recovery of the hydrogen sulfide, which is fed to a conventional recovery unit for conversion to elemental sulfur. There is also the possibility of recovering other organic coproducts, eg, vitamins (qv) and steroids (qv). The rate at which bacteria reduce gypsum to hydrogen sulfide is quite slow, necessitating many large reaction tanks. A 300-t/d plant is estimated to need ten 3785-m^3 (10^6 gal) reactor tanks.

An organism of the *Desulfovibrio* genus has been used to make hydrogen sulfide in early experiments. Subculturing and selectively reisolating active organisms could lead to a strain with higher activity. By such techniques, a 1000-fold increase in activity after 40 generations was achieved in the 1960s (31). With recent achievements in biotechnology, further advances might be expected (see Genetic engineering).

Production

Principal sulfur projects that are in operation or are expected to be in operation are listed in Table 3. Sulfur sources in the United States are expected to undergo significant change during the 1980s. The Frasch process (ie, mines) supplied 53% of the sulfur in the United States in 1980. Recovered sulfur (see Sulfur recovery) was 34%. In 1990, the expected sources will be 29% Frasch, 60% recovered, and 11% miscellaneous (33).

Table 3. Sulfur Projects and Their Capacities[a]

Location	Process type[b]	Capacity, 10^3 t/yr	On-stream date	Developer
Astrakhan, USSR	sour gas	2000[c]	1985	USSR government
Whitney Canyon, Wyo.	sour gas	438	mid-1982	Standard Oil Co. of Indiana
Carter Creek, Wyo.	sour gas	365	1983	Standard Oil Co. of California
Robb-Hanlan, Alberta	sour gas	300	1982	Gulf Resources and Chemical Corp.
Brown Creek, Alberta	sour gas	350	1983–1984	Standard Oil Co. of Indiana
Skopanie, Poland	Frasch	1000	1985	Ministry of the Chemical Industry
Baranow, Poland	Frasch	1200[d]	not determined	Ministry of the Chemical Industry
Coachapa, Mexico	Frasch	250	July 1981	Azufera Panamericana

[a] Ref. 32.
[b] See Sulfur recovery for discussion of sour gas.
[c] Capacity to reach 6×10^6 t/yr by 1990.
[d] Estimated for 1981. Capacity to reach 6.8×10^5 t/yr by 1983.

Economic Aspects

At the beginning of the 20th century, the world's sulfur demand of ca 2×10^6 metric tons was met by sulfur recovered from elemental deposits in Italy and Sicily and from pyrite mined on the Iberian Peninsula. By 1980, sulfur was recovered in more than 70 countries in several forms. Frasch sulfur began to supply the U.S. requirements and displace Italian brimstone in other markets. Disruptions of shipping during World War I caused a conversion in U.S. sulfuric acid plants from the use of pyrites to Frasch sulfur. As a result of the disruption of sulfur production in Europe during World War II and the increasing use of brimstone rather than pyrite for producing sulfuric acid, the United States became a main supplier of sulfur. United States exports more than doubled after the war.

At the time foreign consumers were becoming more dependent on U.S. sulfur, domestic consumption also rose. By the time of the Korean conflict, supplies of sulfur were inadequate to meet world demand. The impetus of the shortage brought about the exploration and development of new sources, measures to conserve sulfur during processing, and the reuse and regeneration of spent sulfuric acid. With the change from short supply, the United States removed controls on sulfur use in 1952 and ceased allocations on international shipments in 1953. United States Frasch producers brought on stream seven new mines by the mid-1950s and Frasch mines were opened in Mexico in 1954.

During this time, the recovery of elemental sulfur from natural-gas processing and petroleum refining was beginning to increase. Output grew from a few tons in 1950 to 4×10^6 t by 1980 in the United States, 6×10^6 t in Canada, and 2×10^6 t in France; recovery processes were also used in more than 40 countries.

By the late 1950s, industrial activities in many of the world's industrial nations had slowed and sulfur supplies outstripped demand. Inventories increased and many of the small or marginal Frasch mines were closed and exploration ceased. At the beginning of the 1960s, however, industrial activity began to recover and demand for sulfur increased, inventories were reduced, and the growing shortage stimulated search for new sources of sulfur. Some of the Frasch mines that had been closed were reopened and economically marginal deposits were developed. In addition, the recovery of elemental sulfur from natural gas and petroleum increased as new projects were brought on stream.

During 1960–1969, the apparent consumption of elemental sulfur in the United States rose from 4×10^6 t to 8.2×10^6 t, which implies an annual growth rate of ca 7.4%. Production of brimstone increased from 5.1×10^6 t to 7.9×10^6 t and inventories dropped almost 4×10^4 t to 3.4×10^6 t.

By 1980, U.S. production of elemental sulfur had grown to 10.4×10^6 t and apparent consumption of elemental sulfur was 12.2×10^6 t; the annual growth rate was 3.7% during the 1970s. Production of brimstone rose to a peak of 10.7×10^6 t in 1974, which also was a peak year for Frasch sulfur production at 8×10^6 t. Frasch sulfur output declined to 5.6×10^6 t in 1978 as overall world demand for sulfur declined in the early 1970s.

Inventories of elemental sulfur held by U.S. producers peaked at 5.6×10^6 t in 1977. In Alberta, Canada, stocks held by producers were 20.6×10^6 t at the end of both 1978 and 1979.

In the later part of the 1970s, world demand for sulfur began to rise, primarily to meet increased use of sulfur as sulfuric acid in processing phosphatic fertilizers. At the same time, problems in shipping sulfur from Canada and Poland, which are both major suppliers, to world markets and the war between Iran and Iraq brought about a worldwide shortage of sulfur. Much of this need was met by shipments of sulfur from Frasch producers' inventories in the U.S. Gulf Coast and, as shipping problems were corrected in Canada, by increased shipments from producers' inventories in Alberta. By June 1981, inventories in the United States had declined to 3.1×10^6 t and in Alberta inventories had declined to 17.6×10^6 t.

In addition to domestic production of Frasch and recovered elemental sulfur, U.S. requirements for sulfur are met by by-product sulfuric acid from copper, lead, molybdenum, and zinc smelting operations and imports from Canada and Mexico. In 1980, Frasch sulfur accounted for 54% of domestic production, recovered elemental sulfur accounted for 34%, and by-product sulfuric acid for 9%. The remainder of production was in the form of pyrites, sulfur dioxide, and hydrogen sulfide (Table 4).

The United States has been a net importer of sulfur since 1975. These imports were essentially Frasch sulfur from Mexico and recovered elemental sulfur from Canada and they met almost 19% of the U.S. apparent consumption in 1980. Apparent U.S. consumption of sulfur from 1976 to 1980 is shown in Table 5.

Table 4. U.S. Production of Sulfur in All Forms, 1000 t Sulfur Content[a]

Property	1976	1977	1978	1979	1980
Frasch	6,365	5,915	5,648	6,357	6,390
recovered elemental from natural gas	1,298	1,426	1,753	1,760	1,730
recovered elemental from petroleum[b]	1,890	2,198	2,309	2,310	2,316
by-product sulfuric acid	957	960	1,103	1,167	1,003
pyrites	297	169	301	400	322
other forms[c]	78	59	61	107	78
Total	*10,885*	*10,727*	*11,175*	*12,101*	*11,839*

[a] Ref. 34.

[b] Includes a small quantity from a coking operation and, since 1978, from utility plants.

[c] Hydrogen sulfide and liquid sulfur dioxide.

Table 5. U.S. Apparent Consumption, 1000 t[a]

Sulfur	1977	1978	1979	1980	1981
Frasch					
shipments	6,030	5,736	7,507	7,400	5,910
imports	781	993	1,229	990	856
exports	1,088	827	1,963	1,673	1,392[b]
Net	*5,723*	*5,902*	*6,773*	*6,717*	*5,374*
recovered					
shipments	3,627	4,088	4,108	4,115	4,207
imports	1,228	1,185	1,265	1,533	1,666
exports from the Virgin Islands	109	39	81	109	
Net	*4,746*	*5,234*	*5,292*	*5,539*	*5,873*
pyrites					
shipments	169	301	400	322	307
by-product sulfuric acid					
shipments	960	1,103	1,167	1,003	1,159
other forms					
shipments[c]	59	61	107	78	72
Total all forms	*11,657*	*12,601*	*13,739*	*13,659*	*12,785*

[a] Crude sulfur or sulfur content, ref. 34.
[b] Includes exports from the Virgin Islands.
[c] Hydrogen sulfide and liquid sulfur dioxide.

Specifications and Standards

Specifications and standards for commercial grades of sulfur are given in Table 6.

Analysis

Elemental sulfur in either its ore or its refined state can generally be recognized by its characteristic yellow color or by its generation of sulfur dioxide when burned in air. Its presence in an elemental state or in a compound can be detected by heating the material with sodium carbonate and rubbing the fused product on a wet piece of silver metal. A black discoloration of the silver indicates the presence of sulfur. The test is quite sensitive. Several other methods for detecting small amounts of elemental sulfur have been developed (35).

Quantitatively, sulfur in a free or combined state is generally determined by oxidizing it to a soluble sulfate, by fusion with an alkali carbonate if necessary, and precipitating it as insoluble barium sulfate. Oxidation can be effected with such agents as concentrated or fuming nitric acid, bromine, sodium peroxide, potassium nitrate, or potassium chlorate. Free sulfur is normally determined by solution in carbon disulfide, the latter being distilled from the extract. This method is not useful if the sample contains polymeric sulfur.

Generally, crude sulfur contains small percentages of carbonaceous matter. The amount of this impurity is usually determined by combustion, which requires an exacting technique. The carbonaceous matter is oxidized to carbon dioxide and water; the carbon dioxide is subsequently absorbed (18). Automated, on-stream determination of impurities in molten sulfur has been accomplished by infrared spectrophotometry (36).

Table 6. Specifications and Standards for Commercial Grades of Sulfur

Grade	Specifications and standards; comments	Uses
amorphous sulfur	see insoluble sulfur	
bright sulfur	crude sulfur free of discoloring impurities; bright yellow	
brimstone	see crude sulfur	
broken rock sulfur	refined sulfur broken and sold as a mixture of lumps and fines	see refined sulfur
broken sulfur	solid crude sulfur crushed to <2.38 mm (−8 mesh)	
collodial sulfur	finely divided sulfur suspended in water; can be prepared by physical or chemical means	mainly pharmaceutical
crude sulfur	commercial nomenclature for elemental sulfur; may be bright or dark but is free of arsenic, selenium, and tellurium	
dark sulfur	crude sulfur discolored by minor quantities of hydrocarbons with up to 0.3 wt % carbon content	
dusting sulfur	finely divided crude or refined sulfur prepared for pesticidal uses	
elemental sulfur	processed sulfur in the elemental form produced from native sulfur or combined sulfur sources, generally with a minimum sulfur content of 99.5 wt %	
flour sulfur	crude sulfur ground through 50–74 μm sieves (300–200 mesh), depending on the brand	rubber vulcanization, dyes, gun powder, agricultural dusts, dusting, and wettable sulfurs
flowable sulfur	synonymous with collodial sulfur but used more frequently to describe agricultural sulfurs	mainly pesticidal
flowers of sulfur (sublimed sulfur)	a powdered form of sulfur produced by sublimation; may contain up to 30% of the amorphous allotrope	rubber vulcanization, agricultural dusts, pharmaceuticals, stock feeds
Frasch sulfur	elemental sulfur produced from native sulfur sources by the Frasch mining process	
insoluble sulfur (Crystex)	produced by extraction of flowers of sulfur by CS_2 or by fast quenching of liquid or vaporized sulfur and extraction by CS_2 to remove soluble sulfur allotropes	rubber vulcanization, rubber cements, cutting oils, high pressure lubricants
lac sulfur	precipitated from polysulfide solutions by sulfuric acid; contains up to 45 wt % calcium sulfate	almost entirely pharmaceutical
liquid sulfur	see molten sulfur	
molten sulfur	crude sulfur in the molten phase	
native sulfur	sulfur that occurs in nature in the elemental form	
precipitated sulfur	precipitated from polysulfide solution by hydrochloric acid and washed to remove all calcium chloride	almost entirely pharmaceutical

Table 6 (*continued*)

Grade	Specifications and standards; comments	Uses
prilled sulfur	solid crude sulfur in pellets; produced by cooling molten sulfur with air or water	
recovered sulfur	elemental sulfur produced from combined sulfur sources by any method	
refined sulfur	elemental sulfur produced by distilling crude sulfur; purity not less than 99.8%	when sulfur is burned in small quantities, refined sulfur provides sulfur dioxide for fumigation, sugar and starch refining, preserving, and bleaching
roll sulfur	refined sulfur cast into convenient sizes	see refined sulfur
rubbermaker's sulfur	ground sulfur of various fineness with special specifications for low acid, ash, and moisture contents	
run-of-mine sulfur	mined by the hot-water process; solid shipments may contain 50 wt % fines, with lump diameters up to 20 cm or more	
screened commercial sulfur	run-of-mine sulfur with particle size determined by screening (generally 9.5 and 1.68 mm (2 and 12 mesh)), plus associated fines	
slated sulfur	solid crude sulfur in the form of slatelike lumps; produced by allowing molten sulfur to solidify in a moving belt	
specialty sulfurs	prepared or refined grades of elemental sulfur that include amorphous, collodial, flower, precipitated, wettable, or paste sulfur	
wettable sulfur	powdered sulfur that has been treated so that it can be dispersed easily in water	

The moisture content of crude sulfur is determined by the differential weight of a known sample before and after drying at ca 110°C. Acid content is determined by volumetric titration with a standard base. Nonvolatile impurities or ash are determined by burning the sulfur from a known sample and igniting the residue to remove the residual carbon and other volatiles.

Health and Safety Factors

Solid Sulfur. Solid elemental sulfur is virtually nontoxic and can be taken internally without injury. Although the inhalation of sulfur dust is not known to cause systemic poisoning, it can irritate the eyes and mucous membranes of the respiratory passages. Certain predisposed individuals may develop an eczematous reaction to sulfur dust. When sulfur burns in air, it forms sulfur dioxide, an irritant gas which readily affects the eyes and respiratory tract. A concentration of 8–20 ppm causes coughing and eye irritation. The generally accepted maximum concentration that can be tolerated for several hours without causing physiological symptoms is 10 ppmv. Exposure to 50–100 ppm can be tolerated for 0.5–1.0 h, and exposure to 500 ppm or more is instantly dangerous.

Although sulfur is nontoxic, dust-tight respirators and dust-tight goggles are recommended for personnel who handle solid sulfur. Because of the low ignition temperature of sulfur and its tendency to develop static charges, sulfur presents a fire and explosion hazard. Additional information on safe practices for handling, storing, and processing of solid sulfur is given in refs. 28, 30–31.

Liquid Sulfur. Sulfur becomes a liquid at ca 100°C. The primary hazards when handling liquid sulfur include its relatively low ignition point, the possible presence of hydrogen sulfide, and the fact that it is a hot liquid capable of producing severe burns. Hydrogen sulfide is not only toxic, but it also has a low explosive limit. The maximum acceptable concentration of hydrogen sulfide averaged over a normal working day and recommended by the ACGIH is 10 ppmv. The OSHA standard is a ceiling concentration of 20 ppm and a peak concentration of 50 ppm for 10 min. Exposure to 100 ppm for 1 h produces symptoms of local irritation of the eyes and respiratory tract. At 500 ppm, inhalation for 15–30 min causes severe irritation of the eyes and respiratory tract with the risk of pneumonia or serious injury to the lungs. Hydrogen sulfide in concentrations of 1000 ppm or more causes immediate unconciousness and results in death unless artificial respiration is applied immediately (see also Sulfur compounds).

At room temperature, the lower explosive limit for hydrogen sulfide in air is 4.3 vol %. At the normal storage temperature for liquid sulfur (132°C), the lower explosive limit for hydrogen sulfide in air is 3.4 vol %. Sulfur storage tanks should be equipped with submerged fill lines to minimize agitation of the sulfur and the release of dissolved hydrogen sulfide. Sulfur storage tanks should also have adequate roof vents to provide natural-draft ventilation of hydrogen sulfide gas, which may form from liquid-sulfur. All sources of ignition should be excluded from the immediate vicinity of liquid-sulfur storage tanks. Personnel engaged in liquid-sulfur operations should be completely protected by clothing including safety boots, safety glasses with side shields, and heat-resistant gloves. An additional full face shield should be worn when disconnecting piping used for transporting liquid sulfur.

Because it is a hot material, liquid sulfur can cause severe burns when it contacts the skin. First-aid treatment should include immediate cooling of the affected area with cool running water, immersion in water, or application of cold compresses. The sulfur crust should not be removed because it serves as a sterile dressing. Further treatment should be administered by a physician. Additional information on safety practices for the handling and storage of liquid sulfur is given in refs. 34, 37–40.

Uses

Sulfur is unusual as compared with most large mineral commodities in that the largest portion of it is used as a chemical reagent rather than as a component of a finished product. Its predominant use as a process chemical generally requires that it first be converted to an intermediate chemical product prior to its initial use in industry. In most of the ensuing chemical reactions between these intermediate products and other minerals and chemicals, the sulfur values are not retained in the final product. Rather, they are more often discarded as the component of a waste product.

Sulfuric acid (qv) is the most important of these intermediate products. More than 85% of sulfur consumed in the world is either converted to sulfuric acid or produced directly as such (see Sulfuric acid and sulfur trioxide). Worldwide, well over half of the sulfuric acid is used in the manufacture of phosphatic fertilizers and ammonium sulfate for fertilizers. The sulfur source may be voluntary elemental, ie, from the Frasch process; recovered elemental from natural gas or petroleum; or sulfur dioxide from smelter operations.

In recent years, the largest demand for sulfur has been for agricultural purposes, the principal use being for processing of phosphatic fertilizers (see Table 7). Other uses have been in petroleum refining; leaching of copper and uranium ores; production of organic and inorganic chemicals, paints and pigments, pulp and paper, and synthetic materials; and in many industrial uses.

Agriculture. Use of sulfur in agriculture accounts for more than half of the sulfur demand. The principal individual requirement is for the manufacture of phosphatic fertilizers, in which sulfuric acid is an essential intermediate. The wet process, ie, the phosphoric acid process, is the most important of those used for this purpose. Phosphate rock is digested with sulfuric acid to produce phosphoric acid and an impure waste-gypsum product. The phosphoric acid is an intermediate product used for the production of a wide range of high grade fertilizers, eg, diammonium phosphate and triple superphosphate. Phosphate fertilizers can be and have been produced in limited quantities by the electric-furnace elemental-phosphorus method without acid or by the use of nitric or hydrochloric acid as a substitute for sulfuric acid. However, the economics of such processes are unfavorable except in certain limited areas of the world.

Another fertilizer use is in the production of ammonium sulfate by reaction of sulfuric acid with ammonia. Some of this is deliberate, but the greater part is by-product material resulting from coke-oven operations; synthetic-fiber manufacture, ie, of caprolactam; and to a lesser extent, utility stack-gas scrubbing (see Ammonium compounds).

In addition, a small amount of sulfur in the elemental form as ammonium thio-

Table 7. U.S. Demand for Sulfur, 1000 t

Property	1976	1977	1978	1979	1980
agriculture	6,164	7,225	7,713	8,142	9,133
soaps, detergents, and water treatment	197	237	301	323	228
plastics and synthetic products	314	455	353	328	330
paper products	276	345	283	372	348
paints and pigments	229	324	330	343	325
metal mining and processing	656	701	773	869	687
petroleum refining	686	879	876	875	1,023
iron and steel production	97	130	305	288	103
storage batteries	33	44	58	51	34
other	2,299	1,317	1,607	2,148	1,424
Total	*10,951*	*11,657*	*12,599*	*13,739*	*13,635*

sulfate and polysulfide or as gypsum is used as a soil amendment and plant nutrient. The latter use of sulfur is growing in importance because the higher grade phosphatic fertilizers being produced by the wet phosphoric acid process do not have the required sulfur content of the older, lower grade fertilizer processes.

Plastics and Other Synthetic Products. Sulfur is used in the production of a wide range of synthetics including cellulose acetate, cellophane, rayon, viscose products, fibers, and textiles. These may account for 2% of sulfur demand in developed countries. The sulfur intermediates involved in their manufacture are equally divided between carbon disulfide and sulfuric acid.

Paper Products. Paper (qv) products account for ca 3% of sulfur demand, with the largest single segment of demand being in the manufacture of wood pulp by the sulfite process (see Pulp). In this process, the main sulfur intermediate is sulfur dioxide, which is generally produced at the plant site by burning elemental sulfur; although some sulfur dioxide is produced as a by-product at smelter operations, purified and liquefied, and shipped to the pulp mills. The sulfur dioxide is converted to sulfurous acid, and the salt of this acid is a principal component of the cooking liquor for the sulfite process.

Paints. Paints account for perhaps 3% of sulfur consumption. The main sulfur use is for the production of titanium dioxide pigment by the sulfate process. Sulfuric acid reacts with ilmenite or titanium slag and the sulfur remains as a ferrous sulfate waste product. Difficulties with this process have led to the development of the chloride process (see Pigments, inorganic; Titanium compounds).

Nonferrous Metal Production. Nonferrous metal production, which includes the leaching of copper and uranium ores with sulfuric acid, accounts for ca 6% of U.S. sulfur consumption and probably about the same in other developed countries. In the case of copper, sulfuric acid is used for the extraction of the metal from deposits, mine dumps, and wastes, in which the copper contents are too low to justify concentration by conventional flotation techniques or the recovery of copper from ores containing copper carbonate and silicate minerals which cannot readily be treated by flotation processes. The sulfuric acid required for copper leaching is usually the by-product acid produced by copper smelters (see Extractive metallurgy).

Sulfuric acid is the most commonly used reagent for the recovery of uranium from ores, and vanadium is often recovered as a coproduct. The sulfuric acid used is either the by-product sulfuric acid produced at smelters or sulfuric acid produced from elemental sulfur.

Petroleum Refining. Petroleum refining includes not only petroleum refining as such but associated chemical processes where process streams may serve both the refinery and the chemical complex. Sulfuric acid requirements for these processes account for ca 6% of sulfur demand. About 60% of the sulfuric acid used in petroleum refining is returned as spent acid for reclaiming; therefore, the demand for new sulfuric acid is ca 3% of the total sulfur demand. The principal use for sulfuric acid is as a catalyst for alkylation, a process by which liquid high-octane gasoline components with very good stability are produced from a combination of gaseous streams. Sulfuric acid and hydrofluoric acid are competing catalysts in this process. Sulfuric acid for refinery processes is manufactured from recovered sulfur produced at the refinery and from contaminated acid (acid sludge) returned to the acid plants for reconstitution (see Petroleum).

Iron and Steel Production. Consumption of sulfur in the form of sulfuric acid accounts for ca 2% of sulfur demand. The sulfuric acid is used as a pickling agent to remove mill scale, rust, dirt, and grease from the surface of steel products prior to further processing (see Metal surface treatments, cleaning, pickling, and related processes). The sulfuric acid pickling process faces increasing competition from hydrochloric acid pickling, largely because of the problem of disposing of the ferrous sulfate waste product. Although there are advantages and disadvantages in the use of hydrochloric acid, it appears that it will largely replace sulfuric acid over the long term. There are no well-defined sources of sulfuric acid for steel pickling, since it is generally obtained from merchant sulfuric acid plants that use the cheapest form of sulfur available in the area where it is produced (see Iron; Steel).

Plant Nutrient. Sulfur, like nitrogen, phosphorus, potassium, and several other elements, is essential for the growth of crops. It performs numerous functions in plant metabolism, one of the most important being its role in the synthesis of certain essential sulfur-containing amino acids and, hence, in the formation of protein. It also plays a role in the formation of vitamins, the fixation of nitrogen by leguminous plants, and other essential functions in plant metabolism (see Mineral nutrients).

Sulfur deficiencies appear to be most widespread in Canada, the United States, South and Central America, Africa, certain parts of the Far East, Australia, New Zealand, and in the South Pacific. Almost all of the important crops require sulfur. Dramatic increases in crop yields result from the application of sulfur-containing fertilizers on soils deficient in this element (41–43).

Soil and Water Treatment. Agricultural soils and waters in the irrigated, arid regions of the world frequently benefit by the application of sulfur and its compounds. These benefits result from the improvements in water penetration and movement and from the increased availability of phosphorus and certain micronutrients which are otherwise rendered unavailable because of the high soil pH. There are extensive areas of such soils in the western provinces of Canada and in many areas west of the Mississippi River in the United States. Throughout the remainder of the world, large tracts of arid, irrigated soils exist, notably in the Middle East, northern Africa, Australia, and many places in both Eastern and Western Europe (44–45).

Animal Nutrition. Sulfur in the diets of ruminant animals is beneficial to the animals' growth (see Pet and livestock feeds). Sulfur increases feed intake, cellulose and dry matter digestion, and the synthesis of microbial protein. This results in increased meat, milk, and wool production (46). The special uses for sulfur in agriculture demonstrate a significant and continuing need for increased use of sulfur. This is estimated to total ca 5×10^6 t/yr for Canada and the United States (47).

Highway Construction. During the past century, many papers describing the preparation of sulfur–asphalt materials have been published. In the 1930s, asphalt (qv) was easily available and priced lower than sulfur. In 1969, The Sulphur Institute concluded an agreement with Iowa State University to reopen the investigation of sulfur–asphalt paving materials and a comprehensive report on the study was published in 1971 (48–49). Four different types of sulfur paving materials are described below.

Replacement for Asphalt. Sulfur-extended asphalt (SEA) binders are formulated by replacing some of the asphalt in conventional binders with sulfur. Binders have been used with sulfur–asphalt weight ratios as high as 50:50, but most binders contain ca 30 wt % sulfur. Greater latitude in design is possible with SEA paving materials, which are three-component systems, whereas conventional asphalt paving materials are two-component systems. Introduction of sulfur can provide some substantial benefits. At temperatures above 130°C, SEA binders have lower viscosities than conventional asphalt. The lower viscosity enables the plant to produce and compact the mix at lower temperatures than with conventional mix. SEA materials have flatter viscosity curves than asphalts and are more viscous than asphalt at summer road temperatures. This causes the pavement to be more resistant to rutting and deformation during hot weather. SEA binders have increased stability as demonstrated by the Marshall and Hveem design methods (50), and sulfur-extended asphalt is very resistant to gasoline, diesel fuel, and other solvents.

During 1974–1982, ca 150 sulfur–asphalt roads were constructed in the world, half of which were in the United States. All U.S. SEA experimental sections that were designed and constructed according to standard practices and with standard materials are performing as well as the control AC (asphalt concrete) sections in these experimental projects (51).

The physical properties of elemental sulfur can be modified by its reaction with various organic and inorganic compounds. Many of the resulting sulfur products tend to have properties similar to paving asphalt (52–54). During 1980–1981, small road trials were organized in seven states. However, it is too early to forecast the performance of these materials.

Sand–Asphalt–Sulfur (SAS). Sulfur is being utilized as a means of upgrading poorly graded aggregates. This technology was developed by Shell Canada Resources, Ltd., under the tradename Thermopave. Sand–asphalt–sulfur does not provide an appreciable reduction of asphalt content of the paving material, but the utilization of sand rather than crushed stone is economically important. Thermopave is a high quality paving material (55–58). After being tested in various provinces of Canada, SAS roads were constructed in the United States.

Recycling of Commercial Asphalt Materials. For many years, repair and maintenance of roads and highways has consisted of putting a layer of asphalt concrete over the existing pavement. In many areas, the road level is at the maximum height for proper drainage and to satisfy safety considerations. As a result, more and more roads must have their surfaces removed before additional paving can take place. The increasing cost of both asphalt and aggregate has presented a greater incentive to reuse rather than to discard the removed material. Many existing roads fail because the asphalt becomes stiff and brittle. If the materials are too stiff, additives that lower the viscosity must be used. The feasibility of using sulfur to soften or reduce the viscosity of the oxidized binder in recycled pavements was successfully demonstrated by the U.S. Bureau of Mines and others (59–60).

Sulfur Concrete. Sulfur concrete is a mixture of sulfur with fine and coarse aggregates. These materials are heated to ca 140°C, placed, and then allowed to cool and solidify into a rigid, concretelike material. Concretes prepared with sulfur as the binder have mechanical properties comparable to portland cement concretes. Methods of preparing sulfur concretes that are not subject to rapid deterioration have been developed recently. A modified sulfur binder, ie, sulfur containing 3–5 wt % organic material, has been developed for aggregates and the resultant sulfur concrete products maintain substantially all their initial strength, even after 300 freeze–thaw cycles.

Sulfur concretes are used in many specialty areas where portland concretes are not completely satisfactory. Since sulfur concretes can be formulated to be resistant to acid and salt corrosion, they are being tested for tanks, electrolytic cells, thickeners, bridged decking, industrial flooring, and pipe (61) (see Cement).

Coatings. Sulfur coating formulations have been used on a concrete warehouse in which potash is stored (62). Adhesion to this nonporous concrete wall has been related to the amount and type of plasticizer. Glass and other fibers, fillers, and fine aggregates affect strength and abrasion resistance but not adhesion. The use of a sulfur spray coating to stabilize a hilltop which was eroding and sliding onto a road and work area below has been described (63). Approximately 8100 m^2 on top of the hill was spray-coated to stabilize the hilltop. The coating has prevented further erosion.

Mortar Substitute. In the early 1960s, a researcher conceived the idea of using sulfur coatings to replace mortar in construction (64). Bricks, blocks, and similar materials are stacked one upon the other and, since the blocks are dry, they can be moved and adjusted until the desired wall configuration is achieved. The blocks are bonded by applying the coating to the wall surfaces only. The coating mixture is applied at 120–150°C and contains 90–95 wt % sulfur with small percentages of fiber and additives. The coating solidifies almost immediately to form a hard, impervious surface. In 1963, the first building of this type was constructed at the Southwest Research Institute. The building has been in continuous use and the coating has not presented any problems.

In 1972, the Agency for International Development (AID) cooperated with The Sulphur Institute to prepare a program to evaluate the feasibility of using the sulfur surface-bond technique in developing countries. Afterwards AID awarded a contract to Southwest Research Institute to construct houses in South America and Africa using this technique. Under the contract, ca 60 houses have been completed (65).

Foams. Sulfur can be foamed into a lightweight insulation which compares favorably with many organic foams, and other insulating materials used in construction. It has been evaluated as thermal insulation for highways and other applications to prevent frost damage (66) (see Foamed plastics; Insulation, thermal).

Sulfur Impregnation. Impregnation of bonded, abrasive grinding wheels with sulfur improves their strength and provides lubricating and coolant qualities during grinding operations. Sulfur-impregnated wheels are well-suited for grinding tough materials, eg, stainless steel, brass, bronze, and nickel. The impregnated wheels cut faster and prevent the welding of metal chips. In more difficult jobs, eg, gear grinding and surface grinding, the sulfur-impregnated wheels have four to eight times the life of nonimpregnated wheels (67) (see Abrasives).

Sulfur can be used to impregnate ceramic tile. Sulfur impregnation reduces water absorption and makes the tile frost-resistant when used on exterior surfaces, eg, floors, roofs, or entrances to buildings (68). Investigators at the Institute of Paper Chemistry described the production of sulfur-impregnated paperboard for use in temporary housing and techniques for manufacturing sulfurized, corrugated container board suitable for self-sustaining shipping containers (69–70). Sulfurized container board is shock-resistant and maintains its strength under hot, humid conditions which cause normal boxboard to lose over half of its strength (70) (see Paper).

Other. Other uses include intermediate chemical products. Overall, they account for 15–20% of sulfur consumption, largely in the form of sulfuric acid but also some elemental sulfur which is used directly, as in rubber vulcanization. These miscellaneous uses, especially those involving sulfuric acid, are intimately associated with practically all elements of the industrial and chemical complexes in developed and developing countries.

BIBLIOGRAPHY

"Sulfur" in *ECT* 1st ed., Vol. 13, pp. 358–373, by F. L. Jackson, E. C. Thaete, Jr., L. B. Gittinger, Jr., L. A. Nelson, Jr., and H. Blanchet, Freeport Sulphur Company; "Sulfur" in *ECT* 2nd ed., Vol. 19, pp. 337–366, by P. T. Comiskey, L. B. Gitlinger, Jr., L. F. Good, F. L. Jackson, L. A. Nelson, Jr., T. K. Wiewiorowski, Freeport Sulphur Co., and M. D. Barnes, Covenant College.

1. P. W. Shenk and V. Thummler, *Z. Elektrochem.* **63,** 1002 (1959).
2. B. Meyer, *Chem. Rev.* **76,** 367 (1976).
3. T. K. Wiewiorowski and F. J. Touro, *J. Phys. Chem.* **70,** 3528 (1966).
4. M. Schmidt, *Angew. Chem.* **12,** 445 (1973).
5. B. Meyer, *Sulfur, Energy, and Environment*, Elsevier Publishing Co., New York, 1977.
6. Von R. Steudel, *Z. Anorg. Allg. Chem.* **478,** 139 (1981).
7. Von R. Steudel and H. J. Mäusle, *Z. Anorg. Allg. Chem.* **478,** 156 (1981).
8. H. J. Mäusle and Von R. Steudel, *Z. Anorg. Allg. Chem.* **478,** 177 (1981).
9. G. Preuner and W. Schupp, *Z. Phys. Chem.* (*Leipzig*) **68,** 129 (1909).
10. H. Braune, S. Peter, and V. Neveling, *Z. Naturforsch Teil A* **6,** 32 (1951).
11. J. Berkowitz and J. R. Marquart, *J. Chem. Phys.* **39,** 275 (1963).
12. W. A. West and A. W. C. Menzies, *J. Phys. Chem.* **33,** 1880 (1929).
13. G. Fouritier, *Compt. Rend.* **218,** 194 (1944).
14. K. Neumann, *Z. Phys. Chem.* **A171,** 416 (1934).
15. R. Farelli, *J. Am. Chem. Soc.* **72,** 4016 (1950).
16. E. W. Washburn, ed., *International Tables of the Numerical Data of Physics, Chemistry and Technology* (*ICT*), Vols. 1–7, McGraw-Hill Book Co., New York, 1926–1930.
17. P. C. L. Thorne and A. M. Ward, *Inorganic Chemistry*, 3rd ed., Nordeman Publishing Co., Inc., New York, 1939, p. 6.
18. W. N. Tuller, ed., *The Sulphur Data Book*, McGraw-Hill Book Co., New York, 1954.
19. J. W. Mellor, *A Comprehensive Treatise on Inorganic and Theoretical Chemistry*, Vol. 10, John Wiley & Sons, Inc., New York, 1930, p. 27.
20. A. Senning, *Dic Schwefelurg Organischer Verbindungen*, Chemisches Institut der Universitat Aarhus, Denmark.
21. D. B. Mason, *Ind. Eng. Chem.* **30,** 740 (1938).
22. R. B. Thompson and D. MacAskill, *Chem. Eng. Prog.* **51,** 3691 (1955).
23. T. P. Forbath, *Trans. AIME* **196,** 811 (1953)
24. T. Kaier, *Eng. Min. J.* **155**(7), 88 (1954).
25. H. R. Potts and E. G. Lawford, *Trans. AIME* **58,** 1 (1949).
26. *Eng. Min. J.* **155,** 9, 142 (1954).
27. G. O. Argall, Jr., *World Min.*, 18 (March 1967).
28. *Chem. Eng.*, 31 (Dec. 5, 1980).
29. *Sulphur* **80** (1969).
30. *Sulphur* **147** (1980).
31. *Chem. Eng. News*, 21 (March 20, 1967).
32. E. P. Johnson, *Chem. Eng.*, 34 (Nov. 2, 1981).
33. *Chem. Eng. News*, 20 (May 24, 1982).
34. *U.S. Bureau of Mines Yearbook*, 1980.
35. F. Feigl, *Spot Tests in Inorganic Analysis*, Elsevier Publishing Co., New York, 1958, pp. 372–375.
36. R. F. Matson, T. K. Wiewiorowski, and D. E. Schof, Jr., *Chem. Eng. Prog.* **61,** 9, 67 (1965).
37. *Handling Liquid Sulfur*, Data Sheet 592, National Safety Council.
38. *Handling and Storage of Solid Sulfur*, Data Sheet 612, National Safety Council.
39. *Prevention of Sulfur Fires and Explosions*, NFPA No. 655, National Fire Protection Association.
40. *Properties and Essential Information for Safe Handling and Use of Sulfur*, Chemical Safety Data Sheet SD-74, Manufacturing Chemists Association (CMA).
41. J. D. Beaton, S. L. Tisdale, and J. Platou, *Crop Response to Sulphur in North America*, Technical Bulletin 18, The Sulphur Institute, Washington, D.C., 1971.
42. P. A. Gallagher, *Proc. R. Dublin Soc. Ser. B* **2**(20) (1969).
43. M. Braud, *Sulphur Inst. J.* **5**(4), 3 (1969–1970).
44. S. L. Tisdale, *Sulphur Inst. J.* **6**(1), 2 (1970).

45. L. K. Stromberg and S. L. Tisdale, *Treating Irrigated Arid-Land Soils with Acid-Forming Sulphur Compounds*, Technical Bulletin 24, The Sulphur Institute, Washington, D.C., 1979.
46. S. L. Tisdale, *Sulphur in Forage Quality and Ruminant Nutrition*, Technical Bulletin 22, The Sulphur Institute, Washington, D.C., 1977.
47. J. D. Beaton and S. L. Tisdale, *Potential Plant Nutrient Consumption in North America*, Technical Bulletin 16, The Sulphur Institute, Washington, D.C., 1969.
48. H. L. Fike, *Some Potential Applications of Sulphur, Sulphur Research Trends, Advances in Chemistry Series 110*, American Chemical Society, Washington, D.C., 1972.
49. D. Y. Lee, *Modification of Asphalt and Asphalt Paving Mixtures by Sulphur Additives*, Interim Report, Engineering Research Institute, Iowa State University, Ames, Iowa, 1971.
50. *Mix Design Methods for Asphalt Concrete (MS-2)*, The Asphalt Institute, March 1974.
51. S. P. LaHue and F. B. Botelho, *Civ. Eng. (N.Y.)*, 57 (May 1981).
52. A. C. Ludwig, B. B. Gerhardt, and J. M. Dale, *Material and Techniques for Improving the Engineering Properties of Sulphur*, FHWA/RD-80-023, U.S. GPO 1980-625-892/2032, 1980.
53. H. J. Lentz and E. A. Harrigan, *Laboratory Evaluation of Sulphlex-233: Binder Properties and Mix Design*, FHWA/RD-80/146, U.S. GPO 1981-0-725-620/1163, 1980.
54. W. C. McBee, T. A. Sullivan, and B. W. Jong, *Modified-Sulfur Cements for Use in Concretes, Flexible Pavings, Coatings, and Grouts*, RI 8545, U.S. Bureau of Mines, Washington, D.C., 1981.
55. R. A. Burgess and I. Deme in James R. West, ed., *Sulphur in Asphalt Paving Mixes, New Uses of Sulphur, Advances in Chemistry Series 140*, American Chemical Society, Washington, D.C., 1975.
56. I. Deme, "Basic Properties of Sand–Asphalt–Sulphur Mixes," *International Road Fed. World Mtg.*, Munich, Oct. 7, 1973.
57. T. A. Sullivan, W. C. McBee, and K. L. Rasmussen, *Studies of Sand–Sulphur–Asphalt Paving Materials*, RI 8087, U.S. Bureau of Mines, Washington, D.C., 1975.
58. D. Saylak and co-workers, *Beneficial Uses of Sulfur in Sulfur–Asphalt Pavements, Advances in Chemistry Series 140*, American Chemical Society, Washington, D.C., 1975.
59. W. C. McBee, T. A. Sullivan, and D. Saylak, *Recycling Old Asphalt Pavements with Sulfur*, ASTM STP622, American Society for Testing Materials, Philadelphia, Pa., 1978.
60. D. L. Strand, *Rural Urban Roads* **18**(7), 42 (1980).
61. J. S. Platou, ed., *Sulphur Research and Development*, The Sulphur Institute, Washington, D.C., 1981.
62. H. L. Fike, *Proceedings of the Symposium on New Uses for Sulphur and Pyrites*, Madrid, Spain, The Sulphur Institute, Washington, D.C., 1976, pp. 215–230.
63. J. E. Paulson, J. Simic, R. W. Campbell, and J. W. Ankers, *Sulfur Composites as Protective Coatings and Construction Materials, Advances in Chemistry Series 165*, American Chemical Society, Washington, D.C., 1978.
64. U.S. Pat. 3,306,000 (1967), M. D. Barnes (to Research Corp.).
65. H. L. Fike and M. Conitz, *Surface Bond Construction Materials*, Envo Publishing Co., Lehigh Valley, Pa., 1976.
66. G. L. Woo, E. P. Antoniades, C. B. Olson, J. W. Ankers, and R. W. Campbell, *Sulfur Foam and Commercial-Scale Field Application Equipment, Advances in Chemistry Series 165*, American Chemical Society, Washington, D.C., 1978, pp. 227–240.
67. U.S. Pat. 3,341,355 (Sept. 12, 1967), T. P. Gallager (to Fuller-Merriam Co.).
68. U.S. Pat. 3,208,190 (Sept. 25, 1965), M. J. Kakos and J. V. Fitzgerald (to Tile Council of America).
69. J. A. Van der Akker and W. A. Wink, *Pap. Ind. Pap. World* **30**(2), 231 (1948).
70. U.S. Pat. 2,568,349 (Sept. 18, 1951), R. C. McKee.

D. W. BIXBY
H. L. FIKE
The Sulphur Institute

J. E. SHELTON
U.S. Bureau of Mines

T. K. WIEWIOROWSKI
Freeport Minerals Co.

SULFUR COMPOUNDS

Carbon Sulfides

The only commercial carbon sulfide is carbon disulfide (qv) [*75-15-0*], CS_2. There are several unstable carbon sulfides.

Carbon subsulfide [*12976-57-2*], C_3S_2, is a red liquid (mp −0.5°C, bp 60–70°C at 1.6 kPa (12 mm Hg)) and is produced by the action of an electric arc on carbon disulfide (1–3). The structure has been shown to be S=C=C=C=S on the basis of its reactions to form malonic acid derivatives and on the basis of physical measurements. It is unstable and decomposes in a few weeks at room temperature; it decomposes explosively when heated rapidly at 100–120°C with formation of a black polymeric substance $(C_3S_2)_x$ (4–5). Dilute solutions in CS_2 are fairly stable, but photochemical polymerization to $(C_3S_2)_x$ occurs.

Carbon monosulfide [*2944-05-0*], CS, is an unstable gas produced by the decomposition of carbon disulfide at low pressure in a silent electrical discharge or photolytically (1–2). It decomposes with a half-life of seconds or minutes to a black solid of uncertain composition (1–2). The monosulfide can be stabilized in a CS_2 matrix at −196°C, and many stable coordination complexes of CS with metals have been prepared by indirect means (6).

Incompletely-Characterized Carbon Sulfides. A poorly characterized black solid, known as carsul, occurs as a residue in sulfur distillation or as a precipitate in molten Frasch sulfur (7–8). Although this material may approach the composition of a carbon sulfide, it is more likely also to contain some chemically bound hydrogen and possibly other elements. Carbon–sulfur surface compounds of the formula C_xS, where x is greater than 4, are prepared by reaction of carbonaceous materials with a sulfur-containing gas, eg, sulfur dioxide at 400–700°C. They are useful as cathodes in electrical cells (9). Other poorly characterized, thermally stable black solids with sulfur contents ranging as high as 46 wt % have been reported as products of high temperature reactions of sulfur with charcoal, cellulose, and various carbon compounds. These have been considered as carbon sulfides by some investigators, but they may contain other elements as well (10–11).

Carbonyl Sulfide. ***Physical Properties.*** Carbonyl sulfide [*463-58-1*] (carbon oxysulfide), COS, is a colorless gas that is odorless when pure; however, it has been described as having a foul odor. Physical constants and thermodynamic properties are listed in Table 1 (12–13). The vapor pressure has been fitted to an equation, and a detailed study has been made of the phase equilibria of the carbonyl sulfide–propane system, which is important in the purification of propane fuel (14–15). Carbonyl sulfide can be adsorbed on molecular sieves as a means for removal from propane (16). This approach has been compared to the use of various solvents and reagents (17).

Chemical Properties. Reviews of carbonyl sulfide chemistry are given in refs. 13, 18–19. Carbonyl sulfide is a stable compound and can be stored under pressure in steel cylinders as compressed gas in equilibrium with liquid. At ca 600°C, carbonyl sulfide disproportionates to carbon dioxide and carbon disulfide; at ca 900°C, it dissociates to carbon monoxide and sulfur. It burns with a blue flame to carbon dioxide and sulfur dioxide. Carbonyl sulfide reacts only slowly with water to form carbon dioxide and hydrogen sulfide. Much technology has been developed for hydrolysis of carbonyl

Table 1. Physical and Thermodynamic Properties of Carbonyl Sulfide

Property	Value	Ref.
mol wt	60.074	12
mp, °C	−138.8	13
bp, °C	−50.2	13
ΔH, fusion at 134.3 K, kJ/mol[a]	4.727	13
ΔH, vaporization at 222.87 K, kJ/mol[a]	18.57	13
density at 220 K, 101.3 kPa (= 1 atm), g/cm^3	1.19	13
sp gr gas at 298 K (air = 1)	2.10	13
critical temperature, °C	105	13
critical pressure, kPa[b]	5946	13
critical volume, cm^3/mol	138	13
triple point, K	134.3	13
$\Delta G°$ formation at 25°C, kJ/mol[a]	−169.3	2
$\Delta H°$ formation at 25°C, kJ/mol[a]	−142.1	2
$S°$ formation at 25°C, J/(mol·K)[a]	231.5	2
C_p° at 25°C, J/(mol·K)[a]	41.5	2
autoignition temperature in air, °C	ca 250	13
flammability limits in H_2O-saturated air at 17.9°C, vol %		
upper limit	9.6	
lower limit	33.2	13
solubility in water at 101.3 kPa (= 1 atm), vol %		
0°C	0.356	13
10°C	0.224	
20°C	0.149	

[a] To convert J to cal, divide by 4.184.
[b] To convert kPa to psi, multiply by 0.145.

sulfide in gas streams to permit the more ready removal of the sulfur content as hydrogen sulfide. The commercial processes are reviewed in detail in ref. 20. Typical catalysts for the gas-phase hydrolysis are based on copper sulfide, chromium oxide, chromia–alumina, and platinum.

Alternative means for removal of carbonyl sulfide for gas streams involve hydrogenation. For example, the Beavon process for removal of sulfur compounds remaining in Claus unit tail gases involves hydrolysis and hydrogenation over cobalt molybdate catalyst resulting in the conversion of carbonyl sulfide, carbon disulfide, and other sulfur compounds to hydrogen sulfide (20).

Carbonyl sulfide reacts slowly with aqueous alkali metal hydroxides, which can therefore be used to free carbonyl sulfide from acidic gases. The half-life at pH 12 (0.01 *N* KOH) is 3 min at 22°C. The product initially formed from carbonyl sulfide and alkali is the thiocarbonate [*534-18-9*] which then breaks down to carbonate and sulfide:

$$\mathrm{COS + 2\ NaOH \longrightarrow NaS\overset{\overset{\displaystyle O}{\|}}{C}ONa + H_2O}$$

$$\mathrm{NaS\overset{\overset{\displaystyle O}{\|}}{C}ONa + 2\ NaOH \longrightarrow Na_2CO_3 + Na_2S + H_2O}$$

To effectively remove carbonyl sulfide from a gas stream, special alkaline scrubbing liquors are used. These contain sodium aluminate or sodium plumbite, or they are made of alkalies with a hydrolysis catalyst based on Zn, Fe, Ni, or Cu. Diethanolamine, diglycolamine, or other alkanolamines (qv) mixed with water remove carbonyl sulfide from sour, ie, acid-gas-containing, gas streams (20–21) (see Carbon dioxide).

Carbonyl sulfide reacts with chlorine forming phosgene and sulfur dichloride [*10545-99-0*] and with ammonia forming urea and ammonium sulfide [*12135-76-1*]. Carbonyl sulfide attacks metals, eg, copper, in the presence of moisture and is thought to be involved in atmospheric sulfur corrosion (22–23). Its presence in propane gas at levels above a few ppm may cause the gas to fail the copper-corrosion test.

It reacts with alcohols in the presence of base to form monothiocarbonate esters:

$$ROH + COS + NaOH \longrightarrow RO\overset{\overset{\displaystyle O}{\|}}{C}SNa + H_2O$$

Carbonyl sulfide reacts with amines to form thiocarbamates:

$$2\ R_2NH + COS \longrightarrow [R_2N\overset{\overset{\displaystyle O}{\|}}{C}S]^-[NH_2R_2]^+$$

With amines such as those used in gas sweetening, it reacts forming mainly ureas:

$$COS + 2\,HOCH_2CH_2OCH_2CH_2NH_2 \rightarrow (HOCH_2CH_2OCH_2CH_2NH)_2CO + H_2S$$

The photolysis of carbonyl sulfide is a laboratory method for the production of monoatomic sulfur, a short-lived species (24).

Occurrence and Preparation. Carbonyl sulfide is formed by many high temperature reactions of carbon compounds with donors of oxygen and sulfur. A principal route is the following reaction:

$$CO + S \rightleftharpoons COS$$

This equilibrium favors COS up to ca 500°C. At higher temperatures, COS dissociates increasingly, eg, to 64% at 900°C. The reaction may be run at 65–200°C to produce carbonyl sulfide if an alkaline catalyst is used (25). A Rhone-Poulenc patent describes the manufacture of carbonyl sulfide by the reaction of methanol with sulfur at 500–800°C (26).

Other important reactions yielding carbonyl sulfide are the following:

Reaction	*Remarks*
$CO_2 + H_2S \rightarrow COS + H_2O$	in Claus furnace and over various catalysts (27)
$CO_2 + CS_2 \rightleftharpoons 2\,COS$	at ca 500°C, catalyzed by silica
$CS_2 + SO_3 \rightarrow COS + SO_2 + S$ [*7446-11-9*] [*7446-69-5*]	above 400°C as gases, at low temperatures as liquids (28)
$HSCN + H_2O \rightarrow COS + NH_3$ [*13129-90-6*]	in H_2SO_4 at 20°C
$FeS_2 + CO \rightarrow FeS + COS$ [*1309-36-0*] [*12305-96-3*]	fast above 800°C

Carbonyl sulfide occurs as a by-product in the manufacture of carbon disulfide and is an impurity in some natural gases, in many manufactured fuel gases and refinery gases, and in combustion products of sulfur-containing fuels (20). It tends to be concentrated in the propane fraction in gas fractionation; an amine sweetening process is needed to remove it.

Carbonyl sulfide is overall the most abundant sulfur-bearing compound in the earth's atmosphere: 430–570 parts per trillion (10^{12}); although it is exceeded by H_2S and SO_2 in some industrial urban atmospheres (22). Carbonyl sulfide is believed to originate from microbes, volcanoes, and the burning of vegetation, as well as from industrial processes. It may be the main cause of atmospheric sulfur corrosion (23).

Production, Shipment, and Specifications. Carbonyl sulfide is available in 97% min purity in cylinders up to 31.8 kg contained weight. It is shipped as a flammable gas. There appears to be no full-scale commercial production of carbonyl sulfide in the United States.

Analytical Methods. Detection of carbonyl sulfide in air can be done by gas chromatography or by combustion to sulfur dioxide and determination of the latter. Where hydrogen sulfide and carbonyl sulfide occur together, the carbonyl sulfide can be determined by combustion after hydrogen sulfide is absorbed by lead acetate, which does not absorb carbonyl sulfide (29).

Health and Safety Factors. Carbonyl sulfide is dangerously poisonous, more so because it is practically odorless when pure. It is lethal to rats at 2900 ppm. Recent study shows an LD_{50} (rat, intraperitoneal) of 22.5 mg/kg. The mechanism of toxic action appears to involve breakdown to hydrogen sulfide (30). It acts principally on the central nervous system with death resulting mainly from respiratory paralysis. Little is known regarding the health effects of subacute or chronic exposure to carbonyl sulfide; a 400-$\mu g/m^3$ max level has been suggested until more data are available (19).

Uses. There may be some captive use of carbonyl sulfide for production of certain thiocarbamate herbicides (qv). A Japanese patent describes the reaction of diethylamine with carbonyl sulfide forming a thiocarbamate salt, which is then alkylated with a 4-chlorobenzyl halide to produce *S*-(4-chlorobenzyl) *N,N*-diethylthiocarbamate [*28249-77-6*], ie, benthiocarb.

Thiophosgene

Physical Properties. Thiophosgene [*463-71-8*] (thiocarbonyl chloride), $CSCl_2$, is a malodorous, red-yellow liquid (bp 73.5°C, d_{20}^{15} 1.509, n_D^{20} 1.5442). It is only slightly soluble with decomposition in water, but it is soluble in ether and various organic solvents.

Chemical Properties. Thiophosgene is more resistant to hydrolysis than its oxygen analogue, phosgene, but it is slowly hydrolyzed to carbon dioxide, hydrogen sulfide, and hydrochloric acid. It can be oxidized to a lacrimatory thiophosgene *S*-oxide (31). Its utility in organic synthesis has been reviewed (32–33).

Thiophosgene reacts with alcohols and phenols to form chlorothionoformates or thiocarbonates. The most studied reactions of thiophosgene are with primary amines to give isothiocyanates and with secondary amines to give thiocarbamyl chlorides:

$$RNH_2 + CSCl_2 \longrightarrow RN{=}C{=}S + 2\ HCl$$

$$2\ R_2NH + CSCl_2 \longrightarrow R_2N\overset{\overset{\displaystyle S}{\|}}{C}Cl + R_2NH.HCl$$

Most of the reactions of thiophosgene involve the expected chemistry of an acid chloride, in which the chlorine atoms are replaceable by various nucleophiles. A reaction involving the C═S bond is the Diels Alder addition:

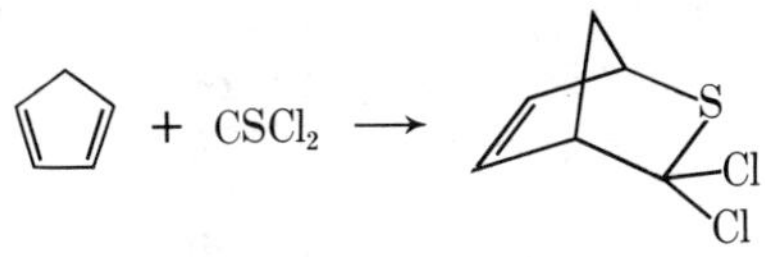

[*1499-71-4*]

Preparation. Thiophosgene forms from the reaction of carbon tetrachloride with hydrogen sulfide, sulfur, or various sulfides at elevated temperatures. Of more preparative value is the reduction of trichloromethanesulfenyl chloride [*594-42-3*] by various reducing agents, eg, tin and hydrochloric acid, stannous chloride, iron and acetic acid, phosphorus, copper, sulfur dioxide with iodine catalyst, or hydrogen sulfide over charcoal or silica gel catalyst (32–33).

There is no full-scale U.S. commercial production of thiophosgene, but it is available in glass ampuls from laboratory reagent suppliers. Thiophosgene is produced in Israel as an intermediate for tolnaftate (2-naphthyl *N*-methyl-*N*-*m*-tolylthiocarbamate) [*2398-96-1*], which is an antifungal drug.

Health and Safety Factors. Thiophosgene has an LD_{50} (rat, oral) of 929 mg/kg and an LC_{50} (inhalation, rat) of 370 mg/m^3 (34). It has both irritant and systemic toxic properties.

Trichloromethanesulfenyl Chloride

Physical Properties. Trichloromethanesulfenyl chloride [*594-42-3*] (perchloromethyl mercaptan, a misnomer but used as the common commercial name), CCl_3SCl, is a strongly acrid, pale-yellow liquid; boiling at 149°C with some decomposition at atmospheric pressure, 68°C at 6.93 kPa (52 mm Hg), and 25°C at 0.8 kPa (6 mm Hg); sp gr (20°C/4°C) 1.6996; n_D^{23} 1.541 (35–36). It slowly hydrolyzes and is soluble in most organic solvents.

Chemical Properties. A detailed review of the chemistry of trichloromethanesulfenyl chloride has been published (37). It is stable for prolonged periods at ambient temperature but decomposes slowly at its atmospheric boiling point forming sulfur monochloride, carbon tetrachloride, carbon disulfide, and polymeric oils. Storage in contact with iron causes it to decompose to carbon tetrachloride and sulfur monochloride. It is hydrolyzed only slowly by water at room temperature in a complex series of steps, proceeding by way of the unstable sulfenic acid and an isolable thiophosgene *S*-oxide ($Cl_2C{=}S{=}O$) [*24768-49-8*] yielding a complex product mixture including hydrochloric acid, phosgene, thiophosgene, trichloromethanesulfonyl chloride, chlorocarbonylsulfenyl chloride [*2757-23-5*], and bis(trichloromethyl) disulfide [*15110-08-4*] (38). At 160°C, hydrolysis is rapid and leads ultimately to the formation of carbon dioxide, hydrochloric acid, and sulfur. Trichloromethanesulfenyl chloride reacts rapidly with sodium hydroxide forming sodium dichloromethanesulfinate [*36829-83-1*], $CHCl_2SO_2Na$ (39).

The oxidation of trichloromethanesulfenyl chloride by nitric acid or oxidative chlorination in the presence of water yields trichloromethanesulfonyl chloride [*2547-61-7*], CCl_3SO_2Cl, which is a lacrimatory solid (mp 140–142.5°C), and is surprisingly stable to hydrolysis and can be steam-distilled. Trichloromethanesulfenyl chloride can be reduced to thiophosgene by metals in the presence of acid and by various other reducing agents. The sulfur-bonded chlorine of trichloromethanesulfenyl chloride is most easily displaced by nucleophilic reagents, but under some conditions, the carbon-bound chlorines are also reactive (40).

$$CCl_3SCl + 2\ RNH_2 \rightarrow CCl_3SNHR + RNH_2.HCl$$

$$CCl_3SCl + RONa \rightarrow CCl_3SOR + NaCl$$

$$CCl_3SCl + RSH \rightarrow CCl_3SSR + HCl$$

$$CCl_3SCl + C_6H_5N(CH_3)_2 \rightarrow p\text{-}CCl_3SC_6H_4N(CH_3)_2.HCl$$

$$CCl_3SCl + 4\ RONa \rightarrow (RO)_4C + 4\ NaCl + (S)$$

$$CCl_3SCl + KF \rightarrow \underset{[2712\text{-}93\text{-}8]}{CFCl_2SCl} + KCl$$

Trichloromethanesulfenyl chloride can be added across double bonds under either free radical or cationic-initiated conditions (41–42):

$$CCl_3SCl + R_2C{=}CR_2 \rightarrow CCl_3SCR_2CClR_2$$

Trichloromethanesulfenyl chloride is useful in the preparation of various series of sulfur-containing heterocycles. A commercially significant example is the preparation of 1,2,4-thiadiazoles.

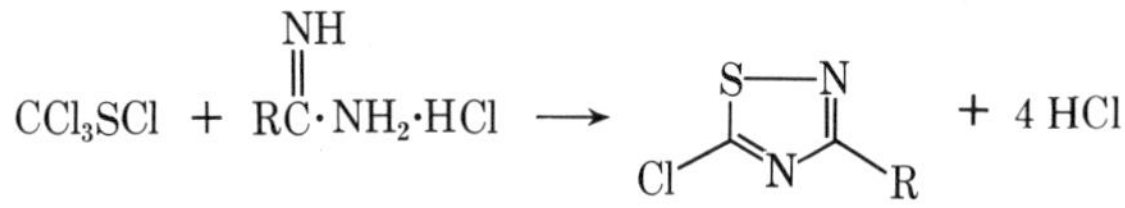

Manufacture. Trichloromethanesulfenyl chloride is made commercially by chlorination of carbon disulfide with the careful exclusion of iron or other metals, which catalyze the chlorinolysis of the C—S bond to produce carbon tetrachloride. Various catalysts, notably iodine and activated carbon, are effective. The product is purified by fractional distillation to a min purity of 95%. Continuous processes have been recently described, wherein carbon disulfide chlorination takes place on a granular-charcoal column (43–44). A series of patents describes means for yield improvement by chlorination in the presence of difunctional carbonyl compounds, phosphonates, phosphonites, phosphites, phosphates, or lead acetate (45).

Shipment and Storage. Much of the trichloromethanesulfenyl chloride that is produced is used captively. However, it can be shipped in drums. Standard materials for storage and handling are nickel, monel, and glass. Trichloromethanesulfenyl chloride is not corrosive to nickel and monel when dry, but care must be taken to prevent its contact with moisture.

Economic Aspects. The U.S. production of trichloromethanesulfenyl chloride according to the 1980 TSCA inventory was 4,500–22,000 metric tons. In the United States, it is produced by the Calhio Company in Perry, Ohio. Most of the usage of trichloromethanesulfenyl chloride is captive. The 1981 price was ca $3.30/kg.

Analytical Methods. A method has been described for gas chromatographic analysis of trichloromethanesulfenyl chloride as well as of other volatile sulfur compounds (46). A method has been recommended for determining small amounts of trichloromethanesulfenyl chloride in air or water on the basis of a color-forming reaction with resorcinol (47).

Health and Safety Factors. Trichloromethanesulfenyl chloride is extremely toxic, having an LC_{50} (mouse) of 9 ppm and a TLV of 0.1 ppm (provisional value) (48). The LD_{50} (rabbit) for an individual skin application is 1410 mg/kg (49). The primary hazard is from inhalation of vapor. Severe local irritation can result from contact of the liquid or vapor with the skin, eyes, mucous membranes, and upper respiratory tract. The irritating nature of the vapors may serve as a warning of its presence.

Uses. The principal commercial application for trichloromethanesulfenyl chloride is as an intermediate for the manufacture of fungicides, the most important being captan (*N*-(trichloromethylthio)-4-cyclohexene-1,2-dicarboximide) [*133-06-2*] and folpet (*N*-(trichloromethylthio)phthalimide) [*133-07-3*] (see Fungicides):

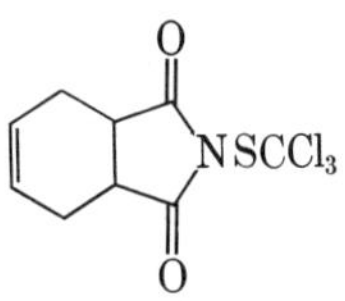

captan
[*133-06-2*]

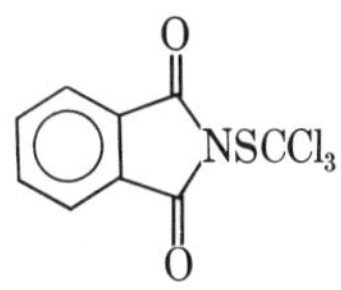

folpet
[*133-07-3*]

A commercial soil or seed treatment fungicide, 5-ethoxy-3-trichloromethyl-1,2,4-thiadiazole [*2593-15-9*] (etridiazole) is prepared from 2-chloro-3-trichloromethyl-1,2,4-thiadiazole [*5848-93-1*], which is a product of trichloromethanesulfenyl chloride and trichloroacetamidine (50).

A fluorinated analogue, $CFCl_2SCl$, is made in Europe from trichloromethanesulfenyl chloride and is used for production of the fungicides dichlorofluanid and tolylfluanid (51):

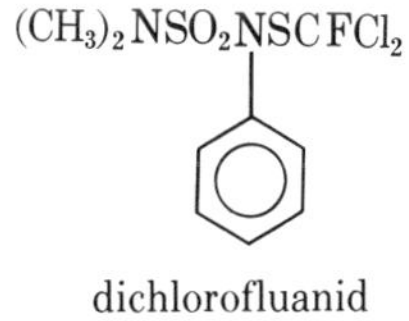

dichlorofluanid
[*1085-98-9*]

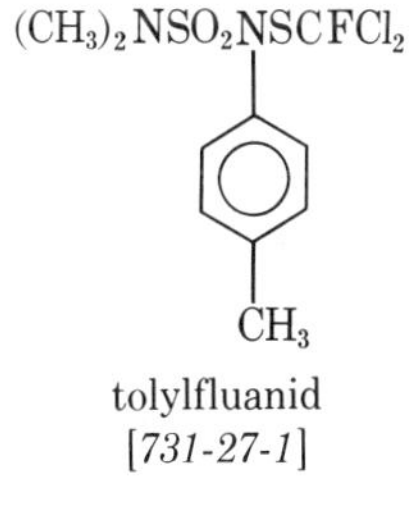

tolylfluanid
[*731-27-1*]

Trichloromethanesulfenyl chloride has also been used for the preparation of lubricant additives (see Lubrication and lubricants). Some higher homologues and analogues of trichloromethanesulfenyl chloride have been reported, ie, trichlorovinylsulfenyl chloride [*19411-15-5*], 1,1,2,2-tetrachloroethanesulfenyl chloride [*1185-09-7*], 1,2,2,2-tetrachloroethanesulfenyl chloride [*920-62-7*], and pentachloroethanesulfenyl chloride [*5940-94-3*] (52–56). A commercial fungicide, Captafol [*2425-06-1*], is the reaction product of 1,1,2,2-tetrachloroethanesulfenyl chloride with tetrahydrophthalimide (57).

Hydrogen Sulfide

Hydrogen sulfide [*7783-06-4*] is present in the gases from many volcanoes, sulfur springs, undersea vents, swamps, and stagnant bodies of water. Bacterial reduction of sulfates and bacterial decomposition of proteins forms hydrogen sulfide. The role of hydrogen sulfide in the natural sulfur cycles is discussed in ref. 58. Of greater importance as a source of sulfur are sour gases, which occur in large amounts in several locations. The concentrations of hydrogen sulfide are as follows (59):

Location	H_2S, *wt %*
France (Lacq)	16.0
FRG (Varnhorn)	22.4
Canada (Harmattan, Alberta)	53.5
Canada (Bearberry, Alberta)	90.0
Canada (Panther River, Alberta)	70–80
United States (Smackover, Miss.)	25–45
USSR (Astrakhan)	22.5
People's Republic of China (Zhaolanzhuang)	60–90

Hydrogen sulfide is a by-product of many industrial operations, eg, coking and the hydrodesulfurization of crude oil and of coal. Hydrodesulfurization is increasing in importance as the use of high sulfur crude oil becomes increasingly necessary (see Petroleum, refinery processes). A large future source of hydrogen sulfide may result if coal liquefaction attains commercial importance.

Physical Properties. Hydrogen sulfide, H_2S, is a colorless gas having a characteristic rotten-egg odor. The physical properties of hydrogen sulfide are given in Table 2.

Detailed studies of the hydrogen sulfide–nitrogen system are reviewed in refs. 64–66 and those of the hydrogen sulfide–water system in refs. 63 and 67. At low temperatures and high pressure, hydrogen sulfide forms a crystalline hexahydrate [*66230-40-8*]. Hydrogen sulfide is soluble in certain polar organic solvents, notably methanol, acetone, propylene carbonate, sulfolane, tributyl phosphate, various glycols, and glycol ethers (20,68). *N*-Methylpyrrolidinone is an exceptionally good solvent, dissolving 49 mL/g at 20°C at atmospheric pressure (69). Hydrogen sulfide is less soluble in nonpolar solvents; for example, the solubility of hydrogen sulfide gas at 20°C is 8.9 mL/g in hexane and 16.6 mL/g in benzene (58). Hydrogen sulfide is very soluble in alkanolamines, which are used as scrubbing solvents for removal of hydrogen sulfide from gas streams (20,70–72). The dissolution of hydrogen sulfide in amines results in the formation of a salt, which generally dissociates upon heating. Detailed studies have been published on the systems hydrogen sulfide–ethanolamine, –diethanolamine, –Sulfinol (a mixture of alkanolamines, sulfolane, and water) and –diglycolamine (73–76) (see Sulfur recovery).

Because of its low dielectric constant, liquid hydrogen sulfide is a poor solvent for ionic salts, eg, NaCl, but it does dissolve appreciable quantities of anhydrous $AlCl_3$, $ZnCl_2$, $FeCl_3$, PCl_3, $SiCl_4$, and SO_2. Liquid hydrogen sulfide or hydrogen sulfide-containing gases under pressure dissolve a large amount of sulfur. The equilibria among H_2S_x, H_2S, and sulfur have been studied (77–78).

Table 2. Physical and Thermodynamic Properties of Hydrogen Sulfide

Property	Value	Ref.
mol wt	34.08	60
mp, °C	−85.60	60
bp, °C	−60.75	60
ΔH fusion, kJ/mol[a]	2.375	60
ΔH vaporization, kJ/mol[a]	18.67	60
density at −60°C, g/cm³	0.993	60
sp gr, gas (air = 1)	1.19	61
critical temperature, °C	100.4	60
critical pressure, kPa[b]	9020	60
critical density, g/cm³	0.3681	61
$\Delta G°$ formation, kJ/mol[a]	−33.6	12
$\Delta H°$ formation at 25°C, kJ/mol[a]	−20.6	12
$S°$ formation at 25°C, J/(mol·K)[a]	205.7	12
C_p°, J/(mol·K)[a]	34.2	12
autoignition temperature in air, °C	ca 260	62
explosive range in air at 20°C, vol %		
upper limit	46	62
lower limit	4.3	
vapor pressure, kPa[b]		
−60°C	102.7	63
−40°C	256.6	
−20°C	546.6	
0°C	1033	
20°C	1780	
40°C	2859	
60°C	4347	
solubility in water at 101.3 kPa (= 1 atm)		61
total pressure, g/100 g soln		
0°C	0.710	
10°C	0.530	
20°C	0.398	

[a] To convert J to cal, divide by 4.184.
[b] To convert kPa to psi, multiply by 0.145.

Chemical Properties. Although hydrogen sulfide is thermodynamically stable, it can dissociate at very high temperatures. The decomposition thermodynamics and kinetics have been reviewed and the equilibrium constant for the reaction has been determined (79–80):

$$2\,H_2S \rightleftarrows 2\,H_2 + S_2$$

Above ca 850°C, the experimental hydrogen yield agree well with that calculated by thermodynamic studies and a decomposition enthalpy of 16 kJ/mol (38 kcal/mol) is obtained and is in close agreement with theory. Below 850°C, equilibrium occurs quite slowly without a catalyst. With silica as catalyst and, to a greater degree, with cobalt molybdate or sulfided platinum as catalyst, greater hydrogen production occurs at 450–850°C than is theoretically predicted on the basis of the above equation. For example, at 477°C, 3.6 wt % H_2 results at equilibrium over cobalt molybdate. These enhanced hydrogen yields are explained on the basis of the formation of sulfur species other than S_2 (80).

The dissociation of hydrogen sulfide has been proposed as a hydrogen source for refineries, which normally need hydrogen and have ample hydrogen sulfide. Recent progress has been made in processes involving transition-metal catalysts, eg, Mo, W, or Ru sulfides, and processes involving the shifting of the equilibrium by removal of the sulfur (78,81). In the presence of a specially designed photoelectrolytic catalyst, visible light cleaves hydrogen sulfide to hydrogen and sulfur (82).

Hydrogen sulfide is oxidized by a number of oxidizing agents, as shown in Table 3. The actual products formed where alternatives are indicated are functions of the quantity of oxidant as well as of reaction conditions.

Certain of the above reactions are of practical importance. The oxidation of hydrogen sulfide in a flame is one means for producing the sulfur dioxide required for a sulfuric acid plant. Oxidation of hydrogen sulfide by sulfur dioxide is the basis of the Claus process for sulfur recovery (qv). The Claus reaction can also take place under milder conditions in the presence of water, which catalyzes the reaction (83). However, the oxidation of hydrogen sulfide by sulfur dioxide in water is a complex process leading to the formation of sulfur and polythionic acids, the mixture known as Wackenroeder's liquid (84).

Oxidation of hydrosulfide salts to sulfur by water-soluble quinones is the basis of the Stretford and the Takahax processes used in sulfur recovery. Oxidation of hydrosulfide to sulfur by sodium vanadate, which can be regenerated by air, is a key reaction of the Beavon process, also used commercially in sulfur recovery. Oxidation of hydrogen sulfide until no odor is detectable proceeds with air over a cobalt molybdate catalyst (85). The reaction of hydrogen sulfide with I_2 is quantitative and is used for analytical determination of H_2S. The oxidation of hydrogen sulfide by hydrogen peroxide in aqueous sodium hydroxide is a means for abatement of hydrogen sulfide in steam from geothermal operations or steam condensate from power plants, sewage, and industrial waste systems (86) (see Geothermal energy). Iron salts are effective catalysts for this reaction.

Table 3. Oxidation Reactions of Hydrogen Sulfide

Oxidizing agent	Conditions	Sulfur-containing products
O_2	flame;	SO_2, some SO_3;
	flame or furnace, catalyst (Claus process);	sulfur;
	aqueous soln of H_2S, catalyst	sulfur
H_2O_2	neutral; alkaline solution	sulfur; $S_2O_3^{2-}$, SO_4^{2-}
Na_2O_2	dry, elevated temperature	Na_2S, Na_2S_x
O_3	aqueous	sulfur, H_2SO_4
SO_2	elevated temperature, catalyst;	sulfur;
	aqueous soln	sulfur, polythionic acids
H_2SO_4	concentrated acid	sulfur, SO_2
HNO_3	aqueous	H_2SO_4
NO	silica gel catalyst	sulfur
NO_2^-	aqueous at pH 5–7	sulfur, NO;
	aqueous at pH 8–9	sulfur, NH_3
Cl_2	gaseous, excess Cl_2;	SCl_2
	gaseous, excess H_2S;	sulfur
	aqueous soln, excess Cl_2	H_2SO_4
I_2	aqueous soln	sulfur (quantitative)
Fe^{3+}	aqueous soln	sulfur, FeS

Anhydrous gaseous or liquid hydrogen sulfide is practically nonacidic, but aqueous solutions, ie, hydrosulfurous acid, are weakly acid; the K_a for the first hydrogen is 9.1 $\times 10^{-8}$ at 18°C and the K_a for the second hydrogen is 1.2×10^{-15}. Reaction of hydrogen sulfide with one molar equivalent of sodium hydroxide gives sodium hydrosulfide; with two molar equivalents of sodium hydroxide, sodium sulfide forms. Hydrogen sulfide reacts with sodium carbonate to produce sodium hydrosulfide and sodium bicarbonate, a reaction which is reversible upon heating. Carbon dioxide can liberate hydrogen sulfide from sodium hydrosulfide. Various metal oxides and hydroxides react with hydrogen sulfide forming sulfides. A useful application is the removal of hydrogen sulfide by reaction with iron oxide (20):

$$Fe_2O_3.H_2O + 3\ H_2S \rightarrow \underset{[12063\text{-}27\text{-}3]}{Fe_2S_3} + 4\ H_2O$$

Anhydrous hydrogen sulfide does not react at ordinary temperatures with metals, eg, mercury, silver, or copper. However, in the presence of air and moisture, the reaction is rapid, leading to tarnishing in the case of silver and copper.

$$4\ Cu\ (\text{or } Ag) + 2\ H_2S \rightarrow \underset{[22205\text{-}45\text{-}4]\ (\text{or } [21548\text{-}73\text{-}2])}{2\ Cu_2S\ (\text{or } Ag_2S)} + 2\ H_2O$$

Hydrogen sulfide causes the precipitation of sulfides from many heavy metal salts. The classical qualitative-analysis scheme depends on precipitation of the sulfides of Hg, Pb, Bi, Cu, Cd, As, Sb, and Sn under acid conditions and the sulfides of Co, Ni, Mn, Zn, and Fe under ammoniacal conditions.

Hydrogen sulfide reacts with molten sulfur and depresses the viscosity of the latter, particularly at 130–180°C; infrared spectral studies show that polysulfanes, H_2S_x, form. The average chain length of these polysulfanes is shorter than the equilibrium chain length of molten sulfur alone at the same temperature; consequently, the viscosity of the molten sulfur is markedly reduced (77–78).

Hydrogen sulfide reacts with olefins under various conditions forming mercaptans and sulfides (87–88). With unsymmetrical olefins, the direction of addition can be controlled by the choice of either a free-radical initiator, including actinic light, or an acidic catalyst:

$$H_2S + CH_3CH{=}CH_2 \xrightarrow[\text{initiation}]{\text{free radical}} \underset{[29255\text{-}31\text{-}0]}{CH_3CH_2CH_2SH}$$

$$H_2S + CH_3CH{=}CH_2 \xrightarrow[\text{catalyst}]{\text{acidic}} \underset{[75\text{-}33\text{-}2]}{(CH_3)_2CHSH}$$

When free-radical initiation is used, cocatalysts, eg, phosphites, increase the conversion (89). If olefins with electron-withdrawing substituents are involved, the addition can be conducted with a basic catalyst:

$$H_2S + CH_2{=}CHCOOR \xrightarrow{\text{base}} HSCH_2CH_2COOR$$

$$H_2S + 2\ CH_2{=}CHCOOR \xrightarrow{\text{base}} S(CH_2CH_2COOR)_2$$

Hydrogen sulfide reacts with alcohols under acid-catalyzed conditions, usually with solid acidic catalysts at elevated temperatures:

$$CH_3OH + H_2S \rightarrow CH_3SH \xrightarrow{CH_3OH} CH_3SCH_3$$

[74-93-1] [78-18-3]

At 550–600°C, hydrogen sulfide reacts with chloroaromatic compounds forming thiophenols and diaryl sulfides (90):

$$H_2S + C_6H_5Cl \rightarrow HCl + C_6H_5SH \text{ (mainly)} + (C_6H_5)_2S$$

[108-98-5] [139-66-2]

It also reacts with epoxides forming 2-hydroxyalkyl thiols and bis(2-hydroxyalkyl) sulfides; both products are made commercially:

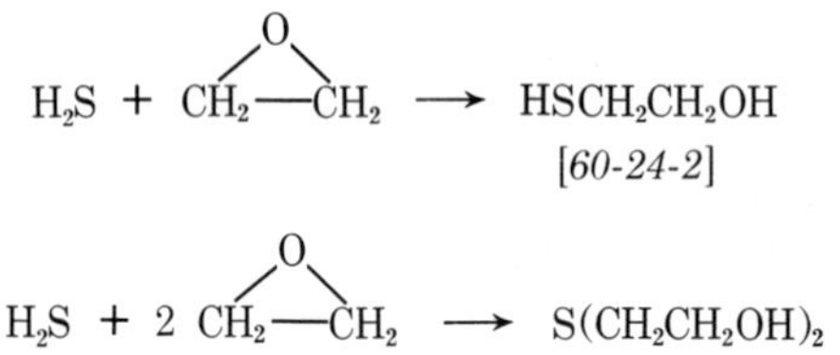

[60-24-2]

[111-48-8]

Hydrogen sulfide reacts with nitriles in the presence of a basic catalyst forming thioamides. A commercial example is its addition to cyanamide with the formation of thiourea:

$$H_2S + NH_2C{\equiv}N \longrightarrow NH_2C(=S)NH_2$$

[62-56-6]

Manufacture. Small cylinders of hydrogen sulfide are readily available for laboratory purposes, but the gas can also be easily synthesized by action of dilute sulfuric or hydrochloric acid on iron sulfide, calcium sulfide [*20548-54-3*], zinc sulfide [*1314-98-3*], or sodium hydrosulfide [*16721-80-5*]. The reaction usually is run in a Kipp generator, which regulates the addition of the acid to maintain a steady hydrogen sulfide pressure. Small laboratory quantities of hydrogen sulfide can be easily formed by heating at 280–320°C a mixture of sulfur and a hydrogen-rich, nonvolatile aliphatic substance, eg, paraffin. Gas evolution proceeds more smoothly if asbestos or diatomaceous earth is also present.

Commercial-scale processes have been developed for the production of hydrogen sulfide from heavy fuel oils and sulfur as well as from methane, water vapor, and sulfur. The latter process can be carried out in two steps: reaction of methane with sulfur to form carbon disulfide and hydrogen sulfide followed by hydrolysis of carbon disulfide (91).

Hydrogen sulfide has been produced in commercial quantities by the direct combination of the elements. The reaction of hydrogen with sulfur vapor proceeds at ca 500°C in the presence of a catalyst, eg, bauxite, an aluminosilicate, or cobalt molybdate. This process yields hydrogen sulfide that is of good purity and is suitable for preparation of sodium sulfide and sodium hydrosulfide (see Sodium compounds). Most hydrogen sulfide used commercially is either a by-product or is obtained from sour natural gas.

Recovery from Gas Streams. The crude oil refined in the United States contains varying amounts of sulfur, eg, 0.04 wt % in Pennsylvania crude to ca 5 wt % in heavy Mississippi crude. Hydrodesulfurization is becoming increasingly important as a refinery operation. More than 90% of the sulfur in crude oils is accounted for in the gas–oil and coke-distillate fractions. The sulfur compounds are mostly acyclic and cyclic sulfides; many of the latter are highly stable aromatic types, eg, benzothiophenes. Their removal is accomplished by passing the sulfur-rich fractions through a fixed-bed catalyst with hydrogen, which is generally by-product hydrogen from catalytic reforming. Besides conversion of 80–90% of the sulfur compounds to hydrogen sulfide, the hydrocarbon saturation is increased. The hydrogen sulfide is usually converted by Claus process (qv) oxidation to sulfur (see Sulfur recovery).

A process was operated commercially by Stauffer for making purified hydrogen sulfide from a crude refinery by-product stream by compression, drying, condensation, and fractional distillation.

Corrosivity. Anhydrous hydrogen sulfide has low corrosivity towards carbon steel, aluminum, Inconel, Stellite, and 304 and 316 stainless steels. However, hard steels, if highly stressed, are susceptible to hydrogen embrittlement by hydrogen sulfide. The risk of embrittlement can be avoided by use of coatings, eg, Teflon, or of 316 stainless or age-hardened Inconel. At elevated temperatures, hydrogen sulfide can produce severe sulfidation of metals, and the resultant sulfide scale is of little protective value against further attack.

Wet hydrogen sulfide is quite corrosive to carbon steel; corrosion rates are as high as 2.5 mm/yr. High strength steels may suffer stress corrosion cracking. Satisfactory steels for wet hydrogen sulfide include 316 stainless steel containing 2.5 wt % Cr and 1 wt % Mo, 4–6 wt % Cr and 0.5 wt % Mo, 2–S or 3–S aluminum alloy, or 18–8 chrome–nickel stainless. 316 Stainless steel can be used to handle hydrogen sulfide vapor to ca 260°C, and 310 stainless steel is useful at higher temperatures. Wet hydrogen sulfide corrodes copper and brass. An extensive compendium of hydrogen sulfide corrosion information has been published, and standards for sulfide-resistant metals for oil field use have been proposed (92–93). High nickel stainless steels and nickel-based alloys are the most resistant to sulfide attack under simulated deep oil-well and gas-well environments (94).

Economic Aspects. Reliable price quotations are not available since most hydrogen sulfide is made and used captively or sold by pipeline at prices which are highly variable, depending on locality. Production in the United States exceeds 1.1×10^6 t/yr. It has recently been estimated that 2.4×10^6 t/yr of sulfur are recovered from H_2S-containing refinery streams and 1.8×10^6 t/yr of sulfur are recovered from H_2S-containing natural gas (95). Annual sulfur production increases of 7% and 12%, respectively, from these two sources are projected for 1978–1990.

Analytical Methods. A method recommended by NIOSH for measurement of hydrogen sulfide in air involves drawing a known volume of air through a desiccant tube containing sodium sulfate to remove water vapor. A tube with a molecular sieve is then used to trap hydrogen sulfide, which is subsequently desorbed thermally and analyzed by gas chromatography. This method is useful for analysis to 15–60 mg/m^3 (96). An alternative method involves aspirating a measured volume of air through an alkaline suspension of cadmium hydroxide. Cadmium sulfide precipitates, and the collected sulfide is determined by spectrophotometric measurement of the methylene blue produced by reaction of the sulfide with an acid solution of *N*,*N*-dimethyl-*p*-phenylenediamine and ferric chloride. This method is useful to 8.5–63 mg/m^3 (97).

A simple laboratory method for detecting hydrogen sulfide is the use of lead acetate paper, which darkens upon formation of lead sulfide. Instruments containing a moving tape treated with lead acetate with an optical device for measuring the reflectance of the tape have been devised for monitoring the presence of hydrogen sulfide. Detector tubes, eg, Draeger tubes, are available for fast, semiquantitative measurement of hydrogen sulfide in air. More recently, electronic devices based on solid-state sensors have become available for continuous monitoring of hydrogen sulfide (98).

Health and Safety Factors. Hydrogen sulfide has an extremely high acute toxicity and has caused many deaths both in the workplace and in areas of natural accumulation, eg, cisterns and sewers (99–100). Hydrogen sulfide can be hazardous to workers in the gas, oil chemical, geothermal energy, and viscose rayon industries and who work in sewer systems, tanneries, mining, drilling, smelting, animal-waste disposal, and on fishing boats. Brief exposure to hydrogen sulfide at a concentration of 140 mg/m^3 causes conjunctivitis and keratitis (eye damage), and exposures at above ca 280 mg/m^3 cause unconsciousness, respiratory paralysis, and death. There is no conclusive evidence of adverse health effects from repeated long-term exposure to hydrogen sulfide at low concentrations, although there is some evidence pointing to nervous-system, cardiovascular, gastrointestinal, and ocular disorders. Symptoms of low level exposure can include headache, nausea, insomnia, fatigue, and inflammation of the eyes and mucous membranes. Irritation of the eyes and respiratory system has been reported at concentrations below 1 ppm.

Hydrogen sulfide is especially dangerous when it occurs in low lying areas or confined workspaces or when it exists in high concentrations under pressure. Because of corrosion and embrittlement, hydrogen sulfide lines and fittings are prone to develop leaks. Hydrogen sulfide is fast-acting and the exposed person may become unconscious quickly, with no opportunity to escape the contaminated space. Moreover, hydrogen sulfide has a deceptively sweet smell at 30–100 ppm and it deadens the sense of smell above this range; therefore, its odor is an unreliable indicator of dangerous concentrations. Self-contained breathing apparatus or a supplied-air respirator is necessary for workers entering areas known or suspected to contain toxic levels of hydrogen sulfide. A summary of the effects of hydrogen sulfide exposure on humans is given in Table 4.

OSHA's acceptable ceiling concentration is 20 ppm. The acceptable maximum peak concentration above the acceptable ceiling concentration for an 8-h shift is 50 ppm for a single maximum duration of 10 min but only if no other measurable exposure occurs.

Protective measures involve prompt detection and adequate ventilation. Continuous monitoring is recommended to signal an evacuation alarm if the workplace concentration exceeds 70 mg/m^3 (50 ppm) and a warning alert if it is present at 15–70 mg/m^3 (10–50 ppm).

The environmental occurrence of hydrogen sulfide is 15–5000 parts per trillion (10^{12}) in urban and industrial areas and 5–50 parts per trillion (10^{12}) in remote areas. Air pollution (qv) by hydrogen sulfide is not a widespread urban problem; it is generally localized near emission sources, eg, kraft paper mills, industrial-waste-disposal ponds, sewage-treatment plants, petroleum refineries, coke ovens, and geothermal-power stations. A primary air quality standard for hydrogen sulfide has not yet been formulated by the EPA. Various states have set limits ranging from 0.168 mg/m^3 for 30 min in Texas to 0.0042 mg/m^3 for 30 min in New Mexico. The Canadian province of

Table 4. Effects of Hydrogen Sulfide Inhalation on Humans[a]

Number of people exposed	Concentration, mg/m^3	Duration of exposure	Effect
1	17,000		death
1	2,800–5,600	<20 min	death
10	1,400	<1 min	unconsciousness, abnormal ECG; death of one person
342	1,400–2,800	<20 min	hospitalization of 320, death of 22 including 13 in the hospital, residual nervous system damage in 4
5	1,400	instant	unconsciousness, death
1	1,400	<25 min	unconsciousness, low blood pressure, pulmonary edema, convulsions, hematuria
4	400–760		unconsciousness
1	320	20 min	unconsciousness, arm cramps, low blood pressure
78	20–35		burning eyes in 25, headache in 32, loss of appetite in 31, weight loss in 20, dizziness in more than 19
6,500	15–20	4–7 h	conjunctivitis
population of Terre Haute, Indiana	0.003–11	intermittent air pollution episodes over a 2-mo period	nausea in 13, headache, shortness of breath in 4, sleep disturbance in 5, throat and eye irritation in 5

[a] Ref. 99.

Alberta, which has a major sulfur industry, permits the long-term presence of 0.004 mg/m^3. Various states have passed emission standards for effluent air or gas from stationary sources. A good summary of these regulations as of 1979 is given in ref. 100. In addition to processes to control emissions by scrubbing hydrogen sulfide from gas streams, considerable progress has been made in controlling smaller hydrogen sulfide emissions from sewage systems and industrial waste systems by use of hydrogen peroxide as an oxidant.

Uses. Most of the hydrogen sulfide recovered as a by-product is converted to elemental sulfur by the Claus process or to sulfuric acid (qv) where there is a market for the acid near the source of the hydrogen sulfide. Hydrogen sulfide is also used to prepare various inorganic sulfides, notably sodium sulfide and sodium hydrosulfide, which are used in the manufacture of dyes (qv), rubber chemicals (qv), pesticides, polymers, plastics additives, leather (qv), and drugs. A large amount of sodium hydrosulfide or sodium sulfide is used and largely recycled in kraft pulping; hydrogen sulfide can be used for replenishing the sulfide content (see Pulp). An important application in organic synthesis is the reaction of hydrogen sulfide to produce thiols (qv) (mercaptans).

In metallurgy, hydrogen sulfide is used to precipitate copper sulfide from nickel–copper-containing ore leach solutions in Alberta, Canada, or to precipitate nickel and cobalt sulfides from sulfuric acid leaching of laterite ores in Moa Bay, Cuba (95). Hydrogen sulfide is also used in the production of heavy water for the nuclear industry (63).

Hydrogen Polysulfides (Sulfanes)

Individual hydrogen polysulfides have been characterized from H_2S_2 [*13465-07-1*] up to at least H_2S_8 [*12026-49-2*] (101). They are of no commercial utility by themselves, although sodium and calcium polysulfides, which are made by addition of sulfur to the corresponding monosulfides, are used commercially. The atmospheric boiling point of H_2S_2 is 70.7°C and the boiling point of H_2S_3 [*13845-23-3*] is 69°C at 0.3 kPa (2 mm Hg). The higher hydrogen polysulfides have been separated by fractional distillation at 13 mPa (10^{-4} mm Hg). They are all unstable. Hydrogen polysulfides are extremely unstable, even to traces of alkalies, and they must be kept in HCl-treated glass or silica vessels, since ordinary glass is too alkaline. A mixture of hydrogen polysulfides can be prepared by addition of an alkali polysulfide to acid (102).

The sulfanes (H_2S_x) are the subject of several reviews (101,103). Except for hydrogen sulfide, they have had no practical utility. Sodium tetrasulfide [*12034-39-8*] is available commercially as a 40 wt % aqueous solution and is used to dehair hides in tanneries, as an ore flotation agent, and in the preparation of sulfur dyes and metal sulfide finishes.

The practical importance of the higher sulfanes relates to their formation in sour-gas wells from sulfur and hydrogen sulfide under pressure and their subsequent decomposition which causes well plugging (104). The formation of high sulfanes in the recovery of sulfur by the Claus process also may lead to persistance of traces of hydrogen sulfide in the sulfur thus produced (78). Quantitative determination of H_2S_x and H_2S in Claus process sulfur requires the use of a catalyst, eg, PbS, to accelerate the breakdown of H_2S_x (105).

Sulfur Halides and Oxyhalides

Sulfur forms several series of halides with all of the halogens except iodine. The fluorides, including the commercially important sulfur hexafluoride [*2551-62-4*], are discussed elsewhere (see Fluorine compounds).

Sulfur Monochloride. ***Physical Properties.*** Sulfur monochloride [*10025-67-9*], S_2Cl_2, is a yellow-orange liquid with a characteristic pungent odor. It was first discovered as a chlorination product of sulfur in 1810. The typical commercial sulfur monochloride has a melting point of −76°C and the pure material has one of −76.1°C. Other constants are bp 137.8°C, sp gr 1.68 at 20°C, refractive index (n_D^{20}) 1.670, flash point 130°C (Cleveland open-cup) (106–107). The vapor pressure is 0.479 kPa (3.59 mm Hg) at 0°C, 1.513 kPa (11.35 mm Hg) at 21.1°C, and 3.5 kPa (26 mm Hg) at 37.7°C. Sulfur monochloride is insoluble in water but is soluble in many organic solvents. The heat of formation at 25°C of liquid sulfur monochloride is −32 kJ/mol (−7.7 kcal/mol) and −58 kJ/mol (−13.9 kcal/mol) for the gas (108–109).

Chemical Properties. The chemistry of the sulfur chlorides has been reviewed (110–111). Sulfur monochloride is stable at ambient temperature but undergoes exchange with dissolved sulfur at 100°C, indicating reversible dissociation. When distilled at its atmospheric boiling point, it undergoes some decomposition to the dichloride, but decomposition is avoided with distillation at ca 6.7 kPa (50 mm Hg). At above 300°C, substantial dissociation to S_2 and Cl_2 occurs. Sulfur monochloride is noncombustible at ambient temperature, but at elevated temperatures it decomposes to chlorine and sulfur (107). The sulfur then is capable of burning to sulfur dioxide and a small proportion of sulfur trioxide.

Sulfur monochloride is hydrolyzed at a moderate rate by water at room temperature but rapidly at higher temperatures. In the vapor state, the hydrolysis rate is slow and involves disproportionation of the primary hydrolysis products:

$$S_2Cl_2 + 2\ H_2O \rightarrow 2\ HCl + SO_2 + H_2S$$

In solution, the hydrogen sulfide and sulfur dioxide [*7446-09-5*] thus formed react to producing polythionic acids and elemental sulfur.

Various reducing agents, eg, hydrogen iodide, can abstract chlorine from sulfur monochloride leaving elemental sulfur:

$$S_2Cl_2 + 2\ HI \rightarrow I_2 + 2\ HCl + 2\ S$$

This reaction is useful analytically.

Some metallic oxides and sulfides are chlorinated by sulfur chlorides:

$$ZnS + S_2Cl_2 \rightarrow ZnCl_2 + 3\ S$$

Sulfur trioxide [*7446-11-9*] reacts with sulfur monochloride to produce pyrosulfuryl chloride [*7791-27-7*], $ClSO_2OSO_2Cl$.

Hydrogen sulfide can react with S_2Cl_2 to produce, depending on conditions, a mixture of sulfanes (H_2S_x where $x > 1$) or dichlorosulfanes (S_xCl_2 where $x > 2$). These compounds tend to be unstable at ambient temperatures.

Numerous organic reactions of sulfur monochloride are of practical and commercial importance. Of particular importance is the reaction of sulfur monochloride with olefins to yield various types of addition products (111). With ethylene, the severe vesicant bis(2-chloroethyl) sulfide [*505-60-2*] or mustard gas forms with elemental sulfur and polysulfides (see Chemicals in war). Propylene reacts similarly:

$$CH_2{=}CH_2 + S_2Cl_2 \rightarrow (ClCH_2CH_2)_2S + (S)$$

Higher olefins appear to yield mainly bis(2-chloralkyl) disulfides.

The reaction of unsaturated natural oils with sulfur monochloride gives resinous products known as Factice, which are useful as art-gum erasers and rubber additives (112). The addition reaction of sulfur monochloride with unsaturated polymers, eg, natural rubber, produces cross-links and thus serves as a means for vulcanizing rubber at moderate temperatures.

With sulfur trioxide plus chlorine or with sulfur dioxide plus chlorine, sulfur monochloride yields thionyl chloride [*7719-09-7*], $SOCl_2$. Various nucleophilic reactions can displace the chlorine atoms of sulfur monochloride:

$$S_2Cl_2 + 2\ RONa \rightarrow ROSSOR + 2\ NaCl$$

$$S_2Cl_2 + RSH \rightarrow RSSSSR + 2\ HCl$$

$$S_2Cl_2 + 4\ R_2NH \rightarrow R_2NSSNR_2 + 2\ R_2NH.HCl$$

The reaction with morpholine is used to make a commercial vulcanizing agent.

Reaction with anilines yields the Herz compounds:

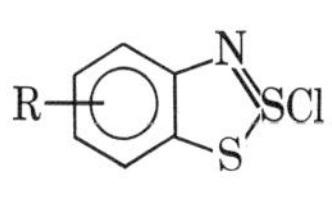

which are intermediates for certain sulfur dyes (qv).

The reaction of S_2Cl_2 with aromatic compounds can yield disulfides or mixtures of mono-, di-, and polysulfides.

$$S_2Cl_2 + 2\,C_6H_6 \rightarrow (C_6H_5)_2S_x$$

A similar reaction with phenols is employed to make commercial vulcanizing agents and antioxidants (see Antioxidants and antiozonants).

Sulfur monochloride can act as a chlorinating agent in some reactions. For example, *O,O*-dialkyl phosphorodithioic acids are converted to *O,O*-dialkyl phosphorochloridothioate:

$$\underset{[756\text{-}80\text{-}9]}{(CH_3O)_2\overset{S}{\overset{\|}{P}}SH} + S_2Cl_2 \longrightarrow \underset{[2524\text{-}03\text{-}0]}{(CH_3O)_2\overset{S}{\overset{\|}{P}}Cl} + HCl + S$$

Carboxylic acids can be converted to acid chlorides or anhydrides:

$$RCOOH + S_2Cl_2 \longrightarrow R\overset{O}{\overset{\|}{C}}Cl$$

$$2\,RCOOH + S_2Cl_2 \longrightarrow (R\overset{O}{\overset{\|}{C}})_2O$$

The reaction of sulfur monochloride with carbon disulfide yields carbon tetrachloride and sulfur:

$$CS_2 + 2\,S_2Cl_2 \rightarrow CCl_4 + 6\,S$$

Manufacture. Sulfur monochloride is made commercially by direct chlorination of sulfur and usually in a heel of sulfur chloride from a previous batch. The chlorination appears to proceed stepwise through higher sulfur chlorides (S_xCl_2, where $x > 2$). If conducted too quickly, the chlorination may yield products containing SCl_2 and S_xCl_2 as well as S_2Cl_2. A catalyst, eg, iron, iodine, or a trace of ferric chloride, facilitates the reaction.

The commercial manufacture of carbon tetrachloride by chlorination of carbon disulfide yields sulfur monochloride.

$$CS_2 + 3\,Cl_2 \rightarrow CCl_4 + S_2Cl_2$$

The sulfur monochloride formed in this reaction can be treated with additional carbon disulfide in the presence of a catalyst to yield carbon tetrachloride and sulfur. Alternatively, the sulfur monochloride and carbon tetrachloride can be separated by careful distillation in an inert gas stream (113). Modern processes utilize continuous introduction of chlorine into liquid sulfur at ca 240°C and 101 kPa (1 atm). The commercial product is assayed for purity by specific-gravity determination. A typical commercial specification is 98.0 wt % sulfur monochloride minimum, 51.4 wt % total chlorine.

Shipment and Storage. Sulfur chlorides are noncorrosive to carbon steel and iron when dry. If it is necessary to avoid discoloration caused by iron sulfide formation, stainless steel should be used. They are shipped in tank cars, tank trucks, and steel drums. When wet, they behave like hydrochloric acid and attack steel, cast iron, aluminum, stainless steels, copper and copper alloys, and many nickel-based materials. Alloys of 62 Ni–28 Mo and 54 Ni–15 Cr–16 Mo are useful under these conditions.

Economic Aspects

The price of sulfur monochloride was \$0.127/kg in 1970, \$0.165/kg in 1975, and \$0.26/kg in 1980. Over 45,000 t/yr of sulfur monochloride is produced in the United States.

Health and Safety Factors. Sulfur monochloride is highly toxic and irritating by inhalation, and it is corrosive to skin and eyes. The OSHA permissible-exposure limit is 1 ppm (6 mg/m^3). Pulmonary edema may result from inhalation. Since its vapor cannot be tolerated even at low concentrations, its presence serves as a warning factor. Sulfur monochloride is not highly flammable, having flash points of 118°C (closed-cup) and 130°C (open-cup) and an ignition temperature of 234°C.

Uses. The principal commercial uses of sulfur monochloride are in the manufacture of lubricant additives and vulcanizing agents for rubber (112,114–115) (see Lubrication and lubricants; Rubber chemicals). The preparation of additives for wear and load-bearing improvement of lubricating oils is generally carried out in two steps, and the technology is described in numerous patents (115) (see Sulfurization and sulfur chlorination).

Sulfur Dichloride. ***Physical Properties.*** Sulfur dichloride [*10545-99-1*], SCl_2, is a deep red, fuming liquid, which decomposes in moist air with the evolution of hydrogen chloride (116). Pure sulfur dichloride is unstable and is supplied commercially as a 73.1–79.4 wt % SCl_2 mixture, with sulfur monochloride comprising the remaining percentage. The melting point is reported in the range −121.5 to −61°C, and the boiling point with decomposition is 59°C. The sp gr is 1.621 at 15°C. Distillation of sulfur dichloride with minimal decomposition and improved storability requires the addition of a stabilizer, eg, PCl_3.

The solubility of sulfur dichloride in water is not meaningful since it reacts rapidly. It is slightly soluble in aliphatic hydrocarbons and very soluble in benzene and carbon tetrachloride. The heat of formation is −22 kJ/mol (−5.3 kcal/mol) for the gas at 25°C (108).

Chemical Properties. Sulfur dichloride in the liquid state at ambient temperature is in equilibrium with sulfur monochloride and dissolved chlorine:

$$2\ SCl_2 \rightleftharpoons S_2Cl_2 + Cl_2$$

The equilibrium constant is 0.013 at 18°C. Sulfur dichloride reacts violently with water, forming hydrogen chloride, sulfur dioxide, hydrogen sulfide, sulfur, and a mixture of thionic acids.

Sulfur dichloride is oxidized by sulfur trioxide or chlorosulfuric acid [*7790-94-5*] (qv) to form thionyl chloride:

$$SCl_2 + SO_3 \rightarrow SOCl_2 + SO_2$$

$$SCl_2 + ClSO_3H \rightarrow SOCl_2 + SO_2 + HCl$$

Economic Aspects. The price of sulfur dichloride was \$0.12/kg in 1970, \$0.18/kg in 1975, and \$0.29/kg in 1980. Over 9500 t/yr of sulfur dichloride is produced in the United States. Sulfur dichloride reacts with an excess of sulfur trioxide forming pyrosulfuryl chloride:

$$SCl_2 + 3\ SO_3 \rightarrow S_2O_5Cl_2 + 2\ SO_2$$

Sulfur dichloride undergoes many of the same reactions with organic compounds as described for sulfur monochloride. Addition to olefins affords a route to bis(2-

chloroalkyl) sulfides and, in certain cases, heterocyclic sulfides (117–118):

$$RCH{=}CH_2 + SCl_2 \longrightarrow (RCHClCH_2)_2S$$

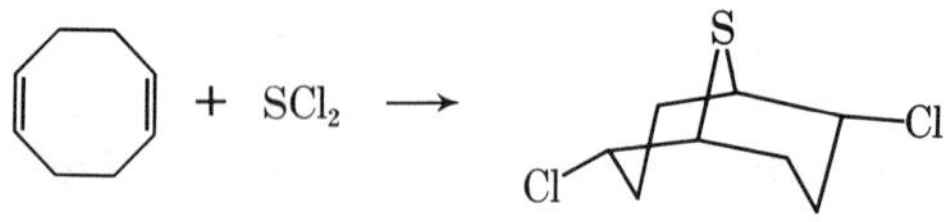

[6522-21-0]

With a few olefins, the addition can yield sulfenyl chlorides:

$$CHCl{=}CHCl + SCl_2 \rightarrow Cl_2CHCHClSCl$$

[2441-27-2]

Manufacture. The manufacture of sulfur dichloride is similar to that of sulfur monochloride, except that the last stage of chlorination proceeds slowly and must be conducted at temperatures below 40°C. The preparation of a high assay sulfur dichloride requires special techniques, eg, continuous chlorination during distillation or distillation with traces of phosphorus trichloride or phosphorus pentasulfide [*1314-80-3*] (119–121). Crude product containing 80 wt % sulfur dichloride to which is added 0.1 wt % phosphorus trichloride can be distilled to yield a 98–99 wt % pure sulfur dichloride, which can be stored for weeks at room temperature without appreciable change.

Shipment and Storage. Sulfur dichloride, if kept dry, is noncorrosive at ambient temperatures; thus, carbon steel and iron can be used in the construction of tanks, piping, and drums. However, when water or humidity is present, materials resistant to hydrochloric acid must be used, eg, Teflon, titanium, Hastelloy C, or possibly a chemically resistant, glass-reinforced polyester. Threaded pipe joints should be assembled with Teflon tape. Hoses should be constructed with a Teflon inner lining with the outer tube constructed of Neoprene or braided 316 stainless steel protected by an adequate thickness of Teflon. Sulfur dichloride should be stored away from heat and away from the direct rays of the sun.

Uses. Sulfur dichloride is used as a chlorinating agent in the manufacture of parathion [*56-38-2*] insecticide intermediates (see Insect control technology):

$$(CH_3O)_2\overset{\overset{\displaystyle S}{\|}}{P}SH + SCl_2 \rightarrow (CH_3O)_2\overset{\overset{\displaystyle S}{\|}}{P}Cl + 2\,S + HCl$$

It is also useful in the rapid vulcanization of rubber, eg, in the preparation of thin rubber goods by coating molds or fabrics with rubber latex. Rubber substitutes, such as white Factice, are made from unsaturated oils and sulfur dichloride (122). The cross-linking ability of sulfur dichloride is also utilized to modify drying oils for varnishes and inks.

An insecticide intermediate, 4,4′-thiobisphenol [*2664-63-3*] has been prepared from sulfur dichloride and phenol:

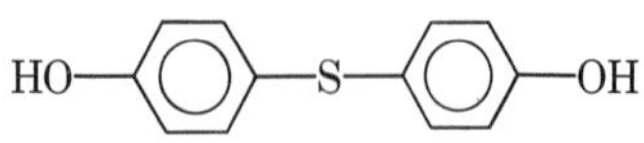

Antioxidants are also made by reaction of sulfur dichloride with phenols:

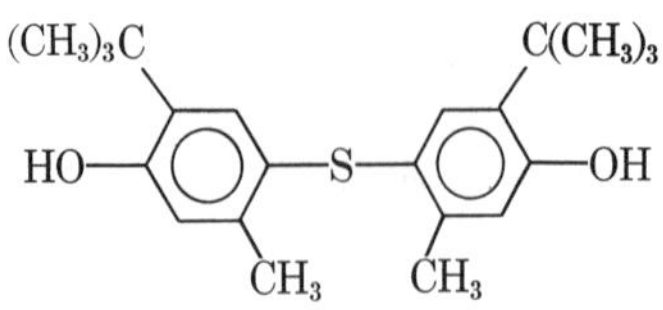

[3818-54-0]

The addition of sulfur dichloride to 1,2-dichloroethylene yields 1,2,2-trichloroethanesulfenyl chloride which, after further chlorination to 1,1,2,2-tetrachloroethanesulfenyl chloride [*1185-09-71*], is used to produce the fungicide captafol [*2425-06-1*] (Difolatan). Lubricating oil additives of types similar to those discussed regarding sulfur monochloride are a significant application for sulfur dichloride.

Oxyhalides

Thionyl Chloride. ***Physical Properties.*** Thionyl chloride [*7719-09-7*], $SOCl_2$, is a colorless fuming liquid with a choking odor (123–125). Selected physical and thermodynamic properties are listed in Table 5. Thionyl chloride is miscible with many organic solvents including chlorinated hydrocarbons and aromatic hydrocarbons. It reacts quickly with water to form HCl and SO_2. Thionyl chloride is stable at room temperature; however, slight decomposition occurs just slightly above its boiling point, so prolonged refluxing should be avoided. It decomposes fairly rapidly above 150°C and completely decomposes at 500°C to chlorine, sulfur dioxide, and sulfur monochloride. Thionyl chloride is nonflammable.

Table 5. Physical and Thermodynamic Properties of Thionyl Chloride

Property	Value	Ref.
mol wt	118.97	12
mp, °C	−104.5	123
bp, °C	76	123
density at 25°C, g/cm^3	1.629	123
molar heat capacity, J/(mol·K)[a]	120.5	123
latent heat of vaporization, kJ/mol[a]	31.3	123
viscosity, mPa·s (= cP)		
0°C	0.801	124
38°C	0.545	124
refractive index, n_D^{20}	1.517	124
dielectric constant at 20°C	9.25	124
electrical conductivity at 25°C, S/cm (= mho/cm)	2×10^{-6}	124
vapor pressure, kPa[b]		
−20°C	1.5	126
0°C	4.5	
5°C	5.9	
20°C	11.6	
26°C	14.7	
$\Delta H°$ formation, kJ/mol[a]		
25°C, liquid	−246	12
25°C, gas	−213	12
$\Delta G°$ formation at 25°C, gas, kJ/mol[a]	−198	12
$S°$ formation, J/(mol·K)[a]	309	12
$C_p°$, J/(mol·K)[a]		
25°C, liquid	121	12
25°C, gas	66.5	12

[a] To convert J to cal, divide by 4.184.
[b] To convert kPa to mm Hg, multiply by 7.5.

Chemical Properties. Thionyl chloride chemistry is reviewed in refs. 127–129. Significant inorganic reactions of thionyl chloride include its reactions with sulfur trioxide to form pyrosulfuryl chloride and with hydrogen bromide to form thionyl bromide [*507-16-4*]. With many metal oxides, it forms the corresponding metal chloride plus sulfur dioxide and, therefore, affords a convenient means for preparing anhydrous metal chlorides.

The reactions of thionyl chloride with organic compounds having hydroxyl groups are very important. Alkyl chlorides, alkyl sulfites, or alkyl chlorosulfites form from its reaction with aliphatic alcohols, depending on reaction conditions, stoichiometry, and the alcohol structure:

$$ROH + SOCl_2 \longrightarrow RCl + SO_2 + HCl$$

$$ROH + SOCl_2 \longrightarrow RO\overset{\overset{O}{\|}}{S}Cl + HCl$$

$$2\ ROH + SOCl_2 \longrightarrow RO\overset{\overset{O}{\|}}{S}OR + 2\ HCl$$

These reactions can be catalyzed by bases, eg, pyridine, or by Lewis acids, eg, zinc chloride. In the case of asymmetric alcohols, steric control, ie, inversion, racemization, or retention of configuration at the reaction site, can be achieved by the choice of reaction conditions (128–130). Some alcohols dehydrate to olefins when treated with thionyl chloride and pyridine.

With phenols, thionyl chloride forms the aryl chlorides only in exceptional cases, eg, with trinitrophenol (picric acid). The reaction of thionyl chloride with primary amines produces thionylamines.

$$RNH_2 + SOCl_2 \rightarrow RN{=}S{=}O + 2\ HCl$$

Sulfamic acids yield sulfamoyl chlorides, which are useful intermediates for herbicide synthesis (131):

$$RNHSO_3H + SOCl_2 \rightarrow RNHSO_2Cl + HCl + SO_2$$

The reaction of thionyl chloride with carboxylic acids yields either acid chlorides or, in some cases, anhydrides depending on the stoichiometry:

$$RCOOH + SOCl_2 \longrightarrow R\overset{\overset{O}{\|}}{C}Cl + HCl + SO_2$$

$$2\ RCOOH + SOCl_2 \longrightarrow (R\overset{\overset{O}{\|}}{C})_2O + 2\ HCl + SO_2$$

This route to acid chlorides is often preferred over the alternative use of phosphorus trichloride, since the by-products (SO_2 and HCl) are gaseous and easily removed. On the other hand, the use of phosphorus trichloride yields phosphorous acid as a by-product, less convenient on a laboratory scale, but salable as a valuable by-product on a commercial scale. Another advantage of thionyl chloride is its low or negligible reactivity toward many other functional groups.

Manufacture. Thionyl chloride may be made by any of the following reactions:

$$SCl_2 + SO_3 \rightarrow SOCl_2 + SO_2$$

$$SCl_2 + SO_2 + Cl_2 \rightarrow 2\ SOCl_2$$

$$SCl_2 + SO_2Cl_2 \rightarrow 2\ SOCl_2$$

The sulfur dichloride can be fed as such or produced directly in the reactor by reaction of chlorine with sulfur monochloride.

In a batch process, the reaction is carried out in a glass-lined, jacketed iron vessel, fitted with an agitator, and with a condenser arranged either for reflux or for distillation (132). The vessel is charged with either sulfur monochloride or sulfur dichloride and about 1% of antimony trichloride as a catalyst. Chlorine is introduced into the reactor near the bottom. Liquid oleum is added to the reactor at such a rate that the temperature of the reaction mass is held at ca 25°C by the use of cooling water in the jacket of the reactor. When the batch is completed, a slight excess of oleum and chlorine is added to reduce to a minimum the residual SCl_2. Since thionyl chloride combines readily with sulfur trioxide to form the relatively stable pyrosulfuryl chloride, it is necessary to maintain the concentration of sulfur trioxide in the reaction mass at a low level; hence, the addition of oleum to sulfur chloride rather than the reverse. Correspondingly, low temperatures are employed to minimize formation of the higher oxides. When all of the reactants are added, heat is applied to the jacket of the reactor, and the batch is refluxed until most of the sulfur dioxide, hydrogen chloride, and chlorine are eliminated. The thionyl chloride is then distilled from the reactor (leaving the high boiling oxides and other impurities) and further purified if necessary.

In another process sulfur monochloride, sulfur dioxide, and chlorine are allowed to react at 200°C in the presence of an activated carbon catalyst (133):

$$S_2Cl_2 + 2\ SO_2 + 3\ Cl_2 \rightarrow 4\ SOCl_2$$

Using a 0–10% excess of chlorine and 100% excess of sulfur dioxide, conversions of around 50% are obtained. The liquids in the reaction product are condensed and separated, the sulfur mono- and dichloride are returned for further reaction, and the excess gases are also recycled, producing an ultimate yield near 100% on all reactants.

At present, thionyl chloride probably is produced commercially by the continuous reaction of sulfur dioxide or sulfur trioxide with sulfur monochloride or sulfur dichloride mixed with excess chlorine (all compounds in the gaseous phase) at elevated temperature over activated carbon (134). Unreacted sulfur dioxide is mixed with the stoichiometric amount of chlorine and allowed to react at low temperature over activated carbon to form sulfuryl chloride, which is liquefied under the process conditions and fed back to the main thionyl chloride reactor.

A number of processes have been devised for purifying thionyl chloride. A recommended laboratory method involves distillation from quinoline and boiled linseed oil. Commercial processes involve adding various high boiling olefins such as styrene to react with the sulfur chlorides to form adducts that remain in the distillation residue when the thionyl chloride is redistilled (135). Alternatively, sulfur can be fed into the top of the distillation column to react with the sulfur dichloride (136).

A laboratory method is described for purification of thionyl chloride by redistillation from triphenyl phosphite (127).

Commercial thionyl chloride has a purity of 98–99.6% minimum with sulfur dioxide, sulfur chlorides, sulfuryl chloride and usually some iron compounds as impurities.

Storage and Shipping. Thionyl chloride is classified as a corrosive material; it is shipped and stored in glass-lined steel tanks, zinc-coated (galvanized) steel drums, or special plastic drums having a DOT permit. Stainless steel (304L, 316L) may be used if all traces of moisture are excluded.

Thionyl chloride stored in tanks or drums should be kept cool and away from direct sunlight and water.

Economic Aspects. The price of thionyl chloride was \$0.53/kg in 1975 and \$0.82/kg in 1980. Until recently, the only U.S. producer was Hooker Chemical Company in Niagara Falls, N.Y. A new 6700 t/yr plant was started up by Mobay Chemical Company at Baytown, Texas, in 1981. The Mobay production is said to supplant imports. Hooker also started up a new plant in Taft, La., in 1979, thereby approximately doubling that company's previous capacity. The reason for the new plant was growth in demand for thionyl chloride in the manufacture of pesticides, herbicides, and engineering thermoplastics and as a specialty chlorinating agent.

Health and Safety Factors. Thionyl chloride as a reactive acid chloride can cause severe burns to the skin and eyes and acute respiratory-tract injury upon vapor inhalation. The hydrolysis products, ie, hydrogen chloride and sulfur dioxide, are believed to be the primary irritants. Depending on the extent of inhalation exposure, symptoms can range from coughing to pulmonary edema.

Uses. A principal use of thionyl chloride is in the conversion of acids to acid chlorides, which are employed in many syntheses of herbicides, surfactants, drugs, vitamins, and dyestuffs (see Surfactants and detersive systems; Vitamins). Possible larger scale applications are in the preparation of engineering thermoplastics of the polyarylate type made from iso- and terephthaloyl chlorides, which are made from the corresponding acids plus thionyl chloride (137) (see Engineering plastics).

The reactive intermediate, $(C_2H_5)_2NCH_2CH_2Cl.HCl$, which is used to produce cationic starch, is made by the reaction of $(C_2H_5)_2NCH_2CH_2OH$ with thionyl chloride. An analogous reaction is used to make fluchloralin herbicide (Basalin) (138):

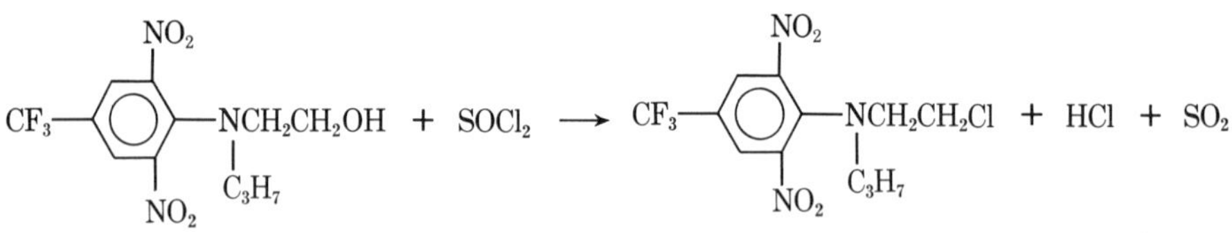

fluchloralin

An intermediate for the herbicide Basagran [*25057-89-0*] can be prepared by the following reactions (129):

$$(CH_3)_2CHNH_2 + SO_3 \longrightarrow (CH_3)_2CHNHSO_3H$$

[*42065-76-9*]

$$(CH_3)_2CHNHSO_3H + SOCl_2 \longrightarrow (CH_3)_2CHNHSO_2Cl$$

[*26118-67-2*]

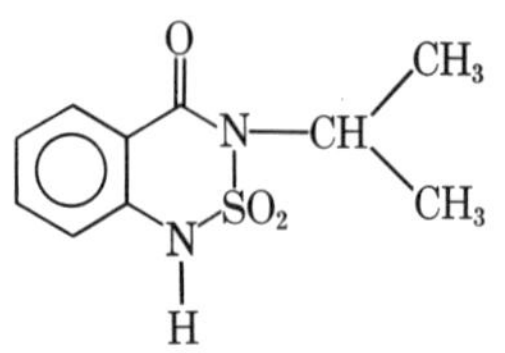

Basagran

An example of a sulfite ester made from thionyl chloride is the commercial insecticide endosulfan [*115-29-7*]. A stepwise reaction of thionyl chloride with two different alcohols yields the commercial miticide, Omite [*2312-35-8*] (139). Thionyl chloride also has potential large-scale applications as a coreactant in sulfonations and chlorosulfonations. A patent describes the use of thionyl chloride in the preparation of a key intermediate, bis(4-chlorophenyl) sulfone [*80-07-9*], which is used to make a commercial polysulfone engineering thermoplastic (see Polymers containing sulfur, polysulfone resins) (140). The sulfone group is derived from chlorosulfonic acid; the thionyl chloride may be considered a coreactant which removes water (see Sulfolane and sulfones).

A new family of high energy-density batteries has been developed and is based on lithium and thionyl chloride. These batteries are used in backup power sources for missile and telephone systems (141–142).

Sulfuryl Chloride. ***Physical Properties.*** Sulfuryl chloride [*7791-25-5*], SO_2Cl_2, is a colorless liquid with an extremely pungent odor. Physical and thermodynamic properties are listed in Table 6. Sulfuryl chloride dissolves sulfur dioxide, bromine, iodine, and ferric chloride. Various quaternary alkylammonium salts dissolve in sulfuryl chloride to produce highly conductive solutions. Sulfuryl chloride is miscible with acetic acid and ether but not with hexane (143–144).

Table 6. Physical and Thermodynamic Properties of Sulfuryl Chloride

Property	Value	Ref.
mol wt	134.968	12
mp of last crystal point, °C	−54	143
bp, °C	69.5	143
density at 25°C, g/cm^3	1.6570	144
latent heat of vaporization, kJ/mol[a]	27.97	143
surface tension at 23.5°C, mN/m (= dyn/cm)	35.26	144
viscosity, mPa·s (= cP)		
0°C	0.918	143
37.8°C	0.595	143
refractive index, n_D^{20}	1.443	143
vapor pressure, kPa[b]		
0°C	5.45	
18°C	12.69	
68.7°C	99.3	
coefficient of expansion, 0–38°C, °C^{-1}	0.0012	143
electrical conductivity, S/cm	3×10^{-8}	143
$\Delta H°$ formation, kJ/mol[a]		
25°C, liquid	−394	12
25°C, gas	−364	12
$\Delta G°$ formation 25°C, gas, kJ/mol[a]	−320	12
$S°$ 25°C, gas, J/(mol·K)[a]	311	12
C_p°, J/(mol·K)[a]		
25°C, liquid	134	12
25°C, gas	77	12

[a] To convert J to cal, divide by 4.184.

[b] To convert kPa to mm Hg, multiply by 7.5.

Chemical Properties. The chemistry of sulfuryl chloride has been reviewed (145). It is stable at room temperature but readily dissociates to sulfur dioxide and chlorine when heated; the equilibrium contant has the following values (144):

T, °C	30	40	50	102	159	191
$K = \frac{[pSO_2][pCl_2]}{[pSO_2Cl_2]}$	2.92	5.13	8.48	240	902	1330

where p = kPa (divide by 101.3 to convert to atm).

The decomposition of sulfuryl chloride is accelerated by light and it is catalyzed by aluminum chloride and charcoal. Many of the reactions of sulfuryl chloride are explainable on the basis of its dissociation products. Sulfuryl chloride reacts with sulfur at 200°C or at ambient temperature in the presence of aluminum chloride producing sulfur monochloride. It liberates bromine or iodine from bromides or iodides. Sulfuryl chloride does not mix readily with water and hydrolyzes rather slowly.

$$SO_2Cl_2 + 2\ H_2O \rightarrow H_2SO_4 + 2\ HCl$$

The reaction of sulfuryl chloride with a stoichiometric amount of sulfuric acid produces chlorosulfuric acid (chlorosulfonic acid) [*7790-94-5*]:

$$SO_2Cl_2 + H_2SO_4 \rightarrow ClSO_3H$$

This latter reaction is reversible; sulfuryl chloride can be fractionally distilled from boiling chlorosulfonic acid in the presence of a catalyst, eg, a mercuric salt.

Iodine reacts with sulfuryl chloride in the presence of aluminum chloride as catalyst forming iodine chlorides:

$$I_2 + SO_2Cl_2 \rightarrow 2\ ICl + SO_2$$

$$ICl + SO_2Cl_2 \rightarrow ICl_3 + SO_2$$

Sulfuryl chloride reacts with anhydrous ammonia yielding a series of sulfamides of the general formula $NH_2SO_2(NHSO_2)_nNH_2$, where $n \geq 0$. A cyclic compound of the formula $(SO_2NH)_3$ [*13954-94-4*] is also produced.

The organic chemistry of sulfuryl chloride involves its use in chlorination and in sulfonation (qv) (131,146–147). As a chlorinating agent, sulfuryl chloride is often more selective than elemental chlorine. The use of sulfuryl chloride as a chlorinating agent often allows more convenient handling and measurement and better temperature control, because of the lower heat of reaction, as compared with chlorine. Sulfuryl chloride sometimes affords better selectivity than chlorine in chlorination of active methylene compounds (148–150):

$$SO_2Cl_2 + CH_3COCH_2COOC_2H_5 \rightarrow CH_3COCHClCOOC_2H_5 + HCl + SO_2$$

In the presence of a free-radical initiator, alkylaromatic compounds are chlorinated in the side chain. With alkanes, chlorosulfonation can take place along with chlorination:

$$SO_2Cl_2 + RCH_3 \rightarrow RCH_2SO_2Cl + HCl$$

This reaction is usefully applied to the manufacture of chlorosulfonated polyethylene. With alkenes, sulfuryl chloride can act as a chlorinating agent in the presence of a free-radical initiator or, in the presence of pyridine, it can undergo addition across the double bond forming a 2-chloroalkanesulfonyl chloride:

$$RCH{=}CH_2 + SO_2Cl_2 \xrightarrow{\text{pyridine}} RCHClCH_2SO_2Cl$$

However, when sulfur monochloride is used as the catalyst, chlorosulfites form

$$RCH{=}CH_2 + SO_2Cl_2 \xrightarrow{S_2Cl_2} RCHClCH_2O\overset{\overset{O}{\|}}{S}Cl$$

With alkylamines, sulfuryl chloride can produce alkylsulfamoyl chlorides, which are useful intermediates in herbicide syntheses (131).

$$RNH_2.HCl + SO_2Cl_2 \rightarrow RNHSO_2Cl + 2\ HCl$$

Manufacture. The preparation of sulfuryl chloride is carried out by feeding dry sulfur dioxide and chlorine into a water-cooled steel vessel containing a catalyst, eg, activated charcoal. Alternatively, chlorine is passed into liquefied sulfur dioxide at ca 0°C in the presence of a dissolved catalyst, eg, camphor, a terpene hydrocarbon, an ether, or an ester. The sulfuryl chloride is purified by distillation; the commercial product is typically 99 wt % pure.

Storage. Dry sulfuryl chloride can be handled in iron or steel equipment but some corrosion may occur; lead, glass, 316 stainless steel, or nickel containers are preferable. Lead resists corrosion by sulfuryl chloride fairly well, even in the presence of moisture. It is important to prevent access of atmospheric moisture and to vent any hydrogen chloride pressure that builds up. Containers of sulfuryl chloride should be kept away from light, heat, and water.

Economic Aspects. The price of sulfuryl chloride was $0.40/kg in 1975 and ca $0.66/kg in 1980. There is only one U.S. producer, Hooker Chemical Company. Production in the United States exceeds 23,000 t/yr. A substantial part of this is apparently used captively (see Uses).

Health and Safety Factors. Sulfuryl chloride is both corrosive to the skin and toxic upon inhalation. The TLV suggested by the manufacturer is 1 ppm. The vapors are very irritating to the eyes and upper respiratory tract, causing prompt symptoms ranging from coughing to extreme bronchial irritation and pulmonary edema.

Uses. Uses of sulfuryl chloride include the manufacture of chlorophenols, eg, chlorothymol for use as disinfectants. It is also used in the manufacture of alpha-chlorinated acetoacetic derivatives, eg, $CH_3COCHClCOOC_2H_5$, which are precursors for important substituted-imidazole drugs, phosphate insecticides, and heterocyclic fungicides (148–149). Other herbicide synthesis uses of sulfuryl chloride, in which compounds with SO_2N linkages are produced, are reported in ref. 131. Sulfuryl chloride is used captively by DuPont on a large scale in the manufacture of chlorosulfonated polyethylene [*9008-08-6*] (see Elastomers, synthetic—chlorosulfonated polyethylene).

Pyrosulfuryl Chloride. ***Physical Properties.*** Pyrosulfuryl chloride (disulfuryl chloride), $S_2O_5Cl_2$, is a colorless liquid having the following properties: mp, −37°C; d_4^{20}, 1.837; n_D^{19}, 1.449; heat of vaporization, 147 J/g (35.1 cal/g); bp, 152°C.

Chemical Properties. At room temperature, pyrosulfuryl chloride is fairly stable. However, at temperatures over 180°C it decomposes into sulfur dioxide, sulfur trioxide, and chlorine. At 360°C, its decomposition is virtually complete. Pyrosulfuryl chloride is hydrolyzed exothermically to sulfuric and hydrochloric acids. If there is excess pyrosulfuryl chloride, some chlorine and sulfur dioxide also form. Pyrosulfuryl chloride can be used as a chlorinating and sulfonating agent (151).

Manufacture and Use. Pyrosulfuryl chloride can be prepared by the chlorination of sulfur trioxide with a wide range of chlorides, eg, S_2Cl_2, $SOCl_2$, PCl_5, $POCl_3$, $SiCl_4$, Si_2OCl_6, NaCl, and CCl_4. However, pyrosulfuryl chloride cannot be prepared by refluxing sulfuryl chloride and sulfur trioxide. Pyrosulfuryl chloride can also be obtained by dehydrating chlorosulfuric acid with phosphorus pentoxide or pentachloride.

The usual method for preparing pyrosulfuryl chloride is the addition of oleum to hot carbon tetrachloride,

$$H_2SO_4 + SO_3 + 2\ CCl_4 \rightarrow S_2O_5Cl_2 + 2\ HCl + 2\ COCl_2$$

The reaction mass is treated with dry sodium chloride to remove any unreacted chlorosulfuric acid as sodium chlorosulfurate [*14989-05-0*]. It is then fractionally distilled to obtain a refined pyrosulfuryl chloride.

Another process used for the production of pyrosulfuryl chloride involves the reaction between sulfur monochloride and sulfur trioxide at 35°C. The batch is then fractionally distilled:

$$S_2Cl_2 + 5\ SO_3 \rightarrow S_2O_5Cl_2 + 5\ SO_2$$

A third process involves the reaction of sulfur trioxide vapor with sulfur dichloride in a heel of pyrosulfuryl chloride held at ca 110°C (152). The reactor is fitted with a reflux condenser to return any reactants or intermediates which may have vaporized from the reaction mass. The pyrosulfuryl chloride produced is collected in the reactor and is fractionally distilled:

$$SCl_2 + SO_3 \rightarrow SOCl_2 + SO_2$$

$$SOCl_2 + 2\ SO_3 \rightarrow S_2O_5Cl_2 + SO_2$$

Pyrosulfuryl chloride is useful as a reagent for drug and dyestuff manufacture, in particular as a chlorinating agent (153).

Sulfur Nitrides

The sulfur nitrides have been the subject of recent reviews (154–156). Although no commercial applications have as yet been developed for these compounds, interest has been stimulated by the discovery that polymeric sulfur nitride is electroconductive (156–157) (see Polymers, conducting; Inorganic high polymers).

Tetrasulfur Tetranitride. Tetrasulfur tetranitride [*28950-34-7*], N_4S_4, is a bright orange crystalline solid (mp 178° with subl). It is explosive and detonates when struck. It decomposes slowly in water. When thermolyzed in vacuum at 300°C or over silver wool at 220°C, N_4S_4 produces disulfur dinitride [*25474-92-4*], N_2S_2; an explosion can occur during this process.

The usual satisfactory synthesis of tetrasulfur tetranitride involves first saturating a solution of sulfur monochloride in carbon tetrachloride with chlorine and subsequently treating it with dry ammonia. However, this method is dangerous because of the risk of explosion.

Disulfur Dinitride. Disulfur dinitride, N_2S_2, is a colorless liquid which, when kept at 20°C, gradually polymerizes to polythiazyl [*56422-03-8*] $(NS)_x$.

Polymeric Sulfur Nitride. Polymeric sulfur nitride (polythiazyl), $(NS)_x$, is a bronze, crystalline substance with metallic luster. It exhibits anisotropic electrical conductivity, which is greatest along one crystal axis. At temperatures approaching absolute zero (<0.26 K), superconductive properties are observed (see Superconducting materials).

Sulfur Oxides

Numerous oxides of sulfur have been reported and those that have been characterized are SO [*13827-32-2*], S_2O [*20901-21-7*], S_nO (n = 6–10), SO_2, SO_3, and SO_4 [*12772-98-4*]. Among them, SO_2 and SO_3 are of principal importance. Sulfur oxide chemistry is reviewed in refs. 158–160.

Sulfur monoxide can be prepared as a gas, which is a biradical and has a half-life of less than 2 ms, even at pressures of <130 Pa (1 mm Hg). It is best made by passing sulfur vapor and sulfur dioxide through an electric arc at low pressure or by decomposition of an episulfoxide (161). Its decomposition proceeds probably through $(SO)_2$ [*53238-43-0*] with the formation of S_2O and SO_2; S_2O is not stable in the condensed phase.

Disulfur trioxide polymer [*39421-39-1*] $(S_2O_3)_x$, a blue-green solid prepared by adding elemental sulfur to liquid sulfur trioxide at 15°C, is probably a mixture (162).

Octasulfur monoxide [*35788-51-3*], S_8O, has been prepared in poor yield from H_2S_7 [*12026-48-1*] and $SOCl_2$ and in a much more practicable way by oxidation of S_8 by trifluoroperoxyacetic acid (163–164). Octasulfur monoxide is an orange-yellow crystalline solid (mp 78°C), which decomposes at 20°C in a few hours to SO_2 and polymeric sulfur. Other members of the series S_nO, where n = 6–10, have been similarly made and characterized; they are less stable than S_8O (163).

Sulfur tetroxide SO_4, is a white solid (mp 3°C dec) and is formed by subjecting a cold gaseous mixture of sulfur dioxide and oxygen to a silent electric discharge. The substance is a powerful oxidizing agent. Its identity as a compound is in question.

Sulfur Dioxide. ***Physical Properties.*** Sulfur dioxide, SO_2, is a colorless gas with a characteristic pungent, choking odor. Its physical and thermodynamic properties are listed in Table 7. Heat capacity, vapor pressure, heat of vaporization, density, surface tension, viscosity, thermal conductivity, heat of formation, and free energy of formation as functions of temperature are presented graphically in ref. 165. A detailed discussion of the sulfur dioxide-water system is also given in ref. 167.

Liquid sulfur dioxide expands by ca 10% when warmed from 20°C to 60°C under pressure. Pure liquid sulfur dioxide is a poor conductor of electricity, but high conductivity solutions of some salts in sulfur dioxide can be made (168). Liquid sulfur dioxide is only slightly miscible with water. The gas is soluble to the extent of 36 volumes per volume of water at 20°C, but it is very soluble (several hundred volumes per volume of solvent) in a number of organic solvents, eg, acetone, other ketones, and formic acid. The use of sulfur dioxide as a solvent and reaction medium has been reviewed (168).

Inorganic Chemistry. Sulfur dioxide is extremely stable to heat, even up to 2000°C. It is not explosive or flammable in admixture with air. The oxidation of sulfur dioxide by air or pure oxygen is a reaction of great commercial importance and is commonly conducted at 400–700°C in the presence of a catalyst, eg, vanadium oxide. In certain sulfur dioxide recovery processes, oxidation of sulfur dioxide is carried out on activated carbon, which also serves as an adsorbent for the sulfur dioxide and the sulfuric acid produced (169). The oxidation of sulfur dioxide to sulfuric acid and sulfates in the atmosphere is important with regard to air pollution studies (170–172). Radicals, eg, HO, HO_2, and CH_3O_2, appear to be the principal species responsible for the homogeneous oxidation of sulfur dioxide in the atmosphere, which can occur at rates as high as 4%/h (172).

Table 7. Physical and Thermodynamic Properties of Sulfur Dioxide

Property	Value	Ref.
mol wt	64.06	12
mp, °C	−72.7	165
bp, °C	−10.02	158
ΔH fusion, kJ/mol[a]	7.40	158
ΔH vaporization at −10.0°C, kJ/mol[a]	24.92	158
vapor density at 0°C, 101.3 kPa (= 1 atm), air = 1	2.263	158
liquid density at −20°C, g/cm³	1.50	165
critical temperature, °C	157.6	165
critical pressure, kPa[b]	7911	165
critical volume, cm³/g	122.0	165
$\Delta G°$ formation, gas at 25°C, kJ/mol[a]	−300.19	12
$\Delta H°$ formation, gas at 25°C, kJ/mol[a]	−296.82	12
$S°$ formation, gas at 25°C, J/(mol·K)[a]	248.1	12
$C_p°$ (gas at 25°C), J/(mol·K)[a]	39.9	12
dielectric constant at −16.5°C	17.27	158
dipole moment at 25°C, C·m[c]	3.87×10^{-30}	158
vapor pressure, kPa[b]		
10°C	230	166
20°C	330	
30°C	462	
40°C	630	
solubility in water at 101.3 kPa, g/100 g H_2O		167
0°C	22.971	
10°C	16.413	
20°C	11.577	
30°C	8.247	
40°C	5.881	

[a] To convert J to cal, divide by 4.184.
[b] To convert kPa to psi, multiply by 0.145.
[c] To convert C·m to D, divide by 3.336×10^{-30}.

Sulfur dioxide can also be reduced by hydrogen to hydrogen sulfide. A commercial process for this was developed by Shell Oil Company. The oxidation of SO_2 catalyzed by nitrogen oxides was important in the early processes for manufacture of sulfuric acid (qv). Sulfur dioxide reacts with chlorine or bromine forming sulfuryl chloride or bromide [*507-16-4*].

$$SO_2 + Cl_2 \rightarrow SO_2Cl_2$$

Reduction of sulfur dioxide to sulfur includes an industrially important group of reactions (173). Hydrogen sulfide reduces sulfur dioxide even at ambient temperature in the presence of water but, in the dry state and in the absence of a catalyst, a temperature of ca 300°C is required.

$$SO_2 + 2\,H_2S \rightarrow 3/x\ S_x + 2\,H_2O$$

This reaction is catalyzed by silica, bauxite, and various metal sulfides. The usual catalyst is activated alumina.

The Claus process, which involves the reaction of sulfur dioxide with hydrogen sulfide to produce sulfur in a furnace, is important in the production of sulfur from sour natural gas or by-product sulfur-containing gases (see Sulfur and sulfur recovery).

When the Claus reaction is carried out in aqueous solution, the chemistry is complex and involves polythionic acid intermediates (84, 159). Sulfur dioxide can also be reduced by hydrogen to hydrogen sulfide.

A Shell process uses hydrogen or a mixture of hydrogen and carbon monoxide to reduce sulfur dioxide, carbonyl sulfide, carbon disulfide, and sulfur mixtures that occur in Claus process off-gases to hydrogen sulfide over a cobalt molybdate catalyst at ca 300°C (174). Reformed natural gas reduces sulfur dioxide in a process developed by Asarco (175).

Reduction of sulfur dioxide by methane is the basis of an Allied process for converting by-product sulfur dioxide to sulfur (176). The reaction is carried out in the gas phase over a catalyst. Reduction of sulfur dioxide to sulfur by carbon in the form of coal has been developed as the Resox process (177). The reduction, which is conducted at 550–800°C, appears to be promoted by the simultaneous reaction of the coal with steam. The reduction of sulfur dioxide by carbon monoxide tends to give carbonyl sulfide [*463-58-1*] rather than sulfur over cobalt molybdate, but special catalysts, eg, lanthanum titanate, have the potential to direct the reaction toward sulfur (178).

Equilibria such as

$$SO_2 + 2\,HCO \rightleftharpoons 1/x\ S_x + 2\,CO_2$$

$$SO_2 + 2\,COS \rightleftharpoons 3/x\ S_x + 2\,CO_2$$

$$SO_2 + CS_2 \rightleftharpoons 3/x\ S_x + CO_2$$

are probably involved in the preparation of sulfur from roasting gases.

With hot metals, sulfur dioxide usually forms both metal sulfides as well as metal oxides. In aqueous solution, sulfur dioxide is reduced by certain metals or by borohydrides to dithionites:

$$Zn + 2\,SO_2 \rightarrow \underset{[7779\text{-}86\text{-}4]}{ZnS_2O_4}$$

$$NaBH_4 + 8\,NaOH + 8\,SO_2 \rightarrow 4\,\underset{[18299\text{-}48\text{-}4]}{Na_2S_2O_4} + NaBO_2 + 6\,H_2O$$

Sulfur dioxide dissolves in water, forming a weak acid solution of sulfurous acid (H_2SO_3); however, the pure substance H_2SO_3 has never been isolated. At 101.3 kPa (1 atm) sulfur dioxide pressure, the solubility of sulfur dioxide is 18.5% at 0°C and 5.1% at 40°C. Lower solubilities are observed when other diluent gases, eg, air, are present. At 25°C, the hypothetical H_2SO_3 has a first ionization constant of 1.72×10^{-2} and a second ionization constant of 1.1×10^{-9}. At low temperatures, concentrated sulfurous acid affords a crystalline hydrate, $SO_2.6H_2O$ [*24402-69-5*].

The absorption of sulfur dioxide in alkaline aqueous solutions affords sulfites (M_2SO_3) and bisulfites ($MHSO_3$) and, at higher temperatures and concentrations, metabisulfites ($M_2S_2O_5$). Sodium sulfite [*7757-83-7*], Na_2SO_3, is produced on a large scale for use in pulp and paper mills. It is also used in boilers and oil wells as an oxygen scavenger and in the reduction of dyes. Calcium [*13780-03-5*] and magnesium bisulfite [*13774-25-9*] are produced from lime and magnesia for use in pulping (see Paper; Pulp).

The chemistry of the interaction of sulfur dioxide with alkaline substances, either in solution, slurry, or solid form, is also of great technological importance in connection with air-pollution control and sulfur recovery (qv) (20,173,179–185).

The reaction of sulfur dioxide with anhydrous ammonia has no technical applications and is very complex. The primary product appears to be HNSO, which disproportionates and polymerizes (186). The reaction of sulfur dioxide with aqueous ammonia produces a solution of ammonium sulfite [*10192-30-0*]. This reaction proceeds efficiently, even with a gas stream containing as little as 1 wt % sulfur dioxide. The sulfur dioxide can be regenerated at a high concentration by acidulation or by steam stripping of the ammonium sulfite solution, or the sulfite can be made to precipitate and the ammonia recovery by addition of lime (187). The process can also be modified to product ammonium sulfate for use as fertilizer (188) (see Fertilizers). In a newer variant of this process, the use of electron-beam radiation catalyzes the oxidation of sulfur dioxide in the presence of ammonia to form ammonium sulfate (189).

Organic bases, notably xylidine and dimethylaniline, have also been used to absorb sulfur dioxide in commercial smelter gas scrubbers. In the case of dimethylaniline, the sulfur dioxide adduct is thermally dissociated to recover the base and sulfur dioxide (188).

Organic Chemistry. The organic chemistry of sulfur dioxide, particularly as it relates to food applications, is discussed in ref. 190. Although no reaction takes place with saturated hydrocarbons at moderate temperatures, the simultaneous passage of sulfur dioxide and oxygen into an alkane in the presence of a free-radical initiator or light affords a sulfonic acid. This is the so-called sulfoxidation reaction (191):

$$2\,C_6H_{14} + 2\,SO_2 + O_2 \rightarrow 2\,C_6H_{13}SO_3H + H_2O$$

[*13595-73-8*]

Simultaneous treatment of an alkane with sulfur dioxide and chlorine affords a sulfonyl chloride and is referred to as sulfochlorination or the Reed reaction (191–192) (see Sulfurization and sulfurchlorination):

$$C_6H_{14} + SO_2 + Cl_2 \rightarrow C_6H_{13}SO_2Cl + HCl$$

[*14532-24-2*]

Isomer mixtures are generally obtained. Chlorosulfonation is used to produce chlorosulfonated polyethylene, a curable thermoplastic.

The reaction of sulfur dioxide with olefins under free-radical-catalyzed conditions produces copolymers which, in most cases, are of an alternating 1:1 type (193–194):

$$SO_2 + CH_3CH{=}CH_2 \longrightarrow \left(\overset{\displaystyle CH_3}{\overset{|}{C}}HCH_2SO_2 \right)_x$$

[*30475-44-6*]

Although these polymers are inadequately stable for ordinary applications as plastics or fibers, their photochemical lability has afforded them some utility as photodecomposable coatings in fabrication of integrated electronic circuits.

Certain olefin–sulfur dioxide copolymers have shown promise of biomedical applications, inasmuch as their films allow passage of oxygen and carbon dioxide and do not cause blood clotting (195).

Sulfur dioxide undergoes the Diels-Alder reaction with acylic dienes, and this reaction is conducted on a commercial scale with butadiene. The initial adduct, sul-folene [*77-79-2*] is hydrogenated to a solvent, sulfolane [*126-33-0*] (qv), which is useful for selective extraction of aromatic hydrocarbons from a refinery stream and as a reaction solvent. Sulfur dioxide reacts with Grignard reagents forming sulfinic acid salts.

$$2\,RMgCl + 2\,SO_2 \rightarrow (RSO_2)_2Mg + MgCl_2$$

The reaction of inorganic sulfites with aldehydes and less-hindered ketones proceeds readily to produce hydroxy sulfonic acid salts.

$$R_2C{=}O + NaHSO_3 \longrightarrow R_2C(OH)SO_3Na$$

Inorganic sulfites also add to carbon–carbon double bonds, under base-catalyzed conditions in the case of olefins substituted by electron-withdrawing substituents and under free-radical-initiated conditions with other olefins. The products are alkylsulfonates.

$$RCH{=}CH_2 + NaHSO_3 \xrightarrow[\text{initiator}]{\text{radical}} RCH_2CH_2SO_3Na$$

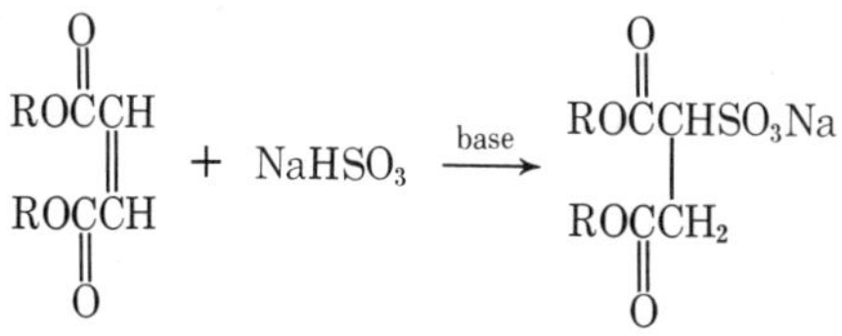

Bisulfite can be used as a reducing agent for nitroaromatics, quinones, diazo compounds, and disulfides. In some cases, the incorporation of a sulfonic group into the reaction product occurs. A review of this chemistry is given in ref. 190.

Manufacture. *Combustion of sulfur.* For most chemical process applications requiring sulfur dioxide gas or sulfurous acid, sulfur dioxide is prepared by the burning of sulfur or pyrite, FeS_2 [*1309-36-0*]. A variety of sulfur and pyrite burners have been developed for the sulfuric acid and wood-pulp industries (see Paper; Pulp). These industries produce and immediately consume ca 90% of the captive sulfur dioxide produced in the United States. The technology of sulfur burners is discussed in refs. 196–200.

The production of sulfur dioxide gas by the combustion of Frasch-process sulfur and sulfur recovered from natural gas or oil-refinery gases is relatively simple and is the preferred method, except in cases in which economic considerations favor the use of pyrite, such as in some European countries. For most applications, with the exception of sulfuric acid manufacture, as high a sulfur dioxide content as possible is desired. For sulfur burners utilizing air, the theoretical maximum is 21 vol % SO_2, and under satisfactory conditions 14–20 vol % can be achieved. The various sulfur burners commercially available are rated by their manufacturers as capable of producing gas in the range of 5–18 wt % sulfur dioxide. In the upper range of sulfur dioxide concentration, theoretical flame temperatures for the combustion of sulfur in air are 1200–1600°C. One kilogram of sulfur requires one kilogram of oxygen for its complete combustion. This is the amount contained in 3.35 m^3 air. The reaction produces 4500 kJ/kg (1942 Btu/lb) of sulfur dioxide produced from sulfur. The selection or design of sulfur burners is determined by several factors: primarily the rate and concentration of sulfur dioxide to be produced and the quality of sulfur to be burned. Continuity of operation and the required degree of operator attention and maintenance are important secondary considerations.

Large sulfuric acid plants now commonly are based on spray or spinning-cup sulfur burners (199). The largest sulfuric acid plants burn up to 39 t/h of sulfur with multiple spray burners or one or two spinning-cup burners. Rotary-kiln burners were used in chamber process sulfuric acid plants, but such plants have been almost entirely replaced by contact-process sulfuric acid plants, which generally have larger capacities. Both the rotary and spray types can be used with most qualities of sulfur and are useful over a wide range of sulfur dioxide concentrations and rates. However, rotary burners are rarely used in current sulfuric acid plant practice.

Spray burners can be operated intermittently and can be used at higher rates than rotary burners. Carbonaceous matter in dark sulfurs poses no problems for spray burners and is readily handled in rotary burners that are properly designed to burn or remove the resulting scum. The ash content of sulfurs can be a problem in sulfur dioxide from high rate spray burners, since ash can be carried into the product gas stream. Pan and cascade burners are generally more limited in flexibility and are useful only where low sulfur dioxide concentrations are desired. Gases from sulfur burners also contain small amounts of sulfur trioxide; hence, the moisture content of the air used can be important to achieving corrosion-free operation. Continuous operation at temperatures above the condensation point of the product gases is advisable where exposure to steel surfaces is involved.

The spray burners in use have to a large extent replaced the earlier pan or retort type. The best known and most widely used of the rotary burners is the Glens Falls rotary sulfur burner, manufactured by Glens Falls Machine Works, Inc. The burner consists of a horizontal steel cylinder that slowly rotates to keep the interior surfaces wet with molten sulfur, which is maintained at close to overflow level, either by mechanical solid feed or preferably by liquid feed from an external sulfur melter. Air is drawn through the burner by a draft applied to the discharge end. At all but the lowest drafts and capacities, the cylinder is followed by a brick-lined afterburner or combustion chamber into which more air can be admitted through dampers. Rated capacities are 4.5–900 kg/h of sulfur, although capacities as high as 2700 kg/h of sulfur have been reported for brick-lined burners. Several improvements have been proposed for the brick rotary burner, to permit the use of bituminous sulfurs (201). These include tilting the burner cylinder for a continuous overflow of carbonaceous scum into the combustion chamber, use of fins within the burner to increase surface, and insulating to increase the temperature for increased capacity and for better combustion of bituminous matter. Regulation of the air supply is the most important factor in the burning of sulfur to form sulfur dioxide. Excess air, particularly if it is moist and if operating temperatures are too low, leads to the formation of sulfur trioxide, and ca 2–3 wt % of the sulfur is usually lost in this way. A lesser difficulty is the formation of sublimed sulfur if the burner is overheated or if too little air is used once the burner has arrived at the proper temperature (200). In an appropriately operated afterburner, this source of loss is practically eliminated.

Several other types of sulfur burners are available to meet various needs. A pan type of continuous sulfur burner with capacities of 11–610 kg/h is made by Acme

Process Equipment Company and can be operated either under pressure or under vacuum. Pressure atomizing-spray burners, which are particularly suitable when high capacities are needed, are offered by the designers of sulfuric acid plants. Molten sulfur is pumped, at pressures up to or somewhat above 700 kPa (100 psig) through a spray gun into a brick-lined combustion chamber, and air is supplied under pressure by a blower. A waste-heat boiler utilizes the heat of combustion to make steam, part of which may be used for melting and pumping the sulfur. Quick starting and stopping and ease of regulation are some of the advantages claimed for spray burners. Capacities are 0.045–36 t/h of sulfur; the higher amount represents the use of multiple spray guns.

Air-atomizing sulfur burners can be based on smaller combustion chambers at the expense of requiring more power to compress the air. Air is preheated in a jacket, permitting faster combustion. The combustion volume can be as low as 0.085 m^3 (3 ft^3) to burn 900 kg/d of sulfur (202). Such high mass-velocity burners are offered by Chemipulp Process, Inc. (Stebbins Engineering) in Watertown, N.Y., and by A. B. Celleco in Tumba, Sweden. These burners can deliver high strength gases of 17–19 vol % SO_2 and are used in plants that produce liquid SO_2 and in pulp mills. A small version of the Celleco burner has recently been introduced (200).

The spinning-cup sulfur burner is suitable for very large capacities, ie, up to 39 t/h of sulfur in a single spinning cup. It is similar to spinning-cup oil burners, except that it is modified to prevent solidification of sulfur. The spinning cup provides centrifugal motion to the sulfur, which is then atomized by a flow of air in an annular space surrounding the cup. This burner also sprays the sulfur into a brick-lined combustion chamber. Spinning-cup sulfur burners are available in the United States from W. N. Best & Co. and in Europe from Saacke Co.

Other burners are used for low capacity operations. A cascade or checker burner, in which molten sulfur flows down through brick checkerwork countercurrent to a flow of air, is being used in small units with a sulfur trioxide converter to condition gases entering electrostatic precipitators at boiler plants operating on low sulfur coal. A small pan burner, which is fed with solid, low carbon sulfur, is used to produce sulfur dioxide for solution in irrigation water to control the pH and to maintain porosity in the soil. The same type of burner is used to disinfect wastewater; sulfur dioxide is used instead of chlorine.

Burning pyrites. The burning of pyrite is considerably more difficult to control than the burning of sulfur, although many of the difficulties have been overcome in mechanical pyrite burners; Herreshoff and Wedge burners are two well known types (198). The pyrite is burned on multiple trays, which are subject to mechanical raking. The theoretical maximum SO_2 content is 16.2 wt %, and levels of 10–14 wt % are generally attained. As much as 13 wt % of the sulfur content of the pyrite can be converted to sulfur trioxide in these burners. The spent pyrite may contain up to 4 wt % sulfur. In most applications, the separation of dust is necessary when sulfur dioxide is made from pyrite. Several methods can be employed for this, but for many purposes the use of water-spray towers is the most satisfactory. The latter also remove some of the sulfur trioxide and cool the gas. For most applications, burner gases need no further treatment other than cooling to permit their absorption either by water or by alkaline solutions or slurries. Lead pipes in flowing water are generally used; ca 1.3 m^2 (14 ft^2) of cooling surface per metric ton of sulfur dioxide per day is sufficient to cool the gas to near ambient temperatures.

Many modern plants burn pyrite and other metallic sulfides in fluid-bed roasters. The pyrite is fed onto a grate where it and its calcines are kept as a fluid by an upward flow of air through openings in the grate (see Fluidization). Steam-generating coils are installed above the grate to remove the excess heat and to produce steam while maintaining the fluidized bed at the desired temperature. Such roasters are offered by Dorr-Oliver, Inc. in Stamford, Conn., and by Lurgi in Frankfurt, FRG. Use of oxygen instead of air in a specially designed fluidized-bed apparatus makes possible the manufacture of high strength sulfur dioxide from sulfide ore (203).

A technique known as flash roasting was developed in the 1920s. Finely ground pyrites or other sulfide ores fall through a chamber in which air is introduced countercurrently from the bottom. The particles burn at 1000–1100°C yielding sulfur dioxide free of sulfur trioxide at a concentration suitable for making sulfuric acid or for sulfite pulping (204).

A basic research study on combustion of sulfur has led to the recent postulation that sulfur trioxide may actually be the primary combustion product and that sulfur dioxide may then be produced by the further reaction of sulfur trioxide with sulfur vapor in the oxygen-deficient region of the flame (205).

Recovery from flue gases. Recovery of sulfur dioxide from flue gases is described in refs. 20, 72, 172, 179. The stack gas from smelting often contains sufficient sulfur dioxide (ca 6 wt %) for economic conversion to sulfuric acid, whereas the lower concentration in power plant stack gases generally requires sulfur dioxide concentration.

Oleum. Production of SO_2 from oleum was developed at Stauffer Chemical Co. and is used commercially to produce liquid sulfur dioxide where it can be integrated with an existing oleum operation or with a concentrated sulfuric acid-consuming operation. The chemistry is as follows:

$$2\,SO_3 + S \rightarrow 3\,SO_2$$

A flow diagram of the plant is shown in Figure 1. Approximately 30 wt % oleum is pumped into a reboiler, 2, and SO_3 evolves. The resultant 20 wt % oleum is returned to the oleum towers of an acid plant. For heat conservation, the oleum streams are interchanged in a heat exchanger, 1. The SO_3 stream passes through molten sulfur in reactor, 3, where the reaction takes place, and molten sulfur is continuously fed to reactor, 3. Periodically, sludge is withdrawn from the reactor; this occurs every six to eight months if bright sulfur is used. The SO_2 vapors pass through a sulfuric acid tower for removal of trace SO_3. Cooling, 9, gas cleaning, 10, compressing, 11, and water condensation, 13, complete the process of producing liquid SO_2 for storage. The reactor, 3, is cooled and/or heated for startup by a closed water circulation loop, 4–6. The sulfur dioxide produced is 99.99% pure. Stauffer Chemical Company operates three commercial plants of this type. Development of gas-cleaning techniques and proper economical materials of construction has resulted in over 96% on-stream time.

Spent sulfuric acid. Spent sulfuric acid recovered from petrochemical and refinery processes can be fed to a high temperature furnace at 870–1260°C, where it is transformed into sulfur dioxide, water, and other gaseous products. After suitable scrubbing and drying, the gases are passed to a conventional contact sulfuric acid plant (207).

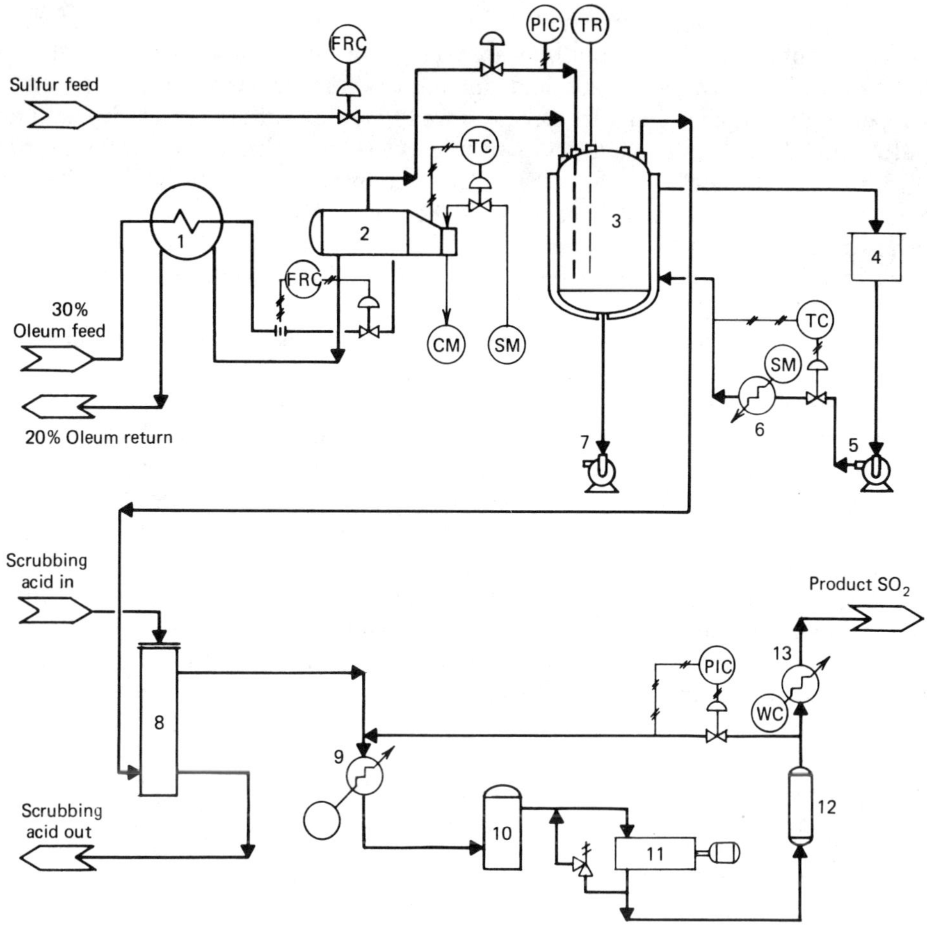

Figure 1. Flow diagram of production of sulfur dioxide from oleum. 1, 30% Oleum exchanger; 2, SO_3 vaporizer; 3, reactor; 4, coolant surge tank; 5, coolant circulating pump; 6, coolant exchangers; 7, sludge and acid pump; 8, scrubber; 9, SO_2 cooler; 10, gas cleaner; 11, SO_2 compressor; 12, pulsation damper; 13, SO_2 condenser; CM, condensate; FRC, flow recording controller; PIC, pressure indicating controller; SM, steam; TC, temperature recorder; WC cooling water (206).

Corrosivity. Almost all common materials of construction are resistant to commercial dry liquid sulfur dioxide, dry sulfur dioxide gas, and hot sulfur dioxide gas containing water at above the dew point (208–209). These include cast iron, carbon steel, copper, brass, and aluminum. Where hot gases or hot solutions are involved, the temperature resistance, particularly for plastics and resins, and the resistance to thermal shock, particularly for ceramic, glass, and stone, should be taken into account. The latter materials are inert to the wet gas, sulfurous acid, and sulfite solutions. Carbon, graphite, and impregnated carbon are suitable for practically all types of sulfur dioxide service. Lead is also resistant to sulfur dioxide and sulfites under most conditions. Aluminum is resistant under a variety of conditions and is favored in some food-industry applications involving wet sulfur dioxide. Organic coatings are generally resistant but may fail if gas diffusion through the film is appreciable. Among the organic materials, hard rubber has been satisfactory in sulfurous acid at moderate temperatures, and butyl rubber probably performs similarly.

Iron, steel, nickel, copper–nickel alloys, and Inconel Ni–Cr–Fe are satisfactory for dry or hot sulfur dioxide, but they are readily corroded below the dew point or by wet sulfur dioxide gas, sulfurous acid, and sulfites. Inconel is especially resistant to very hot sulfur dioxide gas. Metals best suited to a wide variety of wet, dry, and hot sulfur dioxide, sulfurous acid, and sulfite service are nickel–chromium alloys, eg, Worthite and Durimet 20, and several of the austenitic stainless steels. Type 304 may be satisfactory for mild conditions but types 316 and 317 are usually required for more severe applications at high temperatures. When sulfuric acid is also present, the 20-grade stainless steels may be needed. A new Crucible SC-1 stainless steel is said to be useful for sulfur dioxide scrubbers (210).

Liquid sulfur dioxide discolors iron, copper, and brass at ca 300 ppm moisture and produces light scale at ca 0.1 wt % moisture and serious corrosion at ca 0.2 wt % or higher moisture content. Copper and brass can be used to handle wet sulfur dioxide where some corrosion can be tolerated, or where the moisture level is low. Wooden tanks are widely used for sulfurous acid preparation, handling, and storage. Sulfite pulp digestors are made of steel lined with acid-resistant brick. Thermosetting phenolic and furan resins and various inorganic acid-resistant cements have been used for digestor mortar joints exposed to hot bisulfite liquors.

Shipment and Storage. Liquid sulfur dioxide is shipped in tank cars, 1-t drums, and cylinders. All shipping containers are governed by regulations of the DOT, which classifies sulfur dioxide as a nonflammable gas. The various containers are made of specified steel, they must pass hydrostatic pressure tests, and they are protected by safety devices. Single-unit tank cars are available in 1450-, 18,000-, 27,200-, 50,000-kg capacities. Sulfur dioxide tank cars are lagged and provided with a spring-loaded pressure-relief value. Purchasers of tank-car quantities of sulfur dioxide are required to have adequate storage facilities for prompt transfer from tank cars. Tank trucks with 18.2-t capacities are also used for sulfur dioxide shipment. Drums containing 907 kg of sulfur dioxide are shipped on multiunit railroad cars holding 15 drums. These drums are provided with fusible safety plugs, which melt at 74°C. Since the purchaser ordinarily retains the drums until they are empty, they serve as their own storage containers. The common cylinder size for sulfur dioxide is the 68-kg container, although other cylinder sizes are available for refrigeration-grade material. Cylinders are also provided with fusible safety plugs and are affixed with the green label for nonflammable gases.

All shipping containers for liquid sulfur dioxide are arranged so that withdrawal or transfer of the contents can be effected either as a gas or as a liquid. In general, the pressure of the sulfur dioxide in its container is used for the transfer. The most convenient method for attaining the required pressure differential in the case of cylinders and drums is heating. Since these containers are equipped with fusible plugs, it is recommended that the temperature never be allowed to exceed 52°C either in storage or during heating for transfer. In order to minimize the danger of a container of sulfur dioxide becoming full of liquid and failing as a result of the development of hydrostatic pressure with rising temperature, the maximum allowable sulfur dioxide capacity by weight of any container has been set by DOT at 1.25 times the water capacity by weight. It is common practice to fill containers to only 1.15 times the water capacity.

Economic Aspects. Demand for liquid sulfur dioxide was ca 218,000 metric tons in 1980, with the total U.S. capacity being ca 278,000 t. There are eight U.S. producers; the four largest are Stauffer Chemical Co., with three plants totaling 71,000 t of production; Virginia Chemicals, with two plants totaling 82,000 t capacity; Asarco, with one 45,000-t plant in Tacoma, Wash.; and Cities Service, with one 35,000-t plant associated with a smelter at Copperhill, Tenn. (211). Virginia, Stauffer, and Cities Service have recently expanded their capacities.

In 1969–1979, growth was 6.5%/yr; growth through 1984 is projected at 3.5%/yr (212). The price in the period 1952–1979 was $0.076–0.020/kg; the 1980 price was $0.20–0.22/kg.

The use pattern is estimated as 42% in the manufacture of hydrosulfites and other chemicals; 23% in manufacture of paper; 15% in agricultural products; 12% in metal and ore refining; 3% in oil refining; and 5% in miscellaneous uses (212). The market in paper processing and oil refining is expected to decline. However, a new use in manufacture of high fructose corn syrup, which totaled 16,000 t in 1979, is rapidly increasing (see Sugar, special sugars).

In addition to domestic production of sulfur dioxide, growing quantities of imports from Canada are entering the United States despite an import duty.

Grades and Specifications. The main grade of liquid sulfur dioxide is known as the technical, industrial, or commercial grade. This grade contains a minimum of 99.98 wt % sulfur dioxide and is a water-white liquid free of sulfur trioxide and sulfuric acid. It contains only a trace at most of nonvolatile residue. Its most important specification is the moisture content, which is generally set at 100 ppm maximum. The only other grade sold is the refrigeration grade of liquid sulfur dioxide, which is a premium grade having the same purity and specifications as the industrial grade, except for the moisture content, which is specified as 50 ppm maximum. At least one manufacturer sells a single grade for which specifications have been established as follows: color, APHA 25 max; nonvolatile residue, 25 ppm max; moisture, 50 ppm max.

Analytical Methods. The official NIOSH-recommended method for determining sulfur dioxide in air consists of drawing a known prefiltered volume of air through a bubbler containing hydrogen peroxide, thus oxidizing the sulfur dioxide to sulfuric acid. Isopropyl alcohol is then added to the contents in the bubbler and the pH of the sample is adjusted with dilute perchloric acid. The resultant solution is then titrated for sulfate with 0.005 M barium perchlorate, and Thorin is used as the indicator.

There is an end-point color change from yellow or yellow-orange to pink. This method is useful over the range 6.6–26.8 mg/m^3 and is useful for compliance with an OSHA standard of 13 mg/m^3 (5 ppm) (213).

A method that is suitable for analysis of sulfur dioxide in ambient air and is sensitive to 0.003–5 ppm involves aspirating a measured air sample through a solution of potassium or sodium tetrachloromercurate, with the resultant formation of a dichlorosulfitomercurate. Ethylenediaminetetraacetic acid (EDTA) disodium salt is added to this solution to complex heavy metals, and can interfere by oxidation of the sulfur dioxide. The sample is also treated with 0.6 wt % sulfamic acid to destroy any nitrite anions. Then the sample is treated with formaldehyde and specially purified acid-bleached rosaniline containing phosphoric acid to control pH. This reacts with the dichlorosulfitomercurate to form an intensely colored rosaniline-methanesulfonic acid. The pH of the solution is adjusted to 1.6 ± 0.1 with phosphoric acid, and the absorbance is read spectrophotometrically at 548 nm (214).

The Reich test is used to estimate the sulfur dioxide content of a gas by measuring the volume of gas required to decolorize a standard iodine solution (215). Equipment has been developed commercially for continuous monitoring of stack gas by measuring the near-ultraviolet absorption bands of sulfur dioxide (216–218). The determination of sulfur dioxide in food is conducted by distilling the sulfur dioxide from the acidulated sample into a solution of hydrogen peroxide, followed by acidimetric titration of the sulfuric acid thus produced (219).

Health, Safety, and Environmental Factors. Several review are available on the health, safety, and environmental aspects of SO_2 (171,190,220–224). Sulfur dioxide is an irritant gas; concentrations above 20 ppm have a marked irritant, choking, and sneeze-inducing effect. Acute exposures to concentrations of ca 50 ppm promptly irritate the nose and throat and cause coughing and nasal discharge; these symptoms are sufficiently disagreeable that most persons would not tolerate them for more than ca 15 min. Acutely toxic levels can cause suppurative bronchitis and asthmalike and influenzalike symptoms. At very high levels of sulfur dioxide, asphyxia leading to death can result or a chemical bronchopneumonia may develop which can be fatal after several days (223–226). Chronic exposure to sulfur dioxide has been widespread in certain industries, notably smelting and paper manufacture.

Even at levels of 2–5 ppm, symptoms of upper respiratory irritation have been reported. Acclimatization to concentrations at 20–30 ppm has been observed (223). Animal and human studies suggest that sulfur dioxide has only a low degree of chronic toxicity (223–224).

Sulfur dioxide occurs in industrial and urban atmospheres at 1 ppb–1 ppm and in remote areas of the earth at 50–120 parts per trillion (10^{12}) (22). Natural sources include volcanoes, decaying organic matter, and solar action on seawater. Sulfur dioxide has been blamed for contributing seriously to corrosion but part of this corrosive action may be caused by carbonyl sulfide (23). Assessment of the human health effects of sulfur dioxide in polluted atmospheres is complicated by the simultaneous presence of particulates, sulfuric acid, sulfates, and other pollutants (220–224,227).

Green plants are far more sensitive to sulfur dioxide than are people and animals. Injury to vegetation may occur at ground level concentrations below 1 ppm. Alfalfa, barley, cotton, and wheat are particularly sensitive, and injury occurs at as low as 0.15–0.20 ppm. Potatoes, onions, and corn are relatively resistant (226). Because sulfur is an essential nutrient for plants, a certain level of sulfur dioxide in the atmosphere may be desirable and may have a positive effect on plant growth. As a result of air-pollution control measures limiting sulfur dioxide in air, farmers may find it necessary to apply supplementary sulfur to crops (226).

Sulfur dioxide and nitrogen oxides are converted to sulfuric acid and nitric acid, respectively, by oxidation processes in the atmosphere. Acidity of rain contributed to by sulfuric and nitric acid may have caused environmental damage, particularly diminishment of aquatic life in fresh-water lakes; however, the magnitude of the problem, the sources of the pollutants, and their quantitative trends are much disputed as are the preventative measures to be taken. Discussions of the complex topic of acid rain are given in refs. 58, 171, 225, 228–232. Estimated sulfur dioxide emissions by source in the United States for 1975 are given in ref. 227 (see Air pollution).

In an estimate made in 1980, 75% of sulfur dioxide emissions is attributed to the combustion of sulfur-containing fuels and 25% to industrial processes (232). The reduction of sulfur dioxide emissions from sulfur-containing fuels can be approached by a number of principal strategies, some of which are outlined in refs. 233–235. An overview of the problem in the context of U.S. energy needs is available in ref. 236.

The national primary air quality standard, judged by the EPA to be necessary to protect public health and promulgated in 1971, is 80 μg/m^3 (0.03 ppm) calculated as an annual arithmetic mean and 365 μg/m^3 (0.14 ppm) computed as a max 24-h concentration that is not to be exceeded more than once per year. Secondary standards, which protect the public from any known or anticipated adverse effects, allow 0.02 ppm as an annual arithmetic mean, 0.10 ppm as a 24-h maximum not to be exceeded more than once per year, and 0.50 ppm as a 3-h maximum not to be exceeded more than once per year. Individual states are expected to implement the Federal standards (224). In two reviews of the chronic toxicology of sulfur dioxide, it is concluded that more stringent standards than the present ones are not warranted by the available data (224,228).

No data were found by NIOSH implicating sulfur dioxide as a primary carcinogen (223). Although certain *in vitro* tests, eg, the Ames test on *Salmonella*, indicate possible weak mutagenicity there are no definitive mutagenicity results for higher organisms. Reference 237 is an extensive review of the mutagenicity data.

Sulfur dioxide is on the FDA's approved list for a number of food applications, eg, fruit or grain fumigation to retard spoilage, as a bleaching agent in the refining of sugar, and in the processing of fruits and grains. The safety of such treatments has been reviewed (238); likewise, the biochemistry of the interaction of sulfur dioxide with biomolecules and living organisms has been surveyed (190,238–239).

Uses. The dominant use of sulfur dioxide is as a captive intermediate for production of sulfuric acid. There is also substantial captive production in the pulp and paper industry for sulfite pulping, and it is used as an intermediate for on-site production of bleaches, eg, chlorine dioxide or sodium hydrosulfite (see Bleaching agents). There is also a substantial merchant market for sulfur dioxide in the paper and pulp industry. Sulfur dioxide is used for the production of chlorine dioxide at the mill site by the reduction of sodium chlorate in sulfuric acid solution and also for the production of sodium dithionite by the reaction of sodium borohydride with sulfur dioxide (240–241). This last application is growing rapidly in North America.

A smaller but important use for sulfur dioxide is for stabilization of pulp brightness after hydrogen peroxide bleaching of mechanical pulps. Sulfur dioxide neutralizes the alkalinity and destroys any excess hydrogen peroxide, which if left in would cause the pulp to lose brightness (242).

In food processing, sulfur dioxide has a wide range of applications as a fumigant, preservative, bleach, and steeping agent for grain (190): dried fruit, eg, apricots, are preserved by sulfur dioxide; cherries are both bleached and preserved; and the browning of peeled and sliced apples is prevented. In the manufacture of wine (qv), a small amount of sulfur dioxide is added to the must to destroy bacteria, molds, and wild yeasts without harming the yeasts that produce the desired fermentation. Wine casks are also sterilized by sulfur dioxide. The formation of nitrosamines in beer is prevented by sulfur dioxide treatment in the malting process (243). In molasses manufacture, sugar refining, and soy protein processing, sulfur dioxide is used for bleaching as well as for the prevention of microbial growth (244). In making high fructose corn syrups, sodium bisulfite from sulfur dioxide is added to the enzymatic isomerization step to prevent undesired microbial action. In the earlier stages, where corn is steeped in preparation to wet milling, a 0.1–0.2 wt % sulfur dioxide solution is used to swell and soften the kernels and to prevent the growth of microorganisms. Sulfur dioxide is used throughout corn syrup processing to prevent microbial growth and to retard the nonenzymatic development of color by the Maillard reaction. Corn syrups produced in the United States usually contain 15–40 ppm of sulfur dioxide, although those syrups intended for fruit canning must contain less than 10 ppm to avoid corrosion of the cans (249) (see Food processing; Syrups).

In water treatment, sulphur dioxide is often used to reduce residual chlorine from disinfection and oxidation (245). This technology is used in potable-water treatment, in sewage treatment, and especially in industrial waste-water treatment (see Water, industrial water treatment; Water, municipal water treatment; Water, sewage). Although sodium metabisulfite [*7681-57-4*] and sodium sulfite [*7757-83-7*] may be used, the use of liquid sulfur dioxide has convenience and cost advantages and lends itself to automation. Therefore, the larger plants tend to use sulfur dioxide rather than sulfite or bisulfite salts. Sulfur dioxide is also employed as a reducing agent for conversion of chromates to less toxic chromic compounds in plating-shop effluents.

In petroleum technology, sulfur dioxide most commonly as sodium sulfite is used as an oxygen scavenger (246). This use is particularly important in secondary and tertiary oil recovery processes involving flooding of underground oil formations with water or with aqueous solutions (see Petroleum, chemicals for enhanced recovery). To prevent corrosion in piping and storage systems, it is important to lower the oxygen concentration of the water to 50–100 ppb. The rate of oxygen reduction is increased by adding a cobalt compound as catalyst and by maintaining a pH of 8.5–10 (247). Sulfur dioxide and sodium sulfites are also used in oil refining and other industrial processes (notably in boiler waters) as oxygen scavengers.

Another sulfur dioxide application in oil refining is as a selective extraction solvent in the Edeleanu process, wherein aromatic components are extracted from a kerosene stream by sulfur dioxide, leaving a purified stream of saturated aliphatic hydrocarbons which are relatively insoluble in sulfur dioxide. This is an old process but it is probably still practiced in at least one U.S. refinery.

Sulfur dioxide acts as a cocatalyst or catalyst modifier in certain processes for oxidation of *o*-xylene or naphthalene to phthalic anhydride (248–249).

In mineral technology, sulfur dioxide and sulfites are used as flotation depressants for sulfide ores. In electrowinning of copper from leach solutions from ores containing iron, sulfur dioxide prereduces ferric to ferrous ions to improve current efficiency and copper cathode quality. Sulfur dioxide also initiates precipitation of metallic selenium from selenous acid, a by-product of copper metallurgy (250).

In kaolin (clay) processing, sulfur dioxide reduces colored impurities, eg, iron compounds. In the bromine industry, sulfur dioxide is used as an antioxidant in spent brine to be reinjected underground. In agriculture, especially in California, sulfur dioxide is used to increase water penetration and the availability of soil nutrients by virtue of its ability to acidulate saline–alkali soils (251). It is also useful for cleaning ferric and manganese oxide deposits from tile drains (252).

In magnesium casting, sulfur dioxide is employed as an inert blanketing gas. Another foundry application of SO_2 is as a rapid curing catalyst for furfuryl resins in cores. Surprisingly, in view of the many efforts to remove sulfur dioxide from flue gases, there are situations where sulfur dioxide is deliberately introduced. In power plants burning low sulfur coal and where particulate stack emissions are a problem, a controlled amount of sulfur dioxide injection affords improved particulate removal.

Sulfur dioxide is useful as a solvent for sulfur trioxide in sulfonation. This technology is practiced by Cities Service, which has the production capacity for 2700–4500 t/yr of linear alkylbenzenesulfonate; the sulfur dioxide presumably is made captively (253).

In glass-container manufacture, sulfur dioxide is used as a surface alkali neutralizer. The treatment improves the resistance of bottles to scratching during handling and prepares leach-resistant bottles for medicinals, blood plasma, and detergents.

Sulfur Oxygen Acids and Their Salts

Sulfuric acid, H_2SO_4, is the most important commercial sulfur compound (see Sulfuric acid and sulfur trioxide). Peroxymonosulfuric acid [*7722-86-3*] (Caro's acid), H_2SO_5, is discussed elsewhere (see Peroxides and peroxy compounds, inorganic). The lower-valent sulfur acids are not stable species at ordinary temperatures. Dithionous acid [*15959-26-9*], $H_2S_2O_4$, sulfoxylic acid [*20196-46-7*], H_2SO_2, and thiosulfuric acid [*13686-28-7*], $H_2S_2O_3$ are unstable species. A discussion of efforts to isolate and characterize the unstable sulfur acids is given in ref. 254.

Sodium Sulfite. ***Physical Properties.*** Anhydrous sodium sulfite [*7757-83-7*], Na_2SO_3, is an odorless, crystalline solid and the commercial grades are colorless or off-white (255–261). It melts only with decomposition. The specific gravity of the pure solid is 2.633 (15.4°C). Sodium sulfite is quite soluble in water; it has a maximum solubility of 28 g/100 g sol at 33.4°C; at higher and lower temperatures, it is less soluble in water. Below this temperature, the heptahydrate crystallizes; above this temperature, the anhydrous salt crystallizes (260). Sodium sulfite is soluble in glycerol but insoluble in alcohol, acetone, and most other organic solvents.

Chemical Properties. Anhydrous sodium sulfite is stable in dry air at ambient temperatures or at 100°C, but in moist air it undergoes rapid oxidation to sodium sulfate [*7757-82-6*]. On heating to 600°C, sodium sulfite disproportionates to sodium sulfate and sodium sulfide [*1313-82-2*]. Above 900°C, the decomposition products are sodium oxide and sulfur dioxide. At 600°C, it forms sodium sulfide upon reduction with carbon (256).

Aqueous solutions of sodium sulfite are alkaline, and have a pH of ca 9.8 at 1 wt %. The solutions are oxidized readily by air. The redox potential is a function of pH, as would be expected from the equation

$$SO_3^{2-} + 2\,OH^- \rightarrow SO_4^{2-} + H_2O + 2\,e \qquad E_0 = +0.90\text{ V}$$

Although the usual product of oxidation of sulfite by air is sulfate, at high pH substantial amounts of the strong oxidant Na_2SO_5 can form (261).

Sodium sulfite undergoes addition of sulfur to form sodium thiosulfate. At acidic pH, the chemistry of sodium sulfite is that of bisulfite, metabisulfite, and sulfur dioxide.

Manufacture. In a typical process, a solution of sodium carbonate is allowed to percolate downward through a series of absorption towers, through which sulfur dioxide is passed countercurrently. The solution leaving the towers is chiefly sodium bisulfite of, typically, 27 wt % combined sulfur dioxide content. The solution is then run into a stirred vessel where aqueous sodium carbonate or sodium hydroxide is added to the point where the bisulfite is fully converted to sulfite. The solution may be filtered if necessary to attain the required product grade. A pure grade of anhydrous sodium sulfite can then be crystallized above 40°C, since the solubility decreases with increasing temperature.

In a recently patented process, a stirred suspension of sodium sulfite is continuously treated with aqueous sodium hydroxide and a sulfur dioxide-containing gas at 60–85°C, and 96% pure anhydrous sodium sulfite is removed by filtration (260). In another continuous one-step process, substantially anhydrous sodium carbonate and sulfur dioxide are concurrently introduced into a saturated solution of sodium sulfite at pH 6.5–7.6 and above 35°C with continuous removal of sodium sulfite (262).

Various processes have been disclosed wherein moist solid sodium pyrosulfite [*7681-57-4*] is stirred in a steam-heated vessel with sodium carbonate. The exothermic reaction at 80–110°C results in the drying of the product. A continuous fluid-bed version of this process has been described (263). A lower grade of sodium sulfite is produced commercially in the United States as a by-product of the sulfonation–caustic cleavage route to resorcinol (257).

Shipment and Storage. Anhydrous sodium sulfite is supplied in 22.7- and 45.4-kg moistureproof paper bags or 45.4- and 159-kg fiber drums. Most sodium sulfite is shipped by rail in hopper cars. Sodium sulfite should be protected from moisture during storage; when dry, it is quite stable, but when wet, it is oxidized by air.

Economic Aspects. The five main producers and their annual production capacities are Koppers, 45,000 t (by-product of resorcinol production by caustic cleavage of benzenesulfonic acid [*98-11-3*]); Reichhold, 72,000 t; and Stauffer (Southgate, Calif.), 10,000 t. Total U.S. capacity is 146,000 t (264). In 1981, Virginia Chemicals put a 34,000-t plant in Portsmith, Va., on stream (265). U.S. demand for sodium sulfite was ca 136,000 t in 1980 and is projected to be 152,000 t in 1985; thus, a 2.3%/yr growth rate is expected from 1980 to 1985. In the 1970s, demand declined 9.3%/yr because of increased recycling, especially by pulp manufacturers who were obliged to comply with procedures mandated by environmental regulations (264). The price of technical-grade sodium sulfite was $0.099/kg in 1970, $0.24/kg in 1975, and $0.37/kg in 1980. The photographic grade was $0.51–66/kg in 1980.

The use pattern is 60% in sulfite pulping, 15% in water treatment (dechlorination, deoxygenation), 12% in photography, and 13% in miscellaneous uses. The decline in pulp usage is believed to be at an end; high alkali and sulfur prices are expected to cause pulp makers to change from making their own sodium sulfite to purchasing it. Boiler-water deoxygenation, photography, and oil production are expected to be growth areas (264).

Grades and Specifications. The commercial grades of sodium sulfite available in the United States are the following (255–258):

photographic grade = white crystalline material, typically 97.0 wt % minimum Na_2SO_3 assay, 0.5 wt % maximum insoluble matter, 0.15 wt % maximum alkalinity as Na_2CO_3, 20 ppm maximum iron, 20 ppm maximum heavy metals (as Pb), 0.01 wt % maximum thiosulfate (as $S_2O_3^{2-}$); anhydrous technical grade = white granular material, typically 95.0–97.5 wt % Na_2SO_3 minimum assay; *anhydrous Food Chemicals Codex grade:* 95 wt % minimum (typically 98.6 wt %) Na_2SO_3 assay, 3 ppm mixture arsenic, 10 ppm maximum heavy metals (as Pb), 30 ppm maximum selenium; *anhydrous technical (by-product) grade* = pink to red-brown powder, 89 wt % Na_2SO_3 minimum, 5 wt % sodium sulfate, and ca 3 wt % sodium carbonate (This grade is used in sulfite pulping and oxygen scavenging from boiler and process water); and *aqueous solution* = 22 wt % minimum Na_2SO_3, pH 8.8, specific gravity (25°C/25°C) 1.27. The aqueous solution is a convenient form for use under conditions where there are no facilities for handling and dissolving the solid salt.

Analytical Methods. A classical and still widely employed analytical method is iodimetric titration. This is suitable for determination of sodium sulfite, for example, in boiler water. Standard potassium iodate–potassium iodide solution is commonly used as the titrant with a starch or starch-substitute indicator (255). Sodium bisulfite occurring as an impurity in sodium sulfite can be determined by addition of hydrogen peroxide to oxidize the bisulfite to bisulfate, followed by titration with standard sodium hydroxide (255).

Health and Safety Factors. Although sodium sulfite has no detectible odor, its dust and solutions are irritating to the skin, eyes, and mucous membranes. The ingestion of sodium sulfite causes gastric irritation resulting from the liberation of sulfurous acid. Large doses can cause violent colic and diarrhea as well as disturbance of the circulatory and nervous system. The lowest reported lethal value for humans is 500 mg/kg. Various LD_{50} data reported are rat (oral), greater than 1000 mg/kg (256); rat (intravenous), 115 mg/kg; mouse (intravenous), 130 mg/kg; rabbit (intravenous), 65 mg/kg; rabbit (oral, lowest lethal dose), 2825 mg/kg (34). The oral ingestion of subacute amounts of sodium sulfite, in view of the low gastric pH, amounts to ingestion of sulfur dioxide, which has a low order of toxicity.

Uses. *Pulp manufacture.* Sodium sulfite is utilized in neutral semichemical pulping, in acid sulfite pulping, in high yield sulfite cooling, and in some kraft pulping processes (266). Many pulp mills prepare their own sulfite and recycle as much as possible, but use of merchant sodium sulfite by pulp mills is substantial. Much of the by-product sodium sulfite from resorcinol manufacture goes into pulp applications as well as a substantial fraction of the lower assay manufactured sodium sulfite.

Water treatment. Sodium sulfite is an agent in the reduction of chlorine or oxygen in water. Dissolved oxygen in boiler water tends to enhance pitting and other types of corrosion. In boilers operated at below 4.82 MPa (700 psi), a residual concentration of 30 ppm of sodium sulfite is generally effective. Catalytic amounts of cobalt or copper salts are often added to accelerate the reaction of oxygen with sulfite (246–247) (see Water, industrial water, treatment).

In waterflooding of oil fields, oxygen dissolved in the water can contribute to pitting of pipes and to clogging of injection wells and the oil-bearing rock with iron corrosion products. A residual concentration of ca 10 ppm sodium sulfite is usually effective in preventing these problems (255). Sodium sulfite is also used to remove oxygen from drilling muds (see Petroleum, drilling fluids).

In removing excess free chlorine from municipal or industrial water and from wastewater, sodium sulfite competes with bisulfite or sulfur dioxide. Other commercial applications of sodium sulfite in wastewater treatment include the reduction of hexavalent chromium to the less toxic Cr^{3+} salts as well as the precipitation of silver and mercury.

Photography. Sodium sulfite is useful as a reducing agent in certain photographic fixing baths, developers, hardeners, and intensifiers (258). However, the principal use is as a film preservative and discoloration preventative (see Photography).

Miscellaneous. In ore flotation, sodium sulfite functions as a selective depressant. In textile processing, sodium sulfite is used as a bleach for wood and polyamide fibers and as an antichlor after the use of chlorine bleach. Synthetic applications of sodium sulfite include production of sodium thiosulfite by addition of sulfur and the introduction of sulfonate groups into dyestuffs and other organic products.

Sodium Bisulfite. Sodium bisulfite, $NaHSO_3$, is not a stable compound in the solid state. The sodium bisulfite of commerce consists chiefly of sodium metabisulfite, $Na_2S_2O_5$. Aqueous sodium bisulfite solution with 26–77 wt % SO_2 (sp gr 1.36), is a commercial product.

Sodium Metabisulfite. ***Physical Properties.*** Sodium metabisulfite (sodium pyrosulfite, sodium bisulfite (a misnomer)), $Na_2S_2O_5$, is a white granular or powdered salt (specific gravity 1.48) and is storable when kept dry and protected from air. In the presence of traces of water, it develops an odor of sulfur dioxide and in moist air it decomposes with loss of part of its SO_2 content and by oxidation to sodium sulfate. Dry sodium metabisulfite is more stable to oxidation than dry sodium sulfite. At low temperatures, sodium metabisulfite forms hydrates with 6 and 7 mol of water. The solubility of sodium metabisulfite in water is 39.5 wt % at 20°C, 41.6 wt % at 40°C, and 44.6 wt % at 60°C (267). Sodium metabisulfite is fairly soluble in glycerol and slightly soluble in alcohol.

Chemical Properties. The chemistry of sodium metabisulfite is essentially that of the sulfite–bisulfite–metabisulfite–sulfurous acid system. The relative proportions of each species depends on the pH. The pH of a sodium bisulfite solution obtained by dissolving 10 wt % sodium metabisulfite in water at 20°C is 4.9; at 30 wt %, the pH is 4.4 (267).

Manufacture. Aqueous sodium hydroxide, sodium bicarbonate, sodium carbonate, or sodium sulfite solution are treated with sulfur dioxide to produce sodium metabisulfite solution. In one operation, the mother liquor from the previous batch is reinforced with additional sodium carbonate, which need not be totally in solution, and then is treated with sulfur dioxide (268). In some plants, the reaction is conducted in a series of two or more stainless-steel vessels or columns in which the sulfur dioxide is passed countercurrent to the alkali. The solution is cooled and the sodium metabisulfite is removed by centrifuging or filtration. Rapid drying, eg, in a steam-heated shelf dryer or a flash dryer, avoids excessive decomposition or oxidation to which moist sodium metabisulfite is susceptible.

Shipment and Storage. Sodium metabisulfite can be stored under air at ambient temperatures, but under humid conditions the product cakes and the available SO_2 content decreases as a result of oxidation. Therefore, storage should be under cool, dry conditions. The product is shipped in 22.7- and 45.4-kg polyethylene-lined (moistureproof) bags and 45.4- and 181-kg fiber drums. Steel drums are used for export. The solution is shipped in tank cars and trucks. Dry sodium metabisulfite can be handled in iron or steel equipment. Sodium bisulfite solutions can be handled in 316 and 347 stainless steels, lead, rubber, wood, Haveg, or glass-reinforced polyester equipment containers.

Economic Aspects. U.S. production of sodium metabisulfite in 1980 was ca 45,000 t. The principal U.S. manufacturers are Virginia Chemicals and Allied Corp. The price was $0.13/kg in 1970, $0.31/kg in 1975, and $0.44/kg in 1980.

Grades and Specifications. Sodium metabisulfite is available in photographic, food and NF, and technical grades (269). Typical analyses for the food and NF grade are 99 wt % $Na_2S_2O_5$, 0.6 wt % Na_2SO_4; 0.4 wt % Na_2SO_3, 3 ppm iron, 0.2 ppm arsenic, 0.1 ppm lead, and 2 ppm selenium. Typical analyses for the technical grade are 98 wt % $Na_2S_2O_5$, 1.1 wt % Na_2SO_4, 0.8 wt % Na_2SO_3, and 4 ppm iron. An aqueous solution assaying 38 wt % $NaHSO_3$ is also produced commercially.

Health and Safety Factors. Sodium metabisulfite is nonflammable but, when strongly heated, it releases sulfur dioxide. It has a rather low degree of acute toxicity; the LD_{50} (rat, oral) is 0.5–5 g/kg (270), although very large doses can cause colic, diarrhea, and death. The solid product and its aqueous solutions are mildly acidic and irritate the skin, eyes, and mucous membranes. The solid material when moist generates the pungent, irritating odor of sulfur dioxide. Food-grade sodium metabisulfite is permitted in those foods that are not recognized as sources of vitamin B_1, with which sulfur dioxide reacts (267) (see Vitamins, thiamine).

Uses. Sodium metabisulfite is extensively used as a food preservative and bleach in the same applications as for sulfur dioxide. Because sodium metabisulfite is most effective at low pH, the active agent is probably sulfur dioxide or sulfurous acid (267). Sodium bisulfite (sodium metabisulfite) is used in photography, as a reductant and as a preservative for thiosulfate fixing baths. Other reducing-agent applications include reduction of chromate in plating effluents to less toxic chromium salts, which can then precipitate upon addition of lime (271–272). Sodium metabisulfite is also used to reduce chlorine in industrial process water and wastewater. In the textile industry, sodium bisulfite containing metabisulfite is used as a bleach, especially for wool; as an antichlor after bleaching of nylon, for reducing vat dyes; and in rendering certain other dyes soluble. It is a less powerful reductant than sodium dithionite [*7775-14-6*].

In tanneries, sodium bisulfite is used to accelerate the unhairing action of lime. It is also used as a chemical reagent in the synthesis of surfactants. Addition to alpha olefins under radical-catalyzed conditions yields sodium alkylsulfonates, which are wetting agents. The addition of sodium bisulfite under base-catalyzed conditions to dialkyl maleates yields the sulfosuccinates (see Surfactants).

The reversible addition of sodium bisulfite to carbonyl groups is used in the purification of aldehydes. Sodium bisulfite also is employed in polymer and synthetic fiber manufacture in several ways. In free radical polymerization of vinyl and diene monomers, sodium bisulfite or metabisulfite is frequently used as the reducing component of a so-called redox initiator (see Initiators). Sodium bisulfite is also used as a color preventative and is added as such during the coagulation of crepe rubber.

Sodium Dithionite. ***Physical Properties.*** Sodium dithionite (sodium hydrosulfite, sodium sulfoxylate), $Na_2S_2O_4$, is a colorless solid and is soluble in water to the extent of 22 g/100 g of water at 20°C. The commercial product is an anhydrous white powder, assaying 88–90 wt % $Na_2S_2O_4$, with the remainder comprising sodium metabisulfite, sodium thiosulfate, and in some cases sodium formate and sodium carbonate (273).

Chemical Properties. Sodium dithionite is a combustible material which decomposes exothermically in the presence of heat, moisture, or air. Sulfur dioxide is given off violently when the dry salt is heated above 190°C. Aqueous solutions of dithionite are stable for several days at room temperature if the pH is adjusted to 8–9 and if air is excluded. Solutions decompose rapidly above 50°C. The decomposition proceeds by the following equation (273):

$$2\,S_2O_4^{2-} + H_2O \rightarrow 2\,HSO_3^- + S_2O_3^{2-}$$

The decomposition of dithionite is accelerated by thiosulfate, polysulfide, and acids. The addition of mineral acid to a dithionite solution produces first a red color which turns yellow on standing; subsequently, sulfur precipitates and evolution of sulfur dioxide takes place (274). Sodium dithionite is stabilized by sodium polyphosphate, sodium carbonate, and sodium salts of organic acids (275).

Sodium dithionite is most stable and effective as a reducing agent in alkaline solutions, although with excess strong alkali the following reaction occurs:

$$3\ Na_2S_2O_4 + 6\ NaOH \rightarrow 5\ Na_2SO_3 + Na_2S + 3\ H_2O$$

Dithionite is a stronger reducing agent than sulfite. Many metal ions, eg, Cu^+, Ag^+, Pb^{2+}, Sb^{3+}, and Bi^{3+}, are reduced to the metal, whereas TiO^{2+} is reduced to Ti^{3+} (274). Dithionite readily reduces iodine, peroxides, ferric salts, and oxygen. Some of the decolorizing applications of dithionite, eg, in clay bleaching, are based on the reduction of ferric iron.

Addition of sodium dithionite to formaldehyde yields the sodium salt of hydroxymethanesulfinic acid [*79-25-4*] ($HOCH_2SO_2Na$), which retains the useful reducing character of the sodium dithionite although somewhat attenuated in reactivity. The most important organic chemistry of sodium dithionite involves its use in reducing dyes, eg, anthraquinone vat dyes, sulfur dyes, and indigo, to their soluble leuco forms (see Dyes, anthraquinone). Dithionite can reduce various chromophores that are not reduced by sulfite. Dithionite can be used for the reduction of aldehydes and ketones to alcohols (276). Quantitative studies have been made of the reduction potential of dithionite as a function of pH and the concentration of other salts (277–278). The emf of an alkaline solution of dithionite is 1.13 V (273).

Manufacture. A review of the older manufacturing processes is given in ref. 279. Dithionites can be manufactured by the reduction of sulfites, bisulfites, or sulfur dioxide with reductants, eg, sodium or zinc. A typical older process involved the reduction of liquid sulfur dioxide in a continuous reactor with an aqueous slurry of zinc dust (280). During the 1970s, an alternative process was commercialized and was based on the reaction of sodium formate with sodium hydroxide and sulfur dioxide (281–283). This process is less expensive and environmentally preferable to the zinc powder method. The plants are simpler, more compact, and require less labor. Sodium formate is readily available as a by-product from pentaerythritol manufacture. Sodium dithionite made by the formate process is free of zinc which minimizes water pollution problems. A process involving electrolytically produced sodium amalgam to reduce sodium bisulfite also can produce zinc-free sodium dithionite (284–286); however, the formate process is the predominant one in the United States.

Solutions of sodium dithionite can be economically made at the users' location by a commercially important method that was developed recently (240–241). A solution of sodium borohydride and sodium hydroxide (Borol) reduces sulfur dioxide or sodium bisulfite to sodium dithionite. This process has gained in importance partly because of the availability of low cost sodium bisulfite or sulfur dioxide at many paper mills and because of the convenience and safety of handling the stable liquid reagents.

Shipment and Storage. Sodium dithionite is shipped as well-sealed, water-tight, 113-kg, lacquer-lined steel drums bearing the DOT flammable solids label. Sodium dithionite should be stored in a cool, dry place in tightly closed containers. Under these storage conditions, it can be stored with a negligible loss of activity over a prolonged period. Solutions should be prepared by adding the salt to water, not water to hydrosulfite; if the latter is inadvertently done, enough heat may be generated to cause ignition. The use of commercial aqueous sodium dithionite solution obviates the possibility of fire hazard but entails greater problems of storage stability and possible corrosivity resulting from decomposition products.

Economic Aspects. U.S. production of merchant sodium dithionite (100% $Na_2S_2O_4$ basis) was 38,000 t in 1970, 45,000 t in 1975, and 50,000 t in 1980 (287–288). Dithionite made at the location of use by sodium borohydride reduction of sulfur dioxide is probably not included in these figures. The price was \$0.53/kg in 1970, \$0.90/kg in 1975, and \$1.45/kg in 1980.

Virginia Chemicals is the largest U.S. producer and uses the formate process. Royce Chemical Co. produces smaller amounts through the zinc process. Olin Chemical Co. produces only the solution product and does so by the sodium amalgam process.

Grades. Sodium dithionite is sold as an 88–90 wt % active anhydrous solid. It is also available as a 70 wt % blend of sodium dithionite with sodium carbonate as a stabilizing agent. An aqueous, ca 10 wt % solution that is stabilized by alkali is marketed for pulp and paper, textile-dyeing, and clay bleaching applications.

Analytical Methods. Various analytical methods involve titration with oxidants, eg, hexacyanoferrate (ferricyanide), which oxidizes dithionite to sulfite. Iodimetric titration to sulfate in the presence of formaldehyde enables dithionite to be distinguished from sulfite, since aldehyde adducts of sulfite are not oxidized by iodine. Reductive bleaching of dyes can be used to determine dithionite, the extent of reduction being determined photometrically. Reviews of methods for determining mixtures of dithionite, sulfite, and thiosulfates are given in refs. 289–290.

Health and Safety Factors. Dry sodium dithionite, exposed to moist air, heats and can ignite spontaneously. In case of fire, the burning material must be deluged with water, as too little water may be worse than none at all. Carbon dioxide and dry extinguishers are ineffectual, because dithionite provides its own oxygen for combustion. Large amounts of sulfur dioxide are liberated, further complicating fire-fighting efforts; self-contained breathing units should be worn by all fire fighters.

Sodium dithionite is considered only moderately toxic. As with sulfites, fairly large doses of sodium dithionite can probably be tolerated, because oxidation to sulfate occurs. However, irritation of the stomach by the liberated sulfurous acid is expected. As a food additive, sodium dithionite is generally recognized as safe (GRAS) (291).

Uses. Dyeing consumes ca 50% of the U.S. production of sodium dithionite. An important application in recent years has been the indigo dyeing of cotton denims. Sodium dithionite is an important reductant for vat dyes and certain sulfur dyes. Bleaching of mechanical pulp and thermomechanical pulp comprises ca 20% of U.S. production. Miscellaneous uses include reductive bleaching of clay, glue, gelatin, soap, oils, and food products. Textile uses include dye stripping from fabric, cleaning of dyeing equipment, and oxygen scavenging in water used for high pressure boilers or for synthetic rubber polymerization. Since the toxicity of sodium dithionite is only slight, it is permitted in various food-contact applications. In recent years, users of dithionites have largely converted from zinc to sodium dithionite, to avoid water-pollution problems.

Zinc Dithionite. Zinc dithionite [*7779-86-4*], ZnS_2O_4, is a white, water-soluble powder. Although it exhibits somewhat greater stability in aqueous solution compared to sodium dithionite at a given temperature and pH, its use has declined sharply because of regulatory constraints on pollution of water by zinc.

Sodium and Zinc Formaldehyde Sulfoxylates. Although free sulfoxylic acid [*20196-46-7*], H_2SO_2, has not been isolated and its salts are in doubt, organic derivatives, which may be viewed as adducts of sulfoxylic acid, are commercially made. The latter are mainly sodium formaldehyde sulfoxylate [*149-44-0*], $HOCH_2SO_2Na$ (commercially sold as the dihydrate) and zinc formaldehyde sulfoxylate [*24887-06-7*] (279). These compounds are water-soluble reducing agents with uses similar to the dithionites but are more stable. They can be used in reducing and bleaching applications at lower pH values and at somewhat higher temperatures than the dithionites. For example, the formaldehyde sulfoxylates are useful at pH 3.4 as compared to an optimum pH of ca 6 for zinc dithionite and ca 9.5 for sodium dithionite. The most stable compound of this series is basic zinc formaldehyde sulfoxylate, which can be used as a reductant at ca 100°C, compared to ca 50°C for sodium dithionite.

In addition to applications in dyeing, sodium formaldehyde sulfoxylate is used as a component of the redox system in emulsion polymerization of styrene–butadiene rubber recipes.

Thiocyanic Acid and Its Salts

Free thiocyanic acid [*463-56-9*], HSCN, can be isolated from its salts, but it is not an article of commerce because of its instability, although dilute solutions can be stored briefly. Commercial derivatives of thiocyanic acid are principally ammonium and sodium thiocyanates as well as several organic thiocyanates. The chemistry and biochemistry of thiocyanic acid and its derivatives have been reviewed extensively (292–293).

Ammonium Thiocyanate. ***Physical Properties.*** Ammonium thiocyanate [*1762-95-4*], NH_4SCN, is a hygroscopic crystalline solid which deliquesces at high humidities (294–295). It melts at 149°C with partial isomerization to thiourea. It is soluble in water to the extent of 65 wt % at 25°C and 77 wt % at 60°C. It is also soluble to 35 wt % in methanol and 20 wt % in ethanol at 25°C. It is highly soluble in liquid ammonia and liquid sulfur dioxide, and moderately soluble in acetonitrile.

Chemical Properties. Ammonium thiocyanate rearranges upon heating to an equilibrium mixture with thiourea: 30.3 wt % thiourea at 150°C, 25.3 wt % thiourea at 180°C (293,294). At 190–200°C, dry ammonium thiocyanate decomposes to hydrogen sulfide, ammonia, and carbon disulfide, leaving guanidine thiocyanate [*56960-89-5*] as a residue. Aqueous solutions of ammonium thiocyanate are weakly acidic; a 5 wt % solution has a pH of 4–6.

Thiocyanates are rather stable to air, to oxidation and to dilute nitric acid. Of considerable practical importance are the reactions of thiocyanate with metal cations. Silver, mercury, lead, and cuprous thiocyanates precipitate. Many metals form complexes. The deep red complex of ferric iron with thiocyanate, $[Fe(SCN)_6]^{3-}$, is an effective indicator for either ion. Various metal thiocyanate complexes with transition metals can be extracted into organic solvents.

The organic chemistry of thiocyanates is notably that of nucleophilic displacement of alkyl halides by thiocyanate anion to form alkyl thiocyanates:

$$RCH_2Cl + SCN^- \rightarrow RCH_2SCN + Cl^-$$

In certain instances, the thiocyanates can rearrange to isothiocyanates:

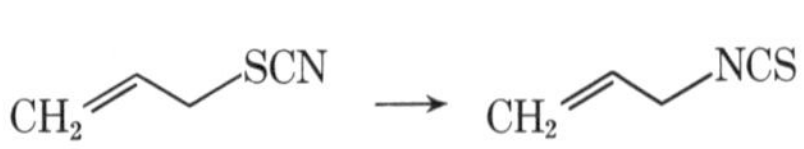

[*764-49-8*] [*57-06-7*]

Ammonium thiocyanate reacts with amines yielding thioureas:

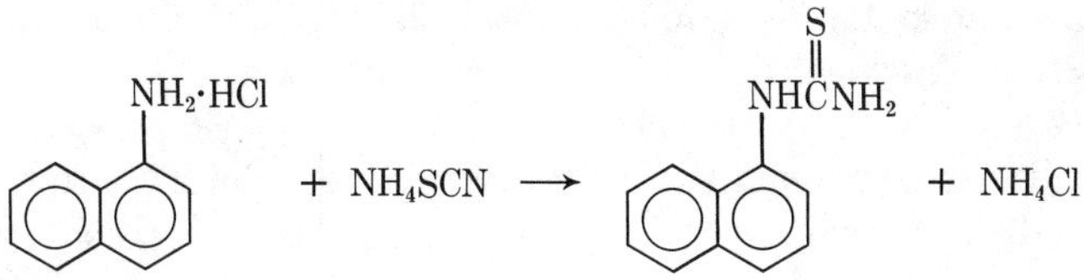

[*86-88-4*]

However, the commercial production of thiourea [*62-56-6*] is probably not by this route, but rather by the reaction of hydrogen sulfide with cyanamide.

Manufacture. An extensive technology was developed initially in the 1930s for isolation of ammonium thiocyanate from coke-oven gases, but this technology is no longer practiced in the United States (292). However, such thiocyanate recovery processes are used industrially in Europe. Likewise, the direct sulfurization of cyanides to thiocyanates is not practiced commercially in the United States. The principal route used in the United States is the reaction of carbon disulfide with aqueous ammonia, which proceeds by way of ammonium dithiocarbamate [*513-74-6*]. Upon heating, the ammonium dithiocarbamate decomposes to ammonium thiocyanate [*1762-95-4*] and hydrogen sulfide.

$$CS_2 + 2\ NH_3(aq) \longrightarrow NH_2\overset{\overset{\displaystyle S}{\|}}{C}SNH_4 \longrightarrow NH_4SCN + H_2S$$

In a typical batch operation, carbon disulfide is added to four molar equivalents of 25–30 wt % aqueous ammonia in a stirred vessel, which is kept closed for the first hour or two. The reaction is moderately exothermic and requires cooling. After 2–3 h, when substantially all of the disulfide has reacted, the reaction mixture is heated to decompose dithiocarbamate and trithiocarbonate and vented to an absorption system to collect ammonia, hydrogen sulfide, and any unreacted carbon disulfide.

This reaction can also be run in a continuous fashion. In the initial reactor, agitation is needed until the carbon disulfide liquid phase reacts fully; the solution can then be vented to a tower where ammonia and hydrogen sulfide are stripped countercurrently by a flow of steam from boiling ammonium thiocyanate solution. Ammonium sulfide solution is made as a by-product. The stripped ammonium thiocyanate solution is normally boiled to a strength of 55–60 wt %, and much of it is sold at this concentration. The balance is concentrated and cooled to produce crystals, which are removed by centrifugation.

Shipment and Storage. The crystalline material is shipped as a nonhazardous material, in polyethylene-lined fiber drums. The solution can be shipped in drums or in bulk. Suitable materials of construction for handling ammonium thiocyanate are aluminum, 316 stainless steel, rubber, poly(vinyl chloride), and glass-reinforced epoxy. Steel, 304 stainless steel, and copper alloys should be avoided (294–295).

Economic Aspects. Ammonium thiocyanate capacity in the United States is well in excess of 9000 t/yr on a 100% solids basis, but current production is substantially less than this capacity. It is made by two suppliers: J. T. Baker and Argus. By-product material from coke ovens is no longer available in the United States, although it may be used in Europe. Production growth is small. The price of ammonium thiocyanate on a 100% basis was $1.06/kg in 1975 and $1.69/kg in 1980.

Grades and Specifications. A technical crystal of 90% minimum assay with ca 2 wt % maximum water content is commercially available as well as a 50–55 wt % aqueous solution (294–295). The latter is the predominant product.

Analytical Methods. Thiocyanate is quantitatively precipitated as silver thiocyanate, and thus can be conveniently titrated with silver nitrate. In the presence of a ferric salt, a red-brown color, which is produced by the ferric thiocyanate complex, indicates the end point.

Health and Safety Factors. Ammonium thiocyanate has a low acute toxicity; the LD_{50} in mammals is 500–1000 mg/kg (294–295). Exposure of the skin, eyes, and mucous membranes should be avoided. Skin eruptions and nasal discharge are symptoms of overexposure. Prolonged exposure may also produce cramps, nausea, and nervous-system disturbances. Thiocyanates are destroyed readily by soil bacteria and by biological treatment systems in which the organisms become acclimatized to thiocyanate. Pyrolysis products and combustion products can include toxic hydrogen cyanide, hydrogen sulfide, sulfur oxides, and nitrogen oxides.

Uses. Ammonium thiocyanate is a chemical intermediate for the synthesis of several proprietary agricultural chemicals, mainly herbicides. Its use directly as a nonselective herbicide is obsolete, although it is still employed as an adjuvant for aminotriazole in nonselective herbicides, which are used principally for brush and perennial weed control. It also is used in photography as a stabilizing agent. It makes undeveloped silver halide substantially insensitive to light. An old use that is still practiced is the incorporation of ca 0.1 wt % ammonium thiocyanate in ammoniacal nitrogen fertilizer solutions to inhibit the corrosion of steel. It is also used in various rustproofing compositions. Ammonium thiocyanate is used in the separation of hafnium from zirconium. Hafnium is extracted as a thiocyanate complex into an organic phase.

There are many smaller specialized uses for ammonium thiocyanate, notable ones being stabilization of glue formulations, as an ingredient in antibiotic fermentations, and as an adjuvant in textile dyeing and printing. A rapidly developing newer use is as a tracer in oil fields. The flow pattern of injected water in enhanced oil recovery operations can be followed by taking water samples from producing wells and analyzing them for thiocyanate colorimetric measurement of the red ferric complex.

Sodium and Potassium Thiocyanates. ***Physical and Chemical Properties.*** Sodium thiocyanate [*540-72-7*], NaSCN, is a colorless deliquescent crystalline solid (mp 323°C). It is soluble in water to the extent of 58 wt % NaSCN at 25°C and 69 wt % at 100°C. It is also highly soluble in methanol and ethanol, and moderately soluble in acetone. Potassium thiocyanate [*333-20-0*], KSCN, is also a colorless crystalline solid (mp 172°C) and is soluble in water to the extent of 217 g/100 g of water at 20°C and in acetone and alcohols. Much of the chemistry of sodium and potassium thiocyanates is that of the thiocyanate anion (292–294).

Manufacture, Shipment, and Analysis. In the United States, sodium and potassium thiocyanates are made by adding caustic soda or potash to ammonium thiocyanate, followed by evaporation of the ammonia and water. The products are sold either as 50–55 wt % aqueous solutions, in the case of sodium thiocyanate, or as the crystalline solids with one grade containing 5 wt % water and a higher assay grade containing a maximum of 2 wt % water. In Europe, the thiocyanates may be made by direct sulfurization of the corresponding cyanide.

Shipping, analysis, and health factors are similar to those of ammonium thiocyanate, except that the alkali thiocyanates are more thermally stable.

Sodium thiocyanate is best handled in 316 stainless steel. At room temperature where some iron contamination can be tolerated, type 304 can be used. Aluminum corrodes more rapidly than stainless steel, but some alloys can be used below 60°C. Some but not all rubber equipment is satisfactory.

Economic Aspects. Capacity for sodium thiocyanate in the United States is substantially the same as that for ammonium thiocyanate, since both products can be made in the same plants, but currently production is estimated at only slightly over 1000 t. The rate of growth is slight. The price on a 100 wt % basis in 1975 was $1.10/kg and in 1980 it was $1.74/kg. Most sodium thiocyanate is sold as the solution. Potassium thiocyanate is a much lower volume product. Its price was $2.86/kg in 1975 and $6.60/kg in 1980.

Uses. The largest use for sodium thiocyanate is as the 50–60 wt % aqueous solution, as a component of the spinning solvent for acrylic fibers (see Acrylic and modacrylic fibers; Acrylonitrile polymers). Other textile applications are as a fiber swelling agent and as a dyeing and printing assist.

Sodium thiocyanate and other thiocyanate salts are used to prepare organic thiocyanates, eg: methylene dithiocyanate [*6317-18-6*] (Stauffer N-948), $CH_2(SCN)_2$, a broad-spectrum industrial biocide used as a slimicide in paper production, also for controlling microbial growth in cooling water, cutting oils and fluids, hides, oil-field injection waters, drilling muds, adhesives, paints, and other polymer emulsions; thiocyanomethylthiobenzothiazole [*21564-17-0*] (Busan 30, Buckman Laboratories, Inc.),

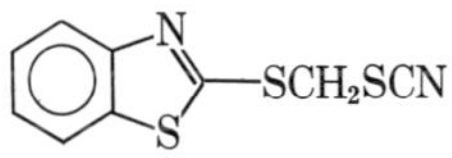

a fungicide used as a seed and bulb treatment and as a preservative in paint; and 2-(2-butoxyethoxy)ethyl thiocyanate [*112-56-1*] (Lethane 384, Rohm & Haas Co.). The latter is sold as a contact insecticide and knockdown agent for mosquito control and for household and industrial insecticide sprays (see Industrial antimicrobial agents; Insect control technology).

Lesser amounts of sodium thiocyanate are used in color toning photographic paper, as a stabilizer in rapid film development, and as a sensitizing agent in color negative-film emulsions.

BIBLIOGRAPHY

"Sulfur Compounds—Structures" and "Sulfur Compounds, Inorganic" in *ECT* 1st ed., Vol. 13, pp. 374–430, by P. Macaluso, Stauffer Chemical Co.; "Sulfur Compounds" in *ECT* 2nd ed., Vol. 19, pp. 367–424, by P. Macaluso, Stauffer Chemical Co.

1. G. Gattow and W. Behrendt in A. Senning, ed., *Topics in Sulfur Chemistry*, Vol. 2, Georg Thieme Verlag, Stuttgart, 1977, pp. 197–206.
2. *Gmelins Handbuch der Anorganischen Chemie*, 8th ed., System No. 14, Part D4, Springer-Verlag, Berlin, Heidelberg, 1977, pp. 1–28.
3. W. Stadlbauer and T. Kappe, *Chem. Ztg.* **101**(3), 137 (1977).
4. A. O. Diallo and J. H. Dixmier, *C. R. Acad. Sci. Ser. C* **263**, 375 (1966).
5. A. P. Ginsberg, J. L. Lundberg, and W. E. Silverthorn, *Inorg. Chem.* **10**, 2079 (1971).
6. M. Herberhold, *Nachr. Chem. Tech. Lab.* **29**, 365 (1981).

7. J. B. Hyne, *Understanding Sulphur*, paper presented at Sulphur 81, International Conference, Calgary, Alberta, May 25–28, 1981.
8. A. R. Shirley, Jr., and R. S. Meline, *Adv. Chem. Ser.* **140,** 38 (1975).
9. U.S. Pat. 4,143,214 (March 6, 1979), C. H. Chang and J. M. Longo (to Exxon Research and Engineering Co.).
10. J. P. Wibaud, *Recl. Trav. Chim. Pays-Bas* **41,** 153 (1922).
11. R. Ciusa, *Gazz. Chim. Ital.* **55,** 385 (1925), and references therein.
12. D. D. Wagman, W. H. Evans, V. B. Parker, I. Halow, S. M. Bailey, and R. H. Schumm, *Selected Values of Chemical Thermodynamic Properties*, NBS Technical Note 270-3, National Bureau of Standards, U.S. Department of Commerce, Washington, D.C., 1968.
13. *Gmelins Handbuch der Anorganischen Chemie*, 8th ed., System No. 14, Part D5, Springer-Verlag, Berlin, Heidelberg, 1977, pp. 2–140.
14. D. B. Robinson and N. H. Senturk, *J. Chem. Thermodyn.* **11,** 461 (1979).
15. R. D. Miranda, D. B. Robinson, and H. Kalra, *J. Chem. Eng. Data* **21**(1), 62 (1976).
16. M. B. Mick, *Hydrocarbon Process.* **55**(7), 137 (1976).
17. S. Weber and G. McClure, *The Cosden/Malaprop Process for Light Hydrocarbon Desulfurization*, Annual Meeting Paper AM-81-49, National Petroleum Refiners' Association, Washington, D.C., 1981.
18. R. J. Ferm, *Chem. Rev.* **57,** 621 (1957).
19. *Carbon Disulfide, Carbonyl Sulfide: Literature Review and Environmental Assessment*, NTIS Report No. PB-257947, report for Environmental Protection Agency, U.S. Department of Commerce by Stanford Research Institute, Menlo Park, Calif., 1976.
20. A. Kohl and F. Riesenfeld, *Gas Purification*, 2nd ed., Gulf Publishing Co., Houston, Texas, 1974, pp. 607–633.
21. K. Nelson and L. Wolfe, *Oil Gas J.*, 183 (July 27, 1981).
22. P. J. Crutzen, L. E. Heidt, J. P. Krasnec, W. H. Pollock, and W. Seller, *Nature (London)* **282,** 253 (1979).
23. T. E. Graedel, G. W. Kammlot, and J. P. Franey, *Science* **212,** 663 (1981).
24. O. P. Strausz in A. Senning, ed., *Sulfur in Organic and Inorganic Chemistry*, Vol. 2, Marcel Dekker, New York, 1972, pp. 1–12.
25. U.S. Pat. 3,235,333 (Feb. 15, 1966), E. A. Swakon and E. Field (to Standard Oil Co. of Indiana).
26. U.S. Pat. 4,007,254 (Feb. 8, 1977), R. Buathier, A. Combes, F. Pierrot, and H. Guerpillon (to Rhône-Poulenc Industries).
27. A. Attar, *Fuel* **57,** 201 (1978).
28. R. Steudel, *Z. Anorg. Allg. Chem.* **346,** 255 (1966).
29. R. G. Confer and R. S. Brief, *Am. Ind. Hyg. Assoc. J.*, 843 (Nov. 1975).
30. C. P. Changelis and R. A. Neal, *Toxicol. Appl. Pharmacol.* **55,** 198 (1980).
31. B. Zwanenburg, L. Thijs, and S. Strating, *Tetrahedron Lett.* **51,** 4461 (1969).
32. H. Tilles in N. Kharasch and C. Y. Meyers, eds., *The Chemistry of Organic Sulfur Compounds*, Vol. 2, Pergamon Press, Oxford, 1966, pp. 311–336.
33. S. Sharma, *Synthesis*, 803 (Nov. 1978).
34. *Registry of Toxic Effects of Chemical Substances*, U.S. Department of Health and Human Services, Public Health Service, Center for Disease Control, National Institute for Occupational Safety and Health, Cincinnati, Ohio, Sept. 1980.
35. *Perchloromethyl Mercaptan*, Product Report, Stauffer Chemical Company, New York, 1966.
36. *Perchloromethyl Mercaptan*, Stauffer Chemical Company, Agricultural Chemical Division, Chemical Intermediates Data Sheet, Westport, Conn., Nov. 1978.
37. F. A. Drahowzal in N. Kharasch, ed., *Organic Sulfur Compounds*, Vol. 1, Pergamon Press, Oxford, 1961, pp. 361–374.
38. J. Silhánek and M. Zbirovský, *Collect. Czech. Chem. Commun.* **42,** 2518 (1977).
39. J. Silhánek and M. Zbirovský, *Int. J. Sulfur Chem.* **8,** 423 (1973).
40. E. Kuhle, *Synthesis*, 561 (Nov. 1970).
41. H. Kloosterziel, *Q. Rep. Sulfur Chem.* **2,** 353 (1967).
42. H. Senning, *Chem. Rev.* **65,** 385 (1965).
43. U.S. Pat. 3,673,246 (June 27, 1972), G. Meyer and H. Mägerlein (to Glanzstoff A.-G.); H. Mägerlein, G. Meyer, and H. Rupp, *Synthesis*, 478 (Sept. 1971).
44. U.S. Pat. 3,808,270 (April 30, 1974), H. Rupp, G. Meyer, H. Zengel, and H. Mägerlein (to Akzo N.V.).
45. U.S. Pat. 4,092,357 (May 30, 1978), C. C. Greco and E. N. Walsh (to Stauffer Chemical Co.); U.S. Pat. 4,093,651 (June 6, 1978), M. L. Honig, C. C. Greco, and E. N. Walsh (to Stauffer Chemical Co.); U.S.

Pat. 4,100,190 (July 11, 1978), V. C. Martines and R. G. Campbell (to Stauffer Chemical Co.); U.S. Pat. 4,101,446 (July 18, 1978), M. L. Honig (to Stauffer Chemical Co.).
46. L. Kremer and L. D. Spicer, *Anal. Chem.* **45,** 1963 (1973).
47. A. Hellwig and D. Hempel, *Z. Chem.* **7,** 315 (1967).
48. *Hazard Information Review No. IR-216,* prepared by Enviro Control, Inc. for TSCA Interagency Testing Committee, Washington, D.C., Nov. 25, 1980.
49. E. H. Vernot, J. D. MacEwen, C. C. Haun, and E. R. Kinkaid, *Toxicol. Appl. Pharmacol.* **42,** 417 (1977).
50. M. Sittig, *Pesticide Manufacturing and Toxic Materials Control Encyclopedia,* Noyes Data Corp., Park Ridge, N.J., 1980, pp. 399–400.
51. F. Kuhle, E. Klauke, and F. Grewe, *Angew. Chem.* **76,** 807 (1964).
52. U.S. Pat. 3,296,302 (Jan. 3, 1967), E. D. Weil, K. J. Smith, and E. J. Geering (to Hooker Chemical Corp.).
53. U.S. Pat. 3,200,146 (Aug. 10, 1965), E. D. Weil, E. J. Geering, and K. J. Smith (to Hooker Chemical Corp.).
54. Belg. Pat. 613,887 (Feb. 28, 1962), (to California Research Corp.).
55. U.S. Pat. 3,259,653 (July 5, 1966), E. D. Weil, E. J. Geering, and K. J. Smith (to Hooker Chemical Corp.).
56. U.S. Pats. 3,144,482; 3,144,483 (Aug. 11, 1964), R. B. Flay (to California Research Corp.).
57. U.S. Pat. 3,178,447 (April 13, 1965), G. K. Kohn (to California Research Corp.).
58. B. Meyer, *Sulfur, Energy, and Environment,* Elsevier Scientific Publishing Co., Amsterdam, 1977.
59. J. B. Hyne, *Recent Developments in Sulfur Production from Hydrogen Sulfide Containing Gases,* paper presented at 181st National Meeting, ACS, Atlanta, Ga., March 29–April 3, 1981.
60. M. Schmidt and W. Siebert in J. C. Bailar, Jr., H. J. Emeleus, R. Nyholm, and A. F. Trotman-Dickenson, eds., *Comprehensive Inorganic Chemistry,* Pergamon Press, Oxford, 1973, pp. 826–837.
61. *Gmelins Handbuch der Anorganischen Chemie,* 8th ed., System No. 9, Part B1, Verlag Chemie, Weinheim/Bergstrasse, 1953, pp. 1–125.
62. *Hydrogen Sulfide, Product Safety Information Sheet,* Stauffer Chemical Co., Industrial Chemical Division, Westport, Conn., 1973.
63. H. J. Neuberg, J. F. Atherley, and L. G. Walker, *Girdler-Sulfide Process Physical Properties,* Atomic Energy of Canada Report AECL-5702, Chalk River Nuclear Laboratories, Chalk River, Ontario, 1977.
64. G. J. Besserer and D. B. Robinson, *J. Chem. Eng. Data* **20,** 157 (1975).
65. H. Kalra, T. R. Krishnan, and D. B. Robinson, *J. Chem. Eng. Data* **21,** 222 (1976).
66. D. B. Robinson, G. P. Hamaliuk, T. R. Krishnan, and P. R. Bishnoi, *J. Chem. Eng. Data* **20,** 153 (1975).
67. M. P. Burgess and R. P. Germann, *AIChE J.* **15,** 272 (1969).
68. D. K. Judd, *Hydrocarbon Process.* **57,** 122 (April 1978).
69. N. L. Yarym-Agaev, V. G. Matvienko, and N. V. Povalyaeva, *J. Appl. Chem. USSR* **53,** 1810 (1980).
70. C. Ouwerkerk, *Hydrocarbon Process.* **57,** 89 (April 1978).
71. F. C. Vidaurri and L. C. Kahre, *Hydrocarbon Process.* **56,** 333 (Nov. 1977).
72. B. L. Crynes, *Chemical Reactions as a Means of Separation,* Marcel Dekker, Inc., New York, 1977, pp. 155–197.
73. J. I. Lee, F. D. Otto, and A. E. Mather, *J. Chem. Eng. Data* **20,** 161 (1975).
74. J. D. Lawson and A. W. Garst, *J. Chem. Eng. Data* **21,** 20 (1976).
75. E. E. Isaacs, F. D. Otto, and A. E. Mather, *J. Chem. Eng. Data* **22,** 317 (1977).
76. J. L. Martin, F. D. Otto, and A. E. Mather, *J. Chem. Eng. Data* **23,** 163 (1978).
77. T. K. Wiewiorowski and F. J. Touro in A. V. Tobolsky, ed., *The Chemistry of Sulfides,* Interscience Publishers, a division of John Wiley & Sons, Inc., New York, 1968.
78. M. E. D. Raymont and R. K. Kerr, *Alberta Sulphur Research Quarterly Bulletin* **12**(4), 38 (1976).
79. M. E. D. Raymont, *Alberta Sulphur Research Quarterly Bulletin* **8**(1), 24 (1971).
80. M. E. D. Raymont, *Alberta Sulphur Research Quarterly Bulletin* **10**(4), 1 (1974); M. E. D. Raymont and J. B. Hyne, *Alberta Sulphur Research Quarterly Bulletin* **12**(2), 3 (1975); **12**(3), 1 (1975).
81. U.S. Pat. 3,962,409 (June 8, 1976), Y. Kutera, N. Todo, and N. Fukuda (to Agency of Industrial Science and Technology).
82. M. Grätzel, *Chem. Eng. News* **59**(30), 40 (July 27, 1981).
. N. G. Vilesov, *J. Appl. Chem. USSR* (Engl. trans. ed.) **53,** 1763 (1981).

84. D. Lyons and G. Nickless in G. Nickless, ed., *Inorganic Sulphur Chemistry*, Elsevier Publishing Co., Amsterdam, 1968, pp. 509–533.
85. R. A. Ross and M. R. Jeanes, *Ind. Eng. Chem. Prod. Res. Dev.* **13**(2), 102 (1974).
86. H. M. Castrantas, *J. Pet. Technol.*, 914 (May 1981).
87. E. E. Reid, *Organic Chemistry of Bivalent Sulfur*, Vol. 1, Chemical Publishing Co., Inc., New York, 1958, pp. 18–21.
88. A. Schöberl and A. Wagner in *Houben-Weyl, Methoden der Organischen Chemie*, 4th ed., Vol. 5, Georg Thieme Verlag, Stuttgart, pp. 20–21.
89. Can. Pat. 902,113 (June 6, 1972), D. J. Martin and E. D. Weil (to Stauffer Chemical Co.).
90. M. G. Voronkov and E. N. Deryagina, *Phosphorus Sulfur* **7**, 123 (1979).
91. R. F. Bacon and E. S. Boe, *Ind. Eng. Chem.* **37**, 469 (1945).
92. R. N. Tuttle and R. D. Kane, eds., *H_2S Corrosion in Oil and Gas Production*, National Association of Corrosion Engineers, Houston, Texas, 1981.
93. *Sulfide Stress Cracking Resistant Metallic Material for Oil Field Equipment*, NACE Standard MR-01-75, 1980 revision, Technical Practices Committee, National Association of Corrosion Engineers, Houston, Texas, 1980.
94. A. I. Asphahani, *Corrosion (Houston)* **37**, 327 (1981).
95. M. C. Manderson and C. D. Cooper, *Sulfur Supply and Demand and Its Relationship to New Energy Sources*, paper presented at 181st National Meeting of the American Chemical Society, Atlanta, Ga., April 2, 1981.
96. *NIOSH Manual of Analytical Methods*, Method No. P & CAM 296, Publication No. 80-125, Vol. 6, U.S. Department of Health and Human Services, Washington, D.C., 1980.
97. *NIOSH Manual of Analytical Methods*, Method No. 54, Vol. 2, U.S. Department of Health and Human Services, Washington, D.C., 1977.
98. *Sulphur* **124**, 47 (May/June 1976); **128**, 43 (Jan./Feb. 1977).
99. *Occupational Exposure to Hydrogen Sulfide*, Criteria Document, Department of Health, Education, and Welfare (NIOSH) Publication No. 77-158; National Technical Information Service Publication No. PB274196, National Institute of Occupational Safety and Health, May 1977.
100. *Hydrogen Sulfide*, Report of the Subcommittee on Hydrogen Sulfide, Committee of Medical and Biologic Effects of Environmental Pollution, Division of Medical Science, National Research Council, University Park Press, Baltimore, Maryland, 1979.
101. K. W. C. Burton and P. Machmer in ref. 84, pp. 336–366.
102. H. F. Meissner, E. R. Conway, and H. S. Mickley, *Ind. Eng. Chem.* **48**, 1347 (1956).
103. E. Muller and J. B. Hyne, *Can. J. Chem.* **46**, 2341 (1968).
104. J. B. Hyne, *Chem. Can.*, 26 (May 1978).
105. T. H. Ledford, *Anal. Chem.* **53**, 908 (1981).
106. *Sulfur Chlorides*, Bulletin B-11317-75, Stauffer Chemical Co., Industrial Chemical Division, Westport, Conn., 1975.
107. *Gmelins Handbuch der Anorganischen Chemie*, Vol. 9, Part B(3), Verlag Chemie, Weinheim/Bergstrasse, 1963, pp. 1748–1790.
108. G. A. Takacs, *J. Chem. Eng. Data* **23**, 174 (1978).
109. K. C. Mills, *Thermodynamic Data for Inorganic Sulphides, Selenides, and Tellurides*, Butterworths, Ltd., London.
110. L. A. Wiles and Z. S. Ariyan, *Chem. Ind. (London)*, 2102 (Dec. 22, 1962).
111. M. Muhlstadt and D. Martinetz, *Z. Chem.* **14**, 297 (1974).
112. W. Hofmann, *Vulcanization and Vulcanizing Agents*, Palmerton Publishing Co., New York, 1965, p. 79.
113. U.S. Pat. 3,884,985 (May 20, 1975), K. P. Hoffman (to Stauffer Chemical Co.).
114. E. R. Braithwaite, ed., *Lubrication and Lubricants*, Elsevier Publishing Co., New York, 1967.
115. Representative patents are the following: U.S. Pat. 2,708,199 (May 10, 1955), H. H. Eby (to Continental Oil Co.); U.S. Pat. 3,346,549 (Oct. 10, 1967), J. F. Ford and E. S. Forbes (to British Petroleum Co., Ltd.); U.S. Pat. 3,471,404 (Oct. 7, 1969), H. Myers (to Mobil Oil Corp.); U.S. Pat. 4,204,969 (May 27, 1980), A. G. Papay and J. P. O'Brien (to Edwin Cooper, Inc.); U.S. Pat. 4,225,488 (Sept. 30, 1980), A. G. Horodysky and P. S. Landis (to Mobil Oil Corp.); U.S. Pat. 4,240,958 (Dec. 23, 1980), M. Braid (to Mobil Oil Corp.).
116. *Sulfur Dichloride*, Product Safety Information Data Sheet, Stauffer Chemical Co., Industrial Chemical Division, Westport, Conn., 1973.

117. E. J. Corey and E. Block, *J. Org. Chem.* **31,** 1663 (1966); E. D. Weil, K. S. Smith, and R. Gruber, *J. Org. Chem.* **31,** 1669 (1966); F. Lautenschlaeger, *J. Org. Chem.* **31,** 1679 (1966).
118. F. Lautenschlaeger in ref. 77, pp. 73–81.
119. U.S. Pat. 3,219,413 (Nov. 23, 1965), K. E. Kunkel and D. S. Rosenberg (to Hooker Chemical Corp.).
120. R. J. Rosser and F. R. Whitt, *J. Appl. Chem.* **10,** 229 (1960).
121. U.S. Pats. 3,071,441 and 3,071,442 (Jan. 1, 1963), J. H. Schmadebeck (to Hooker Chemical Corp.).
122. C. M. Blow, ed., *Rubber Technology and Manufacture*, CRC Press, Cleveland, Ohio, 1971, p. 200.
123. *Thionyl Chloride*, Product Information Bulletin, Mobay Chemical Corp., Inorganic Chemicals Division, Pittsburgh, Pa., June 1981.
124. *Thionyl Chloride*, Product Information Bulletin, Farbenfabriken Bayer A.-G., Leverkusen, FRG, June 1962.
125. *Thionyl Chloride*, Data Sheet 718-H, Process Chemicals Division, Hooker Chemicals & Plastics Corp., Niagara Falls, N.Y., 1981.
126. Ref. 107, pp. 1791–1802.
127. L. F. Fieser and M. Fieser, *Reagents for Organic Synthesis*, John Wiley & Sons, Inc., New York, 1967, pp. 1158–1163.
128. M. Davis, H. Skuta, and A. J. Krusback, *Mech. React. Sulfur Compd.* **5,** 1 (1970).
129. S. S. Pizey, *Synthetic Reagents*, Vol. 1, John Wiley & Sons., Inc., New York, 1974, pp. 321–357.
130. T. G. Squires, W.W. Schmidt, and C. S. McCandlish, Jr., *J. Org. Chem.* **40,** 134 (1975).
131. G. Hambrecht, K. Konig, and G. Stubenrauch, *Angew. Chem. Int. Ed. Engl.* **20,** 151 (1981).
132. U.S. Pat. 2,362,057 (Nov. 7, 1944), J. P. Edwards (to Hooker Electrochemical Co.).
133. U.S. Pat. 2,431,823 (Dec. 2, 1947), A. Pechukas (to Pittsburgh Plate Glass Co.).
134. Ger. Pat. 939,571 (Feb. 23, 1956), H. Jonas and P. Lueg (to Farbenfabriken Bayer A.-G.).
135. U.S. Pat. 3,156,529 (Nov. 10, 1964), D. S. Rosenberg and H. Flaxman (to Hooker Chemical Corp.).
136. U.S. Pat. 3,155,457 (Nov. 3, 1964), K. E. Kunkel (to Hooker Chemical Corp.).
137. U.S. Pat. 4,229,565 (Oct. 21, 1980), H. C. Gardner and M. Matzner (to Union Carbide Corp.).
138. U.S. Pat. 3,854,927 (Dec. 17, 1974), K. Kiehs, K. H. Koenig, and A. Fischer (to BASF A.-G.).
139. U.S. Pat. 3,272,854 (Sept. 13, 1966), R. A. Covey, A. E. Smith, and W. L. Hubbard (to Uniroyal Corp.).
140. U.S. Pat. 4,172,852 (Oct. 30, 1979), W. F. Ark, J. H. Kawakami, and U. A. Steiner (to Union Carbide Corp.).
141. *Elec. Eng. Trans.*, 29 (July 24, 1978).
142. C. R. Schlaikjer, F. Goebel, and N. Marincic, *J. Electrochem. Soc.*, 513 (April 1979).
143. *Sulfuryl Chloride Data Bulletin*, Hooker Chemical Co., Niagara Falls, N.Y., undated.
144. Ref. 108, pp. 1802–1821.
145. I. Tabushi and H. Kitaguchi in J. S. Pizey, ed., *Synthetic Reagents*, Vol. 4, Ellis Horwood, distributed by John Wiley & Sons, Inc., Chichester, UK, 1981, pp. 336–396.
146. Ref. 127, pp. 1128–1131.
147. R. Stroh and W. Hahn in *Houben-Weyl, Methoden der Organischen Chemie*, Vol. 5/3, Georg Thieme Verlag, Stuttgart, 1962, pp. 873–890.
148. U.S. Pat. 3,399,214 (Aug. 27, 1968), M. Kulka, D. S. Thiara, and W. A. Harrison (to Uniroyal, Inc.).
149. T. F. Kozlova, G. B. Shaklova, V. F. Belugin, V. G. Zhelonkin, and N. V. Sedov, *Sov. Chem. Ind.* **6,** 394 (1971).
150. D. Masilamani and M. M. Rogić, *J. Org. Chem.* **46,** 4484 (1981).
151. F. Muth in E. Müller, ed., *Houben-Weyl, Methoden der Organischen Chemie*, Vol. 9, Georg Thieme Verlag, Stuttgart, 1955, p. 515.
152. U.S. Pat. 2,530,410 (Nov. 21, 1950), M. Sveda (to E. I. du Pont de Nemours & Co., Inc.).
153. Brit. Pat. 949,714 (Feb. 19, 1964), (to Farbenfabriken Bayer A.-G.).
154. H. W. Roesky, *Angew. Chem. Int. Ed. Engl.* **18,** 91 (1979).
155. M. Becke-Goehring in *Gmelins Handbuch der Anorganischen Chemie*, 8th ed., Vol. 32, Part 1, Springer-Verlag, Berlin, 1977.
156. M. M. Labes, P. Love, and L. F. Nichols, *Chem. Rev.* **79,** 1 (1979).
157. M. J. Cohen, A. F. Garito, A. J. Heager, A. G. MacDiarmid, C. M. Mikulski, M.S. Saran, and J. Kleppinger, *J. Am. Chem. Soc.* **98,** 3844 (1976).
158. P. W. Schenk and R. Steudel in ref. 84, pp. 369–418.
159. Ref. 60, pp. 859–868.
160. P. W. Schenk and R. Steudel, *Angew. Chem. Int. Ed. Engl.* **4,** 402 (1965).

161. R. M. Dodson and R. F. Sauers, *Chem. Commun.* 1189 (1967); R. M. Dodson and J. P. Nelson, *Chem. Commun.* 1159 (1969).
162. R. T. Gillespie and P. K. Ummat, *Inorg. Chem.* **11,** 1674 (1972).
163. R. Steudel and M. Rebsch, *Z. Anorg. Allg. Chem.* **413,** 252 (1975).
164. R. Steudel and J. Latte, *Angew. Chem. Int. Ed. Engl.* **13,** 603 (1974).
165. C. L. Yaws, K. Y. Li, and C. H. Kuo, *Chem. Eng.*, 85 (July 8, 1974).
166. N. B. Vargaftik, *Tables on the Thermophysical Properties of Liquids and Gases*, Halsted Press, a division of John Wiley & Sons, Inc., New York, 1975, p. 503.
167. Ref. 108, pp. 1131–1293.
168. D. F. Burow in J. J. Lagowski, ed., *Chemistry of Nonaqueous Solvents*, Vol. 3, Academic Press, New York, 1970, pp. 137–185.
169. G. N. Brown, S. L. Torrence, A. J. Repik, J. L. Stryker, and F. J. Ball in *Sulfur Dioxide Processing*, reprint manual published by American Institute of Chemical Engineering, New York, 1975.
170. E. Altwicker, *Adv. Environ. Sci. Eng.* **3,** 80 (1980).
171. J. O. Nriagu, ed., *Sulfur in the Environment*, John Wiley & Sons, Inc., New York, 1978.
172. J. G. Calvert, F. Su, J. W. Bottenheim, and O. P. Strausz, *Atmos. Environ.* **12,** 197 (1978).
173. J. B. Pfeiffer, ed., *Sulfur Removal and Recovery*, Advances in Chemistry Series 139, American Chemical Society, Washington, D.C., 1975.
174. C. D. Swaim, Jr. in ref. 173, pp. 111–119.
175. J. M. Henderson and J. B. Pfeiffer in ref. 173, pp. 35–47.
176. W. D. Hunter, Jr., J. C. Fedoruk, A. W. Michener, and J. E. Harris in ref. 173, pp. 23–34.
177. P. Steiner, H. Juntgen, and K. Knoblauch in ref. 173, pp. 180–191.
178. J. Happel, M. Hnatow, L. Bajars, and M. Kundrath, *Ind. Eng. Chem. Prod. Res. Dev.* **14,** 155 (1975).
179. *Sulphur* **153,** 41 (March/April 1981).
180. R. B. Engdahl and H. S. Rosenberg, *Chemtech*, 118 (Feb. 1978).
181. C. G. Cornell and D. A. Dahlstrom, *Chem. Eng. Prog.* **69**(12), 47 (1973).
182. W. J. Osborne and C. B. Earl in ref. 174, pp. 158–163.
183. L. Korosy, H. L. Gewanter, F. S. Chalmers, and S. Vasan in ref. 173, pp. 192–211.
184. U.S. Pat. 3,911,093 (Oct. 7, 1975), F. G. Sherif, J. S. Hayford, and J. E. Blanch (to Stauffer Chemical Co.).
185. J. P. Hayford, L. P. VanBrocklin, and M. A. Kuck, *Chem. Eng. Prog.* **69**(12), 54 (1973).
186. B. Meyer, B. Mulliken, and H. Weeks, *Phosphorus and Sulfur* **8,** 281 (1980).
187. Ref. 173, pp. 14, 16, 108–110.
188. Ref. 20, pp. 267–278.
189. K. Kawamura, S. Aoki, H. Kimura, K. Adachi, T. Katayama, K. Kengaku, and Y. Sawada, *Environ. Sci. Technol.* **14,** 288 (1980).
190. L. C. Schroeter, *Sulfur Dioxide*, Pergamon Press, New York, 1966.
191. F. Asinger, *Paraffins—Chemistry and Technology*, Pergamon Press, Oxford, 1968, pp. 645–668, 483–571.
192. D. Estel, K. Mateew, W. Pritzkow, W. Schmidt-Renner, *J. Prakt. Chem.* **323,** 262 (1981).
193. E. M. Fettes and F. O. Davis in N. G. Gaylord, ed., *Polyethers, III, Polyalkylene Sulfides and other Polythioethers*, Interscience Publishers, New York, 1962, pp. 225–270.
194. C. S. Marvel and E. D. Weil, *J. Am. Chem. Soc.* **76,** 61 (1954), and references therein.
195. *Sulphur* **122,** 51 (Jan./Feb. 1976).
196. E. H. Conroy in W. W. Duecker and J. R. West, eds., *The Manufacture of Sulfuric Acid, ACS Monograph No. 144*, Reinhold Publishing Co., New York, 1959, pp. 51–60.
197. *Pulp and Paper Manufacture, Vol. 1: The Pulping of Wood*, 2nd ed., Technical Association of the Pulp and Paper Industry of U.S. and Canada, Atlanta, Ga., 1968, pp. 288–295.
198. *Sulfur Manual*, Section 7, Texas Gulf Sulphur Co., New York, 1963. The information from this manual was updated to reflect the state of the technology as of 1981 by A. Gloster, Consulting Engineer, Texasgulf Inc., to whom the present author is indebted.
199. *Sulphur* **151,** 35 (Nov./Dec. 1980).
200. *Sulphur* **155,** 28 (July/Aug. 1980).
201. A. Lippman, Jr., *Ind. Eng. Chem.* **42,** 2215 (1950).
202. J. W. DeVos, *TAPPI* **39,** 883 (1956).
203. U.S. Pat. 3,632,312 (Jan. 4, 1972), W. W. Jukkola (to Dorr-Oliver, Inc.).
204. C.D. Jentz, *Pulp Paper Can.*, 101 (1941).

205. A. Urbanek and M. Burgiell in *Proceedings of the 5th International Sulphur Conference, Nov. 1981*, British Sulphur Corp., Ltd., London, 1982.
206. *Process for Production of Liquid Sulfur Dioxide*, Bulletin of Stauffer Chemical Co., Licensing Administrator, Westport, Conn., 1970.
207. U.S. Pat. 4,177,248 (Dec. 4, 1979), W. T. Richard (to Stauffer Chemical Co.).
208. N. E. Hamner, ed., *Corrosion Data Survey*, 5th ed., National Association of Corrosion Engineers, Houston, 1974.
209. *Mater. Eng.*, 14 (May 1979).
210. S. L. Sakol and R. A. Schwartz in *Sulfur Dioxide Processing*, reprint manual published by American Institute of Chemical Engineers, New York, 1975.
211. *Chem. Mark. Rep.* 1 (Oct. 1, 1980).
212. *SRI International Chemical Economics Handbook*, Stanford Research Institute, Menlo Park, Calif., Dec. 1979, p. 789.0002J.
213. *NIOSH Manual of Analytical Methods*, Method No. S308, Vol. 2, U.S. Department of Health and Human Services, Public Health Service, 1977.
214. *NIOSH Manual of Analytical Methods*, Method No. P & CAM 160, Vol. 1, U.S. Department of Health, Education, and Welfare, Public Health Service, 1977.
215. *Sulphur* **131,** 52 (July/Aug. 1977); J. R. Donovan, E. D. Kennedy, D. R. McAlister, and R. M. Smith, *Chem. Eng. Prog.*, 89 (June 1977).
216. R. S. Saltzman, *Challenges in Stack Sulfur Emissions Measurements*, paper presented at 30th Annual Petroleum Mech. Eng. Conference, Tulsa, Okla., Sept. 21–25, 1975.
217. C. J. Lang, R. S. Saltzman, and G. G. DeHaas, *TAPPI Environmental Conference*, 1975 preprints, Technical Association of the Pulp and Paper Industry, Atlanta, Ga., May 1975, pp. 129–145.
218. D. G. Barrett and J. R. Small, *Chem. Eng. Prog.* **69**(12), 35–38 (1973).
219. *Food Chemicals Codex*, 2nd ed., National Academy of Sciences, Washington, D.C., 1972, 964–965.
220. *Sulfur Oxides*, Committee on Sulfur Oxides, Board of Toxicology and Environmental Health Hazards, National Research Council, National Academy of Sciences, Washington, D.C., 1978.
221. *Sulphur and Its Inorganic Derivatives in the Canadian Environment*, Publication No. 15015, Environmental Secretariat, National Research Council of Canada, Ottawa, 1977.
222. *Sulphur* **153,** 46 (Mar./Apr. 1981).
223. *Criteria for a Recommended Standard. Occupational Exposure to Sulfur Dioxide*, U.S. Department of Health, Education and Welfare, Public Health Service, 1974.
224. F. W. Weir, *J. Occup. Med.* **21,** 281 (1979).
225. P. S. Farber, ed., *J. Air Pollut. Control Assoc.* **30,** 1089 (1980).
226. *Sulphur* **118,** 20 (May/June 1975).
227. D. S. Shriner, C. R. Richmond, and S. E. Lindberg, *Atmospheric Sulfur Deposition—Environmental Impact and Health Effects*, Ann Arbor Science, Publishers, Woburn, Mass., 1980.
228. R. M. Perhac, *Min. Congr. J.*, 19, 58 (Aug. 1981).
229. B. H. Svensson and R. Soderland, eds., *Nitrogen, Phosphorus, and Sulfur—Global Cycles*, SCOPE Editorial Service NRF, Stockholm, Sweden, 1976.
230. L. R. Ember, *Chem. Eng. News*, 20 (Sept. 14, 1981).
231. J. H. Oxley, S. G. Bloom, M. T. Dana, R. B. Engdahl, S. Globe, D. R. Jackson, H. S. Rosenberg, J. R. Schorr, C. W. Spicer, and P. R. Sticksel, *Combustion* **52**(4), 41 (1980).
232. A. J. Buonicore, *Chem. Eng.*, 80 (June 30, 1980).
233. *Sulphur* **134,** 42 (Jan./Feb. 1978).
234. *Sulphur* **148,** 66 (May/June 1980).
235. *Chem. Week*, 24 (Sept. 16, 1981).
236. R. S. Livingstone, T. D. Anderson, T. M. Rosmann, M. Olszewski, A. M. Perry, and C. D. West, *Toward a Desirable Energy Future*, Report ORNL/PPA-81/6, Oak Ridge National Laboratory, Oak Ridge, Tenn., July 1981, and references therein.
237. R. Shapiro, *Mutat. Res.* **39,** 140 (1977).
238. L. F. Green, *Food Chemistry* (England) **1**(2), 103 (1976).
239. D. H. Petering in S. D. Lee and B. Peirano, eds., *Biochemical Effects of Environmental Pollutants*, Ann Arbor Science Publishers, Inc., Ann Arbor, Mich., 1977, pp. 293–306.
240. R. G. Guess, *Pulp Paper* **53**(6), 74, 81 (1979).
241. *Stauffer Sulfur Dioxide in the Production of Sodium Hydrosulfite (Sodium Dithionite)*, Technical Information Bulletin, Stauffer Chemical Co., Industrial Chemical Division, Westport, Conn., May 9, 1980.

242. *Sulfur Dioxide in Pulp Bleaching*, Technical Information Data Sheet, Stauffer Chemical Co., Industrial Chemical Division, Westport, Conn., Jan. 1980.
243. R. A. Scanlon, *CRC Crit. Rev. Food Technol.* **5,** 357 (1975).
244. W. R. Junk and H. M. Pancoast, *Handbook of Sugars*, AVI Publishing Co., Inc., Westport, Conn., 1973, p. 110ff.
245. *Sulfur Dioxide in Water Dechlorination*, Technical Information Bulletin, Stauffer Chemical Co., Industrial Chemical Division, Westport, Conn., Dec. 1977.
246. *Sodium Sulfite Solution 25-9 in Oxygen Scavenging*, Technical Information Bulletin, Stauffer Chemical Co., Industrial Chemical Division, Westport, Conn., Feb. 19, 1980.
247. R. L. Miron, *Mater. Perform.*, 45 (June 1981).
248. Brit. Pat. 1,186,126 (April 2, 1970), J. M. Maselli (to W. R. Grace & Co.).
249. C. N. Satterfield, *Heterogeneous Catalysis in Practice*, McGraw-Hill Book Co., New York, 1980, p. 203.
250. F. Habashi, *Sulphur Inst. J.* **12**(3–4), 15 (1976).
251. *Treating Irrigated Arid-Land Soils with Acid-Forming Sulphur Compounds*, Technical Bulletin No. 24, The Sulphur Institute, Washington, D.C., 1979.
252. L. R. Grass and A. J. MacKenzie, *Sulphur Inst. J.* **6**(1), 8 (1970).
253. *Chemical Economics Handbook*, Stanford Research Institute, Menlo Park, Calif., Aug. 1980, p. 583.5001 ACE.
254. *Gmelins Handbuch der Anorganischen Chemie*, 8th ed., System No. 9, Part B2, Verlag Chemie, Weinheim/Bergstrasse, 1960, pp. 373–613.
255. *Technical Grade Sodium Sulfite—Uses and Applications*, Technical Brochure, Allied Chemical Co., Morristown, N.J., 1980.
256. *Sodium Sulfite Anhydrous*, Technical Data Sheet, Virginia Chemicals, Inc., Portsmouth, Va., undated.
257. *Sodium Sulfite, Technical Anhydrous, Product Data*, Koppers Industrial Chemicals, Koppers Company, Inc., Pittsburgh, Pa., July 1981.
258. *Sodium Sulfite*, Product Data and Product Safety Information Sheets, Stauffer Chemical Co., Industrial Chemical Division, Westport, Connecticut, 1979.
259. K. A. Kobe and K. C. Hellwig, *Ind. Eng. Chem.* **47,** 1116 (1955).
260. Ger. Pat. 2,940,697 (Feb. 5, 1981), A. Metzger and G. Muenster (to Hoechst A.-G.).
261. E. A. P. Devuyst, V. A. Ettel, and M. A. Mosoiu, *Chemtech*, 426 (July 1979).
262. U.S. Pat. 4,003,985 (Jan. 18, 1977), R. J. Hoffman, S. L. Bean, P. Seeling, and J. W. Swaine, Jr. (to Allied Chemical Corp.).
263. Ger. Pat. 900,211 (Dec. 21, 1953), H. Schmidt and J. Schaafhausen (to Farbwerke Hoechst).
264. *Chem. Mark. Rep.*, 9 (April 13, 1981).
265. *Chem. Eng. News*, 18 (Sept. 7, 1981).
266. A. Wong, *Pulp Paper*, 74 (Nov. 1980).
267. *Sodium Metabisulfite*, Food Ingredients Technical Bulletin 1-250, Monsanto, St. Louis, Missouri, 1964.
268. J. A. Kent, ed., *Riegel's Handbook of Industrial Chemistry*, Van Nostrand Reinhold Co., New York, 1974, pp. 141–142.
269. *Sodium Bisulfite Anhydrous (Sodium Metabisulfite)*, Product Data Sheet, Allied Chemical Corp., Industrial Chemicals Division, Solvay, N.Y., 1978.
270. *Sodium Bisulfite*, CHRIS Hazardous Chemical Data, Coast Guard, U.S. Department of Transportation, data sheet revised 1978.
271. *Treatment of Chromium Waste Liquors with Sodium Metabisulfite*, Allied Chemical Corp., Industrial Chemicals Division, Solvay, N.Y., 1980.
272. J. Litt, *Metal Finishing*, 52, 61 (Nov. 1971).
273. *Sodium Hydrosulfite*, Technical literature, Virginia Chemicals, Inc., Portsmouth, Va., 1978.
274. D. Lyons and G. Nickless in ref. 84, pp. 199–239.
275. L. W. Codd and co-workers, eds., *Chemical Technology, An Encyclopedic Treatment*, Vol. 1, Barnes & Noble Books, New York, 1968, p. 461.
276. J. G. deVries and R. M. Kellogg, *J. Org. Chem.* **45,** 4126 (1980).
277. G. Valensi, *Cent. Belge. Etude Corros. Rapp. Tech.* **121,** 207/1-207/22 (1973).
278. A. Teder, *Acta Chem. Scand.* **27,** 705 (1973).
279. A. Janson and F. Scholtz in W. Forst, ed., *Ullmanns Encyklopädie Technischen Chemie*, 3rd ed., Vol. 15, Urban & Schwartzenberg, München-Berlin, 1964, pp. 480–488.

280. Brit. Pat. 1,145,824 (April 21, 1965), (to Electric Reduction Co. of Canada, Ltd.).
281. Ger. Pat. 2,748,935 (May 3, 1979), W. Ostertag, G. Ertl, G. Wunsch, V. Kiener, E. Voelkl, and S. Schreiner (to BASF A.-G.).
282. G. Ertl, V. Kiener, W. Ostertag, G. Wunsch, and K. Bittler, *Angew. Chem.* **91,** 333 (1979).
283. U.S. Pat. 4,218,429 (Aug. 19, 1980), L. C. Bostian and C. E. Winslow, Jr. (to Virginia Chemicals, Inc.).
284. U.S. Pat. 3,523,069 (Jan. 29, 1969), C. W. Oloman (to British Columbia Research Council).
285. M. S. Spencer, P. J. Carnell, and W. J. Skinner, *Ind. Eng. Chem. Proc. Res. Dev.* **8,** 184 (1969).
286. C. Oloman, *J. Electrochem. Soc.* **117,** 1604 (1970).
287. *Chemical Economics Handbook*, Stanford Research Institute, Menlo Park, Calif., 1978, pp. 780.4000 B–C.
288. *Chemical Purchasing*, 24 (Dec. 1979).
289. L. V. Hoff in J. H. Karchmer, ed., *The Analytical Chemistry of Sulfur and its Compounds*, Part 1, Interscience, a division of John Wiley & Sons, Inc., New York, 1970, pp. 245–247.
290. E. Blasius, G. Horn, A. Knöchel, J. Münch, and H. Wagner in ref. 84, pp. 199–239.
291. *Fed. Regist.*, 6117 (Jan. 25, 1980).
292. A. A. Newman, ed., *Chemistry and Biochemistry of Thiocyanic Acid and Its Derivatives*, Academic Press, New York, 1975.
293. Ref. 13, pp. 148–209.
294. *Ammonium Thiocyanate and Sodium Thiocyanate Product Bulletin*, J. T. Baker Chemical Co., Technical Service Department, Phillipsburg, N.J., 1969.
295. *Ammonium Thiocyanate Product Bulletin*, Argus Chemical Division, Witco Chemical Co., Brooklyn, N.Y., 1981.

EDWARD D. WEIL
Stauffer Chemical Co.

SULFUR DIOXIDE, SO_2. See Sulfur compounds.

SULFUR DYES

Sulfur dyes are used mainly for dyeing textile cellulosic materials or blends of cellulosic fibers with synthetic fibers such as polyacrylates, polyamides, and polyesters. They are also used for silk and paper in limited quantities for specific applications. The use of solubilized sulfur dyes on certain types of leather is gaining favor.

From an application point of view, the sulfur dyes are between vat, direct and fiber-reactive dyes. They give good to moderate lightfastness and good wetfastness at low cost and rapid processing.

Traditionally, these dyes are applied from a dyebath containing sodium sulfide. However, over the years, development in dyeing techniques and manufacture has led to the use of sodium sulfhydrate, sodium polysulfide, sodium dithionite, thiourea dioxide and glucose as reducing agents. In the reduced state, the dyes have affinity for cellulose and are subsequently exhausted on the substrate with common salt or sodium sulfate and fixed by oxidation.

The range of colors covers all hue classification groups except a true red. As a rule, the hues are dull compared with other dye classes; black is the most important, followed by blues, olives and browns (see also Dyes and dye intermediates).

The first sulfur dyes were produced in 1873 by heating organic cellulose-containing material, such as wood sawdust, humus, bran, cotton waste, and waste paper, with alkali sulfides and polysulfide (1). These dyes were dark and hygroscopic and had a bad odor. Composition varied and they were easily soluble in water giving greenish dyeings, both from alkali and alkali sulfide baths. The dyes were fixed on cotton fiber and eventually turned brown on exposure to air or chemical oxidation with bichromate solution.

The valuable tinctorial properties of these dyes, in addition to their low price, led to their use in the cotton-dyeing industry. They were sold under the name of *Cachou de Laval*. The main constituent was lignin, which even today is used to produce CI Sulfur Brown 1.

Since then the quality of sulfur dyes has greatly improved, mainly because well-defined organic compounds are used as starting materials. Owing to better techniques, current sulfur dyes have good storage stability and fastness properties, and are generally easily adaptable to modern dyeing methods. An overview of the development of sulfur dyes is given in Table 1.

Chemical Properties

Classification and Structure. Little is known about the structure of sulfur dyes, and therefore, they are classified according to the chemical structure of the starting materials.

The process of sulfurization is usually carried out by a sulfur bake, in which the dry organic starting material is heated with sulfur between 160 and 320°C; a polysulfide bake, which includes sodium sulfide; a polysulfide melt, in which aqueous sodium polysulfide and the organic starting material are heated under reflux or under pressure in a closed vessel; or a solvent melt, in which butanol, Cellosolve, or dioxitol are used alone or together with water. In the last two methods, hydrotropes may be added to enhance the solubility of the starting material. The hydrotropes improve yield and quality of the final dyestuff.

The temperature and duration of heating have a marked effect on both the shade and properties of the dye and conditions have to be carefully controlled in order to achieve uniformity. The precipitation by air oxidation must also be controlled to prevent variations in shade.

Sulfur Bake. To this group belong the yellow, orange and brown sulfur dyes. They are usually made from aromatic amines, diamines and their acyl and nuclear alkyl derivatives. These may be used in admixture with nitroanilines and nitrophenols or aminophenols to give the desired shade. The color formed is said to be due to the formation of the thiazole chromophore (**1**).

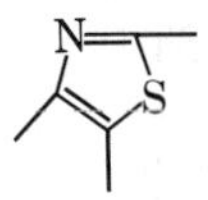

(**1**)

Investigation into the structure of the sulfur-bake dye Immedial Yellow GG, CI Sulfur Yellow 4 (CI 53160) [*1326-75-6*], by chemical degradation of the dye and confirmatory synthesis of the postulated structures showed that a mixture of four dyes (**2–5**) was obtained when benzidine and 4-(6-methyl-2-benzothiazolyl)aniline were baked with sulfur:

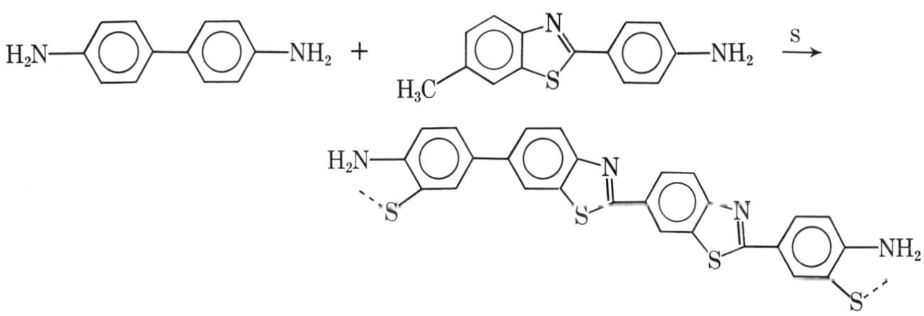

(main part)

(**2**)

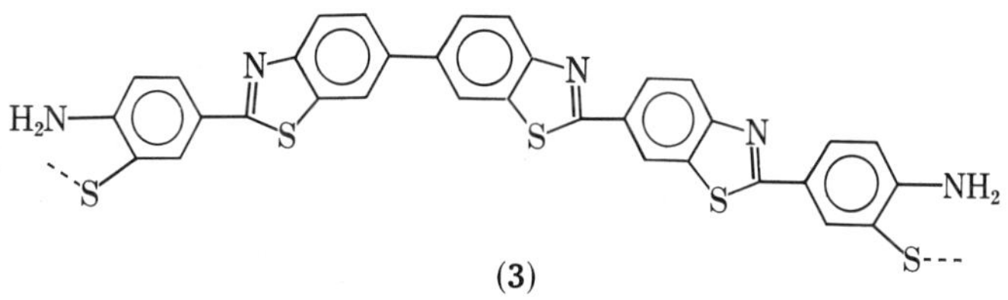

(**3**)

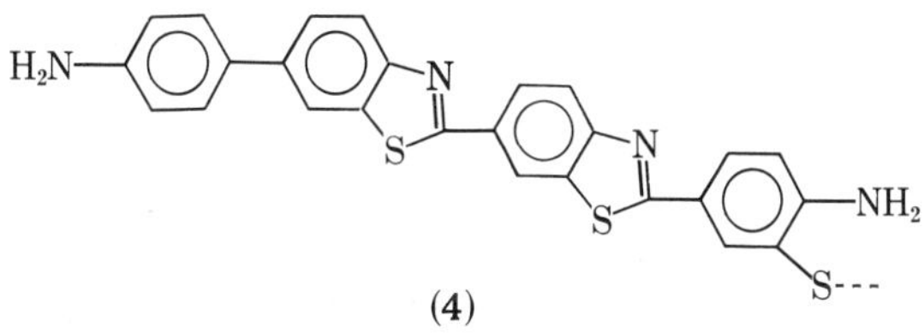

(**4**)

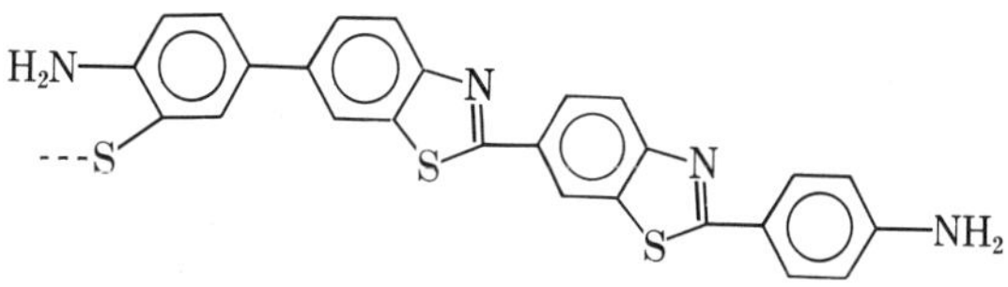

(**5**)

Table 1. Historical Events in the Development of Sulfur Dyes

Year	Development	Product	CI designation		CAS	Ref.
			Name	Number	Registry No.	
1873	heating of organic wastes with alkali sulfides or alkali polysulfides	*Cachou de Laval*	Sulfur Brown 1	53000	[*1326-37-0*]	1
1893	heating of benzene and naphthalene derivatives with alkali polysulfides	Vidal Black	Sulfur Black 3	53180	[*1326-81-4*]	2
1897	sulfurization of diphenylamine derivatives	Immedial Black V	Sulfur Black 9	53230	[*1326-97-2*]	3
1899	inexpensive black from 2,4-dinitrophenol by use of new reflux method of sulfurization	Sulfur Black T	Sulfur Black 1	53185	[*1326-82-5*]	4
1900	use of leucoindophenols	Immedial Pure Blue	Sulfur Blue 9	53430	[*1327-56-6*]	5
	first reddish (bordeaux) dyes from 8-amino-2-hydroxyphenazine	Immedial Maroon B	Sulfur Red 3	53710	[*1327-84-0*]	6
1902	commercial orange and yellow dyes from toluene-2,4-diamine	Immedial Orange C	Sulfur Orange 1	53050	[*1326-49-4*]	7
1904	clear green dye from phenyl-peri acid indophenol	Thionol Brilliant Green GG	Sulfur Green 3	53570	[*1327-73-7*]	8
1908–1909	chlorine fast blue dyes from carbazole and *N*-ethylcarbazole indophenols	Hydron Blue R	Vat Blue 43	53630	[*1327-79-4*]	9
		Hydron Blue G	Vat Blue 42	53640	[*1327-81-7*]	
1926	chlorine fast black dye from *p*-(2-naphthylamino)-phenol	Indocarbon CL	Sulfur Black 11	53290	[*1327-14-6*]	10
1932–1934	clear violet-to-blue dyes from substituted trichlorophenoxazones	Immedial New Blue 5RCF	Sulfur Blue 12	53800	[*1327-96-4*]	11
		Immedial Bordeaux 3BL	Sulfur Red 7	53810	[*1327-97-5*]	

1935–1938	fast browns from decacyclene (13) and its polynitro derivatives, using S_2Cl_2 as sulfurization agent or baking with sulfur	Immedial Katechu 4RL Immedial Yellow Brown GL Immedial Brown GGL	Sulfur Brown 52 Sulfur Brown 60 Sulfur Brown 51	53320 53325 53327	*[1327-18-0]* *[1327-20-4]* *[1327-22-6]*	12
	ready-to-dye liquid dyes from commercial powders or press cake by solution in a mixture of sodium sulfide and sodium sulfhydrate	Sodyesul Black 4GCF	Leuco Sulfur Black 1	53185	*[66241-11-0]*	13
1940–1941	synthetic methods introduced for conversion of organic pigment-type dyes to sulfide soluble dyes	Thionol Ultra Green B	Sulfur Green 14 Leuco Sulfur Red 14		*[12227-06-4]*	14–15
1945	sulfurization with aluminum chloride–sulfur monochloride complex of organic dyes and pigments					16
1948	water-soluble dyes by conversion of conventional sulfur dyes with alkali sulfites or bisulfites to thiosulfonic acid derivatives	Thionol Black BM	Solubilized S Black 1	53186	*[1326-83-6]*	
1959	condensation of intermediates containing NH_2 groups with cyanuric chloride; further conversion to thiol groups					17
1960 1966	Inthion and Dykolite synthetic dyes; Bunte salts (alkyl- and arylthiosulfato dyes) of the nonsulfurized type	Inthion Brilliant Blue I3G Dykolite Brilliant Orange 3G	Condense S Blue 2 Condense S Orange 2	18790	*[12224-49-6]*	
1975	fast brown dyes from Indanthrene-type vat dyes by sulfurization with sulfur in chlorosulfonic acid					18

The dye was degraded by a fusion with caustic potash and the degradation products were identified as various *o*-anilinyl mercaptans. They were identified and characterized by condensation with monochloroacetic acid to give the thioglycolic acids which, on acidification, were converted to well-defined crystalline lactams (**6–8**) together with a small amount of *p*-aminobenzoic acid (**9**).

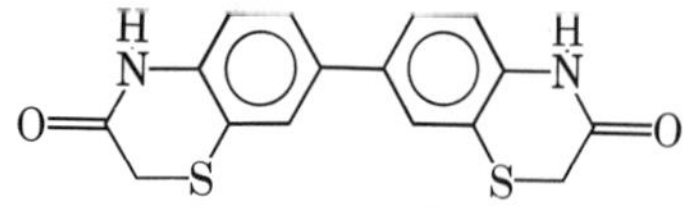

(**6**)

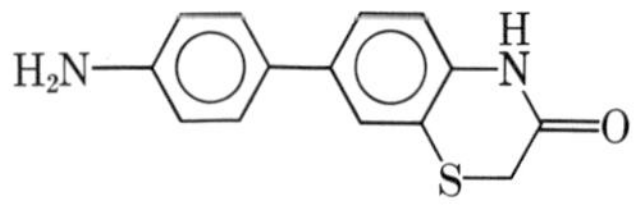

(**7**)

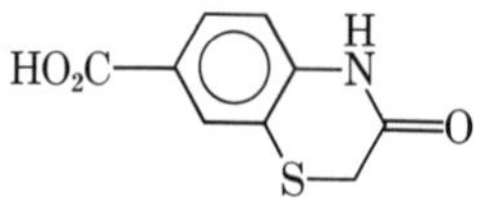

(**8**)

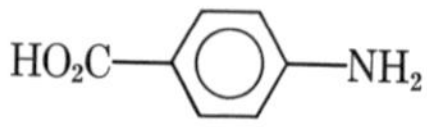

(**9**)

Synthesis of the four components (**2–5**), when mixed in the proportions found, gave a dye whose properties closely resembled that of Immedial Yellow GG. Similarly, investigation of Immedial Orange C, CI Sulfur Orange 1 (CI 53050) [*1326-49-4*], produced from 2,4-diaminotoluene (MTD) revealed ca 6–8 mols of MTD linked by thiazole rings. Whether the linkage is linear (**10**) or branched (**11**) was not ascertained with certainty but again showed that the thiazole chromophore (**1**) is present.

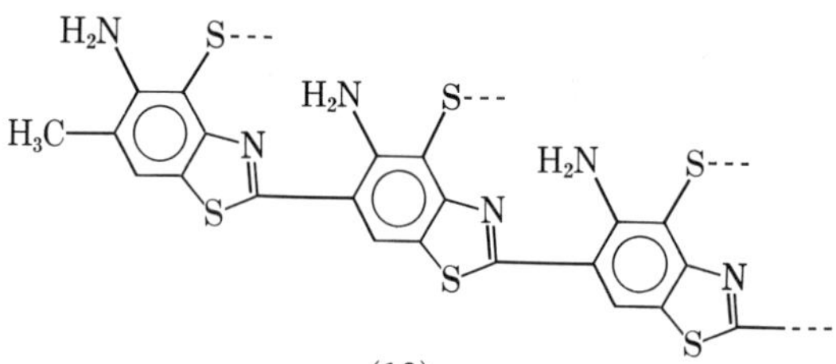

(**10**)

The solubility in alkali sulfide solutions stems from the presence of disulfide groups —S—S— in ortho position to the terminal amino groups. These disulfide groups are reduced to the mercapto groups —SH, which are soluble in alkali.

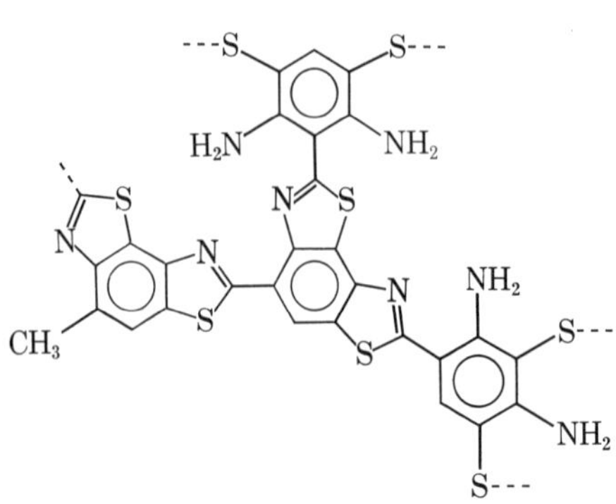

(**11**)

Polysulfide Bake. Although most dyes made formerly by this method are now made by a polysulfide melt, some dyes where certain nitro and phenolic bodies prevented color formation in the sulfur bake are still made by this method. Prereduction of the nitro and polynitro bodies substantially reduces the explosion hazard associated with heating these compounds with sulfur alone. Included in this class are also the exceptionally lightfast sulfur dyes derived from the polynitrodecacyclenes. Decacyclene (**13**) itself is made by a sulfur bake of acenaphthene (**12**) (19). Decacyclene is sulfur-baked to give CI Sulfur Brown 52. However, this color cannot be obtained from acenaphthene in a one-step sulfur bake since a significant amount of diacenaphtho[1,2-*b*; 1′,2″-*d*]thiophene (**14**) is formed which, on sulfur baking, gives a dark brown dye of inferior lightfastness. The structure of the sulfur dyes derived from these polynuclear compounds is unknown.

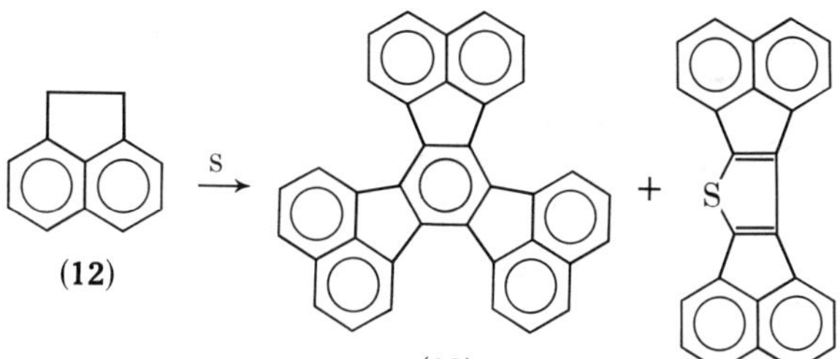

(**12**) (**13**) (**14**)

Polysulfide Melt. CI Sulfur Black 1 (CI 53185) [*1326-82-5*], derived from 2,4-dinitrophenol, is the most important dye in this group which also includes the indophenol-type intermediates. The latter are applied in the stable leuco form. The derived dyes are usually confined to violet, blue and green shades. Other members of this group are intermediates capable of forming quinoneimine (**15**) or phenazone structures (**16**) that produce red-brown or Bordeaux shades:

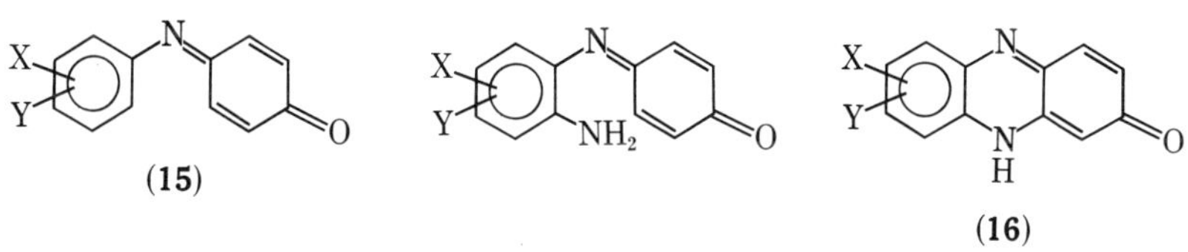

(**15**) (**16**)

With the exception of the phenazones, these compounds lead to formation of the thiazinone ring structure (**17**) which closely resembles the phenazone ring system.

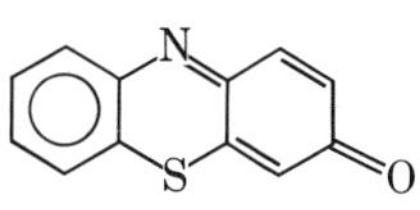

(**17**)

The structure of indophenol-type sulfur dyes was studied (20) on CI Sulfur Blue 9 made from the indophenol (**18**). This compound was purified and brominated to a tetrabromo derivative (**19**) identical with that obtained on bromination of Methylene Violet (**20**). The latter is easily prepared by hydrolyzing Methylene Blue (**21**) with boiling alkali:

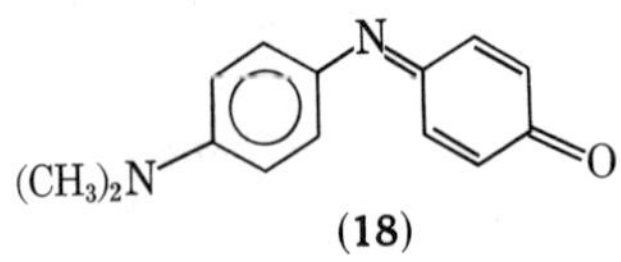

(**18**)

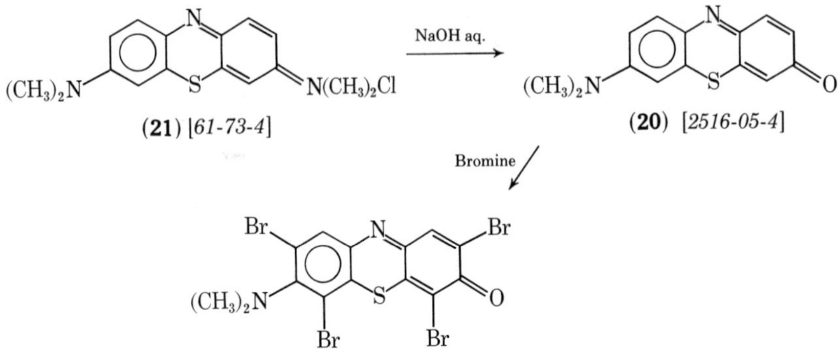

(**21**) [*61-73-4*] (**20**) [*2516-05-4*]

(**19**)

Thus, the presence of a thiazine ring in CI Sulfur Blue 9 was conclusively proved. The thiazine ring is the fundamental chromophore that accounts for the high color value of both the sulfur dye and Methylene Blue, including their ability to form pale yellow leuco forms on reduction.

This work supported the suggestion that sulfur dyes are complex thiazines derived generally from diphenylamine units (21). Depending on the conditions of the thionation, two or more thiazine units could be linked by sulfur to produce a typical sulfur blue (**22**):

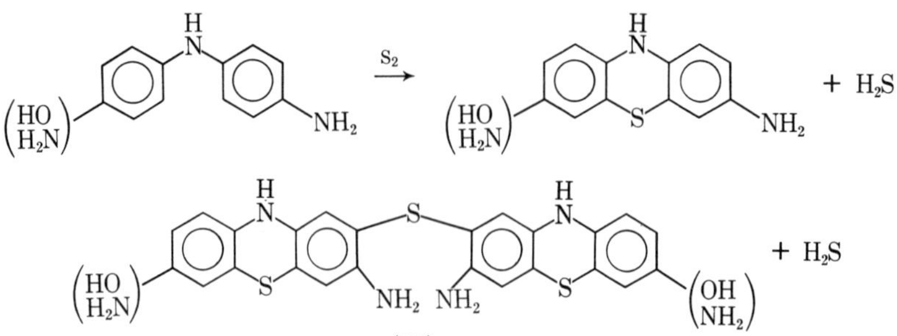

(**22**)

On more severe thionation, a third thiazine ring is formed to give a sulfur black. However, if hydroxyl groups instead of amino groups are attached at positions 2 and 2′, no ring closure would take place and the blue dye would be stable to heat. These formulas are general expressions for the nuclear structures of the blue to black sulfur dyes; they do not take into consideration the quinonoid formation of each dye and other aspects.

Further work elucidated the structure of the dyes in this group by synthesizing trichloroarylthiazinones (**23**) from an *o*-aminoaryl mercaptan (**24**) and chloranil (**25**); treatment with sodium disulfide replaced the halogens by mercapto groups (**26**):

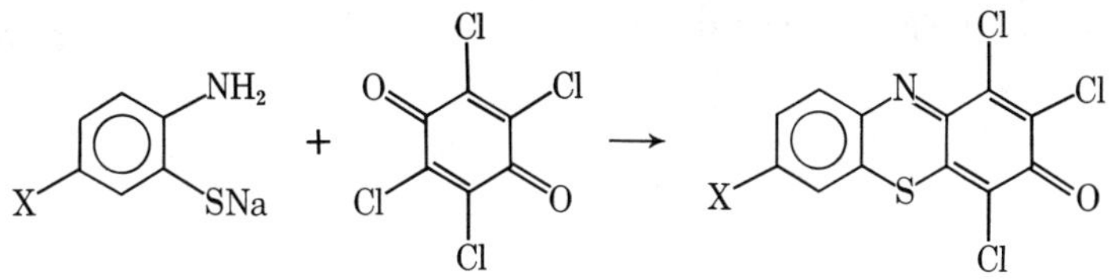

(**24**) (**25**) (**23**)

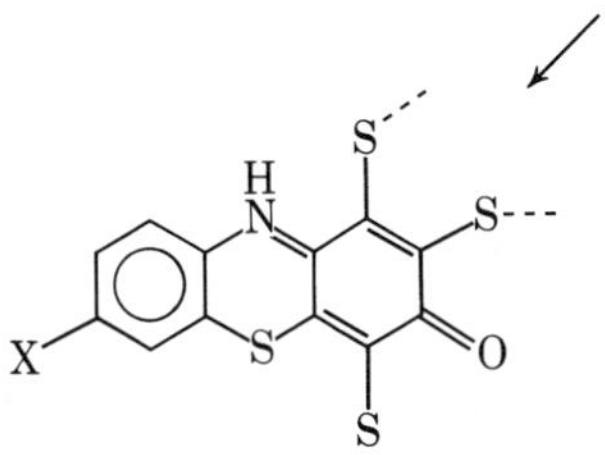

(**26**)

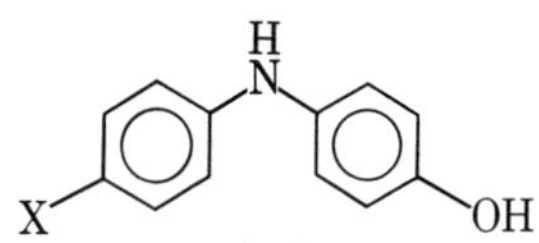

(**27**)

By this method, sulfur dyes derived from 4-hydroxydiphenylamine (**27**) (X = H) are seen to be essentially identical. Similarly, the sulfur dye from 7-anilino-9-methyl-1,2,4-trichlorophenothiazin-3-one (**28**) and the sulfur dye from 4-anilino-4′-hydroxy-2-methyldiphenylamine (**29**) were compared and found to be almost identical, as was the related trichlorophenothiazinone derivative (**30**) from carbazole when compared with Hydron Blue indophenol, CI Vat Blue 43 (CI 53630) (**31**).

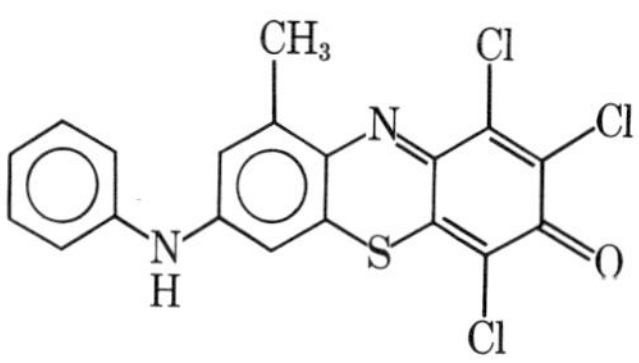

(**28**)

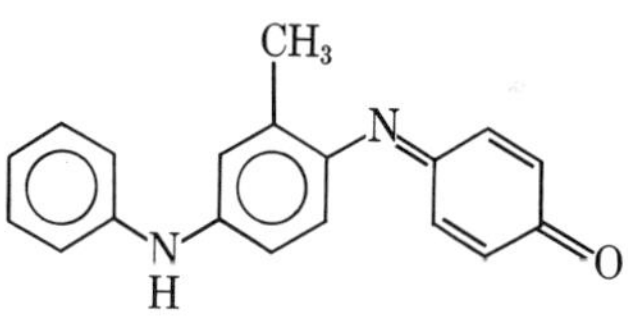

(**29**)

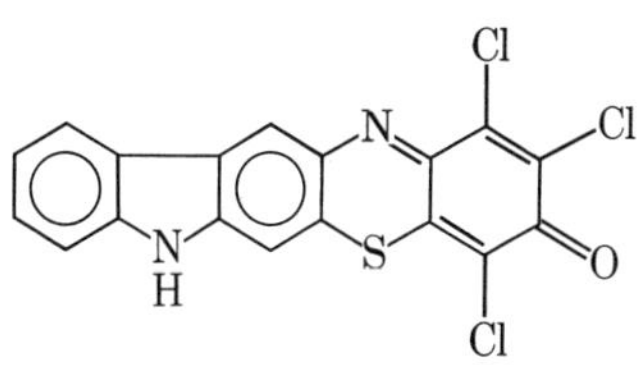

(**30**)

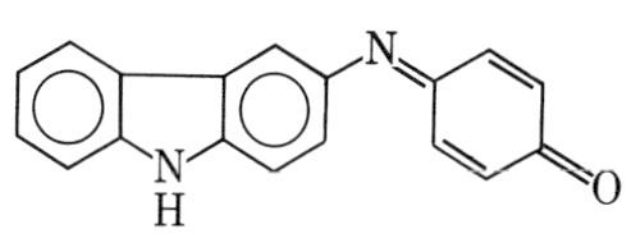

(**31**)

These formulas do not account for all the sulfur or oxygen found in the elemental analysis of the above dyes and formulas have been suggested whereby loosely bound sulfur is depicted as thiozonide

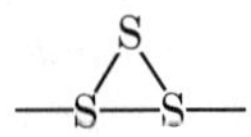

or polysulfide sulfur. Oxygen is contained in sulfoxide groups

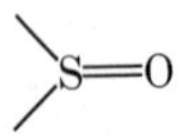

For example, formula (**32**) has been suggested for CI Vat Blue 43 (22), formula (**33**) for a green dye from 4-amino-4′-hydroxydiphenylamine (23) and formula (**34**) for CI Sulfur Black 11 from 4-hydroxyphenyl-2-naphthylamine (24).

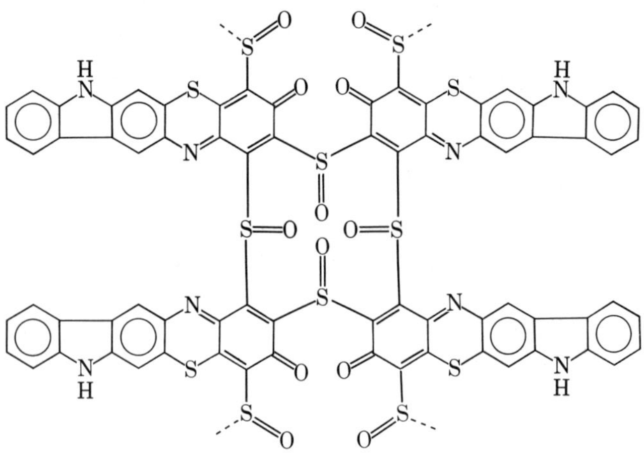

(**32**)

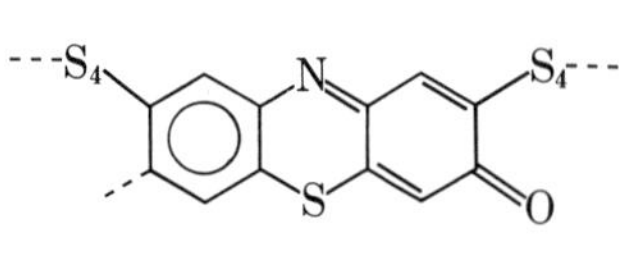

(**33**)

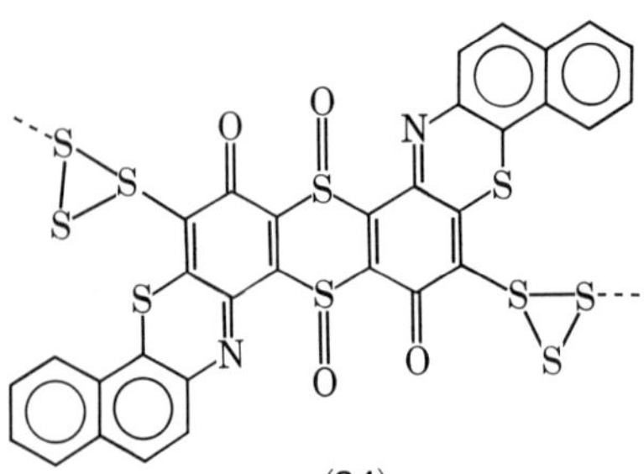

(**34**)

Although the structure of CI Sulfur Black 1, the most important of the sulfur dyes, has been investigated by a number of workers, its definite structure has not yet been established. Starting with 2,4-dinitrophenol, it was found that in color formation one mol ammonia is split off from 3 mol dinitrophenol, indicating that the structure consists of three benzene rings, an important factor that appears to be overlooked in the structural formulas published. Although numerous patents cover the preparation of CI Sulfur Black 1, the products obtained are dissimilar in properties especially with regard to shade and solubility. It is therefore not surprising that structural concepts are at variance and that the products resultant from the thionation are varied and complex mixtures. Examples of the suggested structures are formulas (**35**) (25), (**36**) (26), and (**37–38**) (27), depending on the temperature of the thionation. At 103–105°C, (**37**) is obtained and above 110°C, (**38**), in which sulfur enters at positions X and X′. The bridging sulfur S_n varies from 2 to 7 atoms.

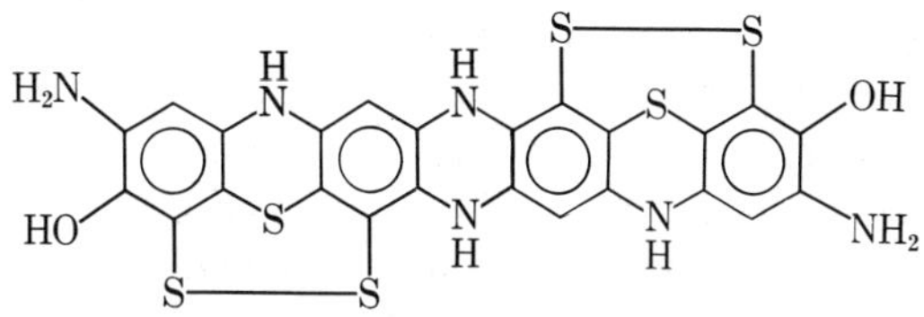

(**35**)

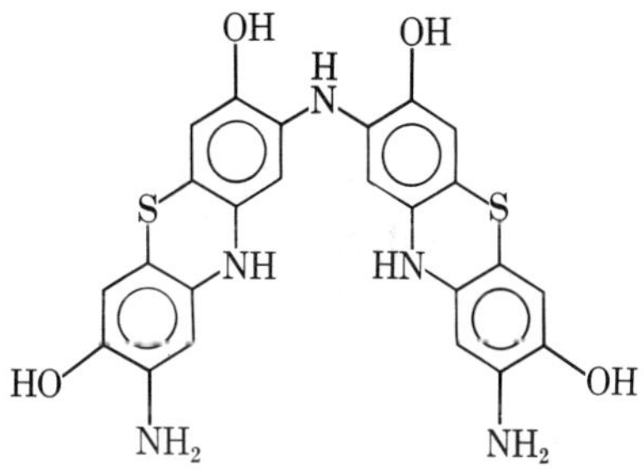

(**36**)

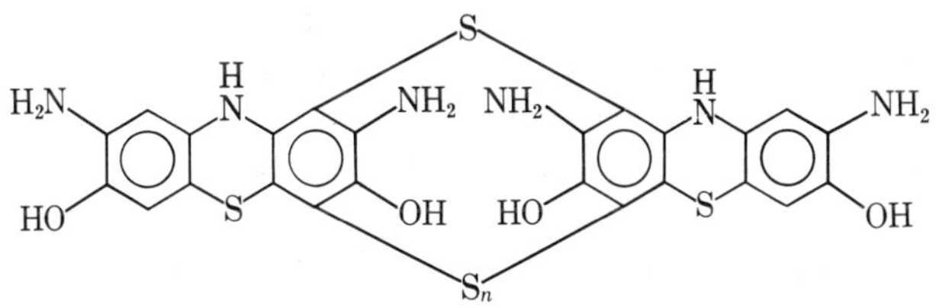

(**37**)

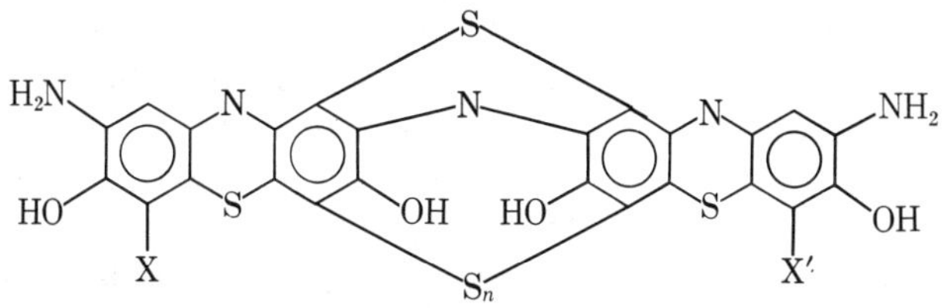

(**38**)

According to one investigator, none of these formulas hold since the ratio of ammonia evolution per mole dinitrophenol is far greater than found (28).

Sulfurized Vat Dyes. These dyes occupy an intermediate position between the true vat colors and sulfur dyes since, like vat dyes, they are dyed preferentially from a sodium dithionite–caustic soda bath. However, some dyes of this class can also be dyed from an alkali sulfide bath or a combination of the two, depending on the dyeing method used and the nature of the substrate to be dyed. This has led to some confusion since CI Vat Blue 42 and 43 are listed in the constitution section of the *Colour Index* under sulfur dyes. Although inferior to true vat dyes in fastness properties, they offer the advantage of better fastness, especially to chlorine, than conventional sulfur dyes.

In dyehouses where sulfide effluent is a problem, sulfur dyes of good chlorine fastness that dye satisfactorily from a dithionite–caustic alkali bath offer an advantage. Included in this group are CI Sulfur Black 11, CI Sulfur Red 10 (CI 53228) [*1326-96-1*], CI Sulfur Brown 96 (CI 53228) [*1326-96-1*], CI Vat Blue 42, and CI Vat Blue 43. The shades of these dyes can be brightened by the addition of vat dyes thus increasing the fastness of the resulting dye.

The structure of the sulfurized vat dyes is uncertain but in the presence of amino or methyl groups, thiazole-ring formation (1) is possible (see Thiazole dyes). Examples have been confined to the dyes whose intermediates have been subjected to the conventional sulfur or polysulfide bake (see Tables 2–5).

Manufacture

Sulfur Bake. Thionation of the various aromatic amino or nitro compounds with sulfur was formerly carried out in iron pans fitted with agitators and heated by gas fire. However, some bakes begin to stiffen up as the reaction progresses preventing further agitation; consequently, the baking process is not uniform throughout. Baking pans have been replaced by more efficient iron cylinder rotary bakers. The bakers, usually of 500–1000 kg capacity, are heated directly by gas jets or hot flue gases from a fire source and rotated at 2–10 rpm on hollow-end trunnions supported by self-aligning bearing rollers. The off-gases from the reaction are led through a catch pot to a scrubber unit containing caustic soda solution. The spent H_2S is converted to Na_2S and NaHS for recycling. The bakers are rotated until the raw dye is ground to a powder. It is discharged and either standardized to strength and shade purified by solution in caustic soda or sodium sulfide. Insoluble matter is removed by filtration. The liquors are evaporated to dryness on a steam-heated rotating single- or double-drum dryer or the dye is precipitated by the addition of acid or sodium bisulfite or by blowing air into the alkaline brew. The H_2S generated is absorbed in caustic soda solution. The precipitated product is filtered, washed, and blown air-dry before discharge. The final drying is usually carried out in fan-assisted steam-heated air ovens.

Fluid Polysulfide Melt. Thionation of the various indophenols is carried out in covered, jacketed, stainless-steel kettles fitted with agitator, thermometer, reflux condenser and facilities for removal and absorption of the hydrogen sulfide gas liberated during thionation. When flammable solvents are employed in the melt, flameproof equipment is used. For bright shades, pure intermediates are used and iron is excluded. The sodium polysulfide may be screened to remove its main impurity, ferrous sulfide, which interferes when copper salts are added to the melt; it may also have a dulling effect on the final shade. Shades are brightened by water-miscible solvents such as ethanol, butanol, glycerol, ethylene glycol, and Cellosolve. In some

Table 2. Sulfur-Bake Dyes

Intermediates	Shade	CI designation Name	Number	CAS Registry No.
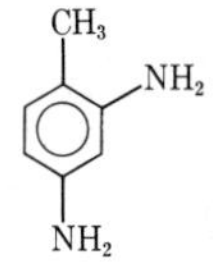	orange	Sulfur Orange 1	53050	[*1326-49-4*]
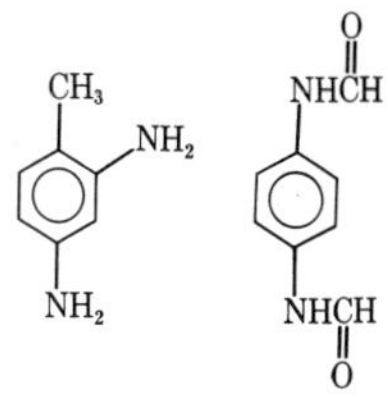	yellowish brown	Sulfur Brown 26	53090	[*1326-60-9*]
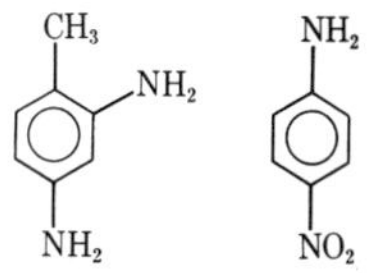	yellowish brown	Sulfur Brown 12	53065	[*1327-86-2*]
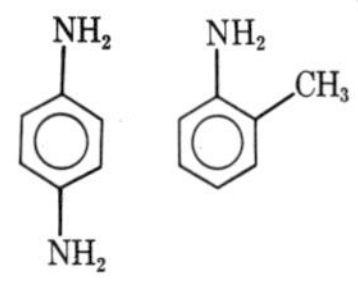	olive	Sulfur Green 12	53045	[*1236-48-3*]
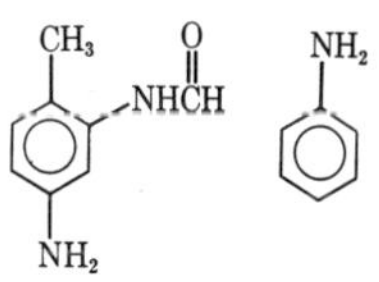	reddish yellow	Sulfur Yellow 1	53040	[*1326-47-2*]
	brown	Sulfur Brown 52	53320	[*1327-18-0*]

cases, the thionation is carried out entirely in solvent after first distilling off any water contained in the raw materials. In order to keep the melt fluid, hydrotropic substances such as urea, thiourea, or xylene sulfonates are added. Hydrotropic substances shorten thionation time and increase yield and brightness.

Metallic additions to the melt, usually in the form of copper sulfate, brighten the shade of certain dyes, such as the bordeaux range made from phenazones and the greens made from the indophenols; the metal forms a complex with the dye. However, copper-containing dyes cannot be applied to material that requires vulcanization.

The composition of the polysulfide varies: it is generally applied in excess with the result that the final dye is wholly or partly in solution in leuco form. At this stage, it can be converted to a liquid form by suitable dilution and the addition of sodium hydrosulfide or diluted, if necessary filtered, and then precipitated by blowing air through the liquor or adding dilute acid. When stronger dyes are required, precipitation may be carried out by adding sodium nitrite directly to the final melt. The excess polysulfide is oxidized and ammonia liberated. Air blowing is usually continued in order to obtain the required brightness of shade. In some cases, caustic soda is added to partly dissolve the dye and air blowing is continued until a satisfactory shade is obtained. The dyestuff is then isolated by filtration, washed, blown air-dry, discharged and dried in a heated air oven.

Table 3. Polysulfide-Bake Dyes

Intermediate(s)	Shade	CI designation Name	CI designation Number	CAS Registry No.
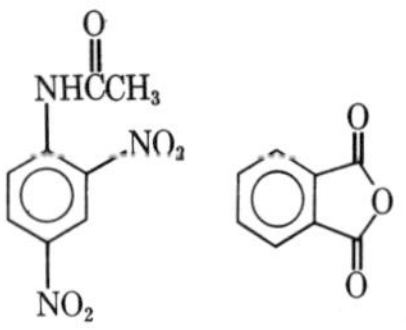	yellow	Sulfur Yellow 9	53010	[*1326-40-5*]
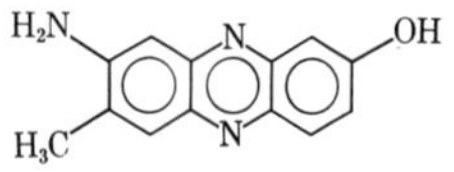	dull reddish brown	Sulfur Brown 56	53722	[*1327-87-3*]
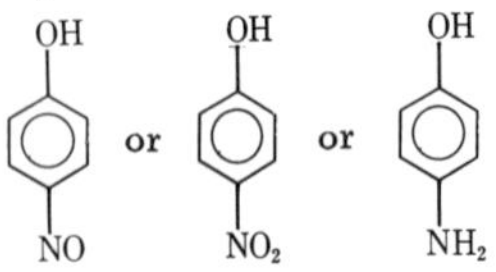 (+ $CuSO_4$)	olive	Sulfur Green 11	53165	[*12262-52-1*]
	dull green	Sulfur Green 1	53166	[*1326-77-8*]
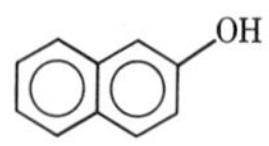	brown	Sulfur Brown 31	53280	[*1327-11-3*]
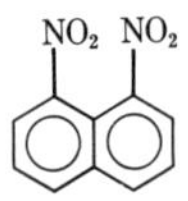	dull reddish brown	Sulfur Brown 7	53275	[*1327-10-2*]
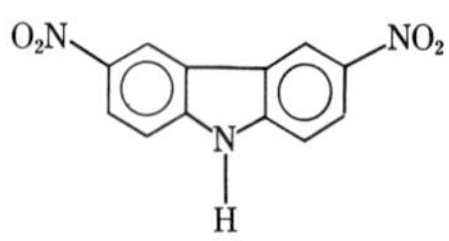	brown	Sulfur Brown 6	53335	[*1327-25-9*]
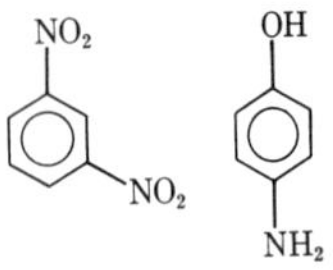	dull green	Sulfur Green 9	53005	[*1326-39-2*]
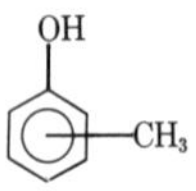	brownish olive	Sulfur Brown 4	53210	[*1326-90-5*]

Application

Sulfur dyes are applied in leuco form. In this form, the dye has affinity for the fiber. After the dye is completely absorbed by the fiber, it is reoxidized *in situ.* In dyes, such as the bright blues, which contain quinonimine groups, further reduction takes place in a manner similar to the reduction of the keto group in vat dyes.

In the third edition of the *Colour Index* 1975, revised 1982, the sulfur dyes are classified according to application method and the structure of the intermediates, into ordinary or conventional dyes, leuco or prereduced dyes, and thiosulfonic derivatives of conventional dyes (solubilized sulfur dyes) (29).

The reducing agent traditionally employed with sulfur dyes is sodium sulfide, but sodium sulfhydrate (NaHS), together with a small quantity of alkali such as sodium carbonate or sodium hydroxide is also widely used. The dyebaths prepared in this way

Table 4 Polysulfide-Melt Dyes

Intermediates	Shade	CI designation Name	Number	CAS Registry No.
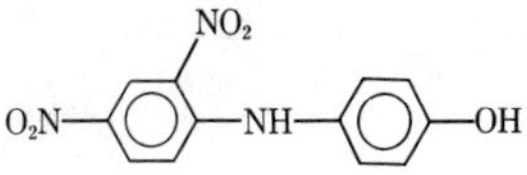	bluish black	Sulfur Black 9	53230	[*1327-56-6*]
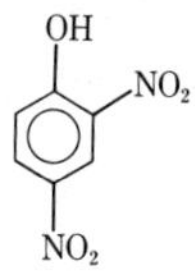	greenish black	Sulfur Black 1	53185	[*1326-82-5*]
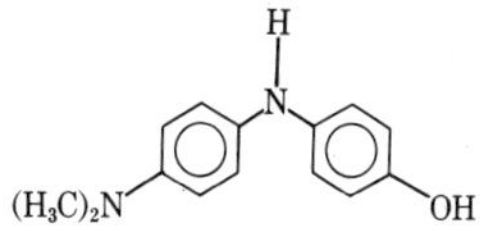	bright blue	Sulfur Blue 9	53430	[*1326-97-2*]
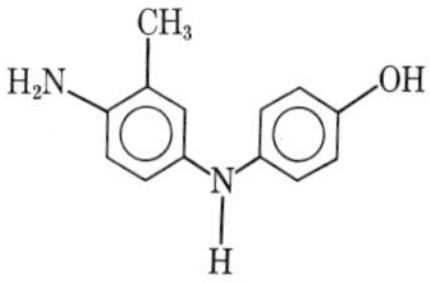	reddish blue to bluish violet	Sulfur Blue 7	53440	[*1327-57-7*]
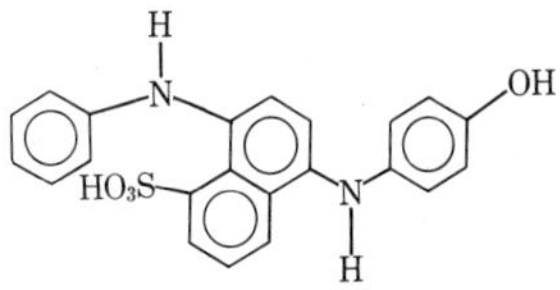	green bluish green	Sulfur Green 3 Sulfur Green 2	53570 53571	[*1327-73-7*] [*1327-74-0*]
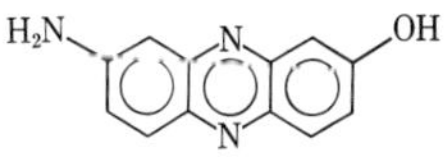	dull Bordeaux	Sulfur Red 3	53710	[*1327-84-0*]
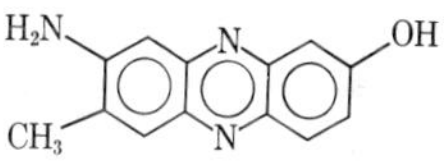	Bordeaux	Sulfur Red 6	53720	[*1327-85-1*]
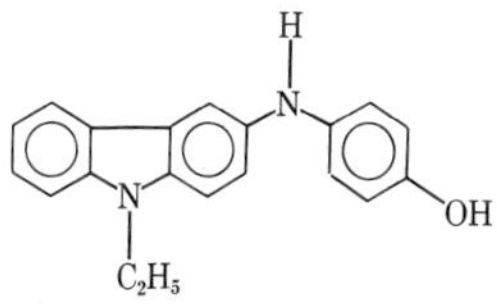	blue to reddish navy	Vat Blue 42	53640	[*1327-81-7*]
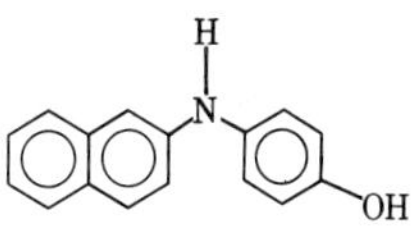	bluish black	Sulfur Black 11	53290	[*1327-14-6*]
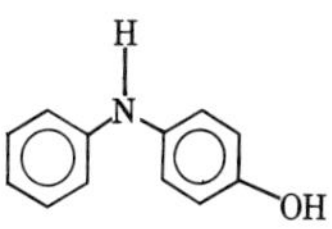	Bordeaux brown	Sulfur Red 10 Leuco Sulfur Brown 96	53228	[*1326-96-1*]
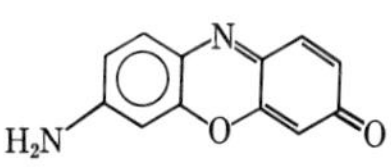	reddish blue	Sulfur Blue 12	53800	[*1327-96-4*]
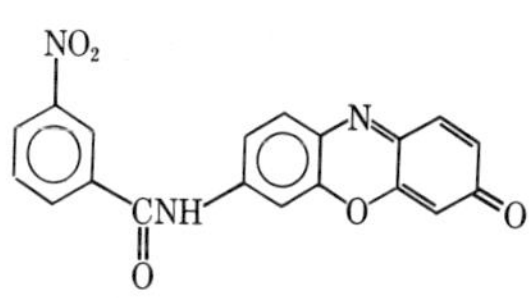	dull bluish red	Sulfur Red 7	53810	[*1327-97-5*]

Table 5. Sulfurized Vat Dyes

Intermediate	Shade	CI designation		CAS
		Name	Number	Registry No.
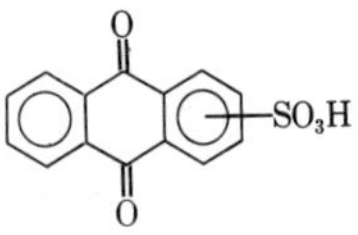 (mono or di)	yellowish brown	Vat dye	58820	[*1328-11-6*]
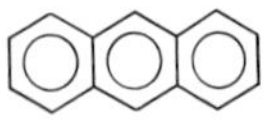	olive	Vat Green 7	58825	[*1328-12-7*]
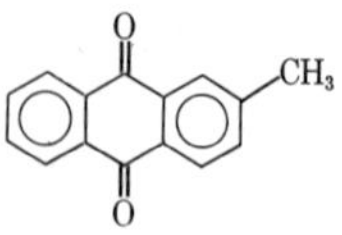	yellowish orange	Vat Orange 21	69700	[*1328-39-8*]
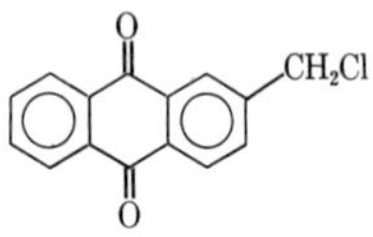	yellow	Vat Yellow 21	69705	[*1328-40-1*]
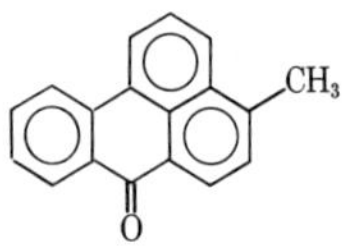	dull greenish blue green	Vat Blue 7 Vat dye	70305 70310	[*6505-58-4*]

are less alkaline than those with sodium sulfide alone which facilitates the rinsing of the dyed goods. Effluent control has resulted in a search for alternative reducing agents. Alkaline sodium dithionite (sodium hydrosulfite) can be used with some sulfur dyes, particularly the blues, but over reduces red-browns such as CI Sulfur Brown 12 [*1327-86-2*], CI Sulfur Red 6 [*1327-85-1*] and similar types. Alkaline sodium formaldehyde sulfoxylate has also been employed with blues, but has the same drawbacks. More recently glucose and sodium hydroxide in almost boiling solution have been proposed for solubilized or dispersed sulfur dyes. Small amounts of sodium sulfide or sodium dithionite together with glucose and sodium hydroxide assist the reduction, but the pH should be maintained above 10.5 throughout the dyeing.

2–5 g/L of sodium polysulfide solutions, as antioxidants, prevent bronzing. Glucose has been used for a similar purpose and sodium borohydride effectively prevents bronzing when dyeing sulfur black. Other dyebath additives, such as wetting and antifoam agents, may not be compatible with the reducing agents, so must be carefully selected.

The conventional sulfur-dye powder is made into a paste with a small amount of soft water and an alkali-stable wetting agent. Boiling for a few minutes in a strong solution of sodium sulfide reduces the dye. The dissolved dye is diluted to the required dyebath volume. When dyeing pale shades, the final bath should contain at least 5 g sodium sulfide (60%) per liter, irrespective of the amount used to dissolve the dye.

The leuco sulfur dyes are in a prereduced form. At present they are produced in liquid form by dissolving paste from the parent sulfur dye in a reducing agent containing sodium sulfhydrate and alkali and, occasionally, other hydrotropic agents. These products were first introduced into the United States in 1936 and are currently available mainly from five manufacturers. They are true solutions requiring only to be diluted with water. Small amounts of reducing agents should be added depending on the dyeing method and the depth of shade required. Solubilized sulfur dyes have no real affinity for the fiber until they are converted into their leuco compounds by a reducing agent. After exhaustion on the fiber, rinsing and oxidation is effected in the same way as for the other types. Solubilized sulfur dyes can be used in aqueous solution to prepad the fabric which can then be treated with reducing agent, with intermediate drying in order to give more level dyeings. Cross-wound packages of yarn may be dyed by this two-stage method. The dyes may be applied in the conventional way by using a leuco dyebath from the start, but it is recommended that they be first dissolved in water only and that the reducing agent is added to the dyebath separately before dyeing.

In the pad–steam method, the padding liquor is prepared by adding the required quantity of prereduced liquid sulfur dye to water at 25–40°C together with additional chemicals and auxiliaries, depending on the dye and the depth of shade desired. The padding liquor is fed into the trough of the dye padder which is fitted with a constant-levelling device and temperature control. A padding mangle squeezes the well-prepared fabric evenly to provide a pick-up of 50–80% of its own weight. The steamer, which should be free from air when operating, is generally set at 102–104°C. A dwell time of 60 s is usually adequate. The remainder of the equipment consists of washing and aftertreatment boxes through which the fabric passes successively before being dried. The continuous operation used in this process is depicted in Figure 1. A typical padding-liquor recipe for a sulfur navy-blue shade on 100% cotton drill is given in Table 6.

Blends of polyester with cotton or viscose are first dyed with disperse dyes, then with sulfur dyes. Disperse and sulfur dyes can also be applied simultaneously in a

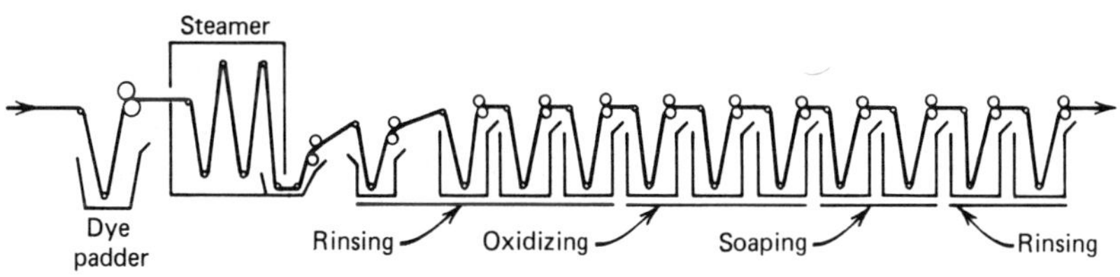

Figure 1. Continuous operation of pad/steam dyeing.

Table 6. Padding-Liquor Recipe at 30°C

Ingredient	g/L[a]
Sulphol Liquid Dark Blue QL (JR)	150
Leucad 71 sodium sulfhydrate, 35%	5
caustic soda 100%	0.5
anionic wetting agent	2.0[b]
EDTA-based sequestrant	0.5

[a] Unless otherwise stated.
[b] ml/L.

pad–dry–thermofix/chemical-reduction pad/steam sequence. In this case, the sulfur dyes cannot be used in their reduced form because of the effect of the sodium sulfide on the disperse dye. Therefore, this method is confined to the solubilized sulfur dyes or sulfur dyes in the dispersed form.

Solubilized sulfur dyes can also be applied without reducing agents (30). The dye, together with urea and thiourea or similar compounds, is padded on 100% cotton, then dried, and thermofixed at 150–175°C. In the case of polyester–cotton, suitable disperse dyes can be added to the padding liquor and thermofixed at higher temperatures together with the solubilized sulfur dye (31).

Before the pressurized package-dyeing machine became established, sulfur dyes were often employed for dyeing cotton warps in rope form, referred to as ball warps. The yarn in ropes or cables, which had been doubled into a W form, was passed several times through a series of baths of reduced dye liquor and baths of rinsing water with intermediate squeezing. The depth of shade was gradually built up; the method resulted in a dyeing of excellent uniformity of shade across a woven fabric. Today, this method is rarely used.

A variation is applied where warps for denims are first dyed with sulfur black or blue and then with indigo. The sulfur dye in its reduced form is added to the prewetting or scouring bath. The yarn is passed through and rinsed before passing to the part of the machine where the indigo is applied by the traditional method of successive stages of impregnation and air oxidation. The sulfur dye improves the appearance and reduces the quantity of indigo required.

Aftertreatment. Since the dye is applied by reduction and oxidation, many methods are available to obtain the correct hue. Air oxidation takes place gradually after the residual reducing agent has been rinsed away, but in general chemical oxidation is faster. The traditional oxidizing agents include sodium or potassium bichromate mixed with acetic acid; addition of copper sulfate slightly improves lightfastness. However, because of ecological restrictions on bichromates, other oxidizing agents are coming into use, such as hydrogen peroxide, sodium perborate, and products based on potassium iodate or sodium bromate mixed with acetic acid. Sodium chlorite is used in alkaline solution together with detergent. These reagents give different effects and are more suitable under some conditions than others. Generally speaking, hydrogen peroxide and sodium perborate give dyeings that are slightly less fast to wet processing than other products.

Aftertreatments include resin finishes, which improve fastness properties, and dye-fixing agents of the epichlorhydrin–organic amine type. These agents react with the dye to give condensation products that are not water soluble and hence more difficult to remove.

Tendering Effects. Cellulosic materials dyed with sulfur black have been known to suffer degradation by acid tendering when stored under moist warm conditions. This effect may be due to the liberation of small quantities of sulfuric acid which occur when some of the polysulfide links of the sulfur dye are ruptured. A buffer, such as sodium acetate, or a dilute alkali in the final rinse, especially after oxidation in acidic conditions, may prevent this occurrence. Copper salts should never be used with sulfur-black dyes because they catalyze sulfuric acid generation. Few instances of tendering with sulfur dyes other than black occur and the problem is largely confined to cotton.

Economic Aspects

The low cost of sulfur dyes, coupled with their good fastness properties and the ease with which they can be applied continues to ensure a high consumption. The number of manufacturers has fallen since the 1960s and the production is now mainly confined to the United States, UK, FRG, and Spain with one principal producer in each of these countries. There are other, less well-known manufacturers in the USSR, People's Republic of China, South Korea, Japan, and Brazil (see Table 7).

Since 1966, when sulfur dyes represented 9.1% of total U.S. dye production and 15.8% of the dyes made for use on cellulosic fibers, few production figures are available, since publication is now regarded as proprietary.

World production is estimated at 110,000–120,000 metric tons per year and at this figure is the highest percentage of any group of dyes. In terms of value, however, the picture is very different.

Table 7. Principal Sulfur Dye Manufacturers and Products

Range name	CI classification	Manufacturer	Country
Asathio	sulfur	Asahi Chemical Co.	Japan
Cassulfon	leuco sulfur, liquid	American Hoechst and Hoechst A.-G.	United States FRG
Diresul	leuco sulfur, liquid	S.A. Cardoner (affiliated to Sandoz S.A.E.)	Spain
Diresul Tina	vat, sulfurized vat dyes, liquid	S.A. Cardoner (affiliated to Sandoz S.A.E.)	Spain
Endurol	vat, sulfurized vat dyes	James Robinson & Co., Ltd.	United Kingdom
Episol	solubilized sulfur[a]	James Robinson & Co., Ltd.	United Kingdom
Hydron	vat, sulfurized vat dyes	Hoechst A.-G.	FRG
Hydrosol	solubilized sulfur	American Hoechst Corp.	United States
Immedial	sulfur	Hoechst A.-G.	FRG
Indocarbon	sulfur, leuco sulfur	Hoechst A.-G.	FRG
Kayaku Sulfur	sulfur	Nippon Kayaku Co., Ltd.	Japan
Kayaku Homodye	sulfur, dispersed form	Nippon Kayaku Co., Ltd.	Japan
Pirocar	sulfur	S.A. Cardoner (affiliated to Sandoz S.A.E.)	Spain
Pirosol	solubilized sulfur		
Sodyeco Liquid	sulfur, sulfurized vat dyes, liquid	Sodyeco Division, Martin Marietta Chemicals and	United States
Sodyesul	sulfur	Rioquim S.A.[b]	Brazil
Sodyevat	vat, sulfurized vat dyes, dispersed form		
Sodyesul Liquid	liquid sulfur		
Solane	sulfur or vat, sulfurized vat dyes	P.C.U.K.[c]	France
Sulfanol	sulfur	P.C.U.K.[c]	France
Sulfanol S	leuco sulfur, powder and liquid	P.C.U.K.[c]	France
Sulphol	sulfur	James Robinson & Co., Ltd.	United Kingdom
Sulphol Liquid	leuco sulfur, liquid	James Robinson & Co., Ltd.	United Kingdom
Sulphosol	solubilized sulfur	James Robinson & Co., Ltd.	United Kingdom
Youhaodron	vat, sulfurized vat	China National Chemicals Import and Export Corp.	People's Republic of China
Youhao Sulphur	sulfur		

[a] For leather.
[b] Formerly Naegeli S.A.
[c] P.C.U.K. = Produits Chimique Ugine Kuhlman; formerly Francolor.

Table 7 lists the principal manufacturers, their range names and the type of dye.

Commercial Forms of Sulfur Dyes

Powders. Powders have, in the past, been the principal form in which sulfur dyes were sold. In general, they are made from the dried press cake, finely ground, and standardized with common salt, sodium sulfate, or soda ash. They are prepared for dyeing by making a paste with water, which is dissolved by boiling with the necessary amount of reducing agent and further addition of water.

Prereduced Powders. These are usually made from press-cake paste to which a reducing agent has been added, such as sodium sulfide, sodium hydrosulfide, or sodium dithionite which solubilize the dye in water. Before drying, the dye paste may be mixed with dispersing and stabilizing agents to aid application.

Grains. Grains are usually prereduced powders. The quantities of sodium sulfide, hydrosulfide, and mineral salts are adjusted to give a grainy product when the paste slurry is dried on a steam-heated drum dryer. Grains offer the advantage of not being dusty.

Dispersed Powders. The principal use of dispersed powders is in pad–dry–chemical pad–steam dyeing techniques. They are normally made from press cake by ball or bead milling to microparticle size in the presence of dispersing agents. The drying is strictly controlled and is carried out in the presence of anticoagulants to prevent aggregation of the dispersed dye particles.

Dispersed Pastes. The milled pastes vary in strength but for ease of handling, the consistency of the dispersed pastes generally permits pouring from the container. Since there is no drying step, dispersed pastes are usually cheaper than the dispersed powders.

Liquids. Some liquid dyes are made directly from the thionation melt by additions of caustic soda and sodium hydrosulfide. Hydrotropic substances are sometimes added, either at the initial thionation stage or after the polysulfide melt is finished in order to keep the reduced dye in solution. Partly reduced liquids are also available. They are usually more concentrated than fully reduced liquids, thus saving packaging and transportation costs. However, they require a further addition of reducing agent to the dyebath in order to obtain full color value. On the other hand, fully reduced liquids are ready to use, since the amount of reducing agent for each dye has been carefully controlled in order to obtain maximum stability on storage and maximum color yield in use. They are not affected by low temperatures as are the dispersed pastes.

Water-Soluble Brands. Water-soluble sulfur dyes occur in both powder and liquid form. The liquid product is cheaper since no drying step is required. These dyes are the Bunte salts, or thiosulfuric acid derivatives of the sulfur dyes. They are made by warming the polysulfide-free pastes with sodium sulfite or bisulfite until they are water soluble. They are salted out from solution or isolated by drum drying the liquor.

Aqueous solutions of these dyes show little or no affinity for cellulosic fiber until a reducing agent has been added and, therefore, they penetrate tightly woven materials. They give level dyeings with less tendency to bronzing.

Health and Safety Factors

During the last 20 years, much more emphasis has been placed on health and safety aspects within the chemical and dyestuff industries. As a consequence, benzidine and β-naphthylamine, which are known carcinogens, have been banned in many countries. In some cases, alternatives to these intermediates have been found in order to retain dyestuffs with similar shades and properties. For example, bright yellow to orange sulfur dyes are obtained from a sulfur bake of 4-(6-methyl-2-benzothiazolyl)aniline, *N*,*N*′-diformyl-*p*-phenylenediamine, and 2,2′- or 4,4′-diformyldiaminodiphenyl disulfide (32). These shades were previously obtainable only from mixtures containing benzidine. The handling of hazardous chemicals such as nitroanilines, dinitro- and diaminotoluenes, nitro- and dinitrophenols and chlorodinitrobenzenes is kept to a minimum. In all cases, suitable protective gear is employed. Hydrogen sulfide is one of the chief hazards of sulfur-dye manufacture and is usually removed by caustic soda scrubber units. Accidental escape of H_2S can be detected by a system that incorporates detector heads specifically for hydrogen sulfide. Process sheets should list all the hazards associated with each operation.

The effluent from the dye manufacture and textile-dyeing industry is usually treated by aeration in large tank farms. The aeration equipment may be a floating rotary blade beater or vertical helix columns through which compressed air is passed or alternatively dissipated through venturi pipes. By these means, ca 10–12% of the oxygen in the air can be effectively utilized in the degradation of the sulfur-dye waste, rendering it acceptable to municipal sewage treatment farms. Small dye houses utilize spent flue gases or hydrogen peroxide as well as ferrous salts to treat plant effluent. However, such an operation is more costly and mixing with spent bleach liquors may reduce costs. Manganese sulfate catalyst assists aeration (33) but the sulfur dyes contained in the effluent also accelerate the oxidation process, thus rendering the inclusion of manganese salts or other catalysts in sulfur-dye effluent seemingly unnecessary (34).

Uses

The sulfur dyes are widely used in piece dyeing of traditionally woven cotton goods such as drill and corduroy fabrics (see Textiles). The cellulosic portion of polyester/cotton and polyester/viscose blends is dyed with sulfur dyes. Their fastness matches that of the disperse dyes on the polyester portion, especially when it is taken into account that these fabrics are generally given a resin finish.

Yarn is dyed with sulfur dyes, although raw stock dyeing has declined in recent years. The dyeing of knitted fabrics, both 100% cotton or blends of cotton with synthetic fibers, is increasing. Although the problems of premature oxidation of the dyebath should not be underrated, sulfur dyes are successfully applied to this type of fabric on open and closed winches and on modern jet-dyeing machines (35). Piece dyeing on the jig uses considerable quantities of sulfur dyes. Although conventional methods are used, special care is necessary to ensure that the dyes are completely dissolved, preferably in a separate vessel, permitting the boiling of dye and reducing agent together.

Cotton yarn is dyed in package machines and the dye exhausted by increasing the temperature and adding salt. The dye must be completely dissolved when preparing the dyebaths to avoid contamination with undissolved dye in the yarn package. The increased availability of the prereduced liquid dyes and the improved quality of sodium sulfide over recent years have reduced this problem. Incorrectly dissolved dye was previously the cause of most faulty dyeings.

Continuous dyeing of piece goods by pad–steam methods is one of the principal outlets for sulfur dyes today, mostly in prereduced liquid form. Nontextile uses for sulfur dyes are limited. Sulfur Black is used for dyeing paper, particularly for lamination applications. More recent developments include the dyeing of leather with the solubilized sulfur Episol dyes. These dyes give well-penetrated dyeings with far better wet fastness than is usually obtained with acid or direct dyes. The dyes are applied to the neutralized leather at pH 7.5–8.0 without reducing agents, and are fixed in a manner similar to the fixation of acid dyes with formic acid at pH 3.8.

BIBLIOGRAPHY

"Sulfur Dyes" in *ECT* 1st ed., Vol. 13, pp. 445–458, by J. J. Ayo, Jr., and E. Kuhn, General Aniline & Film Corp.; "Sulfur Dyes" in *ECT* 2nd ed., Vol. 19, pp. 424–441, by D. G. Orton, Southern Dyestuff Co., a division of Martin Marietta Corp.

1. E. Croissant and L. M. F. Bretonnière, *Bull. Soc. Ind. Mulhouse* **44,** 465 (1874); Brit. Pat. 1489 (April 24, 1873).
2. DR Pat. 84,632 (March 22, 1893); DR Pat. 85,330 (Dec. 10, 1893); Brit. Pat. 19,980 (1893); U.S. Pat. 53,248 (1893), R. Vidal (to Société Anonym Matières Colorantes).
3. DR Pat. 103,861 (Oct. 24, 1897); U.S. Pat. 610,541 (1897): Brit. Pat. 25,234 (1897), G. Kalischer (Cassella & Co.).
4. DR Pat. 127,835 (Dec. 7, 1899), Priebs and O. Kaltwasser (to Akt.-Ges. Berlin).
5. DR Pat. 134,947 (Aug. 19, 1900); U.S. Pat. 693,633 (Aug. 24, 1900), A. Weinberg and R. Herz (to Cassella & Co.).
6. DR Pat. 126,175 (Aug. 11, 1901); U.S. Pat. 701,435 (Jan. 5, 1901), A. Weinberg (to Cassella & Co.).
7. DR Pat. 139,430 (Jan. 12, 1903); DR Pat. 141,576 (May 7, 1902); DR Pat. 152,595 (May 7, 1902), A. Weinberg (to Cassella & Co.).
8. DR Pat. 162,156 (May 25, 1904), M. Boninger (to Sandoz).
9. DR Pat. 218,371 (Dec. 29, 1908), Brit. Pat. 2198 (Feb. 6, 1909), L. Haas and R. Herz.
10. DR Pat. 261,651 (Oct. 24, 1911), (to Aktien-Gesellschaft Für Aniline Fabrikation, Berlin).
11. Brit. Pat. 411,431 (Nov. 29, 1932), W. Zerweck (to I.G. Farbenindustrie).
12. DR Pat. 653,675 (Nov. 30, 1937); DR Pat. 655,487 (Jan. 17, 1938), W. Hagge and K. Haagen.
13. U.S. Pat. 2,130,415 (Sept. 20, 1938), A. J. Buchanan (to Southern Dyestuffs Corp.).
14. Brit. Pat. 547,853 (March 12, 1941), N. H. Haddock (to ICI).
15. Brit. Pat. 541,146 (May 13, 1940), N. H. Haddock (to ICI).
16. U.S. Pat. 2,369,666 (Feb. 20, 1945), A. L. Fox (to E. I. du Pont de Nemours & Co., Inc.); Brit. Pat. 573,831 (Dec. 7, 1945) (to ICI).
17. H. Hiyama, *Synthetic Sulfur Dyes*, Osaka Municipal Technical Research Institute, Japan, 1959.
18. Ger. Offen. 2,347,537 (1975), E. Krusche and co-workers (to Cassella & Co.).
19. DR Pat. 693,862 (July 19, 1936), R. Rieche and B. Schiedk (to I.G. Farbenindustrie); P. Bachmann, *Ber.* **36,** 965 (1903).
20. R. Gnehm and F. Kauffler, *Ber.* **37,** 2619, 3032 (1904).
21. *Mon. Scient.* **11,** 655 (1897); *Mon. Scient.* **17,** 427 (1903).
22. K. H. Shah, B. D. Tilak, and K. Venkataraman, *Proc. Indian Acad. Sci. Sect. A* **28,** 111 (1948).
23. W. N. Jones and E. E. Reid, *J. Am. Chem. Soc.* **54,** 4393 (1932).
24. H. E. Fierz-David and E. Merian, *Abr. der Chem. Tech. der Textilfasern* **146** (1948).
25. I. Chmelnitzkaja and V. Werchowskaja, *Anilinokrasochnaya Prom.* **5,** 67 (1935); *Chem. Abstr.* **29,** 4943 (1935).
26. T. Kubota, *J. Chem. Soc. Jpn.* **55,** 565 (1934).
27. H. Hiyama, *J. Chem. Soc. Jpn. Ind. Chem. Sect.* **51,** 97 (1948).
28. A. Weinberg, *Ber.* **63**(A), 117 (1930).

29. *J. Soc. Dyers Colour.* **89,** 416, 420, 421.
30. C. Heid, *Z. ges Textilind.* **70,** 626 (1968).
31. Brit. Pat. 1,321,453 (Feb. 24, 1970), H. Smithson (to J. Robinson & Co., Ltd.).
32. Jpn. Appl. 71 9309 (Feb. 26, 1971), S. Yoshioka (to Nippon Kayaku & Co., Ltd.).
33. D. A. Bailey and F. E. Humphreys, *J. Soc. Leather Trades' Chem.* **51,** 154 (1967).
34. Y. Maeda and co-workers, *Mizu Shori Gijutsu* **14,** 1253, 60 (1973).
35. H. M. Tobin, *Am. Dyest. Rep.* **68**(9), 26 (Sept. 1979).

General References

H. E. Fierz-David, *Künstliche Organic Farbstoffe*, Berlin, Julius Springer, 1926.

O. Lange, *Die Schwelfelfarbstoffe, Ihre Herstellung und Verwendung*, 2nd ed., Spamer, Leipzig, 1925.

H. A. Lubs, *The Chemistry of Synthetic Dyes and Pigments*, Reinhold Publishing Corp., New York, 1955, Chapt. 6.

J. F. Thorpe and M. A. Whiteley, eds., *Thorpe's Dictionary of Applied Chemistry*, 4th ed., Vol. 11, Longmans, Green & Co., London, 1954.

K. Venkataraman, *The Chemistry of Synthetic Dyes and Pigments*, Vol. 2, 1952, Chapt. 35; Vol. 3, 1970, Chapt. 1; Vol. 7, 1974, Chapt. 1, Academic Press, London and New York.

M. L. Crossley and co-workers, *BIOS (British Intelligence Objectives Committee) Misc.* 55, H.M. Stationery Office, London, UK, 1946.

J. Avery and co-workers, *BIOS 983*, 1946; *BIOS 1155*, D. A. W. Adams and co-workers, H.M. Stationery Office, London, UK, 1947.

FIAT (Field Information Agency Technical) Microfilm 764, reels 82CC, 92AA, and 186C, 1946.

FIAT 1313, Vols. 2 and 3, H.M. Stationery Office, London, UK, 1948.

C. Heid, K. Holoubek, and R. Klein, *Melliand Textilberichte*, International Textile Reports **54,** 1314 (1973).

W. Prenzel, *Chemiefasern + Text. Anwendungstech./Text. Ind.* 24/76, 293 (1974).

Colour Index, 3rd ed., Published by The Society of Dyers and Colourists, Bradford, U.K. and The American Association of Textile Chemists and Colorists, Research Triangle Park, N.C., 1975, revised edition 1982.

W. E. Wood, *Rev. Prog. Color. Relat. Tap.* Society of Dyers & Colourists **7,** 80 (1976).

R. A. Guest
W. E. Wood
James Robinson & Company, Ltd.

SULFURIC ACID AND SULFUR TRIOXIDE

Sulfuric acid [*7664-93-9*], H_2SO_4, is a colorless, viscous liquid with a specific gravity of 1.8357 and a boiling point of 270°C. Its anhydride, SO_3 [*7446-11-9*], is also a liquid with a specific gravity of 1.857 and a boiling point of 44.8°C. Sulfuric acid is by far the largest volume chemical commodity produced and is sold or used commercially in a number of different concentrations including 78 wt % (60°Bé), 93 wt % (66°Bé), 96 wt %, 98–99 wt %, 100%, and as various oleums (fuming sulfuric acid, $H_2SO_4 + SO_3$). Stabilized liquid SO_3 is also an item of commerce.

Sulfuric acid has several desirable properties that lead to its use in a wide variety of applications. It typically is less costly than any other acid; it can be readily handled in steel or common alloys at normal commercial concentrations; and it is available and readily handled at concentrations >100 wt % (oleum). Sulfuric acid is a strong acid; it reacts readily with many organic compounds to produce useful products, and it will form a slightly soluble salt or precipitate with calcium oxide or hydroxide, the least expensive and most readily available base. Concentrated sulfuric acid is also a good dehydrating agent and under some circumstances functions as an oxidizing agent.

History

Sulfuric acid has been an important item of commerce for at least 250 years and has been known and used since the Middle Ages. In the eighteenth and nineteenth centuries, it was essentially entirely produced by the chamber process, in which oxides of nitrogen (as nitrosyl compounds) are used as homogeneous catalysts for the oxidation of sulfur dioxide. The product made by this process is of rather low concentration (typically 60° Baumé, or ca 77–78 wt % H_2SO_4), not high enough for many commercial uses.

During the first part of the twentieth century, the chamber process was gradually replaced by the contact process, and today the former is considered obsolete. However, there has been one recent installation of a modified chamber plant to process very dilute SO_2 gases generated by molybdenum ore roasting (1).

Primary impetus for development of the contact process came from a need for high strength acid and oleum to make synthetic dyes and organic chemicals. The contact process employing platinum catalysts began to be used on a large scale for this purpose late in the nineteenth century. Its development accelerated during World War I to provide concentrated mixtures of sulfuric and nitric acid for explosives production.

In 1875, a paper by Winkler awakened interest in the dormant contact process which had first been patented by Phillips in 1831. Winkler's process was employed by a number of firms; the earliest commercial production was by Emil Jacob at the Kreuznack (Germany) works. Other German firms were using pyrite burner gas as a source of sulfur dioxide, and a tremendous amount of time and money was expended in research, most of it directed toward purification of the roaster gas to prevent contamination and poisoning of the catalyst. This work culminated in Knietsch's classic lecture before the German Chemical Society in 1901 (2) which revealed some of the many investigations that had been carried out by the Badische Anilin- und Soda-

Fabrik. Patents were also taken out between 1898 and 1902 on the Mannheim process, in which a first stage of conversion utilized a shaft filled with ferric oxide, followed by a final stage employing platinum; the Schroder-Grillo process, employing platinum-impregnated magnesium sulfate as catalyst; and the Tentelew Chemical Company process, which introduced many novel features of apparatus design (3).

In 1915, an effective vanadium catalyst for the contact process was developed and used by Badische in Germany. This type of catalyst was employed in the United States starting in 1926 and gradually replaced platinum catalysts over the next few decades. Vanadium catalysts have the advantages of superior resistance to poisoning and of being relatively abundant and inexpensive compared to platinum. After World War II, the typical size of individual contact plants increased dramatically in the United States and around the world to supply the rapidly increasing demands of the phosphate fertilizer industry. This trend has culminated in the recent construction of a single-train sulfur-burning unit rated at 2820 metric tons of acid per day (4).

Another significant change in the contact process occurred in 1963, when Bayer AG announced the first large-scale use of the double-contact (double-absorption) process and was granted several patents (5–8). In this process, SO_2 gas that has been partially converted to SO_3 by catalysis is cooled, passed through sulfuric acid to remove SO_3, reheated, and then passed through another one or two catalyst beds. By this means, overall conversions can be increased from ca 98% up to 99.5–99.8%, thereby reducing emission of unconverted SO_2 to the atmosphere. Because of worldwide pressures to reduce SO_2 emissions, most new plants now utilize double-absorption, and many existing plants have been modified to use it. An early U.S. patent (9) disclosed the general concept but apparently was not reduced to practice at that time.

Physical Properties

Sulfur Trioxide. Sulfur trioxide [*7446-11-9*] at room temperature and atmospheric pressure is a colorless liquid that fumes in air. Trace amounts of water or sulfuric acid can catalyze the formation of polymers. However, polymerization apparently proceeds at a negligible rate so long as the liquid is maintained free of solids (ie, above its freezing point). Once solid polymers are present, temperatures of ca 50–75°C (vs a normal boiling point of 44.8°C) are required to fully convert polymers back to liquid monomer.

There is some controversy about the nature of polymeric sulfur trioxide. Apparently, polymers can have various molecular weights and degrees or types of cross-linking. The literature generally reports that solid sulfur trioxide can exist in three trimorphic phases: alpha-, beta-, and gamma-SO_3 with melting points of 62.3, 32.5, and 16.8°C, respectively. The alpha and beta forms melt to give liquid gamma-SO_3. The alpha-SO_3 phase has a polymeric cross-linked structure that forms crystals resembling ice needles; beta-SO_3 consists of polymeric molecules that form white asbestos-like crystals with a silky luster; and gamma-SO_3 is a colloidal form that consists of cyclic trimer and monomer molecules. An excellent study on the thermodynamic properties of the three SO_3 phases is given in reference 10.

Table 1 presents a summary of thermodynamic properties of pure sulfur trioxide with corresponding references. Significantly lower values have been reported for the heat of fusion of α, β, and γ phases of SO_3 (24.05 kJ/kg (5.75 kcal/kg) for γ, 12.84 kJ/kg (3.07 kcal/kg) for β, and 31.67 kJ/kg (7.57 kcal/kg) for α phase SO_3) (22), as have slightly different critical temperature, pressure, and density values (13).

Table 1. Properties of Sulfur Trioxide

Property	Value	Ref.
critical temperature, °C	217.8	11
critical pressure, kPa[a]	8,208	11
critical density, g/cm^3	0.630	11
triple point temperature (γ phase), °C	16.8	10
triple point pressure (γ phase), kPa[a]	21.13	12
normal boiling point temperature, °C	44.8	12
melting point (γ phase), °C	16.8	12
transition temperature, °C	−183.0	10
liquid density (γ phase at 20°C), g/cm^3	1.9224	13
solid density (γ phase at −10°C), g/cm^3	2.29	14
liquid coefficient of thermal expansion (at 18°C), per °C	0.002005	15
liquid heat capacity (at 30°C), kJ/(kg·°C)[b]	3.222	13, 16
heat of formation of gas (at 25°C), (MJ·kg)/mol[b]	−395.76	17
free energy of formation of gas (at 25°C), (MJ·kg)/mol[b]	−371.07	17
entropy of gas (at 25°C), (MJ·kg)/(mol·°C)[b]	0.25666	17
heat of dilution, MJ/kg[b]	2.109	13
heat of fusion, kJ/kg[b]		
(α)	324.0	13
(β)	151.6	13
(γ)	94.07	13
heat of sublimation, MJ/kg[b]		
(α)	0.8518	10
(β)	0.7269	10
(γ)	0.7029	10
heat of vaporization (γ liquid), MJ/kg[b]	0.5843	10
diffusion in air (at 80°C), m/s	0.000013	18
liquid dielectric constant (at 18°C)	3.11	19
electric conductivity	negligible	20–21

[a] To convert kPa to psi, multiply by 0.145.
[b] To convert J to cal, divide by 4.184.

Figure 1 presents the density of sulfur trioxide as a function of temperature. Curve B, given by Horvath (24), is a composite of data taken from refs. 25–28. The vapor pressures of sulfur trioxide α, β, and γ phases are presented in Figure 2. Different values of SO_3 vapor pressure for α, β, and γ phases have been reported in refs. 10 and 13 (see Table 2).

Table 3 summarizes the thermodynamic properties of sulfur trioxide, and Table 4 summarizes the thermodynamic properties of the oxidation reaction of sulfur dioxide. Thermodynamic data from Boreskov (30) are believed to be more accurate than those of Evans and Wagman (29) at temperatures below ca 435°C.

Sulfuric Acid. Sulfuric acid is a dense, colorless liquid at room temperature, with specific gravity as shown in Figure 3. In the past it was popular to report the concentration of sulfuric acid as specific gravity in degrees Baumé. In the United States, the Baumé scale is calculated utilizing the following formula:

$$°Bé = 145 - (145/\text{sp gr})$$

In Germany (FRG) and in France the Baumé scale is calculated using 144.3 as the constant. The Baumé scale only includes the sulfuric acid concentration range of 0–93.19% H_2SO_4. Higher acid concentrations are not covered because of the great

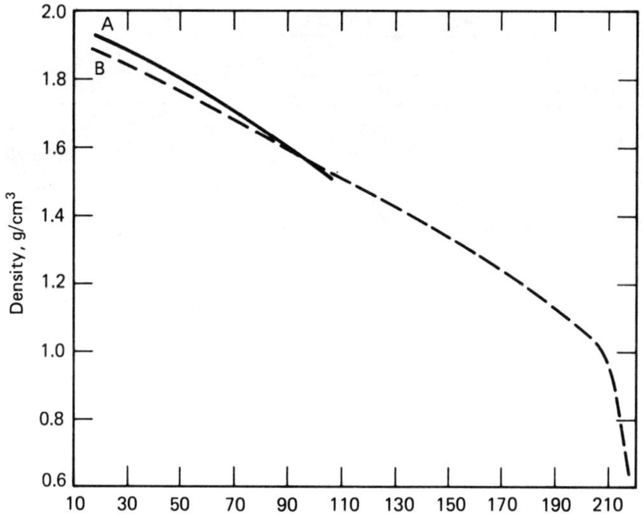

Figure 1. Density of liquid sulfur trioxide. A = ref. 23; B = ref. 24.

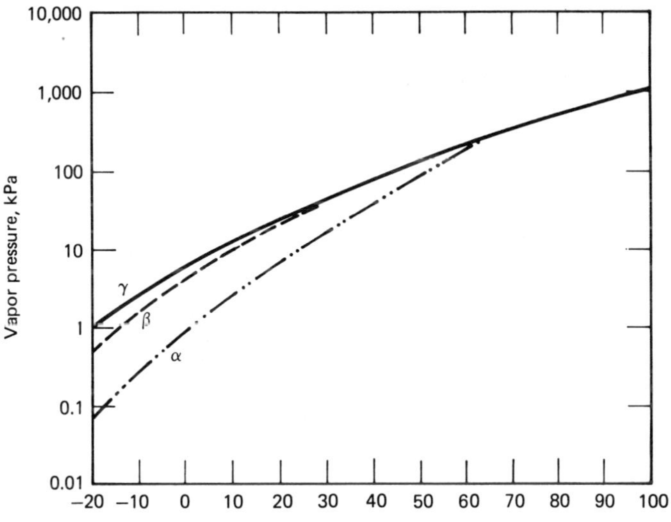

Figure 2. Vapor pressure of sulfur trioxide (23). To convert kPa to psi, multiply by 0.145.

difficulty in differentiating between acid concentrations in the range of 93–100% H_2SO_4 by specific gravity measurements. Data on the density of sulfuric acid solutions and oleum as a function of temperature have been published (31–32).

Figure 4 presents the electrical conductivity of concentrated sulfuric acid solutions (33). For sulfuric acid solutions in the 90–100% H_2SO_4 concentration range, the electrical conductivity measurements reported by Record (33) are believed to be the best values; other conductivity data are also reported (34–36).

The viscosity of sulfuric acid solutions is plotted in Figure 5; other viscosity data may be found in refs. 36–42. Surface tension of sulfuric acid solutions is presented in Figure 6 (43). Surface tension of selected concentrations of sulfuric acid as a function

Table 2. Vapor Pressure of SO_3, α, β, and γ Phases[a]

Temperature, °C	Vapor pressure, Pa[b]		
	α	β	γ
0	773	4,266	5,999
25	9,732	45,860	57,730
50	86,660	126,700	126,700
75	400,000	400,000	400,000

[a] Ref. 13.
[b] To convert Pa to psi, divide by 6895.

Table 3. Thermodynamic Properties of Sulfur Trioxide[a]

Temperature, K	ΔH_f°, kJ/mol[b]	ΔF_f°, kJ/mol[b]
600	−460.2	−359.9
700	−459.6	−343.3
800	−458.9	−326.7
900	−457.9	−310.2
1000	−456.7	−293.9
1100	−455.5	−277.7
1200	−454.1	−261.5

[a] Ref. 29.
[b] To convert J to cal, divide by 4.184.

Table 4. Thermodynamic Properties of $SO_2 + \frac{1}{2} O_2 \rightarrow SO_3$[a]

Temperature, K	ΔH_T, kJ/mol[b]	ΔF_T, kJ/mol[b]	Kp, $Pa^{-1/2}$ [c]
600	−97.99	−41.59	13.13
700	−97.36	−32.30	0.7445
800	−96.57	−23.05	0.1006
900	−95.69	−13.97	0.02033
1000	−94.60	−4.94	0.005686
1100	−93.51	4.02	0.002025
1200	−92.30	12.84	0.000867

[a] Ref. 13.
[b] To convert J to cal, divide by 4.184.
[c] To convert $Pa^{-1/2}$ to $atm^{-1/2}$, multiply by 318.32.

of temperature up to the boiling point is presented in Figure 7 (44); other data can be found in refs. 40–41, 45–47.

The index of refraction of sulfuric acid solutions is presented in Figure 8 (44). Additional data may be found in ref. 48. Solubility data for oxygen in sulfuric acid solutions are presented in Figure 9 (49). The solubility of sulfur dioxide in concentrated sulfuric acid is shown in Figure 10 (50).

Data on chemical properties such as self-dissociation constants for sulfuric and dideuterosulfuric acid can be found in refs. 42,47,51–52. Reference 53 includes an excellent graphical representation of physical-property data of 100% H_2SO_4. Critical temperatures of sulfuric acid solutions are presented in Figure 11 (32).

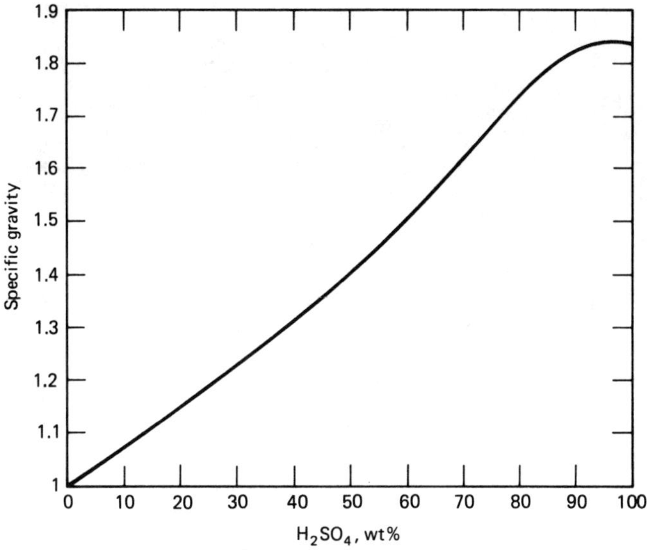

Figure 3. Specific gravity of sulfuric acid solutions, 15/4°C.

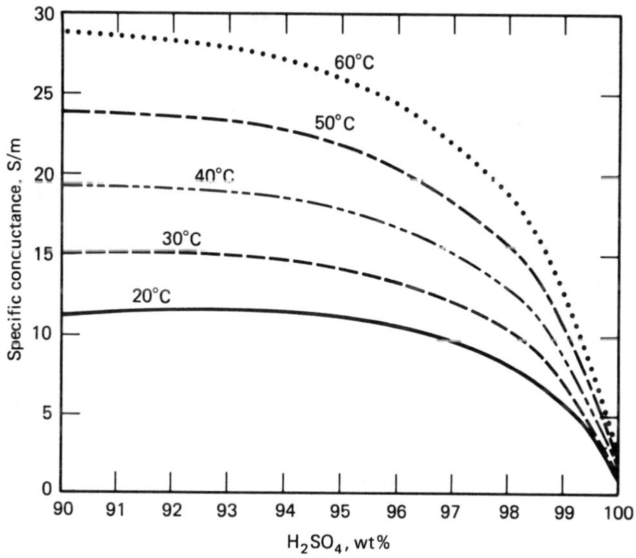

Figure 4. Specific conductance of sulfuric acid (33).

Boiling points of sulfuric acid are given in Figure 12, with some uncertainty existing close to 100% H_2SO_4. Freezing points also are not well established, in part because acid purity and cooling rates significantly affect the observed freezing points. Acid impurities lower the freezing point, and cooling rates may be such that subcooled liquid sulfuric acid is produced. Figure 13 shows the freezing points reported by Malin (55) which do not appear elsewhere in readily available literature. Additional freezing-point data are given in ref. 56 and the best discussions in refs. 41 and 57.

Figure 14 presents the azeotropic composition of sulfuric acid as a function of pressure (58). The vapor pressure exerted by sulfuric acid solutions below the azeotrope is primarily from water vapor; above the azeotropic concentration, SO_3 is the primary

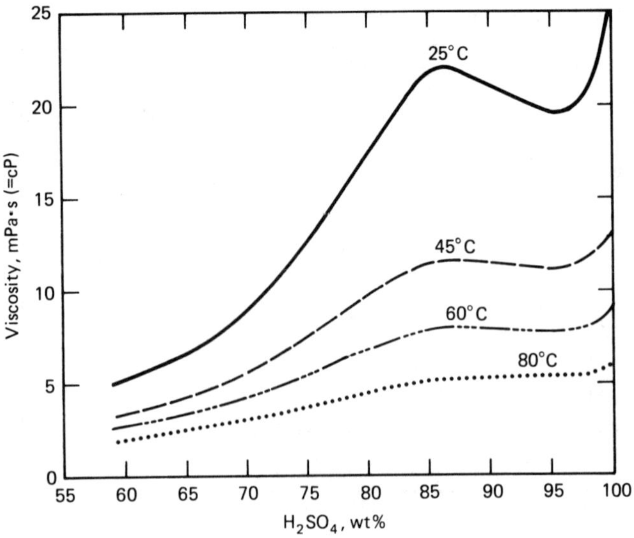

Figure 5. Viscosity of sulfuric acid (37).

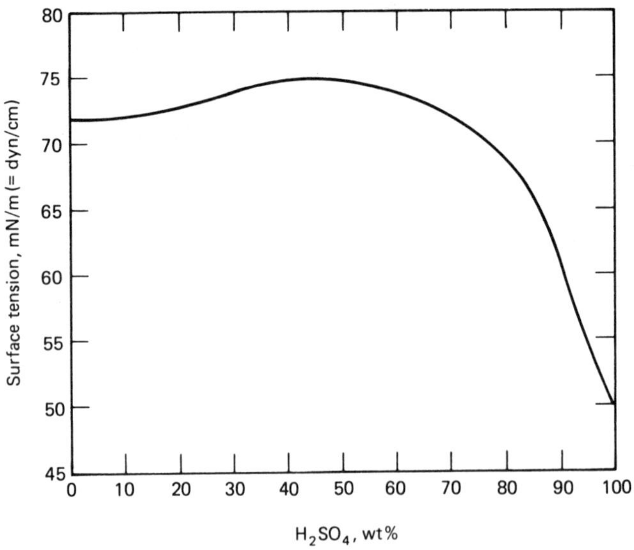

Figure 6. Surface tension of sulfuric acid solutions at 25°C (43).

component of the vapor phase. At the boiling point, sulfuric acid solutions containing <85% H_2SO_4 evaporate water exclusively; those containing >35% free SO_3 (oleum) evaporate exclusively sulfur trioxide. The vapor of sulfuric acid solutions between 85% H_2SO_4 and 35% free SO_3 is a mixture of sulfuric acid, water, and sulfur trioxide vapors. An excellent compilation of vapor-pressure data of concentrated sulfuric acid solutions in the low concentration region is given in ref. 59.

Figure 15 plots consistent total vapor pressures for 0 to 100% H_2SO_4. Table 5 summarizes the partial pressures of sulfuric acid, water, and sulfur trioxide for sulfuric acid solutions at their respective boiling points (60). Partial vapor pressures for sulfuric acid solutions of 85–100% H_2SO_4 at temperatures below the boiling point are presented

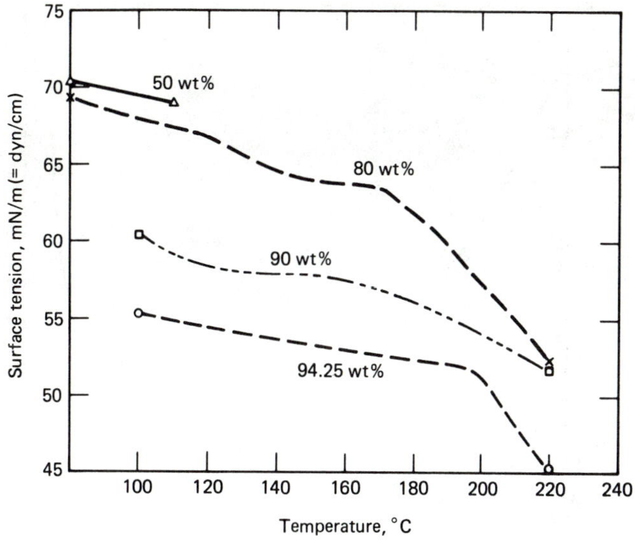

Figure 7. Surface tension of sulfuric acid solutions versus temperature (44).

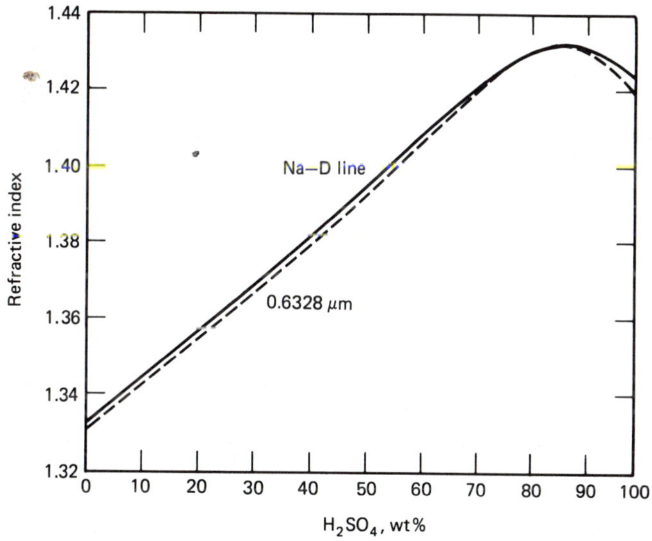

Figure 8. Refractive index of sulfuric acid solutions, Na-D and 0.6328 μm (44).

in Table 6. An excellent recent study on vapor–liquid equilibrium data has been published (61).

Figure 16 presents a composite curve of sulfuric acid heat of mixing from data reported by Kunzler and Giauque (62) and the National Bureau of Standards (63).

Oleum. The density of oleum at 15°C has been reported (31). Oleum concentrations of 30–60% SO_3 are in a crystalline state at 15°C; therefore, their densities are not included. The temperature dependency of oleum density for varying oleum concentrations is presented in Figure 17 from smoothed data reported in ref. 64.

The boiling points of oleum are presented in Figure 18 (65) and freezing points in Figure 19 (56,66). An excellent discussion on the crystallization points of oleum is

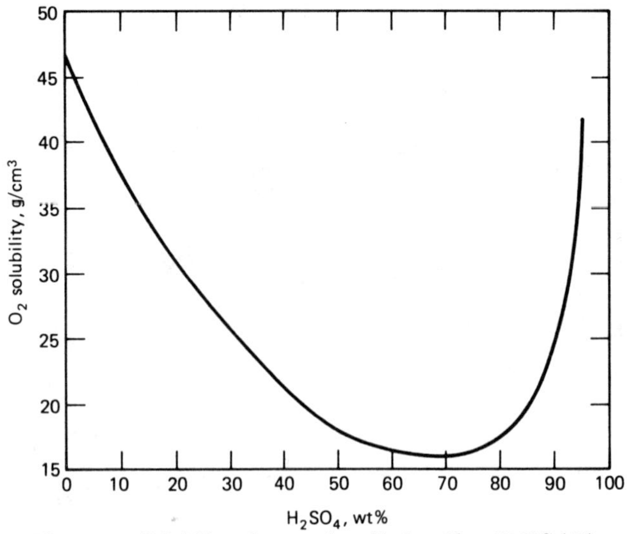

Figure 9. Solubility of oxygen in sulfuric acid at 15.6°C (49).

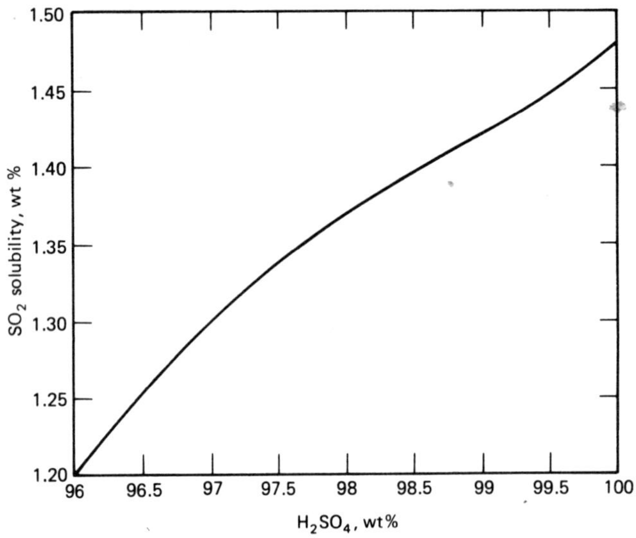

Figure 10. SO_2 solubility in sulfuric acid at 60°C, 101.3 kPa (1 atm) (50).

included in ref. 51. The solubility of sulfur dioxide in oleum is reported (50). Viscosity of oleum is summarized in Figure 20 (37); additional viscosity data are reported in ref. 31.

A composite curve of heat of infinite dilution of oleum from data reported in refs. 67–70 is presented in a compiled form in ref. 71, where heats of formation of oleums from liquid or gaseous SO_3 are also reported (see Tables 7 and 8). Heat of vaporization data are reported in ref. 77. Oleum heat-capacity data are presented in Figure 21 (73).

A smoothed vapor-pressure curve for oleum at 39.35°C from data reported by refs. 10, 60, 65, 74–75 is drawn in Figure 22. Oleum partial vapor-pressure data at the boiling point are presented in Table 9 (60). Data on partial vapor pressures at selected temperatures other than boiling are presented in Table 10.

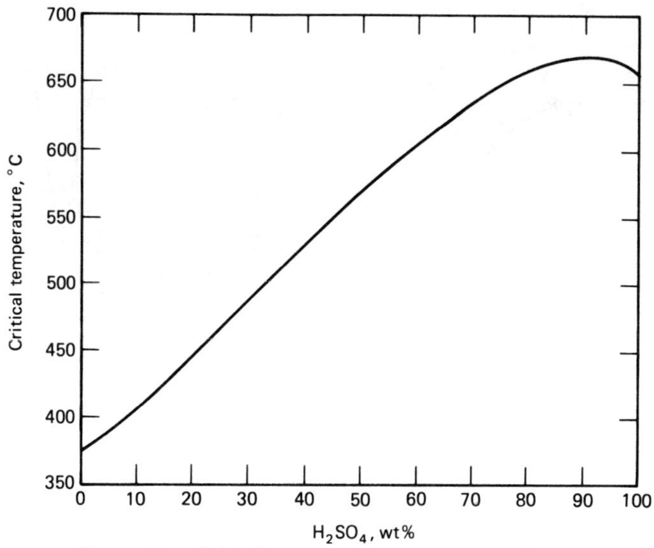

Figure 11. Critical temperatures of sulfuric acid (32).

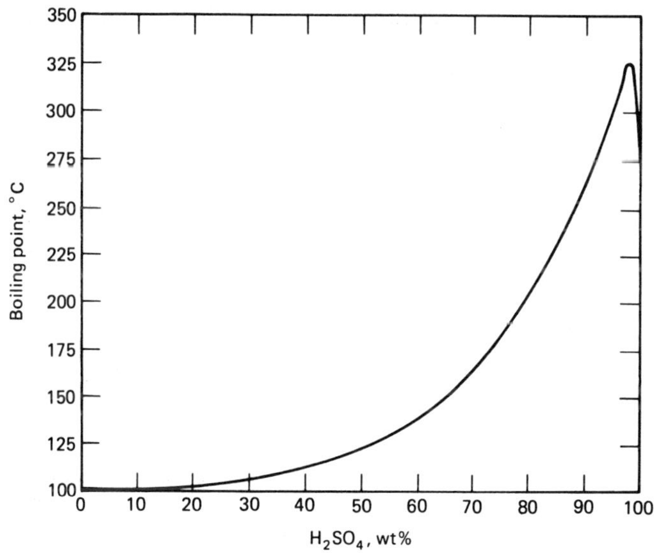

Figure 12. Normal boiling point of sulfuric acid (54).

Manufacture

The contact process produces sulfuric acid from a wide range of sulfur-bearing raw materials by several different process variants, depending largely on the raw material used. In some cases sulfuric acid is made as a by-product of other operations, primarily as an economical or convenient means of minimizing air pollution or disposing of unwanted by-products. Economics favor production of sulfuric acid close to the point of use, since truck or rail freight costs are substantial compared to production costs.

The principal direct raw materials used to make sulfuric acid are elemental sulfur,

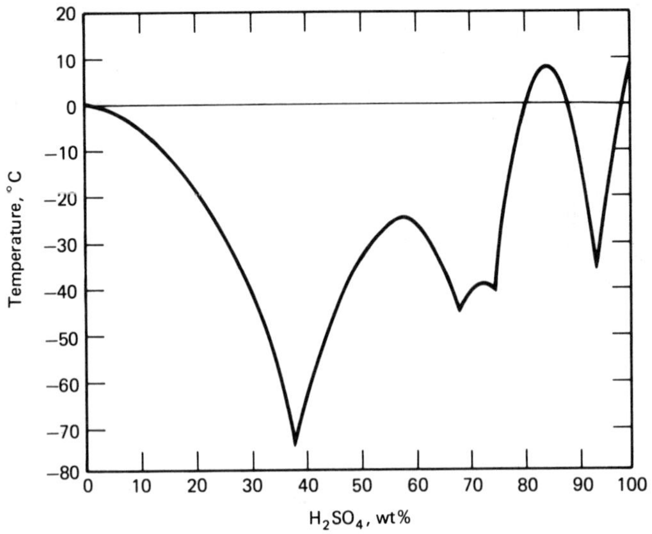

Figure 13. Freezing points of sulfuric acid (55).

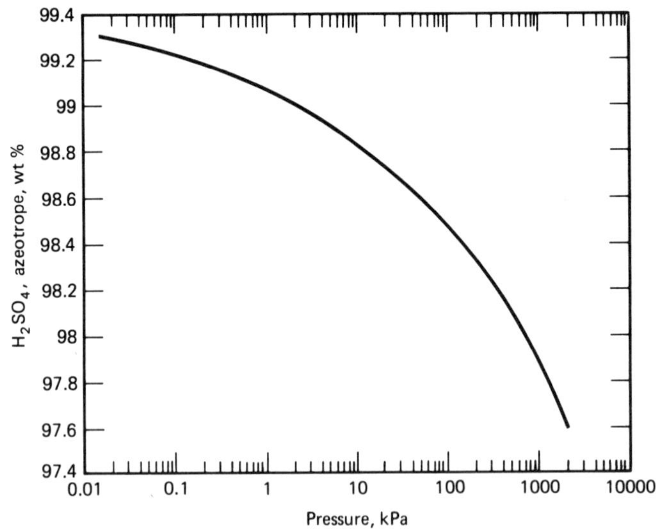

Figure 14. H_2SO_4 azeotrope composition vs pressure (58,61). To convert kPa to psi, multiply by 0.145.

spent (contaminated and diluted) sulfuric acid, and hydrogen sulfide. Elemental sulfur is by far the most widely used. In the past, iron pyrites or related compounds were often used; but this type of raw material is not common today except in Spain and the USSR. A large amount of sulfuric acid is also produced as a by-product of nonferrous metal smelting, ie, roasting sulfide ores of copper, lead, molybdenum, nickel, zinc, or others.

In all types of contact plants, the first steps in the process have the objective of producing a reasonably continuous, contaminant-free gas stream containing appreciable sulfur dioxide and some oxygen. Desirably, the gas stream is dry, but some plants

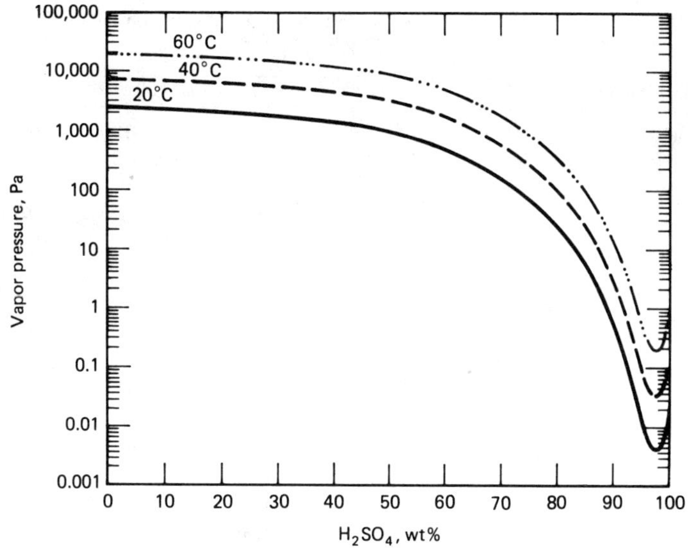

Figure 15. Total vapor pressure of sulfuric acid at 20, 40, and 60°C (60). To convert Pa to psi, divide by 6895.

Table 5. Partial Pressures of Sulfuric Acid Solutions at Their Boiling Points[a]

Sulfuric acid, wt %	Partial pressures, Pa[b] H_2SO_4	H_2O	SO_3
85.0	0.267	101,000	
88.0	13.3	101,000	0.0667
90.0	86.7	101,000	1.067
92.0	680	101,000	24.0
94.0	3,500	97,600	253
95.0	7,800	92,400	1,080
96.0	17,250	80,400	3,720
97.0	31,200	58,600	11,500
98.33	41,900	31,000	28,400
98.4	42,000	29,200	30,100
99.0	39,900	10,700	50,700
100.0	22,900	787	77,600

[a] Ref. 60.
[b] To convert Pa to psi, divide by 6895.

burning H_2S successfully process a gas that contains no more water vapor than required to combine stoichiometrically with the amount of sulfur trioxide formed. If the initial oxygen concentration of the process gas is low, additional air or oxygen must be added prior to (or during) catalytic oxidation to ensure that there is an excess over stoichiometric needs for conversion of SO_2 to SO_3.

The gas stream containing sulfur dioxide is either dried before passing to the catalytic oxidation step, or is oxidized in the presence of water vapor with subsequent acid condensation and removal, or when produced from elemental sulfur, the air used for sulfur burning is predried. In all cases, typical plant designs use sulfuric acid from the process as a drying agent. "Wet" catalytic oxidation is relatively uncommon.

Table 6. Partial Pressures in Pascals[a] for Sulfuric Acid Solutions at Selected Temperatures[b]

		Sulfuric acid, wt %					
Temp, °C	Component	85	90	94	96	98.3	100
20	H_2SO_4				0.0013	0.004	0.0293
	H_2O	5.33	0.533	0.032	0.0053		
	SO_3						0.0173
40	H_2SO_4			0.004	0.012	0.0267	0.1867
	H_2O	25.33	2.933	0.2133	0.04	0.004	
	SO_3					0.0013	0.1467
60	H_2SO_4		0.00267	0.0227	0.0667	0.1733	0.9333
	H_2O	97.33	13.33	1.2	0.2533	0.0267	0.0013
	SO_3					0.0107	0.9333
80	H_2SO_4		0.012	0.1067	0.3333	0.7999	4.00
	H_2O	323.97	51.996	5.333	1.267	0.16	0.0107
	SO_3				0.004	0.0667	4.933
100	H_2SO_4		0.0533	0.4666	1.3332	3.3331	14.665
	H_2O	941.26	170.65	19.998	5.333	0.7999	0.0533
	SO_3			0.0013	0.0213	0.400	21.332
120	H_2SO_4	0.00267	0.1733	1.600	4.800	11.999	46.663
	H_2O	2,480	503.96	67.994	18.665	3.066	0.2666
	SO_3			0.0093	0.1067	1.600	77.33

[a] To convert Pa to psi, divide by 6895.
[b] Ref. 60.

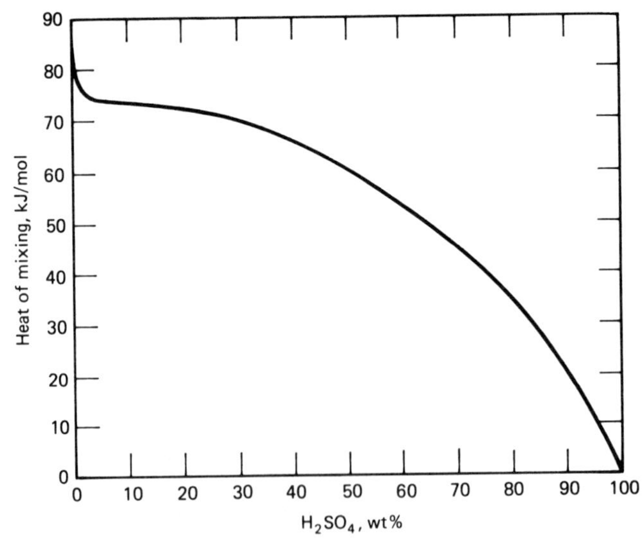

Figure 16. Heat of mixing sulfuric acid from H_2O and H_2SO_4 at 25°C (62). To convert J to cal, divide by 4.184.

The sulfur trioxide produced by catalytic oxidation is absorbed in a circulating stream of 98–99% H_2SO_4 that is cooled to ca 70–80°C. Water or weaker acid is added as needed to maintain acid concentration. Sulfuric acid of ca 98.5% concentration is used because it is the azeotrope having minimum total vapor pressure, ie, the sum of SO_3, H_2O, and H_2SO_4 partial pressures. At acid concentrations <98.5% H_2SO_4, rela-

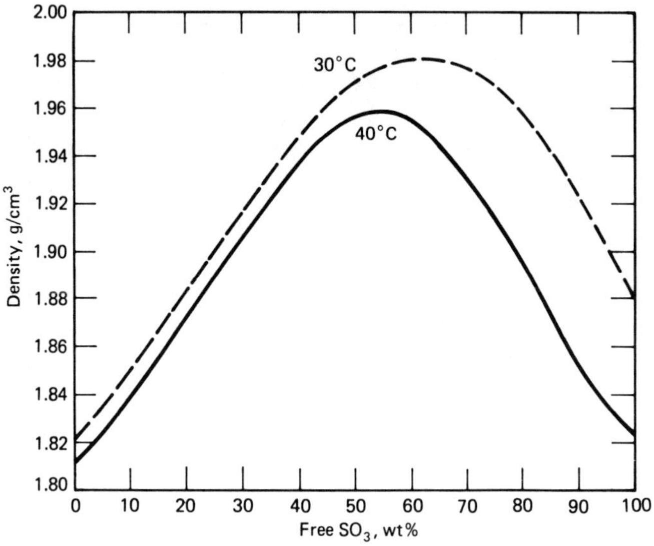

Figure 17. Density of oleums vs temperature (64).

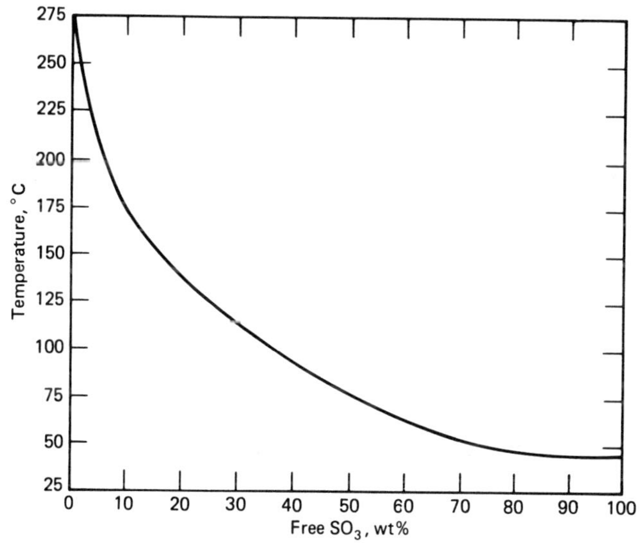

Figure 18. Boiling points of oleums (65).

tively intractable aerosols of sulfuric acid mist particles are formed by vapor phase reaction of SO_3 and H_2O, and at higher acid concentrations, the partial pressure of SO_3 becomes significant.

The catalytic oxidation of SO_2 to SO_3 is highly exothermic and, as expected, equilibrium becomes increasingly unfavorable for SO_3 formation as temperature increases above 410–430°C. Unfortunately, this is about the minimum temperature level required for commercial catalysts to function. Consequently, plant catalytic reactors (converters) are typically designed as multistage adiabatic units with gas cooling between each stage. Typical SO_2 concentrations in the gas stream are ca 4–11 vol %, lower or higher concentrations occasionally being handled by special process or plant modifications.

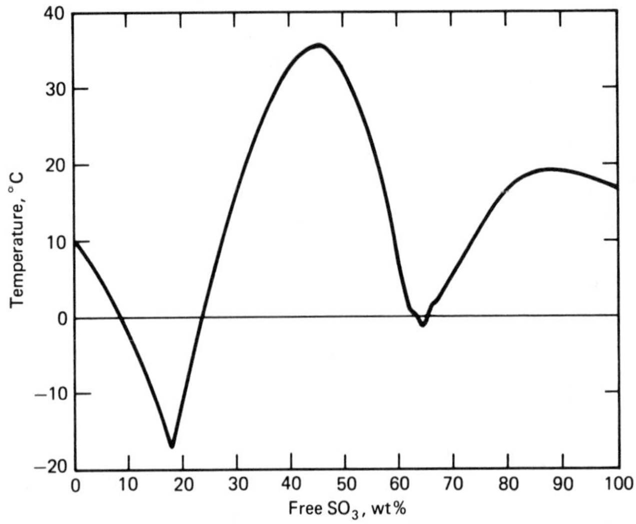

Figure 19. Freezing points of oleums (66).

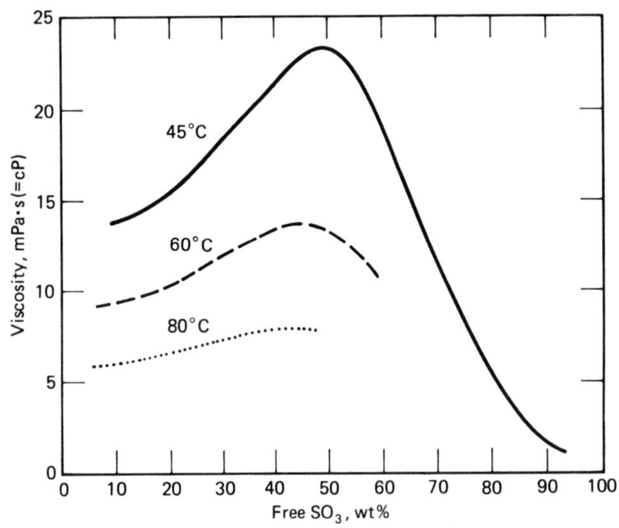

Figure 20. Viscosity of oleums (37).

In early years the contact process frequently employed only two or three catalyst stages (passes) to obtain overall SO_2 conversions of ca 95–96%. Residual sulfur dioxide was vented to the atmosphere as a dilute stream. Later, four-pass converters were commonly used to obtain conversions of ca 97–98% with high SO_2 concentrations or, alternatively, three-pass converters for low SO_2 concentrations.

In recent years, air-pollution requirements have caused adoption of the "double-contact" or "double-absorption" process, which provides overall conversions of ca 99.5–99.8%. The double-absorption process employs the principle of intermediate removal of the reaction product (SO_3) to obtain favorable equilibria and kinetics in later stages of the reaction. A few single-absorption plants are still being built in some areas of the world, or where special circumstances exist, but most industrialized nations

Table 7. Heat of Formation of Oleums by the Reaction H_2O (Liq) + xSO_3 (Liq) = $H_2SO_4.(x-1)\ SO_3$ (Liq)[a]

	Liquid phase	
Free SO_3, %	SO_3, MJ/kg[b]	H_2O, MJ/kg[b]
0	1.107	4.919
10	1.009	5.086
20	0.9067	5.260
30	0.8022	5.434
40	0.6978	5.632
50	0.5905	5.841
60	0.4808	6.061
70	0.3684	6.317
80	0.2535	6.654
90	0.1333	7.106
100	0	

[a] Ref. 71. x = mol total SO_3/mol H_2O.
[b] To convert J to cal, divide by 4.184.

Table 8. Heat of Formation of Oleums by the Reaction H_2O (Liq) + xSO_3 (Gas) = $H_2SO_4.(1-x)\ SO_3$ (Liq)[a]

	Gas phase	
Free SO_3, %	SO_3, MJ/kg[b]	H_2O, MJ/kg[b]
0	1.645	7.311
10	1.547	7.803
20	1.445	8.384
30	1.340	9.080
40	1.236	9.986
50	1.129	11.17
60	1.019	12.84
70	0.9067	15.56
80	0.7918	20.79
90	0.6716	35.83
100	0.5383	

[a] Ref. 71. x = mol total SO_3/mol H_2O.
[b] To convert J to cal, divide by 4.184.

now have emission standards that require double-absorption plants (or tail-gas scrubbers).

Plants producing oleum or liquid SO_3 typically have one or two additional packed towers irrigated with oleum ahead of the normal SO_3 absorption towers. Partial absorption of SO_3 occurs in these towers, and sulfuric acid is added to maintain desired oleum concentrations. Normally, oleum up to ca 35 wt % free SO_3 content can be made in a single tower; two towers are used for 40 wt % SO_3. Liquid SO_3 is produced by heating oleum in a boiler to generate SO_3 gas, which is then condensed. Oleums containing SO_3 >40 wt % are usually produced by mixing SO_3 with low concentration oleum.

Where elemental sulfur or hydrogen sulfide is used as raw material, considerable heat is evolved during the initial combustion. Additional heat is generated by catalytic

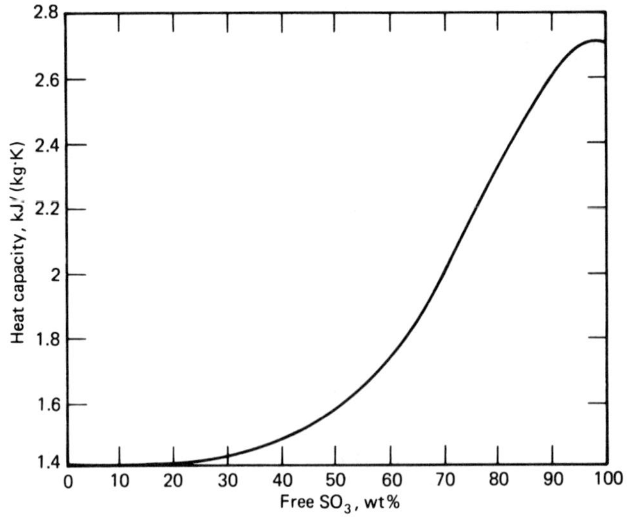

Figure 21. Heat capacity of oleum at 20°C (31). To convert J to cal, divide by 4.184.

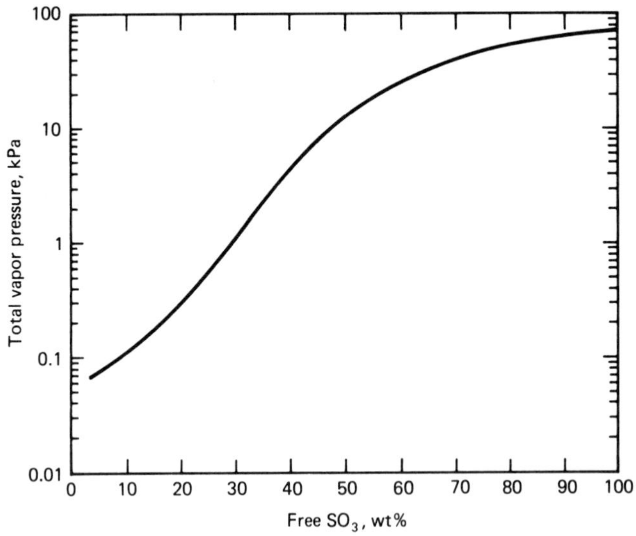

Figure 22. Total pressure of oleum at 39.35°C (10,60,74–75). To convert kPa to psi, multiply by 0.145.

oxidation to SO_3 and by the reaction of SO_3 and H_2O to form H_2SO_4. In such plants, much of the heat is typically used to produce steam, which can be utilized either for heating requirements in other processes or to generate power via turbines. In many cases, large plants of this type are essentially coproducers of steam or power and sulfuric acid; both products have significant economic value. Where spent acid is used as raw material, it usually is decomposed in furnaces fired by gas, oil, or other fuels (sometimes H_2S), and the high temperature gas from such furnaces can also generate steam or power (see Power generation).

In general, plants using SO_2 gas derived from metallic sulfides, spent acids, or gypsum–anhydrite purify the gas stream before drying it by cold- (or wet-) gas puri-

Table 9. Partial Pressures of Oleum at the Boiling Point[a]

	Partial pressures, Pa[b]		
Free SO_3, %	H_2SO_4	H_2O	SO_3
0	22,920	786.6	77,620
2	15,000	240.0	86,090
5	6,639	26.7	94,660
8	3,573	6.00	97,750
10	1,707	1.47	99,620
12	1,013	0.533	100,300
15	466.6	0.107	100,900
18	213.3	0.0227	101,100
20	128.0	0.0080	101,200
22	74.66	0.00267	101,200
25	33.33		101,300
28	16.00		101,300
30	8.67		101,300
32	4.00		101,300
35	1.867		101,300

[a] Ref. 60.
[b] To convert Pa to psi, divide by 6895.

Table 10. Partial Pressures for Oleum at Selected Temperatures, Pa[a,b]

		Free SO_3 in oleum, %					
Temp, °C	Component	5	10	15	20	25	30
20	H_2O						
	H_2SO_4	0.0267	0.0227	0.0160	0.0120	0.0067	0.0040
	SO_3	0.2000	1.600	7.999	31.997	106.7	269.3
40	H_2O						
	H_2SO_4	0.1733	0.1467	0.1067	0.0667	0.0467	0.0267
	SO_3	1.467	9.333	42.66	155.99	478.63	1,125
60	H_2O						
	H_2SO_4	0.8666	0.7333	0.5333	0.400	0.240	0.1333
	SO_3	7.999	46.663	191.98	641.28	1,800	3,953
80	H_2O	0.0013					
	H_2SO_4	4.000	3.066	2.266	1.600	0.9999	0.600
	SO_3	37.33	194.65	725.27	2,240	5,826	12,026
100	H_2O	0.0067	0.0013				
	H_2SO_4	13.33	10.67	7.999	5.333	4.000	2.133
	SO_3	145.32	686.61	2,373	6,779	16,545	32,331

[a] Ref. 60.
[b] To convert Pa to psi, divide by 6895.

fication. In this operation, the gas passes first through scrubbers irrigated with weak sulfuric acid at relatively low temperatures and then through electrostatic precipitators. Many years ago vessels filled with granular coke were used as an alternative or supplement to electrostatic precipitators, but such use is now rare.

Plants that burn good quality elemental sulfur or H_2S gas frequently have no facilities for purifying SO_2 gas but may instead use filters on incoming air and sulfur or H_2S feed streams. Occasionally, hot-gas purification is used, in which the SO_2 gas stream near catalytic reaction temperatures (ca 420–440°C) is passed through beds of granular solids to filter out fine dust particles.

The need for hot-gas purification became borderline with the advent of relatively pure Frasch or recovered sulfur. However, it was frequently used when sulfur was shipped as a solid because of dirt or scale contamination of the sulfur during shipping and handling. In the last few decades, most sulfur used in the United States and Europe has been shipped and handled as a liquid containing very low ash concentrations, typically $<0.005\%$. With this type of raw material, hot-gas purification is not essential, and it is now rarely used. In areas of the world where solid sulfur is still handled, molten sulfur is frequently filtered prior to use as an alternative to hot-gas purification.

Contact plants of the single-absorption type can be designed to cool gases between catalyst passes either entirely by heat exchangers or by a combination of heat exchangers and air dilution. In some cases, a bypassed stream of unconverted feed-SO_2 gas can be used for partial gas cooling, but this practice is uncommon in the United States.

Tail-gas scrubbers are sometimes used in conjunction with single-absorption plants to meet SO_2 emission requirements, but this design has its most common application in existing plants rather than in new plants. Ammonia scrubbing is most popular, but to achieve good economics the ammonia value must be recovered as a usable product, typically ammonium sulfate for fertilizer use. A number of other tail-gas scrubbing processes are available, including use of hydrogen peroxide scrubbing, active carbon for wet oxidation of SO_2, molecular-sieve adsorbents, and the absorption and subsequent release of SO_2 from a sodium bisulfite solution (see Molecular sieves).

Small amounts of sulfuric acid mist (or aerosol) are always formed in sulfuric acid plants whenever gas streams are cooled (or SO_3 and H_2O react) below the sulfuric acid dew point. The dew point varies with gas composition and pressure but typically is ca 140–170°C. Higher and lower dew-point temperatures are possible under varying SO_3 concentration and moisture content of the gas. Such mists are objectionable because of both corrosion in the process and stack emissions.

In recent years, sulfuric acid mists have been satisfactorily controlled by passing gas streams through equipment containing beds or mats of small diameter glass or Teflon fibers. Such units are called mist eliminators (see Air-pollution control methods). This type of equipment was introduced by Monsanto Company and has been a significant factor in making the double-absorption process economical and in reducing stack emissions of acid mist to tolerably low levels. As an alternative, porous ceramic filter tubes or electrostatic precipitators were sometimes used in earlier years, but such installations are now quite rare.

Generation of Sulfur Dioxide Gas. ***Sulfur Burning.*** With the trend to very large single-train plants, current practice is to use horizontal, brick-lined combustion chambers with dried air and atomized molten sulfur introduced at one end. Atomization typically is accomplished either by pressure spray nozzles or by mechanically driven spinning cups. Because the degree of atomization is a key factor in producing efficient combustion, sulfur nozzle pressures have reached 1.14 MPa (150 psi) in recent years. Sulfur burners are typically designed as proprietary items by companies specializing in acid-plant design and construction. Some designs contain baffles or secondary air inlets to promote mixing and effective combustion.

In any procedure involving the handling of molten sulfur, the lines and spray nozzles must be steam jacketed and steam pressure must hold the molten sulfur within the range of ca 135–155°C where its viscosity is at a minimum. Above 160°C the viscosity rises sharply and at 200°C, its viscosity is 3000 times that at 150°C.

The self-sustaining ignition temperature of pure sulfur is ca 260°C but may be slightly higher for "dark" sulfur containing organic impurities. Consequently, a source of ignition is not required if the combustion chamber is preheated to ca 400–425°C before sulfur is admitted. When burning sulfur in air, SO_2 concentrations in the range of ca 3–4 vol % to 13–14 vol % can be produced by the burners described. Special burners capable of producing higher SO_2 concentrations are also available (see Sulfur compounds).

The temperature of gas leaving the sulfur burner is a good indication of SO_2 concentration, even though thermocouples frequently read somewhat lower than true temperatures because of radiation and convection errors. A temperature of ca 970°C corresponds to ca 10.0 vol % SO_2, 1050°C to 11.0 vol % SO_2, 1130°C to 12.0 vol % SO_2, with other temperatures in proportion.

At high flame temperatures, small amounts of nitrogen may react with oxygen to form nitric oxide or N_2O_3. Burner gases are usually cooled rapidly in a subsequent steam boiler, thereby minimizing decomposition of the nitrogen oxides. Consequently, it is not uncommon to find traces of niter either in product acids or somewhat larger amounts in condensed acid collected at mist eliminators.

Sulfur burners are normally operated at moderate pressures, in the range of ca 135.8–170.3 kPa (5–10 psig), using air supplied by the main blower for the plant.

Spent-Acid or H_2S Burning. Burners for spent acid or hydrogen sulfide are generally similar to those used for sulfur with a few critical differences. Special types of nozzles are required both for H_2S, a gaseous fuel, and for corrosive and viscous spent acids. In a few cases, spent acids may be so viscous that only a spinning cup can satisfactorily atomize them. Because combustion of H_2S is highly exothermic, careful design is necessary to avoid excessive temperatures.

Spent-acid burning is actually a misnomer, for such acids are decomposed to SO_2 and H_2O at high temperatures in an endothermic reaction. Any organic or ammonia compounds present in the spent acid are oxidized to produce heat, and excess water is vaporized, which is another endothermic process. To supply the additional heat necessary for spent-acid decomposition, auxiliary fuels, eg, oil or gas, must be burned. Where available, H_2S is an excellent auxiliary fuel.

Relatively high temperatures are required to decompose spent acids at reasonable burner retention times, typically ca 980–1200°C, depending on the type of spent acid. A wide variety of spent acids can be processed in this way, but costs escalate rapidly when the sulfuric acid concentration in spent acid (impurity-free basis) falls below ca 65–75%.

Since large amounts of water vapor are produced by combustion of H_2S or spent acids, atmospheric air, not dried air, is supplied to the burners. In some cases, burners are operated at pressures slightly below atmospheric to pull in outside air, and in other cases preheated combustion air at low pressure may be supplied by ducts.

A few relatively uncontaminated spent acids can be reused without decomposition by evaporating their excess water in concentrators or by mixing them with fresh sulfuric acid of high concentration. A great many literature articles describe various special processes for reclaiming iron-contaminated spent acids from steel-pickling or titanium-pigment production, but such processes are generally an economic burden and are not widely used (76–77).

Ore Roasting, Sintering, or Smelting. Generation of SO_2 at nonferrous metal smelters is determined primarily by the needs of the various metallurgical processes and only incidentally by requirements of the sulfuric acid process. In general, three main types of equipment are used to generate SO_2 gas from metallic sulfides, ie, roasters, sintering machines, and converters. Reverberatory furnaces are also widely used in the copper industry, but the SO_2 gas they generate is too weak and low in oxygen to be handled economically in sulfuric acid plants (see Extractive metallurgy).

By-product SO_2 gas produced at smelters typically contains less oxygen than is found in sulfur-burner gases because oxygen in atmospheric air is partially consumed to form metallic oxides (or sulfates). However, the metal "blow" cycle in batch converters typically forms metal, not oxides; hence, the off-gas from this phase of the operation has an oxygen concentration close to 21%. In addition, many smelters now use oxygen to replace or supplement air, and so produce higher oxygen concentrations than formerly.

Roasters and sintering machines generally operate continuously with fairly uniform concentrations of off-gas. On the other hand, many existing converters are batch units which include both slag and metal blows and considerable downtime for charging and emptying. Hence, gas from these units fluctuates widely in both flow rate and composition, a fact that cause difficulties at acid plants. Some smelters are now adopting continuous-converter processes which provide fairly uniform gas flows at higher SO_2 concentrations than available from batch converters. This trend is beneficial to sulfuric acid plant operation.

Process Details and Flow Sheets

The following equation presents the stoichiometric relation between reactants and products for the contact process.

$$SO_2 + \tfrac{1}{2}\, O_2 \rightleftarrows SO_3$$

There are three important characteristics of this reaction: it is exothermic and reversible, and it shows a decrease in molar volume in the direction of the desired product. To improve equilibrium or driving force for the reaction, the sulfuric acid industry has attempted one or a combination of the following process-design modifications: increasing concentration of SO_2 in the process-gas stream; increasing concentration of O_2 in the process-gas stream by air dilution or oxygen enrichment; increasing the number of catalyst beds; removal of the SO_3 product by interpass absorption (double-absorption process); lowering catalytic-converter inlet-operating temperatures (better catalysts); and increasing the catalytic-converter operating pressure (pressure plants).

Single-Absorption Sulfur-Burning Plants. Single-absorption plants were standard in the industry for many years, using either relatively low strength (8 vol %) SO_2 gas without air dilution, or air-dilution designs. Air dilution is a common design option utilized to improve conversion at the final converter pass by increasing oxygen concentration and reducing equivalent sulfur dioxide concentration of the process gas. Air dilution also eliminates the need for gas-cooling heat exchangers prior to the last one or two converter passes, and has other economic advantages.

Figure 23 shows a typical flow scheme for a four-pass, single-absorption sulfur-

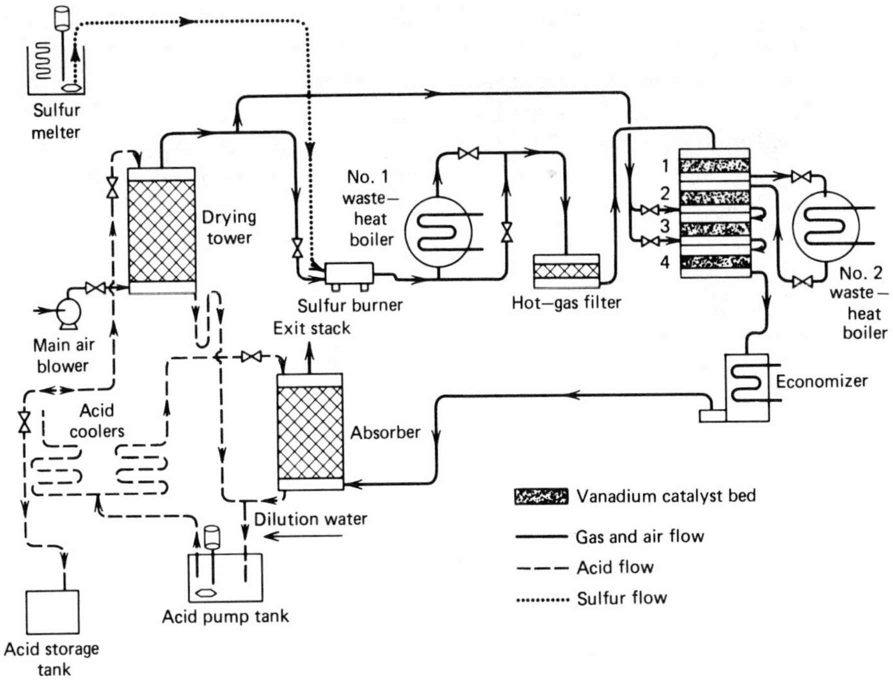

Figure 23. Typical flow sheet for air-dilution sulfur-burning acid plant. The following table of optimum catalyst temperatures at the inlet and exit of each pass is considered normal for ca 9% SO_2 gas when the plant and catalyst are operating efficiently at rated load. If greater or smaller temperature differentials occur, it indicates that the plant or catalyst may not be functioning properly.

Converter pass	*1*	*2*	*3*	*4*
inlet, °C	415–425	440–450	435–440	425–430
outlet, °C	600	500	450	430–435
ΔT, °C	175–185	50–60	10–15	5

burning, air-dilution plant. Plants of this design burn sulfur to generate a process-gas stream of ca 9–10% SO_2. The process-gas stream is diluted with air after the second and third converter passes to lower the gas-stream temperature to design inlet temperatures of the third and fourth converter passes. Air dilution results in a final equivalent process-gas strength of ca 7–8% SO_2.

Process air in sulfur-burning plants is dried by contacting with 93 wt % or 98 wt % sulfuric acid in a countercurrent packed tower. Dry process air is used to minimize sulfuric acid formation in the sulfur burner or subsequently, thus reducing corrosion problems of downstream equipment and stack emissions of sulfuric acid mist.

The heat of combustion from the sulfur burner is removed in a boiler, which reduces the process-gas temperature to the desired converter inlet temperature. Typically, the inlet temperature range to the first converter pass is 410–430°C as dictated by catalyst performance. Commercial vanadium pentoxide catalysts do not show sustained catalytic activity at temperatures <400–410°C, although it is well documented that new catalysts may have initial-reaction ignition temperatures as low as 375°C. The catalyst ignition temperature represents an operating condition below which substantial catalytic conversion (approaching equilibrium) cannot be sustained in any given bed or pass.

SO_2 gas is catalytically oxidized to SO_3 in a fixed-bed reactor (converter) which operates adiabatically in each catalyst pass. The heat of oxidation raises the temperature of process gas in the first converter pass to ca 600–615°C (see Fig. 23). Typically, carbon-steel and cast-iron construction limitations do not allow significantly higher exit temperatures than this. Fortunately, higher temperatures are not usually a problem because the reaction is equilibrium limited. The temperature of hot gas exiting the first pass is lowered to the desired second-pass inlet temperature (430–450°C) by removing the heat of reaction in a steam superheater or second boiler.

In converter passes downstream of the first pass, exit temperatures are limited by thermodynamic equilibrium conversions to ca 500°C or less, a level at which materials of construction limitations are not a significant problem. To obtain desired reaction kinetics and SO_2 conversions, the heats of reaction at the succeeding converter passes are removed by a combination of air dilution and a final "economizer" or gas cooler. It should be noted that the temperature rise of the process gas is almost directly proportional to the SO_2 converted in each pass, even though SO_3 and O_2 concentrations can vary widely.

Process gas leaving the last converter pass is normally cooled to ca 190–210°C prior to SO_3 absorption, the heat being recovered in the steam system by boiler feedwater heaters (economizers). The process gas is not cooled to a lower temperature to avert the possibility of corrosion problems associated with condensing sulfuric acid originating from traces of water in the gas stream.

Gas leaving the economizer or gas cooler flows to a packed tower where SO_3 is absorbed. Concentrated sulfuric acid circulates in the tower and cools the gas to about the acid inlet temperature. The typical acid inlet temperature range for 98.5% sulfuric acid absorption towers is ca 70–80°C. The 98.5% sulfuric acid exits the absorption tower at 100–125°C, depending on acid circulation rate. Acid temperature rise within the absorption tower is attributable both to heat of hydration of sulfur trioxide and to absorbed sensible heat from the process gas. The hot product acid exiting the SO_3 absorption tower is cooled in heat exchangers before being recirculated or pumped into storage tanks.

As discussed, acid circulated over SO_3 absorbing towers is maintained at ca 98.5% concentration to minimize its vapor pressure. Where lower concentration product acid is desired, it is made either in separate dilution facilities, or in drying-tower circuits operated at 93–96% H_2SO_4.

The conversion efficiency of any sulfuric acid plant is better understood by looking at it as an equilibrium-stage process. Figure 24 presents the equilibrium-stage diagram of a single absorption-plant process, such as the one presented in Figure 23.

The horizontal line segments in the diagram represent gas cooling and are horizontal because no appreciable conversion occurs when the process gas is not in contact with catalyst. The diagonal lines in Figure 24 represent adiabatic temperature rise of the process gas within the different converter passes. The slopes of the adiabatic temperature-rise lines are directly proportional to the specific heat capacity of the process gas, which to a reasonably close approximation, can be assumed constant for any degree of conversion.

The two curves in Figure 24 represent SO_2 equilibrium conversions vs temperature for two different initial SO_2 and O_2 gas concentrations. Each initial SO_2 gas concentration has its own characteristic equilibrium curve. For a given gas composition, the adiabatic temperature-rise lines can approach the equilibrium curve but never cross

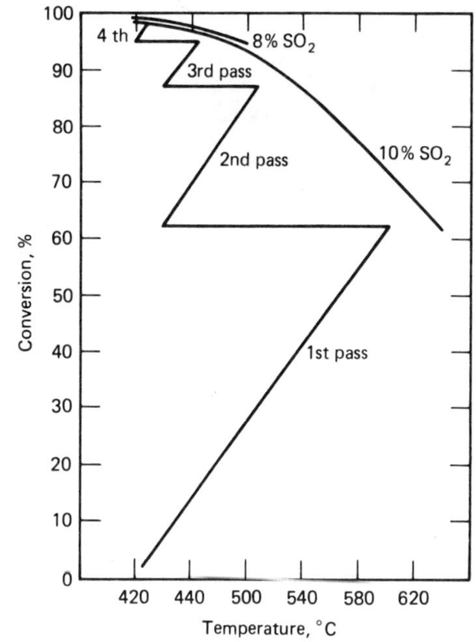

Figure 24. Air-dilution equilibrium-stage diagram.

it. After the third converter pass, the 10% SO_2 equilibrium curve is no longer the limiting curve because the gas composition has been altered by air dilution. In the fourth pass, the 8% SO_2 equilibrium curve is limiting; therefore, the adiabatic temperature-rise operating line can cross the 10% SO_2 equilibrium line.

Double-Absorption Plants. In the United States, new sulfur-burning acid plants are required to limit SO_2 stack emissions to 8 kg of SO_2 per metric ton of 100% acid produced (8 lb/short ton), which is equivalent to a sulfur dioxide conversion efficiency of 99.7%. This high conversion efficiency is not economically achievable by single-absorption plants with current catalyst performance but can be attained in double-absorption plants when the catalyst is not seriously degraded.

Typical double-absorption plant design uses intermediate SO_3 absorption after the second or, more commonly, the third converter pass. Current double-absorption plants usually contain a total of four catalyst passes. Figure 25 presents a typical flow diagram for a sulfur-burning double-absorption plant.

Approximately 90–95% of total sulfur trioxide produced by the double-absorption process is absorbed in the interpass absorption tower; the balance of sulfur trioxide produced in subsequent converter passes is absorbed in the final absorbing tower. Sulfuric acid irrigation rates are adjusted so that acid temperatures exiting the absorbing tower of single-absorption plants and the interpass absorbing tower of double-absorption plants are 100–125°C. The smaller amount of sulfur trioxide absorbed in the final absorption tower of double-absorption plants typically raises its irrigation-acid temperature only to ca 105°C or less. Interpass and final absorbing towers of double-absorption plants are very similar in size since tower diameter is dependent on total gas throughput, not sulfur trioxide concentration.

Another large difference between single- and double-absorption processes is that, after interpass absorption, the process gas has to be reheated from ca 80°C to ca 425°C

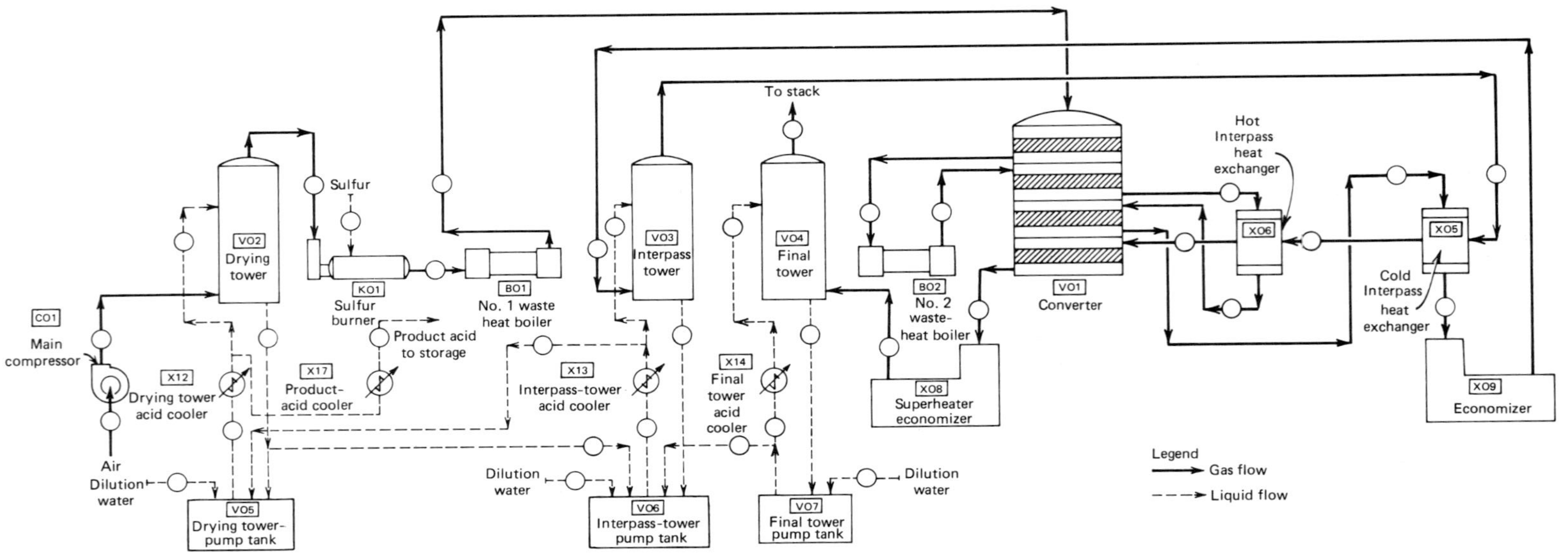

Figure 25. Flow diagram for a double-absorption sulfur-burning plant.

before reentering the converter. Reheating the process gas is accomplished in gas-to-gas-heat exchangers (see Fig. 25), which are a principal cost item. These exchangers simultaneously cool hot gas leaving the converter before it enters the interabsorption tower. All other plant operations are very similar to the corresponding single-absorption processes.

Oleum Manufacture. Production of fuming sulfuric acid (oleum) is accomplished by absorbing sulfur trioxide in one or more special absorption towers irrigated by recirculated oleum. Typically, only 30–60% of the sulfur trioxide present in the gas is absorbed because of oleum vapor-pressure limitations. Since absorption of sulfur trioxide is incomplete, gas leaving the oleum absorption tower or towers must be processed in a subsequent absorption tower, which is one of the normal 98.5% H_2SO_4 towers previously described.

The absorption of sulfur trioxide for oleum production is carried out over a relatively narrow temperature range. The upper temperature is set to provide a reasonable partial pressure driving force for the oleum concentration used. The lower practical temperature limit is the freezing point of oleum, 17°C, which is high enough to be a problem in shipping and handling as well. For many oleum uses it is practical to add small amounts of HNO_3 as an antifreeze (78).

Sulfur Trioxide. The anhydride of sulfuric acid, SO_3, is a strong organic sulfonating and dehydrating agent which has some specialized uses (see Sulfonation and sulfation). Its principal applications are in production of detergents and as a raw material for chlorosulfuric acid (qv) and 65% oleum. In recent years, SO_3 gas has been added to cooled combustion gases at many coal-burning power plants to improve dust removal in electrostatic precipitators (see Air pollution control methods).

Liquid SO_3 is a difficult material to handle because of its relatively low boiling point (44.8°C) and high freezing point (17°C) plus high reactivity with almost all organic substances and water. It reacts explosively with water, in part because of a highly exothermic heat of reaction. In addition, trace amounts of water or sulfuric acid act as polymerization catalysts for liquid SO_3 and produce a series of high molecular weight polymers with elevated melting points. When polymerization occurs, attempts to melt the solids can cause explosions because high vapor pressures (exceeding atmospheric pressure) develop before melting occurs. Polymerization is promoted by freezing the liquid. It can be inhibited by adding small amounts of various patented stabilizers to the liquid, such as 0.3% dimethyl sulfate with 0.005% boric oxide (37). Stabilized liquid SO_3 is an item of commerce, and instructions for its storage and use are available from suppliers.

Liquid SO_3 is usually produced by distilling SO_3 vapor from oleum and condensing it. This operation is normally carried out at a sulfuric acid plant where the stripped oleum can readily be refortified or reused. Eliminating all traces of sulfuric acid from the SO_3 vapor stream is important to minimize polymerization of the liquid condensate. When this is done, it is frequently possible to utilize unstabilized liquid SO_3 in an adjacent process or operation if precautions are taken to prevent its freezing before use. At some plants, gaseous 100% SO_3 is utilized directly instead of producing liquid.

The use of SO_3 to produce sulfonate detergents is a highly specialized application which requires precautions to prevent oversulfonation or charring of the organic compounds used. Where liquid SO_3 is used directly, special high turbulence cooled reactors are employed. In other cases, SO_3 is vaporized and reacts as a gas. Several equipment suppliers offer small-scale SO_3-generating equipment that produces 8–10% SO_3 gas in air for sulfonation or other uses. This equipment is essentially a small-scale contact-acid plant, minus SO_3 absorption towers and acid-handling facilities.

Equipment Details

Absorption and Drying Towers. Towers are typically carbon-steel vessels lined with acid-proof brick and mortar and packed with ceramic rings or saddles. Acid distribution over the packing normally is accomplished by weir troughs or pipes with holes. Well-designed absorption or drying towers operate at absorption efficiencies >99.5%, typically 99.8 to >99.9%. The cooled acid that is used to irrigate tower packing allows trough distributors to be built of cast iron without excessive corrosion. Pipe distributors usually require somewhat more resistant materials or special designs to minimize corrosion of distribution holes.

Drying towers are typically designed with a single- or double-woven wire-mesh pad at the gas exit to remove acid spray and large mist particles. As a more efficient alternative, special glass-fiber mist eliminators are occasionally installed. Glass or Teflon fiber-bed mist eliminators are standard in interpass and final absorption towers where high efficiency collection of acid mist is more critical.

Acid Coolers. Prior to 1970, cast-iron strong-acid coolers were the industry standard, using either "trombone" pipe arrangements or special radiator sections irrigated externally with water. Cast-iron coolers were frequently plagued by gasket leaks, external scaling or corrosion, and other inconveniences and were a significant source of unscheduled plant downtime and maintenance costs. A more detailed discussion of cast-iron acid coolers can be found in ref. 80.

In 1970, anodically passivated stainless-steel shell-and-tube acid coolers became commercially available and proved to have significant maintenance savings and other advantages. This type of cooler is now widely used and is considered standard design.

Anodic passivation uses an impressed voltage from an external electrical power source to reduce metal corrosion. The subject of anodic passivation and its application to sulfuric acid equipment such as stainless-steel acid coolers and carbon-steel storage tanks is well covered in refs. 81–83.

Air cooled, stainless-steel acid coolers, which frequently employ anodic passivation, have been built in the FRG and Canada but to date have been used in only a few plants in cool climates. Anodically protected plate coolers have also been developed in Sweden (84), but full-scale experience is limited. Stainless-steel coolers without anodic protection have a maximum acid temperature limit of ca 80–88°C in 98% sulfuric acid service which limits their use.

Two other commercially available acid-cooler designs are Teflon and graphite coolers. Teflon or so-called spaghetti coolers have the advantage that Teflon tubing can handle a wide range of acid concentrations and contaminants. The Teflon acid-cooler disadvantages are low operating-pressure restrictions, small single-unit capacities requiring multiple units in large acid plants, and potential tube pluggage because of the small diameter tubing utilized (2.5–3.5 mm OD). Larger diameter tubes are sometimes used, but this adversely affects economics.

Impervious (impregnated) graphite coolers are utilized in acid service in which the sulfuric acid concentration is ca 90–93% or less. When operated properly, graphite coolers provide excellent service in weak acids. They are not recommended for acid services over ca 93% concentration because, even though graphite is relatively inert, the graphite-impregnating agent is attacked by concentrated sulfuric acid. The main disadvantage of graphite coolers is that they are quite brittle and subject to damage by mechanical or hydraulic shocks.

Catalysts. Commercial sulfuric acid catalysts typically consist of vanadium and potassium salts supported on silica, usually diatomaceous earth (see Diatomite). Catalyst pellets are available in various formulations, shapes, and sizes depending on the manufacturer and the particular converter pass in which they are to be used. The differences are designed to accommodate the various different service environments within a converter; a study of these factors is presented in ref. 85. Catalyst pellets in first- and second-pass service are subjected to high temperature and dust fouling. Catalysts in subsequent converter passes operate at low temperatures and with high SO_3 and low SO_2 and O_2 concentrations.

In former years, most catalysts were supplied as solid cylindrical extrudates or pellets ranging from ca 4–10 mm dia. Currently, ring-shaped catalysts are beginning to be used, primarily as a means of saving energy via reduced gas-pressure drop. The ring-shaped pellets also have greater resistance to dust fouling.

Increased catalyst-bed pressure drop caused by dust fouling reduces production of acid and significantly increases energy consumption by the plant's blower. To avoid these problems, first-converter-pass catalyst pellets are typically screened annually. Second-pass catalyst pellets are screened less frequently since the first-converter-pass catalyst bed acts as a hot-gas filter for the rest of the converter. Typical screening losses range from ca 4–12% of the catalyst bed per screening. Screening losses depend on screen mesh size and catalyst pellet hardness as well as on screening rate.

Under normal operating conditions, vanadium catalysts in first- and second-pass service slowly lose catalytic activity over time. Catalyst aging is a combination of a loss of catalytically active material from catalyst pellets and irreversible changes within the pellets. It is well documented that at operating temperatures the active catalytic ingredient is a molten salt that migrates from catalyst pellets into adjacent dust. Catalyst aging is accelerated by increasing temperatures and temperature cycling.

Typical sulfuric acid plant operations require that first-pass catalyst beds of good quality catalysts be replaced completely about every five to ten years. Some catalysts of low durability may need to be replaced after one to three years of first-converter-pass service. Catalyst beds subsequent to the first usually operate for many years if not abused, since temperatures and dust loadings are significantly lower than in first-pass service.

Catalytic Converters. Converters vary in design with the different engineering design firms. Most of the traditional converter designs use steel shells (sometimes partly or wholly brick lined) with cast iron and alloy internals. In the high temperature converter passes, carbon steel is protected from hot gases by brick lining or a sprayed aluminum coating. To reduce converter maintenance costs, some newer converter designs are based on stainless-steel construction at a small premium cost over carbon-steel converters. Both carbon- and stainless-steel converters are insulated to reduce heat losses from the exothermic oxidation reaction of SO_2. Essentially all designs use horizontal catalyst beds arranged one over another with gas flowing down through the catalyst.

Gas–Gas-Heat Exchangers. Standard gas–gas heat exchangers in double-absorption plants are built of carbon steel and, in special cases, of stainless steel. Tube diameters range from ca 65–90 mm, and the tubes run vertically in a typical shell-and-tube arrangement. To reduce corrosion and scaling, tubes in high temperature service are Alonized, which is a proprietary aluminum-alloy coating vapor diffused into carbon steel. Alonizing significantly increases the life of carbon steel with the added benefit of no loss of heat transfer from accumulation of corrosion scale.

Mist Eliminators. Modern mist eliminators use glass-fiber beds enclosed or supported by stainless steel or alloy-20 wire mesh or the like. Some mist-eliminator designs utilize Teflon fiber beds.

The aerosol-collection mechanisms of acid-mist eliminators are inertial impaction, interception, and Brownian motion. Inertial impaction works well for aerosols with particle diameters larger than 3 μm; Brownian motion and interception work well with aerosols having smaller particle diameters. Glass-fiber-bed mist eliminators can be designed to operate with almost any desired particle-collection efficiencies, depending on allowable pressure drop and cost.

New patented bicomponent mist eliminators allow improved drainage of coalesced mist collected by the glass fibers, thus essentially eliminating particle reentrainment. Reduced reentrainment allows these units to operate with the same collection efficiency as single-component beds with 20% increase in gas throughput. A discussion of sulfuric acid mist generation, control, and mist-eliminator design is given in ref. 86.

Oleum Equipment and Piping. The traditional material of construction for oleum is carbon steel. Relatively low oleum velocities must be used in steel piping to prevent excessive corrosion. The corrosiveness of oleum decreases with increasing SO_3 concentration. For oleum concentrations <5% SO_3, carbon steel is not recommended because of excessive corrosion. Steel is borderline from 5% to ca 15% SO_3, depending on temperature.

Cast iron is not a recommended construction material for oleum. It is well documented that cast iron fails catastrophically by cracking in oleum service. The mechanism or the cause of cracking is not well understood, but it is agreed that the failures are due to a buildup of internal stresses within the cast iron.

Standard stainless steels have significantly greater corrosion resistance to oleum than carbon steel, but their higher price may make them less economical except for special services such as valves, oleum boilers, etc.

Sulfuric Acid Concentrators. Concentrators for increasing the strength of dilute sulfuric acid by removing water have been used since the early days of the industry. A two-volume text on this subject was published in 1924 (77), in which the topic was subdivided into concentration by hot gases and concentration in heated vessels. More recent discussions of the subject are given in references 13, 77, 88–90.

The need for acid concentrators exists because many uses of sulfuric acid do not consume the material but instead dilute and partially degrade or contaminate it. As a general rule, when sulfuric acid is concentrated, most of its inorganic contaminants become relatively insoluble and can be removed, at least to some degree. In addition, small amounts of organic contaminants are frequently partially oxidized and sometimes can be largely removed by treating with small amounts of hydrogen peroxide or nitric acid to accelerate oxidation (91).

Fired-pot concentrators were one of the earliest types and many are still in use, particularly in Europe. The pots are generally cast iron, for which reason acid concentration within a pot must not be less than ca 80 wt %. Because the pots can operate at acid concentrations up to 98 wt % H_2SO_4 at relatively high temperatures, they are particularly well-suited for processing dark (organic-contaminated) acid. Platinum vessels have also been used in the past for some small-scale applications.

Chemico drum concentrators have been widely used, particularly in the petroleum refining and nitration industries. This type of concentrator uses direct contact with heated air in several staged vessels of lead- and brick-lined construction. The maximum acid concentration obtainable is ca 93–94 wt % H_2SO_4, starting with ca 40–50 wt % (or higher) H_2SO_4.

Vacuum evaporation is also widely used, its main justification being the ability to produce relatively high product-acid concentrations at low operating temperatures, thus reducing corrosion. Control of acid-mist emissions is also easier than with other types of concentrators. The most widely used equipment was formerly designated as Simonson-Mantius. An interesting feature is the use of tantalum-bayonet heating elements (see Evaporation).

In recent years, concentrators that use Pyrex or fused-silica tubes or columns have become available from several different companies (76).

Sulfuric Acid Piping. The traditional material of construction for handling concentrated sulfuric acid (78 wt % and >90 wt %) for most of the past century has been gray cast iron. Gray cast iron was the preferred material because of its low price and availability, coupled with tolerable corrosion. The expected service life of gray cast-iron pipe is ca 7–15 yr (or more) depending on the temperature, concentration, and velocity of the acid. However, gray cast-iron pipe and fittings should not be utilized immediately adjacent to anodically-protected coolers. Gray iron pipe and fittings within 3–4 m of anodically-protected acid coolers will fail catastrophically in three months to a year by spontaneous cracking similar to that observed in oleum service.

In 1977 the U.S. water-supply industry and the centrifugally cast iron-pipe foundries made a change from gray cast iron to ductile cast-iron pipe as the preferred material. This trend forced the sulfuric acid industry to adopt centrifugally cast, ductile iron pipe. Tests have shown that the expected life of ductile iron pipe in concentrated sulfuric acid service is about the same as gray cast-iron pipe with the possible exception of high velocity (>2.5 m/s) or other special services. In contrast to gray cast iron, ductile iron does not suffer from spontaneous catastrophic cracking in acid service near anodically protected equipment.

Cast and ductile iron pipes for sulfuric acid service utilize threaded and flanged joints. Thread taper and thread cuts for flanging typically remove ca 50% of the nominal pipe-wall thickness at the flange face. Because of the larger effective pipe-wall thicknesses of large diameter pipes, acid velocities for small diameter pipes are designed to be considerably lower than for large pipes to reduce corrosion.

Small diameter pipe <75 mm is typically stainless steel or alloy 20, depending on concentration and temperature. Cast-iron pipe is available only in diameters of ≥75 mm.

In metallurgical sulfuric acid plants, the weak-acid scrubbing circuit is typically handled with plastic (FRP or fiber-reinforced plastic) pipe. The contaminants in weak acid usually vary too greatly to allow use of an economical alloy that will give adequate service.

Carbon steel is not normally a suitable piping material for concentrated sulfuric acid because of high corrosion rates in flowing acid. However, where temperatures and flow rates are low, heavy-wall steel pipe is sometimes used for transferring product acid. Carbon steel is used in concentrated sulfuric acid storage tanks because in quiescent and low temperature conditions its corrosion rate is acceptable. Carbon steel is not suitable for handling sulfuric acid in concentrations less than ca 90% even under quiescent conditions, unless passivating agents are present.

Materials of Construction

Resistance of alloys to sulfuric acid corrosion increases with increasing chromium, molybdenum, copper, and silicon content. The corrosiveness of sulfuric acid solutions is highly dependent on concentration, temperature, acid velocity, and acid impurities. An excellent summary of sulfuric acid corrosion is given in ref. 92. Materials of construction for sulfuric acid service are discussed in ref. 73 and nickel-containing alloys in particular, in ref. 93. One of the more resistant alloys available in wrought form is alloy 20, which is designated as CN-7M or Durimet 20 in cast form. Other highly resistant cast alloys are Lewmet 44, Lewmet 55, and Lewmet 66 (Chas. S. Lewis Co., St. Louis, Mo.) (91); Duriron; and Hastelloy D.

Special precautions should be observed when utilizing isocorrosion charts as guidelines for alloy selection. Metal-skin temperature, not bulk-acid temperature, should be used as the criterion of selection since metal-skin temperatures may be significantly higher or lower than bulk-acid temperatures. This is particularly important when designing heat-transfer equipment; heating coils have failed catastrophically because alloy selection was based on acid bulk temperatures instead of the higher metal-skin temperature of the heating coil. In the case of acid cooling, an unnecessarily more expensive higher alloy may be chosen on the basis of bulk-acid temperature.

Oxidizing contaminants, eg, nitric acid and ferric ions, may significantly alter the performance of alloys in sulfuric acid. For example, Hastelloy B-2 performs extremely well in concentrated sulfuric acid but corrodes rapidly in the presence of ferric ions, nitric acid, or free SO_3 in sulfuric acid. For this reason, Hastelloy B-2 is not recommended for oleum service.

Tantalum has excellent corrosion resistance to concentrated sulfuric acid at high temperatures but very poor corrosion resistance to oleum. Tantalum's high price prevents it from becoming a common material of construction in concentrated sulfuric acid service. Zirconium has excellent corrosion resistance to sulfuric acid solutions up to the boiling point in the concentration range of 0–65 wt % H_2SO_4. Duriron (a cast high-silicon iron) has excellent corrosion resistance to sulfuric acid at all concentrations up to the boiling point. Unfortunately, it is very brittle; its use is limited because of its propensity to fracture when subjected to thermal or mechanical shock.

An excellent compilation of the relatively scarce literature data on corrosion of alloys in liquid sulfur trioxide and oleum is given in ref. 93.

In the past, lead was widely used as a material of construction for sulfuric acid concentrations less than ca 80 wt % and occasionally for concentrations up to 93 wt % (at temperatures <45°C). Because of lead's low physical strength and loss of strength at high temperatures, extensive external supports are normally required for it. Consequently, modern practice uses various acid-resisting plastic materials instead of lead whenever practical. Many plastics do not resist concentrated acid well (>50–60 wt

% H_2SO_4), and their properties can vary widely depending on exact composition, degree of polymerization, etc. Hence, specific data or information should be obtained from experts. Teflon is the only common plastic that resists all acid concentrations, although Kynar, fully vulcanized Hypalon, and various proprietary baked phenolic resins are reasonably effective up to 94–95 wt % H_2SO_4 at temperatures near ambient.

Special Plant Designs

Pressure Plants. Catalytic oxidation is normally carried out at absolute pressures of 122–152 kPa (1.2–1.5 atm). Theoretically, higher pressures are advantageous both from thermodynamic equilibrium and kinetic standpoints. However, the practical and economic difficulties of building and operating plants at higher pressures are such that only one large plant of this type has been built to date in Lyon, France (95–96). For sulfuric acid plants operating at pressures of 405–811 kPa (4–8 atm), double absorption is required to meet the EPA mandated SO_2 conversion efficiency of 99.7%. Published information (97) indicates that it may be possible to use single absorption under pressure by operating the final SO_3 absorption step at a low temperature, thus also absorbing part of the unconverted SO_2 and recycling it to the front end of the plant. To date, the economics of pressure plants is not clearly defined, but it appears that pressure plants are at best a break-even proposition when compared to standard plants. A few small units that generate SO_3 gas for sulfonations or other special uses have been operated at higher than normal pressures.

Cement Plants. Calcium sulfate in the form of anhydrite, gypsum ($CaSO_4.2H_2O$), or by-product gypsum from phosphate fertilizers is occasionally used to produce sulfuric acid and cement. Approximately one kilogram of portland cement is produced for each kilogram of sulfuric acid. Because the capital requirement for such installations is ca six to eight times the cost of an elemental-sulfur-burning plant, they are uneconomical except under special circumstances. Plants of this type are currently operating in Austria, Eastern Europe, and South Africa; former units in the UK have been shut down (see Cement).

Oxygen-Enriched Processes. Typically, the cost of obtaining oxygen at high concentrations makes its use uneconomical at sulfuric acid plants unless special circumstances exist. Although proposed processes for oxygen enrichment have been published, perhaps only one plant of this type operated commercially for a sustained period; that plant, remodeled by Consolidated Mining and Smelting Co. (Cominco) to use 25 vol % SO_2, 30 vol % O_2, balance, nitrogen (98–99), is no longer in existence. Concentrated SO_2 was obtained by absorbing low concentration SO_2 from metallurgical operations in ammonia, treating the solution of ammonium bisulfite with concentrated sulfuric acid to form almost pure SO_2 and a solution of ammonium sulfate. Pure oxygen was available as a by-product from electrolysis of water to produce hydrogen for ammonia synthesis, thus providing a low cost supply.

The oxygen-enriched acid process was designed for 240 t/d capacity and operated with two catalytic converters, each with three passes. The reported conversion efficiency of this plant was 99.5%. This oxygen-enriched process also had a second unique feature, namely, most of the process gas was recycled after SO_3 absorption with only a small purge to the atmosphere.

Energy-Efficient Plants. The dramatic increases of energy costs in the past decade have led to significant plant-design changes in the sulfuric acid industry (100). The current trend is to reduce power consumption of the process, to increase heat recovery in the steam system, and to use steam-driven turbines to generate electric power. The following references describe in detail the economics of energy recovery in sulfuric acid plants (4,101–103). To reduce the power consumption of a sulfuric acid plant, three main options are available: increased efficiency of rotating equipment; reduced plant pressure drop on the gas side; and reduced gas volume processed per ton of acid produced.

Increasing the efficiency of rotating equipment has resulted in significant reductions of power consumption in recent sulfuric acid plants. A 5% increase of the main blower's steam-turbine efficiency in a large plant (1820 t/d) typically allows an additional 230–240 kW to be extracted as electric power from a steam-condensing turbine. In water cooling tower operations, the use of variable-speed fans alone conserves up to 2.9×10^8 kJ/yr (800,000 (kW·h)/yr).

Because the main blower is the largest energy-consuming equipment in an acid plant, significant gains in energy efficiency can be made by increasing sizes of gas-handling equipment to reduce overall plant pressure drop. The current price of energy in the United States allows an incremental capital investment for this purpose equivalent to $60–80/Pa ($15,000–20,000/in. water column) of gas-pressure-drop reduction.

To further reduce plant pressure drop, ring-shaped catalysts are currently being adopted. Ring-shaped catalyst pellets reduce gas pressure drop across a catalyst bed by ca 30–60% compared to conventional catalyst-pellet geometries and also reduce pressure-drop increases from dust fouling.

New types of mist eliminators have also been developed which can reduce pressure drop ca 20% as compared to previous designs.

To summarize, former double-absorption plant gas-side pressure drop ranged from ca 44.8–59.8 kPa (180–240 in. of water). Energy-efficient acid plants with the options mentioned above can be designed for 27.4–39.9 kPa (110–160 in. of water) total plant pressure drop.

A second very effective alternative to decrease plant power consumption is to reduce the volume of gas handled by the blower per ton of acid produced. Increasing SO_2 concentration from the 10.5 vol % formerly used to 12.0 vol % not only significantly reduces the power required to produce acid, it also increases high pressure steam production 2.9% by reducing heat losses. Heat that would normally be lost from the process gas to sulfuric acid in the absorption towers or out the stack is reduced by decreasing the amount of inert gas handled. Note that achieving SO_2 conversions of $\geq$99.7% with 12% SO_2 requires more catalyst and significant converter modifications as compared to 10.5% SO_2. At least one patent application of this subject has been filed, and a plant using the process is now ready to start operations.

To increase heat recovery in the steam system, several additional design changes are being adopted. A large improvement in heat recovery comes from the development of new types of steam economizers and boiler feedwater heaters. The key to these designs is improved control to permit operating closer to the sulfuric acid dew point in gas streams and to permit better understanding of corrosion by hot sulfuric acid.

Many past sulfur-burning-plant designs placed the main blower upstream from the drying tower to protect the blower from potential acid mist carry-over, thereby losing the heat of compression to drying-tower acid. New high performance mist eliminators developed recently now allow the blower to be placed after the drying tower without imperiling the blower with acid mist carry-over. Placing the blower after the drying tower allows the heat of compression to be recovered as high pressure steam in the boilers. It should be noted that blowers located after the drying tower have often been used in the past when 93–96 wt % drying acid was employed but were not considered good practice with 98.5 wt % drying acid, as used in sulfur-burning plants.

Higher steam pressures and superheats can increase the sulfuric acid plant steam-system efficiency. At current energy prices, steam pressures of 5.9–6.3 MPa (850–900 psi) at 450–480°C are justifiable. Steam systems of this type in large capacity sulfuric acid plants produce enough steam to operate an electrical turbogenerator. The economics of electrical power cogeneration are such that very attractive paybacks are obtained, particularly where favorable government regulations exist, as in the United States. Where low pressure steam is not required for other processes, maximum efficiency is obtained by condensing turbine exhaust steam under vacuum (104–105)(see Power generation).

As shown in reference 101, the capital cost of an energy-efficient plant plus turbogenerator system is \$30,000,000 vs \$21,000,000 for a typical 1970s design (1800 t/d, 1981 costs). Even with the incremental capital cost for a turbogenerator, the projected simple payout from 15,000 kW of export power is 1.5 yr. These attractive payouts make cogeneration of power attractive for large sulfur-burning plants.

To summarize, sulfur-burning plants that traditionally produced 1.1–1.2 kg of high pressure steam per kilogram of acid produced now can be designed to produce >1.3 kg of high pressure steam per kg of acid. Cogeneration of electric power by sulfur-burning sulfuric acid plants is attractive, aided by U.S. Federal legislation which encourages connections to utility electrical-distribution systems. These advances in energy recovery and conservation give somewhat more expensive plants, but they can readily be justified economically.

Economic Aspects

It has been said that consumption of sulfuric acid is a good measure of a country's degree of industrialization and also serves as an index of general business conditions. Until recent years this statement was valid, but current heavy usage of sulfuric acid by the phosphate fertilizer industry has changed matters. Of total U.S. sulfuric acid production (3.69×10^7 t in 1981 (106)), over two thirds now goes into phosphate fertilizers, other uses have either grown slowly or declined. This trend will undoubtedly continue.

In the production of phosphate fertilizers, the primary role of sulfuric acid is to convert phosphate rock to phosphoric acid and solid calcium sulfates which are removed by filtration. Formerly, considerable normal superphosphate was produced, which is a material containing both mono- and dicalcium phosphates and unseparated calcium sulfates. However, the trend is toward higher analysis fertilizers containing very little calcium sulfates because of savings in freight and handling. Some sulfuric acid is also used to produce ammonium sulfate fertilizer (see Ammonium compounds; Fertilizers).

Other uses of sulfuric acid are in production of inorganic pigments, textile fibers, explosives, petroleum products, alcohols, pulp and paper, detergents, other chemicals, and as a leaching agent for ores, a pickling agent for iron and steel, and a component of lead storage batteries. Use in production of pigments has been relatively static or declining because of a trend to chloride processes for titanium oxide pigments. Iron- and steel-pickling usage is also declining because of the increased use of hydrochloric acid, which can be more easily regenerated on site at large pickling installations (see Metal surface treatments).

Production and consumption trends in the United States are shown in Tables 11 and 12.

Exports and imports are a relatively small percentage of total consumption; imports typically exceeding exports by a ratio of 2:1 or more. Imports originate chiefly from smelter-based plants in Canada and are typically ca 0.5–1.0% of consumption.

Over 70% of production is currently (1981–1982) not sold as such but is used directly by producers to make other materials. This represents a steady upward trend

Table 11. U.S. Sulfuric Acid (100% H_2SO_4) Production by Sulfur Source[a], 1000 Metric Tons

Sulfur source	1960	1965	1970	1975	1980
elemental sulfur	11,180	16,658	20,531	23,532	32,986
hydrogen sulfide	599	344	611	363	90
copper smelter gas	375	335	679	1,619	2,099
zinc smelter gas	699[b]	873[b]	990[b]	646	560
lead smelter gas	na	na	na	118	410[c]
pyrites	1,547	1,252	1,134	572	777
spent acid	1,100	1,642	1,694	1,442	1,569
fortified acid	724	942	1,148	1,066	1,672
Total	*16,224*	*22,046*	*26,787*	*29,358*	*40,163*

[a] Ref. 107.
[b] Includes lead-smelter gas.
[c] Includes molybdenum-smelter gas.

Table 12. U.S. Consumption of Sulfuric Acid (100% H_2SO_4)[a], 1000 Metric Tons

Use	1960	1965	1970	1975	1980
phosphoric acid	3,306	7,225	11,581	17,300	25,500
normal superphosphate	2,057	1,808	1,088	786	740
ammonium sulfate	1,029	1,696	1,472	1,467	1,390
other fertilizers	307	406	386	415	413
petroleum refining	1,542	1,987	1,996	1,870	1,850
copper leaching	na	na	725	1,325	1,160
alcohols	1,571	1,889	1,764	838	1,070
titanium dioxide	1,583	1,585	1,577	1,027	741
hydrofluoric acid	470	629	906	872	770
uranium and vanadium ore processing	306	169	305	355	810
aluminum sulfate	435	524	591	650	610
cellulosics	621	770	625	316	363
iron and steel pickling	973	871	467	254	130
surface active agents	na	316	346	320	152
batteries	86	104	114	122	150
phenol	157	128	181	69	6
all other	na	na	4,338	2,542	5,298
Total	*17,454*	*23,869*	*28,462*	*30,600*[b]	*41,200*[b]

[a] Ref. 107.
[b] Totals rounded.

from 50% in 1960. Almost all large fertilizer plants make their own sulfuric acid on site, and by-product steam from these sulfur-burning plants is generally used for concentrating phosphoric acid in evaporators. Most of the fertilizer plants are located in Florida, Georgia, Idaho, Louisiana, and North Carolina.

Worldwide sulfuric acid production figures are shown in Table 13.

Future growth is expected to be relatively slow in Western Europe and higher than average in the Middle East and in Africa (particularly North Africa, which has extensive phosphate rock deposits). The USSR and People's Republic of China need more fertilizer than they now produce, but it is difficult to predict exactly how rapidly sulfuric acid production will increase in those countries.

Since sulfuric acid has its greatest use in fertilizers, trends in that industry largely dictate what happens in the sulfuric acid business. A considerable portion of U.S. phosphate fertilizer production is exported, but it is likely that exports will gradually diminish (at least on a relative basis) as new fertilizer capacity is developed in other parts of the world. Price is also an important consideration in fertilizer use; high prices reduce consumption by farmers.

The cost and price of sulfuric acid depend in large part on raw material cost and on freight costs. By-product raw material (SO_2) costs at smelters are essentially zero, but the remote locations of many smelters make it very difficult to sell their acid at a profit because of freight costs. In many areas, the delivered cost of sulfur is the most important factor affecting sulfuric acid pricing.

Sulfuric acid prices are usually quoted per ton of equivalent 100% H_2SO_4, even though actual assays may be less or more than 100%. On this basis, concentrations less than 100% sell for ca \$82–104/t (\$75–95 per short ton) as of late 1981/early 1982 depending upon location within the United States, to give a projected total commercial value of $\$3.25 \times 10^9$ for 1982 (108). Oleum typically sells at a premium of ca \$3–5 per metric ton (or a higher premium for concentrations >30% free SO_3). U.S. prices increased an average of ca 5.5%/yr from 1975 through 1978, but subsequently increased more rapidly, averaging ca 20%/yr in 1980 and 1981. This large rate of increase was caused both by high general inflation and by additional upward pressure on sulfur prices caused by increased demand vs available supply. Small acid shipments or spot orders typically command a premium vs quoted contract prices in bulk.

Analysis and Specifications

Analytical methods and specifications for sulfuric acid vary rather widely between different producers and consumers and for different grades of acid. Very little has been

Table 13. World Production of Sulfuric Acid (100% H_2SO_4)[a], 10^6 Metric Tons

Location	1960	1965	1970	1975	1980
North America	18.0	25.0	30.5	34.1	46.8[b]
Central and South America	0.6	1.1	2.0	2.0	3.8
Western Europe	15.4	20.0	24.4	25.1	27.8
Eastern Europe	8.0	13.2	18.8	28.1	32.6
Africa	na	na	2.6	3.8	8.7
Asia	na	na	12.1	13.6	19.6
Oceania[c]	na	na	2.3	2.4	2.9
Total world	*na*	*74 (est)*	*92.7*	*109.1*	*142.2*

[a] Ref. 107.

[b] U.S. production in 1981, 36.9×10^6 t, vs 40.2×10^6 t in 1980 (106).

[c] Primarily Australia and New Zealand.

done to establish industrywide standards in the United States, except for development of the ASTM analytical methods, designated as E 223-68 and summarized in Table 14 (109).

Typical specifications for several common types or grades of acid are shown in Table 15.

Similar limits are generally used for other sulfuric acid concentrations, with the exception of turbidity values for high strength acids (and oleum) and SO_2 and niter values in oleums. Because iron sulfate is relatively insoluble in concentrated acids, the turbidities of 98–99% H_2SO_4 and oleum may be higher than shown, even at acceptable total iron concentrations.

Sulfur dioxide concentrations in oleum are rarely specified or measured, but typical values are considerably higher than in acids of ≤99 wt % concentrations. This occurs because oleum is produced at relatively low temperatures in the presence of appreciable SO_2 in the gas phase, thus leading to high solubility. It is not possible to

Table 14. ASTM Analytical Methods for Sulfuric Acid[a]

Sections	Properties	Principle
1–5	general, reagents	
6–14	total acidity	NaOH titration using phenolphthalein
15–24	Baumé gravity	hydrometer measurement
25–31	nonvolatile matter	evaporation and weighing
32–41	iron	reduce, measure colorimetrically as orthophenanthroline complex
42–49	sulfur dioxide	remove by air sweep, absorb in alkali, add excess iodine, titrate
50–59	arsenic	evolve as arsine, absorb in pyridine and diethylthiocarbamate, measure color

[a] Ref. 109.

Table 15. Typical Sulfuric Acid Specifications[a]

Property	Acid type: Electrolyte[b] (Fed. Spec. O-S-801C) Class 1[c]	Technical (Fed. Spec. O-S-809B) 60°Bé (78%)	Technical (Fed. Spec. O-S-809B) 66°Bé (93%)	Technical (industry typical) 66°Bé (93%)
% H_2SO_4 (min)	93.2	77.5	93.0	93.2
sp gr, 15.5/15.5°C	1.8354	1.7040	1.8347	1.835–1.837
nonvolatiles (max), %	0.03	0.05	0.025	0.02–0.03
As (max), ppm	1.0	2.1	2.5	
SO_2 (max), ppm	40			40–80
iron (max), ppm	50			50–100
nitrate (max), ppm	5.0			5–20
APHA turbidity (max)	nil			100–150
APHA color (max)	per test			100–200

[a] Ref. 110.
[b] Limits are also specified for platinum, organics, copper, zinc, antimony, selenium, nickel, manganese, ammonium, and chloride (see Table 16).
[c] Three additional grades (classes) of lower H_2SO_4 content are also specified.

strip SO_2 from oleums by air blowing, a technique that is frequently applied to product acids of ≤99% concentration.

Measurement and specification of nitrates or other nitrogen oxide compounds in sulfuric acid is a complex subject that has caused many problems for producers and consumers. The difficulty occurs because nitrogen oxides are usually present both as nitrous and nitric compounds, predominantly in the nitrous form. Hence, analytical procedures specific for nitrates only do not give a complete analysis.

A procedure used by Monsanto Company to measure both types of nitrogen oxide compounds at the same time involves development of a pink color by mixing $FeSO_4$ with sulfuric acid followed by measurement or comparison of color intensity. This general type of procedure and possible alternatives are suggested in refs. 111–112 and in UK standards for sulfuric acid.

Although color and turbidity of acid products are important properties, there is very little standardization in such measurements. A frequently used procedure is to determine color and turbidity by comparison with standards originally developed by the American Public Health Association (APHA) for examination of water (113).

A number of different grades of acid are produced for specialized uses, eg, reagent grade, food grade, rayon grade, photographic grade, etc. In addition, some producers offer special premium-priced grades that contain little or no turbidity and color, or in some cases a maximum iron concentration of 10 ppm. Certain objectionable elements such as arsenic, lead, mercury, and selenium are not commonly specified for technical-grade acid, but some producers are attempting to hold each of these at ≤1 ppm (or ≤2–5 ppm in the case of lead) to minimize possible problems. Selenium usually is not present except at a few metallurgical-type plants or at plants using volcanic sulfur as raw material.

Metallurgical (smelter) plants and spent-acid decomposition plants usually produce acid of good (low) color because their SO_2 feed gases are extensively purified prior to use. However, in some cases, particularly at lead smelters, sufficient organic flotation agents are volatilized from sulfide ores to form brown or black acid. Such acid can be used in many applications, particularly for fertilizer production, without significant problems.

The EPA has analyzed a number of sulfuric acid samples from various smelter-based plants to determine if objectional amounts of metals (or selenium) might be present (114). Results are summarized in Table 16.

Descriptions of sulfuric acid analytical procedures not specified by ASTM are given in references 13 and 115. It should also be noted that concentrations of 78 wt % and 93 wt % H_2SO_4 are commonly measured indirectly by determining specific gravity. Higher acid concentrations are normally determined by titration with a base, or by electrical conductivity or other physical properties for plant control.

Health and Safety

Sulfuric acid is injurious to the skin, mucosa, and eyes. Dangerous amounts of hydrogen may develop in reactions between weakened acid and metals. Sulfuric acid at high concentrations reacts vigorously with water, organic compounds, and reducing agents. Oleums and liquid SO_3 frequently react with explosive violence, particularly with water.

Those engaged in handling sulfuric acid should obtain detailed information on

Table 16. Metals in Sulfuric Acid from Smelters, ppm[a,b]

	Max	Min	Mean	Electrolyte (Federal Spec. O-S-801C)
antimony (Sb)	0.18	0.003	0.02	1 (max)
arsenic (As)	0.5	0.02	0.1	1 (max)
beryllium (Be)	<0.001	<0.001	<0.001	
cadmium (Cd)	0.11	0.001	0.005	
chromium (Cr)	3.3	0.28	0.8	
copper (Cu)	0.47	<0.06	0.2	50 (max)
lead (Pb)	2.8	<0.07	0.4	
mercury (Hg)	2.3	0.003	0.05	
magnesium (Mg)	1.7	0.03	0.2	
manganese (Mn)	1.4	0.06	0.3	0.2 (max)
nickel (Ni)	1.7	0.07	0.2	1 (max)
selenium (Se)	0.48	<0.04	0.04	20 (max)
zinc (Zn)	1.83	0.05	0.1	40 (max)

[a] Ref. 114.
[b] Iron concentrations were also measured but varied widely.

safe handling practices. Most large acid producers can supply such information. Safety data sheets dated 1977 are available from the National Safety Council and were also published for many years by the Manufacturing Chemists Association (now Chemical Manufacturer's Association). Although these MCA sheets are now out-of-print, they are still useful as general guides.

The Code of Federal Regulations (CFR) includes detailed rules for packaging and shipping. Pertinent sections are 49 CFR 171.15–171.17, hazardous material incidents and discharges; 49 CFR 172.101–172.102, hazardous material tables; 49 CFR 173.244, stabilized SO_3 requirements; 49 CFR 173.248, spent-acid requirements; and 49 CFR 173.272–183.273, packaging. Requirements for transport in ships or barges are outlined in Section 46 of the CFR.

Sulfuric acid, oleum, and sulfur trioxide are classified as corrosives for labeling and freight purposes and must carry a Department of Transportation (DOT) corrosive label as well as a product label. Small bottles, pails, or drums must be placed inside outer boxes or containers as specified by the above regulations.

Steel tank cars are usually employed for high concentrations of sulfuric acid, often lined to minimize iron contamination. Bottom outlets or valves are not allowed (nor are internal steam coils), and tank contents must be unloaded via standpipe. Using air pressure to unload is not recommended for safety reasons, but if used, gauge pressures should be held at <0.21 MPa (30 psi).

The following general handling precautions should be observed:

Do not get in eyes, on skin, on clothing. Wear chemical goggles, face shield, rubber gloves and full protective clothing whenever there is danger of exposure. Wash thoroughly after handling.

Avoid breathing mist or vapors. Handle only in areas with sufficient ventilation to prevent irritation, or wear a suitable respirator.

Keep containers closed. Do not allow water to enter containers because of violent reaction. When diluting, add the acid slowly with agitation to the surface of the aqueous solution to avoid violent spattering, boiling, and eruption.

Handling containers and pipelines also requires special precautions. An emptied container retains vapor and product residue. Observe all labeled safeguards until the container is cleaned, reconditioned or destroyed. Drums should be periodically vented to prevent accumulations of hydrogen (if not self-vented). To avoid hydrogen explosions when welding, any vessel that has contained sulfuric acid must be thoroughly purged and tested for explosive conditions before welding commences. In dismantling lines and equipment, it should always be assumed that a spray of acid may occur and suitable precautions taken. Iron or other solid sulfates may plug lines or retain pockets of acid. Tightening flange bolts on pipes filled with acid is dangerous because of the possibility of mechanical failures.

In case of contact, immediately flush eyes or skin with plenty of water for at least 15 minutes while removing contaminated clothing and shoes. Call a physician. Wash clothing before reuse. Destroy contaminated shoes.

In case of inhalation, remove to fresh air. If necessary, give artificial respiration, preferably mouth-to-mouth. If breathing is labored, give oxygen. Call a physician.

In case of spill or leak, keep people away and upwind of the spill. If it is necessary to enter the spill area, wear self-contained breathing apparatus and full protective clothing including boots. Dike the area with sand or earth to contain the spill. Remove acid by vacuum truck, and flush the spill area with water. Neutralize washings with lime or soda ash. Notify pollution-control authorities of any runoff into streams or sewers and of any air-pollution incidents.

Safety showers, protected against freezing and with deluge heads, should be readily available at appropriate locations in any plant producing or using sulfuric acid.

Sulfuric Acid Toxicity. Sulfur trioxide does not exist in the atmosphere except in trace amounts because of its affinity for water. It rapidly combines with moisture in air to form sulfuric acid mist. Sulfuric acid aerosol (mist) is a significantly more powerful pulmonary irritant than sulfur dioxide on a sulfur-equivalent basis (116–118).

Researchers have found that physiological responses to sulfuric acid mist inhalation are highly dependent on the particle size of the aerosol (116,118–120). For a constant sulfuric acid aerosol concentration, the irritant action of the aerosol increases with aerosol particle size. Other factors affecting the physiological response of inhaling sulfuric acid mist are humidity, temperature, and previous exposure.

Studies of prolonged exposure to sulfuric acid fumes have been performed on workers in plants manufacturing lead–acid batteries. From these studies it is well documented that prolonged exposure to mineral acid fumes will cause the teeth of the exposed subject to deteriorate (121–123). Overexposure to sulfuric acid aerosols will result in pulmonary edema, chronic pulmonary fibrosis, residual bronchiectasis, and pulmonary emphysema. Additional information is available in refs. 124–125.

The threshold limit value (TLV) of sulfuric acid mist for humans agreed upon by the ACGIH, OSHA and NIOSH is one milligram per cubic meter of air. Sulfuric acid aerosols below the TLV are commonly not detected by odor, taste, or irritation. A TLV of 1 mg/m^3 is recommended by the ACGIH to prevent pulmonary irritation and injury to the teeth at particle sizes likely to occur in industrial situations (126). The following data summarizes human response to various levels of concentration of sulfuric acid aerosols:

Concentration, mg H_2SO_4/m^3 of air	*Response*
0.5–2.0	barely noticeable irritation
3.0–4.0	coughing, easily noticeable
6.0–8.0	decidedly unpleasant, marked alterations in respiration

BIBLIOGRAPHY

"Sulfuric Acid and Sulfur Trioxide" in *ECT* 1st ed., Vol. 13, pp. 458–506, by B. M. Carter and Gregory Flint (Sulfur Trioxide), General Chemical Division, Allied Chemical & Dye Corporation; "Sulfuric Acid and Sulfur Trioxide" in *ECT* 2nd ed., Vol. 19, pp. 441–482, by T. S. Harrer, Allied Chemical Corporation.

1. *Sulphur*, No. 157, 34 (Nov.–Dec. 1981).
2. K. Knietsch, *Ber.* **34,** 4069 (1902).
3. F. D. Miles, *Manufacture of Sulfuric Acid (Contact Process)*, Vol. 4 of A. C. Cumming, ed., *Lunge Series on the Manufacture of Acids and Alkalis*, D. Van Nostrand Co., New York, 1925, p. 348.
4. *Chem. Week* **130**(6), 40 (Feb. 10, 1982).
5. *Chem. Eng. News* **42**(51), 42 (Dec. 21, 1964).
6. *Sulphur*, No. 54, 30 (Oct. 1964).
7. U.S. Pat. 3,142,536 (July 28, 1964), H. Guth and co-workers (to Bayer A.G.).
8. U.S. Pat. 3,259,459 (July 5, 1966), W. Möller (to Bayer A.G.).
9. U.S. Pat. 1,789,460 (June 20, 1931), C. B. Clark (to General Chemical Corp.).
10. J. H. Colwell, *The Physical Properties of SO_3*, Ph.D. thesis, Dept. of Chemistry, University of Washington, Seattle, Wash., 1961.
11. J. F. Mathews, *Chem. Rev.* **72**(1), 71 (1972).
12. A. Smits and P. Schoenmaker, *J. Chem. Soc.*, 1108 (1926).
13. W. W. Duecker and J. R. West, eds., *The Manufacture of Sulfuric Acid*, Reinhold Publishing Co., New York, 1959; reprinted by Robert E. Krieger Publishing Co., Huntington, N.Y., 1971, pp. 136, 447–448.
14. R. Westrick and C. H. MacGillavry, *Rec. Trav. Chim.* **60,** 794 (1941).
15. D. M. Lichty, *J. Am. Chem. Soc.* **34,** 1440 (1912).
16. *Gmelins Handbuch der Anorganischen Chemie*, Teil B, Lief. 1, Verlag Chemie GmbH, Weinheim, FRG, 1953, pp. 337–353.
17. D. R. Stull, Dir., *JANAF Thermochemical Tables, 1965–1968*, National Bureau of Standards, U.S. Department of Commerce, Washington, D.C.
18. C. D. Spangenberg, *Chem. Eng.* **58**(9), 170 (1951).
19. A. A. Maryott and E. R. Smith, *Table of Dielectric Constants of Pure Liquids*, Circular 514, National Bureau of Standards, Washington, D.C., Aug. 10, 1951, p. 44.
20. R. Popiel, *J. Chem. Eng. Data* **9**(2), 269 (1964).
21. P. Walden, *Z. Anorg. Chem.* **25,** 209 (1900).
22. R. C. Brasted, *Comprehensive Inorganic Chemistry*, Vol. 8, D. Van Nostrand Co., Inc., Princeton, N.J., 1961, p. 135.
23. K. M. Malin and co-workers, *Technology of Sulfuric Acid*, Goskhimizdat, Moscow, USSR, 1950, p. 533.
24. A. L. Horvath, *Physical Properties of Inorganic Compounds*, Crane, Russak & Co., Inc., New York, 1975, p. 316.
25. A. Berthoud, *Helv. Chim. Acta* **5,** 513 (1922).
26. A. Berthoud, *J. Chim. Phys.* **20,** 77 (1923).
27. E. Lax and C. Synowietz, eds., *Taschenbuch für Chemiker und Physiker*, 3rd ed., Springer-Verlag, Berlin, FRG, 1967.
28. R. Schenck, *Lieb. Ann.* **316,** 1 (1901).
29. W. H. Evans and D. D. Wagman, *J. Res. Natl. Bur. Stand.* **49,** 141 (1952).
30. G. K. Boreskov, *Catalysis in Sulfuric Acid Production* (Russian), Goskhimizdat, Moscow, USSR, 1954.
31. Ref. 23, pp. 529–559.
32. J. E. Stuckey and C. H. Secoy, *J. Chem. Eng. Data* **8**(3), 386 (July 1963).
33. R. G. H. Record, *Instrum. Eng.* **4**(7), 131 (1967).
34. H. E. Darling, *J. Chem. Eng. Data* **9**(3), 421 (July 1964).
35. R. Haase, P. F. Sauerman, and K. H. Duecker, *Z. Phys. Chem.* **48,** 206 (1966).
36. N. N. Greenwood and A. Thomson, *J. Chem. Soc.*, 3485 (1959).
37. N. F. Bright, H. Hutchinson, and D. Smith, *J. Soc. Chem. Ind.* **65,** 385 (1946).
38. *Engineering Sciences Data Unit, Approximate Data on the Viscosity of Some Common Liquids*, Item No. 66024, Royal Aeronautical Society, London, July 1966.
39. W. Tartakowskaja, J. Bondarenko, and L. Jameljanowa, *Acta Phys. Khim. USSR* **6,** 609 (1937).
40. M. Usanovic, T. Sumarokova, and V. Udovenko, *Zh. Obsh. Khim.* **9,** 1967 (1939).

41. M. Usanovic, T. Sumarokova, and V. Udovenko, *Acta Phys. Khim.* **11,** 505 (1939).
42. R. J. Gillespie and E. A. Robinson in T. C. Waddington, ed., *Non-Aqueous Solvent Systems*, Academic Press, Inc., New York, 1965, pp. 123–128.
43. J. L. R. Morgan and C. E. Davis, *J. Am. Chem. Soc.* **38,** 555 (1916).
44. J. E. L. Maddock, *A Light Scattering Study of Growth of Sulfuric Acid Aerosol Particles*, PhD thesis, Dept. of Chemical Engineering and Chemical Technology, Imperial College of Science and Technology, London, June 1974.
45. L. Sabinina and L. Terpugow, *Z. Phys. Chem.* **173A,** 237 (1935).
46. P. Walden, *Z. Phys. Chem.* **65,** 129 (1909).
47. M. Liler, *Reaction Mechanisms in Sulfuric Acid and Other Strong Acid Solutions*, Academic Press, Inc., New York, 1971, p. 5.
48. E. E. Remsberg, D. Lavery, and B. Crawford, *J. Chem. Eng. Data* **19**(3), 264 (1974).
49. A. Seidell, *Solubilities of Inorganic and Metal Organic Compounds*, 3rd ed., D. Van Nostrand Co., New York, 1940.
50. F. D. Miles and T. Carson, *J. Chem. Soc.*, 786 (1946).
51. J. W. Mellor, *A Comprehensive Treatise on Inorganic and Theoretical Chemistry*, Vol. X, Longmans, Green & Co., New York, 1930, pp. 351–360, 384–425.
52. W. H. Lee in J. J. Lagowski, ed., *The Chemistry of Non-Aqueous Solvents*, Vol. II, Academic Press, Inc., New York, 1967.
53. Ref. 24, pp. 313–335.
54. Ref. 13, 1971, p. 411.
55. Ref. 23, pp. 534–535.
56. C. M. Gable, H. F. Betz, and S. H. Maron, *J. Am. Chem. Soc.* **72,** 1445 (1950).
57. A. Seidell, *Solubilities Inorganic and Metal-Organic Compounds*, Vol. I, 4th ed., D. Van Nostrand Co., Princeton, N.J., 1958, pp. 1161–1162.
58. J. E. Kunzler, *Anal. Chem.* **25,** 93 (1953).
59. C. H. Greenewalt, *Ind. Eng. Chem.* **17**(5), 522 (1925).
60. G. P. Luchinskii, *Zh. Fiz. Khim.* **30**(6), 1207 (1956).
61. H. Lennartz and co-workers, *Vaporization Equilibrium of the Water-Sulfuric Acid System*, Rep. Eur. 6783, Commission of European Communities, Hydrogen Energy Vector 60–70, Europe, 1980.
62. J. E. Kunzler and W. F. Giauque, *J. Am. Chem. Soc.* **74,** 3472 (1952).
63. F. D. Rossini and co-workers, *Selected Values of Chemical Thermodynamic Properties*, Circ. 500, National Bureau of Standards, U.S. Department of Commerce, Washington, D.C., 1952.
64. Ref. 23, pp. 541–542.
65. F. D. Miles, H. Niblock, and G. L. Wilson, *Trans. Faraday Soc.* **36,** 345 (1940).
66. Ref. 31, pp. 535–536.
67. F. R. Bichowski and F. D. Rossini, *Thermochemistry of Chemical Substances*, 1st ed., New York, Reinhold Publishing Co., 1936.
68. H. Howard, *J. Soc. Chem. Ind.* **29,** 3 (1910).
69. R. Knietsch, *Ber.* **34,** 4069 (1901).
70. H. P. J. J. Thomsen, *Thermochemische Untersuchungen*, Vol. III, Barth, Leipzig, GDR, 1883.
71. C. V. Herrmann, *Ind. Eng. Chem.* **33,** 898 (June 1941).
72. F. D. Miles, H. Niblock, and D. Smith, *Trans. Faraday Soc.* **40,** 281 (Sept. 1944).
73. Ref. 23, p. 546.
74. A. G. Amelin, *Sulfuric Acid Technology* (Russian), Khimiya, Moscow, USSR, 1971, pp. 451–452.
75. J. C. D. Brand and A. J. Rutherford, *J. Chem. Soc.* **10,** 3916 (1952).
76. *Sulphur* (152), 37 (Jan.–Feb. 1981).
77. J. M. Connor, *Sulphur* (131), 39 (July–Aug. 1977).
78. A. M. Fairlie, *Sulfuric Acid Manufacture*, Reinhold Publishing Co., New York, 1936.
79. E. E. Gilbert, *Ann. Gen. Chim.* **3,** 42 (1967).
80. T. J. Browder in A. I. More, ed., *Making the Most of Sulfuric Acid, Proceedings of the British Sulphur Corporation's Fifth International Conference*, London, Nov. 1981, pp. 183–205.
81. J. D. Palmer, *Can. Chem. Process.* **60**(8), 35 (Aug. 1976).
82. P. D. Nolan, *Can. Chem. Process.* **61**(5), 40 (May 1977).
83. D. Fyfe and co-workers, *International Corrosion Forum: Corrosion '75, Paper No. 63, Toronto*, National Association of Corrosion Engineers (NACE), Houston, Texas, 1975.
84. P. Cross and co-workers in ref. 80, pp. 157–179.
85. J. R. Donovan, R. M. Smith, and J. S. Palermo, *Sulphur* (131), 46 (July–Aug. 1977).
86. D. R. Duros and E. D. Kennedy, *Chem. Eng. Prog.* **74**(9), 70 (Sept. 1978).

87. P. Parish and F. C. Snelling, *Sulfuric Acid Concentration*, 2 Vols., Ernest Benn Ltd., London, 1924.
88. G. M. Smith and E. Mantius, *Chem. Eng. Prog.* **74**(9), 78 (Sept. 1978).
89. U. Sander and G. Daradimos, *Chem. Eng. Prog.* **74**(9), 57 (Sept. 1978).
90. I. Rodger, *Chem. Eng. Prog.* **78**(2), 39 (Feb. 1982).
91. H. R. Kueng and P. Reimann, *Chem. Eng.* **89**(8), 72 (Apr. 19, 1982).
92. M. G. Fontana and N. D. Green, *Corrosion Engineering*, 2nd ed., McGraw-Hill, Inc., New York, 1978, pp. 223–241.
93. *Resistance of Nickel and High Nickel Alloys to Corrosion by Sulfuric Acid*, Corrosion Engineering Bulletin CEB-1, The International Nickel Co., New York, 1975.
94. G. E. McClain and W. A. Mueller, *Chem. Eng. Prog.* **78**(2), 48 (Feb. 1982).
95. R. A. Bauer and B. P. Vidon, *Chem. Eng. Prog.* **74**(9), 68 (Sept. 1978).
96. *Sulphur*, No. 105, 51 (Mar.–Apr. 1973).
97. G. M. Cameron, *Chem. Eng. Prog.* **74**(9), 47 (Sept. 1978).
98. A. F. Snowball, *Can. Chem. Process Ind.* **31**(12), 1110 (Dec. 1947).
99. S. D. Kirkpatrick, *Chem. Eng.* **55**(4), 96 (Apr. 1948).
100. R. Remírez, *Chem. Eng.* **88**(26), 34 (Dec. 28, 1981).
101. P. R. Minbiole and D. G. Winkler, *Sulfuric/Phosphoric Acid Plant Operations*, CEP Technical Manual, AIChE, New York, 1982, pp. 1–5.
102. L. J. Friedman, *Chem. Eng. Prog.* **78**(2), 51 (Feb. 1982).
103. *Sulphur*, (147), 32 (Mar.–Apr. 1980).
104. J. M. Kovacik and J. C. Frankling, *Chem. Eng. Prog.* **78**(2), 58 (Feb. 1982).
105. R. L. Davis, *Chem. Eng. Prog.* **78**(3), 46 (Mar. 1982).
106. *Chem. Eng. News*, 11 (May 3, 1982).
107. *Sulfuric Acid, SRI Chemical Economics Handbook*, Stanford Research Institute, Menlo Park, Calif., Aug. 1979 and later addenda.
108. *Chem. Eng. News*, 19 (Jan. 25, 1982).
109. *ASTM E 223-68*, Vol. 30 of 1981 Standards, American Society for Testing and Materials, Philadelphia, Pa., 1981, pp. 821–832.
110. *Sulfuric Acid, Electrolyte for Storage Batteries*, Specification O-S-801C, Oct. 20, 1978; *Sulfuric Acid, Technical*, Specification O-S-0809B, Sept. 28, 1967; *Interim Amendment-1*, Jan. 11, 1968, Federal Supply Service, General Services Administration, Washington, D.C.
111. G. Norwitz, *Analyst* **87**(1039), 829 (1962).
112. A. Irudayasamy and A. R. Natarajan, ref. 111, pp. 831–832.
113. *Standard Methods for the Examination of Water, Sewage and Industrial Wastes*, 10th ed., American Public Health Association (APHA), Inc., New York, 1955, pp. 207–211.
114. W. H. Hedley and co-workers, *Determination of Hazardous Elements in Smelter-Produced Sulfuric Acid*, EPA-650/2-74-131, issued as PB-240343/4ST by NTIS, Washington, D.C., Dec. 1974.
115. O. T. Fasullo, *Sulfuric Acid Use and Handling*, McGraw-Hill, Inc., New York, 1965.
116. *Air Quality Criteria for Sulfur Oxides*, National Air Pollution Control Administration, U.S. Dept. of Health, Education, and Welfare, Washington, D.C., Jan. 1969, pp. 89–102.
117. M. O. Amdur, L. Silverman, and P. Drinker, *Am. Med. Assoc. Arch. Ind. Hyg. Occup. Med.* **6,** 306 (Oct. 1952).
118. V. M. Sim and R. E. Pattle, *J. Am. Med. Assoc.* **165,** 1908 (Dec. 14, 1957).
119. R. E. Pattle, F. Burgess, and H. Cullumbine, *J. Pathol. Bacteriol.* **72,** 219 (1956).
120. M. O. Amdur, *Arch. Ind. Health* **18,** 407 (1958).
121. D. Malcolm and E. Paul, *Br. J. Ind. Med.* **18,** 63 (1961).
122. J. B. Lynch and J. Bell, *Br. J. Ind. Med.* **4,** 84 (1947).
123. H. J. Bruggen Cate, *Br. J. Ind. Med.* **25,** 249 (1968).
124. S. Chaney, W. Bloomquist, K. Muller, and G. Goldstein, *Arch. Environ. Health* **35**(4), 211 (July–Aug. 1980).
125. S. M. Horvath and co-workers, *Effects of Sulfuric Acid Mist Exposure on Pulmonary Function*, EPA-600/S1-81-044, issued as PB81-208977 by NTIS, Washington, D.C., June 1981.
126. *Documentation of the Threshold Limit Values*, 4th ed., American Conference of Governmental Industrial Hygienists, Inc., Cincinnati, Ohio, 1980, p. 377.

J. R. Donovan
J. M. Salamone
Monsanto Enviro-Chem Systems, Inc.

SULFURIC AND SULFUROUS ESTERS

Sulfuric and sulfurous acids form a series of esters analogous to those from other acidic materials. The hydrogen of the acid is replaced by a carbon-containing group. Since two hydrogens are present in the sulfur-based acids, there are two series of esters. Replacement of one hydrogen results in an acid ester. If both hydrogens are replaced, whether with the same, with different, or with bifunctional substituents, symmetrical and unsymmetrical diesters form. The two series are represented by the following general formulas:

where R is a carbon group and R′ is a carbon group, hydrogen, or metal cation.

In the acid ester series, there are compounds in which another group, eg, chlorine or amine, is present in place of the hydroxy group. These compounds can be used as intermediates in making diesters. The carbon groups most commonly present are those from short- and long-chain alcohols and from hydroxyaromatic and heterocyclic compounds. In this review, the nomenclature is primarily that used in commercial practice, rather than the more rigorous CA nomenclature by which they are called sulfuric and sulfurous acid organic esters. Table 1 illustrates the variety of known compounds and some of their properties.

Of these, dimethyl sulfate, diethyl sulfate, and long-chain monoalkyl alkali-metal sulfates are the compounds of practical interest. Since ca 1960, many publications and patents have described the activity of the dialkyl sulfates in introducing a methyl or ethyl group into a wide variety of organic compounds. The alkylation chemistry of the lower alkyl hydrogen sulfates or their salts is of far less importance, but is closely related to that of dialkyl sulfates in that the monoalkyl esters are intermediate alkylating agents in those instances where both alkyl groups of the dialkyl sulfate are usable for alkylation. The long-chain monoalkyl hydrogen sulfates, in the form of their alkali-metal or alkylammonium salts, are valuable as surface-active agents and are used commercially. The short-chain monoalkyl hydrogen sulfates are intermediates in commercial conversion routes from olefins to alcohols (see Ethanol; Propyl alcohols, isopropyl alcohol; Sulfonation and sulfation; Surfactants and detersive systems).

Physical Properties

The physical properties are best understood as resulting from the blending of the polar contribution of an acid or neutral sulfate group with the contribution of the usually nonpolar carbon group. The neutral sulfate group of esters is much less polar than the acid sulfate group ($-OSO_3H$). With lower alkyl groups, the polar effects dominate, whereas with higher alkyl groups, the nonpolar interactions of the alkyl dominate.

The lower alkyl hydrogen sulfates are moderately viscous liquids. Higher alkyl

Table 1. Some Sulfuric and Sulfurous Acid Esters

Ester	CAS Reg. No.	Bp$_{kPa^a}$, °C	Mp, °C
Sulfates, $(RO)_2SO_2$			
R			
Open-chain			
methyl	[77-78-1, 75-93-4]	188.8$_{101.3}$, 69.70$_{1.33}$	
ethyl	[64-67-5]	208, 89$_{1.20}$	
n-propyl	[598-05-0]	95$_{0.670}$	
isopropyl	[2973-10-6]	80$_{0.530}$	
n-butyl	[625-22-9]	103$_{0.200}$	
ethyl *n*-butyl	[5867-95-8]	117$_{2.40}$	
chloromethyl	[73455-05-7]	97$_{1.87}$	
2-chloroethyl	[5411-48-3]	150$_{0.940}$	
n-decyl	[66186-16-1]		37.7
n-tetradecyl	[66186-19-4]		58.0
phenyl	[4074-56-0]	194.6$_{2.66}$	
Cyclic			
ethylene	[1072-53-3]		99
1,3-propylene	[1073-05-8]		63
methylene (dimer)	[20757-83-9]		155
Sulfites, $(RO)_2SO$			
R			
Open-chain			
methyl	[616-42-2]	126–127$_{101.3}$	
ethyl	[623-81-4]	159–160$_{101.3}$	
n-propyl	[623-98-3]	82$_{2.00}$	
isopropyl	[4773-13-1]	78$_{2.67}$	
n-butyl	[626-85-7]	110$_{1.73}$	
3-chloropropyl	[83929-99-1]	162$_{1.73}$	
phenyl	[4773-12-0]	185$_{2.00}$	13–16
Cyclic			
ethylene	[3741-38-6]	80$_{3.72}$	
1,2-propylene	[1469-73-4]	84$_{3.92}$	
pentaerythritol,di	[3670-93-7]		154
Organo hydrogen sulfates, $(RO)SO_3H$			
R			
methyl	[75-93-4]	130–140$_{101.3}$ (dec)	
n-decyl	[142-98-3]	(liquid)	
n-dodecyl	[151-41-7]		25–27
n-tetradecyl	[4754-44-3]		37–39
n-hexadecyl	[143-02-2]		40–42
n-octadecyl	[143-03-3]		51–52
phenyl	[937-34-8]		
Organo halosulfates, $(RO)SO_2X$			
R, X			
methyl,chloro	[812-01-1]	134–135$_{101.3}$, 42$_{2.13}$	
methyl,fluoro	[421-20-5]	92$_{101.3}$, 45$_{21.4}$	
ethyl,chloro	[625-01-4]	72$_{15.6}$, 58$_{2.66}$	
ethyl,fluoro	[371-69-7]	113$_{100}$, 21$_{1.60}$	
n-propyl,chloro	[819-52-3]	53$_{1.33}$	
isopropyl,chloro	[36610-67-0]	50$_{0.67}$ (dec)	
Organo halosulfites, $(RO)SOX$			
R, X			
methyl,chloro	[13165-72-5]	35$_{8.00}$	

Table 1 (*continued*)

Ester	CAS Reg. No.	Bp_{kPa^a}, °C	Mp, °C
ethyl,chloro	[*6378-11-6*]	$50–53_{8.00}$	
n-propyl,chloro	[*22598-38-5*]	$78_{10.0}$, $42_{1.69}$	
isopropyl,chloro	[*22598-56-7*]	$71–73_{10.0}$, $55_{5.23}$	

[a] To convert kPa to mm Hg, multiply by 7.5; 101.3 kPa = 1 atm.

hydrogen sulfates are hygroscopic, low melting solids; their hygroscopic nature results from the acid group, and their solid state results from the alkyl group orientation. They do not have boiling points, but decompose on heating.

Stability or storage depends mostly on the purity, with purer materials having longer shelf lives (1). For the higher alkyl groups, the anhydrous compounds are soluble and the monohydrates are insoluble in ether. Solutions in water are strongly ionized and acidic. The lower dialkyl sulfates are liquids with faint but pleasant odors (see Health and Safety Factors); *n*-nonyl and higher normal aliphatic and cyclic sulfates are solids.

The diesters are moderately polar and, therefore, are miscible with most common organic solvents. Solubility in water is low-to-insoluble, with dimethyl sulfate having a water solubility of 2.8 g/100 mL at 18.0°C (2).

The sulfites have some laboratory use, but are not commercially important and are less known. Monoesters of sulfurous acid are quite unstable, although salts have been identified. The diesters of sulfurous acid are mostly liquids with boiling points somewhat less than those of the corresponding sulfates.

Dialkyl sulfites can exist as geometric isomers. The configuration of the three oxygen atoms and the electron pair about the sulfur, which has a minimum of electrostatic energy, is closely tetrahedral. In addition, thermal inversion does not appear to occur in contrast to the behavior of amines. Several isomeric systems have been reported (3–4).

Many esters of halosulfurous acid and of halosulfuric acid are known, ie, halosulfites and halosulfates, respectively. With the smaller alkyl groups, the compounds are high boiling liquids with a penetrating odor and a lacrimatory action. Esters of halosulfuric acid are more stable to moisture when stored than those of halosulfurous acid.

Selected properties of dimethyl and diethyl esters of sulfuric acid are listed in Table 2.

Table 2. Selected Physical Properties of Dimethyl and Diethyl Sulfates

Property	$(CH_3O)_2SO_2$	$(C_2H_5O)_2SO_2$
bp (at 101.3 kPa = 1 atm), °C	188.8	208 (dec) 143 (13.3) (88 (1.33)
mp, °C	−31.8	−24.4
sp gr_{20}^{20}	1.328	1.1803
n_D^{20}	1.3874	1.3396
solubility in water, wt %	2.8 at 18°C	0.7 at 20°C
flash point (open-cup), °C	116	113
autoignition temp, °C	188	
heat of formation (liquid), kJ/mol[a,b]	687	757

[a] To convert J to cal, divide by 4.184.
[b] Ref. 5.

Chemical Properties

Sulfates. The chemistry of alkyl sulfates is dominated by two fundamental process types: reaction with nucleophiles and reaction as acids. Reaction with nucleophiles results in alkylation.

Alkylation. In alkylation, the dialkyl sulfates react much faster than do the alkyl halides, since the monoalkyl sulfate anion ($ROSO_3^-$) is more effective as a leaving group than a halide ion. The high rate is most apparent with small primary alkyl groups, eg, methyl and ethyl. Some leaving groups, such as the fluorinated sulfonate anion, eg, the triflate anion, $CF_3SO_3^-$, react even faster in ester form (6). Against phenoxide anion, the reaction rate is methyl triflate [*333-27-7*] ≫ dimethyl sulfate ≫ methyl *p*-toluenesulfonate [*23373-38-8*] (7). Dialkyl sulfates, as compared to alkyl chlorides, lack chloride ions in their products; chloride corrodes and requires the use of a gas instead of a liquid. The lower sulfates are much less expensive than lower bromides or iodides, and they also alkylate quickly.

Although the first alkyl group is the most reactive, the second alkyl group on the intermediate anion can also alkylate. The temperature must be higher than needed for the first alkyl, and conditions favoring hydrolysis of sulfate ester, eg, the presence of water and alkali, should be minimized. As a very general rule, alkylations by alkyl chloride are intermediate in rate between those of the corresponding dialkyl sulfates and alkyl hydrogen sulfates. In most reported reactions, only one alkyl group reacts. If the alkylated molecule acquires a positive charge, the monoalkyl sulfate anion becomes the negative counterion. This is illustrated in the common reaction of quaternization of a tertiary amine, eg, pyridine.

$$C_5H_5N + CH_3OSO_2OCH_3 \rightarrow C_5H_5NCH_3^+ + CH_3OSO_3^-$$

The reactions of these nucleophilic processes are usually S_N2 rather than S_N1. The reaction rate is methyl > ethyl > isopropyl, as with the alkyl halides. As the species to be alkylated becomes more nucleophilic, alkylation becomes faster, eg, a sulfur-containing anion alkylates more quickly than a phenolic anion.

Dimethyl sulfate and diethyl sulfate react with almost all types of compounds with unshared electron pairs, whether the reaction center is at oxygen, nitrogen, carbon, sulfur, phosphorus, some metals, or most other heteroatoms from groups VA and VIA. Alkylation at oxygen can occur with alcohols, phenols, or acids. Methylation of phenol to give the methyl ether or anisole is carried out commercially. With phenol alone, the reaction is slow with dimethyl sulfate at 100–120°C, giving anisole and some by-product dimethyl ether and sulfonation products (8). Reaction is much faster in basic medium, with phenolate ion as the reactant. The first methyl group reacts at 45–60°C upon addition of dimethyl sulfate to a mixture of molten phenol and sodium hydroxide containing very little water. Continuing the reaction by heating to 100°C results in the reaction of the second methyl group and a 95% yield of anisole (9–11). Ethylation of phenolate ion with diethyl sulfate requires higher temperatures, ie, ca 50–55°C, for the first ethyl group and 145°C for the second ethyl group. Numerous substituted phenols have been alkylated in either aqueous or alcoholic alkaline solutions, or dimethylformamide solution with K_2CO_3 (11–13). Thiophenols react faster than phenols in alkaline solution (14).

Alcohols react readily in alkaline solution. Conditions for pentaerythritol are typical (15). Use of alkyl alcohol or *t*-butanol with KOH in dimethyl sulfoxide gives fair-to-good ethyl ether yields with diethyl sulfate at 50–55°C (16). One hydroperoxide is reported to methylate in high yield in aqueous base (17).

Alkylated oxygen acids include carboxylic, sulfonic, and carbonic acids. Carboxylic acids react with the first alkyl group at 120°C to give esters in high yield, whereas the second alkyl group requires higher temperatures, ie, 200°C (18). The use of carboxylate salts of inorganic or organic bases gives fast esterification under neutral conditions (19). A highly branched acid, eg, trimethylacetic acid, converts readily to the methyl ester; dimethylformamide is a good medium (13,20). Good conditions for diethyl sulfate with lower acids have been determined (21). Dimethyl or diethyl carbonate can be made by heating the corresponding sulfate and alkali-metal carbonate at 150–210°C, which illustrates the high activity of the sulfates (22). Sulfonic acids and their salts alkylate at 100°C (23–24).

The nitrogen of aliphatic and aromatic amines is alkylated rapidly by alkyl sulfates yielding the usual mixtures. Most tertiary amines and nitrogen heterocycles are converted to quaternary ammonium salts, unless the nitrogen is of very low basicity, eg, in triphenylamine. The position of dimethyl sulfate-produced methylation of several heterocycles with more than one heteroatom has been examined (25). Acyl cyanamides can be methylated (26). Metal cyanates are converted to methyl isocyanate or ethyl isocyanate in high yields by heating the mixtures (27–28).

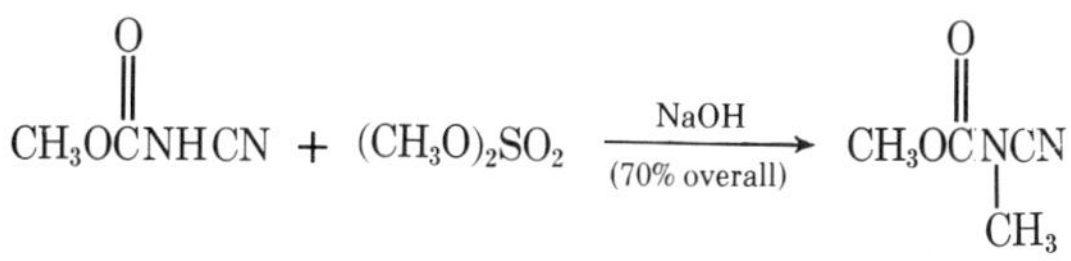

Carbon is alkylated in the form of enolates or as carbanions. The enolates are ambident in activity and can react at an oxygen or a carbon. For example, refluxing equimolar amounts of dimethyl sulfate and ethyl acetoacetate with potassium carbonate gives a 36% yield of the O-methylation product, ie, ethyl 3-methoxy-2-butenoate, and 30% of the C-methylation product, ie, ethyl 2-methyl-3-oxobutanoate (29). Generally, only one alkyl group of the sulfate reacts with beta-diketones, beta-ketoesters, or malonates (30). Factors affecting the O:C alkylation ratio have been extensively studied (31). Reaction in the presence of solid Al_2O_3 results mostly in C-alkylation of ethyl acetoacetate (32).

Carbanions in the form of phenyllithium, sodium naphthalene complex, sodium acetylide, or aromatic Grignard reagents react with alkyl sulfates to give a C-alkyl product (33–36). Grignard reagents require two moles of dimethyl sulfate for complete reaction.

N,N-Dialkylcarboxamides react with dialkyl sulfates at the oxygen to give an addition product (a Vilsmeier-type compound), which is the prototype of a variety of reactive compounds.

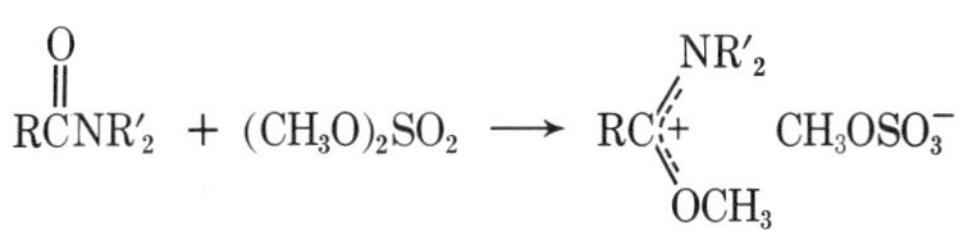

(1)

Most work has been done using dimethylformamide, which gives the complex $[HC(OCH_3)N(CH_3)_2]^+CH_3OSO_3^-$ [*34643-89-5*] (1) (R = H, R′ = CH_3) (37). Further reaction with alcoholates, mercaptides, or amines gives orthoamides from dimethylformamide (DMF) $HC(OCH_3)N(CH_3)_2G$, where G is —OR, —SR, —NR_2 (38). Reaction of the DMF complex with carboxylic acids gives the corresponding methyl esters of the acid (39). The complex from longer-chain amides (R = alkyl) can react with strong bases, eg, NaOR and NaH, to lose a proton and form ketene acetal-type products [$RCH{=}CN(CH_3)_2G$] (40). The intermediate from dimethylformamide is used to convert carbonyl compounds to acetals or ketals (41). Similarly, the complex from *N,N*-dimethylbenzamide (1) (R = C_6H_5, R′ = CH_3) [*611-74-5*] has also been used (42). If an amide of a primary amine is employed, the *O*-alkylation product can be isolated (43) (see Ketenes and related substances).

$$R_2CO + 2\,R'OH + (1) \rightarrow R_2C(OR')_2 + (CH_3)_2NCHO + CH_3OSO_3H$$

Sulfur is reactive in many forms. Mercaptides are alkylated to thioethers, and thioethers react further to give sulfonium salts ($R_3S^+\ CH_3OSO_3^-$). A phosphorothiolate salt reacts similarly (44):

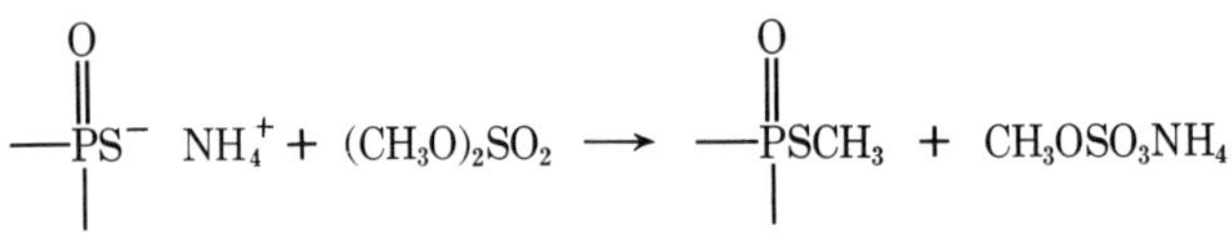

Suitable thiones also alkylate. Thioacridone (45) and thiourea (46) are examples: the first gives the alkylmercaptoacridine and the second gives the isothiourea.

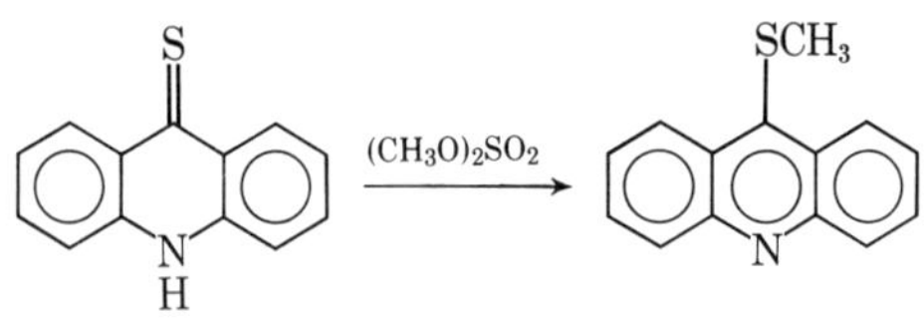

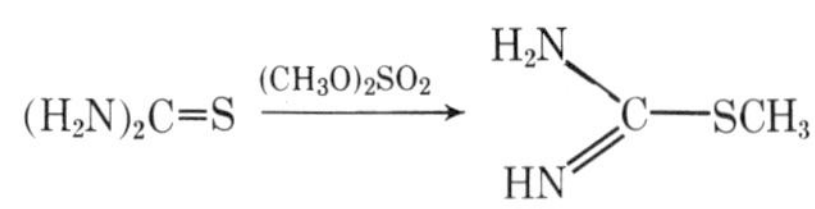

Several inorganic atoms have also been methylated by dimethyl sulfate. Sodium iodide gives methyl iodide, organoxytitanium dichloride gives several highly colored organoxytitanium methyl sulfates, titanium tetrachloride yields titanium chlorosulfates or methyl sulfates, and mossy tin gives dimethyltin sulfate (47–50).

These reactions involve mostly dimethyl and diethyl sulfate. Cyclic sulfates are also reactive, and several have been compared by determining reaction rates with a substituted pyridine or with water (51). In both cases, 1,2-ethylene sulfate is more reactive than 1,3-propylene sulfate or dimethyl or diethyl sulfates.

Some work, although not recent, is reported for alkylation by alkali salts of monoalkyl sulfates, including alkylation of nitrite, cyanide, mercaptide, and carboxylate (52–55). An extensive table of reaction rates of substituted phenolates with mostly methyl sulfate anion at 100°C is given in ref. 56. Commercially, 2-aminoethyl hydrogen sulfate [*926-39-6*] cyclizes at 120–140°C under pressure to give ethyleneimine (aziridine) (57) (see Imines, cyclic). The sulfate [*3840-67-3*] of 3-hydroxypropionamide, $HO_3SOCH_2CH_2CONH_2$, reacts with amines by carbamoylethylation to give 3-aminopropionamides (58). Diethanolamine as the bis(hydrogen sulfate) ester [*15777-78-3*] forms a diazoamino compound with aromatic diazonium compounds. This cyclizes to a 1,3-disubstituted 1,2,3-triazolium compound (59):

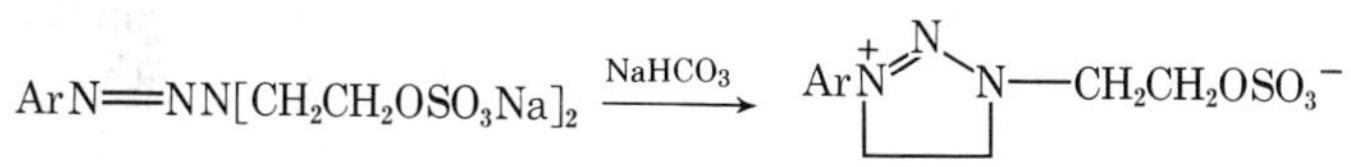

Hydrolysis. The hydrolysis of dialkyl and monoalkyl sulfates is a process of considerable interest commercially. Successful alkylation in water requires that the fast reaction of the first alkyl group with water and base be minimized. The very slow reaction of the second alkyl group results in poor utilization of the alkyl group and gives an increased organic load to a waste-disposal system. Data has accumulated since 1907 on hydrolysis in water under acid, neutral, and alkaline conditions, and best conditions and good values for rates have been reported and the subject has been reviewed (60–70).

Sulfates with alkyl groups from methyl to pentyl have been examined. With methyl as an example, the hydrolysis rate of dimethyl sulfate increases with the concentration of the sulfate. Typical rates in neutral water are first order and are 1.66×10^{-4} s^{-1} at 25°C and 6.14×10^{-4} s^{-1} at 35°C (65–66). Rates with alkali or acid depend on conditions (61,67). Rates for the monomethyl sulfate [*512-42-5*] are much slower, and are nearly second order in base. Values of the rate constant in dilute solution are 6.5×10^{-5} L/(mol·s) at 100°C and 4.64×10^{-4} L/(mol·s) at 138°C (63). At 138°C, first-order solvolysis is ca 2% of the total. The Arrhenius rate at 125–230°C is $k_2 = 5.55 \times 10^6\, e^{-8936/T}$ L/(mol·s) with 50% hydrolysis in 1 min at 200°C and 15% hydrolysis in 5 min at 125°C (68). Hydrolysis of the monoester is markedly promoted by increasing acid strength and it is first order. The rate at 80°C is 3.65×10^{-4} s^{-1} (64). Alkaline solvolysis has been studied by a calorimetric method (69). Heat of hydrolysis of dimethyl sulfate to the monoester under alkaline conditions is 106 kJ/mol (25 kcal/mol) (71).

Cyclic esters show accelerated hydrolysis rates. Ethylene sulfate compared to dimethyl sulfate is twice as fast in weak acid (first order) and 20 times as fast in weak alkali (second order) (70). Catechol sulfate [*4074-55-9*] is 2×10^7 times faster than diphenyl sulfate in alkaline solution (72). Alcoholysis rates of several dialkyl sulfates at 35–85°C are also known (73).

Studies of reaction mechanisms in ^{18}O-enriched water show the following: cleavage of dialkyl sulfates is primarily at the C—O bond under alkaline and acid conditions, and monoalkyl sulfates cleave at the C—O bond under alkaline conditions and at the S—O bond under acid conditions (64,74). An optically active half ester (*sec*-butyl sulfate [*3004-76-0*]) hydrolyzes at 100°C with inversion under alkaline conditions and with retention plus some racemization under acid conditions (75). Effects of solvent and substituted structure have been studied, with moist dioxane giving marked rate enhancement (63,76–77). Hydrolysis of monophenyl sulfate [*4074-56-0*] has been similarly examined (78).

Reactions other than those of the nucleophilic reactivity of alkyl sulfates involve reactions with hydrocarbons, thermal degradation, sulfonation, halogenation of the alkyl groups, and reduction of the sulfate groups. Aromatic hydrocarbons, eg, benzene and naphthalene, react with alkyl sulfates when catalyzed by aluminum chloride to give Friedel-Crafts-type alkylation product mixtures (79). Isobutane is readily alkylated by a dipropyl sulfate mixture from the reaction of propylene in propane with sulfuric acid (80).

Pyrolysis. Thermal stability of the dialkyl sulfate is of interest when using solutions in inert media, eg, hydrocarbons, or on storage of the neat compound. Heating dimethyl sulfate at ca 200°C produces dimethyl ether, SO_3, and other products (81). Higher alkyl sulfates, eg, the diethyl sulfates, decompose to give olefin and oxidation products rapidly at 220°C and slowly at 140°C (82). Heating higher monoalkyl hydrogen sulfates generally gives olefins also, but the monomethyl ester at 130–140°C gives dimethyl sulfate and sulfuric acid.

Sulfonation. Sulfonation is a common reaction with dialkyl sulfates, either by slow decomposition on heating with the release of SO_3 or by attack at the sulfur end of the O—S bond (83). Reaction products are usually the dimethyl ether, methanol, sulfonic acid, and methyl sulfonates, corresponding to both routes. Reactive aromatics are commonly those with higher reactivity to electrophilic substitution at temperatures >100°C. Triphenylamine, diphenylmethylamine, anisole, and diphenyl ether exhibit ring sulfonation at 150–160°C, 140°C, 155–160°C, and 180–190°C, respectively, but diphenyl ketone and benzyl methyl ether do not react up to 190°C. Diphenylamine methylates and then sulfonates as described previously. Catalysis of sulfonation of anthraquinone by dimethyl sulfate occurs with thallium(III) oxide or mercury(II) oxide at 170°C. Alkyl interchange also gives sulfation: ethylene glycol diacetate and dimethyl sulfate at 130°C yield ethylene sulfate (84).

Miscellaneous. Halogenation of the methyl group of dimethyl sulfate by chlorine and by fluorine has been described (85–86). Reduction of dimethyl sulfate by hydriodic acid occurs in a way that is comparable to that shown by sulfuric acid and results in reduction to sulfur (87).

Sulfites. The literature concerning dialkyl sulfites is extensive, although less than for sulfates. Reactions involving alkylation are similar to those of sulfates. Sulfites also undergo elimination, transesterification, and isomerization. The last two parallel reactions of phosphites.

In alkylation, phenols and amines are alkylated by sulfites in high yield and quaternary salts readily form (88). Ethylene sulfite reacts yielding hydroxyethyl derivatives with SO_2 elimination, corresponding to its activity as an ethylene oxide precursor (89).

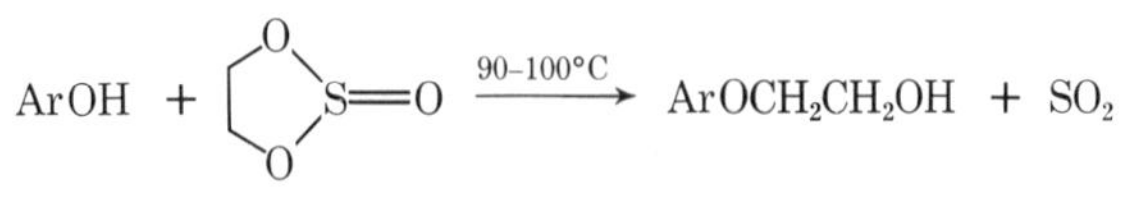

With catechol as the phenol, the cyclic ether forms.

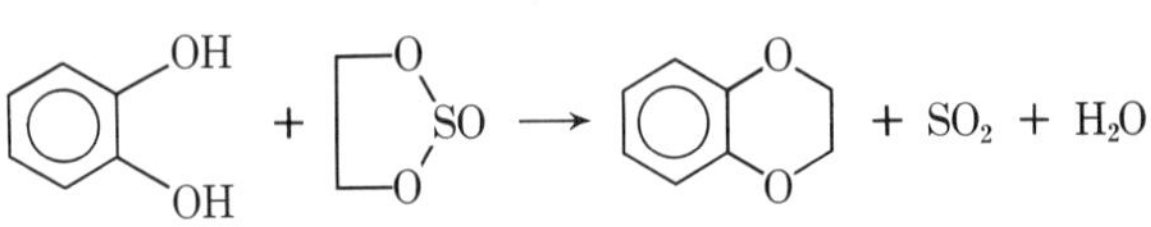

Reaction of carboxylate ion with *p*-nitrophenyl sulfites gives the carboxylate *p*-nitrophenyl esters. If the *p*-nitrophenyl sulfite is unsymmetrical ($O_2NC_6H_4O$-S(O)OR, where R is ethyl or phenyl), carboxylate attacks the *p*-nitrophenyl side (90). Some amino acids react with methyl and benzyl sulfites in the presence of *p*-toluenesulfonic acid to give methyl and benzyl esters of the amino acids as *p*-toluenesulfonate salts (91). With alcohols, the conversion of benzil to a monoacetal upon addition of sulfuric acid to the benzil in methanol and dimethyl sulfite proceeds in high yield (92).

Hydrolysis of dialkyl sulfates under acidic and alkaline conditions, which is followed by the use of $^{18}OH_2$, proceeds by attack at sulfur to give S—O cleavage (93). The rate of hydrolysis is generally faster for cyclic and aryl sulfites than for dialkyl sulfites (94). Activation parameters of hydrolysis are known for some sulfites, and the increased rate for ethylene sulfite results from a reduced entropy of activation which results from a rigid ring structure (95).

The elimination processes usually occur upon heating sulfites, and they yield olefins, ethers, or other oxygenated products. Pyrolysis of 2,3-butylene sulfite [*4440-90-8*] gives some 2,3-butene oxide, but the 1,2-glycol sulfites give ketones from the cis isomers and aldehydes from the trans isomers with associated rearrangement of the carbon skeleton (96–97). Sulfites of 1,3-glycols give mostly tars on pyrolysis, but pentaerythritol disulfite gives some oxetane (98).

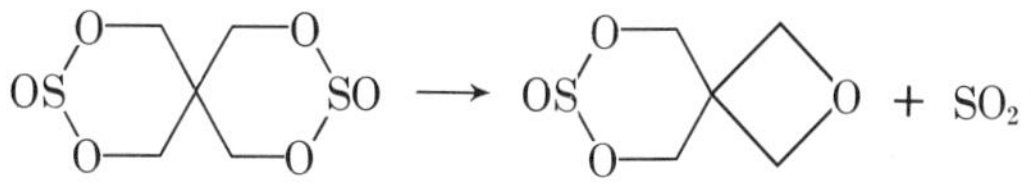

Unsymmetrical sulfites, when saturated with dry HCl at room temperature, give symmetrical sulfites, alkenes, SO_2, and some alkyl chlorides (99). A cyclic sulfite, eg, ethylene sulfite, on reaction with triphenylphosphine gives an olefin (100).

The transesterification reactions occur on mixing alcohol with sulfite and are analogous to hydrolysis. Usually the lower boiling alcohol is distilled, leaving the higher boiling sulfite as the product (101). Tertiary alcohol does not exchange. Ethylene glycol and dimethyl sulfite give ethylene sulfite (102). Exchange occurs at 120°C or is catalyzed by acid or base. The ester exchange can extend to the formation of dialkyl selenites from dialkyl sulfites, SeO_2, and traces of the alcohol (103).

Isomerization of dialkyl sulfite to alkyl alkanesulfonate requires a catalyst. Tertiary amines initiate the process, but alkanesulfonate anion is the chain carrier, and yields are as high as 70% (104).

Halosulfates and halosulfites are used primarily as intermediates in the preparation of symmetrical and unsymmetrical esters. Tertiary bases react with halosulfates even at dry-ice temperatures (−80°C) with decomposition (105).

$$ROSO_2Cl + \text{base} \rightarrow RCl + \text{base}.SO_3$$

This rules out the customary use of such bases as acid acceptors, unless the R—O bond has low susceptibility to nucleophilic attack, eg, when R is R_fCH_2— or Ar—. Hydrolysis rates of some $ROSO_2Cl$ compounds have been given and their mechanisms assessed (106). With a cyclic ether, eg, tetrahydrofuran or tetrahydropyran, the halosulfates yield oxonium-ion addition products (107). Stabilities have been examined and related to the activity in promoting polymerization of tetrahydrofuran (see Polyethers, tetrahydrofuran and oxetane polymers).

Manufacture

Monoester Hydrogen Sulfates. The hydrogen sulfates are prepared by the action of a sulfating agent on the corresponding alcohol or phenol. The following reagents are used for sulfation: sulfur trioxide, sulfuric acid, chlorosulfuric acid, sulfur trioxide–amine complexes, and sulfamic acid (see Sulfonation and sulfation). The latter two produce the amine salt of the sulfate ester directly. The reaction of sulfuric acid with olefins also gives hydrogen sulfates. These reactions are commercially important for the preparation of ethyl, 2-propyl, 2-butyl, and longer-chain sodium alkyl sulfates. The last are useful as detergents (see Alcohols, higher aliphatic, manufacture). With the exception of ethylene, the olefins give esters of secondary alcohols.

Methyl hydrogen sulfate is obtained in yields of 98–99% upon treating sulfur trioxide with methanol below 0°C (108). Ethyl hydrogen sulfate [*540-82-9*] is obtained by the sulfation of ethylene with sulfuric acid to the mixed sulfate, followed by the addition of water to the reaction mixture and hydrolysis of any diethyl sulfate formed. It is usually carried out with 96–98 wt % sulfuric acid at 70 and 80°C and 500–1500 kPa (5–15 atm). The kinetics of this reaction have been thoroughly investigated (109). However, at 150°C and higher pressures with only 70 wt % acid followed by addition of just enough water for hydrolysis, the otherwise necessary reconcentration of the spent sulfuric acid is eliminated (110). Because of the severe corrosion under these conditions, tantalum equipment is necessary. Ethyl hydrogen sulfate can also be prepared in 87% yield by heating sodium hydrogen sulfate with ethanol; a 74–86% yield is obtained by the reaction of ethanol with sulfur trioxide in liquid sulfur dioxide solution or by treating diethyl sulfate with ethanol and distilling the ether *in vacuo* (111–112).

$$(C_2H_5O)_2SO_2 + C_2H_5OH \rightarrow (C_2H_5)_2O + C_2H_5OSO_3H$$

Propylene and butylene require much milder conditions for their sulfation with sulfuric acid. Butylene is sulfated at 30–50°C and 300–600 kPa (ca 3–6 atm) with 30–60 wt % sulfuric acid, and propylene is sulfated at 10–30°C and 500 kPa (ca 5 atm) with 65–85 wt % sulfuric acid. The rate of sulfation of propylene increases sharply with increasing pressure (113). It can also be increased by the addition of kerosene, which raises the concentration of olefin in the liquid phase (114).

Detergents have been manufactured from long-chain alkenes and sulfuric acid, especially those obtained from shale oil or cracking of petroleum wax. These are sulfated with 90–98 wt % acid at 10–15°C for a 5-min contact time and at an acid–alkene molar ratio of 2:1 (115). Dialkyl sulfate initially forms when 96 wt % acid is added to 1-dodecene at 0°C, but it is subsequently converted to the hydrogen sulfate in 80% yield upon the further addition of sulfuric acid. The yield can be increased to 90% by using 98 wt % sulfuric acid and pentane as the solvent at −15°C (116).

Long-chain alcohols, such as are obtained by the hydrogenation of coconut oil, polymerization of ethylene, or the oxo process (qv), are sulfated on a large scale with sulfur trioxide or chlorosulfuric acid to acid sulfates; the alkali salts are commercially important as surface-active agents (see Surfactants). Poly(vinyl alcohol) can be sulfated in pyridine with chlorosulfuric acid to the hydrogen sulfate (117).

Some substituted alkyl hydrogen sulfates are readily prepared. For example, 2-chloroethyl hydrogen sulfate [*36168-93-1*] is obtained by treating ethylene chlorohydrin with sulfuric acid or amidosulfuric acid. Heating hydroxy sulfates of amino alcohols produces the corresponding sulfuric monoester (118).

Since phenolic compounds are easily sulfonated, their sulfation must be accomplished with milder sulfating agents, eg, complexes of sulfur trioxide or chlorosulfonic acid with trimethylamine, dimethylformamide, pyridine, or dimethylaniline, in anhydrous or aqueous medium below 100°C (119–122).

Salts of alkyl hydrogen sulfates that are free of sulfuric acid can be prepared from crude hydrogen sulfates containing $ROSO_3H$ and sulfuric acid in a two-step reaction. First, the crude product reacts in aqueous solution with the hydroxide or carbonate of a metal whose sulfate is insoluble in water. After filtration of the insoluble metal sulfate, the necessary amount of a water-soluble metal sulfate is added to the filtrate, which precipitates the insoluble metal sulfate, leaving the desired salt of the alkyl hydrogen sulfate in solution.

Diorgano Sulfates. Dialkyl sulfates up to octadecyl can be made from the alcohols by a general method involving the following reactions (123):

$$ROH + SO_2Cl_2 \rightarrow ROSO_2Cl + HCl$$

$$2\ ROH + SOCl_2 \rightarrow (RO)_2SO + 2\ HCl$$

$$ROSO_2Cl + (RO)_2SO \rightarrow (RO)_2SO_2 + RCl + SO_2$$

Another method involves the reaction of dialkyl sulfites with sulfuryl chloride (124):

$$2\ (RO)_2SO + SO_2Cl_2 \rightarrow (RO)_2SO_2 + 2\ ROSOCl$$

Mixed esters are synthesized by the reaction of one alkyl chlorosulfate with a different sodium alkoxide or a different dialkyl sulfite as follows (125):

$$ROSO_2Cl + NaOR' \rightarrow ROSO_2OR' + NaCl$$

$$ROSO_2Cl + (R'O)_2SO \rightarrow ROSO_2OR' + R'Cl + SO_2$$

The only commercially important dialkyl sulfates are dimethyl sulfate and diethyl sulfate. Estimated worldwide production in 1980 for dimethyl sulfate was 60,000 metric tons per year. Dimethyl sulfate was initially made by vacuum pyrolysis of methyl hydrogen sulfate:

$$2\ CH_3OSO_3H \rightarrow (CH_3O)_2SO_2 + H_2SO_4$$

Later it was synthesized in a batch process from dimethyl ether and sulfur trioxide (126). In 1947, this combination was adapted for continuous operation (127). Gaseous dimethyl ether was bubbled at 15.4 kg/h into the bottom of a tower 20 cm in diameter and 365 cm high and filled with the reaction product dimethyl sulfate. Liquid sulfur trioxide was introduced at 26.5 kg/h at the top of the tower. The mildly exothermic reaction was controlled at 45–47°C, and the reaction product (96–97 wt % dimethyl sulfate, sulfuric acid, and methyl hydrogen sulfate) was continuously withdrawn and purified by vacuum distillation over sodium sulfate. The yield was almost quantitative, and the product was a clear, colorless, mobile liquid. A modified process is described in ref. 128. Properties are listed in Table 3.

Diethyl sulfate can be prepared by a variety of methods. When ethyl hydrogen sulfate is heated with sodium chloride to 80°C, hydrogen chloride is liberated. The resulting reaction mixture is then distilled at 1.33–2.00 kPa (10–15 mm Hg) at a maximum kettle temperature of 190°C to give diethyl sulfate in 90% yield (129).

$$2\ C_2H_5OSO_3H + NaCl \rightarrow C_2H_5OSO_3H + C_2H_5OSO_3Na + HCl \rightarrow (C_2H_5O)_2SO_2 + NaHSO_4$$

Table 3. Properties of a Commercial Dimethyl Sulfate

Property	Typical value
acidity as H_2SO_4, %[a]	0.11
bp (at 101.3 kPa = 1 atm), °C	
at 95%	190.4
at 97%	192.0
at 1–97%	5.0 range
color[b]	passes 0.0001 *N* iodine, max
sp gr $^{15.6}_{15.6}$	1.334

[a] Determined by nonaqueous titration.
[b] This test is a visual comparison of the color of dimethyl sulfate with that of a 0.0001 *N* iodine solution. Commercial dimethyl sulfate should be lighter in color.

Passing a stream of nitrogen at 95–100°C through a reaction mixture of ethyl ether and 30 wt % oleum prepared at 15°C results in the entrainment of diethyl sulfate. Continuous operation provides a >50% yield (130). The most economical process for the manufacture of diethyl sulfate starts with ethylene and 96 wt % sulfuric acid heated at 60°C. The resulting mixture of 43 wt % diethyl sulfate, 45 wt % ethyl hydrogen sulfate, and 12 wt % sulfuric acid is heated with anhydrous sodium sulfate under vacuum, and diethyl sulfate is obtained in 86% yield; the commercial product is >99% pure (131). In a solvent-extraction route, dilution of the ethylene–sulfuric acid concentrate with water and extraction gives a 35% yield (132).

Lower dialkyl sulfates were made from the alcohols in earlier work; the reaction mass at ca 100°C was stripped by a recirculated inert-gas stream, and the product was recovered by passage through a partial condenser. Yields of 90% for diethyl sulfate and 85% for dimethyl sulfate were reported (133).

In the reaction of ethylene with sulfuric acid, several side reactions can lead to yield losses. These involve oxidation, hydrolysis–dehydration, and polymerization, especially at sulfuric acid concentrations >98 wt %; the sulfur trioxide can oxidize by cyclic addition processes (134).

$C_2H_4 + 2\ SO_3 \longrightarrow$ 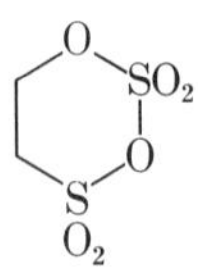

[503-41-3]

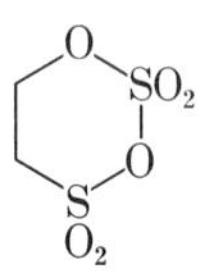 $+ H_2O \longrightarrow HO_3SOCH_2CH_2SO_3H$

[461-42-7]

$HO_3SOCH_2CH_2SO_3H + H_2O \rightarrow H_2SO_4 + HOCH_2CH_2SO_3H$

[107-36-8]

$HOCH_2CH_2SO_3H \rightarrow H_2O + CH_2{=}CHSO_3H$ + polymer

[1184-84-5] *[26101-52-0]*

Dimethyl and diethyl sulfate are available in a variety of containers from 0.5-kg glass bottles to tank cars. The price of dimethyl and diethyl sulfate in tank-car quantities in Dec. 1981 was \$0.93/kg and \$1.26/kg, respectively.

Diethyl sulfate is somewhat less toxic than dimethyl sulfate and is considered noncorrosive, but dimethyl sulfate is classified as a corrosive liquid and ICC regulations must be observed. Mild steel to 306 stainless steel is used for large-volume storage. Dimethyl sulfate is manufactured by E. I. du Pont de Nemours & Co., Inc., and diethyl sulfate by Union Carbide Corp.

Cyclic sulfates can be prepared by a variety of methods. Ethylene sulfate is obtained in low yield from ethylene oxide and sulfur trioxide (135). Methylene sulfate is produced from formaldehyde and sulfur trioxide as follows (136):

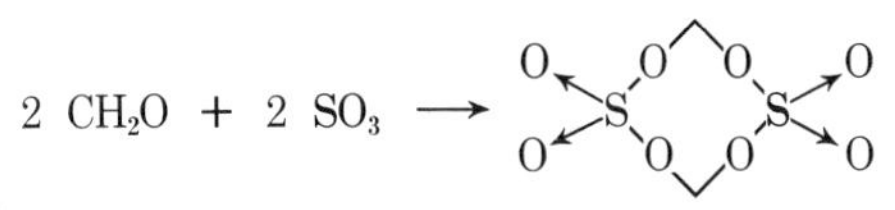

Oxidation of cyclic sulfites with permanganate in acetic acid solution gives cyclic sulfates (137). Heating monohydroxyalkyl hydrogen sulfates with thionyl chloride causes ring closure, as shown below (138):

$$HOCH_2CH_2OSO_3H + SOCl_2 \rightarrow \underline{OCH_2CH_2OSO_2}$$

Acidolysis of cyclic sulfites with sulfuric acid and ester interchange with dimethyl sulfate produce cyclic sulfates (84,139).

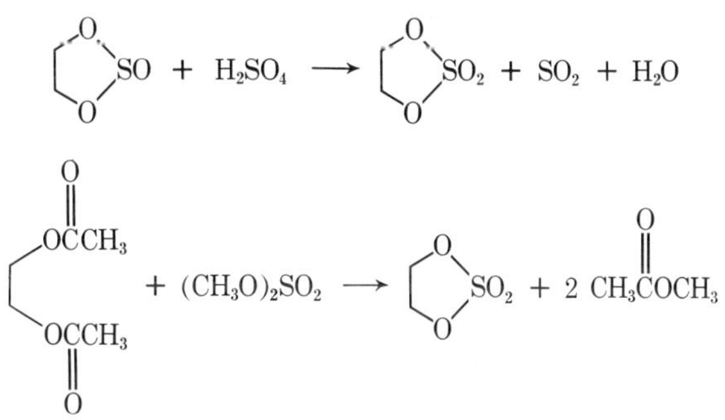

Diorgano Sulfites. Symmetrical or mixed dialkyl sulfites are prepared by the stepwise reaction of thionyl chloride either with two molecules of an alcohol or with stoichiometric quantities of two alcohols in pyridine (140).

$$ROH + SOCl_2 + C_5H_5N \rightarrow ROSOCl + C_5H_5N.HCl$$

$$R'OH + ROSOCl + C_6H_5N \rightarrow R'OSOOR + C_5H_5N.HCl$$

An alternative route is the reaction of iodine or bromine in pyridine with liquid sulfur dioxide at 20°C, which gives good-to-high yields of sulfites (141). The longer chain di-*n*-alkyl sulfites, with a total carbon count of at least eight, give adducts with urea and are suitable for purification (142) (see Clathration).

Heating the adduct of ethylene oxide and sulfur dioxide with primary alcohols in the presence of alkali hydrides or a transition-metal halide yields dialkyl sulfites

(143). Another method for the preparation of methyl alkyl sulfites consists of the reaction of diazomethane with alcoholic solutions of sulfur dioxide (144).

$$ROH + CH_2N_2 + SO_2 \rightarrow ROSOOCH_3 + N_2$$

Cyclic sulfites can be prepared from the glycols and thionyl chloride in the presence of pyridine; this route is analogous to the preparation of dialkyl sulfites (145). Cyclic sulfites are also obtained when the polymerization product of alkylene oxides with sulfur dioxide is decomposed by heating (146). Ester exchange proceeds smoothly with dimethyl sulfite and glycerol or its chlorohydrins to yield the cyclic sulfites (147). Glycerol gives poor results with thionyl chloride.

Aromatic polysulfites can be produced if bisphenols, eg, 4,4′-(1-methylethylidene)bisphenol, are heated with diphenyl sulfite in the presence of lithium hydride (148).

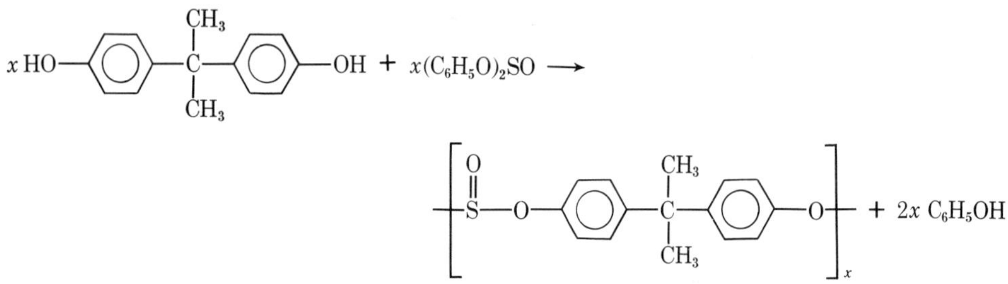

Halosulfates and Halosulfites. A general method for the preparation of alkyl halosulfates and halosulfites is the treatment of the alcohol with sulfuryl or thionyl chloride at low temperatures while passing an inert gas through the mixture to remove hydrogen chloride (149).

$$ROH + SO_2Cl_2 \rightarrow ROSO_2Cl + HCl$$

$$ROH + SOCl_2 \rightarrow ROSOCl + HCl$$

This method is also used with alcohols of the structure $Cl(CH_2)_nOH$ (150). Haloalkyl chlorosulfates are likewise obtained from the reaction of halogenated alkanes with sulfur trioxide or from the chlorination of cyclic sulfites (151–152).

$$ClCH_2CH_2Cl + SO_3 \rightarrow ClCH_2CH_2OSO_2Cl$$

[*13891-58-2*]

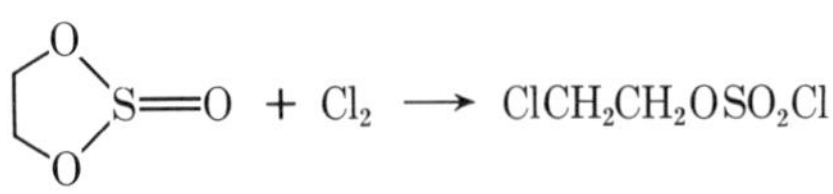

Chlorosilanes form chlorosulfate esters when treated with sulfur trioxide or chlorosulfuric acid (153).

$$(CH_3)_2SiCl_2 \xrightarrow{SO_3} (CH_3)_2Si(Cl)OSO_2Cl \xrightarrow{SO_3} (CH_3)_2Si(OSO_2Cl)_2$$

[*18090-25-0*] [*18090-26-1*]

Another approach to halosulfates is based on the addition of chlorosulfuric or fluorosulfuric acid to alkenes in nonpolar solvents (154).

Health and Safety Factors

The most commonly used dialkyl sulfate is dimethyl sulfate. This is also the most hazardous in liquid and vapor forms. The hazard arises from its toxicity, high reactivity, and to some extent, combustibility. Dimethyl sulfate is corrosive and poisonous, and its effects may be either acute or chronic. Because it has an analgesic effect on many body tissues, even severe exposures may not be immediately painful. Dimethyl sulfate is particularly dangerous to the eyes and respiratory system. It causes severe burns, but symptoms may be delayed. Exposed workers must be immediately and properly treated to avoid either permanent injury to eyes and lungs or even death. Skin burns can also be severe. Ingestion causes convulsions and paralysis, with later damage to the kidneys, liver, and heart (155–157). Genetic effects have been extensively reviewed, and its mutagenicity is correlated with carcinogenicity (158–159).

According to the U.S. Department of Labor (OSHA), exposure to dimethyl sulfate shall not exceed an eight-hour time-weighted average of 1 ppm in air (155). Since both liquid and vapor can penetrate the skin and mucous membranes, control of vapor inhalation alone may not be sufficient to prevent absorption of an excessive dose. Dimethyl sulfate is listed as an industrial substance with suspected carcinogenic potential in man (155). Thus, the ACGIH recommends a time-weighted average threshold limit value of 0.1 ppm, based on tests with laboratory animals.

Dimethyl sulfate lacks warning properties. It looks like water when spilled and has no distinctive odor; these characteristics mandate careful handling, but its high boiling point and careful attention to procedure permit safe usage. CAUTION: *No work should be undertaken with this or diethyl sulfate until the worker has studied and understood the precautions, procedures, and exposure effects given in the material safety data sheets, several government publications, and the product bulletins for the materials* (156–157,160–163). Protective equipment is essential. Clothing and gloves should be made of butyl rubber. All operations should take place in a hood or in an area with good ventilation. Concentrations in air of 10 ppm or higher are immediately dangerous to life and health and self-contained breathing apparatus must be used (163). Exposure to 97 ppm for 10 min can be fatal. In the laboratory, only mechanical pipetting is safe. Protective sleeves, aprons, and face shields are used. If the liquid should contact the skin, the exposed areas should be flushed immediately with water and then soaped and rinsed thoroughly. The victim should receive medical attention. Treatment for other exposures is described in refs. 156 and 157.

Because dimethyl sulfate looks like water, operations are preferably not performed when water is present (eg, wet floors or rain). Any spills or leaks should not be left unattended; they should be contained, and runoff to sewers should be avoided. Minor spills should be flooded with water to dilute and hydrolyze the dimethyl sulfate. The area should then be covered with a dilute (2–5 wt %) caustic solution or a dilute (2–5 wt %) ammonia solution, or soda ash may be sprinkled over the neat liquid and the mix wetted with a gentle spray of water. The neutralizing agent should remain on the affected area for 24 h and then should be washed away. Only personnel wearing protective equipment should perform these operations. The product bulletins should be consulted for procedures to be followed for more severe spills. Concentrated ammonia should not be used with neat dimethyl sulfate, since explosions have resulted after their contact (164).

In the laboratory, excess reagent in a product should be destroyed before workup.

Addition of diluted aqueous ammonia is the most effective practice, if ammonia is otherwise acceptable. Combustibility is a minor problem. The open-cup flash point of 116°C for dimethyl sulfate is well above normal handling temperatures. Flammable, toxic vapors are given off at elevated temperatures.

The acute toxicities of various sulfates have been reported (162). Generally, dimethyl sulfate (LD_{50} 440 mg/kg in rats) is more toxic than diethyl sulfate (LD_{50} 880–1412 mg/kg in rats), which is more toxic than dibutyl sulfate (lowest observed toxic dose, 9500 mg/kg in rats). Ethylene sulfate is more toxic than dimethyl sulfate (51).

The monoalkyl derivatives in salt form appear to have low toxicity. The monomethyl sulfate sodium salt has an approximate oral lethal dose greater than 5000 mg/kg of body weight for rats (165). Monododecyl sulfate sodium salt is widely marketed as a detergent and shampoo ingredient (oral LD_{50} 1268 mg/kg for rats) (162). Both dimethyl sulfate and monomethyl sulfate occur in the environment in coal fly-ash and in airborne particulate matter (166).

Uses

The sulfuric acid esters as compared to the sulfurous esters are the most widely used. In nature they appear as solubilizing groups in detoxification-excretory mechanisms and as sulfated carbohydrate groups in modified proteins. The significant uses are alkylation, formation of long-chain alcohol monosulfates as surfactants, and formation of intermediates in preparation of some lower alcohols. Alkylation (qv) involves primarily dimethyl and diethyl sulfates in the preparation of a wide variety of intermediates and products, especially in the fields of dyes, agricultural chemicals, drugs, and other specialities. In particular, dimethyl sulfate is a powerful reagent yielding quaternary salts in the form of the methosulfates. The use of dimethyl sulfate instead of methyl iodide can be useful when alkylations are carried out in dimethylformide (167). Manufacture of the long-chain alcohol sulfates consumes ca 270×10^6 kg/yr of alcohols (168). Some quaternary ammonium salts are also surfactants and fabric softeners (169) (see Quaternary ammonium compounds).

Minor uses include those as dyes, intermediates, stabilizers, and some specialty polymers. It is not always clear which uses are commercial and which are only alleged. One type of fiber-reactive dye has a pendant alcohol sulfate group. This group reacts with cotton in an alkaline pad both to produce an ether–cellulose bond and thereby a washfast dye, eg, Remazol dyes (Hoechst). Methyl hydrogen sulfate was once used as a solvent in the bromination of indigo (170).

Ethylenesulfonate salt formation and ethyleneimine (aziridine) formation are the most significant uses of the esters as intermediates. The conversion of 2-aminoethanol by the alcohol sulfate to ethyleneimine was carried out for many years (see Imines, cyclic). It is used less because of its toxicity. The ethylenesulfonate is prepared by treating ethyl hydrogen sulfate with SO_3 at 50°C to give ethionic acid [*461-42-7*], which with caustic converts to sodium ethylenesulfonate in 80% yield (134). The product vinyl monomer in salt, ester, or amide form is copolymerized with other monomers to give modified polymers, which exhibit water solubility or basic dye acceptance. The monomer also sulfoethylates cellulose (171). Dimethyl sulfate is also used with boron compounds to stabilize liquid sulfur trioxide (172). Modified polymers, mostly in the form of polyquaternary salts, have some use as coagulants for slimes, suspensions, and emulsions and as antistatic agents (qv) (173) (see Flocculating agents).

Proposed uses of the esters are numerous, but their applications in the modification of cotton, in catalysis, and as solvents are most prominent. The introduction of the aminoethyl group into cotton by the reaction of aminoethyl sulfate on alkali-swollen cotton is well-known (174). The basic group provides a dye site for acid dyes. Preparing methylated and ethylated celluloses with dialkyl sulfates reportedly results in desirable property improvements and some selectivity of reaction; however, such use has been questioned (175–176) (see Cellulose derivatives). Catalyst activity has been reported for ionic, organometallic, and free-radical routes. Acidic activity of monoalkyl sulfates is used to promote hydrolysis of poly(vinyl acetate) and to quench the base-catalyzed polymerization of epoxides (177–178). Some pyrolysis reactions are catalyzed by the presence of dimethyl sulfate, eg, in the formation of ketene from the thermal cracking of acetic acid or alkyl acetate and in the formation of 1,3-dihydroxypropene ether and acrolein acetals from 1,1,3-trihydroxypropane ethers at 200–450°C (179–180). The alkylation of phenol with isobutylenes is promoted by catalytic amounts of dimethyl sulfate (181). Some care is needed in interpreting the catalytic effect of dialkyl sulfates. Many reactions in which activity exists are probably susceptible to acid catalysis. The sulfate can cause the formation of acids either by hydrolysis or by pyrolysis. For example, dimethyl sulfate is a curing agent for furfuryl alcohol condensates, because the neutral ester is a source of sulfuric acid when heated to 180°C (182). Cationic activity in catalysis is shown in the promotion of tetrahydrofuran polymerization by ethyl chlorosulfate and in the initiation of polymerization of several heterocycles (107,183) (see Polyethers, tetrahydrofuran and oxetane polymers).

Organometallic usage is shown in the preparation of titanium- or vanadium-containing catalysts for the polymerization of styrene or butadiene by the reaction of dimethyl sulfate with the metal chloride (184). Free-radical activity is proposed for the quaternary product from dimethylaniline and dimethyl sulfate and for the product from 1,1,4,4-tetramethyl-2-tetrazene and dimethyl sulfate (185–186).

Several solvent uses have been proposed. Dimethyl sulfate has been used as a solvent for the study of Lewis acid–aromatic hydrocarbon complexes (187). It also is effective as an extraction solvent to separate phosphorus halide–hydrocarbon mixtures and aromatic hydrocarbons from aliphatics, and it acts as an electrolyte in electroplating iron (188–191). The toxicity of dimethyl sulfate precludes its use as a general-purpose solvent.

A mixture of dimethyl sulfate with SO_3 is probably dimethyl pyrosulfate [*10506-59-9*], $CH_3OSO_2OSO_2OCH_3$, and, with chlorobenzene, it yields the 4,4′-dichlorodiphenylsulfone (192). Trivalent rare earths can be separated by a slow release of acid into a solution of rare earth chelated with an ethylenediaminetetraacetic acid agent and iodate anion. As dimethyl sulfate slowly hydrolyzes and pH decreases, each metal is released from the chelate in turn and precipitates as the iodate, resulting in improved separations (193).

A number of mixed alkyl and aryl sulfites have been patented as insecticides and biocides (194) (see Insect control technology). Other proposed uses include a polysulfite $\leftarrow\!OSOOCH_2CH_2\!\rightarrow_x$ [*26101-52-0*] as a sensitizer in photographic emulsion, diethyl sulfite for the removal of catalyst fragments in polypropylene, and ethylene sulfite as an accelerator for aminoplastic molding compositions (195–197) (see Photography;

Amino resins). A mixed, cyclic sulfite-carbonate ester of pentaerythritol has been polymerized to high molecular weight solids, which can be used in preparing laminates (198). Use of dimethyl sulfate as an electrolyte in high energy-density batteries has also been described (199) (see Batteries).

BIBLIOGRAPHY

"Sulfuric and Sulfurous Esters" in *ECT* 1st ed., Vol. 13, pp. 506–513, by James M. Straley, Eastman Kodak Co.; "Sulfuric and Sulfurous Esters" in *ECT* 2nd ed., Vol. 19, pp. 483–498, by Julius Fuchs, E. I. du Pont de Nemours & Co., Inc.

1. E. W. Maurer, A. J. Stirtan, and J. K. Weil, *J. Am. Oil Chem. Soc.* **37,** 34 (1960); U.S. Pat. 3,133,946 (May 19, 1964), E. W. Maurer, A. J. Stirtan, and J. K. Weil (to U.S. Dept. of Agriculture).
2. *Dimethyl Sulfate, Properties, Uses, Storage and Handling,* bulletin, E. I. du Pont de Nemours & Co., Inc., Jan. 1981.
3. Y. Shuto and S. Higashimori, *Kogyo Kagaku Zasshi* **67,** 371 (1964).
4. P. B. D. de la Mare and co-workers, *J. Chem. Soc.,* 1813 (1956).
5. H. Mackle and W. V. Steele, *Trans. Faraday Soc.* **65,** 2053 (1969).
6. E. S. Lewis, S. Kukes, and C. D. Slater, *J. Am. Chem. Soc.* **102,** 303, 1619 (1980).
7. E. S. Lewis and S. Vanderpool, *J. Am. Chem. Soc.* **99,** 1946 (1977).
8. L. J. Simon and M. Frejaqueous, *Compt. Rend.* **176,** 900 (1923).
9. H. F. Lewis and co-workers, *Ind. Eng. Chem.* **22,** 34 (1930).
10. E. Y. Wolford, *Ind. Eng. Chem.* **22,** 397 (1930).
11. G. H. Green and J. Kenyon, *J. Chem. Soc.,* 1589 (1950).
12. R. S. Mathur and R. H. Common, *Steroids* **10,** 547 (1967); J. Lille and co-workers, *Tr. Nauch Isslied. Inst. Slantsev (USSR)* (18), 113 (1969); *Chem. Abstr.* **73,** 87567u (1970); O. P. Vig, *J. Indian Chem. Soc.* **52,** 442 (1975).
13. M. Pailer and P. Bergthaller, *Monatsh. Chem.* **99,** 103 (1968).
14. C. M. Suter and H. L. Hanson, *J. Am. Chem. Soc.* **54,** 4400 (1932).
15. L. Orthner and G. Freyss, *Lieb. Ann.* **484,** 146 (1930).
16. D. R. Benedict, T. A. Bianch, and L. A. Cate, *Synthesis,* 428 (1979).
17. Czech. Pat. 160,548 (Oct. 15, 1976), O. Leminger; *Chem. Abstr.* **85,** 62800Z (1976).
18. L. J. Simon, *Compt. Rend.* **176,** 583 (1923).
19. F. H. Stodda, *J. Org. Chem.* **29,** 2490 (1964); A. Werner and W. Seybold, *Chem. Ber.* **37,** 3658 (1904); U.S. Pat. 1,924,615 (Aug. 29, 1933), S. A. Merley (to Doherty Research Co.); N. C. Jameison and D. F. Loncrini, *Chem. Ind.,* 522 (1979).
20. M. H. Richard, *Ann. Chim. Phys.* **21,** 336 (1910).
21. B. K. Zeindov and I. A. Panteeva, *Vses. Soveshch. Sin. Zhirozamen, Poverkhnostnoaktiv. Veschestuam Moyushch. Sredstvam, 3rd, Shebekino,* 225 (1965); *Chem. Abstr.* **66,** 30233h (1967). *Maslozhir. Prom.* **33,** 40 (1967); *Chem. Abstr.* **67,** 111401 (1967).
22. Hung. Teljes 11221 (Mar. 28, 1977), I. Weisz and co-workers; *Chem. Abstr.* **88,** 104702p (1978).
23. A. Etienne, G. Lonchambon, and R. Garreau, *Bull. Soc. Chim. France,* 483 (1977).
24. A. Werner, *Lieb. Ann.* **321,** 269 (1902).
25. V. M. Reddy and K. K. Reddy, *Indian J. Chem.* **17B,** 353 (1979); I. Y. Shirobokov, *Zh. Org. Khim.* **16,** 788 (1980); A. McKillop and R. J. Kobylecki, *J. Org. Chem.* **39,** 2710 (1974).
26. U.S. Pat. 3,941,824 (Mar. 2, 1976), J. J. Fuchs (to E. I. du Pont de Nemours & Co., Inc.).
27. Ger. Pat. 2,828,259 (Jan. 10, 1980), G. Giesselmann, K. Guenther, and W. Fuenten (to Deutsche Gold und Silber-Scheideanstalt); Jpn. Pat. 72-37,615 (Sept. 22, 1972), H. Kadowaki, M. Kametani, and T. Nakamura (to Nitto Chem. Ind. Co.).
28. M. H. Slotta and H. L. Gerhart, *Chem. Res.* **58B,** 1320 (1925).
29. P. S. Clezy, *Tetrahedron Lett.,* 741 (1966).
30. J. U. Nef, *Lieb. Ann.* **309,** 187 (1899).
31. E. M. Arnett and V. M. De Palma, *J. Am. Chem. Soc.* **99,** 5828 (1977); G. Bram, F. Guibe, and P. Sarthov, *Tetrahedron Lett.,* 4903 (1972); Y. Hara and M. Matsuda, *Bull. Chem. Soc. Jpn.* **49,** 1126 (1976); A. Brandstrom, *Acta Chem. Scand.* **30B,** 203 (1976).
32. G. Bram, T. Eillebeen-Kahn, and N. Geraghty, *Synth. Commun.* **10,** 279 (1980).
33. K. K. Anderson and S. W. Fenton, *J. Org. Chem.* **29,** 3270 (1960).

34. D. Lipkin, E. F. Jones, and F. Galiano, *Am. Chem. Soc. Div. Pet. Chem. Repr.* **4**(4), B14 (1959).
35. A. K. Kranzfelder and F. J. Sowa, *J. Am. Chem. Soc.* **59,** 1490 (1937).
36. C. M. Suter and H. L. Gerhart, *J. Am. Chem. Soc.* **57,** 107 (1935).
37. H. Bredereck, G. Simchen, and H. Huffman, *Chem. Ber.* **106,** 3725 (1973).
38. Ger. Pat. 2,215,954 (Nov. 23, 1972), H. Gross, L. Haase, and I. Keitel (to Deutsch. Akad. der Wissenschaften zu Berlin).
39. W. Kantlehner and B. Funke, *Chem. Ber.* **104,** 3711 (1971).
40. H. Bredereck, F. Effenberger, and H. P. Beyerlin, *Chem. Ber.* **97,** 3081 (1964); H. Bredereck, F. Effenberger, and G. Simchen, *Chem. Ber.* **98,** 1078 (1965).
41. W. Kantlehner and H. D. Gutbrud, *Lieb. Ann.* **522,** 1362 (1979); *Synthesis*, 975 (1979).
42. S. Havassian and E. Moralioglu, *Can. J. Chem.* **50,** 233 (1972).
43. U.S. Pat. 4,025,634 (May 24, 1977), G. B. Payne (to Shell Oil Co.).
44. U.S. Pat. 3,917,755 (Nov. 4, 1975), H. G. Franke (to Chevron Research).
45. M. Vlassa, M. Kezdi, and I. Goia, *Synthesis*, 850 (1980).
46. U.S. Pat. 3,896,230 (July 22, 1975), H. L. Klopping (to E. I. du Pont de Nemours & Co., Inc.).
47. R. F. Weinland and K. Schmid, *Chem. Ber.* **38,** 2327 (1905).
48. C. Gopinathan and J. Gupta, *Indian J. Chem.* **3,** 231 (1965).
49. *Ibid.*, **11,** 948 (1973).
50. U.S. Pat. 3,711,524 (Jan. 16, 1973), J. R. Leebrick (Lucille Coon, part interest).
51. G. W. Fischer, R. Jentzsch, and V. Kasanzewa, *J. Prakt. Chem.* **317,** 943 (1975).
52. P. C. Ray and P. Neigi, *J. Chem. Soc.* **89,** 1900 (1906).
53. J. Wade, *J. Chem. Soc.* **81,** 1598 (1902).
54. H. LeB. Gray and G. O. Gutschunst, *J. Am. Chem. Soc.* **42,** 856 (1920).
55. Fr. Pat. 718,762 (1931), (to I. G. Farben).
56. *Tables of Chemical Kinetics, Homogeneous Reactions, Class 212.460*, Circular 510, Suppl. 1, National Bureau of Standards, Washington, D.C., 1956.
57. H. Wenker, *J. Am. Chem. Soc.* **57,** 2328 (1975); U.S. Pat. 3,326,897 (June 20, 1967), R. G. Dunning (to Chemirad Corp.).
58. A. A. Michurin and E. N. Zil'berman, *Zh. Org. Khim.* **1,** 711 (1965).
59. R. Mohr and H. Hertel, *Chem. Ber.* **96,** 114 (1963).
60. R. Kreman, *Monatsh. Chem.* **28,** 13 (1907).
61. H. F. Lewis, O. Mason, and R. Morgan, *Ind. Eng. Chem.* **16,** 811 (1924).
62. G. H. Green and J. Kenyon, *J. Chem. Soc.*, 1389 (1950).
63. G. M. Calhoun and A. L. Burnell, Jr., *J. Am. Chem. Soc.* **77,** 6441 (1955).
64. B. D. Batts, *J. Chem. Soc. B*, 551 (1966).
65. R. E. Roberston and S. E. Sugamon, *Can. J. Chem.* **44,** 1728 (1966).
66. V. A. Kolesnikov and co-workers, *Kinet. Katal.* **18,** 875 (1977); *Chem. Abstr.* **87,** 167141t (1977).
67. J. Kaniewski and co-workers, *Pol. J. Chem.* **52,** 587 (1978); *Chem. Abstr.* **89,** 23395j (1978).
68. J. De Stefano, DuPont, Wilmington, Del., private communication, 1977.
69. Tn. N. Motorova and co-workers, *Khim. Sredstva Zashchity Rast., M.*, 146 (1980); *Chem. Abstr.* **95,** 186298t (1981).
70. C. H. Bamford and C. F. H. Tipper, *Chemical Kinetics*, Vol. 10, Elsevier North-Holland, Inc., New York, 1972, p. 39.
71. E. T. Kaiser, M. Panar, and F. H. Westheimer, *J. Am. Chem. Soc.* **85,** 602 (1963).
72. E. T. Kaiser, I. R. Katz, and T. F. Wulfers, *J. Am. Chem. Soc.* **87,** 3781 (1965).
73. V. A. Kolesnikov, R. V. Efremov, and S. M. Danov, *Kinet. Catal.* **20,** 671 (1979).
74. I. Lauder, I. R. Wilson, and B. Zerner, *Aust. J. Chem.* **14,** 41 (1961).
75. R. L. Burwell, Jr., *J. Am. Chem. Soc.* **74,** 1462 (1952).
76. S. Burstein and S. Liebeman, *J. Am. Chem. Soc.* **80,** 5235 (1958).
77. B. D. Batts, *J. Chem. Soc. B*, 547 (1966).
78. J. L. Kice and J. M. Anderson, *J. Am. Chem. Soc.* **88,** 5242 (1966).
79. H. L. Kane and A. Lowry, *J. Am. Chem. Soc.* **58,** 2605 (1936); J. Epelberg and A. Lowy, *J. Am. Chem. Soc.* **63,** 101 (1941).
80. U.S. Pat. 3,665,050 (May 23, 1972), L. J. McGovern, C. L. West, and O. Webb (to Stratford Energy Corp.).
81. B. C. Lawes, DuPont, private communication, 1981; C. D. Hurd, *The Pyrolysis of Carbon Compounds*, Chemical Catalog Co., New York, 1929, p. 144.
82. J. U. Nef, *Lieb. Ann.* **318,** 43 (1901).

83. E. E. Gilbert, *Sulfonation and Related Reactions*, Interscience Publishers, a division of John Wiley & Sons, Inc., New York, 1965, pp. 23–24.
84. J. Brunken, G. Gluckner, and E. J. Poppe, *Veroeff. Wiss. Photolab. AGFA (Wolfan)* **9,** 61 (1961); *Chem. Abstr.* **57,** 16027e (1962).
85. M. Volmer, *Bull. Soc. Chim. France* **27,** 681 (1920).
86. L. A. Harmon and R. J. Legon, *J. Chem. Soc. Perkin I*, 2675 (1979).
87. A. R. Vasudeva Murthy, *Proc. Indian Acad. Sci.* **47,** 11 (1953).
88. U.S. Pat. 3,168,546 (Feb. 2, 1969), A. Ballauf and co-workers (to Farbenfabriken Bayer).
89. W. W. Carlson and L. H. Cretcher, *J. Am. Chem. Soc.* **69,** 1952 (1947).
90. U.S. Pat. 2,917,502 (Dec. 15, 1959), S. R. Schwyzer (to Ciba Pharmaceutical Company).
91. J. M. Theabald, M. W. Williams, and G. T. Young, *J. Chem. Soc.*, 1927 (1963).
92. Ger. Pat. 2,365,497 (Apr. 24, 1975), J. Bruenisholz and A. Kirchmayr (to CIBA GEIGY).
93. D. Kerr and I. Lauder, *Aust. J. Chem.* **15,** 561 (1962).
94. R. E. Davis, *J. Am. Chem. Soc.* **84,** 599 (1962); P. B. D. De La Mare, J. G. Tillett, and H. F. Van Woerden, *J. Chem. Soc.*, 4888 (1962).
95. P. A. Bristow and T. G. Tillett, *Chem. Commun.*, 1010 (1967).
96. L. Denivelle, *Compt. Rend.* **208,** 1024 (1939).
97. C. C. Price and G. Berti, *J. Am. Chem. Soc.* **76,** 1211 (1954).
98. S. Wauzonek and J. T. Loft, *J. Org. Chem.* **24,** 641 (1959).
99. C. Liberman, *Compt. Rend.* **250,** 2223 (1966); **252,** 1483 (1961).
100. M. A. Shaw and co-workers, *J. Chem. Soc. C*, 2795 (1968).
101. R. C. Mehrotra and S. N. Mathus, *J. Indian Chem. Soc.* **44,** 651 (1967).
102. P. A. Bristow and J. G. Tillett, *Tetrahedron Lett.*, 901 (1967).
103. R. C. Paul, K. K. Bhasin, and R. D. Sharma, *Indian J. Chem.* **13,** 292 (1975).
104. A. J. W. Brooks and R. K. Robertson, *J. Chem. Soc. B*, 1161 (1971).
105. J. Charalambous, M. J. Frazer, and W. Gerrard, *J. Chem. Soc.*, 5480 (1964).
106. E. Buncel, J. P. Millington, and J. F. Wiltshire, *Can. J. Chem.* **55,** 1401 (1977); E. C. F. Ko and R. E. Robertson, *Can. J. Chem.* **50,** 434 (1972).
107. S. Kobayuski and co-workers, *Bull. Chem. Soc. Jpn.* **46,** 3214 (1973).
108. K. A. J. Chamberlain and co-workers, *BIOS Final Report No. 1482*, 1946, p. 6.
109. H. G. Harris and D. M. Himmelblau, *J. Phys. Chem.* **67,** 802 (1963); H. G. Harris, Jr., and D. M. Himmelblau, *J. Chem. Eng. Data* **9**(1), 61 (1964).
110. Ger. Pat. 1,035,632 (Aug. 7, 1958), H. G. Van Raay (to Farbwerke Hoechst A.G.).
111. U.S. Pat. 3,024,263 (Mar. 6, 1962), M. Letherman (to U.S. Dept. of the Navy).
112. D. S. Breslow, R. R. Hough, and J. T. Fairclough, *J. Am. Chem. Soc.* **76,** 5361 (1954).
113. M. S. Nemtsov, *Khim. Promst. Kiev*, 633 (1960); *Chem. Abstr.* **55,** 15070 (1961).
114. G. R. Schultze, J. Moos, and K. D. Ledwoch, *Erdoel Kohle* **11,** 12 (1958); *Chem. Abstr.* **52,** 7677 (1958).
115. G. D. Inskeep and A. Mussard, *Ind. Eng. Chem.* **47,** 2 (1955); D. Steward and E. McNeill, *Chem. Age (London)* **63,** 48 (1950); S. F. Birch, *J. Inst. Pet.* **38,** 69 (1952); K. L. Butcher and G. M. Nickson, *Trans. Faraday Soc.* **54,** 1195 (1958).
116. E. Clippinger, *Ind. Eng. Chem. Prod. Res. Dev.* **3,** 3 (1964).
117. Ger. Pat. 1,086,434 (Aug. 4, 1960), J. Szita (to Farbenfabriken Bayer A.G.).
118. E. Cherbuliez and co-workers, *Helv. Chim. Acta* **48,** 830 (1965); U.S. Pat. 3,194,826 (July 13, 1965), A. Goldstein and F. A. Nowak, Jr. (to Chemirad Corp.).
119. J. Parrod and L. Robert, *Compt. Rend.* **230,** 450 (1950).
120. A. Butenandt and co-workers, *Z. Physiol. Chem.* **321,** 258 (1960).
121. G. N. Burkhart and A. Lapworth, *J. Chem. Soc.*, 684 (1926).
122. E. J. Fendler and J. H. Fendler, *J. Org. Chem.* **33,** 3852 (1968).
123. C. Barkenbus and J. J. Owen, *J. Am. Chem. Soc.* **56,** 1204 (1934); R. Levaillant, *Compt. Rend.* **200,** 940 (1935).
124. Ger. (East) Pat. 32,780 (Jan. 5, 1965), W. Lugenheim, E. Carstens, and H. Fuerst.
125. G. A. Sokol'skii, *Zh. Org. Khim.* **2**(6), 951 (1966).
126. J. Avery and co-workers, *BIOS Final Report No. 986*, pp. 175, 227, 1946–1948.
127. *U.S. Dept. of Commerce Office Technical Service PB Report 75*, Washington, D.C., 1947, p. 463.
128. Czech. Pat. 157,946, M. Sadilek and M. Soulek; *Chem. Abstr.* **84,** 30420b (1976).
129. Brit. Pat. 774,384 (May 8, 1957), E. Roberts (to Ministry of Supply).
130. U.S. Pat. 2,816,126 (Dec. 10, 1957), R. Evans and L. T. Hogarth (to United States of America).

131. Fr. Pat. 1,006,211 (Apr. 21, 1952), (to Société Anon. des Manufactures des Glaces et Produits Chimiques de Saint-Gobain, Chauny et Cirey).
132. Rus. Pat. 139,664 (Nov. 14, 1960), B. K. Zeinalov and R. R. Paschajev.
133. Brit. Pat. 581,115 (Oct. 1, 1946), (to Nicholas Proprietary Ltd.).
134. D. S. Breslow, R. R. Hough, and J. T. Fairclough, *J. Am. Chem. Soc.* **76,** 5361 (1954); D. S. Breslow and R. R. Hough, *J. Am. Chem. Soc.* **79,** 5000 (1957); A. A. Goldberg, *J. Chem. Soc.*, 716 (1942).
135. U.S. Pat. 3,100,780 (Aug. 13, 1963), D. L. Klass (to Pure Oil Co.).
136. U.S. Pat. 2,805,228 (Sept. 3, 1963), J. L. Smith (to Eastman Kodak Co.).
137. J. Lichtenberger and J. Hincky, *Bull. Soc. Chim. France*, 1495 (1961); W. Baker and B. F. Burrows, *J. Chem. Soc.*, 2257 (1961); Brit. Pat. 944,406 (Dec. 11, 1963), L. F. Wiggins, C. C. Beard, and J. W. James (to Aspro-Nicholas Ltd.).
138. Ger. Pat. 1,049,870 (Feb. 5, 1959), J. Brunken and E. J. Poppe (to VEB Filmfabrik Agfa Wolfen).
139. Ger. (East) Pat. 18,485 (Apr. 4, 1960), J. Brunken and E. J. Poppe (to VEB Filmfabrik Agfa Wolfen).
140. W. Gerrard, *J. Chem. Soc.*, 99 (1939); L. Denivelle, *Compt. Rend.* **208,** 1024 (1939); Ger. Pat. 1,133,367 (July 19, 1962), H. F. Wilson (to Rohm & Haas Co.).
141. S. Hasegawa, M. Nojima, and N. Takura, *J. Chem. Soc. Perkin I*, 108 (1976).
142. Ger. Pat. 1,207,380 (Dec. 23, 1965), K. Stuerzer (to Th. Goldschmidt A.G.).
143. Ger. Pat. 1,200,807 (Sept. 16, 1966), K. Stuerzer (to Th. Goldschmidt A.G.).; Ger. Pat. 1,212,072 (Mar. 10, 1966), K. Stuerzer (to Th. Goldschmidt A.G.).
144. H. Hesse and S. Majumdar, *Chem. Ber.* **93,** 1129 (1960).
145. S. Hauptmann and K. Dietrich, *J. Prakt. Chem.* **19,** 174 (1963).
146. U.S. Pat. 3,022,315 (Feb. 20, 1962), W. A. Rogers, Jr., J. E. Woehst, and R. M. Smith (to Dow Chem. Co.); Ger. Pat. 1,217,970 (June 2, 1966), H. Distler and G. Dittus (to Badische Anilin- und Soda-Fabrik A.G.); Ger. Pat. 1,188,610 (Mar. 11, 1965), H. Hoefermann and H. Springmann (to Chemische Werke Huels A.G.).
147. H. F. van Woerden, C. F. van Valkenburg, and G. M. van Woerkam, *Rec. Trav. Chim. Pays-Bas* **86,** 601 (1967).
148. Ger. Pat. 1,213,612 (Mar. 31, 1966), K. Stuerzer (to Th. Goldschmidt A.G.).
149. R. Lavaillant, *Ann. Chim.* **6,** 459 (1936); W. Voss and E. Blanke, *Lieb. Ann.* **485,** 258 (1930).
150. Fr. Pat. 965,161 (Sept. 5, 1950), (to Société Anon. d'Innovations Chimiques dite: Sinnova ou Sadic).
151. F. G. Bordwell and G. W. Crosby, *J. Am. Chem. Soc.* **78,** 5367 (1956); U.S. Pat. 2,860,123 (Nov. 11, 1958), R. V. Jones (to Phillips Petroleum Co.).
152. U.S. Pat. 2,684,977 (July 27, 1954), M. J. Viard (to Société Anon. des Manufactures de Glaces et Produits Chimiques de Saint-Gobain, Chauny et Cirey).
153. M. Schmidt and H. Schmidbauer, *Chem. Ber.* **95,** 47 (1962).
154. U.S. Pat. 1,510,425 (1925), W. Traube; W. Traube and R. Justh, *Brennst. Chem.* **4,** 150 (1923).
155. G. D. Clayton and F. E. Clayton, eds., *Patty's Industrial Hygiene and Toxicology*, 3rd ed., Vols. 2A and 2B, Wiley-Interscience, New York, 1981, pp. 2094, 2892, 2928, and 2930.
156. *Dimethyl Sulfate-Properties, Uses, Storage and Handling*, bulletin, DuPont, Wilmington, Del., Jan. 1981.
157. *Dimethyl Sulfate-Product Safety Bulletin*, bulletin, DuPont, Wilmington, Del., Mar. 1980.
158. G. R. Hoffmann, *Mutat. Res.* **75,** 63 (1980).
159. R. F. Newbold and co-workers, *Nature (London)* **283,** 596 (1980).
160. *Dimethyl Sulfate, Material Safety Data Sheet*, DuPont, Wilmington, Del., 1980.
161. *Diethyl Sulfate, Material Safety Data Sheet*, Union Carbide, New York, 1976.
162. *Registry of Toxic Effects of Chemical Substances*, U.S. Department of Health and Human Services, 1979.
163. *NIOSH/OSHA—Pocket Guide to Chemical Hazards*, U.S. Department of Health and Human Services, U.S. Department of Labor, Sept. 1980.
164. H. Lindler, *Angew. Chem. Int. Ed.* **2,** 262 (1963).
165. Haskell Laboratory, E. I. du Pont de Nemours & Co., Inc., private communication, 1975.
166. D. J. Eatough and co-workers, *Environ. Sci. Technol.* **15,** 1502 (1981).
167. R. Kuhn and H. T. Trischmann, *Chem. Ber.* **96,** 284 (1963).
168. *Chem. Times Trends*, 29 (Oct. 1980).
169. *McCutcheon's Detergents and Emulsifiers*, North American ed., MC Publishing Co., N.J., 1979, annual.

170. K. A. J. Chamberlain, A. M. North, and D. C. Wilson, *BIOS Final Report No. 1482.6*, 1946.
171. U.S. Pat. 2,580,351 and U.S. Pat. 2,580,352 (Dec. 25, 1951), V. R. Grassie (to Hercules Powder Co.).
172. U.S. Pat. 3,160,474 (Dec. 8, 1964), W. G. Schnoor and A. W. Yodis (to Allied Chemical Corp.).
173. U.S. Pat. 4,137,164 (Jan. 30, 1979), A. T. Coscia and M. N. D. O'Connor (to American Cyanamid Co.); U.S. Pat. 3,993,615 (Nov. 23, 1976), S. B. Markofsky (to W. R. Grace & Co.); U.S. Pat. 2,723,256 (Nov. 8, 1955), M. Hayek (to E. I. du Pont de Nemours & Co., Inc.).
174. U. A. Reeves and J. D. Guthrie, *Text. Res. J.* **23,** 522 (1953).
175. U.S. Pat. 3,660,012 (May 2, 1972), J. G. Roberts (to USDA); F. S. H. Head and G. E. Hadfield, *J. Polym. Sci.* **7A,** 2517 (1969); Can. Pat. 968,915 (June 10, 1975), L. R. B. Hervey, J. R. Starr, and R. P. Tschirch (to Cotton, Inc.).
176. N. G. Gaylord and co-workers, *Encyclopedia of Polymer Science and Technology*, Vol. 3, Interscience Publishers, New York, 1965, p. 427.
177. Ger. Pat. 1,915,222 (Oct. 2, 1969), H. Nakamura and A. Saito (to Japan Synthetic Chemical Industries Co., Ltd.).
178. Ger. Pat. 2,656,727 (June 29, 1978), R. Gehm, K. H. Baumann, and W. Harder (to BASF A.G.).
179. K. K. Georgieff, *Can. J. Chem.* **30,** 332 (1952).
180. Brit. Pat. 695,789 (Aug. 19, 1953), R. H. Hall and E. S. Stern (to Distillers Co., Ltd.).
181. U.S. Pat. 3,116,336 (Dec. 31, 1963), J. L. Van Winkle (to Shell Oil Co.).
182. Ger. Pat. 911,659 (May 17, 1954), A. Schmidt (to Chemische Werke Huels A.G.).
183. T. Fujisawa, Y. Yokota, and T. Mukaiyama, *Bull. Chem. Soc. Jpn.* **40,** 147 (1967); Belg. Pat. 666,828 (Nov. 16, 1965), M. H. Litt, A. J. Levy, and T. G. Bassiri (to Allied Chemical Corp.); Ger. Pat. 1,206,585 (Dec. 9, 1965), W. Seeliger (to Chemische Werke Huels A.G.).
184. U.S. Pat. 3,389,128 (June 18, 1968), J. W. Bayer and W. C. Grinonneau (to Owens-Illinois, Inc.).
185. T. Otso and Y. Takemura, *Bull. Chem. Soc. Jpn.* **43,** 567 (1970).
186. K. Sugiyama, *Kinhi Daigako Kogahuba Kenkyu Hokoku* **13,** 27 (1979); *Chem. Abstr.* **93,** 168706d (1980).
187. W. I. Aalbersberg and co-workers, *J. Chem. Soc.*, 3055 (1959).
188. U.S. Pat. 2,801,957 (Aug. 6, 1957), G. C. Ray (to Phillips Petroleum Co.).
189. U.S. Pat. 2,776,327 (Jan. 1, 1957), A. Schneider (to Sun Oil Co.).
190. P. Pascal and M. L. Quinet, *Ann. Chim. Anal. Chim. Appl.* **23,** 5 (1941).
191. V. Y. Rybkovskii and co-workers, *Tr. Kishinev S-kh. Inst.* **123,** 95 (1974); *Chem. Abstr.* **88,** 80954u (1978).
192. U.S. Pat. 2,971,985 (Feb. 14, 1961), R. Joly, R. Bucourt, and C. Fabignon (to UCLAF, Paris); U.S. Pat. 3,355,497 (Nov. 28, 1967), E. Budnick (to Plains Chemical Co.).
193. F. H. Firsching and co-workers, *J. Inorg. Nucl. Chem.* **36,** 1655 (1974).
194. Jpn. Pat. 63-8245 (June 7, 1963), M. Nagasawa and F. Yamamato (to Ihara Agricultural Chemical Co.); U.S. Pat. 3,179,682 (Apr. 20, 1965), A. Covey, A. E. Smith, and W. L. Hubbard (to United States Rubber Co.).
195. Belg. Pat. 615,408 (Apr. 13, 1962), P. P. Chiesa, J. R. Dann, and J. W. Gates, Jr. (to Kodak Soc. Anon.).
196. Brit. Pat. 903,077 (Aug. 9, 1962), (to "Montecatini" Societa Generale per l'Industria Mineraria e Chimica).
197. Brit. Pat. 866,440 (Apr. 26, 1961), C. P. Vale, S. Gutter, and W. Wilson (to British Industrial Plastics Ltd.).
198. U.S. Pat. 3,251,857 (May 1, 1966), F. Hostettler and E. F. Cox (to Union Carbide Corporation).
199. N. P. Yao, E. D'Orsay, and D. N. Bennion, *J. Electrochem. Soc.* **115,** 999 (1968); S. C. S. Wang and D. N. Bennion, *J. Electrochem. Soc.* **128,** 1827 (1981).

General References

E. E. Gilbert, *Sulfonation and Related Reaction*, Wiley-Interscience, New York, 1965, Chapt. 6.
C. M. Suter, *Organic Chemistry of Sulfur*, John Wiley & Sons, Inc., New York, 1944, Chapt. 1.

W. B. McCormack
B. C. Lawes
E. I. du Pont de Nemours & Co., Inc.

SULFURIZATION AND SULFURCHLORINATION

Sulfur reacts with unsaturated compounds forming products that are usually dark and often viscous cross-linked mixtures of mono-, di-, and polysulfides. Sulfur monochloride, S_2Cl_2, also reacts with unsaturated materials forming products that are cross-linked by sulfur but that also contain chlorine (see Properties). For many years, sulfurized (sulfurated) and sulfurchlorinated, unsaturated fatty oils have been added to mineral oils to make lubricants that provide antiwear, low friction, increased load-carrying ability, and improved oxidation resistance (see Lubrication and lubricants).

The first hundred years of industrial production and applications of sulfurized and sulfurchlorinated fatty oils, starting in 1849, is reviewed in reference 1. Most of the early products were resinous materials used in protective coatings. As early as 1918, a patent was issued for the preparation of sulfurized fat and its use as a cutting-oil compound for metals (2) (see also Fats and fatty oils; Vegetable oils).

Patents issued in 1939 describe an important discovery that led to the standard of quality of lubricant additives for ca 30 yr (3). Sulfurized sperm-whale oil was much more soluble in paraffinic oils and more stable than the sulfurized lard oil previously used in gear lubricants. Under the heavy loads in worm-driven, hypoid-gear automotive differentials, extreme-pressure (EP) agents with good stability are needed. The performance advantages of sulfurized sperm oil also held true in many other lubricant uses. Because of its cost effectiveness, it also competed with sulfurized lard oils in metalworking applications (4). Banning the importation of sperm oil since 1971 and the change to using more highly paraffinic mineral oils generated increased interest in development of new sulfurized and sulfurchlorinated products (5).

Sulfurized triglycerides, mainly from hog fat, have provided excellent performance when used in lightly refined, highly aromatic mineral oils. These oils are good solvents and keep the sulfurized triglycerides in solution without problems of solidification and precipitation. The possibility of the aromatic oils being toxic to humans, however, caused a shift to the use of highly refined oils with low aromatic content, which decreases the solubility of sulfurized triglycerides in the oil; thus, modifications were necessary to provide adequate solubility. One approach, which has led to large volumes of commercial products that impart EP properties to more highly paraffinic lubricants, was sulfurization of triglycerides in the presence of esters made from alkyl alcohols and unsaturated fatty acids (6–7). Good pour depressancy is also provided by this type of product. Another important approach involves the cosulfurization of triglycerides with various olefins or interchange with sulfurized olefins (8–13). Both approaches result in improved solubility of sulfurized triglycerides by taking advantage of sulfur linkages to attach olefins or unsaturated esters, ie, triglyceride-S_x-olefin or triglyceride-S_x-ester to make adequately soluble sulfurized fatty oils. Sulfurchlorination of triglyceride–olefin mixes has also been described (11).

Sulfurized and sulfurchlorinated sperm-oil-based extreme-pressure additives have been compared with replacement products (14). Although their composition has not been specifically defined, the replacement products have been described as combinations of materials selected from lard oils, methyl esters and higher esters, and other functionally substituted compounds. In laboratory tests of oils, including Falex and

Table 1. Chemical and Physical Properties of Sulfurized and Sulfurchlorinated Fats and Oils[a]

Product designation	Sulfur, wt %	Active sulfur, wt %	Copper strip tarnish test, 10 wt % in oil	Viscosity mm^2/s (= cSt) 40°C	100°C	Density (at 25°C), g/cm^3	Total acid number, mg KOH/g	Pour point, °C
Base 10-L[b]	10.0	0	1b	1100	80	0.98	25	18
Base 14-L[b]	13.5	3	3b	1700	100	0.99	26	27
Base A-92[b]	16.5	6	4b	625	50	1.00	18	21
Base L-66[b,c]	6.0	0	1b	1000	100	0.99	5	10
Base 10-SE[d]	8.5	0	1b	20	4	0.94	5	10
Base 12-SE[d]	13.0	2	4a	30	6	0.96	4	10
Base 18-SE[d]	16.5	6	4a	40	9	0.98	7	10
Base 99[c,d]	8.0	0	2b	75		1.02	8	10
Sul-perm 10[e]	10.0	0	1b	400	45	0.98	8	16
Sul-perm 12[e]	12.0	1	1b	500	50	0.98	12	18
Sul-perm 18[e]	16.5	6	4a	600	55	1.02	12	16
Sul-perm 88[c,e]	6.5	0	2b	4000	330	0.99	7	18
recommended ASTM procedure	D 129	D 1662	D 130	D 445	D 445	D 1298	D 974	D 97

[a] The products listed in the table and their properties have been submitted by Keil Division of Ferro Corp.
[b] Based on lard.
[c] Sulfurchlorinated fats and oils have chlorine %s similar to their sulfur %s.
[d] Based on methyl ester of lard.
[e] Based on mixture of animal and vegetable oils plus synthetic fatty esters.

Timken tests, and in evaluations of compounded leaded lubricants and greases, data were obtained that showed adequate performance of replacement products. Commercial acceptance for these replacement products has been reported (15). Certain alkyl esters were also reported to be good sperm-oil replacements (16).

Other commercial products have been made that are based on sulfurization or sulfurchlorination of unsaturated hydrocarbons alone. Products of this type usually have very high sulfur contents, very good solubility in all types of mineral oils, and good antioxidant and antiwear properties (see Antioxidants and antiozonants). Large quantities are used for cutting-oil applications, where the low friction type of lubricant action contributed by triglyceride or ester is normally not needed. Sulfurized terpene products are also used commercially (see Terpenoids). Patents on sulfurization of olefins have included the use of cyclic olefins, eg, dicyclopentadiene (17–18). Other patents describe the superatmospheric-pressure sulfurization of a wide variety of olefins, including branched olefins, eg, isobutylene and diisobutylene, and straight-chain alpha olefins, eg, 1-octene, in the presence of a mixture of hydrogen sulfide, sulfur, and a catalyst, eg, tributylamine or other amines (19–20). A feature which gives the products of this process special application is that they are much lighter in color than the usual sulfurized products.

Products have long been made by the simple addition of S_2Cl_2 to olefins. The presence of chlorine gives extra EP activity, and products of this type, when the potential corrosiveness of chlorine is not a problem, are generally used in the more severe, lower speed metal-cutting operations.

Properties

Properties of typical sulfurized fatty oils are listed in Table 1. Sulfur reacts easily with most unsaturated materials in the liquid phase simply by mixing somewhat above its melting point so that a solution–dispersion is present. There have been conflicting and inconclusive reports as to the mechanisms of sulfurization, ie, whether they are predominately free radical or polar. The products of reaction of sulfur with olefins have been researched and discussed in detail in terms of examples with various monoolefins and diolefins (21). Sulfurization has also been reviewed with emphasis upon relatively simple olefins; product distribution has been reported with proposed mechanisms related to the structure of olefins (22). The reaction of sulfur with 1-octene at 140°C gives mainly polysulfides, which are a mixture of the following:

$C_6H_{13}CH(CH_3)S_xCH_2CH{=}CHC_5H_{11}$

(1)

average $x = 6.7$
25%

$C_8H_{17}SC_8H_{17}$ and/or (1) ($x = 1$)

(2)

15%

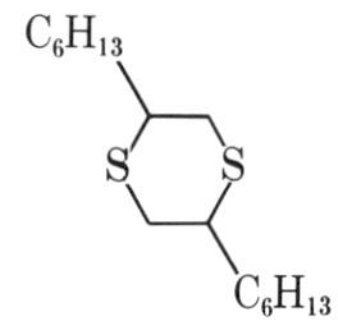

+ $C_8H_{17}{-}S{-}C_8H_{16}{-}S{-}C_8H_{17}$

(3)

15%

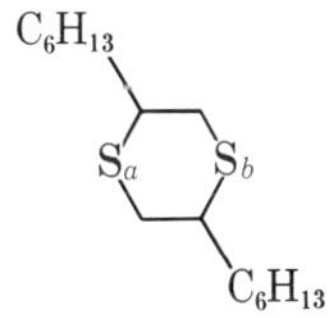

(4)

30%
average $a + b = 6.7$

$C_8H_{17}S_aC_8H_{16}S_bC_8H_{17}$

(5)

15%
average $a + b = 4.7$

Reaction of sulfur with branched and cyclic olefins gives higher amounts of polysulfide (1).

Since most commercial sulfurized products are made from partially unsaturated triglycerides, their chemistry is very complex because of the variety of carboxylic acid moieties present and the multiple choices of sites for reaction and cross-linking of sulfur. This complexity has made the analysis of such sulfurized products difficult. Extensive work was undertaken to determine the characteristics of sulfurization and sulfurchlorination of jojoba oil (23). Sulfurized jojoba oil, which performs very well as a lubricant additive replacement for sperm oil, is composed of monoesters. Although triglycerides are the main raw material used for sulfurized products, this jojoba work is important, not only for work on characterization of the structures of sulfurized and sulfurchlorinated fatty-acid-based material, but for a summary of previous mechanism–structure work on sulfurization and sulfurchlorination. Jojoba oil was treated with S_2Cl_2 in solvent in a manner so that up to ca 8 wt % sulfur in the product was obtained. The reaction took place by addition to the double bond, and it was concluded, based on nmr, that a cross-linked structure of the following type was obtained:

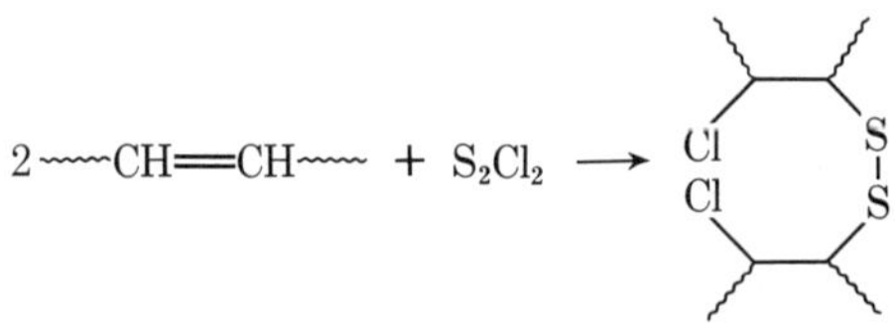

When jojoba oil was treated in solvent with sulfur so that up to ca 8 wt % sulfur was obtained, the nmr spectrum did not allow clear determination of the degree of bridging, but it appeared that the structure was

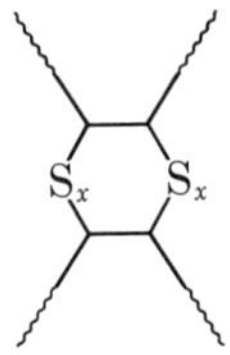

where $x = 1$ when relatively low amounts of sulfur were used. In sulfurization of fatty materials, however, there is almost always considerable evolution of hydrogen sulfide. This indicates hydrogen abstraction and the presence of more complicated mechanisms and products, including formation of a mixture of sulfides and polysulfides of the type described for simple olefins. The relative amount of polysulfides and the value of x depend upon reaction conditions and upon the relative quantity of sulfur charged. Lower quantities of sulfur and higher temperatures cause formation of higher amounts of monosulfide and disulfide products.

As a general rule, the maximum amount of sulfur that can react with an unsaturated material, so that the product does not contain active sulfur, corresponds to one atom of sulfur per double bond, eg, in the formation of the product with the empirical formula for dialkyl disulfide, RS_2R'. The total sulfur that can react, giving stable polysulfide active sulfur products without later precipitation of sulfur, is generally ca 2.5 atoms per double bond, eg, for RS_5R'.

Manufacture

Sulfurization. Manufacture of a sulfurized product is normally carried out in a mild steel, batch-reaction vessel equipped with an agitator. A typical method for sulfurization of triglycerides or of mixtures of triglyceride with other unsaturated materials involves heating to ca 140°C followed by addition of sulfur, which melts, partly dissolves, and starts to react (see Fig. 1). The mixture is heated further, but the reaction is exothermic. Therefore, heating is discontinued as the temperature reaches ca 150°C; the exotherm increases the temperature to ca 180°C. The temperature is usually maintained at 175–200°C for 1–4 h. Depending upon the material sulfurized,

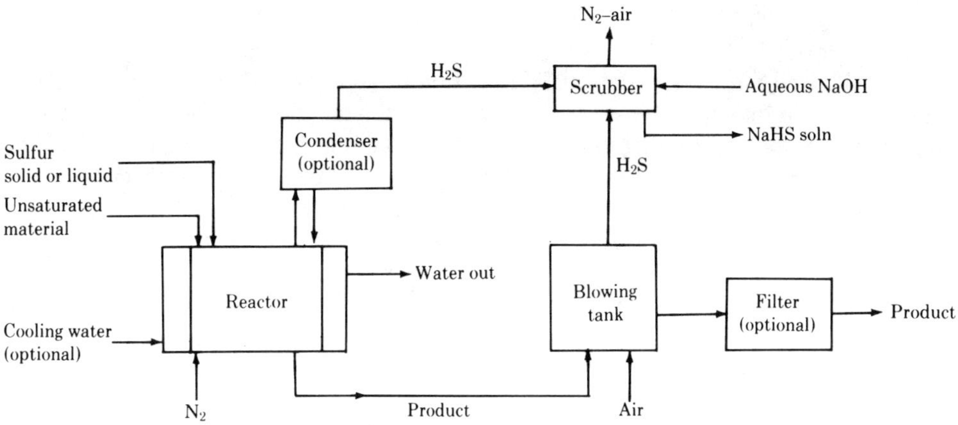

Figure 1. Flow chart for sulfurization.

catalysts, eg, trialkylamines, may be used to help bring about sulfurization at somewhat lower temperatures. Throughout the process, ca 5–12 wt % of the sulfur is lost as hydrogen sulfide, depending upon the material sulfurized and whether the product is to contain active sulfur or is to be an inactive sulfur product. The temperature of the material is allowed to drop to ca 170°C while being transferred with nitrogen purging to a second vessel equipped with a sparging system. The material is then air-blown to remove dissolved hydrogen sulfide and to convert any residual thiols to disulfides, giving a product with a bland odor. The temperature during the blowing period of ca 6 h is slowly lowered to ca 100°C. The viscosity of the product increases during air blowing. Throughout the entire process, hydrogen sulfide passes into a packed-tower caustic scrubber and is recovered as sodium bisulfide solution. Sulfurized products are stored and shipped in simple steel containers.

Sulfurchlorination. Manufacture of sulfurchlorinated products is carried out, preferably in a glass-lined reaction vessel equipped with a sparging system and cooling coils or jacket. The process is schematically similar to that of sulfurization except that a separate vessel for air blowing is normally not used. Dichloro disulfide reacts with most unsaturated materials rapidly and with strong exotherm at relatively low temperatures. In a typical method of production, S_2Cl_2 is added at a controlled rate to the substrate, which must be dry, and the temperature is held lower than ca 50°C during the first half of the addition. It is then added to completion at higher temperatures, ie, up to 80°C, depending upon the material. The temperature is then raised to ca 80°C, and air blowing is carried out for ca 6 h or until the product is bland. The off-gas includes HCl, H_2S, and residual S_2Cl_2 and is passed into a scrubber system containing caustic solution. After air blowing, a small amount of stabilizer, eg, epoxidized soybean oil, is usually added. Most sulfurchlorinated products are stored and shipped in ordinary steel containers, but if long-term storage is desired for relatively high sulfur and chlorine products, eg, those containing 12–18 wt % S and 12–18 wt % Cl, the use of polypropylene-lined drums is advisable.

Economic Aspects

The United States International Trade Commission's figure for the manufacture of sulfur compounds for U.S. sales in 1980 was 76,480 metric tons (24). These figures do not include phosphorodithioates or sulfurized or sulfurchlorinated phenols. The value was $103,205,000 (average unit value $1.35/kg). The companies associated with that report, except those who asked for confidentiality, are Thiokol, Carstab Div.; Chevron Chemical; Elco Corp., a subsidiary of Detrex Chemical Industries, Inc.; Ferro Corp., Keil Chemical Div.; Pennwalt Corp.; Quaker Chemical; Texaco Inc.; White and Bagley Co.; and Mobil Oil Corp. Some of the categories include aliphatic hydrocarbon sulfides, diisobutylene polysulfide, di-*tert*-nonyl polysulfide, sulfurized lard oil, and sulfurized sperm-oil substitutes (see Table 2). Sulfurchlorinated sperm oil was also listed, but is probably no longer available because of the Endangered Species Act and its subsequent effect on sperm-whale oil availability; however, sulfurchlorinated sperm-oil substitutes are available. Of these categories, only sulfurized lard oil was reported separately at 5933 metric tons.

Specifications and Standards

A key property to consider for sulfurized and sulfurchlorinated additives is sulfur activity (see Table 1). Although sulfur content is important for economic reasons, it is the activity of the sulfur which causes the additive to form successfully a metal sulfide sacrificial film under the existing conditions of temperature, pressure, and sliding speed.

Sulfur Activity. If the ratio of sulfur to sites of unsaturation is 1:1 or less, the sulfurized product is usually inactive. This is important because copper, brass, and bronze, which are widely used in bushings and gears, are corroded by active sulfur. ASTM D 130, which describes the method of measuring the corrosion of a copper strip in a 10 wt % concentration of the sulfur-containing additive in oil, is widely used to test for corrosion. After 3 h of immersion at 100°C, the stain on the copper strip is compared

Table 2. 1982 Prices and Manufacturers of Some Sulfurized Products[a]

Additive type	Typical 1982 price, $/kg	Manufacturer
sulfurized lard	0.86	Thiokol, Carstab Div.; Ferro Corp., Keil Chemical Div.; Quaker Chemical; White and Bagley Co.
sulfurized sperm-oil replacements	1.10	Thiokol, Carstab Div.; Elco Corp.; Ferro Corp., Keil Chemical Div.
di-*tert*-nonyl polysulfide (*tert*-nonyl sulfides)	2.10	Pennwalt Corp.
diisobutylene polysulfide (sulfurized 2,4,4-trimethylpentene)	1.66	Texaco Inc.
aliphatic hydrocarbon sulfides	1.40	Elco Corp.; Ferro Corp., Keil Chemical Div.
sulfurchlorinated fats	1.00	Thiokol Carstab Div.; Ferro Corp., Keil Chemical Div.

[a] Ref. 25.

to a set of ASTM standards and graded 1a or 1b for no stain or light tarnish; 2a, 2b, 2c, 2d, or 2e for various light stains or moderate tarnish; 3a or 3b for various heavy stains or dark tarnish; and 4a, 4b, or 4c for black stains or corrosion.

Most fatty oils with iodine values of 65–80, when sulfurized to 8–10 wt % sulfur, produce a copper strip of 1a or 1b. The 14 wt % sulfurized fats are typically rated 3a or 3b, and the 17 wt % sulfurized fats, 4a, 4b, or 4c. The most active sulfurized additives are 2–4 wt % sulfurized hydrocarbon oils which can stain copper at room temperature in a few minutes. Although high sulfur activity is detrimental in lubricants, it is essential in cutting oils where slow sliding speeds and high unit loadings are common.

Sulfurchlorinated fats are active at lower sulfur levels, with 6 wt % sulfur products giving 1b–2b copper strips and 10 wt % sulfur products giving 4+ copper strips. Direct sulfurization of olefins, eg, diisobutylene, yields products with ca 40 wt % sulfur. These are used in cutting oils and are highly active to copper strips. Sulfurchlorinated diisobutylene reacts with sodium polysulfide, thereby removing chlorine and adding more sulfur. The resulting product contains 40 wt % sulfur. The activity to copper is eliminated by the use of copper deactivators and the final product gives a copper strip of 1a or 1b. It is then suitable for use in lubricants that must contact copper-containing metals.

Specific Gravity. The specific gravity of sulfurized fats is generally 0.96–1.03 and is measured with a hydrometer. The very high sulfur products eg, 40 wt % sulfurized olefin, have specific gravities of 1.05–1.15.

Pour Point. Pour points are important if the additives are purchased in drum quantities during the winter months; also, by monitoring the pour points of fatty-based products, correlation with the starting fatty mixes can be made. High tallow contents cause high pour points (20–30°C), whereas the use of low titer lards result in lower values (0–20°C).

Acid Value. Acid values are useful for determining the type of lard used in the starting mix for sulfurized lard oils. A #1 lard oil has 13–16% free fatty acid (ffa), and extra winter-strained lard oil has 2–4% ffa. The acid value is determined by titration with a base to a color indicator or potentiometric end point.

Flash Point. Flash points of sulfurized synthetic sperm oils and lards are ca 220°C with fire points being 30°C higher. The flash and fire points of sulfurized methyl esters are ca 50°C lower, but they are not lower than the mineral oil that the additives are blended in. The most commonly used method for determining flash and fire points is the Cleveland open-cup tester.

Viscosity. Because of the importance of viscosity in lubrication, sulfurized additives used in this industry are closely monitored for this property so that the viscosity of the finished lubricant conforms to specification. Viscosity ranges that can be used for industrial lubricants have been established by ASTM (see Table 3). Viscosities are determined by ASTM D 445 in terms of kinematic viscosity determined by capillary means. The viscosity is reported as mm^2/s (= cSt) at 40°C. Because of past use of Saybolt Universal seconds (SUs), much data exists that was generated at 37.8°C either by capillary or by measuring flow through a Saybolt orifice. Viscosity data should be reported in mm^2/s at 40°C or 100°C to conform with D 445 and to be comparable to the viscosity ranges defined in D 2422.

Table 3. ISO Viscosity Grades[a]

Grade no.	Viscosity range, mm^2/s (= cSt)
2	1.98–2.42
3	2.88–3.52
5	4.14 5.06
7	6.12–7.48
10	9.00–11.0
15	13.5–16.5
22	19.8–24.2
32	28.8–35.2
46	41.4–50.6
68	61.2–74.8
100	90–110
150	135–165
220	198–242
320	288–352
460	414–506
680	612–748
1000	900–1100
1500	1350–1650

[a] ISO = International Standards Organization viscosity grade as defined by ASTM D 2422.

Health and Safety Factors

Sulfurized and sulfurchlorinated fatty oils and esters have had a long history of manufacture and usage without incident of hazard to workers. Of the newer type of products, sulfurized and sulfurchlorinated sperm-oil substitutes and olefins have been subjected to standard animal testing. These materials are classified as nontoxic by oral ingestion and are not primary skin irritants, corrosive materials, nor eye irritants. For example, the acute oral LD_{50} values in all cases are greater than 5.0 g/kg body weight (26).

Uses

Lubrication as related to sulfurized fats and oils is described in ref. 27. Hydrodynamic lubrication is characterized by the presence of a continuous fluid. Under conditions of higher loads, however, hydrodynamic lubrication fails and boundary lubrication must be present in order to prevent contact of metal surfaces with each other. Under conditions of boundary lubrication when the appropriate lubricant additives are present, an extreme-pressure (EP) lubrication effect takes place, with chemical reaction of additives occurring at the point of metal–metal contact. Sulfurized and sulfurchlorinated fatty materials participate in all of these types of lubrication, at first by contributing to a layer of adequate thickness by increasing the viscosity, then by forming a surface film by affinity of the polar part of the molecule for the metal, and then by the reaction of sulfur and chlorine under conditions of heat and pressure. Sulfurized and sulfurchlorinated olefins participate mainly by means of the EP mechanism.

The main uses of sulfurized and sulfurchlorinated products are in lubricating oils, cutting oils for metalworking, gear oils, and greases. The ability of sulfurized fats to reduce the coefficient of friction and act as an antiscuffing agent has encouraged their use in almost all areas of lubrication. In gear oils, they can be used at the 2–6% level with 1–3% lead naphthenate. However, recently this has become an unpopular formulation approach because of the toxicity of lead-containing compounds. Sulfurized or phosphosulfurized fats at the 0.1–0.5% level are used with sulfurized isobutylene and other phosphorus-containing additives to make EP gear compounds. A typical performance criterion is the ability of the gear oil to pass a 27-kg Timken OK load as part of the performance testing to qualify it for a U.S. Steel EP 224 gear-oil compound.

There has recently been a resurgence of interest in sulfurized products in automotive applications. In automotive lubricants, zinc dialkyl dithiophosphates (ZDPs) are the typical antiwear agents. These are made by the reaction of alcohols with P_2S_5 and subsequent formation of the zinc salt. Sulfurized fats are effective friction reducers and antiscuff agents when used with ZDPs in crankcase oils. They are also used in automatic-transmission fluids. A typical motor-oil formulation may contain 1.5 wt % ZDPs and 0.75 wt % of a sulfurized fat as well as additives that impart detergency, dispersancy, and viscosity modification. Although engine oils are given final evaluation by means of engine tests, because of cost, bench tests are used to screen choices for engine testing. A motor oil containing the preceding levels of ZDPs and sulfurized fat on a four-ball wear test at a 40-kg load yields a wear scar of 0.3–0.4 mm after 1 h. Typical test temperatures are ca 120°C. In engine testing, reduced coefficients of friction have resulted in greater efficiency of gasoline usage by ca 2–4% when 1–2 wt % of a sulfurized additive is present. Other benefits include improved resistance to oxidation and to consumption of oil. Recent patents describing benefits of sulfurized products in automotive oils include the use of sulfurized amides and esters of fatty acids with alkanolamines, a sulfurized alkylphenol–olefin reaction product, and improvement by addition of a cosulfurized triglyceride-ester mix (28–30). Up to ca 1% improvement in fuel economy could be obtained by use of 1 wt % sulfurized ester in multigrade gear oil (31).

In metalworking applications, both sulfurized and chlorosulfurized materials are used extensively. For tapping and threading mild steel, high sulfur-containing olefins and 17 wt % sulfurized fats are used. Alloy steels require cutting oils with sulfurchlorinated additives for most metalworking applications. A typical test criterion for a combination of active sulfurized and sulfurchlorinated additives is that the test oil perform favorably on a torque-tapping tester to an established standard. Test results are usually expressed as a percentage of the performance against the standard. Examples of cutting oils made from high sulfur and sulfurchlorinated materials are listed in Table 4.

Unique products for metalworking have been patented, including sulfurization of polyol esters, eg, sorbitan monooleate, in mixes with fats and alkyl esters and sulfurization of unsaturated fatty esters and olefins in the presence of mixed alkyl–alkenyl acid phosphate esters (32–33). Additives that provide improved lubricity of aqueous oil-well drilling fluids have also been patented (34) (see Petroleum, drilling fluids). These are prepared by sulfurization of polyethoxylated unsaturated fatty acids, followed by mixing with polyethylene glycols.

New products made by variations of the sulfurchlorinated route are used in

Table 4. Tapping Torque of Sulfurized and Sulfurchlorinated Cutting Oils[a]

Additive description	Sulfur in additive, wt %	Sulfur in test oil, wt %	Tapping efficiency, %
reference test[b]	38	1.9	100
sulfurized decene	32	1.6	95
di-*tert*-nonyl polysulfide	37	1.85	97
sulfurized diisobutylene (2,4,4-trimethyl-pentene)	40	2.0	95
gear-oil base[c]	40; <1 wt % Cl	2.0	95
sulfurchlorinated octene	17; 19 wt % Cl	0.85	98

[a] Values were determined on a Falex (Faville-Levally Corp., Aurora, Ill.) Model 8 torque-tapping tester. The test oils contained 5 wt % of the additive dissolved in an ISO VG 22 base oil and were used to tap 1117 steel.
[b] Keil's Base 380.
[c] Sulfurchlorinated diisobutylene reacted with sodium polysulfide.

metalworking and in gear oils. This route has been adapted by dehydrochlorination and other methods to the manufacture of high sulfur, low chlorine products from relatively low molecular weight olefins (35–38). The products are lubricant additives with good antiwear, extreme-pressure, antioxidant, and solubility properties but without the corrosion problems associated with the usual sulfurchlorinated olefins.

BIBLIOGRAPHY

"Sulfurization and Sulfurchlorination" in *ECT* 2nd ed., Vol. 19, pp. 498–506, by Raymond S. Dalter and Ingenuin Hechenbleikner, Carlisle Chemical Works, Inc.

1. G. Schiemann and F. Vohwinkel, *Farbe Lack* **64,** 485 (1958).
2. Brit. Pat. 129,132 (July 15, 1981), (to E. F. Houghton & Co.).
3. U.S. Pats. 2,179,060, 2,179,061, and 2,179,066 (Nov. 7, 1939), H. G. Smith (to Gulf Oil Corp.).
4. B. W. Hotten, *NLGI Spokesman* **37,** 174 (Aug. 1973).
5. *Fed. Regis.*, 18320 (Dec. 2, 1970).
6. U.S. Pat. 3,740,333 (June 19, 1973), R. B. Hutchison and K. P. Kammann (to Emery Industries, Inc.).
7. U.S. Pat. 3,850,825 (Nov. 26, 1974), P. Vienna and M. J. Den Herder (to Standard Oil Co., Ind.).
8. U.S. Pat. 3,455,896 (July 15, 1969), M. J. Den Herder and A. Borg (to Standard Oil Co., Ind.).
9. U.S. Pat. 3,825,495 (July 23, 1974), T. D. Newingham, A. D. Recchuite, J. P. Griffith III, and M. W. Haseltine, Jr. (to Suntech, Inc.).
10. U.S. Pat. 4,166,795 (Sept. 4, 1979), A. D. Recchuite (to Suntech, Inc.).
11. U.S. Pat. 4,166,796 and U.S. Pat. 4,166,797 (Sept. 4, 1979), A. D. Recchuite (to Suntech, Inc.).
12. U.S. Pat. 4,149,982 (Apr. 17, 1979), D. A. Lee and J. A. Boslett (to Elco Corp.).
13. U.S. Pat. 4,188,300 (Feb. 12, 1980), H. E. Sanson III and W. R. Hartman, Jr. (to Mayco Oil and Chemical Co.).
14. P. F. Thompson, *NLGI Spokesman* **36,** 208 (Sept. 1972).
15. *Ibid.*, **38,** 20 (Apr. 1974).
16. R. L. Peeler and L. M. Hartmann, *NLGI Spokesman* **37,** 17 (Apr. 1973).
17. U.S. Pat. 4,147,640 (Apr. 3, 1979), G. J. J. Jayne, D. R. Woods, and J. P. O'Brien (to Edwin Cooper and Co., Ltd.).
18. U.S. Pat. 4,188,297 (Feb. 12, 1980), G. J. J. Jayne and D. R. Woods (to Edwin Cooper and Co., Ltd.).
19. U.S. Pat. 4,119,549 (Oct. 10, 1978), K. E. Davis (to Lubrizol Corp.).
20. U.S. Pat. 4,119,550 (Oct. 10, 1978), K. E. Davis and T. H. Holden (to Lubrizol Corp.).
21. L. Bateman and C. G. Moore in N. Kharasch, ed., *Organic Sulfur Compounds*, Vol. 1, Pergamon Press, Inc., New York, 1961, Chapt. 20, pp. 210–228.

22. W. A. Pryor, *Mechanisms of Sulfur Reactions*, McGraw-Hill Book Company, Inc., New York, 1962, pp. 97–116.
23. J. Wisniak and H. Benajahu, *Ind. Eng. Chem., Prod. Res. Dev.* **14**(4), 247 (1975).
24. *Synthetic Organic Chemicals, United States Production and Sales, 1980*, USITC Publication 1183, U.S. Government Printing Office, Washington, D.C., 1981.
25. *U.S. International Trade Commission*, Washington, D.C., 1980.
26. Keil Chemical Division, Ferro Corporation, Hammond, Ind., unpublished data.
27. C. L. Hermann and J. J. McGlade, *J. Am. Oil Chem. Soc.* **51,** 88 (1974).
28. U.S. Pat. 4,201,684 (May 6, 1980), R. E. Malec (to Ethyl Corp.).
29. U.S. Pat. 4,228,022 (Oct. 14, 1980), W. Lowe and T. V. Liston (to Chevron Research Co.).
30. U.S. Pat. 4,282,106 (Aug. 4, 1981), L. Schaap, J. Udelhofen, and S. Harris (to Standard Oil Co., Ind.).
31. Y. Ikemoto, *SAE Tech. Paper Series, No. 811207*, Fuels and Lubricants Mtg., Tulsa, Okla., Oct. 1981.
32. Ger. Offen Pat. 2,842,138 (Apr. 10, 1980), I. Kreutzer, W. Schuele, W. Krueger, and O. Wilken (to Rhein-Chemie Rheinau G.m.b.H.).
33. Ger. Offen. Pat. 2,166,893 (Aug. 12, 1976), (to Rhein-Chemie Rheinau G.m.b.H.).
34. U.S. Pat. 4,172,300 (Oct. 30, 1979), T. O. Walker (to Texaco, Inc.).
35. U.S. Pat. 3,697,499 (Oct. 10, 1972), H. Myers (to Mobil Oil Corp.).
36. U.S. Pat. 4,132,659 (Jan. 2, 1979), B. W. Hotten (to Chevron Research Co.).
37. U.S. Pat. 4,204,969 (May 27, 1980), A. G. Papay and J. P. O'Brien (to Edwin Cooper, Inc.).
38. U.S. Pat. 4,194,980 (Mar. 25, 1980), and U.S. Pat. 4,240,958 (Dec. 23, 1980), M. Braid (to Mobil Oil Corp.).

General References

R. K. Springborn, ed., *Cutting and Grinding Fluids: Selection and Application*, American Society of Tool and Manufacturing Engineers, Dearborn, Mich., 1967.

J. P. Friedrich in E. H. Pryde, ed., *Fatty Acids*, American Oil Chemists Society, Champaign, Ill., 1979, Chapt. 30, pp. 591–607.

H. E. Kenney, E. T. Donahue, and I. Schmeltz, *Lubr. Eng.* **30,** 394 (1974).

E. W. Bell, L. E. Gast, and F. L. Thomas, *J. Am. Oil Chem. Soc.* **54,** 259 (1977).

E. W. Bell, J. C. Cowan, L. E. Gast, and R. E. Koos, *J. Am. Oil Chem. Soc.* **53,** 511 (1976).

H. J. Nieschlag and I. A. Wolff, *J. Am. Oil Chem. Soc.* **48,** 723 (1971).

J. J. O'Connor and J. Boyd, *Standard Handbook of Lubrication Engineering*, McGraw-Hill, Inc., New York, 1968.

U.S. Pat. 4,036,769 (July 19, 1977), R. L. Zipf (to Werner G. Smith, Inc.).

F. A. Faville and W. A. Faville, *Lubr. Eng.* **24,** 349 (1968).

A. L. Mills in G. D. Hobson, ed., *Modern Petroleum Technology*, 4th ed., Applied Sciences Publications Ltd., Barking, UK, 1973, pp. 526–572.

U.S. Pat. 3,514,480 (May 26, 1970), E. K. Fields (to Standard Oil Co., Ind.).

U.S. Pat. 4,161,475 (July 15, 1979), K. E. Davis (to Lubrizol Corp.).

R. W. Mould, H. B. Silver, and R. J. Syrett, *Wear* **19,** 67 (1972).

U.S. Pat. 3,775,322 (Nov. 27, 1973), B. W. Hotten (to Chevron Research Co.).

U.S. Pat. 4,012,368 (Mar. 15, 1977), B. W. Hotten (to Chevron Research Co.).

A. G. Papay and D. W. Dinsmore, *Lubr. Eng.* **32,** 233 (1976).

U.S. Pat. 4,228,021 (Oct. 14, 1980), A. L. P. Lenack (to Exxon Research and Engineering Co.).

H. Okabe and T. Kanno, *ASLE Trans.* **24,** 459 (1981).

M. Kagami and co-workers, *ASLE Trans.* **24,** 517 (1981).

H. Okabe, M. Masuko, and K. Sakurai, *ASLE Trans.* **24,** 467 (1981).

U.S. Pat. 4,207,195 (June 10, 1980), A. G. Horodysky (to Mobil Oil Corp.).

U.S. Pat. 2,646,349 (July 21, 1953), D. A. Wagner (to Sinclair Refining Co.).

U.S. Pat. 2,851,421 (Sept. 9, 1958), A. Manteuffel and W. D. Gilson (to Pure Oil Co.).

S. R. Sprague and R. G. Cunningham, *Ind. Eng. Chem.* **51,** 1047 (1959).

U.S. Pat. 2,993,856 (July 25, 1961), T. C. Heising and B. Corrigan (to Texaco Inc.).

F. T. Barcroft, *Wear* **3,** 440 (1960).

G. Schiemann, H. Duering, and H. Koerner, *Dtsch. Farben-Z.* **17,** 408 (1963).

Fred G. Rounds, *Am. Soc. Lubr. Eng. Trans.* **7,** 11 (1964).

U.S. Pat. 3,280,028 (Oct. 18, 1966), H. J. Matson (to Sinclair Research, Inc.).
U.S. Pat. 3,318,396 (Dec. 31, 1967), R. J. Tailleur (to Gulf Oil Corp.).
U.S. Pat. 3,970,568 (July 20, 1976), B. W. Hotten (to Chevron Research Co.).
U.S. Pat. 3,817,971 (June 18, 1974), N. A. Carlson and B. H. Garth (to DuPont).
R. Buechner and D. Hoetzl, *Schmierungstechnik* **7,** 207 (1976).
H. Gisser, J. Messiner, and D. Chasan, *Wear* **34,** 53 (1975).
U.S. Pat. 3,840,460 (Oct. 8, 1974), D. B. Sheldahl and J. H. Striegler (to Atlantic Richfield Co.).
U.S. Pat. 3,498,915 (Mar. 3, 1970), L. E. Coleman (to Lubrizol Corp.).
U.S. Pat. 4,218,338 (Aug. 19, 1980), A. W. Schwab, L. E. Gast, and H. E. Kenney (to U.S. Dept. of Agriculture).
A. W. Schwab, W. K. Rohwedder, L. E. Gast, *J. Am. Oil Chem. Soc.* **55,** 860 (1978).
U.S. Pat. 4,177,153 (Dec. 4, 1979), W. Lowe (to Chevron Research Co.).
U.S. Pat. 4,161,451 (July 17, 1979), W. Lowe (to Chevron Research Co.).
J. W. Hagemann, J. A. Rothfus, and M. A. Taylor, *Lubr. Eng.* **37,** 145 (1981).
U.S. Pat. 4,236,898 (Dec. 2, 1980), B. T. Davis and J. B. Retzloff (to Ethyl Corp.).
T. K. Miwa and J. A. Rothfus, *J. Am. Oil Chem. Soc.* **56,** 765 (1979).
W. T. Brannen and D. A. Lee, *NLGI Spokesman*, 41 (May 1971).
U.S. Pat. 3,963,692 (June 15, 1976), H. D. Ivery, Jr. (to Lubrication Co. of America).
U.S. Pat. 4,137,187 (Jan. 30, 1979), W. Lowe (to Chevron Research Co.).
U.S. Pat. 4,218,331 (Aug. 19, 1980), J. D. Bacha and R. W. Hill (to Gulf Res. and Dev. Co.).
A. G. Papay, *NLGI Spokesman*, 191 (Sept. 1975).
F. L. Ewald, *Metal Progr.* **90,** 133 (1966).
E. M. Lown and co-workers, *J. Am. Chem. Soc.* **89,** 1056 (1967).
N. Lozac'h and J. Vialle in N. Kharasch and C. Meyers, eds., *Organic Sulfur Compounds*, Vol. 2, Pergamon Press, Inc., New York, 1966, Chapt. 10, pp. 257–285.
J. R. Nooi and G. J. Huijben, *J. Appl. Chem.* **18,** 84 (1968).
B. Shaughnessy and M. Heming, *Fette Oele Tenside Waschmittel* **107,** 307 (1981).
J. G. Mills, *Lubrication Fundamentals*, Marcel Dekker, Inc., New York, 1980.
U.S. Pat. 4,176,072 (Nov. 27, 1979), P. F. Jackish (to Ethyl Corp.).

KARL KAMMANN
GUY E. VERDINO
Keil Chemical Division
Ferro Corporation

SULFUR RECOVERY

Sulfur is generally categorized as elemental sulfur (brimstone) or sulfur in other forms. Elemental sulfur is further classified by its method of production. Frasch sulfur is produced from salt domes or sedimentary deposits by injecting superheated water into the formation, melting the sulfur in place, and pumping out the molten sulfur. Native sulfur is produced by conventional mining methods, and recovered sulfur is generally produced as a by-product of sour natural gas and petroleum by which hydrogen sulfide is stripped from the hydrocarbon and converted to elemental sulfur by the Claus process (see Gas, natural; Petroleum, refinery processes).

Production of recovered sulfur has increased rapidly in the last 25 yr and accounts for about one half of the world production of sulfur in the elemental form. Discovery and development of large sour natural-gas fields in the United States, Canada, France, the FRG, the Middle East, and the USSR have been main factors in this rapid growth. Increased processing of sour crude oil and more strict pollution control has forced most petroleum refineries to recover the sulfur content of the crude in their refineries. Based on the nature of the process or the economics of a particular situation, sulfur is recovered also as a salable sulfur compound.

Processing of Sour Natural Gases and Sour Crude Petroleums

There are many methods for treating gases in order to remove hydrogen sulfide from them, and many factors affect the selection of the gas-treating process including the volume of gas to be processed and its temperature and pressure, the desirability of sulfur recovery, the selectivity required, the types of impurities present, the concentrations of the impurities, air pollution regulations, and specifications to be met (1). In order to sell the processed or sweet gas, the gas processor often must meet certain specifications and standards, eg, maximum hydrogen sulfide content, 5.72 mg/m^3 (0.25 grains/100 SCF); maximum mercaptan content, 22.9–114 mg/m^3 (1–5 grains/100 SCF); maximum carbon dioxide content, 1–3 vol %; maximum water content, 64–112 mg/m^3 (4–7 lb/10^6 SCF); minimum gross heating value, 35.4 MJ/m^3 (950 Btu/SCF); maximum hydrocarbon dew point, −9.4°C; minimum delivery pressure, 4.93 MPa (700 psig); and oxygen content, 0–0.4 vol % (2).

Recovery of Acid Gases. Of the processes in which the reagent is regenerated and the hydrogen sulfide is recovered, the amine reagents in aqueous solution are the most widely used. Three amines are available industrially: monoethanolamine (MEA), diethanolamine (DEA), and triethanolamine (TEA) (see Alkanolamines). The first two are used widely in gas treating. Other amines, as well as nonamine compounds, are also used (see also Carbon dioxide).

Mono- and Diethanolamine Processes. Monoethanolamine in aqueous solutions is used to absorb hydrogen sulfide and carbon dioxide from sour natural gases in many plants. The basic engineering design includes an absorber in which a lean water solution of MEA absorbs the acid gases (hydrogen sulfide and carbon dioxide) from the sour natural gas and a stripper in which heat, usually in the form of steam, separates the acid gases from the pregnant amine solution for recycle. A simplified flow diagram of the conventional MEA cycle is illustrated in Figure 1.

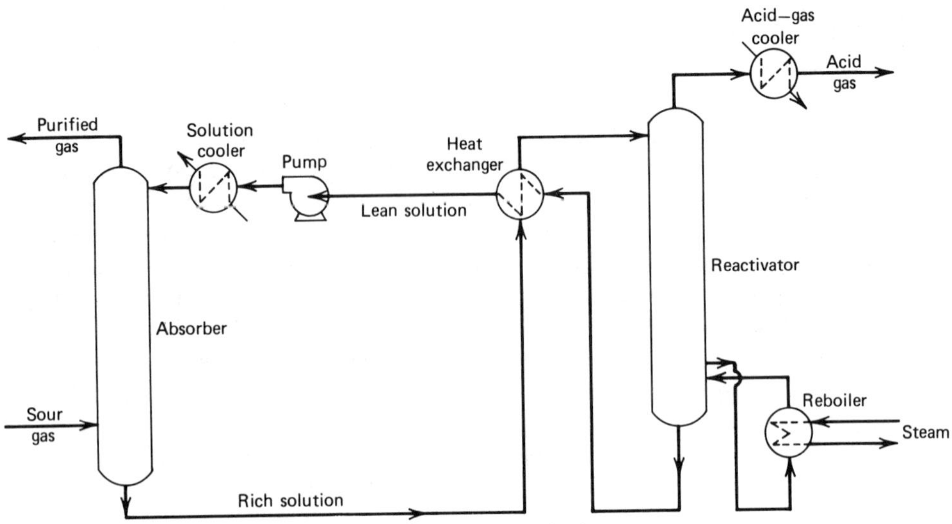

Figure 1. Flow diagram of standard amine process.

Engineering modifications include the split-stream cycle, the combined amine and glycol cycle to sweeten and dehydrate the gas simultaneously, and the absorber–heat-exchanger cycle for more concentrated solutions. In the split-stream cycle, only a portion of the rich amine is stripped to a low concentration of acid gas for recycle with resultant energy savings. Figure 2 is a simplified flow diagram of a split-stream cycle. Monoethanolamine reacts with hydrogen sulfide and carbon dioxide according to the following reversible chemical reactions.

$$2\ HOCH_2CH_2NH_2 + H_2S \rightarrow (HOCH_2CH_2NH_3)_2S \quad (1)$$

$$(HOCH_2CH_2NH_3)_2S + H_2S \rightarrow 2\ (HOCH_2CH_2NH_3)SH \quad (2)$$

$$2\ HOCH_2CH_2NH_2 + CO_2 + H_2O \rightarrow (HOCH_2CH_2NH_3)_2CO_3 \quad (3)$$

$$(HOCH_2CH_2NH_3)_2CO_3 + CO_2 + H_2O \rightarrow 2\ (HOCH_2CH_2NH_3)CO_3H \quad (4)$$

$$2\ HOCH_2CH_2NH_2 + CO_2 \rightarrow HOCH_2CH_2NHCO_2(NH_3CH_2CH_2OH) \quad (5)$$

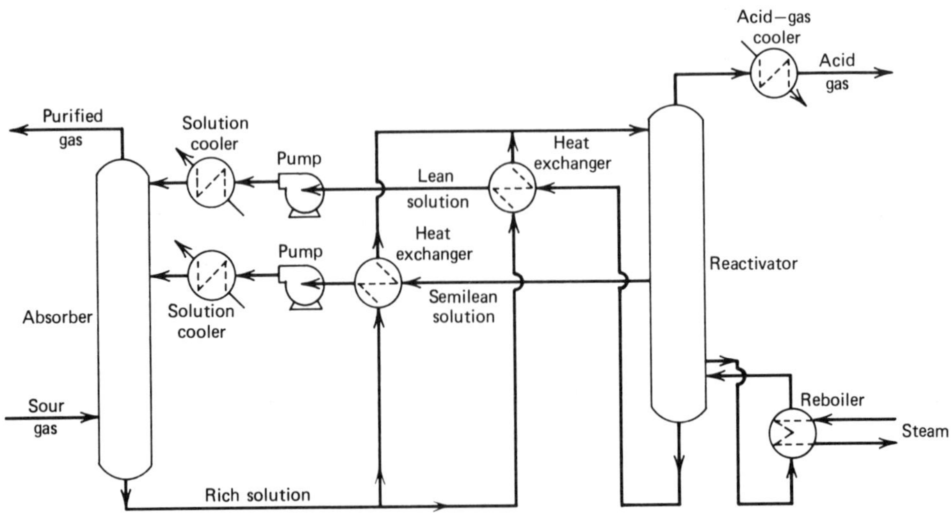

Figure 2. Flow diagram of split-stream amine process.

The reversibility of these chemical reactions is the basis for the cyclic MEA gas-scrubbing processes (see also Gas cleaning).

Certain impurities in the raw or feed gas degrade MEA. The normal impurities in sour natural gases that may cause degradation are carbon dioxide, carbonyl sulfide, and carbon disulfide (3–4). Carbon dioxide forms heat-stable salts, which are unaffected by the amine-regeneration step. As these salts accumulate, the concentration of MEA decreases, and thus it is necessary to add fresh MEA solution from time to time and to bleed a portion of the recycling solution in order to keep the heat-stable salts at a reasonable concentration. Because of environmental constraints, the bleed stream must be treated by some acceptable procedure. The MEA may be recovered by distillation (5). Degradation of MEA by carbonyl sulfide and carbon disulfide can be minimized by adding caustic soda to the cycling solution; the caustic also removes mercaptans (4,6). The following factors must be considered in the design of plants using aqueous MEA as the absorbent: the purity of sweet gas, the equilibrium in the system H_2S–CO_2–MEA, corrosion of construction materials, losses of solution, solution loading, the concentrations of H_2S and CO_2 in the inlet gas, the pressure of the system, and the ratio of gas flow to liquid flow (7–10).

Basically, the diethanolamine process is also a cyclic, heat-regenerative process. The process equipment is essentially identical to that used in an MEA plant. Diethanolamine treatment of sour gas has been reviewed in considerable detail (11). For removal of impurities, 5–10% of the circulation rate may be passed through a precoated leaf-type filter with activated carbon added to the filter aid to remove the degradation products. Other plants have successfully used reclaimers (12). In France, the Société Nationale des Pétroles d'Aquitaine (SNPA) began developing its own DEA system. These activities led to the SNPA DEA process (13–14). The solution strength for DEA is 20–30 wt %. In plants based on the SNPA DEA process, 0.9–1.3 mol of amine per mole of acid gas is used with 30 wt % solution and the plants are operated at a high absorption temperature. Other plants are designed on the basis of 1.5–2 mol of amine per mole of acid gas.

Hot Carbonate Process. The U.S. Bureau of Mines first studied the hot carbonate process for the removal of CO_2 from synthesis gas, then later applied the process to the removal of H_2S and CO_2 from industrial gases (15–18). Others built plants based on the hot carbonate process to remove hydrogen sulfide and carbon dioxide from natural gases (19–23). The hot carbonate process involves the same basic flow schemes as the MEA process. It is best suited for bulk removal of H_2S and CO_2 from gases, with a second process that enables the sweet gas to meet specifications. Attempts have been made to stage the process in order to obtain a gas of higher purity; however, staging increases the capital and operating costs.

Simple chemistry is one advantage of the process. The reactions are

$$K_2CO_3 + H_2S \rightarrow KHS + KHCO_3 \quad (6)$$

$$K_2CO_3 + CO_2 + H_2O \rightarrow 2\,KHCO_3 \quad (7)$$

The reactions are readily reversible. Both COS and CS_2 are at least partially hydrolyzed according to the following equations:

$$COS + 2\,K_2CO_3 + 2\,H_2O \rightarrow 3\,KHCO_3 + KHS \quad (8)$$

$$CS_2 + 3\,H_2O + 3\,K_2CO_3 \rightarrow 4\,KHCO_3 + 2\,KHS \quad (9)$$

The presence of the HCO_3 ion is necessary for the regeneration of potassium hydrosulfide in solution. Hot carbonate plants are subject to corrosion by carbon dioxide. The hot carbonate process has the advantages of low capital investment, low operating costs, and low heat requirements. Corrosion appears to be the main disadvantage.

Sulfinol Process. In order to avoid the degradation losses suffered by MEA solutions, some plants have switched to the relatively new solvent Sulfinol (24–25). The Sulfinol plants are based on the same basic process flow schemes as the MEA plants. The Sulfinol solution usually consists of 40–45 wt % sulfolane (qv) (tetrahydrothiophene dioxide), 40–45 wt % diisopropanolamine, and the balance water. As compared with treating gases with aqueous MEA solutions, Sulfinol provides greater gas-treating capacity, lower solvent circulation, lower heat requirements, and lower rates of solvent degradation. Diisopropanolamine but not sulfolane degrades in service. Carbon steel is an adequate construction material for Sulfinol solutions, but it undergoes normally expected corrosion. A plant that had been running on aqueous MEA solution designed for a loading of 0.4 mol acid gas per mol amine, or 33.5 volumes acid gas per volume solvent, was switched to Sulfinol. The Sulfinol solvent provided a comparable loading of 45 volumes acid gas per volume solvent. A comparison of the heat duties for aqueous MEA and Sulfinol is shown in Table 1, and the amine degradation values for the two solvents are listed in Table 2.

Sulfinol also offers the following favorable characteristics: low foaming tendency, pipeline specification gas, low vapor pressure in use, very low tendency to fouling, and no expansion on freezing. Sulfinol, however, does have the following disadvantages: absorption of heavy hydrocarbons from the feed gas, especially aromatics; Sulfolane and diisopropanolamine are expensive; royalty for use of process; and tendency to escape from threaded piping systems (26–29).

Table 1. Heat Duties of Aqueous MEA and Sulfinol, kJ/m^3 gas[a,b]

Service	Aqueous MEA	Sulfinol
lean solvent cooler	413	279
lean–rich exchanger	339	249
regenerator reboiler	800.4	376

[a] Ref. 27.
[b] To convert kJ/m^3 gas to 10^6 Btu/10^6 SCF gas, divide by 37.23.

Table 2. Amine Degradation by Aqueous MEA and Sulfinol, kg/1000 m^3 CO_2[a,b]

Plant location and type	Aqueous MEA	Sulfinol
Person, Texas (COS, 7 ppm)		
commercial plant	0.961	0.12
East Texas (COS, 86 ppm)		
commercial plant	4	
pilot plant	4	<0.56
Canada		
pilot plant		<0.56

[a] Ref. 27.
[b] To convert kg/1000 m^3 CO_2 to lb/10^6 SCF, multiply by 62.4.

Selexol Process. Allied Corporation developed the Selexol process, which is based on the physical solubility properties of the dimethyl ether of polyethylene glycol (DMPEG) for H_2S, CO_2, COS, and mercaptans (see Fig. 3). The solvent can also be substituted for other liquid absorbents in existing plants (30). By 1980, 26 plants had chosen to use the Selexol process (31). The Selexol process has been successfully employed in the following applications: bulk removal of CO_2 from natural gas with trace amounts of H_2S; selective absorption of H_2S and COS from natural gases both low and high in H_2S, while keeping CO_2; and selective absorption of H_2S in gasification units and in refinery-gas treatment plants. Gas dehydration is also achieved in the absorption step (32).

It is claimed that the gas-treatment cost of the Selexol process is low for the following reasons: solvent regeneration relies on pressure reduction rather than steam stripping; there is a low circulation rate because of high solvent loading (the solvent is pure, rather than in aqueous solution); any plant can handle substantial variability in feed-gas composition; the solvent does not corrode mild steel; and the low vapor pressure and nondegradation minimize solvent losses. The main disadvantage appears to be appreciable hydrocarbon solubility. A number of plants are operating based on Selexol and others are being built (33–35).

Methyldiethanolamine Process. Aqueous solutions of methyldiethanolamine selectively absorb H_2S from gas streams containing both H_2S and CO_2 (36). The flow system of the process is similar to that of the MEA process. The solution has been tried as a replacement for DEA with sour refinery gas. Like DEA, it is degraded somewhat. Its cost, which is higher than the other amines, has prevented its use (37). However, renewed interest has been shown in the process (38–40). Pennwalt is promoting its use based on the following claimed advantages: selectivity for H_2S, energy savings because of lower desorption temperature and lower heat of reaction as compared to MEA and DEA, less corrosion than with MEA and DEA, low vaporization loss, and nondegradation by CS_2 and COS (41–42).

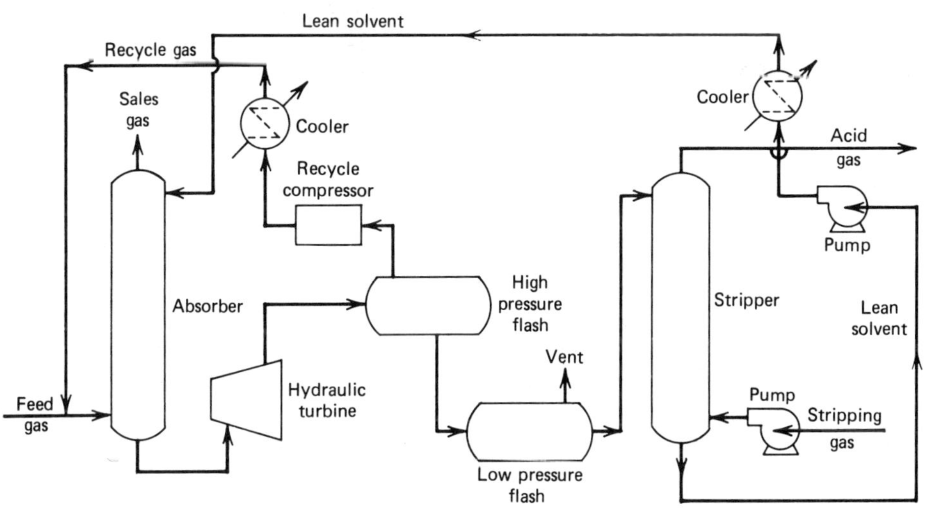

Figure 3. Typical Selexol process flow diagram. Courtesy of Allied Corporation.

Tripotassium Phosphate Process. Although no longer used for sour natural gases, the Shell tripotassium phosphate process has the following advantages for high temperatures, treatment of liquid hydrocarbons, and purification of refinery and synthesis gases: nonvolatility of tripotassium phosphate, insolubility in liquid hydrocarbons, and nonreactivity with COS and other minor impurities. Also, the aqueous solution of tripotassium phosphate has some selectivity for H_2S. Plants using aqueous tripotassium phosphate solutions are designed with a cyclic heat-regenerative cycle similar to those in MEA and other cyclic heat-regenerative processes (43). With CO_2 in the gas stream, the solution concentration is limited to 35 wt % in order to avoid the precipitation of $KHCO_3$. An excellent flow diagram is given in ref. 44.

Sodium Phenolate Process. Koppers Co., Inc. developed the sodium phenolate process for absorbing hydrogen sulfide from gases (45). However, other processes have replaced the sodium phenolate process (46).

Alkazid Process. The Alkazid process, which is of the cyclic heat-regenerative type, was developed in the FRG. The Dik solution, containing the potassium salt of *N,N*-diethylglycine or *N,N*-dimethylglycine, selectively absorbs H_2S in the presence of CO_2 and small amounts of CS_2 and HCN. The M solution, containing sodium alanine, absorbs H_2S and CO_2 simultaneously. The S solution, containing sodium phenolate, is used for gases containing HCN, NH_3, CS_2, and mercaptans, as well as H_2S and CO_2. Aluminum and special alloys are used to prevent corrosion (47–48).

Vacuum Carbonate Process. The vacuum carbonate process permits recovery of hydrogen sulfide in a concentrated form in either potassium or sodium carbonate solutions. Regeneration of the solution is accomplished by distillation under vacuum. A flow diagram of the cyclic process is given in ref. 49. The process has been applied mainly to coke-oven gases because it can handle HCN and other impurities that may cause difficulties in other adsorption systems (50–51).

Water Washing. Water washing of acid gases from sour natural and refinery gases has no industrial importance. Factors adversely affecting the water-washing process are the usually stringent specifications for sweet gas, the loss of hydrocarbons, and the impracticality of air stripping (52–53).

Direct Recovery of Elemental Sulfur. ***Dry-Box Process.*** Coke-oven gases, manufactured gases, and natural gases with trace amounts of H_2S can be treated by the dry-box process, in which the hydrogen sulfide reacts with iron oxide according to the following simplified reaction mechanism (54).

$$Fe_2O_3.H_2O + 3\ H_2S \rightarrow Fe_2S_3 + 4\ H_2O \qquad (10)$$

The revivification of the oxide may be represented as:

$$2\ Fe_2S_3 + 3\ O_2 + 2\ H_2O \rightarrow 2\ Fe_2O_3.H_2O + 6\ S \qquad (11)$$

In some systems the elemental sulfur after revivification of the oxide is recovered by such means as solvent extraction (55–57).

The dry-box process has many disadvantages: the plants are large and require considerable ground area; there is much handling of large quantities of oxide; the activity of the oxide varies; pressure decreases through oxide boxes in series; hydrates form; and the working conditions are unpleasant. A gas containing an appreciable amount of CO_2 that must meet a stringent heating-value specification is another disadvantage. Some of the advantages are selectivity towards H_2S and mercaptans; excellent removal of H_2S; insensitivity to pressure; and a minimum of attention by personnel is required.

Thylox Process. The Thylox process, developed by the Koppers Co., has been used throughout the world to purify industrial gases and particularly coal gases. It is not used to treat natural and refinery gases (58–63).

Giammarco-Vetrocoke Process. The Giammarco-Vetrocoke process consists of three processes: one for CO_2, one for H_2S, and one for acid gases (64). The chemistry of the process is rather complex, involving sodium or potassium carbonate solutions activated by salts of multivalent metals, eg, arsenic, or by certain organic compounds (65–66). The removal of H_2S and CO_2 from gases is represented by the following chemical equations:

$$KH_2AsO_3 + 3\ H_2S \rightarrow KH_2AsS_3 + 3\ H_2O \quad (12)$$

$$KH_2AsS_3 + 3\ KH_2AsO_4 \rightarrow 3\ KH_2AsO_3S + KH_2AsO_3 \quad (13)$$

$$3\ KH_2AsO_3S \rightarrow 3\ KH_2AsO_3 + 3\ S \quad (14)$$

$$6\ KH_2AsO_3 + 3\ O_2 \rightarrow 6\ KH_2AsO_4 \quad (15)$$

$$6\ CO_2 + 2\ K_2AsO_3 + 3\ H_2O \rightarrow 6\ KHCO_3 + As_2O_3 \quad (16)$$

$$CO_2 + K_2CO_3 + H_2O \rightarrow 2\ KHCO_3 \quad (17)$$

$$6\ KHCO_3 + As_2O_3 \rightarrow 2\ K_3AsO_3 + 6\ CO_2 + 3\ H_2O \quad (18)$$

$$2\ KHCO_3 \rightarrow K_2CO_3 + CO_2 + H_2O \quad (19)$$

The H_2S and CO_2 are separated upon regeneration because the sulfide is oxidized to elemental sulfur. The elemental sulfur is recovered either by filtration or by flotation. Arsenic contamination of the sulfur is one drawback.

The Giammarco-Vetrocoke process is a more complex cyclic system than the MEA cycle. Adjusting the solution composition results in the removal of hydrogen sulfide and carbon dioxide or the selective absorption of hydrogen sulfide (67). The process is said to be different from the Thylox process because the monothioarsenate is inert to oxygen and carbon dioxide. No undesirable side reactions occur, and no arsenic sulfide precipitates.

Stretford Process. The original Stretford process was developed in the UK by The Clayton Aniline Company and the North Western Gas Board. It is a cyclic process based on the following chemical steps (68–69):

$$H_2S + Na_2CO_3 \longrightarrow NaHS + NaHCO_3 \quad (20)$$

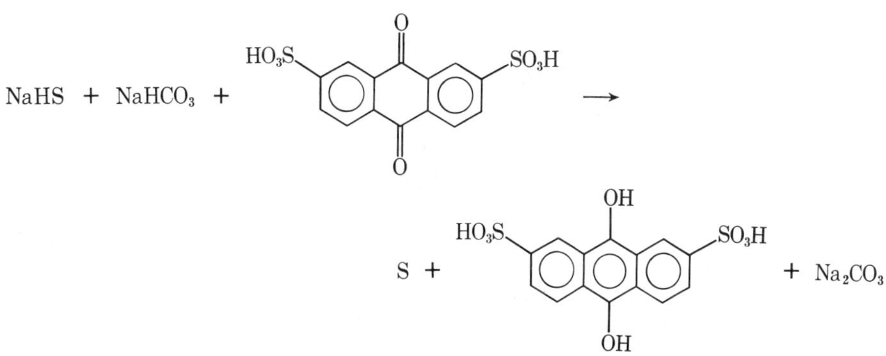

(21)

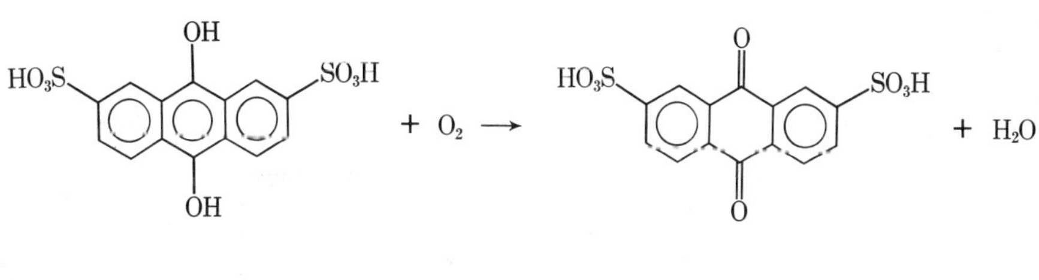

(22)

Reactions 20 and 21 occur in the absorber, and reaction 22 takes place in the regenerator. Sulfur is filtered from the absorbent. The process was improved by the addition of sodium metavanadate ($NaVO_3$) to the absorbent. The vanadium promotes the rapid oxidation of NaHS, thus helping to prevent the formation of undesirable $Na_2S_2O_3$ (70–71).

W. C. Holmes & Company, Ltd., in the UK, built many Stretford process plants. Because W. C. Holmes & Company developed proprietary adjuncts to the Stretford process in the areas of feed-gas pretreatment and effluent treatment, the process is sometimes called the Holmes-Stretford process. The process is claimed to be noncorrosive to mild steel. Its licensor in the United States claims the following advantages: low capital and operating costs, nonplugging absorber, zero effluent discharge, and adaptability to a variety of gas compositions (72–73).

Takahax Process. In Japan the Tokyo Gas Co. devised the Takahax gas-sweetening process, and the process has been used in over 60 plants in Japan. It is particularly well-suited to treating coke-oven gas, waste-gas streams, and refinery tail gases. The Takahax process, which is similar to the Stretford process, operates on a cyclic basis and requires no heat. The absorbent is an aqueous solution of sodium carbonate and sodium-1,4-naphthoquinone-2-sulfonate. In the absorber the following reactions occur:

$$Na_2CO_3 + H_2S \longrightarrow NaHS + NaHCO_3 \quad (23)$$

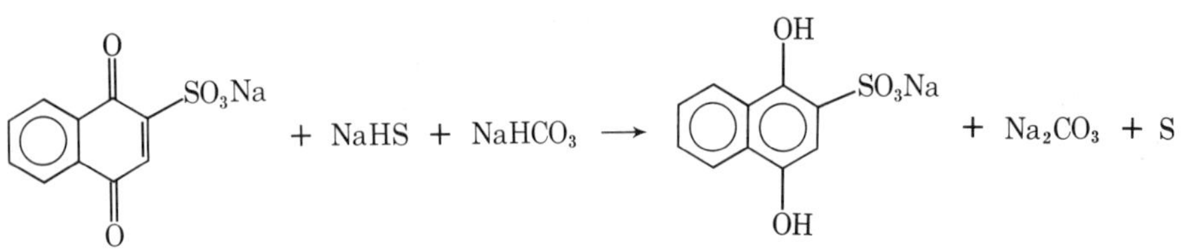

(24)

In the regenerator, air reoxidizes the naphthohydroquinonesulfonate to the naphthoquinonesulfonate. The solution is then ready to be recycled to the absorber. The air strips any CO_2 from the liquor so that excess air and CO_2 leave the top of the regenerator. A bleed stream of the regenerated liquor is filtered to remove sulfur at the same rate as it is produced, and then it is returned to the absorber. Hydrogen cyanide in the sour gas is absorbed and is then oxidized to sodium thiocyanate. Any sulfur dioxide or sulfur trioxide is absorbed and converted to sodium sulfate. The sodium thiocyanate and sodium sulfate do not regenerate but increase in concentration. A bleed stream is needed to keep these impurities at an acceptable level. The Nittetsu Chemical Engineering Co. Ltd. of Japan has developed a method of reducing the wastage of scrubbing liquor by decomposing thiocyanates. Part of the scrubbing solution is concentrated and pyrolyzed. The resulting sodium sulfide and bisulfide are recycled to the regenerator. Gases from the decomposer are scrubbed with the feed liquor to the decomposer (74–75).

Molecular-Sieve Process. There arose a need to remove H_2S selectively from natural-gas streams containing substantial amounts of CO_2, and in ca 1960 molecular sieves (qv) were tried for this separation. Certain molecular sieves separate the hydrogen sulfide from natural gases. The molecular sieve is regenerated with gaseous SO_2, which yields elemental sulfur and water (76). A pilot plant using the flow diagram illustrated in Figure 4 has been built. Sour gas is passed through a tower filled with molecular-sieve beads. After the absorption tower, the gas passes through a gas cooler, a second tower filled with molecular-sieve beads on the cooling cycle, and then through another cooler before transmission as product gas.

About one third of the produced sulfur is burned. The resulting burner gas is sent through another tower filled with molecular-sieve beads loaded with H_2S in order to regenerate the molecular-sieve beads. The Claus reaction takes place and sulfur vapor, water vapor, and noncondensable gases are produced in the regeneration tower. The sulfur vapor is condensed as the product. Switching incoming gas from tower to tower maintains a cyclic operation (77). Other aspects of the process are described in ref. 78.

Molecular sieves have been applied to gas processing in Canada in a variety of ways (79). Attention has been accorded to the absorption and desorption of H_2S (80). Reaction characteristics of molecular sieves in removing small amounts of H_2S from natural gas have been investigated (81) (see Adsorptive separation, gases).

Iron Cyanide Complexes. Processes involving iron cyanide complexes for hydrogen sulfide were developed in Europe in the 1930s and 1940s. Differences between the processes existed mainly in the iron complexes used and the method of regeneration. Typical processes were the Fischer, the Staatsmijnen-Otto, and the Autopurification processes. No large industrial plants have been built (82–83).

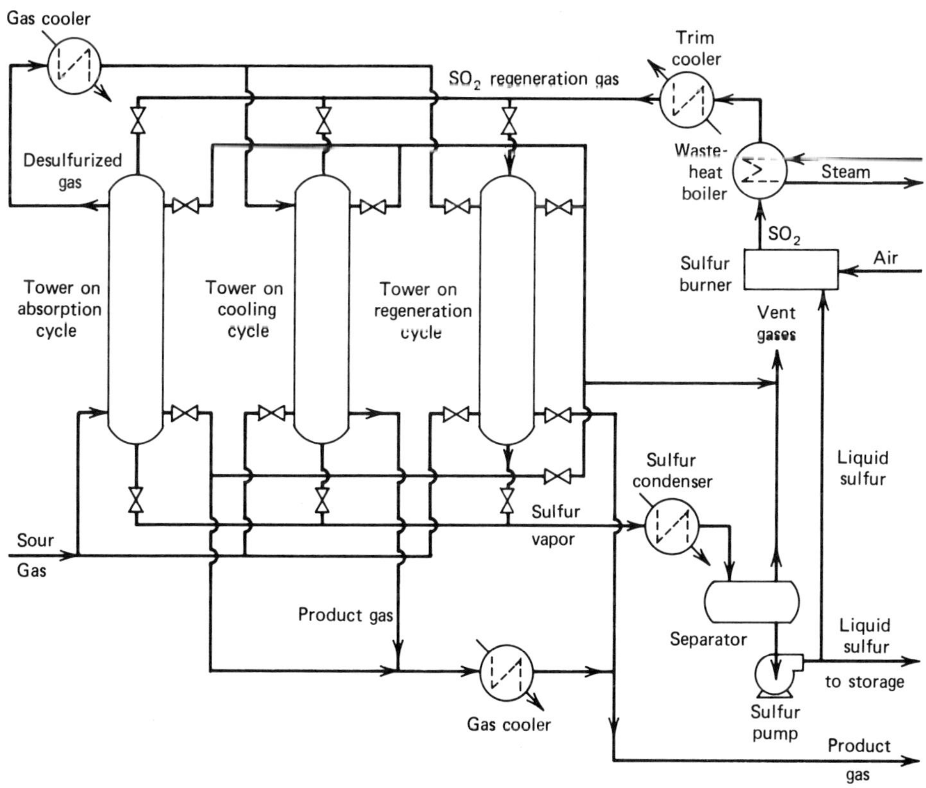

Figure 4. Flow diagram of molecular-sieve process.

Perox Process. The Perox process involves the absorption of H_2S in aqueous ammonia containing a catalyst, eg, hydroquinone. Air oxidizes the ammonium hydrosulfide to elemental sulfur, which is separated by filtration. Impurities, eg, thiosulfates and thiocyanates, may accumulate and must be removed. Some makeup ammonia and hydroquinone also are needed (84–86).

Claus Process. The basic Claus process was invented in 1883. Its large-scale use began in the 1950s in the United States, and there are two principal process variations. The first is the straight-through process, in which all of the air and the acid gas pass through the combustion zone. The second is the split-flow process, in which all of the air and at least one third of the acid gas pass through the combustion zone with the remaining acid gas being sent to the first catalytic reactor (87–88).

Because of air-pollution regulations, many modern Claus plants are equipped with a combustion stage, one-to-four catalytic converters, and a tail-gas clean-up unit. There are several ways of preheating or reheating and of cooling the gases. Condensation of the sulfur vapor has been carried out by various means (89–90). Because of the impurities in the acid gas, eg, ammonia and hydrocarbons, and the low concentrations of H_2S in the acid gas, modified processes have been developed (91–100). A simplified flow diagram of a typical Claus sulfur-recovery plant is shown in Figure 5.

The principal chemical reactions of the Claus process are

$$2\,H_2S + 3\,O_2 \rightarrow 2\,SO_2 + 2\,H_2O \tag{25}$$

$$2\,H_2S + SO_2 \rightarrow 3\,S + 2\,H_2O \tag{26}$$

Many side reactions are possible, eg,

$$H_2S + CO_2 \rightarrow COS + H_2O \tag{27}$$

$$CO + S \rightarrow COS \tag{28}$$

$$2\,CO + 2\,S \rightarrow CO_2 + CS_2 \tag{29}$$

$$COS + H_2S \rightarrow CS_2 + H_2O \tag{30}$$

$$2\,COS + 3\,O_2 \rightarrow 2\,SO_2 + 2\,CO_2 \tag{31}$$

$$CS_2 + SO_2 \rightarrow 3\,S + CO_2 \tag{32}$$

$$CH_4 + 4\,S \rightarrow CS_2 + 2\,H_2S \tag{33}$$

$$CH_4 + SO_2 \rightarrow COS + H_2O + H_2 \tag{34}$$

The thermodynamics of the formation of sulfur from hydrogen sulfide are explained in ref. 101. The thermodynamics of the Claus process was studied for an acid gas containing 25 vol % H_2S and 75 vol % CO_2, and for an acid gas containing 100 vol % H_2S and 0 vol % CO_2 (102). The following conclusions were reached:

	H_2S/CO_2, vol %	
Sulfur formation	*25/75*	*100/0*
theoretical optimum recovery	99.2%	99.3%
practical optimum recovery	97.5%	98.7%

The thermodynamics of sulfane production was also studied (103). The kinetics of the Claus process are incompletely understood. Studies are reported in refs. 104–107.

The energy consumed and liberated by the Claus process has been evaluated.

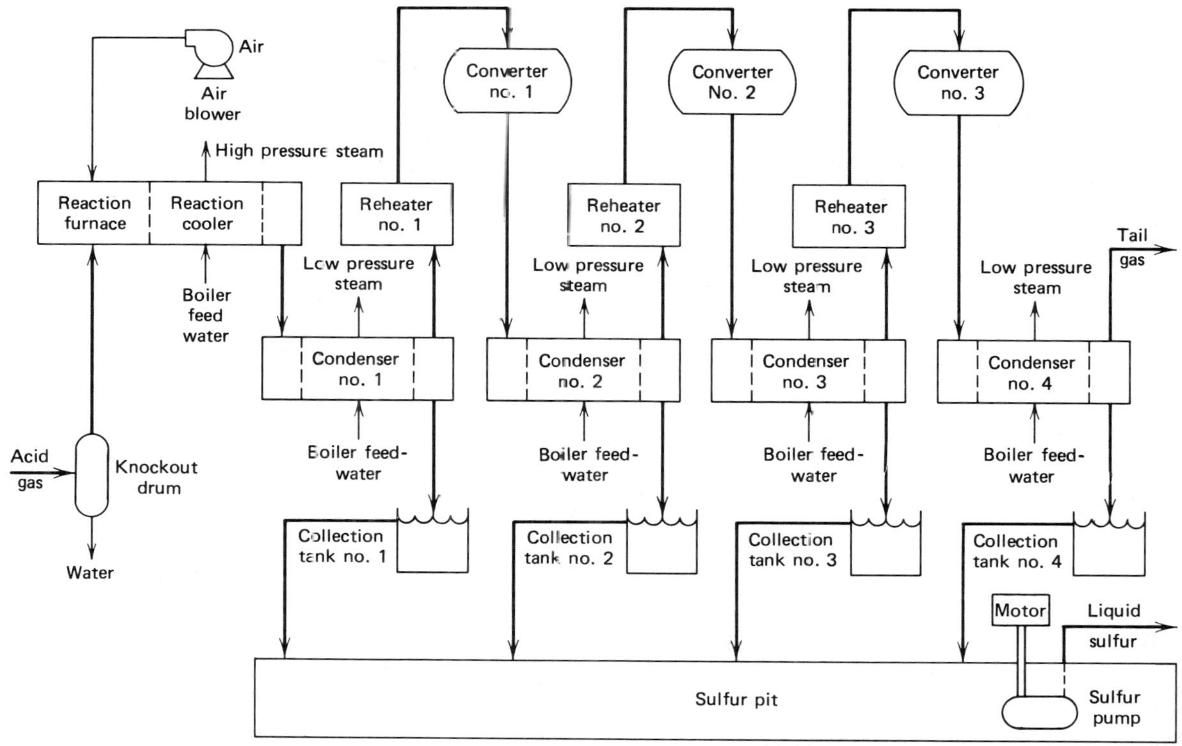

Figure 5. Flow diagram of typical Parsons Claus sulfur-recovery plant. Courtesy of The Ralph M. Parsons Co.

Energy is consumed by the air blower, incinerator, boiler feed water, and reheating of reactor feeds. Energy is liberated in steam, the liquid sulfur product, steam condensate, boiler blowdown, stack gas, and losses (108). The amount of air fed to a sulfur plant depends on the concentration of H_2S in the acid gas and the quantity of combustible impurities. Both the heat balance and the recovery of sulfur are affected by the feed to the plant.

The effect of temperature on the oxidation of hydrogen sulfide is described in ref. 101. The conversion passes through a minimum at 525–625°C. Above this minimum temperature range, the oxidation requires no catalyst, whereas below this range, a catalyst is required to achieve an industrially satisfactory oxidation rate. In the thermal region, ie, where no catalyst is needed, conversion increases with increasing temperature contrary to the catalytic region where the conversion decreases with increasing temperature. Claus sulfur plants operate at nearly atmospheric pressure. Increased pressure lowers the conversion in the thermal region and has the opposite effect in the catalytic region. Water vapor has no effect on the oxidation of hydrogen sulfide but does affect adversely the equilibrium of the reaction of H_2S and SO_2 over the catalyst. The standard catalyst used for many years in Claus sulfur plants was bauxite, although many others were tried (109). Other and specialty catalysts are described in refs. 110–117.

During the oxidation of hydrogen sulfide, the sulfur dew point of the gases increases as sulfur vapor is produced. For example, at ca 100 kPa (14.5 psi), condensation of 70% of the sulfur vapor decreases the dew point from 280°C to 235°C. Fog formation can also be a problem during condensation of the sulfur vapor (118).

In the early Claus sulfur plants, the ratio of H_2S:SO_2 was controlled by wet chemical methods of analysis, eg, the Tutweiler method, of the exit stack gas (119). The gradual imposition of air-pollution controls and the need for economy in operations made it necessary for more complex instrumentation to be installed in the new plants (120–125).

Safety becomes most important in plants handling and processing hydrogen sulfide gas, which is extremely toxic and quickly paralyzes the sense of smell. Therefore, procedures should exist for the safe handling of H_2S and for the treatment of poisoning by H_2S (126).

Claus Plant Tail-Gas Treatment. In the early Claus sulfur plants, either the tail gases were exhausted to the atmosphere through a stack without any treatment, or the gases were incinerated upon leaving the last converter and the SO_2-containing tail gas was passed through a tall stack (127). Air-pollution controls imposed on the Claus sulfur-recovery plants are described in refs. 128–134. Most Claus sulfur plants cannot meet existing or proposed air-pollution regulations without addition to the original design or inclusion of some method of reducing or eliminating the sulfur content of the gas leaving the plant and entering the atmosphere (128). Tail-gas clean-up processes have been proposed and some are in use; most of the known tail-gas processes are listed in Table 3.

The Sulfreen process converts H_2S and SO_2 in the tail gas to sulfur at temperatures near the melting point of sulfur by the classic Claus reaction using a special activated alumina as the adsorbent and catalyst in the reactors. Multiple beds of alumina are cycled through adsorption–reaction cycles and desorption of deposited sulfur. The process operates in batch or semicontinuous modes. An overall Claus conversion of 99% has been claimed (136–140).

Table 3. Tail-Gas Clean-Up Processes[a]

Dry-bed processes	Wet-scrubbing processes
Sulfreen	SCOT
Amoco cold-bed adsorption	Beavon
SNPA/Topsöe	IFP
Lucas	Davy-Powergas/Wellman-Lord
Westvaco	Cleanair
Shell flue-gas desulfurization	ATS
	B&R/Selectox I
	Chiyoda 101
	Clauspol 150
	Citrate
	Aquaclaus
	Trencor-M
	Townsend
	ASR Sulfoxide
	UCAP
	Wiewiorowski

[a] Ref. 135.

In the Amoco cold-bed adsorption (CBA) process, gas from the Claus plant's final sulfur condenser reacts at 125–150°C in an adsorption reactor containing a regular catalyst, eg, alumina. Deposition of sulfur on the catalyst shifts the equilibrium toward higher conversion. The process requires multiple reactors. While adsorption and reaction of the gases occur in one reactor, the hot gas from the first Claus reactor is passed through a second reactor in order to vaporize the condensed sulfur and reactivate the catalyst (128,137,141).

The Lucas process consists of three stages: incineration of the final tail gas from the Claus plant; treatment of the incinerated gas with hot coke; and removal of SO_2 from the gas leaving the coke bed and returning the SO_2 to the beginning of the Claus process. The following reactions occur in the hot coke:

$$C + O_2 \rightarrow CO_2 \tag{35}$$

$$2\ SO_3 + C \rightarrow 2\ SO_2 + CO_2 \tag{36}$$

$$2\ H_2S + 3\ O_2 \rightarrow 2\ SO_2 + 2\ H_2O \tag{37}$$

$$2\ NO + C \rightarrow N_2 + CO_2 \tag{38}$$

The SO_2 is absorbed in aqueous disodium hydrogen phosphate as follows:

$$Na_2HPO_4 + H_2O + SO_2 \rightarrow NaH_2PO_4 + NaHSO_3 \tag{39}$$

On regeneration, reaction 39 is reversed. An overall Claus yield of 99.9% is claimed (140,142–143).

In the Westvaco process, the Claus tail gas is incinerated and diluted with air to bring the O_2 content to ca 8.5 vol %. Then the gas flows through a continuous, countercurrent, three-stage, fluidized-bed adsorber in which the SO_2 is absorbed from the gas as sulfuric acid. The treated tail gas is said to contain <200 ppm SO_2 by volume. In the regenerator, the SO_2 is released and returned to the first stages of the Claus process (140,144).

The Société Nationale des Pétroles d'Aquitaine (SNPA) and Haldor Topsöe have developed a catalytic oxidation process to treat tail gases from a Claus plant. By incineration, all sulfur compounds in the tail gas are converted to SO_2. After the incinerated gas is cooled to 420°C, it is passed over a vanadium catalyst to give SO_3. The cooled converter exit gas passes through an acid concentrator and then an adsorber in which the SO_3 is absorbed forming 80 wt % H_2SO_4. This weak acid flows to the concentrator in which 94 wt % H_2SO_4 is produced. Only one such plant was built at Lacq, France (140).

The Shell flue-gas desulfurization process (SFGD) includes a fixed bed of copper oxide on alumina to adsorb SO_2 from incinerated tail gases of a Claus plant at ca 400°C. Multiple beds are used on a swing-bed scheme of adsorption and regeneration. The SO_2 is desorbed at ca 400°C by a hot reducing gas and is returned to the beginning of the Claus process. One plant is operated by the Showa-Yokkaichi Oil Company in Tokyo, Japan (140).

Shell Internationale Petroleum Maatschappij BV developed the SCOT process for the incineration of Claus plant tail gases. The incinerated tail gas reacts on a cobalt–molybdenum catalyst supported on alumina with a reducing gas, which may be H_2 or a mixture of CO and H_2. Almost all of the SO_2 is reduced to H_2S, and any COS and CS_2 is hydrolyzed. After cooling, the reduced gas is absorbed in aqueous diisopropanolamine (DIPA), which has high selectivity for H_2S versus carbon dioxide. The DIPA solution is regenerated with heat, and the H_2S from the regenerator is returned to the front of the Claus process. A number of these tail-gas treatment plants have been built. About 99.8% overall Claus conversion should be expected. A SCOT unit may cost as much as the Claus plant (128,135,137,145–149).

The Beavon process, offered by The Ralph M. Parsons Co. and developed by Union Oil Co., consists of two processes: the Mark I and the Mark II Beavon sulfur-removal processes (128,135,137,149–152). The Mark I Beavon sulfur-removal process is more complex and is claimed to have >99.6% overall efficiency. The Mark II Beavon sulfur-removal process is intended for situations where an efficiency of 98–99% is acceptable. In the Mark I process, the sulfur-plant tail gas flows to a reducing gas generator where it is heated to 315–400°C. The mixed gases pass to a reactor containing a cobalt–molybdate catalyst developed by Union Oil Co. In the reactor, the SO_2 and sulfur are converted to H_2S, and CS_2 and COS are hydrolyzed to H_2S. After cooling, the gases are treated by the Stretford process. The exit gas is discharged to the atmosphere with no further processing (see Fig. 6).

The Mark II Beavon sulfur-removal process includes the initial hydrogenation–hydrolysis step of the Mark I process. The wet gas is cooled to reduce the water content to one-seventh the value leaving the reactor. The dehumidified gas is sent to a final Claus converter containing a special catalyst developed by Union Oil Co. The overall efficiency is claimed to be ≥98% for the Claus plant with the Mark II process.

The Institut Français du Pétrole of France developed the IFP process. The Claus plant tail gas enters a vertical packed-tower reaction vessel. The gas passes from bottom to top countercurrently to a proprietary solvent–catalyst liquid solution at ca 125°C. Catalyst–solvent solution and steam condensate are added to the circulating solution entering the top of the reactor tower. Polyethylene glycol containing a metal salt catalyst is reported to be the solvent–catalyst. The treated tail gas leaves the top of the reactor at ca 125°C and passes to an incinerator equipped with a stack; liquid sulfur is taken from the bottom of the tower. The basic Claus reaction occurs in the tower. Since the IFP process does not affect CS_2 and COS, other means must be used

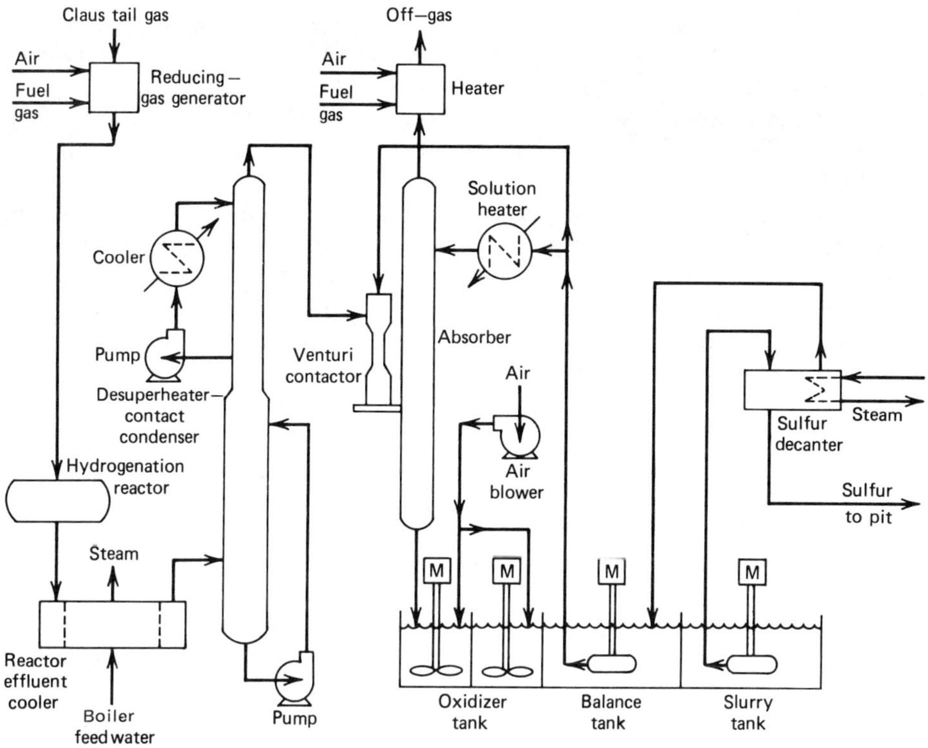

Figure 6. Beavon sulfur-removal process for Claus plant tail gas. Courtesy of The Ralph M. Parsons Co. M = motor.

for their removal. About 99% overall recovery can be achieved (135,149,153–154) (see also Incinerators).

The Davy Powergas/Wellman-Lord process was developed for removing SO_2 from stack gases. At a Claus plant, the incinerated tail gas is cooled, quenched, and fed to the absorber where it is scrubbed countercurrently by a solution of sodium sulfite. The SO_2 in the incinerated gas reacts with the sodium sulfite to form sodium bisulfite. The pregnant solution is treated at 110°C in an evaporator–crystallizer, and SO_2 and water vapor evolve. The SO_2 is returned to the beginning of the Claus plant after the water has been removed. The liquid from the evaporator–crystallizer is returned to the absorber after solution of the sodium sulfite crystals. A bleed stream must be treated to take out sodium sulfate. Claus plants in Japan have used this tail-gas treatment process (135,144).

The Cleanair process of J. F. Pritchard and Co. has three stages. The tail gas from the Claus unit enters stage 3 in which COS and CS_2 are removed. Next the gas flows to stage 1, in which ca 50 wt % of the sulfur is recovered. In stage 1, all of the SO_2 and some H_2S is converted to elemental sulfur. Then the gas passes to stage 2, which is the Stretford process for the removal of H_2S (135,144,149,155).

The ATS Claus tail-gas clean-up process produces an ammonium thiosulfate fertilizer (see Fertilizers). Colorado Interstate Gas Company had the first commercial plant at Table Rock, Wyo. In the ATS process, SO_2 is absorbed from incinerated Claus tail gas in aqueous ammonia to produce ammonium sulfite and ammonium bisulfite in solution. The solution passes to a converter where it flows countercurrent to H_2S from the Claus plant. In the converter, ammonium thiosulfate forms. The ammonium thiosulfate solution is concentrated to 60 wt % and is sold (156).

The B&R/Selectox I sulfur-recovery process and the Beavon Mark II process appear to be identical. The first B&R/Selectox I industrial unit to be built operates in the FRG (157).

Chiyoda Chemical Engineering & Construction Company Ltd. of Yokohama, Japan, has developed the Chiyoda thoroughbred 101 flue-gas desulfurization process which is designed to treat Claus-plant tail gases in Japan. Incinerated tail gas from a Claus plant is absorbed in a circulating stream of dilute sulfuric acid. The treated gas goes from the absorber to a stack. The acid containing dissolved SO_2 flows to an oxidizing tower in which air rises countercurrently to the acid which contains an oxidizing catalyst. Part of the acid then goes back to the absorber while the rest passes to a crystallizer. Limestone is added to the crystallizer to produce gypsum, which is sold commercially in Japan (135,149).

The Clauspol 150 process, developed as an improvement of the IFP process by Institut Français du Pétrole, involving scrubbing of SO_2 from incinerated Claus tail gas with aqueous ammonia. Then the ammonium sulfite solution reacts with H_2S in the IFP reactor as in the IFP process (144).

Other processes for which plants have not been built include the UCAP, Sulfox, Citrate, Trencor-M, Aquaclaus, Townsend, ASR sulfoxide, and Wiewiorowski processes (135,137,141,144,149,158–162).

Recovery from Metallurgical Operations

In early metallurgical operations, sulfur recovery was not practiced (163). Sulfur dioxide in the air may be helpful at times to farmers but can also cause ecological damage and political controversy (164–167). In locations where the economics are favorable, sulfur is recovered as sulfuric acid in metallurgical sulfuric acid plants (see Sulfuric acid and sulfur trioxide). Newer metallurgical processes have been and are being developed. One aim of these processes is to recover sulfur as the element or as a useful compound, eg, ammonium sulfate.

To avoid the formation of SO_2 in winning, ie, recovering, metals from sulfide ore concentrates, some investigators turned to hydrometallurgy. Depending on the process, the sulfide sulfur appears as ammonium sulfate, sulfuric acid, or elemental sulfur. Leaching agents include sulfuric acid, nitric acid, hydrochloric acid, perchloric acid, ammonia, ferric chloride, etc, usually in aqueous solution. Elevated temperatures and pressures were used to speed the reactions or to provide sulfide sulfur in the desired final form. Air and oxygen were used in some of the processes (168–172) (see Extractive metallurgy).

Among the early leaders in the commercial hydrometallurgical treatment of metallic sulfide concentrates were F. A. Forward and Sherritt Gordon Mines Ltd. (173–174). Sheritt Gordon Mines Ltd. built a commercial plant to treat a sulfide concentrate containing ca 12 wt % nickel, 0.5 wt % cobalt, 1.5 wt % copper, 30 wt % iron, and 28 wt % sulfur (175). The finely divided concentrate is repulped in an ammonia solution and sent to a pressure-leaching autoclave. A continuous flow of air reacts with the sulfide minerals at elevated temperature and pressure. The sulfide sulfur is oxidized to sulfate sulfur to form ammonium sulfate. The final solution of ammonium sulfate holds the metals as complex amine salts. Iron as hydrated ferric oxide and the gangue materials are separated from the solution by thickeners and filters. Ammonia is recovered from the spent leach gases as aqueous solution. After recovery of the metals, the sulfur is recovered as fertilizer-grade ammonium sulfate from the metal-free solution.

F. A. Forward and Sherritt Gordon Mines, Ltd. developed a process for pressure-leaching zinc sulfide concentrates, based on the following chemical reaction, to produce elemental sulfur:

$$2\ ZnS + 2\ H_2SO_4 + O_2 \rightarrow 2\ ZnSO_4 + 2\ S + 2\ H_2O \quad (40)$$

A slight excess of sulfuric acid causes the reaction to proceed almost quantitatively at 110–115°C under an oxygen pressure of ca 140 kPa (20 psi). The reaction goes to completion in 2–4 h. The sulfur is recovered by flotation (qv) or pelletizing (see Size enlargement). The sulfur then is melted by raising the pulp temperature at the end of leaching to form liquid sulfur globules. The pellets, which form on cooling below 100°C, are separated from the leach liquor and other solids by screening on a 100-μm (150-mesh) screen (176).

Subsequently, Sherritt Gordon Mines, Ltd. improved the process of pressure-leaching of zinc sulfide concentrates by raising the temperature to 150°C and the oxygen pressure to ca 200 kPa (29 psi). The time was reduced to <2 h, and the molar ratio of H_2SO_4 to Zn became 0.66:1.0 (excess of concentrates). If the zinc concentrates contain no iron, an iron addition as pyrrhotite or ferric hydroxide amounting to at least 5 wt % of the concentrates is needed. The final leach liquor contains in solution zinc and the following impurities (if they are present in the concentrates): cadmium, copper, nickel, and cobalt. Impurities, eg, arsenic, antimony, bismuth, germanium, and tin, precipitate with the iron. Any precious metals remain with the leach residue. The solids are separated from the leachate by filtration. The gangue materials, lead sulfate, and iron compounds are separated from the elemental sulfur and unreacted zinc concentrates by flotation. The elemental sulfur is recovered by filtration of the liquid sulfur. The unreacted zinc concentrates and the sulfur entrapped therein during the filtration are returned to the autoclave (177). An improvement of this process is being installed in 1981 at the electrolytic zinc plant of Texasgulf Inc. at Timmins, Ontario, Canada, in order to expand the leaching capacity for zinc sulfide concentrates, but no elemental sulfur will be recovered.

Many hydrometallurgical processes for sulfide copper concentrates are being developed (178–194). Among these processes are the Arbiter, Cymet, Sherritt-Cominco Copper, cyanidation, chlorination, Lurgi, Anatread, RLE, Lime-RLE, and other ferric chloride processes. Miscellaneous ores also are being processed with reagents not commonly used in hydrometallurgical processing. Most of these processes convert sulfide sulfur to elemental sulfur (195–200). In the Cymet process, aqueous ferric chloride solution leaches the copper sulfide concentrates. With chalcopyrite, elemental sulfur forms (201–202).

As another way of overcoming pollution of the atmosphere with SO_2, other metallurgists produced concentrated SO_2 for sale as liquid sulfur dioxide. Notable among these attempts is the oxygen flash-smelting process of The International Nickel Company of Canada, Ltd. (INCO) (203). In the oxygen flash-smelting process, the dried sulfide concentrates and flux are injected into the smelting furnace with 95 wt % oxygen from an oxygen plant. The exit gas, containing ca 75 wt % SO_2, is scrubbed and treated by a wet Cottrell precipitator before delivery to a liquid SO_2 plant operated by Canadian Industries Ltd. INCO has made modifications and expanded the plant (204–205). At the same time, Outokumpu Oy in Finland undertook flash smelting of sulfide concentrates. Because of the difference in the economies between Canada and Finland, the Finns used preheated air in the original process and made sulfuric acid from the SO_2 (206–207). Outokumpu Oy improved its process and added a process to reduce the sulfur dioxide to elemental sulfur (208–209).

In other processes, the sulfur dioxide is absorbed in organic liquids to produce liquid sulfur dioxide for sale. American Smelting and Refining Co. developed the dimethylaniline (DMA) process for producing liquid SO_2 and built plants in Selby, Calif., and Tacoma, Wash. Another DMA plant for liquid SO_2 was installed in Copperhill, Tenn., by Tennessee Corp. (210–212). I. G. Farben in Wolfen, Germany, had a liquid SO_2 plant which absorbed the SO_2 from the gas into xylidine (213). Early processes for adsorbing SO_2 from smelter gas are described in refs. 214–221.

Other metallurgists in addition to those employed by Outokumpu Oy attempted to reduce the SO_2 in the smelter off-gases to elemental sulfur. The Boliden Company in Sweden operated a commercial plant to reduce SO_2 to elemental sulfur with coke (214,222). Others in Europe investigated the pyrometallurgical reduction of SO_2 smelter gases to elemental sulfur (223).

At Trail, British Columbia, Can., The Consolidated Mining and Smelting Co. of Canada, Ltd. (now Cominco Ltd.) built a plant to reduce concentrated SO_2 to elemental sulfur. The smelter off-gas is adsorbed in a solution of ammonia to make a concentrated solution of ammonium bisulfite. This solution reacts with sulfuric acid forming $(NH_4)_2SO_4$ and SO_2. Ammonium sulfate crystallizes from the solution as fertilizer-grade material. After drying, the 100% SO_2 gas is mixed with oxygen and blown through incandescent coke to produce elemental sulfur. The gases are cooled and the sulfur vapor is condensed. The liquid sulfur is solidified and stored (224–227).

American Smelting and Refining Co. operated a semicommercial plant at Garfield, Utah, to reduce a gas containing 5–7 vol % SO_2 and 5–8 vol % O_2 with natural gas to elemental sulfur. The main reactions taking place in a furnace at 1250°C were as follows:

$$CH_4 + 2\,O_2 \rightarrow CO_2 + 2\,H_2O \tag{41}$$

$$2\,SO_2 + CH_4 \rightarrow CO_2 + 2\,S + 2\,H_2O \tag{42}$$

Because of the high temperature, side reactions took place and it was necessary to cool the exit gases and send them through catalyst beds so that the following reactions could occur:

$$2\,COS + SO_2 \rightarrow 2\,CO_2 + 3\,S \tag{43}$$

$$2\,H_2S + SO_2 \rightarrow 2\,H_2O + 3\,S \tag{44}$$

The product gases were cooled and pure sulfur was condensed (228). No commercial plant was ever constructed. As is apparent from the reactions, equivalent amounts of SO_2 and O_2 consume equivalent amounts of methane. At the present price of natural gas, oxygen becomes a very undesirable constituent of the smelter gas.

Between 1958 and 1960, Texas Gulf Sulphur Co. (now Texasgulf Inc.) operated a pilot plant at Sudbury, Ontario, Can., on roaster gas supplied by The International Nickel Co. of Canada Ltd. In this pilot unit, the SO_2 in the roaster gas was reduced to elemental sulfur catalytically with natural gas. No details other than the patent were published but the project was not continued because of unfavorable economics (229–230).

Orkla Grube Aktiebolag developed a process to recover sulfur as a by-product of copper smelting. A plant was installed at Lokken, Norway, to smelt copper-bearing iron pyrites, coke, silica, and limestone in water-jacketed shaft furnaces. A matte containing ca 12 wt % copper and gas containing ca 300 g of sulfur vapor per cubic meter at ca 450°C left the furnace. After cleaning, the gases were passed through a catalyst and were then cooled to 130°C in waste heat boilers, where most of the sulfur was condensed. The process has been licensed in Spain, Portugal, and the USSR (231–232).

Allied Corporation operated a commercial plant to reduce SO_2 with CH_4 to elemental sulfur near Sudbury, Ontario, Can. Roaster gas was taken from a nickel-bearing pyrrhotite being processed by Falconbridge Nickel Mines Ltd. The methane was partially reformed so that reactions took place at lower temperatures. The dry analysis of the roaster gas was 12–13 vol % SO_2 and 1–1.5 vol % O_2. The gas was cooled and purified in a train of equipment similar to that used in a metallurgical sulfuric acid plant. The gas then went through a regenerator-reactor system. The chemistry can be summarized in terms of the following two reactions:

$$CH_4 + 2\,SO_2 \rightarrow CO_2 + 2\,H_2O + 2\,S \qquad (45)$$

$$2\,CH_4 + 3\,SO_2 \rightarrow 2\,CO_2 + 2\,H_2O + 2\,H_2S + S \qquad (46)$$

The temperature was 815–1315°C. The hot gases were cooled and ca 40 wt % of the total sulfur was condensed. Enough H_2S was generated to give the needed mole ratio of $2H_2S{:}SO_2$ for the Claus part of the plant. Next, the gases were sent to a conventional two-stage Claus plant where the classic Claus reaction of H_2S and SO_2 took place. The sulfur from the Claus unit was condensed. Falconbridge Nickel Mines Ltd. closed the plant because of difficulties in the metallurgical operations (233–235). Allied's process requires more natural gas than is required in reaction 41.

Recovery from Stack Gases

Attempts to recover sulfur from the stack gases emitted from burning sulfur-containing fuels resulted from the environmental laws passed in countries such as Canada, Japan, and the United States (see Air pollution). Some foresaw a glut of sulfur by 1980 with the passage of clean-air laws. But scrubbing of flue gases with lime or limestone, which produces a calcium sulfite–sulfate waste for dumping, has dominated the flue-gas desulfurization (FGD) systems market. These systems were used because they had been proven effective and because the equipment and the scrubbing materials were well-known and available. Other ways to meet air-pollution requirements are by selection of low sulfur fuels, conversion of high sulfur coal to gas or oil with removal of the sulfur during the processing, removal of the SO_2 in the boiler, and use of tall stacks.

The lime–limestone systems dominate the FGD field at present; ca 10 wt % of slaked lime or pulverized limestone is used (see Lime and limestone). The SO_2 and O_2 in the flue gas react forming calcium sulfite and calcium sulfate. The amount of oxygen determines the amount of calcium sulfate. The solids are separated from the liquid by thickening and filtration. The liquid from the thickeners is returned to the scrubbing circuit; the means of disposal of the solids depends on local environmental regulations. If gypsum is desired as a marketable by-product, such as may be the case in Japan, gypsum formation is forced by oxidation in the scrubbing loop. No efforts to use the sulfur content have been recorded. The chemical process types of FGD systems in the United States based on megawatts of capacity are listed in Table 4 (236–242).

The addition of magnesium sulfate to the lime–limestone scrubbing slurry represents a significant modification to the chemistry of the lime–limestone FGD systems. Magnesium sulfate stops scaling and increases SO_2 removal efficiency. Dravo Corporation offers thiosorbic lime, which is a lime solution enriched with 5.6 wt % MgO for use with any existing lime-based FGD system (243–244).

There are several versions of the double-alkali process. Basically, however, SO_2 is scrubbed from the stack gas with a sodium-based solution. The liquor is regenerated by mixing the spent scrubbing solution with a calcium-based solution to form calcium sulfite. Again, there is no sulfur recovery (245–247).

The sodium carbonate process absorbs SO_2 from stack gases in sodium carbonate solution. The spent liquor is pumped to settling basins. No reagent is regenerated and no sulfur recovery is attempted. This process must operate only in arid regions where evaporation outstrips rainfall and spent liquor production (248).

In the aqueous carbonate process, SO_2 is removed from the flue gas by passing the gas through a spray dryer filled with a mist of sodium carbonate solution. The flow of sodium carbonate mist is controlled to maintain the temperature in the dryer at 65–85°C. In this temperature range, dry sodium sulfite powder forms and is collected. The sodium sulfite can either be discarded or be treated by a process familiar to the pulp and paper industry. The spent absorbent is heated with coal or coke at 1000°C to reduce the sulfite and any sulfate to sodium sulfide. After the sodium sulfide melt is quenched and cooled, the solids are separated from the green liquor, which is fed to a carbonation unit for which CO_2 is taken from the reduction step. Sodium carbonate for reuse and hydrogen sulfide form. The latter is sent to a Claus plant to make elemental sulfur (248) (see Pulp).

In the Wellman-Lord process, sodium sulfite solution is used to scrub SO_2 from flue gas with the formation of sodium bisulfite. The sodium bisulfite is heated to re-

Table 4. Process Types and Capacities of U.S. FGD Systems[a]

Process	Capacities, MW Operational	Being constructed	In planning	Total
limestone	7,426	9,127	10,163	26,716
lime	6,110	5,050	4,461	15,621
lime–limestone	20	0	660	680
sodium carbonate	375	509	0	884
magnesium oxide	120	0	726	846
Wellman-Lord	429	486	940	1,855
double alkali	0	1,102	0	1,102
aqueous carbonate	0	500	0	500
citrate	0	60	0	60
Totals	*14,480*	*16,834*	*16,950*	*48,264*

[a] Ref. 237.

generate the scrubbing solution and concentrated SO_2 gas, which can be made into liquid SO_2, sulfuric acid, or elemental sulfur. The scrubbing liquor must be purged of sodium sulfate, which forms upon reaction with oxygen in the flue gas. Allied Corp. and Davy McKee have joined to form the Wellman-Lord/Allied process to produce elemental sulfur by reducing the SO_2 with natural gas (235–236,248–250).

In the citrate process, an aqueous solution of sodium citrate and citric acid is used to scrub SO_2 from stack gases. Hydrogen sulfide gas is passed through the SO_2-rich solution to make elemental sulfur, which is separated from the liquor before the liquid is returned to the absorber. It is necessary to purge sodium sulfate which forms in the flue gas. With the citrate process, it is necessary to have a source of H_2S or to use two thirds of the produced elemental sulfur to make H_2S by reaction with a reducing gas (236,251–252).

In the magnesium oxide process, the flue gas is scrubbed with a slurry of magnesium oxide. The magnesium oxide and SO_2 react to form magnesium sulfite. The magnesium sulfite is centrifuged from the liquid, dried, and calcined to regenerate magnesium oxide and SO_2 which is made into sulfuric acid (236,253). There are many other processes that have been proposed or are being developed to remove SO_2 from stack gases, eg, dry nahcolite (236,254–255), potassium thiosulfate (236), dry cupric oxide (236,256), dry scrubbing (236), Chiyoda thoroughbred 101 (256–257), FW-BF (258–260), Saarberg-Holter (236), Purasiv S (261), Cat-Ox (262), MHI and Chiyoda thoroughbred 102 (263–264), IFP Stackpol 150 (265–266), molten carbonates, alkalized alumina, Reinluft, Wuk, SFGD, and carbon (267–273).

The throwaway processes present a disposal problem in connection with the liquid, semiliquid, or solid wastes. Several methods have been proposed to keep them from being an environmental hazard. The Poz-O-Tec system, marketed by IU Conversion Systems, Inc., of Philadelphia, Pa., involves sulfate–sulfite alkaline-earth compounds, which react to produce fibrous gypsum, and cementing reactions including pozzolanic reactions between fly ash, lime compounds, and sulfur oxide salts. Dravo Corporation of Pittsburgh, Pa., offers Calcilox which turns sludge into an earth-type material. Research-Cottrell, Inc. offers a process involving dewatering of the sludge and mixing it with dry fly ash. Chemfix, Inc. in Pittsburgh has a method based on soluble silicates and silicate setting agents to produce solids from sludge and liquid wastes (248,274–277).

Recovery from Coal

Many of the coals in the United States contain too much sulfur for the combustion products to meet existing or pending air-pollution requirements. However, coals high in sulfur content can be burned if the sulfur is removed or recovered from the flue gases as discussed earlier, the sulfur is removed during the combustion process, the coal is desulfurized before combustion, or the coal is converted to gas or oil (see Coal). Coals contain 40–60 wt % pyritic sulfur and 40–60 wt % organic sulfur. Part of the pyritic sulfur can be removed physically by means such as grinding, heavy media, flotation (qv), centrifugal separation (qv), electrostatic separation, high intensity magnetic separation (qv), and by combinations of these methods (278–284) (see also Air-pollution control methods; Gravity concentration).

The Meyers process chemically removes pyritic sulfur from coal. The chemistry is outlined below.

$$Fe_2(SO_4)_3 + FeS_2 \rightarrow 3\ FeSO_4 + 2\ S \quad (47)$$

$$7\ Fe_2(SO_4)_3 + 8\ H_2O + FeS_2 \rightarrow 15\ FeSO_4 + 8\ H_2SO_4 \quad (48)$$

$$S\text{-}coal \rightarrow S + coal \quad (49)$$

$$4\ FeSO_4 + 2\ H_2SO_4 + O_2 \rightarrow 2\ Fe_2(SO_4)_3 + 2\ H_2O \quad (50)$$

The sulfur can be recovered by solvent extraction or vaporization. The Ledgemont process resembles the Meyers process except that the pyritic sulfur is oxidized only to soluble sulfates which are washed out of the coal with water (285).

A process has been proposed for leaching the organic sulfur from coal with *para*-cresol. Enough of the leaching solvent is vaporized to yield a high sulfur solid residue equal to the sulfur in the coal being processed (286).

The EPA funded a study to demonstrate sulfur removal and recovery by pressurized combustion in a fluidized bed of limestone. The limestone is regenerated to lime with SO_2 production for possible reduction to elemental sulfur (287–288).

Approximately 20 coal-liquefaction processes and ca 30 coal-gasification processes are being developed. Coal-liquefaction processes include pyrolysis, extraction–hydrogenation, and indirect liquefaction. Coal gasification includes fixed-bed, moving-bed, fluid-bed, and entrained-bed as well as molten-bath, rotary-kiln, and *in situ* processes (289–291) (see Fuels, synthetic).

Regardless of the process, some or all of the sulfur appears as H_2S. Associated with the H_2S are other chemical compounds which may cause problems in the processes used to treat sour natural and refinery gases. The process chosen to remove sulfur compounds may have to be modified in order to handle some of the compounds, eg, thiophenes, carbon disulfide, ammonia, etc.

Recovery from Oil Shale

Recovery of oil from shale containing organic matter has been investigated and, in some countries, carried out (292–294) (see Oil shale). Generally, oil from shale can be recovered by mining and processing the shale on the surface or by *in situ* processing. The principal retorting processes are described in the literature (295). The sulfur in the shale passes into the oil vapors and into the gas. The principal sulfur compound in the gas is usually H_2S as the SO_2, which forms during combustion, reacts with CaO and MgO in the shale (296). *In situ* retorting of oil shale also has been actively studied (297–304). Investigations on gasifying oil shale have been made (305–306). The sulfur appears as H_2S. No sulfur is recovered commercially from processing oil shale.

Economic Aspects

In the early 1930s, the world sulfur market was dominated by Frasch sulfur produced in the United States. Gradually Frasch sulfur produced abroad and recovered sulfur reduced the dominant position of the U.S. producers of Frasch sulfur in the marketplace. By 1980, as shown in Table 5, recovered sulfur production was greater than Frasch sulfur production. The substantial changes in U.S. sulfur prices between Jan. 1, 1970 and Oct. 1, 1981 are listed in Table 6 (see also Sulfur).

The total capital investment in a sour-gas sulfur-recovery plant depends on its location. The primary factors affecting the total capital investment are the gas-gathering system, the initial treatment required, the gas-treating system, the utilities, the

Table 5. 1980 Estimated World Elemental Sulfur Production, 10^6 metric tons[a]

Country	Frasch	Oil refinery	Natural gas	Other	Total
United States	6.4	2.3	1.7		10.4
Canada		0.2	5.9	0.3	6.4
Mexico	1.7	0.2	0.2		2.1
other Western Hemisphere		0.2	0.2	0.1	0.5
France		0.2	1.8		2.0
FRG		0.3	0.8		1.1
other Western Europe		0.5			0.5
Iraq	0.5				0.5
Japan		1.2			1.2
other Middle East, Asia, and Oceania		0.2	0.7		0.9
Poland	4.5			0.3	4.8
USSR	1.0	0.3	1.2	1.0	3.5
other East Bloc		0.2	0.1	0.4	0.7
Total production	*14.1*	*5.8*	*12.6*	*2.1*	*34.6*

[a] Courtesy of Market Research Department, Texasgulf Inc.

Table 6. U.S. Sulfur Price, fob Tampa, Fla.[a]

Effective date	Price, $/t	Effective date	Price, $/t
Jan. 1, 1970	31.48	May 1, 1978	67.40
Feb. 1, 1970	29.52	Jan. 19, 1979	72.07
Nov. 1, 1970	26.57	June 13, 1979	76.99
Mar. 10, 1971	24.60	Sept. 5, 1979	84.37
July 7, 1973	27.55	Dec. 6, 1979	93.97
Sept. 10, 1973	30.50	Feb. 1, 1980	104.30
May 1, 1974	35.91	Mar. 15, 1980	114.14
July 1, 1974	41.32	July 16, 1980	125.45
Aug. 14, 1974	49.20	Dec. 1, 1980	137.75
Nov. 1, 1974	50.18	Oct. 1, 1981	145.13
Dec. 20, 1974	56.08	Nov. 1, 1982	132.87
Apr. 16, 1975	63.96		

[a] Courtesy of Market Research Department, Texasgulf Inc.

Claus plant, and the tail-gas treatment required. The capital investment for a Claus plant depends on the composition of the feed to the plant and the particular flow scheme adopted. In 1970, a Claus sulfur plant with a rated capacity of ca 400 t/d of sulfur might have a battery limits cost of $\$2 \times 10^6$; Such a plant may have cost as much as $\$10^7$ in 1981. The direct operating cost of a Claus sulfur plant is usually low and depends on the costs of labor and utilities. The direct cost of treating the tail gas from the Claus plant also depends on the cost of labor and utilities.

BIBLIOGRAPHY

1. R. N. Maddox and M. D. Burns, *Oil Gas J.* **65**(33), 131 (Aug. 14, 1967).
2. B. G. Goar and T. O. Arrington, *Oil Gas J.* **26**(26), 160 (June 26, 1978).
3. L. D. Polderman, C. P. Dillon, and A. B. Steele, *Oil Gas J.* **54**(2), 180 (May 16, 1955).
4. E. M. Berlie, J. W. Estep, and F. J. Ronicker, *Chem. Eng. Prog.* **61**(4), 82 (Apr. 1965).
5. *Oil Gas J.* **50,** 149 (Aug. 25, 1952).

6. W. W. Williams, *Hydrocarbon Process. Pet. Refiner* **43,** 121 (July 1964).
7. F. Zapffe in *Proceedings of the Gas Conditioning Conference*, University of Oklahoma, Norman, Okla., Mar. 5–6, 1958, pp. 17–24.
8. H. G. Muhlbauer and P. R. Monaghan in *Proceedings of the Gas Conditioning Conference*, University of Oklahoma, Norman, Okla., Mar. 6–7, 1957, pp. 69–89.
9. K. J. Fitzgerald and R. A. Richardson, *Hydrocarbon Process.* **45,** 125 (July 1966).
10. J. C. Dingman, *Hydrocarbon Process. Pet. Refiner* **42,** 189 (Sept. 1963).
11. A. L. Kohl and F. C. Riesenfeld, *Gas Purification*, McGraw-Hill Book Co., New York, 1960, pp. 19–20, 34–36, 42–43, 45–46, 49–50, 54, 65–66, 70, 75, 78–81.
12. *Chem. Eng.* 40 (Mar. 4, 1963).
13. L. W. Dailey, *Oil Gas J.* 136 (May 4, 1970).
14. C. J. Wendt, Jr., and L. W. Dailey, *Hydrocarbon Process.* **46,** 155 (Oct. 1967).
15. H. E. Benson, J. H. Field, and W. P. Haynes, *Chem. Eng. Prog.* **52,** 433 (Oct. 1956).
16. J. S. Tosh, J. H. Field, H. E. Benson, and R. B. Anderson, *U.S. Bur. Mines Rep. Invest. 5622*, 1960.
17. J. H. Field, G. E. Johnson, H. E. Benson, and J. S. Tosh, *U.S. Bur. Mines Rep. Invest. 5660*, 1960.
18. D. Bienstock, J. H. Field, and J. G. Meyers, *U.S. Bur. Mines Rep. Invest. 5979*, 1962.
19. B. O. Buck and A. R. S. Leitch, *Pet. Refiner* **37,** 241 (Nov. 1958).
20. F. Zapffe, *Oil Gas J.* **56**(36), 100 (Sept. 8, 1958).
21. *Oilweek* **18,** 31 (Apr. 30, 1962).
22. J. P. Bocard and B. J. Mayland, *Hydrocarbon Process.* **41,** 128 (Apr. 1962).
23. H. W. Manley and G. H. Lewis, *Oil Gas J.* **60**(33), 163 (Aug. 13, 1962).
24. C. L. Dunn, E. R. Freitas, J. W. Goodenbour, H. T. Henderson, and M. N. Papadopoulos, *Hydrocarbon Process. Pet. Refiner* **43,** 150 (Mar. 1964).
25. C. L. Dunn, E. R. Freitas, E. S. Hill, and J. E. R. Sheeler, Jr., *Hydrocarbon Process.* **44**(4), 137 (Apr. 1965).
26. *Sulphur* **62,** 29 (Feb. 1968).
27. C. L. Dunn, E. R. Freitas, E. S. Hill, and J. E. R. Sheeler, Jr., *Oil Gas J.* **63,** 89 (Mar. 29, 1965).
28. *Can. Pet.* **7,** 56 (Nov. 1966).
29. B. G. Goar, *Oil Gas J.* **67**(26), 117 (June 30, 1969).
30. J. W. Sweny and J. P. Valentine, *Chem. Eng.* 54 (Sept. 7, 1970).
31. J. P. Valentine, private communication, Jan. 1982.
32. R. T. Clare and J. P. Valentine, *Energy Process. Can.* **67**(6), 38 (July–Aug. 1975).
33. W. Woelfer, *27th Canadian Chemical Engineering Conference*, Calgary, Alberta, Can., 1977.
34. R. Van Deraerschot and J. P. Valentine, *2nd International Conference on the Control of Gaseous Sulphur and Nitrogen Compound Emission*, Salford, England, Apr. 6–8, 1976.
35. D. K. Judd, *Hydrocarbon Process.* **57,** 122 (Apr. 1978).
36. H. D. Frazier and A. L. Kohl, *Ind. Eng. Chem.* **42**(11), 2288 (Nov. 1950).
37. F. E. Miller and A. L. Kohl, *Oil Gas J.* **51**(51), 175 (Apr. 27, 1955).
38. F. C. Vidaurri and L. C. Kahre, *Hydrocarbon Process.* **56,** 333 (Nov. 1977).
39. R. L. Pearce, *57th Annual Gas Processors Association Convention*, Mar. 20–22, 1978.
40. B. G. Goar, *The Gas Conditioning Conference*, Norman, Okla., Mar. 3–5, 1980.
41. *Pennwalt Product Information, Methyldiethanolamine (MDEA) for Gas Sweetening*, Philadelphia, Pa., S-282, 1980.
42. *Pennwalt Technical Data Specification Sheet 02146*, Philadelphia, Pa., July 1, 1980.
43. H. N. LaCroix and L. J. Coulthurst, *Refiner Nat. Gasoline Manuf.* **18**(8), 334 (Aug. 1939).
44. Ref. 11, p. 131.
45. A. R. Powell, *Nat. Pet. News* **29**(24), 50 (June 16, 1937).
46. J. W. Estep, G. T. McBride, Jr., and J. R. West in J. J. McKetta, Jr., *Advances in Petroleum Chemistry and Refining*, Vol. 6, Interscience Publishers, a division of John Wiley & Sons, Inc., New York, 1962, p. 350.
47. R. Leunddemann, G. Noddes, and H. G. Schwarz, *Oil Gas J.* **57**(32), 100 (Aug. 3, 1959).
48. Ref. 11, pp. 134–137.
49. Ref. 11, p. 123.
50. F. H. Garner, R. Long, and A. Pannell, *J. Appl. Chem.* **8,** 325 (May 6, 1958).
51. H. A. Gollmar and W. W. Hodge, *Proceedings of the Tenth Industrial Waste Conference*, Purdue University Engineering Extension Series No. 89, March 1956, pp. 35–48.
52. Ref. 11, pp. 168–171.

53. H. R. Froning, R. H. Jacoby, and W. L. Richards, *Hydrocarbon Process.* **43**(4), 125 (Apr. 1964).
54. S. M. Ali, M. Younis, and B. Ahmad, *Pak. J. Sci. Res.* 8(4), 239 (1965).
55. S. M. Ali and M. Amin, *Sci. Ind.* (*Karachi*) **5,** 92 (1967).
56. L. G. Townsend, *paper presented to the Manchester and District Association of Gas Engineers*, Manchester, England (Jan. 26, 1949).
57. F. A. Burden and W. B. S. Newling, *Publication No. 397*, The Institution of Gas Engineers, London, Nov. 1951.
58. Ref. 11, p. 295.
59. H. A. Gollmar, *Ind. Eng. Chem.* **26**(2), 130 (Feb. 1934).
60. A. R. Powell, *Joint Committee Conference of the Production and Chemical Committees*, American Gas Association, Inc., New York, May 25–27, 1936, pp. 3–15.
61. G. E. Foxwell and A. Grounds, *Chem. Ind.* 163 (Feb. 25, 1939).
62. A. R. Powell, *Ind. Eng. Chem.* **31**(7), 789 (July 1939).
63. F. A. Toynbee, *J. Applied Chem.* **5**(1), 19 (Jan. 1955).
64. F. C. Riesenfeld and J. F. Mullowney, *Pet. Refiner* **38**(5), 161 (May, 1959).
65. *Chem. Eng.* **67**(9), 166 (Sept. 19, 1960).
66. R. N. Maddox and M. D. Burns, *Oil Gas J.* **66,** 90 (June 3, 1968).
67. F. C. Riesenfeld and J. F. Mullowney, *Oil Gas J.* **57**(20), 86 (May 11, 1959).
68. T. Nicklin and E. Brunner, *Hydrocarbon Process.* **40**(12), 141 (Dec. 1961).
69. *Chem. Ind.* 883 (May 19, 1962).
70. P. Ellwood, *Chem. Eng.* **71,** 128 (July 20, 1964).
71. *Sulphur* **53,** 28 (Aug. 1964).
72. A. J. Moyes and S. Vasan, *Oil Gas J.* **72,** 26 (Sept. 2, 1974).
73. *The Holmes-Stretford Process for Removal of Hydrogen Sulfide from Gas Streams*, Peabody Engineered Systems, 1974.
74. *Sulphur* **101,** 26 (July–Aug. 1972).
75. K. Mitachi, *Chem. Eng.* **80**(24), 78 (Oct. 15, 1973).
76. J. C. Fails and W. D. Harris, *Oil Gas J.* **58**(28), 86 (July 11, 1960).
77. *Chem. Eng.* **67**(25), 86 (Dec. 12, 1960).
78. H. W. Haines, Jr., G. A. Van Wielingen, and G. H. Palmer, *paper presented to American Institute of Chemical Engineers*, New Orleans, La., Feb. 26–Mar. 1, 1961.
79. R. J. Kulperger, *Oilweek* **19**(7), 25 (Apr. 1, 1968).
80. F. Pollio and R. Kunin, *I&EC Prod. Res. Dev.* **7**(1), 62 (Mar. 1968).
81. M. R. Cines, D. M. Haskell, and C. G. Houser, *Chem. Eng. Prog.* **72**(8), 89 (Aug. 1976).
82. Ref. 11, pp. 303–308.
83. Ref. 11, pp. 359–360.
84. H. Brommer and W. Luhr, *Eng. Dig.* 32 (July 1956).
85. Ref. 11, pp. 308–309.
86. Ref. 46, p. 359.
87. Ref. 46, p. 361.
88. *Oil Gas J.* **76,** 92 (Aug. 7, 1978).
89. *Chem. Eng.* **70,** 38 (Apr. 1, 1963).
90. D. K. Beavon and D. J. Cameron, *Meeting of the Canadian Natural Gas Processing Association*, Edmonton, Alberta, Can., Feb. 27, 1970.
91. B. G. Goar, *Hydrocarbon Process.* **53**(7), 129 (July 1974).
92. *Sulphur* **115,** 38 (Nov.–Dec. 1974).
93. D. K. Beavon and J. E. Leeper, *Sulphur* **133,** 35 (Nov.–Dec. 1977).
94. D. K. Beavon and J. E. Leeper, *27th Canadian Chemical Engineering Conference*, Calgary, Alberta, Can., Oct. 23–26, 1977.
95. T. S. Royan and C. E. Loiselle, *paper presented to the Canadian Natural Gas Processing Association*, Calgary, Alberta, Can., June 14, 1978.
96. P. Grancher, *Hydrocarbon Process.* **57**(7), 155 (July 1978).
97. A. G. Maadah and R. N. Maddox, *Hydrocarbon Process.* **57**(8), 143 (Aug. 1970).
98. T. S. Royan and C. E. Loiselle, *Oil Gas J.* **77,** 168 (Jan. 29, 1979).
99. T. S. Royan and C. E. Loiselle, *Energy Process. Can.* **71**(3), 34 (Jan.–Feb. 1979).
100. D. K. Deavon, B. Kouzel, and J. W. Ward, *181st Annual Meeting of the ACS*, Atlanta, Ga., Mar. 29–Apr. 3, 1981.
101. B. W. Gamson and R. H. Elkins, *Chem. Eng. Prog.* **49**(4), 203 (Apr. 1953).

102. R. K. Kerr, *Energy Process. Can.* **68**(6), 28 (July–Aug. 1976).
103. R. K. Kerr and E. M. Berlie, *Energy Process. Can.* **69**(6), 48 (July–Aug. 1977).
104. Ref. 46, pp. 364, 369.
105. R. S. Cunluffe, *Alberta Sulphur Res. Ltd. Quarterly Bull.* **VI**(4), 1 (Jan.–Mar. 1970).
106. R. K. Kerr, H. G. Paskall, and L. C. Biswanger, *Energy Process. Can.* **68**(4), 32 (Mar.–Apr. 1976).
107. R. K. Kerr, H. G. Paskall, and N. Ballash, *Energy Process. Can.* **69**(1), 66 (Sept.–Oct. 1976).
108. J. H. Barnhart, *Chem. Eng. Prog.* **74**(5), 58 (May 1978).
109. Ref. 46, p. 365.
110. I. G. DallaLana, *Canadian Natural Gas Processors Association Research Seminar*, Calgary, Alberta, Can., Nov. 24, 1972.
111. M. J. Pearson, *Gas Process. Can.* 22 (May–June 1973).
112. M. Graulier and D. Papee, *Energy Process. Can.* **66**(4), 32 (Mar.–Apr. 1974).
113. R. A. Ross and M. R. Jeanes, *Prod. Res. Dev.* **13**(2), 102 (June 1974).
114. I. G. DellaLana, *Energy Process. Can.* **66**(6), 38 (July–Aug. 1974).
115. M. J. Pearson, *Energy Process. Can.* **68**(6), 38 (July–Aug. 1974).
116. M. Steijns and P. Mars, *Ind. Eng. Chem., Prod. Res. Dev.* **16**(1), 35 (1977).
117. M. J. Pearson, *Hydrocarbon Process.* 99 (Apr. 1978).
118. H. F. Johnstone, M. D. Kelley, and D. L. McKinley, *Ind. Eng. Chem.* **42**(11), 2298 (Nov. 1950).
119. Ref. 46, pp. 441–444.
120. J. S. Nader and J. L. Molphin, *J. Air Pollut. Control Assoc.* **8**(4), 336 (Feb. 1959).
121. H. C. McKee and W. L. Rollwitz, *J. Air Pollut. Control Assoc.* **8**(4), 338 (Feb. 1959).
122. D. J. Fraade, *Process Control Autom.* 242 (June 1963).
123. *Taylor Technol.* **4**(2), 21 (Fall 1951).
124. P. C. Opeker and B. G. Goar, *Hydrocarbon Process.* **45**(6), 181 (June 1966).
125. R. S. Saltzman and E. B. Hunt, Jr., *ISA Trans.* **12**(2), 103 (1973).
126. E. M. Berlie, *Hydrocarbon Process.* **42**(4), 115 (Apr. 1963).
127. Ref. 46, p. 367.
128. A. R. Laengrich and W. L. Cameron, *Oil Gas J.* **76,** 158 (Mar. 20, 1978).
129. M. H. Bailleul, P. Berthier, and G. Guyot, *Proc. Annu. Conv., Nat. Gas Process Assoc. Tech. Pap.* **49,** 89 (1970).
130. *Can. Chem. Process.* **53**(8), 56 (Aug. 1969).
131. W. D. Saulmon, *Oil Gas J.* **75**(19), 53 (May 9, 1977).
132. P. N. Cheremisinoff, Y. H. Habib, and K. B. Rao, *Ind. Wastes* **18**(5), 22 (Nov.–Dec. 1972).
133. Canadian Petroleum Association, *Submission to Environment Conservation Authority of the Province of Alberta, Canada on Environmental Effects of the Operation of Sulfur Extraction Gas Plants*, Sept. 29, 1972.
134. R. F. Klemm, *Environmental Effects of the Operation of Sulfur Extraction Gas Plants*, submitted to Environment Conservation Authority of the Province of Alberta, Can., Sept. 1972.
135. *Oil Gas J.* **76**(35), 160 (Aug. 28, 1978).
136. *Oil Gas J.* **70,** 85 (June 26, 1972).
137. *Sulphur* **110,** 35 (Jan.–Feb. 1974).
138. L. C. Cameron, *Oil Gas J.* **72**(25), 110 (June 24, 1974).
139. M. M. Morin and Y. M. Philardeau, *Energy Process. Can.* **69**(5), 30 (May–June 1977).
140. *Oil Gas J.* **76**(37), 88 (Sept. 11, 1978).
141. J. E. Nobles, J. W. Palm, and D. K. Knudtson, *Hydrocarbon Process.* 143 (July 1977).
142. A. Doerges, K. Bratzler, and J. Schlauer, *Hydrocarbon Process.* **55**(10), 110 (Oct. 1976).
143. *Sulphur* **128,** 41 (Jan.–Feb. 1977).
144. *Sulphur* **111,** 44 (Mar.–Apr. 1974).
145. J. E. Naber, J. A. Wesselingh, and W. Groenendaal, *Can. Pet.* **14**(8), 32 (Aug. 1973).
146. J. E. Naber, *Energy Process. Can.* **66**(1), 32 (Sept.–Oct. 1973).
147. J. E. Naber, J. A. Wesselingh, and W. Groenendaal, *Chem. Eng. Prog.* **69**(12), 29 (Dec. 1973).
148. C. D. Swaim, Jr., *Paper No. 122 presented at 167th ACS Meeting*, Los Angeles, Calif., Apr. 4, 1974.
149. B. G. Goar, *Energy Process. Can.* **57,** 26 (Nov.–Dec. 1975).
150. The Ralph M. Parsons Co., *Sulphur Recovery Plants and Tail Gas Purification Units*, Bulletin No. 210, Dec. 1976.
151. D. K. Deavon and R. P. Vaell, *Proc. Div. Refining, Am. Pet. Inst.* **52,** 267 (1972).

152. D. K. Beavon and J. C. Brocoff, *Second International Conference, 170th Event of The European Federation of Chemical Engineering*, University of Salford, UK, Apr. 6–8, 1976.
153. *Oilweek* **22**(32), 19 (Sept. 27, 1971).
154. Y. Barthel, S. Frankowiak, P. Renault, A. Deschamps, P. Bonnifay, and J. Andrews, *53rd Annual Gas Processors Association Convention*, Denver, Colo., Mar. 1974.
155. G. L. Thomas and E. Plum, *71st National AIChE Meeting*, Dallas, Texas, Feb. 20–23, 1972.
156. A. Zey, S. White, and D. Johnson, *Chem. Eng. Prog.* **76**(10), 76 (Oct. 1980).
157. D. K. Beavon, R. H. Hass, and B. Muke, *Oil Gas J.* **77,** 76 (Mar. 12, 1979).
158. *Chem. Eng. News* **48**(18), 68 (Apr. 27, 1970).
159. J. S. Hayford, *Hydrocarbon Process.* **52**(10), 95 (Oct. 1973).
160. F. S. Chalmers, *Hydrocarbon Process.* **53**(4), 75 (Apr. 1974).
161. *Sulphur* **117,** 40 (Mar.–Apr. 1975).
162. C. M. Yon, G. R. Atwood, and C. D. Swaim, Jr., *Hydrocarbon Process.* **58**(7), 197 (July 1979).
163. *U.S. Bur. Mines Rep. Invest. 3339*, May 1937.
164. R. S. Dean and R. E. Swain, *U.S. Bur. Mines Bull. 453*, 1944.
165. M. Katz, *Ind. Eng. Chem.* **41**(11), 2450 (Nov. 1949).
166. B. Meyer, *Sulfur, Energy, and Environment*, Elsevier Scientific Publishing Company, New York, 1977, pp. 163–166, 257, 331.
167. L. R. Ember, *Chem. Eng. News* **59**(37), 21 (Sept. 14, 1981).
168. V. N. Mackiw, T. W. Benz, and D. J. I. Evans, *Metall. Rev.* **11,** 143 (1966).
169. R. E. Doane, *Colo. Sch. Mines Res. Inst. Min. Ind. Bull.* **15**(6), 1 (Nov. 1972).
170. N. Arbiter, *IGT Symposium*, Chicago, Ill., Dec. 1974.
171. M. L. Jansen and D. A. Mulligan, *J. Metals* **27**(1), 13 (Jan. 1975).
172. *Sulphur* **140,** 37 (Jan.–Feb. 1979).
173. F. A. Forward, *Trans. Can. Inst. Met.* **LVI,** 373 (1953).
174. F. A. Forward and V. N. Mackiw, *J. Metals*, 457 (Mar. 1955).
175. M. A. Schoening, D. B. Smith, and N. Colvin, *Chemical Engineering Conference of the AIChE and The Chemical Institute of Canada*, Montreal, Can., Apr. 22, 1958.
176. F. A. Forward and H. Veltman in G. R. St. Pierre, ed., *Physical Chemistry of Process Metallurgy*, Pt. 2, Interscience Publishers, New York, 1961, pp. 1255–1276.
177. H. Veltman and P. T. O'Kane, *97th Annual Meeting of the A.I.M.E.*, New York, Feb. 25–29, 1968.
178. R. J. Roman and B. R. Benner, *Min. Sci. Eng.* **5**(1), 3 (Jan. 1973).
179. J. C. Paynter, *J. S. Afr. Inst. Min. Met.* **74**(4), 158 (Nov. 1973).
180. G. M. Swinkels and R. M. G. S. Berezowsky, *CIM Bull.* **71**(790), 105 (Feb. 1978).
181. W. W. Harvey and F. O. Dudas, *Trans. Soc. Min. Eng. AIME* **262,** 46 (Mar. 1977).
182. G. J. Snell and M. C. Sze, *Eng. Min. J.* **178**(10), 100 (Oct. 1977).
183. N. Arbiter, *BNF Conference*, Barcelona, Spain, 1974.
184. A. G. Fonseca, *CIM Bull.* **67**(747), 105 (July 1974).
185. M. C. Kuhn, N. Arbiter, and H. Kling, *CIM Bull.* **67**(742), 62 (Feb. 1974).
186. T. Biegler, H. R. Skewis, and D. A. Swift, *AIME Annual Meeting*, Chicago, Ill., Feb. 25–Mar. 1, 1973.
187. F. P. Haver and M. M. Wong, *J. Metals* 25 (Feb. 1971).
188. I. H. Warren, *Aust. J. Appl. Sci.* **9**(1), 36 (Nov. 1958).
189. C. Chamberlain, J. Newton, and D. Clifton, *Eng. Min. J.* **170**(10), 90 (Oct. 1969).
190. J. K. Gerlach, E. D. Gock, and S. K. Ghosh, *Second International Symposium on Hydrometallurgy*, Chicago, Ill., Feb. 26–Mar. 1, 1973.
191. G. Bjorling and F. Lesidrenski, *Paper No. A68-1, TMS-AIME Annual Meeting*, New York, Feb. 25–29, 1968.
192. W. Kunda, J. P. Warner, and V. N. Mackiw, *Trans. CIM Bull.* **LXV,** 21–25 (Jan. 1962).
193. P. H. Yu, C. K. Hansen, and M. E. Wadsworth, *Second International Symposium on Hydrometallurgy*, Chicago, Ill., Feb. 26–Mar. 1, 1973.
194. J. Dasher, *CIM Bull.* **66**(733), 48 (May 1973).
195. O. Sitnai and P. K. Peeler, *Proc. Australas. Inst. Min. Metall.* (261), 21 (Mar. 1977).
196. T. Okabe and H. Ito, *Technol. Rep. Tohoku University* **XXIX**(1), 157 (1964).
197. E. Peters and F. Loewen, *Paper No. A68-33, TMS-AIME Annual Meeting*, New York, Feb. 25–29, 1968.
198. R. Shantz and T. M. Morris, *Eng. Min. J.* **175**(5), 71 (May 1974).
199. R. A. Meyers, J. W. Hamersma, and M. L. Kraft, *Environ. Sci. Technol.* **9**(1), 70 (Jan. 1975).

200. E. P. Foster, D. R. Taschler, and D. F. Lowe, *J. Metals* **31**(7), 23 (July 1979).
201. J. H. McNamara, W. A. Ahrens, and J. B. Franek, *AIME Annual Meeting*, Denver, Colo., Feb. 26, 1978.
202. P. R. Kruesi, E. S. Allen, and J. L. Lake, *Hydrometallurgical Conference of the CIMM*, Halifax, Nova Scotia, Can., Aug. 27–30, 1972.
203. *Trans. CIM Bull.* **LVIII,** 158 (1955).
204. R. Saddington, W. Curlock, and P. Queneau, *J. Metals*, 440 (Apr. 1966).
205. S. Merla, C. E. Young, and J. W. Matousek, *AIME Annual Meeting*, San Francisco, Calif., Feb. 20–24, 1972.
206. P. Bryk, *British Nonferrous Metals Research Association Misc. Publ. No. 435*, London, UK, Dec. 1951.
207. F. Benitez, *Eng. Min. J.* **154**(10), 76 (1953).
208. *Sulphur* **50,** 33 (1964).
209. *Eng. Min. J.* **169,** 168 (Mar. 1968).
210. E. P. Fleming and T. C. Fitt, *Ind. Eng. Chem.* **42**(11), 2253 (Nov. 1950).
211. *Chem. Eng.* **60**(4), 274 (Apr. 1953).
212. J. M. Henderson and J. B. Pfeiffer, *Min. Eng.* **26**(11), 36 (Nov. 1974).
213. A. H. Loveless, *B.I.O.S. Final Report No. 863, Item No. 22*, London, UK, 1946.
214. M. P. Applebey, *J. Soc. Chem. Ind.* **56,** 139 (May 1937).
215. H. F. Johnstone and A. D. Singh, *Eng. Exp. St. Bull. Ser. No. 324, University of Illinois Bull.* **XXXVIII**(19), 1 (Dec. 31, 1940).
216. H. E. Newell, *Paper No. 55-APC-3, First International Congress on Air Pollution of ASME*, New York, Mar. 1–2, 1955.
217. T. Kennaway, *J. Air Pollut. Control Assoc.* **7**(4), 266 (Feb. 1958).
218. H. F. Johnstone, *Pulp Paper Mag. Can.* **53**(4), 105 (Mar. 1952).
219. S. R. Craxford, A. Poll, and W. J. S. Walker, *J. Inst. Fuel*, 13 (Jan. 1952).
220. J. P. V. Woollam and A. Jackson, *Trans. Inst. Chem. Eng.* **23,** 43 (1945).
221. H. F. Johnstone, H. J. Read, and H. C. Blankmayer, *Ind. Eng. Chem.* **30**(1), 101 (Jan. 1938).
222. R. E. Swain, *Ind. Eng. Chem.* **31**(11), 1358 (1939).
223. D. D. Howat, *Chem. Age* 273 (Dec. 14, 1940).
224. R. Lepsoe and W. S. Kirkpatrick, *Trans. CIMM* **XL,** 399 (Aug. 1937).
225. R. Lepsoe, *Ind. Eng. Chem.* **30**(1), 92 (Jan. 1938).
226. S. D. Kirkpatrick, *Chem. Met. Eng.* **45**(9), 483 (Sept. 1938).
227. R. Lepsoe, *Ind. Eng. Chem.* **32**(7), 910 (July 1940).
228. E. P. Fleming and T. C. Fitt, *Ind. Eng. Chem.* **42**(11), 2249 (Nov. 1950).
229. P. Queneau, E. H. Bracken, and D. Kelly, *J. Metals* **10**(8), 527 (Aug. 1958).
230. U.S. Pat. 3,199,955 (Aug. 10, 1965), J. R. West and E. H. Conroy, Jr., (to Texas Gulf Sulphur Co.).
231. H. R. Potts and E. G. Lawford, *Bull. Inst. Min. Met.* (509), 1 (Apr. 1949).
232. T. Kiaer, *Eng. Min. J.* **155**(7), 88 (July 1954).
233. W. D. Hunter, Jr., and J. P. Wright, *Chem. Eng.* **79**(2), 50 (Oct. 1972).
234. J. P. Wright, *Sulphur* **100,** 72 (May–June 1972).
235. R. G. Bierbower and J. H. Van Sciver, *Chem. Eng. Prog.* **70**(8), 60 (Aug. 1974).
236. *Sulphur* **141,** 34 (Mar.–Apr. 1979).
237. B. A. Laseke and T. W. Devitt, *Chem. Eng. Prog.* **75**(2), 37 (Feb. 1979).
238. A. V. Slack, *Chem. Eng. Prog.* **74**(2), 71 (Feb. 1978).
239. H. Idemura, T. Kanai, and H. Yanagioka, *Chem. Eng. Prog.* **74**(2), 46 (Feb. 1978).
240. J. Battelheim and A. C. Dollins, *Chem. Ind.* (3), 122 (Feb. 1, 1975).
241. M. Epstein, L. Sybert, S. C. Wang, C. C. Leivo, and F. T. Princiotta, *Chem. Eng. Prog.* **70**(6), 53 (June 1974).
242. S. S. Strom and W. Downs, *Chem. Eng. Prog.* **70**(6), 55 (June 1974).
243. W. J. Raymond and A. G. Sliger, *Chem. Eng. Prog.* **74**(2), 75 (Feb. 1978).
244. J. G. Selmeczi and D. A. Stewart, *Chem. Eng. Prog.* **74**(2), 41 (Feb. 1978).
245. C. R. LaMantia, R. R. Lunt, and I. S. Shah, *Chem. Eng. Prog.* **70**(6), 66 (June 1974).
246. C. G. Cornell and D. A. Dahlstrom, *Chem. Eng. Prog.* **69**(12), 47 (Dec. 1973).
247. R. L. Gall and E. J. Piasecki, *Chem. Eng. Prog.* **71**(5), 72 (May 1975).
248. W. D. Brown and E. C. Brown, *CIM Bull.* **71**(795), 82 (July 1978).
249. *Sulphur* **132,** 42 (Sept.–Oct. 1979).
250. E. L. Mann, *Electr. World*, 70 (Nov. 1, 1972).

251. L. Korosy, H. L. Dewanter, F. S. Chalmers, and S. Vasan, *167th ACS National Meeting*, Los Angeles, Calif., Apr. 5, 1974.
252. S. Vasan, *Chem. Eng. Prog.* **71**(5), 61 (May 1975).
253. G. R. Koehler, *Chem. Eng. Prog.* **70**(6), 63 (June 1974).
254. J. M. Genco, H. S. Rosenberg, M. Y. Anastas, E. C. Rosar, and J. M. Dulin, *J. Air Pollut. Control Assoc.* **25**(12), 1244 (Dec. 1975).
255. J. M. Genco and H. S. Rosenberg, *J. Air Pollut. Control Assoc.* **26**(10), 989 (Oct. 1976).
256. R. S. Engdahl and H. S. Rosenberg, *Chemtech* **8**(2), 118 (Feb. 1978).
257. A. Tamaki, *Chem. Eng. Prog.* **71**(5), 55 (May 1975).
258. W. F. Bischoff and Y. Habib, *Chem. Eng. Prog.* **71**(5), 59 (May 1975).
259. *Sulphur* **119,** 24 (July–Aug. 1975).
260. *Environ. Sci. Technol.* **9**(8), 712 (Aug. 1975).
261. J. J. Collins, L. L. Fornoff, K. D. Manchanda, W. C. Miller, and D. C. Lovell, *Chem. Eng. Prog.* **70**(6), 58 (June 1974).
262. W. E. Miller, *Chem. Eng. Prog.* **70**(6), 49 (June 1974).
263. H. Idemura, *Chem. Econ. Eng. Rev.* **6**(8), 22 (Aug. 1974).
264. *Sulphur* **125,** 45 (July–Aug. 1976).
265. *Ibid.*, p. 43.
266. *Environ. Sci. Technol.* **11**(1), 22 (Jan. 1977).
267. C. Grimm, J. Z. Abrams, W. W. Leffmann, I. A. Raben, and C. LaMantia, *Chem. Eng. Prog.* **74**(2), 51 (Feb. 1978).
268. R. A. McIlroy, G. A. Atwood, and C. J. Major, *Environ. Sci. Technol.* **7**(11), 1022 (Nov. 1973).
269. D. H. McCrea, J. H. Field, and E. R. Bauer, Jr., *Paper 68-WA/FU-3, ASME Annual Meeting*, New York, Dec. 1–5, 1968.
270. *Sulphur* **55,** 24 (Dec. 1964).
271. D. Bienstock, J. H. Field, S. Katell, and K. D. Plants, *J. Air Pollut. Control Assoc.* **15**(10), 459 (Oct. 1965).
272. R. J. Sappok and P. L. Walker, Jr., *J. Air Pollut. Control Assoc.* **19**(11), 856 (Nov. 1969).
273. W. D. Lovett and F. T. Cunniff, *Chem. Eng. Prog.* **70**(5), 43 (May 1974).
274. R. J. Evans and W. A. Duvel, Jr., *Pollut. Eng.* **6**(10), 44 (Oct. 1974).
275. J. M. Dulin and E. C. Rosar, *Environ. Sci. Technol.* **9**(7), 627 (July 1975).
276. *Environ. Sci. Technol.* **9**(7), 622 (July 1975).
277. W. H. Lord, *Pollut. Eng.* **8**(6), 40 (June 1976).
278. Research and Education Association, *Modern Energy Technology*, Vol. II, Research and Education Association, New York, 1975, pp. 1413–1422, 1430–1450.
279. J. A. Cavallaro, M. T. Johnston, and A. W. Deurbrouck, *U.S. Bur. Mines Rep. Invest. 8118*, 1976.
280. A. W. Deurbrouck and E. R. Palowitch, *U.S. Bur. Mines I.C. 8282*, 1966.
281. K. J. Miller, *U.S. Bur. Mines Rep. Invest. 7822*, 1973.
282. W. T. Abel, M. Zulkoski, G. A. Brady, and J. W. Eckerd, *U.S. Bur. Mines Rep. Invest. 7732*, 1973.
283. J. W. Leonard and D. R. Mitchell, *Coal Preparation*, 3rd ed., The AIME, Inc., New York, 1968, Chapts. 9–11, pp. 9-1-11-20.
284. T. A. Sladek and C. H. Cox in Editors of Chemical Engineering Progress, *Coal Processing Technology*, Vol. 5, AIChE, New York, 1979, pp. 19–23.
285. J. C. Agarwal, R. A. Giberti, P. F. Irminger, L. J. Petrovic, and S. S. Sareen, *167th National ACS Meeting*, Los Angeles, Calif., Apr. 1974.
286. Ref. 278, pp. 1425–1426, 1429.
287. R. C. Hoke, M. S. Nutkis, L. A. Ruth, and H. Shaw, *NTIS Report PB-23/374*, National Technical Information Service, Washington, D.C., Jan. 1974.
288. R. C. Hoke, L. A. Ruth, and M. Ernst in A. J. Engel, S. M. Slater, and J. W. Gentry, eds., *Emission Control From Stationary Power Sources: Technical, Economic and Environmental Assessments*, AIChE Symposium Series No. 201, Vol. 76, AIChE, New York, 1980, pp. 16–21.
289. A. L. Conn, *Chem. Eng. Prog.* **77**(5), 11 (May 1981).
290. Ref. 278, pp. 943–944.
291. J. W. White and M. Larson, *Symposium Papers—Clean Fuels from Coal*, Institute of Gas Technology, Chicago, Ill., Sept. 10–14, 1973.
292. P. L. Russell, *Q. Colo. Sch. Mines* **71**(4), 173 (Oct. 1976).
293. N. DeNevers, *Sci. Am.* **214**(2), 24 (Feb. 1966).
294. R. F. Cane, *Hydrocarbon Process.* **44** 189 (Sept. 1965).

295. T. A. Hendrickson, *Q. Colo. Sch. Mines* **69**(2), 45 (Apr. 1974).
296. P. Schmalfeld, *Q. Colo. Sch. Mines* **70**(3), 142 (July 1975).
297. B. Grant, *Oil Gas J.* **62**(20), 84 (July 20, 1964).
298. R. D. Ridley, *Q. Colo. Sch. Mines* **69**(2), 21 (Apr. 1974).
299. C. Y. Cha and D. E. Garrett, *Q. Colo. Sch. Mines* **70**(3), 31 (July 1975).
300. R. D. Ridley and R. T. Chew, III, *Q. Colo. Sch. Mines* **70**(3), 123 (July 1975).
301. H. E. McCarthy and C. Y. Cha, *Q. Colo. Sch. Mines* **71**(4), 85 (Oct. 1976).
302. R. J. Gumerman, *Chem. Eng. Prog.* **75**(9), 85 (Sept. 1979).
303. K. L. Berry, *Chem. Eng. Prog.* **75**(9), 72 (Sept. 1979).
304. P. M. Dougan, *Chem. Eng. Prog.* **75**(9), 81 (Sept. 1979).
305. C. L. Tsaros and H. L. Feldkirchner, *Preprint 30D, 58th Annual Meeting of AIChE*, Philadelphia, Pa., Dec. 5–9, 1965.
306. F. C. Schora, Jr., H. L. Feldkirchner, P. B. Tarman, and S. A. Weil, *Hydrocarbon Process.* **53,** 89 (Apr. 1974).

General References

P. A. Ferguson, *Hydrogen Sulfide Removal from Gases, Air, and Liquids*, Noyes Data Corporation, Park Ridge, N.J., 1975.

H. W. Wainwright, G. C. Egleson, C. M. Brock, J. Fisher and A. E. Sands, *U.S. Bur. Mines Rep. Invest. 4891*, Oct. 1952.

W. T. Abel, F. G. Shultz, and P. F. Langdon, *U.S. Bur. Mines Rep. Invest. 7947*, 1974.

F. G. Sawyer, R. N. Hader, L. K. Herndon, and E. Morningstar, *Ind. Eng. Chem.* **42,** 1938 (Oct. 1950).

Yu. F. Makogon and G. A. Sarkisyants, *National Research Council of Canada Technical Translation No. 15476*, Ottawa, Canada, 1976.

S. Katell and L. T. Fan, *U.S. Bur. Mines Inf. Circ. 7886*, 1959.

J. M. Genco and S. S. Tam, *EPA Report EPA-450/3-74-055*, Washington, D.C., June 1974.

C. A. Rohrmann and H. T. Fullam, *EPA Report EPA-650/2-74-085-b, NTIS No. PB237928*, Washington, D.C., Sept. 1974.

U.S. Bur. Mines Inf. Circ. 8527, 1971.

A. G. Starliper and H. Kenworthy, *U.S. Bur. Mines Rep. Invest. 6554*, 1964.

F. P. Haver and M. M. Wong, *U.S. Bur. Mines Rep. Invest. 8006*, 1975.

M. B. Shirts, J. K. Winter, P. A. Bloom, and G. M. Potter, *U.S. Bur. Mines Rep. Invest. 7953*, 1974.

R. B. Worthington, *U.S. Bur. Mines Rep. Invest. 7705*, 1973.

F. P. Haver and M. M. Wong, *U.S. Bur. Mines Rep. Invest. 7474*, 1971.

F. P. Haver, K. Uchida, and M. M. Wong, *U.S. Bur. Mines Rep. Invest. 7185*, 1968.

M. H. Stanczyk and C. Rampacek, *U.S. Bur. Mines Rep. Invest. 6808*, 1966.

J. D. Spagnola, *U.S. Bur. Mines Rep. Invest. 6662*, 1965.

P. Queneau, ed., *Extraction Metallurgy of Copper, Nickel and Cobalt*, Wiley-Interscience, New York, 1975. pp. 503–534.

D. Bienstock, L. W. Brunn, E. M. Murphy, and H. E. Benson, *U.S. Bur. Mines Inf. Circ. 7836*, 1958.

L. A. Hass, *U.S. Bur. Mines Inf. Circ. 8608*, 1973.

R. J. Evans, *U.S. Bur. Mines Inf. Circ. 8633*, 1974.

D. A. Martin and F. E. Brantley, *U.S. Bur. Mines Rep. Invest. 6321*, 1963.

J. I. Paige, J. W. Town, J. H. Russell, and H. J. Kelly, *U.S. Bur. Mines Rep. Invest. 7414*, 1970.

L. A. Hass, T. H. McCormick, and S. E. Khalafalla, *U.S. Bur. Mines Rep. Invest. 7483*, 1971.

J. W. Town, J. H. Russell, and H. J. Kelly, *U.S. Bur. Mines Rep. Invest. 7478*, 1971.

L. A. Hass, T. H. McCormick, and S. E. Khalafalla, *U.S. Bur. Mines Rep. Invest. 7647*, 1972.

J. B. Rosenbaum, W. A. McKinney, H. R. Beard, L. Crocker, and W. I. Nissen, *U.S. Bur. Mines Rep. Invest. 7774*, 1973.

A. V. Slack and G. A. Hollinden, *Sulfur Dioxide Removal from Waste Gases*, 2nd ed., Noyes Data Corporation, Park Ridge, N.J., 1975.

R. Brandland, *Proceedings of the Symposium on Recovery of Pulping Chemicals*, The Finnish Pulp & Paper Institute, Helsinki, Finland, 1969.

P. Marier and T. R. Ingraham, *Mines Branch Research Report R222*, Canadian Department of Energy, Mines, and Resources, Mar. 1970.

B. L. Crynes, ed., *Chemical Reactions as a Means of Separation—Sulfur Removal*, Marcel Dekker, Inc., New York, 1977.

J. O. Nriague, ed., *Sulfur in the Environment—Part I: The Atmospheric Cycle*, John Wiley & Sons, Inc., New York, 1978.
J. O. Nriague, ed., *Sulfur in the Environment—Part II: Ecological Impacts*, John Wiley & Sons, Inc., New York, 1978.
National Academy of Engineering, *Pipeline-Quality Gas*, Pt. I of *Evaluation of Coal Gas Technology*, Superintendent of Documents Stock No. 2414-00057, National Research Council, Washington, D.C., 1973.
P. J. LaRosa and H. J. Michaels, *Study of Sulfur Recovery from Coal Refuse*, Superintendent of Documents Stock No. 5501-0137, EPA, Washington, D.C., Sept. 1971.
T. Fraser, W. L. Crentz, and A. L. Bailey, *U.S. Bur. Mines. Bull. 483*, 1950.
Editors of Chemical Engineering Progress, *Coal Processing Technology*, Vol. 7, AIChE, New York, 1981; see also earlier volumes.
J. Cooperman, J. D. Davis, W. Seymour, and W. L. Ruckes, *U.S. Bur. Mines Bull. 498*, 1951.
A. Matzik, R. O. Dannenberg, J. R. Ruark, J. E. Phillips, J. A. Lankford, and B. Guthrie, *U.S. Bur. Mines Bull. 635*, 1966.
S. Katell and P. Wellman, *U.S. Bur. Mines Technical Prog. Rep. 44*, Oct. 1971.
U.S. Bur. Mines Open File Rep. 32-72 and *33-72*, July 1972.
E. L. Burwell, T. E. Sterner, and H. C. Carpenter, *U.S. Bur. Mines Rep. Invest. 7783*, 1973.
S. Katell and P. Wellman, *U.S. Bur. Mines Technical Prog. Rep. 81*, Sept. 1974.
T. A. Sladek, *Colo. Sch. Mines Min. Ind. Bull.* **18**(2), (1975); see also earlier Bulletin.
L. K. Barker, *U.S. Bur. Mines Rep. Invest. 8011*, 1975.
U.S. Dept. Energy Technol. Information Center, *A Bibliography of Publications Dealing with Oil Shale and Shale Oil from U.S. Bur. Mines, 1917–1974, and ERDA Energy Research Center, 1975–1976*, LERC/R1-77/7, Sept. 1977.
E. L. Burwell, H. C. Carpenter, and H. W. Sohns, *U.S. Bur. Mines Technical Prog. Rep. 16*, June 1969.
A. E. Barak, L. Dockter, and H. C. Carpenter, *U.S. Bur. Mines Technical Prog. Rep. 30*, Jan. 1971.
E. L. Burwell and I. A. Jacobson, Jr., *U.S. Bur. Mines Technical Prog. Rep. 85*, Nov. 1974.
J. R. East, Jr., and E. D. Gardner, *U.S. Bur. Mines Bull. 611*, 1964.
C. O. Hawk, M. D. Schlesinger, H. H. Ginsberg, and R. W. Hiteshue, *U.S. Bur. Mines Rep. Invest. 6548*, 1964.
J. B. Pfeiffer, ed., *Sulfur Removal and Recovery from Industrial Processes*, American Chemical Society, Washington, D.C., 1975.
Sulphur **96,** 22 (Sept.–Oct. 1971); **97,** 23 (Nov.–Dec. 1971); **99,** 30 (Mar.–Apr. 1972); **111,** 36 (Mar.–Apr. 1974); **112,** 35 (May–June 1974); **115,** 38 (Nov.–Dec. 1974); **116,** 21 (Jan.–Feb. 1975); **117,** 40 (Mar./Apr. 1975); **129,** 25, 30 (Mar.–Apr. 1977); **132,** 28 (Sept.–Oct. 1977); **133,** 41 (Nov.–Dec. 1977); **134,** 26 (Jan.–Feb. 1978); **140,** 34 (Jan.–Feb. 1979); **142,** 26 (May–June 1979); **143,** 24 (July–Aug. 1979).
J. H. Gary, ed., *Fourteenth Oil Shale Symposium Proceedings*, Colorado School of Mines Press, Golden, Colo., Aug. 1981; see also earlier Proceedings.

JAMES R. WEST
Texasgulf Inc.

SUNSCREEN AGENTS. See Cosmetics.

SUPERACIDS. See Antimony compounds.

SUPERALLOYS. See High-temperature alloys.

SUPERCONDUCTING MATERIALS

Superconductivity, discovered by Onnes in 1911 (1), is a phenomenon occurring in many metals and alloys at very low temperatures, near the boiling points of liquid helium and hydrogen. The term superconductivity was derived from its primary characteristics, namely, the complete absence of electrical resistance below the superconducting transition temperature T_c. Onnes was well aware of the possible technological implications of his discovery and, in 1913, constructed a lead coil in an attempt to generate a magnetic field by passing a current through the lossless wire (2). The results were disappointing, however, since only very low magnetic fields could be generated before the coil returned to the normal conducting state. More than 40 yr passed before a magnetic induction of ca 0.71 T (7.1 kG) was obtained by wrapping a niobium wire on an iron core (3). In 1960, a solenoid was constructed without the iron core, and 0.4 T (4.0 kG) was obtained at the center of the coil (4). Subsequently, the superconducting solenoid became a common tool in research laboratories for the generation of intense magnetic fields. The Nb–Zr alloys initially used were replaced by Nb–Ti alloys which were true engineering materials. In 1969–1970, the bronze process was discovered and used to fabricate multifilamentary wires of Nb_3Sn and V_3Ga. These are intermetallic compounds capable of generating much higher magnetic fields than the Nb–Ti alloys. At this point, the possible large-scale applications of superconductivity began to be considered in earnest. Simultaneously, the search continued for economical ways to produce superconducting materials with even more desirable properties (see also Magnetic materials).

Superconductivity was first applied on a small scale in 1964 when quantum interference effects were observed between two Josephson junctions located in a superconducting ring (5–6). The device potential of this phenomenon was quickly realized and soon a very sensitive magnetometer known as a superconducting quantum interference device (SQUID) was fabricated. Josephson junctions have since been used in mixers, detectors, and as circuit elements for digital computers. The material problems are somewhat different in this case. For example, fabrication procedures for computer-circuit elements must be compatible with integrated-circuit technology. Common to both large- and small-scale applications is the necessity of finding materials with higher superconducting transition temperatures and good mechanical properties (see Integrated circuits).

Theory

The original data of Onnes on the resistivity of mercury are shown in Figure 1 (1). Below the superconducting transition temperature T_c (ca 4.25 K), the resistivity was observed to drop suddenly to zero. (Recent experiments have shown that the resistivity is so small that a current in a superconducting ring would require $>10^5$ yr to decay.) Onnes observed similar transitions in lead (T_c = 7.2 K) and tin (T_c = 3.7 K). In 1913, Onnes constructed a lead coil and attempted to use it at 4.2 K to generate a magnetic field. The field in the coil, however, was quenched to the normal state after generating a magnetic field of only a few hundred Gauss. The phenomenon of superconductivity has now been observed in nearly half of the elements of the periodic table and in many thousands of alloys and compounds. In some cases it has been necessary

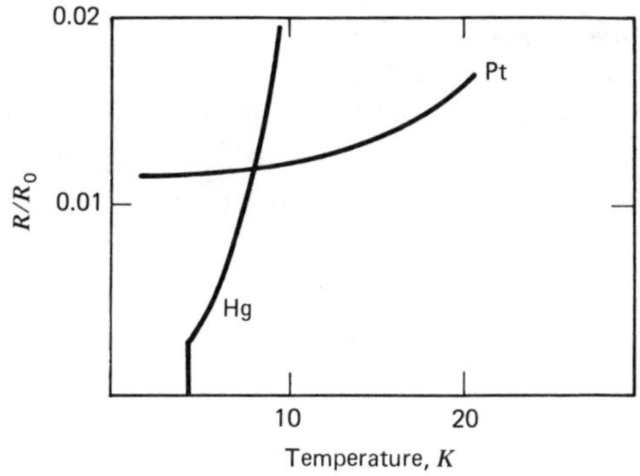

Figure 1. Relative resistance of mercury and platinum vs absolute temperature (1).

to apply very high pressures or to reduce the temperature to values very close to absolute zero in order to observe the phenomenon. Nevertheless, superconductivity has proved to be quite general in nature.

Despite the development of a microscopic theory of superconductivity in 1957 (7), there are no general rules for predicting whether a given material is superconducting or the value of the transition temperature T_c. However, certain trends are evident that furnish the basis for empirical rules for the occurrence of superconductivity. For example, superconductors with high T_c values tend to be poor conductors in the normal state. Furthermore, the occurrence of magnetism usually precludes the occurrence of superconductivity. Perhaps one of the most useful of the empirical rules is the Matthias rule, illustrated in Figure 2 (8). The transition temperature T_c is plotted as a function of the valence electron per atom (e/a) ratio for the 4d transition metals and their alloys. Pronounced maxima occur in T_c for e/a ca 4.8 and e/a ca 6.4, whereas a deep minimum occurs for e/a ca 5.7. This behavior was deduced from experimental results and has been employed to predict new high T_c alloys.

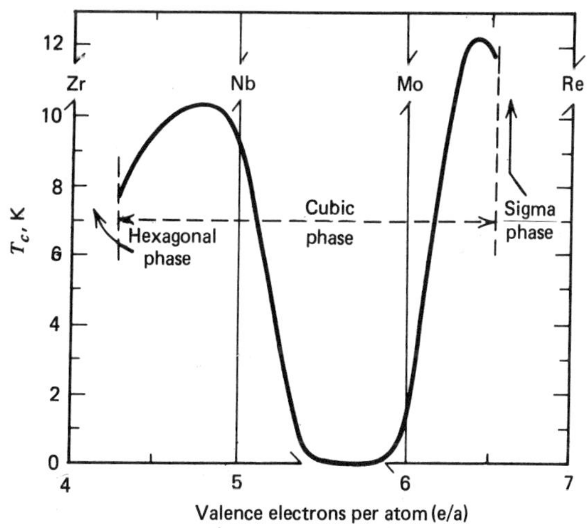

Figure 2. The variation of the superconducting transition temperature T_c of transition metals and alloys vs electron per atom ration (e/a), illustrating Matthias' rule (9).

Critical Magnetic Fields and Critical Currents. Early discussions of superconductivity were dominated by the phenomenon of the infinite conductivity of the material. However, if a superconductor is cooled below T_c in a low applied magnetic field, that magnetic field is expelled from the interior of the superconductor (10). If a superconductor were simply a material whose conductivity was infinite below T_c, then the external magnetic field applied before the material was cooled below T_c would not be expelled but would be kept constant by eddy currents induced on the surface of the sample. Thus, it was shown that the superconducting state is a thermodynamic equilibrium state, resulting from a phase transition to a state with lower free energy than the normal conducting state (10). Exclusion of the external field costs a certain amount of free energy. For a certain value of the external field $H = H_c$ (the critical field), the induced currents can no longer exclude the field and the superconductor makes the transition (quenches) to the normal state. Considering a long cylinder of a pure simple metal, eg, Pb, Sn, etc, oriented parallel to the external field, total exclusion of the external field below the critical field H_c occurs, that is, the magnetic induction $B = H + 4\pi M = 0$ for $H < H_c$ (where M = magnetization). This is referred to as type-I behavior. A different kind of magnetic behavior (type II) is observed in transition-metal alloys. In that case, total flux exclusion ($B = 0$) occurs up to an external field H_{c1}, the lower critical field. Thereafter, flux enters the material in quantized bundles, $\Phi_0 = hc/2e = 2.07 \times 10^{-11}$ T·cm^2 (2.07×10^{-7} G·cm^2), up to an external field H_{c2}, the upper critical field. Although the superconductor does not totally exclude flux between H_{c1} and H_{c2}, the electrical resistivity continues to be zero (below a critical current) up to H_{c2}. Thus, the useful electrical properties of type-II materials persist up to H_{c2}, and it is common to display the useful region on a graph of H_{c2} vs T as shown in Figure 3, for some of the highest critical-field and critical-temperature materials.

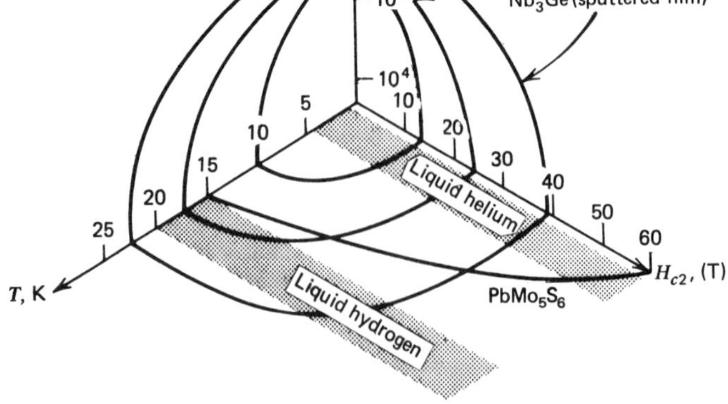

Figure 3. The critical current density J_c (on a log scale) is shown as a function of the critical upper magnetic field H_{c2} and the temperature T for three readily available superconductors. A plot of H_{c2} vs T is also shown for the ultrahigh field compound $PbMo_5S_6$ (11–14). To convert T to G, multiply by 10^4.

The field region between H_{c1} and H_{c2} is commonly referred to as the mixed state. In the mixed state, the quantized bundles of flux or fluxoids arrange themselves in a periodic lattice with hexagonal symmetry, the flux-line lattice. Most superconductors of technological interest are type-II materials and, henceforth, this discussion assumes type-II behavior.

The application of a magnetic field is not the only way to destroy superconductivity. The passage of a large current through the material may also result in quenching to the normal state. The critical current density J_c, which is a function of the temperature, magnetic field, and most importantly, the metallurgical history of the material, is a primary factor in determining the technological utility of a material. The passage of a transport current through a superconductor in the mixed state results in a Lorentz force being exerted on a bundle of flux lines. As long as the flux lines are pinned by some crystalline imperfection or impurity, they remain stationary and no power is dissipated in the material. The current density for which flux-line motion begins is referred to as the critical current density J_c. From a technological viewpoint it is important to note that J_c can be increased by creating lattice imperfections or by inserting impurities to pin the flux lines.

Bardeen-Cooper-Schrieffer Theory. It was recognized as early as 1950 that the electron–phonon interaction is probably very important for superconductivity (15). However, early attempts to arrive at a description of the superconducting state via a perturbation of the normal state were unsuccessful. An important step was the demonstration that the attractive electron–electron interaction, due to exchange of phonons between two electrons with opposite spins and momenta, could lead to a bound pair no matter how weak the interaction was (16). In 1957 Bardeen, Cooper, and Schrieffer (BCS) used this idea of the bound pair (Cooper pair) to write a wave function for the ground state of a superconductor in which all electrons were paired. After adjusting the paired wave functions to minimize the free energy, the paired state was used as the basis of a complete microscopic theory of superconductivity. The successes of the BCS theory are many, including the idea of a temperature-dependent energy gap, which opens up in the electronic energy spectrum at the Fermi energy for $T < T_c$. Another important result is an expression for T_c in terms of the fundamental parameters of the system:

$$T_c \approx \theta_D \exp\,[-1/NV] \tag{1}$$

In equation 1, θ_D is the Debye temperature, N is the density of states at the Fermi energy for one-spin direction, and V represents the effective strength of the attractive phonon-mediated interaction between the electrons. Thus, for a fixed V, higher values of T_c are expected when the density of states at the Fermi level is large. The BCS theory is the fundamental microscopic theory from which all other phenomenological treatments are derived.

Certain characteristic lengths which first appeared in more phenomenological treatments are further clarified by the BCS theory. For example, the coherence length ξ was introduced to explain nonlocal behavior in the electromagnetic properties of superconductors (17). In the phenomenological theory, it is essentially the shortest distance over which the superconducting wave function may vary (18). In the BCS theory this distance is also recognized as the correlation length of two electrons in a bound Cooper pair. The other important length is the penetration depth λ for the external magnetic field. This quantity was first introduced in a discussion of magnetic-field penetration into superconductors and represents a characteristic distance for decay of an external field into a superconductor (19).

Tunneling. The phenomenon of tunneling is a consequence of the wave nature of matter and was discovered early in the history of quantum mechanics. If two strips of metal are separated by a thin (1–10 nm) oxide barrier, a certain probability exists that an electron can pass from one strip to another. Tunneling can occur between two superconductors, with a nonlinear current–voltage curve as shown in Figure 4 (20). The sharp rise in the tunneling current occurs at twice the gap parameter Δ, when the superconductors are equal and at the sum of the gap parameters when the gap parameters are different. This kind of tunneling, involving the passage of one electron across the barrier, is known as quasi-particle (normal-electron) tunneling. Aside from obtaining the value of the gap parameter Δ, the current–voltage characteristics also contain a small structure at higher voltages that can furnish information about the electron–phonon interactions in the superconductor.

Another kind of tunneling was predicted in 1962 involving the tunneling of bound pairs from one superconductor through an oxide barrier to another superconductor. The superconductors are coupled on both sides of the barrier. The resulting supercurrent depends upon the phase difference of the two wave functions and may be present in the absence of an applied voltage as shown in Figure 4. In the presence of microwave radiation, steps appear on the I–V curve of Figure 4 due to the a-c Josephson effect in which the phase difference between the two superconductors is modified by the presence of the microwave field. The d-c Josephson effect is now used in very sensitive magnetometers and in computer memory elements. The a-c Josephson effect is used for very precise determinations of the fundamental constant ratio h/e and in sensitive infrared detectors.

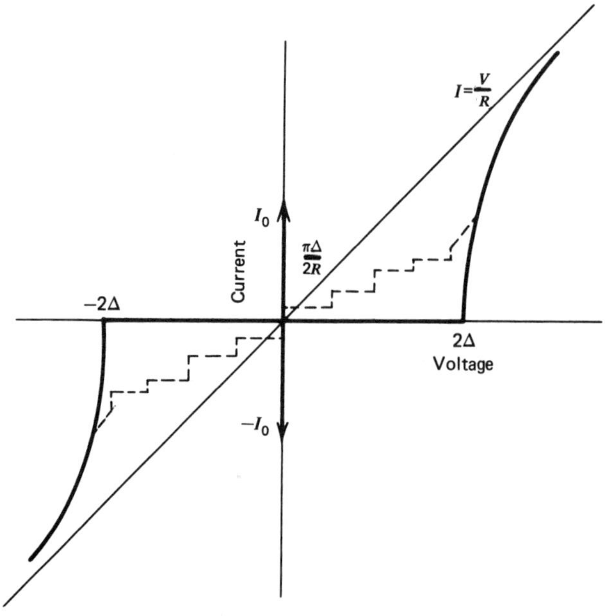

Figure 4. Josephson tunneling current of pairs, quasi-particle tunneling current, and tunneling current when both superconductors are normal. The dotted curve shows the Shapiro steps which are induced by the presence of a microwave field (21). (I = current; V = voltage; R = resistivity.)

Ginzburg-Landau-Abrikosov-Gor'kov Theory. In 1950, Ginzburg and Landau proposed a phenomenological theory of superconductivity based upon an expansion of the free energy in terms of a superconducting order parameter or wave function ψ (18). The square modulus of ψ, $|\psi|^2$, was interpreted as being proportional to the number of superconducting electrons in the superconducting state. Thus, $\psi = 0$ for the normal state ($T > T_c$), whereas $\psi \neq 0$ for $T < T_c$. This theory is most valid for small values of the order parameter, near T_c. The equilibrium value of ψ is that which minimizes the free energy leading to a nonlinear equation for ψ. For certain simple geometries, this equation can be easily solved, and the characteristic lengths ξ and λ appear naturally in this theory. The resulting equations, known as the Ginzburg-Landau (GL) equations, are particularly useful for treating problems involving spatial inhomogeneities. In fact, Abrikosov applied the GL equations to study the penetration of a magnetic field into a superconductor in the mixed state (22). His prediction of a hexagonal fluxlike lattice was subsequently verified experimentally. Although the theory was successful in dealing with situations involving spatial inhomogeneities, it received little attention in the West because of its phenomenological nature. However, in 1959 Gor'kov showed that the GL theory could be obtained from the microscopic theory, correctly reformulated in terms of Green functions (23). The condition for the validity of the theory was shown to be that the temperature be close to T_c. Furthermore, the order parameter ψ was identified with the complex spatial-gap parameter $\Delta(\vec{r})$ of the microscopic theory. Thus, the GL theory, which is so useful for treating type-II superconductors that are of technological interest, had been placed on a firm theoretical footing. This theory, referred to as the Ginzburg-Landau-Abrikosov-Gor'kov (GLAG) theory, has led to many useful results. For example, one familiar consequence of the GLAG theory, which is useful for large values of the ratio λ/ξ, relates the zero-temperature upper critical field in the absence of normal-state paramagnetism $H^*_{c2}(0)$ to the normal resistivity ρ_n and the specific-heat parameter γ:

$$H^*_{c2}(0) = 3.1 \times 10^4 \, \rho_n \gamma T_c \tag{2}$$

where the field is given in gauss or 10^{-4} tesla, ρ_n in Ω·cm, and γ in erg/(cm^3·K^2) or 0.1 J/(m^3·K^2). This expression neglects the Pauli paramagnetism of the electrons in the normal state. Thus it cannot be used for quantitative predictions of upper critical fields for many superconducting materials. Equation 2 helps to explain qualitatively some of the factors influencing the maximum upper critical field.

Materials

At present, the largest commercial applications of superconductors rely upon the property of zero electrical resistance in the superconducting state. That is, for temperatures below T_c, for magnetic fields below H_{c2} (or H_c, as in the case of type-I materials), and for current densities below J_c, useful materials exhibit no electrical resistance. Practical materials for large-scale applications must also possess good mechanical properties. The commercially available superconducting materials can be roughly divided into ductile solid-solution alloys such as those of niobium and titanium, and intermetallic compounds, such as Nb_3Sn or V_3Ga, which are relatively brittle. At this time, NbTi, containing 45–50 wt % niobium, is the most widely used ductile material for superconducting magnet windings. For example, the Fermilab

Tevatron was planned to have 774 bending magnets, each 7 m long, as well as 240 quadrupole magnets, each 1.35 m long. All of these are to be constructed of NbTi alloy, as are the magnets planned for the Isabelle accelerator at Brookhaven National Laboratory: 732 dipole magnets, with lengths up to 4.75 m, and 348 quadrupole magnets with lengths up to 1.7 m. For higher field applications, acceptable Nb_3Sn wires have been fabricated and a Nb_3Sn magnet will be included in the Large Coil Task (LCT) project at the Oak Ridge National Laboratory for magnetic plasma confinement (see Fusion energy).

Consideration of the configuration in which these materials are used clarifies the processing procedure. Between the upper and lower critical fields, type-II superconductors contain flux-line vortices. The passage of a current through the superconductor results in a Lorentz force on the flux lines and may lead to their motion, thereby making the conductor resistive. To prevent this development and improve the current-carrying capability of the wire, the flux lines are pinned by microstructural defects, eg, dislocations and grain boundaries. In spite of the pinning, disturbances such as a small heat pulse caused by friction or a sudden change in the field can result in the motion of the vortices known as flux jumps which leads to resistance. In the earliest superconducting wires, these flux jumps would cause the magnet to revert to normal, often in a catastrophic fashion. To stabilize the magnet against damage caused by flux jumps, the conductor must be modified. The methods of stabilization depend upon the close proximity of the superconductor to some other material of high conductivity, eg, copper or aluminum (24).

Cryostatic stabilization depends upon the superconductor being bonded to a high conductivity normal (nonsuperconducting) metal. When the magnet reverts to normal, the current flows through normal material. The ohmic heating, however, is less than the cooling at the surface and allows the superconductor to return to the state of zero resistance.

Adiabatic stabilization is achieved by embedding fine superconducting wires in a metal with high thermal diffusivity, eg, copper. In this multifilamentary wire, the heat released during a local flux jump can be dissipated by the copper before an instability develops. In NbTi, for example, flux jumps should not occur for a wire with diameter <0.1 mm.

Dynamic stabilization refers to the manner in which the high conductivity of the normal material prevents the growth of a local thermal disturbance.

At present, one or more of these techniques may be employed in the fabrication of a particular conductor. Thus, the most common superconducting wire is a multifilamentary wire embedded in a Cu matrix. The superconducting part has been heavily cold-worked to create the dislocations necessary for pinning the vortices to achieve high critical currents.

Niobium-Titanium Alloys. Some measurements of the superconducting transition temperature (defined as the midpoint of the resistivity or magnetization transition) T_c are shown in Figure 5 (25–29). The T_c peaks at ca 10.1 K for alloys with high Nb content (17–30 wt % Ti). The T_c is slightly lower, 9.0–9.3 K, for Nb alloys that are used in technological applications (45–50 wt % Ti). For the Nb-rich alloys, the materials are single phased and T_c is relatively unambiguous. With increasing Ti content, T_c depends upon the composition of the bcc matrix after precipitation of the α-Ti and upon the presence of the martensitic transition (see phase diagram of Figure 6).

The upper critical field H_{c2} is shown in Figure 7 as a function of composition for

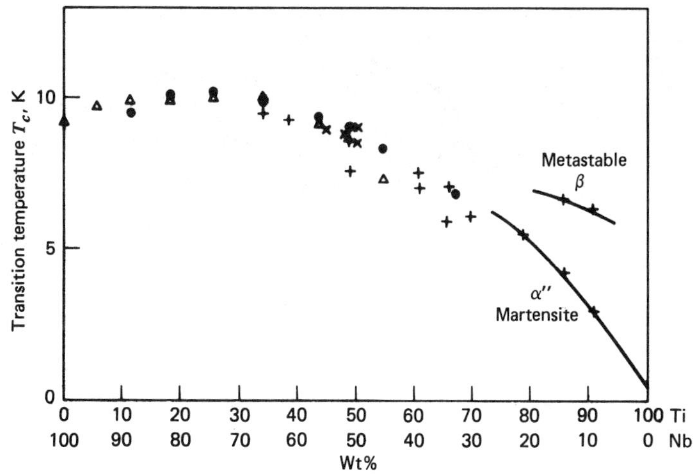

Figure 5. Superconducting transition temperature of Nb–Ti alloys.

the Nb–Ti system (26–29). There are a number of possible ambiguities in determining H_{c2}. For example, when determining H_{c2} from a resistive transition, it is necessary to use a very low current density (0.05–5 A/mm^2) to eliminate any dependence of the measured value upon the measuring current. Furthermore, the surface associated with dislocations may have higher critical fields than the cell interiors. Thus, the amount of cold working should be specified when specifying H_{c2}. It is generally accepted that the highest values of H_{c2} at $T = 4.2$ K are ca 11 T (110 kG) and occur for alloys with 45–50 wt % Ti. For this reason and because this range is most favorable for obtaining high J_c values, practical materials are made with these compositions. The standard Nb alloy used in the United States contains 46.5 wt % Ti, a composition also used in the United Kingdom and by certain Japanese manufacturers. Vacuumschmelze in the FRG and Furukawa in Japan use Nb–Ti (1:1) as standard.

Other ternary or quaternary alloys based on the Nb–Ti system have also been examined with the aim of obtaining higher critical fields and possible additional flexibility in processing in order to obtain high J_c values with alternative processing techniques. For example, Nb–Zr alloys were the first ductile high field alloys to be produced commercially. However, Nb–Ti alloys are more easily drawn, are more compatible with copper, and have higher critical fields, and therefore replaced the Nb–Zr materials. Even so, Nb–Ti–Zr and Nb–Ti–Ta–Zr alloys are produced commercially in Japan and Nb–Ti–Ta materials are produced commercially in the United States. Enhancements of H_{c2} over the best binary Nb–Ti alloys have been reported for Nb–Ti–Ta, Nb–Ti–Ta–Zr, and Nb–Ti–Ta–Hf systems. However, at 4.2 K these enhancements are rather marginal (ca 0.3 T or 3 kG). The differences become significant at low temperatures. The upper critical field of an Nb alloy with 43 wt % Ti–25 wt % Ta reaches 15.5 T (155 kG) at 2 K, which is 1.3 T (13 kG) larger than the best binary alloy. This alloy has been produced commercially and will be used in a 12-T (120-kG) prototype fusion-test coil (31). The addition of Ta or other materials with higher atomic number (Z) favors increased spin-orbit scattering which counteracts the Pauli paramagnetic limitation of the upper critical field.

Since it is necessary to create dislocations in the Nb–Ti alloy in order to obtain high values of J_c, some knowledge of the Nb–Ti phase diagram is desirable. The phase diagram in Figure 6 represents the best estimate of the low temperature portion of

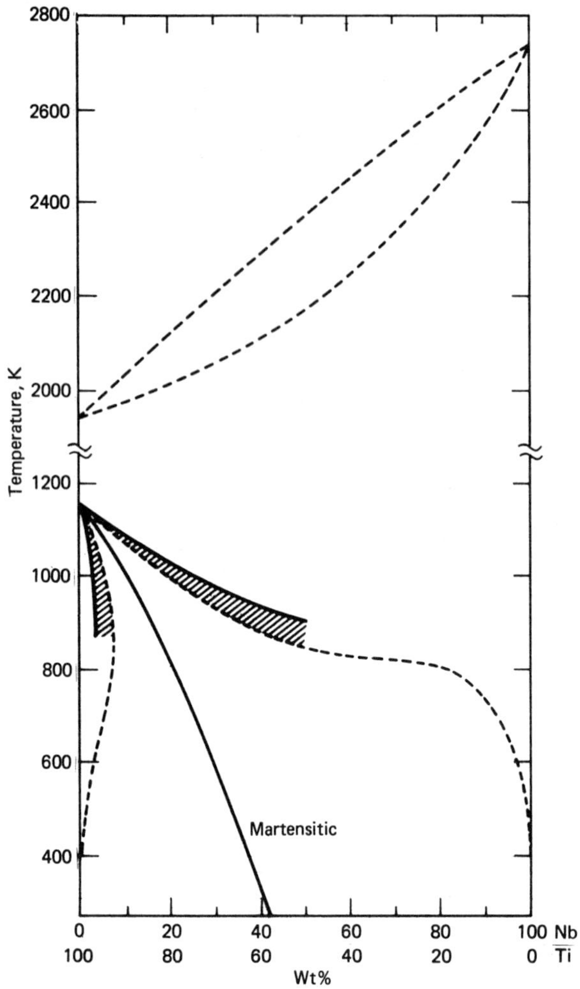

Figure 6. Nb–Ti phase boundaries, present estimates (30).

the phase boundaries of the Nb–Ti system (30). Below 1158 K, solid Ti transforms from the bcc β phase to the hexagonal close-packed (hcp) α phase. Niobium depresses this phase boundary, but there is some difficulty in determining the exact position of the $\beta \rightarrow \alpha + \beta$ boundary because diffusion is very slow at these temperatures (T/T_M ca 0.4 for the Nb–50 wt % Ti alloy). Other metastable phases α', α'', and ω may also occur.

The presence of the precipitated second phases in the β matrix leads to high J_c values. Many details of the microstructure of high J_c commercial Nb–Ti alloys are discussed in ref. 32. In particular, the nature and density of the α-Ti precipitates (T_c = 0.39 K), produced by heat treatments at or near the final wire size after large area reductions, are important. Heat treatments are generally performed at 2–6 times the final wire diameter, at 350–400°C, and sometimes for as long as ca 100 hours. In the Ti-rich region of the phase diagram, the β to α transition takes place martensitically. For Nb–50 wt % Ti, special nucleating methods are required for precipitating α-Ti, and no α-Ti precipitates have been detected above 60 wt % Nb. This is a second factor

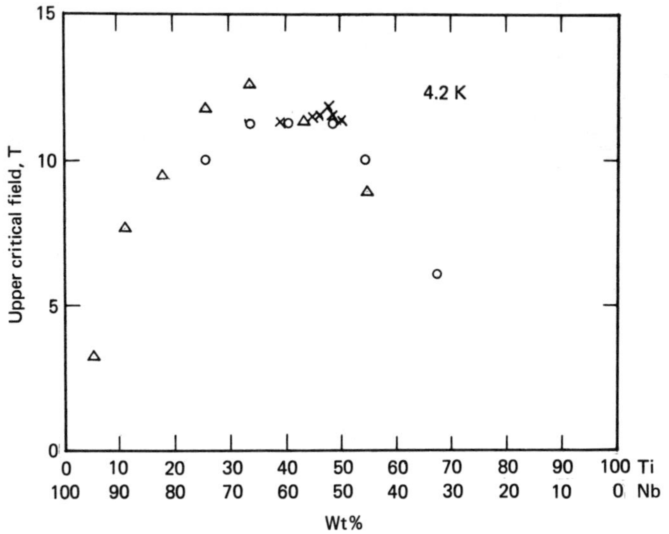

Figure 7. Upper critical field of Nb–Ti alloys at 4.2 K. To convert T to G, multiply by 10^4.

limiting technological alloys to the 45–50 wt % Ti composition range. For these alloys, precipitate densities of 10^{15}–10^{16} particles per cm^3 are observed with particle diameters of 3–20 nm. The resultant critical current densities for some commercial composites are shown in Figure 8. At 4.2 K and a magnetic field of 5 T (50 kG), J_c varies from 1500 to 2500 A/mm^2. Typical Brookhaven and Fermilab conductors have J_c values of ca 2000 A/mm^2, whereas some specially processed materials possess a J_c of ca 3000 A/mm^2. The highest J_c values are obtained by reducing the precipitates to particle sizes of 5–15 nm dia and increasing their densities to ca 10^{17} particles per cm^3.

Among the ternary alloys, the Hitachi X Nb alloy (42 wt % Zr–6 wt % Ti develops high J_c values with little cold work (33). In this alloy, J_c values of up to 2800 A/mm^2 (at 5 T = 50 kG, 4.2 K) have been observed with T_c = 10.3 K and H_{c2} = 10.5 T. However, this high Zr alloy is not ductile and is not produced in multifilamentary form. The Nb–43 wt % Ti–25 wt % Ta alloy mentioned previously has been produced in multifilamentary form with J_c ca 1800 A/mm^2.

The conductor is fabricated in accordance with the intended application which also determines the necessary stabilization procedures. The important parameter in the case of cryostatic and dynamic stabilization is the electrical resistivity of the normal conductor at the application temperature. Generally, high purity materials with high resistivity ratios are used. For example, OFHC-copper (oxygen-free high conductivity) has resistivity ratios of 200–300 in the as-cast material and 100–200 in the final conductor. In applications where eddy current losses must be kept to a minimum, a high resistivity material like Cu–Ni may also be added to the conductor.

For the simple case of a monocore conductor, the Nb–Ti bar is codrawn in a copper tube. The superconductor may continuously bond to the normal metal if the surfaces are clean. This procedure establishes a diffusion bond between the Cu and the NbTi which is absolutely necessary if the ductile Cu is to show the same area reduction as the NbTi. Without a diffusion bond, the copper–superconductor area decreases continuously and failure occurs. Multicore conductors are usually fabricated by an extrusion process. For low filament numbers, a single-step extrusion process is used,

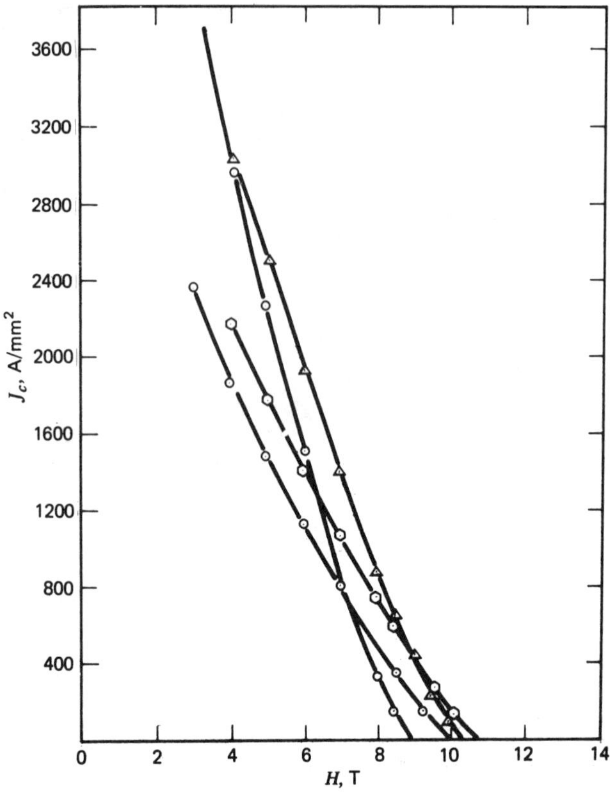

Figure 8. Critical current density vs applied field H at 4.2 K for four commercial filamentary composites (30). ⬢, Nb–46.5 wt % Ti; ▲, Nb–50.4 wt % Ti; ○, Nb–53 wt % Ti; ●, Nb–41 wt % Ti–15 wt % Ta–8 wt % Zr. To convert T to G, multiply by 10^4.

whereas for high filament numbers or for special filament distributions, a two-step extrusion is used. The Nb–Ti composite materials are commonly extruded with conventional commercial equipment. Further discussion of the physical parameters employed for NbTi materials can be found in ref. 34. A single-extrusion conductor with 1500 filaments is shown in Figure 9.

Although extrusion processing is expensive, the price of Nb–Ti wire is determined

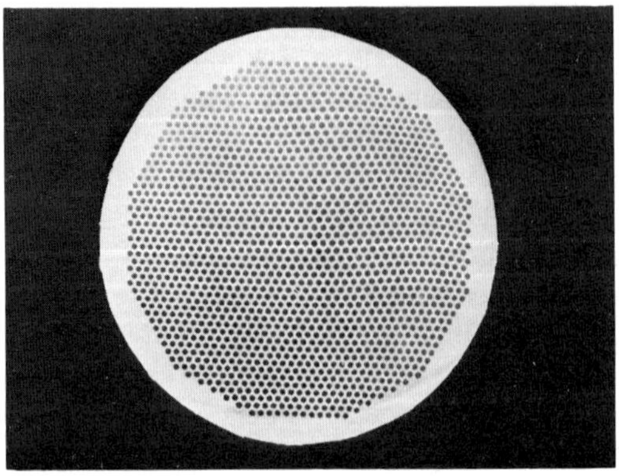

Figure 9. Cross section of a single-step extrusion conductor with 1500 filaments (35).

by the cost of the starting material. In 1980, the price was ca \$200/kg for Nb and \$20/kg for Ti. The importance of these prices can be seen from the Fermilab specifications for Nb–Ti 46.5 wt % given in Table 1. The principal impurities are oxygen and tantalum; the other metallic impurities occur in lower concentrations that can only be obtained by multiple electron-beam melting of the Nb, since the Ti cannot be purified by melting. Thus, high purity Nb is mixed with Ti of lower purity to obtain the starting material for the extrusion process.

The rapid increase in the price of the columbite ore used as the source of Nb has prompted some U.S. manufacturers to use pyrochlore ores. Because of large reserves of pyrochlore ores in North America, the price of Nb should be more stable as a result of this change. In addition, the higher Ta content of the pyrochlore ores may have beneficial effects on the upper critical field.

A-15 Superconductors. In 1954, superconductivity was discovered in the A-15 structure of the intermetallic compound V_3Si (T_c = 17 K) (36). This material was classified as having the β-tungsten structure, a cubic structure type, now known more commonly as the A-15 structure. For a compound A_3B, the B atoms occupy the corners of a cubic cell, whereas two A atoms are positioned on each cube face in such a manner that A atoms on a number of adjacent cubes generate three sets of orthogonal linear chains, as shown in Figure 10. Many early models attempted to explain the properties of A-15 materials in terms of these linear chains. In particular, the high T_c values of these materials were thought to be related to singularities in the density of states, which were, in turn, to have originated with the linear chains (37). The fact remains, however, that all high T_c materials ($T_c > 18$ K) have the A-15 structure. The T_c values for some A-15 compounds are given in Table 2. Shortly after the discovery of V_3Si, Nb_3Sn was fabricated (39). The highest known T_c material, Nb_3Ge, was made ca 19 yr later by sputtering the material onto a heated substrate (40). This metastable material has not been produced in bulk form.

The values given in Table 2 are for ordered materials with the correct stoichiometric composition. Disorder and composition fluctuations depress T_c, as well as the density of states, for the high T_c members of the A-15 materials. Variations in the long range order parameter S, owing to heat treatments, and the resulting changes in T_c have been studied (41). The largest effect occurs in V_3Au where T_c increases from 0.7

Table 1. Fermilab Specifications for Nb–46.5 wt % Ti

Element	Value, ppm max[a]
Ti, wt %	46.5 ± 1.5
O	1000
H	20
C	200
Fe	200
Ta	1000
N	150
Ni	100
Si	100
Cu	50
Al	100
Nb	remainder

[a] Except where otherwise stated.

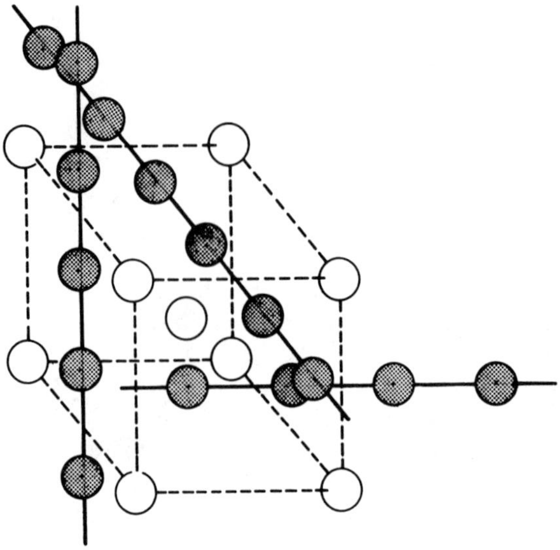

Figure 10. A-15 crystal structure, showing chains of transition-metal atoms (9).

Table 2. Superconducting Properties of Some A-15 Materials [a]

Material	CAS Registry Number	T_c, K	H_{c2} at 4.2 K, T
Nb_3Sn	[*12035-04-0*]	18.3	22.5
Nb_3Al	[*12003-75-7*]	18.9	29.5
Nb_3Ga	[*12024-05-4*]	20.3	33.5
Nb_3Ge	[*12025-22-8*]	23.0	37.0
V_3Ga	[*12024-15-6*]	15.4	22.0
V_3Si	[*12039-76-8*]	17.1	22.5
Ti_3Ir	[*12056-24-5*]	4.3	
Ti_3Pt	[*12038-32-3*]	0.49	
Ti_3Au	[*12006-62-1*]	<0.015	
V_3Pd	[*12504-34-6*]	0.08	
V_3Pt	[*12166-02-8*]	3.2	
V_3Au	[*12256-58-5*]	2.87	
V_3Al	[*12004-01-2*]	16.5	
Cr_3Rh	[*12190-94-2*]	0.07	
Cr_3Ir	[*12410-56-9*]	0.2	
Nb_3Rh	[*12034-99-0*]	2.6	
Nb_3Os	[*12034-90-1*]	1.0	
Nb_3Ir	[*12030-47-6*]	1.6	
Nb_3Pt	[*12034-97-8*]	10.6	
Nb_3Au	[*12006-56-3*]	10.6	
Mo_3Os	[*12163-84-7*]	12.7	
Mo_3Ir	[*12030-45-4*]	8.5	

[a] Ref. 38.

to 3.2 K as S rises from 0.80 to 0.99. On the other hand, radiation-damage-induced disorder has a serious effect on T_c and other superconducting properties of the high T_c members of the A-15 class. This sensitivity to radiation damage is much greater than that for the bcc solid-solution alloys (Nb–Ti). Finally, the T_c values measured in wires are also extremely sensitive to the compressive strains generated by the differential thermal contraction between the A-15 core and its Cu matrix. Since the A-15 compounds are formed by solid-state diffusion near 973 K and are used at 4.2 K, the thermal-contraction strain is generated by a temperature differential of ca 1000 K. For V_3Si, the strained material may have a T_c reduction of ca 2 K, whereas for Nb_3Sn the reduction is ca 1 K.

Values for the upper critical field H_{c2} at 4.2 K are given in Table 2 for a number of high T_c A-15 compounds. The upper critical field also shows a marked sensitivity to disorder and the compressive strains generated by differential thermal contraction, as does the critical current density J_c. The effect of differential contraction on J_c can be seen in Figure 11, where J_c is shown as a function of magnetic field for several values of the matrix-to-core ratio. The effect of strain becomes more important at high magnetic fields.

The first magnet using A-15 material was constructed at Bell Laboratories in 1961 (43). A mixture of tin and niobium powders was packed in the desired proportions in a Nb tube, which was mechanically elongated, coiled into a solenoid, and heated to transform the powders into Nb_3Sn. In addition to low critical current density and lack of stability, it was necessary to cause the material to react after winding the magnet. These disadvantages made the process impractical for wide-scale applications.

The process that first allowed the production of Nb_3Sn and V_3Ga wire in commercial quantities is known as the bronze process (42). The fabrication of V_3Ga wire by this process was reported in 1970 (44–45). It is illustrated in Figure 12(**a**) for Nb_3Sn. A Nb core is inserted into a Sn bronze. After bundling, the composite rods are drawn

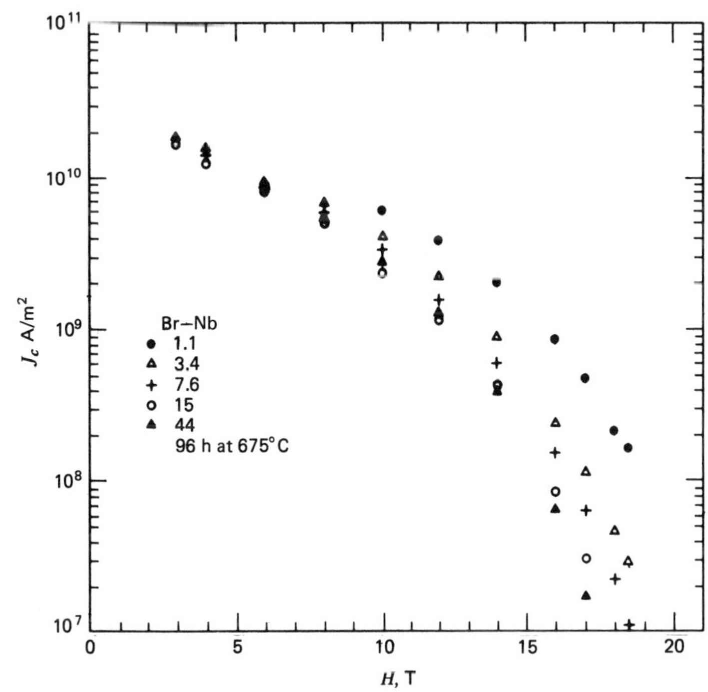

Figure 11. Critical current densities of Nb_3Sn wires with various matrix-to-core ratios as a function of applied magnetic field (42). To convert T to G, multiply by 10^4.

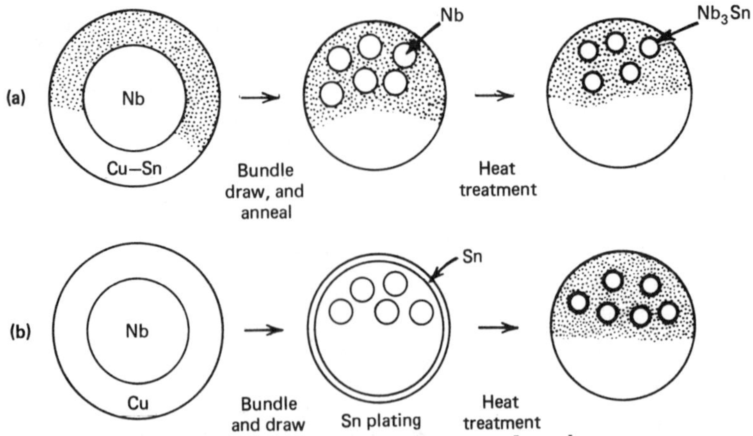

Figure 12. Schematic diagrams for (**a**) the bronze and (**b**) the external diffusion processes for multifilamentary Nb_3Sn and V_3Ga wires (42).

and annealed to final wire size. Heat treatment results in the growth of Nb_3Sn at the interfaces of the Nb cores and the bronze matrix. This process can be understood by considering the relevant phase diagrams. The binary Nb–Sn phase diagram shows that above 1203 K, the only stable compound formed is Nb_3Sn. Indeed, commercial Nb_3Sn tapes are produced by heating Nb tapes in a liquid Sn bath. Above 1203 K, single-phase Nb_3Sn layers are formed. The process is slightly more complicated, however, in the presence of Cu. Ternary phase diagrams show that at 973 K the diffusion path from the Cu–Sn solid solution to the Nb–Sn solid solutions passes through the A-15 phase field. Thus, only the A-15 phase is formed. Furthermore, the T_c of the Nb_3Sn formed is always ca 18 K, indicating that the composition of the material formed is at the stoichiometric composition Nb_3Sn. In practice, however, the bronze process is not completely free of problems. Commercial copper–tin alloys with up to 10 wt % Sn are unsuitable because the tin content of the bronze has to be as high as possible. Usually, a saturated α solution alloy with 13–14 wt % Sn is used which must be fabricated by the same manufacturer who makes the superconducting wire. Furthermore, phosphorus acts as a poison for Sn diffusion in the Nb–bronze system and the presence of oxygen reduces the ductility of the material. Thus, the starting material must be phosphorus-free high Sn content bronze.

The first modification of the bronze process, the external diffusion process, was introduced in an attempt to increase the thickness of the intermetallic layer in V_3Ga by plating the wire with gallium (46). A schematic diagram of this process is shown for Nb–Cu in Figure 12(**b**). The process consists of extruding a Nb–Cu composite and then drawing it to final wire size. The wire is coated with a layer of tin and heated to make a Cu–Sn alloy matrix and, upon further heating, Nb_3Sn filaments. This modification offers the advantage that a Nb–Cu composite can be drawn to a small size without intermediate annealing, whereas the Nb–(Cu–Sn) or V–(Cu–Ga) composites must be continually annealed to remove work-hardening effects. Furthermore, the concentration of Sn is not limited to the 13 wt % Sn as in the original bronze process. The fact that a high Sn concentration in the matrix usually results in a high J_c value offers a considerable advantage. In addition, another element may be added to the matrix to improve the superconducting properties after the Sn diffusion has already taken place. On the other hand, problems are created in the tin-plating process and subsequent diffusion heat treatments by the delamination of the outer layer of the wire when the Sn layer exceeds 5 μm. This limits the wire diameter to 0.25 mm since a larger Sn layer would be necessary to form the Nb_3Sn in thicker wires. The problem is believed to be related to volume expansion of the outer layer by Sn diffusion into the Cu matrix and to the formation of voids. New laminating and heat treatment methods have somewhat reduced these difficulties. Although this process gives wires with higher J_c values than the bronze process, it results in a conductor without a high purity copper-stabilization jacket. In the bronze process, the high purity Cu for stabilization (not shown in Fig. 12) is protected from outward Sn diffusion by the presence of a Ta diffusion barrier. Neither the Ta barrier nor the Cu stabilization material can be obtained with the external diffusion process, and therefore the use of this material is restricted to certain types of braids.

Other more traditional modifications of the Nb_3Sn fabrication process seem to correct some of the deficiencies of the external diffusion method, eg, the internal Sn diffusion process, shown in Figure 13(**a**) (47). In this process Cu-clad Nb bars and Cu-clad Sn–20 wt % Cu alloy are assembled in a Cu tubing with the Sn–Cu alloy at the center. This combination is drawn down to the final size and heat treated to form the Nb_3Sn. As in the external diffusion process, the intermediate annealing steps of the bronze process are eliminated. In addition, the Cu stabilization material and the Ta diffusion barrier can be used since the Sn is at the center of the conductor. The Sn concentration in the matrix is limited. Furthermore, difficulties may be encountered in large-scale applications since the length of the wire is limited because the billet cannot be extruded without melting the Sn–Cu core.

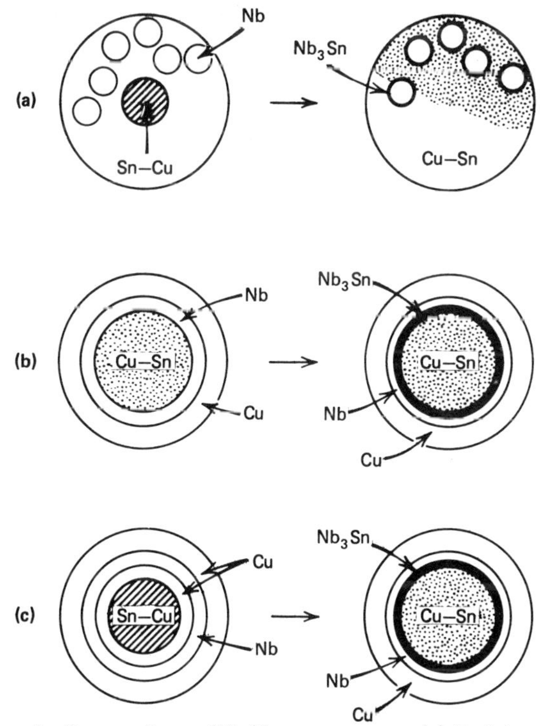

Figure 13. Schematic diagrams for modified bronze processes: (**a**) the internal tin diffusion; (**b**) the internal bronze; and (**c**) the internal Sn–Nb tubing processes. The dotted and shaded areas represent the bronze and Sn–Cu eutectic alloys, respectively (42).

The superconducting Nb_3Sn multifilamentary wires available by these techniques may possess filaments as small as 1–3 μm, whereas their number can exceed 500,000. Individual wires have been combined into cables, braids, or even more complex geometries. Multifilamentary V_3Ga has been produced commercially by these techniques which are also feasible for V_3Si. Aside from these more traditional methods of fabricating Nb_3Sn or V_3Ga conductors, a number of other methods have been developed which attempt to eliminate certain problems of the previous techniques in an attempt to develop multifilamentary composites with higher filament densities and greater ability to resist degradation caused by strains.

The *in situ* technique is based on the superconductivity of Cu containing trace amounts of Nb (48). The broad plateau near 1973 K of the Cu–Nb phase diagram in Figure 14 indicates a strong tendency for phase separation to occur. The solidification path of a Cu–30 wt % Nb alloy involves the massive precipitation of β Nb as the liquidus is crossed. Depending upon the cooling rate, a dendritic structure may be formed. When the temperature reaches 1363 K, the Cu-rich matrix solidifies. In order to produce a filamentary superconductor, this Cu ingot with Nb precipitates is cold swaged and drawn to final wire diameter (ca 0.25 mm). The *in situ* conductor is shown in Figure 15 after the Cu was removed from the composite with acid. Tin is added by either internal or external diffusion. In the latter case, tin is added at the final wire size by electroplating or dipping. This process avoids the intermediate annealing heat treatments and should allow an increase of tin content. However, some of the problems of the external diffusion process can also be present here, in particular, the formation of voids at the Cu–Sn interface. In the internal *in situ* process, Sn is added to the melt before casting. The Cu–Nb–Sn alloy is then swaged and drawn to the final wire size

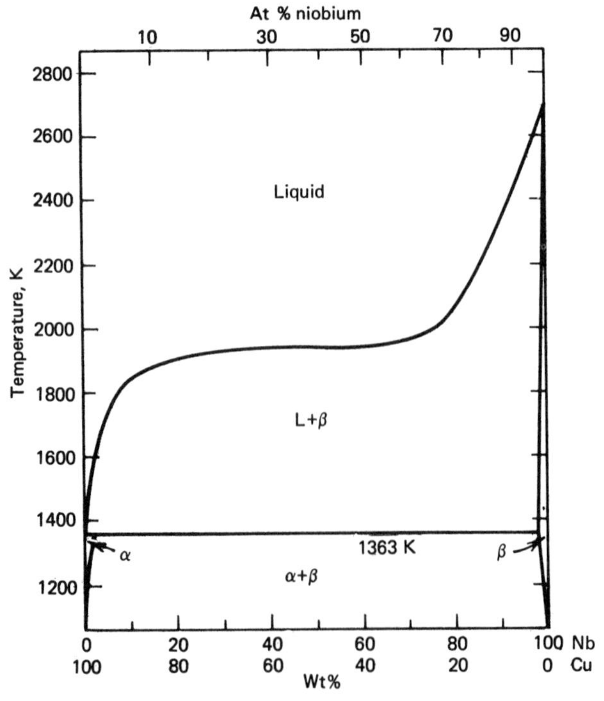

Figure 14. Cu–Nb equilibrium phase diagram (49).

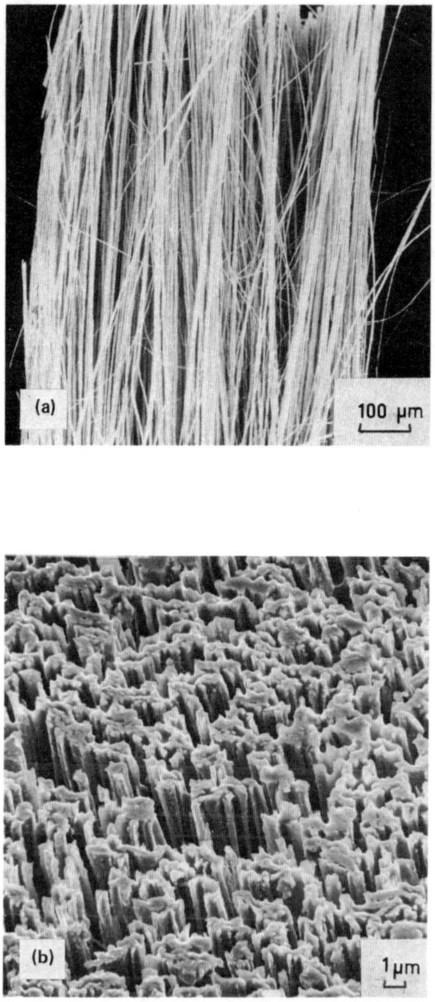

Figure 15. Microstructure of Cu–Nb *in situ* composite after etching to remove Cu matrix: (**a**) longitudinal cross section; (**b**) end view of ribbon-shaped filaments (50).

by numerous annealing heat treatments. As in the bronze process, this process is limited by the upper limit on the Sn content (ca 13 wt % Sn at 773 K) and the need for costly annealings. The main advantage of the *in situ* process is, however, that the multifilamentary structure is easily formed without the necessity of assembling the many rods that ultimately become the filaments. When the conductor in question has a very large number of filaments, the effort saved is considerable. Finally, the J_c values of these materials compare very favorably with Nb_3Sn conductors made by more traditional methods and the strain tolerance of the *in situ* materials is ca 3–4 times better.

Powder metallurgy (qv), commonly used for the preparation of high melting composites or nonequilibrium structures, can be used to produce multifilamentary conductors. Three processes are used to produce multifilamentary conductors via the powder metallurgy method. They involve hot- or cold-working of the mixed powders as well as infiltration of a liquid into a porous rod. Great care must be given to the choice of the starting powders, as well as their sizes and mixing.

In the cold-working technique, the powder size is selected first. The maximum density would be best achieved by a range of sizes. However, to achieve a uniform filament size, it is preferable to restrict the range of powder sizes. Frequently, 40-μm powders are used to obtain submicrometer filaments. The Nb powders are produced by either ball-milling or in a hydride–dehydride process. Since the powders must be highly deformable, the oxygen content of the starting material must be kept to a minimum. Tin may be added either as a powder at the start or may be diffused into the composite at some later point as in the bronze process. When the Sn is added as a final step, the specimen may be cold extruded to half the initial diameter, swaged, and drawn to the final diameter. When the external sheath is a barrier to Sn diffusion, it must be removed. Recently, a central core of Sn powder has been used to introduce the Sn. The J_c values in powder-metallurgy conductors are comparable to those of *in situ* materials in most cases. For a Cu–35 wt % Nb–7 wt % Sn material, critical currents of >10 kA/cm^2 were obtained at 12 T (1.2 kG) (51). Recently the cold-powder process for producing Nb_3Al starting with Nb and Al powders has been developed at MIT. The materials produced had good J_c and T_c values and further developments are expected.

Early attempts at extruding the materials at high temperatures were not successful because of solution hardening of the Nb by oxygen. Recently, however, the extrusion was carried out near the melting point of copper, with a third element, eg, Al, Mg, Ca, Zr, or Hf, added as a getter. The addition of ca 1 wt % Al was sufficient to reduce the oxygen content below 0.1 wt %. This process has been used successfully with Nb–Sn and V–Ga systems.

In the infiltration process, a partially compacted porous Nb rod is infiltrated by molten Sn. The hydride–dehydride Nb powders of ca 40–60 μm size are compacted into rods. After a partial sintering in vacuum at high temperatures, the rod is immersed in liquid tin at 350–400°C. After area reduction, the wire reacts at 930°C to produce the Nb_3Sn. Critical currents in these composites are particularly impressive at high fields: a Nb_3Sn–Nb composite with J_c ca 10 kA/cm^2 at 20 T (200 kG) has been reported (52).

Materials for Tunnel Junctions. Since the discovery of superconducting tunneling and the Josephson effect, various electronic applications of superconductivity have been developed. Measuring devices employing the SQUID have been available for some time, and the tunnel junctions as detectors of infrared radiation have been successfully employed in radio astronomy. However, the potentially largest electronics application of Josephson junctions depends upon their ultrafast, low power-switching characteristics in digital circuits (53). An extensive effort, at IBM and other laboratories, has been made to develop materials for Josephson devices compatible with the thin-film integrated-circuit technology of modern computers (qv). Until recently, IBM emphasized the development of Pb alloys for thin-film devices. Persistent problems with mechanical stability of the Pb-base materials, however, have led to renewed efforts to prepare junctions with refractory Nb alloys (54) (see Magnetic materials, thin films).

Lead-based alloys have a number of desirable properties which makes them suitable for junction-electrode applications. They have relatively high T_c values, are rather insensitive to impurities and film-structure variations, and have reasonably long coherence lengths. These properties facilitate their use in junctions whose characteristics are close to the theoretical predictions and also highly reproducible. On the other hand, pure Pb films have poor mechanical and chemical stability. Devices fabricated with pure Pb electrodes may develop a short after one cycling to cryogenic temperatures. Furthermore, the Pb films can deteriorate during the processing steps involved in preparation of integrated circuits.

For these reasons, IBM has developed a Pb–12 wt % In–4 wt % Au alloy for use as a junction base electrode (55). In its fine grain form, this material exhibits excellent resistance to defects produced by thermal cycling. A layered composite Pb–Au–Pb–In is prepared by deposition onto a substrate held at liquid nitrogen temperature. Upon warming to 350 K for one hour, a two-phase alloy is formed consisting of a superconducting Pb–13 at % In matrix containing $InAu_2$ *[55891-90-2]* particles of low T_c. The $InAu_2$ phase is concentrated in the central region of the film and facilitates the growth of fine-grained films. Furthermore, a protective layer of In_2O_3 is formed epitaxially on the surface of the alloy before removal from the vacuum system. This procedure results in excellent stability during the wet chemical steps of the photolithographic process. Since the In_2O_3 exhibits high tunneling conductivity, there is no need to remove it before forming the junction. For the counterelectrode, a Pb–29 at % Bi alloy has been used. This alloy is in the hcp ϵ phase and has a T_c of ca 8.4 K, higher than that of pure Pb. It is more resistant to damage caused by thermal cycling (56). However, since the Pb–Bi alloy is susceptible to corrosion, it must be protected by a SiO layer before the wet chemical processes can be used (see also Electrical connectors).

Computers employing Josephson junctions would be extremely stable if they could remain at 4.2 K. However, the necessity of making repairs and parts replacements make it inevitable that the computer elements must be cycled to room temperature a number of times during their lifetime. This cycling may result in a rupture of a tunnel-barrier oxide and the formation of a superconducting microbridge between the two junction electrodes. This problem arises because of the stresses induced by differential thermal contraction of the electrode materials and the substrate. If the strain impressed on the film by the substrate exceeds the value it can support elastically, strain relaxation occurs, usually through grain rotation, dislocation glide, or hillock formation. These manifestations of strain relaxation are believed to be responsible for the failure of Pb-alloy junctions.

Josephson junctions with Nb-base electrodes and either Nb or Pb-alloy counterelectrodes offer an alternative to these materials. Stable Nb films with good superconducting properties can be prepared by evaporation onto substrates held at low temperatures (298 K) (see Film deposition techniques). These films, however, have a very high level of intrinsic stress that limits the materials onto which the Nb films can be deposited. For example, photoresist stencils rupture or distort when Nb films are deposited upon them. Thus, Nb films must be patterned after deposition by protecting the wanted regions with photoresist material and etching away the remaining material. Junctions with one or two Nb electrodes have an excellent ability to resist thermal cycling damage. Thus, one group of 16,000 Nb/Pb-alloy junctions has been cycled 1400 times over a period of several years without any failures, a result significantly better than the best results on Pb-alloy junctions (57). The main problem with these Nb/Pb-alloy junctions for digital circuits is their excessively high junction capacitance, with a corresponding decrease in speed of logic and memory elements.

More recently, attempts have been made to fabricate reliable Josephson elements using high T_c materials. Materials with $T_c > 10$ K are technologically important because they can be cooled with small closed-cycle refrigerators which are relatively compact, economical, and reliable. The use of such coolers would greatly expedite the general uses of Josephson junctions. An obvious choice for the high T_c materials would be the A-15 compounds. Indeed, junctions employing V_3Si, Nb_3Sn, and others have been made at Stanford University (58). However, the sensitivity of the superconducting properties of these materials to stoichiometry presents an obstacle to their use.

Another candidate material is NbN (T_c = 16 K), a member of the B1 class (NaCl crystal structure) of superconductors which includes the pseudo-binary $NbN_{0.7}C_{0.3}$ (T_c = 18 K). Early work on these compounds was performed by Meissner and coworkers. The bcc metals readily form NaCl (B1) compounds because the small carbon or nitrogen ions can enter interstitial positions in the bcc lattice and convert it to an NaCl type lattice. The insensitivity of NbN to stoichiometry is shown in Figure 16 where it is compared to other transition-metal nitrides with the B1 structure. The insensitivity of T_c to N content is particularly convenient for this material, since NbN is generally formed by the reactive sputtering of Nb in the presence of N onto high temperature substrates. Tunnel junctions have been prepared by using amorphous Si (a–Si) as a barrier and Nb as a counterelectrode. In this case, the entire NbN/a–Si/Nb layered structure is formed before the device patterning is performed (59).

New Materials. ***Chevrel-Phase Superconductors.*** In 1971, the existence of a new series of ternary molybdenum chalcogenides $M_xMo_6X_8$ (X = S, Se, Te) was reported (11). Soon thereafter it was shown that several of these compounds were superconducting with high T_c values (60). Tremendous interest was generated when measurement of the upper critical field of $PbMo_6S_8$ at 4.2 K gave an H_{c2} of ca 53 T (530 kG) (12). In fact, these materials have the highest critical fields at 4.2 K of any superconductor. Chevrel-phase superconductors and their transition temperatures are given in Table 3.

The basic building block of the system is the Mo_6X_8 unit, which takes the form of a slightly distorted cube. The chalcogen X atoms sit at its corners, whereas the Mo atoms are located slightly outside the middle of the faces of the X cube. These positions result in a hexagonal rhombohedral unit cell with a rhombohedral angle α_{rh} ca 90°. When M is a large cation, ie, Pb, Sn, or a rare earth metal, then x is close to one and

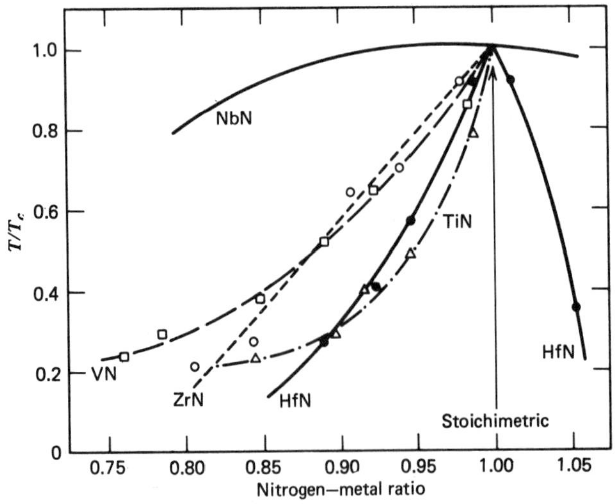

Figure 16. Critical transition temperature vs nitrogen-to-metal ratio for various B1-structure nitrides of the transition metals (9).

Table 3. Some Chevrel Phase Superconductors[a]

Material	CAS Registry Number	T_c, K
$Na_2Mo_6S_8$	*[60581-71-7]*	8.6
$LaMo_6S_8$	*[57485-79-7]*	7.1
$Yb_{1.2}Mo_6S_8$		9.1
$Cu_{1.8}Mo_6S_8$		10.8
$Cu_{1.6}Mo_6S_8$		6.4
$Ag_{1.6}Mo_{6.4}S_8$		9.1
$Sn_{1.2}Mo_{6.4}S_8$		14.2
$PbMo_{6.4}S_8$		15.2
$Y_{1.2}Mo_6Se_8$		6.2
$LaMo_6Se_8$	*[59249-33-1]*	11.4
$PrMo_6Se_8$	*[59249-36-4]*	9.2
$NdMo_6Se_8$	*[57485-82-2]*	8.2
$Sm_{1.2}Mo_6Se_8$		6.8
$Gd_{1.2}Mo_6Se_8$		5.6
$Cu_2Mo_6Se_8$	*[64975-25-3]*	5.9
$Sn_{1.2}Mo_6Se_8$		6.8
Mo_6Se_8	*[57620-26-5]*	6.3
$Mo_6S_6Br_2$	*[64417-64-7]*	13.8
$Mo_6S_6I_2$	*[64417-82-9]*	14.0

[a] Ref. 61.

α_{rh} = 89–90°. When M is a small cation like Cu, Ni, Fe, etc, x can vary over a large range and α_{rh} = 94–95°. The distance between two Mo atoms on neighboring units is larger than that between two Mo atoms of the same unit, resulting in a series of channels that run through the structure along the three rhombohedral axes. The M cations are distributed along these channels. In the case of large cations (Pb, Sn, etc), the M atoms is located at a site of pseudo-cubic symmetry as shown in Figure 17. With decreasing cation size, the M atoms are removed more and more from the center of the pseudo-cubic site. When M = Cu, up to 4 Cu atoms can enter the chalcogen channel leading to a decrease in the Mo–Mo intracluster distances and important changes in physical properties (62).

The high normal-state resistivities of these materials are notable. Measurements on single-crystal $Cu_xMo_6S_8$ and Mo_6Se_8 give ρ values of ca 1–2 mΩ·cm at 300 K, suggesting extremely small electronic mean free paths of ℓ ca 4.0 nm (63). Highly resistive materials are expected to have large upper critical fields H_{c2}. In these materials, the M element plays a limited role in the superconductivity as shown by the fact that nearly all compounds containing magnetic rare-earth elements have been found to be superconducting. However, superconductivity seems to be strongly linked to the electronic density of states. Band-structure calculations for $PbMo_6S_8$ and other chalcogenides suggest that the band structure near the Fermi level is composed of narrow d bands corresponding to the molecular levels of the Mo_6S_8 units (64–66). The width of the d-band at the Fermi level is determined essentially by the intercluster Mo–Mo distance. The charge transfer from the M cation to the Mo_6 cluster determines, in part, the position of the Fermi energy in relation to these narrow d bands and thereby the T_c of the material. The upper critical fields of these materials are generally as high as expected for highly resistive materials. In Figure 18 the upper critical fields

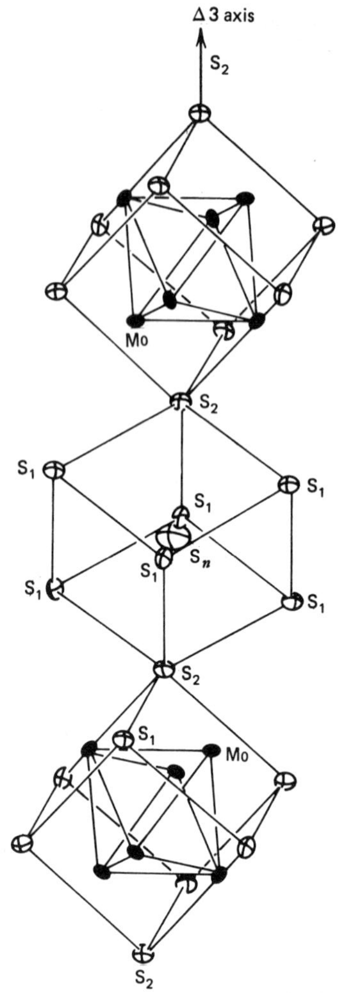

Figure 17. The separation along the ternary axis of two building blocks Mo_6S_8 by a pseudo-cubic site in the Chevral phase (62).

are shown as a function of temperature for several Chevrel-phase compounds. These very high critical field values suggest that these materials are well suited for superconductive applications. Attempts have been made to allow Mo wires and ribbons to react with sulfur and Pb vapor to give $PbMo_6S_8$, but better J_c values have been obtained by drawing powder-filled Ta tubes: J_c (4.2 K) is ca 10 kA/cm^2 at 4 T (40 kG) and 4 kA/cm^2 at 10 T (100 kG). Measured values of J_c on sputtered and evaporated films are comparable to those obtained on bulk materials (67). It has been estimated, using a scaling law for flux pinning, that J_c for $PbMo_6S_8$ may be ca 10 kA/cm^2 at 26 T (260 kG). The promise of these materials to provide conductors for very high field solenoids depends upon the possibility of making a wire or tape that withstands the enormous magnetic forces at these high fields.

Amorphous Superconductors. Amorphous films of simple metals have been studied since 1954 by quenching the vapor phase onto cryogenic substrates (68). The first examples of bulk metallic glasses formed from transition metals were produced

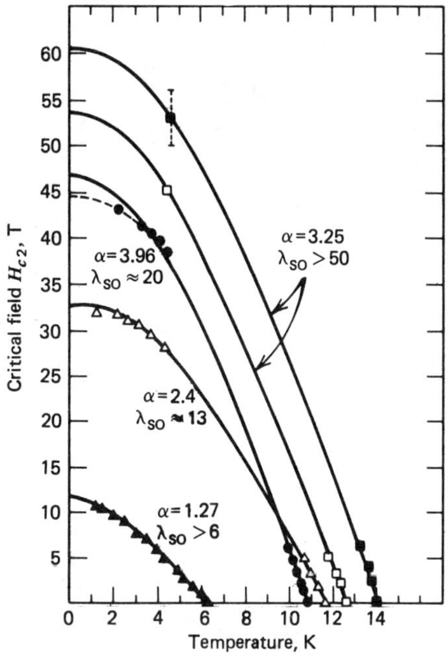

Figure 18. The critical fields of $M_xMo_6X_8$ compounds (62). ■, $Gd_{0.2}PbMo_6S_8$; □, $PbMo_6S_8$; △, $SnMo_6S_8$; ●, $LaMo_6Se_8$; ▲, Mo_6Se_8. To convert T to G, multiply by 10^4.

in 1975. These materials have a number of unusual properties. A most notable feature distinguishing these materials from their crystalline counterparts is the systematics of T_c as shown in Figure 19 (69). The single-peaked, featureless curve of the amorphous materials has a maximum for an electron-to-atom ratio slightly above that of Mo. Indeed, many high T_c amorphous superconductors contain large amounts of Mo. The T_c values of some representative materials are given in Table 4. The maximum T_c value obtained so far is 9.05 K for $Mo_{0.8}B_{0.1}P_{0.1}$. Initially it was believed that the shape of this curve was due to the progressive filling of the amorphous d-band. Recent calculations, however, show that the d-bands of amorphous alloys may have a much more complex structure than supposed and therefore other mechanisms must be invoked (see Amorphous magnetic materials; Glassy metals).

Another property common to these materials is the high resistivity at low temperatures. Actually, the resistivity of amorphous materials is frequently temperature independent or increases slightly with decreasing temperature, contrary to the usual behavior observed in metals. Thus, it is common to encounter low temperature resistivities in the 100–300 $\mu\Omega$-cm range. Large normal-state resistivities, indicative of small electronic mean-free paths, are associated with large upper critical fields. For transition-metal alloys, the field gradient at T_c, $-(dH_{c2}/dT)_{T_c}$ varies from 2.0 to 4.5 T/K (20 to 45 kG/K); the latter value is for amorphous Mo. Curves of H_{c2} vs T are frequently linear, even at low temperatures (71). This behavior has been attributed to quenched-in strains or inhomogeneities in the material (72). The high resistivities are also associated with the atomic disorder quenched into the material while it is being formed in an amorphous state. Quench rates of 10^6 K/s are frequently necessary to obtain the amorphous phase. Such cooling rates are obtained by vapor quenching onto a low temperature substrate, directing a stream of liquid metal onto the surface of a

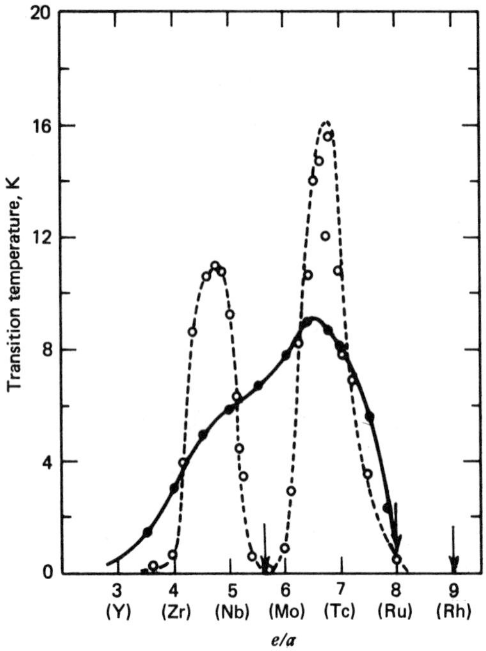

Figure 19. The variation of T_c of transition metals of the 4*d* series in the amorphous and crystalline states (69). ○, Crystalline metals and alloys; ●, Collver-Hammond (amorphous films).

rapidly spinning wheel, or even cooling in a piston-and-anvil apparatus. All of these processes produce foils or films where one spatial dimension is very much smaller than the others.

The possibility of employing amorphous superconductors in large-scale devices depends upon their critical currents. Since amorphous materials tend to be homogeneous and without grain boundaries, their J_c values tend to be rather low. Attempts have been made to increase J_c by creating inhomogeneities. In Figure 20, the J_c values of two samples of $(Mo_{0.6}Ru_{0.4})_{80}B_{10}Si_{10}$ are compared. In the second case, the sample had been heated long enough for partial recrystallization (ca 5% of sample volume) to occur (73). The crystallites have served as flux-pinning sites, and J_c has increased by a factor of more than 10^3 at 2 T (20 kG). In Figure 21 J_c vs H is shown for another promising material $(Zr_{0.7}Hf_{0.3})_{60}V_{40}$. In the amorphous phase this material exhibits great ductility but is not superconducting. After proper heat treatments, the brittle C-15 superconducting phase is formed with reasonable critical currents: $J_c > 100$ kA/cm^2 at 15 T (150 kG) (74). Such a material might be used in a wind-and-anneal fabrication process.

Applications

Magnets for High Energy Physics. Ever since Onnes attempted in 1913 to generate a magnetic field using a resistanceless superconducting Pb coil, the application of superconductivity has been primarily focused on the creation of magnetic fields. First small laboratory-size magnets were constructed. However, the largest superconducting magnet systems have been built for high energy physics. In early applications, superconducting magnets were used to create fields in bubble chambers. Recent projects

Table 4. T_c Values for Some Amorphous Superconductors [a]

Material	CAS Registry Number	T_c, K
$(Mo_{0.8}Ru_{0.2})_{80}P_{20}$		7.31
$(Mo_{0.4}Ru_{0.6})_{80}P_{20}$		4.68
$Mo_{80}P_{10}B_{10}$	*[83847-04-5]*	9.00
$(Mo_{0.8}Re_{0.2})_{80}P_{10}B_{10}$		8.71
$La_{80}Au_{20}$	*[83847-03-4]*	3.5
$La_{80}Ga_{20}$	*[37239-26-2]*	3.84
$Zr_{75}Rh_{25}$		4.3
$Zr_{70}Pd_{30}$	*[57656-11-8]*	2.4
$Zr_{70}Be_{30}$	*[83847-02-3]*	2.80
Nb_3Ge	*[12025-22-8]*	3.6
Nb_3Si	*[12035-03-9]*	3.9
$Mo_{68}Si_{32}$	*[83847-01-2]*	7.4
$Mo_{0.8}N_{0.2}$		8.3
$(W_{0.5}Ru_{0.5})_{80}P_{20}$		4.57
$Zr_{70}Ru_{30}$	*[83847-00-1]*	3.7
$(Mo_{0.6}Ru_{0.4})_{90}B_{10}$		7.15
$(Mo_{0.6}Ru_{0.4})_{82}B_{18}$		6.05
$(Mo_{0.6}Ru_{0.4})_{76}B_{24}$		5.05
$(Mo_{0.6}Ru_{0.4})_{82}Si_{18}$		5.40
$(Mo_{0.6}Ru_{0.4})_{68}Si_{32}$		2.80
$(Mo_{0.6}Ru_{0.4})_{76}As_{24}$		4.60

[a] Refs. 69–70.

were undertaken to save electrical energy and thereby reduce operational and capital costs for large high energy accelerator bending and focusing magnets.

In the United States, two large superconducting a-c magnet rings are in the planning stage: the Fermilab energy doubler at the Fermi National Laboratory and Isabelle at Brookhaven National Laboratory. After construction of a number of magnets, the latter project was reviewed and may be discontinued due to design difficulties. The energy doubler at Fermilab will be installed in the tunnel of the existing conventional 400 GeV machine and will increase the beam energy to ca 1000 GeV. The ring will employ 744 dipole magnets (B = 4.5 T (45 kG) and length = 6 m) as well as 276 quadrupole magnets (length = 1.5–2.0 m). Many of the dipole magnets have been constructed, but there have been some difficulties in achieving the desired field homogeneity. In the FRG, an electron–proton colliding-beam facility, Hadron-Electron Ring Anordnung (HERA), is anticipated. The proton accelerator of this unit will have an energy of 820 GeV and will be equipped with 640 dipole magnets [B = 4.75 T (47.5 kG)] and 164 quadrupoles.

Rotating Machinery. ***Synchronous Generators.*** Most likely, one of the first commercial applications of superconductivity will be in the electric-power industry, where copper conductors and iron flux circuits will be replaced by superconductors (see also Power generation). Large machines operating at 3600 rpm are currently limited to the 1000–1500 MW range. In a conventional three-phase synchronous generator, the diameters of the rotor and the stator are limited by mechanical stresses. If the rotor is longer, axial instabilities become a problem. Superconductors generating high flux densities for the rotor field can substantially reduce this problem. In conventional machines, the flux density is usually limited to the magnetic saturation of the iron [ca

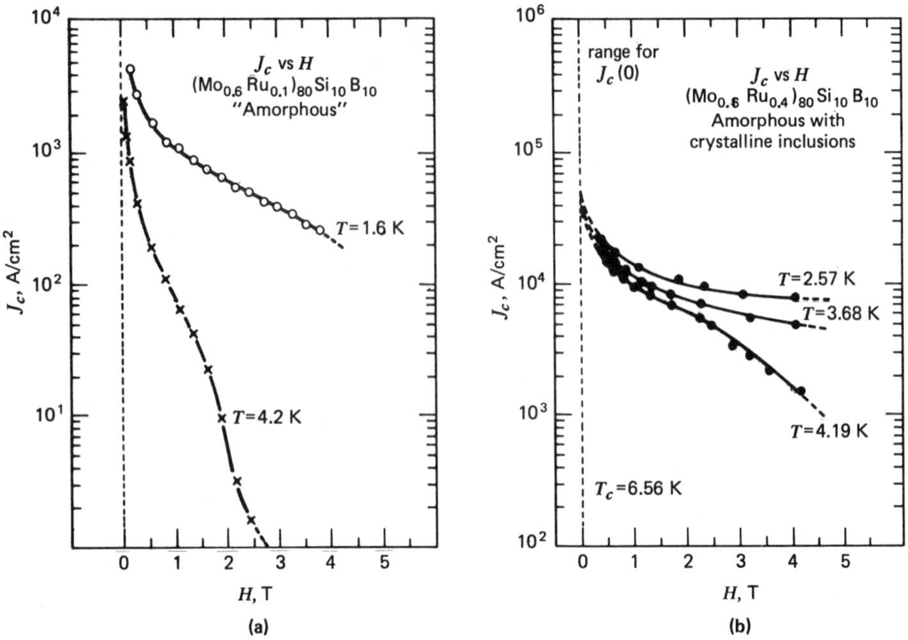

Figure 20. Critical current as a function of magnetic field for (**a**) amorphous $(Mo_{0.6}Ru_{0.4})_{80}Si_{10}B_{10}$, and (**b**) amorphous plus crystalline inclusions $(Mo_{0.6}Ru_{0.4})_{80}Si_{10}B_{10}$ (73). To convert T to G, multiply by 10^4.

2 T (20 kG)], whereas with Nb–Ti superconducting wire, this flux density may easily reach 5 T (50 kG). Furthermore, it can be produced in a much smaller volume, reducing the size of both the rotor and the surrounding armature, which results in improved mechanical stability and reduced cost. It is estimated that weight and volume would be reduced by a factor of ca 2, whereas efficiency in a two-pole machine would increase 0.5–0.8% (75). Reductions of weight and volume make the superconducting generators competitive in the power range of 1000 MW, and if the increased efficiency is taken into account, the break-even point is shifted to values <1000 MW.

The first machine using a rotating superconducting field winding was designed and tested at MIT; it had a rating of 50 kW at 3600 rpm. The rotor produced a dipole field of ca 3.2 T (32 kG), and the machine could be run continuously when connected to a source of liquid helium. The MIT group has completed a second 3-MW generator. This machine has been connected to the Cambridge Electric power line as a synchronous producing condenser drawing no power. Westinghouse has constructed and operated a similar machine with a rating of 5 MW. In the United States, Westinghouse and EPRI (Electric Power Research Institute) have initiated a program to develop a 300-MW 3600-rpm machine, a prototype of the first commercial superconducting a-c generator. It will be installed at TVA before 1985. These programs have generated great enthusiasm for the commercial use of superconductivity. In Europe, studies are in progress on 2-pole generators of 1200–3000 MW. In Leningrad, a 200 MW demonstration generator is planned for 1987.

Homopolar d-c Machines. The main interest in d-c superconducting rotating machines is focused on homopolar machines. In a disk-type homopolar machine, an outer superconducting solenoid produces a large magnetic flux, whereas an inner normal-metal Faraday disk carries a radial current between the inner axis and the outer

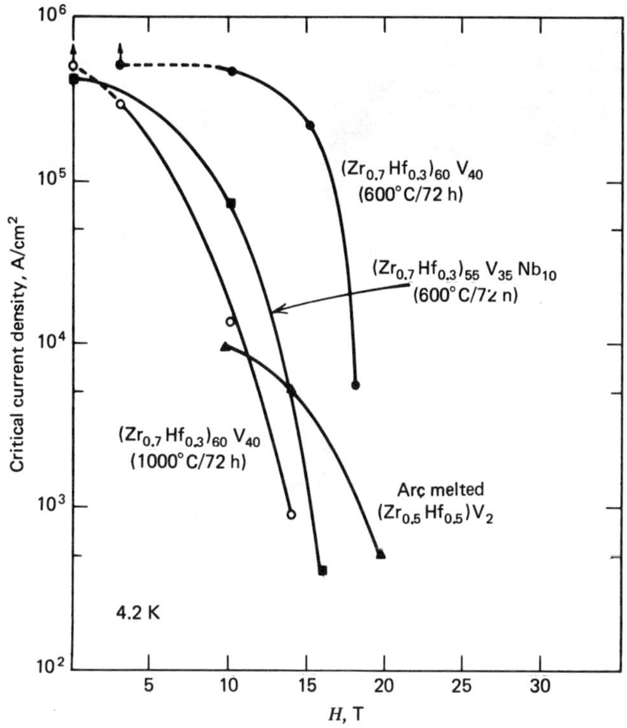

Figure 21. The critical current density for various Zr–Hf–V alloys as a function of magnetic field (74). To convert T to G, multiply by 10^4.

edge. The interaction of the field with the armature current causes a torque on the disk. The current-collecting system between the stationary and rotating parts poses a difficult engineering problem which will be solved either with conventional carbon–copper brushes or rotating liquid metal contacts. Assuming that the low-loss reliable current-collection system can be developed, superconducting homopolar machines can have extremely high torques with good speed control. The first high power superconducting motor, built in 1966 by the International Research and Development Co. (IRD) in the UK, had a power rating of 37.5 kW at 2000 rpm. This company is constructing a 6-MW superconducting motor for the Royal Navy, which will be succeeded by a motor of ca 25 MW.

Fusion Magnets. Probably the most important and widespread use of superconductivity will be in controlled thermonuclear fusion based upon magnetic confinement of a plasma. Although laser and heavy-ion beam-induced implosion of a pellet are being considered as alternatives for fusion, magnetic confinement has made the most progress, and it is only a question of time before feasibility is demonstrated (see Fusion energy). The use of conventional magnets for commercial fusion reactors is ruled out because their power consumption would be about equal to the power output of the reactor. In the case of superconducting magnets, only the power for the He refrigerator need be provided which has been estimated to be ca 0.1% of the reactor output (75).

The two principal types of magnet configuration currently under investigation are the Tokamak and the tandem-mirror arrangement, both shown in Figure 22. The Tokamak arrangement consists of toroidal field coils, supplied by dc, and enclosing

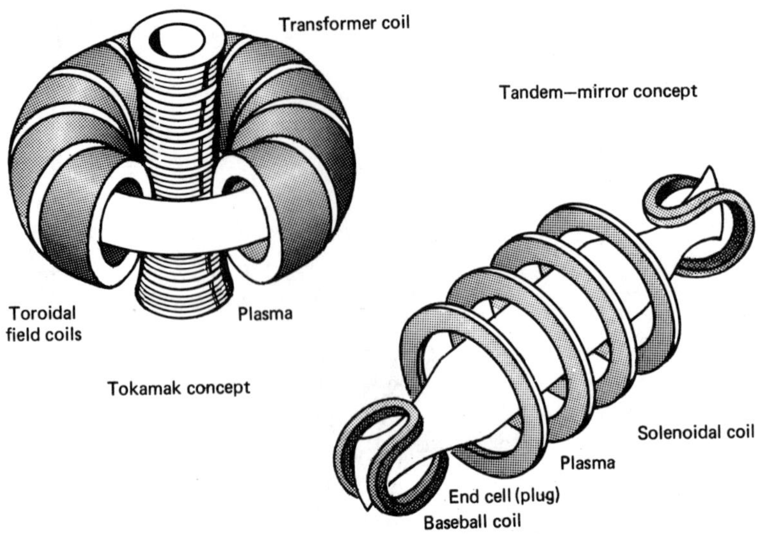

Figure 22. Basic concepts of magnetic confinement systems (75).

a plasma ring and pulsed a-c transformer coils which generate ohmic heating of the plasma by inducing plasma currents. Additional a-c coils are required for plasma stabilization and purification. It is not certain what toroidal field strength is needed for a practical reactor. However, experiments on the high field Tokamak at the MIT Magnet Laboratory, Alcator, indicate that higher confinement fields result in larger values of $n\tau$, the product of density n and confinement time τ, one of the important parameters. In the case of the tandem-mirror concept, mirror cells placed at the ends of a central solenoidal confinement chamber provide longitudinal confinement. In Figure 22 these mirrors are shown as baseball coils. In later tandem-mirror designs, these baseball coils are replaced by Yin-Yang-type coils.

Several years ago a superconducting toroidal magnet system for the T-7 Tokamak experiment was successfully put into operation at the Kurchatov Institute in Moscow (76). A forced-cooled hollow NbTi–Cu composite conductor was used in the magnet. At a current of 6 kA, a flux density of 2.5 T (25 kG) was generated at the torus axis. The next step will be T-15, especially aimed at the generation of high plasma temperatures, ie, (50–70) $\times 10^6$ K, which will be the scientific and technological basis for the thermonuclear Tokamak demonstration reactor (TNTDR). The TNTDR will contain a burning plasma and generate a thermal power of 450 MW (77) (see also Plasma Technology, Supplement Volume).

The U.S. DOE has started a toroidal-coil experiment at Oak Ridge National Laboratory, which is referred to as LCT (Large Coil Task). The toroidal d-c magnet system will be composed of six D-shaped single-disk coils with a 2.5 $\times$ 3.5 m bore. The system is designed to generate a maximum flux density of 8 T (80 kG). This project is being carried out under an international cooperative agreement: three coils are to be fabricated in the United States (General Electric, Westinghouse, General Dynamics), and three are to be supplied by Japan (Hitachi), Switzerland (Braun-Bovery), and Euratom (Siemens) (78).

Studies by several Tokamak and mirror-reactor design groups have pointed out the necessity for magnetic field levels of 11–12 T (110–120 kG). This has given rise to a U.S. 12 T (120 kG) coil program with four coil-project teams, each composed of a national laboratory, an industrial firm with experience in the fabrication of superconducting magnets, and a supplier of superconducting materials.

Magnetohydrodynamic Generators. Power generation by magnetohydrodynamics (MHD), combined with conventional coal burning is attractive in the United States because of the large reservoir of coal and the expected high efficiency of this method (see Coal, magnetohydrodynamics). At present, the open-cycle coal-fired system is favored for MHD electric-power plants (see Fig. 23). As a hot ionized gas passes through the MHD channel at high velocity, the perpendicular magnetic field initiates a force on the gas perpendicular to the gas flow and the magnetic field. The voltage built up between the cathode and anode is capable of driving a d-c current through an external load connected to the MHD generator. The exit gas from the MHD channel is still hot enough to operate a conventional steam-turbine plant. In this way the overall efficiency of the total plant can be increased to 50%, compared with a maximum efficiency of ca 40% for conventional steam plants. As in the case of fusion magnets, only superconducting magnets can be used in commercial plants, since conventional copper magnets would require an amount of energy nearly equal to that produced by the MHD generator. Furthermore, combining MHD and coal burning results in a notable suppression of SO_x and NO_x emission and a reduction in CO_2 generation and thermal pollution.

Many development programs were under way in the United States and the USSR, including some involving large-scale international cooperation. However, all U.S. funding for MHD research programs was halted by executive order in 1981.

Magnets for Nmr Imaging. Recently, computerized axial tomography (CAT, CT) scanning, or x-ray computerized tomography, has been developed to examine the inside of the human body. In this technique, x-ray data are recorded from many different directions and reconstructed mathematically to give cross-sectional anatomical views of selected regions of the body (see X-ray technology). This technique, however, is not sensitive to soft tissue. A new technique is on the threshold of clinical application which does not expose the patient to ionizing radiation and is much more sensitive to soft tissues. This technique, termed nmr imaging, depends upon the well-established ability of nmr to monitor metabolic processes as well as upon the highly advanced new imaging processes (79).

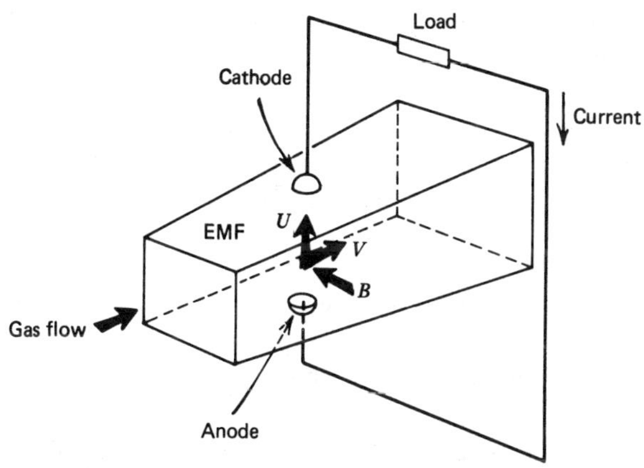

Figure 23. Shape and mode of operation of a MHD generator (75).

The basis of nmr spectroscopy is the fact that transitions between nuclear spin states can be observed during the resonant absorption of electromagnetic radiation by a nucleus in an external magnetic field (see Analytical methods). All nmr images produced so far have been obtained with the nuclei of hydrogen. The fact that the human body is 75% water thus facilitates the nmr imaging process. The magnetic fields involved for H nuclei are small [0.1–0.2 T (1–2 kG)], and therefore, electromagnetic radiation is in the MHz region and penetrates the human body easily. Images thus obtained are related to proton density, but may also be intensified or enhanced by techniques sensitive to differences in T_1, the spin-lattice relaxation time, between neighboring tissues. Although the magnets used generally provide low fields, they must be extremely homogeneous and stable. The magnet itself must possess a homogeneity of 10^{-6} over a 1 m diameter region and this homogeneity is increased by the use of shim coils at room temperature. Furthermore, the temporal stability must be ca 10^{-8}/h. Although both superconducting and conventional magnets are competitive at low fields, the stability of the superconducting coil as well as its ability to be used more economically at higher magnetic fields suggest that the superconducting coil will be preferred for nmr imaging.

Small-Scale Applications. A few years after Josephson's prediction of pair tunneling, quantum interference effects between two Josephson tunnel junctions incorporated in a superconducting ring were observed (5). The implications of this result for applications in measuring devices were soon realized and a number of d-c SQUIDs were developed and used (80). The early devices frequently involved machined Nb screws which were used in point-contact junctions. More recently, thin film tunnel-junction d-c SQUIDs have been constructed with a sensitivity that is several orders of magnitude higher than earlier versions. Another type of SQUID operating with an r-f flux bias, is referred to as the r-f SQUID (81). Both kinds may be used for very sensitive measurements of magnetic induction, voltage, etc. For example, a SQUID magnetometer can be used to take magnetocardiograms of the magnetic fields generated by the periodic electric currents in the heart (82). The signal was typically 5×10^{-11} T (5×10^{-7} G), and a shielded room was necessary. More recently, gradient magnetometers were developed that discriminate against distant sources and therefore can be used without shielding. SQUIDs have been used to obtain magnetoencephalograms (10^{-12} T or 10^{-8} G), and other low field applications are under investigation (83).

Josephson junctions are useful as detectors of radiation and recently much interest has been generated by the possibility of using high T_c materials in conjunction with miniature cryocoolers operating at ca 10 K (see Cryogenics). The junctions may be used as bolometers, and yield an output proportional to the temperature rise ΔT due to the absorption of radiation. On the other hand, junctions exhibit a square-law response to small a-c signals; this property may be employed for video detection. Connected to a local oscillator, the Josephson junction can also be used in a variety of heterodyne circuits. Furthermore, the Josephson effect provides a definition of the international volt in terms of the constant voltage steps induced in the current-voltage characteristics of a Josephson junction irradiated by microwave radiation. These junctions are also used in noise thermometry and frequency measurement (84) (see Noise pollution).

The potentially greatest impact of Josephson junctions could come in the area of large high speed digital computers (53). This application has been under investigation by IBM for a number of years and aims to make Josephson junction elements compatible with large-scale integrated (LSI) circuit technology. Large arrays of logic and memory circuits have been fabricated with standard photolithographic techniques. Logic-2 input OR circuit chains have been operated at an average logic delay as short as 13 ps per stage. Memory cells that are suitable for use in a subnanosecond access-time memory have been operated at a stored energy of ca 0.06 aJ (5.7×10^{-23} Btu). In order to take advantage of the fast switching speed of Josephson circuits, it is necessary to locate the circuits in close proximity to each other to minimize the signal-propagation delays (7 ps/mm). Aside from the materials problems previously mentioned, the ultimate use of Josephson-junction circuit elements requires adequate packaging and integration with room-temperature circuitry.

Economic Aspects

For the last several years the U.S. market for superconducting wire has been ca $\$10 \times 10^6$. Of this market, ca 90% has been sold to the U.S. government for projects like the Fermilab and Isabelle accelerators and the Twin Mirror fusion project at Lawrence-Livermore Laboratory. The remaining 10% represents sales to small companies for custom magnets. Most of the wire has been Nb–Ti materials, but some Nb_3Sn is also included. The near future should show a reduction in government purchases as large accelerator projects are completed. On the other hand, purchases by small manufacturers of custom research magnets are expected to increase over the next few years, so that the total sales to these two sources should decline very slowly. A new factor will be the demand generated by superconducting magnets for nmr whole-body scanners. At this point, magnet manufacturers expect more than half of the machines ordered to be equipped with superconducting magnets. Optimistic estimates indicate that although demand for scanner magnets is virtually nonexistent today, it could represent a market of $\$36 \times 10^6$ in 1985 and $\$119 \times 10^6$ in 1987.

BIBLIOGRAPHY

1. H. K. Onnes, *Leiden Communication*, 124c (1911).
2. *Ibid.*, 133d (1913).
3. G. B. Yntema, *Phys. Rev.* **98,** 1197 (1955).
4. S. H. Autler, *Rev. Sci. Inst.* **31,** 369 (1960).
5. R. C. Jaklevic, J. Lambe, A. H. Silver, and J. E. Mercereau, *Phys. Rev. Lett.* **12,** 159 (1964).
6. B. D. Josephson, *Phys. Lett.* **1,** 251 (1962); *Adv. Phys.* **14,** 419 (1965).
7. J. Bardeen, L. N. Cooper, and J. R. Schrieffer, *Phys. Rev.* **106,** 162 (1957); **108,** 1175 (1957).
8. B. T. Matthias, *Phys. Rev.* **97,** 74 (1955).
9. J. K. Hulm and B. T. Matthias in S. Foner and B. B. Schwartz, eds., *Superconductor Materials Science: Metallurgy, Fabrication and Applications*, Plenum Publishing Corporation, New York, 1981, p. 1.
10. W. Meissner and R. Ochsenfeld, *Naturwissenschaften* **21,** 787 (1933).
11. R. Chevrel, M. Sergent, and J. Prigent, *J. Solid State Chem.* **3,** 515 (1971).
12. S. Foner, E. J. McNiff, Jr., and E. J. Alexander, *Phys. Lett.* **49A,** 269 (1974).
13. B. B. Schwartz and S. Foner, *Physics Today* **30,** 34 (1977).
14. J. R. Gavler, *Appl. Phys. Lett.* **23,** 480 (1973).
15. H. Frohlich, *Phys. Rev.* **79,** 845 (1950).
16. L. N. Cooper, *Phys. Rev.* **104,** 1189 (1956).
17. A. B. Pippard, *Proc. Roy. Soc.* (*London*) **A216,** 547 (1953).
18. V. L. Ginzburg and L. D. Landau, *JETP* **20,** 1064 (1950).
19. F. London and H. London, *Proc. Roy. Soc.* (*London*) **A149,** 72 (1935); *Physica* **2,** 341 (1935).
20. I. Giaever, *Phys. Rev. Lett.* **5,** 147, 464 (1960).

21. B. D. Josephson, *Rev. Mod. Phys.* **46,** 251 (1974).
22. A. A. Abrikosov, *Soviet Phys. JETP* **5,** 1174 (1957).
23. L. P. Gor'kov, *Soviet Phys. JETP* **9,** 1364 (1959).
24. M. N. Wilson in ref. 9, p. 63.
25. J. K. Hulm and R. D. Blaugher, *Phys. Rev.* **123,** 1569 (1961).
26. M. Suenaga and R. M. Ralls, *J. Appl. Phys.* **40,** 4457 (1969).
27. D. G. Hawksworth and D. C. Larbalestier, *Adv. Cryog. Eng.* **26,** 479 (1980).
28. P. H. Bellin, H. C. Gatos, and V. Sadogapan, *J. Appl. Phys.* **40,** 3984 (1969).
29. H. Hillmann and K. J. Best, *IEEE Trans. Mag.* **13,** 1568 (1977).
30. D. C. Larbalestier in ref. 9, p. 133.
31. H. R. Segal, T. A. de Winter, Z. J. J. Stekly, and K. Hemachalam, *IEEE Trans. Mag.* **17,** 53 (1981).
32. I. Pfeiffer and H. Hillmann, *Acta Met.* **16,** 1429 (1968).
33. T. Doi, F. Ishida, U. Kawabe, and M. Kitada, *Trans. Met. Soc. AIME* **242,** 1793 (1968).
34. H. Hillman in ref. 9, p. 275.
35. K. J. Best, H. Hillmann, H. Hoeflich, J. Rudolph, and H. Weber, *Forschungsbericht BMFT NT* 1022, 1979.
36. G. F. Hardy and J. K. Hulm, *Phys. Rev.* **87,** 884 (1953); **93,** 1004 (1954).
37. J. Labbe and J. Friedel, *J. Physique* **27,** 153, 303 (1966).
38. J.-L. Staudenmann, B. DeFacio, L. R. Testardi, S. A. Werner, R. Flukiger, and J. Muller, *Phys. Rev. B* **24,** 6446 (1981).
39. B. T. Matthias, T. H. Geballe, S. Geller, and E. Corenzwit, *Phys. Rev.* **95,** 1453 (1954).
40. J. R. Gavaler, *Appl. Phys. Lett.* **23,** 480 (1973).
41. R. D. Blaugher, R. E. Hein, J. E. Cox, and R. M. Waterstrat, *J. Low Temp. Phys.* **1,** 539 (1969).
42. M. Suenaga in ref. 9, p. 201.
43. J. E. Kunzler, E. Buehler, F. S. L. Hsu, and J. H. Wernick, *Phys. Rev. Lett.* **6,** 89 (1961).
44. K. Tachikawa, *Int. Cryog. Eng. Conf. Berlin* (1970); *Iliffle Sci. Tech. Pub.*, 339 (1971).
45. M. Suenaga and W. B. Sampson, *Appl. Phys. Lett.* **18,** 584 (1971).
46. *Ibid.*, **20,** 443 (1972).
47. Y. Hashimoto, K. Yoshizaki, and M. Tanaka, *Proceedings of the 5th International Conference of Cryogenic Engineering*, Kyoto, Jpn., 1974, p. 332.
48. C. C. Tsuei, *Science* **180,** 57 (1973).
49. C. Allibert, J. Droile, and E. Bonnier, *Compt. Rend. C* **268,** 2277 (1969).
50. R. Roberge in ref. 9, p. 389.
51. R. Flukiger, R. Akihama, S. Foner, E. J. McNiff, Jr., and B. B. Schwartz, *Advances in Cryogenic Engineering*, Vol. 26, Plenum Publishing Corporation, New York, 1980, p. 377.
52. M. R. Pickus, J. T. Holthius, and M. Rosen in M. Suenaga and A. F. Clark, eds., *Filamentary A15 Superconductors*, Plenum Publishing Corporation, New York, 1980, p. 331.
53. J. Matisoo, *IBM J. Res. Dev.* **24,** 113 (1980).
54. R. F. Broom, R. B. Laibowitz, Th. O. Mohr, and W. Walter, *IBM J. Res. Dev.* **24,** 212 (1980).
55. S. K. Lahiri, *J. Vac. Sci. Technol.* **13,** 148 (1976); S. K. Lahiri, and S. Basavaiah, *J. Appl. Phys.* **49,** 2880 (1978).
56. S. K. Lahiri, S. Basavaiah, and C. J. Kircher, *Appl. Phys. Lett.* **36,** 334 (1980).
57. S. I. Raider and R. E. Drake, unpublished results.
58. J. Kwo, R. H. Hammond, and T. H. Geballe, *Bull. Am. Phys. Soc.* **24,** 455 (1979); D. F. Moore, R. B. Zubeck, J. M. Rowell, and M. R. Beasley, *Phys. Rev. B* **20,** 2721 (1979).
59. D. W. Jillie, H. Kroger, L. N. Smith, E. J. Cukauskas, and M. Nisenoff, *Appl. Phys. Lett.* **40,** 747 (1982).
60. B. T. Matthias, M. Marezio, E. Corenzwit, A. S. Cooper, and H. E. Barz, *Science* **175,** 1465 (1972).
61. Ø. Fischer, *Appl. Phys.* **16,** (1978).
62. R. Chevrel in ref. 9, p. 685.
63. R. Flukiger, R. Baillif, and E. Walker, *Mat. Res. Bull.* **13,** 743 (1978); R. Flukiger, H. Devantay, J. L. Jorda, and J. Muller, *IEEE Trans. Mag.* **MAG-13,** 818 (1977).
64. O. K. Andersen, W. Klose, and H. Nohl, *Phys. Rev.* **B17,** 1209 (1978).
65. L. F. Mattheiss and C. Y. Fong, *Phys. Rev.* **B15,** 1760 (1977).
66. T. Jarlborg and A. J. Freeman, *Phys. Rev. Lett.* **44,** 178 (1980).
67. S. A. Alterovitz, J. A. Woolam, L. Kammerdiner, and H. L. Huo, *Appl. Phys. Lett.* **33,** 264 (1978); **31,** 234 (1977).
68. W. Buckel and R. Hilsch, *Z. Phys.* **138,** 109 (1954).

69. W. L. Johnson, *J. Appl. Phys.* **50,** 1557 (1979).
70. W. L. Johnson and A. R. Williams, *Phys. Rev. B* **20,** 1640 (1979).
71. M. Tenhover, W. L. Johnson, and C. C. Tsuei, *Solid State Commun.* **38,** 53 (1981).
72. W. L. Carter, S. J. Poon, J. W. Hull, and T. H. Geballe, *Solid State Commun.* **39,** 41 (1981).
73. B. M. Clemens, W. L. Johnson, and J. Bennett, *J. Appl. Phys.* **51,** 1116 (1980).
74. M. Tenhover, *IEEE Trans. Mag.* **17,** 1021 (1981).
75. G. Bogner in ref. 9, p. 757.
76. D. P. Ivanov, V. E. Keilin, B. A. Stavissky, and N. A. Chernoplekov, *IEEE Trans. Mag.* **MAG-15,** 550 (1979).
77. B. Kadomzew and G. Jelissejew, *Sowjetischer Export* **2,** 30 (1980).
78. *Fusion Technology Development, August 1979,* U.S. Department of Energy, DOE/ET-0116/1 UC-20, A–G.
79. I. L. Pykett, *Sci. Am.,* 78 (May 1982).
80. J. E. Zimmerman and A. H. Silver, *Phys. Rev.* **141,** 367 (1966).
81. J. E. Zimmerman, P. Thiene, and J. T. Harding, *J. Appl. Phys.* **41,** 1572 (1970).
82. D. Cohen, E. A. Edelsack, and J. E. Zimmerman, *Appl. Phys. Lett.* **19,** 16 (1971).
83. D. Cohen, *Science* **175,** 664 (1972); D. Cohen, *Phys. Today,* 34 (Aug. 1975).
84. R. A. Kamper in B. B. Schwartz and S. Foner, eds., *Superconductor Applications: SQUIDS and Machines,* Plenum Publishing Corporation, New York, 1977, p. 189.

General References

P. G. deGennes, *Superconductivity in Metals and Alloys,* W. A. Benjamin, Inc., New York, 1966; physical understanding of superconductivity, especially type II superconductors.

F. London, *Superfluids,* Vol. 1, John Wiley & Sons, Inc., New York, 1950; phenomenological theory of superconductivity before the explosive progress in the 1950 decade. (The famous prediction of flux quantization appears on p. 152.)

E. A. Lynton, Superconductivity, 3rd ed., Menthuen & Co., Ltd., London, 1969; brief survey.

D. Saint-James, G. Sarma, and E. J. Thomas, *Type II Superconductivity,* Pergamon Press, New York, 1969; review of type II superconductors often using a combination of rigorous, phenomenological, and physical viewpoints.

J. R. Schrieffer, *Theory of Superconductivity,* W. A. Benjamin, Inc., New York, 1964; advanced presentation of the theory of superconductivity. (A quantum mechanics and solid state physics background is required.)

L. Solymar, *Superconductive Tunnelling and Applications,* Wiley-Interscience, New York, 1972.

M. Tinkham, *Introduction to Superconductivity,* McGraw-Hill, Inc., New York, 1975; basic physics of superconductivity.

T. Van Duzer and C. W. Turner, *Principles of Superconducting Devices and Circuits,* Elsevier North Holland, Inc., New York, 1981.

S. Foner and B. B. Schwartz, eds., *Superconducting Machines and Devices: Large Systems Applications,* Plenum Press, New York, 1974.

S. Foner and B. B. Schwartz, eds., *Superconductor Materials Science: Metallurgy, Fabrication, and Applications,* Plenum Press, New York, 1981.

R. D. Parks, ed., *Superconductivity,* Vols. I and II, Marcel Dekker, Inc., New York, 1969; specialized chapters on almost all areas of superconductivity theory, experiment, and quantum effects in superconductors.

B. B. Schwartz and S. Foner, eds., *Superconductor Applications: SQUIDS and Machines,* Plenum Press, New York, 1977.

FRANK P. MISSELL
Universidade de São Paulo

BRIAN B. SCHWARTZ
Brooklyn College, CUNY

SUPERCRITICAL FLUIDS. See Supplement volume.

SUPEROXIDES. See Peroxides and peroxy compounds.

SUPERPHOSPHATE. See Fertilizers.

SURFACTANTS AND DETERSIVE SYSTEMS

Surfactants, 332
Detergency and detersive systems, 387

SURFACTANTS

The term surfactant is a contraction of the longer, more awkward term surface-active agent. Coined in 1950, surfactant has become universally accepted to describe organic substances with certain characteristic features in structure and properties. The term detergent is often used interchangeably with surfactant. As a designation for a substance capable of cleaning, detergent can also encompass inorganic substances when these do in fact perform a cleaning chore. More often, detergent refers to a combination of surfactants with other substances, organic or inorganic, formulated to enhance functional performance, specifically cleaning, over that of the surfactant alone. It is so used in this article.

Surfactants are characterized by the following features:

Amphipathic structure. Surfactant molecules are composed of groups of opposing solubility tendencies, typically an oil-soluble hydrocarbon chain and a water-soluble ionic group.

Solubility. A surfactant is soluble in at least one phase of a liquid system.

Adsorption at interfaces. At equilibrium, the concentration of a surfactant solute at a phase interface is greater than its concentration in the bulk of the solution.

Orientation at interfaces. Surfactant molecules and ions form oriented monolayers at phase interfaces.

Micelle formation. Surfactants form aggregates of molecules or ions called micelles when the concentration of the surfactant solute in the bulk of the solution exceeds a limiting value, the so-called critical micelle concentration (CMC), which is a fundamental characteristic of each solute–solvent system.

Functional properties. Surfactant solutions exhibit combinations of cleaning (detergency), foaming, wetting, emulsifying, solubilizing, and dispersing properties.

The presence of two structurally dissimilar groups within a single molecule is the most fundamental characteristic of surfactants. The surface behavior, ie, the surface activity, of the surfactant molecule is determined by the make-up of the individual groups, their solubility properties, their relative size and their location within the surfactant molecule. The term amphipathy was proposed as "the occurrence in a single molecule or ion, with a suitable degree of separation, of one or more groups which have affinity (sympathy) for the phase in which the molecule or ion is dissolved, together with one or more groups which are antipathic to the medium (ie, which tend to be expelled by it)" (1).

Different designations describe the opposing groups within the surfactant molecules, eg, hydrophobic, water hating and hydrophilic, water liking; lipophobic, fat hating and lipophilic, fat liking; oleophobic, fat (oil) hating and oleophilic, fat (oil) liking; lyophobic, solvent hating and lyophilic, solvent liking. The terms polar and nonpolar are also used to designate water-soluble and water-insoluble groups, respectively.

Hundreds of thousands of amphipathic combinations are theoretically possible. Thousands have been described in the literature covering most of the commercially important or theoretically interesting structures.

Surface activity is not limited to aqueous systems. All of the combinations of aqueous and nonaqueous phases are known to occur. However, because water is present as the solvent phase in the overwhelming proportion of commercially important surfactant systems, its presence is assumed in much of the common terminology of industry. Thus, the water-soluble amphipathic groups are often referred to as solubilizing groups.

Until the end of the 19th century, soap (qv), the alkali metal salt of long-chain carboxylic acids, was the only man-made surfactant. With the spread of chemical technology, particularly in the early years of this century, the technical shortcomings of soap, ie, its instability toward water hardness and acidity, were felt particularly acutely in the textile industry. Sulfonated oils were developed as hardness- and acid-stable dyeing and wetting assistants. A critical shortage of soap in Germany during World War I provided further incentive for the development of substitutes for the fat-derived hydrophobe. Further activity in Germany in the years following World War I led to many hundreds of new surfactant structures. These developments can be viewed as the beginning of the modern surfactant industry. The growth of the petrochemical industry after World War II further aided the growth of the surfactant industry in providing high quality, relatively inexpensive starting materials. During the last decade, however, the escalating cost of crude oil and simultaneous improvements in the production of oil-bearing seeds have led to a reexamination of oleochemical feedstocks as potential sources for surfactant manufacture. An extensive shift to oleochemicals is not likely, although some moves in this direction are in evidence.

Surfactants are classified depending on the charge of the surface-active moiety, usually the larger part of the molecule. In anionic surfactants, this moiety carries a negative charge, as in soap:

$$C_{17}H_{35}CO_2^- \; Na^+$$

In cationic surfactants, the charge is positive:

$$(C_{18}H_{37})_2\overset{+}{N}(CH_3)_2 \; Cl^-$$

In nonionic surfactants, as the name implies, there is no charge on the molecule; the solubilizing contribution is supplied by a chain of ethylene oxide groups:

$$C_{15}H_{31}(OC_2H_4)_7OH$$

Finally, in amphoteric surfactants, solubilization is provided by the presence of positive and negative charges in the molecule, as in

$$C_{12}H_{25}\overset{+}{N}(CH_3)_2 \; CH_2CO_2^-$$

In general, the hydrophobic group consists of a hydrocarbon chain containing ca 10–20 carbon atoms. The chain may be interrupted by oxygen atoms, a benzene ring, amide, ester, and other functional groups, and double bonds. A propylene oxide hydrophobe can be considered as a hydrocarbon chain in which every third methylene group is replaced by an oxygen atom. In some cases, the chain may carry substituents, most often halogens. Siloxane chains have also served as the hydrophobe in some surfactants developed in recent years.

Hydrophilic, solubilizing groups for anionic surfactants include carboxylates, sulfonates, sulfates, and phosphates. Cationics are solubilized by amine and ammonium groups. Ethylene oxide chains and hydroxyl groups are the solubilizing groups in nonionic surfactants. Amphoteric surfactants are solubilized by combinations of anionic and cationic solubilizing groups.

The molecular weight of surfactants ranges from a low of ca 200 to a high in the thousands for polymeric structures. A surfactant with a straight-chain C_{12} hydrophobe and a solubilizing group is generally a very effective structure. The optimum can be higher by several C atoms or even slightly lower than 12 depending on the nature of the polar group.

In the application of surfactants, physical and use properties, precisely specified, are of primary concern; chemical homogeneity is of little significance in practice. In fact, surfactants are generally polydisperse mixtures defined as "a preparation containing molecules which are all of the same type, but which vary only in chain length or in some other structural detail . . . (as) distinguished from a mixture of two different molecular species" (2). Examples of such polydisperse mixtures include in the first place the natural fats as precursors of fatty acid-derived surfactant structures. Coconut oil contains glycerol esters of C_6 to C_{18} fatty acids. This range itself varies, depending on the extent of distillative treatment to which the oil has been subjected. Nonionic surfactants of the alcohol ethoxylate type are polydisperse not only with respect to the hydrophobe but also in the length of the ethylene oxide chain. In the case of alkylbenzene, another surfactant raw material of great commercial significance, the alkyl substituent not only contains different chain lengths but also carries the benzene ring at all except the terminal positions along the hydrocarbon chain.

Commercial surfactants, then, are complicated mixtures exceedingly difficult to separate into pure molecular species (3–4). Under these circumstances, the IUPAC and Chemical Abstracts nomenclatures designed to describe pure molecular species are of little utility. The American Oil Chemists' Society (5) recognizes the value of trivial names such as lauryl and myristyl as designating a mixture of chain lengths with a distribution around C_{12} and C_{14}, respectively.

The terms coco or tallow describing a mixture of acids derived from coconut oil and tallow, respectively, are even less systematic but still useful. In analogy with the chemically precise acyl groups, the acid radicals from coconut and tallow fatty acids are sometimes referred to as cocoyl and tallowyl, respectively.

Achieving a useful balance between systematic and common nomenclature is the aim of the *CTFA Cosmetic Ingredient Dictionary* published by The Cosmetic, Toiletry and Fragrance Association, Inc. (6). Chemical names devised for ingredients of cosmetic formulations have attained general acceptance in the United States and are used on labels (see Cosmetics).

In view of the precise specifications of physical and performance properties of polydisperse surfactant mixtures, there is little justification for attempting to separate such mixtures into individual, chemically homogeneous, species. Where the latter have been used as model compounds in basic studies on surface behavior, they were generally synthesized rather than isolated from commercial mixtures.

The abbreviations indicating prominent use properties of the various classes of commercial surfactants are shown in Table 1. The list also includes nonsurfactant properties. Frequently, a combination of surfactant and nonsurfactant properties is useful and salable and hence commercially important.

Most of the properties listed in Table 1 are self-explanatory or are discussed in later sections. A few need explanation: Antimicrobial activity includes germicidal, bactericidal, and bacteriostatic effects. Emolliency describes lubrication or a soft feel which is imparted to skin by surfactants. A hair conditioner is a substantive surfactant applied from aqueous solution to impart a lubricating or antistatic effect. Opacifiers are used to thicken hand-dishwashing products and cosmetic preparations to convey an appearance of high concentration and to retard drainage.

In general, the surfactants given in Tables 3–36 are derived from straight-chain materials and are biodegradable. The relatively small number of surfactant structures that are not expected to be rapidly biodegradable because of extensive chain branching are identified by a footnote.

Table 1. Abbreviations of Properties or Uses of Commercial Surfactants[a]

Property	Abbreviation
antistatic reduction or elimination	A
antimicrobial activity	AM
corrosion inhibition	C
demulsification	dE
defoaming	dF
detergency[b]	D
dispersing efficacy[b]	Di
emulsion polymerization	eP
emulsification[b]	E
emolliency	Em
foaming power[b]	F
foam stabilization	Fs
hair conditioning	H
intermediate for further synthesis	I
lime-soap dispersion	L
opacification	O
oil recovery[c]	OR
penetrating power	P
rewetting effectiveness[d]	rW
solubilizing power[b]	S
softening efficacy[a]	Sp
stabilizing power	St
suspending power	Su
textile assistance[f]	Ta
viscosity building	V
wetting power[b]	W

[a] Compiled from vendors' recommendations.
[b] Key functional property.
[c] Includes enhanced, tertiary.
[d] For paper, textiles.
[e] For fabric, fiber, yarn.
[f] Processing, fiber lubrication, dyeing.

Table 2 lists 73 U.S. companies that market the surfactants discussed in this article. The interests of surfactant chemical producers, marketers, users and of the technical personnel in the surfactant and related disciplines are well served by trade associations, publications, and scientific societies. The Soap and Detergent Association (SDA), the Chemical Specialties Manufacturers' Association (CSMA), and the Cosmetic, Toiletries, and Fragrance Association (CTFA) represent the companies that make and use surfactants and other detergent raw materials. Trade news and practical technology are reported in *Household and Personal Products Industry* (HAPPI), *Soap Cosmetics Chemical Specialties*, *Cosmetics & Toiletries*, and *Chemical Times & Trends*. The *Journal of the American Oil Chemists' Society*, the *Journal of the Society of Cosmetic Chemists*, the *Journal of Colloid and Interfacial Science*, and the appropriate *American Chemical Society* journals provide media for scientific publications.

Physical Chemistry of Interfaces

The usefulness of surfactants stems from the effects that they exert on the surface, interfacial, and bulk properties of their solutions.

Phenomena at Liquid Interfaces. The area of contact between two phases is called the interface; three phases can have only a line of contact, and only a point of mutual contact is possible between four or more phases. Combinations of phases encountered in surfactant systems are given below:

L–G	L–L–G	L–S–G	L–S–S–G
L–L	L–L–L	L–S–S	L–L–S–S–G
L–S	L–L–S	L–L–S–G	

where G = gas, L = liquid, and S = solid. An example of an L–L–S–G system would be an aqueous surfactant solution containing an emulsified oil, a suspended solid, and entrained air (see Emulsions; Foams). The example embodies several conditions that are common to practical surfactant systems. First, since the surface area of a phase increases as particle size decreases, the emulsion, suspension, and entrained gas would each have large areas of contact with the surfactant solution. Next, since interfaces can only exist between two phases, analysis of phenomena in the L–L–S–G system breaks down into a series of analyses, ie, surfactant solution to the emulsion, to the solid, and to the gas. It is also apparent that the surfactant must be stabilizing the system by preventing contact between the emulsified oil and the dispersed solid. Finally, the dispersed phases are in equilibrium with each other through their common equilibrium with the surfactant solution.

Figures 1(**a**) and (**b**) schematically represent typical gas–liquid and liquid–liquid interfaces at equilibrium. Referring to Figure 1(**a**) and assuming that the gas G consists of air and vapor of the liquid L at equilibrium, it is obvious that there is a continuous movement of liquid molecules through the gaseous interfacial region R_G since the rates of evaporation and condensation at the interface I_{GL} are equal. Liquid molecules are also moving continuously into and out of I_{GL} through the liquid interfacial region R_L. R_G and R_L represent nonhomogeneous transitional regions between the homogeneous phases G and L. Systems are known in which R_G and R_L have thicknesses equivalent to two or more layers of molecules but for most analyses the interface I_{GL} can be considered as consisting of a single layer of molecules.

For thermodynamic treatment of surface phenomena, the thickness of the boundary regions can often be ignored or their effect eliminated by selection of a convenient location for the interface I_{GL}. The liquid–liquid interface, I_{LL}, of Figure 1(**b**) is similarly associated with interfacial regions, R_A and R_B, which can be treated like the gas–liquid interface in most analyses. Since few liquids are completely immiscible, mutual saturation is taken as the equilibrium condition.

Surface Energy of Gas–Liquid Interfaces. Each unit of a liquid surface is associated with a quantity of free energy called the free surface energy. Work equivalent to the increase in free-surface energy must be added to a liquid to increase its surface area. The free energy content of a surface is expressed in mJ/m^2 (= erg/cm^2). A pure liquid tends toward the condition of minimum energy, which is the shape of minimum area. Thus, the shape of a free-falling droplet would be spherical under idealized conditions. The tendency of a pure liquid to minimize its surface can be explained on the basis of intermolecular attractive forces. Even though a liquid is a freely flowing fluid mass, the individual molecules are so close together that van der Waals-type forces are the dominant attraction between molecules. These are very short range forces that decrease as the sixth power of the intermolecular distance. Thus, a molecule in the interior of a liquid would not be subject to a net attractive force in any direction because the number of molecules in the sphere surrounding it would, on the average, be the same in all directions. However, a molecule at the surface of the liquid would be subject to a strong inward attraction because the number of molecules in the hemisphere of the vapor close enough to exert an attractive force pulling out of the liquid would be small as compared to the number in the hemisphere of liquid pulling inward from the surface. A reasonable ratio of vapor-to-liquid molecules would be 1:1000. This unbalanced attraction would cause as many molecules as possible to move out of the surface into the bulk liquid, reducing its area to a minimum.

Surface Energy of Liquid–Liquid Interfaces. Each unit area of the interface between two mutually insoluble liquids is associated with a quantity of energy called the interfacial free energy, expressed in mJ/m^2 (= erg/cm^2). Expansion of the interfacial area requires the addition of energy. Conversely, the interface tends to decrease the surface of minimum area. Breaking of an emulsion is an example of a decrease in the area of a liquid–liquid interface. Molecules at a liquid–liquid interface are subject to a more balanced sphere of attraction than molecules at a gas–liquid interface because the number of molecules on both sides of a liquid–liquid interface is of the same order of magnitude. Thus, there is a strong force of attraction between two liquids in contact and the interfacial free energy is always less than the surface free energy of the more energetic liquid.

Surface and Interfacial Tension. Some properties of liquid surfaces are suggestive of a skin that exercises a contracting force or tension parallel to the surface. Mathematical models based on this effect have been used in explanation of surface phenomena such as capillary rise. The terms surface tension and interfacial tension relate to these models which do not reflect the actual behavior of molecules and ions at interfaces. The widespread usage of the terms is perhaps unfortunate. Surface tension is the force per unit length required to create a new unit area of gas–liquid surface. It is expressed as mN/m (= dyn/cm) and is numerically equal to the free surface energy in mJ/m^2 (erg/cm^2). Similarly, interfacial tension is the force per unit length required to create a new unit area of liquid–liquid interface and is numerically equal to the interfacial free energy.

Table 2. Manufacturers or Sources of Surfactant Chemicals

Code	Manufacturer or source
AAC	Alcolac, Inc.
ACI	Aceto Chemical Co., Inc.
ACY	American Cyanamid Co.
AKS	Arkansas Co., Inc.
AMH	American Hoechst Corp., Industrial Chemical Division
AMK	Armak Co., Industrial Chemicals Division
ARD	Ardmore Chemical Co.
CGY	CIBA-GEIGY Corp., Dyestuff & Chemicals Division
CHP	C. H. Patrick & Co., Inc.
CLI	Clintwood Chemical Co.
CO	Conoco, Inc.
CRO	Croda, Inc.
CRS	Carson Chemicals, Inc.
CRT	Crest Chemical Corp.
CTL	Continental Chemical Co.
DEX	Dexter Chemical Corp.
DSC	Diamond Shamrock Corp., Process Chemicals Division
DUP	E. I. du Pont de Nemours & Co., Inc.
EFH	E. F. Houghton & Co.
EK	Eastman Kodak Co.
EMK	Emkay Chemical Co.
EMR	Emery Ind., Chemical Group
FET	Finetex, Inc.
GAF	GAF Corp., Chemical Products
GLY	Glyco Chemicals, Inc.
GWR	W. R. Grace & Co., Organic Chemicals Div.
HAL	C. P. Hall Co.
HCL	Hart Chemical, Ltd.
HDG	Hodag Chemical Corp.
HNC	Henkel Corp., Chemical Specialties Div.
HPC	Hercules, Inc.
HRT	Hart Products Corp.
HXL	Hexcel Corp., Hexcel Chemical Products
HUM	Witco Chemical Corp., Humko Chemical Division
ICI	ICI Americas Inc.
LNZ	Lonza, Inc.
MIL	Milliken & Co., Milliken Chemical Division
MOA	Mona Industries, Inc.
MON	Monsanto Industrial Chemicals Co., Detergents and Phosphate Division
MRS	Mars Chemical Corp.
NLI	NL Industries, Inc.
NLS	Sterling Drug Inc., Lehn & Fink Industrial Products Division
OC	Olin Corp., Olin Chemicals
OXR	Onyx Chemical Co.
PET	Petrochemicals Co., Inc.
PG	The Procter & Gamble Co.
PHC	Philip A. Hunt Chemical Corp., Organic Division
PIL	Pilot Chemical Co.
QCC	Quad Chemical Corp.
QCP	Quaker Chemical Corp.
RCD	Witco Chemical Corp., Richardson Division
RH	Rohm & Haas Co.
RW	Robinson-Wagner Co.
S	Sandoz Colors & Chemicals

Table 2 (*continued*)

Code	Manufacturer or source
SBC	Scher Chemicals, Inc.
SDW	Sterling Drug Inc., Sterling Organics Div.
SFS	Stauffer Chemical Co., Specialty Div.
SHC	Shell Chemical Co.
SHX	Sherex Chemical Co., Inc.
SLC	Soluol Chemical Co., Inc.
SNW	Sun Chemical Corp., Chemicals Division
STP	Stepan Chemical Co.
SYB	Sybron Corp., Chemical Division/Tanatex
TCC	Texaco, Inc., Texaco Chemical Co.
TCM	Intex Products, Inc.
TMH	Thompson Hayward Chemical Co.
UCC	Union Carbide Corp., Ethylene Oxide Derivatives Division
VNC	R. T. Vanderbilt Co.
VND	Van Dyk & Co., Inc.
WTC	Witco Chemical Corp., Organics Division
WVA	Westvaco Corp., Polychemicals Division
WYN	BASF Wyandotte Corp.

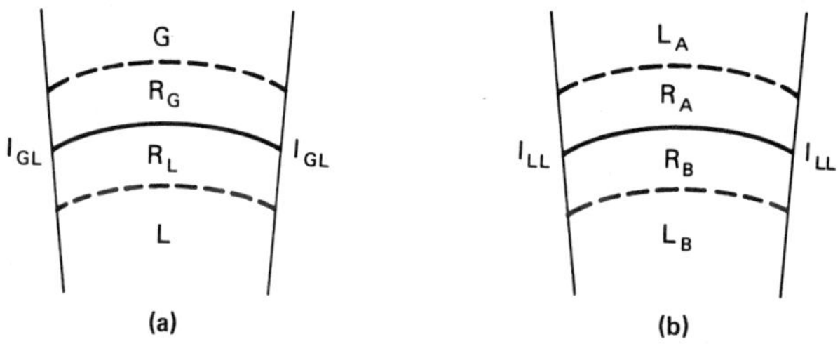

Figure 1. (**a**) Gas–liquid interface. (**b**) Liquid–liquid interface.

Energy of Adhesion. The interfacial energy between two mutually insoluble liquids A and B is equal to the difference in the separately measured surface energies of each phase:

$$\gamma_{AD} = \gamma_A - \gamma_B \tag{1}$$

where γ is free surface or interfacial energy. Liquid A and B are assumed to be mutually saturated with each other. The term γ_{AB} represents the energy that would have to be added to the system to separate the liquids. The rule holds true for most systems. The work of adhesion W_{AB} between two liquids, ie, the energy necessary to increase the interfacial area by one square unit is defined by the Dupré equation:

$$W_{AB} = \gamma_A + \gamma_B - \gamma_{AB} \tag{2}$$

The quantity of energy required to separate the two liquids increases as the interfacial tension between them decreases; the lower the interfacial energy, the stronger the adhesion.

An attraction also exists at the interface between a liquid and an insoluble solid, the interfacial tension is lower than the sum of the surface tensions of the two phases. Dupré's equation is applicable to the solid–liquid interface and has the form

$$W_{SL} = \gamma_{SA} + \gamma_{LA} - \gamma_{SL} \tag{3}$$

However, the surface tension of the solid γ_{SA} and the solid–liquid interfacial tension γ_{SL} cannot be measured directly by simple means. The work of adhesion of the solid to the liquid W_{SL} is usually determined by other techniques.

Contact Angle. The line of contact between the three phases of a G–L–S system is the locus of all points from which the angle of contact between the liquid and the solid can be measured (see Figure 2). The drop of liquid L is resting on the solid S and both of these phases are exposed to the gas G at equilibrium saturation of the liquid in air. The drop is assumed to be small enough for the flattening pressure of gravity to be negligible. The vector X_a is tangent to the liquid at its contact with the solid. The angle θ between the tangent and the surface of the solid is called the contact angle. The equilibrium value of the contact angle is an indicator of the energy relationships between liquid–liquid and liquid–solid interfaces. The contact angle is related to the free surface energies through the Young-Dupré equation:

$$\gamma_{SA} = \gamma_{SL} + \gamma_{LA} \cos \theta \qquad (4)$$

Combination with the Dupré equation for the energy of adhesion of a solid–liquid interface, equation 3, gives

$$W_{SL} = \gamma_{LA}(1 + \cos \theta) \qquad (5)$$

When the contact angle is zero, ie, $\cos \theta = 1$, then $W_{SL} = 2\ \gamma_{LA}$. Under this condition, the attraction of the liquid for the solid is exactly equal to the attraction of the liquid for itself and complete wetting of the solid by the liquid is possible at equilibrium. If the liquid attracts the solid more strongly than it attracts itself, then $W_{SL} > 2\ \gamma_{SL}$ is a possible condition even though the negative contact angle that it implies is impossible. No contact angle or nil is used to distinguish $W_{SL} > 2\ \gamma_{LA}$ from zero for $W_{SL} = 2\ \gamma_{LA}$. The droplet of Figure 2 has a circular line of contact with the solid which is shown in the cross-sectional diagram only at the intercepts, points x. The tangent to the droplet X_a is a vector that represents the tendency of the air–liquid interface toward the spherical shape of minimum energy content. There are two interfaces in the plane of the solid: liquid–solid, and gas–solid. The liquid–solid interface tends to a minimum as represented by the vectors X_1 parallel to the solid surface and directed toward the center of the circle of contact. The surface force represented by the vectors X_s is too weak to effect contraction of the solid surface. Nevertheless, such a force does exist and works toward minimizing the area of the air–solid interface. In the system of Figure 2, the area of the air–solid interface may be reduced by substituting a liquid–solid interface which is the tendency represented by the vectors X_s. For a contact angle of zero, $\cos \theta = 1$, and at equilibrium, $X_s = X_a + X_1$ A contact angle of nil would represent the condition $X_s > X_a + X_1$ in which equilibrium could not be attained and

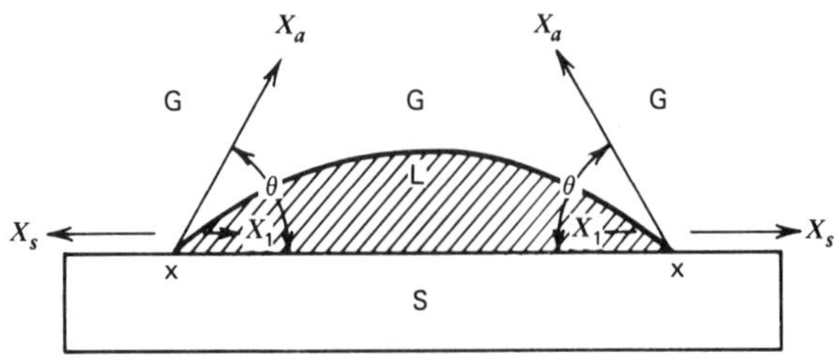

Figure 2. Contact angle at the oil–solid–air boundary line in the roll-back process of oily solid detergency.

the liquid would continue to spread over the solid surface. Contact angles of 90° or greater are associated with very low solid–liquid adhesion. In practical surfactant technology, this condition would be apparent because of the lack of wetting of the solid by the liquid. A contact angle of 180° which would indicate the complete absence of solid–liquid adhesion is not known to exist. A hysteresis effect is nearly always observed in the measurement of contact angles, ie, the advancing angle is greater than the receding angle.

Spreading. The decrease in the free surface energy of a system caused by the spreading of a liquid over the surface of a solid or another liquid is called the spreading coefficient. The total surface energy must be decreased by the spreading process which implies a greater attraction between unlike molecules than between like molecules. For a liquid–solid system, the decrease in the free energy ϕ as a result of the spreading liquid L over solid S is

$$\phi = \gamma_{SA} - \gamma_{LA} - \gamma_{SL} \tag{6}$$

Combining with Dupré's equation for solids, equation 3 gives

$$\phi = W_{SL} - 2\,\gamma_{LA} \tag{7}$$

Thus, the decrease in free surface energy of the total system when the liquid spreads over the solid equals the energy of adhesion W_{SL} minus the energy of cohesion $2\,\gamma_{LA}$. The energy of cohesions is the energy that must be added to a liquid system to separate a column of liquid of unit cross section, thus creating two units of new surface area. If the spreading coefficient ϕ is positive, the liquid spreads over the solid, and the greater the positive value, the more readily spreading occurs. The spreading of a liquid over a plane solid proceeds slowly to equilibrium but in many practical systems capillary action or mechanical agitation speeds the wetting process. The spreading coefficient of a liquid–liquid system is calculated directly by substituting the surface and interfacial tensions in

$$\phi' = \gamma_o - \gamma_w - \gamma_{wo} \tag{8}$$

where γ_o represents the surface tension of the oil phase, γ_w the surface tension of the surfactant solution, and γ_{wo} the interfacial tension. The greater the positive value of ϕ, the greater the ease of spreading.

Effects of Surfactants on the Properties of Solutions. A surfactant changes the properties of a solvent in which it is dissolved to a much greater extent than would be expected from its concentration. This marked effect is due to adsorption at the solutions interfaces, orientation of the adsorbed surfactant ions or molecules, micelle formation in the bulk of the solution, and orientation of the surfactant ions or molecules in the micelles. These effects are caused by the amphipathic structure of a surfactant molecule. The magnitude of these effects depends to a large extent on the solubility balance of the molecule. An efficient surfactant is usually relatively insoluble as individual ions or molecules in the bulk of a solution, eg, 10^{-2}–10^{-4} mol/L.

A schematic representation of adsorption and orientation in a solution of an anionic surfactant is shown in Figure 3. This illustration is intended to show: a relatively greater number of adsorbed molecules in the G–L interface than in the bulk of the solution; a relatively greater number of adsorbed molecules on the S–L interface (beaker walls) than in the bulk of the solution; the adsorbed molecules are oriented with their lyophilic moieties in the solvent and their lyophobic moieties respectively

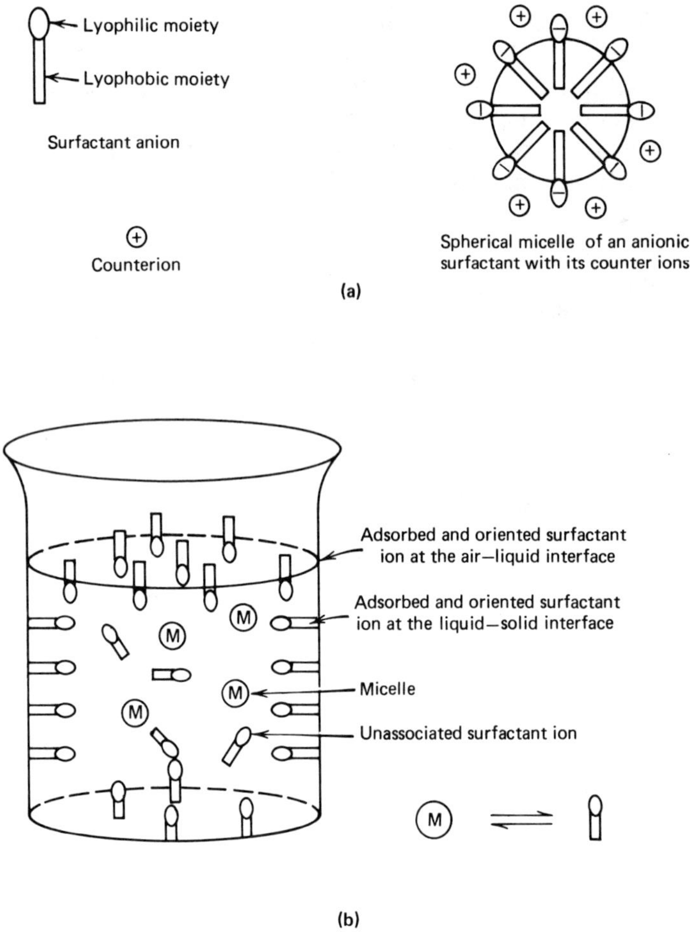

Figure 3. Schematic diagram of an anionic surfactant solution at equilibrium above its critical micelle concentration. (**a**) Components. (**b**) Locations of the components at equilibrium.

in the G–L interface and in contact with the S–L surface; the surfactant molecules aggregated in the micelles are oriented with the lyophilic moieties exposed to the solvent and shielding the lyophobic moieties in the center of the micelles; and the adsorbed molecules are in equilibrium with the micelles through the relatively small concentration of individual solute molecules in the bulk of the solution.

If the G–L interface were replaced by an L–L interface, eg, an oil layer on an aqueous solution, the orientation of the ions in the illustration would not be changed. In practical surfactant systems, L–L interfaces are often the surfaces of emulsified droplets and L–S interfaces the surfaces of suspended solids.

Adsorption at Gas–Liquid and Solid–Liquid Interfaces. Positive adsorption, the concentration of one component of a solution at a phase boundary, results in a lowering of the free surface energy of the solution. Accumulation of a surfactant at a solution interface means that the attractive forces between its molecules or ions are less than the attraction between solvent molecules. As thermal diffusion brings surfactant molecules into the surface, they accumulate there because they cannot reenter the solution against the stronger mutual attraction of the solvent molecules. Negative adsorption occurs when the attraction between solute molecules is greater than the attraction between solvent molecules. This condition exists in concentrated aqueous solutions of inorganic compounds such as NaOH and is associated with a surface tension slightly higher than for pure water.

The quantitative relationship between the degree of adsorption at a solution interface, G–L or L–L, and the lowering of the free surface energy was deduced by Gibbs using thermodynamic methods (7). The resulting equation was rigorous but awkward for calculations. A frequently used approximate form of the Gibbs adsorption isotherm, equation 9, is applicable to dilute binary solutions where the activity coefficient is unity and the radius of curvature of the surface is not too great:

$$\Gamma_s = -\frac{C_s}{RT} \cdot \frac{d\gamma}{dC_s} \tag{9}$$

It states that the surface excess of solute Γ_s is proportional to the concentration of solute C_s multiplied by the rate of change of surface tension, with respect to solute concentration, $d\gamma/dC_s$. The concentration of a surfactant in a G–L interface can be calculated from the linear segment of a plot of surface tension versus concentration and similarly for the concentration in an L–L interface from a plot of interfacial tension. In typical applications, the approximate form of the Gibbs equation was employed to calculate the area occupied by a series of sulfosuccinic ester molecules at the air–water interface (8) and the energies of adsorption at the air–water interface for a series of commercial nonionic surfactants (9).

The adsorbed layer at G–L or S–L surfaces in practical surfactant systems may have a complex composition. The adsorbed molecules or ions may be close-packed forming almost a condensed film with solvent molecules virtually excluded from the surface, or they may be widely spaced and behave somewhat like a two-dimensional gas. The adsorbed film may be multilayer rather than a monolayer. Counterions are sometimes present with the surfactant in the adsorbed layer. Mixed monolayers are known that involve molecular complexes, eg, one-to-one complexes of fatty alcohol sulfates with fatty alcohols (10), as well as complexes between fatty acids and fatty acid soaps (11). Competitive or preferential adsorption between multiple solutes at G–L and L–L interfaces is an important effect in foaming, foam stabilization, and defoaming (see Defoamers).

Adsorption at Solid–Liquid Interface. Practical applications of surfactants usually involve in some manner the adsorption of a surfactant on a solid surface. This adsorption is always associated with a decrease in free-surface energy, the magnitude of which must be determined indirectly. The force with which the adsorbate is held on the adsorbent may be roughly classified as physical, ionic, or chemical. Physical adsorption is a weak attraction caused primarily by van der Waals forces. Ionic adsorption occurs between charged sites on the substrate and oppositely charged surfactant ions. It is usually a strong attractive force. The term chemisorption is applied when the adsorbate is joined to the adsorbent by covalent bonds or forces of comparable strength; it is not an important factor in surfactant technology. The surface condition of a solid adsorbent markedly affects its adsorption characteristics. Important considerations are smoothness, cleanness, particle size, packing of powders, presence of capillary systems, etc. Typical substrates on which adsorption effects are important include metals, glass, plastics, textile fibers, sand, crushed minerals, plant foliage, paper, and the mixed solid dirt that collects on clothing, linen, walls, and floors.

Analysis of the shape of adsorption isotherms has been the technique most widely used in investigations of adsorption mechanisms, but their interpretation has not been established for the numerous combinations of adsorbent–adsorbate and their interactions in multicomponent surfactant systems (see Adsorptive separation, liquids).

Physical and ionic adsorption may be either monolayer or multilayer (12). Capillary structures in which the diameters of the capillaries are small, ie, one to two molecular diameters, exhibit a marked hysteresis effect on desorption. Sorbed surfactant solutes do not necessarily cover all of a solid interface and their presence does not preclude adsorption of solvent molecules. The strength of surfactant sorption generally follows the order cationic > anionic > nonionic. Surfaces to which this rule applies include metals, glass, plastics, textiles (13), paper, and many minerals. The pH is an important modifying factor in the adsorption of all ionic surfactants but especially for amphoteric surfactants which are least soluble at their isoelectric point. The speed and degree of adsorption are increased by the presence of dissolved inorganic salts in surfactant solutions (for textile fibers, see ref. 14).

The state of the adsorbed layer has been studied by electron diffraction, x-ray diffraction, interferometer, and electron microscope. The contact angle of pure liquids on solid surfaces, ie, wetting behavior, has been used effectively in investigations of the properties of adsorbed monolayers (15–17). Adsorption can be demonstrated visually by dipping a clean glass plate, that is readily water wettable, into a solution of a cationic surfactant. On withdrawal the plate is water repellent because of the orientation of the adsorbed layer. When a complete and closely packed film of surfactant molecules or ions is adsorbed on a solid, the substrate no longer exerts any effect on the contact angle. Preferential adsorption or competition between surfactant solutes for the surface of solid substrates is an important effect and the basis of preferential wetting.

Micelles. Surfactant molecules or ions at concentrations above a minimum value characteristic of each solvent–solute system associate into aggregates called micelles. The formation, structure, and behavior of micelles have been extensively investigated. The term critical micelle concentration (CMC) denotes the concentration at which micelles start to form in a system comprising solvent, surfactant, possibly other solutes, and a defined physical environment. If the properties of a surfactant solution are plotted as a function of the concentration of the surfactant, the properties usually vary linearly with increasing concentration up to the CMC at which point there is a break in the curve (18–19). This effect is illustrated schematically for aqueous solutions of sodium lauryl sulfate in Figure 4. Experimental data on this compound have been reported in a similar composite chart. Other properties have also been used for the determination of CMC, eg, refractive index, light scattering, dialysis, and dye solubilization. Although most studies of micellar size and properties have involved aqueous systems, nonaqueous systems have received some attention (20–24). The orientation of surfactant ions in an aqueous solution with the hydrophilic moieties exposed to the water was illustrated in Figure 3. In a nonaqueous system, the orientation is reversed and the lipophilic moieties are exposed to the organic solvent. Although the CMC of surfactants in aqueous solutions depends on the structure of the compounds and the environment, for many anionic surfactants at low electrolyte concentrations and room temperature it is 10^{-3}–10^{-2} mol/L (25); for nonionic surfactants under comparable conditions it is less, 10^{-5}–10^{-4} mol/L (26).

Micelle size is expressed as the micellar molecular weight or, more generally, the

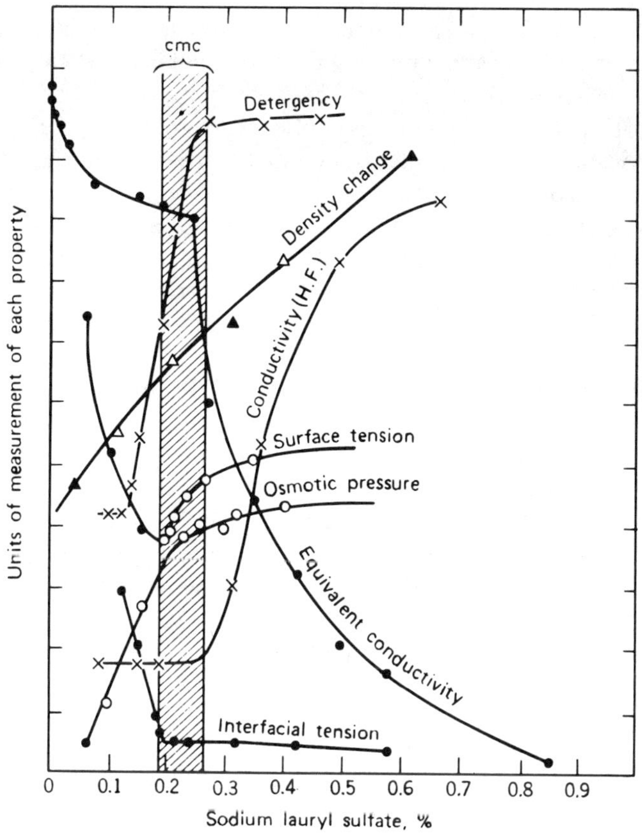

Figure 4. Physical property curves for aqueous sodium lauryl sulfate.

aggregation number, ie, the number of monomers making up the micelles. Micellar aggregation numbers generally lie between 20 and 100, for single-chain anionic and cationic surfactants (27). Large aggregation numbers (>1000) have been reported for nonionic micelles (28), especially as the cloud point is approached. Assuming that the radius of a micelle is just short of the extended monomer chain length, and that the interior of the micelle is filled with the monomer chain, the following relationships were calculated between aggregation number (m) and carbon chain length (n), for surfactants composed of paraffin chains with simple anionic head groups (29):

n	12	14	16	18
m	33	46	60	78

In a surfactant series where the hydrophilic group is constant, but the size of the hydrophobic group is increased, CMC values decrease for both ionic and nonionic surfactants. At constant hydrophobe size, CMC values decrease with decreasing ethylene oxide content of nonionic surfactants. Increasing the electrolyte concentration lowers the CMC for both anionic and nonionic surfactants, especially for the former. The CMC of anionic surfactants increases with temperature, whereas the CMC of the nonionic surfactants decreases with temperature. Since the mid-1950s, micellar properties of homogeneous nonionic surfactants have been studied and in some cases compared with polydisperse nonionic surfactants (30). A theoretical treatment to describe micellization of mixed nonionic systems has been proposed (31). Under the assumption of ideal mixing and a phase separation model, the theory gives excellent prediction for surface tensions and the CMC of the mixtures. The CMC and other micellar properties of homogeneous and polydisperse nonionic surfactants of comparable composition are quite similar. A detailed thermodynamic treatment of solutions of ideal multicomponent micelles has been described (32). In principle, this treatment can be used to predict monomer concentration, CMC, and other micellar properties. However, its practical application presents substantial difficulties. Extensive CMC values for many surfactants under a variety of conditions have been compiled (33). This compilation is especially useful as it contains an evaluation of the reliability of the data presented.

Small micelles in dilute solution close to the CMC are generally believed to be spherical. Under other conditions, micellar materials can assume structures such as oblate and prolate spheroids, vesicles (double layers), rods, and lamellae (34–35). All of these structures have been demonstrated under certain conditions, and a single surfactant can assume a number of structures, depending on surfactant, salt concentration, and temperature. In mixed surfactant solutions, micelles of each species may coexist, but usually mixed micelles are formed. Anionic–nonionic mixtures are of technical importance and their properties have been studied (36–37).

Micellar properties are affected by changes in the environment, eg, temperature, solvents, electrolytes, and solubilized components. These changes include complicated phase changes, viscosity effects, gel formation, and liquefication of liquid crystals. Of the simpler changes, high concentrations of water-soluble alcohols in aqueous solution often dissolve micelles and in nonaqueous solvents addition of water frequently causes a sharp increase in micellar size.

Measurement of Surface Activity. Each surface-active property can be measured in a variety of ways and the method of choice often depends on the characteristics of the substance to be tested. The most frequently determined properties are surface tension, interfacial tension, contact angle, and CMC.

Surface and Interfacial Tension. Theories and experimental methods for measuring these properties are reviewed in ref. 38. The du Nouy Tensiometer, ie, the ring method, is probably the most widely used instrument for the determination of static surface and interfacial tension (39). Other frequently used methods include drop weight (40), pendant drop (41), maximum bubble pressure (42), capillary rise, and sessile drops. Dynamic surface tension can be measured by a speedup of the bubbling rate in the maximum bubble-pressure method (43) or the change in wavelength and form when a jet of liquid issues from an elliptical orifice (44). Dynamic methods provide information on the time required for surfactant molecules to migrate to the interface. Interfacial tensions can also be measured by the spinning-drop method (45). This method is especially effective for the measurement of low interfacial tensions.

Contact Angle. A number of methods for direct determination of the contact angle have been proposed, but these measurements are difficult and the results not always reproducible. The roughness of the solid surface and the difference between the advancing and receding angle lowers the precision of the measurements. The tilted plate is the simplest and most direct method for this determination (46). Contact angles can also be measured by observing single oil drops on solid surfaces (17) or upon single fibers (47).

Critical Micelle Concentration. The rate at which the properties of surfactant solutions vary with concentration changes at the concentration where micelle formation starts. Surface and interfacial tension, equivalent conductance (48), dye solubilization (49), iodine solubilization (50), and refractive index (51) are properties commonly used as the basis for methods of determination of CMC.

Anionic Surfactants

Carboxylate, sulfonate, sulfate, and phosphate are the polar, solubilizing groups found in anionic surfactants. In dilute solutions of soft water, these groups are combined with a 12-carbon-chain hydrophobe for best surfactant properties. In neutral or acidic media, or in the presence of heavy metal salts, the carboxylate group loses most of its solubilizing power.

Of the cations (counterions) associated with polar groups, sodium and potassium impart water solubility, whereas calcium, barium, and magnesium promote oil solubility. Ammonium and substituted ammonium ions provide both water and oil solubility. Triethanolammonium is a commercially important example. Salts of these ions are often used in emulsification.

Higher ionic strength of the medium depresses surfactant solubility. To compensate for the loss of solubility, shorter hydrophobes are used for application in high ionic-strength media.

The U.S. production of anionic surfactants in 1980 amounted to 65.9% of total surfactant production, a reduction from 70.6% a decade ago (see below under Production and Economic Aspects).

Carboxylates. Soaps represent most of the commercial carboxylates. The general structure of soap is $RCOO^-M^+$, where R is a straight hydrocarbon chain in the C_9–C_{21} range and M^+ is a metal or ammonium ion. Interruption of the chain by amino or amido linkages leads to other structures which account for the small volumes of the remaining commercial carboxylates.

Large volumes of soap are used in industrial applications as gelling agents for kerosene, paint driers, and as surfactants in emulsion polymerization (see Soap; Driers and metallic soaps; Emulsions). Soap usage in consumer products is principally in personal washing products, ie, soap bars.

The first laundry powders to be developed were tallow-soap-based laundry powders. They showed excellent detergency in soft water, but have disappeared almost completely from the consumer market. In the presence of polyvalent metal ions, eg, Ca^{2+}, Mg^{2+}, Ba^{2+}, or Fe^{3+}, soap reacts to form lime soap or curds which are unsightly, reduce soap concentration, and affect performance adversely. Incorporation of lime-soap dispersants which, as the name implies, disperse lime soap to the point of near-invisibility, is one approach toward solving this problem. A method for rating lime soap dispersion efficacy has been published (52). Another approach is to build soap powders with sodium carbonate or silicate. Neither approach proved sufficiently effective or economical to prevent replacement of soap powders by so-called syndets, a combination of synthetic surfactants, notably alkylbenzenesulfonate, and builders, notably pentasodium tripolyphosphate. When concern over water pollution resulted in a ban of phosphorus in laundry detergents, the possibility of returning to laundry powders based on a combination of soap and lime-soap dispersants received much attention, particularly by the group at the USDA laboratories in Philadelphia (53).

Mostly for economic reasons, the large number of lime soap dispersants synthesized by this group has not achieved significant commercialization. Lower tallow costs, relative to the cost of derived hydrophobes from petrochemicals, might change this situation.

Carboxylates with a fluorinated alkyl chain are marketed by the 3M company under the trade name of Fluorad surfactants. They include also other functional derivatives of fluorinated and perfluorinated alkyl chains. Replacement of hydrogens on the hydrophobe by fluorine atoms leads to surfactant molecules of unusually low surface tension. This property imparts excellent leveling effectiveness.

Polyalkoxycarboxylates. These surfactants are produced either by the reaction of chloroacetate with an alcohol ethoxylate

$$ROH + n\ \overset{O}{CH_2{-}CH_2} \longrightarrow R{+}OC_2H_4{+}_nOH + ClCH_2CO_2Na \longrightarrow R{+}OC_2H_4{+}_nOCH_2CO_2Na$$

or from an acrylic ester and an alcohol alkoxylate.

$$ROH + n\ \overset{O}{CH_2{-}CHR'} \longrightarrow$$

$$R{+}OCH_2CHR'{+}_nOH + CH_2{=}CHCO_2R'' \longrightarrow \xrightarrow{NaOH} R{+}OCH_2CHR'{+}_nO(CH_2)_2CO_2Na$$

Because of the presence of the ethylene oxide linkage, these products possess a somewhat nonionic character which manifests itself in greater compatibility with cationic surfactants. They find application as lime-soap dispersants, wetting agents, and detergents in cosmetic formulations. Table 3 gives composition and properties of some representative commercial products manufactured by Sandoz Colors & Chemicals.

N-Acylsarcosinates. Sodium N-lauroylsarcosinate [*7631-98-3*] is a good soap-like surfactant. The amido group in the hydrophobe chain lessens the interaction with hardness ions. N-acylosarcosinates have been used in dentifrices (qv) where they are claimed to inactivate enzymes that convert glucose to lactic acid in the mouth (54). They are prepared from a fatty acid chloride and sarcosine:

$$RCOCl + CH_3NHCH_2CO_2Na + NaOH \rightarrow RCON(CH_3)CH_2CO_2Na + NaCl + H_2O \qquad (10)$$

Table 4 gives trade names and properties.

Table 3. Polyethoxycarboxylates, $R{+}OCH_2CH_2{+}_nCH_2CO_2M$

Trade name	R	n	M	Functions[a]
Sandopan DTC	tridecyl[b]	7	Na	L, W
Sandopan KST	cetyl	12	Na	D, L, W
Sandopan TA-10	isostearyl	5	H	Em, S
Sandopan MS-40	nonylphenyl[b]	20	Na	D, S, W

[a] See Table 1.
[b] Branched-chain.

Table 4. *N*-Acylsarcosinates, $RCON(CH_3)CH_2CO_2M$

Trade name	RCO—	M	Concn, wt %	Functions[a]	Manufacturer[b]
Hamposyl C	cocoyl	H	100	C, D, E, eP, Fs	GWR
Hamposyl C-30	cocoyl	Na	30	C, D, E, F, Fs	GWR
Hamposyl L	lauroyl	H	100	C, D, E, eP, Fs	GWR
Hamposyl L-30	lauroyl	Na	30	C, D, E, F, Fs	GWR
Hamposyl O	oleoyl	H	100	C, D, E, eP, Fs	GWR
Maprosyl 30	lauroyl	Na	30	C, D, E, F	OXR
Sarkosyl LC	cocoyl	Na	100	C, D, E, F	CGY
Sarkosyl O	oleoyl	Na	100	C, D, E	CGY

[a] See Table 1.
[b] See Table 2.

Acylated Protein Hydrolysates. These surfactants are prepared by acylation of protein hydrolysates with fatty acids or acid chlorides. The hydrolysates are variable in composition, depending on the degree of hydrolysis. Collagen from leather processing is a common protein source. Acylated protein hydrolysates are mild surfactants recommended for personal-care products (see Table 5) (see also Cosmetics).

Sulfonates. The sulfonate group, —SO_3M, attached to an alkyl, aryl, or alkylaryl hydrophobe, is a highly effective solubilizing group. Sulfonic acids are strong acids and their salts are relatively unaffected by pH. They are stable to oxidation and, because of the strength of the C–S bond, also to hydrolysis. They interact moderately with the hardness ions Ca^{2+} and Mg^{2+}, significantly less so than carboxylates. Modification of the hydrophobe in sulfonate surfactants, by introduction of double bonds or ester or amide groups, either into the hydrocarbon chain or as substituents, gives surfactants that offer specific performance advantages.

Because the introduction of the SO_3H function is inherently inexpensive, eg, by oleum, SO_3, SO_2Cl_2, or $NaHSO_3$, sulfonates are heavily represented among the high-volume surfactants (see also Sulfonation and sulfation).

The first sulfonate was prepared from paraffin and SO_2Cl_2 by the Reed reaction (55–56):

$$RH + SO_2 + Cl_2 \xrightarrow{uv} RSO_2Cl \xrightarrow[H_2O]{NaOH} RSO_3Na \qquad (11)$$

Work-up of the complex reaction mixture was difficult and expensive. As a result, these surfactants were withdrawn from the U.S. market.

Table 5. Acylated Protein Hydrolysates[a]

Trade name	Fatty acid	Cation	Concn, wt %	Functions[b]
Maypon 4C	coco	K	36	D
Maypon 4CT	coco	TEA[c]	38	D
Maypon K	oleic	Na	44	D, E, Ta
Maypon UD	undecylenic	K	35	AM, D

[a] Manufactured by STP (Table 2).
[b] See Table 1.
[c] TEA = triethanolammonium.

In Europe, Hoechst AG produces secondary alkanesulfonates by a similar reaction:

$$RCH_2CH_3 + SO_2 + \frac{1}{2}O_2 \xrightarrow{uv} RCH(CH_3)SO_2OH \xrightarrow{NaOH} RCH(CH_3)SO_3Na \quad (12)$$

These products are marketed under the trade name Hostapur. Hostapur SAS 60, a 60% active paste, is a principal ingredient in European liquid hand-dishwashing products.

Alkylbenzenesulfonates. As a group, the alkylbenzenesulfonates (ABS) rank next to soap in production volume and usage. A product with a linear alkyl side chain is referred to as linear alkylate sulfonate, LAS. In the early 1960s, LAS replaced ABS, a branched-chain alkylbenzenesulfonate which had earlier replaced soap as the principal surfactant in U.S. household laundry products; C_{12} ABS was derived from benzene and a propylene tetramer. Although no longer used in U.S. consumer products, significant volumes (>40,000 t) of ABS are still produced for export.

Alkylbenzenesulfonic acids are light-colored, viscous liquids whose alkali and ammonium salts are soluble in water. The salts are neutral, relatively unaffected by water hardness, and stable toward alkaline and acid hydrolysis. Calcium and magnesium alkylbenzenesulfonates are less water-soluble than sodium or ammonium salts. They are, however, soluble in hydrocarbons and are used extensively in these media. The LAS salts are stable to oxidation and can be incorporated in formulations containing oxidizing agents. Pure alkylbenzenesulfonates have been synthesized for a property study (57–58).

Alkylbenzenesulfonates are effective surfactants, which also respond well to builders and foam boosters in detergent formulations. These properties together with low cost and availability at consistent quality, account for their dominant position in household laundry products.

Alkylbenzenesulfonates are manufactured in a three-step process:

$$RR'CHCl + C_6H_6 \xrightarrow[\text{or } AlCl_3]{HF} RR'CHC_6H_5 + HCl \quad (13)$$

$$R'CH{=}CHR + C_6H_6 \xrightarrow[\text{or } AlCl_3]{HF} R'C(CH_2R)C_6H_5 \quad (14)$$

$$RR'CHC_6H_5 \xrightarrow[\text{or oleum}]{SO_3/\text{air}} RR'CHC_6H_4SO_3H \quad (15)$$

$$RR'CHC_6H_4SO_3H \xrightarrow{NaOH} RR'CHC_6H_4SO_3Na \quad (16)$$

The product from reactions 13 or 14 is a secondary alkylbenzene in which the point of attachment of the benzene ring to the alkyl chain is located randomly along the internal chain positions (see also Alkylation). The relation of performance properties to structure (ie, distribution of the benzene ring along the alkyl chain) has been investigated (4).

The 1980 U.S. production of alkylbenzenesulfonates of 294,197 t showed a slight increase over the 270,340 t produced in 1966. Dodecylbenzenesulfonic acid [*27176-87-0*] and its salts accounted for 80% of total production. Representative commercial alkylbenzenesulfonates are given in Table 6.

Table 6. Alkylbenzenesulfonates, $RC_6H_4SO_3M$

Trade name	R	M	Concn, wt %	Functions[a]	Manufacturer[b]
Calsoft LAS-99	alkyl	H	98	D, E, I	PIL
Conco AAS-45S	dodecyl	Na	45	D, W	CTL
Conco AAS-60S	dodecyl	TEA[c]	60	D, W	CTL
Conco AAS-SPECIAL 3	dodecyl	IPA[d]	98	D, W	CTL
Conco 75S	dodecyl	Ca	65	E	CTL
Conoco C-550	dodecyl	Na	50	D	CO
Conoco C-560	dodecyl	Na	60	D	CO
Conoco C-650	tridecyl	Na	50	D	CO
Conoco SA 597	dodecyl	H	97	D, I	CO
Conoco SA 697	tridecyl	H	97	D, I	CO
DDBSA 99-B	C_{10}–C_{13}	H	97	D, I	MON
Emkal NOBS	nonyl	Na	40	D	EMK
Richonate 60B	dodecyl	Na	60	D	RCD
Richonic Acid B	dodecyl	H	98	D, I	RCD
Surco DDBSA	dodecyl	H	96	D, I	OXR
Surco 60T	alkyl	TEA[c]	70	D	OXR
Witconate 1298	dodecyl[e]	H	95	D, I	WTC
Witconate 1298 Soft	dodecyl	H		D, I	WTC
Witconate P1059	dodecyl[e]	AM[f]	60	D	WTC

[a] See Table 1.
[b] See Table 2.
[c] TEA = triethanolammonium.
[d] IPA = isopropylammonium.
[e] Branched.
[f] AM = ammonium.

Alkylarenesulfonates, Short-Chain. This group of sulfonates includes prominently the sodium and ammonium salts of toluene-, xylene-, and isopropylbenzenesulfonic acids. The alkyl chain in these compounds is too short to confer surface activity. Industrially, however, these materials are important in their function as coupling agents, solubilizers, or hydrotropes. As such, they are widely used in liquid detergent compositions, both for hand dishwashing and laundering, where high concentrations of organic surfactants need to be kept in homogeneous solution in the finished product.

Lignosulfonates. Lignosulfonates are by-products of the paper industry where they are manufactured by reaction of SO_2 and $Ca(HSO_3)_2$ with wood pulp (see also Lignin; Paper; Wood). They are produced in large volume and find application as inexpensive dispersing agents, as ore flotation agents, in oil-well mud drilling formulations and in pelletizing animal feed products.

Like the short-chain alkylarenesulfonates, lignosulfonates are included in production statistics of surfactants but are not further considered here because in dilute solution they do not reduce surface tension to the extent that surfactants do. Moreover, they do not form oriented monolayers or micelles (59).

Naphthalenesulfonates. These surfactants are manufactured by sulfonation and neutralization of naphthalene, tetrahydronaphthalene, alkylnaphthalenes, and formaldehyde–naphthalene condensation products. The alkali metal salts are stable to hydrolysis and oxidation. They are highly water-soluble and hence are not strongly surface active in soft water. Naphthalenesulfonates find wide-spread application as dispersants, wetting agents, stabilizers, and suspending agents for latex, carbon black, pigments, and other disperse systems. Particularly the condensation products find extensive use in the tanning of leather. A selection of commercially available naphthalenesulfonates is given in Table 7.

α-Olefinsulfonates. α-Olefinsulfonates (AOS) are produced by reaction of SO_3–air with an α-olefin followed by intermediate neutralization of sultone intermediates. In its properties, AOS is generally very similar to LAS. It is reported to be somewhat milder to skin and to biodegrade slightly more readily. Because of the simplicity and low cost of the starting materials, α-olefinsulfonates have been of great interest to surfactant producers for many years. Only in recent years, however, has the usage of AOS increased. For 1980, total usage is estimated at 6800 t. The principal applications include liquid soap formulations and cosmetic products.

Commercial AOS is a multicomponent mixture in which the double bond is located at a number of positions along the chain. It contains significant proportions of hydroxyalkanesulfonate isomers in which the hydroxyl group is similarly distributed along the carbon chain.

Representative commercial AOS products are shown in Table 8.

Petroleum Sulfonates. Petroleum sulfonates originally were produced as by-products of petroleum refining. With expanding usage and increasing complexity of their specifications, petroleum sulfonates are now considered coproducts from the refining of certain petroleum fractions. Reaction of high boiling feedstock with oleum

Table 7. Naphthalenesulfonates, $RC_{10}H_6SO_3Na$

Trade name	R or description	Concn, wt %	Functions[a]	Manufacturer[b]
Aerosol OS	isopropyl	75	Di, W	ACY
Alkanol XC, solid		na	Di, Ta, W	DUP
Darvan No 1	CH_2O condensed	100	Di	VNC
Daxad 11	CH_2O condensed	96	Di, E	GWR
Emkal NNS	nonyl	35	D, Di, E, W	EMK
Nekal BA-75	butyl	30	Di, St, W	GAF
Nekal BA-77	isopropyl	65	Di, St, W	GAF
Petro BA	alkyl	50	D, S, W	PET
Tamol SN	CH_2O condensed	93	Di	RH

[a] See Table 1.
[b] See Table 2.

Table 8. α-Olefinsulfonates, $RCH{=}CHSO_3Na$

Trade name	R	Concn, wt %	Functions[a]	Manufacturer[b]
Bio Terge AS-40	C_{10}–C_{12}	40	D, W	STP
Conco AOS-40	alkyl	40	D, W	CTL
Conco AOS-90F	alkyl	90	D, W	CTL
Witconate AOS	C_{10}–C_{12}	40	D, W	WTC

[a] See Table 1.
[b] See Table 2.

yields a crude acid mixture that separates into an upper brown, mahogany layer and a greenish, lower layer. Historically, the names mahogany acids and green acids originate with the color of these layers.

The green acids and the soaps derived from them are high in water-soluble polysulfonic acids of low economic value. They are often burned to recover SO_3. Mahogany acids are predominantly monosulfonated and represent products of considerable value including the salts. When expanding usage led to a shortage of natural petroleum sulfonates, synthetic alkylates such as detergent alkylate bottoms, were sulfonated to provide synthetic petroleum sulfonates. They account for a significant fraction of petroleum sulfonate production.

Petroleum sulfonates are widely used as solubilizers, dispersants, emulsifiers, and corrosion inhibitors. In recent years, they have emerged as the principal surfactant associated with expanding operations in enhanced oil recovery (60). Alkaline earth salts of petroleum sulfonates are used in large volumes as additives in lubricating fluids for sludge dispersion, detergency, corrosion inhibition, and micellar solubilization of water. The chemistry and properties of petroleum sulfonates have been described (61–62).

Precise production figures for petroleum sulfonates are difficult to obtain since they are not generally collected with surfactant statistics and because of the uncertainty in the data which are reported. Principal U.S. producers include Exxon, Shell, Stepan, and Witco.

Sulfonates with Ester, Amide, or Ether Linkages. *Dialkyl sulfosuccinates.* Introduced in 1939 by the American Cyanamid Company under the Aerosol trademark, dialkyl sulfosuccinates have become widely used specialty surfactants (see Table 9). Their properties are described in ref. 9. Sodium di(2-ethylhexyl)sulfosuccinate [*577-11-7*] accounted for nearly 80% of the 11,000 t of sodium dialkyl sulfosuccinates produced in the United States in 1980. Within the limitations in hydrolytic stability imposed by the presence of the ester groups, sulfosuccinates are mild, versatile surfactants used when strong wetting, detergency, rewetting, penetration, and solubilization effectiveness is called for.

The symmetrical diesters are produced by esterification of maleic anhydride

Table 9. Dialkyl Sulfosuccinates, $RO_2CCH_2CH(SO_3Na)CO_2R$

Trade name	R	Concn, wt %	Functions[a]	Manufacturer[b]
Aerosol AY-100	amyl	100	Di, E, W	ACY
Aerosol OT	2-ethylhexyl	100	E, W	ACY
Alrowet D-65	octyl	65	Ta, W	CGY
Crestopen 5X	octyl	60	W	CRT
Emcol 4500	octyl	70	D, S, W	WTC
Monawet MB-45	isobutyl	45	W	MOA
Monawet MT-70	tridecyl[c]	70	Di, eP, W	MOA
Nekal WT 27	octyl	70	E, rW, W	GAF
Schercowet DOS-70	octyl	70	E, W	SBC
Tex-Wet 1001	octyl	60	rW, Ta, W	TCM

[a] See Table 1.
[b] See Table 2.
[c] Branched.

followed by addition of sodium bisulfite across the double bond (see also Succinic acid and succinic anhydride):

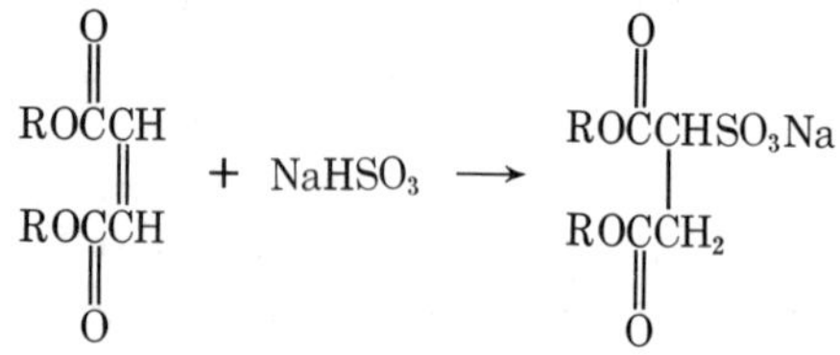

Amidosulfonates (N-Acyl-N-Alkyltaurates). These surfactants are derived from taurine ($H_2NCH_2CH_2SO_3Na$). They are effective surfactants and lime-soap dispersants (see Table 10). Because of high raw material cost, their usage is relatively small, ca 900 t/yr. Technically, they are of interest because they are stable to hydrolysis, are unaffected by hard water, and compatible with soap. They have been used in soap–surfactant toilet-bar formulations. With shorter acyl groups, they make excellent wetting agents.

Sodium *N*-oleoyl-*N*-methyltaurate [*137-20-2*] is produced in three high-yield steps:

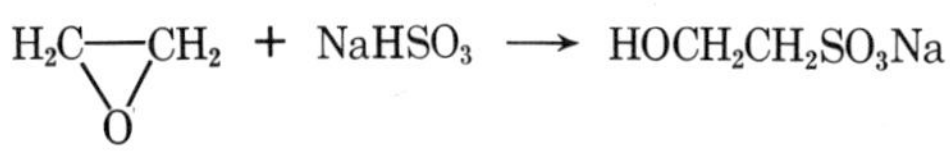

$$CH_3NH_2 + HOCH_2CH_2SO_3Na \longrightarrow CH_3NHCH_2CH_2SO_3Na$$

$$C_{17}H_{33}COCl + HN(CH_3)CH_2CH_2SO_3Na \xrightarrow{NaOH} C_{17}H_{33}CON(CH_3)CH_2CH_2SO_3Na + NaCl$$

2-Sulfoethyl esters of fatty acids (acyl isethionates). Like the *N*-acyl-*N*-alkyltaurates, these compounds are excellent surfactants, mild to skin and unaffected by hardness ions. The presence of an ester linkage limits hydrolytic stability. This limitation, combined with relatively high raw material costs, has prevented wide-spread usage of the 2-sulfoethyl esters in consumer detergent products. The cocoyl derivative, however, is the principal ingredient in a detergent bar for personal use.

Table 10. Sodium *N*-acyl-*N*-alkyltaurates, $RCONR'CH_2CH_2SO_3Na$ (RCO = acyl, R′ = alkyl)

Trade name	RCO—	R′	Concn, wt %	Functions[a]	Manufacturer[b]
Celopon	oleyl	methyl	10	D, Di, W	AKS
Concogel	oleyl	methyl	14	D, Di, W	CTL
Cresterge L	oleyl	methyl	32	D, Ta, W	CRT
Igepon TC 42	cocoyl	methyl	24	D, Di, F	GAF
Igepon TN 74	palmitoyl	methyl	44	D, Di, Su	GAF
Igepon T 77	oleyl	methyl	67	D, Di, F, W	GAF
Igepon TK 32	tall oil	methyl	20	D, Di, Su	GAF
Igepon CN 42	palmitoyl	cyclohexyl	23	D, Di, St	GAF
Tergenol S Liq	oleyl	methyl	32	D, Ta	HRT

[a] See Table 1.
[b] See Table 2.

2-Sulfoethyl esters are produced commercially in two steps:

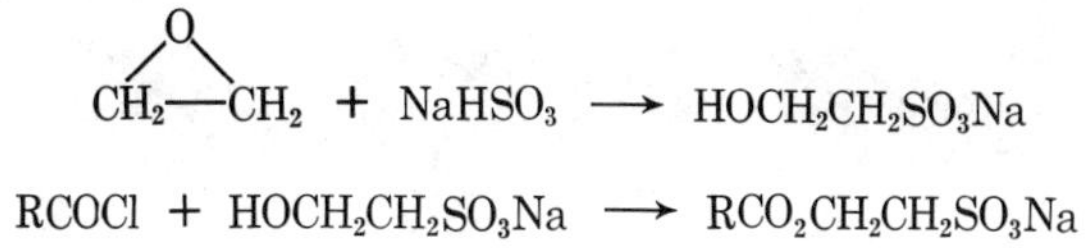

where R = linear long chain.

The viscous fatty acid chloride is mixed with solid sodium isethionate and is heated under vacuum and agitation without water or solvent; HCl is evolved, leaving the product as a finely divided powder. The direct esterification of fatty acid and sodium isethionate has been patented (63).

Sulfonates with ether linkages include ring-sulfonated alkylphenol ethoxylates and a disulfonated alkyldiphenyl oxide. The latter is marketed by the Dow Chemical Co. under the trade name of Dowfax 2A1. This surfactant is characterized by high solubility in salt solutions and high tolerance for strong acids and bases. It is used in industrial and institutional cleaners.

Sulfates and Sulfated Products. The sulfate group, —OSO_3M (where M is a cation) which represents the sulfuric acid half-ester of an alcohol, is more hydrophilic than the sulfonate group because of the presence of an additional oxygen atom. Attachment of the sulfate group to a carbon atom of the hydrophobe through the C–O–S linkage limits hydrolytic stability, particularly under acidic conditions.

Usage of sulfated alcohols and sulfated alcohol ethoxylates has expanded dramatically during the last decade as the detergent industry reformulated consumer products to improve biodegradability and lower phosphate contents.

Alcohol Sulfates. Commercial production of alcohol sulfates started in Europe about 1930 and the products have become well established in specialty markets. The hydrophobes are predominantly of petrochemical origin although at present some shift is evident to oleochemical hydrophobes obtained by reduction of fatty acids and esters. Secondary alcohol sulfates, popular in Europe at one time, have never been used extensively in the United States. These products were marketed under the trademark Teepol by the Shell Oil Co. Alcohol sulfates from straight-chain primary alcohols are similar in performance properties, feel, and emolliency characteristics to soaps of corresponding chain lengths (see also Alcohols, higher aliphatic). With increasing chain length, higher temperatures are required to attain maximum detergency and wetting effects. The physical and interfacial properties of a series of purified alcohol sulfates are described in ref. 64.

Commercial alcohol sulfates are sold as nearly colorless solutions of concentrations 25–40% or as light-colored pastes. Purified sodium lauryl sulfate of activity 90–99% is available, specifically for use in dentifrices (qv).

Alcohol sulfates are stable to hard water. Magnesium lauryl sulfate forms voluminous foam with low water content which is particularly useful in rug shampoos where soil is removed by vacuum pick-up of foam generated by vigorous brushing with a minimum of detergent solution. Sensitivity to hydrolysis in hot alkaline and acidic media is one of the main disadvantages of alcohol sulfates. Nonetheless, their versatility under use conditions is attested to by the variety of different salts offered commercially, as shown in Table 11. Among the principal applications are shampoos, dentifrices, textile processing, and emulsion polymerization.

Table 11. Alkyl Sulfates, $ROSO_3M$

Trade name	R	M	Concn, wt %	Functions[a]	Manufacturer[b]
Avirol SL 1000	lauryl	Na	30	D, Di, E, W	HNC
Avirol SA 4106	2-ethylhexyl	Na	45	E, Di, W	HNC
Avirol SA 4110	2 decyl	Na	30	E	HNC
Conco Sulfate A	lauryl	NH_4	30	D	CTL
Conco Sulfate P	lauryl	K	30	D	CTL
Conco Sulfate EP	lauryl	DEA[c]	35	D, W	CTL
Conco Sulfate WR Dry	lauryl	Na	90	D, W	CTL
Duponol C	lauryl	Na		D	CTL
Emersal 6400	lauryl	Na	30	D, E, F	EMR
Emersal 6434	lauryl	TEA[d]	40	D, F	EMR
Emersal 6462	octyl/decyl	Na	38	F, W, rW	EMR
Emersal 6465	2-ethylhexyl	Na	40	P, W	EMR
Equex S	lauryl	Na	30	D, Di, W	PG
Equex T	lauryl	TEA[d]	40	D, Di, F, W	PG
Maprofix TLS 65	lauryl	TEA	65	D, F	OXR
Maprofix MG	lauryl	Mg	25	D, W	OXR
Richonol A	lauryl	Na	30	D, W	RCD
Richonol AM	lauryl	NH_4	30	D, W	RCD
Richonol T	lauryl	TEA[d]	40	D, W	RCD
Sipex OLS	octyl	Na	31	eP	AAC
Sipex TDS	tridecyl[e]	Na	25	Di, E, W	AAC
Sole Terge TS-2-S	2-ethylhexyl	Na	35	P, W	HDG
Standapol A	lauryl	Na	28	D, F	HNC
Standapol MLS	lauryl	MEA[f]	32	D, F	HNC
Standapol T	lauryl	TEA[d]	40	D, F	HNC
Stepanol WA 100	lauryl	Na	99	D, Di, E, F, W	STP
Stepanol MG	lauryl	Mg	29	D, F, W	STP
Stepanol AM	lauryl	NH_4	29	D, Di, F, W	STP
Witcolate D-510	2-ethylhexyl	Na	40	P, W	WTC

[a] See Table 1.
[b] See Table 2.
[c] DEA = diethanolammonium.
[d] TEA = triethanolammonium.
[e] Branched.
[f] MEA = monoethanolammonium.

Alcohols, Ethoxylated and Sulfated. During the last 15 years, these products have experienced a dramatic growth, from about 1600 t in 1966 to >108,000 t in 1980. Most of the increase has come from expanded usage in consumer detergent products. This growth is due to expanding availability of relatively low-cost primary straight-chain alcohols from petrochemical sources; superior biodegradability of sulfated alcohol ethoxylates compared to the corresponding sulfated alkylphenol ethoxylates; the need for hardness-insensitive surfactants in heavy-duty laundry detergents brought about by the reduction in phosphate content; expansion of reliable, inexpensive sulfation technology; and addition of less expensive ethylene oxide which lowers overall costs.

Ethoxylated alcohol sulfates have several advantages over alcohol sulfates: lower sensitivity to hardness with respect to foaming and detersive effectiveness; less irritating to skin and eyes; and higher water solubility.

Sulfation of alcohol ethoxylates is carried out in high yields with oleum, chlorosulfonic acid, sulfamic (amidosulfuric) acid and gaseous sulfur trioxide. The last is the preferred reagent. It is usually generated *in situ* by burning elemental sulfur (see Sulfonation and sulfation). In particular, SO_3 technology has displaced the use of sulfamic acid, an agent uniquely suited for the preparation of ammonium salts which are widely used in light-duty dishwashing formulations:

$$RO(CH_2CH_2)_nOH + H_2NSO_3H \rightarrow RO(CH_2CH_2)_nOSO_3NH_4$$

Sulfated alcohol ethoxylates generally contain from 10–40% ethylene oxide calculated on the weight of the starting alcohol. They are offered commercially as essentially colorless, odorless liquids containing 35–60% of the active ingredients. At high concentrations, sulfated alcohol ethoxylates become viscous and are diluted with ethanol. The products are stable, compatible with builders in liquid and solid formulations. At very high alkalinity and in acid media, the sulfate linkage is susceptible to hydrolysis, particularly at elevated temperatures. In addition to light-duty liquid formulations, shampoos are widely formulated with sulfated alcohol ethoxylates. They also find application in textile processing and in emulsion polymerization. Table 12

Table 12. Alcohols, Ethoxylated and Sulfated, $R(OCH_2CH_2)_nOSO_3M$

Trade name	R	*n*	M	Concn, wt %	Functions[a]	Manufacturer[b]
Alfonic 1412-17	C_{12}–C_{14} (linear)	3	NH_4	60	D, E, W	CO
Alfonic 1412-5	C_{12}–C_{14} (linear)	5	Na	60	D, E, W	CO
Alipal CD-128	lauryl		NH_4	58	D, Di, E, W	GAF
Avirol SC-3001	lauryl		Na	29	eP	HNC
Avirol SE-3004	tridecyl[c]		Na	32	eP	HNC
Calfoam NEL-60	lauryl		NH_4	60	D, E, F	PIL
Calfoam SEL-60	lauryl		Na	60	D, E, F	PIL
Calfoam SLS-30	lauryl		Na	30	D, E, F	PIL
Conco Sulfate 219	lauryl		Na	60	D, E, W	CTL
Maprofix ESV	lauryl	1	Na	25	D, F	OXR
Maprofix ES	lauryl	3	Na	27	D, F	OXR
Maprofix LES-60C	lauryl	3	Na	58	D, F	OXR
Maprofix LES 60-A	lauryl	3	Na	59	D, F	OXR
Neodol 25-3A	C_{12}–C_{15} (linear, primary)	3	NH_4	56	D, E, F	SHC
Neodol 25-3S	C_{12}–C_{15} (linear, primary)	3	Na	58	D, E, F	SHC
Richonol S-1285C	lauryl		Na	55	D, F	RCD
Richonol S-1300C	lauryl		NH_4	60	D, F	RCD
Richonol S-5260	lauryl		Na	27	D, F	RCD
Sipex EA	lauryl		NH_4	27	D, E, F	AAC
Sipex TDS	tridecyl[c]		Na	25	Di, E, eP, W	AAC
Standapol ES-1	lauryl		Na	25	D	HNC
Standapol EA-3	lauryl		NH_4	59	D	HNC
Standapol EA-40	myristyl		Na	59	D	HNC
Steol CS-460	C_{12}–C_{15} (linear)		Na	60	D, E, F	STP
Steol CA-460	C_{12}–C_{15} (linear)		NH_4	60	D, E, F	STP
Sulfotex PAI	caprylyl/capryl		NH_4	47	D, E, F	HNC
Sulfotex PAI-S	caprylyl/capryl		Na	45	D, E, F	HNC
Witcolate AE-3	C_{12}–C_{15} (linear)	3	NH_4	59	D	WTC

[a] See Table 1.
[b] See Table 2.
[c] Branched.

gives the composition and properties of typical commercial products. The properties of well-characterized sulfated alcohol ethoxylates are described in refs. 64–66.

Alkylphenols, Ethoxylated and Sulfated. These products first appeared on the market about 1950. By 1965, they were widely used in retail detergent products. Because they degrade less readily than the sulfated alcohol ethoxylates, their anticipated expansion failed to materialize. In fact, the 1980 U.S. production of 2526 t showed a significant decrease from the 4354 t produced in 1966. Sulfated alkylphenol ethoxylates are used in industrial products such as hospital cleaning products. They are also used in textile processing and in emulsion polymerization.

Sulfated alkyphenol ethoxylates are sold as colorless, odorless aqueous solutions at concentrations of 30% and higher. The presence of ethylene oxide in the molecule increases resistance to hardness ions and reduces skin irritation. Representative commercial sulfated alkylphenol ethoxylates are given in Table 13.

Sulfated Acids, Amides, and Esters. Given a reactive group, reaction with sulfuric acid may be carried out on fatty acids, alkanolamides, and short-chain esters of fatty acids. The disodium salt of sulfated oleic acid is a textile additive and an effective lime-soap dispersant.

A typical sulfated alkanolamide structure is $C_{11}H_{23}CONHCH_2CH_2OSO_3Na$. Others include the sulfates of mono- and diethanolamines of fatty acids in the C_{12}–C_{18} detergent range. The presence of the amide group confers added stability toward acid and base hydrolysis at elevated temperatures. Sulfated alkanolamides are effective cleaners with good lathering properties and find application as textile additives and in the formulation of industrial cleaners.

The manufacture of sulfated esters is a simple process requiring no complicated equipment. The triglyceride is first transesterified with a low molecular weight alcohol and then sulfated. A typical sulfated ester is $CH_3(CH_2)_7CH_2CH(OSO_3Na)$-$CH_2(CH_2)_5CH_2CO_2R$, where R is methyl, ethyl, propyl, butyl, or amyl. The sulfated esters are marketed as clear, yellow-brown viscous solutions. They are good wetting agents and are used as textile additives. They also possess good detersive and foaming properties.

Among U.S. producers of this group of surfactants are ICI Americas, Inc., Proctor Chemical Co., Diamond Shamrock Corp., Dexter Chemical Corp., and Colgate-Palmolive Co. Total production volume in 1980 was 10,597 t.

Table 13. Alkylphenols, Ethoxylated and Sulfated, $RC_6H_4(OC_2H_4)_nOSO_3M$

Trade name	R	*n*	M	Concn, wt %	Function[a]	Manufacturer[b]
Alipal CO-433	nonyl	4	Na	28	D, Di, E, F, W	GAF
Alipal CO-436	nonyl	4	NH_4	58	D, Di, E, F, W	GAF
Alipal EP-120	nonyl	4	NH_4	30	D, Di, E, F, W	GAF
Cellopal 100	alkyl		TEA	100	D, Di, E, F, W	SYB
Concopal SS	alkyl[c]		NH_4	60	D, Di, E, F, W	CTL
Neutronyx S-60	nonyl	4	NH_4	58	D, F	OXR
Triton X-301	octyl[c]		Na	20	D, Di, F	RH
Witcolate D 51-51	alkyl[c]		Na	30	eP	WTC

[a] See Table 1.
[b] See Table 2.
[c] Branched.

Sulfated Natural Oils and Fats. Sulfated natural triglycerides were the first nonsoap commercial surfactants introduced in the middle of the last century. Since then the sulfates of many vegetable, animal, and fish oils have been investigated (see also Fats and fatty oils). With its hydroxyl group and a double bond, ricinoleic acid, 12-hydroxy-9,10-octadecenoic acid, is an oil constituent particularly suited for sulfation. Its sulfate is known as turkey-red oil. Oleic acid is also suited for sulfation. Esters of these acids can be sulfated with a minimum of hydrolysis of the glyceride group. Polyunsaturated acids, with several double bonds, lead to dark-colored sulfation products. The reaction with sulfuric acid proceeds through either the hydroxyl or the double bond. The sulfuric acid half ester thus formed is neutralized with caustic soda:

$$R'CH{=\!=}CHR + H_2SO_4 \longrightarrow R'CH_2CH(R)OSO_3H \qquad (17)$$

$$R'CH_2CH(R)OH + H_2SO_4 \longrightarrow R'CH_2CH(R)OSO_3H \qquad (18)$$

$$R'CH_2CH(R)SO_3H + NaOH \longrightarrow R'CH_2CH(R)OSO_3Na + H_2O \qquad (19)$$

The first reaction is the more important because double bonds are encountered more frequently than hydroxyl groups. Solubility in water increases with increasing sulfation but so does the danger of hydrolysis of the glyceride linkage.

Sulfated oils are widely used as wetting and penetrating agents, emulsifiers, textile softeners, and lubricants.

The products, offered by more than 30 U.S. manufacturers, are usually sold under descriptions referring to the starting oils, eg, sulfated peanut oil, 50%.

Phosphate Esters. The mono and diesters of orthophosphoric acid:

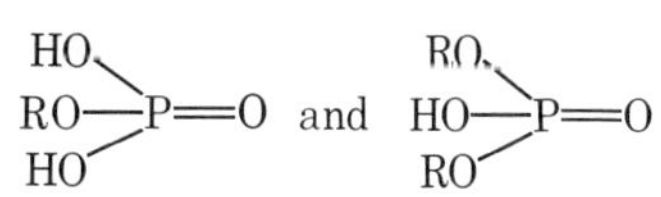

and their salts are useful surfactants which have been known for over 40 years (67). Consumption of phosphate surfactants has grown from about 5,900 t in 1966 to over 17,000 t in 1980 (see under Production and economic aspects for a breakdown).

In contrast to sulfonates and sulfates, the resistance of alkyl phosphate esters to acids and hard water is poor. Calcium and magnesium salts are insoluble. In the acid form, the esters show limited water solubility, although their alkali metal salts are more soluble. The surface activity of phosphate esters is good, although in general it is somewhat lower than that of the corresponding phosphate-free precursors. Thus, a phosphated nonylphenol ethoxylated with 9 mol of ethylene oxide is less effective as a detergent in hard water than its nonionic precursor. At higher temperatures, however, the phosphate surfactant is significantly more effective (68).

Because of their high costs and the limitations noted above, phosphate surfactants find application as specialty surfactants in situations where such limitations are of no concern. As specialty surfactants, phosphate esters and their salts are remarkably versatile. Applications include emulsion polymerization of vinyl acetate and acrylates; dry-cleaning compositions where solubility in hydrocarbon solvents is a particular advantage; textile mill processing where stability and emulsifying power for oil and wax under highly alkaline conditions is called for; and industrial cleaning compositions where tolerance for high concentrations of electrolyte and alkalinity is required. In addition, phosphate surfactants are used as corrosion inhibitors, in pesticide formulations, in papermaking, and as wetting and dispersing agents in drilling mud fluids.

Polyphosphoric acid, P_2O_5, $POCl_3$, and PCl_3 are suitable phosphorylating agents. Reaction of an alkyl sulfate with sodium pyrophosphate has also been reported for preparation of alkyl pyrophosphates (69). In general, phosphorylation leads to a mixture of reaction products that are sold without further separation. Thus, when lauryltri(ethyleneoxy)ethanol reacts with ⅓ mol of P_2O_5 at 50°C and is neutralized with 50% aqueous NaOH, the reaction mixture contains the following products:

$C_{12}H_{25}(OCH_2CH_2)_4OPO(OH)ONa$ [*82679-95-6*]	monoester, largest fraction
$(C_{12}H_{25}(OCH_2CH_2)_4O)_2PO(ONa)$ [*28505-90-0*]	diester, second largest fraction
$(C_{12}H_{25}(OCH_2CH_2)_4O)_3PO$ [*20297-06-7*]	triester, small fraction

and smaller amounts of the phosphate esters of the polyethylene glycol which is present in the ethoxylated alcohol. A trace of NaH_2PO_3 is also present.

Table 14 gives the important commercial phosphate surfactants which are offered commercially under a variety of trade names including Crodafos, Emphos, Findet, Gafac, Hodag PE, Hostaphat, Monafax, Rexophos, Triton, Victawet, and others.

Nonionic Surfactants

Unlike anionic or cationic surfactants, nonionic surfactants carry no discrete charge when dissolved in aqueous media. Hydrophilicity in nonionic surfactants is provided by hydrogen bonding with water molecules (71). Oxygen atoms and hydroxyl groups readily form strong hydrogen bonds, whereas ester and amide groups form hydrogen bonds less readily. Hydrogen bonding provides solubilization in neutral and alkaline media. In a strongly acid environment, oxygen atoms are protonated, providing a quasi-cationic character. Each oxygen atom makes a small contribution to water solubility. More than a single oxygen atom is therefore needed to solubilize a nonionic surfactant in water. Nonionic surfactants are compatible with ionic and amphoteric surfactants. Since a polyoxyethylene group can easily be introduced by reaction of ethylene oxide with any organic molecule containing an active hydrogen atom, a wide variety of structures can be solubilized by ethoxylation.

Polyoxyethylene Surfactants (Ethoxylates). Polyoxyethylene-solubilized nonionics were introduced in the United States as textile chemicals shortly before 1940. The solubility of these compounds derives from recurring ether linkages in a polyoxyethylene chain $—O—CH_2CH_2—O—CH_2CH_2—O—CH_2CH_2—O—$. A single oxyethylene group, $—OCH_2CH_2—$ contributes slightly more to hydrophilicity than a single methylene, CH_2, contributes to hydrophobicity. To effect complete miscibility with water at room temperature, ca 60–75% by weight of polyoxyethylene content is needed on most hydrophobes. With increasing temperature, ethoxylates become less soluble as a result of decreased hydration and increasing micellar size (72–73). The temper-

Table 14. Phosphate Ester Surfactants[a]

Component	Cation	Manufacturer[b]
butyl phosphate	K	DUP
hexyl phosphate	H, K	ICI, SFS
2-ethylhexyl phosphate	H, Na	CHP, DSC, GAF, WTC
octyl phosphate	H, K, A[c]	DUP, DSC, HNK, WTC
decyl phosphate	H	DUP
octyldecyl phosphate	H	DUP
9-octadecenyl phosphate	H	DSC
mixed alkyl phosphate	H, DEA[d]	AMH, CTL, DUP, SFS
hexyl polyphosphate	K	DEX
octyl polyphosphate	H, K	DEX, SNW
other phosphated and polyphosphated alcohols		CHP, EFH, HRT, RCD
glycerol monoester of mixed fatty acids, phosphated		QCP, WTC
butyl alcohol, ethoxylated and phosphated	H	GAF
2-ethylhexanol, ethoxylated and phosphated	H	DSC, PHC
dodecyl alcohol, ethoxylated and phosphated	H	GAF
tridecyl alcohol[e], ethoxylated and phosphated	H	GAF, MIL, SNW, WTC
9-octadecenyl alcohol, ethoxylated and phosphated	H	GAF
mixed linear alcohols, ethoxylated and phosphated	H	AMH, CHP, CRT, CTL, EMR, HNK, HRT, MIL, MOA, QCC, TCC, WTC
polyhydric alcohols, ethoxylated and phosphated	H	DEX, GAF, HDG, HNK
phenol, ethoxylated and phosphated	H	DSC, EMR, GAF, MOA, RH, WTC
octylphenol[e], ethoxylated and phosphated	H, Mg	OXR, RH
nonylphenol[e], ethoxylated and phosphated	H	CRT, CTL, DEX, DSC, GAF, HDG, HNK, HRT, MOA, PHC, QCC, TCC, WTC, WVA
dodecylphenol[e], ethoxylated and phosphated	H	GAF
dinonylphenol[e], ethoxylated and phosphated	H	GAF, MOA, PHC, WTC

[a] Ref. 70.
[b] See Table 2.
[c] A = alkylammonium.
[d] DEA = diethanolammonium.
[e] Branched.

ature at which the appearance of a second phase is observable is referred to as the cloud point, a useful characteristic of nonionics which is independent of concentration in the range of 0.5–10 wt % (74). Small amounts of admixed anionic surfactants may raise the cloud point by several degrees. Surface activity and performance efficiency of nonionics are usually greatest at temperatures just below the cloud point. The surface activity of ethoxylates is not adversely affected by water hardness. High electrolyte concentrations, particularly the presence of sodium ions, decrease the solubility of ethoxylates by a salting-out effect. Hydrochloric acid and calcium ions, on the other hand, increase water solubility. Ethoxylates are capable of solubilizing iodine in aqueous solution through formation of addition complexes. Ethoxylate–iodine complexes maintain the biocidal activity of iodine while reducing its toxicity to humans (75). Ethoxylates are moderate foamers and do not respond to conventional foam boosters. Foaming shows a maximum as a function of ethylene oxide content (76). Low foaming nonionic surfactants are prepared by terminating the polyoxyethylene chain with less soluble groups such as polyoxypropylene groups and methyl groups. Ethoxylates can be prepared to attain almost any hydrophilic–hydrophobic balance. For incorporation into powdered products, they suffer from the disadvantage of being liquids or low melting waxes which complicates the manufacture of free-flowing, crisp powders. Solid products are manufactured with ethoxylates of high ethylene oxide content. The latter, however, are too water-soluble to provide good surface activity.

Nonionic surfactants are often characterized in terms of their HLB (hydrophile–lipophile balance) number (see Emulsions). For simple alcohol ethoxylates, the HLB number may be calculated from

$$\mathrm{HLB} = E/5$$

where E is the weight percentage of ethylene oxide in the molecule. The functionality of nonionic surfactants depends on HLB as follows:

HLB range	*Application*
3–6	water-in-oil, w/o, emulsifier
7–9	wetting agent
8–15	oil-in-water, o/w, emulsifier
13–15	detergent
15–18	solubilizer

Ethoxylation. Base-catalyzed ethoxylation of aliphatic alcohols, alkylphenols, and fatty acids proceeds in two stages: formation of a monoethoxy adduct and addition of ethylene oxide to the monoadduct to form the polyoxyethylene chain. The sequence of reactions is shown in equations 20–22:

$$RO^- + H_2C\overset{O}{\diagdown\!\diagup}CH_2 \xrightarrow{\text{slow}} ROCH_2CH_2O^- \qquad (20)$$

$$ROH + ROCH_2CH_2O^- \xrightarrow{\text{fast}} RO^- + ROCH_2CH_2OH \qquad (21)$$

$$ROCH_2CH_2O^- + H_2C\overset{O}{\diagdown\!\diagup}CH_2 \xrightarrow{\text{fast}} ROCH_2CH_2OCH_2CH_2O^- \qquad (22)$$

Equation 20 is the rate-controlling step. The reaction rate of the hydrophobes decreases in the order primary alcohols > phenols > carboxylic acids (77). With alkylphenols and carboxylates, build-up of polyadducts begins after the starting material has been completely converted to the monoadduct reflecting the increased acid strengths of these hydrophobes over the alcohols. Polymerization continues until all ethylene oxide has reacted. Beyond formation of the monoadduct, the reactivity is essentially independent of chain length. The effectiveness of ethoxylation catalysts increases with their base strength. In practice, ratios of 0.005–0.05:1 mol of NaOH, KOH, or $NaOCH_3$ to alcohol are frequently used.

Alcohol Ethoxylates. During the last decade, these products have emerged as the principal nonionic surfactants in consumer detergent products (see Table 15). Consistent quality, expanding production capacity of relatively inexpensive detergent range straight-chain, highly biodegradable alcohols, and increasing usage, particularly in heavy-duty liquid laundry products, are the principal factors underlying this growth.

Alcohol ethoxylates vary in physical form from liquids to waxes. With increasing ethylene oxide content viscosity increases; a slight hydrophobe odor, present in the lower members of the series, decreases; specific gravity increases from just under 1 to slightly below 1.2; and water solubility increases; ca 65–70 wt % of ethylene oxide content is needed for complete miscibility with water at room temperature.

Alcohol ethoxylates are effective cleaners in domestic and industrial detergent products. Depending on ethylene content, alcohol ethoxylates provide excellent emulsifying, wetting, and dispersing performance.

In the ethoxylation reaction, primary alcohols react most readily, followed by secondary and tertiary alcohols. The rate of reaction of the primary hydroxyl with ethylene oxide is close to that of the ethylene oxide chain build-up reaction. As a result, the ethylene oxide chain is built up before all alcohol has been converted into the monoadduct and unethoxylated alcohol is present in the final ethoxylation product. Alcohol ethoxylates are polydisperse with respect to ethylene oxide chain length. An ethoxylate containing nominally eight ethylene oxide units actually contains significant amounts of other ethoxylates containing from 0–20 ethylene oxide units. The composition of the mixture has been reported as following a Weibull and Nycander distribution more closely than a Poisson distribution. Since the ethoxylates at the extremes of the 0–20 range are less useful than the nominal product, much effort has been expended to peak the distribution more narrowly around the target value. Acid catalysts provide a sharper distribution but can lead to undesired side products such as dioxanes which must be removed. A recent U.S. patent claims a barium hydroxide catalyst promoted with phenolic compounds is capable of yielding a substantially narrow distribution of ethoxylates (78).

Alkylphenol Ethoxylates. In their physical and performance properties, these products are similar to the alcohol ethoxylates. Table 16 indicates that most alkylphenol ethoxylates are derived from alkylphenols carrying branched alkyl side chains, predominantly nonyl and octyl. Chain branching accounts for the lower rate of biodegradation of alkylphenol ethoxylates compared to that of alcohol ethoxylates. On the other hand, chain branching increases water-solubility and contributes to marginally superior detersive properties. Since the stringent biodegradability criteria adopted by the U.S. Soap and Detergent Association apply to consumer products only, it is not surprising that alkylphenol ethoxylates continue to be used extensively in industrial products. Nonylphenol ethoxylates accounted for 73.6% of 1980 U.S. production of alkylphenol ethoxylates.

Alkylphenol ethoxylates are chemically stable and highly versatile surfactants that find application in a large variety of industrial products including acid and alkaline metal cleaning formulations, hospital cleaners, herbicides and insecticides, oil-well drilling fluids, synthetic latices, and many others (see Metal surface treatments; Disinfectants and antiseptics; Herbicides; Insect-control technology; Petroleum, drilling fluids; Elastomers, synthetic).

Commercial alkylphenol ethoxylates are almost always produced by base-cata-

Table 15. Alcohol Ethoxylates, $R\{OCH_2CH_2\}_nOH$

Trade name	R	n	Concn, wt %	Functions[a]	Manufacturer[b]
Alfonic 610-50R	C_6, C_8, C_{10}[c]	3	100	D, E, W	CO
Alfonic 1012-40	C_{10}–C_{12}[c]	2.5	100	E	CO
Alfonic 1012-60	C_{10}–C_{12}[c]	5.7	100	D, E, W	CO
Alfonic 1412-40	C_{12}–C_{14}[c]	3.0	100	E	CO
Alfonic 1412-60	C_{12}–C_{14}[c]	7.0	100	D, E, W	CO
Alfonic 1214-70	C_{12}–C_{14}[c]	10.6	100	D, E, W	CO
Alfonic 1218-70	C_{12}, C_{14}, C_{16}, C_{18}[c]	10.7	100	D, E, W	CO
Antarox BL-214			100	D, rW, Ta	GAF
Antarox BL-225			100	D, W	GAF
Antarox BL-240			100	D, Ta, W	GAF
Antarox BL-330			95	D, W	GAF
Antarox BL-344			90	D, W	GAF
Antarox LF-330			95	D, W,	GAF
Antarox LF-344			90	D, W	GAF
Arosurf 66-E2	isostearyl	2	100	E, Di, S, St	SHX
Arosurf E 20	isostearyl	20	100	E, Di, S, St	SHX
Bio Soft EA-10	C_{10}, C_{12}[c]		100	D, E	STP
Brij 30	lauryl	4	100	E	ICI
Brij 35	lauryl	23	100	E	ICI
Brij 52	cetyl	2	100	E	ICI
Brij 58	cetyl	20	100	E, S	ICI
Brij 72	stearyl	2	100	E	ICI
Brij 78	stearyl	20	100	E	ICI
Brij 92	oleyl	2	100	E	ICI
Brij 98	oleyl	20	100	E	ICI
Brij 700	stearyl	100			ICI
Brij 721	stearyl	21			ICI
Cerfak 1400			85	D, W	EFH
Crestomul L			100	D, E	CRT
Emulphogene BC-420	tridecyl[d]		100	E	GAF
Emulphogene BC-720	tridecyl[d]		100	E, Fs, W	GAF
Emulphogene ON-877	tridecyl[d]		70	Di, E, S	GAF
Examide DA			50	D, E, Ta	SLC
Hyonic PE-360	tridecyl[d]		100	D, W	DSC
Neodol 25-3	C_{12-15}[c]	3	100	Di, I	SHC
Neodol 23-6.5	C_{12-13}[c]	6.5	100	D	SHC
Neodol 25-7	C_{12-15}[c]	7	100	D, W	SHC
Neodol 25-12	C_{12-15}[c]	12	100	D, E, W	SHC
Neodol 45-7	C_{14-15}[c]	7	100	D, W	SHC
Neodol 45-13	C_{14-15}[c]	13	100	D, E, W	SHC
Neodol 91-2.5	C_{9-11}[c]	2.5	100	Di, W	SHC
Neodol 91-6	C_{9-11}[c]	6	100	D, W	SHC
Neodol 91-8	C_{9-11}[c]	8	100	D, W	SHC
Plurafac A-38	alkyl		100	D, E, W	WYN
Poly Tergent J-200	tridecyl[d]		99	D, Di, E, S	OC
Poly Tergent J-500	tridecyl[d]		99	D, Di, E, S	OC
Promulgen 1703	alkyl		100	E, S, St	RW
Renex 30	tridecyl		100	D, Di, W	ICI
Siponic E-3	tallow	6	100	E	AAC
Siponic E-30	tallow	30	100	E	AAC
Siponic L-4	lauryl	4	100	E	AAC
Siponic L-25	lauryl	25	100	E	AAC
Standamul O-5	oleyl	5	100	E	HNC

Table 15 (*continued*)

Trade name	R	*n*	Concn, wt %	Functions[a]	Manufacturer[b]
Standamul B-3	cetyl	30	100	E	HNC
Surfonic JL-80X	alkyl[c]		100	D, Di, E, W	TCC
Surfonic LF-17	alkyl[c]		100	D, W	TCC
Tergitol 25-L-3	C_{12-15}[c]	3	100	Di, I	UCC
Tergitol 25-L-7	C_{12-15}[c]	7	100	D, W	UCC
Tergitol 25-L-12	C_{12-15}[c]	12	100	D, E, W	UCC
Tergitol 15-S-3	C_{11-15}[e]	3	100	I, Ta	UCC
Tergitol 15-S-5	C_{11-15}[e]	5	100	D, E, W	UCC
Tergitol 15-S-7	C_{11-15}[e]	7	100	D, E, W	UCC
Tergitol 15-S-9	C_{11-15}[e]	9	100	D, E, W	UCC
Tergitol 15-S-12	C_{11-15}[e]	12	100	D, E, W	UCC
Tergitol 15-S-15	C_{11-15}[e]	15	100	W, E	UCC
Tergitol 15-S-20	C_{11-15}[e]	20	100	E, Ta	UCC
Tergitol 15-S-30	C_{11-15}[e]	30	100	E, Ta	UCC
Tergitol TMN-3	trimethylnonyl[d]	3	100	Ta, W	UCC
Tergitol TMN-9	trimethylnonyl[d]	9	90	Ta, W	UCC
Trycol DA-4	isodecyl[d]	4	100	E, W	EMR
Trycol DA-6	isodecyl[d]	6	100	I, P, W	EMR
Trycol LAL-4	lauryl	4	100	E	EMR
Trycol LAL-8	lauryl	8	100	D, rW, W	EMR
Trycol LAL-12	lauryl	12	100	D, W	EMR
Trycol LAL-23	lauryl	23	100	S, Ta	EMR
Trycol OAL-23	oleyl	23	100	S, St, Ta	EMR
Trycol SAL-20	stearyl	20	100	E, St, Ta	EMR
Trycol TDA-3	tridecyl[d]	3	100	E	EMR
Trycol TDA-9	tridecyl[d]	9	100	E, I, W	EMR
Trycol TDA-15	tridecyl[d]	15	100	D, Fs	EMR
Trycol TDA-18	tridecyl[d]	18	100	Di, S, Ta	EMR
Volpo 3	oleyl	3	100	E, S	CRO
Volpo 10	oleyl	10	100	E, S	CRO
Volpo 20	oleyl	20	100	E, S	CRO
Witconol 1206	alkyl			W	WTC

[a] See Table 1.
[b] See Table 2.
[c] Linear-primary.
[d] Branched.
[e] Linear-secondary.

lyzed ethoxylation of alkylphenols. Because phenols are more strongly acidic than alcohols, reaction with ethylene oxide to form the monoadduct is faster. The product, therefore, does not contain unreacted phenol. Thus, the distribution of individual ethoxylates in the commercial mixture is narrower, and alkylphenol ethoxylates are more soluble in water.

Carboxylic Acid Esters. In the carboxylic acid ester series of surfactants, the hydrophobe, a naturally occurring fatty acid, is solubilized with the hydroxyl groups of polyols or the ether (and terminal hydroxyl) groups of ethylene oxide chains.

The 1980 production of 105,805 t represents a 59% increase since 1966. Production of all product types increased, except glycerol and propanediol esters, which declined by 5.1 and 65.0%, respectively.

Table 16. Alkylphenol Ethoxylates, $RC_6H_4(OC_2H_4)_nOH$

Trade name	R	*n*	Concn, wt %	Functions[a]	Manufacturer[b]
Conco NI-40	nonyl[c]		100	D, E, W	CTL
Hyonic PE-100			100	D	DSC
Igepal CA-630	octyl[c]		100	E, D, Di	GAF
Igepal CO-210	octyl[c]	1.5	100	D, Di, Fs, I	GAF
Igepal CO-530	nonyl[c]	6	100	E, D, Di	GAF
Igepal CO-710	nonyl[c]	10.5	100	D, W	GAF
Igepal CO-850	nonyl[c]	20	100	D, E, S, W	GAF
Igepal CO-880	nonyl[c]	30	100	D, Di, E, S, W	GAF
Igepal DM-730	dialkyl[c]		100	D, E, S	GAF
Makon 4	nonyl[c]	4	100	D, E, S	STP
Makon 6	nonyl[c]	6	100	D, E, S	STP
Makon 8	nonyl[c]	8	100	D, E, S	STP
Makon 10	nonyl[c]	10	100	A, D, E, S, Ta	STP
Makon 12	nonyl[c]	12	100	A, D, E, S, Ta	STP
Makon 14	nonyl[c]	14	100	D, E	STP
Makon 20	nonyl[c]	20	100	D, E	STP
Makon 30	nonyl[c]	30	100	D, E	STP
Neutronyx 600	nonyl[c]	9	100	D, Di, E, W	OXR
Nonionic E-4	alkyl		100	D, E, W	HDG
Nonionic E-10	alkyl		100	D, E, W	HDG
Poly Tergent B-150	nonyl[c]	4.5	100	D, Di, E, W	OC
Poly Tergent B-200	nonyl[c]	6	100	D, Di, E, W	OC
Poly Tergent B-350	nonyl[c]	9	100	D, Di, E, W	OC
Poly Tergent B-350	nonyl[c]	10.5	100	D, Di, E, W	OC
Poly Tergent B-500	nonyl[c]	15	100	D, Di, E, W	OC
Rexol 25/1	nonyl[c]	1	100	D, Di, I, St	HCL
Rexol 25/507	nonyl[c]	50	70	dE, S, St	HCL
Sterox DF	dodecyl[c]	6.2	99.5	D, E	MON
Sterox DJ	dodecyl[c]	10.3	99.5	D, E, W	MON
Sterox ND	nonyl[c]	4	99.5	E, D	MON
Sterox NJ	nonyl[c]	9.2	99.5	D, Di, E, W	MON
Sterox NK	nonyl[c]	10.3	99.5	D, Di, E, W	MON
Surfonic N-10	nonyl[c]	1	100	D, Di, E, dF	TCC
Surfonic N-31.5	nonyl[c]	3.15	100	D, Di, E, dF	TCC
Surfonic N-40	nonyl[c]	4	100	D, Di, E, W	TCC
Surfonic N-60	nonyl[c]	6	100	D, Di, E, W	TCC
Surfonic N-95	nonyl[c]	9.5	100	D, Di, E, W	TCC
Surfonic N-100	nonyl[c]	10	100	D, Di, E, W	TCC
Surfonic N-120	nonyl[c]	12	100	D, Di, E, W	TCC
Surfonic N-200	nonyl[c]	20	100	D, W	TCC
Surfonic N-300	nonyl[c]	30	100	D, W	TCC
T-DET N-4	nonyl[c]	4	100	D, E	TMH
T-DET N-6	nonyl[c]	6	100	D, Di, E, W	TMH
T-DET N-9.5	nonyl[c]	9.5	100	D, Di, E, W	TMH
T-DET N-15	nonyl[c]	15	100	D, Di, E, W	TMH
Tergitol NP-4	nonyl[c]	4	100	D, E	UCC
Tergitol NP-7	nonyl[c]	7	100	D, Di, E, W	UCC
Tergitol NP-9	nonyl[c]	9	100	D, Di, E, W	UCC
Tergitol NP-40	nonyl[c]	40	100	D, E, W	UCC
Triton N-57	nonyl[c]	7–8	100	D, E	RH
Triton N-111	nonyl[c]	12–13	100	Di, E	RH
Triton X-45	octyl[c]	5	100	D, E	RH

Table 16 (*continued*)

Trade name	R	*n*	Concn, wt %	Functions[a]	Manufacturer[b]
Triton X-102	octyl[c]	12–13	100	D	RH
Triton X-305	octyl[c]	30	70	E, St	RH
Triton X-705	octyl[c]	70	70	E, St	RH

[a] See Table 1.
[b] See Table 2.
[c] Branched.

Glycerol Esters. Commercial glycerol esters, though named after the most abundant species, almost always are mixtures of isomeric mono- and diglycerides (see also Glycerol). Trade names and compositions of typical commercial products are given in Table 17.

Saturated fatty acid mono- and diglycerides are light-colored solids melting between 25 and 85°C. The 1-monoglycerides are higher melting than the corresponding 2-monoglycerides. Glycerides of unsaturated fatty acids are liquid at room temperature. In general, the odor of the glycerides is characteristic of the fats from which they are derived. The two hydroxyl groups in monoglycerides are not sufficiently hydrophilic to carry even an easily solubilized acid like oleic acid into solution. Despite the lack of water solubility, the fatty acid mono- and diglycerides are commercially important and technically interesting surfactants.

Table 17. Glycerol Esters of Fatty Acids

Trade name	Fatty acid	Ester type	Concn, wt %	Function[a]	Manufacturer[b]
Aldo MC	coco	mono	100	E	GLY
Aldo ML	lauric	mono	100	E	GLY
Aldo MR	ricinoleic	mono	100	E	GLY
Aldo MS	stearic	mono	100	E	GLY
Cerasynt 945	stearic	mono	100	E	VND
Emerest 2400	stearic	mono	100	E	EMR
Emerest 2421	oleic	mono	100	E	EMR
Emerest 2426	oleic	di	100	E	EMR
Flexricin 13	ricinoleic	mono	100	E	NLI
Hallco CPH-34-N	lauric	mono	100	E	HAL
Hodag GML	lauric	mono	100	E	HDG
Hodag GMO	oleic	mono	100	E	HDG
Hodag GMR	ricinoleic	mono	100	E	HDG
Hodag GMS	stearic	mono	100	E	HDG
Kessco Ester	oleic	di	100	E	AMK
Kessco Ester	oleic	mono	100	E	AMK
Kessco Ester	stearic	mono, pure	100	E	AMK
Myverol Type 18-30	tallow	mono	100	E	EK
Myverol Type 18-40	lard	mono	100	E	EK
Myverol Type 18-45	cottonseed	mono	100	E	EK
Myverol Type 18-98	safflower	mono	100	E	EK
Witconol MST	stearic	mono	100	E	WTC

[a] See Table 1.
[b] See Table 2.

Most applications involve emulsification, dispersion, solubilization, and lubrication. In bread, cakes, and other bakery products, mono- and diglycerides are almost universally present as emulsifiers, dispersing agents, and lubricants. They are also used in candies, ice cream, yeast, butter, whipped toppings, and icings. Flavor oils for carbonated beverages and for bakery products are emulsified or solubilized by surfactant mixtures that include blends of mono- and diglycerides (see also Flavors and spices). Glycerol monostearate acts as an emulsifier and opacifier in cosmetic formulations. In cutting, drawing, and finishing of metal products, mono- and diglycerides find application as emulsifiers, lubricants, and corrosion inhibitors. They are also used as emulsifiers, dispersants, suspending agents, and grinding aids in the manufacture of paints and polymers.

The reaction of fats with glycerol is the most important industrial method of preparing mono- and diglycerides. In this reaction, the fatty acid groups are redistributed randomly by heating at 180–250°C in the presence of an alkaline catalyst.

Polyoxyethylene Esters. This series of surfactants consists of polyoxyethylene (polyethylene glycol) esters of fatty acids and aliphatic carboxylic acids related to abietic acid (see Resins, natural). They differ markedly from mono- and diglycerides in properties and uses.

Commercial polyoxyethylene fatty acid esters contain varying proportions of monoesters, $RCO\text{+}OC_2H_4\text{+}_nOH$; diesters, $RCO\text{+}OC_2H_4\text{+}_nOOCR$; and polyglycol, $H\text{+}OC_2H_4\text{+}_nOH$. The composition of the mixture can be forced toward the mono- or diester by the ratio of reactants and the process of manufacture. Commercial products are usually described by the acid esterified, the predominant ester species (mono- or di-), and either the molecular weight of the polyglycol or the number of ethylene oxide groups in the polyglycol chain. Composition and functions of typical polyoxyethylene esters are listed in Table 18.

The physical properties of the fatty acid ethoxylates depend on the nature of the fatty acid and even more on the ethylene oxide content. As the latter increases, consistencies of the products change from free-flowing liquids to slurries to firm waxes. At the same time, odor, which is characteristic of the fatty acid, decreases in intensity. Odor and color stability are important commercial properties, particularly in textile applications. Oleic acid esters, though possessing good functional properties, cannot be used since they tend to yellow on exposure to heat and air.

At room temperature, ca 60 wt % ethylene oxide is needed to solubilize the fatty acids. Surface activity of the ethoxylates is moderate and less than that of alcohol or alkylphenol ethoxylates (77). The ethoxylates are low foamers, a useful property in certain applications. Emulsification is the most important function. Its importance is reflected in the wide range of lipophilic solubilities available in the commercial products. Like all organic esters, the fatty acid ethoxylates are susceptible to acid and alkaline hydrolysis.

Fatty acid ethoxylates are used extensively in the textile industry as emulsifiers for processing oils, antistatic agents, softeners, and fiber lubricants, and as detergents in scouring operations. They also find application as emulsifiers in cosmetic preparations, pesticide formulations, etc.

Fatty acid ethoxylates are manufactured either by alkali-catalyzed reaction of fatty acids with ethylene oxide or by acid-catalyzed esterification of fatty acids with preformed poly(ethylene glycol). Deodorization and decolorization steps are commonly incorporated into the manufacturing process.

Table 18. Polyoxyethylene Esters of Rosin, Tall Oil, and Fatty Acids

Trade name	Ester type	Composition			Concn, wt %	Functions[c]	Manufacturer[d]
		Acid	PEG[a]	POE[b]			
Emerest 2620	mono	lauric	200		100	E, dF, V	EMR
Emerest 2622	di	lauric	200		100	E	EMR
Emerest 2625	mono	isostearic	200		100	E, Ta	EMR
Emerest 2534	mono	pelargonic	300		100	Ta	EMR
Emerest 2647	sesqui	oleic	400		100	E, Ta	EMR
Emerest 2652	di	lauric	400		100	E, Ta	EMR
Emerest 2660	mono	oleic	600		100	E, Ta	EMR
Emerest 2672	mono	stearic		40	100	E, S	EMR
Emulphor TO-530		tall oil		10	100	W	GAF
Emulphor TO-639		tall oil		16	97	W	GAF
Ethofat C/15	mono	coco		5	100	D, Di, E	AMK
Ethofat C/25	mono	coco		15	100	D, Di, E	AMK
Ethofat 142/20	mono	tall oil		10	100	D, Di, E	AMK
Ethofat 242/25	mono	tall oil		15	100	D, Di, E	AMK
Ethofat 0/15	mono	oleic		5	100	D, Di, E	AMK
Ethofat 0/20	mono	oleic		10	100	D, Di, E	AMK
Ethofat 60/15	mono	stearic		5	100	D, Di, E	AMK
Ethofat 60/20	mono	stearic		10	100	D, Di, E	AMK
Ethofat 60/25	mono	stearic		15	100	D, Di, E	AMK
Hodag 40-L	mono	lauric	400		100	E, Sp, W	HDG
Hodag 40-R	mono	ricinoleic	400		100	E, Sp, W	HDG
Hodag 42-S	di	stearic	400		100	E, Sp, W	HDG
Hodag 60-L	mono	lauric	600		100	E, Sp, W	HDG
Hodag 60-S	mono	stearic	600		100	E, Sp, W	HDG
Hodag 62-O	di	stearic	600		100	E, Sp, W	HDG
Hodag 150-O	di	oleic	1500		100	E, Sp, W	HDG
Hodag 150-S	mono	stearic	1500		100	E, Sp, W	HDG
Kessco Ester	mono	lauric	200		100	E	AMK
Kessco Ester	di	lauric	400		100	E	AMK
Kessco Ester	mono	lauric	400		100	E	AMK
Kessco Ester	di	oleic	400		100	E	AMK
Kessco Ester	mono	oleic	400		100	E	AMK
Kessco Ester	di	stearic	400		100	E	AMK
Kessco Ester	mono	stearic	400		100	E	AMK
Kessco Ester	mono	lauric	600		100	E	AMK
Myrj 45	mono	stearic		8	100	E	ICI
Myrj 52	mono	stearic		40	100	E	ICI
Nonisol 210		oleic	400		100	E	CGY
Nonisol 300		stearic	400		100	E, St, W	CGY
Nopalcol 4-L	mono	lauric	400		100	E	DSC
Nopalcol 6-R	mono	ricinoleic	600		100	E	DSC
Nopalcol 10-CO		castor	1000		99	E	DSC
Nopalcol 30-S	mono	stearic	1200		100	E	DSC
Pegosperse 400 MS	mono	stearic	400		100	A, dF, E	GLY
Pegosperse 400 MO	mono	oleic	400		100	A, dF, E	GLY
Pegosperse 400 DOT	di		400		100	A, dF, E	GLY
Pegosperse 600 DOT	di		600		100	A, dF, E	GLY
Pegosperse 400 ML	mono	lauric	400		100	A, dF, E	GLY
Pegosperse 600 ML	mono	lauric	600		100	A, dF, E	GLY
Pegosperse 1750 MS	mono	stearic	1750		100	A, dF, E	GLY
Renex 20	mono	fatty and rosin		16	100	D, W	ICI
Surfactant AR150	mono	rosin			100	D, E	HPC

Table 18 (*continued*)

Trade name	Ester type	Composition: Acid	PEG[a]	POE[b]	Concn, wt %	Functions[c]	Manufacturer[d]
Witconol H31 A	mono	oleic	400		100	E	WTC
Witconol L 32-45	di	stearic	150		100	St, V	WTC

[a] Mol wt of esterified polyethylene glycol.
[b] Number of ethylene oxide groups in esterified polyoxyethylene.
[c] See Table 1.
[d] See Table 2.

The principal constituents of rosin (qv) are abietic and related acids. Tall oil (qv) is a mixture of unsaturated fatty and alicyclic acids of the abietic family. Refined tall oil may be high in rosin acids or unsaturated acids, depending on the refining process. Ethoxylates of rosin acids (eg, dehydroabietic acid) are similar to fatty acid ethoxylates in surfactant properties and manufacture, except for their stability to hydrolysis. No noticeable decomposition was observed when a rosin ester of this type was boiled for 15 min in 10% sulfuric acid or 25% sodium hydroxide (79). Steric hindrance of the carboxylate group associated with the alicyclic moiety has been suggested as the cause of this unexpectedly great hydrolytic stability.

Anhydrosorbitol Esters. Fatty acid esters of anhydrosorbitol (see Alcohols, polyhydric) are the second largest class of carboxylic ester surfactants. The important commercial products are the mono-, di-, and triesters of sorbitan and fatty acids (see Table 19). Sorbitan is a mixture of anhydrosorbitols, principally 1,4-sorbitan and isosorbide:

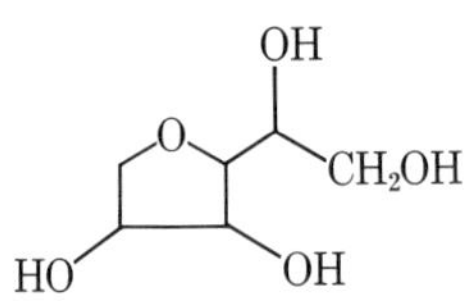

1,4-sorbitan

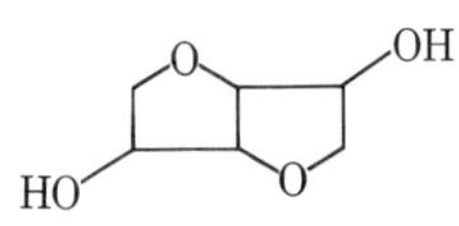

isosorbide

Sorbitan oleates and the monolaurate are pale-yellow liquids. Palmitates and stearates are light-tan solids. Sorbitan esters are not soluble in water but dissolve in a wide range of mineral and vegetable oils. They are lipophilic emulsifiers, solubilizers, softeners, and fiber lubricants which find application in synthetic fiber manufacture, textile processing, and cosmetic products. Sorbitan esters have been approved for human ingestion and are widely used as emulsifiers and solubilizers in foods, beverages, and pharmaceuticals.

Anhydrosorbitol esters are prepared commercially by direct esterification of sorbitol with a fatty acid at 225–250°C in the presence of an acidic catalyst. Internal ether formation and esterification take place under these conditions.

Ethoxylated Anhydrosorbitol Esters. Ethoxylation of sorbitan fatty acid esters leads to a series of more hydrophilic surfactants (see Table 20). All hydroxyl groups of sorbitan can react with ethylene oxide. The structure of the principal component of a nominal polyoxyethylene(20)sorbitan monostearate illustrates the composition of these products:

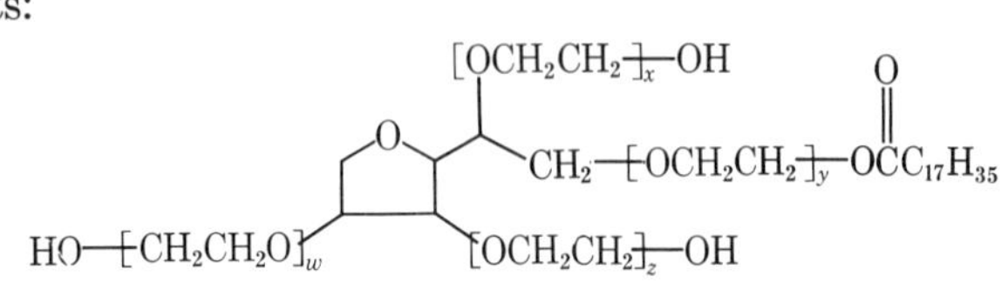

$(w + x + y + z = 20)$

Table 19. Sorbitan Esters of Fatty and Tall Oil Acids

Trade name	Ester type	Acid	Concn, wt %	Functions[a]	Manufacturer[b]
Armotan ML	mono	lauric	100	E, S	AMK
Armotan MO	mono	oleic	100	E, S	AMK
Armotan MS	mono	stearic	100	E, S	AMK
Emsorb 2500	mono	oleic	100	E, C	EMR
Emsorb 2502	sesqui	oleic	100	E	EMR
Emsorb 2503	tri	oleic	100	E, Sp	EMR
Emsorb 2505	mono	stearic	100	E, Ta	EMR
Emsorb 2510	mono	palmitic	100	E, Ta	EMR
Emsorb 2515	mono	lauric	100	E	EMR
Glycomul SOC	mono/sesqui	oleic	100	E	GLY
Glycomul TO	tri	oleic	100	E	GLY
Glycomul TS	tri	stearic	100	E	GLY
Hodag SML	mono	lauric	100	E	HDG
Hodag SMP	mono	palmitic	100	E	HDG
Hodag SMS	mono	stearic	100	E	HDG
Hodag STS	tri	stearic	100	E	HDG
Hodag SMO	mono	oleic	100	E	HDG
Hodag STO	tri	oleic	100	E	HDG
Span 20	mono	lauric	100	E, V	ICI
Span 40	mono	palmitic	100	E, V	ICI
Span 60	mono	stearic	100	E, V	ICI
Span 65	tri	stearic	100	E, V	ICI
Span 80	mono	oleic	100	E, V	ICI
Span 85	tri	oleic	100	E, V	ICI

[a] See Table 1.
[b] See Table 2.

Typical commercial ethoxylated sorbitan fatty acid esters are yellow liquids, except tristearates and the 4- and 5-mol ethylene oxide adducts which are light-tan solids. These adducts, as well as the 20-mol adducts of the triesters, are insoluble but dispersible in water. The monoester-20-mol adducts are water soluble. Ethoxylated sorbitan esters are widely used as emulsifiers, antistats, softeners, fiber lubricants, and solubilizers. In combination with the unethoxylated sorbitan esters or with mono- or diglycerides, they are often used as coemulsifiers. The ethoxylated sorbitan esters are produced by heating sorbitan esters with ethylene oxide at 130–170°C in the presence of alkaline catalysts.

Natural Fats, Oils and Waxes, Ethoxylated. Production of this group of surfactants increased from 1800 to almost 8500 t in the last 15 years. In 1980, ethoxylated castor oil accounted for 53% of the total.

Castor oil is a triglyceride high in ricinoleic esters. Ethoxylation in the presence of an alkaline catalyst to a polyoxyethylene content of 60–70 wt % yields water-soluble surfactants (see Table 21). Since alkaline catalysts also effect transesterification, ethoxylated castor oil surfactants are complex mixtures with components resulting from transesterification and subsequent ethoxylation at the available hydroxyl groups. The ethoxylates are pale-amber liquids of specific gravity just above 1.0 at room temperature. They are hydrophilic emulsifiers, dispersants, lubricants and solubilizers that are used as textile additives and finishing agents, as well as in paper and leather manufacture.

Table 20. Polyoxyethylene Derivatives of Sorbitan Fatty Acid Esters

Trade name	Ester type	Fatty acid	POE[a]	Concn, wt %	Functions[b]	Manufacturer[c]
Armotan PML-20	mono	lauric	20	100	A, E, S, Sp, Ta	AMK
Armotan PMO-20	mono	oleic	20	100	A, E, S, Sp, Ta	AMK
Armotan PMS-20	mono	stearic	20	100	A, E, S, Sp, Ta	AMK
Emsorb 6900	mono	oleic	20	100	Di, E, S	EMR
Emsorb 6901	mono	oleic	5	100	E	EMR
Emsorb 6903	tri	oleic	20	100	E, Sp, Ta	EMR
Emsorb 6905	mono	stearic	20	100	E	EMR
Emsorb 6906	mono	stearic	3	100	E	EMR
Emsorb 6910	mono	palmitic	20	100	E	EMR
Emsorb 6915	mono	lauric	20	100	A, E, Ta	EMR
Glycosperse L-4	mono	lauric	4	100	E	GLY
Glycosperse L-20	mono	lauric	20	100	E	GLY
Glycosperse O-5	mono	oleic	5	100	E	GLY
Glycosperse O-20	mono	oleic	20	100	E	GLY
Glycosperse P-20	mono	palmitic	20	100	E	GLY
Glycosperse S-4	mono	stearic	4	100	E	GLY
Glycosperse S-20	mono	stearic	20	100	E	GLY
Glycosperse TO-20	tri	oleic	20	100	E	GLY
Glycosperse TS-20	tri	stearic	20	100	E	GLY
Hodag PSML-20	mono	lauric	20	100	E	GLY
Hodag PSMP-20	mono	palmitic	20	100	E	GLY
Hodag PSTO-20	tri	oleic	20	100	E	GLY
Tween 20	mono	lauric	20	100	E, S, Ta	ICI
Tween 21	mono	lauric	4	100	E, W	ICI
Tween 40	mono	palmitic	20	100	TE, S, Ta	ICI
Tween 60	mono	stearic	20	100	E	ICI
Tween 61	mono	stearic	4	100	E	ICI
Tween 65	tri	stearic	20	100	E	ICI
Tween 80	mono	oleic	20	100	E, S, Ta	ICI
Tween 81	mono	oleic	5	100	E	ICI
Tween 85	tri	oleic	20	100	E, Ta	ICI

[a] Number of ethylene oxide groups in esterified polyoxyethylene.
[b] See Table 1.
[c] See Table 2.

Table 21. Ethoxylated Castor Oils

Trade name	POE[a]	Concn, wt %	Functions[b]	Manufacturer[c]
Emulphor EL 719	40	96	Di, E, Em, S, Ta	GAF
Emulphor EL 980	200	100	Di, E, S, Ta	GAF
Emulphor EL 719P	40	100	Di, E, S, Ta	GAF
Emulphor EL 620P	30	100	Di, E, Em, S, Ta	GAF
Surfactol-318		100	E, dF, Ta	NLI
Surfactol-380		100	E, St, Ta	NLI
Trylox CO-5	5	100	E, dF	EMR
Trylox CO-30	30	100	E, Sp, Ta	EMR
Trylox CO-40	40	97	E, S, Ta	EMR
Trylox CO-200	200	100	A, Sp, Ta	EMR
Trylox HCO-25[d]	25	100	E, Sp	EMR

[a] Number of ethylene oxide groups in esterified polyoxyethylene.
[b] See Table 1.
[c] See Table 2.
[d] Hydrogenated castor oil.

Table 22. Glycol Esters of Fatty Acids

Trade name	Glycol ester[a]	Concn, wt %	Functions[b]	Manufacturer[c]
Aldo PGHMS	PG monostearate	100	E	GLY
Emerest 2350	EG monostearate	100	O	EMR
Emerest 2355	EG distearate	100	O	EMR
Emerest 2381	PG monostearate	100	O	EMR
Hallco 37-N	EG monostearate	100	E, O	HAL
Hallco-CPH-51-N	PG monostearate	100	E	HAL
Hodag DGL	DEG monolaurate	100	E	HDG
Hodag DGO	DEG monooleate	100	E, Ta	HDG
Hodag DGS	DEG monostearate	100	E, Ta	HDG
Hodag EGS	EG monostearate	100	E	HDG
Hodag PGML	PG monolaurate	100	E	HDG
Hodag PGMS	PG monostearate	100	E	HDG
Kessco Ester	EG distearate		E, O	AMK
Kessco Ester	EG monostearate		E, O	AMK
Kessco Ester	DEG monostearate		E, O	AMK
Pegosperse 50 MS	EG monostearate	100	E	GLY
Pegosperse 50 DS	EG distearate	100	E	GLY
Pegosperse 100 O	DEG monooleate	100	E	GLY
Pegosperse 100 S	DEG monostearate	100	E	GLY
Witconol CAD	DEG stearate	100	E	WTC

[a] EG = ethylene glycol; DEG = diethylene glycol; PG = propylene glycol.
[b] See Table 1.
[c] See Table 2.

Glycol Esters of Fatty Acids. Ethylene glycol, diethylene glycol, and 1,2-propanediol esters of fatty acids together constitute a relatively small group of ester surfactants. The 1980 U.S. production of ca 3400 t was almost the same as in 1966. The mono- and dilaurates of the three glycol types are liquids, the stearates are solids. Glycol esters are strongly lipophilic emulsifiers, opacifiers, and plasticizers that commonly are formulated in combination with hydrophilic emulsifiers. Glycol monoesters can be prepared by the alkali-catalyzed reaction of fatty acids with ethylene or propylene glycol or by esterification of fatty acids with a glycol. As expected, the products are generally mixtures. The principal components of representative commercial glycol esters are given in Table 22.

Lanolin alcohols are obtained by saponification of purified wool grease, a mixture of high molecular esters that is recovered in wool scouring. Ethoxylation of purified lanolin alcohols yields a full series of lipophilic and hydrophilic nonionic emulsifiers whose largest use is in cosmetic preparations. In 1980, 770 t of ethoxylated lanolin was produced in the United States. Manufacturers include Amerchol, Croda, ICI, Emery Industries, and Robinson Wagner.

Carboxylic Amides. Carboxylic amide nonionic surfactants are condensation products of fatty acids and hydroxyalkyl amines. Amides derived from diethanolamine accounted for 78% of the 28,304 t carboxylic amide nonionic surfactants produced in the United States in 1980. This is a decrease from the 42,193 t produced in 1966.

Diethanolamine Condensates. The first fatty acid diethanolamides offered commercially were of the 2:1 or regular type. They are sometimes referred to as Kritchevsky amides after the inventor (80). Regular amides are obtained by heating one mol of fatty acid with two mol of diethanolamine. Nominally, the composition of regular diethanolamides is represented by $RCON(CH_2CH_2OH)_2.HN(CH_2CH_2OH)_2$. The actual composition has been reported as: diethanolamide $RCON(CH_2CH_2OH)_2$,

Table 23. Physical Form and Solubility of Fatty Acid Diethanolamides

Fatty acid	Amine:acid	Physical form	Water solubility
coco	2:1	liquid	soluble
coco	1:1	paste	limited solubility
lauric	2:1	soft paste	soluble
lauric	1:1	paste	limited solubility
oleic	2:1	liquid	dispersible
stearic	2:1	paste	dispersible

50%; diethanolamine $HN(CH_2CH_2OH)_2$, 25%; amide ester $RCON(CH_2CH_2OH)$-CH_2CH_2OOCR, 10%; amine ester $RCOOCH_2CH_2NHCH_2CH_2OH$, 10%; amine soap $RCOO^-H_2N(CH_2CH_2OH)_2^+$, 5% (81).

The second type of diethanolamide is the 1:1 or superamide which contains the components of the reaction of 1 mol fatty acid and 1 mol diethanolamine. A typical superamide composition is: 90% or more of the diethanolamide, 7% unreacted diethanolamine, and 2.5% amine ester and amide ester.

The presence of the large excess of diethanolamide and its derivatives in regular diethanolamides increases the water solubility compared to superamides. The physical form and water solubility of the fatty acid diethanolamides are shown in Table 23. Although the superamides of coco and lauric acids and the regular amides of the higher fatty acids are insoluble in water, they are readily dispersed and solubilized in combinations with more hydrophilic surfactants.

Fatty acid diethanolamides are versatile and widely used surfactants (see Table 24). Foam stabilization and detergency are the most important functional properties. In addition, they are very effective in increasing the viscosity of detergent solutions and provide emolliency, corrosion inhibition, emulsification, wetting, and lime-soap dispersion. Diethanolamides find application in liquid dishwashing detergents, shampoos, janitorial scrub soaps, textile processing, lubricating oils, and drycleaning detergents.

Regular fatty acid diethanolamides are prepared by heating 1 mol fatty acid with diethanolamine at 160–180°C for 2–4 h. Superamides are prepared by heating a fatty acid methyl ester with an equimolar amount of diethanolamine at 100–110°C for 2–4 h; the methanol formed is distilled off. Representative commercial products are described in Table 24.

Monoalkanolamine Condensates. Coco, lauric, oleic, and stearic monoethanolamides and monoisopropanolamides are the principal surfactants in the monoalkanolamide group (see Table 25). As a whole, the group has lost some ground over the last 15 yr. Monoalkanolamides are generally water-insoluble solids that are easily solubilized by hydrophilic surfactants. Except for solubility, their properties and uses are similar to the diethanolamides. Manufacturing processes and yields have been described (82).

Polyoxyethylene Fatty Acid Amides. Ethoxylation of fatty acid amides with one or two mol ethylene oxide yields nominally the same compositions as are obtained by condensing mono- or diethanolamines with fatty acids. Ethoxylation of fatty acid amides with more than 1 mol ethylene oxide yields predominantly secondary amides since the second amide hydrogen is less reactive than the first and also less reactive than the hydroxyl hydrogen. Ethoxylation can be carried out by conventional technology. Although the products of condensation are generally preferred, the longer-

Table 24. Fatty Acid Diethanolamides

Trade name	Fatty acid	Type[a]	Concn, wt %	Functions[b]	Manufacturer[c]
Aminol CA-2	ricinoleic	R	>99	D, Sp	FET
Aminol LA-12	lauric	R	>99	D, Fs	FET
Aminol LNO	linoleic	S	>99	Di, E, V	FET
Aminol OTX	tall oil	R	>97	Di, E	FET
Ardet DC	coco	R	100	D, Fs, E, V	ARD
Ardet DLA	lauric	S	100	Fs, V	ARD
Calamide C	coco	S	100	D, Fs	PIL
Calamide O	oleic	S	100	D, Fs	PIL
Calamide CW-100	coco	R	100	D, Fs	PIL
Clindrol 100C	coco	S	100	E, Fs, V	CLI
Clindrol 100L	lauric	S	100	Fs, V	CLI
Clindrol 100LM	lauric/myristic	S	100	Fs, V	CLI
Clindrol 200-O	oleic	R	100	E, V	CLI
Clindrol 200S	stearic	R	100	Di, E, V	CLI
Condensate PO	coco	R	100	D, E	CTL
Condensate PL	lauric	R	100	Fs	CTL
Emid 6511	lauric	S	100	Fs, V	EMR
Emid 6514	coco	S	100	Fs, V	EMR
Emid 6543	tallow	S	100	E, dF, Sp	EMR
Emid 6544	capric	R	100	Fs, W	EMR
Emid 6545	oleic	S	100	E, dF, Ta	EMR
Gafamide CDD-518	coco	S	100	D, Fs	GAF
Hartamide 9137	oleic	R		A, Fs, V	HCL
Hartamide LDA 70	lauric	S	>90	Fs, V	HCL
Hartamide OS	coco/oleic	S	100	Fs, S	HCL
Marsamide 40	coco	R	100	D, Fs, V	MRS
Marsamide 50	coco	S	100	D, Fs, V	MRS
Monamine ACO-100	lauric	R	100	D, E, V	MOA
Monamine LM-100	lauric/myristic	R	100	D, Fs, V, W	MOA
Monamine T-100	tall oil	R	100	D, Fs, V, W	MOA
Monamid 150-AD	coco	S	100	Fs, V	MOA
Monamid 150-IS	isostearic	S	100	E, C	MOA
Monamid 150-CW	capric	S	100	F, Fs	MOA
Ninol 202	oleic	R		Fs, V	STP
Ninol AA-62 EX	lauric	S	100	D, Fs, V	STP
Onyxol 336	lauric	R	97	Di, F	OXR
Onyxol 345	coco	R	97	Di, Fs, V, W	OXR
Rodea	ricinoleic	R	100	E, S	CLI
Schercomid CDA	coco	R	100	D, Di, Fs	SBC
Schercomid SLL	lauric/myristic	S	100	F, Fs, V	SBC
Standamid CD	capric	R	98	D, E, Fs, St, V	HNK
Standamid KD	coco	S	100	F, V	HNK
Varamide MA-1	coco	S	90	E, Fs, V	SHX
Varamide A-1	coco	R	100	D	SHX
Varamide L-1	lauric	R	100	D, Fs	SHX
Witcamide 82	coco	S	100	V	WTC
Witcamide 5195	lauric	R		Fs, V	WTC

[a] R = regular; S = super.
[b] See Table 1.
[c] See Table 2.

Table 25. Monoalkanolamides of Fatty Acids

Trade name	Fatty acid	Amine	Concn, wt %	Functions[a]	Manufacturer[b]
Ardet LIPA	lauric	isopropanol	93	Fs	ARD
Cerasynt D	stearic	ethanol	100	E, O, V	VND
Clindrol 101LI	lauric	isopropanol	100	D, Fs	CLI
Emid 6500	coco	ethanol	100	Fs, V	EMR
Intermediate 300	tallow/coco	ethanol	100	Fs, St, V	HUM
Intermediate 325	coco	ethanol	100	Fs, St, V	HUM
Monamid LIPA	lauric	isopropanol	100	Fs, V	MOA
Monamid LMA	lauric	ethanol	100	Fs, V	MOA
Monamid S	stearic	ethanol	100	O	MOA
P&G Amide No. 27	coco	ethanol	100	Fs, St, V	PG
Schercomid CME	coco	ethanol	100	D, Fs	SBC
Schercomid SME	stearic	isopropanol	100	D, E, O, V	SBC
Sole-Onic MIO	oleic	isopropanol	100	A, H	HDG
Surco CMEA	coco	ethanol	100	Fs, O, V	OXR
Witcamide 61	oleic	isopropanol	>90	C, E, V	WTC
Witcamide 70	stearic	ethanol	>90	C, O, V	WTC
Witcamide MM	myristic	ethanol	>90	C, O, V	WTC

[a] See Table 1.
[b] See Table 2.

chain ethylene oxide adducts of fatty amides sometimes exhibit special properties. Some representative products are given in Table 26.

Polyalkylene Oxide Block Copolymers. The higher alkylene oxides derived from propylene, butylene, styrene, and cyclohexene react with active oxygens in a manner analogous to the reaction of ethylene oxide. Because the hydrophilic oxygen constitutes a smaller proportion of these molecules, the net effect is that the oxides, unlike ethylene oxide, are hydrophobic. The higher oxides are not used commercially as surfactant raw materials except for minor quantities that are employed as chain terminators in polyoxyethylene surfactants to lower their foaming tendency. The hydrophobic nature of propylene oxide units, —$CH(CH_3)CH_2O$—, has been utilized in several ways in the manufacture of surfactants. Manufacture, properties, and uses of poly(oxyethylene-*co*-oxypropylene) are reviewed in ref. 83.

Poly(oxyethylene-*co*-oxypropylene) Nonionic Surfactants. A great variety of these products is marketed by the BASF Wyandotte Corp under the Pluronic polyol

Table 26. Polyoxyethylene Fatty Acid Amides

Trade name	Fatty acid	POE[a]	Concn, wt %	Functions[b]	Manufacturer[c]
Amidox C-2	coco	2	100	D, E, Fs	STP
Amidox C-5	coco	5	100	D, E, Fs	STP
Amidox L-2	lauric	2	100	D, E, Fs	STP
Amidox L-5	lauric	5	100	D, E, Fs	STP
Ethomid HT/15	hydrogenated tallow	5	100	D, Di, E	AMK
Ethomid HT/60	hydrogenated tallow	50	100	D, Di, E	AMK
Ethomid O/17	oleic	5	100	D, Di, E	AMK

[a] No. of ethylene oxide groups in polyoxyethylene.
[b] See Table 1.
[c] See Table 2.

trademark. The synthesis is shown below:

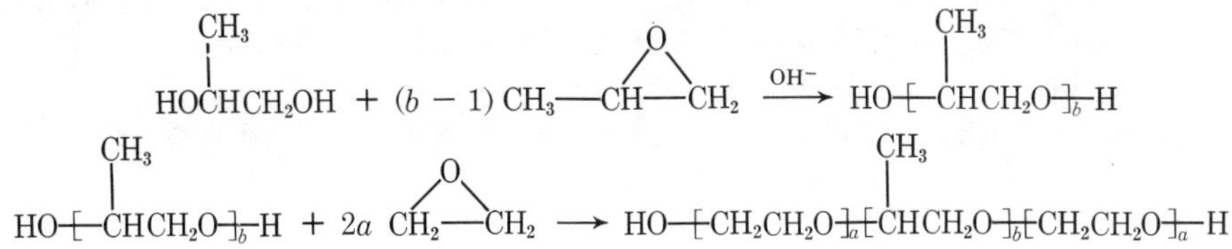

The structure of individual block polymers is determined by the nature of the initiator (1,2-propanediol in the above example), the sequence of addition of propylene and ethylene oxides, and the percentage of propylene and ethylene oxides in the surfactant. Thus, when the order of addition is reversed, a different structure is obtained in which the hydrophobic moieties are on the outside of the molecule. With ethylene glycol as the initiator, the reactions are as follows:

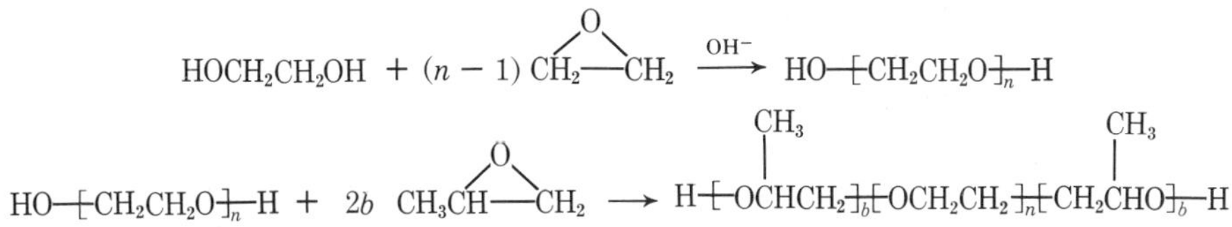

A third series of surfactants can be prepared by simultaneous addition of ethylene oxide and propylene oxide. Initially, propylene oxide predominates in the mixed oxides. In the second reaction step, ethylene oxide predominates. With a trifunctional initiator, eg, 1,1,1-trimethylolpropane, the final surfactant structure consists of three polyoxyalkylene chains each containing blocks of polyoxyethylene and polyoxypropylene, attached to a single carbon (84). Pluronic polyol surfactants vary from mobile liquids to flakeable solids. Solubility in water ranges from slightly soluble to completely miscible at the boil. Pluronic polyols with high ethylene oxide content exhibit no solution cloud point even at 100°C. In general, they are soluble in aromatic solvents but insoluble in kerosene or mineral oil. They exhibit very low toxicity and are nonirritating to the skin and almost tasteless. Chemical stability is comparable to that of the alcohol ethoxylates.

Block polymer nonionic surfactants are not strongly surface active but exhibit commercially useful surfactant properties. Aqueous solutions characteristically foam less than those of other surfactant types. They act as detergents, wetting and rinsing agents, demulsifiers and emulsifiers, dispersants and solubilizers. They are used in automatic dishwashing-detergent compositions, cosmetic preparations, spin finishing compositions for textile processing, metal-cleaning formulations, papermaking, and other technologies.

Block polymer nonionic surfactants are also manufactured by Dow Chemical Co., Union Carbide, and Witco Chemical Corp.

Cationic Surfactants

As the name implies, the hydrophobic moiety of a cationic surfactant carries a positive charge when dissolved in aqueous media. The positive charge resides on an amino or quaternary nitrogen. A single amino nitrogen is sufficiently hydrophilic to solubilize a detergent-range hydrophobe in dilute acidic solution; eg, laurylamine is soluble in dilute hydrochloric acid. For increased water solubility, additional primary, secondary, or tertiary amino groups can be introduced or the amino nitrogen can be quaternized with low molecular weight alkyl groups such as methyl or hydroxyethyl. Quaternary nitrogen compounds are strong bases that form essentially neutral salts with hydrochloric and sulfuric acids. Most quaternary nitrogen surfactants are soluble even in alkaline aqueous solutions. Polyoxyethylated cationic surfactants behave like nonionic surfactants in alkaline solutions and like cationic surfactants in acid solutions.

Cationic surfactants are widely used in acidic aqueous and nonaqueous systems as textile softeners, dispersants, emulsifiers, wetting agents, sanitizers, dye fixing agents, foam stabilizers, and corrosion inhibitors (86). To some extent, the usage pattern mirrors that of the anionic surfactants in neutral and alkaline solutions. The positively charged cationic surfactants are more strongly adsorbed than anionic or nonionic surfactants on a variety of substrates including textiles, metal, glass, plastics, minerals, and animal and human tissue, all of which can often carry a negative surface charge. Substantivity of cationic surfactants is the key property in many applications in which they are uniquely effective. In general, they are incompatible with anionic surfactants. Reaction of the two large, oppositely charged ions gives a salt that is insoluble in water. Ethoxylation moderates the tendency to form insoluble products with anionic surfactants.

Many benzenoid quaternary cationic surfactants possess germicidal, fungicidal, or algicidal activity. Solutions of such compounds, alone or in combination with nonionic surfactants are used as detergent-sanitizers in hospital maintenance. Since biocidal products are classified as economic poisons, their labeling is regulated by the EPA. The 1980 U.S. production of cationic surfactants of ca 141,000 t represented ca 6.4% of total surfactant production. Some of this production is used for the preparation of more highly substituted derivatives (85).

Amines. ***Oxygen-Free Amines.*** Aliphatic mono-, di-, and polyamines derived from fatty and rosin acids make up this class of surfactants which accounted for nearly 32,000 t produced in the U.S. in 1980. This represents a moderate increase from the 27,000 t produced in 1966. Primary, secondary, and tertiary monoamines with C_{18} alkyl or alkenyl chains constitute the bulk of this class. The products are sold as acetates, naphthenates, or oleates. Principal uses are as ore-flotation agents, corrosion inhibitors, dispersing agents, wetting agents for asphalt, and as intermediates for the production of more highly substituted derivatives. For ore flotation, corrosion inhibition and incorporation into asphalt, the amines are generally marketed in proprietary formulations. Physical constants and detailed descriptions of the individual amines are therefore not widely published.

In addition to the mono- and dialkylamines, representative structures of this class of surfactants include:

$$RNHCH_2CH_2CH_2NH_2$$

N-alkyltrimethylene diamine, where the alkyl group is derived from coconut, tallow, and soybean oils; or is 9-octadecenyl

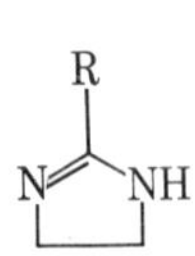

2-alkyl-2-imidazoline, where R is heptadecyl, heptadecenyl, or mixed alkyl, and

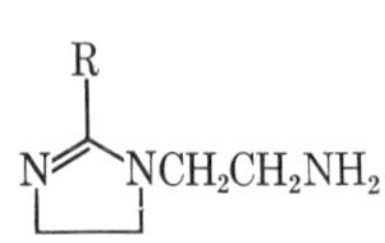

1-(2-aminoethyl)-2-alkyl-2-imidazoline, where R is heptadecyl, 8-heptadecenyl, or mixed alkyl.

Oxygen-Containing Amines, Except Amides. This group includes amine oxides, ethoxylated alkylamines, 1-(2-hydroxyethyl)-2-imidazolines, and alkoxylates of ethylenediamine. It is steadily increasing in economic importance. Production volume of 34,286 t in 1980 is almost twice the 17,463 t produced in the United States in 1966. Amine oxides, alkylamine ethoxylates, and the ethylene diamine derivatives account for 87% of the tonnage in the group; imidazoline derivatives account for the remaining 13%.

Amine oxides. Amine oxides were first patented in Germany in 1939. They were introduced into the United States in 1956 but did not attract widespread attention until 1961, when the first patents describing their use as surfactants were issued (86–87).

The amine oxide group is polar, with the highest electron density around the oxygen atom. Amine oxides exhibit strong hydrogen bonding tendencies and are hygroscopic. They are cationic in acid solutions, and nonionic in neutral or alkaline solution (88):

$$\mathrm{R}\overset{\overset{\mathrm{O}}{\uparrow}}{\mathrm{N}}(\mathrm{CH_3})_2 + \mathrm{H^+} \rightleftharpoons \mathrm{R}\overset{\overset{\mathrm{OH}}{|+}}{\mathrm{N}}(\mathrm{CH_3})_2$$

hydroxyammonium ion

In acidic media, amine oxides and anionic surfactants form precipitates; the CMC is much greater than in neutral or alkaline media. The change in CMC parallels the change from ionic to nonionic form. Amine oxides are stable in formulated detergent products. They do not act as oxidizing agents.

Composition and functions of representative commercial amine oxides are given in Table 27.

Amine oxides have attracted wide-spread interest as replacements for alkanolamides as foam builders in liquid hand-dishwashing compositions. Although considerably more expensive than alkanolamides, amine oxides have maintained their position in hand-dishwashing formulations over the last decade. In addition to foam boosting, amine oxides are effective shampoo surfactants, cotton detergents, wetting agents in concentrated electrolytes, and emulsifiers. They impart softness and body to fabrics and build viscosity in aqueous solutions. They find application in household and industrial detergents, detergent-sanitizers, plating-bath additives, and textile-mill processing formulations.

Amine oxides are marketed as pale-yellow aqueous solutions or light-yellow pastes of concentrations of 30–55 wt %. They are prepared by oxidation of tertiary amines with hydrogen peroxide (89). Yields are above 85%.

Alkyl amine ethoxylates. Ethoxylation is an economical route to obtain the variety of properties required by numerous and sometimes small-volume industrial uses of cationic surfactants. The commercial amine ethoxylates shown in Tables 28 and 29, are derived from linear alkyl amines, aliphatic *t*-alkyl amines, and rosin (dehydroabietyl) amines. Despite the variety of chemical structures, the amine ethoxylates tend to have similar properties. In general, they are yellow or amber liquids or yellowish low melting solids. Their specific gravities at room temperature range

Table 27. Amine Oxides, $R\overset{O}{\overset{\uparrow}{N}}(CH_3)_2$ **or** $R\overset{O}{\overset{\uparrow}{N}}(CH_2CH_2OH)_2$

Trade name	R	$(CH_3)_2$ or $(EO)_2$[a]	Concn, wt %	Functions[b]	Manufacturer[c]
Ammonyx CO	cetyl	$(CH_3)_2$	30	D, E, F, Fs, W	OXR
Ammonyx DMXD-40	lauryl	$(CH_3)_2$	40	D, F, Fs, W	OXR
Ammonyx MO	myristyl	$(CH_3)_2$	30	F, Fs, W	OXR
Ammonyx SO	stearyl	$(CH_3)_2$	25	E, H	OXR
Aromox C/12	coco	$(EO)_2$	50	E, Fs	AMK
Aromox DMCD	coco	$(CH_3)_2$	40	D, Fs	AMK
Aromox DMHTD	hydrogenated tallow	$(CH_3)_2$	40	Fs	AMK
Aromox DM16D	hexadecyl	$(CH_3)_2$	40	Fs	AMK
Aromox T/12	tallow	$(EO)_2$	50	E, Fs	AMK
Aromox 18/12	octadecyl	$(EO)_2$	50	E, Fs	AMK
Barlox 10S	decyl	$(CH_3)_2$	30	D, Em, Fs, V	LNZ
Barlox 12	lauryl	$(CH_3)_2$	30	D, Em, Fs, V	LNZ
Barlox 14	myristyl	$(CH_3)_2$	30	D, Em, Fs, V	LNZ
Conco XAC	cetyl	$(CH_3)_2$	30	D, Em, Fs, V	CTL
Conco XAL	lauryl	$(CH_3)_2$	30	D, Em, Fs, V	CTL
Conco XAM	myristyl	$(CH_3)_2$	30	D, Em, Fs, V	CTL
Conco XAS	stearyl	$(CH_3)_2$	30	D, Em, Fs, V	CTL
Ninox L	lauryl	$(CH_3)_2$	30	D, Em, Fs, V	STP
Ninox M	myristyl	$(CH_3)_2$	30	D, Em, Fs, V	STP
Schercamox DML	lauryl	$(CH_3)_2$	30	D, Em, Fs, V	SBC
Schercamox DMM	myristyl	$(CH_3)_2$	30	D, Em, Fs, V	SBC
Schercamox T-12	tallow	$(EO)_2$	50	Fs, Sp, W	SBC
Standamox O1	oleyl	$(CH_3)_2$	55	AM, Em, Ta, V	HNK
Varox 185E	C_{12-15} oxypropyl	$(EO)_2$	40	D, E, F, Fs	SHX
Varox 188E	C_{8-10} oxypropyl	$(EO)_2$	40	D, E, F, Fs	SHX
Varox 191E	C_{9-11} oxypropyl	$(EO)_2$	40	D, E, F, Fs	SHX
Varox 365	lauryl	$(CH_3)_2$	30	D, E, F, Fs	SHX

[a] $(EO)_2$ = bis(2-hydroxyethyl).
[b] See Table 1.
[c] See Table 2.

from 0.9 to 1.15, and they are soluble in acidic media. Higher ethoxylation promotes solubility in neutral and alkaline media. The lower ethoxylates form insoluble salts with fatty acids and other anionic surfactants. Salts of the higher ethoxylates are soluble, however. Oil solubility decreases with increasing ethylene oxide content but many ethoxylates with a fairly even hydrophilic–hydrophobic balance show appreciable oil solubility and are used as solutes in the oil phase.

Amine ethoxylates are used as emulsifiers, dispersants, antistats, textile softeners and lubricants, and as wetting agents in both acidic and alkaline solutions. As corrosion inhibitors, they are used in aqueous acid solutions, in refined petroleum products, and in formulations for use in the production and refining of petroleum. In addition, they are applied as components in mill-processing formulations for fabrication of metal and textile products, as wetting agents for asphalt (qv) and as frothing agents for ore flotation.

Manufacturing processes and equipment are similar to those employed for the preparation of alcohol ethoxylates. In the absence of steric hindrance, ethylene oxide reacts with both hydrogens of primary amines at relatively low temperatures (90–

Table 28. Aliphatic and Rosin Amine Ethoxylates, $RN\begin{smallmatrix}\nearrow [CH_2CH_2O]_m H \\ \searrow [CH_2CH_2O]_n H\end{smallmatrix}$

Trade name	Amine	$m + n$	Concn, wt %	Functions[a]	Manufacturer[b]
Ethomeen C/12	coco	2	100	A, Di, E, Ta	AMK
Ethomeen C/15	coco	5	100	A, Di, E, Ta	AMK
Ethomeen C/20	coco	10	100	A, Di, E, Ta	AMK
Ethomeen C/25	coco	15	100	A, Di, E, Ta	AMK
Ethomeen S/12	soybean	2	100	A, Di, E, Ta	AMK
Ethomeen S/15	soybean	5	100	A, Di, E, Ta	AMK
Ethomeen S/20	soybean	10	100	A, Di, E, Ta	AMK
Ethomeen S/25	soybean	15	100	A, Di, E, Ta	AMK
Ethomeen T/12	tallow	2	100	A, Di, E, Ta	AMK
Ethomeen T/15	tallow	5	100	A, Di, E, Ta	AMK
Ethomeen T/25	tallow	15	100	A, Di, E, Ta	AMK
Ethomeen 18/12	stearyl	2	100	A, Di, E, Ta	AMK
Ethomeen 18/15	stearyl	5	100	A, Di, E, Ta	AMK
Ethomeen 18/20	stearyl	10	100	A, Di, E, Ta	AMK
Kemamine AS 650	coco	2	100	A, Di, E, Ta	HUM
Kemamine AS 974	tallow	2	100	A, Di, E, Ta	HUM
Kemamine AS 989	oleyl	2	100	A, Di, E, Ta	HUM
Kemamine AS 990	stearyl	2	100	A, Di, E, Ta	HUM
Polyrad 0515	rosin	5	85	C	HPC
Polyrad 1110	rosin	11	90	C	HPC
Sipenol IC 10	coco	10	100	E, eP, I	AAC
Sipenol ISO5	soybean	5	100	E, eP, I	AAC
Sipenol O12	oleyl	12	100	E, eP, I	AAC
Triton RW-20	t-C_{12-14}[c]	2	100	D, E, OR	RH
Triton RW 75	t-C_{12-14}[c]	7.5	100	D, E, OR	RH
Triton RW-150	t-C_{12-14}[c]	15	100	D, E, OR	RH
Trymeen CAM-10	coco	10	100	A, E	EMR
Trymeen OAM-30/80	oleyl	30	80	Ta	EMR
Trymeen SAM-50	stearyl	50	100	A, Ta	EMR
Trymeen TAM-15	tallow	15	100	Ta, Ta	EMR
Trymeen TAM-20	tallow	20	100	A, Sp, Ta	EMR
Trymeen TAM-25	tallow	25	100	A, Ta	EMR
Trymeen TAM-40	tallow	40	100	A, E, Ta	EMR

[a] See Table 1.
[b] See Table 2.
[c] Branched-chain.

120°C) without added catalysts (90). When the nitrogen atom is hindered, as it is in the Triton RW products, only one of the amino hydrogens reacts with ethylene oxide. Once this reaction is complete, a basic catalyst is added and ethoxylation proceeds in the manner of the alcohol-based nonionics. In *N*-alkyl-1,3-propanediamine, all three amino hydrogens are available for reaction with ethylene oxide. *N*-Alkyl-1,3-propanediamines are prepared from fatty monoamines and acrylonitrile, followed by reduction of the resulting 3-cyanoethylalkyl amine.

2-Alkyl-1-(2-Hydroxyethyl)-2-Imidazolines. Compositions and functions of typical commercial products in this series are given in Table 30. 2-Alkyl-1-(2-hydroxyethyl)-2-imidazolines are used in hydrocarbon and aqueous systems as an-

Table 29. Fatty Alkyl-1,3-Propanediamine Ethoxylates, 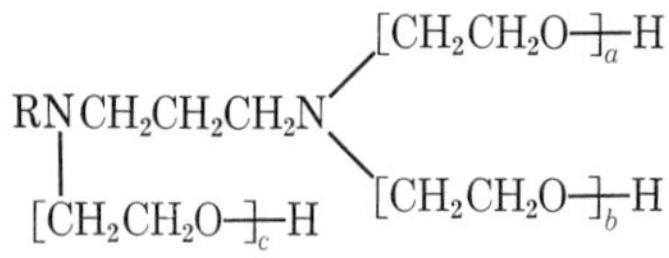

Trade name	R	$a + b + c$	Concn, wt %	Functions[a]	Manufacturer[b]
Ethoduomeen T/12	tallow	2	100	E	AMK
Ethoduomeen T/13	tallow	3	100	Di, E	AMK
Ethoduomeen T/20	tallow	10	100	Di, E, W	AMK
Ethoduomeen T/25	tallow	15	100	Di, E, W	AMK

[a] See Table 1.
[b] See Table 2.

Table 30. 2-Alkyl-1-(2-Hydroxyethyl)-2-Imidazolines, 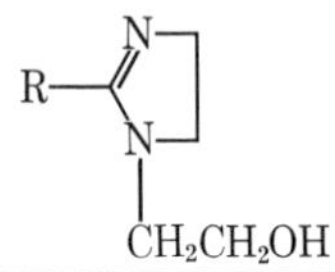

Trade name	Fatty acid source of RC<	Concn, wt %	Functions[a]	Manufac-turer[b]
Finazoline AC	coco	90	A, C, E, Sp, Ta	FET
Hodag Amine C-100-L	lauric	100	C, D, E	HDG
Hodag Amine C-100-O	oleic	100	C, D, E	HDG
Hodag Amine C-100-S	stearic	100	C, D, E	HDG
Monazoline CY	caprylic	90	A, AM, C, D, E, Sp, V	MOA
Monazoline C	coco	90	A, AM, C, D, E, Sp, V	MOA
Monazoline O	oleic	90	A, AM, C, D, E, Sp, V	MOA
Monazoline T	tall oil	90	A, AM, C, D, E, Sp, V	MOA
Varine C	coco	100	A, C, E	SHX
Varine O	oleic	100	A, C, E,	SHX
Witcamine AL 42-12	tall oil	85	A, C, E, I	WTC
Witcamine PA-78B	stearic	100	C, Di	WTC

[a] See Table 1.
[b] See Table 2.

tistats, corrosion inhibitors, detergents, emulsifiers, softeners, and viscosity builders. They are prepared by heating the salt of a carboxylic acid with (2-hydroxyethyl)ethylenediamine at 150–160°C to form a substituted amide; 1 mol water is eliminated to form the substituted imidazoline with further heating at 180–200°C.

Substituted imidazolines yield three series of cationic surfactants: by ethoxylation to form more hydrophilic products; quaternization with benzyl chloride, dimethyl sulfate, and other alkyl halides; and oxidation with hydrogen peroxide to amine oxides.

Ethylenediamine alkoxylates. The reaction 1,2-alkylene oxides with ethylenediamine forms the basis of a series of surfactants of the general structure:

where x and y each varies from ca 4–100. As in the oxygen block polymer surfactants, the polyoxypropylene chain contributes hydrophobicity and the polyoxyethylene moiety hydrophilicity. The two tertiary nitrogens contribute cationic character in the lower molecular weight polymers but are so highly shielded in the higher molecular weight range that the products behave essentially as nonionic compounds.

A number of these structures are offered commercially by BASF Wyandotte Corp. under the trade name of Tetronic polyols. The products are similar to the oxygen block polymers. Although not strongly surface active *per se*, they are useful as detergents, emulsifiers, demulsifiers, defoamers, corrosion inhibitors, and lime-soap dispersants. They are reported to confer antistatic properties to textiles and synthetic fibers.

Amines with Amide Linkages. Representatives of this group are prepared from carboxylic acids and di- and polyamines. The amide linkage connects the amine to relatively inexpensive hydrophobes. Tall oil fatty acid diethylenediamine and polyalkylene polyamine condensates account for 56% of the 1980 U.S. production of 18,000 t. Formulas for typical amide amines are shown below:

$$RCONHCH_2CH_2NHCH_2CH_2NH_2 \text{ or } RCON(CH_2CH_2NH_2)_2$$

where R is derived from coconut, oleic, stearic, or tall oil acids,

$$RCONHCH_2CH_2CH_2N(CH_3)_2,$$
$$RCONHCH_2CH_2CH_2NHCH_2CH_2CH_2NH_2, \text{ or } RCON(CH_2CH_2CH_2NH_2)_2,$$
$$RCONHCH_2CH_2NHCH_2CH_2NHCH_2CH_2NHCH_2CH_2NH_2$$

where R is derived from stearic acid.

Ethoxylation of these structures leads to surfactants more soluble in water. Amide amines are dark-colored liquids that find application in corrosion inhibitors, petroleum demulsifiers, and metal-processing formulations.

Quaternary Ammonium Salts. The quaternary ammonium ion is a much stronger hydrophile than primary, secondary, or tertiary amino groups, strong enough to carry into solution a hydrophobe in the surfactant molecular weight range, even in alkaline media. The discrete positive charge on the quaternary ammonium ion promotes strong adsorption on negatively charged substrates, such as fabrics, and is the basis for the wide-spread use of these surfactants in domestic fabric-softening compositions.

In 1980, ca 43,400 t of quaternary ammonium salts were produced in the United States, slightly more than twice the 1966 production. Bis(hydrogenated tallowalkyl)dimethylammonium chloride, the surfactant used in household softeners, accounted for 56% of the total. The second-largest volume type is represented by alkylbenzyldimethylammonium chlorides (benzalkonium chlorides) with wide-spectrum antimicrobial activity. Representative commercial quaternary surfactants are given in Tables 31–34.

In addition to the structures shown in the tables, commercial quaternary surfactants include a variety of modifications, prepared, eg, by quaternization of simple tertiary amines:

$$RN(CH_2CH_2OH)_2 + R'X \longrightarrow R'R\overset{+}{N}(CH_2CH_2OH)_2\ X^-$$

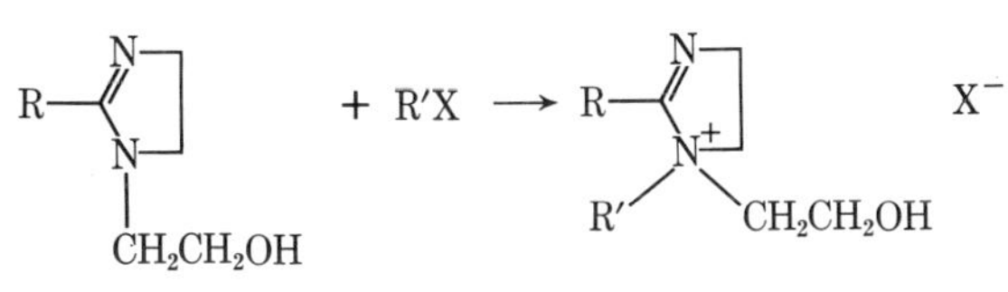

Table 31. Dialkyldimethylammonium Salts, $RR'\overset{+}{N}(CH_3)_2\ A^-$

Trade name	R and R′	A^-	Concn, wt %	Functions[a]	Manufacturer[b]
Adogen 432 CG	C_{12-18}	Cl	65	H, Sp	SHX
Adogen 442	hydrogenated tallow	Cl	75	Sp	SHX
Adogen 462	coco	Cl	75	C, Di, E	SHX
Adogen 470	tallow	Cl	75	Sp	SHX
Arosurf TA100	C_{16-18}	CH_3SO_4	95	H, Sp	SHX
Arquad 2C-75	coco	Cl	75	C, E	SHX
Arquad 2S-75	soya	Cl	75	C, E	SHX
Arquad 2HT-75	stearyl	Cl	75	Sp	SHX
Kemamine Q9702C	hydrogenated tallow	Cl	75	E, H, Sp	HUM
Kemamine Q6502C	coco	Cl	75	C, Di, E	HUM
Kemamine Q2802C	behenyl	Cl	75	H, Sp	HUM
Kemamine Q2982C	erucyl	Cl	75	H, Sp	HUM
Variquat K300	coco	Cl		C, E	SHX
Varisoft 137	hydrogenated tallow	CH_3SO_4	90	Sp	SHX
Varisoft DHT	hydrogenated tallow	Cl	75	H	SHX

[a] See Table 1.
[b] See Table 2.

Table 32. Alkylbenzyldimethylammonium Chlorides (Benzalkonium Chlorides), $R(C_6H_5CH_2)\overset{+}{N}(CH_3)_2\ Cl^-$

Trade name	R	Concn, wt %	Functions[a]	Manufacturer[b]
Barquat MB-50	myristyl	50	AM	LNZ
Barquat OJ-50	oleyl	50	AM	LNZ
BTC 50	$C_{12-14-16-18}$	50	AM	OXR
BTC 824	$C_{14-16-18}$	50	AM	OXR
Ammonyx 4002	stearyl	94	AM	OXR
Ammonyx CETAC	cetyl	25	AM	OXR
Ammonyx KP	oleyl	50	AM	OXR
Dibactol	myristyl	100	AM	HXL
Hyamine 3500	$C_{12-14-16}$	50	AM	RH
Kemamine BQ 2802C	behenyl	75	A, E, H	HUM
Kemamine BQ 9742C	tallow	75	Di, dE	HUM
Roccal II	C_{8-18}	10	AM	NLS
Stedbac	stearyl	100	H	HXL
Variquat 50ME	$C_{12-14-16}$	50	AM	SHX
Variquat B35	hydrogenated tallow	80	E, I	SHX
Varisoft SDC	stearyl	25	AM, E, H	SHX
Zephiran Chloride	alkyl	17	AM	SDW

[a] See Table 1.
[b] See Table 2.

and of amide-amines:

$$RCONHCH_2CH_2CH_2N(CH_3)_2 + R'X \rightarrow RCONHCH_2CH_2CH_2\overset{+}{N}R'(CH_3)_2\ X^-$$

Amphoteric Surfactants

Amphoteric surfactants contain both an acidic and a basic hydrophilic group. These ionic functions may be any of the anionic or cationic groups described in the

Table 33. Alkyltrimethylammonium Salts, $R\overset{+}{N}(CH_3)_3\ A^-$

Trade name	R	A^-	Concn, wt %	Functions[a]	Manufacturer[b]
Acetoquat CTAB	cetyl	Br	95	AM	ACI
Ammonyx CTAC	cetyl	Cl	25	Em, H, Sp	OXR
Bromat	cetyl	Br	100	AM	HXL
Kemamine Q9743C	tallow	Cl	75	AM, dE, Sp	HUM
Mytab	myristyl	Br	100	AM	HXL

[a] See Table 1.
[b] See Table 2.

Table 34. Alkylpyridinium Halides, 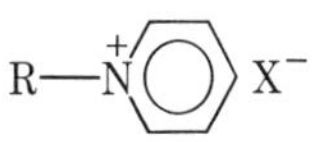

Trade name	R	X^-	Concn, wt %	Functions[a]	Manufacturer[b]
Acetoquat CPC	cetyl	Br	95	AM	ACI
Acetoquat CPB	cetyl	Cl	100	AM	ACI
Dehyquart C	lauryl	Cl	90–94	A, E, H	HNK

[a] See Table 1.
[b] See Table 2.

preceding sections. In addition, ether or hydroxyl groups may also be present to enhance the hydrophilicity of the surfactant molecule. Examples of amphoteric surfactants include amino acids and their derivatives in which the nitrogen atom tends to become protonated with decreasing pH of the solution. Amino acid salts, under these conditions, contain both a positive and a negative charge on the same molecule. In alkylbetaines, $R\overset{+}{N}(CH_3)_2CH_2CO_2^-$, discrete opposing charges are present in the molecule at all pH values.

Amphoteric surfactants are generally considered specialty surfactants. Over the last decade, however, their usage has expanded significantly. Compared to the 1966 U.S. production of 2268 t of amphoteric surfactants, 11,661 t were produced in 1980, of which 64% were imidazolinium derivatives.

Amphoteric surfactants do not irritate skin and eyes; they exhibit good-surfactants properties over a wide pH range; and they are compatible with anionic and cationic surfactants.

A basic nitrogen and an acidic carboxylate group are the predominant functional groups. In a few structures, sulfonate or sulfate provide the negative charge.

Among the simplest amphoteric structures are the alkylbetaines, prepared from alkyldimethylamines and sodium chloroacetate. Relatively inexpensive alkyldimethylamines containing an amide group can be prepared from fatty acids and 3-dimethylaminopropylamine. Reaction with sodium chloroacetate yields acylamidopropylbetaine. Commercial alkyl betaines and their amidopropyl analogues are described in Tables 35–36.

Imidazolinium Derivatives. Amphoteric imidazolinium derivatives are prepared from the 2-alkyl-1-(2-hydroxyethyl)-2-imidazolines and from sodium chloroacetate. The most likely structure of the reaction product is shown below (91):

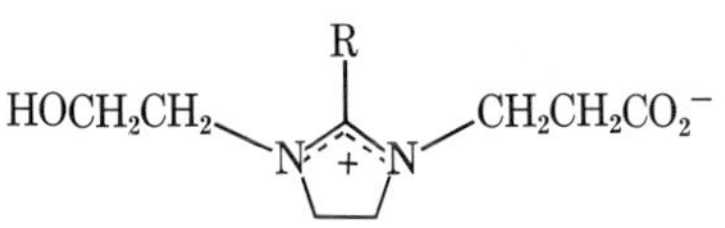

Table 35. Alkylbetaines, $R\overset{+}{N}(CH_3)_2CH_2CO_2^-$

Trade name	R	Concn, wt %	Functions[a]	Manufacturer[b]
Emcol CC 37-18	coco	45	D, E, F, Fs, V	WTC
Lonzaine 10S	decyl	41	E, F, H	LNZ
Lonzaine 12C	coco	32	E, F, H	LNZ
Lonzaine 16S	cetyl	31	E, F, H	LNZ
Product BCO	cetyl		Di, W	DUP
Rewopon Am-DML	lauryl	40	D	EMR
Varion CDG	coco	35	D, E, F, V	SHX
Velvetex AB-45	coco	45	H, V	HNK
Velvetex OLB-50	oleyl	50	H	HNK

[a] See Table 1.
[b] See Table 2.

Table 36. Amidopropylbetaines, $RCONHCH_2CH_2CH_2\overset{+}{N}(CH_3)_2CH_2CO_2^-$

Trade name	RCO	Concn, wt %	Functions[a]	Manufacturer[b]
Aerosol 30	cocoyl	35	F, H	ACY
Amphosol CA	cocoyl	30	D, F, Fs	STP
Carsonam 32	cocoyl	59	D, F, Fs	CRS
Lonzaine CO	cocoyl	35	D, E, F	LNZ
Maprolyte C	cocoyl	35	F, St, W	OXR
Monateric CAB	cocoyl	35	D, E, H, W	MOA
Monateric LMAB	lauroyl	35	D, E, H, W	MOA
Schercotaine CAB	cocoyl	35	D, F, W	SBC
Schercotaine IAB	isostearyl	35	C, Sp, V	SBC
Schercotaine MAB	myristoyl	35	A, D, V, W	SBC
Schercotaine PAB	palmitoyl	35	C, V	SBC
Varion CADG	lauroyl	32	D, E, F, V	SHX
Velvetex BC 35	cocoyl	35	A, D, F, Fs, H	HNK

[a] See Table 1.
[b] See Table 2.

Unlike the 2-alkyl-2-imidazolines, this structure is stable and resistant to hydrolysis. After ring cleavage, reaction with sodium chloroacetate yields linear products:
and

$$RCONHCH_2CH_2NHCH_2CH_2OH + ClCH_2CO_2Na \longrightarrow RCONHCH_2CH_2N(CH_2CH_2OH)CH_2CO_2Na$$

$$RCONHCH_2CH_2NHCH_2CH_2OH + 2ClCH_2CO_2Na \longrightarrow RCONHCH_2CH_2\overset{+}{N}(CH_2CO_2Na)(CH_2CO_2^-)CH_2CH_2OH$$

Commercial imidazolinium derivatives are for the most part not clearly characterized as to their structure and are frequently marketed as proprietary products. As a result, it is difficult to relate properties and uses to structural parameters. Imidazolinium derivatives are recommended as detergents, emulsifiers, wetting and hair conditioning agents, foaming agents, fabric softeners, and antistatic agents. There is some evidence that in cosmetic formulations, certain imidazolinium derivatives reduce eye irritation caused by sulfate and sulfonate surfactants present in these products.

Production and Economic Aspects

Total production of 100% active surfactants of all types amounted to almost 2×10^6 t in 1980, a 48% increase over 1966. Of this total, 892,000 t or 45.7% was either converted into formulated products by the manufacturers or consumed in their processes. The remaining 1.06×10^6 t was marketed for a value of ca $\$1.42 \times 10^9$. By far the greater part of the surfactants sold to industrial customers was formulated by the purchaser and resold at lower concentration of active ingredient but at higher mark-up. Precise statistics on sales and consumption are not easily obtained since different sources include different product categories in their calculations. According to one estimate (92), the total consumption of formulated soap and detergent products in North America in 1980 amounted to ca 6.2×10^6 t which corresponds to a surfactant content of about 17% in the finished products. Another estimate places the value of finished surfactant products, at the manufacturing level, at $\$7.5 \times 10^9$. Table 37 lists the 1980 production and sales of surfactant types on a 100% active basis.

Uses

The literature on surfactant technology tends to be dominated by household and personal products. The sales volume per product in these category is large enough to justify research and investment in large manufacturing facilities. Industrial applications represent smaller sales volumes per product. Research and especially technical service is often conducted by the user or by a service company that supplies a variety of items to the consuming industry.

Household and Personal Products. Detergency is the primary function of household and personal products. In recent years, a secondary function, such as softening in combination with detergency in laundry detergents or conditioning in combination with detergency in shampoos, has been offered as an additional product benefit. In general, products have tended toward functional specialization. An outline is given in Table 38.

Industrial Uses. Surfactants are widely used outside the household for a variety of cleaning and other purposes. Often the volume or cost of the surfactant consumed in industrial processes is small compared to benefit. Although usage of surfactants in industrial applications is smaller than in the large household products market, expansion of enhanced oil recovery activities could consume sufficient surfactant to tip the balance in favor of industrial uses (see Table 38).

DETERGENCY AND DETERSIVE SYSTEMS

The cleaning of a solid object, ie, the removal of unwanted foreign matter from its surface, is accomplished by a variety of methods ranging from simple mechanical separation such as blotting or abrasion to removal by solution or selective chemical

Table 37. U.S. Production and Sales of Surfactants, 1980[a]

Surfactant	Production, 1000 t	Sales: Quantity, 1000 t	Sales: Value, $1000	Average unit price, $/t
anionic				
carboxylic acids and salts	384.2	65.437	67.025	1024
sulfonic acids and salts				
lignosulfonates[b]	399.1			
alkylbenzenesulfonates	294.2	88.020	95.544	1085
alkylarylsulfonates, short chain[c]	34.0			
naphthalenesulfonates	9.7	8.2	12.1	1464
petroleum sulfonates	191.909[d]	156.3	210.7	1348
sulfonates with ester, ether, or amide linkages	34.207	16.4	44.7	2722
all other sulfonates	9.839	9.8	12.4	1269
sulfuric acid esters and salts				
alcohols, sulfated	108.9	24.9	56.0	2253
alcohols, ethoxylated and sulfated	108.4	44.1	48.5	1100
alkylphenols, ethoxylated and sulfated	2.5	2.3	5.4	2372
acids, amides, and esters, sulfated	10.6	8.5	10.6	1246
natural fats and oils, sulfated[e]	5.9	3.9	3.6	922
phosphoric and polyphosphoric acid esters (and salts)		12.621	21.7	1719
alcohols and phenols, alkoxylated and phosphated (and their salts)[f]	9.9			
all other phosphoric and polyphosphoric acid esters (and their salts)	7.1			
all other anionic	31.1	14.9	11.4	763
Total anionic	*1641.5*	*455.3*	*599.6*	*1317*
nonionic				
alcohols, ethoxylated	254.6	204.225	215.3	1054
alkylphenols, ethoxylated[g]	156.7	138.0	150.9	1093
carboxylic esters, ethoxylated		89.1	140.7	1579
glycerol esters	30.8			

polyethylene glycol esters	27.6			
anhydrosorbitol esters	13.2			
ethoxylated anhydrosorbitol and sorbitol esters	11.9			
natural fats and oils, ethoxylated	8.5			
ethylene and diethylene glycol esters	3.4			
propanediol esters	1.1			
all other carboxylic acid esters	16.4			
carboxylic amides, ethoxylated	28.3	22.7	39.7	1743
all other nonionic	53.5	20.7	52.2	1700
Total nonionic[h]	*606.0*	*484.7*	*598.8*	*1235*
cationic				
oxygen-free amines	36.0	33.0	64.930	1950
oxygen-containing amines[i]	34.3	11.9	22.926	1927
amide-linked amines	18.0	13.4	22.547	1680
quaternary ammonium salts	43.4	41.5	72.8	1753
all other cationic	9.3	7.9	16.406	2078
Total cationic	*141.0*	*107.7*	*199.5*	*1852*
amphoteric	11.7	11.2	23.5	2100
Grand total	*2400.2*	*1058.9*	*1421.4*	*1342*

[a] Data based on Tables XII-I and XIV-I of ref. 70 (USITC).
[b] Strictly speaking, not surfactants, but used as dispersing and ore-flotation agents.
[c] Strictly speaking, not surfactants, but used as coupling agents, solubilizers, and hydrotropes.
[d] Estimated to be understated USITC by 56,000 t.
[e] Includes castor, cod, herring, neat's foot oils, and tallow only. Additional sulfated vegetable oils are included under all other anionic surfactants. Precise figures are difficult to establish.
[f] Alkoxylated phenol phosphate esters, 5219 t.
[g] Nonylphenol ethoxylates are 73.6% of this amount.
[h] The total production of nonionics has doubled in the last 15 years. Except for carboxylic amides, which suffered a 30% decline, all groups participated in the growth led by alcohol ethoxylates, which tripled in production.
[i] Including amine oxides.

Table 38. Surfactant Uses

Use	Function
household products[a]	cleansing, abrasion, softening
detergents, scouring powders, hard-surface cleaners, fabric softeners	
personal-care products[a]	cleansing, moisturizing, softening
soaps, shampoos, creams	
agriculture	
phosphate fertilizers	to shorten manufacturing cycle and prevent caking during storage
spray applications of herbicides, insecticides, and fungicides	to wet, disperse, and suspend powdered pesticides and emulsification of pesticide solutions; to promote wetting, spreading, and penetration of the toxicant
building and construction	
paving	to improve the bond of asphalt to gravel and sand, ie, prevent stripping
concrete	to promote air entrainment for control of density, plasticity, insulating properties, etc
elastomers and plastics	
emulsion polymerization	solubilization: monomer and catalyst are dissolved and react in surfactant micelles; emulsification of monomers; stabilization of latexes
foamed polymers	introduction of air and control of cell size
latex adhesive	to promote wetting and thus improve bond strength
plastic articles	antistatic agents
plastic coating and laminating	wetting agents
food and beverages	
food-processing plants	to clean and sanitize walls, floors, and processing equipment
fruits and vegetables	to remove pesticide residues and aid in wax coating
bakery products and ice cream	to solubilize flavor oils, control consistency, and retard staling
beverages	to solubilize flavor oils
crystallization of sugar	to improve washing and reduce processing time
cooking fats and oils	to prevent spattering in frying because of superheating and sudden volatization of water
industrial cleaning	
janitorial supplies	as detergents and sanitizers for walls, floors, and windows
miscellaneous cleaning	as detergents for railroad cars, trucks, airplanes, tunnel walls, engines, etc
descaling	as wetting agents and corrosion inhibitors in acid cleaning of boiler tubes and heat exchangers
soft goods	as detergents for laundry and dry cleaning
wax strippers	to improve wetting and penetration of the old finish
leather	
skins	as detergent and emulsifier in degreasing
tanning	to promote wetting and penetration
hides	as emulsifiers in fat-liquoring
dyeing	to promote wetting, penetration, and level dyeing
metals	
concentration of ores	wetting and foaming, ie, collectors and frothers in ore flotation (qv)
cutting and forming	wetting, emulsification, lubrication, and corrosion inhibition in rolling oils, cutting oils, drawing lubricants, buffing and grinding compounds
casting	as mold release additives
rust and scale removal	wetting, foaming, and corrosion inhibition in pickling (ie, acid cleaning) and electrolytic cleaning

Table 38 *(continued)*

Use	Function
grease removal	emulsification and wetting in emulsion cleaning preparations
plating	wetting and foaming in electrolytic plating baths
mining	
mining and transporting coal and minerals	wetting and dedusting
paper	
pulp treatment	deresinification, pitch dispersion, and washing
paper machine	defoaming, felt washing, color leveling, and dispensing
calender	wetting and leveling in coating and coloring operations
towels and pads	rewetting, ie, improve absorption of moisture
paints and protective coating	
pigment preparation	flushing, ie, promote preferential wetting by the paint vehicle: dispersing and wetting of the pigment during grinding
latex paints	to emulsify the oil or polymer, disperse the pigment, stabilize the latex, retard sedimentation and pigment separation, modify wetting and rheological properties
waxes and polishes	emulsifying waxes, stabilize emulsions and wet substrates in finishes for floor and automobiles; antistat
petroleum production and products	
drilling fluids	to emulsify oils, disperse solids, and modify rheological properties of drilling-and-completion fluids for oil and gas wells
mist drilling	to convert intrusion water to foam in air drilling
(see workover of producing wells)	to emulsify and disperse sludge and sediment in clean-out wells; modify wetting of formation at producing zone
secondary and tertiary (enhanced) recovery	in flooding operations, to release crude oil from the formation surface, ie, preferential wetting
refined petroleum products	as detergent, sludge dispersant, and corrosion inhibitor in fuel oils and turbine oils
textiles	
preparation of fibers and filaments	as detergent and emulsifier in raw-wool scouring; dispersant in viscose rayon spin baths; lubricant and antistat in spinning of hydrophobic filaments
gray-goods preparations	wetting and detergency in slashing and sizing formulations; wetting and detergency in kier boiling and bleaching of cotton and carbonizing of wool; detergency in scouring piece goods; emulsification of processing oils
dyeing and printing	wetting, penetration, solubilization, emulsification, dye leveling, detergency, and dispersion
finishing of textiles	wetting and emulsification in finishing formulations; softening, lubricating and antistatic additives to finishes

[a] Includes an extensive list of consumer products.

action. The term detergency is restricted to cleaning systems in which the following conditions prevail: A liquid bath is present and is the main cleansing constituent of the system. The action of the bath involves more than simple solution or simple hydraulic dislodging of soil although both of these processes inevitably occur and make some contribution to cleaning. The enhanced cleaning effect is caused primarily by the presence in the bath of a special solute, the surfactant, which acts by altering interfacial effects at the various phase boundaries within the system. A typical detersive system thus consists of a solid object to be cleaned, called the substrate; soil or dirt attached to it which is to be removed in the washing process; and a liquid bath that is applied to the soiled substrate. Each of these elements is usually a complex entitity whose property and composition can vary widely. The final cleaning effect is the result of the interaction of these elements and the processing conditions employed, ie, temperature, time, mechanical energy input (agitation) and, in the case of aqueous baths, the presence of hardness ions in the water.

In the cleaning or washing process in a typical detersive system the soiled substrate is immersed in or brought into contact with a large excess of the bath liquor. Sufficient bath is used to provide a thick layer over the whole surface of the substrate. During this stage, air is displaced from soil and substrate surfaces, ie, they are wetted by the bath. The system is subjected to mechanical agitation, either rubbing or shaking, which provides the necessary shearing action to separate the soil from substrate and disperse it in the bath. Agitation also aids the mass-transfer processes in the system, just as in any other heterogeneous chemical reaction. The fouled bath carrying the removed soil is drained, wiped, squeezed, or otherwise removed from the substrate. The substrate is rinsed free of the remaining fouled bath. This rinsing step determines the final cleanliness of the substrate. The cleaned substrate is dried or otherwise brought to the desired finished state.

A rational discussion of detergency requires some definition of clean. In the physicochemical sense, a surface is clean if it contains no molecular species other than those in the interior of the two adjoining phases. Practically, such a state is reached only with great difficulty even under the most exacting laboratory conditions. In a practical sense, a surface is clean if it has been brought to a desired prespecified state with regard to foreign matter present upon it, as judged by some set of agreed-upon tests or standards. Household linen, for example, is judged clean when it is free of visible soil even though it may carry an appreciable burden of starch and softening finish. In the dyehouse of a textile mill, a piece of goods such as this would be rejected as dirty and returned for scouring because these finishes interfere with dyeing. Most standards for cleanness involve a visual or optical judgment as to the presence of foreign matter. In some systems, in the desizing of cotton, for example, the degree of cleanness may be specified by weight percentage of soil on the substrate. In other systems, such as the degreasing of metal, it is the weight of soil per unit area of substrate surface that specifies cleanness. In washing dishes or glass ware, cleanness is often specified by complete water wettability or freedom from water break, as well as by visual clarity (see also Metal surface treatment; Textiles).

Although it is impossible to describe or even to enumerate all the practical detersive systems that might be encountered, a large proportion can be assigned to a relatively small number of classes. This classification disregards surfactant structure and type of substrate (fibrous or hard surface) and is restricted to a consideration of the type of soil present on the substrate, the type and degree of mechanical action employed, the bath ratio, and the nature of the detergent used. Some of the more commonly encountered detersive systems are classified on this basis in Table 39.

Components of Detersive Systems

Substrates. Solid objects to be cleaned by detersive processes, vary widely in chemical composition and surface configuration. With relatively few exceptions, however, they can be divided into fabrics and fibrous materials, and hard surfaces. Fabrics present a highly complex configuration capable of holding soil mechanically even though the soil may be physicochemically removed from the surface. Most fabrics are organic polymeric materials which may be swellable by water or permeable to small molecules and ions dissolved in the bath. Many common polymeric fabrics, cotton, polyester, rayon, nylon, wool and cellulose esters, contain ionogenic or polar centers capable of localizing electric charges. Hard surfaces, on the other hand, are relatively flat and smooth. They cannot mechanically hold soil that has been detached by physicochemical action. In general, they are impermeable to water and water-soluble materials although they vary widely in their wettability and polarity. The most important types are glass, metal, and organic polymeric materials which include painted surfaces, linoleum, and plastic tile.

Soils. Soils vary greatly in composition. They may consist of a single solid or liquid phase but usually two or more phases are present, intimately and randomly mixed and irregularly disposed over the substrate surface. In a surprisingly large number of important detersive systems, the nature of soil and the quantity present are well known. This is the case, for example, in most textile mill operations such as raw wool scouring, the boil-off and scouring of loom-state woven goods, and the soaping of printed cottons. In the cleaning of fabricated metal parts to remove forming and drawing lubricants, buffing compounds, etc, the nature of the soil is well known. The behavior of soils encountered in dishwashing is well characterized.

As a result of many painstaking investigations, the soils on apparel encountered in laundering have been shown to be complex mixtures containing both oily and finely divided solid material (93). The oily material consists largely of fatty acids and polar fatty material but a considerable proportion of neutral nonpolar oil is also present. The solid components vary widely with the locale in which samples are taken and tend to resemble local street dust in composition.

Particle size is one of the most important factors in determining the ease with which solid soil can be removed from a substrate. Particles of >5-μm dia are generally easily removed. Particles of <10-nm dia cannot be removed by ordinary detersive processes once they are attached to a typical textile fabric. Such particles are responsible for the gradual irreversible graying of white goods with continued wear and laundering. Particles in this size range tend to form clumps and clusters before they reach the fiber surface. These clusters behave like individual large particles. Particles or clusters in the range of 100-nm dia resist removal by simple agitation in liquids that are not surface active, but these particles are removable by normal detersive processes. This is the size range of greatest interest.

Soil may include material that is soluble in the bath, such as encrusted sugar residues and molecularly dispersed material such as fruit juice stains. Removal of these soils is an important aspect of cleaning but is not generally considered in discussions of detergency.

Table 39. Detersive Systems

System	Soil	Mechanical action	Bath ratio	Detergent
fabric and fibrous				
textile and allied manufacturing				
raw wool scouring	liquid at operating temperature	very gentle	high	organic surfactant
wool yarn and piece goods	liquid	gentle	moderately high	organic surfactant
scouring gray cottons	mostly solid; waxy and starchy	vigorous	low	built surfactant
scouring manmade fabrics	mixed; oily lubricant and sizing residues	moderate	medium	built surfactant
de-inking paper	mixed; oily and pigment	vigorous	high	heavily built surfactant
laundering				
household	mixed and variable	moderate	high	unbuilt or built surfactant
commercial and industrial	mixed and variable; heavier than household soil	vigorous	high	built surfactant
rug cleaning				
in plant	heavy solid	moderate gentle	medium to low	organic surfactant, may be built
on location	heavy solid	vigorous; superficial brushing	very low; foam bath	organic surfactant foam
hard surfaces				
glass and ceramics				
hand dishwashing	mixed oily and solid organic	moderate to vigorous	high	organic surfactant
machine dishwashing	mixed oily and solid organic	vigorous hydraulic	high	inorganic[a]
commercial bottle washing	light solid organic	vigorous hydraulic	high	inorganic[a]
metals				
prefinishing cleaning	oily	moderate to gentle hydraulic	high	inorganic[a]
postforming emulsion cleaning	oily and mixed	gentle hydraulic	high	mixed inorganic and surfactant
cleaning metal structures and equipment, tanks, dairy equipment, etc	variable; mostly oily or organic solid	usually vigorous rubbing, sometimes hydraulic	usually low in wash cycle; may be high in rinse cycle	inorganic[a], built surfactant
organic surfaces, paint, linoleum, plastic tile	mixed solid, organic solid, and oily	usually vigorous rubbing	low	lightly to moderately built surfactant
cosmetic and personal cleansing				
shampooing	oily	vigorous rubbing	low in wash cycle; high in rinse cycle	organic surfactant

Table 39 *(continued)*

System	Soil	Mechanical action	Bath ratio	Detergent
bathing and washing	mixed, mostly oily	mild to vigorous rubbing	high; sometimes low in wash cycle	organic surfactant

[a] Organic surfactant frequently added to aid wetting and draining.

Baths. The baths discussed here are aqueous solutions. Most nonaqueous cleaning systems, such as metal decreasing operations, depend entirely on solvent action and therefore cannot be considered examples of detersive systems. Some nonaqueous systems, however, are true detersive systems. Modern dry-cleaning baths, for example, contain solutes that are surface active in the conventional hydrocarbon or chlorinated hydrocarbon medium and contribute to soil removal. The physical chemistry of such systems differs considerably from that of aqueous systems (see Drycleaning). Among bath components the solute that is effective in cleaning, usually a mixture of several components, is called the detergent. The term *detergent* is also used frequently in the restricted sense of a surfactant of high detersive power. In many hard-surface systems, however, nonsurfactants such as alkaline silicates and phosphates exert a true detersive effect. They are, in fact, the principal detergents in these systems even in the complete absence of any surfactants. In the cleaning of fabric systems, the most important detersive component in the bath is the surfactant. Nonsurfactant components that augment the cleaning effect of surfactants in these systems are called *builders.* Many materials that act as builders in fabric systems, eg, phosphates and silicates, are primarily detergents in hard-surface systems although there is no significant theoretical connection between the two effects.

Formulation

Detergents are formulated to satisfy the requirements of the soiled substrate to be cleaned and the expected range of washing conditions. Some detergents, the familiar bar or toilet soap, for example, consist essentially of only one component. There are relatively few cleansing systems, however, in which a suitably formulated detergent consisting of several components does not outperform the best single-component system. Detergents for hand dishwashing are essentially mixtures of surfactants and rarely contain builders; ingredients of laundering detergent formulations for fabrics may be divided into the following groups: Surfactant, including soap and the various surfactants discussed in the previous section. Inorganc salts, acids, and bases, including builders, and other compounds that do not contribute detergency but provide other functions, such as regulating density, assuring crispness of powdered formulations, etc. Organic additives that enhance detergency, foaming power, emulsifying power, or soil-suspending effect of the composition. Special purpose additives, such as bleaching agents (qv), fluorescent whitening agents, antimicrobial agents, bluing agents, or starch which provide desirable performance functions but have no direct effect on soil removal (see also Brighteners, optical; Industrial antimicrobial agents).

Fabric detergency formulations for special applications, such as the various specific operations within the textile mill, are frequently much simpler. They tend to contain little if any builder or special-purpose additive. The indispensable ingredient in fabric detergency is the organic surfactant. Formulations for hard-surface detergency, such as those used in automatic machine dishwashing, are also simpler than fabric-washing compositions. An organic surfactant is frequently not needed and inorganic salts are the detersive ingredients.

Surfactants. The most important components of detersive systems are, of course, the surfactants described earlier.

Builders. Builders are substances that augment the detersive effects of surfactants. Most important is the ability to remove hardness ions from the wash liquor and thus to prevent them from interacting with the surfactant. Such interaction reduces detersive effectiveness. Hardness ions are best removed by sequestration to form soluble chelates (see Chelating agents). Less desirable is the formation of insoluble precipitates which may deposit on fabrics and machines and can, over many wash-and-wear cycles, lead to incrustation on machine parts and harshening of fabric. A third mechanism for removing hardness ions from wash liquors is through ion exchange in which sodium ions held by the ion exchanger replace calcium ions in solution. The ion exchanger, in general, is a solid. Unlike precipitated calcium carbonate, however, the particle size of the ion exchanger can be controlled and the problem of the presence of insoluble particles in the wash liquor can be mitigated to some extent.

In general, builders supply alkalinity to the wash liquor and thus function also as alkalies. In addition, they can exert a suspending effect and are therefore effective in keeping detached soil suspended in the wash liquor, particularly builder ions with multiple charges.

Phosphates. Pentasodium triphosphate, sodium tripolyphosphate, STPP, $Na_5P_3O_{10}$, is the most widely used and most effective builder in heavy-duty fabric washing compositions (see also Phosphoric acid and phosphates). It is a strong sequestrant for calcium and magnesium, with a pK_{Ca} of ca 6, and provides excellent suspending action for soils. Because of its high sequestration power, it also finds extensive application in automatic-dishwashing detergents. Sodium tripolyphosphate forms stable hydrates and thus aids in the manufacture of crisp spray-dried laundry powders.

Tetrasodium pyrophosphate ($Na_4P_2O_7$) is another important primary builder and detergent. In sequestration, it is not quite as effective as sodium tripolyphosphate and its usage in heavy-duty laundry powders has declined somewhat in recent years. Functionally, tetrasodium pyrophosphate is both a builder for surfactants and alkali.

Trisodium phosphate (trisodium orthophosphate, Na_3PO_4) is an important constituent of hard-surface cleaners including those for ceramic, metal, or painted surfaces. It may be used with soaps, surfactants, or other alkalies. It precipitates many heavy-metal ions but does not sequester to form soluble chelates. It is thus a precipitant builder and additionally an alkali.

Glassy phosphates (sodium polymetaphosphate, sodium hexametaphosphate) vary in composition, depending upon the manufacturing process. They exert a powerful sequestering and suspending effect combined with a low solution pH, about 6 or 7, and tend to hydrolyze or revert in aqueous solution and heat to pyrophosphates and orthophosphates.

Potassium phosphates, particularly tetrapotassium pyrophosphate, are considerably more soluble than their sodium analogues. They are used as builders in liquid detergents.

In the early 1970s, a number of U.S. jurisdictions, including several states, banned the use of phosphate as detergent builders. Several alternatives to phosphates have therefore been introduced into heavy-duty laundry detergents in the United States.

Sodium Carbonate. Sodium carbonate builds by forming insoluble calcium carbonate with calcium ions in hard water. Build-up of calcium carbonate on machine and fabrics, which can occur with time, is undesirable. Sodium carbonate does not provide any suspending action. It does, however, provide alkalinity to the wash liquor and is an effective alkali.

Silicates. Sodium silicates have been used extensively as soap builders in laundering formulations since well before the advent of synthetic surfactants. Silicates are more effective in sequestering magnesium rather than calcium hardness. Again, they function primarily as alkalies. In addition, they act as anticorrosive agents and prevent deterioration of washing machines, specifically metal pump parts. However, in recent years, many of these machine parts have been replaced by engineering plastics and the anticorrosion function has lost some of its importance. Alkaline silicates act as primary detergents in machine-dishwashing formulations. Commercial alkaline silicates are characterized by the ratio of SiO_2 to Na_2O in the molecule. The silicates used in detergent formulations generally show a ratio >1, usually about 2.4. Thc 1:1 compound, sodium metasilicate, is considered to be too corrosive to be widely used in consumer product formulations.

Zeolites. In recent years, certain zeolites have found application as builders in heavy-duty detergent formulations. The zeolite of choice is a so-called Type A zeolite, of empirical composition $Na_2O.Al_2O_3.2SiO_2.4.5H_2O$ and particle size of the range of 10 μm. This builder functions by ion exchange in which the calcium ions in hard water are replaced by sodium ions released from the zeolite crystal. Its pore size accommodates calcium ions but is not sufficiently large for the highly hydrated magnesium ions. Zeolite A, therefore, is not an effective builder for magnesium hardness. Like sodium carbonate building, the ion exchange process is appreciably slower than soluble chelate formation by strong sequestrants like sodium tripolyphosphate. Type-A zeolite is used principally to replace part of sodium tripolyphosphate in areas where phosphates are limited by law, ie, in the United States, Canada, and certain parts of Western Europe. However, it is not practical as the sole builder in a nonphosphate detergent formulation (see also Molecular sieves).

Clays. Clays (qv), such as kaolin, the montmorillonites, and bentonites, have been recommended and used from time to time as ingredients of washing compositions and other formulations containing surface-active agents. Under certain favorable conditions, particularly in soft water of low dissolved solids content, clay suspensions can have a marked detersive effect on ordinary soiled fabrics. Bentonite also acts as a suspending agent. In addition, sodium bentonite has some water-softening effectiveness by virtue of its ability to sorb calcium ions. However, clays are considerably less effective than Type-A zeolite in water-softening.

In one U.S. laundry powder, a montmorillonite clay serves to adsorb and hold long-chain quaternary ammonium compounds during the washing cycle and prevents interaction with the anionic surfactants. During the spinning step of the laundry cycle the solid clay-cationic particles are filtered onto the washed articles. When the latter are exposed to the heat of the automatic drier, the cationic surfactants melt and provide a softening and antistatic finish. A softening-in-the-wash effect is thus achieved with minimal interference with detersive performance.

Nitrilotriacetic Acid. The trisodium salt of nitrilotriacetic acid, $N(CH_2COOH)_3$, so-called NTA, is a powerful sequestrant builder, comparable to sodium tripolyphosphate. It is therefore noted here, even though it is an organic builder. NTA has been recommended and used as a phosphate replacement in areas where phosphate is banned. However, because of adverse laboratory reports of possible teratogenic effects, NTA was withdrawn in 1970 from consumer products at the suggestion of the U.S. Surgeon General. A recent decision by the EPA permitted reintroduction. Products based on a builder system of NTA and Type-A zeolite are on the market. Because it is a smaller molecule than sodium tripolyphosphate, NTA is theoretically a more effective sequestrant on a weight basis. It is, however, less effective than sodium tripolyphosphate as a suspending agent and is not as easily processed in spray-dried laundry powders.

Alkalies. Caustic soda (sodium hydroxide) is used largely in mechanical bottle washing, glass washing, and metal cleaning. Sodium carbonate, either anhydrous (soda ash) or in hydrated form, has been used as builder or filler in soaps, surfactants, and with inorganic constituents in cleaners for hard surfaces and fabrics. In forms insoluble calcium carbonate with calcium ions in hard water but does not provide any suspending action. Sodium bicarbonate ($NaHCO_3$), sodium sesquicarbonate ($NaHCO_3.Na_2CO_3$), and sodium borate ($Na_2B_4O_7.10H_2O$, borax) are used in place of soda ash when a lower pH is desired. In cases where high solubility is required, the potassium analogues are used.

The alkalies do not sequester heavy-metal ions and have little soil-suspending effect. They are effective in maintaining a high pH and saponify the acidic constituents of soil and thus promote cleaning. In the cleaning of ceramics, glass, and metal surfaces, however, the alkalies act as primary detergents even in the absence of surfactants in these systems.

Neutral Soluble Salts. Sodium sulfate and, to a considerably lesser extent, sodium chloride are the principal neutral soluble salts used in laundering compositions. They are often considered to be fillers although they perform an important standardizing function enabling the formulator to manufacture powders of a desired, controlled density. Sodium sulfate, in addition, lowers the CMC of organic surfactants and thus the concentration at which effective washing can be achieved.

In wool-scouring systems for textile processing that contain nonionic surfactants, sodium chloride acts as a true builder, ie, detergency promoter.

Organic Additives. Certain nonsurfactant organic additives improve the cleaning performance and exhibit other desirable properties.

Such additives are usually present in low percentages and serve one or more of the following specific functions: reduced redeposition of soil from the detergent bath onto the substrate; increased whiteness or appearance of cleanliness; enhanced cleaning effect on specific types of solid and stains; promotion or inhibition of foaming power and stability; increased solubility or other modification of the physical form of the detergent composition; sequestering of heavy-metal ions, both in the concentrated detergent and in the diluted cleansing bath; reduced injurious effects which the detergent may have on the substrate or the washing machine, such as tarnishing of silverware or etching of glassware, corrosion of metals, or irritation of skin in toilet and cosmetic applications.

Antiredeposition agents contribute to the appearance of washed fabrics. Sodium carboxymethylcellulose (NaCMC) is the most widely used, and on cotton fabrics, the most effective. With the advent of synthetic fabrics, other cellulose derivatives, eg, methylcellulose, hydroxybutylcellulose, hydroxypropyl- and mixed methyl and hydroxybutylcellulose ethers have been shown to be more effective than NaCMC (94) (see Cellulose).

Fluorescent whitening agents (see Brighteners, optical) were first disclosed in 1940 in combination with detergents (95). They absorb ultraviolet radiation and subsequently emit some of the radiation energy in the blue part of visible spectrum. As a result, they confer enhanced whiteness to the appearance of washed articles. Highly effective cotton-substantive fluorescent whitening agents were in wide-spread use at relatively high concentrations (ca 0.5%) in the 1950s and 1960s. For synthetic fabrics such as polyester, it has proved to be more effective to prebrighten the fabric by incorporating the fluorescent whitening agent in the spin–melt during manufacture rather than depend on adsorption from the detergent bath. As a result, the usage of fluorescent whitening agents in formulated laundry products has decreased in recent years.

Bluing agents, which are dyes, provide another approach to maintaining fabric whiteness by a mechanism in which a yellow cast of washed fabrics is covered by the blue dye. Since this approach diminishes reflectance, it is inherently less desirable than the use of fluorescent whitening agents which increase reflectance.

Proteolytic enzymes are in general use in European detergent formulations. In the United States, they were used in the late 1960s and early 1970s. They are still present in a few laundry products. Proteolytic enzymes interact with certain proteinaceous stains and contribute to cleanness in addition to the action of the formulated detergent (96).

Bleaching agents (qv), such as sodium perborate ($NaBO_3.4H_2O$), are commonly present in significant amounts in European laundry detergents. At high washing temperatures, sodium perborate effectively bleaches chemical stains such as wine, fruit juices, etc. Because of the lower washing temperature in U.S. machines, sodium perborate is considerably less effective and its usage is restricted to some individual brands of laundry detergents. In automatic-dishwashing formulations, chlorine is needed to remove food stains from dishware. It is present in all U.S. products, either in the form of chlorinated trisodium orthophosphate or as chlorinated isocyanurate.

Foam regulators such as amine oxides and alkanolamides are present in products where high foam volume is functionally or esthetically desirable, mainly hand-dishwashing liquids and shampoos. In automatic dishwashing products, on the other hand, copious foam volumes interfere with the efficiency of the mechanical rotors during operation. In this type of product, a foam depressant is often present.

Organic sequestering agents serve the same purpose as the sequestering phosphates, ie, to remove interfering metal ions from the detergent bath. They are used where the less expensive phosphates are, for one reason or another, not applicable. Nitrilotriacetic acid, discussed earlier, EDTA, and a variety of organic phosphonate structures are commonly used in a wide range of detergent compositions. Certain hard-surface cleaning products such as sanitizing cleaners for hospital use are examples of products generally containing small amounts of organic sequestrants.

Factors Influencing Detergency

Detergency is mainly affected by the concentration and structure of surfactant, hardness and builders present, and the nature of the soil and substrate. Other important factors include wash temperature; length of time of washing process; mechanical action; relative amounts of soil, substrate, and bath, generally expressed as the bath ratio, ie, the ratio of the bath weight to substrate weight; and rinse conditions.

Effect of Surfactant Concentration. A plot of soil removal versus surfactant concentration is generally sigmoid. It starts at the soil removal of water without surfactant, rises slowly, then more steeply until a plateau is reached when detergency is little affected by increase in surfactant concentration. This plateau can be correlated with the CMC of the surfactant, and is generally higher than the CMC. In general, detergency attains its maximum when the CMC of the surfactant is reached, taking into account surfactant adsorption on the soil and substrate. However, with certain surfactants, there is evidence that oily-soil detergency continues to increase above the CMC (97–100).

Surfactant Structure. The chemical structure of the surfactant is an important factor in detersive effectiveness. When relating detersive power to chemical constitution, within limited series certain regularities can be observed, but few if any general principles apply to the whole range of surfactants. Among the homologous fatty acid soaps and the straight-chain alkyl sulfates, optimum detergency under usual washing conditions occurs at a chain length of ca sixteen carbon atoms. Detergency among the ethoxylated nonionic surfactants varies in a regular manner with the length of the ethylene oxide chain as well as with the structure of the hydrophobic group.

Within a series with a fixed hydrophilic head group, detergency increases with increasing carbon chain length, reaches a maximum, and then decreases. This behavior frequently reflects a balance between increased surface activity of the monomer and decreased monomer concentration with increased surface activity. Similar effects are seen in surfactants in biological systems.

The numerous studies of the effect of surfactant structure in detergency include the classical paper on the series of sulfated secondary straight-chain alcohols (101) and various papers on detergent-builder combinations (103); nonionic–anionic mixtures (103–104); the polyethenoxy nonionic series (105–106); detergency of isomeric alkylbenzenesulfonates (107); practical home laundering (108); laundering and detergency effects in seawater (109); *o*- and *p*-alkylbenzenesulfonates with straight and branched chains (110); and studies on alkylbenzenesulfonates (111).

Water Hardness and Builders. The presence of heavy-metal ions, especially calcium and magnesium, has an effect on washing second only to that of the surfactant itself. Distilled water, used in a system where soil and substrate do not contain substantial calcium ions, is a surprisingly effective detergent, when the soils themselves contain fatty acids or sodium soaps. Detergency can be improved by the addition of surfactants in amounts smaller than might be expected. Conversely, the detergency of surfactants is generally decreased by the presence of hardness ions (Ca^{2+} and Mg^{2+}) and hard water alone is a poor soil remover. With traditional formulations containing anionic surfactants or soaps, cotton can be cleaned well by washing only if the calcium concentration is reduced to <0.01 mM. In current synthetic detergent compositions, builders sequester calcium and magnesium ions and thus reduce interaction with surfactants, soils, and fabrics as described above.

The sodium soaps of fatty acid form calcium soaps of such low solubility that they act as their own builders. Initial soap additions precipitate the calcium ion and the soap added thereafter functions in soft water. At high temperatures, the calcium soaps are relatively soluble compared to calcium tripolyphosphate. Thus, STPP (sodium tripolyphosphate) can build soaps in a hot water wash. However, at low temperatures the relative affinity of STPP for calcium decreases so that STPP cannot build soaps in a cold water wash.

Calcium ion enters the system not only in the form of water hardness but also in the form of calcium salts contained in the soil. Other heavy-metal ions such as aluminum and ferric iron may also be present in the soil, and must be removed by an appropriate builder to achieve good soil removal. Effective builders for cotton washing are those for which the calcium dissociation constant, expressed as pK_{Ca}, or $-\log K_{Ca}$ is >4 and preferably >7 (112). Much of the work that led to elucidating the role of builders as calcium sequestering agents in detergency was done in connection with redeposition studies (113).

Legislatively mandated reductions in detergent phosphate concentrations have resulted in numerous attempts to compensate for the attendant cleaning losses. Problems caused by phosphate reduction can be ameliorated by changes in surfactant systems. Thus, calcium-sensitive surfactants such as linear alkylbenzenesulfonates (LAS) can be replaced by calcium insensitive ones such as alcohol ethoxylates (AE) or alcohol ether sulfates (AES). Proper blends greatly reduce surfactant sensitivity to calcium ions. Increasing the amount of surfactants partially compensates for phosphate reduction or elimination. Other calcium-lowering agents include NTA, citrate, zeolite, and carbonate. Citrate and NTA lower calcium levels in solution by sequestration, zeolite by ion exchange, and carbonate by ion-pair formation and precipitation. To date, no entirely satisfactory phosphate replacement has been found. Indeed, some soil-cloth detergency data show that the 1978 non-P products are less effective than the 1969 high P products (114).

In the presynthetic surfactant era, soap was built (and still is) with alkaline salts such as soda ash, silicates, orthophosphates, and borates. These materials buffer the wash solution to a high pH, prevent soap protonation, and thus remain effective. Another type of builder is the neutral inorganic salt such as sodium chloride and sodium sulfate. These materials may improve detergency by increasing the ionic strength and altering the CMC of anionic surfactants.

Antiredeposition Agents. The redeposition effect is best illustrated by agitating a clean swatch of white cotton in a washing bath which has been used and is turbid with dispersed soil. Even though the bath may still be effective and capable of removing soil, the white swatch picks up some soil from the bath and darkens. Any agent that minimizes this redeposition improves net soil removal, as well as decreases the soil accumulation through several use-laundering cycles. Redeposition is in many ways the inverse of soil removal and is influenced by the same factors. Thus, calcium ion promotes redeposition in much the same way that it inhibits soil removal. Surfactants themselves are effective antiredeposition agents. Builders reduce redeposition, independently of their ability to sequester hardness ions.

Sodium carboxymethylcellulose (NaCMC) greatly reduces redeposition in cotton-washing systems based on synthetic surfactants. It is effective at remarkably low concentrations of ca 1% of the standard washing compositions which are used at ca 0.1 to 0.2% in the bath. Thus, ca 0.001–0.002%, or 10–20 ppm NaCMC is sufficient to significantly inhibit redeposition.

A very large number of hydrophilic polymers have been tested with regard to their power of inhibiting redeposition, and it is remarkable how few materials can match NaCMC. Some proteins, polyvinylpyrrolidinones, and poly(vinyl alcohols) with rather specific molecular weight ranges are in the same class as NaCMC. Certain starch derivatives, cellulose sulfates, and cellulose ethers are also effective. Currently, both NaCMC and cellulose ethers are used in home laundry products specifically to control redeposition. The cellulose ethers are used to control oil soil redeposition upon synthetic fibers, especially polyester. Using tracer methods, it was proved that NaCMC is adsorbed on the soil particles and also, under normal laundering conditions, on cotton fiber (115–116). Adsorption on cotton appears to be more important in preventing soil redeposition. This adsorption occurs to a significant extent only in the presence of metallic cations which are furnished as sodium ions from STPP and the filler salts present in the detergent. In the absence of cations, except for the Na^+ present in the NaCMC itself, the adsorption is so slight as to be almost undetectable, and the antiredeposition effect is correspondingly lowered.

Liquid Soil. Liquid soils are frequently removed from fibers by a roll-up mechanism (117). In a study of wool grease removed from single fibers, it was found that the grease layer, originally continuous, rolled up in relatively large globules, which detached themselves from the fiber. The detergent alters the contact angle at the fiber–grease–water interface. In pure water, the contact angle is <90°, measured through the oil. Addition of detergent increases the angle to ca 180°. The receding contact angle, ie, that measured as the oil rolls up, is more important in detergency than is the advancing contact angle. Detergency mechanisms are similar on wool, viscose rayon, and cuprammonium rayon (118). The oil removal depends primarily on the nature of the oil. Those containing free fatty acids are removed more rapidly, followed by neutral glycerides and less polar mineral oils. Similarly, polar oils are more easily removed from wool than nonpolar oils (119). This difference correlates well with differences in contact angle.

Liquid soil can be removed by direct emulsification and solubilization as well as by rollback. Oils that are more readily emulsified or solubilized are removed more rapidly and more completely.

Solid Soil Type and Size. Different solid soils differ greatly in ease of removal and redeposition behavior. These differences can be traced to particle size and soil–substrate bonding. The effect of particle size variation on detergency has been studied with soil removal and redeposition techniques.

In an early systematic investigation of this relationship, a series of carbon-black samples was applied to chopped cotton fibers from an aqueous suspension with vigorous agitation (120). The degree of soiling was estimated by filtering the fibers with suction to form a mat, drying, and measuring the reflectance of the mat. The degree of soiling was found to be directly related to the particle size of the carbon black. As shown in Figure 5, when the primary particles of the carbon black are <50 nm, the soil deposition is severe, whereas at particle sizes >50 nm, the soil deposition is much lighter and relatively independent of particle size. Electron microscope studies of the cotton fiber show that 50 nm is about the upper limit of the crevice width on the surface of cotton fibers. Furthermore, the soil deposition by this procedure using fine-particle carbon blacks is practically irremovable, and the fibers are to all intents and purposes dyed a very fast gray. On this basis and on the basis of other supporting evidence, it was concluded that soiling in this system takes place by microocclusion, ie, that the

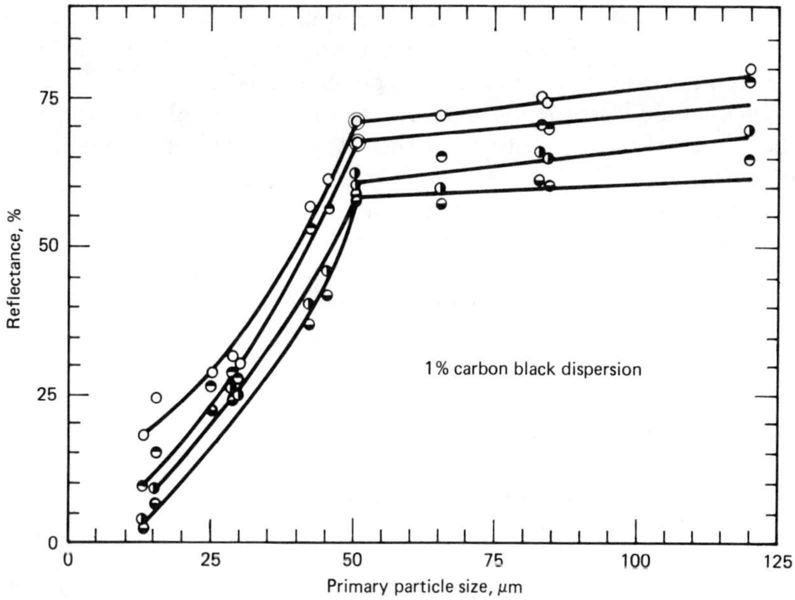

Figure 5. Effect of primary particle size on rate of change in reflectance. ○ = 5 min; ◓ = 20 min; ◑ = 180 min; ◒ = 360 min.

particles of soil are so small that they can be trapped mechanically by the irregularities of the fiber surface, even though these irregularities are scarcely larger than the dimension of a polymeric molecule. This theory accounts well for the observed facts, but the facts could also be accounted for by simply assuming that van der Waals forces control the attachment of the carbon particles to the fiber. Since these forces operate at the surface of the particle, there should be a critical particle size below which the surface forces become more important than the inertial and gravitational forces which are proportional to mass. Electron micrographs of soiled cotton show adherent soil particles in size ranges up to 0.5–2.0 μm clinging to apparently smooth surfaces (121–123). These and other supporting detergency data, indicate that mechanical binding is not the basic mechanism of solid soil binding, even though the binding varies greatly with particle size.

As might be expected, large differences in the removability of solid particulate soil are due to differences in the chemical nature of the particle surface. Thus, iron oxides, lampblacks, and clays, all of the same particle size, differ greatly in their redeposition behavior and the method with which they are removed.

Fabrics and Fibrous Substrates. Certain fibers are easier to wash than others. The hardness of the fiber surface, which varies not only with the basic fiber but also with the surface finish, affects both soilability and soil removal. Generally, soft finishes pick up and retain soil more readily than hard finishes, since soft surfaces permit greater soil–surface contact. The hydrophobicity of the fiber surface also influences soilability and soil removal, and the more hydrophobic fibers show greater soil retention. These effects are explained by the high-energy interface existing between hydrophobic fibers and water (124). Many soil components lower the interfacial energy and therefore locate at the fiber–water interface.

Soils can also penetrate into the fiber. The interior of the cotton fiber, the lumen, is relatively hollow, and soils may collect there. Polyester fibers are solid, but if polyester is washed above its glass-transition temperature, T_g, it becomes relatively fluid. In this case, oils on the fiber surface can mix with the polyester itself. When soils penetrate into the fiber, they are nearly impossible to remove.

Temperature and Mechanical Action. Absorption of external thermal or mechanical energy by a detersive system influences the rate and extent of soil removal. Raising the temperature generally increases the cleaning rate and, therefore, the amount of soil removed during any fixed-time laundering cycle. The effect is only strong at two critical temperatures; one is the temperature where the fatty soil in the system liquefies. As temperature increases through this region, the detergency increases markedly. Typical curves of detergency versus temperature are shown in Figure 6 (125). The second critical temperature is the boiling point. Boiling greatly increases detergency because of the localized mechanical action of steam bubbles forming, expanding, and breaking away at the solid–liquid interfaces. However, in the United States, boiling effects are never encountered in home washing machines. Temperature effects are also important with certain detergent additives. Thus the efficacy of certain bleaches and enzymes is markedly reduced at the low wash temperatures now popular. These effects emphasize the problems inherent in the current trend to lower wash temperatures.

In a study of soil removal as a function of mechanical action, the following relationship is valid within reasonably wide limits:

$$S = K(CFT)^n$$

where S is soil removal, C is detergent concentration, F is a force factor, T is time, and K and n are constants (n = between 0 and unity). This relationship is of limited applicability and can hardly be valid for values of C and T outside of the usual practical ranges. Evaluation of the force factor F is not simple, eg, for washing a wool plug, the curve of detergency versus mechanical action reached a plateau at which further energy input did not increase oily soil removal. Similar results are encountered with normal fabric washing.

As with the case of energy input, detergency generally reaches a plateau after a certain wash time as would be expected from a kinetic analysis. In a practical system,

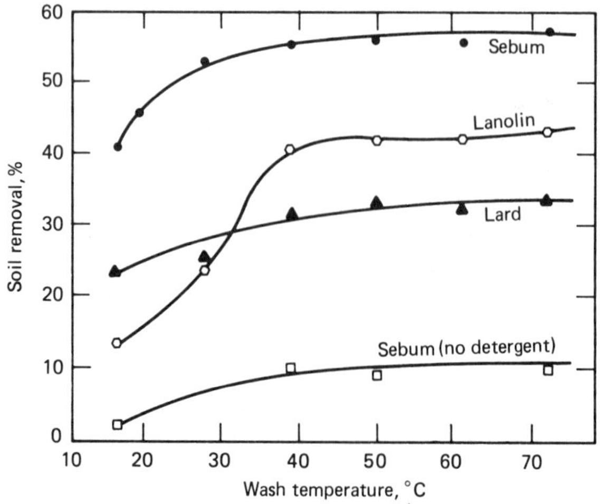

Figure 6. Effect of wash temperature on removal of sebum, lanolin, and lard from cotton (0.25% built detergent) (125).

each of its numerous components has a different rate constant, hence its rate behavior generally does not exhibit any simple pattern. Many attempts have been made to fit soil removal (126) rates in practical systems to the usual rate equations of physical chemistry. The rate of soil removal in the Launder-Ometer could be reasonably well described by the equation of a first-order chemical reaction, ie, the rate was proportional to the amount of removable soil remaining on the fabric (127–128). In a study of soil-removal rates from artificially soiled fabrics in the Terg-O-Tometer, the percent soil removal increased linearly with the log of cumulative wash time.

In detailed studies with a number of different artificial test cloths, first-order kinetics were obtained for soil removal during the first 6–20 min of the cycle (129).

In the washing of fabrics, machine and timing must be carefully adjusted. Too large an energy input for too long a time can injure fabric severely and eventually tear it to shreds. The same effect is quite possible with purely mechanical agitation.

Foam. The relationship of foaming power to detergency has always been of interest, and foaming power has become associated in many consumers' minds with high detersive power (see also Foams). It is generally conceded by detergent technologists, however, that foam has no direct relationship to detergency in ordinary fabric-washing systems, and that it does not improve cleaning in a laundry or home-washing machine. Indeed, excessive foam can inhibit agitation and reduce cleaning. In systems of low bath ratio, on the other hand, foam may play an important role. The individual foam laminae tend to imbibe and segregate particles of both liquid and solid soil that have been removed from the substrate. This prevents redeposition and enables the soil to be easily removed by scraping off or rinsing away the soil-laden foam. This effect is very important in the on-location shampooing of rugs, and to a certain extent in the shampooing of human hair. An excellent series of photomicrographs show the imbibition of liquid–soil droplets by foam films during the washing of oiled glass plates (130). The soil tends to be carried by capillary convection to the low-pressure region in plateau borders where three individual laminae join to form the edge of a foam cell.

Mechanisms

It is evident from the preceding discussion that even the simplest detersive system is surprisingly complex and heterogeneous. It can nevertheless be conceptually resolved into simpler systems that are amenable to theoretical treatment and understanding. These simpler systems are represented by models for substrate–solid soil and substrate–liquid soil. In practice, many soil systems include solid–liquid mixtures. However, removal of these systems can generally be analyzed in terms of the two simpler model systems. Although these two systems differ markedly in behavior and structure, and require separate treatment, there are certain overriding principles that apply to both.

The first principle is that soil systems can be regarded and treated as classical systems of colloid and surface chemistry. A cotton fiber bearing attached small particles of clay or carbon soil, immersed in an aqueous medium, is fully analogous to the sulfur sols or gold sols of classical colloid chemistry (see Colloids, Supplement). Although, the fiber is not of colloidal dimensions, the area over which any individual soil particle contacts the fiber is of colloidal dimensions. Furthermore, the ratio of the soil particle's area to the area of contact and the ratio of the soil particle's mass to the area of contact put this system in the colloidal range. These ratios are the same as the ratios prevailing when one sulfur sol particle makes contact with another. In addition, in the substrate–soil system two different types of surfaces are interacting, whereas in the classical sol systems the particles are all of the same chemical nature and the interacting surfaces are therefore similar. The problem of mixed sols has, however, been explored in classical colloid chemistry to at least a limited extent, and it is fully analogous to the problem of the model soil–detersive system. A fiber or steel plate covered with oil and immersed in an aqueous solution is a good representative of the liquid–liquid–solid system of classical surface chemistry. The well-established concepts of interaction at the liquid–liquid interface and at the three-phase boundary line are fully applicable.

A second principle applying to these model systems is derived from their colloidal nature. With the usual thermodynamic parameters fixed, the systems come to a steady state in which they are either agglomerated or dispersed. No dynamic equilibrium exists between dispersed and agglomerated states. In the solid–soil systems, the particles (provided they are monodisperse, ie, all of the same size and shape) either adhere to the substrate or separate from it. In the liquid–soil systems, the soil assumes a definite contact angle with the substrate, which may be anywhere from 0° (complete coverage of the substrate) to 180° (complete detachment). The governing thermodynamic parameters include pressure, temperature, concentration of dissolved components, and electrical conditions. This concept is at variance with the idea of a dynamic detergency equilibrium which was advanced as a working hypothesis by many of the earlier investigators. It is, however, much more in keeping with the established concepts of colloid chemistry. In the classical theory of lyophobic colloids, no consideration is given to a dynamic equilibrium existing between the peptized state and the agglomerated state. This all-or-nothing behavior has been studied experimentally in model detersive systems by observing with the help of microscopic techniques the behavior of single textile fibers in aqueous carbon black suspensions containing dissolved soap (131). In the absence of soap, the carbon particles deposited on the fiber. When sufficient soap was added, the carbon particles did not deposit on the fiber but remained in suspension. The quantity of soap necessary to maintain the carbon in suspension was proportional to the amount of carbon present in the system. There was a sharp demarcation between the state in which the carbon was suspended and the state in which the carbon was attached to the fiber.

In applying the above concept, the factor of particle size must be continuously borne in mind. A heterodisperse system can reach a steady state where the smaller particles are agglomerated and the larger particles are dispersed, giving the apparent effect of an equilibrium. In ideal monodisperse systems under steady conditions, however, no such effects are noted.

Purely mechanical disturbances (which are not usually considered as thermodynamic variables) may influence the state of aggregation of a colloidal system; for example, floc size in carbon and iron oxide suspensions varies with the degree of agitation being imposed on the system (132). When the agitation is stopped, the flocs revert to their steady-state size. To have any effect, the shear field must be quite high and the particles relatively large. With particles $<0.5\ \mu m$, thermal agitation becomes more important than even vigorous mechanical agitation in determining the state of aggregation.

A final consideration in resolving practical detersive systems into their simpler components relates to soil removal versus redeposition. Superficially, it would appear that the redeposition phenomenon contradicts the all-or-nothing concept that the system must exist in either the agglomerated or the dispersed state. Keeping in mind both the composite nature and the kinetics of a practical system, it is readily shown that no such contradiction exists. The soil particle that redeposits is in a different state from what it was during its initial removal from the substrate. It may have changed its state by becoming detached from adherent oil or from a cluster of similar solid particles. During the time interval between detachment and redeposition, it may have altered its surface character by adsorption or desorption. In this time interval, the state of the substrate surface may also have changed by adsorption or desorption, or by the loss or gain of an oily layer. Additionally, there are probably varying substrate types present in the wash, with varying affinities for the soils present. For example, oily soil could be detached from a cotton fiber, but then deposit on a polyester fiber for which it would have a greater affinity. Thus, the initial group of agglomerated systems, which composes the soil–substrate–bath complex before soil removal, is quite different from the agglomerated system composed of substrate and redeposited soil.

Solid-Soil Detergency. ***Adsorption.*** Many studies have been made of the adsorption of soaps and synthetic surfactants on fibers in an attempt to relate detergency behavior to adsorption effects. Relatively fewer studies have been made of the adsorption of surfactants by soils (133). Plots of the adsorption of sodium soaps by a series of carbon blacks and charcoals show that the fatty acid and the alkali are adsorbed independently, within limits, although the presence of excess alkali reduces the sorption of total fatty acids (134). No straightforward relationship was noted between detergency and adsorption.

In a study of the adsorption of soap and several synthetic surfactants on a variety of textile fibers, it was found that cotton and nylon adsorbed less surfactant than wool under comparable conditions (13). Among the various surfactants, the cationic types were adsorbed to the greatest extent, whereas nonionic types were adsorbed least. The adsorption of nonionic surfactants decreased with increasing length of the polyoxyethylene chain. When soaps were adsorbed, the fatty acid and the alkali behaved more or less independently just as they did when adsorbed on carbon. The adsorption of sodium oleate by cotton has been shown independently to result in the deposition of acid soap (a composition intermediate between the free fatty acid and the sodium salt), if no heavy metal ions are present in the system (135). In hard water, the adsorbate has large proportions of lime soap.

The adsorption behavior of commercial alkylbenzenesulfonates and several other anionic detergents on cotton, wool, and some of the synthetic fibers has been carefully plotted (136). The equilibrium adsorption isotherms, with few exceptions, do not show any startling abnormalities. A number of other studies have attempted to correlate adsorption with detergency (137–139). In general, no significant correlations were found for a number of possible reasons. On porous fibers such as cotton, adsorption may occur on fiber areas that are not beneficial to detergency. The presence of soils on the fiber surface may significantly alter the adsorption behavior. The presence of trace amounts of surface-active agents in the soils themselves could modify adsorption of the active detergent component. For instance, soap present in natural soils could significantly change nonionic adsorption behavior. Finally, and perhaps most importantly, equilibrium adsorption measurements may not realistically reflect adsorption taking place in a 10-min wash cycle.

Adsorption of bath components is a necessary and possibly the most important and fundamental detergency effect. Adsorption is the mechanism whereby the interfacial free energy values between the bath and the solid components (solid soil and substrate) of the system are lowered, thereby increasing the tendency of the bath to separate the solid components from one another. Furthermore, the solid components acquire electrical charges that tend to keep them separated or acquire a layer of strongly solvated radicals that have the same effect. If it were possible to follow the adsorption effects in a detersive system, in all their complex ramifications and interactions, the molecular picture of soil removal would be greatly clarified.

Mass Transfer Near the Substrate Surface. Mechanical action has a great effect on soil removal probably by influencing mass transfer, ie, the diffusion of soluble material away from the immersed fibers (140–141). Mechanical action tends to maintain a high concentration gradient near the fiber, and the resulting increased diffusion causes stronger diffusion currents to flow. These diffusion currents are presumably responsible for carrying away the soil particles that have already been detached or loosened from the fiber surface by physicochemical action. In a Terg-O-Tometer-investigation of mass transfer, using only water soluble substances, the transfer coefficient was found to be directly proportional to agitator speed and stroke angle, inversely proportional to the water-holding capacity of the cloth load, and independent of bath volume (142–143). The soil-removal behavior is very similar to the mass-transfer behavior, supporting the idea that diffusion currents are an important operating factor in soil removal, even though they may play no part in breaking the primary soil–substrate bond. Considering the wide variety of soils and substrates encountered in washing, as well as the wide variation in particle size and fabric geometry, mass-transfer effects could play a principal or minor role of the overall wash process, depending upon the specific system encountered.

Colloidal Stabilization. Surfactant adsorption reduces soil–substrate interactions and facilitates soil removal. For a better understanding of these interactions, a consideration of colloidal forces is required.

The model solid-soil detersive system is advantageously treated as a sol–agglomerate colloid system or, in more general terms, a lyophobic colloid. In the typical lyophobic colloid, consisting of a single disperse phase in an aqueous suspending medium, only one type of liquid–solid interface and one type of solid–solid interface is present. The simplest detersive system, however, has an added degree of complexity in the presence of two types of liquid–solid interface: soil–bath and substrate–bath. Also present are two effective types of solid–solid interface: soil–substrate and soil–soil. The soil–soil interface relates to flocculation of dispersion of soil particles remote from the substrate, and is not of primary concern in the present discussion. The soil–soil interface is, of course, important in practical detergency since soil aggregates can be regarded as large single particles.

There are two general theories of the stability of lyophobic colloids, or, more precisely, two general mechanisms controlling the dispersion and flocculation of these colloids. Both theories regard adsorption of dissolved species as a key process in stabilization. However, one theory is based on a consideration of ionic forces near the interface, whereas the other is based on steric forces. The two theories complement each other and are in no sense contradictory. In some systems, one mechanism may be predominant and in others both mechanisms may operate simultaneously. The fundamental kinetic considerations common to both theories are based on Smoluchowski's classical theory of the coagulation of colloids.

As the particles in a colloidal dispersion diffuse, they colloid with one another. In the simplest case, every collision between two particles results in the formation of one agglomerated particle, ie, there is no energy barrier to agglomeration. Applying Smoluchowski's theory to this system, the half-life, $t_{1/2}$, ie, the time for the number of particles to become halved, is expressed as follows:

$$t_{1/2} = \frac{3\,\eta}{4\,kTN_0}$$

where η is the viscosity of the medium; k, Boltzmann's constant; T, temperature; and N_0, the initial number of particles. For dispersions at moderate concentration, ca 10^9 particles per cm^3, $t_{1/2}$ has a value of the order of a few seconds.

The above expression assumes there is no barrier to collision and every collision is effective. For stable dispersions to exist, an energy barrier, W, is assumed which prevents collision. In this case, the expression for half-life becomes:

$$t_{1/2} = \frac{3\,\eta}{4\,kTN_0}\exp\,(2\,W/kT)$$

If W is 15–20 kT, the half-life is several days and the dispersion is reasonably stable. The problem of colloidal stability thus becomes a problem of the energy barrier, W, that lessens the unifying collisions.

Ionic Stabilization. The quantitative theory of colloid stability based on electrical barriers was developed independently in the USSR by Derjaguin and Landau (144) and by Verwey and Overbeek in Holland (145). It is generally referred to as DLVO theory. There are two opposing forces between particles: a force of attraction that is responsible for agglomeration when the particles approach close enough, and a force of repulsion that prevents close approach. The attraction forces are the London dispersion forces or van der Waals forces. These have the general form:

$$F = K/r^n$$

where r is the distance between particles and n is between 3 and 7, depending on the geometry of the interacting surfaces. The repelling forces are caused by the electrical double layers that surround the particles in aqueous dispersions. The interaction of these two forces is conveniently shown in a plot of potential energy versus distance between particle centers, as in Figure 7. The ordinate in this plot can be either force or potential; for quantitative purposes, they are interconvertible through the definitive relationship that force is the negative gradient of potential. Curve 1 represents the attractive potential caused by van der Waals forces, curve 2 represents the repulsive potential caused by double-layer forces, and curve 3 is the resultant force experienced by the two particles. To separate two agglomerated particles ($r = 2a$), it is necessary to overcome the energy barrier $V + W$, to bring them to distance X_2 at which point repulsive forces predominate. Conversely, to cause agglomeration of two separated particles, it is necessary to overcome the barrier W at X_2.

According to DLVO theory, the shape of curve 2, which controls the shape of the effective curve 3, is determined by the thickness and charge density of the electrical double layer that surrounds each particle. These factors, in turn, are determined by the nature of the ionic species adsorbed at the interface, the degree of adsorption, and the ionic strength of the surrounding medium. The quantitative aspects of the DLVO and the double-layer theory (ie, the structure, formation, and interactions of electrical

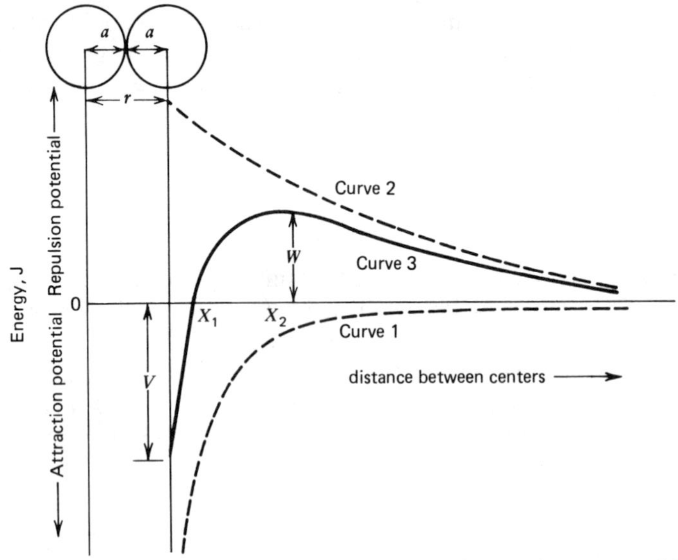

Figure 7. Attraction–repulsion potentials as a function of distance between particle centers.

double layers at phase interfaces) are beyond the scope of this article, and it is recommended that the reader consult the standard publications in this field. Qualitatively, high charge density increases the repulsion forces and thus shifts curve 3 upward, increasing the potential W and decreasing $|V|$. High ionic strength shrinks the double layer, lowering curve 3, lowering W, and tending to increase $|V|$.

Electrical potentials can, of course, be negative or positive. In most detersive systems, the charges acquired by both substrate and solid soil are negative. The adsorption of detergent anions increases the effective density of negative charge and tends to increase the double-layer-repulsion potentials. The counterions, in this case metallic cations, exert the controlling effect on the ionic strength factor, with increasing valence greatly magnifying the effect. Thus, calcium ions have a much greater effect than sodium ions in shrinking the double layer and promoting agglomeration. Aluminum ions have an even greater effect than calcium ions. This is in accord with the well-known Schulze-Hardy rule which was proposed as an empirical finding many years before this quantitative DLVO theory was developed.

The electrokinetic effect is one of the few experimental methods for estimating double-layer potentials. If two electrodes are placed in a colloidal suspension, and a voltage is impressed across them, the particles move toward the electrode of opposite charge. For nonconducting solid spherical particles, the equation controlling this motion is:

$$u = \frac{RFVD\zeta}{6\,\pi\eta}$$

where u = velocity of particles; η = viscosity of medium; V = applied field, V/cm; R = factor for electrical relaxation; D = dielectric constant of medium; F = factor for size of spheres; and ζ = zeta potential.

This equation is a reasonable model of electrokinetic behavior, although for theoretical studies many possible corrections must be considered. Correction must always be made for electrokinetic effects at the wall of the cell, since this wall also carries a double layer. There are corrections for the motion of solvated ions through the medium, surface and bulk conductivity of the particles, nonspherical shape of the particles, etc. The parameter zeta, determined by measuring the particle velocity and substituting in the above equation, is a measure of the potential at the so-called surface of shear, ie, the surface dividing the moving particle and its adherent layer of solution from the stationary bulk of the solution. This surface of shear lies at an indeterminate distance from the true particle surface. Thus, the measured zeta potential can be related only semiquantitatively to the curves of Figure 6. There have been many attempts to correlate detergency and deflocculation with measurements of zeta potential (143); such measurements have been made on both fibers and soil particles using various model detersive baths as the media. Aside from the expected result that high zeta potentials tend to correlate with deflocculation, few direct relationships of theoretical significance have become evident. In an extensive and productive study, it was pointed out that DLVO theory relates the height of the maximum in the potential curve (W in curve 3, Fig. 6) to the square of the zeta potential rather than to its first power (145). From the DLVO theory the surface potential ψ was calculated which is necessary to give a barrier of height (in joules) $W = 15$ kT for soil particles of various size in electrolytes of various concentrations; ζ was measured experimentally. When the ratio ζ^2/ψ^2 was greater than unity, good stability was attained. In calculating soil-substrate interaction, good results were obtained using the ratio of $[(\zeta$ substrate$)(\zeta$ soil$)]/[(\psi$ substrate$)(\psi$ soil$)]$. Application of DLVO theory to detergency has also been made very successfully when the shape factors and factors involving the structure of the double layer were taken into consideration (146).

The strong adverse influence of calcium ions on the stability of lyophobic suspensions is predicted by DLVO theory, and has been demonstrated with many types of simple soils. That calcium ions have an overwhelming effect on the redeposition of carbon soil onto cotton tends to support the idea that DLVO theory is a principal key in explaining detersive action. The redeposition of carbon onto cotton has been correlated quantitatively with the calcium ion content of the system, both in the presence and absence of surfactant (147). The adverse effect of calcium ions on wet soil removal in practical washing has also been well established (148). The effect of calcium in detergency cannot be explained solely, however, by its shrinking of the double layer. Calcium is very strongly adsorbed onto cotton through the carboxyl groups present on the cellulose chains. This adsorption is so strong that cotton rinsed in even slightly hard water transfers a harmful amount of calcium to the next wash liquor. Calcium present in soil particles can act as a bonding agent at the soil–substrate interface, and inhibit soil removal by this secondary mechanism. In one of the most thorough studies of the above effects, it was also demonstrated that calcium promotes the deposition of oil onto cotton (149). The sorption of calcium during washing is complicated by its apparent ability to coadsorb with surfactants and complex phosphate builders, further obscuring the mechanism by which it exerts its adverse effect (150–151).

Despite its successes and its great usefulness as a guide, the limitations of the DLVO theory in explaining detergency are most evident when considering the nature of the soil–substrate bond. The DLVO theory in its simplest form postulates van der Waals bonding as the primary agglomerating mechanism. Even some of the newer theories of double layer interaction, that do not assign such an important role to van der Waals forces, do not encompass agglomerating mechanisms such as polyvalent cation bridges and hydrogen bonding (152–153). There is ample evidence that these latter mechanisms prevail in some important detersive systems.

Steric Stabilization. Double-layer repulsion cannot satisfactorily explain stabilization of lyophobic suspensions by nonionic surfactants nor nonionic detergency in solutions of high salt content. Not only nonionic surfactants but even anionic ones can overcome the agglomerating effects of salt. To explain these phenomena, steric interaction between the surfactant molecules adsorbed at the solid–solution interface must be considered. The adsorbed molecules are oriented with the hydrocarbon tails adherent to the essentially hydrophobic solid surface and the water-soluble heads sticking into the solution. This adsorption and orientation is governed by the same factors governing micelle formation, ie, the amphipathic nature of the surfactant molecule and solution forces. The head groups at both nonionic and anionic surfactants are heavily hydrated. As a result the lyophilic particle is surrounded by an inner layer of hydrophobic surfactant tails and an outer layer of hydrated surfactant heads. In Figure 7, it can be seen that the attractive potential does not greatly increase until the particles are relatively close to each other. The bulk of the hydrated head groups prevents the particles from approaching near enough to each other for the attractive forces to cause agglomeration. In the case of polyoxyethylene nonionic surfactants, the head group is quite large, extending for several nanometers (tens of angstroms) into solution; the head group is associated with a large amount of water of hydration caused by ether–water interaction. The bulk of a hydrated anionic head group is not so great; however, the anionic surfactant retains some ionic stabilization even in salt solution. Agglomeration is achieved only if the particles overcome the energies involved in steric compression of the head groups and desolvation of the head groups. Conversely, if two particles, adherent over a small area of contact, are immersed in a surfactant solution, oriented adsorption on the noncontacting areas, followed by hydration, presumably breaks the adhesive bond and forces the particles apart. This picture would suggest some degree of correlation between the wetting power of a surfactant for a given material and its stabilizing power for sols of the same material. Such a correlation has indeed been found by Lange (154), using paraffin wax with both nonionic and anionic surfactants.

Oily-Soil Detergency. ***Roll-up.*** The principal means by which oily soil is removed is probably by roll-up. The applicable theory is simply the theory of wetting. In briefest outline, a droplet of oily soil attached to the substrate forms at equilibrium a definite contact angle at the oil–solid–air boundary line. This contact angle (see Fig. 2) is the result of the interaction of interfacial forces in the three phase boundaries of the system. These interfacial forces, expressed in mN/m (dyn/cm), or interfacial free energy values expressed in mJ/m^2 (erg/cm^2) are conveniently designated γ_{LA}, γ_{SA}, and γ_{SL}; the subscripts relate to the liquid–air, solid–air, and solid–liquid interfaces, respectively. The equilibrium contact angle at the boundary line θ_{SLA} may be regarded as a thermodynamic quantity since it is functionally related to the free energy values through the Young-Dupré equation:

$$\gamma_{SA} = \gamma_{SL} + \gamma_{LA} \cos \theta_{SLA}$$

When the solid substrate is placed in the bath, the air is displaced by the bath, B. The SA interface is replaced by an SB interface. Similarly, an LB interface replaces the LA interface. The equilibrium free energy values of these new interfaces are not established immediately but gradually through mass transfer (if there is any mutual solubility between L and B; it is assumed that B does not dissolve S) and through adsorption of dissolved components. When these processes have gone to completion the new relationship is:

$$\gamma_{SB} = \gamma_{SL} + \gamma_{LB} \cos \theta_{SLB}$$

In general, γ_{SB} and γ_{LB} are lower than γ_{SA} and γ_{LA}. If B contains a surface-active agent, they tend to become exceptionally low and θ_{SLB} is therefore much greater than θ_{SLA}. As θ_{SLB} approaches the theoretical limit of 180°, the bath theoretically displaces all the soil. If θ_{SLB} is greater than 90°, even though less than 180°, hydraulic action is capable of removing the oil droplet completely from the soil and surface.

Rearrangement of the above equation gives

$$\cos \theta = \frac{\gamma_{SB} - \gamma_{SL}}{\gamma_{LB}}$$

For $\theta > 90°$, there must be $\cos \theta < 0°$. Since the interfacial tension terms are themselves positive, for $\cos \theta < 0$, it follows that $\gamma_{SL} > \gamma_{SB}$. In other words, for effective roll-up, ie, $\theta > 90°$, the interfacial tension between the solid and bath must be less than that between the solid and liquid. As shown in Figure 8, the area of contact can be reduced to zero while maintaining the contact angle and its equilibrium value, and avoiding any necking down and division of the droplet. If θ_{SLB} is less than 90°, it is impossible to separate the oil completely from the surface by hydraulic action alone. As the droplet is withdrawn and the area of contact is reduced, a neck is formed directly above the contact region. If the contact angle is maintained at equilibrium, the drop must divide at this neck, leaving a small quantity of oil adherent to the substrate. This effect is shown in Figure 9.

The value of θ_{SLB} can be estimated on purely theoretical grounds from estimates of the adsorption of surfactant which, in turn, can be estimated from the Gibbs adsorption equation relating adsorption to surface-tension lowering.

Even when the equilibrium value of θ_{SLB} approaches 180°, it is quite difficult in practice to displace all the oil from the substrate by the bath for the following reasons: The contact angle of the oil, as it is rolling back, is a receding contact angle that is considerably smaller than the equilibrium angle. Any surface roughness further in-

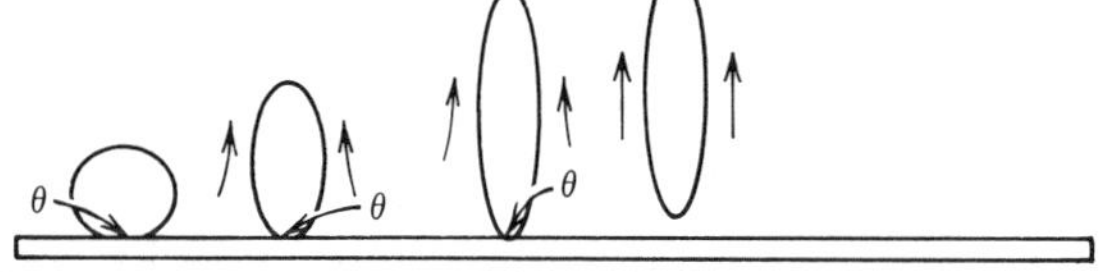

Figure 8. With θ remaining constant at >90°, oil droplet can be removed completely by hydraulic currents (arrows).

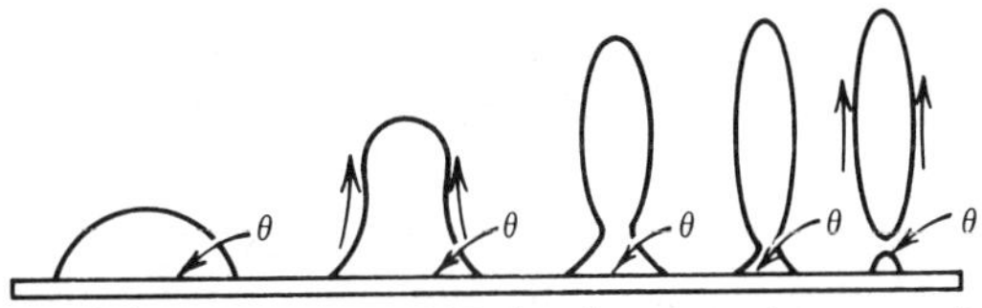

Figure 9. With θ remaining constant at <90°, droplet cannot be removed completely by hydraulic currents (arrows). A small droplet is left attached to the substrate.

creases the hysteresis. Thus, an actual angle of 180° may not be achieved even though the theoretical equilibrium angle has this maximum value. Solid surfaces are notoriously inhomogeneous. Even a small surface area exhibits roughness or spots of high interfacial energy where the moving boundary tends to stick. The hydraulic currents then tear away the bulk of the droplet, leaving some oil at the rough or sticky spot.

In this model, the surface is oil-free under steady-state conditions only when θ_{SLB} is 180°. At any lower value, free-floating oil droplets redeposit and become attached to the substrate, and at equilibrium (on the all-or-nothing basis) all the oil is attached. When θ_{SLB} is high, however, even though it is not 180°, agitation keeps a large proportion of oil separated from the substrate and moving around in the bath. Thus, if the substrate is removed from the agitated bath and rinsed, it is freed of most of its soil burden.

This model for oily soil removal does not apply to systems where the oily soil becomes emulsified. Emulsified oil droplets are characterized by a surface layer that acts either as a physical barrier or as a potential energy barrier against coalescence, or against coming into intimate contact with the substrate. Such droplets must overcome this potential barrier to redeposit on the substrate, whereas the free-floating unemulsified oil droplets of the roll-up model redeposit on contact without surmounting any barrier, energetic or physical. The mathematical model for emulsification can be found in standard treatises on emulsification. It resembles the model for solid-soil dispersion outlined above, although there are many important differences of detail.

Solubilization. The role of micellar solubilization (as the term is used in the physical chemistry of surfactants) in oily-soil removal has been debated for many years. The amount of oily soil that could be present in a normal wash load could not all be removed and held in micellar solution by anionic surfactants. On the other hand, nonionic surfactants could do so, because of their greater solubilizing ability. High solubilizing power is definitely linked with good detergency (155). Thus, a very direct relationship between the solubilizing power of a surfactant for the test dyestuff Orange OT and its ability to remove polar solid from steel surfaces was established (156). In a practical detergency range lying between the surfactant concentration that gives 90% soil removal and a surfactant concentration twice that high, the relationship is expressed by the equation:

$$D = K_1 S + K_2$$

where D is the detergency value, S is the solubilization value, and K_1 and K_2 are constants. Furthermore, in a series of polyoxyethylene nonylphenol nonionic surfactants, the value of K_1 varied linearly with the HLB number of the surfactant. The value of K_2 varied linearly with the log of the interfacial tension measured at the surfactant concentration that gives 90% soil removal. Carrying the correlations still further, it was found that from the detergency equation of a single surfactant with three different polar soils, K_1 was a function of the soil's dipole moment and K_2 a function of the soil's surface tension (156).

Detailed thermodynamic and mechanistic analyses of solubilization are given in refs. 97 and 157. These works showed that under proper circumstances, solubilization can make a significant contribution to oily-soil removal.

Phase Changes at the Soil–Bath Interface. Closely related to solubilization is a phenomenon that involves polar organic soils and surfactant solutions. If a complete phase diagram is plotted for a ternary system containing sodium dodecyl sulfate (or another representative water-soluble surfactant), a polar organic compound (eg, glycerol oleate), and water, several important and unusual features are noted: (*1*) A large area represents a liquid phase consisting of a microemulsion, where the dispersed particles are so small that the system is isotropic, like the familiar soluble oils. (*2*) Over another large area, a liquid crystalline phase is formed, containing all three components. This liquid crystalline phase flows like a liquid, at least in one direction. Flow perpendicular to the oriented planes is accomplished by folding the planes cylindrically, but the physical flow is still of the purely viscous type, with no yield point evident. These two phases, particularly the liquid–crystal phase, play an important part in detergency (158). Furthermore, liquid–crystal formation lowers interfacial tension (159). Although this phenomenon was demonstrated in tertiary oil recovery, the principles could also apply to oily-soil detergency.

When the polar organic component is a solid at ordinary temperatures, the addition of detergent and water markedly lowers the melting point; more specifically as the temperature is raised, a point is reached where surfactant and water penetrate the solid. Thus, the ternary liquid–crystal phase might form spontaneously at room temperature by mixing the components, or, more precisely, an aqueous detergent solution can literally melt and liquefy a relatively large proportion of solid polar fatty matter (see Liquid crystals). What actually happens when these two phases are placed together at a low temperature and slowly warmed is a slow interaction. At a definite critical temperature, however, penetration of the detergent solution into the polar material starts to take place rapidly and the mass soon becomes fluid (97). This interaction is especially favorable in the case of nonionic surfactants which are relatively soluble in polar organic compounds and are the active ingredients of choice when forming microemulsions. In addition, above certain temperatures they display multilayer adsorption isotherms (160–165).

These phenomena are most rapid and easiest to observe in fairly concentrated aqueous detergent solutions, that is, minimally 2–5% detergent solutions. In a practical qualitative way, this is a familiar effect, and there are many examples of the extraordinary solvency and cleaning power of concentrated detergent solutions. Penetration can also be demonstrated at low detergent concentrations. As observed microscopically, the penetration occurs in a characteristic manner with the formation of sheathlike structures, termed myelin; they are filled with isotropic liquid but have a liquid crystalline birefringent skin.

In a detersive system containing a dilute surfactant solution and a substrate bearing a solid polar soil, the first effect is adsorption of surfactant at the soil–bath interface. This adsorption is equivalent to the formation of a thin layer of relatively concentrated surfactant solution at the interface, which is continuously renewable and can penetrate the soil phase. Osmotic flow of water and the extrusion of myelin forms follows the penetration, with ultimate formation of an equilibrium phase. This equilibrium phase may be microemulsion rather than liquid crystalline, but in any event it is fluid and flushable from the substrate surface. This phase change effect explains the detersive behavior of sucrose fatty esters in admixture with alkylarenesulfonates (166).

Measurement of Detergency

The measurement of detergency can be approached from two different points of view. The theoretical approach is concerned with the relative quantity of soil bound to the substrate before and after washing. In this case, measurement is a necessary analytical procedure in the study of the detergency mechanism. The second approach emphasizes the development of reproducible laboratory methods that predict the results of practical cleaning operations. In the development of new household-cleaning compositions, for example, the formulator must know whether his products outperform others under actual use conditions. Realism, or accuracy as it is usually termed, is a prime requisite of any detergency test. It means good correlation between laboratory evaluation and the results of field testing. The practical field evaluation is usually made on the basis of specifications, sometimes implied rather than explicitly expressed, that determine whether or not the cleaning results were satisfactory. In removing spinning lubricants from wool yarns, for example, the most satisfactory cleaning procedure is that which removes most soil from the substrate. In certain metal-cleaning operations, however, the satisfactory outcome is a piece of metal free of solid soil but carrying an even and easily perceivable layer of rust-preventive oil. In judging the cleanness of white fabrics, more interest is usually shown in fabric whiteness rather than in its actual soil content. Thus, most detergency evaluations in which white fabric is the substrate specify cleanness in terms of fabric whiteness or reflectance.

With these limitations in mind, the measurement of detergency in the laboratory requires the following components: A means for measuring or estimating the amount of soil on the substrate or the degree of cleanness both before and after washing; satisfactory substrates and soiling compositions; a means for applying soil to substrate in a realistic manner; and a realistic and reproducible cleaning device. These fundamental requirements apply regardless of the particular type of substrate that is being cleaned. More attention in this area has been centered on textile fabrics than on other substrates but the various substrates and even human skin can be considered from the same point of view.

Fabric Detergency. ***Laundering.*** Reflectance is the most commonly used measurement for the whiteness of fabrics, although the transmittance of light by fabric specimens can also be used. The most commonly used instrument for reflectance measurement is the Gardner colorimeter, although the Zeiss Elrepho is also used. For general detergency, the grayness of the fabric is measured. Color effects can also be measured and fabric yellowing is especially important. It is masked by fluorescent whitening agents (FWA). Special filters are available to eliminate this effect and whitening caused by soil removal can be distinguished from that of FWA deposition.

As would be expected, no single artificial soil or combination of artificial soils can adequately model natural soils in the household laundry. Natural soils vary widely within a single household, between households, and between regions. In addition, natural soiling includes effects of aging and repetitive soiling that are difficult to simulate artificially. Consequently, there is no single set of test conditions that can consistently rate a group of detergents in a completely realistic manner. Limited correlations are usually obtained by suitably adjusting both the laboratory procedure and the practical full-scale procedure.

These problems can be dealt with by using artificial test cloths impregnated with various approximations of natural soils such as vacuum cleaner dust, dirt from air-conditioner filters, clays, carbon black, fatty acids, dirty motor oil, and artificial sebum, either alone or in combination (114,167–171). The soils are applied by spraying, immersion, or padding. If the soils are carefully applied, reproducible results can be obtained. Soil-test cloths can be of great help in detergency studies, when used with an understanding of their limitations.

The device most widely used for laboratory fabric detergency measurements is the Terg-O-Tometer (U.S. Testing Co.), a miniature agitator washer. Just as it is difficult to model natural soils, it is difficult to model a full-scale washing machine. The Terg-O-Tometer consists of four or six small agitator washers in 2-L beakers. Water and detergent are placed into the beakers, and the temperature controlled by a water bath. Standard-soil cloths are added and washed for 5–15 min. The speed as well as the angle of oscillation of the agitators is adjustable. The soil cloths are then rinsed, usually in the Terg-O-Tometer, dried, and the reflectances read. The performance of various detergent mixtures can be compared on the basis of the final reflectance R_f of the washed soil cloths. However, it is generally more useful to express the cleaning as percentage detergency, $\% D$:

$$\% D = \frac{R_f - R_i}{R_o - R_i} \times 100$$

where R_i is the initial reflectance (before the wash) of the soil cloths and R_o the reflectance of the unsoiled fabric used to prepare the soil cloths.

Redeposition of soil can be estimated simultaneously with net soil removal (detergency) by including a white swatch with the soiled swatches. It can also be estimated separately by adding a measured amount of soil to a fresh wash bath, then treating the white swatch in this standard-soil bath. The former method is, superficially at least, more realistic. The latter method, however, is frequently of greater value in development work because both the quantity and composition of the soil are known and can be controlled.

Grayness of a fabric swatch is not directly proportional to its content of black pigment (or artificial soil). A basic formula relating reflectance to the pigment content or concentration can be applied to the evaluation of detergency test swatches (127,172–174). In simple form, it states that the quantity $(1 - R)^2/2R$ (where R is the fraction of light reflected from the sample) is a linear function of the soil content of the sample.

In some cases, it may be impossible, or undesirable, to measure the amount of soil by reflectance. Soil can also be determined by extraction and weighing the cloths, or weighing the ashed cloths.

A number of excellent studies have used a variety of radiolabeled soils to investigate the removal of small amounts of colorless soils such as oils (175–177). Indeed, by proper use of different radiolabels (such as ^3H and ^{14}C), the preferential removal of various components in a soil mixture can be followed.

The above tests all measure the ability of a washing system to remove soil from a fabric. In a single wash however, the detergent or washing machine is often judged by the way the linen appears after several soil–wash cycles. After a series of soilings and washings, the linen acquires an off-white gray or yellowish shade caused by soil accumulation and chemical changes in the soil. Although soil accumulation tests are more tedious than soil removal tests, they give the most realistic results. A laboratory-scale soil accumulation test using vacuum cleaner dirt and small swatches of fabric has been described (178).

Textile Mill Operations. Detergency is important in textile finishing because small quantities of foreign matter on the goods can interfere seriously with dyeing and other finishing treatments. Furthermore, the goods are expected to be uniformly and thoroughly clean when sold. Many detergency tests in this area are of the semipractical type, ie, test swatches are taken from articles that are actually being washed in the plant, and these swatches are analyzed for soil content. This analysis generally consists of gravimetric determination of the soil content either directly from the fabric weight, or by extraction (see Textiles).

Hard-Surface Detergency. Despite the variety of hard-surface objects that are purposefully cleaned at regular intervals, detergency has been studied quantitatively in relatively few cases only. The small-scale user normally judges washing results as satisfactory or unsatisfactory. If satisfactory results are obtained with the amount of detergent and the degree of mechanical action employed, the user is not interested in minor quantitative differences. In those areas where specifications are important and where differences among detergents or mechanical washing equipment are readily perceivable, quantitative methods for measuring detergency have been developed.

In specific cases of metal cleaning where small amounts of residual soil must be detected and are difficult to measure by conventional means, radiotracer methods have been employed (179). Interest in these techniques has been stimulated by the development of methods for decontaminating hard surfaces subjected to atomic fall-out (180).

Quantitative measurements have been obtained for ceramics and glass, metals, and organic surfaces such as painted and plastic tile.

Glassware and Dishwashing. Dishes are washed either by hand or in an automatic dishwashing machine. Hand-dishwashing detergents are generally high foaming compositions containing organic surfactants as the main ingredient. The consumer judges efficiency not only by the cleanness of plates but also by foam persisting throughout the operation. Evaluation of hand-dishwashing products by manufacturers simulates this procedure. The number of plates that can be washed clean, judged visually without or with a color or fluorescence indicator, and the number of plates necessary to kill the foam in the dishpan is taken as the measure of detersive efficiency (181–182).

More objective laboratory methods employ a mechanical device such as a Terg-O-Tometer (183). Food soils are applied to microscope slides or glass tape rather than to actual plates. The soils are tagged with fluorescent materials or with dark pigment to facilitate measurement of residual soil. Reflectance or transmittance may also be read directly (184).

The foam stability of hand-dishwashing compositions can also be measured more directly and more quantitatively using mechanical means to whip up a foam and adding increments of food soil to a predetermined no-foam end-point (185).

The detergents in automatic-dishwashing compositions are largely inorganic but evaluation of residual soil is essentially the same as in hand-dishwashing tests. Low foam avoids inhibition of the free movement of the rotor. Clarity of glass ware is a particularly prized performance feature of home dishwashers and special photometric methods have been reported for measuring freedom from water spots or filming (haze) (186). More frequently, however, the presence of spots and filming is assessed visually in a box in which light is beamed at the interior of inverted glasses in an otherwise black environment.

In restaurant operations, sterilization of dishes is an additional requirement (see Sterilization techniques). Sterilization is determined by the usual swabbing and culturing methods or by employing bacteria tagged with radioactive phosphorus and counting residual radioactivity on the washed dishes (181).

Metal Cleaning. The purpose of cleaning steel is to remove dirt and leave the article in a state where it can be delivered for use without further finishing (see Metal surface treatments). The surface must therefore be covered with a tenacious corrosion-resistant coating as it emerges from the cleaning bath. Many emulsion cleaners remove lubricants and other unwanted dirt while depositing an anticorrosive coating on the metal. The primary test for efficacy in this situation is a corrosive test of the cleaned article.

A more thorough cleaning prepares the surface for further finishing, eg, electroplating or painting, or application of an oxide finish. The basic objective in this case is to ensure maximum adhesion of the finish by freeing the surface of all foreign matter (187). Frequently, the cleaning compositions are straight organic solvents, hydrocarbons or chlorinated hydrocarbons, but in many cases aqueous detergent solutions are used. Regardless of the composition of the cleaning product, detergency test methods involve soiled test specimen, a standard method or device for performing the cleaning operation, and a quantitative method for estimating the extent of soil removal. The test specimens are mostly metal panels or coupons, 10–300 cm^2 in area. They are carefully cleaned, usually by abrasive action, and then soiled. Soils may vary widely but often are based on lubricating oil or asphalt. A dark pigment or fluorescent marker is added if the cleaning result is to be evaluated visually or by photometry. A radioactive tracer, compatible or identical with the soil, may also be added for assessment of residual soil. The cleaning device is often a miniature version of large-scale cleaning equipment which can provide for soaking of the test article and for agitation by hydraulic action or boiling.

When a quantitative estimate of residual soil is not called for and the suitability of a metal surface for further finishing needs to be assessed, the water-break test is used. The term water-break refers to the behavior of a water film on a smooth greasy surface. When the film becomes sufficiently thin by drainage, it suddenly breaks into islands or droplets between which the surface appears dry. On the other hand, when a film drains from a clean water-wettable, nongreasy surface, it becomes progressively thinner and finally disappears by evaporation without ever breaking into droplets. Such a surface is said to be free from water break.

In a similar procedure, the atomizer test, which depends on the behavior of an advancing rather than a receding contact angle, a fine mist of water is applied to the metal surface and the spreading of water is observed. On a clean surface, water spreads to a uniform film. With oleic acid as the test soil, the atomizer test can detect the presence of 10^{-5} mg of soil per cm^2, less than a monomolecular layer (188). For steel which is to be electroplated, the copper dip test is often employed. Steel is dipped into a cupric salt solution and the evenness of the resulting metallic copper deposit is noted.

Correlation of all aspects of the test method with the practical system of interest is always important. The test used for dairy cleaning is an excellent example (189). Milk is used to tag the soil with radioactive ^{45}Ca by an exchange with radioactive $CaCl_2$.

This treatment is applied to stainless steel planchets by suspending the planchets in milk under actual pasteurizing conditions.

Organic Surfaces. Tests for detergency on organic surfaces such as painted walls and plastic tile generally include a rubbing or sponging step corresponding to the manner in which such surfaces are cleaned in practice. The method adopted by the Chemical Specialties Manufacturers' Association includes scrubbing a painted glass surface soiled with a marking pencil in a Gardner Straight Line Washability Machine and assessing detersive efficacy visually on a seven point scale (190). In a similar method, a preconditioned linoleum tile is coated with an oily pigment soil and scrubbed in a Gardner machine; the detersive efficacy is measured by reflectance change (191).

Detergent Manufacture

Liquid Products. The manufacture of liquid detergent products is a straightforward process requiring simple equipment with provisions for metered addition of individual ingredients, agitation, and, if needed, heating.

Spray-Dried Powders. The manufacture of powdered products is more complicated. In the last three decades, high-pressure spray-drying of an aqueous slurry has replaced the earlier process in which a solidified cake of the product had to be broken up mechanically. Spray-drying equipment requires a relatively high capital outlay. The resulting product, however is characterized by several desirable features: a high concentration of surfactant can be accommodated in the finished product; the product consists of hollow beads that dissolve rapidly in the washing solution; powders are considerably less dusty than those produced by the earlier process; powders generally do not lump and most compositions can be packaged in cartons without special liners; and bulk densities are obtainable within the general range of 0.25–0.65 g/L.

First, a slurry is prepared in a crutcher, which is a mixing vessel equipped with heavy-duty agitation and provisions for heating. Solid and liquid components are combined to form a slurry as homogeneous as practicable. In a batch operation, two crutchers are employed; while one crutcher batch is discharged into the spray-drying tower, another batch is prepared in the second crutcher.

Continuous operation can be automated with the aid of load-cell scales and strain-gauge weighing systems. This system offers shorter and more precisely defined residence and hence greater product homogeneity (192). In addition, continuous slurry preparation permits higher solids concentrations and hence reduces the amount of water that must be evaporated. The resulting energy savings can be significant in large installations.

Acids, such as fatty acids and alkylbenzenesulfonic acids, are neutralized with NaOH during slurry preparation to form soap and sodium alkylbenzenesulfonate, respectively.

After preparation, the slurry is transported to an aging vessel. During a residence time of 20–30 min, the neutralization process, hydration of sodium tripolyphosphate and structural changes in the slurry are completed to provide a homogeneous composition. By means of a high pressure pump, the slurry is conveyed to the spray-drying tower under a pressure of ca 10 MPa (100 atm). A representative spray drying tower layout is shown in Figure 10. The slurry, at 80–100°C, is forced through nozzles of 2.5–3.5 mm dia arranged on a nozzle ring. In the tower, the slurry encounters hot air

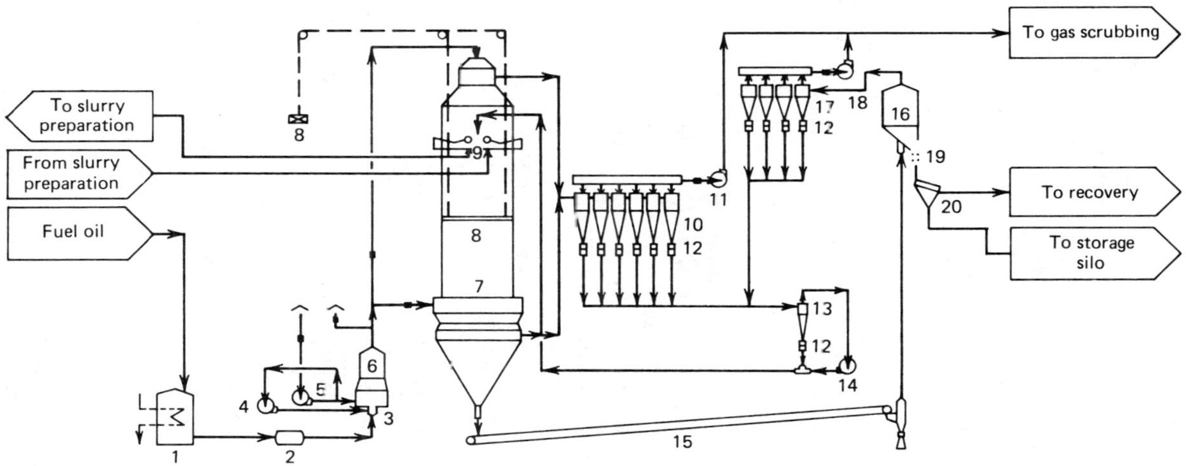

Figure 10. Plant for spray drying of detergents and soaps. 1, Fuel oil day tank; 2, fuel oil pumping station; 3, burner; 4, air combustion; 5, cold air fan; 6, hot air generator; 7, spray-drying tower; 8, cleaning ring; 9, nozzles circuit; 10, tower suction cyclones group; 11, tower suction fan; 12, double flap dischargers; 13, recovery powder cyclone; 14, recovery powder fan; 15, belt conveyor; 16, air lift; 17, air-lift suction cyclones; 18, air-lift suction fan; 19, double flip dischargers; 20, vibrating sieve.

which has entered the tower at 250–350°C. Air flow may be co- or countercurrent. Air circulation is provided by two blowers. A third blower provides suction which carries fines into the collecting cyclones. From there the fines may be reintroduced into the upper part of the tower where they are contacted by wet slurry descending in the tower. Alternatively, the fines may be added to subsequent crutcher batches. Upon exit, the powder is conditioned during passage via a belt conveyor and an airlift to the packaging machinery. The larger spray-drying towers can produce 10–20 t/h.

Because of stringent air-pollution rules the exit gases are wet-scrubbed in brine since high NaCl concentrations reduce foam formation; NaOH scrubs out SO_2 from sulfur-containing fuel. The scrubbing solution saturated with detergent fines is recycled to the water tank for slurry preparation. As shown in Figure 9, the tower is also provided with a cleaning ring which is moved along the tower interior when needed to prevent build-up of powder along the tower walls.

Although spray drying accommodates relatively high content of surfactants, certain types, such as the alkanolamides and some nonionic surfactants are best added to the product after spray drying. Postaddition not only protects the surfactant from the heat of the tower but also prevents the formation of aerosols in the exit gas. Aerosols are more difficult to trap in the scrubbing system than solid fines. They are formed by unsulfonated matter from the manufacture of LAS and nonionic surfactants with short ethylene oxide chains (193).

Dry-Blended Powders. In addition to lower capital outlay, dry blending requires considerably less processing energy. Dry blending of powders provides little control over product density. Final product density, which is usually near unity, depends on the density of the starting materials. Moreover, particle size and product densities are understandably less constant than in spray-dried powders (see Mixing and blending).

Agglomerated Powders. The process of agglomeration is intermediate between spray-drying and dry-blending. Process water concentrations are between 35–45% in a crutcher slurry and essentially zero in dry blending. In agglomeration, a spray of water or other liquid is aimed at dry powder under agitation. For best results, the liquid must be finely divided, liquid and powder must be suspended in space, and both must be kept in motion in order to constantly expose fresh material. Equipment designs include stationary mixers, rotating mixers with spray nozzles, and rotating blenders with liquid dispersion bars, either twin shell or continuous zig-zag (194). Automatic-dishwashing detergents are mostly manufactured by agglomeration (195). Sodium silicate is the usual agglomeration liquid. Other examples of agglomeration processes include the hydration of sodium tripolyphosphate, coloring (speckling) of detergent powders, and the agglomeration of spray tower fines.

Analysis

The literature on analytical methods is voluminous and not easily summarized (200–203). Often the greatest expertise in the analysis of complex detergent mixtures resides with in-house personnel in individual companies who may regard their methods as proprietary information.

An unknown commercial detergent may contain some combination of anionic, nonionic, cationic, and possibly amphoteric surfactants, inorganic builders and fillers as well as some minor additives. In general, the analytical scheme includes separation of nonsurfactant and inorganic components from the total mixture; classification of the surfactants; separation of individual surfactants; and quantitative determination (204).

Health and Safety Factors; Environmental Considerations

As a class, surfactants and detergent products are among the most widely used chemical compositions. Almost everyone is exposed to these products on a daily basis in situations that range from ingestion of food-grade emulsifiers to intimate contact of skin and eyes with personal-care and laundry products. Safety is therefore a matter of great importance (205–206). Ranges of surfactant LD_{50} values are shown in Table 40.

Under conditions of normal use, detergent products are not hazardous to users. Nonetheless, surfactants possess some toxicity and they are mild irritants. Particularly under conditions of misuse, such as accidental ingestion or spillage, they can produce irritation and discomfort in the form of nausea and vomiting as well as irritation to skin and eyes. The long-term effects, however, are minimal (207).

In the last decade, governmental concern with the safety of chemicals has led to a number of legislative acts that regulate the manufacture and sale of chemicals including detergents. The first one was the Food and Drug Act passed in 1938. Food-grade emulsifiers fall under its provisions as well as products containing antimicrobial agents such as deodorant soap bars.

Like other chemicals, new substances introduced into detergent products are regulated by the TSCA of 1977. Since its purpose is to prevent chemicals with long-term deleterious effects from entering the marketplace, the emphasis in testing is on long-term effects such as carcinogenicity, mutagenicity, and teratogenicity.

The manufacture of surfactants and of detergent products is regulated by the OSHA. Dust concentration in detergent plants as well as factory noise levels are the primary areas of relevance, since the individual components in these products are essentially non-hazardous.

Of more immediate concern to the detergent industry is the Federal Hazardous Substances Act (FHSA) of 1960 and the Consumer Products Safety Act (CPSA) of 1972. The FHSA defines specific labeling requirements, such as "Danger" for extremely flammable, corrosive, or highly toxic substances and "Warning" or "Caution" for less hazardous materials.

To assess the degree of hazard, animal acute oral toxicity data are generally relied upon even though such data have limitations in their applicability for predicting possible human effects in the case of a specific exposure. Ranges of detergent LD_{50} values ratings are given in Table 41.

Table 40. Rat Oral LD_{50} Values of Surfactant[a]

Type of compound	Oral LD_{50} (rat)
alkylbenzenesulfonates	700–2,480 mg/kg
alcohol ethoxylates	1,600 to greater than 25,000 mg/kg
sulfated alcohol ethoxylates	7,000 to greater than 50,000 mg/kg
alcohol sulfates	5,000–15,000 mg/kg

[a] Ref. 205.

Table 41. Detergent Animal Acute Oral LD_{50} Versus Probable Lethal Dose for a Human Adult [a]

Animal acute oral LD_{50} (g/kg)		Toxicity rating	Probable lethal oral dose for 70 kg person, g
<0.005	6	super toxic	taste, <0.35 [b]
0.005–0.050	5	extremely toxic	0.35–4.9 (1 tsp)
0.05–0.50	4	very toxic	4.9–28.3 (t tsp–1 oz)
0.5–5.0	3	moderately toxic	28.3–453 (1 oz–1 pt or 1 lb)
5.0–15.0	2	slightly toxic	453–1000 (1 pt–1 qt)
>15	1	practically nontoxic	>1000 (>1 qt, 2.2 lb)

[a] Ref. 208.
[b] 0.35 g = 7 drops.

The range of values for several representative categories of detergent products is given in Table 42.

Because of emesis, it is unlikely that appreciable quantities of most common soaps and detergents could be ingested accidentally.

Methods of testing for eye and skin irritation potential have been reviewed (210). The official FHSA procedure for evaluating ocular irritation potential of detergent products is a modified Draize rabbit eye test (211). Some controversy surrounds this method at present and a search for a procedure less injurious to test animals is in progress. In general, the order of irritation is cationic > anionic > nonionic (212).

Skin irritation potential is assessed by patch tests. More serious than simple irritation is the potential of a product to cause sensitization, ie, to cause a subject to become allergic to even very small amounts of the product.

Several test methods have been devised for sensitization. In the Magnussen-Kligman procedure, guinea pigs are injected just below the skin with a slightly irritating dose of the test substance with and without an antigen. After some days, a patch test is taken on the injection site to stimulate the skin to react. After several weeks, another

Table 42. Acute Oral LD_{50} Ranges for Detergent Products [a]

Detergent type	Albino rat, oral LD_{50}, g/kg
heavy-duty, laundry	
granular	2–7
liquid	2–9
hand-dishwashing, liquid	5–20
automatic dishwashing	2–7
floor-and-wall cleaner	
crystalline	4–6
liquid	8–>16
rug cleaner	7–9
laundry presoak with enzyme, granular	3–11
fabric softener	>10
toilet bar, soap, or synthetic base	7–20
shampoo	
plain	>10
medicated	3–10

[a] Ref. 209.

patch test is taken, this time on an untreated site. A positive reaction is an indication of the sensitizing potential of the test substance (213).

Photosensitization, the potential of a product to cause sensitization in the presence of sunlight, is similarly evaluated by taking a patch test on guinea pigs before and after exposure to uv radiation (214).

Environmental Considerations. The introduction of surfactant products into the environment, after use by consumers or as part of waste disposed during manufacture, is regulated by the Clean Water Act, the Clean Air Act, and the Resource Conservation and Recovery Act. In this respect, surfactants are subject to the same regulations as chemicals in general. There are, however, two areas of specific relevance to surfactants and detergent products, ie, biodegradability and eutrophication.

As early as 1947, incidents of foam blankets on bodies of water were reported. This was at a time when detergents containing synthetic surfactants were in the process of displacing soap as the principal heavy-duty laundry product. It was suspected that they were the cause of foam formation since they were less readily degraded by the microorganisms present in water. In contrast to soap with a straight-chain hydrophobe, alkylbenzenesulfonate, ABS, the predominant surfactant at the time, contained a highly-branched hydrophobe, eg, a propylene tetramer alkyl group. The U.S. surfactant industry, represented by The Soap and Detergent Association (SDA), took the lead in the investigation of the problem, sponsoring a large number of field studies in the development for test methods and analytical procedures for measuring the presence of surfactants in the environment. For anionic surfactants like ABS, the Methylene Blue Active Substance (MBAS) test which depends on the formation of a chloroform-soluble complex of the anionic surfactant with cationic methylene blue, provides the first indication of the extent to which a surfactant has been degraded (215). Nonionic surfactants are assayed by complex formation with cobalt thiocyanate (216).

Among the methods for simulating the fate of the surfactant in the environment, the river die-away test (217) and the shake-flask method (218) have proved acceptable for fast screening and routine use. The semicontinuous activated sludge (SCAS) method is more time-consuming but more accurate and reproducible (219). Determination of biological oxygen demand (BOD) also provides useful data on biodegradation (220).

The extensive investigations led to the conclusion that a branched hydrophobe impedes the rate and extent of degradation of surfactants by microorganisms. The most immediately apparent remedy, therefore, was to replace the propylene tetramer in ABS with a straight hydrocarbon chain giving straight-chain alkylbenzenesulfonate, so-called linear alkanesulfonate (LAS). At the same time, commercialization of the Ziegler process for the oligomerization of ethylene provided another route to straight-chain hydrophobes that could easily be converted to detergent alcohols and straight-chain nonionic surfactants. By 1965, the U.S. detergent industry had completed a voluntary switch from hard to soft surfactants at a cost that has been estimated at ca $\$150 \times 10^6$. In addition to ABS, other surfactants based on propylene oligomers, such as alkylphenol derivatives, have largely disappeared from U.S. consumer laundry products.

Even though the biodegradability problem has been solved for all practical purposes, the subject continues to receive considerable attention. The biodegradation of LAS has been studied intensively and several mechanistic pathways have been identified such as β-, and ω-oxidation as well as reductive and oxidative desulfonation (221). Investigation of the biodegradation of LAS, alcohol ethoxylates, and alkylphenol ethoxylates in the laboratory and under sewage-plant operating conditions showed that LAS and straight-chain alcohol ethoxylates and their sulfates degrade to CO_2 and H_2O (222–223).

Eutrophication. This term, which denotes excessive nutrition or overfertilization, has been applied to the contribution excessive amounts of phosphorus may make to the growth of algae under certain conditions. Phosphorus in water supply originates from run-off of agricultural fertilizers, human excrement, and sodium tripolyphosphate present in detergent formulations. Excessive algal growth in stagnant bodies of water contributes to oxygen depletion which causes starvation of marine life and eventually leads to the death of lakes through silting. Phosphorus can be removed from the effluent of sewage treatment facilities by treatment with materials such as alum. Expansion of sewage treatment has been the method of choice in some areas of the world, notably Sweden, to assure acceptably low phosphorus content in environmental waters (224). In the United States, and later in Western Europe, detergent phosphates were singled out in the early 1960s as the cause of eutrophication and their removal from consumer laundry formulations was proposed as a feasible approach to improvement of environmental water quality. A number of states (Indiana, New York, Minnesota, Vermont, and Michigan) and a number of local jurisdictions, including the city of Chicago, have banned detergent products containing phosphate.

The efforts of the detergent industry toward solution of its part of the eutrophication problem are, at this point, less complete than its response to the biodegradability problem. Soda ash, Na_2CO_3, sodium silicate, and, to a lesser extent, sodium citrate formed the basis of the early formulations marketed in the areas where phosphates were banned. Technically, these substances are considerably less effective than sodium tripolyphosphate. As a precipitant builder, soda ash can lead to undesirable deposits of calcium carbonate on textiles and on washing machines.

At the same time, the industry embarked on an intensive search for phosphate substitutes. Of a very large number of experimental organic builders, a few substances reached commercialization or near-commercialization including trisodium nitrilotriacetate (NTA), trisodium carboxymethoxysuccinate (CMOS), and trisodium carboxymethyltartronate (225). Polymeric builders (polyelectrolytes) proved to be effective calcium sequestrants but failed to satisfy the criterion of acceptable biodegradability. Interestingly, some monomeric polycarboxylates proved to be even more powerful calcium sequestrants that sodium tripolyphosphate but were not sufficiently biodegradable (226).

Trisodium nitrilotriacetate, a sequestrant of effectiveness comparable to sodium tripolyphosphate, reached commercialization in the late 1960s. However, because some laboratory findings suggested potential teratogenicity, it was withdrawn from the market in 1970 at the request of the U.S. Surgeon General. It has continued to be used in Canada and elsewhere. On the basis of many studies supporting the safety of NTA, the EPA has recently dropped its opposition to NTA and this builder, in combination with type-A zeolite, is beginning to appear in some nonphosphate consumer laundry products.

Laboratory assessment of the eutrophication potential of an experimental substance is less clear-cut than that of biodegradation. A frequently used method is the algal assay procedure in which a variety of algal cultures is grown in open shake flasks and the effect of test material on their growth is determined (227). In a second procedure, the MAAP test, one observes the behavior of a microcosm, consisting of bacteria, algae, zooplankton, sediment, and water taken from an oligotrophic lake. The diversity index, an indication of the number of algal species present, is a measure of the nutritive index of the system. A reduction in the diversity index is taken as an indication of the eutrophic potential of the test substance (228).

BIBLIOGRAPHY

"Detergency" in *ECT* 1st ed., Vol. 4, pp. 938–960, by A. M. Schwartz, Harris Research Laboratories; "Detergents" in *ECT* 1st ed., Suppl. 1, pp. 190–223, by A. M. Schwartz, Harris Research Laboratories; "Detergency" in *ECT* 2nd ed., Vol. 6, pp. 853–895, by A. M. Schwartz, Harris Research Laboratories; "Surface Active Agents" in *ECT* 1st ed., Vol. 13, pp. 513–536, by D. M. Price, Oakite Products, Inc.; "Nonionic Surfactants" in *ECT* 1st ed., Suppl. 2, pp. 490–522, by R. L. Mayhew and F. E. Woodward, General Aniline & Film Corporation; "Surfactants" in *ECT* 2nd ed., Vol. 19, pp. 507–593, by C. E. Stevens, Management Research Consultants.

1. G. S. Hartley, *Aqueous Solutions of Paraffin-Chain Salts*, Hermann & Cie, Paris, 1936, p. 45.
2. R. A. Gibbons, *Nature* **200,** 665 (Nov. 1963).
3. B. A. Mulley in M. J. Schick, ed., *Surfactant Science*, Vol. 1, Marcel Dekker, Inc., New York, 1967, Chapt. 13, pp. 422–431.
4. J. Rubinfield, E. M. Emery, and H. D. Cross, *J. Am. Oil Chem. Soc.* **41,** 822 (1964).
5. H. P. Dupuy, American Oil Chemists' Society Nomenclature Subcommittee, *J. Am. Oil Chem. Soc.* **45,** 390A (1968).
6. *CTFA Cosmetic Ingredient Dictionary*, CTFA Information Resources Department, The Cosmetic, Toiletry, and Fragrance Association, Inc., Washington, D.C., 1982.
7. N. K. Adam, *The Physics and Chemistry of Surfaces*, 3rd ed., Oxford University Press, London, 1941, pp. 107–117.
8. C. R. Caryl, *Ind. Eng. Chem.* **33,** 731 (1941).
9. A. M. Mankowich, *J. Am. Oil Chem. Soc.* **43,** 615 (1966).
10. R. Matalon, *J. Colloid Sci.* **8,** 53 (1953)
11. E. D. Goddard and A. Ackilli, *J. Colloid Sci.* **18,** 585 (1963).
12. J. W. McBain and J. C. Henniker, *Colloid Chem.* **7,** 47 (1950).
13. A. S. Weatherburn and C. H. Bayley, *Textile Res. J.* **22,** 797 (1952).
14. L. H. Flett, L. F. Hoyt, and J. Walter, *Am. Dyestuff Rep.* **41,** 139 (1952).
15. H. W. Fox and W. A. Zisman, *J. Colloid Sci.* **7,** 428 (1952).
16. G. J. Kahan, *J. Colloid Sci.* **6,** 571 (1954).
17. E. D. Goddard, M. P. Aronson, and M. L. Gum in T. F. Tadrus, ed., *The Effect of Polymers on Dispersion Properties*, Academic Press, New York, 1981.
18. W. C. Preston, *J. Phys. Colloid Chem.* **52,** 84 (1948).
19. K. Hess, W. Philipoff, and H. Kiessig, *Kolloid Z.* **88,** 49 (1939).
20. C. R. Singleterry, *J. Am. Oil Chem. Soc.* **32,** 446 (1955).
21. S. Kaufman and C. R. Singleterry, *J. Colloid Sci.* **10,** 139 (1955).
22. E. J. Fendler, J. H. Fendler, R. T. Medary, and O. A. El Seoud, *J. Phys. Chem.* **77,** 1432 (1973).
23. O. A. El Seoud, E. J. Fendler, J. H. Fendler, and R. T. Medary, *J. Phys. Chem.* **77,** 1976 (1973).
24. J. H. Fendler, E. J. Fendler, R. T. Medary, and O. A. El Seoud, *J. Chem. Soc. Faraday Trans. 1*, **69,** 280 (1973).
25. R. Goto and co-workers, *J. Chem. Soc. Jpn.* **75,** 73 (1950).
26. M. J. Schick and A. H. Gilbert, *J. Colloid Sci.* **20,** 464 (1965).
27. M. J. Rosen, *Surfactants and Interfacial Phenomena*, John Wiley & Sons, Inc., New York, 1978, pp. 90–92, and references therein.
28. R. R. Balmbra, J. S. Clunie, J. M. Corkill, and J. F. Goodman, *Trans. Faraday Soc.* **58,** 1661 (1962).
29. H. E. Garrett in K. Durham, ed., *Surface Activity and Detergency*, MacMillan and Co., Ltd., London, 1961, pp. 37–38.
30. E. H. Crook, D. B. Fordyce, and G. F. Trebbi, *J. Phys. Chem.* **67,** 1987 (1963).

31. J. H. Clint, *J. Chem. Soc. Faraday Trans. 1*, **71,** 1327 (1975).
32. D. G. Hall, *Trans. Faraday Soc.* **66,** 1351, 1359 (1970).
33. P. Mukerjee and M. J. Mysels, *Critical Micelle Concentration of Aqueous Surfactant Systems*, NSROS-NBS 36, U.S. Department of Commerce, Washington, D.C., 1971.
34. G. J. T. Tiddy, *Phys. Rep.* **57,** 1 (1980).
35. T. A. Bostock, M. P. McDonald, G. J. T. Tiddy, and L. Waring, *Surface Active Agents*, SCI, London, 1979, p. 181.
36. J. M. Corkill, J. F. Goodman, and J. R. Tate, *Trans. Faraday Soc.* **60,** 986 (1964).
37. M. J. Schick and D. J. Manning, *J. Am. Oil Chem. Soc.* **43,** 133 (1966).
38. W. D. Harkins in A. Weissberger, ed., *Physical Methods of Organic Chemistry*, Vol. 1, Interscience Publishers, New York, 1945, Chapt. 6.
39. H. W. Fox and C. H. Chrisman, Jr., *J. Phys. Chem.* **56,** 284 (1952).
40. R. C. Brown and H. McCormack, *Philos. Mag.* **39,** 420 (1948).
41. S. Fordham, *Proc. R. Soc. London Ser. A* **194,** 1 (1948).
42. A. S. Brown, R. U. Robinson, E. H. Sirois, H. G. Thiboult, W. McNeil, and A. Tofias, *J. Phys. Chem.* **56,** 701 (1952).
43. M. Picon, *Ann. Pharm. Fr.* **6,** 84 (1948).
44. K. L. Sutherland, *Aust. J. Chem.* **7,** 319 (1948).
45. J. L. Cayais, R. S. Schecter, and W. H. Wade, *The Measurement of Low Interfacial Tension via the Spinning Drop Technique*, Department of Chemistry and Chemical Engineering, The University of Texas at Austin, Austin, Texas.
46. N. K. Adam, *The Physics and Chemistry of Surfaces*, 3rd ed., Oxford University Press, London, 1941, pp. 180–185, 413.
47. B. J. Carroll, *J. Colloid Interface Sci.* **57,** 488 (1976).
48. A. B. Scott and H. V. Tartar, *J. Am. Chem. Soc.* **65,** 692 (1943).
49. M. L. Corrin and W. D. Harkins, *J. Am. Chem. Soc.* **69,** 679 (1947).
50. S. Ross and J. P. Oliver, *J. Phys. Chem.* **63,** 1671 (1959).
51. H. B. Klevens, *J. Phys. Colloid Chem.* **52,** 130 (1948).
52. H. C. Borghetty and C. A. Bergman, *J. Am. Oil Chem. Soc.* **27,** 88 (1950).
53. W. R. Noble, R. G. Bistline, Jr., and W. M. Linfield, *Soap Cosmet. Chem. Spec.*, 38, 62 (July 1972).
54. U.S. Pat. 2,689,170 (Sept. 14, 1954), W. J. King (to Colgate-Palmolive Co.).
55. U.S. Pats. 2,174,110 (Sept. 26, 1959), and 2,263,312, C. F. Reed (to E. I. du Pont de Nemours & Co., Inc.).
56. W. H. Lockwood, *Chem. Ind. (London)* **62,** 760 (1948).
57. J. M. Davis and E. F. Degering, *Proc. Indiana Acad. Sci.* **56,** 116 (1946).
58. W. E. Truce and J. F. Lyons, *J. Am. Chem. Soc.* **73,** 126 (1951).
59. W. C. Browning in F. Asinger, ed., *International Congress on Surface-Active Substances*, Brussels, 4th, Sect. A, 1964, Vol. 1, Gordon and Breach Science Publishers, New York, pp. 141–154.
60. E. A. Knaggs and J. W. Hodge, *Petroleum Sulfonates—Key Process Chemicals in Micellar Polymer Oil Recovery Systems*, paper presented March 27, 1980, American Chemical Society, Houston Chemical, Marketing & Economics Division.
61. *Petroleum Sulfonates*, Tech Service Bulletin, 5734, Witco Chemical Corp., Division Sonneborn Chemical & Refining Corp., New York, 1973.
62. C. Bluestein and B. R. Bluestein in W. M. Linfield, ed., *Surfactant Science Series*, Vol. 7, Part 2, Marcel Dekker, Inc., New York, 1976, pp. 315–343.
63. U.S. Pat. 3,394,155 (July 23, 1968), A. Cahn and H. Lemaire (to Lever Brothers Co.).
64. J. K. Weil, R. G. Bistline, and A. J. Stirton, *J. Phys. Chem.* **62,** 1083 (1958).
65. J. K. Weil, A. J. Stirton, R. G. Bistline, and E. W. Maurer, *J. Am. Oil Chem. Soc.* **36,** 241 (1959).
66. R. G. Bistline, A. J. Stirton, J. K. Weil, and E. W. Maurer, *J. Am. Oil Chem. Soc.* **34,** 516 (1957).
67. G. A. Kosalapoff, *Organophosphorus Compounds*, John Wiley & Sons, Inc., New York, 1950.
68. R. L. Mayhew and F. Krupin, *Soap Chem. Spec.* **38**(4), 55, 93, 95; (5) 80, 81, 167, 169 (1962).
69. A. M. Schwartz and J. W. Perry, *Surface Active Agents*, Vol. I, Interscience Publishers, Inc., New York, 1949, p. 145.
70. *Synthetic Organic Chemicals, United States Production and Sales, 1980, USITC Publ. 1183*, United States International Trade Commission, U.S. Government Printing Office, Washington, D.C., 1981, pp. 197–227, 247.
71. L. N. Ferguson, *J. Am. Chem. Soc.* **77,** 5288 (1955).

72. H. Arai, *J. Colloid Interface Sci.* **23,** 348 (1957).
73. P. H. Elworthy and C. B. Macfarlane, *J. Chem. Soc.* 311 (1964).
74. J. M. Cross, *Proc. Chem. Specialties Mfrs' Assoc.* 135, 143 (June 1950).
75. U.S. Pat. 2,931,777 (April 5, 1960), H. A. Shelanski (to GAF Corporation).
76. M. N. Fineman, G. L. Brown, and R. J. Meyers, *J. Phys. Chem.* **56,** 963 (1952).
77. A. N. Wrigley, F. D. Smith, and A. J. Stirton, *J. Am. Oil Chem. Soc.* **34,** 39 (1957).
78. U.S. Pat 4,210,764 (July 1, 1980), K. Yang, G. L. Neild, and P. H. Washecheck (to Conoco, Inc.).
79. W. B. Satkowski and W. B. Bennet, *Soap Chem. Spec.* **33**(7), 37 (1957).
80. U.S. Pat. 2,089,212 (Aug. 10, 1937), W. Kritchevsky (to Ninol Laboratories).
81. W. S. Lennon and I. M. Rosenbaum, *Am. Perfum.* **72**(4), 76 (1958).
82. L. J. Garrison and J. H. Pasleau, *Deterg. Age*, 27–29, 98 (Jan. 1968).
83. I. R. Schmolka in ref. 3, Chapt. 10.
84. I. R. Schmolka, *J. Am. Oil Chem. Soc.* **54,** 11 (1977).
85. R. R. Egan, *J. Am. Oil Chem. Soc.* **45,** 481 (1968).
86. U.S. Pat. 3,001,945 (Sept. 26, 1961), H. F. Drew and R. E. Zimmer (to Procter & Gamble Co.).
87. U.S. Pat. 2,999,068 (Sept. 5, 1961), W. Pilcher and S. L. Eton (to Procter & Gamble Co.).
88. K. W. Herrmann, *J. Phys. Chem.* **66,** 295 (1962).
89. D. B. Lake and G. L. K. Hoh, *J. Am. Oil Chem. Soc.* **40,** 628 (1963).
90. H. L. Saunders, J. B. Braunwarth, R. B. McConnell, and R. A. Swenson, *J. Am. Oil Chem. Soc.* **46,** 167 (1969).
91. B. C. Trivedi, A. J. Digioia, and P. J. Menardi, *J. Am. Oil Chem. Soc.* **58,** 754 (1981).
92. J. Weekers, *Tenside Deterg.* **19,** 103 (1982).
93. C. B. Brown, *Research (London)* **1**(10), 46 (Oct. 1947); W. C. Powe and W. L. Marple, *J. Am. Oil Chem. Soc.* **37,** 136 (1960).
94. G. K. Greminger, Jr., and A. S. Teot, *J. Am. Oil Chem. Soc.* **55,** 122 (1978).
95. A. E. Siegrist, *J. Am. Oil Chem. Soc.* **55,** 114 (1978).
96. P. N. Christensen, P. Holm, and B. Sønder, *J. Am. Oil Chem. Soc.* **55,** 109 (1978).
97. M. E. Ginn and J. C. Harris, *J. Am. Oil Chem. Soc.* **38,** 605 (1961).
98. J. A. Shaeiwitz, A. F.-C. Chan, E. L. Cussler, and D. F. Evans, *J. Colloid Interface Sci.* **84,** 47 (1981).
99. K. W. Dillan, E. D. Goddard, and D. A. McKenzie, *J. Am. Oil Chem. Soc.* **56,** 59 (1979).
100. *Ibid.*, **57,** 230 (1980).
101. E. E. Dreger, G. I. Keim, G. D. Miles, L. Shedlovsky, and J. Ross, *Ind. Eng. Chem.* **36,** 610 (1944).
102. H. Stüpel, *Synthetische Wasch-und Reinigungsmittel*, Konradin-Verlag, Stuttgart, 1954.
103. G. E. Barker, *J. Am. Oil Chem. Soc.* **28,** 317 (1951).
104. H. R. Suter and M. G. Kramer, *Soap Sanit. Chem.* **27**(8), 33 (1951).
105. R. L. Mayhew and R. C. Hyatt, *J. Am. Oil Chem. Soc.* **29,** 357 (1952).
106. J. V. Karabinos, G. E. Bartels, and G. E. Kapella, *J. Am. Oil Chem. Soc.* **31,** 419 (1954).
107. F. N. Baumgartner, *Ind. Eng. Chem.* **46,** 1349 (1954).
108. M. S. Furry and E. M. O'Brien, *Am. Dyestuff Rep.* **41,** 861 (1952).
109. R. Bernstein and H. Sosson, *Soap Sanit. Chem.* **30**(1), 46 (1954).
110. F. W. Gray, J. F. Gerecht, and I. J. Krems, *J. Org. Chem.* **20,** 511 (1955).
111. H. A. Ludeman, J. A. Balog, and J. C. Sherrill, *J. Am. Oil Chem. Soc.* **35,** 5 (1958).
112. T. G. Jones in K. Durham, ed., *Surface Activity and Detergency*, Macmillan and Co., Ltd., London, 1961.
113. T. G. Jones and J. P. Parke, *Proceedings of the 4th International Congress on Surface Activity*, 3rd ed., University of Maine Press, 1960, p. 178.
114. B. J. Rutkowski in W. G. Cutler and R. C. Davis, eds., *Surfactant Science Series*, Vol. 5, Part 3, Marcel Dekker, Inc., New York, 1981, p. 1013.
115. S. Stawitz, W. Klaus, and P. Hoepfner, *Fette Seifen Anstrichm.* **60,** 94 (1958).
116. J. W. Hensley and C. G. Inks, *Textile Res. J.* **29,** 505 (June 1959).
117. N. K. Adam, *J. Soc. Dyers Colour.* **53,** 121 (1937).
118. W. Kling, E. Langer, and I. Haussner, *Melliand Textilber.* **25,** 198 (1945).
119. J. C. Stewart and C. S. Whewell, *Textile Res. J.* **30,** 903, 912 (Dec. 1960).
120. J. Compton and W. J. Hart, *Textile Res. J.* **23**(3), 158, 418 (1953); **24**(6), 263 (1954).
121. W. Kling and H. Mahl, *Melliand Textilber.* **35,** 640 (1954).
122. T. H. Shuttleworth and T. G. Jones, *International Congress on Surface Activity, London*, 2nd, Vol. 4, 1957, Academic Press, New York, pp. 52–59.

123. V. W. Tripp, A. T. Moore, B. R. Porter, and M. L. Rollins, *Textile Res. J.* **28,** (June 1958).
124. J. Berch and H. Peper, *Textile Res. J.* **33,** 137 (Feb. 1963); J. Berch, H. Peper, and G. L. Drake, Jr., *Textile Res. J.* **34,** 29 (Jan. 1964).
125. W. C. Powe, *J. Am. Oil Chem. Soc.* **40,** 290 (July 1963).
126. J. C. Dickinson and R. C. Palmer, *J. Textile Inst.* **42,** T510 (Dec. 1951).
127. T. H. Vaughn, A. Vittone, Jr., and L. R. Bacon, *Ind. Eng. Chem.* **33,** 1012 (1941).
128. W. P. Utermohlen, Jr. and E. L. Wallace, *Textile Res. J.* **17,** 676 (1947); L. Loeb, P. B. Sanford, and S. D. Cochran, *J. Am. Oil Chem. Soc.* **41,** 120 (Feb. 1964).
129. H. Schott, *J. Am. Oil Chem. Soc.* **52,** 225 (1974).
130. D. G. Stevenson, *J. Soc. Dyers Colour.* **68,** 57 (1952).
131. I. J. Gruntfest and E. M. Young, *J. Am. Oil Chem. Soc.* **26,** 236 (1949).
132. I. Reich and R. D. Vold, *J. Phys. Chem.* **63,** 1497 (1959).
133. J. C. Harris, *Textile Research J.* **18,** 669 (Nov. 1948); J. C. Harris, *Soap Chem. Spec.* **34**(11), 40 (Nov. 1958); **34**(12), 59 (Dec. 1958); **35**(2), 57 (Feb. 1959).
134. A. S. Weatherburn, G. R. F. Rose, and C. H. Bayley, *Can. J. Res.* **27F,** 179 (1949); **27,** 426 (1949); A. S. Weatherburn and C. H. Bayley, *Textile Res. J.* **22,** 797 (Dec. 1952).
135. K. W. Gardiner and L. B. Smith, *J. Am. Oil Chem. Soc.* **26,** 194 (1949).
136. L. H. Flett, L. F. Hoyt, and J. Walter, *Am. Dyestuff Rep.* **41,** 139 (1952); W. A. Fessler, *Am. Soc. Test. Mater. Spec. Tech. Publ.* **115,** 9 (1951).
137. A. L. Meader, Jr., and B. A. Fries, *Ind. Eng. Chem.* **44,** 1636 (1952).
138. K. McLaren, *J. Soc. Dyers Colour.* **66,** 521 (1950).
139. J. M. Lambert, *Ind. Eng. Chem.* **42,** 1394 (1950).
140. J. Tuzson and B. A. Short, *Textile Res. J.* **30,** 983 (Dec. 1960).
141. B. A. Short in ref. 114, pp. 237–238.
142. B. Derjaguin, *Discuss. Faraday Soc.* **18,** 85 (1954); B. Derjaguin and L. Landau, *Acta Physicochim. URSS* **14,** 633 (1941).
143. E. J. W. Verwey and J. T. G. Overbeek, *Theory of the Stability of Lyophobic Colloids*, Elsevier Publishing Co., Amsterdam, 1948.
144. J. C. Harris, *Textile Res. J.* **28,** 912 (Nov. 1958).
145. K. Durham, *J. Appl. Chem.* **6,** 153 (1956); K. Durham in ref. 122, p. 60.
146. H. Lange, *Kolloid-Z* **154,** 103 (1957); **156,** 108 (1958); *Am. Dyest. Rep.* **50** (12), 25 (June 12, 1961).
147. J. Ross, P. T. Vitale, and A. M. Schwartz, *J. Am. Oil Chem. Soc.* **32,** 200 (1955).
148. P. T. Vitale, *J. Am. Oil Chem. Soc.* **31,** 341 (1954).
149. A. S. Porter in ref. 120, p. 103.
150. W. J. Schwarz, A. R. Martin, and R. C. Davis, *Textile Res. J.* **32,** 1 (Jan. 1962).
151. B. J. Rutkowski and A. R. Martin, *Textile Res. J.* **31,** 892 (Oct. 1961).
152. A. Bierman, *J. Colloid Sci.* **10,** 231 (1955).
153. O. F. Devereux and P. L. De Bruyn, *Interaction of Plane Parallel Double Layers*, M.I.T. Press, Cambridge, Mass., 1963.
154. H. Lange, *J. Phys. Chem.* **64,** 538 (May 1960); *Kolloid Z.* **169,** 124 (1960).
155. G. S. Hattiangdi, W. W. Walton, and J. I. Hoffman, *J. Res. Natl. Bur. Stand.* **42,** 361 (1949).
156. A. M. Mankowich, *J. Am. Oil Chem. Soc.* **39,** 206 (April 1962).
157. R. Mallikarjun and D. B. Dadyburjor, *J. Colloid Int. Sci.* **84,** 73 (1981).
158. A. S. C. Lawrence, *Nature* **183,** 1491 (1959).
159. E. I. Franses, J. E. Puig, Y. Talman, W. G. Miller, L. E. Scriben, and H. T. Davis, *J. Phys. Chem.* **84,** 1547 (1980).
160. J. M. Corkill, J. F. Goodman, and J. R. Tate, *Trans. Faraday Soc.* **62,** 979 (1966).
161. N. A. Klimenko and A. M. Koganovskii, *Kolloidn. Zh.* **36,** 135 (1974).
162. N. A. Klimenko, A. A. Tryasorokova, and A. A. Permilovskaya, *Kolloidn. Zh.* **36,** 624 (1974).
163. N. A. Klimenko, A. A. Permilovskaya, and A. M. Koganovskii, *Kolloidn. Zh.* **36,** 728 (1974).
164. *Ibid.*, **37,** 870 (1975).
165. N. A. Klimenko and co-workers, *Kolloidn. Zh.* **37,** 873 (1975).
166. L. I. Osipow and F. D. Snell, *J. Am. Oil Chem. Soc.* **38,** 184 (1961).
167. F. L. Diehl and J. B. Crowe, *J. Am. Oil Chem. Soc.* **31,** 404 (1954); E. Goette, *Melliand Textilber.* **34,** 534 (1954).
168. C. Z. Draves and O. L. Sherburne, *Am. Dyestuff Rep.* **39,** P771 (1950).
169. H. L. Sanders and J. L. Lambert, *J. Am. Oil Chem. Soc.* **27,** 153 (1950).
170. R. E. Wagg in ref. 120, pp. 35–42.

171. A. R. Martin and R. C. Davis, *Soap Chem. Spec.* **36**(4), 49 (April 1960); (5), 73 (May 1960).
172. O. C. Bacon and J. E. Smith, *Ind. Eng. Chem.* **40,** 2361 (1948).
173. J. C. Harris, M. R. Sullivan, and L. E. Weeks, *Ind. Eng. Chem.* **46,** 1942 (1954).
174. I. Reich, F. D. Snell, and L. I. Osipow, *Ind. Eng. Chem.* **45,** 137 (1953).
175. B. E. Gordon and E. L. Bustin, *J. Am. Oil Chem. Soc.* **45,** 754 (1968).
176. B. E. Gordon and W. T. Shebs, *J. Am. Oil Chem. Soc.* **46,** 537 (1969).
177. J. C. Illman, B. M. Finger, W. T. Shebs, and T. B. Albin, *J. Am. Oil Chem. Soc.* **47,** 379 (1970).
178. A. M. Schwartz and J. Berch, *Soap Chem. Spec.* **39**(5), 78 (May 1963).
179. E. H. Armbruster and G. M. Ridenour, *Soap Chem. Spec.* **31**(7), 47 (1955).
180. G. Segura, C. T. Snell, S. Stigman, and F. D. Snell, *Chem. Ind. (London)*, 1270 (1953).
181. W. G. Mizuno, F. T. Lanners, and J. L. Wilson, *J. Am. Oil Chem. Soc.* **32,** 437 (1955).
182. A. Segesser and H. Stüpel, *Seifen Oele Fette Wachse* **80,** 686 (1954).
183. G. R. Edwards and H. Stüpel, *paper presented at the 4th International Congress on Surface-Active Agents, Brussels, Sept. 1964, Technical Bulletin SC:365-80*, Shell Chemical Co.
184. L. K. Bradacs and W. Schetelig, *J. Am. Oil Chem. Soc.* **41,** 161 (1964).
185. L. E. Weeks, J. C. Harris, and E. L. Brown, *J. Am. Oil Chem. Soc.* **31,** 254 (1954).
186. A. L. Kimmel, H. M. Gadbury, and D. O. Darby, *Soap Chem. Spec.* **37,** 51 (1961).
187. J. C. Harris, W. Stericker, and S. Spring, *ASTM Bull.* **204,** 31 (1955).
188. E. A. Saubestre in ref. 114, Chapt. 19.
189. F. H. Firsching and H. E. Everson, *J. Am. Oil Chem. Soc.* **34,** 547 (1957).
190. *CSMA Designation DCC-04*, Chemical Specialties Manufacturers' Association, Washington, D.C., July 1973.
191. R. R. Adler, R. B. Albin, and B. M. Finger, *J. Am. Oil Chem. Soc.* **46,** 520 (1969).
192. A. Davidsohn, *J. Am. Oil Chem. Soc.* **55,** 134 (1978).
193. J. Kretschmann, H. Latka, and H. Reuter, *Waschmittelchemie*, Henkel & Cie, 1976, pp. 235–252.
194. S. A. Kuti, *J. Am. Oil Chem. Soc.* **55,** 141 (1978).
195. R. J. Fuchs, J. M. Polkowski, and P. P. Carfagno, *Chemical Times Trends*, 37 (1977).
196. G. F. Longman, *The Analysis of Detergents and Detergent Products*, John Wiley & Sons, Inc., New York, 1975.
197. M. J. Rosen and H. Goldsmith, *Systematic Analysis of Surface-Active Agents*, 2nd ed., Wiley-Interscience, New York, 1972.
198. J. Kross in M. J. Schick, ed., *Surfactant Science Series*, Vol. 8, Marcel Dekker, Inc., New York, 1975.
199. J. G. Karabinos, G. E. Kapella, and G. E. Bartels, *Soap Chem. Specialties* **6,** 30, 41 (1954).
200. K. Burger, *Z. Anal. Chem.* **196,** 22 (1963).
201. L. E. Weeks, M. E. Ginn, and G. E. Baker, *Proc. Chem. Specialties Mfrs. Assoc.*, 150 (May 1957).
202. *Standard Methods for the Examination of Water and Wastewater*, 14th ed., American Public Health Association, New York, 1975.
203. R. A. Llenado, *Household Personal Products Industry* **18,** 69, 71, 73 (1981).
204. H. G. Nadeau and S. Siggia in ref. 3, Chapt. 24.
205. *Human Safety and Environmental Aspects of Major Surfactants*, NTIS document PB 301193, a report prepared for the Soap and Detergent Association by Arthur D. Little, Inc., 1977.
206. Ref. 203, *Supplement*, NTIS document PB 1182452, 1981.
207. A. R. Temple, *Detergents—In Depth*, '78, Consumer Information Department, The Soap and Detergent Association, New York, 1978, pp. 36–41.
208. *Cleaning Products and Their Accidental Ingestion*, Scientific and Technical Report No. 5R, The Soap and Detergent Association, New York, 1980, p. 9.
209. *Ibid.*, p. 10.
210. F. N. Marzulli and H. I. Maibach in *Advances in Modern Toxicology*, Vol. 4, John Wiley & Sons, Inc., New York, 1977.
211. *Food, Drug, and Cosmetic Law Reports 233, 440, and 476.*
212. Ref. 210, p. 440.
213. A. H. Gilbert, *J. Am. Oil Chem. Soc.* **58,** 66A (1981).
214. Ref. 210, Chapt. 14.
215. J. Longwell and W. D. Maniece, *Analyst* **80,** 685 (1955).
216. S. L. Boyer and co-workers, *Environ. Sci. Technol.* **11,** 1167 (1977).
217. P. J. Weaver, *Soap Chem. Spec.* **41,** 45 (1965).
218. Subcommittee on Biodegradation Test Methods, *J. Am. Oil Chem. Soc.* **42,** 989 (1965).

219. *Ibid.*, 990 (1965).
220. C. A. Vath, *Soap Chem. Spec.* **40**(2), 56, 182 (1964); **40**(3), 55, 108 (1964).
221. R. D. Swisher, *Surfactant Science Series*, Vol. 3, Marcel Dekker, Inc., New York, 1970.
222. R. J. Larson and G. E. Wentler, *Soap Cosmet. Chem. Spec.* **58**(5), 33 (1982).
223. R. L. Huddleston and A. M. Nielson, *Soap Cosmet. Chem. Spec.* **55**, 34, 36, 37, 44 (1979).
224. H. O. Bouveng, *J. Am. Oil Chem. Soc.* **55**, 178 (1978).
225. M. M. Crutchfield, *J. Am. Oil Chem. Soc.* **55**, 58 (1978).
226. H. C. Kemper, R. J. Martens, J. R. Nooi, and C. E. Stubbs, *Tenside* **12**, 47 (1975).
227. *Algal Assay Procedure: Bottle Test*, National Eutrophication Research Program, Environmental Protection Agency, Corvallis, Ore., 1971.
228. D. Mitchell and J. C. Buzzell, Jr., *J. Sanit. Eng.*, 8333 (1971).

General References

D. J. Shaw, *Introduction to Colloid and Surface Chemistry*, 3rd ed., Butterworths, London, 1980.

A. M. Schwartz in E. Matijević, ed., *Surface and Colloid Science*, Vol. 5, John Wiley & Sons, Inc., New York, 1972.

B. J. Carroll in E. Matijević, ed., *Surface and Colloid Science*, Vol. 9, John Wiley & Sons, Inc., New York, 1976.

K. L. Mittal, ed., *Colloidal Dispersions and Micellar Behavior*, ACS Symposium Series 9, American Chemical Society, Washington, D.C., 1975.

K. L. Mittal, ed., *Solution Chemistry of Surfactants*, Vol. 2, Plenum Press, New York, 1979.

P. Hiemenz, *Principles of Colloid and Surface Chemistry*, Marcel Dekker, Inc., New York, 1977.

K. L. Mittal, ed., *Micellization, Solubilization, and Microemulsions*, Vols. 1 and 2, Plenum Press, New York, 1977.

A. W. Adamson, *Physical Chemistry of Surfaces*, 3rd ed., John Wiley & Sons, Inc., New York, 1976.

P. Becher, *Emulsions, Theory and Practice*, 2nd ed., Reinhold Publishing Corp., New York, 1977.

A. M. Schwartz, J. W. Perry, and J. Berch, *Surface Active Agents and Detergents*, Vol. 2, Interscience Publishers, Inc., New York, 1958.

M. J. Rosen, *Surfactants and Interfacial Phenomena*, Wiley-Interscience, New York, 1978.

A. Hüthig, *Waschmittelchemie*, Henkel & Cie, Heidelberg, 1978.

M. J. Schick, ed., *Surfactant Science*, Vol. 1, Marcel Dekker, Inc., New York, 1966.

K. Shinoda, *Solvent Properties of Surfactant Solutions*, Vol. 2, *Surfactant Science*, Marcel Dekker, Inc., New York, 1967.

E. Jungermann, ed., *Surfactant Science Series*, Vol. 4, Marcel Dekker, Inc., New York, 1970.

W. G. Cutler and R. C. Davies, eds., *Surfactant Science Series*, Vol. 5, Parts 1–3, Marcel Dekker, Inc., New York, 1972, 1975, 1981.

W. M. Linfield, ed., *Surfactant Science Series*, Vol. 7, Marcel Dekker, Inc., New York, 1976.

C. Gloxhuber, ed., *Surfactant Science Series*, Vol. 10, Marcel Dekker, Inc., New York, 1980.

M. Balsam and E. Sagarin, eds., *Cosmetics, Science and Technology*, 2nd ed., Wiley-Interscience, New York, Vol. 1, 1972; Vol. 2, 1972, Vol. 3, 1974.

McCutcheon's Detergents & Emulsifiers, 1981 North American Edition, MC Publishing Company, Glen Rock, 1981.

McCutcheon's 1981 Functional Materials, MC Publishing Company, Glen Rock.

ARNO CAHN
Consultant

JESSE L. LYNN, JR.
Lever Brothers Company

SUTURES

Surgical sutures are sterile filaments used to hold tissues together until they heal adequately for self-support or to join tissues with implanted prosthetic devices. They are normally attached to needles for stitching the edges of wounds or surgical incisions. As ligatures they are used, generally without a needle, to tie off ends of severed tubular structures such as blood vessels and ducts to prevent bleeding or other fluid leakage.

Sutures are characterized according to type of material, physical form, biodegradability, size, surgical use, etc. Among the materials utilized for sutures are natural products, such as surgical gut or silk (qv), many commonly known synthetic fibers, and some fibers from synthetic polymers that have been synthesized specifically for use as sutures (see Fibers, chemical). Sutures may be fabricated as monofilaments or multifilaments; the latter are generally braided, but sometimes twisted or spun, and may be coated with waxes, fluorocarbons, silicones, and other polymers to decrease capillarity and improve handling or functional properties.

A typical appearance of a braided suture, a monofilament, and catgut is shown in Figure 1. The broad range of suture diameters required for widely varying surgical procedures is exemplified in Figure 2 which displays an intermediate size 3-0 monofilament suture and a size 2 product tied with a size 10-0 suture.

An absorbable suture is one that degrades in body tissues and disappears from the implant site, usually within two to six months. A nonabsorbable suture is resistant to biodegradation, becomes encapsulated in a fibrous sheath or is otherwise incorporated by the host, and remains in the tissue as a foreign body unless it is surgically removed (eg, skin sutures) or extruded from the site by the host.

Specific designations are employed to signify areas of surgical use, such as ophthalmic or cardiovascular sutures, or to denote functional use, such as retention or closure sutures. These designations are related to the diameter and properties of the filament as well as to the size and type of attached needle which make it particularly appropriate for a specific surgical application.

For more than 4000 yr, surgeons have employed a variety of natural, often exotic, materials as sutures. In ancient India, beetle heads and the jaws of giant ants held wound margins together. Animal sinews and linen threads were mentioned by the early Egyptians. However, before Pasteur, the frequent development of infection and pus at the suture site, believed to be an inevitable consequence of the materials themselves, led surgeons to rely instead on cautery, boiling oil, compression, and other techniques for the control of hemorrhage or for wound closure. A breakthrough occurred in the 1860s when Lister developed modern aseptic surgery and demonstrated that suturing with a low incidence of infection could be achieved by disinfecting the strands with phenol (carbolic acid) (see Disinfectants and antiseptics). Lister also greatly extended the utility and reliability of catgut and other collagenous materials as absorbable sutures through his discovery that chromic salt treatment afforded longer resistance to tissue digestion. In the following years, a wide range of natural and synthetic fibers as well as metallic wires have been investigated in the continuing search for better suture material.

Suture sales in the United States have been estimated to be $\$(150–250) \times 10^6$; somewhat less than twice that amount is sold in other countries. The market is about

Figure 1. Scanning electron photomicrograph of a braided, a monofilament, and a catgut suture (from top to bottom).

equally divided between the absorbable and nonabsorbable types. In the domestic market, catgut remains the leader with >25% of total suture sales. Synthetic absorbables have made rapid inroads and now hold ca 20% of the market share. Silk has dropped to third place with ca 16% of total sales. A unique aspect of this industry is the requirement for an extensive inventory of products, each of which requires conformance to FDA manufacturing, lot number, record-keeping, and monitoring regulations. With the wide range of available suture materials, and the variety of suture diameters, lengths, colors, number of needles attached, needle types, and packaging variations, a full-line manufacturer may offer as many as 1700 different products to satisfy today's broad range of surgical procedures and techniques.

Suture–Tissue Interactions

As with all surgical implants, some tissue response to sutures is inevitable (1–6). Trauma caused by the passage of needle and thread can be minimized but not eliminated by using materials of optimum design and size, proper knot tension, and other aspects of meticulous technique. The initial tissue reaction, ca 5-d duration, is essentially the same for all sutures. Beyond this point, the tissues respond to the foreign

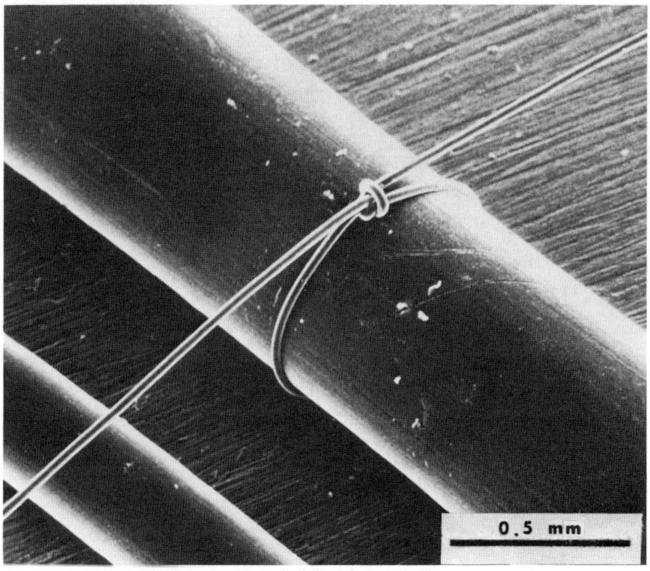

Figure 2. An intermediate size 3-0 monofilament suture, and a size 2 product tied with a size 10-0 suture.

body by absorbing or encapsulating it in an envelope of fibrous connective tissue; the difference in the degree, intensity, and type of cellular reaction depends on the nature of the material used. Catgut and its related collagens evoke the strongest reaction, followed by silk, cotton, and other natural fibers. The synthetic fibers currently used, including polyolefins, polyesters, and polyamides, as well as the newer synthetic absorbable sutures composed of homo- and copolymers of glycolic acid, result in relatively bland reactions as do metallic wires. Obviously, fibrous materials that exhibit gross or localized toxicity are unsuitable as sutures, and special care must be exercised with fibers manufactured principally for textile applications to ensure the absence of undesirable natural impurities or processing materials.

The influence of the biological environment on sutures is an important factor in their effectiveness. Absorbable sutures, whether of natural origin or synthetic, are designed to provide support to tissues during the early stage of healing and then to be eliminated by absorption or metabolic digestion. Nonabsorbable sutures, on the other hand, are intended to provide long-term support to tissues or permanent attachment of tissues to prosthetic devices. It has long been recognized, however, that variation exists in the degree to which various nonabsorbable suture materials retain their strength over long periods of implantation (7).

Mechanical Properties

Tensile strength and knot security are perhaps the two most important mechanical properties of sutures. Since sutures are usually knotted in use, surgeons are concerned with knot-pull, rather than straight-pull tensile strength. This property is generally measured with a tensiometer either by pulling on the free ends of a surgeon's knot (8) or by pulling apart a loop tied with sufficient squared throws to prevent knot slippage (9). As might be expected, metallic sutures are the strongest; natural fibers including linen, silk, catgut, collagen, and cotton are the weakest; and synthetic materials are of intermediate strength.

Knot security, which may be defined as the ability of a suture to resist slippage and untying, is a more elusive property to determine quantitatively than is tensile strength. It appears to be influenced by many factors including the coefficient of friction of the suture material, size, construction, stiffness, type of knot and number of overthrows, moisture absorption, etc (9–15). Metallic sutures, which tend to kink at the knot, show maximum security with minimal throws, followed by uncoated braided sutures of materials such as poly(glycolic acid) or poly(ethylene terephthalate). However, treatment of braids with wax or synthetic coatings, such as is applied to silk, nylon, or polyester, tends to reduce knot security. Catgut, relatively secure when dry, develops a tendency to slip when moistened by tissue fluids (16). In practice, lower knot security is generally rectified by use of additional knot throws as dictated by surgical circumstances or the experience of the surgeon. Sutures with superior tensile strength and knot security permit the use of finer-strand diameters and smaller knots, thus reducing the amount of foreign body placed in the tissues and minimizing the risk of wound disruption and other postoperative complications.

Sterilization

Suture materials must be sterilized when used, and it has become standard practice for manufacturers to sterilize their products before marketing, essentially eliminating a procedure that had mostly been done in the hospital (see also Sterilization techniques). In the United States, the earlier methods of steam and dry-heat sterilization have largely been replaced by ethylene oxide sterilization or ionizing radiation. The choice depends upon economics and the nature of the sutures.

Ethylene oxide is the most widely applicable sterilant as it causes no significant alteration in substrate properties and may be used for materials that are particularly sensitive to steam (eg, absorbable sutures) or ionizing radiation. If metallic foil or other impermeable packaging films are used, open-package sterilization followed by aseptic sealing after degassing is required. Ethylene oxide residues can remain after degassing. Although specific guidelines for residues on sutures have not been set, an interim residue guideline of 250 ppm has been recommended for plastic medicinal implants and devices (17) (see Prosthetic and biomedical devices).

Radiation sterilization offers a fast, continuous, closed-package process with low operating costs if the volume is sufficiently high to justify the initial capital investment (18). The extent of its use for commercial sterilization of sutures is equal to that of ethylene oxide. For sutures, cobalt-60 β-rays with their greater penetrating power are preferred over electron beams. They have been used principally for catgut, nonabsorbable polyester, nylon-6,6, and silk products that withstand sterilization dose levels without unacceptable deterioration. Ethylene oxide is preferred for polypropylene, synthetic absorbable sutures, linen, and cotton (19) (see Sterilization techniques).

In general, sutures are packaged within dual packets designed to protect the sterility of the suture as well as the outer surface of the inner packet which is opened only under aseptic conditions adjacent to a sterile operating field. Figure 3 demonstrates the sequence of events as outer and inner envelopes are opened to permit removal of the suture enclosed in its inner label.

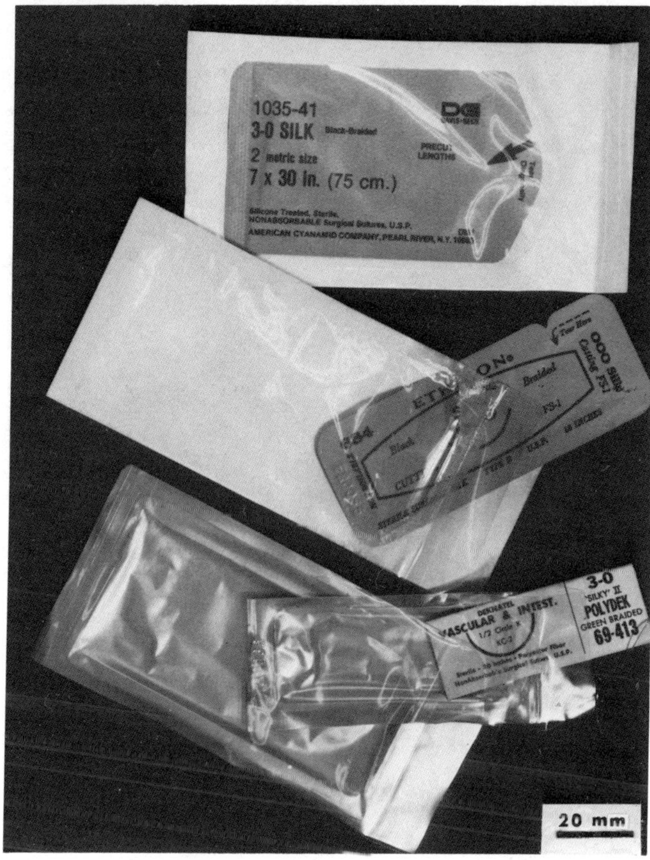

Figure 3. Removal of suture from package.

Needles

Surgical needles may be considered as integral parts of sutures used for stitching. They are generally made from heat-treatable (400 series) stainless steel or from carbon steel which may be nickel-plated to retard corrosion. Wire diameters range from ca 1.55 mm in the large retention or closure needles down to 0.050 mm used for microsurgery. An array of straight and curved shafts is made with blunt, cutting edged, or circularly tapered points; these are specifically tailored for optimum penetration of a variety of soft or tough tissues that may be accessible to the surgeon readily or only with difficulty. Examples of a few of the needle shapes and sizes used in various surgical procedures are displayed in Figure 4.

The traditional eyed needle, threaded in the hospital, has been increasingly replaced by the eyeless types. The latter are sold crimped or swaged to one or both ends of a sterile suture that is fitted into a channel or drilled hole at the end of the shaft. The resultant single strand produces less tissue trauma than the double-stranded end required with the eyed type; tedious and time-consuming needle preparation in the operating suite, particularly troublesome with fine needles, is eliminated.

Figure 4. Surgical needles.

Colorants

Although some surgical procedures require sutures that are not colored, more frequently, dyed or pigmented sutures are used to enhance their visibility while surgery is in progress. Only dyes or pigments are utilized for which Color Additive Petitions have been approved by the Division of Color Technology, Bureau of Foods of the FDA (see Colorants for foods, drugs, and cosmetics). In most cases, the specific dye or pigment lots employed must also be certified by the FDA as meeting the approved chemical specifications, based upon analytical data that have been supplied by the petitioner.

The approved colorant and the sutures for which their use is approved are given in Table 1.

Standards and Test Procedures

For many years the specifications of commercial sutures have been guided principally by standards set forth in The United States Pharmacopeia (USP) (8). In addition to general descriptions of the two classes (absorbable and nonabsorbable), USP specifies methods for measuring length, diameter, knot tensile strength, needle-pullout strength, sterility, extractable color, and packaging and labeling procedures. For collagenous fibers, a test for soluble chromium is given. The USP also permits the use of color additives approved by the FDA.

Diameter ranges and minimum knot strengths are specified for the various suture sizes (see Table 2). Nonabsorbable sutures are subdivided into three USP classes with

Table 1. FDA Approved Colorant and Corresponding Suture Material

Colorant	Colour index	Suture material
chromium–cobalt–aluminum oxide		linear polyethylene [*68071-35-2*]
ferric ammonium citrate/Pyrogallol		plain and chromic catgut
logwood	75290	silk, nylon-6,6 [*32131-17-2*], nylon-6 [*25038-54-4*]
copper(II) phthalocyanine	74160	polypropylene [*9003-07-0*]
FD&C blue no. 2	73015	nylon-6, nylon-6,6
D&C blue no. 6	73000	poly(ethylene terephthalate) [*37353-50-7*], plain or chromic collagen, polypropylene
D&C blue no. 9	69825	cotton or silk
D&C green no. 5	61570	nylon-6,6, nylon-6
D&C green no. 6	61565	poly(ethylene terephthalate), poly(glycolic acid) [*26009-03-0, 26124-68-5*]
D&C violet no. 2	60725	glycolic–lactic acid polyester [*26780-50-7*]

different strength limits: Class I, silk or synthetic fibers, monofilament, twisted, or braided; Class II, cotton, linen, or coated natural or synthetic fibers where the coating forms a casing of significant thickness but does not contribute appreciably to strength; and Class III, mono- or multifilament wire (20).

Diameter ranges for absorbable collagenous sutures are somewhat larger than equivalent sizes of nonabsorbables, a consequence of their tendency to swell in water. The introduction of the Dexon poly(glycolic acid) sutures (American Cyanamid Company, Davis & Geck) in 1971 led to a new USP category, absorbable synthetic sutures; size–diameter range relationships are identical to those of the nonabsorbable classes. Knot-strength specifications are 20–75% higher than for comparable sizes of collagenous absorbables.

Similar standards exist in many foreign countries. The *European Pharmacopeia* (EP) was established for use by the Common Market nations. Joint efforts between the EP representatives and the Revisions Committee of USP led to the addition of the metric size designations in *USP XIX*, which issued in 1975. As a consequence the metric size diameter ranges in EP and USP, with a few small exceptions, are identical.

Regulatory Aspects

Until 1970, only two suture materials, Dexon and Prolene (polypropylene) (Ethicon, Inc.) sutures, had been subjected to a complete premarket clearance as new drugs by the FDA and had received subsequent approval. In 1971, the FDA classified the older suture materials as drugs and specified abbreviated new drug application requirements for manufacture and sale (21). In essence, the manufacture and sale of sutures has since been subject to FDA control like pharmaceutical and biological agents. However, since the enactment of the Medical Device Amendments to the Food and Drug Act in 1976, sutures have been considered devices and are now under the control of the Bureau of Medical Devices (BMD). The introduction of a new polymeric suture material, considered a Class-III product, requires a demonstration of safety and surgical efficacy through controlled laboratory, animal, and clinical studies. On successful completion of such a program, a premarket approval (PMA) is granted by

Table 2. Diameter Ranges and Minimum Knot Strength for Various Suture Sizes[a]

Sutures, USP size	Metric size, gauge no.	Limits on diameter, mm		Limits on knot-pull strength, N[b]		
		Min	Max			
Nonabsorbable[c]				*Class I*	*Class II*	*Class III*[d]
12-0	0.01	0.001	0.009	0.0098		0.0196
11-0	0.1	0.010	0.019	0.0490	0.0392	0.1667
10-0	0.2	0.020	0.029	0.1569	0.1177	0.4904
9-0	0.3	0.030	0.039	0.3531	0.2354	0.5884
8-0	0.4	0.040	0.049	0.5884	0.3923	1.079
7-0	0.5	0.050	0.069	1.079	0.5884	1.569
6-0	0.7	0.070	0.099	1.961	1.079	2.648
5-0	1	0.10	0.149	3.923	2.256	5.296
4-0	1.5	0.15	0.199	5.884	4.511	8.042
3-0	2	0.20	0.249	9.415	6.473	13.34
2-0	3	0.30	0.339	14.12	10.00	17.65
0	3.5	0.35	0.399	21.18	14.22	33.34
1	4	0.40	0.499	26.68	17.75	46.68
2	5	0.50	0.599	34.57	24.91	57.86
3 and 4	6	0.60	0.699	47.86	36.09	89.34
5	7	0.70	0.799	60.41		111.8
6	8	0.80	0.899	71.39		133.4
7	9	0.90	0.999	88.66		155.9
8	10	1.00	1.099			178.5
9	11	1.100	1.199			201.0
10	12	1.200	1.299			223.6
Collagen				*Av, min*	*Individual strand, min*	
	0.01	0.001	0.009			
	0.1	0.010	0.019			
	0.2	0.020	0.029			
9-0	0.3	0.030	0.039			
	0.4	0.040	0.049			
8-0	0.5	0.050	0.069	0.4413	0.2452	
7-0	0.7	0.070	0.099	0.6865	0.5394	
6-0	1	0.10	0.149	1.765	0.9807	
5-0	1.5	0.15	0.199	3.727	1.961	
4-0	2	0.20	0.249	7.551	3.923	
3-0	3	0.30	0.339	12.26	6.669	
2-0	3.5	0.35	0.399	19.61	10.20	
0	4	0.40	0.499	27.17	14.22	
1	5	0.50	0.599	37.27	19.12	
2	6	0.60	0.699	44.23	23.54	
3	7	0.70	0.799	57.86	29.32	
4	8	0.80	0.899	68.65	34.23	
Synthetic					*Av*	
12-0	0.01	0.001	0.009			
11-0	0.1	0.010	0.019			
10-0	0.2	0.020	0.029			
9-0	0.3	0.030	0.039		0.4413	
8-0	0.4	0.040	0.049		0.6865	
7-0	0.5	0.050	0.069		1.373	
6-0	0.7	0.070	0.099		2.452	
5-0	1	0.10	0.149		6.669	
4-0	1.5	0.15	0.199		9.317	
3-0	2	0.20	0.249		17.36	

Table 2 (*continued*)

Sutures, USP size	Metric size, gauge no.	Limits on diameter, mm		Limits on knot-pull strength, N[b]
		Min	Max	
2-0	3	0.30	0.339	26.28
0	3.5	0.35	0.399	38.25
1	4	0.40	0.499	49.82
2	5	0.50	0.599	62.27
3 and 4	6	0.60	0.699	
5	7	0.70	0.799	

[a] Ref. 8. The use of portions of the text of USP XX–NF XV is by permission of the USP Convention. The Convention is not responsible for any inaccuracy of quotation or for any false or misleading implication that may arise from separation of excerpts from the original text or by obsolescence resulting from publication of a supplement.

[b] To convert N to kgf, divide by 9.807.

[c] The limits on knot-pull tensile strength apply to nonabsorbable surgical sutures that have been sterilized. For non-sterile sutures of Class I and Class II, the limits are 25% higher.

[d] The tensile strength of sizes larger than metric size 3 of monofilament Class III (metallic) nonabsorbable surgical suture is measured by straight pull.

the Bureau of Medical Devices. Products considered equivalent to existing commercial sutures and classified as Class I (general controls) or Class II (performance standards) may also be introduced by a new manufacturer, if approved by BMD, following notification by the manufacturer under Section 510(k) of the Act.

Materials

Surgical Gut and Regenerated Collagen. Surgical catgut is essentially composed of the protein collagen and is derived from the small intestines of animals, usually the outer serosal layer in cattle or the submucosal layer in sheep. After food residues and soft tissues are removed by mechanical and chemical stripping, the intestines are slit lengthwise. In order to prolong *in vivo* strength, the ribbons may be treated with a chromic salt solution; this process may be repeated at a later point, or it may be supplemented with an aldehyde cross-linking step. Chromicizing must be controlled to prevent excessive digestion or the development of undesirable mechanical properties. Depending on the size desired, an appropriate number of ribbons are twisted into bonded strands, which are then cut to length, dried, and polished to a uniform diameter. Various regulated heat treatments have been suggested to improve suppleness or for preliminary sterilization, but these require low moisture content in order to avoid tensile-strength loss. Visual and mechanical inspection is carried out at several stages of the process to separate nonuniform segments caused by irregularities present in the intestinal raw material. Since dry catgut is too stiff to be handled, it is packaged in tubing solutions of aqueous ethanol or 2-propanol before sterilization.

Catgut implant studies have been reported for a variety of animal species and tissue sites with inconsistent and often conflicting conclusions regarding tissue reaction and surgical efficacy (2,22). These results are partly due to the variation in techniques and lack of proper controls, but also to the inherent variability of the material itself. In human tissue this observation of nonuniformity has also been made with respect to strength retention (23), and on numerous occasions widely variable absorption rates have been reported (6,24–26).

Manufacturers have made steady improvements in the reliability of this complex natural material, and the extreme ranges originally observed in absorption rate, tissue reaction, and physical properties have been significantly narrowed. A number of approaches have been described to achieve a more uniform material by preparing regenerated collagen strands. These have been made by wet spinning homogenous dispersions of purified collagen obtained from animal tendons. Moderate improvements have been noted in some properties but this material has only limited acceptance as a catgut replacement, principally for fine-sized ophthalmic sutures (19).

The absorption of catgut depends upon the action of proteolytic enzymes, and its rate may differ from one tissue site to another (1). Plain or untreated gut is digested more rapidly and consequently causes a greater degree of early tissue reaction than the chromicized product. Catgut is seldom used for suturing skin because infection often occurs as a sequel to inflammatory tissue response. In addition, infection may accelerate strength loss and absorption, which increase the risk of wound dehiscence in slower-healing sites.

Exposure to aqueous tissue fluids causes catgut to swell and lose ≥20% of its package tensile strength (16); wet catgut also exhibits lower knot security. However, despite its imperfections, catgut has served the surgeon well over the years and is still used more than any other suture material. Until the development of the poly(glycolic acid) sutures, it was the only absorbable suture material available.

Synthetic Absorbable Sutures. The finding that polyester fibers derived from lower α-hydroxy acids had suitable mechanical and biological properties for use as absorbable sutures led to the introduction of Dexon sutures in 1970 (27–28).

Polymers of glycolic acid (**2**) are conveniently prepared by melt polymerization of the dimeric cyclic intermediate glycolide (**1**), preferably with Lewis acid or other metal-salt catalysts (29–30). The use of stannous chloride dihydrate in the presence of small quantities of alcohols has been recommended (31); these combinations function as both initiators and molecular-weight-control agents. The polyester (**2**) has been variously referred to as poly(glycolic acid), polyoxyacetyl, or poly(hydroxyacetic acid).

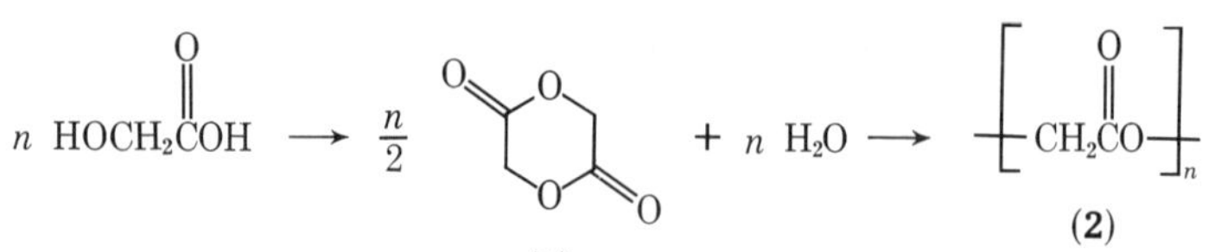

(**1**) (**2**)

Poly(glycolic acid), mp 224–226°C, is converted by conventional melt-spinning and drawing operations into multifilament yarns with high tenacities (0.44–0.88 N/tex or 5–10 g/den) similar to those of the commercial poly(ethylene terephthalate) fibers. Sutures of various sizes are then prepared by braiding yarns of appropriate diameters.

As might be expected with a controllable synthetic material, the physical, absorption, and biological properties of polyester sutures derived from lower α-hydroxy acids are much more uniform than those of catgut (1). Initial knot strengths are significantly higher and animal implant studies indicate that they retain their *in vivo* strength superiority over gut during the first 20 d (32), a time which is beyond the critical period for wound repair in most surgical procedures. The mild reaction to synthetic absorbable sutures led to their use in a wide variety of operative procedures

(33), even in tissues, notably skin and subcuticular, where many surgeons previously employed nonabsorbable sutures to avoid the complications of catgut. They have also been approved for ophthalmic surgery, but their safe use in cardiovascular and neurosurgery has not yet been established. Evaluations in the latter application areas, however, are reported in the literature (34–37).

Absorption of the poly(glycolic acid) sutures is believed to proceed via a hydrolysis reaction that does not require the intervention of enzymes as do the collagenous materials (38). Animal metabolism studies with radiolabeled poly(glycolic acid) sutures have shown that <1% remains at the implant site after 4 mo and that the breakdown products are excreted principally as glycolic acid in the urine and carbon dioxide in the expired air (28).

Homopolymers [*26023-30-3*] of lactic acid and copolymers having lactic acid and minor amounts of other α-hydroxy acids in the chain have been recommended for suture use (39) but apparently have unacceptable absorption characteristics (19). A similar situation exists with the polyester [*26744-04-7*] derived from β-hydroxybutyric acid (40–41), an intracellular component of certain bacteria, which functions as a reserve carbon source. Copolymeric sutures containing 90% glycolic acid and 10% lactic acid units in the polyester chain, eg, Vicryl (polyglactin 910) [*26780-50-7*] (Ethicon, Inc.), have also been introduced (23,42); their properties seem to be essentially the same as those prepared from the homopolymer of glycolic acid.

Both of these synthetic absorbable sutures show handling properties typical of uncoated braids. They are supple and exhibit high knot security but tend to pass through tissue less readily than monofilaments and coated braids. Coated versions, available since 1980 and 1981, have resulted in reduced tissue drag and improved handling properties during knot tying (43–44). The coating applied to Dexon sutures is a poly(oxyethylene-co-oxypropylene) polymer called Poloxamer 188 (Pluronic F-68, BASF Wyandotte Corp.). The coating applied to the Vicryl glycolide–lactide copolymer suture is a mixture composed of equal parts of calcium stearate and a 35–65 copolymer of glycolide–lactide.

Silk. Silk is still the most widely used nonabsorbable suture and is considered by many surgeons to have the best handling and knot-tying characteristics. The silk material is degummed to remove the natural sericin coating and scoured, and mostly used braided. It is usually dyed black and coated with wax or silicone rubber to reduce its capillarity and improve handling. Dermal silk sutures are twisted and treated with an insolubilized gelatin coating. In a specialized ophthalmic application where the fine diameters desired (ca 20–40 μm) are difficult to obtain by braiding, several virgin silk fibers are twisted together and prevented from unraveling by the bonding action of the natural gum. These fibers may be dyed with logwood, or temporary visualization of the fine undyed strands may be achieved by treatment with methylene blue solution before use.

As a natural protein fiber, silk exhibits some variability in tissue reactivity and often loses a third or more of its original strength in the body within six months (1,45). Although classified as a nonabsorbable suture, fragmentation and even complete digestion of implanted fibers have been observed as early as three months after implantation (46).

Cotton. Cotton sutures are prepared by twisting long staple fibers. They handle well and exhibit good knot security but are the weakest of sutures currently used (9). They maintain their *in vivo* strength longer than silk and are similar in tissue reaction, although individual fibers frequently separate from the main suture mass and stimulate a wider response (46).

Linen. Although linen is one of the earliest recorded suture fibers, it is not used extensively today in the United States. It is processed similarly to cotton and is stronger, but control of diameter and strength uniformity is difficult (19). Its principal use is in gastrointestinal surgery.

Polyamides. The nylons developed for textile applications were the first completely synthetic fibers employed as sutures, and after thirty years they are still used extensively, particularly for skin. Monofilaments and braided multifilaments coated with silicone or wax are available; the former predominate. Both nylon-6,6 and nylon-6 are used, eg, as Dermalon and Ethilon. A composite form consisting of an unbraided yarn core enclosed in a relatively thick film sheath of the same polymer is also available (47) but has apparently not gained wide acceptance. Although the nylons exhibit only the usual mild tissue reactions of most synthetic nonabsorbable fibers used in surgery, they undergo a gradual hydrolytic tensile-strength loss after several months in the body (46,48). Therefore, they are not recommended for applications where permanent holding power is required such as the fixation of heart valves (see Polyamides).

Nonabsorbable Polyesters. Poly(ethylene terephthalate) fiber provides the strongest nonabsorbable synthetic suture available (9), and its tensile properties remain relatively constant for long periods of time in tissue (46). These characteristics are also responsible for the fact that tubular prostheses used for arterial replacement are now knitted or woven almost exclusively from this material in preference to nylon, acrylic polymers, and other synthetic fibers. For suture use, poly(ethylene terephthalate) is generally braided to obtain optimum handling properties. Coating of this material with silicone, poly(tetramethylene adipate), or fluorocarbon polymers to facilitate knot tying and passage through friable tissue has greatly increased its popularity, particularly for cardiovascular surgery. These polyester suture products are marketed under trademarks such as Ti·Cron (American Cyanamid Company, Davis & Geck) and Tevdek (Howmedica, Pfizer, Inc.) (see Polyesters).

Polyolefins. Since the early 1960s high-density polyethylene and later isotactic polypropylene monofilaments have been used as sutures (49–52). They are extremely inert and are not subject to the tissue-fluid degradation shown by the polyamides. The knot strength of polyethylene sutures is similar to that of silk sutures (9), and they are less stiff than polypropylene. The latter is closer to nylon in knot strength and exhibits good fatigue resistance and low tissue drag, properties which make it suitable for cardiovascular surgery. Both polyolefin monofilaments have been widely used, particularly in the repair of skin and abdominal incisions (53–55) (see Olefin polymers).

Metals. A wide variety of metallic wires have been used as sutures, ranging from the pure precious metals in ancient times to modern alloys. Aside from relatively small amounts of tantalum [*7440-25-7*] and silver [*7440-22-4*], most wire sutures today are prepared from monofilament or twisted 316 stainless steel [*52013-36-2*]. This material is favored over other metals because of its optimum combination of high strength, low tissue reactivity, inertness, and flexibility. Although it is the strongest suture available, some surgeons dislike it because of handling difficulties; it also tends to kink, may fatigue and fragment, and because of its stiff sharp edges it may occasionally push out through the skin when used for internal closure (1). Sutures and fixation wires of F460 stainless steel provide increased resistance to attack by body fluids. Tantalum wire is particularly recommended for anchoring tantalum reinforcing gauze, bone plates, and other implants to adjacent tissues.

Selection. Selection of the best suture material for a given procedure is difficult if not impossible and is a subject of lively controversy among surgeons (2,25). In some cases any of a variety of materials might perform equally well, and the choice would depend not only on the purpose and particular technique employed but on the training and experience of the user (19). A material that one surgeon can handle well (tie secure knots with ease) may feel uncomfortable in the hands of another.

Braided sutures are often preferred for their better handling and knotting properties, but some surgeons favor the stiffer monofilaments since they pass through tissues with a minimum drag and are more readily removed. It has often been suggested that the interstices present in a braided nonabsorbable material provide permanent potential sites for irritation or persistent bacterial infection which are accessible only with difficulty to leukocytes or antibiotics. Monofilaments have been recommended to minimize this problem (22) although some investigators have disagreed with this recommendation (56).

The use of absorbable sutures minimizes or eliminates this and other permanent foreign-body problems such as irritation or migration caused by tissue movement. In addition, absorbable sutures do not remain to act as a nidus for stone formation in the biliary or urinary tract. It has been suggested that absorbable sutures, particularly the new synthetic polyester sutures derived from the lower α-hydroxy acids, retain *in vivo* tensile strength sufficiently long to allow their use in 90% of all surgical procedures. However, despite the fact that controlled clinical studies have reported their use in abdominal fascia (57–58) and other tissues which provide most of the incisional holding power during healing (33), many surgeons still follow the conservative approach and use nonabsorbable materials. Objective biological data on wound healing in animal and human tissues are given in refs. 39, 53, and 59. These data should help to resolve some of the present uncertainties in this area and allow surgeons to make an optimal selection from the broad range of modern suture materials.

Trends

The most important advance in surgical sutures in the past decade was the introduction of synthetic absorbable material. Further developments, such as braid constructions utilizing more and finer filaments for a given suture diameter and the application of coatings (43–44), have improved handling characteristics. More recent synthetic absorbable materials have been reported that absorb more rapidly (60), as well as monofilaments that exhibit strength retention *in vivo* over a longer period of time (61).

Significant advances have been made in the field of microsurgery, in which surgical repairs are effected in ophthalmic, vascular, nerve, and other tissues while being viewed through an operating microscope, in part owing to the development of increasingly fine-sized needles and both absorbable and nonabsorbable sutures (62–76).

BIBLIOGRAPHY

1. R. W. Postlethwait in D. C. Sabiston, Jr., ed., *Davis-Christopher Textbook of Surgery*, 10th ed., Vol. I, W. B. Saunders Co., Philadelphia, Pa., 1972, pp. 307–318.

2. R. W. Postlethwait in J. E. Dunphy and W. Van Winkle, Jr., ed., *Repair and Regeneration*, McGraw-Hill, Inc., New York, 1969, pp. 263–285.
3. J. Black, *Biological Performance of Materials—Fundamentals of Biocompatibility*, Vol. 8 of the *Biomedical Engineering and Instrumentation Series*, Marcel Dekker, Inc., New York, 1981.
4. S. D. Bruck, *Properties of Biomaterials in the Physiological Environment*, CRC Press, Inc., Boca Raton, Fla., 1980.
5. T. N. Salthouse, *Otolaryngol. Head Neck Surg.* **88,** 658 (1980).
6. R. W. Postlethwait and co-workers, *Ann. Surg.* **181**(2), 144 (1975).
7. R. W. Postlethwait, *Trans. South. Surg. Assoc.* **81,** 299 (1970).
8. *The United States Pharmacopeia XX and the National Formulary XV (USP XX–NF XV)*, The United States Pharmacopeial Convention, Inc., Rockville, Md., 1980.
9. J. B. Herrmann, *Am. Surg.* **37**(4), 209 (1971).
10. H. Becker and M. R. Davidoff, *S. Afr. J. Surg.* **15**(3), 105 (1977).
11. J. G. Thacker and co-workers, *Am. J. Surg.* **133,** 713 (1977).
12. H. Tera and C. Aberg, *Acta Chir. Scand.* **142,** 1 (1976).
13. J. G. Thacker and co-workers, *Am. J. Surg.* **130,** 374 (1975).
14. F. W. Taylor, *Surgery* **5,** 498 (1939).
15. F. W. Taylor, *Ann. Surg.* **107,** 458 (1938).
16. E. L. Howes, *Surg. Gynecol. Obstet.* **137,** 15 (1973).
17. C. W. Bruch, *Proceedings of International Symposium on Industrial Sterilization*, Duke University Press, Durham, N.C., 1973, p. 49.
18. C. Artandi, *Proceedings of the U.S.P. Conference on Radiation Sterilization, Washington, D.C., Oct. 9, 1972*, Jacob Slonim Publishing Ltd., Tel-Aviv, Israel, p. 9.
19. J. C. Forester, *Br. J. Hosp. Med.*, 578 (Nov. 1972).
20. Ref. 8, p. 760.
21. *Fed. Regist.* **36**(218), 21612 (1971).
22. W. Van Winkle, Jr., and J. C. Hastings, *Surg. Gynecol. Obstet.* **135,** 113 (1972).
23. R. W. Postlethwait and B. M. Smith, *Surg. Gynecol. Obstet.* **140,** 377 (1975).
24. H. Laufman, *Surg. Gynecol. Obstet.* **145,** 597 (1977).
25. M. B. Myers, *Am. J. Nurs.* **71**(9), 1725 (1971).
26. A. R. Katz and R. J. Turner, *Surg. Gynecol. Obstet.* **131,** 701 (1973).
27. U.S. Pat. 3,297,033 (Jan. 10, 1967), E. E. Schmitt and R. A. Polistina (to American Cyanamid Co.).
28. E. J. Frazza and E. E. Schmitt, *J. Biomed. Mater. Res. Symp.* **1,** 43 (1971).
29. U.S. Pat. 2,668,162 (Feb. 2, 1954), C. E. Lowe (to E. I. du Pont de Nemours & Co., Inc.).
30. K. Chup and co-workers, *Makromol. Chem.* **100,** 262 (1967).
31. U.S. Pat. 3,442,871 (May 6, 1969), E. E. Schmitt, M. Epstein, and R. A. Polistina (to American Cyanamid Co.).
32. J. B. Herrmann, *Arch. Surg.* **106,** 707 (1973).
33. F. L. Allmann, Jr., *Surg. Gynecol. Obstet.* **136,** 607 (1973).
34. Y. Mii, T. Shimuzu, S. Tamai, A. Fukui, and Y. Hori, *J. Microsurg.* **2,** 42 (1980).
35. W. E. Fee, E. Okamoto, M. Spellman, and P. H. Ward, *Laryngoscope* **88,** 619 (1978).
36. J. C. De Lee, M. T. Smith, and D. P. Green, *J. Hand Surg.* **2,** 38 (1977).
37. D. R. Watts, S. H. Carr, and R. P. Hohf, *J. Biomed. Mater. Res.* **10,** 867 (1976).
38. R. W. Postlethwait, *Surgery* **78,** 531 (1975).
39. U.S. Pat. 3,636,956 (Jan. 25, 1972), A. K. Schneider (to Ethicon, Inc.).
40. U.S. Pat. 3,036,959 (June 29, 1966), J. N. Baptist (to W. R. Grace & Co.).
41. Brit. Pat. 1,034,123 (June 29, 1966), J. N. Baptist and J. B. Ziegler (to W. R. Grace & Co.).
42. J. Conn, Jr., R. Oyasu, M. Welsh, and J. M. Beal, *Am. J. Surg.* **128,** 19 (1974).
43. U.S. Pat. 4,047,533 (Sept. 13, 1977), V. A. Perciaccante and H. P. Landi (to American Cyanamid Co.).
44. J. Conn, Jr., and J. M. Beal, *Surg. Gynecol. Obstet.* **150,** 843 (1980).
45. R. W. Postlethwait, *Ann. Surg.* **171,** 892 (1970).
46. R. W. Postlethwait, *Trans. South. Surg. Assoc.* **81,** 299 (1970).
47. M. Fox and W. Young, *Surgery* **52**(6), 913 (1962).
48. S. A. Boruchoff and P. C. Donshik, *Ophthalm. Surg.* **8,** 42 (1977).
49. A. R. Katz and H. D. Evans, *Am. J. Surg.* **103,** 208 (1962).
50. F. C. Usher and co-workers, *J. Am. Med. Assoc.* **179**(10), 780 (1962).
51. U.S. Pat. 3,359,983 (Dec. 26, 1967), E. H. Northey (to American Cyanamid Co.).

52. U.S. Pat. 3,630,205 (Dec. 28, 1971), G. M. Listner (to Ethicon, Inc.).
53. R. E. Herrmann, *Surg. Gynecol. Obstet.* **138,** 84 (1974).
54. H. Haxton, *Surgical Techniques*, John Wright & Sons, Bristol, UK, 1970, p. 74.
55. H. Haxton, *Brit. J. Surg.* **52**(5), 372 (1965).
56. R. F. Edlich and co-workers, *Ann. Surg.* **177**(6), 679 (1973).
57. A. L. Gallitano and E. S. Kondi, *Surg. Gynecol. Obstet.* **137,** 794 (1973).
58. H. A. Haxton and co-workers, *J. Abdom. Surg.* **16**(9), 239 (1974).
59. R. M. Anderson and R. F. Romfh, *Technique in the Use of Surgical Tools*, Appleton-Century-Crofts, New York, 1980.
60. U.S. Pat. 4,032,993 (July 5, 1977), J. Coquard and co-workers (to Rhone-Poulenc Industries).
61. U.S. Pat. 4,052,988 (Oct. 11, 1977), N. Doddi and co-workers (to Ethicon, Inc.).
62. L. Cherry, *N.Y. Times Mag.*, 12 (Aug. 5, 1979).
63. Sun Lee, *Int. Surg.* **65,** 485 (1980).
64. J. W. Chandler, J. Sugar, and A. R. Gasset, *Arch. Ophthalmol.* **92,** 155 (1974).
65. A. R. Gasset, R. T. Bellows, H. W. Haltmann, and N. Gravenstein, *Ann. Ophthalmol.* **8**(1), 99 (Jan. 1976).
66. T. N. Salthouse, B. F. Matlaga, and M. H. Wykoff, *Am. J. Ophthalmol.* **84,** 224 (1977).
67. *J. Microsurg.* **1**(2), (1979).
68. J. J. Stangl and co-workers, *Fertil. Steril.* **30,** 210 (1978).
69. A. Thiede and co-workers, *J. Microsurg.* **1,** 216 (1979).
70. A. Sugar, *Ophthalm. Surg.* **11,** 22 (1980).
71. Yoshio Mii and co-workers, *J. Microsurg.* **2,** 42 (1980).
72. N. V. Nielsen, J. C. Højbjerg, and E. Westerlund, *Acta Ophthalmol.* **58,** 48 (1980).
73. S. E. Sherman, *Ann. Ophthalmol.* **12,** 1099 (1980).
74. A. Sugar, *Ophthalm. Surg.* **11,** 335 (1980).
75. D. E. Williamson, *Ophthalm. Surg.* **11,** 778 (1980).
76. J. E. Blaydes and J. Berry, *Ophthalm. Surg.* **11,** 790 (1980).

General References

Refs. 54 and 60 are also general references.
D. E. Clarke, *Contemp. Surg.* **17,** 33 (1980).
T. L. Taylor, *J. Am. Podiatry Assoc.* **65,** 649 (1975).
N. M. Bikales, ed., *Encyclopedia of Polymer Science and Technology*, John Wiley & Sons, Inc., New York, Vol. 4, 1966, pp. 1–15, 244–272; Vol. 6, 1967, pp. 505–573; Vol. 8, 1968, p. 516–533; Vol. 9, 1968, pp. 403–440; Vol. 10, 1969, pp. 347–460; Vol. 11, 1969, pp. 1–41; Vol. 12, 1970, pp. 578–585; Supplement Vol. 1, pp. 587–597.
F. Beck and J. B. Lloyd, *Cellular Control Mechanisms Cellular Responses to Environment*, Vol. 4 of *The Cell in Medical Science*, Academic Press, Inc., New York, 1976.
C. Artandi, *Chemtech*, 476 (Aug. 1981).

JAMES B. MCPHERSON
American Cyanamid Company

SWEETENERS

Organic compounds, other than carbohydrates, that taste sweet have been known for centuries (see Alcohols, polyhydric; Sugar). The amino acid, glycine, whose name derives from its taste, is a familiar example (see Amino acids). During the last hundred years, the list of new sweeteners has primarily consisted of synthetic compounds, undoubtedly because of the tremendous advances in the field of synthetic organic chemistry. Although the list is very long, only some of these substances have ever been used and most have fallen by the wayside because of toxicological problems. Any successful sweetener has to fulfill certain minimum criteria (see also Flavor characterization; Flavors and spices; Food additives).

Traditionally, the standard is sucrose, both in terms of quality of taste and taste profile. The closer a compound is to sucrose, the more acceptable it is to the general public. Many substances reported to be sweet have strange, nonsucrose-like flavors. Furthermore, sweetness may coexist with other tastes, eg, licorice or menthol. Some compounds have a lingering bitter flavor or aftertaste, others a persistent long-acting sweet taste. Sucrose solutions have a mouth feel that cannot readily be duplicated.

The sweetening agent should be stable in all applications for which sucrose is currently used. Special handling and formulations could be considered if justified by demand.

The sweetening compound must be water-soluble since water-insoluble substances have no taste. Reasonable solubility and a rapid rate of solution are necessary for some applications. On the other hand, the sweetener should not be hygroscopic since this would impede manufacturing, packaging, distribution, and use.

Ideally, the product should cost less than sucrose on an equivalent sweetness basis. However, interest in low-calorie products is so great that a premium price might be tolerated.

Standards vary from country to country, but generally tend to become more stringent as toxicological testing and analytical methods improve.

No synthetic sweetener has ever met all the above requirements, and it is evident that sucrose cannot be replaced in the foreseeable future. Sucrose, in addition to satisfying the above criteria and being the standard against which new discoveries are measured, adds bulk or structure to solid products such as baked goods, viscosity to beverages and syrups, and functions as a weak antimicrobial agent in jams and jellies by increasing osmotic pressure (see also Food processing; Syrups).

Structure–Activity Relationships

The existence of a wide variety of chemical substances, both carbohydrate and noncarbohydrate, that produce a sweet response at least as one component of their total flavor has prompted much speculation about a possible theory that would correlate sweetness with general structural elements. The impetus came partly from the belief that there was probably only one sweet taste-bud receptor. This theory, however, has been contested (1–2).

An early approach was based on identification of polar or polarizable moieties (3). Because of the observation that in certain amino acids one optical isomer was sweet, whereas the other was not, the theory was expanded to include a third binding site which was thought to be hydrophobic rather than polar (4). Since the discovery of the taste of aspartame, attempts have been renewed to correlate structure and taste (5) with particular application to derivatives of aspartic acid (6–13). The general purpose has always been to develop a model that would account for sweet-tasting compounds and at the same time would explain why closely related analogues failed to be sweet.

Although many of the concepts and detailed analyses are ingenious, the results are unsatisfactory in two ways. First, if the theory is broad enough to encompass diverse structural types, then many compounds, although they fit the theory, are not sweet. Second, when the data of a certain series can be well accommodated, then structures outside the series may not fit the model. In short, these theories and correlations lack useful predictive value. It is not possible at present to design a compound with a structure that has a sweet taste. Even worse, an analogue of a known synthetic sweetener cannot be constructed with a high degree of confidence in predicting the taste, either quality or potency. It is hoped that future research will remedy this situation.

Saccharin. The taste of saccharin [*81-07-2*], 3-oxo-2,3-dihydro-1,2-benzisothiazole 1,1-dioxide (*o*-sulfobenzimide or *o*-benzosulfimide),

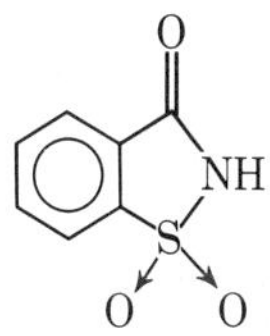

saccharin

was discovered in 1879 during an investigation of the oxidation of *o*-toluenesulfonamide (14). The compound is acidic and not very soluble in water. Therefore, it is normally employed as the sodium [*128-44-9*] or calcium [*6485-34-3*] salt. Both salts are soluble to the extent of ca 1 g in 1.5 mL water. Saccharin was introduced commercially in 1885 and has been more or less in widespread use ever since.

The potency of saccharin relative to sucrose, as with other synthetic sweeteners, is inversely related to sucrose concentration, and ranges anywhere from 200 to 700 times that of sucrose. As normally used, saccharin has a potency of ca 300. Although sweet, saccharin has a bitter, metallic aftertaste which is very difficult to mask. There is probably no one (nontaster) who cannot taste saccharin; some people are more sensitive to the aftertaste than others, but if the concentration is steadily increased, everyone in a test group eventually experiences saccharin's unpleasant flavor. Saccharin is stable to heat, both dry and in aqueous solution, and therefore is suitable for normal food applications.

Saccharin is not metabolized and has no food value. This latter property caused it to be banned in the United States in 1912 because of the fear that people who used saccharin instead of sugar would be deprived of nutritional values (15). Although this ban was lifted, saccharin's sales were low because of its unpleasant aftertaste. The primary users were diabetics; in some countries (eg, France) saccharin was considered to be a drug, sold only in pharmacies and banned, even today, in processed foods. Since 1950, U.S. sales of saccharin have increased because it was discovered that in a mixture of saccharin and the newly introduced synthetic sweetener, cyclamate, the off-flavor and aftertaste of both compounds were significantly suppressed (16). It would be more accurate to say that the dose–response curve of bad taste vs sweetener concentration was displaced to the right since the bitter metallic aftertastes of saccharin and cyclamate can still be detected at high concentrations. Another important factor in the increased usage of saccharin in the United States is the widespread concern over the harmful health effects of overweight and the resulting intense efforts to control food-calorie intake.

Saccharin is manufactured by one of two processes: the first is based on the original discovery (17), whereas the second is designed without *o*-toluenesulfonamide, a suspected carcinogen (18). With modern chromatographic techniques, as many as 15 impurities have been detected in commercial saccharin (19–21).

Process 1:

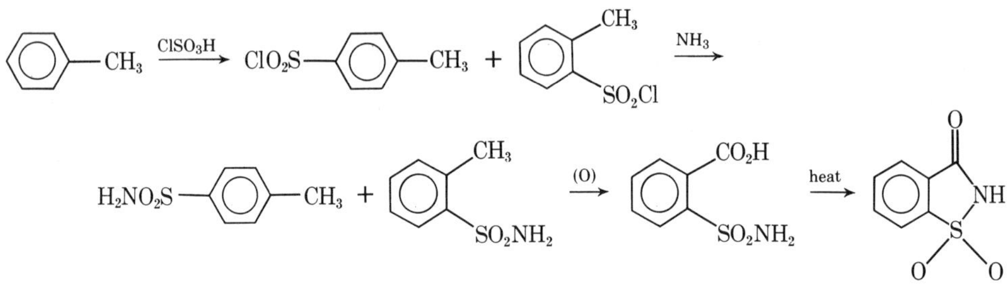

The *o*- and *p*-toluenesulfonamides are separated before oxidation with, eg, chromium, manganese, and iron compounds (see Chelating agents).

Process 2:

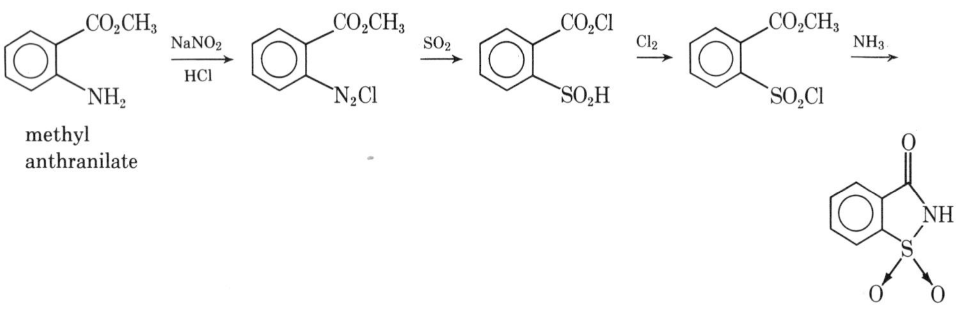

Methyl anthranilate is readily available in three steps from phthalic anhydride.

In recent years, saccharin has come increasingly under fire as a result of a number of toxicological studies (22). In 1969, a study of the effects of a mixture of cyclamate and saccharin (10:1) on rats showed the formation of bladder tumors (23); this was followed by other studies with similar results. This research ultimately led to a ban on the use of cyclamates in the United States. Saccharin, the minor component of the mixture, was allowed to remain on the market. In 1974, the Canadian Health Protection Branch initiated a study that was intended to definitely prove or disprove the carcinogenicity of saccharin. It was concluded in 1977 that saccharin produced bladder tumors in rats and that the effect was not due to the impurity, *o*-toluenesulfonamide, to bladder stones, or bladder parasites. As a result, in Canada saccharin was removed from food and beverages. Since cyclamate had previously been removed from food and beverages in Canada as well as in the United States, this action virtually destroyed the Canadian diet soft-drink industry until the approval of aspartame in 1981. In early 1977, the FDA proposed the same ban on saccharin as Canada. However, because of strong public opposition, the U.S. Congress, in November, 1977, imposed a moratorium against the FDA action which was extended in 1979 and again in 1981. In August, 1983, the question of the regulation of the use of saccharin in the United States comes up again.

A number of saccharin analogues are known. However, analytical data are not available, and these compounds are only characterized as sweet, bitter, or tasteless. It is fairly certain that compounds structurally related to saccharin, but sweeter, have not been found. The one notable exception is thieno[3,4-*d*]saccharin [*59337-79-0*], which is 1000 times as sweet as sucrose without bitter aftertaste (24). *N*-Alkylation removes the taste from analogues retaining the *o*-sulfobenzimide nucleus. Ring substitution in the 4- and 6-positions often give sweet derivatives, whereas substitution in the 5- and 7-positions usually imparts a bitter flavor (25–26). Some examples of sweet saccharin analogues are shown in Table 1. The saccharin moiety may possess a basic bitter taste that is occasionally partially concealed by a sweet component (26). Saccharin is a more important commodity than cyclamate. In the last ten years or so, the People's Republic of China, the Republic of Korea, and the Republic of China have started production, in addition to established manufacturing in Japan. The U.S. market for saccharin is ca 2700 metric tons per year, the European market 1500–2000 metric tons, and the world market 7000–9000 metric tons. The U.S. price of saccharin is $6.60–7.70/kg (1982). However, because of expanding production, special price situations can be found. Allegations of dumping are heard with increasing frequency. The countries in which saccharin and its ammonium, sodium, and calcium salts are

Table 1. Saccharin Analogues

Analogue	CAS Registry No.	Potency[a]
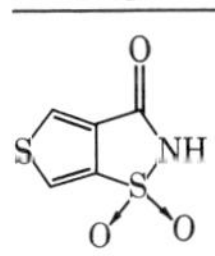	[*59337-79-0*]	1000
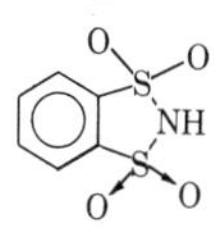	[*4482-01-3*]	sweet
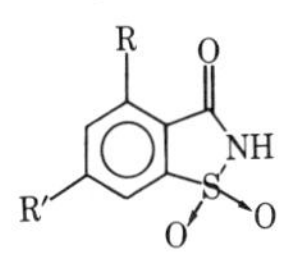		
		sweet
R = NO_2, R′ = H	[*82649-17-0*]	
R = NH_2, R′ = H	[*82649-18-1*]	
R = H, R′ = NO_2	[*22952-24-5*]	
R = H, R′ = Cl	[*46149-10-4*]	
R = H, R′ = F	[*384-45-2*]	

[a] Potency relative to sucrose as 1.

sold include: Argentina, Australia, Austria, Belgium, Brazil, France, FRG, Netherlands, Italy, Japan, Mexico, Portugal, Spain, South Africa, United Kingdom, and the United States.

Cyclamic Acid

The taste of cyclamate was discovered in 1937 at the University of Illinois (27).

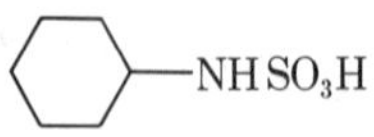

Cyclamic acid [*100-88-9*], cyclohexanesulfamic acid, has a sweet-sour flavor and a very acidic pH of ca 1. It is readily soluble in water at ca 10% and is normally used in the form of its neutral sodium salt, called sodium cyclamate [*139-05-9*] or simply cyclamate. The calcium salt is used in low sodium diets. Cyclamic acid and its sodium salt were patented by DuPont in 1940. Abbott Laboratories introduced cyclamate in 1950 and the product enjoyed a steady increase in popularity and sales until it was banned in the United States in 1970 (28).

Cyclamate is about 30 times as sweet as sucrose, but has an off-flavor or unpleasant aftertaste which can be detected if the concentration is high enough. Because of the off-flavor and other factors, such as feel, rate of onset, and disappearance of taste, it is difficult to determine the sucrose equivalent of cyclamate at relatively high concentrations. As noted above, the discovery of the properties of a 10:1 cyclamate–saccharin mixture established the market for diet products in the United States, particularly carbonated beverages (qv). At corresponding levels of off-taste response in a test population, the mixture is more than twice as sweet as either component alone (16).

Cyclamate is manufactured by sulfonation of cyclohexylamine. A large number of reagents are applicable but commercial processes use sulfamic acid, a sulfamate salt, or sulfur trioxide (29–35). The salt is obtained by neutralization. Before cyclamate was banned, U.S. production amounted to ca 8000–9500 t/yr. The U.S. price in 1969 was ca $1.10/kg. In 1981, the price in Canada was $9.90/kg. The European market is ca 1300 t/yr at a price similar to the Canadian price.

Initially, it was claimed that cyclamate was not metabolized and, in fact, it was placed on the GRAS list by the FDA in 1958. However, in 1966, it was discovered that cyclamate was metabolized to cyclohexylamine and the latter was shown to cause chromosome damage in animals. In 1969, as mentioned above, the finding of bladder tumors in rats after a two-year feeding study of a cyclamate–saccharin mixture was reported (23). A parallel study of cyclohexylamine also turned up a bladder tumor (23). The FDA took cyclamate off the GRAS list and removed it from foods and beverages, but permitted it to be sold in pharmacies. Canada and the U.K. followed suit, among others. About a year later, after a congressional investigation, the FDA totally banned cyclamate from the U.S. market. It is still available in various forms in a number of countries: Argentina, Belgium, Brazil, France, Italy, Netherlands, Japan, Mexico, Portugal, South Africa, and Spain.

As was true with saccharin, analogues of cyclamic acid have not shown much promise. No compound yet reported is significantly sweeter than cyclamate itself. *N*-Alkylation removes sweetness, whereas changing the ring size retains sweetness, especially with 5- or 6-membered rings; the 6-membered ring seems to impart the most sweetness. Some straight- or branched-chain alkanesulfamates are sweet but with very low potency (36).

Aspartame

In 1965, the taste of aspartame [*22839-47-0*; *53906-69-1*] (APM, L-aspartyl-L-phenylalanine methyl ester), was discovered accidentally (37). The compound was an intermediate in the synthesis of the *C*-terminal tetrapeptide of gastrin; the latter was used as a biological standard in connection with an antiulcer project. Aspartame is a white crystalline compound which may be obtained anhydrous or, more usually, as a half hydrate. When heated, it cyclizes to the corresponding diketopiperazine and the melting point of the latter is eventually observed:

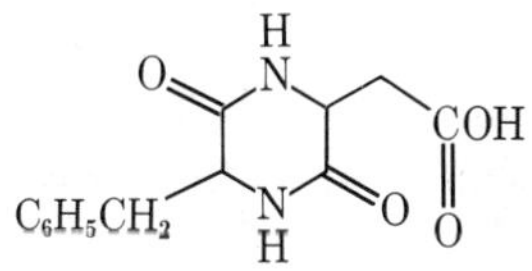

3,6-dioxo-5-benzyl-2-piperazineacetic acid

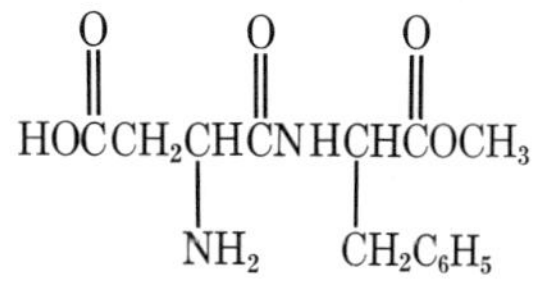

aspartame (APM)

Formation of the diketopiperazine takes place whenever APM is dissolved in aqueous solution; the rate is a function of temperature and pH. Figure 1 shows the overall stability of APM which is satisfactory for most applications. The diketopiperazine is tasteless and at a very advanced stage of conversion, a diminution in sweetness may be noticed. At 25°C and its isoelectric point (pH 5.2), APM gives a 1% solution in water. At pH 2.2, where maximum solubility is achieved, a 10% solution can be obtained. Mixed solvents offer a striking advantage, particularly for synthetic operations. In water or acetic acid, APM gives a 1% solution, and in 70% acetic acid a 30% solution.

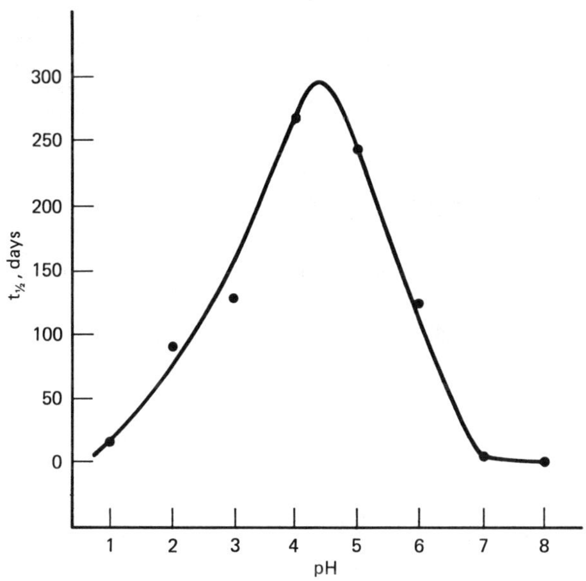

Figure 1. Stability of APM at 25°C.

The stability of aspartame in aqueous systems has been studied in several food preparations under normal conditions of intended use (37). Analytical procedures were developed to determine the diketopiperazine, aspartylphenylalanine, the constituent amino acids and methanol, and unchanged APM. This work is particularly relevant to the use of aspartame in carbonated beverages. In dry foods, stability of aspartame is not of concern unless it is used in combination with hygroscopic ingredients. Dry storage of APM is quite satisfactory for a concentrated table-top sweetener that is a mechanical blend with a carbohydrate and packaged in standard food-service paper packets, and for a table-top sweetener equal to sucrose in volume measure and sweetness, in which APM is spray-dried with partially hydrolyzed corn starch.

The taste of APM could not have been predicted from its constituent amino acids (qv). L-Aspartic acid has a flat taste, whereas L-phenylalanine is bitter (39). When the two are linked together in the proper way and the phenylalanine carboxyl is converted to a methyl ester, the resulting product has a totally unexpected taste. Aspartame is very much like sucrose, and some tasters in paired comparison studies have preferred it to sucrose. The various objectionable taste qualities associated with cyclamate and saccharin are absent in APM, and its taste seems to be invariant. A person to whom sucrose is unpleasant may find that APM tastes good; however, it cannot be called sweet because the word is traditionally associated with sucrose.

It has been found repeatedly that apparent potency of a sweetener relative to sucrose is a function of the sucrose concentration; this relationship is inverse. Some data for APM are given in Table 2 (38). Relative potency is also affected by the presence of other flavors; in other words, it is a function of the test system. In most formulations of practical interest, the potency of APM lies in the range of 150–200 (38) (see Table 3).

A curious property of APM is that untrained tasters are not dramatically affected

Table 2. Aspartame Potency in Water

Sucrose, %	Aspartame, %	Potency
4.3	0.02	215
10.0	0.075	133
15.0	0.15	100

Table 3. Aspartame Content of Typical Foods and Beverages

Product	Sucrose, g/100 g product	Aspartame, g to replace sucrose	Potency of aspartame
dry			
tabletop sweetener	100.0	0.050	200
powdered soft drinks	10.3	0.069	150
instant pudding	17.0	0.085	200
gelatin dessert	18.0	0.113	160
presweetened cereal	37.0	0.205	180
wet			
carbonated soft drinks	10.0	0.055	180
canned fruit cocktail	20.0	0.114	175
flavored yogurt	17.0	0.085	200
frozen dessert	15.0	0.060	250

by concentration changes (38). In coffee, for example, doubling the APM concentration caused the beverage to be judged only slightly oversweetened, whereas halving the original concentration was not even detected. Aspartame shows synergy with other sweeteners, ie, less than the calculated amount of a particular combination is required to give a predetermined degree of sweetness. In addition, APM has been reported to favorably modify other flavors, sometimes at a concentration below that required for the sweetening effect.

Aspartame is a suitable sweetener for a wide variety of foods because of its sucrose-like taste and its ability to blend well with other food flavors. Because of interactions with other flavors, the replacement of sucrose by aspartame requires reformulation and not simple substitution of ingredients.

Aspartame is metabolized into its constituent parts which are then metabolized by the usual body pathways (39). After a long history of regulatory investigation, APM was approved for marketing on August 12, 1981, in Canada (40) and on October 22, 1981, in the United States (41–42). It is also available in other countries. The physical properties of APM suggested that it is suitable for all uses except those requiring extensive heat, eg, baking. Practical experience with APM is in the early stages, but Canadian marketing results clearly show that it is highly satisfactory in carbonated soft drinks (43). Aspartame is a competitor for the \$100 × 10^6/yr (U.S.) table-top sweetener market (44).

The synthesis of aspartame has attracted considerable attention and many routes have been attempted. Because of the probable requirements of large-tonnage production, chemical engineering considerations become crucial and may overshadow the elegance of the organic syntheses. The following synthetic procedures have been tried (45–47):

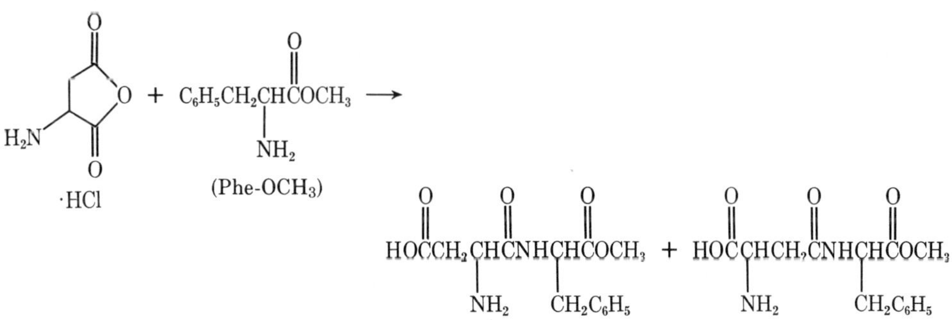

(Asp-Phe-OCH$_3$, α-isomer) [*22839-47-0; 53906-69-7*] (Asp-Phe-OCH$_3$, β-isomer) [*22839-61-8*]

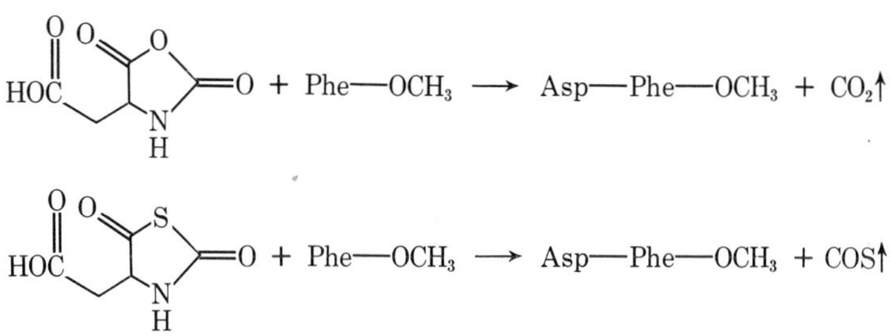

Other synthetic methods are described in refs. 37 and 48–50.

Table 4. Sweetness Potency of L-Aspartic Acid Derivatives

Compounds	CAS Registry No.	Potency[a]
Asp—Phe—OCH_3 (aspartame)	*[22837-47-0]*	150
Asp—Phe(6H)—OCH_3	*[26270-66-6]*	225
Asp—Cys(Bu^t)—OCH_3	*[58336-96-2]*	900
Ama[b]—Phe—OCH_3	*[40203-98-3]*	230
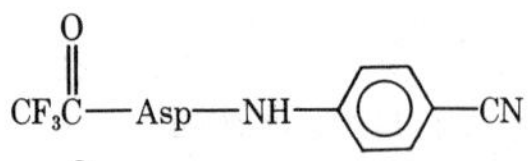	*[39219-30-2]*	3000
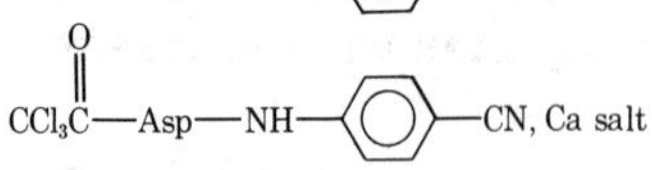	*[41567-02-6]*	3000
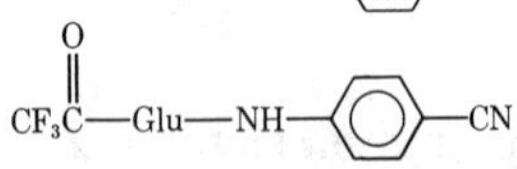	*[77314-15-9]*	3000
Asp—D—Ser—OPr^n	*[52946-03-9]*	320
Asp—D—Ser—OBu^i	*[52946-06-2]*	200
Asp-D-NHCH(CH_3)CH_2OC(=O)R		
R = Bu^t	*[77050-95-4]*	240
R = 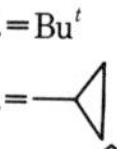	*[77050-92-1]*	200
R = 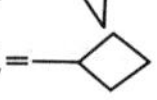	*[77050-96-5]*	220
Asp—Ama[b](OR)—OCH_3		
R = $CH(C_2H_5)_2$	*[38018-03-0]*	360
R = $CH(Pr^i)_2$	*[61016-38-4]*	680
R = 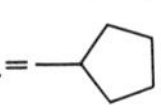	*[38016-18-1]*	600
R = 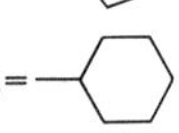	*[38016-05-2]*	880
R = *cis* 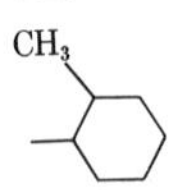	*[51499-60-6]*	4,600
R = *trans* 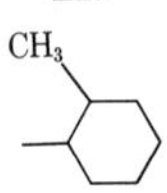	*[61091-20-1]*	7,300
R = 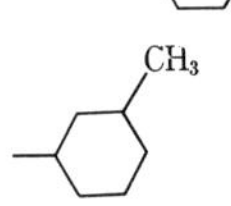	*[61016-39-5]*	600
R = 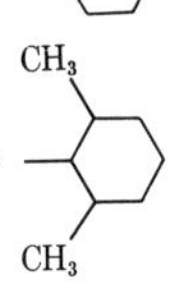	*[61016-41-9]*	5,300
R = 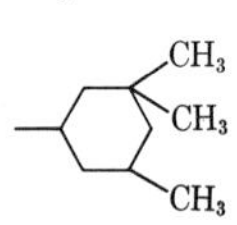	*[61016-42-0]*	320
R = 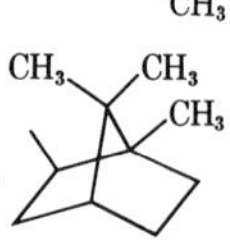	*[61055-87-6]*	760

Table 4 (*continued*)

Compounds	CAS Registry No.	Potency[a]
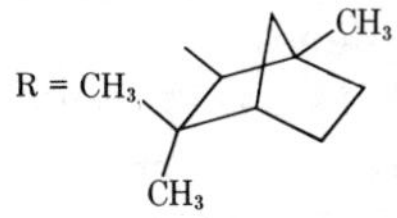	[*61091-21-2*]	33,000
Asp—Ama[b](OR)—OC_2H_5		
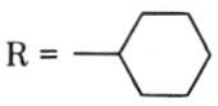	[*61116-59-4*]	280
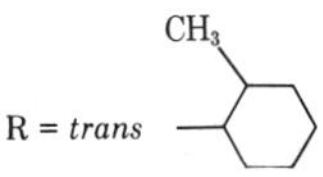	[*61016-48-6*]	640
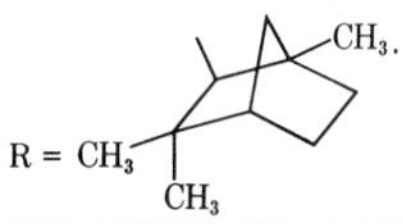	[*61116-60-7*]	5,400

[a] Times sucrose sweetness.
[b] Ama = aminomalonyl.

Aspartame has a novel structure for a synthetic sweetener and the discovery of its taste engendered the synthesis of analogues in a number of laboratories. The analogues are either dipeptides or amino acid amides. In most cases, the analogue was based on aspartic acid.

Aspartame is the L,L-isomer. The other three diastereoisomers (LD; DL; DD) are faintly bitter (37). This is a good example of the rigid steric requirements of a biochemical reaction. None of the natural amino acids so far tested has been able to substitute for aspartic acid in APM. However, aminomalonylphenylalanine methyl ester [*40203-98-3*] is sweet (51). A number of aspartyl dipeptide esters were sweet and, depending on the side chain, the *C*-terminal amino acid could sometimes be the D-isomer (52). The methyl ester was the sweetest in LL compounds, whereas the propyl ester [*83900-00-1*] was sweetest in LD compounds.

Since the various analogues were evaluated in a number of different laboratories, exact comparisons of potency may not be possible. However, only very rarely was a compound claimed to be sweeter than aspartame, except diesters of aspartylaminomalonic acid (53). Some of these derivatives had sweetness potencies one and two orders of magnitude greater than that of APM. The best compounds of this series were sweeter than any previously reported natural or synthetic substance. A possible explanation lies in the highly enolic character of malonic acid. Thus, the *C*-terminal part of the dipeptide is flat and this rigidity may permit a superior fit to the taste-receptor site.

A large number of amides of L-aspartic acid were found to be sweet (54). Aryl, alkyl, and aralkyl amines can be used. The potencies of these compounds were not exceptional and did not in general surpass APM. However, outstanding derivatives with greater sweetness by an order of magnitude were *N*-trichloroacetyl- [*41567-02-6*] (calcium salt) and *N*-trifluoroacetylaspartic acid *p*-cyanoanilide [*39219-30-2*] (55), and *N*-trifluoroacetylglutamic acid *p*-cyanoanilide [*77314-15-9*] (56). Derivatives with sweetness potencies equal to or greater than 200 times sucrose are given in Table 4; the number was chosen arbitrarily. For detailed structure–activity relationships, the original papers should be consulted.

The variety and generality of the above results suggest that a new type of biochemical nucleus has been found. That is, L-aspartic acid although not sweet itself, when derivatized in a certain way, gives products that often have a sweet taste sometimes with astonishing potencies. Such a compound could be called a cryptophore (latent carrier of a property) or, in this case, a cryptoglucophore. In a cryptophore, further structural elaboration exposes or brings to light a property that was hidden in the original molecule. Cryptophores might possibly be discovered in other areas of biological activity involving, for example, particular pharmacological properties (see Pharmacodynamics).

Aspartame's price ($152/kg in 1982) does not appear to restrict its use in food products by several U.S. companies (57).

Other Sweeteners

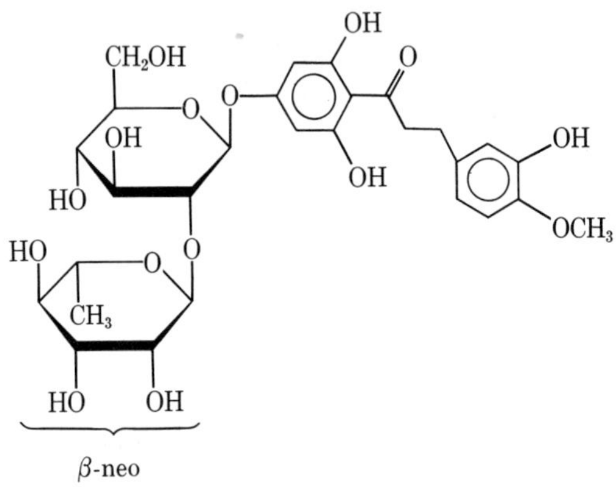

β-neohesperidine dihydrochalcone

β-Neohesperidine Dihydrochalcone. Conversion of the bitter flavanone, naringin, to the dihydrochalcone, β-neohesperidine dihydrochalcone [*20702-77-6*], at the Western Regional Research Laboratories of the USDA resulted in a sweet-tasting product (58). Although the project was concerned with structure–taste relationships, the sweet taste of the dihydrochalcone was totally unexpected. Naringin is obtained from grapefruit rinds and the prospect of utilizing an agricultural waste product was attractive.

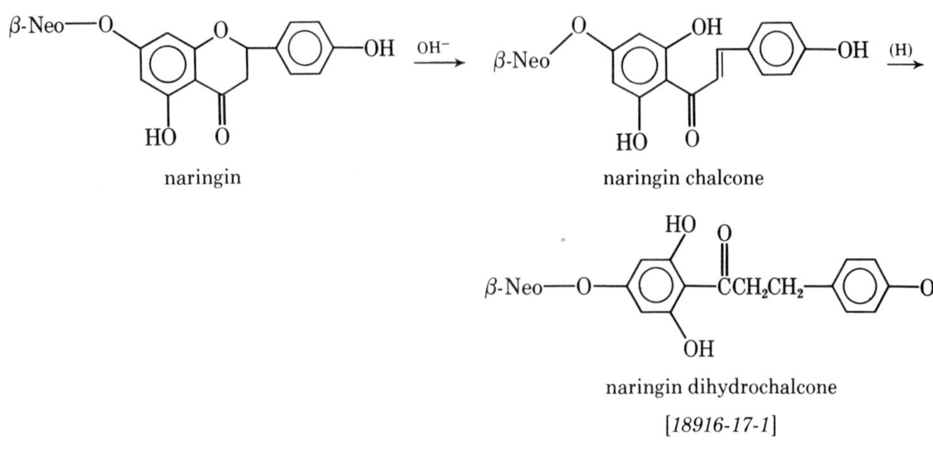

In subsequent work, it was found that neohesperidine from Seville oranges gave a more potent dihydrochalcone. Because Seville oranges are not abundant enough to be a useful raw material, a partial synthesis was developed (59–61). Nevertheless, the dependence on grapefruit rinds probably places an upper limit on the potential availability of this sweetener.

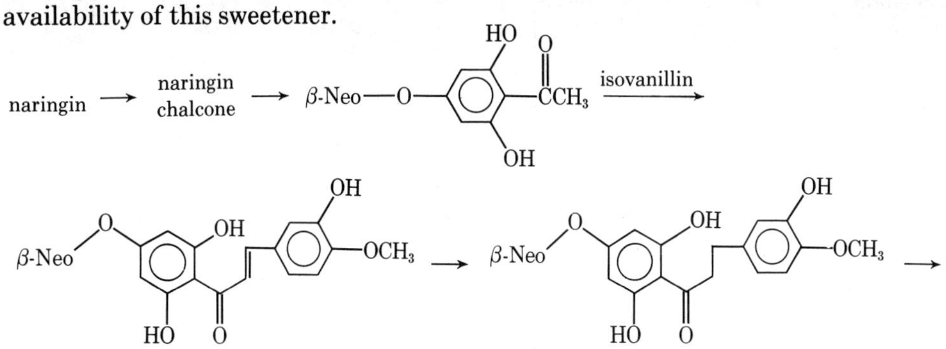

β-neohesperidine
dihydrochalcone

Table 5. Dihydrochalcone Sweeteners[a]

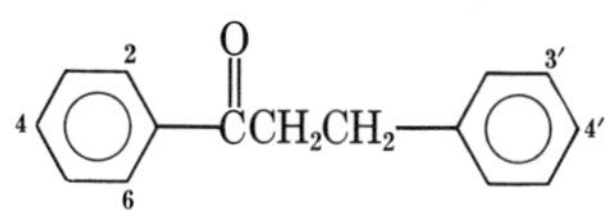

2	4	CAS Registry No.	6	3′	4′	Potency
OH	β-neo	*[20702-77-6]*	OH	OH	OCH_3	340–1500
OH	β-neo	*[19282-00-9]*	OH	OH	OC_2H_5	1000
OH	β-neo	*[19361-37-6]*	OH	OH	OPr	2000
OH	OH	*[35400-60-3]*	OH	OH	OCH_3	850
OH	OCH_2CO_2H	*[61497-77-6]*	OH	OH	OCH_3	300
OH	OCH_2CO_2H	*[70412-94-1]*	OH	OH	OPr	240
OH	$OCH_2COCH_2CO_2H$	*[76799-09-2]*	OH	OH	OCH_3	310
OH	$O(CH_2)_3PO_3H$	*[76799-10-5]*	OH	OH	OCH_3	280
OH	$O(CH_2)_2NHSO_3H$	*[70412-97-4]*	OH	OH	OCH_3	350
OH	$O(CH_2)_2CHOHCO_2H$	*[76799-11-6]*	OH	OH	OCH_3	440

[a] Refs. 62–63.

Table 6. Sweetness Potency of Acesulfam Analogues

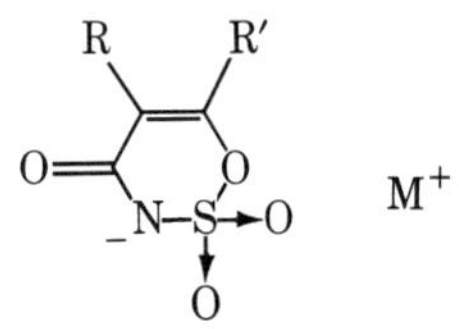

R	R′	M	CAS Registry No.	Potency
H	CH_3	K	*[55589-62-3]*	130
H	C_2H_5	Na	*[82770-53-4]*	150
C_2H_5	CH_3	Na	*[33666-00-1]*	250
Cl	CH_3	K	*[78351-55-0]*	200
H	CH_2Cl	K	*[74530-65-7]*	150
H	CH_2Br	K	*[74530-72-6]*	150
H	CH_3N_3	K	*[78351-54-9]*	150

Table 7. Other Compounds Reported to be Sweet

Compound	CAS Registry No.	Potency	Melting point, °C	Solubility in water, %	Ref.
chloroform	[*67-66-3*]	40	−64	0.5	68
dulcin	[*150-69-6*]	250	173–174	0.12	69
stevioside	[*77-05-4*]	300	198	0.12	70
phyllodulcin	[*26359-43-4*]	400	119–121	0.002	71
SRI oxime V	[*59691-20-2*]	450	91	0.33	72
glycergic acid	[*84215-86-1*]	500	215	crystallizes from water	73
perillartine	[*30950-27-7*]	750	102	0.0083	72
6-chlorotryptophan	[*17808-35-4*]	1300	264–265	crystallizes from water	74
hydrofluorene sweetener	[*34069-54-0*]	1400	296–299	sol	75
thaumatin (190 amino acids)	[*53850-34-3*]	200	na	sol	76
trichlorogalactosucrose	[*56038-13-2*]	2000	syrup	na	77
monellin (94 amino acids)	[*9062-83-3*]	2500	na	sol	78
osladin	[*33650-66-7*]	3000	198–199	na	79
P-4000	[*553-79-7*]	4000	47–48	0.014	80

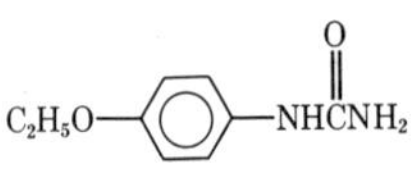
dulcin

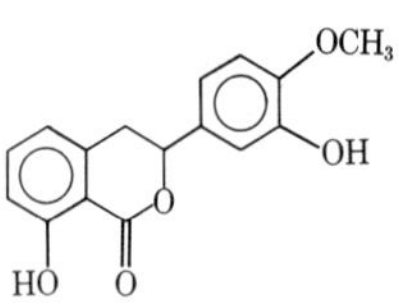
phyllodulcin

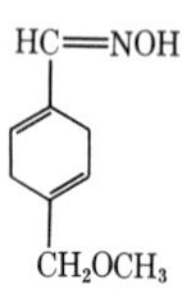
SRI oxime V

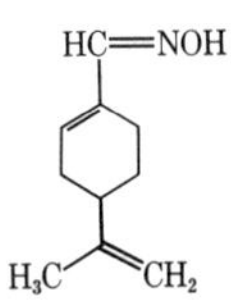
perillartine

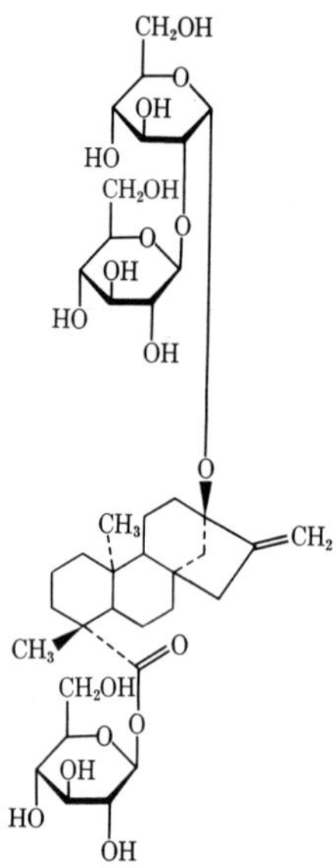
stevioside

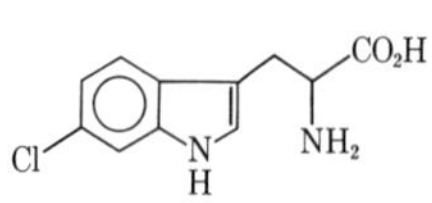
6-chlorotryptophan (D-isomer)

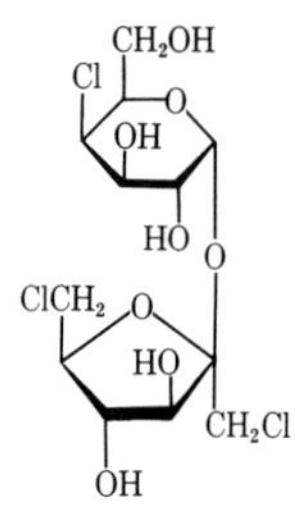
trichlorogalactosucrose

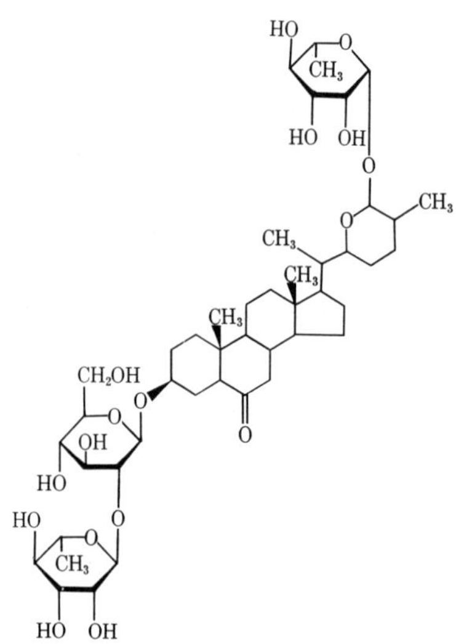
osladin

Table 7 *(continued)*

Compound	CAS Registry No.	Potency	Melting point, °C	Solubility in water, %	Ref.
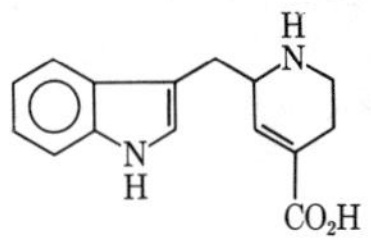 glycergic acid	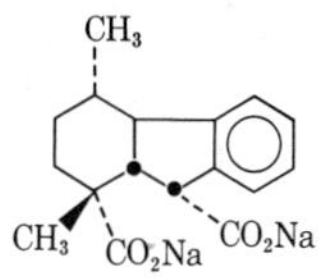 hydrofluorene sweetener			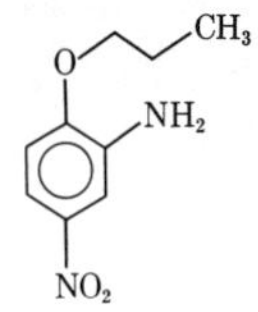 P-4000	

The sweet taste of β-neohesperidine dihydrochalcone is lingering, with a cooling or menthol-like aftereffect. These properties would sharply restrict any possible commercial application. Perhaps for that reason, although extensive toxicological and other studies have been carried out, registration has not been attempted in any country.

The sweetness potency of neohesperidine dihydrochalcone relative to sucrose is reported to be 340–1500. This unusual spread may be due to comparing different sucrose concentrations as well as the lingering taste. The value of 340 is probably the most reliable. It was obtained in an extraordinarily comprehensive study of the taste characteristics of neohesperidine dihydrochalcone and its analogues (62).

Many analogues of dihydrochalcones have been synthesized both with and without sugars. A selection of the more active ones is shown in Table 5. The potencies are not necessarily comparable; furthermore, a bitter taste can be detected alongside the sweet response in certain analogues. In no case has the taste profile been significantly improved in the sense that it is much closer to that of sucrose.

Acesulfam-K. A new class of sweeteners was reported in 1973 from the laboratories of Farbwerke Hoechst A.-G. (64). The structure is reminiscent of saccharin and, in fact, the saccharin homologue containing the oxathiazinone dioxide ring is sweet but also bitter. Acesulfam [*33665-90-6*], 3,4-dihydro-6-methyl-1,2,3-oxathiazine-4-one 2,2-dioxide, is slightly soluble in water but, as a strong acid, has a simultaneous sweet and sour taste which impedes its usefulness. The neutral salts are stable and allow suitable taste comparisons.

The potassium salt, acesulfam-K [*55589-62-3*], has been chosen for development. It gives a 25% solution in water at room temperature. The salt is stable under the usual conditions of handling and storage encountered during processing of foods and beverages although quantitative data have not been published. The taste of acesulfam-K resembles that of cyclamate; it has no metallic aftertaste. The sweetness potency ranges from 100 to 200, depending on whether the comparison solution is 10 or 3% sucrose. This behavior is typical of synthetic sweeteners.

Acesulfam-K is not metabolized and is excreted unchanged; it does not accumulate in the body on repeated administration and has no pharmacological effects. Extensive toxicological studies have been carried out without adverse effects. Applications for approval as a sweetener have been filed in ca 15 countries (65).

Several syntheses of the ring system of acesulfam have been studied. The most versatile in terms of synthesis of analogues involves fluorosulfonyl isocyanate,

$OCNSO_2F$, which is expensive and toxic, and gives the possibility of residual fluoride ion in the product. A procedure using the cheaper and safer chlorosulfonyl isocyanate was developed especially for the synthesis of acesulfam itself (66):

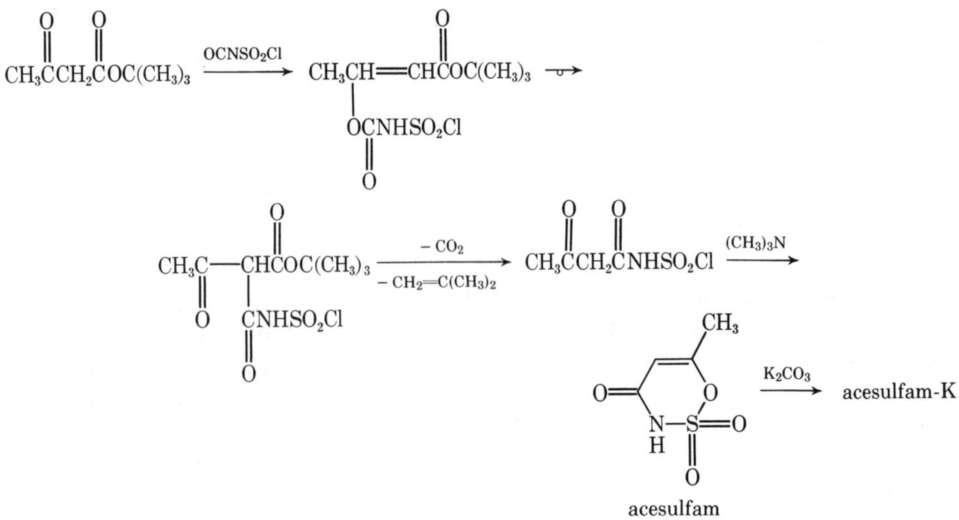

A number of analogues of acesulfam have been prepared; the best is ca twice as sweet. Acesulfam-K, although not the sweetest in the series, was selected because its taste was unusually clean. Furthermore, possible metabolism should lead to innocuous products. Table 6 shows acesulfam-K and other oxathiazinone dioxides with potencies of 150 times sucrose or greater.

Miscellaneous. Many substances other than those discussed above have been claimed to have a sweet taste, eg, so-called left-handed sugars, but development as a marketable product has only been rarely attempted (67). A marketable product must have satisfactory taste, stability, solubility, and low cost, and be safe. To find a compound that offers all these properties is a difficult task. Table 7 lists a small selection of structures reported to be sweet. The potencies refer to sucrose as unity and should be interpreted only as approximate values.

BIBLIOGRAPHY

"Sweeteners, Nonnutritive" in *ECT* 2nd ed., Vol. 19, pp. 593–607, by K. M. Beck, Abbott Laboratories.

1. G. A. Crosby, G. E. DuBois, and R. E. Wingard, *Drug Design*, Vol. 8 of E. J. Ariens, ed., *Medicinal Chemistry*, Academic Press, New York, 1979, p. 215.
2. J. M. Tinti, D. Durozard, and C. Nofre, *Naturwissenschaften* **67,** 193 (1980).
3. R. S. Shallenberger and T. E. Acree, *Nature (London)* **216,** 480 (1967).
4. L. B. Kier, *J. Pharm. Sci.* **61,** 1394 (1972).
5. H. Iwamura, *J. Med. Chem.* **23,** 308 (1980).
6. M. de Nardo, C. Runti, F. Ulian, and L. Vio, *Farmaco Ed. Sci.* **31,** 906 (1976).
7. H. Iwamura, *J. Med. Chem.* **24,** 572 (1981).
8. F. Lelj, T. Tancredi, P. A. Temussi, and C. Toniolo, *Farmaco Ed. Sci.* **35,** 988 (1980).
9. A. van der Heijden, L. B. P. Brussel, and H. G. Peer, *Chem. Senses Flavor* **4,** 141 (1979).
10. M. Goodman and C. Gilon in Y. Wolman, ed., *Peptides 1974*, Halsted Press, New York, 1975, p. 271.

11. F. Lelj, T. Tancredi, P. A. Temussi, and C. Toniolo, *J. Am. Chem. Soc.* **98,** 6669 (1976).
12. L. B. P. Brussell, H. G. Peer, and A. van der Heijden, *Z. Lebensm. Unters. Forsch.* **159,** 337 (1975).
13. A. van der Heijden, L. B. P. Brussel, and H. G. Peer, *Food Chem.* **3,** 207 (1978).
14. I. Remsen and C. Fahlberg, *Am. Chem. J.* **1,** 426 (1879); C. Fahlberg and I. Remsen, *Ber.* **12,** 469 (1879).
15. P. Fabiani, *Ann. Falsif. Expert. Chim.* **74,** 273 (1981).
16. H. C. Vincent, M. J. Lynch, F. M. Pohley, F. J. Helgren, and F. J. Kirchmeyer, *J. Am. Pharm. Assoc.* **44,** 442 (1955); U.S. Pat. 2,803,551 (Aug. 20, 1957), F. J. Helgren (to Abbott Laboratories).
17. U.S. Pat. 319,082 (June 2, 1885), C. Fahlberg (one-half to A. List, Leipzig, Germany).
18. *Chem. Eng. News* **41,** 76 (1963).
19. B. Stavric, R. Klassen, and A. W. By, *J. Assoc. Off. Anal. Chem.* **59,** 1051 (1976).
20. B. Stavric, R. Lacombe, J. R. Watson, and I. C. Munro, *J. Assoc. Off. Anal. Chem.* **57,** 678 (1974).
21. A. M. Szokolay, *J. Chromatogr.* **187,** 249 (1980).
22. C. Lecos, *FDA Consumer*, 8 (Sept. 1981).
23. J. M. Price, C. G. Biava, G. L. Oser, E. E. Vogin, J. Steinfeld, and H. J. Ley, *Science* **167,** 1131 (1970).
24. Ger. Offen. 2,534,689 (March 25, 1976), O. Hromatka and D. Binder (to BASF A.-G.); P. A. Rossy, W. Hoffmann, and N. Mueller, *J. Org. Chem.* **45,** 617 (1980).
25. A. F. Holleman, *Recl. Trav. Chim. Pays-Bas* **42,** 839 (1923).
26. G. H. Hamor, *Science* **134,** 1416 (1961).
27. L. F. Audrieth and M. Sveda, *J. Org. Chem.* **9,** 89 (1944).
28. W. L. Pines, *FDA Consumer*, 19 (Dec. 1974).
29. U.S. Pat. 2,275,125 (March 3, 1942), L. F. Audrieth and M. Sveda (to E. I. du Pont de Nemours & Co., Inc.).
30. U.S. Pat. 2,804,472 (Aug. 27, 1957), D. J. Loder (to E. I. du Pont de Nemours & Co., Inc.).
31. U.S. Pat. 2,804,477 (Aug. 27, 1957), H. S. McQuaid (to E. I. du Pont de Nemours & Co., Inc.).
32. Israel Pat. 18,631 (May 17, 1965) (to Abbott Laboratories).
33. U.S. Pat. 3,060,231 (Dec. 31, 1962), D. Mueller and R. Trefzer (to CIBA Corp.).
34. U.S. Pat. 3,366,670 (Jan. 30, 1968), O. G. Birsten and J. Rosin (to Baldwin-Montrose Chemical Co., Inc.).
35. U.S. Pat. 3,194,833 (July 13, 1965), M. Freifelder and B. Meltsner (to Abbott Laboratories).
36. G. A. Benson and W. J. Spillane, *J. Med. Chem.* **19,** 869 (1976).
37. R. H. Mazur, J. M. Schlatter, and A. H. Goldkamp, **91,** 2684 (1969).
38. C. I. Beck in G. E. Inglett, ed., *Symposium: Sweeteners*, AVI Publishing Co., Inc., Westport, Conn., 1974, p. 211.
39. R. E. Ranney, J. A. Oppermann, E. Muldoon, and F. G. McMahon, *J. Toxicol. Environ. Health* **2,** 441 (1976).
40. *Wall Street Journal*, 25 (July 31, 1981).
41. *Chem. Eng. News*, 6 (July 20, 1981).
42. A. H. Hayes, *Fed. Reg.* **46,** 38284 (1981).
43. A. Ramirez, *Wall Street Journal*, 20 (Sept. 9, 1981).
44. *Chem. Week*, 25 (March 10, 1982).
45. Y. Ariyoshi, T. Yamatani, N. Uchiyama, Y. Adachi, and N. Sato, *Bull. Chem. Soc. Jpn.* **46,** 1893 (1973).
46. Y. Ariyoshi, *Kagakuto Subutsu* **12,** 189 (1974); *Chem. Abstr.* **82,** 1120u (1975).
47. U.S. Pat. 4,256,897 (March 17, 1981), F. J. Vinick (to Pfizer, Inc.); F. J. Vinick and S. Jung, *Tetrahedron Lett.* **23,** 1315 (1982).
48. U.S. Pat. 3,879,372 (April 22, 1975), W. H. J. Boesten (to Stamicarbon); U.S. Pat. 3,933,781 (Jan. 20, 1976), G. L. Bachman, M. L. Oftedahl, and B. D. Vineyard (to Monsanto).
49. A. Yasutake, H. Aoyagi, and N. Izumiya, *Bull. Chem. Soc. Jpn.* **50,** 2413 (1977).
50. Y. Isowa, M. Ohmori, T. Ichikawa, K. Mori, Y. Nonaka, K. Kihara, K. Oyama, H. Satoh, and S. Nishimura, *Tetrahedron Lett.* **20,** 2611 (1979).
51. Brit. Pat. 1,299,265 (Dec. 13, 1972), M. T. Briggs and J. S. Morley (to ICI); M. Fujino, M. Wakimasu, K. Tanaka, H. Aoki, and N. Nakajima, *Naturwissenshaften* **60,** 351 (1973).
52. R. H. Mazur, J. A. Reuter, K. A. Swiatek, and J. M. Schlatter, *J. Med. Chem.* **16,** 1284 (1973).
53. M. Fujino, M. Wakimasu, M. Mano, K. Tanaka, N. Nakajima, and H. Aoki, *Chem. Pharm Bull.* **24,** 2112 (1976).

54. R. H. Mazur, A. H. Goldkamp, P. A. James, and J. M. Schlatter, *J. Med. Chem.* **13,** 1217 (1970).
55. M. Lapidus and M. Sweeney, *J. Med. Chem.* **16,** 163 (1973).
56. M. Kawai, R. Nyfeler, J. M. Berman, and M. Goodman, *J. Med. Chem.* **25,** 397 (1982).
57. *Chem. Week*, 16 (Sept. 15, 1982).
58. U.S. Pat. 3,087,821 (April 30, 1963), R. M. Horowitz and B. Gentile (to U.S. Department of Agriculture).
59. R. Horowitz and B. Gentili in ref. 38, p. 182.
60. R. M. Horowitz and B. Gentili, *J. Agric. Food Chem.* **17,** 696 (1969).
61. G. H. Robertson, J. P. Clark, and R. Lundin, *Ind. Eng. Chem. Prod. Res. Dev.* **13,** 125 (1974).
62. L. Krbechek, G. Inglett, M. Holik, B. Dowling, R. Wagner, and R. Ritter, *J. Agric. Food Chem.* **16,** 108 (1968).
63. G. E. DuBois, G. A. Crosby, and R. A. Stephenson, *J. Med. Chem.* **24,** 408 (1981).
64. K. Clauss and H. Jensen, *Angew. Chem. Int. Ed. Engl.* **12,** 869 (1973).
65. E. Lück, *Ann. Falsif. Expert. Chim.* **74,** 293 (1981).
66. K. Clauss, *Liebigs Ann. Chem.*, 494 (1980).
67. L. O'Brien and R. C. Gelardi, *Chemtech* **11,** 274 (1981).
68. M. Windholz, ed., *Merck Index*, 4th ed., Merck & Co., Inc., Rahway, N.J., 1976, p. 272.
69. Ref. 66, p. 458.
70. Ref. 66, p. 1137.
71. G. A. Crosby and R. E. Wingard, Jr. in C. A. M. Hough, K. J. Parker, and A. J. Vlitos, eds., *Developments in Sweeteners*, Vol. 1, Applied Science Publishers, Ltd., London, 1979, p. 145.
72. E. M. Acton and H. Stone, *Science* **193,** 584 (1976).
73. A. Hofmann, *Helv. Chim. Acta* **55,** 2934 (1972).
74. T. Moriya, K. Hagio, and N. Yoneda, *Bull. Chem. Soc. Jpn.* **48,** 2217 (1975).
75. U.S. Pat. 3,734,966 (May 22, 1973), A. Tahara, Y. Ohtsuka, T. Nakata, and S. Takada (to Rikagaku Kenkyusho).
76. J. D. Higginbotham in ref. 69, p. 112.
77. P. H. Fairclough, L. Hough, and A. C. Richardson, *Carbohydr. Res.* **40,** 285 (1975); Brit. Pat. 1,543,167 (March 28, 1979), L. Hough, S. Phadnis, R. Khan, and M. R. Jenner (to Tate & Lyle, Ltd.).
78. J. D. Higginbotham in ref. 69, p. 98; *Biotechnol. Newswatch*, 2 (Oct 4, 1982).
79. J. Jizba, L. Dolejs, V. Herout, and F. Sorm, *Tetrahedron Lett.*, 1329 (1971).
80. P. E. Verkade, *Farmaco Ed. Sci.* **23,** 248 (1968).

ROBERT MAZUR
GD Searle and Co.

SYNTHETIC AND IMITATION DAIRY PRODUCTS

Dairy products have been staple items of the diet for many centuries and have long been the target for imitation. The origin of most of the imitation dairy products is obscure, but extends back into the 19th century with respect to oleomargarine and filled cheese. The number and complexity of these imitations has increased with increasing technology and knowledge of the functionality of food ingredients. The imitation products require the same technology, processes, and equipment required for dairy products.

The production and marketing of imitation dairy products reflect many facets related in part to technology, economics, legislation, politics, nutrition, and public food habits. The relative significance of these factors to the market varies from country to country and within regions of the same country. Market penetration of imitation products into the sale of dairy products probably is greatest in the United States and the greatest advance of such products has occurred since World War II.

Advances in processing technology since the early 1940s, eg, the widespread use of homogenization, fluid blending, and continuous processes for a wide range of products, have also markedly advanced the development of imitation products. Current processing technology makes it relatively simple to manufacture physically stable products from a wide range of fats, proteins, and carbohydrates. The growth in use of convenience foods and decreased preparation of in-the-home meals have made the U.S. consumer more willing to change food habits and patterns. In almost all cases, the synthetic dairy products are offered at lower cost, and economics is perhaps the most important single factor in the initial acceptance of imitation dairy foods. Because of their improved technical properties, ability to meet consumer requirements, and the fact that many can be produced with an almost unlimited shelf life; many imitation dairy products are no longer competitors for their alternative milk products (1).

Traditionally, the dairy industry has reacted to the introduction of imitation products with restrictive laws that have tended to be self-defeating. Where legal standards exist for synthetic dairy products, the laws are less restrictive and tend to foster rather than stifle creativity and innovation. In addition, restrictive laws tend to create a negative consumer reaction towards the original product.

Definitions

Imitation dairy products are referred to as imitations, simulates, substitutes, analogues, and mimics and are associated with terms such as filled, nondairy, vegetable nondairy, and artificial milk, cheese, etc. The term nondairy has been used indiscriminately to describe both imitation dairy products and products legally defined as not being imitation dairy products. There are no universally accepted definitions of imitation dairy foods. The products can be divided into three types: those in which an animal or vegetable fat has been substituted for milk fat; those that contain a milk component, eg, casein; and those that contain no milk components (see Milk and milk products). The first two types make up most of the substitute dairy products.

A code of principles has been developed for consumer protection and fair practice in the trade of milk and milk products, mainly by ensuring the precise use of the term milk and terms for different milk products; avoiding confusion arising from the mixing of milk and milk products with nonmilk fats and/or nonmilk proteins; and prohibiting the use of misleading names and information for products that are not milk products. Essentially, any product that resembles a dairy product is an imitation or synthetic product. The code has been accepted by 71 countries.

The International Dairy Federation study on Imitation and Synthetic Milk and Dairy Products includes the following terms and definitions:

Imitation products contain at least one main milk constituent, usually skim milk or skim-milk powder, and vegetable fat replaces butterfat. What have been termed filled milk products are included under this definition.

Synthetic products contain no dairy product although some include sodium caseinate.

The following definitions have been proposed:

Filled. Filled imitation dairy products are made in semblence of a dairy product and milk fat has been substituted by a vegetable or animal fat other than milk fat.

Nondairy. Milk-protein-based imitation dairy products are made to resemble a dairy product and contain one or more proteins derived from milk. Nondairy in this case refers only to imitation dairy products that contain proteins derived from milk, since the FDA has ruled that such proteins are chemicals and can not be considered dairy products.

Synthetic. Synthetic imitation dairy products are made to resemble a dairy product but contain no components derived from milk.

Ingredients

The characteristics and stability of imitation dairy products depend largely on the characteristics of the main ingredients, ie, fat, protein, and carbohydrates, and of the minor functional ingredients that stabilize the fat and protein systems. The ingredients generally include fats or oils and the stabilizing emulsifiers, proteins and their stabilizing gums, and salts and carbohydrates (see Fats and fatty oils; Vegetable oils; Gums).

Fats and Oils. In industrial use, fats are considered solid at room temperature and oils liquid at room temperature; the latter is true because of a higher content of unsaturated fatty acids. Hydrogenation of oils converts them to fats. Fats and oils are mainly the triglyceride-rich lipid fraction, which contains small amounts of other lipids, eg, sterols and phospholipids. For imitation dairy products, fats are generally selected to have low and narrow melting-point ranges, generally 32–36°C. The fats and oils are characterized in a number of ways, including iodine number, polyunsaturated fatty acid content, saturated fatty acid content, Wiley melting point, and by solid-fat index at different temperatures. The iodine number is an index of the number of unsaturated bonds in the fatty acid molecules of the fat or oil. The polyunsaturated fatty acid content is given as a percent of the fatty acids with two or more unsaturated bonds. The Wiley melting point is the temperature at which all of the fat or oil is in a fluid state. The solid-fat index is given as the percentage of the fat or oil that is solid at a given temperature, generally 10–40°C. The desired characteristics of a fat or oil for use in imitation dairy foods are bland flavor, low peroxide value and good flavor stability, low level of free fatty acids and resistance to hydrolysis, and a desired solid fat index over the use temperature range of the product.

The common fats used in imitations are hydrogenated coconut, cottonseed, soybean, groundnut, palm kernel, and various blends of these products. Fat systems with lower melting points are generally preferred as they have a better texture mouth feel. Specialized products may contain polyunsaturated oils including corn, rapeseed, safflower, and sunflower oils.

The physical characteristics of a fat or oil for imitation dairy products are not necessarily dictated by the fat being replaced, but by the composition, processing methods, and conditions of use of the substitute product. Thus, the selection of the fat or oil is generally developed experimentally. The chemical and physical nature of the components of the system, their order of addition, shear input, and processing temperature dictate the final interactions and the nature of the product.

A comparison of the characteristics of milk fat and three different fats that are used in satisfactory filled and imitation milks are listed in Table 1 and those that comprise whipped topping, in Table 2.

Emulsifers. An emulsion is a two-phase system consisting of two immiscible liquids, the one being dispersed as fine globules in the other. In food systems, two general types of emulsions exist: oil–water and water–oil. An emulsifier is a product that has a water-soluble and a fat-soluble portion in the same molecule and stabilizes fat-containing food products. Emulsifiers may be natural products, eg, phospholipids and proteins, or chemicals derived from natural products. Most of these products are esters of long-chain fatty acids with a polyhydric alcohol. The types of emulsifiers used in

Table 1. Typical Fat Characteristics for Imitation and Filled Milks[a]

Characteristics[b]	Milk fat	Fat products		
		A	B	C
SFI, %				
at 10°C	34	25	15	15
at 21°C	13	13	11	9
at 27°C	8	9	10	8
at 33°C	0.5	4	6	2.5
iodine value	33	80	95	105
polyunsaturated fatty acid, wt % (cis–cis)	1.6	21	24	36
saturated fatty acids, wt %	92.1	21	20	18

[a] Ref. 2.
[b] SFI = solid fat index.

Table 2. Typical Fat Characteristics for Imitation Dairy Products[a]

Characteristic	Ice cream and whipped topping				Coffee whiteners	
	A	B	C	D	Fluid	Dry
SFI, %						
at 0°C	38	30	24	25	25	61
at 21°C	23	20	13	18	14	50
at 27°C	20	19	10		9	46
at 33°C	12	14	6	4	4	28
at 41°C	4	8	2			
iodine value	68	82	83	85	80	67

[a] Ref. 2.

imitation dairy foods include: mono- and diglycerides, lactic acid and fatty acid esters of glycerol, acetic acid and fatty acid esters of glycerol, polyglycerol esters of fatty acids, sorbitan esters of fatty acids, polyoxyethylene esters of fatty acids, and propylene glycol esters of fatty acids. The emulsifiers should provide for a stable emulsion that is resistant to coalescence and breaking, be bland and contribute no off-flavors, provide an emulsion that is resistant to creaming or separation, and have an optimum hydrophile–lipophile balance for the given system. All imitation dairy foods contain emulsions. Margarine is the only product that is a water-in-oil emulsion; all others are oil-in-water emulsions. Frequently a combination of emulsifiers are used to give the desired characteristics to the final product.

The hydrophile–lyophile balance (HLB) is an empirical system based on the fact that oil–water (o/w) emulsions are best stabilized by water-soluble emulsifiers and water–oil (w/o) emulsions are best stabilized by oil-soluble emulsifiers (3). The method of calculating the HLB and the requirements for HLB application are given in Table 3 (see Emulsions). Table 4 gives HLB values for common emulsifiers. The HLB optimum for a given emulsifier varies with the components of the food system. A coconut oil–water emulsion that shows optimum stability with an HLB of 7–9 shows a shift in requirements for stability with added casein and electrolytes to an optimum stability with an HLB of 3–5. In addition, the stability of an emulsion can be affected by the chemical nature of the emulsifier. The optimum HLB for an emulsifier in a given system is influenced by the other ingredients used. This is illustrated for a model synthetic milk system in Figures 1 and 2.

Generally, emulsifiers are selected to minimize coalescence of globules. However, in the case of high fat whipped products, eg, ice cream and whipped toppings, the emulsifier provides for controlled agglomeration of fat globules, thus giving the desired body to the final product. The agglomeration helps to control air micelle size and stability.

The physical form of the food relates to the selection of the emulsifier. For example, coffee whiteners are marketed as liquids, frozen products, or dry powders. The

Table 3. Hydrophile–Lipophile Balance (HLB) Concept for Emulsifier Selection[a]
Calculation of HLB scale (0–20): HLB = 20 (1 − S/A), where S = saponification number of ester and A = acid number of recovered acid.

Property	HLB value
dispersibility	
no dispersibility in water	1–4
poor dispersibility	3–6
milky dispersion after shaking	6–8
stable milk dispersion	8–10
translucent-to-clear dispersion	10–13
clear solution	13+
application of surfactant	
w/o emulsifier	4–6
wetting agent	7–9
o/w emulsifier	8–18
detergents	13–15
solubilizers	10–18

[a] Ref. 3.

Table 4. Emulsifiers and Their Characteristics[a]

Emulsifier	Melting point, °C	Viscosity (at 25°C), MPa·s (= cP)	HLB rating
mono- and diglycerides from the glycerolysis of edible fats or oils; 52–56 wt % alpha monoglycerides, 61–66 wt % total monoglycerides	57–61		3.2
mono- and diglycerides of fat-forming fatty acids; 47–50 wt % alpha monoglycerides, 54–59 wt % total monoglycerides		ca 150	2.8
mono- and diglycerides from the glycerolysis of edible fats or oils; 40–44 wt % alpha monoglycerides, 48–52 wt % total monoglycerides	46–50		2.8
mono- and diglycerides from the glycerolysis of edible fats or oils; 54–58 wt % alpha monoglycerides, 65–69 wt % total monoglycerides	52–53		3.5
mono- and diglycerides from the glycerolysis of edible fats or oils; 52–56 wt % alpha monoglycerides, 61–66 wt % total monoglycerides	59–62		3.5
mono- and diglycerides from the glycerolysis of edible fats or oils; 54–58 wt % alpha monoglycerides, 65–69 wt % total monoglycerides	54–57		3.5
sorbitan monostearate	53		4.7
polyoxyethylene (20) sorbitan monostearate		600	14.9
polyoxyethylene (20) sorbitan tristearate	ca 33		10.5
polyoxyethylene (20) sorbitan monooleate (polysorbate 80)		400	15.0
mono- and diglycerides from glycerolysis of edible fats or oils (80 wt %) and polysorbate 80 (20 wt %)	54–57		5.2
mono- and diglycerides from the glycerolysis of edible fats or oils (80 wt %) and polyoxyethylene (20) sorbitan tristearate (20 wt %)	54–57		4.3
mono- and diglycerides from the glycerolysis of edible fats or oils (40 wt %), and polyoxyethylene (20) sorbitan tristearate (60 wt %)	54–57		5.9
stearoyl-2-lactylate			4.5

[a] Ref. 3.

optimum emulsifier differs for each system. For liquid products, the emulsifier must impart good stability to the emulsion so that it remains in a uniform state on standing after preparation and prior to sale. It must also prevent oiling off (emulsion breaking) and feathering (formation of unstable white flocs on top that resemble feathers) when used in coffee (qv). For the frozen product, the emulsifier must also contribute to the freeze–thaw stability of the emulsion. In dry products, the emulsifier must contribute to the free-flowing properties and to the dispersal of the product in coffee. The dry

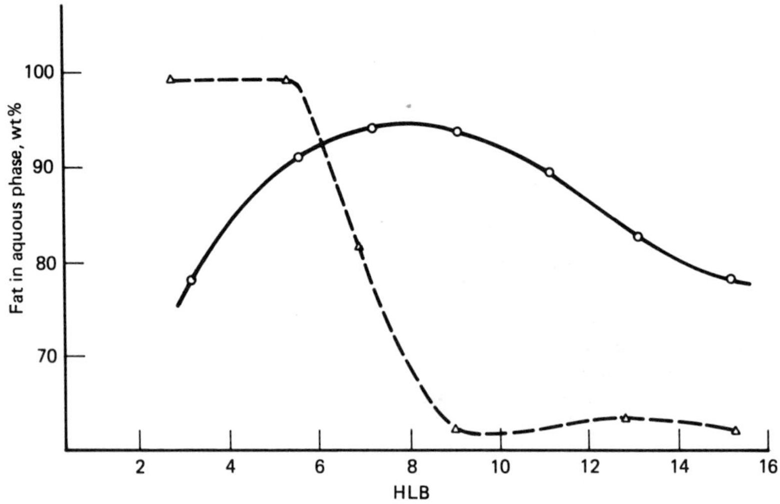

Figure 1. Effect of system components on emulsifier HLB. △: Fat–caseinate–salts–water–emulsifier; ○: fat–water–emulsifier (4).

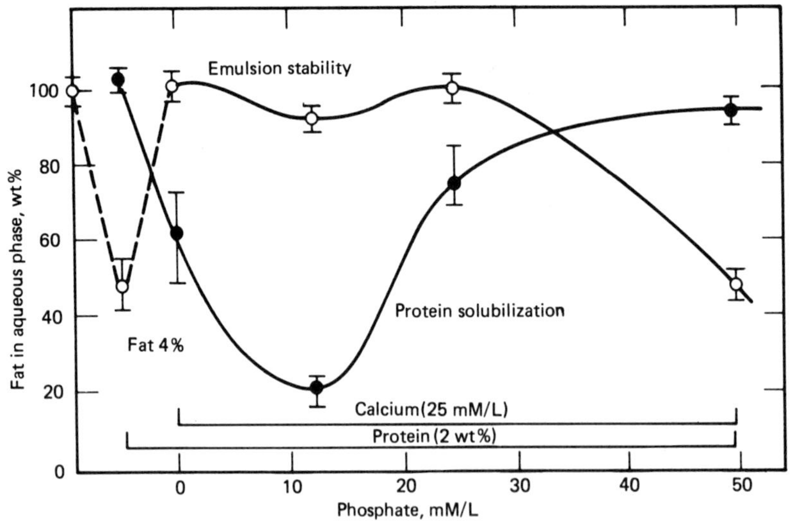

Figure 2. Effect of system components on emulsion and protein stability in an imitation milk (4).

whitener should sink to the bottom of the coffee and disperse rapidly. The selection of emulsifier differs for each product. The lowest HLB emulsifiers are used for liquid products and the highest HLB emulsifiers are used for dry products. Thus, mono- and diglycerides are commonly used in liquid coffee whiteners and blends of the emulsifiers are used for dry products. A typical blend contains mono- and diglycerides, polyoxyethylene, and sorbitan fatty acid esters. The type and level of emulsifier also depend on the type and concentration of protein used in the product. This relates to the fact that the protein also contributes to the emulsification of the fat. A common level of emulsifier in imitation dairy foods is 5% of the fat concentration.

Proteins. Proteins (qv) are especially significant in imitation dairy foods with respect to the nutritional and physical properties to the product (5). The relative significance of the nutritional quality of the protein depends upon the type of product and the extent to which the product contributes to the total protein intake of a given population. Thus, protein quality is extremely important in imitation milk, significant in imitation cheese, and less significant in coffee whiteners and whipped toppings. The role of protein with respect to physical properties varies with the type of product. Proteins contribute to a number of functions in an imitation food, which include emulsification, gelation, melting, and whipping. The desired functionality of proteins in different types of imitation dairy food are shown below:

Imitation	*Desired functionality of protein*
cheese	emulsification, melting
coffee cream	emulsification, whitening, buffering
cultured product	emulsification, gelation
ice cream	emulsification, whipping
milk-based pudding	emulsification, gelation
whipped topping	emulsification, whipping

The factors to be considered in protein selection include: nutritional quality required, specific functionality required, solubility (ease of incorporation into formulation), good flavor quality (contributes no off-flavor to product), and stability to processing conditions (see also Food processing).

A wide number of protein sources are available for use in imitation dairy products. These include: animal proteins, ie, skim milk in liquid, condensed, or dry form (filled products); casein, caseinates, and coprecipitates; whey proteins; fish proteins; and blood proteins. Oil-seed protein sources include soybean protein concentrates and isolates, groundnut protein, cottonseed protein, and sunflower seed, rapeseed, coconut, and sesame seed proteins. Other sources are leaf and single-cell proteins (see Foods, nonconventional). Of these protein sources, milk and soybean proteins are most widely used. Some imitation products include other proteins, eg, whey, groundnut, and cottonseed proteins. Other proteins to be used will be based on economics, functionality, and availability.

Casein and Caseinates. Preliminary definitions for edible casein and edible caseinate have been suggested (5): edible casein is the product obtained by separating, pressing, and drying and lactic- or mineral-acid-precipitated coagulum of skimmed milk, and edible caseinate is the product obtained by drying aqueous solutions prepared by combining dry edible casein or fresh edible casein curd with food-grade alkali. Casein is also made from rennet coagulation of skim milk. Proposed International Dairy Federation (IDF) standards for casein and caseinate are listed in Tables 5 and 6. Casein and caseinates are the most widely used proteins in imitation dairy products, primarily because of their superior functionality and flavor as compared to most vegetable proteins.

Generally, the salt form of casein is used and the two common forms are sodium caseinate and calcium caseinate. Sodium, potassium, ammonium, and calcium caseinates are available. In most cases the sodium salt is preferred for emulsification and the calcium salt is preferred for imitation cheese. Casein and caseinates must be stored carefully and evaluated for flavor before use in products. Improperly manufactured or stored casein–caseinate has a very strong, musty off-flavor. Excessive fat content,

Table 5. IDF Casein Specifications[a]

Property	IDF standard 45, acid-precipitated edible casein	
	Extra grade	Standard grade
flavor and odor	natural	slightly
color	white-to-pale cream	white-to-pale cream
appearance	lump-free	lump-free
protein (dry basis: N × 6.38), wt %	min 95	min 90
moisture, wt %	max 12	max 12
milk fat, wt %	max 1.5	max 2.0
lactose, wt %	max 0.2	max 1.0
ash, wt %		
free acid (mL/g 0.1 *N* NaOH)	max 0.20	max 0.27
scorched particles, mg/g	max 15.0/25	max 22.5/25
foreign matter	none	none
copper, ppm	max 5	max 5
lead, ppm	max 5	max 5
iron, ppm	max 20	max 20
total plate count, g^{-1}	max 30,000	max 100,000
coliforms, 0.1 g^{-1}	neg	max 2
mold and yeasts, g^{-1}	max 50	max 100
thermophilic organisms, g^{-1}	max 5,000	max 5,000

[a] Ref. 6.

Table 6. IDF Caseinate Specifications[a]

Property	IDF standard 72, edible caseinate	
	Extra grade	First grade
color	white-to-pale cream	white-to-pale cream
flavor and odor	bland	bland
moisture, wt %	max 6.0	max 8.0
protein (dry basis N × 6.38), wt %	min 90.0	min 88.0
fat (dry basis), wt %	max 1.5	max 1.5
lactose, wt %	max 0.5	max 1.0
scorched particles, mg/g	max 15.0/25	max 22.5/25
extraneous matter, 50 g^{-1}	free	free
copper, ppm	max 5	max 5
lead, ppm	max 5	max 5
iron, ppm	max 20	max 20
total plate count, g^{-1}	max 10,000	max 100,000
coliforms, 0.1 g^{-1}	neg	neg
mold and yeasts, g^{-1}	max 50	max 50
thermophilic organisms, g^{-1}	max 5,000	max 5,000

[a] Ref. 6.

high lactose and moisture contents, and high storage temperatures contribute to off-flavor development.

Whey Proteins. Whey is the fluid obtained by separating the coagulum from milk, cream, and/or skim milk and is a by-product of either casein or cheese manufacture. Its composition is determined by the method of curd formation, curd handling practices, and methods of handling whey as it separates from the curd. Dried acid whey contains ca 12.5 wt % protein (total nitrogen × 6.38), 11.0 wt % ash, and 59.7 wt % lactose, whereas sweet whey contains 13.5 wt % protein, 1.2 wt % fat, 8.4 wt % ash, and

74 wt % lactose. The composition varies with the type of acid used (7). It has been used in some imitation products but not as a source of protein. Whey proteins have been used in imitation products only in the past few years after the commercialization of ultrafiltration. Membranes are used that retain protein and permit water, lactose, and ash to pass through the membrane as permeate. Protein concentrates are available from both acid and sweet whey and in concentrations of 35–55 wt % protein. The functionality of the protein concentrates varies with whey type and concentration. Table 7 gives compositional data for whey protein concentrates from different sources of whey. They are used in a limited number of products, including ice cream, coffee whiteners, and whipped toppings. A significant problem is the wide variation in properties in the same type of product from different manufacturers and in the same product from the same source. As these products become standardized and more universally available, it is predicted that they will be used widely in imitation products alone and in combination with either caseinates or soybean isolates. They do have a higher protein efficiency ratio (PER) than casein and are equal to egg protein.

Soybean Protein Isolates. Soybean protein isolates are the only vegetable proteins that are widely used in imitation dairy products (1). There has been much research to develop improved soybean protein isolates with good functional properties and improved flavor. Most isolates are derived from isoelectric precipitation, so that the soybean protein isolates have properties that are similar to those of casein, in that they are insoluble at their isoelectric point, have a relatively high proportion of hydrophobic amino acid residues, and are calcium-sensitive. They differ from casein in that they are heat-denaturable and thus heat-labile. The proteins have relatively good nutritional properties and have been increasingly used as a principal source of protein. A main deterrent to their use has been the beany flavor associated with the product. However, flavor is consistantly being improved and use is expected to increase with its improved properties and lower cost as compared to caseinates.

Based on protein content, there are three main groups of products: flakes, flours, and grits with a minimum protein content of 40 wt %; concentrates with a protein content of 70% on a dry weight basis; and isolates containing 90% or more protein on a dry weight basis. For imitation products, the concentrates and isolates are the more

Table 7. Whey Protein Concentrate Compositions from Different Sources (as % Total Solids)[a]

Property	Rennet	Lactic	Sulfuric	Cheese
TN × 6.38	79.06	80.00	80.00	84.84
nonprotein nitrogen (NPN)	0.78	0.35	0.10	0.63
true protein	73.96	77.77	79.36	80.82
lactose	10.70	10.00	9.30	2.33
ash	3.30	2.5	2.5	2.63
fat	3.30	5.0	5.0	7.96
PO_4	0.22	0.60	0.42	0.11
Ca	0.53	0.42	0.37	0.60
SO_4	0.06		0.75	0.15
Mg	0.005	0.010	0.025	0.033
Na	0.28	0.16	0.16	0.58
K	0.69	0.21	0.21	0.17
Cl		0.30	0.24	0.52
lactate		1.50		0.45

[a] Ref. 7.

important products. The flavor quality of the products improves with increased protein content and, at present, only isolates approach a flavor quality sufficient for use in imitation dairy products at the same protein level as the dairy product they resemble. Additional processing steps can further improve flavor but at considerable expense.

Soy protein concentrates are produced from defatted soy flour or flakes (1). These are obtained by leaching oligosaccharides, minerals, and other soluble constituents with acidified water (pH 4.6). Alternatively, the defatted material is denatured by means of moist heat followed by extraction with water. The physical properties differ according to the method used. Concentrates prepared by acid leaching and neutralization in the absence of heat treatment have a higher soluble protein content.

Soy protein isolates are derived from defatted flakes or flours by removing the water-insoluble polysaccharides, oligosaccharides, and other low molecular weight components (1). The flakes or flour receive a minimum heat treatment and are extracted with an aqueous alkaline medium (pH 7–8.5). The clarified extract is adjusted to pH 4.5 with food-grade acid. The precipitated proteins are separated by centrifugation and then are washed, neutralized, and dried as the proteinate. The pH treatment, temperature of precipitation and ionic strength dictate the ratio of the two main protein classes (glycinin and conglycinin) and the functionality of the isolates. Glycinin and conglycinin vary in functionality and variations in the ratios of these products are significant in the behavior of the soy protein isolate in imitation products. Some soy protein isolates are slightly hydrolyzed to improve solubility and emulsifying properties. Blends of soy with either caseinates or whey proteins have been prepared to extend soy protein functionality and to minimize off-flavor problems (see Soybeans and other oilseed proteins).

Compositions of soy protein concentrates and isolates are given in Table 8.

Other Proteins. Groundnut, fish, and cottonseed proteins have been used to a limited degree in imitation products. Their properties are discussed in more detail in an FAO report on imitation milks and imitation milk products (8).

Gum Stabilizers. Gums (qv) are used in imitation dairy products for one or more of the following reasons (3): provide viscosity control and improve mouth feel, improve whipping (aerating) properties of whipped products, provide protective colloid effect to stabilize proteins to heat processing, modify the surface chemistry of fat-globule surfaces to minimize creaming, provide acid stability to protein systems, improve freeze–thaw stability, provide a hard-water resistant coffee whitener, and modify

Table 8. Approximate Analyses of Commercial Soy Protein Concentrates and Isolates[a,b]

	Concentrates			Isolates			
Analysis	A	B	C	A	B	C	D
moisture, wt %	6.7	5.2	3.1	4.7	6.4	7.6	3.7
protein, wt %	66.2	67.3	69.6	92.8	92.2	92.9	94.7
protein (dry basis), wt %	70.9	71.1	72.2	97.4	98.7	100.0	98.4
fat	0.3	0.3	1.2				
crude fiber, wt %	3.5	3.4	4.4	0.2	0.1	0.1	0.2
ash, wt %	5.6	4.8	3.7	3.8	3.5	2.0	2.7
nitrogen solubility index	5.0	69.0	3.0	85.0	95.0		
pH (1:10 aqueous dispersion)	6.9	6.6	6.9	7.1	6.8	5.2	5.5

[a] Ref. 1.
[b] Different products designated by A, B, C, etc.

melting characteristics of imitation cheese. Gums can be classified as neutral and acidic, straight- and branched-chain, gelling and nongelling. Common gums and their characteristics are listed in Table 9. The principal gums and their functions in imitation dairy products are listed in Table 10. A listing of gum applications in dairy foods that apply to imitation products is given in Table 11.

Gums are generally used in concentrations of 0.02–0.5 wt % or 1–6 wt % of the protein level in the food. Because of differing functionalities, combinations of gums provide a better product than an individual gum.

Stabilizing Salts. Citrates and phosphates are used in imitation dairy products for one or more of the following purposes (3): alter buffering capacity of the system, improve the stability of the protein to calcium ions, improve the heat stability of the protein, minimize the age gelation of ultra-high temperature (UHT)-processed imitation products, serve as emulsifying salts in imitation-cheese manufacture, and modify the water-binding capacity of proteins and improve solubility. Common phosphate salts used in imitation foods are monosodium phosphate (MSP), disodium phosphate (DSP), trisodium phosphate (TSP), disodium pyrophosphate, tetrasodium pyrophosphate (TSPP), sodium tripolyphosphate (STP), sodium hexametaphosphate (SHMP), sodium trimetaphosphate, and sodium tetrametaphosphate (3). Although there are similarities in some functional uses, phosphates are more broadly used than citrates in imitation foods.

Phosphates react with calcium to reduce the calcium ion activity and assist in stabilizing calcium-sensitive proteins, ie, caseinate and soy proteinate, during processing. Phosphates react with milk proteins and the extent of reaction depends upon chain length. Casein precipitates upon addition of pyrophosphates, whereas whey proteins do not. Longer-chain polyphosphates cause the precipitation of both casein and whey proteins. The reactions involved are complex and not fully understood. Functions of phosphates in different types of imitation dairy products are summarized in Table 12 (see Food additives).

Composition

The composition of imitation dairy products is highly variable and generally represents the least-cost formulation consistent with consumer acceptance of the product. In the absence of Standards of Identity the imitation products invariably have lower fat and protein levels than the dairy products that they are made to resemble.

The gross compositions of filled milk, mellorine, synthetic milk, sour cream, coffee whiteners, whipped toppings, and cheese are listed in Table 13. Table 14 presents data comparing the composition of selected dairy products and their imitations. A selected listing of recent patents with regard to ingredients, formulations, and processing of imitation dairy foods is given in the following references: cheese (12–29), cheesecake (30), creams (31–35), milk (36–41), sweetened condensed milk (42), milkshake (43–44), milk replacers (calf) (45–48), sour cream (49–51), ice cream (52–53), and low fat margarine (54–57).

Processing

The processes used for imitation dairy products are essentially the same as for their dairy product counterparts (1,5,10,58). The processes common to all fluid products (see Fig. 3) include: ingredient blending, pumping, pasteurization or steril-

Table 9. Common Stabilizing Gums and Their Characteristics[a]

Gum	Source	Sugar components	Solubility	Gelation property	Remarks
alginates	seaweed	D-manaronic acid (1-4), L-glucuronic (1-4)	varies with salt form	gels in presence of divalent cations	viscosity of 1 wt % solution varies with type, 10–2000 mPa·s (= cP)
arabic	tree extract	L-arabinose, D-galactose, L-rhamnose, D-glucaronic acid	very water-soluble	nongelling	low viscosity, unstable in acid
carboxymethyl cellulose	wood pulp	D-glucose (1-4)	varies with molecular weight substitution	nongelling	forms films
carrageenan	seaweed extract	D-galactose, 3,6 anhydro-D-galactose and sulfate acid ester groups	water-soluble at 50–80°C	gels with potassium, calcium, ammonium ion	exists in several forms, κ-type stabilizes casein to heat
guar	seed gum	D-mannose (1-4), D-galactose (1-6) branches	cold-water-soluble	forms nonedible gels with horax	viscosity of 1 wt % solution, 2700 mPa·s
locust bean	seed gum	D-mannose (1-4), D-galactose (1-6) branches	swells in cold water, hot-water-soluble	nongelling	viscosity of 1 wt % solution, 3000–3500 mPa·s gives elastic character to carrageenan gels
tragacanth	tree extract	D-galactose, D-xylose, D-glucuronic acid	swells in cold water, soluble at 50°C in 2 h	nongelling	viscosity of 1 wt % solution, 3400 mPa·s; acid-resistant
xanthan	microbial	D-glucose, D-mannose, D-glucuronic acid	soluble in cold or hot water	nongelling	freeze–thaw-stable, acid-stable

[a] Ref. 3.

Table 10. Gums and Associated Functions in Imitation Dairy Products

Gum	Imitation product use	Function
alginates	ice cream	aeration, reduce whip time
	cheese	texture modifier, prevent oil separation
carrageenan	milk puddings	gelation
	ice cream	prevent wheying off
	infant formula	stabilize proteins to heat
	evaporated milk	stabilize proteins to heat
locust bean gum	ice cream	smooth melt down, freeze–thaw resistance
	cheese spread	texture modifier
guar gum	ice cream	water binding, body control
carboxymethyl cellulose	ice cream	prevent ice crystal growth
	whipped topping	aeration, protective colloid
xanthan gum	sour cream	prevent wheying off

ization, homogenization, cooling, and packaging. Products can be manufactured by batch or by continuous processes based on those developed for dairy products. A large volume of imitation products is made in large plants with continuous processes. Ingredients may be either liquid, dry, or a combination of both. Liquid ingredients can be blended by the use of positive metering pumps to a mixing tank or by in-line blending. Dry ingredients must be hydrated during the blending operations. This is most commonly accomplished by the use of turbine mixers in processing vats or by incorporating the dry material through a high speed, centrifugal pump. The blending temperature depends upon the nature of the ingredients, but it must be above the melting point of the fat and sufficient to fully hydrate gums used as stabilizers. Pasteurization is generally done in high temperature short time (HTST) units, in which the homogenizer is integrated into the pasteurization system. Sterilization is either by retort or by ultra high temperature (UHT) processing, followed by aseptic packaging (see also Sterilization techniques).

Margarine. Margarine or oleomargarine was originally marketed as an imitation butter; however, it now has a recognized identity of its own. Typical fat blends used in margarine manufacture are listed in Table 15. The proportion of the fat blend and other ingredients varies with the type of margarine and with the country of manufacture. Within a given country, the variations in fat blends, aqueous phases, and other ingredients are fixed by legislation. Two margarine compositions are given below (6):

UK margarine
fat blend, 1250 kg
aqueous phase
- 50 L ripened milk
- 100 L water
- 100 L brine (containing 31.7 kg salt)
- 4.42 kg preservative

U.S. margarine
fat blend, 907 kg
aqueous phase
- 208 L ripened milk
- 32 kg salt

additives
- 1.4–3.2 kg vegetable lecithin
- 1760 RE/kg vitamin A concentrate
- 28.3 g diacetyl (flavoring)
- fat-soluble dye, as required

Table 11. Application of Gums in Dairy Products[a]

Product	Agar	Algin	Carrageenan	Furcellaran	Arabic	Xanthan	Tragacanth	Hydroxypropyl cellulose	Locust bean	Guar	CMC[b]	Methyl cellulose	Karaya
ice cream stabilizer	X	X	X	X	X		X		X	X	X	X	X
ice milk	X	X	X	X	X		X		X	X	X	X	X
milkshake	X	X	X				X		X		X		
sherberts	X	X	X	X	X		X		X	X	X		X
chocolate milk drink		X	X	X	X	X	X		X	X	X		
pudding	X	X	X	X	X	X			X	X			X
cream cheese	X	X	X		X				X	X			
cheese spread		X	X		X	X	X	X	X	X	X		X
Neufchâtel process cheese	X	X	X		X		X		X	X			X
whipped cheese		X	X			X		X	X				X
yogurt	X		X		X				X	X			

[a] Ref. 3.
[b] CMC = carboxymethyl cellulose.

Table 12. Functions of Phosphates in Imitation Dairy Products

Imitation dairy food	Phosphate type	Function
milk-based pudding	SHMP, STP, TSPP, TSP	gelation
milk	DSP	heat stability
coffee whitener	DSP, DKP[a], Poly[b], SALP[c]	pH control, reduce feathering, reduce oiling off
whipped topping	DSP, DKP[a], TSPP	pH control, whip stability, overrun improvement
sour cream	DSP, DKP[a], TSP, TSPP	improve body, reduce wheying
ice cream	DSP	protein stability
cheese	MSP, DSP, TSP, TSPP	modify protein, emulsifier

[a] DKP = dipotassium phosphate.
[b] Poly = polyphosphate.
[c] SALP = sodium aluminum phosphate.

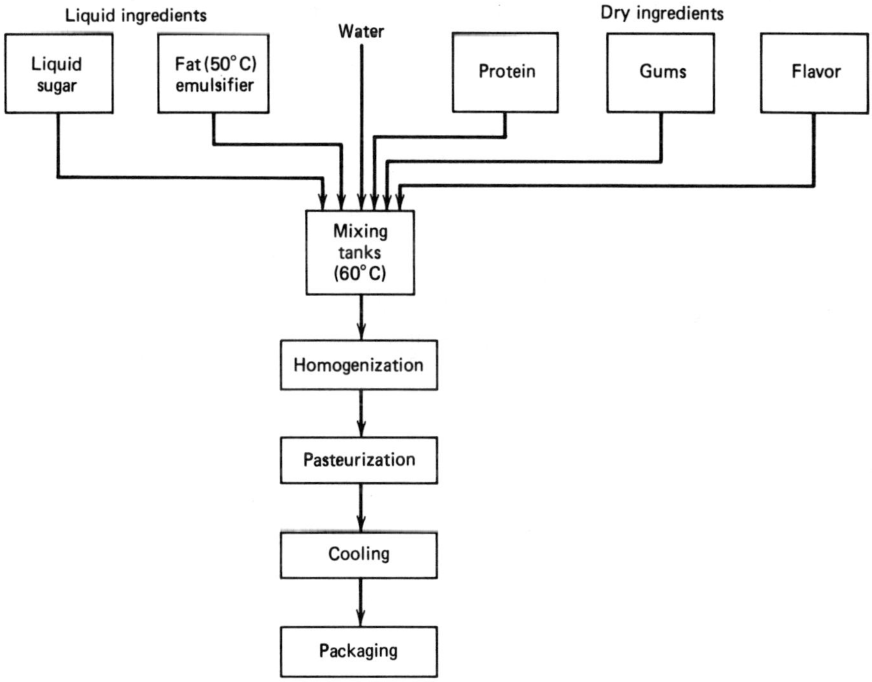

Figure 3. General processing scheme for fluid imitation dairy products.

Traditionally, skim milk was used as the aqueous phase in margarine. Today, margarines may contain caseinates, whey proteins, and soy proteins as the protein component in the aqueous phase, where the laws permit. A milk-free margarine which prevents sticking during frying, because of the incorporation of 0.01–0.1 wt % sodium caseinate in the margarine, has been proposed. Substituting soy protein for milk would have a similar effect.

Table 13. Typical Compositions of Whole and Filled Milk, Mellorines, and Imitation Dairy Products, wt %[a]

Ingredient	Whole milk	Filled milk			Mellorines			Nondairy milk		Sour cream	Liquid coffee-whiteners	Whipped toppings	Caseinate-based cheese
		A	B	C	A	B	C	A	B				
fat	3.8	3.8	3.1	3.4				3.0	3.7				
protein	3.3	4.1	3.3	3.1				1.0	0.9		1.0–3.0		
carbohydrates	4.6	5.3	4.9	5.0				8.0	6.7				
minerals	0.7	0.9	0.8	0.7				0.2	0.5				
vitamins A and D	+[b]	−[b]	+	−				+	+				
riboflavin	+	+	+	+				0	0				
water					62.5	62.0	61.1			82.0–64.0	80.0–85.0	46.0–75.0	
milk solids[c]					13.0	12.5	11.5			6.0–15.0			
sugar					14.0	13.0	12.0					6.0–15.0	
corn syrup solids					6.0[d]	6.0[d]	5.0[d]				5.0–10.0		
vegetable fats					4.0	6.0	10.0			6.0–20.0	8.0–12.0	20.0–32.0	
stabilizer–emulsifier					0.5	0.5	0.4						
total solids					37.5	38.0	38.9						
stabilizers								0.8	+	0.1–1.0	0.02–0.5	0.1–1.5	
emulsifiers								0.4	+	0.3–2.0	0.2–1.0	0.3–1.5	
buffer								0.2	+		0.1–0.5		
sodium caseinate												1.0–5.0	15–35
calcium caseinate													10–35
oil (mp 32–43°C)													1–2
emulsifier salts													0–0.5
gums													
other proteins[e]													0–25
acid (adipic, lactic)													0.1–1.8

[a] Refs. 8–9. wt % = g/100 g.
[b] +: present; −: absent.
[c] Does not include fat.
[d] 42 DE = dextrose equivalent.
[e] May include gluten, soy, egg white, or gelatin.

Table 14. Comparison of Certain Dairy Products and Their Simulated Counterparts[a]

		Gross composition, g/100 g dry matter				Minerals, mg/100 g dry matter				Vitamins		
		Carbohydrates	Fat	Protein	Ash	Calcium	Phosphorus	Sodium	Potassium	A, IU	Riboflavin, μg	Total solids, g/100 g dry matter
milk (fluid), whole		38.9	27.8	27.8	5.6	936	738	396	1142	1112	1350	12.6
white beverages,	A	52.8	31.7	7.4	4.0	25	248					12.1
liquid	B	61.8	27.7	6.8	4.5	136	673	636	3270			11.0
light cream		14.6	72.7	10.5	2.2	353	280	182	331	3020	509	27.5
coffee whiteners	A	48.5	39.9	4.9	2.8	12	718	293	788	110	0	98.9
dry	B	49.1	36.5	5.0	2.7	16	625	258	768	110	0	98.9
	C	46.1	37.2	5.0	2.7	46	561	290	606	440	108	98.9
	D	48.7	35.8	4.9	3.0	12	62	146	1040	200	219	98.5
liquid	A	50.2	47.9	3.0	1.5	23	30	543	6	2170	0	26.7
	B	42.0	48.0	5.0	2.0	70	155	245	300	250	0	20.0
	C	49.1	40.5	8.6	2.7	72	212	496	121	0	0	22.2
	D	37.9	52.6	2.6	1.1	52	57					21.1
whipping (heavy) cream		7.8	85.3	5.6	1.2	190	149	93	134	3510	268	41.0
whipped toppings	A	40.8	43.2	4.6	1.0	17	32	100	48	800	0	99.9
dry	B	40.6	45.4	5.7	0.6	12	46	97	6	1370	0	98.9
liquid	A	25.7	58.5	0.0	0.5	15	3	193	13	1040	0	39.3
	B	31.6	55.3	2.0	0.2	14	20	20	2	2220	0	49.1
aerosols	A	22.0	58.0	10.7	2.3	310	276	255	310	2540	145	35.5
dairy-based	B	20.4	63.3	8.5	2.3	274	267	413	209	2650	189	41.2
nondairy	A	29.2	58.7	0.0	0.5	10	3	243	11	1130	0	39.1
	B	23.9	67.7	7.6	0.5	23	71	173	6	1370	0	39.3

[a] Ref. 11.

Table 15. Typical Composition of Fat Blends of Some Margarines, wt %[a]

Fat	Margarine A	Margarine B	Margarine C	Margarine D	Margarine E	Margarine F	UK wartime standard blend	United States product
palm kernel	32	37	36			55	50 (palm kernel and coconut)	
coconut	37	36	40	50				
palm						15	8.75	
cottonseed								42.5
hydrogenated cottonseed; mp, 34°C				18				57.5[b]
groundnut	20	16			11		25[c]	
hydrogenated groundnut	11	5			69[d]	30[e]	16.25	
soybean		6	24	22	7			
sesame				10	13			
Total	*100.0*	*100.0*	*100.0*	*100*	*100*	*100*	*100*	*100*

[a] Ref. 59 (updated).
[b] Melting point not specified.
[c] Or cottonseed or other similar soft fat.
[d] Mp, 30°C.
[e] Mp, 40°C.

In addition to traditional margarines, the following are available: whipped margarine, margarine containing 15–40 vol % finely and uniformly dispersed gas; liquid margarine, margarine that pours and is made by one of several methods that modify the fat system (9–10,58); and low fat margarine, margarine with a fat content below 80 wt % and generally 30–65%, ie, imitation margarine or spread. By regulation, margarine contains 80 wt % fat. A number of imitation margarines have been introduced with fat contents of 35–65 wt % (60). A number of different processes have been suggested. In most cases, these products contain sodium caseinate, whey proteins, or soybean proteins as the protein portion of the aqueous phase; emulsifiers; and sometimes gums. Emulsifiers of choice are generally mono- and diglycerides. Some low fat spreads are made from a blend of vegetable fat and butter and butterfat. A flow diagram from the production of one such product is given in Figure 4.

Cheese. The development of imitation cheese has been from the substitution of milk components in the development of filled and nondairy cheese and the development of a synthetic cheese based on the Chinese food sufu, which is based on soybean curd.

Filled and Caseinate-Based Cheese. Filled cheese was made traditionally by adding vegetable fat to skim milk, homogenizing at low pressure to make a uniform emulsion, followed by traditional cheesemaking procedures. With the development of cheese processing, filled cheese was made from skim-milk curd to which was added vegetable oil, flavoring, and phosphate and citrate emulsifing salts. The mixture was heated to ca 55–66°C and the melted, processed, and filled cheese was then packaged. The skim-milk curd was frequently treated with proteolytic enzymes to obtain a body similar to that obtained when processing aged cheese. The current caseinate-based imitation cheeses are based on these processes. A variety of approaches have been suggested for making these products. One such product is made by the process shown

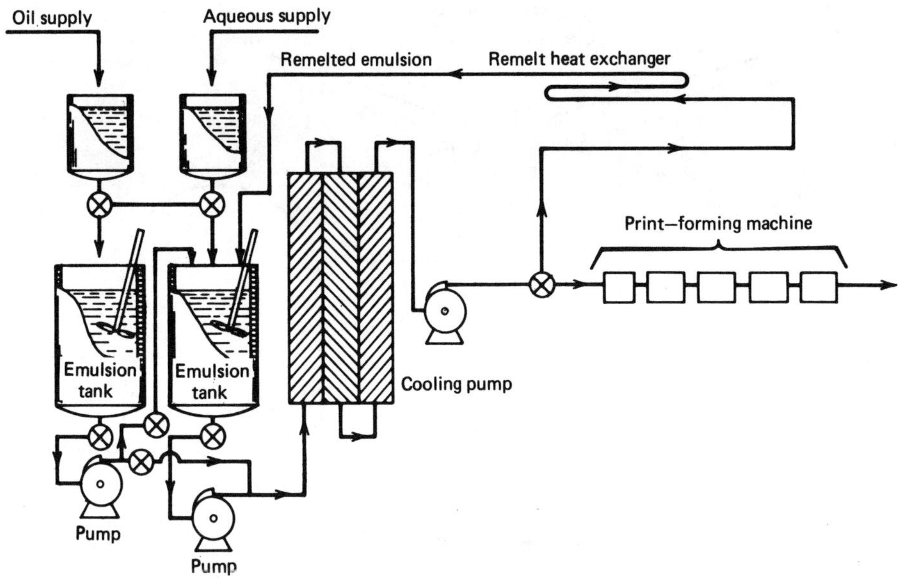

Figure 4. Schematic process flow diagram for a low fat spread (61–62).

in Figure 5 and has the formulation shown below (22):

Dry ingredients, wt %	
calcium caseinate	24.5
tapioca flour	3.0
salt	2.16
adipic acid	0.6
vitamins and minerals	0.1
sorbic acid (mold inhibitor)	0.5
Fat-color blend, wt %	
soybean oil hydrogenated to a Wiley melting point of ca 36°C	21.3
lactylated monoglyceride	0.05
red-orange coloring	0.01
flavoring blend	0.23

The common types of cheese that are imitated are cheddar and mozarella. An imitation cheese containing whey proteins and blood proteins has been reported and the formulation is given in Table 16.

Soybean-Based Cheese. Traditional sufu cheese manufacture is diagrammed in Figure 6. In this process, the soybeans are cooked to give soy milk, which is then formed into a curd with calcium sulfate. This product, which is widely used in the Orient, is unsuited to Western tastes. Interest in developing imitation synthetic cheese gained prominance in the middle 1960s, with sufu as a starting point. Some progress was made using molds and enzyme preparations, but bitter and beany flavors were a deterent to the acceptance of the product. Several U.S. patents report preparation of imitation cheese based on combinations of soy milk or soybean proteins with milkfat and some milk solids. One such product is prepared from specific amounts of natural cheese, pregelatinized starch, a high protein binding agent (preferably soybean), water, and sugar (64). The mixture is heated to 90°C and extruded into the form of small

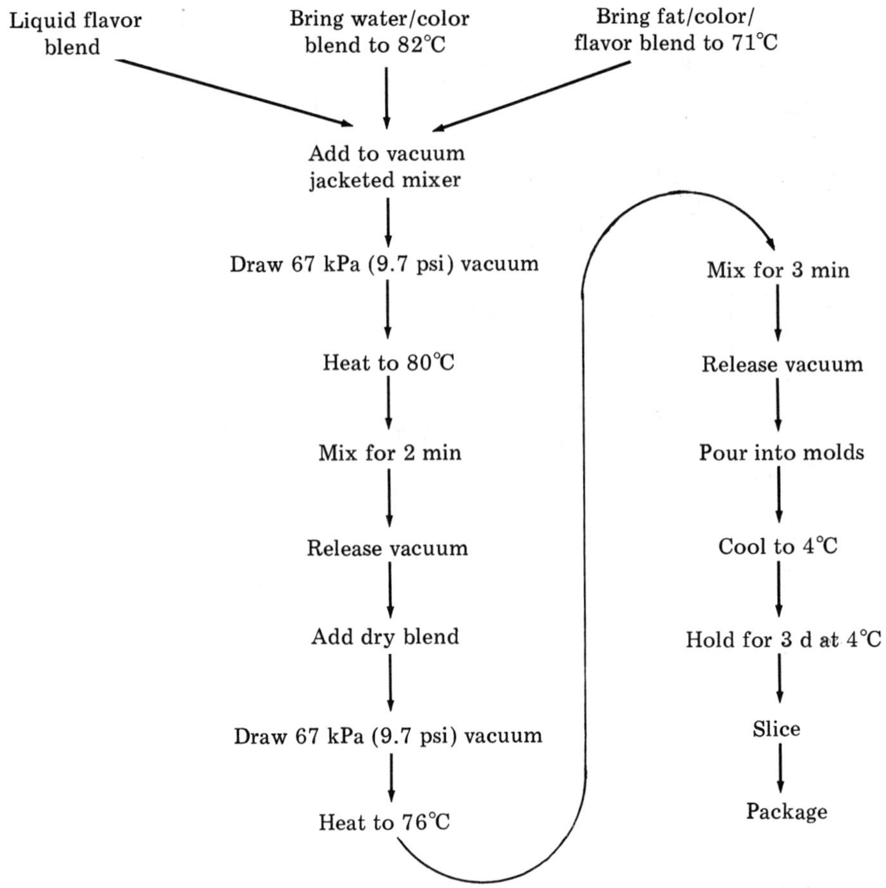

Figure 5. Schematic process flow diagram for an imitation cheese product (21).

strands. More recent approaches have been to combine caseinate and soy proteins (enriched in the glycinin fraction) to make imitation processed cheeses. A limited amount of soy protein can be used at this time to avoid undesirable beany flavors. An imitation cheese spread based on soy protein and caseinate and its method for processing is presented in Table 17.

Coffee Creams/Whiteners. Historically, cream or half-and-half were added to coffee to develop a creamy white color and to modify the flavor characteristics of the coffee. Coffee-cream substitutes have been given the generic name of coffee whiteners and are available in liquid, frozen, and dry forms. The emulsifiers and proteins are key stabilizing factors with respect to the stability of the product in coffee. The type of emulsifier system varies with the type of coffee whitener. In addition, the method of processing has a significant effect on the stability and whitening power of the product. A typical process for a liquid coffee whitener would follow that given in Figure 3. Homogenization is a key process and the desired particle size is 0.2–1.0 μm. Rapid cooling after homogenization assists in producing a stable product. Both caseinate- and soy-protein-based products are marketed. Three typical formulas for coffee whiteners are listed in Table 18 (65–66). There is a relationship between emulsifier and protein amounts with respect to the stability of a soy-based coffee whitener, as shown in Table 19.

Table 16. Formulation and Composition of Cheeselike Emulsions[a]

Property	Wt %
formulation	
globin protein	12.2–17.0
plasma proteins	2.8–6.2
ENR-RO whey solids	5.0
MOR-REX hydrolyzed cereal solids	8.5
butter	8.5
cellulose gum	0.8
composition of processed products	
protein	14.5–18.1
fat	5.9–7.5
ash	4.1–6.7
moisture	54.7–64.3

[a] Ref. 63.

Table 17. Formulation for Snack-Spread Base[a,b]

Ingredient	Wt %
margarine	37.38
sodium caseinate	5.58
maltrin dextrin	5.50
modified food starch	3.52
gelatin	2.46
soy isolate	2.28
whey, sweet	1.66
acid blend	0.86
lactic acid (50 wt %)	(96.33)
citric acid, dry	(3.60)
acetic acid, cone	(0.07)
emulsifier	0.17
salt	0.08
water	40.51
Total	*100.00*

[a] Ref. 60.

[b] Pilot-plant processing: 1. Weigh and blend all dry ingredients thoroughly. 2. Preheat high shear blender with 96–99°C water. Product must be maintained above a minimum temperature of 71°C during mixing. 3. Weigh required hot (96–99°C) water into blender. 4. Activate blender and add dry ingredients into vortex. Maintain vortex throughout addition of dry ingredients; the mixture should be thick, smooth, and creamy. 5. Transfer mixture into a tank that can be heated and cooled and is equipped with a variable-speed agitator. 6. Add margarine and emulsifier and blend thoroughly. Temperature must be above 71°C. 7. Cool mixture to 38°C with slow agitation. 8. Add acid mixture very slowly while agitating to get good mixing.

Ice Cream and Frozen Desserts. Imitation frozen desserts include ice cream, sherbets, and specialty products, eg, milkshake bases. Filled ice cream, called mellorine in the United States, is the dominant product in the world market and is sold both as a soft-serve and as a hardened product. The composition and processing is as for ice cream, except for the substitution of vegetable fat for milkfat. To date, frozen

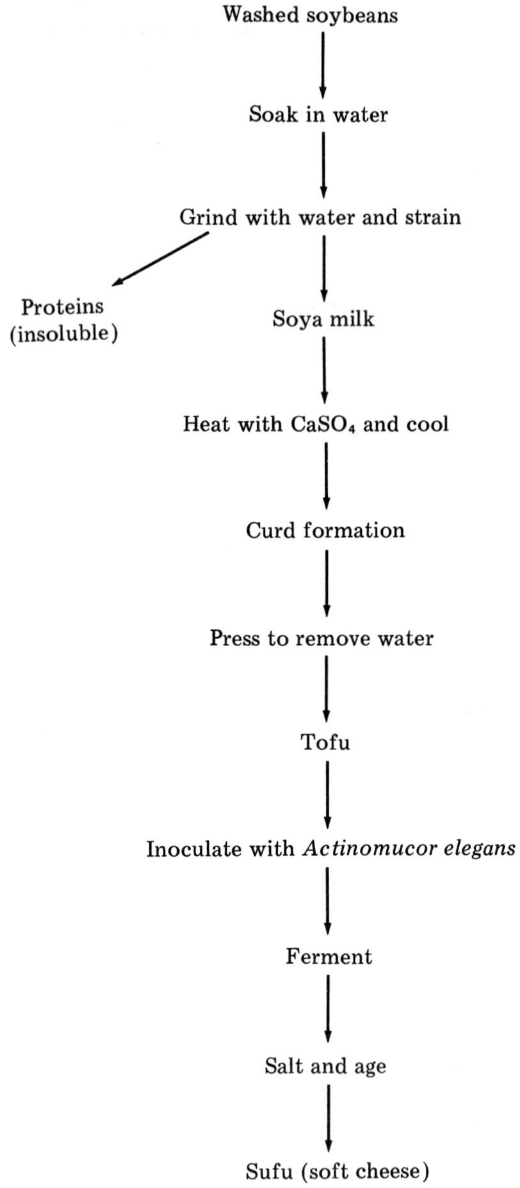

Figure 6. Schematic process flow diagram for the processing of a soybean cheese curd (64).

desserts based on caseinates or whey protein concentrates have not been marketed because there has been no economic advantage to do so. Whey protein has been reported as being more functional than casein in ice cream manufacture. The formulation for a whey-protein-based milkshake product is given in Table 20. Soybean-protein-based ice cream products are also technologically feasible but generally are limited in use because of flavor problems. An acceptable ice cream has been made by replacing 50% of the nonfat milk solids with the remainder of the solids, which is a soybean protein isolate made up of cheese whey (67). Chocolate flavor has been widely used to mask the flavor of soybean proteins in ice cream.

Table 18. Type Formulation of Caseinate and Soy-Based Coffee Whiteners[a]

Ingredient	Sodium caseinate-based, wt %	Soy-protein-based	
		A, wt %	B, wt %
corn syrup solids	7.5	12.4	15.0
vegetable fat	10.0	9.5	10.1
protein	1.5	0.60	0.8
mono- and diglycerides	0.5		0.5
Na stearoyl-2-lactylate		0.1	0.2
sorbitan monostearate		0.2	
polysorbate 60		0.56	0.2
dipotassium phosphate	0.10	0.1	0.2
water	remaining percentage	remaining percentage	remaining percentage

[a] Refs. 65 and 66.

Table 19. Effect of Protein and Emulsifier Variation on Coffee Whitener Stability[a]

Sample	Concentrations of independent variables[b]				Height of feather, mm
	SSL[c]	ATMOS 150[d]	TWEEN 60[d]	Soy isolate	
A	0.10	0.35	0.56	0.70	0
B	0.10	0.20	0.32	0.60	coagulated
C	0.25	0.35	0.56	0.60	1.5
D	0.40	0.35	0.56	0.70	11
E	0.40	0.275	0.44	0.70	coagulated
F	0.40	0.275	0.44	0.60	27
G	0.25	0.275	0.44	0.80	42
H	0.25	0.275	0.44	0.60	23
I	0.25	0.35	0.56	0.70	18

[a] Ref. 66.
[b] The following concentrations were kept constant: hydrogenated coconut oil, 9.5 g/100 g; K_2HPO_4, 0.1 g/100 g; corn syrup, 12.4 g/100 g. Water was the dependent variable.
[c] Sodium steroyl-2 lactylate.
[d] Mfg. by Atlas Co.

Milk. Imitation milks fall into three broad categories: filled products based on skim milk, buttermilk, whey, or combinations of these; synthetic milks based on soybean products; and toned milk based on the combination of soy or groundnut (peanut) protein with animal milk. Few caseinate-based products have been marketed (1,57,68). Milk is the one area where nutrition is a major concern, especially in the diets of the young. Imitation milks are being made for human and animal markets. In the latter area, the emphasis is for products to serve as milk replacers for calves. The composition of milk and filled-milk products based on skim milk have been presented in Table 13. Table 21 gives the composition of a whey/buttermilk-solids-based calf-milk replacer, which contains carboxymethyl cellulose (CMC) for proper viscosity of the product.

Guidelines for fluid milk substitutes have been presented (68). These include recommendations regarding raw materials, methods of preparation, composition, sanitation, control of toxic substances, heat destruction of trypsin inhibitors, and recommendations for the distribution of filled or toned milk and products containing

Table 20. Composition of Strawberry-Flavored Shakes[a]

Ingredients	Wt %
hydrogenated soybean oil	4.0000
sucrose	11.0000
salt	0.0200
dry whey–caseinate blend	10.0000
alginate	0.2600
carboxymethyl cellulose	0.1000
calcium lactate.$5H_2O$	0.0250
anhydrous citric acid	0.1125
hydrolyzed cereal solids	1.0000
FD&C No. 3 (dry) red	0.0005
imitation strawberry flavor (dry)	0.0530
imitation strawberry flavor (liquid)	0.0150
water	73.4140
Total	*100.0000*

[a] Ref. 47.

Table 21. Dry Mass Composition of Milk-Substitute Formulations; Without Vitamins and Trace Elements, Based on Clotted-Cream Buttermilk[a,b]

	Buttermilk, 42%	Clotted whey (pH 1.5), 30%	Sweet whey, 15%		Total, 100%
protein	13.9	3.9	2.0		19.8
lactose	17.6	19.2	11.0		47.9
mineral materials (ash)	3.4	1.8	1.4	0.6	7.2
lactic acid	5.0	4.8	0.6		10.4
Ca	0.59	0.15	0.14	0.25	1.13
Mg	0.09	0.03	0.05		0.17
K	0.76	0.03	0.36		1.15
Na	0.21	0.03	0.11		0.35
total cations	1.65	0.24	0.66	0.25	2.80
PO_4	1.22	0.99	0.33		2.54
Cl	0.42	0.42	0.21		1.05
Ca/P	1.48	0.48	1.30		1.36

[a] Ref. 49.
[b] Fat = 10.5 wt % and CMC stabilizer = 0.5% in all formulations.

noncow-milk components. Most of the cow-milk substitutes are based on either soybean or groundnut (peanut). Special considerations are given to protection against toxic materials from groundnuts and elimination of antinutritional factors in soybean, eg, the trypsin inhibitor and hemaglutinin. Peanuts in particular and soybeans to a lesser degree may be contaminated with *Aspergillus* molds that form aflatoxin. The level of aflatoxin should not exceed 30 ppb on a dry basis in the protein product. Urease enzyme activity is used as an indirect measure of proper heat treatment to destroy antinutritional factors.

A typical process for toned milk is dispersion and hydration of protein in water at 60°C, addition of corn syrup, addition of oil–emulsifier blend, addition of gum, two-stage homogenization at 14 MPa (2000 psi) and then at 3.4 MPa (500 psi), pasteurization, cooling, and packaging (69). The recommended composition of a toned milk is given in Table 22. Toned milk, containing peanut protein, was introduced in India under the name of Miltone.

Soy-Based Beverages. Soy milk is a traditional oriental product and has been the basis for development of products for western markets. Soybean-based beverages may be classified as follows (64): (*1*) traditional soy milks, unfermented, starting with uncomminuted full-fat beans, full-fat or defatted soy flours, or soy protein concentrates; (*2*) traditional fermented yogurtlike products; or (*3*) simulated milks based on soy protein concentrates and isolates, ie, of the dairy type, fluid single strength, concentrated, or sterile, nonfat milk replacers as dry blends of soy protein isolates with dry whey, infant-feeding beverages simulating human milk, and fermented yogurtlike products.

A contemporary process for making a soy milk with minimum beany flavor is presented in Figure 7. The recommended compositions for a fluid and for a dry product are given in Table 23.

Infant Formulas. ***Milk-Based.*** Sometimes referred to as humanized milk, these milk-based products are formulated to resemble the composition of human milk, but contain a lower amount of carbohydrate than cow's milk (71). They are sold in three forms: liquid ready-to-use, liquid concentrates, and dry powders. A typical formulation is given in Table 24; a typical processing scheme is the addition of skim milk to a mixing vat; bring to 60°C; add oil–emulsifier, minerals, and vitamins; pasteurize; homogenize at 21 MPa (3000 psi); standardize composition; package; sterilize; cool; and store.

Vegetable-Protein-Based. Most vegetable-protein-based infant formulas are soybean-based (72). Soybean-based formulas have served as useful replacements for milk in the nutritional management of infants who are allergic to cow's milk. Current formulas tend to contain isolates as the protein source to eliminate or reduce the presence of carbohydrates that are the cause of flatulence and abnormal stools. Care is taken to provide adequate nutrition and to use proteins processed in such a way as to minimize or eliminate any antinutritional factors. The formulation of a typical soy-based infant formula is given in Table 25.

Table 22. Composition of Toned Milk Containing Vegetable Protein[a,b]

Property	Fluid, sterilized	Dry
moisture, wt % max	89	3.5
fat, wt % min	2	18.0
nonfat solids, wt % min	9	
carbohydrates (lactose/sucrose and malt sugar-dextrin), wt %	4.0 min; 8.0 max	35.0 min (lactose, 12.0 min)
vegetable protein (N × 6.25), % min	3.5	20.0
essential fatty acid content, expressed as linoleic acid as wt % of fat, min	2.0	2.0
total ash, wt % max		7.0
acid-insoluble ash, wt % max		0.05

[a] Ref. 69.

[b] The recommended levels of vitamins and minerals are the same as those indicated for the milk substitute.

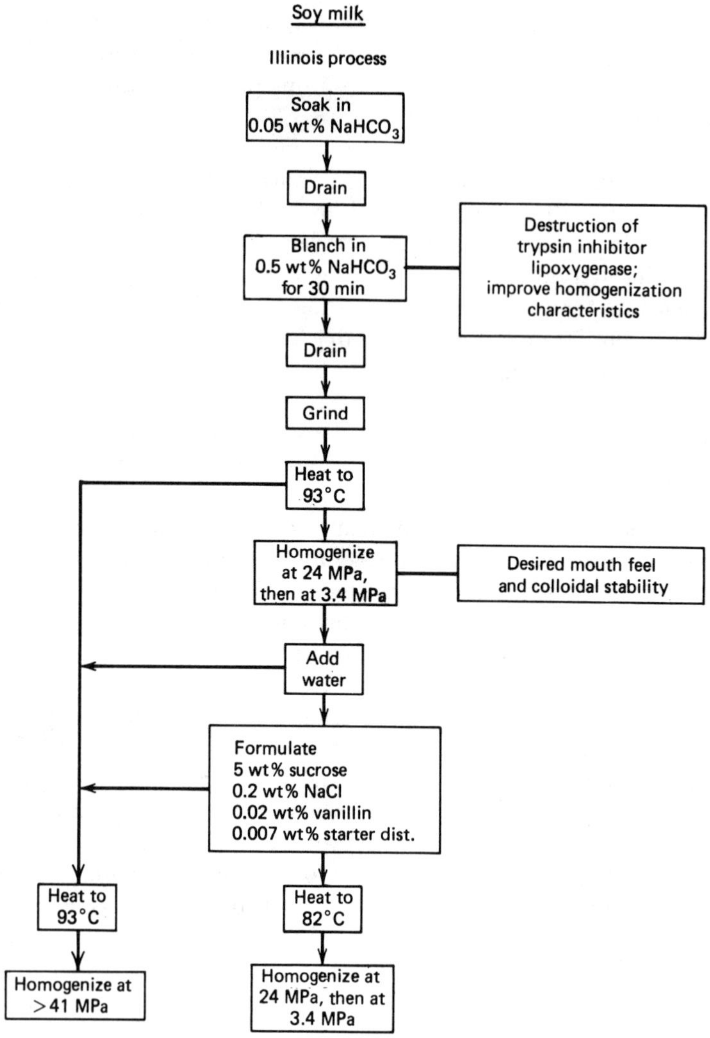

Figure 7. Schematic process flow diagram for a soy milk (70). To convert MPa to psi, multiply by 145.

Whipping Creams and Whipped Toppings. Whipped topping is the generic term used for imitation whipping cream. Products are available as liquids to be whipped, frozen prewhipped products, liquids in aerosol cans that form whips upon release, and dry powdered products that are reconstituted and whipped (65). Toppings were first made from nonfat dry milk or sodium caseinate as the protein source and vegetable fat as the source of fat. Most of these products are heavily dependent upon the gum constituents to impart the desired product characteristics and to provide uniformity in the whipping characteristic of the topping. Generally, the toppings have a lower fat content than traditional whipping cream. Isolated soy proteins have also been used in whipped toppings. Soy-protein-based toppings have a lower protein concentration than caseinated-based toppings. Formulations are adjusted to protein levels, and higher protein levels can result in off-flavors. Typical formulations for a liquid and a frozen, pre-whipped product are given in Table 26.

Table 23. Recommended Chemical Composition for Milk Substitute[a]

Property	Fluid, sterilized	Dry
moisture, wt % max	89.0	3.5
fat, wt % min	2.0	18.0
nonfat solids, wt % min	9.0	
protein (N × 6.25), wt % min	3.5	28.0
essential fatty-acid content, expressed as linoleic acid as wt % of fat, min	2.0	2.0
carbohydrates (sucrose or malt sugar–dextrin), wt % min	4.0	51.5
calcium, mg/100 mL or /100 g, min	100	800
vitamin A, μg/100 mL or /100 g, min	50	450
vitamin D, IU/100 mL or /100 g, min	40	360
iron, mg/100 mL or /100 g, min	0.4	4.0
thiamine, mg/100 mL or /100 g, min	0.15	1.0
riboflavin, mg/100 mL or /100 g, min	0.15	1.3
niacin, mg/100 mL or /100 g, min	1.0	8.0
folic acid, μg/100 mL or /100 g, min	5	50.0
pantothenic acid, μg/100 mL or /100 g, min	200	1600
vitamin B_{12}, μg/100 mL or /100 g, min	0.5	5.0
pyridoxine, μg/100 mL or /100 g, min	200	1600
ascorbic acid, mg/100 mL or /100 g, min	3.0	25.0
total ash, wt % max		7.0
acid-insoluble ash, wt % max		0.05

[a] Ref. 69.

Miscellaneous. Miscellaneous products with current or potential market significance include sour cream, chip dips, milkshake bases, puddings, and yogurt. These products have been formulated from caseinates and soybean proteins.

Market Significance

Most dairy products have been imitated to a greater or lesser extent. In the United States, filled and imitation products for which companies offer dry mixes or bases without fat include: filled buttermilks, filled cheese, filled coffee creams, filled dip products, filled eggnog, filled frozen desserts, filled milk, filled sour cream, filled whipping cream, imitation ice cream, imitation milk and milk products, and imitation coffee cream. Complete dry mixes available are imitation coffee cream, imitation flavored dips, imitation frozen desserts, imitation half-and-half, imitation milk and milk drinks, imitation sour cream, and imitation whipping cream. In addition, a wide variety of filled and imitation products is sold under generic names as: chip dips, coffee whiteners, margarine, spreads, and whipped toppings. These products have identities apart from dairy products and have their own characteristics. Many are sold under fanciful names, for example, coffee whiteners are sold under a wide range of trade names that include: Coffee Light, Coffee Mate, Coffee King, Complement, and Crodamix. Most of these products, including margarine, usually contain a milk component.

Of the filled and imitation products, margarine has made the greatest penetration

Table 24. Examples of Commercially Prepared Milk-Based Formulas Marketed in the United States[a]

components			
protein	nonfat cow milk	demineralized whey and nonfat cow milk	nonfat cow milk and soy protein isolate
fat	vegetable oils	vegetable oils and oleo oil	corn oil
added carbohydrates	lactose or corn syrup solids	lactose	sucrose
main constituents, g/100 mL			
protein	1.5–1.6	1.5	3.6
fat	3.6–3.7	3.6	1.6
carbohydrate	7.0–7.1	7.2	6.6
minerals	0.3–0.4	0.3	0.7
caloric distribution (% of calories)			
protein	9	9	26
fat	48–50	48	27
carbohydrate	41–43	43	47
minerals per liter			
calcium, mg/L	550–600	445	1000
phosphorus, mg/L	440–455	330	800
sodium, meq/L	11–17	7	17
potassium, meq/L	16–28	14	32
chloride, meq/L	12–14	10	29
magnesium, mg/L	40–48	53	85
sulfur, mg/L	130–160	145	310
copper, mg/L	0.4–0.6	0.4	1
zinc, mg/L	2.0–4.2	3.2	4.0
iodine, μg/L	40–69	69	100
iron, mg/L	trace–1.5	12.7	18
vitamins per liter			
vitamin, A RE/L[b]	340–500	530	600
thiamine, μg/L	400–650	710	750
riboflavin, μg/L	600–1000	1060	900
niacin, mg/L	7–8.5	7	10
pyridoxine, μg/L	320–400	423	700
pantothenate, mg/L	2.1–3.2	2.1	5
folacin, μg/L	50–100	32	100
vitamin B_{12}, μg/L	1.5–2.0	1.1	2.5
vitamin C, mg/L	55	58	50
vitamin D, IU/L	400–423	423	400
vitamin E, IU/L	8.5–12.7	9.5	6.3

[a] Ref. 71.

[b] Vitamin A is reported as retinol equivalents per liter (RE = 5 IU = 1.25 μg vitamin A or 3 μg carotene).

into the sale of dairy foods. The replacement of butter has tended to stabilize in recent years in most markets. Data on market penetration from 1955 to 1970 is given in Table 27. In the United States, the market preference for butter and margarine reversed from 1950 to 1970. As indicated below, margarine had ca 72% of the total spread market in the United States in 1980. However, the growth of margarine was relatively slow during the 1970s.

Table 25. Approximate Analysis of a Typical Soy-Protein Infant Formula[a]

Approximate analysis			
Component	Value	Vitamins	Value
protein, g/L	20.0	vitamin A, IU/L	2500
fat, g/L	36.0	vitamin D, IU/L	400
carbohydrate, g/L	68.0	vitamin E, IU/L	15
minerals, g/L	3.8	vitamin C, mg/L	55
calcium	0.70	vitamin B_1, mg/L	0.04
phosphorus	0.50	vitamin B_2, mg/L	0.60
sodium	0.30	vitamin D_6, mg/L	0.40
potassium	0.71	niacin, mg equiv/L	9.0
chloride	0.53	folic acid, mg/L	0.10
magnesium	0.050	vitamin B_{12}, μg/L	3.0
iron	0.012	pantothenic acid, mg/L	5.0
zinc	0.0050	biotin, mg/L	0.15
copper	0.00050	vitamin K_1, mg/L	0.15
manganese	0.00020		
iodine	0.00015		
water, g/L	901.6		
calories per 100 mL	68		

[a] Ref. 72.

Table 26. Formulations for a Liquid Nondairy Topping and a Frozen Prewhipped Topping[a]

Ingredient	Liquid nondairy topping, wt %	Frozen prewhipped topping, wt %
water	55.25	53.40
palm-kernel oil	25.00	26.00
sucrose	18.00	18.00
isolated soy protein	0.40	1.20
sorbitan monostearate	0.30	0.35
flavor	0.30	0.30
lecithin	0.20	
mono- and diglycerides	0.20	
polysorbate 60	0.20	0.30
disodium phosphate	0.15	0.15
guar gum		0.30

[a] Ref. 65.

	Per capita consumption, kg	
Year	*Butter*	*Margarine*
1950	4.8	2.7
1960	3.4	4.3
1970	2.3	4.9
1975	2.2	5.1
1980	2.1	5.3

Countries producing imitation dairy products are listed in Table 28.

Market penetration of imitation dairy products in the United States in 1980 is given in Table 29. Margarine was excluded from this survey (73).

Table 27. Production of Butter and Margarine in 14 Milk-Producing Countries[a,b]

Production	1955	1960	1965	1970
margarine	100	114	117	123
butter	100	112	119	115
Total	*100*	*113*	*118*	*119*
share of total, %				
margarine	50.2	50.7	49.6	51.8
butter	49.8	49.3	50.4	48.2

[a] Ref. 8.
[b] 1960, 1965, and 1970 values are based on the arbitrary 1955 production unit of 100.

Table 28. Countries in which Imitation or Synthetic Products were Available from 1970 to 1972[a,b]

Products	Countries or areas
imitation products	
liquid milk	Australia, UK, United States, Brazil
condensed and evaporated milk	Southeast Asia, United States, Netherlands
milk powder	Australia, UK, United States, Japan
ice cream	Belgium, Brazil, Canada, Israel, Japan, Netherlands, Sweden, Switzerland, UK, United States
cream	Australia, Ireland, Japan, Finland, UK, United States
whipping cream	Israel, United States
baker's cream	Denmark, FRG, Norway, Sweden, UK, United States
synthetic products	
liquid milk	United States
condensed and evaporated milk	UK
cream powder, eg, coffee whiteners, whipping toppings, etc	Australia, Canada, Denmark, Japan, Norway, Netherlands, Sweden, Switzerland, UK, United, States, FRG
ice cream	Israel
infant formula	Australia
desserts	Canada, UK

[a] Ref. 57.
[b] The definitions as defined in ref. apply.

Economic Aspects

The prices of vegetable fats are two to three times lower than milk fat in many countries and this has a direct relationship to the marketing of imitation high fat products, eg, ice cream, whipped toppings, and coffee whiteners (51). Concern over the controversy of diet and cardiovascular disease makes a small contribution to the increased use of vegetable-fat-based imitation products. In addition, the price of caseinate substitutes is ca one-half to two-thirds of the price of caseinate. Substitution of vegetable fat for milk fat is increased in developing countries that produce vegetable fats and import a large portion of their dairy products. Thus, the development of imitation milks in Asian countries has been based on a combination of economics and a desire to use their own raw materials to reduce foreign exchange required for imports.

Table 29. Market Penetration by Dairy Substitutes in the United States Projected for 1980, % [a]

Dairy products	1980 Level, %
fluid whole milk	1
creams	49
low fat milk	1
American and other cheeses	5
evaporated and condensed whole milk	2
ice cream and ice milk	1.3

[a] Ref. 8.

Table 30. Relative Ingredient Costs and Sale Prices of Imitation Dairy Products Compared to Dairy Products [a]

	Percentage of component as compared to dairy product		Imitation product costs and prices as percentages of comparable dairy products		
			Ingredient cost		
Imitation product [c]	Fat	Protein	A	B [b]	Sale price
milk-filled	86	100	70	72	85
coffee whitener	56	80	27	40	100
whipped topping	78	100	33	59	63
ice cream	100	86	52	55	86
cheese	76	89	70	81	67

[a] Ref. 11.
[b] Corrected to same fat and protein level as dairy product.
[c] Products other than filled-milk based on caseinate.

In the Philippines, the sale of filled milk had become 85% of the dairy products market by the early 1970s.

Relative ingredient costs and retail prices for imitation products as a percent of their dairy counterparts are listed in Table 30. Since many of these products contain less fat and protein than the dairy products, the data are based on the relative ingredient costs for equal compositions. A comparison of ingredient costs for 1969 is given in Table 31. Although out of date, the relative costs of ingredients have increased by approximately the same amount and the margin differences are valid today. No comparative study is available for current prices. Except for coffee cream, the average advertised prices for dairy products are 30–60% above the prices of the substitutes. The price differential between cream and coffee whiteners is much less, reflecting convenience as a purchase incentive rather than price. Based on ingredient cost comparisons, the fat cost in dairy products is 40–90% of the total ingredient costs, whereas fat makes up 16–30% of the ingredient cost of substitute products. This is related to both the higher price of milk fat and the higher amount of fat generally in dairy products as compared to their imitations.

Regulatory Aspects

The legal status of imitation dairy products varies widely among countries; the information given below pertains primarily to the legal status of imitation dairy foods

Table 31. Composition and Estimated Ingredient Cost of ca 45.4 kg of Selected Substitutes for Dairy Products in the United States, 1969[a]

Product and estimated ingredient cost	Vegetable fat	Protein agent	Body and sweetening agent	Other ingredients[b]	Water	Total
filled milk:						
composition, kg	1.4[c]	43.9		0.1		45.4
ingredient cost, ¢/kg	0.3	1.7[d]		0.1		2.1
coffee whitener,						
liquid or frozen						
composition, kg	4.6[c]	1.0[e]	3.8	0.3	35.7	45.4
ingredient cost, ¢/kg	1.1	0.5	0.4	0.2		2.3
powder						
composition, kg	16.3[c]	2.3[e]	25.4	0.9	0.5	45.4
ingredient cost, ¢/kg	4.1	1.9	2.5	0.2		7.9
whipped topping:						
composition, kg	11.3[c]	1.1[e]	5.4	0.4	27.1	45.3
ingredient cost, ¢/kg	2.8	0.6	0.6	0.1		4.2
imitation sour cream:						
composition, kg	8.2[f]	1.8[e]	2.3	0.2	32.9	45.4
ingredient cost, ¢/kg	1.4	0.9	0.3	0.3		2.9
imitation ice cream (mellorine):						
composition, kg	4.5[c]	32.9	7.7	0.2		45.3
ingredient cost, ¢/kg	1.1	0.5[g]	0.8	0.2		2.7
imitation ice milk:						
composition, kg	1.8[c]	35.0	8.4	0.1		45.3
ingredient cost, ¢/kg	0.5	0.5[g]	0.9	0.2		2.0

[a] Ref. 11.
[b] Includes buffering, emulsifying, and stabilizing agents.
[c] Coconut oil.
[d] Fluid skim milk Class I price of 8.697 ¢/kg.
[e] Sodium caseinate.
[f] Soybean oil.
[g] Price of fluid skim milk for manufacturing use, 3.34 ¢/kg.

in the United States. The FDA has held that filled products should be nutritionally equivalent to the products they resemble. In the case of imitation milks, the FDA proposed regulations for nutritional equivalency (74). A section of the Food, Drug, and Cosmetic Act defines misbranded foods and the FDA has set up standards of identity for foods under this part of the law (75). In 1978, proposals were suggested for standards of identity for imitation cheese, creams, and milks (76).

The definition of margarine varies in different countries (61). The U.S. Federal Code of regulations on margarine is given in ref. 77. Imitation ice cream called mellorine is defined in ref. 78.

BIBLIOGRAPHY

1. A. H. M. van Gennip, *International Dairy Federation Document 107*, International Dairy Federation, Brussels, Belgium, 1978, pp. 92–93.
2. I. C. Graham, *Proceedings of the Symposium on Margarine and New Edible Oil Products*, Canadian Institute of Food Technology, Toronto, Canada, June 9–12, 1968, pp. 39–54.
3. T. E. Furia, *Handbook of Food Additives*, 2nd ed., CRC Press, Boca Raton, Fla., 1980.

4. K. Sabharwal and D. G. Vakaleris, *J. Dairy Sci.* **55,** 277, 283 (1972).
5. F. Winklemann, *Imitation Milk and Milk Products*, Food and Agricultural Organization of the United Nations, 1974, pp. 1–117.
6. J. Morita, *Jpn. J. Dairy Food Sci.* **29**(1), 26 (1980).
7. M. E. Mathews, *N.Z. J. Dairy Sci. Technol.* **13**(3), 149 (1978).
8. Economic Research Service, *Market Research Report No. 947*, U.S. Department of Agriculture, Washington D.C., 1972.
9. *U.S. Dairy Council, Digest* **39**(2), 7 (1968).
10. T. J. Hedrick, *Dairy Ind.* **34**(3), 127 (1969).
11. H. H. Moede, *USDA Economic Research Service*, MTS-177, U.S. Dept. of Agriculture, Washington, D.C., 1979, pp. 19–23.
12. U.S. Pat. 3,635,737 (Jan. 18, 1972), A. Baron (to Baron).
13. U.S. Pat. 3,741,774 (July 20, 1973), M. P. Burkwall (to Quaker Oats).
14. U.S. Pat. 3,806,606 (April 23, 1974), P. Seiden (to Procter & Gamble).
15. U.S. Pat. 3,922,374 (Aug. 26, 1975), R. G. Bell (to Anderson, Clayton & Co.).
16. U.S. Pat. 3,891,777 (April 28, 1975), R. A. Boyer (to Ralston, Purina Co.).
17. U.S. Pat. 3,891,778 (Nov. 18, 1975), R. A. Boyer (to Ralston, Purina Co.).
18. U.S. Pat. 3,873,729 (Sept. 5, 1975), H. Kubota (to Fuji Oil Co.).
19. U.S. Pat. 3,941,891 (July 28, 1976), R. L. Kasik (to Beatrice Foods).
20. U.S. Pat. 4,016,298 (March 27, 1977), R. L. Kasik (to Beatrice Foods).
21. Brit. Pat. 1,517,956 (July 19, 1978), J. D. Wynn (to Anderson, Clayton & Co.).
22. U.S. Pat. 4,104,413 (Aug. 1, 1978), J. D. Wynn (to Anderson, Clayton & Co.).
23. U.S. Pat. 4,105,803 (Aug. 8, 1978), A. C. Peng (to Ohio Agricultural Research and Development Center).
24. Ger. Pat. 2,834,093 (Jan. 3, 1979), J. A. Moore and R. L. Kasik (to Beatrice Foods).
25. U.S. Pat. 4,166,142 (Aug. 28, 1979), P. K. Chang (to Stauffer Chemical Co.).
26. U.S. Pat. 4,156,018 (May 22, 1979), T. S. Sozzi (to Nestle).
27. Austral. Pat. 500,294 (Aug. 30, 1979), J. P. Wynn and co-workers (to Anderson, Clayton & Co.).
28. U.S. Pat. 4,110,484 (Sept. 10, 1979), C. E. Rule (to SEM Corporation).
29. U.S. Pat. 4,197,322 (April 8, 1980), J. L. Middleton (to Universal Foods Corp.).
30. U.S. Pat. 4,163,806 (Nov. 10, 1979), D. M. Callen (to Callen & Callen).
31. Brit. Pat. 1,469,541 (April 6, 1977), D. G. Gardiner (to General Foods).
32. Brit. Pat. 1,525,526 (Oct. 20, 1978), J. Kruseman and co-workers (to Nestle).
33. Brit. Pat. 1,500,427 (Feb. 8, 1978), H. Kubota, M. Yamaguchi, and T. Taetishi (to Fuji Oil Co.).
34. U.S. Pat. 4,143,174 (March 6, 1979), S. M. Shah (to Beatrice Foods).
35. Brit. Pat. 1,568,834 (June 4, 1980), T. Foley (to Unilever, Ltd.).
36. U.S. Pat. 3,843,828 (Feb. 15, 1974), R. H. Arndt (to Ralston, Purina Co.).
37. U.S. Pat. 3,798,339 (March 19, 1974), A. C. Reng (to Swift & Co.).
38. U.S. Pat. 4,163,069 (July 31, 1979), N. Melachouvis (to Stauffer Chemical Co.).
39. Brit. Pat. 2,030,439A (March 19, 1980), S. M. Irvine, A. Penar, G. B. Wood (to Unilever Ltd.).
40. U.S. Pat. 4,194,018 (March 18, 1980), T. Hotel (to Nestle).
41. Brit. Pat. 2,061,691A (1981), M. Trop (to Israel Bengurian University).
42. U.S. Pat. 4,129,664 (Dec. 12, 1978), S. Kruseman (to Nestle).
43. Ger. (FRG) Pat. 2,811,356 (Sept. 20, 1979), G. Bandash.
44. Brit. Pat. 2,021,921A (Dec. 12, 1979), S. V. Kranenburg (to Unilever, Ltd.).
45. Brit. Pat. 2,022,392 (June 4, 1980), A. Kuipers (to Meggle Industries).
46. Brit. Pat. 1,568,874 (June 11, 1980), C. Glass (to Centrale Veevoeder-Fabrick "Sloten").
47. Brit. Pat. 1,582,597 (Oct. 12, 1981), A. Kuipers (to Meggle Industries).
48. U.S. Pat. 4,031,261 (March 10, 1977), S. R. Durst (to Pillsbury Co.).
49. U.S. Pat. 3,433,643 (March 18, 1969), P. P. Noznich and C. W. Tatter (to Beatrice Foods).
50. U.S. Pat. 3,443,960 (July 11, 1969), P. Noznick (to Beatrice Foods).
51. U.S. Pat. 3,560,220 (Feb. 2, 1791), J. G. Bangert (to Bangert and Halik).
52. U.S. Pat. 3,594,193 (July 10, 1971), A. R. Mishkin (to Nestle).
53. Fr. Pat. 2,387,605 (Dec. 22, 1978), B. M. Davenat (to Davenat).
54. Brit. Pat. 2,011,942 (July 18, 1979), K. Wallgren (to Mjelhcentralen Sweden).
55. Ger. Pat. 2,021,140 (Nov. 28, 1979), D. J. Harvey (to Dairy Developments).
56. U.S. Pat. 2,982,039 (Sept. 21, 1976), G. E. Scihelli (to Stauffer Chemical Co.).
57. Brit. Pat. 1,564,800 (April 16, 1980), J. Bodor (to Unilever, Ltd.).

58. W. J. Harper and C. W. Hall, *Dairy Technology and Engineering*, Avi Press, Newport, Conn., 1976.
59. M. K. Switzer, *Margarine and Other Food Fats*, Leonard Hill, Ltd., London, 1956, pp. 240–273.
60. R. F. Bennet and J. A. Ohren, *Food Prod. Dev.* **8**(5), 15 (1974).
61. M. T. Gillies, *Shortenings, Margarines, and Food Oils*, Noyes Data Corp., Park Ridge, N.J., 1974, pp. 219–220.
62. U.S. Pat. 3,457,086 (July 22, 1969), E. L. Josefowicz, I. Cooper, and D. Melnick (to Corn Products Co.).
63. F. W. Knapp, R. H. Schmidt, W. J. H. Maudlin, and E. H. Ahmed, *J. Food Protection* **41,** 257 (1978).
64. J. J. Jonas, *J. Milk Food Technol.* **39**(1), 39 (1975).
65. C. W. Kolar, I. C. Cho, and W. L. Watrous, *J. Am. Oil Chem. Soc.* **56,** 389 (1979).
66. R. J. Pearce and W. J. Harper, *J. Food Sci.* **47,** 680 (1982).
67. M. D. Wilding, *J. Am. Oil Chem. Soc.* **56,** 392 (1979).
68. *P.A.G. Statement of Milk Substitutes No. 6*, Protein Advisory Group of the United Nations System, New York, 1970.
69. *P.A.G. Guideline No. 15 on Milk Substitutes*, Protein Advisory Group of the United Nations System, New York, June 1972.
70. *Quick Frozen Foods* **34**(9), 100 (1972).
71. S. J. Foman, *Infant Nutrition*, W. B. Saunders Co., Philadelphia, Pa., 1974, pp. 377–382.
72. W. A. B. Thomson, *J. Am. Oil Chem. Soc.* **56,** 386 (1979).
73. G. Agren and A. Oklund, *Bulletin II*(*14*), Protein Advisory Group (PAG) of the United Nations System, New York, 1972, pp. 19–23.
74. *Fed. Reg.* **38,** 20748 (Aug. 2, 1973).
75. Food, Drug, and Cosmetic Act, Section 403.
76. *Fed. Reg.* **43,** 42118 (Sept. 19, 1978).
77. 21 FCR 120, Chapt. 1, Part 45.
78. 21 FCR 120, Chapt. 8, Part 20.

W. JAMES HARPER
Ohio State University

SYNTHETIC LUBRICANTS. See Lubricants and lubrication.

SYRUPS

Dextrose [*50-99-7*] and corn syrups are hydrolysis products of starch (qv) and are generally referred to as corn sweeteners. Although distinct commercial products, they have in common the raw material source, general methods of preparation, and many properties and applications. Dextrose is the common or commercial name for D-glucose, the pure crystalline solid recovered from almost completely hydrolyzed starch. Corn syrups are clear, colorless, viscous liquids prepared by hydrolysis of starch to solutions of dextrose, maltose, and higher molecular weight saccharides. Alternatively, the enzymatic isomerization of dextrose hydrolysate yields high fructose corn syrup (HFCS). Maple syrup, like corn syrup, is a nutritive sweetener produced as a concentrated solution of carbohydrate (sucrose) and is used in food applications. Molasses is a syrup produced as a by-product of sugar manufacture (see Sweeteners; Sugar, special sugars).

Dextrose

Dextrose (D-glucose, corn sugar, starch sugar, blood sugar, grape sugar) is by far the most abundant sugar in nature and occurs either in the free state (monosaccharide form) or chemically linked with other sugar moieties. In the free state, it occurs in substantial quantities in honey, fruits, and berries. As a polymer of anhydrodextrose units, it occurs in starch, cellulose, and glycogen. Sucrose is a disaccharide of dextrose and fructose. Commercial production of dextrose by hydrolysis of starch yields white crystalline sugars that are either anhydrous ($C_6H_{12}O_6$) or hydrated ($C_6H_{12}O_6.H_2O$). Dextrose hydrate [*16824-90-1*], with its one molecule of water of crystallization per molecule of sugar, separates from concentrated solutions at <50°C. Anhydrous α-D-glucose [*26655-34-5*] does not contain water of crystallization and separates at 50–115°C. Another anhydrous form, β-D-glucose [*28905-12-6*], separates if crystallization is carried out at temperatures >110–115°C.

In one of the first attempts to prepare commercial dextrose, grapes were used as the starting material (1). It is generally conceded that Kirchoff's work in 1811 was the forerunner of the starch hydrolysate industry (2). It was first reported in 1815 that acid conversion of starch to sugar is the result of hydrolysis of the starch rather than dehydration, and that the starch sugar is identical with grape sugar (3). It was not until 1842, however, that starch hydrolysis was first practiced commercially in the United States. Crystalline dextrose became a main industrial product when methods of crystallization were developed in the 1920s.

Properties. Physical properties of the three crystalline forms of dextrose are listed in Table 1. Crystallographic data for the α-anhydrous forms are reported in ref. 4. In solution, dextrose exists in both the α and β forms. When α-dextrose dissolves in water, its optical rotation diminishes gradually as a result of mutarotation until, after a prolonged time, an equilibrium value is reached (see arrows, Table 1). At this point, ca 62% of the dextrose is present in the β form. This equilibrium value is not significantly changed over a wide range of temperatures and concentrations. The same equilibrium value apparently exists in anhydrous melts as well as in glassy materials. Pure crystalline β-dextrose is quite sensitive to moisture, and it changes to the more

Table 1. Physical Properties of D-Glucose

Property	α-D-Glucose	α-D-Glucose hydrate	β-D-Glucose
molecular formula	$C_6H_{12}O_6$	$C_6H_{12}O_6.H_2O$	$C_6H_{12}O_6$
mp, °C	146	83	150
solubility (at 25°C), g/100 g soln	62 → 30.2 → 51.2[a]	30.2 → 51.2[a,b]	72 → 51.2[a]
$[\alpha]_D^{20}$	112.2 → 52.7[a]	112.2 → 52.7[a,b]	18.7 → 52.7[a]
heat of soln (at 25°C), J/g[c]	−59.4	−105.4	−25.9

[a] Equilibrium value.
[b] Anhydrous basis.
[c] To convert J to cal, divide by 4.184.

stable α form if exposed to high humidity. At 25°C, α-dextrose monohydrate dissolves fairly rapidly, yielding a solution containing ca 30 wt % dextrose. Very slowly thereafter, further quantities of dextrose dissolve until a saturated solution containing ca 50 wt % dextrose is obtained. The first phase of the dissolving process results from the limited solubility of α-dextrose. The slow, subsequent dissolution is caused by the transformation of part of the dissolved α-dextrose to the more soluble β form. When saturation is finally reached, a mixture of α- and β-dextrose in solution is in equilibrium with solid α-dextrose hydrate. At 25°C, anhydrous α-dextrose dissolves rapidly and beyond the limit of solubility of α-dextrose hydrate. Since the hydrate is the stable form at this temperature, crystallization of the hydrate occurs to its limit of solubility and the pattern then follows that of the hydrate as described above. The rate of attainment of equilibrium is increased by heating or in the presence of acids or bases. Data for the solubility of the equilibrium mixtures, as interpolated from results given in ref. 5, are listed in Table 2. Additional data on the solubility of the three crystalline forms before equilibration and in the metastable regions are given in refs. 5–6.

Dextrose in solution or in solid form exists in the pyranose structural conformation. In solution, a small amount of the open-chain aldehyde form exists in equilibrium with the two cyclic forms (see Fig. 1) and is responsible for the reducing properties of dextrose (7).

Dextrose shows the reactions of an aldehyde, a primary alcohol, a secondary alcohol, and a polyhydric alcohol (see Carbohydrates). In acid solution, either after standing for a prolonged time or after heating, dextrose undergoes polycondensation, ie, dehydration, yielding a mixture of di- and oligosaccharides, most of which are the

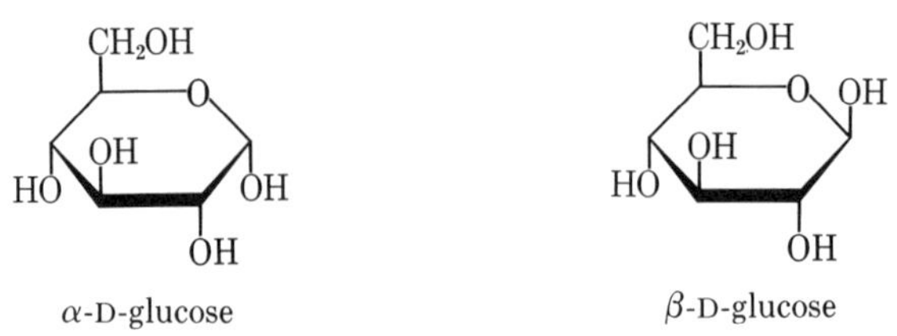

Figure 1. Structures of D-glucose.

Table 2. Solubility of Dextrose in Water

Temperature, °C	Dextrose in soln, wt %	Temperature, °C	Dextrose in soln, wt %
0	34.9	30	54.6
5	38.0	40	61.8
10	41.2	50	70.9
15	44.5	60	74.8
20	47.8	70	78.2
25	51.3	80	81.3

disaccharides gentiobiose and isomaltose. In acid solution and at high temperature, dehydration leads to formation of 5-hydroxymethylfurfural, which is a water-soluble, high boiling, and relatively unstable compound. Polymerization of 5-hydroxymethylfurfural yields dark-colored compounds, and evidence has been presented for the role of this compound as an intermediate in the discoloration of sugar solutions (8). Dextrose decomposition under these conditions also yields levulinic and formic acids.

In mildly alkaline solution, the principal reaction of dextrose is partial transformation, ie, isomerization, to fructose and other ketoses. D-Mannose, saccharinic acids, and other decomposition products form to a lesser extent. In highly alkaline solution and particularly in the presence of atmospheric oxygen, a complex mixture of products of decomposition and rearrangement results (9). Mild oxidation in slightly alkaline solution gives D-gluconic acid in quantitative yield. More vigorous oxidation with nitric acid yields glucaric acid, tartaric acid, oxalic acid, and other compounds resulting from fragmentation of the dextrose molecule. Alkaline Fehling's solution is reduced by dextrose with roughly five atoms of copper being reduced per molecule of dextrose. Electrolytic reduction or catalytic hydrogenation of dextrose is practiced commercially to manufacture sorbitol.

When dextrose is heated with methanol containing a small amount of anhydrous hydrogen chloride, α-methyl-D-glucoside is obtained in good yield and can be isolated by crystallization. Similar reactions occur with higher alcohols, but the reaction products are more difficult to isolate by crystallization. Dextrose reacts with acid anhydrides in the presence of basic catalysts, yielding esters. Complete reaction gives the pentaacylated derivative.

The reaction of dextrose with a nitrogen-containing compound, eg, amino acids or proteins, yields a series of intermediates which form pigments of varied molecular weight (Maillard reaction). Type of pigments produced is dependent on reaction conditions such as pH, temperature, and concentration of reactants (10).

Dextrose is the common intermediary metabolite in carbohydrate metabolism, since other utilizable monosaccharides are converted to dextrose before they are further metabolized. Starch, glycogen, and the common monosaccharides are hydrolyzed enzymatically in the alimentary canal. Dextrose is normally absorbed into the portal-vein blood, by which it is transported first to the liver and then circulated to all parts of the body. Before dextrose or any other monosaccharide can be utilized metabolically, it must be phosphorylated and enter the glycolytic cycle. Other monosaccharides, eg, galactose and fructose, eventually are converted to glucose-6-phosphate and metabolized like dextrose.

Manufacture. Dextrose is manufactured from a variety of starches. Corn starch is used almost exclusively in the United States, whereas starches from wheat, potato, sweet potato, and tapioca, as well as corn are used in other countries (see also Wheat and other cereal grains). Until 1960, commercial dextrose was produced by acid-catalyzed hydrolysis of starch at elevated temperatures (11–14). Because of the formation of degradation products, dextrose yield was limited to ca 86% (15). Improvement in dextrose yield was achieved by partial or complete replacement of acid with one or more enzymes. These processes are referred to as acid–enzyme (A–E) or enzyme–enzyme (E–E), depending on whether initial starch hydrolysis (thinning or liquefaction) is conducted with acid or a bacterial α-amylase. In either case, subsequent hydrolysis to dextrose (saccharification) is achieved with glucoamylase, a fungal enzyme which preferentially cleaves dextrose from the partially degraded starch.

By 1960, the A–E process was in general use and dextrose yields were increased to 92–94% (16). The process involves thinning a 30–35 wt % starch slurry with acid (HCl or H_2SO_4) at a temperature and pressure necessary to achieve a dextrose equivalent (DE) level of 10–20. Dextrose equivalent (DE) is a measure of the reducing-sugar content of a sweetener calculated as dextrose and expressed as a percentage of the total dry substance. Formation of acid-reversion products during thinning limits the dextrose yield attained during saccharification, and therefore, the use of this process is limited (17). Higher dextrose yields are achieved by the E–E process, which was developed in the 1960s. This process is based on the use of a heat-resistant bacterial α-amylase for starch thinning. Lower temperature and neutral pH limits side reactions and results in dextrose yields of 95–97% (15,18). Initial processes were based on α-amylase derived from the bacterium *Bacillus subtilis* (see Enzymes, industrial). In a typical process, 30–40 wt % starch is thinned at 85°C for 1 h followed by a short heat treatment at 140°C; a second enzyme addition is made and reaction at 85°C is continued to complete the liquefaction step (19). The resulting 10–15 DE hydrolysate is then saccharified with glucoamylase. The high temperature heat treatment is required to solubilize insoluble starch particles that are believed to be amylose–fatty acid complexes formed during initial liquefaction (20). Elimination of the heat-treatment step was made possible by a recent improvement in E–E technology and involves a thermostable α-amylase derived from *Bacillus licheniformis* (21–26). Starch slurry is thinned continuously with a steam-injection heater at 30–40 wt % solids, pH 6–6.5, and 103–107°C for 5–10 min. A 1–2 h hold at ca 95°C completes the thinning and yields a 10–15 DE hydrolysate for saccharification. Other types of E–E processes conducted at lower temperatures have been developed for their energy-saving potential (27–31).

Regardless of the type of thinning utilized, dextrose is produced by the action of glucoamylase during saccharification. Glucoamylase is produced from strains of *Aspergillus niger* in submerged fermentation (32). A broth is obtained containing two or more glucoamylase isozymes, α-amylase, transglucosidase, and other enzymes, eg, protease and cellulase. The α-amylase assists in saccharification (33); however, transglucosidase catalyzes the formation of isomaltose and therefore must be removed prior to use. Removal of the enzyme is accomplished by adsorption on a clay mineral or by other techniques (34–40). Alternatively, selection for an *Aspergillus* mutant that does not produce transglucosidase eliminates the removal step (41). The glucoamylase broth is clarified by filtration or centrifugation before evaporation. Glucoamylase is also produced from a *Rhizopus* organism in Japan by surface fermentation; however, enzyme properties are somewhat inferior to those of *Aspergillus* (42). Methods for glucoamylase production are given in refs. 42–44.

Saccharification is conducted by placing thinned starch into a large agitated tank called an enzyme convertor or reactor, adjusting the temperature to 55–60°C and the pH to 4.0–4.5, and adding a specified dosage of glucoamylase. The reaction continues until the maximum dextrose level is reached, usually after 48–96 h depending on enzyme dosage. At the normal saccharification solids of 30–35 wt %, ca 95–96% dextrose is produced. During saccharification, isomaltose and maltose are produced by the reversion reaction, in which two dextrose molecules combine as a result of the reverse action of glucoamylase. Consequently, if the reaction is extended beyond the normal time, the dextrose level decreases because of excessive formation of reversion products. Kinetic and equilibrium data for the reversion reaction are given in refs. 45–47.

Increased dextrose yields can be attained by conducting saccharification at reduced solids content, by addition of a debranching enzyme, eg, pullulanase, or by using a specialized glucoamylase preparation (48–50). Although immobilized glucoamylase technology has been studied extensively, apparently no commercial processes are in use because of the reduced dextrose yield compared to conventional systems (51) (see Enzymes, immobilized). After saccharification, the hydrolysate is clarified by centrifugation or precoat filtration to remove traces of insoluble fat, protein, and starch. Treatment with powdered carbon, granular carbon, or ion-exchange resins is then used to remove residual trace impurities, color, and, in the case of the ion-exchange resins, inorganic constituents. The decolorized liquor is evaporated to 50–55 wt % solids and may be given a second carbon treatment.

Commercial dextrose products include crystalline monohydrate or anhydrous dextrose and liquid dextrose. In addition, the hydrolysate can be refined and evaporated to a high DE corn syrup or dried to a solid product. Compositions are listed in Table 3.

For production of dextrose monohydrate, the decolorized liquor is evaporated under reduced pressure to a syrup containing 70–78 wt % solids and then is cooled and passed to crystallizers. The common form of crystallizer is a horizontal cylindrical tank fitted with a slowly turning agitator, cooling jacket, and cooling coils. A substantial bed of seed crystals comprising 20–25 wt % of the previous batch is left in the crystallizer and the refined syrup, at ca 46°C, is mixed with the seed crystals giving an initial temperature of ca 43°C. The agitated mass is slowly cooled to ca 20–30°C over a period of 3–5 d; by then, ca 60 wt % of the dextrose has crystallized as the monohydrate in a form suitable for separation and washing. The actual values for concentration, initial and final temperature, and rate of temperature drop vary depending primarily on the dextrose content of the hydrolyzate (57–59). The resulting magma in the crystallizers is passed into perforated-screen centrifuge baskets, spun to remove the mother liquor, and sprayed with water while spinning to wash out residual mother liquor. The wet sugar, containing ca 14 wt % total moisture, proceeds to rotary dryers, where it is dried in a stream of warm air until it contains ca 8.5 wt % moisture, which is slightly less than the theoretical 9.1 wt % for one molecule of water of crystallization. The lower final moisture content is a way of minimizing caking tendency by reducing the probability of free water being present in the product.

Mother liquor from the first crop of crystals can be concentrated and crystallized in a similar manner to recover an additional crop of crystals. Depending on its quality, some mother liquor can be partially recycled to the initial crystallization step to in-

Table 3. Saccharide Composition of Dextrose, High Fructose Corn Syrup (HFCS), Corn Syrups, and Maltodextrin

Sweetener type	DE	Amount of constituent, wt %[a] Fructose	Dextrose	DP-2	DP-3	DP-4	DP-5	DP-6	DP > 6	Ref.
dextrose	>99.5	0	>99.5	<0.25	←	←	<0.1	→	→	52
high DE corn syrup	97	0	95	<4	<1	←	←	<1 →	→	52
HFCS		43	52	←	←	5	→	→	→	53
		55	40	←	←	5	→	→	→	54
		90	7	←	←	3	→	→	→	54
acid-converted corn syrup	30	0	9	10	12	9	7	6	47	
	36	0	14	12	10	9	7	6	42	
	42	0	19	14	12	10	8	7	30	
	55	0	31	18	13	10	7	5	16	
high maltose corn syrup	42	0	6	44	13	3	2	2	30	
	48	0	9	52	15	2	2	2	18	
	50	0	3	75	13	←	11	→	→	55
high conversion corn syrup	64	0	37	31	11	5	4	3	13	
	70	0	43	30	7	5	3	2	10	
maltodextrin	10	0	0	3	4	4	4	7	78	56
	15	0	1	5	7	5	5	11	66	56
	20	0	1	8	9	7	7	14	53	56

[a] DP refers to degree of polymerization of dextrose, where DP-2 is disaccharide, DP-3 is trisaccharide, etc.

crease crystal-phase yield as a single crystal crop. The overall yield depends on the dextrose content of the original hydrolysate and the extent to which hydrolysates are refined. The final mother liquor is used by the tanning and fermentation industries and in the manufacture of caramel color.

An exceptionally pure grade of dextrose monohydrate is obtained by dissolving the centrifuge cake obtained as above and recrystallizing in essentially the same manner. This recrystallized dextrose is marketed as a USP grade for special therapeutic purposes, eg, intravenous injection.

Anhydrous α-dextrose is manufactured by dissolving dextrose monohydrate in water and crystallizing at 60–65°C in a vacuum pan. Evaporative crystallization is necessary to avoid color formation at high temperatures and hydrate formation at low temperatures. The product is separated by centrifugation, washed, dried, and marketed as a very pure grade of sugar for special applications (60–61).

A liquid dextrose product at 71 wt % solids is prepared by dissolving crystalline dextrose. Liquid or solid high DE products are prepared by evaporating refined hydrolysate to 70–75 wt % solids or by spray-drying to a moisture content of ca 2 wt %.

Anhydrous β-dextrose can be crystallized from concentrated solutions or melts at high temperatures. With careful control of conditions, a product containing a high percentage of anhydrous β-dextrose can be obtained (61–62).

Dry dextrose is shipped in 45.4-kg bags or in bulk in hopper cars. Liquid products are shipped in insulated railroad tank cars or tank trailers at ca 50°C to prevent crystallization.

Economic Aspects. The price of dextrose monohydrate since 1952 is given in Table 4. Dextrose is generally priced below sucrose and therefore fluctuations in sugar prices result in changes in the price of dextrose (70). Other factors contributing to cost are corn prices, demand, processing capacity, and production costs. Production capacity has not changed significantly in recent years because of the expense of dextrose crystallizers. This factor along with high production costs have caused dextrose prices to increase to levels above those of HFCS and corn syrup. Compared to dextrose monohydrate, anhydrous dextrose sells at a premium; prices in early 1981 were 90¢/kg for anhydrous dextrose and $1.01/kg for USP anhydrous dextrose (see Fine chemicals).

Table 4. Price of Dextrose, HFCS, Corn Syrup, and Maple Syrup in the United States[a]

Year	Dextrose, $/kg[b]	HFCS, $/kg[c]	Corn syrup, $/kg[d]	Maple syrup, $/L[e]
1970	0.21	0.36	0.15	
1975	0.43	0.37	0.31	2.79
1976	0.31	0.24	0.24	2.84
1977	0.29	0.20	0.19	3.07
1978	0.33	0.19	0.18	3.32
1979	0.35	0.21	0.21	3.76
1980	0.59	0.38	0.32	4.11

[a] Refs. 63–64.
[b] In 45.4-kg bags, fob New York, monohydrate basis.
[c] In tank cars, fob New York, 42 wt % fructose, 71 wt % solids.
[d] In tank cars, fob New York, sp gr 1.42 (43°Bé, 80.3 wt % solids).
[e] Refs. 65–69.

Analysis and Specifications. U.S. and international specifications for crystalline dextrose are given in references 71 and 72–74, respectively. High quality anhydrous dextrose produced for the pharmaceutical industry is prepared in accordance with specifications given in ref. 75. Typical product analyses include reducing sugar value, solution color, solution clarity, ash, particle size distribution, and moisture. High pressure liquid chromatography (hplc) is used to measure the concentration of dextrose and other saccharides in solution (76–77).

Health Factors. Dextrose as well as corn syrups are substances that are presumed to be GRAS by the FDA (78). A study of the health aspects of dextrose, fructose, and corn syrups has indicated that these sweeteners are not hazardous at current levels of consumption with the exception of a small contribution to the formation of dental caries (79).

Uses. The main use of dextrose is in the food-processing industry, where it is of value for its physical, chemical, and nutritive properties. Distribution of dextrose to various industries in the United States is shown in Table 5. In the baking industry, it serves primarily as a fermentable sugar which contributes to crust color. In the beverage industry, it is used as a source of fermentables in low calorie beer and as a sweetener in beverage powders. In the canning industry, it supplies sweetness, body, and osmotic pressure and also contributes to better natural color retention in certain products. In the confectionery industry, it is used to supply sweetness and softness control and to regulate crystallization. Dextrose is also used to make tableted products; flavor is often enhanced by the cooling effect obtained when dextrose hydrate dissolves in the mouth. In frozen desserts, it prevents oversweetness and improves flavor.

In many cases, dextrose is used in conjunction with sucrose (80). Although dextrose by itself is somewhat less sweet than sucrose, the combination with sucrose may be as sweet as pure sucrose at the same concentration (81). Dextrose is also used in the pharmaceutical industry for intravenous feeding as well as for tableting and other formulations. In fermentation, dextrose is a raw material for biochemical synthesis

Table 5. Distribution by Industry of Dextrose, HFCS, and Corn Syrup in the United States, 1000 metric tons

Year	Baking	Beverage	Canning	Confectionery	Dairy	Total[a]
Dextrose[b]						
1965	178	8	21	37	6	468
1970	174	8	23	52	7	547
1975	157	18	15	62	7	561
1980	51	66	4	55	2	513
HFCS[c]						
1970	18	39	9	0.5	5	99
1975	129	279	64	4	36	715
1980	365	1039	235	15	140	2659
Corn syrup[d]						
1965	209	36	104	410	159	1211
1970	222	93	98	439	213	1449
1975	315	205	174	416	278	2278
1980	196	384	126	446	241	2201

[a] Includes other applications not listed.
[b] Monohydrate basis.
[c] 71 wt % solids basis.
[d] Sp gr 1.42 (43°Bé, 80.3 wt % solids).

and, chemically, it is a raw material for sorbitol, mannitol, and methyl glucoside production. Yeast fermentation of dextrose to ethanol is of current importance in the production of gasohol (82). The production of other bulk chemicals from dextrose by chemical or biochemical means appears to have promise in future applications (83). Reaction of dextrose with sorbitol and citric acid produces a low calorie dextrose polymer which is used as a bulking agent in reduced-calorie foods (84).

High Fructose Corn Syrups

High fructose corn syrups (HFCS, isosyrup, isoglucose) are concentrated solutions containing primarily fructose and dextrose with lesser quantities of higher molecular weight saccharides. HFCS is produced by partial enzymatic isomerization of dextrose hydrolysates followed by refining and concentration. Commercial products contain 42, 55, or 90% fructose on a dry weight basis.

In nature, fructose (levulose, fruit sugar) is the main sugar in many fruits and vegetables. Honey contains ca 50 wt % fructose on a dry basis. Sucrose is composed of one unit each of fructose and dextrose combined to form the disaccharide. Fructose exists in polymeric form as inulin in plants such as Jerusalem artichokes, chicory, dahlias, and dandelions, and is liberated by treatment with acid or enzyme (85).

Fructose was first isolated in 1847 (86). In 1874, it was recognized that fructose had advantages over sucrose as a sweetener for diabetics (87). Based on the discovery of the alkaline conversion of dextrose to fructose in 1895, a considerable number of investigations were conducted in an attempt to develop a commercial process (88–89). However, because of problems associated with color, off flavor, degradation products, and low fructose yield, a process was never commercialized. Enzymatic conversion of dextrose to fructose with glucose isomerase was first reported in 1957 and patented in 1960 (90–91). Research in this area continued for several years in Japan resulting in a U.S. patent in 1971 and commercial production in Japan in 1966 (92–93). Japanese technology was licensed by Standard Brands and production was initiated in the United States in 1967 by a batch process (94). Initially, a 15 wt % fructose syrup was made followed in 1968 by a 42 wt % fructose product. In 1972, a continuous system was initiated using an immobilized enzyme process (95–96) (see Enzymes, immobilized).

Properties. Fructose, a ketohexose monosaccharide, crystallizes as β-D-fructopyranose [*7660-25-5*], has a molecular weight of 180, and a melting point of 102–104°C. In solution, fructose undergoes rapid mutarotation to a mixture of α-D- [*10489-81-3*] and β-D-fructopyranose and α-D- [*10489-79-9*] and β-D-fructofuranose [*470-23-5*] (see Fig. 2), during which time specific rotation changes from −132° to −92° (97). At equilibrium, ca 60–70 wt % of the fructose is in the β-pyranose form and 20–30 wt %

Figure 2. Structures of α-D-fructose.

is in the β-furanose form. The remainder exists as α-D-fructose (98). Relative amounts of each depend on temperature and pH. At equilibrium, the solubility of fructose in water is 80 wt % (25°C). The crystalline form of fructose is 1.8 times sweeter than sucrose and therefore, can be used in reduced quantities to attain equivalent sweetness. A discussion of the metabolism of fructose is given in ref. 99.

Because of the presence of fructose, HFCS is sweeter than conventional corn syrups, although intensity of sweetness results from many factors, eg, temperature, pH, and concentration. In general, 42 wt % HFCS has the same sweetness as sucrose, 55 wt % HFCS is as sweet as medium invert and 2–5% sweeter than sucrose, and 90 wt % HFCS is 15–30% sweeter than sucrose (100). Other properties of HFCS include high solubility, which reduces the possibility of crystallization during shipment; humectant properties allowing for increased shelf life of bakery products; easier decomposition of fructose during baking resulting in improved color and flavor; and high osmotic pressure for containment of microbial growth. Properties of fructose and HFCS are discussed in refs. 87, 100–101.

Manufacture. HFCS containing 42 wt % fructose is produced commercially by column isomerization of clarified and refined dextrose hydrolysate with immobilized glucose isomerase. Enriched syrup containing 90 wt % fructose is prepared by chromatographic separation and blended with 42 wt % HFCS to obtain 55% HFCS (see Table 3 for product compositions).

Glucose isomerase is produced for commercial use from a variety of bacterial organisms including *Streptomyces*, *Aerobacter*, *Actinoplanes*, *Arthrobacter*, and *Bacillus* (89,102–110). Enzyme properties vary depending on the source, but they are similar in terms of operational pH and temperature. Batch reactions with soluble enzymes require long residence time resulting in the production of color, off flavors, and by-products, eg, psicose which is a nonmetabolizable sugar formed under alkaline conditions. To eliminate these problems and to reduce enzyme cost, glucose isomerase has been immobilized by several different techniques for use in continuous operation. A high concentration of enzyme reduces the reaction time to 0.5–4 h or less, thus substantially eliminating side reactions. In addition, continual use of the enzyme reduces enzyme cost to an economical level.

Several methods of immobilization are used for preparing the enzyme for commercial use. Initial processes involved whole cells which were heat-treated to fix the enzyme within the cells or enzyme-absorbed onto diethylaminoethyl cellulose (95–96). Novo Labs, Inc., markets a whole-cell system in which bacterial cells are homogenized, treated with gluteraldehyde, and dried (111). Other commercial whole-cell systems are described in refs. 102, 112–113. Technology has also been patented for immobilization of isomerase onto porous ceramics (114–116).

HFCS is produced from dextrose hydrolysate which has been clarified, carbon-treated, ion-exchanged, and evaporated to 40–50 wt % dry basis. Magnesium is added as a cofactor to maintain isomerase stability and to prevent inhibition by trace amounts of residual calcium. If the calcium level is below 1 ppm, a magnesium level (added as $MgSO_4$) of 0.0004 M is sufficient, although a higher calcium level requires additional magnesium (117). Hydrolysate is passed through a fixed bed of immobilized isomerase at a controlled flow rate. Process conditions depend on the particular enzyme system used but are generally 60–65°C, pH 7.5–8.5, and a reaction time of 0.5–4 h.

Maximum fructose content at equilibrium is 50–55 wt % dry basis; however, residence time is adjusted to attain 42–45 wt % fructose since a greatly increased reaction time is required to attain higher levels. The enzyme can be used for as long as

10 wk, during which time loss in activity is compensated for by regulating the residence time, ie, the flow rate through the column (118). Enzyme reactors are operated in parallel or in series and are replaced individually as isomerase is inactivated. Isomerized hydrolysate containing ca 42 wt % fructose is adjusted to pH 4–5, carbon-treated to remove color, ion-exchanged to remove salts, and concentrated by evaporation to ca 71 wt % solids. Product shipment is by tank trucks or by rail. Since dextrose content in the HFCS product is 50–52 wt %, storage at 30–32°C is required to prevent crystallization. If crystallization occurs, the syrup is heated to 38°C to dissolve crystals prior to unloading.

Products containing higher levels of fructose are produced by chromatographic separation of 42 wt % HFCS and were first commercially available in limited quantities in 1976. Enrichment is accomplished by passing 42 wt % HFCS through a column of absorbent containing calcium or other cation groups (119). Fructose is retained to a greater degree than dextrose or oligosaccharides, and therefore, a stream containing nonfructose is collected first as raffinate followed by elution of fructose extract with water. In commercial operation, separation is achieved in a batchwise, semicontinuous, or continuous fashion. The continuous procedure involves a simulated moving bed in which feed and desorbent enter the column at different points while fructose and raffinate streams are withdrawn (120–123). Points of entry and withdrawal are changed periodically to correspond to flow through the column, and hence, separation efficiency is maximized. Typical operation involves the addition of 42 wt % HFCS at a dry substance content of 50% to a column at 50–70°C. Columns as large as 6.7-m dia have been used to produce a stream containing 90% of the fructose at a concentration of 94 wt % (120). Raffinate is recycled to isomerization columns for production of 42 wt % HFCS. Enriched HFCS is blended with 42 wt % HFCS to produce a product containing 55–60 wt % fructose. A solids level of 77 wt % is suitable for shipment with minimal crystallization problems. In addition, enriched HFCS at 90 wt % fructose is evaporated to 80 wt % solids for shipment as an essentially noncrystallizable syrup.

Crystalline fructose differs from HFCS in that it is >99% pure and contains negligible dextrose. Although the raw material for crystalline fructose production has been sucrose, a new plant will produce crystalline fructose from HFCS starting in 1981 (124). Estimated initial capacity is 10,000 t/yr (64). A new route to the production of fructose has been reported and involves enzymatic conversion of dextrose to D-glucosone, followed by hydrogenation to fructose (125).

Economic Aspects. Prices of HFCS since 1970 are listed in Table 4. The same factors that affect the price of dextrose also are responsible for fluctuations in the price of HFCS. During the early 1970s, refined sugar prices increased dramatically, resulting in a concomitant increase in the price of HFCS and the construction of new production facilities. However, when sugar prices fell during 1975–1978, HFCS prices also dropped to a level of ca 35% below sugar rather than the traditional 10–20% differential (126). Excess capacity resulted in delaying the startup of some new HFCS plants. However, during the last few years, demand for HFCS has increased significantly, especially in the soft-drink industry (see Carbonated beverages). Consequently, HFCS prices increased dramatically between 1979 and 1980 because of increased demand and higher sucrose prices and has led to a current increase in processing capacity throughout the industry. The rapid increase in U.S. production is shown in Table 5. In 1980, ca 25% of total HFCS sales were the 55 wt % type; this was expected to increase to ca 33% in 1981 (64). For comparison, the 1979–1980 production of HFCS in Japan, Canada, and Europe amounted to 340,000, 75,000, and 45,000 t, respectively (127).

The retail price of crystalline fructose at $2.20/kg is considerably higher than HFCS (64). Estimated world consumption of crystalline fructose for 1980 was 16,000 t, of which 20% was consumed in the United States (128).

Analysis, Specifications, and Health Factors. U.S. and international specifications for HFCS are the same as for corn syrup. Fructose content is determined by high pressure liquid chromatography, dry substance by refractive index, and color by spectroscopy (76,129). Methods of analysis are discussed in ref. 77. The health aspects are the same as discussed for dextrose.

Uses. High fructose corn syrup is used as a partial or complete replacement for sucrose or invert sugar in food applications to provide sweetness, flavor enhancement, fermentables, or humectant properties. It is used in soft drinks, baking, canned fruit, dairy products, and confections (see Table 5). HFCS is used in combination with not only sucrose, but also with other corn sweeteners (80). The main application of 55 wt % HFCS is in soft drinks as emphasized by the rapid increase in use in the beverage industry. The use of a 50–50 blend of 55 wt % HFCS and sucrose was authorized by soft-drink producers in 1980 and use of HFCS by Coca-Cola alone is estimated to constitute a potential market of 590,000 t/yr (130).

HFCS containing 90 wt % fructose is used in low calorie or specialty foods because of its high sweetness and, therefore, reduced usage level and lower caloric value. In Canada, 90 wt % HFCS has been used to replace saccharin in low calorie soft drinks. Other uses are as a liquid table-top sweetener or as a honey-flavored product for use in baked goods and confections. Crystalline fructose is essentially pure and is used at a level that provides sweetness at a lower caloric level than other sweeteners. Uses include health foods and medicines (64).

Corn Syrups

Corn syrups [*8029-43-4*] (glucose syrup, starch syrup) are concentrated solutions of partially hydrolyzed starch containing dextrose, maltose, and higher molecular weight saccharides. Corn syrups are generally sold in the form of viscous liquid products and vary in physical properties, eg, viscosity, humectancy, hygroscopicity, sweetness, fermentability, etc. Production in the United States began in 1842. The syrup is now produced in the United States from corn starch by acid and enzyme processes. Starch sources, eg, wheat, rice, potato, and tapioca, are used elsewhere depending on availability.

Properties. Corn syrups are defined on the basis of reducing sugar content as one of four types having DE of 20–99.4. Lower DE products are classified as maltodextrins [*9050-36-6*] and higher DE products as dextrose. Classifications are

Product	*DE*
maltodextrin	<20
type I corn syrup	20–37
type II corn syrup	38–57
type III corn syrup	58–73
type IV corn syrup (includes HFCS)	73–99.4
dextrose	≥99.5

Syrups may also be classified as acid-conversion, acid–enzyme-conversion, high maltose, etc, syrups. However, as shown in Table 3, the most adequate characterization is with respect to the content of individual saccharides. In many cases, it is the content of individual saccharides or groups of saccharides which determines characteristics of syrups. Functional properties of corn syrups include fermentability, viscosity, humectancy–hygroscopicity, sweetness, colligative properties, and browning reactions.

Fermentability of corn syrups by yeast is important in certain food applications, eg, baking and brewing. The fermentable sugars present in corn syrup are dextrose, maltose, and maltotriose. Fermentability of maltose or maltotriose depends on the specific fermentation process and organism. In general, greater fermentability is obtained at the higher DE levels.

Viscosity of corn syrup is a function of DE value, temperature, and solids concentration. Viscosity decreases with increasing DE and temperature and increases with increasing concentration. For a 43-DE corn syrup, viscosity at 1.42 sp gr (43° Bé) is 56,000 mPa·s (= cP) at 27°C, 14,500 mPa·s at 38°C, and 4,900 mPa·s at 49°C. Corresponding values for a 55-DE syrup at the same density are 31,500 mPa·s, 8,500 mPa·s, and 2,900 mPa·s.

The hygroscopic and humectant properties of corn syrups are of great importance in many applications. Depending on the type of syrup and on the specific conditions of temperature and humidity, the products may either resist or facilitate moisture loss or moisture absorption. The ability to attract moisture or retard its loss increases with increasing DE value. Prevention of moisture pickup is more characteristic of syrups with low DE values.

Sweetness is primarily a function of the levels of dextrose and maltose present and therefore is related to DE. Other properties that increase with increasing DE value are flavor enhancement, flavor transfer, freezing-point depression, and osmotic pressure. Properties that increase with decreasing DE value are bodying contribution, cohesiveness, foam stabilization, and prevention of sugar crystallization. Corn syrup functional properties are described in detail in ref. 131.

Manufacture. Acid hydrolysis of starch is conducted by batch or continuous processes. Batchwise conversion is carried out in large cookers or converters, which are usually built of manganese bronze and have capacities of ca 10 m^3 (10^4 L). A suspension of starch at 35–40 wt % dry solids is passed to the converter, hydrochloric acid is added to a concentration of 0.015–0.02 N, and the converter is steam-heated until a temperature of 140–160°C is reached. The mixture is held at this temperature for a period of time, usually 15–20 min, to produce the desired degree of hydrolysis. Improved process control and therefore better product uniformity is achieved by continuous processes based on indirect heating. Acidified starch slurry is pumped at a constant rate through a series of heat exchangers at reaction conditions similar to those used in batch operations. In either process, hydrolysate is neutralized to pH 4–5.5 by addition of soda ash, clarified by centrifugation or filtration, evaporated to ca 60 wt % solids, carbon-treated to remove color and acid degradation products, and concentrated to 77–85 wt % solids. Sulfur dioxide is added during evaporation to some grades of syrup to reduce color development. Syrups produced by straight acid hydrolysis do not exceed 55 DE because of the formation of difficult-to-remove color and flavor at higher DE levels. Syrup compositions are given in Table 3. Syrups are also produced by acid–enzyme or enzyme–enzyme processes. Starch is first hydrolyzed by acid, as described above, or by enzyme, as described for dextrose, followed by saccharification to the desired composition with one or more enzymes.

Maltose syrup [*69-79-4*], for example, is prepared from ca 20-DE, partially hydrolyzed starch by saccharification with a maltose-producing enzyme at ca 55°C and pH 5. Maltogenic enzymes, eg, β-amylase extracted from germinated barley or fungal α-amylase derived from *A. oryzae*, are used to produce a hydrolysate containing ca 40–50 wt % maltose (see Table 3). Higher levels of maltose, ie, 60–80 wt %, are produced by saccharification with a combination of a maltogenic enzyme and a debranching enzyme, eg, pullulanase (55). Properties of maltose are given in ref. 132. Processes for the production of maltose syrups are given in refs. 133–134.

High conversion syrups of 60–70 DE containing high levels of dextrose and maltose are also produced (see Table 3). These types of syrup are made in the same manner as maltose syrup, except a combination of glucoamylase and maltogenic enzyme is used. Processes are described in refs. 134–135. Maltose and high conversion syrups are refined as described earlier and shipped in rail cars and tank trucks. Syrups are generally heated to ca 38°C to facilitate unloading. Some syrups, particularly those with a limited extent of hydrolysis, are reduced to dry form by spray-drying or roll-drying. These products are commonly called corn-syrup solids and are shipped in moisture-proof bags. Products of <20 DE are referred to as maltodextrins or hydrolyzed cereal solids (see Table 3). These products are bland-tasting, free-flowing, and nonhygroscopic. Maltodextrins are prepared from regular or waxy corn starch by enzyme or acid processes to products of 12–20 DE, clarified, refined, spray-dried to a moisture content of 3–5 wt %, and shipped in 45.4-kg bags. In some cases, a syrup of ca 75 wt % solids is produced.

Economic Aspects. Prices of corn syrup since 1952 are listed in Table 4. In the early 1970s, competition depressed prices to a very low level (12.4 ¢/kg in 1972 (70)). Recovery followed when sugar prices increased and production capacity was reduced resulting from the manufacture of other competitive products. Recent prices have been ca 40% below that of sucrose.

Analysis, Specifications, and Health Factors. U.S. and international specifications for corn syrups and dried corn syrups are given in refs. 136–138. Corn syrups are usually sold with a specification of the Baumé measurement, and these can be related to the solids content (139). The most common value is sp gr 1.42 (43° Bé), corresponding to 80.3 wt % solids for a 42 DE acid-converted syrup. Solids contents for such a syrup at sp gr 1.39, 1.41, 1.44, and 1.45 (41, 42, 44, and 45° Baumé) are 76.3, 78.3, 80.3, 82.3, and 84.3 wt %, respectively. Higher DE syrups have slightly higher solids contents at the same densities. Saccharide composition is determined by high performance liquid chromatography (140). Other analyses include color, iron, and pH. Health and safety aspects are the same as for dextrose.

Uses. Principal uses of corn syrups are shown in Table 5. Corn syrups are often used in combination with sucrose, dextrose, or HFCS (80). The specific type of syrups used depends on the properties desired in the final product. Properties listed above are important to varying degrees in the various products, and changes in formulation can affect the choice of syrup required to supply the most desirable properties. In many cases, corn syrups are used as supplementary sweeteners, with sucrose remaining the primary one.

In the confectionery industry, corn syrups are used extensively in nearly every type of confection, ranging from hard candy to marshmallows. In hard candies, which are essentially solid solutions of nearly pure carbohydrates, corn syrup contributes resistance to heat discoloration, prevents sucrose crystallization, and controls hygroscopicity, viscosity, texture, and sweetness. Maltose syrups, high conversion syrups, and acid-converted syrups (36 and 42 DE) are used for this application.

In the canning and preserving industries, corn syrups are used to prevent crystallization of sucrose, provide body, accentuate true fruit flavors, and improve color and texture.

In the beverage industry, the predominant use is in the beer and malt-liquor areas. High conversion syrups are used to replace dry cereal adjuncts, provide fermentable sugars, enhance flavor, and provide body. These syrups contain controlled amounts of dextrose and maltose for proper fermentation.

Corn syrups used in baking are generally of the high conversion type. They are incorporated into cakes, cookies, icings, and fillings to increase the amount of moisture retained, retard crystal growth of other sugars, enhance tenderness, and increase shelf life. In yeast-raised goods, fermentability is of importance, and therefore, only high DE syrups are used (see Bakery processes and leavening agents, yeast-raised products).

Corn syrups used in ice cream and frozen desserts are generally 36- or 42-DE acid-converted syrups. The syrup serves primarily to provide maximum flexibility in adjusting flavor, texture, body, and smoothness; it also aids in grain control and in the modification of melt-down and shrinkage characteristics of the frozen product.

Syrups of 25–30 DE are used as spray-drying aids in products, eg, coffee whiteners (see Synthetic and imitation dairy products). High conversion syrup, maltose syrup, and 42-DE syrup are used in jams and jellies.

Additional uses of corn syrup include applications in table syrups, baby food, meat packing, breakfast foods, salad dressing, pickles, dehydrated powdered foods, medicinal syrups, textile furnishings, adhesives, and numerous other products and processes.

Maple Syrup

Maple syrup is prepared by evaporating sap from the maple tree to a concentrated solution containing predominantly sucrose. Its characteristic flavor and color are formed during evaporation. Maple syrup is produced from the sap of several varieties of mature maple trees, eg, the sugar maple (*Acer saccharum*) and black maple (*Acer nigrum*). Main producing areas include the northeastern and upper midwestern United States and eastern Canada.

Collection of sap is made sometime between late fall and mid-spring, depending on weather conditions, the best time being when the temperature is ca 7°C during the day and below freezing at night. Sap generally contains 2–3 wt % solids, of which ca 96% is sucrose and the remainder other carbohydrates, organic acids, ash, protein, and ligninlike materials. A taphole is drilled into the tree, the hole is sanitized with a germicidal pellet, and a spout is driven in. Sap is collected in a bucket or bag or alternatively in plastic tubes directed to a centralized collection tank by gravity or vacuum. Evaporation is conducted at atmospheric pressure until a boiling point of 104°C is reached to produce a syrup meeting Federal specifications of at least 66 wt % solids (141). A syrup concentrated to only 65 wt % solids has a thin taste, and one concentrated to ≥67 wt % crystallizes when cooled. A refractometer or hydrometer is used to measure concentration. Final specific gravity should be 1.35 (37.75° Baumé) at 15.6°C. Flavor and color develop during evaporation as a result of reactions occurring between sugar and other components. Syrup is clarified, graded as to color, flavor, and density, and finally packaged in small containers for retail sale as table syrup. Typically the product contains 88–99 wt % sucrose and 0–12 wt % invert sugar. Maple syrup is also used in candy manufacture by blending with sucrose. Other applications include addition to cookies, cakes, ice cream, baked beans, baked ham, and baked apples.

Maple sugar is prepared by evaporating sap to a high solids content, ie, a boiling temperature of 116–121°C, and then allowing the supersaturated solution to crystallize or solidify during cooling. Maple cream or maple butter is made by stirring a supersaturated solution while cooling rapidly to produce a product of creamy texture.

Recent price data for maple syrup are given in Table 4. Annual U.S. production from 1975 to 1980 was $3.5–4.6 \times 10^6$ L, although 1981 production was projected to be 20–30% higher than in 1980 (142).

Molasses

Molasses, another type of syrup, is a by-product of the sugar industry; it is the mother liquor remaining after crystallization and removal of sucrose from the juices of sugar cane or sugar beet and is used in a variety of food and non-food applications

Table 6. Analyses of Molasses Samples

Molasses type	Solids, %	Total sugars as invert, wt %	Crude protein, %	Total ash, wt %
cane (Louisiana)	80.8	59.5	3.0	7.2
cane (refiner's)	75.4	55.9	2.1	8.6
high test (Cuba)	80.4		0.7	1.4
beet (Wisconsin)	78.6	52.7	11.4	9.3
corn	74.9	50.3	0.4	8.9
citrus	71.4	42.4	4.7	4.8

Table 7. Price Comparisons Between Molasses and No. 2 Yellow Corn at Baltimore, Chicago, and Omaha, cents[a]

	Baltimore			Chicago			Omaha		
Year	Corn, 0.035 m^3 (1 bu)	Molasses, 24.6 L	Difference	Corn, 0.035 m^3 (bu)	Molasses, 24.6 L	Difference	Corn, 0.035 m^3 (1 bu)	Molasses 24.6 L	Difference
1977	228.0	178.6	49.4	221.5	209.2	12.3	206.6	223.3	−16.7
1978	241.0	198.6	42.4	233.0	253.5	−20.5	211.2	275.7	−64.5
1979	280.6	346.6	−66.0	265.3	398.5	−133.2	235.2	409.6	−174.4
1980	320.2	405.8	−85.6	306.7	465.7	−159.0	272.2	489.8	−217.6
1981	329.2	377.4	−48.2	315.8	420.0	−104.2	291.2	453.8	−162.6

[a] Ref. 144.

Table 8. Estimated Utilization of Molasses in the United States, 10^6 L [a]

Year	Distilled sprits	Yeast, citric acid, and vinegar	Pharmaceuticals and edible molasses	Total	Mixed feeds, direct feeding and silage	Total utilization
1970	30	379	204	613	2286	2899
1974	61	379	223	663	2169	2832
1977	28	351	97	476	2340	2816
1978	22	352	79	453	1969	2422
1979	33	294	73	400	1964	2364
1980	31	293	71	395	1379	1774
1981 [b]	19	285	72	376	1611	1987

[a] Data for 1970 and 1974 from ref. 145. Data for 1977–1981 from ref. 144.
[b] Estimated.

Table 9. U.S. Molasses Production, Imports, and Exports, 10^6 L [a]

Year	Production					Imports	Mainland exports
	Cane	Beet	Citrus	Hydrol	Total		
1970	731	602	45	83	1461	1446	4
1975	833	734	38	95	1700	1147	23
1979	772	623	13	153	1561	877	73
1980	730	429	24	150	1333	531	90
1981 [b]	830	563	23	74	1490	564	67

[a] Data for 1970, 1975, and 1979–1981 from refs. 145, 66, and 144, respectively.
[b] Estimated.

(see Sugar). Molasses was first produced from sugarcane in China and India centuries ago, later spread into Europe and Africa, and was introduced as the by-product of cane-sugar production into Santo Domingo by Columbus in 1493. During Colonial times, molasses was very important to the American colonies for the production of rum (see Beverage spirits, distilled). In 1733, the British Parliament passed the Molasses Act to tax molasses imported from foreign countries. This attempt to restrict trade was ignored by the colonies and was, in part, responsible for the American Revolution.

Manufacture. Raw sugar is produced from sugarcane by a process that involves extraction of the sugar in water, treatment to remove impurities, concentration, and several crystallizations. After the first crystallization and removal of "first sugar," the mother liquor is called first molasses. First molasses is recrystallized to obtain a second lower quality sucrose (second sugar) and a second molasses. After a third crystallization, the third molasses contains considerable nonsucrose material, and additional recovery of sucrose is not economically feasible. The third molasses is sold as blackstrap, final, or cane molasses. Raw sugar obtained from the above process is mixed with water to dissolve residual molasses and separated by centrifugation. This process is called affination and the syrup is referred to as affination liquor. The sugar is dissolved in water, treated to remove color and impurities, and subjected to several crystallizations to obtain refined sugar. The mother liquor from the final crystallization is combined with affination liquor and crystallized to produce a dark sugar ("remelts") which is recycled to raw sugar. The remaining mother liquor is called refiners molasses and is similar to final molasses but usually of better quality.

Table 10. World Production of Industrial Molasses, 1000 metric tons[a]

Country	1978–1979	1979–1980	1980–1981	1981–1982[b]
Brazil	5,000	4,800	5,400	5,700
USSR	3,306	3,062	2,669	2,749
India	2,564	1,582	2,000	2,600
U.S.[c]	1,916	1,819	2,238	2,340
Mexico	1,410	1,311	1,141	1,252
Thailand	1,059	676	1,029	1,250
Cuba	1,296	1,185	1,111	1,178
China	932	878	970	1,067
France	952	958	983	1,000
Philippines	818	827	835	860
Poland	657	653	469	755
FRG	685	653	656	730
Australia	577	598	713	723
South Africa	674	668	601	700
Argentina	818	789	654	590
Indonesia	455	430	450	500
others	9,195	8,831	8,946	9,710
Total	*32,314*	*29,720*	*30,865*	*33,704*

[a] Ref. 144.
[b] Estimated.
[c] Does not include Puerto Rico.

Table 11. Molasses Price, \$/t[a]

Year	Blackstrap[b]	Beet[c]	Citrus[d]
1960	19.09	31.72	19.46
1965	16.57	29.76	17.79
1970	26.76	40.12	19.29
1975	49.96	61.51	46.26
1979	91.44		
1980	106.37	105.38	114.25
1981	93.59	94.03	113.65

[a] Price fob tank car or tank truck. Based on 715 L for blackstrap and beet molasses and on 731 L for citrus molasses. Data for 1960–1970, 1975, and 1979–1981 from refs. 63, 66, and 144, respectively.
[b] New Orleans, La.
[c] Colorado.
[d] Florida.

In beet-sugar manufacture, the beet juice does not contain reducing sugars (fructose and glucose) which are present in cane juice but may contain raffinose. Because of the absence of reducing sugars, sucrose level in beet molasses is not reduced to the same extent as with cane. Final molasses from beet contains ca 60 wt % sucrose (dry basis) compared to 30 wt % sucrose (dry basis) in cane molasses. Treatment of diluted beet molasses with calcium oxide precipitates sucrose as tricalcium sucrate (Steffen process) which is recycled to the incoming hot beet juice. During reycling, raffinose accumulates in the final molasses and retards crystallization if not removed.

Therefore, a portion of the final molasses is periodically removed and called discard molasses.

High test molasses (invert molasses) is produced from cane sugar when sucrose manufacture is restricted because of overproduction. The cane sugar at ca 55 wt % solids is enzymatically converted to invert syrup to prevent crystallization and evaporated to a syrup. The product is used in the same applications as blackstrap molasses.

Molasses from other sources include citrus and corn sugar (hydrol) molasses. Citrus molasses is produced from citrus waste and contains 60–75% sugars (see also Fruit juices). Corn sugar molasses is the mother liquor remaining after dextrose crystallization and contains a minimum of 43% reducing sugars expressed as dextrose.

Molasses is shipped in drums, barrels, tank trucks, tank cars, barges, and ocean-going vessels. Because of high viscosity, molasses must be heated in some situations to facilitate pumping. However, prolonged heating must be avoided to prevent caramelization.

Composition. Molasses composition depends on several factors, eg, locality, variety, soil, climate, and processing. Cane molasses is generally at pH 5.5–6.5 and contains 30–40 wt % sucrose and 15–20 wt % reducing sugars. Beet molasses is ca 7.5–8.6 pH, and contains ca 50–60 wt % sucrose, a trace of reducing sugars, and 0.5–2.0 wt % raffinose. Cane molasses contains less ash, less nitrogenous material, but considerably more vitamins than beet molasses. Composition of selected molasses products is listed in Table 6. Procedures for molasses analysis are given in ref. 143.

Uses. The primary use of molasses is in animal feed. Molasses provides a carbohydrate source, salts, protein, vitamins, and palatability and may be used directly or mixed with other feeds (see Pet and livestock feeds).

The carbohydrate content of 24.6 L (6.5 gal) of blackstrap molasses is considered to be equal to 0.035 m^3 (one bushel) of corn as measured by the energy produced from 0.035 m^3 of corn and the amount of molasses required to produce the same amount of energy. When molasses is less expensive than corn, sales increase; when the reverse is true, sales decrease. Relative corn and molasses costs are given in Table 7 for 1977–1981.

The second main use for molasses is in various fermentation processes as an inexpensive source of carbohydrate. Molasses is the basic raw material for rum production and is also used for production of yeast (qv) and citric acid (qv) (see Fermentation).

Food applications utilize first and second molasses in baking (bread, cakes, cookies) for the molasses flavor. Molasses is also used in curing of tobacco and meats, in confections such as toffees and caramels, and in baked beans and glazes.

Distribution of molasses to various industries is shown in Table 8.

Production and Economics. Total U.S. molasses production along with imports and exports are shown in Table 9. Principal sources of imported molasses include Brazil, Colombia, Dominican Republic, and Guatemala (144). Principal molasses producers are listed in Table 10.

Molasses prices since 1960 are given in Table 11. Cane and beet molasses prices declined in 1981 due to several factors such as a slight increase in world supplies, higher interest rates, and an overpricing of molasses compared to other feed ingredients.

BIBLIOGRAPHY

"Dextrose and Starch Syrups" in *ECT* 1st ed., Vol. 4, pp. 961–969, by G. R. Dean, Corn Products Refining Company; "Molasses" in *ECT* 1st ed., Vol. 9, pp. 167–180, G. T. Reich, Consultant; "Dextrose and Starch Syrups" in *ECT* 2nd ed., Vol. 6, pp. 919–932, by E. R. Kooi, Corn Products Company; "Molasses" in *ECT* 2nd ed., Vol. 13, pp. 613–633, by Arthur G. Keller, Louisiana State University.

1. H. Wichelhaus, *Der Starkezucker*, Akademische Verlagsgesellschaft, Leipzig, Germany, 1913.
2. G. S. C. Kirchoff, *Acad. Imp. Sci. St. Petersbourg, Mem.* **4,** 27 (1811).
3. T. de Saussere, *Ann. Phys.* **49,** 129 (1815).
4. F. Becke, *Tschermaks Mineral. Petrogr. Mitt.* **10,** 464 (1889).
5. R. F. Jackson and C. G. Silsbee, *Natl. Bur. Stand. (U.S.) Sci. Paper* **437,** 715 (1922).
6. F. E. Young, *J. Phys. Chem.* **61,** 616 (1957).
7. S. M. Cantor and Q. P. Peniston, *J. Am. Chem. Soc.* **62,** 2113 (1940).
8. B. Singh, G. R. Dean, and S. M. Cantor, *J. Am. Chem. Soc.* **70,** 517 (1948).
9. W. L. Evans, *Chem. Rev.* **31,** 537 (1942).
10. D. J. McWeeny in G. G. Birch and L. F. Green, eds., *Molecular Structure and Function of Food Carbohydrate*, John Wiley & Sons, Inc., New York, 1973, p. 21.
11. R. W. Kerr, ed., *Chemistry and Industry of Starch*, 2nd ed., Academic Press, Inc., New York, 1950, Chapt. 14.
12. G. R. Dean and J. B. Gottfied, *Adv. Carbohydr. Chem.* **5,** 127 (1951).
13. W. B. Newkirk, *Ind. Eng. Chem.* **31,** 18 (1939).
14. R. H. Rogge, *Ind. Eng. Chem.* **41,** 2070 (1949).
15. E. R. Kooi and F. C. Armbruster in R. L. Whistler and E. F. Paschall, eds., *Starch: Chemistry and Technology*, Vol. II, Academic Press, Inc., New York, p. 562.
16. A. D. Woolhouse, *Starch as a Source of Carbohydrate Sweeteners*, Report No. CD 2237, Dept. of Scientific and Industrial Research, New Zealand, 1976, p. 20.
17. Ref. 15, p. 563.
18. R. V. MacAllister in R. S. Tipson and D. Horton, eds., *Advances in Carbohydrate Chemistry and Biochemistry*, Vol. 36, Academic Press, Inc., New York, 1979, p. 40.
19. *Dextrose and Starch Sugar*, Brochure B 007g-GB, Novo Industri A/S, Bagsvaerd, Denmark, 1977.
20. R. E. Hebeda and H. W. Leach, *Cereal Chem.* **51**(2), 272 (1974).
21. U.S. Pat. 3,912,590 (Oct. 14, 1975), S. Slott and G. B. Madsen (to Novo Industri A/S).
22. Can. Pat. 964,215 (Mar. 11, 1975), H. Outtrup, O. Andresen, and K. Aunstrup (to Novo Terapeutisk Laboratorium A/S).
23. G. B. Madsen, B. E. Norman, and S. Slott, *Staerke* **25**(9), 304 (1973).
24. N. Saito, *Arch. Biochem. Biophys.* **155,** 290 (1973).
25. J. P. Chiang, J. E. Alter, and M. Sternberg, *Staerke* **31**(3), 86 (1979).
26. P. Rosendal, B. H. Nielsen, and N. K. Lange, *Staerke* **31**(11), 368 (1979).
27. U.S. Pat. 3,922,196 (Nov. 25, 1975), H. W. Leach, R. E. Hebeda, and D. J. Holik (to CPC International Inc.).
28. U.S. Pat. 3,922,197 (Nov. 25, 1975), H. W. Leach, R. E. Hebeda, and D. J. Holik (to CPC International Inc.).
29. U.S. Pat. 3,922,198 (Nov. 25, 1975), E. A. Kuske and D. K. Przybylski (to CPC International Inc.).
30. U.S. Pat. 3,922,199 (Nov. 25, 1975), R. E. Hebeda, D. J. Holik, and H. W. Leach (to CPC International Inc.).
31. U.S. Pat. 3,922,200 (Nov. 25, 1975), R. Walon and E. Berghmans (to CPC International Inc.).
32. U.S. Pat. 3,012,944 (Dec. 12, 1961), F. C. Armbruster (to Corn Products Co.).
33. Ref. 15, p. 558.
34. U.S. Pat. 3,042,584 (July 3, 1962), E. R. Kooi, C. F. Harjes, and J. S. Gilkison (to Corn Products Co.).
35. U.S. Pat. 2,967,804 (Jan. 10, 1961), R. W. Kerr (to Corn Products Co.).
36. U.S. Pat. 3,017,330 (Jan. 16, 1962), R. W. Kerr (to Corn Products Co.).
37. U.S. Pat. 3,067,108 (Dec. 4, 1962), T. L. Hurst and A. W. Turner (to A. E. Staley Manufacturing Co.).
38. U.S. Pat. 3,101,302 (Aug. 20, 1963), G. E. Inglett (to Corn Products Co.).
39. U.S. Pat. 3,108,928 (Oct. 29, 1963), H. R. Kathrein (to Grain Processing Corp.).
40. U.S. Pat. 3,917,513 (Nov. 4, 1975), M. Sternberg (to Miles Laboratories, Inc.).

41. B. E. Norman in R. C. W. Berkeley, G. W. Gooday, and D. C. Ellwood, eds., *Microbial Polysaccharides and Polysaccharases*, Academic Press, Inc., New York, 1979, p. 359.
42. L. A. Underkofler in R. F. Gould, ed., *Cellulases And Their Applications, Advances in Chemistry Series No. 95*, American Chemical Society, Washington, D.C., 1969, p. 343.
43. U.S. Pat. 2,881,115 (Apr. 7, 1959), R. W. Liggett, W. C. Mussulman, D. F. Rentshler, and J. Ziffer (to A. E. Staley Manufacturing Co.).
44. U.S. Pat. 2,893,921 (July 7, 1959), D. P. Langlois and W. Turner (to A. E. Staley Manufacturing Co.).
45. E. J. Hehre, G. Okoda, and D. S. Genghof, *Arch. Biochem. Biophys.* **135,** 75 (1969).
46. J. A. Roels and R. van Tilburg, *Staerke* **31**(10), 338 (1979).
47. G. M. A. van Beynum, J. A. Roels, and R. van Tilburg, *Biotechnol. Bioeng.* **22,** 643 (1980).
48. U.S. Pat. 4,017,363 (Apr. 12, 1977), W. H. McMullen and R. Andino (to Novo Industri A/S).
49. U.S. Pat. 3,897,305 (July 29, 1975), T. L. Hurst (to A. E. Staley Manufacturing Co.).
50. U.S. Pat. 4,254,225 (Mar. 3, 1981), M. Tamura, M. Shimizu, and M. Tago (to CPC International Inc.).
51. M. J. Daniels, *Enzyme Engineering in Food Processing*, Vol. 2 of P. Linko and J. Larinkari, eds., *Food Process Engineering*, Applied Science Publishers, Ltd., London, 1980, p. 103.
52. *Nutrients From Corn. It's Logical*, Corn Products, Englewood Cliffs, N.J.
53. *Buyers Guide*, Corn Products, Englewood Cliffs, N.J.
54. G. E. Inglett, *Food Technol.*, 37 (Mar. 1981).
55. U.S. Pat. 3,565,765 (Feb. 23, 1971), R. E. Heady and F. C. Armbruster (to CPC International Inc.).
56. U.S. Pat. 3,849,194 (Nov. 19, 1974), F. C. Armbruster and E. R. Kooi (to CPC International Inc.).
57. U.S. Pat. 1,508,569 (Sept. 16, 1924), W. B. Newkirk (to Corn Products Refining Co.).
58. U.S. Pat. 1,890,913 (Dec. 13, 1932), W. B. Newkirk (to Corn Products Refining Co.).
59. U.S. Pat. 2,152,874 (Apr. 4, 1939), C. J. Copland (to Corn Products Refining Co.).
60. W. B. Newkirk, *Ind. Eng. Chem.* **16,** 1173 (1924); **28,** 760 (1936).
61. U.S. Pat. 1,693,118 (Nov. 27, 1928), W. B. Newkirk (to Corn Products Refining Co.).
62. U.S. Pat. 2,369,231 (Feb. 13, 1945), A. T. Harding (to Corn Products Refining Co.).
63. *Sugar and Sweetener Report*, USDA, Washington, D.C. **1**(1), (1976).
64. *Ibid.*, **6**(1), (1981).
65. *Ibid.*, **1**(8), (1976).
66. *Ibid.*, **2**(9), (1977).
67. *Ibid.*, **3**(9), (1978).
68. *Ibid.*, **4**(9), (1979).
69. *Ibid.*, **5**(7), (1980).
70. C. R. Keim, *Sugar Azucar Yearbook* **47,** 101 (1979).
71. Office of the Federal Register, General Services Administration, *Code of Federal Regulations*, Sections 168.110 and 168.111, Title 21, Food and Drugs, Pts. 100–169, U.S. Government Printing Office, Washington, D.C., 1980.
72. Codex Alimentarius Commission, *Recommended International Standard for Dextrose Monohydrate*, CAC/RS 8-1969, FAO and WHO, 1970.
73. Codex Alimentarius Commission, *Recommended International Standard for Dextrose Anhydrous*, CAC/RS 7-1969, FAO and WHO, 1970.
74. Codex Alimentarius Commission, *Recommended International Standard for Powdered Dextrose (Icing Dextrose)*, CAC/RS 54-1971, FAO and WHO, 1972.
75. *The United States Pharmacopeia XX (USP XX–NF XV)*, The United States Pharmacopeial Convention, Inc., Rockville, Md., 1980, p. 218.
76. K. M. Brobst and H. D. Scobell, *Cereal Foods World* **26**(5), 224 (1981).
77. R. Bernetti in D. R. Lineback and G. E. Inglett, ed., *Food Carbohydrates*, AVI Publishing Co., Inc., Westport, Conn., 1982.
78. Subcommittee on Review of GRAS list (Phase II), *A Comprehensive Survey of Industry on the Use of Feed Chemicals Generally Recognized as Safe (GRAS)*, DHEW contract no. FDA 70-22, Committee on Food Protection, Division of Biology and Agriculture, National Research Council, National Academy of Sciences, Washington, D.C., 1972.
79. *Evaluation of the Health Aspects of Corn Sugar (Dextrose), Corn Syrup, and Invert Sugar as Food Ingredients*, SCOGS-50, DHEW contract no. FDA 223-75-2004, Life Sciences Research Office, Federation of American Societies for Experimental Biology, Bethesda, Md., 1976.
80. M. M. Godzicki, *Food Eng.*, 11 (Oct. 1975).
81. C. Nieman, *Manuf. Confect.* **40**(8), 19 (1960).
82. S. Elias, *Food Eng.*, 61 (Oct. 1979).

83. *Proceedings from the 1980 Scientific Conference, Basic Chemicals from Carbohydrates*, Corn Refiners Association, Inc., Washington, D.C.
84. A. Torres and R. D. Thomas, *Food Technol.*, 44 (July 1981).
85. M. Kierstan, *Process Biochem.*, 2 (May 1980).
86. T. Doty in P. Koivistoinen and L. Hyvonen, eds., *Carbohydrate Sweeteners in Foods and Nutrition*, Academic Press, London, 1980, p. 259.
87. C. Morris, *Food Eng.*, 95 (May 1980).
88. C. A. Lobry DeBruyn and W. Alberda van Eckenstein, *Rec. Trav. Chim.* **14,** 203 (1895).
89. M. Seidman in G. G. Birch and R. S. Shallenberger, eds., *Developments in Food Carbohydrate*, Vol. 1, Applied Science Publishers Ltd., London, 1977, p. 19.
90. R. O. Marshall and E. R. Kooi, *Science* **125**(3249), 648 (1957).
91. U.S. Pat. 2,950,228 (Aug. 23, 1960), R. O. Marshall (to Corn Products Co.).
92. U.S. Pat. 3,616,221 (Oct. 26, 1971), Y. Takasaki and O. Tanabe (to Agency of Industrial Science and Technology, Tokyo).
93. Y. Takasaki, *Agric. Biol. Chem.* **30,** 1247 (1966).
94. N. H. Mermelstein, *Food Technol.*, 20 (June 1980).
95. U.S. Pat. 3,694,314 (Sept. 26, 1972), N. E. Lloyd, L. T. Lewis, R. M. Logan, and D. N. Patel (to Standard Brands, Inc.).
96. U.S. Pat. 3,788,945 (Jan. 29, 1974), K. N. Thompson, R. A. Johnson, and N. E. Lloyd (to Standard Brands, Inc.).
97. R. V. MacAllister and E. K. Wardrip in M. S. Peterson and A. H. Johnson, eds., *Encyclopedia of Food Science*, AVI Publishing Co., Inc., Westport, Conn., 1978, p. 330.
98. S. J. Angyal and G. S. Bethell, *Aust. J. Chem.* **29,** 1249 (1976).
99. G. L. S. Pawan in ref. 10, p. 65.
100. *Food Prod. Dev.*, 39 (Apr. 1978).
101. B. H. Landis, *Seminar Proceedings, Products of the Corn Refining Industry in Food, May 9, 1978*, Corn Refiners Association, Inc., Washington, D.C., p. 47.
102. U.S. Pat. 3,779,869 (Dec. 18, 1973), M. F. Zienty (to Miles Laboratories, Inc.).
103. U.S. Pat. 3,957,587 (May 18, 1976), F. C. Armbruster, R. E. Heady, and R. P. Cory (to CPC International Inc.).
104. U.S. Pat. 3,813,320 (May 28, 1974), K. K. Shieh and B. J. Donnelly (to Anheuser-Busch, Inc.).
105. U.S. Pat. 3,834,988 (Sept. 10, 1974), K. K. Shieh and B. J. Donnelly (to Anheuser-Busch, Inc.).
106. U.S. Pat. 3,645,858 (Feb. 29, 1972), C. K. Lee, L. E. Hayes, and M. E. Long (to R. J. Reynolds Tobacco Co.).
107. U.S. Pat. 3,979,261 (Sept. 7, 1976), H. Outtrup (to Novo Industri A/S).
108. J. M. Newton, *Res. Dev. Assoc. Mil. Food Packag. Syst.* **29**(2), 83 (1977).
109. W. P. Chen, *Process Biochem.*, 30 (June–July 1980).
110. B. K. Hamilton, C. K. Colton, and C. L. Cooney in A. C. Olson and C. L. Cooney, eds., *Immobilized Enzymes in Food and Microbial Processes*, Plenum Press, New York, 1974, p. 85.
111. U.S. Pat. 3,980,521 (Sept. 14, 1976), S. Amotz, T. K. Nielsen, and N. O. Thiesen (to Novo Industri A/S).
112. U.S. Pat. 3,821,086 (June 28, 1974), C. K. Lee and M. E. Long (to R. J. Reynolds Tobacco Co.).
113. U.S. Pat. 3,838,007 (Sept. 24, 1974), A. G. Van Velzen (to Gist Brocades).
114. U.S. Pat. 3,868,304 (Feb. 25, 1975), R. A. Messing (to Corning Glass Works).
115. U.S. Pat. 3,982,997 (Sept. 28, 1976), D. L. Eaton and R. A. Messing (to Corning Glass Works).
116. U.S. Pat. 3,992,329 (Nov. 16, 1976), D. L. Eaton and R. A. Messing (to Corning Glass Works).
117. *Fructose Syrups*, B 176a-GB, Novo Industri A/S, Bagsvaerd, Denmark, 1978.
118. *Sweetzyme Type Q*, IB 174a-GB, Novo Industri A/S, Bagsvaerd, Denmark, 1978.
119. J. E. Long, *Corn Annual*, Corn Refiners Assoc., Inc., Washington, D.C., 1978, p. 16.
120. D. B. Broughton, H. J. Bieser, R. C. Berg, E. D. Connell, D. J. Korous, and R. W. Neuzil, *Sucr. Belge* **96,** 155 (1977).
121. *Food Process.*, 74 (July 1979).
122. T. Hirota, *Sugar Azucar*, 245 (Jan. 1980).
123. H. W. Keller, A. C. Reents, and J. W. Laraway, *Staerke* **33**(2), 55 (1981).
124. Ref. 87, p. 108.
125. U.S. Pat. 4,246,347 (Jan. 20, 1981), S. L. Neidleman, W. F. Amon Jr., and J. Geigert (to Cetus Corp.).
126. *Forbes*, 67 (June 26, 1978).

127. T. C. Earley, *International Sweetener and Alcohol Conference*, London, 1980, p. 62.
128. C. E. Kean, *Food. Prod. Dev.*, 43 (Apr. 1978).
129. E. C. Conrad and J. K. Palmer, *Food Technol.*, 84 (Oct. 1978).
130. *Chem. Week*, 43 (Apr. 15, 1981).
131. H. E. Horn, *Cereal Foods World* **26**(5), 219 (1981).
132. J. E. Hodge, J. A. Rendleman, and E. C. Nelson, *Cereal Sci. Today* **17**(7), 180 (1972).
133. Ref. 41, p. 350.
134. *Glucose Syrups*, B 006-GB, Novo Industri A/S, Bagsvaerd, Denmark, 1977.
135. Ref. 41, p. 363.
136. Office of the Federal Register, General Services Administration, *Code of Federal Regulations*, Sections 168.120 and 168.121, Title 21, Food and Drugs, Pts. 100–169, U.S. Government Printing Office, Washington, D.C., 1980.
137. Codex Alimentarius Commission, *Recommended International Standard for Glucose Syrup*, CAC/RS, 9-1969, FAO and WHO, 1970.
138. Codex Alimentarius Commission, *Recommended International Standard for Dried Glucose Syrup*, CAC/RS, 10-1969, FAO and WHO, 1970.
139. *Critical Data Tables*, Corn Refiners Association, Inc., Washington, D.C., 1969.
140. L. E. Fitt, W. Hassler, and D. J. Just, *J. Chromatogr.* **187**, 381 (1980).
141. Office of the Federal Register, General Services Administration, *Code of Federal Regulations*, Section 168.140, Title 21, Food and Drugs, Pts. 100–169, U.S. Government Printing Office, Washington, D.C., 1980.
142. *Sugar and Sweetener Report*, USDA, Washington, D.C., **6**(2), (1981).
143. F. Schneider, ed., *Sugar Analysis: ICUMSA Methods*, International Commission for Uniform Methods of Sugar Analysis, Peterborough, England, 1979.
144. *Molasses Market News, Market Summary 1981*, Agricultural Marketing Service, USDA, Denver, Colo.
145. Ref. 63, **1**(11), (1976).

General References

Corn Sweeteners

Refs. 10, 16, 18, 41, 86, and 89 are also general references.

E. R. Kooi and F. C. Armbruster in R. L. Whistler and E. F. Paschall, eds., *Starch: Chemistry and Technology*, Vol. II, Academic Press, Inc., New York, 1967, p. 562.

G. G. Birch, L. F. Green, and C. B. Coulson, eds., *Glucose Syrups and Related Carbohydrates*, Elsevier Publishing Co., Ltd., Amsterdam, The Netherlands, 1970.

D. Howling in G. G. Birch and K. J. Parker, eds., *Sugar: Science and Technology*, Applied Science Publishers Ltd., London, 259 (1979).

Enzyme Technology, Vol. 2 of L. B. Wingard, Jr., E. Katchalski-Katzir, and L. Goldstein, eds., *Applied Biochemistry and Bioengineering*, Academic Press, Inc., New York, 1979.

H. M. Pancoast and W. R. Junk, eds., *Handbook of Sugars*, 2nd ed., AVI Publishing Co., Inc., Westport, Conn., 1980.

S. H. Hemmingsen and B. E. Norman in G. E. Inglett and L. Munck, ed., *Cereals for Food and Beverages*, Academic Press, Inc., New York, 1980, p. 61.

R. V. MacAllister in W. H. Pitcher, Jr., ed., *Immobilized Enzymes for Food Processing*, CRC Press, Inc., Boca Raton, Fla., 1980, p. 81.

Maple Syrup

J. W. White, Jr. and J. C. Underwood in G. E. Inglett, ed., *Symposium: Sweeteners*, AVI Publ. Co., Inc., Westport, Conn., 1974, p. 111.

R. T. Foulds, Jr. in A. H. Johnson and M. S. Peterson, eds., *Encyclopedia of Food Technology*, AVI Publishing Co., Inc., Westport, Conn., 1974, p. 573.

C. O. Willits and C. H. Hills, eds., *Maple Syrup Producers Manual*, Agricultural Research Service, USDA, Washington, D.C., Agricultural Handbook No. 134, 1976.

Molasses

A. Lachmann in A. H. Johnson and M. S. Peterson, eds., *Encyclopedia of Food Technology*, AVI Publ. Co., Inc., Westport, Conn., 1974, p. 608.
G. P. Meade and J. C. P. Chen, *Cane Sugar Handbook*, 10th ed., John Wiley & Sons, Inc., New York, 1977, pp. 359–377.
R. A. McGinnis, ed., *Beet-Sugar Technology*, 3rd ed., Beet Sugar Development Foundation, Fort Collins, Colo., 1982.

R. E. HEBEDA
CPC International

T

TACK. See Rubber compounding.

TACONITE. See Iron.

TALC

The term talc [*14807-96-6*] covers a wide range of natural minerals, most of which are high magnesium silicates. It is a hydrous magnesium silicate, $Mg_3SiO_{10}(OH)_2$, theoretically 31.7% MgO, 63.5% SiO_2, and 4.8% H_2O. The mineral talc is usually, although not always, a main constituent of mineral mixtures offered commercially as talc.

Minerals commonly associated with talc are tremolite ($CaMg_3(SiO_3)_4$), serpentine ($3MgO.2SiO_2.2H_2O$), anthophyllite ($Mg_7.(OH)_2.(Si_4O_{11})_2$), magnesite, mica, and chlorite. The latter is a talclike mineral in which varying proportions of aluminum are substituted for magnesium in the brucite layer. Prochlorite, for example, is $9MgO.3Al_2O_3.5SiO_2.8H_2O$. Other impurities such as dolomite ($Ca,MgCO_3$), calcite ($CaCO_3$), iron oxide, carbon, quartz, and manganese oxide may also be present.

Talc deposits were probably formed by hydrothermal alteration or contact metamorphism of preexisting rocks. The degree of alteration and the nature of the rocks from which the talc was formed largely determined purity and particle form. The whiter, purer talcs derive chiefly from dolomite, dolomitic limestone, and magnesite. Those from ultrabasic igneous rocks usually contain intermediate minerals, eg, serpentine.

Generally speaking, the purest talcs have the greatest commercial value. Those used for cosmetics, pharmaceuticals, selective adsorption, and electrical ceramics approach theoretical purity and are also characterized by a white color.

Talc, soapstone, and pyrophyllite are grouped together by the U.S. Bureau of Mines for reporting purposes, although they differ in physical and chemical properties, as well as in use. Soapstone is a talclike material of varying composition. Impurities prevent its use in applications requiring white color or chemical purity. Pyrophyllite is the aluminum analogue of talc with a theoretical composition $Al_2SiO_{14}(OH)_2$. It has many of the characteristics of talc and may be substituted for it in some applications.

Talc, although not as common as clay, is found in many parts of the world (see Clays). The United States is the world's largest producer (1–2). Vermont, Texas, Montana, New York, and California, in that order, account for >90% of a volume exceeding 10^6 metric tons per year. In recent years, the production in Vermont and Montana has increased, whereas in New York and California it has decreased.

Vermont talcs found near Ludlow, Chester, and Windsor occur in relatively impure form. In order to meet the needs of high grade uses such as cosmetics, Vermont talc producers beneficiate via froth flotation. For lower grade applications such as roofing, roller milling and air classification suffice. In Texas, talc occurs in large deposits in Hudspeth County, ca 160 km east of El Paso. They are dark in color owing to a small content of graphite and also contain significant amounts of dolomite. Their primary use is in ceramics; the carbon is released as oxide in the firing process. Montana talcs from mines near Cameron and Dillon approach theoretical purity, have good color, and are strongly hydrophobic. Although New York talc (St. Lawrence County) is of good white color, it is intimately associated with tremolite and other amphiboles (3). These minerals enhance its value in ceramic wall-tile production but restrict its use to applications where slip, softness, and hydrophilicity are of minor importance. California talcs occur in and around Death Valley and vary widely from very pure to tremolitic, depending on location. The presence of tremolite has adversely affected the sales volumes of both New York and California talcs because of the controversy surrounding tremolite-asbestos as a health hazard (see Asbestos). California production is further restricted by a 1976 law that effectively freezes annual production from mines within Death Valley National Monument to the levels of the mid-1970s.

The next largest producer of talc is the People's Republic of China with very large mines in Liaoning Province; production from Shandong and Guangxi Provinces has become significant in the last ten years. Total Chinese exports exceeded 400,000 metric tons in 1979 with unknown but significant quantities consumed domestically.

The USSR, producing some 500,000 t annually for its own consumption, is the third largest producer of talc. Finland, France, and India, each with 1980 production of ca 300,000 t, occupy fourth, fifth, and sixth positions, respectively. Australia, Austria, Republic of Korea, Italy, and Norway each produce ca 100,000 t/yr.

Most sources, including the U.S. Bureau of Mines, list Japan as a significant producer of talc and pyrophyllite (1–2,4). Although Japan does produce small amounts of both minerals, most of the reported tonnage appears to be a natural mixture of pyrophyllite, kaolin, silica, and sericite, called roseki, which is widely used by that country's ceramic industry (5) (see Ceramics).

Properties

Pure talc mineral is characterized by softness (Mohs scale 1), hydrophobic surface properties, and slippery feel (3,6–7). The crystal form may be foliated, lamellar, fibrous, or massive. Although pure talc is the standard for softness on the Mohs scale, some commercial talcs may be harder because of the presence of impurities and associated minerals, such as dolomite, calcite, tremolite, and quartz. The electron micrograph in Figure 1 illustrates the lamellar structure of a relatively pure Montana talc.

Crude talcs range in color from white to green and brown. After being ground to powder form, pure talcs exhibit varying degrees of white. A typical sample of high quality talc gives a reading of 90–95% on the General Electric brightness scale. High quality talcs in oil or plasticizer suspensions range from 40 to 60% reflectance, and are blue-white to yellowish in cast.

The refractive index of talc is 1.54–1.59, and specific gravity is 2.7–2.8. Pure talc is heat-stable up to 900°C. Differential thermal analysis shows a main peak at 900–1000°C, which represents an endotherm caused by loss of the chemically combined water of the talc mineral. A minor peak at 710°C may be present, corresponding to the loss of CO_2 from a small amount of dolomite.

Figure 2 illustrates the structure of pure talc mineral. It consists of a brucite sheet sandwiched between two silica sheets, to form talc layers that are superimposed indefinitely. Each layer is electrically neutral; adjacent layers are held together only by weak van der Waals forces. The characteristic slippery property of talc results from these layers sliding over one another.

Talc is inert in most chemical reagents, although it exhibits a marked alkalinity (typically pH 9.0–9.5). It is, however, soluble in hot concentrated phosphoric acid. Its cation-exchange capacity is also very low, on the order of 2.4 meq/100 g for a finely ground talc (17.6 m^2/g surface area).

The hydrophobic–organophilic nature of lamellar or platy talcs is of importance in a number of applications.

Figure 1. Lamellar structure of Montana talc.

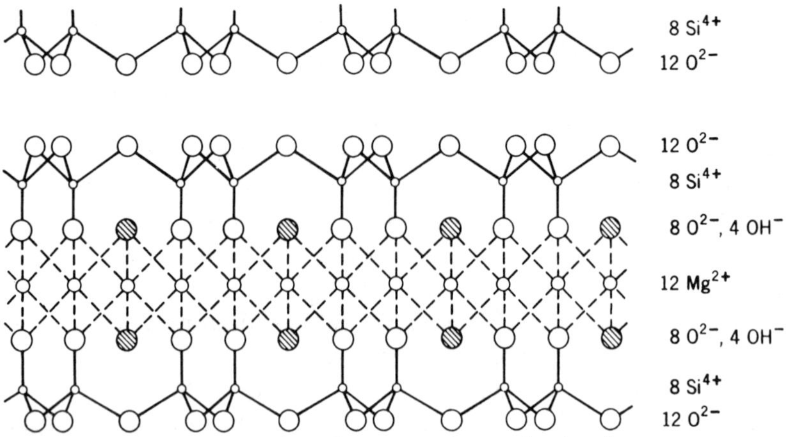

Figure 2. Crystal structure of pure talc mineral.

Mining and Processing

Mining methods vary from straightforward open-pit operations to complicated, carefully timbered undergound operations. Where color and chemical purity are important, screening, washing, and selection by hand or electron beam may be used.

Traditional processing involves crushing with jaw or gyratory crushers; rotary or flash drying, if necessary; and milling with roller-type or ball/pebble mills. Maximum particle size is controlled by air classification. The product is collected in cyclones. For dust control, baghouse dust collectors are used (see Air pollution control technology). These materials are normally ground to a specification of ≥99% through a 74 μm (200 mesh) sieve or ≥99% through a 44 μm (325 mesh) sieve. An increasing proportion of talc is produced to subsieve specification, ie, 30 μm (ca 500 mesh) or below. Some materials are reduced to a maximum particle size of 6 μm (ca 2250 mesh) in fluid-energy equipment using superheated steam. Such products have a mean surface diameter of <1 μm and specific surfaces approaching 20 m^2/g. The term Mistron, and related expressions such as Mistron Vapor and Mistron Frost, are trademarks of Cyprus Mines Corporation that describe such ultrafine talcs.

There are a number of methods that reduce talc to subsieve particle size. In most cases, these depend on high energy input, such as fluid energy using superheated steam or ultrahigh-speed vertical hammer mills. Air classification with or without mechanical rotors is used to remove oversized particles from conventionally milled talc. Classification products may not be the same as ultrafine grinds, since different particle size distribution and specific surface result from the two approaches.

Upgrading deposits of marginal quality to products suitable for high grade uses by froth flotation and acid leaching has increased substantially in the past ten years.

All of Finland's 300,000 t/yr is processed via flotation (qv) to products suited for paper pigments. Similar systems are employed by two Canadian producers. Perhaps the most advanced system is that employed by Windsor Minerals to produce high quality cosmetic talc from relatively low grade Vermont deposits.

Platy talc is relatively easy to float. Separation from platy impurities, such as chlorite and mica, however, may cause problems. The usual solution is larger cell capacity, less agitation, and lower solids content than might be seen in froth-flotation plants processing other minerals.

A typical flow sheet for the processing of platy talc is shown in Figure 3.

Economic Aspects

The world total of talc business may surprise those who think of talc only as face powder base.

The U.S. Bureau of Mines reports world production of the talc group at ca 6.9 $\times 10^6$ t (pyrophyllite included) in 1979 (see Table 1). A good estimate of value of this tonnage would be $434,800,000 (1–2,8–9).

Prices vary widely, depending on chemical purity, fineness of grind, and color. Roofing-grade talc (ground and packed in bags) from Vermont or Texas sells for less than $22/t. Ultrafine talcs from Montana and cosmetic talcs from Italy command more than $220/t at their source.

As indicated earlier, traditional production and market patterns have been affected by the asbestos health controversy to certain types of industrial talc that contain tremolite and other amphiboles as impurities.

Substitution of competitive white minerals such as wollastonite, kaolin, and

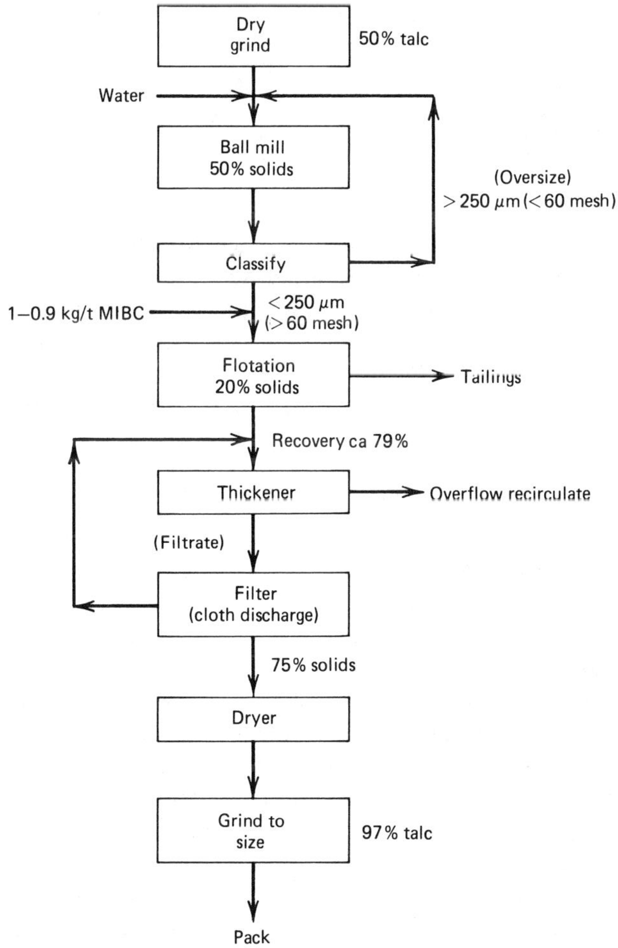

Figure 3. Typical flow sheet for low grade platy talc (MIBC = methylisobutylcarbinol).

Table 1. World Production of Talc and Soapstone

Source	10^6 metric tons[a]	10^6 \$	Principal markets	Applications
North America	1.2	99.9	world	ceramics, paint, paper, cosmetics, plastics
Western Europe	1.0	86.5	world	paper, ceramics, paint, cosmetics
Asia, Oceanic	1.3	105.2	eastern Asia, Europe	paper, paint, ceramics, cosmetics
South America	0.3	18.8	local	roofing, insecticides, rubber
others (including USSR)	0.5	46.6		
pyrophyllite (all countries)	2.6	77.8		
Total	*6.9*	*434.8*		

[a] Approximate values.

calcium carbonate in a number of applications has slowed the growth of U.S. industry. Where talc is an essential ingredient, users have shifted to asbestos-free grades, often at higher costs.

Specifications and Standards

Most talcs have brand names that define the properties of interest to the user. Differences in source and type of raw material, particle shape, impurity limits, fired and unfired color, particle-size distribution, surface area, and oil and water adsorption distinguish various brands. There are a number of more general talc specifications. The *U.S. Pharmacopoeia XX* specifies impurity limits (10). The Toilet Goods Association Specification No. 10 (Talc) has its own version; ASTM Standard D 605-69 has been approved for use by agencies of the Department of Defense for paint (11).

Analytical and Test Methods

Chemical analyses for talc are routinely conducted by means of x-ray fluorescence. X-ray crystallography (diffraction) shows characteristic patterns for talc and its principal impurities. Mineralogical analysis by this method is the principal means of evaluating large numbers of samples such as may be generated by an exploration program. Electron microscopy, including selected area electron diffraction, is often used to identify crystal shape and to determine mineral content of impurities related to asbestos that may be present. Particle-size distribution is normally determined by sedimentation methods; for many applications, Hegman gauge readings provide an excellent and quick measurement of maximum particle size. Specific surface is determined by techniques based on BET nitrogen-gas adsorption.

A wide range of special tests identifies properties of importance to consumers, including oil absorption, color, bulk density (fluffed and packed), abrasiveness, and pitch adsorbency.

Health and Safety Factors (Environmental)

Talc is subject to regulation for respirable dust concentration in work places by OSHA and the Mine Safety and Health Administration. Threshold limit value for pure talc is 2 mg talc per m^3 air (12).

The TLVs for commercial talc containing deleterious impurities such as free silica are lower. Those containing fibrous asbestos impurities, eg, chrysotile, anthophyllite, and tremolite, are regulated as asbestos with a TLV of two fibers of >5 μm in length per cm^3 air (see also Asbestos).

The FDA has jurisdiction over talc used for cosmetic and pharmaceutical purposes, and pure talc is generally regarded as safe for these uses.

There have been numerous investigations of, and publications on, the alleged carcinogenic effects of commercial talcs (13). It appears that most of the controversy is the result of work done on the effects of New York state tremolitic talcs on workers (14–15). Although there are no clinical data to suggest that pure talcs represent any greater occupational hazard other than that of a nuisance dust, they have also been affected by the same controversy.

Since talc appears in cosmetics (qv), pharmaceuticals, and foods, concern has also been expressed about its safety in these uses. The U.S. government's official position was last expressed on Jan. 11, 1979, in a letter by the FDA Commissioner to the Public Citizen Health Research Group (16): ". . . There is to date no conclusive evidence that pure talc is carcinogenic in men or animals."

Environmental Considerations. All open-pit mines require dust-suppression measures such as wet drilling and haul-road watering. Mine-equipment operators are customarily protected from dust by enclosed air-conditioned cabs.

Water quality and disposal is not a problem except in flotation operations, where holding and treatment ponds are used for process tailings. Reclamation of mined areas and waste dumps have become a normal part of good mining practice.

Dry milling operations result in air pollution by discharge of particulate matter unless controls are provided. All large plants provide baghouse dust collection. In some cases, high temperature collectors are required. A well-run talc processing plant has no visible discharge of particulate matter.

Many of the processes used to produce ultrafine talc are energy intensive. Fluid energy mills with superheated steam may require as much as 6 GJ (5.7×10^6 Btu) per ton of finished product.

Uses

The paper industry is the largest consumer of talc (9). In Europe and Japan, talc is a basic raw material, used to enhance opacity and printing properties. In the United States, which has abundant sources of inexpensive kaolin for this purpose, talc is used to control pitch and other oleoresinous impurities in the pulp and papermaking process. Ultrafine, high purity, platy-particle-shaped talcs of high specific surface selectively adsorb these impurities and prevent their deposition on pulp and papermaking machinery (see Paper; Pulp).

The ceramics industry has been the largest consumer in the United States of talc-group minerals for more than three decades. Virtually all of the ceramic wall tile produced in the United States contains a large percentage (up to 65%) of talc (usually tremolitic) or pyrophyllite. Talc prevents delayed glaze crazing, lowers firing temperatures, and reduces fired shrinkage (17).

In Europe and Japan, talc is used in ceramics; electrical ceramics and steatite applications, however, outweigh wall tile. Talcs of high chemical purity and granular particle structure are the bases for low-loss insulators used in a wide variety of electronic and electrical uses (see Ceramics as electrical materials).

The next most important application is in protective coatings. Talc improves exterior durability and controls viscosity and brushing and gloss properties; it also reduces paint-formulation costs.

Other uses of talc are as insecticide carriers, rubber-dusting and textile-filling materials, and as an additive in asphalt roofing compounds. These applications account for large tonnage, but a relatively small proportion of total value, since only low quality talcs are employed.

More recently, the use of talc for plastics filling and reinforcement has grown rapidly (18) (see Fillers). Polypropylene parts reinforced with as much as 40% talc have replaced metal in many automotive applications as manufacturers have strived for weight reduction and improved gasoline mileage. Talc lends resistance to heat distortion and therefore is especially important in under-the-hood uses such as fan shrouds. Polypropylene replacements for metal in domestic applications represents another important new market for talc (see Engineering plastics). It is estimated that polypropylene uses consume >40,000 t of talc per year in the United States, and Europe and Japan each use >50,000 t/yr. Growth should be more rapid than that of the industries served as new applications are found.

Catalytic converters for automotive pollution control have also become important talc consumers since the mid-1970s. The honeycomb substrate requires raw materials of exacting specifications with respect to mineralogical purity and particle-size distribution. Worldwide volume of talc usage in this application is estimated at 20,000 t/yr (see Exhaust control, automotive).

BIBLIOGRAPHY

"Talc" in *ECT* 1st ed., Vol. 13, pp. 566–572, by Hans Thurnauer, American Lava Corporation; "Talc" in *ECT* 2nd ed., Vol. 19, pp. 608–614, by H. T. Mulryan, United Sierra Division, Cyprus Mines Corporation.

1. *Minerals Yearbook*, U.S. Dept. of Interior, Bureau of Mines, U.S. Government Printing Office, Washington, D.C., 1980.
2. R. A. Clifton, *Mineral Facts and Problems*, Bulletin 671, U.S. Bureau of Mines, Washington, D.C., 1980, pp. 1–12.
3. M. Ross, W. L. Smith, and W. H. Ashton, *Am. Mineral.* **53,** 751 (May–June 1968).
4. *Ind. Miner. Mag.*, 40 (Dec. 1977).
5. *Ind. Miner. Mag.*, 25 (July 1977).
6. G. W. Brindley and I. S. Stemple, *J. Am. Cer. Soc.* **43**(1), 34 (1960).
7. J. A. Pask and M. F. Warner, *J. Am. Cer. Soc.* **37**(3), 118 (1954).
8. *Ind. Miner. Mag.*, 40 (Sept. 1979).
9. *The Economics of Talc and Pyrophyllite*, 3rd ed., Roskill Information Services Limited, London, 1981.
10. *The United States Pharmacopeia XX*, The United States Pharmacopeial Convention, Inc., Rockville, Md., 1980.
11. *ASTM D 605-69*, American Society for Testing and Materials, Philadelphia, Pa., 1976.
12. American Conference of Governmental Industrial Hygienists, *Threshold Limit Values (TLVs) for Chemical Substances in Workroom Air Adopted by ACGIH for 1980*, 1980.

13. G. F. Rubino and co-workers, *J. Occup. Med.* **18**(3), 186 (1976); personal communications, G. F. Rubino, Institute of Occupational Medicine, Turin University, Turin, Italy.
14. M. Kleinfeld, J. Messite, O. Kooyman, and co-workers, *Arch. Environ. Health* **14**, 633 (1967).
15. M. Kleinfeld, J. Messite, and M. H. Zaki, *J. Occup. Med.* **15**, 345 (1974).
16. Personal communication, D. Kennedy, Commissioner of Foods and Drugs, FDA to S. M. Wolfe and B. Gordon, Jan. 11, 1979.
17. W. W. Gaskins, *Am. Cer. Soc. Bull.* **31**(10), 392 (1952).
18. J. A. Radosta and N. C. Trivedi, *Handbook of Fillers and Reinforcements for Plastics*, 1978, pp. 160–171.

H. T. MULRYAN
Cyprus Industrial Minerals

TALL OIL

Tall oil [*8002-26-4*] is a natural product of pine trees that is isolated by means of the Kraft pulping process. It was discovered in the late 1800s in Sweden as a product of the acidification of the soapy froth by-product of Kraft pulping of pine wood. Uses for the crude product developed slowly. Its odor and complex nature denied its direct substitution for existing products. Many of the archaic uses for tall oil are detailed in earlier literature (1–2). Current applications are increasingly based on specific chemical functionality of refined rosin and fatty-acid cuts produced by fractional distillation of crude tall oil.

Tall oil is composed of the ether extractable, nonlignin, noncellulosic portion of the pine tree. The two main chemical structures represented are rosin acids and fatty acids. The rosin acids are diterpene carboxylic acids based on an alkyl-substituted perhydrophenanthrene ring structure (see Terpenoids). These acids are nonfood-chain acids and are involved with wound-recovery mechanisms of the tree. The fatty acids (see Carboxylic acids, fatty acids from tall oil) are predominately 18-carbon, straight-chain mono- or diunsaturated fatty acids; oleic and linoleic acids predominate. Approximately 185,000 metric tons of fatty acids was produced in 1980 in the United States. The tall-oil content of a pine species is generally 3% of the dry wood and can be estimated by extraction of the bark-free wood substance of the tree with organic solvents, eg, an alcohol–benzene mixture (3–5).

Composition and Properties

Since its value is minuscule relative to the pulp produced ($6.60–8.80 of tall oil per metric ton of pulp, with pulp sold at $330–550/t), little care is exercised to collect all the tall oil available or to implement changes requiring capital investment to produce a higher quality product. For every metric ton of southern pine pulp produced, ca 30–40 kg of crude tall oil is produced. No crude tall oil is produced from pulping of hardwoods. The southeastern United States produces the most valuable crude tall oil because of its uniformity and composition. Tables 1 and 2 show typical wet analyses for a good southern Kraft tall oil and typical gas liquid chromatographic analysis of the fatty-acid components, respectively.

The tall-oil fatty acids produced exhibit the chemistry expected of fatty acids derived from any vegetable-oil source. These linear, long-chain, unsaturated acids are pale yellow liquids at ambient temperature. They are insoluble in water and soluble in most organic solvents. Their chemistry is centered in the aliphatic carboxyl group and their ethylenic unsaturation. The chemistry is well described in references 8–10.

Physical properties of many of the rosin isomers are given in ref. 9. Commercial tall-oil rosin consists predominately of a mixture of abietic and pimaric acid isomers. Typical tall-oil rosin has a ring structure and a ring-and-ball softening point of 80°C. It is an amber-colored glass and is insoluble in water and soluble in most organic solvents and many nonpolar elastomers and polymers. Typical chromatography analyses of tall-oil rosin are shown in Table 3.

Table 1. Typical Analysis of Southern Crude Tall Oil[a]

Property	Typical value	Range
acid no.	161	158–170
rosin acids, wt %	42	33–47
unsaponifiables, wt %	8.5	6.6–9.7
fatty acids, wt %	48	42–55
water, wt %	0.8	0.04–2.8
toluene insolubles, wt %	0.1	0.04–0.4
color (Gardner)	15	12–18

[a] Ref. 6.

Table 2. Results of a Typical Gas Liquid Chromatography Analysis of ASTM Type II Tall-Oil Fatty Acids[a]

Analysis	Value	Analysis	Value
Fatty acid content, area %		*Wet analysis*	
palmitic	0.1	acid value	197
palmitoleic	0.3	saponification number	199
stearic	2.1	iodine value	130
oleic	48.5	rosin acids, wt %	0.9
linoleic		unsaponifiables, %	1.3
cis-9, *cis*-12	35.3	total fatty acid, %	97.8
cis-9, *trans*-11	4.7		
trans-9, *trans*-11	3.1		
pinolenic (Δ-5,9,12)	3.5		
eicosenoic	1.1		
other	1.3		

[a] Ref. 7.

Table 3. Typical Gas Liquid Chromatographic Analysis of Tall-Oil Rosin

Rosin acid	Area %
9,10-secodehydroabietic acid	0.1
pimaric acids	4.6
tetrahydroabietic acid	0.9
dihydroabietic acid	4.7
palustric/levopimaric acid	10.1
isopimaric acid	5.6
abietic acid	35.4
dehydroabietic acid	27.4
neoabietic acid	2.0
others	9.2

Processing

In practice, tall oil is produced as a by-product of Kraft pulping of pine trees. Pine wood chips are digested in aqueous liquor containing sodium hydroxide and sodium sulfide at 170°C for two hours (11). This digestion delignifies the wood and produces cellulose pulp, sodium soaps of rosin acids, fatty acids, and lignin degradation products. Cellulose is stable under these conditions and remains slurried in the black liquor. When the pulping reaction is complete, pulp is separated from the black liquor and washed. The pulp can be used as such or it is bleached or otherwise upgraded (12).

The black liquor contains valuable inorganic chemicals and must be recovered for economic and environmental reasons. Excess water is evaporated and a curdy soap containing all tall-oil components is salted out and floats to the top. This black-liquor soap is skimmed and is acidulated to produce water-insoluble crude tall oil; an interfacial layer consisting of tall oil, pulp fines, calcium sulfate, and a small percentage of lignin; and a waste acid layer.

Essentially, all of the by-product crude tall oil produced today is refined by fractional distillation to produce tall-oil fatty acids, tall-oil rosin, distilled tall oil, tall-oil heads, and tall-oil pitch. Each of these fractions is used in a wide variety of applications.

The processes of this industry are collection of black-liquor soap, conversion of soap to crude tall oil, fractional distillation to separate rosin and fatty-acid cuts, and upgrading of refinery cuts. Acid refining of crude tall oil, although historically important, now accounts for only 2% of the crude tall oil produced (1–2).

Black-Liquor Soap Recovery. After digestion of pinewood in the mixed sodium hydroxide–sodium sulfide solution, the insoluble cellulose fibers are separated from the black liquor, which contains solubilized lignin, rosin-acid soaps, fatty-acid soaps, unsaponifiables, and inorganic chemicals. Recovery of the latter for recycle and pollution regulations are the economic incentives for evaporation of the liquor to a high solids content so that the residue can be burned. Combustion produces an inorganic sodium carbonate and sodium sulfide smelt. The latter is reconstituted as Kraft cooking liquor.

During the evaporation at ca 23–32 wt % black-liquor solids, the hydrophobic rosin and fatty-acid soaps are salted out. As they separate, they extract oil-soluble unsaponifiables from the liquor and float to the top of the vessel. Approximately 0.3–1.0

m of soap builds up and is collected by skimming (13). Black-liquor soap is soluble in black liquor to ca 0.4 wt %. All soap present in excess of this amount should be collectable. A narrated slide presentation describing recommended soap-collection technology is available from the Pulp Chemicals Association (14).

Since 1970, the market value of black-liquor soap has increased dramatically, and operators of paper mills have become more interested in collecting all available soap. Substantial yield increases have been demonstrated in specific mills by redesign of skimming vessels to reduce turbulence, addition of air to float the soap, and separation of pinewood black liquor from hardwood black liquor which contains no soap (15). Of recent interest is the use of an electric charge to flocculate very small soap particles which have a very low sedimentation rate (16). This equipment increases particle size and, consequently, the rate of rise. At specific mills, this technique works well; at others, it has not been effective.

A good soap-collection system requires good engineering design throughout and close supervision by paper-mill personnel. Once collected, the soap can be acidified at the pulp mill or it can be shipped to a regional facility. Soap quality (value) generally is measured by fatty-acid, rosin-acid, unsaponifiables, insolubles, and black-liquor contents. The composition of a typical southern Kraft black-liquor soap is shown in Table 4.

Essentially no research and development is being directed toward improved soap recovery; collection technology appears to be adequate. Most improvements in collection yields will be accomplished as equipment is updated and existing engineering principles employed.

Research on soap refining prior to acidulation is being conducted. In Scandanavia, solvent extraction with acetone–hexane has been implemented commercially to reduce the high unsaponifiable content (20 wt %) of low grade, mixed birch–pine black-liquor soap. In addition to improvement of the soap value, a valuable sitosterol by-product can be obtained during recovery of the solvents for recycle (17). Application of this technique for upgrading all high unsaponifiables containing crudes is likely by 1995.

Brine washing of crude black-liquor soap is also being studied. Generally, sodium

Table 4. Black-Liquor Soap Composition

	Southern Kraft[a]		Northern Kraft[b]		
	Average	Range	Average	Range	Scandinavian[b]
Based on black-liquor soap					
crude tall oil, wt %	54.2	52–58	52	28–61	
filterable solids, wt %	0.9	0.3–1.1			
black liquor[c], wt %	5.0	5–10			
Based on crude tall oil produced[d]					
acid number	161	147–172	142	122–152	132
unsaponifiables, wt %	9.1	7–12	16.2	11–22	20
rosin acids, wt %	41.5	29–54	26.4	22–49	23

[a] Values based on multiple samples from 6–10 mills, unpublished data.
[b] Mixed pine–birch black-liquor soap (17).
[c] Black liquor is strongly alkaline and requires increased sulfuric acid for acidulation if not properly drained.
[d] Gardner color 14–18.

sulfate brine from crude tall-oil production is mixed with crude black-liquor soap and is separated. The technique reduces the black-liquor, pulp-fines, and soluble-lignin content of the soap, and a more concentrated soap of improved quality is produced. Economics of implementation determine the commercial fate of this technology.

Black-Liquor Soap Acidulation. Black-liquor soap, containing ca 50 wt % tall oil, is converted to crude tall oil by acidification with sulfuric acid. The chemistry is rudimentary. Sodium soaps of rosin and fatty acids react with sulfuric acid producing sodium sulfate and free rosin and fatty acids:

$$2\ RCOONa + H_2SO_4 \rightarrow Na_2SO_4 + 2\ RCOOH$$

where R represents a rosin or fatty-acid alkyl group.

Since the acidified soap is a three-phase mixture, three layers form: a top crude tall-oil layer; an intermediate layer made up of pulp, lignin, tall oil, and a brine emulsion; and a lower brine or waste acid layer. The pH of the final reaction mixture is regulated at ca pH 3 so that all of the carboxylic acids are in the acid form. The commercial process must be carefully controlled if a high yield of uniform product is to be produced.

Although the batch system is still used, the newer commercial systems are continuous (1–2). Drained black-liquor soap from storage is pumped to an in-line mixer where it is combined with dilute (30–40 wt %) sulfuric acid. Either virgin or bleach-plant sulfuric acid may be used. Bleach-plant acid is the residual sulfuric acid from the chlorine dioxide generators in pulp mills. It is normally ca 30 wt % sulfuric acid and contains some impurities, eg, chloride ion from the generation of chlorine dioxide from sodium chlorate. Approximately 200–250 kg (100 wt %) sulfuric acid is required per metric ton of soap neutralized.

The acidification of the sodium salts is essentially instantaneous at 90–95°C. From the mixing section, the reaction mixture is pumped to a 1500-L reactor, where the reaction is completed and mixed calcium and sodium sulfate salts and lignin separate. The reaction mixture overflows to a set of screens or through a filter where the pulp fines and a minor amount of lignin solids are separated, washed with recycled brine, and discarded, typically to landfill.

The tall-oil–sodium sulfate brine mixture then is pumped either to a decanter or to a centrifuge for oil separation. This step is critical since the volume of brine is about three times that of the oil and a loss of 1% oil in waste brine represents 3% of the potential tall oil. Once separated, the crude tall oil is pumped to the washing stage and the waste brine (pH 3–5) is pumped to the paper mill for recycle in a recovery furnace. Alternatively, waste brine can be processed through a sewage plant; however, this increases the cost since no credit is received for heat and chemicals recovered in the paper mill.

The crude tall oil is then washed with hot water or dilute brine for removal of residual acid and inorganic salts. If no washing is done, inorganic materials pass into the distillation step where they may deposit in the pitch stripper or be removed in the pitch.

Continuous soap plants generally process 45–135 t/d of soap. Efficient operation depends on good maintenance, supervision, attention to detail by the operators, and uniformity of the soap used.

Soap-acidulation plants are expected to be continuous when designed to produce 45–135 t/d of soap. The choice of decanter or centrifuge methods for oil–brine separation will probably be equal. Economics of very small plants, ie, those serving one or two mills, favor batch systems. As freight costs and the value of recovered sodium sulfate increase, the economics of small acidulation plants at individual paper mills will improve.

Refining. In the early 1900s, the Scandinavians began to recover minor amounts of crude tall oil (1–2). Initial distillations produced a usable, although dark, fatty-acid fraction with drying properties satisfactory for inexpensive paints and a rosin containing ca 30 wt % fatty acids. Acid refining of crude tall oil (CTO) was also practiced (18). The CTO was dissolved in hydrocarbon solvent and was mixed with sulfuric acid. Very little acid-refined or distilled tall oil is produced today.

Separation of rosin from fatty acids in crude tall oil became more important as increased uses for the separate fatty-acid and rosin fractions developed (see Uses). The first United States distillation units were put on-stream in 1930–1935. In 1949, the first fractional-distillation plant was constructed and operated based on fractionation techniques of the petroleum industry. These resulted in the effective separation of rosin and fatty acids so that each fraction could be used as a chemical entity. Light-colored fatty-acid cuts containing <2 wt % rosin and rosins containing <3 wt % fatty acid were produced. The entire industry rapidly converted to this technology.

Commercial Fractional Steam–Vacuum Distillation. There are eight refiners of crude tall oil in the United States, and their processes are based on fractional distillation (see Table 5). These producers fractionated >6.55×10^5 t of tall-oil fatty acids, 1.9×10^5 t of tall-oil rosin, and 4.7×10^4 t of distilled and acid-refined tall oils. The remainder is the less valuable by-products, pitch and heads. Historically, tall-oil refinery product prices have varied widely with fatty-acid prices being influenced by the price of vegetable oils (soybean oil, etc) and rosin prices being influenced by availability and price of wood and gum rosins. Since tall-oil fatty acids and rosin acids were introduced as low priced substitute materials, tall-oil producers were constantly under pressure to upgrade their products and reduce the costs of their processes. The largest current concern is the cost of energy required for distillation.

Typical tall-oil refineries are described in numerous publications and patents; an excellent overall summary through 1965 is given in ref. 1. The steps of the refining process are removal of pitch in a short tower essentially without fractionation to produce distilled tall oil, gross separation of fatty acids and rosin in a rosin tower to

Table 5. Tall-Oil Fractionators [a]

Company	1980 Capacity[b], 1000 t/yr
Arizona Chemical Company	136
Georgia Pacific	[c]
Hercules	210
Monsanto-Emery	59
Reichhold	90
Sylvachem	90
Union Camp	95
Westvaco	127

[a] Ref. 19.

[b] Excess capacity is ca 20%.

[c] On-stream late 1980, 35,000 t/yr stated capacity.

produce tall-oil rosin and distilled fatty acids, and additional fractional distillations and refining of the fatty acids to produce fractions of the required quality. Distillation is continuous and generally under steam vacuum with plants being operated in a straight-through or blocked (intermediate stored prior to final distillation) fashion.

Modern fractional-distillation towers are constructed of stainless steel and generally are heated with liquid or vaporized Dowtherm liquids. Steam is present throughout the distillation to lower the boiling point of the tall oil and to prevent anhydride formation and decomposition at the high refining temperatures. Towers generally operate at 200–285°C and ca 13.3 kPa (100 mm Hg). A vacuum is generated by barometric condensers.

In a typical system, dry crude tall oil is fed from storage containers through a heat exchanger into the midsection of the pitch tower. On entry to the tower, the crude tall oil vaporizes rapidly. Most of the alcohols present, ie, sitosterol, and C_{22}, C_{24}, and C_{26} wax alcohols, are esterified with fatty acids. Because of the high boiling points of the esters, they condense and are left as a residue or pitch. Lignin, dimerized rosins, and fatty acids also are removed in the pitch fraction. The very highly volatile components, ie, hydrocarbons, sulfur compounds, etc, and noncondensable gases are purged at the top of the tower and the valuable distilled tall oil is removed as a vapor or liquid stream from the pitch tower. Required heat is provided by a reboiler at the base of the pitch tower. Steam can be introduced in the reboiler and to the preheated crude tall-oil feed.

The distilled tall oil produced can be condensed and heated or clay-treated prior to further fractionation, but generally, to conserve energy the vapor or hot distilled tall oil is fed directly to the rosin tower where fatty acids and rosin are roughly separated.

The rosin tower consists of several sections. The lower section serves to separate the highest boiling component rosin. Vaporized feed equilibrates with liquid on the plates, and the rosin descends the tower and is removed as a liquid just above the reboiler or at the base of the tower, depending on the rosin quality desired. If the rosin is removed as a sidestream, the rosin dimer content is minimized, but the product is prone to crystallization.

Above the feed point, fatty acids are effectively separated as vapor from any polymeric material in the feed and are removed near the top of the tower as a distilled fatty acid of good color (Gardner 6 is typical) and low (4–6 wt %) rosin content. The very top or head section of the tower is used as a condensing section which controls the reflux ratio and separates a heads product, ie, the tall-oil heads, which is high in unsaponifiables and low boiling fatty acids.

The distilled fatty acid is then refined further, either in a continuous or a blocked operation. The least pure fatty-acid products, which are commonly sold, go through one more distillation whereas the highest purity materials are distilled as many as three times. After preparation of the distilled fatty acid, a variety of sequences is used, including chemical and physical treatments and clay bleaching as required to produce the fatty acid of the required quality. In these subsequent distillations, an entire tower may be used to remove low boiling impurities as heads, or rosin and polymeric materials as bottoms. These techniques are proprietary.

Standard specifications for tall-oil fatty-acid grades have been established by ASTM (20). Typical characteristics for the various grades of tall-oil fatty acids are shown in Table 6.

Table 6. Tall-Oil Fatty-Acid Grades

	ASTM type[a]	Acid value	Rosin acids, wt %	Unsaponifiables, wt %	Fatty acids, wt %	Gardner color	Iodine value
Acintol FA-1 tall-oil fatty acid[c]	III[b]	194	4.5	2.7	92.8	5	131
Acintol FA-2 tall-oil fatty acid[c]	II[b]	197	0.9	1.3	97.8	≥3	130
Acintol FA-3 tall-oil fatty acid[c]	I[b]	198	0.5	0.7	98.8	≤3	130

[a] Ref. 20.
[b] Essentially all tall-oil producers make products that fit into each of the three ASTM grades. For example, Grade III: Pamak 4, Hercules; Grade II: Unitol AFL, Union Camp Corp.; and Grade I: Aconew 450, Reichhold.
[c] Ref. 7.

Other Distillation Technologies. In addition to the standard steam–vacuum distillation technique, crude tall oil has been processed commercially in a horizontal still called a raining-film still (21). Currently, no tall oil is distilled commercially by this method. A newer technique that is practiced commercially in Europe and was developed by the Luwa Corporation involves the distillation of tall oil without the use of steam under high vacuum (22). This technology is reported to result in reduced energy costs. Equipment limitations and corrosion problems may preclude this process in existing plants.

Government Regulations

The Air Quality Act, which was passed in 1967 with important revisions in 1970 and 1977, requires that air effluents meet certain standards of quality. The requirements have been met by the tall-oil industry with the addition of scrubbers, precipitators, low sulfur fuels, and other air-pollution control equipment. Although a cost factor, air-pollution control is not a technological problem.

Since passage of the Toxic Substance Control Act on Jan. 1, 1977, an inventory has been created and all commercial substances manufactured in the United States have been assigned identification numbers. Toxic Substance Control Act Inventory numbers for the derivatives of the principal tall-oil products and upgraded products are listed in refs. 23–24. This regulation involves additional costs and tends to inhibit innovation (25).

The Resource Conservation and Recovery Act of 1976 initiated regulated handling of all solid wastes. Required hazardous waste management standards, handling, storage, disposal permits, and record keeping are detailed in the Federal Register (26). Production of primary tall-oil products does not generally result in generation of hazardous wastes. Consequently, minimal additional regulatory cost has been added by this legislation.

The Federal Water Pollution Control Act Ammendments of 1972 established the requirements for and issuance of permits to discharge aqueous waste streams into receiving waters (27). Subsequently, the EPA investigated the Gum and Wood Chemicals Manufacturers Industry and issued effluent-limitation guidelines, pretreatment standards, and new source-performance standards (28). Because of the relatively innocuous nature of the tall-oil plant effluents, these guidelines have been met by all producers. In general, biological oxygen demand, total suspended solids, and pH are limited. In certain plants where metal catalysts are used, there are additional limits on the copper, nickel, and zinc contents of the effluents. Recently these guidelines have been cancelled (29).

Analytical Methods and Standards

The tall-oil industry uses many analytical methods from the naval-stores industry and has developed new ones as needed. Standard methods have been developed by ASTM under the auspices of the D-17 Committee on Naval Stores (30). Other standard methods developed by the American Oil Chemists' Society for fatty acids are applied to tall-oil fatty acids (31). The important analytical methods are listed in Table 7.

Uses of Fatty Acids, Rosin, and Refinery By-Products

One unique use of tall-oil fatty acids is for the production of dimer acids; the ratio of unsaturated acids and the low saturated-acid content results in an optimum raw material. Approximately 32,000 t/yr of dimer acids is produced. A large percentage of this is used to produce polyamides used as adhesives (qv). Other uses of these fatty acids are varied with substantial amounts being sold to the mining, rubber, surface-coating, plasticizer, and oil-production industries. The chemistry and utilization of tall-oil rosin is more complex than that of the fatty acids. Many applications depend on properties difficult to measure quantitatively and difficult to correlate to final product use.

Table 7. Standard Methods of Analysis[a]

Analysis	Methods: Tall-oil rosin acid	Methods: Tall-oil fatty acid
acid number	ASTM D 465-59	ASTM D 803-65 AOCS 1a 64T
color	USDA color standards	ASTM D 1544-68
fatty acids, wt %	ASTM D 1585-63	ASTM D 803-65[b]
iodine value	[b]	ASTM D 1959-61 AOCS Tg 1a-64T
rosin acids <15 wt %		ASTM D 1240-54 AOCS TS 1a-64
rosin acids >15 wt %	ASTM D 1585-63	ASTM D 803-65
saponification value	ASTM D 464-59	ASTM D 803-65 AOCS Tl 1a-64T
softening point	ASTM E 28-67	
unsaponifiables	ASTM D 1065-56	ASTM D 803-65 AOCS Ca 66-53
rosin-acid composition	ASTM D 3008-72	
fatty-acid composition		ASTM D 1983-69 AOCS Ce 1-62

[a] Refs. 30–31

[b] Not applicable in presence of high rosin or unsaponifiables.

Tall-oil rosin was introduced in the 1950s as a low cost substitute for gum and wood rosin. Its poor odor and tendency to crystallize prejudiced its use as a direct replacement. For paper size, it was an excellent low cost replacement for wood rosin, especially in fortified sizes (see Papermaking additives). This use has grown dramatically, and it is now the predominant rosin used for paper size. Approximately 5.5 $\times 10^4$ t/yr was sold in the United States to this industry in 1980. As tall-oil rosin properties improved with manufacturing innovation and gum and wood rosin became less available, tall-oil rosin was applied to all the principal uses of these rosins; ca 20.5 $\times 10^4$ t of tall-oil rosin was produced in 1980 in the United States. The ability of rosin to produce tack in other materials led to its use in the rubber and adhesives industries (see Rubber compounding). Upgraded products are expected to be developed for use in the polymer and adhesives industries.

The by-products produced during tall-oil refining are tall-oil pitch, tall-oil heads, and distilled tall oil. Tall-oil pitch is a dark thermoplastic material and is used in asphalt emulsions, low performance mastics, and as a fuel (41.8 MJ/kJ or 18,000 Btu/lb). Tall-oil heads contain lower boiling fatty acids generally with ca 16 carbons. Heads can be used as inexpensive fatty acids where carboxyl functionality is required or as a raw-material source for production of palmitic acid. Distilled tall oil has many uses where its liquid nature and its mixed rosin and fatty-acid content result in favorable product characteristics.

BIBLIOGRAPHY

"Tall Oil" in *ECT* 1st ed., Vol. 13, pp. 572–577, by R. H. Stevens, Herty Foundation; "Tall Oil" in *ECT* 2nd ed., Vol. 19, by Dan C. Tate, U.S. Plywood-Champion Papers Inc.

1. L. G. Zachery, H. W. Bajak, and F. J. Eveline, *Tall Oil and Its Uses*, F. W. Dodge Co., a division of McGraw-Hill, Inc., New York, 1965.
2. J. Drew and M. Propst, *Tall Oil*, Pulp Chemicals Association, New York, 1981.
3. P. Kock, *Utilizing Southern Pines I*, USDA Forrest Service, U.S. Government Printing Office, Washington, D.C., 1972, pp. 202–214.
4. A. H. Connor, M. A. Diehl, and J. W. Rowe, *Wood Sci.* **12**(4), 194 (1980).
5. D. F. Zinkel and D. D. Foster, *Tappi* **63**(6), 137 (1980).
6. Unpublished information, Arizona Chemical Co., Fair Lawn, N.J.
7. *Tall Oil Fatty Acids, Composition and Characteristics*, Arizona Chemical Co., Wayne, N.J., 1980.
8. J. C. Johnson, *Fatty Acid Manufacture, Recent Advances*, Noyes Data Corporation, Park Ridge, N.J., 1980.
9. D. Swern, ed., *Bailey's Industrial Oil and Fat Products*, 4th ed., Vol. 1, Wiley-Interscience, New York, 1979.
10. E. H. Pryde, *Fatty Acids*, American Oil Chemists' Society, Champaign, Ill., 1979.
11. P. Kock, *Utilization of Southern Pines II*, USDA Forest Service, U.S. Government Printing Office, Washington, D.C., 1972, pp. 1431–1436.
12. *Ibid.*, pp. 1454–1460.
13. *Pulp Chemicals Association Special Recovery Symposium*, Nov. 13–14, 1979, Orlando, Fla.
14. Pulp Chemicals Association, 60 E. 42 St., New York, N.Y. 10016.
15. S. R. Boyette in *Pulp Chemicals Association Special Recovery Symposium*, Nov. 13, 1979, Orlando, Fla.
16. Uniflox, Unice Electro Flocculator, Unice Machine Company, San Francisco, Calif.
17. K. Ukkonen, *The Crude Soap Refining Process and the Production of Sitosterols from Unsaponifiables of Crude Sulfate Soap*, Oct. 8, 1980, Pulp Chemicals Association, Washington, D.C.
18. Ref. 1, p. 7.
19. Pulp Chemicals Association, *Statistical Summary 1970–1980*, New York, Apr. 1981.

20. *ASTM Standards*, Vol. 29, American Society for Testing and Materials, Philadelphia, Pa., 1975, p. 270.
21. U.S. Pat. 3,496,071 (Feb. 17, 1970), A. Spence (to Spence and Green Chemical Co.).
22. U.S. Pat. 3,644,179 (Feb. 22, 1972), P. Knoer and G. Piek (to Krems an der Donau).
23. *Toxic Substances Control Act, Candidate List of Chemical Substances, Addendum III, UVCB*, U.S. EPA Office of Toxic Substances, Washington, D.C., Mar. 1978, pp. 43–46.
24. *Toxic Substances Control Act, Chemical Substance Inventory*, Vol. IV, UVCB, U.S. EPA Office of Toxic Substances, Washington, D.C., May 1979, pp. 170–173, 190–193.
25. E. F. Frost, G. V. Cox, and P. B. Hutt, *The First Four Years of the Toxic Substances Control Act*, Chemical Manufacturers Association, Washington, D.C., 1981, p. 94.
26. *Fed. Regist.* **45**(98), 33063 (May 19, 1980); *Fed. Regist.* **45**(229), 27524 (Nov. 28, 1980).
27. *Fed. Regist.* **44**(111), 32854 (June 7, 1979).
28. *Fed. Regist.* **44**(231), 68710 (Nov. 29, 1979).
29. *Fed. Regist.* **46**(80), 23729 (Apr. 27, 1981).
30. *Annual Book of ASTM Standards*, Vol. 20, American Society for Testing and Materials, Philadelphia, Pa., 1973.
31. *Official and Tentative Methods of the American Oil Chemists' Society*, American Oil Chemists' Society, Champaign, Ill., 1980.

HERBERT G. ARLT, JR.
Arizona Chemical Company

TANTALUM AND TANTALUM COMPOUNDS

Tantalum [*7740-25-7*], Ta (at no. 73, at wt 180.948, sp gr 16.6), appears in group VA of the periodic table directly below niobium, Nb, with which it is closely associated in nature and to which it is very similar in properties. Tantalum has only one natural isotope, ^{181}Ta [*7440-25-7*], but several artificial radioactive isotopes have been made. Its thermal-neutron capture cross-section is 21.3×10^{-28} m^2 (21.3 barn). The most common valence of the element is 5. Tantalum is best known as a refractory metal with a combination of unique properties that make it useful in a number of unrelated commercial applications. It is very ductile and can be worked cold into fine wire or thin foil. It is completely inert to strong acids below their boiling points and yet is very reactive to almost all substances at high temperatures. Tantalum is described extensively in refs. 1–4. Tantalum was discovered by Ekeberg in 1802. It was reported in 1844 that there were two similar elements in the earth oxides obtained from columbite: tantalum and niobium (5). In 1866, tantalum and niobium were separated by means of the difference in the solubilities of their complex potassium fluorides (6).

The first ductile tantalum was made in Germany in 1902, and tantalum was first used commercially as a filament in electric lamps. About eleven million tantalum lamps were made before tungsten wire replaced tantalum for this application in 1909. From then until 1922, the uses of tantalum were limited to certain dental and surgical instruments. Tantalum was first produced in the United States in the laboratory of Fansteel Inc., N. Chicago, Ill., in 1922 and has been in continuous commercial production since that time. The production and use of tantalum have increased greatly since 1950, and there is no reason to believe this trend should not continue. Evidence of continued interest in tantalum is demonstrated by the formation and subsequent growth of the Tantalum Producers International Study Center (TIC). Since its charter, the TIC has grown to 61 members, representing miners, merchants, producers, processors, consumers, and users. The First International Symposium on Tantalum, sponsored by the TIC, was held in the FRG in 1978 (7).

Tantalum

Material Availability. ***Natural Occurrence.*** Tantalum occurs in a number of oxide minerals (see Table 1), which almost invariably contain niobium (8–10). It also often occurs in association with titanium and tin and sometimes with the rare earths, uranium, and thorium (10). It does not occur naturally in the free state. The most important tantalum-bearing minerals are tantalite and columbite, which are variations of the same natural compound (Fe,Mn)(Ta,Nb)$_2$O$_6$. The mineral is called tantalite [*1310-23-2*] when the tantalum pentoxide [*1314-61-0*] (Ta_2O_5) content exceeds the niobium pentoxide (Nb_2O_5) content and is usually called columbite [*1310-23-2*] when the niobium pentoxide exceeds the tantalum pentoxide content. Whenever columbite contains 20 wt % or more of Ta_2O_5, it is considered to be a tantalum ore. Tantalite–columbite occurs in some pegmatites in quantities that seldom exceed fractions of a percent. The principal countries producing tantalum mineral concentrates are Canada, Brazil, Australia, Thailand, Zaire, Nigeria, Mozambique, and the USSR (10–11). The tantalum-bearing ores of the placer (alluvium) and pegmatite types can be enriched to concentrates of up to 70 wt % combined Ta_2O_5 and Nb_2O_5 by wet-gravity, magnetic, electrostatic, or flotation methods (12).

Tin Slags. Cassiterite deposits, eg, those in Australia, Brazil, Thailand, Malaysia, Nigeria, and other countries, contain significant amounts of tantalum (10). The bulk of this tantalum is collected in the slags of tin smelters during the smelting operation (12). Over one-half of the tantalum produced in the United States is based on tin slags or a derivative of tin slags as the tantalum feedstock. Most of the world's resources of tantalum occur outside the United States. These resources are largely located in Australia, Brazil, Canada, Egypt, Malaysia, Nigeria, Thailand, and Zaire. The United States is believed to have ca 1.5×10^6 kg of tantalum resources located in identified deposits, but thus far none have been economical to mine in large quantities.

Scrap Recycle. Recycling of various tantalum-bearing scrap materials is a very important secondary source for tantalum. These scrap materials include: powder, foil, wire, and pressed and sintered anodes and capacitors; cemented carbides that contain tantalum carbide as an additive; and various unalloyed tantalum mill product and equipment scrap.

The various metal forms are thoroughly solvent and acid-cleaned, if necessary, to remove impurities. Clean metal is hydrided by heating under a hydrogen atmosphere at ca 900°C, cooled to room temperature, crushed and pulverized to powder, then dehydrided to produce a reclaimed metal powder by heating to ca 900°C in vacuum ($\leq 10^{-2}$ Pa or ca 10^{-4} mm Hg).

Cemented carbide scrap can be treated by several methods. In the Cold-stream process, the carbide scrap is precrushed to coarse particles, which are then accelerated in a high velocity air stream and impinged against a massive stationary carbide target;

Table 1. Tantalum Minerals

Name	CAS Registry No.	Composition	Ta_2O_5 content, wt %	Color and luster	Sp gr	Mohs hardness	Crystal structure	Streak
tantalite–columbite	[*1306-08-7*]	$(Fe,Mn)(Ta,Nb)_2O_6$	0–86	black, metallic	5.2–7.8	6–6½	orthorhombic	brown-to-black
manganotantalite	[*12173-89-6*]	$Mn(Ta,Nb)_2O_6$	40–82	black, metallic	5.0–7.8	6–6½	orthorhombic	brown-to-black
tapiolite	[*1310-29-8*]	$Fe(Ta,Nb)_2O_6$	40–85	black, metallic	6.0–7.8	6–6½	tetragonal	brown-to-black
skogbolite	[*12140-41-9*]	$FeTa_2O_6$	86	black, metallic	7.8	6½	tetragonal	brown-to-black
microlite	[*12173-96-5*]	$Ca_2(Ta,Nb)_2O_6(OH,F)$	60–70	pale yellow-to-light brown, nonmetallic	5.2–6.2	5½	isometric octahedral	light gray
simpsonite or calogerasite	[*12005-43-5*]	$Al_2Ta_2O_8$ + CaO as an impurity	70–72	cream-to-gray, metallic	5.9–6.5	6½–7	hexagonal	white-to-pale yellow
thoreaulite	[*12036-96-3*]	$SnTa_2O_7$ + CaO and Nb_2O_5 as impurities	60–75	white-to-light brown or grey, nonmetallic	7.6–7.9	6	monoclinic	white-to-pale yellow
stibiotantalite	[*12411-68-6*]	$(Sb,Bi)(Ta,Nb)O_4$	35–60	dark to light brown, resinous	5.5–7.5	5–7	orthorhombic	light yellow
yttrotantalite	[*12199-77-8*]	$(Fe,Ca)_2(Y,Er,Ce,U)_2(Ta,Nb)_4O_{15}.4H_2O$	30–45	black, vitreous	5.5–6.8	5–5½	orthorhombic	gray-to-light brown
tanteuxenite	[*12199-43-8*]	$(Y,Er,Ce,U)(Ta,Nb)(Ti)O_6$	20–55	black or brown, resinous	5.5–5.9	5½–6½	orthorhombic	gray-to-light brown
wodginite	[*12178-62-0*]	$Mn_4(Sn > Ta,Ti,Fe)_4(Ta > Nb)_8O_{32}$	50–60	light-to-dark brown	7–7.5	6–6½	monoclinic	brown
struverite	[*12199-39-2*]	$(Ti,Ta,Nb.Fe)_2O_6$	24	iron-black	5.5	6	tetragonal	gray-black
euxenite	[*1317-53-9*]	$(Y,Ca,Ce,U,Th)(Nb,Ta,Ti)_2O_6$		brownish-black	4.6	6½	orthorhombic	yellowish-to-reddish-brown
samarskite	[*1317-81-3*]	$(Fe,Ca,U,Y,Ce)_2(Nb,Ta)_2O_6$	14–28	velvet-black	5.6–5.8	5–6	orthorhombic	dark reddish brown

the particles are shattered to a powder. Since the gas is cooled by expansion from nozzles, oxidation which would normally occur is minimized (13–14).

In the zinc process, the cemented carbide scrap is exposed to molten zinc at >900°C. The zinc alloys with the cobalt binder and expands its volume. Removal of the zinc by vacuum distillation at 950°C results in a porous matte residue, which is subsequently crushed and ground to a metal carbide–cobalt powder suitable for recycling (15–17). In another bloating process, the cemented carbide is heated to >1800°C in a nonoxidizing atmosphere, which makes the mass swell, crack, and become porous. The porous mass is subsequently acid-leached for removal of the binder and is subjected to mechanical reduction to powder (18).

In a process known as leach milling, the cobalt binder in the cemented carbide is leached with acid while the mass is autogenously milled (19). By anodic dissolution of the binder phase under controlled electrochemical potential and current density conditions, a greatly accelerated rate of dissolution of the binder phase is achieved without significant dissolution of the metal-carbide grains (20).

Properties. Some data on physical and mechanical properties and other characteristics of tantalum are listed in Table 2. Certain properties, eg, tensile strength, vary with purity and state of work hardening. More complete physical and mechanical properties are given elsewhere (21–23).

Corrosion Resistance. The presence of a naturally occurring oxide film on the surface of tantalum makes it resistant to corrosion in most severe acid environments and to many other chemicals encountered in industrial applications (see Corrosion and corrosion inhibitors). Tantalum has been tested against hundreds of corrosive substances and the results have been published (24). Regarding industrial usage of tantalum in chemical process applications, among the most important reagents handled are sulfuric, hydrochloric, and nitric acids; organic chemicals; and liquid metals. In general, only those substances that are capable of dissolving the protective oxide film and preventing it from reforming corrode tantalum. It is slowly attacked by strong alkalies, more rapidly by fuming sulfuric acid, rapidly by hydrofluoric acid, and vigorously by a mixture of hydrofluoric and nitric acids. Tantalum is inert to sulfuric acid below ca 150°C and the attack below 175°C is not significant (see Table 3). Corrosion at all temperatures is uniform, ie, without pitting, and hydrogen embrittlement does not occur. This statement is supported without exception by a large number of corrosion tests at several concentrations and by many years of service of heat exchangers installed in sulfuric acid concentrators. However, it has been reported that embrittlement occurred when strips of tantalum were sealed in glass tubes with 80–95 wt % sulfuric acid and heated to the boiling point of the acid in an autoclave (25). Tantalum is inert to hydrochloric acid in all concentrations and at all temperatures up to ca 95°C (see Table 3). Twenty percent acid (constant boiling mixture) does not corrode the metal below 190°C. Tantalum samples tested against concentrated hydrochloric acid at 190°C for 90 h and confined under pressure corrode 150–160 μm/yr and become embrittled (26). Tantalum is not attacked by nitric acid in concentrations up to 70 wt % and at temperatures up to at least 190°C (27–28). Numerous industrial installations with tantalum equipment have been in operation for many years without reported failures caused by corrosion.

In general, tantalum is completely resistant to organic compounds and is used in heat exchangers, spargers, and reaction vessels in several important organic reac-

Table 2. Typical Physical and Mechanical Properties of Tantalum

Property	Value
lattice type	body-centered cubic
lattice constant at 20°C, nm	0.33026
density (at 20°C), g/cm^3	16.6
melting point, °C	2996
boiling point, °C	5425 ± 100
specific heat at 20°C, J/(kg·K)[a]	150.7
linear coefficient of expansion, $°C^{-1}$	6.5×10^{-6}
thermal conductivity (at 20°C), J/(s·cm·K)[a]	0.544
electrical resistivity, $\mu\Omega$-cm	
at 20°C	13.5
at 1000°C	54
at 1500°C	71
temperature coefficient of electrical resistivity (at 0–100°C), $°C^{-1}$	0.00382
thermionic work function, eV	4.12
total emissivity	
at 1500°C	0.21
at 2000°C	0.26
magnetic susceptibility (cgs)	0.93×10^{-6}
tensile strength, MPa[b]	
at room temp	241–483
at 500°C	172–310
at 1000°C	90–117
Young's modulus, GPa[b]	
at room temp	186
at 500°C	172
at 1000°C	152
Poisson's ratio	0.35
ductile-to-brittle transition temp, °C	<-250
working temp	room
recrystallization temp, °C	900–1200
nuclear cross section for thermal neutrons, m^2/atom[c]	21.3×10^{-28}

[a] To convert J to cal, divide by 4.184.
[b] To convert Pa to psi, divide by 6897.
[c] To convert m^2 to barns, divide by 1.0×10^{-28}.

tions, particularly when corrosive inorganics are involved. However, mixtures of anhydrous methanol with chlorine, bromine, or iodine cause pit-type corrosion on tantalum at 65°C (29). This is of particular interest because tantalum is not attacked by methanol, halogens, or the reaction product methyl halide, even at somewhat higher temperatures. Also, pit-type corrosion on tantalum is rare. The corrosive action is apparently caused by the occurrence of the intermediate, haloformic acid.

Tantalum is resistant to several liquid metals and is used in the handling of these materials and in the construction of heat-transfer systems for them.

The occurrence of a galvanic couple or a stray voltage can adversely affect the usefulness of tantalum in a corrosive medium to which it is completely resistant. If tantalum is permitted to become the cathode in an electrolyte, either because it is connected to a metal that is chemically attacked by the electrolyte or because of a stray voltage, it may become embrittled by the atomic hydrogen that is liberated on its surface. In solutions of hydrofluoric acid, which prevents the formation of the pro-

Table 3. Corrosion Rates of Tantalum in Sulfuric and Hydrochloric Acids

Temperature, °C	Corrosion rate, μm/yr	Temperature, °C	Corrosion rate, μm/yr
Sulfuric acid, concentrated		*Hydrochloric acid, 20 wt %*	
175	2.54	190	2.29
200	38.1	200	3.30
250	737	220	22.90
300	8,690	*Hydrochloric acid, 37 wt %*	
Sulfuric acid, fuming		100	1.52
23	2.54	110	4.06
70	234	120	9.14
130	99,060		

tective oxide film, tantalum is less noble than silver, copper, lead, nickel, iron, and niobium, but is more noble than zinc, manganese, aluminum, and zirconium (30).

Tantalum is very reactive at elevated temperatures. It reacts with oxygen or air forming Ta_2O_5; the reaction begins at 260°C and becomes vigorous at ca 800°C. It reacts with molecular hydrogen above 250°C and with atomic hydrogen at room temperature to form the very brittle hydride [*13981-95-8*]. Tantalum containing dissolved oxygen can absorb molecular hydrogen at room temperature when it is undergoing deformation (31). However, all of the hydrogen is released when the hydride is heated above 800°C in a vacuum. All of the halogens react with tantalum forming the metal halides; fluorine at room temperature, chlorine at ca 200°C, bromine at ca 300°C, and iodine beginning at ca 300°C but more rapidly above 340°C. Carbon and boron react directly with tantalum forming tantalum carbide [*12080-06-3*], TaC, and tantalum boride [*12007-35-11*], TaB_2 (32).

Alloys. Tantalum has been alloyed with tungsten to improve its strength with little change in corrosion properties (see Table 4). With the exception of the 2.5 wt % tungsten alloy, which exhibits somewhat improved corrosion properties over that of unalloyed tantalum and is often used as a replacement for tantalum, very little commercial use has been developed for them. Some interest has recently been engendered for higher niobium content (25–40 wt %) as a means of cost reduction (33).

Table 4. Properties of Tantalum–Tungsten Alloys

Alloy	Elongation, %	Tensile strength, MPa[a]	Corrosion resistance in sulfuric acid, μm/yr	
			175°C	200°C
tantalum	42	262	5.08	48.3
tantalum–2.5 wt % tungsten	36	375	5.84	25.4
tantalum–5 wt % tungsten	32	455	5.33	27.9
tantalum–10 wt % tungsten	28	676	5.59	43.2

[a] To convert MPa to psi, multiply by 145.

Two high tantalum alloys with exceptional creep strength have had limited use in the handling of hot liquid metals in nuclear power systems:

	Alloy	
Composition	*T-111*	*T-222*
tungsten, wt %	7–9	9.6–11.2
hafnium, wt %	1.8–2.4	2.2–2.8
carbon, ppm	<40	80–170
tantalum	balance	balance

Tantalum is also used to increase the high temperature strength of nickel- and cobalt-base superalloys, which primarily are used in aircraft gas turbines (see Table 5). The method of addition is either with clean scrap or, preferably, a clean, tantalum-containing thermite alloy of the base metal. Strengthening mechanisms include formation of solid solutions, intermetallic compounds, and carbides (34) (see High temperature alloys).

Manufacture and Processing. The manufacture of tantalum metal is accomplished by extraction of tantalum from the ore or tin slag, separation of the extract of tantalum from other metals present, formation of a pure tantalum compound (usually potassium fluorotantalate [*16924-00-8*], K_2TaF_7), and reduction of the compound to metal powder. The metal powder can be used for electrolytic capacitor applications or it can be consolidated to solid tantalum metal by powder metallurgy (qv) (P/M) or melting methods. These consolidated metal forms are generally further fabricated to mill products. If tantalum carbide is produced, the intermediate tantalum compound is tantalum oxide, which is subsequently carburized to TaC. Niobium that exists concurrently in the ore or tin slag, in association with the tantalum, is recovered by ammoniacal precipitation from the niobium fluoride solution, calcining, and subsequent recovery as niobium oxide, Nb_2O_5. Production steps for tantalum metal, tantalum carbide, and niobium oxide are shown schematically in Figure 1.

Table 5. Typical Nickel- and Cobalt-Base Superalloys Containing Tantalum[a]

	Composition, wt %						
Alloy	Ta	Co	Ni	Cr	Al	C	Other elements
CoTaC[b]	14	balance	10	10		1	
CoTaC 3[b]	13	balance	10	20		1	
MarM-509	7.5	balance	10	24		0.6	W, Ti, Fe
Nitac 13[b]	8.1	3.3	balance	4.4	5.4	0.5	W, Re, V
Nitac[b]	14.9		balance	10	5	1.1	
B-1900	4.3	10	balance	8	6	0.1	W, Mo, Nb, Ti, B, Zr
IN-162	2.0		balance	10	6.5	0.12	W, Mo, Nb, Ti, B, Zr
IN-738	1.7	8.5	balance	16	3.4	0.17	W, Mo, Nb, Ti, B, Zr
MarM-246	1.5	10	balance	9	5.5	0.15	W, Mo, T, B, Zr
MarM-432	2.0	20	balance	15.5	2.8	0.15	W, Fe, Ti, B, Zr
TAZ 8A	8.0		balance	6	6	0.125	W, Mo, Nb, B, Zr
TRW VIA	9.0	7.5	balance	6	5.4	0.13	W, Mo, Nb, Ti, B, Zr

[a] Refs. 34–35.
[b] Directionally solidified eutectic alloys.

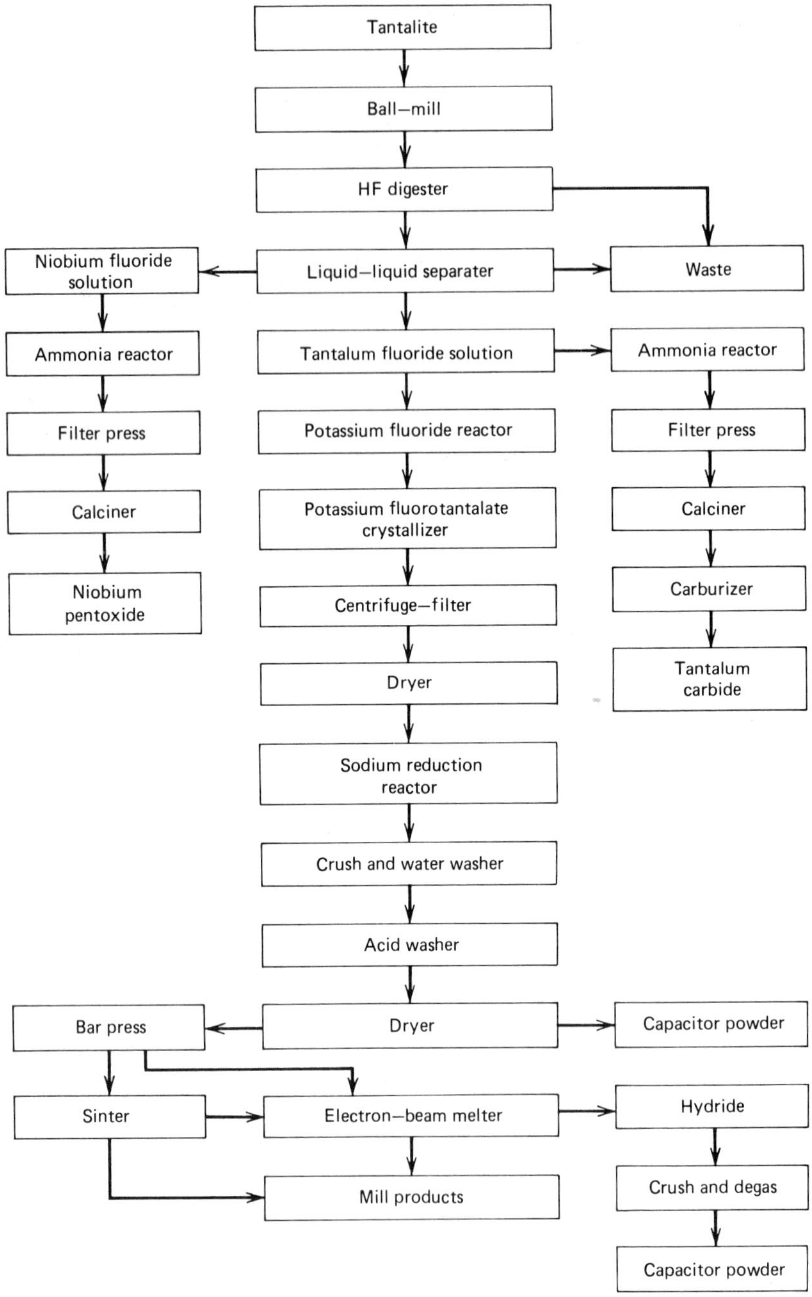

Figure 1. Tantalum production flow sheet.

Extraction. The first successful industrial process used to extract tantalum from tantalite–columbite ores involved alkali fusion to decompose the ore, acid treatment to remove most of the impurities, and Marignac fractional crystallization to separate the tantalum from the niobium and to purify the resulting K_2TaF_7 (6). This method was used to produce all tantalum prior to ca 1957. Since then, liquid–liquid solvent extraction techniques have replaced the Marignac process. The solvent extraction method involves the use of hydrofluoric acid or HF plus H_2SO_4 to dissolve the metal values followed by liquid–liquid separation (36). The technique was thoroughly developed at Ames Laboratory and at the Bureau of Mines (37–40). Subsequent modifications and improvements to the basic process have been made (41–43).

The principles involved in the liquid extraction system are based on the effect of acidity on the solubilities of the fluorides of tantalum, niobium, titanium, zirconium, iron, manganese, etc, in methyl isobutyl ketone (MIBK). When an acid solution of these fluorides is allowed to contact MIBK, the tantalum can enter the organic phase at a relatively low acidity but the niobium only at a high acidity. Under both of these conditions, the other elements remain in the aqueous phase. The difference in the solubilities is sufficient to give clean extractions of pure tantalum and niobium in only a few contacting stages.

The separations are accomplished by either extraction of the tantalum, first by mixing a low acidity feed solution with barren MIBK and then extracting the niobium by increasing the acidity of the feed and mixing it with more fresh MIBK, or by extraction of both tantalum and niobium from a strongly acidic aqueous feed by mixing with fresh ketone and then stripping the niobium from the organic phase by mixing it with a less concentrated acid solution. In either case, the values remaining in the MIBK are recovered by stripping with dilute acid. In industrial practice, the following steps typically are used: Finely pulverized ore is fed slowly into a temperature-controlled tank containing concentrated mixtures of hydrofluoric and sulfuric acids. Digestion is continued for several hours to decompose the minerals completely. The solution is decanted or filtered, adjusted to the required acidity, and then fed into the mixer–settler liquid–liquid system. The resulting aqueous solutions from the liquid–liquid separation containing pure tantalum fluoride or niobium fluoride are converted to hydrated oxides by precipitation with ammonia or to double fluoride crystals through the addition of potassium fluoride.

Slags containing more than 6 wt % Ta_2O_5 can be treated directly by chemical means to recover the tantalum and niobium values (10). However, the application of this direct method is usually uneconomical for lower-grade slags because of the consumption of high cost HF for nonvaluable constituents, eg, SiO_2, TiO_2, and Fe_2O_3, in the slags. Several smelting and subsequent processing methods have been developed to obtain upgraded (higher Ta_2O_5 content) concentrates from lower-grade slags (10,44–48). For example, the tin slag containing recoverable amounts of tantalum and niobium is beneficiated with respect to these elements by first smelting the slag with sufficient carbon in an electric furnace to reduce the iron, tantalum, and niobium components, and these are collected as a high carbon ferroalloy containing the tantalum and niobium in the form of their carbides. This hearth product is subjected to a two-stage oxidation treatment; in the first stage, a metal oxide is used to selectively oxidize carbided components of the slag other than the tantalum and niobium. In the second stage, the carbided tantalum and niobium are reoxidized, and the product can be substituted for tantalum and niobium ores in the recovery of these metals.

Reduction. Several methods have been developed for the reduction of tantalum compounds to tantalum metal, but sodium reduction of K_2TaF_7 to produce tantalum metal powder is the principal industrial reduction method used worldwide. The first pure, ductile tantalum was made in Germany by a modification of the sodium-reduction method. The development of more efficient and advanced techniques for sodium reduction of K_2TaF_7 has been one of the outstanding developments leading to the growth of the tantalum industry. As a general description of the process, a confined molten bath of K_2TaF_7, which often contains a diluent, eg, NaCl, is maintained above 800°C. Molten sodium is fed into the bath, which is continuously agitated. When all of the K_2TaF_7 has reacted, the bath is cooled to ambient temperature and the tantalum metal, which is in the form of powder, is recovered by crushing and washing the solidified mass to remove residual salts. Numerous modifications and improvements have been made on the sodium-reduction method, some examples of which are described in selected patents (49–52).

Other methods that have been used or developed for reduction of tantalum include electrolysis of fused K_2TaF_7 to produce tantalum powder and to electrodeposit coatings of tantalum on base metal substrates (2); carbothermic reduction of Ta_2O_5 with carbon or reaction of TaC and Ta_2O_5; chemical vapor deposition (CVD) from a mixture of $TaCl_5$ and hydrogen to coat tantalum on a metal or nonmetallic substrate (53); reduction of $TaCl_5$ in a hydrogen atmosphere to produce fine particle-size powder (54); Kroll process in which anhydrous $TaCl_5$ is reduced by magnesium under helium or argon (55); chlorination of a ferroalloy in a sodium iron chloride melt, subsequent separation of $TaCl_5$ by distillation, and reduction of the tantalum chloride by hydrogen, magnesium, or aluminum (44); and reduction of K_2TaF_7 by aluminum and copper (56).

Consolidation. Both powder metallurgy (qv) (P/M) and melting methods, and a combination of these are used to consolidate tantalum metal.

Pressing. The consolidation of tantalum begins with the pressing of the metal powder. Two types of pressing equipment, mechanical and isostatic, are used to make the compacts, and pressures of 275–700 MPa (39,900–101,500 psi) are applied. Mechanical pressing can be used when the bars are to be sintered by resistance. The powder is spread evenly in a sturdy, rectangular, compound die set, and the pressure is applied through top and bottom punches and often to the side; thus, a bar with precise dimensions is produced. Mechanically pressed tantalum bars are usually limited to ca 10 kg in weight and 75 cm in length. Hydrostatic (isostatic) compaction allows more freedom in bar dimensions and shape. The metal powder is poured into a rubber or neoprene mold which is supported by an inexpensive steel shell. Bars pressed in this manner are usually sintered or become feedstock for melting furnaces.

Sintering. The sintering process consolidates and purifies tantalum metal at temperatures well below its melting point (2996°C). Pressed bars are sintered for several hours at 2000–2400°C in a good vacuum ($\leq 5 \times 10^{-2}$ Pa or 4×10^{-4} mm Hg). Heating is accomplished either by passing a large electric current through the bar, by induction, or by radiation from a susceptor or resistance elements. Sintering in vacuum removes high vapor pressure impurities, eg, salts, some oxides, and lower melting metals; carbon and oxygen contents are reduced through loss as carbon monoxide. Depending on which of the latter impurities is dominant, Ta_2O_5 or carbon may be added to permit maximum purification through the carbon monoxide reaction. Nitrogen evolves in elemental form. Temperatures in excess of 2300°C are required for essentially complete removal of nitrogen (2). Tantalum made by the sintering method has certain metallurgical characteristics, eg, small grain size, which are useful in the forming and drawing of thin sheet and in the drawing of fine wire.

Vacuum-arc melting. As a result of its high melting point, which is far above the melting point of most other elements, tantalum is readily purified by melting in a vacuum of ($\leq 5 \times 10^{-2}$ Pa or 4×10^{-4} mm Hg), particularly in electron-beam melting. Impurities evaporate during such melting to low levels (<100 ppm). Electron-beam melting is generally used for consolidation of unalloyed tantalum and tantalum alloys containing low vapor pressure alloying metals, eg, tungsten. Consumable-electrode arc melting is used for melting alloys containing certain higher vapor pressure metals and the unalloyed metal. Large ingots can be produced by these melting methods for subsequent fabrication of large tantalum sheet or other mill products. In the arc furnace, a heavy direct-current arc is drawn between the electrode and a pad of tantalum metal on the bottom of a water-cooled copper mold (see Fig. 2). Reverse-polarity direct current (electrode negative) is generally used. The heat of the arc melts the lower end of the electrode (the tantalum feedstock), and the molten metal drops into the mold. A shell of tantalum immediately solidifies on the inside wall of the mold, thereby forming a tantalum crucible which holds a shallow pool of molten metal. The tantalum electrode is continuously advanced as it melts and fills the mold. When the mold is full, the operation is stopped and the ingot is allowed to cool before removal from the furnace. A pressure usually below 10 Pa (0.075 mm Hg) is maintained throughout the entire process. Ingots ≥25 cm in diameter can be made, and ingots 18–23 cm in diameter and 180 cm long are common. A more limited amount of purification occurs during this melting process than when electron-beam melting is used. Melting rates of several hundred kilograms of tantalum per hour are normal.

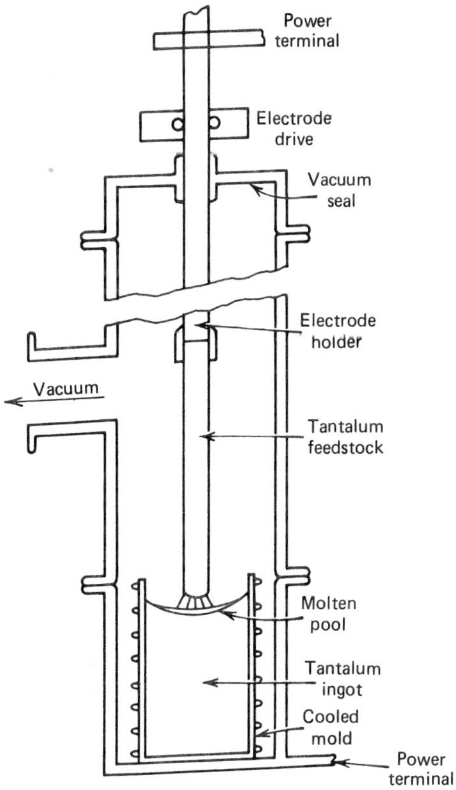

Figure 2. Vacuum-arc melter.

Electron-beam melting. In electron-beam melting, a beam of high velocity electrons, accelerated by a high potential between the cathode and anode in an electron gun, is focused by an electromagnetic field to the end of a formed tantalum bar (the feedstock) and to the surface of a pool of molten metal in the water-cooled copper mold (see Fig. 3). The tantalum feedstock is fed continuously into the furnace as it melts and drips into the pool, and the ingot, as it forms, is progressively lowered down through the mold into the cooling chamber and the liquid level of the ingot near the top of the mold is maintained. This vacuum-melting process effectively removes the gaseous and the more volatile metallic impurities in the feedstock. If further purification is required, the ingot is remelted in the same furnace and the feedstock is lowered vertically into the furnace. Tantalum metal is electron-beam-melted at a rate of ca 100 kg/h, and ingots 20 × 300 cm and weighing more than 1350 kg are in regular production.

Tantalum metal made from electron-beam-melted ingots is noted for its high purity, excellent ductility, and good fabricability and weldability. It is characterized also by its large grain size and lower tensile strength as compared with that produced by P/M sintered bars.

Fabrication. Powder-metallurgy-sintered bars or melted ingots are very ductile and can be rolled cold into sheet and foil. Seamless tubing drawn from sheet is readily available in sizes 0.5–5 cm dia. Capillary-size tubes have also been made. Tantalum wire is made by cold-rolling the bar stock into small rods, oxidizing the surface of the rod to aid retention of the drawing lubricant, and then drawing the coated (oxidized) rod through dies to size. Drawing bare tantalum is not practical because the metal galls.

Ingots, either arc-melted or electron-beam-melted, are prepared for breakdown

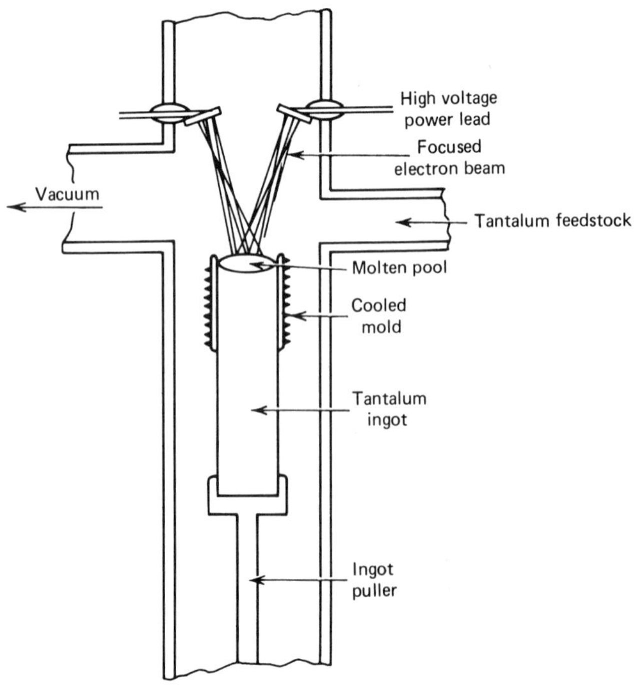

Figure 3. Electron-beam melter.

by machining or grinding to remove the rough surface and provide a smooth, well-blended surface. Ingot breakdown of unalloyed tantalum, especially electron-beam-melted ingots, can be accomplished by press or hammer-forging at ambient temperature despite the coarse, columnar grain structure. However, it is customary to warm the ingots to ca 400–500°C to reduce the energy required for forging. Lean tantalum alloys, eg, Tantaloy 63 metal (Ta–2.5% W–0.15% Nb), also can be forged at similarly low temperatures. However, more highly alloyed, high strength tantalum alloys, eg, Ta–10% W and the T-111 and T-222 alloys, generally are forged or often extruded at >1000°C and sometimes up to ca 1500°C for ingot breakdown to avoid fracture of the ingots.

Unalloyed tantalum and lean tantalum alloy can be reduced 75–90% in cross-section before requiring annealing. The higher alloys should be reduced at least 65% before undergoing a recrystallization heat treatment. Recrystallization is accomplished at ca 900–1200°C, depending on the form and purity, for unalloyed tantalum and at an appropriate higher temperature for the alloys. After ingot breakdown, all of the currently commercially produced grades of tantalum and its alloys can be cold-worked to the final mill products with appropriate intermediate annealings as required. Sheet widths of ≥150 cm and welded coils as long as 100 m have been produced. Table 6 shows the tensile properties of typical P/M and electron-beam-melted tantalum wire and sheet in the as-cold-worked and the recrystallized conditions.

Tantalum is very ductile and can be formed or drawn readily to shapes when adequately lubricated to avoid seizing or galling with the forming tools. It is usually formed in the fully recrystallized condition. Before welding, the tantalum must be chemically clean to minimize contamination of the weld. Preweld cleaning usually requires solvent degreasing followed by chemical cleaning, eg, an acid pickle in an HF plus HNO_3 solution followed by water rinsing and drying (see Metal surface treatments, cleaning, pickling, and related processes). Then it is handled only with clean cotton gloves and is welded as soon as possible. Tantalum must be welded in high purity inert gas or in a high vacuum. Gas tungsten-arc, gas metal-arc, electron-beam, resistance, and laser welding processes are used (57–58) (see Welding). Silver brazing alloys, copper, and a number of specially developed refractory metal brazing alloys can be used to braze tantalum to itself or to dissimilar metals, eg, stainless steels (57) (see Solders and brazing alloys).

Tantalum can be machined with sharp, positive-rake, high speed tool steel and C-2-grade cemented carbide tools. Cutting fluids used include soluble-oil emulsions and cutting oils containing sulfur or chlorine (59).

Table 6. Effects of Recrystallization of Worked Tantalum Sheet and Wire

	Sintered				Electron-beam-melted			
	0.050-cm wire		0.050-cm sheet		0.050-cm wire		0.050-cm sheet	
Property	Worked	Recrystallized	Worked	Recrystallized	Worked	Recrystallized	Worked	Recrystallized
tensile strength, MPa[a]	621	450–520	621	310–345	517	380–450	417	228–310
elongation, %	2	10–20	1	35	2	15–25	2	35–40

[a] To convert MPa to psi, multiply by 145.

Metallography. Tantalum specimens for metallographic examination are prepared by techniques similar to those used for preparing stainless-steel specimens (60). The preparation of a good metallographic surface on recrystallized tantalum requires a little more than the usual care and patience because of the metal's characteristic low hardness and excellent resistance to chemical attack. One procedure is as follows: mount the flat specimen in a plastic or tantalum holder; wet-grind on 240- and then on 400-grit silicon carbide paper; polish on nylon cloth with 30-μm diamond paste and kerosene; repeat previous step using 9-μm diamond paste; polish on microcloth with 1-μm diamond paste; etch in a solution of 15 mL HF, 15 mL H_2SO_4, and 15 mL H_2O activated with 5–20 drops of 30 wt % H_2O_2; repeat last three steps until a satisfactory surface is obtained. The sample should be boiled in a 1 wt % ammonium pentaborate aqueous solution to remove residual HF so that the microscope lens is not etched. Another method for polishing and for polish–etching and recommended etchants for tantalum have been outlined (60).

Economic Aspects. The tantalum content of the world production of ores, slags, etc for 1979 was ca 950 metric tons (61). The U.S. consumption, usually ca 60% of the world total, was 591 t for the same year. The relatively small amount of tantalum in the earth's crust and low concentrations in known deposits keep the cost of concentrates quite high. Although the world reserves of tantalum are ca 65,000 t, which is large relative to the annual consumption, much of the material is low grade and, until improved mining or upgrading technology is available, may also be uneconomical to process.

Increasing demand, irregular production of the concentrates, and warehousing by users and dealers has caused the price to vary considerably yearly and, for spot ore, monthly. The average annual price of tantalum content (concentrate, 60 wt % basis of combined tantalum and niobium pentoxides) is shown for 1958–1980 in Figure 4. Sales to industry from government stockpiles from 1968 to 1975 helped stabilize and perhaps artificially depressed the raw material price. From 1975 to 1980, demand outstripped ore production, the balance of requirements being supplied by depleting producer inventories. The sharp price escalation between 1978 and 1980 was fueled by consumer fears of limited availability. However, this fear and increasing prices also spurred the development of new ore reserves and led to technical innovation, resulting in significant material conservation, eg, higher specific capacity powder (62). For the near term, until 1985 and probably beyond, these factors as well as some softening of demand are expected to stabilize ore price and supply.

Chemical Analysis. The primary method for ore analysis involves ion exchange (63–66). Because this procedure is quite lengthy, faster spectrochemical methods, eg, atomic absorption, plasma emission, and x-ray fluorescence techniques, have been and are being developed (67). The first step for all materials, except x-ray fluorescence, is to dissolve the sample in hydrofluoric acid. For x-ray fluorescence, sample preparation involves forming a glass bead on the ore sample by fusing it with lithium tetraborate. Matrix interference is a problem with spectrochemical procedures and is minimized by developing standards from similar ore bodies. Tantalum alloys are also dissolved in hydrofluoric acid for analysis by plasma emission based on standards prepared by dissolving pure metals that approximate the alloy composition.

Metallic impurities in unalloyed tantalum are determined by computerized optical emission of calcined samples that have been mixed with a halogenated spectrographic carrier for plasma excitation. Oxygen, nitrogen, carbon, and hydrogen in tantalum

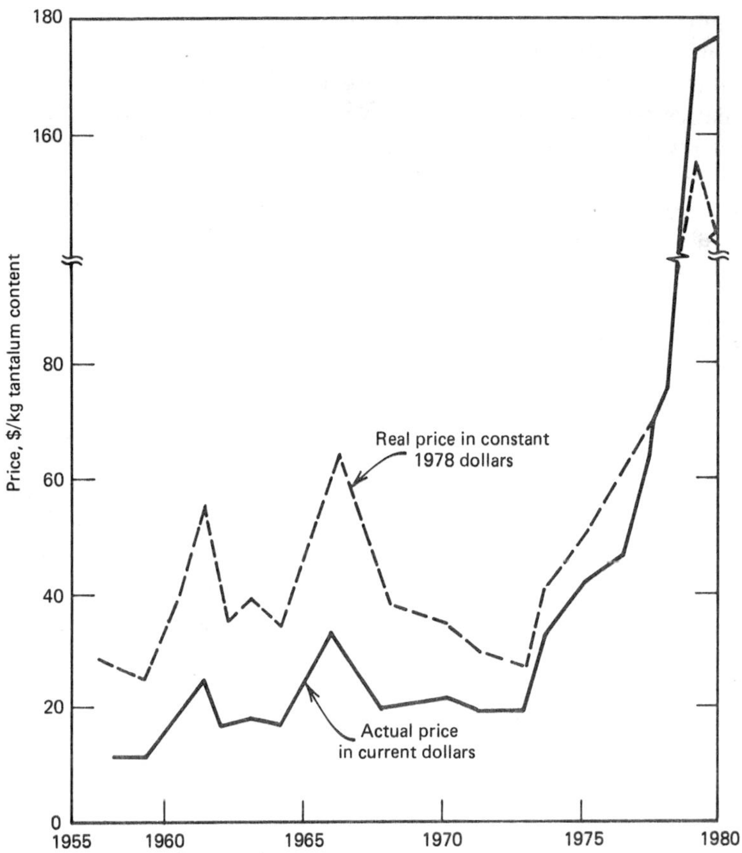

Figure 4. 1958–1980 prices of tantalum in concentrates.

and its alloys are measured by chromatographic methods, as these elements are released from the reaction of the metallic sample with nickel at ca 5000 K. Differential conductivity or infrared emission of elemental nitrogen and hydrogen gases are measured, and oxygen and carbon are determined as carbon dioxide. Analysis for carbon takes place in an oxygen stream and oxygen is measured as it reacts with the graphite crucible.

Toxicology. Tantalum is inert and does not appear to have detrimental affects on the human body, in which it is used in surgical implants (see Prosthetic and biomedical devices). Tantalum, like other reactive metals, burns, particularly if in a finely divided state, eg, powder, when exposed to heat, flame, or a chemical oxidizer. The recommended threshold limit value (TLV) for exposure in workroom air is 5 mg/m^3 (3.8 mg/yd^3) of air (68). Tantalum powder and some compounds have been suspected of causing skin irritation and mild fibrosis of the lungs.

Uses. About 60% of the world's annual production of tantalum is used in capacitors, because of the metal's ability to form a stable dielectric oxide surface film; 27% is used as the carbide, TaC, in cemented carbide cutting tools for steels and special alloys because its presence increases hot hardness and resistance to deformation (see Carbides; Refractory coatings). Its use as a mill product, primarily in corrosion applications, is ca 11%, and ca 8% is used as an alloying addition to high temperature superalloys. A smaller amount is used in other special applications.

Capacitors. During 1981, ca 615 metric tons of tantalum were consumed in noncommunist countries in the form of powder, wire, foil, and cases for the production of 3×10^9 individual capacitors, which represented about one quarter of the total world capacitor market on a dollar basis. Although over one third of the total number of units was produced in Japan, compared to slightly less than one third in the United States, the United States consumes over two thirds of the material used to make capacitors. This reflects the fact that over three quarters of the Japanese market is devoted to the entertainment market, whereas four fifths of the American market for tantalum capacitors is in business machines and military applications, where larger devices are required.

The rest of the capacitor market, on a dollar basis, is made up of inexpensive ceramics (35%), which are used at very low capacitance levels ($<1\ \mu C$ (or $\mu F{\cdot}V$)); aluminum electrolytes (20%), which cover the same general capacitance range as tantalum capacitors but cost about one half as much; and other types (20%), primarily paper and film. Tantalum is a superior capacitor material because of the inertness and stability of its electrolytic oxide film, and it is the standard for reliable performance. There are three basic tantalum capacitor types: foil (Fig. 5), porous-wet-anode in aqueous electrolyte (Fig. 6), and porous-anode, dry electrolyte (Fig. 7).

Foil capacitors are made by rolling two strips of tantalum foil separated by a paper saturated with a glycol-base electrolyte. The anode strip can be etched to increase its surface area prior to electroforming to develop the dielectric oxide. These capacitors have the highest voltage rating because they can be electroformed to more than 500 V and are used with working voltages of 6–500 V. They have the lowest specific capacity, ie, capacitance per unit volume, of the three types being produced. Their production volume is also the lowest.

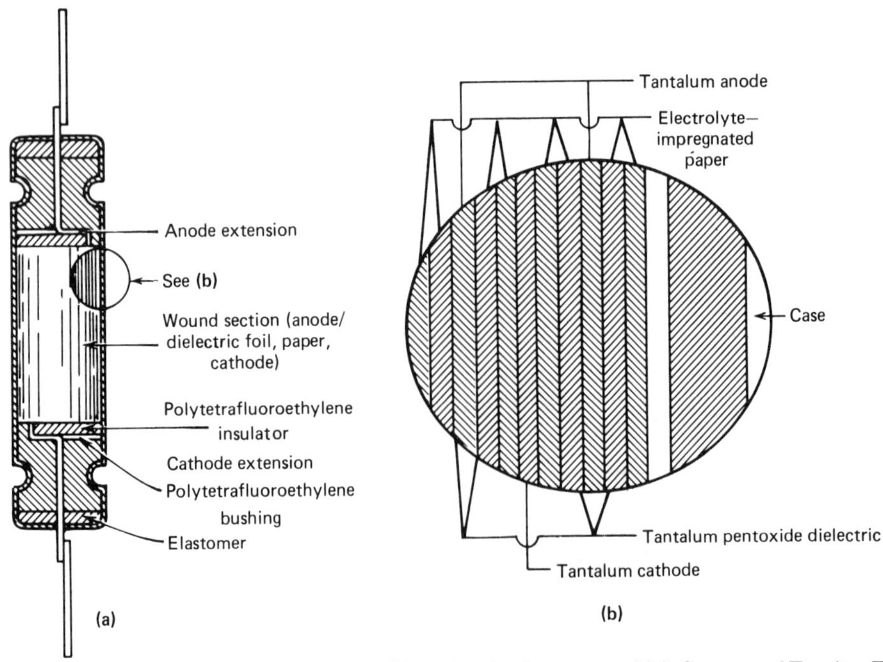

Figure 5. (**a**) Tantalum foil capacitor; (**b**) sectional enlargement of (**a**). Courtesy of Tansitor Electronics, Inc.

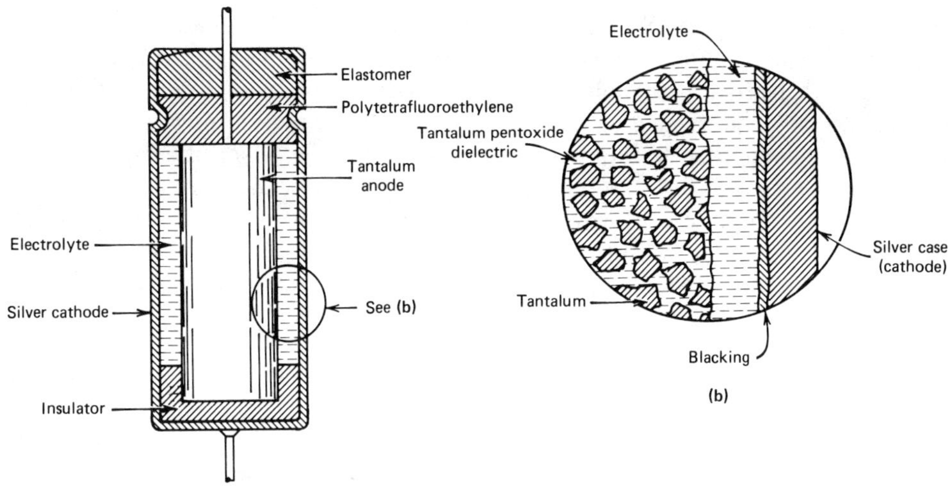

Figure 6. (**a**) Tantalum wet-electrolyte capacitor; (**b**) sectional enlargement of (**a**). Courtesy of Tansitor Electronics, Inc.

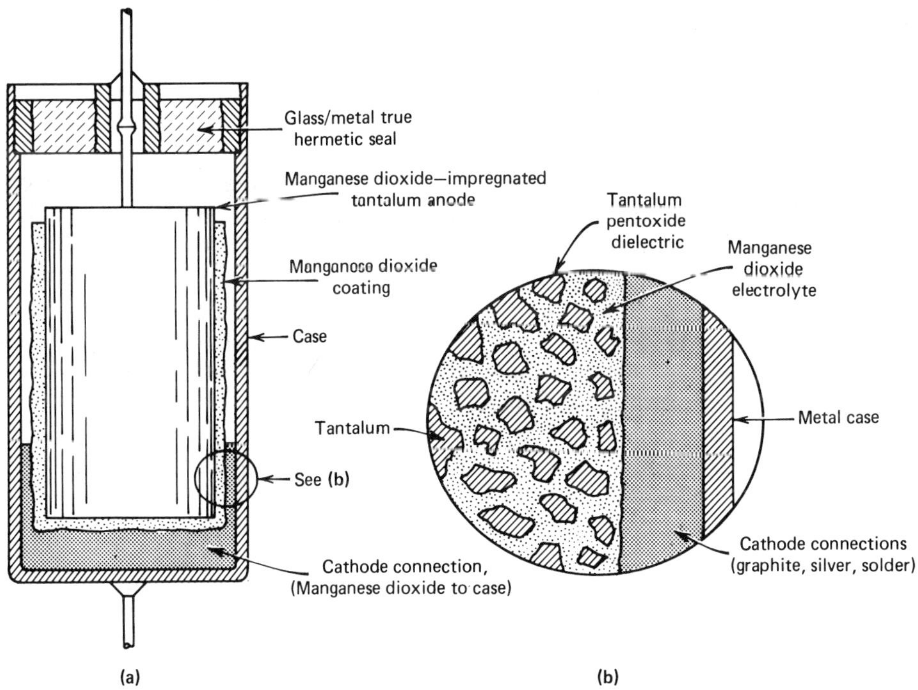

Figure 7. (**a**) Tantalum solid-electrolyte capacitor; (**b**) solid sectional enlargement of (**a**). Courtesy of Tansitor Electronics, Inc.

Wet-electrolyte capacitors are composed of a porous anode, sulfuric acid electrolyte, and a silver or tantalum case. The anode is made by pressing a compact of tantalum powder around a tantalum wire and sintering to weld the particles without closing the voids. Alternatively, the lead wire sometimes is welded to the sintered anode. The procedure produces a very high surface area per unit volume, which results in a high specific capacity. Working voltages range from 6 to 250 V. Porous anodes are made from either sodium-reduced powder or from a powder (commonly called EB powder) made by hydriding, pulverizing, size classifying and dehydriding high-purity arc- or electron- beam-melted tantalum ingots. Both types of powders are often thermally agglomerated to improve flowability, pressability, specific capacity, and other characteristics (69–70). A phosphorus additive may be used to further enhance specific capacity (71). Most low voltage ($\leq$20 V) capacitors are made from sodium-reduced powder, whereas EB powders, which have higher purity and are capable of forming a more perfect Ta_2O_5 dielectric film, are used for higher working voltage ($\geq$35 V) units. Both wet and foil capacitors comprise <1% of the total world tantalum capacitor market.

Solid-electrolyte capacitors have the highest specific capacity, permit the ultimate in miniaturization, and represent 99% of the unit volume in the world's tantalum-capacitor market. They operate well over a wide temperature range, have excellent dielectric properties, and provide good idle and shelf life and long service. Their working voltage range is generally <50 V (100 V max). Both sodium-reduced and EB powders are used; the latter account for ca 20% of the volume, primarily for higher working voltages. Solid-electrolyte, manganese dioxide is incorporated into the electroformed porous anode by repeated impregnation and pyrolysis of a manganese nitrate solution. The anode then is coated with carbon and finally with a silver-rich conducting paint. After soldering of a cathode terminal, the capacitor is encased in a metal or molded plastic container or in a dipped epoxy covering (see Embedding). During 1981, ca 16% of the world's production was of the metal-case type, 18% molded plastic, and 66% dipped epoxy.

A new type of solid tantalum capacitor, the chip, is made by a similar process except the silver coating step is replaced by either a plasma copper spray, an electrolytic coating, or an end cap and silver epoxy cover. Chip capacitors are used unjacketed or jacketed and can be mounted directly onto a thick film hybrid circuit on an Al_2O_3 substrate. The advantage of this construction is that it fits a concept of automatic placement. Wiring and resistors are printed directly on the substrate to which ceramic and tantalum chip capacitors are automatically soldered. Impedance is lowered by this concept because wiring is minimized.

Unit production of capacitors has doubled in the last five years but the weight of the metal used has not increased in proportion, primarily because improvements have been made in the quality and specific capacity of the powder. Since 1950, the average specific capacity in capacitor anodes has increased from ca 1800 to ca 7500 μC/g [(μF·V)/g]. Specific capacities of some production powders are 15,000–20,000 (μF·V)/g, and those of laboratory powders are as high as 30,000 μC/g, depending on sintering conditions. Because further improvements in the efficiency of powder usage are expected, it is unlikely that tantalum demand will parallel increased device production.

Cemented Carbides. Tantalum monocarbide is used in the manufacture of cemented carbides (see Carbides). In commercial grades of complex tungsten carbide-base tools used for steel cutting, the TaC content is ca 5–15 wt %. In lesser concentration, TaC serves as an inhibitor for grain growth of the hexagonal WC. Optimum benefits are obtained in the steel-cutting grades when the tantalum carbide forms a cubic, solid solution phase with the other carbides, WC and usually TiC (72). The effect is to improve the hot hardness and deformation and shock resistance of the cemented carbide and to reduce crater wear. Crater formation is a phenomenon leading to tool failure wherein the metal binder in the cemented carbide, ie, cobalt, diffuses into the steel chip being machined, resulting in the continuous erosion of the tool. Other monocarbides, including NbC and combinations such as (Hf,Nb)C substituting for part of the TaC, are also used (73–74). However, the primary reason for this substitution is economic as there is no evidence that significant improvement in tool properties is achieved (see Tool materials).

Chemical Equipment. Mill products account for about one tenth of the tantalum produced, and the largest use for tantalum sheet is in corrosion-resistant chemical equipment. Fabricated tantalum equipment used in the chemical industry includes lined or clad vessels, bayonet heaters, vapor condensers, multitube heat exchangers, U-tubes, spargers, thermowells, rupture diaphragms, orifices, and many others (57). Tantalum plugs are used to repair perforations in glass-lined steel equipment. Tantalum equipment is widely used in sulfuric acid concentrators, in temperature controllers for chromium-plating systems, in the distillation and condensation of acids and acidic chemicals, and for handling chlorine. It is used in the manufacture of pharmaceuticals and fine chemicals where contamination cannot be tolerated. It is also used for several organic chemical reactions.

Most tantalum equipment can be constructed from sheet less than 0.75-mm thick because of its inertness and adequate strength. Because thermal conductivity is high, efficient heat-transfer designs that minimize equipment cost are possible. Further material economy can also sometimes be obtained by using explosively bonded or roll-clad tantalum on steel, copper, or other substrates (see Metallic coatings). Chemical vapor deposition has been employed to coat tantalum on substrates, especially steel, for applications such as thermowells (53) (see Film deposition techniques).

The availability of larger tantalum sheets from electron-beam-melted ingots has made it practical to line large 2000–8000-L reaction tanks because fewer welded seams are needed. Many such installations are in use and are based on thermal-expansion-compensated, loose-lined tantalum equipment.

Other. Other applications in which lesser amounts of tantalum are used include flame and radiation shields, hardware for vacuum furnaces, boats for evaporation applications, spinnerets for synthetic fibers, and surgical applications, eg, suture wire, cranial repair plates, and wire gauze.

Tantalum Compounds

The refractory nature of tantalum prevents the formation of many ordinary simple compounds of this element. Of those that have been prepared, only a very few are considered important.

Aluminum Tantalide. Aluminum tantalide [*12004-76-1*], Al_3Ta (sp gr 7.02), is a gray refractory powder that forms when aluminum metal is added to potassium fluorotantalate at ca 1000°C. It is quite insoluble in acids and alkalies. It oxidizes very slowly when heated in air above 500°C. It melts at ca 1400°C.

Tantalum Boride. Tantalum boride [*12077-35-11*], TaB_2 (sp gr 11.15, mp ca 3000°C, Mohs hardness ≥8), is a gray metallic powder. It is quite insoluble in acids and alkalies. It can be prepared by heating tantalum and boron in a vacuum at ca 1800°C; by induction-heating a mixture of Ta_2O_5, B_2O_3, and C; or by the electrolysis of fused Ta_2O_5, B_2O_3, CaO, and CaF_2 (42).

Tantalum Carbide. Tantalum carbide [*12080-06-3*], TaC (sp gr 13.9, mp 3880°C, Mohs hardness 9–10), is a crystalline, very fine, gold-to-brown-colored powder. It is insoluble in all acids except a mixture of hydrofluoric and nitric acids. It is an interstitial compound with a face-centered cubic lattice. Stoichiometric tantalum carbide can be prepared by the direct carburization of tantalum powder with the appropriate amount of carbon black or by the reaction of tantalum oxide and carbon at ca 1900°C in an inert atmosphere. The valency of tantalum does not determine the stoichiometric composition of the carbide. Tantalum carbide is converted to the pentoxide when heated to redness in air. It is easily decomposed for analytical purposes by fusion with potassium pyrosulfate or by solution in a mixture of hydrofluoric and nitric acids.

Tantalum Halides. ***Tantalum Pentachloride.*** Tantalum pentachloride [*7721-01-9*], $TaCl_5$ (sp gr 3.68, mp 216.0°C, bp 242°C), is a white (when pure) crystalline salt that hydrolyzes in water but is soluble in absolute alcohol. It also hydrolyzes on exposure to moist air. It forms when tantalum metal is heated in pure dry chlorine above 200°C. It can also be prepared by heating a mixture of tantalum pentoxide and carbon in pure dry chlorine. It decomposes to tantalum and hydrogen chloride when heated above 600°C in hydrogen.

Tantalum Oxytrichloride. Tantalum oxytrichloride [*13597-22-3*], $TaOCl_3$, is a crystalline salt that forms, with some pentachloride, when tantalum pentachloride is distilled in dry air. It hydrolyzes in water.

Tantalum Tribromide. Tantalum tribromide, $TaBr_3$, forms when the pentabromide is reduced with hydrogen at 700°C (76). It reacts with water and alkali, liberating hydrogen and forming $Ta_2O_5.2H_2O$.

Tantalum Pentabromide. Tantalum pentabromide [*13451-11-1*], $TaBr_5$ (sp gr 4.67, mp 265°C, bp 348.8°C), is a yellow crystalline salt that hydrolyzes in water. It can be prepared by heating tantalum metal in pure bromine gas above 300°C. Vapor pressures of the liquid have been reported (75).

Tantalum Pentaiodide. Tantalum pentaiodide [*14693-81-3*], TaI_5 (mp 496°C, bp 543°C), forms a shiny black crystal when tantalum metal is heated to a bright red heat in iodine. It hydrolyzes when exposed to air but does not decompose below its boiling point when protected. It loses iodine when heated somewhat above 1000°C (75). Vapor pressures of the liquid have been reported (75).

Tantalum Pentafluoride. Tantalum pentafluoride [*7783-71-3*], TaF_5 (sp gr 4.74, mp 96.8°C, bp 229.5°C), is a colorless crystalline salt which is soluble in water without hydrolysis. It can be prepared conveniently by the reaction of anhydrous hydrogen fluoride with tantalum pentachloride at room temperature. It forms also when tantalum metal is heated in a stream of fluorine. Tantalum forms a number of double fluorides, of which potassium fluorotantalate is best known.

Potassium Fluorotantalate. Potassium fluorotantalate [*16924-00-8*] (potassium tantalum fluoride), K_2TaF_7 (sp gr 5.24, mp 740 ± 10°C), crystallizes as colorless rhombic needles when potassium fluoride is added to a solution of tantalum fluoride with an excess of hydrofluoric acid. It hydrolyzes when boiled in water containing no free hydrofluoric acid. Its solubility in acidulated water varies from 60 g/100 mL soln at the boiling point to less than 0.5 g/100 mL soln cold. The difference in solubility characteristics of K_2TaF_7 and K_2NbOF_5 was used in the industrial separation of tantalum from niobium. Potassium fluorotantalate is the most important tantalum salt. The recommended TWA maximum workplace exposure for K_2TaF_7 in air is 2.5 mg/m^3 (3.3 mg/yd^3) (fluoride basis). Oral toxicity tests in rats report an LD_{50} value (median lethal dose) of 2500 mg/kg of body weight (68).

Tantalum fluoride forms complexes with the fluorides of the alkali metals, alkaline earth metals, certain heavy metals, ammonia, and pyridine. The following compounds have been described: $2NH_4F.TaF_5$ [*12022-02-5*], $3NaF.TaF_5$ [*17456-13-2*], 2NaF.-$TaF_5.H_2O$, [*19469-74-0*], $2RbF.TaF_5$ [*76056-60-5*], $ZnF_2.TaF_5.6H_2O$ [*36248-24-5*], $CuF_2.TaF_5.4H_2O$ [*36248-23-4*], $2LiF.TaF_5.2H_2O$ [*21729-09-9*], $NaF.TaF_5$ [*19639-20-4*], $CsF.TaF_5$ [*16923-68-5*], $2CsF.TaF_5$, and $C_5H_5N.HF.TaF_5$ [*69454-16-6*] (76). An oxyfluoride complex, $2KF.TaO_2F_3.H_2O$, has also been identified.

Tantalum Hydride. Tantalum hydride [13981-95-8], TaH (sp gr 15.1), is a gray, brittle material having a metallic luster. Tantalum metal absorbs hydrogen by cathodic electrolysis or when heated in hydrogen at ca 450°C. Hydrogen is absorbed to the extent of $TaH_{0.76}$, ie, ca 740 volumes of hydrogen gas per volume of metal (2,77). The absorption is accompanied by an expansion of the body-centered crystal lattice and by an increase in molecular volume of the tantalum. The absorbed hydrogen is liberated when the hydride is heated above 800°C. Tantalum hydride exhibits superconductivity at low temperatures, and it resists attack by acids (78) (see Hydrides).

Tantalum Nitride. Tantalum nitride [*12033-62-4*], TaN (sp gr 13.80, mp 2800 ± 50°C), forms in hexagonal, dull-gray crystals when tantalum is heated in pure nitrogen at ca 1100°C. Nitrogen is liberated if the nitride is heated to 2000°C. Tantalum nitride is insoluble in acids but is easily decomposed by potassium hydroxide with the evolution of ammonia. It is an interstitial compound, like the carbide. It does not exhibit superconductivity above 1.20 K (79) (see Nitrides).

Tantalum Oxides and Tantalic Acid. Tantalum oxides and tantalic acid are described in ref. 80.

Tantalum(II) Oxide. Tantalum(II) oxide [*12035-90-4*] is not easily prepared, but its existence has been reported. Tantalum tribromide decomposes in water to a brown powder, $TaO_2.2H_2O$, which changes to the pentoxide in air at room temperature (3). An impure Ta_2O_4 [*12035-90-4*] probably forms when the pentoxide is partially reduced with carbon at 1900°C.

Tantalum Pentoxide. Tantalum pentoxide [*1314-61-0*] (tantalum oxide), Ta_2O_5 (sp gr 8.2, mp 1800°C), is a rhombic, white powder and is insoluble in water and in most acids and alkalies. It reacts slowly with hot hydrofluoric acid. It can be taken up by fusion with potassium pyrosulfate or alkali-metal hydroxides or carbonates. It is prepared by the ignition in air of tantalum metal, tantalum hydride, or tantalum carbide or by heating tantalic acid. Tantalum oxide reacts with carbon at 1900°C forming the carbide and with S_2Cl_2 at a dull red heat to form the pentachloride [*7721-01-9*].

Tantalic Acid. Tantalic acid [*75397-94-3*], $HTaO_3$ or probably more correctly $Ta_2O_5.nH_2O$, is the name given to hydrated forms of tantalum oxide. It is a white insoluble precipitate which forms when a potassium pyrosulfate fusion containing tantalum is leached with water, or when a solution of tantalum fluoride is made ammoniacal, a solution of alkali tantalates is acidulated with hydrochloric acid, or when a halide of tantalum is hydrolyzed by boiling in water. Tantalic acid forms organic complexes with tannic, oxalic, tartaric, salicyclic, citric, and pyrogallic acids. These complexes are very useful to the analytical chemist. A voluminous, characteristic, yellow precipitate is produced when tantalic acid is formed in the presence of a tannin; it is completely insoluble in water or dilute hydrochloric acid. The oxalic and tartaric complexes of tantalic acid, on the other hand, are soluble in water. The pyrogallol–tantalic acid complex, which is colored, is used in the spectrophotometric determination of tantalum (81). Tantalic acid is very weakly acidic; it readily reacts and dissolves with either hydrofluoric acid or potassium hydroxide. When tantalic acid is ignited, tantalum pentoxide forms.

Tantalic acid forms several series of tantalates. Some of these occur in nature (see Table 1) and many others have been prepared in the laboratory. Potassium tantalate or orthohexatantalate [*12437-78-4*] (4:3 ratio), $4K_2O.3Ta_2O_5.16H_2O$, is a colorless monoclinic salt that is soluble in water. It is the most common potassium salt of tantalum and is prepared by dissolving a fusion of potassium hydroxide and tantalum pentoxide in water and crystallizing the salt by evaporation. Cesium and rubidium form tantalates in this series which has only fourteen molecules of water. Potassium metatantalate [*57126-93-9*] (1:1 ratio), $K_2O.Ta_2O_5$, is a white insoluble salt prepared by leaching calcined potassium hexatantalate with water.

Tantalum Sulfide. Tantalum sulfide [*12138-15-7*], Ta_2S_4, is a black crystalline powder which is stable below 1300°C. It is insoluble in water and is very slowly attacked by hydrofluoric, sulfuric, and nitric acids. It is decomposed by fusion in potassium hydroxide or pyrosulfate. It is formed by heating tantalum metal to a red heat in sulfur vapor.

BIBLIOGRAPHY

"Tantalum and Tantalum Compounds" in *ECT* 1st ed., Vol. 13, pp. 600–613, by D. F. Taylor, Fansteel Metallurgical Corporation; "Tantalum and Tantalum Compounds" in *ECT* 2nd ed., Vol. 19, pp. 630–652, by Donald F. Taylor, Fansteel, Inc.

1. G. L. Miller, *Tantalum and Niobium*, Academic Press, Inc., New York, 1959.
2. F. T. Sisco and E. Epremian, eds., *Columbium and Tantalum*, John Wiley & Sons, Inc., New York, 1963.
3. F. Fairbrother, *Chemistry of Niobium and Tantalum*, American Elsevier Publishing Co., Inc., New York, 1967.
4. C. A. Hampel, ed., *Rare Metals Handbook*, 2nd ed., Reinhold Publishing Corp., New York, 1967.
5. H. Rose, *Pogg, Ann. Phys. Chem.* **53,** 317 (1844).
6. J. C. Marignac, *Ann. Chim. Phys.* **8,** 5 (1866).
7. *First International Symposium on Tantalum*, Rothenburg ob der Tauber, FRG, May 10–12, 1978, Tantalum Producers International Study Center, Brussels, Belgium.
8. E. S. Dana, ed., *System of Mineralogy*, 6th ed., John Wiley & Sons, Inc., New York, 1937.
9. C. S. Hurbut, Jr., *Revision of Dana's Manual of Mineralogy*, 15th ed., 1941.
10. T. S. Jones, *Tantalum*, MCP Mineral Commodity Profiles, Bureau of Mines, U.S. Department of the Interior, Washington, D.C., June 1980.
11. T. S. Jones, *Columbium and Tantalum in 1981*, Mineral Industry Surveys, Bureau of Mines, U.S. Department of the Interior, Washington, D.C.
12. P. Borchers and G. J. Korinek in H. Y. Sohn, O. N. Carlson, and J. T. Smith, eds., *Extractive Metallurgy of Refractory Metals*, conference proceedings, Metallurgical Society of AIME, Feb. 22–26, 1981.
13. M. Shamsudding and H. Y. Sohn in ref. 12.
14. J. Walraedt, *The Coldstream Process*, 7th Plansee Seminar, Reutte, Austria, June 1971, Metalwork Plansee.
15. B. F. Kieffer and E. F. Baroch in ref. 12.
16. Brit. Pat. 582,921 (Dec. 2, 1946), E. M. Trent (to Poderloys, Ltd.).
17. U.S. Pat. 3,595,484 (July 27, 1971), P. Barnard (to U.S., as represented by the Secretary of the Interior).
18. U.S. Pat. 2,485,175 (Oct. 18, 1949), G. J. Trapp.
19. U.S. Pat. 3,438,730 (Apr. 15, 1969), W. M. Shwayder.
20. U.S. Pat. 4,234,333 (Nov. 18, 1980), M. H. Ghanderhari and M. Schussler (to Fansteel, Inc.).
21. F. F. Schmidt and H. R. Ogden, *The Engineering Properties of Tantalum and Tantalum Alloys*, DMIC Report 189, Sept. 13, 1963.

22. *Properties of Selection: Nonferrous Alloys and Pure Metals*, Vol. 2 of *ASM Metals Handbook*, 9th ed., Metals Park, Ohio, pp. 799–804, 1979.
23. *Properties and Selection: Stainless Steels, Tool Materials and Special Purpose Metals*, Vol. 3 of *ASM Metals Handbook*, 9th ed., Metals Park, Ohio, pp. 317–325, 1980.
24. M. Schussler, *Fansteel Corrosion Data Survey on Tantalum*, Fansteel, Inc., N. Chicago, Ill., May 8, 1972.
25. F. S. Badger, *Ind. Eng. Chem.* **50,** 1608 (1958).
26. C. R. Bishop and M. Stern, *Corrosion* **17,** 379, 385 (1961).
27. M. Stern and C. R. Bishop in ref. 2.
28. C. A. Hampel, *Corrosion* **14,** 557 (1958).
29. E. Rabald, *Werkst. Korros.* **12,** 695 (1961).
30. *Corrosionomics* **5**(5), 5 (1960).
31. A. Clauss and H. Forestier, *Compt. Rend.* **246,** 3241 (1958).
32. R. F. Rolsten, *J. Am. Chem. Soc:* **80,** 2952 (1958).
33. D. Lupton, K. Shulze, and F. Aldinger, *Niobium 81 International Conference*, San Francisco, Calif., 1981.
34. C. T. Sims in ref. 33.
35. E. F. Bradley, ed., *Source Book on Materials for Elevated-Temperature Applications*, American Society for Metals, Metals Park, Ohio, 1979.
36. D. F. Taylor, *Chem. Eng. Progr.* **54,** 47 (1958).
37. J. R. Werning and K. B. Higbie, *Ind. Eng. Chem.* **46,** 644, 2491 (1954).
38. H. A. Wilhelm and R. A. Foos, *U.S. Atomic Energy Commission Report ISC-694*, Washington, D.C., Aug. 1954.
39. E. L. Koerner, Jr., M. Smuts, and H. A. Wilhelm, *U.S. Atomic Energy Commission Report ISC-802*, Washington, D.C., Dec. 1956.
40. C. H. Chilton, *Chem. Eng.* **65,** 104 (1958).
41. U.S. Pat. 3,117,833 (Jan. 14, 1964), J. A. Pierret (to Fansteel, Inc.).
42. U.S. Pat. 3,712,939 (Jan. 23, 1973), R. H. Capps and G. S. Harman (to Union Carbide Corporation).
43. W. Ger. Pat. Offen. 29 27 835 (June 4, 1981), H. Meyer (to Hermann C. Starck).
44. G. H. Korinek in ref. 7.
45. U.S. Pat. 3,447,894 (June 3, 1969), R. A. Gustison, F. Gorczyca, and J. A. Cenerazzo (to Kawecki Chemical Company).
46. U.S. Pat. 3,721,722 (Mar. 20, 1973), R. A. Gustison (to Kawecki Berylco Industries, Inc.).
47. U.S. Pat. 4,192,674 (Mar. 11, 1980), B. Krismer and H. Pungs (to Hermann C. Starck).
48. U.S. Pat. 4,211,754 (July 8, 1980), M. C. F. Van Hecke and J. Deweck (to Metallurgie Hoboken-Overpelt).
49. U.S. Pat. 2,950,185 (Aug. 23, 1960), E. G. Hellier and G. L. Martin (to National Research Corporation).
50. U.S. Pat. 3,829,310 (Aug. 13, 1974), T. X. Mahy (to Norton Company).
51. U.S. Pat. 4,149,876 (Apr. 17, 1979), C. F. Rerat (to Fansteel, Inc.).
52. W. Ger. Pat. Appl. P25 17 180.6 (Apr. 18, 1975), R. Hahn and D. Behrens (to Hermann C. Starck).
53. U.S. Pat. 3,767,456 (Oct. 23, 1973), F. A. Glaski (to Fansteel, Inc.).
54. Can. Pat. 702,612 (Jan. 26, 1965), D. R. Brutvan, R. L. Ripley, and H. V. Seklemian.
55. H. A. Johansen and S. L. May, *Ind. Eng. Chem.* **46,** 2499 (1954).
56. U.S. Pat. 2,905,549 (Sept. 22, 1959), D. F. Taylor, R. L. Baughman, and L. F. Yntema (to Fansteel, Inc.).
57. G. D. Corey in ref. 7.
58. *Welding Data Book* issued by editors of *Welding Design and Fabrication*, A/78 and A/79 (1974/75).
59. *Machining*, Vol. 3 of *ASM Metals Handbook*, 8th ed., Metals Park, Ohio, 1967, pp. 433–438.
60. *Metallography, Structures and Phase Diagrams*, Vol. 8 of *ASM Metals Handbook*, 8th ed., Metals Park, Ohio, 1973, p. 109.
61. T. S. Jones, *Bulletin 671*, U.S. Bureau of Mines, Washington, D.C., 1980.
62. *Metals Statistics 1981*, 74th ed., American Metal Market, New York, 1981.
63. K. A. Kraus and G. E. Moore, *J. Am. Chem. Soc.* **71,** 38 (1949); **73,** 2900 (1951).
64. K. A. Kraus and F. Nelson, *ASTM Special Technical Publication*, 195 (1954).
65. J. L. Hague, E. D. Brown, and H. A. Bright, *J. Res. Natl. Bur. Stand.* **53,** 261 (1954).
66. S. Kallmann, H. Oberthin, and R. Liu, *Anal. Chem.* **34,** 609 (1962).

67. *Methods for Emission Spectrochemical Analysis, Optical Emission Spectrographic Analysis of Tantalum Metal*, ASTM Committee E-2, American Society for Testing and Materials, Philadelphia, Pa., 1972.
68. *1979 Registry of Toxic Effects of Chemical Substances*, NIOSH, U.S. Department of Health and Human Services, Washington, D.C., Sept. 1980.
69. U.S. Pat. 3,418,106 (Dec. 24, 1968), J. A. Pierret (to Fansteel, Inc.).
70. U.S. Pat. 3,473,915 (Oct. 21, 1969), J. A. Pierret (to Fansteel, Inc.).
71. U.S. Pat. 4,009,007 (Feb. 22, 1977), S. S. Fry (to Fansteel, Inc.).
72. K. A. J. Brookes, *World Directory and Handbook of Hardmetals*, 2nd ed., Engineer's Digest Ltd., London, 1979.
73. B. Aronsson and L. J. Aschan in ref. 33.
74. P. H. Booker and R. E. Curtis, "Tantalum-Carbide-Free Steel Cutting Grade Tool Alloys," Cutting Tool Engineering, pp. 18–22, Sept./Oct. (1978).
75. K. M. Alexander and F. Fairbrother, *J. Chem. Soc. S*, 223 (1949).
76. C. W. Balke, *J. Am. Chem. Soc.* **27,** 1140 (1905).
77. M. von Parani, *Z. Elektrochem.* **11,** 555 (1905).
78. F. Horn and W. Zweigler, *J. Am. Chem. Soc.* **69,** 92762 (1947).
79. G. F. Hardy and J. H. Hulm, *Phys. Rev.* **93,** 1004 (1954).
80. R. Young and T. Hastings, *J. Am. Chem. Soc.* **64,** 1740 (1942).
81. J. I. Dinnin, *Anal. Chem.* **25,** 1803 (1953).
82. P. G. Melikoff and L. Pissarjewski, *Z. Anorg. Chem.* **20,** 345 (1899).

Robert E. Droegkamp
Mortimer Schussler
John B. Lambert
Donald F. Taylor
Fansteel, Inc.

TAR AND PITCH

Most organic substances, other than those of simple structure and low boiling point, when pyrolyzed, ie, heated in the absence of air, yield dark-colored, generally viscous liquids termed tar and pitch. The differentiation between these terms is not precise. When the by-product is a liquid of fairly low viscosity at ordinary temperature, it is regarded as a tar; if of very viscous, semisolid, or solid consistency, it is designated as a pitch. Thus, in preparative organic chemistry, a tar or pitch is frequently the distillation residue. Some thermal decomposition always accompanies the distillation of vegetable and animal fats and oils resulting in the production of small amounts of vegetable-oil pitch, wool-grease pitch, stearin pitch, and so on. Such products, of ill-defined composition and properties, have little industrial importance and are mainly burned as fuel.

Large amounts of tar or pitch by-products are produced by industrial processes. The distillation of crude petroleum yields a pitch-like residue termed bitumen or asphalt. In the United States, these terms are interchangeable, but in Europe the term asphalt is generally restricted to naturally occurring rock or lake asphalt, whereas the residual product of crude-oil distillation is termed bitumen. Although these are important industrial materials produced in millions (10^6) of metric tons annually, they are not included here (see Asphalt; Petroleum, products).

WOOD TAR

Production

The pyrolysis or carbonization of hardwoods, eg, beech, birch, or ash, in the manufacture of charcoal yields, in addition to gaseous and lighter liquid products, a by-product tar in ca 10 wt % yield. Dry distillation of softwoods, eg, pine species, for the production of the so-called DD turpentine yields pine tar as a by-product in about the same amount. Pine tar was also called Stockholm tar or Archangel tar and was, at one time, imported from the Baltic by European maritime countries for the treatment of cordage and ship hulls; it was an important article of commerce in the 17th to 19th centuries. The small amount produced today is burned as a crude fuel.

Charcoal production from hardwoods, on the other hand, has increased in recent years. Production of charcoal in 1909 was estimated at 508,000 metric tons, decreasing to 213,000 tons in 1947, and is currently between 508,000 and 609,000 t. The value of the charcoal produced had risen from $\$29 \times 10^6$ in 1963 to well over $\$50 \times 10^6$ in 1981.

Composition, Processing, and Uses

There are no statistics available for the amount of wood tar processed, but almost all of it is burned. The commercial by-products from wood carbonization are limited to methanol, denatured methanol, methyl acetate, and acetic acid. These products are derived from the aqueous phase of the condensed products, the so-called pyroligneous acid. On distillation, pyroligneous acid yields wood spirit, acetic acid with small amounts of propionic and butyric acids, and soluble tar. The wood spirit, on refining, yields methanol, methyl acetate, and acetone. The aqueous acid fraction is neutralized with milk of lime to give grey acetate of lime, 82–84% $(CH_3COO)_2Ca$, which is neutralized to give acetic acid or pyrolyzed to give acetone. The so-called soluble tars are mainly condensation products of aldehydes and phenols; they are burned as fuel.

Small amounts of the sedimentation tar, ie, the separated organic layer from the condensed wood-carbonization vapors, are distilled, first at atmospheric pressure to give wood spirit, crude acetic acid, and light wood oils. The first two of these fractions are added to the similar fractions from the pyroligneous acid for further processing. The light wood oil is treated with permanganate and sulfuric acid and used as a solvent or for giving a wood smoke or tar note to perfumes, soaps, and shampoos. Further distillation under reduced pressure yields wood creosote which is used for the preservation of cordage, timber, and Hessian sacks, and as a disinfectant, component of wood stains, and froth-flotation agent.

Chemically, wood tar is a complex mixture that contains at least 200 individual compounds, among which the following have been isolated (1): guaiacol (2-methoxy-phenol), 4-ethylguaiacol (2-methoxy-4-ethylphenol), creosol (4-methylguaiacol), 2,6-xylenol, butyric acid, crotonic acid, acetol (1-hydroxy-2-propanone), butyrolactone, maltol (2-methyl-2-3-hydroxy-4*H*-pyran-4-one), tiglaldehyde (2-methyl-2-propenal), methyl ethyl ketone, methyl isopropyl ketone, methyl furyl ketone, and 2-hydroxy-3-methyl-2-cyclopenten-1-one.

COAL TAR

By far the largest source of tar and pitch is the pyrolysis or carbonization of coal and, generally, the terms tar and pitch are synonymous with coal tar and the residue obtained by its distillation (see Coal, coal-conversion processes, carbonization). The importance of coal tar as an industrial raw material dates back to the first half of the eighteenth century when the carbonization of coal with by-product tar production was expanding rapidly in the United Kingdom. Initially, the crude tar was subjected to a simple flash distillation in pot stills to yield a solvent naphtha, creosote for timber preservation, and a residue of pitch that found an outlet as a binder for coal briquettes. Later, coal tar was the main source of aromatic hydrocarbons, phenols, and pyridine bases needed by the rapidly expanding dyestuffs, pharmaceuticals, and explosives industries. The development of by-product coke ovens and crude benzene recovery at both coke ovens and gas works greatly increased the supplies of crude tar and tar distillates for the recovery of tar chemicals, ie, benzene, toluene, xylenes, phenol, cresols and cresylic acids, pyridine and picolines, naphthalene and anthracene, in addition to the so-called bulk products, eg, creosote, tar paints, road tars, and pitch binders (2–4).

Until the end of World War II, coal tar was the main source of these aromatic chemicals. However; the enormously increased demands by the rapidly expanding plastics and synthetic-fiber industries greatly outstripped the potential supply from coal carbonization and led to production from petroleum. Today, over 90% of the world production of aromatic chemicals is derived from the petrochemical industry whereas coal tar is chiefly a source of anticorrosion coatings, wood preservatives, feedstocks for carbon-black manufacture, and binders for road surfacings and electrodes.

Physical Properties

The physical properties of crude tars vary over a wide range. Investigation has been mainly concerned with establishing correlations between the more readily determined chemical and physical properties of the distillate oils and residual pitch, and other properties. Based on the correlations, other properties can be predicted with an accuracy sufficient for such purposes as plant design (see Table 1).

Viscosity of Coal-Tar Pitch; Change with Temperature. Since pitch is mainly used as a hot-applied binder or adhesive, the viscosity and its change with temperature are important in industrial practice. Some useful correlations, by which the viscosity of pitch at any temperature can be predicted, have been developed. The data on which such correlations are based may be from one of the fixed equiviscous points that characterize a pitch (see Table 2).

The viscosity of a straight-run or fluxed-back pitch can be calculated from the R and B (ring-and-ball) softening point (27):

$$\log \eta_t = -4.175 + \frac{711.8}{86.1 - t_s + t} \qquad (1)$$

Table 1. Correlations for Predicting the Physical Properties of Tars and Tar Products

Property	Applicable to	Correlation expression	Comments and reference
density, d, at 20°C, g/cm^3	coke-oven dry tars and tar oils	$d_{20} = 1.877 \times 10^{-3} M^a + 0.808$ $d_{20} = 7.337 \times 10^{-4} t_b{}^b + 0.890$	5
variation of sp gr with temperature, t, g/cm^3	dry tars	$\frac{\text{sp gr}}{t} = 0.001778 - 0.00098 \times \text{sp gr}$ at 15.6°C	nomograph available; 7
variation of density with temperature, t, g/cm^3	dry tars, tar oils, and pitches	in $d_{t_1} = d_{t_2} - b(t_1 - t_2)$, $b = 0.00068 \pm 0.00005$ for dry tars and $(162.7 - 86.2\, d_{20} \pm 8) \times 10^{-5}$ for tar oils and pitches	8
	dry coke-oven tars and tar oils	in $\beta = \frac{d_0 - d_t}{d_0 t}$, $\beta = \left(-10.5433 + \frac{0.0122}{t_b}\right) \times 10^{-4}$ $\beta = (-12.6114 + 0.0346\, M^a) \times 10^{-4}$	β has negative value; 8
viscosity, η, mPa·s (= cP)	tar oils	$\log \eta_{20} = 0.0078\, t_b - 1.123$, at other temperatures, $\log \eta$ varies linearly with the absolute temperature; at t_b the viscosity of any tar oil is approximately 0.25 mPa·s	nomograph available; 9
specific heat capacity, C, kJ/(kg·K)[c]	tar oils	$C_t = \{(0.7360 + 0.8951 \times d_{20} + 0.00360\, t)/d_{20}\} + (0.00904 - 0.0000221 \times t_b) \times \text{T.A.}^d$	10
	pitches	$C_t = \frac{3.665}{d_{20}} - 1.729 + 0.00389 \times t$	12
thermal conductivity, K, W/(m·K)	tar oils	$K = (1.34 - 0.00084\, t \pm 0.084) \times 10^{-5}$	
	pitches	$K = (1.423 \pm 0.084) \times 10^{-5}$	no significant variation over range 25–105°C, 13
surface tension, S, mN/m (= dyn/cm)	dry tars, tar oils, pitches	$S = 93.8 \times d_{20} - 0.0496 \times t_b - 47.5$ $S_t = 18.4 \times d_t^4$ $S_t = \frac{S_{20} d_t}{d_{20}^4}$	nomograph available (14–15)
latent heat of vaporization, L, kJ/kg[b]	tar oils	at the average boiling point, $L = d_{20}(486.1 - 0.599\, t_b)$	16

[a] Av molecular weight.

[b] Av boiling point defined as the mean of the temperatures in °C at which 10%, 20% . . . 80%, 90% by volume distills in a standard flask distillation. Ref. 6.

[c] To convert J to cal, divide by 4.184.

[d] % Tar acids defined as the percentage by volume extracted by 10% aqueous caustic soda. Ref. 11.

Table 2. Viscosities of Fixed Points for Coal-Tar Pitch

Fixed point	Viscosity Pa·s (10^3 cP)	Difference between fixed point and EVT, °C	Ref.
equiviscous temperature, EVT, °C	25		17
R and B softening point[a], °C	800	−19	19
K and S softening point[b], °C	5,500	−27	22
penetration of 200[c]	2×10^4	−30	23
penetration of 10[c]	10^7	−51	24[d]
ductility point, °C[e]	10^7	−51	25
Fraas brittle point, °C	4×10^7	−65	26
transition to glassy state	10^{12}	−90	

[a] Ring-and-ball method, ASTM D 36-26. In the United States, two other softening point methods are employed: cube-in-air (18) and cube in water (20). Cube-in-air softening point = R and B softening point + 4°C. Cube-in-water softening point = R and B softening point + 10°C.

[b] Krämer and Sarnow (21).

[c] Penetration of 20 or 1 mm using a 100-g weight for 5 s at 25°C.

[d] Ref. 24 gives the relationship between viscosity and penetration (mm × 10): η (Pa·s) = $1.58 \times 10^9/(\text{penetration})^{2.16}$.

[e] When a pitch is tested for ductility, the sample either suffers brittle fracture without elongation or elongates to the maximum distance without breaking. When tested at increased temperatures at a particular point, the ductility point, the behavior changes from the first type to the second.

where η_t is the viscosity at temperature t°C in Pa·s $\times 10^{-1}$ (= P), and t_s is the R and B softening point in °C. This relationship covers a temperature range of 180°C and a viscosity range of 10^8. It cannot be used to predict the viscosity accurately since the softening point itself is only an approximate measure of the viscosity and because there are small differences in the temperature susceptibility of different types of pitch. However, equation 1 provides a useful means of calculating viscosities at any temperature above the softening point with an accuracy sufficient for most design purposes.

The general equation for the temperature function of coal-tar pitch takes the following form (28):

$$\log \log (\nu + 0.8) = K + K_1 \log t \tag{2}$$

where ν is the kinematic viscosity (mm²/s), t is the temperature in K and K and K_1 are constants. K_1 depends on the toluene-insoluble content (TI in weight %):

$$K_1 = 5.8 - 0.03\ \text{TI}$$

This relationship applies to a wide range of straight-run and heat-treated pitches but not to fluxed pitches. If the K and S (Krämer and Sarnow) softening point is taken as a fixed point, equation 2 can be written as follows:

$$\log \log (\nu + 0.8) = 0.08 + 0.04 \log \text{TI} + (5.8\text{–}0.03\ \text{TI})\{\log (t_s + 273) - \log t\}$$

where t_s is the K and S softening point in °C:

Another equation has the form

$$\log \eta = At^{-5} + B \tag{3}$$

where η is the dynamic viscosity in mPa·s (= cP), t is the temperature in K, and A and

B are constants, derived from data relating to a range of pitches and refined tars varying in softening point from 54 to 104°C. The value of A in equation 3 is closely related to the softening point and can be expressed as:

$$A = (0.358\, t_s + 3.74) \times 10^{12} \tag{4}$$

where t_s is the R and B softening point using glycerol as the heating medium or, using water as the heating medium, as shown below:

$$A = (0.358\, t_s + 4.82) \times 10^{12} \tag{5}$$

B in equation 3 is related to the TI content:

$$B = 0.0206 \times \text{TI} - 0.378 \tag{6}$$

Using equation 3, the viscosity of any pitch can be calculated from two measurements in the range 10^4–10^6 mPa·s (= cP) with a precision similar to that which might be expected by direct measurement. By employing equations 3, 4 or 5, and 6, the viscosity of a pitch at any temperature can be calculated from the R and B softening point and the TI content with an accuracy adequate for most engineering purposes.

Chemical Composition

The tars recovered from commercial carbonization plants are not the primary products of the thermal decomposition of coal. The initial products undergo a complex series of secondary reactions. Although pilot-plant tests carried out by the U.S. Bureau of Mines show, that at any given carbonizing temperature the yield of tar is related approximately linearly to the content of volatile matter of the coal, no clear trends could be distinguished between the nature of the coal and the physical properties or chemical nature of the tar produced under the same carbonizing conditions (29).

Recent studies have given a clearer picture. The application of statistical structural analysis based on ir spectrometry and proton nmr (30), a combination of uv and ir spectrometry with gas chromatography (gc) of coal pyrolysis products (30–31), and correlation of refractive index, molecular weights, and densities, have made important contributions (32–33). Evidence suggests that, when coal is heated in the absence of air, relatively small molecules are released that are absorbed in the micropore structure of the coal, including straight-chain and branched-chain alkanes and alkenes with terminal double bonds and aromatic hydrocarbons like benzene, toluene, naphthalene, methyl- and dimethylnaphthalenes, acenaphthene, and fluorene. The principal decomposition occurs at 450–500°C with the volatilization of 15–20% of the coal. This reaction appears to be a depolymerization, probably by breaking the methylene, polymethylene, or conjugated olefin bridges between the aromatic ring clusters to yield free radicals. These radicals are believed to have molecular weights of 300–800 and to consist of systems containing 4–10 fused aromatic or cycloaromatic rings with paraffinic side chains or hydroxy substituents. Some of these free radicals condense to high molecular weight resins which remain in the coke; others are stabilized by hydrogen transfer. The stabilized free radicals undergo secondary cracking reactions as they come in contact with the hot coke.

The nature of the secondary reactions is uncertain. Some believe that the primary tar components are broken down to small free radicals that recombine as they travel toward the retort exit; others suggest that some components remain relatively intact except for the removal of peripheral substituent groups and that the higher molecular weight components of coal tar are, in effect, slightly altered fragments of the original coal structure.

Although these speculations are of academic interest only, it is clear that even tars produced at the lowest commercial carbonization temperatures are very different from primary tars. In fact, low temperature tar, continuous vertical retort (CVR) tar, and coke-oven tar form a series in which the yield of tar decreases, the aromaticity of the tar increases, the content of paraffins and phenols decreases and the ratio of substituted aromatic and heterocyclic compounds to their unsubstituted parent molecules decreases. The reactions accounting for these changes, ie, cracking and cyclization of paraffins, dehydration of phenols, and dealkylation of aromatic and heterocyclic ring compounds, are those that would be expected, on thermodynamic grounds, to occur at the temperatures prevailing in carbonization retorts.

The application of modern methods of analyses such as gc, low-ionization-voltage mass spectrometry, high pressure liquid chromatography (hplc), and nmr has greatly increased knowledge of coal-tar composition without materially altering the qualitative picture.

Coke-oven tar is an extremely complex mixture, the main components of which are aromatic hydrocarbons ranging from benzene and alkylbenzenes to compounds containing as many as twenty or more rings. Heterocyclic compounds containing oxygen, nitrogen, and sulfur, but usually only one heteroatom per ring system are present. Small amounts of paraffinic, olefinic, and partly saturated aromatic compounds also occur. The aromatic and heterocyclic structures occur in both unsubstituted and substituted forms; the main substituents are methyl, ethyl, or hydroxyl. Although longer aliphatic side chains are encountered in small amounts in the lower boiling fractions; they tend, like hydroxyl groups, to disappear in the higher boiling fractions. Molecules containing up to four rings are generally fully condensed, but in the more complex molecules, the degree of condensation is less complete and branched ring-chain structures, in which most of the rings are fused to no more than three other rings, predominate.

Continuous vertical-retort tars differ from coke-oven tars. Whereas the latter contain relatively small amounts of nonaromatic hydrocarbons, CVR tars contain a relatively high proportion of normal straight-chain or slightly branched-chain paraffins; the content of these decreases from ca 20% in the lower boiling fractions to 5–10% in the higher distillate oils.

Coke-oven tars contain, on the average, 3% phenolic compounds, mainly phenol, cresols, xylenols, and naphthols in the fractions distilling up to 300°C. Continuous vertical retort tars, on the other hand, contain 20–30% of hydroxy-substituted hydrocarbons of which some 15% (based on tar) can be extracted from the oils distilling up to 350°C. These phenols are mainly xylenols, methylethyl- and diethylphenols, dihydric phenols (eg, catechol, resorcinol, and their methyl derivatives), hydroxy hydrindenes and their methyl and dimethyl derivatives, naphthols and hydroxysubstituted acenaphthenes, anthracenes, and phenanthrenes.

Furthermore, the aromatic hydrocarbons and heterocyclic structures in CVR tar are predominantly substituted by one or more methyl groups and the content of unsubstituted ring structures, which are the main components of coke-oven tars, is comparatively small.

Low temperature tars contain 30–35 wt % nonaromatic hydrocarbons, ca 30% of caustic-extractable phenols in the distillate oils, and 40–50% of aromatic hydrocarbons; the latter usually contain one or more alkyl substituent groups.

On atmospheric distillation, coke-oven tars yield 55–60% pitch, whereas CVR tars give 40–50% pitch; the pitch yield from low temperature tars is in the 26–30% range.

The number of individual components in coal tar can only be guessed. As many as 461 compounds have been identified (34). The minor components in a neutral fraction of coke-oven naphtha and naphthalene oil were investigated by a combination of high efficiency fractionation, zone refining, and ir spectroscopy. Although the oils analyzed constituted only 13.5% by weight of the tar, 49 components were identified, including all possible C_6–C_9 aromatic hydrocarbons, thiophene and thionaphthene, indan and four methylindans, benzofuran (coumarone) and four methylbenzfurans, benzonitrile and three isomeric tolunitriles, and styrene, α-methylstyrene, *trans*-β-methylstyrene, and indene.

Table 3 gives the average properties of various types of tar and Table 4 lists the amounts of components which have either current or potential industrial possibilities.

Of the total tar bases in UK coke oven and CVR tars, pyridine makes up about 2%, α-picoline about 1.5%, β- and γ-picolines about 2%, and ethylpyridine and lutidines about 6%. Primary bases, aniline and toluidines, account for about 2% of the bases in coke oven and CVR tars and 3.5% of the bases in low-temperature tars. The main basic components in coke-oven tars are quinoline (16–20% of the total), isoquinoline (4–5%), and methylquinolines. These dicyclic bases are less prominent in CVR and low-temperature tars, in which, only a minority of the basic constituents have been identified.

Although gas chromatography has greatly assisted in the elucidation of the structure of tar distillate oils, it is of restricted applicability to pitch. It has, however, been applied to solvent extracts and vacuum distillates of pitch (35). More precise information about the principal components of the lower molecular weight fractions of coke-oven pitch have been obtained by the application of high pressure liquid chromatography, thin-layer chromatography (36), and low voltage mass spectrometric examinations of pitches and pitch solvent fractions at the U.S. Bureau of Mines

Table 3. Properties of Coal Tars

	Coke-oven tars				CVR tars, UK		Low temperature tars, UK av	Lurgi tars, UK av
	UK		FRG	U.S.				
	Av	Range	av	av	Av	Range		
yield, L/t	33.6	16.8–43.2	26.8		70.9	62.7–75.0	95.5	12.7
density at 20°C, g/cm^3	1.169	1.138–1.180	1.175	1.180	1.074	1.066–1.083	1.029	1.070
water, wt %	4.9	3.2–6.5	2.5	2.2	4.0	2.1–6.3	2.2	2.8
carbon, wt %	90.3	86.8–93.5	91.4	91.3	86.0	85.3–87.2	84.0	84.2
hydrogen, wt %	5.5	5.1–6.0	5.25	5.1	7.5	7.2–7.8	8.3	7.7
nitrogen, wt %	0.95	0.73–1.07	0.86	0.67	1.21	0.84–1.65	1.08	1.09
sulfur, wt %	0.84	0.68–0.96	0.75	1.2	0.90	0.67–1.16	0.74	1.39
ash, wt %	0.24	0.03–0.67	0.15	0.03	0.09	0.03–0.24	0.10	0.02
TI[a], wt %	6.7	2.7–7.5	5.5	9.1	3.1	0.9–5.1	1.2	0.7

[a] Toluene-insoluble components.

Table 4. Constituents of Coal Tar

Component, wt % of dry tar	Coke-oven tars, UK Av	Coke-oven tars, UK Range	Coke-oven tars, FRG av	Coke-oven tars, US av	CVR tars, UK Av	CVR tars, UK Range	Low temperature tars, UK av	Lurgi tars, UK av
benzene	0.25	0.12–0.42	0.4	0.12	0.22	0.14–0.26	0.01	0.02
toluene	0.22	0.09–0.35	0.3	0.25	0.22	0.17–0.29	0.12	0.05
o-xylene	0.04	0.02–0.07		0.04	0.06	0.05–0.08	0.05	0.05
m-xylene	0.11	0.08–0.18	0.2	0.07	0.13	0.11–0.18	0.10	0.07
p-xylene	0.04	0.02–0.07		0.03	0.05	0.04–0.07	0.04	0.03
ethylbenzene	0.02	0.01–0.05		0.02	0.03	0.02–0.04	0.02	0.04
styrene	0.04	0.02–0.04		0.02	0.04	0.02–0.06	0.01	0.01
phenol	0.57	0.14–1.15	0.5	0.61	0.99	0.49–1.36	1.44	0.97
o-cresol	0.32	0.10–0.34	0.2	0.25	1.33	0.77–1.53	1.48	1.14
m-cresol	0.45	0.16–1.0	0.4	0.45	1.01	0.58–1.33	0.98	1.83
p-cresol	0.27	0.07–0.70	0.2	0.27	0.86	0.50–1.03	0.87	1.51
xylenols	0.48	0.13–1.30		0.36	3.08	2.70–3.58	6.36	5.55
high boiling tar acids	0.91	0.31–2.09		0.83	8.09	6.57–11.1	12.89	11.95
naphtha	1.18	0.52–2.66		0.97	3.21	2.86–3.84	3.63	3.02
naphthalene	8.94	7.29–11.31	10.0	8.80	3.18	2.15–3.84	0.65	2.01
α-methylnaphthalene	0.72	0.60–0.86	0.5	0.65	0.54	0.46–0.64	0.23	0.63
β-methylnaphthalene	1.32	1.15–1.63	1.5	1.23	0.68	0.62–0.74	0.19	1.05
acenaphthene	0.96	0.42–1.28	0.3	1.05	0.66	0.50–0.80	0.19	0.57
fluorene	0.88	0.46–1.80	2.0	0.64	0.51	0.33–1.23	0.13	0.62
diphenylene oxide	1.50	1.40–2.00	1.4		0.68	0.62–0.74	0.19	0.57
anthracene	1.00	0.52–1.38	1.8	0.75	0.26	0.18–0.30	0.06	0.32
phenanthrene	6.30	2.30–9.80	5.7	2.66	1.75	0.40–2.90	1.60	0.28
carbazole	1.33	0.58–1.73	1.5	0.60	0.89	0.14–1.43	1.29	0.22
tar bases	1.77	1.25–2.60	0.73	2.08	2.09	1.81–2.47	2.09	2.50
medium soft pitch	59.8	49.5–63.9	54.4	63.5	43.7	41.0–49.3	26.0	33.1

(37–38). The results of these studies indicate that pitch contains the following high molecular weight constituents: aromatic hydrocarbons with four rings, eg, chrysene, fluoranthene, pyrene, triphenylene, naphthacene, and benzanthracene; five-membered ring systems are represented by picene, benzopyrenes (BaP and BeP), benzofluoranthene, and perylene. The next highest fraction contains, as the main components, six-membered ring systems, such as dibenzpyrenes, dibenzfluoranthenes, and benzoperylenes. Seven-ring systems, eg, coronene, have also been identified. These basic hydrocarbon structures are accompanied by methyl and polymethyl derivatives and, in the case of the pitches from CVR and low temperature tars, by mono- and polyhydroxy derivatives. As in the case of the distillate oil range, heterocyclic compounds are also present.

Above this relatively low molecular weight range, which constitutes ca 40–50% of a medium soft coke-oven pitch, the information concerning the chemical structure of pitch is only qualitative and derived mainly from statistical structural analysis and mass spectra. The same sequence of polynuclear aromatic and heterocyclic compounds appears to continue, reaching molecules with 20–30 rings. With increasing molecular weight, more heterocyclic atoms appear in the molecule, whereas the number and length of alkyl chains decreases and the hydrocarbon structures are not fully condensed. In the lower temperature pitches, some ring structures appear to be partly hydrogenated.

Manufacture and Processing

The largest volume of coal is carbonized in batch coke ovens to produce a hard coke suitable for blast furnaces for the reduction of iron ore. Oven temperatures, as measured in the flues, are between 1250 and 1350°C and residence time varies between 17 and 30 h. The gas made in this process is mainly used as fuel and other applications in the steel works (see Fuels, synthetic).

Until ca 1960, in countries where natural gas was not available, large amounts of coal were carbonized for the production of town gas, and a coke grade which, although unsuitable for metallurgical use, is satisfactory as a domestic fuel in closed stoves. The early cast-iron and silica horizontal retorts used at gas works were replaced by continuous vertical retorts, some of which are still in operation. They operate at flue temperatures of 1000–1100°C; the volatile products are rapidly swept from the retort by the introduction of steam at 10–20% by weight of the coal carbonized.

The passing of the Clean Air Act in the UK in 1956 resulted in a revival of interest in low temperature carbonization to produce a very reactive coke suitable for open fires. In the Coalite process, the coal is heated at 600–650°C for 4 h in small retorts each holding 6–7 metric tons (38). The Rexco process employs large internally heated retorts in which charges of 34 metric tons are heated to 700–750°C for 6 h (39).

In the future, crude low temperature tar may be supplied as by products of synthetic natural gas (SNG) and syncrude from coal substitute. A number of the more advanced SNG processes such as Lurgi, Bi-Gas, and Cogas employ low temperature pyrolysis of coal and yield a by-product tar. The tar obtained from the Lurgi process appears to be similar to Coalite tar. Lurgi gasifiers have been chosen for five large U.S. plants for the conversion of western lignite and subbituminous coals to SNG. In full operation, these plants are expected to produce 2×10^6 metric tons of by-product tar per year. Lurgi gasifiers are also employed to produce synthesis gas for South Africa's two Fischer-Tropsch plants and will be used in the third which is under construction. These three plants will yield ca 250,000 metric tons of tar.

Primary Distillation. As produced, crude coal tar is of value only as a fuel and although formerly large amounts were burned, the practice has largely been abandoned. Today, 99% of the tar produced in the UK and the FRG and 75% of U.S. production is distilled. Most of the crude tar regarded as being burned in the United States is first topped in simple continuous stills to recover a chemical oil, ie, a fraction distilling to 235°C that contains the bulk of naphthalene and phenols.

Although 10–30 t mild-steel or wrought-iron pot stills, some equipped with fractionating columns, are still in use at smaller tar works, continuous stills with daily capacities from 100–700 t are preferred.

The various designs of continuous tar stills are basically all similar. The crude tar is filtered to remove large-sized solid particles, dehydrated by heat exchange and passage through a waste heat coil, then heated under pressure to ca 360°C, and flashed to separate volatile oils from the involatile pitch. The volatile oils are separated into a series of fractions of increasing boiling range by fractional condensation in a side-stream column or a series of columns. The various designs differ in the extent to which heat exchange is used, in the plan of the pipe-still furnace, in the distillation pressure, and whether recycle of pitch or base tar is involved.

The Abderhalden design, employed at some French distilleries and used in the UK for distilling tar to a base-tar residue, is probably the simplest single-pass, atmospheric-pressure design (40). After straining and the addition of alkali as a corrosion inhibitor, the crude tar is heat-exchanged with the distillation side streams and the hot pitch and then passed through an economizer coil in the convection section of the furnace. This furnace is essentially a rectangular chamber lined with refractory brick divided into two sections by a curtain wall which has apertures to enable the hot flue gases to pass from one compartment to the other. The first section serves as the combustion and radiant-heating section in which coke-oven gas, fuel oil, or creosote–pitch fuel is burned at specially designed nozzles projecting into the chamber. The products of combustion are drawn by a fan from the combustion chamber into the second, or convection, chamber before exhaustion to a chimney stack via a waste-heat coil. The coils through which the tar is pumped are, in some designs, set in the walls of the convection chamber only; in other designs, the main heating coil is partly set in the combustion chamber where it is heated by direct radiation.

In the Abderhalden plant, the crude tar leaves the waste-heat coil at ca 150°C under a back pressure and is expanded into the dehydrator. The temperature in this cylindrical baffled vessel is maintained by circulating hot, dehydrated tar through a coil at the base. The moisture and light oils flash off and pass directly to the side-stream fractionating column. The dehydrated tar from the base of the dehydrator is pumped at 400–500 kPa (ca 4–5 atm) through the main heating coil in the radiant section of the furnace. The hot tar at 350–360°C is injected into the pitch flash chamber together with superheated steam. Steam and oil vapors join the dehydrator overhead distillate at the base of the fractionation column from which an overhead stream of water and light oils, side-stream oil fractions, and a heavy oil residue are taken. The overhead stream is separated in a decantor and the oil layer partly returned to the bubble-cap fractionating column for reflux. A similar design differs only in the extent of heat exchange (41).

Several descriptions have been published of the continuous tar stills used in the USSR (42–44). These appear to be of the single-pass, atmospheric-pressure type, but are noteworthy in three respects: They do not employ heat exchange and they incorporate a column with a bubble-cap fractionating section and a baffled enrichment section instead of the simple baffled-pitch flash chamber used in other designs. Both this column and the fractionation column, from which light oil and water overhead distillates, carbolic and naphthalene oil side streams, and a wash oil-base product are taken, are equipped with reboilers.

The original Heinrich Koppers design is still in use at some older plants in the UK and Europe (45). It is a single-pass, atmospheric-pressure design but differs from other designs by employing separate bubble-tray columns for each oil fraction. After straining, alkali doping, and dehydration, the tar is pumped through the radiantly heated main furnace coil under pressure and, at 360°C, injected into the baffled-pitch column where the pitch is separated from the oil vapors; steam is added to the pitch column to assist vaporization. These vapors then pass through four columns operated at atmospheric pressure and progressively lower temperature to yield the various oil fractions as base products. The temperature at the top of each column is automatically controlled by recycling part of the bottom from each column as reflux to the preceding column. The overhead distillate from the final column consists of light oil and steam. It is condensed, the two layers are separated in a decantor, and part of the light-oil product is recirculated as reflux.

The simplest unit employing vacuum fractionation is that designed by Canadian

Badger for Dominion Tar and Chemical Co. at Hamilton, Ontario (46). In this plant, the tar is dehydrated in the usual manner by heat exchange and injection into a dehydrator. The dry tar is then heated under pressure in an oil-fired helical-tube heater and injected directly into the vacuum fractionating column from which a light-oil overhead fraction, various oil fractions as side streams, and a pitch base product are taken.

A unique design is employed in the Clairton refinery of United States Steel Corporation where a sequence of fractionation stages is operated at increasing temperature (47). Each stage consists of a helical-coil heater, a vacuum flash drum, and a vacuum fractionating column. Dehydration and the removal of light oil are carried out at atmospheric pressure. The topped tar is pumped to the first stage where it yields a phenolic oil overhead. The bottoms pass to the second stage, which yields a close-cut naphthalene oil as the overhead product. The third stage separates residual oils from the pitch, whereas in the fourth stage, the residual oils are separated into fractions required for various grades of creosote.

The modern Teerverwertung-Koppers design, based on the experience at the Rütgerswerke refinery at Meiderich, is a single-pass atmospheric- and reduced-pressure unit illustrated in Figure 1 (48). The crude tar is pumped through five heat exchangers to the dehydrator. The light oil and steam from the top of this unit are cooled, condensed, and separated, and some of the oil recycled as reflux. The dehydrated tar is pumped through a heating coil in the tube furnace and, at 300–310°C, injected into the atmospheric pressure fractionation stage, which consists of two fractionating columns and a splitting column. The products taken from this first stage include residual light oils and water, taken overhead from the first column and passed back to the dehydrator, naphthalene oil as overhead distillate from the second column, and wash oil as the base product from the second column. The base tar from the base of the first column is picked up by a pump and passed through a separate coil in the furnace and, at 300°C, injected into the pitch flash chamber which, like the final fractionating column, is maintained under a vacuum of 13.3 kPa (100 mm Hg). The residual oil vapors, separated from the pitch residue, are split into a heavy wash oil and light and heavy anthracene oils in the vacuum column.

The Rütgerswerke plant at Castrop-Rauxel produces closely fractionated concentrates of a number of polynuclear hydrocarbons that are further separated and purified to provide the main source of these chemicals for non-Communist countries (49). The crude coke-oven tar is dehydrated by heat exchange with the dehydrator overhead distillate and passage through a low pressure steam heater. The dehydrated tar is heat-exchanged with the carbolic-acid stream and the pitch and enters the midpoint of the carbolic-oil column at 250°C. This column operates at atmospheric pressure and contains 40 bubble-cap trays; the reflux ratio is 16:1. Carbolic oil is taken overhead and naphthalene oil is taken as a side stream. The latter is upgraded in a 20-tray splitting column to give a fraction containing 84–88% naphthalene, equivalent to 95% of the naphthalene in the crude tar. The bottoms from the carbolic-oil column passes next to a vacuum fractionating stage consisting of a 40-tray column and a 20-tray splitting column to yield a methylnaphthalene concentrate, a light wash oil, and a fraction rich in fluorene and acenaphthene. The residual base tar is heated to 280°C and pumped to the anthracene column which is maintained at 9.3 kPa (70 mm Hg). Anthracene oil and heavy oil are separated from pitch and the overheads further fractionated in a separate vacuum column into an anthracene–phenanthrene con-

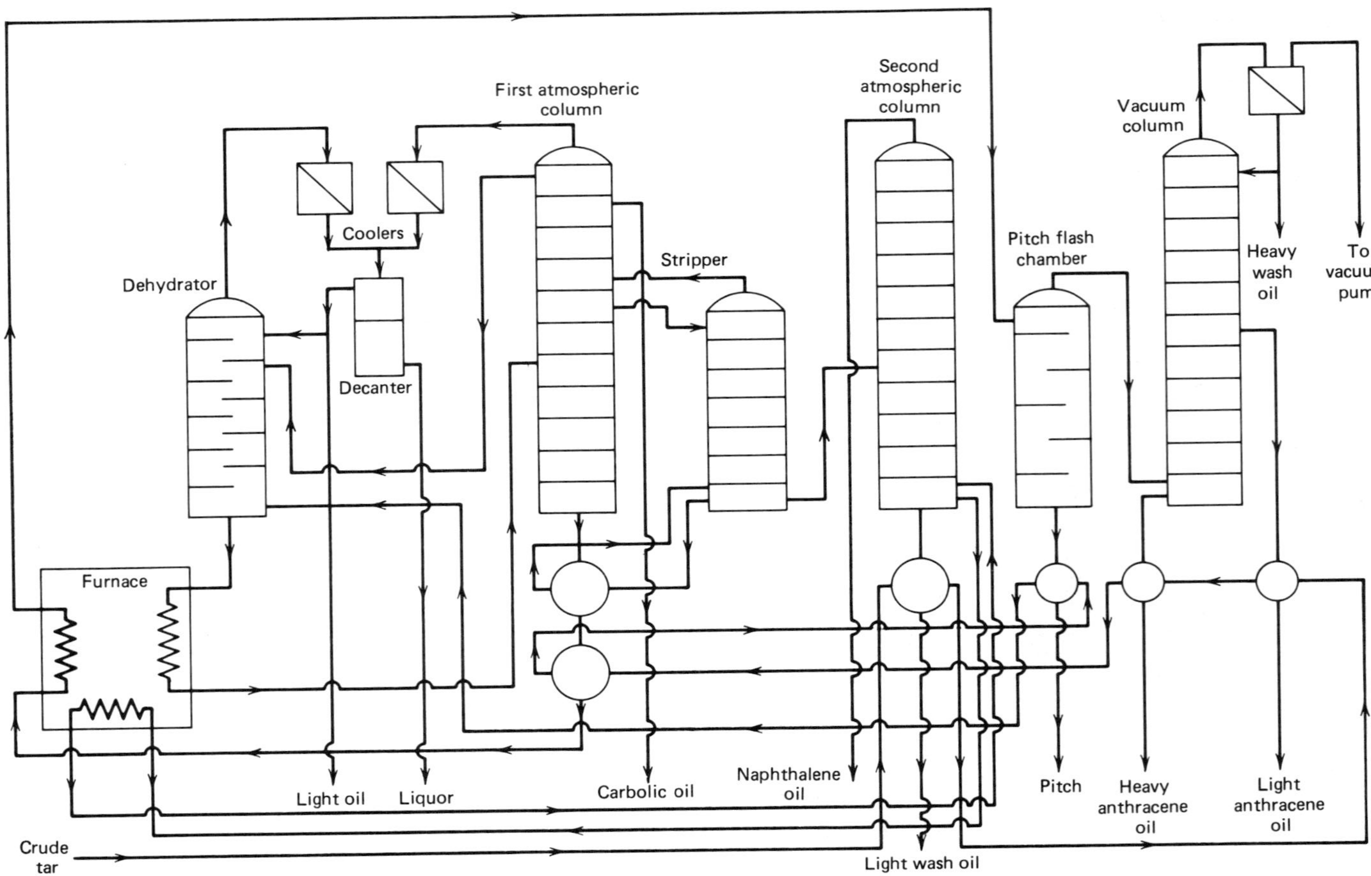

Figure 1. Teerverwertung-Koppers tar distillation plant.

centrate, a fraction containing 95% carbazole, and a heavy oil from which pyrene, fluoranthene, and chrysene can be isolated. Highly efficient columns afford very efficient fractionation. In addition, separate base products from the splitting columns are reboiled by heat from the bottoms products of the main columns.

The most widely used design of continuous coal-tar still is probably the Wilton design supplied by Chemical and Thermal Engineering, Ltd., of Wilmslow, UK. Wilton plants are installed in several refineries in the UK, South Africa, Australia, New Zealand, India, Spain, Argentina, and the Republic of Korea. The popularity of this design, which operates on the recycle principle, is due to good thermal efficiency and flexibility in handling any type of tar over a wide range of throughputs (see Fig. 2). The plant contains a helical-coil furnace, a dehydration column, two valve-tray fractionating columns, and a pitch column divided into an upper stripping section and a lower mixing section. The crude tar is pumped through coarse filters and, after the addition of alkali solution, passed via a heat exchanger to the waste-heat coil in the convection part of the furnace. The heated tar, under a moderate back pressure, enters the baffled dehydration column where the entrained liquor and light oils vaporize, leaving the dehydrated tar which is pumped to the bottom, or mixing, section of the pitch column. Here, it is mixed with a greater volume of hot pitch overflowing from the stripping section. This pitch–dehydrated tar mixture is pumped through the radiantly heated main furnace coil, which is backpressured to prevent vaporization. At 350–360°C, the mixture leaves the furnace and enters the upper part of the pitch column where the distillate oil vapors flash off, leaving the pitch residue; vaporization is assisted by injecting superheated steam at the base of the stripping section. A level of some pitch is maintained at this point, where the main pitch product is drawn off. The remainder of the pitch overflows into the mixing section from which it is recycled together with the dehydrated tar. The total vapors from the pitch column, including those produced in the mixing section by the contact of hot pitch with the dehydrated tar, pass to the base of the bottoms fractionating column. Here, wash oil and anthracene oil are taken as side streams and a heavy oil is taken as base product. Uncondensed vapors from this column, together with the vapors from the dehydrator, are fed to the base of the tops column from which an overhead light oil–water fraction, carbolic oil, and naphthalene–oil side-streams are taken together with a base product which is returned to the bottoms column as reflux. The light oil–water fraction is condensed and separated in a decantor; part of the light oil serves as reflux to the tops column.

There are two serious problems associated with continuous tar distillation. Coal tar contains two types of components highly corrosive to ferrous metals. The ammonium salts, mainly ammonium chloride, associated with the entrained liquor remain in the tar after dehydration and tend to dissociate with the production of hydrochloric acid and cause rapid deterioration of any part of the plant in which these vapors and steam are present above 240°C. Condensers on the dehydration column and fractionation columns are also attacked. This form of corrosion is controlled by the addition of alkali (10% sodium carbonate solution or 40% caustic soda) to the crude tar in an amount equivalent to the fixed ammonia content.

The higher boiling phenols, present in considerable amounts in CVR and low temperature tars, are corrosive to mild steel, especially above 300°C. Cast iron, chrome steel or stainless steel are more resistant. Furnace tubes, the insides of fractionating columns, and the rotors of pumps handling hot pitch and base tar are generally con-

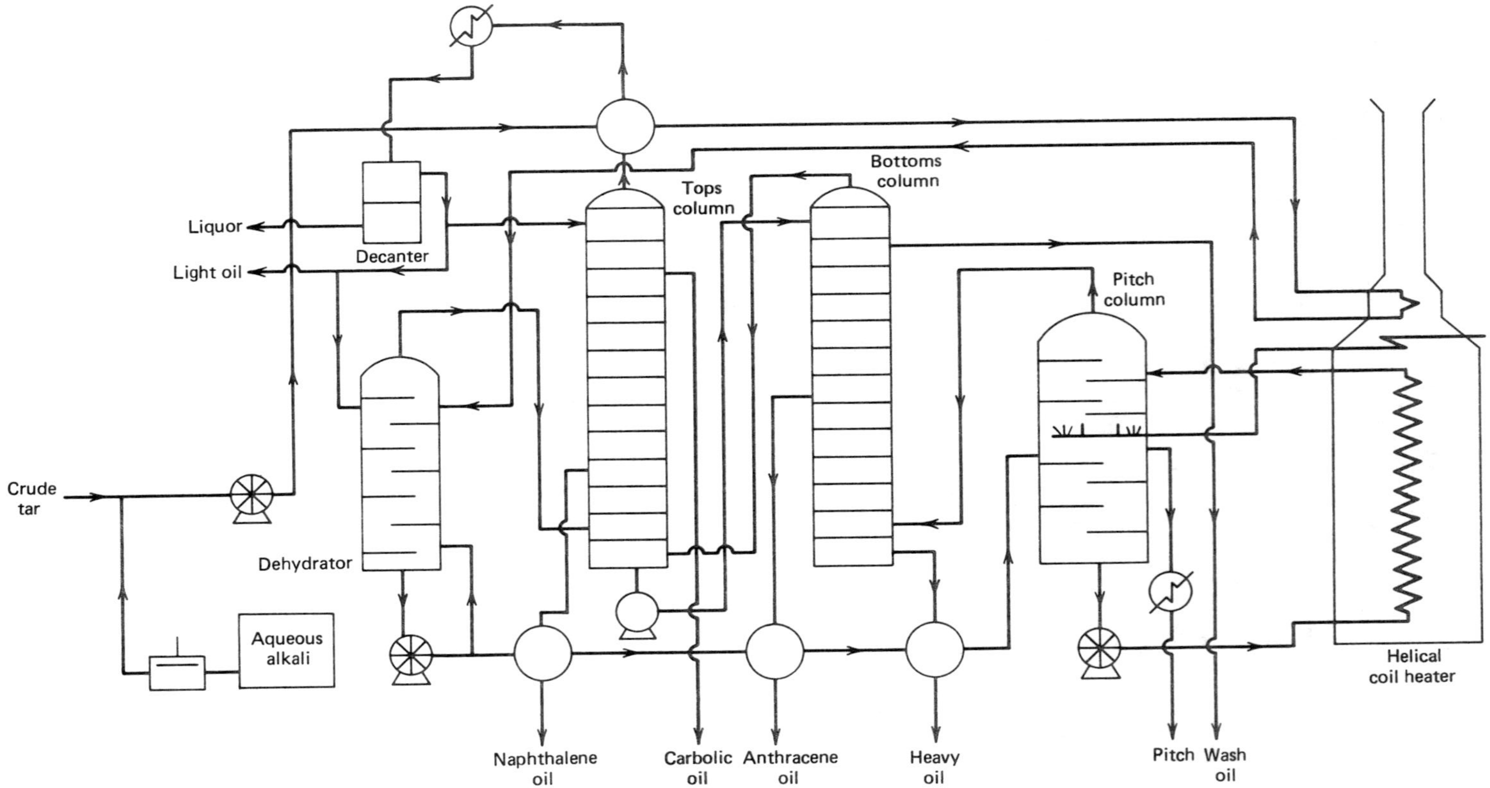

Figure 2. The Wilton tar distillation plant (courtesy of Chemical and Thermal Engineering, Ltd.).

structed of these metals. Nevertheless, to ensure satisfactory furnace-tube life, particularly in plants processing CVR or low temperature tars, the tube temperature should be kept to a minimum.

A more important reason for operating furnaces at the lowest possible temperature is the fact that coal tar is thermally unstable. If heated above a certain temperature, it decomposes, forming coke which rapidly blocks the furnace tubes. The critical temperature for coke formation is lower for vertical retort and low temperature tars than for coke-oven tars. To guard against overheating, different designs offer various options. In the Teerverwertung-Koppers design, by employing vacuum in the second stage, the tar need never be heated above 300–310°C, which is well below the critical temperature. If vaporization is allowed to proceed to any considerable extent in the furnace tubes combined with a high heat flux, tube wall temperatures might rise above the decomposition threshold at the zone where the tube contents change from a foam to a mist. Most plants operate the furnace tubes under a back pressure, which may be as high as 1 MPa (ca 10 atm) to suppress vaporization; the recycle of pitch in the Wilton design also minimizes vaporization and, by increasing the velocity through the tubes, improves the heat-transfer rate.

The main objectives in the primary distillation of crude tar are to obtain a pitch or refined tar residue of the desired softening point, to concentrate in certain fractions components that are subsequently to be recovered, and to yield distillate oils that are suitable for blending. A tar-distilling plant is designed to yield fractions that meet the specifications set by the purchaser and these depend on the product sales pattern. This pattern differs for different companies, with the result that each plant is, to some extent, tailor-made and unique in its details and performance. Although at some refineries as many as seven distillate oil fractions are taken, in most plants the number of primary fractions is smaller. The situation is complicated by the fact that different names are given to identical fractions and, conversely, the same names are given by different distillers to fractions of different boiling ranges.

A typical primary distillation product pattern at a coke-oven tar-processing plant is given in Table 5. At some coke-oven distilleries, only one fraction, designated naphthalene oil, is taken between 180 and 240°C. Two fractions, light creosote or middle oil (230–300°C) and heavy creosote or heavy oil (above 300°C), are taken between the naphthalene oil and pitch.

When using a continuous vertical retort, phenols, cresols, and xylenols are collected in one fraction. A typical range of primary distillation fractions is given in Table 6.

The only valuable components in low temperature tar are the phenols and an oil

Table 5. Primary Coke Oven Tar Distillation

Product	Distillation range, °C	% Dry tar
light oil	up to 180	0.5–1.0
phenolic or carbolic oil	180–205	3–4
naphthalene oil	200–230	10–12
wash oil	240–290	6–8
light anthracene oil	260–310	20–25
heavy anthracene oil	above 300	20–25
medium-soft pitch		53–58

Table 6. Primary CVR Tar Distillation

Product	Distillation range, °C	% Dry tar
light oil	90–170	2–4
carbolic oil	180–240	12
light creosote oil	230–300	14
heavy creosote oil	275–360	11.5
residual oil	300–395	44–46
medium-soft pitch residue		

fraction distilled over the range of 180–310°C, which is collected for tar-acid recovery is taken. A typical primary distillation is given in Table 7.

Secondary Processing of Tar Distillate Oils. ***Light Oils, Refining of Benzene and Naphtha.*** The only processing that light oils might receive at the refinery would be a fractional distillation into crude benzene (formerly called benzol or benzole, now obsolete) distilling up to 150°C, a naphtha fraction distilling from 150 to 190°C, and a creosote residue (see BTX processing; Benzene). Crude benzene from coke-oven tar is normally refined together with crude benzene separated from coke-oven gas. The naphtha is washed with alkali and acid to remove tar acids and bases, then treated with a small amount of sulfuric acid to remove sulfur compounds and olefins and redistilled to give refined solvent naphtha. Alternatively, a special fraction rich in indene can be collected from which indene–coumarone resins are produced as described later.

In the refining of the combined crude benzene, a defronting steam-stripping operation removes the lower boiling components. The defronted benzene may be fractionated in batch or continuous stills to yield a mixture of crude benzene, toluene, and mixed xylenes, and a naphtha residue. Each fraction is purified to meet grade specifications. Alternatively, the crude benzene, after defronting, may be refined and the refined product fractionated. In the refining of crude benzene or its distillate fractions, nonaromatic hydrocarbons and thiophene derivatives are removed. The oil is usually mixed with 2–4% conc sulfuric acid for a short time, washed, and separated from the spent acid. This treatment is followed by a dilute alkali and a water wash, and the product is refractionated.

A number of refineries in the United States, the UK, and Europe employ hydrorefining, in which the vaporized hydrocarbon feedstock and hydrogen are passed over a cobalt molybdate-on-alumina catalyst at 400°C and 3550 kPa (500 psig). The olefins are converted to paraffins or cycloparaffins, and thiophene compounds to hydrogen sulfide. After washing with alkali to remove H_2S, small amounts of nonar-

Table 7. Primary Low Temperature Tar Distillation

Product	Distillation range, °C	% Dry tar
light oil[a]	up to 180	3.5
middle oil	180–310	29
wax oil[b]	300–400	37.5
medium-soft pitch residue		26

[a] Also called fresh oil.

[b] Solidifies at 20°C because of high paraffin wax content.

omatic hydrocarbons are removed by the Udex process, the Shell Sulfolane process, or the *N*-methylpyrrolidinone extraction process of Metallgessellschaft A.G. (see Extraction, liquid–liquid extraction).

Gas works crude benzene or the lower boiling liquid products from low temperature carbonization, so-called fresh oil or low temperature spirit, are too low in aromatic hydrocarbon content to be refined economically to synthesis-grade benzene and toluene. Formerly, these fractions were used as a gasoline additive; currently, they are used as fuel at the refinery.

Coumarone–Indene Resins. These should be called polyindene resins (50) (see Hydrocarbon resins). They are derived from a close-cut fraction of a coke-oven naphtha free of tar acids and bases. This feedstock, distilling between 178 and 190°C, and containing a minimum of 30% indene, is warmed to 35°C and polymerized by adding 0.7–0.8% of the phenol or acetic acid complex of boron trifluoride as catalyst. With the phenol complex, tar acids need not be completely removed and the yield is better. The reaction is exothermic and the temperature is kept below 120°C. When the reaction is complete, the catalyst is decomposed with a hot concentrated solution of sodium carbonate. Unreacted naphtha is removed, first with live steam and then by vacuum distillation to leave an amber-colored resin. It is poured into trays, allowed to cool, and broken up for sale.

Pyridine Bases. Formerly, pyridine bases were recovered from coal-tar light oils (51) but in recent years synthetic pyridine and picolines have replaced the coal-tar products.

First, the tar acids were removed from the naphtha fractions of light oils and, in the case of CVR tars, the carbolic oil. The oils were then mixed with 25–35% sulfuric acid. After separation of the sulfates, the aqueous solution was diluted with water and the resinous material skimmed off. The diluted sulfate solution was boiled to expel any neutral oils, dried by the addition of solid caustic soda or azeotropically with benzene, and fractionated to yield pyridine, α-picoline, and a fraction referred to as 90/140 bases, which consisted mainly of β-picoline, γ-picoline, and 2,6-lutidine. Higher boiling fractions were termed 90/160 and 90/180 bases because 90% of the product distilled at 160 and 180°C, respectively.

Carbolic Oils and Low Temperature Tar Middle Oil, Tar Acids. The fractions of CVR tars and, in some cases, coke-oven tars, distilling in the range of 180–240°C, and the middle oil fraction (180–310°C) from low temperature tars are treated for the recovery of tar acids (52).

The oils are mixed with a slight excess of 10% aqueous caustic soda in stirred vessels, continuous extraction columns, or, at some refineries, by circulating the contents of the mixing vessel by a gear pump. The extraction is carried out at ambient temperature or just above the crystallizing point of the oil. The extraction is best carried out in two stages, using 90% of the alkali to contact the fresh oil and the remainder to complete the removal of the phenols. The crude phenate or cresylate solution is separated, and contains some neutral and basic material which must be removed by extracting with phenol-free light oil or crude benzole and/or treating with live steam.

The purified sodium phenate solution is then decomposed by passing it down ring-packed or coke-filled towers at 80–85°C countercurrent to a gas containing 25–30% CO_2, generated in lime–coke kilns, or a flue gas containing 10–15% CO_2. This operation is called springing. Since the rate of CO_2 absorption depends on the CO_2 partial pressure, the richer gas reduces the number and size of the springing towers required. An upper layer of crude wet tar acids and a lower layer of sodium carbonate solution is obtained. The crude wet tar acids contain ca 20–25% of sodium carbonate solution and, to reduce their moisture content to ca 10%, the crude wet tar acids are passed down an after-carbonation tower countercurrent to a stream of the CO_2-containing gas. Separation of the layers, into which the effluent from the various carbonation tower divides, yields crude tar acids containing ca 10% liquor and a solution of sodium carbonate and bicarbonate.

Caustic soda is recovered from the carbonate/bicarbonate solution by treating with a slight excess of hard-burned quicklime (or slaked lime) at 85–90°C in a stirred reactor. The regenerated caustic soda is separated from the calcium carbonate precipitate (lime mud) by centrifuging or rotary vacuum filtration. The lime mud retains 30–35% liquid and, to avoid loss of caustic soda, must be well-washed on the filter or centrifuge. Finally, the recovered caustic solution is adjusted to the 10% level for recycle by the addition of 40% makeup caustic soda.

Disposal of the washed lime mud poses a problem. Drying and calcining to regenerate lime and CO_2 require an uneconomical expenditure of energy and, at present, lime mud is dumped. The use of sulfuric acid for springing and the partial electrolysis of the sodium sulfate to give a mixture of caustic soda and sodium sulfate as the recycled extracting medium as an alternative to caustic soda extraction and lime recausticization, avoids the production of lime mud. This alternative, although successful in laboratory and pilot-scale tests, has not yet been applied in commercial practice.

In the next stage in the recovery and refining of tar acids, water and pitch are removed from the crude tar acids in a continuous-vacuum still heated by superheated steam or circulating hot oil. The aqueous phenol overhead distillate is recycled, the stream of once-run tar acids is refined, and the phenolic pitch bottoms are burned.

The once-run tar acids are fractionated in three continuous-vacuum stills heated by superheated steam or circulating hot oil. These stills contain 40–50 bubble trays and operate at reflux ratios between 15 and 20:1. The overhead product from the first column is 90–95% phenol; from the second, 90% *o*-cresol; and from the third, a 40/60 *m*-/*p*-cresol mixture. Further fractionation gives the pure products.

At other refineries, only two continuous stills in series are used, but these are of 80–100 plate efficiency and yield pure grades of phenol and *o*-cresol and a base mixture of cresols, xylenols, and higher boiling tar acids. The latter are fractionated batch-wise to various salable grades of cresylic acids.

Fractionation columns in tar-acid refineries are generally operated under vacuum and heated by high pressure steam or circulating hot oil. Calandria in the reboilers, condensers, run-down lines, and receiving tanks are constructed of stainless steel or, in the case of the condensers, of tin or nickel. Cast-iron column shells are satisfactory, but stainless-steel bubble or valve trays are preferred. A flow sheet of a typical tar acid extraction and refining plant is shown in Figure 3.

In the case of low temperature tar, the aqueous liquor that accompanies the crude tar contains between 1 and 1.5% by weight of soluble tar acids, eg, phenol, cresols, and dihydric phenols. Both for the sake of economics and effluent purification, it is necessary to recover these, usually by the Lurgi Phenosolvan process based on the selective extraction of the tar acids with butyl or isobutyl acetate. The recovered phenols are separated by fractional distillation into monohydric phenols, mainly phenol and

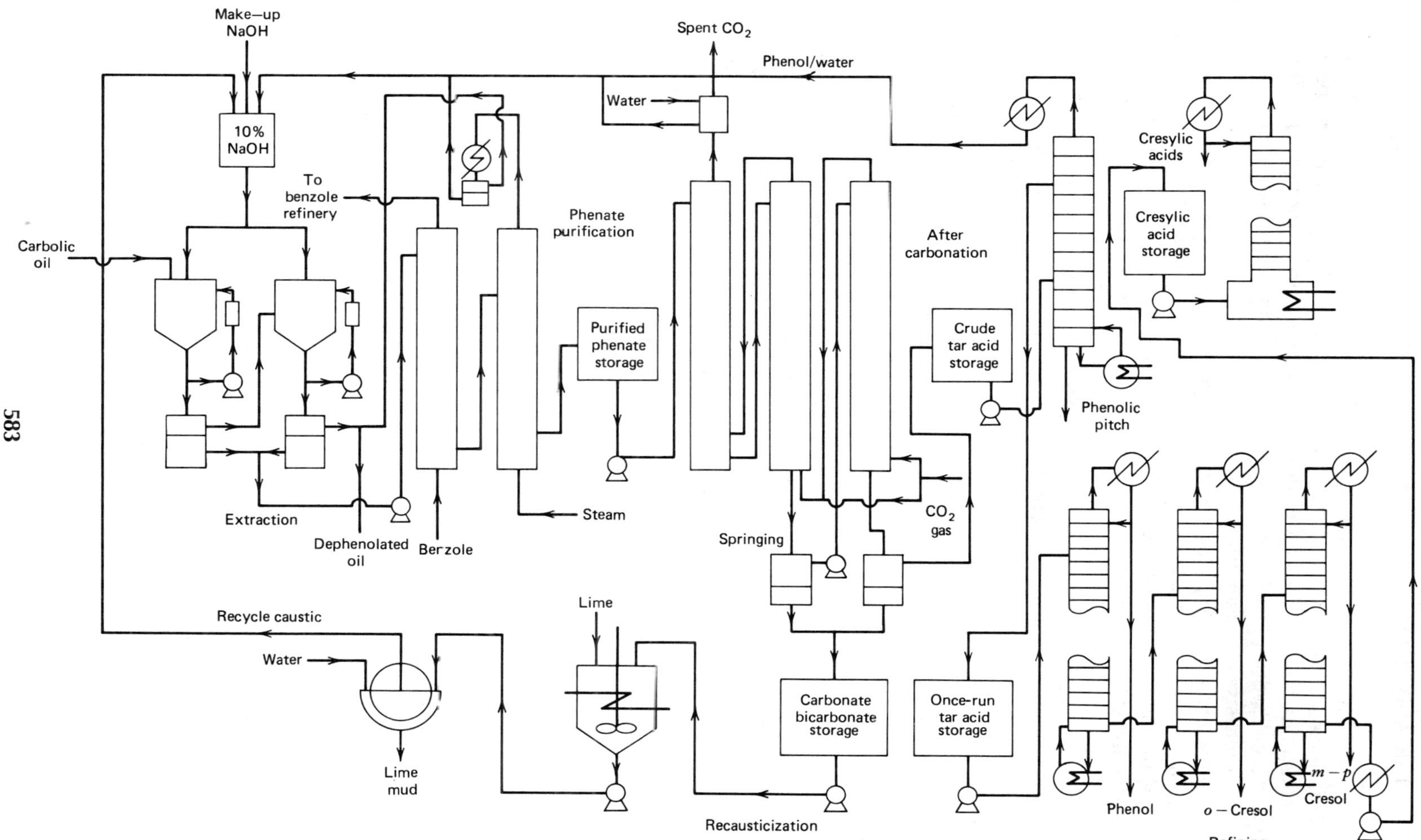

Figure 3. Tar-acid extraction and refining.

cresols, and dihydric phenols, mainly catechol, methyl catechols, and methyl resorcinols (see also Hydroquinone, resorcinol, and catechol). The monohydric phenol fraction is added to the crude tar acids extracted from the tar for further refining, whereas the dihydric phenol fraction is incorporated in wood-preservation creosote or sold to adhesive manufacturers.

Naphthalene Oils. Naphthalene is the principal component of coke-oven tars and the only component that can be concentrated to a reasonably high content on primary distillation. Naphthalene oils from coke-oven tars distilled in a modern pipe still generally contain 60–65% of naphthalene. They are further upgraded by a number of methods.

In the older method, still used in some USSR and East European tar refineries, the naphthalene oil is cooled to ambient temperatures in pans, the residual oil is separated from the crystals, and the crude drained naphthalene is macerated and centrifuged. So-called whizzed naphthalene crystallizes at ca 72–76°C. This product is subjected to 35 MPa (350 atm) at 60–70°C for several minutes in a mechanical press. The lower melting layers of the crystals are expressed as liquid, giving a product crystallizing at 78–78.5°C (95.5–96.5% pure). This grade is satisfactory for oxidation to phthalic anhydride and is referred to as hot-pressed or phthalic-grade naphthalene.

The more modern process adopted in the UK and some European plants (52) is also based on crystallization of the primary naphthalene oil, which is diluted with lower-crystallizing material to give a feedstock crystallizing point at 55°C. This material is cooled in closed, stirred tanks to 30–35°C and the resultant slurry of naphthalene crystals and mother liquor is centrifuged, washed, and spun dried. These operations are automatically timed and controlled.

The crystallization, centrifugation, and washing process is also applicable to the recovery of phthalic-grade naphthalene from CVR naphthalene oils, provided the tar acids are first recovered.

At other European refineries and in the United States, coke-oven naphthalene oil is upgraded by fractionation and further purification.

Several refineries in the UK and Europe employ the Pro-Abd refiner to upgrade whizzed naphthalene to phthalic-grade quality or to convert the latter into the purer chemical grade. The device consists of a rectangular tank fitted with a nest of coils through which either steam or water circulates. The tank is filled with the feedstock which is crystallized by circulating cold water in the coils. When the contents of the tank have solidified, a tap at the base is opened and hot water is circulated until the temperature is just below the desired crystallizing point of the product. This condition is maintained until no more oil drains from the base. The bottom tap is then closed and the contents of the tank are melted by steam circulation and drained.

In a novel naphthalene-purification method, oil containing ca 70% of naphthalene is charged to tall cylindrical tanks equipped with internal heating or cooling coils and cooled to 35°C. Mother liquor is allowed to drain from the base of the tank and, when drainage is complete, the tank is filled with methanol until the solvent level is above the surface of the crystallized naphthalene. After standing for 1 h, the methanol is drained. This procedure is repeated. Finally, the purified naphthalene is melted by passing steam through the internal coils and the liquid product discharged to storage. This product is pure enough for the manufacture of 2-naphthol.

Brodie crystallizers have been installed in the UK and in France for upgrading phthalic-grade naphthalene to 99% purity or better. This apparatus, developed by Union Carbide Corporation, Australia, for separating *o*- and *p*-dichlorobenzene, has been adapted for naphthalene refining (53) (see Naphthalene).

Wash Oils. No tar chemicals can be extracted commercially from tar oils distilling in the range 250–300°C. Although the wash-oil fraction of coke-oven tars, distilling mainly in the range 250–280°C, is employed at coking installations to scrub benzene from coal gas, most oils in this boiling point range are used in creosote blends.

Anthracene Oils. In the UK and Europe, but not in the United States, crude anthracene is isolated from coke-oven anthracene oils. In some cases where the anthracene oil gives only a small residue at 360°C, the oil is diluted with half its volume of drained naphthalene oil or a light wash oil and this blend is cooled to 35°C. The resulting solid–liquid slurry is filtered or centrifuged to give a crude anthracene containing 40–45% anthracene. With other samples of oil, it is necessary to recrystallize the first cake obtained by cooling the undiluted anthracene oil to 30–35°C and separating the crystals by filtration or centrifugation. This first crop of crystals generally contains 12–15% of anthracene and is recrystallized. The anthracene concentrate, termed 40 s anthracene is sold as such.

Shipment, Storage, and Handling

Crude Tar and Tar Products. Where the tar distillery is sited close to the carbonizing plant, the crude tar is transferred directly by pipeline from the tar–liquor separating vessels on the by-product recovery unit to the storage tanks. Otherwise, it is shipped in rail or road tankers or by barge. Crude tar is stored in mild-steel tanks maintained at 40–50°C by steam coils.

Liquid tar products, eg, light oils, cresols, cresylic acids, creosote oil, and road tars, are generally transported in bulk in insulated mild-steel road or rail tankers. They are loaded at a temperature sufficiently high to ensure delivery at the desired viscosity. Small quantities are generally delivered in drums which may have to be steam heated to ensure complete liquidity before discharge.

Pitch. Pitch used to be stored in solid form at the tar distillery in open bays, from which it was removed by small explosive charges. Loading of the lump pitch by mechanical shovel created a dust hazard both at the tar installation and at the customer's, where the lumps had to be ground before use. Today, pitch is stored in tanks heated by superheated steam or circulating hot-oil coils and transported in liquid form in insulated rail or road tankers. When transport as a hot liquid is not feasible, eg, for distances >500 km, or export by ship in bulk, or for small amounts, the pitch is converted into a dust-free particulate form (short rods termed pencils, pastilles, or flakes).

Several plants employ cooled-belt flakers. These consist of flexible steel belts ca 1 m wide and up to 50 m long with short rubber skirting at the edges. Molten pitch flows from a thermostatically controlled tank over a weir to give a flat thin sheet on the belt, which is cooled from below by water sprays. At the end of the belt, the solid pitch is broken up by rotating tines. The pitch flakes are drained and transported to a covered storage silo by belt conveyor; during this time the surface moisture evaporates.

Direct water cooling is also employed in the pitch-pencilling plant at Rütgerswerke. Pitch at ca 50°C above its softening point is pumped through nozzles at a rate and pressure to give a series of almost horizontal jets. These jets project a short distance into horizontal tubes through which a cocurrent streamline flow of cooling water is maintained. The tubes extend for some 25 m and are bent into a semicircle at their exit end. The stream of pitch solidifies into solid rods about 10–20 mm thick which break up onto short lengths as they are forced round the bends. The pitch pencils are delivered into a cooling pond situated underneath the pipe assembly. A conveyor belt picks up the pitch from the pond and transports the pencils to a chute from which they fall to another conveyor belt which transports them to the storage silo. Surface drying is accomplished by warm-air jets.

Another type of pitch-pencilling plant, designed by Chemical and Thermal Engineering, Ltd., employs a vertical cooling tank. Liquid pitch at ca 150°C, held in a thermostatically controlled tank, is circulated round a ring main by a centrifugal pump. A horizontal section of the main contains an extrusion manifold consisting of a number (usually 6–10) 9–12-mm-dia nozzles set 200 mm apart. The circular streams of molten pitch from the nozzles fall by gravity into a deep rectangular tank containing water at 40–50°C where the pitch hardens into solid rods.

Economic Aspects

Crude Tar. Current world output of crude tar is estimated at ca 18×10^6 metric tons; 14×10^6 metric tons are of coke-oven origin (see Table 8).

Table 9 gives the U.S. crude-tar production from 1960 to 1980 (55) and Table 10 lists UK tar production from 1955 to 1981 (56).

Table 8. Estimated World Production of Crude Coke-Oven Tar, Thousand Metric Tons[a]

Country	1975	1980	1981
Benelux	390	280	240
Czechoslovakia	250	na	na
Canada	na	na	223
GDR	1000	na	na
FRG	1350	1120	1089
France	460	430	426
Italy	300	308	298
Japan	2000	na	na
Poland	760	na	na
Spain	200	163	162
South Africa	na	382	292
UK	719	450	403
United States	na	2375	na
USSR	3500	na	na

[a] Partially from ref. 54.

Table 9. U.S. Crude-Tar Production

Year	10^6 t	Year	10^6 t
1960	3.112	1976	2.837
1970	3.397	1978	2.407
1974	3.020	1980	2.375

Table 10. UK Crude-Tar Production, Thousand Metric Tons

Year	Coke ovens	Gas works	Low temperature ovens	Total
1955	1101.3	1930.1	41.9	3073.3
1960	1187.2	1555.3	56.4	2798.9
1965	1076.5	1250.8	119.5	2446.8
1970	935.5	342	100.6	1378.1
1975	718.8	1.6	202.9	923.3
1977	628.7	0	170.1	798.8
1979	523.5	0	134.8	658.3
1980	449.1	0	153.8	602.9
1981	403.1	0	158.1	561.2

Crude-coal-tar production has declined considerably in recent years, especially in the UK. Although there has been some increase in low temperature tar from the Coalite and Rexco operations, the amount of crude tar available for processing in the UK in 1981 is less than one fifth of the 1955 production. Tar chemicals production is given in Table 11.

Creosote oil production in the United States (57) and the UK (56) is given in Table 12.

Table 11. Tar Chemicals Production[a]

Property	10^3 t
benzene	
U.S. total	5191
from coal carbonization	51
UK total[b]	ca 900
from coke ovens	73.62
Western Europe	4000[c]
toluene	
U.S. total	5082
from coke ovens	ca 12
UK total	170.13
from coal carbonization	ca 6.8
phenol	
U.S. total	1120
from coal tar	<22[c]
cresylic acid	
U.S. total	53.5
from coal tar	15.0
naphthalene	
U.S. from coal tar	132.0[d,e]
UK from coal tar	7.5[d]
40s anthracene, UK	2.8

[a] 1980, unless otherwise stated.
[b] Including benzene from toluene dealkylation.
[c] Estimated.
[d] 1981 production.
[e] Used almost entirely for internal phthalic anhydride production.

Table 12. Creosote Oil Production, Thousand Metric Tons

Year	United States	UK
1975	468.0	279.6
1977	440.9	253.7
1979	430.7	207.4
1980	609.9[a]	205.4
1981	510.0[b]	172.5

[a] Uncorrected.
[b] Estimated.

Recent statistics on U.S. pitch production are not available; Table 13 gives earlier data. Production for 1972–1975 was ca 42% of the crude tar production, from which pitch production after 1975 can be calculated. UK production is given in Table 14.

Tar chemicals and tar bulk products compete in the marketplace with the products of the petrochemical industry. Except for anthracene, production from petroleum sources is much larger than from coal carbonization and the prices for these products are lower. Creosote competes in many of its uses with petroleum fuel oils and aromatic residues, whereas road tar competes with road asphalt and bitumen. Only in uses for which the tar product is superior or cannot be replaced does it command a premium. Recent tar bulk products prices are given in Table 15.

Specifications

Tar bulk products are covered by both national specifications and specifications formulated by users.

For instance, creosote for timber preservation is governed by the American Wood

Table 13. U.S. Production of Coal Tar Pitch[a]

Year	10^3 t
1960	1964.5
1970	1786.7
1972	1390.0
1974	1262.3
1975[b]	1016.3

[a] From ref. 55.
[b] Estimated.

Table 14. UK Pitch Production[a]

Year	10^3 t
1975	344.7
1977	278.5
1979	248.1
1980	230.7
1981	195.0

[a] Ref. 56.

Table 15. Tar Bulk Prices, 1982

Product	$/t	$/L
United States		
wood-preservation creosote		
U.S. bulk		0.31
U.S. drums		0.33
electrode pitch	281–295	
aluminum	305–325	
graphitized steel	325–340	
impregnation pitch for refractories	340	
pitch for fiber pipes and coal briquette	235	
roofing pitch[a]	290	
United Kingdom[b]		
road tar	198–216	
carbon-black feedstock and coal-tar fuels	171	
Western Europe		
indene-coumarone resins	810	
wood-preservation creosote	252–261	0.25–0.26

[a] Conforming to Federal specification R-P 81.
[b] Dollar prices based on $1.80/£.

Preserving Association (AWPA) Specifications P1/65 and P13/65 and ASTM D-350-67; in the UK, it is covered by British Standard Specifications (BSS) 144/73 and 3051/71. Creosote specification AWPA P1/65 is intended for the treatment of timber for land and fresh-water use, and the heavier grade specified in AWPA P13/65 for the preservation of marine piling and timber; BSS 144/73 specifies two grades of creosote from coke-oven tar; BSS 3051/71 is intended for tar-oil wood preservatives from lower-temperature tars. These specifications include specific gravity, maximum water content, maximum values for toluene- or benzene-insoluble material and maximum amounts distilling at 230°C, 270°C, 315°C, and 355°C. In the case of the AWPA specifications, there are minimum limits to the specific gravities of each of the distillate fractions.

Other national specifications for wood preservation creosote are CAM 972 (Argentina), AS T505/1965 and AS 1143/1978 (Australia), NBN 439 (Belgium), DS Rec. M314/TP (Denmark), IS 218-1961 (India), DGN R21-1952 (Netherlands), SASS 17-1943 (South Africa), and GOST 2770-59 (USSR). These are generally similar to the U.S. and UK specifications.

Tar-based road binders are the subject of ASTM Specification D 1753-76 (United States), BSS 76/1974 (UK), AS 63-1947 (Australia), DIN 51-551 (1961) (FRG), IS 215-1961 (India), JIS K2472 (Japan), and GOST 4641-49 (USSR). These cover a range of road tars of different viscosity ranges and generally include maximum values for moisture content, phenols, toluene-insoluble material, and, in some cases, naphthalene content. They also give the permissible ranges for specific gravity, for the amount of oils distilling between 200°C and 270°C, between 270°C and 300°C and for the softening point of the residue above 300°C.

A few national specifications cover electrode-binder pitch, eg, GOST 10200-62 (USSR), but in general, this product is supplied to specifications set by the aluminum or steel-manufacturing companies. The requirements for the binder for Söderberg electrodes differ from those for prebaked electrodes and include a minimum specific gravity or density, a narrow softening point range (95–100°C (R and B) for Söderberg electrodes and 100–110°C for prebaked electrode binder), a minimum C/H ratio, a minimum coking value, maxima for the permissible ash content, moisture content and amount of volatile matter at 360°C, a range for quinoline-insoluble matter (8–13% for Söderberg electrode binder and 5–10% for prebaked electrode binder), and a minimum of ca 20% for the amount of toluene-insoluble/quinoline-soluble (β-resin) content. However, in addition to the published specification requirements, aluminum-smelting companies demand that the pitch binder should pass certain preacceptance tests involving flow measurement at various temperatures of the pitch–petroleum coke electrode paste and measurement of strength, porosity, electrical resistance, and reactivity of test electrodes made by slowly carbonizing the electrode paste under compression to 900°C (see Carbon and artificial graphite).

Coal-tar pitch for other uses is similarly subject to a few national specifications but mainly sold to users' specifications. Pitch intended for roofing, damp-proofing, and waterproofing is the subject of ASTM specification D 450-71 and Federal specification R-P-381; hot-applied tar-based coatings (pipeline enamels) are the subject of BSS 4164/1980 and, in the United States, of AWWA (American Water Works Association) specification C203-73.

Analytical Methods

Tar. Before the development of gas chromatography, the quantitative analyses of tar distillate oils involved tedious high efficiency fractionation and refractionation, followed by identification or estimation of individual components by ir or uv spectroscopy. Today, the main components of the distillate fractions of coal tars are determined by gc (58).

The analytical procedures included in the specifications for tar bulk products are given in ASTM standards (6).

Pitch. For the solvent analysis of pitch, a number of methods have been proposed. The solvent may be used sequentially or a fresh sample may be used with each solvent; the least or more powerful solvent may be used first; and the ratio of solvent to pitch or pitch fraction and the temperature and time of extraction vary.

In the Broche-Nedelmann procedure, the pitch is separated into material insoluble in benzene (α-component), material soluble in benzene but insoluble in petroleum ether (β-component), and material soluble in petroleum ether (γ-component). A modification of this procedure is now most widely used in the United States and the UK. Separate samples of the predried pitch are extracted with 100–120°C-boiling range-petroleum ether (5 g of pitch is extracted with 100-mL portions of solvent until the extract is colorless), toluene (1 g of pitch is extracted with 100 mL of toluene under reflux for 30 min and the residue is washed with hot toluene until the washings are colorless), and quinoline (1 g of pitch is extracted with 25 mL of quinoline at 70–80°C for 20 min and the residue is washed with hot quinoline). The residues are dried and weighed and the analysis calculated to give the m/m percentages of the following four fractions: *1*. Petroleum ether soluble (crystalloids or γ-fraction). Generally contains 4–50% of a medium soft coke-oven pitch. Molecular weight range is 175–300. *2*. Toluene soluble–petroleum ether insoluble (β-resins). Generally contains 20–30% of a medium soft coke-oven pitch. Molecular weight range is 300–700. *3*. Quinoline soluble-toluene soluble (C-2 fraction or resins). Generally contains 8–15% of a medium soft coke-oven pitch. Molecular weight range is 1000–2000. *4*. Quinoline insoluble (C-1 fraction). Amount in coke-oven pitches varies from 5 to 15% but rarely exceeds 5% in CVR pitches or 1% in low temperature pitches. It appears to be similar to carbon black. In an earlier UK procedure, the crystalloid fraction was recovered by evaporation of the petroleum ether extract and, after drying and weighing, reextracted with boiling benzene until the solution was colorless; the β-resin fraction was recovered by evaporating the benzene extract. Finally, the benzene-insoluble residue was treated with boiling pyridine to give the pyridine-insoluble material; the pyridine extract was evaporated to obtain the C_2-fraction. The extractions can be carried out in a Soxhlet extraction apparatus.

The Mallison solvent-analysis method is still used in Europe (59). A sample of the dry pitch is extracted with a mixture of anthracene oil and pyridine to leave a residue of pyridine-insoluble material, the so-called H-resins. This fraction is roughly equivalent to the C-1 fraction obtained by the methods described earlier. A second sample of pitch is extracted on the water bath with benzene and the fraction insoluble in this solvent isolated, dried, and weighed. This residue is termed the H + M-resins (high and medium molecular weight resins) and, by subtraction, the M-resins (roughly equivalent to the C-2 fraction) are obtained. A third pitch sample is extracted with hot methanol to give an insoluble fraction regarded as the H + M + N-resins (high, medium, and low molecular weight resins) and, again by subtraction, the N-resins are obtained. The methanol extract is diluted with ammonium chloride solution which precipitates a fraction termed the m-oils, whereas the fraction soluble in the diluted methanol is termed the n-oils.

Health and Safety Factors

The volatile components of coal tar, ie, mononuclear aromatic hydrocarbons, phenols, and pyridine bases, are toxic when ingested, inhaled, or absorbed through the skin and the usual precautions must be taken when crude benzene or tar light oils are handled. Most polynuclear aromatic compounds are primarily skin and eye irritants but are tolerated internally. Naphthalene was at one time prescribed as an internal antiseptic and antihelminthic in daily doses of 0.1–0.5 g; the probable lethal dose is between 5 and 15 g. Anthracene passes through the gastrointestinal tract mainly unchanged, as do most higher polynuclear hydrocarbons. The lowest lethal oral dose for coal-tar creosote is given as 140 mg/kg body weight in the 1974 TSCA list.

The Carcinogenic Hazard of Tar and Tar Products. The main health hazard usually associated with coal tar and its products is carcinogenicity. Although this hazard undoubtedly exists, it is by no means as serious as some reports suggest and is constantly being reduced by improving working conditions. The use of tar products in road surfacing, preservation of telegraph and transmission poles, pitch fiber or coal-tar enamels for water pipes constitutes no danger whatever to the general public.

Cancer was first recognized as an occupational hazard in 1775 when the prevalence of scrotal cancer among London chimney sweeps was noticed. The chemical origin of this form of cancer was not universally accepted until 1922 when it was demonstrated that tumors could be induced on mouse skin with an ethereal soot extract (60). Skin cancer was also noted to be an occupational hazard of workers exposed to pitch dust in the coal-briquetting industry (61) and to workers exposed to crude tar (62). In 1915, tumors were produced in rabbits' ears by prolonged application of crude coal tar (63).

However, it was shown that only the higher boiling fractions were carcinogenic and that the spectra of these fractions resembled those of benzo[*a*]anthracene (64). This work culminated in the isolation of a highly carcinogenic compound, benzo[*a*]pyrene (BaP) from coal-tar pitch (65). Cancerous skin lesions of workers exposed to pitch dust undoubtedly support the belief that these lesions are caused by polynuclear aromatic hydrocarbons, although it has not been possible to demonstrate their carcinogenic action in animals such as monkeys more closely related to humans.

Although high temperature BaP-containing coal-tar creosote causes skin cancer in mice (66), skin cancer among workers in the timber-creosoting industry is not documented. However, protective clothing should be worn to avoid skin irritation. Although epitheliomas can result from prolonged exposure to creosote (67), there is no evidence that the large volume of creosote constitutes a hazard to the workers or the public. Similar large amounts of coal-tar pitch and tar oils are employed for the construction and maintenance of roads. Again, no evidence links any health hazard either to road-working personnel or the general public.

The main hazard is long-term, continual contact of finely divided solid pitch with the skin. In a relatively small, but statistically significant proportion of persons exposed in this manner, premalignant pitch warts appear, usually around the scrotum, hands, and face, particularly around the nostrils. These lesions, if untreated, develop into malignant epitheliomas. Such skin cancers are readily curable if diagnosed early; premalignant warts usually heal spontaneously after excision, but malignant epitheliomas must be treated by radiation. A total surface treatment of 50 Gy (5000 rad) is generally enough to destroy the malignant cells. The condition clears up ca one month after treatment, leaving some alteration in skin pigmentation; there are no reports of recurrence (68). Skin cancer caused by coal-tar pitch may take many years to manifest itself; epidemiological studies at creosoting plants are inconclusive.

A more serious hazard might be thought to exist in the pitch-roofing and road-tar industries where some personnel are continually exposed to the inhalation of pitch fumes, which contain BaP and other carcinogens (69). However, epidemiological studies of such workers have not revealed any increased risk of cancer of the lungs, esophagus, or internal organs (70–71).

The hazard of skin cancer by contact with pitch dust has, in recent years, been reduced by the transport and handling of pitch as a liquid or as dust-free flakes or pencils. Nevertheless, in handling coal-tar products, certain precautions should be taken; these are now obligatory in tar distilleries and plants using pitch or creosote.

Where the lower-boiling products are handled, effective ventilation must be ensured. Protective clothing, including eye protection and PVC gloves, must be worn, suitable respirators must be available, and regular medical checkups must be carried out on all personnel who are at risk of exposure to benzene in significant concentrations. A high standard of personal hygiene must be maintained. Barrier creams, properly formulated against aromatic hydrocarbons, should be provided and used.

Uses

Coumarone–indene resins have outlets in paints, as tackifiers in rubber compounding, and as adhesives in the manufacturing of flooring tiles (see Hydrocarbon resins).

Cresylic Acids. The higher boiling cresylic acids are mixtures of cresols or xylenols with higher boiling phenols (see Phenol; Alkylphenols). Their main uses are in phenol–formaldehyde resins, solvents for wire-coating enamels, as metal-degreasing agents, froth-flotation agents, and synthetic tanning agents (see Table 16). Statistics do not distinguish between the various grades or between cresylic acids derived from coal tar or petroleum or made synthetically (see Phenolic resins).

Naphthalene. Until the 1960s, the principal outlet for naphthalene was the production of phthalic anhydride; however, more recently, *o*-xylene has replaced naphthalene as the preferred feedstock (see Phthalic acids). Today, only ca 21.3% of the phthalic anhydride made in the United States is derived from naphthalene; in Western Europe, ca 10%; and in Japan, ca 30%.

In Europe, the main uses for naphthalene are for the manufacture of β-naphthol and for dyestuff intermediates (see Dyes and dye intermediates).

Creosote. In coal-tar refining, the recovery of tar chemicals leaves residual oils, including heavy naphtha, dephenolated carbolic oil, naphthalene drain oil, wash oil, strained anthracene oil, and heavy oil. These are blended to give creosotes conforming to particular specifications.

Creosote oils are by far the most widely used timber preservatives (see Wood). This use dates back to 1850. The 1981 U.S. consumption is estimated at <500,000 t.

For the treatment of railway ties and marine pilings, the Bethell or Full-cell process is preferred. The timber to be treated is charged to a pressure cylinder, which is evacuated to extract the air from the wood cells. The cylinder is then filled with hot creosote and the pressure increased to 0.8–1 MPa (ca 8–10 atm) to force the oil into the cells. When uptake of the oil has ceased, residual creosote is drained from the cylinder and the treated timber is briefly subjected to a vacuum before discharge.

For the treatment of telegraph and transmission poles, fence posts, and farm buildings, where a clean outer surface is desired, the empty-cell process is generally used. The Rueping process employs an impregnation cycle in which the timber is first subjected to a moderate air pressure to compress the air in the cells before the cylinder is charged with hot creosote. The pressure in the cylinder is increased to force the oil into the wood cells. When no more oil is absorbed, the pressure is released, allowing the air inside the timber to expand and force the creosote out; a protective film remains on the cell wall. After discharging surplus creosote, the treated timber is evacuated

Table 16. U.S. Distribution of Cresol and Cresylic Acids Uses, 1976[a]

Use	%
phosphate esters	20
solvents	
wire-coating enamels	15
antioxidants	15
phenolic resins	15
cleaning compounds	6
froth-flotation agents	6
miscellaneous	13
export	10

[a] Ref. 72.

for a short time to complete the removal of oil before the timber is removed from the cylinder. The Lowry process is similar but omits the initial pressurization.

Timber-preservation creosotes are mainly blends of wash oil, strained anthracene oil, and heavy oil with minor amounts of oils boiling in the 200–250°C range.

Coal-tar creosote is also a feedstock for carbon black (see Carbon, carbon black). In 1964, this usage accounted for 6.6% of the creosote produced in the UK and in 1981, it accounted for 36.5% of the total production of 172,500 t. Almost any blend of tar oils is suitable for this purpose, but the heavier oils are preferred. Other small markets for creosote are for fluxing coal tar pitch and bitumen in the manufacture of road binders (16.5% of 1981 output), and for the production of horticultural winter wash oils and disinfectant emulsions (about 1% of the total in 1981).

Pitch. The principal current outlet for coal-tar pitch is as the binder for the electrodes used in aluminum smelting. These are of two types. Older plants employ Söderberg furnaces which incorporate paste electrodes consisting of a mixture of about 70% graded petroleum coke or pitch coke and 30% of a medium-hard coke-oven pitch. This paste is added periodically to the top of the monolithic electrode as it is consumed. The more modern smelters employ prebaked electrodes requiring less binder (ca 18%).

The specification requirements for electrode binder pitch, eg, high C/H ratio, high coking value, and high β-resin content, effectively rule out pitches from gasworks or low temperature tars. The crude tar is distilled to a medium-soft pitch residue and then hardened by heating for several hours at 385–400°C. This treatment increases the toluene-insoluble content with only a slight increase in the quinoline-insoluble material.

Coke-oven tar distilleries are usually equipped for the heat treatment of pitch, either batch or continuous. In some cases, pot stills, arranged in cascade, are used. The more sophisticated plants employ one or more carbon steel or cast-iron vessels heated electrically and equipped with temperature controls for both the bulk liquid and the vessel walls; contact time is usually 6–10 h.

The demand for electrode binder pitch has grown as aluminum output has expanded. Consumption of binder averages 122.5 kg per metric ton of primary metal produced. Based on the U.S. output of aluminum in 1979 of 4.5×10^6 t, consumption of electrode pitch in that year is estimated at 550,000 t. West European consumption in that year was of the order of 440,000 t of which the UK supplied 155,000 t (65,000 t for UK smelters and 90,000 t for export); most of the remainder came from the FRG.

A smaller volume of coal-tar pitch of similar quality is used as the binder for graphitized electrodes used in electric-arc steelmaking converters.

In Europe, the production of coal briquettes by pressing a mixture of powdered bituminous coal with 8–10% of medium soft coke-oven pitch, heated to 90°C in a pug-mill, into blocks or ovoids, was formerly a big outlet for pitch in Europe. Today, this market is mainly satisfied by imports from Eastern Europe. Smokeless precarbonized briquettes, using indigenous pitch, are made in the UK and France; the UK product, known as Phurnacite, is made from steam-coal fines briquetted with medium soft coke-oven pitch and subjected to surface carbonization in a continuous vertical retort. In 1981, this product used 56,000 t of pitch (see Size-enlargement).

In North America and the Netherlands, coal-tar pitch is used as the adhesive in membrane roofs (see Adhesives; Roofing materials). The concrete or timber decking of flat-topped buildings is first covered with a layer of soft coke-oven pitch. This base, while the pitch is still fluid, is covered with tar-impregnated paper on which a second layer of molten pitch is poured. This sandwich construction is repeated until the required number of pitch–paper layers (3–5) have been installed. The top paper membrane is coated with molten pitch and covered with slag chippings or gravel. It is estimated that the 1981 U.S. consumption of pitch for this purpose amounted to ca 200,000 t.

Pitch Coke. The manufacture of pitch coke provides a large tonnage outlet for coke-oven pitch in Japan, the USSR and, until recently, the FRG (73–74). Production in the USSR is on the order of 150,000 t and FRG production was, until 1980, of the same order; Japanese output of pitch coke is about 85,000 t. Pitch coke is used, either alone or mixed with petroleum coke, as the carbon component of electrodes, carbon brushes, and shaped carbon and graphite articles.

In the USSR, pitch coke is made by carbonizing a hard coke-oven pitch in modified coke ovens. The hard pitch has an R and B softening point of 140–150°C and is made by air-blowing a mixture of medium-soft pitch and recycled coking oils. This feedstock is charged in the molten state over a period of 5 h and coked for 17–18 h at 1250–1300°C. The coke yield is 70%, oils, which are recycled and amount to 20% by weight of the pitch fed; the gas yield (80% hydrogen) is 10%.

A more modern process used in Japan, and a unit in the FRG employs the Lummus delayed-coking process (75). The feedstock is a soft coke-oven pitch from adjacent coke ovens. This flows into the bottom section of the so-called combination tower where it mixes with the vapors from the active coking drum after these have passed through the empty drum for preheating. The lighter components of the feedstock and the vapors are fractionated into fuel gas and light distillates in the upper part of the combination tower and the liquid from the base is heated to its incipient coking temperature and discharged into the preheated drum. The exothermic coking reaction proceeds for about 20 h. At the end of this period, the contents of the drum are steamed and cooled, and the green coke discharged by high-pressure water lances. The empty drum is steamed, the pressure is tested, and it is then ready to be preheated in preparation for the next charge. The green coke is produced in ca 61% yield and has an apparent density of 960–1088 kg/m^3 and a volatile matter content of 7.5–9.5%. It is calcined in a separate kiln to reduce the volatile matter to the 0.5% level required for electrode carbon. The flow sheet in Figure 4 illustrates the Lummus delayed-coking plant installed at the refinery of the Nittetsu Chemical Industries at Tobata, Japan.

Other uses for coal-tar pitch include the production of pitch-fiber pipes, as a binder for foundry cores, as a sealant for dry batteries, and in the manufacture of clay pigeons.

Pitch-fiber pipes are made by pressure-impregnating dried paper tubes with a soft (65–70°C R and B softening point) pitch at 165°C. The process is similar to the full-cell timber-impregnation process except that the impregnation cylinder is operated vertically. The white pipes are stacked upright in the pressure cylinder which is first evacuated. It is then opened to the heated reservoir of molten pitch and, when the level of pitch has risen above the pipes, the cylinder is closed and a pressure of 800–900 kPa (8–9 atm) applied until no more pitch is absorbed. The surplus pitch is then transferred to the pitch reservoir and the impregnated pipes are cooled by immersing them in cold water. Coke-oven pitch with a low quinoline-insoluble content is preferred. Pitch-fiber

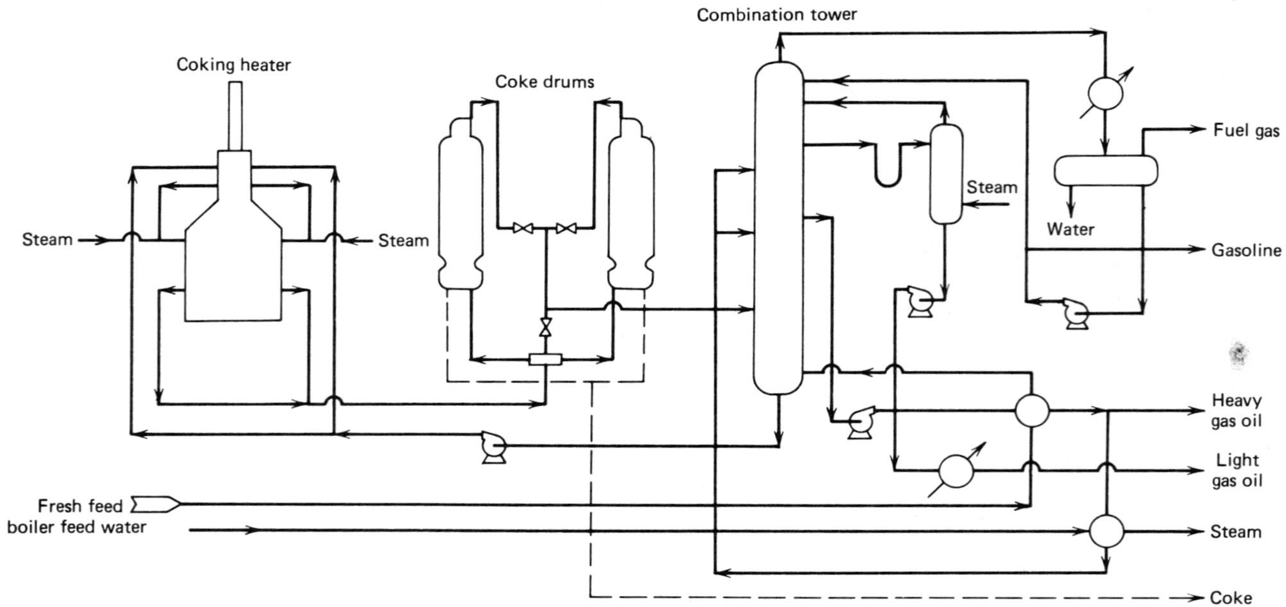

Figure 4. Lummus delayed-pitch coking process (courtesy of the Lummus Co., Ltd.).

pipes are used for the transport of sewage, industrial effluents, and irrigation. This usage of coal-tar pitch is a declining one; in the UK, for example, it has fallen from 30,000 t in 1964 to 2,750 t in 1981.

Pelleted pitch used as the binder in foundry cores, is a hard pitch supplied as spherical granules which are formed by a spray-cooling process. Clay pigeons consist of disks molded from a mixture of hard pitch and a mineral filler such as clay or limestone dust.

Fluxed Pitches and Refined Tars. ***Road Tars.*** In the United States, which has a large supply of bitumen, tar is little used in road construction or maintenance, but in Europe road binders still constitute an important, though declining, market for tar bulk products. Road tars consist essentially of medium-soft pitch fluxed back with higher boiling tar oils. They are produced in a range of viscosities to suit the particular application and the climatic conditions under which they are to be used. They are employed as binders in road bases and base courses and, in tar macadam. For road maintenance, they are used as inexpensive surface dressing to restore the riding qualities of a road surface. The road surface is first swept free of dust, the potholes are filled, and a layer of hot tar sprayed on the surface. While the tar is still fluid, the surface is covered with an evenly applied layer of stone or slag chippings which are rolled into the new surface.

As a road binder, tar has advantages and disadvantages. Since, as seen in Table 2, the temperature interval between the R and B softening point and the brittle point is only 46°C, tar tends to flow in hot weather and suffer brittle fracture in cold weather. Its internal cohesive strength is low. On the other hand, it adheres well to stone and the bond is impervious to water or petroleum-based oils. The disadvantage of the high temperature coefficient of viscosity of road tars can be overcome without sacrificing the favorable properties by blending it with petroleum bitumen, and tar–bitumen blends are now the preferred surface dressing binders in the UK. By adding 1–2% of PVC or certain types of synthetic rubber to road tar, a useful degree of elasticity is imparted. For surface dressing of heavily traveled roads, polymer-modified road tars are popular, particularly in France.

The most widely used binder for the hot-rolled asphalt or asphaltic concrete-wearing courses of heavily traveled roads is a penetration-grade petroleum bitumen.

Table 17. Distribution of Coal-Tar Products, %

Use	United States	United Kingdom
electrode binders	32.5	26.0
wood preservation oils	26.5	12.5
roofing pitch	15.0	
surface coatings	5.0[a]	3.5
carbon black feedstock	4.5	18.0
refractory binders and impregnants	1.5	3.0
pitch-fiber pipes	1.5	0.5
road binders	1.0	15.0[b]
miscellaneous outlets	4.0	2.0
fuels	8.5	7.0
briquetting pitch		12.5

[a] Including pipe-coating enamels and pitch–epoxy and pitch–polyurethane coatings.
[b] Including tar oils used to flux road bitumen.

Such surfacings tend to polish under traffic and lose their resistance to skidding. Addition of 20–25% of coal-tar pitch improves and maintains the skid resistance. Asphalt road surfacings containing pitch–bitumen binder are extensively used in Europe.

The market for tar-based road binders has declined considerably in recent years for a variety of reasons. Less crude tar is available and the profits of electrode pitch and wood-preservation creosote or creosote carbon-black feedstock are higher than road tar profits. In most industrial countries, road construction in recent years has been concentrated on high-speed motorways. Concrete, petroleum bitumen, or lake asphalt are used in their construction. In the UK, for example, the use of tar products in road making and maintenance has fallen from 330,000 t in 1960 to 100,000 t in 1975 and has now reached a fairly stable level of 35,000 t per year.

Low temperature pitch is not suitable for electrode or briquetting binders, but is perfectly satisfactory as the basis for road binders and this outlet currently provides the main market for the 45,000 t of low temperature pitch produced in the UK.

Surface Coatings. Tar-based surface coatings range from the so-called black varnishes, which consist of a soft pitch fluxed back to brushing or spraying consistency with coal-tar naphtha, to pipe–coating enamels and pitch–polymer coatings. The black varnishes, occasionally mixed with small amounts of chlorinated rubber or powdered coal to improve film elasticity, are still used to some extent for the protection of industrial steel work and timber buildings and as antifouling marine paints. The pipe-coating enamels are used for the corrosion protection of buried gas, water, or oil pipes. They are made from a coke-oven pitch fluxed back with strained anthracene oil as the basis and have a softening point of 105–125°C. To ensure good adhesion, the pipe is first primed with chlorinated rubber and chlorinated paraffin wax or chlorinated diphenyl dissolved in a mixture of toluene and trichloroethylene, and the coated pipe is wrapped with glass fiber impregnated with coal tar while the enamel coating is still hot and plastic.

The demand for pipe-coating enamels tends to fluctuate, depending on the success of oil and gas exploration. In 1981, ca 4,000 t of pitch was used for ca 19,000 t of enamel in the UK; most of the enamel was exported to the Middle East. The U.S. production of pipe-coating enamels is believed to exceed 100,000 t per year.

Refined tars are used as extenders for epoxy and polyurethane resins to give surface coatings which have developed a sizeable and growing market (76). These formulations harden into tough abrasion-resistant, waterproof films. They are now widely used for coating underground storage tanks, structural steel items, water storage tanks, marine pilings, and bridge decks. The refined tar supplied for this purpose, over 10,000 t per year in the UK, must meet stringent specifications requiring high aromaticity, low phenolic and base contents and low quinoline-insoluble components. Such properties are secured by fluxing a pitch, made by distillation of a centrifuged coke-oven tar to a relatively low softening point residue, with coke-oven strained anthracene or heavy oils which have been treated for the removal of phenols and bases.

The increasing use of carbon-bonded and carbon-impregnated refractories is providing another growing market for refined tars. These vary from a soft pitch to a refined tar of as low as 30°C EVT (equiviscous temperature) (see Refractories).

Liquid fuels were formerly important outlets for coal-tar pitch–oil blends, and topped tar which could not be disposed of more profitably (77). However, reduced tar supplies phased out this usage.

Distribution of Tar Bulk Products. The 1981 U.S. production of crude tar was ca 2×10^6 t; the bulk products, pitch and creosote, are estimated to amount to ca 1.35×10^6 t. The 1981 distribution of U.S. and UK production is given in Table 17.

BIBLIOGRAPHY

"Tar and Pitch" in *ECT* 1st ed., Vol. 13, pp. 615–632, by E. O. Rhodes, Koppers Company, Inc.; "Tar and Pitch" in *ECT* 2nd ed., Vol. 19, pp. 653–682, by D. McNeil, The Coal Tar Research Association.

1. A. W. Goos and A. A. Reiter, *Ind. Eng. Chem.* **38,** 132 (1946).
2. D. McNeil, *Coal Carbonization Products*, Pergamon Press, Oxford, 1966, Chapt. 3.
3. W. G. Adam, *J. Junior Inst. Engrs.* **55,** 325 (1945).
4. D. Chandler and A. D. Lacey, *The Rise of the Gas Industry in Britain*, British Gas Council, London, 1949.
5. B. S. Gurevich, S. C. Sabirova, and V. M. Rednov, *Coke Chem. USSR* (2), 33 (1973).
6. *Standard Methods of Testing Tar and Its Products*, 7th ed., Serial No. C.O. 3-79, Standardization of Tar Products Test Committee, British Carbonization Research Association, Wingerworth, UK, 1979, p. 477.
7. G. E. Mapstone, *J. Appl. Chem.* **5,** 582 (1955).
8. D. K. H. Briggs, *Report No. 0126*, Coal Tar Research Association, Gomersal, UK, 1955.
9. D. K. H. Briggs, *Report No. 0308*, Coal Tar Research Association, Gomersal, UK, 1963.
10. D. K. H. Briggs and F. Popper, *J. Appl. Chem.* **7,** 401 (1957).
11. Ref. 6, Serial No. 7-79, p. 480.
12. D. K. H. Briggs and W. D. Drake, *Chem. Ind.*, 666 (1957).
13. D. K. H. Briggs and F. Popper, *Fuel* **33,** 222 (1954).
14. D. K. H. Briggs and K. E. Speak, *J. Appl. Chem.* **16,** 137 (1966).
15. D. K. H. Briggs, *Fuel* **43,** 439 (1964).
16. D. K. H. Briggs and F. Popper, *Trans. Inst. Chem. Eng.* **35,** 369 (1957).
17. Ref. 6, Serial No. R.T. 3-79, p. 194.
18. *American Society for Testing and Materials Annual Book of Standards*, Designation No. D 2319 76, Part 15, ASTM, Philadelphia, Pa., 1977.
19. Ref. 6, Serial No. P.T. 3-79, p. 543.
20. Ref. 18, Designation No. 61-75.
21. G. Krämer and S. Sarnow, *Chem. Ind.* **26,** 55 (1903); "Serial No. P2-67," *Standard Methods for Testing Tar and Its Products*, 5th ed., Standardization of Tar Products Test Committee, Gomersal, UK, 1967.
22. Ref. 6, Serial No. P.T. 2-79, p. 504.
23. Ref. 18, Designation D 5-61.
24. R. M. Saal and G. Koens, *J. Inst. Pet. Tech.* **19,** 1976 (1933).
25. Ref. 18, Designation D 113-44.
26. A. Fraas, *Asphalt Teer Strassenbautech.* **13,** 367 (1930).
27. E. H. Binns, *Report No. 0167*, Coal Tar Research Association, Gomersal, UK, 1956.
28. H. G. Franck and O. Wegener, *Brennst. Chem.* **39,** 195 (1958).
29. J. G. Walters, C. Ortuglio, and J. Glaenzer, *U.S. Bureau of Mines Bulletin No. 643*, 1967.
30. R. Chauvin and R. Deelder, *Bull. Soc. Chim. Fr.* (11), 3916 (1969).
31. C. Karr, Jr., and J. R. Comberiati, *U.S. Bureau of Mines Bulletin No. 636*, 1966.
32. M. Vahrman, *Fuel* **49,** 5 (1970).
33. M. Vahrman and R. H. Watts, *Fuel* **51,** 131 (1972).
34. K. F. Lang and L. Eigen, *Fortschr. Chem. Forsch.* 8(1), 91 (1967).
35. R. Ferrand, *Goudron Routes* (3), 5 (1964).
36. D. W. Grant, *Carbonization Research Report No. 23*, British Carbonization Research Association, Chesterfield, UK, 1976.
37. J. L. Schultz, R. A. Friedel, and A. G. Sharkey, Jr., *Fuel* **44,** 55 (1965).
38. G. S. Pound, *J. Inst. Fuel* **24,** 61 (1951).
39. *Gas J.* **291,** 273 (1957).
40. P. Le Roux, *Goudron Routes* **12,** (June 1958).
41. P. van Loyen and O. M. Stuhrmann, *Erdoel Kohle* **13,** 758 (1960).

42. L. S. Elkind, *Coke Chem. USSR* (6), 36 (1962).
43. P. Y. Kuleshov and co-workers, *Coke Chem. USSR* (7), 38 (1960).
44. A. M. Kaznacheev, *Coke Chem. USSR* (3), 44 (1969).
45. *Gas World*, 3 (April 1959).
46. *Chem. Eng. Prog.* **55**(6), 84 (1959).
47. H. J. Turnbull, *Blast Furn. Steel Plant*, 411 (May 1962).
48. H. G. Franck and C. Collin, *Steinkohlenteer: Chemie, Technologie, und Verwendung*, Springer Verlag, Berlin, 1968, p. 35.
49. Ref. 48, p. 36.
50. R. C. Peter, *Ind. Chem.* **31,** 141 (1955).
51. Ref. 1, p. 109.
52. Ref. 1, p. 92.
53. J. G. D. Molinari and B. V. Dodgson, *Chem. Eng.* **287,** 460 (1974).
54. *Conference International du Goudron Statistics*, Paris, 1981.
55. D. L. Richards, *Chem. Mark. Rep.* **209**(9), 3, 54 (1976); and later statistics published in *Coke and Coal Chemicals*, U.S. Bureau of Mines Mineral Survey, Washington, D.C.
56. Statistics published by British Tar Industry Association, London.
57. Statistics published by U.S. International Trade Commission.
58. D. McNeil in M. A. Elliot, ed., *High Temperature Coal Tar*, John Wiley & Sons, Inc., New York, 1981, Chapt. 17, p. 1013.
59. H. Mallison, *Bitumén Teere Asphalte Peche* **1,** 313 (1950).
60. R. D. Passey, *Br. Med. J.* **2,** 1112 (1922).
61. A. Manouvirez, *Ann. Hyg. Publique* **45,** 459 (1876).
62. R. Volkmann, *Beitrage zum Chirurgie*, Breitkopf and Hartel, Leipzig, 1875, p. 370.
63. K. Yamagiwa and K. Ichikawa, *Milt. Med. Fak. Tokio* **15,** 295 (1915).
64. E. L. Kennaway, *J. Ind. Hyg.* **5,** 462 (1924).
65. J. H. Cook, C. L. Hewitt, and I. Hieger, *Nature (London)* **130,** 926 (1932).
66. W. E. Poel, *J. Natl. Cancer Inst.* **18**(1), 41 (1957).
67. H. I. Cookson, *Br. Med. J.* **1,** 368 (1924).
68. F. Pierre and co-workers, *Arch. Mal. Prof. Med. Trav. Secur. Soc.* **T26,** 475 (Sept. 1965).
69. E. Sawicki, F. T. Fox, W. C. Elbert, T. R. Hansen, and J. Meeker, *Am. Ind. Hyg. J.* **23,** 482 (1963).
70. E. C. Hammond, I. J. Selikoff, and P. J. Lawther, *N.Y. Acad. Sci.* **271,** 116 (1976).
71. V. P. Puzinawkas and L. W. Corbett, *Research Report 78-1*, The Asphalt Institute, Maryland, 1978.
72. *Chem. Mark. Rep.* 5 (May 31, 1976).
73. S. Sarkar, *J. Mines Met. Fuels* **10,** 16 (1962).
74. K. H. Osthaus, *Koppers-Mitt.* **3,** 115 (1960).
75. R. Ramirez, *Chem. Eng.* **76,** 74 (1969).
76. P. H. Pinchbeck, *Aust. Oil Colour Chem. Assoc. Proc. News* **14**(6), 10; **14**(7), 6 (1977).
77. W. H. Huxtable, ed., *Coal Tar Fuels*, The Association of Tar Distillers, London, 1960.

DONALD MCNEIL
Technical Consultant
Bradford, UK

TAR SANDS

Tar sands (also known as oil sands and bituminous sands) are sand deposits impregnated with dense, viscous petroleum. Tar sands are found throughout the world, often in the same geographical areas as conventional petroleum. The largest deposits are in the Athabasca area in the northeastern part of the province of Alberta, Can., and in the Orinoco region of east central Venezuela. The Athabasca deposit along with the neighboring Wabasca, Peace River, and Cold Lake deposits contain $>1.86 \times 10^{11}$ m^3 (1.17×10^{12} bbl) bitumen (1). The Venezuelan deposits contain $>1.60 \times 10^{11}$ m^3 (1.0×10^{12} bbl) bitumen (2). Together these deposits represent a significant potential supply.

Deposits of tar sand, each containing $>3 \times 10^6$ m^3 (20×10^6 bbl) of bitumen, have also been located in the United States, Albania, Italy, Madagascar, Peru, Romania, Trinidad, Zaire, and the USSR (see Table 1), a total of ca 450×10^9 m^3 (2.8×10^{12} bbl).

The bitumen can be separated from the tar sands by a wide variety of methods (3–9). Although tar-sand deposits have been known for many years, they have only recently assumed technical importance. The first venture for manufacturing synthetic crude from tar sands, Great Canadian Oil Sands, Ltd. (now Suncor), came on stream in 1967 with a design capacity of 7200 m^3/d (45,000 bbl/d). Suncrude, the second commercial venture, began in 1978 and by the end of 1981 was approaching its rated capacity of 20,500 m^3/d (129,000 bbl/d) of synthetic crude.

Although the Venezuelan tar-sands deposit and several U.S. deposits have been briefly investigated, the bulk of the available data refers to the Athabasca deposit, first recorded in 1788. Serious mapping and exploration were initiated by the Federal Canadian Department of Mines and Technical Surveys in 1913. From that time until ca 1945, except for the continuing research program of the Research Council of Alberta, little interest was expressed in the Athabasca deposit. During World War II, with the possibility of an oil shortage in Canada and the United States, an extensive program

Table 1. Tar-Sand Deposits Listed in Order of Reserve Size[a]

Country	Number of deposits	Original oil in place, 10^6 m^3 [b]	Ref.
Canada	7	186,752.5	1
Venezuela	4	158,989.2	2
USSR	10	88,559.7	2
United States	53	5,467.6	2
Madagascar	2	3,942.9	2
Italy	4	2,250.5	2
Zaire	1	306.0	2
Albania	1	39.0	2
Peru	1	10.5	2
Trinidad and Tobago	1	9.5	2
Romania	1	4.0	2

[a] Ref. 2.
[b] To convert m^3 to bbl, multiply by 6.289.

of exploration was undertaken by the Canadian Government. Between 1940 and 1950, pilot plants were sponsored by both the Federal Canadian Government and the Government of the Province of Alberta. During 1948–1949, the hot-water separation process was demonstrated on pilot scale, and in 1950 the Blair report (10) was published. This engineering study concluded that production of synthetic crude oil from the Athabasca tar sands was technically and economically feasible. Since 1955, industrially sponsored research has been steadily increasing, leading to six formal proposals for commercial ventures (see Table 2).

Among the *in situ* methods for recovering bitumen are fire floods, steam drive and stimulation, and electric heating processes. Processes operating on mined tar sand include direct coking and hot-water, cold-water, and solvent processes. Bitumen itself is too dense, sp gr = 1.00–1.029 (6–10°API), and viscous, 5–50 kPa·s ((5–50) × 10^4 P), to be either salable or transportable. Proposed upgrading processes include coking (partial, conventional delayed, or fluid with catalytic hydrodesulfurization of the coker distillate) and direct catalytic hydrovisbreaking. Commercial ventures for bitumen recovery from the Athabasca tar sands have been proposed by Shell Canada, Ltd. (steam-drive *in situ* recovery process followed by thermal cracking and hydrotreating); Cities Service Athabasca, Ltd. and Great Canadian Oil Sands (GCOS), Ltd. (mining, hot-water extraction, coking, and hydrotreating); Syncrude Canada Ltd. (mining, hot-water extraction, fluid coking, and hydrotreating); Esso Resources Canada Limited (steam stimulation followed by flexicoking and hydrotreating); and Alsands Energy (mining, hot-water extraction, fluid coking, and hydrotreating).

Occurrence and Reserves

Various definitions have been applied to energy reserves (21). Proved reserves are reserves recoverable under current conditions of technical and economic feasibility. They give an accurate, short-term estimate of the working inventory of a particular resource (see Petroleum, resources; Fuels, survey).

Ultimate reserves, or original oil in place, are defined much more broadly as the largest reasonable estimate of the total amount of a particular resource. Not only discovered, but also discoverable (based on reasonable geological extrapolations), resources are included. The estimate is not limited to today's economic conditions or recovery technology. Because of the differences in definition, reserves cannot be compared. Although tar sands are not a principal energy reserve, they certainly are significant with regard to projected energy consumption over the next several generations.

Proved U.S. crude oil reserves remained approximately constant during the 1970s. Nevertheless, the reserve production ratio decreased constantly and perceptibly during the decade. This trend has, particularly recently, caused increasing concern among oil companies as to their long-term positions on crude oil. At this time, several large petroleum companies are committed to the development of synthetic fuels, ie, fuels made from sources other than conventional crude oil, eg, tar sands, oil shale, and coal. Tar sands are clearly the front runner; ie, production from tar sands is today a commercial reality. The next most likely prospect may be oil shale. However, a commercial oil-shale venture seems no closer today than it was 10 yr ago (see Oil shale). Although the technology of coal liquefaction is available now, the economics point 5–10 yr into the future (see also Fuels, synthetic).

Table 2. Commercial Tar-Sand Ventures

Project	Shell Canada, Ltd.	Cities Service Athabasca, Ltd.[a]	Great Canadian Oil Sands, Ltd.[b]	Syncrude Canada, Ltd.[c]	Alsands[d]	Esso Resources Canada Ltd.
Project size						
bitumen, m^3/d[e]	21,000	21,300	9,200	24,100	27,900	25,400
synthetic crude m^3/d[e]	15,900	15,900	7,200	20,500	22,200	22,300
investment, 10^6\$		274[f]	230[g]	2,160[g]	5,000[f]	12,000[f]
Processing sequence						
mining	*in situ*	bucket wheels (two)	bucket wheels (two)	draglines (four)	draglines	*in situ*
bitumen recovery	API separator	dense-phase process[h]	hot water	hot water	hot water	separator
bitumen upgrading	electrical dehydrator	thermal dehydrator	centrifuge	centrifuge	centrifuge	electrical dehydrator
refining	thermal cracking hydrodesulfurization	fluid coking hydrodesulfurization	delayed coking hydrodesulfurization	fluid coking hydrodesulfurization	fluid coking hydrodesulfurization	flexicoking hydrodesulfurization
status	application 1962 withdrawn 1968	application deferred	started 1962 operation Sept. 1967	approval 1972 construction 1973 startup mining 1977, upgrader 1978	application 1979 project terminated 1982	application 1979 project suspended 1982
references	11–12	13–14	15–17	18	19	20

[a] In association with Imperial Oil, Ltd., Richfield Oil Corp., and Royalite Oil Co., Ltd.

[b] A subsidiary of Sun Oil Company, Ltd.

[c] Consortium of Esso Resources Canada Ltd. (31.25%), Canada Cities Service (22%), Gulf Oil Canada Ltd. (16.75%), Petro-Canada (15%), Government of Alberta (10%), Government of Ontario (5%).

[d] Consortium of Shell Canada Resources, Ltd. (25%), Shell Explorer, Ltd. (20%), Petro-Canada (17%), Dome Petroleum Limited (12%), Amoco Canada Petroleum, Ltd. (10%), Chevron Standard, Ltd. (8%), and Gulf Canada (8%).

[e] To convert m^3 to bbl, multiply by 6.289.

[f] Estimated.

[g] Actual.

[h] A hot-water extraction process.

Canadian (Athabasca) Tar Sands. The Alberta tar sands, actually four deposits, are located in the northeast part of the province. Figure 1 gives the geographical location; Figures 2 and 3 give a geological cross section of three of the deposits. The Athabasca tar-sand deposits are the only deposits currently undergoing large-scale commercial exploitation. They are also the only large deposits known to have been investigated scientifically in detail.

Evidently the Cree Indians, native to the Athabasca region of northeastern Alberta, have known about the tar sands for well over 200 yr, as documented by an itinerant New England fur trader who reported that the Indians used "a sticky substance oozing from the river banks to waterproof their canoes" (22).

The first scientific interest in tar sands was taken by the Canadian Government in 1890, and in 1897–1898, the sands were first drilled at Pelican Rapids on the Ath-

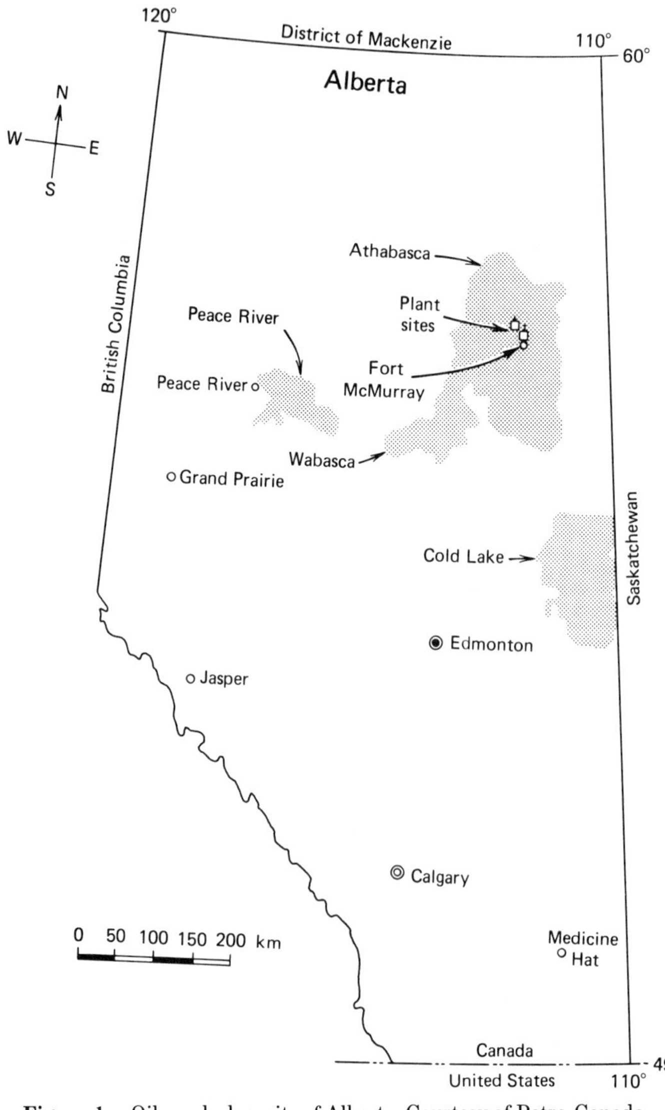

Figure 1. Oil-sands deposits of Alberta. Courtesy of Petro-Canada.

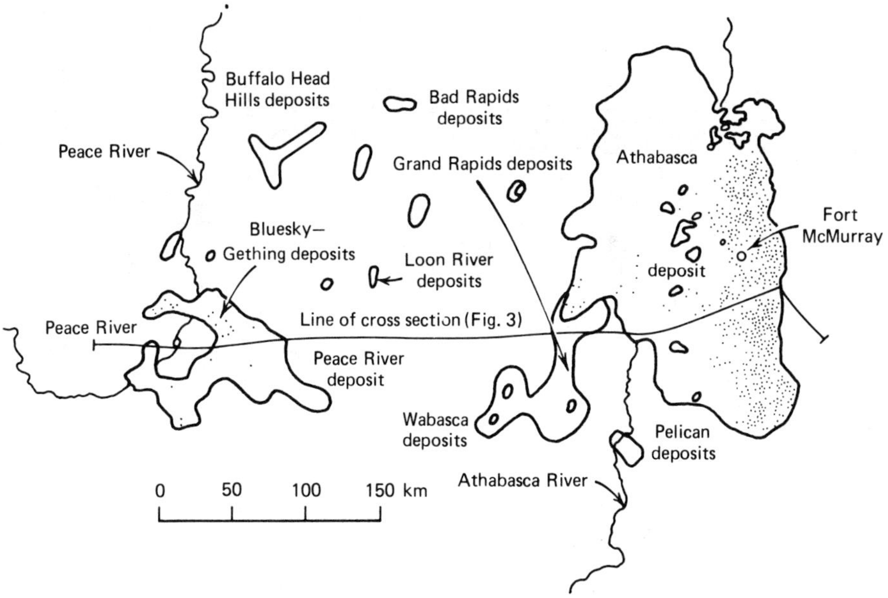

Figure 2. Location of evaluated tar-sands deposits of northern Alberta (22).

abasca River, 129 km southwest of Fort McMurray. Up to 1960, the development was mainly small commercial enterprises. Between 1957 and 1967, three extensive pilot-plant operations were conducted in the Athabasca region, each leading to a proposal for a commercial venture. Shell Canada, Ltd. tested *in situ* recovery of bitumen; a group headed by Cities Service Athabasca, Ltd. operated a pilot plant at Mildred Lake based on a hot-water extraction process; and Great Canadian Oil Sands, Ltd. operated a pilot plant at the present commercial plant site (see Fig. 1) near Mildred Lake. In addition, Mobil Oil Company has conducted tests on *in situ* recovery, and Petrofina established a pilot facility across the Athabasca River from the site of the GCOS plant. The operations of the latter two pilot plants remain proprietary.

There are two schools of thought regarding the origin of the oil. One is that the oil was formed locally and has neither migrated a great distance, nor been subjected to large overburden pressures (23). Since under these conditions the oil cannot have been subjected to thermal cracking, it is geologically young and therefore dense and viscous. The opposing theory assumes a remote origin for the oil, both geographically and in geological time (24). The oil, originally like a conventional crude, is assumed to have migrated into the sand deposit, which may originally have been filled with water. After the oil migrated, the overburden pressures were relieved, and the light portions of the crude evaporated, leaving behind a dense viscous residue. A more recent theory also assumes a remote origin for the oil but postulates that the light hydrocarbons were destroyed by bacteria carried into the petroleum reservoirs in oxygenated, meteoric waters (25). The remote-origin theory would explain the water layer surrounding sand grains in the Athabasca deposit. However, because the metals and porphyrin contents of bitumen are similar to those of some conventional Alberta crude oils of Lower Cretaceous age and because Athabasca bitumen has a relatively low coking temperature, the bitumen may be of Lower Cretaceous age. This is the age of the McMurray formation, which is geologically young. This evidence supports the

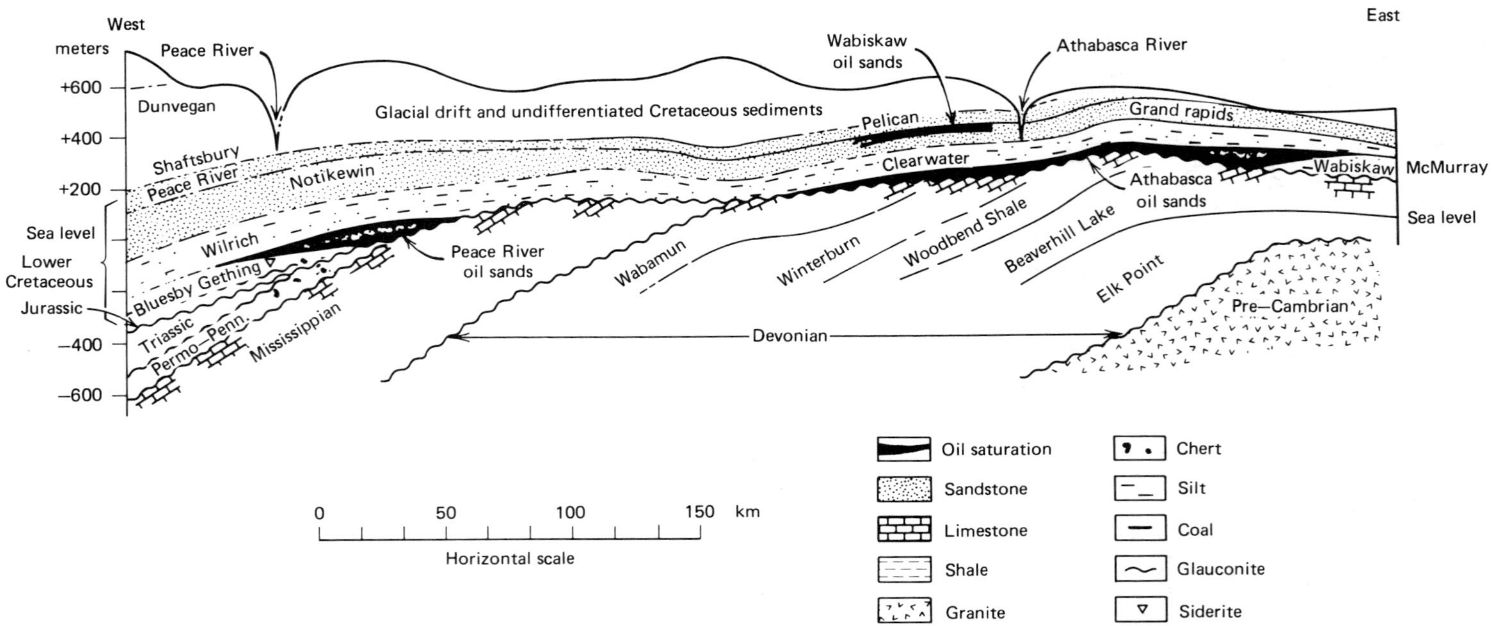

Figure 3. Schematic geological east–west cross section showing the geological setting of the Athabasca tar sands (22).

theory that the oil was formed *in situ* and is a precursor, rather than a residue of some other oil. The issue remains unresolved.

Recoverable Reserves. The two conditions of vital concern for the economic development of tar-sand deposits are the concentration of the resource, or the percent bitumen saturation, and its accessibility, usually measured by the overburden thickness. Information of this type for the Athabasca deposit is summarized in Figures 4 and 5. Recovery methods are based either on mining combined with some further processing or operation on the oil sands *in situ*. The mining methods are applicable to shallow deposits, characterized by an overburden ratio (ie, overburden depth to thickness of tar-sand deposit) of ca 1.0. Since the Athabasca tar sands have a maximum thickness of ca 90 m and average ca 45 m, the data in Figure 3 would indicate that no

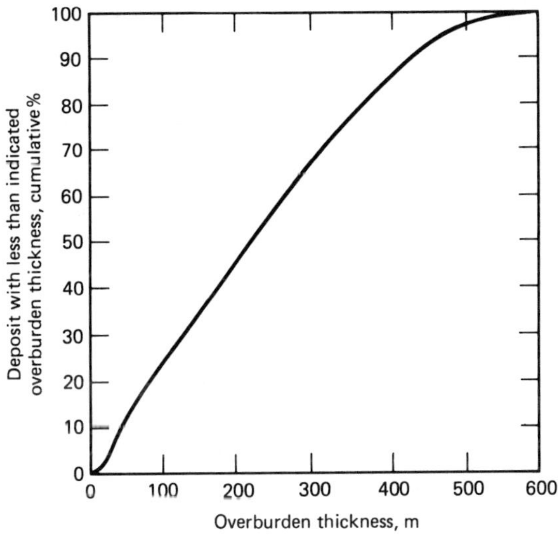

Figure 4. Overburden thickness of the Athabasca tar sands (22).

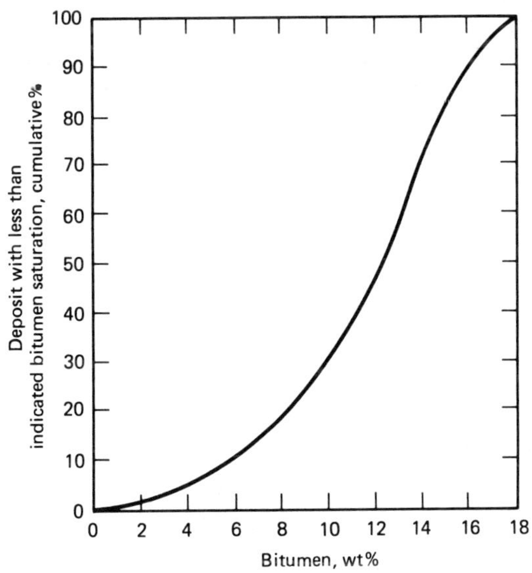

Figure 5. Bitumen saturation of the Athabasca tar sands (22).

more than 10% of the in-place deposit is minable within current concepts of the economics and technology of open-pit mining. This 10% portion may be considered as the proved tar-sand reserves.

In situ processes depend on injecting a heating-and-driver substance into the ground through injection wells and recovering bitumen through production wells. Such processes need a relatively thick layer of overburden to contain the driver substance within the formation between injection and production wells. In current pilot projects, overburden layers could be as thin as 75 m, which means that most of the Athabasca deposit may be suitable for *in situ* recovery.

For tar sands of 10 wt % bitumen saturation, 12.5 metric tons of tar sand must be processed to recover 1 m^3 (6.3 bbl) of bitumen. If the ore contains only 5 wt % bitumen, twice the amount of ore must be processed to recover this amount. Thus, it is clear that below a certain bitumen concentration, tar sands cannot be processed economically. For a mining process, the lower limit may be ca 10 wt %, and therefore, as Figure 5 shows, ca 30% of the Athabasca tar-sand deposit would be excluded from consideration. For an *in situ* process, the lower limit is somewhat less. Nevertheless, considering the limitations both of overburden thickness and bitumen content, and the limitation of recovery efficiency for an *in situ* process, it is clear that when proved reserves are calculated for a given deposit, this figure is much smaller than the total in-place reserves.

Venezuelan Tar Sands. The Venezuelan tar sands are located in a belt 50–100 km wide extending east–west for >700 km immediately north of the Orinoco River (26). The precise limits of the deposit are not well defined because exploration efforts in the past were concentrated on light and medium crude accumulations. An exploration effort is now underway in the Orinoco deposit to define this vast accumulation.

The geological setting of the Orinoco deposit (27) is very complex, having evolved through three cycles of sedimentation. The oil is contained by both structural and stratigraphic traps, depending on the location, the age of the sediment, and the degree of faulting. The tar sands are located along the southern flanks of the eastern Venezuelan basin where three distinct zones are apparent from north to south: a zone of tertiary sedimentation; a central platform with transgressive overlapping sediments; and a zone of Proterozoic erosional remnants covered by sediments. The deposit also contains three systems of faulting. All the faults are normal and many are concurrent with deposition.

United States Tar Sands. Tar sands in the United States are contained in ca 550 separate deposits in 22 states, as shown in Figure 6. Since many of these deposits are small, information on most is limited. The oil-in-place data shown in Table 1 are based on the 53 large accumulations found in California, Kentucky, New Mexico, Texas, and Utah. Development has occurred primarily in California, New Mexico, Texas, and Utah (30).

Properties

Bulk. Tar sand has been defined as sand saturated with a "highly viscous crude hydrocarbon material not recoverable in its natural state through a well by ordinary production methods" (22). Technically, the material should perhaps be called bituminous sand rather than tar sand since the hydrocarbon is bitumen (ie, a carbon disulfide-soluble oil). In petroleum refining, the term tar is reserved to mean the residue

Figure 6. Tar-sand occurrences in the United States (28).

of a thermal process. The term oil sand is used in allusion to the synthetic crude oil that can be manufactured from the bitumen.

Tar sand is a mixture of sand, water, and bitumen (see Fig. 7). The sand component is predominantly quartz in the form of rounded or subangular particles, each of which is wet with a film of water. Surrounding the wetted sand grains and somewhat filling the void among them is a film of bitumen. The balance of the void volume is filled with connate water plus, sometimes, a small volume of gas. Usually the gas is

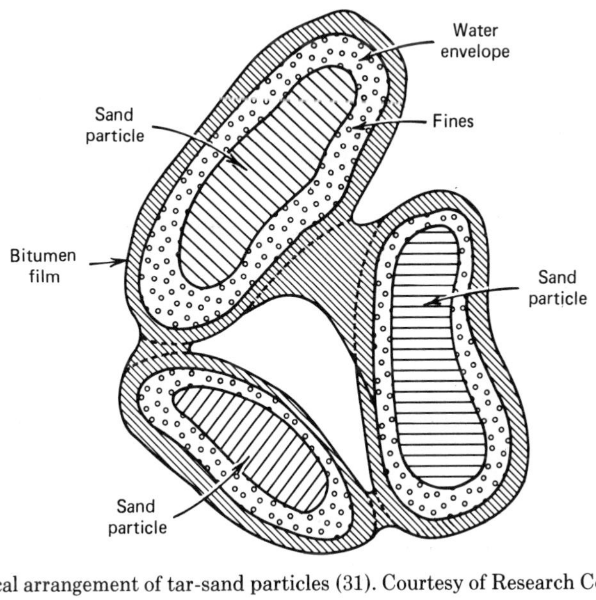

Figure 7. Typical arrangement of tar-sand particles (31). Courtesy of Research Council of Alberta.

air; however, methane has been reported from some test borings in the Athabasca deposit. The sand grains are packed to a void volume of ca 35%, corresponding to a mixture of ca 83 wt % sand; the remainder is bitumen and water. In fact, with considerable regularity bitumen and water constitute ca 17 wt % of the tar sands.

Hot-water processing is based on percent bitumen and percent fines. The former may be determined analytically by extraction with toluene, arranged in such a way that both the water and bitumen contents are measured directly. In this article, percent fines refers to the weight percentage of the dried, extracted tar-sand mineral that passes a fixed mesh size such as 44 μm (325 mesh). The percent fines may be conventionally determined by screening.

Average bulk properties of samples from seven Canadian locations, two Venezuelan locations, and U.S. deposits are given in Table 3. The large number and wide distribution of these deposits result in a very wide range of property values.

Tar-Sand Minerals. Usually, >99% of the tar-sand mineral is composed of quartz sand and clays. In the remaining 1%, more than 25 minerals have been identified, mostly calciferous or iron based. Particle sizes range from large grains (99.9% is finer than 1000 μm) down to 44 μm (325 mesh), the smallest size that can be determined by dry screening. The size between 44 and 2 μm is referred to as slit; sizes below 2 μm (equivalent spherical diameter) are clay.

Clays (qv) are aluminosilicate minerals, for some of which definite chemical compositions have been established. However, as used in this article, clay is only a size classification and is usually determined by a sedimentation method. According to the previous definition of fines, the fines fraction equals the sum of the silt and clay fractions. The clay fraction over a wide range of fines contents is a relatively constant 30% of the fines (see Size measurement).

Bitumen. Although there are wide variations both in the bitumen saturation of tar sand (0–18 wt % bitumen) and in the particle-size distribution of the tar-sand mineral, the chemical composition of Athabasca bitumen is relatively constant over a wide geographical area. Bitumen properties of samples from a number of geographical locations are given in Table 4. Density at 15°C is slightly greater than that of water. About 50% of the bitumen is distillable without cracking. Elemental analyses are relatively constant; sulfur and nitrogen averages are 4.5–5.0 and 0.4–0.5 wt %, respectively. Viscosity values reported for samples from Bitumount, Abasand, and the Mildred-Ruth Lakes area (GCOS) are given in Figure 8; the variation of bitumen density with temperature is shown in Figure 9. A curve for water crosses the bitumen curve twice. The maximum density difference occurs at 70–80°C, near the operating temperature of the hot-water bitumen-extraction process.

Refined Products. Properties of synthetic crude oils made from Athabasca bitumen by various processes are given in Table 5.

Recovery Technology

The tar sand may be mined and transported to a processing plant where the bitumen is extracted and the sand is discharged. Alternatively, the bitumen may be separated *in situ*. The latter process is similar to tertiary recovery of crude oil. Several additional processes of minor importance are described in patents (4). Since the scale of operations described varies from laboratory batch experiments to a 10-t/h pilot plant, comparisons of the relative merits of the various processes must be made with

Table 3. **Bulk Properties of Tar Sands**

Property	Canada, Lower Cretaceous[a]							Venezuela[a]		United States, various deposits[b]
	Athabasca, Wabiskaw/McMurray	Peace River, Bluesky-Gething	Cold Lake A, Grand Rapids	Cold Lake B, Clearwater	Cold Lake C, McMurray	Wabasca A, Grand Rapids	Wabasca B, Wabiskaw	Morichal	Jobo	
proved in-place reserve, km^3	132,000	15,000	26,000	6,400	5,400	10,500	8,000	800	1,400	
depth, m	0–610	450–790	300	410	450	90–330	200–670	1,060–1,220	1,030–1,170	0–975
av pay thickness[c], km	27	14	11	11	9	12	7	68	14–52	4–150
reservoir temperature, °C	10	17		13	10	10		56	54–64	
reservoir pressure, kPa[d]		3,650[e]		3,030				9,410		
initial gas–oil ratio, m^3/m^3								12		
initial oil saturation, %	44–98	45–88	76	73	69	69	64	83	76–88	35–82
porosity, %	39	25–31	34–40	38–46	30–36	27–30	19–32	30–31	30–31	17.5–35
permeability, μm^2[f]	≤500	≤100		≤300		≤350		50–400	150	1.2–600

[a] Ref. 32.
[b] Ref. 28.
[c] 3% bitumen cutoff for Canadian fields.
[d] To convert kPa to psi, multiply by 0.145.
[e] Referenced to 53.3 m Subsea.
[f] To convert μm^2 to darcy, multiply by 1.013.

Table 4. Bitumen Properties

	Canada[a]									
Property	Athabasca, Wabiskaw/McMurray	Peace River, Bluesky-Gething	Cold Lake A, Grand Rapids	Cold Lake B, Clearwater	Cold Lake C, McMurray	Wabasca A, Grand Rapids	Wabasca B, Wabiskaw	Venezuela[a] Morichal	Venezuela[a] Pilon	United States[b], various deposits
sp gr (°API)	1.00–1.014 (8–10)	1.007–1.014 (8–9)		0.986–1.00 (10–12)		0.979–1.014 (8–13)	0.946–1.029 (6–18)	1.061 (11.9)	1.011 (8.4)	0.96–1.12 (−5–16)
Viscosity, Pa·s[c]										
15°C	5×10^3	200		100		8×10^3				
21°C							0.225			
38°C							0.097	3.2	62.26	10×10^3–200×10^3
60°C	3.7							0.59	5.53	
100°C	1.0	0.1		0.1		2.0				
pour point, °C	10							0	20	
Elemental analysis, wt %										
carbon	83.1	82.2		83.7		83.0				
hydrogen	10.6	10.0		10.5		10.3				
sulfur	4.8	5.6		4.7		5.5		2.1	3.7	
nitrogen	0.4	0.1		0.2		0.4		0.53		
oxygen	1.1	2.1		0.9		0.8				
C:H ratio	7.8	8.2		7.9		8.1				
mol wt	570–620	520		490		600				

Hydrocarbon type, wt %						
saturated	22		33			
aromatic	21		29			
resins	39		23			
asphaltenes	18	19.8	15	18.6	10.8	8.6
Metals, ppm wt						
vanadium	250		240	210	250	390
nickel	100		70	75	65	106
Conradson carbon	13.5		12.6		14.0	14.1
Distillation temperature, °C						
initial bp,	260		170	275		
5 vol %	285		255	330		
10 vol %	320		300	370		
30 vol %	425		405			
50 vol %	530		555			

[a] Ref. 32.
[b] Ref. 28.
[c] To converter Pa·s to P, multiply by 10.

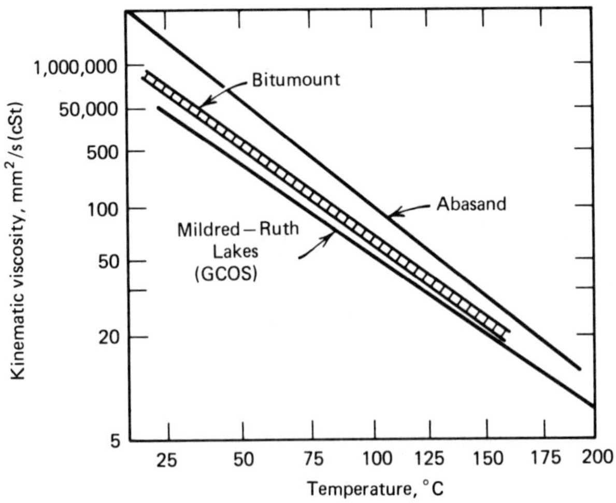

Figure 8. Bitumen viscosity.

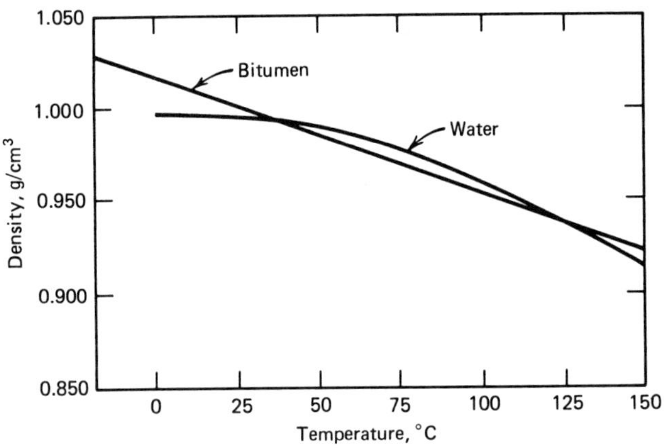

Figure 9. Density of bitumen and water.

great care. The term yield, or bitumen recovery, is subject to so many definitions and possible methods of calculation that comparisons may be meaninglesss. In addition, practically no economic data are available. Despite these shortcomings, process flow sheets may readily be compared. Fundamental process requirements, such as heat, can also be calculated and compared.

***In Situ* Processes.** Conventional crude oil is produced from the oil-bearing formations by drilling wells into the formation. The oil is driven from the formation up through the wells (production wells) by energy stored in the formation, such as the pressure of dissolved natural gas. If this natural energy of the formation is expended or, as in the case of the Athabasca tar sands, if it was never present, then energy must be injected into the formation in order to stimulate production.

Combustion. In *in situ* combustion or fire flooding, energy is generated in the formation in the form of heat. The heat is supplied by actually igniting oil in the formation and sustaining it in a state of combustion or partial combustion. The high temperatures generated decrease the viscosity of the oil and make it more mobile. The high temperatures also cause some cracking of the oil, and an upgraded product rather

Table 5. Properties of Synthetic Crude Oil Refined from Athabasca Bitumen

	Source of samples							
	GCOS[a]		Cities Service	Shell	Direct fluid coking		UOP[b]	Feed for hydro-
Property	Pilot plant	Commercial	Athabasca	Canada, Ltd.	Abasand	Bitumount	pilot plant	treating tests
sp gr	0.837	0.833	0.875	0.861	0.962	0.961	0.901	0.965
(°API)	(37.6)	(38.3)	(30.2)	(32.8)	(15.6)	(15.8)	(25.5)	(15.5)
Distillation temperature, °C								
initial bp	99	72			80	83	48	
5%	126	105			170	211	83	
10%	199	123	129		238	250	118	
30%	193	209	235	204	332	310	226	
50%	252	264	207	282	361	343	288	
90%	360	324		474			399	
end pt	445	379					404	
recovery, %	99						97	
Elemental analysis, wt %								
carbon	86.3							84.4
hydrogen	13.4							11.2
nitrogen[c]	0.02		0.096	0.09				0.24
oxygen								0.10
sulfur[c]	0.03	0.022	0.182	0.29	4.0	3.9	3.6	4.1
heavy metals[d], ppm	0.04							
Hydrocarbon type, wt %								
asphaltenes	0							
resins	0							
aromatic oil	21							
saturated oil		79						
UOP[b] K process							11.0	
	delayed coking: hydrotreating		fluid coking: hydrotreating	thermal cracking hydrotreating	direct coking of tar sand 1790°C		delayed coking	destructive batch distillation
reference	15	16	13	11	33	33	34	35

[a] Mildred-Ruth Lakes area.
[b] Universal Oil Products Co.
[c] Can be removed to ppm concentrations.
[d] Vanadium, nickel, and copper.

than bitumen itself is the fluid recovered from the production wells. These recovery processes are referred to as forward combustion or reverse combustion, depending on whether the combustion front moves with or counter to the direction of air flow. In either case, burning occurs at the interface where air contacts hot, unburned oil or, more likely, coke. Thus, if the flame front is ignited near the injection well, it propagates toward the production well (forward combustion). However, if the front is ignited near the production well, it moves in the opposite direction (reverse combustion). In forward combustion, the hydrocarbon products released from the zone of combustion move into a relatively cold portion of the formation. Thus, there is a definite upper limit of the viscosity of the liquids that can be recovered by a forward-combustion process. On the other hand, since the air passes through the hot formation before reaching the combustion zone, burning is complete; the formation is left completely cleaned of hydrocarbons. In reverse combustion, some hydrocarbons are left in the formation. The theoretical advantage of reverse combustion is that the combustion products move into a heated portion of the formation and therefore are not subject to a strict viscosity limitation. However, most attempts to implement reverse combustion in field pilot installations have been unsuccessful. In many cases, the failure resulted from the onset of secondary combustion at the production well.

Fire flooding is being evaluated in field pilot tests by Amoco in Athabasca and British Petroleum in the Cold Lake deposit in Alberta (36). These projects are at the experimental stage, and neither has reached the degree of development necessary to proceed to commercial application.

Steam Stimulation. In steam stimulation, heat and drive energy are supplied in the form of steam injected through wells into the tar-sand formation. In most instances, the injection pressure must exceed the formation fracture pressure in order to force the steam into the tar sands and into contact with the oil. When sufficient heating has been achieved, the injection wells are closed for a soak period of variable length and then allowed to produce, first applying the pressure created by the injection and then using pumps as the wells cool and production declines.

This process is being tested in field pilot installations by numerous operators in the Alberta tar-sands region (36). The process was pioneered by Imperial Oil (now Esso Resources Canada) in Cold Lake. Starting from a four-well facility in 1964, current activities include more than 150 wells with oil production >1500 m^3/d (9500 bbl/d). By 1979, the process had been developed to a sufficient degree for Esso to apply to the regulatory authorities for approval to implement a commercial recovery scheme utilizing steam stimulation.

Steam-stimulation pilot projects have also been carried out in tar-sands deposits in the Orinoco area in Venezuela (37), California (38), and Utah (38).

Steam Drive. Steam can also be injected into one or more wells with production coming from other wells. This technique is very effective in heavy-oil formations but has found little application in the Alberta tar sands because of the difficulty in connecting injection and production wells. However, the Shell operation in the Peace River deposit in Alberta uses this process (39).

Much of the Peace River deposit has an unusual thin zone of high water saturation at the base of the tar-sand formation. Shell utilizes this zone to carry steam from the injection to the production wells. Once the flow path has been heated, the steam pressure is cycled, alternately moving steam up into the oil zone, then allowing oil to drain down into the heated flow channel to be swept to the production wells. A large pilot project to evaluate this concept was put into operation in 1979 by Shell, the Alberta Oil Sands Technology and Research Authority, and a number of other partners.

Since reservoir temperatures in the Venezuelan tar-sands deposits are higher than in Alberta, the oil is mobile and communication between injection and production wells is easy. Lagoven is evaluating steam drive in a pilot installation in the central Venezuelan Jobo field (37); similar tests are planned by other companies.

Emulsion–Steam-Drive Process. If the viscous bitumen in a tar-sand formation can be made mobile by admixture of either a hydrocarbon diluent or an emulsifying fluid, a relatively low temperature secondary recovery process is possible. If the formation is impermeable, communication problems exist between injection and production wells. However, it is possible to apply a solution or dilution process along a narrow fracture plane between injection and production wells (see also Petroleum, chemicals for enhanced recovery).

Shell Canada, Ltd. tested an emulsion process in field trials between 1957 and 1962 (40). Water emulsification was preferred to a hydrocarbon diluent which is more expensive. The viscosity of a bitumen-in-water emulsion (20–30% bitumen) is essentially the viscosity of water.

Field trials were also conducted with a proprietary nonionic surfactant in water, as well as with a combination of caustic solution (sodium hydroxide in water) and steam-drive technique. For this test, a five-spot pattern of wells was drilled (four injection wells at the four corners of a square with a producing well located in the center). Shell concluded that, in commercial operation, 70% displacement of the in-place oil would be possible from that fraction of the ore body that was wet, or swept, by the caustic solution.

Mining. The alternative to *in situ* processing is to mine the tar sands, transport them to a processing plant, extract the bitumen value, and dispose of the waste sand. In order to support synthetic crude production in a plant of economically viable size, an immense mining operation is needed. For instance, the Suncor project, which produces 7000 m^3/d (44,000 bbl/d) of synthetic crude, requires a mining rate of ca 100,000 t/d of tar sand (ore) only, exclusive of the overburden which must be stripped away. Because of the large scale of mining involved, only open-pit methods have been proposed for the Athabasca tar sands. As noted above, an important parameter in an open-pit mining scheme is the overburden ratio (the thickness of overburden that must be removed to expose unit thickness of the ore body). For the Suncor project, the overburden ratio is ca 0.4 (15). The mining of deposits with overburden ratios as high as 2.5–3.5 has been discussed (41), but not seriously proposed.

Because of the relatively low unit value of tar sand as an ore, mining and transportation costs must be kept down, eg, the feed must be handled minimally between mine and processing plant. On the other hand, continuous units must be designed to operate with a relatively steady round-the-clock feed rate. Although such processing is common in petroleum refineries, it is the exception in mining.

Mining of the Athabasca tar sands presents two additional problems: in-place tar sand requires very large cutting forces and is extremely abrasive to cutting edges, and both the equipment and pit layouts must be designed to operate during the long Canadian winters at temperatures as low as −40°C.

Basically there are two approaches to open-pit mining of tar sands. The first uses a few mining units of custom design, which are necessarily very expensive, eg, bucket-wheel excavators, dredges (both hydraulic and bucket-ladder), and supersized drag lines. In the second approach, a multiplicity of smaller mining units of conventional design is employed at relatively much lower unit costs. Scrapers and truck-and-shovel operations have been considered. Each method has advantages and risks (41). The first approach has been adopted by Suncor (15,17) and Syncrude Canada, Ltd. (42).

In the Suncor pit design, the ore body is divided into two layers, or benches, each nominally 23 m high. The pit floor and the dividing plane between the upper and lower bench are roughly horizontal; two 8100 t/h bucket-wheel excavators are employed for mining. Tar sands loosened from the face of each bench by the bucket wheels are discharged onto a series of conveyors. The overburden is stripped by an electric shovel that discharges to trucks for removal of the overburden material.

Syncrude utilizes a single-bench design with four, 60-m^3 capacity electric draglines that pile the tar sands in windrows along the edge of the pit; four 64,000-t/h bucket wheels transfer the tar sands to a system of trunk conveyor belts that move the material to the extraction plant.

Alsands is considering a scheme very similar to Syncrude's (19,43).

Bitumen Recovery. ***Direct Coking.*** Direct coking of tar sands with a fluidized-bed technique has been tested by the Canadian Department of Mines and Technical Surveys, Mines Branch (33). Tar sand is fed to a coker or still, where the tar sand is heated to ca 480°C by contact with a fluidized bed of clean sand from which the coke has been removed by burning. Volatile portions of the bitumen are distilled. Residual portions are thermally cracked, resulting in the deposition of a layer of coke around each sand grain. Coked solids are withdrawn down a standpipe, fluidized with air, and transferred to a burner or regenerator (operating at ca 800°C) where most of the coke is burned off the sand grains. The clean, hot sand is withdrawn through a standpipe. Part (20–40%) is rejected and the remainder is recirculated to the coker to provide the heat for the coking reaction. The products leave the coker as a vapor, which is condensed in a receiver. Reaction off-gases from the receiver are recirculated to fluidize the clean, hot sand which is returned to the coker. The main product is the condensate from the receiver, a raw heavy synthetic crude oil (see Fluidization).

Other direct coking schemes in varying stages of development are represented by the Lurgi (44) and Taciuk (45) processes. The former uses a short contact-time coking vessel resembling a double-screw conveyor, with a separate coke combustion chamber, similar to that in the approach described above.

The Taciuk system employs a double-walled rotary kiln for both the coking and coke-burning steps. Countercurrent flow of cold-feed tar sand and reject hot sand also provides for recovery of reject heat directly into the feed.

Direct coking of tar sands is a simple, direct treatment, combining both the extraction and primary upgrading steps. Feeds of high fines content offer no particular difficulty. However, direct coking has two disadvantages. Large amounts of sand must be circulated relative to the oil throughput. The sand is likely to be abrasive, causing handling problems. Furthermore, the discharged sand is hot (up to 800°C), which represents a significant heat loss. The recovery of such a large amount of heat presents a challenge.

Anhydrous Solvent Extraction. An anhydrous solvent extraction process for bitumen recovery was developed by Cities Service Athabasca Limited (30) during 1959–1961. The solvent used in the experimental work was not specified. Presumably it was a light hydrocarbon.

The anhydrous process has four steps. In the mixer step, fresh tar sand is mixed with recycle solvent that contains some bitumen and small amounts of water and mineral. Solvent-to-bitumen weight ratio is adjusted to ca 0.5. The drain step consists of a three-stage countercurrent wash. Settling and draining time is ca 30 min for each stage. After each extraction step, a bed of sand is formed and the extract is drained through the bed until the interstitial pore volume of the bed is emptied. From time to time, the bed is plugged with fine mineral or emulsion. In these cases, the drainage rate is essentially zero and the particular extraction stage is ineffective. The last two steps of the process are devoted to solvent recovery. Stripping of the solvent from the bitumen is straightforward. The solvent recovery from the solids holds the key to the economic success of an anhydrous process.

Cold-Water Separation Process. So far, two cold-water bitumen separation processes have been developed to the point of small-scale continuous pilot plants. The process, developed by the Mines Branch of the Canadian Department of Mines and Technical Surveys in 1949–1950, uses a combination of cold water and solvent and was demonstrated in a 90-kg/h continuous pilot plant (46). The first step is carried out in a pebble mill, where the charge is disintegrated and mixed with water, diluent, and reagents. The diluent is kerosene and is added in a ca 1:1 weight ratio to the bitumen in the feed. The total water added is two to three times the weight of tar sands. The pH is maintained at 9–9.5 by the addition of wetting agents and ca 0.77 kg of soda ash per ton of tar sand. The pebble-mill effluent is mixed with more water, and in a raked classifier the sand is settled from the bulk of the remaining mixture. The water and oil overflow the classifier and are passed to thickeners where the oil is concentrated. Clay in the tar-sand feed has a distinct effect on the process; it forms emulsions that are hard to break and are wasted with the underflow from the thickeners. The experimental work on this process was carried out with Athabasca tar sands. The product has the following nominal composition by weight: oil (bitumen diluted 1:1 with kerosene), 73%; mineral, 2%; and water, 25%.

The sand-reduction process is a cold-water process without solvent (47). In the first step, the tar-sand feed is mixed with water at ca 20°C in a screw conveyor in a ratio of 0.75–3 t/t tar sand (the lower range is preferred). The mixed pulp from the screw conveyor is discharged into a rotary-drum screen, which is submerged in a water-filled settling vessel. The bitumen forms agglomerates that are retained by an 840-μm (20-mesh) screen. These agglomerates settle and are withdrawn as oil product. The sand readily passes through the 840-μm (20 mesh) screen and is withdrawn as waste stream. The process is called sand reduction because its objective is the removal of sand from the tar sand to provide a feed suitable for a fluid coking process; ca 80% of sand is removed. Nominal composition of the oil product is 58 wt % oil (bitumen), 27 wt % mineral, and 15 wt % water.

A process called spherical agglomeration closely resembles the sand-reduction process (48–49). Water is added to tar sands and the mixture is ball-milled. The bitumen forms dense agglomerates of 75–87 wt % bitumen, 12–25 wt % sand, and 1–5 wt % water.

Hot-Water Process. The essentials of the hot-water process (13,15,50–51) are shown in Figure 10. In the conditioning step, also referred to as mixing or pulping, tar-sand feed is heated and mixed with water to form a pulp of 60–85% solids at 80–90°C. First the lumps of tar sand as mined are reduced in size by ablation, ie, successive layers of lump are warmed and slough off revealing cooler layers. The pulp so formed

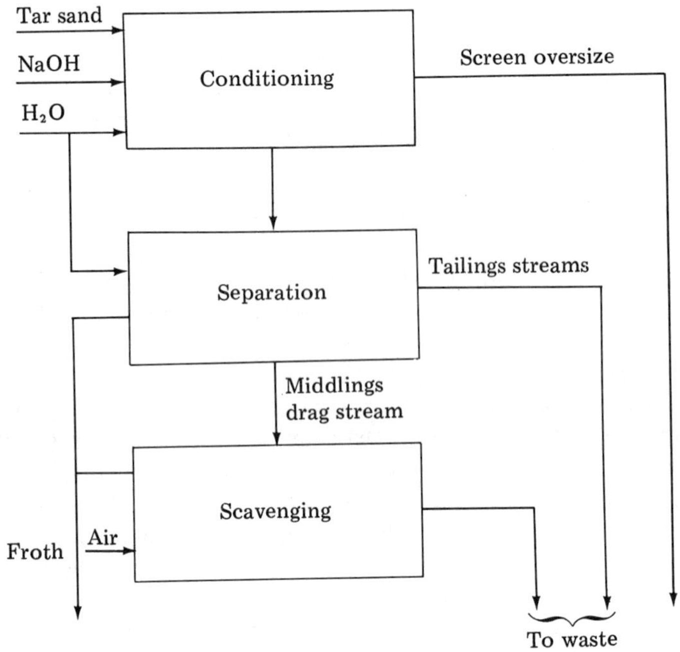

Figure 10. Hot-water bitumen recovery process.

is mechanically mixed, reacts with any chemicals added, is further heated to the process temperature, and is conditioned by open-steam heating in a horizontal rotating drum. With regard to equipment scale-up, conditioning is essentially a heat-transfer process. The effluent from the conditioning drum is screened to remove tramp material or lumps that were not sufficiently reduced in size. The screened pulp is mixed with any added water, adjusted to the proper consistency for pumping as described below, and sent to the separation cell.

The separation cell acts like two settlers, one on top of the other. In the lower settler the sand settles down, whereas in the upper settler the bitumen floats. The bulk of the sand in the feed is removed from the bottom of the separation cell as tailings. A large portion of the feed bitumen floats to the surface of the separation cell and is removed as froth. A middlings stream consists mostly of water with some suspended fine minerals and bitumen particles. A portion of the middlings may be returned for mixing with the conditioning-drum effluent in order to dilute the separation-cell feed for pumping. The remainder of the middlings is called the drag stream, which is withdrawn from the separation cell to be rejected after processing in the scavenger cells. Tar-sand feed contains a certain portion of fine minerals that, if allowed to build up in concentration in the middlings, increases the viscosity and eventually disrupts settling in the separation cell. The drag stream is required as a purge in order to control the fines concentration in the middlings. The amounts of water that can enter with the feed and leave with the separation-cell tailings and froth are relatively fixed. Thus, the size of the drag stream determines the makeup water requirement for the separation cell. The separation cell is an open vessel with straight sides and a cone bottom. Mechanical rakes on the bottom move the sand toward the center for discharge. Wiper arms rotating on the surface push the froth to the outside of the separation cell where it overflows into launders for collection.

The third step in the hot-water process is scavenging. Depending on the drag-stream size and composition, enough bitumen may leave the process in the drag stream to make another recovery step economical. Froth flotation with air is usually employed. The scavenger froth is combined with the separation-cell froth to be further treated and upgraded to synthetic crude oil. Tailings from the scavenger cell join the separation-cell tailings stream and go to waste. Conventional froth-flotation cells are suitable for this step (see Flotation).

The hot-water separation process involves extremely complicated surface chemistry with interfaces among various combinations of solids (including both silica sand and aluminosilicate clays), water, bitumen, and air (52). The control of pH seems to be critical; preferred range is 8.0–8.5, achievable with any of the monovalent bases. Polyvalent cations must be excluded because they tend to flocculate the clays and thus raise the viscosity of the middlings in the separation cell.

A process termed hot-water extraction has been applied on a laboratory scale by the U.S. Bureau of Mines (USBM) to U.S. tar sand (53). In the conditioning step, a fuel oil (0.86 sp gr (33°API)), containing 20–25 vol % aromatic compounds, is added as solvent in a 1:3 weight ratio based on the feed-bitumen content. The product resembles closely that of the Canadian Mines Branch cold-water process. The composition of froth recovered from Edna, Calif., tar sands was 73 wt % diluted bitumen, 6 wt % mineral, and 21 wt % water.

Froth Cleanup. Froth from the hot-water process may be mixed with a hydrocarbon diluent, eg, coker naphtha, and centrifuged. The Suncor process employs a two-stage centrifuging operation (16); each stage consists of multiple centrifuges of conventional design installed in parallel. The bitumen product contains 1–2 wt % mineral (dry bitumen basis) and 5–15 wt % water (wet diluted basis). Syncrude also utilizes a centrifuge system with naphtha diluent. In the Cities Service Athabasca, Ltd. process, the bulk of the mineral is removed by water washing followed by thermal dehydration. The Shell proposal features electrical dehydration, which was also proposed by GCOS in their original application to the Canadian Oil and Gas Conservation Board (1960–1962). It is similar to the electrostatic desalting of conventional crude oil.

Bitumen Upgrading

Bitumen is a hydrogen-deficient oil that is upgraded by carbon removal (coking) or hydrogen addition (hydrocracking) (see also Petroleum, refinery processes).

The overall upgrading process is generally accomplished in two steps. The primary upgrading process improves the hydrogen-to-carbon ratio by either carbon removal or hydrogen addition, cracking the bitumen to lighter products which are more easily processed downstream. Secondary upgrading, usually by further hydrogenation, may achieve further cracking and also removes sulfur and nitrogen. The upgraded or synthetic crude can then be refined to consumer goods such as gasoline, jet fuel, and home heating oil by conventional means (see Gasoline and other motor fuels).

Partial Coking. A partial coking or thermal deasphalting process provides a minimal upgrading of bitumen (54). In partial coking, the hot-water process froth is distilled at atmospheric pressure, and mineral and water are removed (see also Coal, coal conversion processes). A dehydrated mineral-free bitumen product is obtained that contains most of the asphaltenes and Ramsbottom carbon content. The process has been carried out batchwise in laboratory tests over periods ranging from 30 min to 4 h. Thermal cracking begins as the liquid temperature passes 340°C. The distillation is continued to 370–450°C. With slow heating (10°C temperature rise per hour), the coke production rate is ca 1 wt % of feed per hour. As the coke forms about the entrained mineral particles, 1–4 wt % coke up to 50 vol % of the feed is recovered as distillate. After this treatment the residue may be filtered to yield an essentially ash-free production suitable for applications such as metallurgical coke or production of bituminous paints, for which the original mineral content would have disqualified it.

Coking. Coking meets the combination of upgrading needs efficiently and economically. It is carried out in a number of ways, including direct coking, delayed coking, Fluid Coking, Flexicoking, and destructive batch distillation of bitumen. In each case the charge is converted to distillate oils, coke, and light gases. The coke fraction and low energy-content gases of the product satisfy the need for fuel. The coker distillate is a partially upgraded material in itself and is a suitable feed for hydrodesulfurization to produce a sweet synthetic crude. Coking followed by hydrodesulfurization is the commercial process used in the Athabasca tar sands by Suncor and Syncrude. Delayed coking as used by Suncor, is a semibatch process in which feed bitumen is heated before being fed to coking drums that provide sufficient residence time for the cracking reactions to occur.

Both pilot-plant and commercial-yield data are available for Athabasca bitumen (33). Sulfur is distributed throughout the boiling range of the delayed coker distillate, as with distillates from direct coking. Nitrogen, which averaged 5000 ppm in the bitumen feed, is more heavily concentrated in the higher boiling fractions; nevertheless, it is present in all distillate fractions. Raw coker naphthas contain significant olefins and diolefins which must be saturated by downstream hydrotreating. The gas oil has a high aromatic content typical of coker gas oils.

A more recent variant of delayed coking is the Eureka process (55) which uses steam stripping to enhance yield and produce a heavy pitch rather than coke by-product.

Fluid coking (56) is a continuous process employing two vessels with fluidized coke. It provides a better yield of overhead products than delayed coking. Feed oil flows to the reactor vessel where cracking and formation of coke occur; coke is combusted in the burner. Fluid transfer lines between these vessels provide the coke circulation necessary for heat balance. The proportion of coke burned is just sufficient to satisfy heat losses and provide the heat for the cracking reactions.

In flexicoking (57), a gasifier vessel is added to the system in order to gasify excess coke with a gas–air mixture to a low heating-value gas which can be desulfurized and used as a plant fuel.

Coking processes give a comparatively low overall liquid product resulting from conversion of a significant portion of the feed to coke. This may be less of a problem in tar-sand plants with a large fuel demand; however, coke is still far from a premium fuel. Furthermore, secondary upgrading is required for hydrogen-deficient products that have high sulfur and nitrogen content (35,58) (see also Petroleum refinery processes).

In order to take advantage of optimum operating conditions for various distillate fractions, the Suncor coker distillate is treated as three separate fractions: naphtha, kerosene, and gas oil. Suncor uses the Unifining process (licensed by Universal Oil

Products Company). In the operation proposed by Shell Canada Ltd., similar to that of Syncrude, feed is to be treated in two fractions, naphtha and mixed gas oils.

Hydrogenation. The overall liquid yield of direct hydrogenation or hydrocracking of bitumen is substantially higher than that of coking, and significant amounts of sulfur and nitrogen are removed. Currently, however, large quantities of external fuel or hydrogen-plant feedstock are required.

Most processes start with an upflow reactor system in which the 524°C material is cracked or converted. To prevent coking, the processes operate at high pressure with direct contact between bitumen feed and circulating hydrogen. A fluidized additive or catalyst may be employed as an aid to operability and removal of sulfur and nitrogen. This is not necessary with Athabasca bitumen obtained by hot-water extraction since this material contains fine residual mineral matter, which has a beneficial effect in this type of processing.

After the primary conversion, other fluid-bed or fixed-bed catalytic reactors may be added downstream to achieve further hydrogenation as well as sulfur and nitrogen removal. These may be integral with the primary hydrocracker unit or separate downstream hydrotreaters operating on specific fractions.

Hydrocracking processes include H-Oil (59) and LC-Fining (60), which are not used commercially for bitumen feedstocks; second-generation processes include Canmet hydrocracking (61–62), Vebe (63), and Chiyoda ABC (64). Good yields of light products are obtained from bitumen. Table 6 shows typical performance values for the Canmet hydrocracking process.

Hydrocracker products, as expected, have higher hydrogen and lower sulfur and nitrogen contents than those from the coking route and require less secondary upgrading.

Disadvantages of the hydrogen route include relatively high hydrogen con-

Table 6. Upgrading of Athabasca Bitumen Vacuum Bottoms (≥524°C) by the Canmet Hydrocracking Process[a]

Yield	Value, wt % on feed
Feed	
oil + additive	100
hydrogen	2.80
Total	*102.80*
Product	
CO	0.04
CO_2	0.03
H_2S	4.70
NH_3	0.38
light ends, C_1–C_4	9.94
C_5 naphtha, 204°C	20.37
light gas oil, 204–343°C	29.17
heavy gas oil, 343–534°C	28.36
pitch, 524°C	9.81
Total	*102.80*
pitch conversion, wt %	90
sulfur removal, wt %	72
hydrogen consumption, m^3/m^3	350

[a] Courtesy of Petro-Canada.

sumption and high pressure operation. Processes that use conventional (eg, Co—Mo or Ni—Mo) catalysts are susceptible to metals poisoning, which may limit applicability to, or economics of, operation on feeds high in metals such as bitumen.

Economic Aspects

The principal product from tar sands may be either a heavy, high specific gravity (low API gravity) crude oil or a light, upgraded synthetic crude. Crude oil is priced relative to a reference crude which is a light, clean product. Prices for other crudes are discounted depending on their specific (API) gravity, sulfur content, etc. Most of the tar-sands production, apart from the Canadian surface-mining operations, is heavy crude that is priced near the low end of the scale, ie, discounted for gravity and often further discounted for high sulfur and high metals content. The upgraded, desulfurized synthetic crude from surface-mining operations, on the other hand, is a premium product and attracts a top price. Prices for other products from tar-sands plants such as sulfur (see Sulfur recovery) and specialty petroleum products depend on location and market conditions.

Development activity in tar sands occurs in commercial production and field pilot installations. The principal commercial production originates from the surface-mining operations in Canada, currently 3.6×10^5 m^3/yr (2.3×10^6 bbl/yr). Commercial production has also been reported for Cuba, Italy, Mexico, the United States, and Venezuela. Total annual production from these countries is ca 1.2×10^6 m^3 (7.8×10^6 bbl), with the United States accounting for ca 7.3×10^4 m^3 (4.63×10^5 bbl) annually (2).

Field pilot activity is devoted to both *in situ* and mining recovery processes. Most of these projects are located in the United States, Canada, and Venezuela. Although the United States deposits represent only a small proportion of the world's tar sands, they do represent a significant resource. This factor combined with the large number and widespread distribution of the deposits has led to a high level of pilot activity. In 1981, 44 pilot projects were reported (28), 28 related to *in situ* recovery and 16 evaluating mining and plant extraction. In Canada in June 1982, 33 active *in situ* projects and 2 mining–extraction projects were recorded in tar-sand deposits. In addition, ca 50 projects have been completed (65). Venezuela also has a number of active pilot projects. Work in the past has been concentrated on *in situ* recovery of heavy conventional crude (66).

Outlook. Activity in tar-sands development in the United States has increased significantly in the past five years. The current worldwide economic slowdown and depressed world market for crude oil may have a short-term negative effect on the level of activity, but the longer term outlook is for further increases.

Pilot activity in Canada is also at an all-time high. Again, the economic slowdown may have a short-term impact, but the long-term outlook is for increased efforts. Both Syncrude and Suncor have initiated programs to expand their production facilities. New commercial development is, however, at a standstill with all planned projects either abandoned or suspended because of high capital requirements (\$12–15 $\times 10^9$), the current oversupply of crude oil with the consequent reduction in projected prices, and the inability of the companies and governments to reach agreement on terms relating to taxes and royalties. An end to the worldwide economic slowdown will bring a return to demand for crude oil, and further negotiations can resolve the tax and royalty problems. The capital requirements for large projects is not expected to decrease, and consequently, commercial planning is now directed toward smaller-scale developments of 3000–8000 m^3/d ((2–5) $\times 10^4$ bbl/d) rather than the 25,000 m^3/d (1.5 $\times 10^5$ bbl/d) projects.

The most aggressive program for commercial development of tar sands is under way in Venezuela. Lagoven, one of the affiliate companies of Petroleas de Venezuela, has initiated design work and pilot tests aimed at producing 20,000 m^3/d (1.25 $\times 10^5$ bbl/d) of synthetic crude by 1988 in the Orinoco deposit. The other affiliates have similar plans with an overall objective of 1.69 $\times 10^5$ m^3/d (10^6 bbl/d) from the area by the end of the century (66). With the Venezuelan oil industries' heavy dependence on exports, the current crude oversupply will undoubtedly effect these plans in the short term, but is unlikely to do more than delay the implementation of the programs.

Health and Safety Factors

The health and safety factors associated with oil-sands mining are similar to other large-scale surface-mining operations involving large equipment and the movement of huge quantities of material. The principal environmental consideration relating to the mining process is the land reclamation following the completion of the mining. The aspects of greatest concern in the separation process is the disposal of the large quantities of oily, high-pH wastewater. Both air and liquid effluents are subject to controls. More details on the relevant Canadian environmental regulations relating to oil-sands mining are given in ref. 67; environmental factors in Venezuela are discussed in ref. 68.

Health and safety factors in *in situ* operations are associated with high temperature, high pressure steam, or high pressure air. The environmental considerations relate to air and water quality and surface reclamation (67–70). In some environmentally sensitive areas such as the oil-sands deposits in Utah, environmental considerations may make development unfeasible.

BIBLIOGRAPHY

"Tar Sands" in *ECT* 1st ed., Vol. 13, pp. 633–645, by K. A. Clark, Reserach Council of Alberta, and G. B. Shea (U.S. Deposits), U.S. Department of the Interior; "Tar Sands" in *ECT* 2nd ed., Vol. 16, pp. 682–732, by F. W. Camp, Sun Oil.

1. *Alberta's Reserves of Crude Oil, Natural Gas Liquids and Sulphur*, Dec. 1980, Energy Resources Conservation Board, Calgary, Alb., Can., 1981, ERCB 81-18.
2. R. F. Meyer and P. A. Fulton, *Toward an Estimate of World Heavy Crude Oil and Tar Sands Resources*, preprint from *Second International Conference on Heavy Crude and Tar Sands*, UNITAR, Caracas, Venezuela, Feb. 1982.
3. *Oil Sands Fuel of The Future*, Canadian Society of Petroleum Geologists, Calgary, Alb., Can., Sept. 1974.
4. M. A. Carrigy, comptroller, *Athabasca Oil Sands Bibliography (1789–1964)*, prelim. rep. 65-3, Research Council of Alberta, Edmonton, Alb., Can., 1965.
5. *Proceedings of the First Athabasca Oil Sands Conference*, King's Printer, Edmonton, Alb., Can., Sept. 1951.
6. M. A. Carrigy, ed., *Athabasca Oil Sands: A Collection of Papers Presented to K. A. Clark on the Seventy-Fifth Anniversary of his Birthday*, Research Council of Alberta, Edmonton, Alb., Can., Oct. 1963.
7. *Proceedings of the 7th World Petroleum Congress*, Vol. 3, Elsevier Publishing Co., Amsterdam, The Netherlands, 1967, panel discussion no. 13, pp. 549–655.

8. D. A. Redford and A. G. Winestock, ed., *The Oil Sands of Canada—Venezuela*, The Canadian Institute of Mining and Metallurgy, Montreal, Can., 1978.
9. R. F. Meyer and C. T. Steele, eds., *The Future of Heavy Crude Oils and Tar Sands, Proceedings of the First International Conference, UNITAR*, McGraw-Hill, Inc., New York, 1981.
10. S. M. Blair, *Report on the Alberta Bituminous Sands*, Government of the Province of Alberta, Edmonton, Alb., Can., 1950.
11. A. D. Hoskins, *Proceedings of the American Petroleum Institute*, 1964, Sect. III44, pp. 426–429.
12. Application by Shell Canada Ltd., to the Alberta Oil and Gas Conservation Board for approval of a scheme of operation for recovery of oil or a crude hydrocarbon product from Athabasca oil sands, Sept. 6, 1962.
13. R. A. Given, *National Petroleum Refiners Association Meeting*, paper no. 63-15, Apr. 1963.
14. Application by Cities Service Athabasca, Inc., to the Alberta Oil and Gas Conservation Board for approval of a scheme of operation for the recovery of oil or a crude hydrocarbon product from oil sands located in the Mildred Lake area of the Athabasca tar sands, May 9, 1962: amended to Nov. 15, 1962.
15. J. V. D. Fear and E. D. Innes in ref. 7, pp. 633–650.
16. G. F. Andrews, E. W. Dobson, and H. M. Lewis, *Am. Chem. Soc., Div. Petr. Chem., Prepr.* **13**(2), F5 (1968).
17. *Report to the Lieutenant Governor in Council with Respect to the Application of Great Canadian Oil Sands, Ltd.*, Nov. 17, 1960; *Supplemental Report to the Lieutenant Governor in Council with Respect to the Application of Great Canadian Oil Sands, Ltd.*, Sept. 19, 1962.
18. Application to the Alberta Oil and Gas Conservation Board, under Part VI-a of the Oil and Gas Conservation Act by Atlantic Richfield Co., Cities Service Athabasca, Inc., Imperial Oil, Ltd., Royalite Oil Co., Ltd., May 9, 1962: amended to May 3, 1968.
19. *Alsands Fort McMurray Project*, ERCB Report 79-4, Energy Resources Conservation Board, Alberta, Can., Dec. 1979.
20. Application of Esso Resources Canada, Ltd., under Section 43 of the Oil and Gas Conservation Act and Section 27 of the Coal Conservation Act, ERCB Report 79-E, Energy Resources Conservation Board, Alberta, Can., Oct. 1979.
21. D. C. Ion, *Proceedings of the 7th World Petroleum Congress*, Vol. 1B, Elsevier Publishing Co., Amsterdam, The Netherlands, 1967, pp. 25–36.
22. *A Description and Reserve Estimate of the Oil Sands of Alberta*, Oil and Gas Conservation Board of the Province of Alberta, Calgary, Alb., Can., 1963.
23. J. J. Fitzgerald, *Black Gold and Grit*, Gray's Publishing Ltd., Sidney, B.C., Can., 1978.
24. G. W. Hodgson, E. Peake, and B. L. Baker in ref. 6, pp. 75–100.
25. T. A. Link in ref. 15, pp. 55–65.
26. I. A. Berger in ref. 9.
27. G. Fiorillo, *Exploration of the Orinoco Oil Belt, Review and General Strategy*, preprint from ref. 2.
28. R. Valera in ref. 9.
29. L. C. Marchant and C. A. Koch, *U.S. Tar Sand Oil Recovery Projects*, preprint from ref. 2.
30. *Chem. Week* 22 (Oct. 13, 1982).
31. J. H. Cotrell in ref. 6, pp. 193–206.
32. J. Starr and J. M. Prats in ref. 9.
33. W. S. Peterson and P. E. Gishler in ref. 5, pp. 207–236.
34. M. J. Sterba in ref. 5, pp. 257–270.
35. A. R. Aitken, W. H. Merrill, and M. P. Pleet, *Can. J. Chem. Eng.* **42**, 234 (Oct. 1964).
36. "Oil Sands Report," *Oil Week*, 22 (Mar. 22, 1982).
37. C. J. Borregales in ref. 9.
38. L. C. Marchant, C. Q. Cupps, and G. Stosur in ref. 9.
39. R. G. Gorrill, P. Kitzan, and D. P. Komery in ref. 9.
40. T. M. Doscher, R. W. LaBelle, L. H. Swatsky, and R. W. Zwicky in ref. 6, pp. 123–141.
41. F. K. Spragins, *J. Petr. Technol.* **19**(10), 1337 (1967).
42. *Energy Process. (Canada)*, 20 (Sept.–Oct. 1978).
43. *Energy Process. (Canada)*, 22 (Sept.–Oct. 1978).
44. R. W. Rammler, *Can. J. Chem. Eng.* **48**, 552 (Oct. 1970).
45. W. Taciuk, *Energy Process. (Canada)*, 27 (Sept.–Oct. 1981).
46. L. E. Djinghezian in ref. 5, pp. 185–199.
47. J. A. Bichard, C. W. Bowman, R. M. Butler, and J. L. Tiedje in ref. 6, pp. 171–191.

48. J. R. Farnard, H. M. Smith, and I. E. Puddington, *Can. J. Chem. Eng.* **39,** 94 (Apr. 1961).
49. *Can. Chem. Process.* **47**(10), 82 (Oct. 1963).
50. K. A. Clark, *Annual Report 1922*, Report 8, Research Council of Alberta, Edmonton, Alb., Can., 1923, pp. 42–58.
51. K. A. Clark, *Trans. Can. Inst. Mining Metall.* **47,** 257 (1944).
52. C. W. Bowman in ref. 7, pp. 583–604.
53. G. B. Shea and R. U. Higgins, *Report of Investigation 4246*, U.S. Bureau of Mines, Washington, D.C., 1948; *Report of Investigation 4871*, U.S. Bureau of Mines, Washington, D.C., 1952.
54. D. S. Pasternack in ref. 6, pp. 207–229.
55. R. Takahashi and K. Washimi, *Hydrocarbon Process.*, 93 (Nov. 1976).
56. *Petr. Process.*, 135 (Mar. 1956).
57. "Flexicoking Passes Major Test," *Oil Gas J.* **73,** 53 (Mar. 10, 1975).
58. M. A. O'Grady and B. I. Parsons, *The Hydrogenation of Alberta Bitumen over Cobalt-Molybdena Catalyst*, Report R-194, Canadian Department of Energy, Mines, and Resources, Mines Branch, Ottawa, Can., Sept. 1967.
59. W. Mounce and R. S. Rubin, *Chem. Eng. Progr.*, 81 (Aug. 1971).
60. R. P. VanDriesen and co-workers, *Hydrocarbon Process.*, (May 1979).
61. M. A. Menzies, A. E. Silva, and J. M. Denis, *Chem. Eng.*, 46 (Feb. 23, 1981).
62. G. Lunin, A. E. Silva, and J. M. Denis, *Chem. Can.*, (Mar. 1981).
63. U. Graeser and K. Niemann, *Oil Gas J.*, (Mar. 22, 1982).
64. T. Ishikawa and R. G. Keister, *Hydrocarbon Process.*, 109 (Dec. 1978).
65. *Active Projects Alberta*, Energy Resources Conservation Board, Calgary, Alb., Can., June 1982.
66. C. J. Borregalas, *Venezuelan Experience and Recent Development in Heavy Oil Exploration*, preprint from ref. 2.
67. H. W. Thiessen in ref. 9.
68. P. Wenzel and A. Herrera, *Environmental Considerations for Integrated Orinoco Heavy Oil Upgrading and Energy Supply*, preprint from ref. 2.
69. A Goodley in ref. 9.
70. H. Trujillo, *Orinoco Oil Belt: Environmental Protection, Mitigation and Monitoring in the Lagoven-DSMA Project*, preprint from ref. 2.

General References

Refs. 2–9 are also general references.

DONALD TOWSON
Petro-Canada

TARTARIC ACID. See Hydroxy dicarboxylic acids.

TEA

The term tea refers to the plant *Camellia sinensis*, the dried, processed leaf manufactured from it; extracts derived from the leaf; and beverages prepared from leaf or extract of this specie (1). Tea is the most widely consumed of all beverages aside from water. Worldwide per capita consumption, although geographically far from uniform, is 0.1 L per day. Apart from its extensive usage, it is of interest to the food technologist as well as to the biochemist and organic chemist, because of the dependence of the final product characteristics on the unusual chemical composition of the starting material, and because of the complex series of biochemical and organochemical reactions which occur during processing (see Food processing).

Although the plant is now generally considered as a single specie, many botanical variations occur. At least two well-defined varieties are recognized: *assamica*, a large-leaved (15–20 cm) plant, and *sinensis*, a smaller-leaved (5–12 cm) variety (2).

Tea is thought to have originated in East Asia in the area of the People's Republic of China, Burma, Laos, and Vietnam. Its first recorded use dates from the 4th century AD in China. A flourishing trade eventually developed which led to its cultivation in Japan. Tea reached Europe in the 16th century. By the latter part of the 17th century, it had become a popular beverage served in numerous tea houses in London. Tea was discovered in Assam Province of India in 1823 and cultivation in that country followed rapidly. The plant was then introduced into Indonesia and into Sri Lanka after the destruction by disease of its coffee plantings in the 1870s (3). Cultivation of tea on a large scale subsequently spread to many other parts of Asia, the USSR, Africa, and South America.

Tea production was carried out on a small commercial scale in South Carolina around the beginning of the 20th century. At its height, the Pinehurst Tea Farm in Sommerville produced about 4500 kg of tea per year which was sold locally. Commercial operation ceased in 1915 (4).

Tea grows best in tropical and semitropical areas where adequate rainfall (200 cm), good drainage, and slightly acid soils prevail. The small-leaved varieties are generally more frost resistant. In tropical areas, tea quality is improved by planting at higher altitudes, as practiced in south India and Sri Lanka where tea is cultivated at elevations up to 2000 m. The plant is maintained by pruning as a low shrub (1–1.5 m) at a density of 5,000–10,000 plants per ha (5).

Tea propagated from seed is not uniform and botanical differences are visually apparent. Yield and organoleptic properties also vary. During the last few decades, tea has been primarily propagated from leaf cuttings. Planting material for the establishment of new clones is selected on the basis of beverage quality, yield, ease of establishment, pest resistance, frost resistance, and whatever other criteria local considerations might dictate. Selection for yield has greatly improved productivity. Yields from large established fields of the best clonal selections in north India, for example, have reached 4500 kg of manufactured tea per ha. The average yield is only 1500 kg/ha, reflecting the preponderance of seed tea from older plantings (6). Maximum yield is achieved in ca 6 yr. The plant is productive under favorable conditions for many decades and producing 100-year-old plants from seed are not uncommon. It is not known how long clonal tea will be productive since only a few plants of this type are more than 30 yr old. The absence of a tap root and the genetic uniformity encountered in vegetatively propagated plants lead to some risk of decreased resistance to unusual drought, new disease, or new pest infestations unless plantings include a variety of clones.

The use of irrigation, fertilizers, mulches, herbicides, fungicides, and other pesticides has been extensively researched in many tea-growing areas and results are reported in the journals of the various tea research institutes of the world (7). It is significant to note that pesticides banned in the United States are no longer approved for use on tea (8). Tea estates range in size from peasant holdings of 1 ha or less to large establishments of up to 800 ha. Larger estates usually include housing, schools, hospitals, and other facilities for the workers employed in this labor-intensive operation. Labor utilization is often at the level of three persons per 10,000 m^2. Leaf harvesting (plucking) engages more than half of the labor employed. Large estates must also include a factory for the immediate conversion of freshly plucked leaf to manufactured tea. Leaf from small holdings is generally processed at central factories operated by private, government, or cooperative concerns.

New growth (flush) is harvested at intervals of 6–12 days, depending on the climatic conditions. Growth is most rapid during warm weather and heavy rainfall. In some areas, such as north India (Assam), Japan, and the USSR, a period of dormancy occurs during the cold season. In south India, Sri Lanka, Indonesia, and Africa, production continues year-round, permitting more efficient use of labor and manufacturing facilities. The length of the growing season, however, has little effect on annual yield.

The apical bud and several new leaves below it constitute the flush. Typically, two (leaves) and a bud has represented the ideal plucking target, as shown in Figure 1. Very often, however, three or even four leaves are taken. Hand plucking is the most prevalent harvesting mode and the most selective.

In Japan, shears are used as well as hand-held cutters powered by back-carried gasoline engines. Small, self-propelled harvesters which straddle dome-contoured rows are also employed (9).

Figure 1. Two and a bud.

Mechanical harvesting with heavy-duty complex equipment is carried out on a large scale in the USSR (10). This method is also used to a lesser extent in Argentina. Mechanical harvesting is not feasible on the extremely steep slopes planted in parts of India and Sri Lanka. A typical planting for mechanical harvesting is illustrated in Figure 2.

Composition of Fresh Leaf

Like all plant-leaf matter, the tea leaf contains all the enzymes, biochemical intermediates, and structural elements normally associated with plant growth and photosynthesis (2,11–12). In addition, it contains substances that are known to be responsible for the unique properties of tea. A representative analysis of fresh leaf is given in Table 1.

Flavanols. The outstanding chemical characteristic of tea flush is its very high concentration of polyphenolic material. Although amounts vary depending on the factors influencing plant metabolism, such as light, rainfall, temperature, nutrient availability, leaf age, and genetic make up, a small group of flavanols usually constitutes 20–30% of the dry matter. The flavanols of tea belong to the group known as catechins (see Fig. 3).

The tea flavanols are water-soluble, colorless substances with an astringent taste. They oxidize readily, especially in an alkaline environment, and form complexes with many other substances (13). (−)-Epigallocatechin gallate (**4**), which constitutes a very high proportion of the fresh green leaf, has not been reported in other plant matter. Tea beverage quality is positively correlated with fresh-leaf flavanol concentration (14), especially gallated flavanols which decrease as the leaf ages (15). As discussed

Figure 2. Tea under mechanical harvesting.

Table 1. Composition of Tea Flush[a]

Components	% of dry wt
flavanols	25
flavonols and flavonol glycosides	3
polyphenolic acids and depsides	5
other polyphenols	3
caffeine	3
theobromine and theophylline	0.2
amino acids	4
organic acids	0.5
monosaccharides	4
polysaccharides	13
protein	15
cellulose	7
lignin	6
lipids	3
chlorophyll and other pigments	0.5
ash	5
volatiles	0.1

[a] Ref. 11.

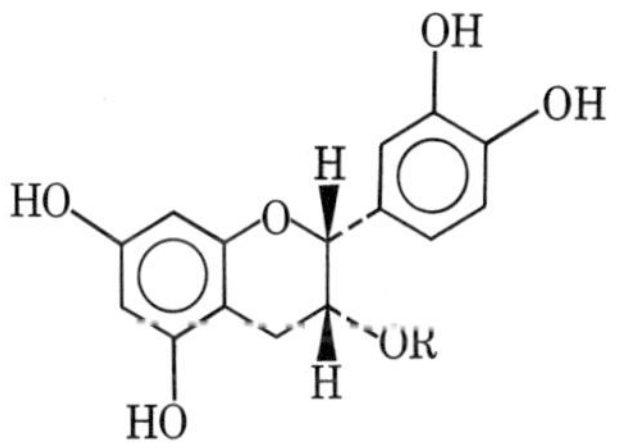

(**1**) (−)-epicatechin;
R = H; 1–3% of dry wt
(**2**) (−)-epicatechin gallate;
R = 3,4,5-trihydroxybenzoyl;
3–6% of dry wt

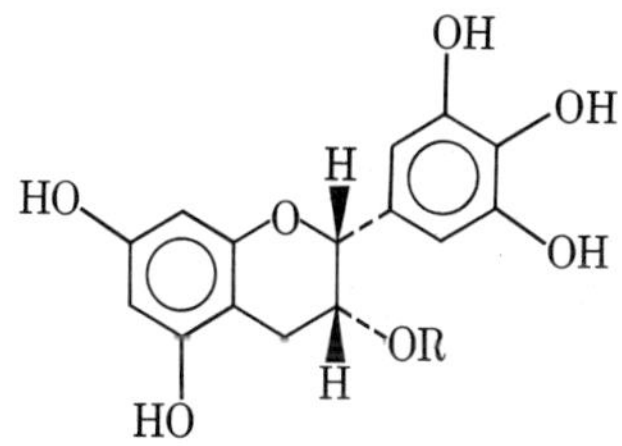

(**3**) (−)-epigallocatechin;
R = H; 3–6% of dry wt
(**4**) (−)-epigallocatechin gallate;
R = 3,4,5-trihydroxybenzoyl;
9–13% of dry wt

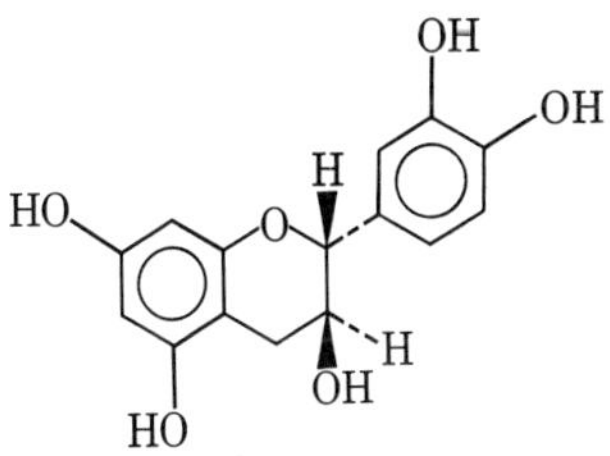

(**5**) (+)-Catechin; 1–2% of dry wt

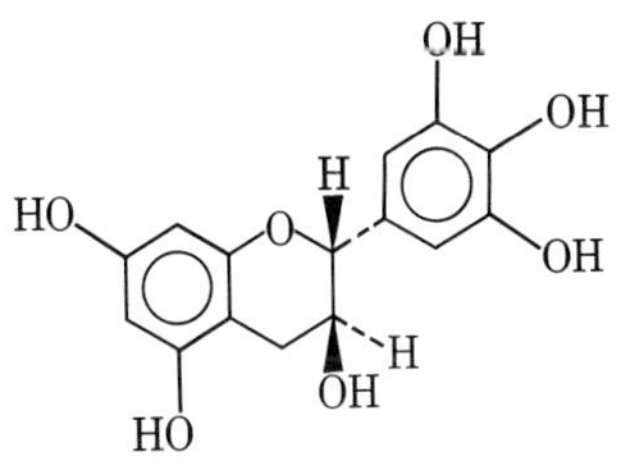

(**6**) (+)-Gallocatechin; 3–4% of dry wt

Figure 3. The flavanols occurring in tea (adapted from ref. 2).

later, this group of compounds plays the most significant role of all substances in tea manufacturing. Tea flavanols occur in the cytoplasmic vacuoles of leaf cells.

Other Polyphenols. These include gallic acid; flavonols, such as quercetin, kaemferol, myricetin, and their glycosides; and depsides such as chlorogenic acid and *p*-coumarylquinic acid (see Fig. 4). Theogallin (3-galloylquinic acid) is unique to tea,

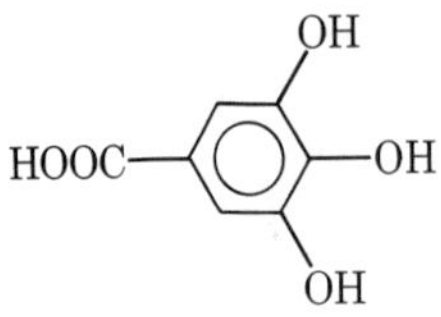

(7) gallic acid

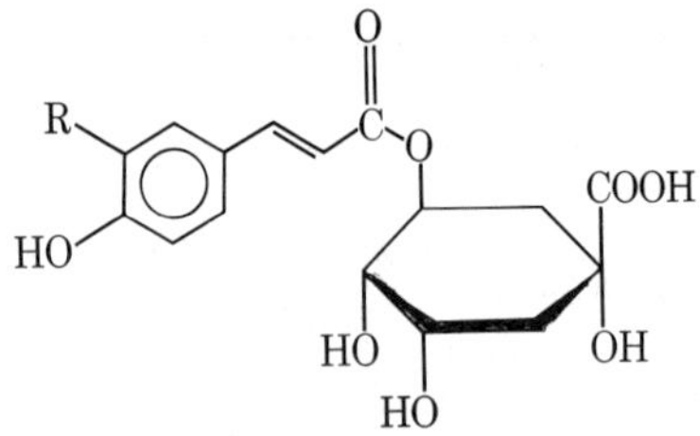

(8) chlorogenic acid; R = OH
(9) *p*-coumarylquinic acid; R = H

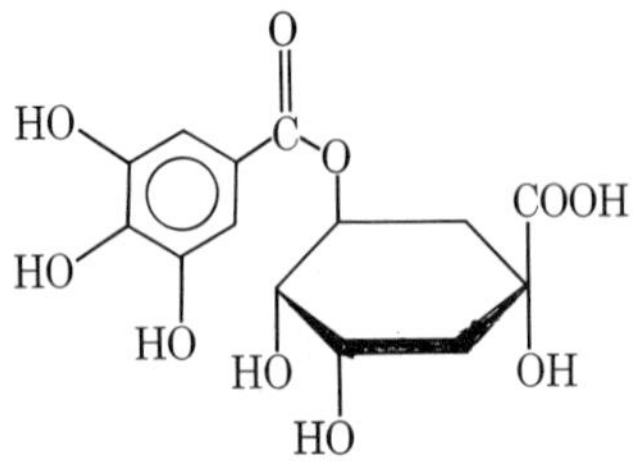

(10) theogallin

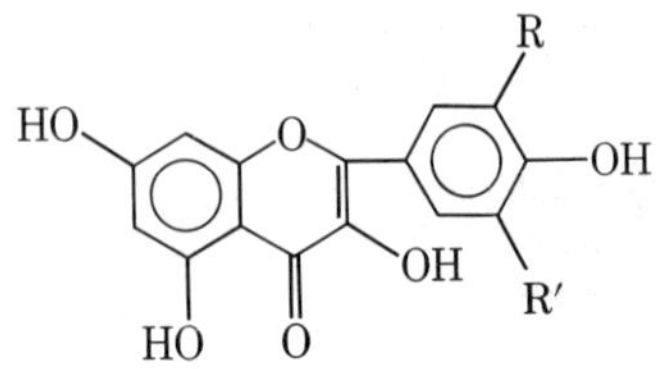

(11) kaempferol; R = R′ = H
(12) quercitin; R = OH, R′ = H
(13) myricetin; R = R′ = OH

Figure 4. Other polyphenols occurring in tea.

occurs at relatively high concentration (1% of dry wt), and is said to correlate with tea quality (16). Several of these polyphenols participate in the important reactions which occur during manufacture.

Caffeine and Other Xanthines. Tea flush contains 2.5–4.0% caffeine and much smaller amounts of the related substances, theophylline and theobromine. Caffeine is responsible for the mild stimulatory properties of tea. Since approximately 45 cups of tea beverage are obtained from 100 g of dry leaf, the caffeine content of tea is usually around 30–50 mg per cup or less than half that of coffee. Its tendency to form complexes with polyphenolic material affects the behavior of tea beverages as well as the technology of instant-tea manufacture.

Theanine and Other Amino Acids. Theanine (5-*N*-ethylglutamine) (**14**) is an amino acid unique to tea (17). The other amino acids found in tea are common to most plants. The amino acids of tea leaf play a role in aroma formation during the manufacturing process:

$$HO_2CCH(NH_2)CH_2CH_2CONHC_2H_5$$

(**14**) theanine

Carotenoids. Carotenoid compounds constitute only 0.03–0.06% of tea flush but are important in aroma formation. These compounds include β-carotene, lutein, zeaxanthin, violaxanthin, and several others (18).

Chlorophyll. This normal component of leaf matter is a precursor of pigments which impart color to manufactured tea leaf (2).

Volatiles. About 300 volatile compounds have been identified in manufactured tea but only about 60 have been reported to occur in fresh leaf (19) (see below).

Lipids. The presence of unsaturated fatty acids derived from phospholipids is of significance to aroma generation.

Enzymes. All enzymes implicated in plant-leaf metabolism undoubtedly occur in tea leaf. The enzymes of primary interest are those involved in the conversion of fresh leaf to the manufactured teas of commerce. The single most important of these is the tea leaf polyphenolase which catalyzes the oxidation of the catechins. It contains copper (0.32%) and exists as a mixture of several isoenzymes. Its main component has a molecular weight of ca 140,000 (20). Alcohol dehydrogenase (21) and leucine α-ketoglutarate transaminase (22) contribute to the development of aroma components during leaf manufacture. Other enzymes of significance are 5-dehydroshikimate reductase (23) and phenylalanine ammonia lyase (24), which mediate polyphenol formation in the growing leaf.

Qualitatively, samples of tea leaf of different age, geographical origin, climatic environment, and genetic constitution show compositional similarity. Quantitative differences, however, are large and account for a wide disparity in the properties, use, and value of products made from them.

Manufacture

The manufacturing process converts freshly harvested leaf to products of commerce. Black tea is the most widely consumed form of tea. It is the result of promoting the oxidation of fresh leaf catechins by atmospheric oxygen through catalysis by tea polyphenolase. The oxidized product is then dried. Green tea, consumed mostly in Japan, the People's Republic of China, North Africa, and the Middle East, is processed in a manner designed to prevent the enzymic oxidation of catechins before drying. Oolong tea is partially oxidized tea, manufactured primarily in the People's Republic of China and the Republic of China. Instant tea, usually a powder, is generally prepared by the aqueous extraction of black tea followed by concentration and drying. It is also possible to make instant tea from green tea or from oxidized leaf before the drying step.

Chemistry of Tea Fermentation; Oxidation. The most significant change that occurs during the manufacture of black tea is the conversion of the colorless catechins to a complex mixture of orange-yellow to red-brown substances and the development of a large number of volatile compounds. This process is incorrectly termed fermentation. Organoleptically, it is reflected in an amber-colored beverage, less astringent than that derived from fresh leaf, with an exceedingly complex flavor profile.

Catechin Oxidation. The first step in tea fermentation is the enzymic oxidation of the catechins to form highly reactive quinones (25):

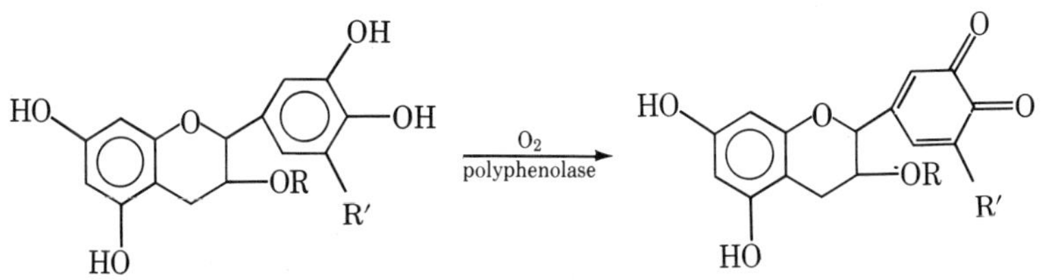

(1)

R = H or 3,4,5-trihydroxybenzoyl
R′ = H or OH

All catechins undergo this oxidation step but the gallocatechins (**3**) and (**4**) are preferentially oxidized. Quinone formation is the primary driving force in fermentation and no other significant mechanism has been validated for the initiation of the reactions which result in the production of the polyphenolic components of black tea.

Theaflavins. A quinone derived from a simple catechin, (**1**), (**2**), or (**5**), reacts with a quinone derived from a gallocatechin, (**3**), (**4**), or (**6**), to form a compound with a benztropolone ring known as a theaflavin:

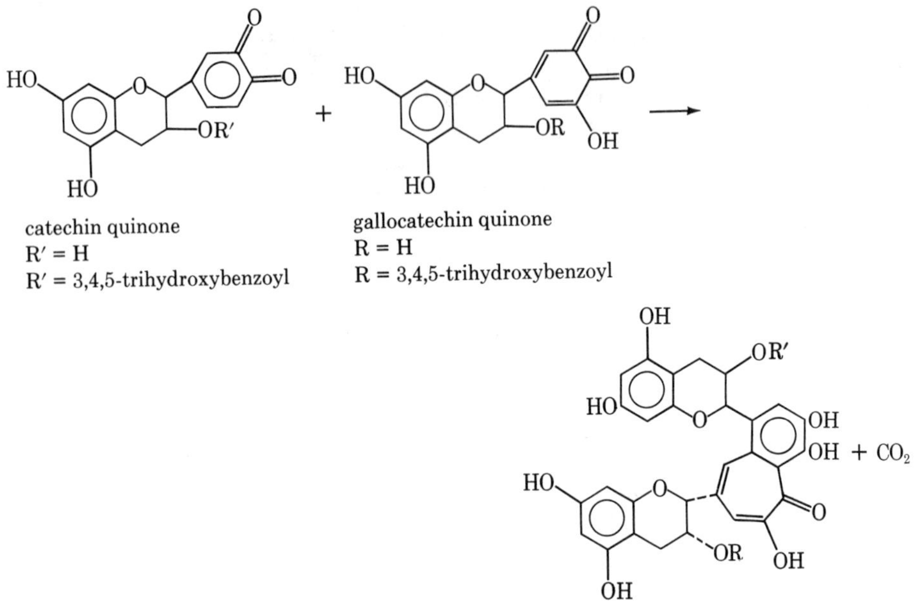

catechin quinone
R′ = H
R′ = 3,4,5-trihydroxybenzoyl

gallocatechin quinone
R = H
R = 3,4,5-trihydroxybenzoyl

(2)

(**15**) theaflavin; R = R′ = H
(**16**) theaflavin gallate A; R = 3,4,5-trihydroxybenzoyl, R′ = H
(**17**) theaflavin gallate B; R = H_1 R′ = 3,4,5-trihydroxybenzoyl
(**18**) theaflavin digallate; R = R′ = 3,4,5-trihydroxybenzoyl

Theaflavin structures were recently established (26). The four theaflavins theoretically derivable from (**1**), (**2**), (**3**), and (**4**) have been isolated and characterized as well as one other form known as isotheaflavin (27).

Theaflavins exhibit a bright orange-red color in solution and are important in the determination of brightness, a desirable attribute of tea beverages. It should be noted that many of the named attributes of tea arise from the tea taster's lexicon and are difficult to define objectively. The theaflavins are distinctive enough in color to be determined spectrophotometrically in a tea brew (28).

Total theaflavin concentration in black tea does not exceed 2% and can be sometimes as low as 0.3% (29). Prolongation of fermentation decreases theaflavin content (30), although the exact fate of these compounds is not known. At most, only 10% of the original catechin content of tea flush is accounted for as theaflavins in black tea.

Theaflavic Acids. Although gallic acid (**7**) is not directly oxidized by tea polyphenolase, it is converted to gallic acid quinone by catechin oxidation products (31). Further reaction with the oxidized form of (**1**), (**2**), or (**5**) gives theaflavic acids:

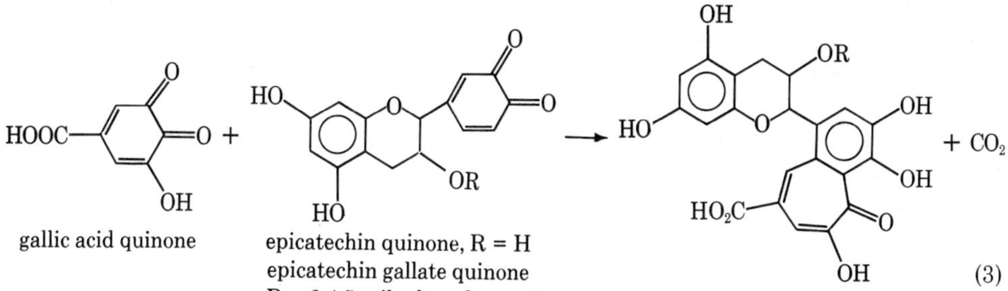

gallic acid quinone

epicatechin quinone, R = H
epicatechin gallate quinone
R = 3,4,5-trihydroxybenzoyl

(**19**) epitheaflavic acid, R = H
(**20**) epitheaflavic acid-3′-gallate; R = 3,4,5-trihydroxybenzoyl

(3)

The theaflavic acids are bright red, acidic substances present only in very small quantities in black tea probably due to their reactivity (11).

Bisflavanols. Reaction of oxidation products derived from equations 3 and 4 results in a series of colorless substances known as bisflavanols (26):

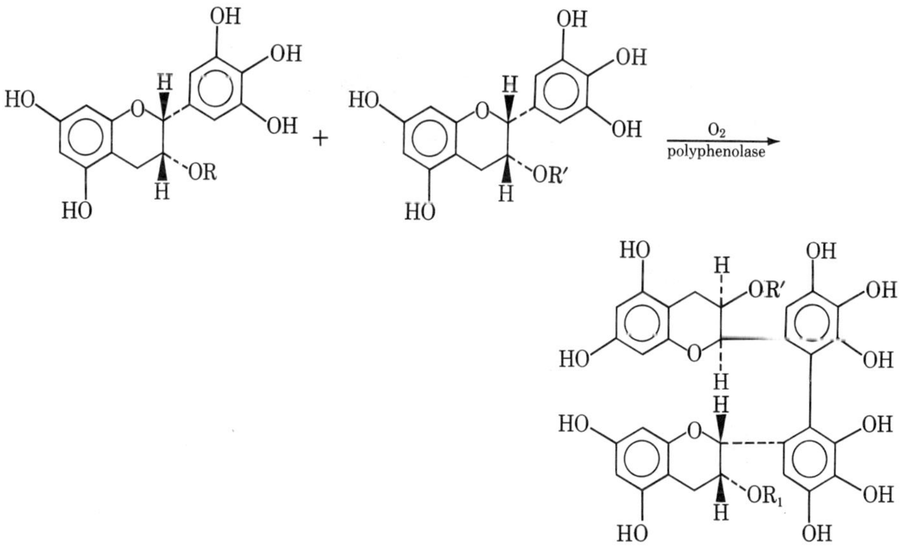

(4)

(**21**) bisflavanol A; R = R′ = 3,4,5-trihydroxybenzoyl
(**22**) bisflavanol B; R = 3,4,5-trihydroxybenzoyl; R′ = H
(**23**) bisflavanol C; R = R′ = H

Bisflavanols occur only in very small amounts in black tea as they are also highly reactive.

Thearubigens. During the formation of black tea, ca 15% of tea catechins remain unchanged and ca 10% are accounted for by theaflavin, theaflavic acid, and bisflavanol formation. Eventually, ca 75% of the catechins are converted to a complex, poorly defined, incompletely separated group of substances known as thearubigens because of their red-brown color. Investigations of thearubigen chemistry have resulted in conflicting data and conclusions, attesting to the complexity of the mixture.

Only partial separation of these compounds has been achieved. Molecular weight determinations indicate a range of 700–40,000 (12) but this must be considered in context with the known increase in molecular weight which occurs when tea infusions are aged (32).

Acid hydrolysis of thearubigen fractions results in the formation of cyanidin (**24**) and delphinidin (**25**) which are related to (**1**) and (**3**):

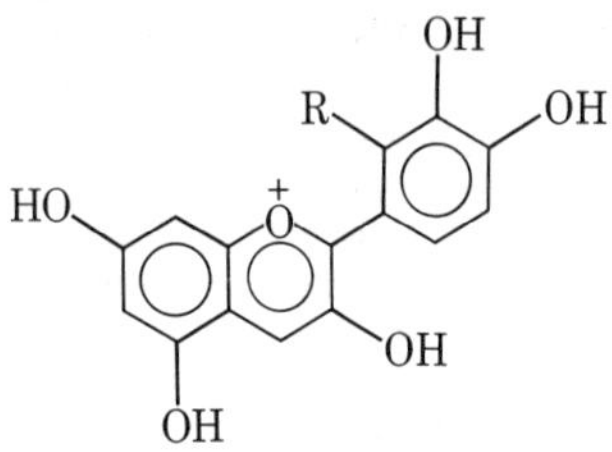

(**24**) cyanidin; R = H
(**25**) delphinidin; R = OH

Reductive hydrolysis of thearubigens yields small quantities of all of the catechins found in tea. Evidence exists for the presence of polymeric proanthocyanidins resulting from the production of new C–C bonds between oxidized catechin residues, but these structures account only for a small fraction (33).

Decrease in the contents of (**15**)–(**23**) during fermentation suggests that these compounds are further oxidized by the catechin quinones and enter into the formation of the thearubigen complex (12). It is also likely that catechin quinones oxidize (**8**), (**10**), and glycosides of (**11**), (**12**), and (**13**) leading to incorporation of their oxidation products into thearubigen (26). Although thearubigens constitute the largest group of compounds in black tea (up to 20% of dry weight) and contribute significantly to color, strength, and mouthfeel of the beverage, they are the least well known of all components. The oxidized polyphenol complex of black tea is frequently referred to as tannin. It has no relationship, however, to tannic acid which is a pentadigalloyl ester of glucose and does not occur in tea.

Black Tea. Until the 1930s, the changes that take place in the black tea manufacturing process, which has been evolved over hundreds of years, were little understood. The process consists of several steps known as withering, rolling, fermenting, firing, and sorting, although several of these are misnomers.

Withering. Plucked leaf arrives at the estate factory within a few hours of its harvest. It is handled carefully to prevent bruising and to dissipate heat generated as a result of continued respiration. In order to reduce the moisture from 75–80 to 55–65%, leaf is distributed in thin layers on open-meshed fabric netting or, more efficiently, in troughs designed for effective air flow through deep beds of leaf (up to 30 cm). This procedure allows more rapid loss of water without heat build-up detrimental to quality.

Reduction of moisture converts the turgid leaf to a flaccid material which is easily handled without excessive fracture. This process takes place over a period of 6–18 h, depending on factory equipment and weather conditions. It is now recognized that other significant changes occur during withering: cell membrane permeability is increased which allows easier disruption of cell contents during the next stage of manufacture; amino acid, caffeine, and organic acid concentrations increase; polyphenolase activity is enhanced. The series of changes that takes place over a period of time independently of water loss has been termed chemical withering (34). The degree of wither is often estimated by leaf texture or may be determined by checking the weight of an isolated portion of the leaf.

Rolling. The next step is the establishment of proper conditions for the enzymic oxidation of the flavanols by molecular oxygen. This is accomplished by disrupting the cell structure to effect enzyme–substrate contact. The leaf mass must also be maintained in a physical state favorable to oxygen diffusion. So-called rolling meets these needs (35). It is primarily a cutting and squeezing process which also includes a twisting action to enhance leaf appearance. The conventional roller consists of a circular table, 1–1.3 m in dia, equipped with battens. Above this, a smaller circular sleeve moves eccentrically over the surface of the table. Withered leaf is charged into the sleeve and a pressure cap lowered into the sleeve. The result depends on the pressure and the rate of movement. Intermittently, leaf sufficiently rolled and reduced in size is separated by sieving and the remainder is rerolled. At the end of the rolling process, leaf juices are well distributed on the surface of the cut, twisted leaf.

More modern equipment is coming into use in most of the tea growing areas. The McTear Rotorvane consists of a cylinder (20 or 37.5 cm in dia) lined with vanes through which a vaned rotorshaft propels withered leaf. Constriction at the end applies pressure and greatly increases the work done on the leaf. Rotating cutter blades are sometimes added to the shaft. From the Rotorvane, tea is frequently passed through a CTC machine (crush, tear, curl) which consists of a pair of ridged cylindrical rollers revolving at different speeds with a narrow clearance between rolls.

The Rotovane–CTC combination provides efficient and uniform maceration and is well suited to the establishment of a continuous manufacturing process. Macerated leaf is fluffed up by passage through a high speed Rotorvane without back pressure to eliminate matting and facilitate air diffusion. Leaf temperature should be kept below 35°C during maceration to prevent flavor deterioration. The Legg cutter, originally designed for cutting tobacco, is widely used in north India for macerating unwithered leaf.

Fermentation. The term fermentation arose from the erroneous concept of black-tea production as a microbial process (36). The factory conversion of green leaf to black tea became recognized as an oxidative process initiated by tea-enzyme catalysis around 1901 (37). The process, which actually starts at the onset of maceration, is allowed to continue under ambient conditions. Leaf temperature should not exceed 25–30°C. Lower temperatures (15–25°C) improve flavor (30). Temperature control and air diffusion are facilitated by distributing macerated leaf in layers 5–8 cm deep, sometimes on the factory floor, but more often on racked trays in a fermentation room maintained at a high rh and at as low a temperature as is feasible. Depending on the nature of the leaf, maceration techniques, ambient temperature, and the style of tea desired, the fermentation time can vary from 45 min to 3 h. More highly controlled systems depend on the timed conveyance of macerated leaf on mesh belts for forced air circulation. If the system is enclosed, humidity and temperature control are improved (38).

During the oxidation process, leaf color changes from green to coppery and a pleasant, characteristic aroma develops. In most instances, the proper termination point is determined by the skill of the process supervisor (tea maker) on whose judgment the value of the final product is highly dependent. Some attempts to control a suitable end point by instrumental techniques have been made. The fermentation step is terminated by firing (drying).

Aroma. The aroma of tea is one of its most important characteristics and to a large extent determines its value. Aroma is influenced by genetic, cultural, and manufacturing variables. The small-leaved varieties growing in areas subject to cool periods are most likely to develop the best aroma. The changes in aroma brought about during fermentation include those initiated by catechin quinones which oxidize amino acids, carotenes, and unsaturated lipids (39).

Aldehydes corresponding to Strecker degradation products of glycine, alanine, valine, leucine, isoleucine, methionine, phenylalanine, and theanine are formed in this way (see also Amino acids). Phenylacetaldehyde (from phenylalanine) is a significant black-tea aroma component.

Ionones, terpene alcohols, terpene aldehydes, and their oxidation products such as theaspirone and dihydroactinidiole result from oxidation by quinones of the carotenoids present in flush. Many of these are believed to be important in black-tea aroma. Similarly, the formation of *trans*-2-hexenal from linolenic acid in fresh leaf reduces tea quality. Other enzymic mechanisms for the formation of volatiles during fermentation have been demonstrated (2,40–41). Important changes in aroma also occur during the firing stage of tea manufacture (40).

In all, over 300 components of black-tea aroma have been identified, but in aggregate they account for only 0.1% of dry weight. Most of the mechanisms of formation are not understood. The most important aroma components include terpenes, terpene alcohols, lactones, ketones, esters, and spiro compounds (41).

Firing. Fermentation is terminated by firing in a hot-air oven with forced circulation in a countercurrent mode. Time–temperature profiles are controlled to result in drying in 18–20 min with little energy wastage.

During the firing process, leaf moisture is reduced to 2.5–3.0%. Many organochemical processes are accelerated during this period, as are enzymic reactions before thermal inactivation. Most noticeable are the leaf-color changes caused by the conversion of chlorophyll to phenophytins and pheophorbide, which are black and dark brown, respectively (2), and marked changes in the complex aroma (40). During firing, many of the lower-boiling aroma components are volatilized. Oxidation of alcohols and carbonyl compounds also occurs. Reactions that generate cyclic nitrogen compounds, such as pyrazines, pyridines, and quinolines, take place. The aroma of fired black tea is therefore markedly different from that of wet fermented leaf.

Grading. Fired tea is graded with a series of oscillating screens of appropriate size. The frequently produced grades in descending particle size are: orange pekoe (OP), pekoe, broken orange pekoe (BOP), broken orange pekoe fannings (BOPF), fannings, and dust. Dust, debris, and some fiber are removed by winnowing. Stalky materials are generally somewhat more moist than the fired leaf and may be removed with an electrostatic sorter. Tea has been traditionally packed in foil-lined, plywood chests measuring 40 × 50 × 60 cm holding ca 60 kg.

Process Variations. Several departures from the conventional techniques described above have been adopted, mainly to achieve greater degrees of automation and control. They include withering modification (42), different types of maceration equipment (43), closed systems for fermentation (44), and fluid-bed dryers (45). A thermal process has been described which utilizes decreased time periods for enzymic reactions but depends on heat treatment at 50–65°C to develop black-tea character (46). It is claimed that tannin–protein complexation is decreased and, therefore, greater tannin extractability is achieved. Tea value is believed to be increased through use of this process.

Green Tea. Green tea is made by rolling and firing without enzymic oxidation. Fresh leaf is subjected to a rapid steaming process in a rotating cylindrical drum for ca 20 s. Steamed leaf is cooled and rolled lightly in a hot-air environment. Partially dried leaf is then rolled more vigorously, dried further, and the process repeated until moisture is reduced to 3%. Drying temperatures are lower than for black tea which causes further differences in both the volatile and nonvolatile fractions. The aroma components of green tea have been extensively studied and compared with those of fresh leaf and black tea (47). A high proportion of catechins is retained in green tea, and the amino acid content is much higher. In the People's Republic of China, inactivation of enzymes is usually effected by dry heat.

Green or black tea can be prepared from any fresh leaf, but in practice different varieties are used for each product. Horticultural techniques effect further modifications. Plants grown for green tea production usually have smaller leaves, and lower catechin and higher amino acid contents (2). Green tea beverage is pale yellow-green, slightly more astringent than black tea, and has a brothy characteristic imparted by theanine and other amino acids.

Other Types. ***Oolong Tea.*** This type is produced in the People's Republic of China and the Republic of China. Withered leaf is lightly rolled and only partially fermented. Enzymic action is terminated by a heating process. More vigorous rolling to achieve desired appearance followed by additional heat treatment and final firing complete the process. In some respects, the beverage characteristics are intermediate between those of green tea and black tea, but oolong tea cannot be duplicated by a mixture of green and black tea since the latter contains oxidation products not formed during oolong manufacture.

Minor Varieties. Brick teas are prepared in the USSR and parts of the People's Republic of China (46). These products are often cooked as a soup with butter or other fats.

The addition of flavorants during processing may also be carried out. Jasmine flowers impart a characteristic flavor. Oil of bergamot is used to prepare Earl Grey tea. Other flavors are added to manufactured black tea (see also Flavors and spices).

Blending, Packaging, Forms of Consumption

Blending. The tea taster plays an important role in purchasing and blending tea. His goal is usually the establishment and maintenance of a chosen standard of tea under rapidly changing conditions. It becomes necessary to include teas from many countries and many gardens within a country in a single blend to ensure constancy in flavor, color and price over a long period of time. The tea taster's jargon includes terms such as bold, tippy, brisk, bright, dull, pungent, flavory, all of which need careful definition for use (48).

Packaging. In most countries, tea is sold in packets. In the United States, more than 90% of leaf tea is sold in tea bags frequently packed at 200 bags per lb (454 g). Larger, family-sized tea bags are also coming into use, as are 1-oz (28-g) food-service bags. An important consideration is paper quality which must be selected so as to avoid flavor contamination, allow efficient infusion, and retain tea fines. Successful packaging protects against light and moisture absorption. Polypropylene or coated cellophane outer wraps for tea packages provide a barrier to loss of tea aroma and retard access of oxygen and foreign aromas. Low temperature improves storage stability.

Forms of Consumption. In the United States, about 78% of all tea consumed is in the form of iced tea, a beverage hardly known in the rest of the world. Tea is prepared from tea bags by infusing in hot water at a ratio of ca 100:1 water to tea. The beverage achieves a solids concentration of ca 0.3% in ca 3 min. Hot black tea is generally consumed with milk in most parts of the world, but often with lemon in the United States; sugar is frequently added.

Economic Aspects

Many new growing areas have been developed in the last few decades, and world production of tea has been steadily increasing. This is particularly true for Africa and South America. Even in most traditional areas, output is increasing because of higher-yielding plants and more scientific cultural procedures. This trend will probably continue since new plantings have not yet reached their maximum yield potential and older plantings are being steadily replaced (49). Total world production for 1980 was estimated at 1.8×10^6 metric tons, distributed as shown in Table 2.

Tea is offered for sale in the producing countries at auctions held in Calcutta, Cochin, Colombo, Jakarta, and Mombassa. In addition, large quantities are shipped to London for auction. Teas are also offered in bulk in Rotterdam for direct sale. Private and government contract sales can be obtained; ca 50% of exported tea is sold through auctions. Per capita consumption of tea in a group of selected countries is shown in Table 3.

Prices. Tea prices rose sharply in 1977 when coffee prices rose and then declined (49–50) (see Table 4). It is convenient to consider London auction prices since annual averages for all lots of tea sold are available. These include teas of all grades and from all of the producing countries.

Tea prices are sensitive to the normal factors influencing supply and demand of agricultural crops but also to the price of coffee and sugar. It is said that tea consumption in India decreases when sugar prices are high. Attempts have been made to predict prices by econometric modeling (51).

The establishment of an international pact to control tea production and prices has been sought on several occasions but has never had adequate support for implementation (52). The need for foreign exchange and the need to employ agricultural labor tend to stimulate increased production. Depressed prices could eventually trigger such an agreement since it is believed that 1981 prices do not cover costs of production in some areas (53).

Table 2. World Production of Tea, Thousand Metric Tons, Dry[a]

Area	1970	1975	1980
Asia			
India	419	487	577
Sri Lanka	212	214	191
People's Republic of China	na	na	290
Japan	91	105	102
Indonesia	44	57	79
Turkey	33	55	100
Iran	20	23	20
Bangladesh	31	29	40
Republic of China	28	26	25
Vietnam	6	6	5
Malaysia	3	3	3
Africa			
Kenya	41	57	90
Malawi	19	26	30
Mozambique	17	13	19
Tanzania	8	14	17
Zimbabwe	4	6	10
Rwanda	1	4	7
South Africa	1	3	6
Zaire	8	7	6
Mauritius	3	3	4
other countries	20	21	5
South America			
Argentina	27	28	34
Brazil	6	8	10
other countries	2	4	5
Papua New Guinea	1	4	8
USSR	67	86	115
Total	*1112*[b]	*1289*[b]	*1798*

[a] Ref. 49.
[b] Exclusive of the People's Republic of China.

Instant Tea; Decaffeination

Instant tea is manufactured in several countries but production and consumption in the United States is greater than in the rest of the world. The 1980 U.S. production was ca 8000 metric tons of soluble tea solids derived from 20–25% of total leaf-tea imports. The basic process for manufacture of instant tea as a soluble powder from dry tea leaf includes extraction, concentration, and drying. In practice, the process is considerably more complicated because of the need to preserve the volatile aroma fraction and to allow the use of the final product to prepare iced tea (54).

Extraction. Tea leaf is extracted with hot water either in columns or kettles (55). To maintain a relatively low water-to-leaf ratio and achieve full extraction, a countercurrent system is usually employed. High pressure and temperature increase extractability, and 35–45% of the components of black tea are recovered. The volatile aroma components are vacuum-stripped from the extract (56) or steam-distilled from the leaf before extraction (57). In either case, aroma may be preserved for addition at a later point.

Table 3. Annual Per Capita Consumption of Tea, Average 1977–1979[a]

Country	kg	Country	kg
Algeria[b]	0.28	Japan	0.97
Australia	1.66	Kenya	0.74
Belgium	0.11	Morocco	0.98
Canada	0.86	Netherlands	0.65
Chile	0.96	New Zealand	2.38
Denmark	0.43	Poland	0.53
Egypt	0.77	South Africa	0.72
France	0.12	Sri Lanka	1.50
FRG	0.21	Sweden	0.36
India	0.50	Turkey[c]	1.84
Iran[b]	1.05	UK	3.1
Ireland	3.59	United States	0.37
Italy	0.06	USSR	0.55

[a] Ref. 49.
[b] 1976–1978.
[c] 1975–1977.

Decreaming. Extract is separated from leaf by centrifugation and cooled to precipitate a fraction known as cream. This material is a complex formed between caffeine and a portion of the oxidized catechins of black tea. Its composition varies, depending on the concentration and temperature of the extract from which it is obtained. It is always characterized by a high caffeine content and a deep color. This material must be removed in order to obtain a product that is soluble in cold water. Several techniques have been described for the solubilization of this fraction to permit the reincorporation of the desirable substances into the product (58).

Concentration. The extract is generally concentrated under vacuum to the solids content desired for drying. Freeze concentration has been described (59). Preserved aroma and the solubilized cream fraction may be added before drying.

Drying. Freeze drying of tea concentrates has been described (60), although spray drying is the usual process employed. Low density products are generally obtained ($0.1\ g/cm^3$).

Table 4. London Tea Prices[a]

Year	Prices (U.S., $/kg)
1960	1.42
1965	1.29
1970	1.10
1975	1.39
1977	2.74
1979	2.17
1980	2.24

[a] Average prices are shown since differences in quality cause prices to vary over a wide range.

Other Methods. Instant tea may also be prepared by directly extracting fermented leaf before the drying stage. These processes are only suited for practice in the tea-growing countries. They have been carried out on a small scale in Sri Lanka, India, and Kenya (61).

Instant Tea Mixes. Instant tea is sold as such, but may also be blended with flavors and sugar or noncaloric sweeteners. Lemon is by far the predominant flavor used.

Decaffeination. Small quantities of decaffeinated tea have begun to appear on the market. It is not yet known whether this represents a significant trend. Solvent extraction of caffeine from tea leaf is described in several patents (62).

BIBLIOGRAPHY

"Tea" in *ECT* 1st ed., Vol. 13, pp. 656–666, by G. F. Mitchell, General Foods Corporation; "Tea" in *ECT* 2nd ed., Vol. 19, pp. 743–755, by E. Hainsworth, Tea Research Institute of East Africa.

1. T. Eden, *Tea*, 3rd ed., Longman Group, Ltd., London, 1976, p. 16.
2. R. L. Wickremasinghe in C. O. Chichester, ed., *Advances in Food Research*, Vol. 24, Academic Press, New York, 1978, pp. 229–286.
3. W. H. Ukers, *All About Tea*, The Tea and Coffee Trade Journal Co., New York, 1935, pp. 1–9, 23–48, 109, 135, 179 (a good reference for the history of tea).
4. G. F. Mitchell, *The Cultivation and Manufacture of Tea in the United States*, Bulletin No. 234, USDA Bureau of Plant Industry, 1912.
5. C. R. Harler, *The Culture and Marketing of Tea*, 3rd ed., Oxford University Press, London, 1964 (a good reference for cultural practices).
6. N. K. Jain, *Two Bud* **26,** 57 (1979).
7. *Annual Reports, Tea Research Institute of East Africa*, Kericho, Kenya; *Tea Q.*, Journal of the Tea Research Institute of Sri Lanka, Talawakelle, Sri Lanka; *Two Bud*, Tocklai Exp. Sta., Jorhat, Assam, India; *Annual Reports, United Planters Association of South India*, Cinchona P.O., Coiombatore District, South India.
8. B. Banerjee, *Two Bud* **27,** 80 (1980).
9. F. Nakano, *Study of Tea* **42,** 72 (1971).
10. T. A. Cheyshvili, *Sakartvelo Tea Picker*, Ganatleva Press, Tiblisi, USSR, 1968.
11. G. W. Sanderson in V. C. Runeckles, ed., *Structural and Functional Aspects of Phytochemistry*, Academic Press, New York, 1972, p. 251.
12. D. J. Millin and D. W. Rustidge, *Process Biochem.* **2,** 9 (1967).
13. E. A. H. Roberts in T. A. Geissman, ed., *The Chemistry of Flavonoid Components*, The Macmillan Company, New York, 1962, pp. 471–479.
14. R. L. Wickremasinghe, *J. Natl. Sci. Counc. Sri Lanka* **1,** 111 (1973).
15. I. S. Bahtia and M. R. Ullah, *J. Sci. Food Agric.* **19,** 535 (1968).
16. G. V. Stagg and D. Swaine, *Phytochemistry* **10,** 1671 (1971).
17. R. A. Cartwright, E. A. H. Roberts, and D. J. Wood, *J. Sci. Food Agric.* **5,** 597 (1954).
18. S. Venkatakrishna, B. R. Premachandra, and H. R. Cama, *Agric. Biol. Chem.* **40,** 2367 (1976).
19. T. Yamanishi, *Nippon Nogei Kagaku Kaishi* **49,** 63 (1975).
20. R. P. F. Gregory and D. S. Bendall, *Biochem. J.* **101,** 569 (1966).
21. A. Hatanaka and T. Harada, *Phytochemistry* **12,** 2341 (1973).
22. R. L. Wickremasinghe, *Phytochemistry* **13,** 2057 (1974).
23. H. A. Stafford, *Annu. Rev. Plant Physiol.* **25,** 459 (1974).
24. K. Iwasa, *Jpn. Agric. Res. Q.* **10,** 89 (1976).
25. E. A. H. Roberts, *Biochem. J.* **33,** 218 (1939). The further work of Roberts is reported in a series of articles through 1963.
26. Y. Takino, *Jpn. Agric. Res. Q.* **12,** 94 (1978); D. J. Cattell and H. E. Nursten, *Phytochemistry* **16,** 1269 (1977).
27. D. T. Coxon, A. Holmes, and W. D. Ollis, *Tetrahedron Lett.* **No. 60,** 5241 (1970).
28. E. A. H. Roberts and R. F. Smith, *Analyst (London)* **86,** 94 (1961).
29. E. A. H. Roberts and R. F. Smith, *J. Sci. Food Agric.* **14,** 689 (1963).

30. J. B. Cloughley, *J. Sci. Food Agric.* **31,** 911 (1980).
31. J. E. Berkowitz, P. Coggon, and G. W. Sanderson, *Phytochemistry* **10,** 2271 (1971); D. T. Coxon, A. Holmes, and W. D. Ollis, *Tetrahedron Lett.* **60,** 5247 (1970).
32. D. J. Millin, D. S. Sinclair, and D. Swaine, *J. Sci. Food Agric.* **20,** 303 (1969).
33. A. G. Brown, W. B. Eyton, A. Holmes, and W. D. Ollis, *Phytochemistry* **8,** 2333 (1969).
34. G. W. Sanderson, *Tea Q.* **35,** 146 (1964).
35. C. R. Harler, *Tea Manufacture*, Oxford University Press, London, 1963, pp. 37–50.
36. *The Tea Cyclopedia*, Whittingham, London, 1882, pp. 211–2.
37. Ref. 3, p. 68.
38. Brit. Pat. 1,268,231 (March 22, 1972), (to The Tea Research Association).
39. G. W. Sanderson and H. N. Graham, *J. Agric. Food Chem.* **21,** 576 (1973).
40. K. L. Bajaj, *R. Ital. Essenze Profumi, Piante Aromi Saponi Cosmet. Aerosol* **61,** 123 (1979).
41. T. Takeo, *Phytochemistry* **20,** 2145 and 2149 (1981); O. G. Vitzthum, P. Werkhoff, and P. Hubert, *J. Agric. Food Chem.* **23,** 999 (1975).
42. Brit. Pat. 2,026,668 (Feb. 6, 1980), D. W. Brooks (to Hambro Machinery, Ltd.).
43. J. B. Cloughley and R. T. Ellis, *J. Sci. Food Agric.* **31,** 924 (1980).
44. Brit. Pat. 1,274,002 (May 10, 1972), A. C. K. Krishnaswami and C. Hariprasad.
45. Brit. Pat. 1,484,540 (Sept. 1, 1977), D. Kirtisinghe and D. P. Ranasinghe (Tea Research Institute).
46. M. A. Bokuchava and N. I. Skobeleva in T. E. Furia, ed., *CRC Critical Reviews in Food Science and Nutrition*, CRC Press, Boca Raton, Fla., 1980, p. 303.
47. T. Yamanishi, *Nippon Nogei Kagaku Kaishi* **49,** 1 (1975).
48. Ref. 35, pp. 102–105.
49. *Annual Bulletin of Statistics*, International Tea Committee, London, 1981.
50. G. P. Tyler, *Oxford Agrarian Studies* **5,** 99 (1976).
51. F. G. Adams and J. R. Behrman, *Econometric Models of World Agricultural Commodity Markets*, Ballinger, Cambridge, Mass., 1976.
52. *Report of the Third Session of the Intergovernmental Group on Tea*, FAO, Rome, April 20–30, 1976.
53. *Financial Times, London*, p. 27 (Sept. 2, 1981).
54. N. D. Pintauro, *Tea and Soluble Tea Products Manufacture*, Noyes Data Corp., Park Ridge, N.J., 1977 (a good summary of tea processing patents).
55. U.S. Pat. 2,902,368 (Sept. 1, 1959), E. Seltzer and F. A. Saporito (to Thomas J. Lipton, Inc.); U.S. Pat. 3,451,823 (June 24, 1969), A. R. Mishkin, W. C. Marsh, A. W. Fobes, and J. L. Ohler (Afico, S.A.).
56. U.S. Pat. 2,927,860 (March 8, 1960), E. Seltzer and F. A. Saporito (to Thomas J. Lipton, Inc.).
57. Brit. Pat. 946,346 (Jan. 8, 1964), (Afico, S.A.).
58. U.S. Pat. 3,151,985 (Oct. 6, 1964), A. Fobes (Afico, S.A.); U.S. Pat. 3,787,590 (Jan. 22, 1974), B. Borders, H. Rivkowich, and W. C. Rehman (Tetley, Inc.); U.S. Pat. 3,163,539 (Dec. 29, 1964), W. E. Barch (to Standard Brands, Inc.); Brit. Pat. 1,249,932 (Oct. 13, 1971), (to Brooke Bond, Ltd.).
59. U.S. Pat. 3,598,608 (Aug. 10, 1971), N. Geniaris (to Struthers Scientific and International Corp.).
60. Brit. Pat. 1,367,616 (Sept. 18, 1974), A. R. Mishkin and J. L. Ardem (to Nestlé Products Ltd.).
61. U.S. Pat. 3,649,297 (March 14, 1970), D. J. Millin (to Tenco Brooke Bond, Ltd.); U.S. Pat. 3,392,028 (July 9, 1968), L. Vuataz (Afico, S.A.).
62. U.S. Pat. 4,167,589 (Sept. 11, 1979), O. Vitzhum and P. Hubert.

General Reference

An extensive collection of literature concerning all aspects of tea is maintained by Thomas J. Lipton, Inc., Englewood Cliffs, N.J.

HAROLD GRAHAM
Thomas J. Lipton, Inc.

TECHNETIUM. See Radioactive tracers.

TECHNICAL SERVICE

Technical service, as practiced in the last 20 or 30 years, has evolved from concepts associated with sales and marketing. The early history of the chemical industry focuses, for the most part, on production and manufacturing of heavy chemicals, with research revealing an occasional dramatic discovery such as nylon. Following World War II, the emergence of the petrochemical industry unleashed a variety of new compounds available in tonnage quantities, and emphasis turned toward finding uses for these materials. Potential customers had to be approached, educated in their need for such materials, and advised how to use them effectively. As a consequence, marketing departments became more and more important and often determined the direction of sales promotion. The service aspect of selling new compounds became a part of this marketing effort. In advising customers, technical-service personnel discovered additional needs and transmitted this information to research and development departments for investigation. Even before this marketing activity began, technical service became almost inherently a part of the industry's need to communicate information ranging from raw materials to products. An early discussion of the term service (with technical implied, though not stated) can be found in ref. 1, in a publication directed at users of industrial boilers. Then, as now, service to a customer was emphasized, rather than the sale of a product. Examples of service included analysis of operating conditions; chemical analyses of water and steam taken from various points in the system; and periodic rechecking (1).

After 1920, the petroleum industry spent considerable effort on lubrication technology, including the development of additives, eg, antioxidants, boundary lubricants, metal deactivators, pour-point depressants, oil detergents, and similar materials, supported with technical information and testing (2) (see Lubrication and lubricants). For a number of years, Monsanto displayed the slogan: Serving industry which serves mankind. Although that slogan was dropped, the emphasis was still on service to industry throughout the world. To achieve this goal, a large part of Monsanto's effort is devoted to application research; product and process improvement; development of alternative raw materials; and development of new uses for existing products (3).

In the 1960s, an example of the transition of one aspect of marketing into technical service was given by Union Carbide by naming their Tarrytown, New York, facility a technical-service laboratory. Technical service was described as a marketing revolution, with a technical salesman backed by a customer-service specialist offering technology to go with a product. The laboratory was designed for the technical study of customer problems; to provide basic application knowledge and information for handling, storage, and processing of chemicals; to provide information to Union Carbide's research and development department for new product development; and to provide training facilities for technical sales personnel. The laboratory staff was organized into 19 different industry or functional areas, eg, oil and gasoline additives, plasticizers, detergents and household products, paint formulation, and synthetic lubricants, among others (4). Originally a part of Union Carbide's marketing department, the laboratory became, by 1965, a part of R&D (research and development) itself, although it was still an integral part of the marketing–technical-service support activity (see also Process research and development).

Definitions

The meaning of the term technical service can be interpreted in several ways. A study of the impact of communication factors on R&D performance divides R&D functions into basic and applied research, development, and technical service. The study defined the latter as tasks meeting specific local needs of an organization such as improving performance or reducing cost in existing products or operations. Opening new markets for existing products was included, and emphasis was placed on applying existing knowledge (5).

The term applied rather than basic research is pertinent for industry (6). In this definition, applied research has specific commercial objectives either related to products or processes; development projects apply the research results to technology. Within the framework of project work, technical service is included within the classification of improving marketing operations. The importance of balancing research with technical-service or administrative activities is emphasized to further individual and group creativity (6).

In recent years, a number of studies have tried to define or distinguish among the various types of research and development activities practiced in the United States (7). In particular, the definitions for basic and applied research seem to vary widely, particularly when surveyed from an industrial point of view. A straightforward definition of technical service was made by the Industrial Research Institute as work performed by R&D personnel, usually requested by other areas of a company, such as manufacturing, marketing, or corporate management. It involves the application of existing technology to company problems in the existing business. Others agree with this approach by including technical service under defensive research activities and suggesting it extends product or process improvements to other internal departments (see Research management).

These definitions may be relevant when looking inward, within a company framework. Assistance by R&D personnel to other departments has always been an important function of most R&D departments. Straight analytical support of production, be it routine quality checks or definition of problems, may frequently be the responsibility of R&D. Information on product properties, performance, and application are generated within R&D and transferred to a marketing, sales, or engineering department for formal preparation and customer distribution. However, an outward approach, as seen by existing or potential customers, may be preferable. The service itself is important, regardless of whether it is supplied by R&D, sales, marketing, or engineering. Technical service may not involve the company's technology itself. There should be no hesitation within the boundaries of legality, propriety, and perhaps impact on current or future profitability, to provide guidance, direction, information, or assistance on subjects outside the technology of a company. A technical-service engineer must be "... an organizer, salesman, safety expert, psychologist, accountant, labor-relations expert—even mechanic, truck driver and floor sweeper, if need be" (8).

An advertisement detailed a technical-service chemist's responsibilities to include: finding new product applications; interfacing with other departments; and traveling at a 10–15% rate for direct customer contact and support (9).

The three main R&D areas are research, development, and technical service. A product is conceived and born in research; investigated, analyzed, and brought to readiness in development; and supported in commercial application by technical service. This definition, from a product viewpoint, with its inherent problem-solving aspect, is the central theme of technical service. From a customer viewpoint, it may not be broad enough. Although many aspects are technical as contrasted to administrative, they may not involve problem-solving applications technology in the traditional sense. Indeed, the narrow definition of the 1960s is being broadened for regulatory reasons. The distribution and handling of chemicals has come under increasingly stringent and numerous federal, state, and local regulations, and providing information to customers on toxicity, disposal, effluent limitations, etc, has become an important function of technical service.

In some fields and in some companies, the line between sales and technical service is blurred. In water treatment, for example, several companies depend on the field salesperson to be the first-line technical-support person as well. In addition to closing a sale, the sales agent is also a troubleshooter and can solve problems that the customer may have with a product's application or the system's performance. Equipped with analytical test kits, a fieldman utilizes knowledge of the system chemistry and recommends a course of action or explains the problems encountered. For instance, testing the fresh water going into a demineralizer exhibiting short runs and finding that the total hardness has increased dramatically over that in previous testing. Perhaps on investigation, a piping error has caused some cooling-tower blowdown water, with concentrated hardness, to enter into the water supply for the demineralizer system. Frequent analysis of a system's water to determine if the concentration of the product is appropriate is another task for the sales agent. For example, the field personnel of Drew Chemical Co. are familiar with the chemistry and engineering of pretreatment and effluent water systems, along with cooling systems and boiler water systems. This company supports its sales–technical representatives with an in-field technical-service senior person and with in-depth technology from their home office's consulting and technical-services group. The latter can initiate research investigations in the laboratory or at the customer's site (10). Other water-treatment companies, such as Betz, Nalco, and Calgon, provide similar support functions (see also Water, industrial water treatment). The Ferroalloys Division of Union Carbide trains its sales agents extensively over an 18-month period, including assignment to a technical-service operation in order to learn how customers use products and what problems they encounter (11).

A gray area is the applications research that results from general problems uncovered in the field, but internal work on specific customer problems qualifies as technical-service work. Analytical data obtained on samples from a customer's system, for example, would be a part of this internal support since it is reported back to the customer or prospect as part of a technical evaluation or recommendation. Nalco and Betz, for instance, have analytical-service laboratories throughout the United States as part of their technical-service support to industrial and municipal customers.

Functions. A number of functions are basic to technical service; these are provided by one person, a group, or a department and include: development of specific technical recommendations for customers; information on products, including use, properties, application, etc; trial or start-up of products at a customer's location; assistance in training customer's or the company's own sales personnel in product application and monitoring; solving application problems; laboratory and testing service; investigation of customer complaints; the gathering and preparing of technical data and information regarding the product line; and information concerning government regulations of the sale and use of chemicals.

Technical Service and the Chemical Business

From the viewpoint of technical service in the chemical industry, the level of service provides the distinction between commodity chemicals and chemical specialties. Inherently, commodities like acetone or sodium phosphate require minimal technical-service support; generally, only standard information is required by a customer. Specific problems of quality may occur, which create a flurry of analytical activity and may lead to a process correction or the identification of a handling problem. The customer does not need assistance or direction in how to use the material; it is defined, for instance, as a raw material or carrier solvent and can be obtained from a variety of sources. Such materials are defined as undifferentiated products (12). They may differ perhaps in content specifications. Technical-service effort in these areas usually focuses on cost reduction or process improvement. Differentiated products, on the other hand, require varying degrees of technical-service support. These materials are generally made by different suppliers to performance specifications rather than to content specifications. Examples include functional chemicals such as pesticides or dyes and formulated products such as quenchant fluids or cooling-water treatments. Differentiated products may include pseudocommodities, ie, large-volume products such as plastics or surfactants, and specialty chemicals. Products in this latter classification are invariably designed to solve customer problems. Arthur D. Little does not use volume as a criterion in defining chemical specialties, but rather the term differentiation as a distinguishing characteristic. Arthur D. Little defined two classes: performance chemicals that benefit a customer through a superior value in use and proprietary chemicals that are made to perform a particular function. The latter are differentiated by the supplier's individual performance capability, as contrasted to the chemical or formulation itself. In this aspect, technical service demonstrates the superior capability of one supplier versus another. Oil-field chemicals, industrial cleaning materials, and water-treatment chemicals are good examples of cases where product effectiveness may be virtually indistinguishable, but active service support in terms of frequent system monitoring, chemical analyses, operational recommendations, etc, may give one supplier a competitive advantage.

Economic Aspects

Sales for specialty chemicals can only be estimated (see Table 1). Another source gives a range for 1978 sales of $(20–40) \times 10^9$ (14). A large variety of industries and many specialties areas within large companies are involved. A recent estimation of growth rates and market size for 21 specialty chemicals is shown in Figure 1. In 1980, the total market for specialties was estimated at $(30–35) \times 10^9$, ca 20–25% of total chemical industry sales (15). It has been stated with regard to new specialty chemicals that "For specialty chemical organizations, product application technology is critical. Scientific and technological competence in solving a customer's problems is the key . . ." (16). In addition, specialty chemicals are useful and difficult to replace. They have a large benefit-to-cost ratio, providing significant performance at low cost compared to the final product of a customer. Furthermore, the user of a specialty product has

Table 1. Estimated Size and Outlook for Selected Chemical-Specialty Sectors, 1978–1983[a]

Industrial and institutional products, sector and sales range	Maturity[b]	Profitability[c]	Sales growth[d]
By market			
Over $500 × 10⁶			
food additives[e]	M/A	M	L
lube-oil and fuel additives	G	H	H
metal-treating compounds	M	M	M
plastics additives[f]	G	M	H
textile chemicals	A/M	L	L
water-treatment chemicals	G	H	H
$(100–500) × 10⁶			
concrete additives	M/A	L	M
cosmetics additives	M	M	M
electronic chemicals	E/G	M	H
graphic arts chemicals	G/M	H	M
laboratory chemicals	M	M	M
leather chemicals	A	L	L
oil-well chemicals	G	H	H
paint additives	M/A	M	L
paper chemicals	A/M	L	L
photographic chemicals	M	H	M
rubber chemicals	A	L	L
<$100 × 10⁶			
foundry chemicals	M	M	L
mining chemicals	M	M	L
By function			
>$10⁹			
cleaning chemicals	M	M	L
adhesives[g]	M	M	M
$500 × 10⁶–$10⁹			
catalysts	M	M	M
diagnostic aids	G	H	H
flavors and fragrances	M/G	H	M
plasticizers	G/M	M	H
specialty adhesives	G/E	M	H
$(100–500) × 10⁶			
corrosion inhibitors	M	M	M
flame retardants	M	M	M
functional fluids	G/M	M	M
fungicides–biocides	M	M	M
industrial enzymes	M/G	M	M
medicinals	M	M	M
specialty industrial coatings	M/G	M	M
specialty gases	G/M	M	M
specialty pigments	M	M	M
specialty polymers	G	M	H
specialty surfactants	M/G	M	M
<$100 × 10⁶			
friction-reduction compounds	E	M	H

[a] Ref. 13.
[b] E = embryonic; G = growth; M = mature; A = aging.
[c] % Net profit on equity: H = >16%; M = 12–16%; L = <12%.
[d] %/Year in current dollars (assumes 7–8%/yr inflation rate): H = >15%; M = 12–15%; L = <12%.
[e] Exclusive flavors.
[f] Exclusive plasticizers.
[g] Exclusive structural adhesives.

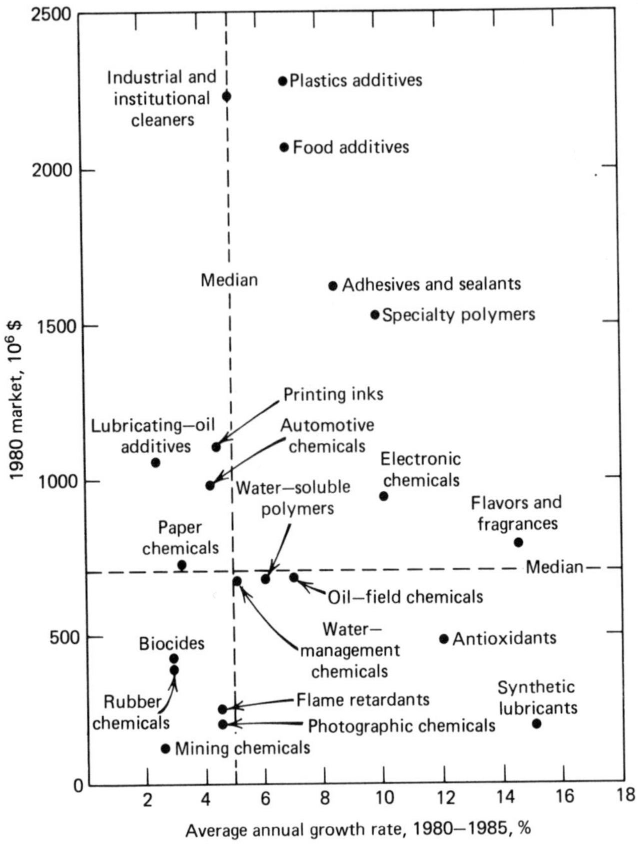

Figure 1. Market size and growth rates for a group of specialty chemicals (courtesy of McGraw-Hill and SRI International).

less technical knowledge than the supplier with regard to the product, and needs technical assistance to use it (14).

Referring to Figure 1, examples for each of the 21 listed specialties could be given to illustrate the above definitions. Water-soluble polymers are used in many areas, including municipal waste treatment. Solid matter that has been coagulated and settled is screened and dried for eventual disposal through incineration or land fill. In order to make the sludge compact, while maintaining draining and drying efficiency, certain types of water-soluble polymers, particularly polyacrylamides, are used as so-called drainage or dewatering aids (see Acrylamide polymers; Resins, water-soluble). Employees of a municipality would generally not know what type of polymer, molecular weight, and concentration should be applied under varying conditions (see Flocculating agents; Wastes, industrial).

Another indication of the market size and nature of various chemical specialties is shown in Table 2, which includes estimates of profitability and future potential and of the maturity of each product line. Chemical specialties are not inherently a growth industry, but like commodities, follow a life cycle. Mature and aging product lines are as prevalent as with emerging chemicals. Several of these cycles are illustrated in Figure

Table 2. Forecast U.S. Merchant Sales of Specialty Chemicals, 1977–1985[a]

Specialty chemical	10^6 \$ 1977	10^6 \$ 1985	%/yr
functional chemical products			
pesticides, formulated	2,850	4,370	5.5
industrial and institutional cleaners	1,485	2,030	4.0
plastics and resins	1,050	2,510	11.5
automotive chemicals	825	1,220	5.0
water-management chemicals	600	1,110	8.0
diagnostic aids	500	1,195	11.5
catalysts	365	480	3.5
synthetic elastomers	365	700	8.5
adhesives and sealants	310	665	10.0
oil-field chemicals	225	355	6.0
laboratory chemicals	170	190	1.5
refinery chemicals	110	140	5.0
Total	*8,855*	*14,965*	
additives			
for petroleum	1,200	1,460	2.5
for plastic additives	945	1,750	8.0
textile specialties	905	1,145	3.0
for food	600	820	4.0
photographic chemicals	470	935	9.0
rubber-processing chemicals	370	545	5.0
for paint	365	480	3.5
metal plating and finishing chemicals	350	460	3.5
for cosmetics	300	480	6.0
for paper	265	470	7.5
printing chemicals	110	150	4.0
foundry chemicals	100	120	2.5
flotation reagents	95	120	3.0
Total	*6,075*	*8,935*	
selected multipurpose additives			
flavors and fragrances	475	980	9.5
biocides	350	535	5.5
others[b]	555	820	5.0
Total	*1,380*	*2,335*	
Total	*16,310*	*26,235*	*avg 6.1*

[a] Ref. 17.
[b] Includes chelating agents, organic pigments, enzymes, and others.

2, ranging from the embryonic electronic chemicals to the aging leather-chemical sector. As suggested in Figure 2, early in a life cycle, technology is the key factor. Education of the customer in application and use is paramount. As a product ages, its function becomes well-defined and higher marketing efforts are necessary to promote the product. Although technical service may be a significant and necessary part of the marketing efforts in mature and aging sectors, in the early phases of a product's development it is mandatory. Even within a growth sector like water-treatment chemicals, individual chemicals are at different points in their own life cycles (see Fig. 3). The water and wastewater chemical market is estimated to be 2×10^9 by 1985, with

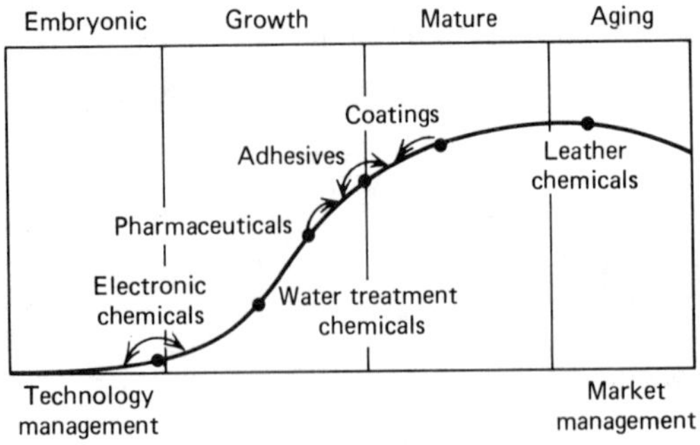

Figure 2. Life-cycle of selected chemical specialties.

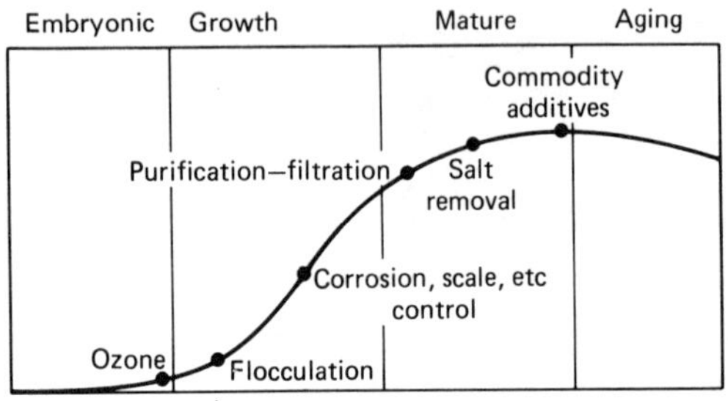

Figure 3. Life-cycle of selected water-treatment chemical business areas.

commodities (or mature and aging chemicals) contributing roughly one third of the total. Overall real growth should be 8–9% per year through 1985 (18).

Examples of Technical-Service Activities

A highly successful chemical-specialty company, Lubrizol, Inc., dominates ca one third of the petroleum additive market, competing against large oil companies like Exxon, Socal, and Amoco. Lubrizol has emphasized research and development and particularly technology in their production of detergents, corrosion inhibitors, and viscosity improvers. Thorough testing of new products is essential both in well-controlled laboratory engines and on the road (19). Exacting test procedures given are in ASTM's Special Technical Publication 315 and in the Society of Automotive Engineers handbook (20). Extensive data on physical properties showing multiviscosity grade qualifications in many base-stock oils may also be essential to motor-oil blenders and marketers. Such thorough testing procedures are expensive and therefore not usually employed by small businesses. Results of testing are generally presented to potential customers by technical personnel, rather than by the sales force. A close working relationship between supplier and customer often guides experimentation.

An example of the development of a new product involving a high degree of customer contact and cooperation was Skydrol, Monsanto's nonflammable hydraulic fluid for aircraft. Developed just after World War II, this product filled a specific need, ie, replacement of flammable oil. Hydraulic systems were actuated, lubricity and viscosity measured, and stability checked under various conditions. Different fluids were tested in Douglas aircraft under close supervision of both parties. In fact, many patents concerning the formulation of alkyl aryl phosphates into fire-resistant fluids were issued to the Douglas Aircraft company (21) (see also Flame retardants; Hydraulic fluids).

Technical-service for hydraulic fluids is divided into three categories by one author (22). In the first, accommodations for applications, it is proved to a customer that the proposed fluid will work properly. A hydraulic excavator is brought on-site and tested first with the original oil, then with the new oil in question. Measurements of temperature and time to complete certain operations show the equivalency of the materials. The second category is solving field problems, such as seal failures, entrained air problems, corrosion, and contamination from unusual sources. Often a site visit is necessary to find the source of a problem, even after analytical tests. The third category is the development of a new product resulting from a customer's special needs, eg, a hydraulic fluid for low temperature operation. Cooperation at all points is essential so that simulated laboratory testing will result in acceptable performance.

In most technical-service situations, customer and supplier cooperate in the investigation of a problem. However, the severity of the problem generally dictates the degree of active customer participation. In general, several suppliers compete and apply their own backgrounds and expertise to the problem. Depending on the particular situation, a customer may permit several potential suppliers to try their approaches in the actual running system for a time, which may be a week or several months. Monitoring the trial would be the responsibility of the supplier, but success should be obvious. For instance, system efficiency may not change but a smaller amount of a chemical-specialty product might have been used, thus offering economic advantages to the customer. In another instance, troublesome deposits on system parts might be removed.

In some situations, industrial systems are difficult or impossible to simulate accurately in a laboratory or pilot-plant setting, and the product may have to be formulated at the customer site. A prime example of this situation is the application of defoamers (qv) in the pulp and paper industry. A brown stock washer, actually a series of mechanical washers, cleans the pulp after the digestor and removes, through dilution, the chemicals used in the digestion process. Surface-active materials cause a great deal of foam in this process unless kept under control with chemical additives. Because of the constantly changing nature of the wood, temperature, pH variations, etc, a standard defoamer may not suffice. Frequently, a researcher is able to blend a variety of raw materials on a small scale and test the mixture immediately on a hot, fluid sample taken from the system. Scale-ups based on this approach have generally been quite successful on a cost–performance basis. Thus, on-site technical-service laboratory studies are often essential early in the sales negotiations.

In providing polyester fibers for the nonwoven textile industry, Hoechst Fibers Industries recognized the need for strong customer collaboration with a technical-services department located at their plant site in Spartanburg, South Carolina. This department assists customers in the initial selection of fibers and the chemical and physical analysis of the fibers, and in the solution of fiber- or machine-related problems (23). Because of the high technology in this rapid growth area, basic commodity manufacturers like Hoechst need to be familiar with textile-industry processes. The acquisition of this familiarity is a critical factor in dealing with potential customers.

The plastics industry has consistently used the technical-service approach in working with customers. Knowledge of molding characteristics and incorporation of dyes or pigments into a plastic are requisite responsibilities of the specialty chemical supplier rather than the fabricator of plastic goods. For example, Sandoz, Ltd., offers a range of services such as color matching or advice on processing techniques to obtain rapid and even distribution of color. The wide variety of polymers, processes, and uses demands a great deal of time, effort, and capital investment by Sandoz into areas other than straight production of pigments (24).

An excellent example of an analytically oriented technical problem is outlined in ref. 25. Although the problem was handled internally, its solution could have been asked of a supplier. The source of the trouble was a corrosion inhibitor fed to the light-hydrocarbon overhead of a crude-oil distillation tower. The inhibitor was a complex mixture of cyclic amines in an aromatic solvent. By experimentally evaluating its thermal stability, a brownish-black tar was produced, similar to the black, spongy material found in the partially plugged reboiler furnace tubes. A great deal of analytical work pinpointed the diagnosis of corrosion caused by sulfur and hydrochloride salts originating from the corrosion inhibitor. Complicated analytical techniques were used. Although this problem was solved by a large oil company, most customers would not be able to deal with such complexities. The skilled specialty-chemical supplier, however, can utilize technical-service expertise to solve the problem.

In the late 1970s, stringent air-pollution limitations led to an effort by suppliers to improve the efficiencies of electrostatic precipitators (see Air-pollution control methods). A number of chemical-specialty manufacturers chose chemical additives to lower the resistivity of fly-ash particles. Typically, proprietary formulations are tailored for specific systems and customers. Companies active in this area include Apollo, Dearborn, Nalco, and Betz Laboratories (26). Investigation of the customer site is necessary for R&D and sales personnel to define the system, conditions, additive-injection point, feed rate, etc. Because of the dynamic nature of such systems, the supplier's technical staff must closely monitor the treatment program in the early stages.

The disposal of wastewater has been strongly affected by legislation and regulation. Oil contaminants typically have to be removed before discharge into a municipal treatment system; limits might be 5–20 ppm and no visible presence of surface oil. Traditionally, these requirements have been met by industry by breaking emulsions with inexpensive commodity chemicals, which permit the oil to be skimmed mechanically from the surface. Acid followed by the addition of alum has been generally effective. Rising oil prices have increased the importance of reclamation of high quality oil (see Recycling, oil). Specialty chemicals, with their inherent technical-service support, offer improved methods of emulsion breaking and coagulation. Cationic polymers, with their higher charge density, for example, are more efficient than alum in neutralizing the repulsive charges between particles and precipitating the emulsifying agents. However, because of the wide variety of oily wastewaters present in industry, conditions, concentrations, nature of the fluid, etc, vary greatly and no one product works effectively in all systems, or in one system all the time. Metalworking, oil refineries, and steel mills are prime examples of areas with different conditions (27). The technical-service personnel of specialty-chemical suppliers work closely with industry and local governments to determine the optimum treatments and methods of application.

Technical Information

With the help of technical-service personnel, new technical information is transmitted to customers, eg, ref. 28. Such publications (29–30) attempt to educate the reader in the many facets of industrial water problems and approaches to solve them. Chemical reactions, system diagrams, test methods, and treatment alternatives are outlined, with minimal reference to proprietary products.

Publications, such as ref. 31, focus on a product line and introduce some basic technology of demulsifiers, scale inhibitors, drilling-fluid additives, and the like (31). Although such a brochure is primarily a sales tool, the efforts of R&D personnel are reflected in the technical information presented. Another popular example of manufacturer's literature is ref. 32. This booklet centers on one product and illustrates its many uses. Specific properties, preparation of water dispersions, and eight formulations for different uses are given. A potential customer is encouraged to try samples in his own application. A next step, of course, would be to work with technical personnel on specific problems. Generally, in the chemical industry, a small brochure serves for basic product and application information. In addition, safety-data handling sheets are prepared for every product.

Formal reports for customer use can also be prepared and presented by technical-service personnel. In the authors' experience, tailored reports on viscosity-index-improver blending studies were prepared for specific oil companies. Laboratory data were obtained using the oil or blends of oil supplied by the potential customer, plus specific motor-oil additives and the proprietary polymeric viscosity-index improver. Formulations were tailored to meet the specific viscosity requirements of, for instance, a multigrade motor oil. Appropriate physical properties were measured and tabulated, and comparisons made with the cost–performance of competitive materials. Presentations would typically be to the R&D representatives of the oil company responsible for developing new motor oils. Sales or marketing personnel from the supplier would accompany the technical-service representative in order to answer questions on pricing, delivery, etc. In a similar vein, technical-service personnel may give a seminar to a prospective customer's R&D staff to acquaint them with the chemistry and theory behind a particular product or product line. Many chemical companies, particularly in the chemical specialty area, provide this type of information service.

Organization

The first approach to a formal technical-service function, whatever its designation, might be within the sales or marketing organization of which customer contact is an integral part. This arrangement offers the best opportunity for intimate knowledge of the customer. Although under any structure, contact through the sales force is primary, a sales orientation to the technical-service group ensures rapid response; up-to-date knowledge of all matters that could affect the customer's reception of the technical-service effort; awareness of the sales situation; and priority setting within the department itself. However, laboratory and other interdepartmental contacts and activities to support a technical-service effort are decreased; there is pressure on the type, amount, and method of data to be presented; and sales details and activities are attended to rather than technical functions.

Technical-service activities within a laboratory organization are more prevalent in the chemical industry. Emphasis in the past was on autonomous laboratories, working both with sales and marketing personnel and with R&D staff. Today, the technical-service responsibility appears to rest with development or applications group personnel within R&D. These groups are usually organized around a product or product line, occasionally subdivided by industry classifications.

Individuals within a group may specialize in a certain area. Assignments might include product improvement; evaluation and development of new products; establishment of application procedures; field evaluation of new products; and direct support of field sales.

In one investigation, it was found that the communications of technical-service project members to those outside of R&D, but within the company studied, averaged three times in two days (33), considerably more external contact than research or development workers had. Performance improves when "the manager is more controlling and structures internal technical communication to a greater degree" (34). It appears that the manager, rather than the workers in technical-service, needs to communicate with other departments, especially marketing, and within the laboratory environment. Development projects that had a high degree of communication with other departments other than R&D were highly successful. It is the conclusion of this investigation that the three R&D functions, research, development, and technical service, are not alike and should not be managed in the same way.

Computer Application

Computer usage can replace experienced personnel who normally would travel extensively to customer locations. Today, data and information are obtained, transmitted to a central location, and analyzed. Results or recommendations are returned almost instantly. A recent development along these lines includes Calgon's Calguard system for cooling-water systems (35). A salesman can dial-up a central computer from on-site and initiate an interactive program that inquires about cooling-tower conditions such as water volume, flow rate, temperatures, pH, and analytical data on the water itself, etc. Using algorithms developed over years of laboratory testing, a particular water-treatment program is designed for using proprietary products and displayed at the plant site for customer review. Betz Laboratories uses a saturation–equilibria (SATEQ) computer program to assess the thermodynamic tendencies of cooling-system water to precipitate insoluble scaling species (36). The advent of programmable hand-held calculators and portable computers has meant an enormous amount of potential power is available for on-site use by technical-service personnel.

BIBLIOGRAPHY

1. *The Betz Indicator*, W. H. & L. D. Betz, Philadelphia, Pa., Jan. 1932.
2. W. A. Zisman in R. C. Gunderson and A. W. Hart, eds., *Synthetic Lubricants*, Reinhold, New York, 1962, pp. 11–30.
3. D. J. Forrestal, *Faith, Hope and $5,000—The Story of Monsanto*, Simon and Schuster, New York, 1977, pp. 175, 238.
4. Union Carbide Chemicals Company Technical Service Laboratory, Union Carbide, Tarrytown, N.Y. (undated brochure).
5. T. J. Allen, D. M. S. Lee, and M. L. Tushman, *IEEE Trans. on Eng. Mgmt.* **EM-27**(1), 2 (Feb. 1980).
6. R. T. Holzmann, *CHEMTECH*, 84 (Feb. 1972).

7. H. K. Nason, *Res. Manage.* **24**(3), 23.
8. C. C. Hager, *Chem. Eng.* **85**(1), 79 (1978).
9. *Philadelphia Inquirer*, 17E (Sept. 20, 1981).
10. E. Q. Belote and G. A. Cappeline, *Chem. Eng. Prog.*, 22, 23 (Jan. 1979).
11. *Sales Management* **108**(4), 38 (1972).
12. C. Kline, *CHEMTECH*, 110 (Feb. 1976).
13. *Strategic Business Assessment of the Chemical Industry*, Report No. R790503, Arthur D. Little, Inc., Cambridge, Mass., 1979, p. 21.
14. R. D. Stewart, *CHEMTECH*, 410 (July 1981).
15. D. J. Deutsch, *Chem. Eng.* **88,** 49 (June 29, 1981).
16. Ref.14, p. 411.
17. J. R. Hickey, *papers presented at the Chemical Marketing Research Association*, San Francisco, Calif., Feb. 7–8, 1979.
18. *AIR/WATER Pollution Report*, Business Publishers, Inc., Silver Spring, Md., Sept. 15, 1980, p. 367.
19. W. Worthy, *Chem. Eng. News*, 12 (March 14, 1977).
20. C. C. Colyer *ASTM Standardization News* **9**(10), 8 (1981).
21. R. E. Hatton, in R. C. Gunderson and A. W. Hart, ed., *Synthetic Lubricants*, Reinhold, New York, 1962, p. 133.
22. D. F. G. Hampson, *Hydraulic Pneumatic Mechanical Power* **20,** 260 (1974).
23. H. D. Stroud, *Supplying Polyester for the Nonwoven Industry*, Hoechst Fiber Industries, Spartanburg, S.C., undated report.
24. M. Cotton, *PRT Polym. Age.* **5**(11), 314 (1974).
25. S. A. Schmidt and V. F. Gaylor, *Anal. Chem.* **48,** 972A (1976).
26. *Chem. Eng. News*, 14 (Oct. 9, 1978).
27. L. C. Hale and D. Bauer, *Plant Eng.* **31**(6), 131 (1977).
28. *Betz Handbook of Industrial Water Conditioning*, 8th ed., Betz Laboratories, Inc., Trevose, Pa., 1980.
29. F. N. Kemmer, ed., *The NALCO Water Handbook*, McGraw-Hill, Inc., New York, 1979.
30. *Principles of Industrial Water Treatment*, Drew Chemical Corp., Boonton, N.J., 1977.
31. *Surfactants for the Petroluem Industry*, Bulletin 250, Witco Chemical Corp., New York, 1977.
32. *The Versatile Ingredient VEEGUM*, Bulletin No. 15, R. T. Vanderbilt, Norwalk, Conn., undated booklet.
33. Ref. 5, p. 4.
34. Ref. 5, p. 11.
35. *Chem. Week* **123**(20), 54 (1978).
36. L. Chen, D. T. Freese, and W. R. Snyder, *The Role of Computer Calculated Supersaturation Ratios in Assessing Scaling Tendencies*, paper to be presented at Corrosion/82, NACE, Houston, Texas, March 22–26, 1982.

General References

P. S. Bender, *Design and Operation of Customer Service Systems*, AMACON, New York, 1976.
W. H. Bluel, *Product Service Planning*, AMACON, New York, 1980.
E. P. McGuire, *Technical Service Handbook*, Padric Publishing Company, Mountainside, N.J., 1962.

ROBERT J. ZIEGLER
HILLEL LIEBERMAN
Betz Laboratories

TELLURIUM AND TELLURIUM COMPOUNDS

Tellurium [*13494-80-9*], Te, at no. 52, at wt 127.61, is a member of the sixth main group (VIA) of the periodic chart located between selenium and polonium and between antimony and iodine. The configuration of its six outer electrons is $5s^25p^4$, and its four inner principal shells are completely filled. Tellurium is more metallic than oxygen, sulfur, and selenium, yet it resembles them closely in most of its chemical properties. Whereas oxygen and sulfur are nonmetals and electrical insulators, selenium and tellurium are semiconductors, and polonium is a metal. Tellurium forms inorganic and organic compounds superficially similar to the corresponding sulfur and selenium compounds, and yet dissimilar in properties and behavior. The valence states assigned to the central atom in tellurium compounds are −2, 0, +2, +4, +6.

Tellurium was discovered in 1782 in Transylvanian gold ore. Its name is derived from the Latin *tellus*, meaning earth (1).

At 0.047 atoms per 10,000 atoms Si, tellurium is about the fortieth element in the order of cosmic abundance. Along with platinum, palladium, and ruthenium, it ranks about the seventy-first in the order of crustal abundance. The average amount in crustal rocks is 0.01 ppm. The chalcogen elements, sulfur, selenium, and tellurium, are primary components of intrusive and extrusive magmas and volcanic gases, and hence of volcanic sulfur deposits. Nevertheless, selenium and tellurium are not essential components of the common igneous rock-forming minerals. Tellurium is widely distributed in the earth's crust in deposits of many different types, from magmatic and pegmatitic to hydrothermal, especially where these deposits are associated with epithermal gold and silver deposits (2–4).

Small concentrations but large quantities of tellurium are present in copper porphyries, massive pyritic copper sulfide and nickel sulfide deposits, and frequently in lead sulfide deposits. In pyritic deposits, tellurium is concentrated chiefly in pentlandite, $(FeNi)_9S_8$; chalcopyrite, $CuFeS_2$; and pyrite, FeS_2 (in decreasing order); and least in sphalerite, ZnS, and pyrrhotite, FeS. Other tellurium-bearing deposits are copper–molybdenum, lead–zinc, gold, tungsten–bismuth, uranium, and mercury–antimony. The S:Se:Te ratio varies widely among the deposits and within the same deposit. At times, tellurium concentration exceeds that of selenium, and often it increases with depth. Little is known about tellurium in sedimentary rocks, though some shales contain 0.1–2 ppm; manganese nodules from the Pacific and Indian Ocean floors contain 0.5–125 ppm (see Ocean raw materials).

Like selenium, tellurium usually forms binary minerals; but whereas selenium isomorphously replaces a part of the sulfur in sulfides, tellurium occurs only as discrete tellurium minerals of microscopic size. Tellurium is a main component of ca 40 mineral species, including 24 tellurides, two tellurates, native tellurium, and a selenium–tellurium alloy. It is a minor constituent of an undetermined number of minerals. It occurs in combination with oxygen, sulfur, and ten other elements with high atomic numbers. At least ten tellurium minerals occur with gold and silver, ten with bismuth, and six with iron. Some of the better known minerals are hessite [*12002-98-1*], Ag_2Te; petzite [*1317-73-3*], Ag_3AuTe_2; calaverite [*37043-71-3*], $AuTe_2$; sylvanite [*1301-81-1*], $AuAgTe_4$; altaite [*12037-86-4*], PbTe; tetradymite [*1304- 78-5*], Bi_2Te_2S; rickardite [*12134-39-3*], Cu_4Te_3; nagyágite [*12174-01-5*], $Au(Pb,Sb,Fe)_8(Te,S)_{11}$; and tellurite [*14832-87-2*], TeO_2.

Like selenium, tellurium minerals, although widely disseminated, do not form an ore. Hence, there are no deposits that can be mined for tellurium alone, and there are no reserves. Resources estimated at millions of metric tons, however, are found in the base-metal sulfide deposits mined for copper, nickel, gold, silver, and lead, where the recovery of tellurium, like that of selenium, is incidental.

Physical Properties

At least 21 tellurium isotopes are known, with mass numbers from 114 to 134. Of these, eight are stable (120, 122–126, 128, 130). The others are radioactive with lifetimes from 2 min to 154 d; the heaviest six (131m, 131, 132, 133m, 133, 134) are fission products (see Radioisotopes). Tellurium illustrates the rule that elements with even atomic numbers have more isotopes than elements with odd atomic numbers. The neighboring antimony has two isotopes, and iodine only one. Because of the preponderance of heavy isotopes, tellurium has a higher atomic weight than iodine.

At ordinary temperature and pressure, solid tellurium, unlike sulfur and selenium, exists in one modification only. It crystallizes in a hexagonal lattice with parameters $a = 0.4457$ nm, $b = 0.5929$ nm, bond angle 103.2°, and trigonal symmetry. The structure is a network of relatively weakly bound spiral chains. The so-called amorphous variety contains very small hexagonal crystals. Single crystals are readily damaged by slips, fractures, deformation, and dislocation. Many physical properties, such as compressibility, strength, thermal expansion, refraction of polarized light, electrical conductivity, and galvanomagnetic properties, are markedly anisotropic. The structural diagram of tellurium shows five solid phases at 300–475°C up to 7 MPa (ca 70 atm).

Tellurium forms a continuous range of isomorphous solid solutions with gray selenium. These solutions contain chains with more or less randomly alternating Se and Te atoms. Tellurium solidified from the liquid is crystalline, grayish white with a metallic luster resembling antimony, rather brittle, and tarnishing somewhat on exposure to air. The crystals are easily cleaved and pulverized. Thin films of tellurium are reddish brown to purple. They are opaque to visible light and transparent to infrared. Finely divided tellurium is black when powdered mechanically, grayish-brown when precipitated with oxygen from alkaline solution or with sulfur dioxide from acidified solutions, and black when precipitated by reduction with zinc, tin, or iron. Colloidal tellurium prepared by reducing aqueous telluric acid with hydrazine is red.

Physical properties are given in Table 1. The vapor pressure between 511 and 835°C is given by the equation (5): $\log p_{kPa} = 6.7249 - (5960.2/K)$ (to convert kPa to mm Hg multiply by 7.5).

The golden-yellow tellurium vapor absorbs light in the visible and ultraviolet regions and contains species from monatomic Te to Te_5. The Te_2 species is predominant.

The electrical resistance of polycrystalline tellurium, as with any semiconductor, varies greatly with the amount and kind of impurities and decreases markedly with increasing temperature, and especially with compression. The intrinsic electrical resistivity ρ at 25°C is about 1 Ω-cm. The photoelectric effect is only approx 0.01% as strong as in gray selenium. The thermal emf is 50–600 $\mu\gamma$ per °C, depending on the

Table 1. Physical Properties of Tellurium

Property	Value
specific gravity[a] at 18°C	
crystalline	6.24
precipitated	6.0–6.2
hardness, Mohs[b]	2.0–2.5
modulus of elasticity, MPa[c]	4140
Poisson ratio at 30°C	0.33
heat capacity at 25°C, kJ/mol[d]	25.70
entropy at 25°C, J/K[d]	49.70
heat of fusion, kJ/mol[d]	17.87
mp, °C	450
viscosity at mp, mPa·s (= cP)	1.8–1.95
bp, °C[e]	990
heat of formation[f], kJ/mol[d]	171.5
heat of vaporization, kJ/g[d]	46.0
thermal conductivity at 20°C[g], W/(m·K)	0.060
Te–Te bond energy, kJ/mol[d]	138
covalent radius, nm	0.137
electronegativity[h]	ca 2.1
first ionization potential, eV	9.01
electron affinity, eV	
first	2.3
second	ca 3.0
volume shrinkage on solidification, %	5–7

[a] Increases under pressure.
[b] Anisotropic.
[c] To convert MPa to psi, multiply by 145.
[d] To convert J to cal, divide by 4.184.
[e] Extrapolated.
[f] Te(g) atom to Te_2(g) molecule.
[g] Polycrystalline material; in single crystals, it is anisotropic and affected by impurities and lattice imperfections.
[h] Pauling scale.

kind and concentration of impurities. Tellurium has a diamagnetic susceptibility, which, like the Hall effect, is anisotropic and affected by impurities. The Hall effect varies irregularly with temperature and shows two sign reversals in the solid state and one in the liquid state. Tellurium and many of its alloys and intermetallic compounds have semiconductor and thermoelectric properties.

Chemical Properties

Although chemically tellurium resembles sulfur and selenium, it is more basic, more metallic, and more strongly amphoteric. Its behavior as an anion or a cation, depending on the medium, eg:

$$TeO_2 + 2\,KOH \rightarrow \underset{[7790\text{-}58\text{-}1]}{K_2TeO_3} + H_2O$$

$$TeO_2 + 4\,HCl \rightleftarrows \underset{[10026\text{-}07\text{-}0]}{TeCl_4} + 2\,H_2O$$

explains the three types of redox reactions in the acid and the alkaline mediums. In an acid medium, the following reactions take place:

$$Te - 4\,e \rightarrow Te^{4+} \qquad Te^{4+} + 4\,e \rightarrow Te \qquad Te^{4+} - 2\,e \rightarrow Te^{6+}$$

In an alkaline medium:

$$Te - 4\,e + 3\,H_2O \rightarrow TeO_3^{2-} + 6\,H^+$$

$$TeO_3 + 4\,e + 6\,H^+ \rightarrow Te + 3\,H_2O$$

$$TeO_3 - 2\,e + H_2O \rightarrow TeO_4^{2-} + 2\,H^+$$

Tellurium forms ionic tellurides with active metals, and covalent compounds with other elements. The valence states are −2 (H_2Te [*7783-09-7*], Na_2Te [*12034-41-2*], CuTe [*12181-15-6*]), +4 (TeO_2 [*7446-07-3*], $TeBr_4$ [*10031-27-3*], $TeOCl_2$ [*16984-34-2*]), and +6 (TeO_3 [*13451-18-8*], H_6TeO_6 [*7803-68-1*], TeF_6 [*7783-80-4*]). Solid crystalline tellurium tarnishes slightly in the air, and more rapidly and to a greater degree in the powdered state. Moist precipitated tellurium oxidizes on drying, especially above 100°C. Molten tellurium is readily oxidized to tellurium dioxide, which can be volatilized by blowing air through the melt. Tellurium reacts with halogens and halogenating agents, and mixes in all proportions with sulfur and selenium. Oxidation with nitric acid, and ignition of the resulting $2TeO_2.HNO_3$ [*23624-18-2*], yields very pure TeO_2. Tellurium dioxide can be used as an oxidant for acetoxymethylation reactions (6):

$$ArH \xrightarrow[\text{LiBr, 120 or 160°C}]{TeO_2,\ CH_3CO_2H} ArCH_2OC(=O)CH_3$$

Tellurium reacts with concentrated, but not with dilute, sulfuric acid to form tellurium sulfite [*84074-47-5*]:

$$Te + H_2SO_4 \rightleftharpoons TeSO_3 + H_2O$$

Dilution with water reverses the reaction, and heating the solution liberates sulfur dioxide. When added to a solution of tellurides, tellurium forms colored polytellurides. Unlike selenium, tellurium is not soluble in aqueous sodium sulfite. This difference offers a method of separating the two elements. Like selenium, tellurium is soluble in hot caustic alkalies, with the exception of ammonium hydroxide. Cooling reverses the reaction. Since tellurium forms solutions of anions (Te^{2-}) and cations (Te^{4+}), tellurium films can be deposited on inert electrodes of either sign.

Elemental tellurium reduces chlorides such as $AsCl_3$, $AuCl_3$, and $PbCl_4$ to the element; it reduces $FeCl_3$ to $FeCl_2$, and SO_2Cl_2 to SO_2. Oxidation of metals by tellurium gives metallic tellurides. Tellurium itself is oxidized by strong reagents, eg, $Na_2Cr_2O_7$, $KMnO_4$, $Ca(OCl_2)_2$, and $HClO_3$. Tellurium dioxide and tellurous acid and its salts are readily reduced to the element with $SnCl_2$, Na_2AsO_3, H_2S, and $Na_2S_2O_4$ and are oxidized to the Te (VI) state with PbO_2, CrO_3, Cl_2, and $KMnO_4$. Selenium, H_2S, and HCl reduce TeO_4^{2-} ions to TeO_3^{2-}. Solid tellurium oxides can be reduced by heating with hydrogen, carbon, carbon monoxide, and sulfur.

The stability of organic chalcogen compounds decreases mostly in the order sulfur > selenium > tellurium.

Manufacture

Recovery. The percentage of tellurium recovered from copper ore is indeed small. About 90% is lost in flotation concentration, and from 20 to 60% in each metallurgical operation, eg, roasting smelting, converting, fire-refining, and slimes treatment (7–18) (see also Copper). Losses are also high in the metallurgical treatment of lead and zinc ores, by-products of which are sometimes sent to a copper smelter. A diagram of the distribution of tellurium during the treatment of various nonferrous sulfide ores and in sulfuric acid manufacture is shown in ref. 19.

Most tellurium of commerce is recovered from electrolytic copper refinery slimes, in which it is present from a trace to 8% (see also Selenium).

When the copper is removed from slimes by aeration in dilute sulfuric acid, oxidative pressure-leaching with dilute sulfuric acid, or digesting with strong acid followed by water-leaching, much of the tellurium is dissolved. This tellurium is recovered by cementing (precipitation on metallic copper), leaching the cement mud with dilute caustic soda, and neutralizing with sulfuric acid. The precipitate from the neutralization contains tellurium as tellurous acid suitable for recovery. Some tellurium is lost during these operations.

The copper-free slimes are treated by one of the following methods: refining with soda ash in a doré or cupeling furnace (10–12); combined oxidation and alkalization by roasting or baking a slimes–soda ash mix; removal of selenium by roasting and caustic soda leaching; or boiling the slime with caustic soda (13–14). The soda slag from the soda refinement or the roasted product of the oxidation is leached with water to extract sodium tellurite [*10102-20-2*]. The insoluble sodium tellurate [*10101-83-4*] in the leached slag is returned to the copper-anode furnace. The liquor obtained from the selenium removal and the boiling with caustic soda contains lead. In all cases, the solution contains selenium and impurities. Whatever the method, the liquor is neutralized to pH 6–6.2 with sulfuric acid to precipitate impure tellurous acid as tellurium mud (15), which contains also lead sulfate, silica, and other impurities. The mud is purified by redissolving in caustic soda and reprecipitating. Impurities such as lead are removed by careful precipitation from the caustic solution with sodium sulfide. Fractional neutralization of the initial impure caustic solution yields tellurous acid of a purity acceptable for reduction to the metal.

Tellurium is recovered from the precipitated tellurous acid by three methods: (*1*) By direct reduction by heating with flour under a cover of borax (heavy fuming of tellurium dioxide and the formation of organic decomposition products are drawbacks; reduction with sulfur is rapid and leaves a clean melt, but the heavy fumes are obnoxious); (*2*) dissolving in hydrochloric acid or in sulfuric acid to which crude common salt is added, followed by precipitation of tellurium with sulfur dioxide, and subsequent filtration, washing, drying, and melting (a drawback is that some oxidation of the precipitate takes place during drying); and (*3*) dissolving in caustic soda to give a solution containing sodium tellurite and free caustic soda, electrolyzing the solution in a cell with stainless-steel electrodes, followed by removing, washing, drying, and melting the cathode deposit (16); the last is the most satisfactory method.

Electrolysis conditions at CCR Division of Noranda Mines, Ltd., Canada (18) are given in Table 2.

Purification. The high boiling temperature of tellurium precludes purification by atmospheric distillation, but low pressure distillation is feasible. Heavy-metal impurities (iron, copper, tin, lead, antimony, bismuth) remain in the still residue.

Table 2. Electrolysis Conditions in Tellurium Recovery[a]

Property	Value
Te, g/L	
in fresh electrolyte	150–200
in spent electrolyte	90–140
NaOH, g/L	
in fresh electrolyte	40
in spent electrolyte	80
electrolyte temperature, °C	40–45
flow rate, L/h	60
no. of anodes	15
no. of cathodes	14
cathodic current density, A/m^2	160
cell voltage, V	2.0–2.5
deposition time, days	3
current efficiency, %	ca 90

[a] From ref. 18.

Volatile selenium is a persistent contaminant, and may be as high as 500 ppm in the distilled tellurium.

In a rather laborious chemical purification method, tellurium is dissolved in strong nitric acid, hydrolyzed to white $2TeO_2.NO_3$ [*23624-18-2*], precipitated by diluting and boiling, separating, and washing (redissolving and rehydrolyzing, if desired), dissolved in hydrochloric acid, and reduced with sulfur dioxide.

Ultrahigh-purity tellurium is prepared by zone refining (qv) in a hydrogen or inert-gas atmosphere. Single crystals of tellurium, tellurium alloys, and metal tellurides are grown by the Bridgman and Czochralski methods (see Semiconductors).

Commercial Products. *Tellurium dioxide*, TeO_2 (79.9% Te theoretical), is made by drying precipitated tellurous acid at 100°C. The acid is purified, if necessary, by redissolving in caustic soda solution and neutralizing with sulfuric acid.

Sodium tellurate, Na_2TeO_4 (53.7% Te theoretical), is made by oxidizing sodium tellurite solution with hydrogen peroxide. The reaction is exothermic.

Ferrotellurium, iron telluride [*12125-63-2*], which usually has a Te content near the theoretical 69.5%, is made by melting iron and tellurium powders in ca stoichiometric proportion.

Tellurium diethyldithiocarbamate [*20941-65-5*], $[(C_2H_5)_2NC(S)S]_4Te$, is made by the reaction of diethylamine, carbon disulfide, and tellurium dioxide in an alcoholic solution.

Metal tellurides for semiconductors are made by direct melting, melting with excess tellurium and volatilizing the excess under reduced pressure, passing tellurium vapor in an inert gas carrier over a heated metal, and high temperature reduction of oxy compounds with hydrogen or ammonia.

Economic Aspects

Tellurium has been produced commercially in the United States since 1918 and in Canada since ca 1934 (17–18). The principal world producers are given in Table 3. In addition, tellurium is produced in Bulgaria, the GDR, India, Yugoslavia, and probably the People's Republic of China.

Production and trade figures are published by the U.S. Bureau of Mines (4,20),

Table 3. Principal Tellurium Producers, 1979[a]

Location	Company
Canada	Noranda Mines, Ltd., CCR Division
	International Nickel Co. of Canada, Ltd.
Fiji	Emperor Mines, Ltd.
FRG	Norddeutsche Affinerie
Hong Kong	Metal Refiners (Asia), Ltd.
Japan	Mitsubishi Metal Corp.
	Mitsui Mining & Smelting Co.
	Sumitomo Metal Mining Co.
	Nippon Mining Co.
Peru	Centromin-Peru, S.A.
United States	Asarco, Inc.
	AMAX, Inc.
USSR	several producers

[a] From ref. 17.

the American Bureau of Metal Statistics (21), and the Canadian Department of Energy, Mines, and Resources (22). The production figures are few and incomplete, because some are not available, some are withheld to avoid disclosing proprietary data, and some are estimated.

The estimated world production of tellurium in 1978 was 350 metric tons. Canada, Japan, and the United States accounted for most of the production in noncommunist countries; Japan has become an important producer. Most of the Canadian production is exported, a large part of it to the United States. The 1979 production in Canada was 47.2 t, down from 53.0 t in 1976. In 1978, almost one half of the U.S. demand was met by imports. The main applications were in ferrous metallurgy, 57%, and nonferrous metallurgy, 14%, for improvement of machinability; chemicals, 25%; and miscellaneous, 4%. The uses in catalysts, semiconductors, thermoelectric alloys, cast iron, and special glass formulations are important, although they require only small amounts.

The United States and Canada are self-sufficient in tellurium. The metal and its compounds are not included in the U.S. Government National Stockpile list of Critical and Strategic materials. Imports are subjected to a 4% duty on the metal and a 5% duty on the compounds and salts.

Because of unsettled business conditions, market quotations have not been published in 1981. The most recently quoted price was \$31.00/kg (Nov. 82).

Specifications and Grades

There are no official U.S. specifications, but some producers publish their own standards. Some users specify the weight and shape of the pieces, a screen analysis of the powder, and a maximum content of certain impurities.

The grades marketed in The United States and Canada are given in Table 4. The commercial grade is available in lumps (≥7.5 cm dia), blocks (450 g, 900 g, and 2.7 kg), sticks (450 g and 900 g), powder (74 μm or <200 mesh), and tablets (1, 2, and 3 g).

The official USSR specifications for technical tellurium are given in Table 5.

Table 4. Tellurium Grades

	Commercial or refined	Pure	Highly pure
tellurium	99.0–99.4%	99.95%[a]	99.999%
selenium	0.5%	500 ppm	
lead	0.3%	25 ppm	
iron	0.1%		
copper	0.02%		
arsenic		15 ppm	
zinc		10 ppm	
other impurities		15 ppm[b]	
impurities, max			10 ppm

[a] Some manufacturers claim 99.99%.
[b] Total of iron, copper, nickel, chromium, antimony, bismuth, and mercury.

Table 5. USSR Specifications for Technical Tellurium

Brand	Min, % Te	Max impurity							
		Se	Pb	Cu	S	Na	Si	Al	Fe
T00	99.9	0.05	0.006	0.004	0.02	0.006	0.001	0.003	0.006
T0	99.8	0.1	0.02	0.004	0.04	0.01	0.002	0.006	0.01
T1	99.0	0.1	0.2	0.05	0.04	0.05	0.05	0.04	0.1
T2	96.0	1.5	1.0	0.3	0.2	0.3	0.2	0.1	0.15

Analytical Methods

Tellurium is usually detected by the formation of a black precipitate with sulfur dioxide in dilute hydrochloric acid, or a reddish-brown tellurium sulfide with hydrogen sulfide (23–25). Elemental tellurium and tellurides, but not tellurites or tellurates, give a purplish-red color when warmed with concentrated sulfuric acid.

Depending on the tellurium content and the nature of the material being analyzed, tellurium can be determined by a number of gravimetric (26), volumetric, polarographic, potentiometric (27), emission-spectrographic, x-ray-fluorescent, and paper-chromatographic methods. In addition, atomic-absorption-spectrophotometric procedures are widely used.

Gravimetric methods are not entirely satisfactory. Precipitation as the element with sulfur dioxide (preferably with the addition of hydrazine) is carried out by coprecipitation of other elements and by oxidation of the precipitate on drying. Tellurium can be precipitated from Te(IV) solutions with ammonia, pyridine, and hexamethylenetetramine, but heavy metals and other elements are coprecipitated, unless complexed with EDTA (28), citric acid, or tartaric acid.

In a common volumetric determination, the sample is dissolved in a 1:1 nitric–sulfuric acid mixture; addition of potassium bisulfate and repeated fuming with sulfuric acid volatilizes the selenium. The tellurite is dissolved by boiling with 10% (v/v) sulfuric acid, followed by threefold dilution with water and titration with potassium permanganate:

$$5\,TeO_3^{2-} + 2\,KMnO_4 + 3\,H_2SO_4 \rightarrow 5\,TeO_4^{2-} + K_2SO_4 + 2\,MnSO_4 + 3\,H_2O$$

The titrant must be standardized against pure telluric acid.

When the tellurium concentration is 0.01%, the sample is concentrated and examined spectrophotometrically or by atomic absorption. A direct spectrographic method is used when the concentration is several ppm.

The determination of tellurium in organic material is discussed in ref. 29.

Health and Safety Factors

Elementary tellurium and the stable tellurides of heavy nonferrous metals are relatively inert and are not thought to be a health hazard (30–34). All other tellurium compounds, however, should be handled with caution, including reactive tellurides and volatile and soluble compounds, such as hydrogen telluride, tellurium hexafluoride, and the organic compounds. Some of these compounds can enter the body by absorption through the skin or inhalation and ingestion of dust, fumes, vapors, or sprays. Hydrogen telluride is a very toxic gas, although contrary to published statements its toxicity does not exceed that of hydrogen sulfide. The soluble tellurites are more toxic than the selenites and arsenites. The toxic effects of the compounds arise on reduction to the relatively harmless elemental tellurium and to dialkyl (chiefly methyl) tellurides. Tellurium compounds and metabolic products are present in exhaled breath, sweat, urine, and feces. The elimination is relatively slow. Continuous exposure to tellurium may result in some accumulation. No definite pathological effects have been observed, but acute industrial exposure to tellurium dioxide fumes and dust, and short exposure to hydrogen telluride, may cause a dryness and metallic taste in the mouth, as well as an unpleasant garlic odor in the breath, suppression of sweat, headache, and somnolence. Not all individuals are susceptible to the same degree.

Unlike selenium, tellurium is not known to be a biologically essential trace element (see Mineral nutrients).

As with selenium, industrial precautions include the commonsense measures of good housekeeping, adequate ventilation, personal cleanliness, and frequent changes of clothing. Gloves and safety glasses should be worn at all times, and dust masks and chemical goggles should be used where needed. Oral administration of ascorbic acid, recommended at one time to alleviate the unpleasant breath, is no longer in favor, because by reducing tellurium compounds, ascorbic acid may enhance their toxic effects. No ambient air standard has been established in the United States; the accepted TLV is 0.01–0.1 mg/m^3.

Tellurium and its commercial products do not present a fire hazard. They should be stored in a dry, ventilated area. There are no shipping restrictions.

Inorganic Compounds

Tellurium forms inorganic and organic compounds very similar to those of sulfur and selenium. The most important tellurium compounds are the tellurides, halides, oxides, and oxyacids (5). Techniques and methods of preparation are given in refs. 35–36. The chemical relations of tellurium compounds are illustrated in Figure 1 (37).

Tellurides. Most elements form compounds with tellurium. Binary compounds of tellurium with 69 elements and alloys of two others have been described (5,38); 58 are metals, three are metalloids, and ten are nonmetals. Most tellurides are prepared by direct reaction, varying from very vigorous with alkali and some alkaline-earth

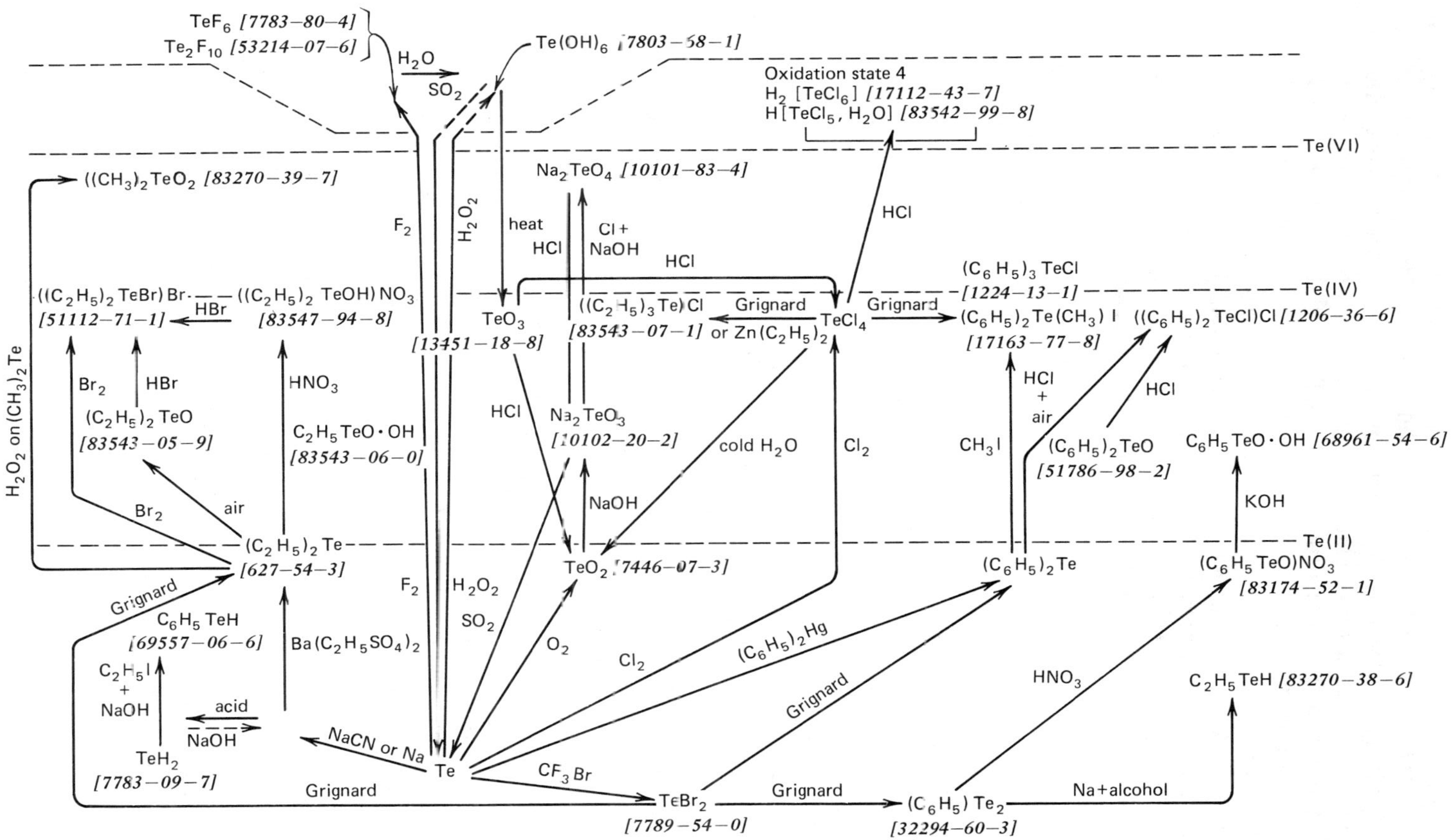

Figure 1. Chemical relations of the compounds of tellurium. Courtesy Longmans, Green & Co., Ltd., London (7).

metals to sluggish and requiring a high temperature with hydrogen. The alkali and alkaline-earth tellurides are colorless ionic solids rapidly decomposed by air, especially in the presence of atmospheric moisture. In solution, they hydrolyze less than sulfides and selenides but are very easily oxidized. The solutions dissolve excess tellurium, forming dark-red tellurides. The alkali–metal tellurides are strong reductants. Hydrotellurides, such as sodium hydrotelluride [*23624-18-2*], NaHTe, are also known. Some metals form more than one telluride, and some metal tellurides show nonstoichiometry. Many tellurides possess semiconductor properties.

Hydrogen Telluride. Hydrogen telluride [*7783-09-7*], H_2Te, is a colorless, toxic gas with an odor resembling arsine. The liquid and solid states are also colorless. Hydrogen telluride is prepared by the action of water on aluminum telluride in the absence of air, or by electrolysis of cold 15–20% sulfuric acid with a tellurium cathode. Hydrogen telluride and its aqueous solution decompose slowly in air. Unless perfectly dry, the compound also decomposes when exposed to light. Hydrogen telluride is a strong reductant and is rapidly oxidized. It reacts with alkalies and the solutions of many metal salts to form the corresponding tellurides. Hydrogen telluride solutions are weakly acidic and ionize to HTe^- and Te^{2-}.

Tellurium Sulfide. In the liquid state, tellurium is completely miscible with sulfur. The Te–S phase diagram shows a eutectic at 105–110°C when the sulfur content is 98–99 atom % (94–98 wt %). Tellurium–sulfur alloys have semiconductor properties (see Semiconductors). Bands attributed to tellurium sulfide [*16608-21-2*], TeS, molecules have been observed.

Tellurium Selenides. Tellurium selenides or selenium tellurides, are unknown. The molten elements are miscible in all proportions. The mixtures are not simple solid solutions but have a complex character. Like the sulfides, the selenides possess semiconductor properties.

Carbon Sulfotelluride. Carbon sulfotelluride [*10340-06-4*], CSTe, exists as a yellow-red liquid with a garlic odor. It is decomposed by light, even at −50°C, to carbon disulfide, carbon, and tellurium.

Carbonyl Telluride. Little is known about carbonyl telluride [*65312-92-7*], COTe. It is formed in poor yield by passing carbon monoxide over tellurium at a high temperature. It is less stable than the selenide.

Tellurium Nitride. Tellurium nitride [*12164-01-1*], Te_3N_4, is an unstable, citron-yellow solid that detonates easily when heated or struck, but it can be kept under dry chloroform. It is said to explode on contact with water, possibly because of the heat of wetting.

Tellurium Halides. Tellurium forms the dihalides $TeCl_2$ and $TeBr_2$, but not TeI_2. However, it forms tetrahalides with all four halogens. Tellurium decafluoride and hexafluoride can also be prepared. No monohalide, Te_2X_2, is believed to exist. Tellurium does not form well-defined oxyhalides as do sulfur and selenium. The tellurium halides show varying tendencies to form complexes and addition compounds with nitrogen compounds such as ammonia, pyridine, simple and substituted thioureas and anilines, and ethylenediamine, as well as with sulfur trioxide and the chlorides of other elements.

Tellurium Tetrafluoride. Tellurium tetrafluoride [*15192-26-4*], TeF_4, forms white, hygroscopic needles melting at 129.6°C. It decomposes at 194°C to TeF_6 and is readily hydrolyzed. Tellurium tetrafluoride attacks glass, silica, and copper at 200°C, but it does not attack platinum below 300°C.

Tellurium Decafluoride. Tellurium decafluoride [*53214-07-6*], Te_2F_{10}, is a stable, volatile, colorless liquid, melting at −33.7°C and boiling at 59°C.

Tellurium Hexafluoride. Tellurium hexafluoride, TeF_6, melts at −38°C and sublimes at −39°C, forming a colorless gas. It hydrolyzes slowly to orthotelluric acid. It is reduced by tellurium to TeF_4.

Tellurium Dichloride. Tellurium dichloride [*10025-71-5*], $TeCl_2$, is a black, hygroscopic solid, melting at 208°C to a black liquid; it boils at 328°C to a bright-red vapor. The solid is stable when pure. It disproportionates in organic solvents and is decomposed by acids and alkalies. It is hydrolyzed to H_2TeO_3, Te, and HCl, and decomposed by HCl to Te and H_2TeCl_6 [*17112-43-5*]. Air oxidizes tellurium dichloride to TeO_2 and HCl.

Tellurium Tetrachloride. Tellurium tetrachloride [*10026-07-0*], $TeCl_4$, forms white, hygroscopic crystals which melt at 225°C to a dark-red liquid. The vapor (bp ca 390°C) is also red. It is prepared by the action of chlorine on tellurium, or of SCl_2, CCl_4, or $AsCl_3$ on tellurium or TeO_2. The crystals are soluble in benzene, toluene, and ethyl alcohol, but not in ethyl ether. It hydrolyzes in cold water to TeO_2, and in hot water gives a clear solution. With dilute HCl, it yields H_2TeO_6 [*56367-33-0*], TeO_2, $HTeCl_5$ [*22742-12-7*], and H_2TeCl_6, and with concentrated HCl, $TeCl_6^{2-}$. On evaporation of the acid solution to dryness, $TeCl_4$ forms $H_2TeCl_6.2H_2O$ [*12192-31-3*], which decomposes at 350°C to TeO_2. With phenylmagnesium chloride, it yields $Te(C_6H_5)_3Cl$ [*1224-13-1*]. Tellurium tetrachloride forms addition compounds with various ligands.

Tellurium Dibromide. Tellurium dibromide [*7789-54-0*], $TeBr_2$, must be prepared very carefully, as it is unstable and has a strong tendency to disproportionate to $TeBr_4$ and Te.

Tellurium Tetrabromide. Tellurium tetrabromide [*10031-27-3*], $TeBr_4$, forms yellow hygroscopic crystals which decompose above 280°C and melt at 363°C under bromine vapor. It boils at 414–427°C, dissociating into $TeBr_2$ and bromine. It is soluble in ether and chloroform but not in CCl_4, and is readily hydrolyzed.

Tellurium Tetraiodide. Tellurium tetraiodide [*7790-48-9*], TeI_4, forms gray-black volatile crystals which, above 100°C, decompose into the elements. They melt at 280°C in a sealed tube.

Tellurium Oxychlorides. Tellurium oxychlorides, $Te_6O_{11}Cl_2$ [*12015-54-2*], and $TeOCl_2$ [*16981-34-2*], have been reported.

Tellurium Oxydibromide. Tellurium oxydibromide [*66461-30-1*] has been prepared.

Tellurium Oxides, Oxyacids, and Salts. The white crystals of tellurium dioxide, TiO_2, melt at 733°C to a clear, dark-red liquid which vaporizes at 790–940°C. Tellurium dioxide is made by dissolving tellurium in strong HNO_3 and decomposing the resulting $2TeO_2.HNO_3$ at 400–430°C; by heating Te in air or oxygen; or by thermal dehydration of H_6TeO_6. It is slightly soluble in water with formation of H_2TeO_3, and readily soluble in halogen acids, forming complex anions, such as $TeCl_6^{2-}$. It dissolves in solutions of alkali-metal carbonates only on boiling. The standard redox potential is intermediate between those of SO_2 and SeO_2. In solution, TeO_2 is a weaker oxidant than SeO_2 for SO_2. Tellurium dioxide is amphoteric, with minimum solubility at pH 3.8–4.2, its isoelectric point. It forms tellurites, $MTeO_3$, with alkali metals, but not with NH_3. With some acids it forms basic salts such as $TeO_2.2HCl$ [*83543-00-4*], $TeO_2.3HCl$, $TeO_2.2HBr$ [*83543-02-6*], $2TeO_2.HNO_3$, $2TeO_2.SO_3$ [*12065-84-8*], and

$2TeO_2.HClO_4$. Heating with metal oxides yields tellurites, but with higher oxides, such as PbO_2, or with oxidants such as KNO_3 or $KClO_3$, tellurates are formed. Heating with metals such as Al, Cs, and Zn, with C (at red heat), or with S, ZnS, HgS, or PbS, reduces TeO_2 to Te. Hydrogen does not reduce it completely even at high temperatures, and H_2O_2 oxidizes aqueous TeO_2 suspension to H_4TeO_6 [*41673-77-2*].

Tellurium Trioxide. Tellurium trioxide, TeO_3, exists in two modifications, the yellow-orange α-form and the grayish β-form. The α-form, sp gr 5.07, decomposes above 360°C to TeO_2. It is prepared by dehydrating H_6TeO_6 at 300–360°C. α-Tellurium trioxide is a strong oxidant; it reacts violently with metals such as Al or Sn, and with nonmetals such as C, P, and S. Although insoluble in water, dilute mineral acids, and dilute alkalies, it dissolves in hot concentrated alkalies with the formation of tellurates. α-Tellurium trioxide is reduced to TeO_2 and $TeCl_4$ by boiling with concentrated HCl (with the liberation of Cl_2). The β-form, a grayish solid, sp gr 6.21, is obtained by heating orthotelluric acid or α-TeO_3, alone or with H_2SO_4, for 10–12 h at 300–350°C. It is soluble in concentrated Na_2S solution. Less reactive than the α-form, it is reduced by hydrogen at 400°C. The β-form does not react with water at 150°C in the presence of acid or alkali. Most likely, it is a polymer, but there are no structural data for either modification.

Tellurous Acid. Tellurous acid [*10049-23-7*], H_2TeO_3, is an unstable white solid that dehydrates readily to TeO_2. It is prepared by the acidification of tellurite solution with HNO_3 or by cold hydrolysis of a tetrahalide. Tellurous acid is a much weaker acid than H_2SeO_3. Both normal and acid tellurites are known. Alkali–metal tellurites are obtained from alkali hydroxides and TeO_2; other tellurites are prepared by double decomposition with metal tellurites, or by fusing TeO_2 with metal oxides or carbonates. The alkali–metal tellurites are water soluble, the alkaline–earth tellurites are less so. Other tellurites are insoluble. Tellurites are oxidized to tellurates by heating in air or by treating a solution with H_2O_2 and other oxidants. In acid solution, the salts are reduced to Te by SO_2, Sn, Zn, Cu, and some organic compounds.

Orthotelluric Acid. The white crystals of orthotelluric acid, H_6TeO_6, are sparingly soluble in cold water and easily soluble in hot water and mineral acids, with the exception of HNO_3. Orthotelluric acid is made by oxidizing Te or TeO_2 (for example by refluxing with H_2O_2 in concentrated H_2SO_4). It exists in a cubic α-form and a monoclinic β-form, which is stable at room temperature. The β-form is obtained from the α-form by heating. Orthotelluric acid tends to polymerize. It is weakly dibasic, forming salts such as KH_5TeO_6 [*15855-68-2*], $Li_2H_4TeO_6$ [*20730-51-2*], and $Na_2H_4TeO_6$ [*20730-46-5*]. An exception is Ag_6TeO_6. Heating dehydrates H_6TeO_6 to TeO_3; further heating gives TeO_2 and oxygen. The acid is a fairly strong oxidant; it liberates iodine from KI and oxidizes HCl and HBr. It is reduced by H_2S, SO_2, Zn, Fe^{2+}, and hydrazine. It forms complex acids and salts with other elements.

Polymetatelluric Acid. Polymetatelluric acid $(H_2TeO_4)_n$ (where n = ca 10), is a white, amorphous, hygroscopic solid obtained by partial dehydration of orthotelluric acid at 100–200°C in air. Polymetatelluric acid always contains some of the orthoacid and reverts to it in solution. It forms esters and salts. Orthotelluric and polymetatelluric acids are the only two telluric acids known.

The so-called allotelluric acid [*13520-55-3*], a colorless syrup, is a mixture of orthotelluric and polymetatelluric acids. It is prepared by heating the ortho acid in a sealed tube at 305°C.

Tellurates. The water-soluble alkali-metal and alkaline-earth tellurates are prepared by chlorinating alkaline solutions of the tellurites or by heating solid tellurites with KNO_3, $KClO_3$, or PbO_2. The insoluble tellurates are made by double decomposition of alkali-metal salt solutions.

Other Inorganic Compounds. Alkali-metal and ammonium telluropentathionates $(Te(S_2O_3)_2)^{2-}$, have been prepared. The S_2O_3 group can be replaced by ethyl xanthate or diethyl dithiocyanate.

Tellurium pseudohalides, such as the dicyanide [*14453-24-8*], $Te(CN)_2$, the dithiocyanate [*83543-04-8*], $Te(SCN)_2$, and thiourea complexes with $Te(SCN)_2$, have been prepared. They are similar to the halides in properties.

Basic tellurium nitrate, $TeO_2.NO_3$, is made by dissolving Te in HNO_3. Thermal decomposition begins at 190°C and is complete at 300°C.

Basic tellurium sulfate [*12068-84-8*], $2TeO_2.SO_3$; selenate, $2TeO_2.SeO_3$; and tellurate, $2TeO_2.TeO_3$, are known. The sulfate is made by slow evaporation of a TeO_2 solution in H_2SO_4. It is stable up to 440–500°C and hydrolyzed slowly by cold, and rapidly by hot water.

Tellurium perchlorate, iodate, and methylthiosulfate, as well as Te(IV) salts of aliphatic and aromatic acids, have been prepared.

Organic Compounds

The chemical properties of organosulfur, organoselenium, and organotellurium compounds are markedly similar. Because bond stability with carbon decreases with the increasing atomic number of the element, thermal stability decreases whereas oxidation susceptibility increases to such an extent that alkyl tellurides are oxidized rapidly by air at room temperature. As a result, less has been written concerning the chemistry of organotellurium than organoselenium compounds. Nevertheless, a sizable literature exists (29,39–42).

Organotellurium compounds range from the simple carbon sulfotelluride to complex heterocyclic compounds and organotellurium ligands (43). Various types of tellurium compounds and specific examples are listed in Table 6.

Not included here are the many organic complexes of inorganic tellurium compounds and their ions.

For detailed descriptions of the laboratory preparation of organotellurium compounds, see refs. 40, 44–46.

Uses

Free-Machining Steels. Tellurium has been shown to be the best additive for improving machinability in several types of ferritic steel (47–49). Its effectiveness in this area has caused a considerable increase in tellurium consumption. The Inland Steel Co., East Chicago, Indiana, developed what is currently the largest single outlet for tellurium as an additive to plain-carbon, leaded, and leaded resulfurized steels (50). The presence of 0.023–0.057% tellurium in cold-drawn steels, containing 0.06–0.09% carbon, 0.28–0.37% sulfur, and 0.15–0.22% lead, increases their machinability by 30–50% and increases the feed rates. The machined surface finish is superior, the life of the cutting tool is lengthened, the metal chips are more uniform, and there is less buildup on the tool edges. Tellurium addition affects the mechanical properties very

Table 6. Isolated Organic Tellurium Compounds[a]

Compounds	CAS Registry No.	Formula	Mp, °C	Bp, °C
tellurols or tellanes		RTeH		>90
ethanetellurol	[*83270-38-6*]	C_2H_5TeH		
benzenetellurol	[*69577-06-6*]	C_6H_5TeH		
alkyl, aryl, and cyclic tellurides				
dimethyl telluride (methyl telluride)	[*593-80-6*]	$(CH_3)_2Te$		82
diphenyl telluride	[*1202-36-4*]	$(C_6H_5)_2Te$	53.4	182
tetraphenyl telluride	[*64109-07-5*]	$(C_6H_5)_4Te$	104–106	
tetrahydrotellurophene	[*3465-99-4*]	$TeCH_2(CH_2)_2CH_2$		
ditellurides		R_2Te		
diphenyl ditelluride	[*32294-60-3*]	$C_6H_5Te{-}TeC_6H_5$	53–54	
alkyltellurium trihalides		$RTeX_3$		
methyltellurium tribromide	[*20350-53-2*]	CH_3TeBr_3	dec 140 C	
dialkyltellurium dihalides		R_2TeX_2		
dimethyltellurium dichloride	[*24383-90-3*]	$(CH_3)_2TeCl_2$	92(α), 134(β)	
tellurium salts		R_3TeX		
trimethyltellurium iodide	[*18987-26-3*]	$(CH_3)_3TeI$		
telluroxides		RTeO		
diphenyl telluroxide	[*51786-98-2*]	$(C_6H_5)_2TeO$	185	
tellurones		R_2TeO_2		
dimethyl tellurone	[*83270-39-7*]	$(CH_3)_2TeO_2$		
telluroketones		R_2CTe		
dimethyl telluroketone	[*83270-40-0*]	CH_3CTeCH_3	63–66	
tellurinic acid		RTeO.OH		
phenyltellurinic acid	[*83270-41-1*]	$C_6H_5Te.OH$	211	
heterocyclic compounds				
1,4-oxatellurane	[*5974-87-8*]	$OCH_2CH_2TeCH_2CH_2$		
3,5-telluranedione	[*24572-07-4*]	$CH_2COCH_2TeCH_2CO$		

[a] Ref. 39.

little, raises the recrystallization temperature only slightly (without affecting the heat-treating properties), and has a minor effect on grain refining. In addition, it is less detrimental than sulfur in lowering the corrosion resistance. Tellurium can also be added to low and high alloy steels that are normally difficult to machine (51). Tellurium also improves the machinability of powder-metallurgy steels (52).

Chilled Castings. Since about 1940, tellurium has been used to control the chill depth of iron castings (see Iron). Hard, wear-resistant, chilled surfaces were originally produced on gray iron castings by pouring molten iron against a metal surface. A thin layer of white iron formed on rapid cooling. By the addition of less than 0.1% Te to the melt, chilled surfaces could be obtained, with higher yields and extended mold life.

Tellurium is 100–150 times as effective as chromium in producing a given depth of chill. It is a powerful carbide stabilizer and minute quantities convert gray iron to white. Since tellurium is highly volatile, it is added to the ladle just before pouring. It may be added as pellets, tablets, powder (in aluminum or copper containers), ferrotellurium or copper telluride [*12181-15-6*] granules, or any other convenient form. Alternatively, the mold is treated by coating the desired areas of the mold surface with a tellurium-bearing wash, surfacing the sand mold with a layer of tellurium-rich sand, or using tellurium-containing tape (53).

The result is a hard, abrasion-resistant surface, important in many applications of cast iron. The depth of the chill may be controlled by regulating the amount of tellurium added. The casting shows a sharp demarcation line between the chilled and unchilled regions; there is no intermediate or mottled zone. Yet, the chilled portion shows excellent resistance to spalling from thermal or mechanical shock. Tellurium-treated iron is more resistant to sulfuric and hydrochloric acids than is untreated, unchilled gray iron. The amount added ranges from 0.005 to 0.1%; ca 60% is lost by volatilization. Excessive addition causes porosity in the castings.

Tellurium-chilled iron has been used in mining, automotive, railroad, and other equipment. Tellurium as a coating of molds and cores, so-called corewashes, eliminates troublesome localized shrinkage in castings. These corewashes may contain ca 25% tellurium; the remainder consists mainly of silica and some bentonite.

Other Ferrous Metal Uses. Tellurium shows an extremely high surface activity in liquid metals (48,54). As a result, addition of tellurium sharply decreases the rate of nitrogen absorption in liquid iron and steel (55). This allows the use of cheaper, lower-purity oxygen in pneumatic or basic oxygen steelmaking, and decreases the nitrogen pickup during supplementary scrap melting.

In 0.1% and smaller additions, tellurium acts as a mild deoxidizer of liquid steel, improves the soundness of castings, refines the grain size, and counteracts the detrimental effect of sulfur on the ductility (56). Tellurium also improves the transverse properties of wrought structural steels (by globulizing the sulfides), electrical steels (by aiding the development of magnetic anisotropy), malleable cast irons (57), and spheroidal (graphitic) cast irons (see also Metal surface treatments).

Copper Alloys. A 99.5% Cu–0.5% Te alloy has been on the market for many years (58). The most widely used is alloy No. CA145 (number given by Copper Development Association, New York), nominally containing 0.5% tellurium and 0.008% phosphorus. The electrical conductivity of this alloy, in that annealed state, is 90–98%, and the thermal conductivity 91.5–94.5% that of the tough-pitch grade of copper. The machinability rating 80–90, compares with 100 for free-cutting brass and 20 for pure copper.

Unlike lead and other additives, tellurium has no adverse effect on the hot-working properties of copper, and does not cause segregation and firecracking. Although the alloy is somewhat less ductile at room temperature than pure copper, it may be extensively hot- and cold-worked; intermediate annealing is required when the cross-section reduction is >40%. Machining methods are the same as for free-cutting brass. Because the copper telluride particles are harder than the lead particles in free-machining brass, carbide-tipped cutting tools are recommended.

A copper alloy, containing 0.02–0.04% tellurium, 0.002–0.015% phosphorus, and 0.002–0.05% oxygen is recommended for use in automobile radiators (59). A continuously cast Amtel copper alloy, containing tellurium 0.4–0.6%, phosphorus 0.007–0.012%, sulfur 0.02% max, and a copper–silver–tellurium–phosphorus alloy (99.90% max) was developed by AMAX Base Metals R&D, Inc. (60).

Tellurium–copper alloys are recommended for situations demanding a high production rate with no significant sacrifice in conductivity. These alloys can be soldered, brazed, or welded without incurring embrittlement. They are used in vacuum applications, forgings, screw-machine parts, welding-torch tips, transistor bases, semiconductor heat sinks, electrical connectors (qv), motor and switch parts, and nuts, bolts, and studs. Addition of tellurium significantly improves the surface of machined parts.

Copper–lead–tellurium alloys have high wear resistance in sliding contacts. In copper–zinc alloys, the benefits of tellurium decrease with increasing zinc content and almost disappear when the zinc content exceeds 35%.

Telnic bronze (CDA No. 191), an alloy that hardens with age, contains 98.3% Cu, 1.1% Ni, 0.5% Te, and 0.2% P. It was developed by Chase Division, Kennecott Copper Corp., about 1950. The machinability rating is 80. The hot-working properties are comparable to those of a high copper commercial bronze and are superior to lead-containing free-machining copper alloys. The tensile strength is high, and the electrical and thermal conductivities are similar to those of nickel bronze CDA No. 190, and ca one half those of copper (see Copper alloys).

Lead Alloys. A tellurium–lead alloy containing 0.02–0.1% tellurium, with or without antimony, was introduced in 1934 (61) as tellurium lead or Teledium. The standard alloy contains about 0.05% tellurium and 0.06% copper. It shows very uniform elongation and retains fine structure to its breaking point. This is attributed to the film of lead telluride [*1314-91-6*] which is insoluble in the lead matrix and prevents the growth of lead grains over a wide temperature range. The properties that distinguish the alloy from common lead are higher tensile, fatigue, and creep strengths, higher recrystallization temperatures and corrosion resistance, and a significantly longer time to soften at 25°C after cold work.

The addition of tellurium has a deoxidizing action and confers on lead useful work- and precipitation-hardening properties, and improves resistance to wear, vibration, and mechanical breakdown. These properties along with improved corrosion resistance are utilized in the sheathing of power communication and marine cables, in chemical equipment (especially that exposed to sulfuric acid), or where resistance to fatigue is important. Tellurium successfully replaces tin (1–3% Sn) in the sheathing. Thinner sheaths can be extruded, and the extrusion pressure for 0.05% Te lead is about the same as for a 3% Sn alloy.

Adding tellurium to lead and to lead alloyed with silver and arsenic improves the creep strength and the charging capacity of storage battery electrodes (see Batteries). These alloys also have been suggested for use as insoluble anodes in electrowinning.

Other Metals. Tellurium has been added to copper-base, lead-base, and tin-base bearing alloys. In babbit-type alloys, tellurium controls the structure and improves uniformity and fatigue resistance by restraining the tendency to segregation (see Bearing materials).

Adding 0.05–1% tellurium to tin and tin-base bearing alloys improves the work-hardening properties, tensile strength, and creep resistance.

Tellurium has been recommended as an additive to magnesium to increase corrosion resistance (see Corrosion and corrosion inhibitors). The addition is highly exothermic but can be controlled by adding one tellurium tablet at a time to a sufficiently large bath of liquid magnesium. The addition to tellurium and chromium improves the stress-corrosion resistance of aluminum–magnesium alloys.

In permanent-magnet alloys of the high cobalt–titanium type, tellurium significantly increases the coercive force.

Metal Coatings. Tellurium chlorides, as well as tellurium dioxide in hydrochloric acid solution, imparts a permanent and attractive black antique finish to silverware, aluminum, and brass. Anodized aluminum is colored dark gold by tellurium electrodeposition. A solution containing sodium tellurate and copper ions forms a black or blue-black coating on ferrous and nonferrous metals and alloys. Addition of sodium tellurite improves the corrosion resistance of electroplated nickel. Tellurium diethyldithiocarbamate is an additive in bright copper electroplating (see Electroplating).

Pigments and Glass. Tellurium has served as base for ultramarine-type, cadmium sulfotelluride (62), and cadmium telluride pigments (63) (see Pigments). In addition, small amounts of tellurium have been used in glass and ceramics to produce blue to brown colors.

Catalysts. Tellurium, tellurium oxides, tellurium tetrachloride, and tellurium tetrabromide are necessary or optional components of catalysts used in petroleum refining and the organic chemical industry. These catalysts promote such diverse reactions as oxidation, ammoxidation, hydrogenation, dehydrogenation, halogenation, dehalogenation, and phenol condensation (64,68). Tellurium passivates nickel, iron, and vanadium in cracking catalysts (see Catalysis).

Lubricants. Tellurides of titanium, zirconium, molybdenum, tungsten, and other refractory metals are heat- and vacuum-stable. This property makes them useful in solid self-lubricating composites in electronics, instrumentation, and aerospace fields (see Lubrication and lubricants).

Organic tellurides are antioxidants in lubricating oils and greases.

Rubber. At one time, the rubber industry was the largest single consumer of tellurium, but has since been replaced by ferrous metal uses. Powdered tellurium is used as a secondary vulcanizing agent in natural rubber and in styrene–butadiene rubbers. In many cases, addition of 0.5% Te increases the rate of vulcanization and improves the aging and mechanical properties of sulfurless and low sulfur stocks. Tellurium is particularly effective when used with tetramethylthiuram disulfide, (bis(dimethylthiocarbamoyl) disulfide), $((CH_3)_2NCSS)_2$, and with selenium diethyldithiocarbamate in sulfurless cures. It is used frequently to eliminate porosity in thick-molded sections. Tellurium dioxide improves the heat stability and aromatic fuel resistance in cured polysulfide elastomers, such as Thiokol. Tellurium rubber is extremely resistant to heat and abrasion. It is used in all-rubber-jacketed portable cables in mining, dredging, welding, etc, and in conveyor belts for special application (see Rubber chemicals).

Tellurium diethyldithiocarbamate, in combination with mercaptobenzothiazole, with or without tetramethylthiuram disulfide, is the fastest known accelerator for butyl rubber. It is used extensively in butyl tubes for buses and similar vehicles and in other butyl applications (see Elastomers, synthetic; Rubber, natural).

Tellurium is also useful for the coupling of rubber to metals.

Explosives. The use of tellurium in delay electric-blasting caps was phased out several years ago. Sodium tellurite is claimed as a jelling promoter in explosive compositions that can be readily poured or pumped into drillholes (65) (see Explosives).

Medical and Biological. A very small, yet very important application of tellurium is in organic derivatives and radioactive isotopes for use as biological tracers, x-ray-contrast agents, and diagnostic aids, and for the treatment of thyroid diseases (see Radioactive drugs; Radioactive tracers). Some inorganic and organic tellurium compounds are effective bactericides, fungicides, and parasiticides.

Thermoelectric Materials. Tellurium is a component of most thermoelectric materials designed for operation at normal temperatures. Alloys based on bismuth telluride and lead telluride have the highest thermoelectric figures of merit, which reflect the efficiency of the reversible heat–electricity conversion (69). Thermoelectric materials are useful in various cooling or heating devices that have no moving parts and are operated with direct current. Alternatively, thermoelectric devices can convert heat to electricity. The high expectations of 2–3 decades ago for a large outlet of tellurium in this field have not materialized, in part because of relatively low efficiency. Nevertheless, there are many special applications, such as small cooling boxes, water coolers, limited remote-power generators, solar-energy converters, and small-scale control apparatus (see Thermoelectric energy conversion).

Other useful thermoelectric materials are copper telluride, silver telluride [*12249-68-2*], and manganese telluride [*12032-88-1*].

Semiconductors. Tellurium rectifiers similar to selenium rectifiers have been developed, but their manufacture and use are not known. Small amounts of tellurium are used in semiconductor lasers and as dopants in semiconductor alloys.

Tellurium dioxide and tellurium are constituents of a number of glasses, compounds, and alloys used in infrared optics, photodiodes, photodetectors (qv) (including infrared and nuclear radiation detectors), piezoelectric devices, photovoltaic cells, and semiconductor films. Tellurium is used in amorphous magnetic alloys and amorphous memory devices. Solar cells based on cadmium telluride [*1306-25-8*], cadmium copper telluride, and cadmium sulfotelluride have been developed (see also Solar energy).

Tellurium is an important dopant, alloying constituent, or film in electrophotographic materials and photoreceptors.

A zinc cadmium telluride–indium telluride combination is useful in solid-state image-conversion devices. Tellurium compounds are used in light emitting devices and phosphors in television tubes (see Light-emitting diodes; Luminescent materials).

Optical fibers have been developed using a cadmium telluride core or tellurium (see Fiber optics).

Photography and Thermography. The use of organotellurium compounds and silver-free tellurium(II) metal complexes and compounds in photoemulsions, photosensitization, thermography, and similar applications has been patented (66–67) (see also Photography). Organotellurium compounds are useful used in thermal recording materials, imaging materials, and information recording.

BIBLIOGRAPHY

"Tellurium and Tellurium Compounds" in *ECT* 1st ed., Vol. 13, pp. 666–676, by E. M. Elkin, Canadian Copper Refiners Ltd., and J. L. Margrave, University of Wisconsin; "Tellurium and Tellurium Compounds" in *ECT* 2nd ed., Vol. 19, pp. 756–774, by E. M. Elkin, Canadian Copper Refiners, Ltd.

1. M. A. Weeks and H. M. Leicester, *Discovery of the Elements*, 7th ed., *J. Chem. Educ.*, Easton, Pa., 1968.
2. B. Mason, *Principles of Geochemistry*, 3rd ed., John Wiley & Sons, Inc., New York, 1971.
3. N. D. Sindeeva, *Mineralogy and Types of Deposits of Selenium and Tellurium*, Interscience Publishers, a division of John Wiley & Sons, Inc., New York, 1964.
4. A. M. Lansche, *Selenium and Tellurium—a Materials Survey*, U.S. Bureau of Mines Information Circular 8340, U.S. Government Printing Office, Washington, D.C., 1967.

5. W. A. Dutton in W. C. Cooper, ed., *Tellurium*, Van Nostrand Reinhold Co., New York, 1971, pp. 100–183.
6. J. Bergman and L. Engman, *J. Org. Chem.* **47,** 5191 (1982).
7. J. H. Schloen and E. M. Elkin, *Trans. AIME* **188,** 764 (1950); A. Butts, ed., *Treatment of Electrolytic Copper Refinery Slimes*, Reinhold Publishing Corp., New York, 1954, Chapt. 11.
8. T. N. Greiver, I. G. Zaitseva, and V. M. Kosover, *Selen i Tellur (Selenium and Tellurium) Metallurgiya*, Moscow, 1977.
9. L. A. Soshnikova and M. M. Kupchenko, *Pererabotka Mednelektrolitnykh Shlamov (Treatment of Copper Electrolysis Slimes) Metallurgiya*, Moscow, 1978.
10. F. Benard, *Trans. AIME* **106,** 369 (1933); **159,** 521 (1944).
11. A. I. Gaev, T. S. Dolzhenko, and B. V. Trutnev, *Tsvetn. Metall.* **11**(10), 64 (1938).
12. R. H. Waddington, *Can. Metals Met. Ind.* **3,** 153, 190, 199 (1940).
13. M. Heberlein, *Metall. u. Erz* **30,** 363 (1933).
14. J. H. Schloen and E. M. Elkin, *Trans. Can. Inst. Min. Metall.* **49,** 143 (1946).
15. U.S. Pat. 2,349,697 (May 23, 1944), N. R. Brierly (to The American Metal Co.).
16. U.S. Pat. 2,258,963 (Oct. 14, 1941), W. G. Woll and R. T. Gore (to International Smelting & Refining Co.).
17. *Proceedings of the Symposium on Industrial Uses of Selenium and Tellurium*, Toronto, Canada, Oct. 1980, Selenium–Tellurium Development Association, Darien, Conn.
18. R. Bresee, D. Vleeschhouwer, and J. Thiriar in ref. 17, pp. 31–49.
19. Rcf. 8, pp. 36–45.
20. *Minerals Yearbook*, U.S. Dept. of the Interior, Bureau of Mines, Government Printing Office, Washington, D.C.
21. *American Bureau of Metal Statistics*, New York, annual issues.
22. *Canadian Minerals Yearbook*, Canada Department of Energy, Mines, and Resources, Ottawa, Canada.
23. W. C. Cooper in N. H. Furman, ed., *Standard Methods of Chemical Analysis*, 7th ed., Vol. 1, D. Van Nostrand Co., Inc., Princeton, N.J., 1962, pp. 925–949; and in ref. 5, pp. 281–312.
24. T. W. Green and M. Turley in I. M. Kolthoff and P. J. Elving, eds., *Treatise on Analytical Chemistry*, Vol. 7, Part 2, Interscience Publishers, a division of John Wiley & Sons, Inc., New York, 1961, pp. 137–205.
25. D. I. Ryabchikov and I. I. Nazarenko in I. P. Alimarin, ed., *Analysis of High Purity Materials*, Israel Program for Scientific Information, Jerusalem, 1968.
26. C. Duval and U. M. Doan, *Anal. Chim. Acta* **5,** 569 (1951).
27. P. W. Bennett and S. Barabas, *Anal. Chem.* **35,** 139 (1963).
28. K. L. Cheng, *Anal. Chem.* **33,** 761 (1961).
29. K. J. Irgolic, *The Organic Chemistry of Tellurium*, Gordon and Breach Science Publishers, Inc., New York, 1973.
30. G. D. Clayton and F. E. Clayton, eds., *Patty's Industrial Hygiene and Toxicology*, 3rd ed., Vol. 2, John Wiley & Sons, Inc., New York, 1981.
31. N. J. Sax, *Dangerous Properties of Industrial Materials*, 5th ed., Van Nostrand Reinhold Publishing Corp., New York, 1979.
32. T. Devlin, *A Review of the Toxicology of Tellurium and its Compounds*, Sandia Laboratories, Albuquerque, N.M., and Livermore, Calif., 1975.
33. V. B. Vouk in *Toxicology of Metals*, Vol. 2, Environmental Health Effects Research Seris PB 268 324, U.S. Environmental Protection Agency, Research Triangle, N.C., reproduced by National Technical Information Service, May 1977, pp. 370–383.
34. J. R. Glover and V. Vouk in L. Friberg and co-workers, eds., *Handbook on the Toxicity of Metals*, Elsevier/North Holland Biomedical Press, Amsterdam, 1979, Chapt. 35, pp. 587–598.
35. *Inorganic Syntheses*, McGraw-Hill Book Co., New York, published at irregular intervals.
36. G. Brauer, *Handbook of Preparative Inorganic Chemistry*, Vol. 1, Academic Press, New York, 1963.
37. P. J. Durrant and B. Durrant, *Introduction to Advanced Inorganic Chemistry*, Longmans, Green & Co., London, 1962, p. 881.
38. D. M. Chizhikov and V. P. Shchastlivyi, *Tellurium and the Tellurides*, Collets Publishers, Ltd., London and Wellingborough, 1970.
39. K. W. Bagnall, *The Chemistry of Selenium, Tellurium, and Polonium*, Elsevier Publishing Co., Am-

sterdam, 1966; *Comprehensive Inorganic Chemistry*, Vol. 2, Pergamon Press, Oxford, 1973, Chapts. 23–24.
40. J. Newton Friend, *A Text-Book of Inorganic Chemistry*, Vol. 11, Part 4, Charles Griffin & Co., London, 1937, pp. 166–260.
41. E. Krause and A. von Grosse, *Die Chemie der Metallorganischen Verbindungen*, Borntraeger, Berlin, 1937 (through ref. 9).
42. R. A. Zingaro and K. Irgolic in ref. 10, pp. 184–280.
43. H. J. Gysling, *The Synthesis of Organotellurium Ligands*, Kodak Laboratory Chemicals Bulletin, Vol. 53, No. 1, Eastman Kodak Co., Rochester, N.Y., 1982.
44. *Organic Syntheses*, published annually since 1921, John Wiley & Sons, Inc., New York.
45. J. Houben, ed., *Die Methoden der Organischen Chemie (Weyls Methoden)*, Georg Thieme Verlag, Stuttgart, 1925–1941 (reproduced by Edwards Bros., Inc., Ann Arbor, Mich.).
46. V. Migrdichian, *Organic Syntheses*, Vols. 1 and 2, Reinhold Publishing Corp., New York, 1957.
47. E. S. Nachtman in ref. 5, pp. 373–388.
48. R. J. Raudebaugh in ref. 17, pp. 201–221.
49. D. Bhattacharya in ref. 17, pp. 222–243.
50. U.S. Pats. 3,152,889 and 3,152,890 (Oct. 13, 1964), M. O. Holowaty (to Inland Steel Co.); U.S. Pat. 3,169,857 (Feb. 16, 1965), A. E. Rathke and A. T. Morgan (to Inland Steel Co.).
51. U.S. Pats. 2,009,713–2,009,716 (July 30, 1935), F. R. Palmer (to Carpenter Steel Corp.).
52. P. W. Taubenblat and W. E. Smith in ref. 17, pp. 254–266.
53. U.S. Pat. 2,979,793 (April 18, 1961), R. L. Wilson, F. B. Herlihy, and T. J. Wood (to American Brake Shoe Co.).
54. R. H. Aborn in ref. 5, pp. 389–409.
55. U.S. Pat. 3,134,668 (May 26, 1964), R. D. Pehlke and M. Weinstein.
56. U.S. Pat. 2,258,604 (Oct. 14, 1941), A. P. Gagnebin (to General Motors Corp.).
57. U.S. Pats. 2,250,488 and 2,250,489 (July 29, 1941), C. H. Lorig and D. E. Krause (to Battelle Memorial Institute); U.S. Pats. 2,253,502 (Aug. 26, 1941) and 2,331,886 (Oct. 19, 1943), A. L. Boegehold (to General Motors Corp.); Brit. Pat. 552,390 (April 3, 1943), E. Morgan and E. Hinchcliffe (to British Cast Iron Research Association).
58. U.S. Pat. 2,027,087 (Jan. 13, 1936), H. L. Burghoff and D. E. Dawson (to Chase Co., Inc.).
59. Ref. 17, p. 217.
60. P. W. Taubenblat, G. C. Van Tilburg, and F. Gruss in ref. 17, pp. 267–276.
61. W. Singleton and B. Jones, *J. Inst. Metals* **51**, 71 (1933); Brit. Pat. 441,524 (June 8, 1934), W. Singleton, W. Hulme, and B. Jones (to Goodlass Wall and Lead Industries, Ltd.).
62. U.S. Pat. 3,012,899 (Dec. 12, 1961), A. Giordano (to Harshaw Chemical Co.).
63. Brit. Pat. Appl. 2,033,418 (May 21, 1980), B. J. Chase and co-workers (to Johnson, Matthey & Co.).
64. V. Kollonitsch and C. H. Kline, *Hydrocarbon Process. Petrol. Refiners* **43**(6), 139 (1964).
65. U.S. Pat. 4,032,376 (June 28, 1977), K. R. Fossan and G. O. Ekman (to Nitro-Nobel A.B.).
66. U.S. Pat. 4,148,659 (April 10, 1979), J. L. Vuyts and co-workers (to AGFA-Gevaert).
67. U.S. Pats. 4,154,503 (May 10, 1979), 4,155,760, and 4,155,761 (May 22, 1979), M. Lelenthal and J. H. Gysling (to Eastman Kodak Co.); U.S. Pat. 4,239,846 (Dec. 16, 1980), J. H. Gysling and S. A. Gardner (to Eastman Kodak Co.); Eur. Pat. Appl. 11,754 (June 11, 1980), U.S. Pats. 4,188,218 (Feb. 12, 1980) and 4,258,128 (March 24, 1981), J. H. Gysling (to Eastman Kodak Co.).
68. W. C. Cooper in ref. 5, pp. 414–416; A. B. Stiles in ref. 17, pp. 286–290.
69. C. H. Champness in ref. 5, pp. 322–372.

General References

J. Newton Friend, *A Text-Book of Inorganic Chemistry*, Vol. 7, Part 2, Charles Griffin & Co., London, 1931.

J. W. Mellor, *A Comprehensive Treatise on Inorganic and Theoretical Chemistry*, Vol. 11, Longmans, Green & Co., Inc., London, 1931.

Gmelins Handbuch der Anorganischen Chemie, 8th ed., System-Nummer 11, Verlag Chemie, Berlin, 1940.

N. V. Sidgwick, *The Chemical Elements and Their Compounds*, Vol. 2, Oxford University Press, London, 1950.

P. Pascal, *Nouveau Traité de Chimie Minérale*, 10th ed., Vol. 13, Masson & Cie., Editeurs, Paris, 1960.

A. F. Wells, *Structural Inorganic Chemistry*, 3rd ed., Oxford University Press, London, 1962.

A. A. Kudryavtsev, *The Chemistry and Technology of Selenium and Tellurium*, Collets Publishers, Ltd., London and Wellingborough, 1974.

E. M. ELKIN
Noranda Mines, Ltd.

TEMPERATURE MEASUREMENT

Sources for devices mentioned in this article are listed in Table 1.

Thermodynamic Kelvin Temperature Scale

Temperature is a measure of the hotness of something. For a measure to be rational, there must be agreement on a scale of numerical values for hotness and upon devices for realizing and displaying these values. The only temperature scale with an absolute basis in nature is the thermodynamic kelvin temperature scale (TKTS), which is based on functions that can be deduced directly from the first and second laws of thermodynamics (qv). Since the lower limit of the TKTS is absolute zero and since it is linear by definition, only one nonzero reference point is required to establish its slope. The reference point of 0°C was chosen because it is the equilibrium temperature of water in its liquid and solid phases at atmospheric pressure. In 1954, the present TKTS was established by changing the reference point to the equilibrium temperature of liquid and solid water under its own vapor pressure, ie, the triple point of water. This point was assigned the value 273.16 kelvins, which is 0.01° in Celsius. The unit of temperature on the TKTS is the kelvin, abbreviated K. Since the temperature interval 1 K is equal to the temperature interval 1°C, K may also be used to indicate an interval, but not a temperature, of the Celsius scale and is so used in this article. The symbol for the TKTS temperature is T; for a temperature on any other scale it is t.

Accurate measurements on the TKTS present substantial experimental difficulties. Most easily used thermometers are not based on functions on the first and second laws of thermodynamics. Thus, most practical thermometers depend upon some function that is a repeatable and single-valued analogue of temperature, and they are used as interpolation devices of practical and utilitarian temperature scales which are themselves artifacts. The main purpose for the study of the TKTS is to establish relationships between the thermodynamic scale of nature and the practical temperature scales of the laboratory, so that measurements made on nonthermodynamic scales can be translated into TKTS terms.

The vapor pressure of an ideal gas is a thermodynamic function, and the instrument generally used for investigations of the TKTS is the helium-gas thermometer. The transfer function of this thermometer may be chosen to be the change in pressure with a change in temperature at constant volume or the change in volume with a change in temperature at constant pressure. Because it is easier to measure pressure accurately

Table 1. Temperature-Measurement Devices

Device	Manufacturer	Price information
cells for producing the triple point of water	Jarrett Instrument Co., Wheaton, Md.	ca $500
cells, furnaces, and associated equipment for metal freezing-point calibration	Yellow Springs Instrument Co., Inc., Yellow Springs, Ohio	$2,000–5,000
standard platinum resistance thermometers	Yellow Springs Instrument Co., Inc., Yellow Springs, Ohio	$3,000–5,000, depending upon calibration required
	Rosemount Engineering Co., Minneapolis, Minn.	
	Heraeus GmbH, Hanau, FRG	
industrial-grade resistance thermometers	Yellow Springs Instrument Co., Inc., Yellow Springs, Ohio	$50 to several hundred dollars, depending upon configuration
	Rosemount Engineering Co., Minneapolis, Minn.	
	Minco Corporation, Minneapolis, Minn.	
	Leeds & Northrup Co., North Wales, Pa.	
precision thermistors		bare thermistor, ca $20; price increases depending upon housing, etc
disk and super-stable thermistors	Yellow Springs Instrument Co., Inc., Yellow Springs, Ohio	
bead thermistors	Thermometrics, Inc., Edison, N.J.	
standard-grade thermocouples	Leeds & Northrup Co., North Wales, Pa.	≥$1,000, depending upon calibration
industrial-grade thermocouples	Leeds & Northrup Co., North Wales, Pa.	a few dollars and up, depending upon housing
	Omega Engineering Co., Norwalk, Conn.	
optical pyrometers	Leeds & Northrup Co., North Wales, Pa.	≥$2,000, depending upon features
quartz thermometers	Hewlett-Packard, Mountain View, Calif.	several thousand dollars and up
fiber-optic thermometers	Luxtron, Mountain View, Calif.	$6,500
liquid-in-glass thermometers	many suppliers and distributors	a few dollars to several hundred dollars
resistance bridges, current comparators, potentiometers	Leeds & Northrup Co., North Wales, Pa.	$5,000–15,000
	Guildline Instruments, Elmsford, N.Y., and Smith Falls, Ont.	$18,000
automatic bridges	Automatic Systems, Ltd., Leighton Buzzard, UK	$15,000–25,000
	Neil Brown Systems, Cataumet, Mass.	$25,000
Novel temperature instruments		
quartz thermometer	Hewlett-Packard, Mountain View, Calif.	na
nmr thermometer	Yokagawa Electric Co., Tokyo, Jpn.	na
fiber-optic thermometer	Luxtron, Mountain View, Calif.	$6,500

than it is to measure volume, constant-volume gas thermometers are more common than constant-pressure gas thermometers.

A rudimentary gas thermometer is shown in Figure 1. Its operation will be illustrated by using it to show that the temperature of the ice point is 273.15 K. The figure shows a spherical bulb of constant volume, connected by constant-volume tubing to a U-tube manometer. A second connection to the U-tube manometer leads to a reservoir of mercury, which contains a plunger P, which varies the height of the mercury in the manometer. The bulb and rigid tubing contain an ideal gas. The bulb is first surrounded by an equilibrium mixture of ice and water (C, in Fig. 1(**a**)). When the gas A is in thermal equilibrium with the water and ice in the bath, the pressure of the gas is adjusted by actuating the plunger so that both columns of mercury in the U-tube manometer are at the same height, corresponding to the index mark 1.

Next, the ice bath is removed and replaced with a bath containing an equilibrium mixture of liquid and vapor water, ie, boiling water. For this mixture to be at 100°C, it is necessary that the atmospheric pressure on the surface of the water be 101,325 Pa or 1 standard atmosphere. As the gas is heated to the temperature of the boiling water it expands, and the mercury in both legs of the manometer is displaced. The mercury in the left leg is readjusted to the index mark 1 by changing the position of the plunger in the reservoir, thereby restoring the requirement of constant volume. This condition is shown in Figure 1(**b**). However, the mercury in the right leg is not now at the index mark 1. The difference between the heights of the mercury legs is a measure of the new pressure of the gas, which is 1.366099 times the initial or 0°C pressure. From the expression (100°C − 0°C)/(1.366099 − 1) = 273.15, the kelvin or the degree Celsius can be defined as 1/273.15 of the pressure ratio change between 0°C and 100°C. Thus, if the temperature is reduced below 0°C by 273.15 kelvin or Celsius

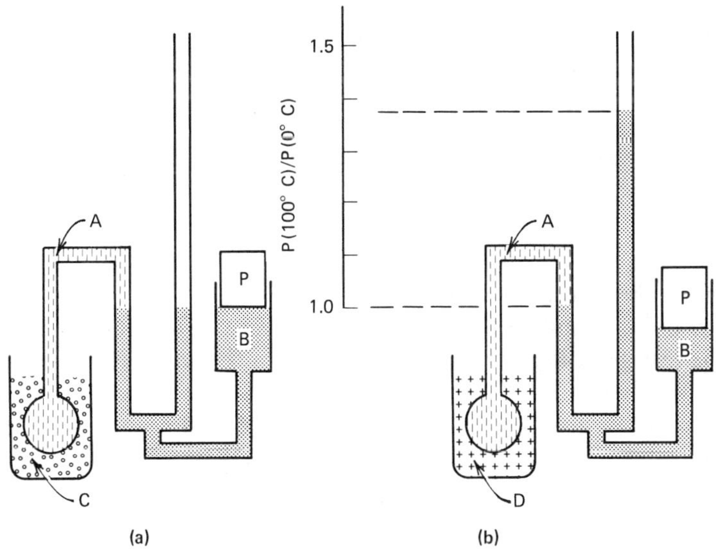

Figure 1. Gas thermometer. A, helium gas; B, mercury, C, water and ice (0°C); D, water and steam (100°C); P, plunger for adjusting height of mercury column.

units, an absolute zero that is the zero on the TKTS is reached. The zero of the Celsius scale is therefore 273.15 K and

$$T = t + 273.15 \tag{1}$$

The temperature interval and the zero of the TKTS is defined without reference to the properties of any specific substance. Real measurements with real gas thermometers are much more difficult than the example suggests.

Practical Temperature Scales

Many properties of materials change with temperature, eg, the expansion and contraction of solids and liquids, the electrical properties of conductors and semiconductors, and the color and brilliance of light emitted from a very hot source. Any of these properties can be used to make a thermometer; all nonthermodynamic ones require the construction of a concensus scale to relate the properties of a prescribed interpolation device to temperature.

Although only one fixed point is necessary to define the TKTS, practical temperature scales require the specification of a sufficient number of fixed points to allow the coefficients of the characteristic equations of the stipulated interpolation devices to be determined. The international temperature scale (ITS) of 1927 was revised in 1948 to the international practical temperature scale of 1948 ($IPTS_{48}$), which was amended in 1954 and in 1960. The zero of the scale is that temperature that is exactly 0.01°C lower than the triple point of water (1). Earlier specification as a fixed point of the boiling point of sulfur was replaced by the more accurately attainable freezing point of zinc. A number of secondary reference points were assigned values on the $IPTS_{48}$.

Between −182.97 to 630.5°C (the normal boiling point of oxygen and the freezing point of antimony, respectively), the interpolation instrument was specified to be a platinum resistance thermometer constructed from platinum of such purity that the ratio of its resistance at the normal boiling point of water to that at the ice point is ≥1.3920. From 0 to 630.5°C, the characteristic equation of such a thermometer is

$$R_t = R_o(1 + At + Bt^2) \tag{2}$$

which requires that the thermometer be calibrated at three fixed points to determine its coefficients. From −182.97 to 0°C, the equation is modified to

$$R_t = R_o\,[1 + At + Bt^2 + C\,(t - 100)\,(t^3)] \tag{3}$$

which requires a fourth calibration point.

Between 630.5°C and 1063°C, the respective freezing temperatures of antimony and gold, the interpolation device was specified as a thermocouple, one leg of which was pure platinum and the other an alloy of platinum and 10 wt % rhodium (see Thermoelectric energy conversion). The characteristic equation for thermal emf E was given as

$$E = A + Bt + Ct^2 \tag{4}$$

Above 1063°C, an interpolation device was not specified, but it was specified that temperature be defined in terms of the ratio of the intensity J of visible monochromatic radiation from a blackbody at temperature T to the intensity J_{Au} of radiation of the same wavelength at the temperature T_{Au} of freezing gold. The interpolation equation is Planck's law:

$$\frac{J(T)}{J(T_{Au})} = \frac{\exp\left[\frac{C_2}{\lambda(T_{Au})}\right] - 1}{\exp\left[\frac{C_2}{\lambda(T)}\right] - 1} \quad (5)$$

where C_2 is the second radiation constant, 0.01438 meter-degrees, and λ is in meters. The interpolation instrument used for the measurement is commonly a radiation pyrometer.

In 1968, an extensive revision of the scale was made (2). The motivation for the revision of the scale was to extend the scale downward from the normal boiling point of oxygen to 13.81 K (−259.34°C) and to bring it into closer conformance with the then best-known values of fixed points on the TKTS. The interpolation instruments of the scale did not change, except the requirement that the platinum-resistance-thermometer ratio be not less than 1.39250, but the platinum-resistance-thermometer interpolation equations were changed to accommodate to changes in the values of all fixed points except those of the triple point and boiling point of water. Table 2 shows the values assigned to the defining fixed points of the ITS, the $IPTS_{48}$, and the $IPTS_{68}$.

Over the platinum-resistance-thermometer range of 0–630.74°C, in place of the simple quadratic of the earlier scales, $IPTS_{68}$ temperatures are defined by the function

$$t_{68} = t' + M(t') \quad (6)$$

where t' is a temperature that is neither t_{48} nor t_{68}, and $M(t')$ is a factor that makes $t' = t_{68}$ at certain fixed points

$$M(t') = 0.045\left(\frac{t'}{100}\right)\left(\frac{t'}{100} - 1\right)\left(\frac{t'}{419.58} - 1\right)\left(\frac{t'}{630.74} - 1\right) \quad (7)$$

Table 2. Defining Fixed Points of the Three International Temperature Scales

Defining fixed point[a]	International temperature scale (ITS 1927), °C	International practical temperature scale (IPTS 1948), °C	International practical temperature scale (IPTS 1968)	
			°C	K
tp e H_2[b]			−259.34	13.81
bp e H_2, 2.5/7.7 MPa[b,c]			−256.108	17.042
nbp e H_2[b]			−252.87	20.28
nbp Ne			−246.048	27.102
tp O_2			−218.789	54.361
nbp O_2	−182.97	−182.97	−182.962	90.188
ice point	0.000	0		
tp H_2O		0.01	0.01	273.16
nbp H_2O	100.000	100	100	373.15
fp Sn			231.9681	505.1181
fp Zn		419.505	419.58	692.73
nbp S	444.60	444.6		
fp Ag	960.5	960.8	961.93	1235.08
fp Au	1063	1063	1064.43	1337.58

[a] tp = triple point, bp = boiling point, nbp = normal boiling point, and fp = freezing point.
[b] e H_2 = *eka* H_2.
[c] To convert MPa to atm, divide by 0.101.

and

$$t' = \frac{1}{\alpha}\left(\frac{R(t)}{R(o)} - 1\right) + \delta\left(\frac{t'}{100}\right)\left(\frac{t'}{100} - 1\right) + M(t') \tag{8}$$

Between −259.34°C and 0°C, the platinum thermometer is divided into four distinct ranges, each represented by a polynomial and each requiring the determination of coefficients by calibrating the thermometer at some of the new low temperature fixed points of the $IPTS_{68}$. The definition below 0°C is

$$W(T) = W'(T) + \Delta W(T) \tag{9}$$

where $W(T) = R(t)/R(o)$ (the ratio of the observed resistance of a specific platinum resistance thermometer at temperature t to its resistance at 0°C), and $W'(T)$ is a reference function of the resistance ratio given by

$$T_{68} = A_o + \sum_{i=0}^{20} A_i \,[\ln W(T_{68})]^i \text{ in } K \tag{10}$$

The 21 constants A_i are stipulated in the text of the scale, and the calculation must be done iteratively with a large number of decimal digits. The National Bureau of Standards has provided a table of the reference function at small temperature intervals (3). Because the manner of carrying out the platinum-resistance-thermometer computations is not clear from the text of the scale and because there apparently is no step-by-step method available in the literature, a detailed discussion of practical computation is included at the end of this article.

The $IPTS_{68}$ is lacking in many respects. The assignment of new values to familiar fixed points makes it vital, when referring to historical data, to know or specify what scale was meant. It cannot be assumed that literature published after 1968 is based on the $IPTS_{68}$. Figure 2 illustrates the differences between $IPTS_{48}$ and $IPTS_{68}$ over much of the platinum-thermometer range. Many industries have rejected the $IPTS_{68}$.

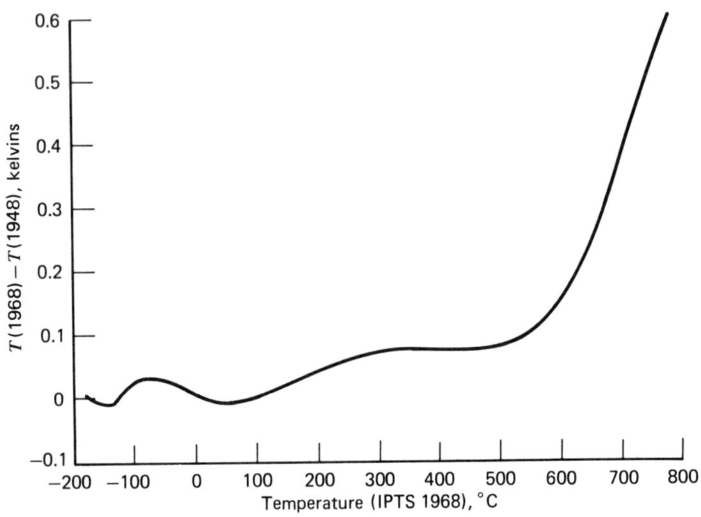

Figure 2. Difference between the IPTSs of 1948 and 1968.

Furthermore, the objective of bringing the IPTS into closer conformance with the TKTS was not achieved. Determinations of thermodynamic temperatures of the fixed points result in experimental values that are subject to change as better experiments are carried out. It has been shown that the normal boiling point of water (which was not changed) is actually 99.97°C, whereas the thermodynamic values of the tin and zinc points (which did change) are closer to the 1948 than the 1968 values. Since a desirable virtue of an artificial scale is permanence, the revision was unfortunate.

Another revision of the IPTS is contemplated in 1987. One change which will almost certainly be made is an increase in the range of the platinum resistance thermometer. At the transition from the platinum resistance thermometer to the standard thermocouple, more than an order of magnitude of accuracy is lost. New developments in platinum resistance-thermometer construction suggest that a platinum thermometer of special construction may be used reliably to the silver point or even to the gold point. The thermocouple could then be deleted. Such a platinum resistance thermometer would most likely have an ice-point resistance of 2.5 Ω.

Fixed Points of the Scale

The fixed points are the basis of the scale and they are of two kinds: defining and secondary or reference fixed points. Defining fixed points are two- and three-phase equilibria of very pure materials to which temperature values are assigned. They are chosen to be as few in number as is consistent with the requirements of the interpolation equations. Secondary or reference fixed points are two- and three-phase equilibria of very pure materials, whose values of equilibrium temperature are not assigned but are established by measurements made with interpolation instruments that have been calibrated at the defining fixed points. Except at very high and low temperatures, fixed points are generally not difficult to obtain, and they offer primary calibration assurance. Equipment for producing some of them is available commercially.

Two-phase equilibria may be solid–liquid, liquid–vapor, or solid–vapor. As is evident from the phase rule of Gibbs, two-phase equilibria are pressure-dependent

$$P + V = C + 2 \tag{11}$$

where P is an integer equal to the number of phases present, C is the number of components present (for an ideally pure material, $C = 1$), and V is an integer giving the number of degrees of freedom. Phase equilibria involving a vapor phase are much more pressure-dependent than liquid–solid equilibria. Equilibria in which all three phases are present (triple points) are independant of external pressure.

The triple point of water is the most important of the fixed points, because it is the one point that the TKTS and the IPTS have in common and because it can be realized with great accuracy. A water triple-point cell is shown in Figure 3. The glass envelope has a well for a thermometer. The cell is almost filled with very pure water of defined isotopic content. The slight headspace contains water vapor, whose pressure at the triple point is 600 Pa (4.5 torr). The integrity of the cell can be checked by inverting it. A sharp click will be heard as the water, uncushioned by air, strikes the opposite wall of the cell; as the bubble of water vapor is caught in the curved portion of the handle, it is compressed until it is almost invisible. These checks are all that is necessary to confirm the integrity of the cell.

In use, a mantle of ice is frozen onto the outer surface of the thermometer well.

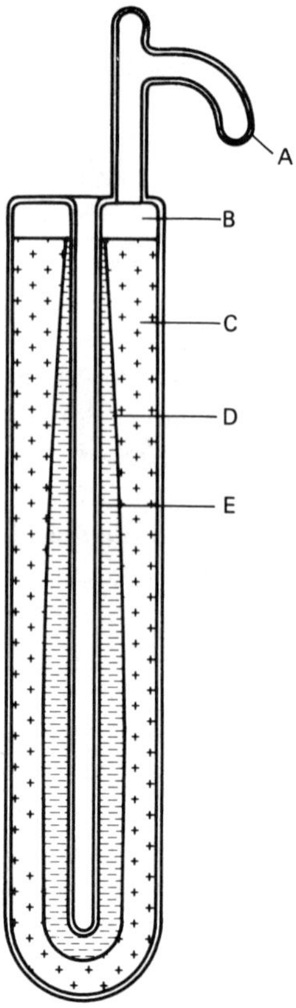

Figure 3. A water triple-point cell. A, handle; B, water vapor; C, liquid vapor; D, solid water (ice mantle); E, thermometer well.

A common way to do this is to fill the well with crushed dry ice until the mantle achieves a good thickness. Descriptions of how to freeze and maintain a cell are given in several publications (3–4). The temperature of the triple point of water is 0.01°C exactly, by definition. The temperature of the cell can be realized within ca 0.00015 K of the definition. In contrast, a bath of ice and water for producing the temperature 0°C is difficult to use with accuracy better than 0.002 K.

Any laboratory requiring precise temperature measurements should be equipped to produce the triple point of water, if no other fixed point. In addition to its status as a primary definition of a temperature, it is an important check upon the laboratory's platinum resistance thermometers. When such thermometers shift, they tend to shift at all temperatures. If the calibration is correct at the water triple point, the probability is very high that it has not changed at other temperatures. With the high cost of recalibration, some laboratories never recalibrate their platinum resistance thermometers unless there is a change at the water triple point.

Metal freezing points are an important class of fixed points. Their pressure dependence is of small order (for tin, for example, it is 3.0×10^{-8} K/Pa (0.003 K/atm)) and unlikely to be a problem in all but the most critical measurements. The metals used in fixed-point cells are usually better than 99.9999% (6 N) pure. Figure 4 is one design for such a cell. The metal is contained in a crucible of purified graphite with a graphite cover and sleeve. The crucible is contained in an envelope of quartz, the

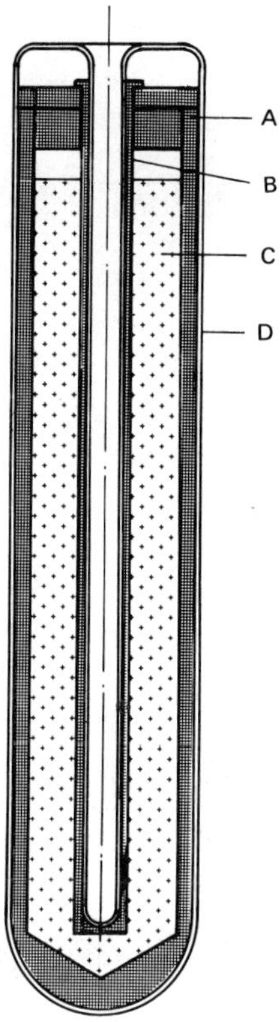

Figure 4. Metal freezing-point cell. A, pure graphite crucible; B, thermometer well; C, highly pure metal; D, quartz envelope.

quartz extending into the graphite sleeve to provide a thermometer well; thus, the graphite is completely enclosed. The cell is charged with pure metal, filled with argon or another inert gas at 101 kPa (1 atm) at the freezing temperature, and sealed. Thus, it is protected against contamination and contains an atmosphere of gas at the correct pressure. Details of the construction of such cells have been published and they are commercially available, as are furnaces suitable for melting and freezing the metal (5). Sealed cells of this type have shown no measurable change in freezing temperature after 15 years of use. Table 3 shows a partial list of useful metal freezing points and their freezing temperatures and pressure coefficients.

To melt these metals, the cells are heated in furnaces that do not respond quickly (see Fig. 5). The furnace control is first set ca 10 K above the melt temperature. The onset of melting is indicated by the monitoring thermometer as a melt arrest, during which temperature does not change, and the completion of melting is indicated by a rise to the controlled temperature. The furnace temperature is then reduced to slightly below the melt temperature. The technique for initiating freezing, for managing supercooling, etc, depends upon the metal and is described in the literature.

Liquid–vapor fixed points or boiling points are the least satisfactory fixed points because of their pressure dependence; at the boiling point of oxygen, 107 Pa of pressure difference (0.8 torr) = 0.01 K. However, the boiling point of oxygen defines the lower end of one platinum resistance-thermometer range, and thus it is still used. The triple point of argon is only 5 K below the oxygen boiling point, and may well replace the oxygen point in the IPTS revision of 1987.

Table 3. Metal Freezing Points

Metal	Freezing point (at 101.3 kPa[a]), C°, IPTS$_{68}$	Pressure coefficient K/atm[a]	Pressure coefficient K/cm of liquid height
mercury	−38.862	0.0054	0.000071
gallium	29.7714 melt point	−0.00198	−0.000012
indium	156.634	0.0049	0.000033
tin	231.9881	0.0033	0.000022
bismuth	271.442	−0.0035	−0.000034
cadmium	321.108	0.0062	0.000048
lead	327.502	0.0080	0.000082
zinc	419.58	0.0043	0.000027
antimony	630.74	0.00085	0.000005
aluminum	660.37		
silver	961.93		

[a] 1 atm = 101.3 kPa.

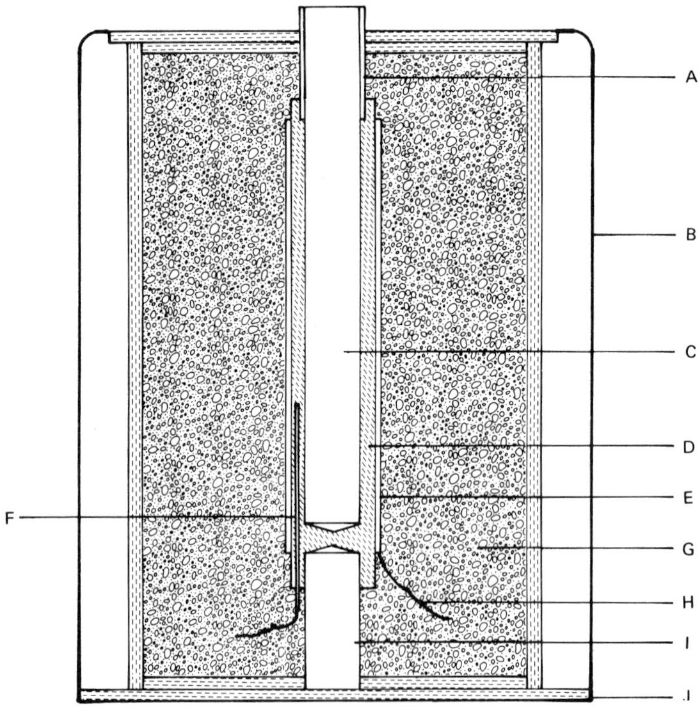

Figure 5. Metal freezing-point furnace. A, ceramic guide tube; B, metal outer shell; C, cell space; D, bronze or nickel core; E, furnace winding; F, control thermometer; G, Fiberfax insulation; H, power leads; I, ceramic standoff; J, Transite sheet and pipe inner shell.

Except at laboratories such as the NBS, the calibration of thermometers in boiling oxygen is usually done by comparison with a known thermometer. In effect, pressure considerations are removed by measuring the temperature of the boiling oxygen with an NBS-calibrated thermometer or by considering the average measured value from several such thermometers.

A list of secondary reference points is given in the text of the scale. A more extensive list, which does not include an evaluation of the quality of the work reported, is given in reference 6. The reason for interest in the development of secondary reference points is illustrated by the gallium point (7). In enzyme reactions, which are of interest in clinical chemistry and other disciplines, the reaction rates may vary by as much as 10%/K. Generally, temperature control of spectrophotometer cuvettes is inadequate to eliminate temperature as a source of error. However, cuvettes are too small to admit thermometers other than tiny thermistors or thermocouples, and these require frequent calibration verification. The gallium liquid–solid equilibrium temperature (29.7714°C) is very near the standard enzyme reaction rate temperature (30°C) and permits calibration of the small thermometers immediately before and after the reaction. Calibrations at 0°C and 100°C are not adequate because these temperatures are generally off-scale of the narrow-range thermometers. Since the development of the gallium point and its acceptance by BIPM (Bureau International des Poids et Mesures) as a reference point of the scale, important other uses for it have emerged in biological, ecological, environmental, and thermal-energy thermometry.

Resistance Thermometers

Resistance thermometers are divisible into three general types: high precision platinum, industrial-grade metallic, and nonmetallic. High precision platinum resistance thermometers are suitable for use as interpolation instruments of the IPTS and are frequently referred to as SPRTs (standard platinum resistance thermometers). These thermometers generally are useful over the entire platinum-thermometer range of the IPTS, ie, from −182.962°C to 650°C, and a special capsule form of this thermometer can be used at −259.34°C to 232°C. The purity of the platinum is characterized by a ratio of resistance at 100°C to that at 0°C which is not less than 1.39250. The thermometer construction is designed to hold the element wire in a strain-free manner. Almost without exception the resistance at 0°C is 25.5 Ω. This ice-point resistance permits a very rough but quick estimation of temperature, since a 1 K change is very nearly equivalent to a 0.1-Ω change.

SPRTs are always provided with two current and two potential leads extending from the measuring element, so that the element can be measured as a four-terminal resistor, and the effects of lead-wire resistances eliminated. The thermometers are generally 7 mm in diameter, ca 40 cm long, and must be immersed at least 20 cm in the medium to be measured. They are usually made with a quartz protection tube around the element and the internal lead wires. They are fragile in the sense that they are easily broken, and they can be strained and the calibration consequently altered by even a slight mechanical shock. They are costly, and so is the ancillary resistance-measuring equipment. A typical SPRT is shown in Figure 6.

Industrial-grade metallic resistance thermometers are wound of metal wire, usually high purity platinum, but also copper, nickel, and alloys. The temperature-sensitive element is typically protected by a covering of ceramic tubing or ceramic cement. Glass is sometimes used, but otherwise suitable glasses are inadequate insulators above 500°C and cause shunt resistance, and strains of the wire caused by semifluid and incomplete recovery of the glass from thermal expansion and contraction produce a thermal hysteresis measurement.

Industrial-grade thermometers are smaller than SPRT elements and may be as

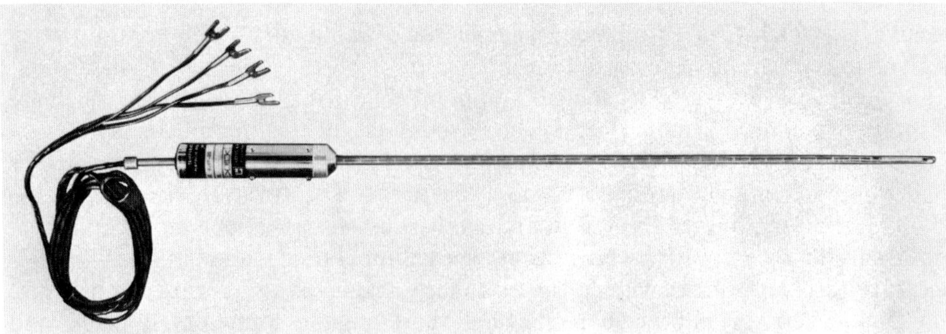

Figure 6. Laboratory standard platinum resistance thermometer (SPRT).

small as 2.5 mm in diameter and 12 mm in length. They are available in various ice-point resistances, including 100 Ω, 200 Ω, and 500 Ω, and are usually interchangeable within ±0.25 K at the ice point; closer interchangeability tolerances are also available. They are sold as unsheathed elements and in a wide variety of sheath and closure configurations, both standard and custom, and they are relatively inexpensive. Several common industrial patterns are shown in Figure 7. Most industrial platinum thermometers are designed to withstand abuse from vibration, shock, etc without significant change in calibration, which argues against their being ideally strain-free, and they are not considered suitable as interpolation standards of the IPTS.

Since the early 1970s, development work has been reported on thick-film, thin-

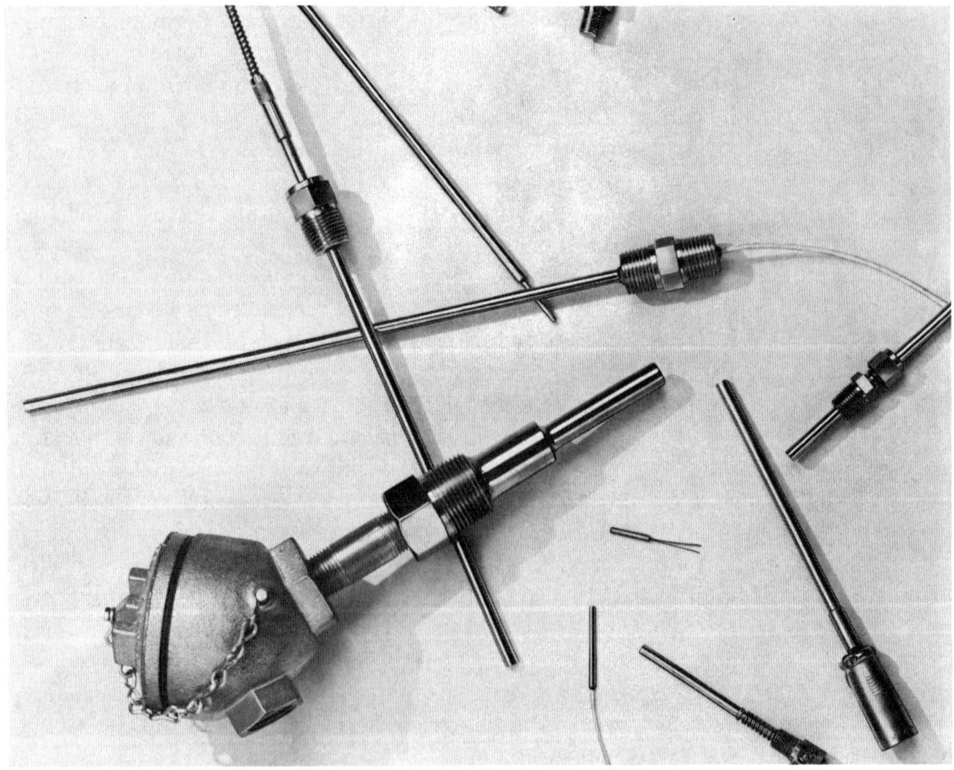

Figure 7. Industrial resistance-thermometer configurations.

film, deposited, and various other resistance thermometers made with materials other than wrought wire. Such thermometers have certain advantages of size, response time, etc. To date they do not have the stability and interchangeability of wrought-wire elements. In general, manufacturers of industrial resistance thermometers do not report to the user, except upon request, the calibrations of individual thermometers, but instead publish tables of resistance-vs-temperature characteristics of a class of thermometers and tolerance charts that show the deviations of various classes from a nominal calibration. A typical tolerance for a precision-class, industrial resistance thermometer is

$$\text{deviation from nominal} < (0.1 \pm 0.002\, t)\ ^\circ\text{C} \tag{12}$$

The tables are computed based on the true $IPTS_{68}$ values for the defining fixed-point temperatures and, for the interpolation, the simpler mathematics of $IPTS_{48}$ are used (see eqs. 2 and 3). The result is a quasi-1968 table, whose departures from strict consonance with $IPTS_{68}$ are nowhere more than 0.01 K, which is industrially acceptable.

Nonmetallic resistance thermometers or thermistors are generally temperature-sensitive semiconducting ceramics (see Ceramics as electrical materials). The variety of sizes, shapes, and performance characteristics is large. One manufacturer of thermistors catalogues a choice of characteristics ranging from 100 Ω at 25°C to 1 mΩ at 25°C. The thermistor material is usually a metal oxide, eg, manganese oxide. Dopants, eg, nickel oxide or copper oxide, may be added to obtain a variety of resistance and slope characteristics. The material can be sintered into a disk or bead with integral or attached connecting wires. Figure 8 shows one series of processes for forming a thermistor.

Bead thermistors are formed by placing two platinum wires in close proximity and bridging them with a drop of oxide slurry, which is then sintered into a hard bead and encapsulated in protective glass. Such thermistors are quite stable, in some cases approaching, over a narrow temperature range, the stability of industrial metallic thermometers. However, the resistance tolerances may vary from unit to unit by as much as ±20% unless careful matching is used. Beads can be made quite small, which sometimes dictates their use, eg, in temperature probes which are mounted intravenously.

Disk thermistors generally are not as stable as beads, but they can be produced to close limits of interchangeability. One manufacturer offers disk thermistors that are interchangeable to ±0.1°C and to ±0.05°C. Disks cannot be made as small as beads;

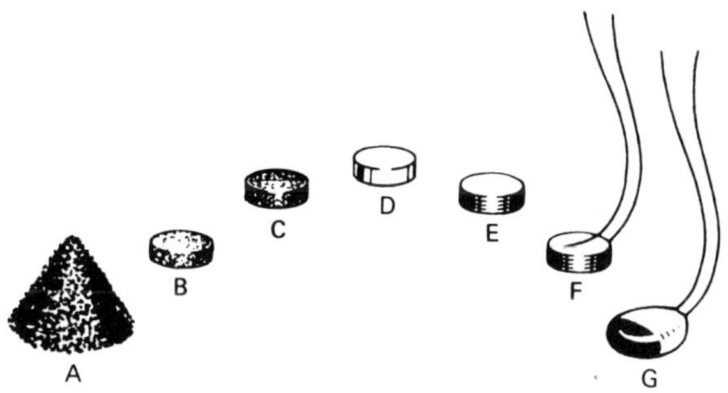

Figure 8. Steps in the manufacture of a disk thermistor. A, ball-milled powder; B, pressed disk; C, sintered disk; D, silvered disk; E, edge-ground disk; F, lead wires attached; G, epoxy-coated.

a 2-mm dia seems an approximate practical limit. Disks are commonly coated with epoxy resins, and it is probable that the lower stability of disks results from a less mechanically compressive coating. Recent developments indicate the possibility of a thermistor having the interchangeability of the best disks and the stability of the best bead thermistors.

Three characteristics of thermistors are distinctly different from those of metallic resistance thermometers. For all but a few special types, the temperature-resistance characteristic is negative; that is, resistance decreases with increasing temperature. All metallic resistance thermometers exhibit a positive characteristic. The temperature-resistance characteristic is very nonlinear. One typical commercial thermistor changes resistance 376 Ω/K at 0°C and 13.5 Ω/K at 70°C. Combinations of two or three thermistors and simple resistive networks can be made to provide approximate linearity over relatively narrow temperature ranges, with some reduction in sensitivity of output vs temperature change. Over favorable portions of the nonlinear characteristic, the change in resistance can be more than an order of magnitude larger than that of metallic resistance thermometers. The use of thermistors is thus generally confined to the near-ambient temperature ranges, eg, the biological range, where linearity is not an objective and where the high $\Delta R/\Delta t$ is a decisive advantage.

Many attempts have been made to develop an equation explaining the semiconducting transfer function of thermistors (see Semiconductors). These have generally been of the form

$$p^{-1} = f(T) \exp(-\Delta E/2kT) \tag{13}$$

where ΔE is the energy gap between the valence-band energies, T is the kelvin temperature, and k is Boltzmann's constant. None of these efforts has provided a simple explicit law that fits the calibration data. The equation generally used is an empirically derived polynomial (8):

$$T^{-1} = A + B\,(\ln R) + C\,(\ln R)^3 \tag{14}$$

Instrumentation. The principal considerations for instrumentation to measure the resistance at temperature of resistance thermometers are adequate sensitivity, elimination of the effects of lead-wire resistance, and avoidance of the effects of self-heating. For the measurement of 0.001 K with an SPRT, the minimum necessary resolution and sensitivity of the resistance-measuring instrument is 100 μΩ. Sensitivity that is an order of magnitude higher is highly desirable. The application of industrial resistance thermometers and thermistors dictates the sensitivity required. Often, for example in the presence of electrical noise, the highest sensitivity does not necessarily produce the best measurement.

With regard to lead resistance, the temperature-sensitive element of a thermometer is connected both internally and externally to a readout instrument by means of conductors that are resistances and whose values probably vary with ambient temperature. Various schemes have been developed to eliminate, compensate for, or minimize these effects.

It is particularly important to eliminate lead resistance effects in high precision measurements with SPRTs. The element of an SPRT is designed as a four-terminal resistor. The traditional measuring instrument is the Mueller Bridge, shown schematically in Figure 9. The ratio arms of such a bridge must be equal, and they usually can be balanced. Two three-terminal measurements are made and averaged. Lead

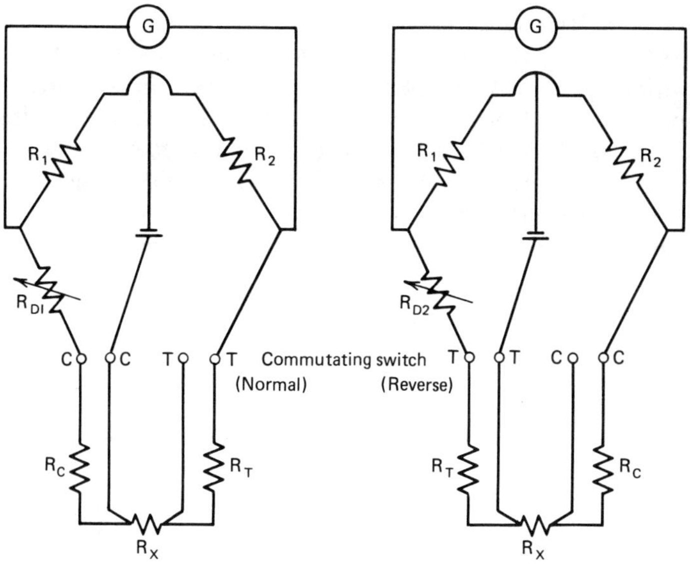

Figure 9. The Mueller bridge for resistance thermometry. R_1 and R_2, equal-arm resistors; R_{D1}, multiple-dial decade resistors; R_C and R_T, resistance of thermometer lead wires; R_X, platinum thermometer resistance; G, galvanometer.

resistance must not change between measurements, and so it is customary to average four measurements based on the sequence NRRN, where N is the normal position of the bridge commutator and R is the reverse.

It is also possible to measure an SPRT potentiometrically with reference to a standard resistor. Perhaps the most elegant way to measure an SPRT as a true four-terminal resistor is with a precision current comparator. The many precision wire-wound resistors of the Mueller Bridge are replaced by tapped transformer turns. Null is determined by the absence of flux in a pick-up coil, and so the comparator is a d-c device from the perspective of the thermometer. Such instruments are available with linearity and long-term stability on the order of 2 ppm. A schematic is shown in Figure 10.

The next generation of inductive bridges is automated, and this trend can be expected to continue as automatic test systems are developed to improve calibration cost-effectiveness. Such a bridge may be a fully automatic kelvin a-c double bridge based on an inductive voltage divider (9–10). By using a-c phase-locked bridge ener-

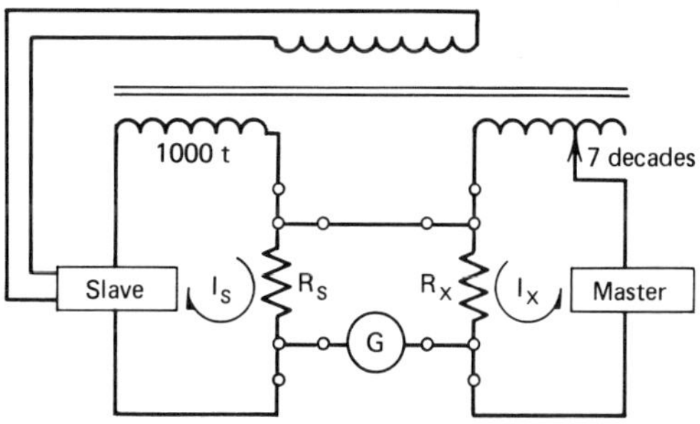

Figure 10. The precision current comparator. Master and slave are precision current sources; R_s, calibrated standard resistor; R_x, platinum thermometer resistor; G, galvanometer.

gizing and detection techniques, great sensitivity is achieved without need for manual balances. Class accuracies of ±100 μK and temperature sensitivities of a few microkelvins are routinely obtained. No special environmental controls are required nor is recalibration. Automated bridges are especially useful for monitoring long-term thermal equilibria and high volume thermometer calibration. All such bridges, whether manually or automatically balanced, are available with analogue and digital interfaces for data logging and computer control. SPRTs are designed as d-c resistors, and problems such as reactance and quadrature must be met and resolved when they are used in a-c circuits.

Alternating-current circuitry is insensitive to accidentally generated thermoelectric voltages, which are d-c. These can appear wherever there are junctions between dissimilar materials and the junctions are at different temperatures. Such effects appear in d-c circuits as offset voltages, which are detected by reversing the polarity of the bridge battery. They are minimized in SPRTs by making the element and internal lead wires of pure platinum and joining the platinum internal leads to the copper external leads in the thermometer head, which is usually cool and isothermal. In industrial thermometers, where the cost of platinum internal lead wires is prohibitive, an attempt is made to position junctions so that they are isothermal.

Most industrial thermometers do not operate at bridge null but indicate temperature from bridge imbalance. Both active and passive circuitry has been developed to minimize the effects of lead resistance and lead-resistance change resulting from temperature. Figure 11 shows some passive circuit configurations which, with certain constraints upon the choice of circuit components, reduce these effects.

The absolute resistance of thermistors is generally high enough relative to lead resistance that lead resistance can be ignored. When this is not true, considerations appropriate to metallic resistance thermometers apply.

Much commercial instrumentation is available, ranging from the simple to the complex, ie, digital ohmmeters, signal-conditioning electronics, data-loggers, and computers, for industrial temperature measurement and control. All such devices must be chosen with consideration of the errors introduced by self-heating (see also Instrumentation and control).

A resistor is a passive element. For its measurement, a current must be passed through it, and thus it becomes an I^2R device which generates some heat. The question

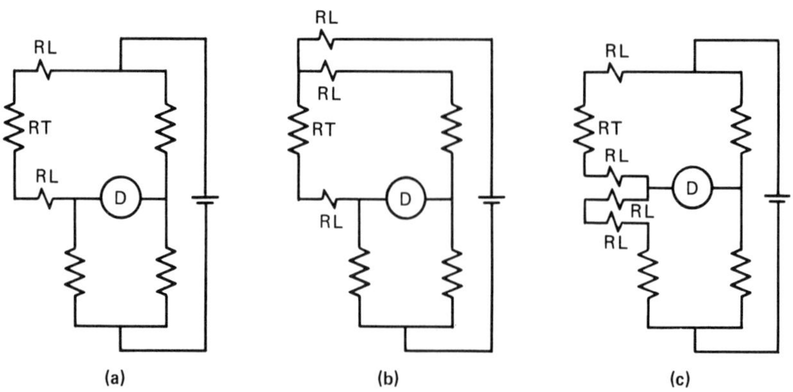

Figure 11. Industrial resistance-thermometer circuits for the suppression of lead resistance effects (RL). (**a**) 2-wire; (**b**) 3-wire; (**c**) 2-wire and 2-wire compensating loop.

is whether the heat generated is significant in terms of the accuracy with which the temperature is to be measured. Resistances of SPRTs are measured at the NBS with a thermometer current of 1 mA. In the standards calibration environment at fixed points, the heat generated results in errors of probably not more than 500 μK, because the dissipative coupling to the medium is good. Calibrations of higher accuracy can be made by measuring the resistance at three current levels, eg, 1, $\sqrt{2}$, and 2 mA, and calculating by means of a determinant in resistance and power (not current) what the resistance would be with the current source removed. This is called zero-power resistance. For SPRTs calibrated at the NBS, the user should be careful to duplicate the 1 mA current level. An SPRT should not be subjected to currents higher than 5 mA, even momentarily.

A thermometer measures no temperature except its own. Heat caused by the measuring current should be dissipated as much as possible by close coupling to the medium. The surface area of an SPRT is large compared to that of industrial thermometers and thermistors, and so the dissipation is larger. For an industrial thermometer with an ice-point resistance of 100 Ω, a measurement current of 1 mA generates a perceptible but perhaps not objectionable thermal offset. Conditions of flow and the nature of the medium into which the thermometer is inserted affect the offset also. A thermometer in a protective thermowell may present special problems. The basic consideration is that the error caused by self-heating is a function of the heat generated minus the heat dissipated.

The problem of self-heating is particularly important in thermistor thermometry, since with negative temperature coefficient, the effect of self-heat is to lower the resistance, which causes an increase in impressed power, and a condition of thermal runaway can occur. Power levels of 8 mW are probably a safe maximum for 2.5-mm dia disk thermistors.

Seebeck-Effect Thermometers

Seebeck-effect thermometers or thermocouples are junctions of two wires of dissimilar materials that accomplish a net conversion of thermal energy into electrical energy with the appearance of an electric current (see Thermoelectric energy conversion). Unlike resistance thermometers, whose output is proportional to temperature, the output of thermocouples is proportional to temperature difference; that is, the magnitude of the current is proportional to the difference of the temperatures of the two junctions; Figure 12 illustrates this principle.

There are three fundamental rules regarding thermocouples: no thermal emf is produced if heat is applied to a circuit comprising a single homogeneous conductor; the thermal emf of a circuit comprising any number of conductive materials at uniform temperature is zero; and the thermal emf developed by a pair of homogeneous but dissimilar conductors, having junctions at T_1 and T_3, is the same as the thermal emf developed by an arrangement of the same conductors having junctions at T_1 and T_2 and at T_2 and T_3. Figure 13 illustrates this latter principle.

It is a common misconception that the generation of the thermal emf is a junction

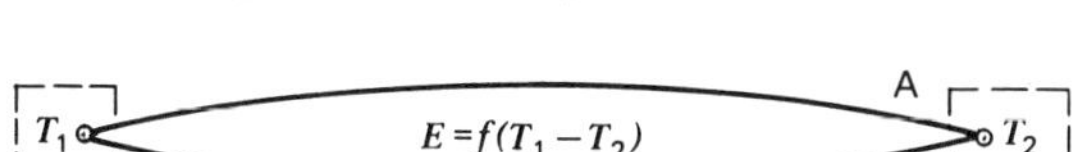

Figure 12. Basic thermocouple circuit. A and B are wires of different materials.

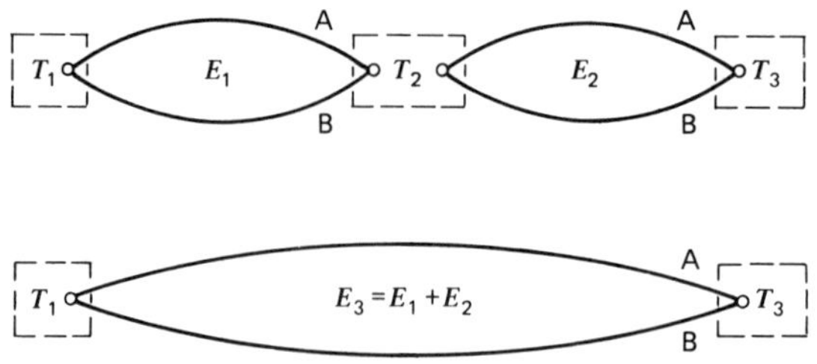

Figure 13. Rule 3 of thermocouples. A and B, the two legs of the thermocouple, are wires of dissimilar materials. Junctions are at temperatures T_1, T_2, and T_3. E_1, E_2, and E_3 are thermal emfs. The thermal emf at T_3 is the same for both circuits.

effect; rather it is a bulk effect of the material. Electrons in metals have an electrical charge and kinetic energy. Along a thermal gradient, charge does not change but kinetic energy does; therefore, a transfer of heat must occur. The thermal emf results from a slight rearrangement of electrons in the conductor to respond to thermal transfer. The function of the junction is to assure electrical continuity between the conductors, and a consequence of this is that it does not matter how the junction is made, eg, by welding, brazing, clamping, etc, as long as electrical conductivity is preserved.

The assumption of homogeneity of the first and third rules is a condition that is never met in practice, and that, if it were met, could not be assumed to be preserved throughout the working life of the thermocouple. For precise work, it is important to verify the performance of the thermocouple by calibration at appropriate intervals.

Standard (IPTS) Thermocouple

At 630–1064.43°C (the freezing temperatures of antimony and gold, respectively), the specified interpolation standard of the $IPTS_{68}$ is a thermocouple, one conductor of which is pure platinum and the other an alloy of platinum with 10 wt % rhodium. The thermal emf required for such a thermocouple is stipulated in the text of the scale. The coefficients of the characteristic quadratic of this thermocouple (eq. 4) are determined by measurements at the two fixed points, which are the high and low ends of its range, and at one intermediate fixed point, the freezing point of silver, 961.93°C. In a letter system for the designation of thermocouple pairs, the Pt–90 wt % Pt–10 wt % Rh thermocouple is called type S.

Type S thermocouples intended for use as IPTS interpolation standards are usually protected by an enclosure which minimizes contamination and provides an oxidizing atmosphere. It is particularly important to avoid reducing atmospheres, metallic vapors such as lead and zinc, nonmetallic vapors such as phosphorus, sulfur or arsenic, or easily reduced oxides. The immediate protective sheath of the thermocouple should never be metallic. The positive element 90 Pt–10 Rh can be affected by neutron flux, since rhodium gradually transmutes to palladium. The negative element Pt forms stable isotopes and is relatively immune to transmutation, but it can be work-hardened by fast neutron bombardment. Calibration values are valid only in the fully annealed state.

Although the scale limits the use of the type S thermocouple as an interpolation standard to the freezing point of gold, such thermocouples may be used as thermometers continuously to 1400°C and intermittently to the melting point of platinum, ie, 1767.6°C. However, such exposure reduces useful life because of diffusion of impurities and propagation of grain growth. Also, sudden quenching from an elevated temperature to room temperature affects the physical properties of the wire and, consequently, the calibration.

Type S thermocouples have low thermal emfs. The nominal emf at 630°C is 554.58 μV and at 1064°C it is 10329.25 μV; ie, the average sensitivity is 11 μV/K.

For reasons of inhomogeneity, contamination, grain growth, alteration of physical state, etc, more than an order of magnitude of accuracy is lost at the transition between the platinum resistance thermometer and the Pt–90 Pt–10 Rh thermocouple. It is this consideration that has stimulated the effort to eliminate the standard thermocouple in the 1987 version of the scale.

Working Thermocouples

Properties of various types of working thermocouples are shown in Table 4. Additional property information is given in refs. 11–12.

The wires of thermocouple pairs must be electrically insulated from each other and protected from the environment. Manufacturers' catalogues indicate the sorts of insulation and protection available. Insulations range from bare wires separated by double-bore ceramic beads or tubing, through servings of various insulating fibers, to enclosure in metal tubing filled with compacted ceramics. Both common and exotic materials are available, and manufacturers' recommendations should be sought. Essentially, the enclosure should do the following: provide mechanical protection, but not contribute to mechanical strain of the wire, eg, if possible, closely match the thermal expansion coefficient of a metallic sheath to that of the thermocouple wires, which should closely match each other; provide protection against contamination and not contribute to it by means of the migration of impurities from the insulation to the wire, via metallic inclusions, the evolution of contaminating vapors, etc; and maintain the insulation and not degrade it, as would be the case, eg, of an unsealed hygroscopic ceramic insulation. To all of these potential alterations of state except the last, the susceptibility of thick wires may be less than that of thin wires.

Another potential source of error, particularly in type K thermocouples used at 200–500°C, is reversible change of metallic state. Such changes are thought to be lattice ordering–disordering. Reversible changes are particularly subtle, may not be readily discoverable by calibration, and may seriously compromise measurements made with

Table 4. Properties of Various Thermocouple Pairs

ASTM type	Materials	Range[a], °C	Seebeck[b] coefficient, μV/C°
J	iron–constantan	0–760	64.3
K	chromel–alumel	0–1260	36.5
R	Pt–87 Pt–13 Rh	0–1450	13.8
S	Pt–90 Pt–10 Rh	0–1450	11.8
T	copper–constantan	−183–375	53.0
E	chromel–constantan	0–875	78.5
B	70 Pt–30 Rh–94 Pt–6 Rh	870–1700	11.6

[a] Wire of small diameter may require closer range.

[b] Emf is given for example at the high temperature end of the range; many thermocouples are nonlinear.

some thermocouple types. Also, although the ASTM letter designations for thermocouple pairs are specified thoroughly in the United States, the nominal calibrations designated by the same letters may be different in other countries and in international standards. It is important to determine the specific characteristics of a specific lot of wire, particularly when making replacements of thermocouples or thermocouple extension wire.

In industrial applications, it is not uncommon that the thermocouple element must be connected to the readout instrument or controller by a long length of wire. It is evident from the differential nature of the thermocouple that, to avoid unwanted thermal junctions, this wire should be thermocouple wire of the same type (ie, for a type J thermocouple the extension wire must also be type J) and be connected properly with respect to positive and negative legs. The junctions between the thermocouple leads and the extension wires should be made in an isothermal environment and protected (eg, by a housing) from drafts, etc, which would upset thermal equilibrium. The junctions must have the same electrical integrity as the thermal junctions. Because of the low voltage, enclosure in a shield or grounded conduit should be considered. When the thermocouple is of a noble or exotic material, the cost of extension wire may prohibit the use of identical materials. Manufacturers of thermocouple wire may suggest less costly alternatives and discuss the effect of the compromise.

Since the thermal emf of the thermocouple is a function of the temperature difference between the hot and the cold ends (the latter are usually located at the readout instrument), the measurement can be no more accurate than the isothermal condition of the thermocouple pair at the cold end and the accuracy with which this temperature is known. Three common methods for assuring these are used:

The cold-junction temperature can be established by immersing the junction in some known thermal environment, often an ice bath. For more precise measurements, a properly maintained water triple-point cell is more accurate. A temperature-controlled oven at above ambient temperature may be used.

The cold junction may be attached to or contained within a relatively isothermal mass, eg, a large block of copper, whose temperature is measured by some independent means, eg, an embedded thermistor.

Cold-junction compensation can be provided by a network, which includes a voltage source and a temperature-sensitive bridge, to provide an offset voltage that is proportional to the temperature sensitivity of the thermocouple and is of opposite sign.

Despite problems in their use, and because of their generally low cost, their temperature ranges, the possibility of making them very small, etc, use of thermocouples may be the only way to solve a temperature-measurement problem. However, in considering cost, the total system cost including extension wire and instrumentation must be included.

Thermopiles are diagrammed in Figure 14. In the circuit arrangement of Figure 14, the thermal emf proportional to the temperature difference between a single hot and cold junction is multiplied by the number of junctions, and higher emfs are generated. When the hot junctions are not all at the same temperature and the cold junctions are maintained at the same temperature, the thermal emf is proportional to the average of the hot-junction temperatures multiplied by the number of junctions minus the cold-junction temperature. This arrangement is useful for finding the mean temperature of a generalized space. When the hot junctions are placed close together

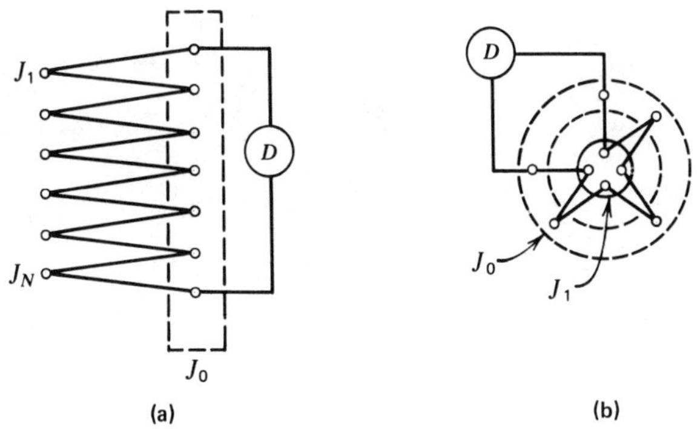

Figure 14. Thermopiles. (**a**) The thermoelectric voltage at D is the average of the thermal emfs developed by $J_1 - J_0 \ldots J_N - J_0$ multiplied by N; all junctions J_0 are maintained at the same temperature. (**b**) The thermoelectric voltage at D is the thermal emf developed by $J_1 - J_0$, multiplied by the number of junctions. Junctions J_1 are mounted on a radiation sink; junctions J_0 are shielded from incident radiation.

and perhaps affixed to a heat-collecting and equilibrating small target, the thermopile forms the sensor of a sensitive radiometer.

Not both elements of the thermocouple need be wire. For example, if a copper pipe conducts a fluid whose temperature is to be measured, a constantan wire attached to the pipe will form a T thermocouple. Such arrangements are difficult to calibrate and require full understanding of the possible inherent problems. Many ingenious arrangements, mountings, and applications of thermocouples are described in refs. 13–14.

Measurement of Emitted Radiation

Temperatures of materials that are too hot to be placed in physical contact with a thermometer, either because the temperature to be measured would be affected by such contact or because the thermometer's characteristics would be changed, may be determined by measuring the radiation emitted by the materials.

Optical Pyrometer. Temperatures above the freezing point of gold (1064.43°C) are defined on the $IPTS_{68}$ by a thermodynamic function, and an interpolation instrument is not stipulated. The interpolation instrument universally used is an optical pyrometer. In the event that the 1987 revision of the IPTS extends the range of the platinum resistance thermometer to the silver point but not higher, the optical pyrometer will be used downward to the silver point. Of the three instruments used for producing the scale, only the optical pyrometer responds directly to a thermodynamic function.

The main elements of a disappearing-filament optical pyrometer are shown in Figure 15. An optical pyrometer consists of an optical system for viewing the radiating target, a lamp filament, filters, an eyepiece, and a calibrated means for varying the lamp current and, consequently, the filament brightness. The pyrometer is sighted on the radiating source. The lamp-filament current is varied until the superimposed image of the glowing filament is of such brilliance that its separate image is indistinguishable from the source brilliance which fills the optical field. The lamp current is then noted. If a second radiating source also causes the image to be indistinguishable at the same lamp current, the temperature of the second source is said to be the same as the temperature of the first.

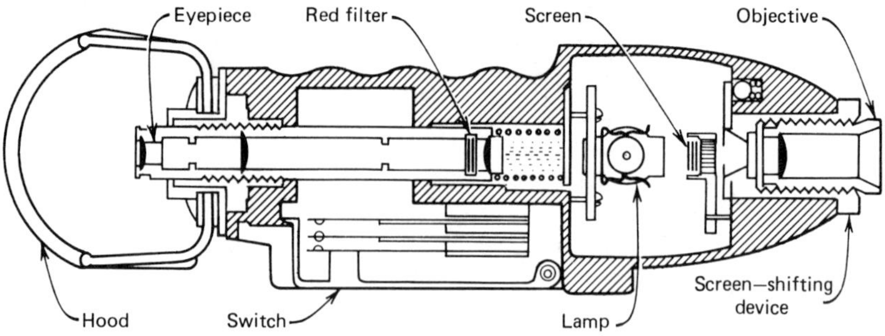

Figure 15. Optical pyrometer.

The radiation pyrometer maintained at the NBS is calibrated by sighting upon a blackbody at the freezing temperature of gold. Secondary pyrometers submitted for calibration are calibrated with respect to the NBS pyrometer by sighting both on strip-lamp filaments.

Blackbody. Materials at any temperature both emit and absorb radiation. In the range of temperature in which the pyrometer is commonly used, the radiation is visible as light. However, if detectors other than the human eye are used, it is not necessary that the radiation be in the visible spectrum.

The ideal blackbody is a closed cavity whose walls are at uniform temperature. At every point of the interior wall, radiation is emitted and absorbed so that emission and absorption are in equilibrium. In this condition there is no net loss, and the internal energy is a function of temperature only. For the sake of seeing into such a cavity, a small hole may be provided which, if it is the proper size and shape, does not disturb the equilibrium condition in a significant way. Under these circumstances, the radiation emerging from the viewing port has the same radiant energy as the interior of the cavity, which corresponds to the temperature of the blackbody.

A simplified drawing of a blackbody and its furnace is shown in Figure 16. A

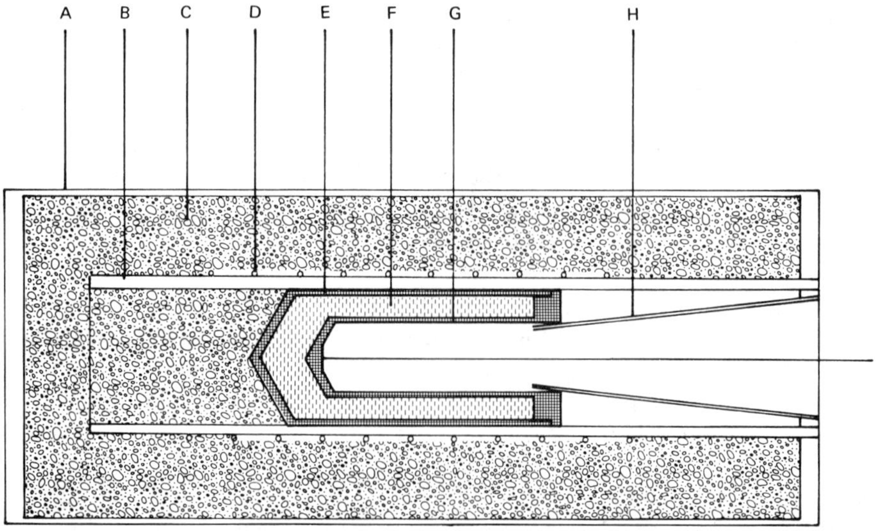

Figure 16. Blackbody and furnace. A, furnace shell; B, furnace tube; C, thermal insulation; D, heater winding; E, outer wall of graphite crucible; F, pure metal (gold or copper); G, inner wall of graphite crucible; H, sighting cone; I, target.

graphite cavity is surrounded by a graphite crucible, so that the cavity can be surrounded by a shell of gold. The NBS gold blackbody contains 1.8 kg of ≥99.999 wt % gold. Because of the high price of gold and the scarcity of high purity gold, copper (with a freezing point of 1084.5°C) is an attractive alternative. One end of the cavity is open and leads to a conical viewing port. The other end provides a target upon which to sight. The furnace is provided to melt and freeze the gold. The NBS gold blackbody has a temperature reproducibility of 0.02°C and an effective emissivity of 0.99999, ie, it departs from the ideal blackbody by 10 ppm. The blackbody and furnace design shown in the figure is intended as a calibration device for use by industrial standardizing laboratories at the copper point (15).

The total energy radiating from a blackbody of unit area is given by the law of Stefan-Boltzmann:

$$E = \sigma T^4 \tag{15}$$

where T is the absolute temperature in kelvins, and σ is the Stefan-Boltzmann constant, 5.6697×10^{-12} W/cm^{-2} T^{-4}. This expression contains no term for frequency or wavelength, which is distributed over a large range, with most of the energy appearing in the infrared.

If the spectral radiance $N_{b\lambda}$ is defined as the energy of that portion of the blackbody radiance which is at wavelength λ, then

$$N_{b\lambda} = \frac{C_1\lambda^{-5}}{\exp\,(C_2/\lambda T) - 1} \tag{16}$$

where C_1 and C_2 are the first and second radiation constants, respectively. The term -1 in the denominator was added by Planck when he observed that Wien's equation did not fit experimental data at high temperatures. To explain this stroke of pure intuition, it was necessary for him to invent the quantum theory, for which he received the Nobel Prize in 1918. This equation permits the evaluation of the energy emitted by a blackbody at a specific wavelength and its comparison to the energy emitted by a second blackbody at the same wavelength but at a different temperature. Letting T_0 be a reference temperature,

$$\frac{N_b\,(\lambda,T)}{N_b\,(\lambda,T_0)} = \frac{\exp\,(C_2/\lambda T_0) - 1}{\exp\,(C_2/\lambda T) - 1} \tag{17}$$

Strip Lamp. A strip lamp is used for the calibration of secondary pyrometers (see Fig. 17). The notched portion of the tungsten strip is the target. A pyrometer that has been calibrated against a radiating blackbody is sighted on the target, and the strip-lamp current is adjusted to radiate at the intensity of the blackbody as transferred by the primary pyrometer. The secondary pyrometer is then substituted for the first, and the current required to raise its lamp filament to the brilliance of the target of the strip lamp is noted.

Measurement of Radiation from Real Objects

Real objects in surroundings generally hotter or cooler than themselves radiate, absorb, and reflect energy. The portion radiated is called the emissivity ϵ. If the portion reflected is r, then

$$\epsilon + r = 1 \tag{18}$$

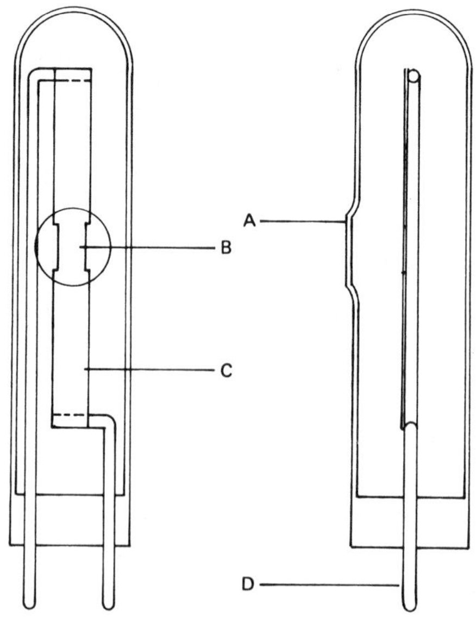

Figure 17. Strip lamp. A, plane window; B, target area of strip; C, strip; D, base pins.

No object can radiate more energy than can a blackbody at the same temperature, since a blackbody in equilibrium with a radiation field at temperature T radiates exactly as much energy as it absorbs. Any object exhibiting surface reflection must have an emissivity of less than unity. Since pyrometers are usually calibrated with respect to blackbodies, this causes one of the most serious problems encountered in the use of radiation pyrometers. The emissivities of some common materials are listed in Table 5.

The filter and screen of the pyrometer shown in Figure 15 require specific mention. From equation 17 it is evident that the observed radiation must be limited to a narrow bandwidth. Also, peak intensity of radiation does not occur at the same wavelength at different temperatures. The pyrometer is fitted with a filter (usually red) having a sharp cutoff, usually 620 nm. The human eye is insensitive to light of wavelengths longer than ca 720 nm. The effective pyrometer wavelength is ca 655 nm.

Table 5. Emissivities of Common Materials at $\lambda = 0.65\ \mu m$

Material	Emissivity	
	Solid	Liquid
carbon	0.8–0.9	
chromium	0.34	0.39
copper	0.10	0.15
gold	0.14	0.22
iron	0.35	0.37
manganese	0.59	0.59
molybdenum	0.37	0.40
nickel	0.36	0.37
platinum	0.30	0.38
silver	0.07	0.07
steel	0.35	0.37
cast iron	0.37	0.40

It is also necessary when measuring temperatures to reduce the intensity of the radiation admitted into the pyrometer, because pyrometer lamp filaments should not be subjected to temperatures >1250°C. Calibration stability may be affected by higher filament temperatures. This reduction is accomplished with a screen or screens. These may be rotating sectored disks, but in manually operated secondary pyrometers, they are usually neutral-density filters.

Several types of secondary pyrometer are available. In addition to those that measure temperature by varying lamp current, some pyrometers maintain the lamp at constant current but interpose a wedge of graduated neutral density, whose position is the measure of temperature. Also, automatic pyrometers, in which the eye is replaced by a detector and the hand on the lamp-filament rheostat by a servo system, are available. Such automatic pyrometers can be used unattended and can actuate recorders, controllers, etc. In general, the accuracy of automatic pyrometers is somewhat lower than that achieved manually by a skilled operator.

The problem of emissivity from real materials has stimulated the study of pyrometers that measure radiation at two wavelengths. The principle of the two-color pyrometer is that the energy radiating from a source at one wavelength increases with temperature at a different rate from that radiated at another wavelength. Thus temperature can be deduced from the ratio of the intensities at the two wavelengths, regardless of emissivity considerations. Two-color pyrometers are not as widely used as single-color instruments.

Other Electrical Thermometers

Many special-purpose electrical thermometers have been developed, either for use as practical thermometers or as research devices for study of temperature and the temperature scale. There are research thermometers based on thermal noise and thermometers based on the change in the speed of sound as a function of absolute temperature.

A novel and useful thermometer based on the change in the resonant frequency of a quartz crystal is produced by Hewlett-Packard. The crystal in the sensor probe is a precisely cut quartz crystal ca 8 mm in diameter. Each quartz sensor has a unique temperature-vs-frequency characteristic, and each sensor is individually calibrated with its associated electronic readout. Sensitivities of 100 μK are obtained at −50°C to 150°C and digital readouts are given. The input signal to the conditioner is very low, and the long-term stability is such that frequent adjustment of zero against reference standards, eg, the triple point of water or the gallium point, is advisable. However the direct readout in units of temperature and the great sensitivity make this thermometer system a useful and practical device in many disciplines.

Yokogawa Electric Works has developed a thermometer based on the nuclear quadrupole resonance of potassium chlorate and usable over the range from −183°C to 125°C. This thermometer involves the fundamental properties of the absorption frequency of the ^{35}Cl nucleus, and its calibration is in itself a constant of nature.

Conventional electrical thermometry is difficult or impossible or involves unacceptable hazards in some measurement situations. These include measurements where the process would interfere with the measurement, eg, in microwave and dielectric or inductive heating, high voltage–high frequency testing, and studies of the biological effects of electromagnetic radiation; measurements where the presence of an electrical current is unacceptable, eg, with highly flammable or explosive materials; or where the presence of any electrical field would perturb the quantity to be measured. A significant discipline is patient monitoring during electrocautery or electrosurgery, during which the insulating leads of electrical thermometers may become capacitive conductors. A new thermometer by Luxtron is based on rare-earth phosphors at the end of a thin fiber-optic line and resolves many of these problems (see Fiber optics). The phosphors are excited by ultraviolet illumination (see Luminescent materials). The fluorescent emissions at different wavelengths have different temperature dependences, but the intensities have a linear dependence upon intensity of excitation. Thus the ratio of emissions is independent of excitation and dependent upon temperature.

Nonelectrical Thermometers

Liquid-in-Glass. The thermal expansions of liquids are reliable analogues of temperature. Liquids are suitable for this use between their freezing and boiling points; mercury and colored alcohol are common fluids. The typical liquid-in-glass thermometer includes a thin-walled glass bulb, which is attached to a capillary stem, partially filled with a visible liquid, and sealed against the environment. That portion of the capillary that is not filled with liquid is commonly filled with a dry gas under pressure to avoid column separation. A scale is etched or fused into the stem. If the scale does not include 0°C, a reference mark is usually provided at 0°C to provide a calibration check at the ice point. A contraction chamber may be included below the main capillary to avoid the need for a long stem before the beginning of the scale.

The main problems with the use of liquid-in-glass thermometers are errors in functions of design and materials. Pressure errors are one type of problem. The change in liquid column height in response to temperature change is a function of the volume of the bulb compared to the volume of the capillary. Consequently, an external pressure or vacuum which tends to alter the bulb volume causes an error of indication. Such errors may be small when the thermometer is subjected to normal barometric variations, but they must be considered when, for example, installing a thermometer in an autoclave or pressure vessel.

Immersion errors also should be considered. The glass as well as the liquid has a temperature coefficient of expansion. Three types of thermometer, each having a different immersion requirement, are common. Complete-immersion thermometers are calibrated for use with the entire thermometer immersed. Total-immersion thermometers are calibrated for use with that portion of the thermometer immersed which contains the liquid column. Partial-immersion thermometers are calibrated for use with the thermometer immersed to a specific index mark on the stem. Most liquid-in-glass thermometers used for accurate measurements are of the total-immersion type. However, it may be difficult to read a thermometer whose mercury column is at or below the surface of the medium. In practice, the thermometer is immersed to a depth that permits the meniscus to be seen. The emergent portion of the liquid is then subject to temperature gradients. In this case or in the case of partial-immersion thermometers, it is necessary to make a correction for the temperature of the emergent stem by measuring the stem temperature with a second thermometer and calculating the correction by means of an equation, eg,

$$C_s = KN\,(t_1 - t_2) \qquad (19)$$

where C_s is the correction for stem emergence and is added algebraically to the temperature indicated by the thermometer, K is a coefficient representing the difference in expansion between the glass and the thermometric fluid (for mercury and usual glass this is 0.00016 for Celsius scales), N is the length of the liquid column extending above the medium expressed in degrees of the scale, t_1 is the temperature of the thermometer bulb, and t_2 is the average temperature of the liquid column extending above the medium.

Glass-creep errors are also encountered. The liquid-in-glass thermometer should always be used to measure temperatures in ascending order. If the thermometer is stored at room temperature, a temporary ice-point depression is produced when the thermometer is heated above room temperature. The error may be as large as 0.01 mK for 10 K of difference between the temperatures. If the thermometer is used to measure a temperature and it is then necessary to measure a lower temperature, the thermometer should be stored at a still lower temperature for at least 3 d prior to the measurement to assure recovery of bulb dimensions.

Parallax errors are occasioned by failure to read the thermometer tube from a position exactly normal to its axis. They can be reduced by taking the mean of the observations of several skilled observers but can be eliminated only by using a filar telescope mounted so as to move along an axis parallel to the axis of the thermometer stem.

Mercury thermometers are subject to separations of the mercury column or to inclusion of bubbles of the fill gas. These can often result from shipping or handling, and they cause a scale offset, which can usually be recognized by visual inspection and always recognized by a 0°C verification.

Frictional errors result from the failure of the liquid to wet perfectly the small thermometer bore. They can usually be removed by lightly tapping the thermometer, eg, with a wooden pencil, before reading.

Temperature scales are seldom identified on liquid-in-glass thermometers.

Despite these difficulties, liquid-in-glass thermometers represent a means for inexpensively making temperature measurements of moderate accuracy. Many types of liquid-in-glass thermometer are available. Among them are: thermometers for general-purpose measurement in industrial and laboratory work; thermometers constructed for the performance of specific tests, eg, ASTM materials tests; thermometers constructed for very short ranges, eg, the NBS SRM (standard reference material) 933 and SRM 934 thermometers for use in clinical chemistry; thermometers for the measurement of small differentials, eg, the Beckman thermometer; thermometers with fixed or variable built-in electrical contacts for alarm and control functions; and maximum, minimum, or maximum–minimum reading thermometers, which show the extreme temperatures reached during the period since reset.

Miscellaneous. Other nonelectrical thermometers are bimetal, filled-system, and pyrometric-cone thermometers. In bimetal thermometers, two metals of differing expansion characteristics are welded to form a strip. If one end of such a strip is fixed, the strip bends as a function of temperature as the metal of higher expansion produces a longitudinal force. If the strip is formed into a helix, the free end rotates. The displacement force actuates a pointer, a switch, or the wiper of a potentiometer. The filled-system thermometer is a system comprised of a rigid bulb, a connecting metal capillary, and a flexible bulb, eg, metal bellows or diaphragm filled with a liquid or gas of known thermal expansion properties. As temperature causes this material to expand, the flexible bulb is displaced, and the displacement is sensed as a mechanical motion and is seen as a pointer displacement, switch actuation, etc. Pyrometric cones are geometrical shapes (commonly conical) and show a physical distortion, eg, slumping or collapse, at some usually high temperature. A main use is in estimating peak kiln temperatures for firing ceramic artifacts.

Computation of Platinum Resistance Thermometers on $IPTS_{68}$

The following are step-by-step instructions for the computation of the resistance-temperature function of platinum resistance thermometers on the $IPTS_{68}$. The mathematical structure of the Scale is different for the range 0–630.74°C and −182.962°C to 0°C. For the first range, the requisite data is measured resistance at three temperatures: the triple point of water (0.01°C), the freezing point of tin (231.9681°C), and the freezing point of zinc (419.58°C). For the range −182.962°C to 0°C, the same data is required as well as a measurement at the normal boiling point of oxygen.

0–630.74°C. For the temperature range 0–630.74°C, the $IPTS_{68}$ requires the computation of an intermediate temperature t', which is neither an $IPTS_{48}$ nor an $IPTS_{68}$ temperature, but expresses the difference between the IPTS temperature and the platinum temperature. $R(t_{68})$ and $R(t')$ are, however, identical, and always represent the same hotness.

By Scale definition:

$$t_{68} = t' + M(t') \tag{20}$$

where

$$M(t') = 0.045\left(\frac{t_{68}}{100}\right)\left(\frac{t_{68}}{100} - 1\right)\left(\frac{t_{68}}{419.58} - 1\right)\left(\frac{t_{68}}{630.74} - 1\right) \tag{21}$$

t' cannot be calculated directly, but it can be calculated by a rapidly converging iterative calculation, in the first step of which it is assumed that $t' = t_{68}$, and in each successive step of which the t' which is used on the right-hand side of the equation is the t' which was obtained on the left in the previous iteration

$$t' = t_{68} - 0.045\left(\frac{t_{68}}{100}\right)\left(\frac{t_{68}}{100} - 1\right)\left(\frac{t_{68}}{419.58} - 1\right)\left(\frac{t_{68}}{630.74} - 1\right) \tag{22}$$

$$t' = t_{68} - 0.045\left(\frac{t'}{100}\right)\left(\frac{t'}{100} - 1\right)\left(\frac{t'}{419.58} - 1\right)\left(\frac{t'}{630.74} - 1\right) \tag{23}$$

and so on.

With the values in ohms of resistance obtained from calibrations at the water triple point and the freezing points of tin and zinc, the value of resistance at 0°C is calculated. This is

$$R_o = R(\text{tp})[1 + At' + B(t')^2] \tag{24}$$

but the coefficients A and B are not yet known. It causes no significant error if B is taken to be zero and A is taken to be 0.00398458; the latter the average value of A for SPRTs calibrated by the NBS over a number of years. Then the calculation of R_o reduces to

$$R_o = R(\text{tp})/1.0000398458 \tag{25}$$

Using this R_o, the ratios of the resistances obtained at the tin and zinc freezing points to the resistance R_o are determined:

$$W_{Sn} = R_{Sn}/R_o \tag{26}$$

$$W_{Zn} = R_{Zn}/R_o \tag{27}$$

With these values of ratio W, the quadratic coefficients A and B are calculated by the simultaneous solution of

$$W_{Sn} = 1 + A(t'_{Sn}) + B(t'_{Sn})^2 \tag{28}$$

$$W_{Zn} = 1 + A(t'_{Zn}) + B(t'_{Zn})^2 \tag{29}$$

The required value for t'_{Sn} may be determined from the iterative equation for t', and the value of t'_{Zn} is identical to the value of t_{68Zn}.

From the A and B coefficients, α and δ are determined directly:

$$\alpha = A + 100\,B \tag{30}$$

$$\delta = -10^4\,B/\alpha \tag{31}$$

Finally, $W = R_t/R_o$ is calculated for the temperature of interest from

$$W = R_t/R_o = \alpha\left\{t' - \delta\left(\frac{t'}{100}\right)\left(\frac{t'}{100} - 1\right)\right\} + 1 \tag{32}$$

The inverse of this equation is

$$t_{68} = \frac{1}{\alpha}\left(\frac{R(t')}{R_o} - 1\right) + \delta\left(\frac{t'}{100}\right)\left(\frac{t'}{100} - 1\right) \tag{33}$$

−182.962°C to 0°C. Over the range −182.962°C to 0°C, the IPTS temperature is defined by

$$W(t) = W'(t) + \Delta W(t) \tag{34}$$

where $W(t) = R_t/R_o$. $W'(t)$ is a reference function representing the calibration of a fictitious, nonideal platinum resistance thermometer, whose value for any t is given by

$$T_{68} = \sum_{i=0}^{20} A_i\,\{\ln W(T_{68})\}^i \tag{35}$$

where the constants A_i are provided in the text of the Scale. The need to evaluate this cumbersome Chebychev polynomial can be avoided by the use of a table of the reference function $W(t')$, which is provided in ref. 3.

For −182.962°C to 0°C,

$$\Delta W(t) = W(t) - W'(t) = A_4 t + C_4 t^3\,(t - 100) \tag{36}$$

and, from the definition of α, it follows that

$$\alpha' = \{W'(100) - 1\}/100 \tag{37}$$

$$\alpha = \{W(100) - 1\}/100 \tag{38}$$

from which

$$\Delta W(100) = W(100) - W'(100) = 100\,(\alpha - \alpha') \tag{39}$$

The text of the Scale gives $\alpha' = 0.0039259668$. Then, substituting with $t = 100°C$:

$$A_4 = \alpha - \alpha' \tag{40}$$

and solving for C_4:

$$C_4 = \frac{\Delta W(t) - A_4 t}{t^3 (t - 100)} \tag{41}$$

The table of resistance vs temperature is calculated from

$$W(t) = W'(t) + A_4 t + C_4 t^3 (t - 100) \tag{42}$$

Except for the need to insert $W'(t)$ from the NBS table for each value of t, the calculations for $IPTS_{68}$ fit within the memory range of programmable calculators, eg, the HP 67 and TI 59 (see Programmable pocket calculators).

BIBLIOGRAPHY

"Temperature Measurement" in *ECT* 1st ed., Vol. 13, pp. 677–699, by P. H. Dike, Leeds and Northrup Co.; "Temperature Measurement" in *ECT* 2nd ed., Vol. 19, pp. 774–802, by William T. Gray, Leeds & Northrup Company.

1. H. F. Stimson, *Natl. Bur. Stand. (U.S.) Monogr.* **37,** 1962.
2. H. Preston-Thomas, *Metrologia* **12,** 1 (1976).
3. J. L. Riddle, G. T. Furukawa, and H. H. Plumb, *Natl. Bur. Stand. (U.S.) Monogr.* **126** (1972).
4. H. E. Sostman, G. E. Smith, and D. J. Jones, *Three Contributions to the Triple Point of Water*, Yellow Springs Instrument Co., Yellow Springs, Ohio, 1979.
5. H. E. Sostman, *Rev. Sci. Instrum.* **48**(10), 1351 (1977).
6. L. Crovini, R. E. Bedford, and A. Moser, *Metrologia* **13,** 4 (1977).
7. H. E. Sostman, *Rev. Sci. Instrum.* **48**(2), 127 (1977).
8. J. S. Steinhart and S. R. Hart, *Deep-Sea Res.* **15,** 497 (1968).
9. J. J. Hill and A. P. Miller, *Proc. IEEE* **109B** (1962).
10. *Ibid.*, **110**(2), (1963).
11. *ASTM STP 470B*, American Society for Testing and Materials, Philadelphia, Pa., 1982.
12. R. L. Powell and co-workers, *Natl. Bur. Stand. (U.S.) Monogr.* **125** (1974).
13. P. A. Kinzie, *Thermocouple Temperature Measurement*, John Wiley & Sons, Inc., New York, 1973.
14. H. D. Baker, E. A. Ryder, and N. H. Baker, *Temperature Measurement in Engineering*, John Wiley & Sons, Inc., New York, 1961.
15. R. D. Lee, *Natl. Bur. Stand. (U.S.) Tech. Note* **483** (1969).

General References

R. P. Benedict, *Fundamentals of Temperature, Pressure and Flow Measurement*, John Wiley & Sons, Inc., New York, 1969.

H. B. Sachse, *Semiconducting Temperature Sensors and Their Applications*, John Wiley & Sons, Inc., New York, 1975.

J. F. Swindells, *Natl. Bur. Stand. (U.S.) Monogr.* **90** (1965).

F. G. Brickwede, ed., *Temperature, Its Measurement and Control in Science and Industry*, Vol. 3, Reinhold Publishing Co., New York, 1962.

H. H. Plumb, ed., *Temperature, Its Measurement and Control in Science and Industry*, Vol. 4, Instrument Society of America, 1972.

J. F. Schooley, ed., *Temperature, Its Measurement and Control in Science and Industry*, Vol. 5, Instrument Society of America, 1982.

HENRY E. SOSTMAN
Yellow Springs Instrument Co., Inc.

TERBIUM. See Rare-earth elements.

TEREPHTHALIC ACID, p-$C_6H_4(COOH)_2$. See Phthalic acid and other benzene polycarboxylic acids.

TERGITOL. See Surfactants and detersive systems.

TERNE PLATE. See Metallic coatings.

TERPENES AND TERPENOIDS. See Terpenoids.

TERPENOIDS

Terpenes are widely distributed in nature, and occur in nearly all living plants. They are generally regarded as derivatives of isoprene, wherein the isoprene units are arranged in a head-to-tail fashion, although there are some exceptions to this arrangement. The terpenes are therefore classified according to the number of isoprene units in their carbon skeletons, with a single terpene unit being regarded as two isoprene units. Terpenes are classified as in Table 1.

The terpenes are further categorized as acyclic (open chain), monocyclic (one ring), bicyclic (two rings), tricyclic (three rings), etc. Although the term terpenes might seem to be appropriate only for unsaturated hydrocarbons, it is generally recognized to apply not only to isoprene oligomers, but also to their saturated or partially saturated isomers, as well as to their derivatives, eg, alcohols, aldehydes, esters, etc. The term terpenoids properly designates such derivatives. Terpenoids, especially the oxygenated derivatives, are important flavor and perfume materials. In general, their mammalian toxicity is relatively low. Many are listed as GRAS (generally recognized as safe) as food additives and flavorings (1–2). Numerous monographs on terpenes and their derivatives

Table 1. Classification of Terpenes

Isoprene units	Carbon atoms	Classification
1	5	hemiterpene
2	10	monoterpenes
3	15	sesquiterpenes
4	20	diterpenes
5	25	sesterterpenes
6	30	triterpenes
8	40	tetraterpenes
>8	>40	polyterpenes

as fragrance materials and on their biological aspects have been published (see Food additives; Flavors and spices) (3–4).

In referring to individual compounds, terpene chemists use trivial names almost exclusively because the systematic names are frequently unmanageably long and complicated and often obscure the terpenic nature of a compound. Trivial names are used throughout this article with a few exceptions. Structural formulas, trivial names, Chemical Abstracts Service (CAS) names, CAS Registry Numbers, and properties of selected terpenes and terpenoids are listed in Tables 2 and 3. Many of the compounds have multiple registry numbers designating specific stereo chemical forms.

Chemists dealing with volatile terpene mixtures rely heavily on gas–liquid chromatography (glc) in conjunction with mass spectrometry and nmr, infrared, and uv spectroscopy. Fractional distillation is the most commonly used method for separating terpene mixtures into their components, but on a smaller scale, liquid–liquid and liquid–solid chromatography are also employed, especially to separate close-boiling components and less volatile compounds. Use of these modern analytical methods has relegated boiling points, melting points of derivatives, refractive indexes, and other conventional physical properties to a position of secondary importance, although they are frequently used in commercial product specifications.

In nature, the two most abundant natural sources of terpenes are turpentine [*8006-64-2*] and other essential oils, and nearly all commercially important terpenes were obtained from these sources until relatively recently (see Oils, essential). Since about the late 1950s, synthetic methods have been developed for manufacturing most of the industrially important monoterpenes, and these synthetics have taken over a large share of the market. Essential oils are still used in substantial quantities for their flavor and aroma characteristics, which largely depend on their terpene content.

Monoterpenes

Production. ***From Turpentine.*** The southeastern United States is the largest turpentine-producing region in the world. Nearly all industrial syntheses of monoterpenes from turpentine involve sulfate turpentine rather than gum or wood turpentine because of its availability and low price, although some wood turpentine is used, especially in the production of dipentene, camphene, and terpineol (pine oil). In 1980, U.S. production of crude sulfate turpentine was ca 98.6×10^6 L, nearly all of which was used in the chemical processing industry (5). Typically, southeastern U.S. sulfate turpentine contains 60–70 wt % α-pinene, 20–25 wt % β-pinene, and 6–12 wt % other components, including *p*-menthadienes, oxygenated terpenes, anethole [*104-46-1*], methyl chavicol [*140-67-0*], and the sesquiterpene β-caryophyllene [*87-44-5*]. The oxygenated terpenes and the *p*-menthadienes are sometimes referred to as natural pine oil. Wood turpentine from the Southeast is rich in α-pinene but contains <5 wt % β-pinene. Turpentine from the western part of the United States contains 10–20 wt % 3-carene for which no significant chemical uses exist, but this turpentine is useful for its α-pinene content.

Industrially, turpentine is separated into its components by high efficiency vacuum fractionation, as indicated in Figure 1.

The chemical uses of α-pinene and β-pinene are intertwined. An equilibrium between the two is easily obtained over a supported palladium catalyst (6) and lies heavily in favor of the α-isomer (α:β = ca 95:5). There is one commercial plant for the

Table 2. Some Properties of Selected Monoterpene Hydrocarbons and Cineoles

Common name	CAS Reg. No.	Bp, °C at 101.3 kPa[a]	Bp, °C at 13.33 kPa[a]	Mp, °C	Density, d^{20}, g/cm^3	n_D^{20}	$[\alpha]_D$	Gas-chromatography retention time relative to α-pinene on Carbowax 4000
tricyclene	[*508-32-7*]	152	85	+65			0	0.96
α-pinene	[*80-56-8*]	156	89	−50; −75, (±)	0.8595	1.4658	±51	1.00
α-fenchene	[*471-84-1*]	157	91.5		0.8697	1.4740	±44	1.22
camphene	[*79-92-5*]	158	91	+49			±108	1.28
β-pinene	[*127-91-3*]	165	98	−50; −61, (±)	0.8722	1.4790	±22	1.58
myrcene	[*123-35-3*]	167			0.7880[b]	1.4680[b]	0	1.70
cis-pinane	[*6876-13-7*]	168	101	−53	0.8575	1.4629	±23	1.25
p-8-menthene	[*6252-33-1*]	169			0.8142	1.4554	0	*trans*, 1.35 *cis*, 1.60
trans-2-*p*-menthene	[*1124-26-1*]	169			0.8100	1.4500	±132	1.17
p-3-menthene	[*500-00-5*]	169	102		0.8129	1.4519	±141	1.18
trans-*p*-menthane	[*1678-82-6*]	170	103		0.7928	1.4367	0	0.97
3-carene	[*13466-78-9*]	170	104		0.8617	1.4742	±16	1.82
cis-*p*-menthane	[*6069-98-3*]	172	105	−90	0.8002	1.4431	0	1.10
1,4-cineole	[*470-67-7*]	172	105.5	−46	0.8986	1.4446	0	2.09
1,8-cineole	[*470-82-6*]	174	108	+1	0.9245	1.4574	0	2.57
α-terpinene	[*99-86-5*]	175	108		0.8315[b]	1.4755[b]	0	2.09
carvomenthene	[*499-94-5*]	176	110		0.8246[c]	1.4563[c]	±118	1.72
p-4(8)-menthene	[*31105-55-0*]	176	110			1.4689	0	1.71
limonene	[*5989-27-5*]	176.5	110	−74; −89, (±)	0.8411	1.4730	±124	2.29
p-cymene	[*25155-15-1*]	177	110	−73	0.857	1.4905	0	3.15
γ-terpinene	[*99-85-4*]	182	116		0.8455[b]	1.4715[b]	0	2.88
p-3,8-menthadiene	[*586-67-4*]	183.5	117.5		0.851	1.4876	±140	3.12
p-2,4(8)-menthadiene	[*138-86-3*]	185	120		0.8535[b]	1.5030[b]	±49	3.66
terpinolene	[*586-62-9*]	186	120		0.8620	1.4861	0	3.51

[a] To convert kPa to mm Hg, multiply by 7.5.
[b] Temperature = 25°C.
[c] Temperature = 18°C.

isomerization of α-pinene to β-pinene, and it has an annual capacity of 2300 metric tons. The optical activity of the β-pinene produced is the same as that of the starting α-pinene. Southeastern U.S. turpentine contains α-pinene which is ca 35 wt % of the optically active (+)-isomer. For many uses of β-pinene, optical activity is not important, but in cases where it is, natural β-pinene, which contains 92–95 wt % of the optically active (−)-isomer, must be used.

One of the chief uses for β-pinene has been as a starting material for the manu-

Table 3. Some Properties of Selected Oxygenated Terpenes

Name	CAS Reg. No.	Bp, $°C_{kPa}$ at 101.3 kPa[a]	Bp, $°C_{kPa}$ at 13.33 kPa[a]	Mp, °C	Density, d^{20}, g/cm³	n_D^{20}	$[\alpha]_D$
fenchone	[*1195-79-5*]	193	122	+5	0.9452	1.4628	±70
linalool	[*78-70-6*]	199			0.8607	1.4616	±22
α-fenchol	[*1632-73-1*]	200	133	48; 39, (±)	0.935[b]	1.4734	±12.5
citronellal	[*106-23-0*]	203			0.851	1.4467	±12
terpinen-1-ol	[*586-82-3*]	208	$106_{3.33}$[a]		0.9171[c]	1.4701[c]	
camphor	[*76-22-2*]	209		179			±45
trans-β-terpineol	[*7299-40-3*]	209	139	33	0.919	1.4712[d]	0
trans-dihydro-terpineol	[*3901-93-7*]	209		35		1.4630	0
cis-dihydro-terpineol	[*3901-95-9*]	210		47	0.8962[e]	1.4664	0
trans-menthone	[*89-80-5*]	210	138	−6	0.8903[c]	1.4500	±29
terpinen-4-ol	[*562-74-3*]	211	$123_{6.66}$[a]		0.9259	1.4762	±29
neomenthol	[*491-01-0*]	212		−22; 53, (±)	0.8917[d]	1.4604	±20
borneol	[*507-70-0*]	212		209; 210, (±)			±38
isoborneol	[*124-76-5*]	214		212 *dl*			±34
menthol	[*1490-04-6*]	216		43; 38, (±)	0.8911[d]	1.4615	±50
γ-terpineol	[*586-81-2*]	218	$132_{6.66}$[a]	70	0.9412	1.4912	0
α-terpineol	[*98-55-5*]	219	149	40; 35, (±)	0.9336	1.4831	±112
citronellol	[*106-22-9*]	224			0.8550	1.4559	±7
nerol	[*106-25-2*]	225	$128_{3.33}$[a]		0.8735[c]	1.4736[c]	0
citral	[*141-27-5, 106-26-3*]	228			0.8972[f]	1.4891	0
geraniol	[*106-24-1*]	230	$131_{3.33}$[a]		0.8770[c]	1.4756[c]	0
carvone	[*99-49-0*]	231	157		0.9550[c]	1.4990	±62
hydroxycitronellal	[*107-75-5*]		$116_{0.67}$[a]		0.9220	1.4494	±10.5
terpin (ordinary)	[*80-53-5*]	258		105; 117, hydrate			0
α-ionone	[*6901-97-9*]	258			0.9309	1.4971	±400
β-ionone	[*14901-07-6*]	271		−35	0.9461	1.5202	0
nerolidol	[*40716-66-3*]	276			0.8778[g]	1.4795	±15

[a] To convert kPa to mm Hg, multiply by 7.5.
[b] Temperature = 40°C.
[c] Temperature = 25°C.
[d] Temperature = 30°C.
[e] Temperature = 45°C.
[f] Temperature = 15°C.
[g] Temperature = 22°C.

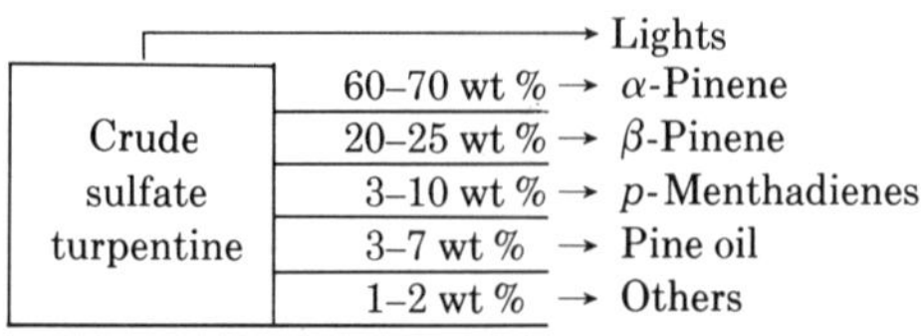

Figure 1. Distillation of crude sulfate turpentine.

facture of geraniol, nerol, and linalool. The first step in the process is the thermal rearrangement of β-pinene to myrcene, which is then hydrochlorinated in the presence of a cuprous chloride catalyst to obtain a mixture of geranyl chloride [*5389-87-7*], neryl

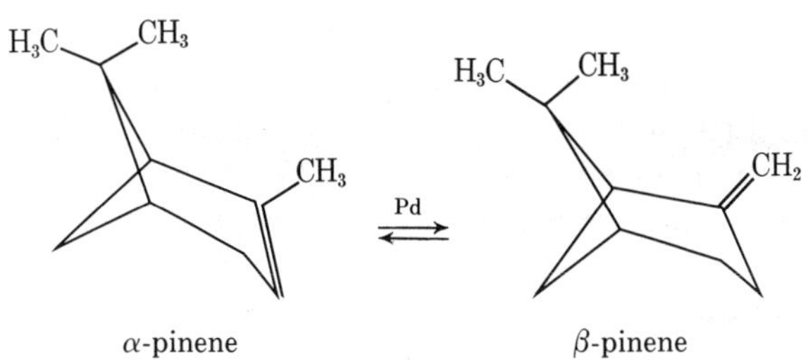

chloride [*20536-36-1*], and linalyl chloride [*471-10-3*], in ratios of ca 50:40:10 (7–8). These chlorides are best converted to alcohols via their acetate esters, which are produced by displacement of the chloride ion by reaction with sodium acetate in the presence of an amine catalyst (9). Saponification of the acetates then produces the alcohols, which are separated by efficient fractionation. If predominately linalool is desired, displacement of the chloride ion can be effected through a solvolytic process involving, eg, acetic acid (10).

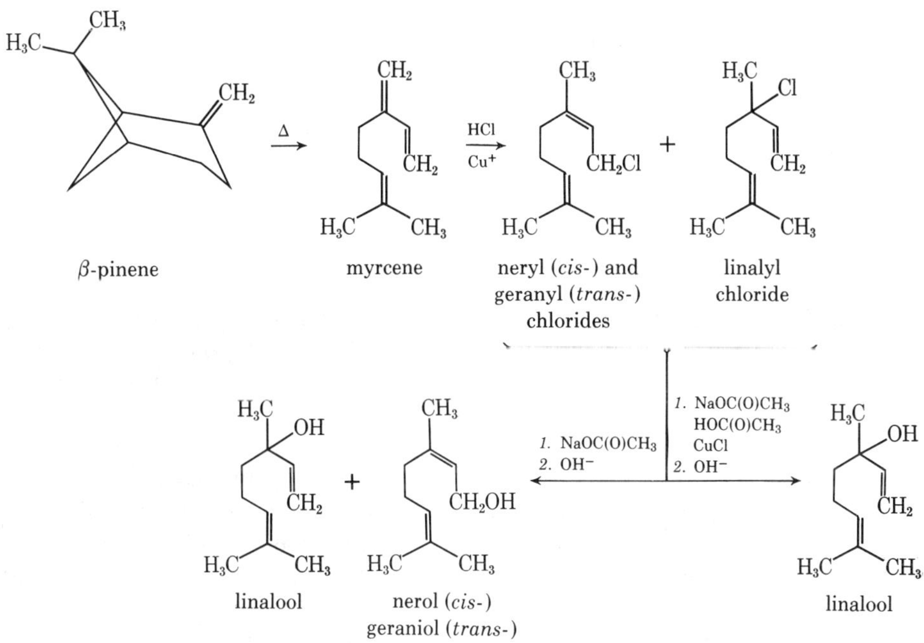

Both the esters and alcohols have rosy, floral, fruity-type odors and are used widely in perfumery (see Perfumes).

The same alcohols are also produced from α-pinene. In this process, α-pinene is selectively hydrogenated to *cis*-pinane (ca 97% selectivity) in the presence of a partially poisoned nickel catalyst (11). Any *trans*-pinane [*33626-25-4*] formed is worthless inasmuch as it does not participate in the rest of the process. The *cis*-pinane is converted to pinane hydroperoxide [*28324-52-9*] (75 wt % *cis*, 25 wt % *trans*) by reaction with air or oxygen in the presence of a free radical initiator. Reduction of the hydroperoxide is accomplished either chemically or by catalytic hydrogenation to produce *cis*-2-pinanol [*4948-21-1*] and *trans*-2-pinanol [*4948-29-2*]. Pyrolysis of 2-

pinanol [*473-54-1*] then produces linalool. The cis/trans isomers can be pyrolyzed together or can be separated by high efficiency fractional distillation and the individual isomers pyrolyzed. If (−)-α-pinene is the starting material, *cis*-2-pinanol gives (+)-linalool and *trans*-2-pinanol gives (−)-linalool (12). A subtle difference in the odor of the two enantiomers exists, some perfumers preferring that of the (+)-isomer (see Perfumes). Conditions of pyrolysis are critical. Linalool cyclization products are formed in substantial amount by excess pyrolysis and are very difficult to separate from linalool by fractional distillation. The cyclization products have no known use.

Linalool is isomerized to geraniol–nerol by use of a vanadium catalyst (13–15). It is proposed that orthovanadate esters transesterify with the linalool, and these rearrange forming the geranyl–neryl esters which, in turn, transesterify with free linalool to produce free geraniol–nerol. The equilibrium mixture contains ca 65 wt % linalool and 35 wt % geraniol–nerol. The amount of geraniol–nerol can be enhanced by carrying out the isomerization on a borate ester of linalool or by forming borate esters as the isomerization proceeds (15). Yields of geraniol–nerol are as high as 90%.

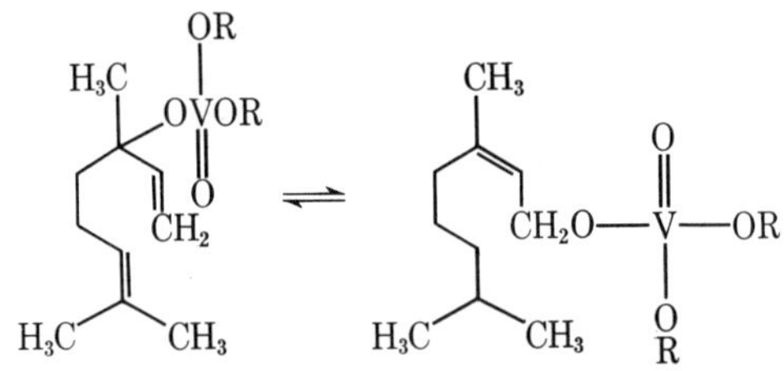

linalool orthovanadate ester — geraniol–nerol orthovanadate ester

From Acetylene and Acetone. Whereas the synthesis of oxygenated monoterpenes from turpentine involves transformations of molecules already containing 10 carbon atoms, the so-called acetylene–acetone route represents a total synthesis of monoterpenes. Industrially, acetone reacts with sodium acetylide in liquid ammonia, producing 2-methyl-3-butyne-2-ol (methylbutynol) (16). Methylbutynol is partially hydrogenated over a Lindlar catalyst to obtain methylbutenol (2-methyl-3-buten-2-ol) (16–18). The acid-catalyzed reaction of methyl isopropenyl ether with methylbutenol gives β-methylheptenone [*110-93-0*] (6-methyl-5-hepten-2-one), which on subsequent reaction with sodium acetylide produces dehydrolinalool [*29171-20-8*] (3,7-dimethyl-6-octene-1-yn-3-ol) (19). Dehydrolinalool is readily hydrogenated to linalool. The intermediate β-methylheptenone has also been made from methylbutenol by its reaction with diketene followed by thermal rearrangement, and by reaction with ethyl acetoacetate via the Carroll reaction (16,18,20–22).

From Isobutylene. Another route to the key intermediate methylheptenone is a very direct one-step process. Isobutylene, formaldehyde, and acetone under high pressure and relatively high temperature react forming methylheptenone (23). This is the isomeric methyl-α-heptenone [*10408-15-8*] (6-methyl-6-hepten-2-one). Methyl-α-heptenone, however, is easily converted to methyl-β-heptenone [*10408-15-8*] over a palladium or metal carbonyl catalyst (24–25). Although the conversion to methyl-α-heptenone is not extremely high, unreacted isobutylene and acetone can be recovered by distillation and recycled to the process, and yields are sufficiently high for commercial use of the process (26–27). That the yields are as good as they are is

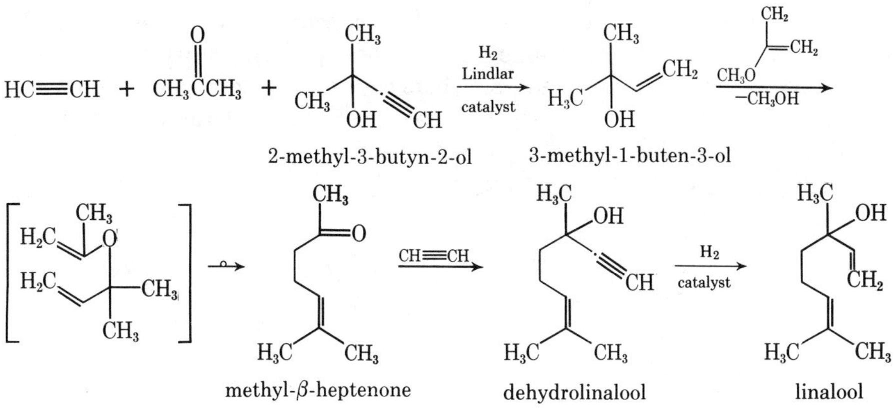

rather surprising considering that the reactants can react with themselves and with each other under the rather drastic conditions of the process. Variations of this method are described elsewhere (28–31). The methyl-β-heptenone of this process can be converted to dehydrolinalool by reaction with sodium acetylide and to linalool by subsequent hydrogenation.

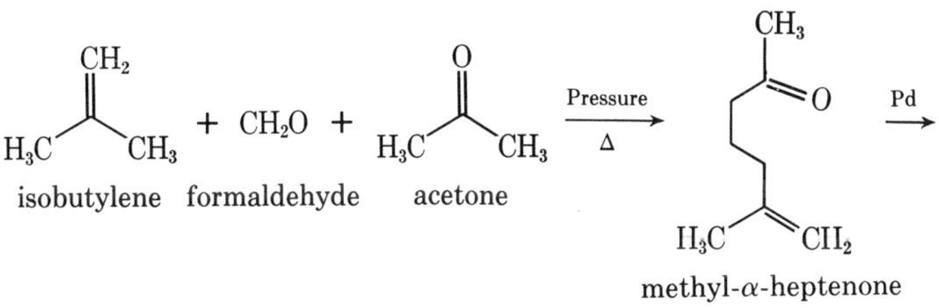

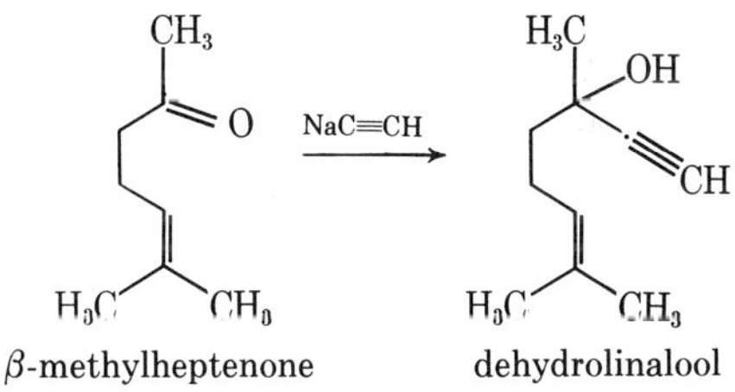

From Isoprene. The hemiterpene isoprene [*78-79-5*] is a very high volume petrochemical used principally in the manufacture of *cis*-polyisoprene rubber. The first step is the production of prenyl chloride [*503-60-6*] (1-chloro-3-methyl-2-butene) by continuous hydrochlorination of isoprene by either gaseous or aqueous HCl; any isomeric 3-chloro-3-methyl-1-butene formed as a by-product can be eliminated by recycling (32). Prenyl chloride next reacts with acetone in the presence of a quaternary ammonium salt catalyst and solid sodium hydroxide at modest temperatures to produce β-methylheptenone (33–35). Again, dehydrolinalool is obtained by reaction of the ketone with acetylene. A continuous ethynylation at atmospheric pressure in the presence of NaOH and *N*-methylpyrrolidinone has been achieved (36).

α-Pinene. ***Reactions.*** Historically, the chief industrial reactions of α-pinene have been acid-catalyzed reactions to produce pine oil and camphene. Recently, however, two additional large-scale uses have been developed commercially. These are the isomerization of α-pinene to β-pinene and the conversion of α-pinene to linalool, geraniol, and nerol (37). Considerable study has been given to the ozonolysis of α-pinene to produce pinonic and pinic acids, but there has been no significant commercialization (38–40). Air oxidation of α-pinene has also received considerable attention (41).

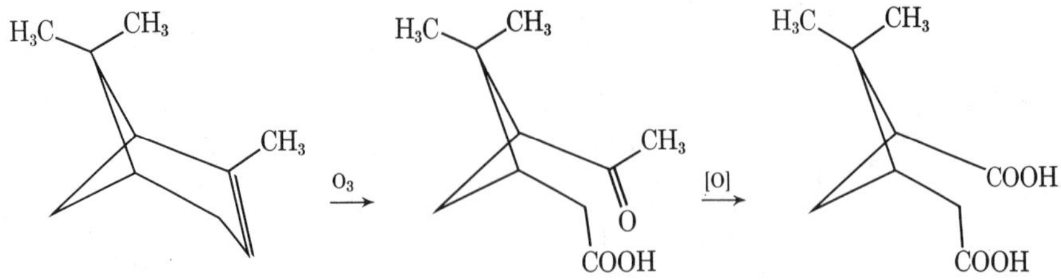

α-pinene pinonic acid pinic acid

Among the products produced by this process are α-pinene oxide [*1686-14-2*], 3-pinene-2-ol [*43119-66-0*], verbenol [*473-67-6*], verbenone [*80-57-9*], myrtenol [*515-00-4*], and myrtenal [*564-94-3*]. The multiplicity of products and the fact that the reaction can be taken only to 20–25% conversion (extensive decomposition and polymerization occur above this level) account for this process not having been commercialized even though some of the products are potential precursors for more valuable products, eg, menthol.

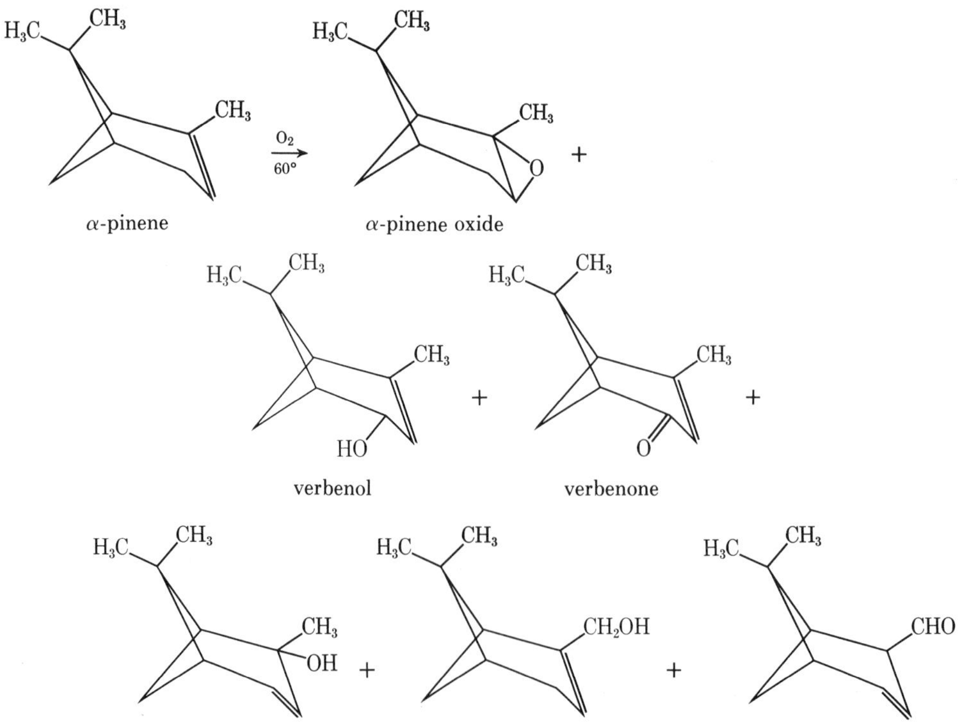

α-pinene α-pinene oxide

verbenol verbenone

3-pinene-2-ol myrtenol myrtenal

Pyrolysis of α-pinene produces dipentene [*138-86-3*] and alloocimene [*673-84-7*] (7). A significant amount of ocimene [*29714-87-2*] (a primary pyrolysis product) can be obtained on a small scale by quick quenching, as for example by immersing a hot (ca 500°C) wire in a large amount of α-pinene maintained at not more than 100°C, but the method is not useful commercially (42).

Manufacture, Storage, and Price. α-Pinene can be obtained readily by isomerization of β-pinene over a variety of catalysts with retention of optical activity (6,43). For most industrial uses, α-pinene from sulfate turpentine is desulfurized before further processing. Desulfurization techniques involve fractionation, absorption, hypochlorite treatment, treatment with metals, or a combination of techniques (44–51). Commercial α-pinene is typically 90–95% pure and is usually shipped in tank cars, tank trucks, or deck tanks. Storage conditions should preclude autoxidation. About 47,000 t of α-pinene were produced in the United States in 1980 (52). The bulk price of the technical-grade product is ca $0.80/kg (53).

Uses. α-Pinene is an important starting material for large-scale manufacture of a variety of other terpenic hydrocarbons, alcohols, aldehydes, and ketones. It also is used in the manufacture of terpene resins. High purity α-pinene is used to a small extent in flavor and perfume applications.

β-Pinene. ***Reactions.*** β-Pinene undergoes many of the same reactions as α-pinene but is seldom used for these reactions when the less costly and more plentiful α-pinene is available. Some of these reactions, for example, are hydration to pine oil, acid rearrangement to camphene, isomerization to *p*-menthadienes by prolonged aqueous acid treatment, hydrogenation to pinanes, dehydrogenation to *p*-cymene, and equilibration to a 95:5 mixture of α-pinene and β-pinene. β-Pinene often reacts under milder conditions, being the higher energy isomer, and sometimes gives somewhat different product ratios and yields. One of the two largest uses of β-pinene depends on its ability to polymerize readily, either alone or with other terpenes, in the presence of Lewis acids to produce terpene resins.

The second large use of β-pinene is in chemical manufacturing. For the chemical manufacturer the most important reaction of β-pinene is its thermal isomerization to myrcene, which is an important intermediate for the production of a variety of terpenic aroma and flavor chemicals and vitamin intermediates (7).

The first commercially significant aroma compounds to be made from β-pinene were nopol [*128-50-7*] (54) and certain of its esters, especially the acetate [*128-51-8*]. Nopol results when β-pinene reacts via a Prins-type reaction with formaldehyde in the presence of an acidic catalyst under pressure at up to ca 225°C.

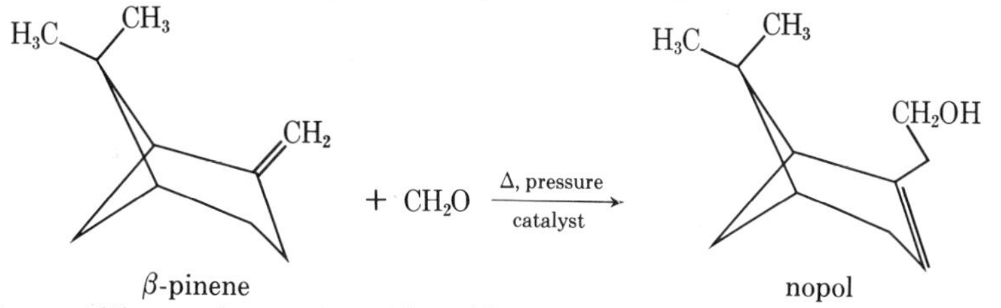

β-Pinene, like α-pinene, is oxidized by air or oxygen, but no commercial application of this reaction has been made for the same reasons as in the case of α-pinene. Ozonization of β-pinene produces nopinone [*24903-95-5*] and formaldehyde, but this reaction has no commercial significance.

Manufacture, Shipping and Storage, and Price. β-Pinene is obtained by fractionation of turpentine and to a much lesser extent by the isomerization of α-pinene. No totally synthetic method for its production exists. Purities of the commercial grades are ca 80–95%, depending on their intended use. β-Pinene is shipped in tank cars, tank wagons, and deck tanks; very high purity β-pinene for use in flavor and perfume applications is shipped in lined drums. Prolonged storage requires conditions precluding autoxidation and polymerization. Much of the β-pinene produced is used captively by the chemical industry. Like α-pinene, β-pinene occurs in minor amounts in many essential oils, but these do not serve as a source of the hydrocarbon. U.S. production of β-pinene in 1980 was ca 21,000 t (52). The resin grade sells for ca $1.00/kg (53).

Uses. The two most important uses of β-pinene are as a starting material for manufacturing aroma and flavor compounds and as a monomer for terpene resins (see Terpene Resins). A minor amount of highly purified β-pinene is used in certain flavor and aroma applications.

3-Carene. (+)-3-Carene, like the pinenes, is obtained by fractional distillation of turpentine. Turpentine from southeastern United States contains negligible amounts of 3-carene, but western U.S. and Canadian turpentines contain 20–25 wt % of this hydrocarbon. Many European turpentines contain appreciably higher amounts, and turpentines from the India–Pakistan area contain ca 55–65 wt % (55). Production volume of this hydrocarbon is unknown.

Reactions. None of the reactions used for α- and β-pinene to produce commercial products has been successfully applied to (+)-3-carene. Much of the industrial research on (+)-3-carene has been directed towards taking advantage of the fact that this hydrocarbon is essentially fully optically active. Two important targets have been the synthesis of (−)-menthol [*2216-51-5*] and (+)-*trans*-chrysanthemic acid [*4638-92-0*].

The potential commercial (−)-menthol syntheses from (+)-3-carene mostly depend on the initial conversion of (+)-3-carene to (+)-2-carene [*4497-92-1*] (formerly called (+)-4-carene) over a wide variety of isomerization catalysts (56–57). (+)-2-Carene can be converted by pyrolysis to (+)-*trans*-isolimonene [*5133-87-1*] via a 1,5 sigmatropic shift. Selective hydrogenation of (+)-*trans*-isolimonene produces (+)-*trans*-2-*p*-menthene [*5113-93-9*], which can be epoxidized to 2,3-epoxymenthane [*5856-74-6*], which is convertible to a mixture of (−)-menthol isomers and carvomenthol [*499-69-4*] isomers. No appreciable use for the carvomenthol isomers exists. Variations of the route from (+)-*trans* isolimonene to optically active menthol exist and are discussed elsewhere (55–58). These involve conversion of (+)-*trans*-isolimonene to (+)-3-menthene or (+)-*p*-2,4(8)-menthadiene [*43124-56-7*] as intermediates.

(+)-*trans*-Chrysanthemic acid [*827-90-7*] is the important acid moiety of pyrethrins (esters), both synthetic and natural; the (−)- and (±)-isomers of the acid are much less effective in synthetic pyrethrins. Pyrethrins are highly desirable insecticides because of their effectiveness at low concentrations, their low mammalian toxicities, and their nonpolluting character (see Insect control technology). Examination of the configuration of (+)-3-carene suggests that it would be a good starting material in the production of chrysanthemic acid of the required chirality. The various routes to chrysanthemic acid from carene are long and arduous and are described elsewhere (55,59–68). A commercial process for manufacturing (+)-*trans*-chrysanthemic acid from (+)-3-carene has not been established.

Fragrance chemicals from 3-carene have also been the subject of many studies. A review of some of these studies has recently been published (69), but as yet none of the fragrance materials described have become large-scale items of commerce.

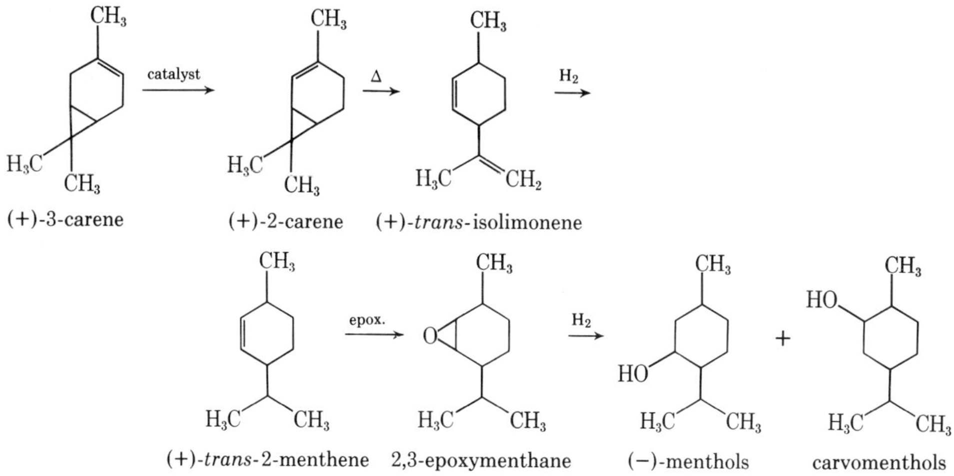

3-Carene can be converted to *m*-menthadienes, usually with appreciable quantities of *p*-menthadienes. The latter have very close boiling points; thus, separation of the meta derivatives is very difficult if not impossible by usual fractionation techniques. A mixture of carvestrene–dipentene has been prepared by isomerization of 3-carene over η-alumina and is useful in the preparation of terpene-type resins, but the process has not been commercialized (70). Some of the chemistry of the *m*-menthadienes and *m*-menthanes has been reviewed recently (71).

Hydrogenation of 3-carene gives chiefly *cis*-carane [*18968-24-6*] under mild conditions, with some 1,1,4-trimethylcycloheptane [*2158-55-6*]; the latter can be made to predominate under more severe conditions (72). 3-Carene autoxidizes readily, producing a variety of products, but this reaction is not industrially important (73). Pyrolysis of 3-carene at 560°C gives a mixture of nine identified hydrocarbons containing the benzene nucleus (74). 3-Carene is not a good resin former.

3-Carene has little industrial use other than as a solvent.

Camphene. ***Reactions.*** One of the most important commercial reactions of camphene is chlorination. Toxaphene [*8001-35-2*], a potent insecticide that is especially effective against the cotton boll weevil, is produced in tonnage quantities yearly by chlorination of camphene to a chlorine content of ca 66–68 wt % (75). The second most important commercial reaction depends on the readiness with which camphene rearranges to the isobornyl structure. When camphene reacts with acetic acid in the presence of a strong acid catalyst, isobornyl acetate [*125-12-2*] (a camphor intermediate) is produced. Acid-catalyzed reaction of camphene with alcohols or glycols produces isobornyl ethers or hydroxyethers, which are notable for their woody, cedarlike odors.

Manufacture. Camphene is produced commercially by treatment of α-pinene with acidic catalysts in the absence of water. It can be produced similarly from β-pinene, but seldom is because β-pinene is more expensive. Tricyclene [*508-32-7*] is always obtained with camphene as an equilibrium product. The isomerization of α-pinene is generally carried out at reflux temperature in the presence of titania or activated clays. Preparation of the catalyst has a great influence on product yield and composition. Many other diverse acidic catalysts have been evaluated but apparently have not achieved commercial importance. The chief by-products of the isomerization are

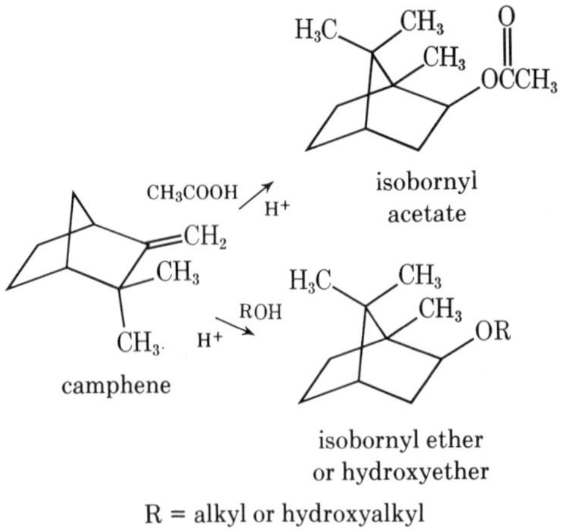

p-menthadienes, which are collectively referred to as dipentene in the trade. These are removed from the camphene–tricyclene product by fractional distillation, and then the camphene–tricyclene is distilled at reduced pressure to produce cuts containing varying amounts of the two products which are blended to suit the intended end use. Commercial camphene contains a camphene:tricyclene ratio of ca 4:1 and has a minimum melting point of ca 39–46°C. In most applications, tricyclene undergoes the same chemical reactions as camphene. Camphene is shipped in tank cars, deck (ship) tanks, and drums. Its price is ca $1.20–1.30/kg (53).

The isomerization of α-pinene to camphene is a classical example of the Wagner-Meerwein rearrangement and is depicted as follows:

Uses. The most important uses of camphene are in the production of toxaphene, isobornyl acetate, isoborneol, and camphor. Comparatively small amounts are used in producing fragrance chemicals by reaction with phenols; typically such compounds have woody odors. Small amounts of camphene are used in fragrances and synthesized essential oils.

p*-Menthadienes and *p*-Cymene.** ***Reactions. *p*-Menthadienes are equilibrated by either acid or base. The equilibrium compositions and rate constants have been determined and discussed (76).

In the acid isomerization, small amounts (2–5 wt %) of *p*-cymene and *p*-menthenes form by disproportionation. After thorough washing of the equilibrium mixture, the three main components can be separated into technically pure products. Although they do not join in the equilibrium, limonene and terpinolene can be used as starting materials to attain the equilibrium. Limonene isomerizes rather slowly to terpinolene, but the irreversible isomerization of terpinolene to α-terpinene is very rapid.

All of the *p*-menthadienes can be hydrogenated to produce a mixture of *cis*- [*6069-98-3*] and *trans*-*p*-menthane [*1678-82-6*], the hydroperoxide of which is useful as a polymerization initiator in the rubber industry (see Rubber chemicals). Conversely, the *p*-menthadienes can be dehydrogenated to produce *p*-cymene or ultimately α-methyl-*p*-methylstyrene [*1195-32-0*].

p-Menthadienes have been disproportionated to produce substantial amounts of *p*-3-menthene, which has been used to produce (±)-menthol commercially (77). Nonconjugated *p*-menthadienes having one site of unsaturation at the 4(8) or 8 po-

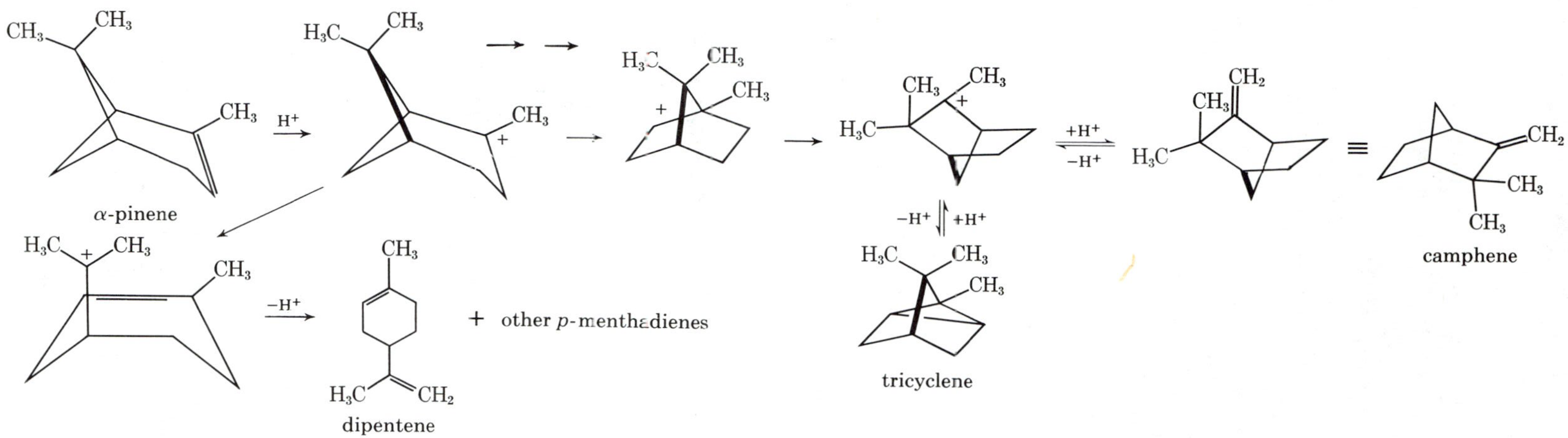
α-pinene
H⁺
−H⁺
dipentene
+ other p-menthadienes
+H⁺
−H⁺
tricyclene
camphene

sition have also been hydrogenated selectively to form *p*-3-menthene in the presence of a homogeneous ruthenium catalyst complex (78). *p*-3-Menthene of 40–50% optical activity has been obtained by hydrogenation of *d*-isolimonene prepared from *d*-3-carene (56–57,79).

Citrus (+)-limonene is the only important starting material for the production of (−)-carvone [*6485-40-1*] and proceeds via the nitrosochloride and oxime (80–81). Air oxidation of limonene has been investigated and leads to a variety of products, but commercialization is not practical (82).

Manufacture, Storage, and Price. The *p*-menthadienes are obtained as by-products from the manufacture of synthetic pine oil and camphene and from extraction of Southeastern pine stumps. Relatively small amounts are obtained on fractionation of sulfate turpentine. They can also be produced by acid treatment of α- or β-pinene, but this is rarely done. *d*-Limonene [*5989-27-5*], a large volume *p*-menthadiene, is a by-product of the citrus industry.

Commercially, most *p*-menthadienes are sold as mixtures called dipentene. Important components present in varying amounts are terpinolene, α-terpinene, camphene, tricyclene, 1,4- and 1,8-cineole, α-pinene, *p*-cymene, *p*-2,4(8)-menthadiene, α- and β-phellandrene, and γ-terpinene; minor amounts of other terpene hydrocarbons also form. Each supplier's product is somewhat different and frequently mixtures of the various types of dipentene are sold. There is a Federal specification (TT-D-376 D, 1975) for technical dipentene for use as a thinner and antiskinning agent in organic coatings. This specification is based solely on physical properties without regard to composition. ASTM D 801-57 establishes procedures for sampling and testing dipentene and likewise does not specify composition. One high quality commercially available dipentene has the following typical properties: n_D^{20} 1.472–1.485, sp gr (15.5°C/15.5°C) 0.863–0.873, color (APHA) 50 max, flash point (TCC) 43–55°C. Kauri-butanol value 113, distillation range (5–95%) 170–190°C. Dipentene is usually shipped in carbon-steel tank cars or in drums. It tends to darken on prolonged storage in black iron; antioxidants are not of much value in preventing this. Production figures are not available, but volumes are large; the price for high quality dipentene is ca $1/kg.

Uses. In addition to being an especially good solvent for paints, varnishes, and enamels that contain synthetic resins, particularly phenolic resins, commercial dipentene serves as an antiskinning agent and as a wetting agent in the dispersion of pigments (see Pigments). The solvency of dipentene for rubber and its swelling and softening properties make it useful in rubber reclaiming and in the processing of natural and synthetic rubbers. Dipentene is also formulated into a variety of cleaners similar to pine-oil cleaners. It is sometimes used as a diluent for pine oil in flotation (qv) applications.

Over 90% pure (+)- and (±)-limonene have been extensively used in the manufacture of terpene resins (see Hydrocarbon resins). In addition to being used as a starting material for the manufacture of (−)-carvone, citrus (+)-limonene is used to a small extent in flavor and fragrance compositions.

Since the early 1950s, *p*-cymene has been used in the manufacture of *p*-cresol by autoxidation followed by decomposition of the resulting hydroperoxide to *p*-cresol and acetone (see Alkylphenols). This process is no longer used on a large commercial scale in the United States, but it is still in use in Japan. *p*-Cymene is used in the synthesis of more complicated molecules and as a heat-transfer medium (see Heat exchange technology).

Myrcene. ***Reactions.*** Hydrohalogenation is the single most important reaction of myrcene, since it is the first step in the industrial synthesis of the important large-volume terpene alcohols geraniol, nerol, and linalool. The halides obtained vary with the reaction conditions. Hydrochlorination of myrcene in the presence of cuprous chloride at ca 0°C gives a product containing geranyl [*5389-87-7*], neryl [*20536-36-1*], and linalyl chlorides [*471-10-3*] in the ratio of ca 50:40:10 with small amounts of myrcenyl [*26492-09-1*] and α-terpinyl chlorides [*39864-10-3*] (8). Similar results can be obtained by use of a cupric chloride catalyst (83). It is this unpurified chloride mixture that is converted industrially to geraniol, nerol, and linalool by acetate formation and subsequent hydrolysis. The mixed chlorides prepared without the use of a copper catalyst give on solvolysis predominately linalool and α-terpineol. Reaction of myrcene with hydrogen bromide, preferably at 75–105°C and in the presence of a free-radical initiator, gives satisfactory yields of mixed geranyl [*6138-90-5*], neryl [*25996-10-5*], and linalyl bromides [*16912-98-4*], but hydrogen bromide is not used industrially for economic considerations (84).

If myrcenyl chloride is desired, the hydrochlorination of myrcene should be carried out in the presence of a Lewis-acid catalyst. In this case, very little geranyl, neryl, or α terpinyl chloride forms (85). Myrcenyl chloride is used as an intermediate in the synthesis of myrcenol and its derivatives.

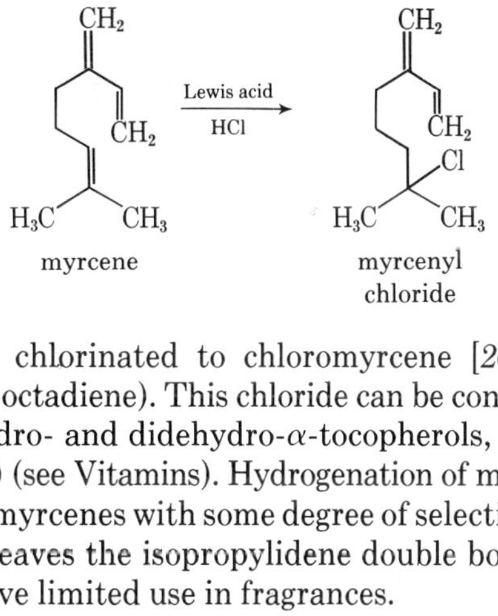

Myrcene has been chlorinated to chloromyrcene [*26492-09-1*] (3-chloro-6-methylene-2-methyl-1,7-octadiene). This chloride can be converted to vitamin E (qv) intermediates and dehydro- and didehydro-α-tocopherols, but the process has not been commercialized (86) (see Vitamins). Hydrogenation of myrcene can be controlled to give isomeric dihydromyrcenes with some degree of selectivity (87). Reaction with two moles of hydrogen leaves the isopropylidene double bond untouched. Some of these hydromyrcenes have limited use in fragrances.

Dialkylamines add to myrcene in the presence of the lithium or sodium salts of the dialkylamines producing *N,N*-dialkylgeranylamines in yields up to 85% with selectivity up to 89% (88). Dialkylnerylamines can be produced from isoprene by reaction with a secondary amine in the presence of a catalyst prepared from the reaction of a secondary amine, a conjugated diene or a polycyclic aromatic compound, and a lithium salt and sodium or potassium metal (89–90). A lithium or sodium salt of a dialkylamine can also be used as the catalyst. Reaction of myrcene hydrochloride [*4490-10-2*] with a secondary amine gives a mixture of dialkylgeranylamine and dialkylnerylamine.

Dialkylgeranylamines and dialkylnerylamines are versatile compounds and can be used to obtain a variety of fragrance compounds and vitamin intermediates. *N,N*-Diethylgeranylamine [*40267-53-6*], when treated with acetic anhydride at reflux temperature in the presence of anhydrous sodium acetate, gives geranyl acetate [*105-87-3*] in 59% yield with only 2 wt % neryl acetate [*141-12-8*] in the product (91).

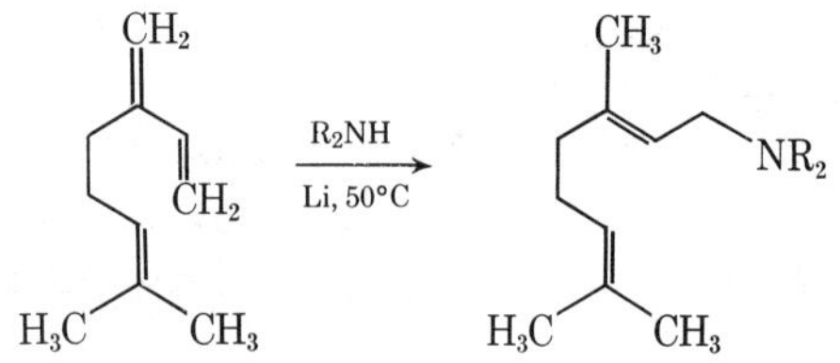

myrcene dialkylgeranylamine

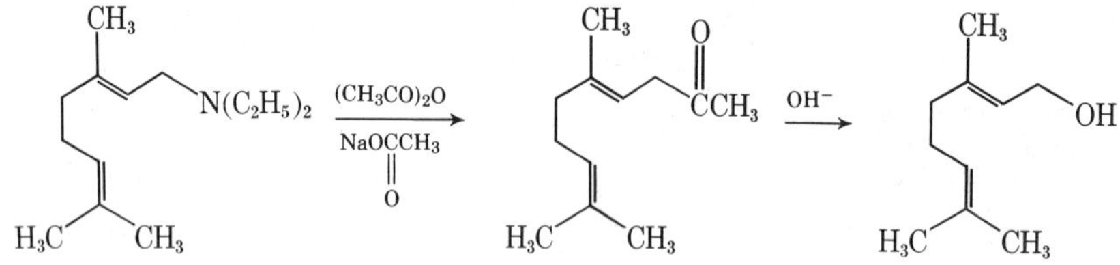

N,*N*-diethylgeranylamine geranyl acetate geraniol

N,*N*-Dipropylnerylamine [*63343-67-9*] reacts with ethyl chloroformate forming neryl chloride, which can be converted to neryl acetate by treatment with potassium acetate in acetonitrile. Saponification then gives nearly pure nerol (92).

Dialkylgeranyl(neryl)amines are hydrated under acid conditions, and the hydrate rearranges over Pd–C to an enamine which, on acid hydrolysis, produces hydroxycitronellal [*107-74-4*]. *N*,*N*-Diethylnerylamine [*40137-00-6*] reacts similarly (93).

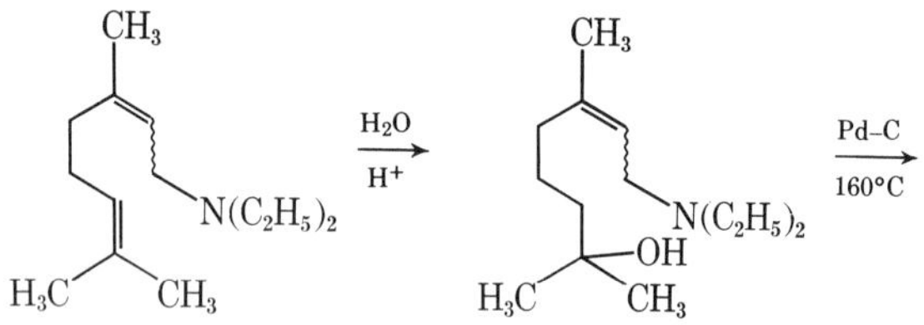

N,*N*-diethylgeranyl(neryl)amine hydrate [*57745-79-6*]

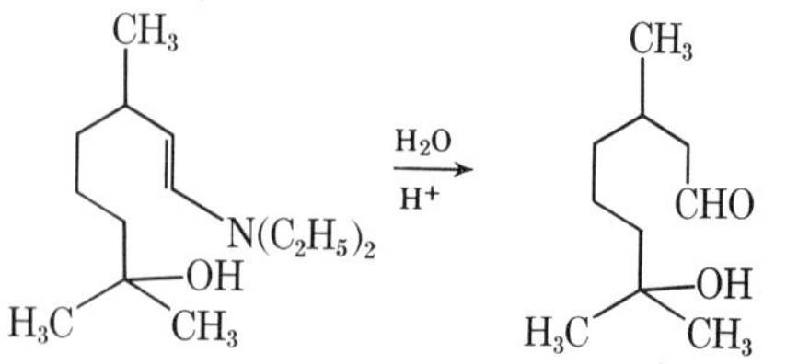

enamine [*1135-22-4*] hydroxycitronellal

N,*N*-Diethylgeranyl(neryl)amine can be isomerized to its 1,3-dienamine over a strong base, eg, an alkali metal 2-aminoethylamine in ethylenediamine. The dienamine on hydrolysis yields dihydrocitral [*57069-90-6*], a vitamin E intermediate (94)..

Reaction of diethylnerylamine with hydrogen peroxide produces amine oxides in high yield (90,95). On heating, these oxides rearrange to *O*-neryl-*N*,*N*-diethylhy-

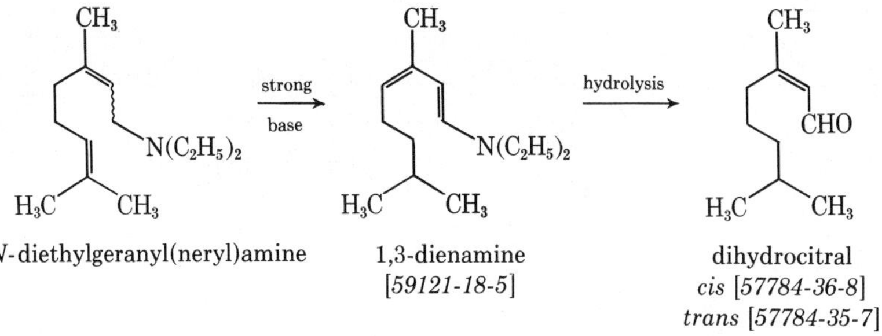

droxylamine [*57816-30-5*], which can be isolated. This hydroxylamine, on further heating, rearranges to *O*-linalyl-*N*,*N*-diethylhydroxylamine [*57784-09-5*], which is separated by distillation.

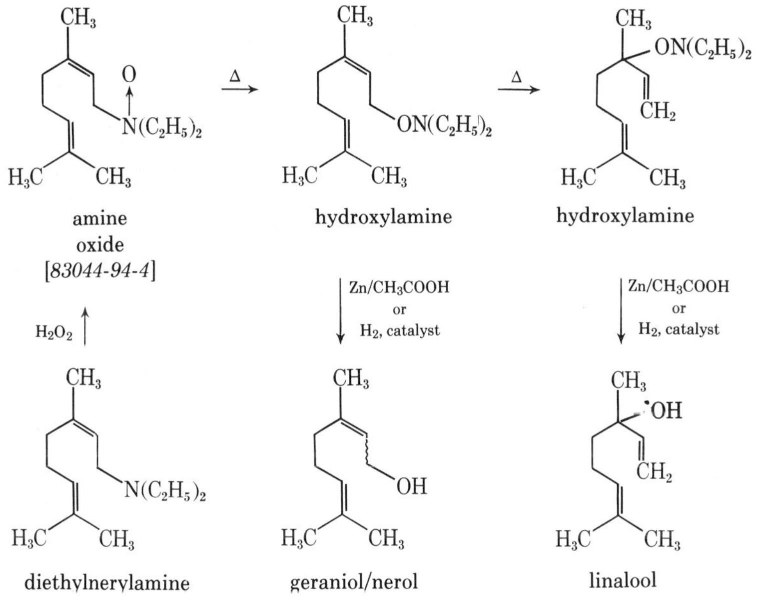

Treatment of the hydroxylamines with zinc in acetic acid gives the corresponding alcohols (90,96). The reduction can also be accomplished by catalytic hydrogenation (97).

Formation of a quaternary ammonium halide of *N*,*N*-dialkylgeranyl(neryl)-amine, acid hydration, and conversion of the resulting product to the corresponding quaternary ammonium hydroxide followed by pyrolysis can be used to prepare a mixture of myrcenol [*543-39-5*] and *cis*- [*7643-59-6*] and *trans*-ocimenols [*7643-60-9*] via a Hofmann elimination (98). If a dialkylgeranylamine is the starting material, the resulting ocimenol is essentially the *cis*-isomer. If the starting dialkylamine is a mixture of the geranyl and neryl forms, a cis–trans mixture of ocimenols with myrcenol results (99).

Myrcene hydrochloride reacts with trimethylamine forming a geranyl- (or neryl) trimethylammonium salt, which converts to, eg, β-cyclogeranyltrimethylammonium chloride on heating. Treatment of the aqueous solution of this salt with aqueous sodium

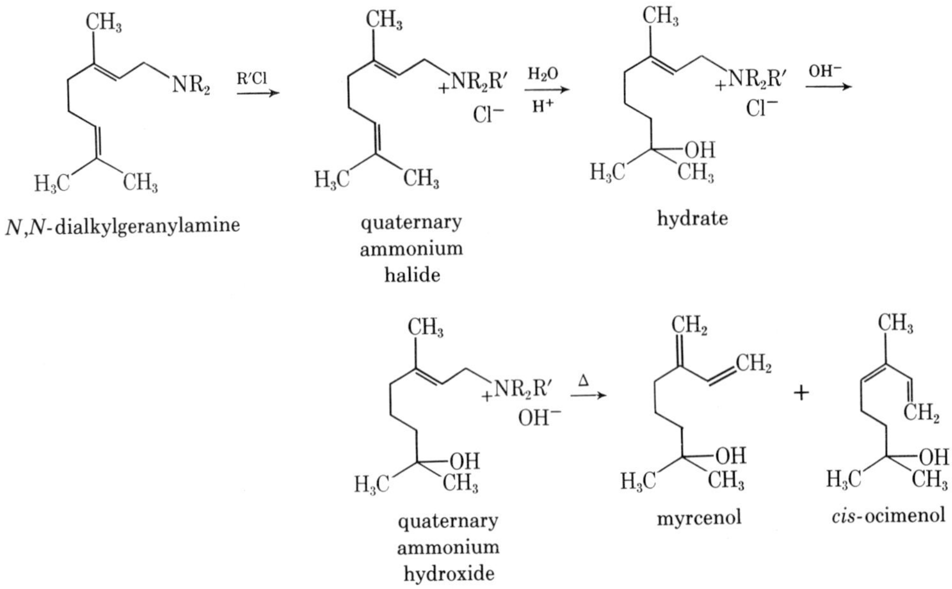

aqueous sodium hydroxide gives γ-pyronene [*514-95-4*] and δ-pyronene [*60714-17-2*] in the approximate ratio of 2:1 (100).

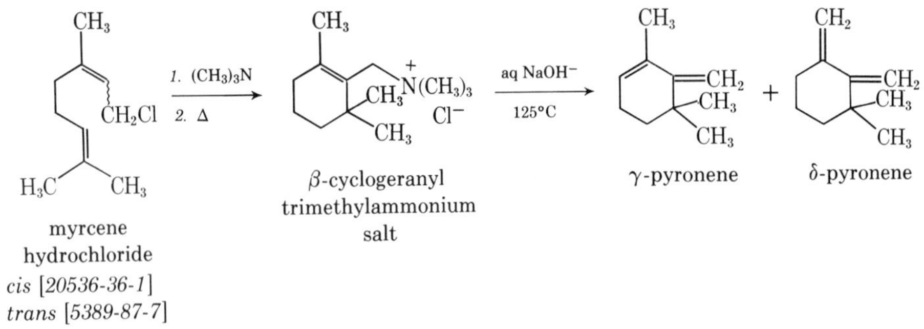

Reaction of myrcene hydrochloride with dimethylamine gives geranyl(neryl)-dialkylamines which, on heating with hydrochloric acid, produce β-cyclogeranyldimethylamine [*78098-31-4*]. Reaction of this amine with acetic anhydride yields β-cyclogeranyl acetate [*68406-89-3*]; saponification of the acetate gives β-cyclogeraniol [*472-20-8*], a potential intermediate for β-ionone (101–102).

Manufacture and Storage. Myrcene occurs to a minor extent in many essential oils, but the only important commercial source is its manufacture by pyrolysis of β-pinene at 550–600°C (103). A typical crude pyrolyzate contains 75–77 wt % myrcene, 9 wt % limonene [*138-86-3*], and a few percent of ψ-limonene [*499-97-8*] and minor cracking products. The reaction probably proceeds through an allylic diradical (104). These diradicals are not typical free radicals, in that they rearrange intramolecularly because of their proximity to one another; hence, little polymerization occurs during the pyrolysis.

The crude pyrolyzate is used without purification in the manufacture of the important alcohols, geraniol, nerol, and linalool. Myrcene can be brought to high purity

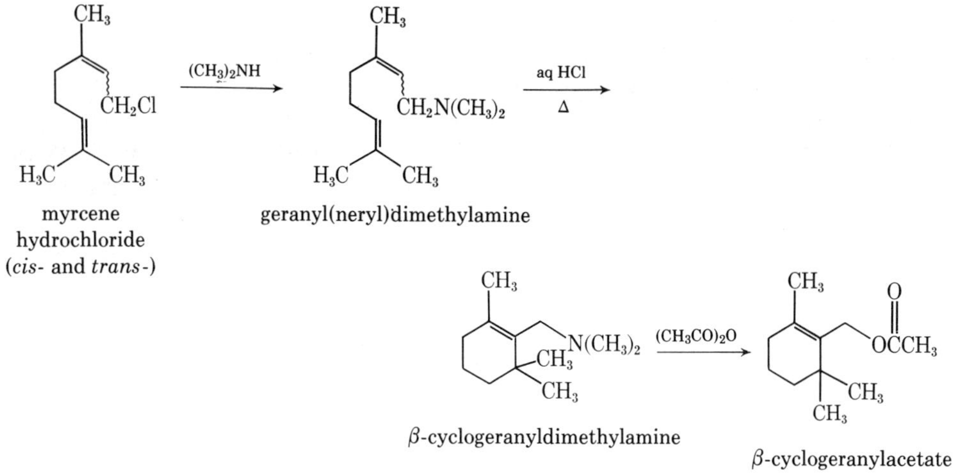

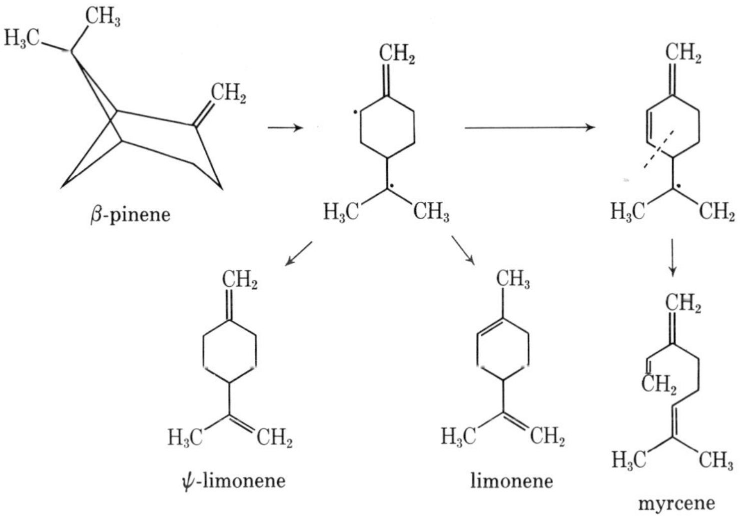

by fractional distillation if extreme precautions are taken to prevent polymerization. Myrcene dimerizes readily on prolonged heating at moderate temperatures.

During shipment or on extended storage, a polymerization inhibitor should be added to either crude or high purity myrcene. Commercial use of high purity myrcene is minimal, and is confined mostly to flavor and fragrance applications (see Flavors and spices). Reliable production and cost figures from crude myrcene are unavailable, but its production volume is large.

Uses. By far the biggest use of myrcene is as an intermediate in the commercial production of terpene alcohols, namely geraniol, nerol, and linalool. These, in turn, serve as intermediates for the production of other large-volume aroma and flavor chemicals, eg, citral, citronellol, citronellal, hydroxycitronellal, *dl*-menthol, the ionones, and others. Large volumes of the alcohols are used to manufacture intermediates for the commercial production of vitamins A and E (see Vitamins). Myrcene is also used in appreciable quantities in the manufacture of specialty aroma compounds, eg, myrcenol and its derivatives.

3,7-Dimethyl-1,6-Octadiene. ***Reactions.*** Under acid catalysis, 3,7-dimethyl-1,6-octadiene [*2436-90-1*] (citronellene, dihydromyrcene) adds water or an alcohol across the 6,7 double bond to produce the hydroxy or alkoxy derivative (105–106). These derivatives can be used in the production of hydroxycitronellal, methoxycitronellal [*3613-30-7*], dihydromyrcenol [*30385-25-2*], and methoxyelgenol [*41890-92-0*]. Hydrogen halides and lower alkyl carboxylic acids add to the diene in a similar manner, and the resulting products are useful in the synthesis of citronellol and dihydromyrcenol (107–108). The diene also reacts with aluminum hydrides, yielding trialkenyl aluminum derivatives which can be converted to citronellol (109).

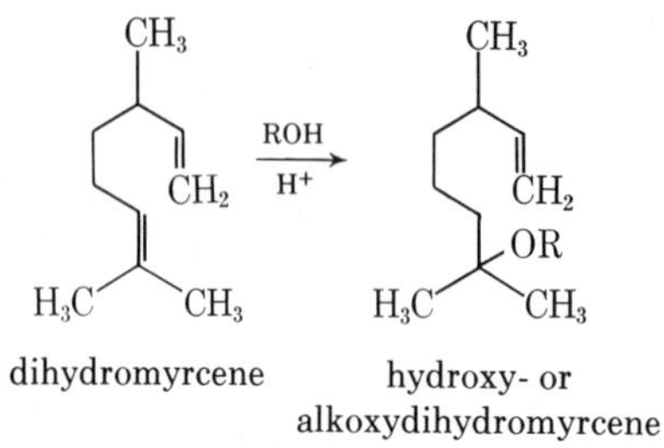

R = H or alkyl

Manufacture. 3,7-Dimethyl-1,6-octadiene is produced by the pyrolysis of pinane, preponderantly the *cis* isomer [*4755-33-3*], which can be made by hydrogenation of either α- or β-pinene or mixtures thereof (11). If the starting material is optically active, the resulting diene is optically active (110).

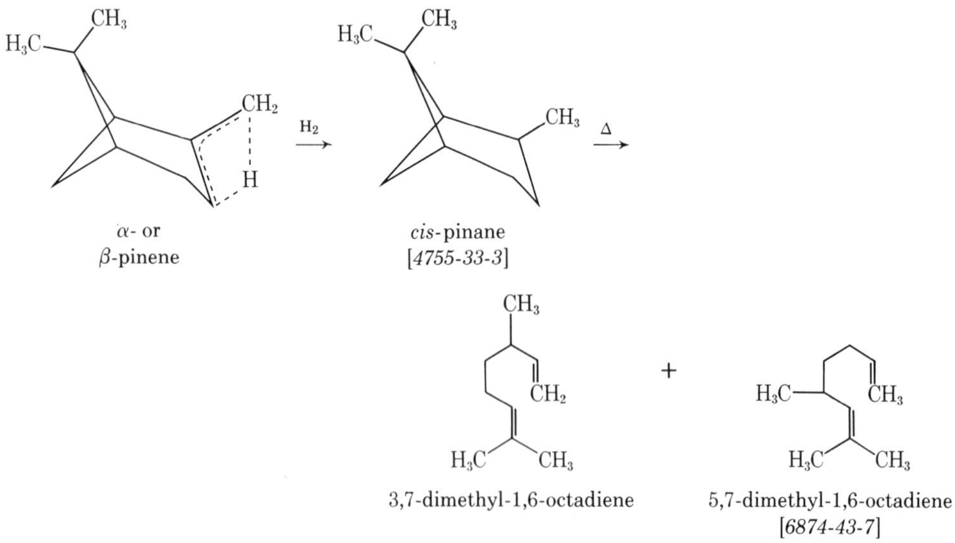

Optimum pyrolysis temperature is ca 600°C. 5,7-Dimethyl-1,6-octadiene is obtained as a by-product, but no use has been developed for this isomer.

Uses. The chief use for 3,7-dimethyl-1,6-octadiene is as a starting material for the synthesis of acyclic perfumery aldehydes and alcohols.

Alloocimene. ***Reactions.*** If alloocimene [*673-84-7*] is hydrochlorinated under the conditions used for the hydrochlorination of myrcene, only a resin is obtained because of polymerization. However, if the reaction is carried out in a continuous fashion with efficient cooling and dilution of anhydrous hydrogen chloride with an inert gas, as much as a mole of hydrogen chloride can be added to alloocimene without significant polymerization (111).

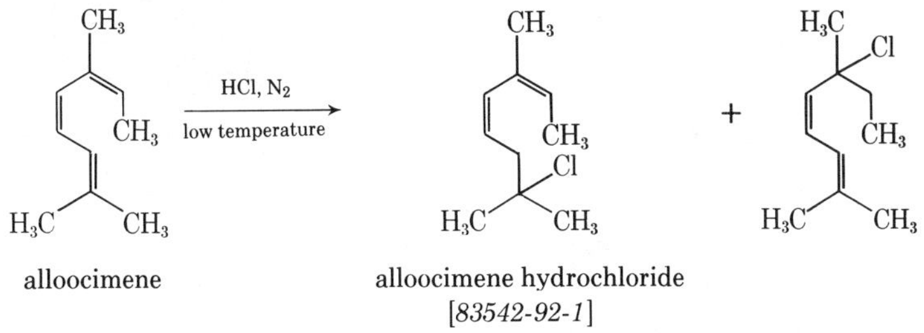

Alloocimene hydrochloride [*83542-92-1*] is quite unstable and polymerizes in less than an hour at ambient temperatures. A substantial amount of the hydrochloride is used in the commercial production of alloocimenol, a mixture of two isomers, by reaction with dilute aqueous sodium hydroxide in the presence of an emulsifying agent. This hydrolysis must be carried out immediately after the hydrochlorination.

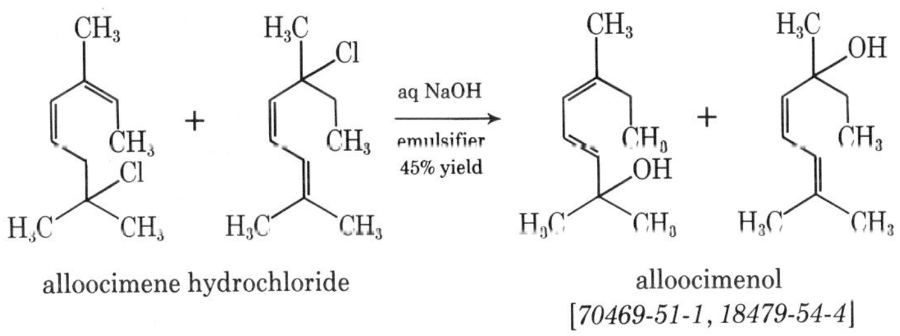

The shift in the double bonds of the one isomer has been attributed to the resonance of a postulated carbonium ion intermediate (111). The alcohol mixture is used in perfumery for its floral scent. The alcohols are also hydrogenated to obtain a mixture of tetrahydromyrcenol [*18479-57-7*] and tetrahydrolinalool [*78-69-3*], which is also sold for its floral scent.

Manufacture and Shipment. Alloocimene is produced by the thermal isomerization of α-pinene with an approximately equal quantity of dipentene and smaller quantities of α- and β-pyronenes (7). The pyrolysis is generally carried out at ca 375°C, and the yields of alloocimene and dipentene are 40–45% each; the combined pyronenes amount to 10–15%. The pyronenes have not been of commercial value; increasing the pyrolysis temperature produces better yields of them but decreases the yield of alloocimene. Although there are four geometrical isomers of alloocimene, only two are obtained in significant quantities from the pyrolysis of α-pinene. These are commonly referred to as alloocimene A [*7216-56-0*] (the 4-*trans*, 6-*cis* isomer) and alloocimene B [*3016-19-1*] (the 4-*trans*, 6-*trans* isomer). The mechanistic pathways for alloocimene formation have been elucidated (112–113).

Alloocimene is shipped with an inhibitor in tank cars, deck tanks, and drums. Production data are not available.

Autoxidation of alloocimene under nonacidic conditions produces a polymeric peroxide that can be thermally depolymerized to produce alloocimene diepoxide [*3765-28-4*] (114).

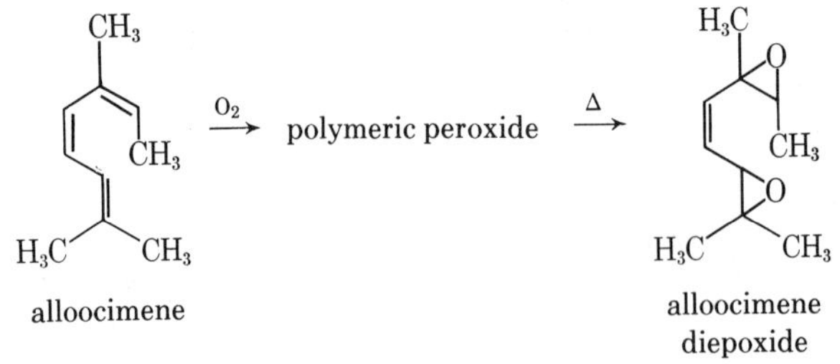

The diepoxide is useful in resin manufacture (114–115). Alloocimene is useful in resins and epoxy modifiers, as a scavenger for decomposition products of halogenated solvents, and as a chemical intermediate for perfumery alcohols.

Oxygenated Monoterpenes

Geraniol and Nerol. ***Reactions.*** In general, geraniol and nerol undergo the same reactions, although sometimes at different rates and under slightly different conditions. The most important industrial reactions are hydrogenation, dehydrogenation (oxidation), and rearrangement. Geraniol–nerol can be converted to citronellol by hydrogenation over a copper chromite catalyst (116). The catalyst probably first dehydrogenates the geraniol to citral. The conjugated double bond of citral becomes activated and is immediately hydrogenated to citronellal, which is then hydrogenated to citronellol. In the absence of hydrogen and under reduced pressures, citronellal can be obtained by rearrangement (117). Geraniol–nerol can be completely hydrogenated, for example over a nickel catalyst, to produce 3,7-dimethyloctan-1-ol [*106-21-8*], a perfumery compound. In the hydrogenation of geraniol–nerol to citronellol, care must be taken to minimize formation of this alcohol. Geraniol–nerol can be isomerized to linalool (13).

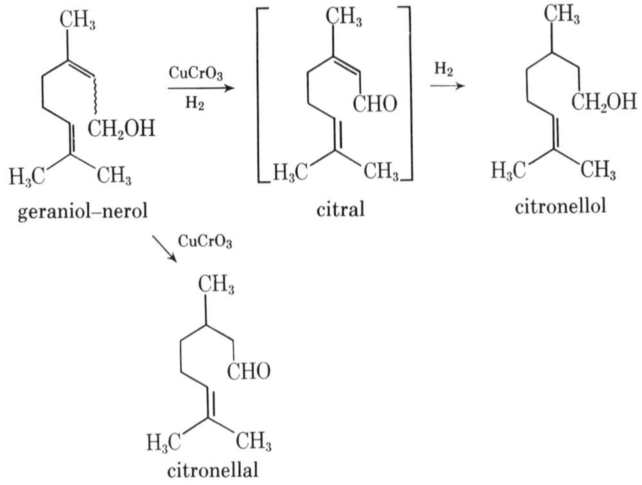

Geraniol–nerol in the vapor phase is converted to citral by oxidative dehydrogenation by an oxygen-containing gas in the presence of a copper catalyst (118–119). The same transformation can be accomplished by oxidation with chromic acid, but industrially this method is too expensive and polluting (120). Oppenauer oxidation of geraniol–nerol catalyzed by aluminum isopropoxide and with furfural as a hydrogen acceptor produces citral in good yield (121). An advantage of this method is that the coproduct produced is furfuryl alcohol which is more valuable than furfural (see Furan compounds). Another method for oxidizing geraniol–nerol to citral involves activated nickel oxide as the oxidant (122). The active oxide is produced by reaction of nickel salts with aqueous sodium hypochlorite.

Manufacture, Shipment, and Price. The chief natural source for geraniol is citronella oil, wherein it occurs with citronellal and citronellol and from which it is separated by fractional distillation. Citronella oil contains only small amounts of nerol. Much greater amounts of geraniol and nerol are produced synthetically than are isolated from natural sources. Geraniol (*trans*) and nerol (*cis*) are geometrical isomers which can be separated by high efficiency fractionation into high purity individual isomers. By far the largest volumes of the synthetic products are sold and used as isomer mixtures; equilibrium mixtures are ca 60% trans and 40% cis. Geraniol–nerol mixtures are commonly referred to as geraniol, but manufacturers' specifications usually give the isomer ratio as determined by gas chromatography.

Shipment is generally made in drums, tank trucks, or deck tanks. U.S. sales of geraniol are ca 1200 t/yr, but much of the geraniol produced is used captively. The sales price is ca $7.70–22.00/kg, depending on the source, quality, and volume (53).

Uses. Geraniol and nerol are used in perfumery for their rosy scents. By far their largest use is as starting materials for the synthesis of citral, citronellol, citronellal, hydroxycitronellal, the ionones, vitamins A and E, and carotenoids.

Linalool. ***Reactions.*** Linalool can be esterified to produce linalyl acetate [*115-95-7*] by reaction with acetic anhydride. Care must be taken to prevent rearrangement or dehydration of linalool during this reaction. To prevent this, a base can be added to neutralize the acetic acid formed by the reaction. Another method is to carry out the reaction in a fractionating column and remove the acetic acid overhead as it forms (123). About 500 metric tons of linalyl acetate was produced in 1980 (52).

Linalool can be converted to geranyl acetone [*3796-70-1*] by the Carroll reaction (21). It is transesterified with ethylacetoacetate, and the resulting ester rearranges thermally with loss of carbon dioxide. Geranyl acetone is an important intermediate in the synthesis of isophytol [*505-32-8*] (used in the manufacture of vitamin E), farnesol [*106-28-5*], and nerolidol [*40716-66-3*].

Linalool can be converted to citral by oxidation with chromic acid, but this reaction is not used commercially because of associated cost and pollution problems. Linalool can be hydrogenated to produce dihydrolinalool (terminal-double-bond hydrogenation) or tetrahydrolinalool, both of which are used to a small extent in perfumery.

Manufacture, Shipment, and Price. The most important natural source of linalool is bois de rose oil from which it is separated by fractionation, but much greater quantities of linalool are produced synthetically (124). Some perfumers still prefer the odor of natural linalool, which is slightly different from that of the synthetic because of minor impurities entrained in the natural product during fractionation.

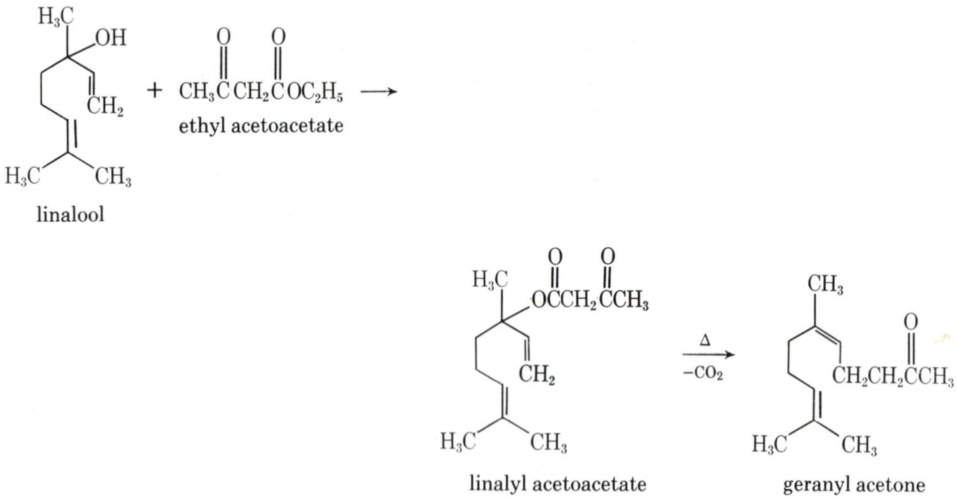

Linalool is shipped in cans, drums, and deck tanks. U.S. production of linalool in 1980 was ca 1200 t (52). The price of synthetic linalool in 1981 was $6.60–8.80/kg, depending on grade and quantity (53).

Uses. Linalool is used in multitonnage quantities for the production of geraniol–nerol and as an intermediate in the synthesis of vitamin E and vitamin A. Substantial quantities of linalool and its acetate are used in perfumery.

Dehydrolinalool. Dehydrolinalool [*29171-20-8*] is not obtainable from a natural source but is manufactured commercially. Production data are not available, but quantities produced are large.

Reactions. Dehydrolinalool is converted to linalool by semihydrogenation of its triple bond over a Lindlar-type catalyst (17). The linalool so produced is optically inactive. Dehydrolinalool is an important precursor for the synthesis of pseudoionone. The transformation has been accomplished in a number of ways. Reaction of dehydrolinalool with diketene gives dehydrolinalyl acetoacetic [*22414-87-5*] which, on pyrolysis, undergoes Carroll rearrangement to a mixture of geometrical isomers of pseudoionone [*141-10-6*] accompanied by loss of carbon dioxide (16,18,21–22).

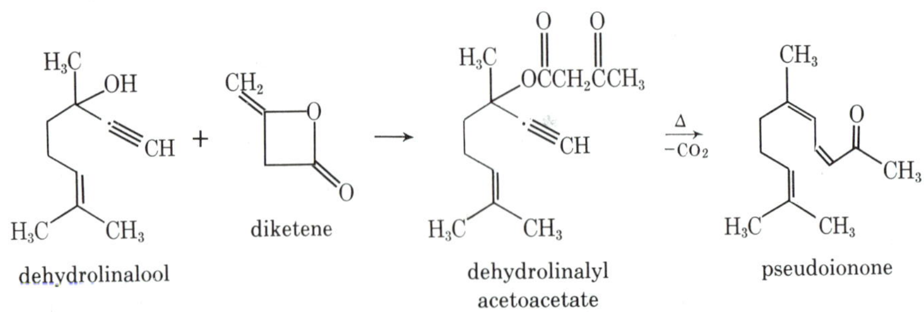

Alkyl esters of acetoacetic acid are commonly used in place of diketene, in which case an alcohol is expelled in the formation of dehydrolinalyl acetoacetate (124–125). Catalytic amounts of a weak acid are sometimes used. A continuous process involving acetoacetic esters with a minimum of by-product formation has been developed (126).

The Claisen-type acid-catalyzed reaction of dehydrolinalool with methyl isopropenyl ether produces an allenic ketone [*16647-05-5*], which can be readily isomerized to pseudoionone by treatment with catalytic amounts of alkali (19,127). A related method involving the dimethyl acetal of acetone has been developed, but it gives somewhat poorer yields (128).

Dehydrolinalool is isomerized to citral by heating with vanadic ester catalysts (129–130).

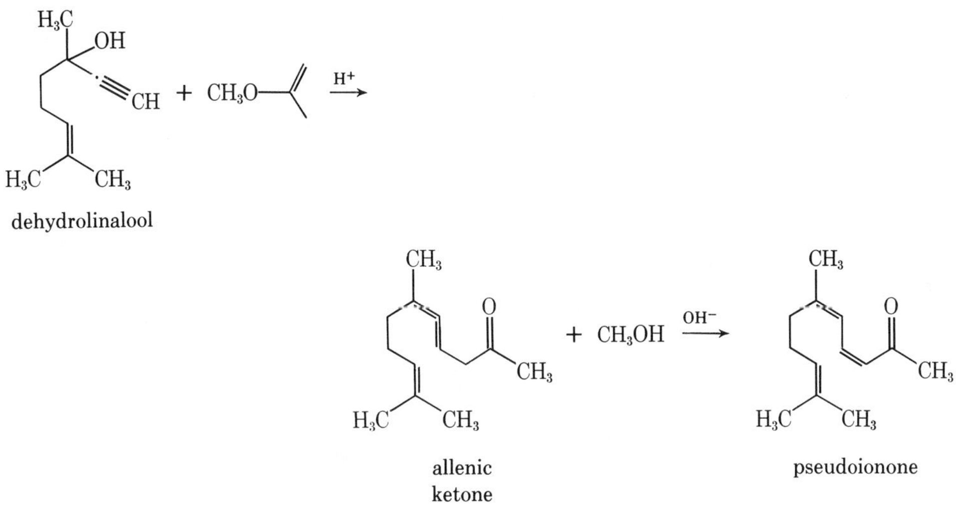

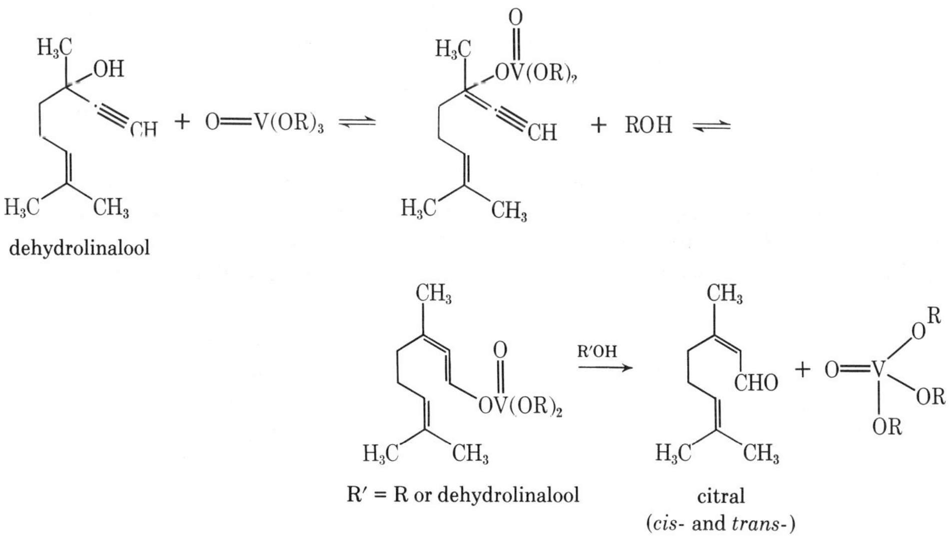

When R represents either an alkoxy or a siloxy group, the degree of conversion of dehydrolinalool is low (20–35%), but selectivity is acceptable (60–80%). An improvement in the method is obtained by using tris(triphenylsilyl)vanadate with small amounts of lauric or benzoic acid, whereby almost complete conversion of dehydrolinalool is achieved and the yield of citral is high (85–90%) (131–132). Triphenylsilyl moieties, wherein the phenyl is substituted in one or more positions with an electron-withdrawing group, eg, halogen, are also claimed to be efficient. The mechanism of the reaction is discussed in ref. 133. Another method for producing citral from dehydrolinalool starts with acetylation of the alcohol, usually by reaction with acetic anhydride (134). The acetate is rearranged by heating with acetic acid in the presence of silver or copper catalysts whereby a mixture consisting primarily of citral diacetate [*41494-93-3*] and an allenic acetate [*41494-92-2*] is obtained. Saponification of this mixture produces citral in good yield.

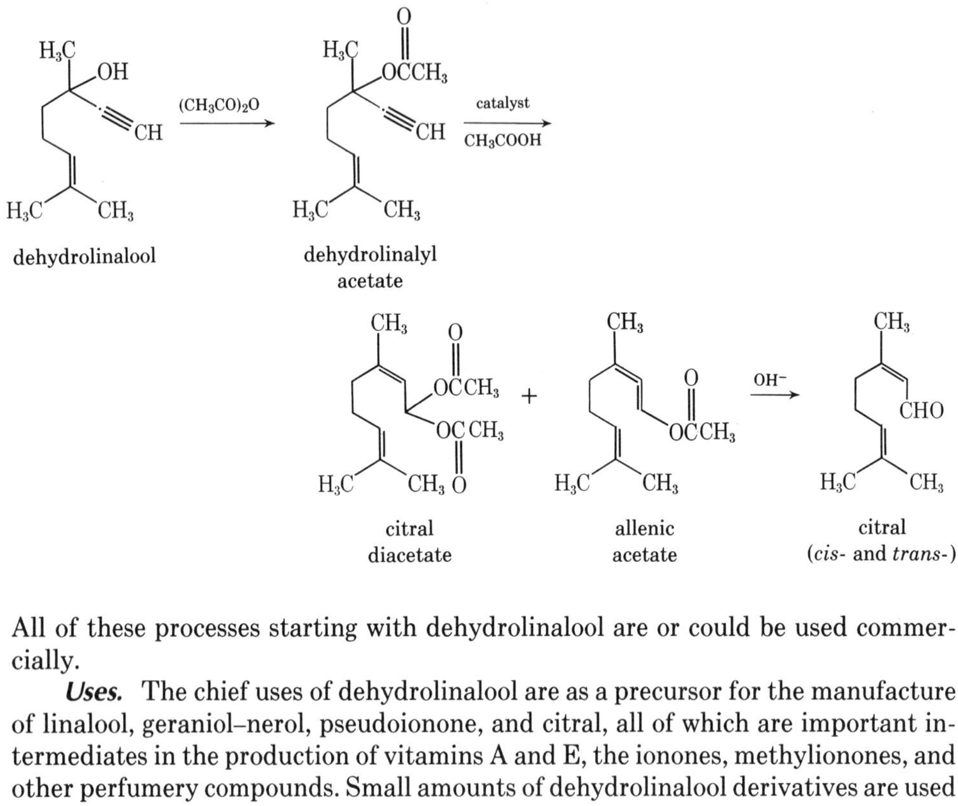

All of these processes starting with dehydrolinalool are or could be used commercially.

Uses. The chief uses of dehydrolinalool are as a precursor for the manufacture of linalool, geraniol–nerol, pseudoionone, and citral, all of which are important intermediates in the production of vitamins A and E, the ionones, methylionones, and other perfumery compounds. Small amounts of dehydrolinalool derivatives are used in perfumery.

Citral. ***Reactions.*** The most important commercial reactions of citral depend on base-catalyzed aldol condensation with ketones. If the ketone is acetone, *cis-* [*3796-54-1*] and *trans-*pseudoionones [*13927-47-4*] are obtained in about the same geometrical ratio as geraniol and nerol were in the starting citral.

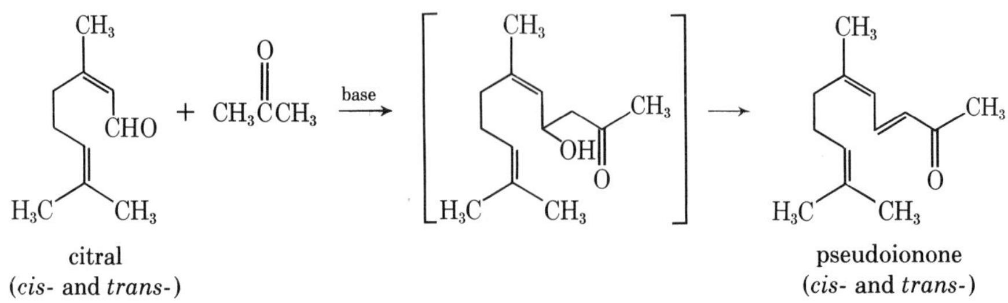

If methyl ethyl ketone is used, pseudomethylionones result. In this case, the condensation occurs involving either a methyl or methylene group yielding a mixture of pseudo-*n*-methylionones [*26651-96-7*] and pseudoisomethylionones [*1117-41-5*]. Some selectivity can be obtained in the condensation by variation of reagents and reaction conditions.

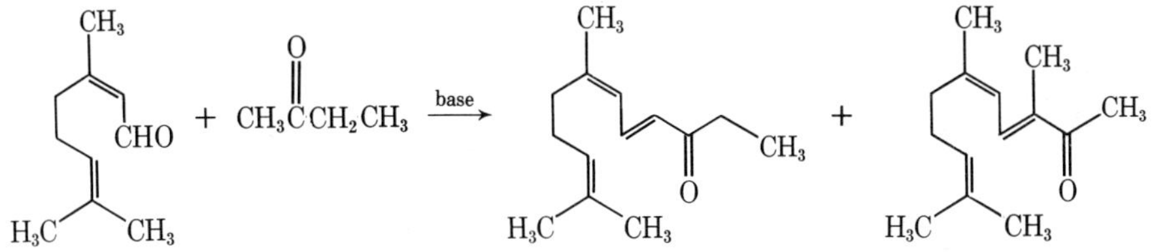

citral (*cis*- and *trans*-) — pseudo-*n*-methylionone (*cis*- and *trans*-) — pseudoisomethylionone (*cis*- and *trans*-)

Although more geometrical isomers are theoretically possible for pseudoionone and pseudomethylionones, the spatial arrangement around the double bond nearest the carbonyl group appears to be fixed (*trans*) or nearly so; thus only two significant isomers are observed in each case. It is not necessary to use highly purified citral for manufacture of pseudoionones. Citral can be purified to a flavor-grade product through its addition compound with sodium sulfite (135). Purification of citral by distillation is very difficult because of its instability.

Hydrogenation of citral can be controlled over a chromium-promoted Raney nickel catalyst to produce mixtures of citronellal and citronellol which, on further hydrogenation, give 3,7-dimethyloctan-1-ol [*106-21-8*], but there are more economical routes (136).

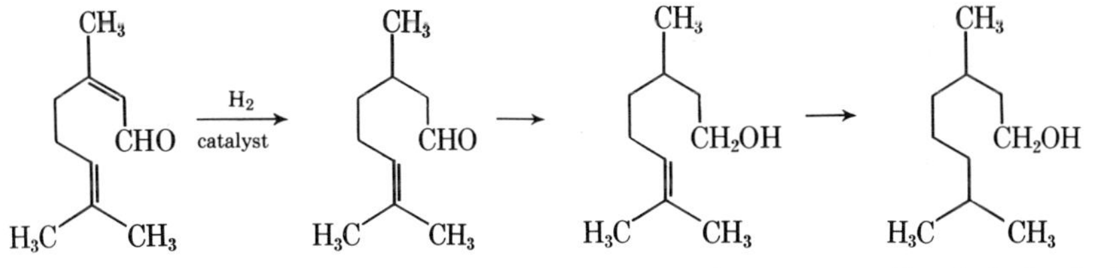

citral (*cis*- and *trans*-) — citronellal — citronellol — 3,7-dimethyloctan-1-ol

Manufacture. Historically, lemongrass oil was the chief source of citral with *Litsea cubeba* oil more recently becoming commercially significant. Both oils contain ca 70–80 wt % citral. Multitonnage quantities of synthetic citral are produced annually; much of it is 70–80% pure. Shipment is usually made in lined drums, pails or aluminum cans.

Citral can also be produced from myrcene hydrochloride. In one such method (Sommelet-type reaction), myrcene hydrochloride reacts with hexamethylenetetramine (urotropine), forming a quaternary ammonium salt which is hydrolyzed in the presence of formaldehyde yielding citral (137). An advantage of the method is that the quaternary ammonium salt can be isolated and purified, which allows high purity citral to be obtained. This method probably is not used commercially because of unfavorable economics and associated pollution problems.

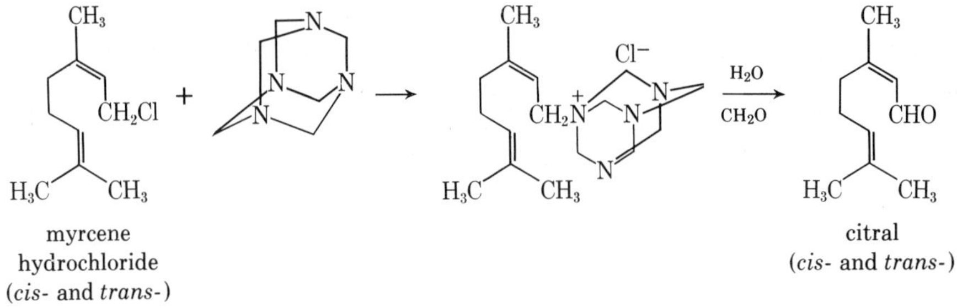

Another method involves the reaction of myrcene hydrochloride with an alkali metal salt of 2-nitropropane to form a nitronate. The ester is unstable and hydrolyzes readily (138). The citral formed by this method is difficult to free of chlorides and other impurities.

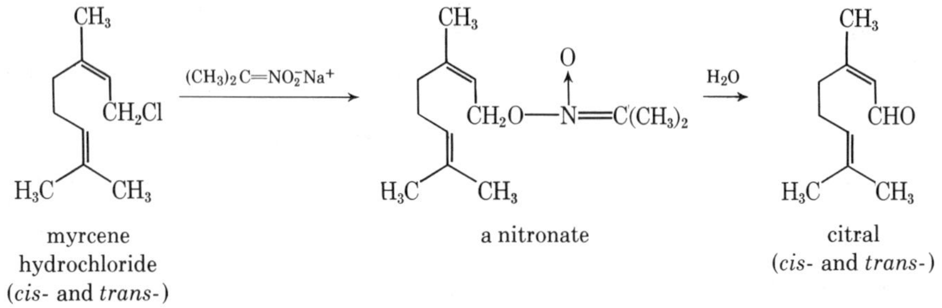

Uses. Multitonnage quantities of pseudoionone are produced each year from citral for use in the manufacture of vitamins A and E. For use in vitamin-E manufacture, pseudoionone is first hydrogenated to the hexahydro derivative [*1604-34-8*]; vitamin-A manufacture proceeds through β-ionone. Appreciable quantities of pseudoionone and the pseudomethylionones are used for producing perfumery ionones and methylionones. A modest amount of high purity citral is used in the flavor and fragrance industries.

Ionones and Methylionones. ***Reactions.*** The only important large-scale commercial reactions of β-ionone are those associated with the manufacture of vitamin A and carotenoids. α-Ionone can be isomerized to β-ionone under acid conditions. The methylionones have no important industrial reactions.

Manufacture. Although the ionones have been identified in small amounts in certain essential oils, no practical commercial natural source exists (139). The largest volume ionone is β-ionone, upon which are based all industrial syntheses of vitamin A.

Pseudoionone (ionone precursor) is generally made industrially from citral. A commercially significant synthesis from mesityl oxide and prenyl chloride has been developed (140–142). In this synthesis, the reaction of mesityl oxide with prenyl chloride produces a ketodiene, which reacts with sodium acetylide yielding 4-isopropenyl-3,7-dimethyl-6-octen-1-yn-3-ol [*64708-52-7*]. Selective synthesis of pseudoionone is achieved by an acetylenic oxy-Cope rearrangement of the dienynol in *N*-methyl-2-pyrrolidinone (NMP) containing a trace of halogen catalysts. When pseudoionone is treated with acid, cyclization occurs to produce a mixture of α- and β-ionone. Common acids used are phosphoric, sulfuric, and Lewis acids. A mixture of the

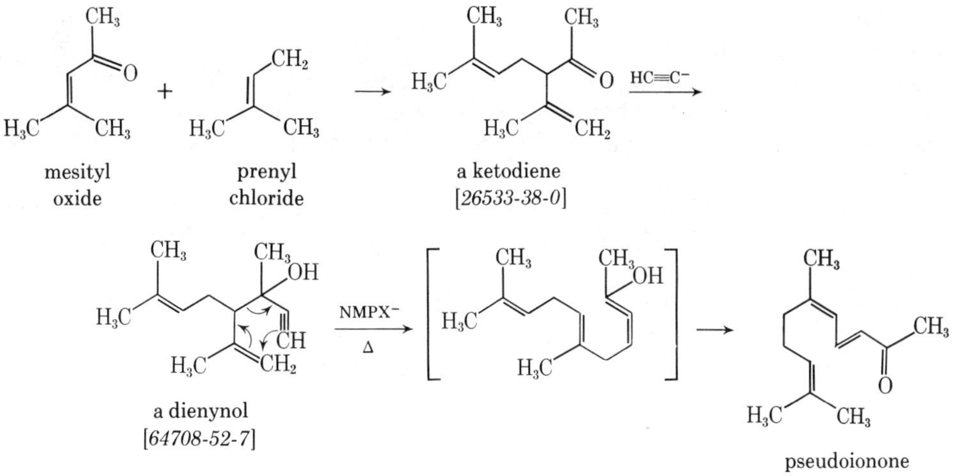

α- and β-isomers is always obtained, but one isomer or the other can be favored by proper selection of the acid and reaction conditions. Inasmuch as the β-isomer is thermodynamically favored, stronger acids and more vigorous reaction conditions enhance the amount of this isomer in the mixture. γ-Ionone [*79-76-5*], the isomer with an exocyclic double bond at the single-ring methyl group is formed in small amounts. The geometrical isomer ratio of the starting pseudoionone is of little importance in the cyclization reaction.

Acid-catalyzed cyclization of a pseudomethylionone cis,trans mixture produces four methylionone isomers.

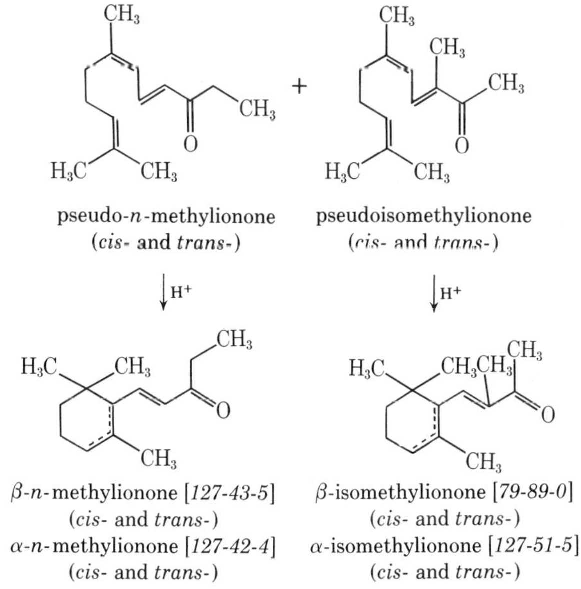

The same general considerations apply for the cyclization of the pseudomethylionones as for the cyclization of pseudoionones.

About 335 t of methylionones was produced in 1980; the average price was ca \$16/kg (52).

Uses. All of the ionones are used extensively in perfumery with α-ionone and α-isomethylionone being the most highly valued. β-Ionone is manufactured in the largest volume of any of the ionones because of its use in vitamin-A manufacture. The volume of β-ionone used for this purpose is not available, but it is large. Perfumery usage of α- and β-ionone combined is ca 90 t/yr with prices averaging ca $18/kg (3,52).

Citronellol. ***Reactions.*** Citronellol can be oxidized or dehydrogenated to produce citronellal or hydrogenated to obtain 3,7-dimethyloctan-1-ol. Esterification of citronellol proceeds readily.

In the presence of acid catalysts, including ion-exchange resins, citronellol is hydrated to hydroxycitronellol [*107-74-4*], a precursor for hydroxycitronellal [*107-75-5*]. If an alkanol is used in place of water, an alkoxycitronellol is obtained.

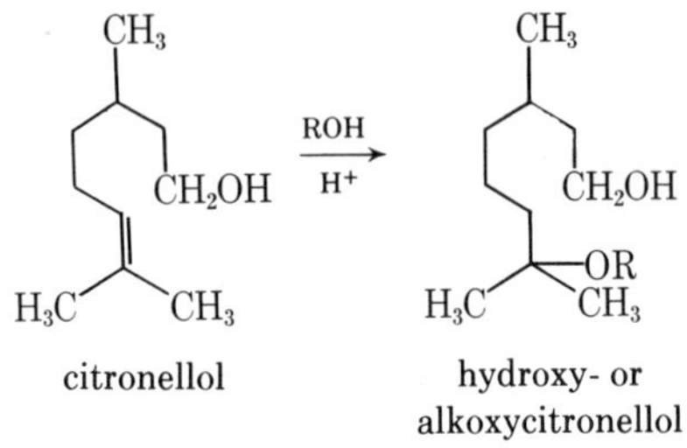

R = H or alkyl

Manufacture, Shipment, and Price. Citronellol is widely distributed in nature in both optical forms. Before the advent of synthetic citronellol, this alcohol was obtained from certain oils of the *Rosaceae* family or by hydrogenation of citronellal obtained from citronella oil.

Citronellol is manufactured on a commercial scale by the hydrogenation of geraniol–nerol or citronellal. It has also been produced from α- or β-pinene. The pinene is first hydrogenated to pinane, which is pyrolyzed to 3,7-dimethyl-1,6-octadiene. The 6,7- double bond of the octadiene is protected by addition of hydrogen chloride, and the resulting product reacts with hydrogen bromide in anti-Markovnikov fashion in the presence of a free-radical catalyst to provide a chlorobromide. The bromide is displaced with acetate ion under catalysis of a quaternary ammonium salt produced *in situ* from triethylamine (143). The 6,7- double bond is regenerated during the displacement step. Hydrolysis of the resulting acetate produces citronellol. If the starting pinene is optically active, optically active citronellol is obtained. For example, (−)-β-pinene or (−)-α-pinene gives (+)-citronellol.

3,7-Dimethyl-1,6-octadiene can also be converted to optically active citronellol by conversion to an aluminum alkyl, followed by oxidation and hydrolysis (109). If the octadiene is produced from (−)-β-pinene or (−)-α-pinene, (+)-citronellol again results.

Citronellol is generally shipped in lined drums, deck tanks, or pails. U.S. production in 1980 was ca 1,140 t at an average price of $9.80/kg (52).

Uses. The chief use of citronellol is in fragrances for the soap and detergent industries for its natural rosy scent. It also serves as an intermediate in the synthesis of certain fragrance and flavor compounds. Esters of citronellol are commonly used in fragrance compositions.

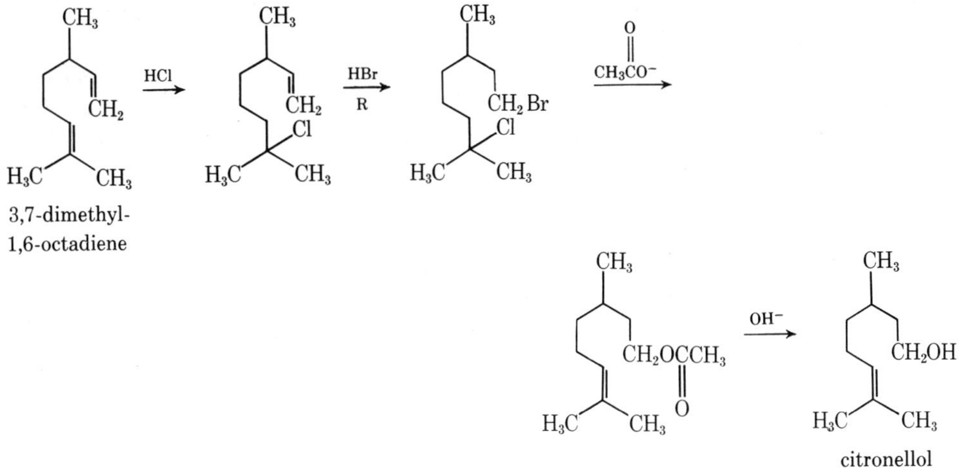

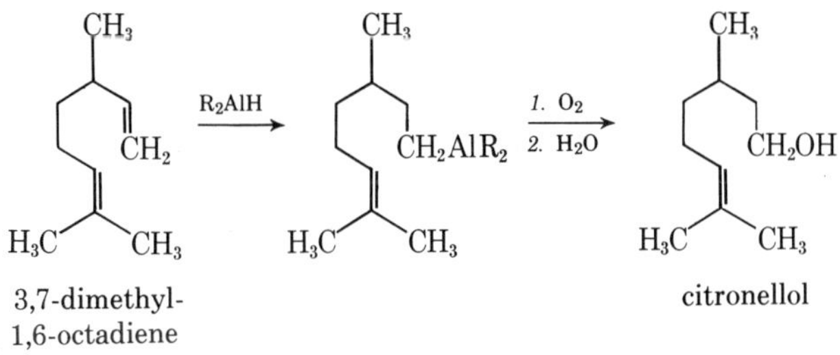

Citronellal. ***Reactions.*** Citronellal undergoes cyclization to isopulegol [*89-79-2*] under acid catalysis (144). If the citronellal is optically active, the isopulegol is optically active; this has been important in the synthesis of (−)-menthol from (+)-citronellal. With sodium bisulfite, citronellal forms an adduct that is useful in protecting the carbonyl group while the molecule undergoes further reactions, eg, hydration and alkoxylation. It can also be used in the purification of citronellal. The aldehyde group of citronellal undergoes the usual carbonyl reactions.

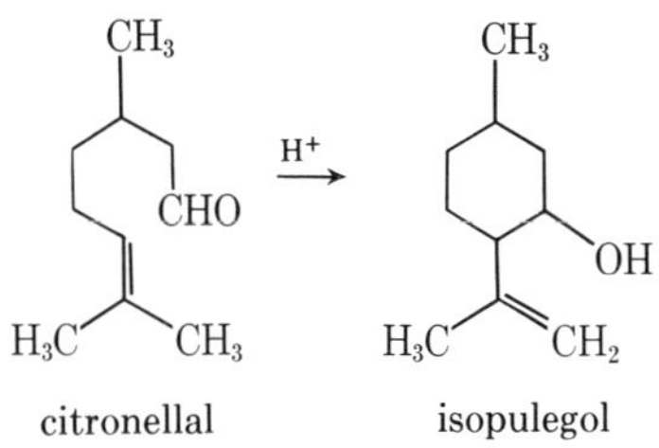

Manufacture. The most common natural sources of citronellal are citronella oil and *Eucalyptus citriodora.* It is produced commercially by rearrangement of geraniol–nerol or by dehydrogenation of citronellol. It can also be made by Oppenauer oxidation of citronellol, but this method is not used commercially because of its high cost. Citral can be hydrogenated into partially optically active citronellal in the presence of a catalytic complex of a rhodium derivative and a chiral phosphine (145) (see also Pharmaceuticals, optically active).

Uses. Citronellal has limited use in fragrances and flavors because of its ease of oxidation and instability to alkalinity. Its chief uses are in the production of hydroxycitronellal, methoxycitronellal, and menthol.

Hydroxy- and Methoxycitronellal. ***Reactions.*** Hydroxycitronellal has no important commercial reactions, except that a small amount is converted to its dimethylacetal to impart stability. An important reaction of methoxycitronellal is its aldol condensation with dialkyl-3-methylglutaconates to produce an intermediate for the synthesis of certain juvenile hormone mimics, eg, methoprene [*40596-69-8*] (146).

Manufacture, Shipment, and Price. Neither hydroxy- nor methoxycitronellal are obtained from natural sources. Hydroxycitronellal is produced by hydration of citronellal under strong acid conditions, but the aldehyde group must be protected before the hydration. The oldest and most common method of protection involves the hydration of the bisulfite adduct and regeneration of the carbonyl group by treatment with base (147).

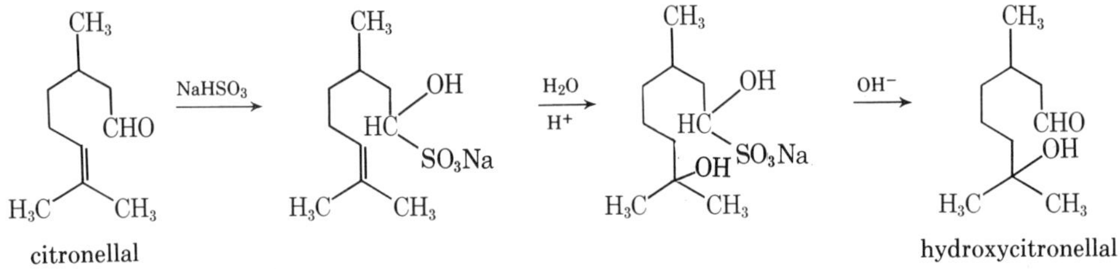

Other methods of protecting the aldehyde group include formation of an intermediate enol acetate, enamine or imine (148–149). In the enamine route, regeneration of the carbonyl group can be accomplished by treatment with water. Hydroxycitronellal is also produced by dehydrogenation of hydroxycitronellol [*107-75-5*] over a copper chromite catalyst (105). Hydroxycitronellol can be produced by direct hydration of citronellol or by hydrogenation of the epoxide of dihydromyrcenol [*30385-25-2*]. Hydrogenation of the epoxide [*40454-18-0*] produces a coproduct, 2,7-dihydroxy-2,6-dimethyloctane [*15340-95-1*], which has no commercial use; this method is not used industrially.

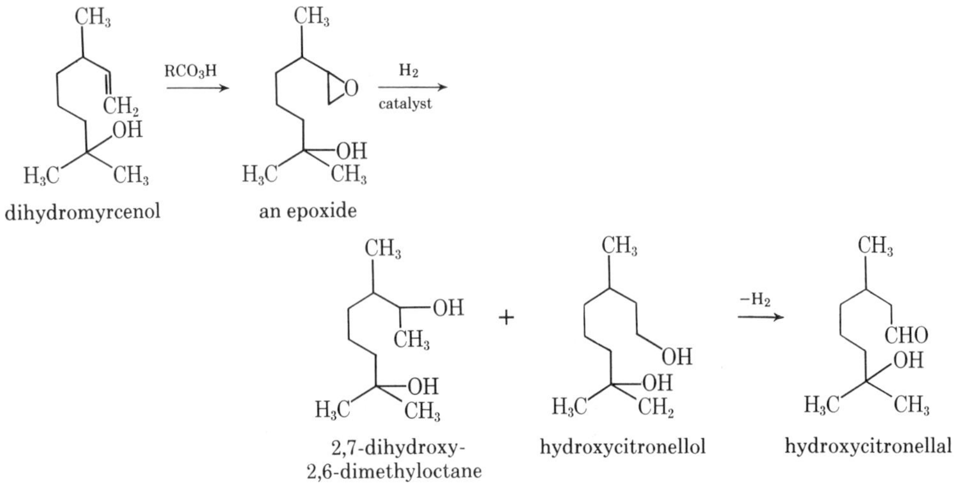

Methoxycitronellal [*3613-30-7*] is produced by the same reactions as hydroxycitronellal, except that in the hydration step, methanol is used instead of water. In this process, the coproduct, 2-methoxy-7-hydroxy-2,6-dimethyloctane [*41890-92-0*] (methoxyelgenol) is useful as a perfumery material (106).

An industrially significant method for producing hydroxycitronellal starts with the reaction of methyl-α-heptenone with methyl Grignard reagent to obtain 2,6-dimethyl-6-heptene-2-ol [*32779-58-1*]. This alcohol, on reaction with carbon monoxide and hydrogen and in the presence of a rhodium catalyst complex under moderately high pressure at 100°C, gives a 90% yield of hydroxycitronellal (150).

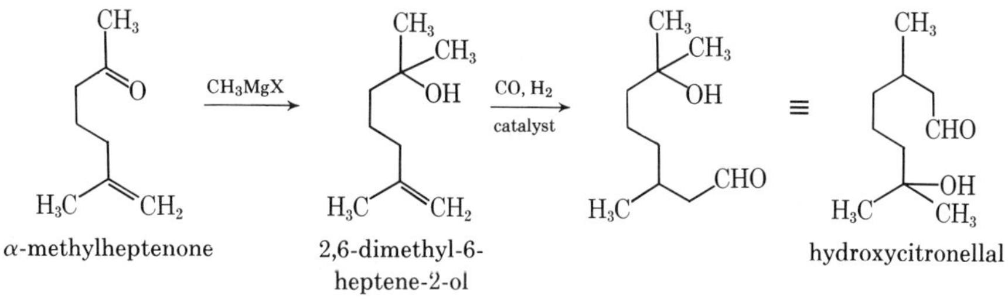

U.S. production is ca 350 t/yr and the price of hydroxycitronellal is ca \$15–20/kg (53). Shipment is usually made in lined drums or pails.

Uses. Hydroxycitronellal is valued for its delicate lily-of-the-valley fragrance by the perfumery industry, which uses large volumes of this compound. Methoxycitronellal also is used in perfumery, but to a much smaller extent. Its chief use is as a starting material for the synthesis of juvenile hormone mimics.

Myrcenol and Dihydromyrcenol. ***Reactions.*** The most important chemical reaction of myrcenol has been its Diels-Alder reaction with acrolein to produce a mixture of two isomeric aldehydes, 3- [*51414-25-6*] and 4-(4-hydroxy-4-methylpentyl)-3-cyclohexene-1-carboxaldehyde [*31906-04-4*], which are used in perfumery (151). It is claimed that the compound having the side-chain substitution at *C*-4 has the more desirable fragrance, and its formation can be enhanced by use of a $ZnCl_2$ catalyst (152).

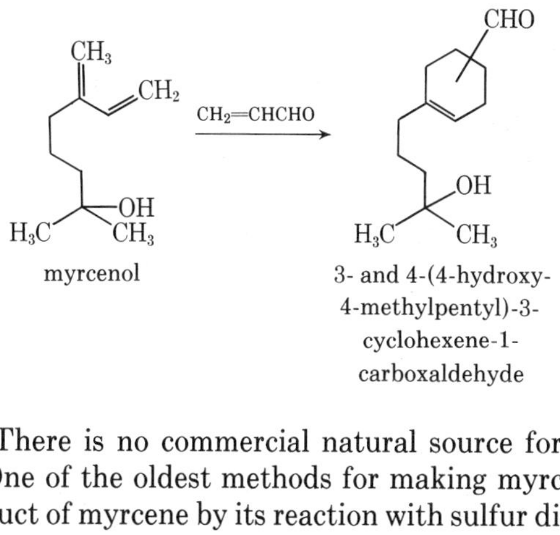

Manufacture. There is no commercial natural source for either myrcenol or dihydromyrcenol. One of the oldest methods for making myrcenol involves initial formation of an adduct of myrcene by its reaction with sulfur dioxide. The adduct is hydrated with aqueous sulfuric acid, and the hydrated product is thermally decomposed to produce myrcenol (153).

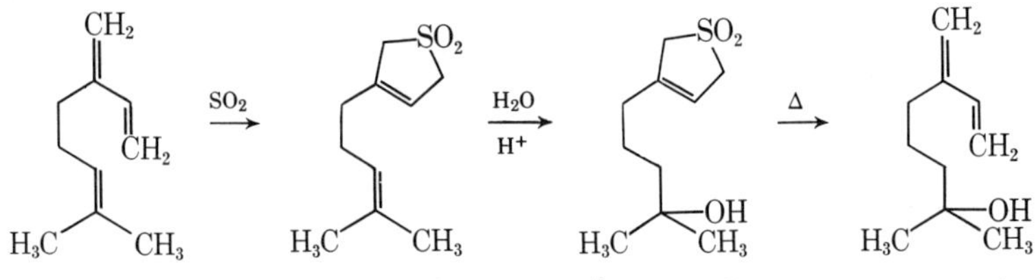

Myrcene can be hydrochlorinated in the presence of a Lewis acid to obtain myrcenyl chloride. Hydrolysis of this chloride with aqueous alkali yields myrcenol (154). Myrcenol can also be prepared from geranyl(neryl)amines.

Dihydromyrcenol can be synthesized by hydrochlorination of 3,7-dimethyl-1,6-octadiene, followed by treatment with dilute caustic (107). It can also be made by reaction of the diene with a mixture of formic and sulfuric acids to produce a formate [*25279-09-8*] which, on subsequent hydrolysis, yields dihydromyrcenol (108).

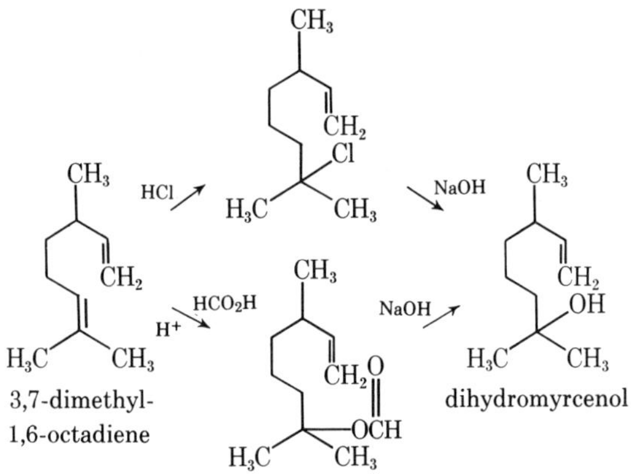

Uses. The primary use of myrcenol is in the production of the aldehyde mixture discussed above. Myrcenol is of limited use in perfumery because of its lack of stability. Some myrcenol is converted to its acetate and propionate, but these too are unstable. Dihydromyrcenol and its esters exhibit somewhat better stability as perfumery materials, but they nevertheless are used sparingly.

Menthol. ***Reactions.*** Menthol undergoes all of the typical chemical reactions of a cyclic secondary alcohol. Probably the two most important reactions commercially are dehydrogenation or oxidation to menthone [*10458-14-7*] and isomenthone [*491-07-6*] and esterification to menthyl acetate [*16409-45-3*] and other esters, but the amounts manufactured are very small compared to the amount of menthol produced.

Manufacture and Price. Of the menthol isomers, only (−)-menthol [*2216-51-5*] and (±)-menthol [*15356-70-4*] are of large-scale commercial importance. Most (−)-menthol is obtained by crystallization from *Mentha arvensis* oil, but the synthetic product is becoming increasingly popular. All (±)-menthol is obtained by synthetic methods.

There are four geometrical isomers (diastereoisomers) of menthol, each containing three asymmetric carbon atoms. These isomers are called menthol, neomenthol [*491-01-0*], isomenthol [*490-99-3*], and neoisomenthol [*491-02-1*].

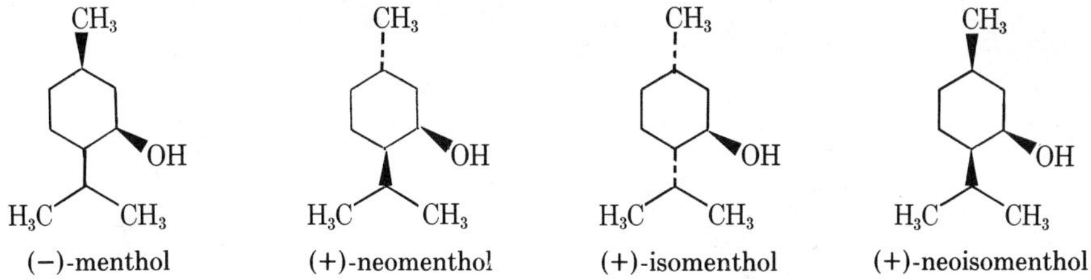

These geometrical isomers also exist in the (+) and (−) forms, making a total of four pairs of enantiomorphic forms and four *dl* forms. The menthols can be divided into two families, the (−)-menthols and the (+)-menthols. The menthols of a given family can be interconverted (equilibrated) in the presence of sodium mentholates without appreciable loss of optical activity or with hydrogenation catalysts, eg, copper chromite, in the presence of hydrogen, but some racemization occurs (155–156). The equilibrium mixture obtained contains ca 62 wt % menthol, 23 wt % neomenthol, 12 wt % isomenthol, and 3 wt % neoisomenthol. These isomers can be separated by high efficiency distillation or crystallization techniques, so stereospecific synthesis of any of the geometrical isomers of the (−)-menthols can constitute a synthesis of (−)-menthol.

(−)-Menthol can be produced by stereospecific synthesis or by resolution of (±)-menthol. An early method of synthesis began with (+)-citronellal [*2385-77-5*], isolated from citronella oil. When (+)-citronellal is treated with an acidic reagent, it cyclizes to produce (−)-isopulegol. Any undesired isomers of (−)-isopulegol can be converted back to (+)-citronellal by pyrolysis (145). (−)-Isopulegol gives (−)-menthol on hydrogenation.

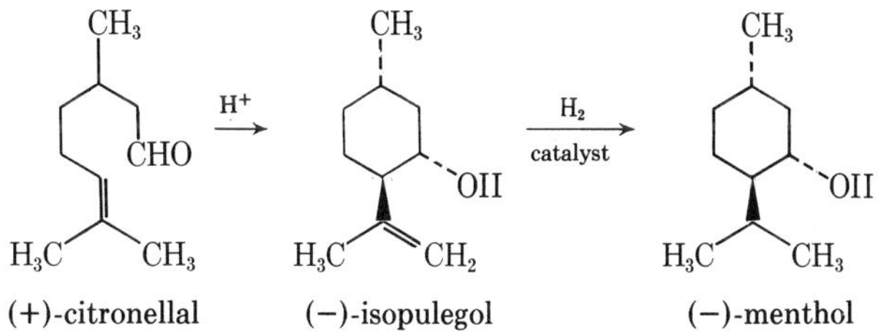

(+)-Citronellal [*2385-77-5*] can also be obtained by dehydrogenation of (+)-citronellol [*1117-61-9*].

In another commercial process, (−)-*β*-phellandrene [*6153-17-9*] (obtained from turpentine) is hydrochlorinated to produce chiefly piperityl chloride (157). The latter on hydrolysis with aqueous base gives a mixture of geometrically isomeric 2-menthen-1-ols [*619-62-5*], (+)-*trans*-piperitol [*65733-27-9*], and (−)-*cis*-piperitol [*65733-28-0*]. Esterification of the mixture with formic acid in the presence of a buffer, followed by hydrolysis, produces an equilibrium mixture rich in (+)-*trans*-piperitol and (−)-*cis*-piperitol which can be separated from the mixture and from each other by efficient fractionation (158). The (−)-*cis*-piperitol can either be hydrogenated and racemized for production of (±)-menthol or isomerized to the equilibrium mixture as described above. Selective hydrogenation of (+)-*trans*-piperitol gives (+)-isomenthol, which can be isomerized to the equilibrium mixture containing (−)-menthol. Efficient fractionation then produces USP (−)-menthol; the other fractions can be recycled to equilibration.

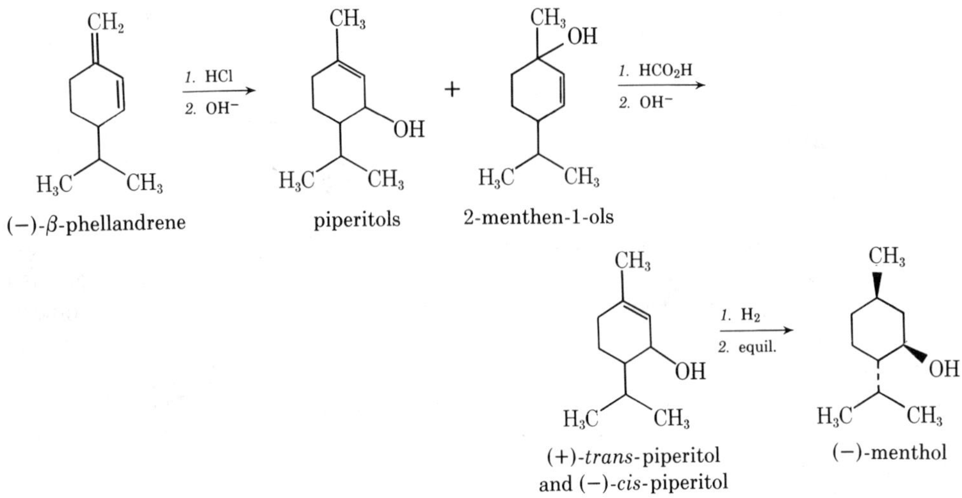

(−)-Menthol is also produced by resolution of (±)-menthol. This has usually been accomplished by fractional crystallization of an ester of (±)-menthol made from an optically active acid, eg, (+)-menthoxyacetic acid [*40248-63-3*]. Optically active (−)-menthol can be obtained by esterifying (±)-menthol with an optically inactive acid, eg, benzoic acid, forming a supersaturated solution or a supercooled melt of the ester and inoculating it with crystals of the (−)-menthyl ester of the same acid (159).

3-Menthene has been used as a starting material for the commercial production of (±)-menthol (160). 3-Menthene is treated with a peracid to obtain a mixture of 3,4-epoxy-*p*-menthane [*4184-02-5*] and 3,4-dihydroxy-*p*-menthane [*4053-64-9*], which is converted to menthone by boiling with dilute sulfuric acid. Mixed menthol isomers are produced by hydrogenation. The menthol isomers are then equilibrated to increase the (±)-menthol content to ca 60%, and the (±)-menthol is isolated by high efficiency fractionation.

If necessary, the *dl*-menthol can be further purified by crystallization. Other isomers are recycled to equilibration.

Thymol [*89-83-8*] serves as the starting material for another method of manufacturing (±)-menthol (161). Thymol is manufactured by isopropylating *m*-cresol with propylene or another isopropylating agent (162). Hydrogenation of thymol produces a mixture of menthol stereoisomers directly. As in the process starting with 3-menthene, the menthol isomers are then equilibrated, and the (±)-menthol is isolated by efficient fractionation.

World production of all menthols is ca 3500 t/yr. The 1981 price of natural (−)-menthol is ca \$15–16/kg; (±)-menthol sells for ca \$12/kg (53).

Uses. (−)-Menthol is used primarily for its physiological cooling effect and minty characteristics. It is widely used in cigarettes, dentrifrices (qv), cosmetics (qv), and pharmaceuticals (qv). It is also used in compounding artificial mint flavors. (±)-

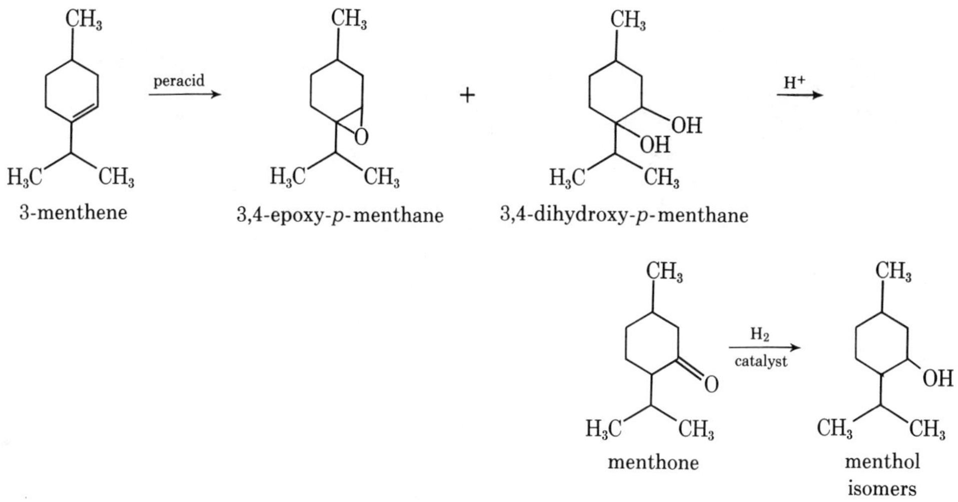

Menthol has only about half of the cooling effect of (−)-menthol and is used to a much smaller extent, chiefly in skin-care products, eg, shaving creams and lotions, as well as in pharmaceutical rubs.

Carvone. ***Reactions.*** Carvone undergoes a variety of reactions, but most are commercially insignificant. Some dihydrocarvone [*7764-50-3*] and carveol [*99-48-9*] are produced by selective reduction of carvone, but the amounts are small.

Manufacture and Price. (−)-Carvone is the most abundant constituent of spearmint oil from which it can be separated by efficient fractionation. If extremely high purity is desired, further purification can be effected by crystallization of derivatives, followed by regeneration of the carvone. (+)-Carvone is the most abundant component of caraway- and dill-seed oils and can be isolated from them by similar procedures. The odors and tastes of the isomers are distinctly different and are characteristic of the oils from which they were separated.

A substantial amount of (−)-carvone is produced synthetically from (+)-limonene. No other method is commercially important. (+)-Limonene reacts with nitrosyl chloride producing limonene nitrosochloride [*31891-52-8*], which is dehydrochlorinated in the presence or absence of a base to obtain carvoxime [*31198-76-2*].
Hydrolysis of the oxime gives (−)-carvone (80–81). The hydrolysis reaction is an equilibrium reaction and it is helpful to add a compound, eg, acetone, to tie up to a large extent the hydroxylamine which is also produced. Acidity must also be controlled to prevent the formation of carvacrol (2-methyl-5-isopropylphenol) [*499-75-2*]. (+)-Carvone can be produced by the same method if (−)-limonene is used as the starting material. (−)-Limonene is expensive, however, and the demand for (+)-carvone is relatively small. Worldwide production of (−)-carvone is ca 270–365 t/yr. The 1981 price was $18–22/kg (53).

Uses. Most (−)-carvone is used in compounding or enhancing mint flavors and aromas for use in dentrifrices, confections, pharmaceuticals, and odorants.

Camphor. ***Reactions.*** Camphor undergoes most of the reactions typical of a cyclic ketone, but few of these are commercially significant (163). Treatment of camphor with sulfuric acid in the presence of acetic anhydride produces camphorsulfonic acid, which is used as a resolving agent if the starting camphor is optically active, ie, the

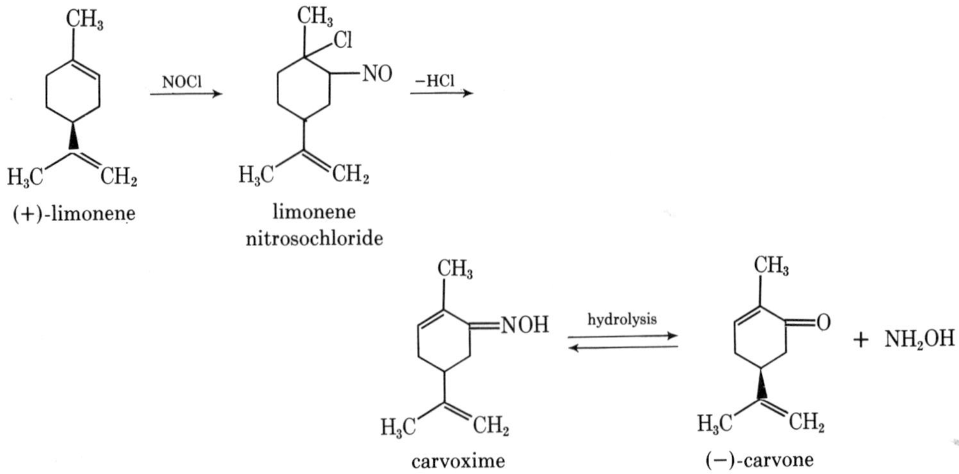

natural form. Camphor is readily brominated alpha to the carbonyl group yielding 3-bromocamphor [*76-29-9*]. (+)-3-Bromocamphor [*1925-58-2*] is used as a topical counterirritant; again, the starting camphor must be optically active.

Manufacture and Price. Camphor is obtained both naturally and synthetically. Natural camphor is obtained from the wood of the camphor tree (*Cinnamomum camphora*), which grows chiefly in the People's Republic of China, the Republic of China, and Japan. The camphor is isolated by combination of steam distillation, filtration, distillation, and sublimation (164). Natural camphor is the optically active dextrorotatory form, whereas synthetic camphor is racemic. Both forms are recognized by the USP.

Most synthetic camphor is produced from camphene made from α-pinene. Camphene reacts with acetic acid in the presence of an acid catalyst, eg, sulfuric acid, to obtain isobornyl acetate [*125-12-2*], which, on saponification, gives isoborneol. The alcohol is then converted to camphor by oxidation with acids, eg, chromic or nitric, or by catalytic dehydrogenation over a catalyst, eg, copper chromite.

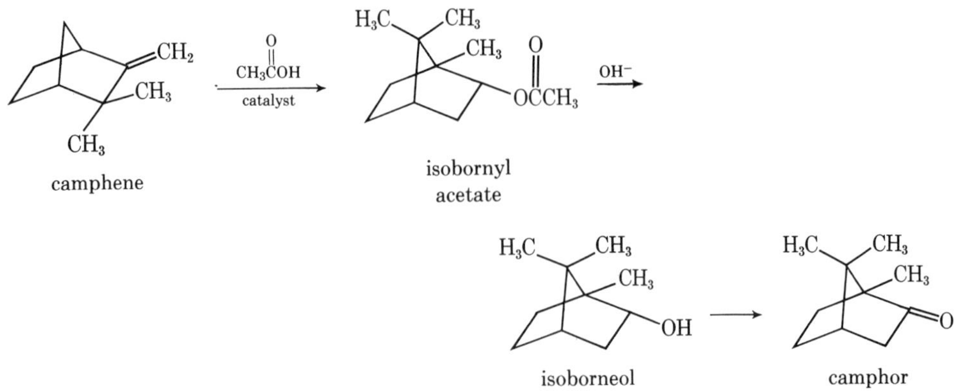

Purification of the camphor can be accomplished by crystallization, sublimation, or distillation.

Other methods for the synthesis of camphor exist but are not commercially important. The 1981 price of the synthetic product is ca \$4–6/kg (53).

Uses. Camphor is used chiefly as an ingredient in pharmaceutical preparations, as a moth repellent (see Repellents in Supplement Volume), a religious incense, and an odorant–flavorant in a variety of consumer products.

Pine Oil. Except for acetylation and hydrogenation of α-terpineol-rich fractions isolated from it, pine oil undergoes no important commercial reactions.

Manufacture and Price. There are three types of pine oil: steam-distilled, sulfate, and synthetic. Steam-distilled pine oil, a product of the wood naval-stores industry, is obtained by distillation of the extract from aged pine stumps in the southeastern United States. Sulfate pine oil is obtained by the fractionation of crude sulfate turpentine. Synthetic pine oil is produced by the acid-catalyzed hydration of pinene, usually α-pinene, followed by distillation. Mineral acids are commonly used in strengths of 20–40 wt % and at ca 30–50°C. The hydration is carried out to produce a maximum of alcohols and a minimum of hydrocarbons and cineoles by controlling the reaction variables of time, temperature, and acid strength. The reaction proceeds through carbonium ion mechanisms. Formation of the monocyclic products is depicted below. Some of the reactions are reversible under the acid reaction conditions.

The chief oxygenated components of pine oils are α-terpineol, β-terpineol [*138-87-4*], γ-terpineol, α-fenchol, borneol, isoborneol, camphor, terpinen-1-ol, terpinen-4-ol, dihydroterpineol [*498-81-7*], methyl chavicol [*140-67-0*], anethole [*104-46-1*], and 1,4- and 1,8-cineole. Not all of these components are present in all pine oils, but all pine oils contain α-terpineol as the main oxygenated component. In addition to the oxygenated components, all pine oils contain amounts of *p*-menthadienes, the most predominant ones being limonene (dipentene), terpinolene, and α-terpinene, with smaller amounts of 2,4(8)-*p*-menthadiene and γ-terpinene; small amounts of pinene and cymene may also be present.

Many grades of pine oil are available commercially and differ according to the source, the efficiency and type of distillation, and the way the various distillation cuts are combined. A high grade synthetic pine oil has the composition shown in Table 4, as determined by capillary gas chromatography.

Synthetic pine oils usually have higher terpinolene contents than the other pine oils, and only trace amounts of camphor are present. They contain no appreciable amounts of sesquiterpenes and do not contain anethole and methyl chavicol, which are usually in the other types of pine oil. Steam-distilled pine oil is relatively high in borneol and fenchol. Commercial pine oils are specified by physical properties and by alcohol content, as determined by either ASTM D 802-57 or gas chromatography. A pine oil, such as described above, typically has the properties listed in Table 5.

Lower grades of α-terpineol for odorant use can be obtained by distillation from pine oil, but these always contain small amounts of borneol, γ-terpineol, or both. High grade perfumery α-terpineol is made by partial dehydration of terpin hydrate [*2451-01-6*] under weakly acidic conditions (165). A substantial portion of α-terpineol produced is converted to its acetate [*80-26-2*] or dihydro-α-terpineol and its acetate which are also widely used in perfumery. U.S. production of synthetic pine oil in 1980 amounted to 20,000 t at an average price of \$1.20/kg (53).

Uses. By far the largest use of pine oil is in the manufacture of cleaners and disinfectants; it is effective against gram-negative enterobacteria but not against gram-positive organisms. Lower grades, eg, pine-oil sulfate, are used as frothers in separating

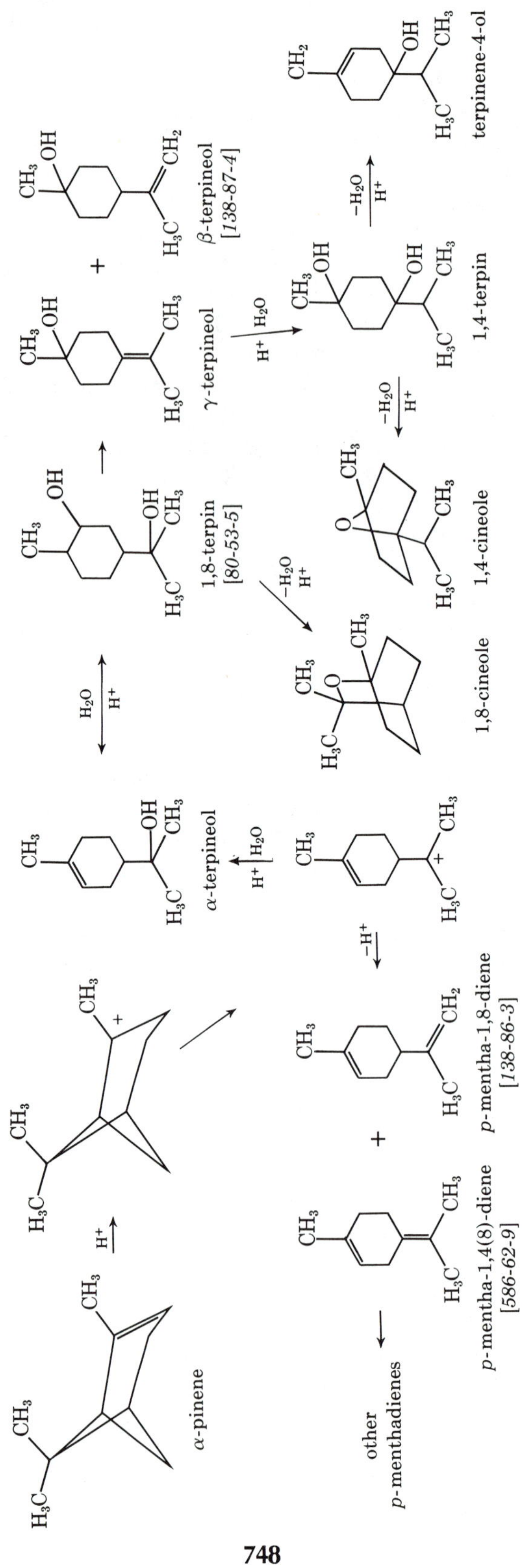

Table 4. Composition of Synthetic Pine Oil

Alcohols	Wt %	Others	Wt %
α- and γ-terpineol	67.0	terpinolene	6.7
terpinene-1-ol	3.9	limonene	2.1
cis-β-terpineol	2.8	*p*-2,4(8)-menthadiene	1.7
isoborneol	2.2	α-terpinene	1.1
borneol	2.1	1,8-cineole	0.8
terpinene-4-ol	2.0	camphor	0.7
fenchol	1.9	1,4-cineole	0.6
trans-β-terpineol	1.3	γ-terpinene	0.3
trace alcohols	2.5	*p*-cymene	0.3

Table 5. Physical Properties of Synthetic Pine Oil

Property	Value
specific gravity (15.5/15.5°C), min	0.930
terpene alcohols[a], wt %, min	85
moisture, wt %, max	0.5
color (APHA)	10
refractive index, n_D^{20}	1.479–1.489
acid number	<1
flash point, °C	75
Kauri butanol value	>500
distillation range, %	
<195°C, max	5.0
<230°C, min	95.0

[a] As measured by gas chromatography.

minerals by flotation. Some of the attributes of pine oil that make it useful in the manufacture of glues and adhesives (qv) are its dispersive and solvent power, its surface activity, and its preservation and plasticizing properties. Another use of pine oil is in the textile industry as a penetrant, dispersing agent, wetting agent, and inhibitor of bacterial growth in wet-processing of cotton, silk, rayon, and woolen goods. It also is used in the protective-coatings, agricultural-chemicals, packing-house, and leather industries (see Disinfectants).

Terpene Resins. Large quantities of terpene hydrocarbons are used in the manufacture of terpene resins. The most commonly used terpenes are β-pinene, α-pinene, limonene, or a combination of these. Other terpenes have been used, but to much less extent. The cationic polymerization is typically accomplished by reaction in a solvent, eg, toluene, and catalyzed by a Lewis acid, $AlCl_3$ being the most commonly used. The resins are thermoplastic and generally of light color (Gardner 2–5). Terpenes have also been copolymerized with a variety of nonterpene monomers. The largest use of terpene resins is in hot-melt and pressure-sensitive adhesives, although other significant uses exist.

Sesquiterpenes

Sesquiterpene hydrocarbons contain 15 carbon atoms usually comprised of 3 isoprene units, although commonly these are rearranged. There are acyclic, monocyclic, bicyclic, tricyclic, and tetracyclic sesquiterpenes. Their structures can be simple or complex. It has been estimated that at least 2000 sesquiterpenes and their derivatives have been discovered in nature (166). Some of the technologically important sesquiterpenes and their sources are listed in Table 6.

Reactions. The two most important commercial reactions of sesquiterpenes are esterification and oxidation to obtain aroma products; sesquiterpene acetates are especially valued for this application. Acetylation of longifolene and hydroxylation of caryophyllene yield products useful in perfumery.

An interesting antiulcer drug, geranyl farnesyl acetate [*51-77-4*] (Gefarnate), has been developed (167–170). A synthesis of this compound involves reaction of farnesyl bromide [*6874-67-5*] with diethyl malonate to produce, after hydrolysis, farnesylmalonic acid [*62088-08-8*] which is decarboxylated to obtain farnesylacetic acid [*23224-49-9*]. Esterification of the acid with geraniol gives geranyl farnesylacetate (171). Heating nerolidol with digeranyl malonate [*67779-10-6*] in dimethylformamide or dimethyl sulfoxide in the presence of aluminum isopropylate also produces Gefarnate (172). Other syntheses exist.

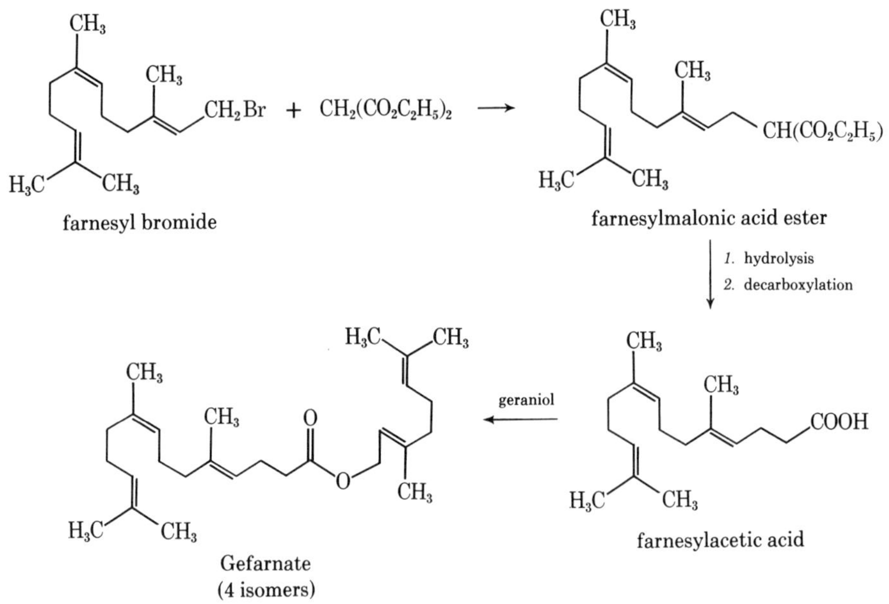

Manufacture. With a few exceptions, the limited number of commercially available sesquiterpenes have been isolated from natural sources, primarily essential oils to which they impart characteristic organoleptic properties. Isolation is usually accomplished by extraction, fractionation, crystallization, or a combination thereof. In general, the commercial products are not highly pure compounds unless they are amenable to crystallization, eg, cedrol.

The literature on the synthesis and transformations of sesquiterpenes is volu-

Table 6. Some Sesquiterpenes from Natural Sources

Sesquiterpene	CAS Registry No.	Structure	Essential oil
cedrol	*[77-53-2]*	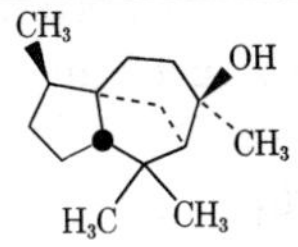	cedarwood
α-santalol	*[115-71-9]*	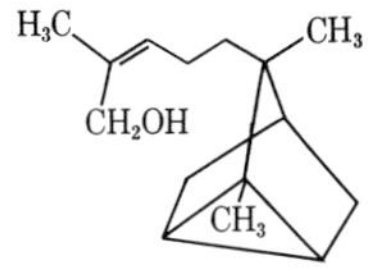	sandalwood
β-santalol	*[77-42-9]*	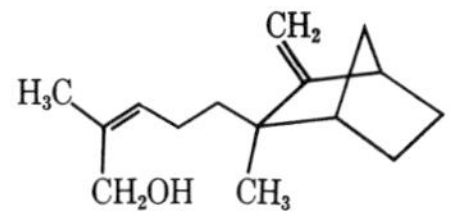	sandalwood
patchouli alcohol	*[5986-55-0]*	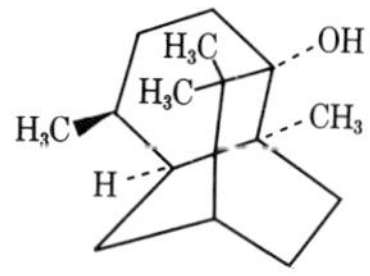	patchouli
guaiol	*[489-86-1]*	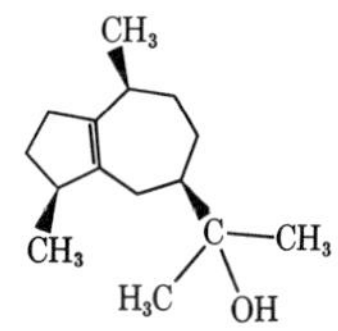	guaiac wood
α-cedrene	*[469-61-4]*	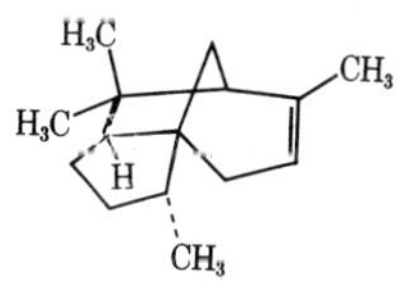	cedarwood
caryophyllene	*[87-44-5]*	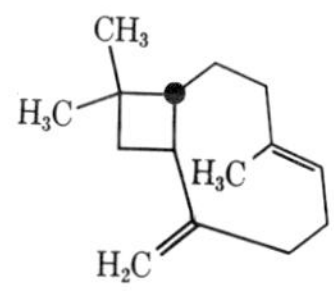	clove
longifolene	*[475-20-7]*	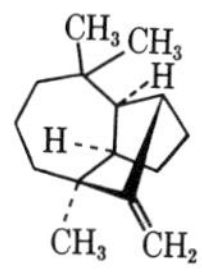	Indian turpentine
α-cadinene	*[24406-05-1]*	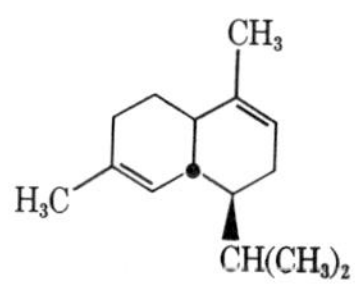	cubebs

minous and mostly of academic interest. An excellent comprehensive review of efforts at the total synthesis of sesquiterpenes has been published (173). The two most important sesquiterpenes synthesized commercially are nerolidol [*7212-44-4*] and farnesol [*4602-84-0*]. The starting material for nerolidol is (geranyl–neryl) acetone [*3796-70-1*, *3879-26-3*]. When geranyl–neryl acetone reacts with acetylene, dehydronerolidol [*2387-68-0*] forms. Hydrogenation of the latter over a Lindlar-type catalyst produces a mixture (ca 60:40 trans:cis) of the geometrical isomers of nerolidol (174). Geranyl–neryl acetone also reacts with vinyl Grignard reagent producing nerolidol directly (175). In both cases, the trans:cis ratio of the nerolidol formed depends on the trans:cis ratio of the geranyl acetone.

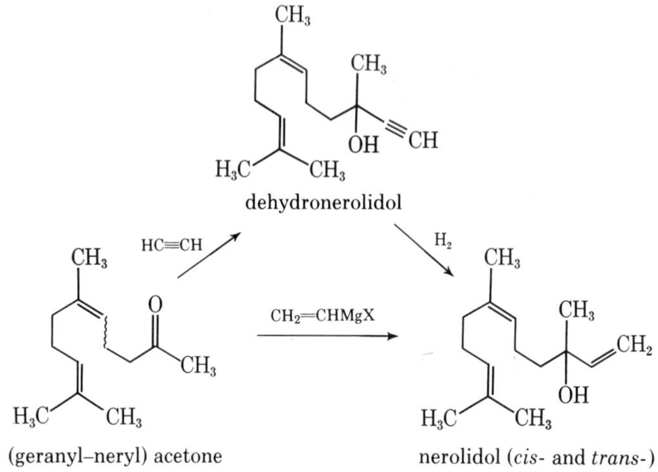

Farnesol can be produced from nerolidol by isomerization over a vanadium catalyst in essentially the same way linalool can be converted to geraniol–nerol (13,15).

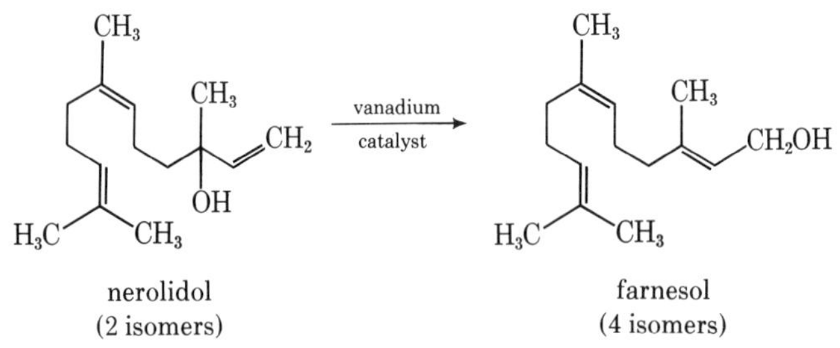

An older method for transforming nerolidol to farnesol involved treatment of nerolidol with anhydrous halogen acid to produce farnesyl halide which, on saponification via the corresponding acetate, yields farnesol (176). Other methods of interconverting linalool and geraniol–nerol are generally applicable to nerolidol and farnesol.

Uses. Sesquiterpenes are used almost exclusively in perfumery; very small amounts are used in flavors. Interest in sesquiterpenes as pharmaceutical intermediates is increasing.

Higher Terpenes and Terpenoids

Diterpenes. Diterpenes contain 20 carbon atoms. The resin acids and vitamin A are the most commercially important diterpenic substances. Gibberellic acid [*77-06-5*] is produced commercially by fermentation processes and is used as a growth promoter for plants, especially seedlings (see Plant growth substances). It is approved by the FDA for increasing the enzymatic activity of malt in brewing applications.

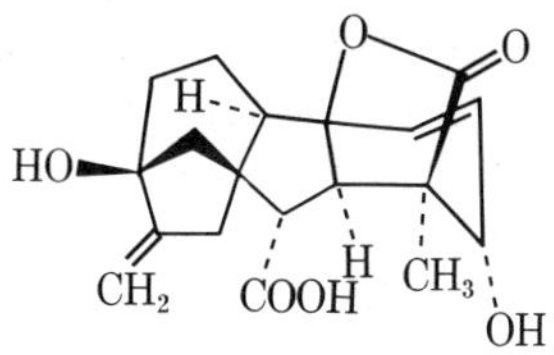

gibberellic acid

The leaves of a Paraguayan shrub, *Stevia rebaudiana* Bertoni, contain a diterpenoid glycoside called stevioside [*57817-89-7*], which is said to be 300 times sweeter than ordinary sucrose (177). Stevioside is noncaloric, tolerant to heat and acidity, and noncariogenic (see Sweeteners). The *Stevia* shrub has been valued as a sweetener for centuries by the Paraguayan natives. The shrub has relatively recently been cultivated quite extensively in the People's Republic of China, the Republic of Korea, and Japan. A crude extract containing mostly stevioside is approved for limited use in food in Japan (178). Approximately 50 t of stevioside was refined in 1979 from 6×10^5 t of *Stevia* leaves for commercial purposes in Japan (179). It is not approved for food use in the United States or Europe. At present, the safety of stevioside is still considered uncertain by some researchers. The possibility that stevioside is converted to the aglycone steviol [*471-80-7*] in vivo has resulted in the hypothesis that it might exhibit significant acute toxicity. The development of stevioside analogues that are nondegradable to steviol is being pursued (180).

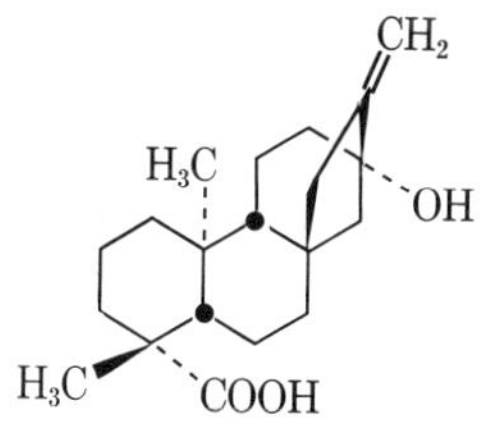

steviol

Isophytol [*505-32-8*] and phytol [*150-86-7*] are important moieties used in commercial syntheses of vitamin E and vitamin K. There is a variety of synthetic methods for their preparation. Chlorophyll [*479-61-8*] is a phytyl ester.

Sesterterpenes. Sesterterpenes contain 25 carbon atoms and were first isolated from and identified in insect protective waxes and fungal sources in the mid-1960s (see Waxes). They form a rare and complicated group of natural products and are not commercially significant (181). A review of their occurrence, structure elucidation, and biosynthesis appears in ref. 182.

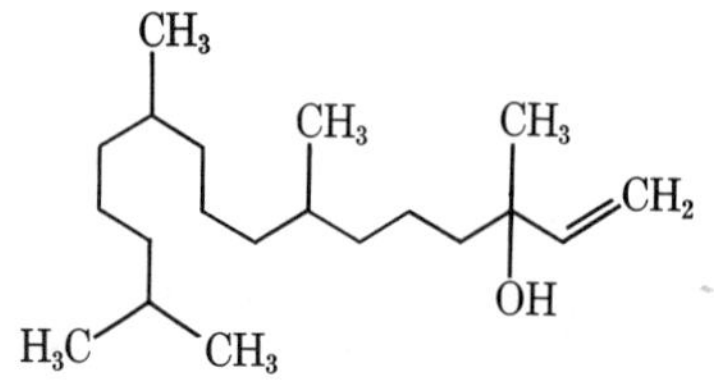

isophytol

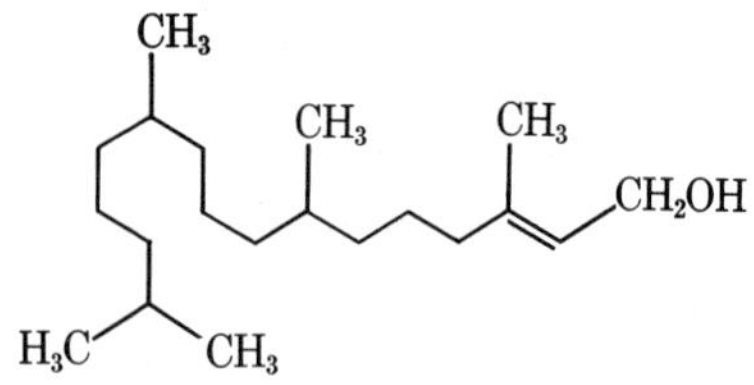

phytol

Triterpenes. The triterpenes contain 30 carbon atoms and are widely distributed in nature, especially in plants, both in the free state and as esters or glycosides. A smaller but important group, including lanosterol [*79-63-0*], occurs in animals. The triterpene acyclic hexaene hydrocarbon, ie, squalene [*111-02-4*], occurs in the liver oils of certain fish, especially those of sharks. Mostly, the triterpenes have four or five rings, which fuse in the same general pattern. Structurally, chemically, and biogenetically, many of the triterpenes have much in common with steroids (qv) (183). It has been substantially verified experimentally that *trans*-squalene [*111-02-4*] is the precursor in the biosynthesis of all triterpenes through a series of cyclization and rearrangement reactions (181,183). Squalene is not used much in cosmetics and perfumery formulations because of its light, heat, and oxidative instability, but its completely hydrogenated derivative, squalane [*111-01-3*], has rather wide usage as a skin lubricant and fixative.

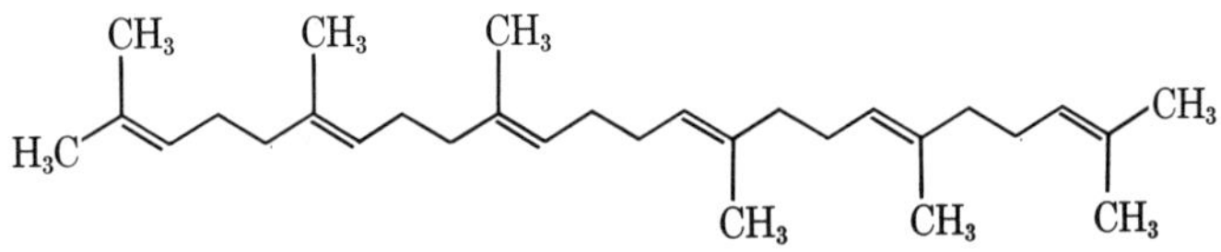

squalene

Squalane is produced synthetically by the coupling of two molecules of geranyl acetone [*3796-70-1*] with diacetylene, followed by dehydration and complete hydrogenation (184). Squalane can also be made by other methods, eg, the dimerization of dehydronerolidol followed by dehydration and hydrogenation (185).

Tetraterpenes. Carotenoids comprise the most important group of C_{40} terpenes and terpenoids, although not all carotenoids contain 40 carbon atoms. They are widely distributed in plant, marine, and animal life. It has been estimated that nature produces ca 100×10^6 t/yr of carotenoids; synthetic production amounts to several hundred tons per year (186–187). Carotenoids are a class of hydrocarbons (carotenes) and their oxygenated derivatives, eg, xanthophyll [*127-40-2*], consisting of eight isoprenoid units joined in such a manner that the arrangement of isoprene units is reversed at the center of the molecule so that the two central methyl groups are in a 1,6 positional relationship and the remaining nonterminal methyl groups are in a 1,5 positional relationship. The class also includes some rearranged and degraded compounds but not retinal and related C_{20} compounds. Carotenoids may be acyclic, monocyclic, or bicyclic. The respective parent compounds for these categories, lycopene [*502-65-8*], γ-carotene [*472-93-5*], and β-carotene [*7235-40-7*], are indexed by *Chemical Abstracts* as ψ,ψ-carotene, β,ψ-carotene, and β,β-carotene, respectively. The geometrical configuration around double bonds is generally trans. The prefix neo is often used to des-

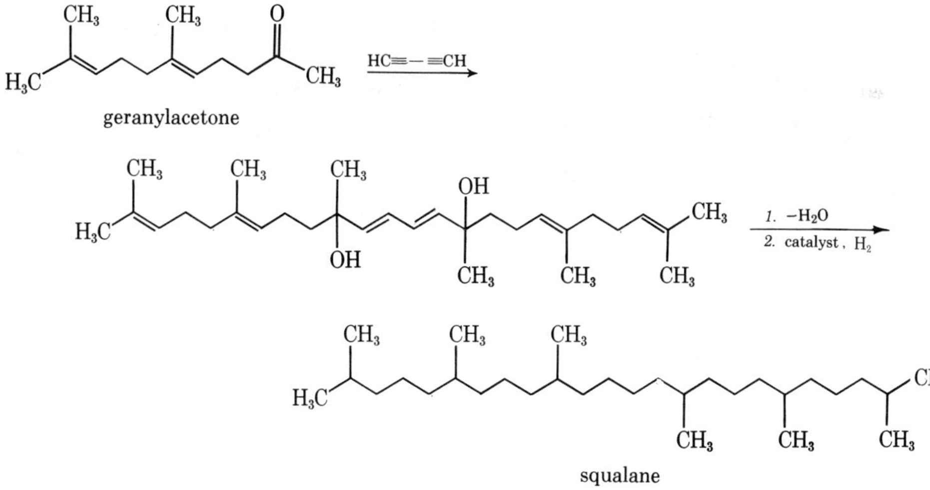

ignate isomers containing at least one cis configuration. The prefix apo indicates carotenoids that are oxidative-degradation products retaining more than half of the carotene structure. There are ca 400 naturally occurring carotenoids of known structure (188–191).

Carotenoids were usually isolated from natural sources by solvent extraction. The extracts were used primarily for coloring foods. At present there are five synthetic carotenoids available commercially. These are β-carotene, canthaxanthin [*514-78-3*], ethyl β-apo-8′-carotenoate [*1109-11-1*], β-apo-8′-carotenal [*1107-26-2*], and citranaxanthin [*3604-90-8*] (192). There is a close relationship between the chemistry involved in their synthesis and the synthesis of vitamin A. A basic overview of the commercial technology has been published (187). All syntheses start with β-ionone. Additional information on industrial syntheses of carotenoids is described in refs. 186–187, 192–198.

Crystalline carotenoids produced by chemical synthesis are of high purity and uniform color. Carotenoids crystallize in a variety of forms, the crystal color varying from red through violet to almost black. The size of the crystals influences the color of the specific carotenoid. Melting points are fairly high and tend to increase with increasing molecular weight and functional groups. The melting ranges of β-apo-8′-carotenal, β-carotene, and canthaxanthin are 136–142°C, 176–182°C, and 207–212°C, respectively (199). Ethyl β-apo-8′-carotenoate melts at 134–138°C, whereas citranxanthin melts at 155–156°C (200–201).

An important function of certain carotenoids is their provitamin A activity. Vitamin A may be considered as having the structure of half of the β-carotene molecule with a molecule of water added at the end position. In general, all carotenoids containing a single unsubstituted β-carotene half have provitamin A activity, but consequently their activity is only about half that of β-carotene. Provitamin A compounds are converted to vitamin A by an oxidative enzyme system present in the intestinal mucosa of animals and humans. This conversion apparently does not occur in plants.

Because of the multiple conjugated double-bond structure in the molecule, pure crystalline carotenoid materials are very sensitive to light and air oxidation and must be stored in sealed containers under vacuum or inert gas to prevent rapid degradation. Thus, commercial utilization as food colorings was initially limited; however, stable marketable forms were developed, eg, emulsions, oil solutions and suspensions, and spray-dried forms (see Colorants for foods, drugs, and cosmetics).

Carotenoids are used as pigments and dietary supplements in animals and poultry feedstuffs (see Pet and livestock feeds). They are added to pharmaceutical products to provide a form of control during manufacturing and to distinguish one product from another; they also enhance aesthetic aspects of the products (202).

β-Carotene is prescribed in the treatment of the inherited skin disorder erythropoietic protoporphyria (EPP) to reduce the severity of photosensitivity reactions in such patients. A recent review considers the essential theoretical background relevant to the role of carotenoids as photoconductors (203). β-Carotene has also been used as a photoconductor in recording-media film (see Electrophotography).

BIBLIOGRAPHY

"Terpene Resins" in *ECT* 1st ed., Vol. 13, pp. 700–704, by W. J. Roberts and A. L. Ward, Pennsylvania Industrial Chemical Corporation. "Terpenes and Terpenoids" in *ECT* 1st ed., Vol. 13, pp. 705–771, by R. S. Ropp, Hercules Powder Company; J. E. Hawkins, University of Florida; E. G. Reitz, Chicago City Colleges, Wright Branch; P. de Mayo, Birkbeck College, University of London; G. C. Harris, Hercules Powder Company. "Terpenes and Terpenoids" in *ECT* 2nd ed., Vol. 19, pp. 803–838, by Albert B. Booth, Hercules Incorporated; J. S. Autenrieth and Albert B. Booth (Terpene Resins), Hercules Incorporated. "Camphor" in *ECT* 2nd ed., Vol. 4, pp. 54–58, by Gastão Etzel, Camphor & Allied Products Limited.

1. 21 CFR Sect. 172–182.
2. National Research Council, *Food Chemicals Codex*, 3rd ed., National Academy Press, Washington, D.C., 1981.
3. D. L. J. Opdyke, ed., *Monographs on Fragrance Raw Materials*, Research Institute for Fragrance Materials, Pergamon Press, Inc., New York, 1979; monographs are also published as a regular feature in *Food Cosmet. Toxicol.*, Pergamon Press, Oxford.
4. R. J. Lewis, Sr. and R. L. Tatken, eds., *Registry of Toxic Effects of Chemical Substances*, 1979 ed., Vols. 1–2, U.S. Dept. of Health and Human Services, Public Health Service, Center for Disease Control, National Institute for Occupational Safety and Health Service, Cincinnati, Ohio, Sept. 1980.
5. *U.S. Department of Agriculture Crop Report*, Pulp Chemicals Association, 1980 statistical summary.
6. U.S. Pat. 3,278,623 (Oct. 11, 1966), J. M. Derfer (to The Glidden Company).
7. L. A. Goldblatt and S. Palkin, *J. Am. Chem. Soc.* **63**, 3517 (1941).
8. U.S. Pat. 2,882,323 (Apr. 14, 1959), R. Weiss (to Van Ameringen-Haebler, Inc.).
9. U.S. Pat. 3,031,442 (Apr. 24, 1962), R. L. Webb (to The Glidden Company).
10. U.S. Pat. 3,076,839 (Feb. 5, 1963), R. L. Webb (to The Glidden Company).
11. U.S. Pat. 4,018,842 (Apr. 19, 1977), L. A. Canova (to SCM Corp.).
12. G. Ohloff and E. Klein, *Tetrahedron* **18**, 37 (1962).
13. U.S. Pat. 3,925,485 (Dec. 9, 1975), P. Chabardes, C. Grard, and C. Schneider (to Rhone-Poulenc S.A.).
14. U.S. Pat. 4,006,193 (Feb. 1, 1977), Y. Ninagawa, T. Nishida, and K. Itoi (to Kuraray Co., Ltd.).
15. U.S. Pat. 4,254,291 (Mar. 3, 1981), B. J. Kane (to SCM Corp.).
16. W. Kimel, J. D. Surmatis, J. Weber, G. O. Chase, N. W. Sax, and A. Ofner, *J. Org. Chem.* **22**, 1611 (1967).
17. H. Lindlar, *Helv. Chim. Acta* **35**, 446 (1952); U.S. Pat. 2,681,938 (Dec. 31, 1954), H. Lindlar (to Hoffmann-La Roche, Inc.); U.S. Pat. 3,674,888 (July 4, 1972), M. Derrien and J. F. Le Page (to Institut Français du Pétrole, des Carburants et Lubrifiants).
18. W. Kimel, N. W. Sam, S. Kaiser, G. G. Eichmann, G. O. Chase, and A. Ofner, *J. Org. Chem.* **23**, 153 (1958).
19. G. Saucy and R. Marbet, *Helv. Chim. Acta* **50**, 2091 (1967).
20. U.S. Pat. 3,023,246 (Feb. 27, 1962), H. Pasedach and M. Seefelder (to BASF A.-G.).
21. M. F. Carroll, *J. Chem. Soc.*, 704 (1940).
22. *Ibid.*, 507 (1941).

23. Ger. Pat. 1,259,876 (Feb. 1, 1968), H. Pommer, H. Mueller, and H. Overwien (to Badische Anilin- & Soda-Fabrik A.-G.).
24. U.S. Pat. 3,670,028 (June 13, 1972), H. Mueller, H. Koehl, and H. Pommer (to Badische Anilin- & Soda-Fabrik A.-G.).
25. Ger. Pat. 1,643,668 (June 9, 1971), H. Pommer, H. Mueller, H. Koehl, and H. Overwien (to Badische Anilin- & Soda-Fabrik A.-G.).
26. H. Pommer and A. Nürrenbach, *Pure Appl. Chem.* **43**(3–4), 527 (1975).
27. W. Reif and H. Grassner, *Chem. Ing. Tech.* **45**(10a), 646 (1973).
28. Ger. Pat. 973,089 (Dec. 3, 1959), W. Friedrichsen (to Badische Anilin-& Soda-Fabrik A.-G.).
29. Ger. Pat. 1,268,135 (May 16, 1968), H. Pommer, H. Mueller, and H. Overwien (to Badische Anilin- & Soda-Fabrik A.-G.).
30. Ger. Pat. 1,286,020 (Jan. 2, 1969), H. Pommer, H. Mueller, and H. Overwien (to Badische Anilin- & Soda-Fabrik A.-G.).
31. U.S. Pat. 3,686,321 (Aug. 22, 1972), H. Mueller and H. Overwien (to Badische Anilin- & Soda-Fabrik A.-G.).
32. Fr. Pat. 1,548,516 (Oct. 28, 1968), M. Pichou (to Société des Usines Chimiques Rhone-Poulenc).
33. Can. Pat. 766,787 (Sept. 5, 1967), W. C. Meuly and P. Gradeff (to Rhodia Inc.).
34. Fr. Pat. 1,384,137 (Jan. 4, 1965), W. C. Meuly and P. Gradeff (to Rhone-Poulenc S.A.).
35. U.S. Pat. 3,983,175 (Sept. 28, 1976), Y. Tamai, T. Nishida, F. Mori, Y. Omura, M. Tanomura, T. Hosogai, Y. Ninagawa, and K. Itoi (to Kuraray Co., Ltd.).
36. Fr. Pat. 1,573,026 (July 4, 1969), E. Monrier (to Société des Usines Chimiques Rhone-Poulenc).
37. *Chem. Eng. News*, 5 (Nov. 22, 1982).
38. G. S. Fisher and J. S. Stinson, *Ind. Eng. Chem.* **47,** 1569 (1955).
39. J. S. Stinson and R. V. Lawrence, *J. Org. Chem.* **19,** 1047 (1954); J. B. Lewis and G. W. Hedrick, *J. Org. Chem.* **24,** 1870 (1959).
40. F. Holloway, H. J. Anderson, and W. Rodin, *Ind. Eng. Chem.* **47,** 2111 (1955).
41. U.S. Pat. 2,863,882 (Dec. 9, 1958), J. P. Bain, A. B. Booth, and E. A. Klein (to The Glidden Company).
42. U.S. Pat. 3,281,485 (Oct. 25, 1966), R. L. Blackmore (to Roberts, A. Boake & Company Ltd.).
43. G. Egloff, G. Hulla, and V. I. Komarewsky, *Isomerization of Pure Hydrocarbons*, ACS Monograph No. 88, American Chemical Society, Washington, D.C., 1942.
44. U.S. Pat. 3,655,803 (Mar. 11, 1972), F. L. Miller (to SCM Corp.).
45. U.S. Pat. 3,778,486 (Dec. 11, 1973), C. B. Hamby, Jr. (to SCM Corp.).
46. U.S. Pat. 3,325,553 (June 13, 1967), J. M. Derfer (to The Glidden Company).
47. U.S. Pat. 3,660,512 (May 2, 1977), C. B. Hamby, Jr., C. W. Barrett, and J. M. Derfer (to SCM Corp.).
48. Can. Patent 973,206 (Aug. 14, 1975), O. Prochazka (to Anglo Paper Products, Ltd.).
49. U.S. Pat. 3,359,342 (Dec. 19, 1967), J. M. Derfer (to The Glidden Company).
50. U.S. Pat. 3,360,581 (Dec. 26, 1967), J. M. Derfer (to The Glidden Company).
51. U.S. Pat. 3,420,910 (Jan. 7, 1969), C. Bordenca, J. M. Derfer, and C. B. Hamby, Jr. (to SCM Corp.).
52. *Synthetic Organic Chemicals, U.S. Production and Sales, 1980*, USITC Publication 1183, U.S. International Trade Commission, Washington, D.C., pp. 143–144, 261, 265.
53. *Chem. Mark. Rep.*, 52, 55, 57 (Oct. 12, 1981).
54. U.S. Pat. 2,340,294 (Feb. 1, 1944), J. P. Bain (to Nelio Resin Processing Corp.).
55. S. Dev, *11th IUPAC Int. Symp. Chem. Nat. Prod.* (4), 433 (1978).
56. U.S. Pat. 3,407,241 (Oct. 22, 1968), A. B. Booth (to Hercules, Inc.).
57. U.S. Pat. 3,407,242 (Oct. 22, 1968), A. B. Booth (to Hercules, Inc.).
58. U.S. Pat. 2,851,481 (Sept. 9, 1958), R. L. Webb (to The Glidden Company).
59. M. Matsui, H. Yoshioka, Y. Yamada, H. Sakamoto, and T. Kitahara, *Agric. Biol. Chem.* **29**(8), 784 (1965).
60. H. Yoshioka, M. Matsui, Y. Yamada, and H. Sakamoto, *Int. Ole Chim. Ind.* **III,** 890 (1967).
61. Y. P. Volkov and L. N. Khachatur'yan, *J. Org. Chem. USSR* **4**(8), 1347 (1968).
62. T. Sasaki, S. Eguchi, M. Ohno, and T. Oyobe, *Bull. Chem. Soc. Jpn.* **42,** 3582 (1969).
63. M. Matsui, H. Yoshioka, H. Sakamoto, Y. Yamada, and T. Kitahara, *Agric. Biol. Chem.* **31**(1), 33 (1967).
64. W. Cocker and H. St. J. Lauder, *J. Chem. Soc. Chem. Commun.*, 684 (1972).
65. W. Cocker, H. St. J. Lauder, and P. V. R. Shannon, *J. Chem. Soc., Perkin 1*, 332 (1975).

66. R. Sobti and S. Dev, *Tetrahedron* **30,** 2927 (1974).
67. Y. Gopichand, A. S. Khanra, R. Mitra, and K. K. Chakravarti, *Indian J. Chem.* **13,** 433 (1975).
68. M. J. De Vos and A. Krief, *Tetrahedron Lett.* (21), 1891 (1979).
69. J. Verghese, *Perfum. Flavor* **4**(4), 23 (1979).
70. U.S. Pat. 3,466,267 (Sept. 9, 1969), J. M. Derfer (to SCM Corp.).
71. J. Verghese, *Perfum. Flavor.* **5**(3), 47 (1980).
72. W. Cocker, P. V. R. Shannon, and P. A. Staniland, *J. Chem. Soc. C*, 41 (1966).
73. B. A. Arbuzov, Z. G. Isaeva, and V. V. Ratner, *Dokl. Akad. Nauk S.S.S.R.* **134,** 583 (1960).
74. W. Cocker, *J. Soc. Cosmet. Chem.* **22,** 249 (1971).
75. U.S. Pat. 2,565,471 (Aug. 28, 1951), G. A.Buntin (to Hercules Powder Co.).
76. R. B. Bates, E. S. Caldwell, and H. P. Klein, *J. Org. Chem.* **34,** 2615 (1969).
77. U.S. Pat. 2,933,542 (Apr. 19, 1960), J. P. Bain and W. Y. Gary (to The Glidden Company).
78. U.S. Pat. 4,204,080 (May 20, 1980), J. O. Bledsoe, Jr., and C. G. Cárdenas (to SCM Corp.).
79. U.S. Pat. 4,249,028 (Feb. 3, 1981), J. O. Bledsoe, Jr., and C. G. Cárdenas (to SCM Corp.).
80. C. Bordenca, R. K. Allison, and P. H. Durstine, *Ind. Eng. Chem.* **43,** 1196 (1951).
81. U.S. Pat. 3,293,301 (Dec. 20, 1966), J. M. Derfer, B. J. Kane, and D. G. Young (to The Glidden Company).
82. U.S. Pat. 3,014,047 (Dec. 19, 1961), J. P. Bain, W. Y. Gary, and E. A. Klein (to The Glidden Company).
83. U.S. Pat. 3,016,408 (Jan. 9, 1962), R. L. Webb (to The Glidden Company).
84. U.S. Pat. 3,400,162 (Sept. 3, 1968), M. G. J. Beets and W. Meerburg (to International Flavors & Fragrances, Inc.).
85. U.S. Pat. 3,412,364 (Nov. 26, 1968), J. H. Blumenthal (to International Flavors & Fragrances, Inc.).
86. U.S. Pat. 4,168,271 (Sept. 18, 1979), C. G. Cárdenas and Z. U. Din (to SCM Corp.).
87. U.S. Pat. 2,919,290 (Dec. 29, 1959), R. L. Webb (to The Glidden Company).
88. Jpn. Kokai 76 70707 (June 18, 1976), G. Hata and M. Tanaka (to Toyo Rayon Co.).
89. U.S. Pat. 4,186,148 (Jan. 29, 1980), A. Murata, S. Tsuchiya, A. Konno, and J. Uchida (to Nissan Chemical Industries, Ltd.).
90. K. Takabe, T. Katagiri, and J. Tanaka, *Tetrahedron Lett.* (34), 3005 (1975).
91. T. Fujita, K. Suga, and S. Watanabe, *Aust. J. Chem.* **27,** 531 (1974).
92. K. Takabe, T. Katagiri, and J. Tanaka, *Chem. Lett.* (9), 1025 (1977).
93. Jpn. Kokai Tokkyo Koho 78 95,909 (Jan. 31, 1977), H. Kumobayashi (to Takasago Perfumery Co., Ltd.).
94. M. Tanaka and G. Hata, *Chem. Ind.*, 202 (1977).
95. U.S. Pat. 4,247,480 (Jan. 27, 1981), A. Murata, S. Tsuchuja, A. Konno, F. Arima, and H. Ikeda (to Nissan Chemical Industries, Ltd.).
96. K. Takabe, T. Katagiri, and J. Tanaka, *Chem. Lett.* (10), 1031 (1975).
97. U.S. Pat. 4,107,219 (Aug. 15, 1978), A. Murata, S. Tsuchuja, H. Suzuki, and H. Iketa (to Nissan Chemical Industries, Ltd.).
98. U.S. Pat. 4,266,087 (May 5, 1981), A. J. Chalk and S. A. Magennis (to Givaudan Corporation).
99. U.S. Pat. 3,932,539 (Jan. 13, 1976), B. J. Kane and R. A. Von Genk (to SCM Corp.).
100. U.S. Pat. 4,179,468 (Dec. 18, 1979), B. J. Kane and R. A. Von Genk (to SCM Corp.).
101. U.S. Pat. 4,244,890 (Jan. 13, 1981), B. J. Kane and R. A. Von Genk (to SCM Corp.).
102. U.S. Pat. 4,267,366 (May 12, 1981), B. J. Kane and R. A. Von Genk (to SCM Corp.).
103. U.S. Pat. 2,420,131 (May 6, 1947), L. A. Goldblatt and S. Palkin (to the free use of the people in the United States).
104. R. L. Burwell, Jr., *J. Am. Chem. Soc.* **73,** 4461 (1951).
105. U.S. Pat. 3,028,431 (Apr. 3, 1962), R. L. Webb (to The Glidden Company).
106. U.S. Pat. 3,963,648 (June 15, 1976), H. R. Ansari, B. G. Jaggers, J. F. Janes, and B. N. Jones (to Bush Boake Allen Ltd.).
107. U.S. Pat. 2,902,510 (Sept. 1, 1959), R. L. Webb (to The Glidden Company).
108. U.S. Pat. 3,487,118 (Dec. 30, 1969), J. H. Blumenthal (to International Flavors & Fragrances, Inc.).
109. U.S. Pat. 2,961,452 (Nov. 22, 1960), R. A. Raphael (to The Glidden Company).
110. U.S. Pat. 3,277,206 (Oct. 4, 1966), J. P. Bain (to The Glidden Company).
111. U.S. Pat. 2,867,668 (Jan. 6, 1959), E. T. Theimer (to Van Ameringen-Haebler, Inc.).
112. K. J. Crowley, *J. Org. Chem.* **33,** 3679 (1968).
113. K. J. Crowley and S. G. Traynor, *Tetrahedron* **34,** 2783 (1978).

114. U.S. Pat. 3,278,530 (Oct. 11, 1966), W. C. Doyle, Jr. (to Tenneco Chemicals, Inc.).
115. U.S. Pat. 3,468,854 (Sept. 23, 1969), E. E. Royals (to Tenneco Chemicals, Inc.).
116. U.S. Pat. 3,346,650 (Oct. 10, 1967), B. J. Kane (to The Glidden Company).
117. USSR Pat. 118,498 (Mar. 10, 1959), V. N. Kraseva, A. A. Bag, L. L. Malkina, O. M. Khol'mer, and I. M. Lebedev.
118. Ger. Offen. 2,338,291 (Feb. 21, 1974), L. M. Polinski, I-Der Huang, and J. Dorsky (to Givaudan, L., et Cie. S.A.).
119. K. Kogami and J. Kumanotani, *Bull. Chem. Soc. Jpn.* **41,** 2508 (1968).
120. K. E. Harding, L. M. May, and K. F. Dick, *J. Org. Chem.* **40,** 1664 (1975).
121. U.S. Pat. 4,055,601 (Oct. 25, 1977), W. J. Ehmann (to SCM Corp.).
122. U.S. Pat. 4,005,031 (Jan. 25, 1977), J. D. Surmatis (to Hoffmann-La Roche Inc.).
123. U.S. Pat. 3,661,978 (May 9, 1972), P. S. Gradeff and B. Finet (to Rhodia Inc.).
124. Fr. Pat. 1,132,659 (Mar. 14, 1957), P. Teisseire (to Établissements Roure-Bertrand Fils and J. Dupont).
125. W. Hoffmann, H. Pasedach, H. Pommer, and W. Reif, *Justus Leibigs Ann. Chem.* **747,** 60 (1971).
126. U.S. Pat. 3,745,189 (July 10, 1973), W. Hoffmann, H. Pasedach, and R. Fischer (to Badische Anilin- & Soda-Fabrik Aktiengesellschaft).
127. R. Marbet and G. Saucy, *Chimia (Switzerland)* **14,** 362 (1960).
128. T. Ishihara, T. Kitahara, and M. Matsui, *Agric. Biol. Chem.* **38**(2), 439 (1974).
129. Brit. Pat. 1,204,754 (Sept. 9, 1970), P. Chabardes and Y. Querou (to Rhone-Poulenc S.A.).
130. U.S. Pat. 3,920,751 (Nov. 18, 1975), P. Chabardes and Y. Querou (to Rhone-Poulenc S.A.).
131. H. Pauling, D. A. Andrews, and N. C. Hindley, *Helv. Chim. Acta* **59,** 1233 (1976).
132. U.S. Pat. 3,981,896 (Sept. 21, 1976), H. Pauling (to Hoffmann-La Roche Inc.).
133. P. Chabardes, E. Kuntz, and J. Varagnat, *Tetrahedron* **33,** 1775 (1977).
134. G. Saucy, R. Marbet, H. Lindlar, and O. Isler, *Helv. Chim. Acta* **42,** 1945 (1959).
135. A. Russell and R. L. Kenyon in E. C. Horning, ed., *Organic Syntheses, Vol. III,* John Wiley & Sons, Inc., New York, 1955, pp. 747–750.
136. U.S. Pat. 4,029,709 (June 14, 1977), R. S. De Simone and P. S. Gradeff (to Rhodia, Inc.).
137. K. V. Leets, A. K. Shumeiko, A. A. Rozenoer, N. V. Kudryasheva, and A. I. Pilyavskaya, *J. Gen. Chem. USSR* **27,** 1584 (1957).
138. U.S. Pat. 2,902,515 (Sept. 1, 1959), M. Montavon and G. Saucy (to Hoffmann-La Roche Inc.).
139. Y.-R. Naves, *Riv. Ital. Essenze Profumi Piante Off. Aromi Saponi Cosmet. Aerosol* **58**(10), 505 (1976).
140. Y. Fujita, T. Onishi, K. Hino, and T. Nishida, *Tetrahedron Lett.* **21**(14), 1347 (1980).
141. T. Onishi, Y. Fujita, and T. Nishida, *Chem. Lett.* (7), 765 (1979).
142. Ger. Offen. 2,705,602 (Aug. 25, 1977), T. Onishi, Y. Fujita, F. Wada, T. Nishida, Y. Omura, F. Mori, T. Hosogai, and S. Aihara (to Kuraray Co., Ltd.).
143. U.S. Pat. 3,005,845 (Oct. 24, 1961), J. P. Bain (to The Glidden Company).
144. Fr. Pat. 1,374,732 (Nov. 9, 1964), (to A. Boake Roberts & Co. Ltd.).
145. U.S. Pat. 4,237,072 (Dec. 2, 1980), P. Aviron-Violet and T.-P. Dang (to Rhone Poulenc Industries).
146. C. A. Henrick, W. E. Willy, J. W. Baum, T. A. Baer, B. A. Garcia, T. A. Mastre, and S. M. Chang, *J. Org. Chem.* **40,** 1 (1975).
147. U.S. Pat. 2,235,840 (Mar. 25, 1941), W. C. Meuly (to E. I. du Pont de Nemours & Co.).
148. U.S. Pat. 3,869,517 (Mar. 4, 1975), P. S. Gradeff and C. Bertrand (to Rhodia Inc.).
149. U.S. Pat. 3,852,360 (Dec. 3, 1974), M. Vilkas and G. Senechal (to L'Air Liquide Société Anonyme Pour Étude et Exploitation des Procedes Georges Claude).
150. U.S. Pat. 3,919,324 (Nov. 11, 1975), W. Himmele, W. Hoffmann, H. Pasedach, and W. Aquilla (to Badische Anilin- & Soda-Fabrik Aktiengellschaft).
151. U.S. Pat. 2,947,780 (Aug. 2, 1960), R. W. Teegarden and L. Steinbach (to International Flavors & Fragrances, Inc.).
152. U.S. Pat. 4,007,137 (Feb. 8, 1977), J. M. Sanders, W. L. Schreiber, and J. B. Hall (to International Flavors & Fragrances, Inc.).
153. U.S. Pat. 3,075,003 (Dec. 31, 1963), J. H. Blumenthal (to International Flavors & Fragrances, Inc.).
154. U.S. Pat. 3,413,364 (Dec. 31, 1968), J. H. Blumenthal (to International Flavors & Fragrances, Inc.).
155. B. D. Sully, *Chem. Ind. (London)*, 263 (1964).
156. T. Yoshida, A. Komatsu, and M. Indo, *Agric. Biol. Chem.* **29**(9), 824 (1965).
157. U.S. Pat. 2,827,499 (Mar. 19, 1958), J. P. Bain, A. B. Booth, and E. A. Klein (to The Glidden Company).

158. U.S. Pat. 2,868,845 (Jan. 13, 1959), R. L. Webb (to The Glidden Company).
159. U.S. Pat. 3,943,181 (Mar. 9, 1976), J. Fleischer, K. Bauer, and R. Hopp (to Haarman and Reimer Gesellschaft mit beschrankter Haftung).
160. U.S. Pat. 2,866,826 (Dec. 31, 1958), H. E. McLaughlin, J. H. Stump, Jr., and M. C. Cleere (to Heyden Newport Chemical Corp.).
161. J. C. Davis, *Chem. Eng.* **85**(12), 62 (May 22, 1976).
162. M. Nitta, K. Aomura, and K. Yamaguchi, *Bull. Chem. Soc. Jpn.* **47,** 2360 (1974).
163. J. L. Simonsen and L. N. Owen, *The Terpenes*, 2nd ed., Vol. II, Cambridge University Press, Cambridge, England, 1949, pp. 373–418.
164. E. Guenther, *The Essential Oils, Vol. IV*, D. Van Nostrand Company, Inc., New York, 1950, pp. 256–328.
165. P. Z. Bedoukian, *Perfumery and Flavoring Synthetics*, 2nd ed., Elsevier Publishing Company, New York, 1967, pp. 336–341.
166. G. Ohloff, *Fortschr. Chem. Org. Naturst.* **38,** 431 (1978).
167. E. Adami, E. Marazzi, and C. Turba, *Med. Exp.* **7,** 171 (1962).
168. E. Adami, *Experientia* **18,** 461 (1962).
169. Belg. Pat. 617,994 (Sept. 17, 1962), (to Istituto de Angeli Societa per Azioni).
170. C. Cardoni, B. Cavaleri, and E. Adami, *J. Med. Chem.* **6,** 457 (1963).
171. G. Pala, A. Mantegani, T. Bruzzese, and G. Sekules, *Helv. Chim. Acta* **53,** 1827 (1970).
172. Jpn. Kokai 78 53614 (May 16, 1978), M. Yagi, N. Yamata, T. Taketomi, and T. Matsui (to Takasago Perfumery Company, Ltd.).
173. A. F. Thomas and Y. Bessière in J. ApSimon, ed., *The Total Synthesis of Natural Products*, Vol. 4, John Wiley & Sons, Inc., New York, 1981, pp. 451–591.
174. O. Isler, R. Rüegg, L. Chopard-dit-Jean, H. Wagner, and K. Bernhard, *Helv. Chim. Acta* **39,** 897 (1956).
175. A. Ofner, W. Kimel, A. Holmgren, and F. Forrester, *Helv. Chim. Acta* **42,** 2577 (1959).
176. I. N. Nazarov, B. P. Gusev, and V. I. Gunar, *J. Gen. Chem. USSR* **28,** 1498 (1958).
177. G. G. Birch, L. F. Green, and C. B. Coulson, eds., *Sweetness and Sweeteners*, Applied Science Publishers, London, UK, 1971, p. 35.
178. *Food Chem. News*, 11 (Feb. 22, 1982).
179. *Jpn. Chem. Week*, 6, (Mar. 27, 1980).
180. G. E. DuBois, P. S. Dietrich, J. F. Lee, G. V. McGarraugh, and R. A. Stephenson, *J. Med. Chem.* **24,** 1269 (1981).
181. A. A. Newman, ed., *Chemistry of Terpenes and Terpenoids*, Academic Press, Inc., New York, 1972.
182. G. A. Cordell, *Phytochemistry* **13,** 2343 (1974).
183. W. Templeton, *An Introduction to the Chemistry of Terpenoids and Steroids*, Butterworth Publishers & Co., Ltd. Inc., London, 1969.
184. Y. Fujita and T. Nishida, *Kukagaku* **29**(11), 814 (1980).
185. Y. Fujita, Y. Ninagawa, T. Nishida, and K. Itoi, *Yuki Gosei Kagaku Kyokaishi* **37**(3), 224 (1979).
186. O. Isler, R. Rüegg, and U. Schwieter, *Pure Appl. Chem.* **14,** 245 (1967).
187. O. Isler, *Pure Appl. Chem.* **51,** 447 (1979).
188. T. W. Goodwin, *The Comparative Biochemistry of the Carotenoids*, Chapman & Hall, London, 1952.
189. T. W. Goodwin, ed., *Chemistry and Biochemistry of Plant Pigments*, 2nd ed., Academic Press, Inc., New York, 1976.
190. O. Isler, ed., and H. Gutman and U. Solms, co-eds, *Carotenoids*, Birkhäuser Verlag Basel, Switz., 1971.
191. O. Staub, *Key to Carotenoids, Lists of Natural Carotenoids*, Birkhäuser Verlag Basal, Switz., 1976.
192. F. Kienzle, *Pure Appl. Chem.* **47,** 183 (1976).
193. O. Isler, H. Kläui, and U. Solms in W. H. Sebrell, Jr., and R. Harris, eds., *The Vitamins*, 2nd ed., Vol. 1, Academic Press, Inc., New York, 1957, pp. 101–113.
194. O. Isler and P. Schudel in R. A. Raphael, E. C. Taylor, and H. Wynberg, eds., *Advances in Organic Chemistry*, Vol. IV, Interscience Publishers, a division of John Wiley & Sons, Inc., New York, 1963, pp. 115–224.
195. O. Isler, M. Montavon, and R. Rüegg, *Ind. Chim. Belge* **32**(3), 853 (1967).

196. O. Isler, W. Guex, H. Lindlar, M. Montavon, R. Rüegg, G. Ryser, G. Saucy, and U. Schweiter, *Chimia* **12,** 89 (1970).
197. H. Pommer, *Angew. Chem. Int. Ed. Engl.* **16,** 423 (1977).
198. M. Rosenberger, P. McDougal, G. Saucy, and J. Bahr, *Pure Appl. Chem.* **51,** 871 (1979).
199. Ref. 2, pp. 25, 73–74, 67–68.
200. H. Kläui and J. C. Bauernfeind in J. C. Bauernfeind, ed., *Carotenoids as Colorants and Vitamin A Precursors, Technological and Nutritional Applications,* Academic Press, Inc., New York, 1981, pp. 47–317.
201. H. Yokoyama and M. J. White, *J. Org. Chem.* **30,** 2481 (1965).
202. K. Münzel in ref. 200, pp. 781–813.
203. J. O. Williams in ref. 200, pp. 787–813.

General References

Ref. 173 is also a general reference.

S. Arctander, *Perfume and Flavor Chemicals (Aroma Chemicals),* Vols. I–II, published by the author, Montclair, N.J., 1969.

P. Z. Bedoukian, *Perfumery and Flavoring Synthetics,* 2nd rev. ed. Elsevier Publishing Co., New York, 1967.

P. Z. Bedoukian, "Annual Review of Perfumery and Flavor Materials" in *Perfum. Flavor.,* Allured Publishing Corporation, Wheaton, Ill.; published as a regular feature.

S. Coffey, ed., *Rodd's Chemistry of Carbon Compounds,* 2nd ed., Vols. IIb and IIc, Elsevier Publishing Co., New York, 1968 and 1969.

S. Dev, *CRC Handbook of Terpenoids, Monoterpenoids,* Vols. I–II, CRC Press, Inc., Boca Raton, Fla., 1982.

T. K. Devon and A. I. Scott, *Terpenes,* Vol. II of *Handbook of Naturally Occurring Compounds,* Academic Press, Inc., New York, 1972.

T. E. Furia and N. Bellanca, eds., *Fenaroli's Handbook of Flavor Ingredients,* 2nd ed., Vol. 2, CRC Press, Inc., Cleveland, Ohio, 1975.

E. Guenther, *The Essential Oils,* Vols. 1–6, D. Van Nostrand Co., Inc., Princeton, N.J., 1948–1952.

F. F. Gunstone, senior reptr., *Aliphatic and Related Natural Product Chemistry,* Vol. 1, The Chemical Society, Burlington House, London, 1979.

J. R. Hanson, senior reptr., *Terpenoids and Steroids,* Vols. 6–9, The Chemical Society, Burlington House, London, 1976–1979.

A. A. Newman, ed., *Chemistry of Terpenes and Terpenoids,* Academic Press, Inc., London, 1972.

A. R. Pinder, *The Chemistry of the Terpenes,* Chapman & Hall, London, 1960.

J. L. Simonsen and co-workers, *The Terpenes,* Vols. 1–3, 2nd ed. rev., Cambridge University Press, Cambridge, New York, 1953 and 1957.

W. Templeton, ed., *An Introduction to the Chemistry of Terpenoids and Steroids,* Butterworths & Co., Ltd., London, 1969.

W. Treibs and co-eds., *Gildemeister and Hoffmann: Die Aetherischen Oele,* 4th ed., Vols. I, II, IIIa, b, c, d, IV, V, VI, VII, Akademie Verlag, Berlin, FRG, 1956–1961.

Physical Properties and Spectral Data of Terpenes and Terpenoids

J. G. Grasselli and W. M. Ritchey, *CRC Atlas of Spectral Data and Physical Constants of Organic Compounds,* 2nd ed., CRC Press Inc., Boca Raton, Fla., 1975.

A. A. Swigar and R. M. Silverstein, *Monoterpenes, Infrared, Mass,* ^{1}H *NMR, and* ^{13}C *NMR Spectra, and Kováts Indices,* Aldrich Chemical Co., Inc., Milwaukee, Wis., 1981.

W. Jennings and T. Shibamoto, *Qualitative Analysis of Flavor and Fragrance Volatiles by Glass Capillary Gas Chromatography,* Academic Press, Inc., New York, 1980; also includes retention indexes and mass spectral data.

C. J. Pouchert, *The Aldrich Library of Infrared Spectra,* Aldrich Chemical Co., Inc., Wilwaukee, Wisc., 1974.

C. J. Pouchert and J. R. Campbell, *The Aldrich Library of NMR,* Vols. 1–11, Aldrich Chemical Co., Inc., Wilwaukee, Wisc., 1974.

E. Breitmaier, G. Haas, and W. Voelter, *Atlas of Carbon-13 NMR Data,* Vols. 1–2 and index, Heyden and Son Ltd., London, 1979.

E. Stenhagen, S. Abrahamsson, and F. W. McLafferty, *Registry of Mass Spectral Data*, Vols. I–II, John Wiley & Sons, Inc., New York, 1974.
EOA Book of Standards and Specifications, Essential Oils Association (succeeded by the Fragrance Materials Association of the U.S., Washington, D.C.), 1976; includes information on essential oils, aromatic chemicals, isolates, general tests, spectra, and suggested apparatus adopted by the scientific section of the EOA through 1975.
First Supplement to the EOA Book of Standards and Specifications, Fragrance Materials Association of the U.S., Washington, D.C., Jan. 1979; includes infrared spectra and gas–liquid chromatographic curves for aromatic chemicals and isolates.

JOHN M. DERFER
MARIAN M. DERFER
SCM Corporation

TETRACYCLINES. See Antibiotics, tetracyclines.

TEXTILES

Survey, 762
Finishing, 769
Testing, 802

SURVEY

Textile materials are among the most ubiquitous in society. They provide shelter and protection from the environment in the form of apparel, they provide comfort and decoration in the form of household textiles such as sheets, upholstery, carpeting, drapery, and wall covering, and they have a variety of industrial functions as tire reinforcement, tenting, filter media, conveyor belts, insulation, etc. Textile materials are produced from fibers (finite lengths) and filaments (continuous lengths) by a variety of processes to form woven, knitted, and nonwoven (felt-like) fabrics. In the case of woven and knitted fabrics, the fibers and filaments are formed into intermediate continuous-length structures known as yarns, which are then either interlaced by weaving or interlooped by knitting into planar flexible sheetlike structures known as fabrics. Nonwoven fabrics are formed directly from fibers and filaments by chemically or physically bonding or interlocking fibers that have been arranged in a planar configuration (see Nonwoven textile fabrics; Tire cords).

Textile fibers may be classified into two main categories and into a number of subcategories, as indicated in Table 1. The generic names of man-made fibers are defined and controlled by the Federal Trade Commission (1). With the exception of glass and asbestos fibers and the specialty metallic and ceramic fibers, textile fibers are formed from organic polymers. Cellulose (qv) and proteins (qv) are the only important natural polymers in naturally occurring fibers (see Biopolymers). Cellulose

Table 1. Classification of Textile Fibers

Naturally occurring fibers
vegetable (based on cellulose), cotton, linen, hemp, jute, ramie
animal (based on proteins), wool, mohair, vicuna, other animal hairs, silk
mineral, asbestos
Man-made fibers
based on natural organic polymers
rayon, regenerated cellulose
acetate, partially acetylated cellulose derivative
triacetate, fully acetylated cellulose derivative
azlon, regenerated protein
based on synthetic organic polymers
acrylic, based on polyacrylonitrile (also modacrylic)
aramid, based on aromatic polyamides
nylon, based on aliphatic polyamides
olefin, based on polyolefins (polypropylene)
polyester, based on polyester of an aromatic dicarboxylic acid and a dihydric alcohol
spandex, based on segmented polyurethane
vinyon, based on polyvinyl chloride
based on inorganic substances
glass
metallic
ceramic

is also the only polymer used in the production of regenerated or derivative man-made fibers. At this time, there are no commercial regenerated protein (azlon) fibers. Attempts at developing and commercializing azlon fibers from either vegetable or animal proteins (corn, peanuts, milk) have proven largely unsuccessful. Since currently used regenerated or derivative fibers (rayon, acetate, triacetate) are based on cellulose, this category of fibers is frequently referred to as "cellulosics." Synthetic polymers, as listed in Table 1, are widely used in fiber production (see Fibers, chemical; Fibers, elastomeric; Fibers, vegetable).

The 1980 total world production of textile fibers amounted to ca 30.6×10^6 metric tons, distributed among the principal types of fibers as shown in Table 2. Cotton (qv) maintains its position of preeminence, although synthetic polymer fibers continue to grow in importance. Although the rate of growth of these fibers has been significantly lower in the late 1970s than in the 1960s, it is reasonable to expect a continuing growth in consumption of these fibers on a worldwide basis. Raw-cotton production is an important agricultural activity in many parts of the world, as indicated in Table 3.

Table 2. Trends in World Fiber Production, 10^6 Metric Tons[a]

Fiber type	1974	1975	1976	1977	1978	1979	1980
natural fiber							
cotton	14.05	11.76	12.48	13.92	12.97	14.27	14.14
wool	1.53	1.54	1.49	1.49	1.53	1.57	1.58
silk	0.04	0.05	0.05	0.05	0.05	0.06	0.06
man-made fibers							
cellulosics	3.53	2.96	3.21	3.28	3.32	3.37	3.24
synthetic polymer	8.20	8.06	9.45	10.09	11.04	11.66	11.56

[a] Ref. 2.

Table 3. World Raw-Cotton Production by Region, 10^6 Metric Tons[a]

Country	1978	1979	1980[b]	1981[b]
Americas				
United States	2.47	3.36	2.54	3.54
Brazil	0.58	0.60	0.65	0.61
Mexico	0.36	0.34	0.37	0.37
other	0.85	0.76	0.71	0.71
Western Europe	0.19	0.16	0.18	0.18
Asia				
USSR	2.74	2.96	3.24	3.08
People's Republic of China	2.27	2.32	2.83	2.95
India	1.41	1.43	1.40	1.47
Pakistan	0.50	0.79	0.75	0.82
other	1.05	1.03	1.02	1.12
Africa				
Egypt	0.46	0.51	0.55	0.53
other	0.70	0.71	0.68	0.73
Total	*13.58*	*14.97*	*14.92*	*16.11*

[a] Ref. 2.
[b] Preliminary data.

Cotton production in any given year is dictated as much by weather conditions as by market demands. The United States, the USSR, and the People's Republic of China are the largest raw-cotton producers, although some cotton is grown in just about every country where rainfall and temperature are adequate. In the United States, ca 80% of the 1980 crop was grown in Arizona, California, Mississippi, and Texas. The high production of raw cotton in the western states is made possible by irrigation and highly mechanized farming practices.

Man-made fiber production is concentrated in the highly developed industrialized regions of the world, as shown in Table 4. The United States and Western Europe account for ca 50% of world production of these fibers. The consumption of the different types of fibers by the U.S. textile industry in the manufacture of various apparel, household, and industrial products is shown in Table 5. Cotton, nylon, and polyester are the dominant fiber types in the United States; polyester is the most widely used since 1977 (see Polyamides; Polyester fibers).

Table 4. 1980 World Man-Made Fiber Production by Region, 10^6 Metric Tons[a]

Country	Cellulosics	Synthetic polymer	Total (%)
Americas			
United States	0.37	3.63	4.00 (26.9)
other	0.15	0.83	0.98 (6.6)
Europe			
Western	0.74	2.58	3.32 (22.3)
Eastern	1.14	1.24	2.38 (16.0)
Japan	0.40	1.45	1.85 (12.5)
all other	0.45	1.88	2.33 (15.7)
Grand total	*3.25*	*11.61*	*14.86 (100.0)*

[a] Ref. 2.

Table 5. Fiber Consumption by the U.S. Textile Industry, 10^6 Metric Tons[a]

Fiber	1971	1976	1980
cotton	1.81	1.54	1.41
wool	0.14	0.04	0.04
rayon	0.45	0.27	0.23
acetate	0.23	0.14	0.14
polyester	0.91	1.50	1.59
nylon	0.77	0.95	1.04
acrylic	0.23	0.27	0.27
polyolefin	0.14	0.23	0.32
glass	0.23	0.32	0.41
Total	*4.91*	*5.26*	*5.45*

[a] Ref. 2.

The cellulosic fibers were the first man-made fibers to become significant commercial materials. They were introduced in the decade following World War I, and they grew in importance until the 1960's when the synthetic polymer fibers began to make their major impact (see Cellulose acetate and triacetate fibers; Rayon). The first synthetic polymer fiber, nylon, introduced in 1940, was the commercial result of pioneering research on condensation polymerization reactions by Carothers and his associates at the DuPont Company (3). Polyester fibers, also based on condensation polymerization reactions, were introduced in the 1950s (4). Addition (vinyl) polymerization reactions, studied extensively in those years, eventually yielded the acrylic and olefin fibers (qv) (5) (see Acrylic and modacrylic fibers).

Man-made fibers are formed by extrusion processes known as melt, dry, or wet spinning (6). In melt spinning, molten polymer is forced under pressure through a metallic dye, known as a spinneret, containing many tiny holes. The emerging fluid jets solidify in a cooling zone to form filaments, which are then wound on containers. The diameter of the filament is determined primarily by the rates of extrusion and take-up. The cross-sectional shape of the filaments can be controlled by the shape of the holes in the spinneret. Melt spinning is the most economical filament extrusion process, but it requires polymers that are thermally stable under melt conditions. Nylon, polyester, and olefin filaments are produced by this method, and glass filaments are produced by an essentially equivalent process. In dry spinning, used for acetate, triacetate, and acrylics, the polymer is dissolved in a suitable solvent, and the solution is forced through a spinneret into a heated zone where the filament solidifies as the solvent evaporates. In wet spinning, the polymer or a chemical derivative of the polymer is dissolved in a suitable solvent, and the solution extruded into a liquid bath where the filament is formed by a combination of precipitation, coagulation, and regeneration mechanisms. Rayon (qv), spandex, and some acrylics are formed by this method.

The spinning or extrusion of filaments is normally followed by the operation known as drawing (6). In this step, the newly formed filaments are irreversibly extended and stabilized by setting or crystallization processes. The combination of spinning and drawing is used to develop a molecular organization characterized by a high degree of orientation and crystallinity. In general, both man-made and natural fibers can be described as having a regular and well-developed molecular organization with considerable structural anisotropy, which is reflected in large differences between axial and transverse physical properties of fibers. In this regard, fibers differ from other polymeric materials such as rubbers and plastics where less well-organized molecular structure predominates.

An important new development in synthetic polymer fiber manufacture is high speed spinning (7–8). In this technology, the spinning (extrusion) and drawing steps are combined into a single operation. The extremely high take-up speed (up to 6000 m/min) induces the high degree of orientation and crystallinity necessary for adequate fiber physical properties. The favorable economics for high speed spinning over the traditional two-step spin-and-draw process suggests that this method will become increasingly dominant, especially for polyester fibers.

With the exception of silk (qv), naturally occurring fibers have finite lengths and generally require several cleaning and purification steps prior to processing into yarns and fabrics (9–10). Man-made fibers, on the other hand, are produced in the form of continuous filaments and can be processed directly into multifilament yarns and subsequently into fabrics. Alternatively, continuous man-made filaments may be cut into discrete lengths to form staple fiber, which can subsequently be processed into yarns and fabrics. Silk is a continuous filament formed by solidification of a protein fluid extruded by the cultivated silkworm, *Bombyx mori* (11). It can be used as a continuous filament or cut for staple fiber. In general, yarns formed from staple fibers (natural, man-made, or blends) are bulkier, softer and have a rougher surface texture than continuous multifilament yarns. In recent years, a number of texturing processes have been developed to impart bulk and softness to continuous multifilament yarns to simulate the characteristics of typical staple-fiber spun yarns (12–13).

The manufacture of textile fabrics from fibers is based on processes that were developed to accommodate the special geometric and physical properties of naturally occurring fibers (9–10). Thus, there is a cotton system of yarn manufacture, a wool system, and a worsted system. Wool and worsted differ in that worsted yarns are manufactured from longer wool fibers that have been drawn parallel by combing operations. Similarly, there are combed-cotton yarns and carded-cotton yarns. In combed-cotton yarns, some of the short fibers have been removed by the combing, and the fibers in the yarn are more parallel, producing a smoother, more lustrous yarn. Carded-cotton yarns are somewhat rougher with fiber ends protruding from the yarn surface. The operations to which fibers are subjected during yarn manufacture are typically opening and cleaning, picking, carding, combing (optional), drawing (referred to as gilling in the case of worsted yarns), roving, spinning, twisting, plying, and winding. Yarns produced by these operations are referred to as spun or staple yarns (13–14). Today, cotton-, woolen-, and worsted-type yarns may be manufactured from staple man-made fibers or from mixtures of natural and man-made fibers. Such blended yarns have found increasing importance since it is possible to combine the desirable properties of man-made and natural fibers.

Yarns used for the production of woven fabrics are classified either as warp yarns or as filling yarns. During weaving, warp yarns are held in the loom under considerable tension whereas filling yarns are inserted and interlaced in various patterns at right angles to the warp. Warp yarns are more highly twisted and less extensible than filling yarns. Yarns used for the manufacture of knit fabrics are, in general, even less twisted and more extensible to produce soft and rather bulky knit fabrics.

After the fabric-formation process, textiles are subjected to either dyeing or printing, and to a variety of finishing operations that may be mechanical or chemical in nature. The specific nature of these dyeing and finishing operations depends largely on the fiber type and on the intended use of the finished fabric.

Dyeing is normally conducted in an aqueous solution of dyestuff (or mixture of dyestuffs) and involves a partition or distribution of the dye from the aqueous phase to the solid fiber phase. Although in most instances dyeing is performed on fabrics, it may also be performed on raw stock (staple fiber) or on yarns. The dye is generally absorbed by the fiber and held by secondary forces, but there are a number of chemically reactive dyes that bond covalently to the fiber polymer. In order to achieve reasonable rates of dyeing, commercial dyeing is normally conducted at 100°C, and in the case of some fibers, particularly polyester, dyeings are conducted under pressure to allow temperatures of approximately 130°C to be used (15). There are many different classes of dyes, including acid, direct, basic, sulfur, vat, disperse, and reactive, each of which has special affinity for certain fiber types. Through the use of dye mixtures and blended yarns and fabrics, it is possible to achieve colorful patterns by dyeing. However, printing, utilizing insoluble pigments instead of dyes, is used more extensively to achieve multicolor patterns and designs. The printing process is much more rapid and requires considerably less energy than dyeing (16) (see also Printing processes).

Mechanical finishing is primarily designed to produce special surface effects. Typical operations in this category are napping, shearing, embossing, and calendering. Chemical finishing operations may involve either the addition of chemicals or the chemical modification of some or all of the fibers comprising the fabric. It is not always possible to draw a clear distinction between additive chemical finishing and chemical modification. The primary purpose of chemical finishing is to confer special functional properties. Typically, procedures may be designed to impart dimensional stability, wrinkle resistance, wash-and-wear or durable-press characteristics, soiling resistance, water repellency, and flame retardance.

Chemical finishing procedures, other than scouring and possibly bleaching, are performed principally on those fabrics that are composed either entirely or in part of cellulosic fibers (ie, cotton and rayon). These fibers have chemically reactive sites (OH groups) that lend themselves to chemical modification and additive chemical treatments. Of particular importance in this regard are treatments of rayon- and cotton containing fabrics with difunctional reagents that are capable of chemically cross-linking cellulose chains (17). Such reactions improve fiber resilience, thereby enhancing wrinkle resistance and recovery. The same reactions can be used to impart wash-and-wear and durable-press characteristics. Wool-containing fabrics are also subjected to chemical-finishing treatments (see Wool). The main purpose of these treatments is to control felting shrinkage when wool fabrics are subjected to mechanical agitation at elevated temperatures in the wet state, conditions that occur in normal washing procedures. Felting shrinkage is caused by an inward migration of fibers in a yarn or fabric (18). This, in turn, is caused by a unique scalar structure of wool, and also of other animal fibers, that causes the frictional coefficient to be significantly higher in the against-scale direction than in the with-scale direction (see Felts). The chemical-finishing treatments are designed to minimize this differential friction by inactivating the scale edges. This is achieved by a variety of treatments, including chlorination, oxidation with permanganate solutions, polymer deposition, and combinations of these (18). Wool fabrics can also be chemically after-treated to impart wash-and-wear and durable-press properties. Fabrics composed of synthetic polymer fibers are frequently subjected to heat-setting operations. Because of the thermoplastic character of some such fibers, for example, nylon, polyester, olefin, and triacetate, it is possible to set such fabrics into desired configurations. These heat treatments involve recrystallization mechanisms at the molecular level, and are thus permanent unless the fabrics are exposed to more severe thermal conditions than were used in the heat-setting process.

Although textiles are usually associated with apparel and household products, industrial uses of fibers are of increasing importance. Of particular significance in this regard is the use of fibers to reinforce plastic materials, producing high performance fiber-reinforced composites (see Composite materials; Laminated and reinforced plastics). Such materials are increasingly used in construction and in the automotive, aerospace, and sports equipment industries. The use of textiles in civil engineering applications, such as road construction, erosion control, and heat-exchange devices, is also becoming more widespread. The term geotextiles has been adopted to describe fabrics used in these types of applications.

BIBLIOGRAPHY

1. Textile Products Identification Act.
2. *Textile Organon*, Vol. 51, Textile Economics Bureau, Inc., Roseland, N.J., 1981.
3. W. H. Carothers in H. Mark and G. S. Whitby, eds., *Collected Papers*, Interscience Publishers, New York, 1940.
4. J. R. Whinfield, *Text. Res. J.* **23,** 289 (1953).
5. P. J. Flory, *Principles of Polymer Chemistry*, Cornell University Press, Ithaca, N.Y., 1953.
6. H. F. Mark, S. M. Atlas, and E. Cernia, eds., *Man-Made Fibers Science and Technology*, Vols. 1–3, Interscience Publishers, a division of John Wiley & Sons, Inc., New York, 1967, 1968, and 1969.
7. U.S. Pat. 4,134,882 (Jan. 16, 1979) and U.S. Pat. 4,195,051 (March 25, 1980), H. R. E. Frankfort and B. H. Knox.
8. H. M. Heuvel and R. Huisman, *J. Appl. Polym. Sci.* **22,** 2229 (1978).
9. M. L. Rollins and D. S. Hamby, eds., *The American Cotton Handbook*, John Wiley & Sons, Inc., New York, 1965, 1966.
10. W. von Bergen, ed., *Wool Handbook*, Vols. I–II, John Wiley & Sons, Inc., New York, 1963, 1968.
11. R. H. Peters, *Textile Chemistry*, Vol. I, Elsevier, New York, 1963, Chapt. 11, pp. 302–314.
12. D. K. Wilson, *Text. Prog.* **9**(3), (1977); **10**(3), (1978); see also S. Backer, *Sci. Am.* **227**(6), 47 (1972).
13. B. C. Goswami, J. G. Martindale, and F. L. Scardino, eds., *Textile Yarns: Technology, Structure, and Applications*, John Wiley & Sons, Inc., New York, 1977.
14. P. R. Lord, *The Economics, Science, and Technology of Yarn Production*, North Carolina State University, Raleigh, 1979.
15. D. M. Nunn, ed., *The Dyeing of Synthetic Polymer and Acetate Fibres*, Dyers Co. Pub., Trust (Society of Dyers and Colourists), Bradford, UK, 1979.
16. L. W. C. Miles, ed., *Textile Printing*, Dyers Co. Pub. Trust (Society of Dyers and Colourists), Bradford, UK, 1981.
17. T. F. Cooke, J. H. Dusenbury, R. H. Kienle, and E. E. Linekin, *Text. Res. J.* **23,** 1015 (1954).
18. K. R. Makinson, *Shrinkproofing of Wool*, Marcel Dekker, Inc., New York, 1979.

LUDWIG REBENFELD
Textile Research Institute

FINISHING

Finishing of textiles includes, in its broadest sense, any process used to improve a knitted, woven, or bonded textile fabric for apparel or other home or industrial use. The principal classes of textile fibers and fabrics include cellulosic, synthetic, protein, glass, and blends. Because of the great differences in the chemical and physical nature of these fibers, finishing processes vary widely among classes and, to a lesser extent, within each class. Mechanical, heat-related, and chemical treatments have been developed over the years since the first attempts to convert fibers into yarns and then into a fabric.

Animal and plant fibers have been used for thousands of years, but the introduction of synthetic fibers into consumer use after World War II caused the most dramatic change in the finishing of natural fibers. Prior to this time it was generally necessary, after washing and drying, to pull or press a garment back into its original shape before wearing. Consumers accepted these disadvantages because there was no real alternative until the introduction of new fibers and their fabrics. When laundered with care, these new fabrics require little or no ironing and retain their shapes because shrinkage and stretching are controlled within narrow limits. These advantages came to be demanded of the garments made from natural fibers. As cotton and rayon were the primary fibers in use, a massive effort was put forth to develop new finishing processes to give cotton and rayon garments the same ease of care and versatility as garments made from synthetic fibers. Similarly, improvements were sought in finishing processes for the synthetics.

Thus, the general topic of textile finishing is very broad, for there are at least ten commercially important textile fibers today, and finishing processes enhance a wide variety of properties including crease retention, ease of care, dimensional stability, abrasion resistance, flame retardance, smolder resistance, odor control, soil release, variable moisture contents, and control of static charges, stretch, water repellency, waterproofing, and resistance to weathering (including sunlight and microorganisms), as well as improving the hand (or feel) and surface appearance of fabrics. Many of these specific topics are presented elsewhere in this Encyclopedia under headings that relate to a specific property, eg, Waterproofing and water repellency or Flame retardants for textiles; to a chemical or a chemical system, eg, Amino resins or Formaldehyde; or to a specific textile fiber, eg, Cotton, Wool, Polyesters, Rayon, etc.

Some of the more important mechanical finishing processes are compressive shrinkage (Sanforized process) and calender finishes, which include schreinering (to produce a silklike appearance), chintz finishing (usually a glazed fabric with bright prints), and embossing. These finishes are semidurable but are made more durable, especially if the base fabric is cotton or rayon, by including a treatment with a conventional N-methylol agent (see Amino resins).

There are many types of mechanical finishing processes employing many types of rollers, but unfortunately there is relatively little discussion of them in the published literature. Trends in fashion have reduced the demand for calendered fabrics, but the processes are still of considerable interest. Permanent effects with fabrics of thermoplastic material are obtainable if the heat and pressure are sufficient to soften the material. Calendering operations have been used to a considerable extent to improve the surface appearance of polyester and nylon fabrics and their blends (see general references).

Chemical Treatments

Generally, chemical treatments are divided into processes involving topical treatments and those involving formation of a chemical bond with the substrate. Most chemical treatments of synthetic fibers are topical because of the general inertness of the substrate. Because the cellulosics and wool are reactive, these fibers may readily form chemical bonds with the finish, as well as act as substrates for nonreactive topical treatment.

Durability of the topical treatments to the various laundering conditions varies widely, but some treatments possess good durability. The properties obtained on synthetics are more permanent when the chemical can penetrate and dissolve in the substrate. Durability of a topical treatment is often improved if the chemicals increase in molecular weight during processing and become entangled with the substrate. Those treatments that involve formation of a chemical bond with the substrate usually produce permanent finishes. However, some of these bonds are broken by hydrolysis during laundering. Some amount of hydrolysis of the finish can be tolerated, and the reactant systems are designed to minimize this reaction. Emphasis in this article has been placed on chemical treatments for the most widely produced commercial products, especially those that involve chemical bond formation with the substrate.

The expression ease of care includes a number of related properties when applied to apparel and related textile goods. The expressions wash-and-wear and durable press generally refer to the performance of a garment after machine washing and line or tumble drying. Since different countries use different washing and drying equipment and conditions, the evaluation for performance (ie, acceptance or rejection by the consumer in a specific country) is varied. What is acceptable in Europe may not be acceptable in the United States, and vice versa. The U.S. market is large, and for the most part, its requirements are of primary concern in this review.

Knitted fabrics generally shed their wrinkles more quickly and completely than woven fabrics regardless of the method of laundering. However, dimensional stability (shrinkage or stretching) of knits is more difficult to control than in woven fabrics. Fortunately, most of the processes that improve the smooth-drying appearance of a fabric also improve its dimensional stability. For example, the cross-linking that occurs during finishing of cellulosic fabric to make it dry smoothly also produces a fabric that resists shrinkage or stretching. Often fabrics are cross-linked with a low level of cross-linking agent to stabilize the fabric. Denim fabrics, for example, are often treated in this manner because a high level of smooth-drying properties is not required. For polyester–cellulosic blends, a heat-setting process is combined with the cross-linking process to achieve maximum performance.

The conventional finishing process for cellulosics involves three steps commonly referred to as pad–dry–cure (Fig. 1). In the first step, the fabrics are immersed in an aqueous finishing formulation and then squeezed to remove excess formulation and to assure even distribution of the reagents in the fabric. The pickup in weight at this point is carefully controlled but usually varies from 60–100% depending upon the process. Interest is growing in new padding technology (1) wherein the increase in weight during padding is even lower than with prior methods. In this new technology,

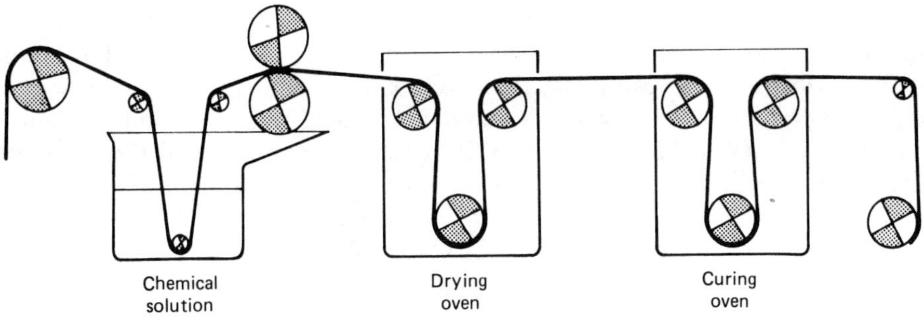

Figure 1. Pad–dry–cure textile finishing.

which includes use of foam or minimum application techniques, the pad bath is more concentrated since the concentration of solids on the dry fabric must be of the same magnitude regardless of the padding technology used.

A finishing pad bath usually consists of a cross-linking agent, a curing catalyst, and hand modifiers. For conventional durable-press processes of cellulosics and cellulosic blends, using a 60–100 wt % wet pickup, the bath contains ca 5–9 wt % cross-linking agent, enough acidic catalyst to effect a cure (0.4–3.5 wt %, depending upon catalyst strength), and 1–3 wt % emulsified polymer which generally acts as a fiber lubricant or a stiffening agent. A topical treatment for a synthetic fabric generally consists of only 1–3 wt % emulsified polymers.

Except for special applications, the drying process simply involves removal of water from the flat fabric on a continuous range. Often, drying and curing are performed in one continuous process at rates that often exceed 100 m/min. On the other hand, the dried fabrics may be made into garments prior to the curing process. This, of course, requires different curing equipment than that used for a flat, continuous length of fabric.

The padding and drying steps are especially important in the conventional treatment of cellulosics. The fabric must be prepared in such a state that, in general, the reagents readily penetrate the fibers to achieve an even treatment within the fiber. Migration of the reactants to the surface of the fiber during drying and curing is a problem and, in an extreme case, leads to a stiffened, unsatisfactory product. Research continues on this subject (2).

After drying, the padded fabric may be allowed to lose some heat or it may be placed in a roll immediately. Depending upon the formulation used in the pad bath, some reaction with cellulose starts during the drying process and continues if the dry roll is hot. The handling of this hot, dry fabric is important for future processing.

In addition to dimensional stability and smooth-drying appearance, ease of care includes crease retention. In general, adequate creases are obtained by chemical finishing of cellulosics and heat setting of polyester and many other synthetics. Blends of cellulosics and polyester are usually subjected to both processes for optimum performance.

To many in the trade, textile finishing refers primarily to chemical finishing (or resin finishing, as it is called in much of the older literature). The primary chemicals used in the early finishing processes were aminoplasts capable of cross-linking the cellulose molecule but also capable of homopolymerization. Not only were these chemicals capable of improving the appearance and the dimensional stability of the fabric, but they also retained pigments used in printing, provided protection for outdoor exposure, and conveyed other related properties desirable in textiles. A rather long and spirited debate developed over the mechanism involved in developing a smooth-drying cellulosic fabric. By the mid or late 1950s, most textile chemists had decided that cross-linking of cellulose and not polymer formation within or about the cellulose molecule was primarily responsible for the desired smooth-drying properties. However, even today the role of polymer formation with aminoplast resins in promoting smooth-drying properties is not completely discarded.

Fabric Preparation. A number of processes are performed after the fabric has been woven or knitted; these are not considered finishing steps. Scouring, mercerization, and bleaching are usually referred to as fabric preparation. Chemicals applied prior to dyeing are generally removed to avoid problems in the dyeing procedures. If the fabric is to be colored, the dyeing or printing process is not considered part of the finishing process. On the other hand, fluorescent brightening agents are often applied during a finishing process (see Brighteners, fluorescent; Dyes, application). However, because of energy costs and other considerations, combination of some of these processes with finishing procedures is receiving much attention from both production and research personnel.

Fabric preparation is very important for the quality of the subsequently dyed or printed fabric. A properly prepared fabric is white and free of contaminating chemicals, and it accepts dyes and other chemicals readily. Also, the fabric should not lose appreciable strength during either the preparation or the dyeing steps (3). Some fabric strength is lost during finishing, but most of this loss is due to embrittlement of the fiber because of cross-linking. Care must be exercised during chemical finishing in the selection of a dye–finishing agent combination to avoid an unwanted change in hue during processing or after exposure to sunlight.

Results of Cross-Linking Cellulosic Fabrics. Essentially any bifunctional compound, except those that cannot penetrate the fiber or that have widely separated reactive groups, will, on reaction with cotton cellulose, produce a fabric with improved smooth-drying appearance.

$$2\,\text{CellOH} + \text{y} - \text{X} - \text{y} \rightleftharpoons \text{CellOXOCell} + 2\,\text{Hy} \qquad (1)$$

Fabric finished in this manner is said to possess a memory. For example, a fabric cross-linked in the flat, dry state is expected to return to that state after washing when it is given an opportunity to shed its wrinkles in a tumble dryer. Similarly, a fabric cross-linked in the flat, wet state returns to that state, ie, sheds its wrinkles, when immersed in similar media. This concept was developed using pad–dry–cure and wet-cure processes with formaldehyde as the cross-linking agent (4).

$$2\,\text{CellOH} + \text{CH}_2\text{O.H}_2\text{O} \overset{\text{H}^+}{\rightleftharpoons} \text{CellOCH}_2\text{OCell} + 2\,\text{H}_2\text{O} \qquad (2)$$

Cross-linking a creased or wrinkled fabric likewise causes retention of the crease or wrinkle for the life of the cross-link. Most cross-linking reactions are reversible, and therefore, the properties imparted by cross-linking are reversible by removing the cross-link through hydrolysis.

Some undesirable properties are imparted by cross-linking cellulose. The most objectionable of these are reductions in breaking strength, tearing strength, and abrasion resistance. In addition, the ability of the fabric to retain or regain moisture is generally reduced by cross-linking. Moisture-regain properties can be retained somewhat by use of swelling agents during cross-linking or, occasionally, by selection of the cross-linking agent (5).

Cross-linking of cellulosic fibers also reduces the accessibility of the fiber. In other words, the cross-linking reaction in a heat-cure process takes place after the moisture is removed while the fiber is in a collapsed state. Therefore, dyeing of fabrics is generally performed prior to cross-linking by a heat cure because the dye cannot penetrate the cross-linked fiber (6). As noted below, this property can be used to advantage.

The appreciable loss of strength and abrasion resistance when cellulosics are cross-linked can be minimized, but this is usually accompanied by a loss in smooth-drying performance. One solution to this problem of loss of strength and abrasion resistance in cellulosics is to blend polyester fibers into the yarns that make up the fabric. The cellulosic portion of the fabric still loses strength because of cross-linking, but the effect is masked by the polyester.

Early Cross-Linking Agents. Formaldehyde, urea–formaldehyde, and melamine–formaldehyde were the first cross-linking systems used in commercial finishing or studied extensively in the early days of chemical finishing. They are still being used today, but other reactant systems have replaced them in general use.

Formaldehyde is the simplest and most economical cross-linking agent known. Despite much research (6–8), formaldehyde has had only limited use in practical heat-cure or vapor-phase applications. It produces a durable, inert cross-link that resists hydrolysis and is unaffected by washing, heat, bleaching, sunlight, or other conditions encountered by a finished textile. A 100% cotton fabric finished with formaldehyde, however, has reduced strength and abrasion resistance compared to a fabric finished with an *N*-methylol agent to a comparable level of durable-press performance. Loss of a great deal of formaldehyde in a conventional pad–dry–cure treatment is another disadvantage, but the primary reason for not using this agent is that it has never been satisfactorily scaled up for continuous industrial production. One of the main failings is the inability to produce a good product consistently. This finish is so stable that errors are essentially impossible to correct.

Urea–formaldehyde or melamine–formaldehyde reactants are resin formers for a large number of products. They have been used as polymers to coat and react with cellulose and as monomeric or oligomeric cross-linking agents. A reactant that coats a fiber generally stiffens it. In the case of a limp fabric, this can be desirable; however, too much resin on the surface will produce a "board." A stiffened, cross-linked fabric usually does not dry smoothly after laundering unless the surface polymer is worked mechanically during washing or in a machine that disturbs the coating. Melamine resins are still used as pigment binders and in binding other additives to cellulosic fabrics. However, during the 1950s both these early resins were rejected as cross-linking agents for cellulosic fabrics primarily because of discoloration and loss of strength upon hypochlorite bleaching. The problem precipitated research into chemical finishing of rayon and cotton aimed at development of new cross-linking systems that would resist the effects of hypochlorite bleach.

Chemistry of *N*-Methylol Cross-Linking Systems. The urea–formaldehyde reactant system is not dimethylolurea, as depicted in equation 3, but generally a methylolated urea in which the ratio of formaldehyde to urea is 1.3–1.8:1. This ratio provides a reasonably stable pad bath, and the reactant is capable of cross-linking cellulose, polymerization, or both. Catalysis is usually provided by metal salts or selected amine

$$HOCH_2NH\overset{\overset{\displaystyle O}{\|}}{C}NHCH_2OH + 2\ CellOH \rightleftharpoons CellOCH_2NH\overset{\overset{\displaystyle O}{\|}}{C}NHCH_2OCell + 2\ H_2O \qquad (3)$$

hydrochlorides.

During bleaching with hypochlorite, some of the N—H groups in the finish are converted to N—Cl. When the bleached fabric is ironed, hydrogen chloride is formed by a series of reactions and the polymer chain of the cellulose substrate is hydrolytically decomposed. The effects can be extreme, ranging from slight discoloration generally accompanied by a slight loss of strength to the fabric falling apart during ironing. Other effects, including removal of part of the finish by the bleaching process, are also encountered (see Bleaching agents).

The chemical mechanisms in chlorine retention and the decomposition of resultant chloroamides have been studied in detail (9–10), and the relation between molecular structure of the agent and its chlorine uptake and release was established. Chloroamides present in finished goods decompose at different rates. Unfortunately, in urea–formaldehyde finishes the chloroamide forms and decomposes readily, producing a heavily damaged fabric (11).

One way of reducing or eliminating chloroamide formation is to completely methylolate the agent and thus limit the number of N—H groups in the finished fabric. The amide–aldehyde reaction is reversible and can approach completion in some cases (eqs. 4 and 5) with excess formaldehyde.

$$—\overset{\displaystyle O}{\overset{\|}{C}}NH_2 + HCHO \rightleftharpoons —\overset{\displaystyle O}{\overset{\|}{C}}NHCH_2OH \quad (4)$$

$$—\overset{\displaystyle O}{\overset{\|}{C}}NHCH_2OH + HCHO \rightleftharpoons —\overset{\displaystyle O}{\overset{\|}{C}}N(CH_2OH)_2 \quad (5)$$

Methylolations are generally carried out under alkaline, aqueous conditions because acidic catalysis usually leads to formation of polymers or to methylenebisamides.

$$—\overset{\displaystyle O}{\overset{\|}{C}}NHCH_2OH + —\overset{\displaystyle O}{\overset{\|}{C}}NH_2 \rightleftharpoons (—\overset{\displaystyle O}{\overset{\|}{C}}NH)_2CH_2 + H_2O \quad (6)$$

Methylolations of primary or unsubstituted amides that approach completion (eq. 5) include those of formamide, carbamates, and melamines. Urea, which has four N—H groups, forms only the trimethylol derivative readily.

Addition to secondary or monosubstituted amides is usually difficult to complete (eq. 7) and forms a mixture of products.

$$—\overset{\displaystyle O}{\overset{\|}{C}}NHR + CH_2O \rightleftharpoons —\overset{\displaystyle O}{\overset{\|}{C}}N\begin{matrix} \diagup R \\ \diagdown CH_2OH \end{matrix} \quad (7)$$

Thus, *N,N′*-dimethylol-*N,N′*-dialkylamides resist formation, and reaction mixtures containing these agents are not effective cross-linking systems for cellulose. On the other hand, cyclic ureas containing secondary NH groups generally methylolate readily, and these agents have been the basis for the improved systems developed to replace urea–formaldehyde.

Glyoxal has been used to produce substituted *N*-methylol reactants capable of reaction with cellulose. Factors that limit the addition of amides to formaldehyde and glyoxal have been studied for imides, carboxylic acid amides, carbamates, and ureas. In general, complete substitution of all amino hydrogens available is limited by steric effects or by low amide acidity in some cases. Complete substitution of all amido hydrogens becomes more general if cyclization occurs during addition to the aldehyde, or if a cyclic amide with nitrogen in the ring, rather than an *N*-alkylamide, acts as nucleophile (12).

Formation of a fully methylolated cross-linking agent does not assure that the finish will be immune from hypochlorite bleaching damage. The curing step does not proceed to completion, even for the most effective cross-linking agent, and some N—H groups (formed by loss of formaldehyde) and some N—CH_2OH groups are found in the finish. Furthermore, laundering conditions can cause hydrolysis of a link to cellulose to form the easily hydrolyzable N—CH_2OH group. Thus, continued washing and bleaching with hypochlorite can alter the original finish and produce one that is susceptible to fabric damage. Most laundry washes are alkaline, but commercial laundering also includes an acid stage. A satisfactory finish must be able to withstand both types of laundering.

$$-\overset{\overset{O}{\|}}{C}NHCH_2OCell + H_2O \rightleftharpoons -\overset{\overset{O}{\|}}{C}NHCH_2OH + CellOH \qquad (8)$$

Reaction Mechanisms for Formation and Hydrolysis of Finishes. It was in the early 1960s that research on these reaction mechanisms, as applied to chemical finishing of cellulosic fabrics, commenced. The formation and hydrolysis (acid) reactions were shown to pass through the same intermediate immonium–carbonium ion (13).

$$-\overset{\overset{O}{\|}}{C}NHCH_2OH + H^+ \rightleftharpoons -\overset{\overset{O}{\|}}{C}NHCH_2\overset{H^+}{O}H \qquad (9)$$

$$-\overset{\overset{O}{\|}}{C}NHCH_2\underset{+}{\overset{H}{O}}H \rightleftharpoons -\overset{\overset{O}{\|}}{C}NHCH_2^+ + H_2O \qquad (10)$$

$$-\overset{\overset{O}{\|}}{C}NHCH_2^+ \leftrightarrow -\overset{\overset{O}{\|}}{C}\overset{+}{N}H{=}CH_2 \qquad (11)$$

$$-\overset{\overset{O}{\|}}{C}NHCH_2^+ + CellOH \rightleftharpoons -\overset{\overset{O}{\|}}{C}NHCH_2OCell + H^+ \qquad (12)$$

Based on the principle of microscopic reversibility, it was reasoned (14) that a highly reactive agent would form an easily hydrolyzable product. This proved true in practice, and the industry rapidly adopted the phrase that "easy on means easy off" (15).

Hydrolysis resistance of a finish is also important during storage of cured goods and during the final pressing and creasing after cured fabrics have been made into garments. Cotton and rayon fabrics regain moisture on standing; the amount regained depends on the blend level and the finishing process (16–17). Since cured, flat fabrics or finished garments are rarely washed prior to sale to the consumer, the acid curing catalyst remains on the fabric until it is laundered by the consumer. This combination of moisture and acidic conditions has considerable time to act upon the finish and thus, again, the resistance of the finish to hydrolysis is of utmost importance. Errors made in processing are often corrected by removal of the finish by acid hydrolysis or "stripping." Stripping must be accomplished under conditions that do not cause excessive strength loss of the cellulosic substrate.

All these factors attest to the importance of an understanding of the acid-catalyzed hydrolysis reaction for the practical development of a superior commercial product. This knowledge is also of great theoretical importance because it is the best and, at times, the only way to determine the chemical structure of the finish. Equation 1 represents the ideal situation in which X remains unaltered during finishing, garment manufacture, and storage. It has been suggested that the true chemical nature of modified cellulose is not properly represented by a simple cross-link (18–19).

A number of standardized conditions for acidic hydrolysis have been used by industry and researchers. Dilute hydrochloric acid is effective but requires careful control if consistent results are to be obtained. Hydrolysis conditions of 30 min at 80°C in an aq soln containing 1.5% phosphoric acid and 5% urea has been found to give reproducible results with a large number of cured fabrics.

In a recent study (20), more than 50 fabrics were hydrolyzed with this urea–phosphoric acid solution. The fabrics were prepared from mono- or polyfunctional *N*-methylol agents under similar curing conditions, and the level of nitrogen content on the fabric before and after hydrolysis was determined. It was concluded that the resistance to hydrolysis was determined primarily by chemical factors, ie, electronic effects of substituent groups on the intermediate immonium–carbonium ion (eqs. 11 and 13).

$$R\underset{\underset{\displaystyle R'}{|}}{\overset{\overset{\displaystyle O}{\|}}{C}}NH\overset{+}{C}HR'' \leftrightarrow R\underset{\underset{\displaystyle R'}{|}}{\overset{\overset{\displaystyle O}{\|}}{C}}\overset{+}{N}H{=}CHR'' \qquad (13)$$

Agents used included formaldehyde and glyoxal adducts of ureas, melamines, carbamates, formamides, acetamides, hydrazides, pyrrolidone, and similar amides. These and related experiments demonstrated that most *N*-methylol and similar agents react with cotton with little or no modification of their structure during curing. Loss of formaldehyde (the reverse reaction in eq. 7) from *N*-methylol groups can be appreciable for some reactant systems (21). Likewise, polymer formation in conjunction with cross-linking occurs in some systems, but this can be controlled (22).

Caution must be used in comparing hydrolysis resistance of cellulose finishes prepared from *N*-methylol agents. A small change in degree of cure, ie, the overall strength of the curing conditions, can make a large difference in the resistance to hydrolysis of a specific finish (23). A trace of base remaining in the fibers after fabric preparation, a change in catalyst type or amount, or variation in cure time or temperature can result in a finish with vastly different resistance to hydrolysis. The reason for this difference in hydrolysis resistance is presently under investigation. However, it has been shown that the variation occurs in finishes prepared from both monofunctional and polyfunctional agents. These differences, especially for a finish prepared from a monofunctional agent, are counter to results obtained from similar determinations performed in homogeneous systems.

Alkaline conditions found in laundering have generally not been strong enough to cause significant hydrolysis of conventional finishes. In research, cross-links are often designed to resist an acid rinse without taking their stability to alkaline conditions into consideration (24). The alkaline hydrolysis of alkoxymethylureas (25) is initiated by removal of a proton from the nitrogen atom followed by the elimination of the alkoxy group.

$$H_2N\overset{O}{\overset{\|}{C}}\overset{-}{N}CH_2OR \rightleftharpoons H_2N\overset{O}{\overset{\|}{C}}N{=}CH_2 + RO^- \qquad (14)$$

The mild conditions required for hydrolysis of alkoxymethylureas were considered surprising, and ethers of methylol compounds of carboxylic acid amides were found to be comparatively stable.

The base-catalyzed cross-linking of cotton with dimethylolurea and the synthesis of bis(methoxymethyl)urea from dimethylolurea and methanol using alkaline catalysis (26) support the above conclusions and the suggested mechanism involving a series of reversible steps involving eq. 14. Further support is provided by the susceptibility of similar finishes to alkaline hydrolysis (27) and their tendency to release considerable amounts of formaldehyde after a relatively strong alkaline home wash (28–29).

Leaving-Group Effects. The first leaving-group effect to be noticed in textile finishing was a minor deactivating effect of the methoxy group in comparison to the hydroxyl (30). This finding later provided a way of improving the resistance of highly reactive *N*-methylol agents to hydrolysis (31). The factors or principles involved are illustrated in the following equations:

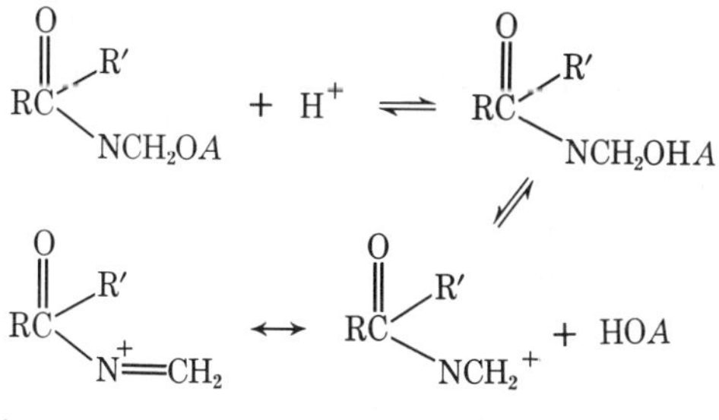

(15)

Ion stability is determined largely by the electronic and steric effects produced by R and R′. Activation by the leaving group depends upon basicity of the leaving group. The activation by an alkoxy leaving group for formaldehyde-based or glyoxal-based agents (32–34) produces at least a (3–5):1 reaction rate ratio for isopropoxy compared to methoxy. Partially isopropylated agents have been used commercially to produce a better cure at lower cure temperatures or times (23,32). Use of acetoxy derivatives produces even a greater activation (as much as 30:1 when compared to the same methoxylated reagent) (31,33).

Leaving-group effects should help explain the durability to acidic hydrolysis developed by strong curing conditions (35). The anhydroglucose unit in cellulose contains one primary and two secondary hydroxyl groups. On reaction with an *N*-methylol compound, these groups in part determine the resistance to hydrolysis of the bond to cellulose. In other words, —OCell contains groups that can function as either poor or good leaving groups.

Curing Catalysts. Many acids have been used successfully in finishing formulations to produce a durable-press finish. Selection of the specific catalyst is often determined by secondary factors, such as effluent standards, influence on dyes, formaldehyde release, discoloration of the fabric, effect on chlorine retention, formation of odors, and storage conditions. Many of the references cited include discussions on the selection of a catalyst for the specific reactant system under investigation. The chemical supplier generally specifies a certain catalyst or a selection of catalysts for use in its durable-press formulation. Frequently, the exact content of the formulation is unknown to the finisher.

Curing catalysts have been classified as mineral or organic acids and as latent acids such as ammonium salts, amine salts, and metal salts (36). Metal salts such as zinc nitrate and magnesium chloride have been the most widely used. Magnesium chloride, used by itself, is a very mild catalyst and is not recommended for curing some agents. Triazone and melamine-based reactants, for example, require a strong catalyst because of their buffering action. Mixed catalyst systems, eg, magnesium chloride with glycolic acid or citric acid, are often used to increase production rates by decreasing the curing time or lowering the curing temperature. Acetic acid and stronger acids, ie, those with a dissociation constant $\geq 10^{-6}$ at room temperature, are suitable (37).

The catalytic activity of metal salts is a complex result of several contributing factors, including acidity and oxidation state of the cation, nature of the anion, ionic size and ion mobility, hydrolysis reactions which the salts undergo in aqueous solution, covalent character, and temperature effects (38). Both Lewis-acid and Brønsted-acid mechanisms operate concurrently to different extents with different catalysts.

One of the greatest disappointments for researchers in this field has been the inability to isolate and characterize a complex of an *N*-methylol reactant and a Lewis acid salt. It is assumed (32,39) that coordination of the metal ion with the methylol oxygen (possibly by transfer from the proximate carboxyl oxygen) would be sufficient to initiate the mechanism shown in equations 9–12.

Physical Nature of Chemically Modified Cellulose. All cross-linking systems applied to 100% cellulosic fabrics under controlled conditions cause similar high losses of strength and abrasion resistance. To date, most finishing research directed at producing an acceptable smooth-drying fabric with improved strength properties involves oligomer and/or polymer formulation with limited cross-linking. Despite numerous independent efforts (40–43), only small advantages have been attained with 100% cellulosics.

Observations with the electron microscope have proven useful in delineating the regions of the cotton cellulose fiber affected by a variety of treatments, including dyeing, mercerization, acid and enzymatic hydrolyses, derivative formation, and resin impregnation (44). Methods to differentiate between impregnation and reaction within the cotton fiber or reaction on the surface of the fiber because of migration to the surface or a failure of the resin to penetrate the fiber have been described (44). Knowledge of this type is of primary importance in understanding the results of experiments designed to distribute reactants to certain portions of the fiber and to determine their mode of reaction, ie, polymerization, reaction with cellulose, or both.

Since cross-linking of cotton reduces strength, abrasion resistance, and other desirable qualities, the relationship between cotton morphology and textile performance continues to be a key research area. Some of the main approaches to improve the physical properties of chemically modified cotton fabrics are examined in ref. 45. The results from conventional pad–dry–cure cross-linking, cross-linking the fiber in the swollen state, simple substitution, and polymer deposition are considered. Attention is given to the pretreatments and the condition of the cotton fiber prior to finishing, eg, ammonia or caustic mercerization and effects of restretching of these pretreated fabrics, the never-dried cotton fiber, the dried fiber conventionally used in commerce, and fibers reswollen with strong alkalies or amines.

Chemically modified cotton fabrics having crease-resistance properties not anticipated on the basis of previous theories (40–41) have been cited (46). These inconsistencies include experimental data obtained from cotton treated with inert solvents, epoxy compounds capable of reacting with fibers preswollen in aqueous base, long-chain monobasic acid chlorides in selected nonaqueous media, and monomers or prepolymers capable of cross-linking or being deposited or grafted. Factors that should be considered in the development of a new theory consistent with all experimental facts include the hydrogen-bonded structure of cotton cellulose, the hydrogen-bonded structure of water, and the interactions between cotton or chemically modified cottons and water.

Further Development of *N*-Methylol Agents. Urea– and melamine–formaldehyde resins, regardless of how they are prepared, are unstable and difficult to keep in solution in a pad bath unless they are alkylated (usually methylated).

$$>NCH_2OH + CH_3OH \overset{H^+}{\rightleftharpoons} >NCH_2OCH_3 + H_2O \qquad (16)$$

These alkylated resins are often used in combination with other resin systems. Differences between these cross-linking systems are chemical in nature because one commercial *N* methylol cross-linking system produces essentially the same set of physical test data for a finished fabric as another.

Melamines and Other Amino-*s*-Triazines. These finishes are unlike urea–formaldehyde finishes in that hypochlorite does not cause fabric degradation, but it does yellow the fabric (47). The *N*-chloromelamines are relatively thermally stable, but a dilute solution of sodium bisulfite will remove color from the fabric. Conventional melamine agents are poor from the standpoint of loss of methylol groups, and even a fully methoxymethylated melamine or amino-*s*-triazine resins will, after processing and laundering, develop sites for color formation.

A large number of *N*-substituted melamines have been prepared, and only those in which no NH_2 group can occur produce nonyellowing finishes. Examples of these types of finishes have been prepared from the formaldehyde addition products of *N*,*N*′,*N*″-tris(2-hydroxyethyl)melamine (48). It is believed that *N*-substituted melamines of this type have not been used in textile finishing. The research demonstrates that the yellow chromophore results from the attack of chlorine on NH_2 groups in the finish and that the color formed is not directly related to the amount of chlorine retained in the finish.

The commercial melamine derivatives used in finishing since the 1940s have generally been mixtures. A partially methylated di- or trimethylolmelamine is an excellent polymer former and also an effective cross-linking agent. Its shelf life is poor, however, and the solution thickens on standing. On the other hand, hexakis(methoxymethyl)melamine is a waxlike crystalline solid that, unlike products with lower levels of methylolation, resists polymer formation and acts almost completely as a cross-linking agent. Some effective melamine cross-linking systems used for durable-press finishing were highly methylolated (ca 5 mol formaldehyde per melamine molecule), partially methylated, and contained methoxymethylated ureas. Resins of this type produced smooth-drying fabrics with good hand, little or no susceptibility to strength losses due to chlorine bleaches, and only a slight yellowness from chlorine after multiple washes.

Specially prepared nonmethylated dry methylolmelamines have been commercially available, but these have not enjoyed wide usage as finishing agents. In addition to their use with cellulosics for durable-press finishing and for regulating the stiffness or hand of these fabrics, melamine resins have been used to shrinkproof wool, rotproof cellulosics, as a binder for pigments, and in numerous other applications.

Guanamines have been suggested for many of the uses discussed above, but it is unlikely that substantial interest will develop because their chemistry is similar and yellowing from hypochlorite is a problem (see Cyanamides).

Alkyleneureas, Triazones, and Urons. These cyclic ureas were the first commercial *N*-methylol products offered as cross-linking agents for reducing or eliminating the hypochlorite bleaching problem. They are still available commercially and are used throughout most of the world. The products are similar in structure and, with a slight excess of formaldehyde, the fully methylolated agent is obtained. The systematic names for the 1,3-bis(hydroxymethyl) compounds and their common names are as follows: 2-imidazolidinone or ethyleneurea, tetrahydro-2-(1*H*)-pyrimidinone or propyleneurea, and tetrahydro-1,3,5-triazin-2(1*H*)-one or triazone.

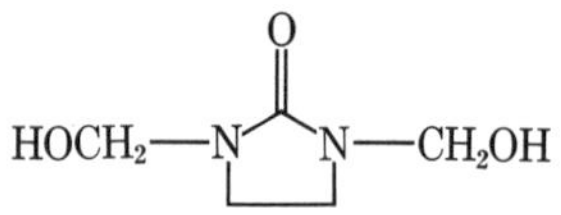

1,3-dimethylolethyleneurea (DMEU)

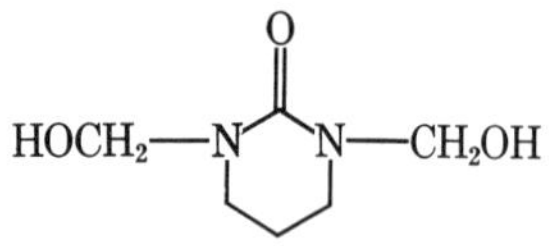

1,3-dimethylolpropyleneurea (DMPU)

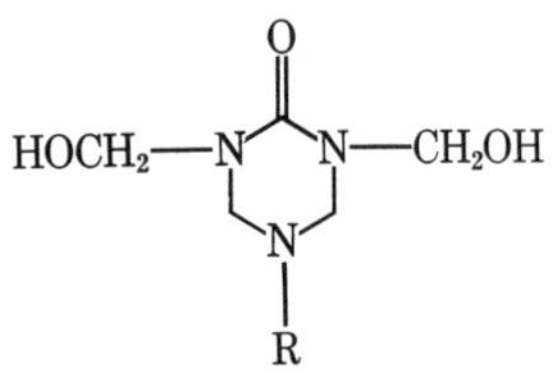

1,3-dimethylol-5-alkyltriazone

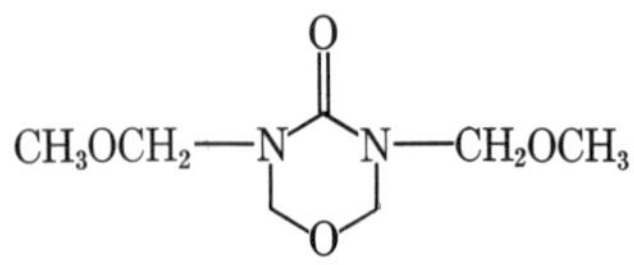

1,3-bis(methoxymethyl)uron

In many cases these cyclic ureas were mixed with urea–formaldehyde or melamine–formaldehyde and used in a number of different formulations. In general, preparation of pure triazone and uron resins is too difficult and too expensive. The main impurity is urea–formaldehyde, but because of the basic nature of the amino nitrogen in the triazone addition, urea–formaldehyde can usually be added to the crude triazone and the formulation can be used to produce finishes that retain strength in hypochlorite bleaching. Similarly, melamine resins can be added to crude uron solutions and the resulting formulations used for commercial finishing. Published work with uron and triazone finishing agents usually describes a method of preparation or states that the uron or unmethylolated triazone was purified before use. Unfortunately, the literature contains much work on formulations of unspecified composition. For this reason a brief review of selected references dealing with the synthesis and chemical finishing with relatively pure materials is given below.

The development of triazone synthesis and use in chemical finishing is described from the synthesis of the first member of its class in 1874 until its development into a generally used finishing agent in 1957 in ref. 49. Methylol derivatives of pure 5-ethyl-, 5-(2-hydroxyethyl)-, and 5,5′-ethylenebistriazinone were compared against four commercial formulations believed to be triazones, a urea–formaldehyde, a melamine–formaldehyde, and a dimethylolethyleneurea formulation. On the positive side, the triazones displayed greater resistance to damage from hypochlorite bleach and to acid hydrolysis.

The longer chain 5-alkyltriazones (from C_8 to C_{18}) are not obtained by the usual methods of synthesis because of formation of a formaldehyde–amine condensate rather than the triazone (50). However, the use of a high boiling solvent and removal of water (by azeotroping with benzene) produces the desired C_8, C_{12}, and C_{18} triazones. None of these triazones penetrates cotton cellulose well, and only the C_8 reactant produced a significant improvement in wrinkle recovery. The C_{18} reactant, on the other hand, appeared to attach itself to the surface of the fiber and to act only as a lubricating agent. These results help provide limits of molecular size and hydrophobic character of a reactant for cellulosic fibers.

The synthesis of triazones by the methods of Burke (51) or Paquin (52) or as noted previously (50) appears to be fairly general. It is a type of Mannich reaction in which the product may form either from the addition of dimethylolurea to the primary amine or from addition of the amine–formaldehyde condensate to urea. Evidently, systems capable of forming stable amine–formaldehyde condensates, eg, substituted hydrazines, resist triazone formation.

Triazones prepared from ethylamine have received the most commercial interest (53). Mixed triazone–urea–formaldehyde resins are said to be cheaper and more effective, and to develop less discoloration on curing at high temperatures than the pure triazone resins. The odor problems noted with triazones could be blamed on either formaldehyde or a volatile amine. Pure triazones are odorless, white, crystalline materials that have been kept for years at room temperature without decomposition or odor development. A solution to the odor problem was thorough washing of the cured fabrics; this procedure has been recommended for all fabrics finished with nitrogenous agents (53).

1,3-Dimethylolhydroxyethyltriazone or 1,3-dimethylol-5-(2-methoxyethyl)-triazone are considered to minimize or eliminate the main failings of triazones. These agents could also be used without an afterwash (54). The pure triazone resins were used alone or with urea–formaldehyde, and no odor other than formaldehyde was noticed.

Possibly the main problem with the triazones and the other agents mentioned in this section is the level of resistance to acid hydrolysis. As is noted below, a relatively low acid stability for a finish correlates with a relatively high level of formaldehyde release. Triazones have not been used generally for many years in the United States, but finishers have not discounted their future use.

The chemistry of urons and their use in textile finishing has been well studied. They are of interest because they represent a potentially excellent cross-linking system based on low-cost starting materials, ie, urea, formaldehyde, and methanol. Both crude reaction mixtures and a fairly pure 1,3-bis(methoxymethyl)uron were developed for commercial use. The unpurified product, expected to contain large quantities of methylated methylolureas, imparts smooth drying properties to cotton fabrics, but the finished fabric is susceptible to strength losses during hypochlorite bleaching and ironing. The pure product, which is readily obtained by vacuum distillation, is an effective cross-linking agent for cotton, and the finished fabric is resistant to chlorine damage even through repeated washes (55).

The uron ring is subject to cleavage during finishing by hydrolysis (eq. 17), but there is strong evidence that this does not occur on finished fabrics (55–57). The

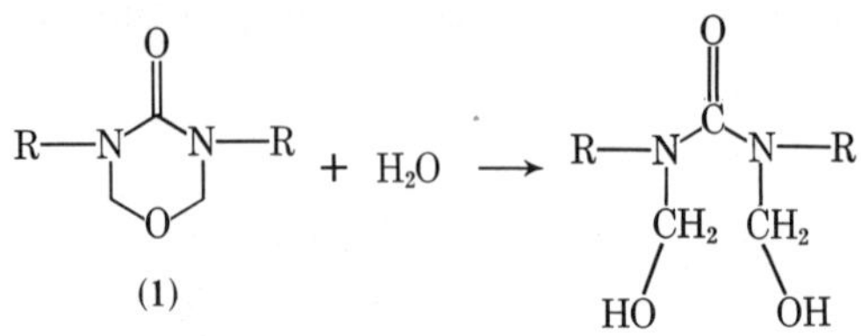

(17)

normally expected precursors ((1), R = H and (1), R = CH_2OH) are difficult to isolate, but both have been purified and characterized (56–57). An alternative method of preparing this compound has been suggested (58). Because of its low cost as a crude product, (1) (R = CH_2OCH_3) continues to be of commercial interest in the United States but is not widely used.

Of the alkyleneureas, DMEU had wide usage in the United States as a replacement for urea– and melamine–formaldehyde, whereas DMPU was used more by the European industry. These materials are still in limited use today, but they have been replaced for the most part with systems that produce a finish with greater resistance to acid hydrolysis.

The ease of hydrolysis of a DMEU-treated fabric has been used to produce bicolored cotton fabrics (59). This was accomplished by applying a thickened DMEU solution to the pile of the fabric, curing the resin, and dyeing the fabric. The pile resists dyeing because of the presence of the cross-links. The bicolored effect is obtained by removing the DMEU cross-links with a mild hydrolysis and then overdyeing the entire fabric.

The cyclic ureas solved the problems caused by hypochlorite bleaching, but complaints about formaldehyde odors in garment-manufacturing plants and in piece-goods stores and demands for a more permanent finish placed an additional requirement on the textile-finishing industry. This new demand grew in strength during the early 1960s, and a significant change in cross-linking systems occurred. The industry stopped talking about wash-and-wear and shifted to products referred to as "durable press" or "permanent press." These products required a finish with greater resistance to hydrolysis and lower formaldehyde release than those in use at that time.

The selected listing of references on the preparation, physical, and chemical properties and pertinent information relative to uses for ethyleneurea is given in reference 60. Some of the firms or research organizations obtaining patents include American Cyanamid Co., J. Bancroft and Sons, Co., BASF, Bayer, Celanese Corp., CIBA, Cities Service Chemical Corp., Courtaulds, E. I. du Pont de Nemours, Gagliardi

Res. Corp., Heberlein, ICI, Monsanto Chem. Co., Norwich Pharm. Co., Rohm and Haas, Sumitomo Chem. Co., Ltd., Sun Chemical Corp., Tootal Broadhurst and Lee Co., Union Carbide, and USDA.

Most commercially prepared DMEU was not only an excellent cross-linking system, but it was also the first general-purpose nitrogenous agent that was not a mixture of materials. Thus, it constitutes an excellent system for basic research because of its relatively high chemical purity and its pertinence to commerce. Nevertheless, finishing research has labored over the years because of the difficult problem of understanding small chemical changes in a complex physical and chemical substrate. Questions concerning cross-link chain length, reasons for chlorine retention, competitive reactions, etc, have been studied, but the conclusions reached have differed (61–64). DMEU may represent the ideal reactant for finishing research, but it has not yet yielded completely satisfying answers to these questions.

Only the chemical cross-linking of prepared cellulosic fabrics with little or no chemical modification have been considered in this article. Similar treatments have been applied to the chemically modified celluloses, but there are relatively few reports of these results. Pretreatments of cotton cellulose, such as etherification of the hydroxyl groups to methyl, hydroxyethyl, cyanoethyl, and carboxymethyl, cause different responses to cross-linking with DMEU. The effect of replacing cellulosic hydroxyl groups depends on the nature of the substituent and the tendency of the etherified cotton to swell (65). Chemically modified celluloses of these types are of interest for a number of practical uses, but they have not been used commercially as substrates for cross-linking to form ease-of-care fabrics.

Delayed Cure and Permanent Press. In the early and mid 1960s, many different processes and many cross-linking agents were used, but none became dominant. For a brief period, two-step cross-linking reactions were used commercially, but more widely in Europe than in the United States. In general, a wet cross-linking treatment was followed by a dry cure. Numerous variations have been tried and continue to be researched, but despite a number of seemingly early successes no commercial process of this type made significant impact in the United States (see general references).

As performance levels increased and special processing conditions involving swelling treatments and two-set applications failed to solve the strength–abrasion problem with cotton, the switch to blends of cotton–nylon or cotton–polyester began in earnest. At high performance levels, the usual points of failure in cotton garments are collar tips, the edge of sharp creases, seams, and the points or other portions of cuffs, depending upon how they were fabricated.

With the emphasis on high performance or permanent press, the curing step is often delayed. Fabric is padded as before with a cross-linking formulation, dried, and fabricated into the garment, which is then cured. In a popular variation of this process, fabrication of the garment follows curing, and creases are formed by high pressure, high temperature presses. The presses in this case are curing the cross-linked fabric a second time by breaking cross-links formed in the initial cure of the flat fabric and allowing them to reform in the fabric in the press. If the fabric in the press is creased, then the reformed cross-links keep the fabric in this state. The cellulosic fibers and yarns in the blended fibers suffer great strength loss in the process, but the synthetic fibers, generally polyester, provide the needed strength for a satisfactory garment. There is, of course, a limit to the severity of conditions that are used in the cure or recure of cellulose–polyester fabrics.

By 1965, at least six permanent- or durable-press processes were recognized and the patent situation on processing was confused. Those in the industry recognized an entirely new set of practical problems to be solved involving such detailed items as preparation of uncured fabric for shipping, type of curing catalyst, time and temperature involved in shipping, cutting and sewing procedures, effects on dyes, etc. Detailed information on these practical problems is provided from the viewpoints of chemical suppliers, researchers, a licensor, and cutters in a series of publications (66–67).

The enthusiastic customer acceptance of permanent press not only forced renewed effort to improve the abrasion resistance of cross-linked cotton, but it prompted a shift from the cyclic ureas, ie, alkyleneureas, urons, and triazones, that had solved the hypochlorite bleach problem to a cross-linking system that was also low in formaldehyde release and more hydrolysis resistant. Some interesting situations resulted. High performance permanent-press garments with durable properties do not normally require even touch-up ironing. As a result, the level of resistance to chlorine damage of the permanent-press garment did not have to be as high as the earlier garments that required some touch-up. On the other hand, since finishers were now only drying or partially curing fabrics, the garment maker was handling uncured goods and completing the cure in buildings heretofore used only for cutting, sewing, and touch-up pressing. This shift necessitated drastic reductions in the amount of formaldehyde released from all types of treated fabrics.

Formaldehyde Analysis and Sources of Formaldehyde in Finished Fabrics. In the finishing of textiles, there are three main areas where various types of formaldehyde analyses are of interest: formaldehyde in the air; free formaldehyde in cross-linking reagents and in the pad baths prepared from these reagents; and free and releasable formaldehyde in finished fabrics and garments. Because of recent studies on the toxicity of formaldehyde, the allowable maximum concentration of formaldehyde in the atmosphere of working places is being reconsidered. A comparison of allowable limits in various countries has been published (68). Concentrations significantly greater than one part per million are believed unusual in areas where finishing workers are confined. Great strides have been made in reducing the free formaldehyde content in finishing formulations and this factor, coupled with the use of improved ventilation and cross-linking systems and a genuine concern for reducing exposure to formaldehyde vapors, has reduced the concentrations in air found in finishing areas.

Free formaldehyde is generally considered to consist of formaldehyde, hydrated formaldehyde, and low molecular weight oligomers. Thus, free formaldehyde relates directly to the odor of a fabric or a padding solution or to the amount of formaldehyde transferable from a treated fabric to a wearer, assuming that the wearer is not exposed to conditions that would cause degradation of the finish (28). Since all finishes degrade somewhat, factors such as extractable formaldehyde and releasable formaldehyde must be considered. The real exposure is difficult to measure, a fact reflected in the number of test procedures used throughout the world.

Extraction tests are used in Japan and Europe, and a release test is used in the United States; standard tests have been compared based on the sources of formaldehyde present in a finished fabric (28,68–70). Finished fabric contains both free formaldehyde and formaldehyde released from the following groups: unreacted $>NCH_2OR$, where R = H, CH_3, or $CH(CH_3)_2$, depending upon the type of agent in the pad bath; $>NCH_2OCell$; $>NCH_2N<$; and $>NCH_2OCH_2N<$. With the types of cross-linking agents in current use, the last two groups are expected to be at very low or zero levels.

The analytical methods (28,71) generally used for free formaldehyde are based on a low temperature (0–5°C) titration that involves the reaction of sodium sulfite and formaldehyde.

$$HCHO + Na_2SO_3 + H_2O \rightarrow HOCH_2SO_3Na + NaOH \quad (18)$$

Performed at the specific temperature and pH ranges, the analyses can be carried out in the presence of *N*-methylol groups. On fabric, the formaldehyde bisulfite compound is decomposed by addition of excess sodium carbonate, and the liberated sulfite is titrated with 0.1 or 0.01 *N* iodine solution (28). The limits of this method have not been studied extensively, but values of free formaldehyde on laundered fabric are generally <25 ppm and confidence in numbers below this level is not good. On the other hand, commercial fabrics are generally first washed and dried by the consumer, and the free-formaldehyde contents start at ca 50 ppm and may go up to several hundred ppm, depending on finishing and storage conditions.

The determination of free formaldehyde in commercial reagents and pad-bath formulations presents a similar situation in that the usual conditions of analysis allow hydrolysis of the *N*-methylol groups. Formaldehyde content usually ranges from less than one percent to several percent. This analysis, also performed at 0–5°C, involves the quantitative determination of NaOH formed (eq. 18) with hydrochloric acid using an adaptation of the Morath-Woods method (71).

Because of an incomplete reaction of sulfite with free formaldehyde, these low temperature methods are believed (72) to analyze only part (usually 80–90 wt %) of the free formaldehyde present in solution. Technique becomes very important and there is a possibility of either a low or a high error. The low temperature titrations on fabric are used primarily for research and in trouble-shooting experiments.

Methods using extractions at higher temperatures are used in a number of countries. Unfortunately, the values obtained are referred to as free formaldehyde content even though part of the formaldehyde reported originates in the nitrogenous finish or unreacted cross-linking reagent on the treated fabric. Presumably, the purpose of these methods is to reduce the exposure of the consumer and the worker to formaldehyde as a skin allergen. For the most part, these test methods have originated with industry; however, in Japan a legal regulation effective in 1975 resulted. The mandatory standards involve an extraction at 40°C. An industrial method used in Europe requires extraction at 22°C followed by a determination of the total formaldehyde content of the extract (73). Use of this method identifies finished fabrics with relatively high levels of unreacted cross-linking agents.

A number of researchers have noted the release of formaldehyde by a chemically treated fabric under prolonged hot, humid conditions (74–77). A method, developed in the United States and used extensively for ca 20 yr, measures the formaldehyde release as a vapor from a fabric stored over water in a sealed jar for 20 h at 49°C or 4 h at 65°C. Results from this test have been used by industry in the U.S. to eliminate the less stable finishes. In this respect the method, AATCC Test Method 112 (78), has been very effective and has led to a new set of finishing agents to replace the alkyleneureas, triazones, and urons.

A urea-glyoxal formaldehyde adduct, 1,3-dimethylol-4,5-dihydroxyethyleneurea (DMDHEU) (79), has been the primary cross-linking agent used in the United States since the combination of formaldehyde release, as measured by the AATCC Test Method 112, and an emphasis on durable properties, ie, hydrolysis resistance, essentially eliminated the other standard agents from general use. Another finish, based on *N,N*-dimethylolcarbamates (80), is the only other durable low-formaldehyde treatment developed industrially, and its use is only about one-fifth that of DMDHEU.

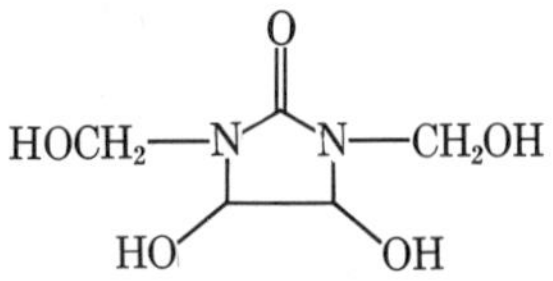

DMDHEU

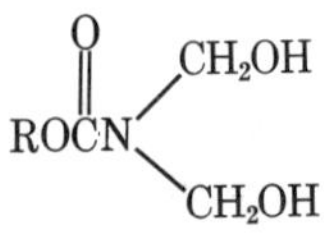

carbamates

Prior to 1965, it was not unusual for unwashed, finished fabrics to release 3000–5000 ppm formaldehyde when tested by AATCC Test Method 112. Free-formaldehyde contents of these fabrics were not determined, but cotton cellulose is capable of retaining quantities of free formaldehyde in excess of these figures. A reduction in the release to ≤2000 ppm was achieved largely through use of DMDHEU and dimethylolcarbamates. The level of formaldehyde release was later reduced to 1000 ppm by some organizations, and today levels of ca 100 ppm are achievable with additives and modifications of the DMDHEU reactant system. It is believed that very little fabric is finished commercially at the 100 ppm level, and it is likely that considerable yardages of fabric are currently being finished at levels >1000 ppm release.

The changes that helped reduce formaldehyde release included the following factors (not necessarily listed in the order of their implementation):

Attention to certain factors in synthesis, formulation, and processing caused a large drop in release values. Factors such as careful pH control, degree of cure, degree of methylolation, free formaldehyde content, and catalyst selection provide significant reductions, even at high resin usage. Residues left from fabric preparation can be neutralized by corrections in pad-bath formulation.

As attention focused on details, the amount of resin solids on the weight of the fabric came under scrutiny and levels began to drop. Performance may have dropped in some cases, but the common practice of applying as much resin to a 50% cotton–50% polyester fabric as is applied to a 100% cotton fabric is unnecessary (81). The lower level of resin solids produces a similar lower level of formaldehyde release.

Nonformaldehyde additives were found to reduce the losses in performance that accompany changes in resin formulation. A number of polymer systems have been applied with varying degrees of success. Those based on acrylate systems appear to be the most widely used. A drop in performance level of durable-press properties may not be noticed immediately, but the use of significantly lower resin contents and reliance on additives usually result in noticeably poorer appearance after repeated tumble or line drying.

Alkylation of *N*-methylol cross-linking agents lowers the formaldehyde-release values significantly (21,82–84). Generally, DMDHEU is methylated to ca 25–50% of theory, considering four hydroxyl groups available for methylation. An optimized treatment with partially methylated DMDHEU at relatively low solids produces a value of 200–400 ppm releasable formaldehyde from an unwashed fabric. A further reduction to ca 100 ppm may require the use of a number of techniques including a scavenger to tie up free-formaldehyde or free-*N*-methylol groups in the finish (84–88).

It has been suggested (89) that these new finishes cannot be satisfactorily monitored for formaldehyde release by procedures that use the Nash reagent or chromotropic acid because of the low values, ie, ≤200–300 ppm, being reported. For release values in the range of 15–175 ppm, a colorimetric procedure based on the use of 3-methyl-2(3*H*)-benzothiazolone is recommended. The question is far from settled because the Nash reagent is specific for formaldehyde and considered accurate to 10 ppm (90). Further, the significance of these low values has not been established and as noted previously, the AATCC method for release does not relate to odor of the fabric.

Once the fabric is in the hands of the consumer and has been laundered, a new situation exists for formaldehyde control. Release values do not correlate with the transport of formaldehyde (28), and thus these values do not correlate with the consumer's exposure to formaldehyde. Should formaldehyde transport from fabric to consumer come under control, then a method of testing for free-formaldehyde content and easily released formaldehyde in the fabric would be more appropriate. However, free-formaldehyde values are very low, even in many unwashed commercial fabrics. In normal consumer garments (finished by today's technology) that have been laundered several times, it is questionable whether the transportable formaldehyde content is high enough to be measured by conventional chemical means.

Despite the encouraging results and conclusions discussed above, it is still considered (91) that formaldehyde release values using the AATCC test procedures can be reduced to below the limit of 500 ppm currently imposed on textile finishers by some members of the garment industry in the United States.

Comparisons of results from the analytical methods used in the United States, Japan, and Europe (28,68) on a single fabric demonstrate that the strength of the hydrolysis conditions used in the test determine the numerical value of the results. Therefore, most formaldehyde release should occur in the washer, in the dryer, or on line drying. The effort to develop new test methods may result in the development of new technologies for finishing. Also being considered is the development and use of nonformaldehyde-containing cross-linking agents (90).

DMDHEU and Carbamates. Unlike triazone and uron cross-linking agents, DMDHEU can be prepared commercially from starting materials in an almost pure state, and no purification steps are necessary. Commercial glyoxal, however, is a mixture of products (92), and its shelf life is limited. Therefore, early commercial DMDHEU agents often varied appreciably in content, but finishing results were generally consistent except for some extreme cases. The yellowing of the fabric that can occur upon curing has been controlled by use of additives (93–94).

Since protonation of either pendant methylols or ring hydroxyls in DMDHEU should lead to the formulation of resonance-stabilized immonium–carbonium ions, it followed that the ring hydroxyls alone could be used to cross-link cellulose (24,95–97). The agent, 1,3-dimethyl-4,5-dihydroxyethyleneurea (DMeDHEU), has been used in Japan since 1974 and has also some limited use in the rest of the world as a non-formaldehyde cross-linking agent for cellulose.

A number of possibilities for finishing with various dihydroxyethyleneureas are shown below.

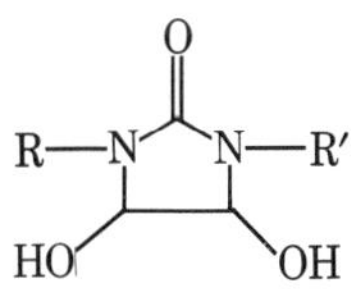

DMDHEU: R = R′ = CH_2OH
MDHEU: R = H, R′ = CH_2OH
DHEU: R = R′ = H
MeMDHEU: R = CH_3, R′ = CH_2OH
DMeDHEU: R = R′ = CH_3

These agents were studied and compared to DMEU with regard to performance (crease recovery of treated fabric) and durability of the finish to hydrolysis (27). All these agents are capable of producing cross-linked fabrics with improved crease resistance. However, not only does the presence of NH groups in the finish result in susceptibility to fabric damage from chlorine bleaching, but these finishes are also susceptible to alkaline hydrolysis. Finishes prepared from DHEU (no formaldehyde present) are readily removed by normal alkaline home washing conditions; the one (for MDHEU) or two moles (for DMDHEU) of added formaldehyde reduces this susceptibility to alkaline hydrolysis. Addition of one or two *N*-methyl groups to DHEU lowers the resistance of the finish to acid hydrolysis. Thus, finishes prepared from DMEU and DMeDHEU are readily removed by acidic solutions whereas DMDHEU finishes resist removal under the same conditions. DMEU represents an upper limit of reactivity for useful finishing. In general, agents should fall within the range of reactivity of DMEU and DMDHEU.

A properly cured DMDHEU finish is permanent and is not removed by standard acid stripping solutions. It can, of course, be removed by stronger acid stripping solutions, but these conditions weaken the cellulose. Similarly, an agent appreciably less reactive than DMDHEU requires curing conditions that lower the strength of the fabric because of hydrolyses or degradation of cellulose.

Thus, DMDHEU represents at present the optimum finishing agent. Pendant and ring hydroxyls appear to have equal reactivity toward small alcohols; however, larger alcohols such as isopropyl alcohol and cellulose are sterically hindered in approaching the 4,5- or ring positions, so that only monosubstitution at these positions is accomplished with ease. As a result, DMDHEU is expected to favor reaction with cellulose through its pendant dimethylol groups with active competition occurring from the 4,5-hydroxyls until one group is substituted (98).

As noted in equations 4 and 5, the amide–aldehyde reaction can approach completion in some cases if excess formaldehyde is used. In practice, a finishing bath cannot tolerate the large excess of formaldehyde needed to force completion, and therefore the reactants present in the practical case are the *N*-methylol and the *N,N*-dimethylol amide and unreacted formaldehyde. Generally, little or no free amide is present, but it is expected to form by demethylolation during drying or curing. A mixture of this type is expected to form the following cross-links:

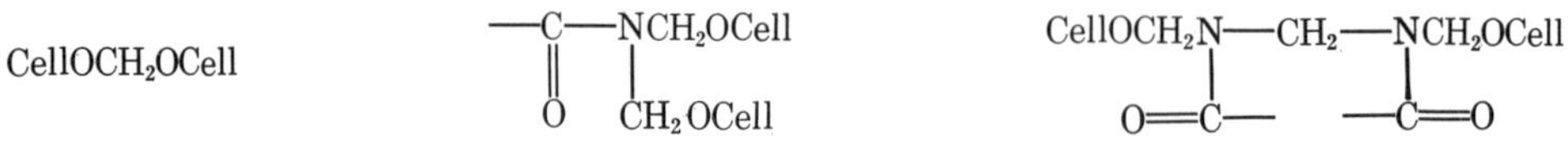

and the following "grafts":

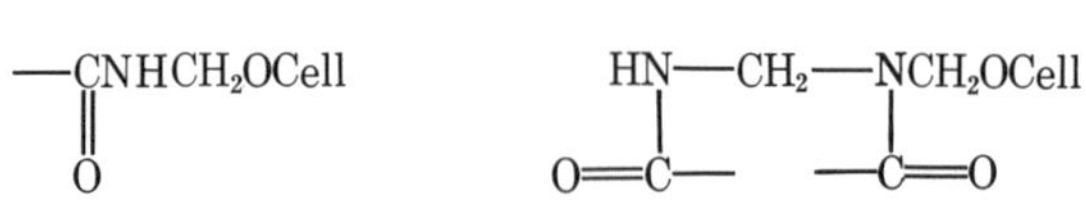

The grafts are sites for *N*-chloroamide formation and subsequent loss of strength on ironing. Because the level of free formaldehyde in the pad bath is relatively high compared to most cyclic urea reactants, the level of formaldehyde cross-links can also be relatively high, thereby weakening the finished fabric.

The utility of this class of agents was first demonstrated with formamide (99) but only the carbamates (24,80) gained wide commercial usage. Because of the thermal stability of the *N*-chlorocarbamates, properly applied carbamate finishes, even though they form *N*-chloro derivatives on bleaching, do not cause fabric damage (100). Methyl, ethyl, isopropyl, isobutyl, hydroxyethyl, and methoxyethyl carbamates are among those that have been used commercially. Mixtures of alkyl carbamates have also been sold in large quantities. Often the manufacturer does not specify the actual carbamate in the formulation. Use of ethyl carbamate is thought to be minimal because of toxicity considerations.

At present, the main advantage of carbamates is price. They are generally a few cents per kilogram less than DMDHEU and are often preferred for some specific uses, eg, sheeting fabrics. The agents may be used in a two-stage curing process (101) or a delayed cure (102). Like other agents, the carbamates can have an adverse effect on the color fastness of dyed fabrics to light, and these effects vary from one carbamate to another. Manufacturers and formulators are the best source of information on lightfastness of specific formulations.

Carbamate finishing agents suffer from the many different types of reaction with cellulose they can produce as noted by the above structures. As a result, the best carbamate crease-recovery and the durable-press ratings are not as high as those obtained from the urea-based agents. In addition, the possibility of formaldehyde cross-links and the subsequent additional loss of strength has, for the most part, limited use of carbamates to cotton–polyester blends.

It is unlikely that carbamate finishing agents will replace DMDHEU-based finishing agents as the principal finishing agent, unless there are unusual quality or price changes in glyoxal manufacturing.

Other *N*-Methylol Agents. Cost has been an important factor in the choice of finishing system. Once an economical formulation is optimized and proven in production, a finisher usually resists change unless another factor enters the picture. The use of different dyes or finishing of white fabrics generally causes a minor change in a formulation. However, as noted before, the susceptibility of a finish to chlorine damage, to hydrolysis, or to release formaldehyde are the main factors that have resulted in significant changes in finishing systems. Further changes might be necessitated by results of toxicity testing of the agents themselves or by the introduction of new dyes. If this should happen, there are a number of other *N*-methylol-type agents that should be considered.

Dimethylolpropyleneurea (DMPU), as noted previously, has been used in other parts of the world but has had only limited use in the United States. Other propyleneureas, including the 5-hydroxy and the 4-methoxy-5,5-dimethyl derivatives, have been developed and provide a wide selection of chemical reactivity comparable to that obtained with ethyleneurea and substituted ethyleneurea reactant systems. A comparison of these systems with respect to reaction mechanisms, rates, structure, and properties is given in reference 103. The same author has provided a comprehensive review of methods of synthesis of cyclic ureas (104).

The characteristics of internally buffered *N*-methylol reagents and their reactions with cotton have been reviewed (105). Included are methylol derivatives obtained from the Michael addition of ammonia or amino acids to acrylamide, triazones containing an additional tertiary amino group, or carbamates containing a tertiary amino group. These reagents in combination with conventional triazones provide a wide range of cross-linking systems containing basic substituents.

In addition to DMeDHEU, a number of other glyoxal-based cross-linking systems with little or no formaldehyde have been evaluated (106). Work continues on the optimization of linear and cyclic compounds containing $>N—CH(OH)—CH(OH)—N<$ groups. The vicinal hydroxyls have been found to possess varying reactivities based on the substituents on the nitrogen atoms and on the size of the nucleophile. These glyoxal-based agents have proven capable of cross-linking cellulose through their vicinal hydroxyls; however, the efficiency of the cross-linking reaction has not been as high as desired, even for the linear *N,N'*-dihydroxyethylenebisamides which should be free to rotate and relieve the steric hindrance.

Miscellaneous Finishing Processes

Finishing Synthetics and Blends. The changing needs and desires of the consumer are met primarily by the fiber producer and the finisher. The primary natural fibers, cellulosics and wool, possess chemical or physical properties that are amenable to chemical modification and allow the finisher to make substantial changes in the physical and chemical nature of fabrics containing them. On the other hand, the synthetic fiber producer develops a fiber that is designed to require little or no chemical finishing and is chemically inert except to strong or unusual conditions. Modern technology has developed many fiber blends and utilizes many finishing techniques to improve textile products.

The finishing of synthetic fibers is limited primarily to heat setting, soil release, hand or comfort effects, and antistatic agents (qv). The main advantages for synthetic fibers are their greater strength, abrasion resistance, chemical inertness, and thermoplasticity. Some of their disadvantages are caused by their oleophilic character, which results in low moisture content, soiling or staining, buildup of static charges, and difficulty in dyeing. These factors differ for specific synthetic fibers and for specific blends of these fibers with natural or other synthetic fibers (see general references).

Heat setting of synthetic, thermoplastic fibers can occur at almost any stage of processing and is often applied during fabric preparation (107–108). Conditions may include either a dry atmosphere or steam; either procedure improves dimensional stability, pilling, and durable-press performance. Stresses introduced into a thermoplastic fiber during its manufacture are reduced by heat treatments. The combined effects of heat treatments in processing a synthetic fabric are a very important factor in determining the properties of the finished fabric, and the finisher must be aware of the extent of these treatments. Similarly, cellulosic fabrics have been durably set with elevated temperatures and mechanical pressure, but the mechanism is not understood (109).

The strong acceptance of durable press around 1961 led to rapid growth in the use of polyester fibers, particularly in comparison to the growth rates of polyamides and acrylics. In blends with cellulosic fibers, polyester has been most widely used but nylon has also been used as the reinforcing fiber. However, polyester's success with cellulosic blends has tended to mask its shortcomings in other areas, eg, soft yarns, knits, and wool blends. These problems and solutions have been reviewed (110).

Pilling is a problem associated with fabrics made from staple yarns and long associated with wool, but one that has increased with the introduction of the new, stronger synthetic fibers. The unsightly surface effect is generally the result of an abrasive action that forms surface fuzz and entanglement of the fuzz into balls. Some of the pills may be removed during wearing of the garment or during laundering, depending on the abrasion resistance and tenacity of the fibers (111). Electron micrographs show that polyester fiber skin flakes cause the entanglement of fibers. In a cotton–polyester blend most of the fibers in the pill are polyester, and the cotton fibers are only passive participants (112). Heat setting is most important in reducing pilling in fabrics that contain polyester staple, regardless of fabric construction.

Topical treatments, often in conjunction with heat setting, are applied to synthetic fabrics to impart hydrophilic properties (113). One of the first polymers used is a condensation product of dimethyl terephthalate, ethylene glycol, and polyethylene glycol. It worked well with polyester but not with nylon or acrylics. Another hydrophilic polymeric finish is based on poly(acrylic acid). Polyester has also been treated with ionizing radiation from a variety of sources to generate free radicals on the surface and then treated with a selected monomer such as acrylic acid (see Initiators; Radiation curing). Because of costs, it is unlikely that radiation-initiated grafting will find much use in the textile industry. Increased hydrophilicity is also obtained by surface hydrolysis of polyester using sodium hydroxide, but this process is thought to have only limited commercial use at present.

The textile industry's experience with commercial oily-soil release treatments for cotton–polyester fabrics has been described in detail. Also, it was noted that the introduction of liquid ammonia mercerization in fabric preparation should reduce the severity of soil-release problems because it involves less cross-linking and no polyester is required for some fabrics (114).

The hydrophilic treatments discussed above are intended to improve soil-release, antistatic, and comfort properties of fabrics containing synthetic fibers. Other types of topical finishes, including fluorochemical, silicone, acrylate, and hydrocarbon systems, have been applied to textiles to obtain similar properties; to alter the hand of fabric; to improve appearance, strength, and abrasion properties; or to provide water repellancy. In the past, tacky topical finishes have resulted in adsorption of particulate matter which produced a gray or dirty look on freshly laundered fabrics. Some of these additives have been evaluated under several conditions involving conventional, wet, and moist curing (115) (see also Surfactants and detersive systems).

Obtaining Properties of Synthetic Fibers in Cotton. Some of the finishing treatments described above are intended to make polyester more like cotton or other natural fibers. On the other hand, cotton and cotton–polyester blends have been chemically modified (116) to accept disperse dyestuffs and thereby rival polyester and other synthetics as a substrate in transfer printing. Successful treatments include use of glycols with cross-linking agents, use of melamine–formaldehyde resins, and acetylation. Attempts to improve cotton and cotton–polyester resistance to wrinkling using polymers or polymer formers have been extensive and are discussed in some of the general references. Film properties of some of the elastomers used in these attempts have been studied and compared to the results obtained from application of these elastomers to cotton (117). For achieving durable-press performance, no elastomer developed equals the conventional *N*-methylol cross-linking reactants. These systems are useful because they can be used to replace part of the *N*-methylol reactants and thereby reduce formaldehyde release in the treated fabric.

Elastic properties needed for stretch textile fabrics are inherent or readily developed in most synthetic, thermoplastic fibers. Considerable production of cotton stretch fabrics for apparel was noted in 1964, and stretch cottons are also used as a backcoating for coated fabrics. The process involves slack mercerization with shrinkage, removal of sodium hydroxide, and setting the yarn crimp by conventional cross-linking. The finishing of cotton stretch fabrics has been reviewed (118), and theoretical aspects including limitations in obtaining stretch properties in cotton have been developed (119–120).

Outdoor Fabrics. Water repellency as a finishing process is discussed in many of the general references at the end of this article and in ref. 121 (see Waterproofing and water repellency). Some fabrics are sufficiently hydrophobic to be water repellent without addition of a finish. For example, polyalkylene and similar synthetic fabrics repel water readily (see Olefin fibers). Similarly, tightly woven fabrics from synthetic or natural fibers that retain natural or processing oils have significant repellent action. However, most consumer goods are treated with a topical finish to provide adequate and fairly durable properties. The level of treatment depends primarily on the nature and construction of the fabric being treated.

The earliest waterproofing treatments were coatings of a continuous layer impenetrable by water. In later improvements the fabrics were made water repellent, but also permeable to air and moisture, thus enhancing the comfort of the wearer. Aluminum and zirconium salts of fatty acids, silicone polymers, and perfluoro compounds are examples of water repellents that have been applied to both synthetic and natural fibers. The water repellency of a fiber increases with the contact angle of water on the surface of the fiber. Hydrophobic fibers exhibit higher contact angles than cellulosics but, in general, they still require a finish (122).

Outdoor fabrics must withstand the combined effect of mildew and rot, sunlight, airborne acids, alternate wetting and drying, and abrasion. Cotton has been widely used, and reference 123 reviews resistance to weathering of such products as awnings, tarpaulins, and tents as of 1960. The lighter, stronger, microbial-resistant synthetic fabrics have made large inroads into markets held by cotton even though the service life of outdoor cottons has been significantly increased since 1960 (see also Fungicides; Industrial antimicrobial agents).

The microorganisms that grow on cotton textiles cause several undesirable effects. Loss of strength is the most serious effect, but they may also stain the fabric or cause defective dyeing. The problem is one that can be greatly aggravated by climatic conditions. Warm, moist conditions found in the tropics can cause unprotected cotton fabrics in contact with the ground to deteriorate in one to two weeks; the same fabric should last for a considerably longer period of time in the northern countries.

Three different types of agents or processes are used to protect cotton against microorganisms. In general, they poison the microorganism, block its action with a protective layer, or change the fiber chemically so that it cannot be utilized by microorganisms. The protective finish must be nontoxic to human beings, durable, colorless, and inexpensive. In addition, the finish must be stable to light, must not accelerate the photochemical degradation of the cotton fabric, and must not adversely affect the physical properties native to cotton. These are exacting specifications and no agent in present use can satisfy them completely. Concerns of a similar nature apply to both synthetic and cotton outdoor fabrics.

Certain phenols, organometallic compounds, and inorganic salts are toxic to microorganisms. However, the phenols generally accelerate photochemical degradation, thereby weakening the fabric. Screening agents are often successfully used with phenols. Many of the highly effective organometallic compounds (particularly those of mercury) are highly toxic to humans, and improved replacements have been developed (124). Copper salts have been widely used (125), particularly copper 8-quinolinolate, but because it is a colored material, its use is limited.

Many of these inorganic and organometallic compounds can be readily applied to cotton using zirconium compounds such as zirconyl acetate and zirconyl ammonium carbonate (126–127). The antimicrobial agents become water soluble by reaction with one of the zirconium compounds. Cotton fabrics are padded with this aqueous solution and heated to 145°C for three to five minutes. This forms a stable, insoluble product with zirconium that resists leaching. A variety of colors can be obtained.

Other processes used to protect cotton against microorganisms include the blocking mechanism and chemical modification of cellulose. These effects are difficult to separate in some instances. However, the protection obtained by cyanoethylation and acetylation of cotton is obviously the effect of chemical modification and not blocking with a polymer. Resin impregnation by *N*-methylolamides appears to involve both effects; however, there is some controversy as to which is more important (128–130). In one process an insoluble melamine–formaldehyde resin is deposited inside and outside the fibers in the presence of water. The cure, catalyzed by hydrogen peroxide, must be complete before the fabric is dried. Curing is obtained at room temperature or in steam (relatively faster). Because this treatment does not prevent the growth of mildew, fungicides (qv) must be included in the process. A similar process also utilizes a melamine–formaldehyde adduct with a strongly acidic catalyst and a dry cure at elevated temperatures. Thiourea is added to reduce cross-linking by a competitive reaction. This treatment is said to limit strength losses and to develop a durable finish.

Flame Retardants. Flame-retardant finishes provide protection to cotton during weathering, and there has been a trend to develop flame-retardance standards for outdoor fabrics. A comparision of experimental and commercial finishes has been made (131). This subject has been covered in detail in other publications (132–136) (see Flame retardants for textiles; Flame retardants).

Flame retardance of textiles has been defined in many different ways. The problem to the researcher is to develop a process that meets the need defined by military, political, economical, consumer, and other considerations. Most textiles are organic polymers and regardless of the treatment will ignite and burn under some condition. Thus, criterion is not complete fire resistance but whether a specific test or standard is met. Smoldering, which can be introduced by a cigarette or another ignition source, is an entirely different process from flaming combustion and will not be considered in detail in this review. Resistance to smoldering appears to have little to do with flame retardance, but more with the ability of the textile to "avoid" the heat source. Thermoplastic fibers that melt away from a heat source provide a satisfactory textile for smolder resistance, but in practice these materials may be the fuel source for an eventual conflagration.

Similarly, thermoplastic fibers that melt at relatively low temperatures are considered flame retardant because the melted polymer drips away from the flame. The present Children's Sleepwear Standard allows many untreated polyester and nylon fabrics to be used for this purpose. Therefore, the finishing of textiles for flame retardancy is limited to a few products such as tentage and work clothing in some selected instances.

Tris(2,3-dibromopropyl) phosphate, generally referred to as Tris, is an effective chemical retardant for polyester, acetate, and triacetate fabrics. In the finishing process it is generally padded onto fabric and forced into the fiber using a thermal treatment. Many millions (10^6) of meters of fabrics especially for children's sleepwear were treated with Tris before the Consumer Product Safety Commission banned its use in 1977. A lowering of the standards followed this ban, and the large-scale flame-retardance finishing of children's sleepwear that existed in 1977 essentially ceased. Other retardants similar to Tris have been evaluated for childrens sleepwear, and some have had limited commercial use.

Tris is not an effective retardant in finishes for cellulosic fabrics, but a number of other phosphorus compounds have been used effectively to flameproof cellulosics for use in garments. Currently, the finishes based on tetrakis(hydroxymethyl)phosphonium chloride (THPC) or the sulfate (THPS) appear to be the most widely used. Formulations of THPS or THPC either as is or partially neutralized and in combination with the dimethyl ether of trimethylolmelamine and urea are padded, dried, and cured to produce a durable finish. A number of variations have been successfully applied using other melamine-based reactants or treatments that eliminate the urea or melamine components and optimize the total formaldehyde content of the systems (see Flame retardants).

Neutralization of these phosphonium salts produces a reactive methylol phosphorus system referred to as THPOH. A comparable reactant system can be prepared from tris(hydroxymethyl)phosphine and formaldehyde. THPOH can be used in a number of pad–dry–cure processes with methylolureas or methylolmelamines, but the most widely used process with cotton uses ammonia gas to form a polymer with THPOH. In this process the polymer is deposited within the cotton fiber, little or no reaction occurs with cellulose, and the strength of the fabric is not affected.

As a general rule, lightweight fabrics (70 g/m^2) are more difficult to flameproof than heavier-weight fabrics. The lighter fabric generally requires a higher add-on. If the finish fixes too much resin-type retardant, the hand of the treated fabric often becomes too stiff for use as a garment. Washing or other mechanical handling of the treated fabric may restore a desirable hand. Cellulosic fabrics treated with phosphonium-based reactants require an oxidative wash to stabilize the finish (P(III) is converted to the oxide, P(V)) and to remove odors.

Tentage fabrics are generally treated with an antimony oxide–halocarbon topical finish. Some of the preferred halocarbon systems include emulsifiable, film-forming, chlorocarbons, or appropriate bromocarbons with a binder. Some finishes of this type do not require a cure and are effective on all types of tentage fabrics. Stiffening of the fabric may occur, but the hand is not as important as for garments. A number of states have enacted flame-retardant tent standards, and the Canvas Products Association International has had success in obtaining uniformity through its efforts.

The level of activity of flame-retardance research on textiles is relatively low and appears to be decreasing. The intensive search for improved finishes began in the late 1960s and lasted ca 10 yr. Some inherently flame-resistant fibers, eg, wool, modacrylics, and aramids, have specialized uses, but only limited consumer appeal. The popular cotton–polyester-blend fabrics, especially the 50–50 blend, has proven extremely difficult to retard with retention of other acceptable properties (137).

Finishing of Wool. Wool (qv) competes for markets primarily with the noncellulosic, synthetic fibers where warmth and wrinkle recovery are desired. Its disadvantages are fuzzing, shrinkage, and difficulty in creasing. Chlorination, modification of the disulfide cross-links, and polymer treatments appear to be chemical reactions of choice.

Untreated fabrics of both wool and cellulosic fibers shrink when immersed in water because of a tendency of the fibers to swell and also because of relaxation of stresses developed during earlier mechanical processing. The degree of shrinkage by this mechanism is greater for cellulosics, but wool with its scaly structure also shrinks because of felting. Mechanical action, eg, washing, causes fibers to move about and become entangled. The roughness of the scales reduces or prevents the fibers from returning to their original configuration, which results in permanent shrinkage. The surface appearance is also altered and appears matted (see Felts).

The problem of felting has been solved by altering the scaly structure of wool. For 100 years or more, commercial processes have generally used chlorination or some other form of oxidation for this purpose. Many different processes of this type have been developed to achieve a more uniform surface and to prepare the surface for additional treatments. The chlorination step is often followed by application of a polymer to form a film on the surface of the wool. Various urea–formaldehyde and melamine–formaldehyde resin systems, vinyl and acrylic polymers and copolymers, silicones, and other polymers have been used for this purpose. Numerous procedures have been used with the ultimate objective of obtaining a smoother fiber without stiffening or discoloring it.

Wool fabrics in general resist creasing and are flame retardant. The disulfide cross-links are readily broken with reducing agents such as thioglycolic acid, sodium bisulfite, and formaldehyde. Such processes, similar to those used in permanent waving of hair, are used to set pleats or creases in wool (138) (see Hair treatments).

Recent efforts continue to utilize ozone (qv) to reduce shrinkage (139), and blends of wool, cotton, and polyester have been finished to provide improved flame-retardancy, durable-press, and shrinkage properties (140). Fabrics from the latter study were intended for uniforms and occupational apparel. The development of wool and wool-blend fabrics has been examined from the point of view of the protection offered in a variety of industrial situations (141).

Hand, Comfort, and Bioactivity. The aesthetic appeal of a garment can be the most important factor in the consumer's purchase or rejection of a finished article. A judgement is made on such properties as hand, appearance, levels of protection, and comfort. The industry is aware of these needs, and chemicals or fibers that irritate the skin must be avoided or neutralized.

Factors involving the thermal and moisture-transport aspects of comfort are possibly the simplest to evaluate and are considered in finishing of textiles, especially the synthetics. Stiffness of a fabric is of primary concern and this has been covered in previous discussions. The combination of both objective and subjective aspects makes the subject of comfort of textiles a complex one (142–145).

A variety of chemical products and finished fabrics said to have antibacterial finishes have been or are on the market. Claims range from statements about preventing odors to the prevention of spread of infections (146). Certainly, this would be a highly desirable product, but no finish appears to have gained appreciable acceptance. Two promising new finishes of this type are based on an organosilicon quaternary ammonium chloride compound (147) or on peroxide-containing complexes of zinc acetate (148).

Economic Aspects

The consumption of textiles (149–150) is usually considered on the basis of the general uses, ie, apparel, household, and industrial. In the lower-income countries, apparel uses account for almost all the total fiber consumption, whereas in the United States, apparel uses account for less than half of the total market. Despite a rapid growth in the use of synthetic fibers, cotton remains the world's number one fiber in terms of total production. The percentage of cotton used in the lower-income countries is generally higher than that used in developed countries. However, it is not unusual for finished fabrics and garments of synthetics or blends of fibers to be a main export item for the lower-income countries. In the past, the fluctuating price of raw cotton and temporary shortages of petroleum for manufacture of synthetic fibers have caused concern.

Costs and Volume. Coloring a textile, either by dyeing or printing, is generally more costly than finishing the textile. Dyes and textile chemicals are sold in the United States by many companies (151–152), generally through a smaller firm, which supplies the user with the desired mix and often with technical service. Many of the larger mills have their own chemical supplier, but it is estimated that their share of the market, even with sales to other mills, is <20%. About 300 companies supply textile chemicals. The small supplier, ie, with annual sales <5×10^6, controls ca 80% of the market (152).

The markets for textile chemicals and for finishing agents have been estimated in 1968 and again in 1978 (152–156). Estimates for these markets have been projected for 1980, 1983, and 1988 (153). The current general business slump and increased import of finished textiles will undoubtedly result in a revision of these figures unless the trends are reversed (157).

Total sales were 457×10^6 in 1978 (153), which represents 852,000 metric tons (most products contain >50% water) with an average price of 52.8¢/kg. Backcoating products, which are used primarily on carpets and upholstery fabrics, represented ca 36% of this dollar volume. Other products include nonwoven binders (16%), with water and stain repellents (13%), durable-press resins–reactants (8%), and softeners (8%). Price per kilogram for specific product types (21 listed) ranges from 26¢ (50% active) for catalysts for durable press to $9.23 (30% active) for soil-release finishes. Many of the leading products and their suppliers are identified (154). A buyer's guide for dyes, pigments, and resin-bonded colors and textile chemical specialties is published yearly by the American Association of Textile Chemists and Colorists (AATCC).

The amount of cross-linking agent used in a pad bath generally adds 2–5 wt % to the cured and washed cotton-containing textile. Efficiency of the reaction is generally ca 85%, but this value can vary. Storage of unwashed goods can result in considerable formaldehyde release and hydrolysis of cross-links, thereby reducing the efficiency of the cure.

Durable, low formaldehyde agents are currently selling at ca 44–88¢/kg for a 45% aqueous solution of tank-truck quantities. The chemical cost of a nonformaldehyde cross-linking system not widely used is ca three times that of a conventional finish. The chemical cost for the catalyst and the auxiliaries for a conventional finish is roughly the same as for the cross-linking agent. However, this cost ratio can vary widely because of special properties desired. For example, some finishers are using relatively high amounts of auxiliaries and lower amounts of cross-linking agent to reduce formaldehyde release in the finished textile.

Elimination of cross-linking agent has been suggested for a specific polyester fiber blended with high wet-modulus rayon or combed cotton in an 80% polyester–20% cellulosic blend (158). Increasing polyester content of blends reduces the need for a chemical finish, but it does not meet the demands for the selection of goods from the stylist or the consumer. A wide range of useful blend levels is available to consumers in the developed countries. The lower limit of polyester in cotton–polyester blends for full-performance treated fabrics has been set at 25% (159).

Another way to eliminate or reduce the use of cross-linking agents has been provided by ammonia mercerization. The first commercial treatment of fabrics in the United States was in 1975 and was based on the efforts of the Sanforized Company. The process is generally considered a pretreatment for chemical finishing, but often the level of smooth-drying properties and dimensional stability obtained is sufficient to eliminate the cross-linking treatment for some fabrics. The practical and theoretical aspects of ammonia mercerization have been the subject of numerous papers (45,160–163).

Analysis of Finishes or Residues

Analytical methods discussed in this review have generally not been concerned with finish or residue analysis. The previously discussed analysis for free formaldehyde on fabric (a residue) is the lone exception, and it is not applicable in all cases because of lack of precision and interference by impurities. The textile chemist uses many analytical methods for fabric analysis at all stages of preparation and finishing. These methods have been gathered in an excellent monograph by the AATCC (164).

Health and Safety Factors

Textiles, especially those used in apparel, are likely to be washed several times during processing, but rarely after entering the finishing plant and before reaching the consumer. It is unlikely that a general health-related problem will show up among consumers because textile workers and garment makers are exposed to much higher levels of obnoxious or potentially toxic materials than the consumer.

The principles of toxicology for textile dyeing and finishing have been reviewed (165). Hundreds of substances are used; the LD_{50}s for common finishing chemicals range from 800 for formaldehyde to 7070 for glyoxal. Many of the polymeric materials used to coat textiles are durably bonded to the fiber and are essentially inert even if removed.

The Consumer Product Safety Commission (CPSC) has been active in a number of textile-related matters involving potential carcinogenicity of a chemical. The most notable of these, ie, the ban on Tris and the transfer of formaldehyde from durable-press and other fabrics, have been discussed previously. The latter subject is currently under review by the CPSC. Some members of the industry concerned with the outcome of the matters have expressed their views (166–167).

Contact textile dermatitis has been traced to dyes, formaldehyde, unreacted monomer from synthetic fibers, and an early flameproofed cotton fabric (168). Skin irritations have also been related to the harshness of the fiber (142–145). A recent review of the literature demonstrates a continuing interest in formaldehyde resins (169), a decrease in the incidence of contact sensitivity to formaldehyde related to government controls in Japan (170), and contact dermatitis on synthetic fibers due to spin finishes (oils, surfactants, and possibly biocides) for polyester (171). Apparently, dermatitis from textiles is generally limited to a small proportion of individuals with apparent high sensitivities or to those (often industrial workers) exposed to chemicals that remove protective oils and fats from their skins. The percentages are low but at times the discomfort is high.

BIBLIOGRAPHY

"Fire-Resistant Textiles" under "Textile Technology" in *ECT* 2nd ed., Supplement Vol., pp. 944–964, by George L. Drake, Jr., U.S. Dept. of Agriculture; "Soil-Release Finishes" under "Textile Technology" in *ECT* 2nd ed., Supplement Vol., pp. 964–973, by Samuel Smith, Minnesota Mining and Manufacturing Company.

1. H. B. Goldstein and H. W. Smith, *Text. Chem. Color.* **12,** 49 (1980).
2. N. R. Bertoniere, W. D. King, and S. P. Rowland, *Tex. Res. J.* **51,** 242 (1981).
3. J. Varghese and D. M. Pasad, *Text. Res. J.* **47,** 802 (1977).
4. W. A. Reeves, R. M. Perkins, and L. H. Chance, *Text. Res. J.* **30,** 179 (1960).
5. R. W. Liggett, H. L. Hoffman, A. C. Tanquary, and S. L. Vail, *Text. Res. J.* **38,** 375 (1968).
6. D. D. Gagliardi and A. C. Nuessle, *Am. Dyest. Rep.* **39,** P12 (1950).
7. S. J. O'Brien and W. J. van Loo, *Text. Res. J.* **31,** 276 (1961).
8. W. A. Reeves, M. O. Day, K. R. McLellan, and T. L. Vigo, *Text. Res. J.* **51,** 481 (1981).
9. L. B. Arnold and co-workers, *Am. Dyest. Rep.* **49,** P843 (1960).
10. H. Petersen in H. Mark, N. Woodling, and S. M. Atlas, eds., *Chemical Aftertreatment of Textiles*, Interscience Publishers, a division of John Wiley & Sons, Inc., New York, 1971, p. 135.
11. S. J. O'Brien, *Am. Dyest. Rep.* **54,** P477 (1965).
12. S. L. Vail and A. G. Pierce, *J. Org. Chem.* **37,** 391 (1972).
13. W. A. Reeves, S. L. Vail, and J. G. Frick, *Text. Res. J.* **32,** 305 (1962).
14. S. L. Vail, *Text. Res. J.* **39,** 774 (1969).
15. H. Petersen, *Text. Res. J.* **41,** 239 (1971).
16. R. A. Gill and R. Steele, *Text. Res. J.* **32,** 338 (1962).
17. R. J. Harper, J. S. Bruno, E. J. Blanchard, and G. A. Gautreaux, *Text. Res. J.* **46,** 82 (1976).
18. R. D. Gilbert and J. H. Rhodes, *J. Polym. Sci. Part C* **30,** 509 (1970).
19. S. P. Rowland, S. M. Stark, V. O. Cirino, and J. S. Mason, *Text. Res. J.* **41,** 57 (1971).
20. S. L. Vail, *Text. Res. J.* **42,** 360 (1972).
21. S. L. Vail and A. G. Pierce, *Text. Res. J.* **43,** 294 (1973).
22. S. L. Vail and G. B. Verburg, *Text. Res. J.* **43,** 67 (1973).
23. S. L. Vail and G. B. Verburg, *Text. Res. J.* **42,** 367 (1972).
24. S. L. Vail, *Chem. Ind.* (*London*), 305 (1967).
25. J. Ugelstad and J. de Jonge, *Acta Chem. Scand.* **10,** 1475 (1956).
26. R. L. Arceneaux and J. G. Frick, *Text. Res. J.* **31,** 1075 (1961).
27. S. L. Vail, G. B. Verburg, and A. H. P. Young, *Text. Res. J.* **39,** 86 (1969).
28. S. L. Vail and R. M. Reinhardt, *Text. Chem. Color.* **13,** 131 (1981).
29. R. M. Reinhardt, B. A. K. Andrews, and R. J. Harper, *Text. Res. J.* **51,** 263 (1981).
30. S. L. Vail, F. W. Snowden, and E. R. McCall, *Am. Dyest. Rep.* **56,** 856 (1967).
31. S. L. Vail, R. M. H. Kullman, W. A. Reeves, and R. Barker, *Text. Res. J.* **40,** 355 (1970).
32. H. Petersen, *Textilveredlung* **8,** 412 (1973).
33. S. L. Vail and H. Petersen, *Ind. Eng. Chem. Prod. Res. Dev.* **14,** 50 (1975).
34. H. Petersen, *Textilveredlung* **5,** 437 (1970).
35. S. L. Vail and W. C. Arney, *Text. Res. J.* **41,** 336 (1971).
36. S. Buchholz, *Am. Dyest. Rep.* **56,** 1025 (1967).

37. U.S. Pat. 3,374,107 (Mar. 19, 1968), J. F. Cotton (to West Point Pepperell, Inc.).
38. A. G. Pierce, R. M. Reinhardt, and R. M. H. Kullman, *Text. Res. J.* **46,** 420 (1976).
39. A. G. Pierce, S. L. Vail, and E. A. Boudreaux, *Text. Res. J.* **41,** 1006 (1971).
40. H. Tovey, *Text. Res. J.* **31,** 185 (1961).
41. J. L. Gardon and R. Steele, *Text. Res. J.* **31,** 160 (1961).
42. J. T. Marsh, *Text. Manuf.*, 157 (1967).
43. A. Hebeish, A. T. El-Aref, E. Allam, and Z. El-Hilw, *Angew. Makromol. Chem.* **80,** 177 (1979).
44. V. W. Tripp, A. T. Moore, I. V. deGruy, and M. L. Rollins, *Text. Res. J.* **30,** 140 (1960).
45. S. P. Rowland, M. L. Nelson, C. M. Welch, and J. J. Hebert, *Text. Res. J.* **46,** 194 (1976).
46. R. R. Benerito, *Text. Res. J.* **38,** 279 (1968).
47. W. F. Herbes, S. J. O'Brien, and R. G. Weyker in ref. 10, p. 319.
48. S. L. Vail, J. G. Frick, and J. D. Reid, *Am. Dyest. Rep.* **51,** P622 (1962).
49. J. D. Reid, J. G. Frick, R. M. Reinhardt, and R. L. Arceneaux, *Am. Dyest. Rep.* **48,** P81 (1959).
50. S. L. Vail, J. G. Frick, P. J. Murphy, and J. D. Reid, *Am. Dyest. Rep.* **50,** P200 (1961).
51. W. J. Burke, *J. Am. Chem. Soc.* **69,** 2136 (1947).
52. A. M. Paquin, *J. Org. Chem.* **14,** 189 (1949).
53. R. L. Wayland, *Text. Res. J.* **29,** 170 (1959).
54. S. L. Vail, C. M. Moran, and J. D. Reid, *Am. Dyest. Rep.* **52,** P712 (1963).
55. R. L. Arceneaux and J. D. Reid, *Ind. Eng. Chem. Prod. Res. Dev.* **1,** 181 (1962).
56. M. T. Beachem and co-workers, *J. Org. Chem.* **28,** 1876 (1963).
57. C. D. Egginton and C. P. Vale, *Text. Res. J.* **39,** 140 (1969).
58. A. A. Eisenbraun, D. S. Shriver, and C. R. Walter, *J. Org. Chem.* **29,** 2777 (1964).
59. E. J. Blanchard, R. J. Harper, J. S. Bruno, and G. A. Gautreaux, *Text. Chem. Color.* **8,** 92 (1976).
60. P. K. Shenoy and J. W. Pearce, *Am. Dyest. Rep.* **57,** 352 (1968).
61. J. G. Frick, B. A. Kottes, and J. D. Reid, *Text. Res. J.* **29,** 314 (1959).
62. P. C. Mehta and J. R. Mody, *Text. Res. J.* **31,** 951 (1961).
63. J. D. Reid, L. W. Mazzeno, R. M. Reinhardt, and A. R. Markezich, *Text. Res. J.* **27,** 252 (1957).
64. J. G. Roberts, *J. Text. Inst.* **58,** P418 (1964).
65. R. M. H. Kullman, J. G. Frick, R. M. Reinhardt, and J. D. Reid, *Text. Res. J.* **31,** 877 (1961).
66. H. B. Goldstein and J. M. May, *Am. Dyest. Rep.* **54,** P738 (1965); R. L. Stultz, p. 744; W. L. Beaumont, p. 746; A. S. Cooper, W. A. Reeves, A. M. Walker, M. J. Hoffman, and H. B. Moore, p. 749; R. Nirenberg, p. 755; J. Midholland, p. 757; R. W. Malburg, p. 759.
67. R. M. Reinhardt, *Text. Bull.* **91**(4), 64 (1965).
68. H. Bille and H. Petersen, *Melliand Textilber.* **57,** 155 (1976).
69. R. M. H. Kullman, A. B. Pepperman, and S. L. Vail, *Text. Chem. Color.* **9,** 195 (1978).
70. K. Schliefer and U. Beines, *Melliand Textilber.* **60,** 960 (1979).
71. J. C. Morath and J. T. Woods, *Anal. Chem.* **30,** 1437 (1958).
72. C. M. Moran and S. L. Vail, *Am. Dyest. Rep.* **54,** 185 (1965).
73. G. Lund, *Shirley Inst. Bull.* **48,** 17 (1975).
74. O. C. Bacon, M. F. Parker, and L. F. Horn, *Am. Dyest. Rep.* **46,** 933 (1957).
75. A. C. Nuessle, *Am. Dyest. Rep.* **55,** P646 (1966).
76. J. D. Reid, R. L. Arceneaux, R. M. Reinhardt, and J. H. Harris, *Am. Dyest. Rep.* **49,** 490 (1960).
77. P. X. Riccobono, R. N. Ring, and A. Roth, *Text. Chem. Color.* **8,** 108 (1976).
78. *Method 112–1975*, technical manual, American Association of Textile Chemists and Colorists (AATCC), Research Triangle Park, N.C., 1975.
79. U.S. Pat. 2,764,573 (Sept. 25, 1956), B. V. Reibnitz, A. Woerner, and H. Scheuermann (to Badische Anilin-and Soda-Fabrik Akt.).
80. R. L. Arceneaux, J. G. Frick, J. D. Reid, and G. A. Gautreaux, *Am. Dyest. Rep.* **50,** 849 (1961).
81. R. J. Harper and J. S. Bruno, *Text. Res. J.* **42,** 433 (1972).
82. B. A. K. Andrews and R. J. Harper, *Text. Res. J.* **50,** 177 (1980).
83. B. A. K. Andrews, R. J. Harper, and S. L. Vail, *Text. Res. J.* **50,** 315 (1980).
84. H. Petersen and P. S. Pai, *Text. Res. J.* **51,** 282 (1981).
85. B. A. K. Andrews, R. J. Harper, R. D. Smith, and J. W. Reed, *Text. Chem. Color.* **12,** 287 (1980).
86. A. Hebeish and K. Schleifer, *Text. Res. J.* **46,** 465 (1976).
87. R. S. Perry, C. Tsou, and C. S. Lee, *Text. Chem. Color.* **12,** 311 (1980).
88. J. D. Turner and N. A. Cashen, *Text. Res. J.* **51,** 271 (1981).
89. W. C. Floyd and S. H. Yoon, *Text. Res. J.* **51,** 276 (1981).
90. S. L. Vail and B. A. K. Andrews, *Text. Chem. Color.* **11,** 14 (1979).

91. P. S. Pai and H. Petersen, *Book of Papers*, 1981 AATCC National Technical Conference, Research Triangle Park, N.C., 1981, p. 205.
92. E. B. Whipple, *J. Am. Chem. Soc.* **92,** 7183 (1970).
93. U.S. Pat. 3,576,291 (Apr. 27, 1971), M. R. Cusano and R. D. Featherston (to Proctor Chemical Company, Inc.).
94. U.S. Pat. 3,765,836 (Oct. 16, 1973), R. L. Readshaw and G. H. Lourigan (to Union Carbide Corporation).
95. U.S. Pat. 3,260,565 (July 12, 1966), M. T. Beachem (to American Cyanamid Company).
96. U.S. Pat. 3,112,156 (Nov. 26, 1963), S. L. Vail and P. J. Murphy (to USA as represented by the Secretary of Agriculture).
97. S. L. Vail, P. J. Murphy, J. G. Frick, and J. David Reid, *Am. Dyest. Rep.* **50,** 550 (1961).
98. S. L. Vail and W. C. Arney, *Text. Res. J.* **44,** 400 (1974).
99. S. L. Vail, J. G. Frick, P. J. Murphy, and J. D. Reid, *Am. Dyest. Rep.* **50,** 437 (1961).
100. A. H. Reine, J. D. Reid, and R. M. Reinhardt, *Am. Dyest. Rep.* **55,** 711 (1966).
101. E. I. Pensa, G. C. Tesoro, R. D. Rau, and P. H. Egrie, *Text. Res. J.* **36,** 279 (1966).
102. R. M. H. Kullman, R. M. Reinhardt, and J. D. Reid, *Am. Dyest. Rep.* **53,** 167 (1964).
103. H. Petersen, *Text. Res. J.* **38,** 156 (1968).
104. H. Petersen, *Synthesis*, 143 (1973).
105. S. P. Rowland, C. P. Wade, and W. E. Franklin, *Text. Res. J.* **44,** 869 (1974).
106. S. L. Vail, *Textilveredlung* **14,** 436 (1979).
107. G. R. Turner, *Text. Chem. Color.* **13,** 246 (1981).
108. E. J. Elliott and R. E. Silva, *Text. Chem. Color.* **13,** 257 (1981).
109. C. G. Tewksbury and B. G. Kidda, *Text. Res. J.* **35,** 347 (1965).
110. E. V. Burnthall and J. Lomartire, *Text. Chem. Color.* **2,** 218 (1970).
111. D. Gintis and E. J. Mead, *Text. Res. J.* **29,** 578 (1959).
112. B. C. Goswami, K. E. Duckett, and T. L. Vigo, *Text. Res. J.* **50,** 481 (1980).
113. B. M. Latta in N. R. S. Hollies and R. F. Goldman, eds., *Clothing Comfort*, Ann Arbor Science Publishers, Inc., Ann Arbor, Mich., 1977, p. 33.
114. B. M. Latta and S. B. Sello, *Text. Res. J.* **51,** 579 (1981).
115. H. E. Bille and H. A. Petersen, *Text. Res. J.* **37,** 264 (1967).
116. E. J. Blanchard, J. S. Bruno, and G. A. Gautreaux, *Am. Dyest. Rep.* **65**(7), 26 (1976).
117. H. R. Rawls, E. Klein, and S. L. Vail, *J. Appl. Polym. Sci.* **15,** 341 (1971).
118. S. L. Vail, *Text. Inst. Ind.* **3,** 309 (1965).
119. A. S. Cooper, H. M. Robinson, W. A. Reeves, and W. G. Sloan, *Text. Res. J.* **35,** 452 (1965).
120. R. L. Colbran and T. M. Thompson, *J. Text. Inst.* **58,** 385 (1967).
121. J. L. Moilliet, ed., *Waterproofing and Water-Repellency*, Elsevier Publishing Company, London, 1963.
122. E. I. Valko in H. F. Mark, S. M. Atlas, and E. Cernia, eds., *Man-Made Fibers*, Vol. 3, Interscience Publishers, a division of John Wiley & Sons, Inc., New York, 1968, p. 499.
123. J. W. Howard and F. A. McCord, *Text. Res. J.* **30,** 75 (1960).
124. C. J. Conner and co-workers, *Text. Chem. Color.* **10,** 70 (1978).
125. A. G. Kempton, H. Maisel, and A. M. Kaplan, *Text. Res. J.* **33,** 87 (1963).
126. C. J. Connor, A. S. Copper, W. A. Reeves, and B. A. Trask, *Text. Res. J.* **34,** 347 (1964).
127. C. J. Conner, G. S. Danna, A. S. Cooper, and W. A. Reeves, *Text. Res. J.* **37,** 94 (1967).
128. R. J. Boyle, *Text. Res. J.* **34,** 319 (1964).
129. A. Ruperti, *Am. Dyest. Rep.* **50,** 762 (1961).
130. W. N. Berard, G. A. Gautreaux, and W. A. Reeves, *Text. Res. J.* **29,** 126 (1959); **30,** 70 (1960).
131. D. A. Yeadon and R. J. Harper, *J. Fire Retardant Chem.* **7,** 228 (1980).
132. V. M. Bhatnagar, ed., *Advances in Fire Retardant Textiles*, Technomic, Westport, Conn., 1975.
133. J. W. Lyons, *The Chemistry and Uses of Fire Retardants*, Wiley-Interscience, New York, 1970.
134. W. A. Reeves, G. L. Drake, and R. M. Perkins, *Fire Resistant Textiles Handbook*, Technomic, Westport, Conn., 1974.
135. R. B. LeBlanc, *Text. Ind.* **143**(2), 78 (1979).
136. *Ibid.*, **144**(2), 82 (1980).
137. R. H. Barker and M. J. Drews, *Development of Flame Retardants for Polyester/Cotton Blends*, NBS-GCR-ETIP 76-22, final report to National Bureau of Standards, Washington, D.C., 1976.
138. H. D. Fedtman and B. E. Fleischfresser, *Text. Inst. Ind.* **6,** 334 (1968).
139. W. J. Thorsen, D. L. Sharp, and V. G. Randall, *Text. Res. J.* **49,** 190 (1979).

140. J. V. Beninate, B. J. Trask, and G. L. Drake, *Text. Res. J.* **51,** 217 (1981).
141. P. N. Mehta, *Text. Res. J.* **50,** 185 (1980).
142. D. G. Mehrtens and K. C. McAlister, *Text. Res. J.* **32,** 658 (1962).
143. N. R. S. Hollies, A. G. Custer, C. J. Morin, and M. E. Howard, *Text. Res. J.* **49,** 557 (1979); N. R. S. Hollies and R. F. Goldman, eds., *Clothing Comfort*, Ann Arbor Science Publishers Inc., Ann Arbor, Mich., 1977.
144. L. Fourt and N. R. S. Hollies, *Clothing Comfort and Function*, Marcel Dekker, Inc., New York, 1970.
145. E. T. Renbourn in J. G. Cook, ed., *Physiology and Hygiene of Materials and Clothing*, Merrow Publishing Co., Ltd., Watford, England, 1971.
146. D. D. Gagliardi, *Am. Dyest. Rep.* **51,** 49 (1962).
147. P. A. Walters, E. A. Abbott, and A. J. Isquith, *Appl. Microbiol.* **25,** 253 (1973).
148. G. F. Danna, T. L. Vigo, and C. M. Welch, *Text. Res. J.* **48,** 173 (1978).
149. M. Stuart, *Textiles (Shirley Inst.)* **10,** 41 (1981).
150. *Cotton Counts Its Customers*, National Cotton Council of America, Memphis, Tenn., 1981.
151. *Mod. Text.* **60**(1), 16 (1979).
152. *Chem. Eng. News*, 10 (May 29, 1978).
153. B. J. Obenski, *Text. Ind.* **144**(9), 116 (1980).
154. *Mod. Text.* **60**(4), 27 (1979).
155. S. C. Stinson, *Chem. Eng. News*, 24 (Oct. 4, 1971).
156. *Am. Text. Rep.* **82**(46), 39 (1968); **83**(26), 17 (1969).
157. *Bus. Week*, 66 (Apr. 9, 1979).
158. F. S. Moussalli and C. L. Browne, *Text. Chem. Color.* **3,** 202 (1971).
159. G. F. Ruppenicker and R. M. H. Kullman, *Text. Res. J.* **51,** 590 (1981).
160. T. A. Calamari, S. P. Schriber, A. S. Cooper, and W. A. Reeves, *Text. Chem. Color.* **3,** 234 (1971).
161. S. A. Heap, *Text. Inst. Ind.* **16,** 387 (1978).
162. B. J. Trask, T. A. Calamari, and R. J. Harper, *Text. Chem. Color.* **9**(10), 33 (1977).
163. S. L. Vail, G. B. Verburg, and T. A. Calamari, *Text. Chem. Color.* **7,** 175 (1975).
164. J. W. Weaver, ed., *Analytical Methods for a Textile Laboratory*, 2nd ed., AATCC Monograph No. 3, Research Triangle Park, N.C., 1968.
165. D. A. Bixler, *Book of Papers*, 1979 AATCC National Technical Conference, Research Triangle Park, N.C., 1979, p. 167.
166. O. P. Beckwith, *Text. Ind.* **144**(2), 84 (1980).
167. H. Bille, *Melliand Textilber.* **62,** 811 (1981).
168. I. Martin-Scott, *Br. J. Dermatol.* **78,** 632 (1966).
169. C. Romaguera, F. Grimalt, and M. Lecha, *Contact Dermatitis* **7,** 152 (1981); contains a number of references on the same subject.
170. T. Sugai and S. Yamamoto, *Contact Dermatitis* **6,** 154 (1980).
171. D. Burrows and H. S. Campbell, *Contact Dermatitis* **6,** 362 (1980).

General References

References 10 and 122 are also general references.

J. T. Marsh, *An Introduction to Textile Finishing*, 2nd ed., Chapman and Hall Ltd., London, 1966; early and current methods of chemical and mechanical finishing.

J. T. Marsh, *Crease Resisting Fabrics*, Reinhold Publishing Corporation, New York, 1962; an unusually thorough review.

Textile Finishing, BASF manual, Ludwigshafen, FRG, 1973.

E. Kornreich, *Introduction to Fibres and Fabrics*, 2nd ed., Heywood Books, London, 1962; fabric structures and industrial equipment are described, and trade names of man-made fibers are included.

J. E. Lynn and J. J. Press, *Advances in Textiles Processing*, Vol. 1, Textile Book Publishers, Inc., a division of Interscience Publishers, Inc., New York, 1961; includes chapters on nonwoven fabrics and on bonding and coating fabrics.

H. C. Speel and E. W. K. Schwartz, *Textile Chemicals and Auxiliaries*, Reinhold Publishing Corporation, New York, 1957; industrial processes and products with special reference to surfactants and finishes.

M. E. Carter, *Essential Fiber Chemistry* in L. Rebenfeld, ed., *Fiber Science Series*, Marcel Dekker, Inc., New York, 1971; review of the ten commercially important fibers.

R. J. Harper, *Durable Press Cotton Goods*, in J. G. Cook, ed., *Merrow Monographs*, Merrow Publishing Co., Ltd., Watford, England, 1971.
R. H. Manly, *Durable Press Treatments of Fabrics—Recent Developments*, Chemical Technology Review No. 68, Noyes Data Corporation, Park Ridge, N.J., 1976; detailed description of treatments for cellulosics and for polyester blends and wool based on U.S. patents issued since 1970.
R. Williamson, *Fluorescent Brightening Agents*, Vol. 4 of *Textile Science and Technology*, Elsevier Scientific Publishing Company, New York, 1980.

SIDNEY L. VAIL
Southern Regional Research Center, USDA

TESTING

The properties of any textile structure, whether made by weaving, knitting, or nonwoven technology, are determined by variables related to the selected fiber, yarn structure, fabric structure, and chemical, thermal, or mechanical finish (see Fibers; Nonwoven textile fabrics). The tests performed on textile structures relate to a measure of use performance and can be categorized as objective or subjective tests. The first type relates to physical performance and includes strength, elastic behavior, shrinkage in care, ironability, colorfastness, and resistance to tear, abrasion, pilling, and degrading media, eg, sunlight, heat, and various chemicals. Subjective test measurements relate more to textile esthetics than to engineering performance and include fabric hand, appearance following laundering or dry cleaning, luster, and comfort.

The properties of the finished textile structure or fabric are an accumulation of the properties of the fiber (or fibers if it is a blend), yarn configuration, construction, and the selected finish. Specific properties are, for fibers: diameter (tex), length, cross-sectional shape, crimp (staple plus textured filament), chemical composition, and fiber structure (molecular orientation and crystallinity); for yarn: fiber properties, fineness or count (number of fibers per cross-sectional area), twist and migration of fibers throughout the bundle cross section, compactness, surface character (hairy or smooth), texture for textured filament yarns, and blend ratio and uniformity of mix (intimate blended yarns); and for fabrics: fiber properties, yarn properties, design (threads per unit length in warp and weft of woven fabric or courses and wales of knitted structure), tension during fabrication, compactness of weave or knit structure, and mechanical or chemical finish. The properties of the textile structure are dictated by interactions of all of the parameters. Details of most of the test methods used by textile evaluators are given in refs. 1–2.

Values obtained in textile testing are only qualitatively related to use performance. For example, a fabric demonstrating twice the resistance to abrasion in a laboratory test compared to another fabric cannot be assumed to last twice as long in the form of a pair of jeans being worn by a teenager. One can only assume that the fabric with the higher abrasion resistance in the laboratory test should be at least equivalent to the second fabric and somewhat superior. The absolute level of abrasion resistance for adequate performance is questionable. A specific value obtained in the laboratory test may or may not represent a satisfactory level of resistance for a particular textile structure in a particular application. The only approach to textile testing, therefore, is to select fabrics of known adequate performance in specific applications and then to evaluate their properties in laboratory tests. Deviations on the low side only suggest that there may be a deficiency in the product relative to the property measured. There are also a number of testing procedures for each property to be measured, eg, abrasion resistance, tear testing, pilling, etc. Depending on particular type of fabric being evaluated, one test method may be superior to another in terms of predicting use properties.

The textile industry and ASTM have developed test standards which, through much experience of relating test properties to actual performance, can be used as guidelines for predicting satisfactory performance for many consumer textile items. A list of these is given in ref. 1. Textile test data are also useful to the buyers and sellers of such products.

The following tests are discussed and interpreted from the engineering standpoint. All tests performed at ambient conditions are performed at 21°C and 65% rh unless otherwise specified.

Fiber Properties

Length. Fibers can be classified as staple or filament. In filament yarns, eg, silk and man-made fibers, with the exception of acrylics which are not available as filament yarns, each filament is essentially infinite in length and is as long as the total length of yarn wound on a package. Thus, the term length has no practical meaning in this form. Staple fibers, for efficient conversion into spun yarns, must have a length-to-diameter (l/d) ratio greater than 10,000. Natural fibers have extremely variable lengths. Fiber lengths of cotton (qv), for example, may be <2.5 cm to ca 7.5 cm, depending on the type of cotton. Wool (qv) fibers may exceed 12.7 cm in length and are extremely variable even if taken from a single sheep. Flax fiber is available in lengths up to 100 cm. Man-made fibers, although originally extruded as continuous filament in extremely large bundles, can be cut to any length to approximate the length of a particular natural fiber for blending purposes or to process readily on equipment designed to convert a fiber of a specific length into a spun yarn. It is possible, though not always desirable, to cut synthetic fibers to extremely uniform lengths. Yarn properties are direct functions of fiber length. The longer the staple length, the better the quality and the higher the strength of the resulting spun yarn, all other factors held constant.

Test methods used to measure fiber length also include measurement of fiber length distribution. ASTM test methods D 1575 (wool) and D 1440 (cotton) include descriptions of techniques for laying out a uniform array of fibers of progressive lengths from the longest to the shortest. Both the average or mean staple length and the distribution of fiber lengths are recorded. ASTM D 1447 describes the use of the Fibrograph to measure the length and length uniformity of cotton fibers. The Fibrograph is a photoelectric optical instrument which scans a combed portion of an appropriately prepared specimen and draws a length–frequency distribution curve from which the upper-half mean and the mean length of the cotton fibers are calculated.

Diameter. Typical apparel and industrial textile fiber diameters are 3–500 μm. Fiber softness or flexibility is a function of diameter, and fineness or quality is particularly significant for the manufacture of lightweight fabrics of adequate strength. Cotton and wool vary greatly in fiber diameter within a given sample, whereas manufactured fibers can be produced with extremely uniform, predetermined cross-sectional areas. Therefore, most test procedures used to determine fineness are designed for cotton, wool, and other animal fibers.

One test procedure for cotton fineness involves the measurement of the resistance to air flow through a plug of fibers; this resistance can be directly related to fiber fineness. ASTM D 1448-79 describes an apparatus for this purpose called the Micronaire, and D 3818-79 describes a similar instrument referred to as the IIC-Shirley Fineness Tester. The fineness of wool fibers can also be examined with the Micronaire, the Port-Air, and the WIRA Fiber Fineness Meter. All of these tests operate on the air-flow principle (ASTM D 1282-77) (1). The diameters of wool fibers and any other fibers with essentially circular cross sections, including man-mades, can be measured directly with a PiNic Image Analyzer. In this test, the diameters of a number of fibers from a test specimen are measured at high magnification. An image-analyzing computer with a vidicon scanner is used to determine diameter and distribution (see ASTM D 31510-76) (1).

A new procedure for measuring the fineness of wool has been developed by Commonwealth Scientific and Industrial Research Organization (CSIRO) in Australia. An instrument called the Fiber-Fineness Distribution Analyzer takes a small snip of wool fibers, disperses them in a liquid, and passes them, one at a time, through a glass cell where they intersect a circular beam of laser light. The scattered light is photoelectrically sensed and a computer translates this pulse into fiber diameter. Therefore, both average diameter and size distribution can be computed. This development is too recent to be accepted as a standard test procedure by ASTM or any other test standard organization. Additional information is given in refs. 3 and 4 on fiber fineness measurement.

Characterization of cotton fibers by cross section, fineness, and maturity grade is discussed in ref. 5, and a simple and rapid method for determining fiber cross-sectional area and length is discussed in ref. 6. A device for measuring the linear densities of cotton in sliver form is described in ref. 7.

Density. The density of fibers can be determined by classical physical means, eg, precise microscopic measurement of length and cross-sectional area, and weight. Specific gravities of the more common fibers are listed in refs. 8–11.

Weight per Length. The relationships between fiber length, diameter, and density result in a common and useful criterion of fiber or yarn size. Both weight-per-length and length-per-weight systems are utilized. The weight per length (or the inverse) of such natural fibers as wool, cotton, and flax is not commonly measured, since staple length and diameter are usually considered sufficient to define fineness. For silk or man-made continuous filaments, the term tex is employed. The tex is the weight in grams of 1000 m of filament. The finer the filament, the smaller the tex value. No consideration of fiber density is made in this measurement of linear density. Another term commonly used is decitex (dtex) which is the weight in grams of 10,000 m of filament.

The preferred method for determining linear density, ie, weight-per-unit length, is that of vibroscoping individual filaments. The vibrating string principle is the basis of this test method, which is the most practical method known for making individual determinations. Single fibers of reasonably high linear density can be weighed individually by means of a highly sensitive torsion balance. The results are reported as weight in grams per 1000 meters. Alternative methods involve weighing bundles of fibers, thereby obtaining average linear densities. Only the individual fiber measurements allow the study of the relationships between properties and individual fiber tex within a given sample group.

The method for determining individual fiber linear density by use of the Vibroscope is described in ASTM D 1577-79 (1). This method also describes the procedure to be used for crimped fibers involving a preloading technique. The use of a preload in linear-density determination helps in the tensile test as well. If the preloading weight is allowed to remain hanging on the fiber as it is inserted in the upper clamp of the testing machine, it helps to align the fiber in the lower clamp and to prevent slack formation. For crimped fibers, the optimum preload essentially removes most of the crimp, thereby minimizing gauge-length corrections necessary because of crimping.

Cross-Sectional Shape. Fibers vary in their cross-sectional shape both naturally and by design (1–2,9). Whereas wool fibers are essentially round, cotton fibers are elliptical. Synthetic fibers prepared by melt spinning can be of any desired shape, depending on the shape of the holes in the spinnerette through which they are extruded. Man-made fibers that are spun from solvents into air (dry spinning) or an aqueous medium (wet spinning) are usually irregular in shape because of the skin-core effect (12). The surface of the fiber coagulates first and surrounds a highly swollen core. As the solvent diffuses out of the fiber center through the coagulated skin, the surface layer is too large for the contracted core and subsequently collapses to the irregular shape. Examples of this type of fiber are acetate, triacetate, rayon, and acrylics. Fibers exhibiting skin-core irregularity are referred to as radially heterogeneous. The properties of the highly oriented skin are different from those of the core. The effect of radial heterogeneity on fiber properties is discussed in detail in ref. 12.

Cross-sectional shape governs the stiffness (flexural rigidity) characteristics of the fiber (13). Cross-sectional shape also influences the tendency of fibers to pack together in yarns. Silk fibers, because of their triangular shape, can pack compactly to give small-diameter, dense yarns. Where lofty yarns of minimum packing are desired, irregularly shaped fibers are often used. However, the position of fibers in spun yarns is normally so random that, to a large extent, yarn density becomes independent of fiber cross-sectional shape. Furthermore, in preparing lofty yarns from either staple or continuous filaments, fiber crimp or waviness is desirable to maintain fiber separation. The crimp effect can far outweigh the effect of cross-sectional shape.

Cross-sectional shape also affects fiber, yarn, and consequently fabric luster. Triangular fibers with essentially flat sides reflect more light, thus yielding a sparkle effect. They are used in carpet fibers where this optical effect results in soil hiding. DuPont manufactures a fiber with a square cross section and containing four continuous holes traveling the length of the fiber (Antron III). This configuration is highly effective in soil obscuration. Carpet fibers of irregularly shaped cross section are referred to as second-generation carpet fibers.

Crimp. Natural fibers that grow in short lengths and are converted to spun staple yarns are very rarely straight. Cotton and wool have a natural spiraling crimp. In addition to producing a certain loftiness to the yarns prepared from these fibers, crimp is essential to conversion from fiber into yarn. Perfectly straight fibers cannot be easily processed through carding, combing, and drafting steps of yarn manufacture. Therefore, even man-made fibers must be crimped in some stage of their manufacture if they are to be processed on equipment used for converting the natural fibers into yarns.

Fiber crimp is also the basic element of textured filament yarns. The convolutions of the individual filaments prevent close packing of the fibers in the yarn, thus allowing increased moisture wicking, which affects air permeability and lessens the frequency of skin contact in fabrics intended for next-to-skin applications. Thus, this fiber crimp enhances the cover or bulkiness and comfort of synthetic filament yarn fabrics made from fibers, eg, polyester, nylon, and olefins.

Fiber crimp is characterized by the frequency and amplitude of the convolutions. ASTM D 3937-80 describes a test for measuring the crimp frequency in manufactured staple fibers and method D 2491 describes measuring crimp in wool. No method is available for measuring the crimp in cotton fibers because of its low frequency and amplitude. None of these test methods consider the amplitude of crimp which, with the frequency and stiffness of the fiber itself, affect the energy or amount of work necessary to remove the crimp from the fiber. All of these factors are also believed to influence processability as well as frequency. A more comprehensive description of the problem in the measurement of fiber crimp energy is discussed in ref. 14.

The amount of fiber crimp in textured filament yarns affects yarn bulk, or covering power and fabric stretch. However, this property is measured in fabric form since the textured yarn develops most of its bulk, stretch, or crimp after hot wet treatment of the finished fabric. A measure of the length of contraction resulting from developed crimp in individual filaments in a textured yarn can be estimated by suitable adaptation of ASTM D 2259-77 (1).

Shrinkage. The shrinkage of staple fibers that contract at least 10% in length in boiling water or hot air can be assessed by applying ASTM D 2101-79 (1). A more reliable technique is to measure the fiber tex by the Vibroscope both before and after subjecting the fiber to the contracting medium, eg, hot water, hot air, solvents, etc. Measurement of fiber shrinkage in the continuous filament yarn form is much less difficult, and the procedure is described in ASTM D 2259-77 (1).

Refractive Index. Most textile fibers exhibit greater molecular orientation or order in their longitudinal direction than in the direction of their transverse axes. Refractive indexes correlate with fiber orientation and, therefore, most textile fibers have different indexes in the longitudinal and transverse directions. The quantitative differences between the two indexes is called birefringence and is a measure of the degree of anisotropy of the fiber. To a considerable degree, birefringence correlates directly with the tensile strength of fibers from a given polymer. Refractive indexes are measurable by classical optical methods (9,15–18).

Tensile Testing of Single Fibers. The tensile testing of single fibers taken from populations of cut staple, tow, and filament yarns requires the utmost precision, since many important decisions are based on the results of these tests (19). Since so much depends on the ability to assess accurately the mechanical properties of fiber products, no effort should be spared in critically assessing every operation in this test to make sure optimum standards of equipment, procedures, and trained personnel are used.

Although there are many types of tensile testers for yarns and fabrics, which work by various principles (pendulum, inclined plane, constant rate of cross-head travel, etc), the field is significantly narrowed with respect to single-fiber testing. The principal reason for this is the high level of sensitivity required to detect the low loads associated with the straining of a single fiber to break. Most textile fibers break at loads of 2–60 g. This rather narrow range in load includes most known fibrous materials used for a range of applications from apparel to tire cord. Variations in load of as little as 0.05 g must be easily detectable. Several manufacturers produce load-measuring devices or load cells with this level of sensitivity.

The following factors should be incorporated in the test equipment and service: versatility, excellent repair service, adaptability to environmental testing, good mechanical construction, good recording apparatus, and electrical and mechanical stability. The Instron tensile tester is an example of an instrument meeting the above requirements.

A constant rate of specimen extension should be used as the straining principle rather than the pendulum or inclined-plane techniques. The speed of response of the recorder often limits the accuracy of extension tests to a maximum extension rate of ca 100%/min for many fibers. This extension rate should not be exceeded for what are normally considered hard fibers. Elastomeric fibers can be tested at extension rates of up to 500–1000%/min. Under these conditions, the response time of the recorder is not exceeded.

Detailed methods for the tensile testing of single fibers and fibers taken from yarns and tows are recorded in ASTM D 3822-79 and D 2101-79 (1). A more detailed analysis of test procedures and their significance is discussed in ref. 14.

It is an extremely lengthy and laborious task to determine strength and tenacity of individual staple fibers, particularly cotton, because of the difficulty of manipulating single fibers and the large number of tests required to obtain a significant average value. Therefore, parallel fiber bundles are tested. ASTM D 1445 (9) describes the Pressley, the Stelometer, and the Clemson bundle tests. These normally show a lower mean strength than the test for single fibers, but they are useful for mill-control testing at considerably less time and expense. The Pressley is the most widely used test.

Errors Introduced in Single-Fiber Testing. Although ASTM D 3822-79 and D 2101-79 adequately describe the clamping system used for the single-fiber test, a very important aspect related to the accuracy of measuring fiber extension is not discussed. Under certain conditions, eg, high strength fibers tested with a soft facing clamp, strains may occur in the volume element of the fiber that is held in the clamp close to the clamp edge. A method for correcting the error introduced by this effect and the complications of fiber crimp in determining fiber extension are discussed in ref. 14.

Testing at Nonambient Conditions. The test methods and discussions cited to date are applicable to the tensile testing of fibers in water and in air at elevated temperatures. Fiber contractions or expansions in nonambient environments must be compensated for in these tests.

Stress–Strain Curve. In most applications, products are discarded for reasons other than simple tensile failure of single fibers within the item. The stress developed at 10% elongation in the fiber to be blended with cotton has far more effect on the fiber's blending efficiency than the ultimate breaking properties of that fiber.

Other than the necessity for the adequate tensile strength to allow processability and adequate finished fabric strength, the performance characteristics of textile items of commerce are governed by properties of fibers measured at relatively low strains (up to 5% extension) and by the change in these properties as a function of varying environmental conditions (20). Thus, the whole stress–strain behavior of fibers from zero to ultimate extension should be studied, and various parameters should be selected to identify certain characteristics that can be related to performance.

Typical patterns of stress–strain behavior are shown in Figure 1 and the relationship to molecular motion is discussed in refs. 14, 21–23. In the low regions of strain (Fig. 1, *a–b*), often referred to as the initial modulus region, the fiber exhibits high resistance to stretch because of the interference of molecular chains as well as intermolecular secondary bonding between chains. Thus, chains or chain segments cannot react independently of their neighbors to the developed stress. In this region of the stress-strain curve, no major displacement of one chain with respect to its neighbors occurs. The small motions that do occur are the results of diffusion of short chain segments into available holes in the fiber microstructure. The extent to which the fiber resists deformation in this low strain region is usually referred to as the initial modulus of the fiber and relates to the hand, crispness, or drape of fabrics made from these fibers, since these fabric properties are observed at relatively minor deformations of the component fibers. The change of fiber modulus with temperature in water can also relate to the wash-and-wear behavior of the final fabrics (14).

At a higher level of stress, individual molecules or at least segments of individual molecules acquire sufficient potential energy to slip by one another as a result of intermolecular attraction being overcome by the applied stress.

The softening behavior allows considerable fiber deformation for relatively small increases in stress (*b–c*); hence, the rate of stress increase per unit strain is less than that observed in the initial modulus region of the stress-strain curve (*a–b*). This behavior has been characterized by several parameters usually referred to as yield point and postyield modulus.

In some fibers, typified in Figure 1 by acetate, the process of dislocation of chain segments with respect to one another continues under additional stress until the fiber breaks. In others, typified by polyester, reinforcement occurs at high levels of extension (*c–d*). Deformation occurs principally in the amorphous regions of the fiber. Load elongation curves for most textile fibers are shown in refs. 8–9.

Numerical parameters. At times, it becomes desirable to characterize the behaviors mentioned above numerically so that a large amount of information can be condensed and many fibers exhibiting different behaviors can be compared. Procedures for measurement of these parameters are described in ASTM D 3822-79 and D 2101-79, and in ref. 14; they are depicted in Figure 2.

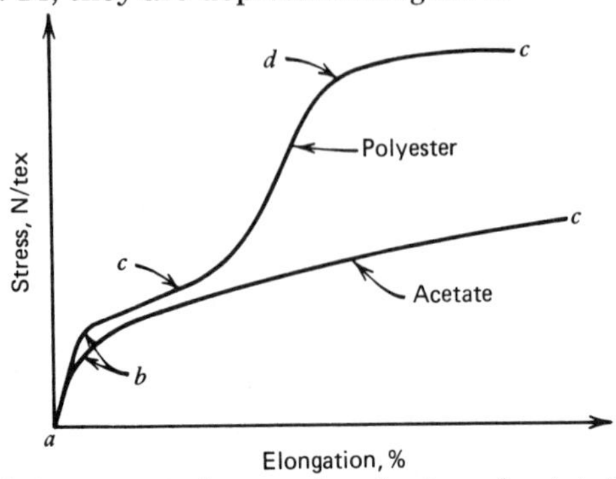

Figure 1. Typical relative stress–strain curves for polyester and acetate fibers (11). To convert N/tex to gf/den, divide by 0.08826.

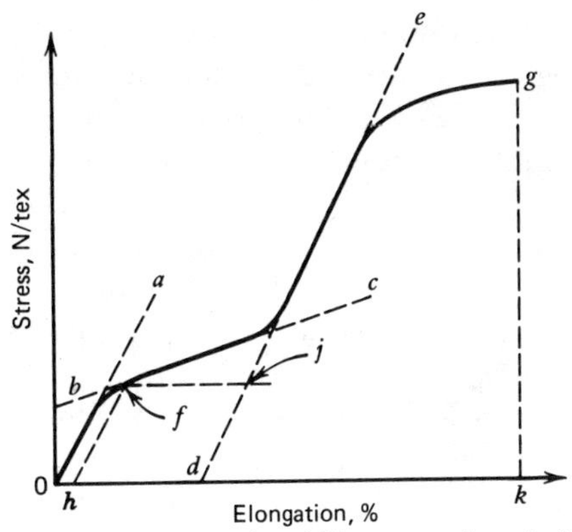

Figure 2. Stress–strain curve parameters; f = yield stress and strain ($0h$ = 1% extension); slope 0–a = initial modulus; slope b–c = postyield modulus; slope d–c = reinforcement modulus; f–j (1% extension) = reinforcement modulus displacement; area $0gk$ = energy to rupture. To convert N/tex to gf/den, divide by 0.08826.

Fabric performance. Stress-strain curve parameters depicted in Figure 2 are related to fabric performance, as indicated in Table 1. These properties should be measured on fibers in the state as nearly representative of their working form as possible. For example, fibers whose ultimate applications are in wearing apparel should be tested in the so-called boiled-off condition to simulate their actual structure after dyehouse wet-finishing.

Stress–strain behavior of various fibers. Load-elongation diagrams for most textile fibers are shown in refs. 8–9. By chemical and physical means, certain fiber-property modifications can be made either to increase or decrease elongation with a concomitant inverse decrease or increase in strength. For example, by chemical swelling techniques, fiber elongation can be increased but usually with an accompanying decrease in the tensile strength. Conversely, synthetic fibers can be drawn by differing amounts to create a myriad of combinations of strength and elongations.

Effect of strain rate on stress–strain behavior. One of the most important factors in testing is the duration of an individual test or the time scale of the experiment. Only by standardization of strain rate can reliable data be obtained. The factors governing the standardization of strain rate largely concern instruments to be used, rather than the particular deformation rates that fibers may encounter in commercial usage. Exceptions to this rule, ie, where some attempt has been made to simulate actual use behavior, include several tests involving stress-and-strain cycles of several minutes.

Table 1. Relationship of Stress–Strain Parameters to Fabric Performance

Parameter	Fabric property
modulus (slope 0–a)[a]	drape, hand, some aspects of wash-and-wear
yield (point f)	wrinkling performance, elastic behavior, shape retention
postyield modulus (slope b–c), reinforcement modulus (slope d–e), and displacement (% extension f–j)	relates to reinforcing efficiency in blended fabrics and ease of large-scale deformability of textile structures
tensile rupture properties (g)	fiber processibility, fabric strength
energy to rupture (area $0gk$)	fabric toughness; resistance to fatigue, pilling, snagging

[a] See Figure 2 for values in parentheses.

On the other extreme, there are tests of extremely short duration simulating high speed impact, eg, ballistic testing for military body armor (24–25). Between the two extremes of testing time scales is a wide region of deformation rates that are particularly significant to the fiber producer and mill representative. These deformation rates approximate the strain history encountered by a fiber in processing from fiber to finished fabric. Although normal laboratory tensile testing speed is ca 2.5 cm/min, and ballistic impact velocities can reach 300 m/s, processing velocities are 1.5–50 m/s. For example, carding involves a speed of motion of ca 20–50 m/s, normal shuttle speeds in weaving are ca 10–15 m/s, and winding and warping subjects fibers to speeds of 1.5–20 m/s.

The particular fiber gauge length, ie, the effective fiber length that is accelerated, must also be considered. For example, a velocity of 1.5 m/s on an effective fiber length of 2.5 cm results in a strain rate of 360,000%/min. If the effective gauge length is 1.25 cm, the strain rate doubles at the same velocity. Since the effective fiber length may be extremely short in certain processing procedures, it is possible to conceive of strain rates of up to several million (10^6) percent per minute being achieved in the deformation of fibers in processing.

An apparatus designed for testing single fibers at strain rates of an order of magnitude of those in textile processing is described in ref. 26. The strain rates available with this apparatus are between those of quasistatic laboratory tests and ultrahigh speed ballistic velocities and are from 2×10^4 to 2×10^6%/min. Other equipment has been constructed for testing fibers and yarns in this high speed range (27–28). Examples of the effect of strain rate on fiber properties for a number of commercial fibers are given in ref. 26.

A detailed study of the stress–strain performance of a material over a complete range of strain rates cannot be done on any single testing machine. Conventional tension testers can be operated up to speeds of ca 125 cm/min; pneumatically operated testers extend the range to several hundred meters per second; and high speed impact up to ca 300–600 m/s is usually achieved by firing missiles at the specimen (28). It is important for each material and each application to consider carefully the range of strain rates over which reliable stress–strain data are needed. Conclusions about performance which are based solely on data obtained at one speed of testing can be entirely misleading.

Tensional repeated stress properties. The ability of a fiber to deform under below-rupture loads and to return to its original configuration or dimension upon load removal is an important performance criterion. Thus, the repeated stress or strain characteristics are of the utmost significance in predicting or evaluating functional properties.

A perfectly elastic material is one that obeys Hooke's law; that is, the stress–strain diagram is linear and the unloading curve superimposes the loading curve. A completely elastic material has a stress–strain diagram that is nonlinear and a recovery curve that does not necessarily superimpose its loading curve but whose recovery, ie, restoration to original dimension, is complete. An imperfectly elastic material is one whose stress–strain diagram is nonlinear, whose recovery curve does not superimpose its loading curve, and whose recovery is incomplete.

When a fiber is stressed, the instantaneous elongation which occurs is defined as instantaneous elastic deflection. The subsequent delayed additional elongation which occurs with increasing time is creep deflection. Upon stress removal, the instantaneous recovery which occurs is called instantaneous elastic recovery and is approximately equal to the instantaneous elastic deflection. If the subsequent creep recovery is 100% (that is, equal to the creep deflection), the specimen exhibits primary creep only and is thus completely elastic. In such an instance, the specimen has probably not been extended beyond its yield point. If after loading and load removal, the specimen fails to recover to its original length, the portion of the creep deflection that is recoverable is still called primary creep; the portion that is nonrecoverable is called secondary creep. Engineers normally call this nonrecoverable elongation permanent set.

It is generally accepted that, all other things being equal, the lower the secondary creep, the better the fiber in terms of wear, shape retention, and crease resistance. This does not mean that glass, which has no secondary creep, is better in abrasion resistance than high tenacity viscous rayon, which has secondary creep, since their respective energy absorption capacities, exclusive of secondary creep, are not equal. Nor does it mean that fibers that exhibit secondary creep are of no value. For fabrics to meet the requirements of wear and crease resistance and shape retention, care must be taken that the load and extension yield points of such fibers are not exceeded in use. For certain use and, particularly, processing requirements, secondary creep may be necessary in order that the fiber be permanently deformed to fit a given state of aggregation in the textile structure. For most uses, however, even if a low region in the stress–strain diagram is required, it is advantageous to obtain this flow region by primary rather than secondary creep.

Elastic performance coefficient (EPC) is also used as a criterion of repeated stress performance. This value reflects the effect of immediate elastic, primary, and secondary creep deflections and the ability of the material repeatedly to absorb and return energy during repeated stressing. The importance of energy-absorbing ability is evident, since to resist destruction the specimen must be capable of absorbing energy imparted to it upon stress application and of releasing this energy upon stress removal without failure. The EPC is a normalized index that expresses degree of perfect elasticity. A material that exhibits identical elastic properties, after repeated stressing with those that it had originally (in one-time loading), has an EPC of 1.0. Similarly, a perfectly viscous material with no recovery has an EPC of zero.

The more quickly and completely a fiber recovers from an imposed strain, the more nearly perfectly elastic it is. The ratio of the instantaneous elastic deflection to the total deflection may be used as a criterion of elasticity (29). The integrated divergences from a theoretical graph of perfect elasticity versus elongation also is used as a criterion for determination of the elasticity index.

A method of testing immediate elastic recovery, delayed recovery, and permanent set of fibers is given in ASTM D 1774-79 (1). A discussion of the effects of strain rate and temperature on the elastic properties of nylon, polyester, and rayon is presented in ref. 14.

Other Fiber Deformations. Textiles must withstand deformations other than tensile ones, and the bending and torsion properties of fibers may become equally important to textile performance. There are five simple types of elastic deformation: simple tension, compression, shear, torsion, and bending.

In order to be useful, textile fibers must be flexible. The stiffness of an ideal cylindrical rod is proportional to the square of the tex. Since the tex, in turn, is proportional to the square of the diameter, stiffness increases as the fourth power of the filament diameter. The shape of the filament cross section also must be considered.

The crush resistance in carpets is achieved to a great extent by the use of a large diameter, high tex fiber.

The problem of shear usually is of much greater significance in brittle, high modulus materials where, although absolute shear strengths are higher than for low modulus materials, inability to deform causes the development of extremely high stress concentrations at certain points (for example, at the crossover of a warp and filling yarn), and eventual failure. This is a disadvantage of glass. For textile purposes and when flexibility is requisite, shear bending and torsional stresses are relatively minor factors when compared with tensional stresses.

Shear moduli and torsional rigidity can be determined by means of a torsion pendulum (30). For example, for polymeric materials, it is possible to ascertain the shear modulus of elasticity and the damping or energy dissipation. The ratio of the tensile modulus to the shear modulus (usually ⅓ for isotropic materials) can be used as an index of fiber anisotropy. The damping is usually measured in terms of the log decrement or tan Δ, which indicates the relationship between the viscoelastic response and the elastic response of the material.

For a complete understanding of the torsional behavior of single filaments, it is necessary to use an instrument capable of rupturing the specimen by twisting while monitoring the torque developed in the specimen. Two instruments are available: a comparator torsion meter and a single-fiber torque meter (30). In the former, the torque developed in a single-fiber specimen is counterbalanced by an oppositely twisted, calibrated reference wire, and the balancing torque is related to the twist in the specimen to give a torque-twist curve. This instrument is extremely accurate, but its sensitivity decreases somewhat for fine fiber dimensions. To overcome this and make the measurement of single-fiber torque-twist properties a continuous procedure, the single-fiber torsion meter was developed. The specimen is suspended from a precision galvanometer and the torque produced by twisting the specimen is electrically counterbalanced. The balancing current flowing through the galvanometer coil can be calibrated in terms of torque and passed to a recorder giving an indexed torque-twist curve. For both test instruments, it is important that the correct tensile loading be used to prevent filament buckling resulting from twisting. The development of torque in biaxial rotation fatigue testing of fibers is discussed in ref. 31.

Loop Strength and Knot Strength. The loop strength of textile fibers is described in ref. 32. There is a direct relationship between tensile elongation to rupture and loop-strength efficiency (8). Knot-strength values are given in refs. 33–34. Test methods for determining the loop and knot strengths of single fibers are presented in ASTM D 3217 (1).

Friction. Fabric properties that are governed in part by frictional effects are fabric hand, strength, elongation, abrasion resistance, dimensional stability, and seam slippage. The conversion of staple fibers into yarns depends upon friction, as do many of the subsequent manufacturing operations, especially procedures employed in the wet finishing of wool. Particular emphasis is given in the literature to wool and animal hairs, since they are probably the only fibers that exhibit two frictional effects, depending upon whether the relative motion is with or against their scales.

Fiber friction can be determined by physical methods. Several investigators use the belt-friction formula, wherein a belt of yarn or fiber slips over a cylindrical surface. The tension t_1 and t_2 at each end of the belt and the angle of contact θ being known, the friction coefficient μ can be determined according to the formula:

$$\frac{t_1}{t_2} = e^{\mu\theta}$$

The friction coefficient can also be measured in terms of the force required to pull two entwined fibers apart. Another test involves an apparatus wherein one or a series of parallel fibers are mounted across a small bridge similar to a violin bridge. This is then pressed against a surface that may be another fiber or some other material, and the fibers alternately slip and stick as they slide across each other. The friction coefficient is not constant with normal pressure but varies with the load in hyperbolic fashion.

In polymeric materials, the friction coefficient can vary with surface speed. This becomes important in the high speed deployment of lines, parachute shrouds, etc, and may cause excessive heat buildup that is sufficient to fuse certain thermoplastic fibers. Thus, it is important that the frictional properties be measured under simulated or use conditions.

Since fiber frictional properties are so important in the conversion of staple yarns to spun yarns, a test method (ASTM D 2612-78) has been designed to measure the cohesive force encountered in the drafting or fiber alignment of sliver or top. This frictional force is affected by surface lubrication, linear density, surface configuration, fiber length, and crimp.

Moisture Absorption. The amount of moisture present in a mass of fibers or a yarn or fabric is calculated as moisture regain or moisture content. Moisture regain is the loss in weight of water upon bone-drying at 105°C as a percentage of the resulting dry fiber weight. Moisture content is water weight loss as a percentage of the combined fiber and water weight. The normal method for determining these values involves weighing, bone-drying, weighing, and calculating. Moisture-regain values of fibers at standard testing conditions of 65% rh and 21°C and at higher relative humidities are given in refs. 9 and 35–37. ASTM D 2654-76 is a test method for measuring moisture regain. Fibers that are capable of absorbing water, when exposed to progressively increasing and then decreasing relative humidities, show regain-versus-relative-humidity curves which do not coincide. The area between the two curves or amount of hysteresis generally directly correlates with the moisture regain determined at standard conditions. Thus, a hydroscopic fiber brought to equilibrium from the wet side always has a higher moisture regain than the same fiber brought to equilibrium from the dry side.

Since the mechanical properties of the hydrophilic fibers are critically dependent upon moisture regain, it is vital that such materials be tested under constant conditions of temperature and humidity. The textile industry has selected 65% rh and 21°C as standard for all testing (1–2,11,38). ASTM lists accepted commercial moisture regain values in the buying and selling of fibers (1,9).

Swelling. When fibers absorb water, they swell radially and longitudinally. Since most fibers are anisotropic, ie, they possess longitudinal orientation of their molecules, the radial swelling is of much greater magnitude than the length increase. As the fibers and, subsequently, the yarn expand radially relative to any increased length, the helical paths in woven and knitted fabrics increase and, consequently, the fabric shrinks.

As a general qualitative rule, the greater the ability of the fiber to swell, the greater the moisture regain. The amount of moisture regain and swelling is a function of the ability of the fiber to absorb water and depends upon the amorphous regions of the fibers. Little water enters the crystalline regions. Swelling mechanics and the properties of fibers are described in refs. 15, 39–41.

Effect of Moisture on Mechanical Properties. Since textiles are sized, bleached, dyed, and finished, they must have tension, torsion, and bending properties to withstand the required wet-finishing operations. For example, the yield point must not be reduced by wetting to a value such that the resulting wet fiber or fabric cannot be processed without becoming permanently and irrevocably distorted. Fabric wetting, however, is necessary in many finishing operations. The ability of a fiber to be rendered plastic by water, whereupon it can be molded into a required geometric configuration and then set in that configuration by redrying, is the basis of many finishing operations. Similarly, if the fiber is to be used in any environment where it is apt to become wet or subjected to high humidities, it must be able to withstand such conditions; for example, in laundering at elevated temperatures.

As might be expected, there is little difference between the wet and dry stress–strain diagrams of the hydrophobic fibers, ie, nylon, acrylic, polyester, etc. The hydrophilic protein fibers and regenerated celluloses exhibit lower tensional moduli on wetting out, that is, their elongations increase and their strengths diminish. The hydrophilic natural cellulosic fibers, ie, cotton, linen, and ramie, are stronger when wet than when dry (see Fibers, vegetable).

Thermal Conductivity and Heat Capacity. Most fibers have similar thermal conductivities and heat capacities. The insulating characteristics of textiles are more related to fabric geometry than dependent on fiber thermal characteristics.

Thermal Stability. Textile fibers may be classified into two main groups with respect to their reaction to heat: thermoplastic and nonthermoplastic. The natural fibers, ie, wool, silk (qv), cotton, linen, and jute, and the synthetic regenerated rayons and proteins are nonthermoplastic, ie, they do not have a fixed melting point (see Rayon). Upon heating they char and decompose. Most of the synthetic fibers, eg, nylon-6,6, polyamides (qv), polyesters (qv), acrylics, and acetate, soften and melt with increasing temperature (see also Acrylic ester polymers; Cellulose acetate and triacetate fibers). Many thermoplastic fibers have no precise melting points because they have two-phase structures, ie, crystalline and amorphous; the amorphous region melts at a lower temperature than the crystalline region. It is customary to report, as the melting temperature, the crystalline melting point as determined by x-ray observation at elevated temperatures.

A group of high-temperature-resistant fibers have been prepared from highly aromatic compounds, eg, Kevlar (DuPont), Nomex (DuPont), and PBI (Celanese's polybenzimidazole) (42–43). These materials exhibit no clear melting point and degrade only at very high temperatures. Their great thermal stability is also responsible for their resistance to burning (see Heat-resistant polymers).

Fiber degradation from heat is affected by temperature, time of exposure, relative humidity, and air circulation. Air circulation may be important because many aging or heat-degradation processes involve oxidation, and the greater the air circulation, the more oxygen there is available to the fiber. With respect to ironing and pressing, pressure and heat transfer also affect the amount of degradation. The effect of heat on fibers is described in refs. 9 and 44–47. Contraction temperatures and melting points of thermoplastic fibers are listed in ref. 48.

In thermal degradation or any form of degradation that involves chain cleavage, fiber molecular orientation affects the observed strength loss. The more the molecular chains that are oriented in the fiber direction (the assumed direction of stress), the greater the strength loss for any given number of chain cleavages. In a highly oriented system, every chain break takes place in a load-bearing element, whereas in a less oriented fiber, the same number of breaks affect a significant number of molecules whose transverse position relative to the fiber axis prevents them from participating in the load-bearing capacity of the fiber. Therefore, these cleavages do not significantly lower the fiber strength.

Stability to Environmental Conditions. The criterion usually used to measure degradation is breaking strength. However, other criteria are also used, depending upon the performance requirements. General practice for the evaluation of sunlight aging of textile materials is given in ASTM G 23-25 (1). These methods involve accelerated testing as well as direct exposure to sunlight and exposure under window glass. The direct sunlight methods, although reliable if exposures of test samples take place simultaneously, fail to take into account the shifting concentration of the short wavelength ultraviolet light from season to season. A procedure for correcting this seasonal shift in the most damaging element of sunlight, thereby allowing test results obtained at different times of the year to be compared directly, is given in refs. 49–50.

Electrostatic Properties. Static electricity is normally generated when two different materials are rubbed together. If the resulting charges are retained and accumulated, a measurable potential is generated. If the material under consideration is a conductor and is grounded, the charge is removed as fast as it is deposited and there is no static electricity. However, if the material is a dielectric, charge builds up to the point where it may interfere with textile processing (inability to card, spin, etc) or be objectionable to a person wearing the material because of its clinging, sparking, etc.

Four criteria are used to study static electricity: surface resistivity, volume resistivity, rate of charge buildup and release, and maximum charge capable of being retained. The AATCC lists the following methods pertaining to the assessment of static in textiles: electrical resistivity of yarns (AATCC 84-1960), electrical resistivity of fabrics (AATCC 76-1964), and electrostatic clinging of fabrics (AATCC 115-1965). Another test method for static is described in ref. 51.

The synthetic hydrophobic fibers have high dielectric strengths and thus are excellent electrostatic generators. The natural and synthetic hydrophilic fibers, because of their higher moisture regains, present less of a processing and use problem, particularly at high relative humidities; however, at low relative humidities these fibers also present electrostatic problems. Although electrostatic properties are influenced by intrinsic fiber properties and moisture regain, these two parameters do not completely explain the electrostatic ranking of the various fibers (see Antistatic agents).

Yarn Properties

Size or Number. Contrary to the linear density method for measuring fibers in which weight is recorded per unit length, yarns are usually characterized by count or the units of length per unit weight. This does not exclude the use of tex for yarns, but generally most spun yarns are sized by the count method. The count of a yarn of a given cross-sectional diameter varies depending upon the particular spinning process that is used to produce it. Listed below are the most common forms of yarn count; they apply for any fiber or blend of fibers processed by the particular system.

System	*Count, m/kg (yd/lb)*
cotton	1695 (840)
English worsted	1130 (560)
U.S. woolen	3225 (1600)

ASTM has published a conversion table for all units of yarn linear density, including the tex system (D 2260). Continuous filament yarn sizes are always designated by the tex numbering system. Determination of yarn count or number is made by weighing specific lengths of yarns and the method is described in ASTM D 1907 (1).

Twist. In staple spun yarns, twist is required to force the fibers into contact with one another, thereby developing the yarn strength. Without twist, the fibers in the yarn would slide past one another easily and, consequently, the untwisted structure would have no practical value as a textile material. In filament yarns, each filament is as long as the yarn being tested; thus, twist has only a secondary effect on yarn properties, ie, the tucking of loose filaments into the yarn bundle to prevent snagging. This can be as easily accomplished by a twist substitute. Various processes have been developed for entanglement of filaments in continuous filament yarns, eg, DuPont's Rotoset. Two methods are used for determining the twist in conventional ring spun yarns. ASTM D 1423 and ASTM D 1422 describe the direct-counting method and the untwist and retwist method, respectively. In the first test, the number of convolutions per centimeter is measured directly. In the latter test, a twisted yarn is untwisted, causing an increase in length, and then retwisted in the opposite direction to the original twist, ie, until it has reached its original twisted length. The number of turns recorded divided by two is an estimate of the original twist level. Twist test results are discussed in ref. 52.

Measurement of twist in open-end spun yarns (a faster process for yarn spinning as compared to conventional ring spinning) is more difficult. Optical observation of the surface fibers has been suggested as the most reliable method for measuring twist in open-end yarns (53). The selection of twist level is very important not only in establishing the surface characteristics of the yarn (eg, low twist for soft fuzzy yarns, high twist for compact smooth yarns), but in determining its strength. Yarn strength as a function of twist level is schematically shown in Fig. 3.

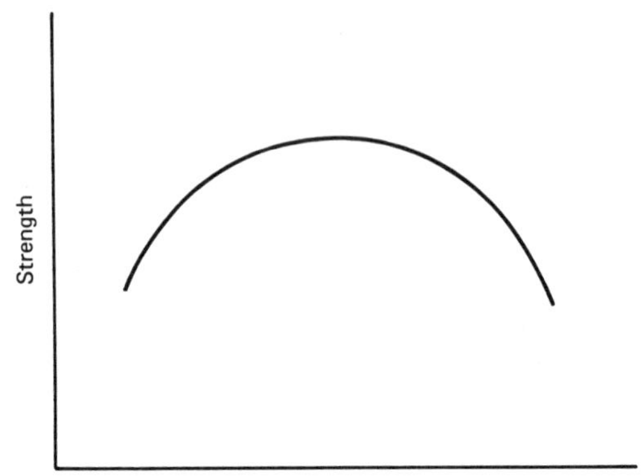

Figure 3. Yarn strength versus twist level.

There is an optimum twist level in terms of strength development in spun yarns. As twist increases, the frictional forces holding the fibers together increase. However, twist in excess of the optimal development of frictional forces causes the fibers to become disoriented with respect to the yarn axis by way of an increased helix angle of the individual fibers. This effect is analogous to the dependence of fiber strength on the orientation of the individual molecules with the fiber axis. Only in the most oriented position do they bear the maximum stress in a tensile test. Therefore, the textile technologist must balance twist level in terms of the yarn strength required for a given application and the desired yarn surface characteristics.

Since a given number of turns per centimeter generates a greater helix angle in a thicker, ie, low count, yarn than in a thinner one, it is customary to express twist in the form of twist multiplier or twist factor. These terms take into account the turns per unit length required to give a constant helix angle in yarns of differing count:

$$\text{Twist multiplier} = \frac{\text{turns per inch (cm)}}{\sqrt{\text{yarn number}}}$$

$$\text{Twist factor} = \text{turns per centimeter}\sqrt{\text{yarn tex}}$$

If this unit of twist measurement is substituted for the horizontal axis in Figure 3, then it is possible to determine the optimum twist levels for maximum yarn strength for any size yarn of a given fiber composition.

Filament yarns require much lower twist factors for optimum strength than spun yarns. These yarns are often used as supplied by the manufacturer and contain producer twist, which may be as low as 0.25 turns per centimeter. Air-entangled yarns are also often used directly without further twisting.

Construction. In addition to single-twisted yarns, many textile threads are a combination of twisted yarns placed together, ie, plies. Several plies twisted together form cables. A method for analyzing yarn construction and the nomenclature used to define this construction is detailed in ASTM D 1244-69.

Fiber Migration. Of importance in determining yarn properties is the desirability for individual fibers in spun yarns to alter their radial positions along their lengths in the yarn bundle. This enhances the interlocking of the fibers in the yarn structure and increases the number of contact points between each fiber. A comprehensive discussion of the effects of fiber migration or lack of it on fabric drape, hand, pill resistance, and abrasion resistance is given in ref. 54. Also described are various methods for measuring fiber migration, including a colorimetric method.

Strength. Yarn strengths are important not only in governing the strengths of the resultant textile structures, but in terms of the ease of conversion of the yarns to these structures. Knitting and weaving processes impart significant stresses on the component yarns. Weak yarns cause difficulties in the fabrication process, resulting in loss of efficiency and fabric defects.

The translation of fiber strength to yarn strength is complex even in the case of filament yarns. One might expect that the yarn strength would be equal to the fiber strength multiplied by the number of filaments in the bundle. However, the yarn strength is usually significantly lower than the sum of the individual fiber strengths. This is explained by the weak-link theory which argues that there is a distribution of individual fiber strengths in the yarn bundle. If all filaments are stressed uniformly, the weakest fiber will break at a combined stress level less than $1/n$th of the average fiber strength, where n = number of filaments in the bundle. The total stress in the yarn then must be shared by the remaining $n - 1$ fibers which causes stress surge in the remaining filaments, thus causing them to fail at some combined stress level below that calculated from the average fiber strength multiplied by the number of filaments in the yarn bundle. The larger the bundle size, the lower the yarn strength relative to the predicted value obtained from the average fiber strength, since the opportunity for the presence of a weak filament is enhanced by the larger number of fibers present.

In spun staple yarns, the yarn strength is only a fraction of that predicted by the individual fiber strengths and the number of fibers in the yarn cross section. In addition to fiber strength, twist, fiber migration, and frictional forces must be considered. Yarns containing two or more fibers of differing stress–strain behavior fail at a stress governed by a combination of the breaking stress of the low elongation fiber and the stress development in the high elongation fiber at the breaking extension of the low elongation fiber. This phenomenon has encouraged the synthetic fiber industry to develop special fiber variants for blending with cotton and wool to ensure compatibility of the stress–strain curves of both blended fibers.

Test methods for measuring yarn strengths are detailed in ASTM D 2256 for single strands and D 1578 for skeins. A method for testing individual yarns, eg, tire cord (qv), is given in ASTM D 885 (1). Recently, equipment has been developed for continuously monitoring the strengths of yarns during their spinning. Such a device is described in ref. 55.

Evenness. Unevenness of yarn diameter, which is more prevalent in spun yarns than in filament yarns, gives rise to poor strength and a nonuniform appearance in fabric form, ie, thick and thin spots. Yarn evenness is usually measured by passing a length of yarn through a chamber that measures the capacitance of the yarn on a continuous basis. Fluctuations in the tester's capacitance readings are proportional to yarn unevenness. This method is detailed in ASTM D 1425. Recent experience with evenness testers for man-made filament yarns is reported in ref. 56. The various methods for measuring yarn evenness are described in ref. 57.

Shrinkage. The length contraction of yarns in air or water at elevated temperatures is measured by exposing yarn skeins to these elements (ASTM D 2259).

Miscellaneous. Many of the tests described for single fibers and filaments apply to yarns. The appropriate methods given by ASTM should be consulted (1). A series of tests is designed for the evaluation of elastomeric yarns, eg, Spandex, which require some modifications in test procedure because of their high extensibility.

Evaluation of Textured Yarn Performance. ASTM does not provide tests for textured yarns. In general, these yarns are evaluated as regular filament yarns, primarily because the texture does not usually develop until a fabric made from them is wet-finished. However, a discussion of the available power of contraction of these yarns in fabric form is given in ref. 58 as are test methods for the measurement of this property. Characterization of textured yarns is usually in terms of the fabric. Of particular importance are test methods to determine fabric shrinkage under hot wet conditions, fabric stretch, and resistance to snagging. The preparation for testing textured yarns is discussed with respect to their initial tensioning in ref. 59.

Fabric Properties

Yarn and Fabric Geometry Effects. Theoretical analyses of ideal geometric structures and empirical investigations of the effects of geometry on physical properties have been made (9,29,60–65). Many inherent and geometric form properties are measurable in quantitative terms on the basis of scientific knowledge; that is, they are objective. There are certain equally important subjective performance characteristics that are evaluated at least in part by personal reaction. These can be expressed in qualitative terms, although quantitative physical parameters in part govern the subjective measurements. Fabric hand or handle, drape, luster, and comfort are of the second category (8–9).

Objective Tests. ***Construction.*** Woven fabrics are characterized in terms of their length, width, weight per square meter, yarn count (threads per centimeter in warp and filling direction), and fabric crimp and takeup. The methods for each of these construction parameters are given in ASTM D 1910-64. In addition to these construction parameters, the weave is usually specified, eg, plain weave, 2/1 twill, basket weave, etc, and yarns are removed from warp and filling for yarn characterization.

Knit fabric construction characterization is discussed in ASTM D 3887. Characterization of knit fabrics includes yield (area per weight), width, length, and yarn count (number of courses and wales per unit length). Type of knit is also specified, eg, raschel, warp knit, double knit. Yarn analysis is difficult and may be impossible because of the problems in removing individual yarns from some knit structures, especially double knits. Since construction has such a significant role in the properties of fabrics, care must be taken to ensure accurate measurement of construction parameters and the evaluator is cautioned to take constructional differences into account when comparing the test data from fabrics of differing constructions.

Thickness and Air Permeability. Thickness and air permeability are essential in measuring the fabric bulk or cover. Two fabrics of identical weight per square meter may exhibit widely differing thicknesses or air permeabilities. Fabrics with high thicknesses and low air permeabilities have high cover. A good example of this concept is the difference between a textured and a nontextured polyester yarn. Fabrics from both forms may be identical in weight, but the fabric from the textured yarn is thicker and exhibits lower air permeability. The measurement of the air permeability of textiles made from textured yarns and exhibiting biaxial stretch is discussed in ref. 66.

A method for determining fabric thickness is given in ASTM D 1777-75. The air permeability test method is described in ASTM D 737 and involves measuring the rate of air flow through a material under a differential pressure between the two fabric surfaces. This parameter is especially important in certain uses, eg, parachute fabrics, boat sails, warm clothing, rainwear and other water-impermeable fabrics, and industrial air filters.

Breaking Strength and Elongation. The breaking strength of woven fabric is used as a criterion for quality control and as a performance standard. For industrial or other purposes, where the fabric is subjected to tension, breaking strength is measured. In cases where the fabric is not called upon to withstand tensile forces, breaking strength is probably the most popular method for maintaining quality control. Breaking strength is an easy test to conduct and is quick and reliable. However, there are many textile applications where strength is of little consequence, and its use as a standard or quality index may be meaningless or far outweighed by other important criteria.

In some cases, the maximum or minimum elongation produced at a selected stress level below rupture is specified as a criterion, eg, percent stretch in stretch fabrics. Two types of fabric strength tests are commonly used: the ravel-strip and grab methods (ASTM D 1682). There are several modifications of these tests.

In the ravel-strip method, a specimen is cut in either the warp or filling direction. Edge yarns in the longer direction are unraveled and removed from the fabric until its width becomes 2.54 cm; a fringe 0.32 cm thick on each side of the fabric thus results. The test strip is clamped lengthwise in the flat jaws of the testing machine. Each jaw is 5 cm wide so that all yarns within the fabric are held and broken during the test. If the fringe is not present and the tensile force progressively increases, the edge yarns pop out of the fabric. While they are held by both jaws, a combination of yarn and fabric is tested with the popped yarns not usually sharing their full burden of the load. The fringe prevents the yarns from popping out and hence the fabric is tested as a fabric.

The strip tensile test is an excellent method for studying the effects of chemical finishes, heat aging, weathering, microbial deterioration, and the like on tensile properties. Most modern electronic tensile-testing machines produce a record of the load-elongation diagram from which strength, elongation, and energy absorption can be determined.

The ravel-strip test requires slow, careful, and often tedious preparation of the specimen and therefore is considered uneconomical by many mills. In addition, it is criticized on the basis that it does not simulate the way fabrics are used. The grab test is also popular, particularly in testing heavy industrial fabrics. A specimen 10 × 15 cm is cut so that the longer dimension is in the direction to be tested. Flat jaws 2.5 cm on a side and 7.5 cm apart at the start of the test are used to clamp the fabric. When the jaws are separated, the yarns clamped in both jaws are subjected to tension. In addition, because of the interaction of warp and filling yarns with attendant frictional forces, a fabric assistance effect develops, and yarns adjacent to those actually clamped in the jaws help to support the load. Thus, grab-strength tests usually produce higher strength values than strip tests.

Empirical attempts have been made to relate strip and grab test values, particularly for cotton fabrics, so that if one strength is known, the other can be calculated. The relationship is complex, depending upon fiber strength and modulus, yarn size and crimp, yarn-to-yarn friction, fabric cover factor, weave, weight, and probably other factors (9).

Bursting Strength. The fabric tensile tests described in the previous section are not applicable to knitted fabrics. The strengths of knitted fabrics are measured by determining their bursting strength. ASTM D 3786 describes the testing of knits of low-to-intermediate extensibility to failure. The fabric is held over an expanding diaphragm, and the hydraulic pressure in the diaphragm at the instant of fabric rupture is reported as the burst strength.

For knitted fabrics exhibiting high extensibility to failure, the ball burst test is recommended (ASTM D 3787). A polished steel ball is pressed onto a rigidly held circle of knit fabric. The force on the ball required to effect fabric rupture is reported as the fabric breaking strength.

Tear Strength and Energy. For all flat, sheetlike materials, eg, woven fabric, films, paper, and leather, the breaking strength of the material in tension is usually greater than its tear resistance. Although it may be difficult to induce a tear in any of these materials, once started the tear can usually be propagated at a relatively low load.

Three types of tear tests are widely, if not always properly, used. The tongue and trapezoid methods involve a standard tensile testing machine for recording the force necessary to tear. The Elmendorf test involves a specially designed pendulum tester which measures tear energy. These methods are presented by ASTM in D 2261, D 2262, and D 1424 (1).

In the tongue tear test, a specimen 20 × 7.5 cm is cut so that the yarns to be ruptured during the tear lie in the shorter dimension. A cut 7.5 cm in length is made along the longer center line of the fabric. This cut thus produces two 7.5 × 3.8 cm tongues, which are placed in the upper and lower jaws of the testing machine. As the jaws move apart, each cross yarn is subjected to progressively increasing tension which occurs in the "del," ie, the triangular-shaped distortion at the active region of tearing. The cross yarns fail singly, doubly, or even triply, depending upon the fabric weave, crimp distribution, yarn strength, elongation properties, etc. The tearing action is manifested on an electronic tensile tester recorder as a diagram of progressively increasing and then sharply decreasing loads. For a proper tear diagram, it is necessary to have a recorder which responds quickly and accurately to changes in load; the modern electronic strain-gauge tensile testers are excellent for this purpose. The purely mechanical pendulum-type tensile testers have too much inertia to give proper results.

In the Elmendorf tester, a pendulum applies energy sufficient to tear through a fixed length of fabric; the tear is of the tongue type. The tester consists of a sector-shaped pendulum with a clamp which is aligned with a fixed clamp when the pendulum is in the raised starting position at maximum potential energy. A rectangular specimen is fastened in the clamp and a slit is made to start the tear. The pendulum is released and the specimen is torn as the moving jaw separates from the fixed jaw. The energy required to tear is the potential energy difference at the beginning and end of the pendulum swing, as measured by the position of the pendulum at the end of the test. A graduated scale attached to the pendulum indicates the percentage of original potential energy available that was consumed in tearing the test specimen. Tear energy is recorded as (N·m)/m (1 N = 102 gf) of fabric torn. The average tearing force can be calculated by dividing the energy by twice the length of tear, because during the test the jaws separate a distance equal to twice the length of the tear.

Much published literature reports tear strength based on the trapezoid tear test. This procedure is actually a modified strength test and most knowledgeable evaluators recognize that the results obtained do not apply to actual tear-type failures occurring in product application. ASTM has discontinued the use of this test method.

Tear resistance is more a function of fabric construction than yarn and fabric strength. Loose open weaves resist tear, whereas tight weaves tear easily. For example, cheese cloth is difficult to tear whereas a nylon taffeta tears more easily. In the open weave, the concentrated force field at the point of tear is dissipated by the compliance of the fabric structure to accommodate the stress field.

Snag Resistance. Although knits do not tear easily, they are prone to snagging on sharp, pointed objects, eg, nails, tree twigs, staples, etc. ASTM test method D 3939 describes a test to quantify the tendency for fabrics to snag. It can be used for woven and knitted fabrics, particularly those made from textured filament yarns. A fabric specimen in tubular form is mounted on a rotating cylinder, and a mace bounces randomly against the rotating specimen. As the mace bounces over the test sample, snags may occur to a degree affected by a variety of fabric factors. The degree of fabric snagging is evaluated by comparison of the tested specimens with visual standards exhibiting a range of snagging resistance. The observed resistance to snagging is reported on a scale ranging from no. 5 (no snagging) to no. 1 (very severe snagging). Like all tests performed in simulated situations, the results are meaningful only if an established correlation exists between performance in this test and in wear situations for the particular types of fabrics being evaluated. The acceptable level of performance in the test should not be selected arbitrarily but should be established in actual wear studies. A study comparing the various snag testing procedures available for testing both knit and woven fabrics is reported in ref. 67.

Pilling Resistance. With the advent of high strength synthetic fibers, fabrics containing these fibers, especially in blends with weaker natural fibers, exhibit a propensity for pill formation in varying degrees. The mechanism of pill formation involves the breaking away of the weak fibers and the retention of the broken lengths by the stronger fibers. The rubbing action causes these fibers to form small balls adhering to the fabric surface. ASTM D 3511, D 3514, and D 3512 are for the laboratory evaluation of the pilling tendencies in fabrics. The three approaches to pilling measurements are the brush pilling tester, elastomeric pad method, and the random-tumble pilling tester. In all three methods, the sample to be tested is exposed to rubbing against a mildly abrasive surface. This is followed by a visual comparison of the fabric surface appearance with a set of standards developed by ASTM.

To a certain degree, fiber producers have minimized the tendency for pilling of the various fibers by advanced technology. Included in the solutions to the problem are a reduction in polymer molecular weight, thereby yielding a lower strength fiber; and notching or etching of the fiber surface either before or after incorporation into fabric form. This increases the tendency for the stronger fiber, which normally retains the pill on the fabric surface, to break off more readily. Most fiber producers offer special low pilling variants.

Abrasion and Wear Resistance. Most textile technologists confine the term abrasion to a simple rubbing action like that which might occur with a laboratory abrasion tester. The term wear is broader in scope and includes the combined effect of many additional factors besides abrasion, eg, laundering, drycleaning, ironing, and wearing of apparel. Prediction of general wear resistance from a laboratory abrasion test is very difficult. The six laboratory tests most commonly used in the United States are listed below:

Oscillatory cylinder method (ASTM D 1175). Abrasion resistance is measured by subjecting the sample to unidirectional rubbing under known conditions of pressure, tension, and abrasive action. It is used for woven textile fabrics.

Uniform abrasion method (ASTM D 1175). Abrasive action is applied uniformly in all directions in the plane of the surface of the specimen about every point in it. The test may be run dry or wet and is used for a wide range of woven and knitted textile fabrics.

Rotary-platform, double-head method (ASTM D 3884). A specimen is abraded by rotary rubbing action under controlled conditions of pressure and abrasive action. The sample, mounted on a platform, turns on a vertical axis against the sliding rotation of two abrading wheels. The resulting abrasion marks form a pattern of crossed arcs. This test is used for a wide range of textile fabrics.

Flexing and abrasion method (ASTM D 3885). A specimen is subjected to unidirectional reciprocal folding and rubbing over a bar having specific characteristics and under known conditions of pressure and tension. This test is used for woven fabrics only. Fabrics may be tested wet or dry.

Inflated-diaphragm method (ASTM D 3886). A specimen is abraded by rubbing either unidirectionally or multidirectionally against an abradant having specific surface characteristics. The specimen is held in a fixed position and supported by an inflated rubber diaphragm which is held under constant pressure. This test is used for woven and knitted fabrics and the fabrics may be tested wet or dry.

Impeller tumble method (AATCC 61). An unfettered fabric specimen is driven by a rotor in a random path so that it repeatedly impinges the walls and abradant liner of the abrading chamber. Although the specimen is continuously subjected to extremely rapid, high velocity impacts, the zigzag orbit it takes is generally circular. The specimen is subjected to flexing, rubbing, shock, compression, stretching, and other forces. Frictional abrasion is produced by the rubbing of surface against surface, and surface against abradant. This method is used for woven and knitted fabrics alike.

A great deal of latitude is offered in terms of evaluation techniques of the abraded fabrics. Although one obvious result of the experiment is the appearance of a hole in the fabric, it is easily recognized that, in wear patterns on textile consumer products, unsightliness preceeds this stage. Alternative techniques are the measurement of weight loss or measure of remaining fabric strength following varying degrees of exposure to the abrading media. However, even these results exceed the degree of fabric surface damage objectionable to the consumer. For example, in blended fabrics containing two fibers of slightly varying shades of color, preferential wear of one fiber causes a shade change in the abraded area. This wear is referred to as frosting. It is therefore recommended that, in addition to more severe test results, visual observations be made frequently in the course of the experiments and that detectable appearance changes be noted at their earliest stage of development. Five requirements for high abrasion resistance are low modulus of elasticity, large immediate elastic deflection, high ratio of primary to secondary creep, high magnitude of primary creep, and high rate of creep recovery.

There are three methods of fiber abrasion: frictional wear, surface cutting, and fiber rupture or slippage. Thus, a fiber can be abraded by frictional wear such as occurs when smooth metal surfaces are rubbed together; subjecting the fiber to a surface-cutting process such as occurs in metal cutting or grinding (a type of wear which results when a very fine abrasive or emery surface is rubbed across a fabric), and is applicable where surface projections of the abrading surface are small relative to fiber diameter; and plucking the fiber, which may develop when the surface protuberances of the abradant are large compared to the fiber diameter and the normal forces between the abradant and the cloth planes are high (29).

Fabric geometry is of great significance with respect to the abrasion resistance of fabrics. The following factors affect abrasion resistance: weave; geometric area of contact between fabric and abradant; local pressures or stress concentrations which develop on specific yarn points or area; threads per cm; crown height, ie, the extent of deformation out of the plane of the fabric resulting from the intersection of warp and filling yarns; yarn size; fabric thickness; yarn crimp; float length; yarn cohesiveness; compressible compliance, ie, compressional resilience of the fabric and backing; fabric tightness; cover factor; direction of abrasion; and magnitude and direction of tensions developed during abrasive action (60).

In addition to fiber and fabric influences on abrasion resistance, chemical finishes also must be considered. Many thermosetting resins used to impart wash-and-wear characteristics to cellulosic fibers and fabrics reduce their resistance to abrasion as a result of fiber embrittlement.

Laundering Shrinkage. When textile fabrics are wet-laundered, they often shrink. Shrinkage is of three types: relaxation, swelling, and felting. Relaxation shrinkage occurs when a fabric that has been finished in a stretched state is exposed to heat and moisture while relaxed. Swelling shrinkage may occur when a fabric that is composed of hydrophilic fibers is soaked in water and is the result of the increase in fiber diameter relative to the length change. Felting shrinkage is associated exclusively with wool and other animal fibers, which are feltable in the practical sense. It is characterized by a continued reduction in fabric area and a continued increase in fabric density as a result of application of heat, moisture, and mechanical motion occasioned by water washing. The laundering of wool fabrics in domestic washing machines and the related test methods and assessments of results are described in ref. 1.

Since the hydrophobic fibers do not swell, they do not shrink by this mechanism when wet-laundered. They may, however, exhibit relaxation shrinkage if they have not been adequately set during finishing. In addition, some hydrophobic fibers are susceptible to thermal shrinkage, and the thermal effect of the hot water used in laundering may therefore cause fabric shrinkage. AATCC offers five separate methods for determining dimensional stability of fabrics and garments (2). Each one is based on differing types of laundering and fiber composition: dimensional changes in automatic home laundering of woven or knit fabrics (AATCC 135-1978), a test for fabric shrinkage in home laundering; dimensional changes in automatic home laundering of garments (AATCC 150-1879), a test for garment shrinkage; dimensional changes in laundering of woven and knitted fabrics except wool (AATCC 96-1980), a test to evaluate dimensional stability in commercial laundering; changes on dry cleaning in perchloroethylene: machine method (AATCC 158-1979), a test for determining shrinkage during dry cleaning; and dimensional changes (shrinkage), felting and relaxation of woven or knitted wool textiles (AATCC 99-1975), a test for determining the shrinkage of wool fabrics specifically.

In all of these tests, temperature and time are programmed variables, so that data may be collected under a variety of washing or cleaning and drying conditions corresponding to the care label recommendations for garment maintenance (see also Felts).

Compressional Resilience. Resilience does not always denote complete and instantaneous recovery after deformation. ASTM lists a series of terms that relate to the hand of fabrics, including some which refer to properties influenced by compressional resilience. Compressional resilience and other deformation-recovery characteristics of textile fabrics are important to such effects as wrinkle resistance and retention of shape, drape, and hand, including softness, stiffness, harshness, limpness, liveliness, boardiness, and springiness; all depend upon the resilience of the structure. Resilience is an important parameter for evaluating blankets and wearing apparel in which warmth is a factor, permanency of pile fabrics including carpets, and bulk fiber utilization in mattresses, cushions, etc. The general method for determining compressional resilience is to compress and unload the material cyclically and then plot a repeated stress diagram, ie, compressive force versus thickness of fabric.

Thermal Transmission. Thermal conductivity or insulating ability must be distinguished from the subjective feeling of warmth or coolness by a human subject. The former is defined in strict engineering terms and is one of the factors involved in warmth. Three methods are normally used to measure thermal conductivity: the fabric is placed between two plates (usually metal) at different temperatures, and the rate of heat flow is determined; a hot surface is wrapped with the fabric and its rate of cooling is measured; or a hot surface is wrapped with the fabric and the amount of heat necessary to maintain temperature is measured (9).

Since the thermal conductivity of air is lower than that of fibers, it follows that the greater the entrapment of dead air space within a textile structure, the better the insulating ability. There is an almost linear relationship between thermal insulating value and fabric thickness and the chief, but not the sole, factor that determines the thermal insulation value is thickness. There is almost no correlation between thermal insulation and weight per unit area.

Given that fiber conductivity has little effect upon a fabric's thermal transmission nor on the conductivity–thickness relationship, the state of aggregation of the fiber in the fabric structure is important. As long as the required thickness is maintained, concomitant insulation results. However, the ability to maintain thickness under normal use conditions of compression, tension, bending, laundering, dry cleaning, and wear must not be overlooked. One additional factor is fabric density. Two fabrics may have the same thickness and, hence, the same thermal conductivity, but to obtain such equal thickness, different weights of fiber may be needed. Thus, aggregates of different fibers may have different bulk densities, and on this weight basis one fiber may exhibit a thermal insulating advantage over another.

Fabric structure rather than fiber properties governs the movement of air within the fabric. Air moves within the interstices of yarn and fabric rather than within the fiber. Thus, air permeability is influenced by the void spaces within the textile structure. Since the same void space is influential in controlling thermal conductivity, it follows that thermal conductivity is influenced by air permeability. However, the movement of air through a textile structure can dissipate heat by convection because the air is no longer dead or still air. Thus, thermal insulation falls as air velocity rises. Also, relative humidity has a negligible effect upon heat transmission.

Moisture Transmission. Water-vapor permeability is related to the suitability of fabrics for apparel purposes. Two methods for measuring the diffusion of moisture through a fabric barrier include the absorption-cup method and the evaporation method. The absorption-cup method involves ascertaining the amount of water that diffuses through the test fabric and into a cup of drying agent; the test fabric acts as the cover of the vessel containing the desiccant (see Drying agents). The other side of the fabric is exposed to an atmosphere of constant humidity, temperature, and air velocity. The amount of water absorbed per unit area per unit time can be calculated by determining the increase in weight of the desiccant at selected time intervals. The evaporation method consists of placing the test fabric over a carefully calibrated cup filled with water to a prescribed depth. Again, the cup and fabric assembly is exposed to an atmosphere of constant humidity, temperature, and air velocity, and the rate of loss in water weight in the cup is determined. Although these methods give valid results for impermeable materials, eg, plastic films, and may be useful for comparing fabrics, they do not give precise results for open fabrics of high moisture transmission. As the water is absorbed into the desiccant or evaporates, the relative humidity of the air adjacent to the fabric may not remain constant. A modified evaporation method involves placing a permeable cover fabric 0.5 cm above the test fabric specimen (68). This causes a constant humidity to develop above and below the specimen.

If it is assumed that a moisture differential exists between two sides of a fabric, water vapor diffuses through the fabric at a rate that depends primarily upon the relative amount of air contained within the structure (69). The resistance of the fiber substance to the passage of water vapor is as much as several orders of magnitude higher than that of air. Since normal fabrics contain 50–90 vol % of air, a negligible amount of water vapor passes through the fibers, and differences between fibers are much less important than differences between fabric geometries. Very tightly woven fabrics have higher resistance to moisture vapor transfer than more open fabrics, and may, for this reason, be less comfortable to wear.

Water Repellency and Water Resistance. Water repellency is the ability of a textile fiber, yarn, or fabric to resist wetting; water resistance is a general term denoting the ability of a fabric to resist wetting and penetration by water (2). AATCC lists the following test methods for measuring water repellency and water resistance: rain test, hydrostatic pressure test, impact penetration test, spray test, static absorption test, and dynamic absorption test. These are described by AATCC in methods 35-1980, 127-1980, 42-1980, 22-1980, 21-1978, and 20-1978 (2) (see Waterproofing and water oil repellency).

A waterproof fabric is one in which the pores or interstices between fibers and between yarns are filled with appropriate substances which cause the fabric to have a continuous surface and very little air permeability. A water-repellent fabric is one whose fibers are usually coated with a hydrophobic-type compound and whose pores are not filled in the course of the treatment. The latter types of fabrics are quite permeable to air and water vapor (70).

Basic methods for measuring water repellency or water resistance are measurement of the pressure required to force water through a fabric, measurement of the amount of surface wetting or penetration by falling droplets, measurement of water absorption upon immersion, or measurement of fiber or fabric wettability in terms of the contact angle of the water droplets. Detailed descriptions of all test methods are given in refs. 8–9.

Stretch. The recent interest in comfort stretch in sportswear and slacks has initiated test methods to determine both the amount of stretch in a fabric and its rate and extent of recovery. ASTM D 2584 describes such a test for knitted stretch fabrics. To determine stretch recovery, a specimen is extended to a specific percent stretch and held for a period of time. It is then allowed to recover under zero load. During the recovery period, the length of the specimen is measured at various time intervals. The amount of growth at each time interval is represented as a percentage of the original unloaded specimen length.

To determine the amount of stretch, a specified load is applied to a fabric of known length. The length of the specimen after load application is then related to its original length as percent stretch.

ASTM has not yet developed a similar test for stretch-woven fabrics. However, it is believed that the method for knitted fabrics can be adapted to yield meaningful data concerning the stretch and recovery properties of woven fabrics. To determine the suitable load to apply to measure percent stretch in any fabric, it is recommended that the sample fabric be stretched in an electronic tensile (eg, Instron) tester which yields a continuous plot of load versus extension.

Color. Spectrophotometers are commonly used to quantitize color shade and intensity. White light impinged on the surface of a dyed fabric causes a light of specific wavelength to be absorbed or reflected, depending upon the fabric color. The reflected light passes through a prism and is measured by a photoelectric cell, and a plot of reflection versus wavelength relates the shade and intensity. Although this curve is adequate to define quantitatively any color, the transmission of these plots from customer to dyehouse or computer storage and retrieval of this information is difficult. However, it is possible to define a color by comparing the intensities of the three reflected primary colors, ie, red, green, and blue. Thus, a system of tristimulus values allows the storage and transmission of color data. Dyestuff manufacturers list these tristimulus values for each of their manufactured dyes as measured for fabrics of specific fiber content. This allows the dyer to approximate closely the desired color. Final adjustment continues to be made by the human eye skilled in shade matching (see Color).

A test method for performing instrumental color measurements is presented in AATCC 153-1978 (2). Further information regarding the science of color measurement is given in refs. 71–72.

Colorfastness. There are many tests for measuring the fastness or color retention of a dyed cloth exposed to light, weathering, laundering, dry cleaning, etc (2,9). Fabrics are exposed for various periods of time under many prescribed conditions. Evaluation of the results are performed by visual comparison of the exposed samples against the international gray scale developed by AATCC or other forms of visual grading scales. An alternative method of evaluation involves color measurement spectrophotometrically.

Lightfastness. A general method describing techniques for determining the lightfastness of textile products is given in AATCC method 16-1978 (2). The two basic forms of exposure are direct exposure to sun rays (AATCC 16C-1977 and 16B-1977) and accelerated methods involving the carbon-arc lamp (AATCC 16A-1977 and 16D-1977) and the xenon-arc lamp (AATCC 16E-1977 and 16F-1977) (2). Better correlation between accelerated testing and direct weathering is expected from the xenon lamp, since its spectral distribution is more similar to that of sunlight than the carbon-arc lamp. Also included as part of these methods is the option to alternate light exposure with dark periods at elevated humidities in an attempt to simulate more closely actual exposure conditions of products.

Laundering fastness. AATCC 61-1980 describes a series of colorfastness-to-laundering tests, which consist of laundering the specimen sewed to a standard white multifiber test cloth and grading the color change in both the test cloth and the dyed sample. Depending upon test requirements, a selected washing formulation of soap, bleach, or detergent is used at various laundering temperatures. A 45-min test is roughly equivalent to five home or commercial laundering cycles.

Gas fading. Another colorfastness problem is that of gas fading of dyed acetate and other man-mades. Many dyes fade or change color when exposed to oxides of nitrogen present in industrial gases and smoke in urban areas. Gas fading can also occur in some sensitive resin-bonded pigments applied to cotton fabrics. The laboratory test for evaluating gas fading consists of exposing small swatches of the test specimen with a dyed control sample to nitrogen oxide fumes until the control fades a prescribed standard amount. The test specimen is then examined to ascertain if fading has occurred. If it has not, the cycle is repeated until fading is obvious. The specimen is then ranked from class 1 (poor) to class 5 (excellent), depending upon the number of exposures required to produce an appreciable shade change. This method is described in AATCC 23-1975 (2).

Crocking fastness. The crocking test is used to determine the extent to which color is transferred from the surface of a dyed fabric to another surface by rubbing. The test specimen is fastened to the base of a crockmeter and is rubbed under controlled pressure by a standard white cotton print cloth for 20 rubbing cycles. The white standard cloth is then examined for the presence of color. The test is conducted dry or wet (AATCC 8-1977) (2). Another method involves the use of the rotary vertical crockmeter (AATCC 116-1977) (2). This test is particularly useful in measuring dye transfer from an area in a fabric too small for testing with the crockmeter.

Other. Dyed fabrics are tested for their resistance to color change when exposed to the following chemicals or conditions: acids, alkalies, chlorine bleach, peroxide bleach, sulfur dioxide, dry cleaning, dry and wet hot-pressing and heat setting, perspiration, salt water, and soap (2,73).

Soil Redeposition and Soil Release. Hydrophobic synthetic fibers attract oily soil to a greater extent than natural fibers and hydrophilic man-made fibers. This fact has led to the development of a number of soil-release and improved whiteness-retention finishes. Two factors are involved in evaluating the soiling characteristics of fabrics: the amount of soil that can be deposited on a fabric during laundering of dirty clothes and the ease with which the soil can be removed during laundering. Procedure AATCC 130-1977 is used to measure the ability of a fabric to release oily stains during home laundering and is particularly useful in evaluating the effectiveness of soil-release finishes (2). Test procedures AATCC 151-1980 and 152-1980 measure the degree of soil redeposition during laundering; the former with a Launder-Ometer and the later a Terg-o-Tometer for accelerated effects. Evaluation consists of visual grading to standards or the measurement of change in fabric light reflectancy. A test method for the simulation of the soiling of textiles by skin fat and tests of the mechanism of soil adhesion and removal in textile fabrics are described in refs. 74–75.

Carpet soiling. AATCC 123-1979 involves a procedure for comparing the soiling propensities of two or more carpets by tumbling them together with a prepared soil in a laboratory ball mill. Test 122-1979 involves subjecting the carpets to foot traffic. Method 11-1979 describes the visual rating method for evaluation of soiling propensity. Appearance does not always correlate with the amount of dirt present in the carpet. Alteration of fiber cross-sectional shape (to avoid round fibers) and longitudinal voids in the fiber hide soil effectively.

Flammability. Flame-burning tests are used to identify fibers. A series of flame tests is described in ref. 11. These tests differ from determinations of fabric flammability (see Flame retardants in textiles).

Three terms are commonly used to describe the flammability characteristics of fabrics: fireproof (total resistance to destruction by fire), flameproof (does not withstand the action of fire without some change in its physical state, but does prevent self-combustion once the igniting source has been removed), and flame-resistant (does not burn readily).

Treated and industrial fabrics. The fire resistance of industrial fabrics, including duck, tentage, awning, and treated textile fabrics used for decorative or other purposes is tested by mounting a 7 × 2.5 cm test specimen vertically on a frame in a test chamber. The sample is subjected to a standard bunsen burner flame for 12 s in such a manner that the lower end of the specimen is 1.9 cm above the top of the gas burner. The flame is then removed and the specimen is evaluated for afterflaming, afterglowing, and length of char (ASTM 1230).

Clothing textiles. Flammability of clothing textiles is tested as specified in the Flammable Fabrics Act under Commercial Standard CS 191-53 (76). Specimens are prepared by brushing if they have raised fiber surfaces, by dry cleaning or laundering if they have flame-retarding finishes, and by drying. The dried specimen is held in a special apparatus at a 45° angle, a standardized flame is applied to the surface near the lower end for 1 s, and the time required for flaming to proceed up the fabric a distance of 12.7 cm is recorded. Fabrics are classified in three categories depending upon the rate of flame spread. Most apparel is required to pass this test, ie, exhibit a flame spread rate of more than 3.5–4 s. In ASTM D 3411, another aspect of fabric flammability, ie, ignition time, is evaluated. A more severe test is described in ASTM D 3659. The sample is suspended vertically, as fabrics usually burn more readily in this position, to simulate the hem of a skirt or trouser cuffs. Evaluation of fabric flammability is judged on the basis of the extent of area damage of the fabric under the conditions of the test. A comparison of methods for measuring apparel fabric flammability is reported in ref. 77.

Children's sleepwear. Although the only mandatory standard enforced by Federal law for general apparel flammability is CS 191-53, there are special, more stringent requirements enforced by the Consumer Product Safety Commission for children's sleepwear. Initially, two tests were used, one for younger children (sizes 0–6X) and the other for older children (sizes 7–14) (78–79). The tests are vertical ignition tests in which, in the case of 0–6X, minimal melt drip was defined as a criteria for acceptance. Although enforcement of these methods has led to a reduction in clothing-related burn injuries to children, the severity of the test, especially that used for sizes 0–6X, eliminated many desirable children's sleepwear fabrics, unless the chemical Tris [tris(2,3-dibromopropyl) phosphate] was used to meet the criteria of the melt-drip requirement. The banning of Tris left the industry with a poor selection of fabrics exhibiting adequate resistance to flammability. A reconsideration of the safety aspect of the melt-drip requirement has led to the elimination of the melt-drip requirement in the children's sleepwear standard. The revised description of the test and requirements for compliance are given in the 1980–1981 revision of the standard (80).

Floor coverings. Federal standards require that carpeting sold to the consumer must pass the methanamine pill test. A small tablet of this material is placed on the surface pile and ignited and the area of char is evaluated. The method is described in ASTM D 2859 and the Federal standard is described in CFR-1630 FF1-70.

The flammability of contract carpeting, ie, floor coverings used in public buildings and multiple-dwelling residences, is regulated by Federal, state, and local building codes. These codes should be consulted for reference to specific test and standard criteria for resistance to ignition and flame spread. Perhaps the most reliable test in terms of relating results to actual fire hazard is the radiant panel test, by which the heat flux required to sustain burning is measured. This method is described in Government Services Administration Test No. 372 and in ASTM E 648.

Subjective Tests. ***Fabric Hand or Handle.*** Stiffness can be measured by such means as the deflection of a cantilever beam, hardness by measurement of compressional resilience, and roughness by the coefficient of friction. Flexibility can be measured by a cantilever bending length test. A specimen 15 × 2.5 cm is extended lengthwise over an edge, the force of gravity causing it to deflect from the horizontal. From the relationship between the deflected length and accompanying angle, the bending length can be calculated; the stiffer the material, the greater the bending length. The details of this test method are described in ASTM D 1388.

Flexural rigidity is a measure of the resistance to bending as by the fingers. Two fabrics may bend to the same extent under their own weight, but the heavier of the two exerts more resistance to bending and so feels stiffer. Flexural rigidity G is related to bending length c by the equation $G = wc^3$, where w is the fabric weight per unit area.

Other characteristics of fabric hand are thickness, hardness, bending modulus (related to flexural rigidity by thickness), compression modulus, density, and extensibility. Certain test instruments are generally used to measure various stiffness properties: the Gurley, the Clark, and the Olson stiffness testers and the Schiefer flexometer (8–9).

Drape. Whereas hand implies tactile criteria, drape normally denotes a visual reaction to the fabric's appearance as produced, for example, by its tendency to fall into pleasing and graceful folds. However, fabric drape depends to a large degree on the same properties that influence fabric hand, eg, stiffness, flexural rigidity, thickness, and compressibility. Although stiffness is normally measured in a single direction, drape implies bending in all directions.

Two instruments have been developed to measure fabric drape, the MIT Drape-o-meter and the Fabric Research Laboratories' drapemeter (81–82). The MIT Drape-o-meter measures the rate of change in curvature radius occurring when a fabric is bent about a cylinder. With the FRL drapemeter, a horizontal circular specimen, eg, 25 cm dia, is placed concentrically over a supporting disk, eg, 12.7 cm dia. The fabric overlap is deflected downward and assumes a draped configuration, depending upon its multidirectional stiffness and flexibility. Thus, various areas of the specimen are bent in three varying directions relative to themselves. The projected area as a percentage of the theoretical maximum area is a measure of the degree of drape.

Preliminary research upon the subjects of handle and drape shows that people are more capable of tactile ranking of fabrics for stiffness, flexural rigidity, and hand than they are of visually ranking them for drape (83). Despite the objective tests described above, the information obtained does not always correlate with the subjective hand test of the consumer/customer. A more reliable indication of the esthetic acceptability of fabrics to consumer tastes is achieved through controlled comparative tests performed by those skilled in recognizing the desirable hand traits of a wide range of fabrics. A discussion of the procedure for human panel testing and a method for analyzing the results are given in ref. 54 for wool and wool blends. This technique is readily adaptable to a wide range of fabric types containing various fiber compositions. Reference 84 includes a revised theory of the quantitative analysis of hand.

Crease Retention, Wrinkle Resistance, and Durable Press. On bending or creasing of a textile material, the external portion of each filament in a yarn is placed in tension and the internal portion is placed in compression. Thus, the wrinkle-recovery properties must be governed in part by the inherent, tensional elastic deformation, and recovery properties of the fibers which make up the yarns and fabrics. In addition to the inherent fiber properties, the yarn and fabric states of aggregation, ie, yarn and fabric geometry, must also be considered.

Descriptions of various laboratory test methods for measuring wrinkle resistance and crease retention are given in refs. 8 and 85. These range from qualitative random wrinkling by a clenched fist squeeze followed by relaxation to quantitative measurement of recovery angle after a specimen is creased in a prescribed, reproducible manner. The most popular of the latter methods was originally the Tootal-Broadhurst-Lee (TBL) test. This consists of folding a 4 × 1 cm sample in the middle under a prescribed pressure for a given length of time and then allowing it to recover while hanging over a pin. The recovery angle is then measured as a criterion of wrinkle resistance. The basic TBL method has been mechanized (AATCC 66-1978). Because of the relative simplicity of this tester, it has had fairly wide adoption as a standard test instrument (2). Generally, those materials exhibiting recovery angles greater than 135° are considered to exhibit good wrinkle-resistance performance under service conditions. This does not mean that every fabric that tests over 135° automatically gives good wrinkle resistance in service. Since wrinkle resistance is particularly important for summer clothing, the AATCC stipulates that crease-recovery angles be measured at 90% rh, as well as at 65% rh; the former is considered to represent the most humid condition to which apparel textiles may be subjected in summer.

An extension of this principle has led to the development of durable-press or permanent-press clothing. In one process, the finishing plant treats the fabric with uncured resin. The garment maker presses and then bakes the garment at ca 160°C. This cures the resin and sets the garment in the desired configuration and is referred to as the postcure process. In the precure process, the fabric is baked prior to garment manufacture. Although the latter process does not result in permanent pleats, the smoothness of the cured fabric is retained during wear and laundering. The resulting garments have durable permanent pleats, creases, and flat areas that, ideally, are insensitive to creases and wrinkles under both wet and dry conditions, so that such garments can be wet-laundered without wrinkling and therefore need not be ironed. Durable-press-type goods are therefore tested to determine both dry and wet crease-recovery angles, both of which should be above 135° to render proper performance.

A second wrinkling test (AATCC 128-1980) consists of preparing a 15 × 28 cm cylindrical fabric specimen, twisting it to induce wrinkles, allowing it to recover, and subjectively ranking its smoothness. Visual ratings determined by comparison with a standard are assigned. The advantage of this method is that it may be used for knitted and woven fabrics. The effects on wrinkle resistance of fiber staple length, diameter and shape; of yarn construction, twist and diameter; and of fabric weave, weight, and thickness, are discussed in ref. 85.

Appearance After Laundering. The extent to which permanent-press fabrics retain their smoothness after laundering is determined by conducting a series of tests devised by AATCC: AATCC 143-1975, Appearance of Apparel and Other Textile End Products After Repeated Home Laundering; AATCC 124-1978, Appearance of Durable Press Fabrics After Repeated Home Laundering; and AATCC 88C and 88B, Appearance of Wash-and-Wear Items: Creases After Home Laundering, Seams After Home Laundering.

In each of these tests, the specimens are subjected to standard procedures simulating home-laundering conditions. A choice is provided for alternative washing temperatures and drying procedures to conform with care instructions recommended for the fabric or final garment. Evaluation is performed by visual comparison with standards prepared by AATCC in prescribed lighting and five grades from very poor to excellent are given.

Luster. Luster depends upon inherent fiber, yarn, and fabric geometry properties so far as they control the degree of order or regularity of surface from which light can be reflected. Surface irregularities tend to diminish the degree of luster. The greater the degree of parallelization of fibers in yarns and of yarns in fabrics, the greater the amount of luster. Similarly, in fabrics, the longer the float (number of filling yarns, for example, that a perpendicular warp yarn crosses over before actually intersecting the filling yarn) and the smaller the number of warp and filling yarn intersections, the greater the degree of luster; this is why satins are so shiny. Also fiber cross-sectional shapes affect luster.

Six methods for measuring luster are described in ref. 86, but only three are applicable to textiles (87). These are determination of the amount of light which a surface can reflect specularly (light falls on the test sample from some chosen angle, usually between 45° and 80° to the normal, and the intensity of reflected light is measured from the corresponding specular angle); measuring subjectively the contrast between specular and diffuse reflection on the premise that such contrasts contribute most effectively to the visual impression of luster; and goniophotometric methods, the fundamental physical methods for studying the lustrous or light-reflecting properties of surfaces (the intensity of reflected light is observed at all angles in the plane containing the angle of incidence) and a curve or group of curves is obtained, which describe the light-reflecting properties of the material.

Each of the three general methods listed gives some interpretation of the effect obtained if color filters are used. There is no assurance, however, that the instruments thus modified are affected by color in the same manner as human vision. Moreover, none of the methods adequately interprets the texture and binocular effects to which the eye is so sensitive. Therefore, it seems likely that visual observation will continue to be necessary for the final appraisal of textile luster.

Comfort. The evaluation of fabric or garment comfort is as subjective and controversial as any textile property. A partial list of the variables that influence comfort includes warmth, coolness, insulating efficiency, moisture absorption, drying speed, softness, bulk, fabric construction, air permeability, skin-surface contact, weight, etc. However, some of these factors are in opposition one to another, eg, warmth and coolness, moisture absorptivity and speed of drying. No controlled laboratory test methods are available for the measurement of comfort. Most research into this subject entails clothing humans with garments made from various fabrics to be evaluated and subjecting them to carefully programmed physical activities in chambers of varying ambient conditions. Data is collected from changes in pulse rate, extent of perspiration, body temperature, and subjective evaluations.

With the advent of textured filament yarns, which allow moisture wicking and less skin contact, the comfort factors of synthetics are markedly improved. This improvement led workers to demonstrate the lack of importance of moisture absorption on comfort in some cases and its detrimental effects in others (88). More important are constructional factors and fabric softness. This is not to say that synthetics are generally more comfortable than natural fibers. An excellent analysis of the work performed by many researchers in the field of comfort is given in ref. 89.

BIBLIOGRAPHY

"Textile Testing" in *ECT* 1st ed., Vol. 13, pp. 908–927, by E. R. Kaswell, Fabric Research Laboratories, Inc.; "Textile Testing" in *ECT* 2nd ed., Vol. 20, pp. 33–62, by E. R. Kaswell, G. A. M. Butterworth, and N. J. Abbott, Fabric Research Laboratories, Inc.

1. *Book of ASTM Standards*, Parts 32 and 33, ASTM, Philadelphia, Pa., 1980.
2. *AATCC Technical Manual*, American Association of Textile Chemists and Colorists, Durham, N.C.
3. E. R. Schwarz in H. R. Mauersburger, ed., J. M. Matthew's *Textile Fibers*, 5th ed., John Wiley & Sons, Inc., New York, 1947.
4. J. L. Fowler and K. L. Hertel, *J. Appl. Phys.* **11,** 497 (1940).
5. K. Gilhaus, *Luenenschloss, Int. Text Bull. Spinning* (2), 117 (1980).
6. J. C. Abbott, L. B. Jaycox, and G. M. Ault, *TAPPI*, 119 (1979).
7. P. N. Bobrov, S. T. Volkova, and V. A. Kozubenko, *Tekst. Promst.* (*Moscow*) **41**(5), 74 (1981).
8. E. R. Kaswell, *Textile Fibers, Yarns, and Fabrics*, Reinhold Publishing Corp., New York, 1953.
9. E. R. Kaswell, *Wellington Sears Handbook of Industrial Textiles*, Pepperell Co., Inc., West Point, N.Y., 1963.
10. M. A. Sieminski, *Rayon Textile Monthly* **24,** 585 (1943).
11. *J. Text. Inst.* **42,** 525 (1951).
12. M. A. Sieminski, R. W. Singleton, and B. S. Sprague, *Text. Res. J.* **31**(11), (Nov. 1961).
13. H. DeW. Smith, *Am. Soc. Test. Mater. Proc.* **44,** 543 (1944).
14. R. W. Singleton in H. F. Mark, S. M. Atlas, and E. Cernia, eds., *Man-made Fibers—Science and Technology*, Vol. 3, Interscience Publishers, a division of John Wiley & Sons, Inc., New York, 1968.
15. J. M. Preston, *Modern Textile Microscopy*, Emmot & Co., Ltd., London, 1933.
16. J. M. Preston, *J. Text. Inst.* **41,** 679 (1950).
17. C. C. Bunn in J. M. Preston and co-workers, eds., *Fibre Science*, British Textile Institute, Manchester, 1949.
18. N. H. Hartshorne and A. Stuart, *Crystals and the Polarising Microscope*, 3rd ed., Edward Arnold, Ltd., London, 1960.
19. D. R. Gentry, *Textile World*, 131 (May 1981).
20. L. Beste and R. Hoffman, *Text. Res. J.* **10,** 441 (1950).
21. R. Meredith in R. Meredith and J. W. S. Hearle, eds., *Physical Methods of Investigating Textiles*, Interscience Publishers, New York, 1956.
22. W. E. Morton and W. S. Hearle, *Physical Properties of Textile Fibers*, The Textile Institute, Butterworths, Manchester-London, 1961.
23. P. F. Dismore and W. Statton, *J. Polym. Sci. Polym. Symp.*, 133 (1966).
24. J. C. Smith and co-workers, *Text. Res. J.* **32,** 721 (1961).
25. D. R. Petterson, G. M. Stewart, F. A. Odell, and R. C. Maheux, *Text. Res. J.* **30,** 411 (1960).
26. H. Morgan, R. Coskren, R. Singleton, and B. S. Spague, *paper presented at Fiber Society Meeting*, West Point, N.Y., 1960.
27. S. Backer and J. G. Krizic, *High-Speed Testing*, Vol. 2, Interscience Publishers, a division of John Wiley & Sons, Inc., New York, 1961.
28. R. J. Coskren, H. M. Morgan, and C. C. Chu, *High Speed Testing*, Vol. 3, Interscience Publishers, a division of John Wiley & Sons, Inc., New York, 1962, pp. 91–104.
29. S. Backer, *Text. Res. J.* **21,** 453 (1951).
30. G. A. M. Butterworth and N. J. Abbott, *ASTM J. Materials* **2,** 487 (1967).
31. S. F. Calil, B. C. Goswami, and J. W. S. Hearle, *J. Phys.* **13,** 725 (1980).
32. M. J. Coplan, *The Effect of Temperature on Textile Materials*, Document 53-21, Wrighton Development Center, Dayton, Ohio, 1953.
33. F. Maillard, *J. Text. Inst.* **40,** 379 (1949).
34. H. Shiefer, L. Fourt, and R. Kropf, *Text. Res. J.* **18,** 18 (1948).
35. P. W. Carlene, *J. Soc. Dyers Colour.* **60,** 232 (1944).
36. M. Harris, *Handbook of Textile Fibers*, Textile Book Publishers, Inc., New York, 1954.
37. *Textile World*, McGraw-Hill Book Co., Inc., New York, 1962.
38. *Textile Test Methods*, Federal Test Methods Standard No. 191, U.S. Government, Dec. 14, 1968.
39. R. Meredith in J. M. Preston, ed., *Fibre Science*, British Textile Institute, Manchester, UK, 1949.
40. I. H. Welch, *Am. Dyest. Rep.* **50,** 25 (1961).
41. J. M. Preston and M. V. Nimkar, *J. Text. Inst.* **40,** 674 (1949).
42. R. W. Singleton and H. D. Noether, *J. Polym. Sci. Polym. Symp.* (19), 65 (1967).

43. R. W. Singleton and co-workers, *J. Polym. Sci. Polym. Symp.* (9), 133 (1969).
44. W. Newell, *Text. World*, (Sept. 1951).
45. L. Ray, *Text. Res. J.* **22,** 144 (1952).
46. W. G. Wolfgang in J. J. Press, ed., *Man-Made Textile Encyclopedia*, Textile Book Publishers, Inc., a division of Interscience Publishers, Inc., New York, 1959.
47. R. Brown and co-workers, *Ind. Eng. Chem.* **40,** 1773 (1948).
48. J. M. Preston, *J. Text. Inst.* **40,** T767 (1949).
49. R. Kunkel, R. W. Singleton, and P. Cook, *Text. Res. J.* **35** (1965).
50. R. W. Singleton and P. Cook, *Text. Res. J.* **39** (1969).
51. O. Kovalcik and R. Simo, *Makrotest*, 175 (1980).
52. H. J. Vogt and R. Punthe, *Textilbetrieb (Wuerzberg, Ger.)*, (6), 39 (June 1981).
53. P. R. Lord, *Text. Res. J.*, 778 (Sept. 1971).
54. R. W. Singleton, *Textile Res. J.* (Aug. 1980).
55. J. S. Graham, *U.S. Department of Agriculture Journal Vol. U8026*, U.S. Department of Agriculture, Washington, D.C., April 30, 1979.
56. R. Furter, *Fiber Prod.* **8**(2), 46 (April 1980).
57. A. A. Gryadoskii, *Tekhnologiya Tekstil Noi Promysklennosti* **142**(4), 15 (1981).
58. R. W. Singleton and co-workers, *Text. Res. J.* **38,** (1968).
59. S. V. Rogina, *Tekst. Promst. (Moscow)* **41**(7), 61 (1981).
60. S. Backer and S. Tannenhaus, *Text. Res. J.* **2,** 635 (1951).
61. E. V. Painter, *Text. Res. J.* **22,** 153 (1952).
62. F. T. Peirce, *J. Text. Inst.* **28,** T45 (1937).
63. M. M. Platt, W. J. Hamburger, and co-workers, *Textile Series Reports*, Office of the Quartermaster General, Washington, D.C.
64. J. Pollitt, *J. Text. Inst.* **40,** 11 (1949).
65. E. R. Schwarz, *Textiles and the Microscope*, McGraw-Hill Book Co., Inc., New York, 1934.
66. V. V. Veselov, G. V. Kolotilova, and I. M. Stepkov, *Tekhnologiya Tekstil Noi Promysklennosti* **138**(6), 9 (1980).
67. R. T. Cary, *Text. Res. J.* **51,** 61 (1981).
68. M. E. Whelan and L. E. MacHattie, *Text. Res. J.* **25,** 197 (1955).
69. P. Nordon and J. G. Downes, *J. Text. Inst.* **56,** T8 (Aug. 1965).
70. J. Rowen and D. Gagliaidi, *Am. Dyest. Rep.* **36,** 261 (1944).
71. D. B. Judd and G. Wyszecki, *Color in Business*, 3rd ed., Interscience Publishers, a division of John Wiley & Sons, Inc., New York, 1975.
72. F. W. Billmeyer and M. Soltzman, *Principles of Color Technology*, Interscience Publishers, a division of John Wiley & Sons, Inc., New York, 1966.
73. J. R. P. Monkman, *Insight* **2**(10), 2 (June 1981).
74. H. R. Schmid and W. Krucker, *Textilveredlung* **16**(9), 375 (1981).
75. H. J. Jacobasah, *Tenside Deterg.* **17**(3), 113 (1980).
76. Flammable Fabrics Act, Public Law 90-189, 90 Congress S1003, U.S. Government, Washington, D.C., Dec. 14, 1967.
77. L. Stone and I. Block, *Text. Chem. Color.* **12,** 302 (1980).
78. *CFR 16-15, Standard for the Flammability of Children's Sleepwear Sizes 0–6X* (FF3-74), available from U.S. Government Printing Office, Washington, D.C.
79. Ref. 78, *sizes 7–14* (FF5-74).
80. Ref. 78, revised as of 1980(81).
81. E. R. Schwartz, *Text. Res. J.* **9,** 216 (1939).
82. C. Chu, C. Cummings, and N. Teixeira, *Text. Res. J.* **10,** 539 (1950).
83. W. J. Hamburger, M. M. Platt, and C. Chu, *Factors Which Influence the Draping Properties of Cotton Fabrics*, Report 96, Fabric Research Laboratories, Inc., Dedham, Mass.
84. V. L. Alley, *J. Eng. Ind. Trans. ASME* **102**(1), 25 (Feb. 1980).
85. G. Buck and F. McCord, *Text. Res. J.* **19,** 216 (1949).
86. S. Harrison, *Measurement of Gloss*, Printing and Allied Trades Research Association, 1945.
87. Ref. 85, p. 715.
88. F. T. Wallenberger, *Text. Res. J.* **50**(5), (May 1980).

89. N. R. S. Hollies and L. Fourt, *Clothing: Comfort and Function (Fiber Science Series)*, Marcel Dekker Inc., New York, Nov. 1970.

General Reference

R. W. Singleton in H. F. Mark, S. M. Atlas, and E. Cernia, eds., *Man-Made Fibers—Science and Technology*, Vol. 3, Interscience Publishers, a division of John Wiley & Sons, Inc., New York, 1968, pp. 571–634.

ROBERT W. SINGLETON
University of Connecticut

THALLIUM AND THALLIUM COMPOUNDS

Thallium Metal

Occurrence. Thallium [*7440-28-0*] was discovered by Sir William Crookes in 1861 while searching for tellurium in residues from sulfuric acid plants (1). It belongs to Group IIIA of the periodic table with boron, aluminum, gallium, and indium. ^{203}Tl [*14280-48-9*] (29.5%) and ^{205}Tl [*14280-49-0*] (70.5%) are the two stable naturally occurring isotopes, whereas $^{191-202}Tl$ [*32148-21-3; 18235-45-5; 32148-22-4; 18235-46-6; 26683-69-2; 18724-77-1; 14107-52-9; 15743-50-7; 15064-66-1; 15720-55-5; 15064-65-0; 15720-57-7*], respectively, ^{204}Tl [*13968-51-9*], and $^{206-210}Tl$ [*15035-09-3; 14133-67-6; 14913-50-9; 15690-73-0; 13966-01-3*], respectively, are the decomposition products of natural radioactive series.

Thallium is not a particularly rare metal and its abundance in the earth's crust is ca 0.3 ppm. It occurs not only in oxide minerals but also as a chalcophilic element. The metal commonly occurs in potash minerals and in a number of thallium-containing minerals, eg, crookesite [*12414-86-7*], $(Cu,Tl,Ag)_2Se$ (17% Tl), in Sweden; lorandite [*15501-93-6*], $TlAgS_2$ (59% Tl), which is found in Greece; hutchinsonite [*12198-34-4*], $(Tl,Pb)S.(Ag,Cu)_2S.5As_2S_3$, in Switzerland; vrbaite [*12006-31-4*], $TlSbAs_2S_5$ (30% Tl); and avicennite [*12022-82-1*], $7Tl_2O_3.Fe_2O_3$. Of these, crookesite and lorandite are the most important ones.

Properties. Thallium is grayish-white, heavy, and soft. When freshly cut, it has a metallic luster which quickly dulls to a bluish-gray tinge like that of lead. A heavy oxide crust forms on the metal surface when in contact with air for several days. The metal has a close-packed hexagonal lattice below 230°C, at which point it is transformed to a body-centered cubic lattice. At high pressures, thallium transforms to a face-centered cubic form. The triple point between the three phases is at 110°C and 3000 MPa (30 kbar). The physical properties of thallium are summarized in Table 1.

In moist air, thallium slowly oxidizes to thallium(I) oxide [*1314-12-1*]. Steam and air or oxygen react readily with thallium forming thallium(I) hydroxide [*12026-06-1*]. Thallium dissolves only slowly in sulfuric or hydrochloric acid since the resultant salts have low solubilities. It is not soluble in alkaline solutions.

Table 1. Physical Properties of Thallium

Property	Value
atomic weight	204.37
melting point, °C	303
boiling point, °C	1457
density, g/cm^3	11.85
thermal conductance, W/(cm·K)[a]	0.39
specific heat (at 20°C), J/g[b]	0.13
heat of fusion, J/g[b]	21.1
heat of vaporization, J/g[b]	795
atomic volume, W/(C·m)[c]	5.15×10^{30}
first ionization energy, (kJ)/mol[b]	590
electronegativity, Pauling's scale	1.8
covalent radius, nm	0.148
Brinell hardness	2
linear coefficient of expansion	28×10^{-6}
electrical resisvitity, μΩ·cm	18
tensile strength, MPa[d]	9.0
crystal structure	
below 230°C	close-packed hexagonal
above 230°C	body-centered cubic
nuclear spin, ℏ	
^{203}Tl	0.5
^{205}Tl	0.5
nuclear magnetic moment, J/T[e]	
^{203}Tl	1.495×10^{-23}
^{205}Tl	1.509×10^{-23}

[a] To convert W/(cm·K) to (cal·cm)/(s·cm^2·°C), divide by 4.184.
[b] To convert J to cal, divide by 4.184.
[c] To convert W/(C·m) to W/D, divide by 2.998×10^{29}.
[d] To convert MPa to psi, multiply by 145.
[e] To convert J/T to Bohr magneton, divide by 9.274×10^{-24}.

Production and Economic Aspects. Thallium is obtained commercially as a by-product in the roasting of zinc blende and lead sulfide ores in the production of sulfuric acid (2–4) (see Sulfuric acid and sulfur trioxide). The thallium is collected in the flue dust in the form of oxide or sulfate with other by-product metals, eg, cadmium, indium, germanium, selenium, and tellurium. The thallium content of the flue dust is low and further enrichment steps are required. If the thallium compounds present are soluble, ie, as oxides or sulfates, direct leaching with water or dilute acid separates them from the other insoluble metals. Otherwise, the thallium compound is solubilized with oxidizing roasts, by sulfatization, or by treatment with alkali. The thallium precipitates from these solutions as thallium(I) chloride [*7791-12-0*]. Electrolysis of the thallium(I) sulfate [*7446-18-6*] solution affords thallium metal in high purity (5–6). The sulfate solution must be acidified with sulfuric acid to avoid cathodic separation of zinc and anodic deposition of thallium(III) oxide [*1314-32-5*]. The metal deposited on the cathode is removed, kneaded into lumps, and dried. It is then compressed into blocks, melted under hydrogen, and cast into sticks.

Demand for thallium has not increased to any substantial extent since the early 1970s, and the annual production is very small. Although there is no information available on the amount of production or consumption, it is estimated to be 5,000–

15,000 kg/yr. The principal domestic producer is Noah Chemical Division, Noah Industrial Corporation, after American Smelting and Refining Company discontinued its operation in February of 1981. Thallium is available in three grades, 99.9%, 99.99%, and 99.999% pure. The 99.9% material is priced at \$60–80/kg (\$17/kg 1969). From February to December of 1981, the supply of thallium was very small and pricing was unstable because of the decrease in production caused by environmental problems associated with the safe disposal of thallium waste.

Uses. Thallium has limited commercial applications because of its toxic nature. Thallium forms alloys readily with many metals and some of these alloys have unique properties. A number of binary, ternary, and quaternary eutectic alloys are known and have very low coefficients of friction and good resistance to acids (7). These alloys can be used in bearings, eg, Ag–Tl [*83542-95-4*] and Au–Al–Tl [*83547-99-3*]. Alloys of silver and thallium are used in contact points, and an alloy of lead, tin, and thallium is used in the production of anodes. The most important alloy of thallium is the mercury–thallium alloy [*83542-96-5*] which forms a eutectic at 8.7 wt % Tl with a melting point of −60°C. It can be used as a substitute for mercury in switches and seals for equipment used in the polar region or the stratosphere.

Thallium Compounds

Unlike boron, aluminum, gallium, and indium, thallium exists in both stable univalent (thallous) and trivalent (thallic) forms. There are numerous thallous compounds and they are usually more stable than the corresponding thallic compounds. The thallium(I) ion resembles the alkali-metal ions and the silver ion in properties. In this respect, it forms a soluble, strongly basic hydroxide and a soluble carbonate, oxide, and cyanide like the alkali-metal ions. However, like the silver ion, it forms a very soluble fluoride, but the other halides are insoluble. Thallium(III) ion resembles aluminum, gallium, and indium ions in properties.

Thallic salts are readily reduced to the thallous salts by common reducing agents. The standard potentials for the Tl(III)/Tl(I) and Tl(I)/Tl systems in aqueous solutions have been measured (8):

$$Tl^+ \rightarrow Tl^{3+} + 2\,e \qquad E° = -1.252 - 0.0295 \log \frac{[Tl^{3+}]}{[Tl^+]}$$

$$Tl \rightarrow Tl^+ + e \qquad E° = +0.336 - 0.0591 \log\,[Tl^+]$$

Thallium(I) salts are oxidized only by very powerful oxidizing agents, eg, MnO_4^- or Cl_2, in acid solutions. The properties of the more important thallium compounds are listed in Table 2.

Thallous sulfate, thallous nitrate, and thallous and thallic oxide are the main compounds produced in bulk quantities by Noah Chemical, and the November, 1982, price was \$85–185/kg. Consumption figures are not available. Approximately 20 other thallium compounds are also available commercially from Noah Chemical, Cooper Chemical, and Alfa Products, Ventron Division, Thiokol Corp., in research and production quantities. However, demand for thallium compounds is small and is limited to such applications as synthetic or analytical reagents. The prices will probably remain stable and supply is expected to be ample.

Table 2. Properties of Thallium Compounds

Compound	Formula	CAS Registry No.	Color and crystal form	Mp, °C	Density, g/cm^3	Solubility, g/100 g, solvent (°C)	
						Water	Others
thallous carbonate	Tl_2CO_3	*[29809-42-5]*	colorless, monoclinic	272	7.16	5.2 (18), 22.4 (100)	
thallous formate	TlO_2CH	*[992-98-3]*	colorless, needles	101	4.967	500 (10)	methanol
thallous acetate	TlO_2CCH_3	*[563-68-8]*	silky white, needles	131	3.765	very soluble	alcohols
thallous fluoride	TlF	*[7789-27-7]*	colorless, cubic	327	8.23	78.6 (15)	alcohols, HF
thallic fluoride	TlF_3	*[7783-57-5]*	olive green, orthorhombic	550 dec	8.65	insol	
thallous chloride	TlCl	*[7791-12-0]*	colorless, cubic	430	7.0	0.32 (20), 2.38 (100)	
thallic chloride	$TlCl_3$	*[13453-32-2]*	hexagonal plate	155		very soluble	alcohols, ether
thallous bromide	TlBr	*[7789-40-4]*	green–yellow, cubic	456	7.5	0.05 (20), 0.25 (60)	
thallic bromide	$TlBr_3$	*[13701-90-1]*	yellow			sol	alcohols
thallous iodide	TlI	*[7790-30-9]*	yellow, rhombic	440	7.29	0.0006 (20)	
thallous oxide	Tl_2O	*[1314-12-1]*	black	300	9.52	dec	
thallic oxide	Tl_2O_3	*[1314-32-5]*	black	7.7	10.11	insol	
thallous hydroxide	TlOH	*[12026-06-1]*	yellow, needles	139 dec	7.44	25.9 (0), 52 (40)	alcohols
thallous nitrate	$TlNO_3$	*[10102-45-1]*	colorless	206	5.55	8 (15), 594 (104.5)	
thallic nitrate	$Tl(NO_3)_3$	*[13746-98-0]*	colorless			dec	
thallous sulfate	Tl_2SO_4	*[7446-18-6]*	colorless, prisms	632	6.77	4.87 (20), 18.45 (100)	

Carbonates. Thallium forms only thallous carbonate and bicarbonate (9). Thallous carbonate, Tl_2CO_3, is prepared by saturating a hot aqueous solution of thallium hydroxide with carbon dioxide. The carbonate separates upon reduction of the volume and cooling the concentrated solution. It dissolves in water yielding a strongly basic solution because of hydrolysis. It is stable to 175°C in air but loses CO_2 on further heating. Thallous bicarbonate [*65975-01-1*], $TlHCO_3$, forms by the action of carbon dioxide on a dilute aqueous solution of Tl_2CO_3.

Carboxylates. Thallium forms a large number of carboxylate derivatives in both Tl(I) and Tl(III) forms. The Tl(I) salts of carboxylic acids can be prepared from the acid and thallium(I) hydroxide or carbonate. Thus, thallous(I) formate, TlOOCH, is prepared by treating Tl_2CO_3 or TlOH with formic acid followed by evaporation of the solution in vacuum (9). Thallium(I) acetate is prepared similarly. Alternatively, it can be obtained by adding acetic acid to a solution of thallium(I) ethoxide [*20398-06-5*] in ethanol (10). These carboxylates are colorless, light-insensitive, hygroscopic, sharp-melting, and stable solids. A 1:1 thallium formate–thallium malonate [*2757-18-8*] saturated aqueous solution, which is called Clerici's solution [*61971-47-9*], has a density of 4.324 g/mL at 20°C, and has been used used as a heavy liquid in sink–float separation of minerals on a laboratory scale (see Flotation; Gravity separation).

Thallium(III) salts of formic, acetic, and trifluoroacetic acids are prepared from the corresponding acid and thallium(III) oxide (11). Other thallium(III) carboxylates can be obtained from the metathesis of thallium(III) acetate and the carboxylic acid (12). They are colorless, hygroscopic solids.

Halides. Thallous halides are air-stable, light-sensitive, anhydrous salts (9). The water-soluble thallous fluoride is made from Tl_2CO_3 and aqueous hydrogen fluoride. The salt is obtained upon concentrating and cooling the solution. Thallous chloride, bromide, and iodide are prepared by precipitation from aqueous solution of Tl_2SO_4 with dilute solutions containing the corresponding anions. The chloride can be purified by recrystallization from hot water, whereas the bromide is purified by digesting with boiling water. Thallium(I) halides form many double salts and complexes with nitrogen, oxygen, and sulfur ligands.

Thallium(III) fluoride has been prepared by the action of fluorine or bromine trifluoride on thallium(III) oxide at 300°C. It is stable to ca 500°C but is extremely sensitive to moisture. Thallium(III) chloride can be obtained readily as the tetrahydrate [*13453-33-3*] by passing chlorine through a boiling suspension of TlCl in water. It can be dehydrated with thionyl chloride. Thallium(III) bromide tetrahydrate [*13453-29-7*] is prepared similarly, whereas the iodide prepared in this manner is thallium(I) triiodide [*13453-37-7*], $Tl^+I_3^-$.

A large number of halide complexes of thallium(III) have been prepared by precipitation of the complexes from solution with a suitable cation, eg, Tl^+, $(C_2H_5)_4N^+$, $(C_6H_5)_4As^+$, and K^+. Both 4-coordinated $[TlX_4]^-$ and 6-coordinated $[TlX_6]^{3-}$ ions exist in solutions and in solid states.

Hydrides. Unlike boron and aluminum, which form many stable and useful hydrides, there is very little evidence for the existence of TlH and TlH_3 (13). Lithium tetrahydridothallate(III) [*82374-47-8*], $LiTlH_4$, has been obtained from $TlCl_3$ and LiH in ether at −15°C (14). It decomposes rapidly above 30°C to hydrogen, thallium, and lithium hydride (9).

Thallium(I) salts of tetrahydridoborate and aluminate are obtained from a Tl(I) compound, eg, ethoxide, perchlorate, or nitrate, and $LiBH_4$ or $LiAlH_4$ in ether. Thallium(I) tetrahydridoborate [*61204-71-5*], $TlBH_4$, is unstable at 40°C evolving diborane. Thallium(I) tetrahydridoaluminate [*82391-12-6*], $TlAlH_4$, is extremely unstable decomposing to H_2 and Tl at −80°C.

Oxides and Hydroxides. Thallous oxide is most readily obtained by dehydration of TlOH in high vacuum at 50°C. It is black, crystalline, and hygroscopic. It reacts with water forming the hydroxide and dissolves in ethanol yielding the ethoxide (9).

Thallic oxide can be prepared by the reaction of thallium with oxygen or hydrogen peroxide and an alkaline thallium(I) solution. However, it is more easily made from the oxidation of thallous nitrate by chlorine in aqueous potassium hydroxide solution. It is insoluble in water but dissolves in carboxylic acids giving carboxylates.

There is no evidence for the existence of thallic hydroxide, and addition of hydroxide to an aqueous solution of a Tl(III) salt gives Tl_2O_3 instead. Thallous hydroxide can be isolated as yellow needles by the hydrolysis of thallous ethoxide [*20398-06-5*], which is conveniently prepared as a heavy oil by the oxygen oxidation of thallium metal in ethanol vapor. Thallous hydroxide darkens at room temperature and decomposes to Tl_2O and H_2O on warming.

Nitrates. Thallous nitrate, a convenient source of other thallium(I) derivatives, eg, halides, is prepared from the reaction of the pure metal with dilute nitric acid (9). The solid is stable to 300°C and decomposes at 800°C to Tl_2O, NO, and NO_2. Thallic nitrate is obtained as a trihydrate upon dissolving Tl_2O_3 in cold nitric acid. It decomposes to Tl_2O_3 on heating to 100°C or upon hydrolysis. Thallic nitrate is soluble in alcohols and ethers of polyethylene glycols and is often used as an oxidizing agent in organic syntheses.

Sulfates. Thallous sulfate is a commercial product which is produced by reaction of the metal with sulfuric acid and concentration of the solution until crystallization begins (9). It reacts with thallic sulfate yielding $Tl^{I}Tl^{III}(SO_4)_2$ [*37475-01-7*], and with SO_3 forming the pyrosulfate, $Tl_2S_2O_7$ [*82391-11-5*]. Thallic sulfate is extremely unstable and therefore cannot be isolated. Reaction of thallic oxide with dilute sulfuric acid results in $HTl(SO_4)_2$ [*15478-75-8*] or $Tl(OH)SO_4$ [*37205-71-3*], depending on the concentration of sulfuric acid used.

Organometallics. Organothallium compounds have attracted a great deal of interest in recent years mostly because of the new applications of these compounds in organic synthesis. Stable compounds occur in both Tl(I) and Tl(III) oxidation states. Only the cyclopentadienyl ligand is known to stabilize the Tl(I) ion, and the parent compound, cyclopentadienylthallium(I) [*34822-90-7*] is prepared from freshly distilled cyclopentadiene and thallium(I) hydroxide (15). It is a yellow, air-stable solid which is insoluble in most common organic solvents. X-ray and microwave spectroscopic studies in the gas phase indicate C_{5v} symmetry consistent with the half-sandwich structure. The crystal has a polymeric structure with zigzag chains of alternating C_5H_5 unit and Tl atoms. A number of alkylated derivatives of cyclopentadienylthallium(I), $(RC_5H_4)Tl$ where R = CH_3 [*34034-67-8*], C_2H_5 [*28553-52-8*], and C_4H_9 [*28553-53-9*], can also be prepared by the same procedure.

Organothallium(III) derivatives can be classified into three types: R_3Tl, R_2TlX, and $RTlX_2$. The trialkyl derivatives are reactive, unstable compounds, whereas the dialkyl derivatives are among the most stable and least reactive organometallic compounds. The monoalkyl compounds are rather unstable and often cannot be isolated. They are important intermediates in some Tl(III)-promoted organic reactions.

R_3Tl compounds, eg, methyl [*3003-15-4*], ethyl [*687-82-1*], isobutyl [*3016-08-8*], and phenyl [*3003-04-1*] thallium(III), are usually prepared by the reaction between a dialkyl or diarylthallium halide and an organolithium reagent in ether (16):

$$R_2TlX + LiR \rightarrow R_3Tl + LiX$$

or a Grignard reagent in tetrahydrofuran (17):

$$R_2TlX + RMgX \rightarrow R_3Tl + MgX_2$$

However, the most convenient method of preparation is from the reaction of thallium(I) iodide and organolithium, as illustrated below:

$$TlI + 2\,CH_3Li + CH_3I \rightarrow (CH_3)_3Tl + 2\,LiI$$

The intermediate involved in this reaction is thought to be CH_3Tl.

R_3Tl compounds react readily with acids, halocarbons, or sulfur dioxide forming R_2TlX. They also form neutral complexes ($R_3Tl.L$) with Lewis bases (L), eg, amines and phosphines, in a similar manner as the gallium and indium analogues.

R_2TlX compounds are best prepared from thallium(III) halides and an excess of Grignard reagent (18)

$$TlX_3 + 2\,RMgX \rightarrow R_2TlX + 2\,MgX_2$$

However, the yields are usually poor because of the oxidation of the Grignard reagent by thallium(III). Reaction of R_3Tl with acids also affords a convenient route to this type of compound.

Dialkyl and diarylthallium(III) derivatives are stable, crystalline solids which melt at 180–300°C. The dimethylthallium derivatives of CN^-, ClO_4^-, BF_4^-, and NO_3^- contain linear $(CH_3)_2Tl$ cations and the free anions (19). In aqueous solutions, they ionize to the $(CH_3)_2Tl(H_2O)_x^+$ ions, except those derivatives containing alkoxide, mercaptide, or amide anions which yield dimeric structures (20–21).

$RTlX_2$ compounds can be obtained from the reaction of thallic halide or carboxylate with an organoboron compound (22):

$$TlCl_3 + C_6H_5B(OH)_2 + H_2O \rightarrow \underset{[19628\text{-}33\text{-}2]}{C_6H_5TlCl_2} + B(OH)_3 + HCl$$

or with an organomercury compound (23).

$$Tl(O\overset{\overset{O}{\|}}{C}R)_3 + (C_6H_5)_2Hg \rightarrow C_6H_5Tl(O\overset{\overset{O}{\|}}{C}R)_2 + C_6H_5HgO\overset{\overset{O}{\|}}{C}R$$

They can also be prepared by thallation (24)

$$C_6H_6 + \underset{[23586\text{-}53\text{-}0]}{Tl(O\overset{\overset{O}{\|}}{C}CF_3)_3} \rightarrow \underset{[23586\text{-}54\text{-}1]}{C_6H_5Tl(OOCCF_3)_2} + CF_3\overset{\overset{O}{\|}}{C}OH$$

or oxythallation (25)

$$C_6H_5CH{=}CH_2 + \underset{[2570\text{-}63\text{-}0]}{Tl(O\overset{\overset{O}{\|}}{C}CH_3)_3} + CH_3OH \rightarrow \underset{[37011\text{-}27\text{-}1]}{C_6H_5\underset{\underset{OCH_3}{|}}{C}HCH_2Tl(O\overset{\overset{O}{\|}}{C}CH_3)_2}$$

$RTlX_2$ derivatives are covalent compounds and are generally soluble in organic solvents. The aryl and vinyl derivatives are more stable than the corresponding alkyl compounds. This type of compound has been postulated to be an intermediate in many organic synthetic reactions involving thallium(III) species.

Uses. Thallium compounds have limited use in industrial applications. The use of thallous sulfate in rodenticides and insecticides has been replaced by other compounds less harmful to animals (see Insect control technology; Poisons, economic). Thallium sulfide has been used in photoelectric cells (see Photovoltaic cells). A thallium bromide–thallium iodide mixture is used to transmit infrared radiation for signal systems. Thallous oxide is used in the manufacture of glass (qv) that has a high coefficient of refraction. Thallium formate–malonate aqueous solutions (Clerici's solution) have been used in mineral separations. Many thallium compounds have been used as reagents in organic synthesis in research laboratories.

Thallium(I) Compounds. Carboxylic anhydrides can be prepared by the reaction of acyl or aroyl halides in ether with thallium(I) carboxylates (26):

$$R'\overset{O}{\overset{\|}{C}}Cl + TlO\overset{O}{\overset{\|}{C}}R \longrightarrow R'\overset{O}{\overset{\|}{C}}O\overset{O}{\overset{\|}{C}}R + TlCl$$

Thallium(I) carboxylates also react with alkyl iodides to form the esters (27):

$$R'I + TlO\overset{O}{\overset{\|}{C}}R \rightarrow R\overset{O}{\overset{\|}{C}}OR' + TlI$$

Thallium acetate reacts with iodine and an alkene giving the *trans*-iodoacetate derivatives in 90% yield (28):

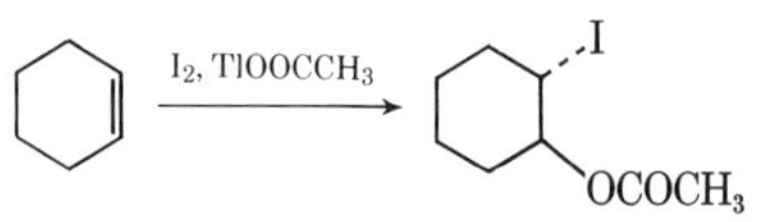

Cyclopentadienylthallium and its alkylated derivatives are used in the synthesis of metallocenes and other transition-metal cyclopentadienyl complexes (29).

Thallium(III) Compounds. Thallium(III) derivatives have been used extensively as oxidants in organic synthesis. In particular, thallic acetate and trifluoroacetate are extremely effective as electrophiles in oxythallation and thallation reactions. For example, ketones can be prepared from terminal acetylenes by means of $Tl(OOCCH_3)_3$ in acetic acid (oxythallation) (30):

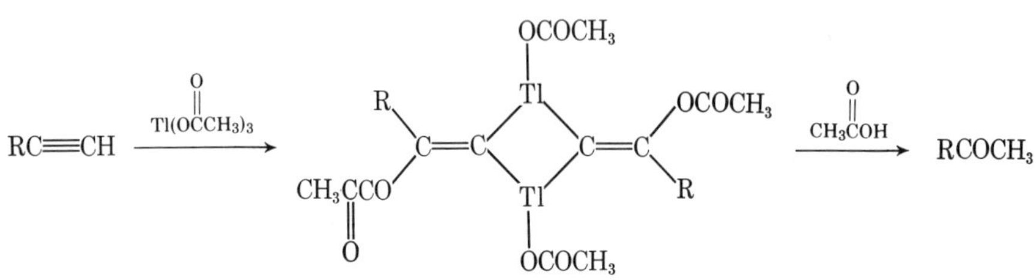

Oxythallation reactions of olefins provide useful synthetic routes to products, eg, glycols, aldehydes, and ketones (31):

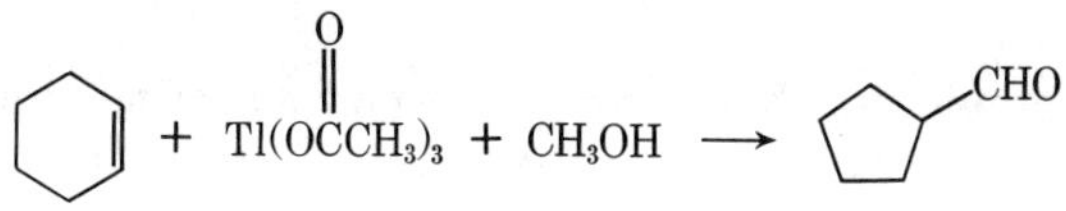

Thallium(III) nitrate in nitric acid and thallium(III) sulfate in sulfuric acid are also effective.

The enhancement of the electrophilic properties of thallium(III) trifluoroacetate makes it a very important thallation reagent. The products of thallation, eg, arylthallium bis(trifluoracetate), undergo a variety of substitution reactions yielding iodides, fluorides, nitriles, thiophenols, phenols, and biaryls.

$$ArH \xrightarrow{Tl(O\overset{O}{\overset{\|}{C}}CF_3)_3} ArTl(O\overset{O}{\overset{\|}{C}}CF_3)_2 \xrightarrow{X^-} ArTlX_2 \rightarrow ArX$$

Thallium(III) trifluoroacetate promotes olefin cyclization reactions and intramolecular coupling reactions (32–33).

The highly ionic thallic nitrate, which is soluble in alcohols, ethers, and carboxylic acids, is also a very useful synthetic reagent. Oxidation of olefins, α,β-unsaturated carbonyl compounds, β-carbonyl sulfides, and α-nitrato ketones can all be conveniently carried out in good yields (31,34–36).

Toxicology

The relative toxicities of thallium compounds depend on their solubilities and valence states. Soluble univalent thallium compounds, eg, thallous sulfate, acetate, and carbonate, are especially toxic. They are rapidly and completely absorbed from the gastrointestinal tract, skin peritoneal cavity, and sites of subcutaneous and intramuscular injection. Thallium is also rapidly absorbed from the mucous membranes of the respiratory tract, mouth, and lungs following inhalation of soluble thallium salts. Insoluble compounds, eg, thallous sulfide and iodide, are poorly absorbed by any route and are less toxic.

Thallium does not occur naturally in normal tissue and it is not essential to mammals but it does accumulate in the human body. Levels as low as 0.5 mg/100 g of tissue suggest thallium intoxication. Based on industrial experience, 0.10 mg/m^3 of thallium in air is considered safe for a 40-h work week (37). The lethal dose for man is not definitely known, but 1 g of absorbed thallium is considered sufficient to kill an adult and 10 mg/kg body weight has been fatal to children. In severe cases of poisoning, death does not occur earlier than 8–10 d but most frequently in 10–12 d. Thallium excretion is slow and prolonged. For example, thallium is present in the feces 35 d after exposure and persists in the urine for up to 3 mo.

The usual symptoms in human thallotoxicosis resulting from acute, subacute, or chronic intoxication are generally the same. Common symptoms include nausea, vomiting, abdominal colic, pain in legs, nervousness and irritability, chest pain, gingivitis or stomatitis, and anorexia. Alopecia (hair loss) does not always occur, especially in cases of mass intake of thallium and low resistance where the patient may die before the occurrence of hair loss. Thallium intoxication during the first trimester of pregnancy can cause skeletal deformities, alopecia, rash, low birth weight, and premature birth (38).

A variety of therapies for thallium poisoning have been suggested by neutralizing thallium in the intestinal tract, hastening excretion after resorption, or decreasing absorption. Recent reports recommended Berlin-Blue (ferrihexacyanate) and sodium iodide in a 1 wt % solution. Forced diuresis hemoperfusion and hemodialysis in combination results in the elimination of up to 40% of the resorbed thallous sulfate (39).

BIBLIOGRAPHY

"Thallium and Thallium Compounds" in *ECT* 1st ed., Vol. 13, pp. 927–932, by K. M. Herstein, Herstein Laboratories, Inc.; "Thallium and Thallium Compounds" in *ECT* 2nd ed., Vol. 20, pp. 63–69, by H. E. Howe, American Smelting and Refining Co.

1. W. Crookes, *Chem. News* **3,** 193, 303 (1861).
2. U.S. Pat. 2,060,453 (Nov. 10, 1936); and 2,011,882 (Aug. 20, 1935), R. Teats (to American Smelting and Refining Co.).
3. R. Kleinert, *Z. Erzbergbau Metallhuettenwes.* **16,** 67 (1963).
4. J. D. Prater, D. Schlain, and S. F. Ravitz, *Recovery of Thallium from Smelter Products*, Rep. Inv. 4900, U.S. Bureau of Mines, Washington, D.C., 1963.
5. T. W. Richards and C. P. Smyth, *J. Am. Chem. Soc.* **44,** 525 (1922).
6. T. W. Richards and J. D. White, *J. Am. Chem. Soc.* **50,** 3290 (1928).
7. M. A. Filyand and E. I. Semenova, *Handbook of the Rare Elements*, MacDonald, London, 1969.
8. W. M. Latimer, *The Oxidation States of the Elements and Their Potentials in Aqueous Solutions*, 2nd ed., Prentice-Hall, Englewood Cliffs, N.J., 1952.
9. G. Brauer, *Handbook of Preparative Inorganic Chemistry*, 2nd ed., Academic Press, New York, 1963.
10. C. M. Fear and R. C. Menzies, *J. Chem. Soc.* **129,** 937 (1926).
11. A. McKillop, J. D. Hunt, M. J. Zelsko, J. S. Fowler, E. C. Taylor, G. McGillivray, and F. Kienzle, *J. Am. Chem. Soc.* **93,** 4841 (1971).
12. J. K. Kochi and T. W. Bethea, *J. Org. Chem.* **33,** 75 (1968).
13. E. Wiberg, O. Dittmann, H. Noth, and M. Schmidt, *Z. Naturforsch. Teil B* **12,** 61, 62 (1957).
14. E. Wiberg, O. Dittmann, and M. Schmidt, *Z. Naturforsch. Teil B* **12,** 60 (1957).
15. H. Meister, *Angew. Chem.* **69,** 533 (1957).
16. H. Gilman and R. G. Jones, *J. Am. Chem. Soc.* **62,** 2357 (1940).
17. O. Yu, Okhlobystin, K. A. Bilevitch, and L. I. Zakharkin, *J. Organomet. Chem.* **2,** 281 (1964).
18. H. Kurosawa and R. Okawara, *Organomet. Chem. Rev.* **2,** 255 (1967).
19. G. B. Deacon, J. H. S. Green, and R. S. Nyholm, *J. Chem. Soc.*, 3411 (1965).
20. M. G. Miles, J. H. Patterson, C. W. Hobbs, M. J. Hopper, J. Overbend, and R. S. Tobias, *Inorg. Chem.* **7,** 1721 (1968).
21. R. C. Menzies and A. R. P. Walker, *J. Chem. Soc.*, 1678 (1936).
22. F. Challenger and O. V. Richards, *J. Chem. Soc.*, 405 (1934).
23. H. J. Kabbe, *Ann. Chem.* **656,** 204 (1962).
24. A. McKillop, J. D. Hunt, and E. C. Taylor, *J. Organomet. Chem.* **24,** 77 (1970).
25. A. McKillop and E. C. Taylor, *Chem. Br.* **9,** 4 (1973).
26. E. C. Taylor, G. W. McLay, and A. McKillop, *J. Am. Chem. Soc.* **90,** 2422 (1968).
27. R. C. Menzies, *J. Chem. Soc.*, 1378 (1947).
28. C. C. Hunt and J. R. Doyle, *Inorg. Nucl. Chem. Lett.* **2,** 283 (1966).
29. R. C. Cambie, B. G. Lindsay, P. S. Rutledge and P. D. Woodgate, *Chem. Commun.*, 919 (1978).
30. S. Uemura, H. Miyoshi, H. Tara, M. Okano, and K. Ichikawa, *Chem. Commun.*, 218 (1976).
31. A. McKillop, J. D. Hunt, E. C. Taylor, and F. Kienzle, *Tetrahedron Lett.*, 5275 (1970).
32. Y. Yamada, S. Nakamura, K. Iguchi, and K. Hosaka, *Tetrahedron Lett.* **22,** 1355 (1981).

33. E. C. Taylor, J. G. Andrade, and A. McKillop, *Chem. Commun.*, 538 (1977).
34. A. McKillop, B. P. Swann, and E. C. Taylor, *Tetrahedron Lett.*, 5281 (1970); A. McKillop, B. P. Swann, and E. C. Taylor, *Tetrahedron* **26,** 4031 (1970).
35. Y. Nagao, K. Kaneko, and E. Fujita, *Tetrahedron Lett.*, 4115 (1978).
36. A. McKillop, D. W. Young, M. Edwards, R. P. Hug, and E. C. Taylor, *J. Org. Chem.* **43,** 3773 (1978).
37. A. Hamilton and H. L. Hardy, *Industrial Toxicology*, 3rd ed., Publishing Science Group Inc., 1974.
38. W. J. Stevens and co-workers, *Acta Clin. Belg.* **31,** 188 (1976).
39. S. Moeschlin, *Clin. Toxicol.* **17,** 133 (1980).

General References

N. V. Sidgwick, *The Chemical Elements and Their Compounds*, Vol. I, Oxford University Press, Ltd., London, 1950.

A. G. Lee, *The Chemistry of Thallium*, Elsevier Publishing Co., Amsterdam, 1971.

K. Wade and A. J. Banister in J. C. Bailar, Jr., and co-workers, eds., *Comprehensive Inorganic Chemistry*, Vol. 1, Pergamon Press, Oxford, 1973.

F. A. Cotton and G. Wilkinson, *Advanced Inorganic Chemistry, A Comprehensive Text*, 4th ed., John Wiley & Sons, Inc., New York, 1980.

A. N. Nesmeyanov and R. A. Sokolik, *The Organic Compounds of Boron, Aluminum, Gallium, Indium and Thallium*, North-Holland Publishing Co., Amsterdam, 1967.

A. McKillop and E. C. Taylor in F. G. A. Stone and R. West, eds., *Advances in Organometallic Chemistry*, Vol. 2, Academic Press, New York, 1973, p. 147.

B. Venugopal and T. D. Luckey, *Metal Toxicity in Mammals, Vol. 2, Toxicity of Metals and Metalloids*, Plenum Press, New York, 1978.

J. F. Reith in J. J. G. Prick, W. G. S. Smith, and L. Muller, eds., *Thallium Poisoning*, Elsevier Publishing Co., Amsterdam, 1955.

Charles Pfizer and Co., Inc., *Thallium Poisoning, Spectrum (N.Y.)* **6,** 558 (1958).

BENJAMIN C. HUI
Alfa Products

THERMAL POLLUTION BY POWER PLANTS

An important by-product of most chemical technologies, in the broad sense, is heat. Few industrial processes are carried out without heat being rejected at some point in the process stream, even if the net energy balance is endothermic. Historically, it was more convenient and less costly to reject such waste heat to the environment than to attempt significant recovery. Although energy scarcities and high costs are stimulating innovative heat-recovery schemes, the low temperatures of waste heat in relation to process requirements often make reuse impractical and disposal the only attractive alternative.

Heat rejected to the environment by most industries is of little consequence. Cooling flows of air or water are deployed over equipment or through heat exchangers and the relatively small quantities of heat are dissipated to the surrounding air. Small cooling towers, often of the evaporative type, are ubiquitous fixtures of an industrial facility.

Concern over heat rejection arose when the quantities at localized sites rose dramatically as the electric utility industry shifted to water-cooled, thermal-electric generating stations of high unit capacity in the 1950s, particularly in the UK and the United States. Concern was further heightened by a planned shift to nuclear fission reactors as the energy source, which entailed both a further increase in localized generating capacity because of economics of scale and a higher percentage of rejected heat compared to useable electrical energy. In the late 1950s and 1960s, the volumes of cooling water projected to be used by the United States electrical supply system and heated to temperatures 10°C or more above intake temperatures were startling. For example, it was estimated that perhaps 25% of the total available freshwater runoff in the entire nation would be needed for cooling purposes by 1980. The term thermal pollution took on fearsome portents among aquatic scientists, fishery managers, and eventually water-pollution control agencies (see Water, water pollution) (1). Directly lethal effects of high temperatures on aquatic life were predicted and, where sublethal temperatures were maintained, effects on reproductive cycles, growth rates, migration patterns, and interspecies competition were hypothesized based on the pervasive nature of temperature as a factor which controls ecological processes (2).

Risks

Power-station cooling is fairly straightforward. Generation of electricity by the steam cycle, the most common method regardless of fuel type, ie, coal, oil, gas, nuclear, solid waste, entails production of waste heat (Fig. 1). Although there are some atmospheric losses in the steam cycle, most of the heat is rejected to flowing water through the heat of condensing steam. For a modern 1200 MW (electrical) light-water nuclear power plant, this release is ca 7100 kJ/(kW·h) (1700 kcal/(kW·h)), ie, approximately twice the thermal equivalent of the generated electric energy. The amount of reject heat is proportionately less in fossil-fueled plants, in which up to 40% of the fuel energy can be converted to electricity. The flowing water for the steam condenser was traditionally pumped from a nearby lake or river and returned to it at an elevated temperature in the once-through or open-cycle system; a 1000-MW nuclear power plant with once-through cooling requires ca 49 m^3 of water per second (ca 13,000 gal/s), which

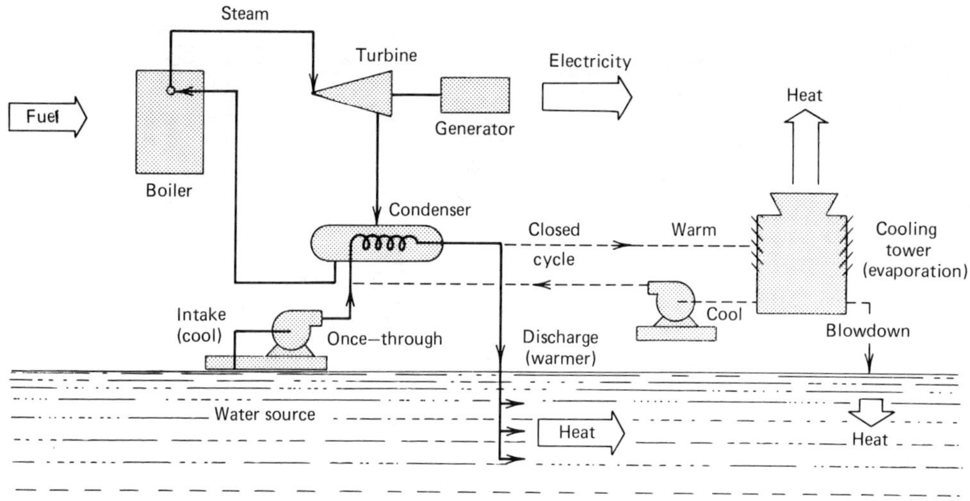

Figure 1. The energy cycle of a thermal electric generating station with two alternative cooling systems: the open-circuit or once-through system (solid lines) and a representative closed cycle, cooling-tower system (dashed lines). Reproduced by permission from ref. 30.

is elevated by 10°C. Screens at the water intake prevent entry of objects larger than ca 1 cm. A biocide, usually chlorine, is used to clean heat-exchange surfaces and water conduits (see Heat-exchange technology).

For purposes of clarifying risks to aquatic life, a clear distinction must be made between heat, which is a quantitative measure of energy that depends upon the mass of an object, in this case the volume of cooling water, and temperature, which is a measure of energy intensity. An amount of heat distributed in two unit volumes of water yields half the elevation of temperature as the same amount of heat distributed in one unit. Although heat is the waste product of electricity generation, temperature is the environmental characteristic to which organisms respond. Thus, the quantity of water used to carry away the load of reject heat is crucial to determining the temperatures created in the environment. A risk is considered here to be the biological or ecological damage that could be done by a certain human alteration of the environment with the likelihood or probability that the specific damaging alteration will occur.

Initially, the source of environmental risk from cooling water was assumed to be the pollutant discharged, ie, heat, in the form of the elevated temperature of the water released from the condensers. Heat is now recognized as only one of several potential risks of power-station cooling (Fig. 2). In this article, the broad view of thermal pollution is taken to encompass each of these ancillary risks.

Generally unheralded until the early 1970s was the physical entrapment and impingement of fish on cooling-water intake screens. Always occurring at chronic low levels at small fossil-fueled plants, the losses of fish and some large invertebrates rose markedly with the startup of new nuclear stations. In particular, plants on the Hudson River began to impinge large numbers of juvenile striped bass, a species of considerable importance locally.

Impingement risks for affected species seem more direct and identifiable than the indirect aspects of temperature change because there are more visible, countable

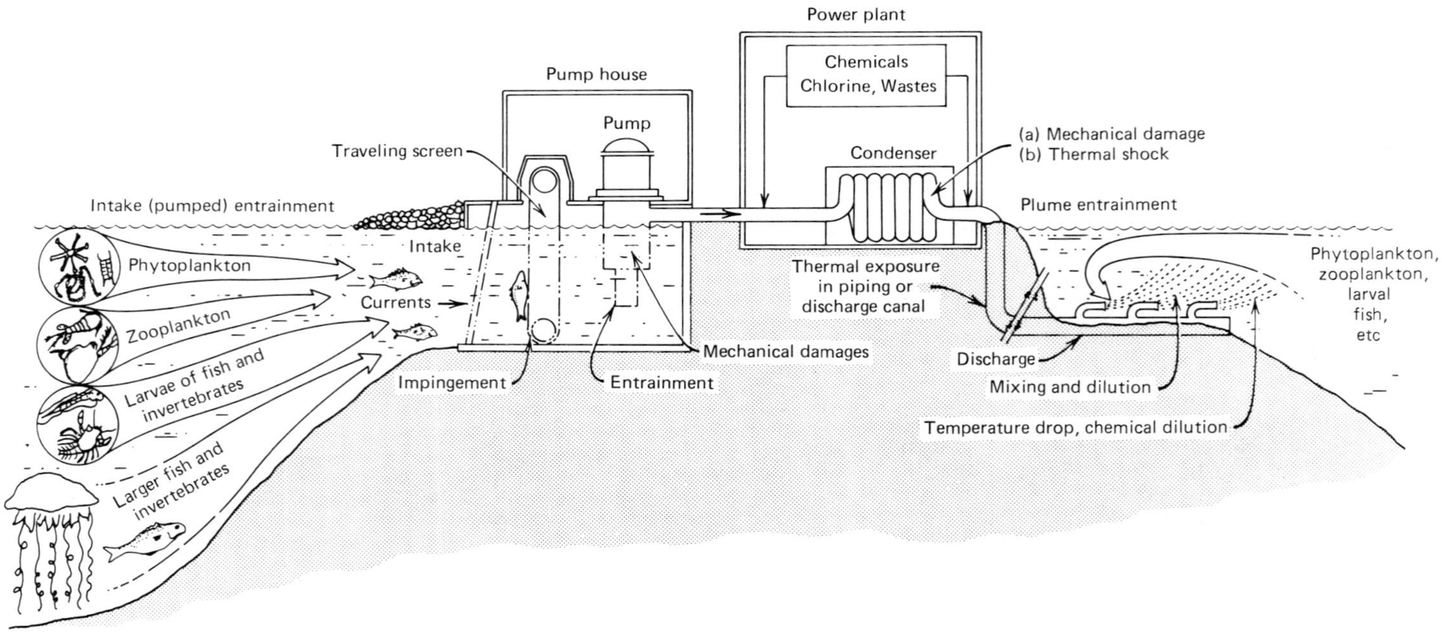

Figure 2. Sources of potential biological damage in the immediate vicinity of a power-station cooling system. Reproduced by permission from ref. 4.

deaths. Even when fish can be removed alive from their attachment to screens, the fatigue and damages suffered generally mean poor chances for survival. The effects of impingement include long periods of futile swimming in screen wells, being held by water currents against the screen mesh (often resulting in suffocation through inability to ventilate the gills), and physical injury from the rotating screen or high pressure water spray used to wash the screen. Beyond risks to individual fish are risks for populations of important species that could be seriously depleted by introducing a new form of chronic mortality into the life cycle. Since impingement acts selectively, affecting some species more than others, ie, generally schooling, pelagic fish, fears of shifts in species composition arose.

Serious analysis of the risks associated with power-plant impingement began when environmental impact statements for nuclear power plants were initiated following the U.S. National Environmental Policy Act of 1969 (NEPA) and the ensuing Calvert Cliffs court decision which assigned responsibility for all cooling-water impacts of nuclear stations to the Atomic Energy Commission (now the Federal Energy Regulatory Commission). The analogous problem of impingement of downstream migrating salmon on screens of irrigation diversion dams had been addressed for many years on the U.S. west coast, but it was not until 1973 that the commonality of both of these problems and of possible engineering solutions was widely recognized (4). Emphasis on solutions was stimulated by the U.S. Federal Water Pollution Control Act Amendment of 1972, which allowed once-through cooling only where intakes minimized losses.

Biocides, principally chlorine, that were used periodically (0.5 hours per day per condenser) for condenser cleaning were identified as toxic risks for organisms in the cooling circuit at the time and for those in the vicinity of the discharge where the biocide dissipates. Concern over residual toxic effects of power-plant chlorination was also largely an outgrowth of NEPA reviews of nuclear stations in the early 1970s, when it coincided with renewed concern over ecological effects of chlorination of treated sewage effluents and evidence of chlorinated organic materials formed in chlorinated water (5–6) (see also Industrial antimicrobial agents; Water, industrial water treatment).

A fourth risk is from combined damages, ie, thermal, physical, and chemical, sustained by small organisms that are pumped through the cooling system, ie, entrainment. These damages were among the earliest to be recognized, but the early emphasis on thermal stresses alone retarded examination of the physical components of the damage, which are now appreciated as the principal risks at many installations (7). During entrainment, any organisms, including phytoplankton, zooplankton, larval fish, invertebrates, and many small fish that cannot swim against the induced current at the cooling-water intake, are drawn into the cooling circuit unless they are large enough to be screened out initially. In the cooling circuit, they receive in rapid sequence (usually <1 min) a series of stresses. These include a pressure drop in front of the pump impeller, risk of physical impact with the impeller or shear stress of a near miss, rapid pressurization downstream of the pump, shear stress as the cooling water is divided among hundreds of condenser tubes ca 2.5 cm in diameter, rapid temperature elevation as heat is transferred to the water through condenser tubes, maintenance of high temperature (usually 8–10°C above ambient) through the discharge system, decreasing pressure in discharge piping (sometimes below atmospheric), followed by turbulent mixing and cooling as the condenser water rejoins the source water body (8). Many entrained organisms do not survive. During periods of biocide treatment to remove heat-transfer-retarding biological slimes from condenser tubes, entrained organisms are also exposed to lethal concentrations of a toxicant, usually chlorine.

Many organisms are exposed to some of the thermal, chemical, and physical stresses of entrainment by being mixed at the discharge with the heated water; this is plume entrainment. The exact number exposed there, which may be many times those pumped through the plant, depends upon the percentage of temperature decline at the discharge that is attributed to turbulent mixing rather than to radiative or evaporative cooling to the atmosphere.

Entrainment was in the forefront of concerns over condenser cooling in the early 1970s as two trends in the planning of power plants developed: gradual decrease in the temperature rise across condensers (ΔT) to reduce thermal effects, which was necessarily accompanied by an increase in volume of water pumped, and increased exploitation by large nuclear stations of estuarine waters, which generally contain far more planktonic eggs and larvae of important fishes and invertebrates than the inland rivers or lakes used previously. Even more than with impingement, the long-term risk with entrainment lies in chronic losses of vulnerable stages to physical stresses which can contribute to instability or decline in populations of valued species. Complex scientific debates over the magnitudes and probabilities of population effects of entrainment mortalities have punctuated power-plant licensing proceedings since 1971, particularly those for the Indian Point Nuclear Power Plant on the Hudson River (9).

Questions also arose over the risks to aquatic life from changes in gas content of the water as a by-product of temperature change; warmer water holds less gas in solution. Dissolved oxygen was hypothesized to decrease below levels necessary for fish by a combination of physical solubility relationships and increased biological demand for oxygen at elevated temperatures. Supersaturation of dissolved gases in water was also viewed, particularly in the northwest United States, as a significant risk arising from heating already saturated river water. Fishery scientists on the Columbia River were especially sensitive to this potential problem, because high gas saturation levels from dam spillways were demonstrably harmful to experimental salmon and trout (10).

Physical changes in habitats near the cooling water intake and discharge structures of power stations were also identified in NEPA analyses as posing some risk or at least potential for change to segments of aquatic communities. Concrete structures, rock jetties, and altered current patterns contribute to habitat modifications that influence suitability of the area for desirable species and can be linked to the overall problem of dissipating rejected heat.

Risks from cooling towers were identified as comparative ecological analyses of cooling-system alternatives became part of the assessment process. Throughout the 1960s, evaporative cooling towers, whether natural or mechanical draft (see Fig. 1), were seen as the panacea: a closed cooling cycle would replace the transgression of the once-through system on natural waters. Cooling towers operating in the United States and abroad, especially in the UK, seemed to justify their immediate implementation in the United States.

The closed cycle is, however, not really closed. Continuous evaporation would gradually accumulate unwanted salts if there were not some flow through in addition to replacement of cooling water for evaporative losses. This flow through or blowdown carries with it numerous chemicals added to the cooling circuit to prevent corrosion, eg, chromates, zinc, and organophosphorus complexes, to slow mineral scaling of heat-exchange surfaces by, for example, acids or to eliminate biological fouling, eg, by chlorine. Blowdown was traditionally discharged untreated into water bodies that supplied the tower intake, and organisms susceptible to the chemicals were placed at risk. The chemicals also escape the towers in drift in the form of small water droplets and aerosols that are dispersed with air flows to terrestrial surroundings. Cooling-tower sludge, the solid material that accumulates in tower basins, must also be discarded. This sludge includes precipitated chemical elements from the water supply, corrosion inhibitors, and airborne dust, pollen, and tree leaves washed from the tower air flow. At the intake, entrainment and impingement are not eliminated by the addition of cooling towers, although water flows can be reduced to 5–10% of that needed for once-through cooling. The presumed advantage of reduction in water volume for minimizing entrainment risks may be illusory, as organisms entrained into cooling-tower circuits are exposed to lethal conditions, whereas many often survive a rapid once-through passage at well-designed open-cycle plants. Among the first environmental risks recognized from cooling towers were increases in fogging and icing in northern climates.

An alternative closed-cycle system that had begun to be used particularly in the Midwest, Southeast, and Texas was the artificial cooling lake. Such lakes exist in a gradient of designs from large multipurpose public reservoirs that provide essentially once-through cooling to privately controlled, diked ponds or canals that rely on evaporative cooling and are often augmented by spray modules. The risks to public resources also vary greatly. Despite the obvious multipurpose benefits of cooling reservoirs where water resources are scarce, these cooling systems were judged in violation of the 1972 Water Pollution Control Act and for several years were not allowed for new power stations. Thus, the number of cooling lakes being constructed has diminished greatly in recent years (11). There has been much debate over cooling lakes, and regulatory preclusion of numerous water and fish management opportunities from their use is viewed as unfortunate.

Changing perceptions of the sources of environmental risk from cooling systems, especially during the period 1968–1973 when thermal concerns were augmented by those of entrainment and impingement, were partly the result of rapid technical improvements in thermal-discharge designs. Spurred by water-temperature standards and limits to sizes of mixing zones imposed by states under the Federal Water Pollution Control Act of 1965, utilities quickly moved away from the traditional discharge schemes. Shoreline canal discharges with plumes of highly elevated temperatures that had been preferred for heat dissipation but which neglected biological requirements gave way to offshore diffusers that exploited turbulent mixing from discharge momentum and buoyancy to quickly diminish temperatures. Such designs had been effective in reducing temperatures rapidly to nondamaging levels at older discharges, such as those in the Columbia River. Risks of thermal damage to aquatic communities were thus diminished, and risks associated with increased cooling-water flow became more apparent.

In the mid-1970s, ecological risks that had commanded the most attention were joined by the risks to human health. The source of risk was the familiar heat additions, but the vector of potential risk, ie, thermophilic microorganisms, was new to most ecological analysts. First to gain attention was the pathogenic amoeba of the genus

Naegleria, which invades the central nervous system and causes a rapidly fatal meningoencephalitis (12). These protozoans have been isolated from both naturally occurring and artificially heated warm, organic-rich aquatic environments. Evidence is rapidly accumulating that another organism, *Legionella pneumophila*, is also well adapted to warm water and can cause epidemics of pneumonia and general systematic infection, eg, Legionnaires' disease, or a nonpneumonic combination of fever, headache, and myalgia, eg, Pontiac fever (13). Common source outbreaks of Legionellosis have occurred in a pattern that suggests airborne spread of aerosol nuclei generated by cooling towers or evaporative condensers that were documented to contain the organism. Another epidemic occurred among workers cleaning a power-plant condenser. Both amoebae and *Legionella* appear to be basically soil or aquatic-sediment microorganisms for which the ecological niche and dispersal mechanisms to man have been expanded by increasing use of heat-rejection systems, particularly cooling towers.

Minimizing the Risks

Selection of the best system involves matching engineering options to the local aquatic system potentially at risk. General principles of aquatic ecology and of the life histories and environmental requirements of species represented locally can be adapted to local water-resource goals and with detailed understanding of the local aquatic setting to achieve site-specific risk prevention.

Thermal Effects. Temperature is the most all-pervasive environmental factor that influences aquatic organisms. There is always an environmental temperature, whereas other factors may or may not be present. Nearly all aquatic organisms, with the exception of marine mammals, are for all practical purposes thermal conformers or obligate poikilotherms. As such, they are not able to exert significant influence on maintaining a certain body temperature by physiological means. Their body temperatures thus fluctuate in close accord with the temperature of the immediate surrounding water; although especially large, active-moving fish, such as tunas, maintain deep-muscle temperatures slightly higher than the surrounding water. Intimate contact between body fluids and water at the gills and the high specific heat of water assure this near-identity of internal and external temperatures. Behavioral thermoregulation or the control of body temperature by selection of water temperature in natural gradients is, however, a common feature of many fish. Behavioral thermoregulation serves an important ecological role in partitioning aquatic habitats among species, and its understanding is a powerful feature for estimating and ameliorating the impacts of thermal discharges.

Thermal effects on aquatic organisms have been given critical scientific review periodically over the past thirty years or more. In addition, annual reviews of the thermal effects literature have been published since 1968 (14). Water temperature criteria for protection of aquatic life were prepared by the NAS in 1972, and these criteria have formed the basis of the EPA recommendations for establishing water-temperature standards for specific water bodies (15–16).

The importance of temperature for organisms lies partly in its fundamental physical–chemical influences. The biochemistry of life is, in general, subject to the basic dictum that a rise of temperature by 10°C results in about a doubling of the rate of chemical reactions. Life also has an upper thermal limit, set partly by the chemical stability of organic molecules, eg, proteins or enzyme systems, and by the balance of food energy inputs and expenditures, that is characteristic of each species and, in some cases, different life stages. Although aquatic organisms must conform to water temperature, many have evolved internal mechanisms other than body-temperature regulation to allow continued functioning as temperature changes occur geographically, seasonally, and daily. Through gradual biochemical adjustments termed acclimation, physiological processes are maintained relatively constant over a prescribed thermal range which is species- or life-stage-specific. Within this range, each species has a zone of optimal physiological performance which, because it tends to coincide with temperatures preferred in gradients, has recently been termed its thermal niche in the environment.

Organisms evolving under annual temperature cycles and in environments with varying temperatures spatially have incorporated thermal cues in reproductive behavior, habitat selection, and certain other features which act at the population level. Thus, the balance of births and mortalities, which determines whether a species survives, is akin to the metabolic balance at the physiological level in being dependent upon the match, within certain limits, to prescribed temperatures at different times of year. At the ecosystem level, relationships among species, eg, predators, competitors, prey animals, and plant foods, are related to environmental temperatures in complex ways. Many of these interactions are poorly understood, and the role of temperature remains speculative in many cases.

Despite the immensity and complexity of known and suspected roles of temperature in aquatic ecosystems, certain thermal criteria have been especially useful in minimizing risks from thermal discharges, at least those short-term ones with which we have the most direct experience. More data have been organized at the physiological level than at higher levels of organization, so that such data have factored prominently in current approaches.

Preventing Mortality. When power stations use small quantities of cooling water, temperatures are raised by large amounts. Organisms entrained in these flows or exposed at the discharge can receive exposures to large temperature rises that, especially in summer, can carry them beyond their upper thermal tolerance. Conversely, organisms that have been nurtured in the warm discharge of a power station in winter can have their cold tolerance exceeded should the power station suddenly shut down. Upper and lower temperature tolerances of aquatic organisms have been well conceptualized over the past 30 yr, and standardized methods are available for determining a species' tolerance ranges under different conditions of thermal history (Fig. 3(**a**)). On a short time scale, mortality is highly dependent on duration of exposure (Fig. 3(**b**)), so that brief exposures to potentially lethal temperatures are not actually lethal. Graphical and mathematical representations of these data for important species at a site can allow a design engineer to tailor temperature elevations and duration of exposure in a plant's piping or in the effluent mixing zone to maximize organism survival (15). This is usually accomplished with detailed mathematical models of cooling-water-effluent dispersion and heat dissipation in the near field where temperatures are highest (18–19). One option used extensively for this purpose in the 1970s was increasing the cooling-water flow and thus decreasing the temperature elevation to below lethal levels. As noted earlier, this tactic had the undesirable result of increasing the numbers of organisms damaged by the physical effects of entrainment and impingement. Thus, there must be a quantitative balance between minimizing direct thermal mortalities and increasing those risks related more to the volume of cooling-water flow.

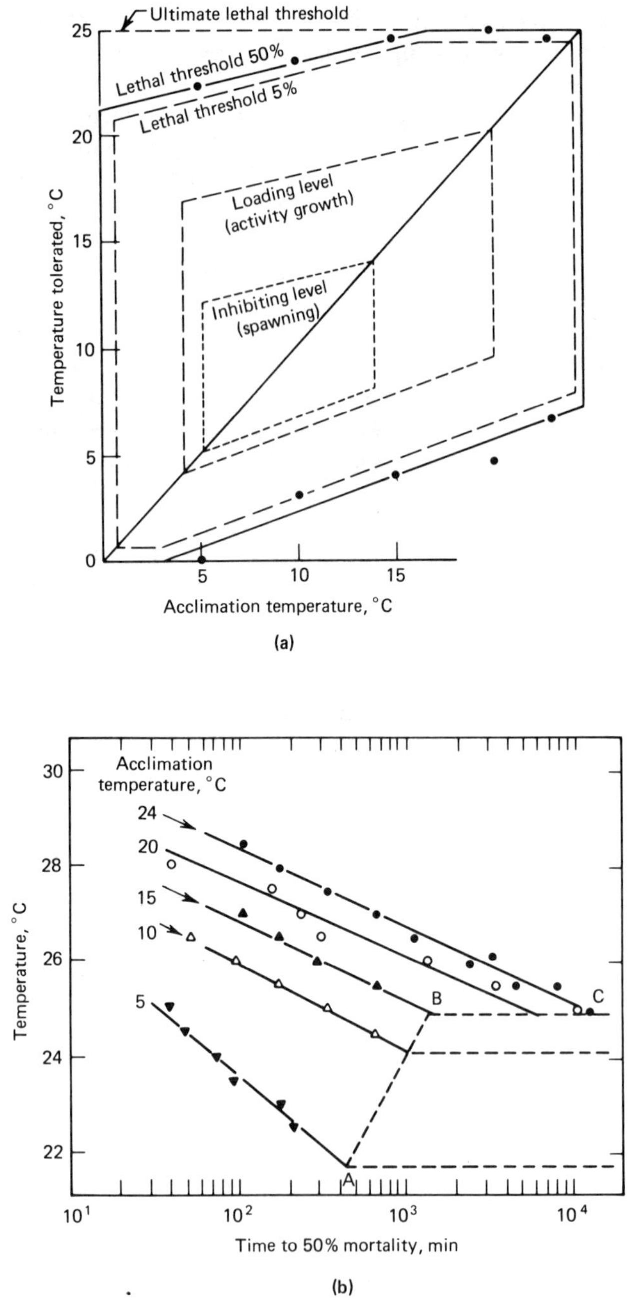

Figure 3. Lethal temperature thresholds for aquatic species. The patterns are general for all species, but exact temperatures are species-specific. (**a**) Tolerance polygon of upper and lower lethal (50%) temperatures for one-week exposures of an example species (juvenile sockeye salmon) which has been held at the acclimation temperature, with more restrictive thresholds indicated as dashed lines; (**b**) time-dependent mortality (50%) of an example species (juvenile chinook salmon) at temperatures above the one-week lethal threshold after holding at different acclimation temperatures. The dashed line ABC indicates transition to less than 50% mortality at lower temperatures and coincides with the upper lethal threshold of this species' tolerance polygon. Reproduced by permission from ref. 17.

Preventing Stressful High Temperatures Over Long Periods. For the long term, ie, greater than 1–2 d, simply preventing mortality is insufficient for protecting aquatic species. All of the physiological functions normally performed must be carried out to maintain healthy individuals that are capable of competing in the natural ecosystem. Physiological functions are sufficiently complex and diverse that no single metabolic process has been adequate to define health over a range of temperatures. However, an aggregate measure, growth rate, has proved useful. Growth is an integrator of all physiological functions and some behavioral ones, eg, feeding rate. Growth occurs only if all other metabolic demands are being met and when sufficient food energy is left over for adding biomass. Typically, many physiological functions of well-fed, cold-blooded organisms proceed optimally over a temperature range in which growth rate is maximal (see Fig. 4). Above the temperatures of maximum growth rate, the rate typically declines steeply to a temperature of zero growth, which often occurs 1–2°C below the temperature at which direct mortalities begin. Intuitively, the healthy fish

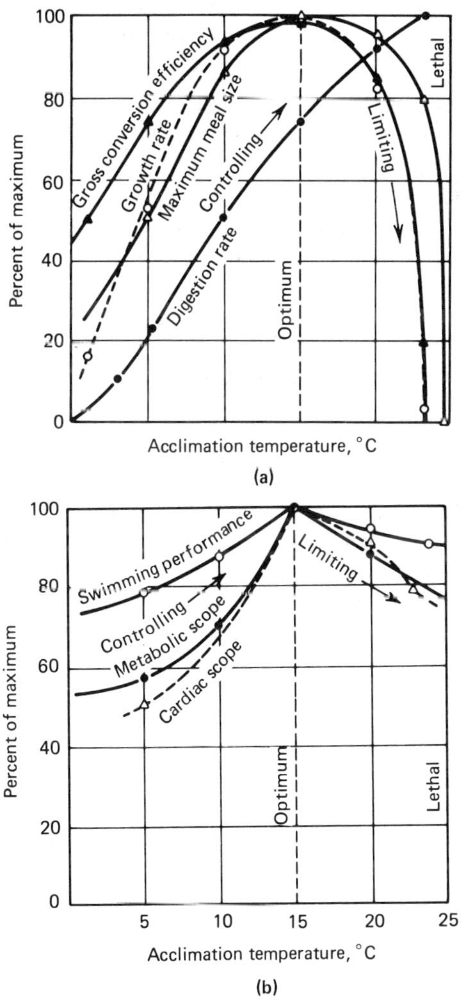

Figure 4. Convergence of numerous physiological functions, including overall growth rate, at an optimum acclimation temperature which is specific for the species, juvenile sockeye salmon. Reproduced by permission from ref. 17.

becomes unhealthy as the long-term temperatures it experiences rise from those that yield maximum growth to that which stops growth. Alternative methods for calculating the upper danger level have been proposed, but each suggests that a long-term decline of growth rate below ca 75% of maximum at high temperatures is unduly risky (15,17).

Standards for upper limits on water temperatures for particular water bodies over periods of about one week or more can be based on species-specific growth rates. With an inventory of important species and life stages in the area during the warm season, the analyst can ascertain the upper temperature limit which does not stress those in the desired aquatic assemblage. Hydrothermal models of heat dissipation in the far field beyond the zone of effluent mixing are important for estimating the zones that may present a long-term risk from elevated temperatures.

Recently, the preferred and avoided temperatures in a gradient have been used as surrogates for optimum growth and upper danger levels for fish. One practical drawback to using growth rate as an index of optimal temperatures is the experimental cost of determining it. As more and more data link temperatures of optimal growth rate with preferred temperatures for a species, there is an increasing tendency to conduct only thermal preference tests (20). There is also accumulating evidence that long-term abundance of species at power-plant sites is generally correlated with preferred and avoided temperatures (Table 1).

Preserving Reproduction Cycles. Organisms in the temperate zone have evolved in concert with seasonal temperature variations. Reproductive cycles are generally keyed to the annual cycle, and temperature is often the controlling factor, although

Table 1. A Comparison of Preoperational (1968–1971) and Postoperational (1972–1981) Catch Rates of Fish in the Wabash River, Indiana, at the Cayuga Electric Generating Station[a]

Common name	Species	Preferred temperature, °C	Avoided temperature, °C	Percent change in catch rate
silver redhorse	*Moxostoma anisurum*	24	26	−23%
shorthead redhorse	*Moxostoma macrolepidotum*	24	26	−17%
mooneye	*Hiodon tergisus*	24	27	−27%
golden redhorse	*Moxostoma erythrurum*	24	26	−31%
sauger	*Stizostedion canadense*	24	28	−22%
smallmouth bass	*Micropterus dolomieui*	24	27	−9%
spotted bass	*Micropterus puctulatus*	24	27	−25%
goldeye	*Hiodon alosoides*	24	27	−48%
shortnose gar	*Lepisosteus platostomus*		33	−49%
buffalofish	*Ictiobus* sp.		33	−56%
blue sucker	*Cycleptus elongatus*			−80%
skipjack herring	*Alosa chrysochloris*		30	90%
longnose gar	*Lepisosteus osseus*	30–33	34	155%
freshwater drum	*Aplodinotus grunniens*	29	31	163%
North River carpsucker	*Carpiodes carpio*		34	178%
gizzard shad	*Dorosoma cepedianum*		31	217%
white bass	*Morone chrysops*	28–30	31	260%
carp	*Cyprinus carpio*	29–32	34	498%
largemouth bass	*Morone salmoides*	27	30	760%
flathead catfish	*Pyladictus olivaris*		32	1647%

[a] Ref. 21.

changing day length is of equal or greater importance for some species. Reproduction success depends in part on the preservation of an annual temperature pattern, although the precise timing is usually not critical. An analyst can obtain dates and temperatures for spawning of many important aquatic species, and thermal discharges can be designed, usually with the help of far-field mathematical models, to assure the necessary thermal periodicity. Unless the cooling system is a very heavily heat-loaded stream or cooling pond, the thermal output of a power-station complex is rarely sufficient to offset the large natural cooling rates of winter. Thus, annual thermal cycles are generally maintained despite anthropogenic heat rejection.

Maintaining Ecosystem Structure and Function. Thermal heterogeneity of water bodies is an important structural feature of the environment and plays a large role in determining the composition and functioning of most aquatic systems. Vertical thermal stratification of lakes, reservoirs, and many estuaries in summer (see Fig. 5) segregates an available habitat into discrete zones with differing thermal and water quality characteristics. In the evolution of aquatic species, these discrete zones have been the basis for partitioning the environment, such that different species or life stages within some species occupy discrete portions based, in part, upon temperature selection behavior. This thermal segregation reduces competition for food resources and spawning sites, and often minimizes predation and cannibalism. The disruption of thermal structure by power-station cooling systems can be more damaging to ecosystem composition and function than the direct killing of many individual organisms.

Particularly significant are thermal refuges during periods of extreme high or low temperatures. Aberrant weather or artificial thermal changes can cause large portions of a water body to exceed the physiological limits of some aquatic species for brief periods. The species usually affected are those at the northern or southern limits of their geographic ranges. These species often survive in isolated zones or refuges that retain suitable temperatures, eg, springs where trout can remain cool in the warmest

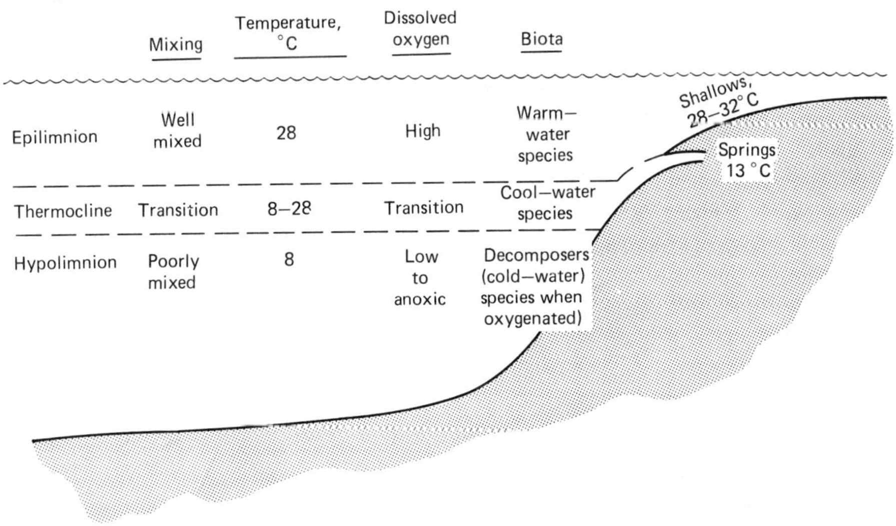

Figure 5. Typical thermal stratification of a lake, reservoir, or poorly mixed estuary in summer which, because of density differences, establishes discrete zones with differing thermal, water quality, and biotic characteristics.

summer months or water bodies where cold-sensitive species, eg, threadfin shad, find relatively warm waters in mid-winter. The demise or survival of such geographically marginal species, which is determined by the availability and sizes of such thermal refuges, can be paramount in establishing the ecological interactions of a water body and ultimately its productivity for sports and commercial fisheries or other uses (22). Power-station discharges can prematurely force desirable cool-water species into refuges in summer and can provide essential warm refuges for desirable warm-water species in winter.

Current thinking is drawing together the previously sketchy and disparate knowledge of thermal heterogeneity of the environment, thermal optima for organism growth rates, and behavioral temperature selection in ways that are useful for predicting ecosystem effects of thermal alterations (20). Thermal-niche concepts can be used to identify critical habitats for important species in ways that focus the potential influences of power-station discharges (22–23). For example, striped bass (*Morone saxatilis*), an East Coast species introduced on the West Coast and in freshwater reservoirs, partitions a water body in summer along thermal gradients among its age or size classes. Young bass prefer and grow optimally at high temperatures near 24–26°C, subadults prefer ca 22°C, and mature adults select temperatures of 20°C or less. Thermal dicharges may benefit growth rates and survival of juveniles, but the adults face a different prospect. Forced to cool water in summer by their genetically based temperature preferences, they may find this habitat severely restricted by thermal additions or compromised by simultaneous depletion of dissolved oxygen as a result, in part, of decomposition of thermally stimulated plankton production. Overcrowding in limited thermal refuges that have sufficient dissolved oxygen has led to starvation, high disease incidence, and abnormally high fishing susceptibility, all of which cause high numbers of deaths.

Altered predator–prey interactions, previously viewed as intractable long-term influences of changing water temperatures, can be estimated at least qualitatively through use of thermal-niche concepts. Adult striped bass that are restricted from access to surface waters by high temperatures seek food among the cool-water species, eg, trout in deep water, rather than among the surface-dwelling shad species which are their normal prey. Trout populations can be decimated, as they were in Lake Mead after introduced striped bass grew to the size and thermal preference of adults.

Careful planning of thermal additions, including creation of new cooling reservoirs, can yield thermal structures which enhance rather than damage desirable aquatic species. Knowledge of the thermal niches of these species and their potential competitors or predators can permit special provisions for thermal refuges, eg, cool summer zones in a heavily loaded cooling pond. Cooling-water circulation can be designed so that the thermal stratification patterns that are essential for some desired species are maintained. From a different perspective, aquatic species introduced to waters used for power-station cooling can be selected so that their thermal niche matches the thermal structure that the facility creates. Multiple species introductions can be evaluated beforehand to estimate whether available thermal patterns will abnormally aggregate or separate potential predators and prey.

Impingement. An explosion of innovative technology was spawned in the 1970s by the deaths of fish and some invertebrates, eg, blue crabs, on power-station intake screens. Unlike the seemingly intractable complexities of thermal effects in a water body, impingement deaths are clearly related to inlet water velocities, lack of escape routes, and retention on screens either in the water or during the screenwash procedures that were designed for handling nonliving trash. Most of these techniques have been reviewed in a series of conference publications (4,24–25). Repellents, eg, sound, electricity, and light, were used to keep susceptible fish away from intakes with only moderate success. More useful were orientations of the inlet structure such that screens flush with the shoreline allowed lateral escape (Fig. 6(**a**)). Where this screen orientation was impossible, guidance systems were developed for intake bays to direct fish from the incoming water flows to alternative escape routes (Fig. 6(**b**)). Guidance devices were also installed at the entrances to unscreened, offshore inlets, where a horizontal velocity cap placed a few feet over a vertical opening proved especially effective in preventing fish entrance (Fig. 6(**c**)). Small-scale modifications to existing rotating trash screens allow many fish to survive as they are raised from the water and deposited in sluiceways for return to the water.

Experience has also shown that deaths of fish on intake screens, although spectacular, often have minor consequences for local populations of many species (26). Most commonly impinged are open-water, schooling species that have high reproductive potential and a propensity for large natural variations in population numbers. The most common season for impingement mortalities, ie, winter, is a time of natural population reduction because of cold stress for several species. Cold-stressed fish may be impinged in large enough numbers to shut down the facility, but evidence suggests that most are moribund and destined not to survive regardless of cooling-water intervention.

Current perspective on minimizing impingement risks focuses on site-specific analyses of potentially vulnerable species and selection of engineering designs which, within acceptable cost limits, keep impingement deaths few. Successfully accomplishing this objective requires intimate cooperation between biologists and engineers.

Biocides. Chlorine and other biocides are used occasionally in cooling water to kill and dispose of organic growths on heat-exchange surfaces and on piping where water flow could be hampered by such growth. Of necessity, organisms passing through the cooling circuit or residing in the effluent area during periodic chlorine injections experience the potentially lethal exposures. The objective of conscientious plant operators is, however, to maximize the intended kill and minimize extraneous damages, particularly in the receiving water. Methods for accomplishing this goal have been developed progressively over the past several years (27).

Chlorine is a toxicant with a typical dose–response pattern for biota (Fig. 7). There is a time-dependent mortality at high concentrations and a low concentration above which long-term chronic effects are shown. Early methods for using the time-dependent effects of high temperature for predicting safe temperatures and exposure times during cooling-water exposures led to a similar approach for chlorine (3). Because chlorine toxicity data were scant, aggregate graphs for all available species were used to identify safe levels. Different aggregates had to be developed for freshwater and marine assemblages because of the markedly different chlorine chemistry in fresh and salt water. The analytical tools are available for minimizing chlorine damages, although species-specific data are urgently needed.

Entrainment. Stresses to small nonscreenable organisms, eg, fish larvae, during passage through the cooling circuit come from a combination of thermal shock, physical abuses, and periodic injections of biocide. Whereas the thermal and biocide compo-

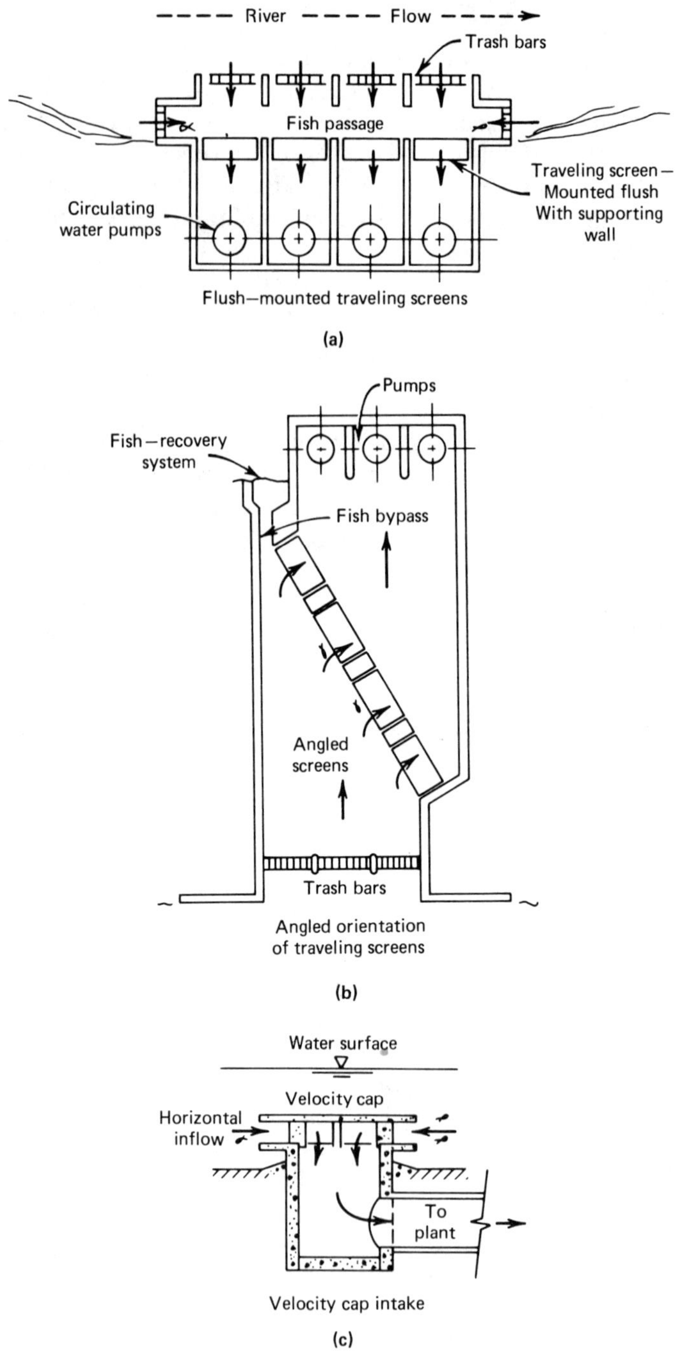

Figure 6. Engineering approaches to minimizing entrapment and impingement of fish and large aquatic invertebrates; eg, blue crabs, on trash screens at intakes. (**a**) An inlet pump house with vertical traveling screens mounted flush with a river shoreline to minimize obstructions to animal movements parallel to shore; (**b**) a fish guidance system composed of vertical traveling screens angled to the direction of water flow to direct fish to a recovery chamber that returns to the water body; (**c**) a velocity cap atop a vertical, offshore inlet; induces a horizontal flow which fish avoid instead of a vertical flow which they do not. Reproduced by permission from ref. 25.

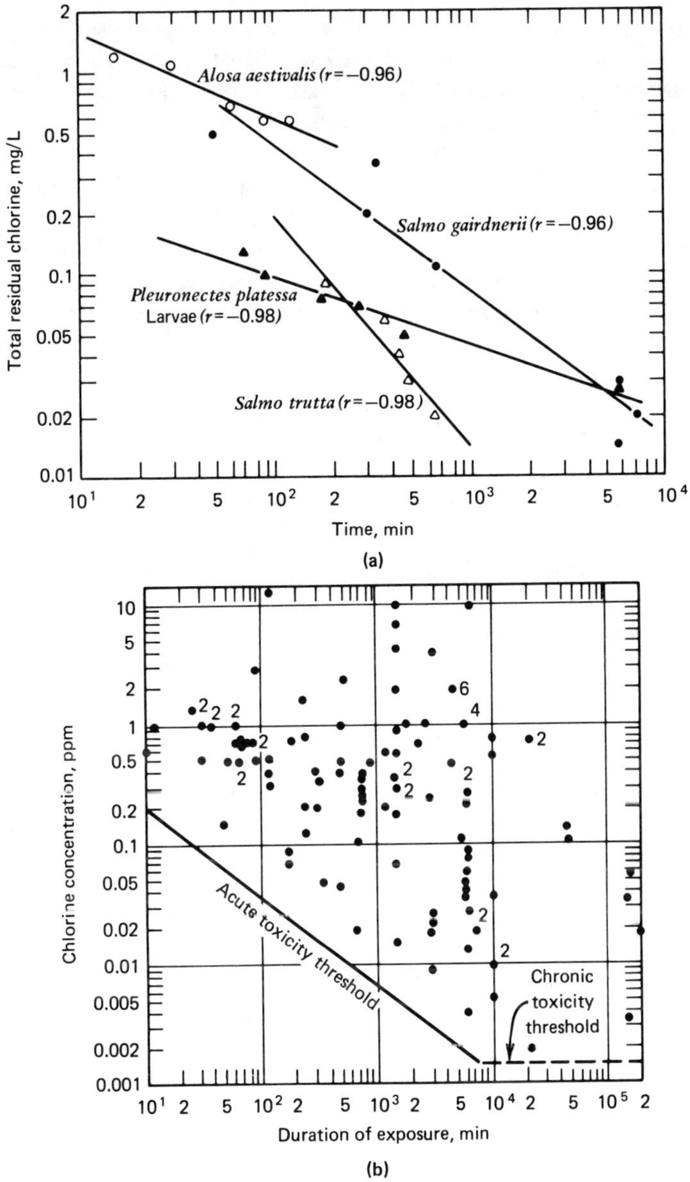

Figure 7. Toxicity of chlorine to aquatic organisms. (**a**) Time-dependent mortality (50%) of four example species in various levels of total residual chlorine in the laboratory. r = correlation coefficient of the curve; (**b**) A summary of chlorine toxicity to freshwater species, indicating overall no-effect thresholds for acute and chronic exposures. Numbers indicate where more than one test yielded the same result. A different summary figure applies to marine organisms because of differences in the chemistry of chlorine in seawater. Reproduced by permission from ref. 27.

nents, ie, the abrupt rise of temperature or chlorine concentration in the condensers followed by a holding period in piping and a decline in the discharge area, are well conceptualized and can be used quantitatively for predicting organism responses, the physical abuses are less well-understood (8). Pressure changes, shear forces, and physical contact with pump impellers or tubing walls are sources of potential damage that vary from system to system. Minor differences in configuration of piping can mean the presence or absence of such devastating features as microcavitation cells. The physical configuration does not lend itself easily to laboratory experimentation, although three such physical simulators have been tried (28–30).

A principal frustration in attempts to minimize entrainment damages has been the contradictory demands of thermal and physical stresses. Thermal stresses can be quantitatively predicted based on dose–response data and minimized by increasing water-flow volumes, which dilute the fairly constant supply of rejected heat. The added volume of cooling water, however, includes proportionately more planktonic organisms, which are subjected to physical stresses. Attempts to minimize thermal damage in new power stations had the demonstrable effect of increasing physical damage. One proposed solution, which is based on the assumption of a high percentage of mortality resulting from physical stresses, is to return to high condenser temperatures at which all of the relatively few organisms entrained would be killed, but those remaining in the waterbody would be assured survival (8).

The assumption of high percentage mortality resulting from physical stresses has recently been criticized. New methods of sampling entrained biota, which involve extensive precautions against damaging stresses during the sampling process, indicate a much higher survival of even delicate fish larvae than had been realized with earlier and more crude sampling methods. It appears that ameliorations of thermal stress with flow increases generally are not cancelled by additional physical mortalities except in cases of exceedingly high flows. An optimization procedure, such as that suggested by the Committee on Entrainment, appears fruitful for identifying on a site-specific and seasonally varying basis the most appropriate cooling-water flow regime (8).

There is also a compromise between entrainment and impingement. One of the engineering methods suggested for minimizing entrainment was replacement of 1-cm intake screens with screens of much smaller aperture. Although fewer organisms are entrained, most of those saved are, in fact, impinged on the smaller mesh of the screen. The probable new impingement mortality may exceed the risks during entrainment.

Gas Balance. Oxygen in solution does not generally change in a cooling circuit unless there is strong aeration of originally saturated water heated maximally, in which case some dissolved gas is lost. In no case does the loss reduce levels to below that required by aquatic life. When intake water is undersaturated, the agitation of the cooling circuit generally yields an increase. Abundant data support these generalities. Polluted receiving water does, however, have an accelerated microbial oxygen demand because of raised temperature, and this microbial deoxygenation can yield exaggerated dissolved oxygen sag zones downstream. The interaction of thermal effluent and waste stabilization in the river must be considered whenever polluted water is used for cooling, and engineering models for predicting oxygen sag zones are generally capable of incorporating temperature changes.

Damaging supersaturation of dissolved gases has occurred in some cooling-water discharges. Damage has been isolated to circumstances, usually in winter, when temperature elevations above the cool ambient are highest based on the assumption that even a large increase in winter results in temperatures below lethal levels or prevalent water quality standards. The practice of winter increases in temperature rise across condensers by cutting back on pumping capacity has either ceased in general practice or the immediate discharge areas have been engineered to prevent long-term residence by susceptible biota. These remedial measures can be completely effective.

Cooling-Tower Chemicals. The new risk posed by changing power-station systems from traditional once-through or open-cycle cooling to cooling towers is from chemicals added to the recirculating water. Blowdown to aquatic systems and drift to the terrestrial landscape carry these chemicals to locations where natural biota can be damaged through direct poisoning or where toxicants can accumulate to potentially detrimental levels by food-chain transfer. Risks from such chemicals can be minimized with fairly straightforward engineering approaches. Airborne drift has been reduced significantly by installation of physical baffles at the air outlets which intercept droplets, coalesce them, and return the water to the cooling-water flow. Blowdown can be treated for removal of chemical constituents, eg, chromates, with the additional benefit of chemical recycling and, thus, cost recovery. Chemical-laden sludges become a more long-term disposal problem; current practices include ponding and landfills. Chemical-recovery processes are also available for sludge treatment. Of significance for all of these processes is the cost which, when added to the initial capital cost of the cooling towers and the operating costs of pumps and fans, can reduce the attractiveness of closed-cycle cooling, especially when less costly mitigative measures exist for the open-cycle system at many locations (see Wastes, industrial).

Human Pathogens. Most studies of the risks to humans from pathogens stimulated by environmental conditions in cooling systems, eg, amoebae, *Legionella*, are preliminary. There is, however, considerable indication that proper designs and good maintenance practices can reduce the potential risks to very low levels. Thermophilic protozoans which thrive in water near human body temperature that is also rich in organic matter find animal tissues similarly favorable, particularly the nasal passages which allow easy penetration to the brain. Field evidence indicates that these organisms are most abundant in stagnant pools and roadside ditches; fatal infections have generally been traced to exposures in such environments. The rapidly flowing water in power-station discharges, although quite warm in summer, generally does not provide either stable substrates or high concentrations of organic material. Additionally, plant security measures generally ensure that bathers do not use the warmest zones. Allowing water contact by the public in discharges, particularly wherever the outlet is a slowly moving canal with abundant vegetation, is risky. The Legionnaires' disease microorganism, recently identified as a common inhabitant of many cooling-tower systems, especially small ones used for building-temperature control, can be held in check by systematic use of biocides in recirculating cooling water. Contact with the infectious organisms can be minimized if care is taken to isolate inlets for ventilating air from the drift aerosols which are emitted from cooling towers.

Maximizing the Benefits

Increasing attention is being given to finding productive uses for power-plant waste heat. The waste of approximately two thirds of the fuel energy during conversion of fossil or nuclear fuels to electricity is increasingly viewed as extravagant. Methods are thus being developed for converting the thermal pollution from power stations into useful heat resources. There are potential physical applications of power-plant rejected heat, eg, industrial heating, and biological applications such as fish culture, soil warming, and heating greenhouses and livestock shelters.

The most important factor determining the possible uses of such heat is the temperature of the heat source. There is a threshold between low grade and high grade heat for engineering uses at ca 100°C. Most waste heat from electric power stations is in the form of low grade heat. This is so low in temperature that disposal to the environment has traditionally been considered the only practical alternative. Discharges from once-through cooling systems of thermal power plants have outlet temperatures of 10–40°C, depending upon the season. Circulating water in closed-cycle cooling systems is only slightly warmer (20–50°C). These low temperatures are the result of careful engineering of power stations to extract the maximum amount of electrical energy from the fuel before the rejected heat is dumped to circulating water in steam condensers. Such engineering has produced high efficiencies compared to most other mechanical systems: 33% in nuclear plants to 40% in fossil-fueled plants is the fuel energy converted to electricity.

A more useful, higher grade heat is in a form that has not been degraded to the low temperatures of most power-plant thermal discharges but which remains at 40–200°C following production of initial amounts of electricity. Low pressure steam can be extracted from the turbine system of a power plant before condensation to waste-heat temperature, thus permitting its use for functions requiring higher temperatures. Although this gives better utilization of the energy remaining in the steam, the efficiency of the low pressure turbine for producing electricity is diminished. Condensers can also be operated with smaller cooling-water flows to yield higher discharge temperatures but with some penalty to generation efficiency because of higher turbine back pressures. The overall efficiency of fuel energy use in such multiple-purpose systems can exceed that of electricity production alone, however.

Ideally, a power station could be a potential supplier of electricity, high grade heat, and low grade heat, even though electricity is the main product and most new power stations are optimized accordingly. Depending upon the desired uses at particular sites, future power stations could be designed to alter the ratios of electricity and different grades of heat to produce the most efficient total energy use. Such multiple use was common at early generating stations in the United States and is still common in Europe.

Aquaculture. Culture of some aquatic species in essentially unmodified thermal effluents of power stations has been attempted both experimentally and commercially (see Aquaculture). The principle behind such culture is the temperature-dependence of growth rates discussed earlier. Use of rejected heat to prolong optimal growth temperatures in cool months can significantly increase the sizes attained in a year. Even though power stations rarely provide high enough temperatures in winter for optimal rates, temperatures can usually be maintained at levels that increase growth rates. Maximum use of available temperatures has recently been made by culturing warm-water shrimp in summer and cool-water trout in winter, each of which can utilize the added temperature imparted by power-station cooling (31).

Open-Field Agriculture. Use of warmed water for field crops and orchards has been tested in several studies. Buried pipes can convey thermal-discharge heat to soils where warming aids plant growth and extends the growing season. Soil warming with open irrigation water has also been studied. There are advantages to spraying thermal-effluent water on crops and orchards to prevent freezing of buds and blooms. Whether or not the heat is used, power-plant cooling water can conveniently be combined with irrigation water systems to provide multiple uses of the water. An extensive demonstration project has been established by Electricité de France at their St. Laurent des Eaux power station.

Greenhouse Agriculture. Greenhouse agriculture is well known for its many advantages over open-field agriculture for certain crops. Yields are greater per land area and year-around culture is possible, which allows matching of crop harvest with high demand and price. One drawback is high expense for heating in winter, which can be the most costly part of greenhouse operation. Current energy shortages have made fuel not only costly but often unavailable. The use of waste heat from steam-electric power plants therefore appears promising as a source of low cost heat for use in greenhouses, especially if the power stations have cooling towers with wintertime operating temperatures of 16°C or higher. The economic analysis is favorable (32).

Experimental greenhouses based on waste heat for temperature control have been operated at several sites and have demonstrated the principle that cooling water can be used to heat a greenhouse in winter. Some have also been based on power-plant effluent for cooling in summer; inexpensive porous packing through which the heated water drips and air circulates is used (Fig. 8). Evaporation provides cooling in summer and sensible heat transfer warms the air in winter. Additional heat and humidity control can be provided with a finned-tube heat exchanger. The French have used pipes and plastic tubes laid in the floor of the greenhouse; warm thermal effluent or water warmed further with heat pumps operating from thermal discharge waters circulates through the pipes and tubes. As with aquaculture systems, greenhouses should affiliate with multiple-unit power stations to ensure continuous warm water supplies.

Animal Shelters. The advantages of temperature control for maximizing weight gain and avoiding animal losses in livestock and poultry are well known. Low grade heat from power station cooling offers the possibility of low cost heating of animal shelters, although few demonstration projects have been developed. Heating animal shelters is a special case of space heating, although capital costs generally must be lower for the system to be economical. The greatest potential for heating animal shelters appears to be in continental climates where exceedingly cold winters can lead to deaths of livestock.

Space Heating. A large percentage of the energy requirements of most countries in temperate zones is for heating and cooling of living and working spaces and for hot water. The historical use of dual-purpose power plants for electricity generation and

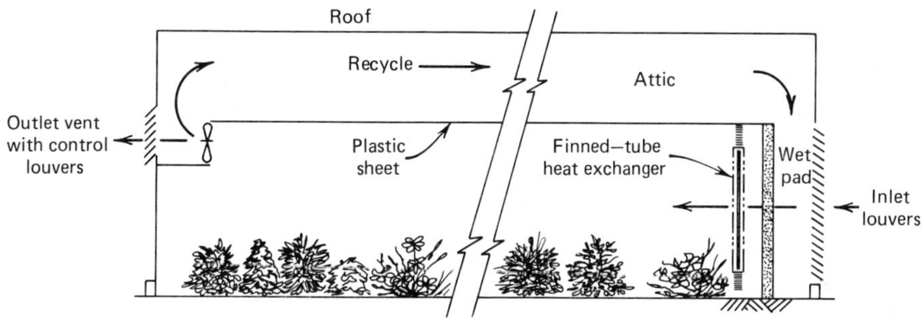

Figure 8. Longitudinal section of an experimental waste-heat greenhouse in which temperature control in all seasons is provided by evaporation and heat transfer as air passes through a fiber pad soaked with power-station cooling water or by heat transfer as air passes through a finned-tube heat exchanger that carries cooling water. A false ceiling provides for recycle of air through the heat-transfer medium. Reproduced by permission from ref. 33.

central district heating in the United States and their current extensive use in such countries as the USSR, Sweden, and the FRG suggests that expansion of this form of waste-heat utilization can contribute significantly to energy conservation and control of concentrated thermal discharges worldwide.

The most economical form of long-distance transport of thermal energy for space heating and cooling by adsorption methods is by heated water rather than steam. Steam had been used from early dual-purpose power plants prior to development of modern water-cooled condensers. It is still used in areas of dense loads, but its range of effective distribution is small because of large pressure drops in distribution systems. Modern experience with dual-purpose power plants (mostly in Europe) and new district heating systems in the United States, eg, in colleges, institutions, and shopping centers, has shown hot water to be superior for dispersed loads, including single-family residences. These dual-purpose stations operate at high thermal efficiency; the Swedish Malmo Plant is researching new technologies for distribution pipes in order to expand economical district heating to dispersed single-family residences.

Economic analyses for the United States indicate that supplying thermal energy to the commercial–residential sector from dual-purpose power stations is more economically competitive in new applications than in cases where existing buildings are to be serviced (34). There is little difference for supplying industrial heating. Such a thermal grid is most competitive with new fossil fuels where there is a high heat-load density and expensive fuel costs.

Industrial Process Heat. Many industries use process heat at 77–110°C. Much of this heat is supplied by combustion of oil and natural gas. Equipment manufacturers are developing industrial heat pumps to capture free industrial-plant waste heat and regenerate it to the desired process-heat temperature, thereby greatly reducing energy costs associated with direct heating. Operations that use heat in the 88–110°C range that can be supplied by new high efficiency heat pumps based on rejected heat sources are washing, blanching, sterilizing, and cleaning operations in food processing; grain drying; metal cleaning and treating processes; distilling operations in the food and petrochemical industries; and industrial space heating. Power-plant rejected heat could be valuable for developing industries that perform these processes (see Steam; Power generation).

Cooling Reservoirs. The most extensively developed productive use for power-plant cooling is in multiple-purpose cooling reservoirs. Small impoundments built especially for heat dissipation have been managed for extensive recreational uses as well. Highly productive fisheries for warm-water species, eg, largemouth bass (*Micropterus salmoides*) and channel catfish (*Ictalurus punctatus*), have made cooling reservoirs highly popular. Such reservoirs have been extensively developed in lake-free areas, eg, Texas and Illinois. Broad-scale ecological research on some of these cooling reservoirs has documented the valuable synergistic relationship between cooling-lake fisheries and power-station heat dissipation (35).

BIBLIOGRAPHY

1. E. W. Moore, *Ind. Eng. Chem.* **50,** 87A (1958).
2. Pennsylvania Department of Health, *Heated Discharges—Their Effects on Streams*, Publication No. 3, Division of Sanitary Engineering, Pennsylvania Department of Health, Harrisburg, Pa., 1962.
3. C. C. Coutant, *Crit. Rev. Environ. Control* **1,** 341 (1970).

4. L. D. Jensen, ed., *Entrainment and Intake Screening*, Publication No. 74-049-00-5, Electric Power Research Institute, Palo Alto, Calif., 1974.
5. J. A. Zillich, *J. Water Pollut. Control Fed.* **44,** 212 (1972).
6. R. L. Jolley, *J. Water Pollut. Control Fed.* **47,** 601 (1975).
7. S. Markowski, *J. Anim. Ecol.* **28,** 243 (1959).
8. J. R. Schubel and B. C. Marcy, eds., *Power Plant Entrainment—A Biological Assessment*, Academic Press, New York, 1978.
9. S. W. Christensen, W. Van Winkle, L. W. Barnthouse, and D. S. Vaughan, *Environmental Impact Assessment Review* **2,** 63 (1981).
10. W. J. Ebel, *Fishery Bulletin* **68,** 1 (1969).
11. J. Z. Reynolds, *Science* **207,** 367 (1980).
12. S. A. Carter, *Environment* **19,** 16 (1977).
13. D. W. Fraser, D. C. Deubner, D. L. Hill, and D. K. Gilliam, *Science* **205,** 690 (1979).
14. *J. Water Pollut. Control Fed.*, annual review issues.
15. *Water Quality Criteria 1972*, Report No. R-73-033, National Academy of Sciences—National Academy of Engineering, U.S. Environmental Protection Agency, Washington, D.C., 1973, pp. 151–171 and appendix.
16. *Quality Criteria for Water*, U.S. Environmental Protection Agency, Washington, D.C., 1976.
17. C. C. Coutant, *Crit. Rev. Environ. Control* **3,** 1 (1972).
18. D. R. F. Harleman, *Report of a Workshop on the Impact of Thermal Power Plant Cooling Systems on Aquatic Environments*, CONF-750.980, EPRI SR-38, Asildmar, Pacific Grove, Calif., Sept. 28–Oct. 2, 1975, pp. 128–135.
19. H. H. Carter, J. R. Schubel, R. E. Wilson, and P. M. J. Woodhead, *Environ. Manage.* **3,** 353 (1979).
20. J. J. Magnuson, L. B. Crowder, and P. A. Medvick, *Am. Zool.* **19,** 331 (1979).
21. J. R. Gammon, unpublished results, Depauw University, Greencastle, Ind., 1981.
22. C. C. Coutant in H. Clepper, ed., *Marine Recreational Fisheries 5*, Sport Fishing Institute, Washington, D.C., 1980, pp. 179–187.
23. C. C. Coutant, *Striped Bass and the Management of Cooling Lakes, Proceedings of the Third Conference on Waste Heat Management and Utilization*, Hemisphere Publishing Corporation, Wash., in press.
24. L. D. Jensen, ed., *Third National Workshop on Entrainment and Impingement*, Ecological Analysts, Inc., Melville, N.Y., 1977.
25. L. D. Jensen, ed., *Fourth National Workshop on Entrainment and Impingement*, Ecological Analysts, Inc., Melville, N.Y., 1978.
26. R. B. McLean, P. T. Singley, J. S. Griffith, and M. V. McGee, *Threadfin Shad Impingement: Effect of Cold Stress*, NUREG/CR-1044, ORNL/NUREG/TM-340, Oak Ridge National Laboratory, Oak Ridge, Tennessee and National Technical Information Service, Springfield, Va., 1980.
27. J. S. Mattice in *Environmental Effects of Cooling Systems*, Technical Report Series 202, International Atomic Energy Agency, Vienna, 1980, pp. 12–26, 148–167.
28. R. J. Kedl and C. C. Coutant in G. W. Esch and R. W. McFarlane, eds., *Thermal Ecology II*, Conf-750425, National Technical Information Service, Springfield, Va., 1976, 394–400.
29. J. M. O'Connor, T. C. Ginn, and G. V. Pase, *The Evaluation and Description of a Power Plant Condenser Tube Simulator*, Report No. 8-0248-956, New York University Medical Center, New York, 1977.
30. J. S. Suffen, *The Physical Effects of Entrainment—Current Research at ORNL*, ORNL/TM-5948, Oak Ridge National Laboratory, Oak Ridge, Tennessee and National Technical Information Service, Springfield, Va., 1977.
31. W. Majewski and D. C. Miller, eds., *Predicting Effects of Power Plant Once-Through Cooling on Aquatic Systems*, Technical Papers in Hydrology 20, United Nations Educational, Scientific and Cultural Organization (Unesco), Paris, 1979.
32. M. Olszewski, S. J. Hillenbrand, and S. A. Reed, *Waste Heat vs. Conventional Systems for Greenhouse Environment Control: An Economic Assessment*, ORNL/TM-5069, Oak Ridge National Laboratory, Oak Ridge, Tennessee and National Technical Information Service, Springfield, Va., 1976.
33. S. E. Beall and G. Samuels, *The Use of Warm Water for Heating and Cooling Plant and Animal Enclosures*, ORNL-TM-3381, Oak Ridge National Laboratory, Oak Ridge, Tennessee and National Technical Information Service, Springfield, Va., 1971.
34. M. Olszewski, *Preliminary Investigations of the Thermal Energy Grid Concept*, ORNL/TM-5786,

Oak Ridge National Laboratory, Oak Ridge, Tennessee and National Technical Information Service, Springfield, Va., 1977.
35. W. Larimore and co-workers, *Evaluation of a Cooling Lake Fishery*, EA-1148, 3 Vols., Electric Power Research Institute, Palo Alto, Calif., 1979.

General References

Environmental Effects of Cooling Systems, Technical Report Series No. 202, International Atomic Energy Agency, Vienna, 1980.
C. C. Coutant, *A Critical Review of Ecological Effects of Power Plant Cooling*, U.S. Department of Energy, Washington, D.C. (in press).
J. R. Schubel and B. C. Marcy, Jr., eds., *Power Plant Entrainment—A Biological Assessment*, Academic Press, New York, 1978.
G. F. Lee and C. Stratton, *Ind. Water Eng.* **9,** 12–16 (1972).
H. Precht, J. Christophensen, H. Hensel, and W. Larcher, *Temperature and Life*, Springer-Verlag, New York, 1973.
C. C. Coutant, *Crit. Rev. Environ. Control* **3,** 1 (1972).
S. E. Beall, C. C. Coutant, M. Olszewski, and J. S. Suffen, *Ind. Water Eng.* **14,** 8 (1977).
L. B. Goss and L. Scott, *Factors Affecting Power Plant Waste Heat Utilization*, Pergamon Press, New York, 1980.

CHARLES C. COUTANT
Oak Ridge National Laboratory

THERMODYNAMICS

The science of thermodynamics deals with energy and its transformations in macroscopic systems, most especially in systems for which temperature is a characteristic coordinate. Every physical and chemical process can be examined thermodynamically.

The coordinates of classical thermodynamics do not include time. Classical thermodynamics does not predict the rates at which processes take place; nor does it provide a macroscopic description of physicochemical systems in the absence of experimental data. It does provide equations for the determination of equilibrium states and for the effect of temperature, pressure, and composition changes on the equilibrium properties of a wide variety of systems. For example, it allows calculation of the maximum or equilibrium yield possible from a chemical reaction for a given feed composition and given final conditions of temperature and pressure, but it implies nothing about the time required to approach this final state.

Thermodynamics is built around two fundamental laws that restrict the behavior of material systems. The first law of thermodynamics affirms the principle of energy conservation; the second law states the principle of entropy increase. In the formal application of these laws, attention is focused on a particular object, quantity of matter, or region of space as the system. All else then constitutes the surroundings. A system may be open or closed to the exchange of matter with its surroundings, or it may be isolated from its surroundings, in which case it can exchange neither matter nor energy. Once isolated, a system is independent of its surroundings and can only change toward a final equilibrium state characterized by properties that have no further tendency to change. Systems not in equilibrium states undergo processes during which the properties of the system change; moreover, the system and surroundings may exchange energy in the forms of heat Q and work W. A process that proceeds so that the system is never displaced more than differentially from an equilibrium state is said to be reversible, because such a process can be reversed in direction at any time by an infinitesimal change in external conditions.

Thermodynamics is covered in depth in refs. 1–10. A list of nomenclature used in the following discussion is given at the end of this article.

First and Second Laws of Thermodynamics

The four basic principles of thermodynamics are as follows: *1.* Associated with systems in equilibrium is an intrinsic property called internal energy U. For closed systems at rest, changes in this property are given by

$$dU = dQ - dW \tag{1}$$

2. First law of thermodynamics: The total energy change of any system together with its surroundings is zero. *3.* Associated with systems in equilibrium is an intrinsic property called entropy S. For reversible processes, changes in this property are given by

$$dS = \frac{dQ}{T} \tag{2}$$

4. Second law of thermodynamics: The total entropy change of any system with its surroundings is positive and approaches zero for reversible processes (11). Thus:

$$\Delta S_{\text{total}} \geq 0 \tag{3}$$

The first and third axioms are prerequisite to the other two, which state the laws of thermodynamics. Each asserts the existence of a thermodynamic property, and each provides an equation connecting the property with measurable quantities. These are not defining equations; they merely provide a means to calculate changes in each property.

For a closed system at rest, the first law for a finite process is expressed as

$$\Delta U = Q - W \tag{4}$$

where ΔU is the energy change of the system and Q and W are heat and work, respectively, passing the boundary between system and surroundings. Equation 1 is the differential form of this equation. These equations require the numerical value of Q to be positive for heat transfer to the system and negative for heat transfer from the system. The opposite convention applies to W: numerical values are positive for work done by the system and negative for work done on the system.

Heat Engines and Heat Pumps

A general equation has been developed for the work produced by a cyclic engine that receives heat Q_H from a heat reservoir at temperature T_H and discards heat Q_C to another heat reservoir at the lower temperature T_C. A heat reservoir is a body that can absorb or reject unlimited amounts of heat without change in temperature. The entropy change of a heat reservoir depends only on its temperature and on the quantity of heat transferred and is always given by the integrated form of equation 2:

$$\Delta S = \frac{Q}{T} \tag{5}$$

Thus, the entropy changes of the two heat reservoirs are given by

$$\Delta S_H = \frac{Q_H}{T_H} \tag{6}$$

and

$$\Delta S_C = \frac{Q_C}{T_C} \tag{7}$$

Since the engine operates in cycles, it experiences no change in properties and, therefore:

$$\Delta S_{total} = \frac{Q_H}{T_H} + \frac{Q_C}{T_C} \tag{8}$$

An expression for the first law with respect to the engine is

$$\Delta U_{engine} = Q - W \tag{9}$$

The heat quantities Q_H and Q_C apply to the engine as well as to the heat reservoirs, but with opposite signs. Thus, for the engine:

$$Q = -Q_H - Q_C \tag{10}$$

Since $\Delta U_{engine} = 0$:

$$W = -Q_H - Q_C \tag{11}$$

Elimination of Q_H between equations 8 and 11 gives

$$W = -T_H\,\Delta S_{total} + Q_C\left(\frac{T_H}{T_C} - 1\right) \tag{12}$$

The work W can range from zero, if the engine is completely ineffective, to some maximum positive value attained for reversible operation. If $W = 0$, then the process degenerates to one of simple heat transfer, for which

$$\Delta S_{total} = Q_C\left(\frac{T_H - T_C}{T_H T_C}\right) \tag{13}$$

At the other limit, $\Delta S_{total} = 0$, and

$$W = Q_C\left(\frac{T_H}{T_C} - 1\right) \tag{14}$$

From this result and if W is to be positive and finite, Q_C must also be positive and finite. Thus, even for reversible operation, a finite amount of heat must be rejected from the engine to the reservoir at T_C. Two additional equations result from combining equations 11 and 14:

$$\frac{Q_C}{T_C} = \frac{-Q_H}{T_H} \tag{15}$$

and

$$\frac{W}{-Q_H} = 1 - \frac{T_C}{T_H} \tag{16}$$

These are Carnot's equations, and they apply to all reversible heat engines operating between a heat source and a heat sink at fixed temperature levels, ie, to Carnot engines. The Qs in these equations refer to the heat reservoirs and this is the origin of the minus signs on Q_H. Common practice is to delete the minus signs and write

$$\frac{|Q_C|}{|Q_H|} = \frac{T_C}{T_H} \tag{17}$$

and

$$\frac{|W|}{|Q_H|} = 1 - \frac{T_C}{T_H} \tag{18}$$

Equation 18 yields the thermal efficiency of a Carnot heat engine, ie, the fraction of the heat taken in that is converted into work. A Carnot engine is reversible and, thus, can be run as a heat pump or refrigerator, as indicated in Figure 1. Equations 17–18 apply to the Carnot heat pump as well as to the Carnot engine, because all quantities are the same for the two cases. The measure of performance of the Carnot heat pump is given by its coefficient of performance:

$$\frac{|Q_C|}{|W|} = \frac{T_C}{T_H - T_C} \tag{19}$$

an equation obtained upon division of equation 17 by equation 18.

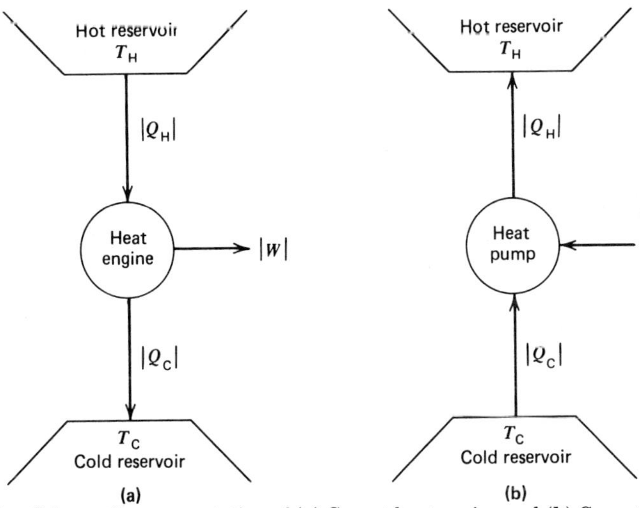

Figure 1. Schematic representation of (**a**) Carnot heat engine and (**b**) Carnot heat pump.

For given temperature levels T_H and T_C, no heat engine operating in cycles can have a thermal efficiency greater than that of a Carnot engine, and no refrigerator can have a coefficient of performance greater than that of a Carnot heat pump. For example, a power plant operating between a boiler temperature of 560 K and a condenser temperature of 310 K typically has a thermal efficiency (work out/heat in) of ca 0.3, whereas a Carnot engine operating between the same temperatures has an efficiency (given by eq. 18) of 0.446.

Fundamental Property Relations for *PVT* Systems

The systems of interest in chemical technology are usually comprised of fluids not appreciably influenced by surface, gravitational, electrical, or magnetic effects. For this large class of fluids, molar or specific volume is observed to be a function of just temperature, pressure, and composition. Systems made up of such fluids are called *PVT* systems. The macroscopic intensive properties of homogeneous fluids existing at equilibrium in *PVT* systems are functions of temperature, pressure, and composition. The macroscopic intensive properties include molar or specific volume V, molar or specific internal energy U, and molar or specific entropy S; the symbols V, U, and S represent these mass-independent or intensive properties. Total-system properties (V^t, U^t, S^t) depend upon system size and are extensive. Thus, if the system contains n moles of fluid, $M^t = nM$, where M is a molar property of the fluid.

Consider a closed, nonreacting *PVT* system containing n moles of a homogeneous fluid mixture. The mole numbers of the individual chemical species sum to

$$n = n_1 + n_2 + \ldots + n_N \tag{20}$$

If this system undergoes a differential change of state in a reversible process, equation 1 can be written as

$$d(nU) = dQ_{\text{rev}} - dW_{\text{rev}} \tag{21}$$

According to equation 2:

$$dQ_{\text{rev}} = Td(nS) \tag{22}$$

and by the definition of work:

$$dW_{\text{rev}} = Pd(nV) \tag{23}$$

These three equations combine to give

$$d(nU) = Td(nS) - Pd(nV) \tag{24}$$

Although derived for a reversible process, equation 24 relates properties only, irrespective of the process. It relates the three thermodynamic properties so far considered and applies to any change in the equilibrium state of a homogeneous, closed, nonreacting fluid system.

According to equation 24:

$$nU = F(nS, nV) \tag{25}$$

and, therefore:

$$d(nU) = \left[\frac{\partial(nU)}{\partial(nS)}\right]_{nV,n} d(nS) + \left[\frac{\partial(nU)}{\partial(nV)}\right]_{nS,n} d(nV) \tag{26}$$

The subscript n indicates that all mole numbers n_i are held constant. Comparison with equation 24 shows that

$$\left[\frac{\partial(nU)}{\partial(nS)}\right]_{nV,n} = T \qquad (27)$$

and

$$\left[\frac{\partial(nU)}{\partial(nV)}\right]_{nS,n} = -P \qquad (28)$$

For the more general case of a homogeneous fluid in which the n_i vary, for whatever reason, it is assumed that nU is a function of the n_i as well as of nS and nV:

$$nU = \mathcal{F}(nS, nV, n_1, n_2, \ldots, n_N) \qquad (29)$$

Then:

$$d(nU) = \left[\frac{\partial(nU)}{\partial(nS)}\right]_{nV,n} d(nS) + \left[\frac{\partial(nU)}{\partial(nV)}\right]_{nS,n} d(nV) + \sum \left[\frac{\partial(nU)}{\partial n_i}\right]_{nS,nV,n_j} dn_i \qquad (30)$$

where the summation is over all species i, and the subscript n_j signifies that all mole numbers are held constant except the ith. Equations 27–28 with the definition of the chemical potential:

$$\mu_i \equiv \left[\frac{\partial(nU)}{\partial n_i}\right]_{nS,nV,n_j} \qquad (31)$$

reduce equation 30 to

$$d(nU) = Td(nS) - Pd(nV) + \sum \mu_i dn_i \qquad (32)$$

This fundamental property relation is the basis for development of all other equations relating the properties of PVT systems.

Alternative forms of the fundamental property relation follow from the definitions of three additional thermodynamic functions:

enthalpy $\qquad nH = nU + P(nV) \qquad (33)$

Helmholtz function $\qquad nA = nU - T(nS) \qquad (34)$

Gibbs function $\qquad nG = nU + P(nV) - T(nS) \qquad (35)$

Taking the total differentials of equations 33–35 and eliminating $d(nU)$ by equation 32 gives

$$d(nH) = Td(nS) + nVdP + \sum \mu_i dn_i \qquad (36)$$

$$d(nA) = -nSdT - Pd(nV) + \sum \mu_i dn_i \qquad (37)$$

$$d(nG) = -nSdT + nVdP + \sum \mu_i dn_i \qquad (38)$$

Equations 32 and 36–38 are equivalent and apply to changes between equilibrium states in any homogeneous fluid system, either open or closed. The choice of which equation to use in a particular application is dictated by convenience. However, the Gibbs function G is of particular importance because of its unique functional relationship to T, P, and the n_i, which are the variables of primary interest in chemical technology. Thus, by equation 38:

$$\left[\frac{\partial(nG)}{\partial T}\right]_{P,n} = -nS \qquad (39)$$

where n as a subscript indicates that all n_i are held constant. This equation may, therefore, be written

$$\left(\frac{\partial G}{\partial T}\right)_{P,x} = -S \tag{40}$$

where subscript x denotes constant composition. Similarly:

$$\left(\frac{\partial G}{\partial P}\right)_{T,x} = V \tag{41}$$

and

$$\left[\frac{\partial(nG)}{\partial n_i}\right]_{T,P,n_j} = \mu_i \tag{42}$$

Equations 32 and 36–38 apply to the special case of 1 mol of a homogeneous fluid of constant composition. Thus, for $n = 1$ and $dn_i = 0$, these equations become

$$dU = TdS - PdV \tag{43}$$

$$dH = TdS + VdP \tag{44}$$

$$dA = -SdT - PdV \tag{45}$$

$$dG = -SdT + VdP \tag{46}$$

Since these are exact differential expressions, the Maxwell equations can be written by inspection; the two most useful ones are derived from equations 45–46:

$$\left(\frac{\partial S}{\partial V}\right)_T = \left(\frac{\partial P}{\partial T}\right)_V \tag{47}$$

$$\left(\frac{\partial S}{\partial P}\right)_T = -\left(\frac{\partial V}{\partial T}\right)_P \tag{48}$$

Enthalpy and Entropy as Functions of *T* and *P*

For a homogeneous fluid of constant composition:

$$H = H(T, P) \tag{49}$$

and

$$S = S(T, P) \tag{50}$$

Therefore:

$$dH = \left(\frac{\partial H}{\partial T}\right)_P dT + \left(\frac{\partial H}{\partial P}\right)_T dP \tag{51}$$

and

$$dS = \left(\frac{\partial S}{\partial T}\right)_P dT + \left(\frac{\partial S}{\partial P}\right)_T dP \tag{52}$$

By definition

$$C_P \equiv \left(\frac{\partial H}{\partial T}\right)_P \tag{53}$$

is the heat capacity at constant pressure. If equation 44 is divided by dT and restricted to constant P:

$$\left(\frac{\partial H}{\partial T}\right)_P = T\left(\frac{\partial S}{\partial T}\right)_P \tag{54}$$

With equation 53, this becomes

$$\left(\frac{\partial S}{\partial T}\right)_P = \frac{C_P}{T} \tag{55}$$

When equation 44 is divided by dP and restricted to constant T:

$$\left(\frac{\partial H}{\partial P}\right)_T = T\left(\frac{\partial S}{\partial P}\right)_T + V \tag{56}$$

In view of equation 48, this becomes

$$\left(\frac{\partial H}{\partial P}\right)_T = V - T\left(\frac{\partial V}{\partial T}\right)_P \tag{57}$$

Combination of equations 51, 53, and 57 gives

$$dH = C_P dT + \left[V - T\left(\frac{\partial V}{\partial T}\right)_P\right] dP \tag{58}$$

Similarly, combination of equations 48, 52, and 55 gives

$$dS = \frac{C_P}{T} dT - \left(\frac{\partial V}{\partial T}\right)_P dP \tag{59}$$

For an ideal gas:

$$PV = RT \tag{60}$$

and

$$\left(\frac{\partial V}{\partial T}\right)_P = \frac{R}{P} = \frac{V}{T} \tag{61}$$

Thus, for an ideal gas, equations 58–59 reduce to

$$dH = C_P dT \tag{62}$$

and

$$dS = \frac{C_P}{T} dT - \frac{R}{P} dP \tag{63}$$

Internal Energy and Entropy as Functions of *T* and *V*

Since $V = V(T, P)$ for a constant-composition fluid, T and V can be considered independent variables; therefore:

$$U = U(T, V) \tag{64}$$

and

$$S = S(T, V) \tag{65}$$

Then:

$$dU = \left(\frac{\partial U}{\partial T}\right)_V dT + \left(\frac{\partial U}{\partial V}\right)_T dV \tag{66}$$

and

$$dS = \left(\frac{\partial S}{\partial T}\right)_V dT + \left(\frac{\partial S}{\partial V}\right)_T dV \tag{67}$$

By definition

$$C_V \equiv \left(\frac{\partial U}{\partial T}\right)_V \tag{68}$$

is the heat capacity at constant volume. Two relations follow from equation 43:

$$\left(\frac{\partial U}{\partial T}\right)_V = T\left(\frac{\partial S}{\partial T}\right)_V \tag{69}$$

and

$$\left(\frac{\partial U}{\partial V}\right)_T = T\left(\frac{\partial S}{\partial V}\right)_T - P \tag{70}$$

With equations 47 and 68, these become

$$\left(\frac{\partial S}{\partial T}\right)_V = \frac{C_V}{T} \tag{71}$$

and

$$\left(\frac{\partial U}{\partial V}\right)_T = T\left(\frac{\partial P}{\partial T}\right)_V - P \tag{72}$$

Combining equations 66, 68, and 72 yields

$$dU = C_V dT + \left[T\left(\frac{\partial P}{\partial T}\right)_V - P\right] dV \tag{73}$$

and

$$dS = \frac{C_V}{T} dT + \left(\frac{\partial P}{\partial T}\right)_V dV \tag{74}$$

For an ideal gas:

$$\left(\frac{\partial P}{\partial T}\right)_V = \frac{R}{V} = \frac{P}{T} \tag{75}$$

and equations 73–74 reduce to, respectively,

$$dU = C_V dT \tag{76}$$

and

$$dS = \frac{C_V}{T} dT + \frac{R}{V} dV \tag{77}$$

Equations 58–59 and 73–74 provide basic relationships which allow calculation of numerical values for the thermodynamic properties U, H, and S from experimental heat-capacity and PVT data.

Energy Equations for Steady-Flow Processes

Industrial production of chemicals is largely by continuous processes in which the rates of inflow and outflow of mass are constant and equal. Moreover, conditions at all points in the process are maintained constant with time. A simple steady-flow process is represented in Figure 2. A process occurs within the fixed control volume between points 1 and 2 that changes the properties of a fluid element as it flows from point 1 to point 2. The system is the fluid in the control volume plus the fluid element of mass δm_1 (Fig. 2**a**) that enters the control volume at point 1 during time $\delta\tau$. After time interval $\delta\tau$, the system appears as in Figure 2**b** and consists of fluid in the control volume plus the fluid element of mass δm_2 that has left the control volume. $\delta m_2 = \delta m_1$ for a steady-flow process, and the properties of the fluid in the control volume are unchanged. The fluid elements δm_1 and δm_2 have properties as measured at points 1 and 2, and these include a velocity u and an elevation above a datum level z. Thus, these masses have kinetic and potential energy as well as internal energy, and the general energy balance for a closed system, as given by equation 4, must be changed to

$$\Delta U + \Delta E_K + \Delta E_P = Q - W \tag{78}$$

where Δ indicates the change from the initial to the final state of the system. As applied to the change shown in Figure 2 during time interval $\delta\tau$, this equation becomes

$$\Delta(U\delta m) + \Delta E_K + \Delta E_P = \delta Q - \delta W \tag{79}$$

where U is internal energy per unit mass and

$$\Delta(U\delta m) \equiv U_2\delta m_2 - U_1\delta m_1 \tag{80}$$

The kinetic and potential energies of the fluid elements at points 1 and 2 are given by

$$E_K = \frac{1}{2}(\delta m)u^2 \tag{81}$$

and

$$E_P = (\delta m)zg \tag{82}$$

where g is the local acceleration of gravity. The energy equation is now:

$$\Delta(U\delta m) + \Delta\left(\frac{u^2\delta m}{2}\right) + \Delta(zg\delta m) = \delta Q - \delta W \tag{83}$$

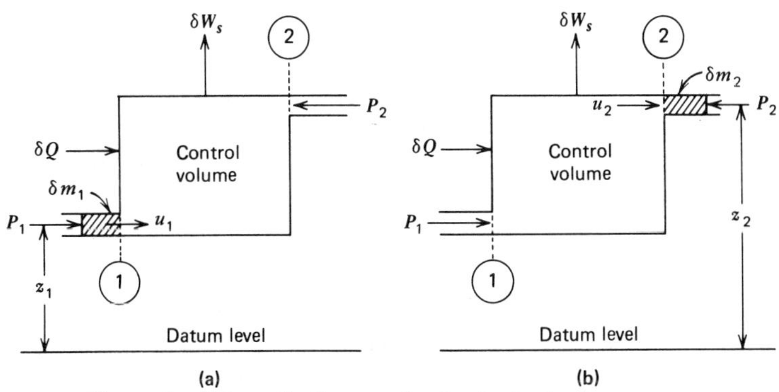

Figure 2. Schematic diagram of a simple steady-flow process.

The quantity δW includes two kinds of work: the shaft work δW_s shown in Figure 2 and the work done by pressures P_1 and P_2 moving through the volumes occupied by the fluid masses δm_1 and δm_2. Thus:

$$\delta W = \delta W_s + P_2(V_2\delta m_2) - P_1(V_1\delta m_1) = \delta W_s + \Delta(PV\delta m) \tag{84}$$

and the energy equation becomes

$$\Delta(U\delta m) + \Delta\left(\frac{u^2\delta m}{2}\right) + \Delta(zg\delta m) + \Delta(PV\delta m) = \delta Q - \delta W_s \tag{85}$$

Factoring δm yields

$$\Delta\left[\left(U + PV + \frac{u^2}{2} + zg\right)\delta m\right] = \delta Q - \delta W_s \tag{86}$$

Since by definition $H = U + PV$, this latter equation becomes

$$\Delta\left[\left(H + \frac{u^2}{2} + zg\right)\delta m\right] = \delta Q - \delta W_s \tag{87}$$

Division of equation 87 by the time interval $\delta\tau$ transforms it into an expression involving rates:

$$\Delta\left[\left(H + \frac{u^2}{2} + zg\right)\dot{m}\right] = \dot{Q} - \dot{W}_s \tag{88}$$

Equations 87–88, developed with respect to a process with one entering and one leaving stream, are applicable for any number of such streams. The Δ simply means the difference between the sum over all leaving streams and the sum over all entering streams. However, in the common case of a single entering and a single leaving stream, $\dot{m}$ in equation 88 must be the same for both; therefore:

$$\Delta\left(H + \frac{u^2}{2} + zg\right)\dot{m} = \dot{Q} - \dot{W}_s \tag{89}$$

Multiplication by the time required for a unit mass of fluid to enter or leave the control volume gives

$$\Delta H + \frac{\Delta u^2}{2} + g\Delta z = Q - W_s \tag{90}$$

where each term is for a unit mass of fluid.

Partial Molar Properties

Since the macroscopic intensive properties of homogeneous fluids in equilibrium states are functions of T, P, and composition, it follows that the total property of a phase, say nM, can be expressed functionally as

$$nM = m(T, P, n_1, n_2, n_3, \ldots) \tag{91}$$

The total differential of nM is, therefore,

$$d(nM) = \left[\frac{\partial(nM)}{\partial T}\right]_{P,n} dT + \left[\frac{\partial(nM)}{\partial P}\right]_{T,n} dP + \sum \left[\frac{\partial(nM)}{\partial n_i}\right]_{T,P,n_j} dn_i \tag{92}$$

or

$$d(nM) = n\left(\frac{\partial M}{\partial T}\right)_{P,x} dT + n\left(\frac{\partial M}{\partial P}\right)_{T,x} dP + \sum\left[\frac{\partial(nM)}{\partial n_i}\right]_{T,P,n_j} dn_i \qquad (93)$$

The derivatives in the summation are called partial molar properties and are denoted by $\overline{M}_i$; thus, by definition:

$$\overline{M}_i \equiv \left[\frac{\partial(nM)}{\partial n_i}\right]_{T,P,n_j} \qquad (94)$$

Equation 93 can now be written

$$d(nM) = n\left(\frac{\partial M}{\partial T}\right)_{P,x} dT + n\left(\frac{\partial M}{\partial P}\right)_{T,x} dP + \sum \overline{M}_i dn_i \qquad (95)$$

Important equations follow from this result through the identities

$$d(nM) = ndM + Mdn \qquad (96)$$

and

$$dn_i = d(x_i n) = x_i dn + n dx_i \qquad (97)$$

Combining these expressions with equation 95 and collecting like terms gives

$$\left[dM - \left(\frac{\partial M}{\partial T}\right)_{P,x} dT - \left(\frac{\partial M}{\partial P}\right)_{T,x} dP - \sum \overline{M}_i dx_i\right] n + \left[M - \sum \overline{M}_i x_i\right] dn = 0 \qquad (98)$$

Since n and dn are independent and arbitrary, the terms in brackets must each be zero; thus:

$$dM = \left(\frac{\partial M}{\partial T}\right)_{P,x} dT + \left(\frac{\partial M}{\partial P}\right)_{T,x} dP + \sum \overline{M}_i dx_i \qquad (99)$$

and

$$M = \sum x_i \overline{M}_i \qquad (100)$$

Differentiation of equation 100 yields

$$dM = \sum x_i d\overline{M}_i + \sum \overline{M}_i dx_i \qquad (101)$$

If equation 99 and equation 101 are both valid:

$$\left(\frac{\partial M}{\partial T}\right)_{P,x} dT + \left(\frac{\partial M}{\partial P}\right)_{T,x} dP - \sum x_i d\overline{M}_i = 0 \qquad (102)$$

This result is known as the Gibbs-Duhem equation. It imposes a constraint on how the partial molar properties of any phase may vary with temperature, pressure, and composition. In particular, at constant T and P it represents a simple relationship among the $\overline{M}_i$ to which measured values of partial properties must conform if they are to be consistent with this thermodynamic requirement.

Equations 94–95, 99–100, and 102 are the general property relations between partial molar properties and solution properties. The symbol M may represent any intensive thermodynamic property other than the system coordinates T, P, and x_i. For example, when $M \equiv H$, the derivatives $(\partial H/\partial T)_{P,x}$ and $(\partial H/\partial P)_{T,x}$ are given by equations 53 and 57. Then equations 99–100 and 102 become

$$dH = C_P dT + \left[V - T\left(\frac{\partial V}{\partial T}\right)_{P,x}\right] dP + \sum \overline{H}_i dx_i \tag{103}$$

$$H = \sum x_i \overline{H}_i$$

and

$$C_P dT + \left[V - T\left(\frac{\partial V}{\partial T}\right)_{P,x}\right] dP - \sum x_i d\overline{H}_i = 0 \tag{104}$$

When $M \equiv G$, the required derivatives are given by equations 40–41. Moreover, the derivative on the left side of equation 42 defines the partial molar Gibbs function $\overline{G}_i$. Therefore:

$$\mu_i = \overline{G}_i \tag{105}$$

and equation 95 reduces to equation 38. Equations 100 and 102, respectively, become

$$G = \sum x_i \overline{G}_i = \sum x_i \mu_i \tag{106}$$

and

$$-SdT + VdP - \sum x_i d\overline{G}_i = 0 \tag{107}$$

If temperature and pressure are constant, equation 107 reduces to

$$\sum x_i d\overline{G}_i = \sum x_i d\mu_i = 0 \qquad \text{(constant } T, P) \tag{108}$$

which is a common form of the Gibbs-Duhem equation.

Fugacity and Fugacity Coefficient

The fugacity and fugacity coefficient are auxiliary functions useful in phase- and chemical-equilibrium calculations; they are defined in relation to the Gibbs function. For a constant-composition mixture, the fugacity f and fugacity coefficient ϕ are defined by the equations:

$$dG = RTd \ln f \qquad \text{(constant } T, x) \tag{109}$$

$$\lim_{P \to 0} (f/P) = 1 \tag{110}$$

and

$$\phi \equiv f/P \tag{111}$$

These equations apply as well to pure species i, a special case of a constant-composition mixture, as indicated by subscripts: G_i, f_i, and ϕ_i. For species i in solution, $\hat{f}_i$ and $\hat{\phi}_i$ are defined by a similar set of equations:

$$d\mu_i = d\overline{G}_i = RTd \ln \hat{f}_i \qquad \text{(constant } T) \tag{112}$$

$$\lim_{P \to 0} (\hat{f}_i / x_i P) = 1 \tag{113}$$

$$\hat{\phi}_i \equiv \hat{f}_i / x_i P \tag{114}$$

The calculation of values for f and ϕ for a constant-composition mixture or a pure species depends on equation 46:

$$dG = VdP \qquad \text{(constant } T,x) \tag{115}$$

Eliminating dG in equation 109:

$$d \ln f = \frac{V}{RT} dP \qquad \text{(constant } T,x) \tag{116}$$

By equation 111:

$$d \ln \phi = d \ln f - \frac{dP}{P} \tag{117}$$

Combining the last two equations and rearranging results in

$$d \ln \phi = \left(\frac{PV}{RT} - 1\right)\frac{dP}{P} = (Z - 1)\frac{dP}{P} \tag{118}$$

where the compressibility factor Z is defined as $Z \equiv PV/RT$. Integration from $P = 0$, where $\phi = 1$, to $P = P$ gives

$$\ln \phi = \int_0^P (Z - 1)\frac{dP}{P} \qquad \text{(constant } T,x) \tag{119}$$

An analogous equation allows calculation of $\ln \hat{\phi}_i$:

$$\ln \hat{\phi}_i = \int_0^P (\overline{Z}_i - 1)\frac{dP}{P} \qquad \text{(constant } T,x) \tag{120}$$

where $\overline{Z}_i$ is the partial molar compressibility factor. Given a value of ϕ or $\hat{\phi}_i$, the value of f or $\hat{f}_i$ is determined from equation 111 or 114. Alternative expressions for $\ln \phi$ and $\ln \hat{\phi}_i$ are given by equations 187 and 195.

In general, the composition dependence of $\hat{f}_i$ must be determined experimentally. However, a useful standard of comparison is provided by the Lewis-Randall rule, a model for solution ideality (12):

$$\hat{f}_i^{\text{id}} = f_i(T,P)x_i \tag{121}$$

where f_i, the fugacity of pure species i, is a function of temperature and pressure.

Property Changes of Mixing and Excess Properties

The thermodynamic properties of mixtures are often expressed in relation to the properties of the pure species through property changes of mixing, defined by

$$\Delta M \equiv M - \sum x_i M_i \tag{122}$$

where the solution property M and the pure-species properties M_i are all at the same T and P. The two property changes of mixing that can be directly measured are the volume change of mixing ΔV and the heat of mixing ΔH. These properties are functions of temperature, pressure, and composition, and equations 93–95, 99–100, and 102 apply when M is replaced by ΔM and $\overline{M}_i$ is replaced by

$$\overline{\Delta M}_i \equiv \overline{M}_i - M_i \tag{123}$$

Evaluation of ΔG and ΔS is indirect. Integration of equation 112 from the state of pure i to a state of i in solution at the same T and P gives

$$\overline{G}_i - G_i = RT \ln \frac{\hat{f}_i}{f_i} \tag{124}$$

Adding and subtracting $RT \ln x_i$ on the right yields

$$\overline{G}_i - G_i = RT \ln \frac{\hat{f}_i}{x_i f_i} + RT \ln x_i \tag{125}$$

The activity coefficient of species i in solution is defined as

$$\gamma_i \equiv \frac{\hat{f}_i}{x_i f_i} \tag{126}$$

Then:

$$\overline{G}_i - G_i = RT \ln \gamma_i + RT \ln x_i \tag{127}$$

Multiplication by x_i and summation over all species gives

$$\sum x_i \overline{G}_i - \sum x_i G_i = RT \sum x_i \ln \gamma_i + RT \sum x_i \ln x_i \tag{128}$$

or, from equation 106:

$$G = \sum x_i G_i + RT \sum x_i \ln \gamma_i + RT \sum x_i \ln x_i \tag{129}$$

and from equation 122:

$$\Delta G = RT \sum x_i \ln \gamma_i + RT \sum x_i \ln x_i \tag{130}$$

Values of $\ln \gamma_i$ are found from phase-equilibrium data. By equations 33 and 35:

$$S = \frac{H - G}{T} \tag{131}$$

and it follows that:

$$\Delta S = \frac{\Delta H - \Delta G}{T} \tag{132}$$

By equation 121 for an ideal solution:

$$\hat{f}_i^{\text{id}} = f_i x_i \tag{133}$$

In this case, equations 126–127 and 129 become, respectively,

$$\gamma_i^{\text{id}} = 1 \tag{134}$$

$$\overline{G}_i^{\text{id}} = G_i + RT \ln x_i \tag{135}$$

and

$$G^{\text{id}} = \sum x_i G_i + RT \sum x_i \ln x_i \tag{136}$$

Subtracting equation 135 from 127 and equation 136 from 129 gives

$$\overline{G}_i - \overline{G}_i^{\text{id}} = RT \ln \gamma_i \tag{137}$$

and

$$G - G^{\text{id}} = RT \sum x_i \ln \gamma_i \tag{138}$$

Differences like those on the left sides of these two equations are called excess properties. Thus, in general:

$$\overline{M}_i^E \equiv \overline{M}_i - \overline{M}_i^{\text{id}} \tag{139}$$

and

$$M^E \equiv M - M^{id} \tag{140}$$

In particular:

$$\overline{G}_i^E = RT \ln \gamma_i \tag{141}$$

and

$$G^E = RT \sum x_i \ln \gamma_i \tag{142}$$

Moreover, by equation 130:

$$G^E = \Delta G - RT \sum x_i \ln x_i \tag{143}$$

The quantities $\overline{M}_i^E$ are partial molar properties with respect to M^E, and equations 93–95, 99–100, and 102 apply when M is replaced by M^E and $\overline{M}_i$ is replaced by $\overline{M}_i^E$. For most properties $\overline{M}_i^{id} = M_i$, $M^{id} = \Sigma\, x_i M_i$, and $M^E = \Delta M$. These relationships hold when M represents Z, V, U, C_V, H, and C_P. Properties G, A, and S are the exceptions. Equations 141–143 apply for G. For the entropy,

$$\overline{S}_i^{id} = S_i - R \ln x_i \tag{144}$$

$$S^{id} = \sum x_i S_i - R \sum x_i \ln x_i \tag{145}$$

and

$$S^E = \Delta S + R \sum x_i \ln x_i \tag{146}$$

In addition, the following relations among excess functions are valid:

$$H^E = G^E + TS^E \tag{147}$$

$$H^E = -RT^2 \left[\frac{\partial(G^E/RT)}{\partial T}\right]_{P,x} \tag{148}$$

$$S^E = -(\partial G^E/\partial T)_{P,x} \tag{149}$$

and

$$V^E = (\partial G^E/\partial P)_{T,x} \tag{150}$$

According to equation 141:

$$\ln \gamma_i = \frac{\overline{G}_i^E}{RT} \tag{151}$$

and, therefore, $\ln \gamma_i$ is a partial molar property with respect to G^E/RT. Because of the primary role played by γ_i in vapor–liquid equilibrium calculations, this relationship is important. However, for it to be useful, an expression for G^E/RT as a function of composition is needed. Of the various empirical expressions that have been proposed to represent this dependence, the modern local-composition equations are of greatest practical use. Notable among these are the Wilson equation, the NRTL equation, and the UNIQUAC equation (13–15). All are inherently applicable to multicomponent systems. Two group-contribution methods for the prediction of activity coefficients have developed from local-composition equations and are commonly used. They are the UNIFAC method and the ASOG method (16–17).

PVT and Phase Behavior of Pure Substances

The volumetric and phase behavior of pure substances is most readily described by reference to schematic PV and PT diagrams. Figure 3, a PT diagram, displays regions of single-phase equilibrium as areas. Curves on this figure represent states of two-phase equilibrium, and pressures represented by the curve are called saturation pressures (P^{sat}).

In Figure 4, a PV diagram, regions of two-phase equilibrium appear as areas. Thus, the region of vapor–liquid equilibrium is delineated by curves representing saturated liquid and saturated vapor. For temperatures less than the vapor–liquid critical temperature T_c, isotherms to the left of the liquid saturation curve represent states of *subcooled liquid*, and isotherms to the right of the vapor saturation curve are for *superheated vapors*. The critical isotherm exhibits a horizontal inflection at the critical state and, thus:

$$\left(\frac{\partial P}{\partial V}\right)_{T;\,\mathrm{cr}} = \left(\frac{\partial^2 P}{\partial V^2}\right)_{T;\,\mathrm{cr}} = 0 \tag{152}$$

For sufficiently large molar volumes V, all isotherms are approximated by the ideal-gas equation $P = RT/V$.

The treatment of pure-component vapor–liquid equilibria is facilitated through the use of property changes of vaporization ΔM_i^{lv}, defined by

$$\Delta M_i^{\mathrm{lv}} \equiv M_i^{\mathrm{v}} - M_i^{\mathrm{l}} \tag{153}$$

where M_i^{l} and M_i^{v} are molar properties of pure i for states of saturated liquid and saturated vapor. The heat of vaporization ΔH_i^{lv} is directly related to the entropy change of vaporization:

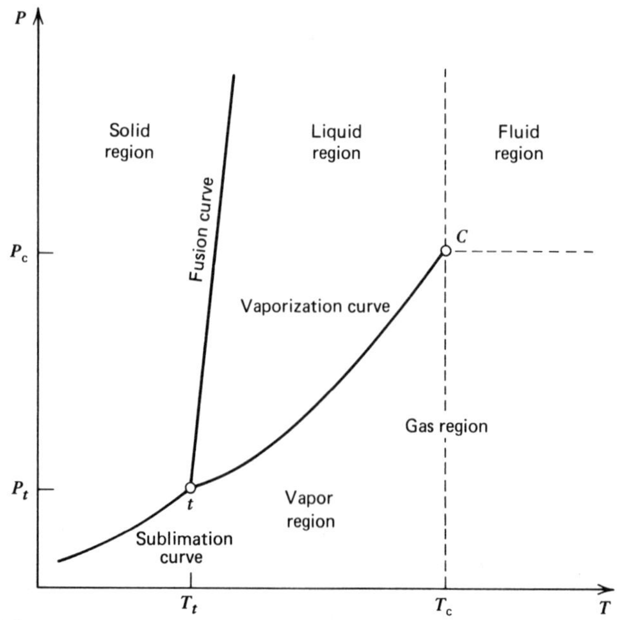

Figure 3. PT diagram for a pure substance that expands on melting (not to scale). For a substance that contracts on melting, eg, water, the fusion curve has a negative slope. Point t is a triple state and point C is the gas–liquid critical state.

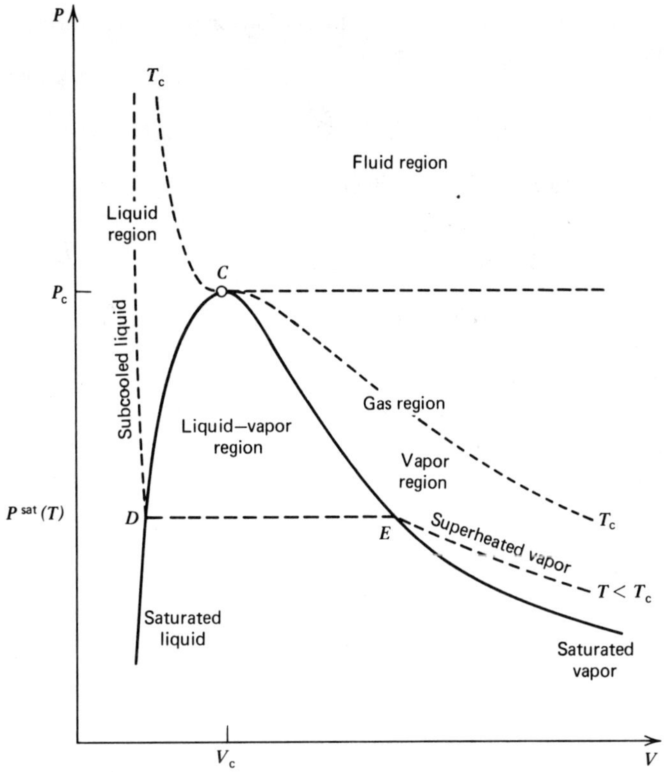

Figure 4. PV diagram for a pure fluid (not to scale). Point C is the gas–liquid critical state, and $P^{sat}(T)$ is the constant pressure at which phase transition occurs at temperature T.

$$\Delta H_i^{lv} = T\Delta S_i^{lv} \tag{154}$$

Moreover, the heat of vaporization is related to the slope of the liquid–vapor saturation curve through the Clapeyron equation:

$$\frac{dP_i^{sat}}{dT} = \frac{\Delta H_i^{lv}}{T\Delta V_i^{lv}} \tag{155}$$

Many empirical expressions have been proposed for the temperature dependence of the vapor–liquid saturation pressure, and they can all be rationalized through equation 155. By far the most popular of these expressions is Antoine's equation:

$$\ln P_i^{sat} = A_i - \frac{B_i}{T + C_i} \tag{156}$$

Numerical values for the three Antoine constants have been tabulated (18).

Residual Functions

A class of thermodynamic properties called residual functions is defined as the difference between a molar property M' that a substance would have were it an ideal gas and the actual value M at the same temperature, pressure, and composition:

$$\Delta M' \equiv M' - M \tag{157}$$

Residual functions reflect the contributions of intermolecular forces to the value of a thermodynamic property. For $\Delta H'$ and $\Delta S'$:

$$\lim_{P\to 0} \Delta M' = 0 \tag{158}$$

It follows from equations 157 and 158 that:

$$\Delta M' = \int_0^P \left[\left(\frac{\partial M'}{\partial P}\right)_T - \left(\frac{\partial M}{\partial P}\right)_T\right] dP \qquad \text{(constant } T,x) \tag{159}$$

Thus, values or expressions for the residual functions $\Delta H'$ and $\Delta S'$ can be calculated from PVT data or from equations of state.

Property changes are readily determined for substances in the ideal-gas state, and these in combination with residual functions can be used to compute property changes of real substances. The computational scheme is suggested in Figure 5 and is based on the identity:

$$\Delta M \equiv \Delta M_1' + (M_3' - M_1') + (M_2' - M_3') - \Delta M_2' \tag{160}$$

Thus, for $M = H$:

$$\Delta H = \Delta H_1' + \int_{T_1}^{T_2} C_P' \, dT - \Delta H_2' \tag{161}$$

and, for $M = S$:

$$\Delta S = \Delta S_1' + \int_{T_1}^{T_2} \frac{C_P'}{T} dT - R \ln \frac{P_2}{P_1} - \Delta S_2' \tag{162}$$

Evaluation of the integrals requires empirical expressions for the temperature dependence of the ideal-gas heat capacity C_P'; these are given in ref. 18.

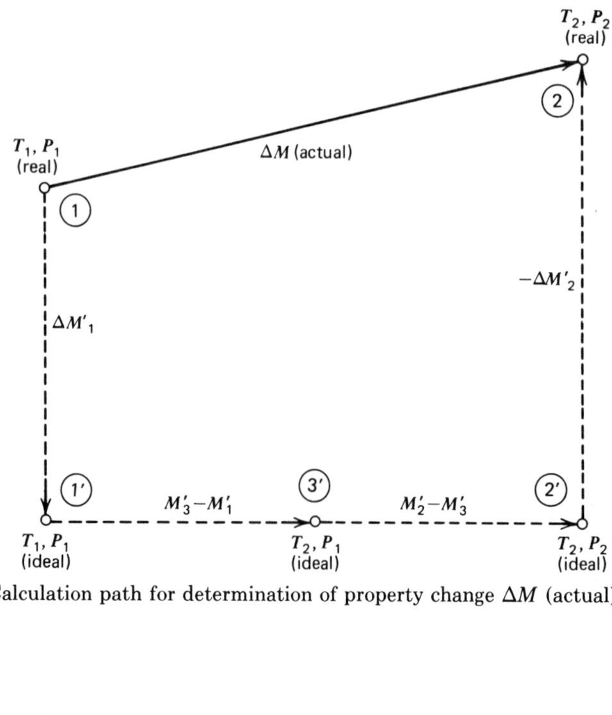

Figure 5. Calculation path for determination of property change ΔM (actual) by residual functions.

Generalized Correlations and *PVT* Equations of State

The volumetric properties of fluids are conveniently represented by generalized correlations or by analytical equations of state. Modern generalized correlations are based on the extended theorem of corresponding states, which in its simplest, three-parameter form asserts that

$$Z = Z(T_r, P_r; \pi) \qquad \text{(all fluids)} \tag{163}$$

Here, Z is a universal function of the reduced temperature T_r ($\equiv T/T_c$), the reduced pressure P_r ($\equiv P/P_c$), and a third corresponding-states parameter π. According to equation 163, all fluids having the same value of parameter π have identical values of Z when compared at the same T_r and P_r. Since equation 163 is presumed valid for all T_r and P_r, then generalized correlations similarly are obtained for those derived properties, eg, the residual functions and the fugacity coefficient, which depend on T and P through Z and its derivatives.

The most popular three-parameter, generalized correlations incorporate Pitzer's acentric factor ω as the third parameter π, where, by definition:

$$\omega \equiv -\log_{10} (P_r^{sat})_{T_r=0.7} - 1 \tag{164}$$

Pitzer correlations for Z are written as

$$Z = Z^0(T_r,P_r) + \omega Z^1(T_r,P_r) \tag{165}$$

which is a special form of equation 163. The linearity of equation 165 implies similar linear Pitzer correlations for certain of the derived properties, eg, $\ln \phi$, $\Delta H'/RT_c$, and $\Delta S'/R$. A comprehensive modern Pitzer-type correlation for Z and for derived properties has been developed (19).

PVT equations of state may be variously classified; the most popular are virial equations, cubic equations, and extended virial equations. The virial equations are infinite series representations of Z, with either molar density ρ ($\equiv V^{-1}$) or pressure P taken as the independent variable of expansion:

$$Z = 1 + B\rho + C\rho^2 + D\rho^3 + \ldots \tag{166}$$

$$Z = 1 + B'P + C'P^2 + D'P^3 + \ldots \tag{167}$$

Parameters $B, C, D, \ldots$, are called density-series virial coefficients, and $B', C', D', \ldots$, are called pressure-series virial coefficients. The second virial coefficients are defined by the recipes:

$$B \equiv \frac{1}{1!}\left(\frac{\partial Z}{\partial \rho}\right)_{T,y;\rho=0} \tag{168}$$

$$B' \equiv \frac{1}{1!}\left(\frac{\partial Z}{\partial P}\right)_{T,y;P=0} \tag{169}$$

Similarly, the third virial coefficients are defined as

$$C \equiv \frac{1}{2!}\left(\frac{\partial^2 Z}{\partial \rho^2}\right)_{T,y;\rho=0} \tag{170}$$

$$C' \equiv \frac{1}{2!}\left(\frac{\partial^2 Z}{\partial P^2}\right)_{T,y;P=0} \tag{171}$$

Higher virial coefficients are defined analogously.

It is clear from equations 168–171 that the virial coefficients depend on temperature and composition only. Moreover, the pressure-series coefficients and the density-series coefficients are related to one another:

$$B' = \frac{B}{RT} \tag{172}$$

$$C' = \frac{C - B^2}{(RT)^2} \tag{173}$$

$$D' = \frac{D - 3\,BC + 2\,B^3}{(RT)^3} \tag{174}$$

etc

Thus, the virial series in pressure, ie, equation 167, can be written

$$Z = 1 + \frac{BP}{RT} + \left[\frac{C - B^2}{(RT)^2}\right] P^2 + \ldots \tag{175}$$

in which the density-series coefficients are used. This form is preferred to equation 167 because the density-series coefficients are the ones normally reported by experimentalists and are the ones for which correlations, in the case of B and C, are available.

In practice, one must work with truncations of the virial equations, and since virial coefficients beyond the third are rarely available, equations 166 and 175 are usually truncated after two or three terms. For low pressure, the two-term truncation of equation 175 is preferred:

$$Z = 1 + \frac{BP}{RT} \tag{176}$$

For more severe conditions, the three-term truncation of equation 166 is used:

$$Z = 1 + B\rho + C\rho^2 \tag{177}$$

Equation 176 should not be used for densities greater than about half the critical value, nor equation 177 for densities exceeding the critical value.

The virial equations are unsuitable for the calculation of properties of liquids or dense gases. The simplest expressions appropriate (in principle) for such applications are equations that are cubic in molar volume. These equations, inspired by the van der Waals equation of state, may be represented by the general formula:

$$P = \frac{RT}{V - b} - \frac{\theta(V - \eta)}{(V - b)(V^2 + \delta V + \epsilon)} \tag{178}$$

where parameters b, θ, δ, ϵ, and η can each depend on temperature and composition.

Special cases of equation 178 are obtained on specification of values or expressions for the various parameters. Because of their generalizability, two-parameter cubic equations are among the most popular. Three modern examples are (20–22)

$$P = \frac{RT}{V - b} - \frac{\theta_S}{V^2 + bV} \tag{179}$$

$$P = \frac{RT}{V - b} - \frac{\theta_{PR}}{V^2 + 2bV - b^2} \tag{180}$$

$$P = \frac{RT}{V - b} - \frac{\theta_H}{V^2 + 3bV - 2b^2} \tag{181}$$

In each of these expressions, parameter θ, different for the different equations, depends on temperature. Numerical values for b and for $\theta(T)$ are determined for a given substance by subjecting the equation of state to the critical derivative constraints of equation 152 and by requiring that the equation reproduce experimental values of the vapor–liquid saturation pressure P^{sat}.

The cubic equations, although simple and able to provide semiquantitative descriptions of real-fluid behavior, are not generally useful for accurately representing volumetric data over wide ranges of T and P. For such cases, more comprehensive expressions with large numbers of adjustable parameters are needed. The simplest of these are the extended virial equations, exemplified by the eight-constant Benedict–Webb–Rubin (BWR) equation of state (23):

$$Z = 1 + \left(B_0 - \frac{A_0}{RT} - \frac{C_0}{RT^2}\right)\rho + \left(b - \frac{a}{RT}\right)\rho^2 + \frac{a\alpha}{RT}\rho^5 + \frac{c\rho^2}{RT^3}(1 + \gamma\rho^2)\exp(-\gamma\rho^2) \tag{182}$$

Modifications to the BWR equation have been made (24–25).

As noted, expressions for derived properties may be determined from PVT equations of state; the most important of these properties are $\Delta H'$, $\Delta S'$, and $\ln \phi$. If an equation of state is explicitly solvable for volume, ie, is volume-explicit, then the following general formulas are appropriate:

$$\Delta H' = RT^2 \int_0^P \left(\frac{\partial Z}{\partial T}\right)_P \frac{dP}{P} \qquad \text{(constant } T,x\text{)} \tag{183}$$

$$\Delta S' = R \int_0^P \left[T\left(\frac{\partial Z}{\partial T}\right)_P + Z - 1\right]\frac{dP}{P} \qquad \text{(constant } T,x\text{)} \tag{184}$$

$$\ln \phi = \int_0^P (Z - 1)\frac{dP}{P} \qquad \text{(constant } T,x\text{)} \tag{119}$$

If an equation of state is solvable explicitly for pressure but not for volume, then alternative formulas must be used:

$$\Delta H' = RT^2 \int_0^\rho \left(\frac{\partial Z}{\partial T}\right)_\rho \frac{d\rho}{\rho} - RT\ln(Z - 1) \qquad \text{(constant } T,x\text{)} \tag{185}$$

$$\Delta S' = R \int_0^\rho \left[T\left(\frac{\partial Z}{\partial T}\right)_\rho + Z - 1\right]\frac{d\rho}{\rho} - R\ln Z \qquad \text{(constant } T,x\text{)} \tag{186}$$

$$\ln \phi = \int_0^\rho (Z - 1)\frac{d\rho}{\rho} + Z - 1 - \ln Z \qquad \text{(constant } T,x\text{)} \tag{187}$$

Although generalized correlations and PVT equations of state are based on data for pure fluids, they are frequently applied to mixtures. In generalized correlations of the form of equation 163, mole fraction is introduced as a variable through empirical recipes for the composition dependence of the corresponding-states parameters T_c, P_c, and π. The simplest such recipes for pseudoparameters are linear in mole fraction:

$$T_c = \sum x_i T_{c_i} \tag{188}$$

$$P_c = \sum x_i P_{c_i} \tag{189}$$

$$\pi = \sum x_i \pi_i \tag{190}$$

However, more complicated expressions have been proposed (18).

The virial equations are unique in that rigorous expressions are available for the composition dependence of the virial coefficients. Statistical mechanics shows that the nth virial coefficient of a mixture is nth order in the mole fractions. Thus, the exact mixing rules are obtained:

$$B = \sum_i \sum_j y_i y_j B_{ij} \tag{191}$$

$$C = \sum_i \sum_j \sum_k y_i y_j y_k C_{ijk} \tag{192}$$

etc

The subscripted coefficients B_{ij}, C_{ijk}, . . ., are functions of T only, and their numerical values are unchanged on permutation of the subscripts. Coefficients with identical subscripts, eg, B_{11}, C_{222}, are properties of pure materials. Those with mixed subscripts, eg, $B_{12} = B_{21}$, $C_{122} = C_{212} = C_{221}$, are mixture properties and are called interaction virial coefficients.

Mixing rules for the parameters in an empirical equation of state, eg, a cubic equation, are necessarily empirical. With the cubic equations, linear or quadratic expressions are normally used. Thus, in equations 179–181, parameters b and θ for mixtures are usually given by

$$b = \sum_i x_i b_i \tag{193}$$

$$\theta = \sum_i \sum_j x_i x_j \theta_{ij} \tag{194}$$

where, as for the second virial coefficient, $\theta_{ij} = \theta_{ji}$.

Given the appropriate mixing rules for the parameters, the mixture properties are computed from a PVT equation of state by exactly the same procedures used for pure components. Partial molar properties of species in solution can also be computed from an equation of state. The most important of the partial properties is $\ln \hat{\phi}_i$ because of its frequent use in phase-equilibrium calculations. For a volume-explicit equation of state

$$\ln \hat{\phi}_i = \int_0^P (\overline{Z}_i - 1) \frac{dP}{P} \qquad \text{(constant } T,x) \tag{120}$$

and for a pressure-explicit equation of state

$$\ln \hat{\phi}_i = \int_0^\rho \left\{ \left[\frac{\partial (nZ)}{\partial n_i} \right]_{T,\rho/n,n_j} - 1 \right\} \frac{d\rho}{\rho} - \ln Z \qquad \text{(constant } T,x) \tag{195}$$

where subscript $\rho/n = 1/V^t$ and indicates constancy of total volume.

Criteria for Phase Equilibria

The following criterion of phase equilibrium can be developed from the first and second laws of thermodynamics. For a closed, multiphase system of constant, uniform temperature and pressure, the equilibrium state is that for which the total Gibbs function is a minimum:

$$G^t(\text{constant } T,P) = \text{minimum} \tag{196}$$

Note that it is presumed in this statement that equilibrium in a multiphase system implies uniformity of T and P among the phases. In certain situations, eg, osmotic equilibrium, pressure uniformity is not required, and equation 196 is then not a useful criterion. Here, however, it suffices.

If the T and P of a multiphase system are constant, then the quantities capable of change are the individual mole numbers n_i^p of the various chemical species i in the various phases p. In the absence of chemical reactions, which is assumed here, the n_i^p may change only by interphase mass transfer (qv) and not, since the system is closed, by the transfer of matter across the boundaries of the system. Hence, for phase equilibrium in a π-phase system, equation 196 is subject to the set of material-balance constraints:

$$\sum_p^{\pi} n_i^p = n_i^t \qquad (\text{all } i) \tag{197}$$

where n_i^t is the constant, total number of moles of i in the system.

Consider the application of equation 196 to the case of multicomponent, two-phase equilibrium. The phases are denoted by α and β, and the following general expression is written for G^t:

$$G^t = n^tG = \sum_k n_k^\alpha \mu_k^\alpha + \sum_k n_k^\beta \mu_k^\beta \tag{198}$$

This expression is minimized subject to the material-balance constraints of equation 197, here written as

$$n_k^\alpha + n_k^\beta = n_k^t \qquad (k = 1, 2, \ldots, N) \tag{199}$$

Equation 199 asserts that only N of the 2 N mole numbers are independently variable. If the independent mole numbers are chosen as n_k^α, it follows from equation 196 that a set of N conditions on G^t at the equilibrium state can be written:

$$\left(\frac{\partial G^t}{\partial n_i^\alpha}\right)_{T,P,n_j^\alpha} = 0 \qquad (i = 1, 2, \ldots, N) \tag{200}$$

If equations 200 and 108 are applied to equation 198, with $\partial n_i^\beta/\partial n_i^\alpha = -1$, then:

$$\mu_i^\alpha = \mu_i^\beta \qquad (\text{all } i) \tag{201}$$

which is the familiar algebraic criterion for two-phase equilibrium.

The approach just illustrated for two-phase equilibrium is easily extended to an arbitrary number of phases π and produces the following generalization of equation 201:

$$\mu_i^\alpha = \mu_i^\beta = \ldots = \mu_i^\pi \qquad (\text{all } i) \tag{202}$$

It is conventional and convenient for the formulation of practical calculations to work with the fugacity $\hat{f}_i$ rather than with the chemical potential μ_i. For such formulations, equation 202 is replaced by the equal-fugacity criterion, which follows from the integrated form of equation 112:

$$\hat{f}_i^{\alpha} = \hat{f}_i^{\beta} = \ldots = \hat{f}_i^{\pi} \qquad \text{(all } i\text{)} \tag{203}$$

Vapor–Liquid Equilibria (VLE). By equation 203, a criterion for vapor–liquid equilibrium is

$$\hat{f}_i^{l} = \hat{f}_i^{v} \qquad \text{(all } i\text{)} \tag{204}$$

Effective use of this general equation requires explicit introduction of the compositions of the phases. This may be done through the activity coefficient γ_i or the fugacity coefficient $\hat{\phi}_i$. Two procedures are commonly used. By the gamma–phi approach, activity coefficients are used for the liquid phase and fugacity coefficients for the vapor phase; equation 204 becomes

$$x_i \gamma_i f_i = y_i \hat{\phi}_i P \qquad \text{(all } i\text{)} \tag{205}$$

Equation 205 is most commonly used for low pressure VLE, in which case it is rewritten as

$$x_i \gamma_i^{+} P_i^{\text{sat}} = y_i \Phi_i P \qquad \text{(all } i\text{)} \tag{206}$$

where P_i^{sat} is the vapor pressure of pure i. Quantity Φ_i is given by the formula:

$$\Phi_i = \frac{\hat{\phi}_i}{\phi_i^{\text{sat}}} \exp\left[-\int_T^{T^+} \frac{\bar{H}_i^E}{RT^2}\,dT + \int_P^{P^+} \frac{\bar{V}_i^E}{RT}\,dP + \int_P^{P_i^{\text{sat}}} \frac{V_i}{RT}\,dP\right] \tag{207}$$

where T^+ and P^+ denote reference conditions at which the γ_i^+ are evaluated, presumably from an expression for G^E.

If the liquid phase is an ideal solution, the vapor phase an ideal-gas mixture, and if liquid properties are independent of pressure, then $\gamma_i^+ = \Phi_i = 1$ and equation 206 reduces to Raoult's law:

$$x_i P_i^{\text{sat}} = y_i P \qquad \text{(all } i\text{)} \tag{208}$$

Of the assumptions inherent in this approximate relationship, that of liquid-phase ideality is usually the least realistic. Therefore, the simplest generally useful special case of equation 206 is the modified Raoult's-law expression:

$$x_i \gamma_i P_i^{\text{sat}} = y_i P \qquad \text{(all } i\text{)} \tag{209}$$

where the γ_i are functions of temperature and composition but are assumed to be independent of pressure. Equation 209, unlike equation 208, can represent vapor–liquid azeotropy.

The second common procedure for VLE calculations is the equation-of-state approach. Here, fugacity coefficients are used for both the liquid and vapor phases and equation 204 becomes

$$x_i \hat{\phi}_i^{l} = y_i \hat{\phi}_i^{v} \qquad \text{(all } i\text{)} \tag{210}$$

Application of equation 210 requires the availability of a single equation of state suitable for both vapor and liquid mixtures. Cubic equations of state are widely used for VLE calculations, mainly for systems comprising nonpolar, nonassociating substances, eg, cryogenic fluids and light hydrocarbons.

The proper design of distillation and absorption columns depends on knowledge of vapor–liquid equilibrium, as do flash calculations used to determine the physical state of streams at given conditions of temperature, pressure, and composition. A detailed treatment of vapor–liquid equilibria is given in ref. 10.

Other Kinds of Phase Equilibria. Equation 203 is the basis for other kinds of phase-equilibrium formulations. Thus, for liquid–liquid equilibria (LLE):

$$\hat{f}_i^\alpha = \hat{f}_i^\beta \qquad \text{(all } i\text{)} \tag{211}$$

where α and β designate the two liquid phases. Eliminating fugacities in favor of activity coefficients gives the gamma–gamma formulation for LLE:

$$x_i^\alpha \gamma_i^\alpha = x_i^\beta \gamma_i^\beta \qquad (i = 1, 2, \ldots, N) \tag{212}$$

For most LLE applications, the effect of pressure on the γ_i can be ignored, and thus equation 212 constitutes a set of N equations relating the equilibrium compositions to each other and to temperature. Solution of these equations for a particular temperature requires a single expression for the composition dependence of G^E suitable for both liquid phases. Not all expressions for G^E suffice, even in principle, because some cannot represent liquid–liquid phase-splitting.

For vapor–liquid–liquid equilibria (VLLE), equation 203 gives

$$\hat{f}_i^\alpha = \hat{f}_i^\beta = \hat{f}_i^v \qquad \text{(all } i\text{)} \tag{213}$$

where α and β designate the two liquid phases. With activity coefficients applied for the liquid phases and fugacity coefficients for the vapor phase, the 2 N equilibrium equations for subcritical VLLE are

$$\left.\begin{aligned} x_i^\alpha \gamma_i^\alpha f_i^\alpha &= y_i \hat{\phi}_i P \\ x_i^\beta \gamma_i^\beta f_i^\beta &= y_i \hat{\phi}_i P \end{aligned}\right\} \text{(all } i\text{)} \tag{214}$$

As for LLE, an expression for G^E capable of representing liquid–liquid phase splitting is required; as for VLE, a vapor-phase equation of state for computing the $\hat{\phi}_i$ is also needed.

Chemical-Reaction Stoichiometry

Consider a system containing N chemical species, any or all of which can participate in r chemical reactions. The r reactions can be represented schematically by the algebraic equations:

$$0 = \sum_i^N \nu_{i,j} A_i \qquad (j = \text{I, II}, \ldots, r) \tag{215}$$

where the A_i represent chemical species and $\nu_{i,j}$ is the stoichiometric number for species i in reaction j. Each $\nu_{i,j}$ has a magnitude and a sign:

$$\text{sign}(\nu_{i,j}) = \begin{cases} - \text{ for a reactant species} \\ + \text{ for a product species} \end{cases}$$

If species i does not participate in reaction j, then $\nu_{i,j} = 0$.

The stoichiometric numbers provide relationships among the changes in mole numbers of chemical species which occur as the result of chemical reaction. Thus, for reaction j:

$$\frac{\Delta n_{1,j}}{\nu_{1,j}} = \frac{\Delta n_{2,j}}{\nu_{2,j}} = \ldots = \frac{\Delta n_{N,j}}{\nu_{N,j}} \tag{216}$$

Since all of these terms are equal, they can be equated to the change of a single quantity ϵ_j, called the reaction coordinate for reaction j, thereby giving

$$\Delta n_{i,j} = \nu_{i,j} \Delta \epsilon_j \begin{cases} i = 1, 2, \ldots, N \\ j = \text{I}, \text{II}, \ldots, r \end{cases} \tag{217}$$

Now the total change in mole number n_i is just the sum of the changes $\Delta n_{i,j}$ resulting from the various reactions. Thus, by the last equation:

$$\Delta n_i = \sum_j^r \Delta n_{i,j} = \sum_j^r \nu_{i,j} \Delta \epsilon_j \qquad (i = 1, 2, \ldots, N) \tag{218}$$

If the initial number of moles of species i is n_{i_0} and if the convention is adopted that $\epsilon_j = 0$ for each reaction in this initial state, then

$$n_i = n_{i_0} + \sum_j^r \nu_{i,j} \epsilon_j \qquad (i = 1, 2, \ldots, N) \tag{219}$$

Equation 219 is the basic equation of material balance for a closed system in which r chemical reactions occur. It asserts the important fact that in such a system there are at most r mole-number-related quantities ϵ_j capable of independent variation. Note that there are no implied restrictions with respect to chemical-reaction equilibria; the reaction-coordinate scheme is merely an accounting procedure, valid for tracking the progress of each reaction to any arbitrary level of conversion.

Property Changes of Reaction

Consider a single-phase system in which there may occur chemical reactions. At constant T and P, the reaction coordinates ϵ_j are the quantities subject to independent variation; thus:

$$d(nM) = \sum_j \left[\frac{\partial(nM)}{\partial \epsilon_j}\right]_{T,P,\epsilon_{k \neq j}} d\epsilon_j \qquad (\text{constant } T,P) \tag{220}$$

where the derivative is a measure of the property change associated with reaction j. Since

$$nM = \sum_i n_i \overline{M}_i \tag{221}$$

it can be shown that

$$\left[\frac{\partial(nM)}{\partial \epsilon_j}\right]_{T,P,\epsilon_{k \neq j}} = \sum_i \nu_{i,j} \overline{M}_i \tag{222}$$

Imagine now that all reactants exist in standard states, states of pure i (real or hypothetical) at the system temperature T and at reference pressure $P°$, and that they react forming products in standard states at the same T and $P°$. Then $\overline{M}_i = M_i^\circ$, and the standard (state) property change of reaction ΔM_j° is defined by analogy to equation 222 as

$$\Delta M_j^\circ \equiv \sum_i \nu_{i,j} M_i^\circ \tag{223}$$

Note that in equation 223 the physical (standard) states may be mixed; any or all of the species may be gases, liquids, or solids. For species present as gases in the actual reactive system, the standard state is the pure ideal gas at pressure P°. For liquids and solids, it is usually the state of pure real liquid or solid at P°. The standard-state pressure P° is conventionally fixed at 101.325 kPa (1 atm).
ΔC_{Pj}°:

$$\Delta G_j^\circ \equiv \sum_i \nu_{i,j} G_i^\circ = \sum_i \nu_{i,j} \mu_i^\circ \tag{224}$$

$$\Delta H_j^\circ \equiv \sum_i \nu_{i,j} H_i^\circ \tag{225}$$

$$\Delta C_{Pj}^\circ \equiv \sum_i \nu_{i,j} C_{Pi}^\circ \tag{226}$$

Function ΔH_j°, ie, the standard heat of reaction, is used to estimate real heat effects, and ΔG_j° to calculate equilibrium compositions. Standard heat-capacity changes of reaction are used for extrapolating ΔH_j° and ΔG_j° with T. Numerical values for ΔG_j° and ΔH_j° are computed from tabulated formation data (26), and ΔC_{Pj}° is determined from empirical expressions for the T dependence of the C_{Pi}° (1,9).

Criteria for Chemical-Reaction Equilibria

The general criterion of chemical-reaction equilibria is the same as that for phase equilibria, namely that the total Gibbs function of a closed system be a minimum at constant, uniform T and P (see eq. 196). If the T and P of a single-phase, chemically reactive system are constant, then the quantities capable of change are the mole numbers n_i. The independently variable quantities are just the r reaction coordinates, and thus the equilibrium state is characterized by the r necessary derivative conditions:

$$\left(\frac{\partial G^t}{\partial \epsilon_j}\right)_{T,P,\epsilon_k} = 0 \qquad (j = \text{I, II}, \ldots, r) \tag{227}$$

subject to the material-balance constraints of equation 219.

Consider the case of a single-phase, multicomponent system undergoing just a single chemical reaction. The total Gibbs function is

$$G^t = n^t G = \sum_i n_i \mu_i \tag{228}$$

and is minimized subject to the constraints

$$n_i = n_{i_0} + \nu_i \epsilon \qquad (\text{all } i) \tag{229}$$

Equation 227 requires that

$$\left(\frac{\partial G^t}{\partial \epsilon}\right)_{T,P} = 0 \tag{230}$$

If equations 230 and 108 are applied to equation 228, with $\partial n_i/\partial \epsilon = \nu_i$, then:

$$\sum_i \nu_i \mu_i = 0 \tag{231}$$

which is the familiar algebraic criterion for single-reaction equilibria.

The approach just illustrated is easily extended to an arbitrary number of independent reactions r and produces the expected generalization of equation 231:

$$\sum_i \nu_{i,j} \mu_i = 0 \qquad (j = \text{I, II}, \ldots, r) \tag{232}$$

Equilibrium Constants

It is convenient to eliminate the μ_i in equation 232 in favor of fugacities. Integrating equation 112 from the standard state of species i to its real state at the same T gives

$$\overline{G}_i = G_i^\circ + RT \ln (\hat{f}_i/f_i^\circ) \tag{233}$$

or

$$\mu_i = \mu_i^\circ + RT \ln (\hat{f}_i/f_i^\circ) \qquad (\text{all } i) \tag{234}$$

Substitution of equation 234 into equation 232 yields, upon rearrangement:

$$\prod_i (\hat{f}_i/f_i^\circ)^{\nu_{i,j}} = K_j \qquad (\text{all } j) \tag{235}$$

where

$$K_j \equiv \exp (-\Delta G_j^\circ/RT) \tag{236}$$

Quantity K_j is the chemical-reaction equilibrium constant for reaction j, and ΔG_j° is the corresponding standard Gibbs function change of reaction (see eq. 224).

Use of equation 236 for actual calculations requires explicit introduction of composition as a variable. As in phase-equilibrium calculations, this is normally done for the gas phase through the fugacity coefficient and for condensed phases through the activity coefficient. Thus, either

$$\hat{f}_i/f_i^\circ = y_i \hat{\phi}_i P/f_i^\circ(P^\circ) \tag{237}$$

or

$$\hat{f}_i/f_i^\circ = x_i \gamma_i f_i^\circ(P)/f_i^\circ(P^\circ) \tag{238}$$

where the notations make explicit the pressures at which standard-state fugacities are evaluated. Equations 237–238 or variations on them are the bases for the various forms into which equation 235 is usually cast. An important special case of equation 235 is obtained when ideal-solution behavior is assumed for a chemically reactive phase. In this event, $\hat{\phi}_i = \phi_i$ in equation 237 or $\gamma_i = 1$ in equation 238, and equation 235 reduces in either case to the law of mass action:

$$\prod_i y_i^{\nu_{i,j}} = K_j' \,(T,P) \qquad (\text{all } j) \tag{239}$$

or

$$\prod_i x_i^{\nu_{i,j}} = K_j''(T,P) \qquad (\text{all } j) \tag{240}$$

where

$$K_j'(T,P) = \frac{K_j}{\prod_i [\phi_i P/f_i^\circ(P^\circ)]^{\nu_{i,j}}} \tag{241}$$

and

$$K_j''(T,P) = \frac{K_j}{\prod_i [f_i^\circ(P)/f_i^\circ(P^\circ)]^{\nu_{i,j}}} \tag{242}$$

The significant feature of the approximate equations 239–240 is that composition effects are separable from the effects of T and P.

Nomenclature

A = Helmholtz function
B, C, D = virial coefficients for the expansion in density (eq. 166)
B', C', D' = virial coefficients for the expansion in pressure (eq. 167)
B_{ij}, C_{ijk} = interaction virial coefficients
C_P = heat capacity at constant pressure
C_V = heat capacity at constant volume
E_K = kinetic energy
E_P = potential energy
f = fugacity of a mixture
f_i = fugacity of pure species i
$\hat{f}_i$ = fugacity of species i in solution
f_i^{id} = fugacity of species i in ideal solution
G = Gibbs function
g = acceleration of gravity
H = enthalpy
K_j = equilibrium constant for reaction j
M = intensive thermodynamic property of a solution (M can represent a variety of thermodynamic functions, including U, C_V, H, C_P, S, A, G, or Z)
M' = ideal-gas value of a thermodynamic property
M_i = intensive thermodynamic property of pure species i
$\overline{M}_i$ = partial molar property of species i in solution
M^{id} = intensive thermodynamic property of an ideal solution
$\overline{M}_i^{id}$ = partial molar property of species i in an ideal solution
$\overline{M}_i^E$ = partial molar excess thermodynamic property
M_i° = standard-state thermodynamic property
ΔM_j° = standard property change of reaction j
ΔM = property change of mixing
$\Delta\overline{M}_i$ = partial molar property change of mixing
$\Delta M'$ = residual thermodynamic property
m = mass
n = number of moles
n_i = number of moles of species i
P = absolute pressure

P_c	= critical pressure
P_i^{sat}	= saturation or vapor pressure of species i
Q	= heat
R	= universal gas constant
S	= entropy
T	= absolute temperature
T_c	= critical temperature
t	= as superscript, denotes a total value
U	= internal energy
u	= velocity
V	= volume
W	= work
W_s	= shaft work for a flow process
x_i	= liquid-phase mole fraction of species i
y_i	= vapor-phase mole fraction of species i
Z	= compressibility factor
z	= elevation above a datum level

Superscripts

E	= excess
id	= ideal solution
l	= liquid phase
lv	= phase transition from liquid to vapor
r	= reduced
t	= total value
v	= vapor phase

Subscripts

C	= cold
H	= hot
t	= triple state

Greek letters

α, β	= as superscripts, identify phases
γ_i	= activity coefficient of species i in solution
μ_i	= chemical potential of species i
$\nu_{i,j}$	= stoichiometric number of species i in reaction j
π	= corresponding-states parameter (eq. 163); number of phases (eqs. 197, 202–203)
ρ	= density
τ	= time
Φ_i	= defined by equation 207
ϕ	= fugacity coefficient of a mixture
ϕ_i	= fugacity coefficient of pure species i
$\hat{\phi}_i$	= fugacity coefficient of species i in solution
ω	= acentric factor

BIBLIOGRAPHY

"Thermodynamics" in *ECT* 1st ed., Vol. 14, pp. 1–37, by J. C. Chu, Polytechnic Institute of Brooklyn; "Thermodynamics" in *ECT* 2nd ed., Vol. 20, pp. 118–146, by H. T. Chen, Newark College of Engineering, and D. F. Othmer, Polytechnic Institute of Brooklyn.

1. M. M. Abbott and H. C. Van Ness, *Schaum's Outline of Theory and Problems of Thermodynamics*, McGraw-Hill, New York, 1972.
2. R. E. Balzhiser, M. R. Samuels, and J. D. Eliassen, *Chemical Engineering Thermodynamics*, Prentice-Hall, Englewood Cliffs, N.J., 1972.
3. K. Denbigh, *Principles of Chemical Equilibrium*, 3rd ed., Cambridge, 1971.
4. M. L. McGlashan, *Chemical Thermodynamics*, Academic Press, New York, 1979.
5. M. Modell and R. C. Reid, *Thermodynamics and its Applications*, Prentice-Hall, Englewood Cliffs, N.J., 1974.
6. J. M. Prausnitz, *Molecular Thermodynamics of Fluid-Phase Equilibria*, Prentice-Hall, Englewood Cliffs, N.J., 1969.
7. I. Prigogine and R. Defay, *Chemical Thermodynamics*, Longmans, Green, London, 1954.
8. S. I. Sandler, *Chemical and Engineering Thermodynamics*, John Wiley & Sons, Inc., New York, 1977.
9. J. M. Smith and H. C. Van Ness, *Introduction to Chemical Engineering Thermodynamics*, 3rd ed., McGraw-Hill, New York, 1975.
10. H. C. Van Ness and M. M. Abbott, *Classical Thermodynamics of Nonelectrolyte Solutions: With Applications to Phase Equilibria*, McGraw-Hill, New York, 1982.
11. R. Clausius, *The Mechanical Theory of Heat*, Macmillan and Co., London, 1879.
12. G. N. Lewis and M. Randall, *Thermodynamics*, McGraw-Hill, New York, 1923.
13. G. M. Wilson, *J. Am. Chem. Soc.* **86,** 127 (1964).
14. H. Renon and J. M. Prausnitz, *AIChE J.* **15,** 785 (1969).
15. G. Maurer and J. M. Prausnitz, *Fluid Phase Equilibria* **2,** 91 (1978).
16. Aa. Fredenslund, J. Gmehling, and P. Rasmussen, *Vapor–Liquid Equilibria Using UNIFAC*, Elsevier, Amsterdam, 1977.
17. K. Kojima and T. Tochigi, *Prediction of Vapor–Liquid Equilibria by the ASOG Method*, Elsevier, Amsterdam, 1979.
18. R. C. Reid, J. M. Prausnitz, and T. K. Sherwood, *The Properties of Gases and Liquids*, 3rd ed., McGraw-Hill, New York, 1977.
19. B.-I. Lee and M. G. Kesler, *AIChE J.* **21,** 510, 1040 (1975).
20. G. Soave, *Chem. Eng. Sci.* **27,** 1197 (1972).
21. D.-Y. Peng and D. B. Robinson, *Ind. Eng. Chem. Fundam.* **15,** 59 (1976).
22. A. Harmens, *Cryogenics* **17,** 519 (1977).
23. M. Benedict, G. B. Webb, and L. C. Rubin, *J. Chem. Phys.* **8,** 334 (1940); **10,** 747 (1942).
24. K. E. Starling and J. E. Powers, *Ind. Eng. Chem. Fundam.* **9,** 531 (1970).
25. E. Bender, *Cryogenics* **13,** 11 (1973).
26. *Selected Values of Properties of Chemical Compounds*, Thermodynamics Research Center Data Project, and *Selected Values of Properties of Hydrocarbons and Related Compounds*, American Petroleum Institute Research Project 44, serial publications of the Thermodynamics Research Center, College Station, Texas; "Thermodynamic Data Banks," *CHEMTECH*, 691 (Nov. 1982).

HENDRICK C. VAN NESS
MICHAEL M. ABBOTT
Rensselaer Polytechnic Institute

THERMOELECTRIC ENERGY CONVERSION

Thermoelectric energy conversion is the interconversion of heat and electrical energy for power generation or heat pumping and is based on the Seebeck, Peltier, and Thomson effects. In the early 1950s, progress in solid-state physics and chemistry led to the development of semiconductor thermoelements with the result that reasonably efficient thermoelectric devices could be constructed. Metallic thermocouples provide only very low efficiencies, the most favorable being combinations of bismuth and antimony, which provide efficiencies of ca 1%; selected semiconductors can provide efficiencies of ca 8–10% (see Semiconductors, fabrication and characterization).

The technique of direct energy conversion is characterized by the independence of size vs efficiency, the absence of moving parts, high reliability, quietness, lack of vibration, low maintenance, simple startup, and absence of pollution problems. Thermoelectric generators have been used increasingly in specialized applications in which combinations of their desirable features outweigh their high cost and low generating efficiencies, which are typically ca 3–7%. Large-scale thermoelectric generators cannot compete with oil-fired central power stations, which operate at efficiencies of 35–40%. The most advanced thermoelectric systems are the radioisotope thermoelectric generators (RTGs), which have been developed for military and space systems under the aegis of DOE. The RTGs most recently operated in space were used to power the Voyager I and II spacecrafts and had conversion efficiencies of 6.7% and specific powers of 4.2 W/kg. Other RTGs have been used for such applications as floating and terrestrial weather stations, cardiac pacemakers, and navigational buoys. Fossil-fired thermoelectric generators have been developed for military and commercial applications. Some of these applications include power for remote navigational lights, communication line repeaters, and cathodic protection, eg, protection of the east–west pipeline across Saudi Arabia by 34 thermoelectric stations.

Thermoelectric heat pumping, like thermoelectric power generation, has had increased application in those areas where the advantages of the thermoelectric conversion process, ie, small space, light weight, high reliability, no noise or pollution, and simple temperature control, can be utilized. Thermoelectric cooling devices have been developed for a variety of military and commercial applications (see Refrigeration). These include submarine air-conditioning systems, small refrigerators and recreational cooler chests, cooling for electronic components, temperature-control units, laboratory instruments, and cooling for electro-optical systems. The state of the art is characterized by individual couples having pumping capacities of 1–4 W. Since the early 1970s, the trend in thermoelectric cooling has been increasingly for commercial applications.

The conversion efficiency of a thermoelectric generator and the coefficient of performance of a thermoelectric refrigerator depends upon the properties of the thermoelectric materials as expressed by their figure of merit. To date, three material technologies are established; these are bismuth telluride, lead telluride, and the Si–Ge alloys. The development of solid-state materials with enhanced figures of merit have not been realized. Therefore, thermoelectric energy conversion provides a unique solid-state technology that complements rather than replaces existing technologies.

Principles

The important events in thermoelectricity are given in refs. 1–12. The period 1800–1850 was characterized by the discoveries of the thermoelectric effects; the period 1890–1920 was characterized by the development of thermoelectric theories and the use of metallic thermoelements in devices; and the period 1950 to the present has been characterized by the development and application of suitable semiconductors (qv) for thermoelectric devices.

Kelvin Relations

The Seebeck effect was reported in 1822 and concerns the net conversion of thermal energy into electrical energy with the appearance of an electric current (2). The Seebeck effect occurs when two dissimilar conductors are joined to form a closed circuit and the junctions are at different temperatures. The Seebeck voltage is the open-circuit emf developed in the circuit. In Figure 1(**a**), the Seebeck voltage V_{AB} develops as a result of the temperature difference ΔT between the junctions of the

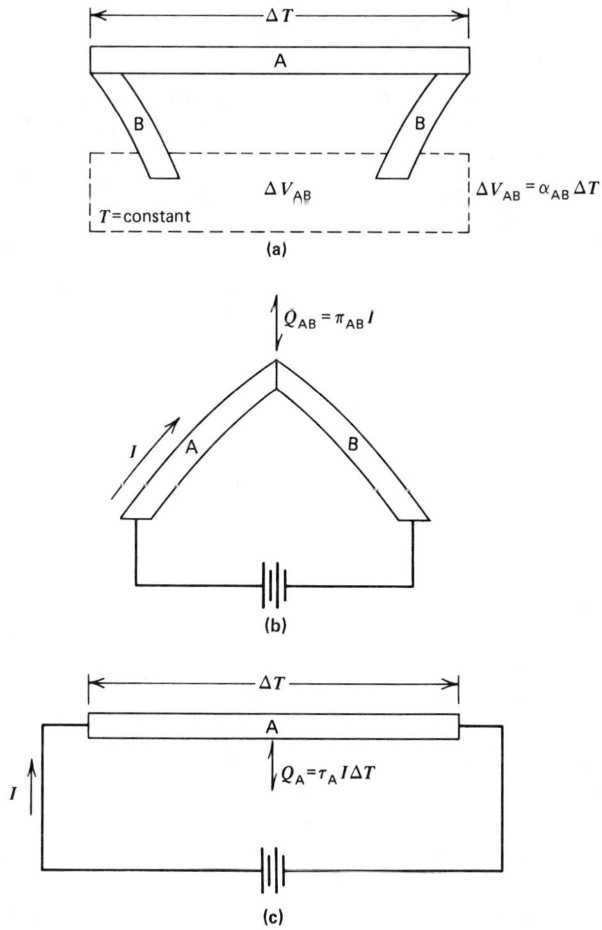

Figure 1. Thermoelectric effects. (**a**) Seebeck effect; (**b**) Peltier effect; (**c**) Thomson effect.

conductors A and B. The Seebeck coefficient α_{AB} is defined by

$$\alpha_{AB} = \lim_{\Delta T \to 0} \frac{\Delta V_{AB}}{\Delta T} \tag{1}$$

The Peltier effect was discovered in 1834 and involves the reversible evolution or absorption of heat that occurs when an electric current is passed through a junction between dissimilar metals, as shown in Figure 1(**b**) (3). The rate of reversible heat generation or evolution Q_{AB} is given by the equation

$$Q_{AB} = \pi_{AB} I \tag{2}$$

where π_{AB} is the Peltier coefficient and I is the current.

The Thomson effect was reported in 1855 and concerns the reversible evolution or absorption of heat that occurs whenever an electric current traverses a region where a temperature gradient exists along a single homogeneous conductor, as shown in Figure 1(**c**) (4). The rate of reversible heat generation or evolution Q_A is given by the equation

$$Q_A = \tau_A I \Delta T \tag{3}$$

where τ_A is the Thomson coefficient, I is the current, and ΔT is the temperature gradient.

The Peltier and Thomson effects are reversible effects and may be positive or negative, depending upon whether heat is absorbed or evolved, and are independent of the irreversible Joule heating effect. For practical thermoelectric devices, materials must be selected with reversible Peltier effects significantly greater than the irreversible Joule heating effects.

The interrelationship of the thermoelectric effects, also called the Kelvin relations, have been derived by both reversible and irreversible thermodynamics (13–14). These relationships are

$$\alpha_{AB} T = \pi_{AB}$$

and

$$T \frac{d\alpha_{AB}}{dT} = \tau_A - \tau_B \tag{4}$$

Theory of Thermoelectric Devices

The thermoelectric effects in a thermoelectric circuit produce useful heating, cooling, and power generation. The efficient operation of devices based upon these effects requires the optimization of both the circuit parameters and the properties of the materials used in the device. Since the efficiency of a thermoelectric generator and the coefficient of performance of a thermoelectric heat pump are independent of the capacity of the unit, these parameters can be derived on the basis of a single thermocouple, as shown in Figure 2(**a**) and 2(**b**). Figure 2(**a**) shows a generalized thermoelectric circuit for describing the operation of a thermoelectric generator, and Figure 2(**b**) shows a circuit for a generalized thermoelectric refrigerator. The circuits consist of two dissimilar thermoelements, n and p, in contact with a heat source at temperature T_2 and a heat sink at temperature T_0. The parameters of the circuits

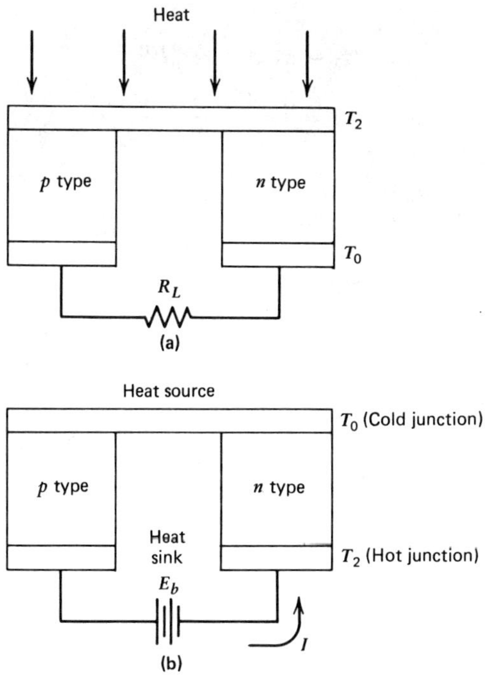

Figure 2. Generalized thermoelectric circuits. (**a**) Operation as a thermoelectric generator. (**b**) Operation as a thermoelectric refrigerator.

are

l_n, l_p	= element length
A_n, A_p	= element cross-sectional area
F_n, F_p	= form factor = A/l
κ_n, κ_p	= specific thermal conductivity
ρ_n, ρ_p	= specific electrical resistivity
κ_t	= total thermal conductance of two parallel branches
R_t	= sum of electrical resistances of branches
α_{np}	= relative Seebeck coefficient
π_{np}	= relative Peltier coefficient
R_L	= external load resistor for power generation
E_b	= external d-c power supply for refrigeration

The basic theory of thermoelectric devices has been thoroughly treated in a number of recent publications (15–16).

Power Generation. In Figure 2(**a**), under steady-state conditions, a temperature difference $T_2 - T_0$ is established across the thermoelements and causes a current I to flow through these branches and the load resistor. This device, therefore, represents the direct conversion of heat into electrical energy with a conversion efficiency η, which is defined as

$$\eta = \frac{\text{power supplied to load } (W_L)}{\text{heat absorbed at hot junction } (Q_2)} \tag{5}$$

The useful power is simply I^2R_L, and if the Thomson heat is neglected, the heat absorbed at T_2 consists of the Peltier heat H_1 and the heat withdrawn from the hot junction H_2:

$$H_1 = \pi_{np} I = \alpha_{np} I T_2 \tag{6}$$

$$I = \frac{\alpha_{np}(T_2 - T_0)}{R_t + R_L}$$

and

$$H_2 = \kappa_t (T_2 - T_0) - \tfrac{1}{2} I^2 R_t \tag{7}$$

The expression for the efficiency is

$$\eta = \frac{I^2 R_L}{\alpha_{np} I T_2 + \kappa_t (T_2 - T_0) - \tfrac{1}{2} I^2 R_t} \tag{8}$$

and the maximum efficiency η_{max} is

$$\eta_{max} = \left[\frac{T_2 - T_0}{T_2}\right] \frac{\sqrt{1 + Z\overline{T}} - 1}{\sqrt{1 + Z\overline{T}} + T_0/T_2} \tag{9}$$

where

$$\overline{T} = \frac{T_2 + T_0}{2}$$

$$Z = \frac{(\alpha_n - \alpha_p)^2}{[\sqrt{(\rho\kappa)_n} + \sqrt{(\rho\kappa)_p}]^2}$$

$$\frac{F_n}{F_p} = \sqrt{\frac{\kappa_p \rho_n}{\kappa_n \rho_p}}$$

The first factor in the expression for the maximum efficiency (eq. 9) is the thermodynamic efficiency of a reversible Carnot cycle. The second factor represents the decrease in this efficiency resulting from irreversible heat conduction along the branches and power dissipation in the form of Joule heat. The maximum efficiency η_{max} is plotted vs temperature difference for a range of values of Z in Figure 3. For maximum efficiency, the factor ZT should be maximized, ie, a high value of Z should be obtained over the widest possible range and at the highest operating temperature.

In practice, the two thermoelements have nearly similar material constants. In this case, the concept of the figure of merit for a single component is given by

$$Z = \frac{\alpha^2}{\rho\kappa} \tag{10}$$

This relationship is useful for comparing the relative thermoelectric efficiencies of various materials. The current state of the art is characterized by materials having figures of merit up to $(3\text{–}3.5) \times 10^{-3}\ K^{-1}$. It should be emphasized that, in actual device applications, there are other heat losses in the system and the efficiency given in equation 9 can never be fully realized (see also Power Generation).

Refrigeration. In Figure 2(**b**), a current I is passed through a thermocouple with one of its junctions in thermal contact with a heat source and the other in contact with a heat sink. Under these conditions the device pumps heat from the heat source to the heat sink, and under steady-state conditions the temperatures are T_0 and T_2, respectively. The parameters of principal importance in evaluating the performance of a refrigerator are the coefficient of performance (COP), the heat-pumping rate, and the maximum temperature difference that the device produces.

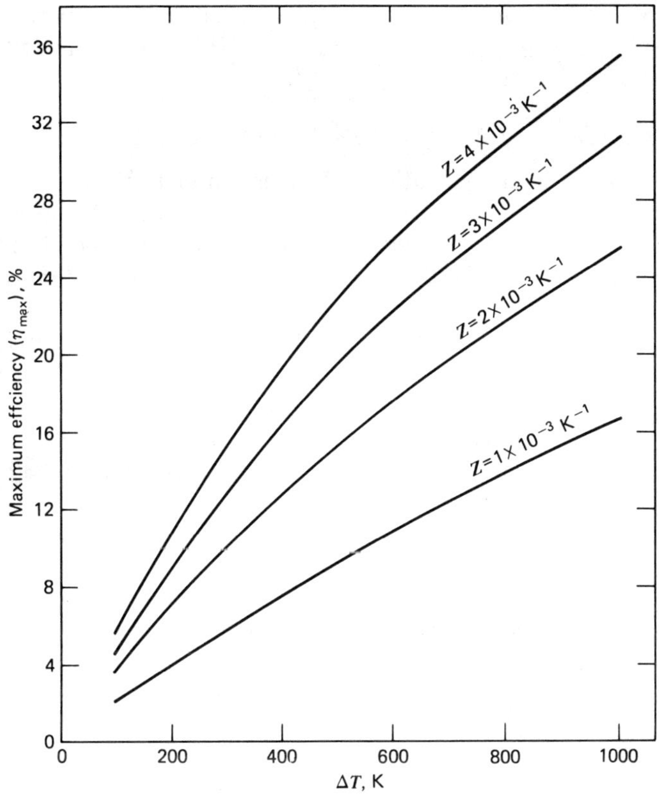

Figure 3. Maximum efficiency of a thermoelectric generator plotted against temperature difference for a cold-junction temperature of 300 K.

The coefficient of performance is

$$\text{COP} = \frac{\text{cooling obtained } (Q_c)}{\text{power input } (E_b I)} \tag{11}$$

At the cold junction T_0, the Peltier heat removed is opposed by the thermal conduction of heat along the thermoelements from the heat sink at temperature T_2 and one half of the Joule heat produced in the thermoelectric circuit. The cooling obtained, therefore, is given by

$$Q_c = \alpha_{np} IT - \kappa_t (T_2 - T_0) - \tfrac{1}{2} I^2 R_t \tag{12}$$

The power input to the thermocouple circuit consists of $\alpha_{np} I(T_2 - T_0)$ to overcome the developed Seebeck voltage and $I^2 R_t$ to overcome the resistance of the thermoelement branches. The power input, therefore, is

$$E_b I = \alpha_{np} I(T_2 - T_0) + I^2 R_t \tag{13}$$

and the COP is

$$\text{COP} = \frac{\alpha_{np} IT_0 - \kappa_t (T_2 - T_0) - \tfrac{1}{2} I^2 R_t}{\alpha_{np} I(T_2 - T_0) + I^2 R_t} \tag{14}$$

Thus, for a given pair of thermoelectric materials and for a given hot- and cold-junction temperature, the COP is a function of the current I, the electrical resistance R_t, and the thermal conductance κ_t. However, R_t and κ_t are not independent and the COP reaches a maximum value when the dimensions of the thermoelements satisfy equation 9 and the current is optimized. The expression for the maximum COP is, therefore,

$$\mathrm{COP}_{\max} = \left[\frac{T_2}{T_0 - T_2}\right] \frac{\sqrt{1 + Z\overline{T}} - T_0/T_2}{\sqrt{1 + Z\overline{T}} + 1} \qquad (15)$$

where $\overline{T}$ and Z are defined in equation 9. The maximum COP of a thermoelectric refrigerator is related to the Carnot efficiency and to a factor containing Z and $\overline{T}$, just as the maximum efficiency of a thermoelectric generator is related to these factors. A graph of $\mathrm{COP}_{\max}$ vs $(T_2 - T_0)$ for various values of the figure of merit Z is given in Figure 4. This graph shows that the maximum difference in temperature occurs under adiabatic conditions when the COP is zero. This graph also shows that, as the figure of merit increases, the $\Delta T_{\max}$ also increases and, at a constant ΔT, the $\mathrm{COP}_{\max}$ is greater for materials with higher figures of merit. A higher COP implies that less power is needed to pump the same quantity of heat. Therefore, the highest performance refrigerator is achieved with thermoelectric materials having the highest figures of merit (see Refrigeration).

Properties

The Seebeck coefficient, electrical conductivity, and thermal conductivity are properties of materials that can theoretically be related to the atomic structure of the

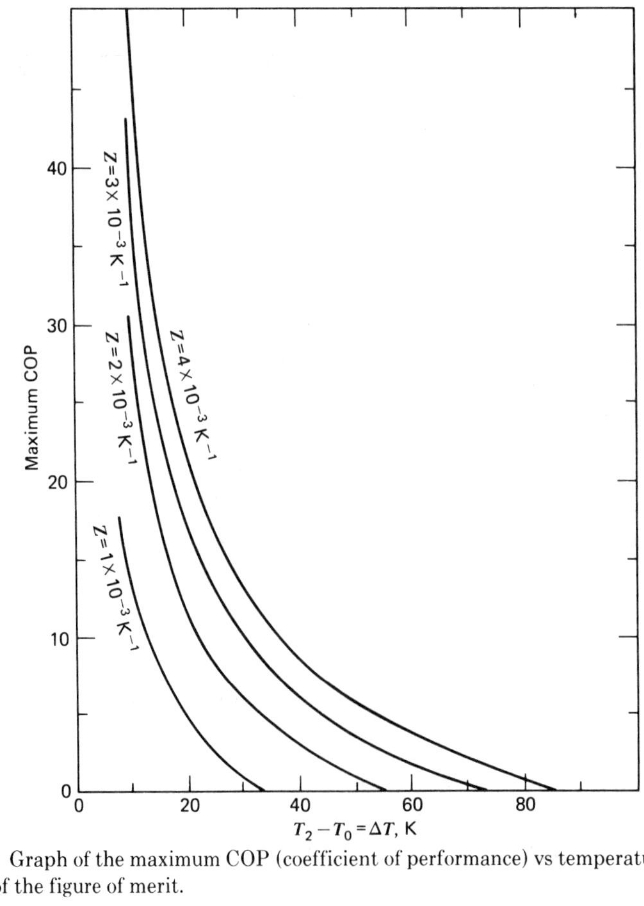

Figure 4. Graph of the maximum COP (coefficient of performance) vs temperature difference for various values of the figure of merit.

materials. However, any real material is much too complicated to analyze completely, although various simplified models of the electronic and thermal processes in metals and semiconductors have been made and are useful for describing the various material parameters (17–18). The thermoelectric properties of some metals and semiconductors at room temperature are given in Table 1 (19). It may be observed that for a metallic system the highest figure of merit is $0.6 \times 10^{-3}\ K^{-1}$, and for a semiconductor system it is $2.3 \times 10^{-3}\ K^{-1}$. It may be seen from Figure 3 that the latter yields ideal efficiencies at least three times greater than those characteristic of metallic systems. Metallic systems are not good sources of thermoelectric power where high power and efficiency are important, whereas semiconductors, since they have significantly higher efficiencies, do offer the potential for greater device utility.

Semiconductors. The Seebeck coefficient, the electrical conductivity, and the thermal conductivity are related to the transport of charge and energy along the thermoelement branch. The electrical conductivity σ is the ratio of the current density in the material and the applied electric field. It is given by

$$\sigma = ne\mu \tag{16}$$

where n is the density of charge carriers (electrons or holes) per unit volume, e is the electronic charge, and μ is the mobility, ie, the drift velocity of the charge carriers in a unit electric field. The thermal conductivity κ is the ratio of the heat-flux density in the material and the temperature gradient. For unipolar conductors, in which a single sign of charge carrier predominates, the total thermal conductivity can be written as the sum

$$\kappa = \kappa_{el} + \kappa_{ph} \tag{17}$$

Here, the electronic thermal conductivity κ_{el} is carried by the charge carriers and is related to the electrical conductivity by the Weidemann-Franz law (17). The lattice or phonon thermal conductivity κ_{ph} is related to the thermal motion of the lattice and may be regarded as independent of the charge-carrier concentration. The Seebeck coefficient α_{np} describes the tendency of a temperature gradient to give rise to a potential difference. If a finite conductor with a uniform distribution of charge carriers is perturbed by a temperature gradient, the charge particles diffuse from the hot to the cold junction. This gives rise to a back emf, which opposes further flow of charge. This counter-emf-per-unit-temperature gradient is the Seebeck coefficient.

Table 1. Thermoelectric Properties of Metals and Semiconductors at Room Temperature[a]

Substance	Seebeck coefficient, $\mu V/°C$	Electrical conductivity, S/cm	Thermal conductivity, W/(cm·K)	Figure of merit, K^{-1}
Cu	2.5	5.9×10^5	3.96	9.3×10^{-7}
Ni	18	1.5×10^5	0.87	5.6×10^{-5}
Bi	75	8.6×10^8	0.08	6.0×10^{-4}
Ge	200	1000	0.636	6.3×10^{-7}
Si	200	500	1.133	1.8×10^{-5}
InSb	200	2000	0.17	4.7×10^{-4}
InAs	200	3000	0.315	3.8×10^{-4}
Bi_2Te_3	220	1000	0.02	2.3×10^{-3}
ZnSb	170	556	0.03	5.8×10^{-4}

[a] Ref. 19.

All the parameters occurring in the figure of merit are functions of the carrier concentration, and their dependence on the carrier concentration is shown in Figure 5. The figure of merit reaches a maximum value at carrier concentrations near 10^{19} cm^{-3}, which corresponds to heavily doped semiconductors.

Seimconductor solid-state theory of electron transport provides a useful description of these macroscopic properties in terms of fundamental semiconductor material constants (18) (see Semiconductors). Analysis of the figure of merit in terms of these fundamental properties has shown that the figure of merit is given by

$$Z \alpha \frac{\mu (m^* T)^{3/2}}{\kappa_{ph}} \tag{18}$$

where the effective mass m^* is a parameter related to the band structure of the material (15,16,19). From equation 18, the optimum thermoelectric material should have a high mobility, a high effective mass, and a low value for the lattice thermal conductivity. This equation allows the evaluation of materials based upon the solid-state parameters μ, m^*, and κ_{ph}. However, detailed considerations of the factors governing the choice of thermoelectric materials have demonstrated that these quantities are interdependent, eg, the relationship between the mobility and the effective mass is governed by the mechanism through which the charge carriers are scattered (15,16,19). Theoretical considerations of the physics and chemistry of semiconductors has not yet advanced to the point where the quantities μ, m^*, and κ_{ph} can be predicted from first principles. Therefore, in order to make predictions about potential thermoelectric materials, it is necessary to rely primarily upon empirical information about these

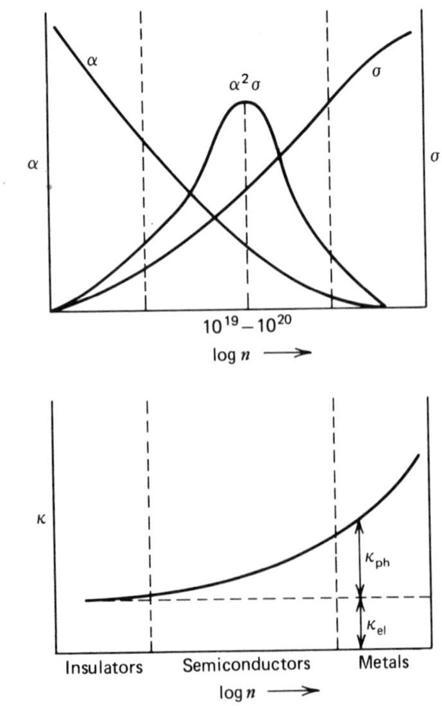

Figure 5. Schematic representation of the dependence of α, σ, and κ on the concentration of free carriers.

quantities. It has been shown that significant improvements in the figure of merit are achieved by solid-solution alloying with an isomorphous element or compound (15). This has the effect of reducing κ_{ph} and increasing the figure of merit, as shown by equation 18.

The optimum thermoelectric material should be a semiconductor doped to 10^{19}–10^{20} carriers/cm^3 and have a Seebeck coefficient of ca 200 μV/°C. For power generation, it should have a large bandgap (ca 1 eV for operation at 800–1000°C), a high melting point, a low vapor pressure, and good mechanical strength, and should be chemically stable.

Thermoelectric Materials. The chemical, physical, and thermoelectric properties of many materials have been examined for possible utility in thermoelectric devices. These materials include elements, alloys, compounds, and mixtures of compounds. A summary of the best thermoelectric materials that have been developed to date is given in Figure 6. A tabular summary of the properties of the best thermoelectric materials is given in Table 2. The highest overall efficiency can only be attained with

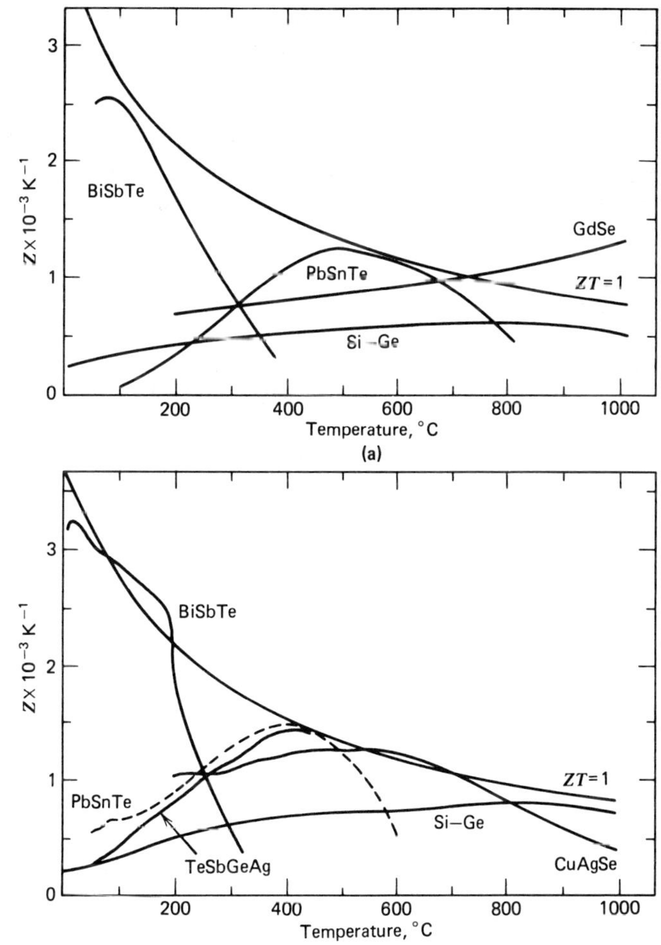

Figure 6. Figures of merit vs temperature for the best thermoelectric materials; (**a**) n-type materials; (**b**) p-type materials (see eqs. 9 and 10).

Table 2. Properties of Thermoelectric Materials

Material	Eg, eV	Mp, °C	Type	Z max, K^{-1}	Z max, T[a]	T max[b], °C
Bi_2Te_3	0.15	585	n or p	2.0×10^{-3}	0.6	250
$BiSb_4Te_{7.5}$			p	3.3×10^{-3}	0.99	250
Bi_2Te_2Se	0.3		n	2.3×10^{-3}	0.69	250
PbSnTe + MnTe	0.3	904	n or p	1.5×10^{-3}	1.00	400
$Si_{78}Ge_{22}$	1.0	1215	n	0.63×10^{-3}	0.70	1000
$Si_{78}Ge_{22}$	1.0	1215	p	0.80×10^{-3}	0.91	1000
TeSbGeAg	0.6	576	p	1.40×10^{-3}	0.56	400
$Cu_{1.97}Ag_{0.03}Se$			p	1.23×10^{-3}	0.95	900
$GdSe_x$			n	1.25×10^{-3}	1.59	1000

[a] Product of maximum figure of merit and temperature, K.
[b] Maximum operating temperature.

various thermoelectric materials over specific temperature ranges. Of all the thermoelectric materials that have been investigated, only bismuth telluride, lead telluride, and the silicon–germanium alloys are established. At this time, the selenide materials, Cu–Ag–Se (p type) and Gd–Se (n type), cannot be considered reliable and developed because of problems associated with using these materials in devices. In some cases, thermal instabilities may limit the useful temperature range over which a material can be used. For example, PbSnTe (PbSnTeMn) cannot be used at >500°C for extended periods because tellurium vaporizes at such temperatures. It may be observed from Table 3 that the best thermoelectric materials are characterized by $Z_{max}T$ values approaching unity.

Fabrication of Thermoelectric Devices

The thermoelectric materials are prepared by either hot pressing, cold pressing and sintering, polycrystalline-casting, or single-crystal techniques. The elements are usually prepared in cylindrical sizes from 1.2 cm OD × 2.5 cm long to 0.6 cm OD × 0.6 cm long. The most general module configuration incorporates the π couple, which is shown in Figure 7(**a**). With Seebeck coefficients of ca 200 μV/°C, a large number of thermocouples must be connected electrically in series to provide a useful output. For bismuth telluride, contacts are usually made with Ni and Cu; for lead telluride, contacts are usually made with W or Fe; and for the Si–Ge alloys, contacts are usually made with Mo or Nb (20) (see also Electrical connectors).

As shown in Figure 6, even the best thermoelectric materials do not exhibit uniformly good thermoelectric properties over the temperature ranges required for obtaining high efficiencies and high heat-pumping characteristics. Therefore, considerable effort has been directed towards the construction of devices based on a combination of materials so that each material operates over its own optimum temperature range (15,21). This has been achieved by using either a cascaded or a segmented arrangement. In a cascaded arrangement, the heat rejection from one stage is used to supply the thermal input to the next stage, as shown in Figure 7(**b**). The cascaded design is more commonly used for heat pumping than for power generation. In the segmented or sandwich arrangement, the thermoelement legs are made of different materials that are integrally bonded together, as shown in Figure 7(**c**). The segmented design is more commonly used for power generation than for heat pumping.

Table 3. Nuclear Thermoelectric Systems

Unit[a]	Power, W_e	Wt, kg	Specific power, W/kg	Thermoelectric material	Heat generator	T_h, °C	T_c, °C	Efficiency	Application
SNAP 1A	85	79.5	1.07	PbTe	isotope ^{144}Ce	538		2.8	demonstration
SNAP 3	2.5	2.3	1.09	PbTe	isotope ^{238}Pu	510		3.75	Department of Defense Satellite
SNAP 7A	10	849	not optimized	PbTe	isotope ^{90}Sr	484	49	6	light buoy
SNAP 7B	60	2088	not optimized	PbTe	isotope ^{90}Sr	484	69	6	lighthouse
SNAP 7C	10	849	not optimized	PbTe	isotope ^{90}Sr	493	60	6	Antarctic weather station
SNAP 7D	60	2088	not optimized	PbTe	isotope ^{90}Sr	516	93	6	ocean weather station
SNAP 7E	7	2724	not optimized	PbTe	isotope ^{90}Sr	482	38	6	deep-sea acoustic beacon
SNAP 9A	25	12.3	2.03	PbTe	isotope ^{238}Pu	493		5	Department of Defense Satellite
SNAP 10A	500	295	1.69	Si–Ge	nuclear reactor	499	324	1.6	space power generator
SNAP 11	25	13.6	1.84	PbTe	isotope ^{244}Ce	524	177	3	NASA Surveyor
SNAP 19	30	13.6	2.21	PbTe	isotope ^{238}Pu	500		5	NASA weather satellite
Pioneer/Viking	35	14	2.5	TeSbGeAg/PbTe	same	540	180	7.1	study of Jupiter, Saturn, and Mars
SNAP 27	63.5	19.4	3.27	PbTe	same	610	275	4.4	Moon missions
MHW	155	36.3	4.27	Si–Ge	isotope ^{238}Pu	1000	300	6.7	Voyager spacecraft

[a] SNAP = systems for nuclear auxiliary power; MHW = multihundred watt.

A number of chemical and metallurgical problems arise from the assembly of thermoelectric materials into modules and the use of these modules in actual devices. These include achieving and maintaining low resistance contacts between the thermoelectric materials and the hot and cold shoes, preventing degradation of thermoelectric performance as a result of inward diffusion of material from the shoes or oxidation and vaporization of the surface of the thermoelectric material, and maintenance of mechanical stability of the module during the cyclic stresses that occur in the actual device. These problems are most prevalent with thermoelectric generators, where high temperatures are involved, and are not generally considered to be limiting factors for thermoelectric cooling devices.

The state of the art for power generation can be characterized by thermoelectric

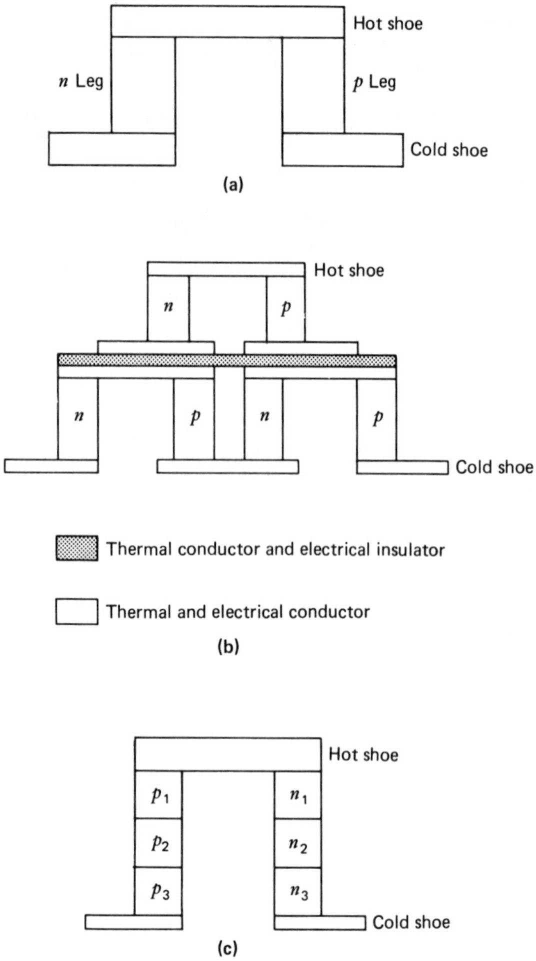

Figure 7. Thermoelectric module arrangements. (**a**) Pi couple. (**b**) Cascaded thermoelectric module. (**c**) Segmented thermoelectric module.

modules operating at 7–10% efficiency at up to 1000°C and average lifetimes of ca 2 yr (12). For thermoelectric cooling devices, the state of the art can generally be characterized by modules operating at ΔT_{max} of ca 70–75°C from 300 K, and having a mean time before failure of 10 yr (22).

Power Generation. A thermoelectric generator converts heat directly from a heat source into electrical energy and is basically a power-limited, low voltage, high current d-c device. Most devices require multivolt inputs, and dc-to-dc converters are used in these instances to raise the voltage to an acceptable level. Therefore, the essential components of a thermoelectric generator are a heat source, a thermoelectric module, and a dc-to-dc converter. Since ca 1960, thermoelectric generators have been built using solar, oceanic temperature gradients and fossil-fuel and nuclear-energy heat sources. These systems have the following inherent advantages in thermoelectric power generation: efficiency independent of size, high reliability, unattended and silent operation, lack of moving parts, resistance to nuclear radiation, and environmental compatibility. On the other hand, the disadvantages are low efficiency and power level, low specific power (power-to-weight ratio), low operating temperature, and high cost (see Nuclear reactors; Solar energy).

Solar Thermoelectric Generators. The earliest attempts at thermoelectric power generation were with generators based on solar heat (8,23). In the early 1960s, several solar thermoelectric generators for space applications were constructed, but because of the low efficiencies involved they do not appear to be potentially useful as an efficient means of space power (24). For underdeveloped countries, solar thermoelectric generators have been developed for powering small irrigation pumping systems, and recently, the use of salty ponds to supply energy has been proposed for powering irrigation pumping systems in the United States (25–26) (see Solar energy).

Fossil-Fuel-Fired Generators. Thermoelectric generators based on fossil-fuel heat sources, eg, propane, butane, natural gas, and JP-5, are used in remote areas where conventional sources of electrical power are not available. The first successful thermoelectric generators involving semiconductors were developed in the USSR and the United States (27–28). A unit has also been developed in the United States to power a gas-burner control that uses heat from a pilot flame.

Both Teledyne Energy Systems and Global Thermoelectrics in Canada have developed a series of fossil-fueled generators based on natural gas, propane, or butane for use in remote locations. The Teledyne system uses bismuth telluride technology and the Global system uses lead telluride. Commercially available units have output powers of 10–100 W, specific powers of ca 1 W/kg, costs of ca $60/W, and efficiencies of 2–4%. Some applications for these units include navigational aids, data systems, communications, and cathodic protection of pipelines, wellheads, and offshore platforms. The advantages of these generators are high reliability, safety, no moving parts, simple operation, operability under all weather conditions, versatility, cleanliness, and mounting versatility.

Fossil-fueled generators have also been developed for certain military applications for field use (29). These generators have power levels up to 500 W, specific powers of ca 17 W/kg, multifuel capability (gasoline, diesel, kerosene, and JP-5), and efficiencies of ca 3%, and they are inaudible beyond 10 m.

Nuclear Thermoelectric Generators. For certain applications requiring electrical power, small size, low weight, and long life, the use of nuclear-energy heat sources provides the optimum method for power generation. The long life of the heat source coupled with its high energy density facilitates the use of thermoelectric generators that can operate unattended for long periods of time. In addition to power for space vehicles, radioisotope thermoelectric generators (RTGs) are used in a variety of medical, terrestrial, and marine applications.

Space Power Systems. Since 1961, the United States has launched 22 NASA and military space systems having all or part of their power requirements supplied by nuclear power sources. Twenty-one of these space systems were powered by RTGs and one by a nuclear reactor (30). Table 3 summarizes the different types of nuclear power systems that have been built and demonstrated. The history of these systems has shown that they can be safely and reliably built and launched to meet a variety of mission objectives.

The RTGs most recently flown in space are the multi-hundred-watt (MHW) units developed by General Electric (31). These 155-W units were used to power the Voyager spacecraft, which sent back new information about Jupiter and Saturn. A photograph of an MHW unit is shown in Figure 8. These units have provided greater-than-pre-

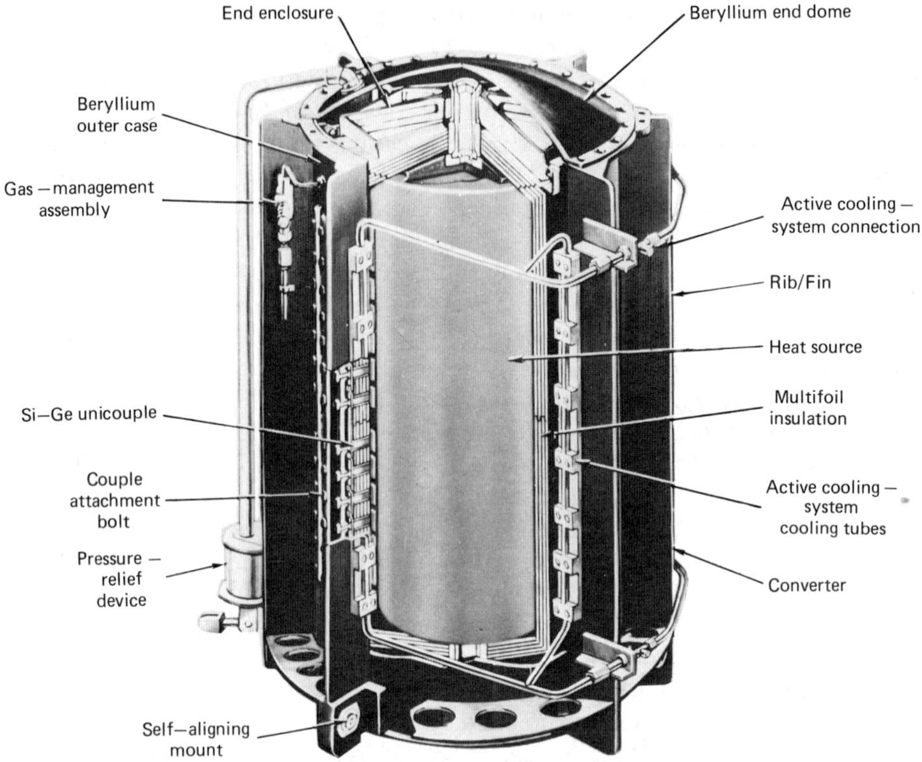

Figure 8. RTG multihundred watt (MHW) thermoelectric generator for space power. Courtesy of General Electric Company.

dicted power for these missions over the 1.5- and 3-yr durations it has taken the spacecraft to reach Jupiter and Saturn.

It can be observed from Table 3 that PbTe has been the mainstay of the U.S. space program, but it now appears that it might be replaced by the Si–Ge alloys, especially in view of the success of the MHW units. For space applications, the state of the art for nuclear-fueled thermoelectric systems is characterized by specific powers of 4 W/kg, conversion efficiencies of 6–7%, and power levels of hundreds of watts.

Terrestrial and Marine Applications. Radioisotope thermoelectric generators are also attractive power systems on earth and marine environments for certain applications, particularly where there are conditions of remoteness, inaccessible or lack of air supply, and need of long-term operation with minimal maintenance. Surface and terrestrial applications include powering light buoys, automatic weather stations, and communications relay stations. Marine applications include power for undersea structures and experiments, deep-sea applications for depths up to 9000 m, acoustic beacons, and oceanographic sensors.

The primary commercial suppliers of these RTGs are General Atomics and Teledyne Energy Systems. The latter have designed a number of Sentinel generators that use TeSbGeAg–PbTe couples and Sr-90 isotopic heat sources, range in power up to 140 W, and have specific powers of 0.01–0.11 W/kg.

Medical Applications. The development of nuclear-powered batteries in the early 1970s provided physicians with power sources to drive implantable devices for their intended period of operation. The radioisotope cardiac pacemaker (RCP) is certainly the best-known application for these nuclear thermoelectric sources. In essence, the RCP is isotope-powered with Pu-238 (half life, 87 yr), has bismuth telluride thermoelements, a dc-to-dc converter, and a pulse generator. The pacemaker controls heart rhythms by delivering electrical impulses to the heart muscle and requires a power level of only ca 3 μW. By 1975, the number of RCP implants was ca 1600 (20) (see Prosthetic and biomedical devices).

Cooling Devices

Since the early 1960s, advances in solid-state chemistry and physics have improved the quality and performance of thermoelectric materials to the point where they are now competitive with existing hardware for certain applications. The advantages of a thermoelectric cooling system are COP independent of size, silent operation, space and weight savings, no moving parts except fans, high reliability, environmental compatibility, simplified maintenance, and modular construction. The disadvantages are low COP and high costs of equipment and power.

In recent years, thermoelectric cooling has been used for a variety of spot-cooling commercial and military applications, eg, cooling electronic components, temperature-control units, medical and laboratory instrumentation, and small volume refrigeration systems. The use of thermoelectric cooling for large-scale commercial-refrigeration and air-conditioning systems is limited by the low efficiencies of the materials involved. Military applications have included cooling systems for fleet submarines and cooling of electronic components for cruise missiles (32–33). Initially, a large number of manufacturing companies explored thermoelectric cooling because of the potential that existed for large-scale commercial and residential markets. However, in the last decade this enthusiasm has decreased, and only Borg-Warner, Melcor, Marlow Industries, and Cambion continue to be active in the field.

In practical use, couples similar to the π couple shown in Figure 7(**a**) are combined in a module where they are connected in series electrically and parallel thermally. Normally, a module is the smallest unit available. The user can tailor the quantity, size, or capacity of the module to fit exact requirements. For operation at or near room temperature, bismuth telluride materials are nearly always used. Modules are available in a variety of sizes, shapes, operating currents, operating voltages, number of couples, and ranges of heat-pumping levels.

Typical operation is shown in Figure 9 for a Melcor CP-2-31-06 module. This figure is for a module containing 31 Bi–Te couples. Each element is 2 mm^2 in cross section and 1.5 mm in length. This figure shows that for a hot-side temperature of 35°C and a cold-side temperature of 10°C (ie, ΔT = 25°C), it will pump ca 16 W of cooling at a current of 10 A and a voltage of 2.8 V. The current (1981) cost of these modules is ca $1.00/W of heat pumped.

Couples have heat-pumping capabilities of 1–4 W. Therefore, large-scale thermoelectric cooling systems require large numbers of couples and result in high costs. However, for applications requiring less than ca 600 W, thermoelectric cooling provides higher COPs than can be obtained by conventional vapor-compression methods. Consequently, the commercial thermoelectric cooling field has been generally restricted

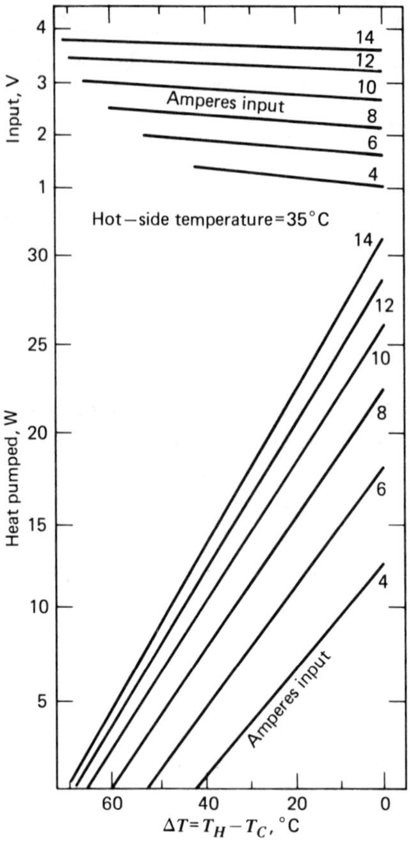

Figure 9. Heat-pumping and electrical characteristics for a 31-couple thermoelectric cooling module with a 35°C hot-junction temperature. Courtesy of Materials Electronics Products Corporation (Melcor).

to coolers for small specialized devices. Some of these applications include coolers for electronic components, eg, infrared detectors, integrated circuits, and nuclear-radiation detectors; temperature-control units, eg, constant-temperature chambers, temperature baths, and environmental chambers; small commercial-refrigeration systems, eg, cream coolers, juice dispensers, and coolers for the recreational outdoors market; and medical and laboratory instruments, eg, cold probes, dewpoint hydrometers, immersion coolers, and cooling for x-ray spectrometers. Borg-Warner has developed a 320-W thermoelectric cooler (Model AHP-1700) for applications in electronics, computers, and machine tools. Thermoelectric coolers are being used increasingly in electrooptical systems. Recent advances in ir-detector technology (see Infrared technology) and thermoelectric cooling have produced effective systems that operate at 170–193 K (34). These lower temperatures are achieved by cascading, as shown in Figure 7(**b**).

Proposals for low temperature thermomagnetic cooling devices have been presented, but no full reductions to commercial applications have taken place. It has been demonstrated that couples based upon this technology can achieve continuous cooling down to 128 K (35). In the 1970s, thermoelectric cooling has been primarily a commercial business. The extensive program in thermoelectric cooling conducted by the U.S. Navy in the 1960s was significantly reduced during the 1970s.

BIBLIOGRAPHY

"Thermoelectric Power Conversion" in *ECT* 1st ed., Suppl. 1, pp. 840–866, by William Arbiter and Richard C. Ross, Nuclear Development Corporation of America; "Thermoelectric Energy Conversion" in *ECT* 2nd ed., Vol. 20, pp. 147–172, by Alfred C. Glatz, National Aeronautics and Space Administration.

1. A. Volta, *Philos. Trans.*, 403 (1800).
2. T. J. Seebeck, *R. Acad. Sci.*, 265 (1822).
3. J. C. Peltier, *Ann. Chim. Phys. (2nd Ser)* **56,** 371 (1834).
4. W. Thomson, *Philos. Trans. R. Soc.*, 1 (1855).
5. U.S. Pat. 511,245 (April 9, 1889), M. Mestern.
6. E. Attenkirch, *Phys. Z.* **10,** 560 (1909).
7. *Ibid.*, **12,** 920 (1911).
8. M. Telkes, *J. Appl. Phys.* **25,** 765 (1954).
9. A. F. Joffe, *Semiconductor Thermoelements and Thermoelectric Cooling*, Infosearch Ltd., London, 1957.
10. A. F. Joffe, *Dokl. Akad. Nauk SSSR* **106,** 931 (1956).
11. C. M. Bhandari and D. M. Rowe, *Contemp. Phys.* **21,** 219 (1980).
12. E. Hampl, J. Hinderman, W. Mitchell, R. Reylek, and D. Wald, *Intersoc. Energy Convers. Eng. Conf.* **10,** 714 (1975).
13. G. N. Hatsopoulos and J. H. Keenan, *J. Appl. Mech.* **25,** 428 (1958).
14. C. A. Domencalli, *Rev. Mod. Phys.* **26,** 237 (1954).
15. H. J. Goldsmid, *Applications of Thermoelectricity*, Methuen & Co., Ltd., London, 1960.
16. R. R. Heikes and R. W. Ure, Jr., *Thermoelectricity: Science and Engineering*, Interscience Publishers, Inc., New York, 1961.
17. F. Seitz, *The Modern Theory of Solids*, McGraw-Hill Book Co., Inc., New York, 1940, pp. 139–194.
18. R. A. Smith, *Semiconductors*, Cambridge University Press, Cambridge, England, 1959, pp. 1–188.
19. D. A. Wright in K. H. Spring, ed., *Direct Generation of Electricity*, Academic Press, Inc., New York, 1965, pp. 285–331.
20. D. M. Rowe, *Proc. IEE* **125,** 1113 (1978).
21. F. D. Rose, E. F. Hockings, and N. E. Lindenblad, *RCA Rev.* **12,** 82 (1961).
22. M. Bohn, D. Benson, and T. Jayadev, *J. Sol. Energy Eng.* **102,** 119 (1980).
23. W. W. Coblentz, *Sci. Am.* **127,** 324 (1922).
24. N. Fuschillo, R. Gibson, F. Eggleston, and J. Epstein, *Adv. Energy Convers.*, 103 (1966).
25. W. Thorp, *Fifth Int. Power Sources Symp.*, Brighton, England, Sept. 20–22, 1966.
26. D. Benson and T. Jayadev, *Third Int. Conf. Thermoelectr. Energy Convers.*, University of Texas at Arlington, Dallas, Texas, Mar. 12–14, 1980.
27. V. Daniel-bek, A. Voronin, and N. Roginskaya, *Radio* **2,** 24 (1954).
28. R. W. Fritts, *Proc. IEEE* **51**(5), 713 (1963).
29. G. Guazzoni, A. Herchakowski, and J. Angello, *Proc. Intersoc. Energy Convers. Eng. Conf.* **13,** 1978 (1978).
30. G. Bennett, J. Lombardo, and B. Rock, *Proc. Intersoc. Energy Convers. Eng. Conf.* **16,** 362 (1981).
31. A. Pitrolo, R. Morrow, and A. Arker, *Proc. Intersoc. Energy Convers. Eng. Conf.* **6,** 492 (1971).
32. R. Strucko and J. Stanton, *Thermoelectric Cooling System and Power Supply for Fleet Nuclear Submarines*, NSRDC Report 3920, Naval Ship Research and Development Center, Washington, D. C., 1973.
33. R. Buist, J. Fenton, and J. Lee, *Proc. Intersoc. Energy Convers. Eng. Conf.* **11,** 1619 (1976).
34. R. Buist, *Proc. Electro-Opt. Syst. Des. Conf.*, West San Francisco, Calif., Nov. 5–7, 1972.
35. W. Yim and A. Smith, *Solid-State Electron.* **15,** 1161 (1972).

General References

Refs. 9, 15, 16, 19, and 20 are also general references.

H. J. Goldsmid, *Thermoelectric Refrigeration*, Heywood-Temple Industrial Publications, Ltd., London, 1964.

ALFRED C. GLATZ
General Foods Corporation

THIAZOLE DYES

Direct dyes, 918
Disperse dyes, 924

DIRECT DYES

Dyes are classified according to their mode of application, eg, acid dyes, direct dyes, disperse dyes, etc, or by their chemical structure, eg, azo dyes (qv), anthraquinone dyes, xanthene dyes (qv). The thiazole nucleus (**1**) is a moiety common to the basic, direct, disperse, cationic, and vat dyes.

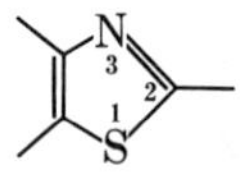

(**1**)

Although these dyes once enjoyed wide usage, they have declined commercially in recent years; only four products were reported by the U.S. International Trade Commission in 1980 as being manufactured in or imported into the United States (Table 1) (see also Dyes and dye intermediates).

Direct dyes are commercially important because of their excellent substantivity to cellulose attributed to the ability of the dye molecule to form hydrogen bonds with the cellulosic substrate. These bonds form most readily when the dye molecule assumes a linear and coplanar conformation. The basic building block for this type of dye is dehydrothio-*p*-toluidine [*92-36-4*] (**4**) [2-(*p*-aminophenyl)-6-methylbenzothiazole], prepared by the fusion of *p*-toluidine with sulfur. The discovery of this reaction in 1887 constitutes the beginning of thiazole dye technology.

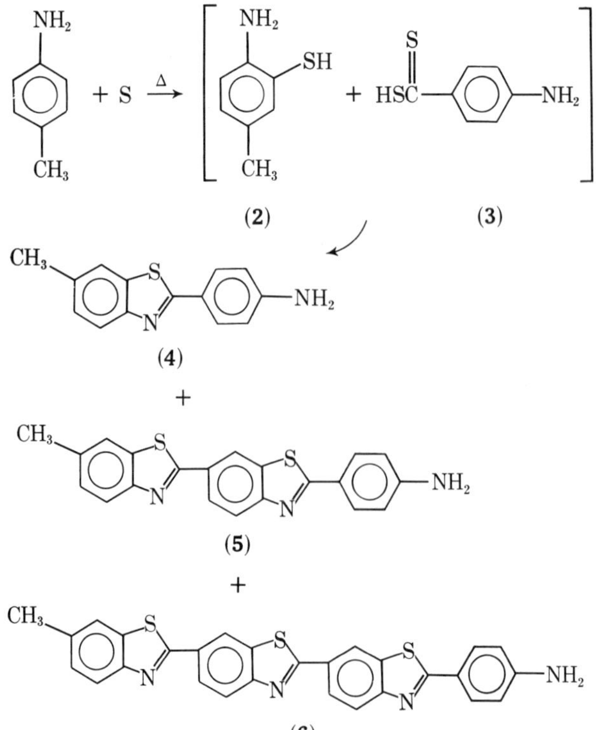

(**6**)

Table 1. U.S. Production and Imports of Thiazole Dyes

Year	Basic Yellow 1, CI 49005 [*2390-54-7*]	Direct Yellow 27, CI 13950 [*10190-68-8*]	Direct Yellow 28, CI 19555 [*8005-72-9*]	Vat Yellow 2, CI 67300 [*129-09-9*]
1976				
U.S. production[a], kg			37,146	297,170
value, $			394,000	686,000
1977				
U.S. production[a], kg			38,052	
value, $			421,000	
imports[b], kg		2,442	679	
1978				
U.S. production[a], kg			25,821	
value, $			313,000	
imports[b], kg		901	453	
1979				
U.S. production[a], kg			12,684	
value, $			218,000	
imports[b], kg		1,389	453	
1980				
U.S. production[a], kg			16,761	
value, $			259,000	
imports[b], kg		951		
U.S. manufacturers[c]	DUP	ATL	ATL, BAS, CK, DUP, and TRC	AC, TRC, and VPC

[a] Ref. 1.
[b] Ref. 2.
[c] DUP, E. I. du Pont de Nemours & Company, Inc.; ATL, Atlantic Chemical Corporation; BAS, BASF Wyandotte Corporation; CK, Crompton and Knowles Corporation; TRC, Toms River Chemical Corporation; AC, American Color and Chemical Corporation; and VPC, Verona Division of Mobay Chemical Corporation (3).

p-Toluidine (1 mol) is heated in a fusion kettle with a slight excess of sulfur (>2 mol technical grade, not crystalline) and a trace of soda ash at 180°C for 1 hour. The temperature is then raised to 190, 200, 210, and 220°C in four-hour stages. Evolution of H_2S ceases at 220°C; the principal products are dehydrothio-*p*-toluidine (**4**) and primuline base [*95-22-7*] (**5**). Dehydrothio-*p*-toluidine is obtained by vacuum distillation at 230–240°C at 266–399 Pa (2–3 mm Hg) with further purification by extraction in boiling ethanol.

The proportions of compounds (**4**) and (**5**) can be controlled by varying the ratio of sulfur to *p*-toluidine; 4:10 favors compound (**4**); 7:10 favors compound (**5**). The thiophenol (**2**) and the dithiobenzoic acid (**3**) are probable intermediates. Continued heating for 18–20 h at ≥220°C may introduce a third thiazole ring (**6**). The bases (**5**) and (**6**) are sulfonated with approximately five times their weight of 20 wt % oleum at 40°C. The sulfonic acids are separated according to the solubilities of the ammonium salts; the sodium salt of the dibenzothiazolesulfonic acid (**7**) [*10360-31-3*] readily dissolves as the ammonium salt, whereas the sodium salt of the monobenzothiazolesulfonic acid [*8005-72-9*] (**8**) is only sparingly soluble.

The classical dyestuff primuline [*8064-60-6*] (Direct Yellow 59, CI 49000) is a mixture of the sodium salts of compounds (**7**) and (**9**) [*10132-80-6*]. Primuline itself, when dyed directly on cotton, gives a greenish-yellow color too fugitive to be of practical value. However, when the amino group is diazotized on the fiber and coupled to aromatic phenols and amines, dyes with good washing fastness are produced. If coupled to resorcinol, an orange shade is obtained; to β-naphthol, a red; to R-Salt (2-naph-

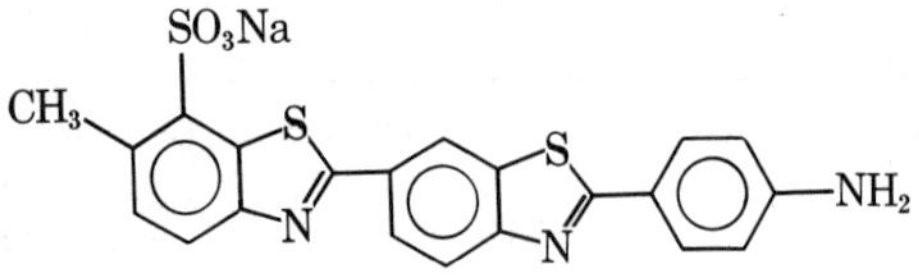

(7)

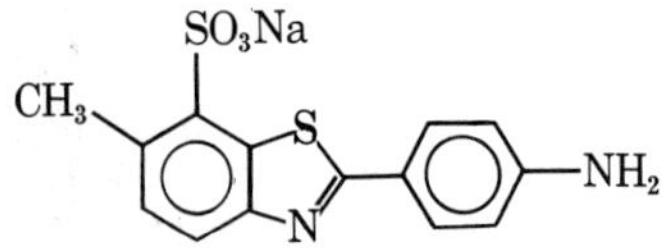

(8)

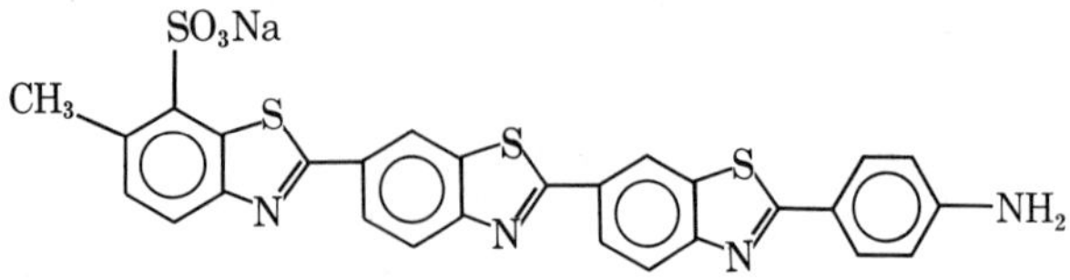

(9)

thol-2,6-disulfonic acid disodium salt), a maroon; to *m*-phenylenediamine, a brown.

Commercially the most important direct dye derived from dehydrothio-*p*-toluidine monosulfonic acid (**8**) is Direct Yellow 28 [*8005-72-9*] (CI 19555). It is obtained by treating an aqueous solution of compound (**8**) with sodium hypochlorite. The dye is a mixture of ca 60% of the azo dye [*10114-47-3*] (**10**) produced by this oxidative coupling and 25% of the corresponding azine product (**11**) [*83487-93-8*].

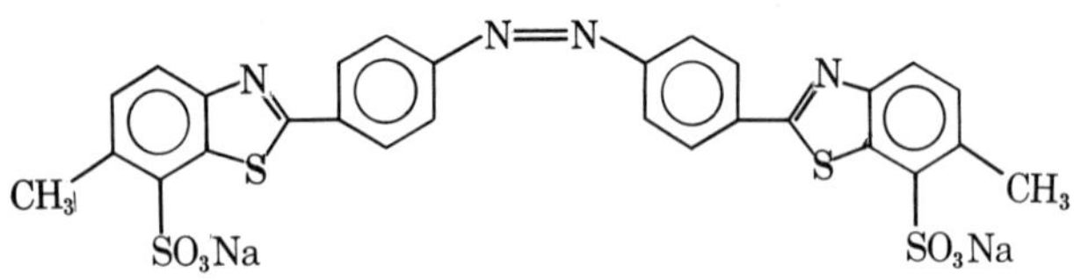

(10)

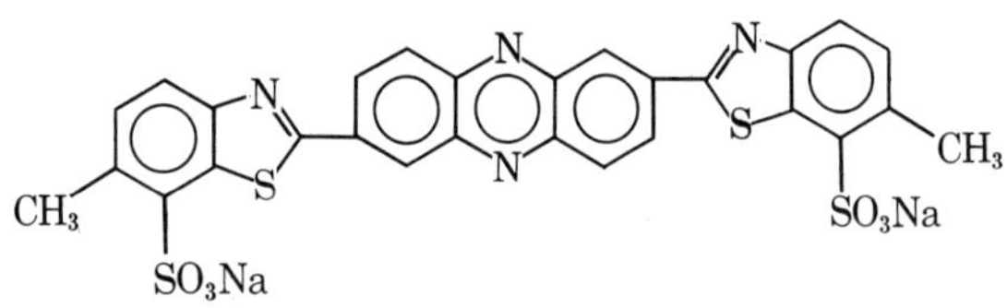

(11)

Other direct dyes obtained from dehydrothio-*p*-toluidine monosulfonic acid are given in Table 2.

Basic Dyes. Basic dyes are synthesized from dehydrothio-*p*-toluidine, as for example, Basic Yellow 1 [*2390-54-7*] (CI 49005) (**17**) by methylation in hydrochloric or sulfuric acid at 160–170°C under pressure. The phosphotungstomolybdic acid salt of compound (**17**) is Pigment Yellow 18 [*1326-11-0*] (CI 49005:1).

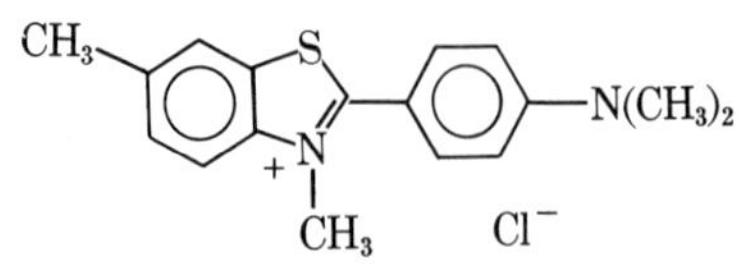

(17)

Vat Anthraquinone Dyes. The only commercially important thiazole vat dye is Vat Yellow 2 [*129-09-9*] (CI 67300) (**18**), prepared by heating 2,6-diaminoanthraquinone, benzotrichloride, sulfur, and cuprous chloride in naphthalene at 180–185°C

Table 2. Direct Azo Dyes from Dehydrothio-*p*-Toluidinesulfonic Acid

Dye	CAS Reg. No.	CI no.	Molecular formula	Structure	Structure no.	Coupling component
Direct Yellow 27	*[10190-68-8]*	13950	$C_{25}H_{22}N_4O_9S_3.2Na$	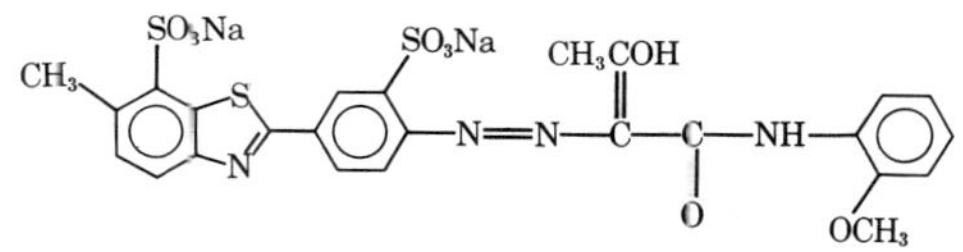	**(12)**	*o*-acetoacetanisidide
Direct Yellow 8	*[10130-29-7]*	13920	$C_{24}H_{20}N_4O_6S_2.Na$	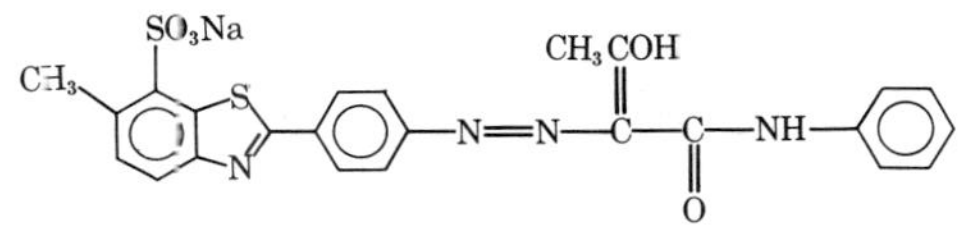	**(13)**	acetoacetanilide
Direct Orange 17	*[6442-00-8]*	19160	$C_{24}H_{17}N_5O_8S_3.3Na$	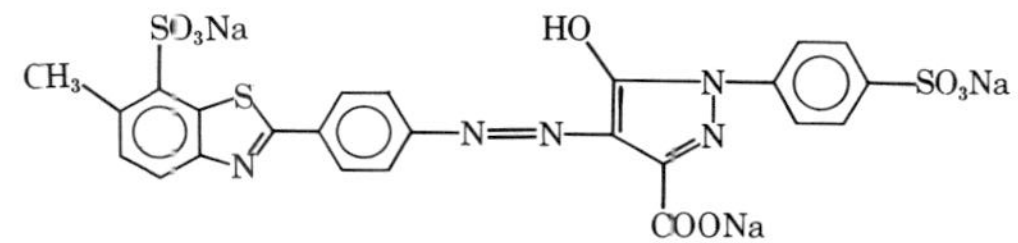	**(14)**	3-carboxy-1-(*p*-sulfophenyl)-5-pyrazolone
Direct Orange 18	*[5863-88-7]*	20215	$C_{26}H_{19}N_5O_5S.Na$	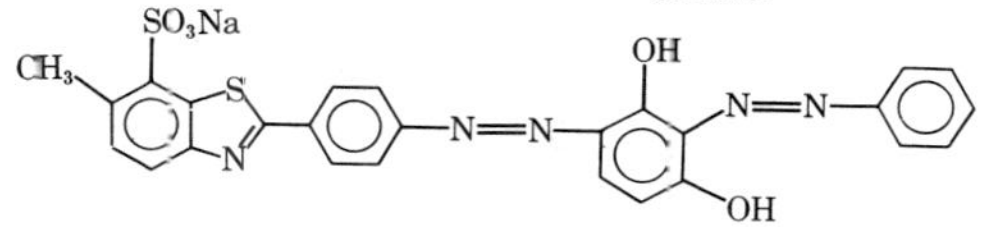	**(15)**	resorcinol[a]
Direct Red 45	*[2150-33-6]*	14780	$C_{24}H_{17}N_3O_7S_3.2Na$	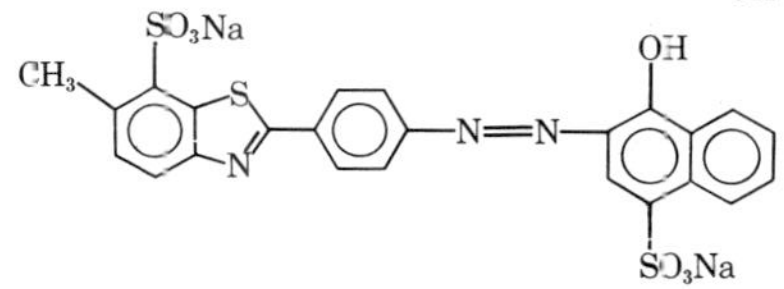	**(16)**	Nevile and Winther's acid

[a] Followed by reaction with aniline.

Table 3. Vat Anthraquinone Dyes Containing the Thiazole Nucleus

Dye	Structure no.	CAS Reg. No.	CI no.	Structure	Intermediates
Vat Yellow 11	(**19**)	[*6451-11-2*]	70405	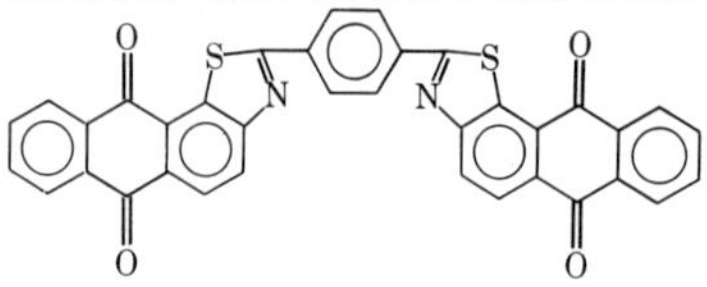	2-amino-1-thiol-anthraquinone and terephthaloyl chloride
Vat Red 20	(**20**)	[*6371-49-9*]	67100	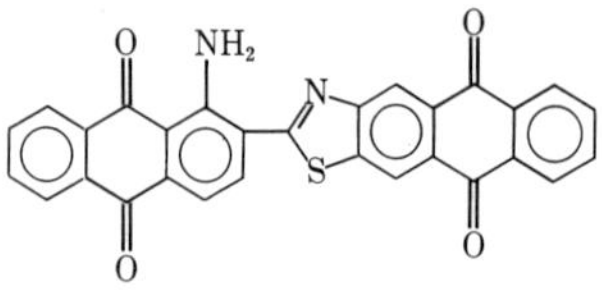	2-amino-3-chloro-anthraquinone and 1-nitro-2-anthraquinone-carboxylic acid
Vat Blue 30	(**21**)	[*6492-78-0*]	67110	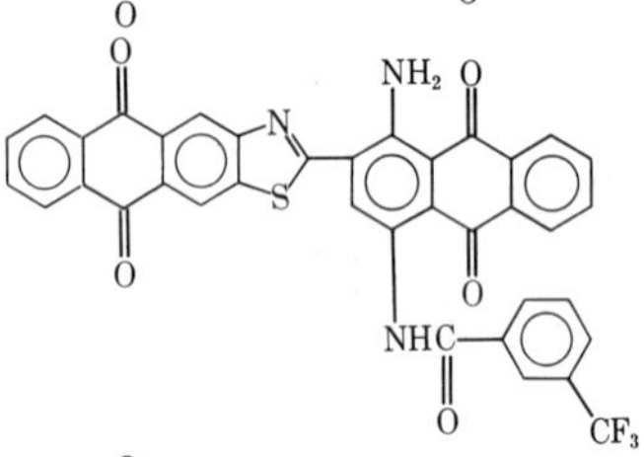	1-amino-4-nitro-2-anthraquinone-carboxylic acid and 2-amino-3-chloroanthraquinone; *m*-trifluoro-methylbenzoyl fluoride
Vat Blue 31	(**22**)	[*6371-50-2*]	67105	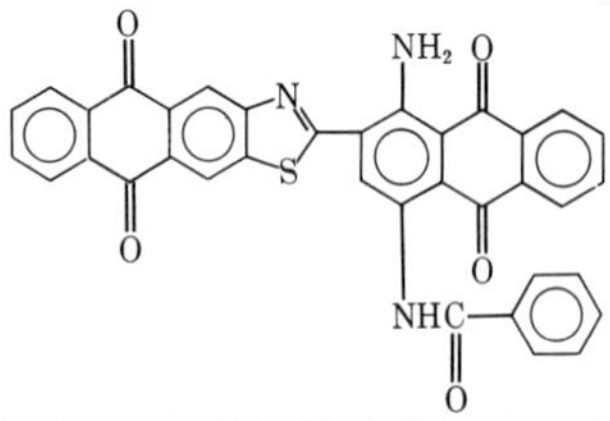	1-amino-4-nitro-2-anthraquinone-carboxylic acid and 2-amino-3-chloroanthraquinone; benzoyl chloride

for six hours. After heating for four more hours at 224°C, the mixture is cooled to 180°C and diluted with more naphthalene. The product is filtered at 130°C on a pressure filter and washed several times with liquefied naphthalene; a 70% yield is obtained. The dye is purified by reprecipitation from sulfuric acid followed by sodium hypochlorite oxidation.

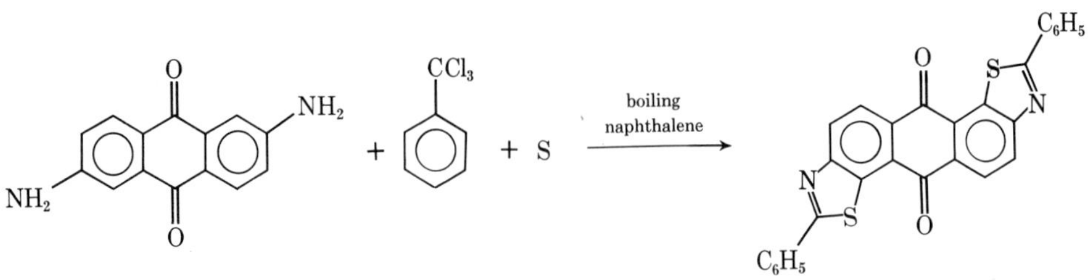

(**18**)

Other vat dyes containing the thiazole nucleus are given in Table 3.

Table 4. Toxicological Characteristics of Some Yellow Dyes

Dye	Rat, oral LD_{50}, g/kg	Rabbit Eye irritation	Rabbit Skin irritation
Direct Yellow 27	>5	negative	slight
Direct Yellow 28	>5	negative	negative
Vat Yellow 2	>15.4	negative	negative

Health and Safety Factors

As a class, dyes derived from the thiazole nucleus have not exhibited properties warranting special health and safety precautions; however, standard chemical labeling instructions are required. Toxicological characteristics of some commercial dyes are given in Table 4.

BIBLIOGRAPHY

"Thiazoles" in *ECT* 1st ed., Vol. 14, pp. 49–58, by Stuart D. Kerr and Martha T. Rathgeb, Cincinnati Chemical Works, Inc.; "Thiazole Dyes" in *ECT* 2nd ed., Vol. 20, pp. 185–191, by R. S. Powell, Allied Chemical Corporation.

1. U.S. International Trade Commission, *Synthetic Organic Chemicals; United States Production and Sales*, U.S. Government Printing Office, Washington, D.C., 1976–1980.
2. U.S. International Trade Commission, *Imports of Benzenoid Chemicals and Products*, U.S. Government Printing Office, Washington, D.C., 1977–1980.
3. Ref. 1, 1980.

General References

Colour Index, 3rd ed., Vol. 4, Society of Dyers and Colourists, Bradford, England and American Association of Textile Chemists and Colorists, Research Triangle Park, N.C., 1971, pp. 4056, 4070, 4132, 4137, 4144, 4441, 4566, 4567, 4584.

H. A. Lubs, *The Chemistry of Synthetic Dyes and Pigments*, Reinhold Publishing Corporation, New York, 1955, pp. 269–272.

K. Venkataraman, *The Chemistry of Synthetic Dyes*, Vol. 1, Academic Press, Inc., New York, 1952, pp. 622–627.

BIOS (British Intelligence Objectives Subcommittee) Final Report 1153, 1946, pp. 349–352 and *Final Report 987*, 1946, p. 59.

H. E. Fierz-David and L. Blangey, *Fundamental Processes of Dye Chemistry*, Interscience Publishers, Inc., New York, 1949, pp. 333–335.

M. Schubert, *Ann.* **558,** 10 (1947).

Fr. Pat. 1,372,691 (Sept. 18, 1964), W. H. Vinton (to E. I. du Pont de Nemours & Co., Inc.).

U.S. Pat. 3,301,860 (Jan. 31, 1967), R. A. Clarke (to E. I. du Pont de Nemours & Co., Inc.).

Ger. Offen. 2,142,947 (Mar. 30, 1972), A. C. Little (to Allied Chemical Corp.).

Ger. Offen. 2,151,795 (Apr. 20, 1972), J. Blackwell (to E. I. du Pont de Nemours & Co., Inc.).

Ger. Offen. 2,206,551 (Aug. 31, 1972), G. Schoefberger (to Sandoz, Ltd.).

RUSSELL E. FARRIS
Sandoz, Inc.

DISPERSE DYES

Many disperse dyes contain a thiazole nucleus. The term disperse describes the manner in which this class of dyes is applied to textile fibers, ie, from a dispersion in water with the aid of a dispersing agent, eg, sodium lignosulfonate (see Dispersants; Emulsions). The thiazole dyes in this class are azo compounds derived from 2-aminothiazole and 2-aminobenzothiazole derivatives, insoluble or sparingly soluble in water (see Azo dyes). Before 1950, these dyes were used to dye acetate and triacetate fibers (see Cellulose acetate and triacetate fibers). Since 1950, their main use has been for polyester fibers. The *Colour Index* gives structures for three dyes of this type: 1,2,3,4-tetrahydro-6-[(6-methoxy-2-benzothiazolyl)azo]benzo[*h*]quinoline-3,7-diol [*6054-52-0*] (**1**) (Disperse Blue 15, CI 11435), 2,2′-[4-[[(6-methoxy-2-benzothiazolyl)azo]phenyl]imino]diethanol [*6373-93-9*] (**2**) (Disperse Red 58, CI 11135), and an ethoxy derivative of (**2**) (Disperse Red 64 [*6358-50-5*], CI 11136):

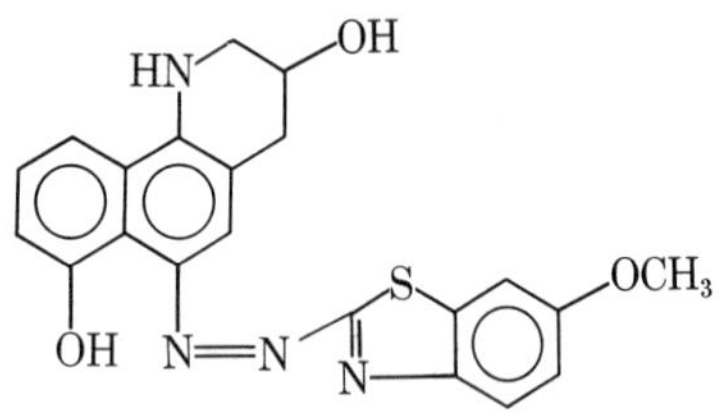

(**1**)

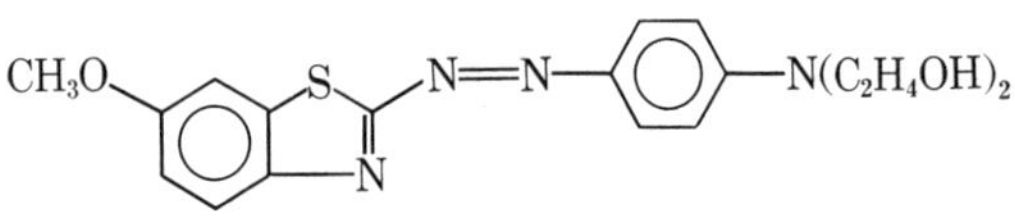

(**2**)

These three dyes no longer have commercial value because of their modest lightfastness on acetate and polyester fibers.

Azo dyes with improved lightfastness on acetate and polyester fibers can be obtained from 2-aminobenzothiazole with electron-withdrawing substituents instead of electron-donating substituents (1). Dyes from 6-acyl- or aroyl-, 6-(carbalkoxy)-, 6-cyano- or 6-cyanothio-, 6-nitro-, 5,6-dihalo-, 4-(trifluoromethyl)-, and 6-(alkylsulfonyl)-2-aminobenzothiazole have been reported (2–8). Also, some new dyes from the previously mentioned 2-aminobenzothiazoles and new dialkylaniline couplers have been reported (9–12).

An important group of disperse azo dyes derived from 2-amino-6-(methanesulfonyl)benzothiazole is shown in Table 1 (8,13–14). This benzothiazole is most conveniently prepared by the thiocyanation of aniline with four equivalents of ammonium thiocyanate to form 4-amino-*m*-phenylene dithiocyanate, which cyclizes under the acidic conditions of the reaction to 2-amino-6-(cyanothio)benzothiazole. This compound is converted to the 6-(methylthio) derivative by treatment with sodium sulfide followed by dimethyl sulfate. The (methylthio)benzothiazole is then oxidized with hydrogen peroxide to 2-amino-6-(methanesulfonyl)benzothiazole. The intermediate,

Table 1. Disperse Azo Dyes from 2-Amino-6-(Methanesulfonyl)Benzothiazole

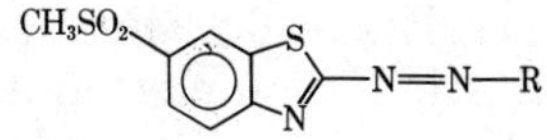

R	CAS Registry No.	Dyed on: Cellulose acetate	Dyed on: Polyester	Color
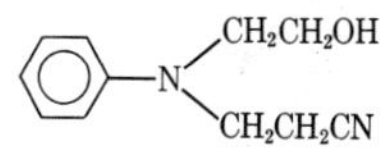	*[24170-48-7]*	X		red
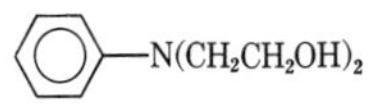	*[63467-02-7]*	X		pink
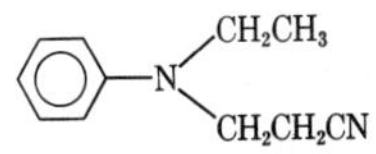	*[63467-01-6]*	X	X	yellowish red
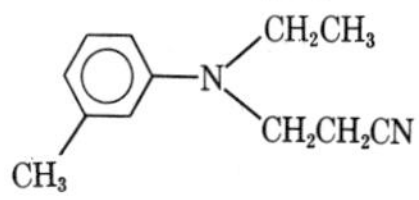	*[16588-67-3]*	X	X	red
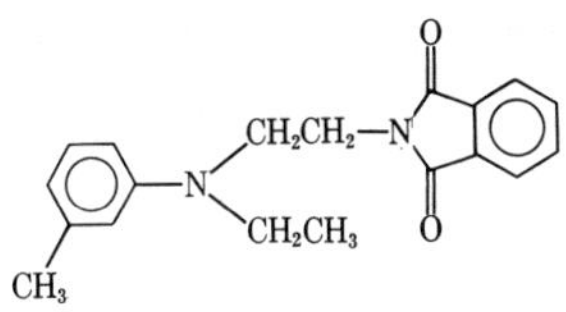	*[2203-01-2]*		X	bluish red
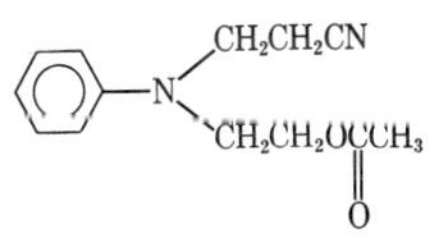	*[54983-30-1]*	X	X	red
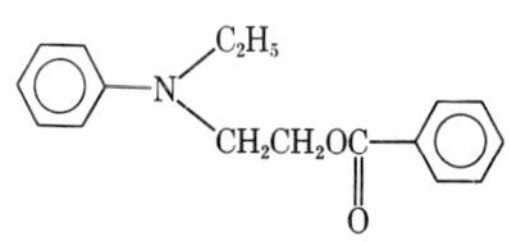	*[43042-04-2]*	X	X	bluish red

2-amino-6-(cyanothio)benzothiazole, is also useful in the preparation of disperse dyes (4):

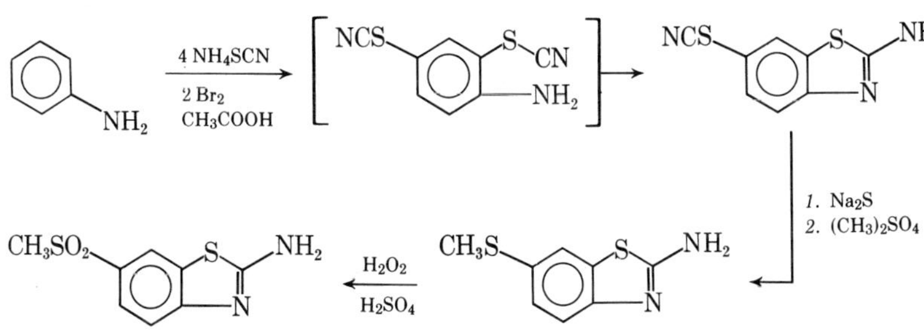

Preparation of a Disperse Azo Dye

Preparation. To a solution of 2.28 g of 2-amino-6-(methanesulfonyl)benzothiazole in 51 g of 50 wt % H_2SO_4 at 5–10°C is added a solution of nitrosyl sulfate (often referred to as nitrosylsulfuric acid by dye chemists) prepared from 0.72 g of $NaNO_2$ and 9 g of conc H_2SO_4. The precipitated sulfate salt gradually dissolves as diazotization proceeds. After ca 1.5 h, the diazotization is complete and the solution is added to an internally ice-cooled solution of 1.90 g of 3-[*N*-(2-hydroxyethyl)anilino]propionitrile in 15 mL of 15 wt % H_2SO_4. After 0.5 h, the coupling is complete, and 400 mL of water is added to the solution with stirring. Neutralization is not required, since the acid salt of the dye dissociates completely when diluted with water. The dye is collected on a funnel, washed with water until acid-free, and dried at 60°C. Since the product is the dye cake, it is necessary to prepare a dispersion before use.

Commercial disperse azo dyes from substituted 2-aminobenzothiazoles and selected dialkylaniline couplers impart bright scarlet to red-violet hues of high tinctorial power to acetate, triacetate, polyester, and polyamide fibers. On these fibers, these dyes exhibit good-to-excellent fastness to light, gas (oxides of nitrogen), and dry heat. The benzothiazole dyes are more bathochromic, have higher tinctorial power, and have better fastness properties than the corresponding azo dyes derived from aniline (1). An excellent illustration of the higher tinctorial power and bathochromicity of the benzothiazole dyes is shown in Table 2 (15).

Three of the more important benzothiazole dyes are Disperse Red 177 [*58051-98-2*], Disperse Red 179 [*58051-98-2*], and Disperse Red 153 [*58051-96-0*]. The first and second are derived from 2-amino-6-nitrobenzothiazole and the third from 2-amino-5,6-dichlorobenzothiazole. In 1977, ca 145.5 metric tons of Disperse Red 177 was produced in the United States (16). The price for this dye was \$13.20–15.40/kg in 1980.

Dyes from 3-amino-2,1-benzisothiazole, which is an isomer of 2-aminobenzothiazole, have recently been described (17–20). These dyes impart reddish- to greenish-blue shades to polyester and acetate fibers and are more bathochromic than the corresponding 2-aminobenzothiazole dyes. 3-Amino-2,1-benzisothiazole is prepared by the addition of hydrogen sulfide to *o*-aminobenzonitriles followed by oxidation (17):

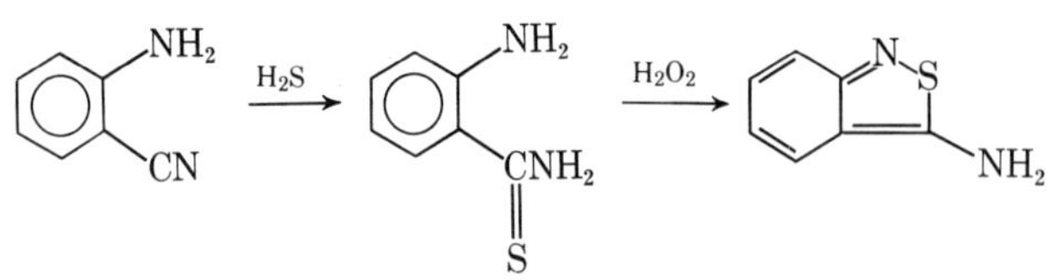

Table 2. Absorption Spectra: Benzothiazolylazo Dyes Versus Phenylazo Dyes

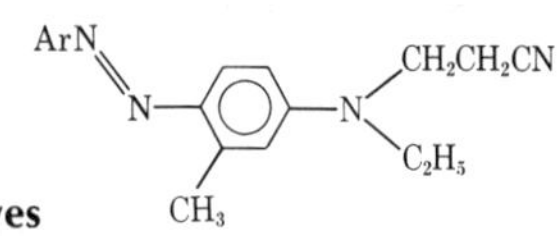

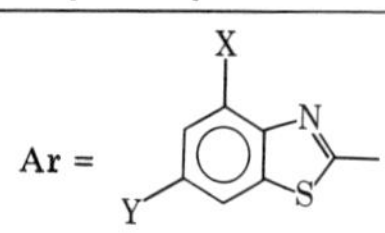

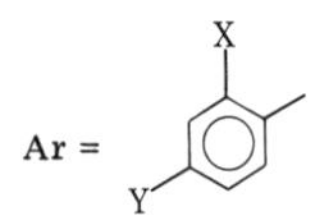

X	Y	CAS Registry No.	λ_{max}, nm	A_{max}	CAS Registry No.	λ_{max}, nm	A_{max}
H	H	[*16588-59-3*]	500	130	[*16588-58-2*]	410	90
H	CH_3O		500	118	[*16588-60-6*]	410	78
H	CH_3SO_2	[*16588-67-3*]	530	124	[*16588-66-2*]	438	78
H	NO_2	[*16586-42-8*]	538	130	[*16586-41-7*]	470	98
Cl	NO_2	[*16586-44-0*]	550	128	[*16586-43-9*]	490	90

Disperse Blue 148 [*61968-29-4*] is derived from 3-amino-5-nitro-2,1-benzisothiazole and is used to dye acetate and triacetate fibers.

Another group of disperse azo dyes is derived from 2-amino-5-nitrothiazole (21–22). Important dyes of this group are listed in Table 3. These dyes are especially noted for their excellent gasfastness, dischargeability, and deep-dyeing properties on acetate fibers. They impart reddish- to greenish-blue shades to acetate fibers and are more bathochromic than the corresponding dyes from 2-aminobenzothiazole. 2-Amino-5-nitrothiazole is prepared by the nitration of 2-aminothiazole and the subsequent rearrangement of the nitroamino intermediate:

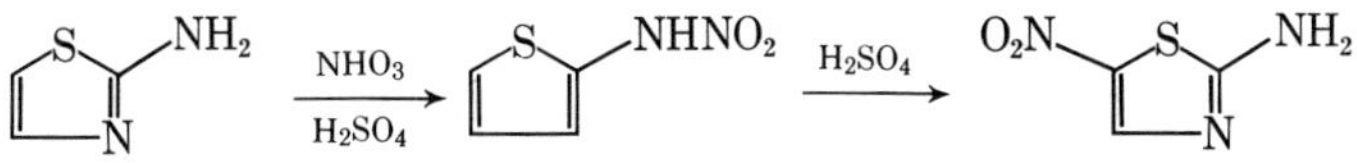

Several disperse azo dyes of the thiazole series are disclosed in recent patent literature (23–26).

Cationic Dyes

The cationic thiazole dyes are characterized by a positive charge on the dye molecule. Acrylic and modacrylic fibers and the more recent basic dyeable polyester and polyamide fibers are produced with acid sites. During the dyeing process, the colored cation of the dye molecules associates with the fiber's acidic sites, the anion. Before 1970, the structures of only two thiazole dyes, Basic Yellow 1 [*2390-54-7*] (CI 49005) (**3**) and Basic Yellow 12 [*4208-81-5*] (CI 48065) (**4**) were listed in the *Colour Index*. These two dyes were originally used to produce bright-yellow shades on cellulose acetate, but they exhibited only moderate fastness. Basic Yellow 1 (**3**) was also used as a paper dye and as a constituent in printing inks (27):

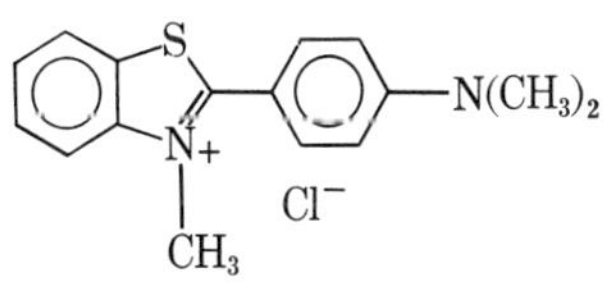

(**3**)

Table 3. Disperse Azo Dyes from 2-Amino-5-Nitrothiazole

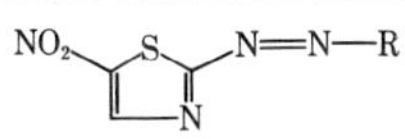

R	CAS Registry No.	Color on cellulose acetate
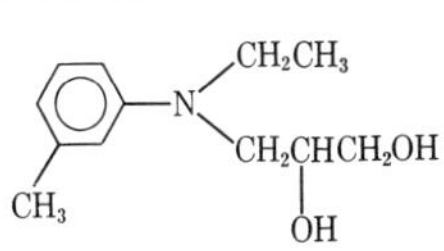	[*69766-79-6*]	reddish blue
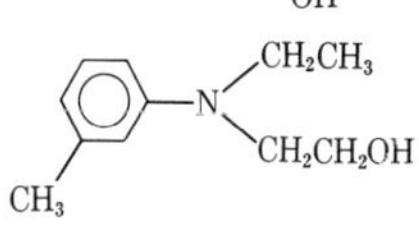	[*68516-81-4*]	reddish blue
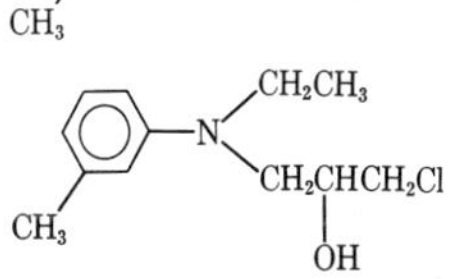	[*70865-21-3*]	reddish blue

With the advent of acrylic and modacrylic fibers, a new use was found for Basic Yellow 12 (**4**). This dye (**4**) is prepared by the condensation of the 1,3,3-trimethyl-Δ2,α-indolineacetaldehyde (Fischer's aldehyde) with 6-amino-2-methylbenzothiazole in an acidic solution at room temperature.

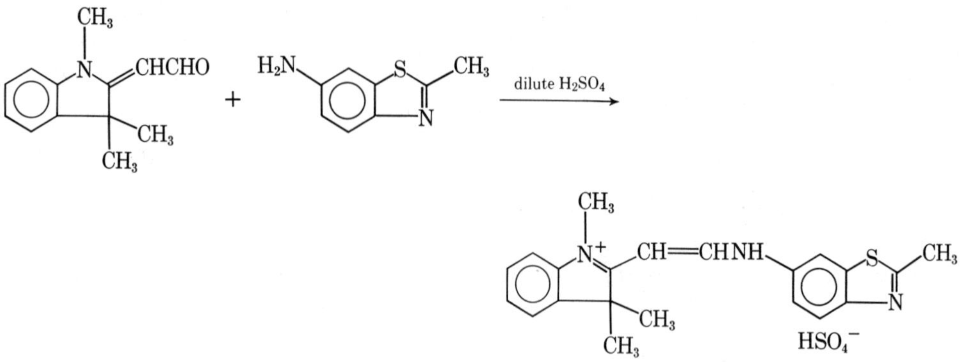

(**4**)

A greenish-yellow cationic dye (**5**) (Basic Yellow 24 [*52435-14-0*], CI 11480), which exhibits excellent fastness properties on acrylic fibers, is prepared by the condensation of 3-methyl-2-benzothiazolinone hydrazone with 2-(dichloromethyl)benzimidazole, followed by alkylation with dimethyl sulfate (28):

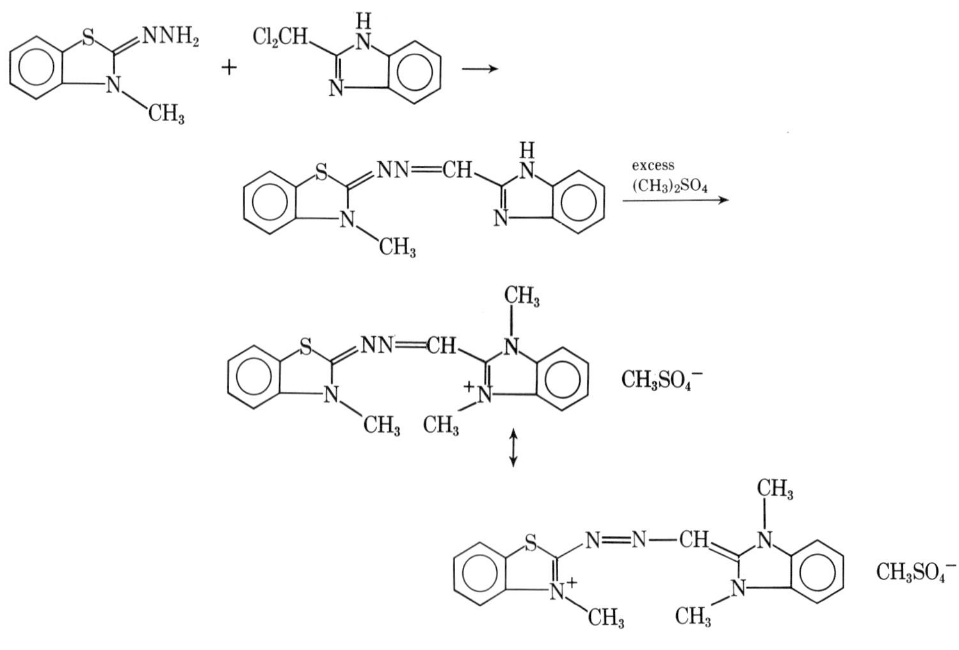

(**5**)

There also is a newer group of cationic dyes containing the thiazole nucleus in the azo dye class. They are similar to azo disperse dyes except for the presence of a

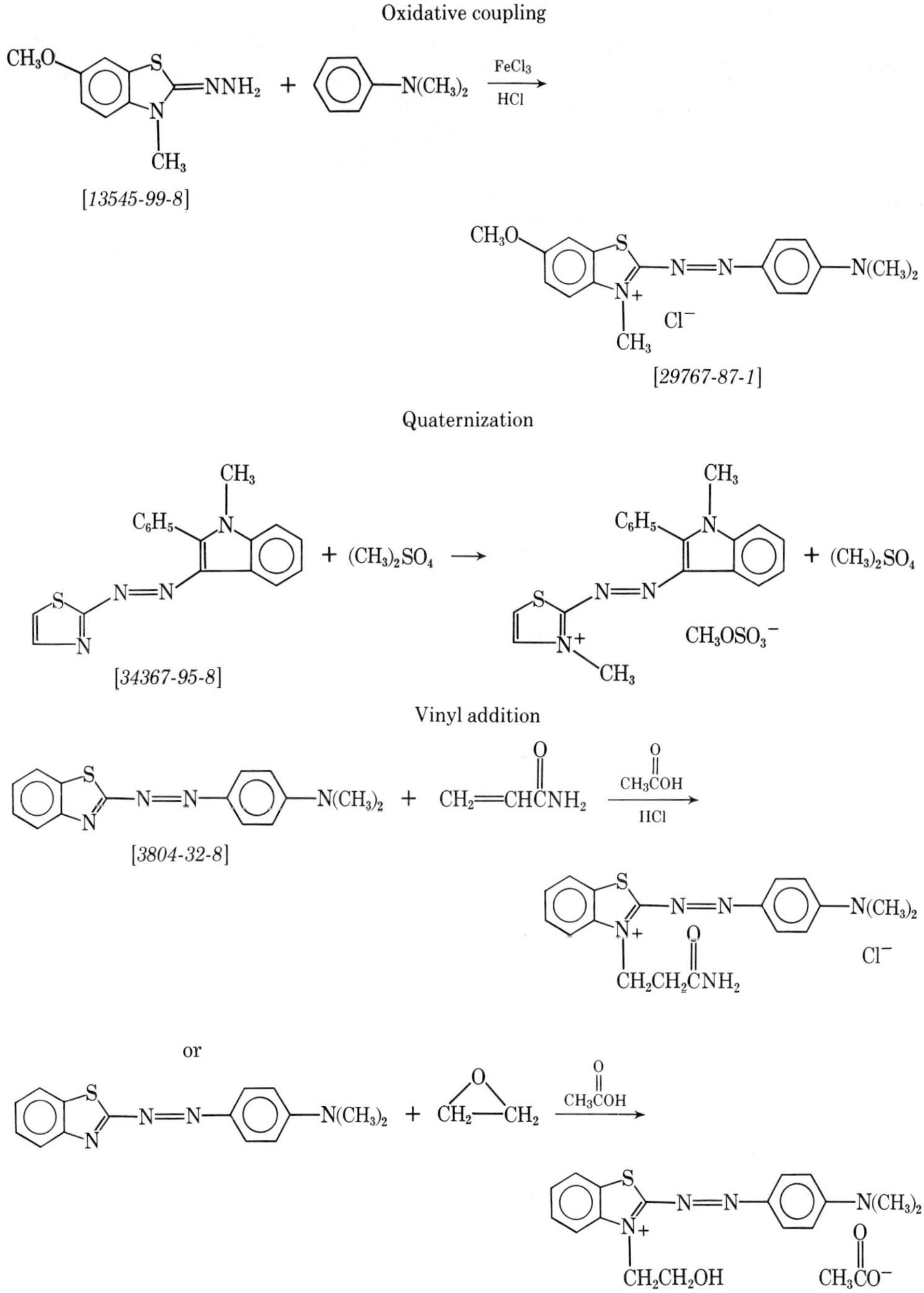

Figure 1. Preparation of quaternized thiazole dyes.

Table 4. Quaternized Azo Thiazole Dyes

CI name	CI no.	Structure	CAS Registry No.
Basic Yellow 24	11480	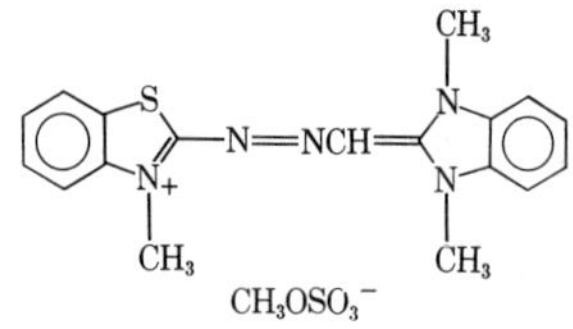	*[52435-14-0]*
Basic Red 29	11460	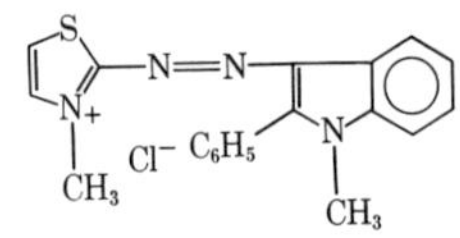	*[42373-04-6]*
Basic Blue 54	11052	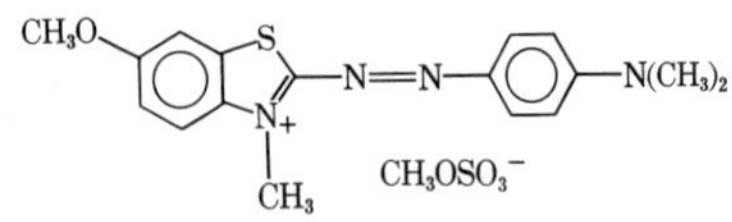	*[15000-59-6]*
Basic Blue 66	11075	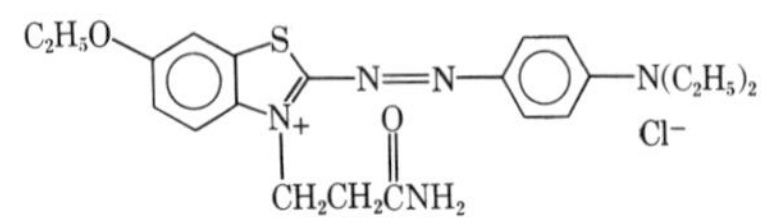	*[12221-38-4]*
Basic Blue 41	11154	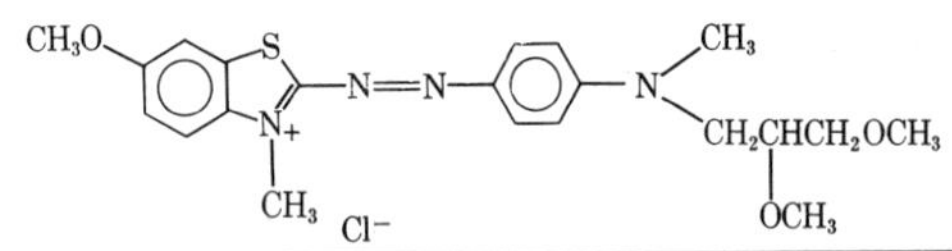	*[12270-13-2]*

quaternized nitrogen atom in the molecule. Cationic azo dyes may be categorized as those in which the ring nitrogen atom is quaternized, eg, (**6**), and those which possess a quaternary nitrogen on the dye coupler, eg, (**7**):

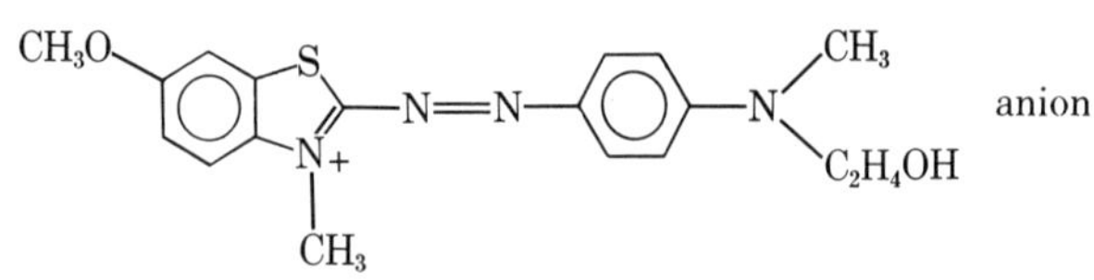

(**6**)

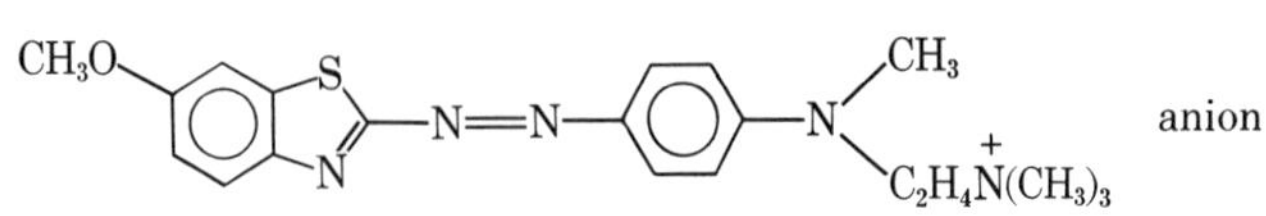

(**7**)

There is a large difference in color between the two types of dyes. Those that have the quaternized thiazole nitrogen (**6**) absorb at 70–100-nm longer wavelengths than the other type (**7**). Thus, in the examples shown, (**6**) is blue and (**7**) is red.

Dyes of type (**6**) were initially prepared by oxidative coupling of a heterocyclic hydrazone in the presence of an aniline coupler (29). These dyes can be prepared commercially by alkylation of the preformed azo dye with an alkylating agent, eg, dimethyl sulfate, or by vinyl addition with either acrylamide or ethylene oxide in the presence of a mineral acid (30–31). These methods are illustrated in Figure 1. An excellent discussion on the preparation of these dyes is given in ref. 32.

The recent patent literature discloses many new dyes of the type shown in Figure 1 (33–41). The structures for some of the cationic dyes for acrylic and modacrylic fibers are shown in Table 4. The fastness properties and colors of these dyes are independent of the anion associated with the dye. The solubility of the dyes in water depends on the anion.

Type (**7**) dyes are derived from coupling the diazotized 2-aminothiazole or benzothiazole with the quaternary-nitrogen-containing coupler. These dyes have greater utility for dyeing acid-modified polyesters and nylon fibers than for dyeing acrylic fibers (39–41).

Health and Safety Aspects

The treatment of effluent from dye and dye intermediate factories and the toxicity of disperse dyes has been described (see Dyes and dye intermediates).

BIBLIOGRAPHY

"Thiazole Dyes" in *ECT* 2nd ed., Vol. 20, pp. 185–197, by J. G. Fisher, Tennessee Eastman Company, Division of Eastman Kodak Company.

1. J. M. Straley in K. Venkataraman, ed., *The Chemistry of Synthetic Dyes*, Vol. 3, Academic Press, Inc., New York, 1970, p. 425.
2. U.S. Pat. 3,329,669 (July 4, 1967), M. F. Sartori (to E. I. du Pont de Nemours & Co., Inc.).
3. U.S. Pat. 4,013,632 (March 22, 1977), J.-P. H. Stiot, J.-M. L. Leroy, and C. M. H. E. Brouard (to Produits Chimiques Ugine Kuhlman).
4. U.S. Pat. 3,084,153 (April 2, 1963), B. R. Fishwick and E. L. Johnson (to Imperial Chemical Industries, Ltd.).
5. Brit. Pat. 896,232 (May 16, 1962), B. R. Fishwick (to Imperial Chemical Industries, Ltd.).
6. U.S. Pat. 3,502,645 (March 24, 1970), D. G. Carmichael (to Martin Marietta Corp.).
7. U.S. Pat. 4,052,379 (Oct. 4, 1977), R. N. Gourley (to Eastman Kodak Co.).
8. U.S. Pat. 2,785,157 (March 12, 1957), J. M. Straley and D. J. Wallace (to Eastman Kodak Co.).
9. U.S. Pat. 3,950,130 (April 13, 1976), E. E. Renfrew and H. W. Pons (to American Aniline Products, Inc.).
10. U.S. Pat. 3,816,388 (June 11, 1974), M. A. Weaver, H. S. Pridgen, and C. A. Coates (to Eastman Kodak Co.).
11. U.S. Pat. 3,639,385 (Feb. 1, 1972), M. A. Weaver and H. S. Pridgen (to Eastman Kodak Co.).
12. U.S. Pat. 3,423,394 (Jan. 21, 1969), M. A. Weaver and D. J. Wallace (to Eastman Kodak Co.).
13. U.S. Pat. 3,161,631 (Dec. 15, 1964), J. M. Straley and D. J. Wallace (to Eastman Kodak Co.).
14. Brit. Pat. 1,324,235 (July 25, 1973), J. F. Dawson (to Yorkshire Chemicals, Ltd.).
15. M. F. Sartori, *J. Soc. Dyers Colour.* **83**, 144 (1967).
16. *Synthetic Organic Chemicals, United States Production and Sales, 1977*, United States International Trade Commission Publication 920, United States Government Printing Office, Washington, D.C., 1978.
17. U.S. Pat. 3,455,898 (July 15, 1969), M. Seefelder, H. Ambrust, and H. G. Wippel (to Badische Anilin- und Soda-Fabrik A.G.).
18. U.S. Pat. 3,573,273 (March 20, 1971), M. Seefelder, E. Hahn, and H. G. Wippel (to Badische Anilin- und Soda-Fabrik A.G.).
19. U.S. Pat. 3,658,784 (April 25, 1972), G. Lange (to Badische Anilin- und Soda-Fabrik A.G.).
20. U.S. Pat. 3,919,188 (Nov. 11, 1975), H. Hagen and G. Hensen (to Badische Anilin- und Soda-Fabrik A.G.).

21. U.S. Pat. 2,659,719 (Nov. 17, 1953), J. B. Dickey and E. B. Towne (to Eastman Kodak Co.).
22. U.S. Pat. 2,683,708 (July 13, 1954), J. B. Dickey and E. B. Towne (to Eastman Kodak Co.).
23. U.S. Pat. 4,016,153 (April 5, 1977), E. E. Renfrew (to American Color and Chemical Corp.).
24. U.S. Pat. 3,515,714 (June 2, 1970), J. I. Dale and M. A. Weaver (to Eastman Kodak Co.).
25. U.S. Pat. 3,997,520 (Dec. 14, 1976), R. Altermatt (to Sandoz, Ltd.).
26. U.S. Pat. 4,252,530 (Feb. 24, 1981), J. Ribka, H. Tappe, D. Roth, and H. Weyer (to American Hoechst Corp.).
27. H. A. Lubs, ed., *The Chemistry of Synthetic Dyes and Pigments*, ACS Monograph 127, ACS, Washington, D.C., 1955, p. 271.
28. U.S. Pat. 2,913,303 (Nov. 17, 1959), H. Baumann, F. Arnemann, J. Eisele, and W. Federkiel (to Badische Anilin- und Soda-Fabrik A.G.).
29. U.S. Pat. 3,043,828 (July 10, 1962), S. Huenig (to Badische Anilin- und Soda-Fabrik A.G.).
30. U.S. Pat. 3,132,132 (1964), M. Suzuki, M. Iizuki, and M. Yamamoto (to Hodogaya Kagaku Kogyo Kabushiki Kaisha).
31. U.S. Pat. 3,634,389 (Jan. 11, 1972), R. Entschel, C. Muller, and H. Siegrist (to Sandoz, Ltd.).
32. D. R. Baer in K. Venkataraman, ed., *The Chemistry of Synthetic Dyes*, Vol. 4, Academic Press, Inc., New York, 1971, pp. 188–200.
33. U.S. Pat. 3,573,272 (March 30, 1971), G. Kaupp, H. Wegmuller, and J. Voltz (to J. R. Geigy A.G.).
34. U.S. Pat. 3,634,389 (Jan. 11, 1972), R. Entschel, C. Mueller, and H. Siegrist (to Sandoz, Ltd.).
35. U.S. Pat. 3,928,311 (Dec. 23, 1975), J. G. Fisher and J. M. Straley (to Eastman Kodak Co.).
36. U.S. Pat. 3,948,877 (April 6, 1976), H. P. Kuhlthau (to Bayer A.G.).
37. U.S. Pat. 4,252,716 (Feb. 24, 1981), H. Kuhlthau (to Bayer A.G.).
38. U.S. Pat. 4,251,440 (Feb. 17, 1981), V. Kaeppeli (to Sandoz, Ltd.).
39. U.S. Pat. 3,414,559 (Dec. 3, 1968), M. F. Sartori (to E. I. du Pont de Nemours & Co., Inc.).
40. U.S. Pat. 3,933,787 (Jan. 20, 1976), H. Moser (to Sandoz, Ltd.).
41. U.S. Pat. 3,945,991 (March 23, 1976), V. Ramanathan and H. W. Liechti (to CIBA-GEIGY A.G.).

J. G. FISHER
G. T. CLARK
Eastman Kodak Company

THIOCYANATES. See Sulfur compounds.

THIOGLYCOLIC ACID

Thioglycolic acid [*68-11-1*] (2-mercaptoacetic acid), $HSCH_2COOH$, is the sulfur analogue of glycolic acid. It is a colorless liquid with a distinct sulfurous odor. It was identified in 1862 when it was prepared by reaction of potassium hydrosulfide with potassium chloroacetate (1). Its primary use is in the form of its salts and esters as a component of hair-waving and depilatory compositions, and the metal mercaptide derivatives of its esters are used as stabilizing additives to prevent deterioration of plastics during processing and use (see Hair preparations; Heat stabilizers).

Properties

Pure thioglycolic acid freezes at −16.5°C and distills unchanged under reduced pressure. Reported constants are $bp_{3.9\ kPa\ (29\ mm\ Hg)}$ 123°C and $bp_{0.8\ kPa\ (6\ mm\ Hg)}$ 90°C, d_4^{20} 1.3253 g/cm^3, and heat of combustion 1446 kJ/mol (345.6 kcal/mol). The acid is miscible with water, methanol, ethanol, acetone, ether, chloroform, and benzene, but it is practically insoluble in aliphatic hydrocarbons. Both the carboxyl group and the mercapto group of thioglycolic acid are acidic. First and second dissociation constants of 3.55–3.92 pK_a and 9.20–10.56 pK_a, respectively, have been reported. At constant ionic strength in 1.0 *M* $NaClO_4$, the first dissociation constant increases with temperature; this is attributed to increased solvation of the anion relative to the unionized acid (2).

Reactions. Thioglycolic acid is stable at room temperature in ≤70 wt % aqueous solutions. In more concentrated solutions, self-esterification takes place to some extent. In 346 d at room temperature, the loss in assay is ca 1% of the initial value in a 74 wt % soln, 2% in an 84 wt % soln, and 9% in 95 wt % thioglycolic acid (3). At 100°C in air, commercial pure thioglycolic acid (98.7 wt % by acid assay and 99.4 wt % by SH-group assay) loses 10% of initial acid assay and 13% of initial SH assay in 24 h (4). A variety of self-esterification products have been reported: *S*-mercaptoacetylthioacetic acid [*99-68-3*], $HSCH_2CSCH_2COOH$, 1,4-dithioglycolide

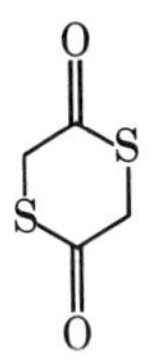

[*4385-42-6*]

and polythioglycolides $HS(CH_2CS)_xCH_2COOH$. Greater loss of SH assay than car-

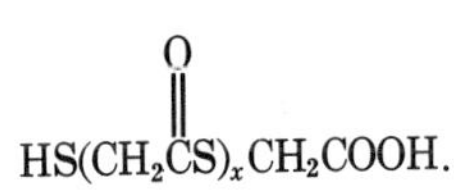

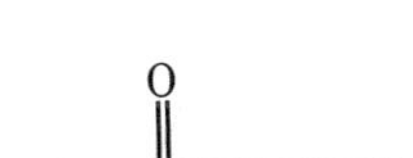

boxyl assay is accounted for by the formation of some ortho-thioester product formulated as

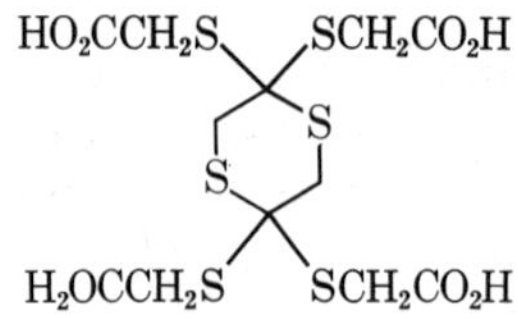

[52959-43-0]

The self-esterification can be reversed in the presence of dilute acids or alkalies, but at ≥150°C, self-esterification is accompanied by some irreversible elimination of hydrogen sulfide.

A further indication of the stability of thioglycolic acid is the appearance of a substantial thioglycolic acid peak (m/e = 92) in the fragmentation pattern of certain esters of thioglycolic acid in the mass spectrometer (6). Thioglycolic acid undergoes reactions typical of carboxylic acids, forming salts, esters, and amides; and of mercaptans, forming thioethers, disulfides, and metal mercaptides. In addition, a number of reactions characteristic of thioglycolic acid involve both the carboxylic acid group and the mercapto group simultaneously.

One of the most widely used reactions of thioglycolic acid is that with disulfides, particularly cystine, and in proteins, eg, wool (qv) and hair (7–8). It reacts reversibly with cystine, especially at elevated pHs, with an equilibrium constant of ca 18 (9) (see Amino acids):

$$\underset{\text{thioglycolic acid}}{2\,HSCH_2COOH} + \underset{\text{cystine}}{HOOC\underset{|}{\overset{}{C}}HCH_2SSCH_2\underset{|}{C}HCOOH} \longrightarrow$$
$$\qquad\qquad\qquad NH_2 \qquad\qquad NH_2$$

$$\underset{\text{cysteine}}{2\,HOOC\underset{\underset{NH_2}{|}}{C}HCH_2SH} + \underset{\text{dithiodiglycolic acid}}{HOOCCH_2SSCH_2COOH}$$

The rate at which equilibrium is established is a function of pH; it is slow at pHs <6 and rapid at pHs of 8–10. The kinetics of this thiol–disulfide interchange have been studied, and the reacting species has been identified as the mercaptide anion (10).

Thioglycolic acid and especially its alkaline aqueous solutions are rapidly oxidized by air to the disulfide, dithiodiglycolic acid [*505-73-7*], or salts thereof in the presence of small amounts of copper, manganese, or iron (11). Mild chemical oxidation, eg, with iodine, also leads to dithiodiglycolic acid, whereas more vigorous oxidative conditions, eg, with dilute nitric acid, produce sulfoacetic acid [*123-43-3*], HO_3SCH_2COOH. In the presence of aqueous alkali or another suitable acid acceptor, thioglycolic acid reacts with organic halides RX giving, after acidification, the thioether carboxylic acid $RSCH_2CO_2H$ (12). When the halide is sodium chloroacetate, the reaction forms thiodiglycolic acid [*123-93-3*], $S(CH_2CO_2H)_2$, which usually forms as a by-product in the manufacture of thioglycolic acid. Alkylene dihalides yield the corresponding *S,S′*-alkylenebis(thioacetic acid).

Thioether carboxylic acids are also obtained by the reaction of thioglycolic acid with an olefin by addition, as in the preparation of carboxymethylthiosuccinic acid [*99-68-3*] from maleic acid; 2-phenylethylthioacetic acid [*58343-10-0*] from styrene; and long-chain alkylthioacetic acids from 1-alkenes (13).

With acetyl chloride, thioglycolic acid yields *S*-acetylthioglycolic acid [*1190-93-8*], CH_3COSCH_2COOH ($bp_{2.3\ kPa\ (17\ mm\ Hg)}$ 158–159°C). Several kinds of products can be obtained from thioglycolic acid with aldehydes and ketones. Acetals with aldehydes (mercaptals), $RCH(SCH_2COOH)_2$, and ketones (mercaptoles), $RR'C(SCH_2COOH)_2$, are obtained by reaction of two moles thioglycolic acid per mole of aldehyde or ketone. Metal chelating agents (qv) having up to eight possible binding sites have been prepared from thioglycolic acid with aliphatic and aromatic dialdehydes (14). Acetals such as mercaptoles have been used for the study of nmr spectra. Under appropriate conditions, both acidic and neutral reaction products of carbonyl compounds with thioglycolic acid in a 1:1 molar ratio can be obtained. Thus, 2,2-pentamethylene-1,3-oxathiolan-5-one, *S*-cyclohex-1-enylthioglycolic acid, and *S,S'*-cyclohexylidenebis(thioglycolic acid) [*4265-54-7*] have been obtained from thioglycolic acid and cyclohexanone; formation of the first involves the SH and carboxyl groups simultaneously, whereas reaction at the mercapto group alone affords the other two derivatives (15).

Frequently, the choice of experimental conditions as well as reactant structure can direct the course of reaction either to a heterocyclic product or to an open-chain derivative. For example, thiocyanic acid and thioglycolic acid react yielding rhodanine (2-thioxo-1,3-thiazolidine-4-one), whereas *N*-substituted organic isothiocyanates with thioglycolic acid at pH 3–7.5 give *N*-substituted rhodanines; at a pH >9.5, the corresponding *N*-substituted thiocarbamoylthioglycolic acids form (16). Similarly, thioglycolic acid can function as a unidentate ligand, a bidentate ligand, or a bridging ligand in forming metal complexes. Complexes studied include metals of periodic table groups IIB, IIIA, IVA, VA, VIA, VIB, VIIB, VIII, as well as the rare-earth metals and uranium. The doubly chelated structure of antimony hydrogen bis(thioglycolate) [*16743-24-1*], $Sb(SCH_2CO_2)_2H$, has been confirmed by x-ray crystallography. Units of four-coordinate antimony with two thioglycolate chelating groups are linked through symmetrical hydrogen bonds into apparently endless chains (17) (see also Coordination compounds).

Thioglycolic acid cleaves polynuclear polyphenolic materials, eg, tannins and lignins, by substitution at the benzylic carbon atom. Both carbon–oxygen and carbon–carbon bonds are cleaved, resulting in substituted *S*-benzylthioacetic acids. These reactions have been used in structure determinations of hemlock bark tannin and to solubilize and remove lignin from wood cellulose (18–19) (see Pulp).

Manufacture, Processing, Shipment, and Storage

Thioglycolic acid is manufactured by the reaction of sodium or potassium chloroacetate with alkali-metal hydrosulfide in an aqueous medium under controlled conditions of pressure, temperature, and pH to give a high yield of thioglycolate salt in solution and to minimize the formation of such by-products as the salts of thiodiglycolic acid, dithiodiglycolic acid, and glycolic acid (20). The reaction mixture is acidified to liberate thioglycolic acid, which is extracted from the solution into an organic solvent and then purified by vacuum distillation or converted directly into derivatives.

Alternative methods have been developed to avoid the formation of the thiodiglycolic acid by-product. Instead of the hydrosulfide, a sulfur compound is used which reacts with chloroacetic acid or sodium chloroacetate to give a carboxymethylthio compound convertible to thioglycolic acid but incapable of further carboxymethylation. Such sulfur compounds, the structure of the carboxymethylation product, and the method by which the latter is converted to thioglycolic acid are shown in Table 1. All methods mentioned result in aqueous solutions of thioglycolic acid and inorganic salts, from which pure thioglycolic acid is obtained by solvent extraction and distillation. The original extraction solvent, ie, diethyl ether, has probably been replaced by less hazardous solvents in most manufacturing plants. Where thioglycolic acid is manufactured mainly as an intermediate for conversion to higher alkyl esters, the alcohol to be esterified can be used to extract and immediately esterify the thioglycolic acid from the aqueous solution (27).

Thioglycolic acid is marketed primarily as 80 wt % aq soln and as neat (95–98 wt %) acid. The ammonium salts and monoethanolamine salts are available as aqueous solutions containing 55–65 wt % and 40 wt % thioglycolic acid, respectively. Isooctyl thioglycolate (the thioglycolic acid derivative used in the largest amount in the United States) and glycerol monothioglycolate [*30618-84-9*] are supplied anhydrous. Thioglycolic acid and the ammonium salt are discolored by traces of iron and are shipped and stored in glass- or plastic-lined containers; the ammonium salt is also shipped and stored in aluminum containers. Under regulations of the DOT, the acid is classified as a corrosive liquid and is affixed with a prescribed CORROSIVE label with the United Nations identification number UN 1940 (28).

Economic Aspects

In 1981, there were probably two manufacturing plants for thioglycolates in the United States, four plants in Europe, and one plant in Japan. The estimated quantities,

Table 1. Sulfur Reactants, Carboxylation Products, and Method of Conversion to Thioglycolic Acid

Sulfur compound	Carboxymethylation product (reaction with chloroacetic acid)	Conversion method to thioglycolic acid	Ref.
$Na_2S_2O_3$	$NaO_3SSCH_2CO_2Na$	hydrolysis and reduction	21
thiourea	$H_2N\overset{NH}{\overset{\|}{C}}SCH_2CO_2H$	hydrolysis, aminolysis	22–23
alkali polysulfide	$(SCH_2CO_2Na)_2$	reduction	24
sodium trithiocarbonate	$S{=}C(SCH_2CO_2Na)_2$	hydrolysis	25
sodium alkyl xanthate	$(CH_3)_2CHO\overset{S}{\overset{\|}{C}}SCH_2CO_2Na$ or $CH_3CH_2O\overset{S}{\overset{\|}{C}}SCH_2CO_2Na$	hydrolysis, aminolysis, or thiohydrolysis	26

calculated as contained thioglycolic acid, of all thioglycolates produced in the United States and the published U.S. prices of thioglycolic acid in 1970, 1975, and 1980 are listed below.

Year	*Production, metric tons*	*Price, $/kg*
1970	1800	1.75
1975	3000	3.50
1980	4000	3.80

Specifications

Specifications and test methods for ammonium thioglycolate have been published (29). These can also be applied to thioglycolic acid and to other salts and esters of thioglycolic acid, as shown in Table 2 (29).

Analytical and Test Methods

Identification. Thioglycolic acid can be identified by its ir spectrum, by tlc of the 2,4-dinitrophenylthioether, and by color reactions (29–30). A violet-purple color forms with sodium nitroprusside, $Na_3[Fe(CN)_5NO].2H_2O$, in dilute ammoniacal solution, and a colored nitrosyl mercaptide forms with nitrous acid. Ferric ion turns a neutral solution blue and an aqueous ammonia solution deep red-violet.

Table 2. Specifications for Ammonium Thioglycolate and Isooctyl Thioglycolate[a]

Property	Ammonium thioglycolate [5421-46-5]	Isooctyl thioglycolate [25103-09-7]
solubilities	miscible in H_2O, C_2H_5OH; immiscible in acetone, benzene, chloroform, ether	miscible in methanol, toluene, carbon tetrachloride
color	colorless to faint pink	colorless to slightly yellow
odor	as specified by buyer	typical of isooctyl ester
identification	close match to Cosmetic, Toiletry, and Fragrances Association (CTFA) ir spectrum	gas chromatogram
pH at 25°C	5.5–6.5	na
assay (iodometric), %	within ±5	97.5 min[b]
acid number, mg KOH/g max	na	1[b]
dithiodiglycolate (as acid), wt % max	2.0	
sulfated ash, wt % max	0.05	
arsenic (as As), ppm max	3	
copper (as Cu), ppm max	1	
iron (as Fe), ppm max	1	
lead (as Pb), ppm max	20	
sp gr at 25°C		0.970 ± 0.002[b]
bp, °C		$125_{2.3\ kPa}$[c]
flash point (Fisher tag closed cup), °C min		115

[a] Ref. 29.

[b] Manufacturer's specification for isooctyl thioglycolate.

[c] To convert kPa to mm Hg, multiply by 7.5.

Assay. Thioglycolic acid is determined by its quantitative reaction with iodine (one mole of iodine oxidizes two moles of thioglycolic acid). Iodine can be used as a standardized solution or it can be generated from iodide ion in the presence of thioglycolic acid. The CTFA method specifies iodide and a potassium iodate reagent (31).

Titration of thioglycolic acid in thioglycolate esters with alkali is complicated by the easy hydrolysis of the esters under the conditions of the titration. Therefore, the acid is determined in a cold-water extract from which the ester has been separated before titration.

In hair-waving and depilatory products, thioglycolic acid has been determined by the CTFA titration method or by gas chromatography with or without derivatization as the methyl ester (32). Thioglycolic acid has been determined in air by condensation in water and by spectrophotometry or by adsorption on silica gel, elution with methanol, and colorimetric measurement as a ferric hydroxamate (33).

Health and Safety Factors, Toxicology

Thioglycolic acid is considered moderately toxic on the basis of short-term animal experiments. The single oral LD_{50} in mice is 250–300 mg/kg body weight; in rats it is 120–150 mg/kg. Toxicological values from the pre-1960 literature for thioglycolic acid, its sodium and ammonium salts, and its methyl and ethyl esters are summarized in ref. 34.

Single doses of thioglycolic acid, isooctyl thioglycolate, and 2-ethylhexyl thioglycolate [*7659-86-1*] were administered to mice by stomach tube, intraperitoneal injection, and inhalation for 6 h. The following LD_{50} values in mg/kg body weight and LC_{50} values in mg/L of air were calculated from the observed results (35).

Compound	*LD_{50} oral, mg/kg*	*LD_{50} intraperitoneal, mg/kg*	*LC_{50} inhalation, mg/L*
thioglycolic acid	530 ± 120	160 ± 5	256
isooctyl thioglycolate	1310 ± 40	1050 ± 37	364
2-ethylhexyl thioglycolate	1920 ± 80	1320 ± 30	>470; air space is saturated

Pure thioglycolic acid is a stronger acid than acetic acid and must be handled with precautions appropriate for strong acids. Chemical splash goggles, a neoprene apron, and rubber or impervious plastic gloves and boots should be worn when working with more than gram quantities of the neat acid. Accidental contact of the skin or eye should be flushed immediately with generous quantities of water; medical attention should be obtained in case of eye contact. Splashed clothing and shoes should be immediately removed and thoroughly cleaned before reuse.

Neutral or slightly alkaline solutions of thioglycolates are less irritating than solutions of the free acid. Hair-waving solutions formulated with up to 9 wt % thioglycolate and with a pH ≤ 10 are considered safe for general use (see Uses). Nevertheless, such solutions, if spilled on the skin or eyes, should be washed away as quickly as possible. Those working professionally with hair-waving solutions should wear gloves to avoid softening and discoloration of fingernails and hands from prolonged contact (36).

After exposure of rabbits' skin to a daily application of 0.6 *N* ammonium thioglycolate (pH 9.3) for 20 d, the extent of absorption and resulting toxic effects depend on the surfactant present. Without a surfactant, ammonium thioglycolate is innocuous at 6.5 mL/(kg·d). When 0.5 wt % surfactant is included, toxicity is noted, with the magnitude of LD_{50} depending on the kind of surfactant. The LD_{50} ranges from 4.1–4.4 mL/(kg·d) with sodium alkylaryl sulfonate, oleate, or stearate down to 2.3–2.8 mL/(kg·d) with sodium lauryl sulfate or a nonionic or quaternary ammonium surfactant. Application of ^{35}S-thioglycolate leads to increased urinary ^{35}S excretion (37). Injection of S-labeled thioglycolic acid into rats and rabbits leads to the excretion of >80 wt % of the label in the urine. The label is rapidly depleted from the blood, with <3 wt % remaining after one hour and ≤0.1 wt % 4–7 h after injection (38). Feeding or injection of thioglycolate depresses the activity of liver and kidney succinoxidase, but not other enzymes.

Sodium thioglycolate was rated noncarcinogenic as a result of a study in which 1 wt % and 2 wt % solutions in acetone were applied to the shaved skin of mice and rabbits twice weekly during the lifetime of the animals. Specific types of tumors found in 38 of 40 positive control animals treated with 7,12-dimethylbenz[*a*]anthracene were entirely absent in untreated (negative control) animals and in animals treated with sodium thioglycolate (39).

Uses

Hair Waving. Salts of thioglycolic acid have been successfully used for the cold waving of human hair since ca 1940. The waving solution, typically containing 5–9 wt % ammonium thioglycolate in water adjusted to ca pH 9, is applied to the hair which is rolled on rods, and is allowed to penetrate or "process" for some time. Shampooing the hair before treatment and addition of surfactants to the waving solution facilitate penetration. During the process time, some of the disulfide linkages of cystine units in the keratin hair protein react with the thioglycolate forming SH-terminated cysteine units and diminishing the number of cross-links between polypeptide chains. As a result, the hair is softened and responds more easily to the shaping stress imparted by being wound on the rods. After processing has taken place to the desired extent, the excess waving solution is rinsed from the hair. The diameter of the rod used largely determines the appearance or tightness of the final hair wave. The curl in the hair is fixed and some permanence is achieved by reforming the cystine disulfide linkages. Usually this is accomplished with an oxidizing treatment before the hair is dried and styled. The oxidizing treatments are formulated to react with remaining SH groups in the treated hair protein and with the alkalinity of any remaining waving solution, thereby restoring as closely as possible the natural condition of the hair in both redox potential and pH. Aqueous and dry neutralizers are supplied, which contain hydrogen peroxide, sodium perborate, or sodium or potassium bromate, as well as citric, tartaric, or phosphoric acids for pH adjustment and wetting agents to help the agent penetrate into the cortex of the hair (40). A detailed formulation of waving solutions and neutralizers and a full description of the procedure are given in ref. 41.

Many efforts have been made to minimize the ammoniacal and sulfurous odor of the standard thioglycolate waving solutions, to avoid damage to the hair resulting from excessive contact, and to enhance the convenience and safety of the process without detracting from its efficacy. The glycerol monoester and the monoethanolamine salt of thioglycolic acid can be used instead of the ammonium salt. Since both of these are less basic than ammonium thioglycolate, more of the effective mercaptide ion is present at any given pH, and effective waving solutions can therefore be formulated at lower alkalinity, typically pH 6.5 (10).

One-package waving solutions that do not require the use of a separate neutralizer have been patented. These are based on combinations of ammonium thioglycolate with its disulfide, diammonium dithiodiglycolate [*68223-93-8*], and are said to be easier to use than conventional waving solutions (42). However, neutralizer solutions are used with disulfide-based products in the United States. The commercial practice of hair waving and much of the patent literature of the 1970s is reviewed in ref. 43 (see also Hair preparations).

Hair Straighteners. The chemical reactions occurring when ammonium thioglycolate is used to straighten excessively curly hair are the same as in hair waving, but the mechanics are different. In hair straightening, the hair is pressed or stretched instead of wound on a rod. Typical products are formulated at pH 9.2–9.6 in a heavy cream that holds the agent to the hair (44).

Depilatories. Thioglycolates adjusted to pH 10–12.5 act on hair with so much greater speed and intensity than waving solutions that the action in this range is used for removing unwanted hair from the skin. Sodium [*367-51-1*], calcium [*814-71-1*], and ammonium salts are used. A laboratory "depilometer" has been designed to simulate practical use conditions for testing products quickly and objectively (45–46). Sodium thioglycolate adjusted to pH 11 provides faster depilatory action than other bases with thioglycolic acid at the same pH.

Intensification of Hair Colors. Production of SH groups linked to the hair protein by treatment of the hair with thioglycolates, as in waving, has been used to make the hair receptive to certain reactive dyes capable of combining with SH groups. The reactive dyes are azo dyes (qv) with a Bunte salt (—SO_3Na) group in the diazonium intermediate (47). The strength of hair color obtained with direct dyes, ie, azo dyes and aminoanthraquinone dyes, is intensified by including in the dye bath a small quantity of water-insoluble thioglycolic acid derivative, eg, isobutyl thioglycolate [*10047-28-6*] (48).

Stabilizers for Vinyl Chloride Polymers. Stabilizers are added to PVC to minimize degradation manifested by color change during processing at 150–210°C, which is caused by irregularities in the structure of the polymer, eg, tertiary chlorine

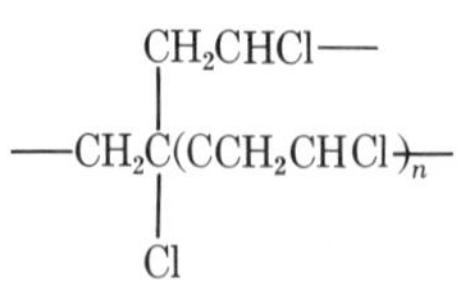

and internal allylic chlorine

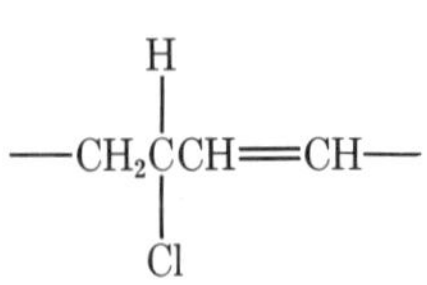

moieties (49). Once initiated, degradation propagates along the polymer chain with the formation of more of these unstable moieties as hydrogen chloride is eliminated. Stabilizers react with the active chlorine of these unstable moieties and form derivatives with reduced tendencies to undergo elimination reactions under PVC processing conditions (50) (see Vinyl polymers).

Most effective stabilizers contain one or more salts of a heavy metal, eg, zinc, cadmium, tin, antimony, or lead. Tin, in the form of alkyltin mercaptides with the characteristic carbon–tin–sulfur–carbon linkage, dominates the field of extruded and molded rigid PVC. By 1953, isooctyl thioglycolate became the preferred sulfur compound used in alkyltin mercaptide stabilizers, eg, dibutyltin bis(isooctyl thioglycolate) [*25168-24-5*] and butyltin tris(isooctyl thioglycolate) [*25852-70-4*] (51).

Additional alkyl thioglycolate stabilizers for PVC include di-*n*-octyltin compounds with controlled levels of octyltin trimercaptide and trioctyltin mercaptide minor ingredients; methyltin, dimethyltin, and antimony mercaptoesters; "ester-tin" (alkoxycarbonylethylenetin) derivatives; and alkyltin or ester-tin alkyl thioglycolates bridged through a sulfide or polysulfide group (52–53). The octyltin and methyltin alkyl thioglycolate compositions have been regulated for use in PVC food packaging and PVC water pipe in food-processing plants by the FDA and regulating authorities in many other countries (54).

Polymerization Catalysts, Accelerators, and Chain-Transfer Agents. The reactivity of mercaptans (RSH), including thioglycolic acid and its esters, with free radicals resulting from the decomposition of organic peroxides and other free-radical sources, is well known (55) (see Thiols). The result of this reaction is the abstraction of hydrogen from the mercaptan group, forming an inactive molecule and a new radical derived from the mercaptan, conveniently symbolized by RS·. In the polymerization of a monomer proceeding by a free-radical mechanism, molecules of the monomer can react competitively with either a growing polymer radical or an RS· radical. The net effect of the presence of mercaptan and resulting RS· radicals in a polymerization depends on the relative rates of RS· formation and monomer reaction with RS· and with the growing polymer radical, and it can represent either acceleration or slowing of the polymerization and influence the molecular weight and molecular-weight distribution of the resulting polymer. Depending on the purpose for which a thioglycolate additive is used in a polymerization, it is termed a catalyst, a promoter or accelerator, or a chain-transfer agent. Small quantities (0.1–7.5 wt %) of the ethylene glycol thioglycolate, alkyl thioglycolates, and thioglycolic acid have been included in recipes for the polymerization of acrylonitrile, methyl methacrylate, and vinyl chloride to regulate the molecular weight and porosity of the polymer, prevent discoloration, and minimize formation of volatiles and the buildup of deposits on the wall of polymerization reactors (56–58) (see Initiators; Polymerization processes).

Other. Thioglycolic acid and its derivatives in small concentrations improve the efficiency of metal cleaners and pickling solutions and reduce the corrosion of metals being treated (see Metal surface treatments). The effect has been noted over a wide pH range, from the pickling of steel with 15 wt % hydrochloric acid and 0.16 wt % thioglycolic acid to the removal of oxide from copper, zinc, cadmium, lead, nickel, and tin and their alloys by a combination of thioglycolic acid and sodium sulfite adjusted to the neutral or alkaline pH appropriate to each metal (59–60). Thioglycolic acid is included in a multicomponent cleaner–inhibitor solution for automobile cooling systems at pH 10.6 (61). Silver cleaning and polishing compositions contain higher alkyl thioglycolates, eg, the hexadecyl ester, to minimize tarnishing (62).

In photographic fixing baths, thioglycolate derivatives prevent deposition of silver metal and improve the receptivity of prints to oily inks by forming silver complexes that are more stable than silver thiosulfate (63) (see Photography). The reactivity of thioglycolates with metals has also been utilized in the extraction of valuable metals, ie, Cu, Ag, Au, and Hg, from dilute aqueous solutions into water-immiscible alkyl thioglycolates (64). Thioglycolic acid has been recommended as a cocatalyst with mineral acid and sometimes minor amounts of dihydric phenols in the preparation of bisphenol A by the condensation of phenol with acetone (65) (see Alkylphenols).

Copper, cobalt, rhenium, vanadium, iron, and nitrite ions have been determined spectrophotometrically after reaction with thioglycolic acid (66). Isooctyl thioglycolate has been used to separate antimony(III) and tin(IV) by selective extraction from hydrochloric acid solution (67).

Derivatives

All derivatives of thioglycolic acid that one or more companies reported having manufactured or imported for a commercial purpose from 1975 through 1977 are listed in ref. 68.

Thioethers. *S*-Substituted thioacetic acids are prepared by treatment of a sodium mercaptide RSNa with sodium chloroacetate or by reaction of thioglycolic acid with an unsaturated hydrocarbon, acid, ester, or amide, or an alkyl halide with an acid acceptor (12–13,69). *S*-Benzylthioacetic acid [*103-46-3*] and *S*-*t*-butylthioacetic acid also result from rearrangements in the attempted preparation of benzyl and *t*-butyl thioglycolate esters by conventional esterification. 2-Naphthylthioacetic acid [*99-21-0*] results from the acid-catalyzed reaction of thioglycolic acid with β-naphthol.

Esters. Thioglycolate esters of most primary alcohols and many glycols and polyhydric alcohols are readily prepared by conventional direct esterification (6,70–71). A continuous esterification procedure for alkyl thioglycolates has been described (72). Special synthesis techniques have been described for obtaining a monothioglycolate of a 600 mol wt polyethylene glycol and for *t*-butyl thioglycolate (73). The tertiary butyl group of *t*-butyl thioglycolate serves as a protective group that can be removed by treatment with trifluoroacetic acid (74).

Amides. Thioglycolic acid amides can be prepared by the reaction of an amide of chloroacetic acid with a sulfur donor reagent, eg, ammonium thiocyanate or sodium thiolbenzoate, and hydrolysis of the intermediate to the mercapto amide. The sequence has been used in the mercaptoacetylation of amino acids (75). Certain amides have also been prepared by direct reaction of thioglycolic acid or an ester with the desired amine, eg, *N*,*N*′-ethylenebis(2-mercaptoacetamide), mp 138–139°C, from ethylenediamine and ethyl thioglycolate and *p*-(2-mercaptoacetamido)-diphenylamine, mp 105–108°C, from thioglycolic acid and *p*-aminodiphenylamine in refluxing toluene (76–77). The latter is an antioxidant for rubber (78) (see Antioxidants).

Amino Acids, Peptides, and Proteins. Thioglycolyl proteins originally were formed by reaction of the ϵ-amino groups in the lysine contained in casein with di- or polythioglycolides in aged samples of thioglycolic acid (79). 2-Mercaptoacetyl derivatives of histidine, glycylhistidine, alanylhistidine, tryptophan, phenylalanine, and glutamic acid have been prepared from the chloroacetylated amino acid and thiolbenzoic acid. The compounds possess antihypertensive activity (80). 2-Mercaptoacetylcysteine has been prepared from cysteine and a dehydration product of thioglycolic acid obtained by the action of *N*,*N*-dicyclohexylcarbodiimide (81).

Metal Complexes. Thioglycolic acid forms a multiplicity of very stable complexes with many metal ions, and its use in analytical masking instead of the traditional but toxic cyanide ion has been recommended (82). Depending on the particular metal and experimental conditions, the thioglycolate complexes can be polynuclear or mononuclear and have the structure of a carboxylate salt in which the SH does not participate, a metal mercaptide in which the carboxy group is free, or a true chelate. Moreover, isolated complexes may or may not retain the carboxyl proton or include an alkali metal or ammonium ion replacement (83). With certain metals, ie, nickel and zinc, complexes having various metal-to-thioglycolate ratios coexist in solution (84). Since ions of transition metals, eg, iron, existing in more than one valence state can oxidize thioglycolic acid directly and catalyze its oxidation by air, the stable complexes obtained when air is excluded are complexes of the metal in its lower valence, eg, Fe(II). Unstable Fe(III) thioglycolate complexes have been obtained in ethanol solution at −75°C (85).

Complexes containing thioglycolate and another ligand have been obtained with zinc, cadmium, mercuric, uranyl, stannous, Co(II), and Co(III) ions (86–91).

BIBLIOGRAPHY

"Thioglycolic Acid" in *ECT* 1st ed., Vol. 14, pp. 78–83, by S. D. Gershon and M. M. Rieger, Lever Brothers Company; "Thioglycolic Acid" in *ECT* 2nd ed., Vol. 20, pp. 198–204, by S. D. Gershon, Lever Brothers Company, and M. M. Rieger, Warner Lambert Company.

1. L. Carius, *Ann.* **124,** 43 (1862).
2. K. J. Ellis and co-workers, *J. Solution Chem.* **6,** 183 (1977).
3. J. F. Mulvaney, *Proc. Sci. Sect. Toilet Goods Assoc.* **5,** 133 (1946).
4. W. Bowers, memorandum, Argus Chemical Corp., 1979.
5. A. Schoberl and G. Wiehler, *Angew. Chem.* **66,** 273 (1954); *Ann.* **595,** 101 (1955); J. F. Blount and co-workers, *J. Org. Chem.* **39,** 2374 (1974).
6. F. Duus and co-workers, *Arkiv Kemi* **28,** 423 (1968).
7. H. Lecher, *Ber.* **53,** 591 (1920).
8. D. Goddard and L. Michaelis, *J. Biol. Chem.* **106,** 605 (1934).
9. T. Bersin and I. Steudel, *Ber.* **B71,** 1015 (1938).
10. A. Fava and co-workers, *J. Am. Chem. Soc.* **79,** 833 (1957).
11. M. Kharasch and co-workers, *J. Biol. Chem.* **113,** 537 (1936).
12. E. E. Reid, *Organic Chemistry of Bivalent Sulfur*, Vol. 1, Chemical Publishing Co., Inc., New York, 1958, pp. 441–442.
13. E. Friedmann and co-workers, *Brit. J. Pharmacol.* **3,** 335 (1948); B. Smith and S. Hernestam, *Acta Chem. Scand.* **8,** 1111 (1954).
14. J. J. Ritter and M. J. Lover, *J. Am. Chem. Soc.* **74,** 5576 (1952); M. M. Jones, *J. Inorg. Nucl. Chem.* **36,** 194 (1974); M. Brink and E. Larson, *Tetrahedron* **27,** 5713 (1971).
15. U.S. Pat. 3,209,012 (Sept. 28, 1965), E. H. Miller and co-workers (to Carlisle Chemical Works, Inc.).
16. U.S. Pat. 3,885,039 (May 20, 1975), N. J. Pinkowski and co-workers (to M and T Chemicals, Inc.); V. Knoppova and co-workers, *Collect. Czech. Chem. Commun.* **39,** 1589 (1974).
17. I. Hansson, *Acta Chem. Scand.* **22,** 509 (1968).
18. K. D. Sears and R. L. Casebier, *Phytochemistry* **2,** 1589 (1970); H. Nimz and K. Das, *Chem. Ber.* **104,** 2359 (1971).
19. U.S. Pats. 3,490,991 and 3,490,993 (Jan. 20, 1970), W. E. Fisher and co-workers (to Owens-Illinois Inc.); K. Kringstad and co-workers, *Nor. Skogind.* **22,** 230 (1968).
20. Jpn. Kokai 76-29,440 (Mar. 12, 1976), Y. Hirose (to Otsuka Chemical Co., Ltd.); U.S. Pat. 3,927,085 (Dec. 16, 1975), H. Zengel and co-workers (to AKZO N.V.); Ger. Offen. 2,711,867 (Sept. 21, 1978), H. Klug (to Hoechst A.G.); Jpn. Pat. 68-17,564 (July 25, 1968), M. Otsuka (to Otsuka Chemical Pharmaceutical Co., Ltd.); Jpn. Kokai 74-75,526 (July 20, 1974), T. Tomioka (to Tomioka Chemical Co., Ltd.).

21. U.S. Pat. 2,413,361 (Dec. 31, 1946), H. M. Martin (to Martin Labs., Inc.); U.S. Pat. 2,594,030 (Apr. 22, 1952), R. J. Coons and co-workers (to Gillette Safety Razor Co.).
22. E. G. Curphey, *Chem. Ind. (London)*, 840 (1951).
23. U.S. Pat. 3,522,295 (July 28, 1970), N. A. Konstantinov.
24. Ger. Pat. 180,875 (Oct. 27, 1905), (to Kalle & Co.); M. Pereira de Almeida and co-workers, *Trib. Farm.* **42,** 40 (1974).
25. Neth. Appl. 6,510,637 (Feb. 14, 1967), (to Dr. Spiess GmbH); T. Taguchi and co-workers, *Tetrahedron Lett.* **41,** 3631 (1969).
26. U.S. Pat. 3,860,641 (Jan. 14, 1975), H. Zengel (to AKZO N.V.); S. Afr. Pat. 7404097 (May 21, 1975), M. F. Werneke (to American Cyanamid Co.).
27. Ger. Offen. 2,005,375 (Aug. 19, 1971), H. Herzog (to Deutsche Advance Produktion GmbH); E. Ger. Pat. 90,542 (June 12, 1972), H. Helbig and co-workers.
28. *U.S. Code of Federal Regulations*, Title 49, sections 172.101 and 173.244.
29. *Ammonium Thioglycolate*, CTFA Specification, rev. May 30, 1971; *Infra-Red Spectrum*, CTFA Method G 1-1, Cosmetic Toiletry and Fragrance Association, Inc., Washington, D.C.
30. E. Thamm, *Seifen, Oele, Fette, Wachse* **98,** 258 (1972).
31. *Determination of Thioglycolic Acid and Dithiodiglycolic Acid*, CTFA Method D9-1, Cosmetic Toiletry and Fragrance Association, Inc., Washington, D.C., May 30, 1971.
32. N. Goetz and co-workers, *Analyst* **104,** 1062 (1979).
33. R. B. Tupeeva, *Gig. Sanit.* **32,** 62 (1967); *Chem. Abstr.* **67,** 5517s (1967); N. D. Rozova, *Gig. Tr. Prof. Zabol.* **14,** 58 (1970); *Chem. Abstr.* **74,** 15543f (1971).
34. H. E. Christensen, ed., *Registry of Toxic Effects of Chemical Substances*, U.S. Department of Health, Education, and Welfare, Rockville, Md., 1976, p. 28; J. A. Norris, *Food Cosmet. Toxicol.* **3,** 93 (1965); Y. S. Rotenberg and co-workers, *Gig. Tr. Prof. Zabol.* **13,** 48 (1969); *Chem. Abstr.* **71,** 2055d.
35. A. W. Sheldon, M & T Chemicals Inc., Rahway, N.J., Oct. 1981.
36. A. H. Powitt, *Hair Structure and Chemistry Simplified*, Milady Publishing Corp., New York, 1972, p. 220–221.
37. L. L. Gershbein, *J. Pharm. Sci.* **68,** 1230 (1979).
38. S. Bakshi and L. L. Gershbein, *Arch. Int. Pharmacodin. Ther.* **197,** 5 (1972); **201,** 77 (1973).
39. F. G. Stenbäck and co-workers, *Food Cosmet. Toxicol.* **15,** 601 (1977).
40. Ref. 36, p. 195–201.
41. U.S. Pat. 3,399,683 (Sept. 2, 1968), A. W. Forbriger and co-workers (to Realistic Co.).
42. U.S. Pat. 3,840,656 (Oct. 8, 1974), G. Kalopissis and co-workers (to L'Oreal).
43. W. W. Edman and co-workers, *Cosmet. Toiletries* **94**(4), 35 (1979); B. Bach, *Cosmet. Toiletries* **94**(4), 43 (1979).
44. Ref. 36, pp. 224–233; U.S. Pat. 3,654,936 (Apr. 11, 1972), T. Wajaroff (to Wella A.G.).
45. M. M. Rieger, *Cosmet. Toiletries* **94**(3), 71 (1979).
46. T. J. Elliott, *J. Soc. Cosmet. Chem.* **25,** 367 (1974).
47. U.S. Pat. 3,415,606 (Dec. 10, 1968), R. E. Randebrock.
48. U.S. Pat. 3,567,355 (Mar. 2, 1971), K. J. Boosen (to Therachemie G.m.b.H.).
49. L. I. Nass in N. M. Bikales, ed., *Encyclopedia of Polymer Science and Technology*, Vol. 12, John Wiley & Sons, Inc., New York, 1970, pp. 738–42,746–8, 759–65.
50. W. H. Starnes, Jr., in D. L. Allara and W. L. Hawkins, eds., *Stabilization and Degradation of Polymers*, American Chemical Society, Washington, D.C., 1978, pp. 309–323; L. R. Brecker, *Pure Appl. Chem.* **53,** 577 (1981).
51. U.S. Pat. 2,641,596 (June 9, 1953), W. E. Leistner and co-workers (to Argus Chemical Corp.); U.S. Pat. 2,648,650 (Aug. 11, 1953), E. L. Weinberg and co-workers (to Metal & Thermit Corp.).
52. U.S. Pat. 3,640,947 (Feb. 8, 1972), C. R. Gloskey (to M & T Chemicals Inc.); U.S. Pat. 3,887,519 (June 3, 1975), L. B. Weisfeld and co-workers (to Cincinnati Milacron Chemicals Inc.); U.S. Pat. 2,684,956 (July 27, 1954), E. L. Weinberg (to Metal & Thermit Corp.); U.S. Pats. 4,080,363 (Mar. 21, 1978) and 4,134,878 (Jan. 16, 1979), J. W. Burley and co-workers (to AKZO GmbH); U.S. Pat. 3,530,930 (Feb. 23, 1971), O. S. Kauder and co-workers (to Argus Chemical Corp.); U.S. Pat. 3,890,277 (June 17, 1975), T. G. Kugele and co-workers (to Cincinnati Milacron Chemicals Inc.); U.S. Pat. 4,113,678 (Sept. 12, 1978), M. Minagawa and co-workers (to Argus Chemical Corp.)
53. U.S. Pat. 3,424,717 (Jan. 28, 1969), J. Gottlieb and co-workers (to M & T Chemicals Inc.); U.S. Pat. 3,507,827 (Apr. 21, 1970), M. W. Pollock (to Argus Chemical Corp.); U.S. Pat. 3,787,357 (Jan. 22, 1974), L. R. Brecker (to Argus Chemical Corp.); U.S. Pats. 4,029,618 (June 14, 1977) and 4,159,261 (June 26, 1979), D. J. Dieckmann (to Dart Industries Inc.).

54. *U.S. Code of Federal Regulations*, Title 21, Sections 178.2650 and 178.2010.
55. R. A. Gregg and co-workers, *J. Am. Chem. Soc.* **70,** 3740 (1948).
56. Jpn. Pat. 71-37,728 (Nov. 6, 1971), S. Negishi (to Teijin Ltd.); E. Ger. Pat. 89,709 (May 5, 1972), G. Siegemund and co-workers.
57. Ger. Pat. 1,645,232 (Sept. 4, 1975), A. Wohnhas (to Roehm GmbH); U.S. Pat. 3,801,541 (Apr. 2, 1974), G. H. Hofmann (to E. I. du Pont de Nemours & Co.); Jpn. Kokai 78-21,288 (Feb. 27, 1978), T. Hamada (to Sumitomo Chemical Co., Ltd.).
58. K. Thinius and K. Buehler, *Plaste Kautschuk* **13,** 333 (1966); U.S. Pat. 3,846,390 (Nov. 5, 1974), I. Ito and co-workers (to Sumitomo Chemical Co., Ltd.); Ger. Offen. 2,748,469 (Apr. 27, 1978), R. J. Stanaback (to Tenneco Chemicals Inc.).
59. Jpn. Kokai 74-96,929 (Sept. 13, 1974), K. Ryu (to Senda, Chisako).
60. U.S. Pat. 3,282,848 (Nov. 1, 1966), J. H. Haslam (to E. I. du Pont de Nemours & Co.).
61. U.S. Pat. 3,962,109 (June 8, 1976), A. W. Oberhofer and co-workers (to Nalco Chemical Co.).
62. U.S. Pat. 3,330,672 (July 11, 1967), H. Kroll and co-workers (to Phillip A. Hunt Chemical Corp.); Ger. Pat. 1,282,414 (Nov. 7, 1968), H. Schlegel and co-workers (to Wuerttembergische Metallwarenfabrik).
63. U.S. Pat. 3,647,439 (Mar. 7, 1972), J. D. Bass (to Eastman Kodak Co.); J. D. Bass, *Res. Discl.* **140,** 33 (1975); G. M. Haist and D. A. Pupo, *Photogr. Sci. Eng.* **20,** 220 (1976).
64. Israeli Pat. 44867 (Mar. 31, 1977), A. Suszer and co-workers.
65. U.S. Pat. 2,468,982 (May 3, 1949), J. L. Jansen (to B. F. Goodrich Co.); Czech Pat. 152,945 (Apr. 15, 1974), Cerny and co-workers; U.S. Pat. 4,052,466 (Oct. 4, 1977), K. K. Sun (to the Upjohn Co.).
66. L. P. Pandey and co-workers, *NML Technol. J.* **18,** 39 (1976); R. S. Ramakrishna and M. Williams, *Talanta* **24,** 639 (1977); S. Banerjee and R. K. Dutta, *Indian J. Technol.* **11,** 185 (1973); E. L. Abramova and co-workers, *Uzb. Khim. Zh.* **12,** 19 (1968); *Chem. Abstr.* **69,** 64377k (1968); E. L. Abramova and co-workers, *Uzb. Khim. Zh.* **12**(3), 3 (1977); *Chem. Abstr.* **87,** 145319s (1977); J. P. Rawat and co-workers, *Indian J. Chem.* **13,** 627 (1975).
67. M. A. Hartwig and K. Rengan, *J. Radioanal. Chem.* **42,** 105 (1978).
68. *Toxic Substances Control Act Inventory of Chemical Substances.* U.S. Environmental Protection Agency, Washington, D.C., May 1979.
69. R. R. Mod and co-workers, *J. Am. Oil Chem. Soc.* **52,** 455 (1975).
70. U.S. Pat. 2,789,963 (Apr. 23, 1957), A. C. Hecker (to Argus Chemical Corp.).
71. Ger. Offen. 2,426,399 (Dec. 11, 1975), H. Schaffner and co-workers (Bayer A.G.); Ger. Offen. 2,448,850 (Apr. 22, 1976), K. W. Kluepfel (to Hoechst A.G.); U.S. Pat. 3,340,236 (Sept. 5, 1967), S. O. Greenlee and co-workers (to CIBA Ltd.).
72. E. Ger. Pat. 127,117 (Sept. 7, 1977), E. Mendow and co workers.
73. Ger. Offen. 2,646,482 (Apr. 28, 1977), A. Dessaint and co-workers (to Ugine Kuhlmann).
74. E. Honkanen, *Acta Chem. Scand.* **24,** 1120 (1970).
75. Y. Sugiura and Y. Hirayama, *J. Am. Chem. Soc.* **99,** 1581 (1977); H. Sakurai and co-workers, *Chem. Pharm. Bull.* **27,** 3022 (1979).
76. T. C. Owen and co-workers, *J. Org. Chem.* **38,** 937 (1973).
77. Brit. Pat. 1,441,621 (July 7, 1976), B. T. Ashworth and co-workers (to Imperial Chemical Ind., Ltd.).
78. A. H. Weinstein, *Rubber Chem. Technol.* **50,** 650 (1977).
79. A. Schöberl, *Angew. Chem.* **A60,** 7 (1948).
80. M. Oya and co-workers, *Chem. Pharm. Bull.* **29,** 63, 940 (1981).
81. Jpn. Kokai 79-63017 (May 21, 1979), J. Iwao (to Santen Pharmaceutical Co., Ltd.).
82. H. F. DeBrabander and co-workers, *Anal. Chim. Acta* **70,** 401 (1974).
83. R. H. Lane and co-workers, *Inorg. Chem.* **16,** 102 (1977); O. Farook and co-workers, *Ann. Chim. (Rome)* **64,** 275 (1974); M. R. Harrison and co-workers, *K. Tek. Hoegsk. Handl.*, 248 (1972).
84. D. D. Perrin and I. G. Sayce, *J. Chem. Soc. A*, 82 (1967); R. Abu-Eittah and G. Arafa, *Indian J. Chem. A* **16A,** 407, 1978.
85. D. L. Leussing and I. M. Kolthoff, *J. Am. Chem. Soc.* **75,** 3904 (1953); A. Tomita, *Inorg. Chem.* **7,** 760 (1968).
86. B. R. Panchal and co-workers, *J. Inorg. Nucl. Chem.* **34,** 3932 (1972); **35,** 1685 (1973); *Indian J. Chem.* **11,** 394 (1973).
87. A. Baranski and co-workers, *J. Electroanal. Chem. Interfacial Electrochem.* **75**(2), 613 (1977); *Chem. Abstr.* **86,** 147656C (1977).
88. P. V. Selvarag and M. Santappa, *J. Inorg. Nucl. Chem.* **38,** 837 (1976); **39,** 119 (1977).
89. R. Woytas, *Rocz. Chem.* **48,** 219 (1974).

90. Ref. 84, p. 25.
91. R. H. Lane and co-workers, *Inorg. Chem.* **16,** 93 (1977).

OTTO S. KAUDER
Argus Chemical Corporation

THIOLS

Thiols are the sulfur analogues of alcohols, ie, the oxygen in the alcohol is replaced with sulfur. Until 1930, these compounds were named as mercaptans and, particularly in commerce, the name mercaptan is still widely used. In 1834, the synthesis and physical properties of ethanethiol were described (1). Since then, numerous thiols have been prepared and characterized. An excellent review of the preparation, properties, and reactions of most of the thiols known up to 1954 is compiled in refs. 2–3 and several more recent reference texts have been published (4–6). This review deals primarily with the simple alkanethiols with limited reference to the aromatic thiols.

Nomenclature

Ethanethiol was originally called mercaptan because of its ability to remove mercury from solutions as a crystalline solid. Mercaptan is still accepted as a descriptive name for compounds containing the —SH group. In the thiol nomenclature system, the compound is named as the hydrocarbon with the longest carbon chain followed by thiol; eg, butyl mercaptan becomes butanethiol. Compounds containing more than one —SH group are named as dithiols, trithiols, etc; for example, $HSCH_2CH_2SH$ is named 1,2-ethanedithiol and is retained in the IUPAC names for compounds with other functional groups. The nomenclature depends on the order of precedence assigned to the functional groups; in *Chemical Abstracts*, $HSCH_2CH_2OH$ is 2-mercaptoethanol [*60-24-2*] and $HSCH_2COOH$ is mercaptoacetic acid [*68-11-1*]. On the other hand, since amine falls below thiol in the order of precedence, $H_2NCH_2CH(SH)CH_3$ is 1-amino-2-propanethiol [*10229-29-5*]. An additional type of nomenclature is also still in use. In this system, thio is inserted into the common or trivial name to designate that oxygen has been replaced by sulfur, ie, C_6H_5SH is called thiophenol [*108-98-5*] and $HSCH_2COOH$ is called thioglycolic acid [*68-11-1*] (qv).

Occurrence

Naturally occurring compounds containing thiol groups and related disulfide (RSSR) linkages are important in biological processes (7). An excellent review of the role of the —SH group in biochemistry is ref. 8. Cysteine [*52-90-4*] and cystine are two common amino acids which can be isolated from protein hydrolysates. They are generally considered to be interconvertible *in vivo*:

$$\underset{\text{cysteine}}{HSCH_2\underset{\substack{|\\ NH_2}}{C}HCOOH} \underset{H_2}{\overset{O_2}{\rightleftharpoons}} \underset{\text{cystine}}{[-SCH_2\underset{\substack{|\\ NH_2}}{C}HCOOH]_2} \quad (1)$$

Although there are only small amounts of cysteine and cystine in living tissue, most of the thiol and disulfide groups in nature come from combined forms of these two acids (9).

Glutathione [*70-18-8*],

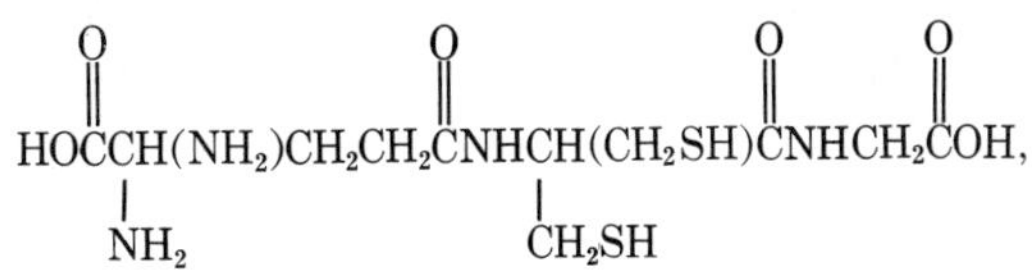

a tripeptide containing a mercaptomethyl group, is widely distributed in animal and plant cells. Another thiol important to life processes is coenzyme A [*85-61-0*], frequently designated as CoA—SH. It is an acyl-transport medium in the synthesis and degradation of fatty acids and functions via the thiol acetate (CoA—$SCOCH_3$); it is also involved in other biochemical conversions.

Some foodstuffs contain trace amounts of methanethiol, eg, cheese, milk, coffee, and oysters (10–13). Small amounts of methanethiol and ethanethiol are components of natural human breath (14). Butanethiol and 2-methylbutanethiol [*1878-18-8*] are contained in the defensive secretion of skunks (15). Methanethiol, ethanethiol, 1-propanethiol, 2-propanethiol, 2-methyl-1-propanethiol, 2-methyl-2-propanethiol, and 1-butanethiol occur in the air and manure gas near domestic animal pens (16).

Most crude oils contain a variety of sulfur compounds (17–19). The amount and type of compounds vary widely depending upon the geographical source. It is generally believed that sulfur compounds in crude oil were formed from the reaction of elemental sulfur generated by microbial action on sulfate in rock with hydrocarbons, resulting in the formation of nonthiophene-type compounds, eg, thiols and sulfides. As the oil ages, these compounds slowly convert to thiophenes (20). In general, thiols are a small portion of the total sulfur compounds present. Forty-four different alkanethiols in the 111–150°C distillate from a Texas (Wasson) crude have been identified. These include all of the possible one-through-six carbon-atom open-chain thiols, cyclopentanethiol [*1679-07-8*], four methylcyclopentanethiols, cyclohexanethiol, and some heptanethiols (21) (see Petroleum, composition).

Physical Properties

Thiols are notorious for their obnoxious odors; this is particularly true for the lower alkanethiols. The odors can be detected at very low concentrations, eg, ethanethiol can be detected by humans in air at less than 0.5 ppb (0.5×10^{-9}). The odor thresholds of various thiols have been measured (22). The physical constants of various alkanethiols are shown in Table 1. The melting points of various long-chain alkanethiols (C_{12}–C_{28} with even-numbered chain lengths) are listed in Table 2.

The lower molecular weight alkanethiols boil at lower temperatures than the corresponding alcohols because of the weaker intermolecular hydrogen bonding in the thiols. Boiling point differences decrease with increasing molecular weight and finally reverse, ie, 1-heptanol and 1-heptanethiol boil very close together and 1-octa-

Table 1. Physical Constants of Some Alkanethiols

Compounds	CAS Registry No.	Boiling point, °C$_{kPa}$[a]	Freezing point, °C	Refractive index, n_D^{20}	d_4^{20}, g/cm^3	Ref.
methanethiol	[*74-93-1*]	5.9	−122.97		0.8665	23
ethanethiol	[*75-08-1*]	35.0	−147.89	1.43105	0.83914	24
2-propanethiol	[*75-33-2*]	52.6	−130.54	1.43554	0.81431	25
1-propanethiol	[*107-03-9*]	67.8	−113.13	1.43832	0.84150	23
2-methyl-2-propanethiol	[*74-66-1*]	64.2	1.11	1.42320	0.80020	24
2-butanethiol	[*513-35-3*]	85.0	−140.14	1.43673	0.82988	25
2-methyl-1-propanethiol	[*513-44-0*]	88.5	−144.86	1.43877	0.83428	23
1-butanethiol	[*109-79-5*]	98.4	−115.67	1.44298	0.84161	25
1-pentanethiol	[*110-66-7*]	126.5	−75.7	1.44692	0.84209	24
1-hexanethiol	[*111-31-9*]	152.6	−80.49	1.44968	0.84242	23
1-heptanethiol	[*1639-09-04*]	176.9	−43.23	1.45215	0.94310	23
1-octanethiol	[*111-86-6*]	199.1	−49.2	1.4540	0.8433	26
1-decanethiol	[*143-10-2*]	240.6	−26	1.4569	0.8443	26
1-dodecanethiol	[*112-55-0*]	142.5_{2}	−9.2	1.4589	0.8450	2
1-hexadecanethiol	[*2917-25-2*]	$123–8_{0.07}$	18–20			26
1-octadecanethiol	[*2885-00-9*]	$188_{0.1–0.2}$	28	14645	0.8475	26
cyclohexanethiol	[*1569-69-3*]	158		1.4921	0.9782	26

[a] To convert kPa to mm Hg, multiply by 7.5.

Table 2. Melting Points of Some Long-Chain 1-Alkanethiols

Compound	CAS Registry No.	Melting point, °C	Ref.
dodecanethiol	[*112-55-0*]	−9.2	27
tetradecanethiol	[*2079-95-0*]	7.0	28
hexadecanethiol	[*2917-26-2*]	18.0	29
octadecanethiol	[*2885-00-9*]	32–32.5	30
eicosanethiol	[*13373-97-2*]	37.5–38.5	31
docosanethiol	[*7773-83-3*]	42.5–43	31
tetracosanethiol	[*16331-24-1*]	51–52	31
hexacosanethiol	[*16331-25-2*]	56.5–57.5	31
octacosanethiol	[*16331-26-3*]	61–62	31

nethiol boils higher than the corresponding alcohol. A comparison of the boiling points of various *n*-alkanethiols and the corresponding alcohols is shown in Table 3.

Highly accurate vapor pressure–temperature relationships for a number of thiols have been determined and, based on these data, constants for the Antoine and Cox

Table 3. Boiling Points of Alcohols and the Corresponding Thiols

Alkyl group	Boiling point, °C Alcohol	Thiol
methyl	64.7	5.9
n-butyl	117.4	98.4
n-heptyl	176.3	176.9
n-nonyl	215.0	220.2[a]

[a] CAS Registry No. [*1455-21-6*].

equations have been derived (32). Ethanethiol, propanethiol, and butanethiol form azeotropes with aliphatic hydrocarbons but not with benzene or toluene (33). The lower molecular weight thiols are substantially less soluble in water than the corresponding alcohols, eg, the solubility of ethanethiol at 20°C is less than 1 part in 100 parts of water.

Some important chemical thermodynamic properties of a number of organic sulfur compounds, including alkanethiols, cycloalkanethiols, and benzenethiol, have been determined and the data have been reviewed (34–39). Calorimetric studies of several alkanethiols have been carried out and the data have been used to calculate their thermodynamic properties (40–43). Spectral studies of general interest include the following: infrared spectra of thiols (44–45), valence force field (46), hydrogen-bonding properties (47), correlation of mass spectral data (48), and interpretation of proton magnetic resonance spectra (49).

Physical properties of a few other thiols and dithiols are given in Table 4.

In the series $HS(CH_2)_nSH$, physical constants have been reported for 1,2-ethanedithiol through 1,12-dodecanedithiol [*33528-63-1*], and the heat of combustion has been determined for several of the lower members of the series (50–51). The *gem*-dithiols, illustrated by 2,2-propanedithiol, are surprisingly stable considering their structure; several alkane and cycloalkane species have been made. Benzenethiol and *p*-benzenedithiol melt much lower than their oxygen analogues; various substituted derivatives in the benzene and naphthalene series have been identified. For the nitrogen heteroaromatics, eg, pyridine-2-thiol, or the 4-isomer, the spectral data indicate predominance of the tautomeric thione form,

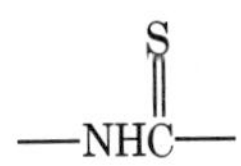

rather than the thiol form (57).

Chemical Properties

Oxidation. Thiols can be oxidized readily to the corresponding disulfides. In the presence of base, the conversion can be accomplished with molecular oxygen (air) (58). Various metal salts, eg, salts of copper, cobalt, or nickel, catalyze the oxidation (59). This process has been reviewed and a mechanism involving the coupling of the thiyl

Table 4. Boiling and Melting Points of Various Thiol Types

Compound	CAS Registry No.	Boiling point, $°C_{kPa}$[a]	Melting point, °C	Refs.
1,2-ethanedithiol	[*540-63-6*]	146	−41.2–49.1	50–51
2,2-propanedithiol	[*1687-47-4*]	57_{13}	4–6	52–53
benzenethiol	[*108-98-5*]	169.1	−14.9	23
p-benzenedithiol	[*624-39-5*]		97.5–98.5	54
pyridine-2-thiol	[*2637-34-5*]		130–132	55
thiophene-2-thiol	[*7774-74-5*]	$54_{0.7}$		56

[a] To convert kPa to mm Hg, multiply by 7.5.

radical (RS·) has been suggested (60). The stoichiometry is shown below:

$$4\,RSH + O_2 \rightarrow 2\,RSSR + 2\,H_2O \qquad (2)$$

Several of the petroleum-refinery sweetening processes, that convert highly odorous thiols to less odorous disulfides involve oxidations of this type. Salts of cobalt phthalocyanine in combination with base are excellent catalysts for these air oxidations (61–64).

Other mild oxidizing agents can be used. Manganese dioxide and some other metals oxides and dimethylsulfonium bromide [*41285-92-1*] give high yields of disulfides (65–66). Heating a thiol with dimethyl sulfoxide is a convenient laboratory method for preparing the corresponding disulfide (67). This method is applicable for converting some terminal dithiols to cyclic disulfides (68):

$$2\,RSH + (CH_3)_2SO \rightarrow RSSR + (CH_3)_2S + H_2O \qquad (3)$$

$$HS(CH_2)_4SH + (CH_3)_2SO \rightarrow \overline{S(CH_2)_4S} + (CH_3)_2S + H_2O \qquad (4)$$

[*1191-08-8*]

Chlorination of thiols under controlled conditions give sulfenyl chlorides, RSCl, although it is usually better to prepare alkanesulfenyl chlorides by cleavage of a disulfide with chlorine. Sulfenyl chlorides are reactive intermediates. Aryl and alkyl derivatives have been investigated extensively (69–70):

$$RSH + Cl_2 \rightarrow RSCl + HCl \qquad (5)$$

$$RSCl + RSH \rightarrow RSSR + HCl \qquad (6)$$

$$RSSR + Cl_2 \rightarrow 2\,RSCl \qquad (7)$$

Vigorous oxidation of thiols, eg, with nitric acid or permanganate, gives sulfonic acids, RSO_3H, as the products. This oxidation can be accomplished in steps to give disulfide, thiosulfinate, thiosulfonate, disulfide trioxide, disulfide tetraoxide, and, finally (with hydrolysis), sulfonic acid (71):

$$\underset{\text{thiol}}{2\,RSH} \longrightarrow \underset{\text{disulfide}}{RSSR} \longrightarrow \underset{\text{thiosulfinate}}{R\overset{O}{\overset{\|}{S}}SR} \longrightarrow \underset{\text{thiosulfonate}}{RSO_2SR} \longrightarrow$$

$$\underset{\substack{\text{disulfide}\\ \text{trioxide}}}{RSO_2SOR} \longrightarrow \underset{\substack{\text{disulfide}\\ \text{tetroxide}}}{RSO_2SO_2R} \xrightarrow{H_2O} \underset{\substack{\text{sulfonic}\\ \text{acid}}}{2\,RSO_3H} \qquad (8)$$

In the case of 1-dodecanethiol, all products except the disulfide trioxide derivative have been isolated and characterized (72).

Addition to Carbon–Carbon Multiple Bonds. Thiols add to olefins to give thioethers. The anti-Markovnikov addition can be promoted by free-radical generators, ie, with peroxide or an azonitrile or with ultraviolet irradiation (73–74). Thus, under these conditions, the addition of ethanethiol to propylene gives ethyl *n*-propyl sulfide. There have been several reviews on free radical additions of this type (75–79):

$$C_2H_5SH \rightarrow C_2H_5S\cdot + CH_2{=}CHCH_3 \rightarrow C_2H_5SCH_2\dot{C}HCH_3 \qquad (9)$$

$$C_2H_5SCH_2\dot{C}HCH_3 + C_2H_5SH \rightarrow C_2H_5SCH_2CH_2CH_3 + C_2H_5S\cdot \qquad (10)$$

The oxidative addition of a thiol to an olefin to give a β-hydroxysulfoxide is a related reaction (80). Hydroperoxide intermediates have been isolated, but they are usually unstable at room temperatures and rearrange to a sulfoxide (60):

$$RSH + CH_2{=}\underset{R'}{\underset{|}{C}}H \xrightarrow{O_2} (RSCH_2CHOOH) \rightarrow R\overset{O}{\overset{|}{S}}CH_2CHR'OH \quad (11)$$

The addition of thiols to olefins can proceed by an ionic mechanism to give normal Markovnikov addition products. Benzenethiol and isobutylene in sulfuric acid give *tert*-butyl phenyl sulfide:

$$C_6H_5SH + CH_2{=}C(CH_3)_2 \xrightarrow{H^+} C_6H_5SC(CH_3)_3 \quad (12)$$

Free-radical-promoted additions of a thiol to butadiene or 2,3-substituted butadienes give the 1,4-adduct exclusively; some conjugated dienes with external substitution, eg, 2,5-dimethyl-2,4-hexadiene, give the 1,2-adduct as the main product (81–82):

$$RSH + H_2C{-}CHCH{=}CH_2 \rightarrow RSCH_2CH{=}CHCH_3 \quad (13)$$

$$RSH + (CH_3)_2C{=}CHCH{=}C(CH_3)_2 \rightarrow RSC(CH_3)_2CH_2CH{=}C(CH_3)_2 \quad (14)$$

Thiols add to acetylenes under conditions generating free radicals (78,82). Reaction with an alkylacetylene generally gives the anti-Markovnikov monoadduct and the 1,2-diadduct, eg, 1-ethylthiopropene (*cis–trans* mixture) and 1,2-bis(ethylthio)-propane are obtained from the reaction of ethanethiol with propyne in the presence of azobis(2-methylpropionitrile) as the initiator (83). With phenylacetylene, monoaddition of *p*-toluenethiol [*106-45-6*] occurs readily, in one instance without an initiator (84). Reaction with excess ethanethiol under extended ultraviolet irradiation gives the terminal diadduct; the geminal diaddition is attributed to the stability of the intermediate benzyl radical, $C_6H_5\dot{C}HCH(SC_2H_5)_2$ (85):

$$3\ C_2H_5SH + 2\ CH_3C{\equiv}CH \rightarrow CH_3CH{=}CHSC_2H_5 + CH_3CH(SC_2H_5)CH_2SC_2H_5 \quad (15)$$

$$CH_3C_6H_4SH + C_6H_5C{\equiv}CH \rightarrow C_6H_5CH{=}CHSC_6H_4CH_3 \quad (16)$$

$$2\ C_2H_5SH + C_6H_5C{\equiv}CH \rightarrow C_6H_5CH_2CH(SC_2H_5)_2 \quad (17)$$

Thiolate anions also add readily to acetylenes, and this reaction is stereoselective. Thus, heating a solution of the sodium salt of *p*-toluenethiol or of methanethiol with phenylacetylene gives the cis monoadduct, $C_6H_5CH{=}CHSR$ (86). This reaction and related nucleophilic reactions of thiols with chloroethylenes have been reviewed (87). The base-catalyzed reaction of a thiol with acetylene can be accomplished to give a vinyl sulfide, $RSCH{=}CH_2$, or the diadduct, $RSCH_2CH_2SR$ (88–89).

Addition to Carbonyls. Mercaptals, which are analogous to acetals, are obtained by the acid-catalyzed reaction of a thiol with an aldehyde. The intermediate carbonyl addition product (a hemimercaptal) is usually not stable enough to be isolated. However, there are exceptions, eg, trichloroacetaldehyde (chloral) and thiols form stable hemimercaptals (90). Mercaptals can also be obtained by an interchange reaction between diphenyl acetal and a thiol (91):

$$RSH + R'CHO \xrightarrow{H^+} R'CH(OH)SR \xrightarrow[H^+]{RSH} R'CH(SR)_2 + H_2O \quad (18)$$

$$Cl_3CCHO + RSH \xrightarrow{H^+} Cl_3CCH(OH)SR \quad (19)$$

$$2\,RSH + R'CH(OC_6H_5)_2 \xrightarrow{H^+} R'CH(SR)_2 + 2\,C_6H_5OH \qquad (20)$$

Thiols also undergo similar additions to a variety of ketones to give mercaptoles, $RR'C(SR'')_2$. Thiol additions to aldehydes and ketones have been investigated extensively and reviewed (92–93). Thiols can be halomethylated by treatment of a thiol–aldehyde mixture with a hydrogen halide. This reaction is particularly useful for the preparation of reactive chloromethyl sulfide (94):

$$RSH + HCHO + HCl \rightarrow RSCH_2Cl + H_2O \qquad (21)$$

1,4-Addition of thiols to quinones occurs readily at room temperature (95). With alkanethiols or benzenethiol and benzoquinone, the hydroquinone product is oxidized by the unreacted benzoquinone and a 2-alkyl or phenylthio-1,4-benzoquinone is the product normally isolated (96–97). Multiple addition gives 2,5- and 2,6- disubstitution (98):

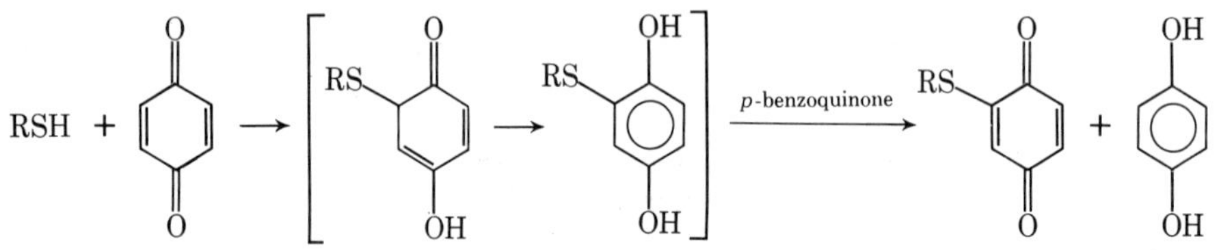

(22)

Formation of Mercaptides. The metal salts of thiols are called mercaptides. Most are insoluble in water and form with a variety of metal cations (3). Many of the mercury mercaptides, $Hg(SR)_2$, have well-defined melting points and have been used for thiol identification. Precipitation of silver mercaptides with silver nitrate has been used for detection and quantitative determination of thiols. The search for methods of removing thiols from sour petroleum distillates led to extensive investigations of mercaptide formation with lead, copper, and other metal ions. One of the oldest sweetening methods is the doctor process, which involves formation of lead mercaptide followed by conversion with free sulfur to organic disulfide and lead sulfide. There are a number of variations of this process; the basic equations are usually written as follows:

$$2\,RSH + Na_2PbO_2 \rightarrow Pb(SR)_2 + 2\,NaOH \qquad (23)$$

$$Pb(SR)_2 + S \rightarrow R_2S_2 + PbS \qquad (24)$$

Conducted on a test-tube scale, these reactions constitute the doctor test for detection of thiols in liquid petroleum products; the product is said to be doctor sweet when no black precipitate, ie, lead sulfide, is present.

The lower molecular weight alkanethiols are soluble in aqueous sodium hydroxide but are readily hydrolyzed:

$$RSH + NaOH \rightleftharpoons RSNa + H_2O \qquad (25)$$

Solubility in water and aqueous caustic decrease with increasing molecular weight. Sodium mercaptides can be prepared in anhydrous media by treating a thiol with metallic sodium:

$$2\,RSH + 2\,Na \rightarrow 2\,RSNa + H_2 \qquad (26)$$

Miscellaneous Reactions. Thiols undergo a variety of other reactions, and most of these involve the thiolate ion. In general, thiolate ions are strong nucleophiles, and this property has been the subject of a number of kinetic studies (99–102). A few examples of the many types of reactions are the following: (27) with alkyl and aryl halides to give thioethers (99,103); (28) with epoxides (base-promoted) to give 2-hydroxyethyl thioethers (104); (29) and (30) with activated carbon–carbon double bonds, eg, in methyl acrylate and acrylonitrile, in the presence of catalytic amounts of base to give the substituted thioethers (105); (31) with carbon disulfide and sodium hydroxide to give trithiocarbonate esters (106); (32) with isocyanates and catalyzed by tertiary amines to give thiourethanes (107–108); (33) with elemental sulfur in the presence of an amine or sodium hydroxide to give S_3–S_5 polysulfides (109–110); (34) with acid chlorides to give thiol esters (a procedure frequently used because of unfavorable equilibria in the thiol–acid reaction) (111); (35) with carbamyl chlorides to give thiolcarbamic acid esters (112); (36) with phosphorus trichloride in the presence or absence of a hydrogen chloride acceptor to give phosphorotrithioite esters (113–114); (37) with sulfur monochloride to give alkyldithiosulfenyl chlorides (115); and (38) with chlorine at low temperatures to give alkylsulfur trichlorides followed by hydrolysis to give alkylsulfinyl chlorides (116–117):

$$RSNa + R'Br \longrightarrow RSR' + NaBr \quad (27)$$

$$RSH + \overset{\overset{O}{\diagup\;\diagdown}}{CH_2{-}CH_2} \longrightarrow RSCH_2CH_2OH \quad (28)$$

$$RSH + H_2C{=\!=}CHCO_2CH_3 \longrightarrow RSCH_2CH_2CO_2CH_3 \quad (29)$$

$$RSH + H_2C{=\!=}CHC{\equiv}N \longrightarrow RSCH_2CH_2C{\equiv}N \quad (30)$$

$$RSH + CS_2 + NaOH \longrightarrow RS\overset{\overset{S}{\|}}{C}SNa + H_2O \quad (31)$$

$$RSH + R'NCO \longrightarrow R'NH\overset{\overset{O}{\|}}{C}SR \quad (32)$$

$$2\ RSH + S_{(x+1)} \longrightarrow RSS_xSR + H_2S \quad (33)$$

$$RSH + R'\overset{\overset{O}{\|}}{C}Cl \longrightarrow R'\overset{\overset{O}{\|}}{C}SR + HCl \quad (34)$$

$$RSH + R'_2N\overset{\overset{O}{\|}}{C}Cl \longrightarrow R'_2N\overset{\overset{O}{\|}}{C}SR + HCl \quad (35)$$

$$3\ RSH + PCl_3 \longrightarrow (RS)_3P + 3\ HCl \quad (36)$$

$$RSH + S_2Cl_2 \longrightarrow RSSSCl + HCl \quad (37)$$

$$RSH + 2\ Cl_2 \longrightarrow RSCl_3 + 2\ HCl$$

$$RSCl_3 + H_2O \longrightarrow R\overset{\overset{O}{\|}}{S}Cl + 2\ HCl \quad (38)$$

Preparation

Commercial Methods. The bulk of the alkanethiols manufactured on a commercial scale are prepared by the reaction of hydrogen sulfide with olefins or with alcohols. Lesser amounts are prepared by converting alkyl halides to the corresponding mercaptans with sodium or ammonium hydrosulfide. Hydrogen sulfide and a tertiary olefin in the presence of an acidic catalyst give a tertiary alkanethiol:

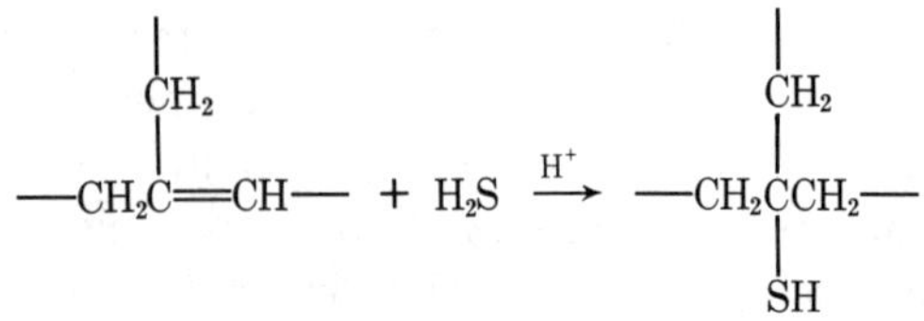
(39)

This reaction is usually carried out in a continuous flow system by passing the reactants over a solid contact catalyst. Low molecular weight olefins, eg, isobutylene to give 2-methyl-2-propanethiol, or higher molecular weight olefins, eg, propylene tetramer to give *tert*-dodecyl mercaptan [*25103-58-6*], are used as feedstocks (118–120). Several Friedel-Crafts catalysts, eg, aluminum chloride and other acid halides, hydrogen fluoride, and boron trifluoride, are described in the patent literature as catalysts for hydrogen sulfide addition to olefins (3). Processes for the production of *tert*-dodecyl mercaptan from propylene tetramer with boron trifluoride and aluminum chloride have been described (121–122).

With the exception of ethanethiol, essentially all of the terminal thiols (primary mercaptans) are made commercially in the United States by the ultraviolet-light-promoted addition of hydrogen sulfide to 1-olefins; ethanethiol is prepared by the vapor-phase reaction of ethylene with hydrogen sulfide. This anti-Markovnikov addition is carried out in the liquid phase and proceeds by a free-radical mechanism (123–124):

$$H_2S \rightarrow H\dot{S} + \dot{H} \tag{40}$$

$$H\dot{S} + CH_2{=}CHR \rightarrow HSCH_2\dot{C}HR \tag{41}$$

$$HSCH_2\dot{C}HR + H_2S \rightarrow HSCH_2CH_2R + H\dot{S} \tag{42}$$

Ultraviolet radiation below 3×10^4 nm, eg, light from a medium-pressure mercury vapor lamp, is an effective radical generator for this reaction. Promoters or sensitizers have been used to increase the reaction rate (123–126). Other free radical generators, eg, azonitriles, can be used to promote the reaction (127). As might be expected, the addition of thiol to olefin to give dialkyl sulfide as a by-product is a competitive reaction. The amount of sulfide formed can be minimized by using excess hydrogen sulfide.

The first catalytic method for the synthesis of thiols involved the vapor-phase reaction of an alcohol with hydrogen sulfide. Thoria or thoria on pumice are effective catalysts for this reaction (128–129):

$$ROH + H_2S \rightarrow RSH + H_2O \tag{43}$$

Later, it was realized that promoted α-aluminas, eg, potassium tungstate on alumina and supported heteropoly acids and their salts, are efficient catalysts

(130–132). Until recently, several terminal thiols, eg, 1-propanethiol, 1-butanethiol, and 1-octanethiol, were produced commercially from the corresponding alcohols. However, at this time in the United States, only methanethiol is manufactured by the alcohol process; the other terminal thiols are obtained by the ultraviolet light process with the exception of ethyl mercaptan, which is produced from ethylene and H_2S over an acid catalyst.

Other methods for the production of alkanethiols described in the patent literature include the vapor-phase reaction of a carbonyl compound with hydrogen and hydrogen sulfide and the cleavage of an organic sulfide with hydrogen sulfide (133–135):

$$\mathrm{R\overset{\overset{\displaystyle O}{\|}}{C}H + H_2 + H_2S \longrightarrow RCH_2SH + H_2O} \qquad (44)$$

$$\mathrm{R\overset{\overset{\displaystyle O}{\|}}{C}R' + H_2 + H_2S \longrightarrow R\overset{\overset{\displaystyle SH}{|}}{C}HR' + H_2O} \qquad (45)$$

$$\mathrm{RSR + H_2S \longrightarrow 2\ RSH} \qquad (46)$$

Benzenethiol is produced on a commercial scale by the red phosphorus reduction of benzenesulfonyl chloride (136). Methods for obtaining this thiol from the high temperature reaction of monochlorobenzene and hydrogen sulfide have been described (137–138):

$$\mathrm{C_6H_5Cl + H_2S \rightarrow C_6H_5SH + HCl} \qquad (47)$$

Laboratory Methods. There are a number of other synthetic routes used primarily for small-scale preparation (3). The first syntheses of thiols were achieved by the reaction of an alkyl sulfate with a metal hydrosulfide. The similar reaction of an alkyl halide or *p*-toluenesulfonate with sodium or potassium hydrosulfide has been used extensively:

$$\mathrm{RX + NaSH \rightarrow RSH + NaX} \qquad (48)$$

The inherent tendency for the formation of dialkyl sulfides in these displacement reactions led to development of other methods. Perhaps the most widely used is cleavage of an isothiouronium salt. The intermediate salts are made from an alkyl halide and thiourea or by reaction of an alcohol with a hydrogen halide and thiourea (139–141). Cleavage is usually with aqueous sodium hydroxide. A modified procedure, which gives good yields of alkanethiols and dithiols, consists of *in situ* cleavage with a high boiling amine (142):

$$\mathrm{RS\overset{\overset{\displaystyle NH}{\|}}{C}NH_2.HX + NaOH \longrightarrow RSH + NaX + H_2NCN + H_2O} \qquad (49)$$

$$\mathrm{RS\overset{\overset{\displaystyle NH}{\|}}{C}NH_2.HX + R'_2NH \longrightarrow RSH + R'_2N\overset{\overset{\displaystyle NH}{\|}}{C}NH_2.HX} \qquad (50)$$

Hydrolysis of thiol esters is also a method of general utility, particularly in making olefin-derived thiols from thioacetates. The addition of thioacetic acid [*507-09-5*] to an olefin proceeds readily under conditions generating free radicals to give the anti-Markovnikov product (27,143–144). The intermediate esters have also been made by the reaction of a thiolacid sodium salt with an alkyl halide or sulfate. The thioacetates are readily hydrolyzed with aqueous base. An alternative cleavage method is the reaction with ammonia, which gives thiol and acetamide (145):

$$RCH{=}CH_2 + CH_3\overset{O}{\overset{\|}{C}}SH \longrightarrow RCH_2CH_2S\overset{O}{\overset{\|}{C}}CH_3$$

$$RCH_2CH_2S\overset{O}{\overset{\|}{C}}CH_3 \xrightarrow{NaOH} RCH_2CH_2SH + NaO\overset{O}{\overset{\|}{C}}CH_3 \quad (51)$$

$$RCH_2CH_2S\overset{O}{\overset{\|}{C}}CH_3 \xrightarrow[NH_3]{} RCH_2CH_2SH + H_2N\overset{O}{\overset{\|}{C}}CH_3 \quad (52)$$

Dithiols of the type $HS(CH_2)_nSH$ have been made by the procedures used for monothiols, eg, from the reaction of dihalides with sodium hydrosulfide, from bis-isothiouronium salts, and from bis-thiolacetates (146). Lithium aluminum hydride reduction of cyclic trithiocarbonates has been used for the synthesis of some *meso* and *dl* isomers of vicinal alkanedithiols (147). Cleavage of bis-trithiocarbonates with acid to give α,ω-dithiols has been reported (148):

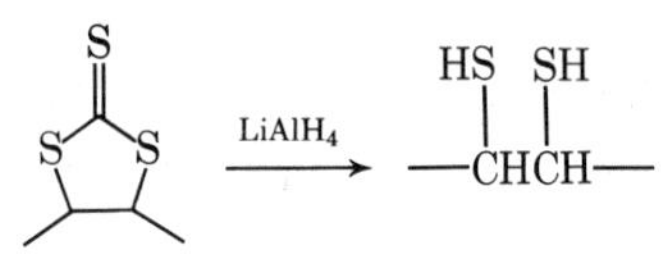

(53)

$$NaS\overset{S}{\overset{\|}{C}}S{-}R{-}S\overset{S}{\overset{\|}{C}}SNa \xrightarrow{H^+} HS{-}R{-}SH + 2\ CS_2 \quad (54)$$

gem-Dithiols were first isolated from reactions in which ketones or aldehydes were warmed with liquid hydrogen sulfide under pressure (52):

$$R\overset{O}{\overset{\|}{C}}R' + 2\ H_2S \longrightarrow RC(SH)_2R' + H_2O \quad (55)$$

Later, these dithiols were made by passing hydrogen sulfide into a solution of a ketimine in dimethylformamide (53,149).

Standard procedures for the synthesis of thiophenols consist of the reduction of sulfonyl chlorides, eg, with zinc and sulfuric acid, or the hydrolysis of an *S*-phenylxanthate formed by replacement of a diazonium group (150–152):

$$C_6H_5SO_2Cl \xrightarrow{3\ H_2} C_6H_5SH + HCl + 2\ H_2O \quad (56)$$

$$C_6H_5N_2Cl + KS\overset{S}{\overset{\|}{C}}OC_2H_5 \longrightarrow C_6H_5S\overset{S}{\overset{\|}{C}}OC_2H_5 \xrightarrow{KOH} C_6H_5SK \quad (57)$$

Procedures have also been developed for reducing aromatic sulfonyl chlorides with phosphorus and the xanthates have been cleaved with lithium aluminum hydride (136,153–154). Benzenedi- and -trithiols can be obtained from bromobenzenes by the formation and reduction cleavage of phenyl alkyl sulfides (54):

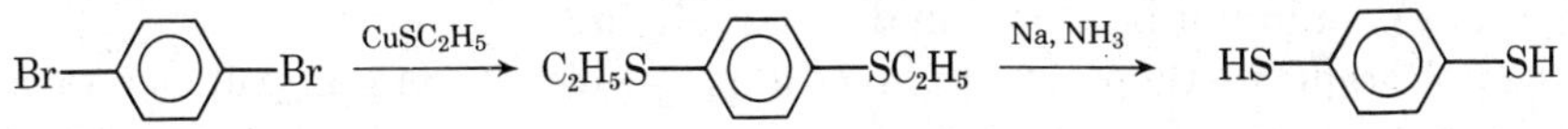

(58)

A method for the synthesis of aromatic thiols from the corresponding phenols or pyridinols has been developed and involves the formation of an *O*-aryldimethylthiocarbamate, thermal rearrangement to the *S*-arylthiolcarbamate, and hydrolysis (155):

$$ArONa + (CH_3)_2N\overset{\overset{S}{\|}}{C}Cl \longrightarrow (CH_3)_2N\overset{\overset{S}{\|}}{C}OAr + NaCl \qquad (59)$$

$$(CH_3)_2N\overset{\overset{S}{\|}}{C}OAr \xrightarrow{\Delta} (CH_3)_2N\overset{\overset{O}{\|}}{C}SAr \xrightarrow{KOH} ArSH \qquad (60)$$

Production and Economic Aspects

Production figures for individual thiols are not available. It is estimated that in 1980 the total annual world capacity for thiol production was $(1.2–1.4) \times 10^5$ metric tons. Of this amount, Western Europe could produce $(3–3.5) \times 10^4$ t/yr. Consumption appears to be growing ca 7%/yr (156). An estimate of thiol consumption in the United States from 1972 to 1979 is shown in Table 5. Mid-1981 prices of thiols were \$1.30–4.20/kg, depending on the specific thiol.

Handling and Storage

Commercially, thiols are shipped in the liquid phase. Methanethiol is shipped in various sizes of pressure vessels; other thiols are shipped in 0.02-m^3 cans, 0.204-m^3 drums, tank trucks, and tank cars according to the DOT regulations pertaining to these containers. Regulations require methanethiol to be labeled as a flammable gas, although it is shipped in the liquid state. Other thiols with flash points below 38°C (in general, C_6 thiols and lower) must be labeled as flammable liquids.

Table 5. Estimated U.S. Thiol Consumption

Year	Consumption, 1000 t/yr
1972	32
1974	43
1976	48
1978	59
1979	61

Metal drums containing thiols should be unloaded and handled carefully to prevent damage, which would permit odors to escape. A leaking drum should be removed to an open area or well-ventilated enclosure where the leak can be stopped or the drum contents transferred. On warm days, drums containing lower molecular weight thiols (C_2–C_4) should be cooled before opening. In summer and in warm climates, thiols should not be stored in direct sunlight.

The preferable method of emptying drums is with a hand pump or electric pump equipped with an explosion-proof motor. Drums can also be unloaded with a dip tube. An inert gas is used to pressurize the drum and force the product out the dip tube. Care should be taken not to allow the pressure in the drum to exceed 136 kPa (5 psig) to prevent accidental rupture of the drum.

Most tank-truck deliveries, particularly deliveries of the more volatile thiols, are made in producer-owned trailers by drivers trained in handling thiols. Larger quantities of thiols are shipped in top-loading liquid-petroleum-gas (LPG) tank cars. The cars are unloaded through liquid eduction valves in the tops of the tank cars.

Process equipment in contact with thiols should be constructed of carbon steel, aluminum, stainless steel, or other copper-free alloys. Thiols stored in carbon steel should be kept dry and blanketed with an inert gas to prevent formation of iron–sulfur complexes. Rubber hoses and gaskets should not be used, because they swell and eventually degrade in the presence of thiols. Fluoroplastic and asbestos are satisfactory materials for gasketing. Stainless steel should be used in the construction of process equipment that is to be in contact with thioalcohols, eg, 2-mercaptoethanol.

Thiols have an obnoxious odor which, unless promptly neutralized, usually leads to employee and neighbor complaints. An effective method for converting the spilled thiol to a relatively low odor disulfide is through chemical oxidation. This can be done by spraying or flooding the spill area with a dilute bleach solution. Sodium hypochlorite, eg, Purex, or calcium hypochlorite solutions may be used. Concentrated oxidizers should never be used, eg, dry calcium hypochlorite powder on a concentrated mercaptan, because sufficient heat may be generated to ignite the mercaptan.

Detailed information on the safe handling of mercaptans may be obtained from the Pennwalt Corporation or Phillips Petroleum Company.

Health and Safety Aspects, Toxicology

In general, most thiols are considered moderately toxic compounds. However, there are two exceptions: 2,4,4-trimethyl-2-pentanethiol (*tert*-octyl mercaptan) and benzenethiol (thiophenol) are much more toxic than other thiols. Since thiols can be detected at very low levels, with the exception of the two highly toxic thiols mentioned above, they adequately warn of their presence long before dangerous concentrations occur. The threshold odor levels and toxicological properties of various thiols are listed in Table 6.

Uses

One of the main uses for aliphatic thiols is as polymerization modifiers in emulsion polymerization systems (see Emulsions; Polymerization mechanisms and processes). They function by controlling the length of the polymer chains to give a product with the desired properties. During World War II, the supply of natural rubber was severely

Table 6. Toxicological Properties of Various Thiols as Determined for Rats

Thiols	Formula	Odor threshold, ppm	LC_{50}[a] inhalation, ppm/4 h	LD_{50}[b] oral, mg/kg	LD_{50}[b] intraperitoneal, mg/kg	Ref.
methanethiol	CH_3SH	0.01	1,664			157–158
ethanethiol	C_2H_5SH	0.0003	4,970	1,034	450	158–159
1-propanethiol	n-C_3H_7SH	0.002	7,300	1,730	515	158–159
2-methyl-2-propanethiol	*tert*-C_4H_9SH	0.0005	24,300	4,729	590	159–161
1-butanethiol	n-C_4H_9SH	0.001	4,460	2,575	679	158–159
1-hexanethiol	n-$C_6H_{13}SH$		1,200	1,580	600	159
2,4,4-trimethyl-2-pentanethiol [*141-59-3*]	*tert*-$C_8H_{17}SH$		50	64		160
1-octanethiol	n-$C_8H_{17}SH$		953[c]	1,600		160
tert-dodecanethiol [*25103-58-6*]	*tert*-$C_{12}H_{25}SH$[d]			7,600		160
1-dodecanethiol	n-$C_{12}H_{25}SH$			7,000	4,225	162
benzenethiol	C_6H_5SH	0.0003	59	46.2	25.2	159, 163

[a] A concentration which is lethal to 50% of the animals tested.
[b] A dose which is lethal to 50% of the animals tested.
[c] Test air saturated with 1-octanethiol.
[d] Mixed C_{12} mercaptans obtained from propylene tetramer.

curtailed (see Rubber, natural). This led to the rapid development of processes to produce large amounts of synthetic rubber (see Rubber chemicals). 1-Dodecanethiol was the first modifier used; it was synthesized from relatively scarce lauryl alcohol, which is derived from coconut oil. This led to the development of a process for making *tert*-dodecyl mercaptan from petroleum-derived olefins (118,164). Currently, both straight-chain thiols in the production of chloroprene and tertiary thiols in the production of styrene-butadiene rubber are used as modifiers. The tertiary C_{12} thiol derived from propylene tetramer has the widest use as a polymerization modifier. However, several commercially available blends of tertiary thiols (C_9 and higher) are claimed to offer certain advantages in some styrene-butadiene rubber processes (165–166). Several types (C_9–C_{16}) and blends of tertiary aliphatic thiols are marketed for use in the manufacture of special-purpose elastomers.

Lower molecular weight alkanethiols, eg, methanethiol, ethanethiol, propanethiol, and 1-butanethiol, are used as intermediates in the manufacture of various agricultural chemicals, including insecticides, acaricides, herbicides (qv), and defoliants (see Insect control technology). In general, the products are organophosphorus compounds, ie, thioates, dithioates, and trithioates, and thiocarbamates (see Phosphorus compounds). An excellent reference describing the commercial processes used to produce various agricultural chemicals has been published (167). Methanethiol is used in the first step of the commercial synthesis of methionine, an amino acid poultry feed supplement (see Pet and livestock feeds). In this step, methanethiol is added to acrolein to give 3-(methylthio)propanal.

Lower molecular weight thiols are used as odorants in natural and liquefied petroleum gas (LPG). In case of leaks, they serve as warning agents so that the gas can be detected before it reaches hazardous concentrations, ie, they are intended to protect against explosions, fires, and other hazards. Ethanethiol in concentrations of 12–24 mg/L is the odorant of choice in LPG. For odorization of natural gas, blends of thiols, eg 2-methyl-2-propanethiol (*tert*-butyl mercaptan), 1-propanethiol (*n*-propyl mercaptan), 2-propanethiol (isopropyl mercaptan), and dimethyl sulfide are used. In the United States for residential and industrial usage, the two most common blends are an 80 wt % *tert*-butyl mercaptan and 20 wt % dimethyl sulfide mixture and a 75 wt % *tert*-butyl mercaptan and 25 wt % mixed propyl mercaptan mixture. The blends are generally added at 0.08–0.12 kg odorant/10^4 m^3 gas (0.5–0.75 lb odorant/10^6 ft^3 gas).

2-Mercaptobenzothiazole [*149-30-4*] and derivatives are an important class of rubber vulcanization accelerators. Mercaptobenzothiazole can also be used as an inhibitor in aqueous systems for preventing corrosion of copper and copper alloys (see Corrosion and corrosion inhibitors). Alkylated and chlorinated benzenethiols are used as peptizers in processing some rubbers and in rubber-reclaiming processes (see Recycling).

Polysulfide polymers with terminal thiol groups, prepared from dihalo compounds and sodium polysulfides, are used in sealant and adhesive applications (see Sealants; Polymers containing sulfur). Other polythiol products are used in epoxy-curing agents and in sealant formulations. These include products, eg, $ROCH_2CH(OH)CH_2SH$ obtained from the reaction of a polyalkylene glycol, epichlorohydrin, and sodium hydrosulfide and

$$R(O\overset{O}{\overset{\|}{C}}R'SH)_n$$

obtained from the reaction of mercaptoacetic acid or 3-mercaptopropionic acid [*107-96-0*] with polyols (168–169). The polymercaptan obtained by esterification of a polyalkylene glycol with a mixture of 3-mercaptopropionic acid and thiodipropionic acid is claimed to give a mercaptan-terminated polymer which can be formulated into a sealant with improved properties (170–171).

The use of alkanethiols and alkanethiol derivatives in the mining industry is a potential large-scale market for thiols. 1-Dodecanethiol, sodium alkyl trithiocarbonates, and dialkyl trithiocarbonates have been reported to be excellent collectors for sulfide minerals (106,172–173). The sodium alkyl trithiocarbonates are obtained from the reaction of sodium hydroxide and carbon disulfide with thiols. Reaction of these salts with an alkyl halide gives dialkyl trithiocarbonates.

$$NaOH + CS_2 + RSH \longrightarrow NaS\overset{S}{\overset{\|}{C}}SR + H_2O \qquad (57)$$

$$NaS\overset{S}{\overset{\|}{C}}SR + R'X \longrightarrow R'S\overset{S}{\overset{\|}{C}}SR + NaX \qquad (58)$$

Ammonium or alkanolamine salts of mercaptoacetic acid (thioglycolic acid) are used in permanent-wave formulations to soften hair (see Hair preparations). The reaction involved in the softening process is reductive cleavage of the disulfide linkage, ie, a thiol–disulfide interchange between the sulfhydryl group of mercaptoacetic acid and cystine units of the keratin. The zinc salts of 2-pyridinethiol-1-oxide [*1121-31-9*] are used as anti-dandruff agents in shampoo.

Mercaptoacid esters and alkanethiols are intermediates in the preparation of organotin sulfides used for polymer resin stabilization. Dibutyltin *S,S′*-(isooctylmercaptoacetate) [*25168-24-5*] has been described as an effective thermal stabilizer for poly(vinyl chloride) resins (see Heat stabilizers).

Medicinal uses for thiols include the following: *o*-mercaptobenzoic acid [*149-93-3*] for the preparation of ethylmercurithiosalicylate [*148-61-8*] (sodium salt), which is the sterilizing agent Merthiolate; ethanethiol for the preparation of 2,2-bis-(ethylsulfonyl)propane (sulfonal) and related hypnotics (see Hypnotics, sedatives, anticonvulsants). 2-Mercaptoethylamine [*107-96-0*] imparts some protection to animals against the effects of ionizing radiation, and it has been the model structure for a considerable volume of government-sponsored research into more effective and less toxic antiradiation drugs (see Radioprotective agents).

BIBLIOGRAPHY

"Mercaptans" in *ECT* 1st ed., Vol. 8, pp. 858–868, by E. E. Reid, Johns Hopkins University; "Thiols" in *ECT* 2nd ed., Vol. 20, pp. 205–218, by S. D. Turk, Phillips Petroleum Co.

1. W. C. Zeise, *Ann.* **11,** 1 (1834).
2. E. E. Reid, *Organic Chemistry of Bivalent Sulfur*, Chemical Publishing Co., Inc., New York, 1958–1962.
3. Ref. 2, Vol. 1, 1958.
4. E. Block, *Reactions of Organic Sulfur Compounds*, Academic Press, New York, 1978.
5. S. Oae, ed., *Organic Chemistry of Sulfur*, Plenum Press, New York, 1977.
6. C. J. M. Stirling, ed., *Organic Sulphur Chemistry*, Butterworths, Wodurn, Mass., 1975.
7. L. Young and G. A. Maw, *The Metabolism of Sulfur Compounds*, John Wiley & Sons, Inc., New York, 1958.
8. P. C. Jocelyn, *Biochemistry of the SH Groups*, Academic Press, New York (1972).
9. Ref. 8, pp. 3–4.
10. L. M. Libbey and E. A. Day, *J. Dairy Sci.* **46,** 859 (1963).
11. D. B. Emmons, J. A. Elliott, and D. C. Beckett, *J. Dairy Sci.* **49,** 1325 (1966).
12. S. Segall and B. Proctor, *Food Technol.* **13,** 679 (1959).
13. A. P. Ronald and W. A. Thomson, *J. Fish. Res. Board Can.* **21,** 1481 (1964).
14. S. Chen, L. Zieve and V. Mohadevan, *J. Lab. Clin. Med.* **75,** 628 (1970).
15. E. E. Beach and A. White, *J. Biol. Chem.* **127,** 87 (1939).
16. T. Janowski, J. Gawlik, and S. Zimnal, *VDI Ber.* (*Ver. Dtsch. Ing.*) **226,** 123 (1975).
17. G. P. Hinds, *Proc. Am. Petrol. Inst. Sect. VIII* **42,** 18 (1962).
18. *Proceedings of the Seventh World Petroleum Congress, Mexico City, 1967*, Vol. 9, Elsevier Publishing Co., Ltd., Essex, England, 1967.
19. R. A. Dean and E. V. Whitehead in ref. 18, pp. 165–175.
20. T. Y. Ho, M. D. Rogers, H. V. Drushel, and C. B. Koons, *Am. Assoc. Pet. Geol. Bull.* **58,** 2338 (1975).
21. H. J. Coleman, C. J. Thompson, R. L. Hopkins, and H. T. Rall, *J. Chem. Eng. Data* **10,** 80 (1965).
22. S. K. Katz and E. J. Talbert, *Intensity of Odors and Irritating Effects of Warning Agents for Inflammable and Poisonous Gases*, Technical Paper 480, U.S. Department of Commerce, Bureau of Mines, U.S. Government Printing Office, Washington, D.C., 1930, pp. 1–37.
23. J. C. Morris, W. J. Lanum, R. V. Helm, W. E. Haines, G. L. Cook, and J. S. Ball, *J. Chem. Eng. Data* **5,** 112 (1960).
24. W. E. Haines, R. V. Helm, C. W. Bailey, and J. S. Ball, *J. Phys. Chem.* **58,** 270 (1954).
25. *Ibid.*, **60,** 549 (1956).
26. R. C. Weast, ed., *Handbook of Chemistry and Physics*, 63rd ed., CRC Press, Boca Raton, Fla., 1982–1983.
27. U.S. Pat. 3,270,063 (Aug. 30, 1966), J. Faith and J. M. Kolyer (to Thompson Chemical Co.).
28. W. V. Smith, *J. Am. Chem. Soc.* **68,** 2064 (1946).
29. B. Flaschentrager and G. Wannschoff, *Chem. Ber.* **67,** 1121 (1934).
30. D. A. Shirley and J. R. Zeitz, *J. Org. Chem.* **18,** 1591 (1953).
31. J. E. Hesse, *J. Chem. Eng. Data* **12,** 266 (1967).

32. A. G. Osborn and D. R. Douslin, *J. Chem. Eng. Data* **11,** 502 (1966).
33. R. L. Denyer, F. A. Fidler, and R. A. Lowry, *Ind. Eng. Chem.* **41,** 2727 (1949).
34. D. W. Scott and J. P. McCullough, *The Chemical Thermodynamic Properties of Hydrocarbons and Related Substances*, Bulletin 595, U.S. Department of Interior, Bureau of Mines, U.S. Government Printing Office, Washington, D.C., 1961.
35. N. Kharasch, ed., *Organic Sulfur Compounds*, Pergamon Press, Inc., New York, 1961–1966.
36. J. P. McCullough, D. W. Scott, and G. Waddington in ref. 35, Vol. 1, 1961, Chapt. 3.
37. D. W. Scott and G. A. Crowder, *J. Chem. Phys.* **46,** 1054 (1967).
38. H. L. Finke, J. P. McCullough, J. F. Messerly, S. B. Guthrie, and D. R. Douslin, *J. Chem. Thermodyn.* **2,** 27 (1970).
39. M. Zaki El-Sabban and D. W. Scott, *The Chemical Thermodynamics of Hydrocarbons and Related Substances*, Bulletin 654, U.S. Dept. of Interior, Bureau of Mines, U.S. Government Printing Office, Washington, D.C., 1970.
40. W. G. Good and B. L. DePrater, *J. Phys. Chem.* **70,** 3606 (1966).
41. J. F. Messerly, S. S. Todd, and G. B. Guthrie, Jr., *J. Chem. Eng. Data* **12,** 426 (1967).
42. W. G. Good, *J. Chem. Eng. Data* **17,** 158 (1972).
43. J. F. Messerly, H. L. Finke, and S. S. Todd, *J. Chem. Thermodyn.* **6,** 635 (1974).
44. C. A. Frenzel, D. W. Scott, and J. P. McCullough, *Infrared Spectra of Organic Compounds in the Region of 15–35 Microns*, Report of Investigation 5658, U.S. Dept. of Interior, Bureau of Mines, Washington, D.C., 1960.
45. G. F. Bolshakov, *Izu. Vyssh. Uchebn. Zaved. Neft Gaz* **23**(1), 39 (1980).
46. D. W. Scott and M. Zaki El-Sabban, *J. Mol. Spectrosc.* **30,** 317 (1969).
47. T. Medda, S. Mukherjee, and S. Palit, *J. Indian Chem. Soc.* **53**(11), 1113 (1976).
48. E. J. Levy and W. A. Stahl, *Anal. Chem.* **33,** 707 (1961).
49. S. A. Marcus and S. I. Miller, *J. Phys. Chem.* **68,** 331 (1964).
50. W. P. Hall and E. E. Reid, *J. Am. Chem. Soc.* **65,** 1466 (1943).
51. M. Mansson and S. Sunner, *Acta Chem. Scand.* **16,** 1863 (1962).
52. T. L. Cairnes, G. L. Evans, A. W. Larcher, and B. C. McKusiek, *J. Am. Chem. Soc.* **74,** 3982 (1952).
53. B. Magnuson, *Acta Chem. Scand.* **17,** 273 (1963).
54. R. Adams and A. Ferretti, *J. Am. Chem. Soc.* **81,** 4939 (1959).
55. A. Albert and G. B. Barlin, *J. Chem. Soc.*, 2394 (1959).
56. W. H. Houff and R. D. Schrietz, *J. Am. Chem. Soc.* **75,** 6316 (1953).
57. A. R. Katritzky and J. M. Lagowski in A. R. Katritzky, ed., *Advances in Heterocyclic Chemistry*, Vol. 1, Academic Press, Inc., New York, 1963, p. 396.
58. J. Xan, E. A. Wilson, L. D. Roberts, and N. H. Horton, *J. Am. Chem. Soc.* **63,** 1130 (1941).
59. J. D. Hopton, C. J. Swan, D. L. Trimm, *Adv. Chem. Ser.*, 216 (1968).
60. A. A. Oswald and T. J. Wallace in ref. 35, Vol. 2, 1966, Chapt. 8.
61. U.S. Pat. 4,127,474 (Nov. 28, 1978), R. R. Frame (to UOP Inc.).
62. U.S. Pat. 4,159,964 (July 3, 1979), R. R. Frame (to UOP Inc.).
63. J. H. Schutten and J. Zwart, *J. Mol. Catal.* **5**(2), 109 (1979).
64. Y. M. Afanas' ev, A. V. Sokolina, and V. A. Borodalvya, *Zh. Prikl. Khim. (Leningrad)* **52,** 405 (1979).
65. T. J. Wallace, *J. Org. Chem.* **31,** 1217 (1966).
66. G. A. Olah, M. Arvanaghi, and D. Yoshwant, *Synthesis*, 721 (1979).
67. C. N. Yiannios and J. V. Karabinos, *J. Org. Chem.* **28,** 3246 (1963).
68. T. J. Wallace, *J. Am. Chem. Soc.* **86,** 2018 (1964).
69. N. Kharasch, S. J. Potempa, and H. L. Wehrmeister, *Chem. Rev.* **39,** 269 (1946).
70. I. B. Douglas in ref. 35, Vol. 1, 1961, Chapt. 30.
71. W. A. Savige and J. A. Maclaren in ref. 31, Vol. 2, pp. 367–371.
72. P. Allen and J. W. Brook, *J. Org. Chem.* **27,** 1019 (1962).
73. F. Ashworth and G. N. Burkhardt, *J. Chem. Soc.*, 1791 (1928).
74. S. O. Jones and E. E. Reid, *J. Am. Chem. Soc.* **60,** 2542 (1938).
75. F. R. Mayo and C. Walling, *Chem. Rev.* **27,** 351 (1940).
76. C. Walling, *Free Radicals in Solution*, John Wiley & Sons, Inc., New York, 1957, Chapt. 7.
77. W. A. Pryor, *Mechanisms of Sulfur Reactions*, McGraw-Hill Book Co., Inc., New York, 1962, Chapt. 3.
78. F. W. Stacey and J. F. Harris in A. C. Cope and co-workers, eds., *Organic Reactions*, Vol. 13, John Wiley & Sons, Inc., New York, 1963, Chapt. 4.

79. G. Sosnovsky, *Free Radical Reactions in Preparative Organic Chemistry*, The Macmillan Co., New York, 1964, Chapt. 3.
80. N. S. Kharasch, W. Nudenberg, and G. J. Mantell, *J. Org. Chem.* **16,** 524 (1951).
81. J. Longfield, R. Jones, and C. Sivertz, *Can. J. Res.* **28B,** 373 (1950).
82. A. A. Oswald and K. Griesbaum in ref. 31, Vol. 2, 1966, Chapt. 9.
83. T. L. Jacobs and G. E. Illingworth, *J. Org. Chem.* **28,** 2692 (1963).
84. E. P. Kohler and H. Potter, *J. Am. Chem. Soc.* **57,** 1316 (1935).
85. A. T. Bloomquist and J. Wolinsky, *J. Org. Chem.* **23,** 551 (1958).
86. W. E. Truce and J. A. Simms, *J. Am. Chem. Soc.* **78,** 2756 (1956).
87. W. E. Truce in ref. 31, Vol. 1, 1961, Chapt. 12.
88. S. A. Miller, *Acetylene*, Vol. 2, Academic Press, Inc., New York, 1966, Chapt. 6.
89. J. J. Nedwick, *Ind. Eng. Chem. Process Des. Dev.* **1,** 137 (1962).
90. R. L. Frank, S. S. Drake, P. V. Smith, Jr., and C. Stevens, *J. Polymer Sci.* **3,** 50 (1948).
91. S. Tanimoto, S. Iwata, T. Imanishi, T. Sugimoto, and M. Okano, *Bull. Inst. Chem. Res. Kyoto Univ.* **57**(3), 235 (1979).
92. Ref. 2, Vol. 3, 1960, Chapt. 5.
93. E. Campaigne in ref. 35, Vol. 1, 1961, Chapt. 14.
94. H. Bohme, H. Fischer, and R. Frank, *Ann.* **563,** 54 (1949).
95. T. Posner, *Ann.* **336,** 85 (1904).
96. J. M. Snell and A. Weissberger, *J. Am. Chem. Soc.* **61,** 450 (1939).
97. R. Lombard and W. Bolchert, *Compt. Rend.* **257,** 1313 (1963).
98. H. S. Wilgus, E. Fraunglass, E. T. Jones, R. F. Porter, and J. W. Gates, *J. Org. Chem.* **29,** 594 (1964).
99. O. R. Quayle and E. E. Royals, *J. Am. Chem. Soc.* **64,** 226 (1942).
100. J. F. Bunnett and G. T. Davis, *J. Am. Chem. Soc.* **76,** 3011 (1954).
101. B. Dmuchovsky, F. B. Zienty, and W. A. Vredenburgh, *J. Org. Chem.* **31,** 865 (1966).
102. J. F. Bunnett and E. Baciocchi, *J. Org. Chem.* **32,** 11 (1967).
103. J. R. Campbell, *J. Org. Chem.* **29,** 1830 (1964).
104. A. Rosowsky in A. Weissberger, ed., *The Chemistry of Heterocyclic Compounds*, Interscience Publishers, a division of John Wiley & Sons, Inc., New York, 1964, Vol. 19, pp. 327–340.
105. G. D. Hurd and L. L. Gershbein, *J. Am. Chem. Soc.* **69,** 2328 (1947).
106. W. A. Schulze, G. H. Short, and W. W. Crouch, *Ind. Eng. Chem.* **42,** 916 (1950).
107. E. Dyer, J. F. Glenn, and E. G. Lendrat, *J. Org. Chem.* **26,** 2919 (1961).
108. J. P. Smith and E. C. Friedrich, *J. Am. Chem. Soc.* **81,** 161 (1959).
109. B. D. Vineyard, *J. Org. Chem.* **32,** 3833 (1967).
110. U.S. Pat. 3,392,201 (July 9, 1968), P. F. Warner (to Phillips Petroleum Co.).
111. Ref. 2, Vol. 4, 1962.
112. U.S. Pat. 2,983,747 (May 9, 1961), R. G. Campbell and G. E. Klingman (to Stauffer Chemical Co.).
113. A. Lippert and E. E. Reid, *J. Am. Chem. Soc.* **60,** 2370 (1938).
114. U.S. Pat. 2,682,554 (June 29, 1954), W. W. Crouch and R. T. Workman (to Phillips Petroleum Co.).
115. U.S. Pat. 2,886,593 (May 12, 1959), R. P. Louthan and C. W. Kruse (to Phillips Petroleum Co.).
116. I. B. Douglas and D. R. Poole, *J. Org. Chem.* **22,** 536 (1957).
117. U.S. Pat. 3,253,028 (May 24, 1966), R. P. Louthan (to Phillips Petroleum Co.).
118. W. A. Schulze, J. P. Lyon, and G. H. Short, *Ind. Eng. Chem.* **40,** 2308 (1948).
119. U.S. Pat. 2,610,981 (Sept. 16, 1952), G. H. Short (to Phillips Petroleum Co.).
120. U.S. Pat. 3,083,231 (March 25, 1963), C. A. Ray (to Phillips Petroleum Co.).
121. J. F. Frantz and K. I. Glass, *Chem. Eng. Prog.* **59,** 68 (1963).
122. P. Bapseres, *Chim. Ind. (Paris)* **90,** 358 (1963).
123. U.S. Pat. 3,050,452 (Aug. 21, 1962), R. P. Louthan (to Phillips Petroleum Co.).
124. U.S. Pat. 4,140,604 (Feb. 20, 1979), D. A. Dimmig (to Pennwalt Corp.).
125. U.S. Pat. 2,411,983 (Dec. 3, 1946), W. E. Vaughan and F. F. Rust (to Shell Development Co.).
126. Eur. Pat. Appl. 5,400 (Nov. 14, 1979), J. Ollivier and G. Souloumiac (to Societe Nationate Elf Aquiltaine, S.A.).
127. U.S. Pat. 2,551,813 (May 8, 1951), P. S. Pinkney (to E. I. du Pont de Nemours & Co., Inc.).
128. P. Sabatier and A. Mailhe, *Compt. Rend.* **150,** 1217 (1910).
129. R. C. Kramer and E. E. Reid, *J. Am. Chem. Soc.* **43,** 880 (1921).
130. H. O. Folkins and E. L. Miller, *Ind. Eng. Chem. Prod. Res. Dev.* **1,** 271 (1962).
131. H. O. Folkins and E. L. Miller, *Proc. Am. Petrol. Inst. Sect. III* **42,** 188 (1962).

132. U.S. Pat. 3,035,097 (May 15, 1962), T. E. Deger, B. Buchholz, and R. H. Goshorn (to Pennwalt Corp.).
133. U.S. Pat. 3,994,980 (Nov. 30, 1976), D. H. Kubicek (to Phillips Petroleum Co.).
134. U.S. Pat. 4,059,636 (May 4, 1976), D. H. Kubicek (to Phillips Petroleum Co.).
135. U.S. Pat. 4,005,149 (Jan. 25, 1977), D. H. Kubicek (to Phillips Petroleum Co.).
136. U.S. Pat. 2,947,788 (Aug. 2, 1960), H. Pitt (to Stauffer Chemical Co.).
137. U.S. Pat. 3,799,989 (March 26, 1947), F. T. Sherk and D. H. Kubicek (to Phillips Petroleum Co.).
138. U.S. Pat. 4,088,698 (May 3, 1977), N. A. Fishel and D. E. Gross (to Monsanto Co.).
139. G. G. Urquart, J. W. Gates, and R. Connor in E. C. Horning, ed., *Organic Synthesis*, Coll. Vol. 3, John Wiley & Sons, Inc., New York, 1955, pp. 363–365.
140. R. L. Frank and P. V. Smith, *J. Am. Chem. Soc.* **68,** 2103 (1946).
141. D. F. Lee, B. Smith, and B. R. Trego, *Chem. Ind. (London)*, 868 (1960).
142. B. C. Cossar, J. O. Fournier, D. L. Fields, and D. D. Reynolds, *J. Org. Chem.* **27,** 93 (1962).
143. R. Brown, W. E. Jones, and A. R. Pinder, *J. Chem. Soc.*, 2123 (1951).
144. J. S. Showell, J. R. Russell, and D. Swern, *J. Org. Chem.* **27,** 2853 (1962).
145. U.S. Pat. 3,086,049 (April 16, 1963), J. J. Godfrey (to California Research Corp.).
146. L. N. Owen in ref. 35, Vol. I, 1961, Chapt. 19.
147. C. G. Overberger and A. Drucker, *J. Org. Chem.* **29,** 360 (1964).
148. D. J. Martin and C. C. Greco, *J. Org. Chem.* **33,** 1275 (1968).
149. C. Djerassi and B. Tursch, *J. Org. Chem.* **27,** 1041 (1962).
150. R. Adams and C. S. Marvel in H. Gilman and A. H. Blatt, eds., *Organic Synthesès*, Coll. Vol. 1, John Wiley & Sons, Inc., New York, 1932, p. 504.
151. D. S. Tarbell and D. K. Fukushima in ref. 139, p. 809.
152. H. C. Godt and R. E. Wann, *J. Org. Chem.* **26,** 4047 (1961).
153. A. W. Wagner, *Chem. Ber.* **99,** 375 (1966).
154. E. E. Campaigne and S. W. Osborn, *J. Org. Chem.* **22,** 561 (1957).
155. M. S. Newman and H. A. Karnes, *J. Org. Chem.* **31,** 3980 (1966).
156. *Chem. Age*, 12 (Feb. 1, 1980).
157. M. Horiguchi, *J. Oska City Med. Center* **9,** 5257 (1960).
158. C. K. Katz, *Technical Paper 480*, U.S. Department of Commerce, Bureau of Mines, U.S. Government Printing Office, Washington, D.C., 1930, pp. 1–37.
159. E. J. Fairchild and H. F. Stokinger, *Am. Ind. Hyg. Assoc.* **19,** 171 (1958).
160. Unpublished data, Phillips Petroleum Co., 1981.
161. A. D. Nevers and W. H. Oister, *A.G.A. Operating Section Conference Paper 65-P-126*, American Gas Association, New York, 1965.
162. M. S. Gizhlaryan, *Biol. Zh. Arm.* **19**(7), 41 (1966).
163. E. A. Blinova, *Hyg. Sanit. (USSR)* **30,** 18 (1965a).
164. C. F. Fryling in G. S. Whitby and co-workers, eds., *Synthetic Rubber*, John Wiley & Sons, Inc., New York, 1954, Chapt. 8.
165. U.S. Pat. 2,549,961 (April 25, 1951), W. W. Crouch and E. G. Marhofer (to Phillips Petroleum Co.).
166. U.S. Pat. 3,188,304 (June 8, 1965), R. H. Goshorn, H. E. Albert, and B. Buchholz (to Pennwalt Corp.).
167. M. Sittig, *Pesticides Process Encyclopedia*, Noyes Data Corp., Park Ridge, N.Y., 1977.
168. U.S. Pat. 3,258,495 (June 28, 1966), G. M. LaFave, F. Y. Hayashi, and A. W. Fradkin (to Diamond Alkali Co.).
169. U.S. Pat. 3,352,810 (Nov. 14, 1967), G. M. Cameron and A. J. Duke (to Ciba, Ltd.).
170. U.S. Pat. 3,549,565 (Dec. 22, 1970), F. B. Jones and B. D. Simpson (to Phillips Petroleum Co.).
171. C. M. Lamb and R. P. Williams, *Adhes. Age* **22**(11), 44 (1979).
172. R. M. Parlman and C. R. Bresson, *New Collectors for Sulfide Minerals, Especially Copper and Molybdenum, Based on Sulfur Chemicals Derived From Petroleum*, Paper No. 80-394, 1980 SME-AIME Fall Meeting, Minneapolis, Minn., 1980.
173. U.S. Pat. 4,211,644 (July 8, 1980), A. Wiechers (to Pennwalt Corp.).

JOHN NORELL
R. P. LOUTHAN
Phillips Petroleum Co.

THIOPHENE AND THIOPHENE DERIVATIVES

Thiophene [*110-02-1*], C_4H_4S, a 5-membered heterocyclic compound (1), is a colorless liquid with an odor similar to benzene, bp 84°C, immiscible with water, and soluble in most organic solvents.

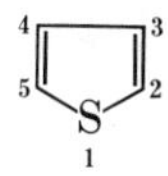

(1)

In general, it undergoes substitution reactions more readily than benzene, almost exclusively in the 2-position. It is highly flammable and moderately toxic and should be handled with proper safety precautions in a well-ventilated area free from sources of sparks and flames. It is available in commercial quantities from Pennwalt Corporation in the United States and from Croda Synthetic Chemicals, Ltd. and Pennwalt-Holland B.V. in Europe. It is used primarily as an intermediate to pharmaceuticals, dyestuffs, and agricultural chemicals.

Thiophene was discovered in 1882 by Victor Meyer when a sample of benzene failed to give the characteristic blue indophenine color test during a laboratory demonstration. Pursuing this anomaly, Meyer found that his assistant had provided him with high purity benzene prepared by pyrolysis of calcium benzoate, instead of the conventional benzene derived from coal tar. Subsequently, Meyer isolated thiophene from coal-tar-derived benzene, showed it to be the component responsible for the indophenine color test, established its structure, and synthesized it by ring closure of succinic acid with phosphorus pentasulfide (1–2).

Thiophene and its derivatives have been found in carbonaceous deposits of lignite, peat, shale, coal, and certain crude oils, but to date its extraction from these sources has not proven commercially feasible.

In 1947 the Socony-Vacuum Company (now Mobil Oil Corporation) started up a semicommercial plant in the United States capable of producing 450 kg of thiophene per day by a continuous, noncatalytic, vapor-phase reaction of butane and sulfur. As a result of its availability and its similarity to benzene, research interest in thiophene was stimulated, particularly by pharmaceutical companies (3–11).

Substituted thiophenes are named similarly to substituted benzenes, using the numbering system shown in compound (1). The important radicals are shown in structures (2), (3), and (4).

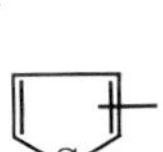

(2) thienyl

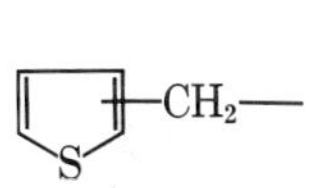

(3) thenyl

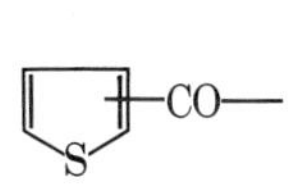

(4) thenoyl

For the hydrogenated thiophenes, *Chemical Abstracts* uses the nomenclature dihydro-

and tetrahydrothiophene, but the names thiolene (**5**) and (**6**) and thiolane (**7**) appear frequently in the literature. The corresponding sulfones, or 1,1-dioxides, are frequently referred to as the sulfolenes (**8**) and sulfolanes (**9**) (see also Sulfur compounds).

(**5**)
2,3-dihydrothiophene
[*1120-59-8*]
(2-thiolene)

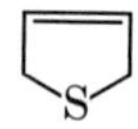

(**6**)
2,5-dihydrothiophene
[*1708-32-3*]
(3-thiolene)

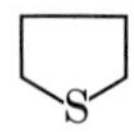

(**7**)
tetrahydrothiophene
[*110-01-0*]
(thiolane)

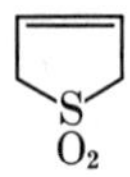

(**8**)
2,5-dihydrothiophene 1,1-dioxide
[*77-79-2*]
(3-sulfolene)

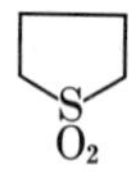

(**9**)
tetrahydrothiophene 1,1-dioxide
[*126-33-0*]
(sulfolane)

The designation thiophene is preferred by *Chemical Abstracts*, as illustrated by compounds (**10**), (**11**), and (**12**).

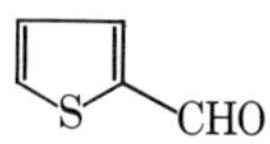

(**10**)
2-thiophenecarboxaldehyde
[*98-03-3*]

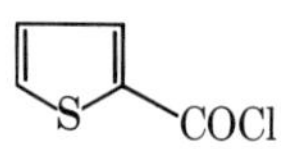

(**11**)
2-thiophenecarbonyl chloride
[*5271-67-0*]

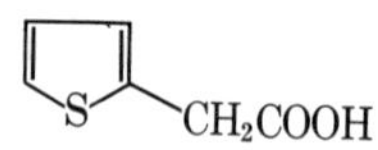

(**12**)
2-thiopheneacetic acid
[*1918-77-0*]

Properties

Physical properties are given in Table 1. For nmr, uv, ir, and mass spectra, see refs. 3 and 24.

Thiophene is aromatic and planar; its degree of aromaticity is between that of furan and benzene (see also Furan derivatives). It undergoes the usual substitution reactions associated with aromatic character and is resistant to addition reactions. Based on bond lengths, resonance energies, and dipole moments, the resonance structures (**13**) were proposed (25) with the classical structure (**13a**) contributing ca 70 wt % to the hybrid, and structures, (**13f**), (**13g**), (**13h**), (**13i**), and (**13j**) involving

Table 1. Physical Properties of Thiophene

Property	Value	Ref.	Property	Value	Ref.
freezing point, °C	−38.3	12	coefficient of cubical expansion, per °C	0.00112	19
boiling point at 101.3 kPa[a], °C	84.16	13	*Critical constants*		
flash point, °C	−1.11	14	temp, °C	306.2	20
density, g/cm^3			pressure, MPa[b]	5.70	20
d_4^0	1.0873	15	volume, mL/mol	219	20
d_4^{25}	1.0573	15	density, g/cm^3	0.385	20
d_4^{50}	1.0285	15	heat of formation at 298.16 K, kJ/mol[c]		
refractive index, n_D^{25}	1.52572	16			
viscosity, abs at 25°C, mPa·s (= cP)	0.621	12, 17	liquid	81.67	13
			gaseous	116.4	13
surface tension at 20°C, mN/m (dyn/cm)	31.34	18	heat of combustion[d] at 101.3 kPa[a], 25°C, kJ/mol[c]	−2791.5	13
vapor pressure, kPa[a]		12–13	heat of fusion at −38.3°C, J/mol[c]	5.31	17
temp., °C					
0	2.86		heat of vaporization at bp, kJ/mol[c]	31.47	13
20.0	8.36				
50.1	31.16		dielectric constant at 20°C	2.7	21–22
71.0	66.66		dipole moment at 25°C in benzene, C·m[e]	1.73×10^{-30}	23
84.16	101.3				
89.7	120		parachor, 20–30°C (surface tension = N/m)[f]	188.8	16
95.9	143.3				
119.8	270.1				

[a] To convert kPa to mm Hg, multiply by 7.5.
[b] To convert MPa to atm, divide by 0.101.
[c] To convert J to cal, divide by 4.184.
[d] C_4H_4S (l) + 13/2 O_2 (g) + 0.7 H_2O (l) → 4 CO_2 (g) + [H_2SO_4.1.7H_2O] (l).
[e] To convert C·m to D, multiply by 3×10^{29}.
[f] Parachor = mol wt × $\gamma^{1/4}$/D − d, where γ = surface tension N/m, and D and d are densities g/cm^3. To convert N/m to dyn/cm, multiply by 1000; to convert parachor (using N/m) to parachor (using dyn/cm), multiply by 5.623.

a 10-electron expansion of the sulfur valence shell by utilization of its available 3*d*-orbitals (26).

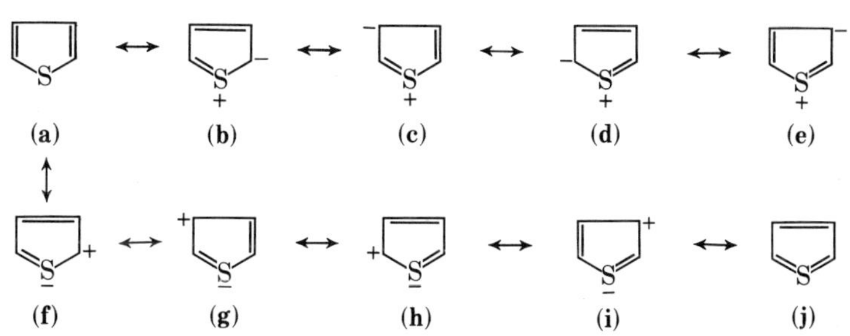

(13)

Reactions

Oxidation. The thiophene ring is readily ruptured by strong oxidizing agents, eg, nitric acid, to give maleic acid. 2,5-Thiophenedione [*6007-87-0*] (**14**) is obtained by passing thiophene and oxygen over a vanadia-phosphorus oxide catalyst at elevated temperatures (27).

(**14**)

Peroxides and peracids attack at the sulfur atom, but the sulfoxide and sulfone of the parent thiophene are unstable and readily dimerize. The dimer (**15**) has been isolated (28–29).

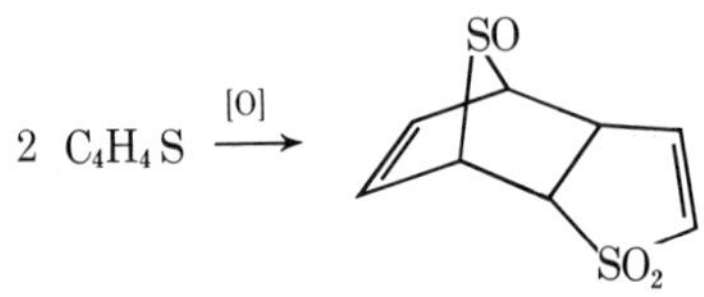

(**15**)

Thiophenes containing two or more halogen or aryl substituents, as well as the dihydro- and tetrahydrothiophenes, form stable 1,1-dioxides (30–31). A chlorinated thiophene 1,1-dioxide is marketed by Diamond-Shamrock Corporation as a slimicide to control bacteria in papermaking and other industrial operations (32). 2,5-Dihydrothiophene 1,1-dioxide (3-sulfolene) (**8**) is produced commercially from butadiene and sulfur dioxide. Hydrogenation yields tetrahydrothiophene 1,1-dioxide (sulfolane) (**9**). Sulfolene and sulfolane, used primarily as specialty solvents in petroleum refining, are available from Phillips Petroleum Company (see Sulfolane and sulfones).

Reduction and Hydrodesulfurization. Thiophene is reduced to the 2,3- and 2,5-dihydrothiophenes (**5**), (**6**) and to tetrahydrothiophene (**7**) by sodium in alcohol or ammonia, or with palladium, cobalt, molybdenum, or rhenium catalysts under carefully controlled conditions. Tetrahydrothiophene (thiolane), an intensely odorous material, is marketed by Pennwalt Corporation as a gas odorant.

Under more rigorous hydrogenation conditions, the sulfur atom is removed, resulting in hydrogen sulfide and butane. This reaction is called hydrodesulfurization (HDS) and is utilized in petroleum refining to remove thiophene and other organosulfur contaminants from crude oil. A wide variety of commercial HDS catalysts is available for this purpose.

Halogenation. Thiophene is readily halogenated in the absence or presence of light, at low temperatures, and with or without catalysts. Substitution occurs first in the 2-, then in the 2,5-positions. At this point, halogen addition becomes a competing reaction, and subsequent dehydrohalogenation by refluxing with alkali is usually required to isolate the 2,3,5- [*17249-77-3*] and 2,3,4,5-substituted [*6012-97-1*] products. The final chlorination product is 2,2,3,3,4,4,5,5-octachlorotetrahydrothiophene [*18043-32-8*].

Alkylation and Acylation. The alkylation of thiophene suffers from poor selectivity. Treatment of propylene or isobutylene with thiophene in the presence of sulfuric acid, for example, gives about equal amounts of the 2- and 3-alkylthiophenes. Another drawback is a tendency for polymerization. Alkyl-substituted thiophenes are best prepared by reduction of the appropriate acylated derivatives.

2-Thiophenecarboxaldehyde [*98-03-3*] (**10**) is prepared from thiophene and DMF (33) or *N*-methylformanilide (34) in the presence of phosphorus oxychloride as a catalyst. Acylation of thiophene with an acid anhydride and phosphoric acid, or an acid chloride and a Lewis acid such as aluminum chloride, stannic chloride, or boron trifluoride gives 2-thienyl ketones.

Sulfonation. 2-Thiophenesulfonic acid [*79-84-5*] is readily produced by treating thiophene with 95% sulfuric acid at ambient temperature. Milder reagents such as sulfur trioxide, pyridine-SO_3, or dioxane-SO_3 are often preferred. Higher temperatures give disulfonic acids, predominantly the 2,5-isomer. Chlorosulfonic acid readily gives 2-thiophenesulfonyl chloride [*16629-19-9*] (**16**). Reaction of the latter with ammonia or amines leads to 2-thiophenesulfonamides. 2-Thiophenesulfonyl chloride undergoes Friedel-Crafts reactions with aromatic compounds to form 2-thienyl sulfones, eg, 2-thienyl phenyl sulfone [*22407-40-5*] (**17**).

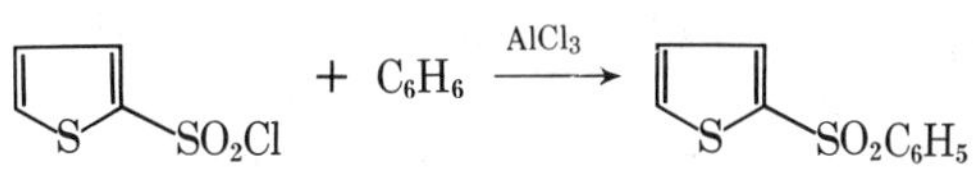

(**16**) (**17**)

Nitration. The conventional mixed-acid procedures used to nitrate benzene are too harsh for thiophene and cause decomposition. In the recommended procedure for the synthesis of 2-nitrothiophene [*609-40-5*], a solution of thiophene in acetic anhydride is slowly added to a solution of nitric acid in glacial acetic acid at 10°C (35). Dinitration gives mixtures of the 2,4- [*5347-12-6*] and 2,5-[*59434-05-8*] isomers. Although the nitro derivatives can be reduced, the free amines are unstable and are best isolated as the stannic chloride double salts. The double salts, but not the free amines, can be diazotized in the usual manner, and the diazonium salts can be coupled with phenols to form dyes (see Dyes and dye intermediates).

Metalation. 2-Thienyllithium [*2786-07-4*] is prepared from thiophene and butyllithium in ether (36), whereas 2-thienylsodium [*83682-19-3*] is readily formed by transmetalation of thiophene with sodium amalgam and an alkyl or aryl halide in refluxing ether (37). The sodium and lithium derivatives and carbon dioxide give 2-thiophenecarboxylic acid; with dimethylformamide, 2-thiophenecarboxaldehyde (**10**) is obtained. Treatment of thiophene with a dilute alcoholic solution of mercuric chloride yields chloro-2-thienylmercury [*5857-39-6*], C_4H_3SHgCl, and the dichloro-2,5-thiophendiyldimercury [*43030-22-4*], $C_4H_2SHg_2Cl_2$.

Polymerization. Thiophene is readily polymerized by acid catalysts, eg, sulfuric acid, hydrogen fluoride, ferric chloride, stannic chloride, aluminum chloride, silica-alumina, or orthophosphoric acid. Self-curing polymers containing up to ca 120 thiophene units (mol wt 10,000) have been prepared with methanesulfonic acid as the catalyst (38). Thiophene condenses with formaldehyde under mildly acidic conditions to form thermoplastic and thermosetting resins (qv). Thiophene–phenol condensation products with formaldehyde can be formed at alkaline pH in the presence of hexamethylenetetramine.

Other Reactions. Thiophene reacts with aqueous formaldehyde and concentrated hydrochloric acid to give 2-(chloromethyl)thiophene [*765-50-4*], a somewhat unstable lacrimator, but nonetheless a useful drug intermediate. 2-Vinylthiophene [*1918-82-7*] is prepared from thiophene and acetaldehyde in concentrated hydrochloric acid, followed by dehydrohalogenating of the chloroethyl intermediate with pyridine (39). 2-Vinylthiophene polymerizes at room temperature in the presence of benzoyl peroxide (40).

Manufacture

In the preferred laboratory synthesis of thiophene, powdered, anhydrous sodium succinate is heated with phosphorus trisulfide to high (gas flame) temperatures under a stream of carbon dioxide. The crude thiophene from the effluent gas stream is condensed, steam distilled, and dried; the yield is 25–30% (41). Alternatively, succinic anhydride is treated with phosphorus trisulfide or pentasulfide.

The Socony-Vacuum (Mobil Oil Corporation) process consisted of a continuous, high temperature (560°C), noncatalytic reaction of butane and sulfur (3,42). It gave a ca 40% yield of thiophene, based on butane, and about an equal amount of highly odorous thiophene tars from which 3-thiophenethiol [*7774-73-4*] has been isolated. The Socony-Vacuum process was licensed and operated by the Pennwalt Corporation in the 1950s, until it was abandoned in the early 1960s because of low yields, corrosion problems, and the difficulty of disposing of the thiophene tars as thiophene production was increased.

Newer processes for thiophene and derivatives are continuous, vapor-phase processes that utilize a variety of C_4 raw materials and sulfur sources, with special metal oxide catalysts to enhance reaction rates, increase yields, and avoid the formation of thiophene tars. In one such process, the vapor-phase reaction of furan and hydrogen sulfide over a heteropolyacid-promoted metal oxide catalyst produces thiophene and water at relatively low temperature (300–400°C) (43). The thiophene and water are separated, and the crude thiophene is continuously distilled; yields are high. The catalyst has a long service life and need not be periodically regenerated. The process can also be used to produce 2-methylthiophene [*554-14-3*] from 2-methylfuran and tetrahydrothiophene from tetrahydrofuran (see Furan and furan derivatives). In another process, C_4 raw materials such as 1-butene, butadiene, *n*-butyl alcohol, or crotonaldehyde react continuously with carbon disulfide over an alkali-promoted metal oxide catalyst at high temperatures (500°C) (44). The thiophene yields vary according to the raw material. Carbon disulfide serves as the sulfur source and is apparently converted to methane or carbon dioxide. Periodic regeneration of the catalyst is required to remove coke from the catalyst surface. This process can also be used to produce 2- and 3-alkyl substituted thiophenes for which the appropriate higher carbon raw materials are available. In a third process, a continuous, high temperature (500–600°C) reaction of butane and sulfur takes place over a mixed metal oxide catalyst which serves to dehydrogenate the butane (45). The hydrogen formed reacts rapidly with the sulfur feed to produce hydrogen sulfide in a highly exothermic reaction. The heat generated by this reaction obviates the need for further external heating and thereby sharply reduces corrosion to the reactor. The hydrogen sulfide reacts with the butane residue to form thiophene in good yields. Periodic air regeneration of the catalyst is required. The process overcomes the main problems associated with the original Socony-Vacuum process, namely, low yields, corrosion, and thiophene tar formation. It can also produce 2- and 3-alkyl-substituted thiophenes from higher hydrocarbons.

Shipping; Specifications

Thiophene is available in tank cars, tank trucks, and 208-L drums from Pennwalt Corporation, and in 180–220 kg drums from Croda Synthetic Chemicals, Ltd. The product specifications for the Pennwalt material are given in Table 2 (19). Specifications for the Croda material are similar, with a 0.3% wt/wt maximum for carbon disulfide.

Thiophene is classified as a flammable liquid and must be shipped with a red label. The U.S. selling price (Sept. 1982) is $6.39/kg.

Health and Safety Factors

Based on the toxicological data available to date, thiophene is considered to be moderately toxic (46). An LC_{50} inhalation value of 9.5 mg/L for mice exposed for a period of two hours and an oral LD_{50} value of 420 mg/kg for mice and 1400 mg/kg for rats have been reported (47). The same report indicates that a daily 6-h exposure to 1.01–1.12 mg/L of thiophene for 144 d caused hyperfunction of the thyroid gland and a decrease in liver detoxification function in rats, rabbits, and mice, and that oral administration of 200 mg/kg to rats produced degenerative changes of unspecified severity in the liver, kidneys, and heart. A considerably higher LC_{Lo} value of 29.9 mg/L has also been reported for inhalation tests on mice (48). In tests carried out by an independent laboratory for Pennwalt Corporation, thiophene gave an LC_{50} value of 9.1 mg/L on mice after one hour of exposure by inhalation (49). An LD_{Lo} value of 830 mg/kg is reported for thiophene administered subcutaneously to rabbits (14,46), and an LD_{Lo} value of 100 mg/kg is reported when it was administered intraperitoneally

Table 2. Specifications and Properties of Commercial Thiophene

	Value or characteristic
Specification	
color	APHA 10 max, water-white
appearance	no turbidity or suspended matter
Engler distillation[a] at 101.3 kPa[b]	
initial bp, min °C	83.0
95% distilled, max °C	85.0
purity[c], wt % min	98.8
Property	
sp gr, 15.6/15.6, °C	1.070
av density, g/cm^3	1.07
net wt, 208-L drum, kg	217.7
flash point, COC[d], °C	−6.6
coefficient of cubical expansion, per °C	112×10^{-5}

[a] ASTM D 1078.
[b] To convert kPa to mm Hg, multiply by 7.5.
[c] By gas chromatography.
[d] Cleveland open cup.

to mice (14). Skin-absorption tests on rabbits at a dose of 1.0 g/kg resulted in no signs of skin irritation or systemic toxicity (49). In *in vitro* mutagenicity assays in *Salmonella typhimurium* (Ames test) (50) and in *Klebsiella pneumoniae* (51), no mutagenic activity was observed.

Uses

The world market for thiophene has grown slowly over the past decade as certain older applications were phased out and new applications appeared. The market is estimated in 1982 at ca 1000–1400 metric tons per year. The principal application for thiophene today is as an intermediate for the manufacture of pharmaceuticals, particularly antihistamines that contain the thiophene moiety (see Histamine and histamine antagonists), and more recently, several potent new cephalosporin-type antibiotics for the treatment of venereal diseases (see Antibiotics, β-lactams in Supplement Vol.). It is also an intermediate in the manufacture of anthelmintics and sheep-dip preparations for veterinary medicine (see Chemotherapeutics, anthelmintics; Veterinary drugs). Other applications for thiophene derivatives reported in the recent literature include pesticides, dyestuffs, optical brighteners, flavorings, and new drugs for the treatment of cardiovascular disorders.

BIBLIOGRAPHY

"Thiophene" in *ECT* 1st ed., Vol. 14, pp. 95–102, by D. E. Badertscher and H. E. Rasmussen, Socony Mobil Oil Co., Inc.; "Thiophene" in *ECT* 2nd ed., Vol. 20, pp. 219–226, by Otto Meth-Cohn, University of Salford, Salford, England.

1. V. Meyer, *Die Thiophengruppe*, Braunschweig, 1888.
2. W. Steinkopf, *Die Chemie des Thiophenes*, Leipzig, 1941 (lithographed by Edward Bros., Ann Arbor, Mich.).
3. H. D. Hartough, *Thiophene and Its Derivatives*, Interscience Publishers, Inc., New York, 1952.
4. H. D. Hartough and S. L. Meisel, *Compounds with Condensed Thiophene Rings*, Interscience Publishers, Inc., New York, 1954.
5. C. D. Hurd, *Q. Rep. Sulfur Chem.* **4**(2–3), 75 (1969).
6. D. T. Clark in D. H. Reid, ed., *Organic Compounds of Sulfur, Selenium, and Tellurium*, Vol. 1, Burlington House, London, 1970, pp. 29–48.
7. S. Gronowitz in ref. 6, Vol. 2, 1972, pp. 352–495.
8. S. Gronowitz in ref. 6, Vol. 3, 1975, pp. 400–491.
9. S. Gronowitz in ref. 6, Vol. 4, 1977, pp. 244–278.
10. S. Gronowitz in D. R. Hogg, ed., *Organic Compounds of Sulfur, Selenium, and Tellurium*, Vol. 5, Burlington House, London, 1979, pp. 247–279.
11. M. Davis in ref. 10, Vol. 6, 1981, pp. 271–281.
12. F. S. Fawcett and H. E. Rasmussen, *J. Am. Chem. Soc.* **67**, 1705 (1949).
13. G. Waddington, J. W. Knowlton, D. W. Scott, G. D. Oliver, S. Todd, W. N. Hubbard, J. C. Smith, and H. H. Huffman, *J. Am. Chem. Soc.* **71,** 797 (1949).
14. N. I. Sax, *Dangerous Properties of Industrial Materials*, 5th ed., Van Nostrand Reinhold Co., New York, 1979, p. 1029.
15. M. Windholz, ed., *Merck Index*, 9th ed., Merck & Co., Inc., Rahway, N.J., 1976, p. 9090.
16. W. E. Haines, R. V. Helm, C. W. Bailey, and J. S. Ball, *J. Phys. Chem.* **58,** 270 (1954).
17. R. D. Obolentsev, *Khim. Seraorg. Soedin.* **6,** 220 (1964).
18. G. H. Jeffrey, R. Parker, and A. I. Vogel, *J. Chem. Soc.*, 570 (1961).
19. *Organosulfur Intermediates*, Pennwalt bulletin S264, Pennwalt Corporation, Philadelphia, Pa., 1973, p. 3.
20. A. P. Kudchadker, G. H. Alani, and B. J. Zwolinski, *Chem. Rev.* **68,** 659 (1968).
21. R. S. Holland and C. P. Smyth, *J. Phys. Chem.* **59,** 1088 (1955).

22. R. Phillippe and M. A. Piette, *Bull. Soc. Chim. Belg.* **64,** 600 (1955).
23. T. J. Barton, R. W. Roth, and J. G. Verkadis, *J. Am. Chem. Soc.* **94,** 8854 (1972).
24. S. Gronowitz in A. R. Katritzky, ed., *Advances in Heterocyclic Chemistry*, Vol. 1, Academic Press, Inc., New York, 1963, pp. 1–124.
25. V. Schomaker and L. Pauling, *J. Am. Chem. Soc.* **61,** 1769 (1939).
26. C. C. Price and S. Oae, *Sulfur Bonding*, The Roland Press Co., New York, 1962, p. 28.
27. U.S. Pat. 3,265,712 (Aug. 9, 1966), T. J. Jennings (to Shell Oil Co.).
28. J. Melles and H. Backer, *Rec. Trav. Chim.* **72,** 314, 391 (1953).
29. W. Davies and F. James, *J. Chem. Soc.*, 15 (1954).
30. C. G. Overberger and J. Whelan, Jr., *J. Org. Chem.* **26,** 4326 (1961).
31. H. Bluestone, R. Bimber, R. Berkey, and Z. Mandel, *J. Org. Chem.* **26,** 346 (1961).
32. *Chem. Mark. Rep.*, 62 (Nov. 10, 1975).
33. U.S. Pat. 2,853,493 (Sept. 23, 1958), A. W. Weston (to Abbott Laboratories).
34. A. W. Weston and R. J. Michaels, Jr., in N. Rabjohn, ed., *Organic Syntheses Collective Volumes*, Vol. 4, John Wiley & Sons, Inc., New York, 1963, p. 915.
35. V. S. Babasmian in A. H. Blatt, ed., *Organic Syntheses Collective Volumes*, Vol. 2, John Wiley & Sons, Inc., New York, 1943, p. 466.
36. R. A. Benkasser and R. B. Currie, *J. Am. Chem. Soc.* **70,** 1780 (1948).
37. J. W. Schick and H. D. Hartough, *J. Am. Chem. Soc.* **70,** 286 (1948).
38. U.S. Pat. 3,557,068 (Jan. 19, 1971), L. K. Huber (to Pennwalt Corporation).
39. W. S. Emerson and T. M. Patrick in ref. 34, p. 980.
40. J. F. Scully and E. V. Brown, *J. Am. Chem. Soc.* **75,** 6329 (1953).
41. R. Phillips in ref. 35, p. 578.
42. H. E. Rasmussen and F. E. Ray, *Chem. Ind.* **60,** 693 (1947).
43. U.S. Pat. 3,197,483 (July 27, 1965), B. Buchholz, T. E. Deger, and R. H. Goshorn (to Pennwalt Corporation).
44. U.S. Pat. 3,822,289 (July 2, 1974), N. R. Clark and W. E. Webster (to Synthetic Chemicals, Ltd.).
45. U.S. Pat. 3,939,179 (Feb. 17, 1976), T. R. Bell and P. G. Smith (to Pennwalt Corporation).
46. E. E. Sandmeyer in G. D. and F. E. Clayton, ed., *Patty's Industrial Hygiene and Toxicology*, 3rd rev. ed., Vol. 2A, John Wiley & Sons, Inc., New York, 1981, p. 2087.
47. I. B. Mikhailets, G. A. Mikhailets, D. G. Pel'ts, and A. V. Valiakhmetov, *Gig. Tr. Prof. Zabol.* **10,** 57 (1966); *Chem. Abstr.* **64,** 20496 (1966).
48. *1980 Registry of Toxic Effects of Chemical Substances*, U.S. Department of Health and Human Services, NIOSH Publication No. 81-116, U.S. Government Printing Office, Washington, D.C., 1981, entry XM735000, p. 716.
49. Unpublished data supplied to Pennwalt Corporation by A. R. Latven, Pharmacology Research, Inc., Darby, Pa., 1965.
50. I. Florin, L. Rutberg, M. Curvall, and C. R. Enzell, *Toxicology* **18,** 219 (1980).
51. C. E. Voogd, *Mutat. Res.* **38,** 117 (1976).

BERNARD BUCHHOLZ
Pennwalt Corporation

THIOSULFATES

The thiosulfate ion, $S_2O_3^{2-}$, is a structural analogue of the sulfate ion where one oxygen atom is replaced by one sulfur atom. The two sulfur atoms are not equivalent and the unique chemistry of the thiosulfate ion is dominated by the sulfide-like sulfur atom which is responsible for the reducing properties and complexing abilities of thiosulfates. Their ability to dissolve silver halides through complex formation is the basis for their commercial application in photography (qv).

Physical Properties

Structure. The structure of pure thiosulfuric acid has not been determined conclusively because of its lack of thermal stability. Radioactive sulfur exchange studies of aqueous solutions have shown the sulfur atoms to be nonequivalent (1). When a sulfite is treated with radioactive sulfur and the resulting thiosulfate decomposed to sulfur and sulfite by acids, the radioactivity appears in the sulfur:

$$^{35}S + SO_3^{2-} \rightarrow {}^{35}SSO_3^{2-}$$

$$^{35}SSO_3^{2-} + H^+ \rightarrow {}^{35}S + HSO_3^-$$

When the silver salt of the radioactive thiosulfate [*83682-20-6*] is decomposed the radioactivity appears in the sulfide:

$$Ag_2{}^{35}SSO_3 + H_2O \rightarrow Ag_2{}^{35}S + H_2SO_4$$

If both sulfur atoms were equivalent, the radioactivity would have been equally divided between the two products.

Calculations from binding-force measurements (2) indicate the limiting structural forms of the thiosulfate ion:

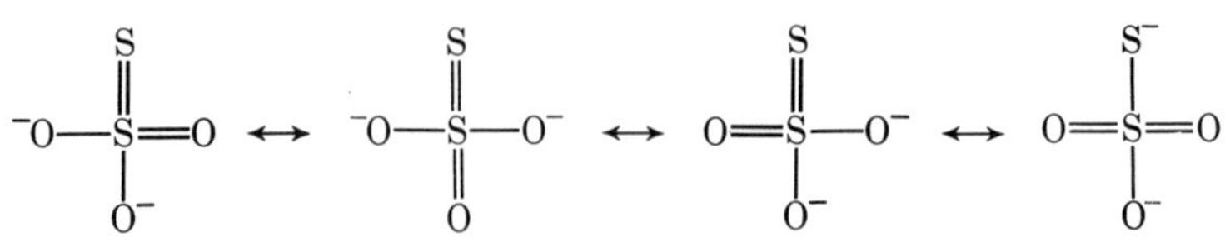

(1) (2)

Structure (**1**) explains the formation of sulfur and sulfite in the presence of acid, and (**2**) the formation of sulfide and sulfate in the presence of heavy metals. The bonding in thiosulfate complexes and the chemistry of thiosulfates are normally explained on the basis of structure (**2**) (see also Sulfur compounds).

The thiosulfate ion has tetrahedral C_{3v} symmetry and the six fundamental modes are both infrared and Raman active. The calculated frequencies (3) are in good agreement with experimental values (4).

Calculated, cm⁻¹	*Experimental, cm⁻¹*
995	1002
669	672
435	451
1123	1125
541	541
335	339

The low value of the S–S force constant compared to the S–O force constant is consistent with the ease of cleavage of the S–S bond. Spectral data indicate that the structure of the thiosulfate ion in solid thiosulfates is the same as that of the ion in solution.

X-ray crystallographic analysis of the sodium thiosulfate pentahydrate [*10102-17-7*] crystal indicates a tetrahedral structure for the thiosulfate ion with S–S bond distance of 0.197 nm and a S–O bond distance of 0.148 nm (5). Neutron diffraction of a barium thiosulfate monohydrate [*7787-40-8*] crystal confirms the tetrahedral structure and bond distances for the thiosulfate ion (6).

Thermodynamic Properties. The heat of formation of the thiosulfate ion, −5.75 kJ/g (−1.37 kcal/g), was determined by studying the equilibrium of the reaction:

$$2\ Ag + S_2O_3^{2-} \rightarrow Ag_2S + SO_3^{2-}$$

and by direct calorimetric experiments (7). The standard free energy of formation is −4.58 kJ/g (−1.09 kcal/g). The partial molal entropy is 62.8 ± 25.1 J/K (15 ± 6 cal/K).

Electrochemical Properties. The oxidation potential for the reaction:

$$2\ S_2O_3^{2-} = S_4O_6^{2-} + 2\ e$$

ranges from 0.2 to 0.4 V in neutral solution, depending on the method of measurement (8).

The electrolytic oxidation of thiosulfate solutions yields tetrathionate, $S_4O_6^{2-}$, as the principal product with HSO_3^- as by-product in acid solutions, and $S_3O_6^{2-}$ in alkaline solutions. The electrolytic oxidation is catalyzed by trace amounts of iodide (9).

Electrolytic reduction with a mercury or platinum electrode produces equimolar amounts of sulfide and sulfite:

$$S_2O_3^{2-} + 2\ e \rightarrow S^{2-} + SO_3^{2-}$$

Chemical Properties

Thiosulfuric Acid. Thiosulfuric acid [*14921-76-7*] is relatively unstable and thus cannot be recovered from aqueous solutions. In laboratory preparation, a lead thiosulfate [*26265-65-6*] solution is treated with H_2S to precipitate PbS or a concentrated solution of sodium thiosulfate [*7772-98-7*] is treated with HCl and cooled to −10°C to crystallize NaCl. Aqueous solutions of thiosulfuric acid spontaneously decompose to yield sulfur, SO_2, and polythionic acids, $H_2S_nO_6$. Thiosulfuric acid is a strong acid comparable to sulfuric acid. Dissociation constants, $K_1 = 0.25$, $K_2 = 0.018$, have been determined from pH measurements using a glass electrode (10).

Pure thiosulfuric acid has been prepared in liquid CO_2 at −50°C (11) or in diethyl ether at −78°C (12). It decomposes at −30°C to $H_2S_3O_6$ [*27621-39-2*] and H_2S, and rapidly at higher temperatures to H_2O, SO_2, and sulfur (13).

Thiosulfates. The ammonium, alkali metal, and alkaline-earth thiosulfates are soluble in water. Neutral or slightly alkaline solutions containing excess base or the corresponding sulfite are more stable than acid solutions. Thiosulfate solutions of other metal ions can be prepared, but their stability depends on the presence of excess thiosulfate, the formation of complexes, and the prevention of insoluble sulfide precipitates.

Acidification of thiosulfate with strong acid invariably leads to decomposition with the formation of colloidal sulfur and sulfur dioxide. The mechanism of this reaction is complex and depends on the thiosulfate concentration and the pH (14). The following reaction explains the formation of the main products:

$$H^+ + S_2O_3^{2-} \rightarrow [HS_2O_3^-] \rightarrow HSO_3^- + S \rightarrow SO_2 + S + H_2O$$

By-products are also formed:

$$3\,S_2O_3^{2-} + 3\,H^+ \rightarrow H_2S + S_4O_6^{2-} + HSO_3^-$$

$$S_2O_3^{2-} + H_2O \rightarrow SO_4^{2-} + H_2S$$

$$5\,S_2O_3^{2-} + 6\,H^+ \rightarrow 2\,S_5O_6^{2-} + 3\,H_2O$$

In dilute aqueous solution, the following equilibrium is established (15):

$$S_2O_3^{2-} + H^+ \rightleftarrows HSO_3^- + S \qquad K = 0.013 \text{ at } 11°C$$

This equilibrium explains the stabilization of thiosulfate solutions using sulfite or bisulfite as one of the components of acid photographic fixing baths.

The existence of anhydrothiosulfuric acid [*83682-21-7*] has been proposed to explain the apparent stability of thiosulfate in concentrated hydrochloric acid solution (16):

$$2\,H_2S_2O_3 \rightleftarrows H_2S_4O_5 + H_2O$$

Reactions. Catalytic amounts of arsenic, antimony, or tin salts promote the formation of pentathionate (16):

$$S_2O_3^{2-} + H_2AsO_3^- \rightleftarrows HSO_3^- + HAsSO_3^{2-}$$

$$S_2O_3^{2-} + 4\,HSO_3^- + 2\,H^+ \rightleftarrows 2\,S_3O_6^{2-} + 3\,H_2O$$

$$S_3O_6^{2-} + HAsSO_3^{2-} + H^+ \rightleftarrows S_4O_6^{2-} + H_2AsO_3^-$$

$$S_4O_6^{2-} + HAsSO_3^{2-} + H^+ \rightleftarrows S_5O_6^{2-} + H_2AsO_3^-$$

Mild oxidizing agents such as hydrogen peroxide in acid solutions produce tetrathionates and trithionates (17):

$$2\,S_2O_3^{2-} + H_2O_2 \rightleftarrows S_4O_6^{2-} + 2\,OH^-$$

$$3\,S_2O_3^{2-} + 4\,H_2O_2 \rightleftarrows 2\,S_3O_6^{2-} + 2\,OH^- + 3\,H_2O$$

The presence of Fe^{2+} promotes oxidation to the sulfate:

$$S_2O_3^{2-} + 4\,H_2O_2 \overset{Fe^{2+}}{\rightleftarrows} 2\,SO_4^{2-} + 2\,H^+ + 3\,H_2O$$

The reaction with iodine in neutral or slightly acid solution:

$$2\,S_2O_3^{2-} + I_2 \rightarrow S_4O_6^{2-} + 2\,I^-$$

is the basis of a volumetric analytical procedure.

Stronger oxidizing agents such as chlorine, bromine, permanganate, chromate, or alkaline hydrogen peroxide oxidize thiosulfate quantitatively to sulfate:

$$S_2O_3^{2-} + 2\,OH^- + 4\,H_2O_2 \rightleftarrows 2\,SO_4^{2-} + 5\,H_2O$$

Hypochlorite, hypobromite, and hypoiodite are also strong enough to oxidize thiosulfate to sulfate:

$$S_2O_3^{2-} + 4\,BrO^- + 2\,OH^- \rightarrow 2\,SO_4^{2-} + 4\,Br^- + H_2O$$

Thiosulfates are reduced to sulfides by metallic copper, zinc, or aluminum:

$$S_2O_3^{2-} + 2\,Cu \rightarrow Cu_2S + SO_3^{2-}$$

The thiosulfate reaction with cyanide to give thiocyanate is the basis for the use of thiosulfate as an antidote in cyanide poisoning:

$$S_2O_3^{2-} + CN^- \rightarrow SO_3^{2-} + SCN^-$$

Thiosulfates form complex ions with a number of metal ions by the coordination of more than one thiosulfate ion. The stability constants for lead thiosulfate complexes have been determined (18):

$$PbS_2O_3 + S_2O_3^{2-} \rightarrow [Pb(S_2O_3)_2]^{2-}$$

$$[Pb(S_2O_3)_2]^{2-} + S_2O_3^{2-} \rightarrow [Pb(S_2O_3)_3]^{4-}$$

Mercury forms a thiosulfate complex by reaction of the oxide with potassium thiosulfate [*10294-66-3*] (19):

$$HgO + 4\,K_2S_2O_3 + H_2O \rightarrow \underset{[83682\text{-}22\text{-}3]}{K_6[Hg(S_2O_3)_4]} + 2\,KOH$$

The stability of the silver complex thiosulfates $[Ag(S_2O_3)_2]^{3-}$, and $[Ag(S_2O_3)_3]^{5-}$ is the basis for the use of thiosulfates to dissolve the residual silver chloride in photographic gelatin coatings. The structure of some thiosulfate complexes has been determined (20).

The addition of thiosulfate to aqueous solutions of silver, lead, and copper precipitates the corresponding thiosulfates, which, on heating, decompose to the sulfides. In this manner, thiosulfate can be used as a reagent for most metals whose sulfides are insoluble. Details of the reactions of other metals with thiosulfate are available (21).

Corrosion. Copper-base alloys are seriously corroded by sodium thiosulfate (22) and ammonium thiosulfate [*7783-18-8*] (23) with corrosion rates exceeding 10 kg/(m^2·yr) at 100°C. High silicon cast iron has reasonable corrosion resistance to thiosulfates with a corrosion rate <4.4 kg/(m^2·yr) at 100°C. The preferred material of construction for pumps, piping, reactors, and storage tanks is austenitic stainless steels such as 304, 316, or Alloy 20. The corrosion rate for stainless steels is <440 g/(m^2·yr) at 100°C (see also Corrosion and corrosion inhibitors).

Preparation

Thiosulfates are normally prepared by the reaction of sulfur and sulfite in neutral or alkaline solution:

$$S + SO_3^{2-} \rightarrow S_2O_3^{2-}$$

Polysulfides react similarly:

$$S_x^{2-} + SO_3^{2-} \rightarrow S_2O_3^{2-} + S_{x-1}^{2-}$$

Sulfides react with sulfur dioxide, sulfite, or bisulfite:

$$2\,S^{2-} + 2\,SO_2 + 2\,HSO_3^- \rightarrow 3\,S_2O_3^{2-} + H_2O$$

$$2\,S^{2-} + 3\,SO_2 + SO_3^{2-} \rightarrow 3\,S_2O_3^{2-}$$

$$2\,HS^- + 4\,HSO_3^- \rightarrow 3\,S_2O_3^{2-} + 3\,H_2O$$

These three methods are employed commercially. In addition, decomposition of polythionates in alkaline solution or their reaction with sulfide or sulfite gives thiosulfates:

$$4\,S_4O_6^{2-} + 6\,OH^- \rightarrow 5\,S_2O_3^{2-} + 2\,S_3O_6^{2-} + 3\,H_2O$$

$$S_3O_6^{2-} + S^{2-} \rightarrow 2\,S_2O_3^{2-}$$

$$S_5O_6^{2-} + SO_3^{2-} \rightarrow S_2O_3^{2-} + S_4O_6^{2-}$$

The high temperature hydrolysis of sulfur in alkaline solutions also produces thiosulfates:

$$S_8 + 6\,OH^- \rightarrow 2\,S_3^{2-} + S_2O_3^{2-} + 3\,H_2O$$

Sodium Thiosulfate

Sodium thiosulfate, either the anhydrous salt, $Na_2S_2O_3$, or the crystalline pentahydrate, is commonly referred to as hypo or crystal hypo. As can be seen from the equilibrium phase diagram shown in Figure 1, when a concentrated sodium thiosulfate solution (50–60 wt %) is cooled to <48°C, the pentahydrate, containing 63.7% $Na_2S_2O_3$, crystallizes in monoclinic transparent prisms. The monohydrate [*55755-19-6*] and the heptahydrate [*36989-91-0*] are also known.

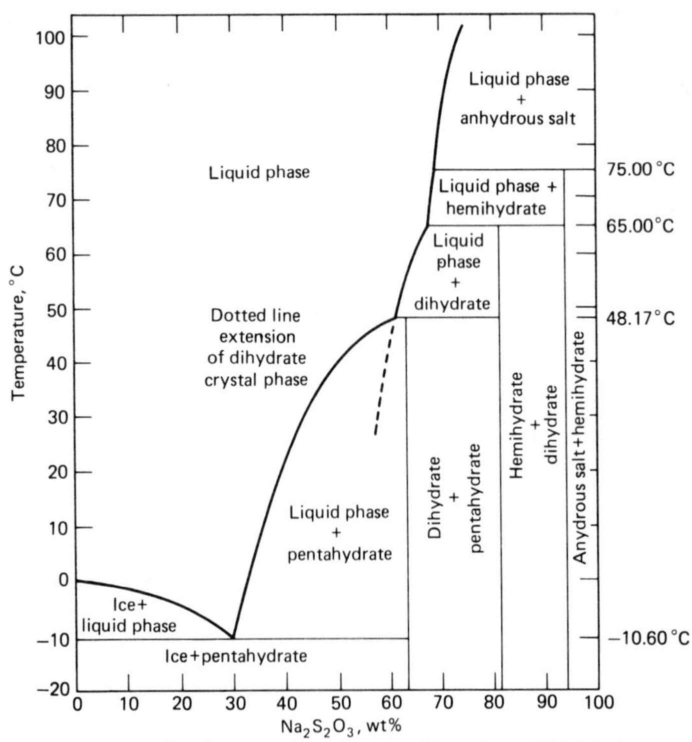

Figure 1. Equilibrium phase diagram, $Na_2S_2O_3$ and H_2O (24).

Although 16 different crystalline modifications have been identified (24–25), the α-pentahydrate is the stable form below 48°C. Solutions of sodium thiosulfate in the absence of seed crystals can be easily supercooled below their normal crystallization temperatures. The dotted line extension of the dihydrate phase in Figure 1 is an indication that, if supercooling takes place below this line, solutions normally giving the pentahydrate may form the dihydrate [*36989-90-9*] instead.

Selected physical properties of sodium thiosulfate pentahydrate are shown in Table 1. The crystals are relatively stable, efflorescing in warm, dry air and deliquescing slightly in moist air. They melt in their water of hydration at 48°C and can be completely dehydrated in a vacuum oven at this temperature, or at atmospheric pressure at 105°C. Anhydrous sodium thiosulfate can also be crystallized directly from a 72% solution above 75°C. It decomposes at 233°C:

$$4\ Na_2S_2O_3 \rightarrow 3\ Na_2SO_4 + Na_2S_5$$

Further heating to 440–500°C gives sodium sulfide and sulfur dioxide (33).

Aqueous sodium thiosulfate solutions are neutral. Under neutral or slightly acidic conditions, decomposition produces sulfite and sulfur. In the presence of air, alkaline solutions decompose to sulfate and sulfide. Dilute solutions can be stabilized by small amounts of sodium sulfite, sodium carbonate, or caustic, and by storage at low tem-

Table 1. Physical Properties of Sodium Thiosulfate Pentahydrate

Property	Value	Ref.[a]
refractive index, n_D^{20}	1.4886	27
density, d_4^{25}, g/cm^3	1.750	28
heat of solution in water at 25°C, J/g[b]	−187	29
heat of formation, kJ/g[b]	−10.48	30
heat of fusion, J/g[b]	200	31
specific heat solid, J/(g·K)[b]	1.84	28
specific heat molten salt, J/(g·K)	2.38	28
dissociation pressure, kPa[c]		28
20°C	0.796	
25°C	1.154	
30°C	1.679	
35°C	2.395	
vapor pressure of saturated solutions, kPa[c]		28
33°C	1.33	
57°C	5.60	
90°C	31.06	
120°C	100.4	
density of aqueous solutions, d_{20}^{20}, g/cm^3		32
wt % $Na_2S_2O_3$		
1.00	1.0083	
10.00	1.0847	
20.00	1.1760	
30.00	1.2762	
40.00	1.3851	

[a] Ref. 26 gives a comprehensive summary of the properties of sodium thiosulfate and its aqueous solutions.

[b] To convert J to cal, divide by 4.184.

[c] To convert kPa to mm Hg, multiply by 7.5.

peratures away from air and light. Oxidation is inhibited by HgI_2 (10 ppm), amyl alcohol (1%), chloroform (0.1%), borax (0.05%), or sodium benzoate (0.1%).

Manufacture. Sodium thiosulfate has been produced commercially by the air oxidation of sulfides, hydrosulfides, and polysulfides as, for example, in the recovery of sodium thiosulfate from the alkali wastes of the Leblanc soda process, according to the following reactions:

$$Ca(SH)_2 + 2\ O_2 \rightarrow CaS_2O_3 + H_2O$$

$$\underset{[10124\text{-}41\text{-}1]}{CaS_2O_3} + Na_2SO_4 \rightarrow Na_2S_2O_3 + CaSO_4$$

The purity of the crystallized product was stated to be 98% $Na_2S_2O_3.5H_2O$. This process offered a means of disposal of Leblanc waste and supplied essentially the entire demand for sodium thiosulfate. When the Solvay ammonia process replaced the Leblanc process, the latter route was abandoned (see also Alkali and chlorine products, sodium carbonate).

Sodium thiosulfate is a by-product of the manufacture of Sulfur Black and other sulfur dyes, where organic nitro compounds are treated with a solution of sodium polysulfide to give thiosulfate. The dyes are insoluble and are recovered by filtration. The filtrate is treated with activated carbon and filtered to obtain a sodium thiosulfate solution. After concentration and crystallization, the final product assays ca 96% $Na_2S_2O_3.5H_2O$ (34) (see also Dyes and dye intermediates).

Other commercial processes are based on the reaction of sodium sulfide or hydrogen sulfide with sulfur dioxide, and caustic or soda ash:

$$2\ Na_2S + Na_2CO_3 + 4\ SO_2 \rightarrow 3\ Na_2S_2O_3 + CO_2$$

$$2\ Na_2S + 3\ SO_2 \rightarrow 2\ Na_2S_2O_3 + S$$

Excess sulfur is filtered before evaporation and crystallization.

In a recent modification, excess sulfur is preground in the sodium sulfide solution and an equivalent amount of sodium sulfite added (35).

Another procedure utilizes a slurry of sodium sulfite, produced by the reaction of soda ash with sulfur dioxide, which is digested with excess sulfur until all of the sulfite is used up:

$$Na_2SO_3 + S \rightarrow Na_2S_2O_3$$

Cationic surface-active agents promote wetting of the sulfur and thereby increase the reaction rate (36). The quality of the product is improved by using photographic-grade sodium sulfite or bisulfite. Excess sulfur is filtered before evaporation and crystallization. Since evaporation is energy intensive, it is important to produce the thiosulfate solution at the highest possible concentration. The purity of the product is typically >99%; sulfite and sulfate are the main impurities.

Economic Aspects. As of 1981, five manufacturers produced sodium thiosulfate in the United States with an estimated total capacity of 30,000 metric tons. Actual production and trade-list prices are shown in Table 2.

Despite declining volume, prices have increased, reflecting increased raw material, energy, and labor costs.

Production outside of the United States is at least 25% of the U.S. production, mainly in the FRG.

Table 2. U.S. Sodium Thiosulfate Capacity, Production, and Prices, 1981[a]

Year	Capacity, t	Production, t	$/t
1960	41,000	31,000	89.79
1965	41,000	27,000	89.79
1970	41,000	22,500	127.89
1975	31,000	14,400	150.56
1978	32,000	16,200	217.68
1979	32,000	17,100	290.24
1980	31,000	16,200	362.80
1981	30,000	16,200	435.36

[a] Pentahydrate basis.

Specifications. Sodium thiosulfate pentahydrate and the anhydrous salt are available in various grades, as shown in Tables 3 and 4.

Table 3. Specifications for Sodium Thiosulfate Pentahydrate

Specifications	ACS reagent grade[a]	USP grade[b]
assay, % min	99.5	99.0
water, %		32–37
insoluble matter, % max	0.005	
sulfate and sulfite, as SO_4, % max	0.1	
sulfide, as S, ppm max	1	
pH of 5% solution at 25°C	6.0–8.4	
arsenic, ppm max		3
heavy metals, ppm max		20
calcium		to pass test
nitrogen compounds, as N, % max	0.002	

[a] Ref. 37.
[b] Ref. 38.

Table 4. American National Standard Specifications for Photographic-Grade Sodium Thiosulfate[a]

Specifications	Crystalline pentahydrate	Anhydrous
assay, %		
min	99.0	97.0
max	101.0	
insoluble matter[b], % max	0.2	0.4
alkalinity, as NaOH, % max	0.02	0.06
acidity, as H_2SO_4, % max	0.01	0.01
sulfide, as S, ppm max	4	6
heavy metals, as Pb, ppm max	10	20
iron, as Fe, ppm max	30	50
appearance of solution	to pass test	to pass test

[a] Ref. 39.
[b] Including calcium, magnesium, and ammonium hydroxides, ppt.

Analytical and Test Methods. An aqueous solution of sodium thiosulfate forms a white precipitate with hydrochloric acid and evolves sulfur dioxide gas which is detected by its characteristic odor. The white precipitate turns yellow, indicating the presence of sulfur. The addition of ferric chloride to sodium thiosulfate solutions produces a dark-violet color which quickly disappears.

Sodium thiosulfate is determined by titration with standard iodine solution (37). Sulfate and sulfite are determined together by comparison of the turbidity produced when barium chloride is added after the iodine oxidation with the turbidity produced by a known quantity of sulfate in the same volume of solution. The absence of sulfide is indicated when the addition of alkaline lead acetate produces no color within one minute.

Health and Safety Factors. The LD_{50} of anhydrous sodium thiosulfate for mice is 7.5 ± 0.752 g/kg (40). Because of its low toxicity, it can be safely used in veterinary medicine. Sodium thiosulfate pentahydrate is affirmed as a GRAS indirect and direct human food ingredient under the Federal Food, Drug, and Cosmetic Act (41) (see Food additives).

It has been reported that humans can consume up to 12 g sodium thiosulfate orally per day with no effect other than catharsis (43).

Uses. The estimated use distribution of sodium thiosulfate in 1981 is given in Table 5.

The principal use for sodium thiosulfate continues to be as fixative in photography to dissolve undeveloped silver halide from negatives or prints (see Photography). In applications where rapid processing is required, such as the processing of x-ray film, it has been largely replaced by ammonium thiosulfate.

Sodium thiosulfate is still used in chrome leather tanning as a reducing agent in two-bath processes to reduce dichromate (hexavalent chromium) to chrome alum (trivalent chromium) (see Leather).

In paper and textiles manufacture, sodium thiosulfate removes the residual bleach before dyeing. It inhibits fermentation in dyeing baths, and is a source of sulfur dioxide in the bleaching of wool. It is also used as a dechlorinating agent for waste streams.

Minor and potential new uses include flue-gas desulfurization (43–44), silver-cleaning formulations (45), thermal-energy storage (46), cyanide antidote (47), cement additive (48), aluminum-etching solutions (49), removal of nitrogen dioxide from flue gas (50), concrete-set accelerator (51), stabilizer for acrylamide polymers (52), extreme pressure additives for lubricants (53), multiple-use heating pads (54), in soap and shampoo compositions (55), and as a flame retardant in polycarbonate compositions (56).

Table 5. Use Distribution of Sodium Thiosulfate

Use	Percent
photography	65
leather tanning	8
paper and textiles	6
flue-gas desulfurization	3
others[a]	18

[a] Dechlorination, secondary oil recovery, thermal-energy storage, pharmaceuticals, and lubricant additives.

Ammonium Thiosulfate

Ammonium thiosulfate, $(NH_4)_2S_2O_3$, commonly referred to as ammo hypo, has displaced sodium thiosulfate in photography. It is normally sold in the United States only as the aqueous solution. In addition, a crystal slurry and anhydrous crystal are available in Europe.

The anhydrous monoclinic crystalline form has a density of 1.679 g/cm^3 (57); no hydrates are known. Solubility in water is given in Table 6. Ammonium thiosulfate solutions decompose slowly below 50°C and more rapidly at higher temperatures. The anhydrous salt decomposes above 100°C to sulfite and sulfur (58):

$$(NH_4)_2S_2O_3 \rightarrow (NH_4)_2SO_3 + S$$

At 230–245°C, the ammonium sulfite decomposes.

Manufacture. Ammonium thiosulfate has been produced by the reaction of ammonium sulfite with sulfur, sulfides, or polysulfides:

$$(NH_4)_2SO_3 + (NH_4)_2S_8 \rightarrow (NH_4)_2S_2O_3 + (NH_4)_2S_7$$

$$(NH_4)_2SO_3 + (NH_4)_2S_7 \rightarrow (NH_4)_2S_2O_3 + (NH_4)_2S_6$$

This series continues until the last polysulfide is ammonium sulfide and the process is completed by the reaction with sulfur dioxide:

$$3\ SO_2 + (NH_4)_2SO_3 + 2\ (NH_4)_2S \rightarrow 3\ (NH_4)_2S_2O_3$$

Ammonium bisulfite can be used in place of the sulfur dioxide. The solution is treated with activated carbon and filtered to remove traces of sulfur. Excess ammonia is added and the solution evaporated if the anhydrous crystalline form is desired. The crystals are dried at low temperature in the presence of ammonia to prevent decomposition (59–61).

Other commercial processes currently practiced are based on the direct reaction of ammonium sulfite with sulfur (62–63):

$$(NH_4)_2SO_3 + S \rightarrow (NH_4)_2S_2O_3$$

Both batch and continuous processes employ excess sulfur and operate at 85–110°C. Trace amounts of polysulfides produce a yellow color which indicates that all the

Table 6. Solubility of Ammonium Thiosulfate

Temperature, °C	Solubility, wt %
−30	53.5
−20	56.5
−10	58.5
0	60.5
10	62.5
20	64.0
30	65.5
40	67.2
50	68.5
60	69.6
70	70.9
80	72.2

ammonium sulfite has been consumed. Ammonium bisulfite is added to convert the last polysulfide to thiosulfate and the excess ammonia to ammonium sulfite. Concentrations of at least 70% $(NH_4)_2S_2O_3$ are obtained without evaporation. Excess sulfur is removed by filtration and color is improved with activated carbon treatment or sodium silicate (64). Upon cooling, the aqueous concentrated solution, ammonium thiosulfate crystallizes.

Agricultural grades of ammonium thiosulfate are prepared by similar processes and contain some excess sulfur. The sulfur can be removed by washing with carbon disulfide. A typical sulfur-free product contains 87% $(NH_4)_2S_2O_3$, 3.4% $(NH_4)_2SO_3$, and 9.6% $(NH_4)_2SO_4$ (65).

Economic Aspects. As of 1981, ten manufacturers produced ammonium thiosulfate in the United States with an estimated capacity of 200,000 t. Actual production was probably 94,000 t. Trade list prices for photographic-grade ammonium thiosulfate in September, 1982, were $479/t in 208-L drums and $235/t in bulk-tank trucks as 60% solution. Prices for agricultural-grade materials are not available.

Specifications. The specifications for photographic-grade ammonium thiosulfate are shown in Table 7. There are no corresponding specifications for the agricultural-grade material.

Analytical and Test Methods. Analysis and test methods are similar to those for sodium thiosulfate. Sulfite is determined by an indirect method based on the titration of the acid liberated when both the sulfite and thiosulfate are oxidized with iodine solution (66).

Health and Safety Factors (Toxicology). The toxicological properties of ammonium thiosulfate are generally considered to be the same as those of sodium thiosulfate and thiosulfates in general (42).

Table 7. American National Standard Specifications for Photographic-Grade Ammonium Thiosulfate Solution

Specifications	Aqueous solution[a]	Crystalline solid[b]
assay as $(NH_4)_2S_2O_3$, %		
min	56.0	97.0
max	60.0	
insoluble matter[c], % max	0.2	0.4
alkalinity, as NH_4OH, %		
min	0.3	
max	1.5	0.4
sulfide, as S, ppm max	20	10
sulfite, as SO_3, % max	0.70	1.4
heavy metals, as Pb, ppm max	10	20
iron, as Fe, ppm max	2.5	50
residue after ignition, % max	0.1	0.2
specific gravity, at 15/15°C		
min	1.310	
max	1.335	

[a] Ref. 66.
[b] Ref. 67.
[c] Including calcium and magnesium, precipitated by ammonium hydroxide.

Uses. The use distribution of ammonium thiosulfate in 1981 was estimated to be photography, 48%; agricultural applications, 50%; and others including dechlorination, 2%.

The principal use of photochemical-grade ammonium thiosulfate continues to be in photography, where it dissolves undeveloped silver halides from negatives and prints. It reacts considerably faster than sodium thiosulfate, and the fixing solutions can be used about twice as long as sodium thiosulfate solutions; the washing period to remove residual thiosulfate is shorter.

Agricultural uses for ammonium thiosulfate take advantage of both the sulfur and ammonium content by blending with other nitrogen fertilizers such as urea (68). Some foliar-spray fertilizers contain ammonium thiosulfate together with other metal micronutrients (69–70). Ammonium thiosulfate or mixtures with ammonium nitrate can also be used as desiccants and defoliants in crop-bearing plants such as cotton, soybean, alfalfa, rice, and peppers (71–72).

Minor and potential new uses for ammonium thiosulfate include flue-gas desulfurization (73–74), removal of nitrogen oxides and sulfur dioxide from flue gases (75–76), converting sulfur in hydrocarbons to a water-soluble form (77), and converting cellulose to hydrocarbons (78–79) (see Sulfur recovery).

Other Thiosulfates

Many other metal thiosulfates, eg, magnesium thiosulfate [*10124-53-5*] and its hexahydrate [*13446-30-5*], have been prepared on a laboratory scale, but with the exception of the calcium, barium [*35112-53-9*], and lead compounds, they are of little commercial or technical interest. Although thallous [*13453-46-8*], silver, lead, and barium thiosulfates are only slightly soluble, other metal thiosulfates are usually soluble in water. The lead and silver salts are anhydrous; the others usually form more than one hydrate. Aqueous solutions are stable at low temperatures and in the absence of air. The chemical properties are those of thiosulfates and the respective cation.

Thiosulfates are generally prepared by treating aqueous solutions of either calcium or barium thiosulfate with the corresponding carbonate or sulfate of the desired metal. The insoluble calcium or barium sulfates or carbonates are filtered and the thiosulfate recovered from the filtrate by vacuum evaporation.

Other metal thiosulfates have been prepared by reaction of suspensions of the metal sulfide with sulfur dioxide. However, these thiosulfates are usually contaminated with polythionates (80).

Some metal thiosulfates are inherently unstable because of the reducing properties of the thiosulfate ion. Ions such as Fe^{3+} and Cu^{2+} tend to be reduced to lower oxidation states, whereas mercury or silver, which form sulfides of low solubility, tend to decompose to the sulfides. The stability of other metal thiosulfates improves in the presence of excess thiosulfate by virtue of complex thiosulfate formation.

The most common form of calcium thiosulfate is the hexahydrate [*10035-02-6*], $CaS_2O_3.6H_2O$, triclinic crystals with a density of 1.872 g/cm^3 at 16°C (81). Heating, however, does not give the anhydrous salt because of decomposition at 80°C. At lower temperatures, dehydration stops at the monohydrate [*15091-91-5*]. The solubility of calcium thiosulfates in water is given below.

°C	0	10	20	30	40
CaS_2O_3, wt %	25.8	29.4	33.0	36.6	40.3

Aqueous solutions decompose on heating as low as 60°C with formation of sulfur.

Calcium thiosulfate has been prepared from calcium sulfite and sulfur at 30–40°C, or from boiling lime and sulfur in the presence of sulfur dioxide until a colorless solution is obtained. Alternatively, a concentrated solution of sodium thiosulfate is treated with calcium chloride; the crystalline sodium chloride is removed at low temperature. Concentrated solutions of calcium thiosulfate are prepared from ammonium thiosulfate and lime; the liberated ammonium ion is recycled to the ammonium thiosulfate process (82).

Calcium thiosulfate is not produced commercially in the United States. Uses include fungicide formulations (83), a noncorrosive concrete-set accelerator (84), and a catalyst for polyolefin manufacture (85).

Complexes and Organic Thiosulfates

Gold thiosulfate complexes of the form $Na_3[Au(S_2O_3)_2].2H_2O$ [*19153-98-1*] are prepared by addition of gold trichloride to concentrated sodium thiosulfate solution (86). The gold is completely reduced and some thiosulfate is oxidized to tetrathionate. This complex has been used in the treatment of rheumatoid arthritis.

Other complex thiosulfates have been prepared to study crystal properties, eg, cadmium ammonium thiosulfates (87), $NaAgS_2O_3.H_2O$ [*37954-66-8*] (88), $K_2Mg(S_2O_3)_2.6H_2O$ [*64153-76-0*] (89), and $(NH_4)_9[Ag(S_2O_3)_4]Cl_2$ [*12040-89-0*] (90).

Organic thiosulfate salts are usually prepared by the reaction of alkyl chlorides with sodium thiosulfate:

$$RCl + Na_2S_2O_3 \rightarrow RSSO_3Na + NaCl$$

Sodium ethyl thiosulfate [*26264-37-9*] is also known as Bunte's salt after the name of its discoverer. Bunte salts may be thought of as esters of thiosulfuric acid (91–93). In essentially all of their chemical reactions, the cleavage is between the divalent and hexavalent sulfur atom. For example, acid hydrolysis produces a thiol and the acid sulfate:

$$RCH_2SSO_3Na + H_2O \rightarrow RCH_2SH + NaHSO_4$$

Bunte salts have bacterial, insecticidal, and fungicidal properties, and are also used as chelating agents or surfactants (94–95). Bunte salts have been tested for preirradiation protection for mammals exposed to lethal radiation doses (96–97) (see Radioprotective agents).

BIBLIOGRAPHY

"Thiosulfuric Acid and Thiosulfates" in *ECT* 1st ed., Vol. 14, pp. 102–115, by R. V. Townend, Allied Chemical Corp.; "Thiosulfates" in *ECT* 2nd ed., Vol. 20, pp. 227–247, by C. A. Wamser, Allied Chemical Corp.

1. T. Moeller, *Inorganic Chemistry*, John Wiley & Sons, Inc., New York, 1959, p. 543.
2. H. Siebert, *Z. Anorg. Allg. Chem.* **275,** 225 (1954).
3. U. Agarwala, C. E. Rees, and H. G. Thode, *Can. J. Chem.* **43,** 2802 (1965).
4. Y. Y. Kharitonov, N. A. Knyazeva, and L. V. Goeva, *Opt. Spektrosk.* **24,** 639 (1968).
5. P. G. Taylor and C. A. Beevers, *Acta Crist.* **5,** 341 (1952).

6. L. Manojlovic-Muir, *Acta Crystallogr. Sect. B*, 135 (1975).
7. H. C. Mel, Z. Z. Hugus, Jr., and W. M. Latimer, *J. Am. Chem. Soc.* **78,** 1822 (1956).
8. *Gmelins Handbuch der Anorganischen Chemie*, 8th ed., Schwefel, Part B, No. 2, Verlag Chemie, G.m.b.H., Weinheim/Bergstrasse, 1960, p. 868.
9. I. P. Chernobaev and M. A. Loshkarev, *Ukr. Khim. Zh.* **33,** 253 (1967).
10. F. M. Page, *J. Chem. Soc.*, 1719 (1953).
11. J. Piccard and E. Thomas, *Helv. Chim. Acta* **6,** 1032 (1923).
12. M. Schmidt, *Z. Anorg. Allg. Chem.* **289,** 141 (1957).
13. M. Schmidt and G. Talsky, *Angew. Chem.* **70,** 312 (1958).
14. Ref. 8, pp. 875–894.
15. F. Foerster and R. Vogel, *Z. Anorg. Allg. Chem.* **155,** 161 (1926).
16. H. Bassett and R. G. Durrant, *J. Chem. Soc.*, 1401 (1927).
17. F. Yokosuka, T. Kurai, A. Okuwaki, and T. Okabe, *Nippon Kagaku Kaishi*, 1901 (1975).
18. G. M. Vol'dman, *Zh. Fiz. Khim.* **44,** 2066 (1970).
19. P. Ray and J. Das-Gupta, *J. Indian Chem. Soc.* **5,** 483 (1928).
20. A. N. Freedman and B. P. Straughan, *Spectrochim. Acta, Part A* **27,** 1455 (1971).
21. Ref. 8, pp. 919–947.
22. *Metals Section, Corrosion Data Survey*, 5th ed., National Association of Corrosion Engineers, Houston, Texas, 1974, pp. 172–173.
23. Ref. 22, pp. 18–19.
24. S. W. Young and W. E. Burke, *J. Am. Chem. Soc.* **26,** 1413 (1904); dotted line extension of dihydrate crystal phase confirmed by Allied Chemical Corp., unpublished results, 1978.
25. S. W. Young and W. E. Burke, *J. Am. Chem. Soc.* **28,** 315 (1906).
26. *Gmelins Handbuch der Anorganischen Chemie*, 8th ed., Verlag Chemie, G.m.b.H., Weinheim/Bergstrasse, 1966, pp. 1162–1174.
27. *Gmelins Handbuch der Anorganischen Chemie*, 8th ed., Verlag Chemie, G.m.b.H., Berlin, 1928, p. 614.
28. Ref. 26, p. 1166.
29. Ref. 26, p. 1168.
30. E. W. Washburn, ed., *International Critical Tables*, Vol. 5, McGraw-Hill Book Co., New York, 1929, p. 201.
31. Ref. 30, p. 131.
32. R. C. Weast, ed., *Handbook of Chemistry and Physics*, 63rd ed., CRC Press, Boca Raton, Fla., 1982–1983, p. D-268.
33. T. Golgotiu and V. Rotaru, *Bul. Inst. Politeh. Iasi*, 37 (1972).
34. F. A. Lowenheim and M. K. Moran, *Faith, Keyes, and Clark's Industrial Chemicals*, 4th ed., John Wiley & Sons, Inc., New York, 1975, pp. 769–773.
35. USSR Pat. 779,297 (Nov. 15, 1980), T. G. Akhmetov, A. A. Murav'ev, E. V. Polyakov, and I. N. Smirnov.
36. U.S. Pat. 2,763,531 (Sept. 18, 1956), G. I. P. Levenson (to Eastman Kodak Co.).
37. *American Chemical Society Specifications—Reagent Chemicals*, 6th ed., American Chemical Society, Washington, D.C., 1981, p. 551.
38. *United States Pharmacopeia XX* (*USP XX–NF XV*), United States Pharmacopeial Convention, Inc., Rockville, Md., 1980, p. 738.
39. *ANSI PH 4.250-1980*, American National Standards Institute, New York, Nov. 16, 1979.
40. G. Plume, *Latv. Lopkopibas Vet. Zinat. Petnieciska Inst. Raksti* **24,** 73 (1970).
41. *Fed. Regist.* **43,** 22937 (May 30, 1978).
42. N. I. Sax, *Dangerous Properties of Industrial Materials*, 5th ed., Van Nostrand Reinhold Company, New York, 1979, pp. 372, 991, 1030.
43. Ger. Offen. 2,700,549 (July 13, 1978), H. Hoelter, H. Gresch, and H. Igelbuescher.
44. W. I. Nissen and R. S. Madenburg, *Inf. Circ. U.S. Bur. Mines*, IC 8806, 1979, p. 16.
45. Can. Pat. 926,554 (May 22, 1973), S. Young and K. Mueller (to Pioneer Marketing Associates).
46. U.S. Pat. 4,280,553 (July 28, 1981), S. L. Bean, J. W. Swaine, Jr., and P. R. Crawford (to Allied Chemical Corp.).
47. U.S. Pat. 4,292,311 (Sept. 30, 1981), S. J. Sarnoff.
48. USSR Pat. 814,930 (March 23, 1981), N. Buchvarov, I. Botev, and N. Teneva.
49. Australian Pat. 461,061 (June 27, 1974), R. Lowe.

50. Jpn. Kokai 76 03,387 (Jan. 12, 1976), T. Yamada, H. Yamamoto, A. Kita, and T. Iwata (to Osaka Soda Co.).
51. Fr. Demande 2,222,327 (Oct. 18, 1974), A. Binick (to Société Française Lanco).
52. U.S. Pat. 3,753,939 (Aug. 21, 1973), H. S. Von Euler-Chelpin (to Kemanord A.-G.).
53. Fr. Pat. 1,555,358 (Jan. 24, 1969), L. M. Niebylski (to Ethyl Corporation).
54. Ger. Offen. 2,917,192 (Nov. 6, 1980), G. Arrhenius (to Kay Laboratories).
55. U.S. Pat. 4,295,985 (Oct. 20, 1981), H. G. Petrow and M. L. Weissman.
56. U.S. Pat. 4,028,297 (June 7, 1977), J. L. Webb (to General Electric Co.).
57. Y. Elerman, A. Aydin Uraz, N. Armagan, and Y. Aka, *J. Appl. Crystallogr.* **11**(6), 709 (1978).
58. L. Erdey, S. Gal, and G. Liptay, *Talanta* **11,** 913 (1964).
59. U.S. Pat. 2,219,258 (Oct. 22, 1940), W. H. Hill (to American Cyanamid).
60. U.S. Pat. 2,412,607 (Dec. 17, 1946), H. V. Farr and J. R. Ruhoff (to Mallinckrodt Chemical Co.).
61. U.S. Pat. 2,586,459 (Feb. 19, 1952), H. V. Farr and J. R. Ruhoff (to Mallinckrodt Chemical Co.).
62. Ger. Offen. 2,635,649 (Feb. 9, 1978), K. H. Henke and G. Weiner (to Hoechst A.-G.); Brit. Pat. 1,175,069 (Dec. 23, 1969), (to Farbwerke Hoechst A.-G.).
63. U.S. Pat. 3,473,891 (Oct. 21, 1969), E. Mack (to Th. Goldschmidt A.-G.).
64. U.S. Pat. 3,890,428 (June 17, 1975), M. D. Jayawant (to E. I. du Pont de Nemours & Co., Inc.).
65. U.S. Pat. 3,524,724 (Aug. 18, 1970), R. L. Every and P. F. Cox (to Continental Oil Co.).
66. *ANSI PH 4.252-1980*, American National Standards Institute, New York, Aug. 15, 1980.
67. *ANSI PH 4.253-1960*, American National Standards Institute, New York, Dec. 21, 1960.
68. Brazil Pedido PI 79 00,040 (Aug. 14, 1979), J. E. Sansing and T. M. Parham (to Allied Chemical Corp.).
69. U.S. Pat. 4,191,550 (March 4, 1980), E. F. Hawkins and T. M. Parham (to Allied Chemical Corp.).
70. U.S. Pat. 4,210,437 (July 1, 1980), L. E. Ott and R. J. Windgassen (to Standard Oil Co., Indiana).
71. U.S. Pat. 3,730,703 (May 1, 1973), R. H. Daehnert (to Allied Chemical Corp.).
72. U.S. Pat. 3,689,246 (Sept. 5, 1972), D. C. Young (to Union Oil Co. of California).
73. A. Zey, S. White, and D. Johnson, *Chem. Eng. Prog.* **76**(10), 76 (1980).
74. U.S. Pat. 4,163,776 (Aug. 7, 1979), M. D. Kulik and E. Gorin (to Conoco, Inc.).
75. Fr. Demande 2,299,070 (Aug. 27, 1976), (to Exxon Research and Engineering Co.).
76. Can. Pat. 1,022,728 (Dec. 20, 1977), A. B. Welty, Jr. (to Exxon Research and Engineering Co.).
77. U.S. Pat. 4,201,662 (May 6, 1980), R. L. Horton (to Phillips Petroleum Co.).
78. U.S. Pat. 3,864,096 (Feb. 4, 1975), P. Urban (to Universal Oil Products Co.).
79. U.S. Pat. 3,864,097 (Feb. 4, 1975), P. Urban (to Universal Oil Products Co.).
80. W. E. Henderson and H. B. Weiser, *J. Am. Chem. Soc.* **35,** 239 (1913).
81. *Gmelins Handbuch der Anorganischen Chemie*, 8th ed., Part B, No. 3, Verlag Chemie, G.m.b.H., Weinheim/Bergstrasse, 1961, pp. 785–793.
82. U.S. Pat. 4,105,754 (Aug. 8, 1978), J. W. Swaine, Jr., W. W. Low, and S. L. Bean (to Allied Chemical Corp.).
83. Jpn. Kokai 77 38,018 (March 24, 1977), Y. Miyahara (to Sankei Chemicals Co.).
84. Jpn. Pat. 73 06,039 (Feb. 22, 1973), H. Tanaka and S. Nakagawa.
85. Ger. Offen. 2,364,170 (July 25, 1974), M. Galliverti, M. B. Ghirga, and B. Calcagno.
86. M. Windholtz, ed., *The Merck Index*, 9th ed., Merck and Co., Rahway, N.J., 1976, pp. 586–587.
87. Z. Gabelica, *Bull. Cl. Sci. Acad. R. Belg.* **59,** 1164 (1973).
88. L. Cavalca, A. Mangia, C. Palmieri, and G. Pelizzi, *Inorg. Chim. Acta* **4,** 299 (1970).
89. Z. Gabelica, *Rev. Chim. Miner.* **14,** 269 (1977).
90. F. Bigoli, A. Tiripicchio, and M. Tiripicchio Camellini, *Acta Crystallogr. Sect. B* **28,** 2079 (1972).
91. B. Milligan and J. M. Swan, *Rev. Pure Appl. Chem.* **12,** 72 (1962).
92. H. Distler, *Angew. Chem. Int. Ed. Engl.* **6,** 554 (1967).
93. D. L. Klayman and R. Shine, *Quarterly Reports on Sulfur Chemistry*, Vol. 3, No. 3, Intra-Science Research Foundation, Santa Monica, Calif., 1968, pp. 191–309.
94. U.S. Pat. 3,364,247 (Jan. 16, 1968), M. H. Gollis (to Monsanto Co.).
95. R. D. Westland, E. R. Karger, B. Green, and J. R. Dice, *J. Med. Chem.* **11,** 84 (1967).
96. J. Tulecki, J. Kalinowska-Torz, E. Musial, H. Nacewicz-Anjedani, L. Senczuk, and D. Skwarski, *Ann. Pharm. (Poznan)* **13,** 139 (1978).
97. R. D. Westland, J. L. Homes, M. L. Mouk, D. D. Marsh, R. A. Cooley, Jr., and J. R. Dice, *J. Med. Chem.* **11,** 1190 (1968).

General References

References 8, 26, 27, and 81 are also general references.
Gmelins Handbuch der Anorganischen Chemie, 8th ed., Natrium, Verlag Chemie, G.m.b.H., Weinheim/Bergstrasse, 1964, pp. 247–252.

JAMES W. SWAINE, JR.
Allied Corp.

THORIUM AND THORIUM COMPOUNDS

Thorium [*7440-29-1*], element number 90, has an atomic weight of 232.038. Before 1940 it was grouped in the Periodic Table with the other tetrapositive metals, titanium, zirconium and hafnium, group IVB. It is now considered to be the second of a 5*f* series of elements, the actinides (qv), by analogy with the 4*f* (rare earth or lanthanon) series in which the second element, cerium, also shows a tetrapositive state. Thorium has no tripositive state to correspond with that of cerium. The electronic ground state of the gaseous atom is $6d^27s^2$, and only at the third level, elevated 7795.270 cm^{-1}, is an *f*-electron indicated, $5f6d7s^2$ ($^3H_4$) (1). Only at spectroscopic Th^{3+} does the ground state (5*f*) contain an *f* electron (2). Trebly charged thorium is not a stable chemical form and Th^{4+}, the stable form, has only the radon core of electrons. Chemically, thorium is less basic than the rare earths, but there are insoluble thorium fluorides, carbonates, hydroxides, oxalates, and phosphates. The crystal radius of Th(IV) (0.095 nm), is also close to that of the rare earths, which leads to their co-occurrence, eg, in the mineral monazite, (Ce,La,Nd)PO_4.

Refractory thorium dioxide [*1314-20-1*], when heated to incandescence by a normally weakly yellow gas flame, emits a brilliant white light. The best effect is achieved with material containing ca 1 wt % cerium oxide. Commercial mantles are made up of a mesh hood of oxide surrounding a flame and are used in decorative exterior lighting, gasoline lanterns for camping, etc. As gas lighting was replaced by electrical lighting after World War I, the commercial importance of thorium declined. However, in the 1940s, it was discovered that the product of neutron absorption by thorium was ^{233}U, which could undergo fission by slow neutrons.

$$^{232}\text{Th}(n,\gamma) \quad ^{233}\text{Th}; \quad ^{233}\text{Th} \xrightarrow{\beta} {}^{233}\text{Pa} \xrightarrow{\beta} {}^{233}\text{U}$$

[*7440-29-1*] [*15720-66-8*]

The work of the Manhattan Project during World War II on the fission chain reaction created tremendous commercial interest in thorium (see Nuclear reactors, fast-breeder reactors) (3–5). This is still the main interest in the element, although small-scale applications of it have been developed in the fields of electronics and catalytic processes. A larger use of thorium and one that has growth potential is that of a strengthening alloying element in metallic magnesium (TD Mg); this application is being extended to alloys such as TD Ni, TD NiCr, and TD NiCrAl. The amount of terrestrial thorium is several times that of terrestrial uranium and several hundred times the amount of the naturally occurring fissile isotope ^{235}U.

Thorium is mildly radioactive, decaying by emission of alpha particles with a half-life of 1.41×10^{10} yr, and it is the progenitor of the $(4n + 0)$ natural radioactive decay series (see Radioactivity, natural). Thorium refined from ores free of uranium would be almost monoisotopic ^{232}Th, ie, the radiothorium (^{228}Th [*14274-82-9*]) of its own $(4n + 0)$ decay chain would be one part in 10^{10}. The presence of uranium in the ore would introduce imponderable amounts of short-lived ^{231}Th [*14932-40-2*] and ^{227}Th [*15623-47-9*] from the decay of ^{235}U, and ^{234}Th [*15065-10-8*] from ^{238}U decay. More ^{230}Th [*14269-63-7*] (ionium) (half-life, 8×10^4 yr) from ^{238}U would be present, but generally less than radiothorium. From practically thorium-free uranium ores, however, thorium preparations rich in ionium can be obtained. One such preparation from an unselected ore contained 26.4 wt % ^{230}Th and 73.6 wt % ^{232}Th (6–7).

Occurrence

Thorium is not a very rare element; it comprises 0.001–0.002 wt % of the earth's crust and is distributed widely at up to 15 ppm in soils (8–9). Thorite [*15501-85-6*], the silicate of the element, and thorianite [*12036-19-0*], a mixed thorium–uranium dioxide, are the minerals highest in thorium content, but they are not significant industrial sources of the element. Uraninite [*1317-99-3*], the high uranium end of the dioxide mixture, does furnish thorium but as a by-product to the extraction of the uranium, which is more valuable. The significant mineral source for thorium is the anhydrous rare-earth phosphate, monazite, from which it is currently obtained as a by-product. Thorium contents of monazite are commonly 5–30 wt % (expressed as oxides). Some sample analyses are given in ref. 10.

Monazites occur in many countries since the chemical resistivity and specific gravity of the material contribute to its accumulation as beach placers. The Travancore deposits in India are historically famous, but Brazil, the Union of South Africa, Australia, Malaysia, the Malagasy Republic, and the United States, among others, have large deposits. Commercially available deposits have changed with time as it became recognized that these materials were valuable as potential sources of nuclear energy.

Recovery from Ores

Chemically stable monazite can be disintegrated successfully by either strong acid or by strong alkali (8,11). Strong acid dissolves monazite by converting phosphate ion to $H_2PO_4^-$ and H_3PO_4. The thorium in the solution can then be separated from rare earths or lanthanons by differential precipitation, ie, precipitation of the thorium, which generally is less soluble than corresponding lanthanon compounds under certain conditions. The simplest precipitation, and one which facilitates further treatment, is that of the hydroxide. If the acid solution is partially diluted and neutralized to ca pH 1 with ammonia, the thorium precipitates with very little entrainment of rare earths. The precipitated thorium solids are then washed with alkali to free the mass from soluble sulfate and phosphate, leaving the residual hydroxide solid to be redissolved in the acid chosen for the purification procedures.

When the monazite is treated with the strong alkali, both thorium and lanthanon hydroxides remain in the insoluble solids, and phosphate is removed in the supernate. In special cases, there may be reason to dissolve the thorium in a differential complexing medium, eg, alkali carbonate. The procedure of choice is dissolving the cake in acid (preferably HCl) and then adding ammonia to cause the thorium to reprecipitate. The final acid solution of the thorium in both cases is best purified by extraction with an immiscible organic solvent. The procedures are summarized in Figures 1 and 2.

Thorianite–uraninite ores are considerably more dilute in their mineral values

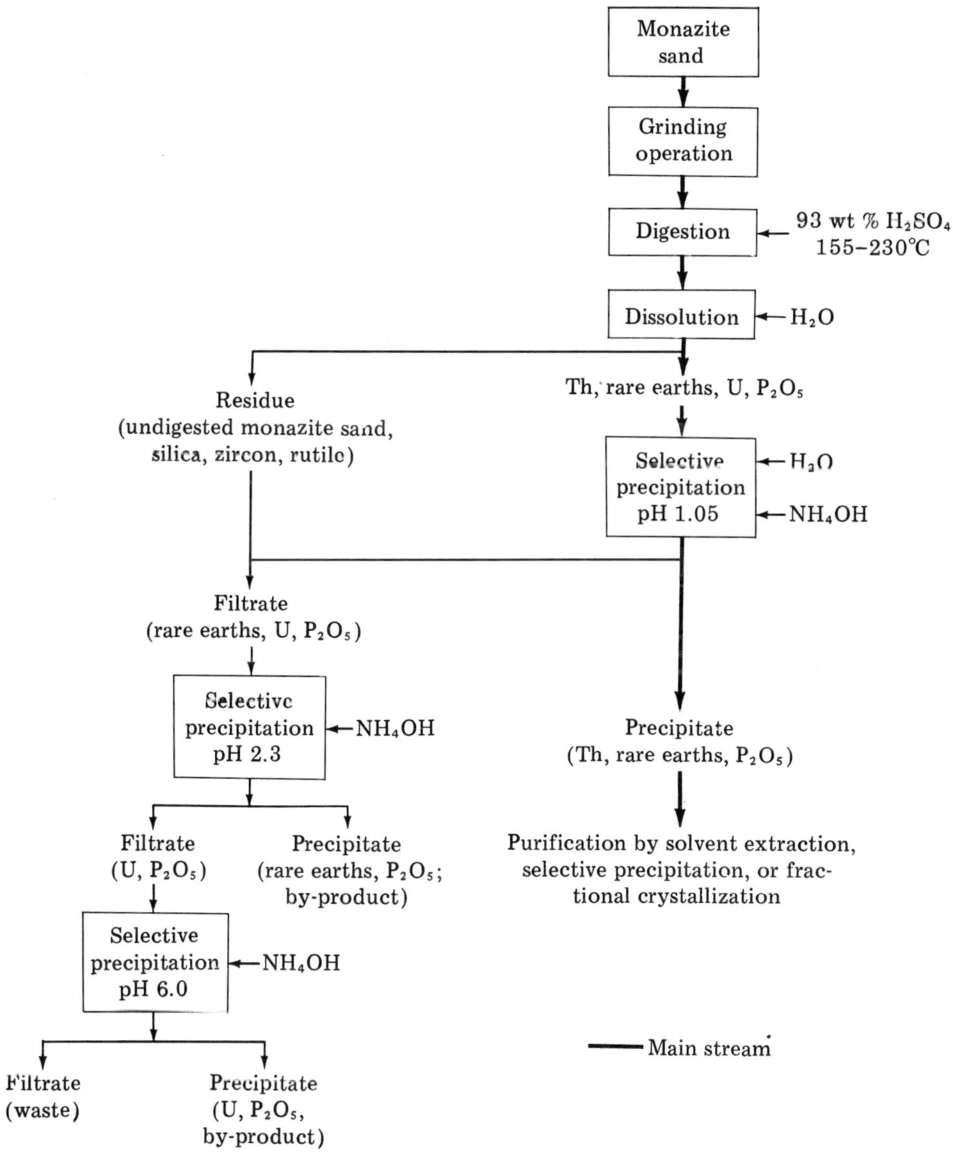

Figure 1. Simplified schematic diagram of sulfuric acid digestion of monazite sand and recovery of thorium, uranium, and the rare earths (12).

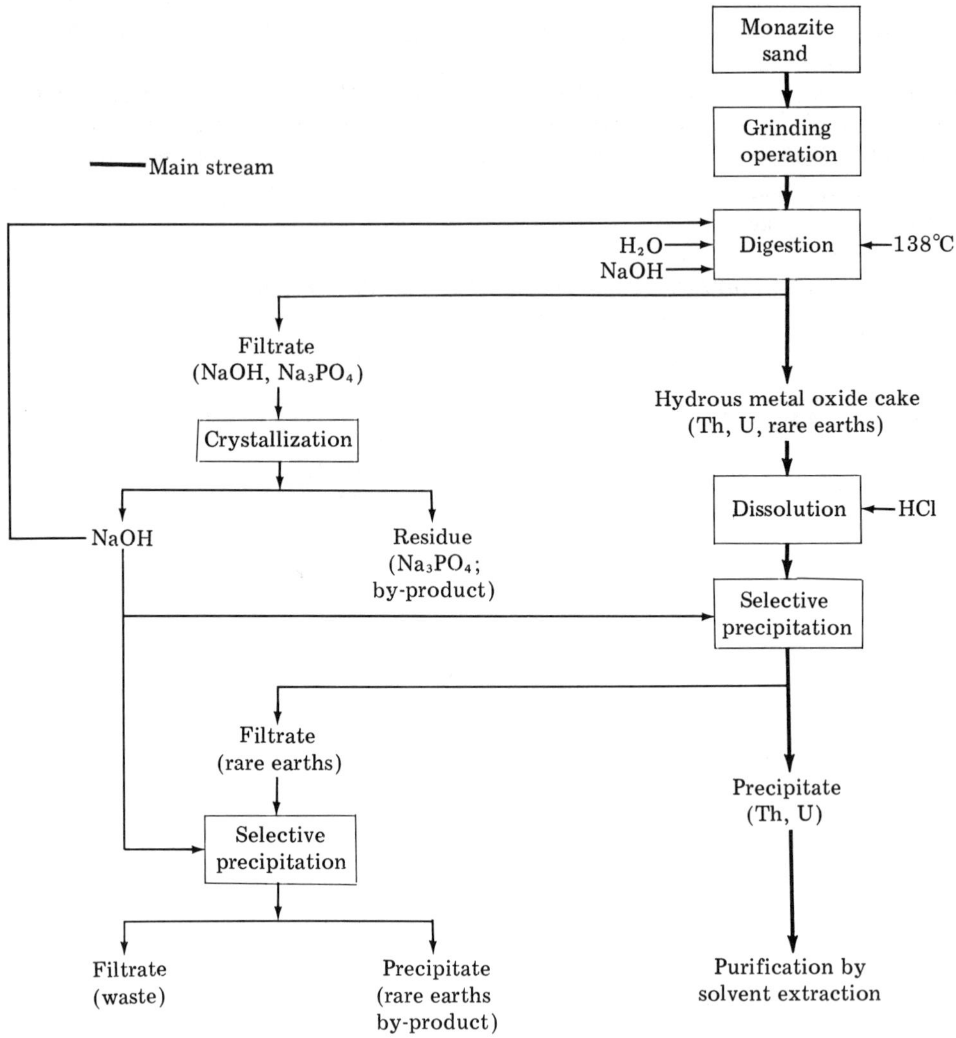

Figure 2. Simplified schematic diagram of caustic soda digestion of monazite sand and recovery of thorium, uranium, and the rare earths (13).

and are processed primarily for their uranium content. A typical ore, as found in the Elliot Lake region of Ontario, Canada, may contain 0.1–0.15 wt % U_3O_8 and one-third that amount of ThO_2. The uranium and thorium are leached from the milled ore with H_2SO_4. The Th(IV) remains cationic and the uranyl ion forms anionic sulfate complexes, which can be removed from the solution on an anion-exchange column. The thorium in solution can then be isolated by an efficient solvent extraction procedure.

There are several different methods for extracting thorium into immiscible organic solvents. These methods are efficient and selective and frequently are used to remove minor concentrations of thorium from mixtures such as leach solutions. One procedure involves tributyl phosphate (TBP) or its analogues dissolved in kerosene. Treatment of the separated organic phase with 10 N H_2SO_4 strips it of the thorium. In the case of the strongly acidic monazite digest, high molecular weight amines dissolved in hydrocarbons may be used to extract the thorium, but this is not very selective. Treatment with amines is equivalent to use of a liquid anion-exchange resin, which carries along any other sulfate-complex forming cations, eg, uranium separation from uraninite (see above). Solvent extraction is very useful in the monazite process, however, for purification of thorium separated by ammonia precipitation. The thorium hydroxide dissolves in nitric acid, and from this solution the thorium may be extracted either by the TBP-kerosene reagent or by fortifying the solution with, eg, calcium nitrate, and extracting it with methyl isobutyl ketone (14–15). A multistage countercurrent system gives optimum results. The basis for the TBP extraction is the formation of a very firm $Th(NO_3)_4.2TBP$ complex, which is strongly hydrocarbon soluble.

Thorium Metal

Properties. Pure thorium is a bright, silvery metal and is dense, very high melting, weaker than the structural metals, softer than steel, and quite reactive chemically (16–17). It darkens on prolonged exposure to air. The gross density is 11.25–11.7 g/cm^3, and the crystallographic density is 11.724 g/cm^3. At the allotropic phase-change temperature of ca 1345°C, the solid changes from face-centered cubic to body-centered cubic. The purest metal (<50 ppm total impurity) melts at ca 1750°C. The heat of sublimation of the solid at 25°C is 597.5 ± 4.6 kJ/mol (142.8 ± 1.1 kcal/mol). The metal is diamagnetic but is superconducting below 1.3–1.4 K (see Table 1).

Thorium alloys readily with iron, cobalt, nickel, copper, gold, silver, boron,

Table 1. Physical Properties of Thorium[a]

Property	Value
atomic diameter, nm	0.3596
melting point, °C[b]	1750
boiling point, °C	ca 3800
density, x ray, g/cm^3	11.72
crystal structure	
in fcc up to 1400°C, a_o, nm	0.5089
in bcc 1400°C to mp, a_o, nm	0.411
heat of fusion, kJ/mol[c]	<19.2
heat of vaporization, kJ/mol[c]	ca 586
heat capacity, 25°C, J/(mol·K)[c]	27.32
thermal conductivity at 200°C, J/(m·K)[c]	38.9
thermal diffusivity at 200°C, cm^2/s	0.28
electrical resistivity, $\mu\Omega$·cm	14
elastic constants	
Young's modulus, MPa[d]	7.24×10^4
shear modulus, MPa[d]	2.76×10^4
Poisson's ratio	0.27
emissivity, 1000–1700°C at 0.667 μm	0.38
work function, eV	3.51
standard electrode potential (V) for the reaction $Th \rightarrow Th^{4+} + 4e$	−1.899

[a] Refs. 18–19.
[b] Higher values have been reported.
[c] To convert J to cal, divide by 4.184.
[d] To convert MPa to psi, multiply by 145.

platinum, molybdenum, tungsten, tantalum, zinc, bismuth, lead, mercury, sodium, beryllium, magnesium, silicon, selenium, aluminum, and perhaps with other elements. Simple eutectics form with chromium and with uranium. The metal is miscible with cerium in both the liquid and the solid states. Data on binary phase diagrams for thorium systems with ca 32 different metallic elements have been published, and many compounds of thorium and other metals have superconducting properties (16–17).

Finely divided thorium metal is pyrophoric in air, and thorium ribbon burns in air to give the oxide. Massive metal, however, is stable at ordinary temperatures, though the surface darkens with time. Under various temperatures, hydrogen, nitrogen, the halogens, and sulfur react vigorously with thorium metal, and binary compounds form upon heating thorium with either carbon or phosphorus.

Aqueous hydrochloric acid attacks thorium metal vigorously; 12–25 wt % of the metal remains undissolved as a dark, hydridic Th(IV) compound, HTh(O)X (X is a mixture of OH^- and Cl^-) (20). A concentration of 0.01–0.05 *M* fluoride or fluorosilicate in the HCl enables the solution process to go to completion. Nitric acid passivates thorium metal, but a small concentration of fluoride or fluorosilicate allows the solution reaction to continue to completion. Dilute hydrofluoric, nitric, or sulfuric acids and concentrated perchloric or phosphoric acids attack thorium slowly with evolution of hydrogen. Thorium is not attacked by alkali hydroxides.

Preparation. Historical preparations of thorium metal by a variety of procedures gave only powdered, impure metal. This was probably caused by the combination of its very high melting point and a chemical reactivity approximating that of magnesium metal. The commercial-scale production process first developed during the Manhattan Project circumvented both problems (21). Reduction of an admixture of zinc halide with thorium tetrafluoride [*13709-59-6*] by metallic calcium in a bomb was the key to the successful preparation. As both halides are prepared in high purity and free of water, side products are minimized. The reduction of the halide mixture produces a great deal of heat, which is largely confined to the bomb, and the metallic zinc produced forms a low melting, low boiling alloy with the thorium metal which cools to a solid metallic button. The zinc is then boiled to evaporation, leaving an excellent yield of quite pure Th-metal sponge. This is melted and recast to a solid metallic billet.

Other modern reduction procedures, eg, the reduction of ThO_2 by calcium in the presence of $CaCl_2$ or the electrolysis of potassium–thorium double fluorides in a fused KCl–NaCl salt bath, yield powdered metal. To make small samples of ultrapure thorium metal, the van Arkel-de Boer process has been used (22–23). Lower purity thorium metal and a small amount of iodine or of thorium iodide are confined in a moderately heated evacuated vessel. Thorium iodide vapors pass to a very hot surface, eg, a tungsten filament, where they decompose and deposit the metal. The iodine vapors recycle to the lower purity thorium and then react forming fresh metal iodide, and the cycle continues.

Compounds

Nonmetallic thorium is in the +4 oxidation state exclusively. Occasional odd compounds have been thought erroneously to represent lower oxidation states. Thorium diiodide [*13779-95-8*], ThI_2, prepared from metal and iodine, is probably $Th(IV)I_2(e)_2$, with the two electrons in a conduction band giving metal-like properties. Certain metalloorganics also may be cited. Th^{4+} is one of the least hydrolyzed tetrapositive ions having a first acid dissociation constant of ca $10^{-3.6}$. With a small ionic radius (0.095 nm), it has a strong tendency to hydrate and to coordinate electron donors. Its coordination numbers are 11, 10, 9, 8, and occasionally 6. In general, the halides and some strongly binding oxygen compounds favor the lower coordination numbers, and weakly basic hydrated compounds favor the higher ones.

Complexes are formed with fluoride, iodate, bromate, nitrate, chlorate, chloride, sulfate, sulfite, carbonate, phosphate, pyrophosphate, and molybdate ions. Complexes with most anions of organic acids, eg, formate, acetate, chloroacetates, oxalate, tartrate, malate, citrate, salicylate, sulfosalicylate, etc, also form. Chelates form very readily, some of which lend themselves especially well to the solvent extraction process.

Pure massive Th metal is made available for chemical synthesis by its disintegration with hydrogen gas (24). Massive metal may require heating to 300–400°C before reaction with hydrogen takes place; powdered or sintered metal reacts immediately and exothermically at room temperature. The induction period of the reaction is a function of the impurity content of the metal. The hydride completely decomposes to the elements at 900°C under vacuum. The metal product is gray to black and powdered, or it is in the form of an easily disintegrated mass. This powdered metal is readily attacked by oxygen, sometimes pyrophorically, forming ThO_2. A number of other elements react with the powdered metal forming binary compounds. The direct reaction of thorium hydride, either with a gaseous element, eg, Cl_2 or Br_2 at 250–350°C, or with the hydrogen compounds, eg, steam, hydrogen halides, hydrogen sulfide, phosphine, or ammonia, gives the binary thorium compound. Methane or carbon dioxide do not react with thorium hydride.

For further information on thorium compounds, see ref. 25.

Hydrides. The two thorium hydrides are ThH_2 [*16689-88-6*] and Th_4H_{15} [*12055-07-1*]. The latter is generally written as $ThH_{3.75}$ and was the first metal hydride to demonstrate superconductivity (26–30). The transition temperature is 7.5–8 K, with metallic conduction to room temperature (see Superconducting materials). Some ternary hydrides have also been prepared: three with palladium as the additional element, $ThZr_2H_{7+x}$, and $ThTi_2H_{6+x}$. The latter two are of potential practical interest, as they combine an extremely high volume density of hydrogen with relatively low equilibrium pressures of hydrogen at elevated temperatures and are stable in air (31) (see Hydrogen energy; Hydrides).

Carbides. Carbon reacts with thorium metal yielding ThC_2 [*12071-31-7*], which exists in three polymorphic forms. Controlling the amounts of material in the reaction leads to the formation of the phase ThC [*12012-16-7*] (32–33). Both carbides are high melting, ie, mp = 2625 ± 25°C and 2655 ± 25°C, respectively. Thorium sesquicarbide [*12151-76-7*] (Th_2C_3) is stable only at very high temperatures and pressures but sometimes occurs at room temperatures in solid solution with carbides of other elements. Thorium dicarbide also forms through heating of thorium dioxide with excess carbon. The dicarbide is reactive, as it burns in air to the dioxide and reacts with sulfur or selenium vapor and with ammonia. The halogens react with thorium carbide yielding the anhydrous halides, and the carbide is readily hydrolyzed by water or moist air. Thorium carbide, ThC, yields methane this way, and ThC_2 gives principally ethane and hydrogen (34–35).

Oxides. Thorium dioxide [*1314-20-0*] (thoria), ThO_2, exists as a single cubic phase of fluorite structure up to its melting point of ca 3390°C. The x-ray density of the crystal at 25°C is 10.001 g/cm^3. Amorphous thoria crystallizes upon fusion with sodium carbonate or potassium orthophosphate. Thoria forms double oxides with the alkaline earths, eg, $ThBaO_3$ [*12230-90-9*]. Double oxides with the alkali metals are of the type Na_2ThO_3 [*12058-67-2*] (36). It also forms oxides with germanium, titanium, vanadium, niobium, tantalum, and molybdenum.

Ignited oxide or oxide fused in large particles is one of the most refractory substances known; even fusion with potassium sulfate may fail to alter it chemically. However, aqueous nitric acid containing 0.01–0.05 M HF or sodium fluorosilicate frequently is a convenient reagent and is easier to use (37–38). Alkali carbonate does not attack thoria, but hot aqueous HF or HF gas at 250–700°C converts thorium dioxide to thorium fluoride [*13709-59-6*]. Thoria acts as a catalyst for reactions involving gases or vapors absorbed onto it. With minor concentration of yttria, its solid-state conductivity gives it important potential application in emf cells for monitoring and controlling high temperature processes.

Addition of hydrogen peroxide to a solution of a thorium salt produces a solid precipitate only when the solution is ≥ 0.3 N in sulfuric acid; all other precipitates are gelatinous (39). The composition of the precipitate is $Th(O_2)SO_4.3H_2O$. The gelatinous peroxide precipitates are highly hydrolyzed, contain more than one peroxide group per thorium, and vary in composition.

Nitrides. Thorium nitride [*12033-90-8*], Th_3N_4, is chestnut to black in color, which may represent variable contamination with Th_2N_2O [*12196-76-8*], which is tan, and ThN [*12033-65-7*], which is golden yellow and is the stable, high temperature decomposition product of all the other nitrides. A phase that analytically is ThCN [*12654-85-2*] contains nitride groups and acetylenic C_2 groups, as in thorium carbide (40). Other double nitrides, eg, Li_2ThN_2 [*11113-24-9*] and $BeThN_2$ [*11112-22-4*], Th_2N_2O and its analogues, compounds containing (NH) groups, etc, have been identified; Th^{4+} and N^{3-} units are involved and charge neutralization is observed. The nitrogen compounds are generally moisture sensitive, yielding ammonia. Thorium fluoride nitride [*12182-88-6*] (ThNF) and to some extent ThNCl are exceptions, requiring heat and acid for hydrolysis. Even though reactions may result with powdered metal at high temperatures, the smoothest synthesis of the binary nitrides proceeds through thorium hydride. Nitrides heated in oxygen generally produce thoria.

Sulfides and Analogues. Binary compounds of thorium with sulfur can be produced by reaction with hydrogen sulfide, sulfur, or carbon disulfide. Thorium disulfide [*12138-07-7*] (ThS_2) is purple or brown-violet and has a density of 7.36 g/cm^3. Heated in vacuum, it loses sulfur to yield Th_7S_{12} [*12286-36-1*] (thorium black), which has sometimes been mislabeled Th_4S_7 (41). Further decomposition gives Th_2S_3 [*12286-35-0*] (thorium brown) (42). The lowest sulfide is ThS [*12039-06-4*] (thorium silver) with a density of 9.56 g/cm^3. All of the sulfides are temperature resistant. Thorium disulfide begins to decompose above 1500°C, whereas ThS sinters at 1950–2050°C with no appreciable vapor pressure at the melting point, ie, >2200°C. The well-sintered solid looks very much like a metal and can be machined, filed, or polished. Thorium brown is perhaps even more useful as a crucible material; it melts at 1900–2000°C with no appreciable volatility. It can be cast, can be machined and filed readily, has metallic luster, and is a fairly good diamagnetic electrical conductor. All the sulfides are attacked by acids and may decompose upon fusion with alkali. Yellow ThOS [*12218-77-8*] can be prepared directly but may form inadvertently when oxygen has been admitted to a synthetic system or has not been removed.

Thorium compounds analogous in composition to the sulfides form with selenium (43). Several tellurium–thorium compounds are analogues of sulfide compositions, but their crystal structures and properties differ from the sulfides and selenides (44). Thorium also forms compounds with phosphorus, arsenic, and bismuth, and there are a number of ternary compounds containing these elements.

Fluorides. Anhydrous thorium fluoride [*13709-59-6*], ThF_4, can be prepared by reaction of the elements or by fluorine or HF with the other halides, thorium hydride, or thorium carbide. Hydrogen fluoride on thorium oxide, hydroxide [*12688-06-1*], oxycarbonate [*61504-43-6*], or oxalate [*2040-52-0*] can also be used to produce ThF_4 (45). Thorium fluoride octahydrate precipitates from aqueous solution. When dehydrated, it gives the tetrahydrate, dihydrate, then $Th(OH)F_3.H_2O$, and at red heat, $ThOF_2$ [*13597-30-3*]. Steam at 900°C is required to remove the last of the fluoride. Anhydrous thorium fluoride produces a large variety of double salts with alkali metal fluorides. Divalent cations give $MThF_6$.

Chlorides. Synthesis of $ThCl_4$ [*10026-08-1*] can be achieved by treating thorium metal, hyride, or carbide with chlorine or by reaction of HCl with the metal or the hydride. A number of chlorinating agents convert ThO_2 to the chloride, and chlorine with ThS_2 also is useful. The oxychloride, $ThOCl_2$ [*13637-74-6*], frequently forms if oxygen has not been excluded or eliminated. However, under high vacuum heating, $ThCl_4$ sublimes, with the oxide disproportionation product remaining (46). The oxychloride forms directly upon hydrolysis of $ThCl_4$, heating stoichiometric amounts of $ThCl_4$ and ThO_2, or treating $ThCl_4$ with oxygen at high temperatures (47). Double salts of thorium chloride with monovalent chlorides are numerous and varied, and compounds form with $BaCl_2$. In crystals, a stable $ThCl_6^{2-}$ entity is important and occurs in some vapor-phase addition molecules. In molten KCl, $ThCl_5^-$ and $ThCl_7^{3-}$ groups exist, and in some vapor species, 5, 8, or 9 chlorides may be coordinated. Hydrated double salts of alkali metals can be isolated from water. Thorium chloride is very soluble in water, and a number of hydrates are known. Solvates also form with electron donors other than water.

Bromides and Iodides. Thorium tetrabromide [*13453-49-1*] is prepared by syntheses similar to those indicated for the other thorium halides. The oxybromide, $ThOBr_2$ [*13596-00-1*], forms through hydrolysis of $ThBr_4$, but it can also be made by the action of Sb_2O_3 on $ThBr_4$ under anhydrous conditions (48). The hydrated salts hydrolyze readily with heating. Solvates with ammonia and a number of organic bases are known, as are double salts. The oxybromide also hydrates. The bromides generally are light sensitive and hydrolyze more easily than the chlorides.

Thorium tetraiodide [*7790-49-0*], ThI_4, can be made by the reaction of iodine with thorium metal or by treating thorium hydride with hydrogen iodide. The basic iodide, $ThOI_2$ [*19173-92-3*], is made by treating ThI_4 with Sb_2O_3 or with ThO_2 (48–49). The iodides are very readily hydrolyzed, even in the solid state, and they yellow on exposure to light. The tetraiodide reacts with thorium metal giving two crystal forms of ThI_2, which are black and golden-metallic, respectively, and a questionable third substance designated ThI_3 [*13779-96-9*] (50–52). It is not certain whether the ThI_2 is truly a Th(II) material.

Hydroxide and Carbonate. Addition of alkali to a solution of a thorium salt generally gives a gelatinous precipitate of hydrated thorium hydroxide, although a vitreous form is claimed to form under certain conditions (53). Dilute acids redissolve the material. Fresh precipitate dissolves in solutions of ammonium oxalate, alkali carbonates, sodium citrate or tartrate, and other complexing anion solutions. Dehydrated $Th(OH)_4$ [*13825-36-0*] seems stable at 260–450°C; at ≥470°C, it converts to the oxide (54).

Thorium hydroxide absorbs carbon dioxide strongly, yielding hydrated $ThOCO_3$, possibly $Th(OH)_2CO_3.2H_2O$ [*12538-66-8*] (55). Reaction with alkali or ammonium carbonate causes the precipitation of similar material from solutions of thorium salts. Raising the concentration of carbonates dissolves the precipitated material. The complexes formed are strong enough to prevent precipitation of thorium by ammonia, fluoride, or acid phosphate. Complexes in solution may contain 4–6 carbonates per thorium, but a wide range of solid double carbonates contain the $Th(CO_3)_5^{6-}$ grouping in a hydrated crystal (56–57).

Nitrates and Perchlorates. Nitric acid dissolves thorium hydroxide yielding thorium nitrate [*13823-29-5*], which is the thorium compound most highly soluble in water. Depending on the acidity, the pentahydrate [*33088-17-4*], tetrahydrate [*33088-16-3*], or the addition product, $Th(NO_3)_4.2N_2O_5$, may crystallize (58–59). If the nitrogen oxides are removed from the latter compound, questionably stable anhydrous nitrate is obtained. In <1 wt % acid, the crystals that form from solution look like those of the pentahydrate but are a partially hydrolyzed material of uncertain stoichiometry. Thorium nitrate is also very soluble in a large number of oxygenated organic solvents, eg, alcohols, ketones, ethers, and esters. Solid solvates with the organic component alone or which contain water and solvent have been identified as have solvates with nitrogen bases. An immiscible organic solvent can extract thorium nitrate from a very concentrated aqueous solution or one with a high concentration of ammonium nitrate added, which is a method for thorium separation from rare earths. For example, tributyl phosphate binds strongly and specifically with the metal ion yielding anhydrous species with two and three phosphate molecules per thorium nitrate (60–61). These species can be extracted into inert fluids in which the thorium nitrate is not otherwise soluble, eg, benzene, CCl_4, and aliphatic hydrocarbons (61–62). Thorium nitrate forms a variety of double salts.

Thorium perchlorate [*16045-17-3*] forms upon the reaction of perchloric acid with thorium hydroxide, and it is very soluble in water. From an acid solution, thorium perchlorate crystallizes as $Th(ClO_4)_4.4H_2O$ (63). At ca 280°C, this converts to $ThO(ClO_4)_2$ and, at ca 355°C, to thoria.

Sulfates. Thorium sulfate forms by metathesis of other thorium compounds. Hydrated sulfate deposits slowly from certain concentrated solutions, eg, thorium nitrate, after addition of strong sulfuric acid. The hydration of the crystals is unpredictable because equilibration is slow and metastable hydrates tend to deposit first. Recognized hydrates have 9, 8, 6, and 4 molecules of water (64). Dehydration is complete at 350–400°C, and at higher temperatures SO_3 dissociates, leaving thoria. Variably hydrated $ThOSO_4$ [*49788-94-5*] can be prepared from solution. The monohydrate is $Th(OH)_2SO_4$ (65). There are a variety of double sulfates, including one with sulfuric acid.

Phosphates. Most thorium phosphate compounds have very low solubilities in water, so that in addition to the normal $Th_3(PO_4)_4$ hydrates there may be acid phosphates, basic phosphates, pyrophosphates, and perhaps a metaphosphate. Thorium phosphate [*14485-31-5*] precipitates as a gelatinous, difficult-to-coagulate solid of uncertain composition. One series of double phosphates contains a $Th_2(PO_4)_3^-$ unit, and another contains a $Th(PO_4)_2^{2-}$ unit.

Oxymetallates and Carboxylates. Thorium oxymetallates tend to be insoluble, and their analytical significance is discussed in refs. 66–67. The chromate, $Th(CrO_4)_2 \cdot 3H_2O$ [*76663-32-6*], forms when dichromate is added to a solution of a thorium salt. Hydrated thorium molybdate [*15414-73-0*], $Th(MoO_4)_2$, precipitates quantitatively from acetic acid medium. Thorium iodate is another precipitate that is used analytically.

Many carboxylates and other organic acids cause the precipitation of thorium from solution, but the solids are generally basic, hydrated, and of indefinite composition. They can serve, however, for analytical ignition to ThO_2; the oxalate has been particularly useful for this purpose. Stoichiometric salts of *m*-nitrobenzoic acid, picrolonic acid, sebacic acid, and the noncarboxylic ferron (7-iodo-8-hydroxyquinoline-5-sulfonic acid) have been prepared.

Chelates. Noncarboxylic chelating agents form stable compounds with thorium. Tetrakis(acetylacetone) thorium [*17499-48-8*] is one the most volatile thorium compounds. A variety of substituted acetylacetones, particularly the trifluor compound, behave similarly (68). Oxine (8-hydroxyquinoline) also binds thorium strongly in tetrakis compounds. An important characteristic of all of these and other chelates is that they can be used to extract thorium from aqueous solutions into nonaqueous solutions.

Metallo-Organics. Prepared thorium metallo-organic compounds seem to be Th(IV) compounds. Sandwich compounds form with cyclooctatetraene ($Th(COT)_2$ [*12702-09-9*]), phthalocyanine ($Th(Pc)_2$ [*12581-78-1*]), and perfluorophthalocyanine (69–73). Nonsandwich thorium tetrabenzyl [*54008-63-8*] has been reported, as have the cyclopentadiene compounds, $(C_5H_5)_3ThR$, where R is an alkyl or aryl group. When R is *n*-butyl, the compound may rearrange in solution to deposit $((C_5H_5)_2Th(C_2H_4))_2$, in which two Th(IV) atoms are bridged by a pair of rings (74).

Economic Aspects

Until the ^{233}U–^{232}Th breeding cycle of nuclear energy becomes commercially significant, the production capacity of the thorium industry will far exceed the demand for thorium. One factor is the continuing processing of monazite for its rare-earth content and that of uraninite–thorianite for uranium. Known world reserves of monazite and thorite are estimated to contain over 700,000 metric tons of ThO_2, with U.S. monazite contributing ca 100,000 t. Consideration of low grade ores may extend this to several million (10^6) tons. The principal commercial thorium compound is thorium nitrate [*13823-29-5*]. A few hundred thousand kilograms are consumed annually and on a worldwide basis for use in gas mantles.

Health and Safety Factors

There are a few potential hazards relating to thorium, although no significant incidents have been reported. One potential hazard is the reactivity of finely divided metal and of the hydrides, with respect to oxygen and halogens. Such a reaction can produce explosive or inflammatory results. Th^{4+} is a heavy element capable of being hazardous when introduced into the body. If injected as the soluble nitrate, for example, it can cause hemolysis. The tendency of Th^{4+} to form compounds of low solubility with organic carboxylates in general, particularly near the pH values of body fluids, tends to localize its effects, though it does form soluble complexes reasonably readily. Similar considerations make its ingestion from the gut lumen problematic. The long half-life of ^{232}Th makes it a minimal radiation hazard. External to the body, the hazards from gamma rays in the accompanying decay chain are small, since their abundance is low, and many are in the x-ray range, ie, <85 keV, and largely self-absorbed in the bulk of the material. There are some residuals in the 240 and 727 keV energy ranges that can lead to radiation from tonnage amounts of material, eg, from monazite beds; such sources merit minimal safety precautions in terms of distance and shielding. Prolonged retention of thorium in the body, particularly in a fixed location, allows the possibility of radiation damage by the energetic alpha particles of some of the short-lived daughter members of the decay chain, of which there are five for every thorium alpha. This probably contributed to stopping the medical use of a radiopaque material based on thorium (see Radiopaques). In the laboratory and in industry, the relevant precaution is against breathing thorium-bearing particles or dusts, which could lodge in the lungs; thoria is prone to dusting, for example. In general, a conventional dust respirator should be sufficient protection.

Thorium compounds are legally classified as source materials for nuclear energy and thus are regulated by various governmental agencies, eg, the NRC. The relevant regulations cover licensing and safety aspects.

BIBLIOGRAPHY

"Thorium and Thorium Compounds" in *ECT* 1st ed., Vol. 14, pp. 116–124, by Howard E. Kremers, Lindsay Chemical Company; "Thorium and Thorium Compounds" in *ECT* 2nd ed., Vol. 20, pp. 248–259, by W. L. Silvernail and J. B. McCoy, American Potash & Chemical Corporation.

1. A. Giacchetti, J. Blaise, C. H. Corliss, and R. Zalubas, *J. Res., U.S. Natl. Bur. Stand.* **A78,** 247 (1974).
2. P. F. A. Klinkenberg and R. J. Lang, *Physica* **15,** 774 (1949).
3. G. T. Seaborg and L. I. Katzin, *Production and Separation of U^{233}: Survey*, National Nuclear Energy Series, Div. IV, Vol. 17A, 1951, TID-5222, AEC.
4. L. I. Katzin, *Production and Separation of U^{233}: Collected Papers*, National Nuclear Energy Series, Div. IV, Vol. 17B, 1952, TID- 5223, AEC.
5. *Proceedings of the International Conference on the Peaceful Uses of Atomic Energy*, Vols. 6–9, United Nations, New York, 1956.
6. E. K. Hyde in ref. 4, pp. 554–557.
7. W. Rall and A. J. Dempster in ref. 4, pp. 639–642.
8. F. L. Cuthbert, *Thorium Production Technology*, Addison-Wesley Publishing Co., Reading, Mass., 1958.
9. C. Frondel, *Systematic Mineralogy of Uranium and Thorium*, Geological Survey Bulletin 1064, U.S. Government Printing Office, Washington, D.C., 1958.
10. J. Strod, *Am. Ceram. Soc. Bull.* **32,** 122 (1953).
11. R. J. Callow, *The Industrial Chemistry of the Lanthanons, Yttrium, Thorium, and Uranium*, Pergamon Press, Inc., New York, 1967.
12. K. G. Shaw and co-workers, *U.S. Atomic Energy Commission Report ISC-407*, Iowa State College, Ames, Iowa, 1954.
13. G. D. Calkins and co-workers, *U.S. Atomic Energy Commission Reports BMI-243 and BMI-243A*, Battelle Memorial Institute, 1950.
14. E. K. Hyde and M. J. Wolf in ref. 4, pp. 197–222.
15. F. H. Spedding, A. Kant, J. M. Wright, J. C. Warf, J. E. Powell, and A. S. Newton in ref. 4, pp. 486–507.
16. H. W. Wilhelm, ed., *The Metal Thorium*, American Society for Metals, Cleveland, Ohio, 1958.
17. J. F. Smith, O. N. Carlson, D. T. Peterson, and T. E. Scott, *Thorium: Preparation and Properties*, Iowa State University Press, Ames, Iowa, 1975.

18. Ref. 8, Chapt. 2.
19. H. A. Wilhelm in C. A. Hampel, ed., *The Encyclopedia of the Chemical Elements*, Reinhold Publishing Corporation, New York, 1968, pp. 711–718.
20. L. I. Katzin, L. Kaplan, and T. Seitz, *Inorg. Chem.* **1,** 963 (1962).
21. F. H. Spedding, H. A. Wilhelm, W. H. Keller, J. E. Iliff, and C. Neher in ref. 4, pp. 428–445.
22. A. E. Van Arkel and J. H. deBoer, *Z. Anorg. Allg. Chem.* **148,** 345 (1925).
23. D. T. Peterson and F. A. Schmidt, *J. Less-Common Met.* **24,** 223 (1971).
24. H. Lipkind and A. S. Newton in ref. 4, pp. 398–403.
25. L. I. Katzin in J. J. Katz, G. T. Seaborg, and L. Morss, eds., *The Chemistry of the Actinide Elements*, Chapman-Hall, London, in press.
26. R. W. Nottorf, A. S. Wilson, R. E. Rundle, A. S. Newton, and J. E. Powell in ref. 4, pp. 350–369.
27. R. E. Rundle, C. G. Shull, and E. O. Wollan, *Acta Cryst.* **5,** 22 (1952).
28. C. B. Satterthwaite and I. L. Toepke, *Phys. Rev. Lett.* **25,** 741 (1970).
29. C. B. Satterthwaite and D. T. Peterson, *J. Less-Common Met.* **26,** 361 (1972).
30. M. Dietrich, W. Gey, A. Rietschil, and C. B. Satterthwaite, *Solid State Comm.* **15,** 941 (1974).
31. R. Van Houten and S. Bertram, *Met. Trans.* **2,** 527 (1971).
32. H. A. Wilhelm and P. Chiotti, *Trans. Am. Soc. Met.* **42,** 1295 (1950).
33. P. Chiotti, F. W. Korbitz, and G. J. Dooley, *J. Nucl. Mater.* **23,** 55 (1967).
34. C. P. Kempter and N. H. Krikorian, *J. Less-Common Met.* **4,** 244 (1962).
35. E. B. Hunt and R. E. Rundle, *J. Am. Chem. Soc.* **73,** 4777 (1951).
36. R. Scholder, D. Rade, and H. Schwarz, *Z. Anorg. Allg. Chem.* **362,** 149 (1968).
37. C. J. Smithells, *J. Chem. Soc.*, 2226 (1922).
38. F. W. Schuler, F. L. Steahly, and R. W. Stoughton in ref. 4, pp. 307–317.
39. J. W. Hamaker and C. W. Koch in ref. 4, pp. 318–338, 339–347.
40. R. Benz, G. P. Arnold, and W. H. Zachariasen, *Acta Cryst.* **B28,** 1724 (1972).
41. E. D. Eastman, L. Brewer, L. A. Bromley, P. W. Gilles, and N. L. Lofgren, *J. Am. Chem. Soc.* **34,** 128 (1951).
42. E. D. Eastman, L. Brewer, L. A. Bromley, P. W. Gilles, and N. L. Lofgren, *J. Am. Chem. Soc.* **72,** 4019–4023 (1950).
43. R. W. M. D'Eye, *J. Chem. Soc.*, 1670 (1953).
44. R. W. M. D'Eye and P. G. Sellman, *J. Chem. Soc.*, 3760 (1954).
45. A. S. Newton, H. Lipkind, W. H. Keller, and J. E. Iliff in ref. 4, pp. 419–427.
46. O. Knacke, F. Miller, and E. van Rensen, *Z. Physik. Chem.* **80,** 82, 91 (1972).
47. B. G. Korshunov and D. V. Drobot, *Zh. Neorg. Khim.* **16,** 556 (1974).
48. K. W. Bagnall, D. Brown, and J. F. Easey, *J. Chem. Soc. A*, 288 (1968).
49. J. D. Corbett, R. A. Guidotti, and D. G. Adolphson, *Inorg. Chem.* **8,** 163 (1969).
50. J. S. Anderson and R. W. M. D'Eye, *J. Chem. Soc.*, (Suppl. 2), S244 (1949).
51. D. E. Scaife and A. W. Wylie, *J. Chem. Soc.*, 5450 (1964).
52. L. J. Guggenberger and R. A. Jacobson, *Inorg. Chem.* **7,** 2257 (1968).
53. M. Guymont, *C. R. Hebd. Seances Acad. Sci. Ser. C* **285**(10), 345 (1977).
54. T. Dupuis and C. Duval, *Compt. Rend.* **228,** 401 (1949).
55. Yu. Ya. Kharitonov, A. K. Molodkin, and T. A. Balakaeva, *Zh. Neorg. Khim.* **14,** 2761 (1969).
56. J. Dervin and J. Faucherre, *Bull. Soc. Chim. France*, 2930 (1973).
57. J. Dervin, J. Faucherre, P. Herpin, and S. Voliotis, *Bull. Soc. Chim. France*, 2634 (1973).
58. J. R. Ferraro, L. I. Katzin, and G. Gibson, *J. Am. Chem. Soc.* **76,** 909 (1954).
59. *Ibid.*, **77,** 327 (1955).
60. J. C. Warf, *J. Am. Chem. Soc.* **71,** 3257 (1949).
61. M. R. Anderson, *U.S. Atomic Energy Commission Dec. ISC-116*, Washington, D.C., 1950.
62. L. I. Katzin, J. R. Ferraro, W. W. Wendlandt, and R. L. McBeth, *J. Am. Chem. Soc.* **78,** 5139 (1956).
63. P. R. Murthy and C. C. Patel, *Indian J. Chem.* **3,** 134 (1955).
64. J. W. Mellor, *A Comprehensive Treatise on Inorganic and Theoretical Chemistry*, Vol. VII, Longmans, Green and Co., Ltd., London, 1941, p. 241.
65. G. Lundgren, *Arkiv Kemi* **2,** 535 (1950).
66. F. S. Grimaldi in I. M. Kolthoff and P. J. Elving, eds., *Treatise on Analytical Chemistry*, Vol. 5, Pt. 2, Interscience Publishers, New York, 1961, pp. 139–210.
67. D. I. Ryabchikov and E. K. Gol'braikh, *The Analytical Chemistry of Thorium*, The MacMillan Co., New York, 1963.

68. G. F. S. Wessels, J. G. Leipoldt, and L. D. C. Bok, *Z. Anorg. Allg. Chem.* **393,** 284 (1972).
69. A. Streitweiser, Jr., and N. Yoshida, *J. Am. Chem. Soc.* **91,** 7528 (1969).
70. J. Goffert, J. Fuger, B. Gilbert, and B. Kanellakopoulos, *Inorg. Nucl. Chem. Lett.* **8,** 403 (1972).
71. F. Lux, D. Dempf, and D. Graw, *Angew. Chem. Intl. Ed.* **7,** 819 (1968).
72. F. Lux, F. Ammentop-Schmidt, D. Dempf, D. Graw, and W. Hagenberg, *Radiochim. Acta* **14,** 57 (1970).
73. F. Lux, D. Brown, D. Dempf, R. D. Fischer, and W. Hagenberg, *Angew. Chem. Intl. Ed.* **8,** 894 (1969).
74. E. C. Baker, K. N. Raymond, T. J. Marks, and W. A. Wachter, *J. Am. Chem. Soc.* **96,** 7586 (1974).

General References

Refs. 3, 4, 8, 11, 16, 17, 18, 25, 66, and 67 are general references.

LEONARD I. KATZIN
Consultant

THULIUM. See Rare-earth elements.